W9-BLT-006

BIOLOGY
THE SCIENCE OF LIFE

BIOLOGY
THE SCIENCE OF LIFE

THIRD EDITION

ROBERT A. WALLACE
UNIVERSITY OF FLORIDA

GERALD P. SANDERS
GROSSMONT COLLEGE

ROBERT J. FERL
UNIVERSITY OF FLORIDA

HarperCollins*Publishers*

Sponsoring Editor: Glyn Davies
Cover Designer: Jaye Zimet

Credit lines for illustrations are placed in the Acknowledgments section at the end of the book, which is to be considered an extension of the copyright page.

Library of Congress Cataloging-in-Publication Data

Wallace, Robert A.
Biology, the science of life/Robert A. Wallace, Gerald P. Sanders, Robert J. Ferl.—3rd ed.
p. cm
Includes bibliographical references.
ISBN 0-673-38044-0 (student edition)
ISBN 0-673-53378-6 (teacher edition)
1. Biology. I. Sanders, Gerald P. II. Ferl, Robert J. III. Title.
QH308.2.W34 1990
574—dc20 89-70114
CIP

90 91 92 93 9 8 7 6 5 4 3 2 1

To
Big John and Leslie,
BBPs both,
and to all who pass through
the portals of
Phoenix Nest Farm

RAW

To
Jenny, Josh, and Jon
for a happier future;
and to Tim, the new kid
on the block.

GPS

To
Mary-B and Evan
with love, admiration,
wonder, and amazement

RJF

PREFACE

GOALS OF THE THIRD EDITION

Someday we would like to revise one of our books. They have all now been through several editions and none has been a simple revision. This one was no exception. In fact, we do not consider this edition a revision at all. There is too much of it that is new or markedly restructured. We literally dismantled the second edition (which was itself, a very successful effort), and reassembled the third.

It is a monumental task to start from the ground and virtually build an entirely new book every time out, but we feel the result makes it all worthwhile. We would like to be able to sometimes say, "It's good enough, let's let it ride." But somehow we never seem to let it ride. One of us will invariably dig a little deeper, think about it, work through it, and see if there isn't a better way. It seems there always is. This "better way" is presented to the others, and then we all go to work to integrate the new material into the flow of the text. The result is what you see here: the third edition of a successful book that really doesn't seem to be a revision at all. We trust that you will agree that it has all been worthwhile.

We set our sights on several specific goals as we began the third edition of *BIOLOGY: THE SCIENCE OF LIFE*. First was our desire to enhance the major themes of evolution and adaptation and to insure that they were strongly woven throughout the text. We strove to dramatically improve the very important illustration program of the text as well as create a more functional and aesthetic design for the book. Up-dating of material, particularly in the areas of genetics and molecular biology was a priority. We aimed at improving the organization where necessary and fine-tuning the writing. Finally, we took a close look at the learning aids with an eye to maximizing their value.

WITH A LITTLE HELP FROM OUR FRIENDS

With so much new in biology, it is obvious that we had to be quite selective in what to present at the introductory level. We made these decisions with the best advice we could get. To begin with we solicited past adopters, those best qualified to tell us about what they wanted to see in the next edition of *BIOLOGY: THE SCIENCE OF LIFE*. They had a lot to tell us, and it was with their input that we began defining our goals. Next we sought the advice of numerous reviewers. In fact, we are convinced that the third edition has gone through one of the most penetrating and searching review cycles possible. But, in addition, it occured to us early on that our efforts might be much more effective if we had a closer working relationship with some of the front-line research biologists, prominent scientists representing the major areas of biology, yet people devoted to teaching and science education. Thus, we were able to put together our Advisory Board: a group of active scientists and educators with whom we would work closely as the new edition emerged. In addition to advising us on new developments in their fields, anticipating new trends in biology, and lending their teaching expertise, our advisors thoroughly reviewed the second edition offering numerous suggestions throughout the process.

THEMES TO WEAVE

At the very essence of life lies its great complexity. We believe, therefore, that any discussions of life and its processes should seek to place this great diversity into some kind of understandable framework by establishing underlying themes. We believe that biology is best understood in terms of its development through *evolution* and *adaptation,* so we have tried to weave our discussions around these twin themes throughout the book.

THE DESIGN AND ILLUSTRATIONS

In a very real sense, and for whatever reasons, learning these days is strongly dependent on the visual presentation of the material. Not only must the presentation be pedagogically effective, but more than ever, it must be attractive and appealing. We believe that we have met both objectives in these pages. The artwork was conceived, developed, revised, improved, and reworked in what sometimes seemed endless cycles. It was painstakingly developed to serve, then, not only as an effective teaching tool, but as a way to invite the student into the pages with a lively, colorful, and appealing presentation. Of particular importance, we feel, is the fact that the design and illustrations were developed as the textual material itself evolved. The result is a particularly coordinated and integrated approach that will enhance learning and make sometimes complex material more easily understandable.

THE UPDATING

A major and vital goal in the preparation of any new edition of a science text is bringing the content up to date. In biology, perhaps the most dynamic field of science, this in itself is always a major undertaking. It is one of the reasons texts seem to get longer and longer. There is so much that is new, and yet, much of what came before seems to be equally important (or maybe we cling to favorite ideas and stories as we ourselves age). If for no other reason, the older material provides basic background and historical context from which to proceed to the new.

Whereas all areas of biology have experienced growth, of particular concern to us were the areas of molecular biology and genetics where, as before, incredible progress, particularly in our understanding of the workings of the gene and of its application in genetic engineering, has come about. Look for a very visible overhaul in this part of the text, much of it brand new input from Rob Ferl (With Rob's enthusiastic consensus, we did retain Jack King's charming asides and musings on Darwin, Mendel, and other greats.). New and exciting chapters have been added covering gene regulation, genetic engineering, and mutation. Rob tells about gene splicing and cloning, and other complexities of genetic engineering in a delightfully familiar manner that somehow makes it seem easy.

Other areas of concern to us included immunology, a subject that stymied most biologists in the past, but one whose molecular mysteries are now being revealed one after another. Look for a brand new chapter on immunology, with some of the latest thinking on cell surface receptors, MHC restriction, clonal selection, the primary immune response, and some new, striking illustrations. Some of the available information is so new that it hasn't had time to really gel, so we had to pick and choose among the various possibilities. For some of us (products of the 60's) there is an element of *deja vu* connected with the rapidly emerging field of immunology. Just as the immense scientific attack launched against cancer in the 60's and 70's led to the resolving of so many cell mysteries, the present marshaling of forces against AIDS is leading to almost daily discoveries on the working of the immune system. Of course, AIDS has become a topic of vital interest to all, and as teachers, biologists have a unique opportunity and responsibility to provide the best available information on this new and terrible scourge. We've devoted considerable space to new discussions of the biology of HIV in our virus chapter, and an equal amount to the clinical aspects, epidemiology and social impact of AIDS in our immunology chapter.

We also focused special attention on neurobiology and developmental biology, fast moving areas whose fundamental molecular mechanisms, like those of immunology, have, until recently, evaded our understanding. We were delighted to see new and promising inroads into the molecular and genetics aspects of development, a much needed connection. We've done two things with animal development: we have expanded the descriptive embryology, and we have focused our attention on newly discovered developmental mechanisms. So look for new content, including homeotic genes, homeoboxes, and patterns of development.

You can also look for our updated and reorganized chapters on plant biology, particularly in the area of regulation and response. The animal behavior chapters have been thoroughly overhauled and we have sacrificed some of our favorite stories for a look at new concepts, trends and theories. Finally, our coverage of community ecology has been heavily overhauled and strengthened as has much of the population ecology.

ORGANIZATION

We hasten to say that our overall organization has been retained. We have, however, made certain internal changes, including changes in chapter sequences here and there. In particular you will find such changes in Part Two: "Molecular Biology and Heredity." We've moved our discussions of cell division and Mendelian genetics ahead of molecular biology. And as mentioned above, molecular biology has been expanded from three chapters to six chapters. We've also reorganized our evolution discussions into four chapters, including a brand new introductory chapter that treats some of the history of evolutionary thought. The lively, intimate visit with Charles Darwin and the introduction to evolution in Chapter 1 has been retained. The subject of immunology, as we've mentioned, has been expanded into a full-blown chapter. Within the plant section, we've divided our lengthy plant reproduction and structure chapter into two, thus making the material a little easier to handle. We have also rearranged chapters in Part VI, animal diversity and function. The neurobiology chapters have been moved up to an earlier position - another recommendation by users and reviewers. At the prompting of our advisors, we have also moved some of our discussion on population ecology into the community ecology chapter. In all, however, people that know the text will see that the general progression of topics in the third edition is still quite familiar.

STUDENT LEARNING AIDS

BIOLOGY: THE SCIENCE OF LIFE has always received very favorable comments for its learning aids, and in particular, its comprehensive chapter summaries and thought-provoking chapter questions. We haven't reinvented the mousetrap here. However, we have added to our list of learning aids a number of *floating glossaries*. You will find such glossaries within key chapters, where critical terminology, those terms that will appear again and again in the chapter and text, is first introduced. Thus when the student reads the phrase, "homozygous recessive" or "chemiosmostic phosphorylation", there is no need to go searching for the bold-faced terms. A concise definition is provided in the floating glossary. Of course, the main glossary, one of the most comprehensive in the industry, is available when needed. We have also retained the other appendix items, that is, the lexicon and classification of organisms. Also, secure in the knowledge that they are quite useful and because of an ongoing devotion to the "if it isn't broken, don't fix it" school, we've kept the geological time table and metric conversions within the front and back covers respectively.

A LAST FAREWELL AND INTRODUCING ROB FERL

Two of us, Bob Wallace and Jerry Sanders, have been working together on various textbook projects for over twenty years. Jack King was our co-author on the first two editions of this book, and with Jack's death we lost not only a good friend but a brilliant colleague. Each of us, we felt, brought certain talents and expertise to the project, and Jack's contributions were in the areas of evolution, genetics, and molecular biology. As we began to think about this edition, we knew that it was time to put the past behind us and look for a new co-author. We needed someone with very special knowledge in Jack's areas and yet someone dedicated to teaching. We knew that our lives (both personal and professional) were tied to this book in a very real sense, so it was critical that we find someone who agreed with our approach to writing about biology and, not incedently, someone we liked. We are delighted to have found such a person.

Rob Ferl is Associate Professor of Food and Agricultural Sciences at the University of Florida and has served as Acting Chairman of the Department of Botany, but neither title tells much about his specialties. Rob is currently the Director of the University of Florida's Biotechnology DNA Sequencing Laboratory. He also serves on advisory panels for both NIH and USDA. Rob teaches general biology for both majors and nonmajors and is Director of the Biological Sciences Teaching Program. He is a committed teacher who recently received the Distinguished Teaching Award and was named Teacher of the Year. We are delighted to have Rob join us and we believe these pages will reflect a commitment by all of us to what has been an exciting and challenging project.

ANCILLARIES

Instructor's Manual Newly written for this edition by Jay Templin of Widener University, the manual includes overviews and outlines for each chapter as well as helpful teaching tips, list of resources, and suggestions for coordination with other supplements.

Student Study Guide Also completely new for this edition, this helpful study aid for students is written by Joe Leverich of St. Louis University. It not only includes objectives and outlines of each chapter but review summaries of the key concepts and key terms as well as self-tests to help the students prepare for examinations.

Laboratory Manual The new *HarperCollins Laboratory Manual* was written by Bill Tietjen of Bellarmine College. Although this lab text can be used independently and in various lab structures, it has been carefully coordinated to complement a lecture course using *BIOLOGY: THE SCIENCE OF LIFE*. All labs have been carefully chosen and class tested. Each lab begins with an introduction to the lab topic. Throughout each exercise students will find cautions that alert them to hazards or items of particular importance. In addition boxes that demonstrate mathematical or statistical material are included when needed. Well-developed art is included for each exercise, helping to clarify the experiment.

Laboratory Manual Instructor's Guide This preparation guide includes hints for preparing labs, methods of building equipment, learning objectives, projected times for each exercise, sample problems, help on caring for organisms, supply sources, and transparency masters.

Testbank A written testbank containing 1500 questions is available free to adopters. All questions are multiple choice. The testbank is co-authored by Karen Morse, Judith Goodenough, William Hixon of University of Massachusetts, Amherst, William Guy of Eastern Michigan University, and Alan Cady of Miami University, Ohio.

Biology Encyclopedia Laser Disk The Biology Encyclopedia Laser Disk, produced in conjunction with Nebraska Interactive Video, Inc., offers the latest in visual technology. It contains transparencies, micrographs, slides, and film and video footage. Over 1500 images were provided by Carolina Biological Supply. The laser disk allows instant access to any image or footage, frame by frame or moving, simply by pushing a few buttons on a hand-held remote. The disk enhances the principles of biology covered in the text much more effectively than transparencies or videos.

Acetate Transparencies A comprehensive set of 125 four-color acetates of art and photo micrographs from the text are available free to adopters.

Testmaster The testbank is available to adopters in a computerized form for your IBM or MacIntosh.

ACKNOWLEDGMENTS

We extend our profound thanks to our collegues in our Advisory Board, who are:

Louis Guillette, ***Department of Zoology, University of Florida*** Lou is probably the most dynamic person we know. Anyone losing their enthusiasm for biology need only spend a few minutes listening to Lou Guillette's ideas. Lou freely provided us with rare insights into reproductive physiology and endocrinology. The two heavily revised (and revitalized) chapters are clear evidence of his influence.

Peter Heywood, ***Division of Biology and Medicine, Brown University*** We are immensely grateful for Peter's invaluable aid in all matters pertaining to cells and their functions, but we also owe him for his memorable story-telling (the aftermath of which nearly got us kicked out of more than one restaurant). It is rare to find someone who has, in addition to an in-depth knowledge of his own current research area, such a broad grasp on the whole of biology.

Carmine Lanciani, ***Department of Zoology, University of Florida*** Carmine is one of the most careful, well-reasoned scientists we know. His contributions were in the areas of ecology and population biology, and his gifts were in keeping us on the leading edges of these fields and in being such a good companion when the day's work was done.

Alan J. Neumman, ***Department of Biology, Texas A&M University*** He doesn't call himself one, but Alan is a real "marine biologist." In fact, Alan may be the only officer in the marine Corps Reserve who holds a PhD in botany. We thank him for his many contributions to our microbiology and plant biology chapters.

Jan Pechenik and Linda Eyster, ***Department of Biology, Tufts University*** Jan and Lindy provided us with the most comprehensive and penetrating review of invertebrate zoology that any of us had ever experienced. We thought we knew all about those crusty and soft-bodied creatures until we met this couple.

Kathleen Scott, ***Department of Biological Sciences, Rutgers University*** We thank this no-nonsense person, whom we very much admire and respect, for her sound advice in reorganizing the evolution section and for the generous way she shared her rich background in vertebrate zoology.

OUR THANKS

This edition of *BIOLOGY: THE SCIENCE OF LIFE* has fallen under the stewardship of a number of publishing groups, for a variety of reasons—finally, we're happy to say, finding a home with the people at HarperCollins.

Among those who worked so diligently on the book in its earlier stages are Susan Schapper, Bess Arends, and Liz Rudder. We are indeed grateful for their efforts. Our editor at the time was Bonnie Roesch. Later, the book fell under the direction of Barbara Schneider and her dedicated group at Scott, Foresman. We are very grateful for the way they held things together and pressed forward during times of transition. The final stages of the book were developed under the direction of our new editor, Glyn Davies. It is indeed exciting to make new plans with him and with Editor in Chief Marianne Russell and Publisher Susan Katz. They are an energetic, cheerful, and determined group, and we're looking forward to working with them on future projects.

OUR REVIEWERS

We said earlier that this edition received the most thorough reviews of any text we have produced. It is the input from reviewers that keeps text writers on the "straight and narrow"! We indeed appreciate their criticism as well as their encouragement. We would like to thank our reviewers, not only for their comments, but in many cases for their friendship. Our reviewers include:

Kenneth B. Armitage, *The University of Kansas*
Karl Aufderheide, *Texas A & M University*
William E. Barstow, *University of Georgia*
Gerald Bergtrom, *University of Wisconsin*
James Blahnik, *Lorain County Community College*
Guy Cameron, *University of Houston*
Nina Caris, *Texas A & M University*
Brenda J. Claiborne, *University of Texas*
David Davis, *University of Alabama*
David DeGroote, Ph.D., *St. Cloud State University*
Lee Drickamer, *Southern Illinois University*
Robert Ebert, *Palomar College*
David L. Edens, *West Virginia State College*
David Eldridge, *Baylor University*
Thomas Emmel, *University of Florida*
Sharon Eversman, *Montana State University*
Ross S. Feldberg, *Tufts University*
Michael Filosa, *Erindale College, University of Toronto*
Susan A. Foster, *Mt. Hood Community College*
David Fox, *Columbia University*
Lawrence Friedman, *University of Missouri, St. Louis*
Elizabeth Godrick, *Boston University*
Judith Goodenough, *University of Massachusetts*
Thomas Griffiths, *Illinois Wesleyen University*
Jeffrey A. Hughes, *Ursinus College*
Margaret C. Jefferson, *California State University, Los Angeles*
Mark Kirkpatrick, *The University of Texas, Austin*
Deana T. Klein, *St. Michael's College*
Daniel Koblick, *Illinois Institute of Technology*
Leila Koepp, *Bloomfield College*
Paul Kugrens, *Colorado State University*
Ruth Logan, *Santa Monica College*
John Makemson, *Florida International University*
Thomas McGrath, *Corning Community College*
Thomas E. McQuistion, *Millikan University*
Neil A. Miller, *Memphis State University*
Herbert Monoson, *Bradley University*
Thomas O'Connor, *Washburn University*
F. Scott Orcutt, *University of Akron*
Wiltraud Pfeiffer, *University of California, Davis*

L. Scott Quackenbush, *Florida International University*
David T. Rogers, Jr., *The University of Alabama*
John T. Romeo, *University of South Florida*
David Sadava, *The Claremont Colleges*
Edward Saiff, *Ramapo College*
Walter Sakai, *Santa Monica City College*
Robert F. Slechta, *Boston University*
John Smarrelli, *Loyola University at Chicago*
Gary Smith, *Tarrant County Community College*
Robert H. Smith, *Skyline College*
Aura Star, *Trenton State College*
Kingsley Stern, *California State University, Chico*
Steve Strand, *University of California, Los Angeles*
Gerald Summers, *University of Missouri, Columbia*
C.J. Swanson, *Wayne State University*
Roger Thibault, *Bowling Green State University*
Frank Toman, *Western Kentucky University*
James Traniello, *Boston University*
Benjamin Tremmel, Ph.D., *Benedictine College*
James E. Urban, *Kansas State University*
Randy Webb, *The Pennsylvania State University*
Terry Webster, *University of Connecticut*
Cherie L.R. Wetzel, *City College of San Francisco*
R. Stimson Wilcox, *State University of New York*

CONTENTS

PART 1 MOLECULES TO CELLS 1

* These elements appear at the end of each chapter.

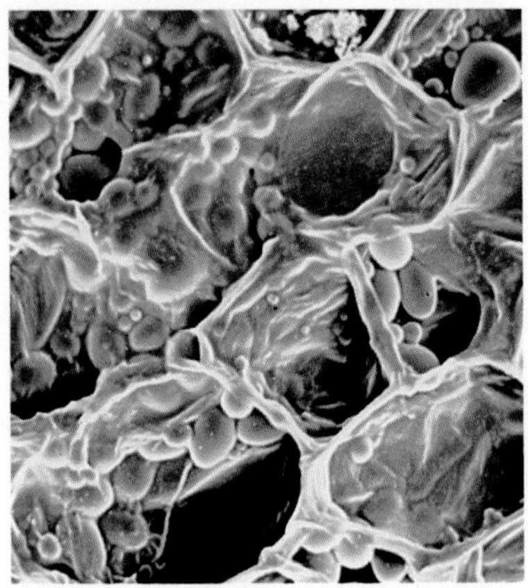

3 THE MOLECULES OF LIFE 50

4 CELL STRUCTURE 76

5 CELL TRANSPORT 110

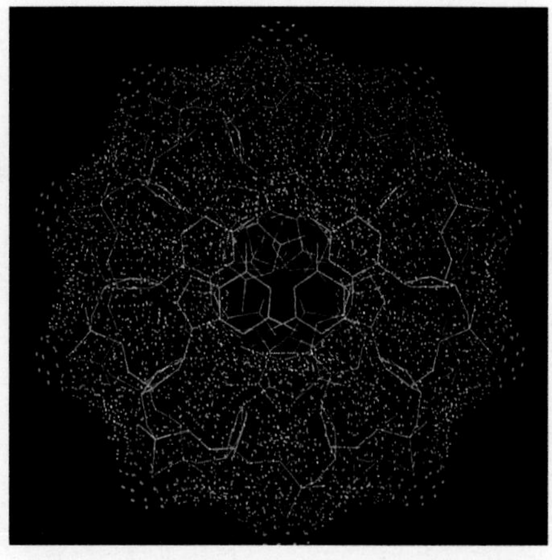

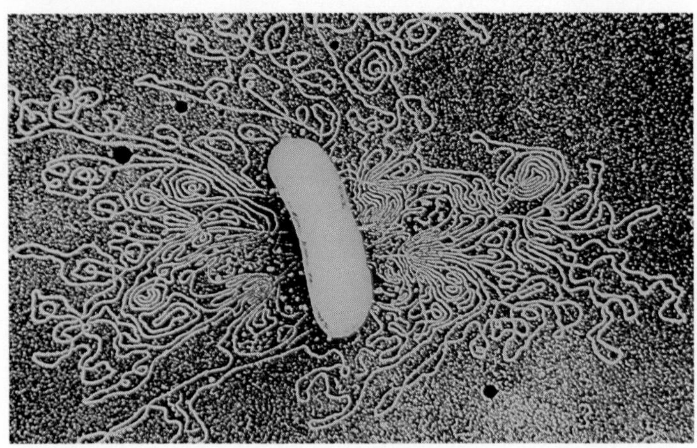

PART 3 EVOLUTION 371

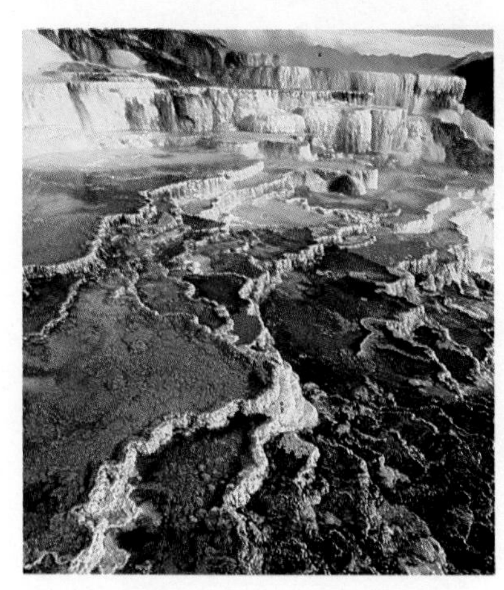

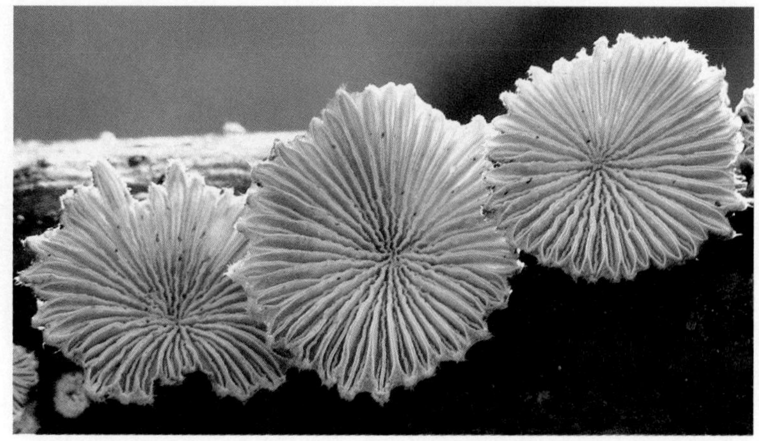

PART 6 DIVERSITY AND FUNCTION: ANIMALS 659

30 INVERTEBRATES I 660

31 INVERTEBRATES II 688

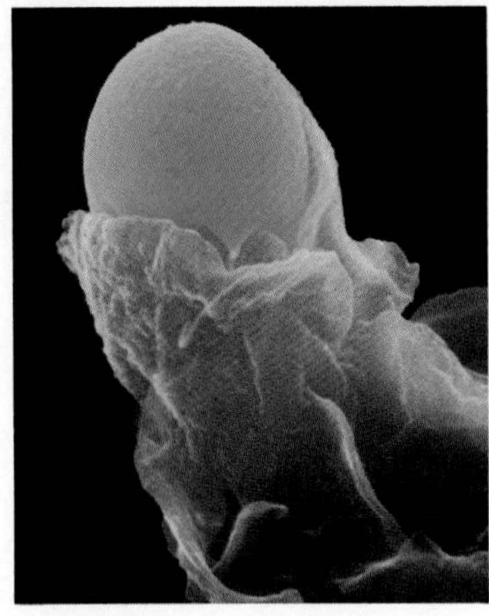

PART 7 BEHAVIOR AND ECOLOGY 1085

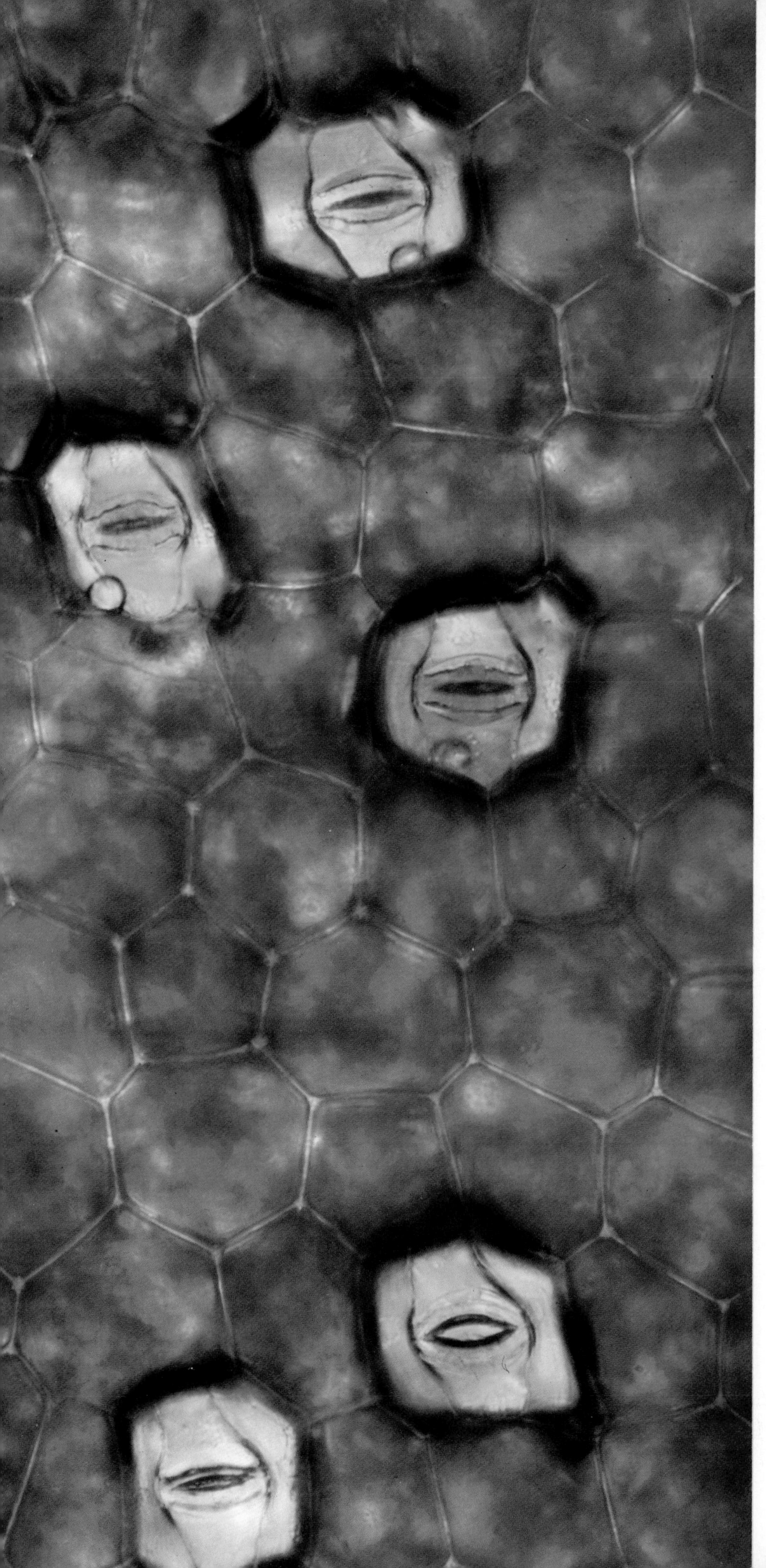

1

MOLECULES TO CELLS

1

MR. DARWIN AND THE MEANING OF LIFE

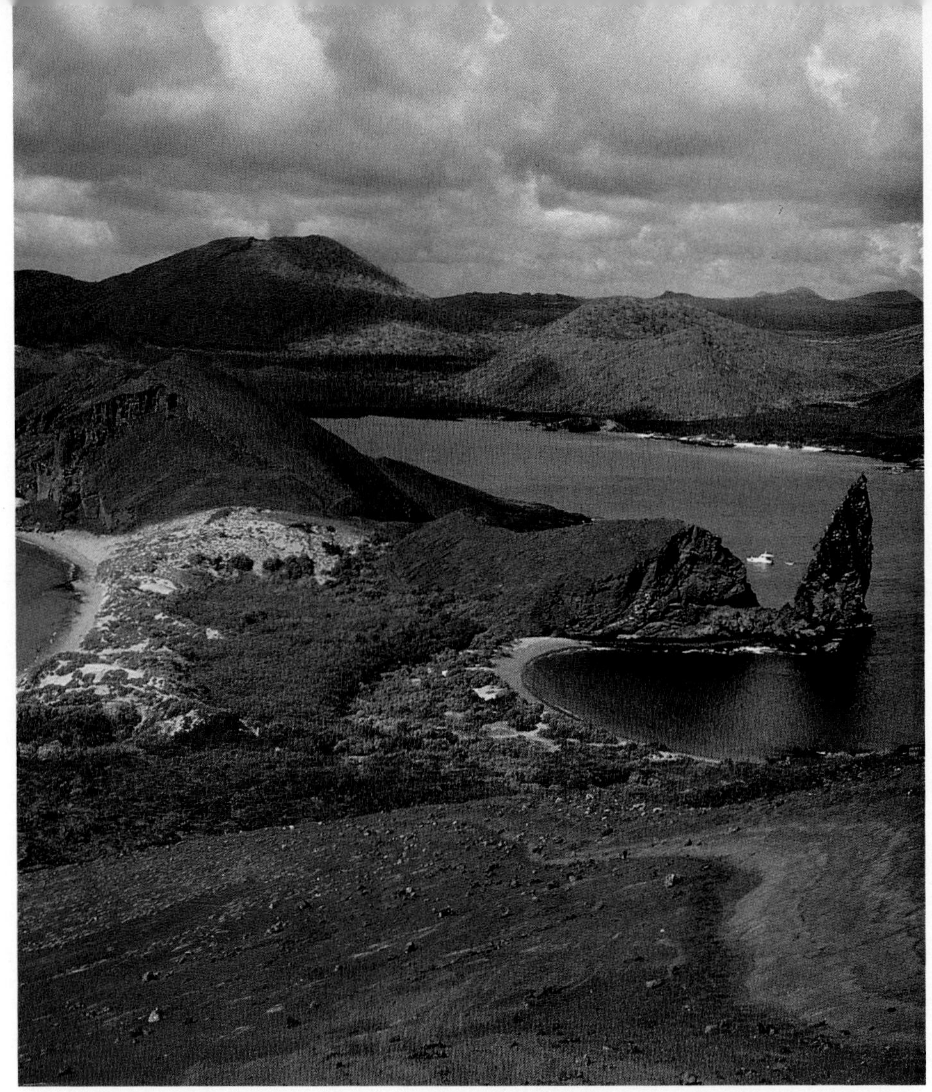

SOMETHING WAS WRONG. HE COULDN'T PUT HIS FINGER ON IT, BUT HE KNEW it just the same. The feeling had nagged at him before, but he had dismissed it. After all, it was hard to be troubled while standing on those grassy hills with the fresh winds of an exotic land brushing across his face.

The good ship *Beagle* lay at anchor off the coast of Argentina, and the young naturalist was 200 miles inland, glad to be ashore, crossing the Argentine pampas on horseback. Life had not always been a joyous adventure for Charles Darwin. Robert, his father, and his grandfather had been among the wealthiest and most famous physicians in England. Charles had been expected to follow in their footsteps and, at the age of 16, was sent off to medical school. But he found that he became ill at the sight of blood, and he nearly fainted upon witnessing his first operation. He saved himself that disgrace only by rushing from the room. At one point Charles was mortified to be told by his father that "You care for nothing but shooting, dogs, and ratcatching, and you will be a disgrace to yourself and all your family."

So Charles tried again—this time law. But he had no aptitude for it and was soon shuttled into training for the clergy. He was duly enrolled in divinity school at Cambridge, where he promptly showed almost as little aptitude for divinity as for medicine. His curriculum included classics, which he loathed, and mathematics, which he couldn't understand. He once wrote a friend about his trouble with mathematics and said, "I stick fast in the mud at the bottom and there I shall remain." Still, his college experience had its pleasant aspects; he could keep up with his insect and rock collecting. At Cambridge he found a friend in one of his teachers, the Reverend John Henslow, who was a botanist as well as a clergyman. Often the two would take long walks in the countryside around Cambridge and discuss the natural history of the area. Darwin was even known by some as "the one who walks with Henslow."

When at last Charles surprised himself and his family by passing his final examinations at Christ's College, he came home shouting, "I'm through! I'm through!" No more school. He was ecstatic. He was now expected to enter the clergy, but he spent his first post-graduate summer happily "geologizing" around the English countryside, away from difficult decisions and away from his family.

When Darwin returned home in the fall he found a letter from Henslow waiting. Henslow had been offered an appointment as naturalist on a British naval survey ship that was to sail around the world. (Actually, the captain desired a companion of his own social standing.) Mrs. Henslow, however, had become so disconsolate at the idea of her husband being gone so long that he had reluctantly refused the offer. He recommended that young Darwin go in his place. Darwin thought it was a great idea.

Unfortunately, Charles' father would have none of it. Darwin was disheartened, but continued to press for permission. Finally his father told him, "If you can find any man of common sense who would advise you to go, I will give my consent." Young Darwin didn't think anyone would give such advice and was prepared to decline the offer once and for all when his uncle Josiah Wedgwood (of pottery fame) said that he thought it would be a splendid thing for his favorite nephew. The two of them together persuaded Dr. Darwin to hold to his word.

On September 5, 1831, Charles was summoned to London to be interviewed by the captain of the *Beagle,* James Fitzroy. Darwin was only a little younger than the 26-year-old captain, who nevertheless had already distinguished himself as a seaman of remarkable abilities. There was an initial awkwardness between them. (Fitzroy thought that the shape of Darwin's nose indicated a weak character.) But soon Fitzroy's doubts about Darwin's nose evaporated, and Darwin was accepted as the *Beagle's* naturalist (Figure 1.1). In fact, Charles shared quarters with the captain himself. There was no salary, and Darwin had to pay for his own room and board throughout the voyage.

Captain Fitzroy's major mission (Figure 1.2) was to chart the waters off South America, but his private mission—his personal passion—was to find evidence that would establish once and for all the literal truth of the biblical account of the creation of the world. For that he needed a naturalist, and this amiable young divinity student who was so hungry for adventure turned out to be just right for the job.

WHERE ARE THE RABBITS?

Now, as Darwin gazed across the lush grassland of South America, he felt that there was something unusual about the place, something not quite right. What was it? Why was he troubled? There was certainly no apparent reason to feel uneasy. The warm breeze gently smoothed the unkept grass, the sky was clear and blue, and his confidence and energy were high. But something was nagging him. What was it? Then it came.

There were no rabbits.

No rabbits. The phrase could be engraved in the consciousness of Western civilization along with $E = mc^2$ and *E Pluribus Unum.* No rabbits; such an innocuous phrase, but in a sense, it signalled the beginnings of a revolution.

Darwin was not thinking about revolution, though. He was thinking about rabbits. Where were they? Darwin knew rabbit country when he saw it, and this place was a rabbit heaven. There was grass for rabbits to eat and bushes to hide in and dirt to dig in—still, there were no rabbits.

Where rabbits should have been, there were other, strange little animals eating grass, digging holes, and hiding in bushes. They had long legs like rabbits, and large ears, and

FIGURE 1.1
CHARLES DARWIN (1809-1882)
This is Darwin at about age 27, shortly after returning from the voyage of the *Beagle.*

FIGURE 1.2
HMS *BEAGLE*
The *Beagle* was one of many British vessels whose primary mission was to chart the oceans and collect oceanographic and biological information. The vessel, just 238 tons in draft, was guided unerringly through a five-year voyage around the world by James Fitzroy, her young captain. Much of the time was spent in South America, where Charles Darwin, the ship's naturalist, made extensive observations and collections.

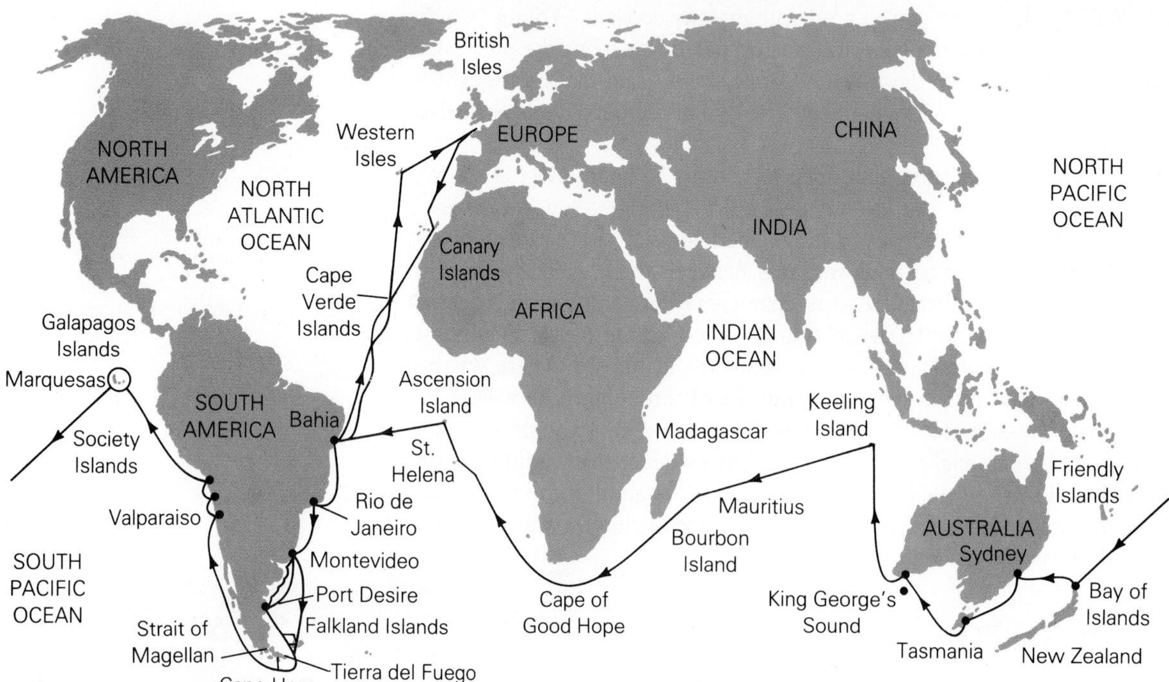

did many things that rabbits did, yet they were clearly not rabbits. They looked more like guinea pigs but in reality, they were Patagonian hares or Mara (see Figure 1.3). But where were the rabbits?

On one level Darwin knew perfectly well why there were no rabbits. There were no rabbits on the Argentine pampas because he was in South America, and rabbits don't live there. That kind of answer usually satisfies almost everyone; however, it is really no answer at all. If the question *Why aren't there any rabbits in South America?* had been pressed, another naturalist might have answered that South America was really not the proper place for rabbits, that the land couldn't support them. But Darwin thought the land *could* support them. He continued to mull over the question and at last a partial answer formed in his mind.

Perhaps there are no rabbits in South America, he thought, *because rabbits can't swim across the Atlantic Ocean.*

That question and its apparently simple answer were to change our perception of the world forever.

An Explanation about Rabbits and Oceans

If Darwin's tentative answer doesn't seem earthshaking, or even particularly interesting, perhaps it is because more than a century and a half have passed since Darwin looked across the lonely pampas. Our view of the world has changed dramatically since that time, largely because of the questions posed during that voyage and the answers Darwin suggested. Of course, rabbits cannot swim across the ocean. But that answer leads to new questions, those to still others, and eventually the whole line of questioning was to produce the logic of evolution. The absence of rabbits on the pampas meant that perhaps animals and plants are where they are, not because they were put there, but because their ancestors either made the journey or they, themselves, originated there.

Darwin also noticed that the rabbitlike rodents of the pampas were very similar to other South American rodents, such as guinea pigs. Why? The conventional wisdom was that South American rodents were similar because that general form of life was well adapted to (or designed for) life in South America. In Darwin's mind, a newer, different answer began to take shape: perhaps South American rodents were similar to each other because they were related.

FIGURE 1.3
THE PATAGONIAN HARE
Although rabbits are not native to South America, their ecological parallel is the Patagonian hare or mara *(Dolichotis patagonum)*. Although the two animals are similar, the mara is a rodent and not at all related to rabbits. The similarity is attributed to convergent evolution—common adaptations of unrelated organisms to similar environments.

It wasn't just the rodents and the rabbits that were stirring Darwin. He had dug up and reconstructed the bones of several extinct mammals, including a gigantic armadillo and some even larger giant ground sloths that were very much like the hippopotamus. Sloths are still found only in South and Central America, but the present-day sloths are small creatures. The extinct giants were clearly similar to–in fact, Darwin had to say, were apparently *related* to—the small burrowing armadillos and the tree-dwelling sloths that were still around. So he had bones of animals that no longer existed and living animals that were so similar that they seemed to be related.

Darwin found evidence that a number of South American mammals had become extinct. He speculated that they were driven to extinction by competition from invading North American species. But why hadn't the South American extinctions occurred earlier? Why had the North American animals taken so long to go south? Darwin thought that perhaps the narrow Isthmus of Panama, which connects the two great continents, had once been under water and that the animals had only advanced southward when a land bridge was formed. This was an excellent guess, and it suggested that conditions on the earth and the relationship of its inhabitants could change.

The Time-Life Connection Darwin had taken with him on the *Beagle* a copy of the newly published first volume of his friend Charles Lyell's revolutionary book, *Principles of Geology*. By the time he had crossed the Atlantic, Darwin was a convert to the new geology. While in South America he received the second volume. Lyell had some rather startling things to say about the physical evolution of the earth. He said that the world was much older than anyone had imagined; that over long periods of time, continents and mountains rose slowly out of the sea; and that they just as slowly subsided again or were washed away. Most importantly, Lyell claimed that the very forces that had so changed the earth in the past were still at work and that the world was still changing.

Darwin's own observations of South American geology seemed to confirm Lyell's position at every hand. In his adventurous climbing of the Andes, he had found fossil clam shells at 10,000 feet. Below them, near an ancient seashore at 8000 feet, he found a petrified pine forest that had clearly once lain beneath the sea because it, too, was interspersed with seashells. In fact, the *Beagle* had arrived in Peru just after a strong local earthquake had destroyed several cities, in some places *raising the ground level by two feet.* The earth had changed and clearly was still changing.

Darwin was excited by his developing idea, but he kept the most revolutionary of his thoughts to himself because he was sometimes uneasy with his ideas and often full of doubts. After all, he had studied for the ministry and had believed in the literal truth of the Bible, but the evidence seemed to contradict the creation account in every detail.

Captain Fitzroy had no such doubts. To him the bones of extinct mammals merely proved the account of the Flood, if one simply allowed that perhaps Noah hadn't been able to round up all the animals. If there were no rabbits in South America, it was because rabbits did not belong in South America. There was a very good reason for everything. Fitzroy believed in laws and rules, and furthermore he knew what the laws and rules were. Darwin, on the other hand, was blessed (or cursed) with an ever-inquiring mind and was always ready to consider an alternative hypothesis. The hypothesis that was forming in his mind now was a dangerous one, and Darwin knew it could lead to trouble.

The Enchanted Isles Darwin's ideas were to be buttressed by his observations during one momentous part of his five-year trip. The *Beagle,* on a dead run from the coast of South America, had reached a peculiar little group of islands, and its anchor clattered into the quiet lee waters of an apparently insignificant island the English called Chatham. Chatham was one of the Galapagos Islands, a recently formed group of volcanic islands that lie astride the equator some 600 miles off the coast of Ecuador. Physically, the islands were quite unlike anything Darwin had seen on the mainland: black, bleak, dry, and hot. And isolated—there were relatively few species of plants and animals. In fact, there were no mammals other than those brought by European ships, and only a few species of birds other than sea birds (see Essay 1.1).

The Galapagos archipelago was, as Darwin would write later, a little world unto itself. Yet Darwin noticed that there was something familiar about the plants and animals. Though most of the species were unique to the islands, Darwin had seen species like them only recently. They were similar to South American species and, as he was to learn, unlike those of Europe, Asia, North America, or Australia. Why should this be?

The usual explanation just wouldn't do. There was nothing in the physical environment to suggest that the islands were somehow appropriate for South American creatures. Indeed, the environment was almost identical to that of the Cape Verde Islands, volcanic islands where the *Beagle* had tied up for nearly a month early in its voyage. The Cape Verde Islands, however, lay off the coast of Africa, and its species were typically African.

Because rabbits can't swim across ocean . . .

Darwin reasoned that it would be difficult, if not impossible, for Asian, Australian, and North American species to cross the Pacific Ocean to settle on these volcanic islands. But perhaps from time to time, a few drifting seeds, a few reptiles on floating logs, and a pair of birds blown off course might have traveled across the 600 miles from the South American mainland. Darwin wondered if, finding a hospitable island free from competition, they could have survived and increased in number.

The Rumblings of Revolution By now Darwin and Fitzroy were engaging in lively, if not heated, discussions about the nature and origins of life. Fitzroy would probably have been chagrined to think he was *helping* Darwin form his "heretical" ideas by providing a sounding board. Like most of his contemporaries, Fitzroy was convinced of the immutability or "fixity" of species, which, in the English vernacular of the nineteenth century, means their inability to change. In his mind species were as they had always been, ever since the Creator placed them here. Darwin probably was hesitant to reveal his true thoughts, for he was beginning to think that life does, in fact, change—that living species are modifications of earlier and different species. Other people, including Darwin's own grandfather, had said the same thing. But where their musings had been unsupported, Darwin was slowly gathering evidence. (Species, as we will see later, refers to any of the millions of unique kinds of organisms, such as humans, red maples, and rainbow trout. Members of a species are similar enough to interbreed successfully—see Chapter 20.)

Darwin saw evidence of changing species, that is, of **evolution**, everywhere. Could his observations prove that evolution is a fact of life on the planet? How are such facts established? For the answer we turn to the inner workings of science.

THE WORKINGS OF SCIENCE

In its broadest sense science might be defined as the way one gets at the truth, or at least as close to the truth as possible. As one prominent scientist put it, "doing one's damnedness with one's mind, no matter what." Nonetheless, even in o[illegible] freewheeling world, there are rules for getting at the truth, and some methods ar[illegible] in greater esteem than others. If we witness a volcano rising from the seabed, we c[illegible] something about how volcanoes are formed. This, then, is a part of how science is done. But simple observation is not enough. Today, the rules are rather formalized for getting one's ideas accepted by the scientific community. Usually, the process involves one or both of two methods, called **inductive reasoning** and **deductive reasoning.**

Inductive Reasoning

Inductive reasoning involves reaching a conclusion based on a number of observations; it moves from the specific to the general. In other words, it begins with observations, and these eventually lead to the formulation of a general statement about what the observations mean.

One of the questions of Darwin's time was, how are coral atolls formed? They often appear as broken circles of islands, formed by the hardened bodies of countless coral animals that make up immense colonies. No one knew just how they came to be, and how the broken circles were formed. The two approaches to the problem involved inductive and deductive reasoning. The inductive approach (Table 1.1) involved a group of statements (observations) that could best be explained by a larger, general statement.

Deductive Reasoning

Unlike the inductive process, deductive reasoning involves drawing specific conclusions from some larger assumption. It leads from the general to the specific, that is, from a broad idea to one or more specific statements. Deductive reasoning leads logically to predictions, which are often described as a form of "if . . . then" reasoning. In Table 1.1 we see that the deductive process begins with a general statement about atolls and, from this, more specific statements are drawn. As you can see, the process is essentially the opposite of inductive reasoning. It is important to understand that neither of these processes, in reality, must exclude the other. The human mind works in complex ways, and in problem-solving, no matter how hard one works to embrace one philosophy, the result is likely to be a product of both.

TABLE 1.1 AN EXAMPLE OF INDUCTIVE AND DEDUCTIVE REASONING

INDUCTIVE REASONING

Observation: Coral atolls usually consist of a circle of islands.
Observation: Coral atolls form from the deposits of living animals.
Observation: Coral animals without direct access to fresh seawater tend to die.
Observation: The interior of an atoll seems to consist of sunken coral.
Generality: Coral atolls are formed as coral animals secrete deposits. The animals in the center, lacking nutrient-laden water, die and sink. This leaves a ring that, in turn, breaks apart to form a circle of islands.

DEDUCTIVE REASONING

Generality: Coral atolls form as coral animals secrete deposits. The animals in the center, lacking nutrient-laden water, die, and sink. This leaves a ring that, in turn, breaks apart to form a circle of islands.
Deduction: Coral atolls will have a sunken center.
Deduction: Coral animals need contact with fresh seawater.
Deduction: Seawater contains something that coral animals need.
Deduction: Coral atolls comprised of more nearly complete rings of land are probably recently formed.

ESSAY 1.1

THE GALAPAGOS ISLANDS

The Galapagos archipelago includes habitats ranging from dry, lowland deserts to wet, species-rich highlands. The islands are home to a bizarre collection of life, from grazing lizards and giant tortoises to a variety of bird life and shore dwellers. Darwin, who despised the place, was to make it famous because he saw it as an experiment of nature in progress.

(*right*) *Opuntia* cactus. The plant is a favorite food of the Galapagos tortoise.

(*above left*) Galapagos tortoise. The species vary according to the island on which they are found.

(*above*) The beautiful Sally Lightfoot crab, named for a dancer because of its quick and graceful movements.

(*far left*) Land iguana, a relatively rare land-bound cousin of the marine iguana.

(*left*) Marine iguana. They graze beneath the sea.

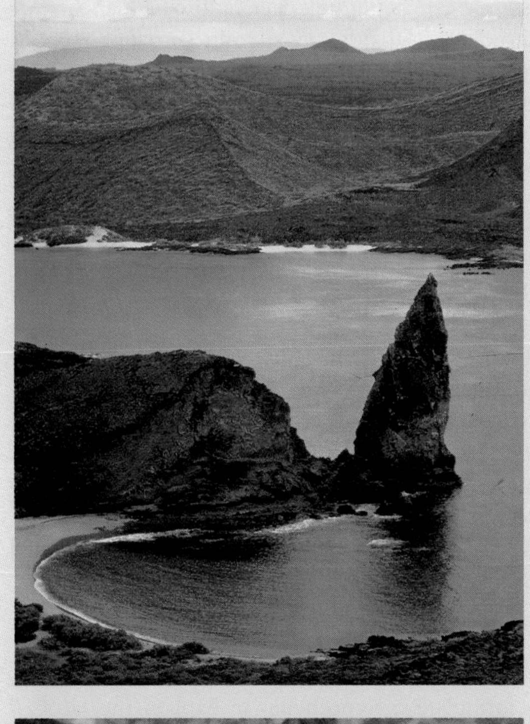

(*left*) Bartolome Island. Sharks are common in these shallow waters.

(*below*) Crown of Thorns, volcanic rocks that were once part of a larger island.

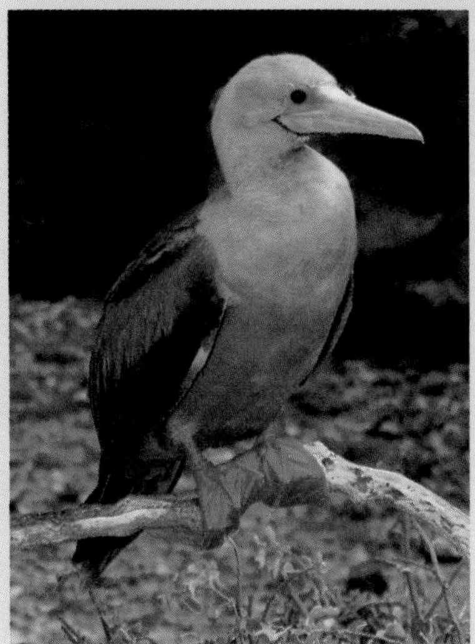

(*above left*) Courting blue-footed boobies.

(*above*) Courting frigate birds. The throat of the male is expanded as a display.

(*middle left*) Waved albatrosses engaged in *billing*, a courtship behavior.

(*right*) Red-footed booby.

Scientific Method

The scientific method has been explained many ways, yet it always eludes precise definition. Essentially, it is the process of establishing new facts and of understanding mechanisms. Though there is no set algorithm—that is, no prescribed set of directions for accomplishing these things, we generally know the scientific process when we see it. The process usually begins and ends with observations about the real world. Between the first observation and the last occurs a fair amount of human mental activity. In a sense, then, science is the interaction of the human mind with the facts of nature. But where does it begin?

There are the initial observations, which are presumably not understood. Darwin, for example, observed that the land birds of the Galapagos are unique species with strong physical similarities to those of mainland South America. But what was he to do with such information?

What happens next is the least known aspect of science, and perhaps the most important. It begins with mulling over the observations and wondering, especially wondering *why*. Almost no one writes down very much at this stage, which may be why it is so little understood. Perhaps most scientists are wondering and mulling over facts most of the time, and most of the time nothing comes of it. Sometimes, though, wondering leads to speculation. Speculations are surmisings about untested ideas. We could no more do without speculation than we could do without facts, but even facts coupled with speculation do not add up to science.

Hypothesis, Theory, and Law

At some stage the investigator must formulate a **hypothesis** to explain the facts. A hypothesis is a possible explanation of some observation. It may have very little evidence to support it, but it usually suggests ways to get at that evidence. The hypothesis represents but one level of scientific activity. At a somewhat loftier level we find the **theory**, which is usually considered to be a stronger explanation of some observation—one that is supported by a considerable amount of evidence.

As more and more evidence comes in that is consistent with a given theory, the explanation may become virtually irrefutable (but not quite) and is called a **law.** Laws enjoy more confidence than theories and, in fact, seem to withstand any kind of testing we can invent. Examples are the law of gravity and the laws of thermodynamics, both universal statements from physics. Biology is notoriously short on laws because life is by nature shifty, elusive, and hard to define. Although biology has its "laws" (such as Mendel's laws and the Hardy-Weinberg law), biological statements wear their titles provisionally and uneasily. Most biological laws are based on mathematical descriptions rather than simple observation.

Hypothesis and Experiment A hypothesis, then, is a tentative explanation of some phenomenon. It can be used to make predictions, which can then be tested. The hypothesis can arrive fully formed, or in very rough outline so that modifications and refinements can be made to fit newer observations. In any case, if a scientific hypothesis is to be useful it must lead to *predictions,* and these must be *testable.*

Testing is usually done through additional observations, although such observations may take place as organized **experiments.** As an example, assume that an investigator has observed certain abnormalities in the growth of bird embryos in an area. He or she suspects that a certain herbicide (plant killing chemical), known to have been used to control plant growth in the area, is responsible. It can never be proved that the herbicide was responsible for past events, but if the idea is tested, the results may support the hypothesis strongly enough to convince people to stop using the herbicide. At any rate, the hypothesis becomes: "Herbicide X can cause abnormalities in bird embryos."

The hypothesis immediately suggests the prediction: "If I administer herbicide X to bird embryos, I should be able to observe abnormalities as they grow." The prediction, as you can see, determines the next observation, which in this case will be an experiment.

In the experiment, the investigator deliberately sets up carefully specified conditions under which certain observations or results can be expected, according to the prediction being tested. In this case, herbicide X will be administered to bird embryos and the effects observed.

Controls The experiment must be set up in such a way that there can be only one explanation for the observations to come; frequently this involves using **controls.** A control is a replica of the experiment in which the special treatment being studied is omitted. In **controlled experiments,** the experimental subjects are arranged in two groups, often labelled "control" and "experimental." Both groups are treated identically in every way except that the special condition under consideration is *not* applied to the control group. The special condition (herbicide X in our example), is often referred to as the **variable,** since it represents a variation from the usual, or normal, condition. The control group may be given an inactive substitute for the variable—a **placebo,** as it is called. Where a chemical is being administered, the placebo may be any inert substance, preferably the solvent used to dissolve the active chemical being tested. Where drugs are being tested on humans, the placebo often turns out to be an innocuous sugar pill. The number of subjects comprising each group, incidentally, is quite important. The greater the number of subjects, the more confidence the experimenter may feel in the results.

The control group, then, represents a standard to which the treated group is compared. Since they are treated identically in every way except for the single variable, any difference in the results can, with confidence, be attributed to that variable. In reality, it is not unusual for unaccountable differences, or "uncontrolled variables," as they are known, to creep into experiments. For that reason, experiments are often repeated after corrective design changes are made. Even the hypothesis itself may need to be revised after the experimenter has had some experience with its testing. For example, as the investigator began the tests of our herbicide X hypothesis, he or she would probably be confronted with the problem of what herbicide concentration to use. Accordingly, the hypothesis may then end up as: "Herbicide X, *in certain environmental concentrations,* can cause abnormalities in bird embryos." Figure 1.4 illustrates some of the problems in our example and how they are addressed through the use of a controlled experiment.

FIGURE 1.4
A CONTROLLED EXPERIMENT
Both experiments are intended to determine the effect, if any, of a certain herbicide on bird development. The first (a), an uncontrolled experiment, involves a number of eggs pierced, treated with the herbicide, and sealed as shown. Note the list of potential errors. The second experiment (b) utilizes three groups that are treated differently and one that is untreated. Suggest reasons for the use of each group and explain how each helps to answer the criticisms listed in (a).

(a) Uncontrolled experiment
Herbicides injected into a number of fertilized eggs. These are incubated and studied at certain stages for undesirable effects.

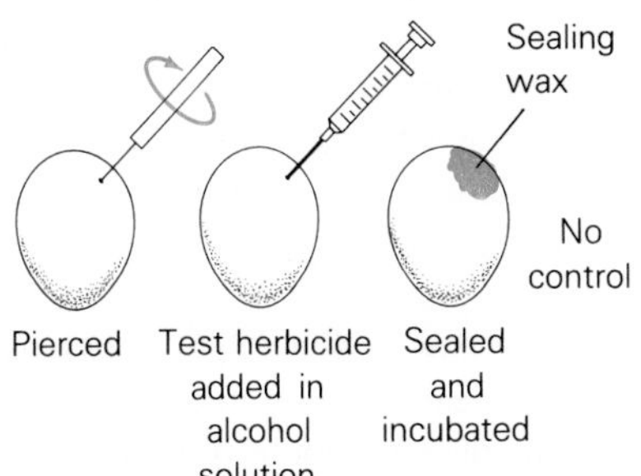

Potential sources of error:
1. Does piercing affect development?
2. Does adding alcohol affect development?
3. Could the result be due to chance?
4. Could there be something wrong with the incubator?

(b) Controlled experiment

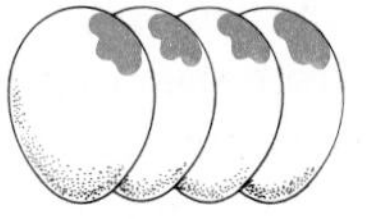

Group 1 Experiment: herbicide in alcohol solution added to a number of eggs

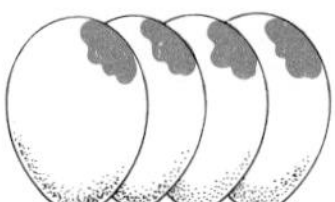

Group 2 Control: alcohol solution added to a number of eggs

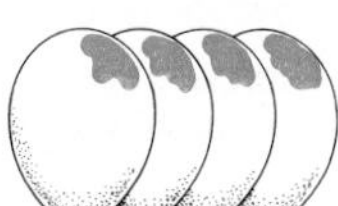

Group 3 Control: a number of eggs pierced but nothing added

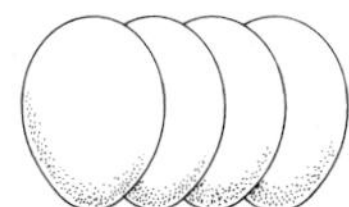

Group 4 Control: a number of eggs untreated but handled in identical ways

All four groups are treated alike in all other ways

Conclusions and Beyond The results of the experiment may or may not support the investigator's predictions. The next step will be to simply accept or reject the hypothesis accordingly. This step, in more formal language, is the investigator's **conclusion.** The conclusion can be tentative or firm, depending on the investigator's confidence in the strength of the evidence. Throughout the entire procedure the investigator must constantly review what others have said and written; someone else may have expressed the same ideas and made the same observations, but tested them and explained them in a different way.

It is important to note that not all experiments are performed in laboratories. There have been many excellent field experiments—that is, those done under natural conditions. An example is a mark-and recapture experiment, in which animals are captured, marked, released, and later recaptured, thus providing information about such things as their movement, growth, or longevity. Hypotheses can also be tested without experiments at all. An ecologist might predict that a certain relationship will be found in nature and then test the prediction by going out into the field to look for it. No matter which method one selects, however, it is still necessary to first make the prediction and then test it.

Finally, publication is an important part of the scientific method. Discovering something and keeping it to oneself is not in keeping with the goals of modern science. As you can see in Essay 1.2, there are rather formal guidelines about how one spreads the word.

In reality, the investigator's behavior is rarely as precise and straightforward as it would appear from reading a published article. For example, we know little about the phase that involves mulling, dreaming, or inspiration. We also know very little about the scientist's own motivation, his or her need to know and tell. The very best scientists are terribly curious, of course, but they also tend to be driving, egocentric, proud enthusiasts who may covet fame, or titles and prizes, or the adulation of students. Some are also of a romantic nature, searching for order and beauty in the natural world. There is also the driving force of the individual's very human need for the acceptance, approval, and admiration of peers.

EVOLUTION: THE THEORY DEVELOPS

Darwin was no exception. Near the conclusion of its voyage, when the *Beagle* arrived at Ascension Island in the South Atlantic Ocean, Darwin found a letter from his sisters waiting for him. In it they mentioned that Adam Sedgwick, an English scientist, had told Darwin's father that on the strength of the letters and specimens that Henslow had already received from the *Beagle,* his son Charles would take a place among the leading scientists. Darwin wrote, "After reading this letter I clambered over the mountains of Ascension with a bounding step and made the volcanic rocks resound under my geological hammer!"

After the Galapagos, the rest of the voyage was somewhat anticlimactic. The *Beagle* made for home as quickly as possible, which in this case meant continuing westward around the world. It called at Tahiti, New Zealand, Tasmania, Australia, Mauritius, South Africa, St. Helena, Ascension, and Brazil. As always when the *Beagle* was under sail, Darwin was wracked with seasickness.

By the time the *Beagle* returned to English waters, it was late in 1836 and Darwin was 27 years old. When the boat tied up, a grateful Darwin leaped ashore and immediately took a carriage home. Since he arrived at a late hour, he decided not to rouse his family and took a room at an inn nearby. The next morning he received a joyous greeting from his family, but he was especially delighted when his dog greeted him and immediately set off down the trail on which they had last enjoyed their morning walk five years before.

Darwin did indeed take his place among the leading scientists. His observations on geology and zoology were published, and he was revealed as a keen observer of nature, undoubtedly one of the greatest natural historians of all time. He was a good storyteller

ESSAY 1.2

HOW SCIENTISTS SPREAD THE NEWS

The process of scientific discovery knows no unbreakable rules and is dominated by intuitive, creative eccentrics "doing one's damnedest with one's mind, no matter what." The conventions of communication among scientists, however, are as formal and intricate as an English country dance. It is not enough just to unravel the universe's little secrets. If new discoveries are to be taken seriously by other scientists, certain rigid criteria must be met. This means, among other things, that the experiments must be done right. Often, as it happens, the discovery comes first, not infrequently by (educated) accident. Still, before the new finding can be published and accepted, the experiment must be repeated and verified according to accepted procedures.

New findings are almost always communicated through a formal **scientific paper** or **journal article.** Scientists may write books, book reviews, and review articles, but these are all based on the all-important research articles. The article may appear in any one of several thousand **scientific journals**—usually one devoted to the narrow specialty of the investigator. The most important new findings may appear in a journal of general interest to all scientists, such as *Science* or *Nature*.

In either case, the article will not be published until it has undergone the scrutiny of the journal's editor and of two or three anonymous volunteer referees. This is one of the extensive safeguards of formal science. The referee system, however, is not without drawbacks. The most important new ideas in science are those that break with established paradigms (world views) to permit fresh, unfamiliar perceptions, but referees may not be ready for fresh, unfamiliar ideas. Many of the most important landmark papers in any scientific field have had to withstand an initial rejection by suspicious referees. Of course, these referees have also prevented the publication of innumerable allegedly grand ideas and supposed paradigm shifts that were in fact, pure hokum.

A standard scientific article consists of six parts: the abstract or summary; introduction; methods and materials; results; discussion; and literature cited. The **Abstract** includes the principal finding, or conclusion, of the experiment being reported. A reader can rapidly skim through a whole pile of journals, reading just the titles and abstracts and delving into the rest of an article only if the abstract seems interesting.

A short **Introduction** reviews any previous relevant work and explains the reasons for proposing the hypothesis that is to be tested. The writing style here, as throughout the article, is usually quite impersonal—maddeningly impersonal to anyone not thoroughly familiar with scientific presentation. One usually learns very little about the investigator's actual thought processes, hunches, dreams, or lucky accidents by reading scientific journals. The experiments themselves are the focus of scientific writing.

The **Methods and materials** section tells exactly how the experiment was conducted. It is written with enough detail and clarity that anyone who is sufficiently interested can repeat the experiment. Repeatability is the only guarantee that the findings are legitimate.

The **Results** section is the key part of most papers. It includes the observations made and experimental data compiled, along with any statistical analysis required to clarify the data. The investigator must accumulate enough data until it becomes extremely unlikely that the results could be due to chance. If the investigator cannot show this, no reputable journal will publish the paper, and other scientists won't take the results seriously.

In the **Discussion** section, the author may be a little less formal and can even indulge in speculation, comparisons, and suggestions for future research. Here ambiguities in the data can be accounted for and potential objections explored, and persuasion and argument are allowed. After all, it doesn't do much good to make a new discovery if you can't convince the rest of the world. To a surprising degree—given the austere formality and pretended impartiality that good scientific manners demand—the ultimate impact of a work of science depends on the skillful presentation of ideas, on the ability to be interesting, and on just plain good writing. Good writing will not make a dull experiment interesting or render an unimportant discovery important; still, our greatest scientists have often been not only great experimentalists and theorists, but great writers as well.

Some papers end with a **summary** that condenses the primary question and the findings. Every scientific paper is sprinkled with numerous parenthetical notes or reference numbers referring the reader to the **Literature Cited,** a list of other journal articles that concludes the presentation. Each item on the list includes the name of the author or authors, year of publication, journal title, volume, and page number, and, sometimes, the title of the cited work. These references alert the reader to other important work in the field, in case he or she wishes to pursue the subject. Just as importantly, they give credit to workers who have made previous contributions. The first person to make a discovery has an eternal claim on all who follow. All humans hunger for recognition, appreciation, approval, and understanding from their peers, and most take pleasure in giving recognition where recognition is due. The formality of scientific citation is a humble acknowledgment that science is a cumulative, cooperative, and, above all, human venture.

as well. His journal, *The Voyage of the Beagle,* became successful popular reading in England and remains a classic today. Of course, he is better known for other writings, particularly those that would seek to explain just why life changes on a changing planet.

A key event in Darwin's voyage aboard the *Beagle* was his visit to the Galapagos Islands. His attention focused on a variety of little finches fluttering about there. These simple, unobtrusive birds were later to give him one of his strongest clues on the evolutionary forces that effect each species.

The Galapagos finches were dark and drab and notably unspectacular. Nevertheless, Darwin collected a number of them to be examined later by specialists in England. Even while on the Galapagos, Darwin had already noticed two things about them: they were all rather similar to species on the South American mainland, and they differed from each other in many critical ways, such as bill size and foraging behavior (Figure 1.5). As Darwin worked out the details of his theory after arriving home, he wondered if the island species could have been somehow modified from an ancestral mainland stock. This idea did not fit with the prevailing idea of how species arose, but it was a cornerstone for Darwin's developing theory. We will come back to the Galapagos later (see Chapter 20).

The Theory of Natural Selection

Beginning in 1837, Darwin started to keep a journal entitled *Transmutation of Species* and was soon making entries referring to "my theory." Reading that journal today, and the journal of the voyage of the *Beagle,* is like reading a detective story after you already know whodunnit—the suspense is in watching the detective sift through clues for the right answer. In his journal entries Darwin began to toy with the idea of **natural selection.** It is fascinating to watch the idea develop in his writings.

Keep in mind that Darwin was a country boy. He knew a lot about agricultural practices, and he knew about livestock breeding. Any good farmer knows that a breed

FIGURE 1.5
THE GALAPAGOS FINCHES
These species all sprang from the same ancestral stock. There are six species of tree foragers (1-8) and eight species of ground foragers (9-14). Their bill sizes and shapes reflect their distinct modes of foraging.

can be improved by *selection,* or, in biological terms, **artificial selection**—selecting the best individuals of each generation for breeding (Figure 1.6). By longstanding folk tradition, the best of the breed (whether cattle, fowl, dog, or cucumber) were honored annually at country fairs and chosen for propagation. The reasoning was simple: like begets like, offspring tend to resemble their parents, and the "best" parents produce the "best" offspring.

But Darwin wondered, could selection operate *without* human intervention? And, if so, how? Agricultural selection involved the conscious choice of the breeder. If selection occurred in nature, who was the selector? This line of thought seemed at first to lead right back to a supernatural factor. Without a conscious selector, it seemed, the inferior individuals were as free to breed as were the most superior, in which case no improvement or change or adaptation would occur.

Did selection have to be conscious, then? Perhaps not. Darwin was impressed with an essay on population that had been written by Reverend Thomas Malthus three decades earlier. Malthus stressed that all species had enormous reproductive capabilities and that their numbers tended to expand rapidly geometrically (2, 4, 8, 16, 32, and so on) unless held in check by starvation or disease. Natural populations, he argued, reached a balance in which all but a few of the young of each generation were forced to perish. *All but a few.* The environment, he thought, could select the individuals that were allowed to breed, and these would be the hardiest. And it all happened through natural means. Darwin's journal shows that by 1838 he had solved the major riddle of evolution. The mechanism of evolution, he said, was *natural selection.* Briefly, natural selection includes (1) overproduction of offspring, (2) natural variation within a population, (3) limited resources and the struggle for survival, and (4) selection by the environment for those with traits that enable the individual to survive and reproduce.

In other words, Darwin noted that living things do not leave as many offspring as they otherwise might because many are killed or their reproduction is curtailed by natural forces. The environment, both the living and nonliving surroundings, determines

(a)

FIGURE 1.6
ARTIFICIAL SELECTION
The results of selective breeding indicate that artificial selection is a powerful mechanism of change in animals and plants. **(a)** The ancestor of the modern horse was probably similar to Przewalski's horse, a hardy little version from the Asian steppes. **(b)** When selectively bred, the wild plant species *Brassica oleracea,* closely resembling the vegetable kale, can be used to produce several common vegetables, each quite different in appearance from the others.

(b)

FIGURE 1.7
SELECTION AT WORK
This splendid male Olympic elk will mate with a number of different females that constitute his personal "harem." He will vigorously attack any other male that approaches, and, being the superb specimen that he is, will usually win. In male elks, then, selection favors physical prowess, and only those males with great strength and fighting ability pass their genes along to the next generation.

(selects) those individuals that are to survive and reproduce. The traits that permit survival, then, will increase in the population as other, less advantageous, traits are weeded out (Figure 1.7).

Darwin's *On the Origin of Species*

Recall that Darwin had, on the east coast of South America, seen fossil remains of extinct species of ground sloths and armadillos and had wondered if they could be related to living species. He had also noticed the similarities of South American rheas and African ostriches, and wondered if they could be related. As Darwin was mulling these ideas over in his mind, he read Malthus's essay, and the grand idea began to come together for him. His theory of evolution by natural selection was developed slowly and methodically, even as he routinely tested each premise in rigorous conversations with a few close scientific allies. He continued gathering his evidence for some 20 years after his return from the voyage of the *Beagle,* yet he made no formal statement, publishing nothing on the topic.

Finally, in 1856, Charles Lyell and the botanist Joseph Hooker persuaded Darwin to set out his argument. Darwin began, almost reluctantly. Keep in mind that Darwin was not shy on matters relating to his science. He was already acknowledged as one of the leading scientists of his day. But for some reason, he was dragging his feet in publishing this theory. In fact, two years later, in 1858, Darwin had written only ten chapters of a work that was to include several volumes.

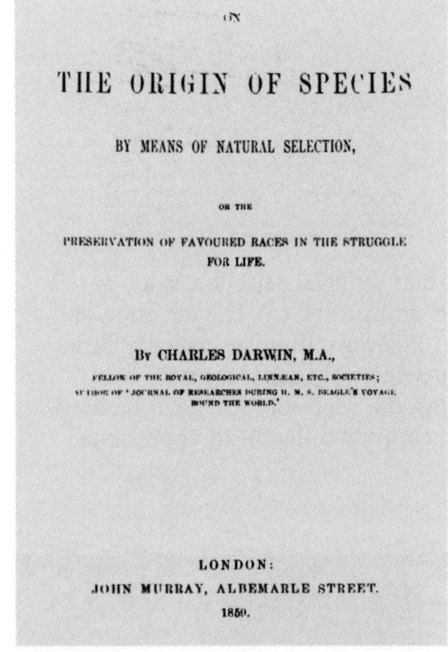

ON

THE ORIGIN OF SPECIES

BY MEANS OF NATURAL SELECTION,

OR THE

PRESERVATION OF FAVOURED RACES IN THE STRUGGLE FOR LIFE.

BY CHARLES DARWIN, M.A.,

FELLOW OF THE ROYAL, GEOLOGICAL, LINNÆAN, ETC., SOCIETIES;
AUTHOR OF 'JOURNAL OF RESEARCHES DURING H. M. S. BEAGLE'S VOYAGE ROUND THE WORLD.'

LONDON:
JOHN MURRAY, ALBEMARLE STREET.
1859.

FIGURE 1.8
DARWIN PUBLISHES HIS THEORY
In November 1859, Charles Darwin laid his ideas open to public scrutiny with the first printing of *On the Origin of Species by Means of Natural Selection.* The book was a sellout! As the printers rushed to restock, the Western world began to marshall forces as the intellectual world polarized into camps that still exist today.

A half a world away, though, things were moving at a different pace. A young biologist named Alfred Russel Wallace was on an expedition to investigate life in the Malay archipelago. He, too, was wondering about the relationship between natural forces and the changing panorama of life. He, too, had read Malthus's essay and wondered how it all fit together. Then, while in bed with fever one night, it all came to him in a burst of insight. He saw that the "fittest would continue the race" after nature had weeded out the rest from among a variable population.

Wallace had corresponded with Darwin before on other matters. So he sent his notes to Darwin, and a shocked Darwin became almost despondent. He was ready to let Wallace accept all the credit for the idea until Lyell and Hooker took it upon themselves to present both papers at a scientific meeting just a few weeks later. Darwin's paper was read first in keeping with his much more substantial evidence, but the whole topic received very little attention that night. A year later, though, the world was alerted to the idea, and when Darwin's *On the Origin of Species* (Figure 1.8) was published, all 1,250 copies sold out the first day.

Testing Evolutionary Hypotheses

Even the brilliant debater Thomas Huxley, Darwin's most pugnacious defender in the 19th century (Figure 1.9) felt that the idea of evolution would remain untested and unproven until someone directly observed the experimental creation of a species. However, Darwin did not believe such a test was possible. He thought that such events simply take too much time. One of the key ingredients in his theory is time—a great deal of time—and a hypothesis cannot legitimately be tested with observations that were made before the hypothesis was formed; thus the theory's power to explain ancient observations did not provide a valid proof. It is not enough to have explanatory power; a hypothesis must have predictive power if it is to be tested and thereby to gain the status of a theory.

Can hypotheses about the past be tested? Yes. Predictive power means only the power to predict the outcome of observations before they are made. For instance, hypotheses and predictions in paleontology can be made and then tested by digging up fossil bones or by analyzing fossil bones that have previously been dug up.

Keep in mind that evolution and natural selection are not the same thing. The theory of evolutionary origins of species by repeated branching is one thing; the theory of natural selection is another, and the latter is exceedingly difficult to test experimentally. Of course, one can simulate natural selection in the laboratory. Darwin himself observed, "To keep up a mixed stock of even such extremely close varieties as the variously coloured sweet-peas, they must be each year harvested separately, and the seed then mixed in due proportions, otherwise the weaker kinds will steadily decrease in numbers and disappear" (from *On the Origin of Species*). An experimental test of natural selection in moths, under more natural conditions, is described in Chapter 18.

The difficulty with the concept of natural selection is not that it predicts too little, but that it explains too much. As soon as it became apparent that natural selection explained the adaptations of organisms to their environment, natural selection began to be used in a lazy way to explain all kinds of biological phenomena. To explain any phenomenon, it began to seem that the speculating biologist need not prove anything new, but needed merely to dream up some halfway plausible way in which the phenomenon might benefit the organism. And if the imagination failed in this, it seemed adequate to state that even if the benefit of the phenomenon to the organism was not obvious, surely there must be one. For instance, if the Indian rhinoceros has one horn and the African rhinoceros has two (Figure 1.10), it might be argued that there must be something about the two environments that makes this arrangement the best one for all concerned.

This extravagant faith in the power of natural selection leads to the benign view that everything is always for the best and there is a reason for everything. Such a viewpoint elevates natural selection to the status of a new, all-powerful deity, and such faith is contradictory to science. It is certainly a contrast to Darwin's own conviction of the role

FIGURE 1.9
DARWIN'S "BULLDOG"
Although Darwin rarely entered into public debate, he was more than adequately represented by Thomas Huxley. Huxley, a powerfully convincing orator who delighted in a hot debate, staunchly defended Darwin's ideas on natural selection.

FIGURE 1.10
NATURAL SELECTION THE ANSWER?
Why does the African rhinoceros have two horns and the Indian rhinoceros have only one? Remember that any explanation would have to be testable. What explanations come to mind? Can they be tested? Is it possible that there is no selective advantage of one horn over two? Or perhaps there is no *longer* a selective advantage. You won't find many biologists willing to say much about questions like this.

of chance and historicity in evolution. Historicity is the notion that things are as they are because of events that occurred in the past. Natural selection is a powerful phenomenon, but it is limited. The adaptations of organisms are marvelous, but they are never perfect, just as evolutionary change is always opportunistic and never predictable.

The Impact of Darwin

Of course, Darwin's ideas were immediately and bitterly controversial. And, unlike the once-controversial ideas of Newton and Einstein, his writings continue to resist resolution. Almost all biologists believe in the central notions of natural selection and evolution, but they still differ greatly among themselves on such questions as what constitutes a species, how species really change, why different species can't mate, and whether most evolutionary change comes in small, continuous increments or in larger and less regular leaps. Darwin, through his works, remains an active participant in the debate.

Darwin went on to publish other major theoretical works and continued his simple but first-rate experimentation. For example, he discovered plant hormones, as we'll see in Chapter 29. *The Expression of the Emotions in Man and Animals* (1872) was the foundation of the modern sciences of ethology and comparative animal behavior (the subject of Chapter 44). *The Formation of Vegetable Mould Through the Action of Worms* (1881) established the importance of earthworms in soil ecology. Because of his work on orchids, climbing plants, and insectivorous plants, modern botanists claim Darwin as one of their own.

But it is *On the Origin of Species* and a related work, *The Descent of Man,* that Darwin's reputation is based, and it is the idea of natural selection that has become the central concept of the science of biology.

There has been a longstanding argument over just how much of an intellectual achievement the theory of natural selection really was. Some biologists and historians of science have maintained, in all seriousness, that the idea is so simple as to be trivial. However, among Darwin's many gifts was an ability to see simple and obvious things that had previously escaped notice and to give them simple and obvious explanations; at least they seem simple and obvious to us, living in a post-Darwinian age. But again, much of science involves clever people pointing out something that, once explained, seems apparent. Our reactions must be somewhat irritating to those involved in developing the explanations. It is interesting that Thomas Huxley, Darwin's contemporary and perhaps his best known supporter, reacted in a blunt and disarming fashion when first presented with the idea. When the *Times* of London sent him a copy of *On the Origin of Species* to review, he is said to have exclaimed to himself, "How extremely stupid not to have thought of that!"

Reducing and Synthesizing

While Darwin was able to focus on such detailed processes as earthworm diggings and snail longevity, he was also a master at seeing the grand scale, the overall picture. His theory of evolution, in fact, could not have been crafted without this ability to generalize and deduce encompassing principles. Today, however, those who seek to test the grand old theory do not take the same approach; rather, they resort to testing fine detail. And so we see that there are different ways of approaching science.

Most of the scientific progress in biology in this century has not been achieved through such grand conceptual breakthroughs as the theory of evolution by means of natural selection, but through what can be called **reductionism.** Reductionism involves the assumption that the properties of biological systems can be understood in terms of physical laws, such as those concerning physics and chemistry. In reductionist science, the questions asked are small ones that can be stated and answered in specific, precisely defined terms. Cause and effect are determined, whenever possible, by eliminating all competing explanations until one is left. Finally, the reductionist seeks to find mechanisms, not reasons, for observed phenomena.

Those dealing with overview and the big picture are called **synthesists.** They seek underlying order in other ways. In general, they seek to show that seemingly unrelated observations can be related after all. Synthesists are the emotional descendants of Darwin, clearly interested in forming grand rules and sweeping generalities. Of course, for science to work well, the reductionist and synthesist approaches should be integrated. The synthesist, after all, is able to generalize because he or she has available so much detailed data produced by the reductionist. Furthermore, the synthesist's vision can be validated only through precise experimentation by the reductionist.

CHARACTERISTICS OF LIFE: THE ULTIMATE REDUCTION

As the reductionists take us spiralling inward to ask the most fundamental questions about life, they are finally confronted with the most basic question of all: What is life? How can we know the living from the nonliving? At many different levels, this is a simple question. It is easy to tell living things from nonliving—until the question falls into the hands of the reductionist. Then it can become most difficult, indeed.

In fact, there are disagreements over the characteristics that mark life, but there are some rather constant features that have been agreed upon, including these six (Essay 1.3):

1. Living things are organized. The world is an unkempt and unruly place that tends to move toward further disorganization. For life to resist this trend it must be orderly and organized.
2. Living things take chemicals and energy from the environment. With these chemicals and energy, they grow and maintain themselves.
3. Living things reproduce. By reproducing, living things pass on a chemical code (usually in the particular arrangements of a molecule called DNA) of their organization to later generations.
4. Living things respond to their environment. Certain characteristics of the environment will be registered and, to some degree, reacted to by living organisms.
5. Living things adapt to the environment. Adaptation involves adjusting in a beneficial way. Individuals can adapt, as when a wolf grows thicker fur in winter. Populations can also adapt, as those individuals in the group with the traits that best fit the environment survive longer and reproduce more. In time, such adaptation by generations of populations results in *evolution*.
6. Living things tend to maintain a balanced, "steady state" (a process called **homeostasis**). Life tends to exist under physical and chemical constancy that it maintains through complex systems of sensitivity and response.

We should not rely too strongly on any such list alone, since there are always exceptions. Further, many nonliving things possess at least one of these traits. Exceptions include such beings as viruses, whose responsiveness is pretty much limited to reproducing when they can and dying when they must. However, under some severe conditions, instead of dying they form crystalline structures. Some people have difficulty believing that anything that could crystallize could be alive. Regarding nonliving (we can't say "dead" because of the implicit assumption that it had once been living) entities mimicking life processes, we see that mineral crystals and oil droplets can grow and seem to reproduce by spontaneously breaking up, rusting iron takes energy from the environment, and many metals are highly ordered. Thus, we must admit that our definitions of life are inadequate and forge ahead. But keep in mind that the inadequacy is not just with our definition. The problem is fundamental. We really don't know much about the basic nature of life (although you may think that seems a bit improbable, considering the size of this book).

BIOLOGY, THE STUDY OF LIFE

Since biology is the study of life, it should be easy to say what biologists do: they study life. That seems simple enough, but the simplicity is deceptive. That's because, as we

ESSAY 1.3

SIGNS OF LIFE

Life is hard to define, but we know that it has certain properties. For example, life takes in and uses energy to retain its own highly organized state. *(right)* Most energy enters the living realm through photosynthesis in green plants.

While all organisms respond to their surroundings, some living things are often extraordinarily responsive to external stimuli. *(below)* The chameleon's *(Chameleo senegalensis)* lightning "tongue-flick" in the presence of a moth clearly supports this assertion.

The aphid seen here *(below left)* is giving "live birth" to an offspring, an attribute of "higher life forms," but the real surprise is that the offspring developed *parthenogenetically*—from an unfertilized egg, just one more twist in diversity. Living things reproduce in an endless variety of ways. *(below right)* The embryos of some terrestrial vertebrates develop in a protective egg, as did this emerging gavial (or gharial, *Gavialis gangeticus*), an endangered species from northern India.

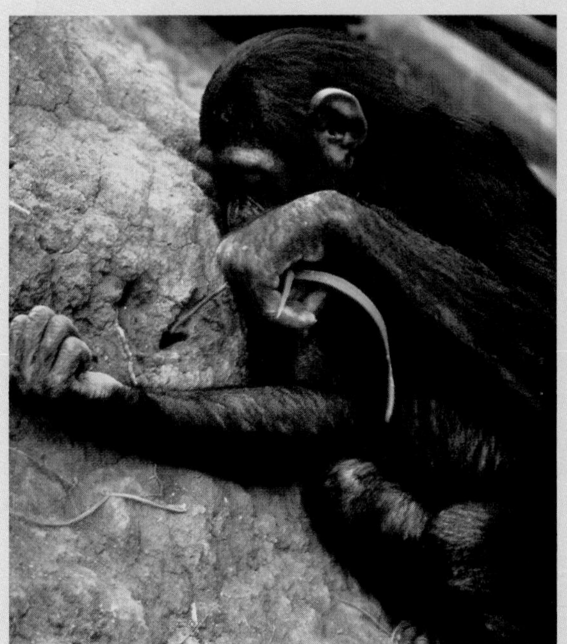

It is important that living things adapt to their environment, and organisms do so in quite surprising ways. *(left)* For instance, chimpanzees are known to make and use simple tools. The chimp carefully strips a slender stem of its leaves and uses it to gently probe into a termite's nest. The stick is carefully withdrawn, and any clinging termites are eaten with relish. *(below)* When not soaring through majestic mountain passes, these bald eagles pass the time freeloading in a garbage dump located on Adak Island in the Aleutians. Such visits occur on a regular basis when the usual food supplies become scarce.

With the physical environment in constant change, living things must also change, or evolve, if they are to remain in harmony with their surroundings. Much of what we understand about evolution comes from the study of fossils, but the fossil record is notoriously incomplete, especially in providing clues to transitional organisms—extinct forms representing steps in the evolution of today's lines. *(right)* The fossil insect seen here *(Sphecomyrma freyi)*, is a rare and exciting find, a transition species representing a stage in the evolution of ants from waspy ancestors. Experts E. O. Wilson and Frank Carpenter state that its head, eyes, abdomen, and sting are typical of wasps, but its thorax (middle body) and waist are decidedly antlike and its antennas contain characteristics of both insects.

Organization is the catchword of life. Even seemingly simple organisms such as the radiolaria *(middle left)* are highly organized. The radiolarian skeleton is essentially of glass, and its sculptured appearance is testimony to how beautifully complex life forms can become.

Life exists within narrow limits, and many forms have devised remarkable ways of regulating their environments. *(below left)* These bumblebees are vigorously fanning the nest, a common practice when temperatures rise.

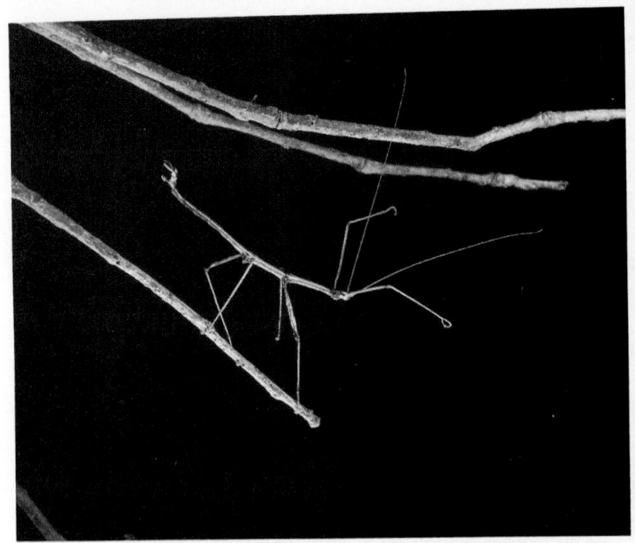

FIGURE 1.11
PRODUCTS OF ADAPTATION
As life adapts to the challenges of the earth, it sometimes produces variations that may seem bizarre.

FIGURE 1.12
HUMAN INQUISITIVENESS
Humans are by their nature inquisitive and have left the marks of this trait in far flung places.

have seen, life itself is notoriously difficult to define. In addition, life is complex—whether one is considering a process, an individual, a population, or a trend. So any study of life, at virtually any level, leads to extreme complexities. Some we understand and some we don't.

Life is also difficult (and fascinating) to study because it is variable and adaptive; it changes in response to the demands of the environment (Figure 1.11). So biologists sometimes find themselves saying something like: "A chameleon is green, except when it isn't." So what can we say about chameleons? Not much about their color at any given time, but life also has broad, adaptive themes, such as its dependence on water and its inheritance encoded in molecules called DNA and RNA. So we can study the general principles of all life, even as some of us are intrigued by the smaller questions, such as a lizard's color. One great advantage is our wealth of simple curiosity, which has historically sent us scattering in countless directions over the planet, probing, peering, counting, measuring, watching, and gathering as we learned more about this place called home. Sometimes we learned, it seems, just by being there. Our quest has been a grand celebration of inquisitiveness, and we have left our marks everywhere (Figure 1.12).

APPLICATION OF IDEAS

1. In explaining scientific and other intellectual achievement, someone once said, "Chance favors the prepared mind." What does this mean to you, and how do you define "prepared"? As you answer, think of the background and experiences of Charles Darwin, and consider how both chance and preparation fit in with what he accomplished.
2. Develop an organizational diagram that illustrates how science might work. Include both the intellectual and technological aspects of science and clearly distinguish these.
3. One important aspect of scientific progress is the rise of new technology. With new inventions and techniques, it is possible to test hypotheses that were heretofore untestable. Yet no technological innovation has replaced the human intellect. Consider the relationship between technology and intellect. How important might an advanced scientific technology have been to Darwin?
4. An antievolutionist derides Darwinian evolution, saying that the notion of new species arising is "unscientific" since such an event has never been observed. How would you counter such an assertion? Are there other, perfectly acceptable, scientific mechanisms that have never been observed?
5. Are humans in modern society subject to natural selection or have we managed to thwart the process as it occurs in other organisms? How might we interfere with the process? Could there be other selective forces at work?

KEY IDEAS

WHERE ARE THE RABBITS?

An Explanation about Rabbits and Oceans

1. Many of Darwin's ideas about evolution arose from observations made during his voyage in the *Beagle*. Many of these were made in South America and neighboring islands.
2. Darwin's observations of life in South America started him wondering about the absence of some species (rabbits) where one would expect them (grasslands). He pondered the meaning of fossils that were different from, yet vaguely similar to living species. He was also challenged by the similarity of different species within a common area.
3. Important to Darwin's growing notions about evolution was the revolutionary work of geologist Charles Lyell. Lyell's theories provided both a substantial time frame and far-reaching geological changes essential to an evolutionary scenario.
4. The prevailing opinion of the 19th century was that species were immutable (unchanging), but Darwin's observations suggested evolutionary descent with change.
5. The isolated Galapagos Islands offered Darwin a veritable laboratory of animals that had originated elsewhere and undergone change as they adapted to conditions there.

THE WORKINGS OF SCIENCE

Scientific reasoning may take two approaches: inductive and deductive reasoning.

Inductive Reasoning

Inductive reasoning proceeds from the specific to the general. Thus, a number of observations lead to a larger, more encompassing statement.

Deductive Reasoning

Deductive reasoning proceeds from the general to the specific. Thus, a more encompassing statement leads to number of deductions or predictions.

Scientific Method

While the **scientific method** can be elusive and difficult to describe, certin elements are standard. Most often, it begins with observations, followed by wondering, mulling over, and speculating about what an observation means.

Hypothesis, Theory, and Law

1. A **hypothesis** is a provisional or tentative explanation of some observed phenomenon. A **theory** is a larger generalization, often based on tested hypotheses but always on a larger body of evidence. A **law** carries even more weight; for most practical purposes, is considered to be irrefutable.
2. To be useful, a hypothesis must lead to predictions and be *testable*. The predictions suggest the test, which is often an experiment, or may simply call for more observations.
3. **Controlled experiments** include both an experimental, or variable, and a controlled element. The **variable** is condition being tested for. The **control** becomes the standard of comparison and is used to prove that any difference between the two is due to the variable being tested and not to chance. The control must be identical to the experimental group to avoid introducing **uncontrolled variables.** In some kinds of experiments, a **placebo** (inactive substitute) is administered to subjects in the control group.
4. The **conclusion** is the investigator's decision about the hypothesis. Based on the experimental or observed results alone, the hypothesis will be accepted or rejected.

EVOLUTION: THE THEORY DEVELOPS

1. Upon the completion of his voyage, Darwin was an immediate success as a scientist. He set to work, mulling over his notes and collections.
2. Darwin was convinced of the idea of evolutionary change, but lacked an explanatory mechanism.

The Theory of Natural Selection

1. A knowledge of livestock breeding and **artificial selection** suggested an evolutionary mechanism to Darwin, but it was the essays of Thomas Malthus that suggested the agent itself. Malthus observed that although more offspring were produced than could survive, natural populations reached a balance.
2. Malthus's ideas suggested that the environment was the agent of selection, and only the fittest individuals lived to reproduce. Darwin called selection by the environment **natural selection.**
3. Natural selection includes (a) overproduction of offspring, (b) natural variation within a population, (c) limited resources and the struggle for survival, and (d) selection by the environment for those with traits that enable the individual to survive and reproduce.

Darwin's *On the Origin of Species*

Darwin's extensive observations, the thinking of Malthus, and Darwin's conversations with other scientists finally came together in a book Darwin began in 1856, setting forth the theory of evolution by means of natural selection. After Alfred Russel Wallace developed the theory independently, Darwin completed and published his book, *On the Origin of Species*.

Testing Evolutionary Hypotheses

1. Evolutionary hypotheses can be tested by first making predictions and then, from the experiments or observations they suggest, accepting or rejecting the hypothesis.
2. While evolution itself can be tested through observation, natural selection is far more difficult to test.

The Impact of Darwin
Darwin's major achievement was his theory of evolution through natural selection, but he also developed many other important ideas in biology.

Reducing and Synthesizing
Science includes two aspects, **reductionism** and **synthesism.** Reductionism involves breaking problems down and investigating the smaller elements. Synthesism consists of putting many smaller ideas together into new grand schemes.

CHARACTERISTICS OF LIFE: THE ULTIMATE REDUCTION

Living things have the ability to

1. maintain a complex organization;
2. take in matter and energy for growth and maintenance;
3. respond to stimuli from the environment;
4. adapt to a changing environment;
5. reproduce their kind;
6. maintain a "steady state" called **homeostasis.**

REVIEW QUESTIONS

1. How did people in the nineteenth century perceive the species on the earth? What was their feeling about fossils that represented extinct life? (6)
2. In what way is the mara like the rabbit? How do biologists explain the similarities? (3-5)
3. List three specific observations that might have been instrumental in Darwin's early thoughts about evolutionary descent. (5, 6)
4. In what ways were Lyell's revolutionary ideas on geology important to the emerging theory of evolution? (5)
5. How does a hypothesis differ from a theory? (10-12)
6. What are the two most important characteristics of a hypothesis? (10)
7. Comment on the scientific validity of the statement, "The experimenter then set out to prove the hypothesis." (12)
8. Suggest how a controlled experiment might be organized to test a new medicine. (11)
9. What is the purpose of an experimental control? (11)
10. Upon what, specifically, must a scientist base his or her conclusions? How are such conclusions usually reached? (12)
11. Was Darwin a careful scientist? Support your answer in detail. (general)
12. List four important elements of the mechanism of natural selection. (14, 15)
13. Why is the explanatory power of natural selection a problem? (17)
14. How can evolution, a gradual process, ever be tested? Be specific. (17)
15. Distinguish between reductionism and synthesism. How is one dependent on the other? (18, 19)
16. List four characteristics of life. Do any of these apply to nonliving entities? Explain. (19)

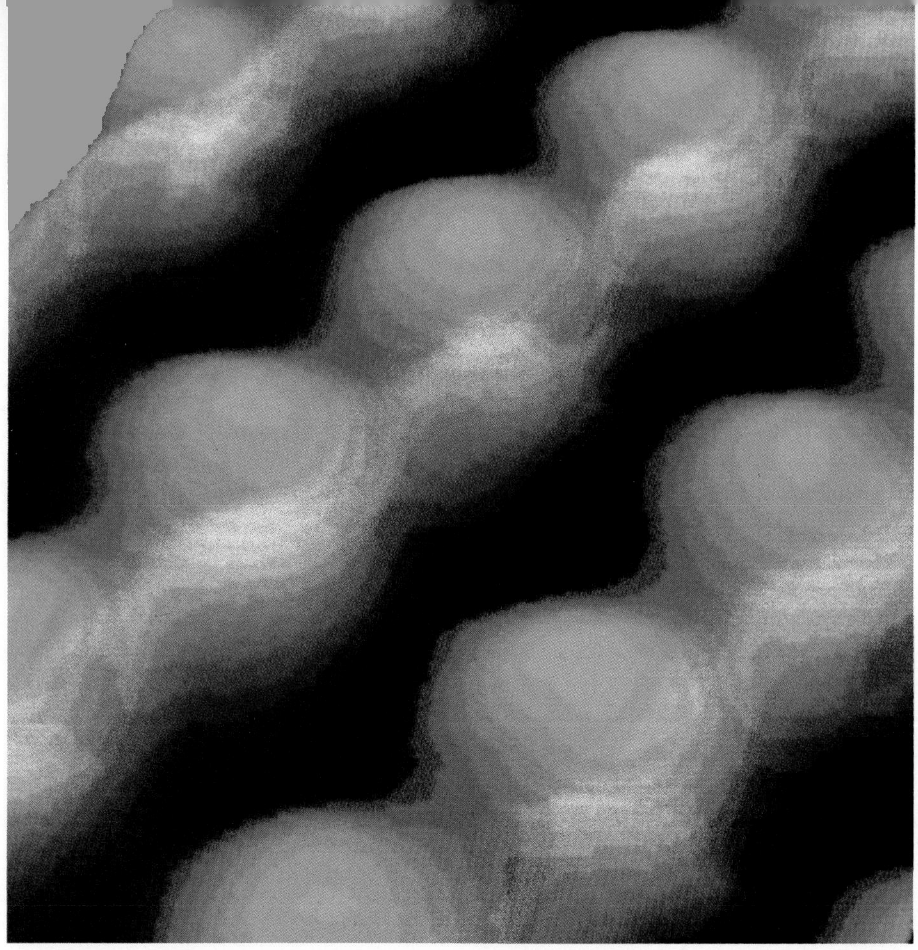

2

SMALL MOLECULES

> In my hunt for the secret of life, I started research in histology. Unsatisfied by the information that cellular morphology could give me about life, I turned to physiology. Finding physiology too complex I took up pharmacology. Still finding the situation too complicated I turned to bacteriology. But bacteria were even too complex, so I descended to the molecular level, studying chemistry and physical chemistry. After twenty years' work, I was led to conclude that to understand life we have to descend to the electronic level, and to the world of wave mechanics. But electrons are just electrons, and have no life at all. Evidently on the way I lost life; it had run out between my fingers.
>
> Albert Szent-Györgi
> *Personal Reminiscences*

SZENT-GYÖRGI'S WRY COMMENT ON HIS SEARCH FOR THE SECRET OF LIFE describes what has been called a reductionist's nightmare. This is because the "meaning of life" cannot be reduced, it seems, to basic understandable processes. It seems that the nature of life will never be grasped as a pure, crystalline gem of truth. If the secret is ever unveiled, it will probably be found somewhere in the very complexity that Szent-Györgi tried to discard. Nonetheless, the reductionist approach is valid; the processes of life depend ultimately on the behavior of lifeless molecules, atoms, and electrons moving mindlessly in space. In order to make sense of what we do know about life, we must know about its components. It is for this reason that we find ourselves immersed, from time to time, in the precise and measured world of biochemistry.

Of course, biology can be "done" without biochemistry. Darwin, for example, had very little knowledge of the subject, and even now, the woods are full of biologists with a love of the outdoors but a hatred of beakers, bases, and balanced equations. However, perhaps their love might even increase if they had a greater appreciation for the complexities of life at the molecular level. It cannot detract from the beauty of a delicately veined leaf to know that it can make food from carbon dioxide and water.

It may be true that some biochemists must periodically be convinced of the existence of the platypus, but only a biochemist can tell us how a fat little hummingbird is able to fly nonstop across the Gulf of Mexico. In this chapter, then, we will briefly enter the biochemist's world. We will begin with some basic information about atoms, inorganic molecules, and small organic molecules. At some point we'll cross the line between the lifeless world of chemicals and the vibrant world of life—but we're not sure where that line is.

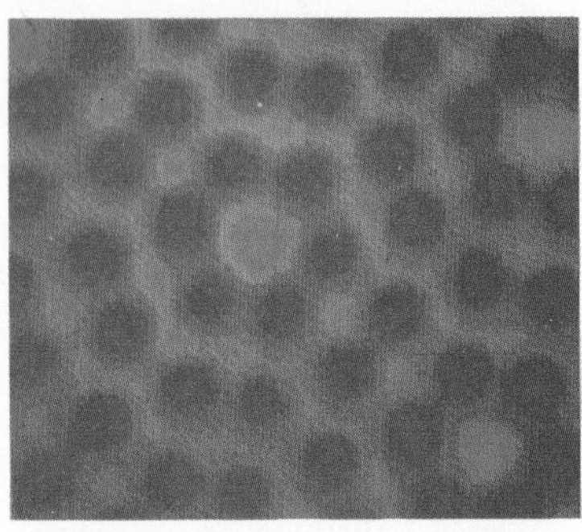

FIGURE 2.1
A LOOK AT ATOMS
Atomic structure is made visible through the scanning tunneling microscope.

ELEMENTS, ATOMS, AND MOLECULES

An **element** *is a substance that cannot be separated into simpler substances by purely chemical means.* To make such separations requires exposing the element to intense energy or bombarding it with high energy particles, treatments that alter the element by bringing about the release of its own subatomic particles and energy. There are 92 naturally occurring elements in the universe. Of these, only six—sulfur, phosphorus, oxygen, nitrogen, carbon, and hydrogen—make up approximately 99% of living matter. SPONCH (from the initial letters of their names) is an acronym that might help you remember these six elements. Of course, these are not the only elements important to life. For instance, plants could not manufacture food without magnesium; and while a pine tree can live with only trace amounts of sodium, you cannot. Sodium is important in the functioning of your nerves and muscles. Table 2.1 lists elements with significant roles in life. Most of the remaining elements are rare in nature and of less interest to biologists.

An **atom** is the *smallest indivisible unit of an element* (Figure 2.1). Atoms of a single type form elements. An atom can be divided into smaller parts, as we'll see shortly, but

TABLE 2.1 ELEMENTS ESSENTIAL TO THE PROCESSES OF LIFE

ELEMENT	% OF SPONCH ATOMS IN HUMANS	SYMBOL	ATOMIC NUMBER	ATOMIC MASS	EXAMPLE OF ROLE IN LIFE
Calcium		Ca	20	40.1	Component of bone; muscle contraction
Carbon	10.50%	C	6	12.0	Constituent (backbone) of organic molecules
Chlorine		Cl	17	35.5	HCl in digestion and photosynthesis
Cobalt		Co	27	58.9	Part of vitamin B_{12}
Copper		Cu	29	63.5	Part of oxygen-carrying pigment of mollusk blood
Fluorine		F	9	19.0	Necessary for normal tooth enamel development
Hydrogen	60.90%	H	1	1.0	Part of water and of all organic molecules
Iodine		I	53	126.9	Part of thyroid hormone
Iron		Fe	26	55.8	Hemoglobin (oxygen-carrying pigment of many animals); cytochromes (electron carriers)
Magnesium		Mg	12	24.3	Part of chlorophyll, the photosynthetic pigment; essential to some enzyme action
Manganese	Mn	25	54.9		Essential to some enzyme action
Molybdenum		Mo	42	95.9	Essential to some enzyme action
Nitrogen	2.47%	N	7	14.0	Constituent of all proteins and nucleic acids
Oxygen	25.60%	O	8	16.0	Molecular oxygen in respiration; constituent of water and nearly all organic molecules
Phosphorus	0.16%	P	15	31.0	Energy transfers
Potassium		K	19	39.1	Generation of nerve inpulses
Selenium		Se	34	79.0	Essential to the workings of many enzymes
Silicon		Si	14	28.1	Diatom shells; glass sponge exoskeleton; arteries
Sodium		Na	11	23.0	Ion balance; nerve conduction
Sulfur	0.06%	S	16	32.1	Constituent of most proteins
Vanadium		V	23	50.9	Oxygen transport in tunicates
Zinc		Zn	30	65.4	Essential to the workings of some enzymes

SPONCH shown in color

if it is, it loses the special properties associated with the element it represents. Each element owes its special characteristics to the specific structure of its intact atoms. In other words, once separated, the subatomic parts no longer represent an element.

A **molecule** *is a unit formed by two or more atoms joined together*. The atoms of a molecule can be of the same kind or of different kinds; that is, they can be formed of the same or different elements. For example, a molecule of the element oxygen consists of two oxygen atoms bound together. The symbol for oxygen is simply O, but the molecule is symbolized by the chemical formula O_2. (The oxygen of the air is molecular oxygen.) Similarly, hydrogen gas consists of molecular hydrogen, H_2. And you are undoubtedly aware that two atoms of hydrogen and one atom of oxygen combine to form one molecule of water, H_2O. (You might want to briefly review here the meaning of the chemical symbols. See below.[1])

Water, by the way, is a compound. And what is a compound? *A* **compound** *is any pure molecular substance in which each molecule contains atoms of two or more different elements in specific proportions.*

Nucleus of hydrogen

1 proton
0 neutrons
1 dalton*

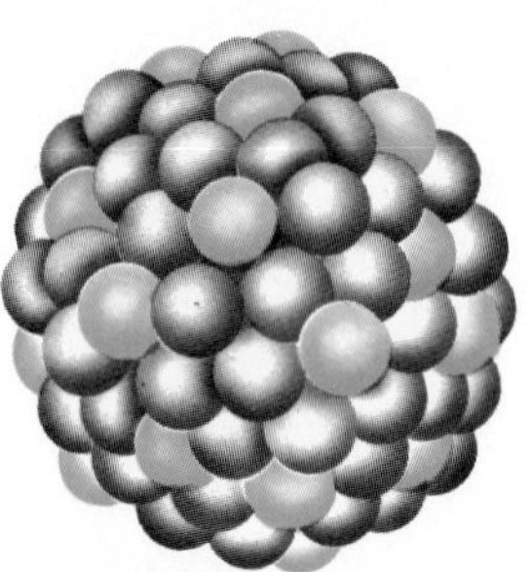

Nucleus of uranium-238

92 protons
146 neutrons
238 daltons

*A dalton is a unit of mass.

FIGURE 2.2
ATOMIC NUCLEI
The comparatively massive nucleus of uranium-238 has 92 protons and 146 neutrons. Hydrogen's nucleus (in the most common form) consists only of a proton.

Atomic Structure

Atoms are made up of about a hundred known kinds of subatomic particles, but most of them are short-lived and play no known role in biology.

The three stable subatomic particles that make up atoms are **neutrons, protons**, and **electrons**, and it is these that will concern us here. Protons and neutrons cluster together to form the **atomic nucleus** (Figure 2.2). Electrons occur outside the nucleus in arrangements we will look into shortly. Neutrons and protons are about equal to each other in mass, and are much heavier than electrons (some 1836 times heavier). Accordingly, we can say that protons and neutrons make up most of the mass of the atom, and indeed, most of the mass of the universe.

The name neutron refers to the neutral electrical state of these particles; that is, neutrons have no electrical charge. However, protons have a positive (+) electrical charge, and electrons have an equally strong negative (−) electrical charge. The electrical charges have important implications for atomic structure. For instance, unlike charges (+ and −) attract each other, and this attraction helps keep the atom together. Where atoms are in their elemental state, that is, where they exist singly, the number of electrons and protons are equal. Thus, the charges cancel each other out.

Whereas unlike charges attract each other, like charges (+ and +, or − and −) exert a strong repelling force. The question arises, then, with all those positively charged protons so close together, why doesn't the nucleus fly apart? We are told by physicists that the nucleus is held together by "strong nuclear forces." These forces at present are poorly understood.

The number of neutrons plus protons determines an atom's **atomic mass** (or **atomic weight,** an older but still common term). Atomic mass is measured in **daltons.**[2] An atom of hydrogen, the lightest element, has a mass of about 1 dalton, while an atom of one of the heaviest elements, uranium-238, weighs just about 238 times as much, and so, has a mass of about 238 daltons (Figure 2.2). And, by the same token, *the* **molecular**

[1]The number appearing *before* a chemical symbol shows how many units there are of whatever follows that number. The small subscript *after* a chemical symbol indicates the number of atoms of the element directly preceding. Thus in H_2O, we find two atoms of hydrogen and one of oxygen, combined to form one molecule of water. The symbol $12H_2O$, then, refers to 12 molecules of water. If a subscript number follows a molecular symbol that is in brackets, it indicates the number of molecules that precedes it and assumes they are part of an even larger molecule. Thus $(CH_2O)_3$ is such a molecule, composed of three CH_2O subunits.

[2]By definition, a dalton is a unit of mass equal to one-twelfth of the mass of an atom of carbon-12, which has six protons and six neutrons in its nucleus. The mass of every other known element is established relative to the mass of this atom.

mass (or molecular weight) *of a molecule is the sum of the atomic masses of the atoms that make up the molecule.* For example, consider water (H_2O):

$$\begin{aligned} \text{Two H} &= 2 \times 1 = 2 \\ \text{One O} &= 1 \times 16 = \underline{16} \\ \text{mass} &= 18 \text{ daltons} \end{aligned}$$

Each element has its own **atomic number,** *which equals the number of protons in the atom.* For instance, the atomic numbers of the six SPONCH elements are 16, 15, 8, 7, 6, and 1, respectively. (Notice that the acronym SPONCH lists the six elements in order of decreasing atomic number.) The number of protons in atoms also equals the number of electrons.

Neutrons, Isotopes, and Biology From what we have just seen, determining the atomic mass of an element seems straightforward—we just add protons and neutrons. Hydrogen has one proton and no neutrons, so its mass is 1. Carbon has six of each, so its mass is 12, and oxygen's, with eight protons and eight neutrons, is 16. But, as it turns out, things aren't quite as neat as we might hope. Atomic mass varies among the atoms within most elements simply because the number of neutrons varies, and such variants are called **isotopes.** More precisely, isotopes are particular forms of an element, differentiated by the number of neutrons in their atomic nuclei.

The nuclei of some isotopes are unstable, or **radioactive.** "Unstable" refers to the decomposition or *decay* of the atomic nucleus, which releases radiation in some form—either an energetic subatomic particle (alpha or beta particles—helium nuclei or electrons, respectively), a highly energetic proton (gamma ray—similar to an X-ray), or some combination of these. In the process of decay, some of the radioactive isotopes or **radioisotopes,** as they are also known, change from one element to another as they lose mass. The new element may or may not be radioactive.

The time required for half of the atoms of any radioactive material to decay is called the isotope's **half-life.** Half-lives can vary considerably. Some of the laboratory-produced radioisotopes have a fleeting half-life of seconds or minutes. On the other hand, most naturally occurring radioisotopes are extremely durable; some have half-lives of billions of years. Uranium-238, for instance, has a half-life of about one billion years, after which half of its atoms would have formed an isotope called lead-206.

Isotopes are important to scientists in a number of ways. The longer-lived ones are often used in dating fossil-bearing samples from the earth's crust, or indeed, the earth itself. Such dating yields important clues to ancient geological and evolutionary events. In medicine, powerful radioisotopes are used in radiation treatment of cancer-ridden tissues. And. of course, scientists are vitally interested in the destructive effects of radiation on all life.

Research biologists also use short-lived, relatively benign, low-energy radioisotopes as **tracers,** to determine where certain chemicals go and how they behave in living cells. For example, the use of radioactive phosphorus, sulfur, and hydrogen has been vital in determining the structure and function of DNA, the gigantic molecules that bear the hereditary information of a species.

ELECTRONS AND THE CHEMICAL CHARACTERISTICS OF ELEMENTS

Biologists are interested primarily in the chemical activity of elements, and this brings them to the behavior of electrons. As we will soon see, all chemical reactions depend on the behavior of these tiny, fast-moving particles. In particular, biologists are concerned with the behavior of electrons on the periphery of the atom, since these are the ones that are most likely to interact with other atoms, yielding molecules. The elements of principal interest are the SPONCH elements and those like them, all of which are comparative lightweights. This is handy for us because the characteristics and behavior of electrons in the lighter elements are fairly simple.

Electron Energy Levels and Shells

Electrons occur in definite spatial arrangements around the atomic nucleus, where they are in constant movement. Each electron, with its negative charge, has a specific amount of energy, and it is this energy that keeps it from being drawn into the positively charged nucleus. Furthermore, the specific amount of energy of an electron affects its arrangement in respect to the nucleus. Electrons with the least amount of energy are closest to the nucleus, while those with more energy are further away. In fact, electrons can be categorized in terms of their energy, with these categories being referred to simply as **energy shells** (Figure 2.3).

Electrons must remain in their shells unless they can somehow gain or lose enough energy to move to another shell. Furthermore, any such change is abrupt or set—there are no electrons at intermediate levels.

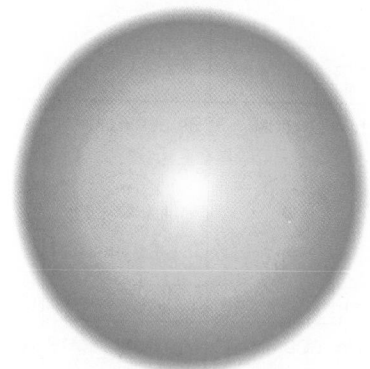

FIGURE 2.3
ENERGY SHELLS
Electrons of varying energy form concentric energy levels or shells. Their energy increases with the distance from the nucleus.

Energy Shells and Orbitals Electrons in each energy shell are precisely arranged, and they move in a highly specific manner. First, most occur in pairs. Second, they move in certain pathways known as **orbitals.** Next, there are specific maximum numbers of electrons in each orbital and specific numbers of orbitals in each energy shell. Let's look at energy shells and electron orbitals, starting with the innermost.

The innermost shell contains one orbital, the **1s orbital,** which contains a maximum of two electrons (a pair). Hydrogen, with its lone electron, is the only element with an unfilled 1*s* orbital. The 1*s* electrons are the innermost in an atom and have the lowest energy. They follow a circular path about the nucleus, but since the plane of the circle is constantly changed, their total movement actually describes a small sphere. The element helium (atomic number 2) has only the 1*s* orbital, filled to capacity with helium's two electrons.

The arrangement of electron orbitals in the second shell also follows strict rules. It will contain no more than eight electrons (although it may contain fewer). The eight electrons occur in four orbitals, each holding no more than one pair. One pair moves in a spherical **2s orbital** (like the 1*s* orbital, but larger), while the other three pairs occur in three dumbbell-shaped **2*p* orbitals.** The 2*p* orbitals are arranged in such a way as to place the electron pairs in each as far from the others as possible. Figure 2.4 illustrates the orbitals of neon, an element that happens to have even numbers of electrons and full outer shells. Note in Figure 2.4d that the axes of the 2*p* orbitals are labelled *x*, *y*, and *z*.

When an element has more than 10 electrons, a third shell must exist, its orbitals resembling those of the second, but designated **3*s*** and **3*p*.** If this is the outer shell, then it too may contain no more than eight electrons. Additional electrons will, as you would expect, fall into a fourth shell, which is as far as we will go. The heaviest element we

FIGURE 2.4
ELECTRON ORBITALS OF NEON
Cloudlike figures represent electron orbitals. **(a)** Neon's two innermost electrons occupy a single spherical 1*s* orbital. **(b)** The next two electrons occupy the 2*s* orbital. **(c)** Each of the 2*p* orbitals is dumbbell-shaped. **(d)** The axes of the three 2*p* orbitals, indicated as *x*, *y*, and *z*, lie as far from one another as possible.

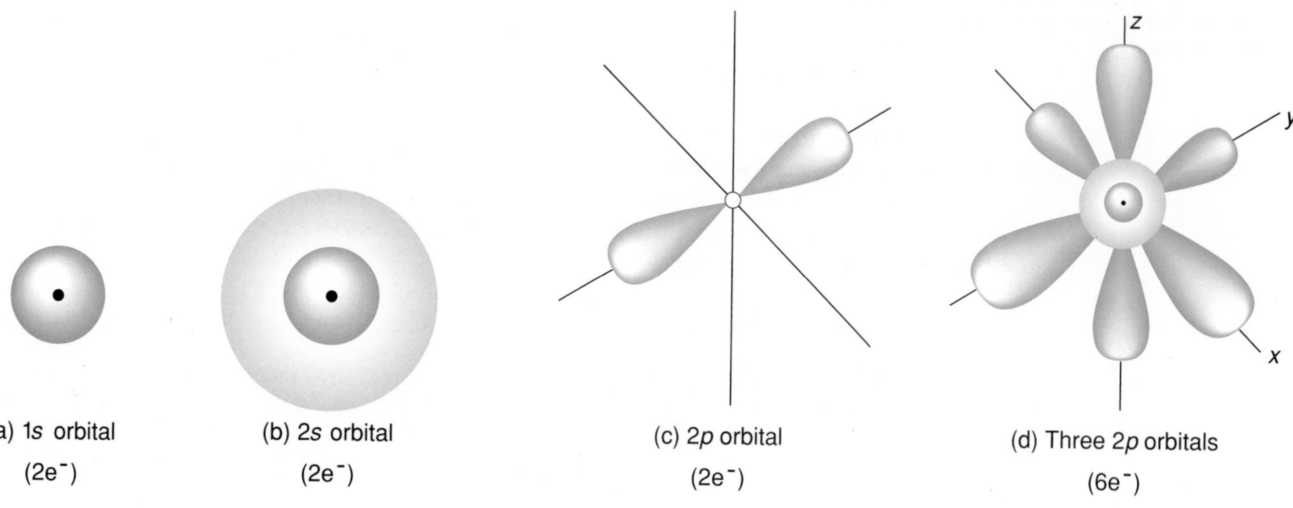

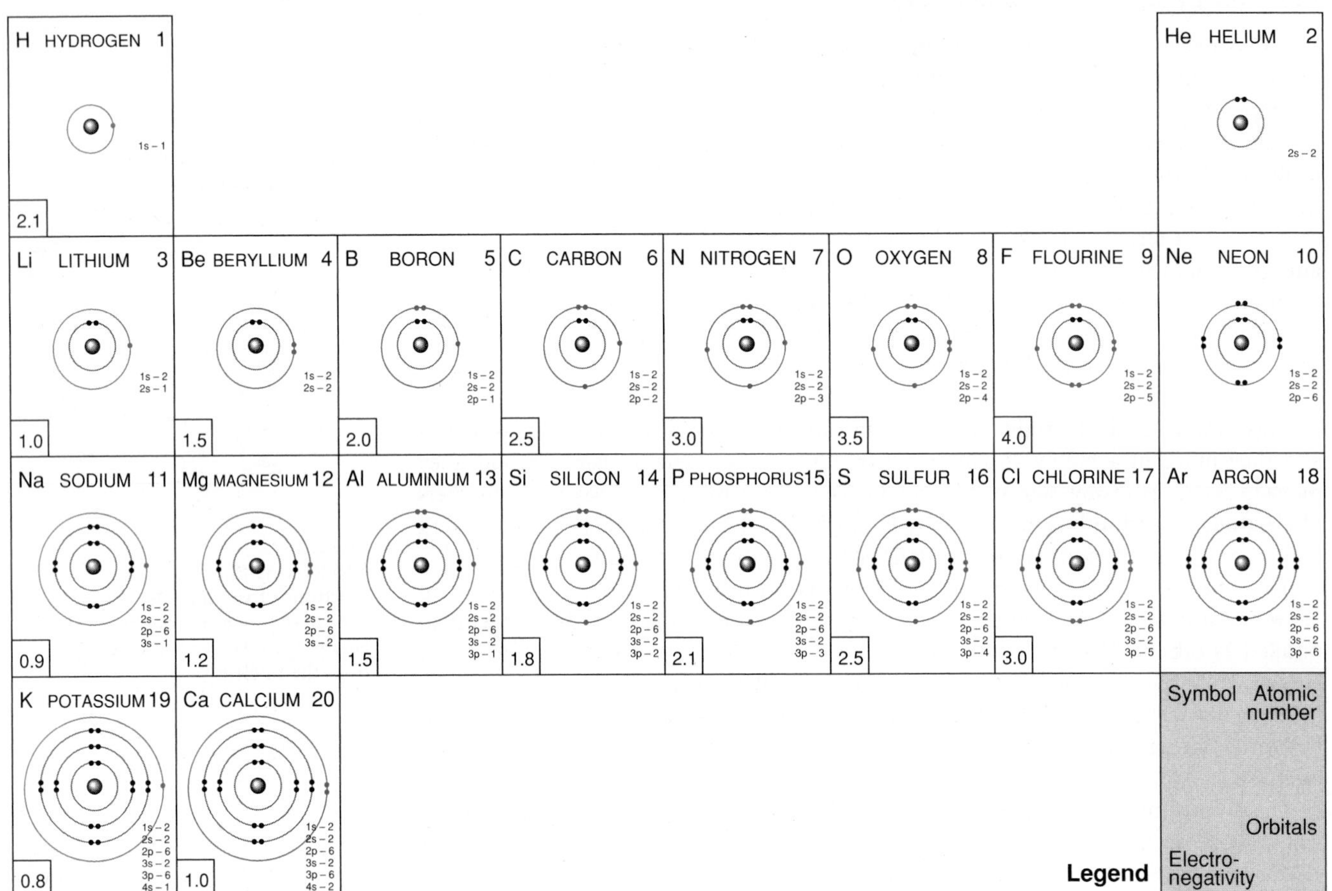

FIGURE 2.5
THE FIRST 20 ELEMENTS
The first 20 elements are arranged in order of increasing atomic number and in vertical columns according to the number of outer shell electrons. As you might expect, elements in a particular column have similar chemical properties. For convenience, the energy shells are shown as flat circles. Each illustration includes the chemical symbol, name of the element, atomic number, electron configuration, and electronegativity (which will be discussed shortly).

will consider here is calcium with 20 electrons, and if our arithmetic is right it should contain just two electrons in its outer shell, both in a **4s orbital**. Figure 2.5 illustrates the orbital arrangement of the first 20 elements.

So atoms have only two electrons in the first shell and, with qualifications,[3] a maximum of eight in each of the next three. Figure 2.6 shows the arrangement for sulfur. Note that the electron energy shells are illustrated in the form of concentric circles. This is an older convention still used as a convenience. (No one wants to draw so many different kinds of orbitals.)

Energetic Tendencies

Electrons behave in precise ways, according to what are called their **energetic tendencies.** To understand these tendencies we should be aware that wherever possible, all matter (molecules, atoms, and their components) follow a universal "downhill" trend, always tending to move towards lower energy states. (Essentially, wherever atoms exist at anything but their lowest state, they will react until that state is reached.) When atoms are at their lowest energy configuration, in order for them to react again, their energy levels must be elevated. With this in mind, let's review three important energetic tendencies in the behavior of atoms.

The first energetic tendency is one referred to earlier, that electrons tend to occur in pairs, a condition that represents their lowest energy state. Since unpaired electrons are

[3]Chemists refer to the eight-electron outer shell maximum as the **octet rule.** The octet rule applies to the second and subsequent shells, but beware of generalities. While we can apply the rule to the 2nd and 3rd shells of the 20 lightest elements, it changes when heavier elements are encountered. To simplify things a bit, think of the numbers 2–8–8 as a way of applying the octet rule to the lighter elements, as we have illustrated for sulfur in Figure 2.6.

not in their lowest energy state, they have the potential to interact with any available electron, forming a new pair. So unpaired electrons will readily pair up with others of any element. As we will see, this is an important trait in the occurrence of chemical reactions.

The second energetic tendency is that elements tend to fill their outer shells. This tendency is naturally met by several elements—for instance those with two, ten, and eighteen electrons. Note the structure of helium, neon, and argon in Figure 2.5. These elements, and a few additional heavier elements, are quite unreactive, and are often referred to as "noble elements." (The idea is that like royalty they remain aloof. The term was obviously coined in an earlier age.)

The noble elements are exceptions, since most elements have unfilled outer shells, and many have unpaired electrons in these shells. Such elements are often extremely reactive. Consider, for example, those elements with "odd" atomic numbers such as hydrogen (1), nitrogen (7), sodium (11), and chlorine (17), each of which has unpaired electrons and unfilled outer shells (see Figure 2.5). Each reacts (forms new chemical combinations) with startling swiftness. Thus the first two energetic tendencies explain much of the ongoing barrage of chemical activity about us.

The third and final energetic tendency is that atoms tend to retain the electrically neutral condition they have as individual atoms (where protons equal electrons in number). But it might occur to you that should either of the first two energetic tendencies, the pairing of unpaired electrons or filling of outer shells, become realized, the balance of charges would be upset. The third energetic tendency, it seems, competes with the first two. Actually, a kind of compromise is possible. It turns out that while the first two tendencies bring about chemical interaction, the third is the product of such events. Let's see what this means.

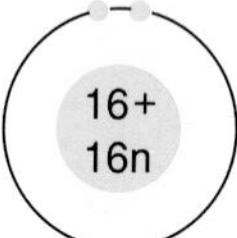

First Shell
2 electrons

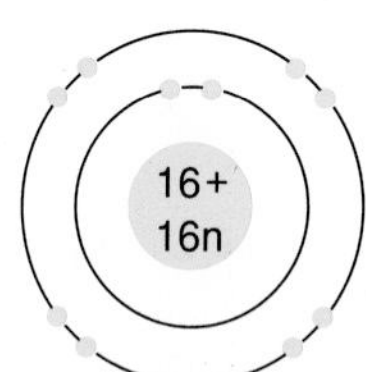

Second Shell
8 electrons

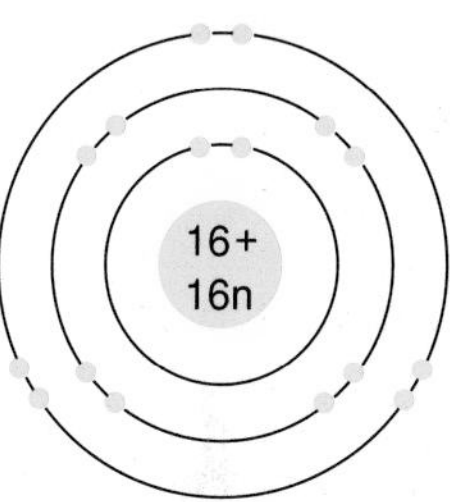

Third Shell
6 electrons

Sulfur: 16 electrons must fill according to the rules

FIGURE 2.6
THE SHELLS OF SULFUR
In this model of the sulfur atom (atomic number 16), the innermost energy shell contains *two* electrons, the next shell, *eight electrons,* and the outer shell, the *six* remaining electrons. Since its outer shell is unfilled, sulfur readily enters into reactions.

Chemical Reactions: Filling the Outer Electron Energy Shell

So if atoms tend to fill their outer shells, just how do they go about it? It turns out that an atom can fill its outer shell in one of only three ways: it can gain one or more electrons from another atom, filling up the hole or holes in its outer shell; it can lose one or more electrons to another atom, stripping its original outer shell bare and leaving a new, full outer shell at a lower energy level; or it can share one or more electron pairs with another atom. Thus, outer shells are filled through the loss, gain, or sharing of electrons.

In each instance, two or more atoms become involved, and through this involvement the atoms engage in what are called **chemical reactions.** Since the loss, gain, or sharing of electrons between or among atoms changes the properties of the elements involved, chemists often define chemical reactions in a more general way. Chemical reactions are chemical changes in which the starting substances—the **reactants**—are chemically changed into new substances—the **products.** Thus, if we choose as reactants two deadly substances—sodium, a silvery metal that reacts violently with water, and chlorine, a heavy, pungent, greenish-yellow gas—these elements will enter a chemical reaction in which the product is sodium chloride, simple table salt. During this chemical reaction, the reactants, you will agree, have undergone a dramatic change in their physical properties.

Whether interacting atoms gain, lose, or share electrons is predictable, and an atom's ability to hold its electrons and to attract more depends on its **electronegativity.** As we see in Figure 2.5, atoms of the different elements differ considerably in this characteristic. The range of electronegativity goes from potassium with a low of 0.8 to fluorine with a high of 4.0. (It may not surprise you to learn that when fluorine and potassium react, fluorine will attract electrons from potassium.)

The difference in electronegativity between interacting elements determines what happens to electrons. As a rule of thumb, if the difference is greater than 2.0, then the atoms of the element with greater electronegativity will completely capture electrons from atoms of an element of lesser electronegativity. If the difference is less than 2.0, then there will be a tendency for interacting atoms to share electrons. (We will see how this sharing is done shortly.) Actually, it's not an either-or situation. Rather, it is a continuum, with the complete loss or gain of electrons at one end and the totally harmonious sharing of electrons at the other. In between there are molecules in which the electron is not evenly shared, but one atom holds the electron an inordinate amount of time.

Where electrons are gained or lost, the formerly electrically balanced atoms take on a net charge. That is, those gaining electrons become negative while those losing electrons become positive (their protons exceed their electrons). Such charged atoms are called **ions.** In situations where sharing occurs but in an unequal manner, the molecule formed has no real net charge on it, but takes on a *polar* nature. That is, the end of the molecule lacking electrons most of the time becomes positive, while the end holding the electrons most of the time becomes negative. (One of the best examples of a polar molecule is water, which, as we will see, owes most of its life-supporting qualities to the very fact that it is highly polar.) Finally, where electrons are equally shared, the molecules are nonpolar. In nonpolar molecules the charges are equally distributed throughout.

The interaction of electrons between and among atoms enables the atoms to form **chemical bonds**—those forces that hold molecules together. Of particular interest to us here are three kinds of chemical bonds: **ionic bonds, covalent bonds,** and **hydrogen bonds**. The first two involve the transfer and sharing of electrons, respectively, whereas the third includes attractions between certain oppositely charged groups in adjacent molecules.

Ions and the Ionic Bond The ionic bond is best demonstrated in the formation of common table salt. Sodium (Na) and chlorine (Cl) differ considerably in electronegativity, with chlorine obviously being the candidate to capture electrons. But, it turns out that energetically, the two elements have complementary needs. As you see in Figure 2.7, sodium has 11 protons. In its elemental form it also has 11 electrons: two in the first shell, eight in the second shell, and only one in the third shell—seven electrons short of a completed outer shell, or one electron too many, depending on how you look at it. And that one extra electron is an energetic, unpaired electron as well. Sodium is only weakly electronegative, so there is no way for it to fill its outer shell—to gain seven electrons—at the expense of chlorine, but if it can get rid of one electron, the already full second shell will become the outer shell. On the other hand, the atom of chlorine, a strongly electronegative element, has 17 protons and 17 electrons in its free atomic state. Its third orbital has seven electrons, which is one short of a full shell—and one of the seven is an unpaired electron. Chlorine, therefore, can fill two of its three energetic needs by accepting one more electron.

Because of their special properties, atomic sodium and atomic chlorine react together very swiftly as an electron passes from sodium, the **electron donor,** to chlorine, the **electron acceptor.** As a result, sodium has only ten electrons to balance its eleven protons, which gives it a net positive charge of +1. But chlorine has 18 electrons as

FIGURE 2.7
IONIC BONDING
When sodium and chlorine atoms meet, sodium's lone outer shell electron is attracted to chlorine. After electron transfer, both atoms have filled outer shells, but since each now has a net charge, they are called ions (Na^+ and Cl^-). Electrostatic attraction between the oppositely charged ions forms an ionic bond, holding the two together.

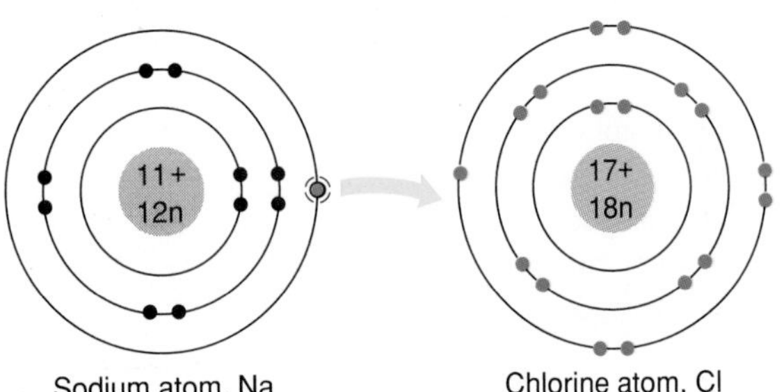

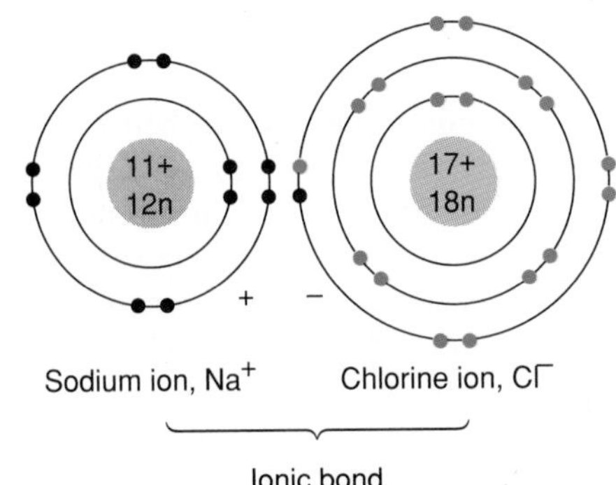

compared to its 17 protons, so it takes on a net negative charge of −1. The opposite charges attract, and sodium joins chlorine to form sodium chloride, common table salt (see Figure 2.7). Note that the name of the negatively charged atom has changed to *chloride* (although sodium is sodium, charged or not). As we mentioned, any charged atom or molecule, such as sodium (Na^+) or chloride (Cl^-), is called an ion.

The electrostatic attraction between the positively charged ion and the negatively charged ion forms the ionic bond. The ionic bond is defined as *a bond formed by electrostatic attraction after the complete transfer of an electron from the donor to the acceptor.*

The ionic bond is responsible for some interesting and complex structures. For example, sodium chloride forms crystals that can range from tiny grains to massive accumulations held together by ionic bonds between individual sodium and chloride ions (Figure 2.8). Since such linkages are quite fragile, if a sodium chloride crystal is placed in water, it will immediately dissociate (separate into its component ions); thus, we say that the salt *dissolves*. The net charge of the resulting *solution* formed remains approximately balanced—you can't get a beaker full of dissolved, positively charged sodium ions without approximately as many negatively charged chloride ions. Incidentally, salt water is an excellent conductor of electricity because of its mobile charged ions.

Before going on to the covalent bond, let's take a moment to consider energetic tendencies and the formation of sodium chloride. What has happened to satisfy the competing demands? The demands are satisfied because

1. in sodium chloride there are no longer unpaired electrons;
2. charges are balanced in the sodium chloride crystal;
3. the outer shells of both sodium and chlorine become filled to capacity.

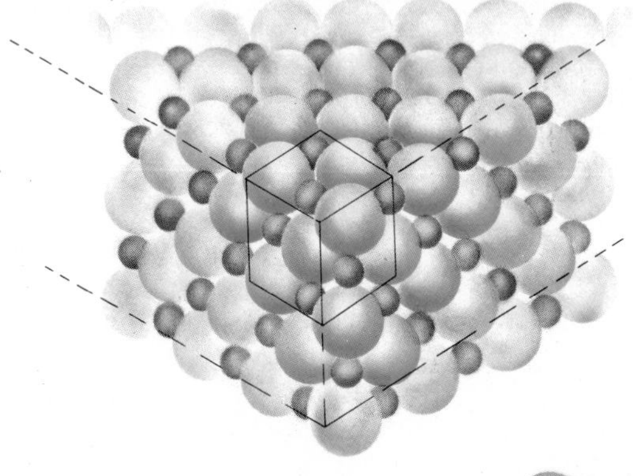

FIGURE 2.8
SALT CRYSTAL
Oppositely charged sodium and chloride ions attract each other in mass, forming a crystalline array of alternating ions. Although the crystals have a very specific geometric form, the potential size range is enormous.

The Covalent Bond: Sharing Electron Pairs The covalent bond forms when one or more pairs of electrons are shared by two atoms. Sharing occurs when the electronegativity of either element is not great enough to allow one element to completely capture electrons from the other. Thus, two such atoms in close proximity can satisfy their shell requirements by simply sharing a pair of electrons. Atoms give up very little by sharing electrons, since such sharing allows them to meet their three energetic tendencies simultaneously. What they do give up, in energetic terms, is the freedom to go their separate ways. The resulting covalent bond holds the atoms together thereby forming a molecule.

Consider the simplest covalent molecule, molecular hydrogen (H_2). It consists of two hydrogen atoms, each comprised of a nucleus with one proton and a single unpaired electron. In forming a molecule, the energetic tendencies of the element are satisfied as follows:

1. The two hydrogen atoms can pool their electrons to make a pair.
2. Charges between protons and electrons in the hydrogen molecule are balanced.
3. By pooling their electrons and then sharing the pair, they simultaneously satisfy the requirement of shell-filling for both atoms (two in the first, or innermost, shell).

Thus we have a molecule of hydrogen, as seen in Figure 2.9. Unlike the ionic bond in sodium chloride, the covalent bond produced in such a way is a rather powerful bond, not easily disrupted. Hydrogen molecules do not dissociate in water.

Since the two hydrogen atoms in the hydrogen molecule share their electrons in an equitable manner, the molecule contains a *nonpolar* covalent bond. Most molecules made up of one element only contain this type of bond. But, since the sharing of electrons often involves different elements with differing electronegativity, the sharing does not always occur on an equal basis. When the sharing is unequal, as in the water molecule, a *polar* covalent bond forms. The main reason water is such an excellent solvent is that it has polarity—a negative end and a positive end that readily interact with other water molecules and with other substances to form solutions.

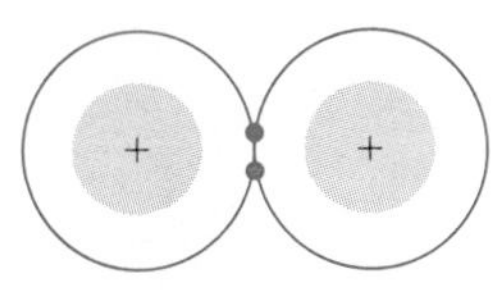

Molecular hydrogen, H_2

Checklist:
Electrons paired? ✓
Charges balanced? ✓
Outer shells filled? ✓

FIGURE 2.9
COVALENT BONDING
In covalent bonding, two or more atoms share electrons. As new molecular orbitals form, a molecule emerges. The simplest example of covalent bonding is seen in the hydrogen molecule (H_2).

CHEMICAL BONDS AND THE SPONCH ELEMENTS

We will now reconsider some of the SPONCH elements and see how they form covalent and ionic bonds. Then we can discuss the hydrogen bond which accounts for some of the amazing characteristics of water.

Carbon and the Covalent Bond

Molecules containing carbon, along with hydrogen and perhaps one or more other elements, are said to be **organic molecules.** There is an enormous range of molecules in this group, and some are incredibly complex. However, we can learn a great deal about the chemistry of carbon compounds through a brief look at *methane,* one of the simplest.

Methane The atomic number of carbon is 6, so free atomic carbon has two electrons in its inner shell and four in its outer shell. How can carbon fill its outer shell and yet retain a balance of charges? Carbon is only weakly electronegative, so it can't capture four electrons. And since it only has six protons in its nucleus to balance the charges, it can't give up its four electrons because a serious imbalance in charges in the opposite direction would result. So, carbon must share electrons to reach stability. Specifically, it forms four covalent bonds. When these four bonds are formed with hydrogen, we have a molecule of methane. **Methane** (CH_4) is a principal component of marsh gas, the waste product of certain primitive bacteria that rot organic material in the depths of oxygen-deficient bogs. More important, methane makes up most of the natural gas that helps fuel the world.

The usual way of writing the formula for methane is CH_4, but this is simply the **chemical formula.** It is often useful to use **structural formulas** to illustrate carbon compounds. Such formulas help indicate the geometric arrangement of atoms and reveal how electrons are shared. The structural formula of methane is written as:

```
     H
     |
 H — C — H
     |
     H
```

The single lines indicate single covalent bonds, but remember that each covalent bond involves two shared electrons. The electron shell diagram is shown in Figure 2.10.

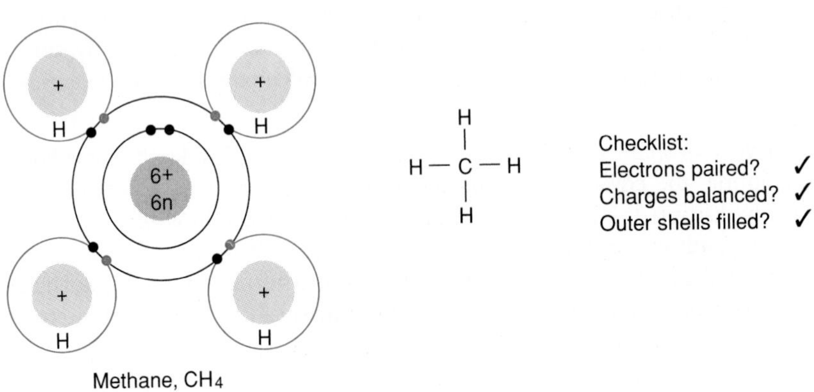

FIGURE 2.10
COVALENT BONDING IN METHANE
In the formation of methane (CH_4), the carbon atom's four outer shell electrons pair up with those of four hydrogen atoms, giving rise to four covalent bonds. As in many hydrocarbons, the electrons are equally shared. Thus, methane is nonpolar.

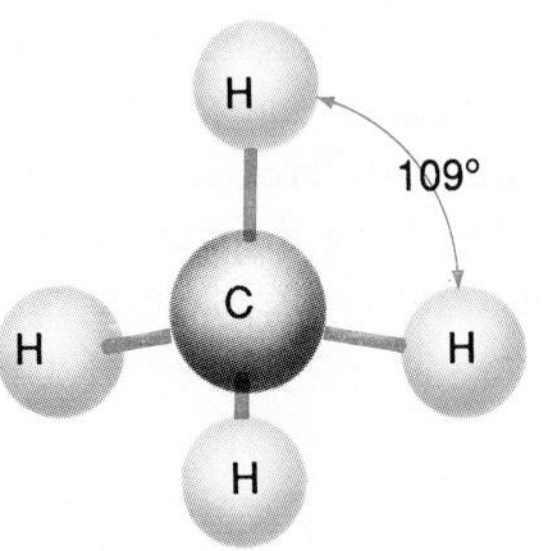

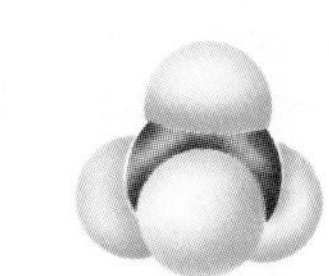
(b) Space-filling model

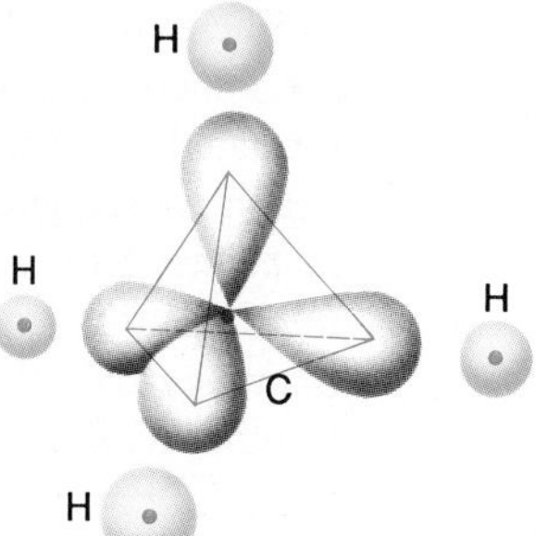

(c) Bond angles of carbon

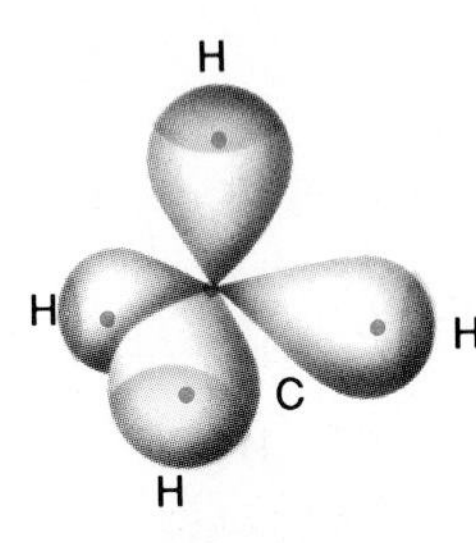

(d) Orbitals

Actually, the electron orbitals of the second shell don't lie in a plane, but in three dimensions. If imaginary lines are drawn from the carbon to each of the four hydrogens, the angle between any two of them—the "bond angle"—is slightly more than 109°, and the figure drawn is a tetrahedron.

Tetrahedrons are not easy to illustrate, but Figure 2.11 includes an attempt. (Tetrahedrons bring similarly shaped "caltrops" to mind. These three-pointed spikes were thrown in the path of advancing armies in ancient times, and in more recent conflicts, were scattered by low-flying aircraft on airstrips used by enemy fighters.) Note that there are several different "models" or ways of representing the methane molecule in three dimensions: the electron cloud or molecular-orbital model, the ball-and-stick model, and the space-filling model.

Carbon can join with itself and with other elements in enormously varied ways as it forms molecules. Because of such versatility, it will become apparent that the chemistry of life is virtually the chemistry of carbon.

(e) Caltrop

FIGURE 2.11
FOUR MODELS OF METHANE
Four representations of methane include **(a)** a ball-and-stick model, useful for visualizing bond angles, **(b)** a more realistic space-filling model, **(c)** the geometric tetrahedron formed by the bond angles, and **(d)** a model that emphasizes molecular orbitals, **(e)** the caltrop.

More on the Magic of Carbon The biochemical magic of carbon, so essential to life, lies in the four electrons of its outer electron shell. As we saw with methane, a carbon atom can form four single covalent bonds. It can also form double bonds, as it does in carbon dioxide. Carbon dioxide contains two sets of **double bonds** (each with two pairs of shared electrons) with the carbon in the middle, double-bonded to an oxygen atom on each side: O ═ C ═ O. (See Essay 2.1 for more on CO_2). Carbon can even form *triple* bonds as in acetylene gas: H—C≡C—H; and in hydrogen cyanide, H—C≡N. Most importantly, carbon can form bonds with other carbon atoms, and the molecules formed can take the form of long chains, rings, and other complex structures. There seems to be no limit to how large a carbon-based molecule can be. A single giant DNA molecule, as we will see later on, can contain up to 50 billion atoms. These giants are aptly called **macromolecules** (*macro-*, "large").

Carbon Backbones Carbon atoms in long chains are often referred to as **carbon backbones** because they support various side groups that extend from the chains like ribs from a backbone. Such side groups include atoms of hydrogen, oxygen, nitrogen, phosphorus, sulfur, and others, forming mixed backbones of chains or rings.

Some of the simpler backbones include the **hydrocarbons.** A hydrocarbon is a compound that consists solely of carbon and hydrogen. We have already looked at the simplest hydrocarbon—methane. Propane and butane are three- and four-carbon chains, respectively. Both are familiar fuels, distilled from petroleum. Butane is seen in Figure 2.12.

Note that the ball-and-stick model of butane shows that the carbon backbone is not really straight. The single covalent bonds operate like little swivels, in which the carbon atoms can rotate freely about the axis of the chain.

The carbon-carbon linkages just mentioned are single covalent bonds, but carbon-carbon linkages can also be double or triple bonds (if two adjacent carbons share four or six electrons, respectively). Double and triple bonds don't swivel, and thus these molecules are more rigid than molecules with only single bonds.

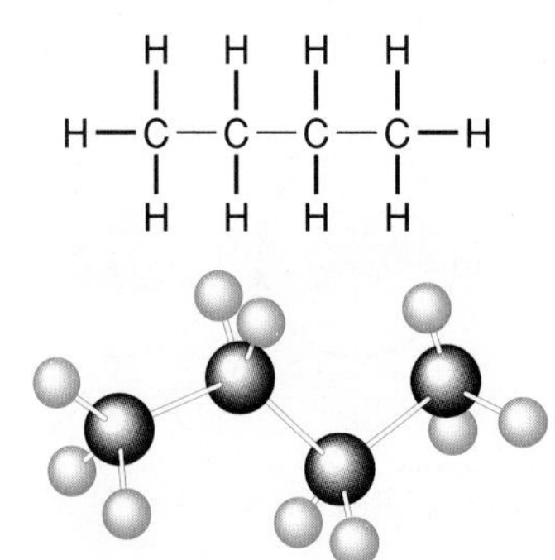

FIGURE 2.12
THE HYDROCARBON BUTANE
Whereas the structural formula of butane (C_4H_{10}) is visually useful, the ball-and-stick model reveals the actual bond angles between carbons.

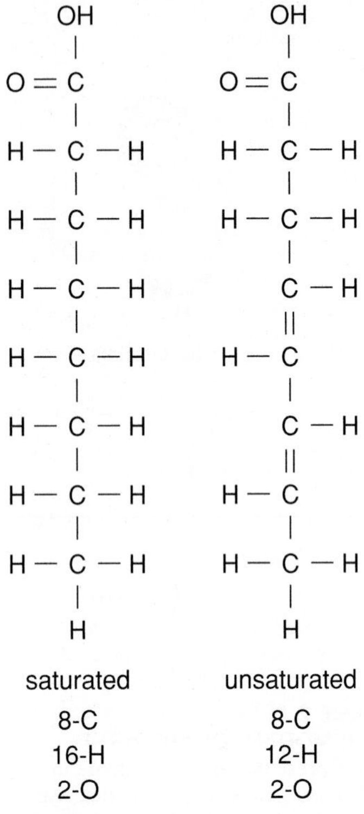

FIGURE 2.13
SATURATED AND UNSATURATED HYDROCARBONS
Saturated and unsaturated fatty acids differ in seemingly minor ways, but such differences are biologically important. Note the angles produced by double bonds in the hydrocarbon tail.

There may be significant biological differences between molecules in which carbons are joined by single bonds or multiple bonds. Consider the dietary implications of saturated fats versus unsaturated fats for example. We are constantly warned of the health consequences of a diet heavy in saturated fats. The chemical difference, as seen in Figure 2.13, is that the fatty acid components of saturated fats have the maximum number of hydrogens possible, whereas the fatty acids of unsaturated fats have fewer than the maximum possible. The terms "saturated" and "unsaturated" apply equally well to other hydrocarbons.

Nitrogen

The atomic number of nitrogen is 7, so, allowing for two electrons in its inner shell, its outer shell has five proton-balancing electrons. Three more electrons are needed to fill the outer shell. Nitrogen usually forms three covalent bonds, as in molecular nitrogen (N_2) and in ammonia (NH_3). In some instances, however, nitrogen forms four covalent bonds, as it does in the ammonium ion (NH_4^+). Four-bonded *(quaternary)* nitrogen always carries a positive charge.

Nitrogen is an essential component not only of proteins, but also of molecules such as the nucleic acids (DNA and RNA), some carbohydrates, some lipids (fats), and a number of other biological molecules. We will return to these classes of compounds in Chapter 3, but at this point, let's learn a few basic points about nitrogen. This is a peculiar element because it is so common (78% of air is molecular nitrogen or N_2), yet so difficult for organisms to obtain in usable forms. This difficulty can present problems. Not only does a lack of suitable nitrogen compounds cause stunted growth in your garden plants, but lack of suitable organic nitrogen molecules results in severe malnutrition in people. The reason for this peculiar shortage in the midst of plenty is that N_2 is very, very stable and tends not to react with other atoms or molecules under most conditions. Let's see why this is.

The nitrogen-nitrogen linkage is a triple covalent bond ($N \equiv N$), involving the mutual sharing of six electrons. It takes a lot of energy, such as lightning, or some very special chemical capability in cells, to break this bond. Therefore, most organisms cannot use molecular nitrogen at all and must depend on other forms of the element such as the soluble ammonium (NH_4^+) or nitrate (NO_3^-) ions (Figure 2.14). However, a few species of bacteria have the chemical capability to *fix* nitrogen from atmospheric N_2, that is, to convert atmospheric nitrogen into forms that can be used in living systems.

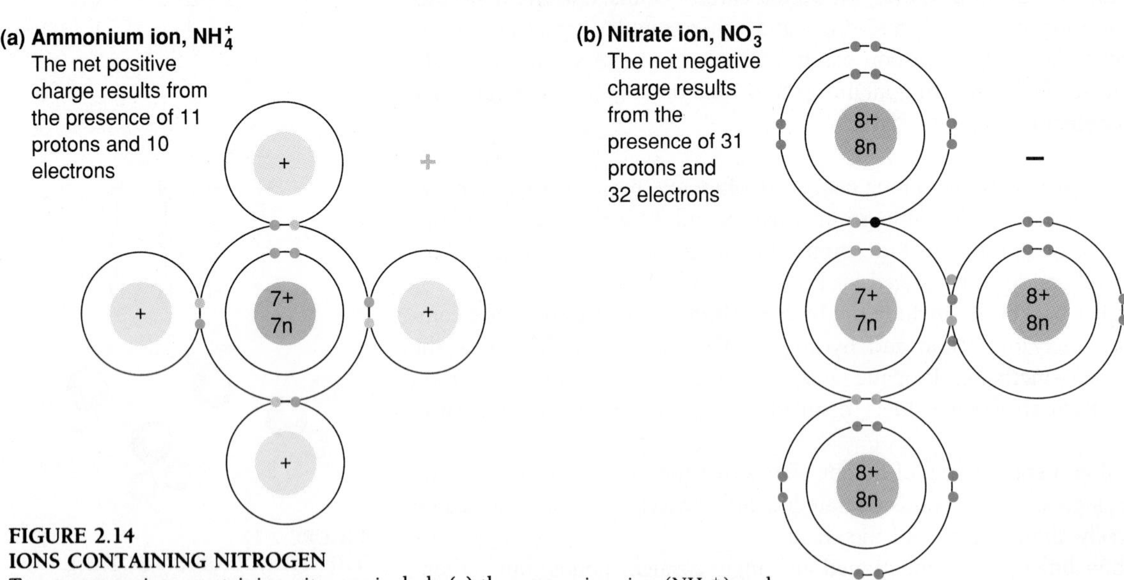

FIGURE 2.14
IONS CONTAINING NITROGEN
Two common ions containing nitrogen include **(a)** the ammonium ion (NH_4^+) and **(b)** the nitrate ion (NO_3^-). Together they form ammonium nitrate, a common but valuable fertilizer.

A few plants—notably the legumes such as peas, beans, peanuts, and alfalfa—meet their own nitrogen needs by harboring mutualistic (mutally beneficial) nitrogen-fixing bacteria within specialized root cells (Figure 2.15) (see Chapters 22 and 47).

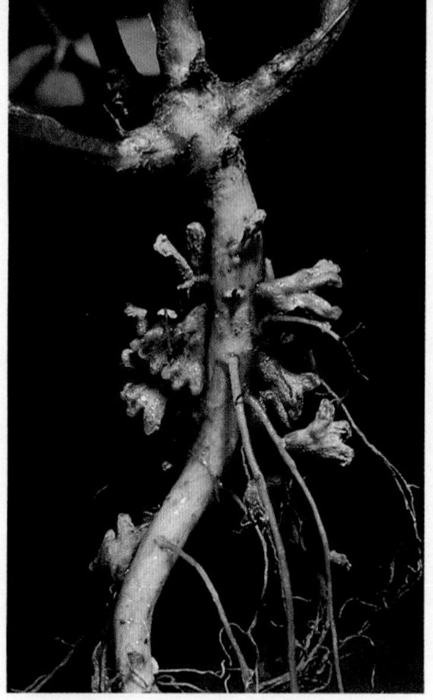

FIGURE 2.15
NITROGEN-FIXING BACTERIA
The swellings throughout this root are called nodules. They harbor nitrogen-fixing bacteria that are capable of converting atmospheric nitrogen (N_2) into nitrogen compounds that are used by both the bacteria and the plant host.

Phosphorus

Phosphorus is another of the essential SPONCH elements. In biological systems it is always combined with oxygen as a **phosphate.** A phosphate can be a free inorganic ion, or it can be combined with a larger organic molecule to form a **phosphate group** (Figure 2.16). In the cellular and extracellular fluids of organisms, phosphate ions exist as HPO_4^{-2}, or as $H_2PO_4^-$ (with one or two hydrogen atoms and two or more negative charges, correspondingly).

In phosphate, the four oxygens are tightly bound to the phosphorus atom, forming the four corners of a tetrahedron with the phosphorus inside. Since the phosphate often occurs as a free ion, it is given a symbol of its own, $\mathbf{P_i}$, which represents **inorganic phosphate.**

Phosphate also occurs in certain fatty substances called **phospholipids,** which make up much of the structure of the many membranes found around and within the cells of organisms. It is also a part of the calcium phosphate that forms the hard matter of bones and teeth.

ATP: The Energy Carrier All living organism use special phosphate-to-phosphate bonds as a means of storing energy and shuffling it around to where it is needed. Such bonds are found chiefly in molecules known as **adenosine triphosphate,** or more familiarly, **ATP,** and its less energetic product, **adenosine diphosphate,** or **ADP.** We'll have much more to say about ATP and ADP in future chapters.

Sulfur

Sulfur, our last SPONCH element, has several roles in the chemistry of life. It appears in some of the **amino acids,** small molecules that, when joined together in chains, form

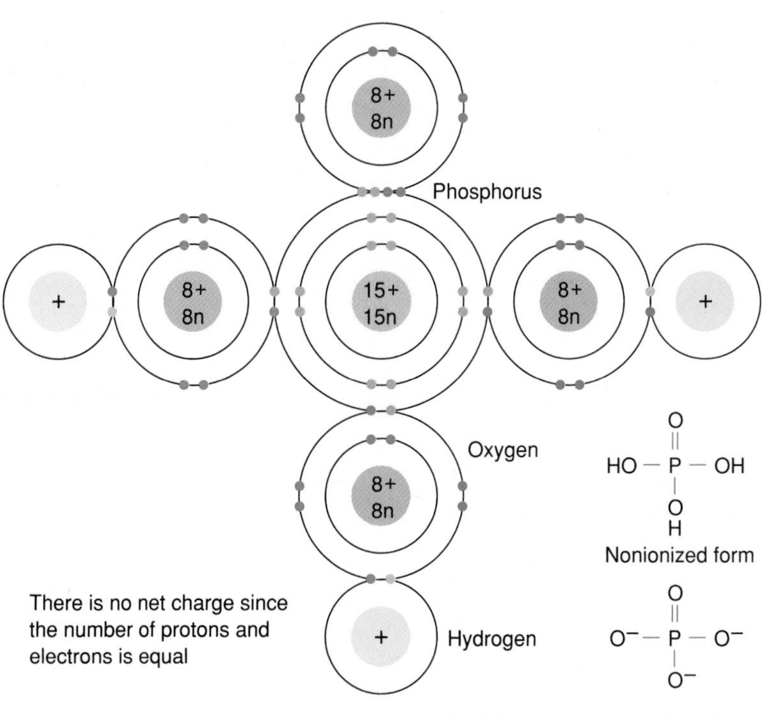

FIGURE 2.16
PHOSPHORIC ACID
(a) In a complex sharing of electrons between phosphorus and oxygen, phosphoric acid is formed. **(b)** In its nonionized or phosph*oric acid* state, it contains three —OH (hydroxide) groups. In its ionized or phosph*ate* state, it loses up to three protons, becoming a negative ion.

the familiar proteins in life. The sulfur occurs in the amino acid as a **sulfhydryl group,** a sulfur atom covalently bonded to a hydrogen atom (—SH). When the amino acids are assembled into protein, pairs of sulfhydryl groups serve as a kind of interlocking "snap" or "hook" that can bond the long strands of protein together, aiding them in forming their special shapes. Two sulfhydryl groups give up their hydrogen and covalently bond together to form a reversible **disulfide linkage**, or **bridge**:

$$\text{amino acid S—H + H—S amino acid} \rightleftharpoons$$
$$\text{amino acid—S—S—amino acid} + H_2$$

The —SH group is only one of several special functional groups or functional side groups, found in organic molecules.

A **functional group** is a specific chemical group that appears rather frequently in the many kinds of organic molecules. They are important because their chemical behavior is pretty much the same, regardless of the kind of molecule to which they are attached. Some important functional groups are the carboxyl and amino groups, found in amino acids; phosphate and sulfhydryl groups (already discussed); and methyl groups, hydroxyl groups, aldehydes, and ketones. Table 2.2 discusses the structure and characteristics of the most familiar functional groups.

TABLE 2.2 SELECTED FUNCTIONAL GROUPS AND THEIR CHARACTERISTICS

X—OH	Hydroxyl group X—O—H	Common in alcohols, slightly polar, tends to form H-bonds
X—NH_2	Amino group X—N(—H)—H	Common in amino acids, weak base in water, accepts a proton: X—N(—H)(—H)—H^+
X—COOH	Carboxyl group X—C(=O)—OH	Common in amino acids and other organic molecules, weak acid in water, releases a proton: X—C(=O)—O^- + H^+
X—CH_3	Methyl group X—C(—H)(—H)—H	Common in organic molecules, nonpolar—rejected by polar molecules, insoluble in water
X—COH	Aldehyde group X—C(=O)—H	Slightly polar, soluble in water, common in sugars
X—SH	Sulfhydryl group X—S—H	Common in protein, where it forms vital covalent linkages
X—CO—X	Ketone group X—C(=O)—X	Slightly polar, water soluble, common in sugars
X—H_2PO_4	Phosphate group X—O—P(=O)(—O—H)—O—H	Polar, weak acid, releases protons in water, common energy carriers of cells, un-ionized and ionized form: X—O—P(=O)(—O^-)—O^-

The letter *X* is just a convenience used to represent an undesignated or unnamed molecule to which the functional group is attached. Some of the functional groups listed above will ionize in water, producing a charged condition, as we saw in the sodium and chloride ions.

FIGURE 2.17
WATER AND THE DESERT
Water has dramatic influence in the desert. Long periods of drought and dry, searing winds leave only the best-adapted plants to dot the landscape. When water is again available, the desert bursts into a short-lived riot of color and new growth.

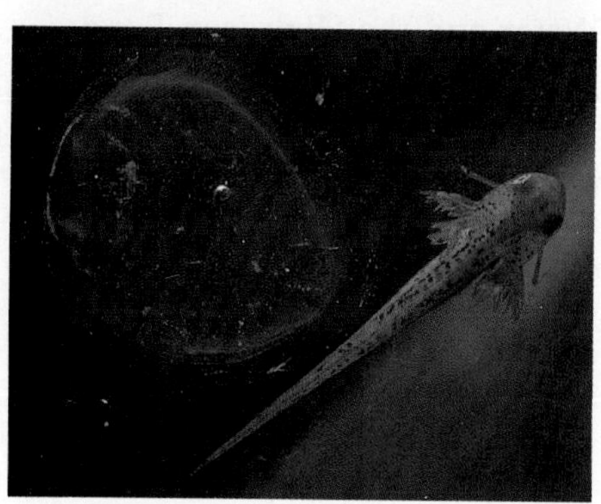

FIGURE 2.18
ANIMALS OF THE WATER
Aquatic animals and their embryos are particularly high in water content. The vulnerability of early growth stages to drying emphasizes the critical role of water in life.

THE WATER MOLECULE AND HYDROGEN BONDING

If you were to walk across some desolate desert, you might see no obvious signs of life for miles. Then, in the distance, you may see green. Coming closer, you find a few cottonwood trees and scattered tall bushes. You know why the plants are there. You might not be able to see it, but you know that it's there, beneath the surface at least: water. Where there's life, there's water. And just about any place where there is water, there is life (Figure 2.17).

On a grander scale, most biologists accept the idea that life began in water, and certainly, the association between water and life endures. Hope of discovering life on Mars faded when we found that water was virtually absent on our celestial neighbor. After all, most of life's chemical reactions occur in aqueous solution; in fact most living organisms are between 50% and 90% water (Figure 2.18). (You are about two-thirds water.)

Thus, water is essential to life as we know it, and on our unique planet water is plentiful. Three fourths of the earth's surface is water, and an enormous volume of water constantly shifts between the earth and its atmosphere in what is called the **hydrologic cycle** (Figure 2.19). The cycle is not simply a physical entity but a process that involves the organisms of the earth as well. Plants, animals, and the many other forms of life use water, incorporating it for a time in their cells as they go about their chemical activities. Eventually, all this water is returned to the cycle of the atmosphere and oceans as the organisms respire or die. In the later chapters, we will see the importance of water to life in many different ways, but for now, let's consider some of the molecular peculiarities of this vital, life-sustaining substance, and why its characteristics encourage chemical reactions.

FIGURE 2.19
THE CYCLING OF WATER
The hydrologic cycle is an ongoing exchange of water through evaporation and precipitation. Living organisms also participate, taking in and releasing sizable quantities of water.

The Polar Nature of Water

In each water molecule, two hydrogen atoms are covalently bonded to one oxygen atom. That is, the two hydrogens share electron pairs in two of the four electron-pair orbitals of the oxygen's outer shell. Figure 2.20a shows an electron shell diagram of water. Notice that the water molecule is quite polar—that is, it has positive and negatively charged sides. Further, the two hydrogen atoms are close together, leaving the molecule a bit lopsided. (In certain views, a diagram of the water molecule reminds some people of Mickey Mouse. Now try looking at it without seeing that.) The polarity and lopsidedness of water are both important to its fascinating properties, as we shall see.

The two hydrogen atoms (the two Mickey Mouse "ears") of water occur at two corners of an imaginary tetrahedron (Figure 2.20b) similar to the one we saw in the methane molecule. No matter which two of the four corners are occupied, the molecule is equally lopsided. More specifically, a pair of lines from the center of the oxygen to the centers of the two hydrogens would form an angle of approximately 105°.

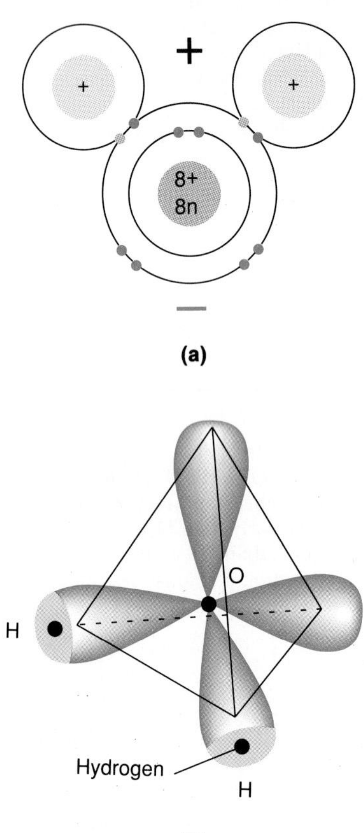

FIGURE 2.20
POLARITY IN WATER
(a) Although electrons are shared between hydrogen and oxygen in the formation of water, the sharing is unequal in favor of oxygen. Thus water molecules are polar, each with positive and negative ends. **(b)** The molecular orbitals of water.

Water and the Hydrogen Bond

Now we will explore the last kind of bonding—the hydrogen bond—particularly with reference to its role in the behavior of water molecules. Hydrogen bonds are defined as *electrostatic attractions between slight positive and negative charges,* thus, they do not involve the transfer or sharing of electrons as seen in ionic and covalent bonding. In general, such electrostatic attractions occur between hydrogen atoms that are attached to oxygen or nitrogen in one polar molecule and certain electrons in oxygen or nitrogen atoms in another polar molecule. In water, the electrostatic attraction occurs between a hydrogen in one water molecule (slightly positive) and the oxygen of another (slightly negative). This attraction tends to hold the two molecules together, even if just for an instant. Thus, the hydrogen bond is quite fleeting in nature.

Hydrogen bonds are quite common in the chemistry of organisms and perform vital roles in helping macromolecules hold their shape and remain intact. One of the best examples of this role is found in the famous DNA double helix, whose two long intertwined strands are held together by nearly countless, very weak, hydrogen bonds. In fact, the very weakness of the hydrogen bond is the key to its usefulness in biological reactions.

Ions, Hydration Shells, and Water as a Solvent Water owes much of its excellent solvent qualitites to its polar characteristics. We've already mentioned that salts such as sodium chloride dissociate into ions when dissolved in water. This dissociation is aided by the tendency of water to form **hydration shells**, which are layers of water molecules that are loosely bound to an ion in solution. (Don't confuse hydration shells with electron shells.) For example, water molecules orient their positive (hydrogen) ends toward a negative ion, such as chloride, surrounding it with a hydration shell as shown in Figure 2.21. This means that the water molecules of the innermost hydration shell have their negative (oxygen) ends pointing outward. This hydration shell, in turn, attracts the positive ends of other water molecules, and so on, forming progressively weaker concentric shells of oriented water molecules. The same kind of thing happens around positively charged ions, except that the orientation of the water molecules is reversed (positive end out). One of our more imaginative colleagues has compared the water molecules that surround an ion to groupies clustering around a highly charged rock star.

Water will form hydration shells around polar molecules as well as charged ions. For instance, sugars contain slightly polar, protruding hydroxl functional groups (—OH; see Table 2.2) with which water can build loose hydrogen bonds and form hydration shells. This keeps the somewhat polar sugar molecules from clumping together; in other words, sugar stays dissolved because hydration shells are formed.

Water and Nonpolar Molecules If you mix a teaspoonful of water in a jar of salad oil, you might notice that the water will quickly form droplets that will coalesce and isolate themselves from the oil. The reason for this is the strong mutual attraction of water molecules. Water has very little attraction for salad oil, which is a nonpolar compound that can't form hydrogen bonds.

If you try to mix a teaspoonful of salad oil in a jar of water, the results will be about the same. This time it is the oil that forms droplets that eventually coalesce. It would appear that the oil molecules too have a strong mutual attraction, but this is not the case. In fact, nonpolar molecules have very little attraction for each other. It is the strong mutual attraction among water molecules that excludes the oil: thus it is effectively isolated. Because the attraction among water molecules is strong, the forces tending to push nonpolar molecules together can be very strong in such a mixed system. Later, we will find that this force is vital to the structure of proteins and of cell membranes.

Hydrophobia, Dishwater, and Mayonnaise Molecules can be **hydrophobic** ("water-fearing") or **hydrophilic** ("water-loving"), although some are actually both. Hydrophobic molecules, such as the oil mentioned above, are nonpolar (uncharged) and are repelled by water. Hydrophilic molecules are polar (having charged regions) and, like ions, readily interact with water. Detergents have both characteristics. They are hydrophobic at one end and hydrophilic at the other, thus they interact with both polar and nonpolar molecules. For example, dishwashing detergents are useful as cleaning agents because they form bridges between the grease left on your dishes and the surrounding water molecules, effectively lifting the grease from the dishes and dispersing it throughout the dishwater. In another example, lecithin, a substance found in large quantities in egg yolk, is a natural detergent that is very useful in forming a bridge between dilute vinegar (which is polar) and salad oil (which is nonpolar). Beat well and the gel that results from your mixture of egg yolk, vinegar, and oil is called mayonnaise. In the next chapter, we will come across molecules called phospholipids, which perform similar linking actions within the membranes of cells.

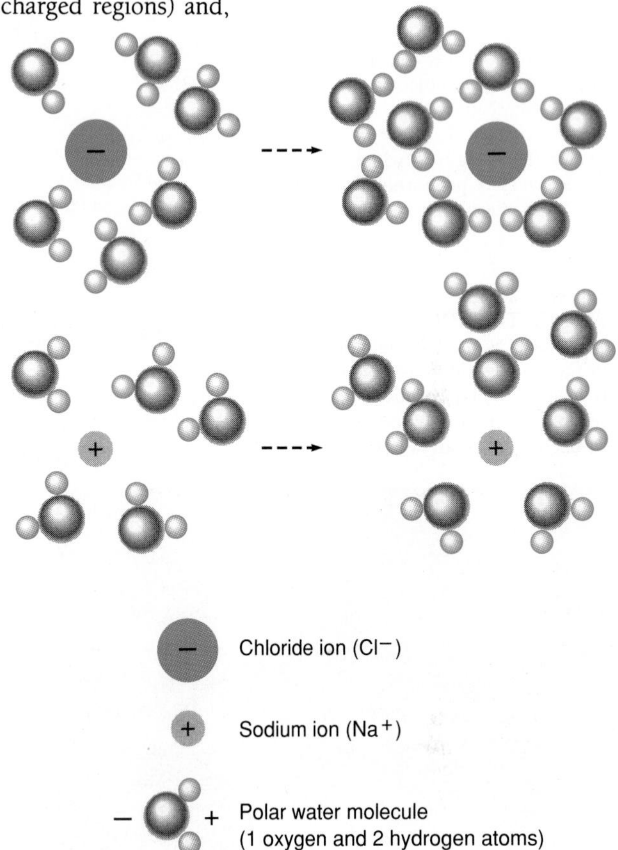

FIGURE 2.21
HYDRATION SHELLS
In its interaction with sodium and chloride ions, water forms hydration shells. Note the opposite orientation of water to the two ions.

Water Is Wet Water tends to get things wet. But what does this really mean? It means that it forms hydrogen bonds with the surface molecules of solid objects, except, of course, with objects made of oily or waxy substances that are composed entirely of nonpolar molecules. This wetting ability is the result of **adhesion**—an attraction between two dissimilar substances. **Cohesion** is the attraction between similar substances; the hydrogen bonds between water molecules give water a considerable cohesion.

The tendency of water to adhere to and spread over solid surfaces is one of its special properties. If a thin glass tube is lowered into a beaker of water, the water wets the inside of the tube and, as it does so, a column of water rises in the tube until it is higher than the water level in the beaker. If glass tubes of different diameters are put into the same beaker, water will rise higher in the tube with the smallest bore. This is called **capillary action**, and it is due in part to the adhesion of water to glass and in part to the cohesion of water to itself. The two forces also explain the peculiar concave bend (meniscus) seen at the top of the water in a graduated cylinder (Figure 2.22).

Similar to capillary action, but on a finer scale, is **imbibition**, the movement of water into porous substances, such as wood or gelatin through **absorption**, the adhesion to surfaces. The substances swell as the water moves in, and in fact, the swelling can generate a startlingly powerful force. Seeds can split their tough coats by the force of

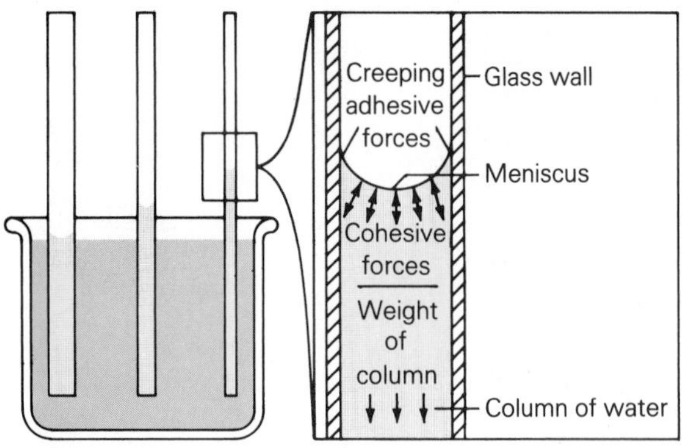

FIGURE 2.22
CAPILLARITY
The combined effects of cohesion and adhesion in water molecules, known as capillarity, can be seen when glass tubes of different diameter are placed in water. The height to which water rises is inversely proportional to the diameter of the tubing.

FIGURE 2.23
SURFACE TENSION
Insects such as the whirligig beetle and the water strider are able to literally walk on water because of surface tension. A surprisingly tough water film occurs at the air-water interface.

imbibition. And it has even been suggested that the great stones used in the construction of the Egyptian pyramids were quarried by driving wooden pegs into holes in the rock face and then soaking the pegs with water.

Water Has High Surface Tension An old party trick involves carefully "floating" a needle in a glass of water. How is this possible—surely a needle cannot float? It works because of **surface tension.** Surface tension is one aspect of the cohesion of water. Where air and water meet, the water molecules at the interface have a much greater attraction to the water next to and beneath them than to the air above them. While hydrogen bonding readily occurs among the surface water molecules, it cannot occur between water and air. The surface thus forms a tough, elastic film of hydrogen-bonded water molecules. The familiar whirligig beetle and water strider are able to walk on this film without breaking it (Figure 2.23). Surface tension is also the force that produces rain drops. Thus spring showers are delicate and pleasurable, rather than terrifying, as they would be if the same amount of water fell in disorganized and ponderous masses.

Water, Heat, and Life One of the most important qualities of water as far as life is concerned is its temperature stability. It takes a large amount of heat to raise the temperature of water measurably, but once heat has been absorbed by water, it is just as stubbornly retained. Chemists are quite interested in the effects of heat on matter and refer to the amount of heat needed to raise the temperature of a given substance as its **specific heat.** Specific heat can be stated in calories; in fact the calorie itself is defined as the heat needed to raise the temperature of one gram of water 1°C. Compared to most other substances, water has a high specific heat. For example, to raise the temperature of lead 1°C requires only 0.03 calories, while a one-degree rise in the temperature of table sugar requires 0.30 calories, and ethyl alcohol, 0.60 calories. But liquid ammonia (NH_3) has a greater specific heat than water, requiring 1.23 calories of heat to raise its temperature 1°C (for this reason it has been used as a refrigerant in refrigerators). Interestingly, molecules of ammonia, like water, tend to join by forming hydrogen bonds.

The high specific heat of both water and ammonia is closely related to the presence of hydrogen bonds. Chemists explain that temperature is a product of the rapid movement of molecules in a substance—more specifically, the average kinetic energy (energy of motion) of molecules. It is important to realize that under ordinary conditions molecules are always in very rapid random motion. But such motion in water and other substances that are subject to hydrogen bonding is greatly retarded. And, when hydrogen bonds are present, much of the heat input is required simply to rupture or counteract these numerous attractive forces. Once this requirement is met, the heat can bring about increased molecular movement, and a thermometer will record this activity as a temperature rise. So, in a real sense, water simply captures and holds much of the heat to which it is exposed.

Incidentally, as you may expect, water resists evaporation better than do most liquids. Chemists (never at a loss for terminology—like biologists) say that water has a "high heat of vaporization." The heat energy required to convert one gram of water to water vapor is 540 calories. This is more than twice that required for ethyl alcohol, and nearly twice that required to vaporize liquid ammonia. The vaporization (or evaporation) of water is also retarded by hydrogen bonds, for the same reasons stated above. Water must absorb a considerable amount of heat before its molecules move fast enough to escape as vapor.

Water's high resistance to vaporization is also significant to animals, such as ourselves, that rely on the evaporation of water from our body surface as a cooling device. It works well simply because evaporating water molecules remove a considerable amount of heat as they escape. On a much grander scale, in the cycling of water into the atmosphere and back, an enormous amount of heat is also carried aloft and redistributed over the earth's surface (see Chapter 46). This is a key factor making most of the earth a habitable place. Further, the waters of the earth—particularly the large bodies—are quite hospitable to life since their temperatures vary only slightly compared to temperatures on land.

The cooling of water reveals even more of its peculiarities. Water freezes slowly because of the great amount of heat that must be withdrawn. As it cools, its molecules at first move closer together, increasing in their density and reaching maximum density at 4°C. The hydrogen bonds become more rigid at this time, and the molecular latticework closes up. Thus, with the arrival of autumn in the temperate regions, layers of cooling lake water sink into the lake depths, helping to create a seasonal revolution or overturn that carries oxygen downward and nutrients upward (see Chapter 47). But as waters approach 0°C, the water lattice opens up and the molecules become widely separated, reaching their lowest density—becoming lightest—as ice crystals form. The separation is apparently consistent with the formation of four hydrogen bonds around each water molecule. (Figure 2.24 portrays the three states of water.) The fact that water at 0°C reaches its least dense state explains the buoyancy of the ice cubes in your drink and why ice skating is more popular than it might otherwise be.

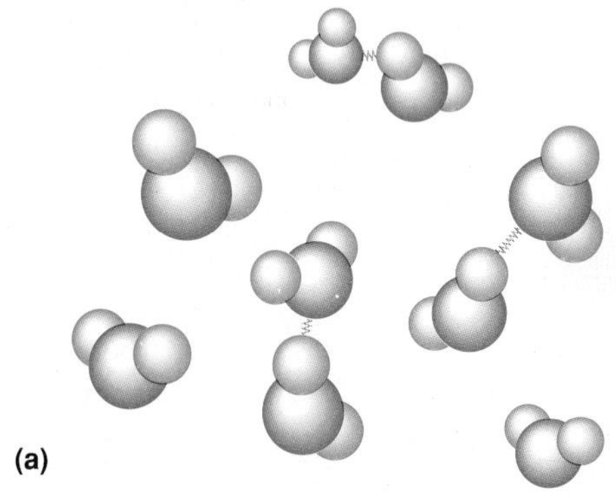

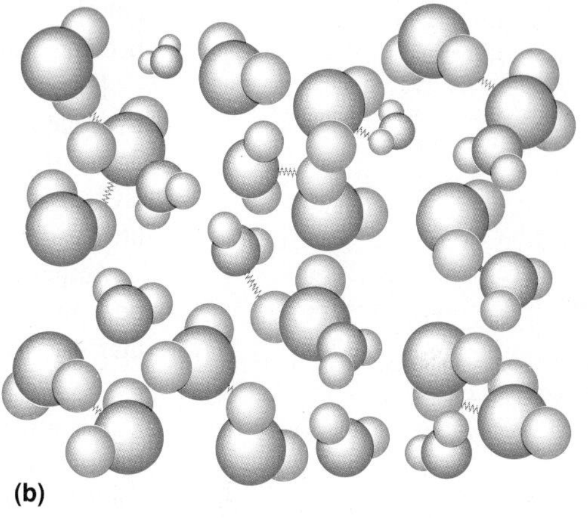

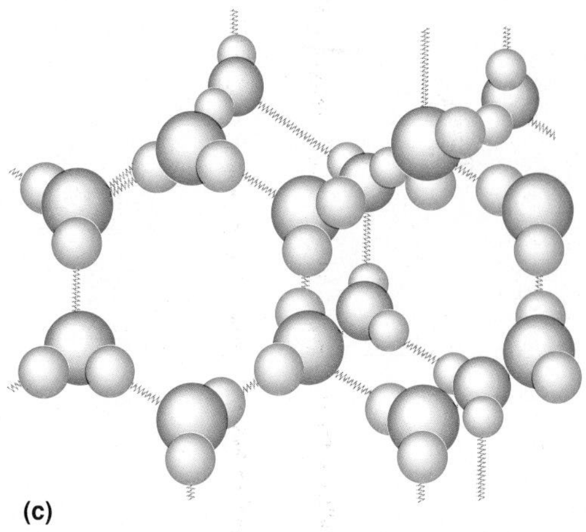

FIGURE 2.24
HYDROGEN BONDING IN WATER
Weak hydrogen bonds attract water molecules together. The degree of attraction depends upon heat energy, which disrupts the hydrogen bond. **(a)** In the gaseous state (above 100° C), molecular movement is rapid and random. **(b)** In the liquid phase (0 to 100° C), molecular movement lessens, and the association between molecules, though firmer, is still quite loose, resulting in sliding rows or lattices. **(c)** Below water's freezing point (0° C), hydrogen bonds hold the water molecules rigidly in place in the expanded, crystalline lattice called ice.

Water: Ionization, pH, and Acids and Bases

Although water usually exists as a covalently bonded molecule, a very tiny fraction of the molecules in a drop of water will briefly and reversibly dissociate into a **hydrogen ion** (H^+) and a **hydroxide ion** (OH^-). That is, water molecules are continually breaking apart into ions, and the ions are continually rejoining to make the neutral molecule again. In pure water, at any one instant, something like 1 molecule in 550 million will be dissociated into a pair of positively and negatively charged ions. An instant later, they will be back together again, but another approximately 1 in 550 million water molecules will dissociate in the meantime.

The *molar concentration* of hydrogen ions in pure water is 0.0000001 *mole per liter*. A **mole** of any substance is the weight *in grams* that equals the molecular mass in daltons of one molecule. Thus, a mole of hydrogen ions weighs 1 gram, and the concentration of hydrogen ions by weight in pure water is 0.0000001 *gram per liter*. The number 0.0000001, or the digit "1" seven places to the right of the decimal point, can also be written as 10^{-7}, which, in scientific notation is the equivalent of 1 divided by 10 million. The molar concentration of hydrogen ions in pure water, then, is 10^{-7} mole per liter, and the same concentration of hydroxide ions is also 10^{-7} mole per liter.

Actually, the hydrogen ion would be a naked proton, and while such things exist, they are not formed during the dissociation of pure water. In reality, the hydrogen nucleus from a dissociating water molecule becomes bound up with another water molecule to make a **hydronium ion** (H_3O^+). Keep in mind that whenever we refer to a hydrogen ion, or write it as H^+, we mean a hydronium ion, H_3O^+. The dissociation and reassociation reaction may be symbolized as follows:

$$2H_2O \rightleftharpoons OH^- + H_3O^+$$

where the longer arrow pointing left indicates that most molecules are in the H_2O form at any one time. In the standard convenient fiction, of course, the same reaction would be written this way:

$$H_2O \rightleftharpoons OH^- + H^+$$

Some substances, when in water, release hydrogen ions in measurable quantities. We call solutions of such substances **acids.** Stomach acid, for instance, is dissolved hydrochloric acid, HCl. Pure HCl is a gas, but when it is dissolved in water, HCl ionizes to become paired H^+ and Cl^- (hydrogen ions and chloride ions).

Just how strong, or *acidic,* an acid solution is depends on the concentration of the hydrogen ions. The molecular weight of hydrochloric acid is 36 (1 for the hydrogen, 35 for the chloride). Thus, 1 mole of pure HCl gas weighs 36 grams, and 36 grams of HCl dissolved in 1 liter of water would create a concentration of 1 mole of HCl per liter, a decidedly strong acid.

Actually, since the HCl ionizes completely, the concentration of *chloride ions* would be 1 mole per liter, and the concentration of *hydrogen ions* would also be 1 mole per liter—a concentration of hydrogen ions that is 10 million times greater than that of pure water. That's very acidic.

A shorthand notation for the strength of acid solutions, or for acidity in general, is the **pH scale.** This scale uses scientific notation to express the H^+ concentration in a solution. Thus, if the acidity of some rather tart orange juice is 0.01 mole of H^+ ions per liter, which is the same thing as 10^{-2} moles of H^+ per liter, that orange juice has a pH of 2. We just leave out the "ten to the minus" part of the number. On the pH scale, then, pure water has a pH of 7. Table 2.3 gives the pH values of various substances. Note that the smaller the pH value, the more acid the solution.

TABLE 2.3 pH VALUES

MOLAR CONCENTRATION OF H^+ IONS	pH	EXAMPLE	MOLAR CONCENTRATION OF OH^- IONS
$1.0 = 10^{0}$	0	1 molar nitric acid	10^{-14}
$0.1 = 10^{-1}$	1	Gastric juices	10^{-13}
$0.01 = 10^{-2}$	2	Lemon juice	10^{-12}
$0.001 = 10^{-3}$	3	Vinegar, Coca Cola	10^{-11}
$0.0001 = 10^{-4}$	4	Tomato juice	10^{-10}
$0.00001 = 10^{-5}$	5	Urine (varies), black coffee	10^{-9}
$0.000001 = 10^{-6}$	6	Saliva, milk	10^{-8}
$0.0000001 = 10^{-7}$	7	Pure water	10^{-7}
$0.00000001 = 10^{-8}$	8	Seawater	10^{-6}
$0.000000001 = 10^{-9}$	9	Baking soda	10^{-5}
$0.0000000001 = 10^{-10}$	10	Soap solution	10^{-4}
$0.00000000001 = 10^{-11}$	11	Ammonia solution	10^{-3}
$0.000000000001 = 10^{-12}$	12	Washing soda	10^{-2}
$0.0000000000001 = 10^{-13}$	13	Oven cleaner	10^{-1}
$0.00000000000001 = 10^{-14}$	14	1 molar sodium hydroxide	10^{-0}

ESSAY 2.1

CARBON DIOXIDE AND THE BICARBONATE ION

Carbon dioxide (CO_2) is an interesting and familiar substance. It makes soft drinks, beer, and champagne tingle, but where does it come from? In champagne and beer, the carbon dioxide is produced as a waste product of yeast metabolism. (Another waste product of yeast is alcohol, of course.) Your own body is loaded with CO_2, which is a waste product of your own metabolism. The carbon dioxide in soft drinks comes from steel cylinders.

There is a tiny amount of CO_2 in the air—not much, about ⅓ of one percent of air by weight, but this small amount is the only source of carbon for plants. Carbon dioxide is a symmetrical, linear molecule, which can be written as O=C=O. (Note the paired bonds when two pairs of electrons are shared. This is called a *double bond.*)

Carbon dioxide dissolves easily in water, and most of the CO_2 in beer and soft drinks occurs as CO_2 in simple solution. But CO_2 also reacts to some extent with the water to form *carbonic acid,* which is a weak acid that further dissociates into a *bicarbonate ion* and a *hydrogen ion* **(a)**. The dissociation (separation) of hydrogen ions is what makes CO_2 solutions acidic (tart). Notice that the arrow is bidirectional. This indicates that the reaction can occur in either direction.

Sodium bicarbonate is a sodium salt of the bicarbonate ion. When you dissolve a spoonful of "bicarb" in a glass of water, some of the bicarbonate ions combine with H^+ ions from the water to reform carbonic acid, with the release of hydroxide ions **(b)**.

In your stomach, sodium bicarbonate neutralizes some stomach acid (hydrochloric acid) and releases CO_2, which takes the form of an unseemly belch. Note that the sodium and chloride ions don't actually enter into the reaction **(c)**.

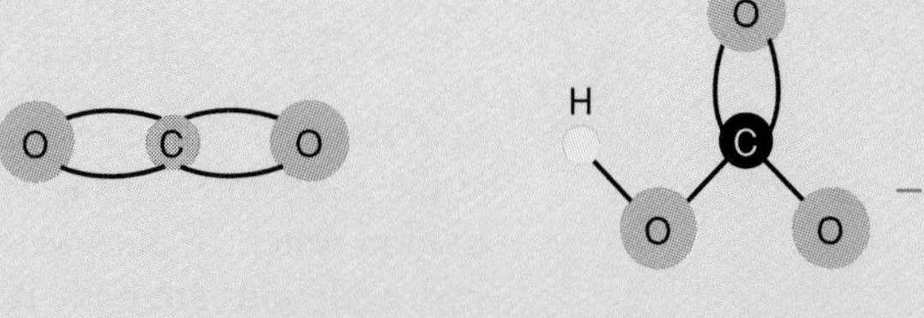

Ball-and-stick model (CO_2)   Ball-and-stick model (HCO_3^-)

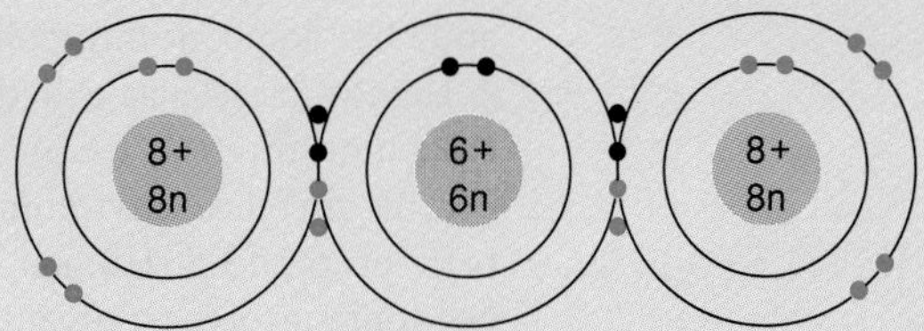

Electron shell diagram (Carbon dioxide, CO_2)

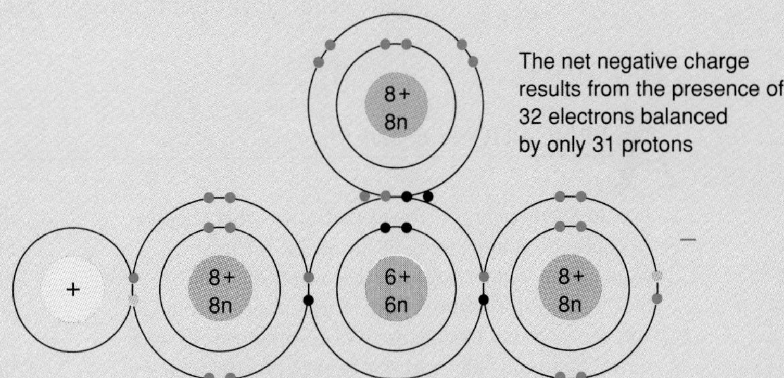

Electron shell diagram (Bicarbonate ion, HCO_3^-)

(a)

$$\underset{\text{Carbon dioxide}}{CO_2} + \underset{\text{Water}}{H_2O} \rightleftharpoons \underset{\text{Carbonic acid}}{H_2CO_3}$$

$$\underset{\text{Carbonic acid}}{H_2CO_3} \rightleftharpoons \underset{\text{Hydrogen ion}}{H^+} + \underset{\text{Bicarbonate ion}}{HCO_3^-}$$

(b)

$$\underset{\text{Water}}{H_2O} + \underset{\text{Bicarbonate ion}}{HCO_3^-} \rightleftharpoons \underset{\text{Carbonic acid}}{H_2CO_3} + \underset{\text{Hydroxide ion}}{OH^-}$$

$$\underbrace{Na^+ + HCO_3^-}_{\text{Sodium bicarbonate}} + \underbrace{H^+ + Cl^-}_{\text{Hydrochloric acid}} \longrightarrow \underset{\text{Sodium ion}}{Na^+} + \underset{\text{Chloride ion}}{Cl^-} + \underset{\text{Water}}{H_2O} + \underset{\text{Carbon dioxide gas}}{CO_2}$$

A **base,** or **alkali,** is a substance that accepts protons and releases hydroxide ions (OH^-) when dissolved in water. Lye, or sodium hydroxide (NaOH), is a familiar example of a strong base. Ammonia is another base; although it contains no hydroxide ion itself, it accepts a proton from water to form an ammonium ion and a hydroxide ion:

$$\underset{\text{Ammonia}}{NH_3} + \underset{\text{Water}}{H_2O} \rightleftharpoons \underset{\text{Ammonium ion}}{NH_4^+} + \underset{\text{Hydroxide ion}}{OH^-}$$

A solution can be acidic, neutral, or basic (alkaline), but it cannot be acidic and basic at the same time. This is because hydroxide ions and hydrogen ions join spontaneously to form water. In pure water, which is neutral, the concentrations of hydrogen ions and hydroxide ions are both 10^{-7} moles per liter. But when the concentration of hydrogen ions rises to 10^{-1} mole per liter, as in gastric juice, the molar concentration of hydroxide ions falls to 10^{-13}, which is a very small number. In general, the exponents of the H^+ and OH^- ion concentrations in a solution always add up to -14. Thus, seawater, which is slightly basic, has a hydroxide ion concentration of 10^{-6} and a hydrogen ion concentration of 10^{-8}: $(-6) + (-8) = -14$. The seawater, then, has a pH of 8.0.

The chemical environment in which most life processes go on has a pH value of between 6 and 8. Human blood, for instance, has a pH of approximately 7.4, and the cell contents of most organisms have similar nearly neutral pH values. Exceptions include the stomach's digestive juices and fluids contained in citrus fruits, both of which are quite acidic. Plant fluids generally range from pH5 to pH7.

APPLICATION OF IDEAS

1. Reread Szent-Györgi's lament that opens this chapter. Actually, he is an extremely successful scientist, a Nobel Prize winner, and the discoverer of many of life's secrets (for example, the structure of vitamin C and much of the biochemistry of respiration). In what sense, then, did "life" run out between his fingers?
2. Elements like hydrogen, sodium, and chlorine are rarely found in their elemental form. Explain why this is true. From a periodic table of the elements, list other elements with the same chemical characteristics and try to determine whether these are ever found in their elemental form. (The organization of the periodic table will tell you what the others are.)
3. It is interesting to compare silicon with carbon, since they both have four electrons in their outer shells. Silicon can form long chains called silicones, which are like the long chains of carbons called hydrocarbons. In addition, silicon combines with oxygen (SiO_2) to form crystals of silica or quartz and readily combines with fluorine to form highly soluble SiF_4. This similarity has not escaped the attention of science fiction writers who have described life based on silicon rather than carbon. What might such life forms be like? What similarities and differences would you expect between silicon- and carbon-based life forms? What might such alien scientists have to say about the "special properties" of silicon?

KEY IDEAS

ELEMENTS, ATOMS, AND MOLECULES

1. An **element** is a substance that cannot be separated into simpler substances by purely chemical means. Examples common to life include sulfur, phosphorus, oxygen, nitrogen, carbon, and hydrogen.
2. An **atom** is the smallest indivisable unit of an element. A **molecule** is two or more atoms chemically joined. A **compound** is a molecule in which two or more different elements occur in specific proportions.

Atomic Structure

1. Atoms consist of positively charged **protons,** uncharged **neutrons,** and negatively charged **electrons.**
2. Protons and neutrons make up the **atomic nucleus,** while electrons—minute particles—are in motion about the nucleus.
3. The combined mass of protons and neutrons account for the **atomic mass (atomic weight)** of an element. **Molecular mass** (molecular weight) is the combined atomic masses of the atoms in the elements making up that molecule.
4. The **atomic number** of an element is the number of protons (or electrons) in its atoms.

5. The number of neutrons in atoms of an element may vary, resulting in a number of **isotopes** of that element.
6. **Radioactive** isotopes are unstable isotopes that give off energy and/or matter in the form of radiation. The rate at which **radioisotopes** disintegrate is measured in **half-lives.**
7. Radioisotopes called **tracers** are used to determine the role of various elements and compounds in living organisms.

ELECTRONS AND THE CHEMICAL CHARACTERISTICS OF ELEMENTS

The behavior of electrons determines the chemical activity of elements.

Electron Energy Levels and Shells

1. The energy of an electron determines its position in respect to the nucleus. Accordingly, electrons occur at specific **energy shells**.
2. Each shell has a set number of electron **orbitals** (pathways), each holding up to two electrons. In the first three shells, beginning with the innermost and least energetic, the orbitals include:
   a. first shell: one 1*s* (2 electrons)
   b. second shell: one 2*s* (2 electrons), three 2*p*'s (2 electrons each)
   c. third shell: one 3*s* (2 electrons), three 2*p*'s (2 electrons each)

Energetic Tendencies

1. All matter tends to occur at the lowest energy state possible.
2. Atoms follow three chemically important **energetic tendencies**:
   a. Orbiting electrons tend to form pairs.
   b. Atoms tend to form full outer shells.
   c. Positive and negative charges tend to balance.
3. Atoms of the lighter elements fill their shells according to the **octet rule**, with the maximum numbers of 2, 8, and 8 in the first three shells. The so-called noble elements (such as helium, neon, and argon) have filled outer shells in their elemental condition and thus are unreactive.

Chemical Reactions: Filling the Outer Electron Shell

1. The tendency to fill the outer shell brings about chemical reactions and allows chemical bonds to form.
2. **Chemical reactions** are events in which **reactants** undergo changes into **products**. Chemical reactions involve losing, gaining, or sharing outer shell electrons.
3. Elements that are strongly electronegative tend to gain electrons when they react, while those less strong lose electrons. When **electronegativity** in the reactants is about equal, sharing of electrons occurs. Where sharing is equal, *nonpolar* molecules form, but with unequal sharing, polar molecules (slightly positive or negative) result.
4. The forces that hold atoms together include **ionic bonds** (loss or gain of electrons), **covalent bonds** (sharing electrons), **hydrogen bonds** (attraction of negative and positive regions).
5. In their chemical reaction, sodium, an **electron donor**, loses an electron to chlorine, an **electron acceptor**. The resulting imbalance in proton and electron charges results in the formation of the positive sodium **ion** and the negative chloride ion.
6. The electrical attraction between negative and positive ions is the **ionic bond**. Ionically bonded substances tend to form crystals of indeterminate size. Ionic molecules separate readily into their ions in water.
7. When outer shells are filled through electron-sharing, a **covalent bond** results. Each atom has its outer shell filled part of the time. When two pairs of electrons are shared, a **double bond** is formed.
8. Covalently bonded molecules do not dissociate into ions in water.
9. Sharing is equal in H_2, so it is nonpolar, but in H_2O, the electrons are more attracted to oxygen, so water is polar, with positive and negative sides.

CHEMICAL BONDS AND THE SPONCH ELEMENTS

Carbon and the Covalent Bond

1. **Organic molecules** are produced naturally by organisms and contain carbon, hydrogen, and other selected elements.
2. In **methane**, a simple hydrocarbon, one carbon atom shares its four outer shell electrons with four hydrogens, forming a molecule of tetrahedral shape.
3. Chemical formulas, structural formulas, ball-and-stick models, and space-filling models are all useful ways of portraying carbon (as well as other) compounds.
4. Carbon can form single, double, and triple bonds, and can form very long, straight or branched chains called **carbon backbones**. The backbones may also take the form of rings. Molecules composed mostly of carbon and hydrogen are called **hydrocarbons**.
5. Fatty acids with only single covalent bonds between carbons are saturated, those with one or more double carbon to carbon bonds are unsaturated.

Nitrogen

1. Nitrogen occurs naturally as stable, triple bonded N_2, making up 78% of the atmospheric gas.
2. Nitrogen gas is unavailable to most forms of life but can be converted to useful ammonia and nitrate by nitrogen-fixing bacteria.
3. Nitrogen occurs in proteins, nucleic acids, and in some carbohydrates and fats.

Phosphorus

1. In life, phosphorus is utilized as phosphoric acid or the **phosphate** ion—designated as HPO^{2-}_4 or as $H_2PO^-_4$ and symbolized as **P$_i$** (inorganic phosphate). It is commonly attached to other molecules as a **phosphate group**.
2. Phosphate is used in **ATP (adenosine triphosphate)** and **ADP (adenosine diphosphate)**, molecules used as energy sources for the chemical reactions of life.

Sulfur

1. Sulfur occurs in **sulfhydryl** (S—H) **groups** of certain amino acids. When such amino acids are incorporated into protein, the S—H groups form important **disulfide linkages** or **bridges** within the molecule.
2. **Functional groups** are specific groups of atoms found in many molecules, with each having its own chemical characteristics. Amino, carboxyl, and phosphate functional groups ionize in water, while others such as hydroxyls, aldehydes, and ketones are slightly polar, and methyl groups are nonpolar.

THE WATER MOLECULE AND HYDROGEN BONDING

1. Water is so essential that without it life as we know it would not exist.
2. Water in the environment cycles between its liquid and gaseous phases in what is known as the **hydrologic cycle**.

The Polar Nature of Water

Water is polar. The unequal sharing of electrons renders one part of the molecule negative, the other positive.

Water and the Hydrogen Bond

1. Water molecules interact, with weak, temporary **hydrogen bonds** forming between their positive and negative regions.
2. Water molecules tend to surround positive and negative ions, forming **hydration shells**. This is the characteristic that makes water a good solvent.
3. Nonpolar molecules are repelled by water, so they are forced together in dense associations.
4. Detergents are **hydrophilic** (water loving) at one end and **hydrophobic** (water fearing) at the other, so they form bridges between polar water and nonpolar oils.
5. **Capillary action** is the tendency for water to "creep" through minute spaces. **Adhesion** and **cohesion** are involved.
6. Water enters minute spaces through **imbibition**, which occurs through **adsorption** (the adhesion of molecules to surfaces).
7. **Surface tension** occurs where air and water interfaces occur.
8. Water has high **specific heat**—the amount of energy needed to raise its temperature. This is attributed to the resistance to molecular movement brought about by the presence of numerous hydrogen bonds. Water's ability to retain heat helps in climate moderation and as a medium for life, providing its inhabitants with relatively constant conditions.
9. Water reaches its greatest density at 4° C, but below this, its density decreases rapidly to a minimum density with 0° C (freezing).

Water: Ionization, pH, and Acids and Bases

1. Pure water itself ionizes, but in minute amounts, spontaneously forming **hydrogen ions** (H^+) (protons or, more accurately, **hydronium ions**) and **hydroxide ions** (OH^-).
2. When H^+ ions outnumber OH^- ions, a solution is **acidic.** When OH^- ions outnumber H^+, a solution is **basic** or **alkaline.** Equal numbers of the ions produce a neutral condition.
3. The strength of acids and bases is represented by a **pH scale** from 0 to 14. Acids range from a pH of 7 down to 0 for the strongest acids, while bases range from 7 up to 14 for the strongest bases.
4. The chemical reactions of life generally occur at a pH near neutral (pH7).

REVIEW QUESTIONS

1. Write the names, symbols, and atomic numbers of the chemical elements abbreviated in the acronym SPONCH. (26)
2. Define the terms *atom, element, molecule,* and *compound.* (26, 27)
3. Explain in terms of atomic structures how one element differs from another. (27, 28)
4. Explain how the atomic number and the atomic mass of an element are determined. (27, 28)
5. What are isotopes? Cite an example of a radioactive isotope with an extremely long half-life. (28)
6. List three energetic tendencies of atoms. What do these have to do with chemical reactions? (30, 31)
7. List the orbitals in the first two energy levels or shells and describe their shapes. (29, 30)
8. State the shell-filling (octet) rule and, using potassium (atomic number 19), illustrate its operation. (30)
9. Using argon as an example, explain why the noble elements are unlikely to react. (31)
10. Define electronegativity and explain what it has to do with the behavior of outer-shell electrons. (31)
11. Using potassium and chlorine (atomic numbers 19 and 17) as examples, show how they would form ions if they reacted together. Explain how an *ionic bond* would form between the two elements. (32)
12. Using a drawing, show in detail what would happen if the ionic compound potassium chloride were to be placed in water. (33)
13. Using two atoms of nitrogen (atomic number 7) as examples, illustrate how covalent bonding occurs. What is unusual about the bonding of the N_2 molecule? (33, 34)
14. List three important uses to which plants put the element nitrogen. Why is using molecular nitrogen such a problem for plants? (36)

15. In what form is the element phosphorus usually found in living things, and what are two important uses? (37)
16. In which of the molecules of life would one look for sulfur? What purposes does it serve there? (37, 38)
17. Draw structural formulas for the following molecules: ethane (C_2H_6), propane (C_3H_8), pentane (C_5H_{12}), and octane (C_8H_{18}). (35)
18. Identify the following functional groups: X—CH_3, X—SH, X—NH_2, X—OH, X—COOH X—O. (38)
19. Using water as an example, explain what a polar molecule is and discuss the nature of hydrogen bonding. (39, 40)
20. What do hydration shells have to do with water's excellent solvent qualities? (40)
21. Define the terms *hydrophilic* and *hydrophobic*. What is the peculiar nature of soap molecules that enables them to interact with grease and water? (41)
22. Explain the meaning of specific heat, and discuss two ways in which the specific heat of water is important to life. (42)
23. What do hydrogen bonds have to do with water's resistance to temperature change? (42, 43)
24. Which of the two ions of water predominates in acidic solutions? Basic solutions? Briefly describe the pH scale and cite examples of strong and weak acids and bases. (43, 44)

3

THE MOLECULES OF LIFE

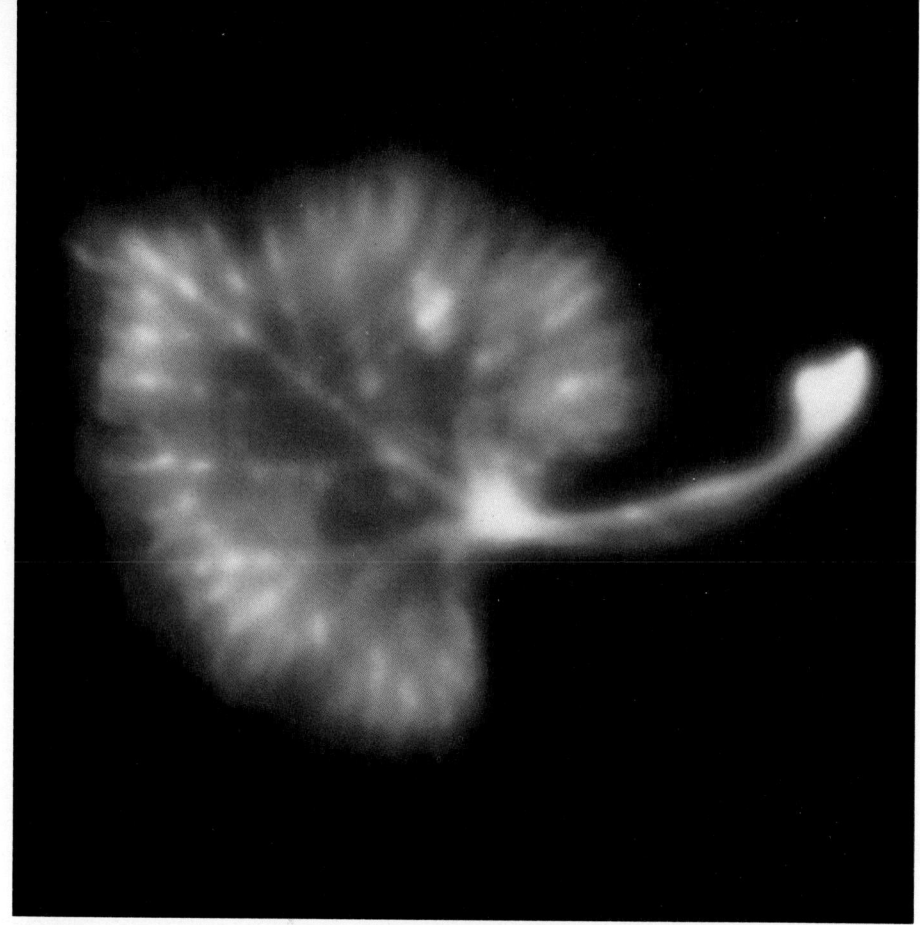

LIFE IS NOT SIMPLE. HOW MANY TIMES WE'VE HEARD THAT, USUALLY FROM weary or worried people trying to live out theirs and hoping their problems are only part of a larger, unpleasant phenomenon. But even those of us who don't focus on life's problems would have to agree that life is complex and often mysterious. In fact, it is complex in a number of ways, including its origins, mechanisms, and structures. Here, then, let's look at the molecular structure of living things, focusing on their building blocks. Specifically, we will concentrate on the nature of the large molecules that comprise living things.

In a sense, we can apply what we learned about small molecules to the study of large ones. Although large molecules can appear dazzlingly complex, they are usually just **polymers**—that is, composed of many identical or similar small molecules, or **monomers**. The monomers, then, are assembled into polymers. The large polymers are also known as **macromolecules** ("large molecules").

In this chapter we will look at structure and function in the four major classes of molecules: the carbohydrates, lipids, proteins, and nucleic acids. Actually, such categorizing is not always accurate because the four classes aren't always completely distinct from each other. For example, along the way we will run into molecules that cross the class lines, such as glycoproteins (sugar-proteins) and lipoproteins (lipid-proteins).

THE CARBOHYDRATES

Carbohydrates are familiar to us as the sugars and starches in our diets. But, as we will see, they are important in other ways as well. Most carbohydrates have the empirical formula $(CH_2O)_n$. (The numbers in empirical formulas are reduced to their simplest terms. Here, the *n* means that there can be any multiple of CH_2O.) Actually, the simplest common carbohydrates are $(CH_2O)_3$, or 3-carbon compounds, but we will primarily be

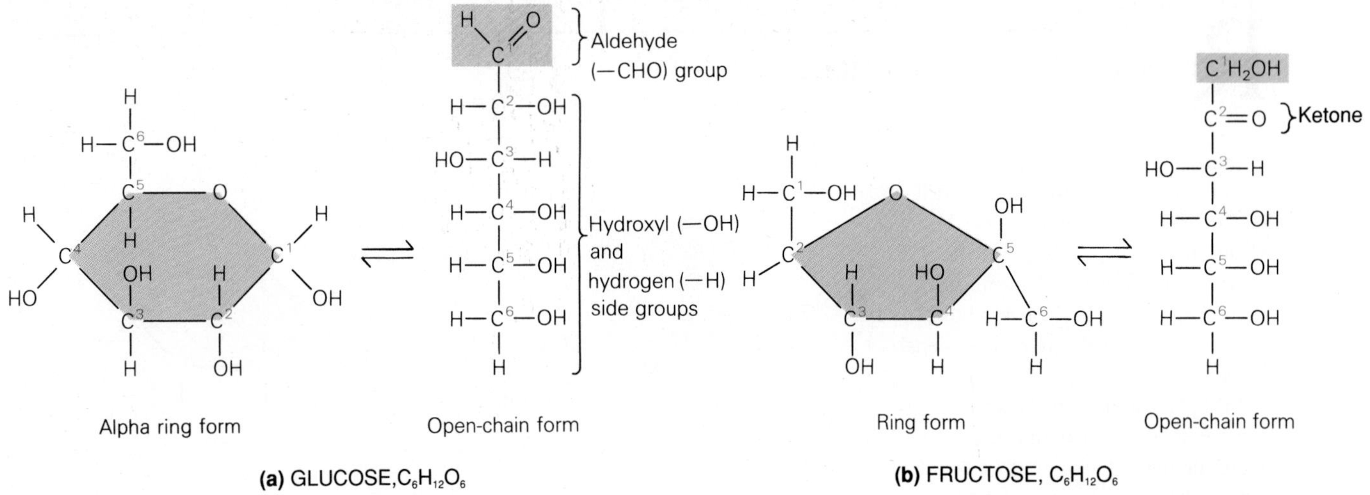

Aldehyde (—CHO) group

Hydroxyl (—OH) and hydrogen (—H) side groups

Ketone

Alpha ring form — Open-chain form

Ring form — Open-chain form

(a) GLUCOSE, $C_6H_{12}O_6$ **(b)** FRUCTOSE, $C_6H_{12}O_6$

FIGURE 3.1
GLUCOSE AND FRUCTOSE
(a) Glucose occurs in both the straight chain and ring forms. (Follow the numbering to see how one form is converted to the other.) The side groups branching from the carbon backbone include one aldehyde group and a number of hydrogens and hydroxyls. **(b)** Fructose has the same chemical formula as glucose, but its geometry differs. Note further that a ketone group is present, rather than an aldehyde.

interested here in 6-carbon carbohydrates and the way they are linked together to form the large polymers. Let's begin by looking at the organization of the "single" and "double" sugars.

Monosaccharides and Disaccharides

All carbohydrates contain single sugars, or **monosaccharides**, as they are called (*mono-*, "one"; *saccharide*, "sugar"). There are many simple sugars, but the most familiar is the 6-carbon sugar, **glucose**. The next most complex type of carbohydrate is the **disaccharide**, which as its name suggests (*di-*, "two") contains two monosaccharides. The most familiar disaccharide is **sucrose**, or table sugar. Then, more complex molecules are formed from longer chains of simple sugars until we reach the **polysaccharides** (*poly-*, "many"), which may contain hundreds or thousands of monosaccharide subunits, covalently linked into chains.

A monosaccharide consists of a short chain or ring of carbon atoms (usually five or six), with nearly every carbon having a hydroxyl (—OH) functional group as well as a hydrogen (—H) side group (Figure 3.1). Note that the empirical formula $(CH_2O)_n$ for carbohydrates suggests one oxygen for every carbon. Often, one carbon at the *end* of each simple sugar forms a double bond with its oxygen, producing an **aldehyde** group:

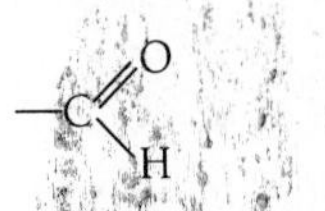

$$\mathrm{-C(=O)H}$$

In other instances a carbon *within* the chain forms a double bond with an oxygen group, yielding what is called a **ketone** group:

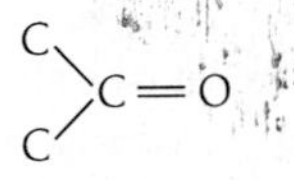

$$\mathrm{(C)(C)C{=}O}$$

Simple sugars, we know, can exist as rings or open chains, but in polysaccharides the simple sugar subunits occur as rings.

As we said, the most common and most important monosaccharide is glucose (an aldehyde), the 6-carbon sugar that is not only fundamentally involved in energy metabolism and photosynthesis, but is the building block from which many complex carbohydrates are built. It is also one of the two subunits of sucrose. Glucose is called blood sugar, corn sugar, or grape sugar, depending on its source. But whatever its source, glucose—or dextrose, as it is also called—is the energy source intravenously introduced into patients in emergencies and after surgery. (See Essay 3.1 for a look at some of the ways glucose can be represented.)

ESSAY 3.1

READING STRUCTURAL FORMULAS

Chemists have several ways of representing organic molecules. We'll use glucose as an example. The chemical formula, $C_6H_{12}O_6$, provides information about the elements and their proportions, but conveys little about the molecule's three-dimensional form.

(a) Space-filling models are the most representative of what an actual molecule might look like if we could see it. They are also useful for viewing the arrangement of molecular orbitals. Recall that the orbitals of each atom change as molecules form.

Glucose is most often represented by ring formulas. In some **(b)**, the bonds are shown in a tapered form in an effort to present a three-dimensional or geometric view. This can also be done in the so called "chair" form **(c)** of the molecule, which goes even further in providing a three-dimensional view.

A somewhat abbreviated structural formula **(d)** might omit the symbol for carbon in the ring but leave the —H and —OH groups in place. Note that the number 6 carbon is shown. A "bare-bones" structural formula **(e)** leaves out all elements except the oxygen of the ring. This form is often used when a number of glucose subunits are shown, such as in the case of a polysaccharide.

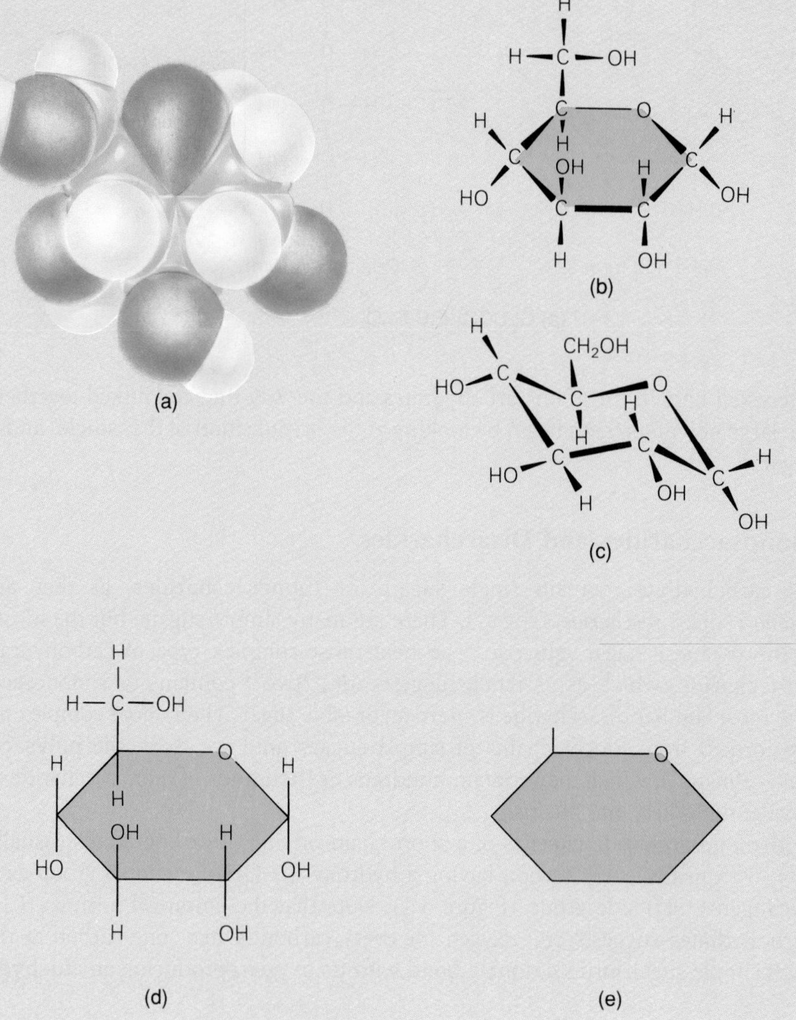

Sucrose, a disaccharide, is made up of one glucose and one fructose subunit (Figure 3.2). Technically, it is a 12-carbon sugar consisting of two 6-carbon sugars (glucose and fructose) linked together in a **dehydration linkage**.

Figure 3.2 illustrates what is meant by the term *dehydration linkage*. Two —OH side groups combine (X_1—OH + HO—X_2) to form an oxygen bridge (X_1—O—X_2) with the liberation of the two hydrogens and one oxygen as H_2O. Such a water-removing chemical reaction, called **dehydration synthesis**, requires the presence of biologically active proteins called **enzymes**. As you might expect of such a chemical reaction, dehydration synthesis requires a source of energy. Many of the synthetic reactions of life are driven by ATP (adenosine triphosphate), a universal energy storage molecule in organisms.

Enzymes abound in the cell and are the catalysts—the activating agents—of most of the chemical reactions of life. (Enzymes are discussed in detail in Chapter 6, but let's just say here that enzymes are molecules than encourage certain chemical reactions within the body.)

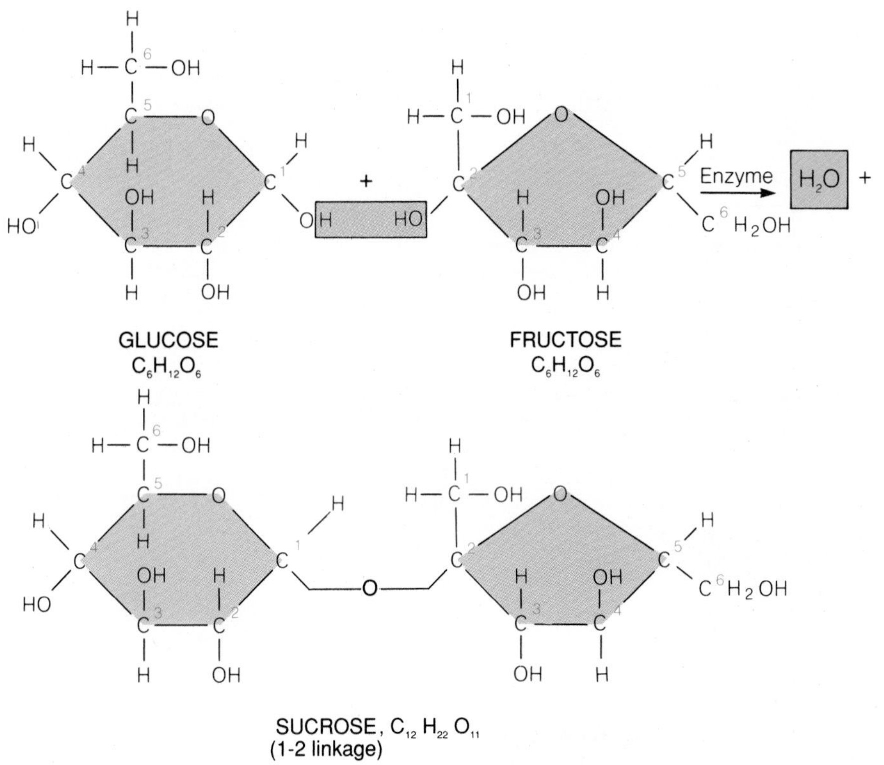

GLUCOSE $C_6H_{12}O_6$ + FRUCTOSE $C_6H_{12}O_6$ —Enzyme→ H_2O +

SUCROSE, $C_{12}H_{22}O_{11}$
(1-2 linkage)

FIGURE 3.2
THE MAKING OF SUCROSE: TABLE SUGAR
The monosaccharides glucose and fructose become linked through dehydration to form the disaccharide sucrose (table sugar). In the enzyme-mediated reaction, a hydrogen (—H) group from one sugar and a hydroxyl (—OH) group from the other combine to form water. The oxygen and carbon to which they were bonded now share electrons, forming a covalent 1–2 linkage.

In your gut (the biologist's indelicate term for gastrointestinal tract), the sucrose is broken back down into glucose and fructose by **hydrolytic cleavage** (*hydro-*, "water"; *lysis*, "rupture"), which is just the opposite of dehydration synthesis. Here, with the assistance of an enzyme, a water molecule is added to the linkage, breaking the sucrose into its component parts. This is necessary because your gut can't absorb sucrose very efficiently, and the cells of your body can't metabolize it. (If you gorge yourself on table sugar, some undigested sucrose may get into your bloodstream, but it will be excreted in your urine.)

The carbons of each sugar are numbered as we saw earlier (see Figure 3.1), so we can precisely describe where the linkages occur. For example, the dehydration linkage between glucose and fructose is a 1–2 linkage, since the number 1 carbon of glucose is linked by an oxygen bridge to the number 2 carbon of fructose. Biochemists also refer to the bond between adjacent monosaccharides as a **glycosidic linkage.**

Another disaccharide of interest is **lactose**, one of the so-called milk sugars. Lactose, in fact, is found *only* in milk (Figure 3.3). Lactose is not as sweet as sucrose, and whereas babies thrive on it, many adults (including nearly all non-Caucasian adults) who consume milk products suffer from **lactose intolerance**. They lack the enzyme that breaks lactose into its two subunits—glucose and galactose—and therefore can't digest lactose. Instead, the lactose, with other milk sugars, passes into the colon (bowel), where it is attacked by gas-forming bacteria, often causing an accumulation of painful gas.

Polysaccharides

Turning to the larger carbohydrates, the polysaccharides, let's again note that they are formed from monosaccharide subunits covalently linked into lengthy chains of **polymers.** The polysaccharides have many forms and serve a large number of functions, among which are food storage and structure. Many familiar foods (potatoes, wheat and corn flour, whole grains, seed vegetables, fruits) contain large amounts of polysaccharides in the form of **plant starches. Glycogen** is another common polysaccharide, formed by animals as a means of storing glucose and often called "animal starch." Then there are **cellulose** and **chitin**, two polysaccharides used, not as storage carbohydrates, but as

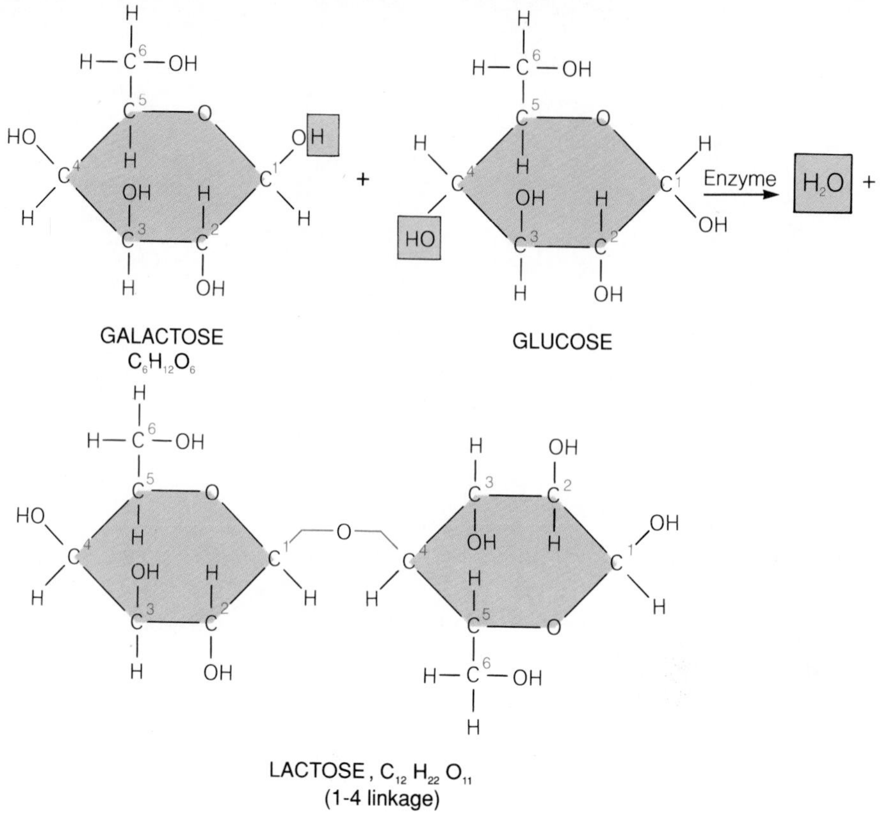

FIGURE 3.3
THE MAKING OF LACTOSE: MILK SUGAR
The monosaccharides galactose and glucose are identical except for the orientation of —H and —OH groups on carbons one and four. Together, they form the disaccharide lactose (milk sugar) through the usual dehydration reaction, but notice that one of the subunits (here we've made it glucose) turns over when the 1–4 linkage forms.

GALACTOSE
$C_6H_{12}O_6$

GLUCOSE

Enzyme → H_2O +

LACTOSE, $C_{12}H_{22}O_{11}$
(1-4 linkage)

structural material. Cellulose is a structural polysaccharide in plant cell walls, and chitin is used by insects to form exoskeletons (skeletons surrounding the body), as well as by fungi to form the walls surrounding their cells.

Plant and Animal Starches Starches form a large part of our diet, perhaps too large a part. Most snack foods ("junk foods"?) are high-calorie foods largely because of their high starch and sucrose content, and, often, a considerable amount of fat. Nevertheless, the starches are vitally important storage carbohydrates, which are readily metabolized for energy. The two most common starches are **amylose** and **amylopectin**—important food reserves for higher plants. All starches are made entirely of glucose subunits.

Amylose Amylose is the simplest starch (Figure 3.4). Its molecules consist of unbranched chains of hundreds of glucose subunits, joined together by 1–4 dehydration linkages. Potato starch is about 20% amylose. In the gut or in a sprouting potato, breakdown (digestion) occurs in three steps: one enzyme breaks amylose into fragments of varying size, attacking the starch at random points; a second enzyme works on the ends of the fragments, cleaving off two glucose units at a time as disaccharides (glucose-1, 4-glucose, or *maltose*); and a third enzyme cleaves the disaccharides into glucose monomers.

Amylopectin The other 80% of potato starch is amylopectin, which is a large, highly branched polymer. Its main chain of 1—4 linked glucose subunits gives rise to side branches that form via 1—6 glucose linkages. Such branches, in turn, produce their own branches, again through 1—6 glucose linkages (Figure 3.5). The digestive enzymes that cleave the 1—4 linkages of amylose also cleave those in amylopectin, but digestion is stalled at the 1—6 linkages until a new enzyme can act there.

Glycogen Glycogen, the storage polysaccharide of animals, is usually a much larger molecule than are amylose or amylopectin, and is highly branched. The predominant

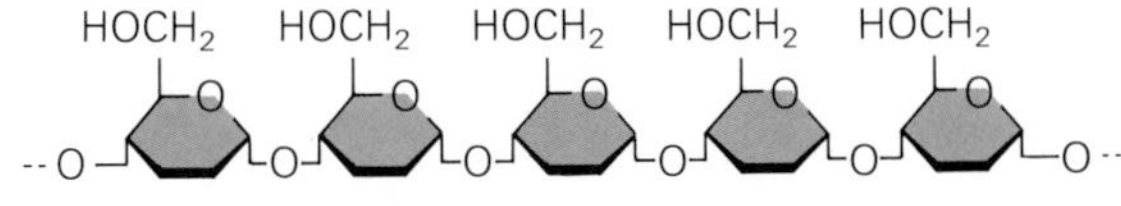

(a) Unbranched chain (amylose)

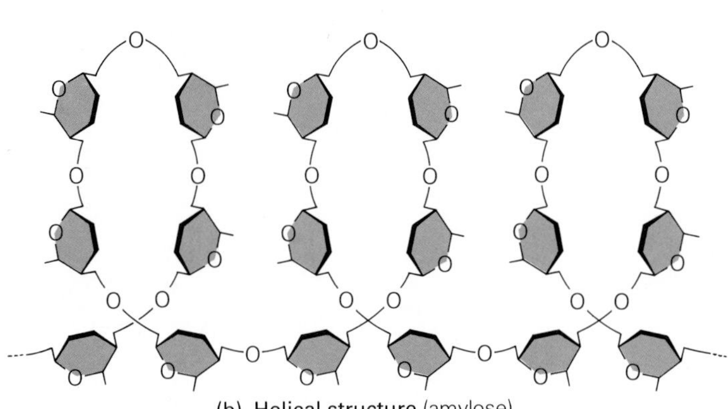

(b) Helical structure (amylose)

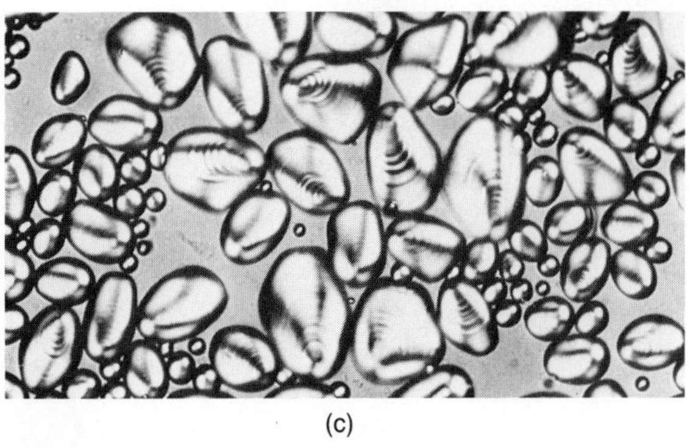

(c)

FIGURE 3.4
THE PLANT STARCH AMYLOSE
(a) The plant starch amylose, a polysaccharide, consists of hundreds of glucose subunits joined through 1–4 linkages. **(b)** When in water, the chain spontaneously forms a helix, and in this form is insoluble. **(c)** In plant cells, such as those of the potato tuber, amylose chains form readily visible starch grains.

linkage is 1–4, but there are many 1–6 linkages as well. Glycogen is a temporary, short-term storage unit in animal cells, particularly prevalent in the liver and muscles of vertebrate animals (Figure 3.6).

Primary and Secondary Structure of Starches Starches can be described according to their structures. The straight- and branched-chain polymers we've just described are called the *primary structures* of the starches. The primary structure of a polymer results from covalent bonding alone, and usually can be shown as a two-dimensional figure. When we referred to amylose as an unbranched chain, we had the primary structure in mind. In reality, amylose is a three-dimensional structure that loops and folds and forms helices (like winding staircases), as shown in Figure 3.4b. Amylose forms helices because

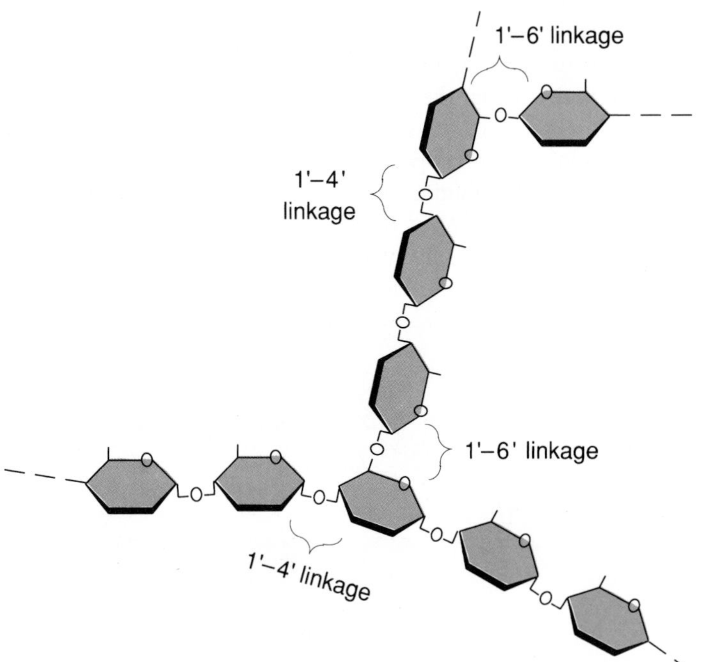

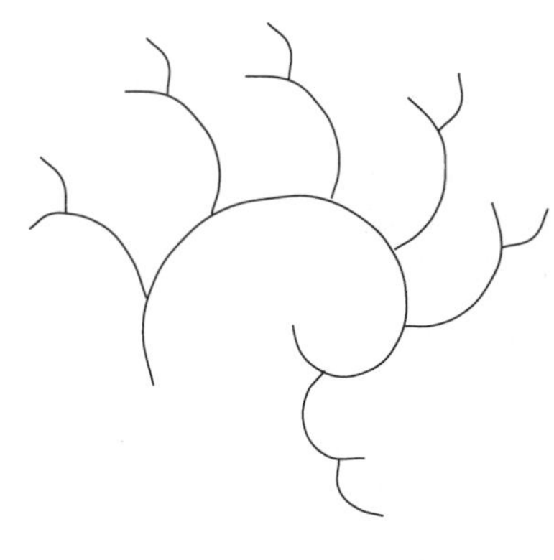

Branching pattern (a 1'–6' branch every 30 glucose units)

FIGURE 3.5
THE PLANT STARCH AMYLOPECTIN
The plant starch amylopectin includes a main chain and its branches, all composed of glucose subunits joined by 1–4 linkages. The branches attach to the main chain through 1–6 linkages.

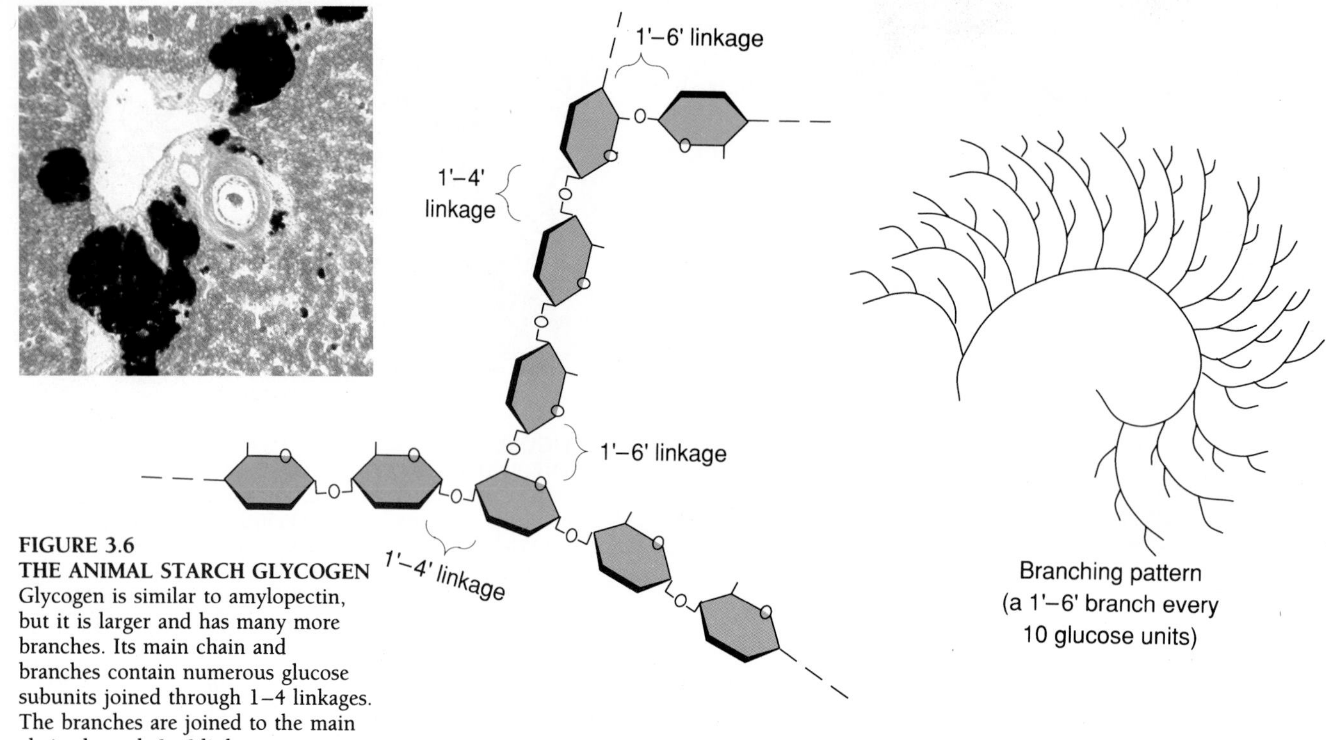

FIGURE 3.6
THE ANIMAL STARCH GLYCOGEN
Glycogen is similar to amylopectin, but it is larger and has many more branches. Its main chain and branches contain numerous glucose subunits joined through 1–4 linkages. The branches are joined to the main chain through 1–6 linkages.

of hydrogen bonding between subunits along its length. These helices make up the *secondary structure* of the molecule. Whereas we are fairly confident about the secondary structure of amylose, the secondary structures of amylopectin and glycogen are less well understood.

Structural Polysaccharides: Cellulose Cellulose is a linear polymer of glucose subunits put together with 1–4 dehydration linkages. If that sounds familiar, it's because amylose is also a linear polymer of glucose subunits put together with 1–4 dehydration linkages. Nonetheless, amylose and cellulose are very different. Starch is fairly soluble, while cellulose is not; cellulose has great tensile (stretch) strength, while starch does not; starch is readily broken down by digestive enzymes, but cellulose is completely indigestible to all but a very few organisms. So, whereas the structural differences in the subunits of starch and cellulose are very slight (Figure 3.7), those slight differences have far-reaching effects. Since cellulose is by far the most abundant polysaccharide on earth, we will want a close look.

The differences between cellulose and other polysaccharides can be traced, ultimately, to the specific ring form of its glucose subunits and the way in which they link up in the polymer. There are actually *two* ring forms of glucose in addition to the straight chain form. Up to now we have concerned ourselves with the alpha ring only (the kind of glucose found in sucrose and in starches). However, there is a beta ring form of

Alpha glucose

Amylose

Beta glucose

Cellulose

FIGURE 3.7
COMPARING AMYLOSE TO CELLULOSE
Like other polysaccharides, cellulose consists of 1–4 linked glucose subunits, but because beta-glucose is the subunit, the structure of cellulose and some of its chemical and physical characteristics are quite different from those of amylose. Note the orientation of —H and —OH side groups on the number 1 carbon of beta-glucose and how this effects the positioning of each beta-glucose subunit in the chain. Note also that there is an absence of branches.

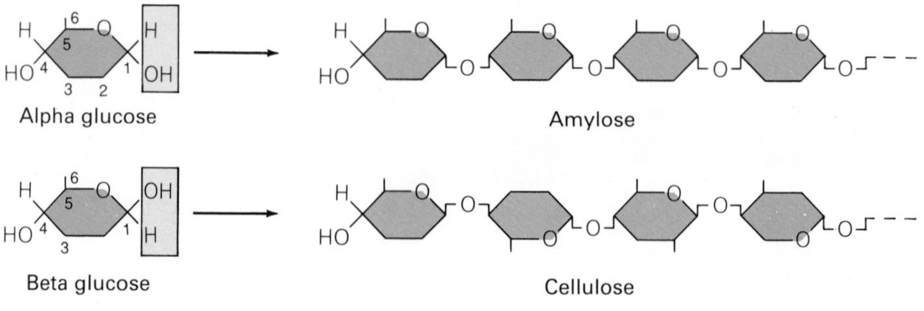

glucose as well. The difference between the two is in the orientation of —H and —OH groups on the number 1 carbon, as seen in Figure 3.7. Consequently, as the 1–4 linkages of cellulose form, every other glucose in the chain is turned over.

Cellulose is difficult to digest precisely because of these linkages. In cellulose, glucose units are linked by what is called a **beta-glycosidic linkage**. Relatively few organisms produce the enzyme **cellulase** necessary to break such a linkage. Garden snails can digest cellulose (as in the leaves of whatever you are planting). Fungi and some bacteria also have cellulase. But what about animals such as grazers and termites? It turns out that they don't digest cellulose themselves, but they harbor helpful intestinal microorganisms that can break it down. The products of cellulose digestion are absorbed (and even the microorganisms themselves are ultimately digested and absorbed). What about the fact that people eat plants? People can eat all the cellulose they want and it will just go right through. In fact, it may go through fairly fast because it irritates the gut lining, causing lubricating fluids to be released. (Whole-bran cereals are rich in cellulose.) Some researchers are convinced that enhancing the expediency of nature's call through increased fiber reduces the time that potential carcinogens are in contact with the bowel, thereby reducing the chance for cancers to develop there.

Digestive considerations aside, the beta-glycosidic linkages of cellulose produce other unusual characteristics. Unlike other polysaccharides, cellulose polymers do not form coils, nor do they cluster into granules. They remain linear, and each polymer attracts others until some 60 to 70 are drawn together into dense cable-like strands called **microfibrils** (Figure 3.8). The attraction is due to hydrogen bonds that occur between oxygen in one polymer and the hydrogen of hydroxyl groups in another. Whereas the individual hydrogen bonds are quite weak, there are a great many in each microfibril. Thus cellulose microfibrils are enormously strong. (Actually, their tensile strength is greater than that of steel wire of a similar diameter.)

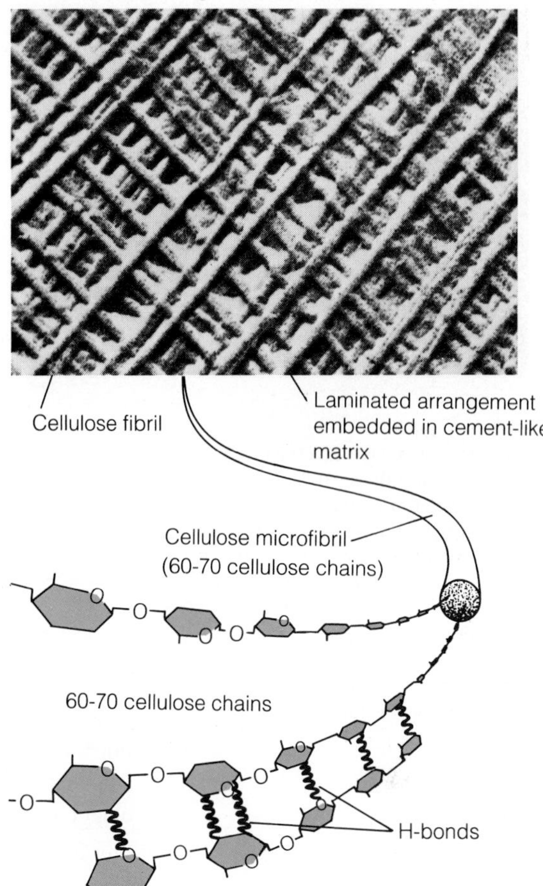

FIGURE 3.8
CELLULOSE MICROFIBRILS
Sixty to seventy hydrogen-bonded cellulose chains form microfibrils, which in turn make up the fibrils seen in the electron micrograph. The fibrils are embedded in cementing materials, forming a laminate that provides great strength to the plant cell wall.

Cellulose microfibrils, when organized into larger strands called **fibrils**, are used in the construction of plant cell walls. As the fibrils are laid down, they take on a laminated and criss-cross arrangement. Then, as growth of the cell wall is completed, cement-like hardening and strengthening agents are added (see Figure 3.8). As a result, plant cell walls are among the strongest of biological structures. If you don't believe this, go throw your best punch at a tree, and report back. The walls of countless non-living cells in a tree trunk, when impregnated with another chemical, **lignin**, become the tough, useful wood of woody plants. But, in spite of its strength, cellulose is also flexible, and under certain conditions it can even be made soft. Your experiences with the paper goods in college restrooms aside, those soft bathroom tissues we hear about on prime time TV are almost pure cellulose.

Other Structural Polysaccharides In the younger tough walls that surround plant cells, cellulose fibers are imbedded in a gluey matrix of **hemicelluloses** and **pectins**. The hemicelluloses, despite the name, are not structurally related to cellulose. Instead, they include polymers of some of the less common 5-carbon sugars. Pectin is the substance that gives jelly its consistency. The strong cell walls of plants can be compared to iron-reinforced concrete, or resin-impregnated fiberglass, or any other materials that owe their strength to fibers imbedded in an amorphous (shapeless) matrix.

Other structural polysaccharides are found in various seaweeds, from which they are extracted and sold to the food industry for such esoteric uses as thickeners in milk shakes. Also, many bacteria secrete slimy protective coats of polysaccharide material.

FIGURE 3.9
CHITIN AND THE EXOSKELETON
Lobsters and other arthropods have an exoskeleton containing the complex carbohydrate chitin, a polymer of N-acetylglucosamine subunits. In many aquatic arthropods (both in fresh and salt water), the exoskeleton is hardened by calcium.

N-acetyl glucosamine

Chitin

Chitin is a principal constituent of the exoskeletons (external skeletons) of insects and other arthropods, including lobsters and crabs (Figure 3.9). Chitin itself is rather flexible and leathery, as is found in the exoskeletons of grasshoppers and cockroaches, but it can become very hard when impregnated with calcium, as in lobsters and crabs. It is a structural polysaccharide similar in many ways to cellulose, except that instead of glucose, the basic unit is **N-acetylglucosamine**. Notice that this carbohydrate contains nitrogen as well as carbon, oxygen, and hydrogen. (The *N*-acetyl part of the name of this compound means that an acetyl group, —$COCH_3$, is attached to the main structure through a nitrogen atom.) Chitin is indigestible to most animals.

Acid Mucopolysaccharides and Proteoglycans

For the record, **proteoglycans** are a class of molecules that contain acid **mucopolysaccharides** complexed with protein. Acid mucopolysaccharides are essentially polysaccharides, but their sugars contain nitrogen, and as their name implies, they also have acidic groups. In vertebrates, proteoglycans form much of the jellylike substance that fills the eye and spaces between cells. Proteoglycans occur in cartilages, tendons, skin, and in the fluids that lubricate skeletal joints. Another important molecule in this class is **heparin**, which is present in the walls of arteries. It has the vital role of inhibiting the clotting of blood within the circulatory system. Of course, where an injury has occurred, heparin's effect is overridden by the blood-clotting mechanism. Heparin is used clinically where the prevention of clotting is required, such as in the treatment of stroke.

THE LIPIDS

The **lipids** are a diverse group of molecules, defined by their solubility rather than by their structure.

Lipids tend to be greasy or oily, and fat-soluble rather than water-soluble. They are generally nonpolar and tend to dissolve in organic solvents, such as gasoline, chloroform, paint thinner, and salad oil. These are all substances that lack charges in their functional groups. Whereas carbohydrates and proteins are hydrophilic ("water-loving"), lipids are hydrophobic ("water-fearing;" see Chapter 2). It may seem odd that an important class of organic compounds is defined solely on the basis of a solubility characteristic, but it does appear that water-insolubility and fat-solubility gives all lipids, no matter how

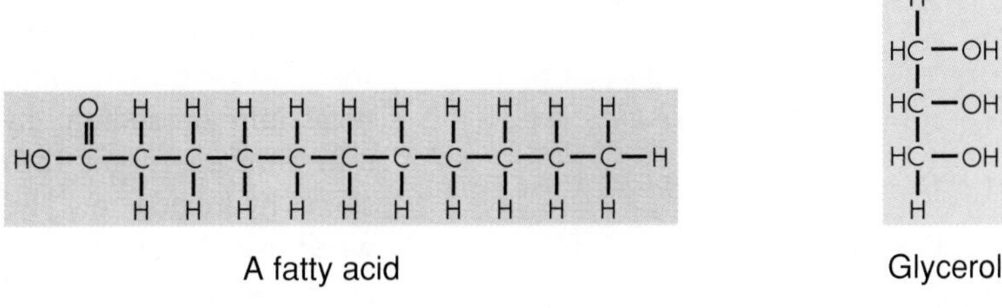

FIGURE 3.10
TRIGLYCERIDE SUBUNITS
Triglycerides consist of one molecule of glycerol and three molecules of fatty acid. Many different fatty acids are known.

different they are structurally, some common characteristics. Lipids may be small molecules, large molecules, monomers, polymers, energy storage molecules, structural molecules, hormones, lubricants, or parts of proteins and carbohydrates.

Important groups within the lipid category are **fats**, **oils**, **sterols**, **waxes**, and **phospholipids.** Most familiar to us are the fats (such as beef tallow) and the oils (such as corn oil, safflower oil, and cod liver oil). All fats and oils are **triglycerides**—that is, compounds consisting of three fatty acid chains attached to a molecule of glycerol (see below). The difference between a fat and an oil is that a fat has a higher melting point; fat is solid at room temperature but oil, with its lower melting point, is liquid at room temperature. But oils, and even fats, also tend to liquefy at the normal temperature of the warm-blooded animals in which they occur. Because of this ambiguity, and to avoid confusion with petroleum oil and with other meanings of the word *fat,* biologists prefer to use the term *triglycerides* to refer to oils and fats.

Triglycerides

Glycerol, formerly known as *glycerine,* is a small molecule. It has three carbons and three hydroxyl side groups. Linked to the glycerol in a triglyceride are three fatty acid molecules. A **fatty acid** molecule consists of a hydrocarbon chain (that is, a chain consisting of a backbone of carbon atoms with only hydrogens as side groups) with a carboxyl group at one end (Figure 3.10). The synthesis of a triglyceride from glycerol and three fatty acids is a dehydration process whereby three molecules of water are formed for each triglyceride produced. Note that enzymes—the organic catalysts of cellular reactions—are required for the reaction to go on (Figure 3.11).

Fatty acids can be of many different lengths and usually contain an even number of carbon atoms, occurring most commonly in chains of 14, 16, 18, or 20. Free fatty acids are not very soluble in water, but in alkaline solutions the carboxyl group becomes ionized, making it strongly polar and thus soluble. Sodium and potassium salts of fatty acids are known as "soap" and are soluble in water. If you can get along without the

FIGURE 3.11
FORMATION OF A TRIGLYCERIDE
In the synthesis of a triglyceride, dehydration linkages form between the three —OH groups of glycerol and the —OH groups of three fatty acids. The products are one molecule of triglyceride and three of water.

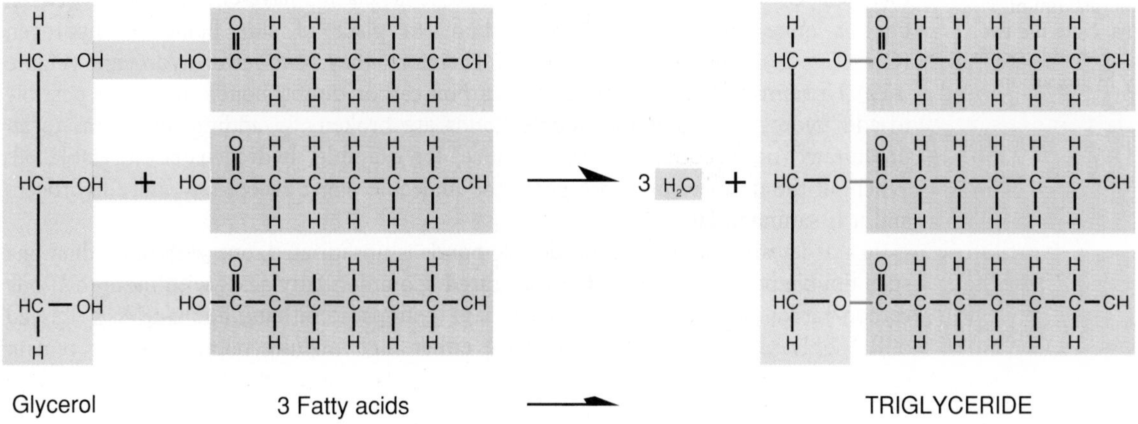

Glycerol + 3 Fatty acids → $3\ H_2O$ + TRIGLYCERIDE

added perfumes and deodorants, you can make your own soap from waste grease from the kitchen and potassium or sodium hydroxide (our forefathers got their potassium hydroxide or "lye" from wood ashes). Just obtain a recipe and be careful with the sodium hydroxide. You can put this soap in the guest bathroom.

Calcium salts of fatty acids, on the other hand, are not soluble in water, which is why soap forms disconcerting curds in hard (calcium-rich) water. Calcium salts also explain the "ring around the tub" and the dullness of hair washed in hard water.

Triglycerides are completely nonpolar because the carboxyl groups of fatty acids are bound up in linkages with the hydroxyl groups of glycerol. A dehydration linkage between a carboxyl group and an alcohol side group is called an **ester linkage**. An **alcohol** is technically any substance with the formula X—C—OH. For example, the ethyl alcohol, or the ethanol of liquor, is CH_3CH_2—OH. Thus, glycerol is an alcohol.

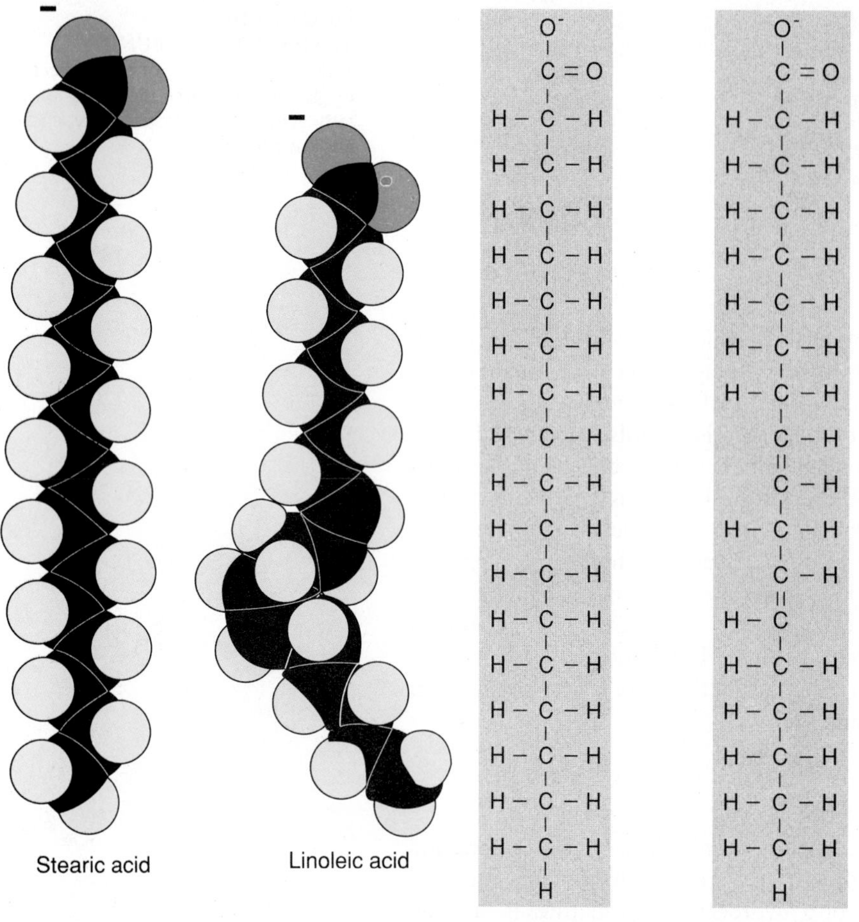

FIGURE 3.12
SATURATED AND UNSATURATED FATTY ACIDS
Stearic acid is a principal fatty acid of animal fat. As in other saturated fats, its carbon chain contains the maximum number of hydrogens. Linoleic acid, abundant in cotton seeds, is an unsaturated fatty acid. Because of its two double bonds, its carbon chain contains less than the maximum number of hydrogens. Note the two sharp bends where the double bonds occur in the lengthy hydrocarbon chain.

Saturated, Unsaturated, and Polyunsaturated Fats

Why all the diet talk about saturated fats? The obvious question here is, saturated with what? You may recall from an earlier discussion that the answer is hydrogen. A **saturated fat** is a triglyceride lacking carbon-carbon double bonds. Therefore it contains all the hydrogens possible.

The carbon backbone of the hydrocarbon tails of fatty acids is composed primarily of carbon-to-carbon single bonds (one pair of shared electrons between successive carbons: C—C—C—C—C, and so on). But some fatty acids have double bonds between one or more pairs of successive carbons in place of side bonds to hydrogen (C = C—C—C—C = C, and so on). Therefore, they have fewer hydrogens (Figure 3.12). **Unsaturated fat** has one or more carbon-carbon double bonds and so it is possible to add hydrogen to it if the double bonds are broken. By adding hydrogens to an unsaturated triglyceride, one can produce, for example, hydrogenated vegetable oil. When all the double bonds are gone, the molecule won't accept any more hydrogen, and it is saturated fat.

Just as fat with carbon-carbon double bonds is unsaturated, one with more than one such double bond is called **polyunsaturated**. Common fatty acids with more than one carbon-carbon double bond are linoleic acid (cottonseed oil) and linolenic acid (linseed oil). Highly saturated fats tend to have rather high melting points (tend to remain solid, like lard, at room temperature), while polyunsaturated fats tend to have low melting points (tend to remain liquid or "oily"). Simple vegetable oils, such as peanut oil, are largely unsaturated, which explains why the oils in old-fashioned peanut butter separate, forming a layer at the surface (which usually manages to run down the outside of the jar). The problem is solved by the food industry through *hydrogenation*—partially saturating the oil with hydrogen, or adding a hydrogenated oil, thus forming a harder fat (cooks call it "shortening") that does not separate at room temperature.

Unsaturated fatty acids are essential in the diet. They are precursors to important molecules we produce, and the membranes of our cells are rich in unsaturated fatty acids. But because chemical processes in humans and other mammals cannot produce double bonds beyond the ninth carbon in a given fatty acid chain, it is necessary to include a small amount of unsaturated fats in the diet. We require both linoleic and linolenic acids as essential fatty acids. Actually, such fatty acids are common in many foods, so lipid nutritional deficiencies rarely arise.

Because of their molecular properties, fats make ideal energy storage molecules. Plants store triglycerides in their seeds, and many kinds of animals store body fat as an adaptation for lean seasons or migration. Humans also store fat under the skin and around internal organs, although agriculture and food storage techniques have largely exempted us from the rigors of seasonal food depletion.

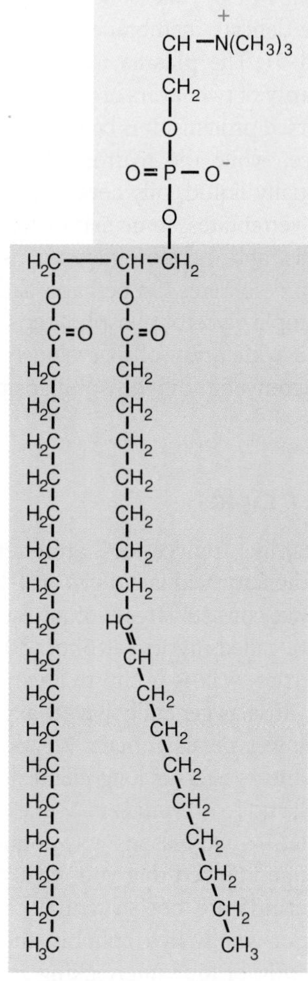

(a) Structural formula of phosphatidylcholine

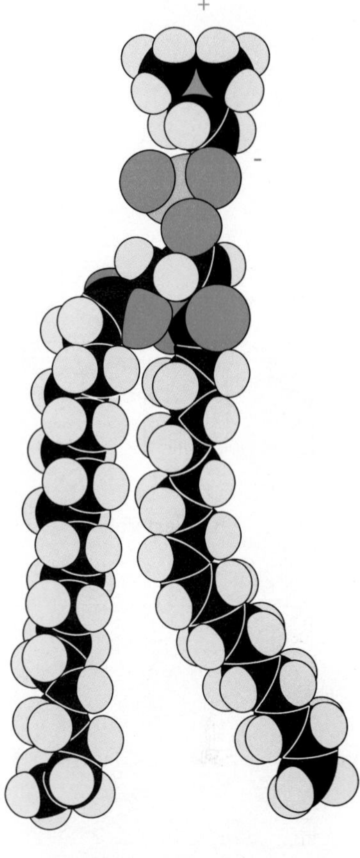

(b) Space-filling model

FIGURE 3.13
THE PHOSPHOLIPID
(a) In phospholipids, two of glycerol's oxygens are bonded to fatty acids (typically one saturated and one unsaturated). The third is linked to a phosphate group, which in turn is joined to one of several possible variable groups, in this example, the entire molecule shown here is called phosphatidylcholine. Phospholipids are common constituents of cell membranes. **(b)** The space-filling model reveals the peculiar bent form of the unsaturated fatty acid.

Phospholipids

While the **phospholipids** are structurally closely related to the triglycerides, there are vital differences between them. In the phospholipids, two fatty acids are linked to a backbone of glycerol by ester linkages. In place of the third fatty acid found in triglycerides, however, phospholipids have a phosphate group, which is commonly linked to still other organic molecules (variable groups) as shown in Figure 3.13. Of the two fatty acids, one is commonly unsaturated and, in that one, the presence of the carbon-to-carbon double bond produces a peculiar but very significant bend in the tail, as seen in the illustration. The bent tail has a lot to do with membrane fluidity, as we will see in Chapter 5. The phosphate group is hydrophilic, so the charged phosphate heads mingle freely with water and other polar molecules. The tails themselves are hydrophobic, so they reject water but mingle freely with nonpolar substances. Actually, this gives phospholipids a detergent property in the sense that they can form bridges between water and oils (see Chapter 2). In addition they reduce the surface tension of water, thus acting as wetting agents.

Phospholipids can occasionally be found as food storage compounds (for example, lecithin in egg yolk), but they are more important in their roles as major components of the delicate membranes that surround and occur within most living cells (Chapters 4 and 5). The **plasma membrane**, as the surrounding membrane is known, consists primarily of two layers of phospholipids, along with glycolipids, glycoproteins, and other dispersed proteins. It is believed that the hydrophilic phosphate ends form the membrane surface, while the hydrophobic fatty acid tails dissolve into one another, giving an essentially liquid, oily core to the membrane.

In vertebrates, some nerves are sheathed in layers of plasma membranes called myelin sheaths, and therefore nervous tissue—especially that of the brain—has a high phospholipid content. (So perhaps "fathead" is an unintentional compliment.) In addition to the simple glycerol phospholipids already mentioned, the brain and other nervous tissue have a wide array of marvelously named and complex phospholipids and glycolipids: sphingomyelin, dihydrosphingosine, cerebrosides, gangliosides, and many more just as poetic.

Cholesterol

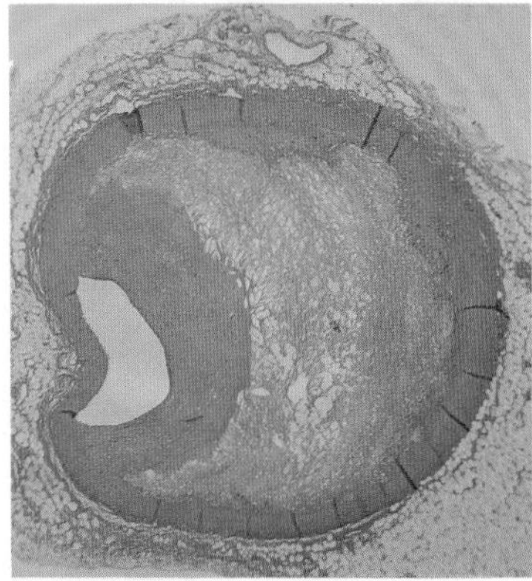

FIGURE 3.14
CHOLESTEROL AND ARTERIAL DISEASE
The steroid cholesterol is implicated in atherosclerosis, a vascular disease in which thickenings and rough spots called plaques form in arterial linings. In addition to restricting blood the roughened linings may trigger blood-clotting, leading to stroke.

Other Lipids

Waxes, like triglycerides, are esters of fatty acids. However, the alcohol that combines with the fatty acid is not glycerol, but a long-chain alcohol with a single hydroxyl group. Beeswax consists largely of an ester of palmitic acid (16 carbons) with a straight-chain alcohol called myricyl alcohol (30 carbons). Because of its hydrophobic, water-repelling properties, wax is useful to a variety of living things, especially those that need to save water. Insects generally have waxy cuticles. So do plants; the waxes of leaves, fruit skins, and flower petals contain very long fatty acids (up to 36 carbons), both as free fatty acids and as esters of long-chain alcohols. Many marine invertebrates manufacture various waxes, and so do your ears. Waxes are generally much harder and even more hydrophobic than fats—two reasons why commercial wax is used as a protectant. Heads would turn if you announced that you were going to "fat" your car.

Steroids are not structurally similar to fatty acid lipids, but since they are either partly or wholly hydrophobic, they qualify as lipids. All steroids contain a peculiar core consisting of four interlocking rings. The differences between various steroids depend on the variation in the side chains. Some steroids are highly hydrophobic, some less so. (By the way, steroids with —OH groups are also called **sterols**.) Let's consider two specific examples of steroids.

Lanolin (or **lanosterol**) is a greasy, almost waxy substance that is commercially refined from sheep's wool. Small amounts of lanolin are found in your own skin and hair, helping to keep them flexible.

Cholesterol is a substance we've all heard of because various food manufacturers have assaulted us over the airways with the message that their products don't contain any. Why should we care? We might be interested because cholesterol (Figure 3.14) has been accused of contributing to "plaque" (thickening) in the lining of our blood vessels and thereby reducing their internal diameter and increasing blood pressure the way one might increase water pressure by placing the thumb over the end of a garden hose. But cholesterol is not all bad. In fact, it has a number of vital functions. For example, like phospholipids, cholesterol is a major constituent of plasma membranes. Also, bile acids, which are necessary for fat digestion, are modified cholesterol. When irradiated with ultraviolet light, a derivative of cholesterol becomes vitamin D, which is necessary for normal bone growth and maintenance. You may have a greater appreciation for cholesterol when you realize that it can also be modified to make the various steroid hormones, including the sex hormones.

THE PROTEINS

For many reasons, proteins are incredibly important molecules. We've seen that carbohydrates function both as food reserves and as structural molecules. Proteins may function in these ways as well, but they also play other, equally vital, roles. Some are chemical messengers, or hormones, whereas others function in oxygen transport in the blood stream. Still others have contractile qualities and are responsible for movement. The antibodies that defend us against invading organisms are also proteins.

Perhaps the most interesting of all are the enzymes, prime movers in chemical reactions of living organisms. We have mentioned the basic role of enzymes, but here let's stress that enzymes are biologically active proteins that work in cells as catalysts—agents that enhance chemical activity and in some instances, bring about reactions that might never occur in their absence. As we proceed into the chemical reactions of cells you will soon develop an abiding respect, perhaps even an awe, for these amazing perpetrators of cellular activity, since they seem to be involved in nearly every conceivable aspect of life. We will return to enzymes later, but putting them aside for now, let's move to a more general consideration of proteins and perhaps pose some basic questions.

First, what *is* a protein? A **protein** is a macromolecule composed of one or more **polypeptides.** A polypeptide is a linear chain of **amino acids** joined together by **peptide bonds** formed, basically, through the enzyme-mediated dehydration process. (This is a huge oversimplification of polypeptide synthesis, as we will see in Chapter 13.) A polypeptide can be sizable, often containing hundreds of amino acids. On a considerably smaller scale is the **dipeptide,** two amino acids joined by a single peptide bond, and the **peptide,** a short chain of amino acids joined together by peptide bonds.

The distinction between polypeptide and protein is not always clear, by the way. Most proteins consist of two or more polypeptides, are biologically active, and take on complex shapes; however, polypeptides may also be active and complex in shape. Thus, some polypeptides are functional proteins, but for now, let's consider polypeptides as simple chains of amino acids. We will take up protein structure shortly.

The range in molecular weight of the various proteins is enormous—from about 5700 for insulin (which has 55 amino acids) to 7–8 million in some of the largest enzymes (those with 70–80 thousand amino acids). Let's begin with the protein subunits, the amino acids.

Amino Acids

Amino acids are the molecular subunits, the building blocks, of proteins. There are just 20 different primary amino acids found in the proteins of living cells, and each amino acid has unique properties that help determine the structure and functional capabilities of the protein in which it occurs. Twenty may seem to be a small number considering how diverse and complex proteins are, but we might bear in mind that this book was written with only 26 letters (some may have been used more than once). Actually, there are more than 20 total amino acids but the additional ones are secondary, that is, they are derived from the primary amino acids.

All amino acids have the same core structure, consisting of a central or **alpha carbon** to which an amino group (NH_2), a carboxyl group (COOH), and an **R group** (variable group) are bonded (along with one hydrogen atom). Figure 3.15 shows the amino acid in its molecular and ionized state and reveals some of its geometry. Amino acids readily ionize under standard physiological conditions, which means, within cells and in extracellular fluids. During ionization, the carboxyl group gives off a proton (which is why the molecule is called an acid), and the amino group takes in a proton (which means that it is also a base). In its ionized state, then, an amino acid has both a negative and a positive charge.

It is in their R groups that the twenty amino acids differ. Figure 3.16 shows the amino acids arranged according to R group properties, emphasizing in particular the R group polarity or lack of polarity and the presence or absence of net charges. The polarity of an R group refers to its tendency to interact with water through the formation of hydrogen bonds. Charges form if R groups ionize. These characteristics are of particular interest to biologists because the cell is a naturally watery environment.

Note in Figure 3.16 that there are eight nonpolar, uncharged amino acids; their R groups are mainly hydrocarbons, so there is no tendency for these groups to form

FIGURE 3.15
AMINO ACID STRUCTURE
(a) Each amino acid contains an alpha carbon to which are attached an amino group (—NH_2), a carboxyl group (—COOH), and one of the 20 different R groups. **(b)** Amino acids commonly ionize, with a proton leaving a carboxyl group and joining an amino group. **(c)** The ball-and-stick model reveals some of the geometry of the amino acid. Note that some of the bonds are quite flexible, permitting free rotation. This flexibility affects the shape of polypeptides.

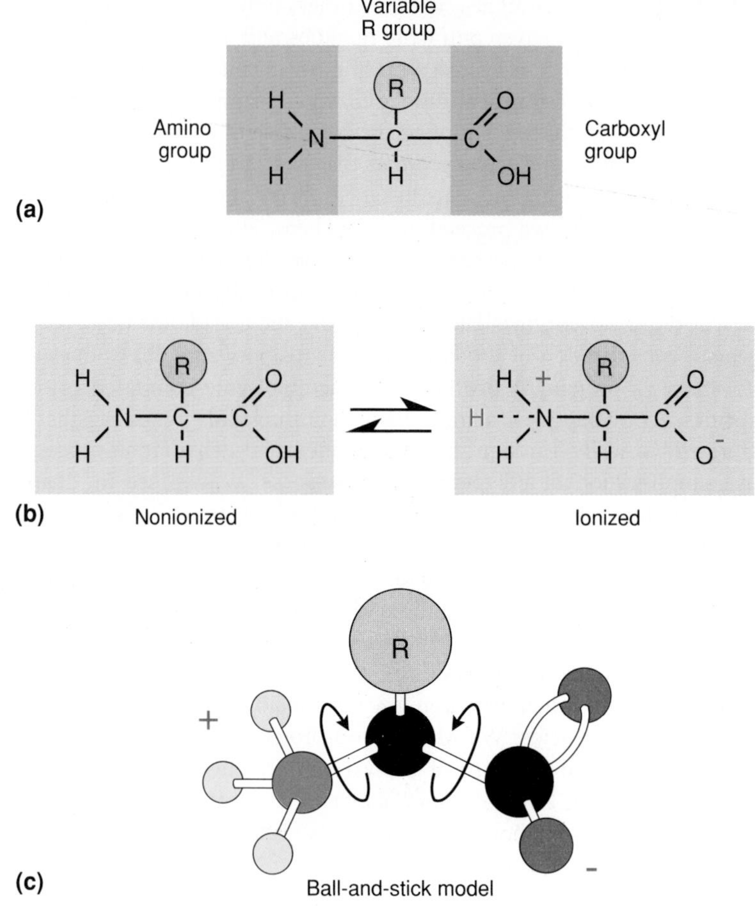

hydrogen bonds or to ionize, thus the nonpolar amino acid R groups are hydrophobic and under the water physiological conditions in the cell, are forced into dense clusters.

The second group of seven amino acids has polar R groups, although these R groups do not ionize and thus have no net charges. The polar R groups readily form hydrogen bonds with water, so the corresponding amino acids are hydrophilic. They are much more soluble than the nonpolar amino acids.

The R groups of the remaining five amino acids ionize in water and thus bear positive or negative charges. Specifically, the R groups of arginine, histidine, and lysine contain an amino group, so these amino acids have a total of two (one in the R group and one attached to the alpha carbon). As a result, when the second amino group ionizes, the amino acid gains a proton and takes on a net positive charge. The R groups of the other charged amino acids, aspartic acid and glutamic acid, contain a carboxyl group. When this group ionizes, it loses a proton. As a result, the last two amino acids bear a net negative charge.

One of the polar, but uncharged amino acids, cysteine (pronounced SIS' tuh'een), has a very special property. It contains a sulfhydryl side group (—SH). Under certain conditions, the sulfhydryl groups of two adjacent cysteines will react, forming a strong covalent bond known as a disulfide bridge (or *disulfide linkage*):

$$R_1\text{—SH} + \text{HS—}R_2 \leftarrow R_1\text{—S—S—}R_2$$

We'll see how important this can be to protein structure in a moment.

FIGURE 3.16 (facing page)
AMINO ACIDS AND THEIR R GROUPS
Proteins are made up of twenty different amino acids. Whereas each (except for proline) has identical amino and carboxyl groups, the R groups differ. The R groups fall into three general types: nonpolar, uncharged polar, and charged (negative or positive). R group characteristics are highly significant to the physical structure of protein.

Polypeptides and Proteins

In polypeptides, the carboxyl group of one amino acid is linked, through the familiar dehydration linkage, to the amino group of the next amino acid; the carboxyl group of

Amino acids with nonpolar R groups (green)

Alanine (Ala)	$CH_3-CH(NH_3^+)-COO^-$
Isoleucine (Ile)	$CH_3-CH_2-CH(CH_3)-CH(NH_3^+)-COO^-$
Leucine (Leu)	$(CH_3)_2CH-CH_2-CH(NH_3^+)-COO^-$
Methionine (Met)	$CH_3-S-CH_2-CH_2-CH(NH_3^+)-COO^-$
Phenylalanine (Phe)	$C_6H_5-CH_2-CH(NH_3^+)-COO^-$
Proline (Pro)	CH_2-CH_2 / $CH_2-\overset{+}{N}H_2$, ring closed through $C(H)-COO^-$
Tryptophan (Trp)	indole ring $(C=CH-NH)$ $-CH_2-CH(NH_3^+)-COO^-$
Valine (Val)	$(CH_3)_2CH-CH(NH_3^+)-COO^-$

Amino acids with uncharged polar R groups (green)

Asparagine (Asn)	$O=C(NH_2)-CH_2-CH(NH_3^+)-COO^-$
Cysteine (Cys)	$HS-CH_2-CH(NH_3^+)-COO^-$
Glutamine (Gln)	$O=C(NH_2)-CH_2-CH_2-CH(NH_3^+)-COO^-$
Glycine (Gly)	$H-CH(NH_3^+)-COO^-$
Serine (Ser)	$HO-CH_2-CH(NH_3^+)-COO^-$
Threonine (Thr)	$CH_3-CH_2(OH)-CH(NH_3^+)-COO^-$
Tyrosine (Tyr)	$HO-C_6H_4-CH_2-CH(NH_3^+)-COO^-$

Amino acids with acid R groups (green)
(negatively charged at pH 6.0)

Aspartic acid (Asp)	$^-O(O=)C-CH_2-CH(NH_3^+)-COO^-$
Glutamic acid (Glu)	$^-O(O=)C-CH_2-CH_2-CH(NH_3^+)-COO^-$

Amino acids with basic R groups (green)
(positively charged at pH 6.0)

Arginine (Arg)	$^+H_3N-C(=NH)-NH-CH_2-CH_2-CH_2-CH(NH_3^+)-COO^-$
Histidine (His)	$HC{=}C-CH_2-CH(NH_3^+)-COO^-$, ring: ^+HN, NH, $C(H)$
Lysine (Lys)	$^+H_3N-CH_2-CH_2-CH_2-CH_2-CH(NH_3^+)-COO^-$

*The portion of the amino acid that makes each amino acid different is colored. Note that some of the amino acids contain more than one amino or acid group, giving them greater basic or acid qualities than others. Cysteine contains a sulfur-hydrogen group at its *R*-terminal. This has special importance in determining the shapes of proteins. The abbreviations given in parentheses are used for convenience in writing protein formulas.

FIGURE 3.17
FORMATION OF A PEPTIDE BOND
Peptide bonds form through the dehydration process. They are linkages that join two amino acids by their carboxyl and amino groups as shown here.

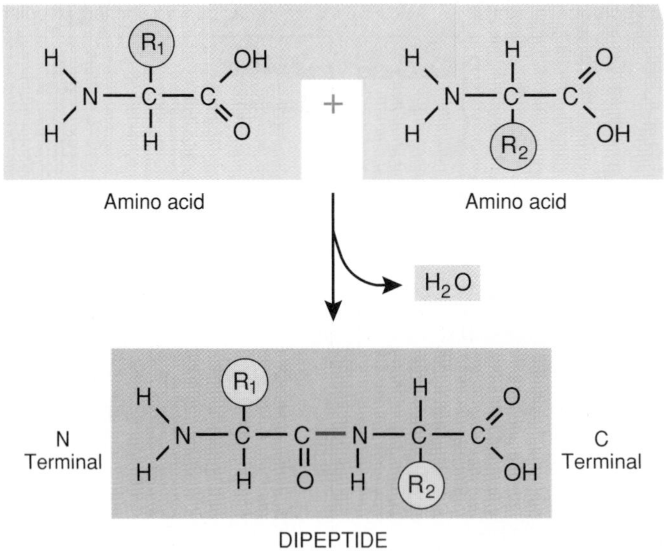

the second amino acid is linked to the amino group of the third; and so on down the line. Each carboxyl-amino linkage is a peptide bond. The complete polypeptide has an **N-terminal,** or amino end (the first amino acid in the chain), and a **C-terminal,** or carboxyl end (the last amino acid). The structure of polypeptides is traditionally written from left to right, with the terminal amino end on the left and the terminal carboxyl end on the right (Figure 3.17).

Levels of Protein Organization Proteins may have as many as four levels of organization, and most have at least three. The first or **primary level** (Figures 3.18 and 3.19a) is determined by the number, kind, and order of the amino acids joined together by peptide bonds to form a simple polypeptide strand. The amino acid sequence, as we will see, is genetically determined, so each kind of protein has a very specific arrangement and number of amino acids. This is highly important to higher levels of organization.

The **secondary level** (Figure 3.19) of organization forms spontaneously, as soon as the polypeptide has been synthesized. In many polypeptides, the chain of amino acids coils, forming what is called an **alpha helix** (right-handed coil). Coiling is possible in a polypeptide because the carbon and nitrogen bonds extending from the alpha carbon of each amino acid are quite flexible and permit rotation. The coiling itself is brought about by hydrogen bonding within the polypeptide strand. This occurs between double-bonded oxygen and the hydrogen of the —NH group. But, whereas each amino acid is involved in such hydrogen bonding, the attractions only occur between the oxygen of one amino acid and the —NH group of a second amino acid located four places away in the strand (see Figure 3.19b). The individual hydrogen bonds in an alpha helix are quite weak, as expected, but there are so many of them that the coil is fairly stable.

Another common secondary structure is a zigzag folding of the polypeptide in what is called a **beta sheet.** Here, the polypeptide does not coil, but remains in an extended form. The extended form is maintained by the formation of hydrogen bonds between adjacent polypeptide strands (see Figure 3.19b). As in the alpha helix, the hydrogen bonds form between double bonded oxygens in one polypeptide and the ═ NH groups in the other. Where an array of parallel polypeptide strands occurs, it often takes on a pleated appearance and is aptly termed a **beta pleated sheet.** The beta sheet is common in structural proteins, including silk. It is the final level of organization in many proteins.

The **tertiary** (third) **level** of protein organization chiefly involves highly specific folding of the polypeptide. It is the attraction of various R groups along the length of the polypeptide that causes the looping and folding associated with tertiary organization.

FIGURE 3.18
PRIMARY STRUCTURE
The primary level of protein organization is simply the sequence of amino acids in the polypeptide chain. Shown here is the primary structure of the polypeptide hormone glucagon, with 29 amino acids.

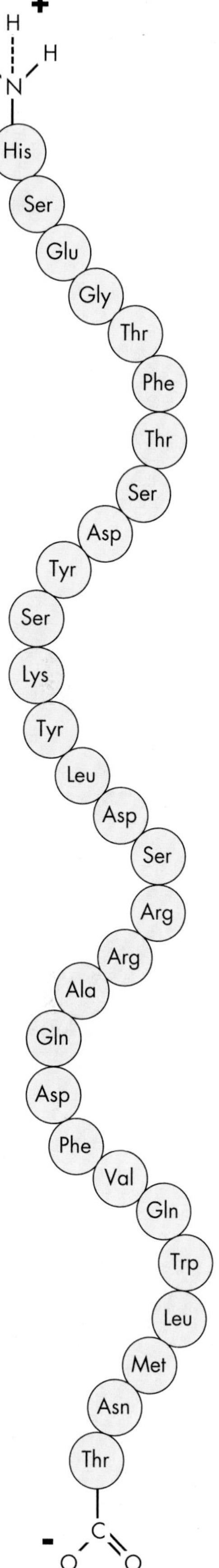

Because the R groups are so specific, each with unique bonding characteristics, the tertiary organization of each protein is, itself, very specific.

There are four principal forces producing folding. First, adjacent polar R groups may simply form *hydrogen bonds* with one another, usually involving opposing —OH and —N groups. Second, *ionic attractions* occur between charged R groups, specifically between those bearing opposite electrical charges. Recall that some amino acids have extra acidic or amino groups, and these take on negative or positive electrical charges, respectively, when ionized. Then, third, where groups of nonpolar amino acids are close together, *hydrophobic interactions* form. Such groups are drawn into tight, water-excluding clusters. Each of the first three forces is individually weak, but when repeated throughout the polypeptide, the combined effect is large.

The fourth interaction involves *covalent bonding,* so here a much stronger force is at work. Covalent bonding occurs between two nearby cysteine amino acids. Recall that the cysteine R group contains a sulfhydryl side group (—SH), and two such side groups react, shedding their hydrogen and forming a disulfide linkage (—S—S—). Obviously those proteins lacking cysteine must rely on the other forces alone for maintaining their tertiary structure. Incidentally, the permanent wave industry is based primarily on the disulfide bonds of keratin, the protein constituent of hair. Permanent wave solution contains an agent that breaks disulfide bonds in keratin, allowing the molecules to literally uncoil, becoming quite limp. Then a second agent is used to bring about the formation of new disulfide bonds, hopefully in places that will produce an aesthetically pleasing look.

There is another aspect to polypeptide folding and the tertiary level of structure. The actual bends or foldings generally occur at flexible "linker regions," places where the alpha helix fails to form. Such regions often begin with the amino acid proline, which as you'll recall, has an unusual ringlike amino group, and thus does not support the formation of a helix. Linker regions are essential because they permit parts of the polypeptide to come together in precise ways. The tertiary level is the final level of organization for proteins containing only a single polypeptide chain.

Some of the globular proteins reach a fourth or **quaternary level** (Figure 3.19d) of organization. In this level, two or more polypeptides join to form the finished protein. Needless to say, quaternary proteins are giants. Among them is the oxygen-carrying blood protein **hemoglobin.** It contains two pairs of interacting polypeptides, attracted to each other by forces similar to those of the tertiary level. The two pairs are designated alpha and beta (not to be confused with the alpha helix and beta sheet). Also present are four very special iron-containing **heme groups,** the actual sites of oxygen transport. The forces involved in protein structure are reviewed in Figure 3.20.

While proteins assume very specific three-dimensional shapes, and their shapes are highly important to function, they can readily lose both shape and function through **denaturation.** One common agent of denaturation is heat. Heat breaks some bonds and causes the random formation of others, and such events are generally irreversible. This is why it is hard to "unfry" an egg. Other denaturing agents are strong acids and bases and certain chemicals. Cold may also have a denaturing effect on protein, as will *urea,* a nitrogen waste produced by many animals. However, in both these instances, the denaturing effects are temporary and reversible, and the protein resumes its shape and function once normal conditions are restored.

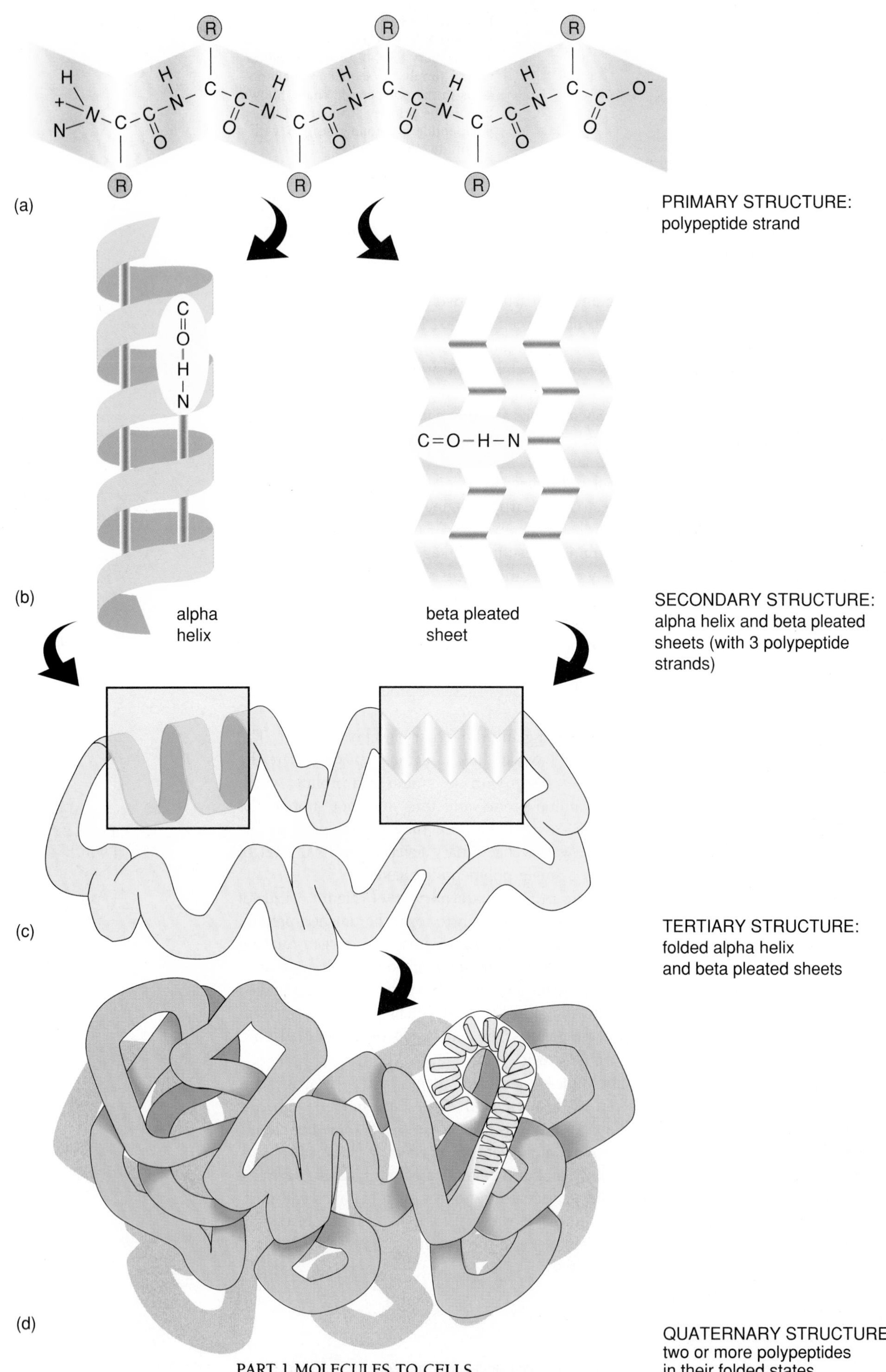

(a)
PRIMARY STRUCTURE:
polypeptide strand
C=O-H-N
(b)
alpha helix
beta pleated sheet
SECONDARY STRUCTURE:
alpha helix and beta pleated sheets (with 3 polypeptide strands)
(c)
TERTIARY STRUCTURE:
folded alpha helix
and beta pleated sheets
(d)
QUATERNARY STRUCTURE:
two or more polypeptides
in their folded states

FIGURE 3.19
LEVELS OF PROTEIN ORGANIZATION
(a) Hydrogen bonding acts on the primary structure shown to produce (b) secondary structure, including the common alpha helix and beta pleated sheet. The alpha helix is produced by hydrogen bonding between amino acids four positions apart along a single polypeptide. The beta pleated sheet is maintained by hydrogen bonding between adjacent polypeptide strands. (c) At the tertiary or third level, hydrogen bonding, ionic attractions, hydrophobic interactions, and covalent bonding (especially disulfide linkages) within the chain produce a highly specific folding in the finished protein. The bending occurs in places where the alpha helix is absent. (d) In the quaternary or fourth level, two or more polypeptide chains are joined through covalent bonding, thus forming a functional protein.

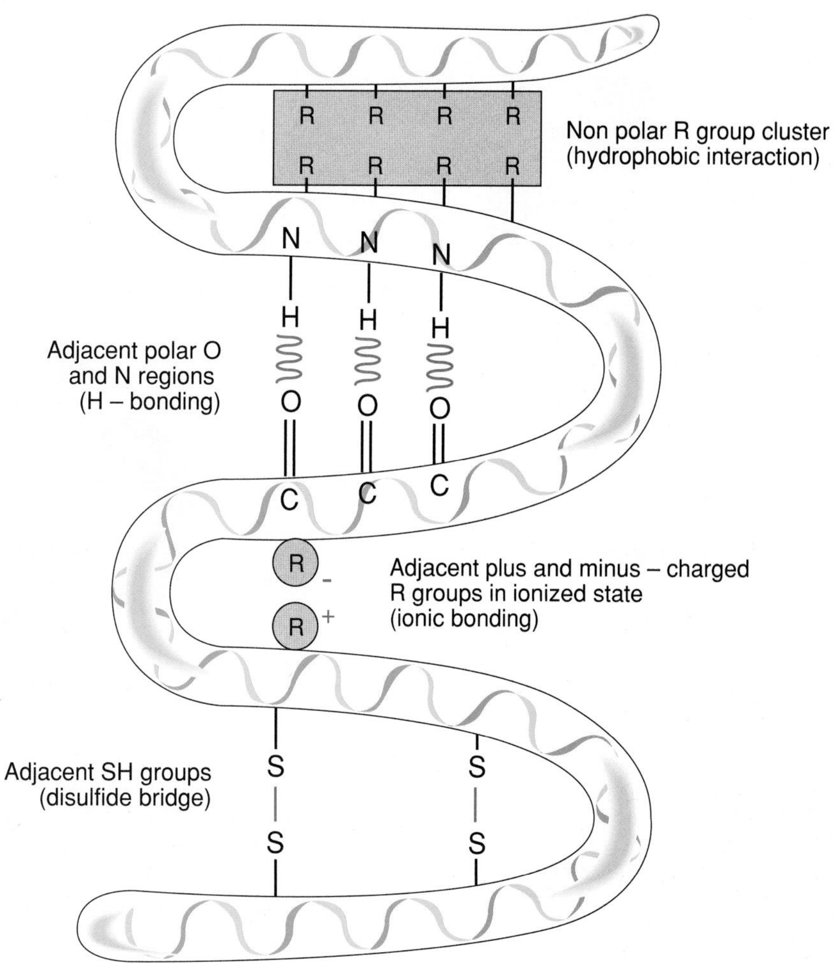

FIGURE 3.20
SUMMARY OF TERTIARY AND QUATERNARY FORCES
The tertiary and quaternary levels of protein structure are brought about by four forces: hydrophobic interaction (among nonpolar R groups), hydrogen bonding (among polar R groups), ionic interaction (among charged R groups), and covalent disulfide linkages (between cysteines).

Conjugated Proteins and Prosthetic Groups

A protein that consists only of amino acids, in one or more polypeptides, is called a **simple protein. Conjugated proteins** have something extra, either covalently bonded to the amino acid side groups or otherwise bound to the polypeptides. The something extra is called the **prosthetic group** (it might help to remember that a wooden leg or a glass eye, is called a *prosthesis,* or added part). Among the conjugated proteins are glycoproteins (the prosthetic group is a carbohydrate), chromoproteins (*chromo-* refers to color, so the prosthetic group is a pigment), and lipoproteins (the prosthetic group is fat-soluble). The hemoglobin molecule is a conjugated protein because each subunit contains a heme group, which in turn contains an iron atom (Figure 3.21). Since heme is deeply colored, hemoglobin is a chromoprotein, although color has nothing whatever to do with its function. Blood just *happens* to be red.

The prosthetic group of a protein often determines function. The catalytic activities of many enzymes, for instance, are dependent on **cofactors,** some of which are small prosthetic groups attached to the enzyme. Neither the cofactor nor the remainder of the enzyme alone has any catalytic properties. While some prosthetic groups are simple metals, others comprise the familiar dietary vitamins.

Binding Proteins

As we've seen, proteins also perform a diversity of nonenzymatic, nonstructural functions. Many of these are related to protein's peculiar ability to form specific structures and

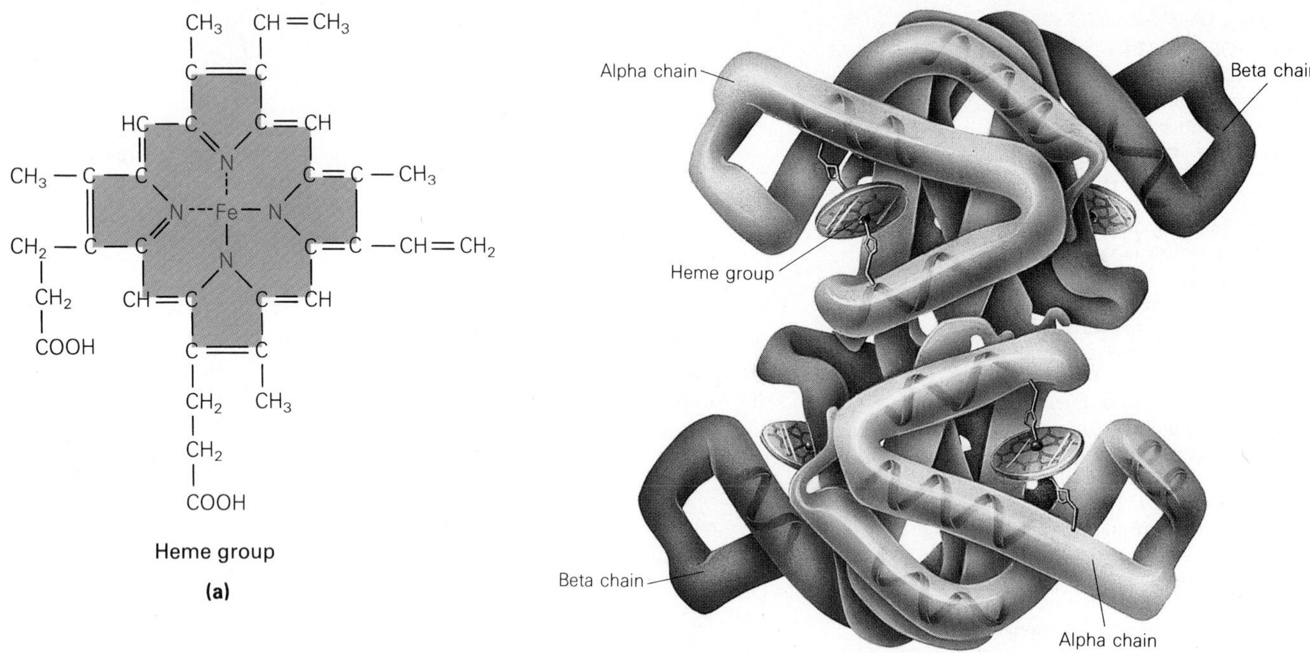

FIGURE 3.21
THE HEME GROUP AND HEMOGLOBIN
(a) Each of the four heme groups of the protein hemoglobin contains a complex hydrocarbon ring that surrounds four nitrogen atoms. At the center of the cluster is one atom of iron. Each heme group in hemoglobin can temporarily bind to one molecule of oxygen. **(b)** Hemoglobin consists of four polypeptide chains and four heme groups.

shapes that enable them to *bind* to other substances. Hemoglobin, for instance, is a carrier protein; it binds oxygen in the lungs and releases it where it is needed in the body, and, to a lesser degree, it binds carbon dioxide in the tissues and releases it in the lungs. There are a number of other circulating binding proteins, such as *transferrin,* which binds iron, and *haptoglobin,* which binds hemoglobin. Egg white has a protein, *avidin,* that binds tightly to *biotin,* an important vitamin; fungi that attempt to invade eggs die of biotin vitamin deficiency.

One of the most critical functions of proteins is their ability to bind to other molecules as part of the antibody response. Some proteins are synthesized by white blood cells for the express purpose of binding with and ultimately destroying or inactivating foreign molecules or particles. (See the discussion of immune responses in Chapter 41.)

We should add that there is good evidence that cells recognize each other during embryonic development through interactions of binding proteins on their surfaces. In addition, binding proteins on cell surfaces lend peculiarities to cells that enable us to smell and taste certain molecules.

Structural Proteins

Structural proteins are important in maintaining the physical form of organisms. They may be *intracellular* (inside cells) or *extracellular* (outside cells). In animals with backbones (vertebrates), common extracellular proteins include **collagen, elastin,** and **keratin.** These are long, fibrous molecules that clump together to make larger fibers. Collagen, the principal component of connective tissue such as tendons, ligaments, and muscle coverings, makes up about 25% of the protein in humans.

Elastin gets its name from its remarkable ability to stretch. Elastin fibers give elasticity to connective tissue, including that of your ears and skin. Ears are very rich in elastin, so if you reach over and tug on your neighbor's ear, it will snap right back to its original form (if all goes well). You can also use pinching as an age test. If you pinch the skin on the back of your hand and it snaps back, you're young. If a ridge stands there, you're not. This is because with aging, elastin loses its elastic properties. The same phenomenon is responsible for the bagginess under the eyes and about the face and neck that awaits all of us.

Keratin is the protein of hair and of the outer layer of skin, as well as that of feathers, claws, nails, horns, antlers, and scales. In contrast to elastin or collagen, it is basically an intracellular protein.

THE NUCLEIC ACIDS

Probably by now almost everyone has heard of the nucleic acids, particularly DNA, simply because some of the recent discoveries and technical breakthroughs in DNA research have been relentlessly and breathlessly touted by the popular press. Understanding the two nucleic acids, DNA and RNA, is so fundamental to biology that we will focus on it in several chapters to come (Chapters 9 and 12–17). Here, we will be brief, introducing the basic structure and making a few comments on how the nucleic acids do their work.

The initials **DNA** and **RNA** refer to **deoxyribonucleic acid** and **ribonucleic acid**, respectively. Both molecules are present, with a very few exceptions, in all cells. DNA is often referred to as the molecular core of life, since it contains the units of heredity we call genes. Essentially, genes are chemical codes that specify the kinds of protein an organism can produce. The DNA in genes does this by specifying the order and arrangement of amino acids in every protein an organism produces. We've seen how the order and arrangement of amino acids determines polypeptide and protein structure or shape through the various levels of protein organization, and that these levels permit the final protein to become functional. Protein structure is quite significant since, directly or indirectly, each organism's proteins determine both its visible structure and its biochemical capabilities. These are often referred to as an organism's genetic characteristics. In humans, for example, skin and eye color are the products of specific pigments that are DNA-specified proteins. In less direct ways, protein structure helps determine other genetic traits, such as body build, facial structure, and hair texture. Of even greater fundamental importance, enzymes are proteins, and enzymes are responsible for promoting the countless chemical reactions that characterize life.

Whereas DNA contains the chemical coding required for the assembly of amino acids into protein, it does not participate directly in the assembly process. Instead, the chemical coding is transcribed into the second kind of nucleic acid, RNA, which assembles amino acids into proteins. So for any of our many thousands of genes to carry out their task, they must first bring about the construction of the RNA molecules that actually do the work.

DNA has another important capability. It is self-replicating. This means it can make copies of itself. This is critical for growth, since organisms grow through cell division. When cells are preparing for division, each DNA molecule goes through replication, and then, at the moment of division, the replicas are properly separated and delivered to the two newly emerging cells. Thus each generation of cells has a complete set of genes. The same holds true for reproduction; DNA must be replicated and copies delivered to developing sperm and egg cells. When egg and sperm join, they bring together the genes of two different individuals.

Nucleic Acid Structure

Each DNA molecule consists of two extremely long polymers held together by hydrogen bonding and coiled into what is known as a **double helix** (Figure 3.22). Each of the two polymers consists of a great number of subunits called **nucleotides**, covalently bonded to form the long strands. Each nucleotide, in turn, consists of three parts: a *phosphate group*, a simple *5-carbon sugar*, and a *nitrogen base*. Nitrogen bases are single- or double-ringed molecules whose rings are made up of nitrogen and carbon. Some of the rings have branching side groups, again of carbon or nitrogen.

Whereas the phosphates and sugars are the same up and down the long DNA molecule, the nitrogen bases differ. There are four different nitrogen bases in DNA, designated as **adenine**, **thymine**, **guanine**, and **cytosine**. We will look into the molecular details of

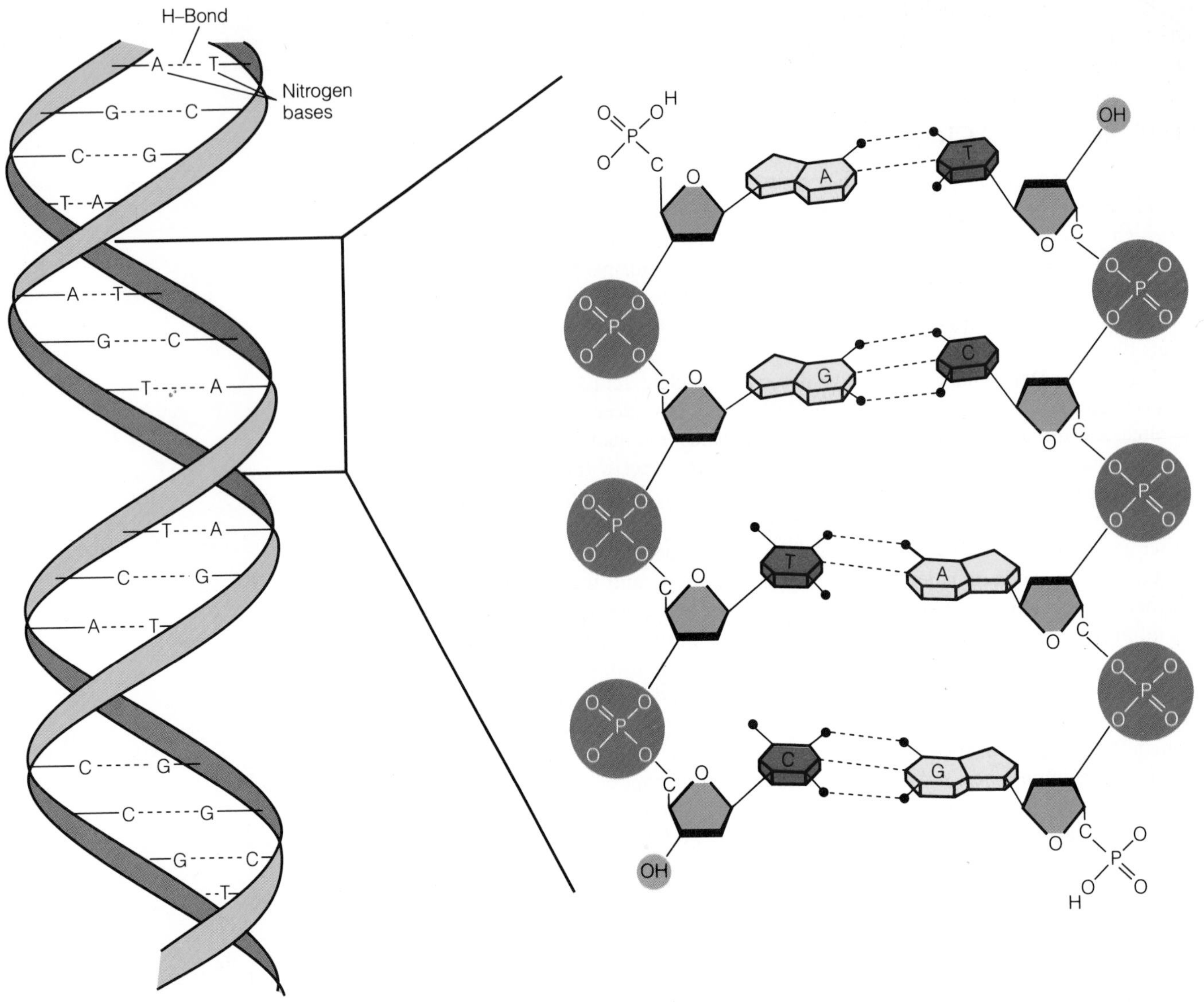

(a) DNA polymer

(c) Molecular structure of the helix

(b) A nucleotide

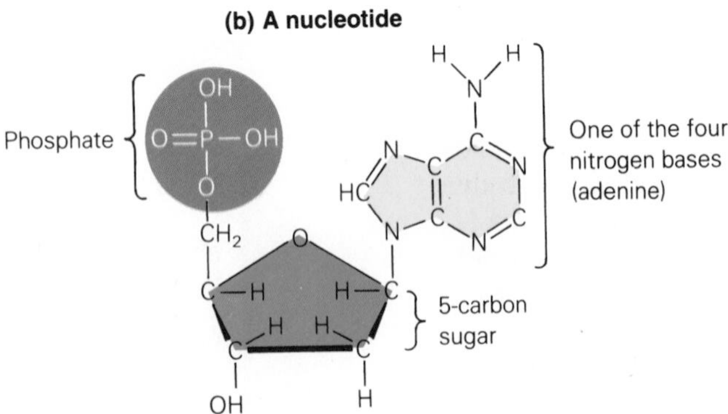

FIGURE 3.22
A FIRST LOOK AT DNA
(a) DNA consists of two very long polymers of nucleotides usually coiled around each other in the form of a double helix. A, T, G, and C are the initial letters of the four nitrogen bases, whose arrangement in DNA makes up its coded message. **(b)** Each of DNA's nucleotide subunits contains the 5-carbon sugar, deoxyribose, to which are linked one of four different nitrogen bases and a phosphate group. **(c)** In the formation of each of DNA's two polymers, nucleotides are assembled through covalent bonding between the phosphate of one nucleotide and the sugar of the next, thus making up a backbone or framework with the nitrogen bases protruding. The two polymers are held together via numerous hydrogen bonds, which attract the nitrogen bases together all along the molecule. The hydrogen bonds also draw DNA into its double-helix configuration.

the nitrogen bases later, but let's note here that the specific, linear arrangement of the bases along the DNA polymer, like signal flags along a rope, makes up the chemical coding (the genetic coding) we mentioned earlier. When read in groups of three, the nucleotides become signals that translate into the placement of one or another of the twenty amino acids at this or that location along the polypeptide. Again, the message is actually read on an RNA molecule that was formed by following the DNA coding. So we see that the genes we all like to talk about at family gatherings are basically an arrangement of nucleotides in our DNA that simply determines what our proteins will be like.

RNA is a single polymer of nucleotides and does not form a continuous helix like DNA. Some forms do, however, have regions where hydrogen bonding brings some nitrogen bases of the single strand together. RNA too has four different nitrogen bases: adenine, **uracil,** guanine, and cytosine. Except for uracil, which replaces thymine, they are identical to the nitrogen bases of DNA. The sugar making up part of each RNA nucleotide is also different from that of DNA.

RNA is produced from DNA through a process called **transcription.** In transcription, the DNA helix opens in places, the two DNA strands separate, and the RNA nucleotides are assembled according to the coding on one of the separated DNA strands. In this way the protein-specifying code is copied into RNA. Since each episode of transcription usually involves just a few genes, or even just one, and there are thousands of genes along each DNA molecule, it follows that each RNA polymer is going to be much shorter than its DNA counterpart. There are several types of RNA required for protein synthesis, but we will look into these and other aspects of DNA structure and activity in future chapters.

APPLICATION OF IDEAS

1. Nucleic acids and proteins are considered to be "informational molecules," but carbohydrates are not, even though they may be extremely large and complex. Why do carbohydrates fail to qualify? Could lipids be "informational molecules"?
2. Animals readily use enzymes to transform glucose and oxygen into water, carbon dioxide, and energy. Since all enzymatic reactions are reversible, can they transform water, carbon dioxide, and energy into oxygen and glucose? Explain your answer and plan to come back to this question after having read Chapters 6, 7, and 8.
3. Discuss the nutritional consequences of various fad diets, such as the all-banana or all-egg diet, a fat-free diet, or a low-carbohydrate, low-fat, all-meat diet. How might such diets affect one's health?

KEY IDEAS

THE CARBOHYDRATES

The empirical formula for **carbohydrates** is $(CH_2O)n$; thus all types of carbohydrates are multiples of this structure.

Monosaccharides and Disaccharides

1. The **monosaccharide glucose** is the subunit for many complex carbohydrates. **Sucrose**, a disaccharide, includes glucose and fructose as its subunits.
2. Functional groups in the simple sugars include **aldehydes** (glucose) and **ketones.**

Polysaccharides

1. Monosaccharides form disaccharides and polysaccharides through **dehydration synthesis**, an enzymatic dehydration process in which **dehydration linkages** (covalent bonds) form between adjacent monosaccharides. Water is also yielded. In the opposing process, **hydrolytic cleavage**, water is restored and the dehydration linkage is broken. In sugars, dehydration linkages are also called **glycosidic linkages.**
2. Plant starches include **amylose** and **amylopectin,** polysaccharides with hundreds of glucose subunits. The first is joined by 1–4 linkages, and the second by 1–4 and 1–6 linkages.
3. **Glycogen**, a highly branched animal starch, also contains both 1–4 and 1–6 linkages.

4. While covalent bonding produces **primary structure** in starches, hydrogen bonding produces a coiled form known as the **secondary structure.**
5. **Cellulose**, composed of beta-glucose subunits, is insoluble and is digestible by only a few organisms that have the enzyme **cellulase.**
6. Cellulose forms strong structural elements (plant cell walls and fibers) because of its fibrous organization. Between 60 and 70 polymers become hydrogen-bonded into microfibrils, which are incorporated into fibrils. The fibrils are strengthened through the addition of **hemicelluloses, pectins,** and **lignins.**
7. **Chitin,** a structural polysaccharide of insect **exoskeletons** and fungal cell walls, contains ***N*-acetyl glucosamine.**

THE LIPIDS

Lipids are fat-soluble, mainly hydrophobic molecules used for energy reserves, membrane components, and hormones.

Triglycerides

1. **Triglycerides** are nonpolar molecules consisting of one **glycerol** bonded to three **fatty acids.** Triglycerides differ from each other in the kinds of fatty acids present.
2. Fatty acids themselves are polar and hydrophilic at their acid end and nonpolar and hydrophobic at their carbon chain.

Saturated, Unsaturated, and Polyunsaturated Fats

1. **Saturated** fats (chiefly the more solid animal fats) have no carbon-to-carbon double bonds and hold the maximum number of hydrogen atoms.
2. **Unsaturated** fats (chiefly liquid oils in plants) contain at least one carbon-to-carbon double bond, with fewer than the maximum number of hydrogens. **Polyunsaturated** fats have two or more carbon-to-carbon double bonds. Unsaturated fats are essential to the diet.

Phospholipids

Phospholipids, components of cellular membranes, contain glycerol bonded to two fatty acids (one saturated, one unsaturated), and a polar phosphate group attached to some variable group. Like fatty acids, the heads of phospholipids are polar and the tails nonpolar. The nonpolar tails form an oily, water-resistant core in cell membranes.

Other Lipids

1. **Waxes** lack glycerol and are more hydrophobic than fats and oils, serving as a waterproofing material in both plants and animals.
2. **Steroids** are lipid-soluble, multiple-ring molecules and include **lanolin** and **cholesterol.**
3. Cholesterol is a constituent of cell membranes, liver bile, and the starting molecule for many steroid hormones and for vitamin D.

THE PROTEINS

1. **Proteins** are important in cell structure, are chemically active as enzymes, are good sources of energy, and serve in movement and as hormones.
2. Proteins are composed of **amino acids,** linked via covalent **peptide bonds** and arranged into **polypeptides.** In their final, active form, one or more polypeptides constitute a protein. The size and weight range of proteins is enormous.

Amino Acids

1. There are 20 primary amino acids in protein, each of which contains an amino and a carboxyl functional group, along with a variable R group, each attached to the **alpha carbon.** In the ionized state, the carboxyl group is negative and the amino group, positive.
2. Amino acid R groups include those that are nonpolar, polar but uncharged, and charged. Nonpolar amino acids are hydrophobic, clustering together in water. Polar but uncharged amino acids are hydrophilic, readily forming hydrogen bonds and interacting with water. Amino acids containing extra carboxyl and amino groups ionize in water, thus becoming charged. Cysteine, a polar amino acid, has a sulfhydryl group (—SH), which interacts with others like it to form disulfide bridges or linkages.

Polypeptides and Proteins

1. Polypeptides are lengthy chains of amino acids. Each polypeptide has an ***N*-terminal** and a ***C*-terminal.**
2. The **primary level** of protein organization is a specified sequence of amino acids.
3. The **secondary level** includes the **alpha helix,** which is produced by hydrogen bonding between one amino acid and another, four positions away, in the same polypeptide. Another common secondary structure is the zigzag **beta sheet,** wherein hydrogen bonding occurs between amino acids in adjacent polypeptide strands.
4. The **tertiary level** is a folding that is maintained by interactions among amino acid R groups. Included are hydrophobic interactions, hydrogen bonding, ionic attractions, and covalent disulfide linkages. Bends occur in linker regions where certain amino acids fail to form the alpha helix.
5. The **quaternary level** of protein organization involves two or more polypeptides, usually linked by disulfide linkages, and maintained by the same forces responsible for tertiary structure.
6. **Hemoglobin,** the oxygen-carrying pigment of many animals, is a quaternary protein. It contains four polypeptides and four iron-containing **heme groups.**

Conjugated Proteins and Prosthetic Groups

Unlike **simple proteins, conjugated proteins** have non-amino acid groups such as carbohydrates, pigments, and lipids. The heme group of hemoglobin is a pigment. Some **prosthetic groups** act as **cofactors,** necessary for some enzymes to function.

Binding Proteins

Binding proteins are those that recognize and join with certain molecules.

Structural Proteins

Structural proteins in vertebrates include **collagen** (in connective tissue), **elastin,** (in flexible tissue), and **keratin** (hooves, nails, and hair).

THE NUCLEIC ACIDS

1. Nucleic acids include **DNA (deoxyribonucleic acid)** and **RNA (ribonucleic acid).**
2. DNA comprises the genes, forming a chemical or genetic coding. It is self-replicating, so copies can be sent from cell to cell and from organism to offspring. RNA is involved in translating the genes into protein products.

Nucleic Acid Structure

1. DNA is double-stranded, consisting of two very long chains of covalently linked **nucleotides** intertwining into a **double helix.**
2. There are four different nucleotides, and their linear order in DNA constitutes the genes and specifies the structure of the many kinds of proteins. While the specifications reside in DNA, they are copied into RNA, which carries out the actual synthesis of protein.

REVIEW QUESTIONS

1. State the empirical formula for carbohydrate. Explain what such a formula means. (50)
2. To what do the terms *monosaccharide, disaccharide,* and *polysaccharide* refer? What do all three have in common? (51)
3. Write the straight chain and ring structural formulas for alpha-glucose, and number the carbons. (51)
4. Explain what is meant by a dehydration linkage or dehydration synthesis. What is the opposite process called? How are both important to most biological molecules? (52, 53)
5. List the important differences between amylose, amylopectin, and glycogen. Where would each be found? (55, 56)
6. Describe the secondary structure of amylose. What force is responsible for it? (56, 57)
7. Compare the primary structure of amylose with that of cellulose. How does the difference affect the use of each carbohydrate as a food? (56, 57)
8. What is it that makes cellulose strong? How is its strength further increased in plant fibers? (57)
9. What is the role of chitin, and how is it different from other carbohydrates? (58)
10. List the essential parts of a triglyceride. Through what enzymatic process are triglycerides joined? What makes one triglyceride differ from the next? (59)
11. Distinguish between an oil and a fat and between a saturated fat, an unsaturated fat, and a polyunsaturated fat. (60)
12. Describe the structure of phospholipids, and explain their arrangement into membranes. (61, 62)
13. List the special uses of steroids and waxes. How do steroids differ structurally from other lipids? (62)
14. What are four significant uses for protein? (63)
15. Carefully define the term *protein.* How is a protein different from a polypeptide? (63)
16. What do all amino acids have in common? How do they differ? (63, 64)
17. Describe the levels of organization possible in a protein, and explain what forces are involved at each level. (66, 67)
18. Discuss the following about the hemoglobin molecule: shape (globular or fibrous), level of organization represented, existence of prosthetic groups if any, size in comparison to other proteins, special function. (69, 70)
19. List the three different types of R groups in amino acids. Which is vital to linking polypeptides? (63)
20. Describe the general characteristics of the DNA molecule. What are its units of structure? (71, 72)
21. What are two roles of DNA and one of RNA? (71)

4

CELL STRUCTURE

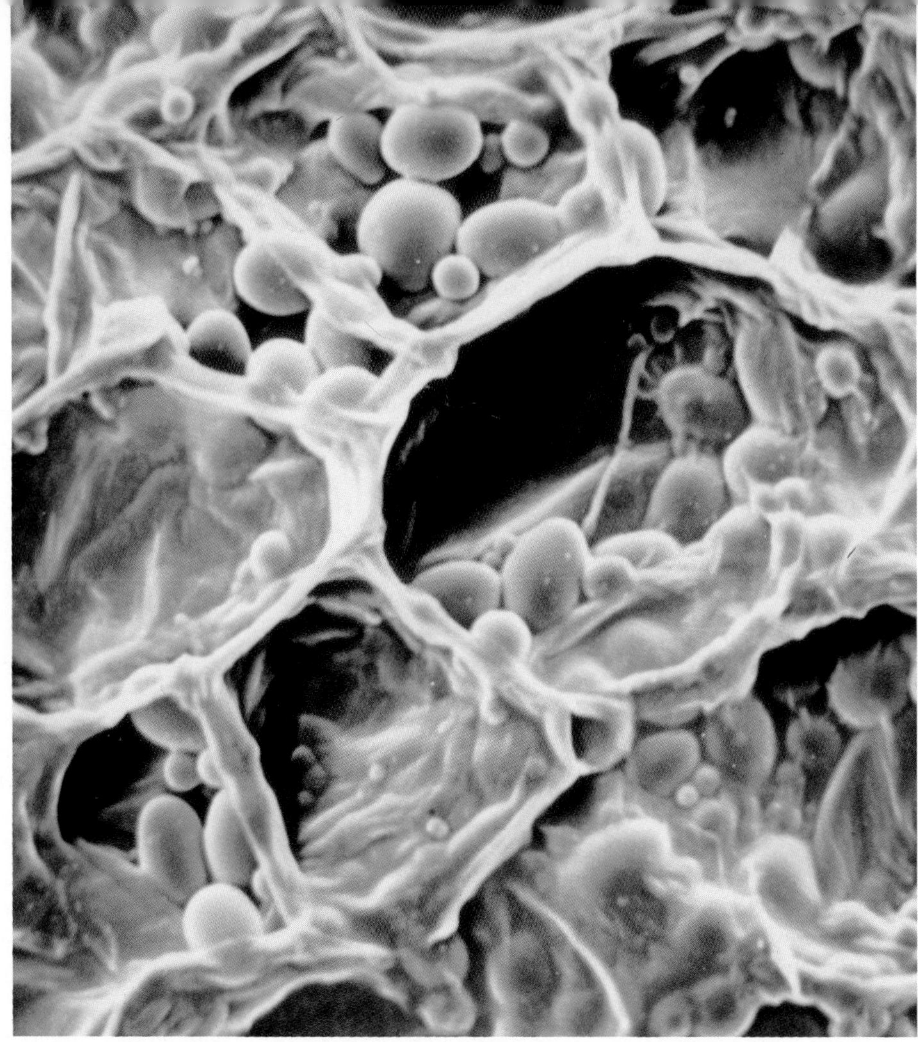

ROBERT HOOKE HAD JUST BEEN APPOINTED CURATOR OF EXPERIMENTS FOR the prestigious Royal Society of London, and he knew he had to come up with something good for the next weekly meeting. He was only too aware that the elite of British science would be there, and he wanted to present a demonstration that would enlighten, entertain, and impress them. It would not be an easy task.

Hooke had an idea. Perhaps he would use an exciting new technology of the seventeenth century—the casting and grinding of glass lenses. The world was buzzing with talk of lenses. With a pair of lenses in a frame the nearly blind were able to see—a miracle come true. Old men who had not been able to read for years had their books and letters restored to them. Earlier in the century, Galileo had pointed a lens to the sky and had started an intellectual revolution. The human eye has a voracious appetite, and Hooke knew it. But Hooke himself was interested in a new use of the lens—to look at very small things. In fact, he had built his own microscope, one of the first in the world.

So for a scientific demonstration, Hooke thought of using the microscope to try to see why cork floats. Cork was a mystery. It appears to be solid, yet it floats. Perhaps it is not so solid after all, Hooke thought. He decided it was worth a look.

Hooke trained his microscope on a cork, and what do you think he saw? Nothing. It turns out that microscopes do not work very well on reflected light. So out of curiosity he took out his pen knife and cut a very thin sliver of cork and shined a bright light *through* it. Then, when he peered into the microscope, he did see something. And it puzzled him. Hooke wrote at the time that the cork seemed to be composed of "little boxes." The little boxes, he surmised, were full of air, and that's why cork floats. He called the little boxes *cells* because they reminded him of the rows of monks' cells in a monastery, and a new scientific field, **cell biology**, was born.

CELL THEORY

The group that week was pleased by Hooke's demonstration, but a full century would pass before the scientific world would grasp the meaning and importance of Hooke's cells. One of the first people to move on the idea was the German naturalist Lorenz Oken, who wrote in 1805 what was to become known as the **cell theory**: "All organic beings originate from and consist of vesicles or cells." But in spite of this rather clear and encompassing statement, the formulation of the cell theory is usually credited to two other Germans, the botanist Matthias Jakob Schleiden and the zoologist Theodor Schwann. In 1839, they published a conclusion that they had reached more or less simultaneously; they said that all living things, from oak trees to violets, and from worms to tigers, were composed of cells. Either they were better at public relations than Oken, or perhaps the world was simply more ready to listen in 1839, but Schleiden and Schwann got the credit. Then, about 20 years after Schleiden and Schwann revealed their ideas, a fourth German, Rudolf Virchow, elaborated upon the proposal to add *"omnia cellula e cellula"*: all cells from cells. Under the conditions on earth today, cells only arise from preexisting cells. This addition was to become known as the biogenetic law, which, like the cell theory, remains a thoroughly respectable idea today. The provision, "under today's conditions," leaves room for a prominent theory proposing that life (cells) arose spontaneously in the early earth, but under much different environmental conditions (see Chapter 21).

We know now that not only must cells come from cells, but virtually every living thing is composed of cells. They are, indeed, the basic units of life. Once Hooke had described his little boxes, the art of microscopy blossomed, and nothing—literally nothing—proved sacred. People were peering through their handmade microscopes at everything. Of course, they went over the human body with avid interest, as they did other animal life and plant life as well—looking at and into everything. What they found were not just the dead cell walls that Hooke had described, but variable, changing cells of all descriptions (Figure 4.1). They could see some of these easily, but others needed to be stained before they were clearly visible. Living cells, they found, were full of a very active, changing, shifting fluid—now called **cytoplasm**—a puzzling finding, indeed. In time, cell researchers would also find that the cells of every creature are unique, differing not only from one species to the next but from place to place within the same individual. We will delve into some aspects of this cell diversity, and you will find that in spite of the apparent diversity, cells are remarkably similar in many quite fundamental ways.

WHAT IS A CELL?

We have said that the cell is the fundamental unit of life, but like most definitions, this one raises more questions than it answers. The phrase "fundamental unit" requires some explanation. What it really means is that the cell contains all of the structures and molecular constituents needed for life. With these entities in place, the cell can *(1) take in raw materials and from these, (2) extract useful energy, (3) synthesize its own molecules, (4) grow in an organized manner, (5) respond to stimuli from its surroundings, and, very significantly, (6) reproduce itself.*

Such a list suggests that even the simplest of cells must have a minimal array of constituents. For instance, a confining membrane, the **plasma membrane,** defines the perimeter of the cell and acts as a barrier, selectively permitting the passage of some substances while summarily rejecting others. Thus, the cell retains its internal integrity—a vital task, since the structure and chemical organization of the cell is vital to life. We'll return to cell organization soon, but first let's consider an aspect of cells that directly involves the surrounding membrane—the size of cells.

The Size of Cells

How big are cells? Not very. Most cells are far too small to be seen with the naked eye. But, of course, there are exceptions: a chicken egg yolk is technically a cell, and a frog

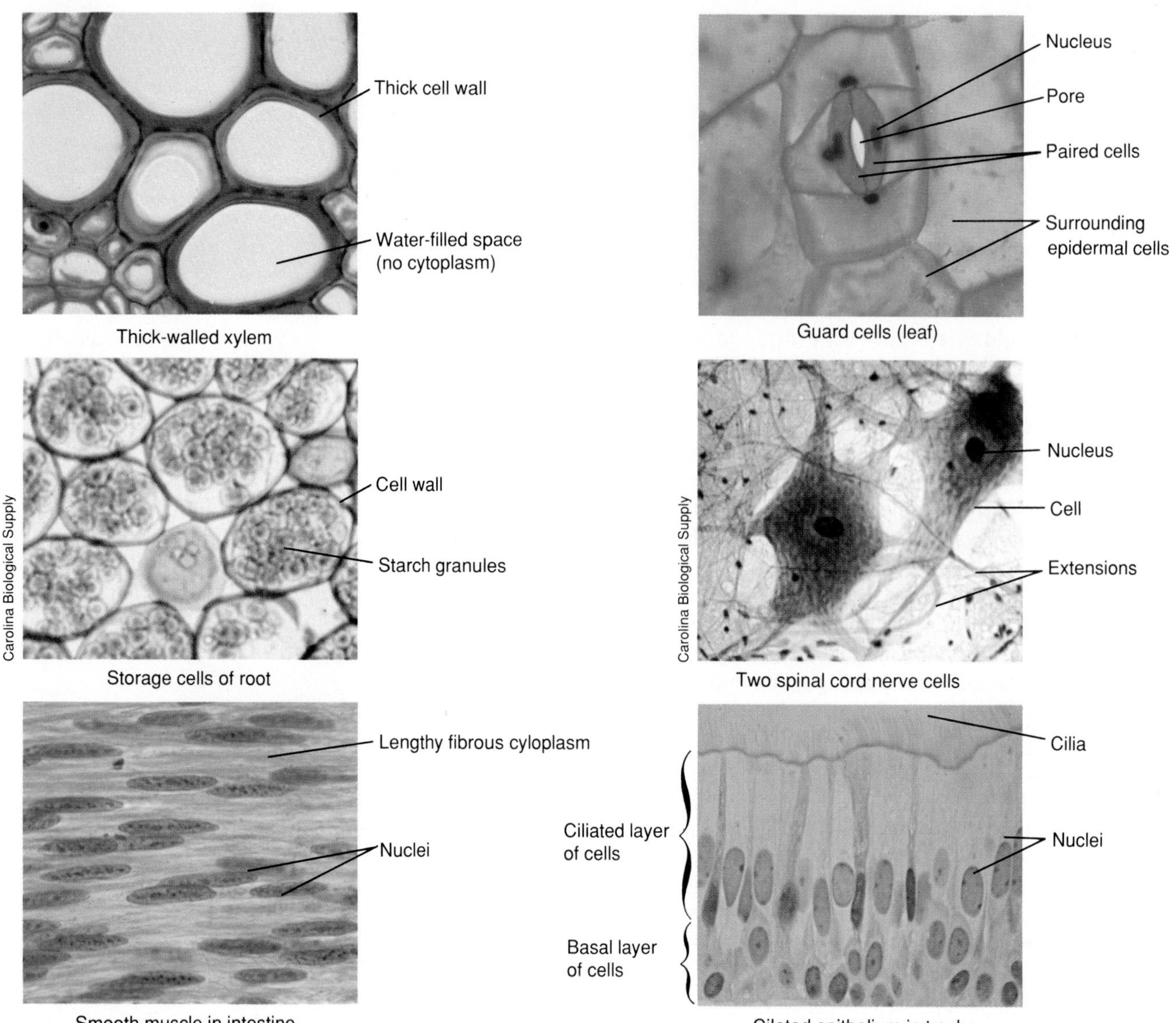

FIGURE 4.1
CELL DIVERSITY
Cells in multicellular organisms vary enormously, each adapted to its own specialized task.

egg is somewhat more convincingly so. Neurons (nerve cells) may be over a meter long (such as those that run down a giraffe's leg), but they are still too thin to be seen. However, apart from such specialized cells, most cells fall within a surprisingly small size range. Very few are smaller than 10 μm (micrometers) in diameter or larger than 100 μm. (See Appendix A for a discussion of metric units of measurement.) Plant cells tend to be somewhat larger than animal cells but perhaps only because they contain large, water-filled internal cavities (or vacuoles); the average amount of cytoplasm is about the same in plant and animal cells. Bacteria, as we shall see, are much smaller than either, seldom exceeding a few micrometers in diameter or length.

Why Are Cells So Small? This all brings up an interesting question. Why are most cells small, and why hasn't cell size increased as organisms became larger? For that matter, why aren't there gigantic single-celled creatures? We have yet to hear of an 800-pound ameba lurking in the old swimming hole, and no one has reported sighting single-celled whales and elephants. The evolutionary trend has been to keep cell size small and simply increase the number to accommodate increases in size. Thus, large organisms are large because they have so many more cells than small organisms. Why?

The Surface-Volume Hypothesis The generally accepted hypothesis explaining the minute cell size has to do with some clear-cut physical problems. The size to which a cell can grow is dictated by its metabolic requirements, and these in turn are largely dependent on the ability of the plasma membrane to serve the cytoplasmic volume within the cell. In other words, perhaps cells remain small because they are restrained by problems of area and volume. This is called the **surface-volume hypothesis,** and its operation would put close restraints on how large cells can be. Let's see why this should be.

The surface-volume hypothesis is based on the observation that as the volume of a body is doubled, its surface area also increases, but not proportionately. If the surface-to-volume ratio at the start is compared to that at the end, we will find that surface loses ground. Figure 4.2 reviews such a situation in a cuboid cell. But, given such a change, why is this important to the maximum size a cell may attain?

Cells, especially growing cells, require an active exchange of substances (chiefly an uptake of food and oxygen and output of wastes and carbon dioxide). Carrying out such an exchange is a function of the plasma membrane surrounding the cell. Since the plasma membrane covers the surface area of a cell, and since the cell contents (the cytoplasm and nucleus) fill its volume, the surface-volume relationship is a critical one. So the factors that restrict cell size can be restated: as a cell increases in size, its volume increases disproportionately to its surface area. Thus, as growth continues, a point is reached at which the membrane can no longer service the needs of the active cytoplasm and nucleus. Typically, before such a point is reached, growth stops, and the cell divides in two: thus a more ideal relationship between surface and volume is established in the new cells.

Certain kinds of cells, we should add, grow large without dividing. But these are often highly specialized cells or those in which special circumstances exist. For example, some plant cells grow quite large because much of their volume is taken up by a large water-filled vacuole (a saclike container). Further, some plant cells exhibit a circulating motion in their cytoplasm, called **cytoplasmic streaming,** which helps in the distribution of materials. Nerve cells can grow extremely long extensions, without large increases in volume. Their massive surface area thus puts them in touch with other cells with which they must function in concert. Certain cells in the small intestine and kidney specialize in the transport of materials across their membranes. They have adapted to this task through changes in shape that drastically increase their surface areas. For instance, cells of the small intestine have highly folded "brush borders," where each fold or projection in the plasma membrane represents a large surface area for absorbing digested foods (Figure 4.3). Ostrich eggs, which are essentially one cell with a lot of stored food, are not limited by the problem of surface and volume, since there is little metabolic activity except at a small region on the yolk surface.

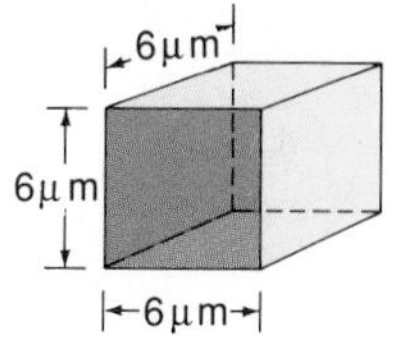

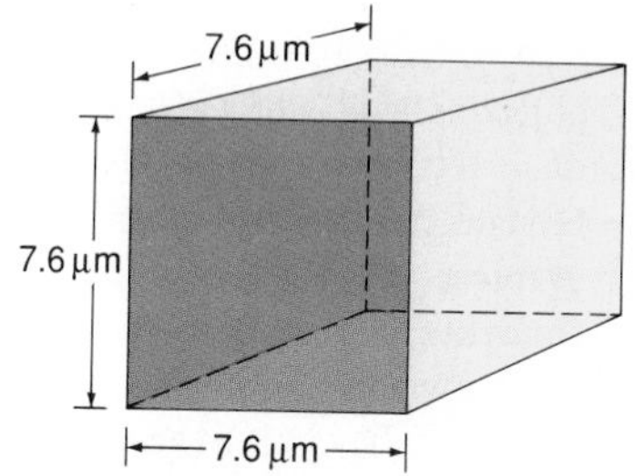

FIGURE 4.2
THE SURFACE-VOLUME RELATIONSHIP
Doubling the size of a cell changes its surface-to-volume ratio disproportionately—the volume increases faster than the surface. If we assume that a certain amount of plasma membrane (surface area) is required to support a specific amount of cytoplasm (volume), then we can predict that when the surface-to-volume ratio becomes critically small, cell division will restore a satisfactory ratio.

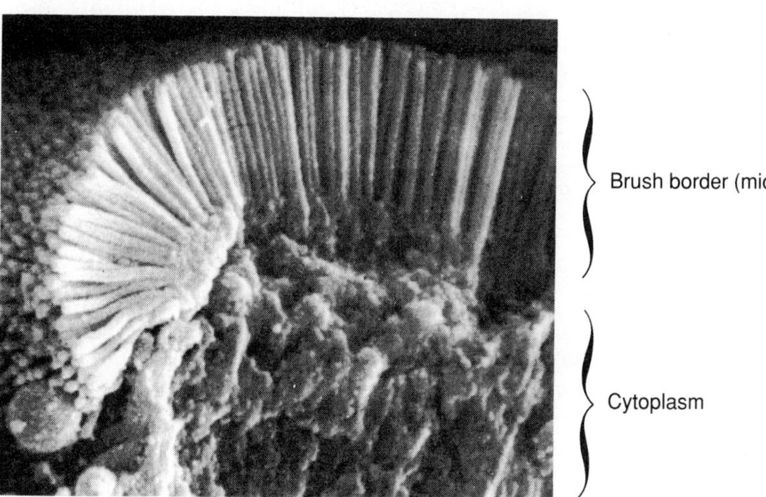

FIGURE 4.3
INCREASING MEMBRANE AREA
The epithelial cells lining the small intestine carry on transport across a surface that is greatly expanded by numerous microvilli, brushlike extensions of the plasma membrane.

ESSAY 4.1

LOOKING AT CELLS

THE LIGHT MICROSCOPE

The light microscope was fully perfected by the start of this century. As a result, almost all cell structures that can be seen with a light microscope had been identified and catalogued before your grandparents were born.

The light microscope, or **compound light microscope**, consists basically of a tube with lenses at its ends. The **objective lens** is brought very close to the object being viewed. The eye is brought close to the **ocular lens** or **eyepiece.** An additional lens, called a **condenser,** lies just above the light source, where it focuses light into a cone over the object. The condenser also contains an **iris diaphragm**, which (like the iris of the eye) can be opened or closed to adjust brightness.

Preparation and Staining of Specimens for the Light Microscope

Small things tend to be transparent, so special **stains** (dyes) must be used to distinguish cellular structures. Stains may be of a general nature, simply darkening the cytoplasm or nuclei of cells, or (depending on their chemistry) stains may be highly selective for certain structures in the cell. Selective staining has been invaluable to the fields of cell biology and **histology** (the study of tissues), and many of the original staining methods are still in use.

Generally, microscope specimens have to be quite small or very thin in order for light to pass through. Small cells such as red blood cells and bacteria can be viewed whole, but most plant and animal tissue must be cut into very thin slices, or **sectioned,** before staining and viewing.

Resolving Power: The Practical Limit

The limits of useful magnification in any microscope is determined by the instrument's **resolving power.** Resolving power is a measure of how close two objects can be and still be distinguished as separate. (The resolving power of the normal, unaided human eye at close range is about one-tenth of a millimeter.) There is a definite limit to the resolving power of even the finest light microscopes, and many important cellular structures are well below this limit. Optical physicists have shown that no system can resolve points that are closer together than half the wavelength of the light used to view them, so the limitation is really light itself. The resolving power of the best light microscopes is approximately 250 nm (0.25 mm), about 500 times the resolving power of the eye. The bottom line is that the limit of useful magnification with the light microscope is about 1400x (1400 times actual size). While this magnification can be exceeded, the resolution will not be improved. The result is called **empty magnification.**

THE TRANSMISSION ELECTRON MICROSCOPE

The **transmission electron microscope,** or **TEM,** has enormous powers of resolution. The object to be viewed is flooded with electrons rather than light waves, and the wavelength of electrons is much smaller than that of the shortest light wave. The TEM's useful magnification can reach 250,000x. Its resolving power, theoretically, is about 0.025 nm, but technically the limits are about 0.2 to 0.3 nm (1000 times the light microscope's).

The principle upon which the TEM works is relatively simple. It makes use of electrons shot from an electron gun similar to the electron gun in the picture tube of a television set. The term "transmission" refers to the formation of an image by those electrons that actually pass through and around the object to be viewed.

Focusing the electron microscope begins when electromagnetic condensers

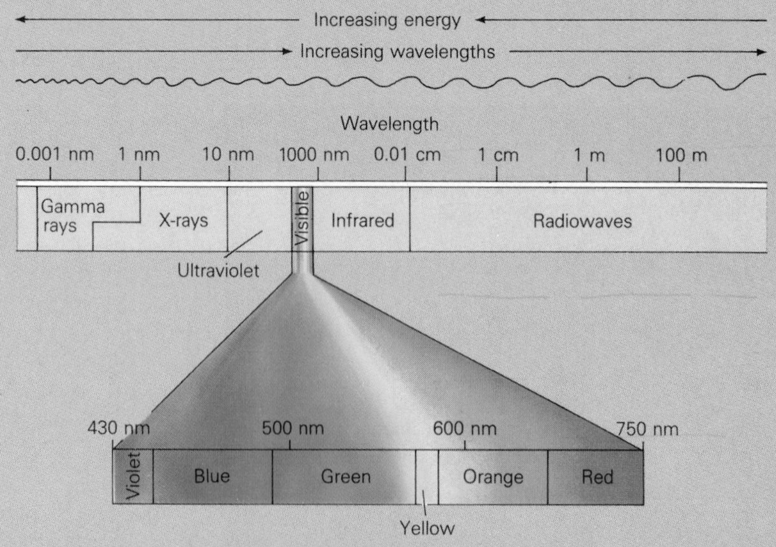

The electromagnetic spectrum. Our eyes are barraged by electromagnetic radiation of a variety of wavelengths, but we see only that portion with wavelengths between 430 and 750 nm. Within that range, light of different wavelengths is discerned as color. The wavelengths in the visible spectrum also restrict microscope resolving power, since objects closer together than 250 nm will interfere with light passing between them and the separation will not be visible. The 250 nm is just about half the shortest wavelength in the visible spectrum.

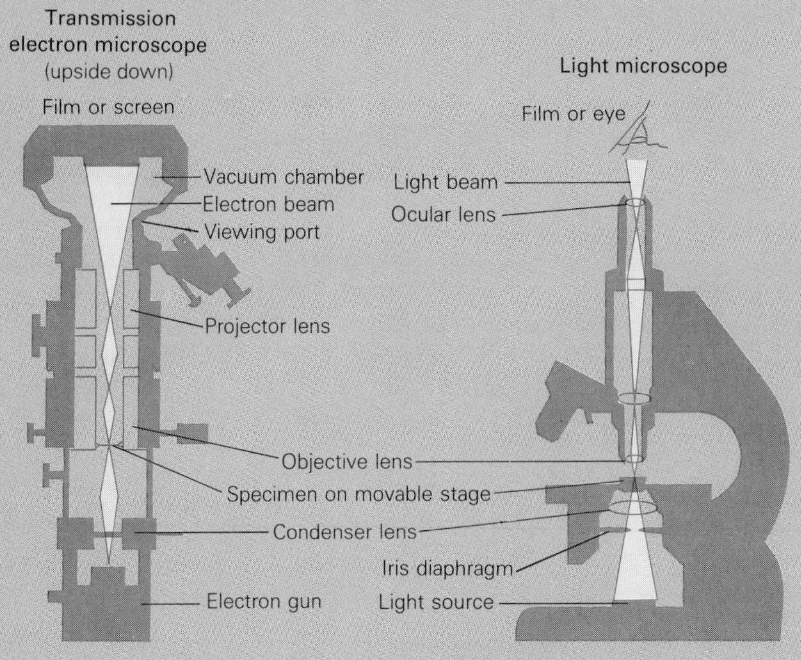

The transmission electron microscope and the light microscope work on very similar principles. In both cases a focused beam, of photons or electrons, is transmitted through a very thin slice of the specimen and is absorbed differentially by different structures of stains. The emerging beam is then refocused to produce an image, either on a fluorescent screen that can be seen by the viewer (electron microscope), directly into the eye (light microscope), or onto film (either system). The light microscope lenses are glass, and the electron microscope lenses are magnetic fields. Here, the electron microscope diagram is shown upside down to emphasize the correspondence between the two systems.

direct the electron beam onto the specimen. Focusing is possible because electrons have negative charges that respond to the surrounding electromagnetic field. The focusing of electrons somewhat resembles the bending of light rays by glass lenses. As the electrons pass into the specimen, some are absorbed by the nuclei of heavier atoms, and others pass through unhindered. Those that get through pass between additional electromagnetic lenses that spread them out. Magnification can be varied by changing the strength of the magnetic lenses.

Finally, the electrons are focused onto a screen for direct viewing or onto special photographic emulsions for preparing electron micrographs.

Preparation of Specimens for the Electron Microscope

In the preparation of tissue specimens for the electron microscope, electron-dense metallic salts are applied. The heavy metal helps solidify the cytoplasmic proteins and combines with them in specific ways. Thus, the metal selectively stains the specimen, its atoms later serving to absorb some electrons and permitting others to pass through. The tissue is embedded in a plastic matrix and then sliced into exceedingly thin sections. Sectioning is necessary because electrons do not pass through thick material very well.

Small objects such as viruses, chromosomes, or the bacterial flagellum can be viewed without slicing. They are first spread onto a thin film of protein that is supported by a wire screen. Then the mounted specimen is placed in a vacuum chamber, and atoms of platinum or gold are spattered onto the specimen at an angle—a procedure

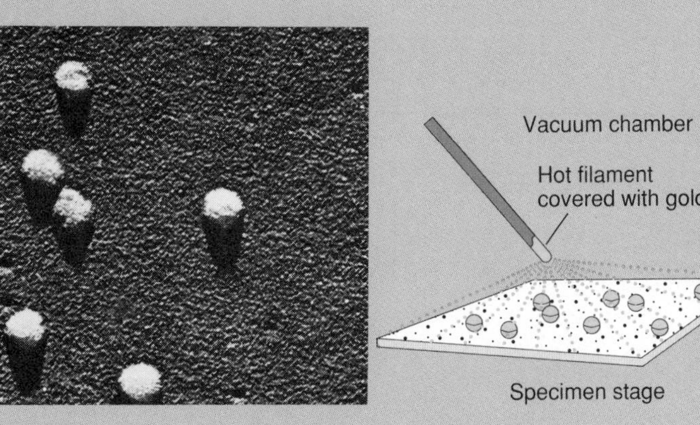

Heavy metal shadow casting. The objects in this photo **(a)** are viruses. The moonscape appearance is attributed to heavy metal shadow casting. Molecules of heavy metals are sprayed at an angle and thus deposited on one side of each virus. The result is a distinct shadowing that helps emphasize the shapes. This is a positive print of the original electron micrograph, so the heavy metal deposits are made to look light, strikingly like reflected sunlight. **(b)** Shadow casting is done by heating and vaporizing a small dab of gold or other heavy metal in a vacuum chamber. The gold "spatters" over the specimen.

ESSAY 4.1

LOOKING AT CELLS—CONT'D

called **shadow casting.** The gold absorbs electrons, and the resulting image is a surprisingly lifelike photograph of three-dimensional bodies.

One of the most specialized methods of specimen preparation is used in the study of cellular membranes. This is a process known as **freeze-fracturing.** Tissue is frozen hard at w100B C and then sectioned. The blade reveals natural weaknesses in the hardened tissue, such as between the phospholipids of the bilayer making up cell membranes. Shearing the membrane in this manner splits it between layers, thereby exposing its inner structure. Such methods have been vital in learning about the protein component of cellular membranes.

A 3-D View: The Scanning Electron Microscope

The **scanning electron microscope** or **SEM** was perfected in the 1960s. While the resolving power of this instrument is far less than the TEM, it has the distinct advantage of producing three-dimensional images with unusually great depth of field. Reflected electrons from a moving beam are captured and used to generate an image electronically on a screen. The specimen can be moved for changing the viewing angle.

The High Voltage Electron Microscope

A recent addition to the cell biologist's growing arsenal of high-tech tools is the **high voltage electron microscope,** a gigantic, three-story version of the TEM. With its enormous energy output (one million electron volts), tissues can be viewed without slicing, and the image produced is three-dimensional. It was the high voltage electron microscope that first revealed details of the intricate cytoskeleton.

Ultracentrifugation

One of the most useful innovations applied to cell study, and an aid to electron microscope technology, has been the **ultracentrifuge.** A centrifuge separates liquid-suspended materials according to density simply by rotating sample tubes at high speed. Modern ultracentrifuges can attain rotating speeds of up to 70,000 revolutions per minute, thereby exerting a force of up to 300,000–400,000 times normal gravity. Such forces are enough to separate various components of cells into layers in the centrifuge tubes. The cells are first disrupted, freeing the components in a liquid suspension. The preparation is placed in a sucrose solution of a known density and subjected to rotational force. Structures and molecules in the suspension soon become distributed according to their sizes and densities, the lighter materials ending up towards the top and heavier materials towards the bottom. As a result, the investigator can remove specific layers and subject them to further biochemical tests or to electron microscope observation.

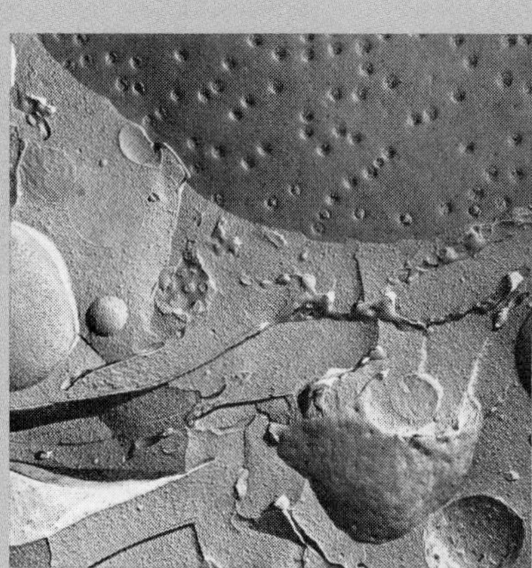

The freeze-fracture technique. The TEM photograph shown here was made of a specimen prepared by freeze-fracture. This technique produces fracture lines within membranes that reveal their inner structure. Here, the nuclear membrane has been cleaved, revealing its pores. This answers some of the long-standing questions about transport in and out of the nucleus.

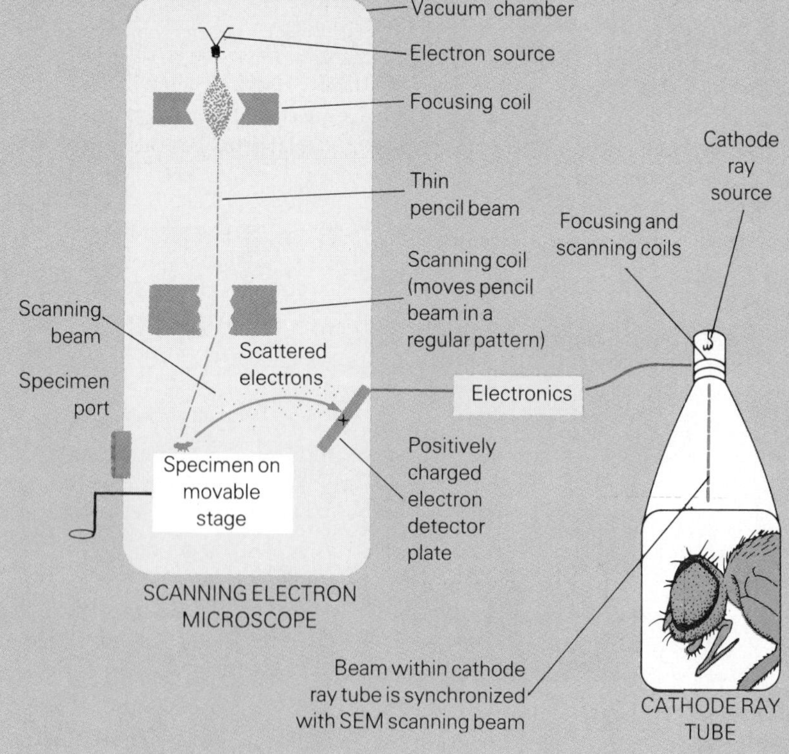

THE PROKARYOTIC CELL

The name **prokaryote** will seem foreign to you until we tell you that prokaryotes are **bacteria** (We'll often use the terms interchangeably.) A detailed treatment of this group will be made in Chapter 22, so here we will be brief.

Prokaryotes are a diverse group of minute, usually single-celled organisms, whose origins can be traced to the most ancient forms of life known. All other forms of life (including human) are **eukaryotes.** You are already quite aware of prokaryotes, for among them are the familiar "germs" that linger in all sorts of places and inspire fascinating TV commercials.

In a more objective light, let's note that in terms of their energy sources, prokaryotes (like other organisms) fall into two major categories, **heterotrophs** ("other-feeding") and **autotrophs** ("self-feeding"). The heterotrophs require complex organic food sources, generally those produced by other organisms. Heterotrophs include the aforementioned "germs," clinically known as **pathogenic** (disease-causing) bacteria. Pathogens are parasites, of course. But most heterotrophic bacteria are from a group known best as **decomposers** (decay bacteria), those that use nonliving organic matter for an energy source. They are often the cause of those pungent odors (or perhaps stenches) associated with body wastes and decay, and while they thrive in our foods and other goods, they generally do not invade our bodies. In fact, decomposers are vital to life because they recycle essential elements such as carbon and nitrogen.

The considerable number of autotrophic prokaryotes use simple inorganic compounds such as carbon dioxide, water, and certain minerals to form their own organic molecules. This of course requires a considerable input of energy. Some bacteria are **photoautotrophic**; like green plants they use light as a source of energy to do their molecule-building. The process itself is called **photosynthesis.** Other bacteria are **chemoautotrophs,** prokaryotes that live deeper in the earth and its waters, where they derive their energy from chemical reactions involving elements such as sulfur and iron.

Whatever their energy source, prokaryotes tend to be structurally simple (Figure 4.4). They generally range in size from one to ten micrometers, only a tenth the diameter of representative eukaryotic cells. One group, the cyanobacteria, is exceptional, often being larger than other bacteria. Cyanobacteria have an elaborate inward extension of their surrounding membrane that contains light-absorbing pigments necessary for photosynthesis. This is the prokaryote version of the eukaryote thylakoid.

In addition to their surrounding membrane, virtually all bacteria have a protective **cell wall.** (So do many eukaryotic cells, but their walls are composed of cellulose or chitin rather than the peptidoglycan and protein of prokaryotic cell walls.) Some bacteria surround the wall with a sheath, and others, with a slimy protective capsule. Within the plasma membrane is the cytoplasm, the living material of the cell—a veritable cauldron of chemical activity (actually, not unlike what goes on in eukaryotic cytoplasm).

The genetic material of bacteria, their DNA, lies free in the cell in an area called the **nucleoid.** (In bacteria there is no organized cell nucleus.) Bacterial DNA occurs as one continuous molecule, referred to as a **chromosome,** which although full of twists and turns, forms a closed loop and can thus be described as circular. Some bacteria have small amounts of DNA incorporated into circular **plasmids** separate from the main chromosome. When prokaryotes approach the time of cell division, the DNA is replicated (copied), and the two copies are separated.

Cell division occurs through a simple process called fission. During fission, some species form a **mesosome**, an inward extension of the plasma membrane that is believed to aid in the separation of DNA replicas. Slender **pili** are used simply to attach the cell to surfaces.

Finally, bacteria have structures that provide movement. Many aquatic and soil species move about through the action of a **flagellum,** which in prokaryotes is a novel, S-shaped structure (see Figure 4.4). The unique prokaryotic flagellum spins on its axis like a propeller and is the only spinning structure seen in cells.

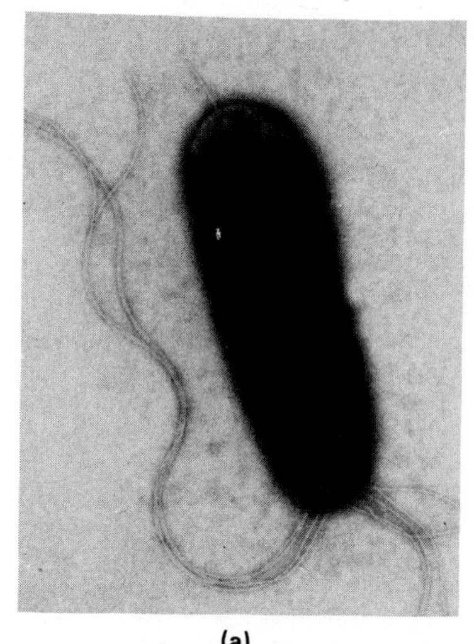

(a)

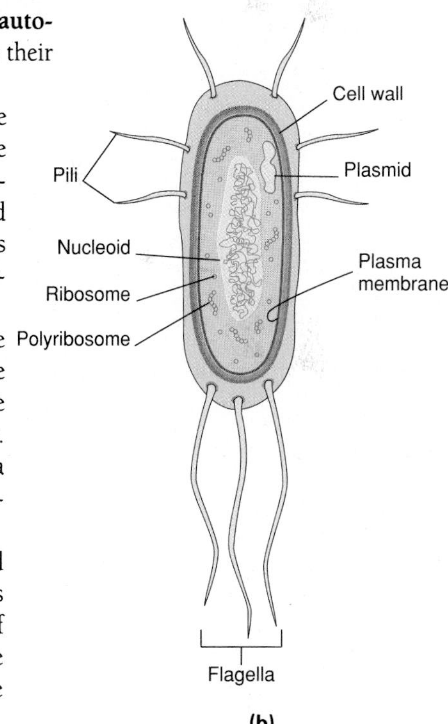

(b)

FIGURE 4.4
THE PROKARYOTIC CELL
(a) The scanning electron micrograph reveals little internal detail in the prokaryotic cell, although a number of flagella are seen. **(b)** Although they are structurally much simpler than most eukaryotic cells, prokaryotic cells carry out many of the same chemical activities.

THE EUKARYOTIC CELL

Eukaryotic cells, including those of plants, animals, fungi, and protists are more complex than those of prokaryotes. However, some structures, many of the molecules of life, and many biochemical processes are similar in the two. For example, eukaryotic cells are surrounded by a plasma membrane, some have dense cell walls, and a few move through the use of a flagellum. But we've already noted that cell walls are chemically different in the prokaryotes and eukaryotes. So are flagella, and from here on there are many other striking differences.

Eukaryotic cells have an extensive array of membrane-surrounded **organelles** ("little organs"), each with its own specific function (Figure 4.5). The most visible organelle is likely to be the **nucleus**, whose membranes (it has two) maintain some separation of the genetic material and the surrounding cytoplasm. The genes, while composed of DNA, are arranged along a number of individual chromosomes that are linear rather than circular. Further, the DNA of eukaryotes is intimately bound to proteins along its length (see Chapter 9).

Although eukaryotic cells share these and other features, they may be quite distinct from each other. The five kingdoms into which living things are classified (**Monera, Protista, Fungi, Plantae,** and **Animalia**; see Table 4.1) are readily distinguished by much of their cell structure. Many eukaryotes in the kingdom Protista are single-celled, and differences among them reflect ways in which these organisms have adapted to their environment through evolution. In multicellular (many-celled) organisms, different kinds of cells have their own specialized interdependent role to play in the more complex organism. Note the differences in size, shape, surfaces and other features in the specialized cells of Figure 4.1. In the discussion to come, we will confine most of our remarks to the cells of plants and animals.

TABLE 4.1 THE KINGDOMS OF LIFE: A PREVIEW

The earth's millions of species are organized into five major groupings known as kingdoms. Assignments to such kingdoms are made on the basis of similarities in structure and function, and biochemical similarities, which, to the biologist, indicate evolutionary relatedness.

KINGDOM	EXAMPLES	MAJOR CHARACTERISTICS
		PROKARYOTES
Monera	Bacteria	Single cells, often in colonies; little internal organization; circular DNA with little protein; includes heterotrophs, photoautotrophs, (photosynthesizers), and chemoautotrophs
		EUKARYOTES
Protista	Protozoans and algae	Single-celled, multicellular, and in colonies; heterotrophs and photoautotrophs
Fungi	Molds, mushrooms, smuts, rusts	Multicellular; simple tubular body form; elaborate reproductive structures; heterotrophs
Plantae	Mosses, ferns, cone-bearers, and flowering plants	Multicellular; well-developed tissues, organs, and systems; photoautotrophs
Animalia	All invertebrates and vertebrates, from sponges to grasshoppers and from fish to humans	Multicellular; well-developed tissues, organs, and organ systems; heterotrophs

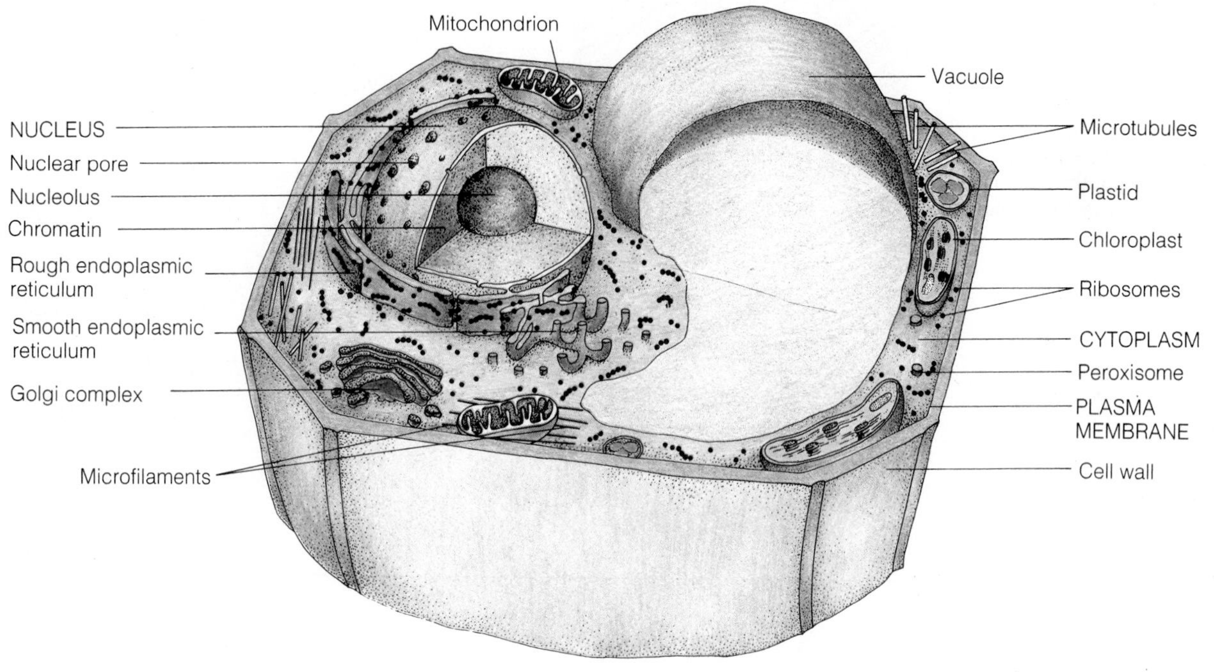

(a) Plant cell

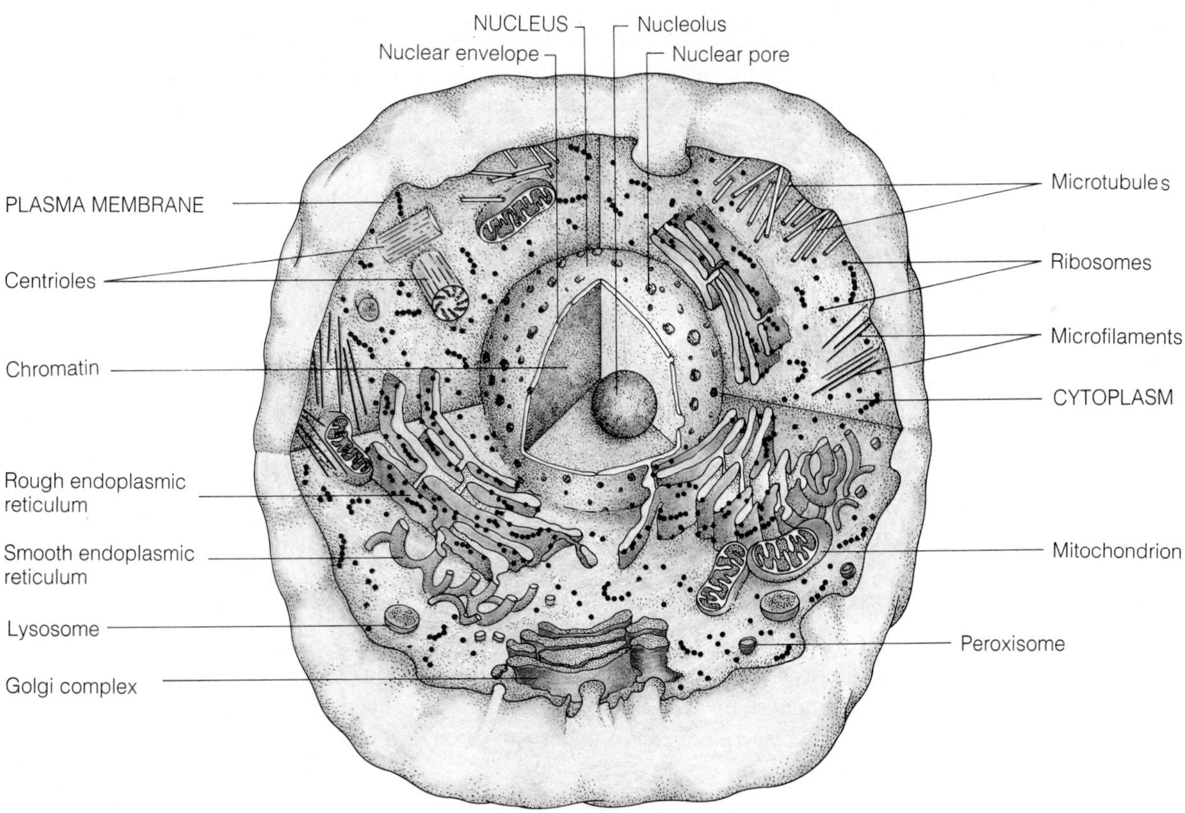

(b) Animal cell

FIGURE 4.5
REPRESENTATIVE CELLS
Few cells of either plants (**a**) or animals (**b**) have all of the structures seen in these composite drawings. Both plants and animals are eukaryotes, which means their cells have a prominent, membrane-surrounded nucleus and several kinds of organelles that are also surrounded by membranes. Plant and animal cells have many features in common, but they also have important differences.

FIGURE 4.6
THE PLANT CELL WALL
Plant cell walls consist of several layers of cellulose microfibrils embedded in a dense matrix (photo). The outermost layer forms first, followed by several added layers. In cells that grow mainly in length, the older microfibrils become oriented longitudinally.

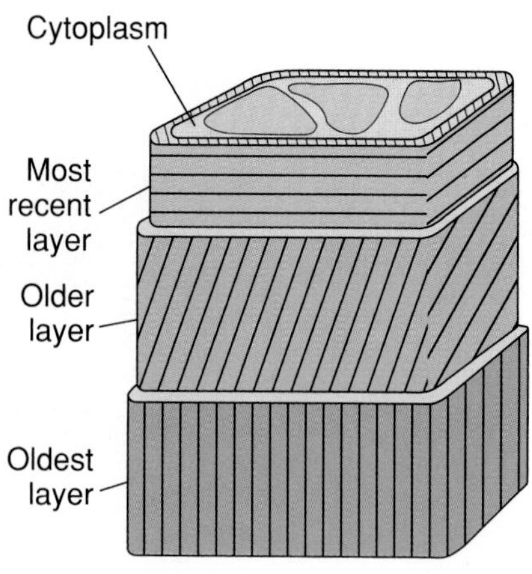

CELL STRUCTURES AND ORGANELLES IN EUKARYOTES

Now that we have some idea about what cells are and how they vary, let's look closely at their structure. In keeping with the characteristics of life discussed earlier, the structures will be grouped according to basic cellular functions. A good place to begin is with those structures that support the cell and form the necessary boundaries that maintain the cell's integrity and provide for transport. From there we will progress to structures or organelles devoted to control and reproduction, synthesis and storage, energy transfer, and finally movement. Keep in mind, though, that there is often some unavoidable overlap among these categories. Cells simply don't fill our neat little categories as well as we might want.

Surface Structures

The Cell Wall As we've already noted, plant cells are surrounded by fairly rigid, nonliving walls composed primarily of cellulose and other polysaccharides (Figure 4.6). Microfibrils of cellulose chains are arranged in layers, with each layer lying at an angle with respect to the one below it. The result is a laminated and very strong but porous covering over the cell (not unlike plywood). Once impregnated with lignin and pectin (hardening substances), the cell wall maintains the shape of the cell and has great strength and resiliency.

Actually, newly formed cell walls can't be too rigid, because each time a plant cell divides to form two cells, the new cells must be able to grow to the size of the original cell. As the cell divides, a new cell wall appears, effectively separating the two new cells. Walls of adjacent cells are cemented together with pectin to form a layer known as the **middle lamella.**

As the plant cells mature and differentiate into specialized tissues, other substances are deposited into the cellulose and pectin matrix. In the stems of woody plants, for instance, lignin, a complex chemical that is technically an alcohol, is continually secreted into the cellulose, forming a hard, thick, decay-resistant wall. **Suberin,** a wax substance, is secreted into the outer layer of stem cells, forming a protective waterproof layer. Waterproofing of the upper surface of leaves, on the other hand, is accomplished by the secretion of **cutin,** a waxy substance added to the epidermal cell wall. Waterproofing in plants, by the way, is intended to keep water in.

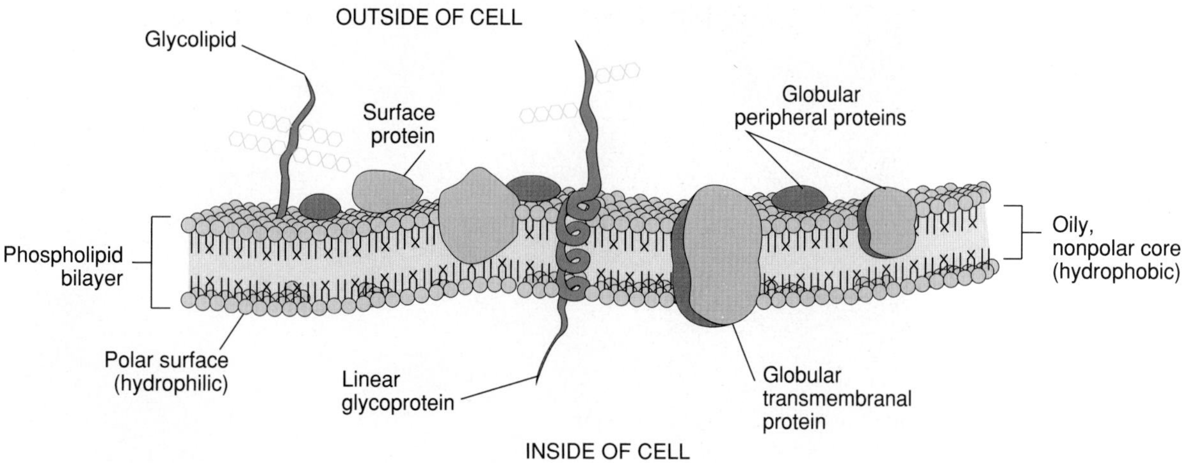

FIGURE 4.7
THE PLASMA MEMBRANE
The plasma membrane consists of a bilayer of phospholipids, their charged or polar heads forming the hydrophilic surface and their uncharged, nonpolar tails forming a hydrophobic oily core. There are numerous protein components, some on the surface, others partly submerged, and still others penetrating the phospholipid bilayer. (Other molecular surface structures are presented in Chapter 5.)

The Plasma Membrane We have already mentioned this vital cell boundary, and vital is the right word because if anything is important to the integrity of the cell, the membrane is. Basically, it keeps the inside of the cell in and the outside out, and it lets through the materials that the cell must exchange with its environment. The membrane's complex structure permits it to accept effortlessly the passage of some substances, utterly reject others, and at times, expend energy to actively assist the transport of still others. Thus, the plasma membrane is not simply a passive envelope, but is highly selective in what it allows to cross.

These important functions, and others, are made possible by the plasma membrane's intricate structure. It is essentially a double lipid film, consisting of two layers of phospholipids. Their fatty acid tails intertwine while their polar heads project outward (Figure 4.7). Interspersed in the lipid film are a number of different kinds of proteins, all essential to the movement of materials and other functions. We will look into plasma membrane function in detail in the next chapter, but we should point out that the basic **phospholipid bilayer** construction is common in cells. It is seen in the nuclear membrane and the membranes that surround organelles throughout the cell. Any difference from one membrane to the next is generally attributed to its protein constituent, which in turn, is related to the membrane's function.

Internal Support: The Cytoskeleton

Up to a few years ago, little was known about structure in the visually clear regions of cytoplasm, those regions surrounding the prominent organelles. It appeared that such places were simply structureless and fluid-filled and were thus designated as the "cell matrix" or "ground substance." But with the advent of fluorescence microscopy and of the high voltage transmission electron microscope, much more about this region is known. We know that the matrix is permeated wth an intricate, mazelike network of supporting fibers which form the **cytoskeleton** (Figure 4.8). Entire cells, or at least thick slices of cells, can now be examined, thus providing a three-dimensional view.

From such images cell biologists have determined that the cytoskeleton is made up of three types of fibrous elements: very fine **microfilaments,** chiefly of the protein actin (6 nm in diameter), **intermediate filaments** (7–11 nm in diameter), and still larger **microtubules** (22 nm in diameter). Their lengths vary widely. The cytoskeletal elements provide a supporting framework for cell organelles, aid in cell movement, anchor the plasma membrane, and provide surfaces upon which chemical activity can occur in an organized manner.

Actin Microfilaments **Actin microfilaments** are composed of two slender chains of actin wound around each other. Microfilaments have a variety of roles, depending upon

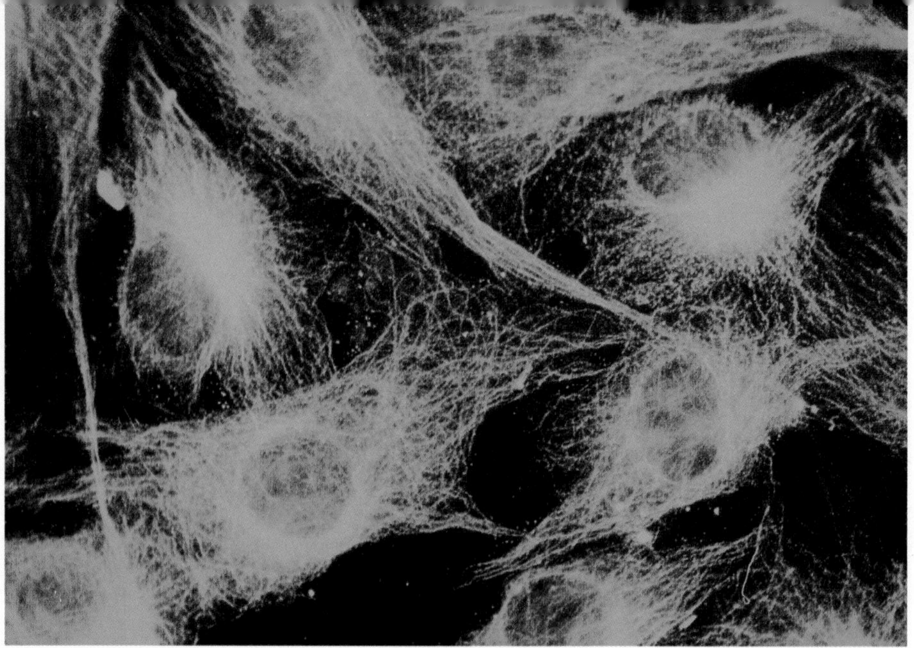

FIGURE 4.8
THE CYTOSKELETON
The fibrous cell cytoskeleton is made visible here through the use of fluorescence microscopy. The procedure involves the use of selective fluorescent dyes that join specific molecular complexes that then fluoresce under UV or shorter wavelength radiations.

other proteins with which they associate. For instance, a network of fibers underlying and crisscrossing the plasma membrane is bound together at intersections by short segments of actin. The strong, flexible fiber network is important since there really isn't much strength in the phospholipid bilayer alone.

Actin microfilaments join other proteins to form a supporting network within the **microvilli** (brush border) of intestinal lining cells. The microfilaments also join intermediate filaments deeper below the surface providing further support (Figure 4.9). In still other protein associations, actin microfilaments become involved in certain types of cell movement. The cytoplasmic streaming mentioned earlier includes an active circular movement of cytoplasm within certain plant cells. The actin microfilaments form a fixed boundary along which the freer cytoplasma moves. Researchers believe that certain proteins, linked to the moving organelles, may interact with actin in a way the promotes the ongoing streaming action. In **ameboid movement**, the organism actually moves itself by squeezing parts of its body into **pseudopods** ("false feet"), tubelike extensions through which its more fluid parts flow. Since the extended and withdrawn states are interconvertible, the cell seems to "pick itself up and follow." Actin microfilaments are also involved in the pinching in of the cell membrane that occurs as animal cells divide (Figure 9.7). We'll also see in Chapter 33 how actin microfilaments interact with the protein myosin during muscle contraction.

Intermediate Filaments Intermediate filaments are the least understood of the three cytoskeletal elements, but they appear in such diverse human cell types as epithelial (lining) cells, most nerve cells, cells of the brain and spinal cord, muscle fibers, and blood cells. The intermediate filaments of epithelial cells are composed of the protein keratin (also found in hair and nails), which lends strength to the important linings they form. Intermediate filaments of keratin are shown as part of the supporting network of microvilli in Figure 4.9.

Microtubules Microtubules, the largest of the cytoskeletal elements, are hollow and rodlike, formed from numerous molecules of **tubulin,** a globular protein. Each tubulin molecule is actually a dimer, composed of two spherical polypeptides. As microtubules form, these doublets are assembled into spiralling rows (Figure 4.10). Cells can rapidly assemble microtubules when needed and just as quickly disassemble them, conserving the tubulin units for future use.

Microtubules, like the other cytoskeletal elements, have highly varied roles. In addition to a purely supporting role in the cytoskeleton, microtubules form **centrioles, basal bodies, cilia,** and flagella. Such structures, as we'll see, are directly or indirectly involved in movement and cell organization. But nowhere is the role of microtubules more dramatically visible than in the dividing cell, where many of them cross the cell, taking

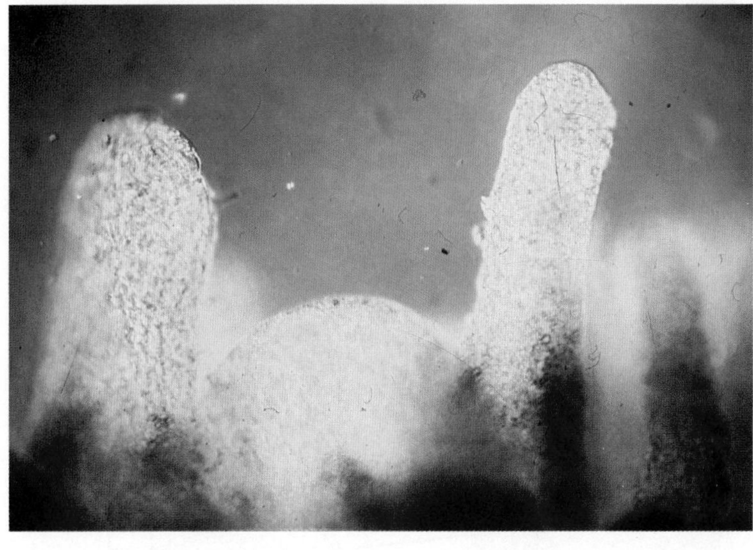

(a)

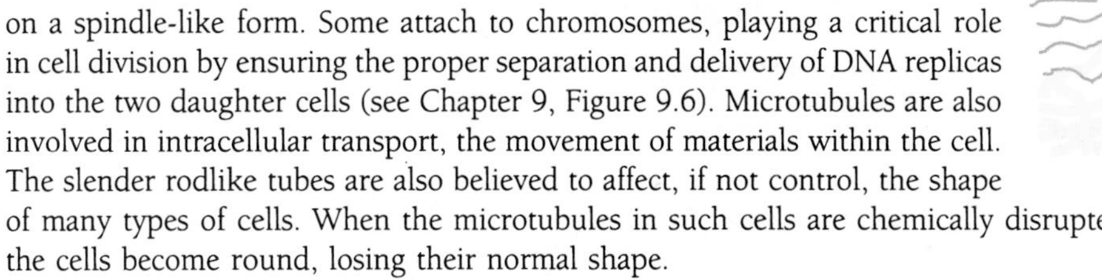

on a spindle-like form. Some attach to chromosomes, playing a critical role in cell division by ensuring the proper separation and delivery of DNA replicas into the two daughter cells (see Chapter 9, Figure 9.6). Microtubules are also involved in intracellular transport, the movement of materials within the cell. The slender rodlike tubes are also believed to affect, if not control, the shape of many types of cells. When the microtubules in such cells are chemically disrupted, the cells become round, losing their normal shape.

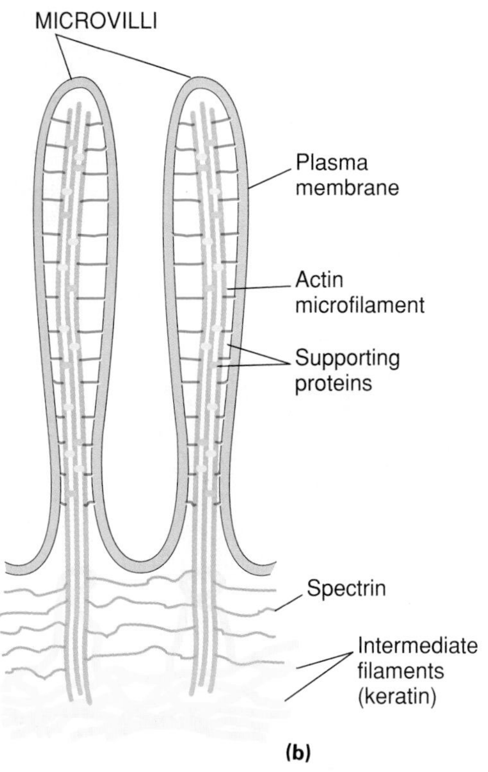

(b)

FIGURE 4.9
MICROVILLI
(a) The cytoskeleton is quite prominent in the delicate microvilli of the intestine. **(b)** Microfilaments extending into each projection form intricate interconnections with other proteins, helping to keep the microvilli erect. Microfilaments emerging from the microvilli interact with a denser maze of keratin-rich intermediate filaments below.

Control and Cell Reproduction: The Nucleus

The significant functions of control and cell reproduction go to the nucleus, the most prominent organelle of the cell. Because the nucleus is easily stained, and therefore easily viewed under the microscope, it was described early in the history of cell study. In fact, the word *nucleus,* which means "kernel," was first used in 1831 by botanist Robert Brown about the time of the emergence of the cell theory. Nuclei have since been found in virtually every type of eukaryotic cell.

The nucleus contains most of the cell's DNA, which, as we've seen, is assembled into a number of linear DNA molecules. Each DNA molecule, in turn, carries a number of genes, or hereditary units, somewhat like beads along a string. Each DNA molecule and its associated proteins are referred to as a chromosome (*chrome,* "color;" *soma,* "body." Chromosomes readily take on certain colored stains). Most of the time there is very little visible detail in the nucleus, although dark-staining bodies called **nucleoli** (singular, **nucleolus**—"little nucleus") are usually seen (Figure 4.11a). Each nucleolus is involved in intense nucleic acid synthesis, specifically nucleic acids of a variety called ribosomal RNA. This nucleic acid, along with certain proteins, is later assembled into protein-synthesizing bodies called **ribosomes.**

The chromosomes are most visible when the cell prepares for cell division, at which time they condense into short sausagelike bodies. Prior to this transformation, each chromosome will have undergone replication, so the replicas first appear as doublets. The replicas will separate, and upon the completion of cell division each new cell will have a complete set of genes.

Whereas most organelles have one surrounding membrane, the nucleus has two, aptly called the **nuclear envelope.** Each membrane consists of the usual phospholipid bilayer along with associated proteins. Unlike other cell membranes, the nuclear envelope contains a great number of **nuclear pores** scattered over its surface (Figure 4.11b). Many materials passing between the nucleus and cytoplasm move through these pores. The pores are not simple holes; under high magnification (Figure 4.11c) they are seen as eight protein clusters surrounding a central cluster.

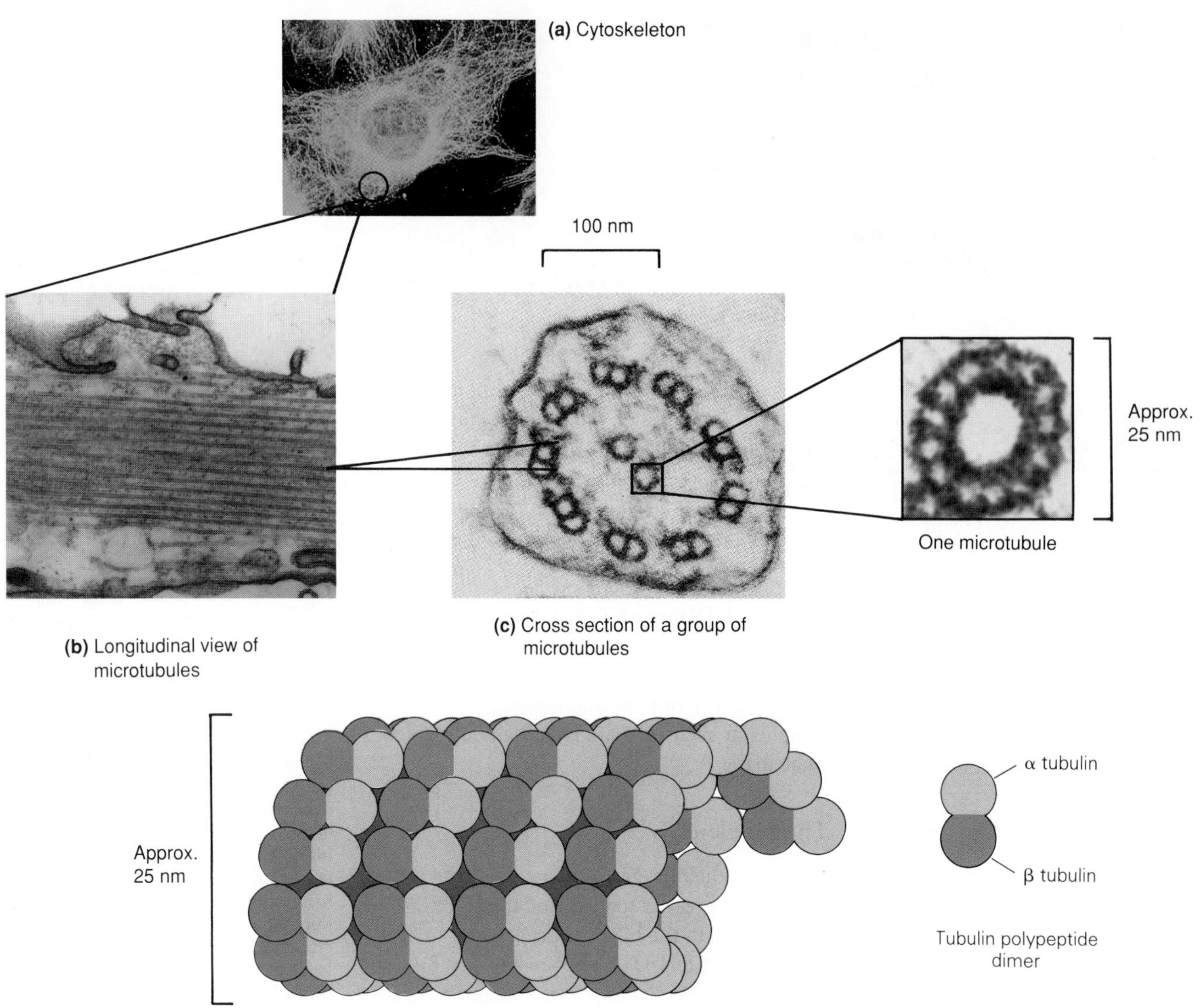

FIGURE 4.10
MICROTUBULE STRUCTURE
(a) Part of the cytoskeleton is made up of microtubules. **(b)** A greatly magnified, longitudinal view shows microtubules as a number of parallel lines, but the enlarged cross section in **(c)** reveals their hollow nature. **(d)** Dimeric tubulin molecules, doublets of spherical alpha and beta tubulin, follow a spiralling pattern as they are assembled into the microtubule.

Organelles of Synthesis, Storage, and Cytoplasmic Transport

There are extensive internal membranes in the eukaryotic cell. Their arrangement establishes a number of regions or compartments within the cytoplasm where very specific chemical functions can be carried out, a situation that increases the efficiency of such processes. The first of these compartments we will consider is formed by the endoplasmic reticulum, an internal membrane.

The Endoplasmic Reticulum The **endoplasmic reticulum (ER)** was unknown in the days when we were restricted to using light microscopes. In fact, in those days the contents of cells were thought of as a formless, soupy "protoplasm" (primary living material). But electron micrographs reveal a complex membrane system that takes up a large part of the cytoplasm of eukaryotic cells, especially those cells that are engaged in significant protein synthesis. The membranes of the ER contain the same phospholipid bilayer seen in the plasma membrane. Although the ER system was first seen in the mid-1940s, it wasn't until 1953 that Keith Porter of the Rockefeller Institute first suggested the name. A year later, Porter and George Palade suggested that the reticulum was a very dynamic, ever-changing structure.

Painstaking electron microscope studies of the ER have led researchers to realize that it is one continuous membrane folded back and forth within the cytoplasm and forming

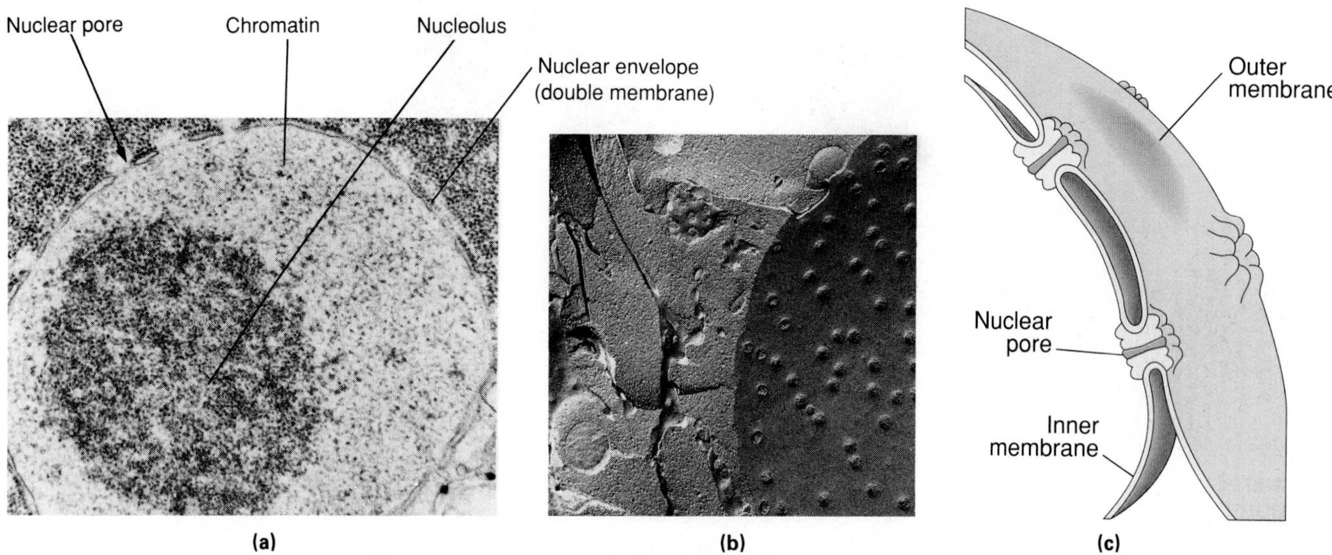

FIGURE 4.11
A STUDY OF THE NUCLEUS
(a) The dark circular area is the nucleolus, where the ribosomal elements are manufactured or gathered prior to assembly in the cytoplasm. The grainy surrounding region contains chromatin (DNA and protein). **(b)** A freeze-fracture preparation exposes the inner nuclear membrane, revealing many nuclear pores. **(c)** Each nuclear pore consists of eight protein granules surrounding one central protein granule.

a closed sac. Its internal space, the **ER lumen,** may account for up to 10% of the entire cytoplasmic volume, and the ER membrane itself, more than half the membrane of an entire cell. It has been clearly demonstrated that the ER is continuous with the outer of the two membranes of the nuclear envelope, so actually the nuclear material is separated from the ER lumen by a single membrane only.

The ER plays a central role in the cell's synthetic activity. Newly synthesized materials such as protein, lipids and carbohydrates are transported within the ER lumen, where they may undergo further processing before being transferred to other organelles.

Rough Endoplasmic Reticulum The endoplasmic reticulum appears in two main forms: rough and smooth. The **rough endoplasmic reticulum (RER)** receives its name from the appearance of ribosomes, dense granules that tightly adhere to the outer sides of the membrane, making it look a little like coarse sandpaper (Figure 4.12). Channels are formed by two such membranes lying side-by-side with their rough surfaces (ribosomes) out.

Rough endoplasmic reticulum occurs most commonly in cells that manufacture proteins destined for secretion outside the cell. Two examples of such proteins are digestive enzymes and certain hormones. As the polypeptides that form these active proteins are synthesized by the ribosomes of the rough endoplasmic reticulum, they enter the lumen, its inner space. Once inside, each linear polypeptide folds into its final shape, and following this other modifications are made. We've mentioned ribosomes in passing, so before going on we will consider the structure and role of these interesting bodies.

Ribosomes: Sites of Polypeptide Synthesis Ribosomes contain sites where amino acids are assembled into polypeptides. In doing this they follow chemical instructions provided by genes situated along DNA (see Chapter 13). While not actually organelles, ribosomes are far too large to be considered molecules. They are large molecular complexes consisting chiefly of **ribosomal RNA (rRNA)** and many kinds of protein. Each ribosome consists of two subunits that come together just before polypeptide synthesis begins (Figure 4.13). The ribosomal subunits in prokaryotes are somewhat smaller and differ chemically from those of eukaryotes. In fact, it is this difference that makes antibiotics such as tetracycline and streptomycin so effective against bacterial infection. The two antibiotics halt the work of bacterial ribosomes, thus stopping protein synthesis. They have no effect on our own eukaryotic ribosomes.

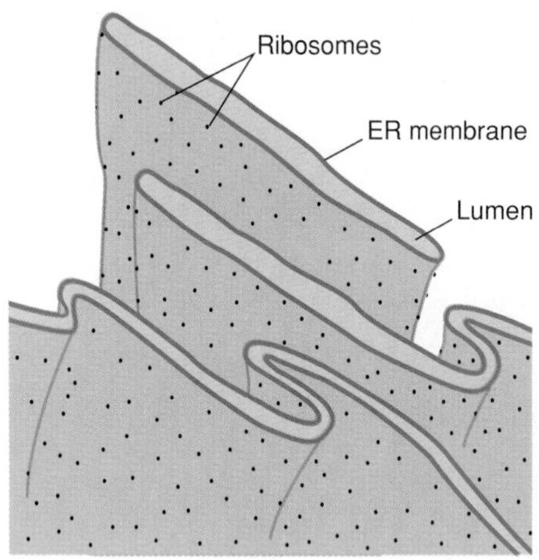

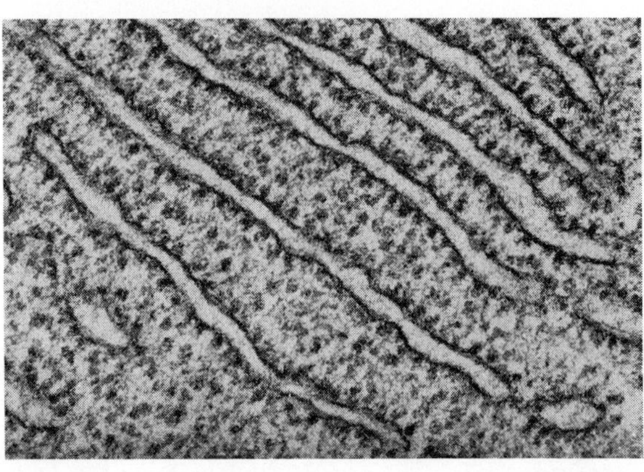

FIGURE 4.12
ROUGH ENDOPLASMIC RETICULUM
The RER consists of parallel membranes that form closed, flattened channels. The membrane is dotted with numerous ribosomes, sites of protein synthesis.

Some ribosomes appear free in the cytoplasm, occurring either singly or in beadlike strands called **polyribosomes** (or just **polysomes**). Others appear to be attached to the endoplasmic reticulum. The latter, called "bound ribosomes," are only temporarily held there while they assemble amino acids into the polypeptides that are processed within the lumen of the endoplasmic reticulum. Such proteins are generally destined to be exported from the cell.

Smooth Endoplasmic Reticulum The **smooth endoplasmic reticulum,** or **SER** (Figure 4.14), occurs in most cells but abounds in cells that synthesize, secrete, and store carbohydrates, steroids, lipids, and other non-protein products. The structure of smooth endoplasmic reticulum resembles an intricate network of tubes and sacs. We find a lot of smooth endoplasmic reticulum in cells of the testis, oil glands of the skin, some hormone-producing gland cells, and absorptive cells lining the small intestine. There is strong evidence that the smooth ER in small intestine cells plays an important role in the way animals take up the products of lipid digestion.

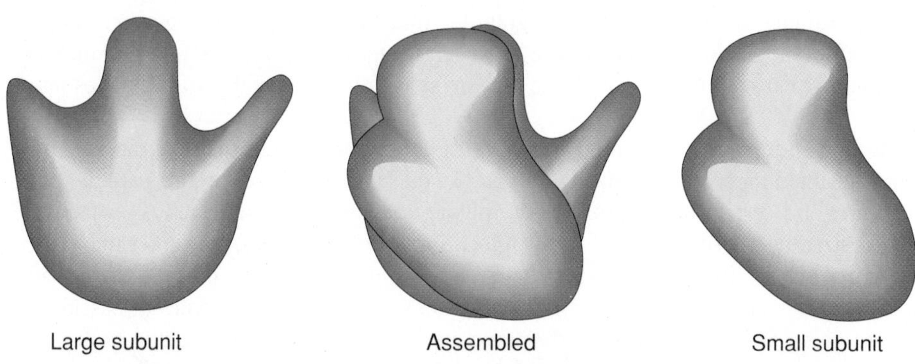

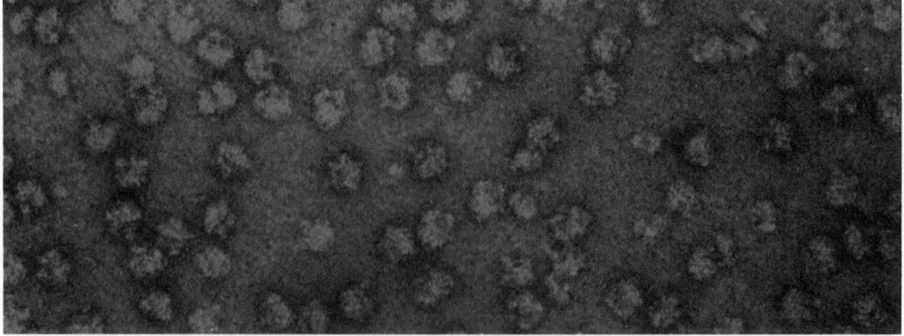

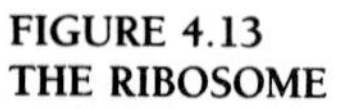

FIGURE 4.13
THE RIBOSOME
Ribosomes are molecular assemblies of RNA and protein. Each consists of an interlocking large and small subunit that join when the ribosome is active in protein synthesis.

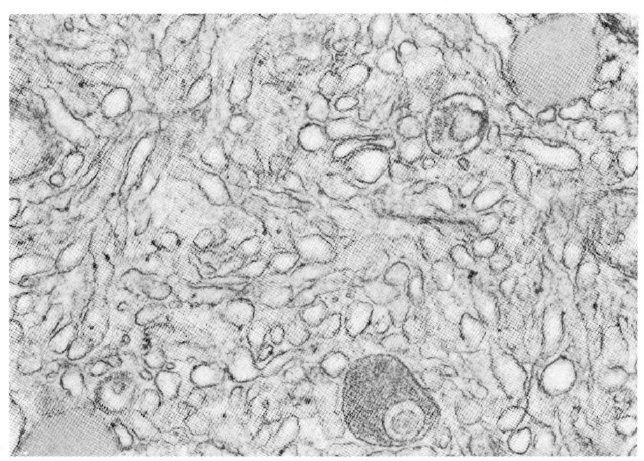

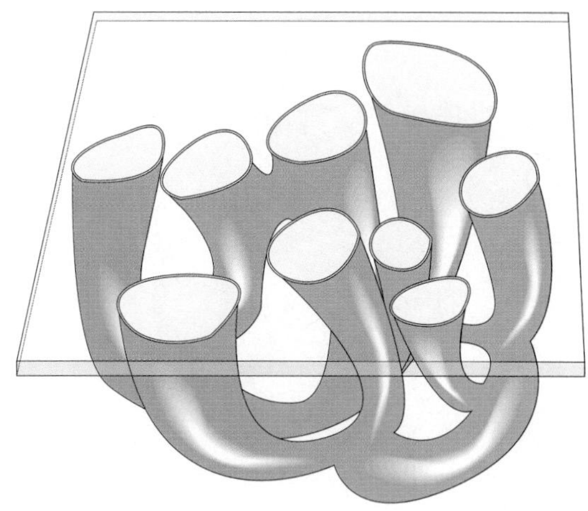

FIGURE 4.14
SMOOTH ENDOPLASMIC RETICULUM
Unlike the broad channels formed by RER, the SER takes the form of a network of interconnected tubelike channels.

Smooth endoplasmic reticulum in liver cells is known to have still another function: it contains oxidizing enzymes that detoxify certain chemicals. Animals injected with large doses of the sedative phenobarbital, for example, revealed a substantial increase in their liver cell smooth endoplasmic reticulum and in the associated enzymes. Also in the liver cells, we find glycogen (animal storage starch) closely associated with smooth endoplasmic reticulum. Studies reveal that the enzymes responsible for one of the steps in the breakdown of glycogen to glucose is present in the SER of liver cells.

The Golgi Complex In 1898, Camillo Golgi, an Italian scientist, tried a tissue-staining technique involving metallic salts. Through its use, Golgi was able to observe structures that were not theretofore visible through the light microscope. For instance, he discovered axons and dendrites—fine, lengthy, communicating extensions of nerve cells. He also observed a peculiar structure that today bears his name, the **Golgi complex.** It was one of the first cytoplasmic structures to be seen, but despite its early discovery, some 70–80 years passed before the function of the Golgi complex was fully appreciated.

Modern electron microscope studies of the Golgi complex show it to be a series of flattened, baglike sacs or **cisternae,** often found near the nucleus. Such a series is referred to as a "Golgi stack." Some biologists refer to such stacks in plant cells as **dictyosomes.**

The Golgi complex is one of the cell's more dynamic organelles ever busy and ever changing. In Figure 4.15, we see that Golgi cisternae actually originate from portions of the nearby rough endoplasmic reticulum known as **vesicles.** The vesicles, laden with products formed in the RER, continually fuse with what is called the ***cis* face** (forming face) of the Golgi complex. The *cis* face, in turn, buds its own vesicles, and these are added to the next cisterna in line. The action repeats, and vesicles are sequentially formed at one cisterna and added to the next progressively through the stack. This is how the Golgi stack is formed. Eventually vesicles contribute to the ***trans* face** (maturing face). Vesicles formed from the *trans* face leave the Golgi complex altogether, headed for various destinations.

So what is all the budding about? Biochemical studies have provided some of the answers. The Golgi stack, as it turns out, is actually an assembly line, involved in the step-by-step modification of a variety of proteins arriving from the rough ER. A battery of enzymes, each specific to its own cisterna in the stack, is responsible for the many reactions of this complex assembly line. Final treatment occurs in the *trans* face, where the different products are sorted out, and each is packaged in its own vesicle. For example, certain powerful digestive enzymes receive their final modification in the *trans* face and enter numerous vesicles that later coalesce to form tough **lysosomes** (our next organelle). Other products are stored temporarily in storage granules, while still others go into secretory vesicles that move directly to the cell membrane, where their contents

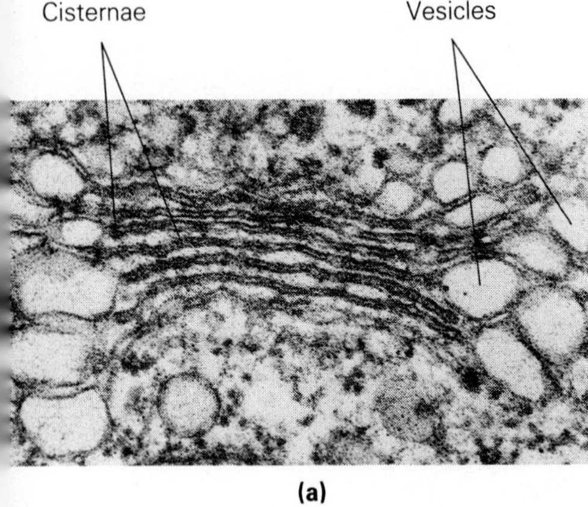

(a)

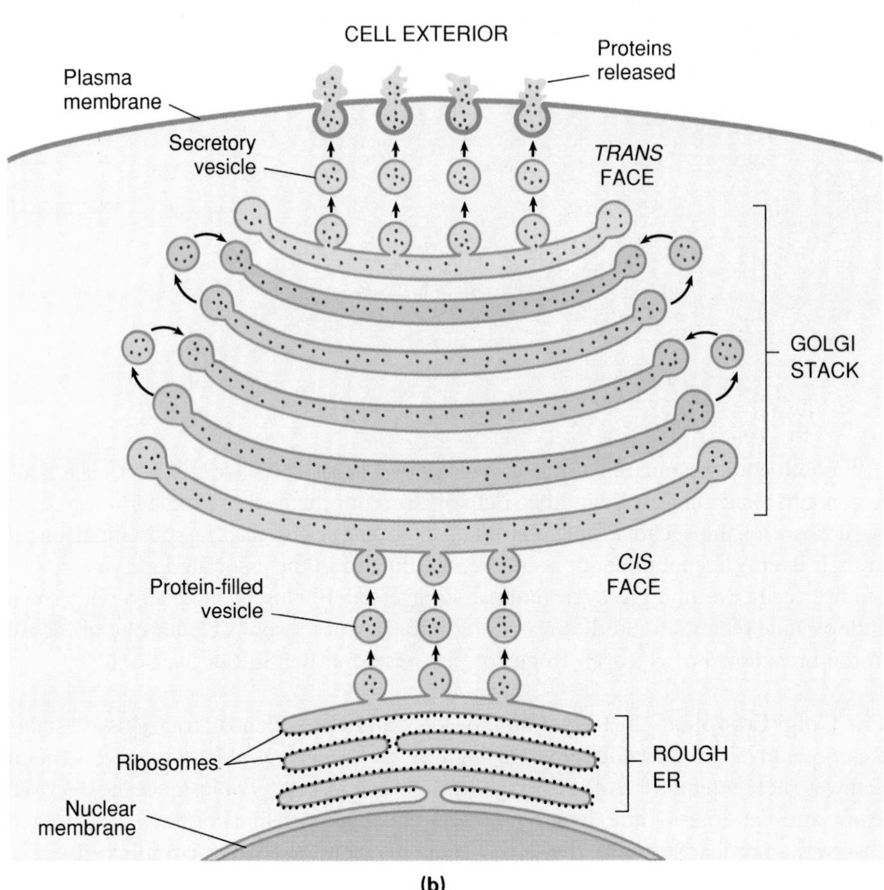

(b)

FIGURE 4.15
THE GOLGI COMPLEX
(a) The Golgi complex includes a number of flattened cisternae also called Golgi stacks. **(b)** The cisternae form continuously as protein-laden vesicles from the nearby rough endoplasmic reticulum fuse at the *cis* (forming) face. The stack is maintained by the continued budding and transfer of vesicles from one cisterna to the next. Molecules moving through the series undergo continuing chemical modification until the finished products leave the stack via vesicles from the *trans* (maturing) face. These vesicles have varying roles according to their contents.

are secreted (expelled) from the cell. In plant cells, newly formed cell wall materials are delivered via such vesicles to the developing cell wall. Cell biologists know a good deal about the Golgi complex, but there are still mysteries. Their latest puzzle: how does the Golgi complex manage the complex sorting process?

Lysosomes Lysosomes have been found in nearly all types of animal cells, although there is some disagreement about whether they exist in plant cells. They are membrane-bounded sacs that are roughly spherical in shape (Figure 4.16a). But the simple appearance of the lysosome belies the rather startling role they play in the life of a cell. Lysosomes are bags of powerful hydrolytic enzymes, synthesized in the rough ER and packaged primarily by the Golgi complex (Figure 4.15b). In all, about 40 different enzymes have been detected in lysosomes. Interestingly, the enzymes require an acidic condition to work efficiently, which helps prevent damage to the cell cytoplasm should the lysosome leak. The lysosome creates its own acidic condition by actively pumping protons (H^+) from the cytoplasm to its interior.

Cell biologist Christian de Duve predicted the presence of lysosomes from biochemical evidence and went looking for them with his electron microscope. When he found them, de Duve described lysosomes as little "suicide bags." His poetic fancy was not entirely unwarranted, since lysosomes are known to engage in **autophagy** (self-eating), a tidying-up process whereby damaged or aged organelles within the cell are taken into a digestive vacuole and hydrolyzed by lysosomal enzymes (see Figure 4.16b). Actually, autophagy is not as destructive as it may seem. Sometimes, in fact, the destruction of cells is a normal part of metabolism. In other instances, lysosomes might destroy a superfluous

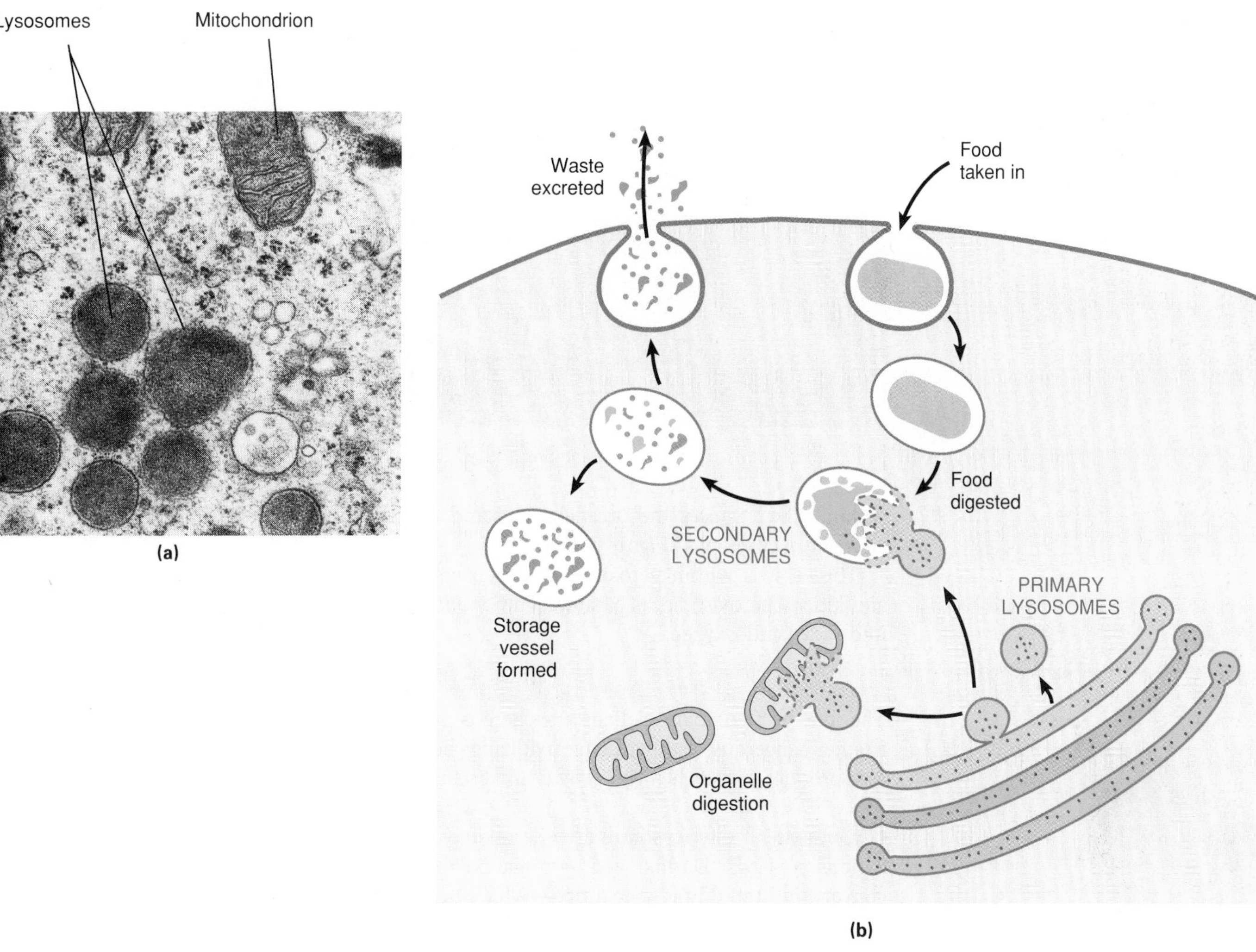

FIGURE 4.16
LYSOSOMES
(a) Lysosomes are dense storage bodies. **(b)** They form continuously from the *trans* face of the Golgi complex, emerging as primary lysosomes. They become secondary lysosomes when they fuse with other bodies. Their varied roles include digesting bulk materials (such as bacteria), modifying and storing molecular material taken in, and destroying aging or damaged cytoplasmic organelles.

cell that is not functioning well, or one in a part of the body that is undergoing reduction as part of a developmental process, such as the tissue between the fingers in a developing hand. Following the action of lysosomes, phagocytic white cells—cells that engage in **heterophagy** (other-eating)—clean up what's left.

Lysosomes also have specific functions in certain healthy, active cells. Such cells take in solid substances and form *digestive vacuoles.* Afterwards, a lysosome fuses with a digestive vacuole, its enzymes aiding in the digestion of whatever has been captured.

There are a number of genetic disorders in which some essential lysosomal enzyme is altered or absent. This means that certain substances in their unreacted, raw form accumulate in the cell, causing so-called **storage diseases.** Most such diseases are fatal in the first five years of human life. Tay-Sachs disease, a genetic disease that is relatively common among infants of European Jewish descent, is one such storage disease. In the disease, the absence of the critical lysosomal enzyme (N-acetylhexosaminidase) results in the buildup of certain lipids called gangliosides in brain cells, producing retardation, blindness, and, eventually, death.

Microbodies

Peroxisomes The **peroxisomes** belong to a class of organelles called **microbodies.** Peroxisomes often lie near mitochondria or chloroplasts (both of which are discussed below). Peroxisomes are found in a great variety of organisms, including plants, animals, and protists. In animals, they are most common in liver and kidney cells, where they are believed to originate from outpocketings of the smooth endoplasmic reticulum. Some peroxisomes appear as very dense bodies with a unique crystalline inclusion (or body)

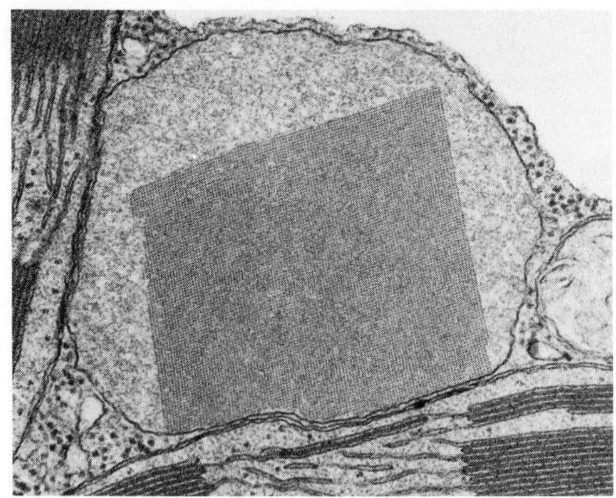

FIGURE 4.17
PEROXISOMES
Peroxisomes are readily identified by the crystalline bodies inside them. This peroxisome is from a plant cell, where it shares borders with two membranous chloroplasts.

within, which causes the organelle to stand out unmistakably when it shows up in electron micrographs (Figure 4.17).

The crystals within peroxisomes are enzymes, and a principal enzyme of the liver and kidney peroxisomes is **catalase**, important in the breakdown of hydrogen peroxide into water and oxygen:

$$2H_2O_2 \rightarrow 2H_2O + O_2$$

The role of the peroxisomal enzymes here is probably protective. Hydrogen peroxide is a rather dangerous, highly reactive oxidizing chemical, produced in considerable quantity as a product of the chemical activity in certain cells.

Glyoxysomes **Glyoxysomes** are microbodies that commonly occur in the lipid-storing regions of seeds. During seed germination (sprouting), enzymes of the glyoxysomes convert the stored lipids to sugars—which help maintain the young seedling just long enough for it to begin synthesizing its own food through photosynthesis.

Vacuoles The term **vacuole** is rather general. It refers merely to any membrane-bound body with little or no inner structure. Vacuoles generally hold something, but their contents vary widely, depending on the cell and the organism.

Plant cells generally have larger vacuoles than do animal cells. In many types of plant cells, a vacuole dominates the central part of the cell, crowding all other elements against the cell wall (Figure 4.18). Such vacuoles are filled with a fluid, the **cell sap**, which is primarily water with various substances in solution or suspension. The solutes may include atmospheric gases, inorganic salts, organic acids, sugars, pigments, or other materials. Sometimes the vacuoles are filled with water-soluble blue, red or purple pigments (the *anthocyanins*), which are responsible for many flower colors.

The large, watery vacuoles of plants serve another curious purpose. While all animals have elaborate excretory systems for getting rid of cellular and metabolic wastes, plants lack such systems. Instead, metabolic wastes and other poisons are sequestered in the central vacuoles, frequently forming crystals. Some plants store large amounts of waste poisons in their vacuoles, and it has been suggested that this serves as a protective device against herbivores. If a herbivore should eat part of the plant and crush the cell, it will get not only food, but stored poisons. Of equal significance, the large, water-filled vacuoles of plants, particularly in cells of the leaves and in the stems and leaves of soft-bodied plants, hold the plant erect and firm.

The Endomembranal System The sharing and transfer of membranes among organelles is quite common in the eukaryotic cell. You've seen that, in places, the nuclear membrane is continuous with the endoplasmic reticulum. Further, you now know that membranes of the endoplasmic reticulum contribute to the Golgi complex, which in

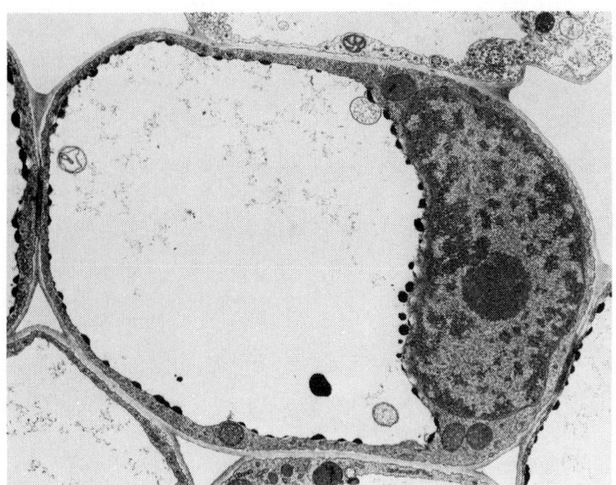

FIGURE 4.18
VACUOLES
Plant cells often have large central vacuoles filled with watery sap. In leaf cells, numerous water-filled vacuoles exert pressure on their cell walls, keeping the leaf rigid and erect.

turn, uses its own membranes to form specialized vesicles. Later, certain vesicles fuse with the plasma membrane, thus contributing their membranes to it, while other vesicles coalesce with vacuoles.

At one time the membrane-bounded organelles were believed to be more or less separate and autonomous, but this view is no longer held, at least for a number of organelles. The ongoing and dynamic formation of one organelle from another has led cell biologists to consider the participants as part of an **endomembranal system.** The endomembranal system, as they see it, is concerned with the synthesis of cell products, with their modification and storage, and eventually with their final disposition. In the coordination of such sequential activities, reactants and products require isolation, and it is the dynamic, sequential formation of endomembranal system organelles that makes this possible.

Energy-Generating Organelles

The energy-generating organelles are the **chloroplasts** and **mitochondria.** Their basic function is to convert energy to a form useful to the cell, and each does this in its own way. Their unique functions make the chloroplast and mitochondrion interesting on their own, but when one considers how these energetic little organelles may have gotten into cells in the first place, they can only be described as fascinating (see Essay 4.2).

Chloroplasts Chloroplasts are found in plants and algae, and their presence is one of the features that distinguishes plants from animals (Figure 4.19a). They are large, as organelles go, and quite complex, with a surrounding double membrane, an intricate inner membranous structure, and, surprisingly, their own circular DNA and ribosomes. With DNA and ribosomes present, chloroplasts are able to reproduce themselves and to carry out the synthesis of some proteins, although many others are taken in from the cell cytoplasm, where they were produced through the action of the nucleus. In the mature cell, chloroplasts always arise from preexisting chloroplasts. This happens through the division of the chloroplast by a simple pinching in two, which looks quite similar to fission in prokaryotes. As you would expect, the division process is preceded by DNA replication, which provides each new organelle with a copy of chloroplast genes. We will find that these characteristics are true of mitochondria as well.

Chloroplasts function in that critical and intricate process called photosynthesis, in which the energy of light is used to convert carbon dioxide and water to glucose and other important molecules. Among the vital participants in this process are active molecules known as the **chlorophylls.** *Chloroplast* means "green form," and the green comes from the photosynthetic pigment chlorophyll. We should note that many bacteria also contain chlorophyll pigments, but the chlorophyll in these prokaryotes has a somewhat different chemical structure.

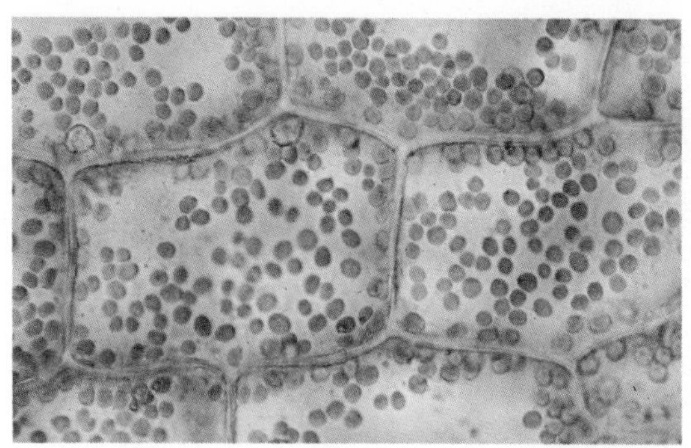

(**a**) CHLOROPLASTS IN PLANT CELLS

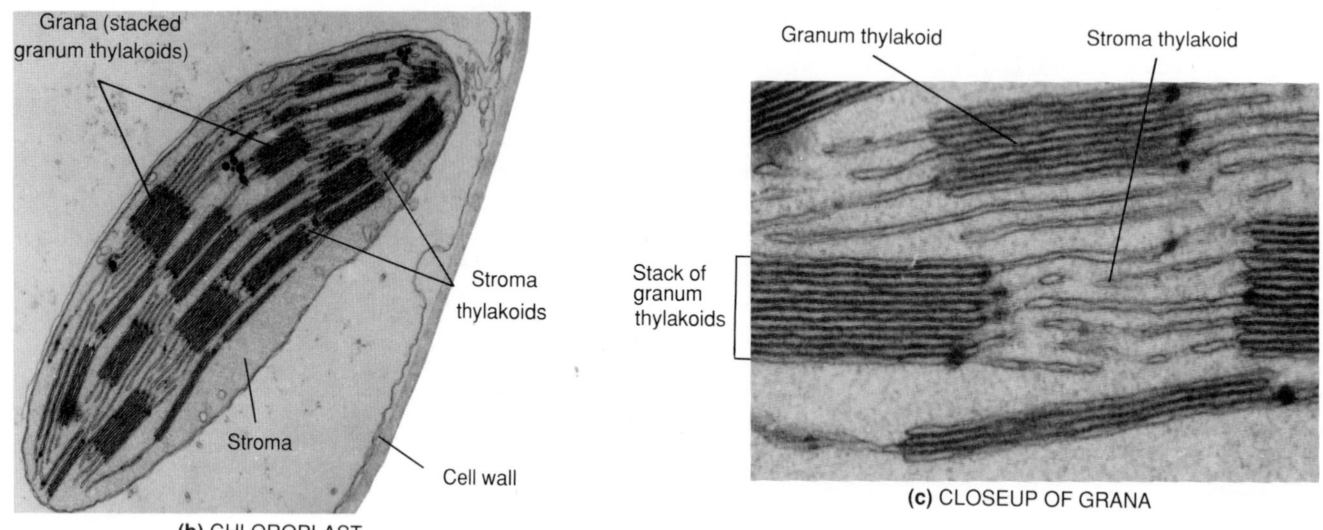

(**b**) CHLOROPLAST

(**c**) CLOSEUP OF GRANA

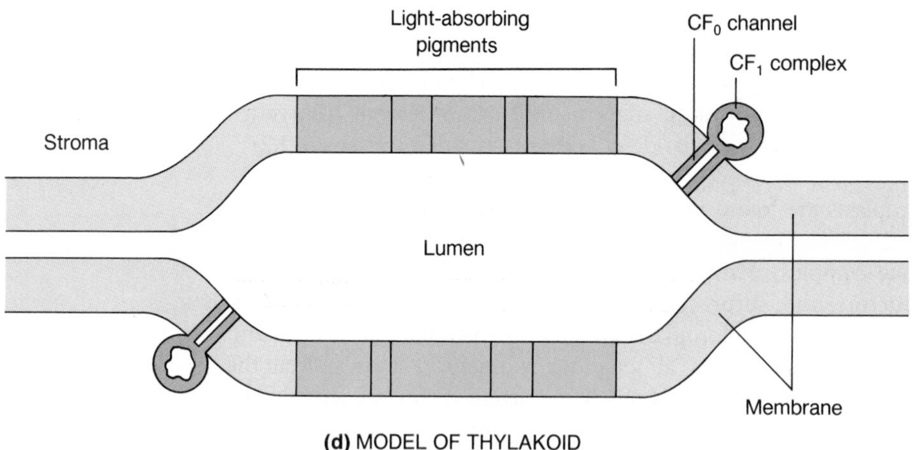

(**d**) MODEL OF THYLAKOID

FIGURE 4.19
A STUDY OF THE CHLOROPLAST
(a) In this light microscope view, numerous chloroplasts fill the plant cells. (b) Each chloroplast is an oval, membrane-surrounded body containing a highly folded, inner membrane system. (c) A closeup reveals individual thylakoids, some of which occur as stacked granum thylakoids and others as connecting stroma thylakoids. (d) Photosynthesis occurs within active elements organized in the thylakoid membrane, as seen in this model. Each thylakoid is disk-shaped, its membranes enclosing an inner chamber, the lumen. ATP is synthesized within bodies called CF$_1$ complexes.

Inside its double membrane, each chloroplast contains a clear, watery area called the **stroma.** It is invaded by an extensive third membrane system that forms saclike vesicles called **thylakoids.** Thylakoids occur in two arrangements (Figure 4.19b and c). First, they are folded into stacks referred to as **grana** (singular, **granum**) where the individual vesicles are called **granum thylakoids.** Second, they occur singly as **stroma thylakoids,** those that cross the open stroma as they form interconnections between grana.

Situated along the thylakoid membranes are numerous granules designated as **CF_0CF_1 complexes.** The CF_0 portion anchors the whole assemblage into the thylakoid membrane and forms a channel into the lumen (Figure 4.19d). The CF_1 complex contains the enzyme ATP synthetase. The latter term should tell you that these bodies are involved in the synthesis of ATP, the cell's energy storage molecule. The entire CF_0CF_1 complex is made up of some nine polypeptides, five of which go to form the bulbous head of the ATP synthetase region (the CF_1 portion).

Other Plastids Chloroplasts are but one of a group of plant **plastids,** all of which are derived from a line of undifferentiated bodies called **proplastids.** Proplastids, present in the plant embryo, differentiate into the various types of plastids as the young plant grows. Apparently, the kind of plastid emerging depends upon clues from the cell's surroundings.

All plastids are surrounded by a double membrane or envelope, a distinct difference from most of the organelles considered so far. In addition to green chloroplasts there are **leucoplasts** ("white plastids") and **chromoplasts** ("colored plastids"). The leucoplasts are starch-storing bodies, present in most plant cells, but most common in storage tissues such as those in onions and apples. Chromoplasts are named for their pigments, which include orange carotenes, yellow xanthophylls, and several pigments that are red. They are highly visible as the bright colors of flowers and fruits.

Mitochondria Mitochondria (sing. mitochondrion) are complex, ATP-generating organelles found in virtually every eukaryotic cell. Like chloroplasts, mitochondria (1) have double membranes, (2) have their own circular DNA, (3) have their own ribosomes and synthesize some of their own proteins, and (4) arise only from preexisting mitochondria (through DNA replication and organelle division). To some extent, chloroplasts and mitochondria do exactly opposite things. Put simply, chloroplasts use energy and raw materials to produce carbon compounds and oxygen, while mitochondria use carbon compounds and oxygen to produce usable energy. We will deal with both these important biochemical processes in Chapters 7 and 8.

We are emphasizing these interesting similarities in order to introduce an important idea. Many biologists believe that both mitochondria and chloroplasts are the descendants of once-independent prokaryotic cells. Such cells were supposedly captured by other very ancient lines of prokaryotes, and they survived as **endosymbionts** (*endo,* "within"; *symbiosis,* "living together"). The idea is embodied in what is called the **serial endosymbiosis hypothesis.** It was through such invasions, the hypothesis maintains, that the membrane-bounded organelles of the eukaryotic cell arose. Although the evidence for serial endosymbiosis is fascinating and compelling, not everyone is convinced. The **autogenous hypothesis** explains the origin of membrane-bounded organelles in a very different manner. Both hypotheses are presented in Essay 4.2.

Mitochondria are considerably smaller than chloroplasts, especially in cross section. In part, this is because chloroplasts tend to be more or less spherical, whereas mitochondria are usually long and slender. In electron micrographs, mitochondria usually appear as oval structures, with the inner of the two mitochondrial membranes appearing as curious folds that extend partway across the inner cavity (Figure 4.20 a, b).

The inner membrane of the mitochondrion is highly folded. These folds are known as **cristae.** The liquid-filled maze within the inner membrane is called the **inner compartment,** or **matrix,** and the space between the two membranes is the **outer compartment.** The folding of the inner membrane greatly increases its surface area. This is important since most of the biochemical work is done on the cristae themselves. They

ESSAY 4.2

EVOLUTION OF THE EUKARYOTIC CELL: AUTOGENY OR ENDOSYMBIOSIS?

The evolutionary history of the eukaryotic cell, with its complex membrane-bounded organelles and organized nucleus, is a subject of considerable controversy. Eukaryotes are thought to have been around as long as 2.5 billion years ago, an idea based on fossil evidence that is still very tentative. It is certain, however, that eukaryotes were well established over one billion years ago, which is still quite ancient. Ideas on the evolution of eukaryotes have changed considerably over the years, and today, two fundamentally different proposals have been brought forward: the **autogenous hypothesis** and the **serial endosymbiosis hypothesis.**

The autogenous hypothesis (*auto-*, "self"; *gen-*, "beginning") is based on the premise that the many membrane-bound structures within the eukaryotic cell were derived from the plasma membrane. The membrane underwent what was probably a gradual and lengthy evolution that began with events in simple and primitive prokaryotes. The idea is that inpocketings in the plasma membrane pinched off, floating free as they commonly do as today's ameboid cells feed. In time some of these structures took on specific functions. They gave rise to specialized organelles, particularly those of the endomembranal system: the nuclear envelope, the endoplasmic reticulum, the Golgi complex, lysosomes and vacuoles.

The development of mitochondria and chloroplasts, according to the autogenous hypothesis, was a bit more complex. Their double membranes would have necessitated more extensive inpocketings of the plasma membrane, or perhaps secondary inpocketings of certain organelles derived from the plasma membrane. The accompanying illustration outlines the hypothetical steps.

In light of what we know about the endomembranal system and the ongoing rise of organelles, the autogenous hypothesis is quite reasonable. Difficulties arise as we try to explain the origin and peculiar characteristics of mitochondria and chloroplasts. As we will see, these difficulties are solved through the next hypothesis, but we might also consider that the two ideas are not necessarily mutually exclusive.

The endosymbiosis hypothesis is an ingenious and reasonably well-accepted alternative. It proposes that the eukaryotes arose principally through a series of events wherein certain prokaryotic cells were engulfed by other, larger prokaryotic cells, those whose membranes had evolved the ability to take in food through phagocytosis (see Chapter 5). Perhaps because the digestive machinery was not yet particularly efficient, some of the ingested organisms would have survived and continued living within the predator, who then became the host. The invaders, the hypothesis goes on, lived somewhat independently at first, but soon an interdependence was established. It was through these incorporations that the newly emerging eukaryote came into possession of mitochondria and chloroplasts, although some biologists add cilia or flagella to the list.

Other eukaryotic structures, those of the endomembranal system, the ER, the Golgi complex, and so on, could have originated autogenously, since, as we've said, the two hypotheses are not mutually exclusive.

In a revival of older versions of the serial endosymbiosis hypothesis, Lynn Margulis of Boston University and others have utilized information from a wide variety of fields to bolster the idea. In her version, Margulis maintains that the earliest eukaryotic cell was derived through at least three separate events that involved the union of four prokaryotic lines, designated simply as **Lines A, B, C,** and **D.**

Line A was a biochemically simple prokaryote capable of only the simplest kinds of metabolism, not unlike anaerobic bacteria that obtain energy from foodstuffs through simple fermentation. (Anaerobic cells cannot utilize oxygen and have a low ATP yield.) However, Line A had one important evolutionary advancement: the ability to engulf other cells, a novel way to feed in those times. There is no such anaerobic cell-engulfing organism alive today.

Line B was aerobic—oxygen using—which means it had a far more efficient energy-generating metabolism. Margulis purposes that its members were frequently engulfed by the line A cells, but some survived within the new host. In time a mutualism developed as each capitalized on what the other offered. The captive cell made use of its host's energy-rich wastes to generate its ATP, and any ATP excesses went to the host. The descendants of the line B cells lost the ability to live outside their hosts and are represented today by mitochondria.

The next event involved cells that Margulis designated as **line C,** those that moved through the action of flagella. If line C attached to the outer surface of the new AB complex, by which we mean line A with its captive "protomitochondria," it might have given rise to flagellated, centriole-containing cells. There are a number of biologists who remain unconvinced regarding this proposed line-C event.

The final symbiotic event was the acquisition of **Line D** cells, photosynthetic prokaryotes (perhaps cyanobacteria). They used sunlight energy along with carbon dioxide and water to produce the organic precursors needed to form the molecules of life. These primitive photosynthesizers were

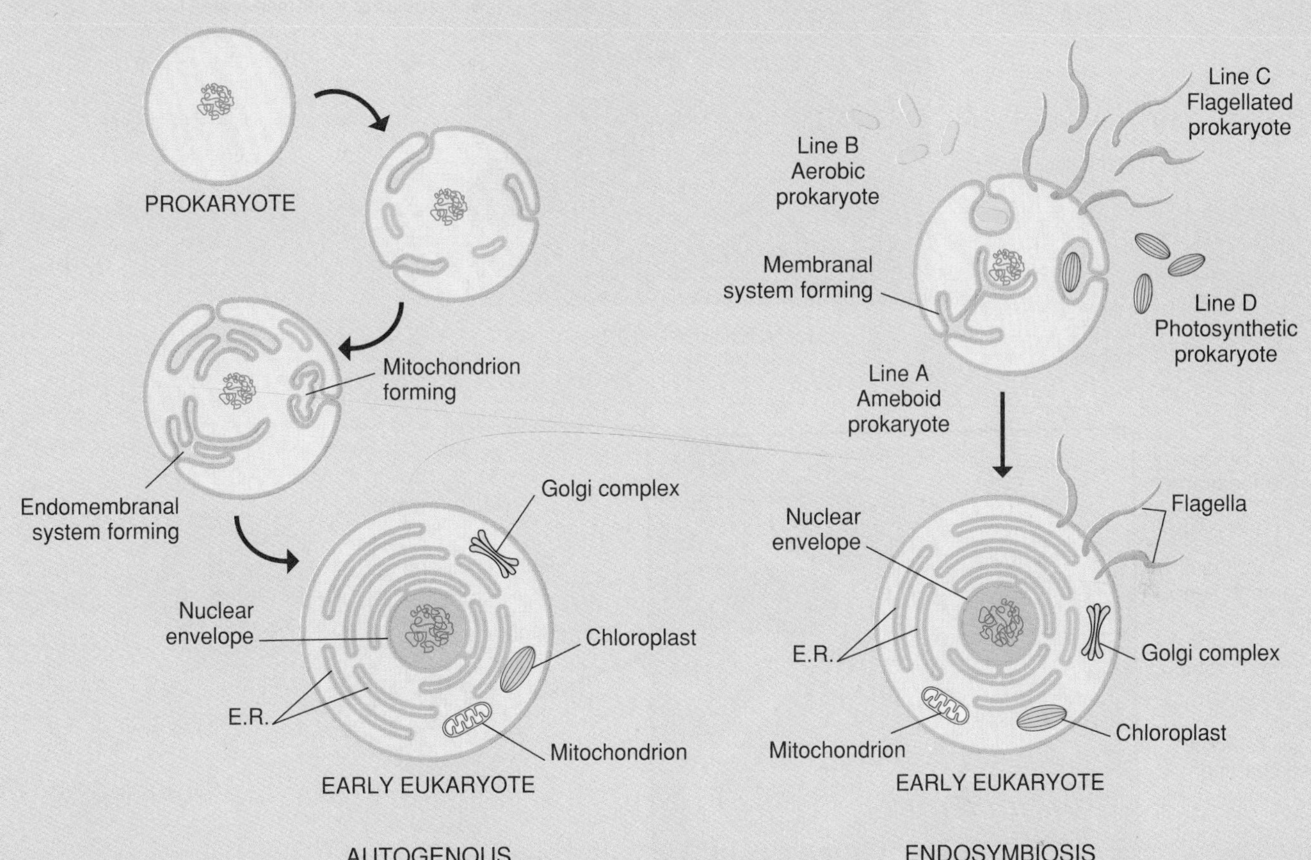

the proposed forerunners of chloroplasts. Once they became permanent symbionts, the emerging eukaryote would thus far have acquired mitochondria, highly efficient organelles for generating ATP, and a light-harvesting organelle that could make use of abundant sunlight energy to convert simple molecules to useful products. The acquisition of a flagellum gave the host added mobility. With the incorporation of chloroplasts, the eukaryote line was well underway.

How strong is the evidence for serial endosymbiosis? The case for mitochondria and chloroplasts is quite strong. Even today, mitochondria retain a circular chromosome, certain types of RNA, and ribosomes that are all remarkably like those of bacteria. The fact that mitochondria arise only from preexisting mitochondria is also heady suggestive evidence of independent origins. Further, at least one mitochondrial protein, *cytochrome c*, is recognizably similar in shape and amino acid sequence to the cytochromes of certain bacteria. Biologists often use such information to estabish evolutionary relationships.

What about the chloroplasts? Photosynthetic endosymbionts are found in many of the simpler marine animals and in many protists, so this mutual relationship is not unusual. There are many similarities between chloroplasts and modern photosynthetic bacteria, including their structure and biochemistry. Also, chloroplasts contain DNA and ribosomes that are quite similar to those found in prokaryotes. Further, both chloroplast and mitochondrial DNA are susceptible to certain antibiotics that do not affect eukaryotic DNA—further evidence of similarity between the organelle and its prokaryote counterpart. And, like mitochondria, chloroplasts arise independently of other organelles from a continuing line of proplastids. We will be looking forward to further developments in the endosymbiosis hypothesis, which is reviewed in further detail in Chapter 23.

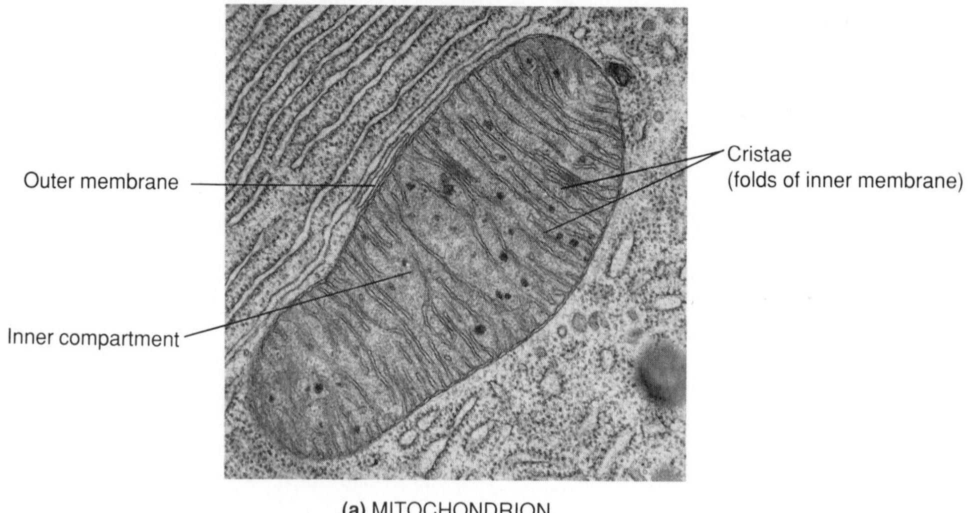

(a) MITOCHONDRION

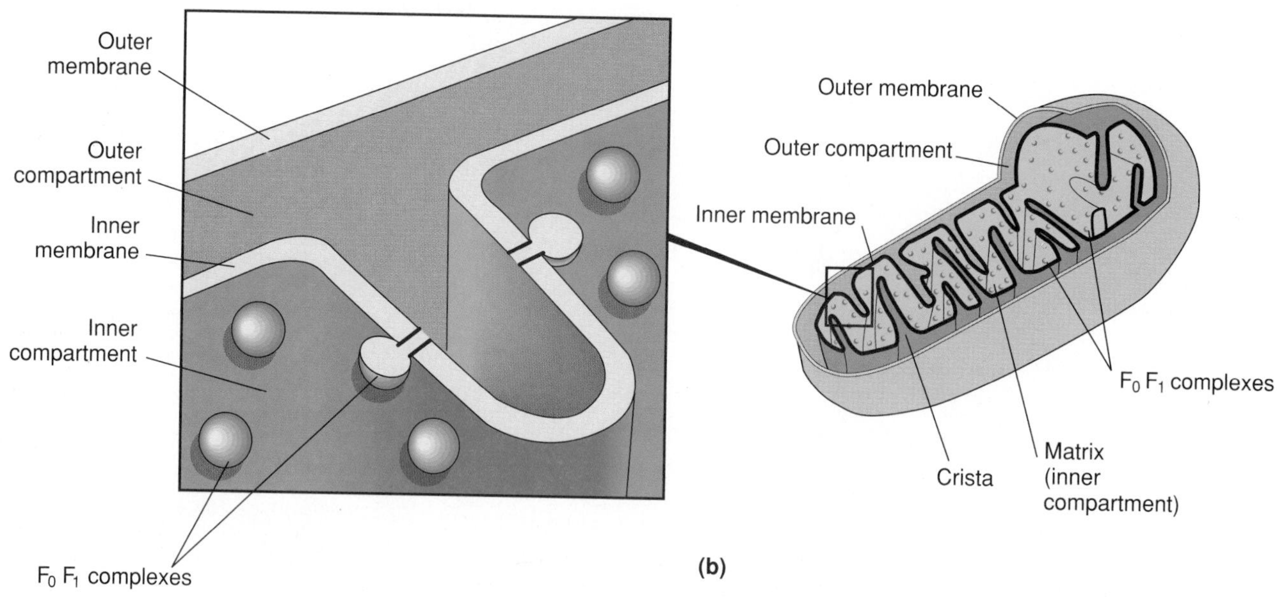

(b)

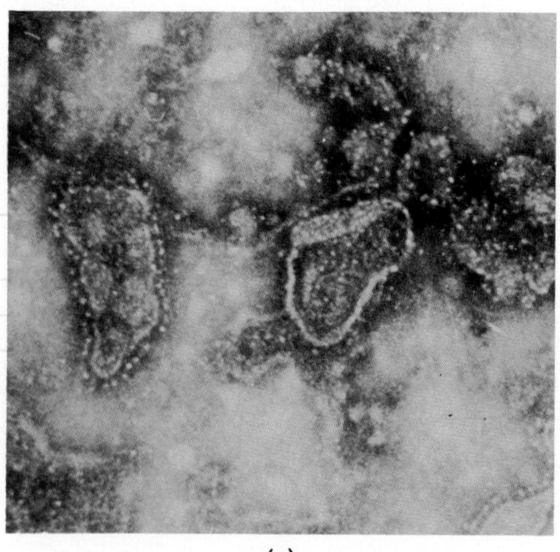

(c)

FIGURE 4.20
A STUDY OF THE MITOCHONDRION
(a) Mitochondria are sausagelike bodies when viewed whole. They have two membranes, which divide the mitochondrion into an outer and an inner compartment (matrix). The inner membrane is highly folded, forming numerous cristae that extend nearly across the organelle. **(b)** Careful reconstruction from photographic studies helps us to visualize the inner and outer compartment and the orientation of the F_1 complexes, each of which has functional importance. **(c)** The inner membrane is studded with F_1 complexes, sites of ATP synthesis. (In this view the inner membrane has been turned inside out by chemical treatment that exposes the F_1 complexes to view.

are far from simple, containing many versatile proteins and other active molecules. Special electron microscope techniques reveal that the tiny shelflike cristae are covered with small, round bodies, very similar to those in the chloroplast (Figure 4.20 c). The spherical granules and their membrane-embedded tails are designated **F_0F_1 complexes.** The spherical F_1 subunits contain ATP synthetases. As in the chloroplast's CF_1 complex, ATP synthetase is responsible for ATP synthesis.

Although mitochondria are found in all eukaryotic cells, there are more in some cells than in others. The number varies in proportion to the cell's respiratory activity—that is, in proportion to the rate at which the cell uses oxygen. This is because mitochondria are specialists in oxygen metabolism, or oxidative respiration. For example, a single hardworking liver cell may contain as many as 1000 mitochondria, but few can be found in a fat storage cell. As you might expect, mitochondria are abundant in muscle cells.

Organelles of Movement

Several organelles are directly or indirectly involved in cell movement. Some create movement within the cell, and others are involved in locomotion, moving the cell through a surrounding fluid medium. Still others move surrounding fluids past the cell.

Centrioles and Basal Bodies A centriole consists of two short cylinders of microtubules, usually seen lying at right angles to each other. Each cylinder contains nine groups of triplet microtubules, arranged in a circle. In cross section an individual cylinder looks a bit like a child's toy pinwheel (Figure 4.21).

Centrioles are not found in all cells. Whereas they exist in the cells of animals and protists, they are not present in fungi and most plants. More specifically, centrioles occur only in eukaryotes that develop cilia or flagella at some point in their lives. Apparently the fungi and most plants (the most recently evolved) have lost the centrioles that were present in the cells of their ancestors.

Two functions of centrioles are known with some certainty. They give rise to **basal bodies,** which in turn produce cilia and flagella. The basal body forms the root of the flagellum, and like the centriole, has nine groups of triplet microtubules. The other role of the centriole is to somehow determine the location of the furrow that forms in animal cells as they divide (see Chapter 9, Figure 9.7). The ever-deepening cleavage furrow (as it is known), is itself produced by contracting actin microfilaments, as we mentioned earlier.

Cilia and Flagella Cilia and flagella are fine, hairlike, movable organelles found on the surfaces of some cells (Figure 4.22). Although cilia and flagella appear to be outside the cell, they really are not; the cell membrane protrudes to cover each cilium or flagellum. Thus, we can consider them to be outgrowths of the cell proper.

FIGURE 4.21
CENTRIOLES
(a) Centrioles are paired cylindrical bodies. The cylinders generally appear at right angles. Here we see two centrioles (four bodies), the products of a recent replication. **(b)** In cross section, the nine sets of triplet microtubules show clearly. **(c)** The three-dimensional illustration reveals the spatial arrangement of microtubules in each triplet.

Centriole

(a)

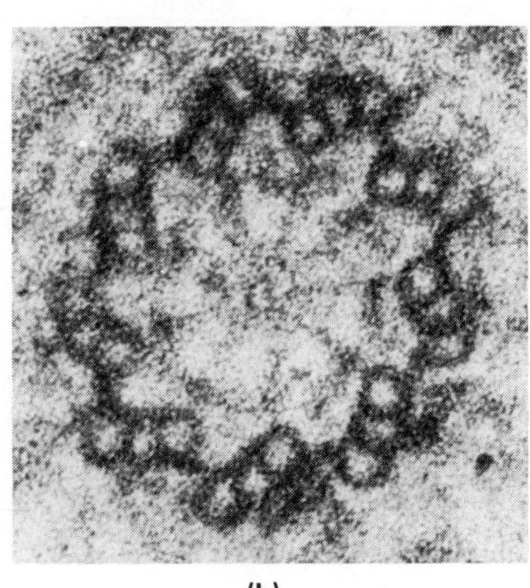

(b)

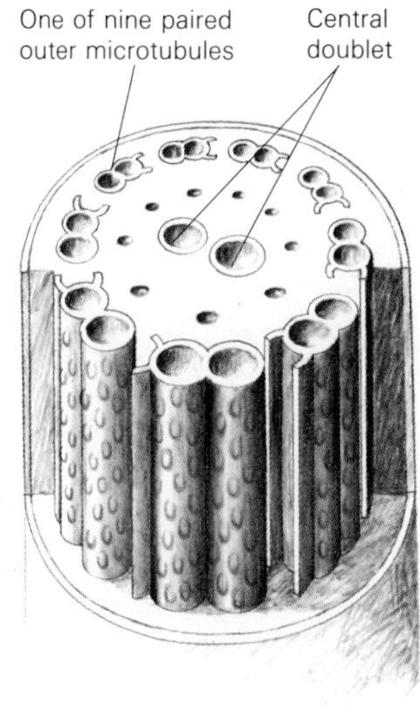

(c)

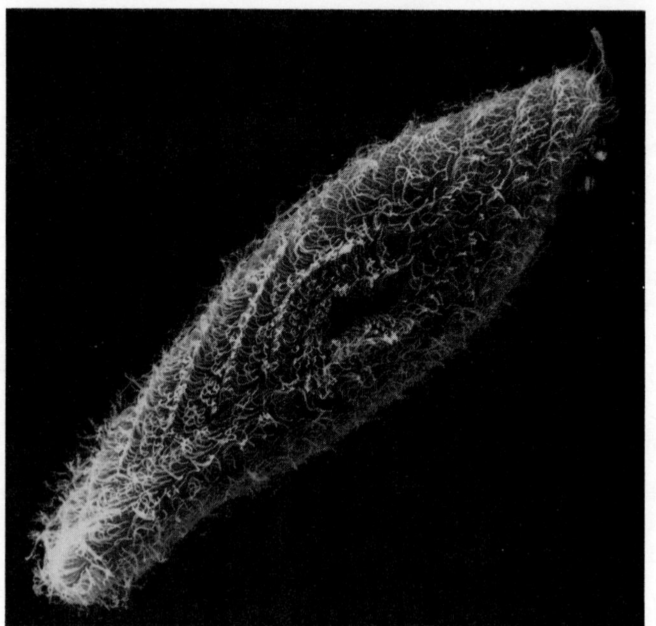

FIGURE 4.22
***PARAMECIUM,* AN ORGANISM COVERED WITH CILIA**
The protist, **Paramecium**, uses its highly coordinated cilia to swim about and to feed.

Both cilia and flagella may either move the cell through some surrounding fluid or move some surrounding fluid past the surface of a stationary cell. For example, a sperm cell swims by undulations (wavelike movements) of its flagellum, a *Paramecium* (single-celled protist) moves by the coordinated beating of rows of cilia, and the cilia of the cells that line the passage of your lungs (unless you have permanently paralyzed them with tobacco smoke) sweep a cleansing film of mucus upward to your pharynx.

TABLE 4.2 KEY DIFFERENCES AMONG THE CELLS OF PROKARYOTES, PLANTS, AND ANIMALS

FEATURE	PROKARYOTIC CELL	PLANT CELL	ANIMAL CELL
Cell membranes	Plasma membrane only	Plasma membrane plus membrane-bounded organelles	Plasma membrane plus membrane-bounded organelles
Cytoskeleton	Absent	Present	Present
Nuclear envelope	Absent	Present	Present
Chromosomes	Single, circular, DNA only	Multiple, linear, complexed with protein	Multiple, linear, complexed with protein
Membrane-bounded organelles	Absent	Present	Present
Ribosomes	Small, free	Larger, some bound to ER	Larger, some bound to ER
Cell wall	Peptidoglycan or protein	Cellulose	None
Flagella or cilia (when present)	Solid, rotating	Rarely present*	Microtubular (9 + 2 pattern)
Ability to engulf solid matter	Absent	Absent	Present, extensive movable membranes
Centrioles	Absent	Absent*	Present

*Although absent in most plants, these features are found in more reproductively primitive plants. Apparently they have been lost in the course of evolutionary change.

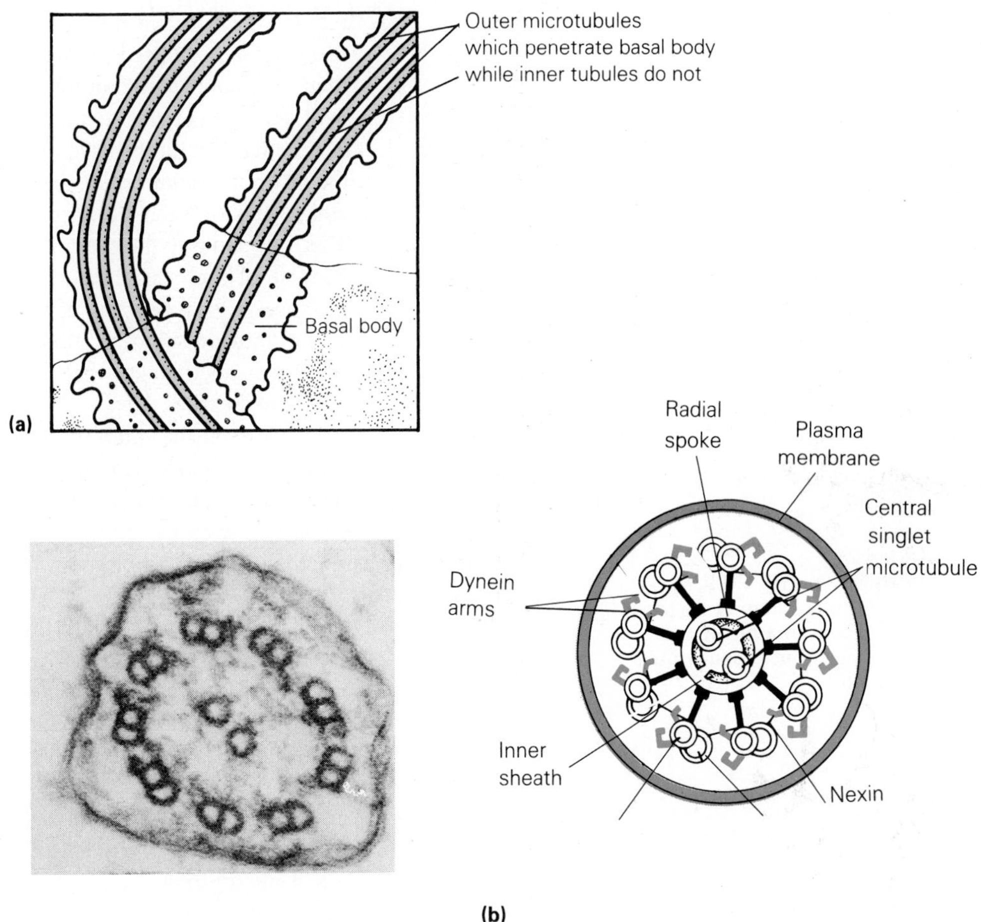

FIGURE 4.23
A STUDY OF THE CILIUM
(a) The longitudinal section of a cilium in an electron micrograph shows that while the outer microtubule doublets penetrate the basal body, the inner pair does not. Note also that microtubules in basal bodies occur as triplets, just as in centrioles (from which they are believed to be derived). **(b)** A cross-sectional electron micrograph view of the cilium reveals the 9 + 2 arrangement of microtubules. The accompanying illustration, reconstructed from numerous electron micrographs, points out that all of the microtubules have definite interconnections with each other and with the central pair.

Cilia and flagella are structurally almost identical. They differ only in length, in the number per cell, and in their pattern of motion. Cilia are short, numerous, and have a highly coordinated, unified rowing motion (like oars, except these "oars" must bend on the return stroke since they stay in the water). Flagella are long, few in number, and move by undulation. More specifically, most cilia are between 10 and 20 μm long, whereas flagella may, in exceptional cases, be several thousand micrometers long—that is, several millimeters long. The flagellum on a *Drosophila* (fruitfly) sperm, for instance, may be longer than the fly itself!

Fine Structure and Movement in Cilia and Flagella Under the electron microscope, the microtubular arrangement in eukaryotic cilia and flagella is similar in cross section to that of centrioles. But, whereas the half-centriole is made up of a circle of nine triplets of microtubules, both cilia and flagella have a "9 + 2" structure: a circle of nine pairs of microtubules with two single microtubules in the middle (Figure 4.23). Each of the nine doublets includes a complete microtubule and one that is only partially complete and fused to the other. Emerging from each complete microtubule are a pair of **dynein arms,** armlike structures composed of the protein dynein. Dynein can make use of energy from ATP, the energy storage molecule of cells. In addition, there are three important accessory proteins. The protein **nexin** forms connections between the microtubular doublets, and extending inward from each of the doublets are the **radial spokes,** which apparently contact an **inner sheath** of protein surrounding the two central microtubules. The entire active core of cilia and flagella—the microtubules, dynein arms, and accessory proteins—are known as the **axoneme** of the organelle. These parts interact to produce the special kind of movement we find in cilia and flagella.

Cilia bend through a process called "dynein-walking," which involves the nine doublet microtubules. Using ATP as an energy source, the dynein arms of one doublet micro-

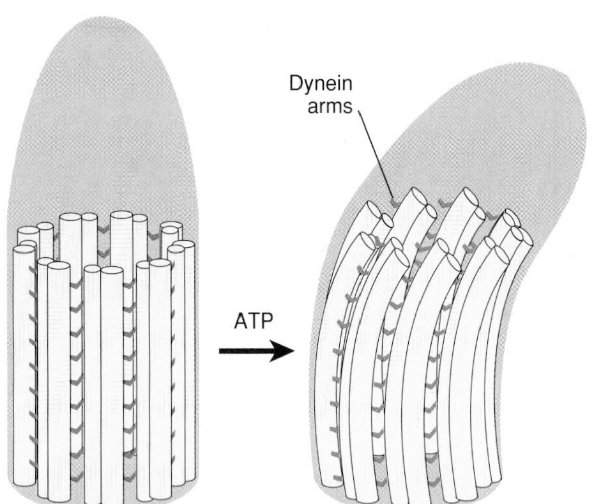

FIGURE 4.24
MOVEMENT IN THE CILIUM
The bending of a cilium or flagellum is brought about by a highly controlled sliding of microtubular units past each other.

tubule attach to and pull against the nearest doublet microtubule. If the action is viewed in microtubules isolated from the axoneme, then one doublet microtubule is seen to actually slide along the next, its dynein arms literally walking up the nearby doublet. But in the intact cilium or flagellum, such sliding is resisted because of the nexin interconnections between doublets and the radial spokes which anchor each doublet to the central microtubules (Figure 4.24). Apparently the central microtubules act as a stationary base. Because of these restraints, the force exerted by the dynein arms produces a local bending of the cilium or flagellum. This bending action, moving smoothly along the length of the structure, first on one side and then on the other, results in the undulating effect mentioned earlier.

While it is the interaction of doublets around the cilium that brings about movement, it is the total organization of the axoneme that makes such movement useful. The idea of one filamentous structure forcefully sliding against another is not unique to cilia and flagella. This is precisely the principle involved in muscle contraction, a process we will look into in Chapter 33.

Each cilium and flagellum terminates beneath the surface of the cell at a basal body, which as we have seen, arises from the centriole. The nine pairs of microtubules of the cilium or flagellum, as they join the basal body, become nine triplet microtubules; the two single microtubules just end blindly. Incidentally, groups of basal bodies are usually joined by interconnecting fibers that presumably enable the cilia to coordinate their movements.

Cilia and flagella also apparently have some sort of primitive cellular sensory function. Many of our own sensory receptors are believed to have evolved from cilia. Amazingly, the 9 + 2 arrangement of microtubules can still be seen in (1) the rods and cones of the retina, (2) the olfactory fibers of the nasal epithelium, and (3) some sensory hairs of the internal ear. It seems, then, that we see, smell, hear, and balance ourselves with highly modified cilia-bearing cells.

APPLICATION OF IDEAS

1. Many exceptions to the cell theory have been noted. In view of this, how can the theory continue to be important and instructive? If you think it is not, offer reasons for this conclusion.
2. A biological principle called complementarity of structure and function holds that the two are very closely related. Cite five or six examples of how complementarity works at the cellular level.

KEY IDEAS

CELL THEORY

1. The **cell theory**, proposed in 1839 by Schleiden and Schwann, maintains that living organisms are composed of cells. Rudolf Virchow added that all cells come from preexisting cells.
2. Cells are extremely diverse, their differences reflecting specialized functions.

WHAT IS A CELL?

1. Cells are the basic units of life. They are capable of taking in materials, extracting energy, synthesizing molecules, growing, responding to environmental stimuli, and reproducing.

The Size of Cells

1. Most cells range in size from 10 to 100 micrometers.
2. As cells increase in size, the volume increases proportionally faster than the surface area.
3. The **surface-volume hypothesis** maintains that cells must remain small in volume in order for the **plasma membrane**, or cell surface, to provide sufficient transport of materials in and out of the cell.
4. Cells can overcome the surface-volume limitations through growth in one dimension only (length or width), or by producing a highly folded plasma membrane, or by dividing so that a more satisfactory surface-to-volume ratio occurs.

THE PROKARYOTIC CELL

1. **Prokaryotes,** or bacteria, are single-celled organisms. Some are **heterotrophs,** living as parasites or as **decomposers,** and requiring organic molecules. Others are **autotrophic,** requiring only inorganic materials. They include **photoautotraphs,** which use light energy for photosynthesis, and **chemoautotrophs,** which derive energy from inorganic matter.
2. Prokaryotic cells have a plasma membrane that, in some, forms infoldings making up a photosynthetic **thylakoid.** A **cell wall** of peptidoglycan or protein surrounds the membrane.
3. Bacterial DNA is continuous, making up one circular **chromosome.** Some DNA may occur in separate **plasmids.**
4. Bacteria reproduce asexually through fission, preceded by DNA replication. The **mesosome** assists in separaton of DNA replicas. Sexual reproduction includes a one-way gene transfer through the **sex pilus,** which forms during **conjugation.** Other **pili** form surface attachments.
5. Prokaryotic movement may be through an S-shaped, rotating **flagellum.**

THE EUKARYOTIC CELL

Eukaryotes include plants, animals, fungi, and protists. They differ from prokaryotes principally in the presence of membrane-surrounded **organelles,** including an organized **nucleus.**

CELL STRUCTURES IN EUKARYOTES

Surface Structures

1. Dense cell walls surround the cells of plants, fungi, and many protists, maintaining shape and providing strength and flexibility.
2. In plants, the cell wall consists of cellulose fibers, laid down in several directions. A **middle lamella** of pectin holds adjacent walls together. **Lignin, suberin,** and **cutin** are secreted into some plant cell walls.
3. The plasma membrane controls the passage of materials into and out of the cell. It is composed of a bilayer of phospholipids and many surface and transmembranal proteins.

Internal Support: The Cytoskeleton

The **high voltage electron microscope** has resolved a detailed picture of the cytoskeleton, showing that the cytoplasmic matrix contains a supporting network. It consists of **actin microfilaments** that help support the plasma membrane, microvilli, and cytoplasm. Microfilaments function in **ameboid movement** and cytoplasmic division. **Intermediate filaments** strengthen epithelial linings, microvilli, and nerve cells. **Microtubules** support the cytoplasm, form **centrioles, basal bodies, cilia,** and **flagella,** and ensure proper chromosome separation during cell division. Microtubules are produced through the assembly of **tubulin** subunits. They are readily assembled and disassembled.

Control and Cell Reproduction: The Nucleus

1. The nucleus is a prominent, spherical body, containing the chromosomes, which are made up of DNA (the genes), and protein.
2. Deep staining **nucleoli** contain ribosomol RNA, which joins protein to form **ribosomes.**
3. The **nuclear envelope** is a complex double membrane containing protein-filled **pores.** Its outer membrane is continuous with the endoplasmic reticulum.

Organelles of Synthesis, Storage, and Cytoplasmic Transport

1. The **endoplasmic reticulum** includes extensive, dynamic, membrane-lined channels through the cytoplasm, enclosing the **lumen.** It also occurs in a *vesicular* form when portions pinch away. The **rough endoplasmic reticulum** is named for the presence of numerous ribosomes (protein synthesizing bodies).
2. Ribosomes are molecular complexes consisting of ribosomal RNA and protein organized into two subunits that join when polypeptides are synthesized. Some are bound to RER; others are free but may occur in strands as **polyribosomes.**
3. The **smooth endoplasmic reticulum** occurs in cells where carbohydrates, lipids, and other nonprotein products are formed.
4. The **Golgi complex** consists of flattened **cisternae** as vesicles from the ER fuse with the ***cis*** face. Its function, the modification and packaging of ER products, occurs sequentially as materials are transferred via vesicles from one cisterna to the next. Products are finally sorted and packaged into vesicles that leave the ***trans*** face. Some materials are stored, whereas others are secreted.
5. **Lysosomes** are membrane-bounded sacs that contain hydrolytic enzymes for the hydrolysis of damaged or aging cell components **(autophagy),** or for the digestion of materials taken into phagocytic vesicles **(heterophagy).** In certain genetic **storage diseases,** critical enzymes are missing, and an abnormal buildup of chemicals occurs in the lysosomes.
6. **Peroxisomes** are membrane-bounded **microbodies** that contain **catalase,** a hydrogen peroxide-metabolizing enzyme. **Glyoxysomes** convert plant lipids to sugars in sprouting seeds.
7. **Vacuoles** are saclike bodies lacking inner structure, containing **cell sap,** water, certain chemicals, and pigments. When water-filled, they give firmness to softer plant structures.
8. The **endomembranal system** includes the above synthetic and storage organelles. In the system, organelles arise from each other in sequence.

Energy-Generating Organelles

1. **Chloroplasts** are complex double-membraned **plastids** that carry on photosynthesis: they convert sunlight energy into chemical bond energy and, using water and carbon dioxide, use this energy to produce carbohydrates. **Chlorophylls,** the light-trapping

photosynthetic pigments, are arranged in membranous **thylakoids.** Enzymatic activity occurs in the clearer **stroma** between **grana**, the stacks of thylakoids. ATP is synthesized in **CF_1 complexes,** bodies located in thylakoids.

2. Plastids include metabolically active chloroplasts, starch-storing **leucoplasts**, and colorful, pigment-filled **chromoplasts.** Plastids differentiate from **proplastids,** present in the plant embryo.
3. **Mitochondria** have double membranes, their own DNA and ribosomes, and divide like bacteria. These and other factors led to the **serial endosymbiosis hypothesis,** which maintains that both chloroplasts and mitochondria originated as free prokaryotes, taken in by an ancient prokaryotic line. The **autogenous hypothesis** maintains that such organelles arose through ingrowths of the plasma membrane.
4. Mitochondria are complex bodies that carry out cell respiration, the conversion of chemical bond energy in cellular fuels to usable energy in the bonds of ATP. Their outer membrane is simple and surrounds an **outer compartment.** The inner membrane is highly folded, forming the shelflike **cristae,** which surrounds an enzyme-rich **inner compartment,** or **matrix.** The cristae contain the **F_1 complexes,** in which ATP is synthesized.

Organelles of Movement

1. Cellular movement includes movement within the cell, movement of the cell through a surrounding fluid, and movement of fluids past the cell.
2. Centrioles are paired, rodlike, self-replicating bodies near the nucleus. Centrioles occur only in eukaryotes that develop cilia or flagella at some point in their lives (generally, animals and protists). Centrioles give rise to basal bodies, which produce cilia and flagella, and determine the location of cleavage furrows in dividing animal cells.
3. Cilia and flagella are involved in the movement of cells. They differ only in tail length, number, and movement pattern. They consist of microtubules in a 9 + 2 arrangement. **Dynein arms,** with their **radial spokes** and interconnecting **nexin,** and an **inner sheath** surrounding the central doublet, form the **axoneme.** The microtubules produce bending action by their energetic movement past each other. Cilia are also believed to be the forerunners of many sensory structures.

REVIEW QUESTIONS

1. Summarize Robert Hooke's contributions to cell biology. (77)
2. Summarize the contributions to cell biology of Schleiden, Schwann, and Virchow. (77)
3. List six fundamental activities of cells. (77)
4. What is the usual size range of cells? Cite two exceptions in which cells are much larger. (78-79)
5. What happens to the relationship between surface area and volume when a cell doubles its size? (79)
6. Summarize the surface-volume hypothesis. How do large cells overcome such limitations? (79)
7. Distinguish between prokaryotic heterotrophs and autotrophs, and describe two variations within each group. (83)
8. Prepare a drawing of a prokaryotic cell, labelling the following: cell wall, plasma membrane, pili, capsule, mesosome, chromosome, and thylakoid. (83)
9. Generally contrast the structure of a prokaryotic and eukaryotic cell. (84)
10. List and explain two important functions of the plant cell wall. (86)
11. What is the basic structure of the plasma membrane? What are its functions? (87)
12. Name the three components of the cytoskeleton and suggest five specific functions carried out by this fibrous network. (87-88)
13. Describe the structure of a microtubule. List three of its uses. (88-89)
14. Summarize two important roles of the nucleus. Be specific. (89)
15. Describe the nucleus and its contents, including the envelope, chromosomes, and nucleoli. (89)
16. Compare the structure and function of the rough endoplasmic reticulum with the smooth endoplasmic reticulum. (91-93)
17. What is the relationship between the endoplasmic reticulum and Golgi complex? List two functions of the Golgi complex. (93-94)
18. Explain how the lysosomes function in autophagy and heterophagy. (94-95)
19. Describe a common chemical reaction in peroxisomes and explain why it is important. (95-96)
20. Discuss two important roles of water vacuoles in plants. (96)
21. Using a simple drawing, illustrate the structure of a chloroplast, including the following: envelope, grana, stroma, thylakoids, CF_0CF_1 complexes. (97-98)
22. Generally discuss the function of chloroplasts, and explain briefly what happens in the thylakoids and stroma. (99)
23. How do evolutionists explain the presence of complex cellular organelles such as chloroplasts and mitochondria? (Essay 4.2)

24. Illustrate the mitochondrion. Include the outer membrane, inner membrane, outer compartment, inner compartment, cristae, and F_0F_1 complex. What is the function of the mitochondrion? (99, 103)

25. List the four characteristics of mitochondria and chloroplasts that suggest prokaryotic origins for these organelles. What is the logic of this idea? (99)

26. Using a drawing, explain the structure of a cilium. Be sure to include the 9 + 2 microtubular arrangement and the components of the axoneme. (105-106)

27. What are two differences between cilia and flagella? (105)

28. Explain the bending action of a cilium or flagellum. (105-106)

5

CELL TRANSPORT

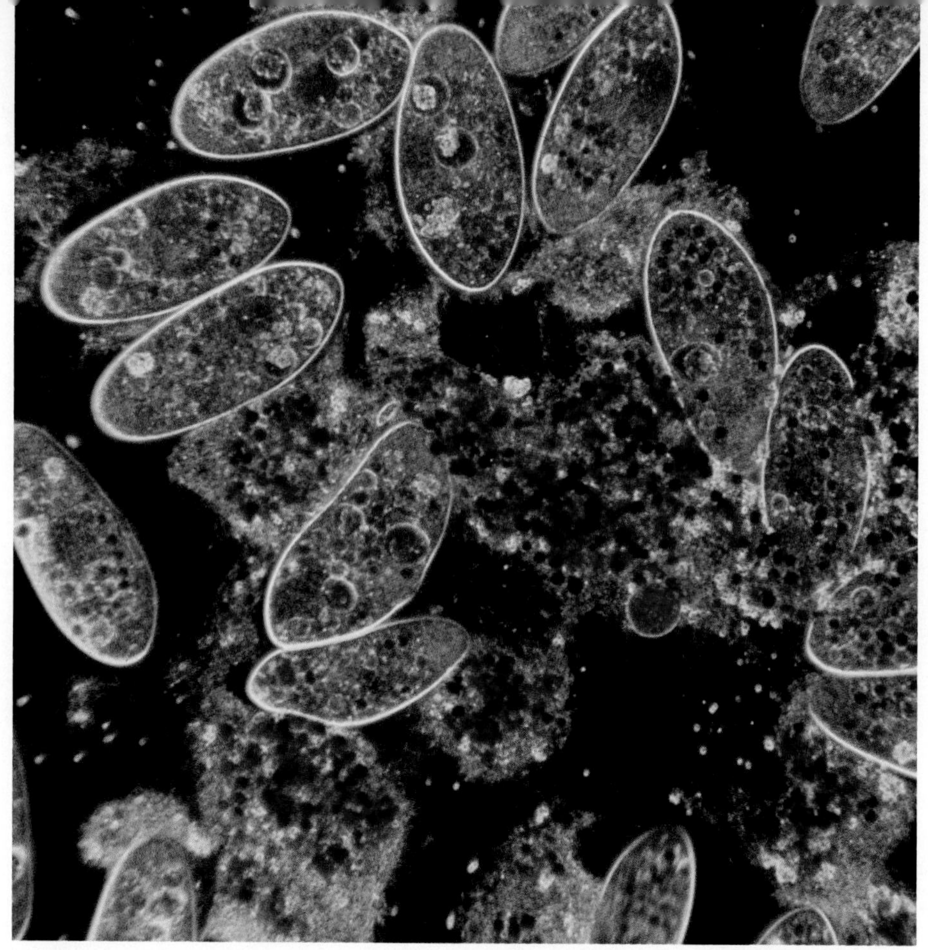

MOST CELLS ARE SO TINY THEY CAN'T BE SEEN, YET WITHIN THEM ARE SYSTEMS so complex that they are not yet understood. One problem that has presented an enduring challenge is the way cells shift critical materials around so that the delicate processes of life can continue. After all, for cells to remain active they require a constant uptake of vital substances and a constant output of metabolic wastes—and, in many instances, the special substances they produce. Obviously, materials moving into or out of cells must pass through that vital envelope called the plasma membrane, so it is here that we can begin our search for the mechanisms of such movement.

THE PLASMA MEMBRANE

The plasma membrane is a critical structure, and rather taken for granted in its role as a living envelope around the cell. But for a long time some people refused to believe it existed. This is understandable; the plasma membrane is so thin that its existence, as well as its structure, was postulated entirely on circumstantial evidence until the advent of the transmission electron microscope (TEM). The circumstantial evidence was gathered from observations that something was regulating the passage of materials into and out of living cells. The circumstantial evidence even suggested the composition of the membrane. For example, it had long been postulated that the membrane has a lipid core simply because lipid-soluble materials and lipid solvents readily pass through. Since water also passes through the membrane, though at a slower rate, scientists believed that the membrane also had water-admitting pores, perhaps surrounded by proteins. As we will see, these ideas were generally on target, although water admitting pores have never been observed.

With the transmission electron microscope we could see for the first time visual evidence that verified much of what the membrane theorists had proposed. Membranes, as viewed through the TEM, look like two dark lines separated by a clear line. The clear line, it turns out, is about 5.0 nanometers (nm) wide, which happens to be twice the average length of the hydrocarbon tails of membrane phospholipids. This suggested a two-layer structure, with the tails fitting neatly back-to-back. The dark lines apparently represent the phosphorus-rich polar heads of phospholipids, along with proteins associated with the membrane.

The Fluid Mosaic Model

Figure 5.1 shows a diagram of what we believe the plasma membrane structure to be. This is called the **fluid mosaic model.** The name "fluid mosaic" is purposefully vague. It refers to the fluidlike qualities of the phospholipid core and the dynamic behavior of proteins that seem to drift on the "lipid sea," some afloat, others partially or fully submerged. The phospholipids are represented as spheres with one straight tail and one bent tail. Recall that phospholipids contain, in addition to phosphate, a variable group, which in the plasma membrane is often nitrogen-containing choline. This phospholipid is called **phosphatidylcholine.** The polar (charged) heads have two hydrocarbon tails each, and these point inward. This arrangement forms a hydrophilic (water-accepting) inner and outer surface and a hydrophobic (water-rejecting) core between the two. These characteristics have a profound effect upon the membrane's interaction with its protein constituent and with materials entering and leaving the cell. A second type of lipid component, cholesterol, is often widely distributed in the plasma membrane of animals. Its presence just below the phospholipid heads is believed to have a stabilizing influence on an otherwise highly fluid bilayer.

The Membranal Proteins The protein component of the membrane is highly varied and includes both globular and linear proteins. The linear proteins occur chiefly as glycoproteins, which make up part of a surface layer called the **glycocalyx** which we will take up shortly. Here we will consider the globular components.

Freeze-fracture preparations of the plasma membrane helped substantiate the existence of globular proteins. When frozen cells are sheared between the two phospholipid layers, individual transmembranal proteins remain projecting from one surface or the other. On the corresponding surface there are indentations, spaces where the protein had fitted in the intact site.

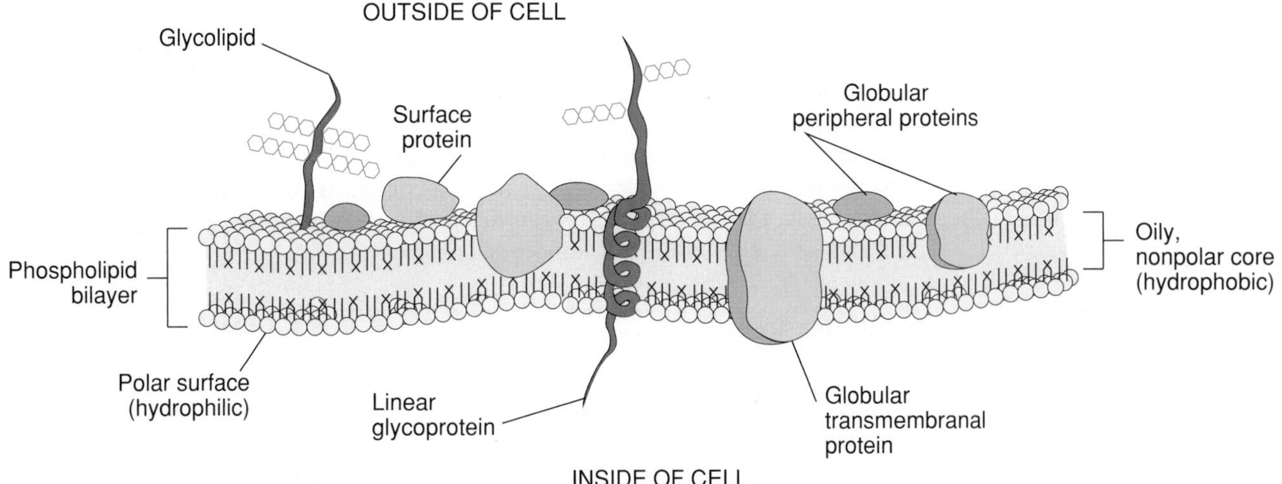

FIGURE 5.1
DETAILED VIEW OF THE PLASMA MEMBRANE
The plasma membrane is basically a bilayer of phospholipids with a varied protein component. Whereas the basic construction was presented in Figure 4.7, a more detailed representation is seen here. Note how the phospholipid bilayer gives rise to hydrophobic and hydrophilic regions. Globular proteins take on different arrangements depending on their hydrophobic and hydrophilic nature. Chainlike glycoproteins have branched sugar groups at the surface, their linear protein component passing through the membrane. The branching sugar groups of glycolipids also occur on the surface, emerging from fatty acid tails that are anchored in the outer phospholipid layer.

Today three arrangements of globular proteins are recognized. **Transmembranal proteins** extend completely through the membrane. Such proteins tend to have hydrophobic midregions and hydrophilic ends (forming a match with the phospholipid bilayer). Smaller **peripheral proteins** have a hydrophobic region but just one hydrophilic region, so they lie partly submerged in the lipid bilayer. The third such component consists of **surface proteins,** those that are entirely hydrophilic. They lie on the surface, not penetrating the hydrophobic core at all.

The transmembranal proteins have several important functions, chiefly transport. Some act as specialized ports, permitting the free passage of selected ions such as those of sodium and potassium. Others, working as carriers, utilize some of the cell's energy reserves to move sodium and potassium actively across the membrane. There are also carriers that enhance the movement of molecules or ions across the membrane without drawing on the cell's energy reserves. Some of the smaller protein carriers are quite mobile, free to pass back and forth through the membrane with their cargo attached.

Some globular membrane proteins are enzymes, carrying out their reactions right in the membrane. Finally, some globular proteins are involved in support. Those on the membrane's inner side may form bridges with elements of the cytoskeleton, which we described earlier (see Chapter 4).

The Glycocalyx: Glycoproteins and Glycolipids The plasma membrane of many animal cells has a fuzzy, indistinct outermost region called the glycocalyx (Figure 5.2). It is made up of complex surface carbohydrates, which are attached to linear proteins or lipids that are anchored in the membrane. You may recall from Chapter 3 that these hybrid molecules are called glycoproteins and glycolipids. The glycolipids have a surface component of branched sugar chains that lead to two fatty acid tails. The fatty tails simply join the phospholipid fatty acid tails that make up the membrane's lipid bilayer. The glycoproteins also have a branched sugar surface component, but their linear protein tails often pass right through the membrane, emerging on the other side. Portions of the tails within the membrane form the familiar alpha helix configuration (see Figure 5.1).

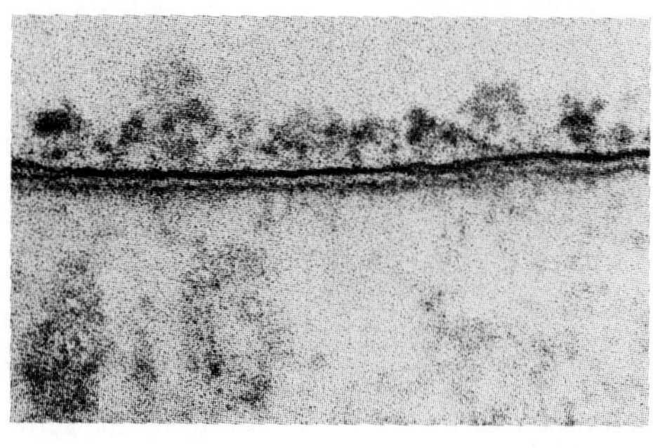

FIGURE 5.2
THE GLYCOCALYX
Glycolipids and glycoproteins form the glycocalyx border seen clearly in the electron micrograph.

Some glycoproteins, glycolipids, and other surface molecules play important roles in cell recognition, whereas others act as receptors for chemical messengers such as hormones. Plant or animal hormones may bind to a specific recognition site. Once the hormone is in place, the hormone/receptor complex sets in motion events that initiate the specific response to that hormone.

Cell recognition in our own bodies (that is, the ability of our cells to identify and interact with other cells) is vitally important to the work of our immune system. The immune system responds to the presence of disease agents or foreign molecules. Cell recognition occurs through the presence of specialized membranal recognition sites that identify specific cell surface proteins, carbohydrates, or other molecules on other cells. For example, surface recognition sites on leukocytes, cells of the immune system, enable them to recognize other body cells as they mount defensive measures against invasion. Interestingly, the same leukocytes can detect the presence of foreign molecules on the cell surface of cells invaded by cancer or by viruses.

Cell Junctions

If we have left the impression that cells in multicellular organisms tend to be isolated from each other, let's correct that now. Communicating channels of several kinds have been found between the closely packed cells of multicellular tissues. Included are the intricate **gap junctions** of animal cells and the slender **plasmodesmata** of plant cells, both of which are very common.

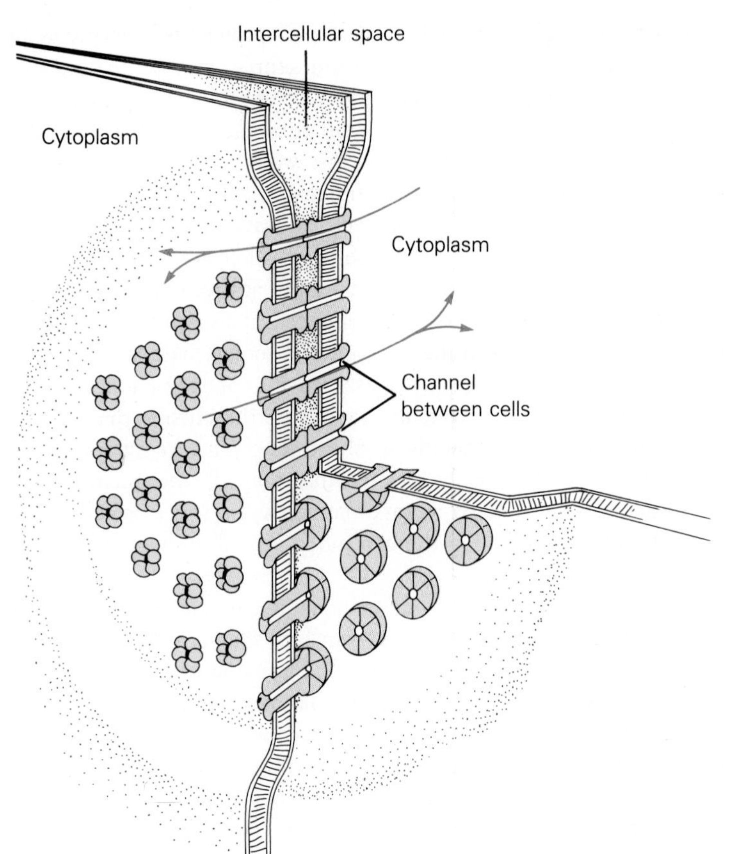

FIGURE 5.3
JUNCTIONS BETWEEN CELLS
Materials move between certain animal cells through pores called gap junctions. Each pore is formed through the special assembly of six transmembranal proteins. The assemblies in each membrane are arranged back to back, creating a minute pipeline connecting the cytoplasm of the two cells and squeezing the two membranes together. The electron micrographs show gap junctions at increasing magnifications.

Gap Junctions Gap junctions are formed by rosettes of protein that firmly hold adjacent membranes together but at the same time form simple, minute pipelines or tubes that make the cytoplasm of adjacent cells effectively continuous (Figure 5.3). The passages are 1.5–2 nm in diameter and will freely admit substances up to a molecular weight of 1000 daltons. This would include most ions, water, amino acids, sugars, vitamins, hormones, and even nucleotides. Gap junctions are believed to be important to heart muscle, where they facilitate the spread of electrical waves, and in the embryo, where they help with the passage of materials through layers of cells.

The protein-lined channels of gap junctions are not simply passive openings but are known to break loose or "uncouple" from the neighboring cells if an injury occurs. This effectively isolates the damaged cell, preventing the loss of essential materials from the continuous cytoplasm. The researchers who first observed this phenomenon also determined that it was probably an influx of calcium ions from the outside that caused the uncoupling. Free calcium ions are generally in short supply in the cytoplasm of cells, but when these ions are experimentally injected into the cell, the junctions are rapidly uncoupled. Eventually, a cell treated in this manner will recover, isolating the calcium ions in its mitochondria, whereupon the gap junctions will be reestablished.

Plasmodesmata Junctions known as plasmodesmata (singular, *plasmodesma*) have been recognized in multicellular plants for some time. In fact, there are very few plant cells that do not communicate through these fine channels. Plasmodesmata are fingerlike cytoplasmic extensions that pass from cell to cell through plasma membrane-lined channels in adjacent cell walls (Figure 5.4). The diameter of plasmodesmata ranges between 20 and 40 nm. Also extending through the porelike openings in adjacent walls are elements of the endoplasmic reticulum (ER) of the two cells. They form slender channels, the **desmotubules,** through the pores, which means that the lumen of the ER in the two cells is also continuous. Researchers have discovered thickenings at either end of the plasmodesmata and have suggested some valvelike role, permitting some control over the passage of cytoplasmic materials between adjacent cells.

If interpretations of the plasmodesmal structure are correct, then it is certain that these cytoplasmic connections are important in intercellular transport. They may be especially significant to the passage of water and other materials through dense tissues and over great distances. One theory of plant transport proposes that the process of *cytoplasmic streaming* (see Chapter 4) and the presence of plasmodesmata may explain how sugar solutions are rapidly transported through living cells in plants. The abundance of plasmodesmata in secretory cells provides further indirect evidence of their transport function.

Creative plant physiologists have also gathered evidence of a more direct nature. For example, when certain dyes that cannot normally cross the plasma membrane are introduced into plant tissues containing plasmodesmata, the dyes spread rapidly from cell to cell. Further, tissues whose cells abound with plasmodesmata will conduct electrical currents more readily and with less loss than those without plasmodesmata. Actually, plasmodesmata, like their counterparts in animals, the gap junctions, are not entirely passive but apparently have considerable control over what passes through. For example, in spite of their generally larger diameter, the plasmodesmata will not admit molecules as large as those that pass through the gap junction, apparently being limited to those lighter than 800 daltons. Nevertheless, the combined effect of countless plasmodesmata in the plant body is to create a virtually continuous cytoplasm, thereby greatly facilitating coordination of the plant's activities.

We have seen that the incredibly thin plasma membrane is a delicate but exquisitely complex structure. Its lipid, protein, and carbohydrate components have precise roles in the membrane's essential barrier, transport, and recognition functions. Now that we have a clearer picture of the nature of the plasma membrane, let's consider how materials may move across it.

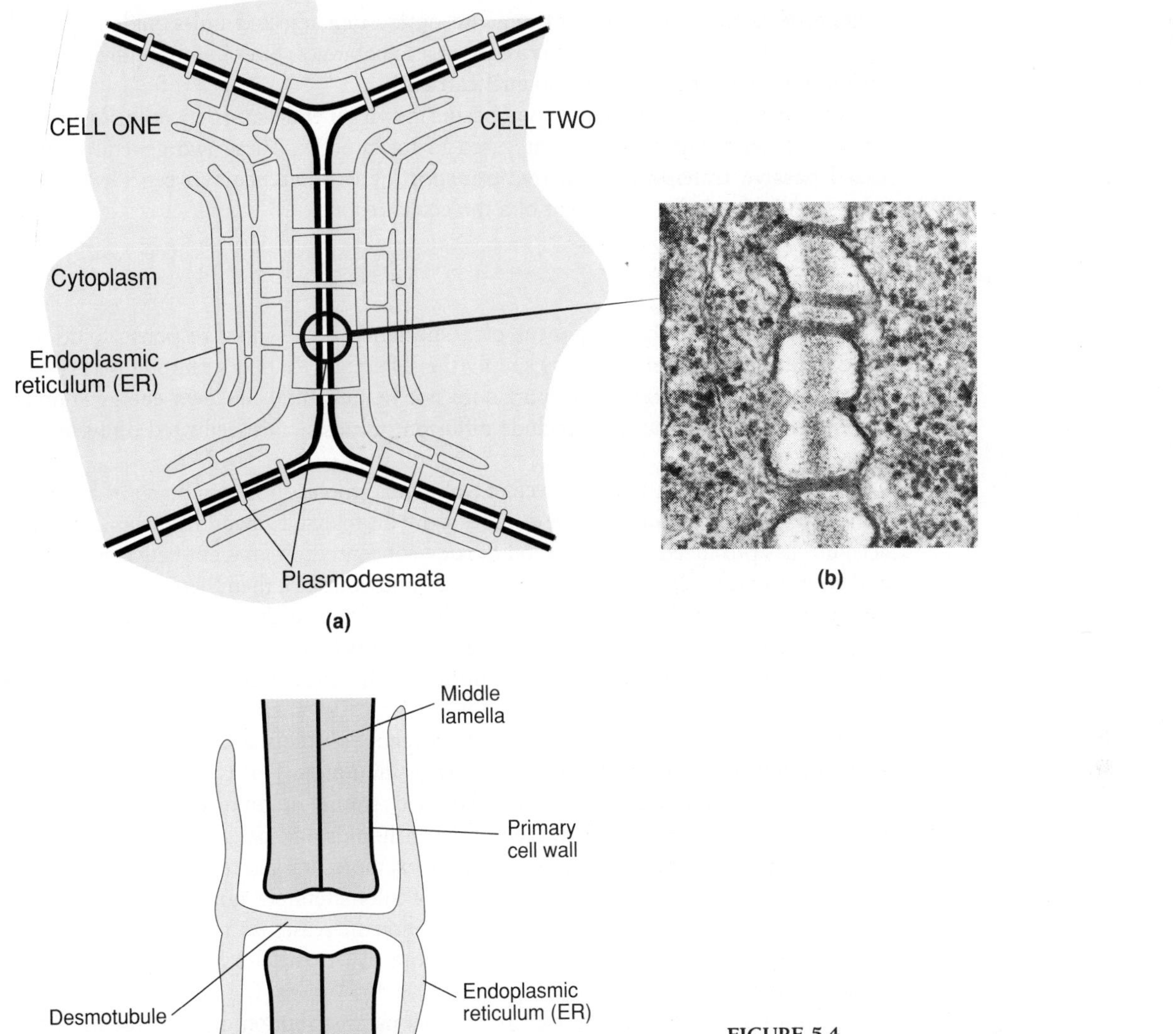

FIGURE 5.4
PLASMODESMATA
(a) The presence of porelike plasmodesmata between plant cells permits the ready passage of materials from one cell to the next. **(b)** The electron micrograph views plasmodesmata in a longitudinal section. **(c)** The cytoplasmic connection often includes desmotubules, tubelike segments of endoplasmic reticulum that pass between the cells.

MECHANISMS OF TRANSPORT

We should begin by noting that the plasma membrane is highly selective in its transport role. Some substances pass readily through with little help or interference, while others can neither enter nor leave the cell. The degree to which substances can pass through a membrane defines the membrane's **permeability.** Because the plasma membrane is more permeable to some substances than others, it is said to be **selectively permeable.** The admission or rejection of substances by the plasma membrane depends on a number of factors, but the most significant are size, polarity, and electrical charge. As a rule, small, nonpolar molecules (such as oxygen and nitrogen) and small, slightly polar, but uncharged molecules (such as water, carbon dioxide, glycerol, and urea) pass rapidly across membranes. Somewhat larger molecules such as glucose and sucrose have great difficulty passing through the membrane on their own, and charged particles, such as sodium, potassium, and calcium ions are quite often rejected unless aided across in some way (Table 5.1). Such ions cross plasma membranes, but only through special ion channels or with the aid of membranal carriers.

We've mentioned that cell transport occurs in two general ways, depending on whether or not the cell's energy reserves are used in the process. These two general ways are called **passive transport** and **active transport.** The difference between the two is in what brings about the movement of a molecule or ion.

Passive Transport

In passive transport the movement of a substance is a simple response to its own **concentration gradient,** which is simply the difference in its distribution over a given space. Importantly, passive transport does not require the cell's own energy supplies. Examples of passive transport include diffusion, osmosis, and facilitated diffusion.

Diffusion **Diffusion** is the movement of atoms, molecules, or ions *down* a concentration gradient. For example, the audible sigh you released as you read this fascinating account, has produced a steep carbon dioxide concentration gradient between your face and your surroundings. There is more CO_2 around your face than there is two feet away. So, the tendency will be for the CO_2 you just expelled to diffuse away, to where it is less concentrated. (It may be helped along by air movement as well.) The carbon dioxide molecules will continue to move away until they are equally dispersed throughout the surroundings.

Diffusion is possible because at temperatures above absolute zero, particles of matter—that is, atoms, ions, and molecules—are always in motion. This is true in gases, liquids, and solids, although such motion in solids may simply amount to particles vibrating against each other. Movement is much freer in liquids and gases. The speed at which gaseous particles can move is considerable. For example, at room temperature a hydrogen atom can reach speeds of several thousands of kilometers per hour, but it only travels a few nanometers in any direction. Such motion increases as temperatures increase. Importantly, the movement of matter is random, with particles forever bumping each other, rebounding, and taking new paths.

For diffusion to occur there must be an unequal distribution of the particles, that is, there must be a concentration gradient. When such a gradient exists, the trend is for

TABLE 5.1 PERMEABILITY OF PHOSPHOLIPID MEMBRANES TO SELECTED SUBSTANCES

HIGHEST	Uncharged molecules arranged by increasing polarity:
	Oxygen
	Nitrogen
	Benzene
	Some amino acids
	Glycerol
	Carbon dioxide
	Water
	Urea
MODERATE TO LOW	Larger, polar, uncharged molecules
	Glucose
	Sucrose
	Some amino acids
LOW	Ions
	Charged amino acids
	H^+, Na^+, HCO_3^-, K^+, Ca^{2+}, Cl^-, Mg^+

the particles to move away from the area of highest concentration. There is simply more freedom of movement in such directions. The trend continues until the substance is randomly distributed in its space.

So we see that diffusion is the *net* movement of the particles of a substance down their gradient, that is, in the direction of fewer particles of that substance, until a random distribution occurs. At this time the particles will have reached a state of **dynamic equilibrium.** This means that movement will continue as before, but there will be *no net movement in any direction.* Diffusion, by definition, will have stopped at that point.

We can also look at diffusion in terms of the energy present in a system. Concentrations of matter represent systems of greater energy than systems whose constituents are in dynamic equilibrium. When randomly distributed, atoms, ions, or molecules of a substance will have reached their lowest energy state as far as diffusion is concerned.

Diffusion is easy to demonstrate (Figure 5.5). If a chunk of brightly colored, soluble material is placed in a glass container filled with water, the material will soon start to dissolve and spread through the water. The dissolving crystal represents a highly concentrated substance and has a steep concentration gradient in its watery surroundings. Soon the color will spread further and further from the source, and given time, all of the substance will enter solution, and the color gradient will have disappeared. What has happened is, the individual atoms, ions, or molecules of the substance have diffused down their gradient and become randomly distributed. As far as directional movement is concerned, the particles will be in a state of dynamic equilibrium.

FIGURE 5.5
DIFFUSION
As the brightly colored, soluble material breaks down, its individual molecules diffuse throughout the water-filled container. The visible color gradient is also a concentration gradient along which a net movement of molecules occurs. Eventually the molecules will reach a state of dynamic equilibrium, in which there is no net movement.

Diffusion and Life Living cells frequently take advantage of diffusion, particularly in their ongoing exchange of gases. As an aerobic, or oxygen-requiring organism, you undoubtedly know that your cells require a continuous supply of oxygen, which is rapidly used up during respiration. Also during respiration, your cells continually produce carbon dioxide, a waste product of this process. Thus, the greatest concentrations of carbon dioxide are found within cells, and the greatest concentrations of oxygen occur outside. In both instances, the net movement of the two gases is in opposite directions, so that the two gases move down their concentration gradients. We are emphasizing gases here because their net movement in cells is usually straightforward. With fluids such as water, the situation is somewhat more complicated.

Osmosis Because water is so crucial to life, there has always been a keen interest in its passage across membranes as it enters and leaves the organism and moves from place to place within. We've seen that water can readily pass through the plasma membrane, so its diffusion is likely to play an important role in life. Biologists refer to the diffusion of water molecules across membranes as **osmosis.** More formally, osmosis is the net movement of water across a membrane from a region of higher water concentration to a region of lower water concentration.

Since living cells are so complex, osmosis is more easily understood through the use of models such as the simple laboratory demonstration shown in Figure 5.6. Here, a thistle tube is filled with a 3.0% sugar solution (3% sugar, 97% water) and its large end, which is covered by a selectively permeable membrane, is immersed into a beaker of distilled (pure) water. The membrane is permeable to water but not to sugar. Let's pose some questions: On which side of the membrane is the concentration of water molecules greater? (On which side is the number of water molecules per cubic millimeter greater?) If you wildly guessed, "in the beaker," you win. The beaker is pure water; thus it is as concentrated as it can be.

But which way, then, would you predict the water will move, into the thistle tube or out of it? Remember, the sugar molecules can't get out—only water can cross the

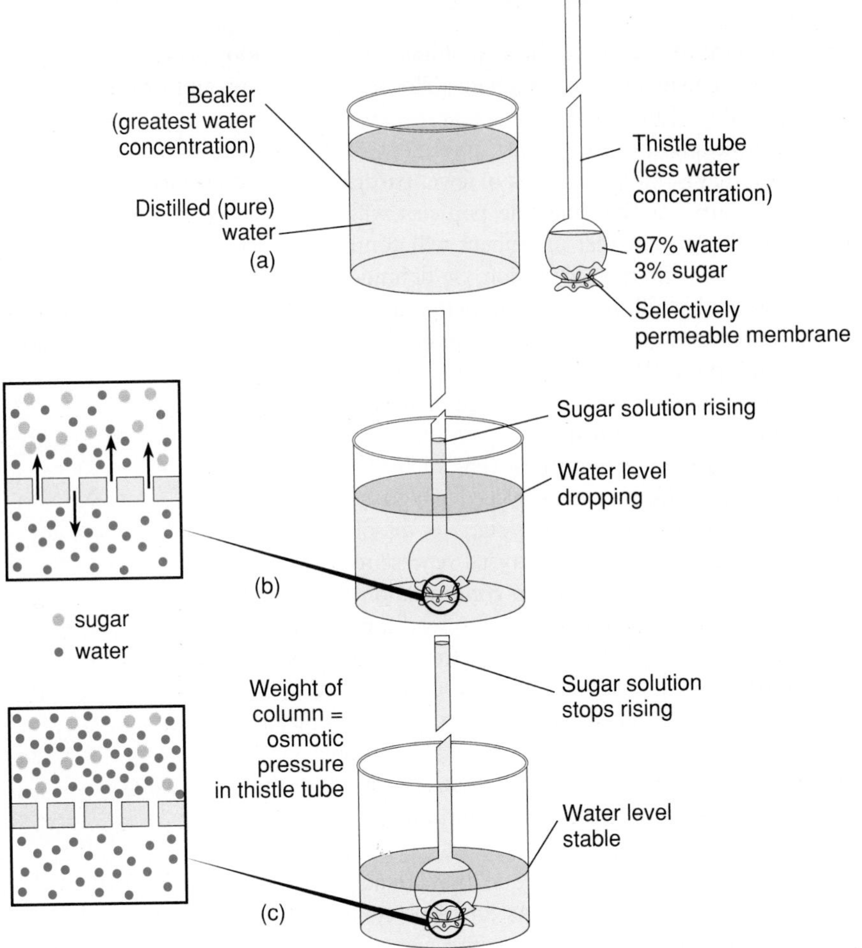

FIGURE 5.6
OSMOSIS
Osmosis, the diffusion of water across a selectively permeable membrane, can be readily demonstrated. The membrane must be permeable to water and impermeable to the solute (sugar). **(a)** Water is in greater concentration in the surrounding beaker than it is across the membrane in the sugar solution. **(b)** Water soon enters the tube and its level rises. **(c)** When the column reaches a critical level, its weight will counteract osmosis and the rise will cease. (Water level changes in the beaker are exaggerated.)

membrane. Also keep in mind the definition of diffusion. That is, particles (atoms, ions, and molecules) tend to move down concentration gradients. Since some of the space in the thistle tube is taken up by sugar molecules, water concentration is lower there and higher in the beaker. Thus, the net movement of water will be out of the beaker, across the membrane, and into the thistle tube. The level of solution in the thistle tube will rise.

How long will the upward movement of water continue? If we assume that the tube is tall enough, we can predict that at some point, the sheer weight of the water column in the thistle tube will force water molecules back across the membrane until, finally, the movement of water in the whole system reaches an equilibrium. This means the net movement will be zero. The force (pressure) that is just enough to impede osmotic movement is a measure of what is called **osmotic pressure.**

More specifically, osmotic pressure is defined as a measure of the tendency for water to cross a selectively permeable membrane from a less concentrated to a more concentrated solution. What determines the osmotic pressure of a solution? Osmotic pressure is directly proportional to the concentration of the solution—the greater the amount of solute, the greater the osmotic pressure. (A solute is any atom, molecule, or ion that will go into solution.) Another way of saying this is that the power of osmosis—the force itself—depends on the difference in the solute concentration on either side of a membrane that is permeable to water.

In summary, then, osmotic systems greatly influence the movement of water in living organisms by virtue of the selectively permeable nature of membranes and the solutes commonly found in cells. The ability of water in osmotic systems to move from one place to another is influenced by solute concentrations and limited by osmotic pressure. These factors will be more readily understood as we see them at work in cells, so let's turn to osmosis in plant and animal cells, where the impact of these physical forces is of interest to us.

FIGURE 5.7
WATER AND STRUCTURAL SUPPORT IN PLANTS
Leaves and young shoots are held erect partly by water pressure within the cells. Should the plant's loss of water exceed its uptake, the cells will lose water. The leaves and shoots will then lose their water-filled, turgid state and wilt.

Osmosis and the Cell Cells are always subject to varying osmotic conditions, and through various adaptations most can exist within some range of these conditions. The two problems confronting both plant and animal cells are the possibilities of water loss and water gain. Plant cells can cope well with water gain by virtue of their tough, resilient cell walls, and they make use of steep water gradients to maintain shape and rigidity in their leaves and young shoots (Figure 5.7). Animal cells, however, lack cell walls and are easily disrupted by excess water intake.

Because of their susceptibility to water intake, animal cells under study in the laboratory must be maintained in what is called an **isotonic** environment. An environment is isotonic if the surrounding solution contains the same concentration of water and solutes as does the cell. Red blood cells, for instance, will remain stable and maintain their shape for a time if they are placed in an isotonic 1.0% salt solution. Should these same blood cells be placed in a **hypotonic** environment—in a solution containing relatively *more* water (less solute) than is found within their cytoplasm—water gain through osmosis will be excessive, and the cell will literally burst. The same life-threatening situation exists for the animal-like protists, tiny single-celled organisms living in fresh water. However, they are provided with mechanisms for getting rid of excess water, as we will see when active transport is discussed. At the other extreme, in a **hypertonic** environment—one in which the water concentration is lower and the solute concentration higher than that of the cell—water is rapidly lost through osmosis. Such a cell will shrink in what is known as **plasmolysis.** Plant cells are also susceptible to plasmolysis in a hypertonic environment (Figure 5.8).

Plants encourage the uptake of water from what is usually a hypotonic environment. They do this by accumulating high solute concentrations in their large cell vacuoles. This, of course, decreases the water concentration there. The subsequent intake of water creates great **hydrostatic pressure** (water pressure) against their cell walls. This pressure partly explains why even spindly tomato plants can stand upright. Botanists refer to the water-swelled condition as a state of **turgor** and often refer to hydrostatic pressure as **turgor pressure.** Turgor is maintained only when the soil water concentration exceeds the water concentration in the plant cells.

But, if your tomato plants are not watered you may soon see the effects of a *shifting* water concentration. As the soil water is lost, the solute concentration in soil water increases, reducing water concentration there. (Adding too much dry fertilizer, which forms solutes in the soil water, can create the same effect.) When the water concentration in the plant cells, particularly those of the root, is greater than that of the soil, the movement of water through osmosis is reversed, and water exits down the new gradient. In a word, the plant wilts. Fortunately for the plant, the massive plasmolytic effects of wilting are reversible, up to a point, and with timely watering it will again become turgid.

Water Potential The movement of water through the plant body involves far more than simple concentration gradients, so physiologists have developed a broader concept that they call **water potential.** Water potential involves energy states. We may think of water potential in terms of the "free energy" state of a system. The free energy of a system is the energy that is available for work (also see Chapter 6). Simply put, a great deal of water concentrated in one place represents great free energy. Thus, a high concentration of water molecules on one side of a membrane, should they be free to

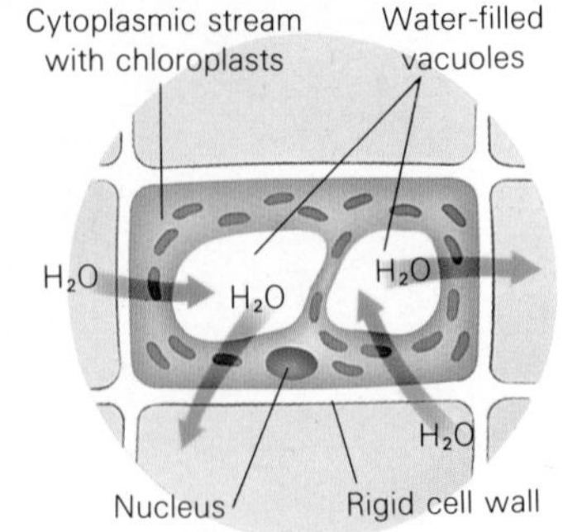

Plant cell in pond water, its usual hypotonic environment, creates a turgid condition, maintaining shape.

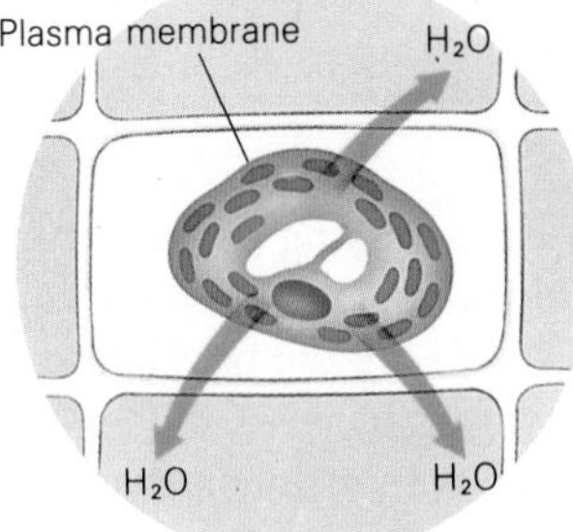

Plant cell in 3% salt solution (97% water), hypertonic to cytoplasm, loses its turgor and shrinks.

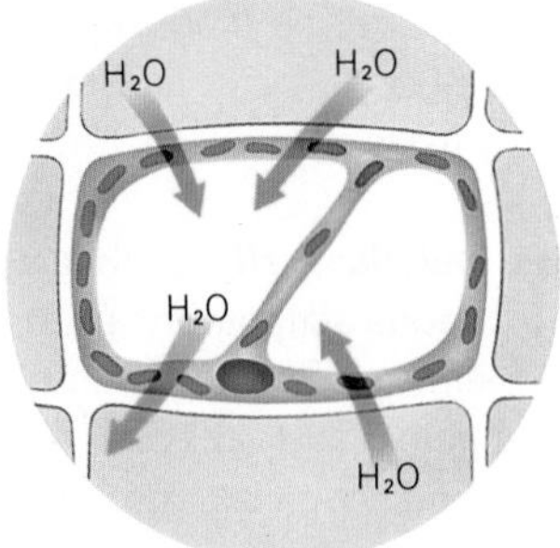

Plant cell in distilled (100%) water—increased turgor with slight bending of walls.

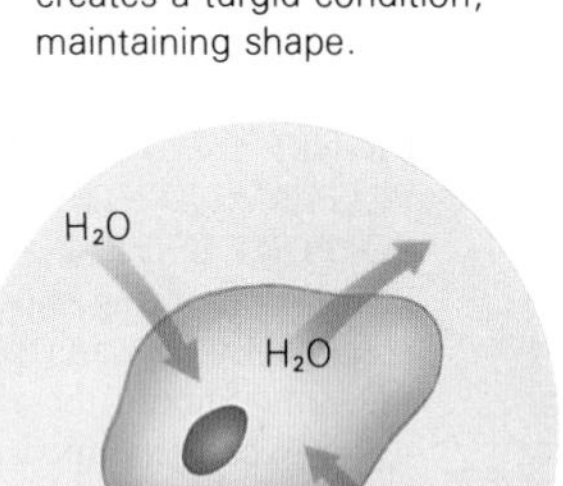

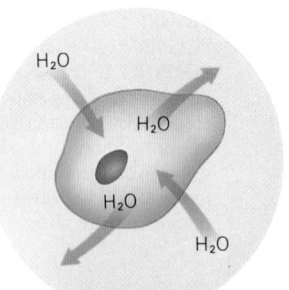

Animal cell in isotonic solution (about 1% solutes). Retains shape.

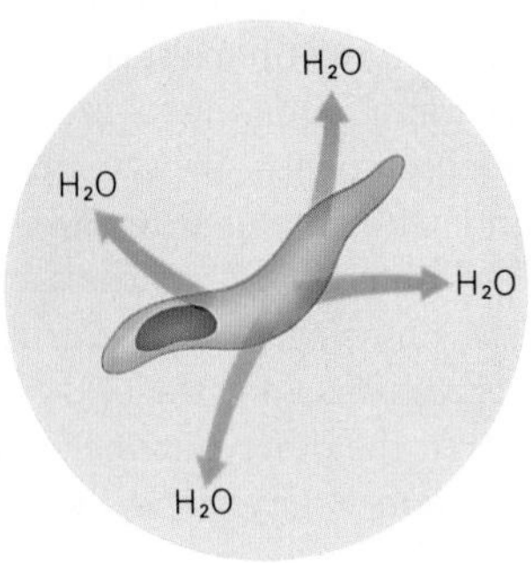

Animal cell in 3% salt solution, hypertonic to cytoplasm, loses water.

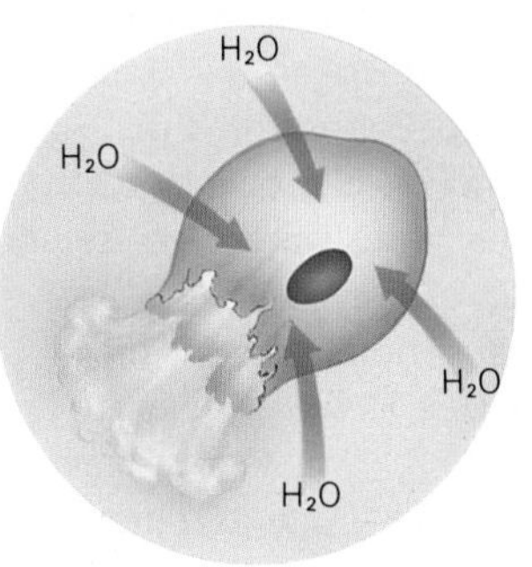

Animal cell in distilled (100%) water, hypotonic to cytoplasm, gains water, bursting (lysing).

FIGURE 5.8
OSMOTIC RESPONSES
Plant and animal cells respond differently to osmotic changes in their surroundings. Although animal cells cannot tolerate severe water gains, plants make use of such gains to maintain turgor.

move, would quickly move to places where the concentration was lower. That is, they would move down their gradient. As dynamic equilibrium is reached, the water will have reached a lower free energy state. (Recall that this is a common tendency in matter: to reach its lowest energy state.) Water potential in any system is determined by comparing it with a standard: pure water at one atmosphere of pressure and at equal temperature. Because pure water is the standard, water potential in the cells and tissues of organisms is almost always less than zero, a negative value.

Not unlike osmosis, water potential has both concentration and solute components. But it goes further, since its components also include factors such as temperature, pressure, and even the surrounding matrix, whether that matrix be cell walls or soil particles. Increases in temperature increase the free energy of water, and plants often have steep temperature gradients between the sun-warmed leaves and the roots. Increased pressure also increases free energy in water, and plants have water pressure gradients throughout. Finally, the surrounding matrix affects water potential, usually by decreasing the free energy. Water adhering to charged clay particles in the soil or adhering to the cellulose of cell walls is effectively removed from the concentration gradient. Adhesion is a major force in water transport in plants, especially in the phenomenal ability of plants to move water from deep in the soil to leaves that may be up to 100 meters above. We'll again come to grips with the components of water potential when we look into transport in plants (Chapter 28).

Facilitated Diffusion and the Role of Permeases Facilitated diffusion, our last example of passive transport, refers to the movement of molecules across a membrane with the assistance of special **carrier molecules** embedded within the membrane. It is similar to simple diffusion in that the energy involved is thermal energy, and the net movement of molecules is always from regions of higher concentration to regions of lower concentration. It differs in that in facilitated diffusion only certain kinds of molecules can cross plasma membranes, and these cross much more readily than do similarly sized, similarly charged competing molecules. How does this happen? Facilitated diffusion is not well understood, but certain carrier molecules, proteins called **permeases** embedded in the membrane, are believed to increase greatly the membrane's permeability to specific substances. Some permeases have even been isolated, but no one knows just how they work. The hypothetical scheme seen in Figure 5.9 shows a permease carrier at work, but the scheme is highly tentative. It relies on the ability of some proteins to change their molecular conformation (tertiary shape).

Active Transport

So far, the various kinds of movements we have considered are all passive. Passive transport does not involve the expenditure of the chemical energy stored by the cells. Active transport, on the other hand, does require an expenditure of the cell's chemical energy, usually in the form of ATP. Active transport may involve individual atoms, ions, or molecules, moved by membranal carriers, or it may involve the uptake or expulsion of materials in bulk, so that a large segment of the membrane becomes involved. In either case the process is characterized by (1) movement of materials regardless of (often against) their concentration gradient, and (2) an expenditure of the cell's chemical energy.

The Work of Membranal Carriers Active transport across membranes commonly involves transmembranal and peripheral proteins that act as carriers. It isn't entirely clear how these proteins move molecules in or out of the cell, but they do have some of the characteristics of enzymes (Chapter 6). Carrier proteins are specific; each type has a site or sites whose shape matches the kind of molecule it transports. Carriers work fastest when a large number of molecules to be transported are present, and they do their work by undergoing conformational (shape) changes. These changes are apparently the key to the movement of some substances in or out of the cell.

The most straightforward membrane carrier proteins, **uniport carriers**, move a single substance—a molecule or ion—across a membrane. Somewhat more complex movement involves **cotransport carriers**, those in which the transport of one substance requires the simultaneous movement of another, either in the same direction or in opposing directions. Examples of two kinds of molecules being moved in the same direction are to be found in the cells lining the digestive tract. Here the transport of glucose and amino acids, products of digestion, is sodium-dependent. It must be ac-

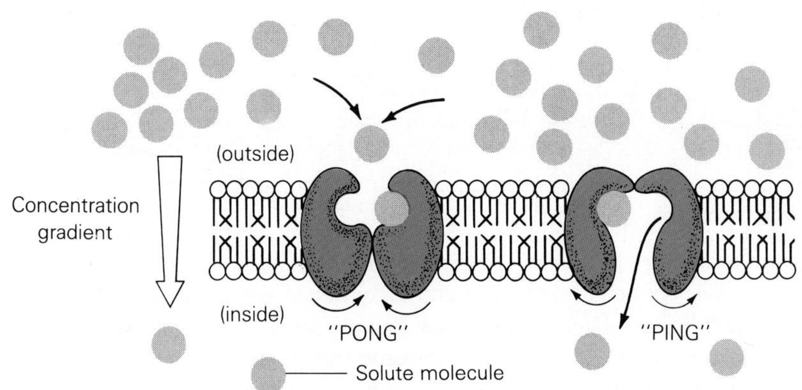

FIGURE 5.9
FACILITATED DIFFUSION
In facilitated diffusion, membrane proteins called permeases utilize conformational changes to accelerate selected molecules or ions down their diffusion gradient. The so-called "ping-pong" model suggests that the entry of a molecule triggers one event (the shape change), and its release triggers another (the change back).

companied by the transport of sodium ions. One of the best examples of cotransport occurring in opposite directions involves the ions of sodium and potassium.

Sodium/Potassium Ion Exchange Pumps One of the most active and common transport mechanisms is the **sodium/potassium ion exchange pump**, or **NA^+/K^+ pump.** (The term "pump" is figurative.) This exchange pump actively transports potassium ions (K^+) into the cell and sodium ions (Na^+) out. It is so important to cells that an estimated one third of an animal's ATP energy reserve is expended in its operation. In nerve cells, where these exchange pumps are most active, the figure exceeds two thirds. This is because neural impulses involve the diffusion of sodium ions into the nerve cell, followed by the diffusion of potassium ions out of the cell. The sodium/potassium ion exchange pumps in nerve cells must then restore conditions, moving sodium out and potassium back in. Considering how much neural activity goes on in an animal, the great expenditure of energy is understandable.

A model of the sodium/potassium exchange pump is seen in Figure 5.10. In the sequence of events, sodium is taken into its entry site, and an ATP molecule reacts with the carrier. ATP thus adds a phosphate to the carrier and provides the required energy. (Note that at this time the potassium site is not receptive to that ion.) The carrier then undergoes a conformational change, releasing the sodium outside the cell and simultaneously closing that sodium site and opening the potassium site. Removal of the phosphate then restores the carrier's original shape, the potassium escapes into the cell cytoplasm, and the pump is set up for a new cycle.

Na^+/K^+ pumps aid other cells in maintaining specific osmotic conditions. Through their continued work in the cotransport of the two ions, the cell can be maintained in a state that is isotonic to its surroundings. There will then be no net movement of water molecules.

In other instances, the net movement of water in or out may be highly desirable. For instance, when the body's water must be conserved, certain cells in the kidney transport ions into their surroundings, the growing solute there producing a hypertonic condition. Water immediately follows, to be channeled into the blood for redistribution to the body (see Chapter 36). So the strategic transport of ions can be used in regulating and maintaining osmotic conditions in the organism and in the management of water.

Plants also influence the movement of materials by altering osmotic conditions. The active transport of ions into root tip cells encourages water to follow. Likewise, the active transport of sugars from leaf cells, where they are produced, to the plant's transport system is followed by the net movement of water into that system. The rising hydrostatic pressure then creates a flowing stream that carries the sugars throughout the plant. Thus we see that in the plant and animal body, active transport can be coupled with passive transport, thereby aiding in the regulation of body processes.

Other Ion Pumps Other pumps of great significance to living animal cells include **calcium ion pumps**, which play important roles in the specialized functioning of nerve and muscle cells. For instance, calcium ions (Ca^{2+}) taken in by nerve cells initiate the movement of chemical-laden storage vesicles to the cell membrane. When such chemicals (called neurotransmitters) are released, they activate other nerve cells, thereby relaying the neural message. In muscle cells, the presence of calcium ions is essential to contraction. But in between contractions, when muscle activity might be counterproductive, calcium pumps actively transport the ions into reservoirs, where they remain until the next contraction (see Chapter 33).

Calcium ion pumps in the cells of the plant root tip aid in the root's growth towards gravity. This ensures better anchorage and a continued supply of water and minerals. It just wouldn't do for plants to send their roots upward by mistake (see Chapter 29).

Finally, hydrogen ion (H^+) pumps, or **proton pumps** as they are better known, play a vital role in cellular energetics. As we'll see in the next three chapters, it is through

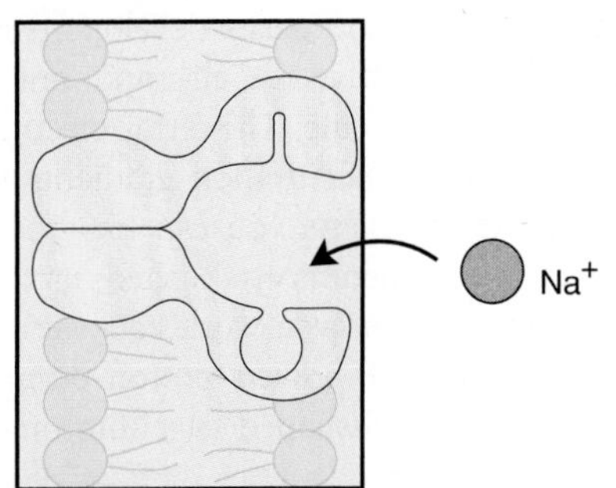

1. The sodium ion site is open to the cytoplasm, but the potassium site is closed.

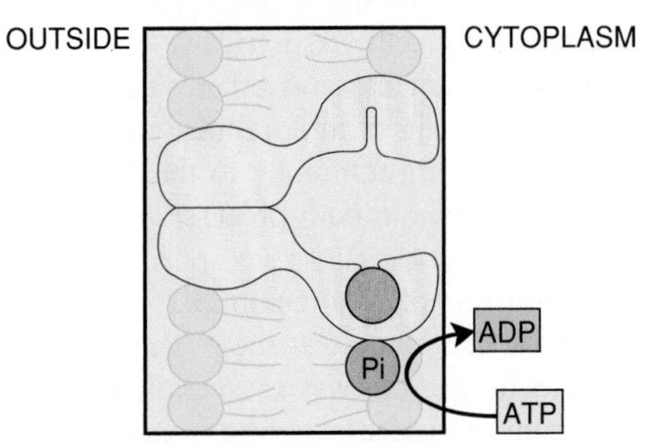

2. Sodium fills its site in the pump and the carrier is phosphorylated by ATP.

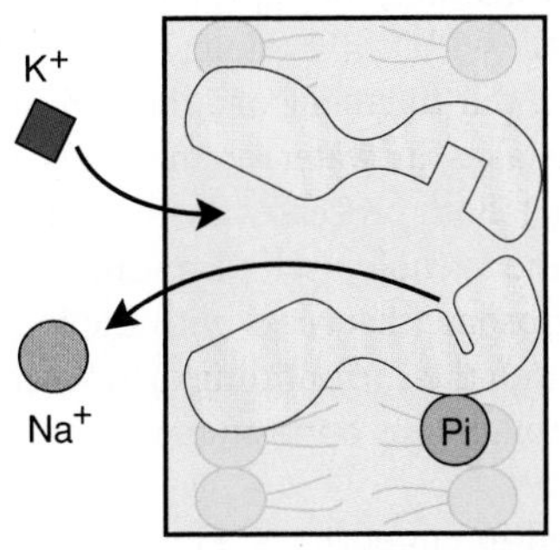

3. Using the energy ATP just provided, the carrier changes shape, releasing sodium to the outside and opening its potassium site.

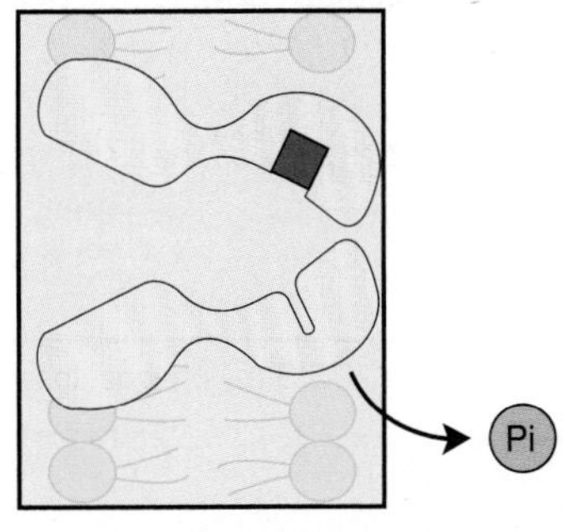

4. Potassium enters its site, and the carrier is dephosphorylated.

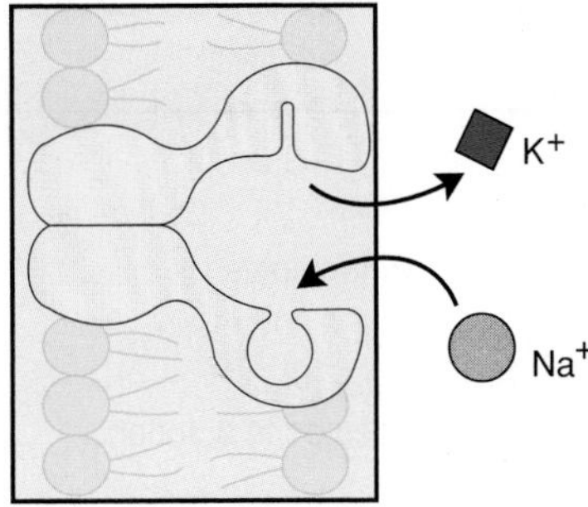

5. Another shape change occurs, releasing potassium to the cytoplasm and once again opening the sodium site.

FIGURE 5.10
A SODIUM/POTASSIUM ION EXCHANGE PUMP
The cotransport of sodium and potassium ions occurs through an ATP-driven sodium/potassium ion exchange pump. Two kinds of conformational changes occur in the protein as the pump works. One involves the sodium and potassium sites, the other the entire carrier.

the action of proton pumps that mitochondria and chloroplasts develop steep *electrochemical gradients* whose free energy can be used to drive the all-important reactions that generate ATP, the cell's main energy-releasing molecule. The term "electrochemical gradient" refers to the formation of steep chemical concentration gradients that have electrical characteristics (separated + and − charges). As you can imagine, the free energy of such a system is considerable. Once the gradient is established, its free energy can be used to produce ATP. Further, such systems can even work backwards, as they do in the case of the prokaryotic flagellum. There, ATP powers proton pumps, which then provide the mechanical force needed to spin the S-shaped flagellum on its axis.

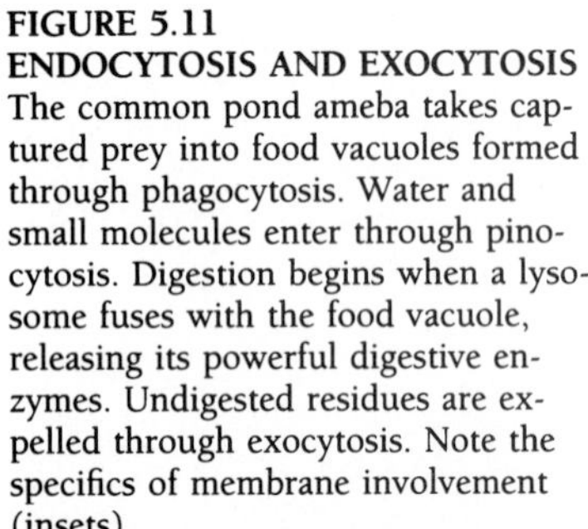

FIGURE 5.11
ENDOCYTOSIS AND EXOCYTOSIS
The common pond ameba takes captured prey into food vacuoles formed through phagocytosis. Water and small molecules enter through pinocytosis. Digestion begins when a lysosome fuses with the food vacuole, releasing its powerful digestive enzymes. Undigested residues are expelled through exocytosis. Note the specifics of membrane involvement (insets).

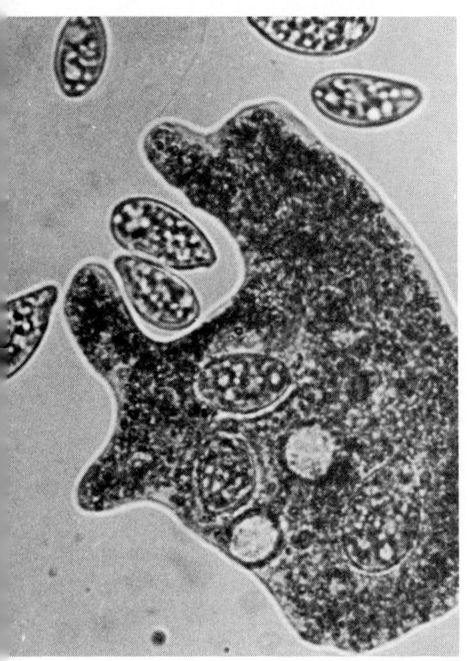

Endocytosis and Exocytosis **Endocytosis** and **exocytosis,** as the terms suggest, refer to the uptake and the expulsion of substances by cells, respectively. They were first observed in the feeding and digestive activity of amebas and other single-celled organisms. They were later seen in certain white blood cells of multicellular animals, as well as in other cells. Both endocytosis and exocytosis are testimony to the versatility of the plasma membrane, which plays a highly active role in both processes.

Endocytosis Endocytosis begins as a depression in the plasma membrane, eventually forming an "inpouching" and finally a spherical, saclike vacuole as the pouch deepens and is finally pinched off from the surface. Vacuoles formed this way are actually portions of the plasma membrane turned inside out. Their inner surface is derived from the outer surface of the original plasma membrane. In this peculiar manner, the captured material that was originally *outside* the cell ends up *inside* the cell, within a vacuole, of course. If the forming vacuole receives a solid object, such as a cell fragment or a bacterium, the process is called **phagocytosis** ("cell eating"). If it takes in water or materials in solution, the process is called **pinocytosis** ("cell drinking"). Both processes are illustrated in Figure 5.11. If digestion is involved it will occur in the vacuole itself. Typically, the enzymes to digest the engulfed food will come from a lysosome that fuses with the food vacuole. As digestion proceeds, the products will diffuse into the surrounding cytoplasm. Undigested residues will be released from the cell through the opposing process of exocytosis.

Actually, most eukaryotic cells are constantly involved in endocytosis. For the most part this involves pinocytosis as cells take in larger molecular entities, but phagocytosis

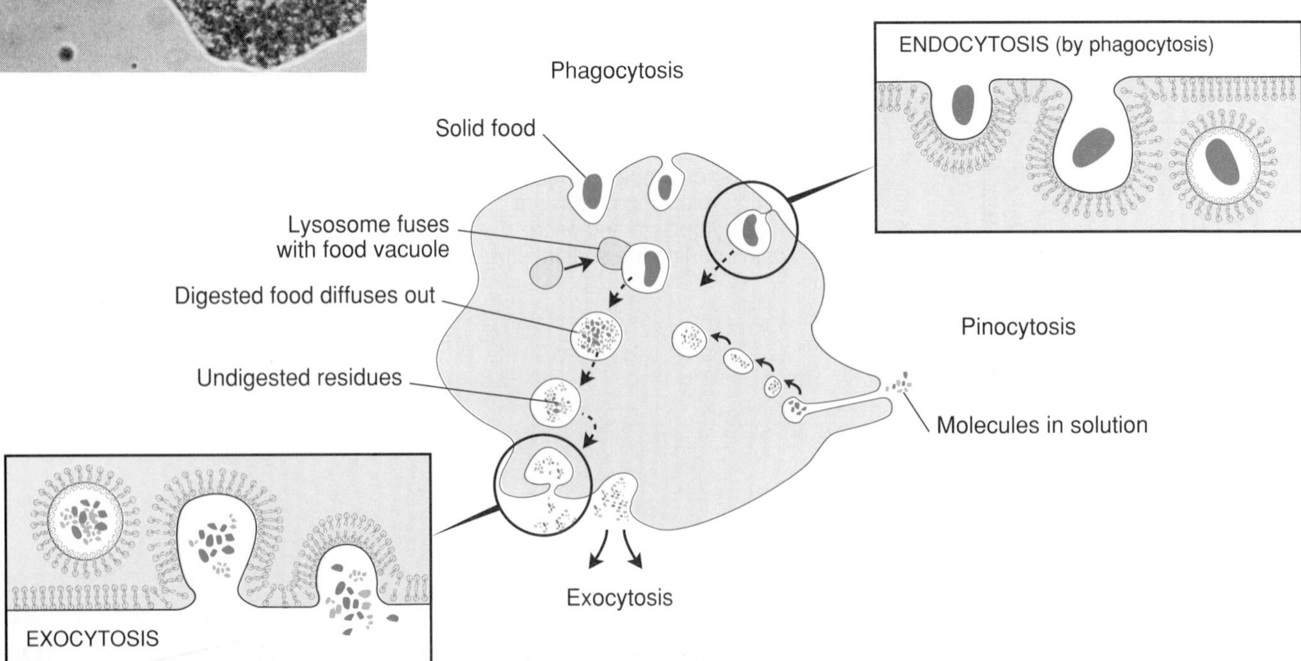

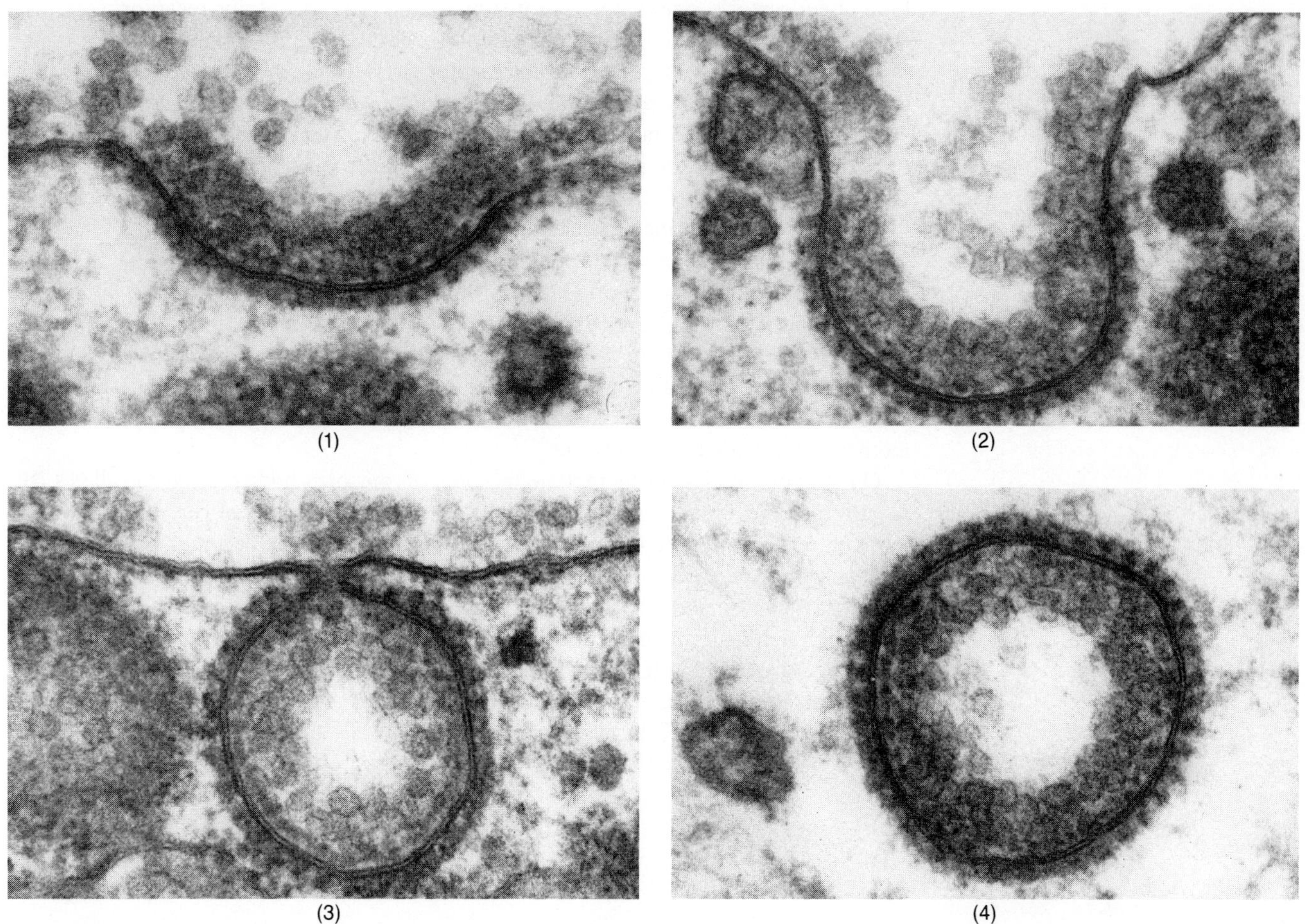

FIGURE 5.12
RECEPTOR-MEDIATED ENDOCYTOSIS
In receptor-mediated endocytosis, receptor sites on the plasma membrane fill with metabolites (substances to be used by the cell). With the filling of the sites, a coated pit develops below, followed by the pinching off of a coated vesicle. Such vesicles lose their coats, which recycle to the surface, and may next be joined by others like them (not shown).

is also common. If you think carefully about the effects of endocytosis, you may recognize a problem: cells constantly lose bits of membrane that end up forming vacuoles. How much do they lose? Some of the larger phagocytic white blood cells are so heavily involved in endocytosis that they can literally use up their plasma membrane in about one hour. However, such cells retain a stable volume and surface area in spite of this loss. Obviously, plasma membrane is replaced as fast as it is used. Typically we find that much of the plasma membrane lost through endocytosis is constantly restored through the opposite process of exocytosis. And as we will see, the plasma membrane is rapidly recycled during receptor-mediated endocytosis.

Exocytosis In exocytosis, materials are expelled from the cell. But exocytosis provides for far more than simply ridding the cell of wastes. While this is the first time we have encountered the specific term, we discussed several instances of exocytosis in the last chapter. For example, it is through exocytosis that the Golgi vesicles of secretory cells expel their contents. For exocytosis to occur, a vacuole containing such material must fuse with the plasma membrane. The fusion is a complex and little-understood process, but studies reveal that in some instances, contractile microfilaments are involved. They apparently become attached to the vacuole and the membrane and draw them together. Then, membranal elements of the two fuse, the site of contact opens up, and as the vacuole is turned inside out, its contents are deposited outside the cell. The vacuole's membrane is thus integrated into the plasma membrane (see Figure 5.11).

Receptor-Mediated Endocytosis More recently, cell biologists have identified another form of endocytosis, one that is **receptor-mediated.** That is, specific receptor molecules in the plasma membrane bind with substances from the external environment for transport into the cell. As more and more receptor sites are filled, a surface indentation, a **coated pit,** forms, followed by a complete inpouching, and finally separation of a **coated vesicle** containing the transported substance (Figure 5.12). The vesicle loses its coat (which is recycled to the plasma membrane) and then takes on the characteristics of a storage vesicle. The cell then acts upon the material taken in, generally through the fusion of the newly formed vesicle with a lysosome.

Receptor-mediated endocytosis was first reported in the early 1960s when Thomas Roth and Keith Porter determined that this process helped bring protein into birds' eggs and into the immature eggs of the mosquito for storage. In the latter, they identified 300,000 pitted sites on the oocyte surface, each of which was capable of forming a protein-laden vesicle. Other substances entering the cell through this process include cholesterol, iron, hormones, and mammalian antibodies being transported from the mother to the fetus late in fetal life.

One important class of membrane receptors in our own bodies is the so-called **LDL receptors** (low density lipoprotein). They occur on the surfaces of certain cells whose job is to remove cholesterol from the blood. When cholesterol becomes bound to LDL receptors, it is taken into the cell and safely metabolized to be stored as simple fat. While this is what normally happens, there are people with too few LDL receptors on their cholesterol-metabolizing cells. This condition leads to *hypercholesterolemia,* the elevation of blood cholesterol levels, resulting in an increased chance of premature death due to the arterial disease, atherosclerosis. Recall that atherosclerosis includes the obstruction of blood vessels by plaques in the arterial walls and the danger of stroke when the roughened plaque surfaces bring on the formation of blood clots (see Chapter 3).

APPLICATION OF IDEAS

1. It was once thought that frogs could not survive in distilled water—that is, until someone actually kept some in a pan of distilled water for months without ill effect. Theoretically, what should have happened to the frogs? Why do you think it did not?
2. Early methods of preserving food included drying, pickling, smoking, salting, and sugar-curing. Offer a physiological explanation, based on osmosis, for the failure of spoilage bacteria and molds to grow readily on food treated in these manners.
3. In an experiment, the leaves of the water plant *Anacharis* are treated in several ways and observed under the light microscope. The cells are normally rectangular in shape, with a large central vacuole and a thin layer of chloroplast-containing cytoplasm between the vacuole and plasma membrane. Explain each microscopic observation and compare it with the others.
   a. Leaves in pond water show no change.
   b. Leaves in distilled water show no change.
   c. Leaves in 1% and 2% glucose solutions show no change, but in 3% glucose, the chloroplasts form a tight sphere near the cell center.
   d. Leaves in 1% NaCl solution show no change, but those in a 1.5% NaCl solution resemble those in the 3% glucose solution.

KEY IDEAS

THE PLASMA MEMBRANE

1. The presence of a plasma membrane and its specific structure at first were only theorized on the basis of permeability studies.
2. The earliest models of the plasma membrane included a double layer of phospholipids and a covering of protein on either side.
3. The sandwichlike structure of the plasma membrane was confirmed by early electron microscope studies.

The Fluid Mosaic Model

1. In the **fluid mosaic model** of the plasma membrane the bilayer of phospholipids of earlier models is retained, but the arrangement of membrane proteins is much different. The two layers of phospholipids are arranged with their polar heads outward and their nonpolar tails intermingling. Some proteins are **transmembranal proteins,** passing through the entire membrane, some are **peripheral proteins,** partly

submerged in the lipid core, and others, the **surface proteins**, appear on the surface only.

2. The proteins serve several functions, including actively transporting substances in and out, recognition and binding with other cells, recognition of messenger molecules, and immune response.
3. Other surface components include **glycoproteins** and **glycolipids**, which make up the **glycocalyx.** Included among these are cell-surface carbohydrates such as those that determine blood type, along with others involved in cell recognition.

Cell Junctions

1. **Gap junctions**, protein-lined pores between adjacent animal cells, permit the movement of materials between cells.
2. **Plasmodesmata**, slender strands of cytoplasm that extend through pores in adjacent plant cell walls, help in the intracellular transport of materials. The pores are lined by the plasma membrane, which is continuous from one cell to the next; portions of the ER—the **desmotubules**—also communicate between cells. Plasmodesmata may be involved in the rapid transport of sugars and water through the plant body.

MECHANISMS OF TRANSPORT

1. Membranes that admit some substances and not others are **selectively permeable.**
2. Factors influencing the membrane's **permeability** include particle size and electrical charges, if any.
3. There are two known mechanisms of transport. **Passive transport** does not require ATP energy output by the cell, while **active transport** does.

Passive Transport

1. **Diffusion** is the net movement of atoms, molecules, or ions, down their **concentration gradient**, from areas of their greater concentration to areas of their lesser concentration.
2. Molecule or ion movement never ceases, but diffusion ends when their arrangement becomes random. At this time, the particles are said to be in **dynamic equilibrium**—there is no net movement in any direction.
3. The diffusion of water down its gradient is referred to as **osmosis.**
4. Osmosis is the movement of water across a membrane from an area of greater concentration to one of lesser.
5. **Osmotic pressure** is a measure of the force that must be applied to stop the osmotic movement of water across a membrane.
6. The net movement of water in and out of cells through osmosis is determined by the comparative amounts of water and solutes on either side of the plasma membrane.
   a. When the solutes are equal, the system is **isotonic**, and no net movement of water occurs.
   b. When the concentration of water outside is greater (and solutes less) than inside, the surroundings are **hypotonic**, and water will move in, sometimes bursting the cell.
   c. When the concentration of water outside is less (and solutes greater) than inside, the surroundings are **hypertonic**, and water will move out, causing **plasmolysis.**
7. When plant cells fill with water, **hydrostatic pressure** and a state of firmness called **turgor** are produced. A loss of hydrostatic pressure results in wilting.
8. **Water potential**, a term used in reference to plants, is a measure of the free energy of water as compared to pure water. Like osmotic potential it is affected by solutes, but unlike it, water potential is also a function of temperature and pressure. Water moves from its higher water potential to its lower.
9. In facilitated diffusion, no ATP energy is used, but special membrane **carrier molecules**, some called **permeases**, speed the passage of specific molecules or ions through the membrane.

Active Transport

1. In active transport, materials move against an established gradient, and ATP is consumed. Where **uniport carriers** move one substance, **cotransport carriers** move two—in the same or different directions.
2. In the **sodium/potassium ion exchange pump**, a transmembranal protein undergoes energy-driven conformational (shape) changes, carrying sodium ions into the cell and potassium ions out. ATP is converted to ADP and P_i.
3. Ion pumps have many functions in cells: influencing of osmosis by altering their ion (solute) concentrations, bringing about neurotransmitter release, supporting gravity responses in plant roots, and producing ATP through the creation of electrochemical gradients.
4. In **endocytosis**, bulk substances are taken in by an inpouching portion of the plasma membrane, which forms a vacuole. Solid material is taken in by **phagocytosis**, and fluids are taken in by **pinocytosis.**
5. In **exocytosis**, a cellular vacuole or vesicle attaches to the cell membrane, whereupon the two membranes fuse and the materials are expelled from the cell.
6. In **receptor-mediated endocytosis**, receptor molecules in the plasma membrane are activated by a substance, and when enough molecules react, a vacuole forms and the material enters the cell. One of the suspected receptors is transferrin, an iron-containing protein.

REVIEW QUESTIONS

1. Prepare a simple drawing of the fluid mosaic model of the plasma membrane. Label the phospholipids and the surface, transmembranal, and peripheral proteins. (111-112)
2. Carefully explain the significance of the phospholipid arrangement and that of the transmembranal proteins.
3. List four functions of membranal proteins. (111-112)
4. What molecules occur in the glycocalyx, and what are their functions? (111-112)
5. Why is it that molecules such as glycerol, oxygen, carbon dioxide, and steroid hormones pass readily through the plasma membrane, while glucose and many essential ions have difficulty? (115-116)
6. What are gap junctions? In what way do they serve the animal cell? (112-114)
7. Plasmodesmata are common in plant cells. Fully describe their structure. What important questions about transport might they help answer? (114)
8. Specify two significant differences between passive and active transport. (116)
9. Write a precise definition of the term *diffusion* and explain how it works in cells. (116)
10. Under what conditions will diffusion begin? When is it over? Why is the term "net" included in discussions of molecular movement in diffusion? (116-117)
11. The term *osmosis* was coined to cover a special case of diffusion. Explain what this case is. (117)
12. Explain what would happen if cells averaging about 1% solute concentration were placed in a 3% salt solution. What term describes the osmotic conditions in the cell? Outside the cell? (119)
13. Distinguish between the terms permeable, impermeable, and selectively permeable. (115)
14. Explain why microorganisms might fail to grow on food that has been preserved by salting. (119)
15. In what way do plants rely on high hydrostatic or turgor pressure in their cells? (119-120)
16. In what way does facilitated diffusion differ from ordinary diffusion? (121)
17. Explain how the sodium/potassium ion exchange pump is thought to work. (122-123)
18. What is the role of the plasma membrane in endocytosis? How does this differ from its role in exocytosis? (124-125)
19. How do phagocytosis and pinocytosis differ? (124)
20. Explain *receptor-mediated endocytosis.* (125-126)

6

CELL ENERGETICS

FLOWERING PLANTS GROWING PEACEFULLY ON A FLOWER-STREWN HILLSIDE are the very picture of beauty and tranquility. However, this picture is deceptive. In their own way, the plants are engaged in a battle for life or death. They must quietly fight for survival as they compete with other plants for sunlight, water, and soil minerals. The fact that they're there at all indicates a certain success in the struggle. But success itself can spell new problems. Once a plant has competed successfully and managed to garner enough energy to store some in its own molecules, it may suddenly have to yield its hard-won gains to some casual grazer who then rearranges the plant's molecules and energy to suit its own needs. The plant eater, in turn, often falls to some sharp-toothed predator, bent on fulfilling its own requirements (Figure 6.1). Some predators serve as prey for yet others, and some do not. But every living thing eventually meets its fate, and then the microbes of decay have their way. In this, the final episode, the simplest of life forms extract the last remnants of energy as they rearrange the molecules of plant and animal corpses.

The name of life's game, then, is energy. Energy, as we will view it here, is shuffled about in molecules where it is boosted, drained, and eventually lost. As we consider the energy of molecules, however, we must not lose sight of the "big picture." After all, these various shifts of molecules and energy all give rise to the more visible and familiar panorama we associate with life.

ENERGY

We can begin with the obvious questions. What is energy? What are its characteristics, and how does it behave? Like life itself, energy is an elusive concept; in fact, we can only perceive the existence of energy through its effect on matter. Nonetheless, let's begin with a functional, if somewhat creaky, definition. **Energy** is the ability to do work. Work, here in its broadest context, means to apply force, to move something, as it were.

FIGURE 6.1
THE GAME IS CALLED "ENERGY"
Energy is transferred from one organism to another in a variety of ways.

Work applies to every aspect of life, from the chemical work of molecule building to the physical work associated with movement and the transporting of molecules "uphill"—that is, against their concentration gradient.

Energy—the ability to do work—occurs in two states, **potential energy** and **kinetic energy.** The first really defines itself, implying that energy is "stored" and not doing anything. For example, the five-ton boulder on the hill above your house represents a considerable store of potential energy. It was raised to that level in some forgotten time when geological forces carried out a great deal of work to get it there. Should the soil holding the rock give way, you may quickly come to grasp the difference between potential and kinetic energy. Kinetic energy, as you probably have determined by now, is energy of motion. We know this because of its effect on matter. Not only does the boulder leave a trail through the brush as its potential energy becomes kinetic, but it might have a substantial effect on the matter that is your house.

Forms of Energy

Energy can take various forms. Thus, we speak of chemical bond energy, electrical energy, magnetic energy, mechanical energy, and radiant energy. The latter includes the familiar electromagnetic radiations, such as infrared, light, ultraviolet rays, and so on. We will soon see that all of these forms of energy are interchangeable; that is, under certain conditions, one form of energy can become another.

The Laws of Thermodynamics

Descriptions of the behavior of energy have been greatly simplified into time-honored principles known as the **laws of thermodynamics.** These laws are based on certain observations about the behavior of matter and energy that are remarkably invariable from one time to the next. Such consistency leads to predictions. What happens, for example, to objects raised above the ground and then released? What happens to an object that is heated and then set aside, away from the heat? Everyone can predict that the first object will fall to the ground and that the second will cool. Such observations, made time and again, have eventually led to the formation of *laws* (or laws of nature). We like to think that these laws cannot be broken under any circumstances. Biology, because of the almost infinite variety of life, is actually rather short of laws. But living systems are made of matter that interacts chemically and physically, and the laws of physics and chemistry are as inviolable in living systems as anywhere else. So like the rest of the known universe, living systems obey the laws of thermodynamics.

FIGURE 6.2
HUMANS RELY HEAVILY ON ENERGY TRANSFORMATIONS
Energy can change from one form to another, sometimes quite spectacularly.

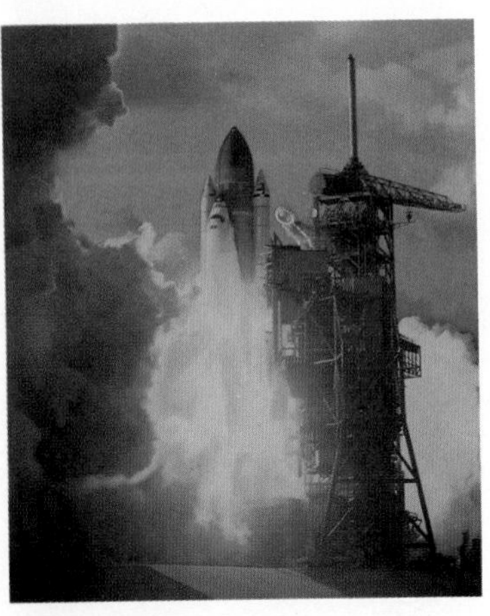

The First Law The **first law of thermodynamics** states that *energy can neither be created nor destroyed.* It is also called "the Law of Conservation of Energy." What this means is that the total amount of energy present in a system remains constant. This constancy, of course, can be maintained only if the system is "isolated"—that is, if energy can neither enter or leave. In any such system, however, the energy within it can readily change from one form to another (Figure 6.2).

Energy changes occur in an often bewildering variety of ways. When you start your lawnmower engine you can begin to appreciate the idea of energy conversion—especially if yours is hard to start. But your own expended energy aside, you might consider that the gasoline in the lawnmower's tank is a veritable storehouse of chemical energy. It is, of course—potential energy, locked away in the chemical bonds that hold the long chains of carbon and hydrogen atoms together.

When the engine is finally started, a mix of compressed gasoline and air encounters an electrical discharge from the spark plug. Chemical bonds in the molecules of fuel break, and chemical energy becomes heat, which leads to the expansion of gases in the engine cylinder. The next energy transformation is to mechanical (kinetic) energy, or

[1] Isolated systems—those in which matter and energy cannot enter or leave—do not really exist (except perhaps for the universe), but are contrived as models by scientists who wish to test their ideas under hypothetical conditions that can be limited and controlled.

energy of motion, which comes about when the gases expand and push against the piston. The piston moves up and down, its connecting rod rotating the crankshaft, which spins the lawnmower blade. At certain points in the piston's movement, valves open, and the expanding gases escape into the surroundings, their energy dissipating as heat. The escaping gases—carbon dioxide and water—are at considerably lower energy than gasoline, and the difference between the two energy levels is to be found in exhausted heat and the energy of motion. The latter largely becomes heat as well, as the moving parts of the lawnmower encounter friction.

What we've seen here is characteristic of energy as it changes form. Energy transformations are accompanied by the formation of heat, and when such heat has dissipated, it is no longer capable of work, at least as far as that system (in this case, the lawnmower) is concerned. The transition of systems with great energy to systems with low energy extends beyond lawnmowers; it is a tendency of the universe at large. Physicists express this general observation as the **second law of thermodynamics.**

The Second Law The second law of thermodynamics, sometimes called the "Law of Entropy," asserts that *energy transfer leads to less organization.* Disorder spontaneously increases, and it can be measured as **entropy.** As energy shifts through the universe, moving from higher levels to lower, entropy (or disorder) increases, and the universe becomes more random. From this statement we can deduce that there is more energy in a more organized system, and that as its energy shifts to lower levels, a system becomes less organized—or more random. Carbon dioxide and water molecules represent a more random state than do molecules of gasoline, and so the lawnmower system has moved to one of greater entropy. Much of the energy itself, once in the chemical bonds of the fuel, is now heat.

The second law further tells us that changes from one form of energy to another are imperfect—that some energy is lost, usually as heat, in each transition. Thus, when chemical energy changes to the energy of motion, some of the energy is lost as heat. The energy of heat is not lost from the system; it is just normally unavailable to do useful work. Dammed-up water roaring through a hydroelectric facility spins the generator turbines, converting energy of motion to electrical energy, but because of friction the transformation is imperfect, and there is an unavoidable escape of energy in the form of heat.

We have said that the laws of thermodynamics apply to the living as well as to the physical world. Since all living organisms are highly ordered, the question arises, how do they resist the inexorable trend towards increased entropy?

FIGURE 6.3
DEATH: MAXIMUM ENTROPY
Organisms maintain their complex molecular state only by taking in energy. When energy input ceases, the molecular state is rapidly simplified and randomized, and the energy dissipated.

Life and the Laws of Thermodynamics Life, in accordance with the second law, requires a constant input of energy. It is this energy that enables living things to remain ordered. If anything interferes with this uptake of energy, or if anything sufficiently alters the ordered state (such as disease or injury), entropy increases as order gives way to randomness (Figure 6.3).

Also according to the second law, transfers of energy in living things are not entirely efficient; energy is lost at each stage. However, living things generally do much better at complete energy transfer than do physical systems. Living cells are particularly efficient at extracting the energy from fuels such as glucose. Oxygen-using organisms can transfer about 40% of the chemical bond energy of glucose to chemical bond energy in ATP, the most common useable form of energy in the cell. Such efficiency is rarely achieved in artificial systems.

Considering the energy loss in each transition, how can life sustain itself? Life goes on because a virtually endless supply of energy bathes the earth—energy from the sun. In fact, there is so much light energy available that living things sustain themselves on less than 1% of the total reaching the earth. As long as this energy source is available, and no one figures out how to make some marvelous new weapon, life on earth should continue. In fact, the second law will not make a final impact for another 10 billion years or so, when the sun is scheduled to burn out. At that point the earth and everything on it will be reduced to simple molecules and heat.

APPLYING ENERGY PRINCIPLES TO LIFE

Now let's take the next step and apply these principles of energy to the chemistry of life. We have noted that complex organic molecules have great potential energy because of their highly organized state. They remain organized because they are held together in precise configurations by bond energy. When such bonds are broken during chemical reactions, some of their energy becomes available for work.

Such available energy is referred to as **free energy** or, more technically, as Gibbs free energy (in honor of Josiah W. Gibbs, a nineteenth-century chemist who clarified many of the ideas we have been discussing). The term "free" has nothing to do with the "cost" of energy, but refers instead to its *availability* for work (as in, "Are you 'free' for work today?"). The free energy of reactants usually changes in a reaction, and such changes are symbolized as ΔG (*delta* meaning "change," *G* for "Gibbs" free energy).

Glucose, or $C_6H_{12}O_6$, when burned or metabolized in the presence of oxygen (such as occurs in aerobic cells) is broken down into $6CO_2$ and $6H_2O$. The difference in free energy between the reactants and the two products turns out to be 686 kcal/mole (kcal = kilocalories or 1000 calories). The reaction is stated as:

$$\underbrace{C_6H_{12}O_6 + 6O_2}_{\text{reactants}} \rightarrow \underbrace{6CO_2 + 6H_2O}_{\text{products}}$$

$$\Delta G = -686 \text{ kcal/mole}$$

The negative sign of ΔG indicates that the reactants have lost free energy.

Exergonic and Endergonic Reactions

There are two basic types of chemical reactions: those in which the products have *less* free energy than the reactants and those in which the products have *more* free energy than the reactants. Reactions of the first type, such as the breakdown of glucose into CO_2 and H_2O, are called **exergonic reactions** ("energy out," see Figure 6.4a). Those of the second type, such as occur when glucose is produced from CO_2 and H_2O in photosynthesis, are called **endergonic reactions** ("energy in," see Figure 6.4b). Whereas exergonic reactions may proceed spontaneously, endergonic reactions do not; they require energy other than that contained in the reactants. In other words, energy from somewhere else must be pumped into the reaction. The formation of glucose in plants illustrates a case from nature.

During photosynthesis (Chapter 7), plants utilize light energy to make glucose from carbon dioxide and water. What occurs in this is a reversal of what occurs in the breakdown of glucose, so $\Delta G = +686$ kcal/mole, a positive number. A chemist would write:

$$\underbrace{6CO_2 + 6H_2O}_{\text{reactants}} \rightarrow \underbrace{C_6H_{12}O_6 + 6O_2}_{\text{products}}$$

$$\Delta G = +686 \text{ kcal/mole}$$

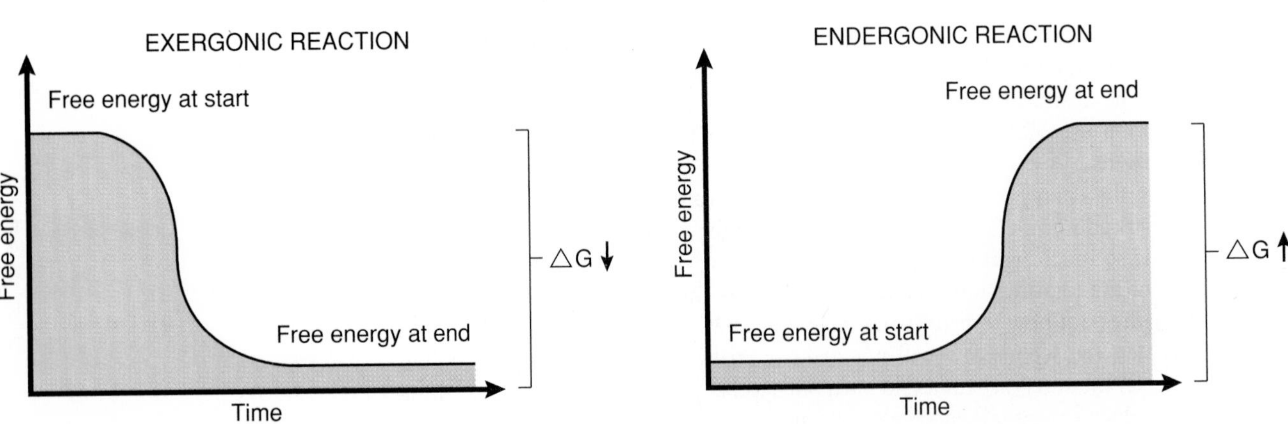

FIGURE 6.4
EXERGONIC VS ENDERGONIC REACTIONS
The products of an exergonic reaction have less free energy than the reactants. The reduction in free energy is expressed as $-\Delta G$. The reverse is true of endergonic reactions, in which the products have more free energy than the reactants, and the difference is expressed as $+\Delta G$.

Chemical Reactions, Free Energy, and Equilibrium

In the reactions we've just seen, the arrows point just one way. This indicates that the reactions are moving in that direction, toward the formation of product. However, in the real world, sometimes the reactions proceed the other way, and some of the products begin to produce the reactants—the starting material. This is possible because molecules oscillate in place. This vigorous movement causes product molecules to collide with enough force to convert their kinetic energy into new chemical bonds, causing the reactants to reform. Thus we find that a small but significant quantity of reactants is generated from products just formed in the reaction, ensuring that some amount of reactant is always likely to be present.

In spite of some small amount of reverse reactions, most proceed in the original direction. That is, the *net* direction is from "left to right" in our previous equations. This net direction of reaction continues as long as the free energy of the reactants exceeds the free energy of the products. But at some point, often near the end of the reaction, the free energy of the two may become equal. At that point the original reactants are formed as fast as are the products, and there is no longer a net direction to the reaction. The concentration of products and reactants now remains constant. At this point, reactions are said to be at **equilibrium**, with reactants and products proceeding in neither direction.

In strongly exergonic reactions, nearly all of the individual molecules of reactant will have formed product before equilibrium is reached. In weakly exergonic reactions, much more of the reactants are left when equilibrium is reached.

In weakly endergonic reactions, those with little energy available to drive the reaction, equilibrium will be reached before much product has formed. At the opposite end of the spectrum, strongly endergonic reactions, those with a rich energy source (such as the sunlight used in photosynthesis), most of the reactant forms product before equilibrium is reached.

Many reactions in the cell may not reach equilibrium at all. (In fact, should life's reactions reach an equilibrium state, we would probably be quite dead!) In such instances, the products have been consumed in other reactions, or perhaps have left the cell as fast as they are formed. When we carry on respiration we exhale CO_2 as it is formed; thus, in an overall sense at least, the respiratory reactions go "to the right." Further, chemical reactions in the cells of organisms do not usually occur as isolated incidents, as they might in a test tube. Instead, one reaction is commonly *coupled* to another. Let's see what this means.

Coupled Reactions The chemical processes of life rarely are isolated and independent. Rather, they are more likely to be part of complex and interacting chains of events. Would it be adaptive for an organism to restrict its exergonic reactions to those that lead only to the escape of free energy as heat? Other than producing body heat (a certain amount of which is adaptive to the organism), this would not be of much value. Instead, exergonic reactions are commonly coupled with endergonic reactions.

The overall process of glucose metabolism during respiration, for example, is an exergonic process yielding CO_2, H_2O, and energy. But a close look at the process reveals that many **coupled reactions** exist along the way. Thus, certain exergonic steps provide the free energy to drive coupled endergonic steps. Specifically, some of the free energy in such coupled steps ends up in new chemical bonds in ATP. For each mole of glucose, some 40 percent of the free energy lost by the reactants ends up in ATP bonds. Of course, the ATP may next enter an exergonic reaction in which its free energy is used to drive still other endergonic reactions. We will frequently come across such coupled reactions as we look further at cellular energetics.

Getting Reactions Started

We sometimes find dramatic examples of the interconversion of energy at work. Consider the case of the airship *Hindenburg*. Back in the 1930s, Germany's Third Reich intended to impress America with this hydrogen-bloated airship, the most famous of the giant zeppelins. It indeed impressed America; it blew up! While landing at Lakehurst, New Jersey, the *Hindenburg's* hydrogen gas combined violently with molecular oxygen of the atmosphere to produce water. The rest is history.

But why was the water formed so explosively? It was a matter of energy states. Water (H_2O) is in a lower energy state than an equivalent amount of H_2 and O_2. We know this because a mixture of hydrogen gas and oxygen gas confined in a space and ignited by a spark will release heat. More specifically, if one mole (see Chapter 2) of H_2 and a half mole of O_2 (the quantities needed to produce one mole of water) are placed in a device called a calorimeter, and a spark is added, one mole of water (actually steam) will be produced, and a heat rise of just about 58 kilocalories (kcal) will occur. On a gargantuan scale, the exploding *Hindenburg* illustrated this same principle. When the *Hindenburg* burned, the chemical mix went quickly to a lower free energy state, producing low-energy water and releasing the excess energy as an enormous fireball of heat and light.

The molecular reaction of the tragedy may be written as

$$2H_2 + O_2 \rightarrow 2H_2O$$
$$\Delta G = -58 \text{ kcal/mole}$$

If you try to impress your little brother by mixing a little oxygen with hydrogen before his very eyes, however, he will be totally unimpressed. Nothing will happen; the molecular hydrogen and molecular oxygen will just sit there with their outer electron shells nicely filled. Your little brother is likely to wander off. What now? Perhaps a spark of intuition.

Energy of Activation It turns out that with a tiny spark, the mixture will blow his hat off and gain you great respect. But why? Why didn't the mixture blow up without the spark? The reason molecular oxygen and molecular hydrogen don't react with each other at room temperature is that before they can, the atoms of each molecule must first be forced apart. The H_2 will then become 2H, and the O_2 will become 2O. This frees the electrons that were tied up in the covalent bonds, permitting individual hydrogen and oxygen atoms to join, forming water. The spark provides the energy required to initiate the process. This energy is referred to as the **energy of activation.**

The energy of activation must exceed the bond energy that originally held the molecules together. Actually, the spark, or energy of activation, only gets things started. Once the first molecules are disrupted, the energy liberated sets off an avalanche of reactions. Actually, we might say the system catches fire. As mentioned, $2H_2O$ is in a lower energy state than a mix of $2H_2 + O_2$, so as that system goes to its lowest energy state, the energy difference will be revealed as heat and light. The heat and light released in the *Hindenburg* disaster were considerable.

ENZYMES: BIOLOGICAL CATALYSTS

Catalysts: Quieter Changes in Energy States Hydrogen and oxygen will combine to form water at low temperatures, but only in the presence of a **catalyst.** Catalysts greatly lower the energy of activation required by chemical reactions. For instance, hydrogen and oxygen will combine readily, even at room temperature, if powdered platinum is present. The hydrogen first combines with the platinum and then with the oxygen, leaving the platinum in its original state. By definition, catalysts lower the energy of activation requirement, but they are not changed by the reactions they facilitate. Biological systems make use of enzymes, a special class of catalysts.

If you were to leave hydrogen and oxygen gases together, some of them would randomly unite to form water, but not to any significant degree. Nevertheless, the process would go on. A piece of wood left on the forest floor will gradually react with oxygen from the air, but not as fast as it would in a forest fire. In both instances, a spark is necessary to speed up processes that would have occurred anyway. However, in a structure as delicate as a cell, the heat necessary to initiate biochemical processes would cause the cell to disappear with a crackle of boiling cytoplasm, and yet, the cell certainly can't wait for reactions to occur spontaneously. They must be encouraged. Thus, the cell's activities are greatly dependent on the presence of enzymes.

Enzymes act as biological catalysts by accelerating chemical reactions. There are a great number of enzymes in any cell's biochemical arsenal, each of which functions in only one kind of reaction. The substance on which an enzyme acts is called its **substrate.**

Enzymes have several important characteristics: (1) Enzymes are proteins. (2) Each enzyme reacts with a specific substrate. (3) Enzymes do not require substantial heat to bring about chemical reactions (Figure 6.5). (4) Enzymes cannot change equilibrium concentrations for a given reaction, but they bring the reactants to that point much faster. (5) Enzymes do not affect the free energy changes (ΔG) of a given reaction. (6) Finally, enzymes emerge unaltered from reactions, ready to act again and again.

Let's also note that the union of substrate and enzyme occurs only because of the rapid, random motion of molecules, so the union is really just a matter of chance. Yet under the right conditions, reactions occur with dazzling speed, with thousands to millions of such events occurring each second. Finally, be mindful that in spite of our frequent references to "the enzyme," these remarkable catalysts are usually present in great numbers; thus, any reaction is a matter of molecular activity on an immense scale.

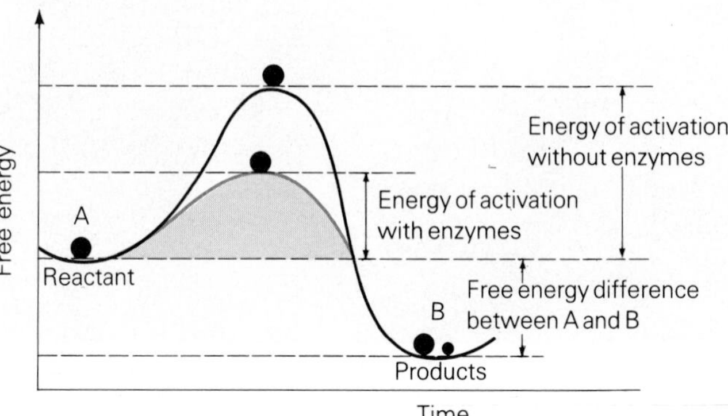

FIGURE 6.5
ENZYMES AND THE ENERGY OF ACTIVATION
Reactions that utilize enzymes need far less energy of activation than do reactions without enzymes.

Cofactors (Vitamins and Minerals)

Enzymes may require the assistance of other agents to carry on their activities. Some, those called **cofactors**, may be permanently bound to the enzyme, or alternatively, may become loosely bound to the substrate. In any case, they are essential for proper enzyme function. Most cofactors are atoms of metals such as zinc, iron, and copper. Interestingly, excesses of the same metals may be highly poisonous. Cofactors may also be organic molecules called **coenzymes.** Most vitamins are coenzymes, or at least their precursors (raw materials from which such molecules are made). Some vitamin excesses are nearly as harmful as metallic cofactor excesses.

The Enzyme-Substrate Complex: A Matter of Fit

One reason proteins make efficient enzyme catalysts is that they assume precise shapes. The importance of shape is best seen in the **active site** (Figure 6.6a), a crucial part of any enzyme. The shape of this site makes it possible for the enzyme to interact specifically with a certain substrate and no other. The enzyme and substrate join briefly at the active site, forming what is called the **enzyme-substrate complex**, seen in Figure 6.6b.

Whereas the fit between substrate and the enzyme's active site is close, it is not as perfect as it was once thought to be. In what biochemists call the **induced fit hypothesis,** it is proposed that as the substrate binds to the active site, the site changes to fit the substrate more closely. In so doing, the substrate is placed under physical stress, an important factor because it helps with the disruption of chemical bonds. Figure 6.6c looks into the actual events at the active site by considering the work of a hydrolytic enzyme.

Influences on the Rates of Enzyme Action

The rates of enzyme-catalyzed reactions are strongly dependent on three conditions:

1. Substrate concentration. As the substrate concentration increases (starting with zero), the probability of an enzyme molecule colliding with a substrate molecule

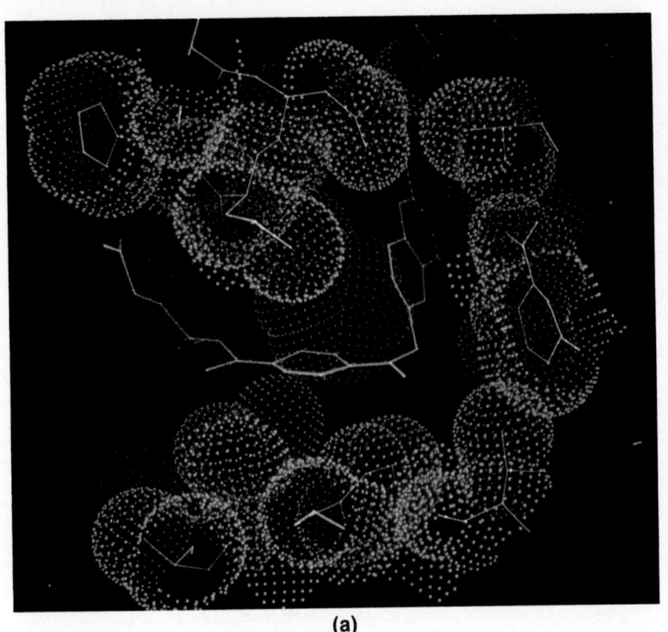
(a)

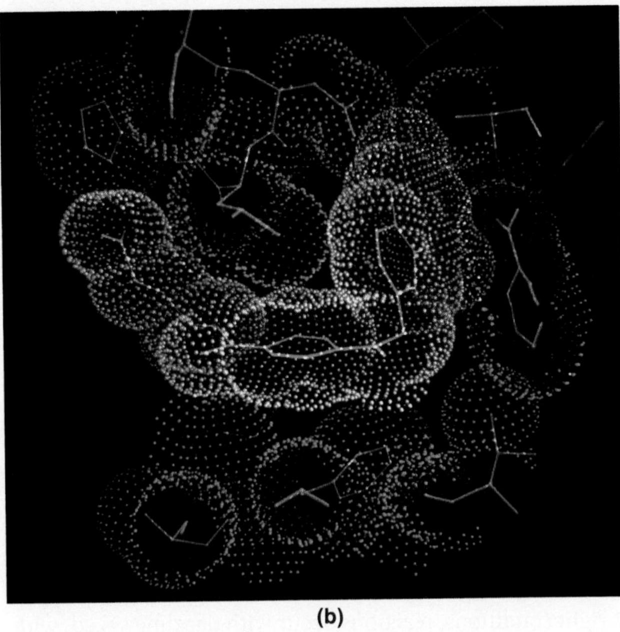
(b)

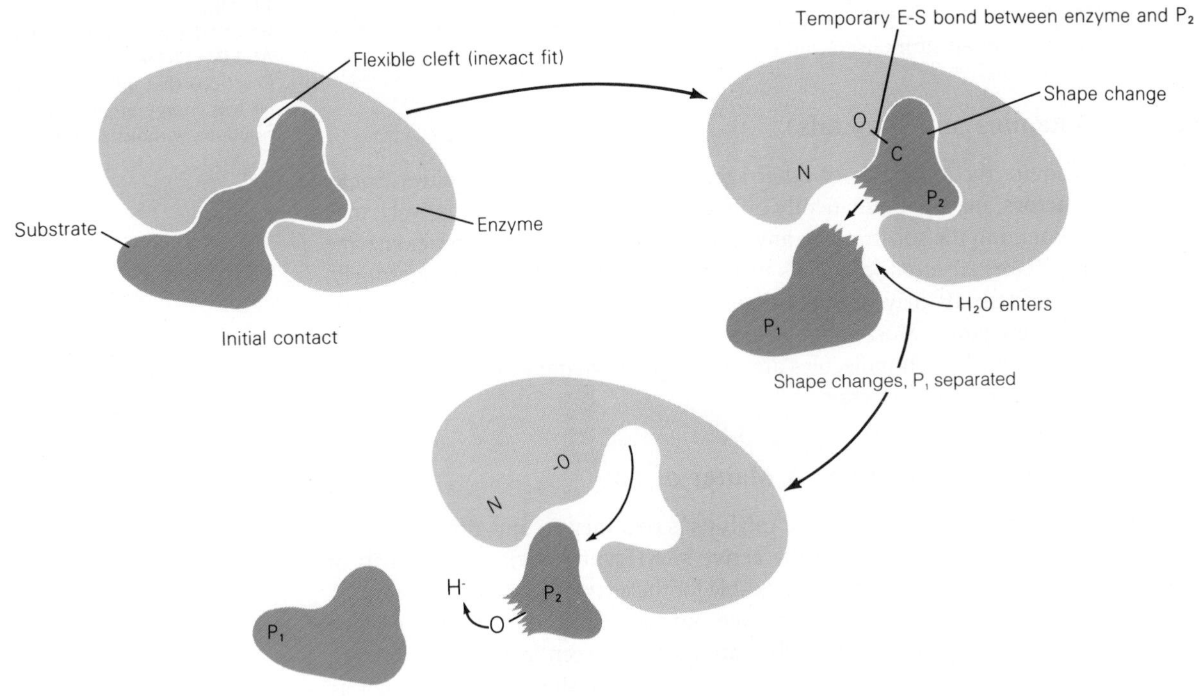

(c)

FIGURE 6.6
ENZYMES AND SUBSTRATES
(a) A three-dimensional computer-generated view of an enzyme reveals its globular shape and a cleft that forms the active site. **(b)** The enzyme's specific substrate fits closely into the active site, where it will react. The combination forms the enzyme-substrate complex. **(c)** This model of a hydrolytic enzyme takes us through a series of steps. First, a substrate enters the active site of the enzyme, forming an imperfect fit and held in place chiefly by hydrogen bonds and other subtle forces. The enzyme, a dynamic molecule, changes shape slightly, placing stress on the substrate at key points and cleaving it into products 1 and 2 (P_1 and P_2). A temporary enzyme-substrate (ES) bond forms between P_2 and a nearby amino acid in the enzyme. Next, water enters, the temporary ES bond breaks, and hydrogen from the water joins the enzyme while the $-OH$ group joins P_2. P_2 is released as the enzyme resumes its inactive shape.

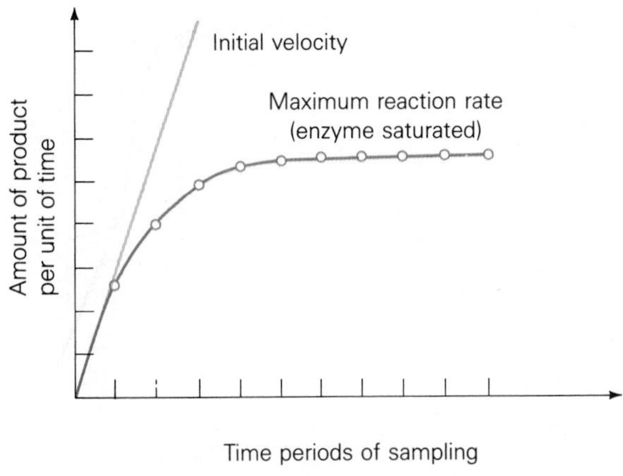

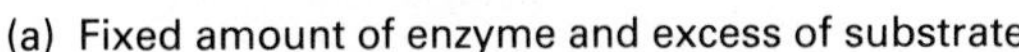

(a) Fixed amount of enzyme and excess of substrate

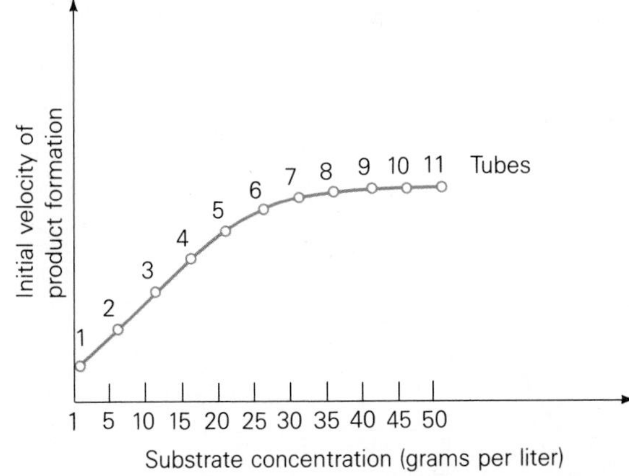

(b) Fixed amount of enzyme, but increasing amounts of substrate carried out in eleven reaction tubes

FIGURE 6.7
SUBSTRATE CONCENTRATIONS AND REACTION RATE
(a) The graph plots the amount of product formed when a fixed amount of enzyme and an excess of substrate are placed in a single reaction tube and product formation is measured over time. The initial velocity of the reaction is steep, but soon slows and levels off, indicating that the enzyme is now working at the turnover rate, the maximum under the conditions present. **(b)** The quantity of enzyme in each tube is the same, but the amount of substrate is increased through the series. The amount of product is then measured after a predetermined time. Increasing substrate concentrations increase the rate of product formation, but only to a point. This indicates that the combination of substrate and enzyme is a random process, and that the more substrate present the better the chance of meeting. But when the turnover point is reached (at about tube number 9), all enzyme molecules are fully employed, and increasing the substrate has no further effect.

increases, and so the overall rate of reaction increases. At some concentration, though, a point called **enzyme saturation** is reached. In enzyme saturation, each active site is engaged, and the reaction rate is then determined by "turnover time," the time required for the individual reaction to occur. Any increase in substrate, without an accompanying increase in enzyme, will not affect the overall reaction rate (Figure 6.7).

2. Effects of pH. Acidity also affects the active site of an enzyme. A certain pH encourages the ionization of critical amino acid R-groups in the enzyme's active site, causing the enzyme and substrate to react more readily. By the same token, excessive acidity or alkalinity can denature the enzyme, rendering it inoperative. The optimal pH condition depends upon the specific enzyme (Figure 6.8).
3. Effects of temperature. As a rule of thumb, an increase of 10°C doubles the rate of most chemical reactions, including enzyme activity (Figure 6.9). This is because heat in a system is expressed as molecular motion; thus an increase in the movement of the substrate and enzyme molecules increases the chances of their collision.

   Of course, there is a limit to how much heat can usefully be added to any system. After all, heat in excess can denature protein—that is, break the fragile bonding forces that maintain the tertiary shape of enzymes.

Teams of Enzymes: Metabolic Pathways

As you might expect, enzyme activity in the cell can become quite complex. A single enzymatic reaction may be just one link in a long sequence of such reactions. Such a chain of reactions is called a **metabolic pathway**, and there are many of these at work in the cell. A metabolic pathway includes teams of different enzymes, each involved in its own chemical change. In effect, the product of one enzyme becomes the substrate of the next, and so on along the pathway. Metabolic pathways are quite like industrial assembly lines, in which each worker makes some change in the product passing by.

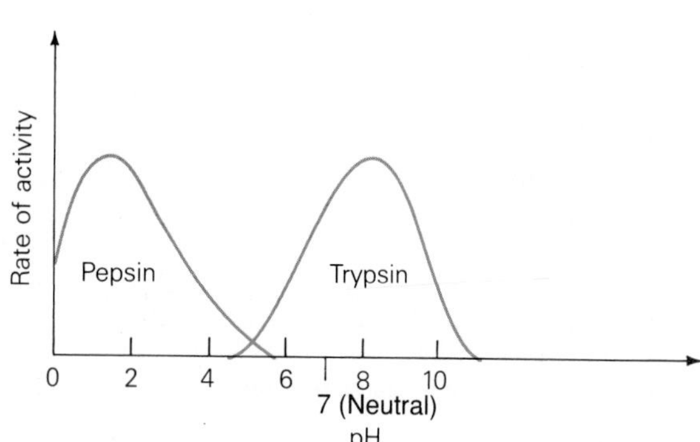

FIGURE 6.8
pH AND ENZYME ACTIVITY
Most enzymes, like trypsin, operate optimally near pH7, the neutral point. Pepsin, a stomach enzyme, is most active in a strongly acidic environment.

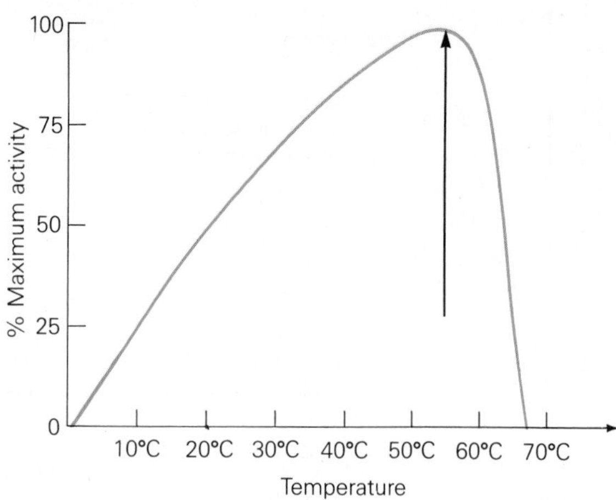

FIGURE 6.9
TEMPERATURE AND REACTION RATE
The addition of heat accelerates the action of enzymes—up to a point. The sudden reduction in activity at 50–60°C indicates that the enzymes have probably been denatured by heat, rendering them inactive.

Quite often, a metabolic pathway includes numerous coupled exergonic-endergonic reactions. As we've seen, in such paired reactions, free energy from the exergonic reaction makes the endergonic reaction possible.

Mechanisms of Enzyme Control

Although the activities of enzymes can be modified by surrounding conditions such as temperature, pH, and substrate concentrations, far more precise controls exist. Such controls, working through activating and inhibiting mechanisms, control the accessibility of the enzyme.

Enzyme-Activating Mechanisms In some instances the enzymes are available but remain in an inactive form until certain chemical signals prompt them into activity. We find a good example of such control in the animal liver cell, which contains enzymes that operate in pathways that help regulate blood sugar levels. When the hormone epinephrine (also called adrenaline) arrives at the surface of a liver cell, it binds to specific recognition sites on the cell surface. These, in turn, activate a key membranal enzyme whose role is to begin a cascade of chemical activity that ends with the activation of certain other enzymes. (So we have enzymes activating enzymes.) The latter are responsible for the breakdown of glycogen, a storage polysaccharide, into free glucose. The glucose leaves the liver cell and enters the blood, bringing the blood sugar level to its optimal concentration.

Enzyme-Inhibiting Mechanisms The opposite end of the spectrum from enzyme activation is enzyme inhibition, where something interferes with the functioning of existing enzymes. Two major ways this negative kind of regulation can occur are called **competitive inhibition** and **noncompetitive inhibition.** In competitive inhibition, a substance similar to the usual substrate fills the active site of the enzyme. The enzyme cannot react with this "lookalike" agent, so its activity is blocked. In noncompetitive inhibition, an agent will bind to the enzyme, but at a locale other than the active site. Such binding alters the enzyme's shape enough to render the active site inoperable.

Inhibition is the mechanism through which certain antibiotics work. Penicillin, a competitive inhibitor, fills the active site of enzymes that synthesize the cell wall materials in certain groups of bacteria and a faulty wall is produced. Although the bacterial cell is healthy enough in other ways, without a proper cell wall it cannot cope with osmotic fluctuations in its environment. Eventually, enough water will diffuse in to burst the cell.

Allosteric Control **Allosteric enzymes** are generally large enzymes that act in metabolic pathways, often at their start. These enzymes are named for their secondary sites (*allo-*,"other"; *stereos,* "space") and are highly sensitive to changing conditions. This is because they respond to two kinds of **modulators:** activators and inhibitors. The binding of modulators to the secondary sites does not involve strong bonds, just weak attractions. This permits the modulator and enzyme to separate easily, another factor that makes the enzyme exquisitely sensitive to changing conditions.

Allosteric enzymes have a quaternary level of structure, that is, they contain two or more polypeptide subunits (see Chapter 3). Each subunit contains the usual active site, but where the polypeptide subunits join, secondary or **allosteric sites** form. It is important to know that when any allosteric site is occupied by a modulator, it affects all active sites on the enzyme.

Allosteric enzymes seem to be restless; they continually switch between active and inactive states, their "engines idling" until a modulator arrives (Figure 6.10). If an activating modulator binds to the allosteric site, it locks the enzyme into its active state, and reactions begin. But when the bound modulator is an inhibitor, the enzyme becomes locked into its inactive state, and no reactions occur. In either case, the presence of the activator or inhibitor affects the shape of the active sites nearby, opening or closing them to substrate, respectively.

In metabolic pathways, the activator molecule may be the very substrate that enters the pathway. When it is present in large quantities, it binds to the allosteric sites of one or more different enzymes in the pathway, and the pathway becomes active. So the initial substrate carries a clear message: "Speed up." Alternatively, an inhibitor molecule may be one of the products further along the pathway, perhaps the final product itself. When such products begin to accumulate, they back up, soon filling allosteric sites of enzymes at the start of the pathway. Thus, the product sends its own message: "Stop."

The inactivation of allosteric enzymes by an end product of the pathway is an example of what biologists call **negative feedback inhibition**, or simply, negative feedback. In negative feedback the product of an action slows or stops (inhibits) that action. This provides a metabolic pathway with a means of automatic regulation. Inhibition goes on until the final product diminishes (is used up or removed), whereupon the product molecules leave the allosteric sites, and the reactions resume. Figure 6.11 reviews an example of allosteric enzyme control through negative feedback.

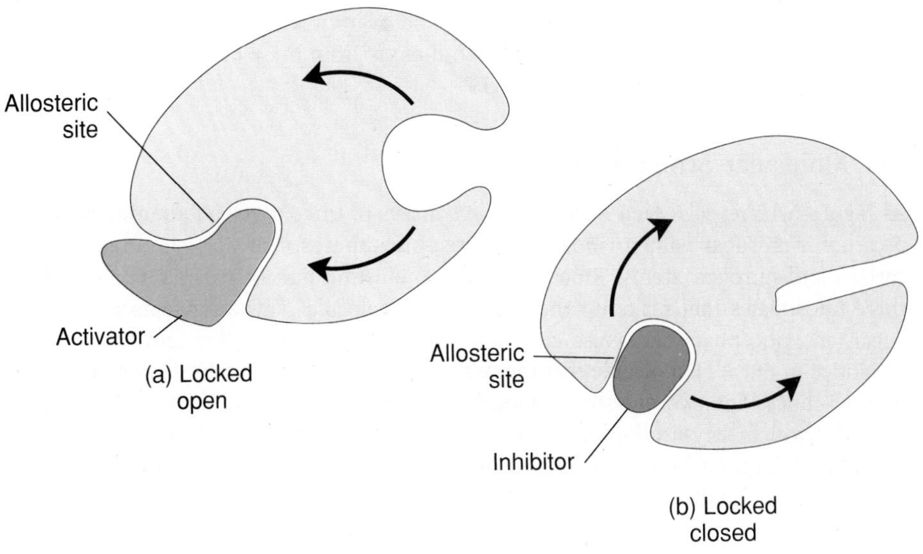

FIGURE 6.10
ALLOSTERIC SITES AND CONTROL
(a) When activating modulators insert into the allosteric site, the enzyme's active site is made receptive to substrate. **(b)** When inhibitory modulators are in place, the active site is altered, making it unreceptive to substrate.

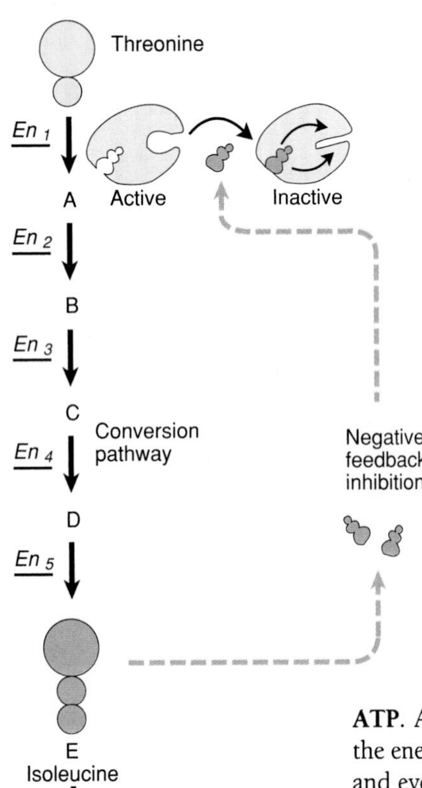

FIGURE 6.11
NEGATIVE FEEDBACK INHIBITION
In the metabolic pathway converting threonine to isoleucine, an excess of the final product forms a negative feedback loop to the first enzyme (E_1), filling its allosteric site and blocking the pathway.

Summarizing the Enzyme

At this point let's review some of the major characteristics of enzymes and their functions:

1. Enzymes are proteins with highly specific shapes.
2. Each kind of enzyme has a specific role, reacting with a specific substrate.
3. Enzymes may have slightly flexible active sites, which fit the substrate's molecular configuration.
4. Enzymes, like other catalysts, emerge unaltered from their reactions and can be used repeatedly.
5. Enzymes lower the energy of activation required for a reaction to occur.
6. The rates at which enzymes function is determined by substrate and product concentrations, heat, and pH.
7. Enzymes may occur in groups that catalyze sequences of metabolic reactions called metabolic pathways.
8. Enzyme activity may be, in large part, controlled by activating and inhibiting modulators.

ATP: THE ENERGY CURRENCY OF THE CELL

Living things have to stay busy or perish. Even the laziest, most sedentary creature on earth is a virtual hotbed of activity, even if all the activity takes place quietly, within its cells. No matter what this creature is, all its activities are intimately tied to one kind of molecule, **adenosine triphosphate**, or **ATP.** ATP is important as a ready source of energy, but we should keep in mind that the energy is used in a variety of ways, such as building molecules, transport, movement, and even in the thought processes of fireflies (at one end) and their glow (at the other).

ATP can accurately be called the "coin" of the cell's energy transactions. Like the penny, our smallest monetary unit, an ATP molecule represents a basic energy unit. When larger amounts of energy are needed, more ATP must be used. The fact that small amounts of energy are available within each ATP molecule is very important to the processes of life. Such amounts permit the expenditure of free energy to match the task at hand. If nothing else, life is parsimonious; wastefulness is soon dealt with by natural selection.

All living cells, by the way, use ATP. It is a universal molecule of energy transfer. If energy is made available by some cellular process, such as photosynthesis or respiration, it is first transferred to ATP. If that energy is then needed, it is released by ATP. If you recall from Chapter 3, carbohydrates, lipids, and even proteins contain energy reserves the cell can use. But energy cannot be taken directly from these molecules. First it must be transferred to the special bonds of ATP.

The Molecular Structure of ATP

As Figure 6.12 reveals, each ATP molecule consists of three parts: an **adenine base**, the 5-carbon ring sugar named **ribose**, and three **phosphates.** Adenine is a double ring of carbon and nitrogen atoms. Ribose forms a connecting link between adenine and the three phosphates that make up the "tail" of the molecule. This arrangement of a base, sugar, and phosphate tail is also common in the nucleic acids, DNA and RNA.

Notice in our ATP molecule that the two outermost phosphates are shown connected by wavy lines. The two outermost phosphates are special because the bonds (the wavy lines) by which they are joined to each other and the rest of the molecule are readily hydrolyzed (broken by the addition of water). The products are **ADP (adenosine diphosphate)** and a free phosphate group (P_i). When the terminal phosphate's bond is

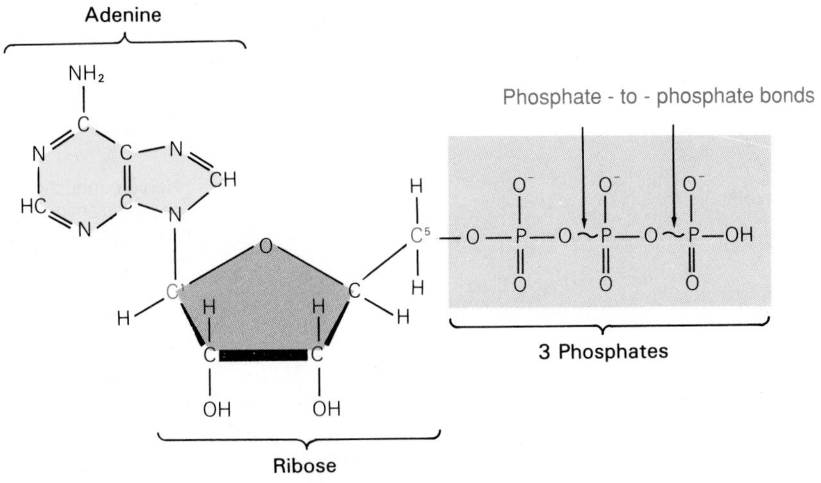

FIGURE 6.12
THE ATP MOLECULE
ATP consists of the nitrogen base adenine, the pentose sugar ribose, and three phosphates. The phosphate bonds (curved lines) are readily cleaved, releasing energy and making the phosphates available for other reactions.

thus broken, the free energy change or ΔG is -7.3 kcal per mole.[2] That is, 7.3 kcal of energy is freed for work. The phosphate-to-phosphate bonds were once called "high energy bonds," but actually several other sorts of cellular molecules yield more energy upon hydrolysis than ATP does. The ATP bonds are instead important in that they are easily broken and therefore are a *readily available* source of energy.

Most of the free energy used in the cell's many chemical reactions is provided by the hydrolysis of ATP. Thus, it is important that ATP continue to react even when in low concentrations. Since it is highly exergonic, the hydrolysis of ATP to ADP and P_i will not reach equilibrium until much of the ATP is used up. In other words, the reaction in which the phosphate bonds are hydrolyzed is clearly towards the right.

$$ATP + H_2O \rightleftharpoons ADP + P_i$$
$$(\Delta G = -7.3 \text{ kcal/mole})$$

The simple hydrolysis of an ATP molecule would do little for the cell except provide some heat and use up some water, except that such exergonic reactions are usually closely coupled with endergonic reactions in which some of the free energy lost by ATP is put to work. Such endergonic reactions generally involve **phosphorylation**—the addition of phosphate to something else. The easy hydrolysis of ATP's phosphate group releases the phosphate to be transferred to other molecules. When another molecule accepts one of the phosphates, it gains new free energy. The energized molecule is then more likely to enter into yet other reactions. For example, temporary phosphorylation of a molecule sets that molecule up for a substitution reaction in which the phosphate is subsequently replaced by another chemical group. In fact, it is precisely through a number of such phosphorylations and substitutions that sugars are formed during photosynthesis (see Chapter 7).

The Cycling of ATP

Since ATP supplies the major share of the cell's energy requirements, you might think that living things would be bulging with such molecules. This is not the case. Quite simply, cells don't need much ATP because they recycle it so efficiently. (We humans, in fact, have only about two ounces of ATP in our entire bodies.)

When ATP is hydrolyzed in the cell, it is usually just the terminal phosphate group

[2]The value 7.3 kcal/mole is determined under specific laboratory conditions where concentrations, temperature, and pressure are set to a certain standard. The free energy change for the hydrolysis of ATP in the cell may be much greater. In one instance, the free energy change of ATP hydrolysis in red blood cells was estimated to be 12.4 kcal/mole.

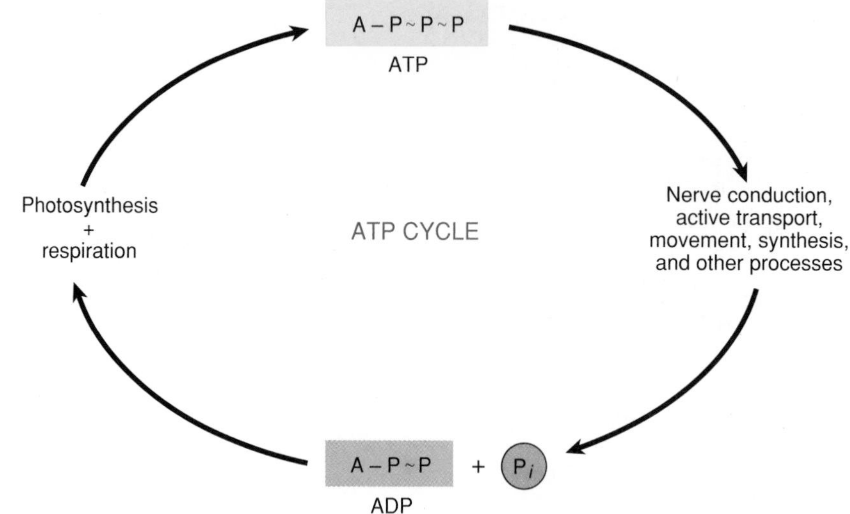

FIGURE 6.13
ATP CYCLE
ATP enters into many exergonic reactions and provides the energy for numerous cellular processes. In so doing it is converted to ADP, which is then recycled through endergonic reactions, during which the terminal phosphate is restored.

that is cleaved, forming adenosine diphosphate or ADP. (Less frequently, the two outermost phosphates are cleaved, and **adenosine monophosphate**, or **AMP**, is the result.) The ADP is then recycled—sent through processes in which the terminal phosphate is restored. As we will see in the next two chapters, most ATP is reconstituted in two major ways, through photosynthesis and cell respiration. During either process, phosphates are joined to ADP or AMP, and ATP is formed (Figure 6.13).

We should not lose sight of the significance of the role of ATP. ATP helps living things to remain organized and delays the inexorable trend toward entropy. In the upcoming sections, we will learn more about the recycling or regeneration of ATP from ADP, but first we'll set the stage by introducing a few additional molecules with related roles.

THE COENZYMES AND OTHER ELECTRON CARRIERS

Coenzymes, you'll recall, work together with enzymes. But unlike enzymes, coenzymes are not proteins. They are much smaller than proteins, and as a matter of fact, the components of some are similar to those of ATP. Coenzymes are involved in many kinds of cellular reactions, but one of their major roles is to participate in reactions that lead eventually to the synthesis of ATP.

The two most common coenzymes are **NAD** and **NADP** (nicotinamide adenine dinucleotide and nicotinamide adenine dinucleotide phosphate). We will also consider a third one, **FAD** (flavin adenine dinucleotide). These are usually pronounced as "nad," "nad-phosphate," and "fad." NADP operates in photosynthesis, and NAD specializes in cell respiration. FAD functions in both processes.

Chemically, NAD and NADP are quite similar, but NADP contains one more phosphate. Like ATP, both contain the nitrogen base adenine, along with ribose and phosphate. But in addition, they contain a compound known as **nicotinic acid** (Figure 6.14), a component of vitamin B. Nicotinic acid is the active component in both coenzymes. FAD is similar to the other two coenzymes, but its active group is isoalloxazine, a subunit of riboflavin (also from the B vitamin).

In general, the three coenzymes carry out similar functions. They team up with enzymes that are involved in the **oxidation** and **reduction** of substrates.

Redox Reactions: Oxidation and Reduction

It may seem strange that so much of life depends on simply passing electrons along (giving it almost an *electrical* quality), but that is precisely what we're about to see. Electron shifting takes place in what are called **redox reactions,** coupled reactions involving both oxidation and reduction. **Oxidation** is the removal of electrons from a

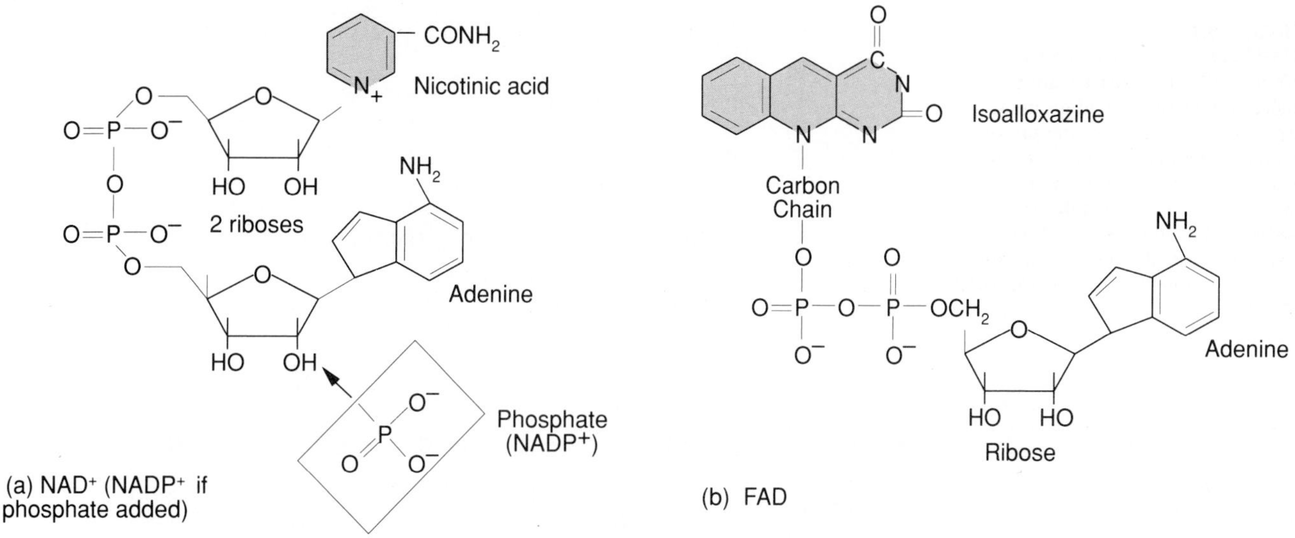

(a) NAD^+ ($NADP^+$ if phosphate added)

(b) FAD

FIGURE 6.14
COENZYMES
(a) The coenzymes NAD^+ and $NADP^+$ differ only in the presence of an added phosphate group in the latter. The nicotinic acid ring is emphasized because it is the group that enters into oxidation and reduction reactions. **(b)** The active group of FAD is an isoalloxazine.

substance, and **reduction** is the addition of electrons to a substance. When something is **oxidized** it loses electrons, and those electrons do not wander around alone for long. They quickly (almost instantaneously) join with another molecule, thereby reducing it. (Because oxidation and reduction are usually paired this way, the two are referred to as redox reactions.) Thus when something is oxidized, something else is reduced. We have already looked into reactions where electrons are lost or gained, such as those involving ionic bonding (see Chapter 2). So now we can review one of our examples using the new terminology.

In the formation of common table salt, atoms of sodium lose an electron to atoms of chlorine. In redox language, sodium is oxidized and chlorine is reduced. Further, because sodium is an electron donor, it can be thought of as a **reducing agent**—here, one that reduces chlorine. It follows, then, that chlorine, the electron receiver, is an **oxidizing agent**, one that accepts electrons from sodium. Chlorine, as you may be aware, is an excellent oxidizing agent, which is why it is effective as a laundry whitener and water purifier, where it removes electrons from stains and bacteria, respectively. In fact, chlorine is such a strong oxidizing agent that iron filings or steel wool will readily burn in its presence.

In some cases, particularly during chemical activity in cells, when electrons are removed during oxidation, protons tag along. Thus, *oxidation becomes the removal of hydrogen,* and *reduction becomes the addition of hydrogen.* (One proton plus one electron equals a hydrogen atom equivalent.) Thus, biologists often define oxidation as the removal of *electrons* or *hydrogen* and reduction as the addition of *electrons* or *hydrogen.* Let's turn to the role of oxidation and reduction in the use of fuels.

Redox reactions in cells often start with a fuel such as glucose. You may not have thought much about this, but gasoline and glucose have a lot in common. They are both fuels and both are loaded with carbon-hydrogen bonds. In fact, people often speak about "burning" foods in their bodies. But in spite of the obvious similarities between the two "fuels," cell respiration differs substantially from combustion, for which we can all be thankful. The main difference is in the speed with which the redox reactions proceed. The complete oxidation of glucose during cell respiration is gradual and highly controlled, utilizing many steps in lengthy metabolic pathways. In this way, much of the energy can be conserved and transferred to ATP, instead of being released as heat, as occurs in combustion.

Coenzymes in Action

Coenzymes have a critical role in the transfer of energy. They temporarily accept electrons (and protons) and then pass them along to other molecules. They thereby change the

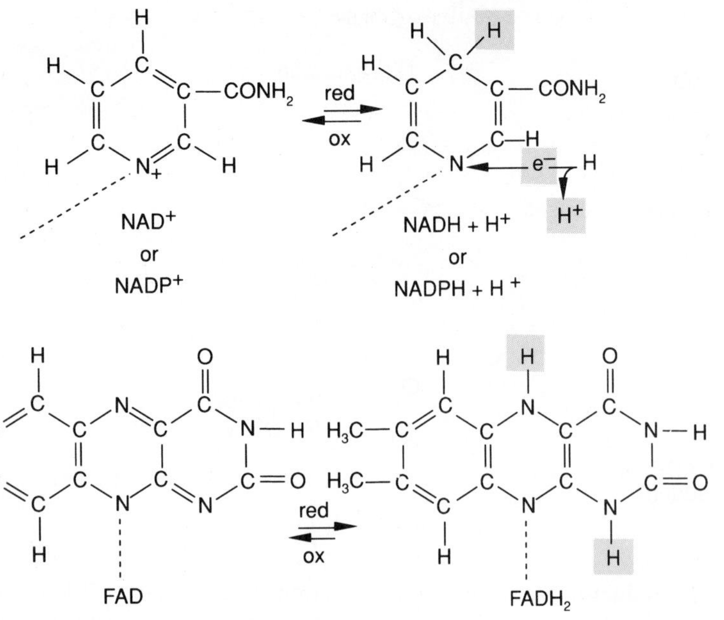

FIGURE 6.15
REDUCTION OF COENZYMES
When NAD^+ and $NADP^+$ are reduced, the nicotinic acid group accepts two electrons and one proton, with an additional proton entering from the surrounding medium. The two coenzymes are then designated as $NADH + H^+$ and $NADPH + H^+$. FAD, however, accepts two electrons and two protons (2H), thus becoming $FADH_2$.

energy levels of both the electron donors and recipients. Because of their role, coenzymes are often referred to as **electron carriers.**

At the pH levels normally found in the cell, oxidized NAD and NADP bear positive charges, and are generally designated as NAD^+ and $NADP^+$. Usually, both NAD^+ and $NADP^+$ can take on two electrons, one of which attracts a proton (H^+) from the surrounding medium, while the other neutralizes the positive charge. Thus, we usually show the reduced forms as NADH and NADPH. Since the reduction of the two coenzymes often involves *two* protons, as well as two electrons, some people prefer to write the reduced forms as $NADH + H^+$ and $NADPH + H^+$, to indicate that the second proton will be found in the surrounding medium (Figure 6.15).

What about FAD, you're probably anxiously muttering—what happens to FAD? You will be glad to hear that FAD doesn't give us the same problems since it readily takes on two electrons and two protons (the equivalent of two hydrogen atoms), so in its reduced state FAD is written $FADH_2$ (see Figure 6.15).

To return to the combined action of an enzyme, a coenzyme, and a substrate, we can symbolize a reaction in this manner:

$$\text{Substrate} + NAD^+ \xrightarrow{\text{Enzyme}} \text{Oxidized substrate} + NADH$$

The enzyme isn't changed in the oxidation reaction, so by tradition it is listed above the arrow.

It is important to note that the reduction of a coenzyme increases its free energy state. As a result, it becomes more reactive. Its usual reaction is to pass its electrons and hydrogens off to some molecule that is in a lower free energy state. The ability to reduce other substances is called **reducing power.**

Whereas NAD, NADP, and FAD are small, mobile carriers, we find electrons are also passed along by large, stationary carriers that are also important in the ATP generating processes.

Stationary Electron Carriers of Mitochondria and Chloroplasts

Stationary electron carriers are found in strategic locations in the membranes within mitochondria and chloroplasts. Actually, we find them in the *inner membrane* of the mitochondrion and in the *thylakoid membranes* of the chloroplast, structures that were discussed in Chapter 4. Most membrane-bound electron carriers are proteins with active prosthetic groups attached. The most common fall into a family of proteins called **cytochromes.** As in the case of the giant oxygen-carrying protein hemoglobin, the prosthetic group of the cytochrome is a heme group, that is, a ring of nitrogen with an iron atom at its center. As you might expect, it is the heme group that is specifically involved with redox reactions. As the iron of the heme group receives and passes electrons, it switches back and forth between two ionic states: Fe^{2+}, the reduced state, and Fe^{3+}, the oxidized state. (When it gains an electron its net positive charge is reduced from +3 to +2, and when it loses that electron it reverts back.) A row of cytochromes carrying out a sequence of redox events would look like this:

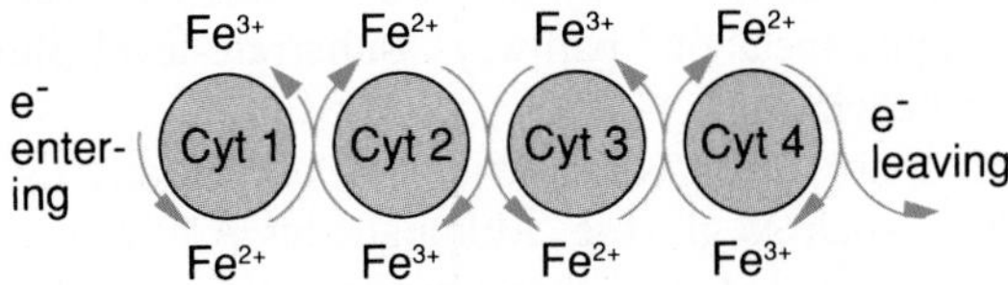

Imagine a series of such carriers arranged in such a manner that they run across a membrane and back (Figure 6.16). Then imagine that, beginning with the first, each carrier in the system, when reduced by an electron, has a slightly greater reducing power than the next in line. Thus carrier 1 can reduce carrier number 2, and number 2 can reduce carrier number 3, and so on through the system. If the first carrier is reduced by a mobile carrier with great reducing power, such as NADH or NADPH, the members can begin to pass electrons, like buckets of water in a bucket brigade, down the system. And "down" is the right word since the sequential action occurs in an energetically downhill direction. In any such system, the free energy of electrons being transported is depleted as they move through the carriers, and, in fact, it is through just such a simple system that ATP is synthesized.

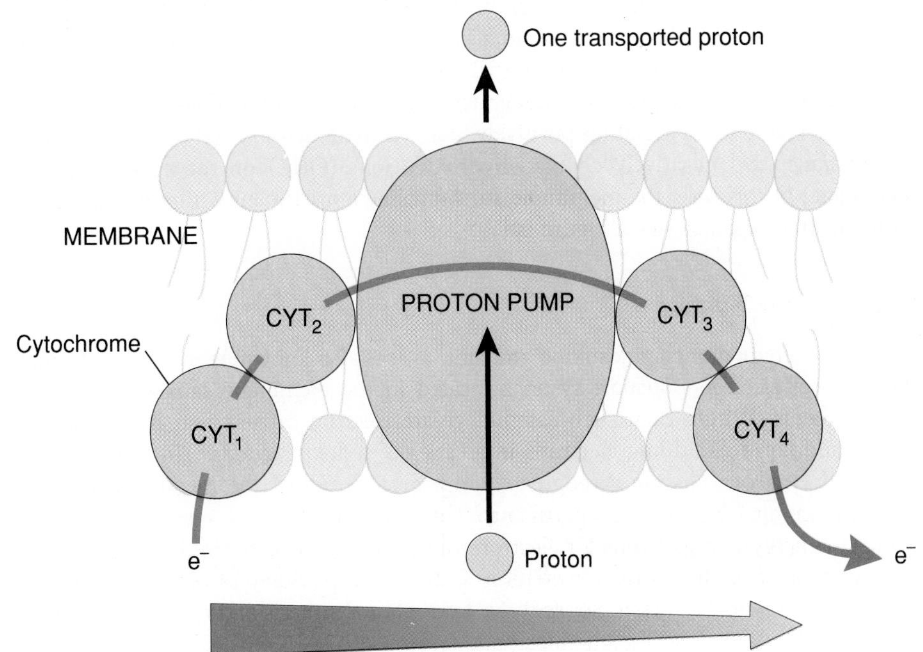

FIGURE 6.16
ELECTRON TRANSPORT SYSTEMS
Electron carriers (cytochromes) in electron transport systems form a sequence within membranes according to their reducing power (those with the greatest reducing power are shown to the left). Energized electrons passing through the electron transport system gradually lose free energy. The free energy is then made available to do work. In this instance the work is carried out by a proton pump, which uses the free energy to transport protons across the membrane.

We have been leading up to just how ATP is made for some time, but it is important that each aspect be in place. Now we can tie together the action of oxidizing and reducing enzymes, the mobile coenzymes, and the membrane-bound carriers of electrons and protons. We will bring these entities together in what is called the theory of chemiosmotic phosphorylation.

THE FORMATION OF ATP IN CELLS

As we consider the production of ATP let's first remind ourselves that ATP is a cycling molecule—that is, the phosphorylation of ADP forms ATP which, when used by the cell, is broken back down to ADP. Now we will see that ADP can be "recharged" to ATP through two distinct metabolic pathways: **substrate-level phosphorylation** and **chemiosmotic phosphorylation.**

Substrate-level phosphorylation is the generation of ATP directly from the chemical bonds of cellular fuels such as glucose. Although this ATP-generating mechanism is commonplace, it delivers far less ATP than chemiosmotic phosphorylation. Substrate-level phosphorylation is thought to be an ancient process, probably used by the earth's first heterotrophic organisms. We will return to the process in Chapter 8; here we will consider some of the principles of chemiosmotic phosphorylation.

Chemiosmotic Phosphorylation

The theory of chemiosmotic phosphorylation, or chemiosmosis as it is also known, was proposed by Peter Mitchell in 1961. Mitchell, it seems, had his own peculiar way of looking at things. While other biochemists were trying to understand ATP synthesis by analyzing each specific component in its manufacture, Mitchell took a broader view and looked at ATP production in terms of the intact chloroplasts and mitochondria. His first clue was a basic and well-substantiated observation: neither chloroplasts nor mitochondria can make ATP unless they are physically intact. From this and other lines of evidence, Mitchell devised the scheme that became known as chemiosmotic phosphorylation. It proposes that ATP synthesis in the mitochondrion (during cell respiration) and chloroplast (during photosynthesis) is dependent upon the creation of steep proton concentration gradients by isolating positively charged hydrogen ions (H^+) on one side of a membrane, and negatively charged hydroxide ions (OH^-) on the other side of a membrane. In this case, the membrane surrounds a compartment within a larger compartment (the organelle) (see Figure 6.17).

Charging the System

The proton gradient a chemiosmotic gradient, as it is also known, is produced through the work of electron transport systems located in the membrane separating the two compartments. Within each such assemblage are electron carriers that have a second role in addition to shuttling electrons in a series of redox reactions. They use the free energy of electrons passing through to pump protons across the membrane into the proton reservoir (the inner compartment). This is a critical step in chemiosmosis. The source of energy-rich electrons for electron transport systems is twofold. Energetic electrons originate from light-driven oxidations in the chloroplast and through the oxidation of fuels in the mitochondrion. So, as we've said, photosynthesis and respiration provide the energy for "charging" the chemiosmotic system.

The Proton Gradient: A System of Great Free Energy

Since an accumulation of H^+ and OH^- ions is involved, the chemiosmotic gradient can be characterized in several ways. It is a concentration gradient (so the ions have a strong tendency to diffuse). It is an electrical gradient (separated positive and negative charges).

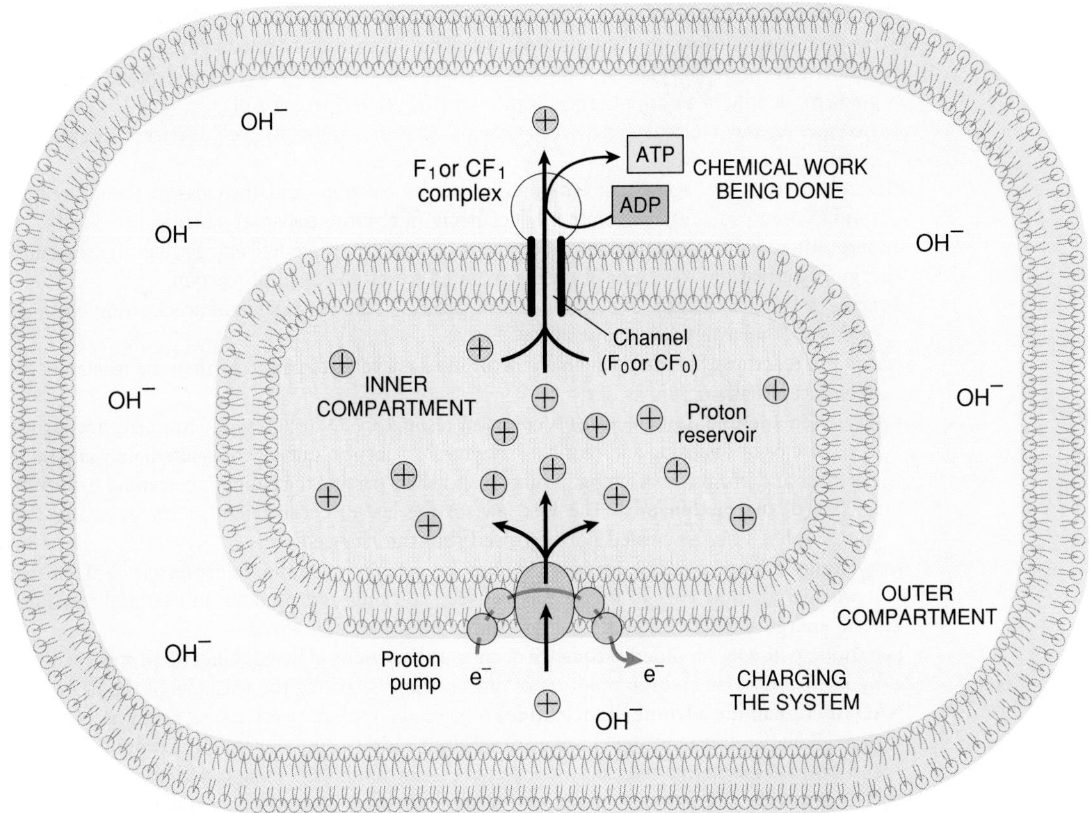

FIGURE 6.17
MODEL CHEMIOSMOTIC SYSTEM
The simplest model of a chemiosmotic system makes use of two membranous sacs, one within the other. Protons, pumped across the inner membrane by an electron transport system (ETS), accumulate within the inner compartment, charging the system. Their presence establishes a steep electrochemical gradient. Next, the protons pass down their gradient across the membrane via a channel (F_0 or CF_0) into an ATP synthesizing complex (F_1 or CF_1). There, ATP synthetase makes use of the free energy to do useful work—that is, to power the conversion of ADP to ATP.

And it is a pH gradient (the protons constituting an acid and the hydroxide ions in a base). Each of these represents a system of great free energy, and free energy can do work. The free energy of such proton gradients isn't imaginary—it has been measured. For example, the electrical potential across such membranes has been measured at 0.14 volts, quite sufficient to power the phosphorylation of ADP.

Needless to say, if the free energy were released in an uncontrolled manner, the result could be disastrous to the cell, perhaps even deadly. But suppose the protons were allowed to trickle down their gradient, the energy being released gradually. Then suppose that such an exergonic event were coupled to an endergonic one, a reaction in which phosphate joins ADP to form ATP. That's essentially what's going on.

The sac-within-a-sac arrangement in nature is represented by membranous compartments in chloroplasts and mitochondria. As you may recall (see Figure 4.19 and 4.20) both have an inner and an outer compartment, thus providing for a proton reservoir. The free energy of the proton gradient of these bodies is tapped by letting the protons escape down the steep concentration gradient and move out of the reservoir.

Tapping the Free Energy of the Gradient

The free energy of the system is released as the protons move down their gradient by passing through special channels that lead into bulbous enzyme complexes. We mentioned earlier that these are called the CF_0CF_1 complexes in the chloroplast and F_0F_1 complexes in the mitochondrion. The enzymes in the bulbous portion are ATP synthetases—those that use the free energy of the gradient in forming ATP. So the proton pump moves protons into the reservoir, and as they move outward through enzyme-laden complexes, their free energy is used to phosphorylate ADP, making ATP (see Figure 6.17).

ATP (adenosine triphosphate) A universal energy carrier molecule in cells, consisting of the nitrogen base adenine, the sugar ribose, and three inorganic phosphate groups.

chemiosmotic phosphorylation A process in which the free energy of a chemiosmotic, or proton gradient, is utilized in the phosphorylation of ADP, thus forming ATP.

coenzyme Organic molecules that act as enzyme cofactors; agents whose presence is essential for certain enzymes to work.

electron carrier Any agent that temporarily accepts electrons and then passes them along to another agent; such agents commonly occurring in electron transport systems.

endergonic reactions Reactions in which the products have more free energy than the reactants.

energy of activation The input of energy required to start a chemical reaction.

enzyme A cellular catalyst. Biologically active proteins that are capable of accelerating chemical reactions at relatively low temperatures.

exergonic reactions Reactions in which the products have less free energy than the reactants (the difference usually occurs as heat).

FAD (flavin adenine dinucleotide) A common respiratory and photosynthetic coenzyme, acting as a cofactor to oxidizing and reducing enzymes or electron carriers, thus alternately accepting electrons and protons (becoming reduced), and later, passing them along (becoming oxidized).

First law of thermodynamics The first law, or the law of conservation of energy, states that energy can neither be created nor destroyed (but can change form).

free energy Energy available for work (Gibbs free energy), commonly expressed as Δ G, that represents the difference in free energy between reactants and products in chemical reactions.

kinetic energy Energy of motion; energy being released or transferred.

metabolic pathway An orderly sequence of chemical reactions in living things; each step catalyzed by its own enzyme and the products of one reaction becoming the reactants of the next.

NAD (nicotinamide adenine dinucleotide) A common respiratory coenzyme, acting as a cofactor to oxidizing and to reducing enzymes or electron carriers, thus alternately accepting electrons and protons (becoming reduced), and passing them along (becoming oxidized).

NADP (nicotinamide adenine dinucleotide phosphate) A common photosynthetic coenzyme, acting as a cofactor to oxidizing and reducing enzymes or electron carriers, thus alternately accepting electrons and protons (becoming reduced), and passing them along (becoming oxidized).

oxidation The removal of electrons and/or protons from a substance by an oxidizing agent, with the substance thus being oxidized and the agent reduced.

potential energy Stored energy or energy at rest, not being expressed in motion.

reduction The addition of electrons and/or protons to a substance by a reducing agent, with the substance thus being reduced and the agent oxidized.

Second Law of Thermodynamics The second law, also called the law of entropy, states that transfer of energy leads to a more random organization or increase in disorder. Further, changes in energy from one form to another results in losses, generally as heat.

APPLICATION OF IDEAS

1. Many inventors have attempted to design "perpetual motion machines." All failed. The reasons for such consistent failure were not understood until the laws of thermodynamics were formulated. Explain how they apply to the problem.
2. When asked in an essay exam to define life, a thoughtful student once wrote an answer totaling six words. "Life is an interruption in entropy." Is this true? Using the laws of thermodynamics for background, comment on this statement.
3. Reviewing what has been presented so far on chemiosmosis in mitochondria and chloroplasts and the electrical differential generated across their membranes, develop an analogy using a rechargeable battery. How are they similar, and where does the analogy fail?

KEY IDEAS

ENERGY

1. **Energy** is often defined as the ability to do work, and work can be defined as the transfer of energy to a body through the application of force.
2. **Potential energy** is energy stored, or energy available for work; **kinetic energy** is energy in motion, or energy being transferred.

Forms of Energy

Energy can take various forms, including chemical bond, mechanical, magnetic, and electromagnetic energy—all of which are interchangeable.

The Laws of Thermodynamics

1. The behavior of all matter in the universe is described by the **laws of thermodynamics.**
2. The *First Law* (the Law of Conservation of Energy) states that energy can neither be created nor destroyed. The sum of energy in an **isolated system** remains the same, but energy can be transformed from one state to another.
3. The *Second Law* (Law of Entropy) states that the free energy of a system is always decreasing. Complex, organized systems tend to reach simplicity and randomness, and a measure of the disorganization and randomness is called **entropy**. Of great biological importance, transfers of energy are always accompanied by a loss of energy.
4. The molecular complexity of life on earth is possible only because of readily available light energy. Without it, the overall state of entropy would soon increase, and life would cease.

APPLYING ENERGY PRINCIPLES TO LIFE

The breaking of chemical bonds releases energy that is called **free energy,** meaning that it is available for work. Free energy changes, designated as ΔG, may be positive or negative.

Exergonic and Endergonic Reactions

Exergonic reactions are those in which the free energy of the product(s) is less than those of reactants (ΔG is negative). In **endergonic reactions** the free energy of the product(s) is greater than that of the reactants (ΔG is positive). Whereas, the breakdown of glucose in cells is exergonic, the manufacture of sugars is endergonic.

Chemical Reactions, Free Energy, and Equilibrium

1. A dynamic **equilibrium** is reached when the free energy of the products equals the free energy of the reactants. The reaction proceeds equally in both directions.
2. Strongly exergonic reactions will not reach equilibrium until most reactant has formed product. Weakly exergonic reactions reach equilibrium sooner. The strength of endergonic reactions depends upon the free energy of reactants and the outside energy source.
3. Exergonic reactions are often coupled with endergonic reactions, the first providing the free energy to drive the second.

Getting Reactions Started

1. The *Hindenburg* disaster illustrates the nature of an exergonic reaction—one in which energy is released. Hydrogen gas, in the presence of a spark, rapidly united with oxygen to form water, and both gases reached a lower energy state. The energy difference was the light and heat of the explosion.
2. At usual temperatures, molecules remain stable unless activated by energy. Such energy is called **energy of activation.** Such energy sets off a chain reaction whereby the energy of one reaction provides the heat of activation for another, etc.
3. **Catalysts** are agents that lower the required heat of activation; thus chemical reactions can occur at lower temperatures than is otherwise possible.

Cofactors (Vitamins and Minerals)

Enzyme function often requires bond or free agents called **cofactors**, commonly in the form of zinc, iron and copper. Cofactors may also be organic molecules called **coenzymes**, some of which are vitamins.

ENZYMES: BIOLOGICAL CATALYSTS

Enzymes are biological catalysts that (1) are proteins, (2) are highly specific for their **substrate**, (3) bring about reactions without the usual heat of activation requirement, (4) do not affect equilibrium or free energy changes, and (5) emerge unaltered from reactions.

The Enzyme-Substrate Complex: A Matter of Fit

1. Enzymes are usually large, globular, quaternary proteins with **active sites** that closely fit the shape of their substrate. When joined they form an **enzyme-substrate complex.**
2. Older ideas of enzyme and substrate shape have been replaced with the **induced fit hypothesis.** Rather than a perfect fit, the substrate fits imperfectly, thus assuming a stressed condition when in the active site.
3. Side groups of enzyme and substrate, along with water, interact chemically. The substrate emerges chemically altered, but the enzyme emerges fully restored.

Influences on the Rates of Enzyme Action

The rate of enzyme activity is strongly affected by several factors.

1. Increasing substrate concentration increases the rate of enzyme activity until saturation is reached, whereupon the rate stabilizes, determined by turnover time.

2. With few exceptions, most enzymatic reactions occur at a near neutral pH, and strong acids or bases slow reactions or denature the enzyme.
3. Heat increases the chances of collision between substrate and active sites, thus speeding up enzymatic reaction. At some point heat denatures the enzyme, and the reaction ceases.

Teams of Enzymes: Metabolic Pathways

Metabolic pathways include teams of enzymes that perform sequential reactions. The product of the first enzyme becomes the substrate of the second, and so on through the pathway.

Mechanisms of Enzyme Control

1. In some instances, hormones set enzyme activating mechanisms in motion.
2. **Inhibition** may be **competitive**, wherein substrate lookalikes block active sites. In **noncompetitive inhibition**, other sites become filled, altering the active site.
3. **Allosteric enzymes** are often first in metabolic pathways, where they are rate regulators. Allosteric enzymes have second sites, which when filled, may affect the active site. Some make use of stimulators that enable the enzyme to work, whereas others make use of inhibitors, whose presence in the **allosteric site** disables the active site.
4. Metabolic pathways may be controlled through **negative feedback inhibition,** wherein an end product of the pathway, when in excess, fills the allosteric site of the first enzyme, where it acts as an inhibitor.

ATP: THE ENERGY CURRENCY OF THE CELL

1. ATP is the universal energy carrier of the cell. ATP's available energy is modest but appropriate for the smallest cellular needs. Greater needs are met simply by reacting greater amounts of ATP.

The Molecular Structure of ATP

1. ATP consists of Adenine, ribose sugar, and three phosphates.
2. The phosphate bonds are readily hydrolyzed with a ΔG of -7.3 kcal/mole. The value of ATP energy is in its ready availability.
3. ATP reactions often end in phophorylation of some molecule, preparing it for further chemical activity.

The Cycling of ATP

1. The hydrolysis of ATP yields energy, ADP and P_i. The products are recycled to endergonic reactions that provide the energy for the resynthesis of ATP. ATP cycling is intense since it is used for most biochemical work in the cell.

THE COENZYMES AND OTHER ELECTRON CARRIERS

1. **Coenzymes** are small, nonprotein molecules that work in cooperation with enzymes.
2. The most familiar coenzymes are **NAD**, **NADP**, and **FAD**, which operate in oxidation-reduction reactions, accepting or transferring electrons and protons (hydrogen ions). They are derived from vitamins—**nicotinic acid** from vitamin B_6 and riboflavin from vitamin B_2.

Redox Reactions: Oxidation and Reduction

1. Electrons are passed from molecule to molecule through redox reactions—coupled oxidation and reduction reactions.
2. **Reduction** involves the loss of electrons and **oxidation** is the gain of electrons, thus when sodium and chlorine react, forming NaCl, chlorine is reduced (gains an e_-) and sodium is oxidized (loses an e_-). Sodium is then a reducing agent and chlorine is an oxidizing agent.
3. Oxidation and reduction can involve protons as well as electrons.
4. Oxidation occurs as glucose gives up its energy to the cell.

Coenzymes in Action

1. When reduced, coenzymes may in turn pass their electrons or hydrogens to other coenzymes, thus called **electron** or hydrogen **carriers.**
2. The coenzymes in their oxidized and reduced forms, respectively, are:
   a. $NAD^+ \rightarrow NADH + H^+$
   b. $NADP^+ \rightarrow NADPH + H^+$
   c. $FAD \rightarrow FADH_2$
3. A reduced coenzyme is at a greater free energy level than an oxidized coenzyme and is said to have **reducing power.**
4. Although the coenzymes above are often mobile, many electron-hydrogen carriers are stationary, or fixed in place, usually within membranes such as those of the thylakoid and the cristae of the mitochondrion.
5. Stationary carriers include **cytochromes**—protein complexes and coenzymes.
6. In their passage through the carriers of electron transport systems, electrons gradually lose free energy, and this energy is harnessed in useful ways.

THE FORMATION OF ATP IN CELLS

ATP is recharged through substrate-level phosphorylation (the generation of ATP directly from the chemical bond energy of fuel substrate) and chemiosmotic phosphorylation.

Chemiosmotic Phosphorylation

For chemiosmotic phosphorylation to occur, chloroplasts and mitochondria must be intact. The process requires the presence of a high concentration of protons (H^+) (a proton gradient) separated by a membrane from a hydroxide ion (OH^-) concentration.

Charging the System

The proton, or chemiosmotic, gradient is produced by

membranal proton pumps powered by the free energy of electrons. Such energy is made available to the pumps as the electrons pass through electron transport systems.

The Proton Gradient: A System of Great Free Energy

1. Proton (chemiosmotic) gradients, systems of great free energy, can be thought of as concentration gradients, electrical gardients, or pH gradients. Their energy is used to phosphorylate ADP forming ATP.
2. The sac-within-a-sac chemiosmotic system is naturally formed by the complex membranes of mitochondria and chloroplasts.

Tapping the Free Energy of the Gradient

The free energy of the chemiosmotic gradient is tapped as protons pass down their gradient through CF_0 or F_0 channels into CF_1 or F_1 complexes. There, phosphorylating enzymes use the free energy of the system to form ATP from ADP and P_i.

REVIEW QUESTIONS

1. Using the terms "work" and "force," define energy.
2. Distinguish between the terms potential, kinetic, and free energy. (130-131)
3. What are some of the forms in which energy is expressed? Cite real-life examples of energy going through several conversions. (130)
4. Using the First and Second Laws, explain why life is improbable in an isolated system. (130)
5. Distinguish between exergonic and endergonic reactions and list an example of each. (132)
6. What factors determine when chemical equilibrium is reached during a reaction? (133)
7. What is the value of coupled reactions? (133)
8. What does *energy of activation* have to do with chemical reactions? How does a catalyst affect this requirement? (134-135)
9. List three important characteristics of enzymes. (135)
10. Summarize the events within the enzyme-substrate complex when a hydrolytic reaction occurs. (136)
11. Compare the "lock and key" analogy with the "induced fit hypothesis" of enzyme action. (135)
12. Summarize the effect of heat on enzymatic reactions. What rule applies, and what are the limits? (137)
13. Using the allosteric site as an example, explain how negative feedback operates. Why is such an automated process valuable and necessary? (139)
14. Carefully describe the structure of ATP and point out specifically why their structure makes these molecules useful energy carriers (140-141).
15. Describe a phosphorylation reaction. Why are these used by the cell? (141)
16. Draw an ATP-ADP cycle, and explain what is happening. (141-142)
17. Carefully define oxidation and reduction and explain the role of a coenzyme in these processes. (142-143)
18. List the three common mobile coenzymes of cells and write their formulas in the oxidized and reduced states. (143)
19. What are "stationary carriers"? What do they carry and what purpose do they serve? (145)
20. What is an electron transport system? What function do such systems serve in the mitochondrion and chloroplast? (145)
21. What is a chemiosmotic differential, and why is it considered both electrical and chemical? (146-147)
22. Suggest how a proton gradient is used in the synthesis of ATP. (147)

7

PHOTOSYNTHESIS

OUR FREQUENT BANTERING ABOUT THE MERITS OF SOLAR ENERGY HAS AN almost touching naivete about it. We seem to treat solar energy as a recent idea, a new concept. We seem to forget that the sun is an ageless source of energy. The fossil fuels—oil, gas, and coal—are simply releasing solar energy stored away in the bodies of long-dead plants and algae. So while we continue to wrestle with the maze of engineering problems associated with harnessing the sun's energy to produce electricity, perhaps we should turn to the real experts for some ideas. And the real experts are likely to be green (Figure 7.1).

It is no secret that plants get their energy from sunlight. But it's less well known that plants long ago evolved the ability to convert **photons** of radiant energy into *electrical* energy and ultimately into *chemical* energy. (Light energy is often described in terms of photons—discrete packets of energy traveling in a wavelike manner.) Light is captured by those tiny, highly organized bodies called chloroplasts. **Chloroplasts** convert sunlight energy into chemical bond energy by first passing this energy to electrons, which, in a new high energy state, move along pathways known as **electron transport systems.** This flow of electrons has many of the characteristics of a current of electricity moving along a conductor. We will return to this point shortly, but first let's step back for a broader look at the process of photosynthesis.

PHOTOSYNTHESIS: AN OVERVIEW

Photosynthesis is the process by which the energy of sunlight is used to bond certain molecules together to produce carbohydrates, commonly, the simple sugar glucose. The raw materials of photosynthesis are carbon dioxide and water. The products are glucose, water, and oxygen. Thus, the reactions are highly endergonic. The process can be expressed in the general equation:

$$6CO_2 + 12H_2O \rightarrow C_6H_{12}O_6 + 6O_2 + 6H_2O\ (\Delta G + 686\ \text{kcal/mole})$$

The reaction, translated, simply tells us that water and carbon dioxide, in the presence of light and chlorophyll, are used to produce glucose, water, and oxygen, with a large gain in free energy.

Obviously, the venerable old equation hides a vast amount of detail. For example, the equation doesn't indicate how something as ethereal as sunlight could possibly power the formation of chemical bonds, nor does it suggest that certain wavelengths of light are more effective than others. There is no suggestion that the water at the right side of the equation is not the same water as that at the left, or that the oxygen at the right comes from the water at the left rather than the carbon dioxide. Finally, the equation hides the fact that before glucose can be synthesized, the energy of light is used to form such energetic substances as ATP and NADPH—which are not even mentioned in the formula. So let's tie up some of these loose ends. Much of what we will consider here represents hard-won information from discoveries made mostly over the past 60 years, and a few from earlier periods. For the historical highlights, see Essay 7.1.

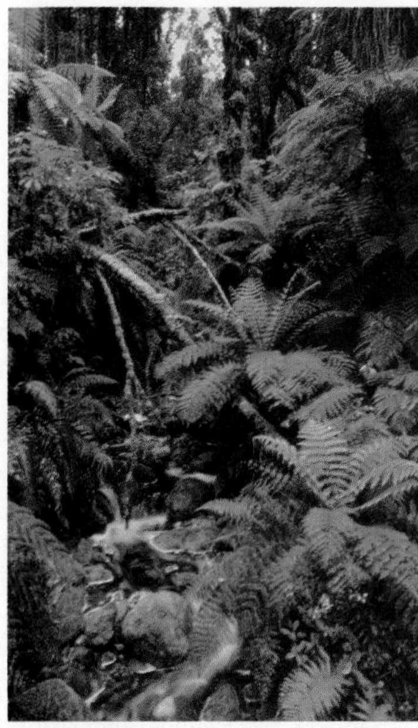

FIGURE 7.1
CAPTURING SUNLIGHT
Plants are specialists in the capture and utilization of solar energy.

The Two Parts of Photosynthesis

The events of photosynthesis can be divided into two parts, those that require light (the **light reactions**) and those that do not: (the **light-independent reactions**). In a sense, the first "charges" the system, and the second lets it run back down (Figure 7.2). As we have emphasized, a system of free energy, when running down, can accomplish a great deal of useful work.

In the light reactions, sunlight energy is used in two ways: to convert ADP to ATP and to reduce $NADP^+$ to NADPH. Both substances will be needed in the light-independent reactions—ATP for its readily available energy and phosphate, and NADPH for its reducing power and hydrogen. The light reaction reads:

$$ADP + P_i + NADP^+ \xrightarrow[\text{chlorophyll}]{\text{light energy}} ATP + NADPH + H^+$$

(Note that we have not attempted to balance the equation. Chlorophyll is the light-absorbing pigment.)

The products of the light reaction become reactants in the light-independent reaction, to yield overall:

$$CO_2 + H_2O + ATP + NADPH + H^+ \xrightarrow{\text{Enzymes}} C_6H_{12}O_6 + ADP + P_i + NADP^+$$

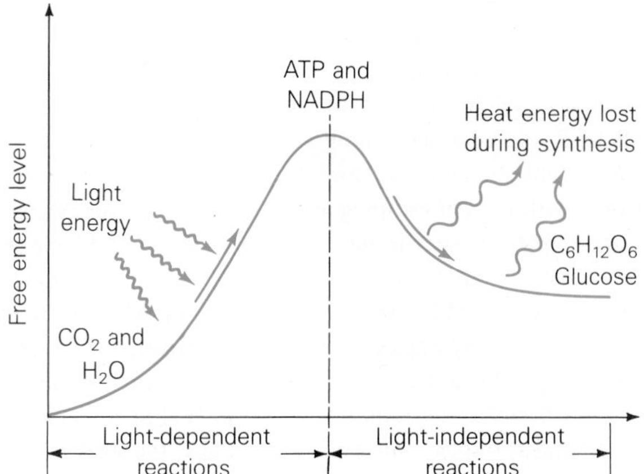

FIGURE 7.2
CHARGING THE SYSTEM AND TAPPING THE ENERGY
The light reactions of photosynthesis charge the system, using light energy to increase its free energy. Then, by allowing the system to run down in the light-independent reactions, the free energy is used in the synthesis of useful products.

ESSAY 7.1

FROM WILLOWS, MICE, AND CANDLES TO ELECTRON MICROSCOPES

Prior to the time of Jan Baptiste van Helmont, a Belgian physician of the 17th century, it was commonly accepted that plants derived their matter from materials in the soil. (Probably, many people who haven't studied photosynthesis would go along with this today.) We aren't sure why, but van Helmont decided to test the idea. He carefully stripped a young willow sapling of all surrounding soil, weighed it, and planted it in a tub of soil that had also been carefully weighed. After five years of diligent watering (with rain water), van Helmont removed the greatly enlarged willow and again stripped away the soil and weighed it. the young tree had gained 164 pounds. Upon weighing the soil, van Helmont was amazed to learn that it had lost only 2 ounces. Van Helmont's conclusion was inescapable: "It's the water!" he cried, but he was only half right. (Carbon dioxide hadn't been discovered yet.)

The next investigative breakthrough occurred late in the eighteenth century, around 1771, when British clergyman Joseph Priestley made an astute observation. In a simple but classic series of experiments, Priestley learned that plants growing in a sealed container would support the burning of a candle and breathing of a mouse longer than either would go on without the plant present. What Priestley had found, in addition to the reciprocity of gases in plants and animals, was that plants produce oxygen.

Priestley's discovery was followed almost immediately by the ambitious effort of a Dutch engineer named Jan Ingenhousz. Ingenhousz carried out some 500 experiments (some say, all within one year), which supported the conclusion that light was essential for photosynthesis. Ingenhousz determined that plants gave off oxygen only in the light and that under darkened conditions, they even produced some carbon dioxide. By the mid-1800s, the basic equation for photosynthesis was well established. Toward the end of the 1800s, Theodore Engleman determined that the chloroplasts were the sites of photosynthesis. He observed the peculiar manner in which bacteria clustered about the ribbon-shaped, oxygen-evolving chloroplasts of certain filamentous algae.

Many of the mysteries in the biochemistry of photosynthesis were resolved in the 1930s, particularly through the efforts of C.B. van Niel. In van Niel's time, it was held that during photosynthesis, light energy broke down carbon dioxide, with oxygen escaping as a gas and carbon joining water to form CH_2O—the empirical carbohydrate. Van Niel knew from his studies of sulfur bacteria that photosynthesis could occur without the production of oxygen. In sulfur bacteria, the source of hydrogen is hydrogen sulfide, and when oxidized, the byproduct is elemental sulfur rather than oxygen. This fact suggested to van Niel that the source of oxygen released during plant photosynthesis was water rather than carbon dioxide. He proposed a general equation for all photosynthesis:

$$CO_2 + 2H_2A \rightarrow (CH_2O) + H_2O + 2A$$

In this equation the H_2A represents the source of electrons, which could be either hydrogen sulfide, water, or some other similar substance—but not carbon dioxide.

While van Niel's arguments were certainly persuasive and logical, persuasion and logic are not proofs. There was, at the time, no way of knowing whether bacterial photosynthesis was similar enough to photosynthesis in algae and plants for van Niel to be certain of his hypothesis. In fact, the final confirmation had to await the development of new research techniques, which in this case turned out to be the use of isotopes of oxygen and carbon, and ways of detecting their presence.

In the early 1940s, Samuel Ruben and Martin Kamen made use of a heavier isotope of oxygen, ^{18}O, whose presence in molecules produced during photosynthesis could be detected by a device called mass spectrophotometer. In one experiment the two researchers raised algae in ordinary carbon dioxide but used water containing the heavy isotope of oxygen ($H_2{}^{18}O$). They found that the oxygen gas given off contained the heavy isotope of oxygen, whereas the carbohydrate formed did not. In a second experiment, ordinary water was used, but the carbon dioxide gas contained the heavy isotope of oxygen ($C^{18}O_2$). In this case the oxygen gas given off was free of the heavy isotope of oxygen, but it was subsequently found to be present in the carbohydrate product of photosynthesis. Van Niel's hypothesis was strongly supported.

The biochemical pathway incorporating carbon dioxide and producing glucose was discovered in the 1940s by Melvin Calvin and his associates. They made use of radioactive isotopes of elements involved in photosynthesis. Radioactive carbon ($^{14}CO_2$) was administered to green algae, which were then exposed to bright light for a few seconds. The algae were quickly killed (to stop any biochemical reactions), and the cells disrupted and searched for new, radioactive, molecular intermediates. Calvin found the radioactive carbon incorporated into 3-phosphoglycerate (3-PG), which turned out to be the first stable intermediate in the pathway. Continued research eventually determined the other intermediates of the carbon-fixing pathway, including the key carboxylating enzyme Rubisco. Calvin and his associates went on to propose a cyclic pathway leading to the formation of glucose, soon dubbed the "Calvin cycle."

In the late 1930s, British biochemist Robert Hill provided strong experimental evidence that also supported van Niel's theorizing. Hill was among the first to isolate successfully and experiment with chloroplasts. From his experiments it was determined that oxygen production in chloroplasts could occur independently of carbon dioxide fixation. In fact, it occurred in the absence of carbon dioxide. All that was required was the presence of an electron acceptor. From these observations, it became clear that photosynthesis occurs in two separate series of reactions: the light reactions involving water and the light-independent reactions involving carbon dioxide fixation. From Hill's discoveries, the experimental focus turned to the role of light in creating a flow of electrons and the pathways taken by the electons. By 1951, the role of $NADP^+$ (first called TPN^+) as an electron acceptor was known. In 1954, Daniel Arnon was able to demonstrate the entire photosynthetic process in isolated chloroplasts. His accomplishment was a milestone in the saga of research in photosynthesis.

Arnon and his associates at the University of California at Berkeley, taking their clues from Hill's brilliant work, used intact chloroplasts as their "experimental organism." Much of the essential material in this chapter has come down from Arnon's initial discoveries. For example, in an early experiment, he determined that if carbon dioxide was withheld, chloroplasts could carry out all of the known photosynthetic reactions except carbohydrate synthesis. In other words, the process yielded only ATP, $NADPH + H^+$, and oxygen—but no glucose. From this observation, Arnon was able to propose that photosynthesis occurred in two distinct phases, the light and "dark" reactions.

Arnon then determined that if he withheld both carbon dioxide and NADP, providing his chloroplasts with ADP, P_i, water, and light only,then the only photosynthetic product would be ATP. There was, as expected, no glucose, no $NADPH + H^+$, and, perhaps, surprisingly, *no oxygen*. The lack of oxygen was significant in that it indicated that a simpler, cyclic process may be going on in the absence of NADP that could yield ATP. We now call this independent process the cyclic light reaction.

In a final experiment, elegant in its conceptual simplicity (although technically very difficult), Arnon placed his chloroplast suspension in the dark for a time, washed away any late-forming products, and waited for synthesis to stop. He then supplied the chloroplasts with CO_2 along with the products normally produced in the light-dependent reactions: ATP and $NADPH + H^+$. Once again, the amazing little organelles became active and went on to produce carbohydrate. As you can see, we owe much of what we understand about photosynthesis to the efforts of Robert Hill and Daniel Arnon and his associates.

Research in photosynthesis has been consistently intense over the years, but recently the interest has moved from the specific chemical reactions to the structures in which they occur. Motivating the switch has been the chemiosmotic theory of Peter Mitchell, the British biochemist. The chemiosmotic theory has become well entrenched in both photosynthetic and respiratory biochemistry today.

Recent studies of the thylakoid membrane, using freeze-fracture techniques, reveal details that help reinforce the chemiosmotic hypothesis. As Mitchell's hypothesis predicted, the CF_0CF_1 complexes are structurally independent of all other membrane proteins. Mitchell had predicted that phosphorylation was a separate process from the photoactivation of electrons and their subsequent transport through electron transport systems. Today's efforts have also revealed an elaborate protein array in the thylakoid membranes, and all biologists are almost certain that they have identified specific proteins and other elements of photosystem I and photosystem II. A recent article by Jan M. Anderson in *Annual Review of Plant Physiology* (vol. 37, 1986) suggests the organization of various elements as seen in the illustration.

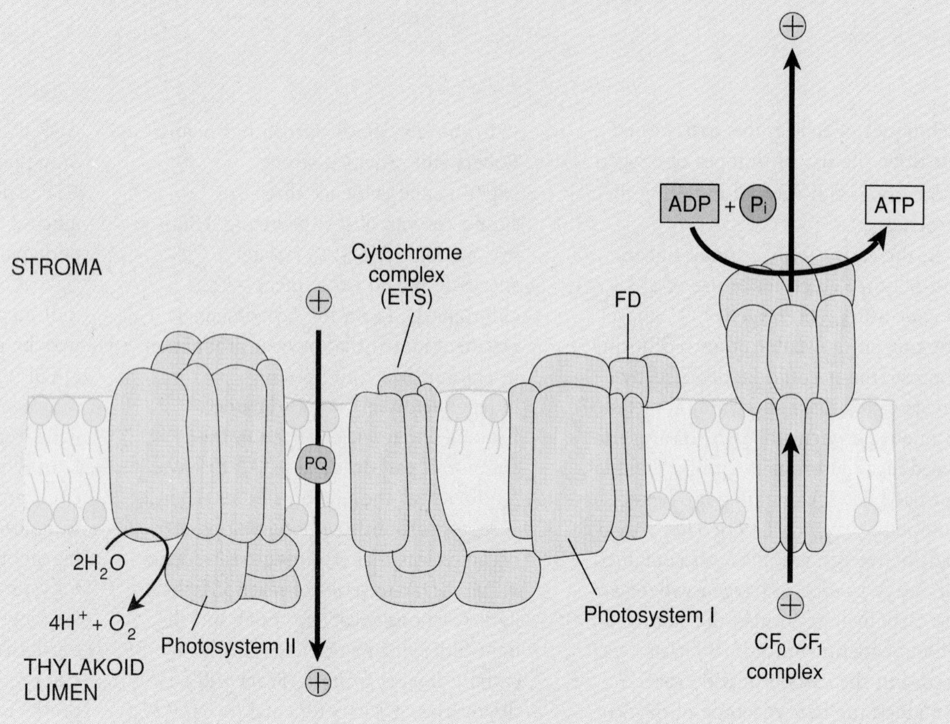

As we can see, after ATP and NADPH make their contributions to glucose-building, they are returned to their original condition (and recycled). We will see some of the many reactions involved in all this shortly.

Now that we are aware of some of the overall reactions of photosynthesis, let's turn to the structures in which the process occurs. Modern research in photosynthesis is focused on the cell, and we are are beginning to see clear associations between biochemistry and structure. The structure of concern here is the chloroplast, with its intricate membranous inner structures, the **thylakoids.**

CHLOROPLASTS

We first considered the organization of chloroplasts in Chapter 4, where we saw that they are relatively large organelles surrounded by a double membrane. Within the chloroplast is a watery **stroma** interrupted by an extensive third membrane system whose foldings form individual closed vesicles or sacs called thylakoids. Where the foldings are extensive, they form stacks made up of **granum thylakoids**, each stack referred to as a **granum**. Here and there, the membranes of granum thylakoids extend out across the stroma and join with other granum thylakoids. These interconnections are called **stroma thylakoids** (Figure 7.3). The thylakoids are at the very heart of photosynthesis, since it is within their membranes that the light reactions—the capture and harnessing of light energy—are carried out.

The Thylakoid

Both the stacked granum thylakoids and the stroma thylakoids contain an extensive, and apparently continuous, channel-like interior known as the **lumen** (Figure 7.4). Since the lumen and the stroma are separated by a membrane, the two are, in effect, separate watery compartments. They form the all important "sack-within-a-sack" arrangement necessary for chemiosmotic phosphorylation. Recall that both chloroplasts and mitochondria utilize this mechanism of ATP formation (see Chapter 6).

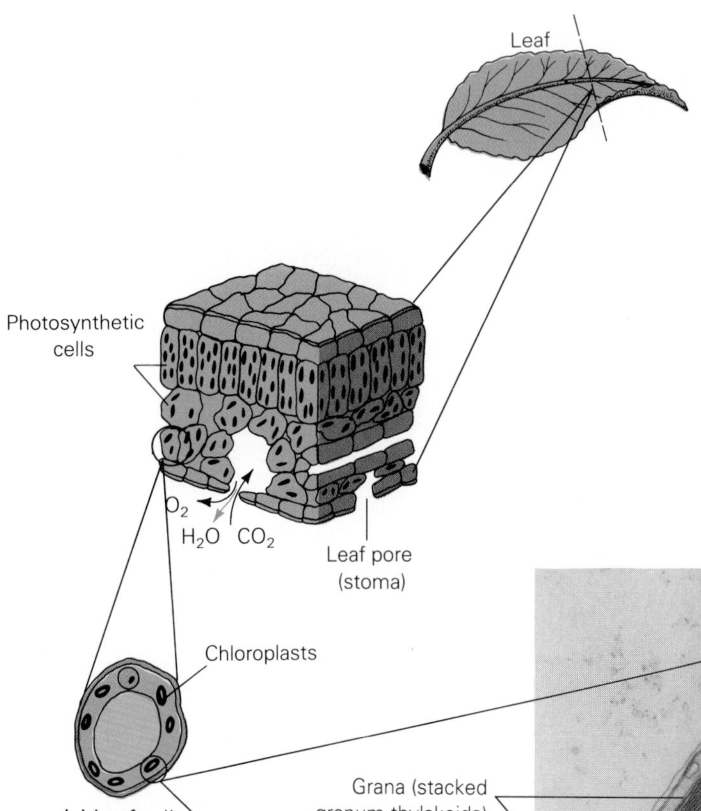

Leaf

Photosynthetic cells

O_2

H_2O CO_2

Leaf pore (stoma)

Chloroplasts

(a) Leaf cell

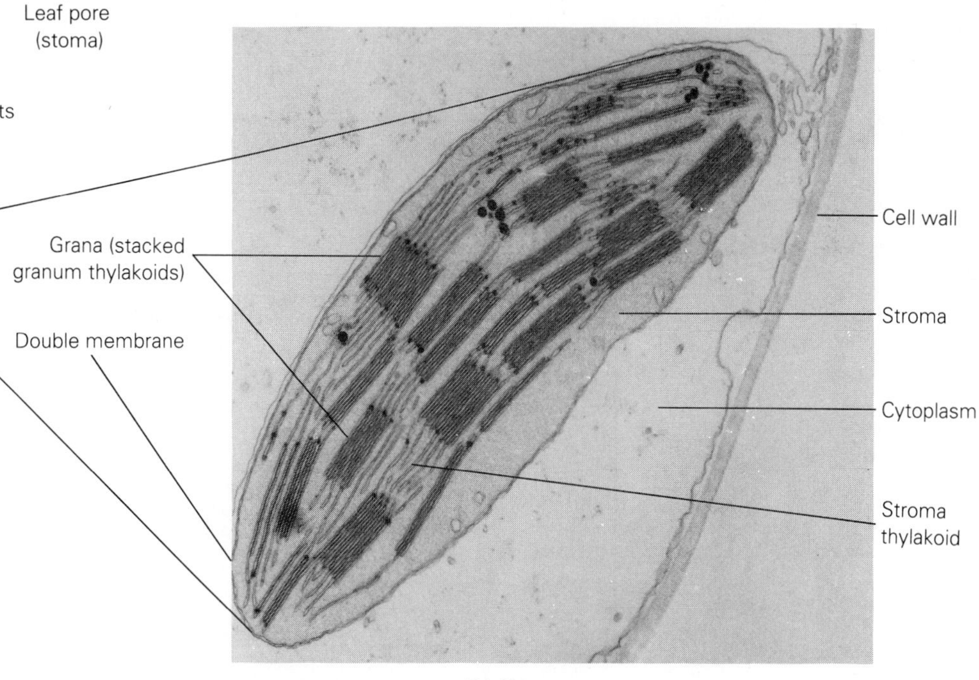

FIGURE 7.3
ORGANIZATION OF THE CHLOROPLAST
The machinery of photosynthesis can be viewed by venturing through several levels of organization. (**a**) The leaf, a photosynthetic structure, contains great numbers of cells, many of which contain chloroplasts. (**b**) Each chloroplast concentrates its photosynthetic pigments in its thylakoid membrane system. (**c**) Through an intricate folding, this membrane system forms grana, stacks of granum thylakoids. Extensions of granum thylakoids form simpler stroma thylakoids.

Grana (stacked granum thylakoids)

Double membrane

Cell wall

Stroma

Cytoplasm

Stroma thylakoid

(b) Chloroplast

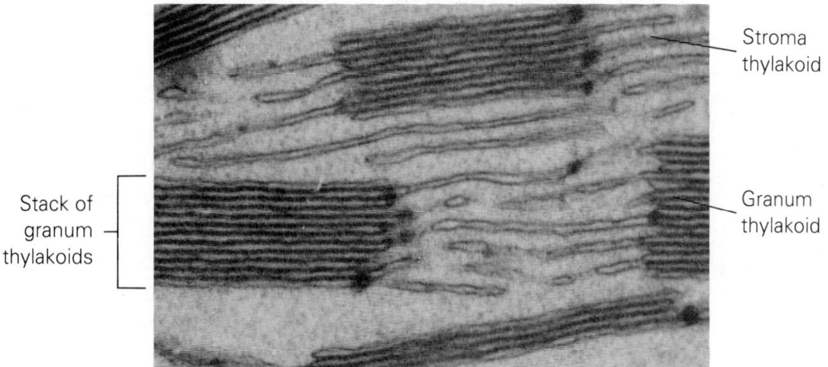

(c) Closeup of grana

FIGURE 7.4
THYLAKOIDS
(a) An exploded 3-dimensional view of the thylakoid membrane system reveals some of its details. Through an intensive folding, the membrane forms stacks of granum thylakoids in some places. Each thylakoid contains a fluid-filled interior or lumen. The lumen of one granum thylakoid may be continuous with the one above or below, and some extend to the next granum via tubular stroma thylakoids. **(b)** A model of the thylakoid membrane system emphasizes the numerous foldings making up the granum thylakoids. Many photosystem I and II complexes (the sites of the light reactions) are integrated into the membrane, as are a number of CF_0CF_1 complexes (the sites of ATP synthesis).

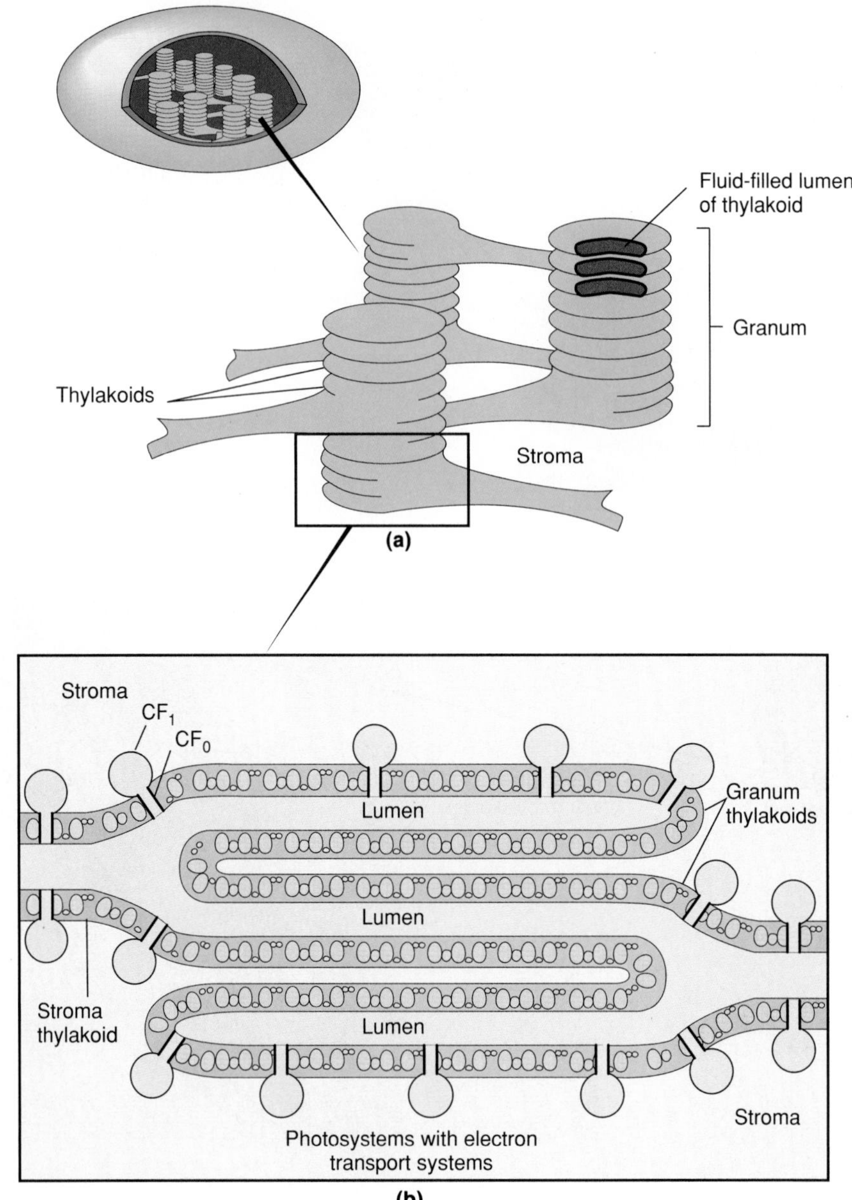

The Two Photosystems The special light-capturing and -harnessing structures of the thylakoid membrane are found in two complex assemblages designated **photosystem I** and **photosystem II (PS I** and **PS II**). Each photosystem includes a **light-harvesting antenna** and **reaction center** (Figure 7.5), and each is closely associated with an electron transport system.

Chlorophyll and the Light-Harvesting Antennas Light-harvesting antennas intercept light and concentrate its energy. They consist of three kinds of pigments, **chlorophyll *a*, chlorophyll *b*,** and a number of **carotenoids.** Although the two chlorophylls are quite similar in their chemical structure, they absorb different light wavelengths. Together they manage to cover the two ends of the visible light spectrum, orange-reds (600–700 nm) and purple-blues (400–500 nm). The major carotenoids of plants, called accessory pigments, absorb not only blue light, but some green as well. Not much green is absorbed, however, and it is this reflected color that we see in the plants that grace our lives. Actually, the relative amounts and kinds of light-absorbing pigments differ among the photosynthetic organisms, providing a wide range of colors in plants and especially in algae. Essay 7.2 presents more on the light-absorbing pigments.

ESSAY 7.2

THE VISIBLE LIGHT SPECTRUM, PIGMENTS, AND PHOTOSYNTHESIS

The earth is constantly bathed in radiation emanating not only from the sun, but from a host of other celestial bodies. Part of the radiation that reaches us is **visible light.** Visible light, however, is only part of an **electromagnetic spectrum** that includes (from lower to higher energy) radio waves, microwaves, infrared radiation, visible light, ultraviolet (UV) radiation, X rays, and gamma rays.

At the low-energy end of the spectrum are very long waves, while at the other extreme are very short, highly energetic waves. Near the middle of this continuum is visible light. Visible light is visible because it interacts with special pigments such as chlorophyll, the molecule that absorbs the energy of light and provides power for photosynthesis.

Light is very hard to describe in nontechnical terms. One reason is that it can be considered in two ways: as particles called photons, or as waves. Arguments have raged over these two concepts for years but today most scientists agree that light is composed of photons that move like waves. This enables us to describe the energy of a photon in terms of its wavelength.

The specific light-absorbing qualities of the chlorophylls and carotenoids can be determined by using a device known as a spectrophotometer. First, the pigments are extracted and dissolved in a solution. Next, light of a known wavelength is passed through the solution, and whatever light is not absorbed is detected on the other side. The wavelength on the entering light can be varied to see which wavelength is most absorbed by the solution. Finally, the data are usually plotted on a graph to form what is called an **absorption spectrum.** Note the absorption spectrum for chlorophylls *a* and *b* on the graph shown here. The peaks represent light that is absorbed by the pigment, while the valleys represent light that passes through the solution. As you can see, the green and yellow hues are least absorbed and are transmitted or reflected, although the yellows are often masked by the darker green color. Thus, these are the colors we see when we look at a chlorophyll solution—or when we look around us at the greenery we treasure.

Certain wavelengths, such as violet-blue and orange-red, are strongly absorbed by chlorophyll. This indicates that these wavelengths are used in

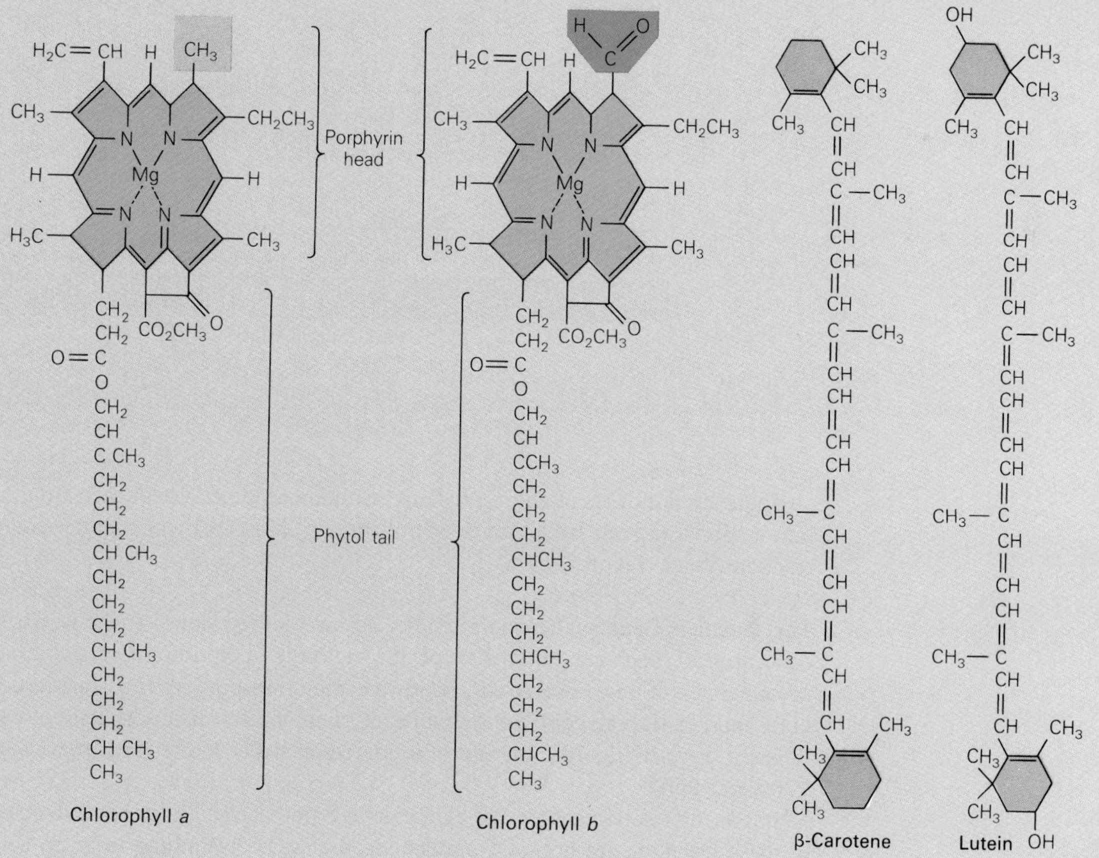

Porphyrin head

Phytol tail

Chlorophyll *a*

Chlorophyll *b*

β-Carotene

Lutein

photosynthesis, but the evidence is circumstantial. It is possible, however, to obtain more direct kinds of evidence. One way is to discover the rate at which some product of photosynthesis is produced when groups of plants are subjected to different wavelengths of visible light. Plants produce oxygen gas at a rate proportional to the rate of photosynthesis: for every glucose produced, six oxygen molecules are released. Measuring the volume of oxygen gas produced under varied wavelengths can determine which wavelengths are most effective in photosynthesis.

With these data we can plot what is known as an **action spectrum** (as shown here). Action spectra for the chlorophylls turn out to be very similar to absorption spectra, indicating that the wavelengths of light absorbed by chlorophyll are the wavelengths that drive photosynthesis.

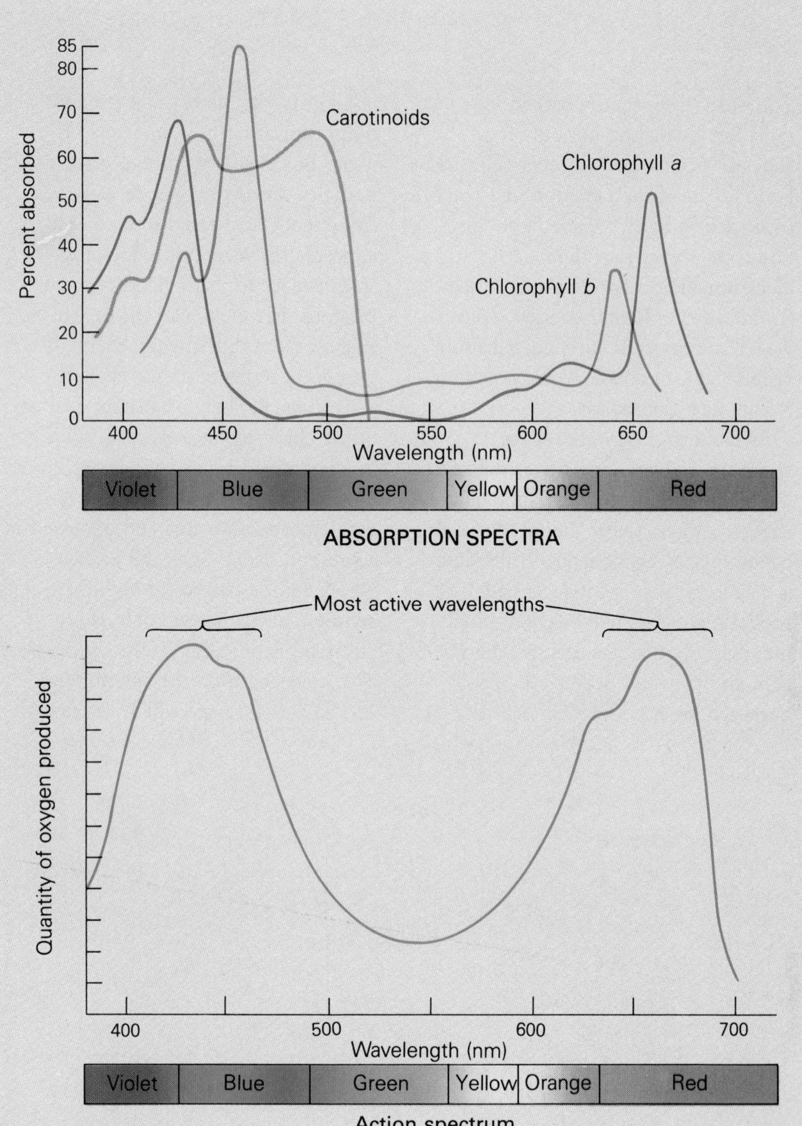

ABSORPTION SPECTRA

Action spectrum

There are hundreds of chlorophyll and carotenoid molecules in each antenna, bound into a special pattern by a framework of protein. Most of them channel energy to the reaction center (Figure 7.5).

The Reaction Center Reaction centers act as "energy sinks" (regions where energy concentrates). Each center consists of one molecule of chlorophyll *a* and an associated protein that is closely linked to the nearby electron transport system. The reaction centers of PS I and PS II, both containing a molecule of chlorophyll *a*, absorb light of a somewhat different wavelength, 700 nm and 680 nm, respectively so they are often referred to as **P700** and **P680**.

It is in the reaction center that light energy is transformed into chemical energy. Many of these reactions are not well understood, but we're beginning to grasp some of the goings-on. When chlorophyll *a* of the reaction center absorbs sufficient energy from the surrounding antenna, an electron from this molecule becomes excited and escapes from its orbital. This is where the transformation of sunlight energy to chemical energy begins.

In the everyday world of the molecule, the ejection of an energized electron is not an unusual event; it happens routinely as the free energy of a molecule changes. Usually, though, an ejected electron simply gives off its absorbed energy and falls back to its lower energy "ground state." When an excited electron of chlorophyll *a* leaves its orbit, however, it does not at once release its newfound energy, nor does it fall back to its ground state—not yet. *Instead, the activated electron is immediately captured by the closely associated electron transport system,* where its new free energy can be put to use doing chemical work. Herein lies the specialness of photosynthesis.

Actually, one photoactivated electron cannot accomplish much photosynthetic work. But considering the lightning speed of such an event, the many photosystems at work in each thylakoid, the great number of thylakoids in a chloroplast, and the number of chloroplasts in a leaf cell, we begin to develop a great respect for what's going on in those unapplauded green leaves.

Electron Transport System Electron transport systems consist of a number of electron carriers precisely arranged in the thylakoid membrane (see Figure 7.5 and Essay 7.1). These assemblages have two roles. First, they use the energy of light-activated electrons to concentrate protons in the thylakoid lumen, where their free energy can be used to generate ATP. Second, the electrons are used to reduce $NADP^+$ to NADPH + H^+. This reduction occurs in photosystem I. Both ATP and NADPH, you will recall, are essential for the light-independent reactions.

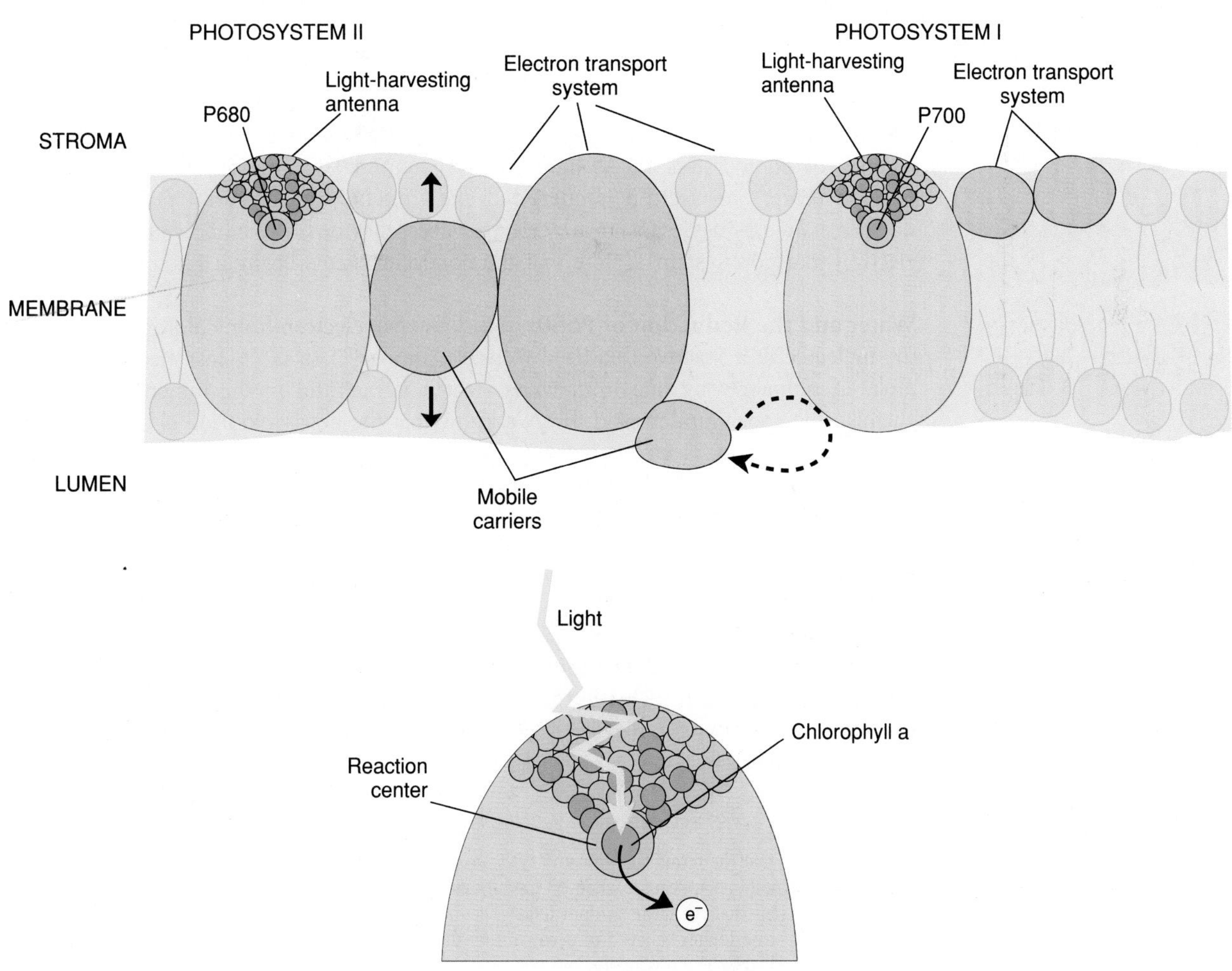

FIGURE 7.5
THE PHOTOSYSTEMS
The thylakoid membrane contains the structures responsible for photosynthesis. Included in photosystems I (right) and II (left), their light-harvesting antenna and associated electron transport systems. Their role in photosynthesis is to capture sunlight energy and use it to power the light reactions. In an enlarged view (bottom), light energy is seen being shunted to the reaction center.

The CF_0CF_1 Complexes In addition to the photosystems, the thylakoid membrane contains a large number of **CF_0CF_1 complexes** (see Figure 7-4b and Essay 7.1). Recall that these protein assemblies play a key role in the ATP-yielding phosphorylation of ADP (see Figure 6.17). The CF_0 complex is a protein-lined channel, originating in the lumen and leading across the thylakoid membrane. Its arrangement permits protons to move down their concentration gradient and into the bulbous CF_1 head. The head contains the important enzyme **ATP synthetase**, which utilizes the free energy of the concentrated protons for the phosphorylation of ADP.

Now let us see how the various elements of the photosystems do their work. The antennas are ready, the electron transport systems are idling, the CF_1 complexes have gathered in ADP and P_i, and the $NADP^+$ is waiting. We also have a supply of water on hand. So what do we need? We need light.

THE LIGHT REACTIONS OF PHOTOSYNTHESIS

Light falling on a green leaf passes into its photosynthetic cells and enters thousands upon thousands of green chloroplasts. Within the many thylakoids, millions of light-harvesting antennas respond. As we take a close look at only one set of photosystems, the events of the light reactions begin to unfold. There are two ways in which photosystems I and II can act: (1) the two can work cooperatively to build the chemiosmotic gradient and reduce $NADP^+$ to NADPH, an effort known as **noncyclic photophosphorylation,**[1] or (2) photosystem I can act alone, simply building the chemiosmotic gradient for ATP synthesis in what is called **cyclic photophosphorylation.** We will refer to these simply as the noncyclic and cyclic events. In both, light-activated electrons, ejected from the reaction centers, pass through electron transport systems, where their free energy is used to accomplish work. Let's look at the details.

The Noncyclic Events

The noncyclic events (Figure 7.6) begin as light is absorbed by the light-harvesting antenna of photosystem II. As light energy is shunted to chlorophyll *a* of the P680 reaction center, an electron is energized and, resisting the powerful pull of its own nucleus, it quickly passes to a nearby electron acceptor (the first member of the associated electron transport system). (Note that the noncyclic events begin at Photosystem II.)

Water and the Reduction of P680 The loss of an electron from chlorophyll *a* leaves the molecule in a reactive, oxidized ("electron hungry") state. Its lost electron will be replaced ultimately by one from water, which is plentiful in the lumen. The actual chemistry is poorly understood, but we know that the newly oxidized chlorophyll *a* is actually reduced by electrons from a nearby manganese-containing protein. Once its missing electron is restored, it is prepared to act again. The manganese-containing protein regains its lost electron by stripping it from nearby water (see Figure 7.6). Although we are considering one electron at a time, in the complete disruption of each water molecule (as seen below) *two* electrons are released, as are two protons and one atom of oxygen:

$$H_2O \rightarrow 2\ e^- + 2\ H^+ + 1/2\ O_2$$

The oxygen atom ($1/2\ O_2$) joins another from a similar disruption of water, and oxygen gas is given off. (This is the oxygen we breathe.) *Importantly, for each electron taken from water, a proton is released directly into the thylakoid lumen, where it helps form the proton gradient.*

[1]We should note that the terms *noncyclic* and *cyclic photophosphorylation* were developed before the chemiosmotic hypothesis appeared, when ATP synthesis was believed to occur during electron transport. It will become obvious that the terms are not totally accurate in describing the light reactions, since phosphorylation is no longer believed to occur during the noncyclic and cyclic events, but is now thought to be a separate process. For this reason, we will refer to "chemiosmotic phosphorylation" in our descriptions of ATP synthesis.

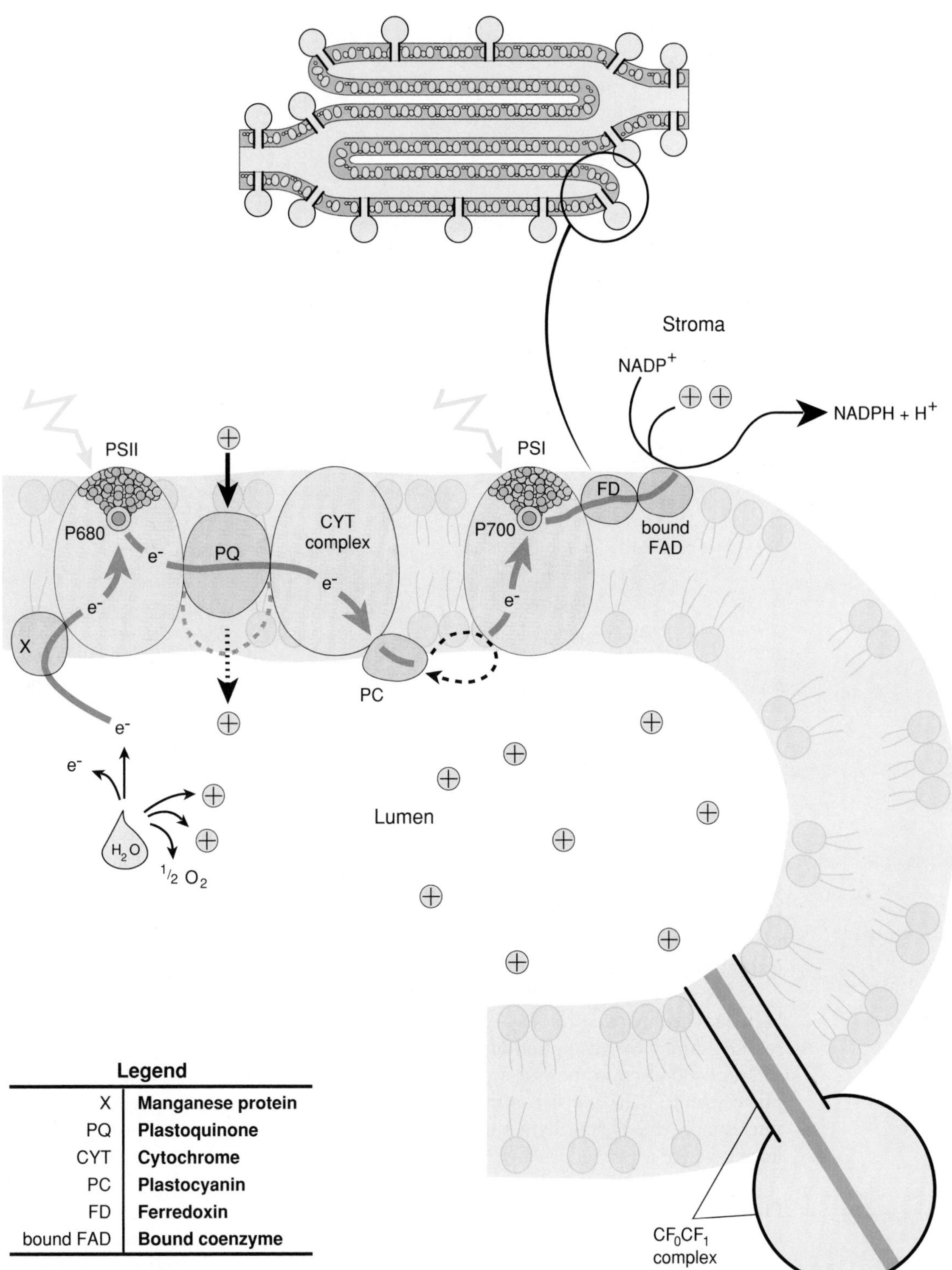

Legend

X	**Manganese protein**
PQ	**Plastoquinone**
CYT	**Cytochrome**
PC	**Plastocyanin**
FD	**Ferredoxin**
bound FAD	**Bound coenzyme**

FIGURE 7.6
THE NONCYCLIC LIGHT REACTIONS
The noncyclic light reactions begin when light is absorbed by the photosystem II light-harvesting antenna. A light-activated electron passes from the P680 reaction center to the electron transport system. The P680 electron is replaced by an electron from the nearby manganese complex (X) that is itself reduced by an electron from water. The complete disruption of each water molecule yields 1/2 of an oxygen molecule, two electrons and two protons, the latter joining the proton gradient. Plastoquinone (PQ) then transports a proton to the lumen. The electron is then passed to a cytochrome complex and then on to plastocyanin (PC), a free surface carrier. PC will transport the electron to photosystem I, replacing an electron lost when that system is oxidized. The excited P700 electron next reduces $NADP^+$ to form NADPH.

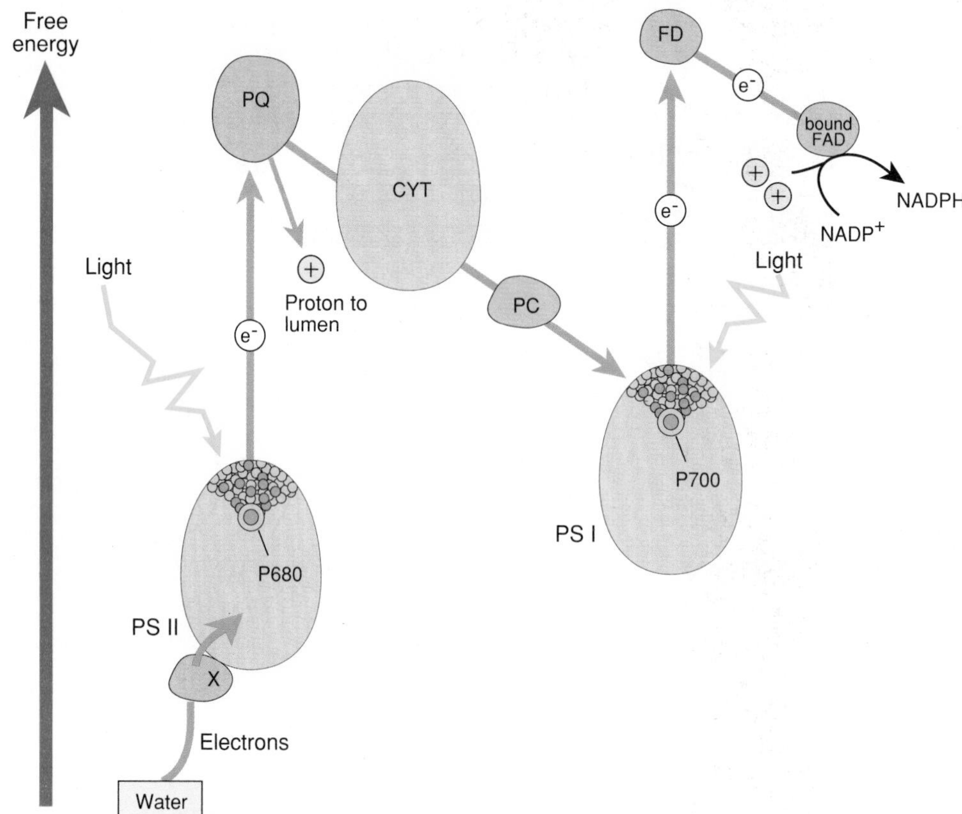

FIGURE 7.7
THE Z-SCHEME
The traditional Z-scheme is arranged to show the relative free energy levels of the participants in the two photosystems. An electron is seen originating in water and passing through the photosystems to $NADP^+$—another way of looking at the noncyclic light reactions. The role of PS I, in this view, appears to be to provide a second free energy boost to the P680 electron. Note that the highest energy level in the system, and the agent with the greatest reducing power, is FD (an iron-sulfur protein), the first carrier to receive electrons from P700. This clearly indicates that NADPH, the final product, has a great deal of reducing power.

Electron Energy and Proton Transport We return to the excited electron, which has now passed out of the P680 reaction center. It is soon transferred to a small, mobile carrier called **plastoquinone (PQ).** PQ has a key role to play, the pumping of a proton. Having taken in the P680 electron, PQ captures a proton from the stroma and releases it to the lumen. After this, PQ gives up its electron to the next carrier, that of a large cytochrome complex.

So we see that through the activation of a P680 electron, the proton gradient has been enriched by two protons—one pumped or shuttled in from the stroma, the other released during the disruption of water.

The P680 electron, now at a considerably lower free energy state, will move on to photosystem I. The transfer is made by another mobile carrier **plastocyanin,** or **PC.** Interestingly, PC is actually a surface protein, apparently free to rove throughout the lumen, where it can reduce any P700 reaction center that has been oxidized.

Photosystem I: the Reduction of $NADP^+$ P680 electrons can only enter P700 when it is ready to receive them, which means that the chlorophyll in that reaction center must be oxidized. This occurs as light is absorbed and an excited P700 electron enters the first acceptor in the photosystem I electron transport system. Thus, while oxidized chlorophyll *a* in the P680 reaction center is reduced by electrons from water, it is the P680 electron that reduces the P700 reaction center.

The transit of P700 electrons through the associated electron transport system is direct (see Figure 7.6), with no pumping of protons. The only purpose of photosystem I in the noncyclic events is to use the energy of excited PS I electrons to reduce $NADP^+$ to NADPH. The reduction actually requires two electrons from the P700 reaction center, along with two protons from the watery stroma. This is the final step in the noncyclic events.

In summary, we have seen that the chemiosmotic proton gradient is increased twice—once during the disruption of water and once in the photosystem II electron transport system. We have also seen how electrons from photosystem I are used to reduce $NADP^+$ to NADPH. Further, the noncyclic events can be viewed as a continuous process, with electrons flowing from water to $NADP^+$, not unlike the flow of electrons in an electrical current.

Traditionally, the shifting energy levels of electrons and carriers and the currentlike flow through the photosystems has been shown somewhat differently, in what is called the **Z-scheme** (Figure 7.7). It may help to look this over and gain a somewhat different perspective on the process. Note that the vertical scale suggests free energy level.

Finally, we will note that the noncyclic events of the light reactions must occur many times to provide enough NADPH and ATP for the synthesis of one glucose molecule in the light-independent reactions.

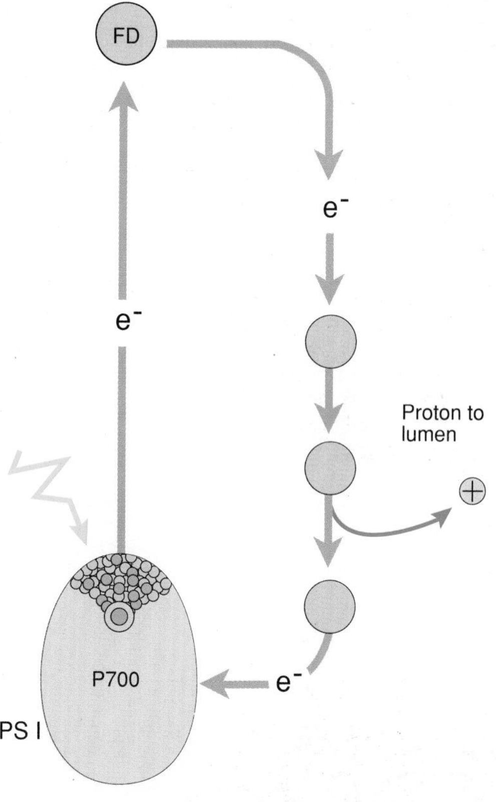

FIGURE 7.8
THE CYCLIC LIGHT REACTIONS
Following the Z-scheme format, we see that a photo-activated electron from P700 is passed to ferredoxin, FD. The exact pathway of electrons in the cyclic reactions isn't clearly known, but the free energy of the electron drops as it passes along from carrier to carrier. At one point, the free energy is used to capture and pump one proton, thus increasing the proton gradient. The electron finally cycles back into P700.

Summing Up the Noncyclic Events Now let's briefly recap what goes on in the noncyclic light reactions. In the order of occurrence, the events are

1. Absorption of photons by P680;
2. Movement of excited P680 electrons into the electron transport system;
3. Oxidation of water, restoring electrons to P680, adding protons to the lumen, and releasing oxygen gas;
4. Use of the free energy of P680 electrons to power proton transport from the stroma to the lumen;
5. Absorption of photons by P700;
6. Replacement of P700 electrons by those of P680;
7. Movement of excited P700 electrons into the electron transport system;
8. Use of P700 electron energy to reduce $NADP^+$ to NADPH.

The most important point to keep in mind is that the light reactions eventually provide ATP and NADPH for the light-independent reactions.

The Cyclic Events

As we indicated earlier, the photosystems can act in a cooperative, noncyclic manner, or photosystem I can act on its own, in a cyclic manner. The term *cyclic* refers to the circular pathway of P700 electrons that, when activated by light, leave P700, pass through an electron transport system, and return to P700. At the moment, we can only make informed guesses about the exact path of cycling electrons, so we will put them in the context of a Z-scheme rather than a membrane scenario to depict the events (Figure 7.8). Unlike the noncyclic events, cyclic activity does not produce NADPH. Instead, the decreasing energy of P700 electrons is used strictly for pumping protons from the stroma to the lumen. Thus, the cyclic events help only in producing ATP.

No one is quite sure how the cyclic events came to be, but there are some prevailing ideas. One of these is that the cyclic events occur when $NADP^+$ is in short supply. In other words, photosystem I goes into "idle," switching to proton pumping, until the light-independent reactions can cycle oxidized NADP ($NADP^+$) back to the light reactions. This effort is certainly not wasted, since any increase in the chemiosmotic gradient means a potential gain in ATP.

Chemiosmotic Phosphorylation

Now that we have seen how the chemiosmotic gradient is built up during the noncyclic and cyclic events, let's turn to chemiosmotic phosphorylation and the production of ATP.

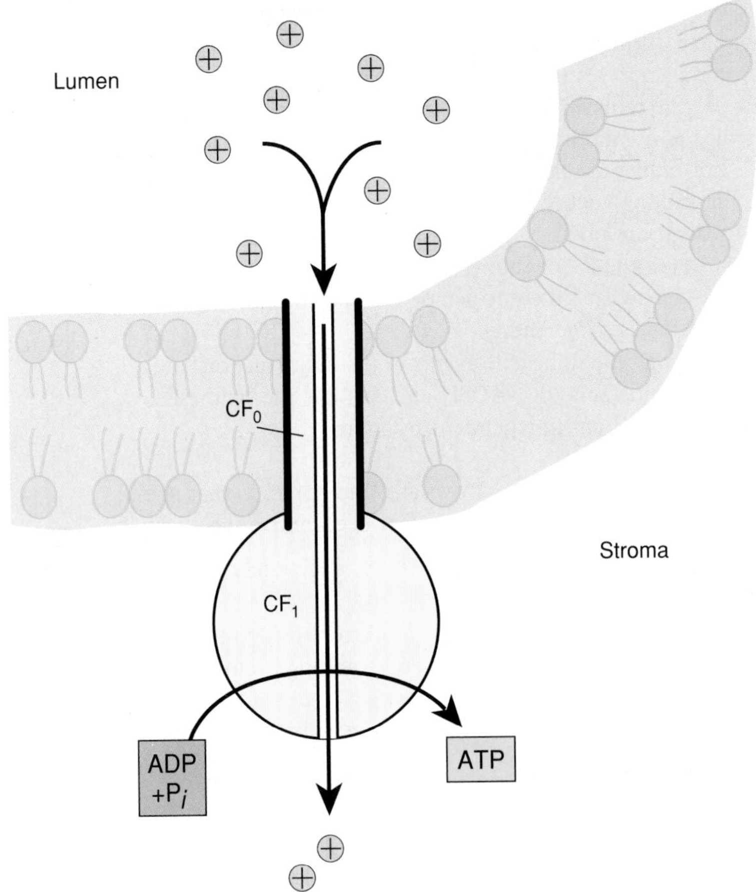

FIGURE 7.9
CHEMIOSMOTIC PHOSPHORYLATION
Protons, concentrated in the acidified lumen, pass down the gradient through a CF_0 channel into a CF_1 complex. There, the free energy of the proton gradient is used in generating ATP from ADP and P_1.

As positively charged protons accumulate in the lumen, they tend to attract negatively charged chloride ions (Cl^-) from the stroma. Chloride ions move rather freely across the thylakoid membranes, and as a result, the lumen becomes a pool of concentrated **hydrochloric acid** (HCl). Further, as protons are pumped out of the stroma, this outer region takes on an alkaline or basic characteristic, simply because of the loss of protons. Thus, the chemiosmotic system in the thylakoid becomes a concentrated acid solution (about pH 4 to pH 5), separated by the thylakoid membrane from an alkaline solution (about pH 8). Such a system has great free energy that can be used to accomplish useful chemical work as the protons are permitted to escape down their gradient (that is, from a place of higher concentration to a place of lower concentration) to join hydroxide ions in the stroma. They can escape only through specific CF_0 channels that direct them into the CF_1 complexes. The CF_1 complexes contain the enzyme ATP synthetase along with supplies of ADP and P_1. It is here that the free energy of the proton gradient is put to work phosphorylating ADP to ATP (Figure 7.9).

We have portrayed the light reactions in a rather mechanical, and slightly unrealistic, manner. The process seems to be a rather plodding event when, in reality, it occurs with dazzling speed and in countless repetitions in innumerable chloroplasts. Perhaps the light reactions are more accurately visualized as a constant flow of electrons and energy through the photosystems. P680 and P700 work in both a "pushing" and "pulling" manner, with P680 providing the push and P700 providing the pull. The overall flow of electrons can best be portrayed as passing in a continuous stream, like an electrical current, but from water to NADPH. Thus, water must be constantly available, entering along with sunlight at one end, with oxygen, ATP, and NADPH leaving at the other.

THE LIGHT-INDEPENDENT REACTIONS

Through the light reactions, the chloroplast builds up a considerable store of potential energy and reducing power in ATP and NADPH. Thus, with an input of carbon dioxide, glucose production can actually begin. The glucose is synthesized in the stroma, the unstructured region of the chloroplast. While there is no membranous organization in the stroma, it does contain a battery of synthesizing enzymes. The essentials of the light-independent reactions can be summarized as:

$$CO_2 + NADPH + H^+ + ATP \xrightarrow{\text{Enzymes}} C_6H_{12}O_6 + NADP^+ + ADP + P_i$$

(The reactants and products have not been numerically balanced.)

Basically, in the second half of photosynthesis, the products of the light reactions and carbon dioxide are used to produce glucose. One would think the process would be simple enough, just a matter of taking six CO_2s and some hydrogen to make a six-carbon sugar. However, carbon dioxide is at a far lower free energy level than is glucose, and therefore would be quite reluctant to form glucose spontaneously. Only the free energy of ATP and NADPH from the light reactions and a battery of synthesizing enzymes make such an event possible.

With the aid of a key enzyme, which we will look at shortly, CO_2 molecules are ushered in, and each is chemically joined to a 5-carbon substance. It is then that the formation of glucose can begin. All of this happens in a circular pathway called the **Calvin cycle.** An abbreviated version is seen in Figure 7.10.

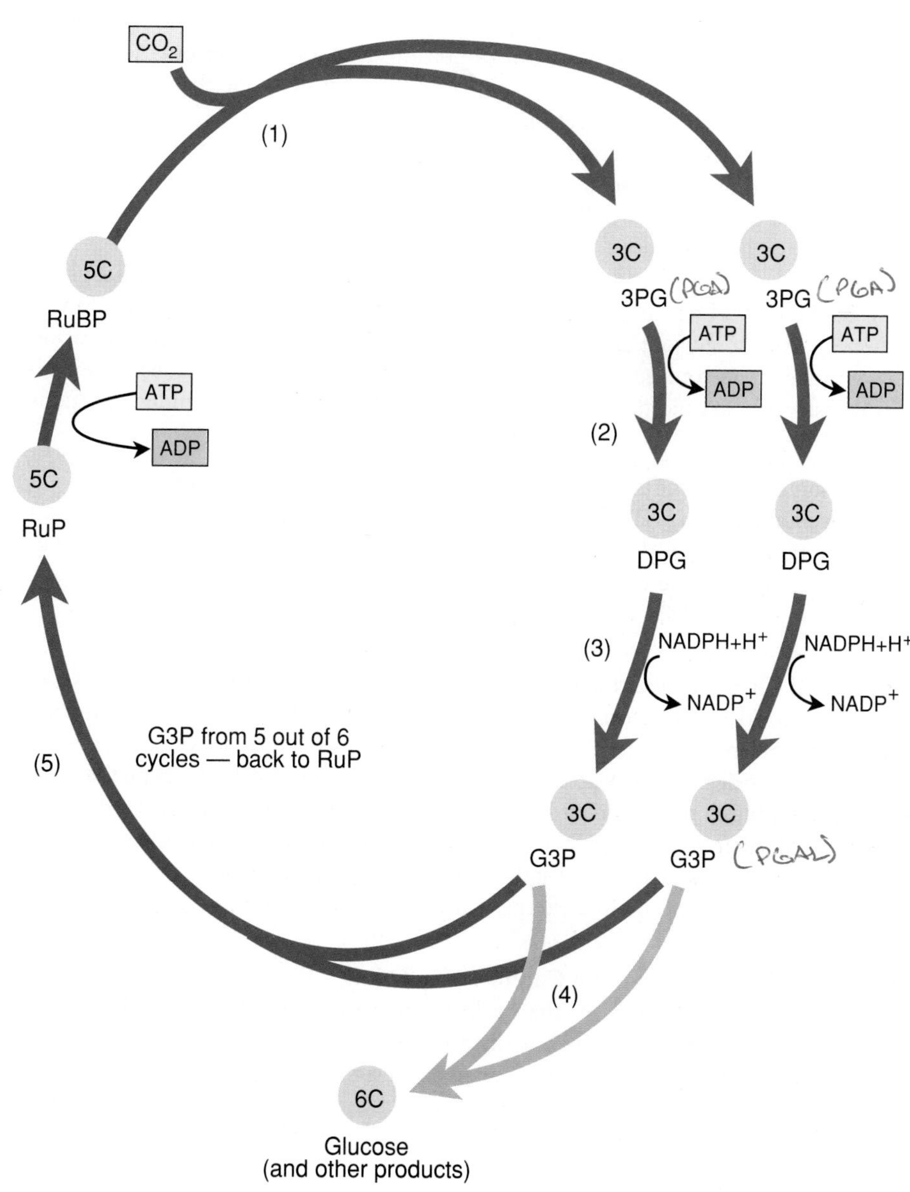

FIGURE 7.10
OVERVIEW OF THE CALVIN CYCLE
Through the use of an abbreviated Calvin cycle, we can readily keep track of the light-independent reactions just by counting carbons. The cycle begins (1) when carbon dioxide enters and is combined with the 5-C resident, which simultaneously splits, forming two 3-C products. The two are phosphorylated by ATP (2), reduced by NADPH (3), and a final 3-C product emerges. This last product is at a crossroads, some (4) continuing on to form glucose and some (5) continuing in the Calvin cycle, to be phosphorylated (6), which restores the original 5-carbon resident.

The Calvin Cycle

The Calvin cycle is named in honor of Melvin Calvin, who was one of its principal discoverers (see Essay 7.1). He described the nature of carbon dioxide incorporation in the late 1940s, an accomplishment that earned him a Nobel Prize.

The Calvin cycle essentially consists of a biochemical pathway in which each enzyme in a team takes its turn at altering the substrate. With each change, the substrate takes on a new form and identity until the work is complete. The cycle has several key events or reactions, which we will describe briefly. The discussion is keyed to the reactions illustrated in Figure 7.11, where further details are presented.

The first key event is **carboxylation**, the addition of carbon dioxide (reaction 1), (also called carbon fixation). As we've noted, one molecule of carbon dioxide is taken in during each complete cycle. The enzyme responsible for reaction 1 is called **ribulose-1,5-bisphosphate carboxylase,** named for its job: adding carbon dioxide to the waiting molecule, **ribulose-1,5-bisphosphate (RuBP)**, forming an unstable 6-carbon intermediate that cleaves into two 3-carbon products.

We won't be discussing each enzyme, but the carbon dioxide adding enzyme-"Rubisco," as biochemists like to call it, is one of the most significant enzymes on earth (and perhaps the most abundant protein). For most life forms it represents the chemical link between the physical and biological worlds. This may sound bold, but the vast majority of the earth's organisms rely, directly or indirectly, on carbon compounds first formed by photosynthesizers in the Calvin cycle.

The second key event (reaction 2) is phosphorylation, the transfer of a high energy phosphate group from ATP to the two 3-carbon products of reaction one. Recall that phosphorylation (see Chapter 6) is generally a preliminary step, one that increases the free energy of the substrate, thus setting it up for further reactions.

The next key reaction (reaction 3), involves reduction. NADPH reduces the two products of reaction two, replacing one of their phosphates with hydrogen. The cleaved phosphates, along with the $NADP^+$s and the ADPs from the previous reaction simply cycle back to the light reactions.

At this point a major division in the pathway occurs. **Glyceraldehyde-3-phosphate (G3P),** the product of reaction 3, can proceed into either of two pathways. One of these (reaction 4), leads to glucose (and other products), whereas the other, the **regeneration pathway**, makes the continuation of the Calvin cycle possible. We will see how both pathways can occur in a moment, but first let's summarize the light-independent reactions so far:

1. Carbon dioxide is added into the Calvin cycle by the enzyme Rubisco.
2. Phosphorylation by ATP sets up the substrate for reduction by NADPH.
3. Reduction by NADPH yields G3P, which can be drawn away to form glucose or recycled to keep the Calvin cycle going.
4. ADP, P_1, and $NADP^+$ recycle to the light reactions.

Glucose and Other Products In the formation of glucose (reaction 4), two G3Ps are first assembled into fructose-1,6-diphosphate, which is then converted to glucose-1-phosphate. From there, free glucose may be formed. Alternatively, glucose-1-phosphates can be dephosphorylated and converted to glucose, which (through the usual dehydration synthesis reaction) can be assembled into starch (see Chapter 3). But G3P can follow still other pathways, where it is modified and used in the synthesis of fatty acids, glycerol, or amino acids. Alternatively, G3P can be further modified and sent into the mitochondrion, where its free energy will be used to produce more ATP.

G3P, Glucose, and Keeping the Calvin Cycle Going It may have occurred to you that drawing the two G3Ps from the cycle to form glucose would, for all practical purposes, end the cycle. This doesn't happen because glucose is only formed when the Calvin cycle reactions have occurred six times. In five out of six turns of the cycle, the

10 G3Ps formed simply enter the regeneration pathway, where they are used to keep the cycle going (reactions 5 and 6). But in the sixth turn, the two G3Ps are drawn off to form glucose. Actually, there are so many cycles going on simultaneously in the stroma that things aren't quite this neat. However, the net result is as we have described it.

If you are keeping tabs on reactants and products as you study Figure 7.11, you may note that in addition to one CO_2, *each complete turn* of the cycle requires 3 ATPs (2 in the main cycle and 1 in the regeneration cycle), along with 2 NADPHs and 2 H^+s.

The synthesis of one molecule of glucose requires six turns, so 6CO_2s, 12 NADPH (and 12 H^+), and 18 ATPs are consumed (12 ATPs in the main cycle and 6 in the regeneration phase). In the next chapter, we will determine how many ATPs are generated when a glucose molecule is completely broken down into CO_2 and H_2O.

PHOTORESPIRATION: TROUBLE IN THE CALVIN CYCLE

If everything went along in the Calvin cycle as we have described it, the chemical efficiency of photosynthesis (captured light energy versus free energy in glucose) would be a respectable 38%. But in the real world things do not always go this way. It seems the carbon dioxide capturing ability of Rubisco depends to a large extent on the amount of carbon dioxide present. When carbon dioxide levels are low, or when it is taken up faster than it can be provided by diffusion, the efficiency of the enzyme falls off drastically. (Keep in mind that CO_2 is not a very abundant gas; it occurs at about 4 parts per 10,000 in air.)

When the concentration of this vital gas falls below minimal levels, plants undergo **photorespiration:** oxygen replaces carbon dioxide in step 1 of the Calvin cycle. Rubisco, it turns out, works with either gas, reacting with oxygen when carbon dioxide levels fall. When oxygen joins ribulose-1,5-bisphosphate, the product goes through several reactions before finally emerging as glycolate. Glycolate is shuttled off to cytoplasmic microbodies, where it is converted to glyoxylate which then enters a mitochondrion where further reactions yield carbon dioxide plus the three-carbon compound **3-phosphoglycerate (3-PG,** as in the Calvin cycle). The carbon dioxide is released, but the 3-PG is utilized as a fuel.

Having some fuel available is nice, but since there is no glucose formed and vital carbon is lost, a strong element of waste seems to be present. Since there seems to be little of value in photorespiration, it remains a scientific enigma. However, natural selection has been at work over the ages, and as a consequence, some plants have a way of avoiding photorespiration.

C4 Carbon Fixation: A Solution

Many tropical and subtropical plants have evolved pathways that overcome the limited availability of carbon dioxide. Plants such as maize, sugar cane, sorghum, and Bermuda grass (Figure 7.12) make use of a special pathway that assures them of an adequate concentration of carbon dioxide for the Calvin cycle. It is noteworthy that such plants live in regions where they are often exposed to intense sunlight, regions where accelerated photosynthesis encourages photorespiration.

In the mid-1960s, M. D. Hatch, C. R. Slack and others discovered the existence of an alternative carbon-fixing pathway involving certain 4-carbon acids, which was soon dubbed the **C4 pathway**. In other plants, those restricted to carbon fixing by means of the Calvin cycle, the first stable molecules formed following carboxylation are 3-carbon compounds, so they are aptly called **C3 plants** (Figure 7.13a).

In **C4 plants**, carbon dioxide diffuses into chloroplasts of **leaf mesophyll cells** (loosely arranged photosynthetic cells) just as it does in C3 plants. But the enzyme it encounters in the C4 chloroplast is not Rubisco, but **PEP carboxylase** (phosphoenolpyruvate carboxylase), an enzyme that works quite efficiently where CO_2 levels are low. The incorporation of carbon dioxide is the beginning of a short 4-carbon metabolic

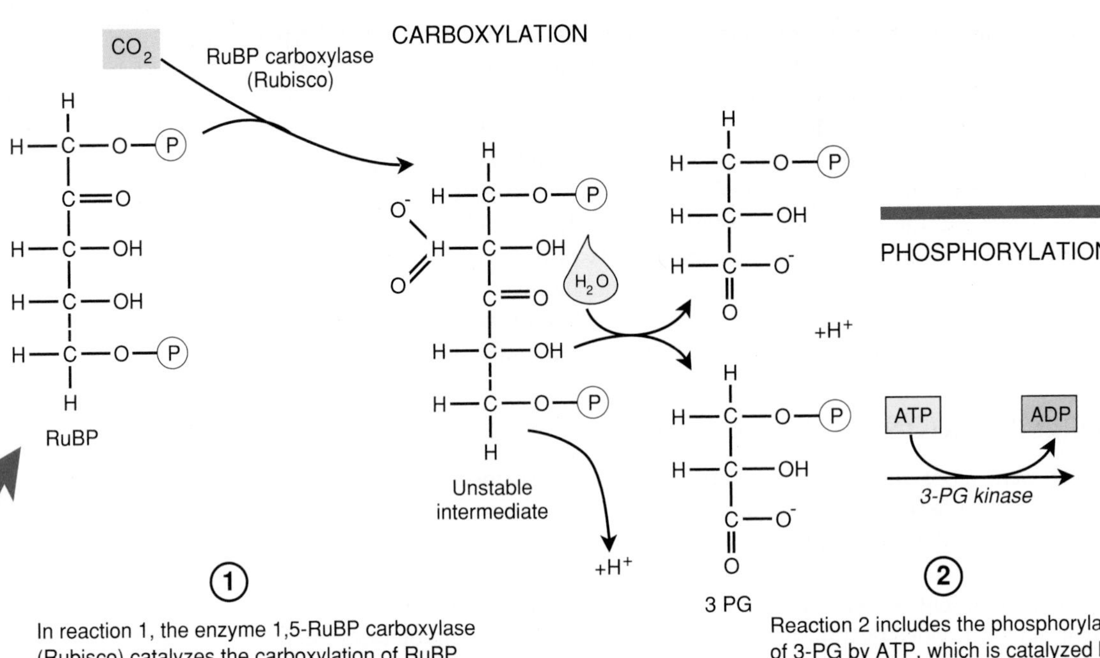

① In reaction 1, the enzyme 1,5-RuBP carboxylase (Rubisco) catalyzes the carboxylation of RuBP, forming a very unstable 6-carbon intermediate that breaks down spontaneously into two 3-carbon molecules of 3-PG. Water is consumed in the process. This is a highly exergonic reaction (ΔG= -12.4 kcal/mole).

② Reaction 2 includes the phosphorylation of 3-PG by ATP, which is catalyzed by the enzyme 3-PG kinase. The products are two 3-carbon DPGs. ADP recycles to the light reactions for phosphorylation into ATP.

CALVIN CYCLE

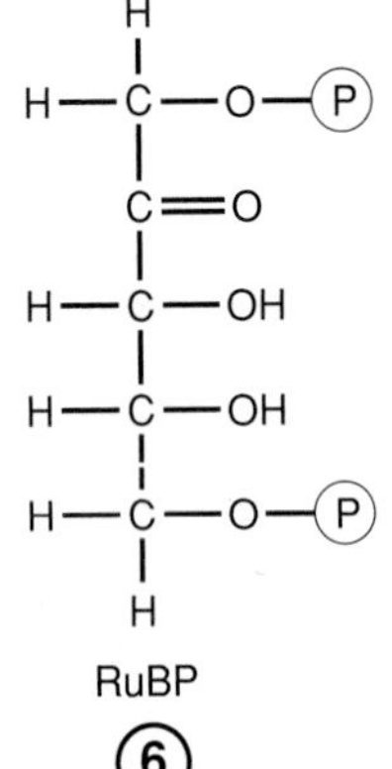

ADP ATP *RuP kinase*

⑥ Reaction 6 requires ATP for its completion. The product is RuBP, the starting molecule of the Calvin cycle. ADP is recycled to the light reactions.

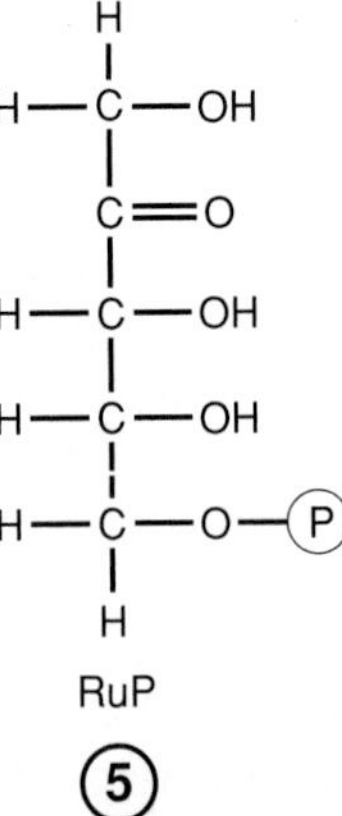

⑤ Reaction 5 represents the regeneration pathway in which G3Ps are utilized in the formation of new RuP. The G3Ps from five out of six turns of the cycle (or 10 out of every 12 G3Ps) must enter the regeneration pathway.

Key

RuP	**Ribulose phosphate**
RuBP	**Ribulose bisphosphate**
3-PG	**3-phosphoglycerate**
DPG	**1,3 diphosphoglycerate**
F-1, 6-BP	**Fructose-1, 6-bisphosphate**
G-1-P	**Glucose-1-phosphate**
G3P	**Glyceraldehyde-3-phosphate**

FIGURE 7.11
THE CALVIN CYCLE
In this expanded version of the Calvin cycle each molecule is shown and each major reaction illustrated.

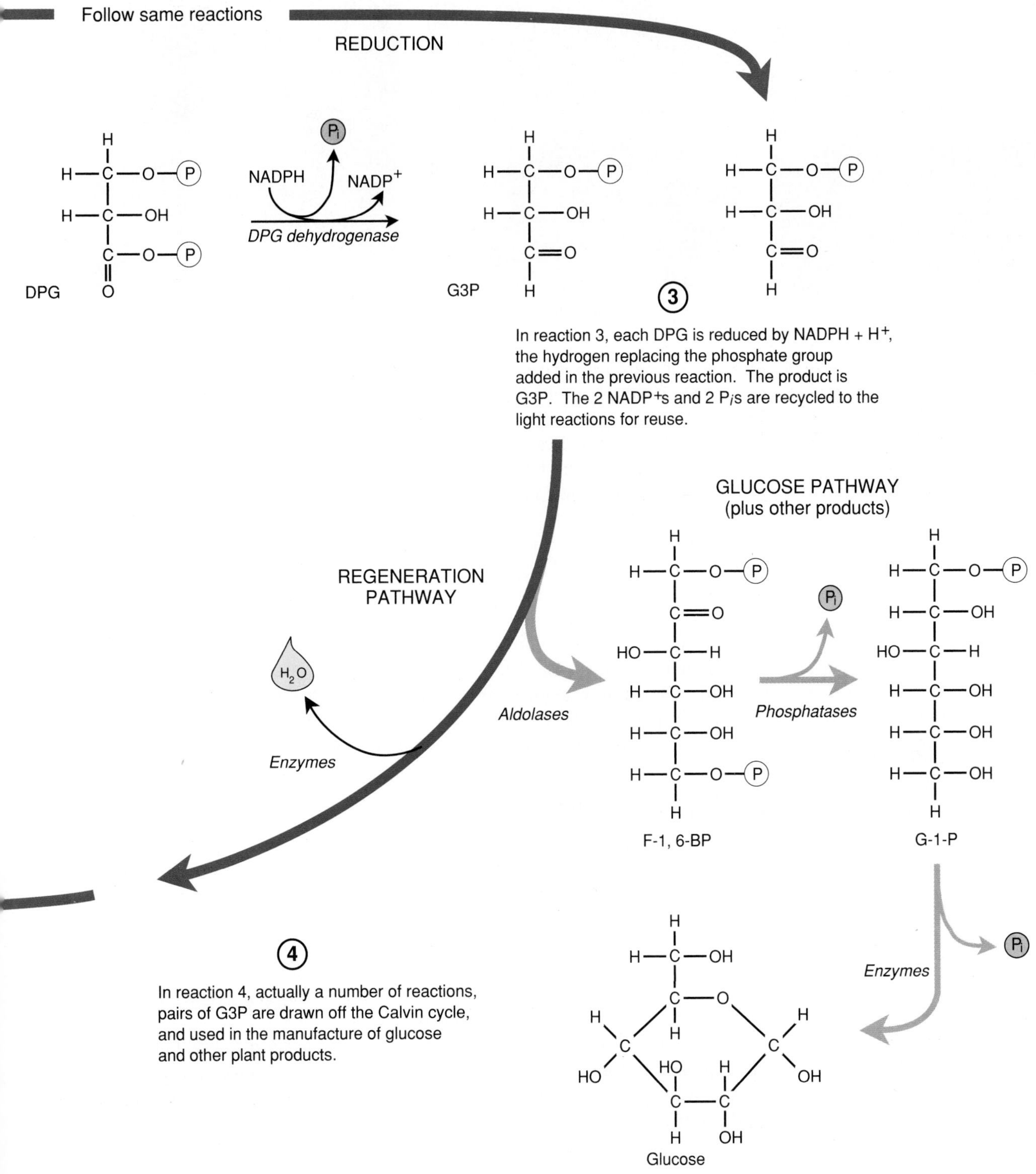

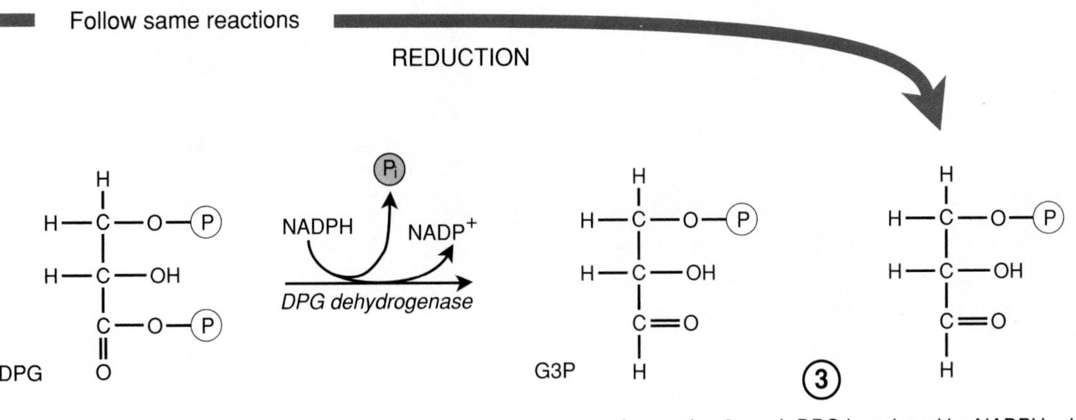

(3)

In reaction 3, each DPG is reduced by NADPH + H^+, the hydrogen replacing the phosphate group added in the previous reaction. The product is G3P. The 2 $NADP^+$s and 2 P_is are recycled to the light reactions for reuse.

GLUCOSE PATHWAY
(plus other products)

REGENERATION
PATHWAY

H_2O

Enzymes

Aldolases

Phosphatases

P_i

F-1, 6-BP

G-1-P

Enzymes

P_i

Glucose

(4)

In reaction 4, actually a number of reactions, pairs of G3P are drawn off the Calvin cycle, and used in the manufacture of glucose and other plant products.

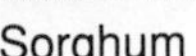

Sorghum

Sugarcane

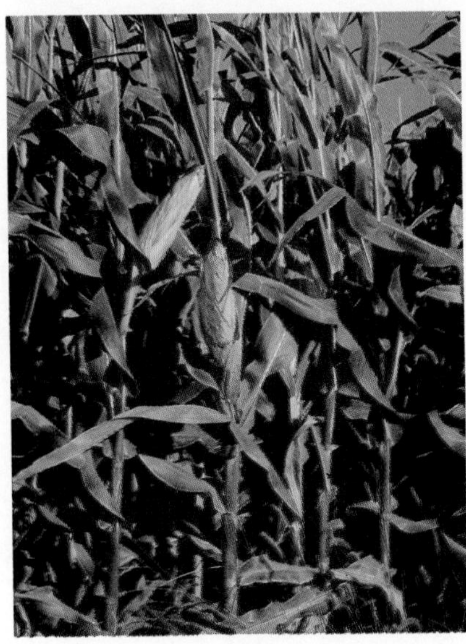

Corn

FIGURE 7.12
C4 PLANTS
C4 plants include over 100 genera distributed in at least 10 families, and the list is growing. C4 plants have made structural and biochemical adaptations to the intense sunlight of the tropics and subtropics.

pathway (Figure 7.13b), with the final product (malate) leaving the leaf mesophyll cells to enter the chloroplasts of adjacent **bundle sheath cells** (cells surrounding leaf veins). Then, other reactions cleave the 4-carbon substrate, yielding a 3-carbon product and free CO_2. Whereas the 3-carbon product recycles to the leaf mesophyll cells to react again, the CO_2 is taken up by Rubisco, which ushers it into the Calvin cycle. So the C4 pathway occurs in the leaf mesophyll cells and the Calvin cycle in the bundle-sheath cells.

Thus we see that the purpose of the C4 pathway is to make use of an efficient carboxylating enzyme and a special biochemical pathway to deliver carbon dioxide to the Calvin cycle. Through the ongoing reactions of the C4 pathway, then—repeated many times and at great speed—the Calvin cycle enzyme, Rubisco, is assured of an adequate supply of carbon dioxide. In this way C4 plants avoid the problem of photorespiration altogether. All of this occurs at no small cost. As you can see in Figure 7.13, the C4 pathway requires an ongoing investment of ATP beyond that usually required by plants using only the Calvin cycle. But apparently, the plants generate sufficient ATP under their brightly lit conditions to afford to splurge.

C4 Plants and Evolution

It would be tidy to say that C4 plants dominate the sunny tropics and deserts, and C3 plants, those regions of the earth where light is less intense. However, no such simple distribution exists. The vast majority of plants, well over 99%, are C3 plants, many of which are alive and well in the tropics. Thus, C3 and C4 plants grow side by side in sunny climates. Further, there are no clear-cut evolutionary divisions between the two. C3 and C4 plants exist in the same plant families, and even within the same plant genera. Evidently, C4 plants represent a rare but interesting evolutionary venture. Whether they are on the increase or not is unknown.

Evolutionary theorists disagree about the origin of the C4 pathway, but most agree that it arose far more recently than the C3. In fact, the C3 pathway is believed to be ancient. Rubisco, as a primeval enzyme, may have been present in the earliest photosynthetic bacteria, those that thrived at a time when the atmosphere was far richer in carbon dioxide, and oxygen was absent or nearly so (see Chapter 21). This may explain Rubisco's inability to discriminate between CO_2 and O_2 at all concentrations. Later, as water-utilizing plants and algae evolved, the oxygen given off accumulated in greater and greater abundance, and since the enzyme simply wasn't able to discriminate between the new gas and CO_2, photorespiration has plagued many plants and algae ever since.

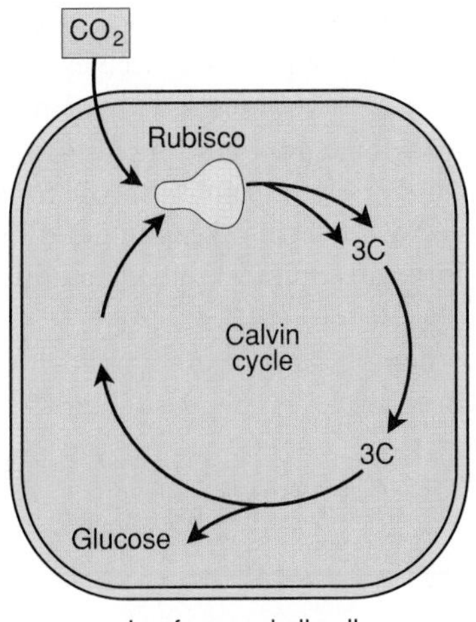

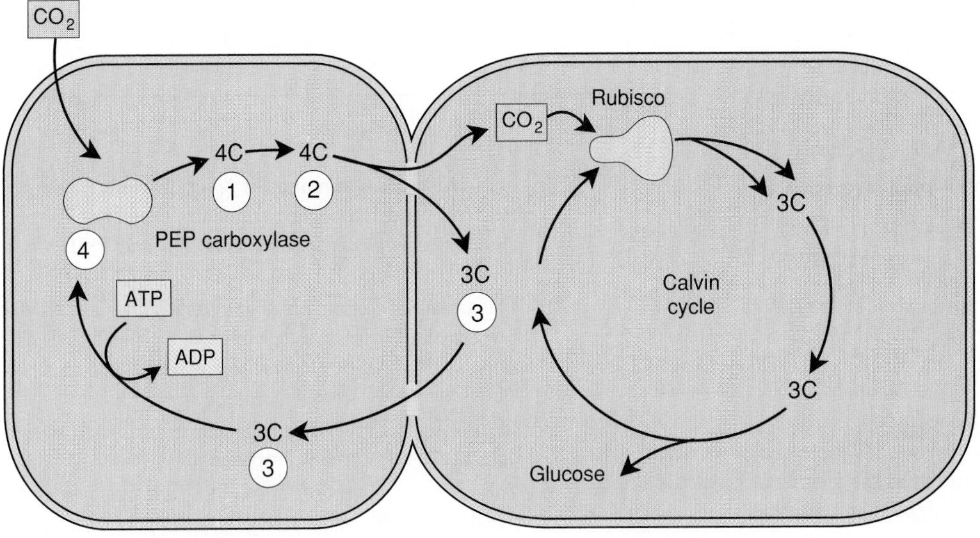

FIGURE 7.13
LEAF CELLS AND THE C4 PATHWAY
(a) C3 and C4 plants differ biochemically and anatomically. The uptake of CO_2 in C3 plants and its use in the Calvin cycle occur in chloroplasts of the leaf mesophyll cells. **(b)** In chloroplasts of the leaf mesophyll cells in C4 plants, the main activity is the uptake of CO_2 and its incorporation into 4-carbon products. The 4-carbon products enter the bundle sheath cells, where the carbon dioxide is released for use in the Calvin cycle.

In Chapter 28 we'll look into another special photosynthetic adaptation in which certain desert plants, the CAM plants, use their C4 pathway to take in and store carbon dioxide during the nighttime hours. This strategy enables the plants to keep their stomatal pores closed tightly during daylight hours, thus avoiding excessive water loss to the superheated, parching, desert winds, while at the same time heading off photorespiration.

HAVE YOU THANKED A PLANT TODAY?

Because many of the mechanisms of photosynthesis are frustratingly complex, as we work our way through them it is easy to get caught up in the details and to lose sight of the tremendous importance of photosynthesis in the overall scheme of things. So let's just take a moment to consider the fate of some of the molecules involved in the marvelous, if maddening, process.

First, we should keep in mind that the sugar made in leaves supplies the entire plant with both energy and the carbon needed to synthesize all of the plant's molecules. Most of the glucose is consumed in respiration, but even so, photosynthesis usually manages to keep ahead of demand, and those molecules not used in respiration can be stored in the form of starch or oils. In addition, a significant amount of the plant's glucose forms cellulose, the most abundant carbohydrate in all of life. We must also keep in mind that the products of photosynthesis can be altered to form any number of other critical molecules, such as proteins and nucleic acids.

Finally, we need only remind ourselves that everything we and all other animals eat was once locked in the molecules of a growing plant or alga.

APPLICATION OF IDEAS

Earlier in this century, plant physiologists determined the precise role of water and carbon dioxide in photosynthesis by using the radioactive isotopes carbon-14, oxygen-18, and tritium (hydrogen-3). Using your new knowledge of photosynthesis, suggest how these isotopes might be used in such determinations, and what you might expect in the results.

KEY IDEAS

PHOTOSYNTHESIS: AN OVERVIEW

The general equation for **photosynthesis** is:

$$6CO_2 + 12H_2O \rightarrow C_6H_{12}O_6 + 6O_2 + 6H_2O$$

Although photosynthesis produces glucose as shown, much of the process is not seen in the general formula (including the roles of light, chlorophyll, enzymes, electron carriers, coenzyme NADP, and ATP).

The Two Parts of Photosynthesis

1. Free energy in the thylakoid is increased in the **light reactions** and decreased as the synthesis of glucose occurs in the **light-independent reactions.**
2. In the light reactions, water, ADP, P_i, and NADP are the reactants, and through the capture of light energy the products oxygen, ATP, NADPH, and H^+ are produced.
3. In the light-independent reactions, ATP, NADPH, H^+, and carbon dioxide are the reactants, while glucose, ADP, P_i, and NADP are the products.

CHLOROPLASTS

Chloroplasts have two surrounding membranes, within which there is an outer fluid compartment, the stroma, and an extensive third membrane whose intricate folds form vesicles called thylakoids. The latter occur as simple **stroma thylakoids** and stacked **granum thylakoids.**

The Thylakoid

1. The space within the thylakoid membrane system is the **lumen,** a second watery compartment that becomes the proton reservoir.
2. Thylakoids have a large number of **photosystems** embedded in their membranes. Each includes **light-harvesting antennas**, a **reaction center**, and an **electron (and hydrogen) transport system** (ETS). Photosystems designated I and II (**PSI** and **PSII**) work in pairs, although PSI also works alone.
3. The antennas contain **chlorophylls *a*** and ***b*** and **carotenoids**. In each, one molecule of chlorophyll *a* along with an associated protein is designated as a reaction center.
4. Chlorophylls *a* and *b*, and the carotenoids each absorb light of slightly different wavelengths.
5. The reaction centers of PSI and PSII are designated **P700** and **P680**, respectively, according to their light absorption.
6. Photosystems shunt energy to their reaction centers, where it is absorbed by electrons of chlorophyll *a*. Instead of energized electrons dropping back to their old energy level, they escape the reaction center to begin reducing carriers of the electron transport system (ETS).
7. The passage of electrons through the ETS yields free energy that is used in pumping protons from the stroma to the thylakoid *lumen*. The electrons also reduce $NADP^+$ to NADPH + H^+.
8. The phosphorylation of ADP occurs in the CF_1 complexes, which abound in the thylakoid membrane.

THE LIGHT REACTIONS OF PHOTOSYNTHESIS

The light reactions include both **noncyclic** and **cyclic photophosphorylation.**

The Noncyclic Events

1. Light absorbed in P680 of PSII excites electrons that pass into the associated ETS.

2. Oxidized chlorophyll *a* in P680 is reduced by electrons from a manganese-containing protein, which is itself reduced by electrons from water molecules within the lumen.
3. As water molecules are disrupted, oxygen gas is liberated, and protons are released into the lumen, adding directly to the chemiosmotic differential.
4. Each electron passing from P680 to P700 powers the transport of a proton from the stroma to the lumen. The proton is shuttled from the stroma to the lumen by carrier **PQ.**
5. Light-excited electrons in P700 leave PSI and pass to NADPH, reducing it to NADPH + H^+. These electrons are replaced by those from P680, transported to PSI by carrier **PC.**
6. The noncycliclight reactions enrich the chemiosmotic gradient and reduce NADPH, thus providing for ATP formation and a supply of hydrogen for glucose synthesis.

The Cyclic Events

Photosystem I also acts independently, capturing light energy and sending its electrons through its ETS, from which they return to chlorophyll *a* (thus water is not involved). Protons are pumped into the lumen, enriching the chemiosmotic gradient.

Chemiosmotic Phosphorylation

1. The chemiosmotic gradient of the thylakoid is a pH differential produced by a buildup of protons and chloride ions in the lumen and an increase in hydroxide ions in the stroma.
2. As protons escape the lumen through the CF_0 channels to the CF_1 complexes, their free energy is used in the phosphorylation of ADP to ATP.

THE LIGHT-INDEPENDENT REACTIONS

In the second half of photosynthesis, ATP and NADPH are used in the reduction, or fixing of carbon dioxide into 3-carbon molecules from which glucose can be made. This occurs in the unstructured stroma, where synthesizing enzymes are located. The synthetic pathway is known as the **Calvin cycle.**

The Calvin Cycle

1. The Calvin cycle is a pathway where several enzymes make subtle changes in resident molecules, bringing in ATP, NADPH, and carbon dioxide to produce carbohydrate.
2. The Calvin cycle includes the following chemical steps:
   a. **Carboxylation:** RuBP + $CO_2 \rightarrow$ two **3-PG**
   b. *Phosphorylation:* (2) 3-PG + two ATP $\rightarrow$ (2) 1,3-DPG + 2 ADP
   c. *Reduction:* (2) 1,3-DPG + 2NADPH + $H^+ \rightarrow$ (2) G3P + $2NADP^+$
   d. *Regeneration:* $NADP^+$, ADP, and P_i are recycled to the light reactions.
   e. *Phosphorylation: RuP* + ATP $\rightarrow$ **RuBP** + ADP
3. The product G3P can be used to produce glucose or other essential molecules, but some must be recycled to keep the Calvin cycle going.
   4. Since only one carbon dioxide is brought in per turn of the cycle, six turns must occur before two G3Ps can be withdrawn to form one glucose. The other 10 G3Ps thus formed are recycled to generate the starting molecule. This recycling keeps the cycle going.

PHOTORESPIRATION: TROUBLE IN THE CALVIN CYCLE

1. Photosynthesis in **C3 plants** becomes inefficient when carbon dioxide gas is in low concentration, a time when the enzyme *1,5-ribulose bisphosphate carboxylase* cannot readily incorporate carbon dioxide.
2. When carbon dioxide is unavailable, **photorespiration** ensues. During this process, RuBP is changed to **glycolate,** which is then converted to glyoxylate, which next forms 3-PG and carbon dioxide.

C4 Carbon Fixation: A solution

1. The problem of photorespiration is avoided in the **C4 plants,** which have an alternative pathway for incorporating carbon dioxide that involves certain 4-carbon acids. This alternative is called the **C4 pathway.** Carbon is incorporated by the enzyme **PEP carboxylase.**
2. Carbon dioxide enters **leaf mesophyll cells,** where it enters the C4 pathway which yields 4-carbon malate.
3. Malate enters the **bundle sheath cells,** where carbon dioxide and 3-carbon pyruvate are formed.
4. The carbon dioxide then enters the Calvin cycle, and the pyruvate sent back to the leaf mesophyll cell when the C4 cycle resumes.
5. C3 plants appear to be adapted to temperate climates, while most C4 plants are desert dwellers.
6. Comparative studies of C3 and C4 plants reveal that in intense sunlight, C3 plants undergo photorespiration, while C4 plants continue photosynthesis.

C4 Plants and Evolution

1. C3 plants dominate, even in tropical and desert environments. The two are not ecologically or evolutionarily distinct since both exist in the same genus and in the same habitats.
2. The C3 pathway is much older and rubisco was probably present in the early photosynthatic bacteria.

Have You Thanked a Plant Today

1. Plants utilize most glucose for cell respiration, storing excesses as starch and oils. Much of the yield ends up in cellulose, amino acids, and nucleic acids.

REVIEW QUESTIONS

1. Write the general formula for photosynthesis and list several factors missing from this simple representation. (152)
2. Briefly explain why disruption of the chloroplast and thylakoids stops photosynthesis. (general)
3. Explain what is meant by this statement: In the light reactions the system is charged up, while in the light-independent reactions it runs back down. (153)
4. Write a detailed equation for the light reactions that includes water, $NADP^+$, ATP, ADP, Pi, NADPH, and oxygen. (153)
5. Write a detailed balanced equation for the light-independent reactions including glucose, water, ATP, and $NADP^+$, ADP, Pi, CO_2, and NADPH. (153)
6. Prepare a large drawing of a chloroplast, labelling grana, grana thylakoid, stroma thylakoids, lumen and stroma. (156)
7. Prepare a simplified drawing of paired photosystems I and II, carefully labelling the following: lumen, stroma, light-harvesting antennas, reaction centers, and ETS. (158-161)
8. Compare the absorption of light among chlorohylls *a* and *b* and the carotenoids. In what way is the combined range of these pigments adaptive? (Essay 7.2)
9. Summarize the events of the noncyclic light reactions, beginning with the absorption of light by P680.
   a. path of electrons
   b. role of water
   c. protons gained in (a) and (b)
   d. final electron acceptor (162-165)
10. What is gained by the noncyclic events? (165)
11. Summarize the cyclic events:
    a. path of electrons;
    b. role of water, if any;
    c. protons gained;
    d. role of NADP, if any. (165)
12. Explain how and when elements of the electron transport system become hydrogen transporters. Be specific. (165)
13. Describe the CF_1 complexes—their location, orientation, structure, and spatial relationship to the photosystems. (166)
14. Using a simple diagram, illustrate the phosphorylation of ADP in a CF_1 complex. (166)
15. Summarize the results of the light reactions. What are the products and what, in general, are they used for in the light-independent reactions? (166)
16. Write a short paragraph summarizing the biochemical pathway known as the Calvin cycle as though you were explaining it to a novice. (167-169)
17. The Calvin cycle contains only six major steps. List these from memory and add one word for each step that defines the kind of reaction occurring (such as phosphorylation). (168-169)
18. What are the roles of carbon dioxide, NADPH, ATP, and enzymes in the Calvin cycle? (167-168)
19. Briefly explain why it requires six turns of the Calvin cycle to generate one molecule of glucose. (168-169)
20. Summarize the steps used in converting G3P to glucose. What alternative products are possible? (168)
21. From a plant's point of view, what is wrong with photorespiration? Under what conditions does it occur? (169)
22. Describe the four steps of the C4 cycle, listing the reactants and products of each. Include the specific cell types in which each reaction occurs. (169, 172)
23. Sucking fluids from desert cacti has quenched the thirst of many a desert traveler. When would you expect the fluids to taste best: in daylight or at night? Why? (general)

8

CELL RESPIRATION

WE HAVE SEEN THAT LIFE CAN ONLY EXIST THROUGH INTENSE AND PRECISE efforts to keep its molecules organized. As soon as the tendency to remain organized ceases, so does life. A corpse is a once-organized entity, gradually becoming disorganized as it decays until finally, its molecules have no more to do with each other than they do with any other molecules. At this point they are behaving randomly. They are no longer organized, and the corpse is gone. Before the final episode, though, the organism lives. And it does so by staying organized. Here, then, we will see how living things make use of energy to maintain that organization.

We should begin by reminding ourselves that the energy available to living things is used in a variety of ways. Energy is required for all cellular work, including such processes as movement, active transport, communication, and molecular synthesis. A key participant in these activities is ATP, the direct source of energy for most of the cell's activities.

With the exception of ATP produced directly through photosynthesis, most of the earth's organisms produce their ATP by using chemical bond energy in organic molecules—carbohydrates, fats, and proteins. We will focus primarily on the carbohydrate called glucose, the most familiar of these "cellular fuels."

Before going on, we should remind ourselves of just where the glucose came from. It was manufactured during photosynthesis. In fact, we will now essentially see the reverse of the process by which glucose was made, so that the energy held within the glucose molecule can be released to do work.

Compare the overall reaction producing glucose and the overall reaction representing its breakdown.

$$\text{Production: (photosynthesis) } 6CO_2 + 6H_2O + \text{energy} \rightarrow C_6H_{12}O_6 + 6O_2$$

$$\Delta G = +\ 686 \text{ kcal/mole}$$

$$\text{Breakdown: (respiration) } C_6H_{12}O_6 + 6O_2 \rightarrow 6CO_2 + 6H_2O + \text{energy}$$

$$\Delta G = -\ 686 \text{ kcal/mole}$$

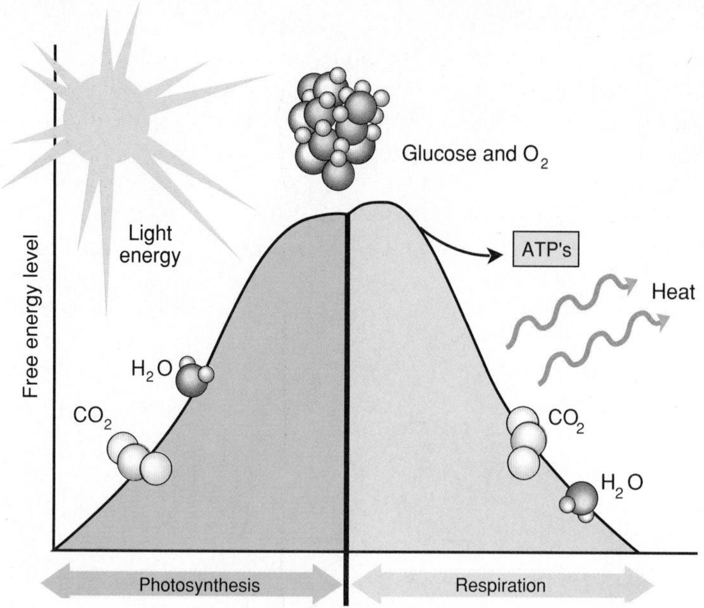

FIGURE 8.1
UPHILLS AND DOWNHILLS IN CELL ENERGETICS
Photosynthesis is portrayed as an uphill (endergonic) process in which sunlight energy is utilized to power the synthesis of glucose from carbon dioxide and water. Respiration is a downhill (exergonic) process in which glucose is broken down to carbon dioxide and water, with some of its free energy conserved in ATP.

The reactions are not simply reversed, though. For example, the energy for photosynthesis originates as photons of light, whereas in the metabolic breakdown reaction (that is, in the catabolism of glucose), energy is extracted primarily from the chemical bonds linking hydrogen to carbon (C—H).

The free energy changes (ΔG) show another important difference. The photosynthetic reaction is essentially an endergonic, or *uphill* process, with the product, glucose, having much more free energy than the reactants—carbon dioxide and water. In the breakdown of glucose, which we will consider next, the glucose and oxygen fall to considerably lower free energy states. Some of the reactions of this *downhill,* or exergonic reaction, though, are coupled with endergonic steps that produce energy-laden ATP. The production and utilization of glucose are represented in Figure 8.1.

THREE WAYS OF UTILIZING GLUCOSE

The fundamental process in cell energetics is the conversion of the great deal of free energy in glucose to the small, more easily managed, energy units of ATP. (It's about like changing a thousand dollar bill for 100 tens—smaller, more spendable units.) In terms of the transfer of glucose energy to ATP, organisms can be divided into three groups: those that utilize **aerobic respiration,** those using **anaerobic respiration,** and those using **fermentation.**

Respiration Begins with Glycolysis

Each of the three begins with **glycolysis,** an initial phase in the metabolism of glucose. We will say more about glycolysis shortly, but here let's just note that glycolysis yields a small amount of ATP through **substrate-level phosphorylation.** Much greater amounts of ATP are produced in chemiosmotic phosphorylation (described in Chapters 6 and 7), but substrate-level phosphorylation is much simpler and more direct. In certain reactions during glycolysis, the fuel molecules are phosphorylated (phosphate is added), greatly increasing their free energy and reactivity. Then at critical stages, the phosphates, along with their valuable energy, are transferred to ADP, yielding ATP. The final products of glycolysis, still energy-rich, are then metabolized in aerobic or anaerobic respiration, or in fermentation.

Aerobic vs. Anaerobic Respiration

Most organisms utilize aerobic respiration, the metabolic pathway that requires oxygen. The products of glycolysis are oxidized and their electrons and protons sent through the electron transport systems. Having passed through these systems, the spent electrons finally combine with oxygen. Thus oxygen becomes a final electron acceptor. As electrons move along these systems, as we saw in Chapters 6 and 7, their free energy is used in creating the proton gradient that powers chemiosmotic phosphorylation. Most of the ADP phosphorylated to form ATP in aerobic respiration comes from the chemiosmotic process.

Some organisms employ anaerobic respiration, a metabolic pathway that does not use oxygen. Most are bacteria, those specifically adapted to anaerobic conditions such as exist in the muddy bottom sediments of anaerobic lakes, or in airless pockets far down in the soil. Many of these bacteria are actually poisoned by oxygen. Most of the ATP formed in anaerobic respiration comes from chemiosmotic phosphorylation, just as it does in aerobic respiration. However, the final electron acceptor cannot be oxygen. Instead, spent electrons from the electron transport system join with inorganic ions such as sulfate or nitrate, or with carbon dioxide. Some final products include nitrite, nitrous oxide, hydrogen sulfide, and methane gas.

Fermentation

As far as glucose utilization goes, the fermenters, organisms that utilize fermentation as a means for generating ATP, are by far the simplest. Fermentation is identical to glycolysis except for the addition of a few final reactions. Thus, fermenters make use of the preliminary glycolytic pathway in which a small amount of ATP is produced through substrate-level phosphorylation. The final products of glycolysis, rather than entering the efficient ATP-generating pathways of respiration, are simply converted to energy-rich organic wastes, and with these the metabolism of glucose ends. So fermenters make do without electron transport systems and chemiosmotic phosphorylation.

Fermenting organisms include several kinds of bacteria and other microorganisms whose waste products are often commercially valuable. These waste products include several common alcohols (isopropyl, butyl, and ethyl) and organic acids (acetic, lactic, propionic, and formic). The most familiar fermenters are probably yeasts, single-celled fungi that brewers use to produce ethyl alcohol. We will return to fermentation later in the chapter.

The Three Parts of Respiration

Our discussion of respiration will be divided into three parts, as summarized in Figure 8.2. The first occurs in the cytoplasm and the other two in the mitochondrion.

Part I, Glycolysis: breakdown of glucose and substrate-level phosphorylation;

Part II, Citric acid cycle: the complete oxidation of fuels to CO_2 and H_2O, yielding protons and electrons;

Part III, Electron transport and chemiosmotic phosphorylation: using the free energy of electrons to produce a proton gradient, and using the proton gradient to form ATP.

GLYCOLYSIS

Since glycolysis is common to most of the earth's organisms, and since it does not require oxygen, it has been suggested that the glycolytic pathway arose early in the evolution of life. This notion agrees with the conviction among scientists that there was little oxygen available in the early atmosphere (see Chapter 21). Cells of today have retained the simple glycolytic pathway, but as we've noted, it is now just a preliminary part of respiration that occurs in the cytoplasmic fluids. The other two parts of aerobic respiration involve mitochondria, structures that appeared later in the evolution of cells.

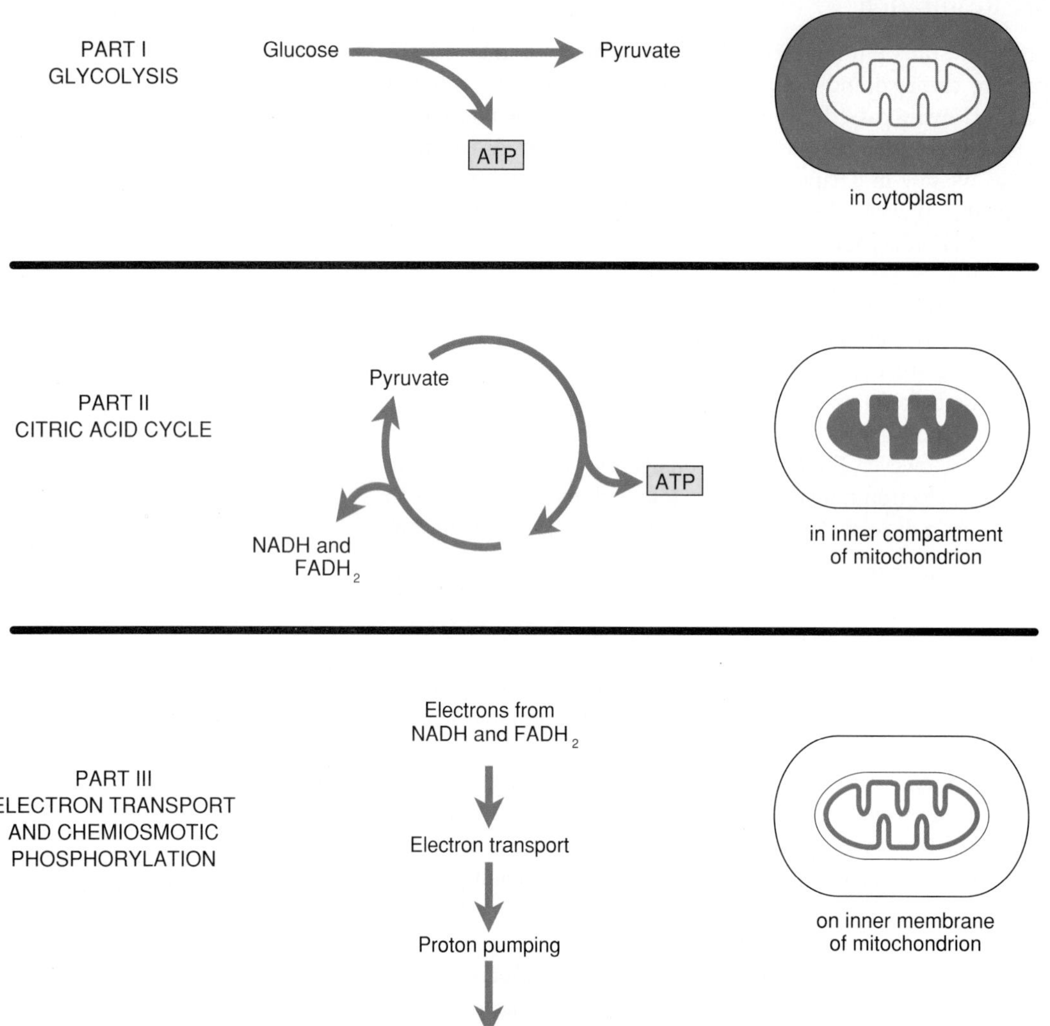

FIGURE 8.2
THREE PARTS OF RESPIRATION
The three parts of glucose catabolism are incorporated into a cell scenario. Part I: Glycolysis occurs in the cytoplasmic fluids, where some ATP is generated. Part II: The products of glycolysis enter the mitochondrion, where they are oxidized in the citric acid cycle. Part III: Electrons from the citric acid cycle are sent through electron transport systems, during which their free energy is used to boost the chemiosmotic proton gradient. Finally, the free energy of the gradient is used to phosphorylate ADP in F_1 complexes, where most of cell's ATP is produced.

In glucose catabolism (metabolic breakdown), it is important for the cell to break the chemical bonds of glucose in such a way as to avoid the sudden release of energy, since this would result in the wasteful (and dangerous) production of heat, an unsatisfactory situation for cells. The transfer of free energy from glucose to ATP proceeds through a gentler, more gradual series of coupled reactions, some losing, others gaining energy.

As an overview we can note that the glycolytic pathway begins with a glucose molecule and ends with two pyruvates (the ions of pyruvic acid). Along the way two coenzyme NAD^+s are reduced to NADH, and four ADPs are phosphorylated, forming four molecules of ATP. Thus, some of the free energy of the original glucose has been conserved in NADH and ATP (Figure 8.3).

The Highlights of Glycolysis

A study of Figure 8.4, the complete glycolytic pathway, will reveal that there are some nine principal reactions in glycolysis, each involving a specific enzyme. The reactions can be conveniently divided into two phases, a "preparatory phase" and a "yielding phase." The first increases the free energy of the substrate (or fuel), thus causing it to become unstable and reactive. In the second phase, the free energy is used in phosphorylating ADP. As we proceed, we'll confine our discussion of glycolysis to just the key reactions in each phase. (The numbers refer to reaction numbers in Figure 8.4.)

The Preparatory Phase: Priming the Fuel The first group of preparatory reactions, in particular reactions 1 and 3, are phosphorylations using phosphates from ATP. While costly, this investment of ATP serves to activate the glucose molecule. That is, it provides the necessary energy of activation, thus priming it for further reactions.

Following this, the doubly phosphorylated fuel is cleaved (reaction 4). Its important products, two glyceraldehyde-3-phosphates (G3P), undergo a complex double reaction that ends the preparatory phase. In reaction 5 the two G3Ps are oxidized and phosphorylated, reactions that greatly elevate their free energy. The oxidations are coupled with the reduction of two NAD^+s to NADH. The newly added phosphate group is in a high energy form, a state that primes it for transfer to another substrate.

The Yielding Phase: Forming ATP The two doubly-phosphorylated products are now fully primed for a key reaction (6). An energy-rich phosphate group from each is transferred to two waiting ADPs. This direct enzyme-mediated phosphorylation of ADP is the substrate-level process we mentioned at the start of this chapter. Thus reaction 6 yields two ATPs.

In the overall glycolytic scheme, the yield of two ATPs so far is really just a "payback," since this replaces the ATPs invested earlier. But a second yield is coming up. This happens in reaction 8, in which a dehydration reaction creates molecular instability. Whereas the removal of H_2O does not increase the molecule's free energy, its distribution become highly uneven. Much of that energy is now centered in the remaining phosphate. When this phosphate bond is cleaved, the ΔG will be large (-14.8 kcal/mole). Part of this energy is conserved as the second pair of ADP phosphorylations occurs. This step produces the net yield from glycolysis: two ATPs for each glucose.

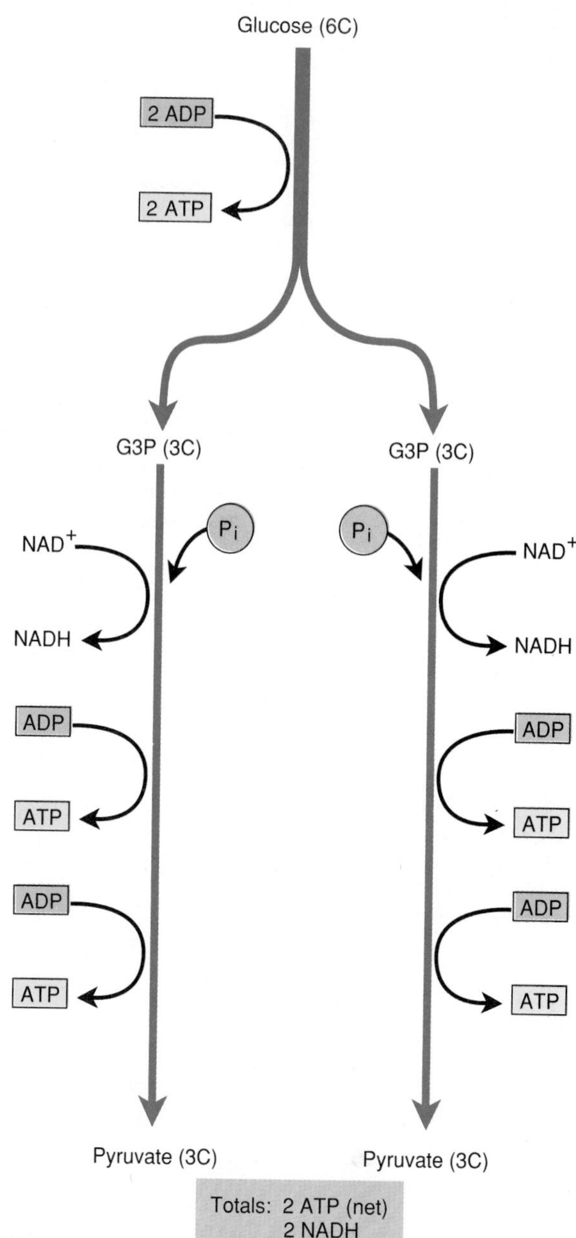

FIGURE 8.3
QUICK VIEW OF GLYCOLYSIS
A simplified view of glycolysis shows glucose entering a pathway, where it is twice phosphorylated by ATP and then split into 3-carbon fuel products. The fuel molecules are oxidized by enzymes and phosphorylated, which prepares them for two substrate-level phosphorylations in which 4 ATP are produced. The net end products of glycolysis are 2 ATP, 2 NADH, and 2 pyruvate molecules.

The Net Energy Yield Using some fairly simple calculations, we can determine the net energy yield in ATP from glycolysis. Such calculations are made on a mole-to-mole basis, as was described in Chapter 6. Recall that when ATP is hydrolyzed to ADP under standard laboratory conditions, the free energy change is -7.3 kcal/mole. At the end of glycolysis, two ATPs were gained, so the *net free energy gain* in ATP, per mole of glucose sent through glycolysis, is 14.6 kcal (Table 8.1)

How much of the free energy originally in glucose does this represent? The free energy of glucose, when the sugar is broken down all the way to CO_2 and H_2O, is 686 kcal/mole. The energy conserved in ATP during glycolysis, then, is only about 2.1% of the total ($14.6/686 \times 100 = 2.13\%$). As we will find, this is a low figure compared to the

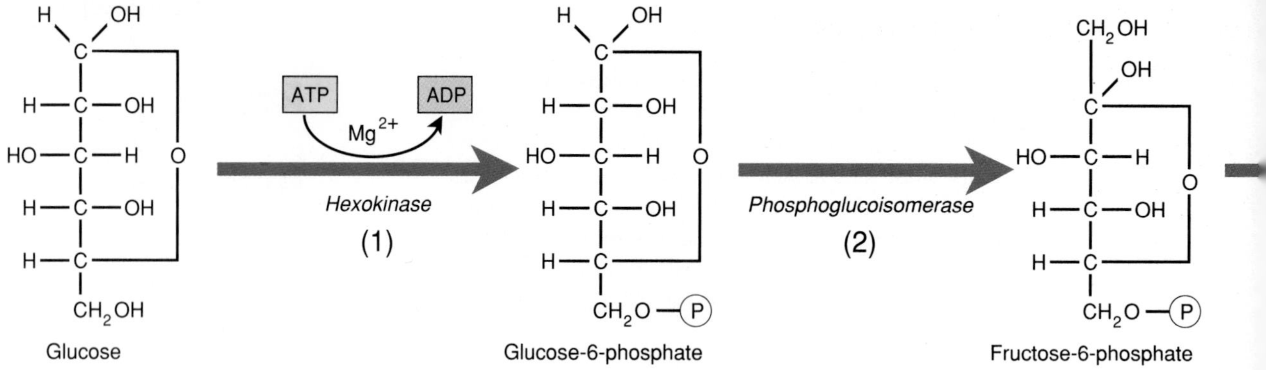

Glucose

ATP → ADP (Mg^{2+})

Hexokinase

(1)

Glucose-6-phosphate

Phosphoglucoisomerase

(2)

Fructose-6-phosphate

Reaction 1 is strongly exergonic. ATP phosphorylates glucose, yielding glucose-6-phosphate. In the process the free energy of glucose is increased by 3.3 kcal/mole (4.0 kcal are lost as heat).

Reaction 2 is slightly endergonic. Fructose-6-phosphate is formed, preparing the fuel for the next phosphorylation by ATP.

GLYCOLYSIS

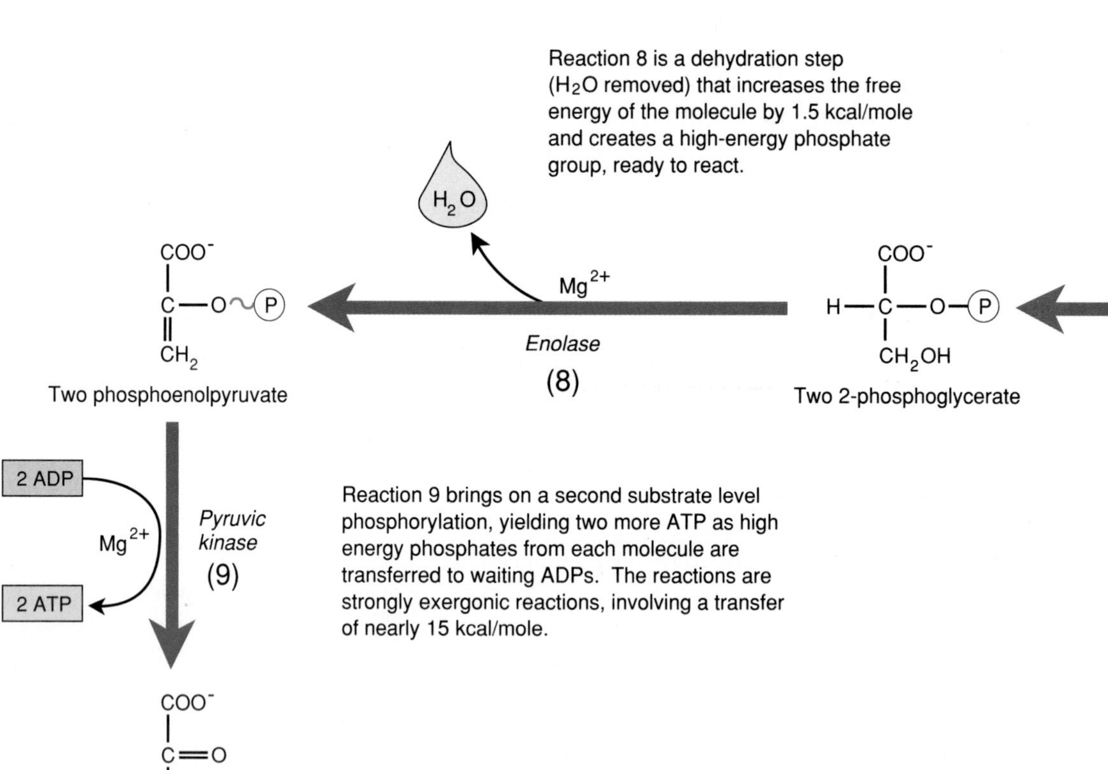

Reaction 8 is a dehydration step (H_2O removed) that increases the free energy of the molecule by 1.5 kcal/mole and creates a high-energy phosphate group, ready to react.
H_2O
Mg^{2+}
Enolase
(8)
Two phosphoenolpyruvate
Two 2-phosphoglycerate
2 ADP
Mg^{2+}
2 ATP
Pyruvic kinase
(9)
Reaction 9 brings on a second substrate level phosphorylation, yielding two more ATP as high energy phosphates from each molecule are transferred to waiting ADPs. The reactions are strongly exergonic reactions, involving a transfer of nearly 15 kcal/mole.
Two pyruvate

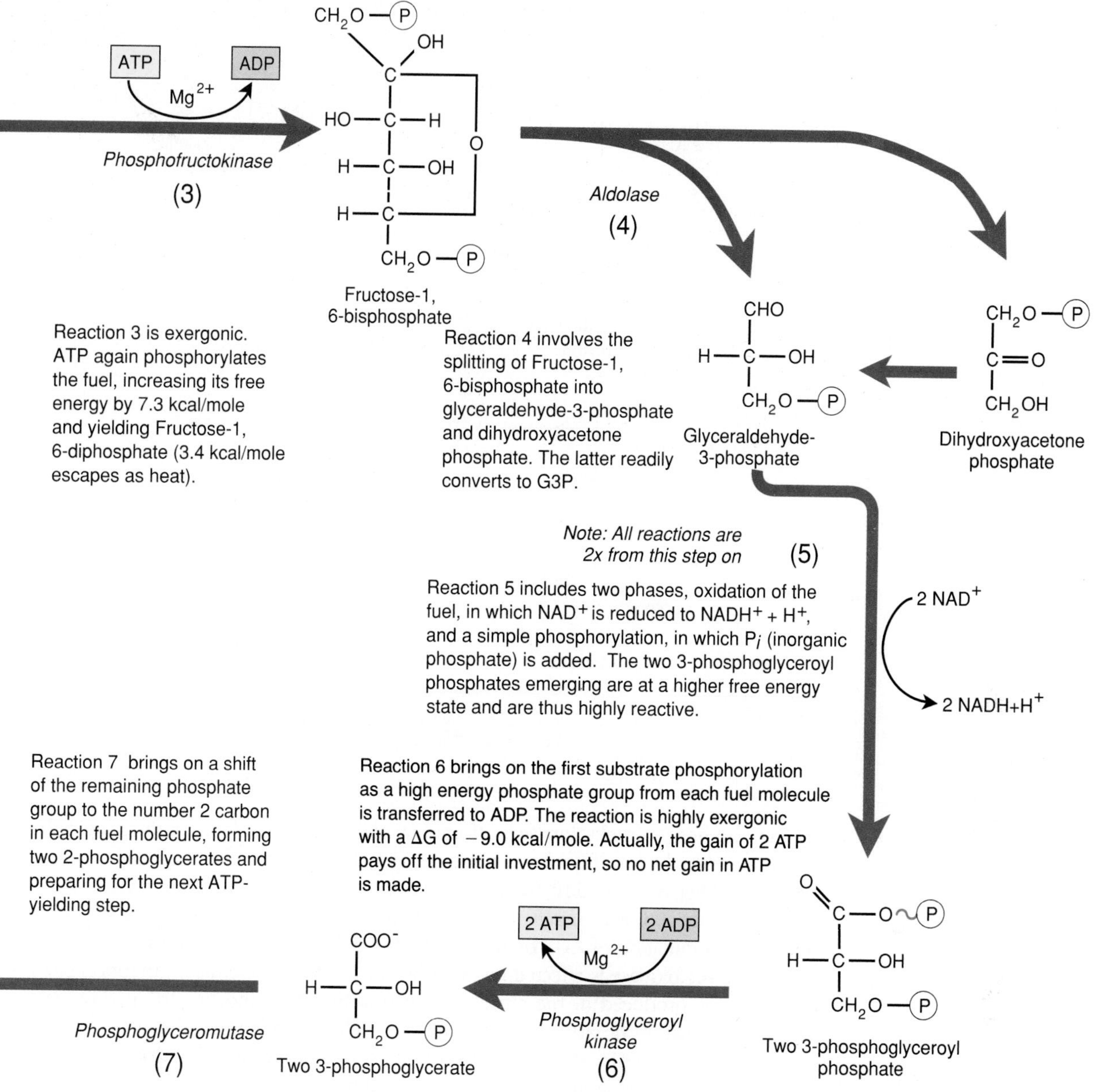

ATP
ADP
Mg 2+
Phosphofructokinase
(3)
Fructose-1,
6-bisphosphate
Reaction 3 is exergonic. ATP again phosphorylates the fuel, increasing its free energy by 7.3 kcal/mole and yielding Fructose-1, 6-diphosphate (3.4 kcal/mole escapes as heat).
Aldolase
(4)
Reaction 4 involves the splitting of Fructose-1, 6-bisphosphate into glyceraldehyde-3-phosphate and dihydroxyacetone phosphate. The latter readily converts to G3P.
Glyceraldehyde-3-phosphate
Dihydroxyacetone phosphate
Note: All reactions are 2x from this step on
(5)
Reaction 5 includes two phases, oxidation of the fuel, in which NAD+ is reduced to NADH+ + H+, and a simple phosphorylation, in which Pi (inorganic phosphate) is added. The two 3-phosphoglyceroyl phosphates emerging are at a higher free energy state and are thus highly reactive.
2 NAD+
2 NADH+H+
Reaction 7 brings on a shift of the remaining phosphate group to the number 2 carbon in each fuel molecule, forming two 2-phosphoglycerates and preparing for the next ATP-yielding step.
Reaction 6 brings on the first substrate phosphorylation as a high energy phosphate group from each fuel molecule is transferred to ADP. The reaction is highly exergonic with a ΔG of −9.0 kcal/mole. Actually, the gain of 2 ATP pays off the initial investment, so no net gain in ATP is made.
2 ATP
2 ADP
Mg 2+
Phosphoglyceromutase
(7)
Two 3-phosphoglycerate
Phosphoglyceroyl kinase
(6)
Two 3-phosphoglyceroyl phosphate

FIGURE 8.4
GLYCOLYSIS
Glycolysis occurs in 9 principal reactions, each involving a specific enzyme. The first (preparatory) phase includes phosphorylations by ATP, cleaving of the fuel into 3-carbon products, and finally oxidation and phosphorylation. The second (yielding) phase includes two substrate-level phosphorylations, with a net yield of 2 ATPs. The final products are ATP, NADH, and pyruvate.

energy yield from the aerobic process to come. Yet, interestingly, the fermenters mentioned earlier get along quite well with this modest yield of ATP from glucose. The free energy left after glycolysis is now present in the two pyruvates and two NADHs.

Control of Glycolysis

Like so many biochemical processes, glycolysis is under a number of delicate control factors. As a general observation, when the cell is using its ATP at a rapid rate, glycolysis speeds up, and conversely, if its ATP use is curtailed, glycolysis slows down. But, what adjusts such rates? Obviously, the availability of glucose is one controlling factor, but there are also far more subtle regulatory mechanisms.

A major regulatory factor is the responsiveness of a key participant, **phosphofructokinase**, an allosteric enzyme responsible for the conversion of fructose-6-phosphate to fructose 1,6-diphosphate, reaction 3 in glycolysis. As an allosteric enzyme (see

TABLE 8.1 ENERGY YIELD IN GLYCOLYSIS (PER MOLE)

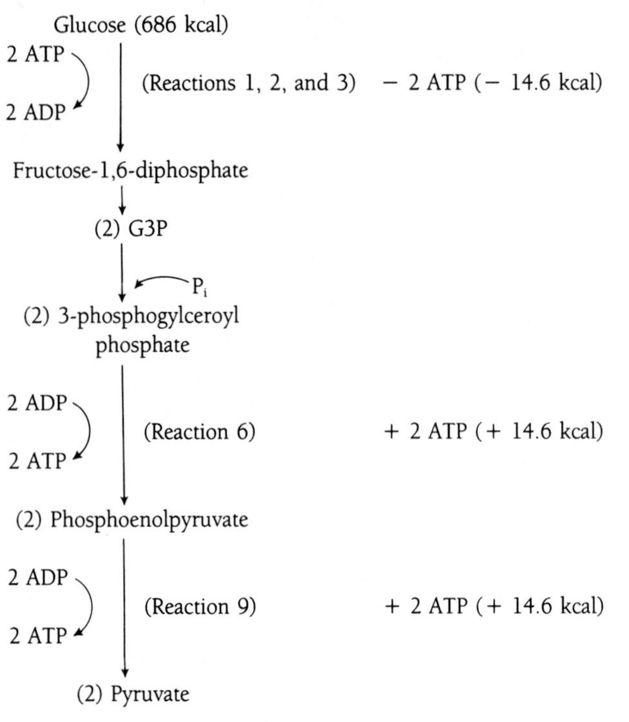

Net gain	+2 ATP (+ 14.6 kcal)
Percentage efficiency	2.1%

Chapter 6), phosphofrucktokinase is affected by both activators and inhibitors. Logically, it is inhibited by high cellular concentrations of ATP and is stimulated by high concentrations of either ADP or AMP. This makes sense, since an accumulation of ATP would indicate that there wasn't much metabolic activity going on in the cell. On the other hand, an accumulation of ADP or AMP or both would suggest the opposite.

Phosphofructokinase is affected by other factors as well. One of its inhibitors is citric acid, which is produced in the citric acid cycle within the mitochondrion (discussed next). The formation of citric acid early in that cycle depends upon the availability of pyruvate from glycolysis. Should citric acid diffuse out of the mitochondrion, it would indicate that reactions within are "backing up" and more pyruvate is being produced than can be removed by the citric acid cycle. So, with citric acid inhibiting phosphofructokinase, a drop in pyruvate soon follows. Thus glycolysis is slowed and adjusted to current needs.

Rates of reaction in this versatile key enzyme can change radically. When muscle is working at a maximum, the activity of phosphofructokinase increases several hundred times. This is important because glycolysis provides much of the ATP utilized in hard-working muscles.

AEROBIC RESPIRATION: THE MITOCHONDRION

Recall that in most organisms, glycolysis is just the beginning of glucose catabolism. Depending on the organism and in some instances on the tissues of that organism, pyruvate, the final product of glycolysis, can enter one of several pathways. The most common is the aerobic respiration process, which occurs in the mitochondrion. There, much of the free energy remaining in pyruvate will be transferred to ATP. We can begin our account of aerobic respiration in the mitochondrion where we left off with glycolysis—with pyruvate and NADH, products that are formed in the cytoplasm.

Pyruvate to Acetyl-CoA

NADH cannot enter the mitochondria, but there is a mechanism, a shuttle carrier as it were, to carry its energetic electrons inside. The pyruvate, however, can enter the mitochondria, where it is altered so that it can take part in the citric acid cycle. It becomes part of an energy-rich substrate called **acetyl-CoA (acetyl-coenzyme A).** This transition turns out to be a complex process involving a gigantic enzyme/coenzyme complex. Coenzymes, incidentally, include derivatives of four vitamins that may be familiar to you: thiamine, riboflavin, pantothenic acid, and nicotinic acid. (So those caring people who pestered us about eating our vegetables all those years were right after all. One wouldn't want to run out of acetyl-CoA.) With so many large interacting molecules, the CoA complex turns out to be almost as large as a ribosome. The intricate details are beyond us, but the overall reaction for each pyruvate entering the complex is:

$$\text{pyruvate} + NAD^+ + \text{CoA} \rightarrow \text{acetyl-CoA} + NADH + H^+ + CO_2$$

As you see, coenzyme A is covalently linked to a two-carbon acetyl group (CH_3CO), forming acetyl-CoA (Figure 8.5).

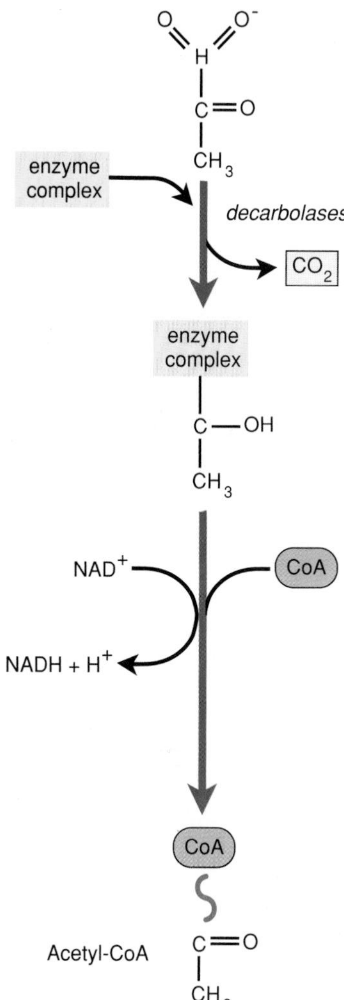

FIGURE 8.5
THE ACETYL-COA STEP
The conversion of pyruvate to acetyl-CoA requires the activity of a large enzyme/coenzyme complex. The products are carbon dioxide, NADH, and a 2-carbon acetyl group bonded to a molecule of coenzyme A.

The Citric Acid Cycle

The **citric acid cycle** (or **Krebs cycle**) was first described in 1937 by Hans A. Krebs. Ironically, although his paper has become a classic in biochemistry, it was rejected by the disbelieving editors of *Nature,* the prestigious British journal to which it was first submitted. However, Professor Krebs was later awarded the Nobel prize for his efforts.

One of the key functions of the citric acid cycle is the oxidation of fuel molecules. The oxidation reactions are coupled to the reduction of coenzymes NAD^+ and FAD, which then transfer captured energetic electrons and protons to the electron transport system. The free energy of the electrons powers proton pumps that produce the all-important proton gradient used in the chemiosmotic phosphorylation of ADP. For an overview of the cycle see Figure 8.6.

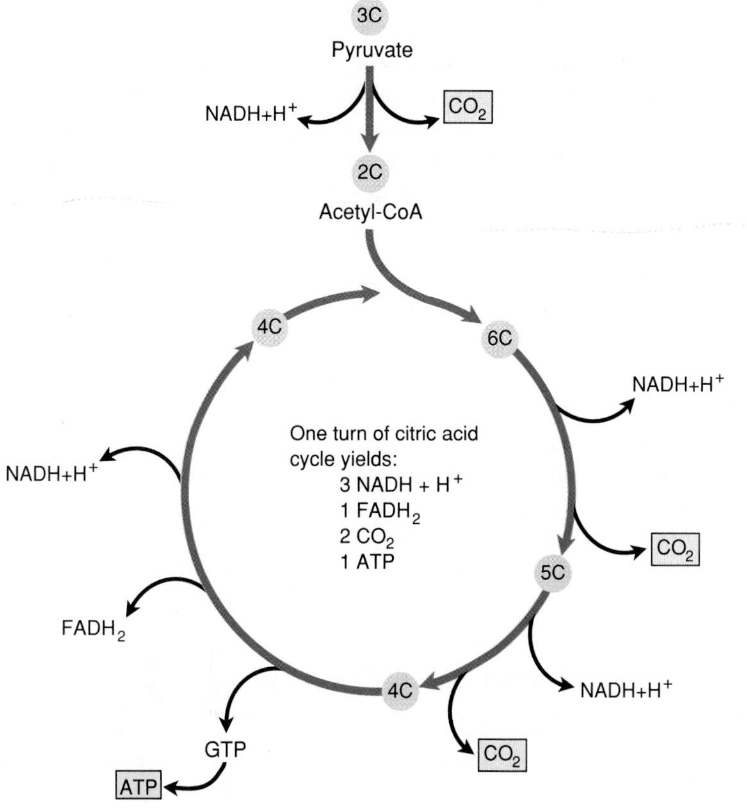

FIGURE 8.6
QUICK VIEW OF THE CITRIC ACID CYCLE
The abbreviated view of a citric acid cycle lets us follow one turn by counting carbons and reduced coenzymes. The cycle begins when the 2-carbon acetyl-CoA and a 4-carbon acid join, forming a 6-carbon product. Four oxidations lead to the yield of 3 NADH and 1 $FADH_2$. One substrate-level phosphorylation produces an ATP, and two decarboxylations (CO_2 removals) account for the loss of two carbons. The final product, a 4-carbon acid, is ready to join another 2-carbon acetyl-CoA and enter another cycle.

Principal Reactions of the Citric Acid Cycle Figure 8.7 reviews each of the nine citric acid cycle reactions, so here we will mention only those steps where key reactions occur.

Enter Acetyl CoA Acetyl-CoA enters the citric acid cycle by reacting with four-carbon **oxaloacetate** (what might be called the "resident molecule," because it must be present for the cycle to begin each time). The enzyme that prompts this stage is called citrate synthetase, and the product is 6-carbon citrate, or citric acid, the same acid found in citrus fruits and the one for which the cycle is named. Because of the great free energy of acetyl-CoA, the reaction occurs quite readily. Once freed, the CoA group simply recycles to join with more pyruvate, generating more acetyl-CoA.

Carbon Dioxide Release Since this pathway is a cycle, oxaloacetate will later form again, bringing the series of reactions back where it began. To proceed from 6-carbon citrate back to 4-carbon oxaloacetate, however, there will have to be two decarboxylations (CO_2 removals; see reactions 4 and 5). Although these reactions do not bring about an immediate energy yield, they are interesting because this is the CO_2 we exhale with each breath—the CO_2 all organisms release as they carry on aerobic respiration.

Oxidations in the Cycle As with other biochemical pathways, some of the reactions result in subtle molecular rearrangements and free energy shifts, each of which prepares the molecule for major reactions to follow. In the citric acid cycle, the major reactions are primarily oxidations, so we'll consider reactions 4, 5, 7, and 9. In each of these, oxidation of the fuel is coupled with the transfer of electrons and protons to coenzyme NAD^+ or coenzyme FAD, reducing them to NADH and $FADH_2$, respectively. The free energy change in these redox reactions is great, with ΔG sometimes in the neighborhood of -7 kcal/mole. But most significantly, the newly reduced coenzymes themselves have the reducing power needed to activate nearby electron transport systems and proton pumps.

Substrate-Level Phosphorylation Another event of special interest is found in reaction 6. Here we see a substrate-level phosphorylation, much the same as occurs in glycolysis. In this reaction, GDP (guanosine diphosphate) is phosphorylated to form GTP (guanosine triphosphate), which in turn phosphorylates ADP to form ATP. GTP is chemically similar to ATP, but with the nitrogen base guanine replacing adenine. The free energy exchange in the reaction generating GTP is high ($\Delta G = -8.0$ kcal/mole). Thus, we see that the key reactions of the citric acid cycle are decarboxylations, oxidations/reductions, and substrate-level phosphorylations.

Glucose Catabolism So Far Let's stand back now and look at the mitochondrial reactions so far. The two important kinds of yields are the reduced coenzymes NADH and $FADH_2$, and the ATP formed at the substrate level. For each pyruvate entering the mitochondrion there is a yield of an ATP, one NADH from the acetyl CoA step, and three NADHs and one $FADH_2$ resulting from oxidations within the citric acid itself. However, since we started respiration with glucose, and each glucose yields two pyruvates, we should take this into account. Thus, in the reactions so far we can note that there are 2 ATPs, 2 NADH formed in the acetyl CoA step, six NADH and 2 $FADH_2$ from oxidations in the citric acid cycle. As we proceed into electron transport and chemiosmotic phosphorylation, you will see the importance of these yields to the ongoing generation of ATP.

ELECTRON TRANSPORT SYSTEMS AND CHEMIOSMOTIC PHOSPHORYLATION

You may recall from Chapter 7 that the light reactions of photosynthesis were dependent on the precise structure of the chloroplast. Chemiosmotic phosphorylation in the mi-

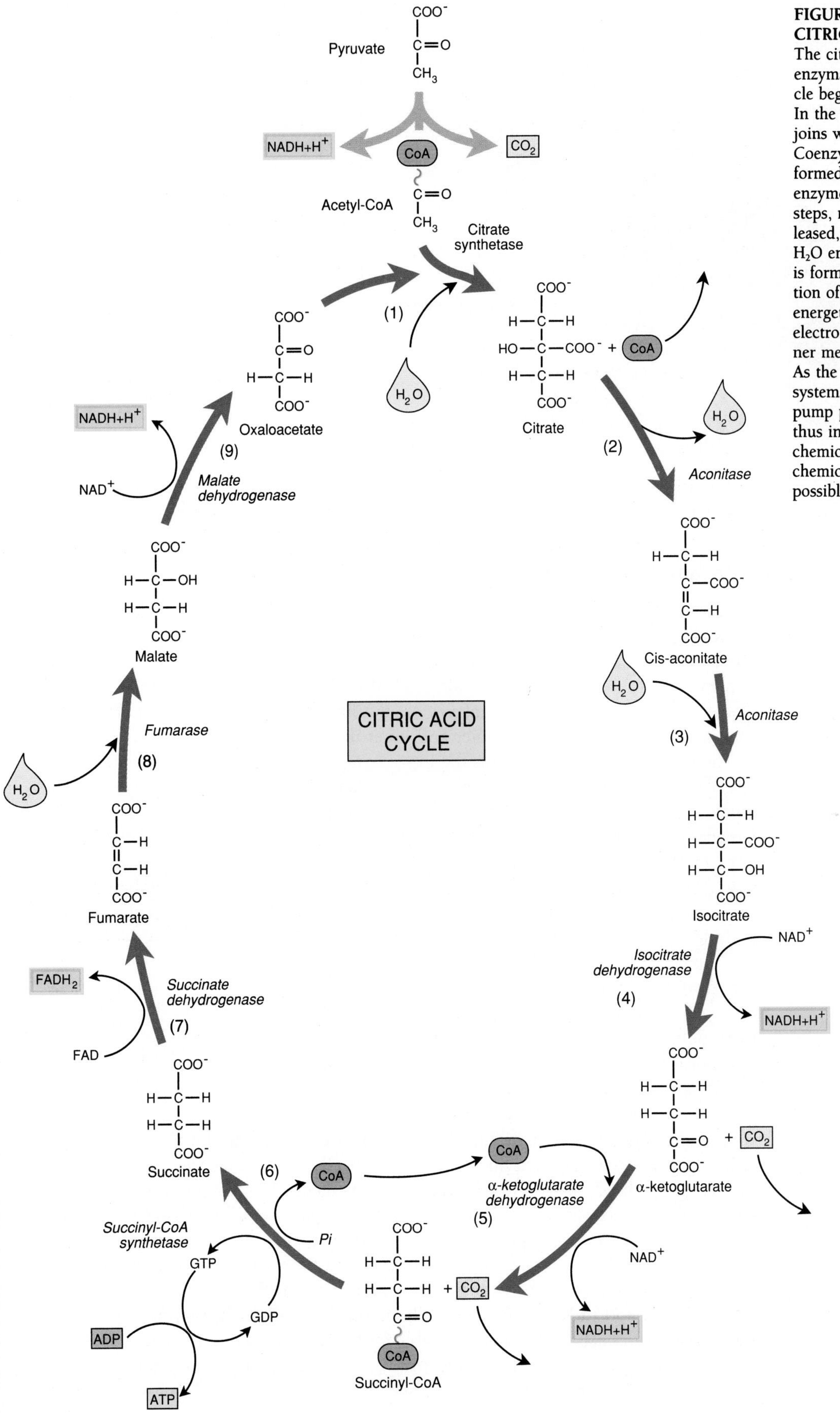

FIGURE 8.7
CITRIC ACID CYCLE
The citric acid cycle has nine major enzymatic steps. Each turn of the cycle begins and ends with oxaloacetate. In the first reaction, oxaloacetate joins with acetyl-CoA and water, and Coenzyme A is released. The product formed is citrate (citric acid). As each enzyme does its job in the succeeding steps, molecules change, CO_2 is released, NAD^+ and FAD are reduced, H_2O enters and leaves, and one ATP is formed. However, the main function of the citric acid cycle is to make energetic electrons available to the electron transport systems in the inner membrane of the mitochondrion. As the electrons pass through these systems, their energy will be used to pump protons across the membrane, thus increasing the free energy of the chemiosmotic gradient, and making chemiosmotic phosphorylation possible.

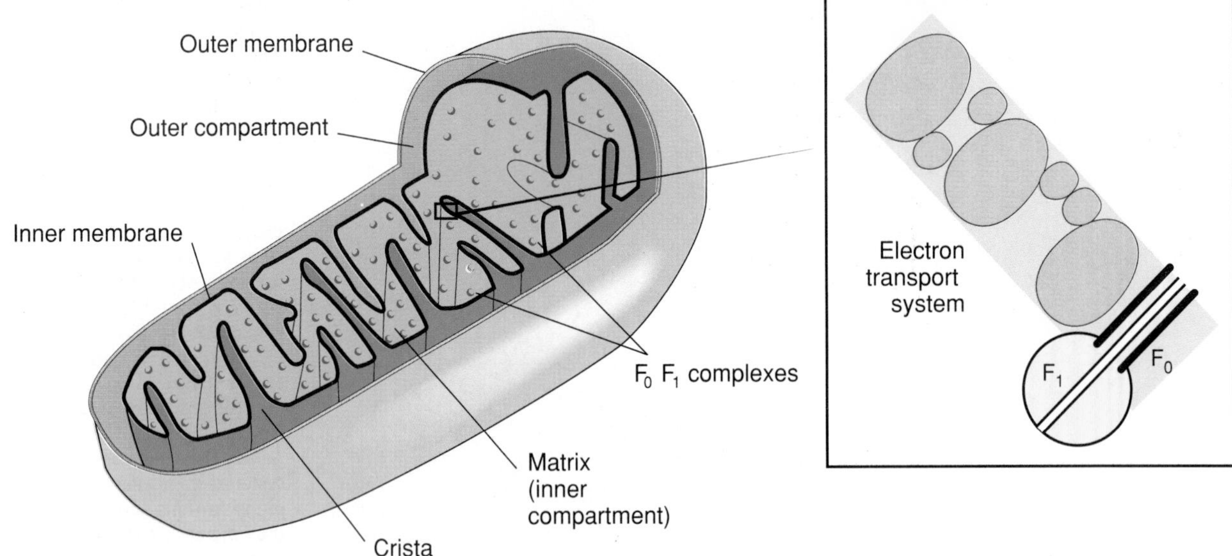

FIGURE 8.8
THE MITOCHONDRION
A three-dimensional view of the mitochondrion provides clues to its functions. Note the two compartments, separated by the inner membrane. Along the inner membrane are numerous aggregations of electron carriers that form highly organized electron transport systems. Also embedded in the inner membrane are F_0F_1 complexes, their channels (F_0) leading from the outer compartment into the bulbous F_1 complexes where ATP is synthesized.

tochondrion is likewise dependent on precise internal structure. In fact, it may occur to you as we proceed that there are many functional parallels between the two organelles.

Mitochondrial Structure

We saw in Chapter 4 that the mitochondrion contains an **outer** and an **inner membrane**—the sac-within-a-sac arrangement typical of chemiosmotic systems (Figure 8.8). The outer membrane is much simpler than the inner membrane. It is readily permeable to many small molecules and ions, although it does contain certain transport proteins that are dedicated to the active transport or shuttling of materials in and out. The **outer compartment** lies between the outer and inner membranes.

The inner membrane surrounds the **inner compartment** or **matrix.** This membrane forms extensive folds or **cristae** that extend deep into the inner compartment. The cristae, you may recall from Chapter 4, vastly increase the surface area of the inner membrane, thus providing an enormous reaction surface. The inner membrane is highly impermeable to most small molecules and ions, but it permits the transit of protons at special sites. There are also special assemblages of proteins in the inner membrane. They are involved in electron transport and the production of ATP. We'll look very closely at these.

Electron Transport Systems and F_0F_1 Complexes The inner mitochondrial membrane contains numerous electron transport systems. Although these systems pass electrons along from one carrier to the next, some of the carriers also serve as proton pumps, transporting protons from the inner to the outer compartment. Most of the carriers are cytochromes, proteins that contain iron. As explained in Chapter 6, it is the iron that receives electrons and passes them along.

The electron carriers are arranged so that each carrier along the route has greater reducing power than the next one in line. Thus as electrons are received by one carrier, they are immediately passed to the next, at an energetically lower level, in an ongoing sequence of reductions and oxidations. Figure 8.9 illustrates the arrangement of the carriers. We'll call them simply sites I, II, and III. The three sites are somewhat distant from each other, but electrons can pass from one to the next via mobile carriers.

Other prominent elements of the inner membrane are the F_0F_1 complexes. The tubular part is the F_0, and the enzyme-laden bulbous part is the F_1. The enzymes, you recall, include ATP synthetase, the one responsible for the generation of ATP. The F_0 channels lead from the outer compartment into the F_1 complexes that protrude into the inner compartment. This arrangement, as we have seen, is important to chemiosmotic phosphorylation.

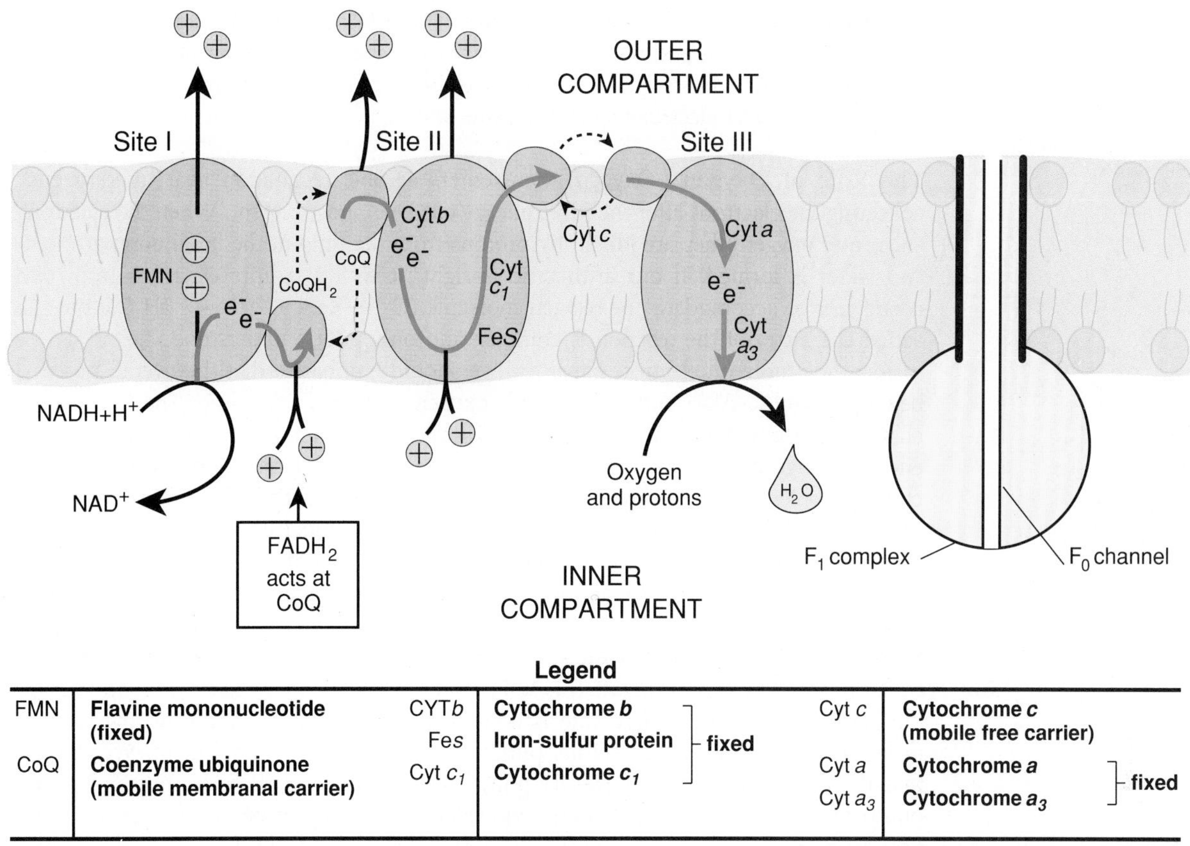

FMN	**Flavine mononucleotide (fixed)**	CYT*b*	**Cytochrome *b*** (fixed)	Cyt *c*	**Cytochrome *c* (mobile free carrier)**
CoQ	**Coenzyme ubiquinone (mobile membranal carrier)**	Fes	**Iron-sulfur protein** (fixed)	Cyt *a*	**Cytochrome *a*** (fixed)
		Cyt c_1	**Cytochrome c_1** (fixed)	Cyt a_3	**Cytochrome a_3** (fixed)

FIGURE 8.9
MITOCHONDRIAL ELECTRON TRANSPORT SYSTEM
The electron carriers within the inner mitochondrial membrane form a sequence starting with FMN at Site I and ending with cytochrome a_3 at Site III. Their arrangement is such that electrons move to the right, as shown here. Note that protons are pumped into the outer compartment at three locations within the ETS. The flow of electrons and pumping of protons in the mitochondrial electron transport system occur as follows: NADH reduces the first carrier FMN, which receives 2 electrons and 2 protons. The 2 protons are pumped into the outer compartment, and the 2 electrons pass into CoQ. CoQ shuttles 2 protons to the outer compartment and passes the electrons along to the central cytochrome complex (Site II). The central cytochrome complex passes the two electrons to cytochrome *c* and shuttles two more protons across to the outer compartment. The electrons are then passed to the final cytochrome complex (Site III). At the last complex, the two electrons join oxygen, the final acceptor, forming water. Note that coenzyme $FADH_2$ reduces CoQ further down the system.

NADH Electrons Build the Chemiosmotic Gradient

We're now ready to see just how electrons and protons from the citric acid cycle, now carried by NADH and $FADH_2$, are put to work building the chemiosmotic gradient. We'll begin with the arrival of NADH at the first carrier of an electron transport system (Figure 8.9).

The flow of electrons begins at Site I when NADH passes two electrons and two protons to the first carrier, **flavin mononucleotide**, or **FMN**, a large transmembranal protein. The protons from NADH pass to the far side of FMN, and are released into the outer compartment, leaving the two electrons behind. Thus we see that FMN represents the first proton pump.

The two electrons, still in a highly energetic state are passed along to a small, mobile carrier called **coenzyme Q**, or simply **CoQ**. The next event isn't entirely clear, but apparently, reduced CoQ captures two protons from the inner compartment, becoming $CoQH_2$ (its reduced form). It then crosses the membrane to the Site II complex and releases its two protons into the outer compartment. Upon the release of the two protons, CoQ's electrons are passed along to the cytochrome *b* in Site II. Then, once more in its oxidized form, CoQ returns back for its next cycle. Thus CoQ acts as a proton shuttle, with each trip enriching the chemiosmotic gradient by two more protons (for a total of 4).

Site II is the third and final proton pumping site, and two more protons are transported into the outer compartment. From Site II, the electrons move on to cytochrome *c*, a small mobile surface carrier between Site II and Site III. Cytochrome *c* passes its electrons to Site III. At Site III the two electrons, now at a very much reduced free energy state, move to the matrix where oxygen, the final electron acceptor, awaits them, forming water.

Let's note that CoQ also has another role. It can pick up electrons and protons from the coenzyme $FADH_2$ and transport the protons to the outer compartment. So $FADH_2$, with less reducing power than NADH, acts further down the chain. From there on $FADH_2$ and NADH electrons work the same way.

The Role of Oxygen Oxygen, then, enters aerobic respiration near its very end, collecting the electrons after their free energy has been mostly spent. When the electrons encounter oxygen, they are joined by protons from the fluid in the inner compartment, and water is formed. If our arithmetic is right, it would require electrons from two coenzymes to accomodate one oxygen molecule ($O_2 + 4e^- + 4H^+ \rightarrow 2H_2O$), but the actual chemistry of the reaction is complex and only poorly understood.

As a final note on electron transport, we should emphasize that the overall process, beginning with NADH and ending with oxygen, is highly exergonic. The free energy change from one end to the other is −53 kcal/mole. Some idea of the nature of this energy change is suggested by Figure 8.10, a representation that is often likened to a child's marbles (the electrons) bouncing down a stairway. Let's keep in mind, however, that the −53 kcal/mole is the difference between the initial free energy state and the final free energy state. It doesn't tell us where the energy went. Of course, we now know that a considerable amount ends up as the free energy of the proton gradient.

Summing up the Proton Yields Two pyruvates entering the mitochondrion (representing one glucose), generate 8 NADHs and 2 $FADH_2$s. (See Figure 8.6 and include the NADH formed when pyruvate is converted to acetyl-CoA.) We can now express these products in terms of protons pumped into the outer compartment. We've seen that the electrons from each NADH pump six protons, and those from each $FADH_2$ pump four. Thus the chemiosmotic equivalent of two pyruvates turns out to be 56 protons pumped into the proton reservoir, $(8 \times 6) + (2 \times 4)$. But there are also the two NADHs from glycolysis to be considered (see Figure 8.3). Of course, we left them in the cytoplasm, and NADH cannot enter the mitochondrion so there's more to tell.

Cytoplasmic NADH reduces a mitochondrial coenzyme, either FAD or NAD^+, via a shuttle system in the outer mitochondrial membrane. The coenzyme reduced depends upon the kind of cell involved, but we'll consider it to be FAD, the coenzyme reduced in the animal muscle mitochondrion.

So we find that two cytoplasmic NADHs generate two mitochondrial $FADH_2$s (and in so doing, free NAD^+ to recycle back to glycolysis, where it is needed). The two

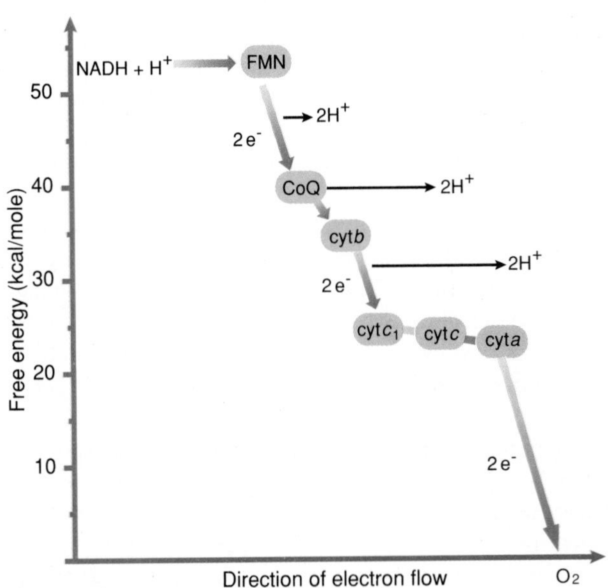

FIGURE 8.10
ENERGETICS OF THE ELECTRON TRANSPORT SYSTEM
Losses of reducing power and free energy by electrons traversing the mitochondrial electron transport system take the form of a descending stairway. Note that there are three steps in which the free energy change is substantial. These changes represent points along the system where protons are pumped, and thus can be equated to energy available for the generation of ATP.

$FADH_2$s account for eight more protons pumped, bringing the total to 64. Keep this number in mind as we turn once again to the chemiosomotic gradient and its use in the phosphorylation of ADP.

Reviewing the Chemiosmotic Gradient and ATP Formation

The chemiosmotic gradient in the mitochondrion has two components; the proton concentration gradient and the electrical charge difference across the inner membrane. The latter is due to the many positive charges (H^+) in the outer compartment and far fewer in the inner compartment. These two components have led to the alternate term "electrochemical gradient," and some biochemists refer to its free energy as the *proton-motive force*. But whatever we call it, what is important is the free energy, and how it is finally put to work.

As we've seen (Chapters 6 and 7), the free energy of the chemiosmotic gradient can be harnessed as its protons move to a lower free energy state. The protons move down their concentration gradient towards the negatively charged medium on the other side of the inner membrane, but due to the impermeability of the inner membrane the protons can only pass through the F_0 channels.

The phosphorylation of ADP to ATP occurs as protons pass down their gradient, from the outer compartment, through the F_0 channels, and into the F_1 complexes. The F_1 complexes contain the enzyme ATP synthetase, and it is here that the free energy of the proton gradient is used to phosphorylate ADP, thus generating new ATP (Figure 8.11).

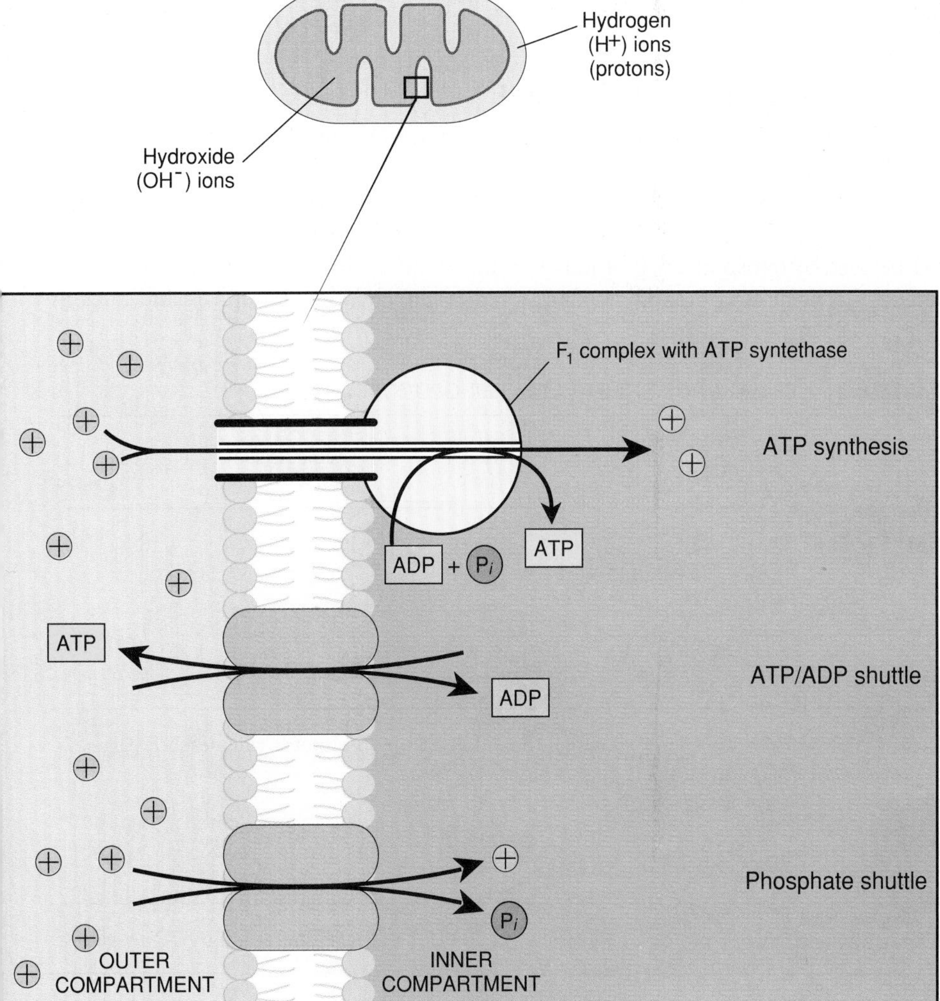

FIGURE 8.11
CHEMIOSMOTIC PHOSPHORYLATION
As protons escape the steep proton gradient of the outer compartment, their free energy is used within the F_1 complexes (bulbous heads) to form ATP from ADP and P_i. Each complex contains ATP synthetase, the phosphorylating enzyme of ATP. Note the two special cotransport carriers, one of which uses the proton gradient to exchange ATP for ADP, and the other which brings in Pi (inorganic phosphate).

Although there is some question about the numerical relationship between escaping protons and ATP produced, it appears that each pair of protons passing into an F_1 complex generates one ATP. This fits well with older calculations, wherein the complete oxidation of glucose was said to lead to the synthesis of 36 ATPs. Actually, this number takes into account both chemiosmotic phosphorylation and substrate-level phosphorylation, and the arithmetic is fairly straightforward (although open to debate). It is generally thought that the total number of ATPs per glucose is:

- Substrate-level phosphorylation
  - Glycolysis .... 2 ATPs
  - Citric acid cycle .... 2 ATPs
- Chemiosmosis
  - 64 protons gained .... 32 ATPs
- Total per glucose .... 36 ATPs

The yields in terms of protons, NADH, $FADH_2$, and ATP are summarized in Table 8.2.

How efficient is the entire catabolism of glucose? In other words, how much of the free energy of glucose ends up in the phosphate groups of ATP? The free energy increase in ADP as it is phosphorylated to ATP is, again, 7.3 kcal/mole. Since there are 36 produced for each glucose, the total kcal per mole value is about 263. As we've seen, the corresponding free energy value of glucose is 686 kcal/mole. Thus the efficiency of glucose catabolism in the cell turns out to be about 38% (263/686 × 100 = 38.3%). (In cells where cytoplasmic NADH shuttles its electrons and protons to mitochondrial NAD^+, the percentage comes out to about 40.) By most standards, 38% represents an efficient energy utilization. The automobile, when tuned to perfection and driven with utmost care, is less than 25% fuel-efficient!

Other Uses for the Proton Gradient

So far we've talked about the energy of the proton gradient being used only in the phosphorylation of ADP, but that free energy can also be tapped for other roles.

Some of free energy can be used to move phosphate (P_i) into the mitochondrion for use in phosphorylation (see Figure 8.11). In addition, chemiosmotic gradient energy can be used to transport calcium ions into the mitochondrion. Calcium is needed in a variety of cellular activities and is often stored in mitochondria. Phosphate and calcium

TABLE 8.2 ATP BALANCE SHEET (ATPs PER GLUCOSE)

	Glycolysis	Acetyl CoA	Citric Acid Cycle	Total per glucose
NADH + H+	(2) ↓ (shuttle)	2	6	8
$FADH_2$	2		2	4
ATP	2		2	4

Each NADH provides the energy to enrich the differential by 6 protons, while each $FADH_2$ provides the energy to enrich it by 4 protons. Therefore:

8NADH = 48 protons
$4FADH_2$ = 16 protons
Total 64 protons

ATP synthesis requires 2 protons: 64 protons = 32 ATP (+ 4 ATP direct) = *36 ATP per glucose.*

ions are transported by special cotransport carriers (see Chapter 5), their movement involving the simultaneous passage of protons.

The transport of ADP into the mitochondrion itself occurs in another way. It enters via a cotransport shuttle (see Figure 8.11). This versatile carrier brings one ADP into the mitochondrion while simultaneously assisting one ATP out. The transport of ADP, phosphate, and calcium is a routine use of the proton gradient, but there is another that is quite unusual, one involving "brown fat" and body heat.

This rather remarkable use of the proton gradient is found in mammals that are born relatively hairless—humans being good examples. A special type of fatty tissue called **brown fat** accumulates in the neck and upper back regions of the young of these species. Its dark color is attributed to a high concentration of cytochromes, contained within numerous mitochondria. Fatty tissue normally doesn't contain a lot of mitochondria, but in brown fat their presence is an adaptation toward generating body heat. Heat is generated as protons from the chemiosmotic gradient escape to the inner compartment, but not through the F_0F_1 complexes. They escape via special pores that lack the phosphorylating enzymes. Without the phosphorylating enzymes to make use of the energy, it is simply released as heat, and this heat is used in keeping the young mammals warm.

THE FERMENTATION PATHWAYS

As we saw earlier, some organisms utilize a relatively simple anaerobic fermentation pathway to obtain energy. There are no electron transport systems or chemiosmotic systems involved. In fermentation, glucose enters a metabolic pathway that is identical to glycolysis, from the phosphorylation of glucose to the emergence of pyruvate, NADH, and ATP. But where glycolysis ends with pyruvate, fermenters carry the reactions a bit further, and the final product, a metabolic waste, is an energy-rich organic molecule.

Fermentation wastes are formed through the reduction of pyruvate by NADH. The coenzyme, having passed its electrons and protons to pyruvate, resumes its oxidized form, NAD^+, and as such it is freed to recycle back to the glycolytic pathway. All cells have only a small amount of NAD^+ available, and so there must be a means of regenerating it after it has reacted in glycolysis. We should note that the recycling of NAD^+ is important in the aerobic respiratory pathways as well. Recall that cytoplasmic NADH hands its protons and electrons off to a mitochondrial membrane carrier, restoring it to NAD^+, its oxidized form.

Earlier we mentioned several fermentation pathways, but here we will look into just two: **alcoholic fermentation** and **lactate fermentation** (see Figure 8.12). The first occurs in yeasts and some bacteria, and the final products are NAD^+, carbon dioxide, and ethyl alcohol. The second occurs in bacteria as well, but is best known as an important pathway in animal muscle tissue, where the products are NAD^+ and lactate (lactic acid).

Alcoholic Fermentation

Ethyl alcohol, is the alcohol of beer, wine, and liquor. Fermentation by yeast is also important to the baking industry, but bakers are primarily interested in the carbon dioxide, which causes bread to "rise." (The ethyl alcohol produced simply evaporates in the oven.)

The transformation of pyruvate into ethyl alcohol occurs in two steps, each mediated by a specific enzyme. The pyruvate is first acted upon by a decarboxylase enzyme, which removes the carboxyl group (COOH or COO^-), converting it to carbon dioxide. The nearly depleted fuel molecule that remains is known as acetaldehyde. Next, with the aid of the enzyme alcohol dehydrogenase, the NADH formed earlier is oxidized. In a closely coupled reaction, its protons and electrons are passed to acetaldehyde, reducing it to ethyl alcohol. In the reaction, NAD^+ is regenerated, whereupon it cycles back into the fermentation pathway, as mentioned.

We've been saying that the fermentation pathway ends with energy-rich wastes, and

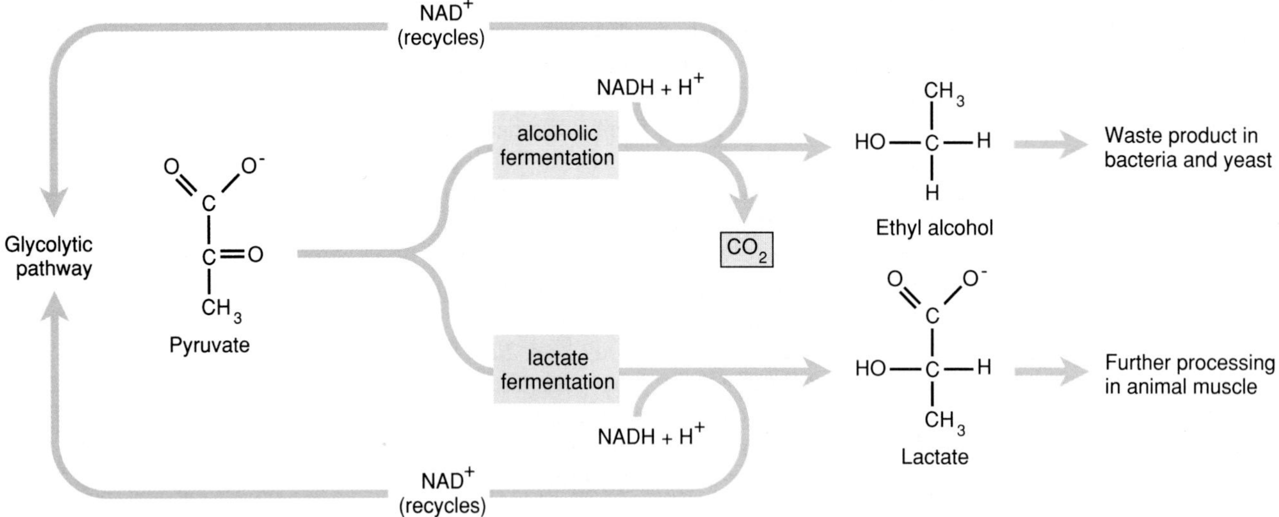

FIGURE 8.12
FERMENTATION PATHWAYS
Alcoholic fermentation is common to yeasts. The products are ethyl alcohol and carbon dioxide. Lactate fermentation, the reduction of pyruvate to lactate, occurs both in certain bacteria and in animal muscles. In both anaerobic processes, NADH is oxidized, thus allowing it to recycle to be used again in glycolysis.

ethyl alcohol (ethanol) certainly qualifies. You should know that alcoholic beverages are loaded with calories, a fact familiar to dieters. Because it has so much free energy, ethyl alcohol can be combined with gasoline, forming "gasohol." (Gasohol is one answer to the limited availability of petroleum.)

Yeasts in general are **facultative anaerobes**; that is, depending on conditions, they can switch the metabolism of pyruvate between anaerobic alcohol fermentation and fully aerobic respiration, where carbon dioxide and water are the end products. The aerobic cell respiration of yeast, by the way, is biochemically identical to human aerobic respiration.

Lactate Fermentation

We have noted that in some cases, the end product of fermentation is lactate, a 3-carbon derivative of pyruvate. Among the lactate fermenters are certain bacteria of genus *Lactobacillus* and *Streptococcus,* whose lactate secretions are used to create the unique flavors of rye breads, yogurt, and some cheeses. And as we mentioned, lactate is also produced in the active muscles of vertebrates as well as many other animals. As in the alcoholic fermentation pathway, the reduction of pyruvate frees NAD^+ to cycle back to glycolysis. In this case, the reduction of pyruvate leads to the formation of lactate (see Figure 8.12). The story of lactate fermentation in our own muscles is interesting, since it explains a lot about exercise, fatigue, and physical conditioning.

Glycolysis and Lactate Fermentation in Muscle Tissue During heavy exertion, a great deal of ATP can be used up quickly. Skeletal muscle can operate aerobically, carrying out oxidative cell respiration, but in larger animals at least, there is no way for the circulatory system to bring in oxygen fast enough for oxidative phosphorylation to replace the ATP being used during such activity. However, muscle cells have two backup systems. The first backup system is a store of high-energy phosphate in a molecule that is abundant in muscles, **creatine phosphate** (or **phosphocreatine,** as it is also known).

Creatine phosphate is a much more compact molecule than ATP, and the cell can store a great deal of readily available energy in this form. Creatine phosphate doesn't provide muscle contraction energy directly, but it can transfer its own energy-rich phosphate to ADP to regenerate ATP:

$$\text{Creatine phosphate} + \text{ADP} \rightarrow \text{ATP} + \text{Creatine}$$

As the creatine phosphate is gradually used up, the muscle tissue falls back to yet another quick energy source—glycolysis. So, in addition to pyruvate, the end products include a modest supply of ATP and NADH. Of course, pyruvate is still energy-rich and can be sent into the mitochondrion to undergo aerobic cell respiration. This would provide a good source of ATP and a continuous supply of recycled NAD^+. But when the oxygen required for aerobic respiration in muscle cannot be supplied fast enough, the continued recycling of NAD^+ becomes threatened, and as we have seen, without it glycolysis cannot proceed. The answer, at least temporarily, is to recycle NAD^+ another way. Using the hydrogen of NADH, pyruvate is reduced to lactate. The conversion to lactate is, in a sense, a dead end, since this product cannot be used directly by muscle.

As a result, lactate accumulates rapidly during intense muscular activity, and sooner or later something must be done with it. Much of the lactate is carried out of the muscles by the blood stream and directed to the liver (and to a much lesser extent to the kidneys). Some of it may remain in the muscles until the amount of oxygen being brought in by the circulation exceeds the muscle's current needs, in which case the lactate is converted back to pyruvate, which then proceeds into the oxidative respiration pathway. There it is broken down to carbon dioxide and water, and its chemical bond energy used to produce ATP.

The lactate sent to the liver has a different fate. It is also converted back to pyruvate, but the pyruvate is then sent through a special biochemical pathway known as **gluconeogenesis.** In several steps, a few of which are clearly a reversal of glycolysis, the lactate is converted to glucose. Some of this same glucose finds its way back to the muscles again. The conversion is not without cost, however, because glucose has greater free energy than lactate. Thus, much of the initial pyruvate produced must be oxidized to provide energy to run the gluconeogenic pathway. (Figure 8.13 summarizes each aspect of muscle activity.)

Oxygen Debt After a period of heavy exertion, the muscle tissues in humans and other vetebrates will be depleted of creatine phosphate, and both the liver and the muscle will be loaded with lactate. Perhaps you have experienced this condition; fatigue hurts, and most people try to avoid it, although a few eccentrics like marathon runners claim to find it exhilarating. The final stage is sheer exhaustion. But marathon running or not, following any strenuous muscle activity, it takes a period of time and a large amount of oxygen and ATP for the lactate to be metabolized and for the creatine to be regenerated as creatine phosphate. During this time you can expect to find yourself continuing to breathe hard, taking in as much oxygen as the lungs can handle. The state of oxygen and creatine phosphate depletion creates what is known as **oxygen debt**, which can be expressed as the amount of extra oxygen needed to restore the system to its preexertion equilibrium. How long it takes to repay the oxygen debt obviously depends on a person's physical condition. Physical conditioning, in turn, involves increasing respiratory and circulatory capacity, as well as increasing the storage capacity and possibly the mass of the muscles themselves. Conditioned runners have enlarged lung capacities, hearts that pump more blood with each stroke, and an increased ability to use oxygen. Some may even have a high tolerance for pain—at least they act as though they do (Essay 8.1).

Interestingly, many smaller animals generate all or most of their muscle ATP aerobically and have no problem with lactate accumulation and oxygen debt. Examples include seemingly tireless migratory birds and small, fast-running mammals. But the great body mass of larger animals does not allow their circulatory systems to keep up with the increased oxygen demands. Nile crocodiles, for instance, are generally sluggish creatures. But when threatened or when stalking prey on land, they can make astonishingly fast charges and can lash their tails about with results that are legendary. Following such outbursts, the giant reptiles must remain still, requiring many hours to repay the oxygen debt.

The late Albert Lehninger of Johns Hopkins University mused that an oversized oxygen

FIGURE 8.13
ENERGY SUPPLIES
(1) The energy available for physical activity comes from limited ATP stores. These stores are replenished **(2)** through the use of creatine phosphate, which regenerates ATP from ADP. **(3)** Leftover creatine is then phosphorylated by ATP, which was produced through glycolysis **(4)**. NADH and pyruvate, the other two products of glycolysis, react **(5)**, yielding NAD^+ and pyrulactate. The NAD^+ recycles to glycolysis and the lactate is sent to the liver **(6)**. During gluconeogenesis, some of the free energy of pyruvate is used to convert the remainder back to glucose.

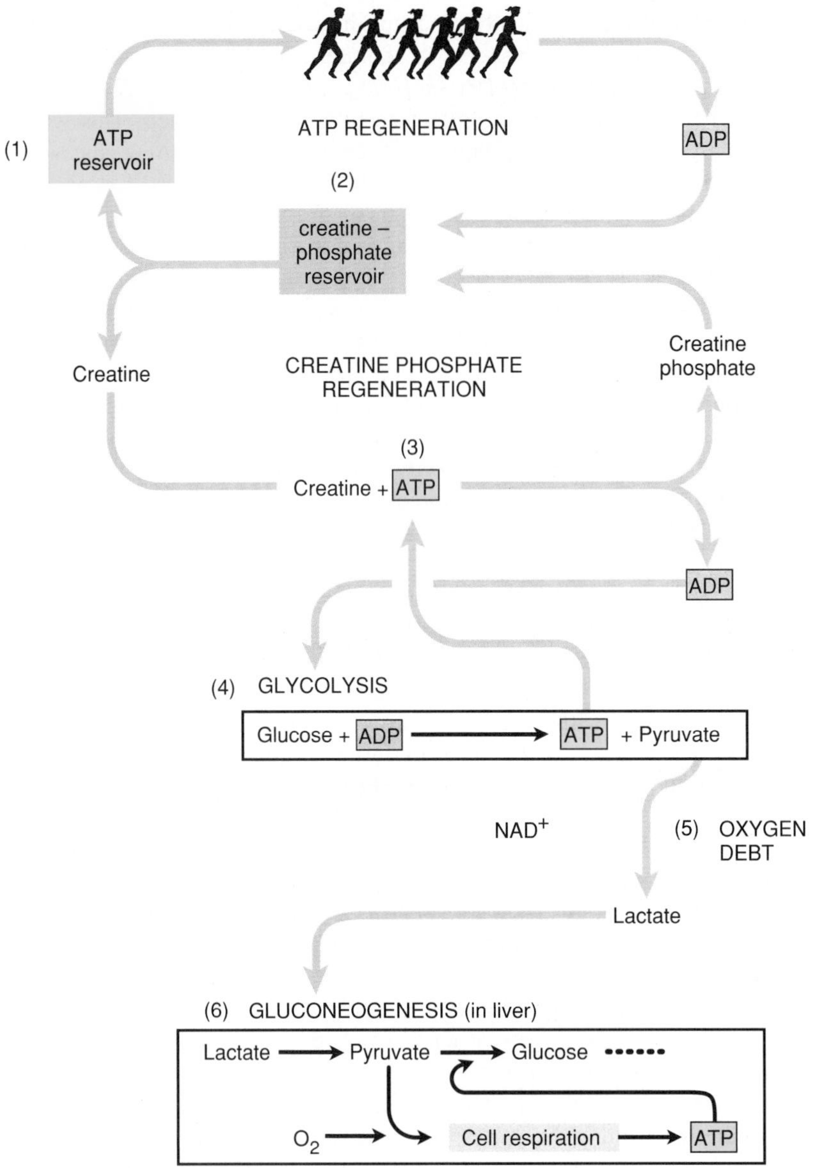

debt may have cost many a giant dinosaur its life, for following any kind of vigorous activity, it would have had to enter a metabolic stupor when it would have been vulnerable to smaller, sharp-toothed predators. (Shall we add the oxygen-debt hypothesis to the long list of explanations for the extinction of these great beasts?)

ALTERNATIVE FUELS FOR THE CELL

Physiologists talk about glucose so much that it might appear that this is the only cellular fuel. While carbohydrates are important and common cellular fuels, they are not the only ones, and quite often they are not even the most important ones. For instance, we all rely heavily on stored fats as a primary energy supply. We maintain a supply of glucose in the form of glycogen, but only enough to maintain us for about a day of normal activity. Following brief periods of fasting our body fats become mobilized, ready for use by cells. By morning, for instance, our bodies have switched to fats, so that most of the ATP energy produced in the mitochondria has come from the oxidation of fatty

ESSAY 8.1

A MARATHON RUNNER MEETS "THE WALL"

There is a a point in the marathon when many runners hit the dreaded "wall." The wall usually looms before them at about mile 18 or 20, and it is here that many flounder--falling prostrate, staggering, or simply walking. It is here that other, better-trained runners begin to pass those whose dust they had been eating, on their way to the conclusion of the 26-mile, 385-yard race. (Some runners say that a marathon is divided into two equal parts: the first twenty miles and the last six.)

There was been a lot of discussion over the years about what the wall really is, but everyone who has encountered it can attest to its reality. Physiologically, it seems to correspond to the period when muscle glycogen, the most readily available reserve of glucose, is depleted. Glucose is normally depleted after two to three hours of slow running, at about an eight-minute-per-mile pace. The runner at this point must switch to matabolizing fat molecules. The blood pH lowers, and the transition is, for some reason, exceedingly stressful. It has also been suggested that women switch to fat metabolism more easily than men, a suggestion supported by their increased relative numbers in 50- and 100-mile "ultramarathons."

Interestingly, when one encounters the wall, sheer willpower has little effect. You simply can't go on. The muscles refuse. Any effort is extremely painful and fatiguing. In addition, blood sugar levels drop, and you may suffer the psychological depression associated with hypoglycemia.

Marathoners have been able able to beat the wall by training so hard that it no longer exists for them. This training includes long runs of 20 miles or more, during which muscle glycogen is repeatedly depleted and then restored in even greater abundance during the following week. Some runners believe they are able to move the wall back by "carbo loading," that is, eating primarily foods high in carbohydrates for the week preceding the race. The idea is that the excess carbohydrates will be stored as glycogen that can then be called on at mile 20.

acids. It should be good news to dieters that body fats (particularly those stored in places we would rather not draw attention to) can also be sent through the respiratory mill. In fact, fats are excellent energy sources, yielding about twice the amount of energy as carbohydrates on a gram per gram basis. Proteins can also be used as energy sources, but their use is complicated by their much more varied composition. Interestingly, fatty acids are the preferred chemical fuel of heart muscle, but they are utterly rejected by the brain, whose chief energy source is glucose.

As all of us are aware, the three basic foods, carbohydrate, lipid (fat), and protein, must be broken down into their chemical subunits to be of use to our bodies as fuels. All of the common molecules—amino acids, lipids, and carbohydrates—can be modified sufficiently to be used in glycolysis and the citric acid cycle. The pathways through which such conversions occur are part of what is called **intermediary metabolism** (Figure 8.14) But for use in respiraiton, even the amino acids and lipid subunits must undergo considerable modification before they can be fully oxidized to yield their energy.

The Metabolism of Fats

Fatty acids must be activated by coenzyme A (CoA) in order to be transported across the mitochondrial membranes to the inner compartment, where they are prepared for the citric acid cycle. Activation takes place on the outer membrane and consists of joining

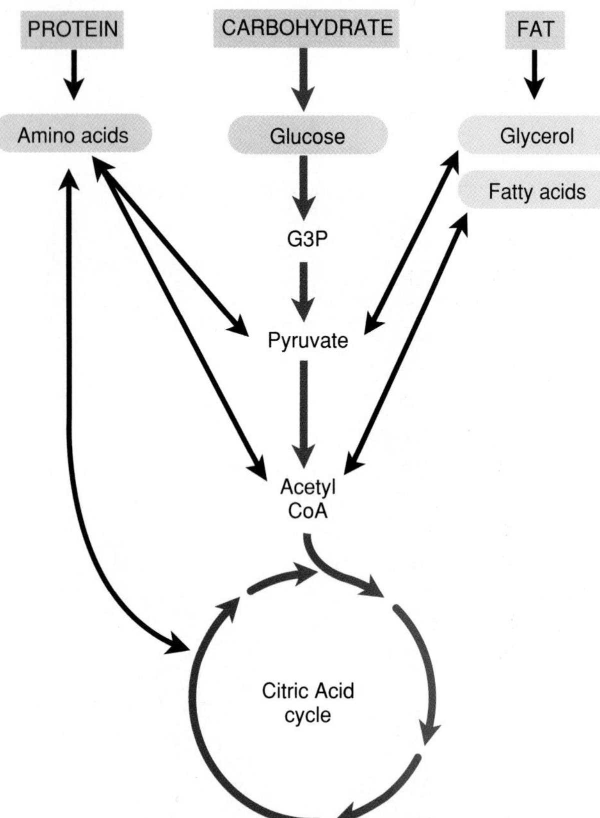

FIGURE 8.14
ALTERNATIVE FUELS
The use of alternative fuels by the mitochondrion is commonplace. Fatty acids are converted to acetyl-CoA with a large gain in NADH. The acetyl-CoA then enters the citric acid cycle. The specific entry of amino acids into the mitochondrial system is varied. Some are converted to pyruvate, others to acetyl-CoA, and still others to acids of the cycle itself.

the fatty acid to coenzyme A. The activated fatty acid is then ferried across the membrane by special carriers, ending up in the matrix. Once there, a 2-carbon acetyl-CoA is split away to enter the citric acid cycle. Then another coenzyme A joins the remaining fatty acid fragment, and the sequence is repeated, freeing a second 2-carbon acetyl-CoA, and so on until the fatty acid chain has been completely fragmented, two carbons at a time, into acetyl CoAs.

For example, **palmitic acid**, a 16-carbon fatty acid, goes through seven reactions yielding eight acetyl-CoAs. But there is a special bonus in all of this: each reaction includes an oxidation and a reduction, so along with the acetyl-CoAs, seven NADH + H^+ and seven $FADH_2$s also form. The reduction products, of course, greatly enrich the chemiosmotic gradient of the mitochondrion, leading to the production of ATP. This explains why fats or lipids are such high-energy food.

The Preparation of Proteins

Proteins are first hydrolyzed by digestive enzymes into the usual 20 amino acids. They then enter the citric acid cycle in several different ways. All must first have their amino groups (NH_3^+) removed, a process called **deamination**, and these groups must be rendered harmless. (Figure 8.15) (Unattended, the amino groups would accumulate as dangerous ammonium ions.) Five of the 20 amino acids are simply converted to pyruvate, which is then converted to acetyl CoA. Six more of the amino acids skip the pyruvate step and enter as acetyl CoA itself. The other amino acids are actually converted into specific acids of the citric acid cycle, joining it directly.

The amount of energy available from each type of amino acid depends, as you might suspect, on where its conversion product joins the citric acid cycle. This makes sense because the point of entry also determines how much NADH is generated. The five amino acids converted to pyruvate have the most to offer, since they yield the maximum number of NADH possible. Those entering as acetyl CoA are next in yield, while those entering as acids of the cycle are, of course, the least productive.

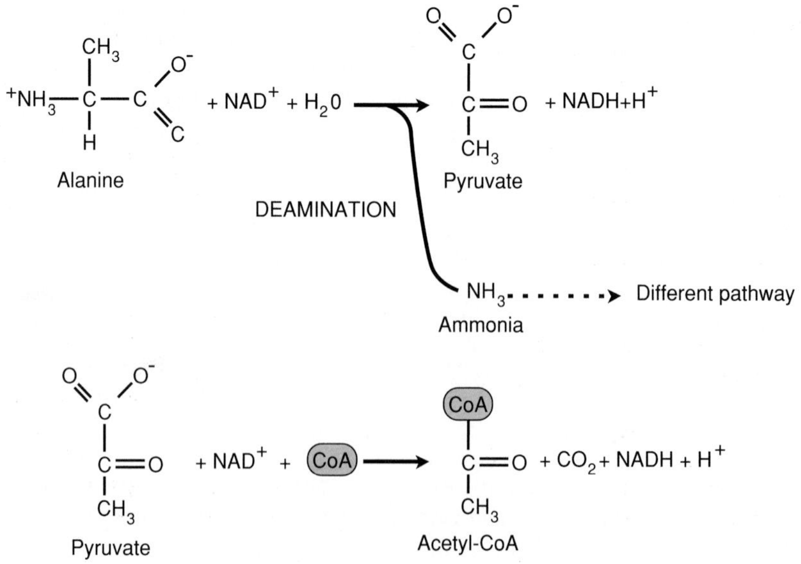

Alanine + NAD^+ + H_2O → Pyruvate + NADH+H^+

DEAMINATION

NH_3 (Ammonia) ········> Different pathway

Pyruvate + NAD^+ + CoA → Acetyl-CoA + CO_2 + NADH + H^+

FIGURE 8.15
DEAMINATION OF AN AMINO ACID
Amino acids prepared for use as energy sources must first have their amino groups removed. These are sent through a separate pathway for processing. The useful deamination product in this case is pyruvate.

Intermediary Metabolism: Biosynthetic Pathways

The citric acid cycle, while central to respiration, serves another purpose. It provides raw materials for other processes. Several of its acids can be diverted from the cycle for use in the synthesis of new molecules, a part of intermediary metabolism called **biosynthesis.**

Notice the double arrows in Figure 8.14. They are meant to indicate that many of the reactions are, in fact, reversible. For example, we see that although fatty acids may enter the cycle as acetyl-CoA, under certain conditions (with energy provided), the reactions can be reversed and the acetyl groups can be converted to fatty acids. Perhaps the greatest versatility is in the biosynthetic pathways involving amino acids. Many of the amino acids can actually be manufactured from certain citric acid cycle intermediates. For example, one of these, alpha-ketoglutarate (see Figure 8.7, reaction 4), can be drawn off and used in the synthesis of glutamic acid, proline, or arginine. Oxaloacetate, the resident molecule, can be converted to aspartate, which in turn, can be used as a precursor for five other amino acids. Conversely, amino acids entering the reactions as acetyl CoA can be turned around and converted directly into fatty acids. (Such information might be important to dieters starting on some high intensity protein or carbohydrate diets—anything we eat can end up as fat). In a real sense, because of such versatility, glucose can then be considered as a starting molecule for many important molecular subunits. Once it has entered respiration, its products can be used in a great variety of ways. But, more to the point, the importance of glycolysis and the citric acid cycle to the cell goes much farther than providing energy.

APPLICATION OF IDEAS

1. What does it mean for a runner to "go anaerobic"? Who depends more on anaerobic respiration, a sprinter or a marathon runner? Explain your answers.
2. Is anaerobic respiration really inefficient? Consider the question in terms of the free energy actually available to dwellers of the anaerobic environment. Also consider the answer in terms of whether or not anaerobic organisms are successful, well-adapted organisms.

KEY IDEAS

Most of the cell's energy needs are fulfilled by ATP.

THREE WAYS OF UTILIZING GLUCOSE

Cells form ATP from the free energy of glucose in (1) **aerobic**, (2) **anaerobic respiration**, and (3) **fermentation**. The first requires oxygen as a final electron acceptor. Both are preceded by **glycolysis,** a preliminary, anaerobic process with a small ATP yield. Anaerobic respiration differs in the final electron acceptor, which is often a sulfate or nitrate ion. Except for the end product, usually lactate or ethyl alcohol, **fermentation** and glycolysis are identical.

The general equation for respiration is the opposite to that of photosynthesis. Overall, respiration is exergonic, whereas photosynthesis is endergonic.

The Three Parts of Aerobic Respiration

The three parts of respiration are glycolysis (cytoplasm), **citric acid cycle** (mitochondrial **matrix**), and electron transport and chemiosmotic phosphorylation (both mitochondrial).

Glycolysis

Glycolysis may be an ancient process. It includes nine principal reactions occurring in the glycolytic pathway.

The Highlights of Glycolysis

1. The preparatory phase includes the phosphorylation of the fuel by 2 ATPs, cleaving of the molecule into 3-carbon fuels, oxidation, and phosphorylation by P_1. Oxidation is coupled to the reduction of NAD^+ to NADH.
2. The yielding phase occurs when high energy phosphates join ADP in two places in the pathway. The net gain is two ATP, which represents about 2.1% of the available energy. The end products are 2 ATPs, 2 NADHs and 2 pyruvates.

Control of Glycolysis

The allosteric enzyme, **phosphofructokinase**, provides a measure of control. It is inhibited by ATP and stimulated by ADP. It is also inhibited by excess citrate from the citric acid cycle.

AEROBIC RESPIRATION IN THE MITOCHONDRION

Pyruvate enters the mitochondrion, where its complete oxidation to CO_2 and O_2 occurs.

Pyruvate to Acetyl-CoA

Pyruvate is acted upon by a large enzyme-coenzyme complex, Coenzyme A. The reactions yield CO_2, NADH, and the 2-carbon acetyl fuel in the form of **acetyl-CoA.**

The Citric Acid Cycle

The important events of the cycle include the reduction of coenzymes by electrons and protons of fuels and the formation of a small amount of ATP.

1. Acetyl-CoA joins 4-carbon **oxaloacetate,** forming 6-carbon citrate with coenzyme A recycling. As the reactions proceed, two carbons are liberated as CO_2.
2. Four oxidations bring about the reduction of three NAD^+s and one FAD.
3. A substrate-level event generates one ATP.

ELECTRON TRANSPORT SYSTEMS AND CHEMIOSMOTIC PHOSPHORYLATION

Mitochondrial structure is essential to chemiosmotic phosphorylation.

Mitochondrial Structure

1. The double membrane provides for an **outer** and **inner compartment,** the first of which is the proton reservoir. The **inner membrane** is highly folded and contains assemblages making up electron transport systems.
2. Many electron carriers are proteins called cytochromes, molecules with iron groups that take in electrons and pass them along.
3. Numerous F_0F_1 complexes dot the inner membrane, their channels leading from the outer mitochondrial compartment.

NADH Electrons Build the Chemiosmotic Gradient

1. Electrons and protons from NADH join the electron transport system at Site I. At this site and in the next two, two protons are pumped into the outer compartment, for a total of 6 per NADH.
2. Electrons and protons from $FADH_2$ enter the electron transport system at **coenzyme Q,** powering the transport of four protons across the inner membrane.
3. Spent electrons join oxygen and, along with protons from the matrix, form water.
4. The large free energy change in the electron transport system ends up as an increased proton gradient.
5. NADH and $FADH_2$ from two pyruvates account for the pumping of 56 protons. Eight more are provided by the reducing power of cytoplasmic NADH, which through a mitochondrial shuttle, reduces mitochondrial FAD.

Reviewing the Chemiosmotic Gradient and ATP Formation

1. The charged mitochondrion has two components, a voltage differential and a proton gradient, their potential energy known as **proton-motive force.**
2. The escape of protons through F_0 channels into the F_1 heads provides the free energy for phosphorylating ADP.
3. The ATP yield from all of respiration is 36 per glucose—2 from glycolysis, 2 from the citric acid cycle, and 32 from chemiosmotic phosphorylation. Thirty-six ATPs represent an efficiency of 38%.

Other Uses for the Proton Gradient

1. The proton gradient also powers the transport of P_i and calcium ions.
2. The uptake of ADP is coupled with the output of ATP through a membrane carrier.
3. In brown fatty tissue, the proton gradient is used in generating body heat.

THE FERMENTATION PATHWAYS

Fermentation is identical to glycolysis, except for its final steps.

Alcoholic Fermentation

In **alcoholic fermentation** pyruvate loses a carbon, as CO_2, and the product, acetaldehyde, is reduced to ethyl alcohol by NADH. NAD^+ recycles. Fermentation occurs in yeasts and bacteria.

Lactate Fermentation

1. In **lactate fermentation,** pyruvate is reduced by NADH to form **lactate.** It occurs in lactic acid bacteria and in animal muscle.
2. During muscle exertion ATP is provided as follows:
   a. A limited reserve of ATP gets the muscle going.
   b. ADP is at first phosphorylated by energy-rich phosphate bonds from **creatine phosphate (phosphocreatine).**
   c. Creatine phosphate is replenished by ATP produced during glycolysis (lactate fermentation).
   d. During glycolysis, lactate is built up in the muscle tissue, but much of it is carried away by the blood to be metabolized in the liver.
   e. When the muscle rests, and oxygen supplies increase, its aerobic respiration begins to catch up; thus it replenishes its ATP and CP reserves.
   f. Lactate in the liver is first converted back to pyruvate. It then undergoes **gluconeogenesis,** where it is reconstituted into glucose.
3. The amount of extra oxygen needed to restore ATP of the resting state is known as the **oxygen debt.**

ALTERNATIVE FUELS FOR THE CELL

In addition to glucose, cells make use of fatty acids and amino acids as fuels. They pass through intermediary metabolic pathways during their preparation for the mitochondrion.

The Metabolism of Fats

Fats are broken down to fatty acids and glycerol with a large NADH yield, entering respiration as pyruvate and acetyl-CoA. On a gram-per-gram basis, they have a much higher ATP yield than other fuels.

The Preparation of Proteins

Proteins are broken down into amino acids, which can then be processed for respiration. Some enter as pyruvate, others as acetyl-CoA, and still others enter the citric acid cycle as one of several intermediaries. The earlier they enter respiration, the greater their energy value.

Intermediary Metabolism: Biosynthetic Pathways

The respiratory pathways also act as biosynthetic pathways wherein the one type of molecular subunit can be used in the synthesis of another. Examples include the use of acetyl-CoA formed from amino acids or pyruvate for the synthesis of new fatty acids. Several citric acid cycle acids can be used in the synthesis of amino acids.

REVIEW QUESTIONS

1. List several similarities and one difference between anaerobic and aerobic respiration. (179)
2. Since neither uses oxygen, how does fermentation differ from anaerobic respiration? (179)
3. Discuss two basic ways in which the chemistry of photosynthesis differs from that of respiration. (177)
4. List two specific processes through which ATP is generated. (178)
5. Describe five things that happen to fuel molecules in the preparatory phase of glycolysis. What does this phase accomplish? (179-181)
6. What is the importance of the high-energy phosphates that form twice during glycolysis? (181)
7. List three products of glycolysis. What eventually happens to each? (181)
8. How much of the free energy of glucose is later represented in the phosphtae bonds of ATP? (181)
9. Briefly explain how ADP, ATP, and citrate affect the operation the enzyme phosphofructokinase. How is this important to the cell? (183-184)
10. Write a word equation that summarizes the acetyl-CoA step of respiration. What happens to each product? (185)
11. What is the overall purpose of the citric acid cycle in respiration? What molecules play the role of intermediary between the cycle and the electron transport system? (185)
12. Determine the total number of NADHs, $FADH_2$s, and ATPs produced when two pyruvates enter mitochondrial respiration. (Table 8.1, p 186)
13. With what molecule does the citric acid cycle begin and end? Six CO_2s are released for each glucose catabolized in respiration. Where are these carbon dioxides formed? (185-186)
14. With a simple line drawing, illustrate the arrangementof the two membranes of a mitochondrion. Explain the significance of the two compartments to chemiosmosis. (188)
15. What does NADH pass along to FMN in the electron transport system? What happens to each? (189)
16. What is the total number of protons pumped per NADH reaching the electron transport system? (190)
17. Where does $FADH_2$ act in the electron transport chain? How does this affect the number of protons pumped? (190)
18. What becomes of the electrons that pass through the electron transport systems? What becomes of much of this product in our own bodies? (190)
19. What is proton-motive force? What are its two components in the mitochondrion? (191)
20. Explain how the free energy of the proton gradient is used in respiration. (191-192)
21. How is the proton gradient put to work in maintaining body temperature in babies? (193)
22. Why is it so important for fermenters to reduce their pyruvate? What are the products of the two main fermentation pathways? (193)
23. Which organisms alternate between alcoholic fermentation and aerobic respiration? (194)
24. List three steps involved in providing ATP for vigorous physical activity in our muscles. (194-196)
25. Specifically, what is the fate of lactate in our own bodies? (195)
26. Explain how an "oxygen debt" arises and how it is paid off. (195)
27. Explain, from a biochemical view, why fats have such a high energy value. (197-198)
28. List three places in which deaminated amino acids may enter the respiratory pathways. What determines their ATP energy value? (198)
29. What is biosynthesis? By using an example, explain how the respiratory pathways are used for biosynthesis (199).

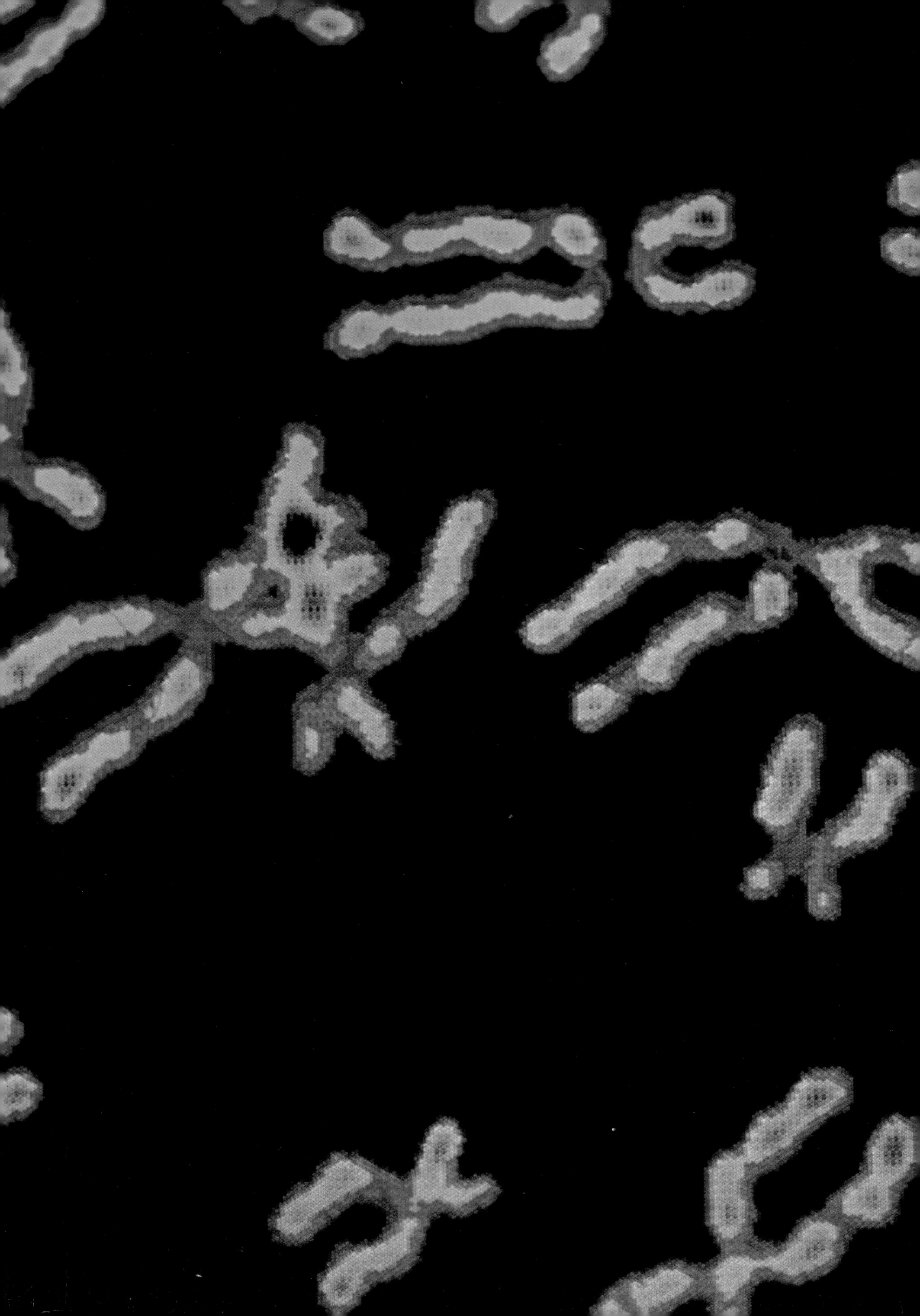

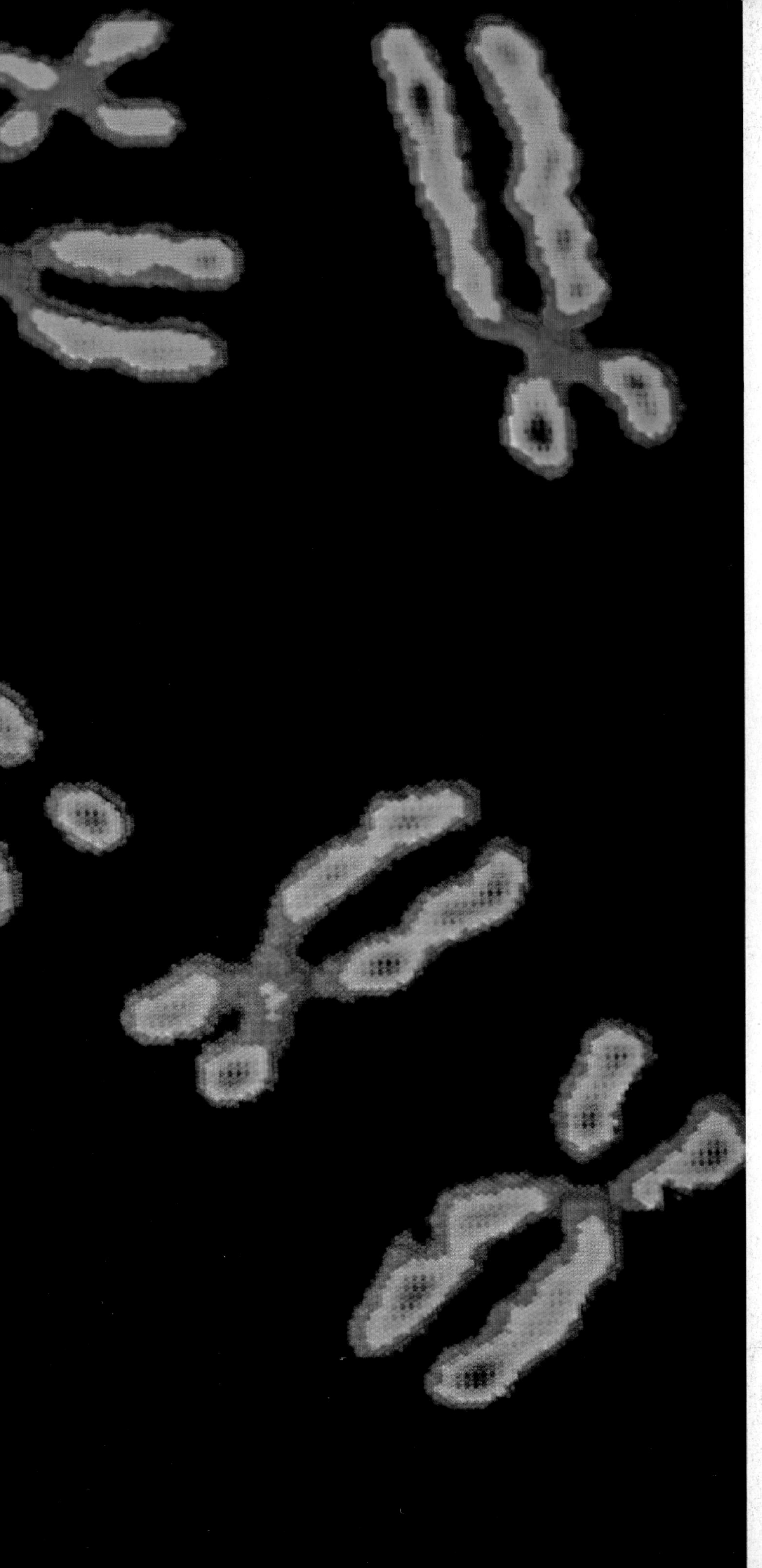

2

MOLECULAR BIOLOGY AND HEREDITY

9

EUKARYOTIC CELL REPRODUCTION

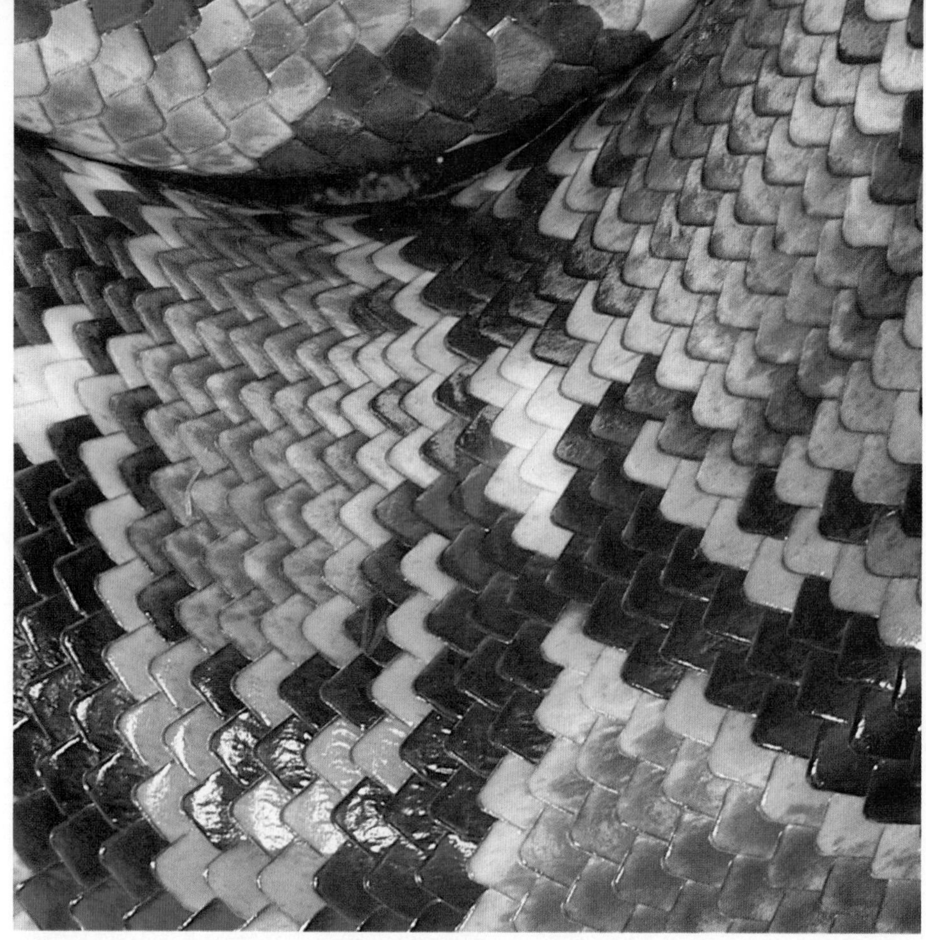

YOU ARE NOT THE SAME PERSON YOU WERE A FEW YEARS AGO. IN FACT, you aren't the same person you were a few seconds ago. This is not to say that these sentences are so profound that you will never be the same for having read them. It's just that even as you sit here reading, your cells are growing, dying, and reproducing. Your body is coping with a rigorous world, and it must change or perish. Some cells, of course, reproduce more rapidly than others. For example, the parts of your body that receive a great deal of friction, such as the palms of your hands or the lining of your small intestine, must constantly replace cells that have died or been worn away. Other parts of your body, however, have very low cell turnover. And some, such as the skeletal muscles and nervous system, are never replaced in adults. Nonetheless, at this very moment, your cells are quietly accomplishing the processes of life, many of them slowly and methodically reorganizing themselves and dividing again and again, duplicating themselves ever so precisely in their unending pageant until the day it all ends.

While this precision rules most of the time to produce the harmonic symphony of life, a dissonant note in this orchestrated performance, just a single miscue, can spell disaster. Thus it is important in our understanding of life to be aware of how cells reproduce so unerringly. Here, then, we will see just how closely regulated and precise the process of eukaryotic cell reproduction really is, beginning with the events in the nucleus.

THE NUCLEUS

The eukaryotic cell nucleus, its structure, and contents are central to our considerations of cell reproduction, so we will begin with a general review of this prominent organelle shown again in Figure 9.1. Recall that the nucleus contains the cell's primary hereditary information, assembled into chromosomes. In addition to chromosomes, the nucleus generally contains one or more prominent nucleoli, dense bodies that produce a special class of RNA called ribosomal RNA. Through most of the life of a cell, the nucleus is surrounded by a protective barrier, the nuclear envelope. It consists of a double membrane that is continuous with the endoplasmic reticulum (see Chapter 4).

The nuclear envelope is a highly selective barrier, readily admitting some substances and summarily rejecting others. Transport of molecules between the nucleus and the cytoplasm is accomplished via **nuclear pores.** Nuclear pores are not simply openings in the nuclear envelope, but are areas with complex protein mechanisms that facilitate the passage of certain macromolecules. Because of this selectivity, materials within the nucleus can be quite different from those of the cytoplasm. During cell reproduction, the nuclear envelope disappears, only to be reestablished after the completion of cell division. With this brief review, then, we can go on to see just what roles the various organelles play in cell reproduction, starting with the behavior of chromosomes.

DNA, Chromatin, and the Eukaryotic Chromosomes

DNA, you recall, is a double helix composed of two extremely long strands of nucleotides. In fact, these molecules can be so long that if the DNA molecule of the largest human chromosome were stretched out, it would be approximately 12 cm (4.7 in) in length. But fortunately for our physical integrity, this never happens. Instead, the DNA is condensed and packaged into manageable bundles. This packaging is accomplished by the formation of **chromatin,** which is DNA that is associated with an array of proteins. These proteins are of two general types—the histones, and others known (logically enough) as nonhistone chromosomal proteins.

Chromatin is formed in several steps (Figure 9.2), each increasing the compactness of the chromosome. First, the negatively charged DNA molecule is wound two and a half times around globules consisting of two copies each of four different kinds of histone protein (called H2a, H2b, H3, and H4). Histones carry a strong positive charge, and

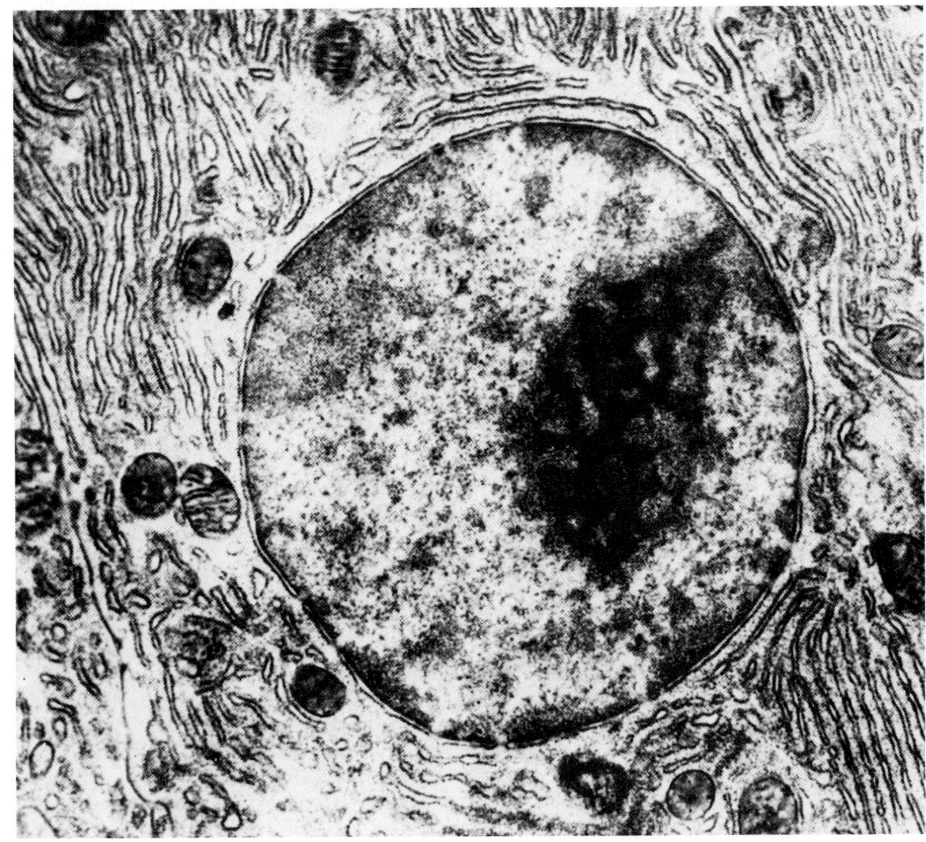

FIGURE 9.1
THE EUKARYOTIC CELL NUCLEUS
The most prominent features of the nucleus are the nucleolus and the chromatin net, both of which stain darkly in typical electron microscope preparations.

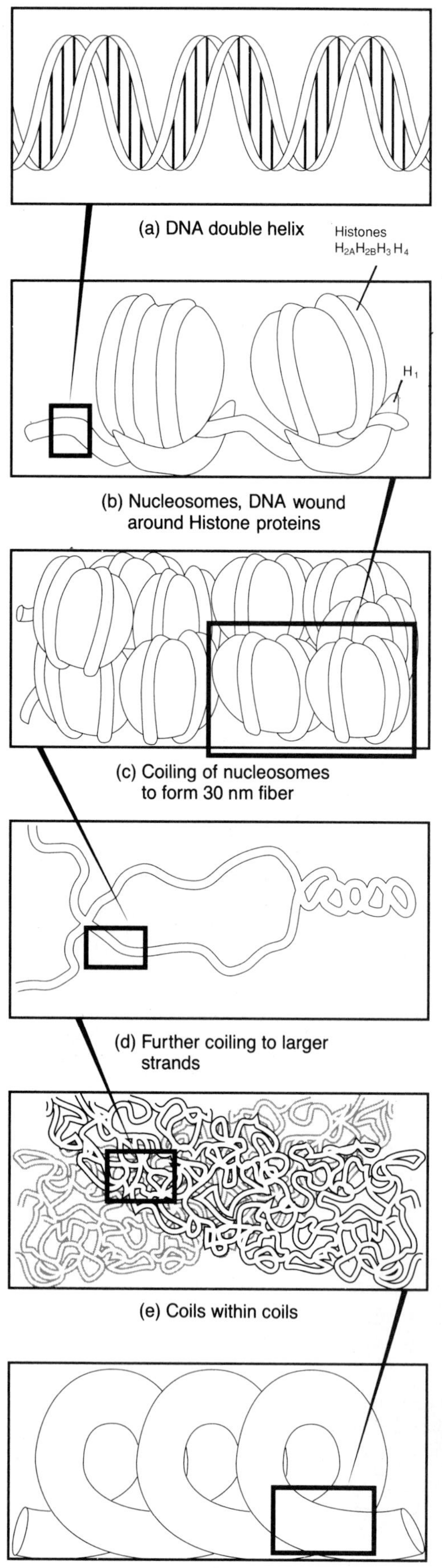

this aids in their association with the negatively charged DNA. The units formed in this way are called **nucleosomes.** The millions of nucleosomes along the length of a chromosome are connected by short stretches of histone-free "linker" DNA to form a "beads-on-a-string"-like structure. This nucleosome-linker-nucleosome structure is the basic unit of DNA packing.

Higher levels of chromosome condensation are accomplished by further packaging of the beads-on-a-string (Figure 9.2). First, an additional histone molecule, called H1, assists in holding the nucleosomes closer together. Then the nucleosomes are stacked into a coil with a width of 30 nm; it is therefore called the 30 nm fiber. Even higher levels of condensation are accomplished with the aid of the non-histone chromosomal proteins by further coiling of the 30 nm fiber, until the familiar eukaryotic chromosome appears, which is condensed (potentially) to 1/10,000 of its original length!

The mass of chromatin in nondividing cells, which resembles a bundle of yarn subjected to the attentions of a demented kitten, is sometimes referred to as the chromatin net because of its netlike appearance under the electron microscope (Figure 9.1). Parts of the chromatin net are thin and diffuse, while other parts are thicker and more tangled. This varied appearance is due to the fact that the chromatin of nondividing cells is found in many different stages of condensation.

The Cell Cycle: Growth, Replication, and Division

A cycle is, by definition, endless, and so the cell cycle is an endless repetition of growth, reproduction, growth, reproduction, etc. (Figure 9.3). However, not all cells cycle continuously. Some cells eventually break out of their cycle, as shown by Option 1 of Figure 9.3. But when they do, they pay dearly. As an example, fingernail cells become filled with keratin (a tough protein) and then die, only to be pushed along by dividing cells further back. The nuclei of red blood cells in birds, reptiles, and fish also are permanently inactivated—turned off, so to speak—so that once formed, these cells will die without undergoing another division. It should be even more apparent that the red blood cells of mammals, which have no nuclei, are also living on borrowed time. Still other cells, most notably some types found in the liver and all nerve cells, cease dividing when we are quite young. Although such cells are said

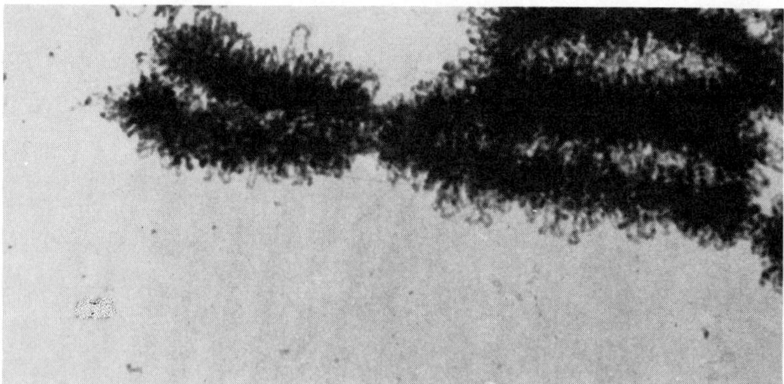

FIGURE 9.2
CHROMOSOMAL ORGANIZATION
The familiar chromosome is the ultimate result of several levels of organization and condensation. Naked DNA (a) associated with histone proteins forms nucleosomes (b) that give a "bead-on-a-string" impression. Coiled packing of nucleosomes forms the 30 nm fiber (c). Further twisting of the 30 nm fiber forms increasingly more dense coils (d,e), which eventually intertwine to form the chromosomal strands (f) that form the arms of the chromosome.

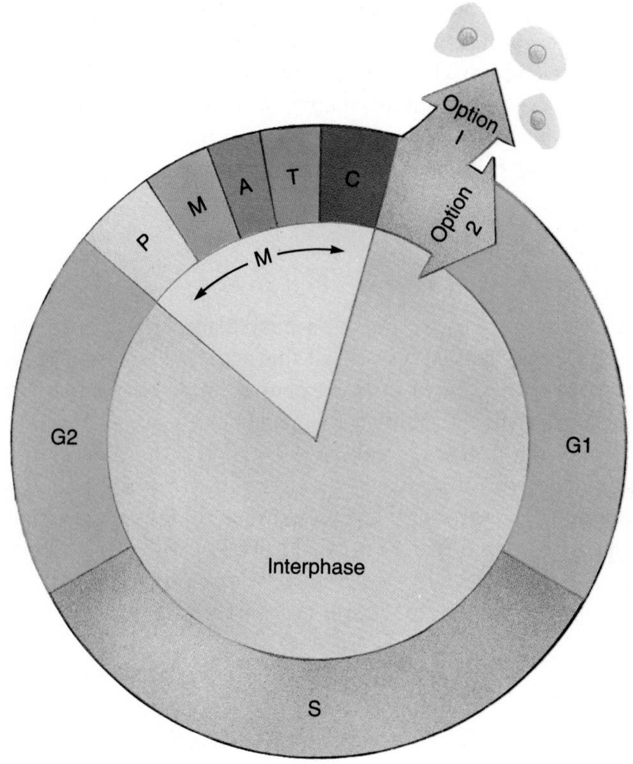

FIGURE 9.3
THE CELL CYCLE
There are four generally recognized parts to the cell cycle; G_1, S, G_2, and M. M, or Mitosis, is further subdivided into prophase, metaphase, anaphase, telophase, and cytokinesis. After mitosis a cell may cease to cycle and permanently become a particular cell type (Option 1); or it may continue to go through the cell cycle (Option 2), generating more daughter cells.

to "last a lifetime," it may, in fact, be the gradual accumulation of their deaths that help determine when we grow old and die. Thus, there are exceptions to cell cycling (biology is full of exceptions), but in this chapter we will focus on cells that cycle.

The cell cycle is traditionally divided into two phases, **interphase** and **mitosis** (also called M-phase). Each of these phases, in turn, is made up of several different parts. Whereas the cell actually spends most of its time in interphase, we will concentrate on the process of mitosis, because it is that phase of the cell cycle that results in the division of one cell into two daughter cells. Nonetheless, we should first see what goes on in interphase.

Interphase may be defined as the stage of the cell cycle between successive mitoses. It includes three different parts, the G_1 phase, the S phase, and the G_2 phase (Figure 9.3). When cells cycle (Option 2 of Figure 9.3), G_1 follows immediately on the heels of mitosis. This is a very active period, a time when the cell synthesizes its vast array of proteins, including the enzymes and structural proteins it will need for growth. In G_1 each chromosome consists of a single molecule of DNA and its proteins, but this soon changes.

During the S phase (or synthesis phase), all of the DNA within the nucleus will be replicated. This means that each chromosome will be faithfully copied so that by the end of S there will be two DNA molecules for each one present in G_1. The importance of this step in the cell cycle is obvious. It ensures that each emerging daughter cell will have the same genetic content as does the mother cell.

G_2 is an important preparatory period preceding mitosis. The proteins of the **spindle** are organized at this time, in preparation for the coming nuclear division. The spindle is an elaborate structure, formed of protein fibers, that is involved in chromosome movement during mitosis. It is constructed anew for each cell cycle, then dismantled after it has been used. (The name "spindle" comes from a fancied resemblance to the spindle of a handloom.) You may recall from Chapter 4 that the mitotic spindle fibers are microtubules, which are long polymers of tubulin subunits.

The relative amount of time the cell spends in interphase varies greatly according to the cell's type and stage of development. The shortest periods generally are found in the rapidly growing embryo. The large, newly fertilized frog egg, for example, will complete a cell cycle every 30 minutes, thereby producing about 8000 cells in 12 hours. In the rapid cell proliferation of early embryos of many species, the G_1 and G_2 phases are very short, so these cells become progressively smaller, up to a point, with each division.

MITOSIS

During interphase the nucleus is, indeed, a busy place. It was once called the resting phase, though, for a very simple reason: all this activity is invisible when viewed with the light microscope. However, in the next phase, mitosis, the activity becomes starkly apparent, as the nuclear material shifts, reorganizes, and moves around.

Mitosis is simply the process of cell division. It is a continuous process with one event leading gradually to the next, yet biologists have described the process in parts: **prophase**, **metaphase**, **anaphase**, **telophase**, and **cytokinesis.** Cytokinesis can also be viewed as a stage of its own, separate from mitosis. We will look briefly into the readily visible events of each phase and then return to selected events for a more detailed discussion. A simplified view of mitosis is seen in Figure 9.4.

Prophase

The onset of mitosis is marked in nearly all cells by the condensation of the chromosomes (Figure 9.4a). Remember that the chromosomes were just replicated in S phase, so that there are now two copies of each chromosome. The two copies are called **chromatids** (or sister chromatids) and are joined together at a region known as a **centromere** (Figure 9.4b). Generally, the individual chromatids, connected by their centromeres, are not visible until late in prophase. As condensation of the chromosomes proceeds, the spindle begins to form, the nucleoli degrade, and in the final stages of prophase the nuclear envelope disappears.

The centrioles (where present, as in animals, many protists, and a few plants—see Chapter 4) migrate to opposite sides of the cell. The centrioles then become surrounded by a cluster of microtubules that radiate outward, taking on a "starburst" form called an **aster.** (Keep in mind that centrioles and asters are not found in the cells of fungi and most plants.)

As prophase continues, the chromosomes begin to move in an agitated manner, presumably being tugged about by the spindle. Following the breakdown of the nuclear envelope, each pair of sister chromatids is attached to **centromeric spindle fibers** from opposite directions. The spindle is attached to the centromere region of the chromatids by a special structure known as the **kinetochore** (Figure 9.5, p. 210). One of the sister chromatids is attached to a spindle fiber emanating from the "north pole" of the dividing cell; the other becomes attached similarly to the "south pole" (Figure 9.6, p. 210). Other microtubules form **polar spindle fibers,** fibers that are not attached to chromosomes, but span the entire distance between the two poles.

Lastly, in late prophase, movement becomes less meandering, as each chromosome heads to the cell's center (Figure 9.4), finally coming to rest at the equatorial plane. There, the pair of chromatids line up across the center of the cell, thus marking the arrival of the next phase, metaphase.

In summary, prophase is an incredibly busy time in which the chromatin condenses, the spindle apparatus forms, the nuclear envelope breaks down, and the chromosomes become arranged across the cell. The key feature is the precise and opposite attachment of centromeric spindle fibers to sister chromatids. This sets the stage for equitable division of the replicated DNA.

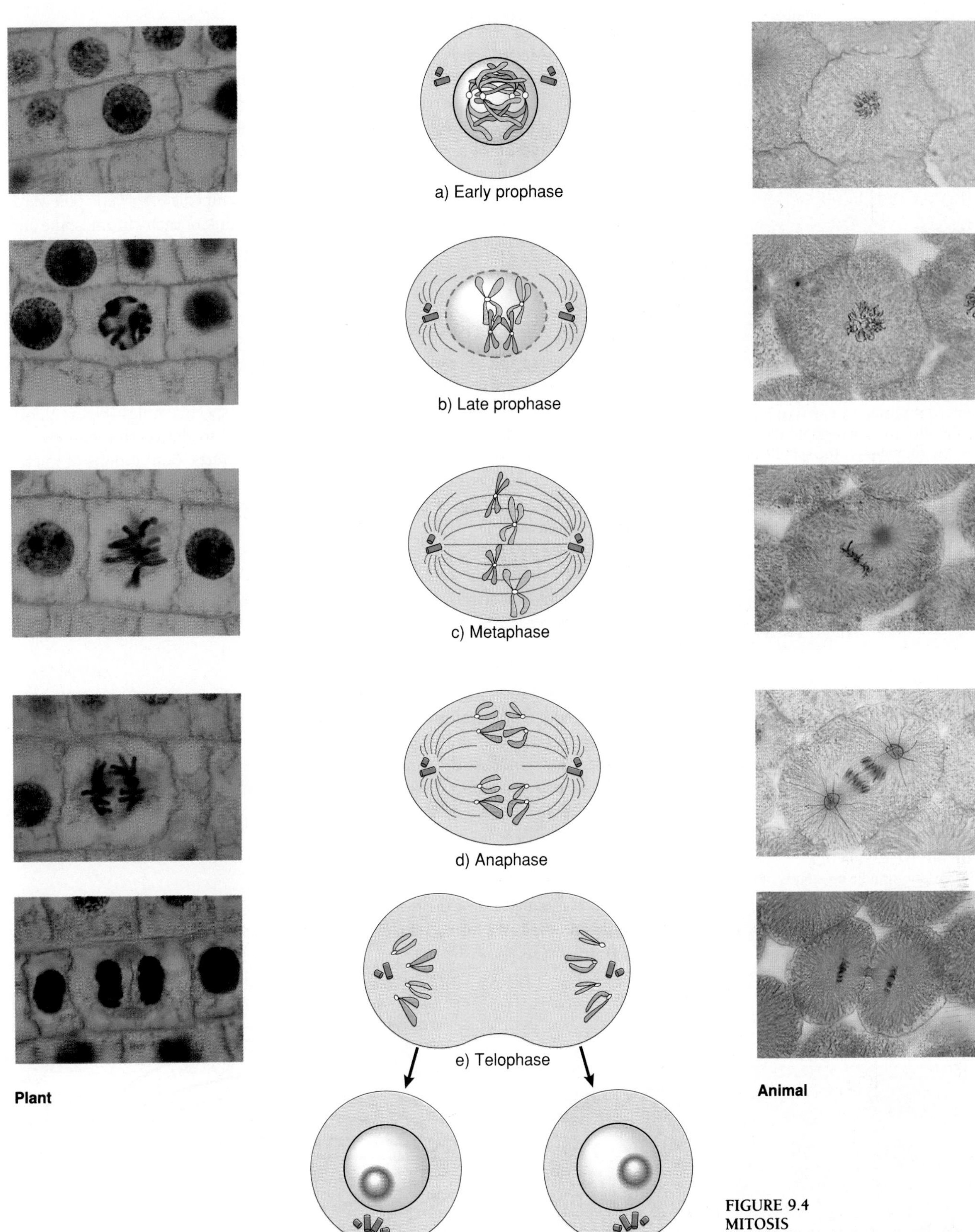

FIGURE 9.4
MITOSIS
The stages of mitosis are diagrammed and aligned with photomicrographs of examples from plants and animals.

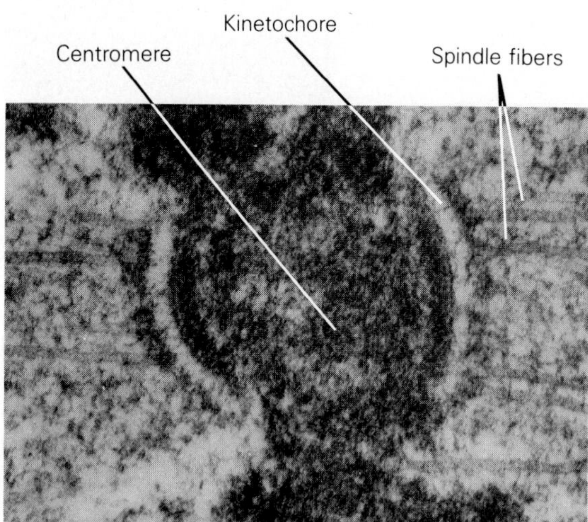

FIGURE 9.5
SPINDLE FIBER ATTACHMENT
In this electron micrograph, the hollow microtubules of the spindle fibers are seen attached to the curved kinetochore, which is physically associated with the centromere of the chromosome.

Metaphase

Metaphase is simply that stage of the process that is marked by the arrival of the chromosomes to the equatorial plane, a region of the spindle known as the **metaphase plate** (Figure 9.4c). Here the chromosomes are momentarily held in place. However, while the chromosomes appear to be at rest, their static position is maintained by strong opposing forces. Observers note that stray cytoplasmic particles fumbling into the metaphase plate are immediately swept poleward. If you've a mind for fanciful analogies, think of each chromosome and its opposing microtubules as a horse roped and held tenuously in place by two struggling but determined handlers.

Anaphase

Metaphase ends abruptly as the centromeres holding the sister chromatids together split apart. The chromatids can now be referred to as chromosomes. They take on rough V-shapes as they are drawn to their respective poles (Figure 9.4d), centromere first, by the spindle fibers, which are attached to the centromeres by the kinetochores. Chromosome movement requires several minutes. Both groups of microtubules are thought to be responsible for this movement, the centromeric fibers literally pulling the chromosomes along and the polar fibers elongating, thereby pushing the poles apart (see Essay 9.1).

It is in anaphase that the precise and equal separation and distribution of chromosomes takes place, even though the cell has yet to divide. One sister chromatid (one copy of each chromosome) has been brought to each end of the cell. This must happen with utterly faithful precision if the two emerging cells are to survive.

Telophase

In telophase, many of the events of prophase are essentially reversed. The chromosomes begin to decondense, their sausagelike form fading into the diffuse chromatin we saw at interphase. The spindle is dismantled, its proteins to be recycled in forming the new cytoskeleton. A new nuclear envelope is formed from remnants of the old, and soon the nucleoli are reestablished.

Cytokinesis

Cytokinesis is the division of the cytoplasm. In other words, this is the point at which two new, separate cells are formed (Figure 9.4e,f). The process is fundamentally different in plants and animals.

FIGURE 9.6
THE MITOTIC SPINDLE
The mitotic spindle apparatus of the rat is shown at metaphase, demonstrating the relationship of the spindle fibers, asters, centrioles, and chromosomes.

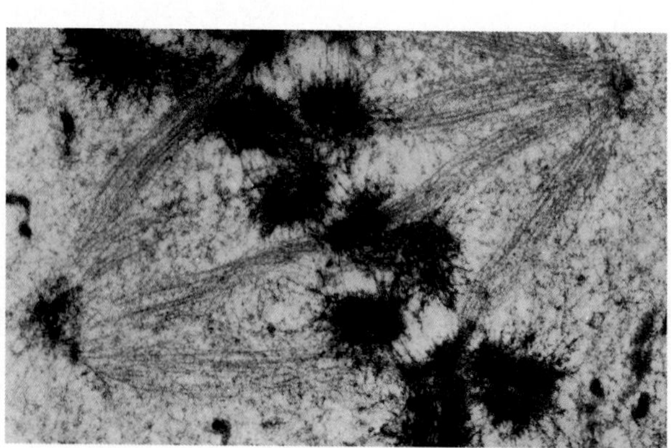

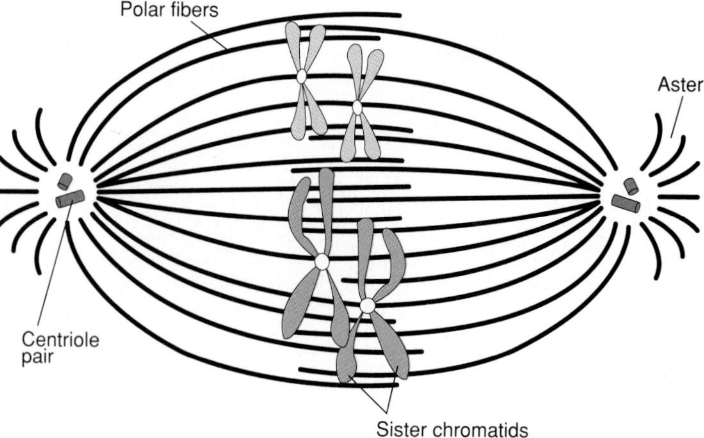

ESSAY 9.1

THE ROLES OF THE CENTRIOLE AND SPINDLE

We have noted that in early prophase, the two centrioles migrate to opposite spindle poles from a position near the nucleus. But what is their role in mitosis and cell division? Until recently, they were ascribed important mitotic roles, but this notion has been discarded. After all, plants and fungi lack centrioles, yet they form perfectly good spindles and go faultlessly through mitosis. It is far more likely that their separation in early prophase is just the cell's way of making sure one pair ends up in each daughter cell. Centrioles, as you learned in Chapter 4, are identical in structure to basal bodies, which in turn organize the formation of cilia and flagella.

Whereas they have no apparent role in chromosome movement, centrioles may influence cytokinesis. Evidence suggests that they play some role in establishing the plane of cell division, at least in animal cells. In experimental situations, anything that affects the position of the centrioles affects the division plane accordingly. Researchers are still trying to learn why.

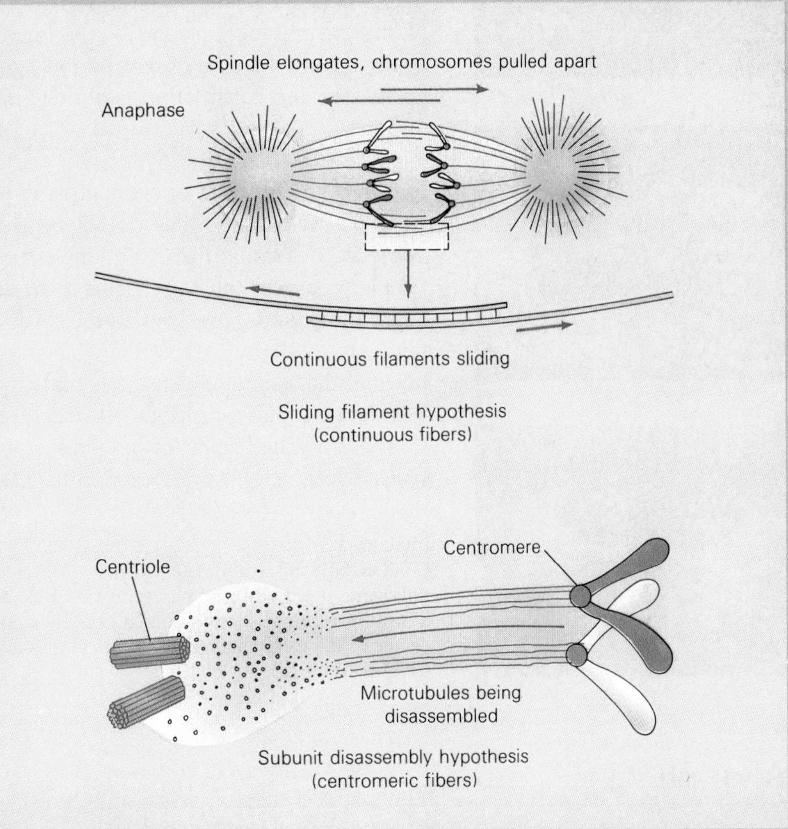

We've seen that the spindle apparatus is architecturally at the very core of mitotic movement. It consists primarily of microtubules composed of the protein tubulin, which is also a component of cilia and flagella. In fact, in the late G_2 phase of the cell cycle, tubulin accounts for as much as 10% of the cell's protein. The assembly of tubulin into the microtubules of the spindle occurs in early prophase, but the details of this assembly are not known. The locale of microtubule assembly is a region at each spindle pole. Curiously, these regions, aptly called microtubule organizing centers (MTOCs) lie directly around the centrioles in animal cells. Yet a functional relationship between the MTOC and centriole has yet to be established.

Centromeric spindle fibers actively separate the chromosomes by pulling on them during anaphase, and then the polar spindle fibers somehow elongate the spindle, apparently aiding in this separation. While we don't know exactly how they work, there are some hypotheses. Let's begin with the microtubules of the centromeric spindle fibers.

Careful observations of the centromeric spindle fibers clearly indicate that as the chromosomes are pulled to the poles, the connecting microtubules get shorter. This observation is explained by the microtubular disassembly hypothesis. Apparently, these spindle fibers are continually disassembled at the microtubule organizing center, disappearing entirely at the end of anaphase. Researchers have found that when substances known to inhibit microtubule disassembly are applied, chromosome movement slows. Conversely, when substances known to accelerate microtubule disassembly are applied, chromosome movement speeds up.

Apparently, the polar spindle fibers operate quite differently. First, let's note that the microtubules do not actually extend from pole to pole, but rather, they extend out from each pole, meeting and overlapping each other. The discovery of this overlapping relationship immediately suggested a mechanism such as that seen in the cilium and flagellum. There the protein dynein hydrolyzes ATP, bringing on a sliding and bending action (see Chapter 4). The application of dynein ATPase inhibitors effectively stops the sliding in polar microtubules. Thus, movement of polar microtubules appears to operate on a very similar principle to those of cilia and flagella.

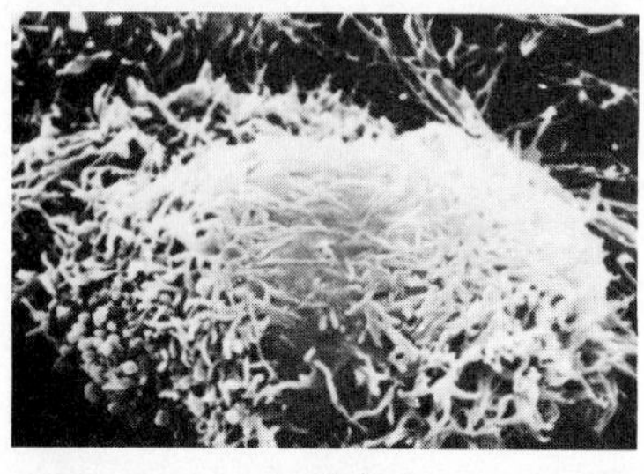

(a)

(b)

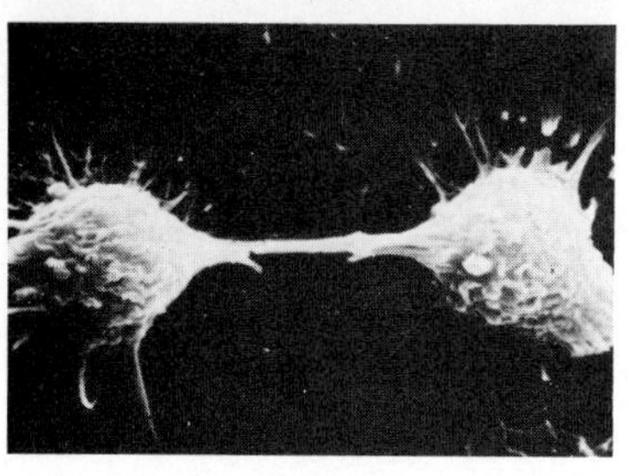

(c)

Animals In animals, cytokinesis generally begins with an equatorial furrow—an indentation, or furrow, in the surface of the cell, usually at about the plane of the metaphase plate. It looks as though someone had looped an invisible thread around the cell and were pulling it tight, pinching the cell in two. For this reason, animal cytokinesis is often referred to as cleavage. The furrow is created by the contraction of a ring of microfilaments (made up of a protein called actin) in the cytoplasm just beneath the plasma membrane. So actually, the furrow is being drawn in from below. The remains of the spindle may be caught by the tightening ring of contracting microfilaments, but no matter, the constriction continues until eventually the cell divides in two (Figure 9.7).

Plants The problem of cytokinesis in plants is complicated by the presence of tough, rigid cell walls. Cell walls cannot bend and flex very much, so cleaving is out of the question. In plants, then, cytokinesis proceeds quite differently. At about the time of telophase, a new cell wall begins to be assembled in the plane of the metaphase plate. Small membranous vesicles formed from the Golgi body accumulate along a plane in the center of the cell. There they join and form a virtually continuous double membrane, but one that contains dense materials that will later be used in cell wall construction. This dense, double-membraned structure is called the **cell plate.**

The cell plate begins forming in the middle of the cytoplasm and then, as more Golgi vesicles join, it grows outward to fuse with the plasma membrane at the periphery of

FIGURE 9.7
CYTOKINESIS IN ANIMAL CELLS
Following the division of chromosomes and their migration to opposite ends of the cell **(a)**, cytokinesis is accomplished in animal cells by a contraction of the cell membrane **(b)** around the center of the cell. This results in the sequestering of the divided chromosomes into separate daughter cells **(c)**.

TABLE 9.1 MITOSIS

INTERPHASE	PROPHASE	METAPHASE	ANAPHASE
Chromosomes are decondensed S phase: DNA and chromosomal proteins synthesized Spindle proteins and other mitotic proteins formed in G_2 phase Cell increases in volume	Chromosomes condense Separate chromatids may become visible Asters and mitotic spindle begin to form Nuclear membrane breaks down Nucleolus disperses Centromeres become attached to the centromeric spindle fibers Chromosomes migrate toward the metaphase plate	Spindle is fully formed Chromosomes are aligned with their centromeres on the metaphase plate Centromeres begin to divide	Centromeres divide Sister chromatids separate to become chromosomes Daughter chromosomes go to opposite poles Centromeric spindle fibers shorten; the spindle as a whole elongates
TELOPHASE	**CYTOKINESIS (CYTOPLASMIC CELL DIVISION) IN ANIMALS**	**CYTOKINESIS (CYTOPLASMIC CELL DIVISION) IN PLANTS**	**INTERPHASE**
Chromosomes begin to decondense Nuclear membranes reform Nucleoli reappear Spindle disappears Cytokinesis (cytoplasmic cell division) usually occurs	Microfilaments associated with the cell membrane form a circular band around the cell Microfilaments contract, pinching apart daughter cells	Small membrane vesicles fuse in the plane of the previous metaphase plate to form the cell plate Cell plate grows to separate daughter cells Cell walls are laid down	Daughter cells are in the G_1 stage of interphase (one DNA molecule per chromosome) after mitosis Chromosomes become diffuse

the cell (Figure 9.8), unlike the animal cell division furrow, which begins at the periphery and extends toward the middle.

A summary of the major features of mitosis is presented in Table 9.1.

Replication without Division There are exceptions to the general pattern of mitosis. For instance, early in the development of many insect larvae, and in many algae and fungi, cells undergo a number of nuclear divisions without cytoplasmic division, thereby becoming multinucleated. In fact the early embryo of the tiny fruit fly ***Drosophila melanogaster*** is a mass of some 6000 nuclei, all in the same single cytoplasm (Figure 9.9a). Only later do the cell membranes form between them to create individual cells.

Other kinds of cells may normally undergo rounds of chromosome replication with neither nuclear nor cytoplasmic division. One such example is found in the giant polytene chromosomes in the salivary glands of flies, including *Drosophila*. Their great size results from repeated chromosomal replication without subsequent separation (Figure 9.9b).

The Advantages of Mitosis

Mitosis, we see, is a critical step in the process that allows cells to divide, to produce two daughter cells where there was one. It is a fundamental property of many kinds of life, but its mechanisms are basically the same wherever it occurs. But what are the advantages of mitosis?

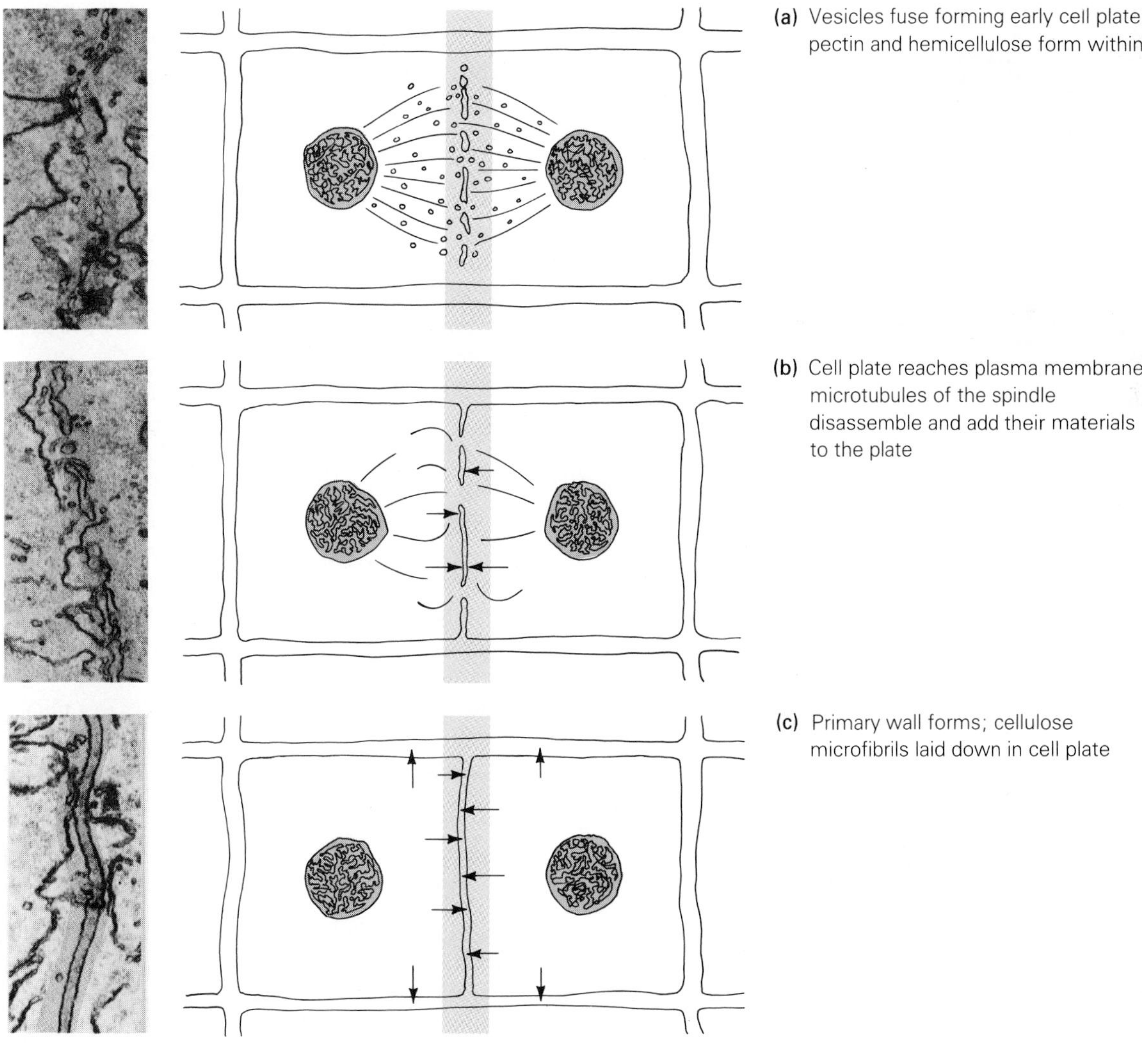
(a) Vesicles fuse forming early cell plate; pectin and hemicellulose form within
(b) Cell plate reaches plasma membrane; microtubules of the spindle disassemble and add their materials to the plate
(c) Primary wall forms; cellulose microfibrils laid down in cell plate

FIGURE 9.8
CYTOKINESIS IN PLANT CELLS
Cytokinesis in plant cells takes place not by a constriction such as that seen in animal cells, but by the building of a new membranes and cell wall between the daughter cells. First membrane vesicles are formed **(a)**, then aligned and fused to form the plasma membrane and then the cell wall **(b,c)**.

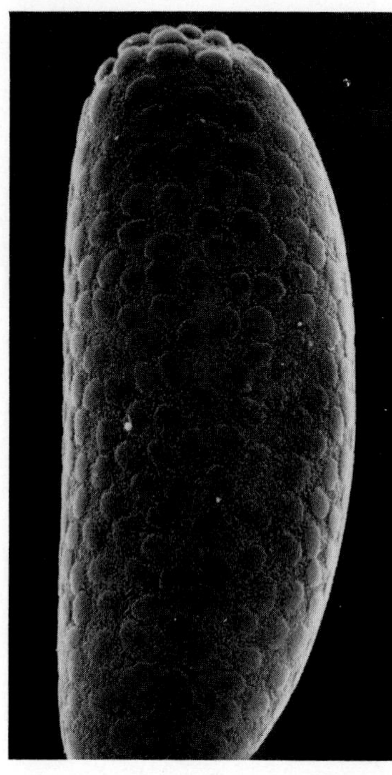

(a)

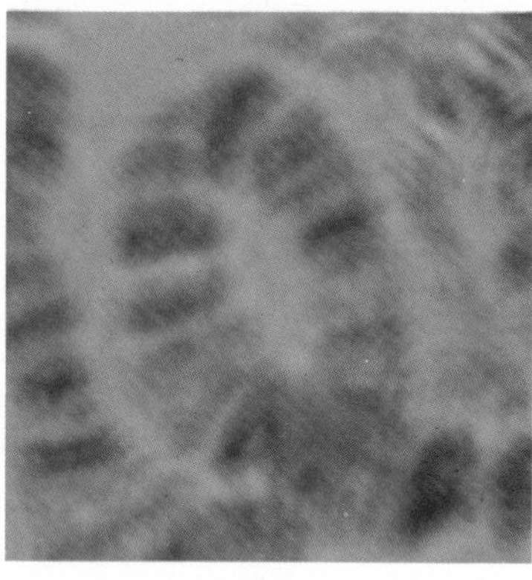

(b)

FIGURE 9.9
REPLICATION WITHOUT DIVISION
The two prominent types of chromosomal replication without accompanying cell division are exemplified in **(a)** the early embryo of the fruit fly *Drosophila melanogaster,* where many nuclei are formed without any cytokinesis, and **(b)** the salivary chromosomes of Drosophila, where chromosome replication has occurred hundreds of times, but the chromosomes remain closely side-by-side within the same nucleus.

In many single-celled organisms, cell division by mitosis is a way of reproducing asexually (without combining genetic material from two sources). Since mating is not involved, the organisms can make copies of themselves at a prodigious rate.

In multicellular organisms, mitosis permits growth through the formation of new cells. In some cases these replace older, worn cells. In humans, for instance, billions of new red blood cells must be produced daily to keep up with the rapid attrition of these cells. New cells can also provide material for cell specialization. Here, the new cells take different developmental pathways that will enable them to perform different tasks.

In some species, mitosis permits very precise wound healing, as damaged cells are replaced; in others, regeneration of lost body parts is possible through complex processes involving mitosis. Plants can reproduce asexually, regenerating whole organisms from small pieces. Obviously, mitosis is a critical and tightly controlled process in the lives of cells. Now, however, we will consider another process, seemingly similar and also tightly controlled, but with an entirely different function.

MEIOSIS

In most plants and animals, individuals are **diploid.** This means that every cell in the organism has two sets of chromosomes, one set of hereditary information from each parent. In sexual reproduction, the two partners produce **gametes** (sex cells, usually eggs or sperm), which join through fertilization to produce an offspring. Thus sexual reproduction ensures that the offspring will receive chromosomes from each of the two parents.

Two related questions come to mind: first, if fertilization joins the genetic material of two parents, why is it that the amount of genetic material is not doubled in every generation? Second, if each parent has two sets of genes and chromosomes, how is it that the offspring receives only one set from each parent?

It turns out that the answer to the second question also answers the first. Gametes are **haploid**—that is, they have only one set of chromosomes. Thus, each gamete (sperm or egg) enters into fertilization with only one set of chromosomes instead of two. Because of this, the number of chromosomes does not double each generation. The problem then becomes, how is the number of chromosome halved during gamete formation? In other words, how do gametes with one set of chromosomes per cell come from cells with two sets of chromosomes? The answer involves a complex and orderly set of steps called **meiosis.** A highly simplified scheme is seen in Figure 9.10.

We can define meiosis as the process whereby a diploid set of chromosomes is reduced to a haploid set of chromosomes in a cell, resulting in the production of a gamete. In diploid organisms, meiosis guarantees (1) that the chromosome number will remain stable from generation to generation and (2) that each sexually reproduced offspring will receive two complete sets of genetic instructions. We will soon see that the meiotic process also virtually guarantees that each gamete will be different from every other gamete. In humans there are 46 chromosomes in nearly every cell of the body—23 originating from the father's sperm (the paternal chromosomes) and 23 from the mother's egg (the maternal chromosomes). It is in meiosis that the 46 chromosomes per cell are reduced to 23 chromosomes per gamete.

It is important to note that most of what we will say about meiosis applies to animals, most plants, and a few protists—specifically to those organisms that spend most of their life as diploids. The fungi and most protists differ in many aspects of sexual reproduction, as we will see in Chapter 24.

Homologous Chromosomes

As we said, the 46 chromosomes in each human diploid cell come in 23 pairs; one member of each pair comes from the father's sperm and one comes from the mother's egg. The two members of each pair are called **homologues,** or homologous chromosomes. The two homologues are functionally equivalent in that they contain the information for the same features, but they may carry different versions of the information for those

FIGURE 9.10
MEIOSIS
The stages of meiosis are diagrammed. Representative stages from Lilly are shown in the photomicrographs.

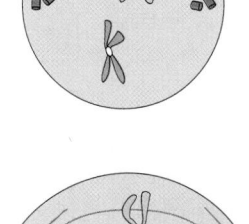

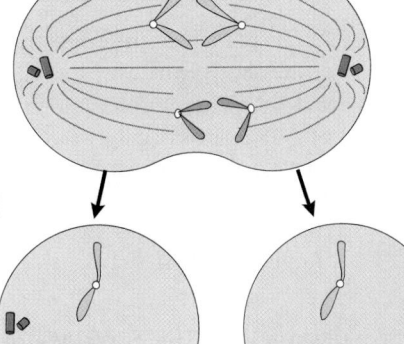

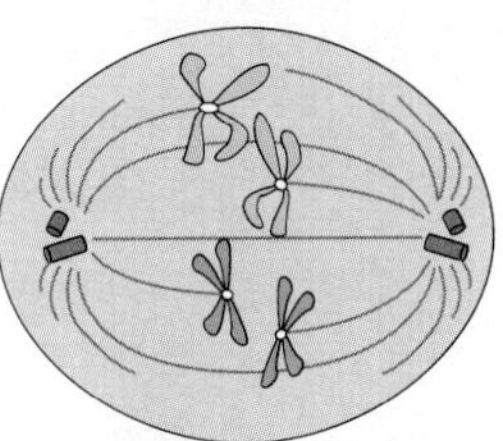
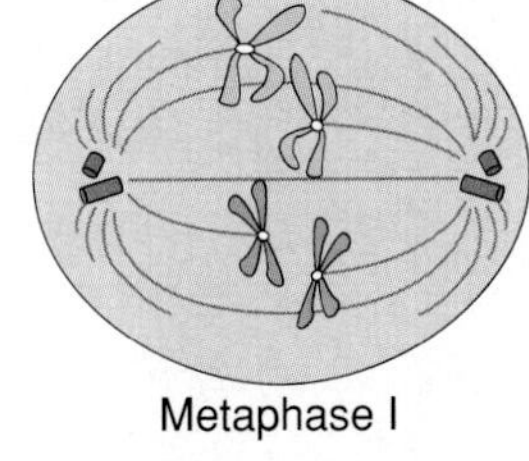

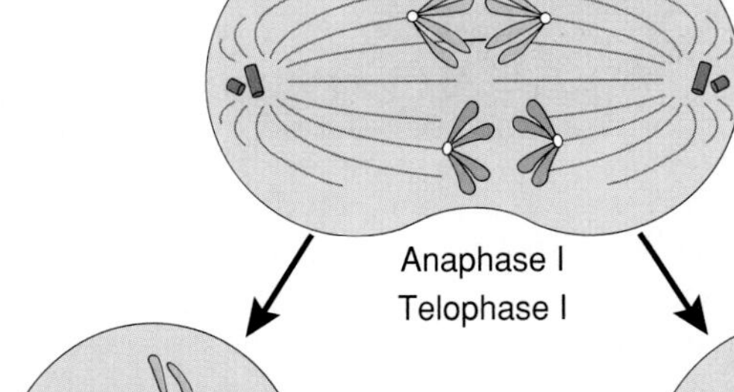

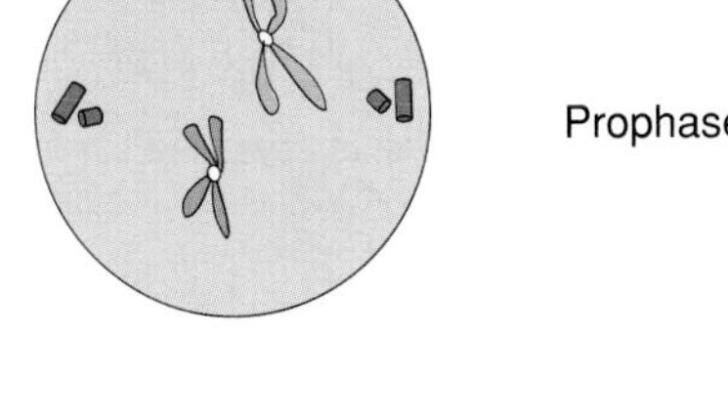
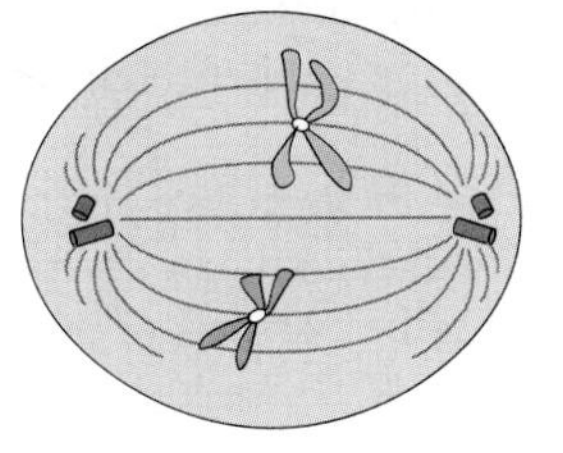
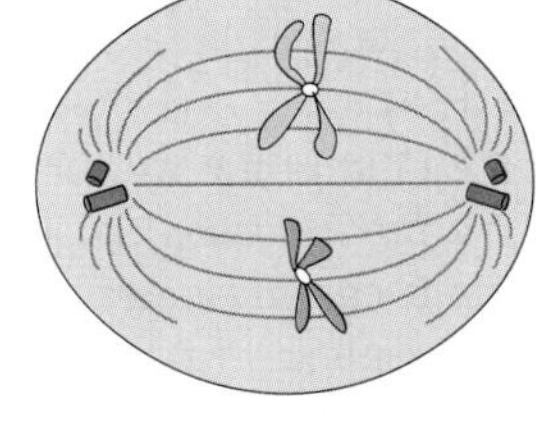

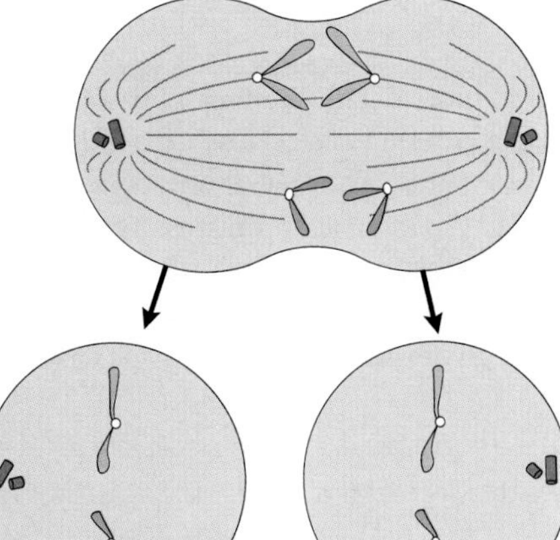
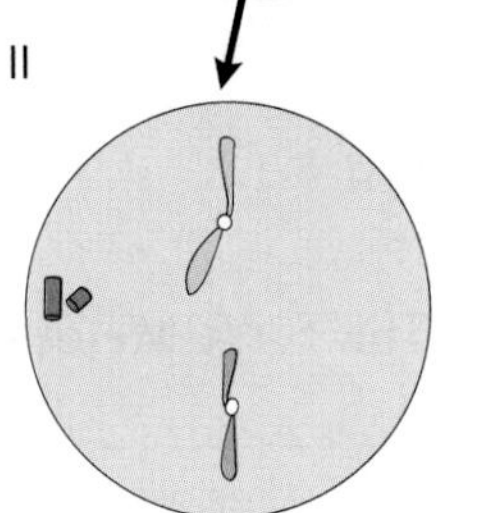

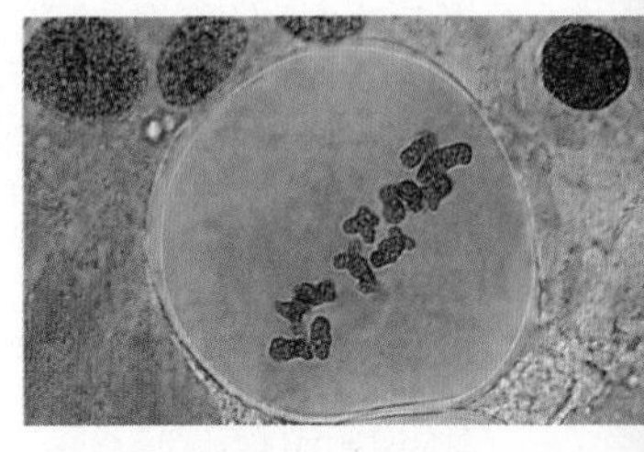
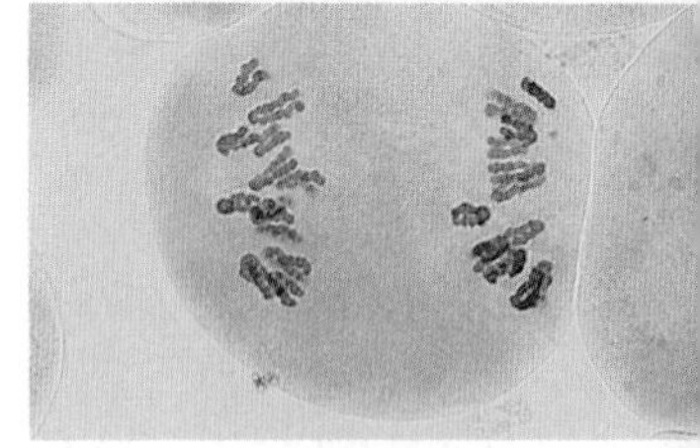
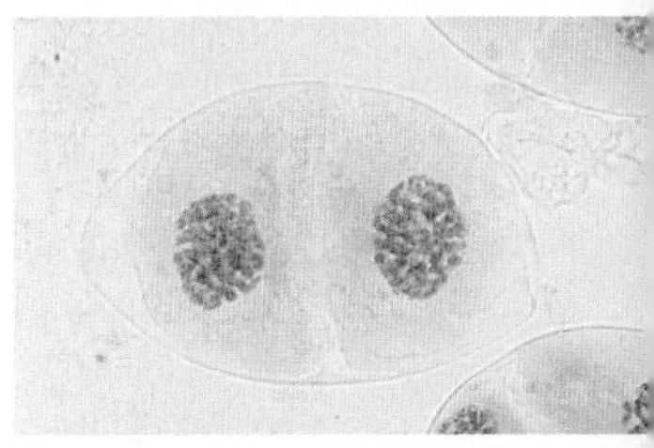
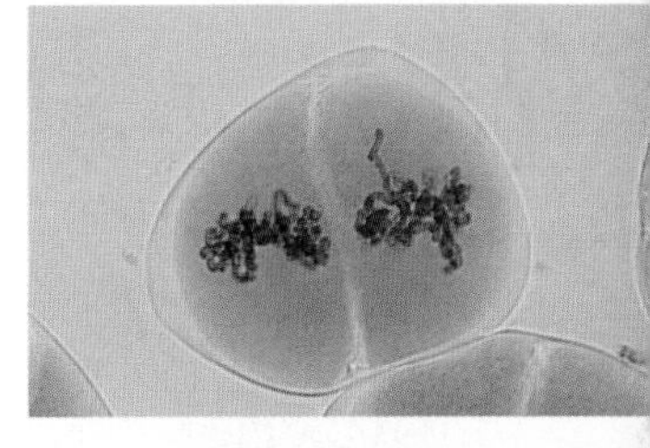
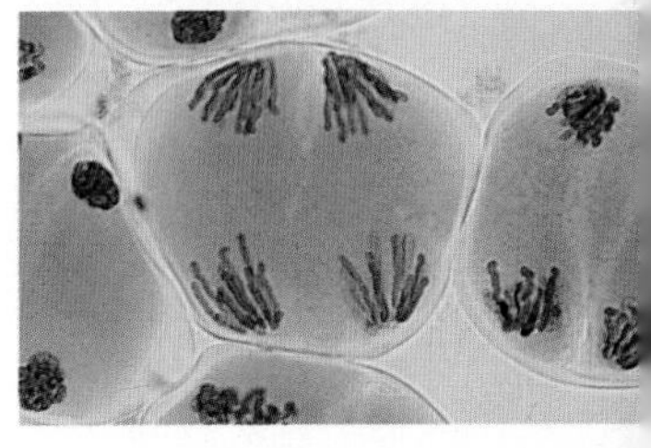
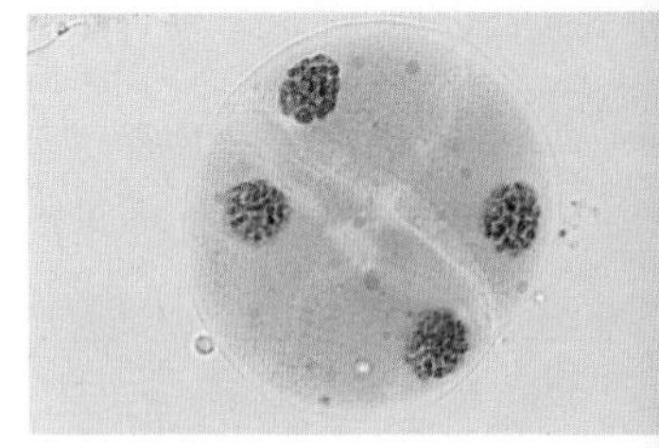

ESSAY 9.2

KARYOTYPING

A **karyotype** is a graphic representation of the chromosomes of any organism in which individual chromosomes are systematically arranged according to size and shape. Each species has its particular karyotype, and so we know the number and kinds of chromosomes found in carrots, fruit flies, and people. Karyotyping in humans is done by a simple and straight-forward method.

A blood sample is drawn and the white cells are separated and transferred to a culture medium. The medium contains not only nutrients, but also chemical agents that first induce mitosis and then stop the process when the chromosomes are at their maximal condensation (the stage at which the chromatids are seen most easily). The cells are then put on microscope slides and stained.

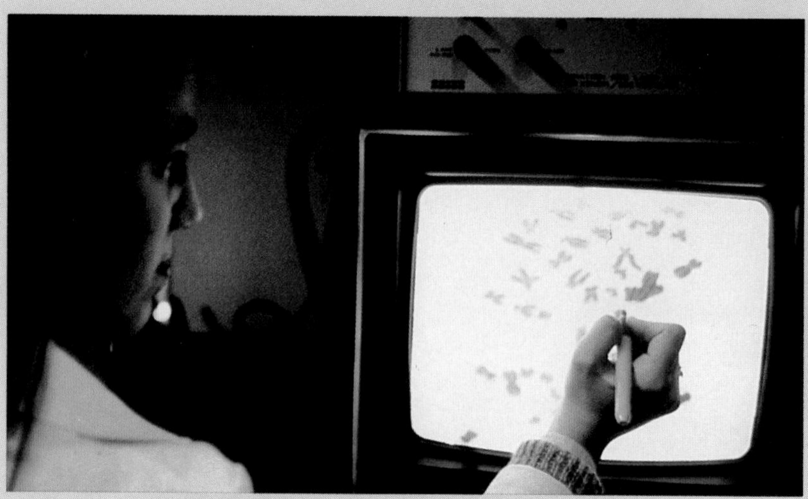

In the (not so) old days, cells showing all of the chromosomes were photographed through a microscope. Then a large print was made and the chromosomes were cut out with scissors, sorted by size and shape, then mounted with rubber cement.

Computer aided image analysis has modernized the process. Now technicians can view the chromosomes, identify them and sort them all electronically to produce the karyotype.

features. For instance, the homologue from the father might carry the information that tends to make eyes blue, and the homologue from the mother might carry information that tends to make eyes green.

Each chromosome within a cell of an individual is distinctive and can be identified through a process called **karyotyping** (Essay 9.2). In the human karyotype, each of the 23 different chromosomes is identified by its size, the position of its centromere, and its unique pattern of dark-staining bands. In Figure 9.10 we identify the homologous chromosomes simply by their size and color.

An Overview of Meiosis

At this point we can review the meiotic process in more detail. We begin by noting that at the start of meiosis, the chromosomes of cells entering meiosis will have doubled, and each chromosome now consists of two chromatids. Thus, in the human meiotic cell, there are actually 92 chromatids at this stage, the same as in mitosis. The cell will then go through two divisions, called meiosis I and meiosis II, in order to produce gametes with 23 chromatids apiece. (In the arithmetic of meiosis, one cell with 92 chromatids produces two with 46 each, which, in turn, produce four cells with 23 chromatids each).

Meiosis I: The First Meiotic Division

This first division in meiosis sets the stage for all that which follows. Remember that the main function of meiosis is to reduce the genetic material in half, so that fertilization can restore the diploid state. In the first meiotic division, meiosis I, the homologous chromosomes undergo **synapsis,** where they physically pair with each other. Then they align at the metaphase plate (as a pair) so that one member of the pair (in this case, two chromatids attached together by the centromere) is sent to each pole. (This is distinct from mitosis, where the chromatids split apart.) Thus, each pole receives one of the homologues—one pole receives the paternal homologue, while the other pole gets the maternal one. It is important to understand that the alignment of homologous pairs is random, so that the homologues arriving at either pole is a random assortment of maternal and paternal chromosomess .

Prophase I Prophase of meiosis I is longer and more complex than that of mitotic prophase. It has been divided into five substages for simplicity's sake, each with its own distinct events. As we see in Figure 9.11, the substages are

1. **Leptotene,** where we find the earliest visible condensation of the chromosomes.
2. **Zygotene,** where homologous chromosomes pair up, or undergo synapsis.
3. **Pachytene,** where the paired chromosomes condense further and can exchange chromosome parts in a process called **crossing over.**
4. **Diplotene,** where the paired chromosomes begin to separate, and the results of crossing over are seen as **chiasmata** (singular, **chiasma**).
5. **Diakinesis,** where the separation of the homologous chromosomes is nearly complete, though the homologues may still be joined at some points.

The process of chromosome condensation during leptotene and zygotene (Figure 9.11a) is quite similar to that of mitosis. The main difference is that the homologues must pair together in synapsis. How do homologues find each other in zygotene? Chromosomes can't move under their own power (as far as we know), and this early movement and pairing occurs before the spindle has even formed, so their movement can't be due to the spindle dragging them along. The only clue we have is that the nuclei of most organisms actually rotate during the early stages of meiosis, so that with time lapse photography they look rather like dryer windows in a laundromat. Pairing or synapsis can be seen to occur when homologous chromosomes touch each other in the right places, but first there seems to be a lot of random groping (see Figure 9.11b). When two homologues happen to touch in the same region, they hold. Then they join at adjacent regions. The result, over a period of time, is that the two homologues seem to come together like the two halves of a zipper. In fact, in synapsis there is a specialized

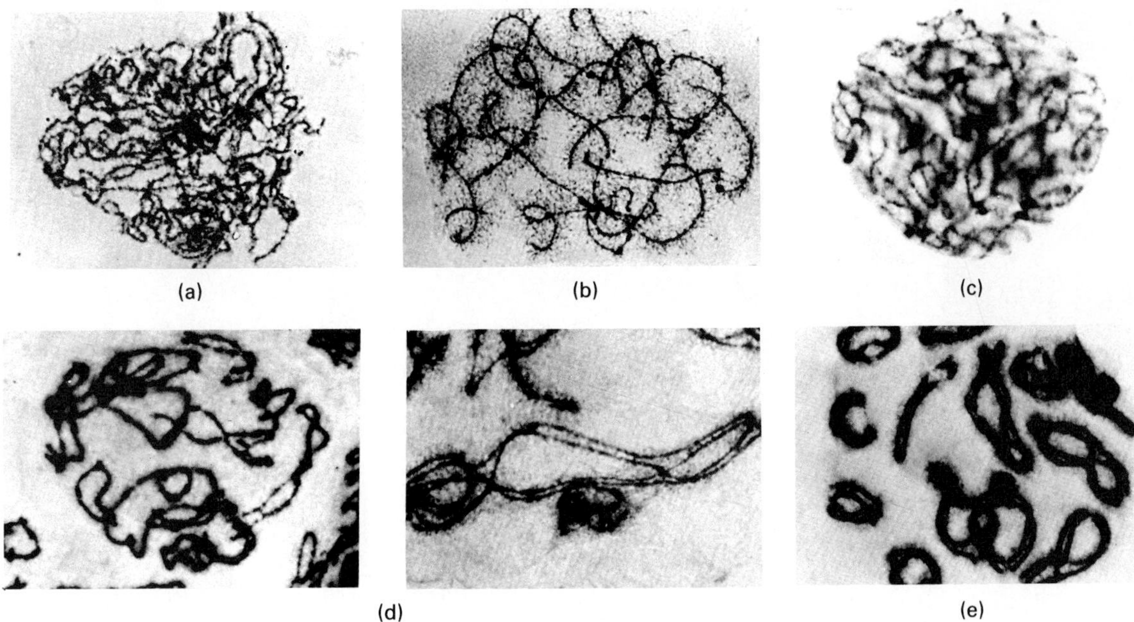

FIGURE 9.11
SUBSTAGES OF MEIOTIC PROPHASE
Meiotic prophase is a continuous process, with each "stage" blending smoothly into the next. These stages have received names based on recognizable events within the process. Some of the stages are particularly clear in these photos of meiotic prophase in the salamander. In **leptotene, (a)** the process of condensation begins, and in **zygotene (b)** the homologous chromosome pairs are joined by the synaptonemal complex. **Pachytene (c)** is the stage where crossing over occurs, though it is difficult to see the crossing-over events. In **diplotene (d)** the homologous chromosomes begin to separate, but remain joined by the chiasmata that result from crossing over during pachytene. **Diakinesis (e)** signals the end of prophase, with the chromosomes condensed in the form in which they will be arranged on the metaphase plate.

structure, the **synaptonemal complex** (made up of protein and RNA), which bridges the two homologues in meiotic prophase, and it even looks like a zipper (Figure 9.12a).

Because of the pairing up of chromosomes in zygotene, each grouping (joined homologues) actually consists of four chromatids called a **tetrad.** The meiotic cell of a human has 23 such tetrads. Each tetrad, remember, has two centromeres, one from each of the two members of the homologous pair.

A Closer Look at Pachytene and Crossing Over In pachytene, the four chromatids are held in close contact by the synaptonemal complex. This sets the stage for a very important process. While they are so closely bound, they are in a position to exchange segments of their chromosomes through a process called crossing over (Figure 9.12b). At any point along the DNA backbone of a chromatid, special multienzyme complexes, forming gigantic recombination nodules, can cause the chromatids to undergo crossing over. This series of events is complex and incompletely understood.

It is quite possible that the synaptonemal complex plays an important role in crossing over. It always seems to be present where crossing over is occurring, and absent where it is not. For example, crossing over has never been detected in the males of *Drosophila melanogaster,* whose paired chromosomes lack a synaptonemal complex. The same can be said for females of the silk moth, *Bombyx,* in which neither the synaptonemal complex nor crossing over has ever been seen.

In human cells, there is an average of about ten crossovers for every meiotic tetrad; in other species, there may be fewer. But one thing is clear: after meiotic prophase, one cannot properly refer to individual chromatids as maternal or paternal. The crossovers produce new kinds of chromatids, and it should be mentioned that this is an important source of genetic variation in a population. This constantly renewed variation is crucial to the evolutionary process because it is from such variable populations that nature will select those individuals whose genes will pass into the next generation.

Evidence of crossing over having occurred is seen first in the diplotene stage (see Figure 9.11d). In this stage, whereas the sister chromatids of each of the two homologues are still held tightly together, the homologous chromosomes begin to separate. Again like a zipper, the synaptonemal complex "unzips," permitting the homologues to begin

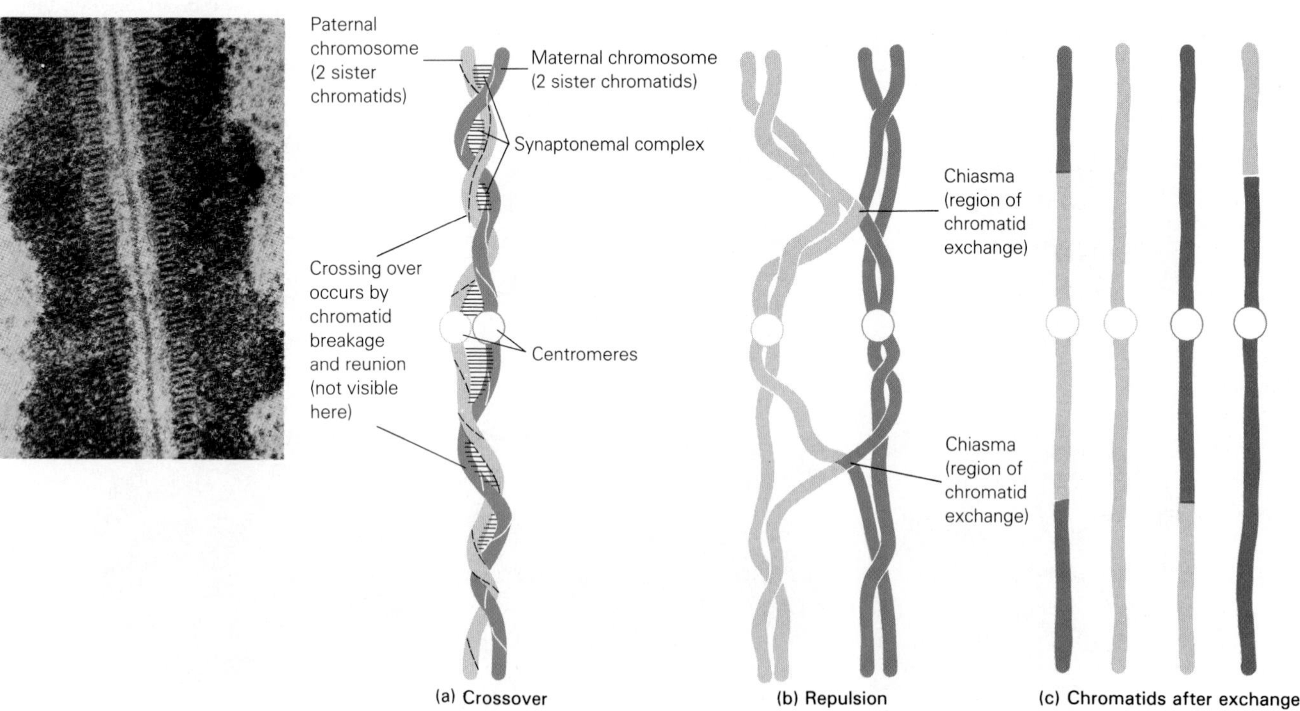

FIGURE 9.12
CROSSING OVER AND CHIASMA FORMATION
In pachytene, the maternal and paternal homologous chromosome pairs are intimately associated and bound together by the synaptonemal complex, shown in the electron microscope and diagrammatically in **(a)**. During this association regions of chromosomes are exchanged between the homologues in the process of crossing over. In diplotene **(b)**, the homologues separate from each other, but are held together at chiasmata, which are the places where crossing over has occurred. If the four chromatids could be unwound and separated (as they will at later stages of meiosis), and if the regions of maternal and paternal origin could be separately indicated, they would appear as shown in **(c)**. Note that any one exchange involves only two of the four chromatids. While we have shown only one exchange in each arm, there are usually many.

parting. But wherever crossing over has taken place, the homologues tend to cling together, forming the cross-shaped configurations we earlier called chiasmata. So for a brief period, the chiasmata reveal where crossing over has occurred.

It is in diakinesis (Figure 9.11e) that the separating chromatids finally become visible as individuals. Recall that this occurred very early in prophase of mitosis. The chromatids of individual chromosomes are joined at their centromeres, just as you expect. However, while the homologous chromosomes move apart along most of their length, they still manage to cling together at the crossover points, just as they did in diplotene. This clinging will continue into metaphase I.

Metaphase I In the first metaphase of meiosis (Figure 9.10), the tetrads are brought to the metaphase plate just as in mitosis, but the appearance and behavior of the chromosomes are quite unlike any mitotic metaphase. In mitosis, you'll recall, centromeric spindle fibers from each pole attached to each centromere. In metaphase I of meiosis, however, centromeric spindle fibers from just one pole attach to each centromere (see Figure 9.10). The importance of this orientation only becomes clear when anaphase I begins. You can see that the centromeres of homologous chromosomes appear to be separated from one another as far as possible, with only the ends of the chromosome arms touching.

Anaphase I Anaphase I follows quickly (Figure 9.10). Because of the way in which the centromeric spindle fibers are attached to the chromosomes, the results are quite different from those of mitotic anaphase. In mitosis, the pull on each chromosome came from two directions, and as the centromeres divided, sister chromatids moved to opposite poles. In metaphase I of meiosis, the pull is from one direction only, and the centromere does not divide. Thus, when migration begins, homologue separates from homologue, and each chromosome, still composed of sister chromatids, is drawn to a pole. In humans, this means that 23 chromosomes go to each pole and that the homologues—the chromosome pairs—are forever separated. So, following the events of anaphase I, we can see that the number of chromosomes arriving in each daughter cell has been reduced by half. For this reason, the first meiotic division is known as the **reduction division.**

Telophase I and Meiotic Interphase Anaphase I is followed by telophase I (Figure 9.10), complete with the reorganization of nuclei, decondensation of chromosomes, and duplication of centrioles, followed in turn by cytoplasmic division. The cell then enters a meiotic interphase.

Meiotic interphase is unlike mitotic interphase in one very important detail: there is no S phase. In other words, there is no DNA replication. Depending upon the species, meiotic interphase may be extremely brief, or it may be exceedingly long.

Meiosis II: The Second Meiotic Division

Meiosis II is much easier to follow than meiosis I because it is essentially the same as mitosis (Figure 9.10).

Prophase II In Prophase II, we immediately see similarities to mitotic prophase. The chromosomes condense, with each containing two chromatids, still attached by their centromeres. This time the centromeric microtubules become attached in standard mitotic fashion: that is, the kinetochores of each chromosome attach to fibers from opposite poles. The chromosomes then move to the metaphase plate.

Metaphase II and Anaphase II In metaphase II the chromosomes line up on the metaphase II plate in preparation for separation. Anaphase II is marked by the division of the centromeres and the separation of the chromatids into individual chromosomes. These chromosomes are then pulled apart as they are in mitotic anaphase, with the same forces and attachments. It is important to remember that the chromosomes that have separated in anaphase II very likely have been rearranged by crossing over.

Telophase II As anaphase II proceeds into telophase II, nuclear envelopes once again form around the four groups of decondensing chromosomes. Unlike the centrioles in mitotic telophase, the newly isolated centrioles do not replicate. (Why do you suppose this is?) The four resulting cells then enter interphase, where they are destined to remain in the G_1 phase. Although cellular activities continue as the egg and sperm develop, DNA replication is over until fertilization triggers new cell cycles.

To sum up, then, we see that each of the four chromatids of the prophase I tetrad (that have potentially been rearranged by crossing over) ends up in one of the four cells that are the products of meiosis. As in mitosis, the amount of genetic material was doubled before the process started (during the S phase), but in two divisions the products were distributed to four cells. Each cell now has one fourth of the normal G_2 amount of DNA. Each of these is now haploid—that is, each cell has only one of each of the different types of chromosomes of the species (for example, 23 chromosomes in human gametes). Meiosis and mitosis are directly compared in Figure 9.13.

There are two major take-home lessons from the story of meiosis. First, meiosis halves the chromosome number of eggs and sperm, making fertilization feasible. Second, meiosis provides a means of shuffling and reorganizing chromosomes, thus increasing genetic variation in offspring. This shuffling and reorganization takes place in two major ways: (1) the prophase I exchange of chromatid parts by crossing over; (2) in the random lining up of homologous chromosomes in metaphase I, so that maternal and paternal centromeres are randomly distributed to the two poles. This is not to mention the somewhat random way two sexual individuals often get together in the first place.

MEIOSIS IN HUMANS

Before we get into the details of meiosis in humans, we should make a few preliminary points. First, meiosis occurs in what is called **germinal tissues** (*germ,* "seed"), tissue that gives rise to gametes. All other tissue is called **somatic tissue.** In animals, the

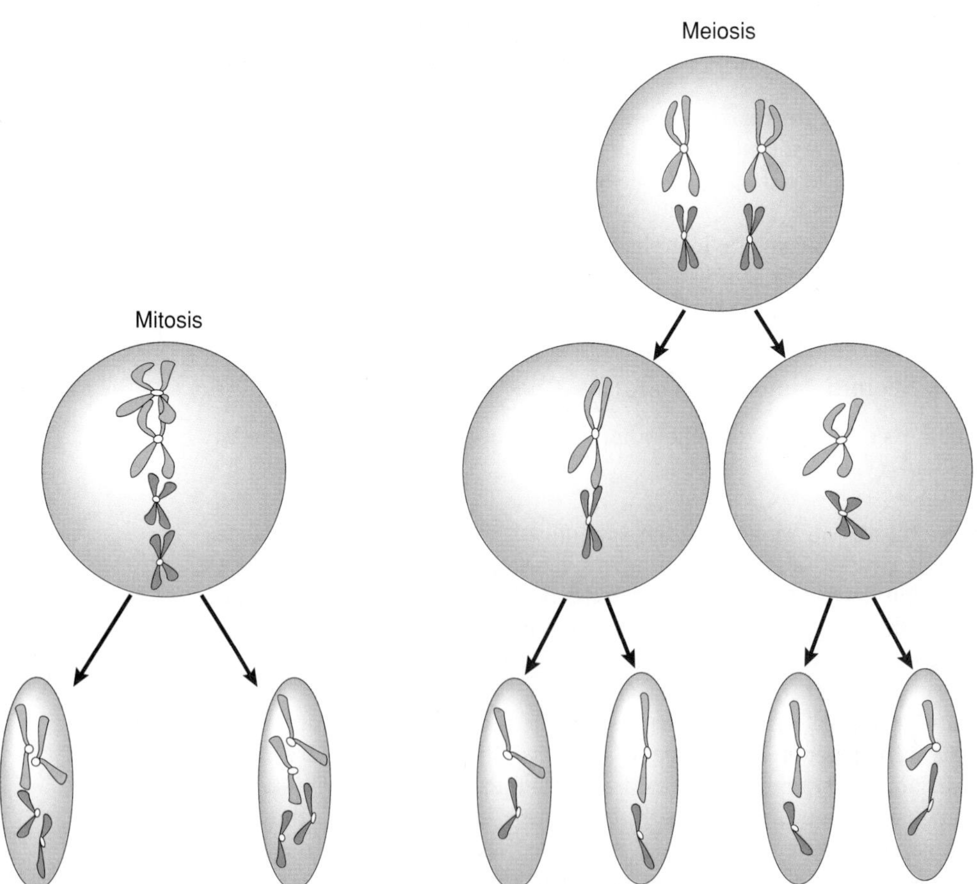

FIGURE 9.13
MITOSIS AND MEIOSIS COMPARED
Simplified schemes of mitosis and meiosis are compared side by side to highlight similarities and differences. It should be clear why the second divison of meiosis is often called a "mitotic" division.

germinal tissue goes through several developmental phases. It begins with a small group of germinal cells in the embryo. These will become the germinal tissue of the gonads, then meiotic cells, and finally the gametes. You might keep in mind that not all the tissues of the gonads are germinal. Some gonad tissue plays a supporting role, but does not undergo meiosis.

The germinal tissue of the gonads (ovaries and testes) is called **germinal epithelium** (Chapter 42). (Epithelium is a general term for surface tissue, either internal or external.) The germinal epithelium of the testes and the ovaries, like any other tissue, undergoes repeated mitosis during development. Meiosis will occur later.

Now on to meiosis in humans. During fetal development in human males, the germinal epithelium of the testes forms the lining of long, convoluted tubes—the seminiferous tubules (Chapter 42), but both mitosis and meiosis in this tissue are suppressed until puberty. Then, the germinal epithelium undergoes a type of asymmetrical mitotic division. With each mitotic division, one of the daughter cells remains a germinal epithelium cell, capable of further mitosis, but the other daughter cell becomes differentiated as a **primary spermatocyte**, a specialized diploid cell ready to begin that once-in-a-life-cycle happening, meiosis. The primary spermatocyte will proceed through two divisions and eventually form four sperm cells (Figure 9.14). Billions of sperm will be produced in the germinal epithelium daily for the rest of the man's life.

Meiosis in human females is quite unlike that in males. During fetal development in human females, germinal epithelium forms in the ovaries. The cells of the germinal epithelium then begin to undergo meiosis (Figure 9.14). There are countless cells in this tissue, but they develop more or less simultaneously during the final months of fetal development, completing most of the first prophase of meiosis. Then meiosis suddenly stops. Thus, when a baby girl is born, her germinal cells, now referred to as **oocytes**, are well along in the process of gamete formation, but frozen in their development. And there they will remain, suspended in this state for 10 to 12 years. Then, when the menstrual cycle begins, they will be released, one or sometimes two at a time, during each monthly cycle for the next 35 years or so. Interestingly, meiosis will not be completed unless the oocyte is fertilized. Although a woman may have several thousand oocytes, only about 400 to 500 will become mature, and, with any luck, most of these will not be fertilized.

Meiosis in Females, a Special Case Since all oocytes are suspended in meiotic prophase in human females, meiotic prophase can last 45 or perhaps 50 years. (Actually, the long prophase I in females is typical of many animals, including all vertebrates.)

In spite of such differences in the beginning stages of meiosis in males and females, once it is underway it proceeds in much the same fashion in both sexes. That is, at least, until meiotic cytokinesis. Here sexual differences again appear.

The primary difference is that whereas each germinal cell in males produces four sperm, each germinal cell in females produces only one viable egg. This happens as the cell constituents are divided unevenly among the products of meiosis, with most ending up in one cell (Figure 9.14). Only that one cell gets to be the egg, and it is, indeed, fat and opulent and swollen with nutrients that will sustain it until it can be fertilized and begin to draw nutrients from the mother's tissues. The unequal division begins about the time of ovulation, where the oocyte's meiotic spindle forms off to one side of the cell, actually just under the surface. A normal first meiotic chromosome separation and cell division occur, but one of the two daughter cells is very small and is called a **polar body.** The other cell is known as the **secondary oocyte.** After fertilization, the second unequal meiotic division occurs, pinching off another tiny cell as a secondary polar body, which emerges just under the first.

The first polar body normally does not bother to complete its own meiosis II division. What would be the use, anyway? In fact, what is the function of polar bodies? Generally the polar bodies can be regarded as no more than convenient little garbage cans into which an unwanted three quarters of the genetic material of the primary oocyte can be dumped, leaving one large oocyte with its haploid complement of chromosomes and the lion's share of its predecessor's cytoplasm and nutrients.

FIGURE 9.14
SPERMATOGENESIS AND OOGENESIS
Spermatogenesis occurs as expected, with one cell yielding four haploid products that will become sperm. However, oogenesis is different. In each of the two divisions, the metaphase plate forms to one side of the oocyte, and unequal cleavage occurs. Thus only one large ovum results from each meiotic event. The ovum in the photo shows a polar body on its right side.

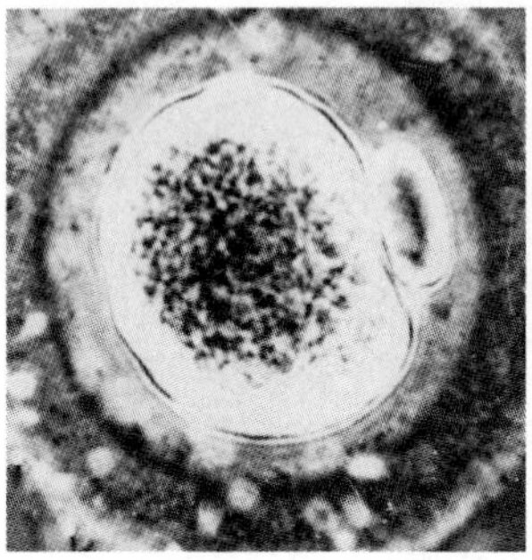

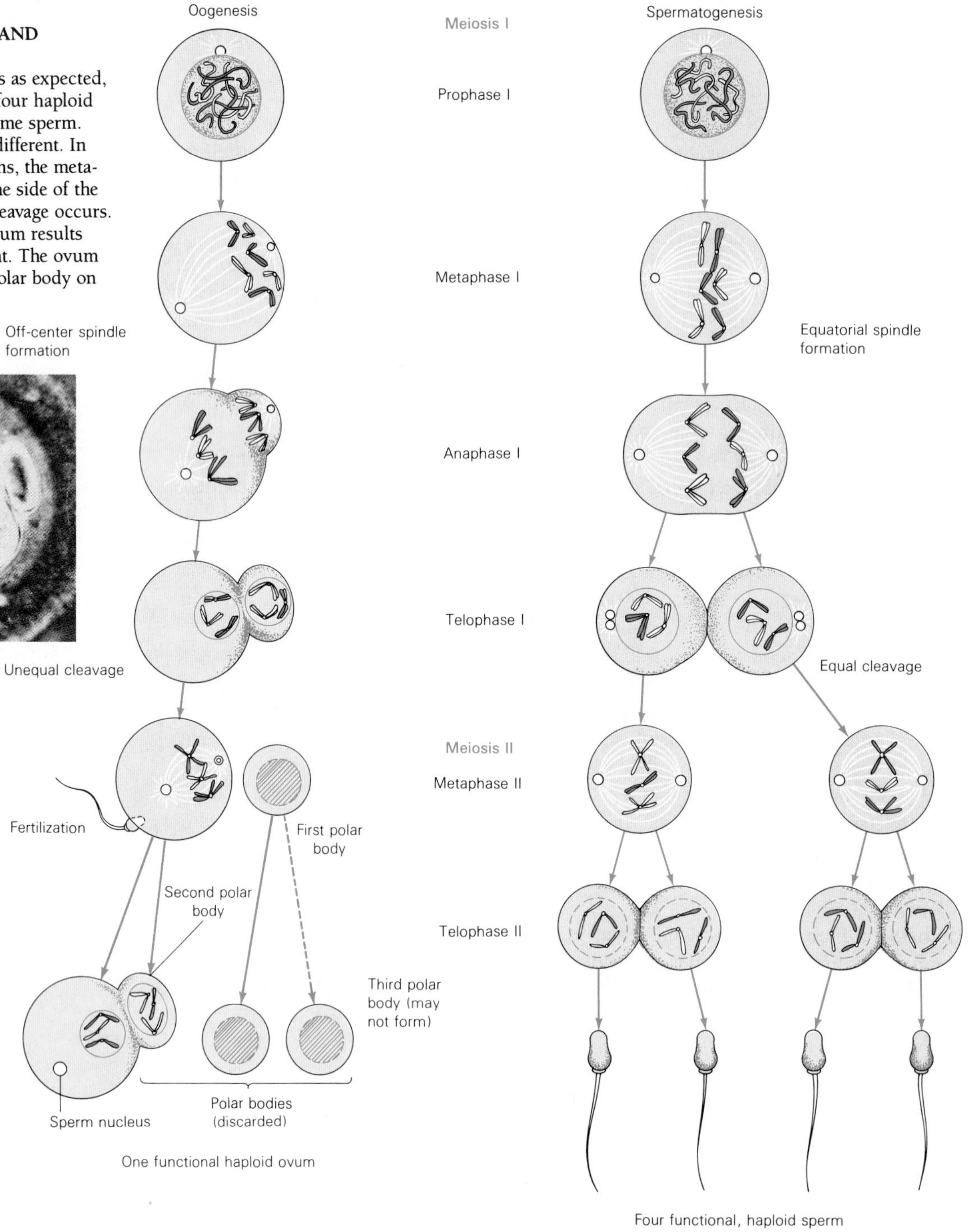

Meiosis, Evolution, and Sex Meiosis, we see, is a complicated process. Furthermore, its complexity means that all too frequently something goes wrong (see Essay 9.3). So why do organisms bother? Maybe because of sex. (Perhaps you haven't noticed that we have been dealing with sex all along.) Meiosis, after all, is part of sex. So the next question is, what is sex good for in the long run? (You probably have an answer about the short run.)

In the long run, of course, we are all dead. But our genes live on. When we leave offspring ("we" includes all us sexual beings), we pass our genetic information on to

ESSAY 9.3

WHEN MEIOSIS GOES WRONG

Meiosis is a much more complicated process than mitosis. When you consider the lengthy prophase with all the subphases of chromosome pairing, crossing over, in addition to the presence of two cell divisions, you shouldn't be surprised that frequently something goes wrong. In humans, for instance, about one third of all pregnancies spontaneously abort in the first two or three months. When the expelled embryos can be examined, it turns out that most of them have the wrong number of chromosomes. Failure of the chromosomes to separate correctly at meiosis is termed nondisjunction.

Not all failures of meiosis result in early miscarriage. There are late miscarriages and stillbirths of severely malformed fetuses. Even worse, about one live-born human baby in 200 has the wrong number of chromosomes, accompanied by severe physical and/or mental abnormalities.

About one baby in 600 has three copies of tiny chromosome 21. Such persons may grow into adulthood, but many have abnormalities. The syndrome is known both as **trisomy-21** and **Down's syndrome,** the latter after the nineteenth-century physician who first described it. Characteristics of the syndrome are general pudginess, rounded features, and a rounded mouth in particular, an enlarged tongue that often protrudes, and various internal disorders. Often a peculiar fold in the eyelids is seen, and in the past this was erroneously equated with the characteristic eye fold of Asians (thus the earlier name "mongoloid").

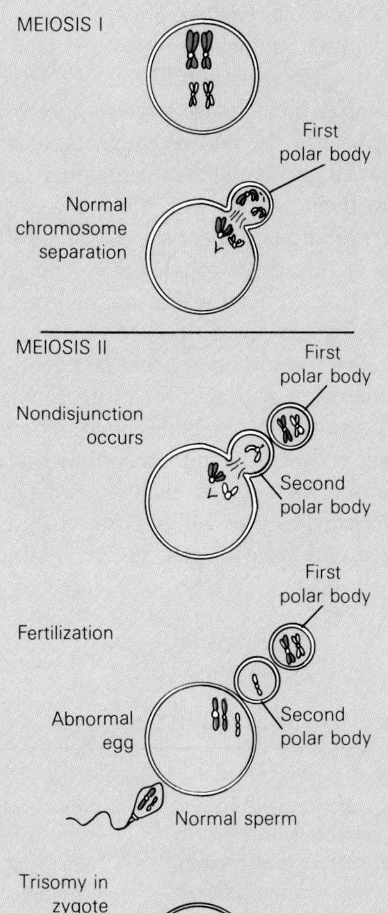

Trisomy-21 individuals also have a characteristic barklike voice and usually happy, friendly dispositions.

Trisomy-21 occurs most frequently among babies born to women over 35 years old. At that age the incidence is about 2 per 1000 births. By age 40, this climbs to about 6, and by age 45, 16 children with Down's syndrome are born for each 1000 births. The age of the father apparently has little if any effect, even though trisomy-21 can result from nondisjunction in the male. We can guess that the much-prolonged prophase I of the human oocyte might have something to do with this.

our descendants. The problem is, our descendants may not find the world to be the same as the one we lived in; the weather changes, new diseases show up, and unexpected competitors eat our food. Logically, no one particular combination of genes, including ours, will be adequate to meet all new situations. So, according to evolutionary theory, sexual reproduction comes to the rescue.

Meiosis and sexual unions keep reshuffling the genes of all the successful individuals in the population so that virtually infinite possible combinations of genes are produced. The reshuffling occurs because crossing over and random assortment of the homologous pairs occurs during meiosis I. Most of the new combinations will be less successful than the old ones because, whereas the old ones are tried and true, the new ones are genetic experiments. In the long run this is the cost of success. The winning evolutionary strategy is to cover as many bets as possible by having highly variable offspring. Then, when the weather changes or the Creeping Purple Flu shows up, perhaps not all of our descendants will be wiped out. Some of them are likely to have the right combinations of genes to be able to live under these conditions.

We can see this principle at work in certain organisms that can reproduce either sexually or asexually. For instance, *Daphnia,* the little water flea, reproduces asexually for generation after generation, females producing only females. In fact, sometimes one can see an asexual female embryo within an asexual female embryo within an asexual adult! The little animals keep up this kind of reproduction as long as conditions are stable. But when their pond begins to dry up, or things otherwise get dangerous and unpredictable, they change strategies. They begin to produce both males and females. These go on to mate and reproduce sexually in the usual way. Thus, in hard times, a few of their highly variable offspring may have that particular combination of genes that will allow them to cope with their new environment.

Some evolutionary theorists do not deny that sexual reproduction may be triggered by stress or trauma or rough times. They note, for example, that intensive sexual reproduction may follow a disaster such as a forest fire. (Interestingly, the number of human births may shoot up after a national disaster, such as war.) The theorists argue, however, that the advantage of sexual reproduction does not lie in producing an increased level of variation within subsequent populations. They note, for example, that some asexual organisms (such as the fungi imperfecti) can also be highly variable.

The silent dances of mitosis and mieosis are indeed intricate pageants in the drama of life, and critical to the evolutionary direction on the planet. An understanding of these mechanisms, we will see, sets the stage for our next story, that of how such mechanisms are used by cells to parcel out the genetic information.

APPLICATION OF IDEAS

1. Single-celled organisms that reproduce asexually through mitosis die in enormous numbers. But potentially, at least in some sense, such organisms are immortal. Multicellular organisms that can reproduce by fission (splitting) or budding (asexually producing miniatures of themselves) are also potentially immortal in the same sense. What do these statements mean? What kind of a tradeoff might there have been in the evolution of sex? Is death more of a reality to sexual beings than to asexual beings?

KEY IDEAS

THE NUCLEUS

1. The nuclear structures include a double membrane called the nuclear envelope and the nuclear pores that selectively admit the passage of materials.
2. The nuclear envelope disappears from sight during much of cell division.
3. The nucleus contains the chromosomes.

DNA, Chromatin, and the Eukaryotic Chromosomes

1. Chromosomes consist of **chromatin,** which includes DNA and nuclear proteins known as histones and nonhistone chromosomal proteins.
2. The DNA of chromosomes is wound about numerous beadlike **nucleosomes,** which consist of histones.
3. In cell division, chromosomes are highly condensed. When not in division the chromosomes are uncoiled and spread out, forming the chromatin net.

The Cell Cycle: Growth, Replication, and Division

1. The cell cycle includes growth, replication, and **mitosis.**
2. Some cell specialization requires that cells leave cycling. Many such cells no longer reproduce so their

life span is usually limited. Others survive the lifetime of the organism.

3. Mitosis includes formation of the mitotic **spindle**, separation of chromosome replicas, and division of the cell into two. The remaining phases constitute **interphase.**
4. **G_1 phase** is a period of protein synthesis, while the **S phase** involves DNA replication, and the **G_2 phase** includes synthesis of the mitotic spindle protein.
5. At the end of G_2, each chromosome consists of two **chromatids** joined at their **centromere.**

MITOSIS

1. Mitosis includes five subphases: **prophase, metaphase, anaphase, telophase**, and **cytokinesis.**

Prophase

1. Chromosome condensation is a matter of coiling and supercoiling in at least three orders.
2. Condensation facilitates the movement of chromosomes during mitosis.
3. In its proper use the term *chromatid* applies only to DNA replicas held together at the centromere in a chromosome. Upon centromeric division, chromatids become chromosomes.
4. Spindle fibers attach to centromeres at a region called the **kinetochore.**
5. The mitotic spindle consists of microtubules made of the protein tubulin.
6. At the onset of mitosis, the centrioles move to opposite spindle poles. Their precise function is unknown, but they are not essential to spindle formation.

Metaphase

1. At metaphase, the chromosomes pause after aligning on the **metaphase plate.** Centromeric microtubules are seen on each side of each centromere.

Anaphase

1. At anaphase, the centromeres divide, and the newly separated chromosomes migrate to opposite poles.

Telophase

1. Telophase is the reverse of prophase, including: chromosome uncoiling, reestablishment of nuclear envelope, and the formation of nucleoli. Newer data suggest that bits of the old nuclear envelope are conserved to be reassembled at this time.

Cytokinesis

1. In animals, cytokinesis, or cleavage, involves the contraction of microfilaments below the cell membrane, forming a cleavage furrow that tightens, dividing the cytoplasm.
2. Because of the cell wall, cleavage does not occur in plants, but cytoplasmic division requires that a new wall form between daughter nuclei.
3. Following telophase in plants, vesicles from the Golgi complex coalesce to form a cell plate between daughter nuclei.
4. Mitosis can occur without cytokinesis, forming multinucleate cells. Repeated replications without mitosis occur in some insects, producing giant polytene chromosomes.

The Advantages of Mitosis

1. Mitosis and cell division are a primary means of asexual reproduction in single-celled organisms. In multicellular organisms, mitosis accommodates growth, cell replacement, repair, and cell specialization.

MEIOSIS

1. **Diploid** organisms must undergo **meiosis** to form **haploid** sex cells. Meiosis reduces the number of chromosomes in half (by the separation of pairs) and permits new gene combinations to arise.

Homologous Chromosomes

1. Chromosomes occur in pairs (in the original fertilized cell, one from each parent), and each pair can be identified through karyotyping.
2. Members of a chromosome pair are called **homologues,** or homologous chromosomes. Each member of a pair has the same array of genes, although there may be a difference in the way each expresses itself.

An Overview of Meiosis

1. Meiosis requires two divisions, meiosis I and meiosis II, which in humans reduce the 23 pairs of chromosomes (actually 46 chromatids because of replication) in a diploid cell to 23 chromosomes in each haploid gamete.

Meiosis I: The First Meiotic Division

1. Meiotic prophase includes five substages:
   a. **Leptotene:** earliest condensation
   b. **Zygotene:** synapsis, or pairing up
   c. **Pachytene:** crossing over
   d. **Diplotene:** X-shaped **chiasmata** form (regions still touching represent points of crossover).
   e. **Diakinesis:** homologue separation
2. During synapsis, homologues are bridged by the **synaptonemal complex** which is made up of protein and RNA. Its function is unknown, but it is absent in homologues that do not cross over.
3. A group of four chromatids joined at a single centromere is called a **tetrad.**
4. In crossing over, enzyme complexes called recombination nodules overlie the homologues. Regions along chromatids break and cross over.
5. Crossovers produce new arrangements of chromatids, genetic recombination, the creation of new gene associations in the chromosomes.
6. Chiasmata reveal where crossovers have occurred.
7. In metaphase I the highly condensed chromosomes are aligned on the metaphase plate. Centromeric spindle microtubules from one pole attach to *only one side of each centromere.*

8. In anaphase I, no centromeric division occurs, and because of the unique spindle attachment, each set of homologues separates, going to opposite poles. This is called **reduction division.**
9. During telophase, the reorganization of the nucleus occurs, followed by cytoplasmic division.
10. There is no DNA replication during meiotic interphase.

Meiosis II: The Second Meiotic Division

1. Meiosis II closely resembles mitosis, and in anaphase II, centromeres divide and single chromosomes (former chromatids) are drawn to opposite poles.

Meiosis in Humans

1. Meiosis occurs in germinal tissues. Such tissues completely differentiate in the embryo of animals, but in plants can arise from simple undifferentiated tissues whenever plants flower.
2. Germinal tissue gives rise to gametes. All other cells are called somatic cells.
3. In the human male embryo, germinal epithelium forms in the seminiferous tubules of the testes, but actual meiosis does not begin until puberty.
4. In the human female embryo, germinal epithelium develops in the ovaries and by the time of birth meiosis has begun in all germinal cells or **oocytes.** Meiosis ceases during prophase I, and remains suspended in each oocyte until it actually undergoes ovulation.
5. While each meiosis in males produces four functional sperm, meiosis in females results in only one oocyte forming. The other cells are lost as **polar bodies** following unequal cleavages in telophase I and II.
6. Meiosis provides variation through (a) random genetic exchange during crossing over, (b) random chromosome lineup at the metaphase plate, (c) random discarding of chromosomes into polar bodies, (d) random selection of oocytes for development and ovulation, and (e) random success of genetically different sperm.

REVIEW QUESTIONS

1. Using the terms *DNA, chromatin, histone,* and *nucleosome,* describe the organization of a chromosome. (205-206)
2. Why is isolation of the nuclear contents of less importance during mitosis than during interphase? (206)
3. What is a chromatin net? Of what significance are the dense and diffuse regions? (206)
4. Mention two kinds of cells that cease cycling and die, and explain how their death makes their function possible. (206-207)
5. List the stages of a cell cycle, and briefly describe the events of each. (206-207)
6. What is the relationship between the chromosome and the chromatid? In what case is the latter term appropriate? (208)
7. In a few sentences, clearly explain the function of mitosis. (204, 213-214)
8. List the phases of mitosis, and summarize the events of each. (208-210)
9. Describe the centromere. What observation suggests that it is a permanent part of the chromosome? Upon which part of a centromere do the spindle microtubules attach? (208)
10. How do the centrioles behave during mitosis? What observation suggests that they are not essential to mitosis? (211)
11. Name and describe the orientation of the three different spindle microtubules. (208)
12. Describe the arrangement of chromosomes and the centromeric spindle fibers at mitotic metaphase, and also explain the significance of the arrangement. (210)
13. Summarize the events of mitotic anaphase. What do these events assure about the chromosomes of daughter cells? (210)
14. Briefly describe cytokinesis in animal cells. What part of the cytoskeleton is involved? (210, 212)
15. Explain how a new cell wall forms between daughter cells in the plant. Why is cleavage impossible? (212)
16. Discuss the importance of mitosis to single-celled organisms such as algae and protozoans. (214)
17. Humans have 46 pairs of chromosomes. How many chromosomes would one find in a cell in G_1, a cell in mitotic metaphase, a sperm cell, a cell in meiotic prophase I, a cell in meiotic telophase I, a cell in meiotic telophase II? (214)
18. Using humans with their 46 chromosomes as an example, explain the number problem solved through meiosis. (214)
19. In a few words, summarize the events in the following: leptotene, zygotene, pachytene, diplotene, diakinesis. Be sure to use the terms *repulsion, synapsis, crossing over, condensation,* and *chiasma.* (217)
20. Compare the attachment of centromeric microtubules in metaphase I of meiosis with that of mitosis. What specific difference will this make in anaphase? (219)
21. Summarize the events of meiosis II, comparing each phase to those of mitosis. (219-220)
22. Briefly summarize the development of germinal epithelium in the human female. How does this development differ in the male? (220-221)
23. Using a diagram, suggest how meiotic cytokinesis in the female vertebrate differs from cytokinesis in the male. Offer a logical reason for the difference. (222)
24. List four ways in which genetic variability is increased through sexual reproduction. (223)

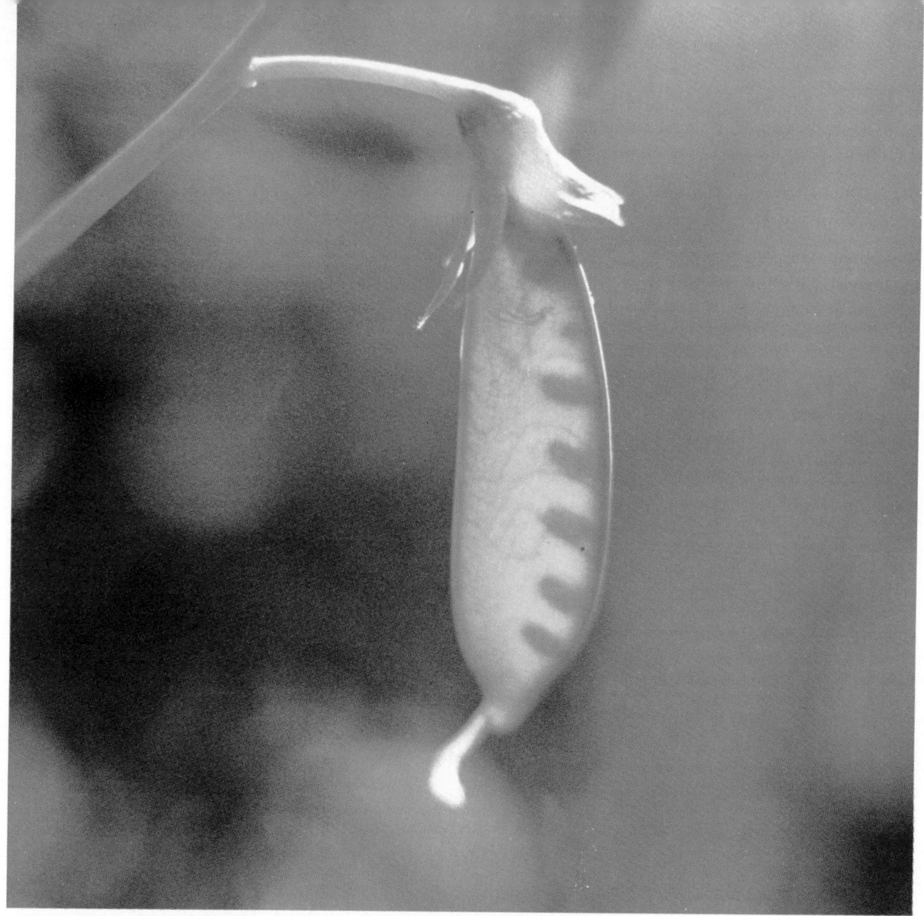

10

MENDELIAN GENETICS

OCCASIONALLY YOU MAY HEAR SOMEONE JUST BACK FROM A TRIP TO SPAIN boasting of having "fought a bull." In fact, that very noun may come to mind when you learn that the "bullfight" was actually with a heifer (a young cow). At certain times of the year, guests are invited to the ranches where fighting bulls are bred. There they can watch the ranchers test the young cows to determine which stock will be used for breeding. As part of the festivities, the guests may be invited to try their hand at caping a heifer. The ranchers seem to think that nothing is quite so festive as watching gringos trying to stand their ground with an angry cow.

Professionally, the ranchers are interested in how the cows fight because it is traditionally accepted in such circles that a fighting bull gets its strength from its father, but its courage from its mother. The origin of such beliefs is lost in tradition, but it is evident that people applied knowledge of genetic phenomena long before they knew anything about genes.

WHEN DARWIN MET MENDEL (ALMOST)

Charles Darwin was probably the nineteenth century's greatest biologist and, as we will see, he developed evolutionary biology as one of the cornerstones of modern science. It has been suggested that his basic knowledge of certain principles of inheritance helped set the tone of his thinking as he worked out his great theory. His knowledge of such matters, though, was no greater than that of any other country person of the time. As the son of a gentleman farmer, Darwin was well aware, quite simply, that offspring tend to resemble their parents. The offspring of heavier animals, he knew, tended toward heaviness; the daughters of higher milk-yielding cows tended to produce more milk. Nonetheless, with this meager understanding of inheritance, Darwin was able to formulate some of the most powerful ideas in modern biology. Even though Darwin came to be known as the nineteenth century's expert on inheritance of traits, most of his ideas about hereditary mechanisms were wrong.

Darwin's chief mistake was to accept the only intellectually respectable theory of his day, which was "blending inheritance." It was believed that the "blood," or hereditary traits, of both parents blended in the offspring, just as two colors of ink blend when they are mixed. The blending hypothesis appeared to work reasonably well for some traits, such as height or weight, that are continuously graded, but it couldn't account for other observations.

It is always surprising, in hindsight, to recognize the degree to which a strongly held belief will blind its proponents to obvious contradictions. The blending theory would predict that the offspring of a white horse and a black horse should always be gray, and that the original white or black color should never reappear if gray horses continued to be bred. But it turns out that the offspring are not always gray, and black and white descendants do appear. Similarly, if a honey-color cocker spaniel were crossed with a black one, the offspring would usually be black or honey, not some kind of golden black. Something obviously was wrong with the theory, but no one seemed to notice, or if they did, they tried to ignore it.

In fact, the blending theory doesn't work even for continuously graded traits like height and weight. If it did, the offspring in every generation would always be less extreme than their parents for every trait. Every time a fat horse mated with a thin horse, the offspring should always have been an "average" horse. In that case the world would by now be full of average, gray horses and average everything else, but in fact, in a group of horses that eat about the same amounts, some will be fat and others thin. Horses are also still found in many different colors, and even expert breeders usually can't predict the color of a foal.

When Darwin finally got around to publishing his theory of natural selection, some of his sharper critics used the weakness of his genetic arguments against him. They pointed out, quite correctly, that natural selection will not work with blending inheritance because any variation will be blended away. Try as he might, Darwin could not come up with an answer for them, and the tired old man, so sickened in his later years, began to backtrack. In later editions of his famous book, *On the Origin of Species,* he began to give more and more weight to the possibility of the inheritance of acquired traits, but his desperate explanations were so bad that we won't bother you with them. The idea of "inheritance of acquired traits," which stated that traits acquired during the lifetime of an individual (muscular development, criminal tendencies, knowledge, and so on) could be passed on to offspring, is actually ancient but remained quite popular and respectable in the nineteenth century. A leading proponent was Jean Baptiste Lamarck, the famous nineteenth-century French naturalist. The basic tenets of Lamarck's hereditary notions have largely been discarded.

The remarkable part of this story is that Darwin's thinking was on the right track and that he came very close, indeed, to solving some of his most vexing problems with explaining the mechanisms of inheritance. Actually, while he was still working on *Origin,* he did some very good experiments to try to solve this nagging puzzle.

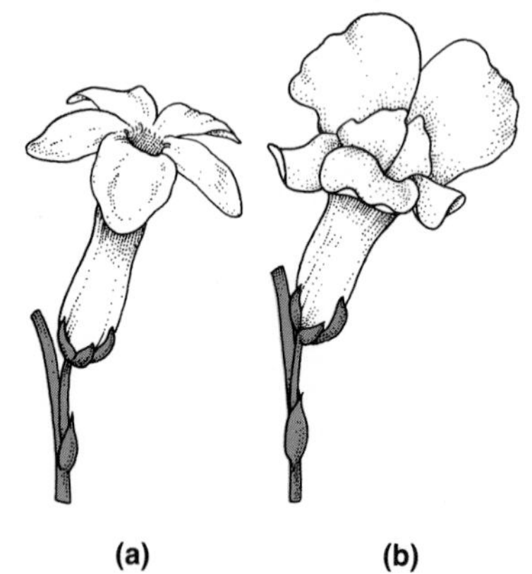

FIGURE 10.1
DARWIN'S SNAPDRAGONS
One of Charles Darwin's excellent experiments included breeding snapdragons. One of his observations was that when true-breeding peloric-flowered snapdragons **(a)** were crossed to true-breeding bilateral flowers **(b)**, the peloric flower trait was lost in the progeny. He termed the dominance of one trait over another prepotence.

In his set of experiments, Darwin crossed two true-breeding strains of snapdragons. (True-breeding means that when individuals of a strain are crossed, the offspring are the same, generation after generation.) The two strains differed in only one respect: one strain had abnormal, radially symmetrical flowers (what botanists called *peloric* flowers), and the other had the normal, bilaterally symmetrical snapdragon flowers (Figure 10.1). Of the progeny of this first cross, Darwin wrote, "I thus raised two great beds of seedlings, and not one was peloric." Darwin called this tendency of one trait in a cross to suppress another *prepotence.* We now call it dominance, and we will consider it in depth shortly.

While Darwin was working on his snapdragon experiment, scientific history was being made across the English Channel, deep in Europe. The self-effacing but incredibly bright and dedicated Gregor Mendel was crossing strains of garden peas. Darwin and Mendel performed almost exactly the same experiments. The two great scientists, unknown to each other, obtained almost exactly the same results. But only Mendel was able to understand what had happened.

The Abbot Gregor Johann Mendel (1822-1884), a member of an Augustinian order in Brunn, Moravia (now part of Czechoslovakia), is often depicted as a kindly old man of the cloth who, while puttering around his monastery garden, somehow stumbled onto important genetic laws. Again, our tendency to embellish the memory of already worthy individuals has led us astray. The real Mendel (Figure 10.2) was remarkable enough in his own right. In many ways he seems more like a twentieth-century biologist somehow displaced into the wrong century. His first published paper is a landmark in its clarity and a model experimental report. Many scientists believe that it is perhaps the best scientific paper ever written.

Early in his life Mendel began training himself, and he became a rather competent naturalist. To support himself during those years, he worked as a substitute high school science teacher. The professors at the school, noting his unusual abilities, suggested that he take the rigorous qualifying examination and become a regular member of the high school faculty. Mendel took the test and did reasonably well, but he failed to qualify. He joined a monastic order instead.

His superiors, confident of his abilities, sent him in 1851 to the University of Vienna for two years of concentrated study in science and mathematics. There, he learned about the infant science of statistics, which was to serve him well.

When Mendel returned to the monastery he began his plant-breeding studies in earnest. He impressed his fellow monks with his intelligence and his vigor, and he applied these qualities to the study of plant breeding. He developed new varieties of fruits and vegetables, kept abreast of the latest developments in his field, joined the local science club, and became active in community affairs.

FIGURE 10.2
GREGOR MENDEL (1822-1884)
Mendel discovered the underlying fundamental principles that govern the inheritance of observed traits. His work, though published in 1866, was essentially unnoticed and unappreciated until the turn of the century.

MENDEL'S CROSSES

Mendel began to experiment by observing the effects of crossing different strains of the common garden pea. To begin with, he based his research on a very carefully planned series of experiments *and* on a statistical analysis of the results. The use of mathematics to describe biological phenomena was a new concept. Clearly, Mendel's two years at the University of Vienna had not been wasted.

The careful planning that went into Mendel's work is reflected in his selection of the common garden pea as his experimental subject. There were several advantages in this choice. Since others had successfully used the garden pea in experimental crosses, he wasn't proceeding entirely on guesswork. Furthermore, pea plants were readily available and fairly easy to grow, and Mendel was able to purchase 34 true-breeding strains. These strains differed from each other in very pronounced ways, so that there could be no problem in identifying the results of a given experiment. Mendel chose seven different characters to work with, each of which has two variations or conditions.

1. seed shape: round or wrinkled
2. seed color: yellow or green
3. seed coat color: white or gray
4. unripe pod color: green or yellow
5. pod shape: inflated or constricted
6. stem length: short (9–18 inches) or long (6–7 feet)
7. flower position: axial (along the stem in leaf junctions) or terminal (at stem tips only)

Mendel's choice of the pea was fortunate, since peas produce large numbers of offspring. He knew from his training in statistics that large numbers would increase the significance and validity of his experiments. Keep in mind that each pea in a pod is a seed—essentially a new plant, an individual with its own genotype and phenotype. An organism's **genotype** is its total combination of genes, and the **phenotype** is the combination of its observable traits. Thus, the phenotype of a plant's offspring—traits such as seed shape and coat color—can easily be determined by simply examining the peas.

The pea, however, is not without its problems. Mendel's artificial crosses relied on tedious manipulation. Left to themselves, most pea plants will simply self-fertilize. To get a cross between two plants or two strains, it was necessary to open the pea flower, remove the pollen-producing anthers (to prevent self-pollination), and apply the foreign pollen with a small paintbrush. This way, Mendel could control his crosses, prevent accidental contamination, and allow self-pollination only when it suited his needs.

Mendel's approach was to cross two true-breeding strains that differed in only one character, such as the seed color. We now call the original parent generation P_1 and designate the first generation offspring the F_1 (F_1 = "first filial," or first family). When the F_1 plants are crossed with each other or are allowed to self-pollinate, the resulting offspring are called the F_2 generation, and so on.

The Principle of Dominance

When Mendel crossed his original P_1 parental plants, he found that any variation of the characters of the two plants didn't blend, as prevailing theory said they should. As shown in Figure 10.3, when plants grown from yellow seeds were crossed with plants grown from green seeds, their F_1 offspring were not intermediate seeds. Instead, all of them were yellow seeds. Mendel termed the trait that appears in the F_1 generation the **dominant** trait, and the one that had failed to appear the **recessive** trait. But he was now left with a vexing question. What had happened to the recessive trait? It had been passed along through countless generations in his purebred lines, so it couldn't have just disappeared.

Mendel then allowed his F_1 pea plants to pollinate themselves. In the offspring of the F_1 generation (which, remember, we call the F_2 generation) Mendel found that roughly one-fourth of the peas were green and that about three-fourths were yellow. The recessive trait had reappeared! He repeated the experiment with other pea strains and obtained similar results. He crossed a round pea strain with a wrinkled pea strain. All of the F_1 peas were round, but in the F_2 generation about one-fourth of the peas were wrinkled again. The ratios did not escape the tenacious Mendel, determined to badger the problem until he could make some sense of it.

One of Mendel's early observations was critical to the development of his genetic model. Mendel found that there were two kinds of yellow peas: the true-breeding kind, like the original parent stock, which would grow into plants that would bear only yellow peas; and another type, which when grown and self-pollinated would produce pods containing both yellow and green peas. At this point we can almost hear Mendel musing,

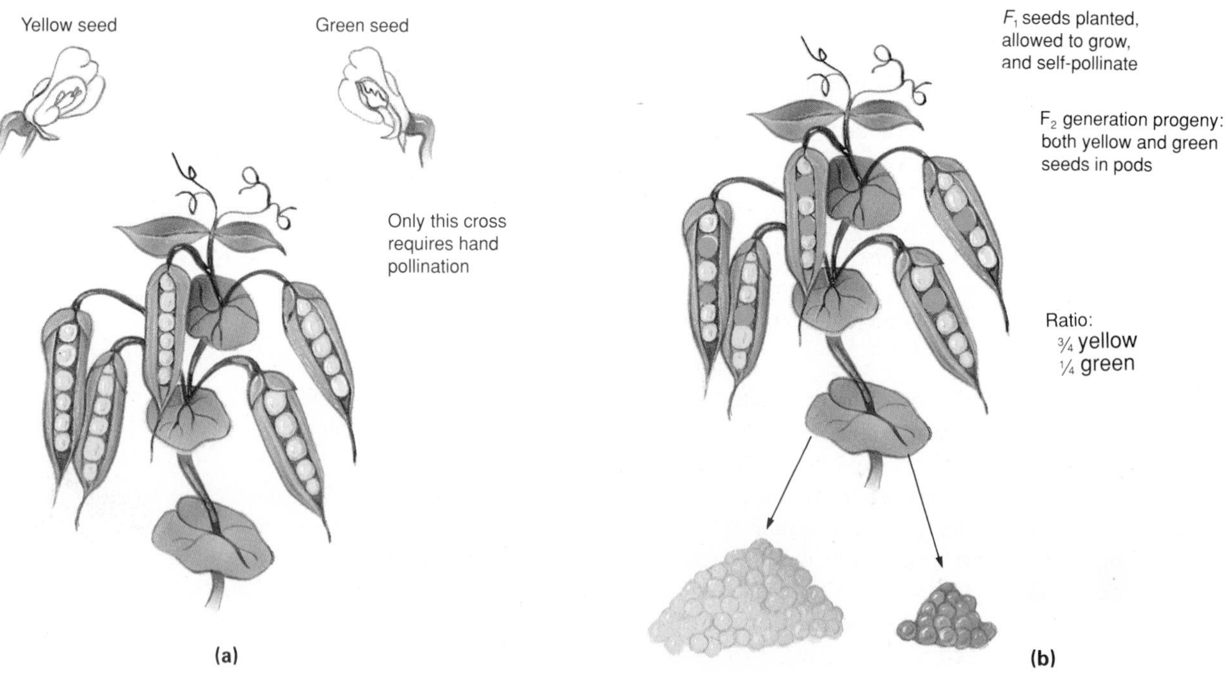

FIGURE 10.3
SINGLE FACTOR CROSS: YELLOW × GREEN PEAS
In one of his early experiments, Mendel crossed pea plants that were true-breeding for yellow peas with plants that were true-breeding for green peas. This cross produced only yellow seeds in the pods of the F_1 generation progeny (**a**). When those yellow seeds were grown and allowed to self-pollinate (**b**), they produced F_2 plants carrying both yellow and green seeds within their pods. When those seeds were counted, three-fourths were yellow and one-fourth green. Or, there was a 3:1 ratio of yellow to green seeds.

". . . two kinds of yellow peas–one pure-breeding, one not." Were there also two kinds of green peas? He found that there were not. When green peas were cultivated and allowed to self-pollinate, they always bore only green peas.

Mendel figured that there were two hereditary "factors" for each character in an individual, one contributed by each of the parents, and that factors might exist in alternate forms or variants. Thus, seed color is determined by the "seed color factor," and there is one variant for green seeds and another for yellow seeds. Mendel's factors have since been renamed **genes,** and the variants have been termed **alleles.** Thus, each gene in an organism is presented by two alleles. If the two alleles are identical, the individual is said to be **homozygous** for that gene. If the individual has two different alleles for the gene, it is referred to as **heterozygous** for that gene.

As we've seen, Mendel realized that there were two kinds of yellow peas in the F_2 generation, homozygous and heterozygous, and that it was impossible to tell the difference between the two kinds of yellow peas unless they were allowed to mature and reproduce. He decided he needed an F_3 generation, which he obtained through a procedure called **progeny testing** (Figure 10.4). This test would allow two seemingly identical yellow seeds to show just how different they really were. The kind we call homozygous would be true-breeding—if allowed to self-pollinate, they would produce one type of pea (yellow in this case). However, those that were heterozygous would produce two types of peas. He planted the yellow peas of the F_2 generation, waited another year for them to grow and bear pods, and then opened the pods to determine which plants were true-breeding. He found that one-third of the F_2 yellow peas had been true-breeding (were homozygous) and that two-thirds had not (and thus were heterozygous).

Mendel was intrigued again by the ratios he observed. He had already determined that three-fourths of the F_2 generation had been yellow and now one third of those yellow F_2 had been true-breeding. One-third of three-fourths is one-fourth. So, one-fourth of the offspring were homozygous and yellow. Further, when he considered all the F_2 peas together, he found the following, a 3:1 ratio of yellow to green.

1/4 of the total F_2 were yellow and true-breeding (homozygous)	3/4 yellow
1/2 of the total F_2 were yellow and not true-breeding (heterozygous)	
1/4 of the total F_2 were green and true-breeding (homozygous)	1/4 green

FIGURE 10.4
THE F_3 GENERATION
Mendel took the piles of green and yellow seeds from the F_2 generation of the single-factor cross of Figure 10.3, grew them up, and allowed them to produce an F_3 generation. He found that all of the green seeds (which was one-fourth of the F_2) were true-breeding, in that they produced only green progeny. However, only one-third of the yellow seeds (again, one-fourth of the F_2) were true-breeding. The remaining two-thirds of the yellow seeds (which amounts to one-half of the F_2) continued to produce both yellow and green seeds.

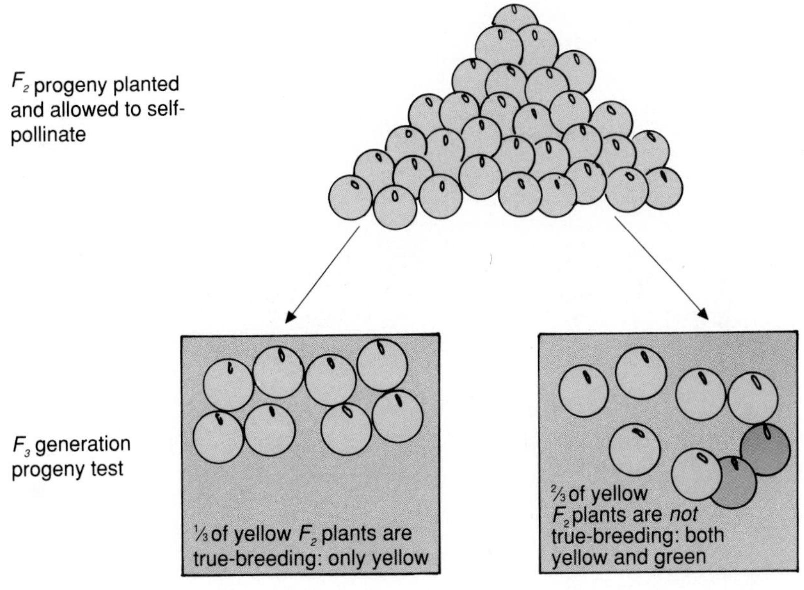
F_2 progeny planted and allowed to self-pollinate
F_3 generation progeny test
⅓ of yellow F_2 plants are true-breeding: only yellow
⅔ of yellow F_2 plants are *not* true-breeding: both yellow and green

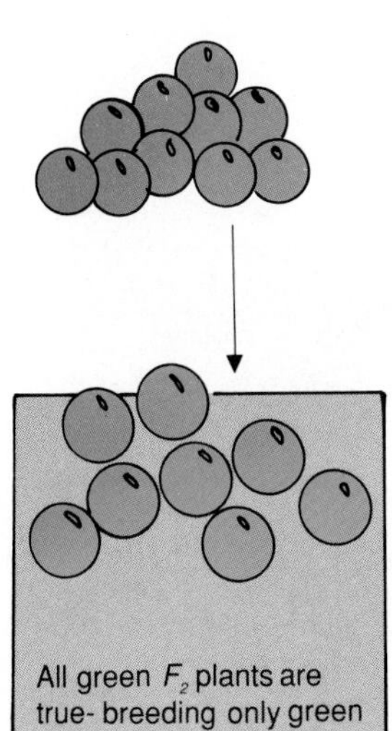
All green F_2 plants are true- breeding only green

The same thing worked for round and wrinkled peas. There were two kinds of round and one kind of wrinkled. Mendel looked at all seven characters (see Table 10.1), and the same principle worked in each case; roughly the same ratios were generated. In every case, one form, bearing what Mendel called the dominant trait, was present in 100% of the F_1 peas or pea plants. In every case, three-fourths of the F_2 peas or pea plants showed the dominant form of the trait (we would say the dominant phenotype), and one-fourth of the F_2 peas or pea plants showed the contrasting, recessive trait.

In every case, a progeny test revealed that in actuality, one-fourth of the F_2 were the dominant form and true-breeding (homozygous), one-half of the F_2 were the dominant form but not true-breeding (heterozygous), and one-fourth of the F_2 were the recessive form and true breeding (homozygous). Clearly, something determining the recessive form was passed down from the true-breeding recessive parental strain, through the hybrid F_1, to the true-breeding recessive F_2.

Mendel thought he could use algebraic symbols to help in his analysis (Figure 10.5). He let **A** represent the allele that determines the dominant form, and he let **a** represent the allele that determines the recessive form. The F_1 hybrid, and in fact all the non-true-breeding plants, must have both alleles present, as represented by the symbol **Aa.**

TABLE 10.1 MENDEL'S F_2 GENERATIONS

The dominant and recessive traits analyzed by Mendel are shown along with the results of F_1 and F_2 crosses. Note the large numbers he worked with. How does a large sample size help with the conclusions? How well do his numbers in the last two columns agree with what probability tells us to expect in the crosses? The proportion of the F_2 generation showing recessive form is in the last column.

Dominant Form	Number in F_2 Generation	Recessive Form	Number in F_2 Generation	Total Examined	Ratio	Proportion of F_2 Generation
Round seeds	5,474	Wrinkled seeds	1,850	7,324	2.96:1	0.253
Yellow seeds	6,022	Green seeds	2,001	8,023	3.01:1	0.249
Gray seed coats	705	White seed coats	224	929	3.15:1	0.241
Green pods	428	Yellow pods	152	580	2.82:1	0.262
Inflated pods	882	Constricted pods	299	1,181	2.95:1	0.253
Long stems	787	Short stems	277	1,064	2.84:1	0.260
Axial flowers (and fruit)	651	Terminal flowers (and fruit)	207	858	3.14:1	0.241

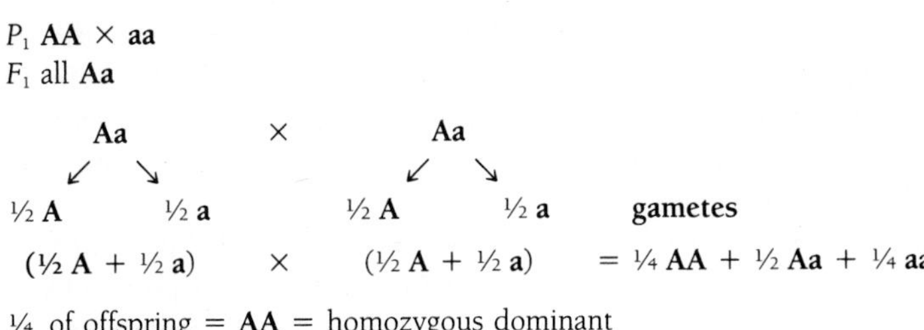

P_1 AA × aa
F_1 all Aa
Aa × Aa
½ A ½ a ½ A ½ a gametes
(½ A + ½ a) × (½ A + ½ a) = ¼ AA + ½ Aa + ¼ aa
¼ of offspring = AA = homozygous dominant
½ of offspring = Aa = heterozygous
¼ of offspring = aa = homozygous recessive

FIGURE 10.5
FOLLOWING ALLELES IN A CROSS
In a generalized scheme for following alleles through a series of crosses, we can see that the outcome of any cross of a homozygous dominant plant with a homozygous recessive plant can only be a generation of heterozygous plants. When the F_1 heterozygous plants are bred, we see that the final outcome of the F_2 generation is strictly dependent upon the possible combinations of gametes.

Since there are two parents, Mendel figured that in the hybrid, **A** comes from one parent and **a** from the other. Mendel determined experimentally that it didn't matter which parental strain bore the peas and which provided the pollen. He let the capital letters signify dominance and the lowercase letters signify recessivity, which merely meant that the **Aa** individuals would look exactly the same as the **A**-bearing parental stock.

If the heterozygous plants get **A** from one parent and an **a** from the other parent, and are symbolized by **Aa,** it makes sense that the true-breeding dominant forms get two **A** alleles—one from each parent—and can be symbolized, **AA.** In the same way, the true-breeding recessive forms get **a** alleles from both parents and can be symbolized by **aa.** We can use **AA** to symbolize the dominant homozygote, **aa** to represent the recessive homozygote, **Aa** to represent the heterozygote, and **A_** to symbolize those plants with the dominant phenotype whose complete genotype is not known.

In summary then:

A = dominant allele
a = recessive allele
AA = homozygous dominant
Aa = heterozygous
aa = homozygous recessive
A_ = dominant phenotype; genotype unknown, **AA** or **Aa**

Now Mendel was ready for some conclusions. As we'll see, they were truly far-reaching, and eventually were codified into what we call Mendel's First Law. This law embraces several principles that you will, by now, recognize.

MENDEL'S FIRST LAW: THE SEGREGATION OF ALTERNATE ALLELES

In the first law, also called the **law of segregation,** Mendel's most fundamental principle proposes that *each genetic character is produced by a pair of alleles, and that alleles segregate* (became separated) during meiotic formation of pollen and ovule (or sperm and egg). Members of a pair of alleles may differ in their expression, with one dominant over the other.

The combination of alleles in the next generation is a matter of chance. Let's look further into these principles through a brief foray into probability itself.

Segregation and Probability

Mendel was the first to realize that when a heterozygote reproduces, its gametes (in plants the sperm-containing pollen and egg-containing ovule) will be of two types in equal proportions.

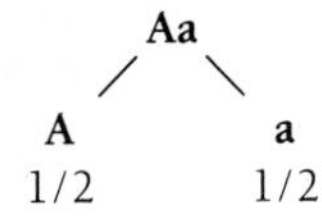

Now, how did he come up with that? He did it by playing with the numbers his experiments were generating. He figured that when the heterozygous plant was allowed to pollinate itself, it would produce half **A** ovules and half **a** ovules. It would also produce half **A** pollen and half **a** pollen. When it self-pollinated, what proportion of all the fertilized ovules would be **aa**?

The probability of any given ovule being **a** is 1/2. The probability of this ovule being fertilized by an **a** pollen is also 1/2. One of the basic laws of probability, the **multiplicative law**, states that the probability of two independent events both occurring is equal to the product of their individual probabilities (Figure 10.6). Since the probability of an ovule being **a** and the pollen being **a** have no influence on each other, the probability that an **a** pollen will meet an **a** ovule during self-pollination of an **Aa** parent is $1/2 \times 1/2 = 1/4$. (If the probability of your passing your next exam is 1/6 and the probability of rain on that day is 1/5, then the probability that both these things will happen is 1/30, unless your test performance influences the weather, or vice versa.)

The same reasoning applies to the quarter of the F_2 progeny that are **AA**: a 1/2 chance of **A** pollen times a 1/2 chance of **A** ovule gives a 1/4 chance of a pea (a newly fertilized ovule) being **AA.**

But half of Mendel's F_2 peas were **Aa.** How do we account for that? We find that there are two different ways that **Aa** zygotes are formed in an **Aa** × **Aa** cross. Either an **a** pollen fertilizes an **A** ovule, or an **A** pollen fertilizes an **a** ovule. The probability of the first combination of events is $1/2 \times 1/2 = 1/4$, and the probability of the second combination of events is also 1/4.

There is no other way of getting an **Aa** zygote from such a cross, and the two possibilities are mutually exclusive (they can't both happen to the same zygote). Furthermore, Mendel had shown that the **Aa** heterozygotes were identical regardless of which parent contributed which allele.

This leads to another basic law of probability, the **additive law:** The probability of either one or another of two mutually exclusive events occurring is equal to the sum of their individual probabilities. (If there is a 1/2 probability of a coin coming up heads, and a 1/2 probability of it coming up tails, then you can be 100% sure [$1/2 + 1/2 = 1$] that it will come up either heads or tails.)

But back to the question: Why were half the F_2 peas **Aa**? We know that the two mutually exclusive possibilities are (1) **a** pollen and **A** ovule, and (2) **A** pollen and **a** ovule. If either one or the other of these events occurs, the zygote will be an **Aa** heterozygote. The probability of a zygote from such a cross being an **Aa** heterozygote is $1/4 + 1/4 = 1/2$.

We have already seen another example of the additive law in probability: the events "zygote is **AA**" and "zygote is **Aa**" are mutually exclusive. The probability of the first in an **Aa** × **Aa** cross is 1/4, and the probability of the second is 1/2. Thus, the probability that an individual will have the dominant phenotype **A_** is $1/4 + 1/2 = 3/4$, which, of course, is what Mendel observed.

Early in the twentieth century, these relationships were put into a graphic form by a fan of Mendel named Reginald Crandall Punnett. Figure 10.7 shows a Punnett square used to keep track of a cross between two F_1 yellow individuals. Rather than use **A** and **a** for everything, Figure 10.7 shows the convention by which a single letter code can be used for a given character. The capital letter is chosen for the dominant trait. In this case **Y** = yellow. The lower case is used to show the recessive trait. In this case **y** = green. Each little square represents the simultaneous occurrence of the event directly above it and to its left. Or we might simply say that gametes from one parent are written at the left and from the other parent at the top (as shown). The symbols within the square represent possible fertilizations. Note that there are two **Yy** squares, which must be added together to account for all the **Yy** individuals in the F_2 progeny, a visual confirmation of the additive law.

We have called Mendel a mathematical biologist not just because he was trained both in mathematics and biology, or because he was among the first biologists to use statistical anaylsis in his work, but also because of the way he arrived at his conclusions. So, we

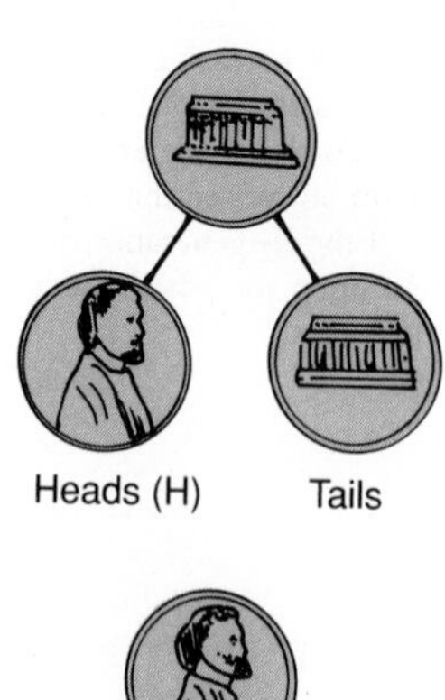

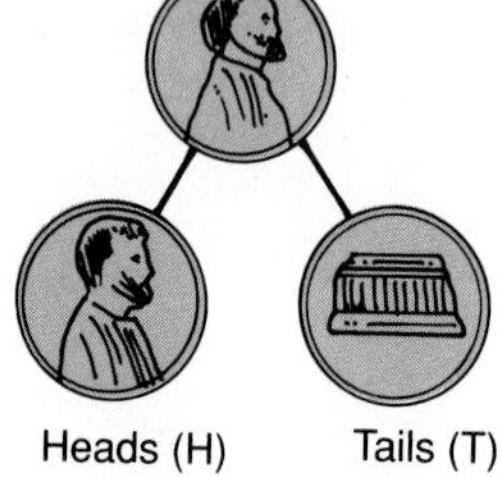

FIGURE 10.6
PROBABILITY AND GENETICS
It is easy to see why Mendel's ratios happen to be such simple fractions. Diploid organisms can have, at most, two different alleles for any one gene, and for nearly all cases meiosis produces an equal number of gametes of, for example, **A** and **a**. The probability of an individual receiving either an **A** or an **a** from any parent is one-half, the same as the probability of betting heads (or tails) from the flip of a penny. When two pennies are flipped together, we can in effect see a simulation of a cross of two **Aa** individuals. Can you see why the probability of getting one head and one tail (a heterozygote is $2 \times ½ \times ½$? And do you see the relationship to the crosses shown in Figures 10.4 and 10.5?

might now ask, what is a mathematical biologist and how does he or she work? Usually, he or she starts with some set of observations. In Mendel's case it was the dominance of one trait in the first generation and the reappearance of the recessive trait in some following generation. By some mental process involving both intuition and logic, the mathematical biologist then constructs a model. The model is an imaginary biological system, based on the smallest possible number of assumptions, and it is expected to yield numerical data consistent with past observations. New experiments are then performed to test further predictions of this model. If the new data don't fit the predictions, the model is discarded or adjusted to fit the new observations so that further experiments can be done. A model, then, is a biological hypothesis with mathematical predictions (although the term is often applied in a nonmathematical manner).

It is important to realize that in science, models and hypotheses cannot be *proven* with experimental data. We can only say that the data are consistent with the model. Mendel did not prove his model of the segregation of alternate traits, but the simplicity of the model and the exellence of the fit enabled him to make new predictions and test them. His success came very close to a proof, at least as far as he was concerned. However, others were unconvinced until after the discovery of chromosomes and meiosis. Have you noticed how well Mendel's findings fit with what you already know about meiosis? Imagine how elated Mendel would have been if meiosis had been discovered in his own lifetime.

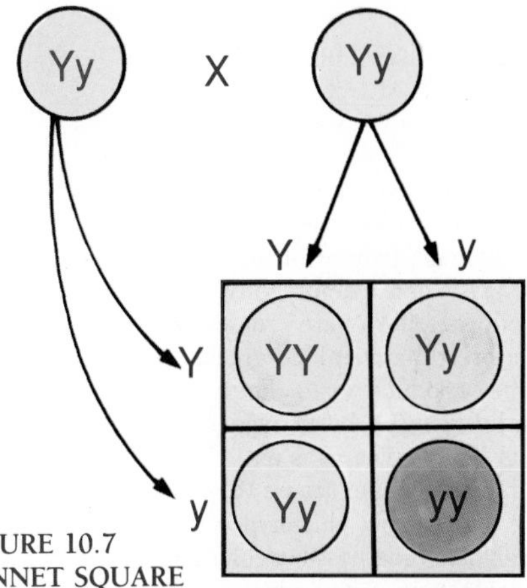

FIGURE 10.7
PUNNET SQUARE
Punnet squares can be used as a means of keeping track of alleles in simple crosses. In this example, two heterozygous yellow individuals (from the F_1 of Figure 10.3) are shown. Each produces gametes that are arranged on the axes of the square. In each quadrant of the square, the intersection of the gametes of the parents determines the genotype (and hence the phenotype) of the offspring. In this example, each square represents one-fourth of the offspring. Compare this picture of an F_1 cross with Figures 10.3, 10.5, and 10.6 to be sure that you see that these are all just different ways to view the same thing.

MENDEL'S SECOND LAW: INDEPENDENT ASSORTMENT

In his second law, the **law of independent assortment**, Mendel states that *the inheritance of a pair of alleles affecting one character occurs independently of the simultaneous inheritance of alleles affecting any other character*. Let's see how he arrived at this.

We have noted that Mendel succeeded in breaking his problem down to its smallest parts, partly by studying only one genetic character at a time. The next step was to study two characters at a time, so he decided to cross a true-breeding strain that bore yellow, round peas with one bearing, green wrinkled peas.

The F_1 offspring were all round and yellow. We can symbolize this as follows:

P_1: **YYRR** × yyrr
F_1: All **YyRr**

where **R** and **r** are symbols for the two alleles of the round-wrinkled gene and **Y** and **y** are symbols for the two alleles of the yellow-green gene. Although we are using Mendel's basic symbols and concepts, we can now introduce another modern term. We defined **allele** earlier as "a particular form of a gene;" now let's note that each gene resides at a specific *locus*. The term **locus** (plural, **loci**) derives from our knowledge that each gene occupies a specific place, or locus, on a chromosome. For Mendel's purposes, it was sufficient to suppose that different features were affected by different loci.

Now let's see what happened in the F_2 generation when Mendel conducted a dihybrid cross—that is, crossed plants that were different in two ways (Figure 10.8). First, as we have said, the F_1 peas were round and yellow. In the F_2, the offspring of $F_1 \times F_1$ (**RrYy** × **RrYy**), Mendel classified 556 peas into four groups (recall that **R_** and **Y_** indicate that although the individual is dominant, the second allele isn't known).

315 round and yellow	R_Y_
101 wrinkled and yellow	rrY_
108 round and green	R_yy
32 wrinkled and green	rryy

Note that 133 peas altogether were wrinkled, and 140 peas altogether were green. Both numbers are close to 139, which is one-fourth of 556.

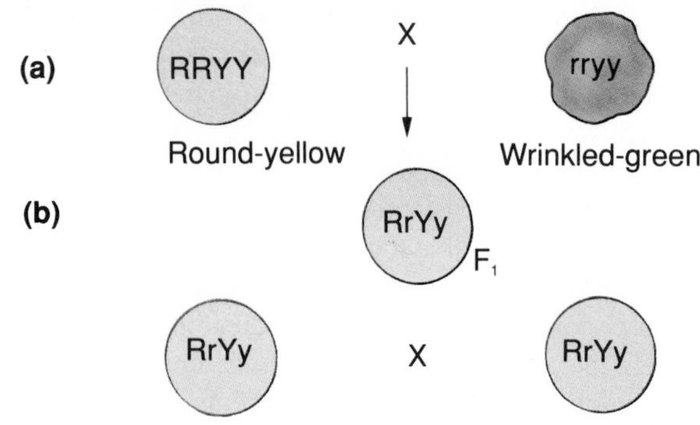

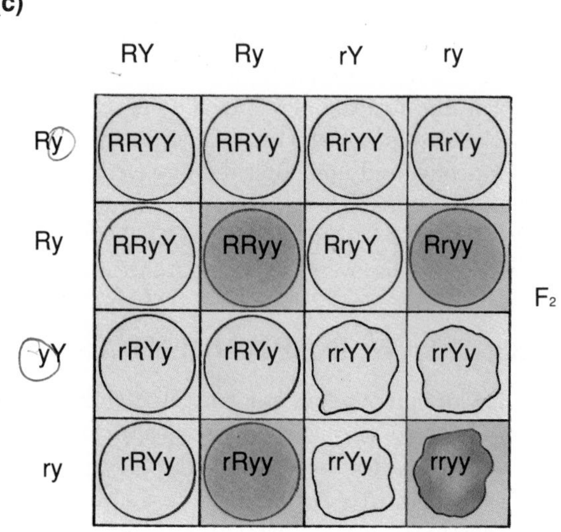

FIGURE 10.8
INDEPENDENT ASSORTMENT OF TWO PAIRS OF ALLELES
Mendel developed his second law by observing the behavior of two traits simultaneously. The traits being considered here are color (yellow or green) and shape (round or wrinkled) of seeds. **(a)** When true-breeding round-yellows (**RRYY**) are crossed with true-breeding wrinkled-greens (**rryy**), they produce an F_1 that is round and yellow **(b)**, heterozygous for both traits (**RrYy**). The F_2 is diagrammed in the Punnet square (c). There are four distinct phenotypes, which include all the possible color and shape combinations. These phenotypes occur in a 9:3:3:1 ratio, but the important thing to remember is that any individual trait (if you just look at yellow versus green, for example) still is represented in a 3:1 ratio.

The results supported Mendel's first law: about one-fourth of the F_2 peas were wrinkled and one-fourth were green, while three-fourths were round and three-fourths were yellow. But the data indicated more than that. Mendel had now shown that the alleles of the two characters were inherited independently. We have already mentioned the multiplicative law of probability, which gives the probability of two independent events both occurring. If the probability of being round is 3/4 and the probability of being yellow is also 3/4, the probability of being both round and yellow is $3/4 \times 3/4 = 9/16$ *if* the two events (seed shape and seed color) are unrelated. Nine-sixteenths of 556 peas is 312.75 peas. When Mendel counted he found 315 round, yellow peas—remarkably close (Table 10.2)! (Researchers rarely reach such good agreement with their hypotheses.)

TABLE 10.2 MENDEL'S PREDICTIONS AND RESULTS FOR F_2 PHENOTYPE

PHENOTYPE OF F_2	FRACTION PREDICTED	RATIO	NUMBER PREDICTED OUT OF 556	NUMBER ACTUALLY OBSERVED
Round and yellow	9/16	9	312.75	315
Wrinkled and yellow	3/16	3	104.25	101
Round and green	3/16	3	104.25	108
Wrinkled and green	1/16	1	34.75	32

TABLE 10.3 MENDEL'S PREDICTIONS AND RESULTS FOR F_2 GENOTYPE (529 F_2 OFFSPRING)

GENOTYPE OF F_2	FRACTION EXPECTED	NUMBER EXPECTED ACCORDING TO HYPOTHESIS	NUMBER ACTUALLY OBSERVED
RRYY	1/16	33	38
RRyy	1/16	33	35
rrYY	1/16	33	28
rryy	1/16	33	30
RRYy	1/8	66	65
rrYy	1/8	66	68
RrYY	1/8	66	60
Rryy	1/8	66	67
RrYy	1/4	132	138

Mendel thought the fit between expected and observed numbers was impressive. He went on to determine which of these peas were true-breeding and which were not. Here again, the fit was good. He expected one-fourth to be **RR**, half to be **Rr**, one-fourth to be **rr**, and one-fourth to be **YY**, half to be **Yy**, and one-fourth to be **yy**, so that:

$1/4 \times 1/4 = 1/16$ should be **RRYY**

$1/2 \times 1/4 = 1/8$ should be **RrYY**

$1/2 \times 1/2 = 1/4$ should be **RrYy**

and so on. Mendel was able to get 529 of his 556 F_2 peas to bear F_3 progeny. The breakdown of their genotypes is listed in Table 10.3.

Again, Mendel felt that the numbers fit his model quite well. He tried combinations of other traits; he even tried three traits together. In each case, the different pairs of alternative traits behaved independently. We know why, of course: maternal and paternal chromosome pairs line up and separate independently during meiosis. (See Figure 10.9, and refer back to Figure 9.10.) But Mendel had never heard of meiosis (except that in a sense, he discovered it). In Mendel's words, "The behavior of each pair of differing traits in a hybrid association is independent of all other differences between the two parental plants . . . ." This then became known as Mendel's second law, or the law of independent assortment. Now we can restate it quite simply: *Each locus will assort independently of other loci.*

What Mendel didn't know was that the law of independent assortment works only for genes that are not on the same chromosome. If **R** and **Y** are on the same chromosome, they are said to be **linked**, and Mendel's law of independent assortment simply doesn't hold for such loci. We will come back to all this in the next chapter. As it happened, none of the gene loci that Mendel used in his multifactor crosses were linked. Perhaps Mendel was just lucky. In fact, Mendel had been very lucky.[1]

[1]In 1936, Mendel's data were reanalyzed by R.A. Fisher, a noted statistician and geneticist. The data fit Mendel's model, all right. But they fit better than they should have by random chance; his data were literally too good to be true. Either Mendel had fudged his data, consciously or unconsciously, or he had presented only his best results and had left out other, less favorable experiments. Or, we could say that the man was absurdly lucky to an unlikely degree. You can take your choice. Perhaps it is worth noting that Mendel recorded several thousand pea and pea plant phenotypes, and, unlike many other researchers, including Darwin with his snapdragons, never once found a pea he felt he couldn't classify one way or another.

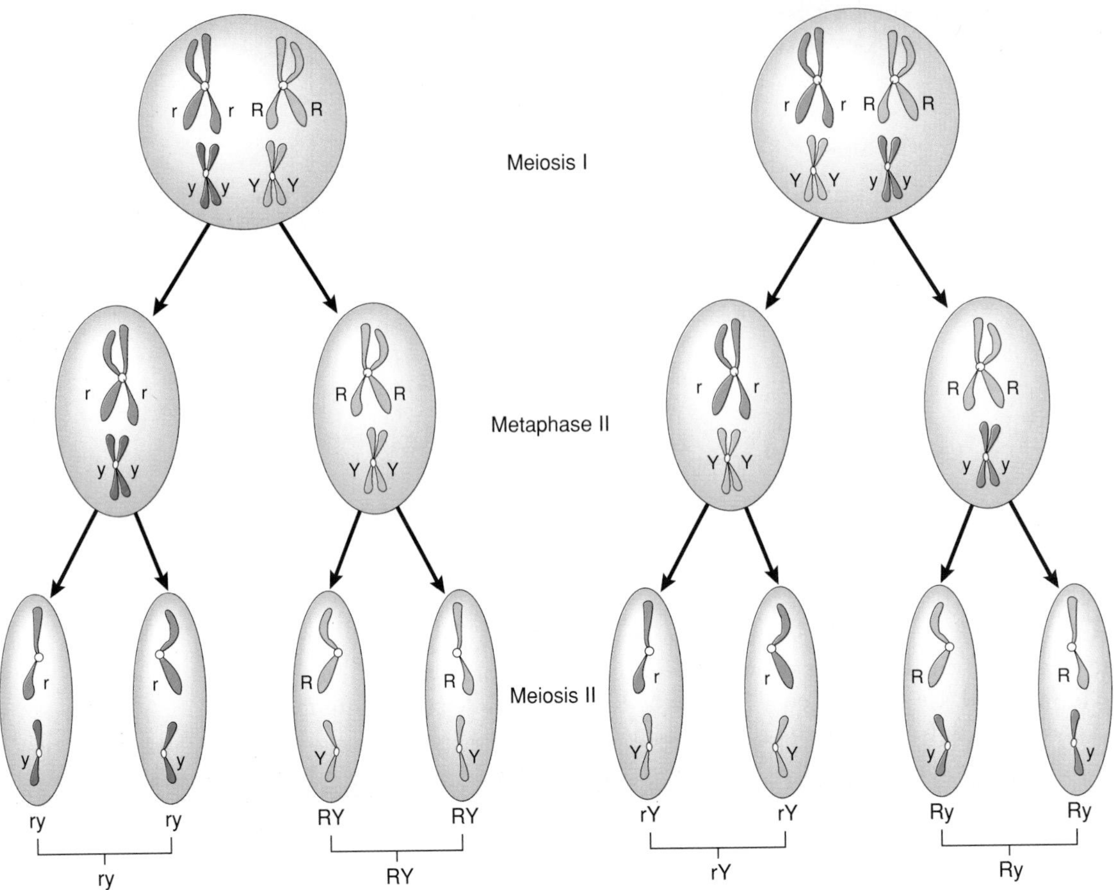

FIGURE 10.9
THE CHROMOSOMAL BASIS OF INDEPENDENT ASSORTMENT
Mendel derived his second law to describe the behavior of traits, but because genes are located on chromosomes, independent assortment also describes the behavior of chromosomes at meiosis. A cell entering meiosis with two pairs of genes on two different chromosomes has only two equally likely alternatives at the first metaphase of meiosis. The important thing to remember, as the name of the law states, is that the segregation of traits on one chromosome is in no way related to the segregation of traits on another chromosome.

Mendel's Testcrosses

We've seen that Mendel was able to determine the genotypes of members of his F_2 generations that were showing the dominant phenotype by permitting the individuals to self-pollinate and then observing the phenotypes of the F_3. Mendel also devised a simpler procedure in which the genotype of any dominant individual could be ascertained. In what is called a **testcross**, such individuals are crossed with subjects that are true-breeding (homozygous) recessives for the trait or traits under consideration. (Remember, the genotype of an individual with the recessive trait is unambiguous, such as **aa**.) For instance, if the phenotype of a seed is "round," its genotype could be either **RR** or **Rr**. Crossing it with a homozygous recessive individual, wrinkled **rr**, would readily solve the problem. If the genotype is **RR**, the cross would be **RR** × **rr**, and all offspring must be **Rr** and thus, round. If the genotype is **Rr**, then the cross is **Rr** × **rr**, and the phenotypic ratio in the offspring is one-half round and one-half wrinkled. The principles of the testcross are demonstrated in Figure 10.10.

Testcrosses can also be used to determine the genotype of dominant individuals where two traits are involved. Such testcrosses, as a matter of fact, greatly aided Mendel in further testing the idea of independent assortment. The individual used as the recessive test parent would, of course, be doubly recessive; thus in our last example, the individual used would be wrinkled and green, or, **rryy**. Any dominant F_2 individual from Mendel's dihybrid cross could have its genotype determined in this manner. For instance, a round, yellow seeded individual, **R_Y_** could be any of the following: **RRYY, RrYY, RrYy**, or **RRYy**.

What testcross results would you predict in each case? You may see the answers intuitively, or, like most of us, you may have to draw Punnett squares and carry out each cross (if all else fails, see the example in Table 10.4). Incidentally, Mendel, never one to be easily satisfied, made it his practice to carry out additional crosses—actually testcrosses on testcross progeny. Such endless probing and checking could explain why, in spite of seemingly endless challenges, his laws are as valid today as they were well over 100 years ago. (It may also explain why it took so long for anyone to become interested in reading his work!)

THE CHROMOSOMAL BASIS FOR MENDEL'S LAWS

While Mendel was completing his work, others were making advances in new directions. One of them was Theodore Boveri, a German who, by the way, discovered the centriole. Boveri, working with sea urchin eggs and sperm, determined that the chromosomes were an essential part of fertilization and development. You can see what fundamental parts of the puzzle were still missing.

Another important advance came from a bright young graduate student at Columbia University. His name was Walter Sutton, and in 1902 he published a paper in which he reported a relationship betwen Mendelian inheritance and meiosis. Keep in mind that Mendel didn't know about meiosis and that those studying it weren't sure how it fit into the big picture. The field was obviously patchy, and one of the most difficult chores of scientists is to merge patches into whole cloth.

The realization that there is a relationship between the physical behavior of chromosomes in meiosis and Mendel's mathematically based laws is rather straightforward for us today (Figure 10.9). However, the initial finding required that investigators observe the actual random lineup of homologous chromosomes on the metaphase plate and their subsequent random separation during anaphase. At first, this may seem easy because, in our figures here, we often show the chromosomes as clearly distinguishable from each other. However, this is a necessary fiction. In the real world under the microscope, homologous chromosomes are not clearly distinguishable from each other. All the same, the researchers needed to actually "see" chromosomes engaged in segregation and in independent assortment.

This is exactly what they did. Investigators found heteromorphic chromosome pairs in the meiotic cells of male grasshoppers. In heteromorphic chromosome pairs, the two homologous chromosomes have some visible difference, such as an extra knob of chromatin on one end, or a dark or faint band here or there. (The sex chromosomes, *X* and *Y*, which are clearly distinguishable in most species, are an example of a heteromorphic chromosome pair.) When two different heteromorphic chromosome pairs were present

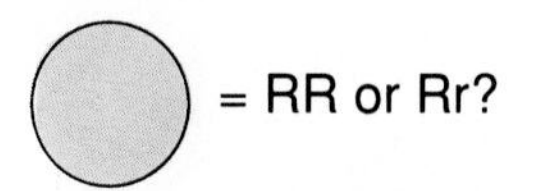

If RR, testcross is RR x rr.

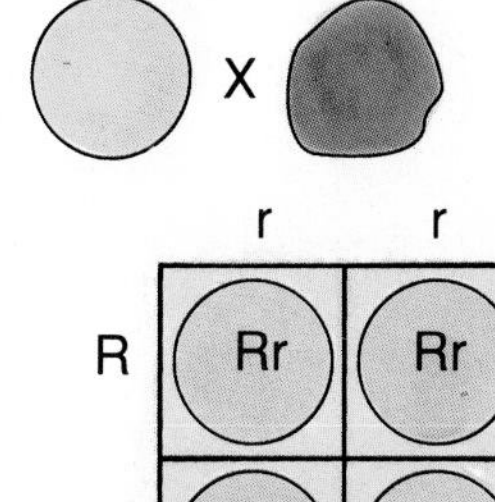

All round offspring

If Rr, testcross is Rr x rr.

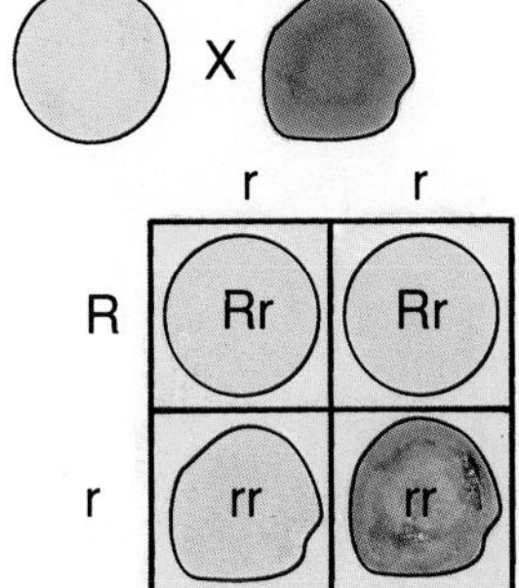

Half round, half wrinkled offspring

FIGURE 10.10
THE TESTCROSS
The two types of round seeds (heterozygous, homozygous) can be distinguished by the results of a cross by homozygous recessive **rr**. The homozygous **RR** produces all round offspring. The heterozygous **Rr** produces 1:1 round:wrinkled.

TABLE 10.4 THE DIHYBRID TESTCROSS ON PLANTS WITH ROUND AND YELLOW SEEDS OF UNKNOWN GENOTYPE (R_Y_)

POSSIBLE GENOTYPE	TESTCROSS	RESULTS
RRYY	**RRYY x rryy**	All offspring round and yellow
RrYY	**RrYY x rryy**	½ round and yellow ½ wrinkled and yellow
RrYy	**RrYy x rryy**	¼ round and yellow ¼ round and green ¼ wrinkled and yellow ¼ wrinkled and green
RRYy	**RRYy x rryy**	½ round and yellow ½ round and green

in the same organism, they were seen to segregate independently in different meiotic cells, giving visible evidence of Mendel's second law (independent assortment: the segregation of one pair of factors has no effect on the segregation of any other pair). Thus, while Mendel drew his inferences from statistical analysis of numbers, the cytologists could actually view alternate segregation and independent assortment in the meiotic process through their light microscopes.

Sutton went on to predict something called gene linkage. He reasoned that since there are only a small number of chromosomes in any one cell, and there are many hereditary factors, each chromosome must carry many genes. But we will get further into that in the next chapter.

PEDIGREES AND HUMAN TRAITS

Mendel's principles are so basic that they can be applied to the analysis of hereditary patterns in virtually any plant or animal. But it is important to know that his analytical methods apply best where large numbers of progeny are forthcoming. Individual pea plants (at least, in the strains Mendel used) produce thirty or so seeds per cross, and even with this seemingly reasonable number, Mendel's records revealed that one of his plants produced 30 yellow seeds and only one green! To obtain his famous 3:1 and 9:3:3:1 ratios, Mendel had to combine the progeny from many identical crosses.

Statisticians today refer to Mendel's problem with the aberrant (30:1) plant "random statistical variation" or more commonly, **sampling error.** Because of such problems, they recommend sample sizes of at least 100. Imagine the impact of this when it comes to determining hereditary patterns in humans, where family sizes are commonly only two to three offspring! As you see, it is virtually impossible to reach valid conclusions on the basis of a single mating.

The approach to this problem is called **pedigree analysis.** A pedigree is a geneticist's way of charting the passage of a trait through many generations. In pedigree charts, darkening the squares and circles (male and female, respectively) indicates the occurrence of the trait in question. The pedigree shown in Figure 10.11 is a hypothetical genetic disease.

The first step in any pedigree analysis is to apply logic in deciding whether the trait is dominant or recessive. Specifically, an individual must be homozygous to express a recessive trait, but need only be heterozygous to express a dominant trait. More importantly, individuals expressing a dominant trait must have at least one parent who also expresses that trait. The disease in this example is clearly not caused by a dominant allele, since the parents of the affected children do not have the trait. It must be recessive.

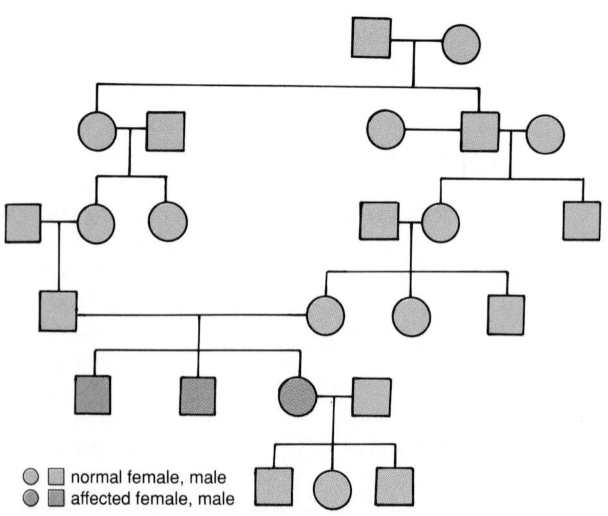

FIGURE 10.11
PEDIGREE OF A HUMAN FAMILY
This hypothetical family has members that suffered a rare disease. It is a recessive condition and only occurred in this family when close relatives were married. Marriages of closely related individuals are termed consanguineous and are indicated in pedigrees by a double horizontal line connecting the couple.

If you are wondering why this trait is so common in this family, consider that the parents of the affected individuals were related, so that both could have gotten the allele from a common ancestor. (The double lines signify consanguineous or "of common blood" marriage.) In this pedigree the disease allele must have occurred originally in one of the two individuals in generation I and passed with a probability of 1/2 to a child in each subsequent generation, hidden or "carried" in the heterozygous condition, until a consanguineous marriage between two heterozygotes resulted in a 1/4 probability of children being homozygous. Actually, most instances of rare recessive diseases are the product of consanguineous mating, and this is the reason that there are laws designed to prohibit marriages within families.

The individual from whom the pedigree was obtained (indicated by the arrow in Figure 10.11) has a strong personal stake in the analysis. A quick look at the pedigree should, however, have been comforting. His mother and father were unrelated, so although he must have gotten the allele from his mother, the chances of his also receiving this very rare allele from his father are minute.

THE DECLINE AND RISE OF MENDELIAN GENETICS

In 1865, after seven years of experimentation (at the very time Darwin was beating his brains out over the enigma of heredity), Mendel presented his results at a meeting of the Brunn Natural Science Society. His audience of local science buffs probably did not understand a word of what they were hearing. After a polite applause, they burst into a vigorous discussion of the hot new idea of the day: natural selection. His single paper was published in the society's proceedings the following year and was actually distributed rather widely. However, the learned scientists of the day were just as baffled and just as uninterested as Mendel's original audience. Eventually, a German botanist included an abstract of Mendel's work in an enormous encyclopedia of plant breeding. Again, the world responded with silence. Apparently, no one had the foggiest notion of the importance of the Austrian monk's experiments and analysis; 1865 minds were just not ready for twentieth-century mathematical biology.

Historians have come up with a small, sad, but remarkable, piece of information. In Darwin's huge library, which is still intact, a one-page account of Mendel's pea work appears in that German encyclopedia of plant breeding. Some relatively obscure work is described on the facing page, and it is covered with extensive notes in Darwin's handwriting. The page describing Mendel's work is clean. Darwin must have seen the paper that would have explained his own snapdragon work and, more important, could have clarified his theory of natural selection, saving him years of agony and uncertainty. But even Darwin was not ready for mathematical biology, and he too failed to grasp Mendel's simple but profound ideas.

Darwin, when he recorded the results of his own crosses, probably did not notice that some of his results approximated simple fractions like one-fourth. Even if he had noticed, it would probably have been of no more significance to him than one-third or one-fifth. Darwin apparently had a very different kind of mind than Mendel. Mendel was not only one of the first mathematical biologists familiar with statistics, but he also knew how to isolate small parts of great problems. Darwin's genius, on the other hand, was in the enormous breadth and scope of his ideas and the ability to fit seemingly unrelated details into a grand scheme.

Mendel's work continued to be ignored—and the ignorance lasted until 1900. In that year three biologists in three different countries, all trying to work out the laws of inheritance, searched through the old literature and all came up with Mendel's paper. All three immediately recognized its importance. Science had changed in 35 years. The obscure monk became one of the most famous scientists of all time. But he had been dead for 16 years.

Later in the twentieth century, Mendelian genetics was applied to the theory of natural selection, and Darwin's reputation, which had faded considerably, ascended to new heights. Darwin had been right all along, if only he had gotten his genetics straight. It is often too easy for those of us who have just had something explained to us to say, ". . . but of course." However, breaking new conceptual ground, even a little, can be stultifyingly difficult. Mendel's work and his reasoning, like Darwin's, are so obvious to us now that we often forget how formidable the monk's task was as he placed those first pea seeds in the carefully cultivated soil of that monastery garden. You may well have found that even when Mendel's careful reasoning is all laid out, it can still be difficult to follow. Biologists can usually be divided into two groups; those who dote on abstract reasoning, who eagerly devour Mendel's work and find that solving genetics problems is fun, and those who don't.

allele one of two or more alternative forms of a gene (**A** and **a**, **R** and **r**)
dominance where the expression of one allele masks the expression of its alternative
dihybrid a cross in which two different pairs of alleles are under consideration
factor Mendel's term for a gene
genotype the combination of genes producing a trait (e.g., **Tt**, **RR**, **gg**, **Gg**)
heterozygous an individual with two alternative forms of a pair of alleles (e.g., **Tt**, **Rr**, **Gg**)
homozygous an individual with identical alleles for a gene (e.g., **TT**, **rr**, **gg**, **GG**)
independent assortment where the segregation of one pair of alleles has no effect on the segregation of another
linked genes genes situated on the same chromosome
monohybrid cross a cross in which one pair of alternative alleles is under consideration
phenotype an individual's visible traits (e.g., **tall**, **round**, **green**)
progeny testing determining the genotype of offspring by self-pollinating or inbreeding
Punnett square a grid used to record the possibilities in a cross
recessivity where an allele's expression is masked by its alternative allele
segregation the physical separation of homologous alleles (those on homologous chromosomes) when meiosis occurs
test cross determining the genotype of a dominant individual by crossing it to a recessive **(T_ × tt)**
true-breeding individuals whose offspring are genetically identical to themselves (homozygous)

PPLICATION OF IDEAS

GENETICS PROBLEMS

To the dismay of many biology students, most educators agree that solving genetics problems is a tremendously instructive means toward truly understanding the ins and outs of Mendelian genetics. It is widely held that ability to do genetics problems is directly related to one's understanding of the principles, AND that a thorough understanding of the principles requires doing genetics problems. Below you will find a series of genetics problems designed around the main points of this chapter. They begin with the obvious and straightforward type, then move onto the more subtle kind that require some sleuthing. If you work them through from beginning to end, yourself, you will see that there is a limited number of possible answers to any particular part of a problem, and that the more difficult problems represent a collection of simple parts. The key is seeing through to those simple parts. The key is based on Mendel's laws.

1. In a cross of two heterozygous **(Aa)** individuals, (a) if they have a child what is the probability that it is homozygous recessive? (b) If they have two children, what is the probability that both will be heterozygous? (c) If they have two children, what is the probability that the first one will be homozygous recessive and the second one will be homozygous dominant? (d) If they have two children, what is the probability that one will be homozygous recessive and one will be homozygous recessive? (e) What is the difference between parts (c) and (d)?

2. In a cross of a homozygous dominant **(AA)** by a homozygous recessive **(aa)**, (a) if they have a child, what is the probability that the child will be heterozygous? (b) What is the probability that the child will be homozygous recessive? (c) If they have two children, what is the probability of the first one being homozygous dominant and the second one being homozygous recessive?
3. In a cross of two homozygous dominant individuals **(AA)**, (a) if they have a child, what is the probability that the child will be heterozygous? (b) If they have a child, what is the probability that the child will be homozygous dominant?
4. In humans, Albinism is caused by a single gene. The recessive allele **(a)**, when homozygous, results in no pigment deposition in the skin and therefore a very pale skin color. Ann Smith is albino. She married Mike Jones, and they had two kids, Sara (who had skin pigment), and Martha (who was albino). (a) Write out the genotypes of Ann, Mike, and their children Sara and Martha. (b) What is the probability that their next child would be albino? (c) What is the probability that if they have two more children, that both children will be normally pigmented? (d) If they have two more children, what is the probability that one will be albino and one will be normally pigmented?
5. Ann and Mike divorce. Ann remarries, this time to Ralph Johnson. They have six normally pigmented children. (a) What would you assume was Ralph's genotype? (b) Assuming, however, that Ralph was heterozygous, what is the probability of their having those six normally pigmented children?
6. Given two separate, independently assorting genes, **A** and **B**, with **A** dominant to **a** and **B** dominant to **b**, (a) what is the probability of getting an **AB** gamete from an individual that is heterozygous at both loci? (b) What is the probability of getting an **Ab** gamete from an individual that is heterozygous at both loci? (c) What is the probability of getting an **AB** gamete from an **AABb** individual? (d) What is the probability of getting an **AABB** child from a cross of two people heterozygous at both loci? (e) From a cross of two people heterozygous at both loci, what is the probability of having a child that expresses the dominant phenotype for both loci?

KEY IDEAS

WHEN DARWIN MET MENDEL (ALMOST)

1. Darwin subscribed to the idea of blending inheritance, which held that traits from both parents are "blended" in the offspring.
2. Critics of natural selection pointed out how blending would destroy variations—a key part of Darwinian evolution.
3. Darwin crossed true-breeding strains of snapdragons, in which he observed **dominant** traits, which he then called prepotent.

MENDEL'S CROSSES

1. Mendel used carefully planned experiments and applied statistical analysis to his data.
2. Mendel reported on experiments with seven different characteristics of garden peas. These included seed form, color of contents, color of coats, color of pods, shape of ripe pods, length of stem, and position of flowers.
3. His approach was to cross true-breeding strains, manipulating pollen by hand to avoid self-pollination.
4. In peas, each pea in a pod has its own **genotype** and **phenotype.** (*Genotype* is total combination of an organism's genes; *phenotype* is the combination of observed or measured traits, generally what is readily visible.)
5. The symbols P_1, F_1 and F_2 are used to designate first and subsequent generations in crosses.

The Principle of Dominance

1. Both Mendel and Darwin perceived the same results in their crosses. There was no blending, but the **recessive** trait disappeared in the F_1 generation and reappeared in the F_2. Darwin called this latency and pursued the problem no further, but Mendel noted that the reappearance of a *recessive* trait occurred with a definite frequency in one-fourth of the F_2.
2. He determined that for each dominant trait there were two kinds (genotypes), those that produced two kinds of offspring **(heterozygous)**, and those that produced one **(homozygous)**.
3. To determine whether an individual with a dominant trait is heterozygous or homozygous requires **progeny testing.** Mendel did this by breeding an F_3 generation from F_2 round peas. He determined that one-third of the round F_2 peas were true-breeding, two-thirds were not. From this he determined that one-fourth of the total F_2 were round and true-breeding; one-half of the F_2 were round and not true-breeding; and one-fourth of the F_2 were wrinkled and true-breeding.
4. From his work so far, Mendel concluded that characteristics were controlled by factors in two forms, dominant and recessive. In modern terms, factor is replaced by gene. The alternate forms of a gene are called **alleles.**

MENDEL'S FIRST LAW: THE SEGREGATION OF ALTERNATE ALLELES

1. Heterozygous **(Aa)** individuals produce two kinds of gametes (sex cells) in equal proportions:

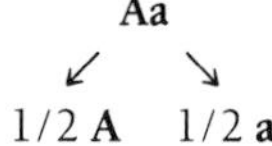

2. Following this, the **multiplicative law** from the laws of probability can be applied to crosses: "The probability of two independent events both occurring is equal to the *product* of their individual probabilities."
3. The F_1 cross **Aa** × **Aa** can be stated (1/2**A** + 1/2**a**) × (1/2**A** + 1/2**a**). Multiplying produces 1/4**AA** + 1/2**Aa** + 1/4**aa.** This 1/2**Aa** is determined algebraically but can be explained genetically. The combination **A** + **a** can occur two ways, **A** + **a** or **a** + **A.**
4. The **additive law** is: "The probability of either one or another of two mutually exclusive events occurring is equal to the *sum* of their individual probabilities."
5. The probability of an **A** pollen and an **a** ovule combining is one-fourth, as is the probability of **a** + **A.** Since they are mutually exclusive events, the probability of a heterozygous F_2 individual is 1/4 + 1/4 = 1/2.
6. Punnett illustrated Mendel's principles using squares:

	A	a
A	AA	Aa
a	Aa	aa

When summed up the results are:
Genotype: 1/4**AA** + 1/2**Aa** + 1/4**aa**
Phenotype: 3/4 dominant + 1/4 recessive

7. Mendel can be considered a mathematical biologist because he constructed a model that yielded numerical predictions consistent with observations.

MENDEL'S SECOND LAW: INDEPENDENT ASSORTMENT

1. Mendel crossed two alleles of the round **locus, R** and **r**, and two alleles of the yellow locus, **Y** and **y** *Locus* refers to a specific gene location on a chromosome.

   RRYY × **rryy**(P_1 cross)
   all **RrYy** (F_1 offspring)
   RrYy × **RrYy** (F_2 dihybrid cross)

2. To predict the results, consider the following:
   a. You know that **Rr** × **Rr** produces 1/4**RR**, 1/2**Rr**, and 1/4**rr.** Likewise, **Yy** × **Yy** produces 1/4**YY**, 1/2**Yy**, and 1/4**yy**.
   b. To predict the results when both are considered simultaneously, follow the multiplicative law for all possible **R** and **Y** combinations.
   These are as follows:

Genotype	Separate proba-bilities	Combined proba-bilities	Grouped into *phenotypes*
RRYY	¼ × ¼ =	1/16	
RrYY	½ × ¼ =	⅛	Round and Yellow
RRYy	¼ × ½ =	⅛	9/16
RyRy	½ × ½ =	¼	
rrYY	¼ × ¼ =	1/16	Wrinkled and yellow
rrYy	¼ × ½ =	⅛	3/16
RRyy	¼ × ¼ =	1/16	Round and green
Rryy	½ × ¼ =	⅛	3/16
rryy	¼ × ¼ =	1/16	Wrinkled and green 1/16
			16/16

3. In addition to being consistent with his expectations, Mendel's findings indicate that two traits, pea shape and color, are inherited independently. If they were not, the multiplicative law wouldn't have worked.
4. Today we know that all of the characters Mendel studied this way were located on different chromosomes. Because of the random way pairs of chromosomes align at metaphase and separate at anaphase, genes separate independently of each other.
5. Mendel's second law can be restated: "If an organism is heterozygous at two unlinked loci, each locus will assort independently of the other."

Mendel's Testcrosses

1. Mendel used testcrosses to determine whether a dominant type was homozygous or heterozygous, and whether assortment was independent. Suspected heterozygotes are crossed with homozygous recessive individuals.
2. Testcrossing can help determine whether the test subject is homozygous or heterozygous.

THE CHROMOSOMAL BASIS FOR MENDEL'S LAW

1. Sutton proposed that the hereditary factors were contained on the chromosomes and that each chromosome carried many factors.
2. Establishing the relationship between Mendel's first and second laws and meiosis required following two heteromorphic pairs of chromosomes through meiosis.

PEDIGREES AND HUMAN TRAITS

1. Pedigree analysis allows insight into genetic effects in situations (as in human genetics) where controlled crosses are impractical.

THE DECLINE AND RISE OF MENDELIAN GENETICS

1. Mendel's findings, which were not understood in his time, were rediscovered about the turn of the century.

REVIEW QUESTIONS

1. What is blending inheritance? How did Darwin's use of this concept open him to severe and justifiable criticism of his natural selection hypothesis? (227-228)
2. Describe Darwin's experiments in the heredity of snapdragons. What did he mean by prepotence? (228)
3. What kind of preparation did Mendel have for mathematical biology? What other characteristics led to his success? (229)
4. Describe Mendel's general procedure from P_1 to F_2. What kinds of crosses did he make? (230-233)
5. What important question confronted Mendel as he observed F_1 generation in each cross? (230)
6. Distinguish between the terms *genotype* and *phenotype*. (229)
7. Carry out Mendel's yellow-versus-green crosses from F_1 to F_2 and verify his ratios in the F_2. State the phenotypic ratio of the F_2. (232)
8. Distinguish between the terms *heterozygous* and *homozygous*. (231)
9. Explain how Mendel used progeny testing to determine the *genotype* of his F_2 peas. What did he learn about the F_2 yellow peas? (231)
10. State Mendel's first law. (233)
11. Using the cross **Aa** × **Aa** and applying the multiplicative law, answer the following:
    a. What is the probability of an **A** sperm fertilizing an **A** egg? Why? (234)
    b. What is the probability of an offspring carrying the **Aa** combination? Explain carefully. (234)
    c. What is the probability of an **a** sperm fertilizing an **A** egg? Why? (234)
    d. How does the *additive law* law apply to predicting the **Aa** offspring? (234)
12. Carry out Mendel's two-character cross from P_1 through F_2. Write the *phenotypic* ratio of the F_2. (234)
13. State Mendel's second law and explain what it has to do with the results of the above cross. (235)
14. Using diagrams of chromosomes to represent the cross **AaBb** × **AaBb**, show how independent assortment works. (Review meiosis in Chapter 11.) *Hint:* There are two ways the homologous pairs of chromosomes can align on the metaphase plate. (237)
15. Show the two types of testcrosses or backcrosses Mendel made for the double heterozygote **RrYy**. What is the purpose of a *testcross?* (238-239)
16. How would a testcross tell you whether a white ram carried a recessive gene for black? Prove your answer. (238)
17. Why do progeny testing if your testcross was to a homozygous dominant individual? (238)

11

GOING BEYOND MENDEL

MENDEL FRANKLY ATTRIBUTED HIS SUCCESS TO HIS DELIBERATE DECISION to work only with factors that always produced large, dramatic effects with clear and distinct phenotypes. He examined such traits, one or two at a time, in highly inbred, genetically unvarying pure strains. Only in such simple systems could he have worked out his famous ratios. Mendel's discoveries were valuable because nearly all genes are, in fact, transmitted according to his principles, at least on a genotypic level. But most visible genetic variation is not so simple at the level of the phenotype because of the complexities of development and gene expression that Mendel had so fortuitously avoided.

In this chapter we will consider those mitigating circumstances that influence many phenotypic ratios, causing them to be something other than a perfect 3:1 (or 9:3:3:1 for dihybrids). The complications fall into the following main classes:

1. **Dominance relationships**, where dominance other than the simple, complete type that Mendel observed in his pea alleles occurs.
2. **Multiple alleles**, where even though a single individual can have only two alleles for a gene, more than two alleles may be involved in the cross.
3. **Gene interactions**, where the expression of one gene is influenced by the expression of another, separate gene.
4. **Conditional gene expression**, where the expression of a gene is dependent upon nongenetic factors such as environmental conditions or age of the individual.
5. **Polygenic inheritance**, where the expression of a trait is governed by the action of more than one gene.
6. **Linkage**, where separate genes are located on the same chromosome.

Even as you look over this list, you can probably see that any of these situations would have serious impact on the ideal ratios generated by Mendel's laws. However, this does not invalidate the laws themselves. Because after all is said and done, essentially all genes in the nucleus behave according to the laws of random segregation and independent assortment.

To illustrate how genes following such clear-cut laws can produce unexpected results, consider the following. Most of the genetic differences we see in our friends—differences in height, weight, body build, skin color, temperament, facial features, athletic ability, intelligence, and hairiness—are due to normal allelic variation (along with the modifying effects of the environment). These normal phenotypic differences, which add so much to human interest, seldom show up in the usual Mendelian ratios, although the genes responsible for them may segregate and assort with faithful Mendelian precision. Even blue and brown eye color, a popular example of Mendelian inheritance in humans, turns out to be quite complex and unpredictable. People do not merely have blue or brown eyes; they may have gray, light blue, deep blue, hazel, flecked, or green eyes. In this chapter we will look beyond Mendel to learn more about how Mendelian genes behave in non-Mendelian ways.

DOMINANCE RELATIONSHIPS

The various ways in which two alleles at one gene locus can affect the phenotype are called dominance relationships. You are aware, of course, that the two alleles come from the two parents. Thus, each parent contributes to the effects controlled by each locus. Such a system is adaptive, in extreme cases, by providing a genetic backup in case one of the alleles should somehow fail or be defective in some other way. More routinely, though, the two alleles will interact through dominance relationships.

What happens when the gene from the father and the corresponding gene from the mother give conflicting information? Say, an allele for brown eyes comes from one parent and an allele for blue eyes comes from the other. What then? Actually, many things can happen, but one of the most common results is that the information from one allele will appear to be ignored. In that case, as we've seen, the other allele is dominant. That's what Mendel saw in his peas, for example. Dominant, as you know by now, means that the phenotype of the heterozygote (the individual with the conflicting genetic instructions) will be exactly like that of one of the homozygotes. The allele that is expressed, no matter what its partner is, is called the dominant allele, and the allele that is suppressed in the heterozygote is called the recessive allele.

But so far we have given only observations and definitions, not explanations. What really happens when two different alleles occur together? How is one suppressed? And how does the organism "choose" between its two sets of information? Actually, there are several different ways that dominance can happen. In most cases the recessive allele isn't expressed because it simply isn't doing anything; its instructions aren't being read. This is clearest in rare medical disorders, where the absence of an enzyme can have a severe and often lethal effect. In a relatively benign example shown in Figure 11.1, albinos lack an enzyme that is necessary to make melanin pigments. They didn't get a functioning gene from their father, and they didn't get one from their mother, so their cells cannot make the pigment. Such individuals are homozygous for recessive alleles, which in this case are alleles that aren't functioning. Heterozygotes for albinism or other enzyme deficiencies, on the other hand, have one working allele and one that doesn't work, and produce only half the usual amount of enzyme; but in most cases half the normal amount of enzyme is still enough to metabolize all the enzyme's substrate, and the phenotype of the heterozygote will be perfectly normal.

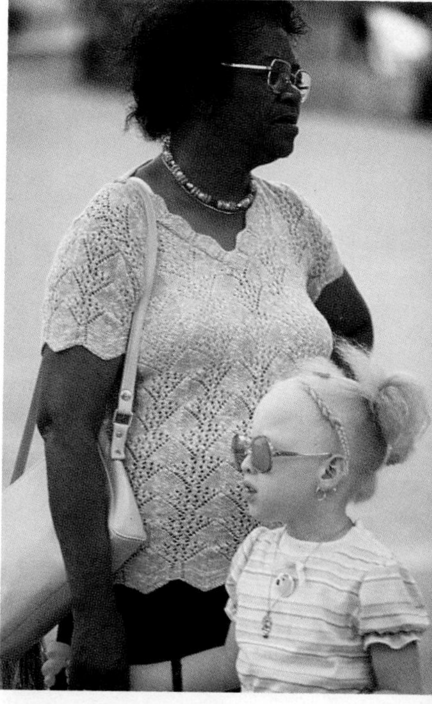

FIGURE 11.1
ALBINISM IN A HUMAN FAMILY
Albinism, the absence of pigmentation in hair and a light skin color, is a striking, simple recessive genetic trait.

In other cases a recessive allele will function a little. The allele, then, is being read, but its product is scant. In such cases, its effect may be swamped by its more active dominant partner, and for all practical purposes it will appear as if it is not functioning at all. Sometimes dominance is in the eye of the beholder. For instance, many alleles of several genes have been found that create changes in the eye color of *Drosophila* (see Figure 11.2). Almost all of them are recessive to the normal alleles, which are involved in the synthesis of one or both of two strong eye color pigments.

In some cases the dominant and recessive alleles interact in another way. Here, the recessive allele is able to perform the normal function, but the dominant allele makes something that prevents that function. For example, true-breeding white leghorn chickens are homozygous for an allele, **I**, that inhibits melanin (color) formation. That's why

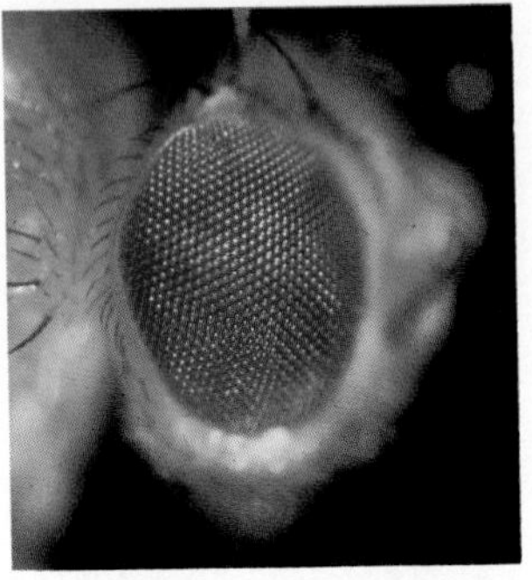
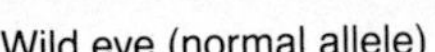

Wild eye (normal allele)

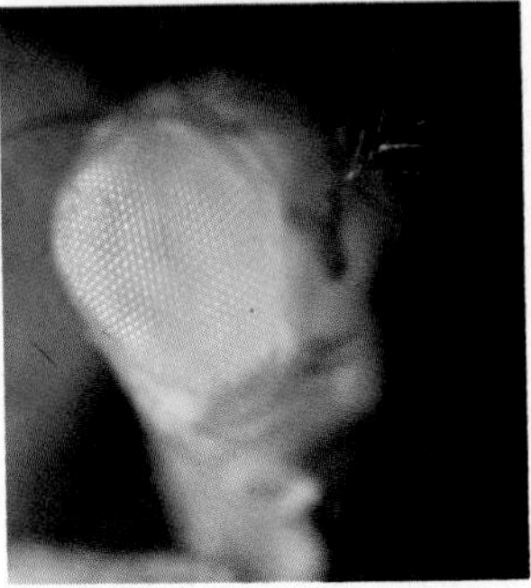

White eye

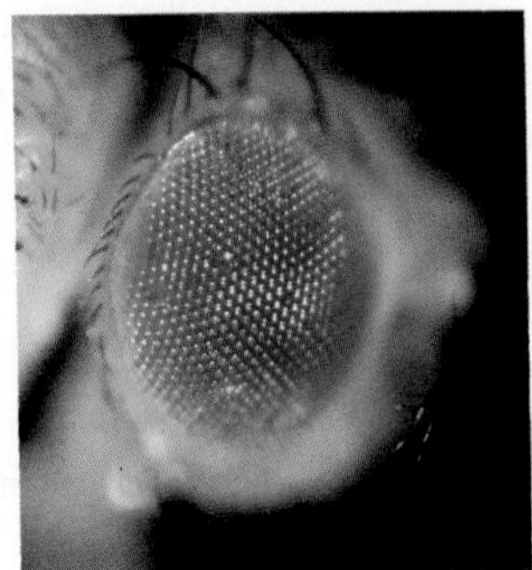

Brown eye

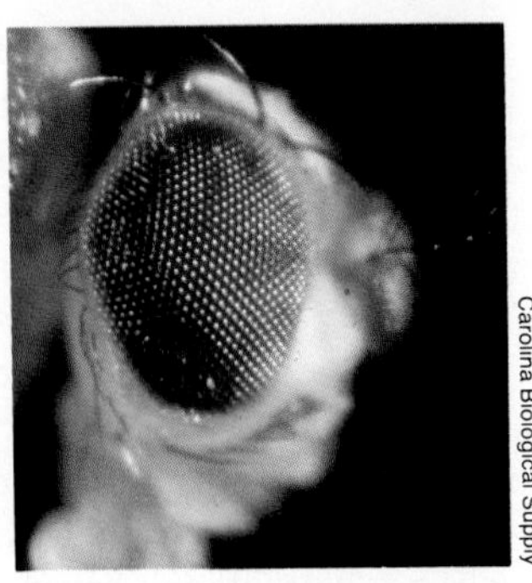

Sepia eye

Carolina Biological Supply

FIGURE 11.2
SOME EYE COLORS IN THE FRUIT FLY
Some of the conspicuous eye color genes and alleles found in the fruit fly are naturally occurring, though most have been created in experiments designed to alter, or mutate, genes involved in eye color.

they're white. Other chicken breeds are homozygous for the alternative allele, **i**, and can be all sorts of colors. In a cross between a white leghorn chicken and another chicken, all the offspring are F_1 heterozygotes:

$$P_1\text{: } \mathbf{II} \times \mathbf{ii}$$
$$F_1\text{: } \mathbf{Ii}$$

And, because the inhibitor allele is dominant, all the offspring of such a cross will be white also. A similar example of dominant color inhibition exists in sheep, where the black recessive allele shows up occasionally in spite of attempts to eliminate the trait (Figure 11.3).

Partial Dominance

Whenever the heterozygote is somewhere between the phenotypes of the two homozygotes, and not exactly like either one of them, we call the relationship **partial dominance** (or, sometimes *incomplete dominance*). A classic example of partial dominance is found in snapdragons, in crosses between strains with red and white flowers. Here, the two alleles can be symbolized by $\mathbf{c^R}$ and $\mathbf{c^W}$. (Historically, incompletely dominant alleles are usually designated by small letters. Hence, we use $\mathbf{c^R}$ and $\mathbf{c^W}$ here, instead of **R** and **r**, which would imply complete dominance.) When the homozygous $\mathbf{c^Rc^R}$ red-flowered snapdragons are crossed with homozygous $\mathbf{c^Wc^W}$ white-flowered snapdragons, the plants in the F_1 generation, the $\mathbf{c^Rc^W}$ heterozygotes, all have pink flowers. If the pink-flowered F_1 plants are self-pollinated to produce an F_2 generation, we have an interesting Mendelian ratio. Instead of a 3:1 or 75:25 ratio, the F_2 snapdragons are 25% red, 50% pink, and 25% white. Since all three genotypes are easily distinguished, the F_2 phenotypic ratio is exactly the same as the F_2 genotypic ratio (Figure 11.4), a diagnostic feature of partial dominance.

Sometimes the phenotype of the heterozygote is not simply a compromise between the two homozygous phenotypes, but has some unique characteristics of its own. Consider Roy Rogers's horse Trigger, for instance. Trigger was a beautiful Palomino horse, with a golden coat and blonde mane and tail. (You can see for yourself. He is stuffed and mounted in the Roy Rogers and Dale Evans Museum in southern California.) All Palomino horses are heterozygotes (Figure 11.5). Crosses between Palomino horses yield brown, Palomino, and white foals in an approximate 1:2:1 ratio.

FIGURE 11.3
DOMINANT COLOR INHIBITION
Apparent white pigmentation is not necessarily due to albinism. There are several examples, such as the sheep shown here, where white coloration is due to a dominant allele that prevents any pigmentation from forming.

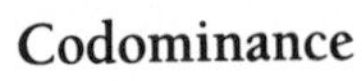

Codominance

Sometimes one homozygote will show one phenotypic trait, the other homozygote will have a different phenotypic trait, and the heterozygote will show both traits. This condition is called **codominance.** In codominance both traits are expressed in the heterozygote, while in partial dominance there is a blending of the two traits. Codominance is actually rather rare among traits that are easily seen, but it is common among genetic traits that can only be measured by biochemical tests. Codominance is often encountered, for instance, in the genetics of blood groups. Let's briefly consider what is known as the MN blood group in humans as an example of codominance.

In the MN blood group system, two codominant alleles, **M** and **N** account for the three genotypes, **MM, MN,** and **NN.** The phenotype is revealed when blood samples are tested with two kinds of antisera, anti-M and anti-N. The two antisera are produced from blood serum removed from animals that have been previously sensitized against **M** or **N** blood. When blood samples react with an antiserum, the red cells agglutinate or clump together, forming visible clusters. Here is how the three genotypes are determined:

GENOTYPE	REACTION WITH ANTI-M	REACTION WITH ANTI-N
MM	+	−
MN	+	+
NN	−	+

Here, " + " means that the red blood cells agglutinate, or clump, and " − " means that they do not. Look at the reaction with anti-M. Considering these three reactions alone, it is clear that allele **M** is dominant. Now look at the reaction with anti-N. For this test, **N** is clearly dominant. When both antisera are used, alleles **M** and **N** are codominant.

FIGURE 11.4
PARTIAL DOMINANCE
Flower color in snapdragons presents a good example of partial or incomplete dominance. The hallmark of partial dominance is that the heterozygotes appear to have a phenotype that is somewhere in between the homozygous dominant and the homozygous recessive. Another distinguishing feature is that the phenotypic ratio of the progeny is always the same as the genotypic ratio. Can you see why?

Recessive Lethals

There are alleles that when homozygous cause the death of the individual, yet have no effect when carried in a heterozygote. These are called **recessive lethal** alleles. A good example is the white seedling allele of maize. Homozygous **ww** individuals fail to produce green chloroplasts; hence the seedlings don't survive. They grow just fine as long as they are small and can live off the stored reserves in the seed (Figure 11.6), but they die when those reserves are spent. Heterozygous **Ww** plants produce normal chloroplasts and are indistinguishable from the homozygous **WW** plants.

Dominant Lethals

Here's a question: if an allele is both lethal and dominant, all individuals carrying the allele (heterozygous as well as homozygous) would die . . . so how would the allele ever be transmitted? The answer is that the individuals carrying the allele must survive at least long enough to reproduce. A famous example is Huntington's chorea, a debilitating and lethal genetic disease in humans. Death is the ultimate result, but the disease does not begin to have its effects until age 40 or so, leaving plenty of time for the allele to be passed to offspring.

FIGURE 11.5
TRIGGER, THE PALOMINO HORSE
The beautiful golden Palomino color, here shown in the most famous of Palomino horses (who is stuffed and on display), is a result of heterozygosity of alleles for both brown and white coat color in horses. Happy trails.

Other Lethals

Lethal alleles can also behave as partial dominants, meaning that you can tell the heterozygotes from the surviving homozygotes. A good example is achondroplastic dwarfism, or achondroplasia. Individuals of **AA** genotype die very early in development. The heterozygous carriers (**Aa**) of the defective allele appear normal at birth, but their arm and leg bones, in particular, stay very short. Affected persons have normal-sized heads, and are generally in good health and often extremely athletic, although they are conspicuously stunted because of their short limbs (Figure 11.7).

If achondroplastic dwarfs marry, and their mates happen to be normal homozygotes (**aa**), about half their children will be dwarfs like themselves

FIGURE 11.6
WHITE SEEDLINGS, A RECESSIVE LETHAL TRAIT
This white seedling trait in corn is recessive, here segregating in a 3:1 ratio. White seedling is also lethal, since the seedlings fail to produce chloroplasts and die when the energy reserves in the seeds are exhausted.

(which is in keeping with Mendel's First Law). When two achondroplastic dwarfs marry each other (**Aa** × **Aa**) they might be expected to produce some children who are homozygous for the dwarf allele. However, they do not, because of the very early stage at which lethality occurs.

MULTIPLE ALLELES

Fortunately for Mendel, he had to deal with only two alleles at any gene locus; otherwise he may have become hopelessly entangled in complex genetic systems. With the wrong choice of plants, or a different selection of traits, he could have been confounded early on. This is because many alleles can occur at a given gene locus (although any one individual can have only two—one from each parent). A Harvard research group recently did an intensive study of a randomly chosen enzyme locus in a North American fruit fly population. They uncovered 37 alleles at one locus, out of a sample of only 146 flies. As far as could be determined, all 37 variants functioned normally.

Let's look at a simple example of multiple alleles. Coat color in rabbits is determined by a single gene, **C** (Figure 11.8). The agouti coloration is the brown mottled color that you find in wild rabbits, and is determined by a dominant allele, **C**. Albino rabbits are white due to a lack of pigment, and are homozygous for the recessive **c** allele. Agouti (**C**) is completely dominant to albino (**c**). There is a third type of coat color that you may be familiar with, called himalayan. It is mostly white, but has black fur on its nose, paws, and ears. The allele that causes the himalayan phenotype is $\mathbf{c^h}$. The $\mathbf{c^h}$ allele is dominant to **c**. Therefore a cross of two heterozygous individuals ($\mathbf{c^hc}$ × $\mathbf{c^hc}$) will produce himalayans and albinos in a 3:1 ratio. But himalayan is recessive to agouti! $\mathbf{Cc^h}$ individuals are agouti. A cross of individuals heterozygous for the agouti and himalayan alleles ($\mathbf{Cc^h}$ × $\mathbf{Cc^h}$) will produce agoutis and himalayans in a 3:1 ratio.

Thus we can see, once again, that dominance is often a question of your point of view, and that even trying to be more precise by the use of uppercase letters to show dominance can be troublesome. More importantly, however, we can see that a population may have more than just two alleles for any given gene. But remember, please, that any individual can have only two alleles for one gene.

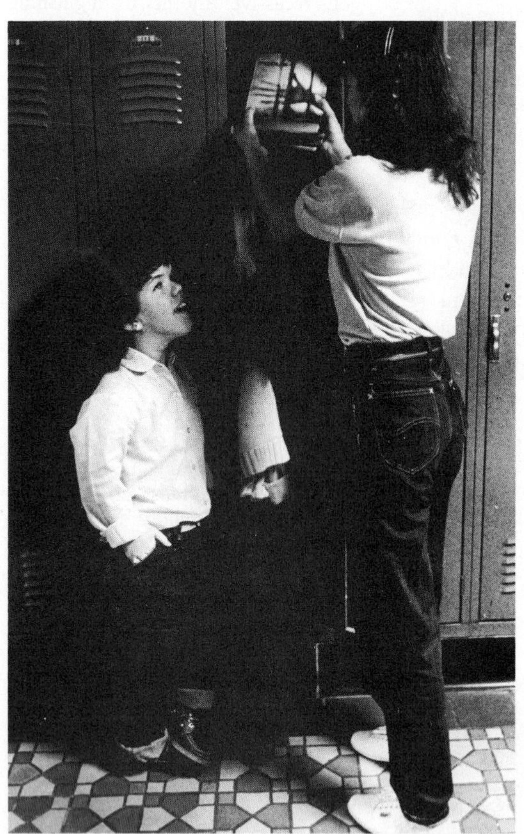

FIGURE 11.7
ACHONDROPLASIA
A person affected with achondroplasia is heterozygous **Aa** and markedly short in stature. The homozygous **AA** condition is lethal. Since you can tell the difference between heterozygous and homozygous dominant, achondroplasia can be considered a partial dominant lethal.

Blood Groups

If you are asked what your blood type is, your answer will probably include at least one of these letters: A, B, or O. This is because the most important and best-known blood group system is the **ABO system**, and the corresponding four blood types which people can have are A, B, O, and AB.

The blood type is determined by antigens on the surfaces of the red blood cells. The term antigen refers to any substance that produces a response in the body's immune system. Part of this response is the production of proteins called antibodies. Antibodies bind to antigens, clumping them together for easier destruction by certain white blood cells. The antigens of the ABO system happen to be polysaccharides associated with the red blood cell membranes. The red blood cells of a type A person, for instance, have A antigens on their cell surface, and the blood cells of a type B person have B cell-surface antigens. The **I** gene (**I** stands for isoagglutinogen, which is another term for antigen) is responsible for the production of these polysaccharides. Type AB people have both A and B antigens, so the two alleles for A and B are codominant. The two alleles are usually written $\mathbf{I^A}$ and $\mathbf{I^B}$ to distinguish the allele that makes the antigen from the antigen itself. Individuals with the fourth blood type, type O, have neither the A antigen nor the B antigen on the surfaces of their red blood cells. A third allele, then, is $\mathbf{I^O}$ (or **i**), which is recessive to both $\mathbf{I^A}$ and $\mathbf{I^B}$. The possible genotypes and their corresponding phenotypes are shown in Figure 11.9.

(a)

(b)

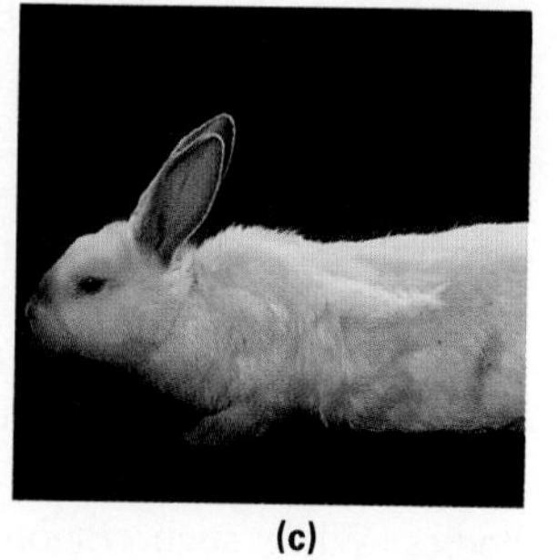
(c)

FIGURE 11.8
MULTIPLE ALLELES AND RABBIT COAT COLOR
There are three alleles that can interact to produce coat colors in rabbits, C, c, and c^h. C_ rabbits are agouti (a), cc rabbits are albino (b), while c^hc^h and c^hc rabbits are himalayan (c). Therefore, C is dominant to c^h and c, c^h is recessive to C but dominant to c, and c is recessive to both C and c^h.

The three alleles can combine into six different genotypes, but since I^AI^A and I^AI^O are both type A, and I^BI^B and I^BI^O are both type B, there are only four ABO blood types or phenotypes.

The ABO genotypes have another, more indirect effect on the phenotype. The immune systems of all individuals, for reasons that are not entirely understood, always produce antibodies against the A and B antigens, whichever is lacking on the individual's own cell surfaces. Thus, type A persons produce anti-B antibodies, and type B persons produce anti-A antibodies; type O persons produce both kinds of antibodies. Since type AB persons lack neither—that is, they have both A and B antigens present—they produce neither the anti-A nor the anti-B antibody.

The antibodies are specific, two-headed binding proteins (see Chapter 41). By two-headed we mean the antibody protein has two active sites. Each antibody is able to bind to two of the specific molecules that it is directed against. The anti-B antibody, for instance, can bind to the B molecule (antigen) on the surface of one red blood cell and at the same time bind to another B molecule on the surface of another red blood cell. Since each red blood cell is covered with these antigens, the antibodies will cause susceptible cells to agglutinate, or clump. Tests of blood types are carried out with drops of blood and antisera on glass microscope slides (see Figure 11.9).

As a result of the anti-A and anti-B antibodies produced by persons of various genotypes, the blood types of donors and recipients have to be very carefully matched. Type A blood transfused into a type B person will be agglutinated by the antibodies of the recipient, forming possibly fatal clumps. In some cases, type O blood can be trans-

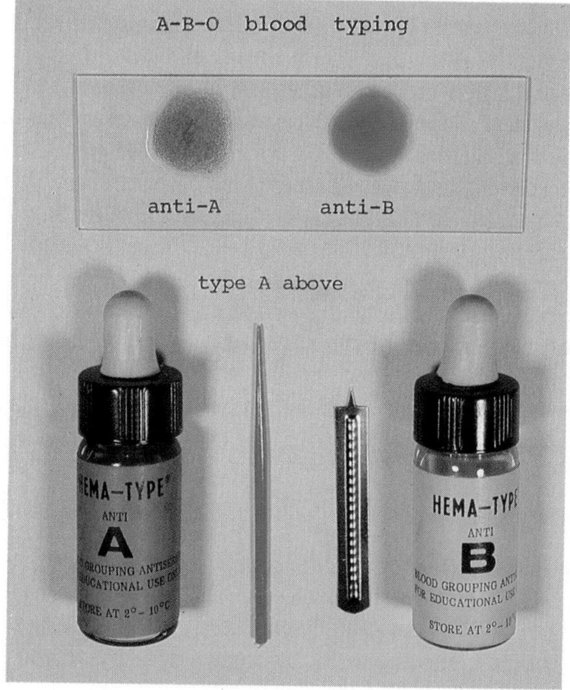

FIGURE 11.9
ABO BLOOD TYPING
The presence of A and B antigens in human blood can be determined in a simple test using anti-A and anti-B test reagents. A positive reaction (agglutination) is characterized by a distinct graininess occurring in the mixture.

Genotype	Agglutination with anti-A?	Agglutination with anti-B?	Blood group	ABO antibodies in serum
I^AI^A	+	–	A	Anti-B
I^AI^O	+	–	A	Anti-B
I^BI^B	–	+	B	Anti-A
I^BI^O	–	+	B	Anti-A
I^AI^B	+	+	AB	None
I^OI^O	–	–	O	Anti-A *and* Anti-B

fused into persons of other blood types, but even this is unwise, because the introduced type O blood brings with it some anti-A and anti-B antibodies, though not usually enough to be harmful.

For years biologists wondered why we sometimes react so quickly and violently to each other's red cells. Actually, most immune reactions occur very slowly, requiring time for the body to recognize a potential antigen and then to initiate the proper response (See Chapter 41). So why is this particular reaction to red blood cells so rapid? Immunologists now think they have the answer. It's a matter of diet. Some of the carbohydrates we take in act as antigens, triggering immune responses and sensitizing our bodies. While such responses quickly subside, the immune system has a long-lived molecular memory, and should the antigen show up again, the antibody response is rapid and effective. As it turns out, these offending dietary carbohydrates are strikingly similar to red cell surface carbohydrates of certain blood types. Should we receive transfusions of such red cells, our previously aroused immune systems go quickly to work, and the antibody response is almost immediate. Our antibodies quickly start the clumping reaction, preparing the offending cells for phagocytosis. While the response itself is normal and typical, the clumping of red cells in our circulatory system may, as we've said, create dangerous blockages in smaller vessels.

The Rh Blood Group Many other blood group loci also have multiple alleles. For example, there is the **Rh** (from *Rhesus,* the monkey in which it was first found) **blood group system**, which has eight fairly common alleles and many rare ones. As in the case of the ABO system, the different alleles can be identified according to how the blood types react to known antibodies in an antiserum. You are probably aware that you're referring to the Rh system when you add the words "positive" or "negative" to your bloodtype. Rh blood groups are classified as Rh positive and Rh negative on the basis of the presence or absence of the most potent Rh antigen.

People don't naturally have anti-Rh antibodies, but transfusing antigen-containing Rh positive blood into an Rh negative person would sensitize that person against the Rh positive factor. The recipient would produce anti-Rh antibodies, but in first exposures the reaction is generally minor. However, should such a transfusion occur again, the next reaction would be rapid and massive, and the effects quite dangerous. Mismatching blood types in transfusions isn't too common, but there is another, more familiar Rh problem—one that confronts mothers.

At one time, Rh negative women could not afford to be kindly disposed toward mates who were Rh positive (Figure 11.10a-d). This is because following a pregnancy in which the baby turns out to be Rh positive, Rh negative women sometimes build up antibodies against the Rh positive antigen. The word "sometimes" should be emphasized because such a reaction requires a leakage of the baby's Rh positive blood across the placenta into the mother's blood, and this seldom happens. But, when it does, Rh negative women can become sensitized against their Rh positive fetuses. This can cause trouble in subsequent pregnancies because the mother's new antibodies can enter the blood of the next fetus shortly before birth and destroy its red blood cells, a condition called erythroblastosis fetalis (Figure 11.10e).

Rh negative women who have just given birth to an Rh positive baby are now routinely given injections of anti-Rh serum, which destroys any fetal red blood cells that may have leaked into the mother's body before her immune system starts building up antibodies against them. This prevents Rh incompatibility problems with any future pregnancies.

GENE INTERACTIONS

You will recall that Mendel found that the F_2 of the round-yellow and wrinkled-green dihybrid cross yielded a 9:3:3:1 phenotypic ratio (see Table 10.1). After Mendel's work was rediscovered, his enthusiastic followers gleefully produced this ratio again and again. Their work is a textbook standard and clearly indicates some of the ways in which genes at different loci can interact to produce different phenotypes.

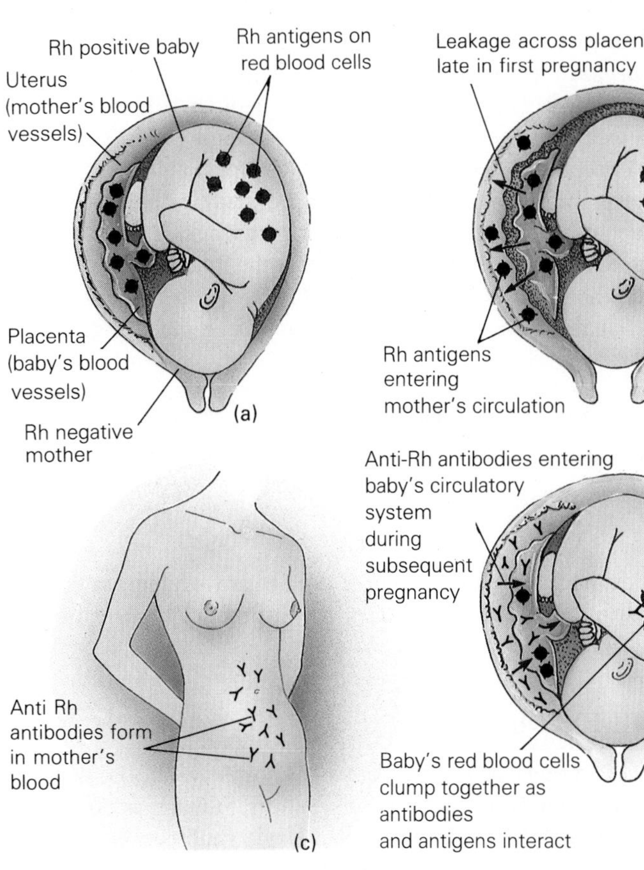

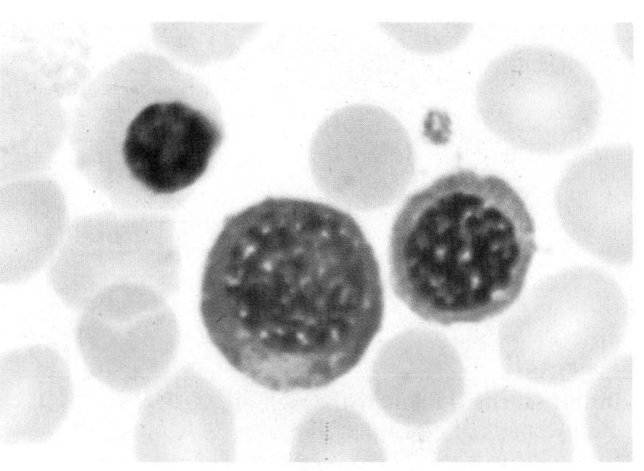

FIGURE 11.10
Rh FACTOR AND INCOMPATIBILITY
Rh incompatibility can arise only when an Rh negative mother gives birth to an Rh positive baby (**a**). During birth (**b**), Rh antigens may leak into the mother's bloodstream, causing her to develop antibodies to the Rh factor (**c**). Then, the mother's antibodies will react with the Rh factors of the *next* Rh positive baby (**d**). If so, the result is erythroblastosis fetalis—destruction of the baby's red blood cells (**e**).

This all worked out very nicely for Mendel and his followers because the round, wrinkled characters do not influence the inheritance of the yellow, green characters, but this is not true of all pairs of gene loci. For instance, it doesn't work for loci that regulate coat color in mice. In this case, we find that at one gene locus, **B** is dominant to **b**, such that **BB** and **Bb** mice are black and **bb** mice are brown. A cross between a homozygous black (**BB**) mouse and a homozygous brown (**bb**) mouse will produce nothing but black heterozygotes (**Bb**). A cross between two heterozygous black mice produces an F_2 generation with three-fourths black mice and one-fourth brown mice. So far so good.

But at another gene locus, **C** is dominant to **c**, such that **CC** and **Cc** mice can make pigment (black or brown) but the **cc** mice cannot, and are thus albinos. The **C** allele, in effect, allows for coat color. Hence, the two gene loci at **B** and **C** control two different steps in the biochemical pathway that produces the pigment normally present in mouse fur. Thus, the **C** locus is said to be epistatic to the **B** locus. **Epistasis** refers to one gene's interfering with the expression of another.

Now consider a mating between a true-breeding white mouse and a true-breeding brown mouse. What would you expect? You might not expect the entire litter to be black. But you shouldn't be too surprised, either, because of the possibility of this cross:

P_1: **CCbb** (brown) × **ccBB** (white)
F_1: **CcBb** (black)

So the black F_1 are heterozygous at two gene loci. The real surprise comes at the next cross, however. If you mate two such double heterozygotes as in Figure 11.11, you will find that in the F_2 generation one-fourth are **cc** (white), regardless of what's happening at the **B** locus. Of the remaining colored mice, three-fourths are black and one-fourth are brown. The phenotypic classes of the F_2 are 9/16 black, 3/16 brown, and 4/16 white (Figure 11.11). This is just the old 9:3:3:1 ratio with the last two terms combined (9:3:4), because once a mouse is white you can't tell whether it might have been brown or black. Perhaps we should put this another way: one cannot distinguish between **BBcc**, **Bbcc**, and **bbcc** without doing a progeny test.

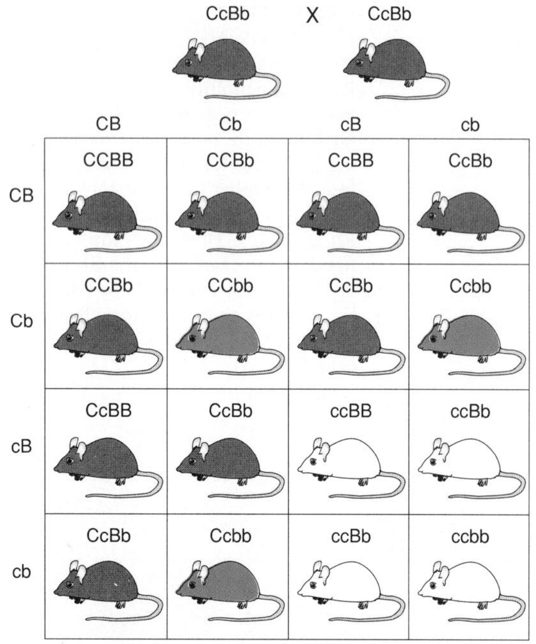

FIGURE 11.11
EPISTASIS
In this scheme we follow the path of the two pairs of genes that influence coat color in mice. The **B** (black) and **b** (brown) alleles occur at one locus, and the **C** (color) and **c** (albino) alleles occur at another locus. The alleles of the two genes assort independently but produce what seem to be strange ratios when two heterozygous mice are inbred. The result is a 9:3:4 ratio of black:brown:white because of the epistatic interaction between the two loci.

Similarly, one can combine two different gene loci that have identical effects. For instance, **pp** mice are also white, while **PP** and **Pp** mice are normally pigmented. These genes control yet another step in the biochemical pathway that produces pigment as its end product. The F_1 offspring of **PPcc** and **ppCC** (both white phenotypes) are **PpCc**, a genotype that gives normally pigmented mice (let's say black). In the F_2 generation, about half are white and about half are black. If we have large enough numbers, we may be able to show that the ratio is not really half black or half white, but 9/16 black to 7/16 white. This is the 9:3:3:1 ratio again, but this time the last three groups are lumped together to give a 9:7.

9/16 **P_C_**	black	9/16 black
3/16 **P_cc**	white	
3/16 **ppC_**	white	7/16 white
1/16 **ppcc**	white	

In other words, once a mouse is white, it can't be any whiter, but as long as it has both a **P** allele and a **C** allele, it will produce a normal black coat.

CONDITIONAL GENE EXPRESSION

We have left the question of conditional gene expression until now so you could develop an understanding of gene action and interaction without having to worry about qualifications. But now it's time to face the realization that a single genotype hardly ever expresses itself in exactly the same way in any two individuals, or for that matter the same way in different tissues of a single individual. We will see, now, that just how a gene is expressed is conditional—it depends on the conditions under which it exists.

Environmental Influences

Environmental influences may be very obvious or very subtle. Let's consider a straightforward example: the Siamese cat. One of the enzymes in its pigmentation pathway is temperature-sensitive; it won't function when it is warm. As a result, pigmentation of the fur occurs primarily in the colder extremities of the cat—the ears, the tail, the feet, and the nose. While these parts are black or dark brown, the rest of the cat is tan or almost white, which is why Siamese cats look like Siamese cats (Figure 11.12a). If you keep your Siamese cat in the warm house, it may grow to be almost white, but if you put it out at night, it will get to be quite dark.

Another strikingly dramatic example of the effect of environmental temperature on phenotype is seen in the *Drosophila* wing mutant called *curly*. Fruit flies raised at temperatures exceeding 16°C lose their ability to fly effectively as the wings curl upwards over the back (Figure 11.12b).

(a)

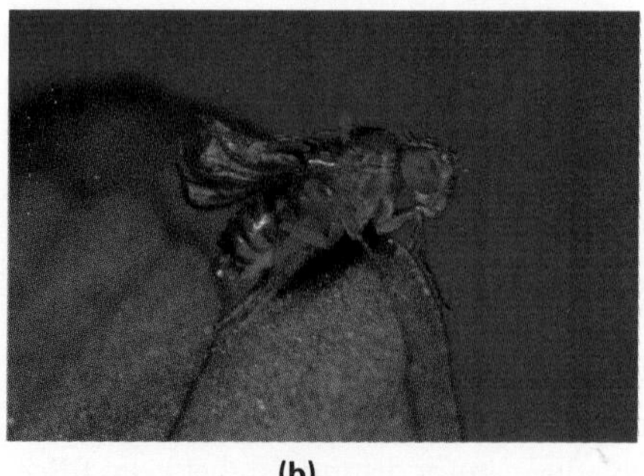
(b)

FIGURE 11.12
CONDITIONAL GENE EXPRESSION IN CATS AND FRUITFLIES
Environmental temperature can have a drastic effect on the expression of coat color genes in the Siamese cat and wing shape genes in *Drosophila melanogaster*. Siamese cats **(a)** have their distinctive color pattern because the enzyme that produces color functions well only in the cooler extremities of the cat. The *curly* mutants of the fruit fly **(b)** suffer a dramatic curling of their wings, but only if they are raised above 16°C.

Incomplete Penetrance and Variable Expressivity

In some cases an individual may have a dominant genotype without showing it, a condition known as **incomplete penetrance.** For instance, the gene that gives one the ability to roll his or her tongue is dominant, yet not all persons with the dominant allele will be able to roll their tongues. But for any one person it is all or nothing: there are no people who can roll their tongue only part way. In fact, in a room full of people with the dominant allele, only about 8/10 will actually be able to roll their tongue. Therefore we say that tongue rolling allele has 80% penetrance.

There is a related situation called **variable expressivity,** where the mutant phenotype will be expressed to varying degrees. For instance, a rare dominant trait in human genetics is polydactyly, the tendency to have extra fingers or toes (Figure 11.13). Persons carrying this dominant allele show variable expressivity in that all four extremities may be affected, or only the one hand or one foot. Both hands and both feet of a given carrier have the same genes and the same environment but may have either normal or abnormal numbers of digits, indicating that some form of developmental chance is at work. Of course, this means that a person carrying the allele may just be lucky enough to have only five toes on each foot and only five fingers on each hand. In such a case, fortune has smiled four times, and the person would be unaware of carrying this allele if it weren't for the fact that unless fortune continued smiling, about half of his or her children would have extra fingers and toes.

Sex-Limited and Sex-Influenced Effects

A dominant gene is known to be responsible for a rare type of cancer of the uterus. Since, needless to say, the gene affects women only, it controls a **sex-limited trait.** A sex-limited trait, then, is a trait that shows up in *only* one sex or the other. **Sex-influenced** traits are another matter. Sex-influenced traits can affect both sexes, but the effect is different. The most common kind of middle-aged male baldness, for example, is caused by a dominant allele that produces only thinning of the hair in women. It doesn't affect eunuchs, either—unless they have been given injections of male sex hormones. (You may have immediately deduced that there *is* a cure for baldness!) Another example is pyloric stenosis, a common malformation of the digestive tract; it runs in families but affects five times as many boy babies as girl babies.

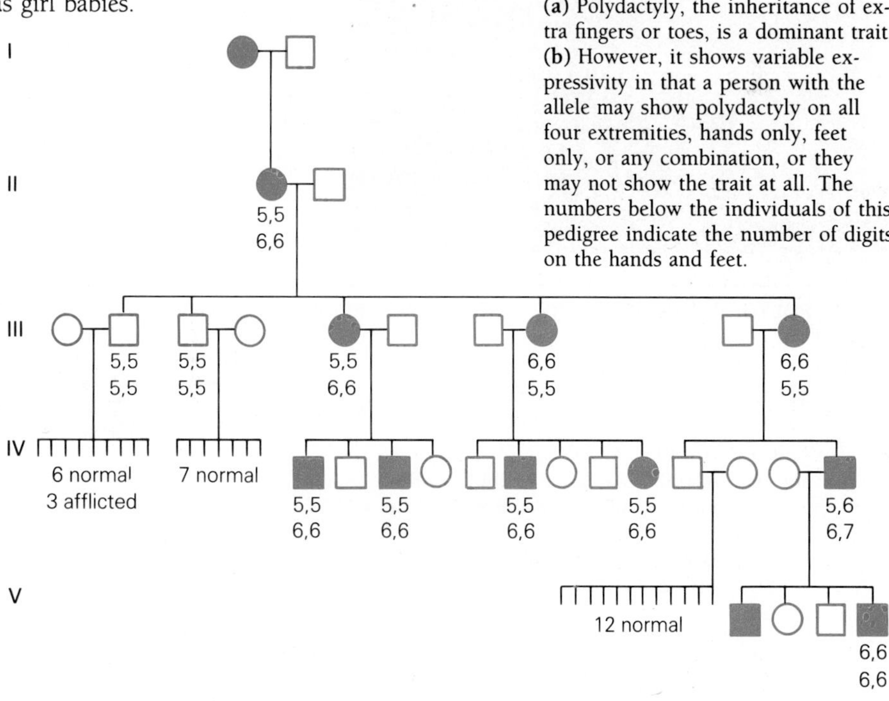

(a)

(b)

FIGURE 11.13
POLYDACTYLY AND VARIABLE EXPRESSIVITY
(a) Polydactyly, the inheritance of extra fingers or toes, is a dominant trait. **(b)** However, it shows variable expressivity in that a person with the allele may show polydactyly on all four extremities, hands only, feet only, or any combination, or they may not show the trait at all. The numbers below the individuals of this pedigree indicate the number of digits on the hands and feet.

Variable Age of Onset

Baldness and muscular dystrophy both have variable ages of onset. Muscular dystrophy can begin at very different ages even in affected brothers, who would have received the same abnormal allele. We have already seen that Huntington's chorea (or Huntington's disease, as it is now known) is lethal after a certain age. However, the exact age of onset is quite variable.

Pleiotropy

Of the thousands of children who show up in hospital emergency rooms with fractured bones, a few become frequent visitors. Some are accident prone, and others may be the victims of physical abuse, but occasionally a sharp-eyed physician, one who paid close attention to his genetics courses, will spot a very rare genetic abnormality called blue sclera-brittle bone disease. Blue sclera refers to a bluish tint in "the white of the eyes," but the brittle bones that accompany the blue sclera are the product of defective calcium metabolism. Both are examples of **pleiotropy** ("many turnings"), the tendency for an allele to be expressed in different ways in different tissues. While this example is rare, pleiotropy is really commonplace.

Perhaps the best example of pleiotropy in humans is a disease of newborns called phenylketonuria, or PKU (see Chapter 12). Because of an enzyme deficiency, PKU victims cannot metabolize the amino acid phenylalanine, so it accumulates in the blood. Its effects are widespread, depending on the tissues. For example, the accumulation causes brain damage leading to mental retardation, and in addition the head fails to grow to normal size. Further, a shortage of pigments normally produced at the end of the phenylalanine pathway can result in light hair and skin color.

CONTINUOUS VARIATION AND POLYGENIC INHERITANCE

Many of the phenotypic traits that are most important to biologists, and especially to plant and animal breeders, do not fit into "either-or" categories (such as round or wrinkled peas). Instead, these traits occur in a gradient, a situation called **continuous variation** (Figure 11.14). Of

FIGURE 11.14
CONTINUOUS VARIATION AND POLYGENIC INHERITANCE
Traits that can be quantitatively measured, such as height in humans, can vary over a considerable and continuous range, as demonstrated by a bell-shaped curve of human height variation. The chart at the bottom presents a model for the inheritance of height in humans, with three loci that each contribute equal increments to the final height.

POLYGENES CONTROLLING HEIGHT: A MODEL OF THREE LOCI

Genotypes	Number of "Tall" Alleles	Number of "Short" Alleles	Height in Centimeters	Distribution
AABBCC	6	0	190	1/64 (1.6%)
AaBBCC, AABbCC, AABBCc	5	1	185	6/64 (9.4%)
aaBBCC, AAbbCC, AABBcc, AaBbCC, AaBBCc, AABbCc	4	2	180	15/64 (23.4%)
aaBbCC, aaBBCc, AabbCC, AaBBcc, AaBbCc, AAbbCc, AABbcc	3	3	175	20/64 (31.3%)
aabbCC, aaBBcc, AAbbcc, AaBbcc, AabbCc, aaBbCc	2	4	170	15/64 (23.4%)
aabbCc, aaBbcc, Aabbcc	1	5	165	6/64 (9.4%)
aabbcc	0	6	160	1/64 (1.6%)

Variation in adult human male height could be explained by supposing three equivalent gene loci, each with two alleles. Each "tall" allele (capital letters) adds five centimeters to height. Many different genotypes may have the same phenotypic effect. The distribution is calculated on the assumption that the "tall" and "short" alleles are equally common in the population. The same distribution would be expected in the F_2 of a hypothetical controlled cross between **AABBCC** and **aabbcc** individuals of, say, corn plants. Although models like this involve many unrealistic assumptions, they nevertheless have proven to have good predictive power.

course, all the alleles for any condition still occur in pairs, and they still segregate and assort according to Mendelian law. The continuous variation exists because alleles at more than one locus are involved. When alleles at more than one locus contribute to the same trait, this is called polygenic inheritance and the trait, a polygenic trait. (How does this differ from multiple alleles and epistasis?) Examples in humans include skin color, foot size, nose length, birth weight, height, and intelligence. Let's look closer at one example.

A great many gene loci determine human height, but to simplify things we'll assume that height is determined by only three loci (three gene pairs at three locations). Also, in reality multiple alleles may be possible at each gene locus, but in our example we'll assume only two alternatives are available: "short" alleles and "tall" alleles. Further, we will assume that the presence of a "tall" allele rather than a "short" allele increases adult height by five centimeters (about 2 inches). People with only the "short" alleles (six in all) grow to be about 160 cm (5′3″), while those with only "tall" alleles (again, six) grow to 190 cm (6′3″). In the middle with three "short" alleles and three "tall" alleles, are the average individuals about 175 cm tall (5′9″).

The chart in Figure 11.14 summarizes the seven height categories possible with this model. We have added the relative frequencies of the seven height categories that would be predicted in the offspring from a large number of heterozygous couples (**AABbCc** × **AaBbCc**). Note that the distribution approximates a "bell-shaped curve" or normal distribution.

LINKAGE

In science, it is often said, the answer to one question gives rise, inevitably, to new questions. And this was certainly the case with early twentieth-century genetics. For example, not long after Sutton and others decided that chromosomes bore the hereditary factors and that many such factors, or genes, were contained in each chromosome, they began to notice that Mendel's principle of independent assortment didn't work out with all genes. The question was, why? They decided that if genes didn't randomly separate, they must be joined together, somehow *linked*. They further decided that the traits were linked together because they are part of the same chromosome, and so these genes moved together as part of a **linkage group**. A linkage group came to be defined as any group of genes that tended to be inherited together because they were on the same chromosome (Figure 11.15).

Such a linkage of genes would, of course, confound the law of independent assortment. And soon enough William Bateson and Reginald Punnett (of Punnett square fame), while trying to confirm Mendel's findings, got some puzzling results that turned out to be due to linkage. They started with two true-breeding strains of sweet peas: one with blue flowers **(BB)** and long pollen grains **(LL)**, the other with red flowers **(bb)** and round pollen grains **(ll)**. The F_1 offspring of this cross had blue flowers and long pollen grains, hence "blue" and "long" were known to be dominants. Their P_1 cross was:

P_1: **BBLL** × **bbll**

F_1: **BbLl**

So far, there were no surprises. Then, following Mendel's now-established procedures, they sought to reconfirm the Law of Independent Assortment by crossing the doubly heterozygous F_1 back to the doubly recessive parental stock, a standard testcross:

BbLl × **bbll**

Mendel's second law predicted that they should get equal numbers of all four possible phenotypes—a 1:1:1:1 ratio of blue-long, blue-round, red-long, and red-round (you may want to confirm this for yourself), but this is not what Bateson and Punnett observed. Their ratio was approximately 7:1:1:7 for the four phenotypes. Their expectations and results are as follows:

FIGURE 11.15
LINKED GENES
If the allele pairs for flower color and pollen shape are on the same chromosome, they cannot assort independently. With this being the case, the phenotypic ratio of the offspring of this testcross should be 1:1, blue-long:red-round. In reality, few genes are linked so very tightly.

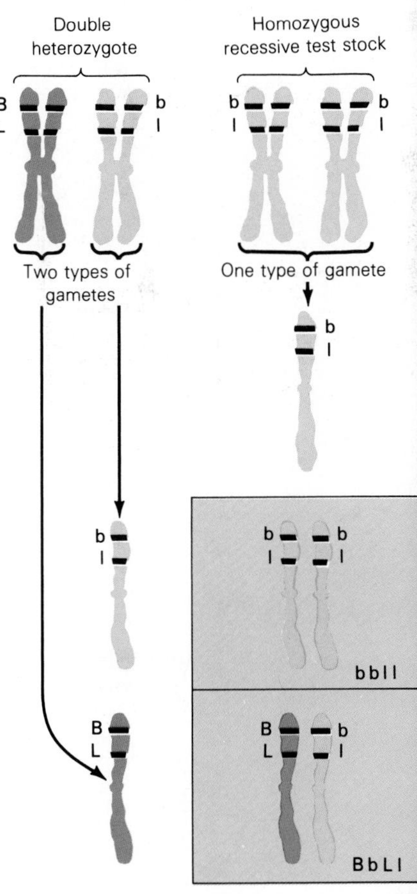

PHENOTYPE	GENOTYPES	EXPECTED	OBSERVED
Blue, Long	BbLl	25.0%	43.7%
Blue, round	Bbll	25.0%	6.3%
red, Long	bbLl	25.0%	6.3%
red, round	bbll	25.0%	43.7%

As you can see, there is significant disagreement with the outcome predicted by Mendel's second law. Perhaps, then, gene linkage could explain the results. But if the genes were linked, with the alleles for blue and long on one chromosome, and the alleles for red and round on its homologue, the results should have been 50% blue-long, 50% red-round (Figure 11.15). Blue flowers would always appear with long pollen grains and red flowers with round pollen grains. Bateson and Punnett's results were inconsistent with either model. They were never able to figure out what was going on. So what was going on? Bateson and Punnett were dealing with linked genes all right, but in a small proportion of meiotic events, the genes had become "unlinked" and represented *genetic recombination*. Crossing over had occurred (Figure 11.16). It's small wonder that Bateson and Punnett gave up in disgust. However, a new generation of geneticists were soon to unravel the mystery. We will leave the history for a time and review some things we already know about crossing over from Chapter 9.

FIGURE 11.16
GENETIC RECOMBINATION
Continuing from Figure 11.15, we look at the same testcross, but with crossing over occurring. Note that with crossing over the types of gametes double. The result is two new classes of offspring, but not necessarily or even generally in a frequency equal to the expected phenotypes.

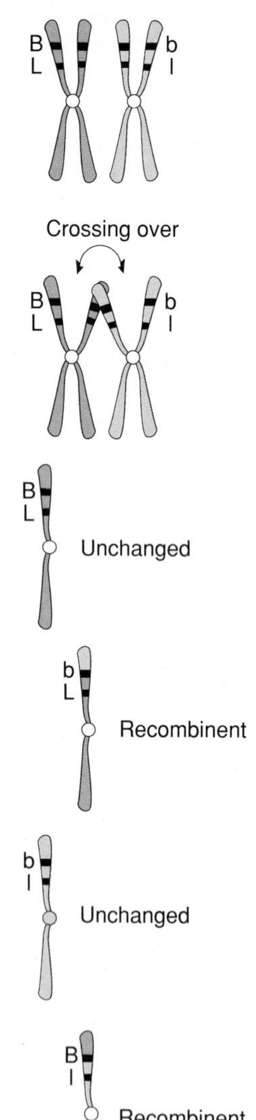

Linkage, Crossing Over, and Genetic Recombination

Crossing over brings about the recombination of alleles of two different genes that are physically parts of homologous chromosomes. This, you will recall, requires the breaking of chemical bonds in the DNA and actual exchange of chromatid regions (see Chapter 9). The early geneticists managed to work out the concept of crossing over using only results from their crosses, testcrosses and progeny counts. Part of the confusion at first was that some pairs of genes showed deviations from the law of independent assortment, but each of these pairs seemed to have its own degree of deviation. Two genes that tended to assort together in a testcross (such as **B** and **L**, and **b** and **l**) were called "linked," but, strangely enough, some gene pairs showed stronger tendencies to "link" than others. In the Bateson-Punnett example, 12.6% of the testcross progeny were recombinants, that is they showed a different linkage of alleles than did the parents. But in other crosses recombinant progeny might represent only a fraction of a percent. Gene pairs that had very low percentages of recombination came to be known as "tightly linked genes;" those with higher percentages, "loosely linked." These testcross percentages had none of the appeal of Mendel's wonderfully precise ratios. Figure 11.17 presents a crossing-over model that explains the Bateson-Punnett results (note that their numbers have been rounded off).

Crossing over, then, allows the alleles of linked gene loci to occasionally change their linkage relationship with each other. *Crossing over is the physical, cytological phenomenon that is responsible for the genetic phenomenon of recombination.* By the way, crossing over can occur without recombination (if the genes of interest on the homologous chromosomes are homozygous, genetically you cannot tell if there is recombination), but recombination cannot occur without crossing over.

Crossing over is a fairly common event. In fact, crossing over is so frequent that genes located far apart on the same long chromosome may seem to obey Mendel's second law. Although they are physically part of the same molecule at the beginning of meiosis, they are so far apart that the probability of the genes ending up in the same gamete is just about the same as if they had been on separate chromosomes to begin with, 50%.

The situation is different with genes that lie close together on a chromosome, since the probability of a crossover event happening between them is correspondingly small. They tend to be shunted around as a unit during meiosis and thus tend to be inherited as a linkage group in testcrosses. For example, let's say that alleles of two genes, **A** and **B** are so close together on a chromosome that there is only a 1.0% chance that they will be separated by crossing over and a 99% chance that they won't. That is, there is a 1% chance of recombination in the region between the two loci.

As an exercise, start with a double heterozygote, with one chromosome carrying **AB** and with its homologue carrying **ab**. You can see that if there were no crossing over at all, half the gametes would carry the **AB** chromosome and half would carry the **ab** chromosome; none would carry **aB** or **Ab**. But in reality, only 99% of the chromosomes are **nonrecombinants**, retaining the linkage arrangement of the parents. Of the **recombinant** chromosomes, half will be **aB** and half will be **Ab**. Thus, we can expect the following kind of gametes:

Nonrecombinants, 99.0%: 49.5% **AB**, 49.5% **ab**

Recombinants, 1.0%: 0.5% **Ab**, 0.5% **aB**

In a testcross, all of the gametes from the double homozygote test stock parent would be **ab**, so the distribution of the testcross progeny should be the same as the distribution of the gametes from the heterozygous parent:

49.5% **AaBb**
49.5% **aabb**
Nonrecombinants = 99.0%

0.5% **Aabb**
0.5% **aaBb**
Recombinants = 1.0%

Therefore, the total recombination between the **A** locus and the **B** locus is 1.0%. In general, let **R** represent the **frequency of recombination** between two gene loci. Then, from the progeny of the testcross **AaBb** × **aabb**, **R** is the number of recombinant offspring (**Aabb** and **aaBb**) divided by the total number of offspring. The answer is .01, which we usually will convert to 1%.

R is related to the probability that crossing over will occur, but it is not exactly equal to this probability. Also, the farther apart two genes are, the more **R** becomes an underestimate of the distance. This is because of double crossing over. If two crossing-over events occur between **A** and **B**, the first would separate **A** and **B**, and the second would put them back together again. The result would be no genetic recombination. Thus, **R** is equal to the probability of an *odd number* of crossing-over events.

Crossing over is an interesting phenomenon in its own right, but it turned out that continued interest in the process led to a much broader understanding of the chromosome.

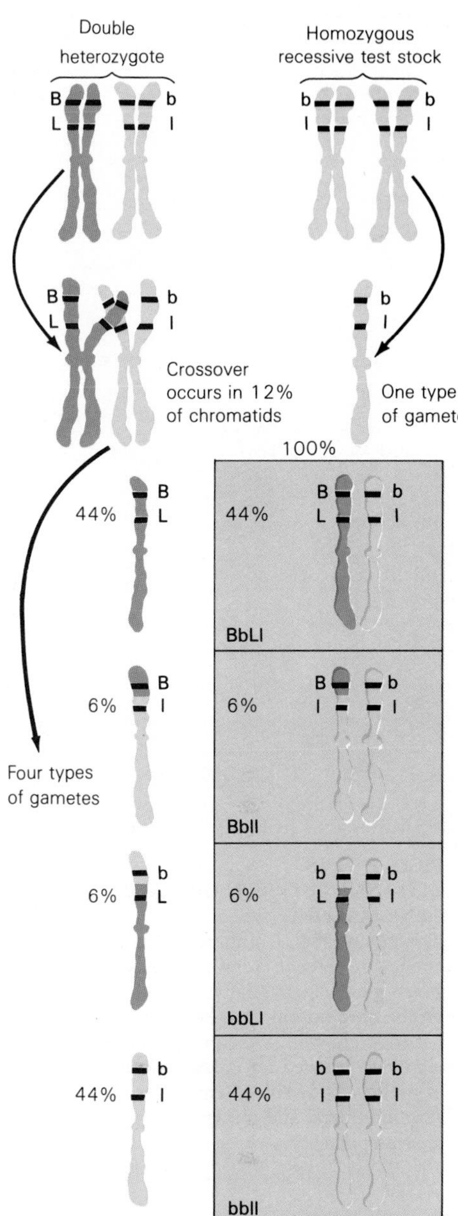

FIGURE 11.17
12% CROSSING OVER
In this example of crossing over, the frequency of occurrence is 12%, which means the noncrossing over frequency is 88%. We still end up with the four types of gametes that we saw in Figure 11.16, but now we must consider the percentage at which each gamete is produced. The end result is near a 7:7:1:1 ratio, which is a far cry from the 1:1:1:1 that would be expected of independent assortment, or the 1:1 ratio of very tight linkage.

Mapping Genes

The laboratory of Thomas Morgan was responsible for bringing the fruit fly, *Drosophila melanogaster,* to its pedestal as the premier eukaryotic organism for genetic study. Morgan and his associates were quite successful in their ongoing search for new *Drosophila* mutants. As these traits were uncovered, crosses were carried out in a variety of ways, often involving pairs of genes. In some, the results conformed nicely to Mendel's law of independent assortment, while in others, gene linkage was clearly established, though for some gene pairs the recombination frequency was low (they were tightly linked), while in other cases the frequency was high.

It didn't take long for Morgan's group to collect data on recombination percentages for a fair number of genes. Upon inspection, they found that each pair of linked genes had its own characteristic recombination fraction; no two linked gene pairs gave the same results. Morgan himself was the first to propose an explanation. Intuitively, he reasoned that the rate of crossing over, the recombination frequency or percentage should be proportional to the distance separating two genes on a chromosome. Then Alfred Sturtevant, a gifted young undergraduate at the time, made a highly logical extension to Morgan's assertion. In Sturtevant's own words, "It would seem . . . that the proportion of crossovers could be used as an index of the distance between any two factors. Then, by determining the distances between A and B, and B and C, one should be able to predict AC. For if proportion of crossovers really represents distance, AC must be approximately either AB plus BC, or AB minus BC."

With this idea in mind, we can follow the lead in Figure 11.18 and construct a map of several genes in *Drosophila*. To begin, we see that the mutant alleles for yellow body and white eyes have a very low recombination (1%), and so they must be very close together. On the other hand, the recombination between the yellow-body and vermilion-eye alleles is 32.2%, and the recombination between white eye and vermilion eye 30.0%. (Note that especially when mapping, recombination percentages are called **map units.**) The three recombinations fit together reasonably well if we begin mapping with vermilion, placing white 30 units away, and yellow 1 unit further still.

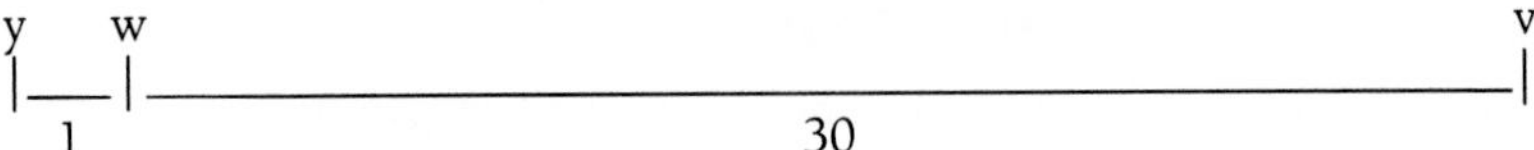

(To see how this works, ask yourself why yellow is placed 1 map unit to the left of white, instead of one unit to the right.)

It follows that miniature wing must be three units to the other side of vermilion—its position is apparently farther from white and yellow than is vermilion's (by 3 map units). So with the addition of the fourth gene, the map becomes:

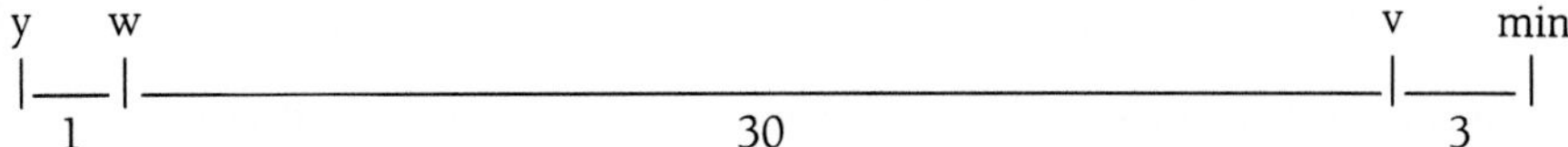

You'll notice that the individual recombination fractions are rather close to additive, as they should be. Geneticists since Morgan and Sturtevant have used this basic method to determine the relative positions of thousands of genes on the chromosomes of *Drosophila*.

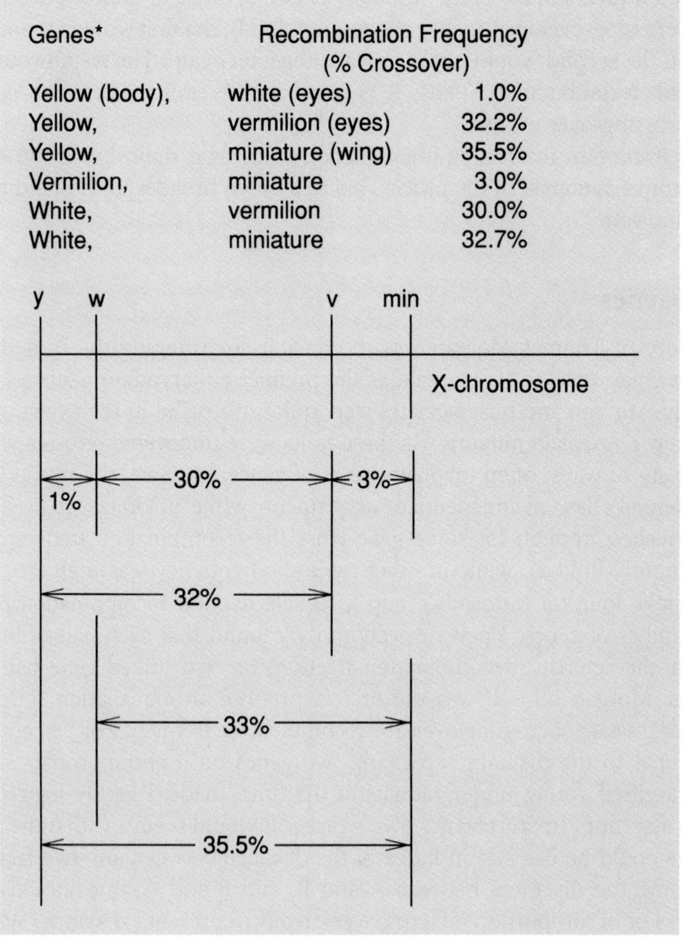

Genes*		Recombination Frequency (% Crossover)
Yellow (body),	white (eyes)	1.0%
Yellow,	vermilion (eyes)	32.2%
Yellow,	miniature (wing)	35.5%
Vermilion,	miniature	3.0%
White,	vermilion	30.0%
White,	miniature	32.7%

FIGURE 11.18
DISTANCE BETWEEN GENES AND GENE MAPPING
By studying the recombination frequency or percent crossing over between genes, their relative locations on the chromosome can be determined. Here we list the percent crossing over observed for crosses involving yellow body color, white eyes, vermilion eyes, and miniature wings in *Drosophila*.

As we see, then, a genetic map yields information regarding not only the order in which gene loci occur on the chromosome, but the distances between the loci as well (Figure 11.19). Mapping enables the discoverer of a new mutant allele to determine whether the discovery is a variant of a known gene locus. And it makes it possible to determine which genes belong to which chromosomes.

FIGURE 11.19 CHROMOSOME MAPS Maps of a *Drosophila* chromosome can take several forms, two of which are shown here. First **(a)**, the positions of the loci can be arranged as a genetic map, where the positions of the loci are listed and positioned according to their recombination relative to each other, and numbered from one end of the chromosome. Second **(b)**, the positions of the genes can be described on the cytological map, a map of the visible bands on the polytene chromosome.

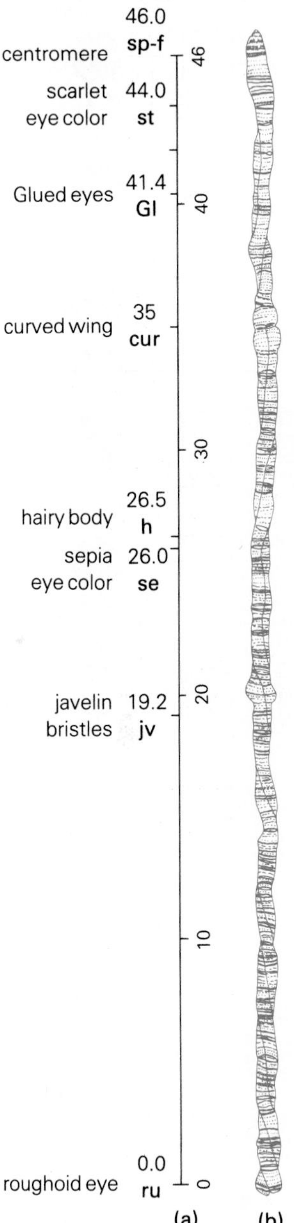

Chromosomes and Sex

When Morgan first began his search for variation in the traits of *Drosophila,* all the flies looked alike, except that males were visibly different from females. However, as Morgan carefully scrutinized each new generation, he eventually turned up one variant. Among a group of flies with normal brick-red eyes, Morgan found a single male with white eyes. He carefully nurtured his little white-eyed specimen and crossed it with several of its red-eyed virgin sisters. (In the laboratory, a fruit fly will mate with any other fruit fly of the opposite sex, if given no other choice. But female fruit flies, if given a choice, prefer to mate with strangers rather than their own brothers, although they will accept their brothers if no other males are around. Male fruit flies don't seem to care one way or another and will even attempt to mate with each other.) From these matings, all the F_1 were red-eyed, to the surprise of none of the "new Mendelians." Obviously, white eyes was a recessive trait.

P_1: white ♂ × red ♀

F_1: all red, ♂ and ♀ alike

The experiments continued, and when the F_1 flies were mated with one another to produce an F_2 generation, sure enough, about one-fourth of the F_2 flies were white-eyed and about three-fourths were red-eyed (Figure 11.20). The actual numbers were not as close to this expected ratio as Morgan had hoped because, as it turned out, the white-eyed flies have a somewhat lower rate of survival than the normal flies. But there was something more peculiar about the F_2 flies. Every single white-eyed fly was a male! In fact, the F_2 ratio approximated;

1/4 red-eyed males: 1/2 red-eyed females: 1/4 white-eyed males

At this point, you may have decided that only males can be white-eyed. You would be wrong. Morgan discovered this when he first did a testcross, mating his original, now-geriatric, white-eyed male to its own red-eyed F_1 daughters. A simple testcross should have provided a 1:1 ratio of dominant to recessive and this one did. In fact, the testcross offspring consisted of approximately equal numbers of red-eyed males, red-eyed females, white-eyed males, and white-eyed females. So females could have white eyes. But even more surprises were in store. When Morgan mated white-eyed females to red-eyed F_1 males, in what is called a reciprocal testcross, again, half of the offspring were red-eyed and half were white-eyed. But now every single one of the males was white-eyed, and every female had red eyes.

Morgan was aware of Sutton's suggestion that a single chromosome may carry a number of hereditary factors. Morgan surmised that the sex-determining factor and the eye-color factor are linked together, since these traits did not follow the law of independent assortment in the F_2 or in the testcrosses. Hence, he reasoned that, as the X chromosome segregates at anaphase, so do the genes on it, including the red-eye or white-eye alleles.

Sex Chromosome in *Drosophila* By this time, Morgan knew something about the chromosomes that determine gender. He knew that male and female *Drosophila* were different with regard to one of their four chromosome pairs, and he guessed that this chromosome difference was the cause and not the result of sex differences. The X chromosome had been named and described earlier by H. Henking, who couldn't figure out what his finding meant (hence the letter "X," for unknown). Female flies have two X chromosomes, while the males have one X and one Y (the same as humans). At

FIGURE 11.20
WHITE EYES AND SEX LINKAGE
In his first cross, Morgan mated his newly discovered white-eyed male with a normal, red-eyed female (**a**). All the F_1 offspring were red-eyed. Inbreeding the F_1 (**b**) produced an F_2 that was three-fourths red-eyed and one-fourth white. But all the white-eyed flies were males. A testcross between the original white-eyed male and a red-eyed female from the F_1 (**c**) showed that the female was heterozygous. Another testcross was done (**d**), with a red-eyed male and a white-eyed female. All the male offspring were white-eyed.

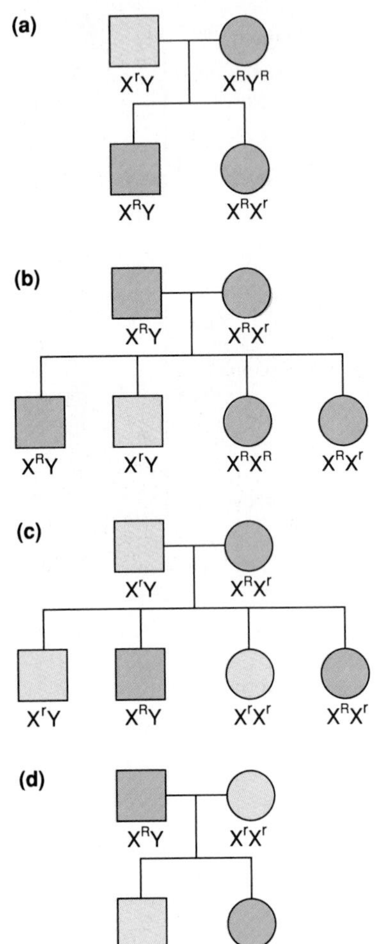

metaphase in *Drosophila,* the X appears as a long, rod-shaped chromosome and the Y, as a shorter, J-shaped chromosome (Figure 11.21). (The human X and Y chromosomes are shown in Essay 9.2.)

We have been saying that you have two of every gene—one from your father and one from your mother. But now we see that this is only partly true. While females are truly diploid, males are only partly diploid. Actually, at about 10% of their gene loci, men have one gene from their mother—period. Although the X and Y chromosomes behave like homologues in meiosis, lining up together, they do not carry the same genes. While the X chromosome has many genes—those determining growth patterns, enzymes, and so on—the little Y chromosome bears almost nothing other than a few genes relating to male sexual development. In *Drosophila,* the Y chromosome carries about six genes, all having to do with male fertility.

A gene on an X chromosome has no homologous gene on the Y chromosome to interact with or yield to. Thus, the Y chromosome behaves as if it has a recessive allele for virtually all the X chromosome loci. However, for a recessive allele to be expressed in a female, it must be present on both of her two X chromosomes. It follows that since a male has only one copy of any X-linked gene, it will express itself, whether it is recessive or dominant in females. For X-linked genes in males, the term **hemizygous** (*hemi-,* "half") is used instead of homozygous or heterozygous.

If you now consult Figure 11.20 for a fresh look at Morgan's crosses, you'll probably have an easier time following the hereditary patterns produced by sex-linked genes. First note that the F_1 offspring clearly established that red eye color is dominant over white. Next, if the eye color gene is carried on the X-chromosome, Morgan's original female must have been $\mathbf{X^RX^R}$, and his white-eyed mutant male would have been $\mathbf{X^rY}$. (Note the use of the superscript on the X to denote alleles located on the X chromosome, and remember that females must be XX, males XY). It follows that each F_1 female would have received an $\mathbf{X^R}$ from her mother and an $\mathbf{X^r}$ from her father, so the F_1 females were all red-eyed heterozygotes. Each F_1 male would have received a Y chromosome from his father and $\mathbf{X^R}$ from his mother, so all F_1 males were red-eyed ($\mathbf{X^RY}$). In the F_2, however, each male would have received a Y from his father and *either* an $\mathbf{X^R}$ *or* $\mathbf{X^r}$ from his mother, the chances of the latter being 50:50. Thus, half the F_2 males were red-eyed, and half were white. Finally, the two test crosses show us that females do express sex-linked traits, but *only if each parent carries the allele in question.* Interestingly, as sex-linked traits pass down through generations they seem to show up in every other one, following a "crisscross" pattern. That is, the traits are most commonly hidden in females, expressed in half their sons, hidden again in the granddaughters, only to be expressed once more in half the great-grandsons. The primary rules regarding these X-linked traits are *that males expressing an X-linked recessive trait got the allele from their mother,* and *daughters expressing an X-linked recessive trait have a father with that trait.*

Human Sex Chromosomes Although normal females are XX and normal males are XY in both *Drosophila* and humans, the physiological mechanisms determining sex is somewhat different in the two species. In *Drosophila,* sex is determined by the number of Xs: two Xs, female; one X, male. Abnormal XXY flies are fully functional females, and XO flies (one X, no Y) are sterile males that look normal. In humans and other mammals, the presence of a Y determines the development of testes, which in turn determines male development.

At about the sixth week of development, a gene on the human Y chromosome becomes active. It is the testis determination factor (TDF gene), and its product stimulates testis formation. The testes subsequently secrete male sex hormone, which prompts further

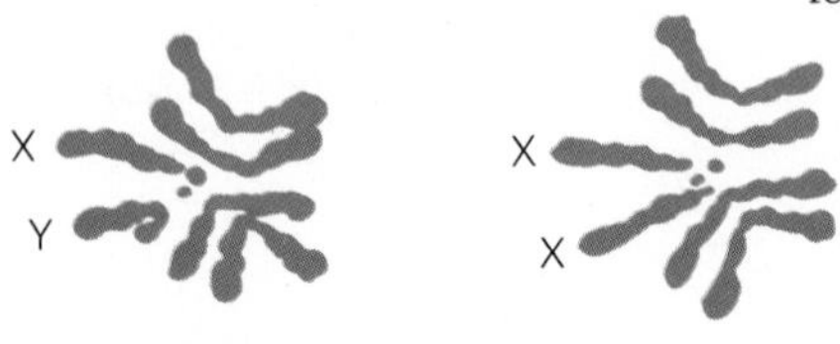

FIGURE 11.21
DROSOPHILA CHROMOSOMES
Drosophila has four pairs of chromosomes. In males only three of the four pairs are homologous. The fourth pair consists of an X chromosome (which is identical to that of the female) and a J-shaped Y (which is not). The Y contains only the genes for male fertility.

male development. Humans having only an X chromosome are designated XO, and are sterile, abnormal females, whereas XXY humans are sterile, abnormal males. You will recall from Essay 9.3 that such abnormalities are the result of nondisjunction during meiosis, although there we were considering autosomal (non-sex chromosome) abnormalities. For more on X and Y chromosome abnormalities see Essay 11.1.

X Chromosomes and the Lyon Effect The "saliva test" is sometimes used in women's athletic competitions as a test of whether a competitor is in fact a woman. Actually, the test has nothing to do with saliva and a great deal to do with X chromosomes. A swab of the inner surface of a person's cheek will pick up a few cells from the mucous membrane lining. When these cells are stained, the cell nucleus in females shows a dark-staining body, the Barr body, which is absent in male cells. The Barr body is actually a condensed X chromosome. The condensed X chromosome can also be viewed microscopically in certain white blood cells, where it forms a characteristic projection from the cell nucleus called a drumstick (Figure 11.22). But what is this about condensed X chromosomes and females?

Since every female has twice as many X-linked genes as are present in male cells, we can presume that there would be a physiological imbalance if all these genes functioned twice as much in one sex as in the other. In any case, evolution has solved the problem by permanently inactivating one of the two X chromosomes in each XX cell during embryonic development; each cell, that is, except for the germ-line cells. Which of the two X chromosomes is inactivated—the one from the father or the one from the mother—seems to be a matter of chance. Every tissue in the adult female is a mosaic of cell lines in which one or the other X has been inactivated. The highly condensed X is replicated normally and is passed from cell to daughter cell, but it is inactive with regard to genetic function. The peculiar behavior of the X chromosome was first established by an English geneticist, Mary Lyon, and is therefore known as the **Lyon effect.**

Incidentally, the Lyon effect probably accounts for the fact that extra numbers of sex chromosomes do not have the disastrous effects that autosome imbalances have. XXX females, for instance, have two Barr bodies and two drumsticks in the cheek and blood cells, but are otherwise normal. As you will see in Essay 11.1, humans with sex-chromosome number abnormalities survive rather well, although they are usually sterile.

FIGURE 11.22
THE CONDENSED X
One of the X chromosomes in each human female somatic cell is represented by permanently condensed heterochromatin. The condensed X chromosome can appear as a simple blob at the side of the nucleus (Barr Body) **(a)**, or in the form of a drumstick **(b)**.

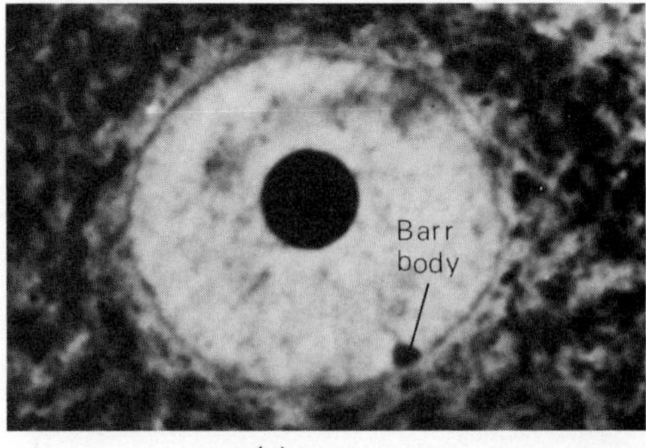

(a)

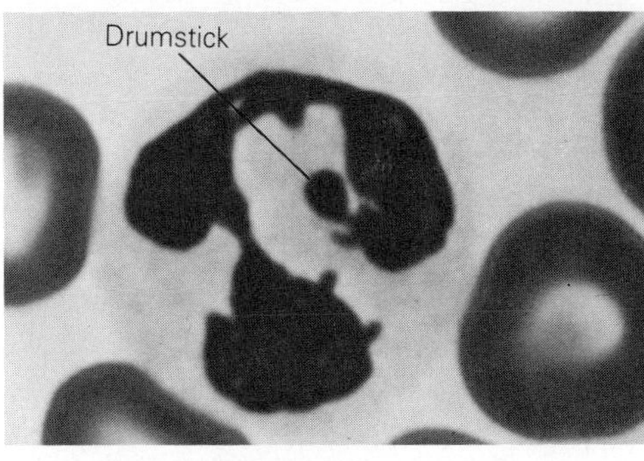

(b)

Colorblindness Colorblindness in humans is usually caused by a recessive allele at either of two closely linked gene loci on the X chromosome. Human color vision depends on the differential sensitivity of three groups of receptors in the retina, called cones. One group of cones is maximally sensitive to blue light, one to red light, and one to green light. Perception of other colors and of subtle hues depends on the relative stimulation of these three types of cones. Persons homozygous (or hemizygous) for a recessive allele at one of the two X-linked loci lack the cones that are more sensitive to green, while homozygotes for recessive alleles at the other X-linked locus lack the cones that are maximally sensitive to red. By the way, the locus controlling the blue-sensitive group is autosomal, and blue-insensitive colorblindness is very rare. Both X-linked defects are called red-green colorblindness (Figure 11.23) because both types of colorblind people have trouble distinguishing many shades of red and green. The defect was first described in a little boy who couldn't learn how to pick ripe cherries. He always brought home a mix of red and green fruit.

Somebody in the British Navy was aware of this situation well over a century ago and set about developing "running lights" on the boats that even colorblind men could tell apart. The green (starboard) side had a touch of blue, and the red (port) side had a touch of orange. When traffic lights were introduced on railroads and city streets, these readily recognizable hues were the logical choice, and we now see them daily.

ESSAY 11.1

SEX CHROMOSOME ABNORMALITIES

Turner's syndrome

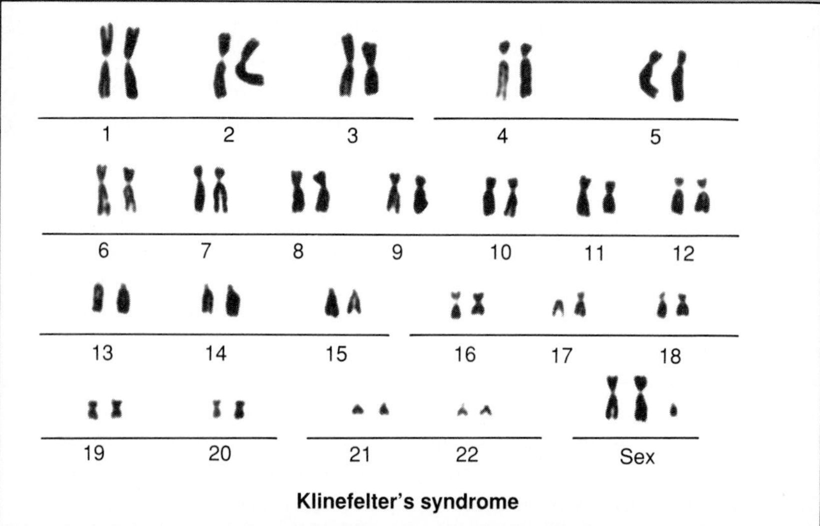

Klinefelter's syndrome

As with trisomy-21, or Down's syndrome, abnormal numbers of sex chromosomes are brought about by nondisjunction—the failure of chromosomes to assort properly during meiosis. A leading factor causing this problem is believed to be aging in the oocytes. While most nondisjunctions in autosomal chromosomes are fatal to the embryo, wrong numbers of sex chromosomes in humans result in live babies and abnormal adults. There are many varieties of sex chromosome conditions. The shorthand designations of the normal individuals and the most common abnormalities are:

- XX Normal female (two X chromosomes)
- XY Normal male (one X, one Y)
- XO Turner's syndrome female (one X, no homologue)
- XXY Klinefelter's syndrome male (two Xs, one Y)
- XYY Extra Y, or XYY syndrome male
- XXX Trisomy-X, or XXX female

In addition to the above, there are many more extreme situations, such as XXYY, XXXY, XXXYY, XXXX, XXXXX, XXXXYY, and so on, each syndrome having its own distinguishing characteristics. However, we can make four generalizations: first, one must have at least one X chromosome to live. Second, the presence of a Y causes the individual to develop as a male, and the absence of a Y causes the individual to develop as a female. Third (and this is probably why these syndromes are not fatal), all but one of the X chromosomes will condense into heterochromatin and be visible as a Barr body when stained, so that XO females lack a Barr body, XXY males have one, XXX females have two, XXXXXYY males have four, and so on. Fourth, the more sex chromosomes a person has, the taller he or she will be, so that XO females are tiny, while on the average XXX, XXY, and XYY individuals are usually much taller than chromosomally normal men and women, and XXYY men are huge.

XO (Turner's) individuals are phenotypically female but do not develop ovaries. They remain sexually immature as adults unless given hormones. XXY (Klinefelter's) males are tall and have small, imperfect testes and low levels of male hormones. They may have femalelike breast development and somewhat feminine body contours. XXX females are tall and frequently sterile but otherwise appear normal. XYY males appear normal except for their extreme height and for a tendency toward severe acne. They are also generally sterile. On the average, they have somewhat reduced IQs and, in common with other low-IQ groups, they average significantly increased criminal arrest records. At one time there was speculation that XYY males had "genetic criminal tendencies," but other analysis suggested that an XYY male is no more likely to be arrested than an XY or XXY male of the same IQ.

About 8% of American men have one form or another of X-linked color blindness—about 6% are hemizygous for a recessive allele at the locus coding for green-sensitive cones, and about 2% are hemizygous for a recessive allele at the locus coding for red-sensitive cones. Women are affected far less often—only about 0.4% of American women are red-green color blind. The reason for the sex difference is that, to be affected, a man need only receive one recessive allele from his mother. But an affected woman must receive recessive alleles from both her mother and her father. (Recall that every daughter expressing a sex-linked recessive trait has a father with that trait.) The chance that both parents will carry the rare allele is much smaller than the chance that just one of them will. Recall our probability discussion in Chapter 10. "The likelihood of two independent events occurring simultaneously is equal to the product of their individual occurrences."

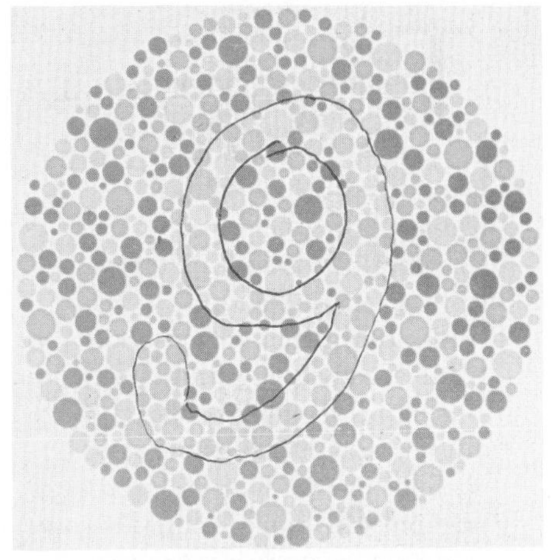

FIGURE 11.23
A COLORBLINDNESS TEST
Color vision is tested using colored plates such as the one shown here. Actually, several plates are required for the complete test. If you are having trouble seeing the number 9 (in red and orange) you may want to take the complete test.

A woman who is heterozygous for color blindness shows no symptoms of the condition. Among her children, however, she can expect half her sons to be colorblind and half her daughters to be carriers like herself—assuming that she marries a man with normal vision. What could she expect if she marries a colorblind man? Figure 11.24 shows the appearance of colorblindness in one family.

Other Sex-Linked Conditions Many sex-linked genetic conditions have had great impact on people's lives. Two of the most common, but also dramatic, of all genetic disorders are hemophilia, the bleeder's disease—in which the blood doesn't clot normally—and muscular dystrophy, in which muscle tissue breaks down in late childhood. The usual forms of both of these are sex-linked. There are actually two common forms of X-linked hemophilia, governed by different X-linked loci. Most hemophiliac males formerly bled to death in their youth. But in recent years, modern medicine has allowed affected hemophiliacs to survive and reproduce, thanks to blood transfusions and to infusions of a blood-derived substance known as antihemophilic factor, which supplies the critical substance missing in hemophiliacs. They now, however, must contend with a risk of contracting AIDS (Chapter 41). So we find adult hemophiliac males, a new situation in human history. In fact, even hemophiliac females occur occasionally (in spite of the fact that homozygous females must receive a recessive allele from each parent). Hemophilia has had interesting implications in European history, as we find in Essay 11.2.

There is, as yet, no effective treatment for muscular dystrophy, so boys with this genetic disease have a fairly short life expectancy. Muscular dystrophy has a delayed age of onset, so that hemizygous boys appear perfectly normal as infants and toddlers, only to begin wasting away some time during their elementary school years. Their heterozygous mothers appear to be free of disease symptoms. Interestingly, however, microscopic tissue samples of females heterozygous for muscular dystrophy show that clusters of muscle cells accounting for about half the total number of muscle cells have atrophied by adulthood. We assume that the affected muscle cells are simply demonstrating the Lyon effect, and that the normal X happened to become condensed. The women do not weaken, because the remaining muscle cells expand in response to the increased load they must carry—another example of how dominance can work.

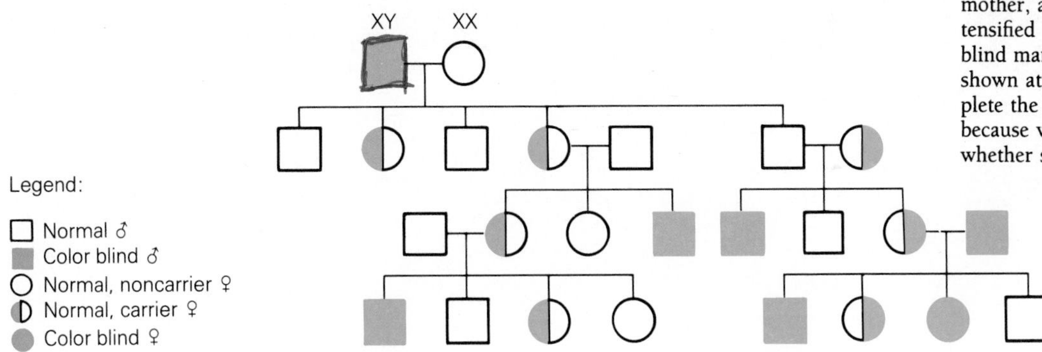

FIGURE 11.24
COLORBLINDNESS IN A FAMILY
In this hypothetical family tree, colorblindness can be traced back to the great-grandfather and great-grandmother, although the problem was intensified by the marriage of a colorblind man and a carrier woman, as shown at the right. We did not complete the great-grandmother's circle because we want you to decide whether she was a carrier.

ESSAY 11-2

THE DISEASE OF ROYALTY

Because it was the practice of ruling monarchs to consolidate their empires through marriage alliances, a highly restricted "royal mating population" was created, and hemophilia was transmitted throughout the royal families of Europe. Hemophilia is a sex-linked recessive condition in which the blood does not clot properly, so that any small injury can result in severe bleeding and, if the bleeding cannot be stopped, in death. Hence, it has sometimes been called the *bleeder's disease*.

The hemophilia of European royalty has been traced back as far as Queen Victoria, who was born in 1819. One of her sons, Leopold, Duke of Albany, died of the disease at the age of 31. Apparently, at least two of Victoria's daughters were carriers, since several of their descendants were hemophilic.

Hemophilia also played an important historical role in Russia during the reign of Nikolas II, the last czar. Alexis, the only son of Nikolas II, was hemophilic, and his mother, the czarina, was convinced that the only one who could save her son's life was the monk Rasputin, known as the "mad monk." Through this hold over the reigning family, Rasputin became the real power behind the disintegrating throne.

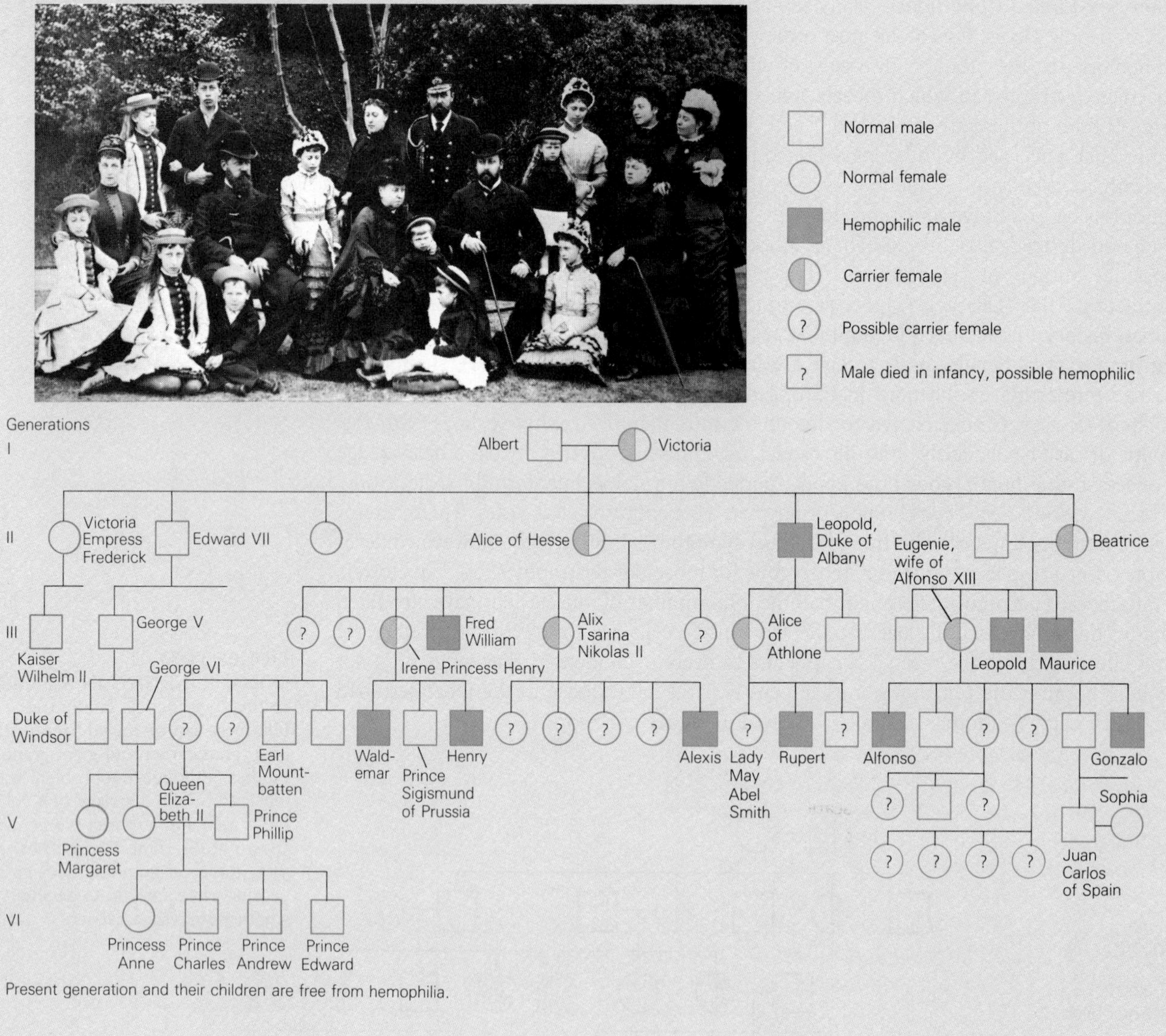

Present generation and their children are free from hemophilia.

Sex-Linked Dominants A few sex-linked mutant alleles are dominant. Dominant brown spotting of the teeth, for instance, is passed from affected men to all their daughters and none of their sons; affected women pass the condition on to half their daughters and half their sons. (Work that one out.) Dominant brown spotting affects about twice as many women as men, since they have two chances of receiving the X chromosome with the dominant brown-spotting allele. Another curious X-linked genetic disease is known as the oral facial-digital syndrome, which involves irregularities of the mouth, face, fingers, and toes. This condition affects only women, who get it only from their mothers. Such women pass the condition on to half their daughters but to none of their sons. It turns out that as a group, affected women have twice as many daughters as sons. We can therefore surmise that the allele in the hemizygous state is lethal.

This chapter has covered a lot of ground, and, it must be admitted, genetics is not the easiest of topics to deal with. Just be sure to keep track of the underlying themes: Mendel's laws. They are the basis for all of these phenomena. In Chapter 10 we covered the law of segregation and the law of independent assortment in their pure form. In this chapter we covered the ways in which those laws seem to be violated or altered. But what we in fact found is that those laws hold up perfectly well, and we merely have to be careful with our interpretation.

APPLICATION OF IDEAS

1. In one breed of chickens, *pea* comb **(PP, Pp)** is dominant to *single* comb **(pp)**. In another breed of chickens, *rose* comb **(RR, Rr)** is dominant to *single* comb **(rr)**. When a homozygous *pea* comb chicken **(PPrr)** is mated with a homozygous *rose* comb chicken **(ppRR)**, the F_1 progeny all have yet a fourth phenotype, *walnut* comb **(P_R_)**. What are the expected phenotypic ratios among the F_2 progeny of two F_1 *walnut* comb **(PpRr)** chickens? In the progeny of an F_1 *walnut* comb chicken **(PpRr)** cross to a *single* comb chicken **(pprr)?**
2. The Japanese geneticist, Hagiwara, crossed two true-breeding strains of Japanese morning glories, both of which had blue flowers. The plants of the F_1 generation all had purple flowers. When Hagiwara crossed the purple-flowered F_1 back to one of the parental strains, the progeny of the backcross were approximately 50% blue-flowered and 50% purple-flowered. When the purple-flowered F_1 was crossed to the other parental strain, the results were the same. But when the F_1 purple-flowered progeny were crossed to each other, the F_2 offspring included purple-flowered, blue-flowered, and scarlet-flowered plants. Hagiwara concluded that two gene loci were involved, an **Aa** locus and a **Bb** locus. The scarlet-flowered plants, then, must have been **aabb.**
   a. What were the genotypes of the parental strains, the F_1 hybrid, and the progeny of the two backcrosses?
   b. List all phenotypes and all genotypes that appear in the F_2.
   c. Below is a list of individual crosses between pairs of F_2 plants, together with the ratios of phenotypes found in the progeny. Determine the genotypes of the parents in each cross.

MALE PARENT	FEMALE PARENT	PROGENY BLUE	PURPLE	SCARLET
purple	scarlet	—	100%	—
purple	scarlet	50%	50%	
purple	scarlet	50%	25%	25%
blue	blue	50%	25%	25%
blue	blue	75%	—	25%
blue	blue	100%	—	—
blue	blue	—	100%	—
blue	blue	50%	50%	—

3. Baur crossed two true-breeding strains of *Antirrhinum majus,* the same snapdragon that Darwin had experimented with. One strain had regular (bilaterally symmetrical) white flowers, and the other had peloric (radially symmetrical) red flowers. The F_1 progeny, which had regular pink flowers, were crossed to each other to produce an F_2 generation. Here is Baur's data on the F_2, in numbers of plants of each flower phenotype:

regular pink	94	peloric pink	28
regular red	39	peloric red	15
regular white	45	peloric white	13

   a. Explain the results.
   b. For the observed total of 234 F_2 progeny, compute the *expected* number of each phenotype, and compare with the observed number.
4. Eleanor Perkins is phenotypically normal, but her family has its share of sex-linked abnormalities. Her husband, Garvey, had the X-linked dominant allele for brown spotted teeth but is otherwise normal. Her brother, Arthur, and her son, Little Chester both suffer

from hemophilia A, the most common type of sex-linked bleeder's disease, Her father, Grandpa, is not hemophilic, but he is colorblind.

a. Draw and label a pedigree diagram of this family.
b. From this information alone, Eleanor knows that she herself is a carrier. Which X-linked mutant gene or genes does Eleanor carry?
c. Once Grandpa came to breakfast wearing one red sock and one green sock and couldn't understand why Eleanor, Garvey, and Arthur were all laughing at him. But when Little Chester didn't get the joke, Eleanor knew that her son was both hemophilic and colorblind, and she suddenly realized what must have happened in one of her own oocytes shortly before her birth. Explain.
d. Eleanor and Garvey are expecting another child. If it is a daughter, what is the probability that it will be hemophilic? Colorblind? Have brown teeth?
e. If their child is a son, what is the probability that he will be hemophilic? Colorblind? Have brown teeth?
f. What is the probability that such a son will be normal—that is, he will have none of the traits mentioned? (Note: This last question requires more information. Haldane and Smith calculated that the recombination frequency between the red-green colorblindness and hemophilia A locus is about 10%. List the expected frequencies of all possibilities for Eleanor's and Garvey's sons, and be sure they add up to one.)

5. Thomas Hunt Morgan, discoverer of X-linkage in fruit flies, later studied inheritance in an insect, *Euchistus.* The bug also proved to have XX females and XY males. In one species, *Euchistus variolarius,* all males have a black spot on the abdomen. In a closely related species, *Euchistus servis,* the males lack the spot, just as do the females of both species. A mating between *E. variolarius* females and *E. servis* males produced numerous F_1 hybrid progeny in which all the males had spots and the females did not. When the F_1 were allowed to inbreed, Morgan found 97 spotless females, 32 spotless males, and 84 spotted males in the F_2 generation.

a. How many gene loci are involved in the spotting phenomenon?
b. Is spotting a dominant or recessive trait?
c. Is the spotting gene on an X chromosome, a Y chromosome, or on an autosome? Explain.
d. If the original cross had been between a spotted male *E. variolarius* and an unspotted female *E. servis,,* what would have been seen in the F_1 and F_2 generations?

6. In many plant species, male and female floral parts are borne on separate plants. Bauer and Shull took pollen from a narrow-leaved male *Lynchnis alba* and dusted it on the flowers of a broad-leaved female plant. In the F_1 both males and females were all broad-leaved. The F_1 male and female plants were crossed to each other, and in the F_2 the females were broad-leaved, but the males were both broad- and narrow-leaved.

a. Explain these results.
b. Is breadth of leaf sex-linked or sex-limited?
c. What do you expect would happen in the F_1 and F_2 of a cross between a male of a true-breeding broad-leaved strain and a female of a true-breeding narrow-leaved strain?

7. T. H. Morgan crossed a white-eyed, yellow-bodied *Drosophila* female with a red-eyed, brown-bodied male. In the F_1 progeny, the females were red-eyed and brown-bodied, while the males were white-eyed and yellow bodied. Allowing these to cross to produce an F_2 generation, Morgan counted the following:

EYES	BODY	SEX	NUMBERS
white	yellow	male	474
white	yellow	female	543
red	brown	male	512
red	brown	female	647
white	brown	male	11
white	brown	female	6
red	yellow	male	5
red	yellow	female	7

a. What happened?
b. Of the total of 2205 F_2 flies, what proportion were white-eyed, Yellow bodied? Male? Recombinant?
c. Do these values depart from the expected 50% of Mendel's first and second laws? Why?
d. What is the map distance between the white and yellow loci?

KEY IDEAS

DOMINANCE RELATIONSHIPS

1. When the information is in conflict, the recessive allele is usually ignored, and the dominant allele is expressed.
2. Recessivity may mean that a gene is simply not functioning. An example of this is albinism, in which a recessive cannot fulfill its role in producing the pigment melanin. Two alleles failing to produce pigment results in an albino individual.
3. Recessive alleles may be functioning, but to a lesser degree than dominant genes.
4. In some instances, dominance is apparent visually, but on closer examination or measurement there is a difference between having one or two genes functioning normally. In fruit flies, for example, there may be less pigment produced by one gene than by two.
5. Some dominant alleles are also known as *inhibitor*

alleles, since they prevent recessive alleles from expressing themselves. The dominant white of white leghorn chickens is an example.

6. Specific **dominance relationships** are **partial dominance**, and **codominance.**
7. Partial dominance in snapdragons is seen when red and white are crossed. The offspring are pink. Phenotypic and genotypic ratios are the same (¼: ½: ¼).
8. Lethality is seen in achondroplasia or achondroplastic dwarfism. Homozygous offspring die as embryos, while heterozygotes survive. For lethality to be expressed, two abnormal alleles must be present.
9. Codominance occurs when one homozygote expresses a trait differently from the other homozygote, but the heterozygote shows both traits.
10. The blood groups M and N are an example of codominance. When tested with antisera, *agglutinations* show that each genotype can be biochemically identified. Both M and N are clearly expressed.

MULTIPLE ALLELES

The term *multiple alleles* means that even though each individual in a population gets two alleles for a trait, more than two different alleles are present in the population's genes.

Blood Groups

1. The blood types of the **ABO system** are determined by cell-surface antigens. Antigens are large molecules that are capable of reacting with specific antibodies. In terms of dominance, $\mathbf{I^A}$ and $\mathbf{I^B}$ are codominant, and both are dominant over $\mathbf{I^O}$. $\mathbf{I^A}$, $\mathbf{I^B}$, and $\mathbf{I^O}$ are multiple allelles.
2. From the alleles $\mathbf{I^A}$, $\mathbf{I^B}$, and $\mathbf{I^O}$ there are six possible genotypes. From these six genotypes there are only four blood groups or phenotypes.
3. The immune systems of individuals produce either anti-A or anti-B antibodies. The type A person produces anti-B, type B produces anti-A, type AB produces neither, and type O produces both.
4. When antibodies meet opposing antigens on red cell surfaces, they bind to these and form bridges to other red cells nearby. The massive binding of these cells is called agglutination, or clumping.
5. Because of antigen-antibody reactions, blood transfusions have to be preceded by careful blood-matching tests.
6. **Rh blood groups** can be divided into Rh positive and Rh negative. The Rh positive antigen is very potent. In blood transfusions, this factor must be considered along with the ABO factors.
7. Rh antibodies are only present when a person receives Rh^+ antigens. The presence of antigen induces an immune reaction, and the antibodies are produced.
8. The immune reaction of Rh^- people to Rh^+ antigens extends into reproduction. Rh^+ fathers pass the antigen-producing gene to their offspring. When Rh^- mothers bear Rh^+ children, there can be mixing of maternal and fetal blood, causing the mother to produce antibodies. Subsequent pregnancies are potentially dangerous, since the antibodies can enter the fetal blood and cause agglutination. The immune response can be clinically suppressed if the incompatibility is known.

GENE INTERACTIONS

1. Mendel's two-factor crosses worked well because the pairs of genes involved did not interact with each other.
2. In many instances the 9:3:3:1 ratio Mendel observed is not produced when other loci or organisms are studied in similar crosses.
3. Sometimes pairs of alleles control different steps in producing the same trait. These are known as **epistatic** genes or alleles.
4. An example of epistatic genes at work is seen in mouse hair color. One pair of alleles produces brown or black, but another pair permits or prevents the presence of color. When the color preventer (a recessive) is homozygous, the mice are white.
5. In the epistatic cross **BbCc** × **BbCc**, the phenotypic ratio in the offspring turns out to be 9:3:4 instead of the classic 9:3:3:1 because every offspring with a **cc** is white.
6. In the epistatic cross **PpCc** × **PpCc**, two color preventors are at work, and the phenotypic ratio in the offspring becomes 9:7 because either **pp** or **cc** in the offspring produces white (or prevents color).

CONDITIONAL GENE EXPRESSION

Many alleles are not necessarily expressed. They may be subject to environmental influence, sexual influence, or those as yet not understood.

1. Dark hair pigment in Siamese cats increases in response to cold.
2. The dominant gene for polydactyly (extra digits), as an example of **incomplete penetrance,** may not be expressed at all, or the degree of expression (number of affected digits) may vary widely.
3. **Sex-limited** traits (for example, a tendency toward uterine or prostate cancer) can only affect one sex, while **sex-influenced** traits (such as baldness) tend to be expressed in one sex and not in the other.
4. Huntington's disease, a neuromuscular disorder, has a variable age of onset, appearing most often near middle age.

CONTINUOUS VARIATION AND POLYGENIC INHERITANCE

1. In many instances a trait is produced by the accumulative effect of genes from different loci. This is called **polygenic** inheritance, and its effect is to produce continuous variation in a trait. Examples include body height and skin color.

2. When polygenic traits are measured and plotted, the data commonly produce normal distributions, which take the form of bell-shaped curves.

LINKAGE

Linkage, the occurrence of genes in **linkage groups**, and crossing over were discovered when expected Mendelian ratios failed to appear. In a testcross with garden peas, Bateson and Punnett found odd ratios and seemingly impossible combinations of traits in the progeny.

Linkage, Crossing Over, and Genetic Recombination

1. In the testcross **AaBb** × **aabb**, the predictable Mendelian ratio in the progeny would be 1:1:1:1, or 25% **AaBb**, 25% **Aabb**, 25% **aaBb**, and 25% **aabb**. If the genes are on different chromosomes, no other combinations are possible.
2. If, however, the genes in the above testcross were linked on the same chromosome, then the ratio in the progeny would be 1:1, or 50% **AaBb** and 50% **aabb.** If they are fully linked, no other combinations are possible.
3. Frequently, even when linkage is established, testcrosses can reveal "impossible" new combinations and strange ratios. For example, the progeny might be 9:9:1:1, or 45% **AaBb**, 45% **aabb**, 5% **Aabb**, and 5% **aaBb.** Apparently, an exchange of alleles between homologous chromosomes has occurred.
4. In the situation above, crossing over in meiosis occurred often enough between the **A** and **B** loci for 10% of the gametes to bear the results of recombination (and did not occur the other 90% of the time). The gametes can be represented as: 90% **nonrecombinants:** 45% **AB** and 45% **ab**, and 10% **recombinants:** 5% **Ab** and 5% **aB.**
5. Actually, more crossovers occur than recombinant genotypes reveal, since double crossovers (or any even number) put the alleles back on their respective homologues.

Mapping Genes

1. **Recombination frequencies** are used to construct genetic maps. Map distances are measured in recombination percentages between loci.
2. Recombination genetic maps identify the chromosome in which a locus is found, and relative locations of genes and distances between them.

Chromosomes and Sex

1. In 1910, Morgan began experimenting with *Drosophila melanogaster,* the fruit fly, which became an important research organism. It is easily maintained, has a short generation time, and produces hundreds of offspring. It has four pairs of readily identifiable chromosomes.
2. Morgan's discovery of a mutant white-eyed male led to the discovery of sex linkage. In his experimental crosses he observed the following:
   a. P_1: White-eyed males × red-eyed females → F_1: all red-eyed.
   b. F_1: Red-eyed males × red-eyed females → F_2: 3/4 red, 1/4 white, but *all white-eyed flies were males.*
   c. Testcross: P_1 white-eyed male × F_1 red-eyed female → 1/2 red, 1/2 white-eyed offspring evenly distributed between males and females.
   d. **Reciprocal testcross:** Red-eyed male × white-eyed female → 1/2 white, 1/2 red, but *all females were red-eyed and all males were white-eyed.*
   e. The solution to the above was the linkage between the white eye allele and the alleles that determined sex: both were on the X chromosome.
3. In *Drosophila* the sex chromosomes are heteromorphic: the X is rod-shaped, and the Y is J-shaped. Females have two X chromosomes, while males have an X and a Y.
4. The *Drosophila* X chromosome carries about 1000 genes, while the Y has only six, and these control fertility. Thus, all X-linked genes are **hemizygous** in males, acting as dominants.
5. In *Drosophila,* sex is determined by the number of Xs; thus, the abnormal XXYs are functional females in the fly, but sterile males in humans *(Klinefelter syndrome).* So, the presence of a Y chromosome in humans produces maleness. Further, the absence of a second sex chromosome (XO) in fruit flies produces a sterile male, while this combination in humans results in a sterile female *(Turner's syndrome).*
6. In each cell of the human female, one X chromosome remains condensed. It is detected as a Barr body in cheek cells and a drumstick figure in the nucleus of certain white cells. The selection of such Xs is random, and a mosaic of maternal and paternal Xs occurs in the tissues, producing the **Lyon effect.**
7. Red-green colorblindness in humans, an abnormality in the cones of the retina, is caused by a recessive allele at either of two closely linked loci on the X chromosome. Colorblindness occurs in 8% of American males and 0.4% of American females.
8. Women heterozygous for a sex-linked trait do not express the trait, although half their sons do.
9. Other sex-linked traits include hemophilia and muscular dystrophy. Because of the mosaic Lyon effect, half the muscle cells of women heterozygous for muscular dystrophy show signs of deterioration.

REVIEW QUESTIONS

1. Explain in terms of pigmentation how dominance works in albinisim. (247)
2. Explain why dominance is not the cut-and-dried concept perceived by Mendel. (247-248)
3. Define and give an example of each of the following:
   a. Partial dominance
   b. Recessive-lethal
   c. Codominance. (247-249)

4. List the responses of the MM, MN, and NN blood groups to anti-M and anti-N antisera. Use the term *agglutination*. (249)
5. Explain what is meant by multiple alleles. (250)
6. List both the blood phenotypes (types) and the genotypes possible in the ABO blood system. (251)
7. Why is blood type O considered to be a universal donor? (251)
8. Review the problem of *Rh* incompatibility between *Rh* positive males and *Rh* negative females. What are the clinical solutions? (252)
9. Carry out the epistatic cross between two **BbCc** mice (as described in the text). From your results, derive the 9:3:4 ratio and explain why the familiar 9:3:3:1 phenotypic ratio did not show up. (253)
10. Review the contributions of Morgan and explain why they were valuable to understanding heredity.
11. List several ways in which *Drosophila* is well suited for the study of heredity. (general)
12. To test your understanding of sex linkage, carry out the calculations that explain the behavior of the white-eye trait in fruit flies. Cross the white-eyed male with a homozygous red-eyed female. Inbreed the F_1 and explain the results in the F_2. (261)
13. Can females express sex-linked traits? Using fruit fly eye color, prove your answer. (261)
14. Compare the shapes, gene complement, and sex-determining behavior of the X and Y chromosome in humans and *Drosophila*. (261-263)
15. Using a Punnett square and the symbols XX (female) and XY (male), determine the probability of male or female offspring. (general)
16. Explain why males always inherit X-linked traits from their mothers and never from their fathers. (262)
17. Explain the statement: Males are hemizygous for X-linked traits. (262)
18. What is the probability of colorblindness in the children of a colorblind carrier woman and her normal-visioned mate? (263)
19. Suggest two reasons why hemophilic females did not appear in the genealogy of European royalty. (266)
20. What observations of women who carry muscular dystrophy suggest that the allele for the condition is really codominant to the normal allele? (265)
21. Briefly describe two examples of how physical conditions in the surroundings determine whether or not a trait will be expressed. (254)
22. How do sex-limited traits differ from sex-influenced traits? (255)
23. How does the expression of the dominant allele for Huntington's disease differ from the usual? (256)
24. Using height as an example, explain what is meant by continuous variation. If two pairs of alleles were responsible, how many different heights would there be? What does your answer tell you about the genetics of this trait? (256-257)
25. How does polygenic inheritance differ from that of multiple alleles? Epistasis? (256-257)
26. What two aspects of their garden pea test-cross data thoroughly confused Bateson and Punnett? (257-258)
27. Using the testcross **GgOo** × **ggoo**, what genotypes in what ratios might one expect if (a) the genes were unlinked, (b) the genes were linked but could not cross over, (c) the genes were linked but crossed over 20% of the time? (257-259)
28. Using a diagram, carefully explain why only odd numbers of crossovers are revealed in recombination frequencies. (259)
29. What is the basis for map distances in genetic or recombination mapping? Explain the logic. (259-260)
30. Construct a simple recombination map from the following alleles and recombination frequencies: Genes **P** and **L**, 30 map units; genes **L** and **X**, 5 units; and genes **X** and **P**, 35 map units. (260)

12

DNA AS THE GENETIC MATERIAL

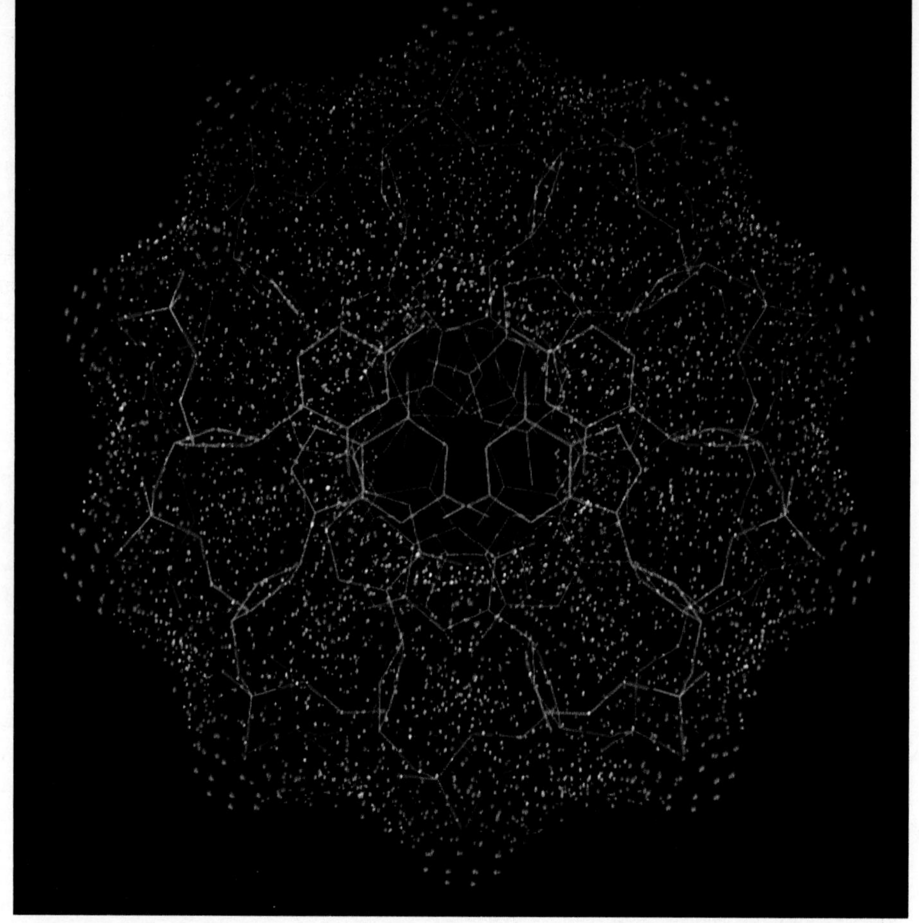

NOT VERY LONG AGO, AS RECENTLY AS THE EARLY 1950s, THE CHEMICAL BASIS of heredity was unknown. In fact, it was widely considered to be so mysterious and complex as to be quite literally beyond human comprehension. Scientists had generally agreed that the gene had a chemical composition, but what genes were, how they were put together, and how they worked seemed hopelessly beyond grasp. As early as the 1940s there was some good evidence that the genetic material was deoxyribonucleic acid (DNA), but at midcentury, most biologists, if they allowed themselves an option at all, would have said that genes were probably made of protein. After all, they knew that enzymes were proteins, and they believed that genes must be much more complex even than enzymes. And before that time, enzymes themselves seemed to be so enormous and complex that their own structure could never be known. (It was not until 1953 that Frederick Sanger determined the amino acid sequence of insulin, one of the simplest proteins.) How could we ever hope to understand genes? It had been pointed out that genes not only had to direct all the life processes of the cell and the development of the organism, but they also somehow had to make exact copies of themselves every cell generation.

The more pessimistic souls must have been shocked when two relentlessly energetic young researchers, James Watson and Francis Crick, showed that the structure of genes and the mechanism of gene function and replication were essentially not so hard to understand after all. In fact, the secrets lay in the simplicity of DNA, a molecule whose presence in the cell had been known for about 80 years. The gene itself turned out to be a surprisingly simple chemical coding of protein-synthesizing instructions. Furthermore, the code contained only four elements. These four elements determined the order in which the 20 amino acids are arranged in any protein. And, after all, it is the order of amino acids that makes one protein different from the next.

One might ask, what does protein synthesis have to do with genes and heredity? The answer is, everything: Remember that proteins form much of the structure of living things, as well as the enzymes and hormones that regulate the processes of life.

As we will see, the story of Watson and Crick, their friends, their personal lives, and their luck and successes, has turned out to be one of the classical, even legendary, stories of biology. In this chapter, we will review the early days of their epic, when they first assembled their ideas on DNA structure and function. What they reveal to us is pleasingly logical and understandable when it is told one step at a time. So let's take it that way, beginning with the structure of DNA. Try to keep in mind that many scientists of the 1950s found it difficult to accept what we are about to say. To many, the mechanical simplicity of these ideas threatened to take away the central mystery of life.

HIGHLIGHTS FROM THE DISCOVERY OF DNA AS THE GENETIC MATERIAL

A fundamental question in biology in the first half of this century was what, precisely, was the material that constituted the factors described by Mendel? Although there were strong indications of ties between chromosomes and inheritance, what were these chromosomes exactly, and how did they hold the genetic information?

The Early Efforts

Surprisingly, DNA itself was discovered while both Darwin and Mendel were working. In 1869, Friedrich Miescher, a Swiss chemist, used the enzyme pepsin to digest the proteins of the nucleus of white blood cells taken from pus and showed that a strange, phosphorus-containing material remained. In a private letter, discovered much later, Miescher actually speculated that this material might be the stuff of heredity. Later, in 1914, a German chemist named Robert Feulgen invented a still widely used staining procedure, Feulgen staining, that is specific for DNA. The Feulgen staining procedure has the advantage of staining cell nuclei more or less strongly according to how much DNA is present. Thus, the amount of DNA present can be calculated by measuring the strength of the color. A few key experiments revealed that virtually every cell nucleus in a given plant or animal has the same amount of DNA, except for gametes (eggs and sperm), which have half the amount.

Once the existence of DNA was known, most biologists still couldn't bring themselves to consider seriously the possibility that it was the hereditary material, for a few very convincing reasons. In the first place, the composition of DNA is very simple: just four different nucleotides are present (refer to Chapter 3). How could anything so simple be the physical basis of anything so wonderful as the gene? How could only four nucleotides produce the complex variations of life, or contain the information for the arrangement of the 20 different amino acids known to be in proteins? In fact, P.A. Levene, who initially described the chemical composition of DNA and first described the nucleotide, set the intellectual tone against DNA as an informational molecule. He mistakenly assumed that because the four nucleotides occurred in approximately equal proportions, DNA was a simple, repeating polymer of the four nucleotides. In the second place, DNA didn't seem to do anything. It just sat there, some scientists said, probably holding the chromosome together, or making it acidic, or doing something even less significant.

But chromosomes also contain proteins. Proteins! Now there was a likely source of variation. Proteins are wonderfully complex and do all kinds of marvelous things. So most bets were on proteins, providing one was willing to believe that genes were chemicals at all. Not every biologist had given up ideas about such things as "vital forces" and other mystical, unexplainable entities. If this all seems absurd to us now, remember that even today no one knows the biochemical basis of such things as thoughts and memories, and indeed, any biochemical basis of such things remains to be shown.

Transformation

In 1928, bacteriologist Fred Griffith conducted what seemed to be an oddball experiment, but one that proved to be a classic. He was studying the virulence (disease-producing capability) of two strains of *Pneumococcus*, a bacterium that causes pneumonia. One strain was dangerous and one harmless. The virulent (disease-producing) strain synthesized a smooth, gummy polysaccharide coat that seemed to protect it from the host's defenses; the harmless strain (a laboratory curiosity) did not. When grown in the laboratory, the virulent strain produced "smooth"-looking colonies; the harmless strain lacked the proper enzymes to coat themselves and produced colonies that appeared "rough."

When Griffith injected smooth-strain bacteria into mice, the mice died. When he injected rough-strain bacteria into mice, the mice did not die. He then killed some smooth-strain bacteria by heating them and injected their bacterial corpses into more mice. The mice did not die. But then Griffith killed some smooth-strain bacteria and mixed them with live rough-strain bacteria (both of which, in separate experiments, had proved to be harmless) and injected the mixture into still more mice. These mice came down with severe pneumonia and died. (At this point, why not pause and make your own best guess about what was happening.) Did the chemical remains of the smooth-strain bacteria help the rough strain do its dirty work? To further confuse things, autopsies showed that the dead mice were full of virulent, living, smooth bacteria! Where did they come from?

Griffith thought that perhaps he had erred in his experimental technique, so he repeated the experiment with great care, again and again. The results were clearly not due to faulty techniques, nor were they due to accidental contamination; the dead smooth-strain pneumococci were indeed dead. As Sherlock Holmes said, when you have eliminated the impossible, whatever is left must be the truth, no matter how improbable. It seemed unlikely, but apparently the living rough-strain *Pneumococcus* had somehow been transformed. That is, they had incorporated hereditary material from an outside source and, in so doing, expressed a new (smooth) trait (Figure 12.1).

Others improved on the experiment, trying to discover what was behind these results. They found that transformation could occur in test tubes, as well as in living mouse

FIGURE 12.1
TRANSFORMATION (1928)
Griffith's experiments clearly laid the groundwork for the discovery of the genetic material. His smooth strain of *Pneumococcus* was virulent and caused pneumonia in mice, that is unless heat was used to kill the bacteria. Griffith's rough strain was harmless, but could be "transformed" into the virulent, smooth strain by the addition of some substance or "factor" from the heat-killed smooth strain. Later work of Avery and his coworkers indicated that DNA might be the factor responsible for transformation.

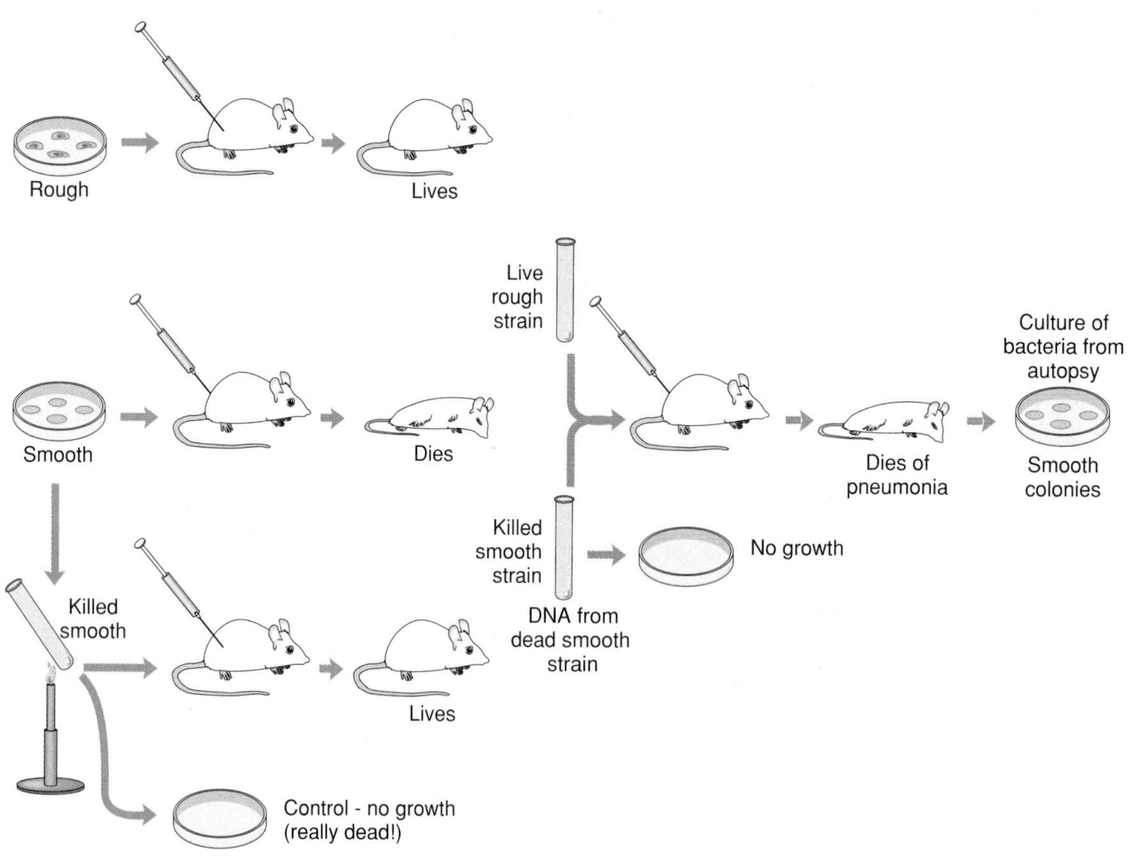

hosts. Various materials from the dead smooth bacteria were isolated and purified and then injected to see if they were the mysterious transforming substance. Not until 1944, and the work of O. T. Avery, C. McLeod, and M. McCarty was it demonstrated that pure DNA extracted from smooth-strain *Pneumococcus* could transform rough-strain bacteria, giving them ability to synthesize the necessary enzymes for making the smooth polysaccharide coat. Avery and his coworkers further showed that purified proteins would not cause transformation, nor would carbohydrates or lipids. After years of research we now know that the harmless rough cells actually take in pieces of smooth-cell DNA. With a low but measurable frequency, the DNA-repairing enzymes of the rough cell insert the deadly smooth-strain DNA fragments into the cell's own chromosome, replacing the rough strain's gene. The few rough bacteria in a mixture that are transformed in this way are then able to synthesize new enzymes and become smooth and able to infect a mouse.

By 1944 it was becoming clear that DNA was the genetic material—clear, that is, to a few of the more visionary biologists. Most researchers were still far from convinced, even after some 30 different instances of bacterial transformation by purified DNA had been reported. But perhaps a bit of caution and a critical attitude aren't all that bad in science.

The Experiments of Hershey and Chase

Alfred Hershey and Martha Chase performed another classic experiment that, in retrospect at least, firmly established that DNA was the genetic material. They used a new tool in genetic analysis, the bacteriophage (called phage for short; see also Chapters 15 and 23).

If you have ever wondered whether a germ can get sick, you will be glad to learn that it can. It can become infected with phage (a kind of virus) and may even die. The phage Hershey and Chase worked with destroys *Escherichia coli,* the common, rod-shaped bacterium that lives harmlessly in your intestine. This phage consists of a DNA chromosome contained in a coat made of protein. Protein and DNA, then, are the only components of this phage. Hershey and Chase set out to see which of these components were passed into the bacterium host and used to produce the next generation of phage particles.

The important point discovered by Hershey and Chase is that only the DNA of the phage enters the host cell; the protein portion of the phage stays outside (Figure 12.2a). Hershey and Chase didn't know this at the time, but discovered it themselves through experimentation. They did this by infecting bacteria grown in two types of media, one containing radioactive sulfur (^{35}S), the other radioactive phosphorus (^{32}P) (Figure 12.2b). Since proteins contain sulfur but no phosphorus, and nucleic acids contain phosphorus but no sulfur, the protein and nucleic acids conveniently labeled themselves by incorporating the radioactive sulfur and phosphorus, respectively.

Hershey and Chase then infected "cold" (normal, nonradioactive) bacteria with their "hot" (radioactive) bacteriophages and allowed enough time for the phages to attach themselves and inject their DNA—but not enough time for the production of new phage. Then they put the phage and bacteria into a kitchen blender. The empty "ghosts" of the bacteriophage were dislodged from the bacterial surface and could be separated by centrifugation. It turned out that nearly all the radioactive sulfur was found in the empty bacteriophage ghosts; nearly all the radioactive phosphorus was found inside the infected bacteria. Thus, Hershey and Chase proved that only the DNA entered the cell, and that DNA would then be responsible for the production of the next generation of phage. More important, they later showed that the purified phage DNA has all the information necessary to enable the host to make new bacteriophage DNA, and new bacteriophage protein as well.

FIGURE 12.2
THE BACTERIOPHAGE: HERSHEY AND CHASE (1950)
Phage viruses are composed of only a protein coat surrounding a DNA core. (**a**) The virus attacks a bacterial cell, causing the cell to mass-produce a new generation of viruses. (**b**) Hershey and Chase utilized the fact that proteins could be uniquely labeled with ^{35}S and DNA could be labeled with ^{32}P, to ask the question, "Is it protein or DNA that is passed into the bacterium to direct the formation of the next generation?" They found that it was DNA.

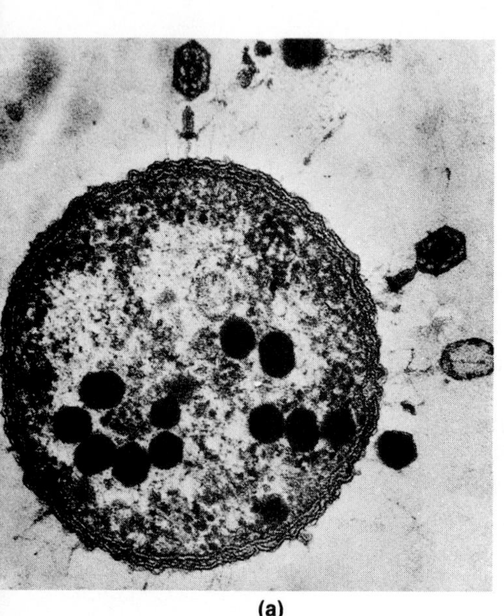

(a)

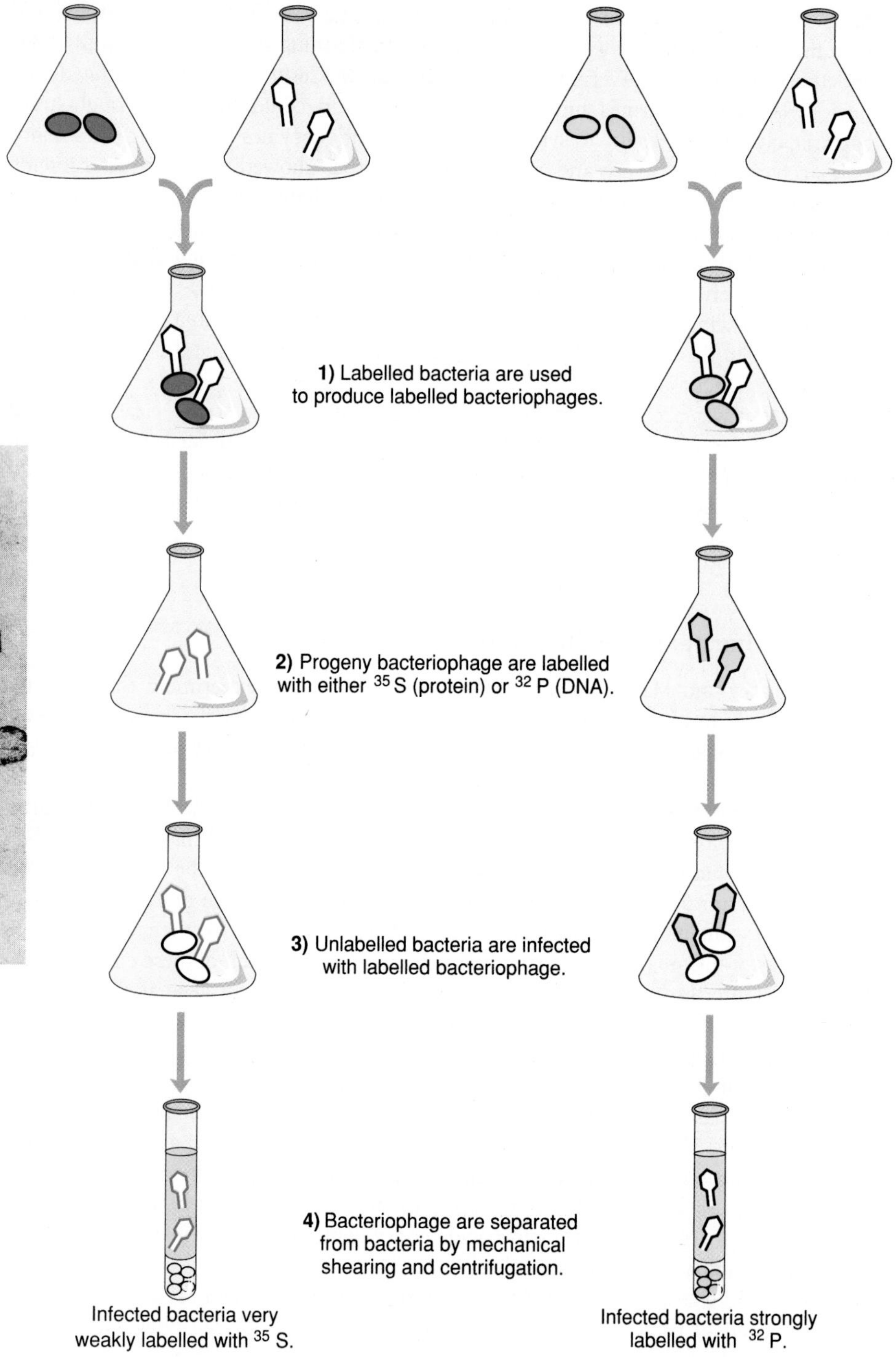

1) Labelled bacteria are used to produce labelled bacteriophages.

2) Progeny bacteriophage are labelled with either ^{35}S (protein) or ^{32}P (DNA).

3) Unlabelled bacteria are infected with labelled bacteriophage.

4) Bacteriophage are separated from bacteria by mechanical shearing and centrifugation.

Infected bacteria very weakly labelled with ^{35}S.

Infected bacteria strongly labelled with ^{32}P.

(b)

Thus, it appeared that DNA was the stuff of which genes were made—at least the genes for smooth coats in *Pneumococcus* and the genes of certain bacterial viruses. Actually, Hershey and Chase were lucky because some viruses use RNA as their genetic material (see Chapters 15 and 23), and in certain viruses, some or all of the protein does in fact enter the host cell. Yet the work of Hershey and Chase stands out as the pivotal set of experiments that began to establish firmly the growing idea that DNA was indeed the genetic material.

Still, no one had very much of an idea of how DNA worked. Genes were known to produce enzymes, which are complex proteins. The big question was, how could a molecule with only four different subunits determine the specificity of proteins, which are composed of 20 different amino acids?

THE STRUCTURE OF DNA

The question of how a simple molecule like DNA could produce complex molecules like enzymes turns out to be based on the molecule's very simplicity.

DNA Nucleotides

We saw in Chapters 3 and 9 that DNA molecules are incredibly long polymers, each containing millions of the structural units called nucleotides. DNA nucleotides, you recall, consist of three parts: phosphate groups, the five-carbon sugar deoxyribose, and one of four different nitrogen bases. The sugar links the phosphate and the nitrogen base:

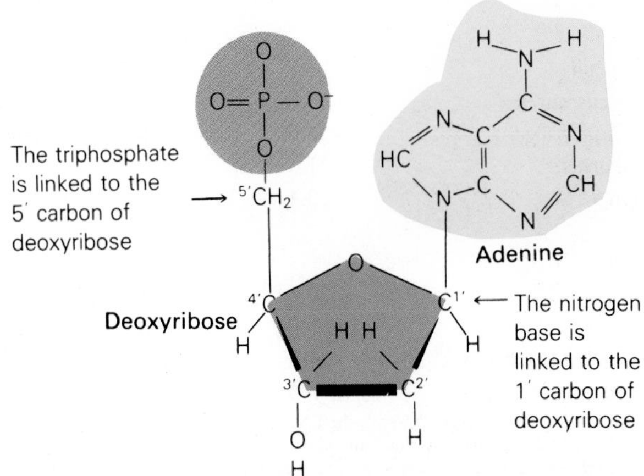

Deoxyadenosine triphosphate (dATP)

Pyrimidines (single ring)

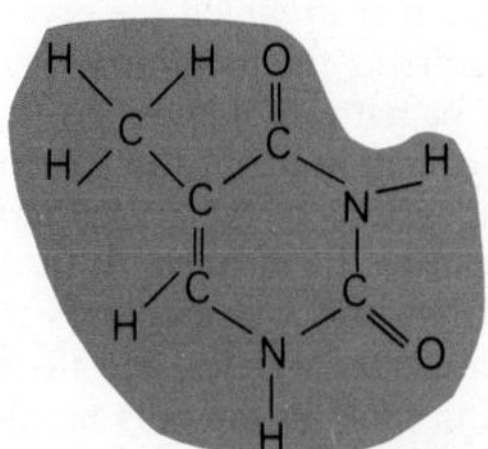
Thymine (T)

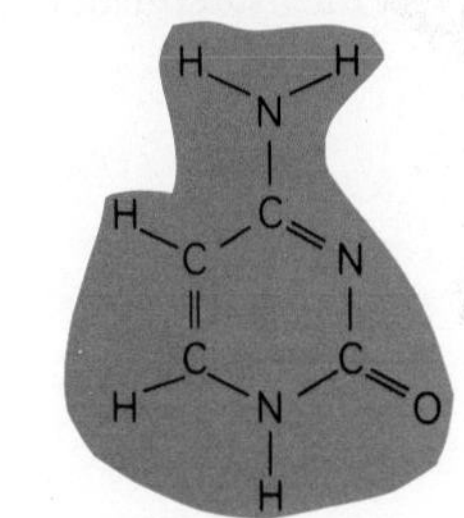
Cytosine (C)

Purines (double ring)

Adenine (A)

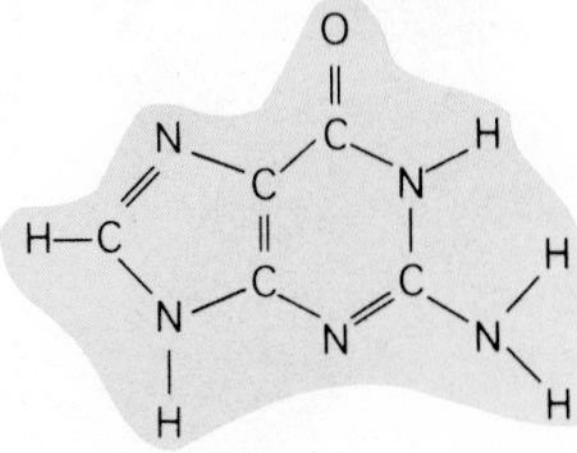
Guanine (G)

The phosphate groups are the familiar acidic groups we saw in ATP (see Chapter 6). When incorporated into the DNA polymer, only the single phosphate group attached directly to the sugar ring will remain. The five-carbon sugar, deoxyribose, is quite similar to ribose, the five-carbon sugar also familiar from our discussion of ATP. As you see, deoxyribose has a simple —H or hydrogen group on its 2′ carbon, instead of the —OH or hydroxyl group seen in ribose. Note that the five carbons of deoxyribose are traditionally numbered as 1′ through 5′ (read, "one prime through five prime").

The four nitrogen bases of DNA are known as **adenine, guanine, thymine**, and **cytosine.** The latter two, thymine and cytosine, called **pyrimidines,** consist of a single six-cornered ring of nitrogen and carbon. Adenine and guanine are more complex, double-ringled molecules, called **purines.** They contain a six-cornered and a five-cornered ring of carbon and nitrogen. (It may help to keep the purines and pyrimidines straight if you recall that the smaller molecules have the longer names.) The colored N-H groups seen in the molecules are sites where the nitrogen bases bond to deoxyribose. In the assembled nucleotide, these groups are covalently linked to the 1′ carbon of deoxyribose, while the phosphate group is covalently linked to the other end of deoxyribose, its 5′ carbon.

Adenine, guanine, thymine, and cytosine are also known simply by the letters A, G, T, and C, respectively. Interestingly, both adenine and thymine are named for the thymus, a gland from which they were first isolated; *adeno* means "gland." Cytosine uses the

TABLE 12.1 CHARGAFF'S RULE (1949–1953),[a]

	BASE COMPOSITION (MOLE PERCENT)			
	A	T	G	C
Animals				
Human	30.9	29.4	19.9	19.8
Sheep	29.3	28.3	21.4	21.0
Hen	28.8	29.2	20.5	21.5
Turtle	29.7	27.9	22.0	21.3
Salmon	29.7	29.1	20.8	20.4
Sea urchin	32.8	32.1	17.7	17.3
Locust	29.3	29.3	20.5	20.7
Plants				
Wheat germ	27.3	27.1	22.7	22.8
Yeast	31.3	32.9	18.7	17.1
Aspergillus niger (mold)	25.0	24.9	25.1	25.0
Bacteria				
Escherichia coli	24.7	23.6	26.0	25.7
Staphylococcus aureus	30.8	29.2	21.0	19.0
Clostridium perfringens	36.9	36.3	14.0	12.8
Brucella abortus	21.0	21.1	29.0	28.9
Sarcina lutea	13.4	12.4	37.1	37.1
Bacteriophages				
T7	26.0	26.0	24.0	24.0
λ	21.3	22.9	28.6	27.2
ϕX174, single strand DNA[b]	24.6	32.7	24.1	18.5
ϕX174, replicative form	26.3	26.4	22.3	22.3

[a] By determining the composition of nitrogen bases in the DNA of a variety of organisms, Chargaff and his contemporaries were able to provide vital information. Pay close attention to the relative quantities of A and T, and G and C here. (But note that the values are not exactly equal due to experimental error.)

[b] Note that this virus has single-stranded DNA, which does *not* follow Chargaff's rule. Why not?

Adapted from A. L. Lehninger, *Biochemistry,* 2d ed. (New York: Worth, 1975).

FIGURE 12.3
X-RAY DIFFRACTION OF DNA
The actual image of an X-ray diffraction pattern is a subtle pattern of light and dark spots, caused by the manner in which the crystal lattice diffracts the incoming X-ray beam. The pattern of spots gives critical information about the structure and dimensions of the crystal lattice that causes the diffraction. In the case of DNA, the X-shaped crossing pattern indicated a helix, and the spacing of the spots in the dots provided clues to the dimensions of the helix and its repeating substructure.

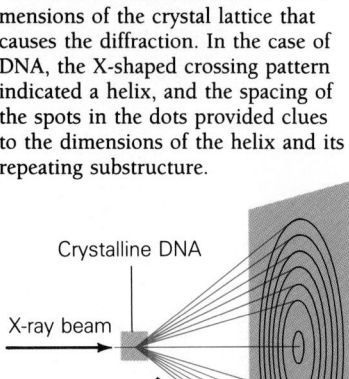

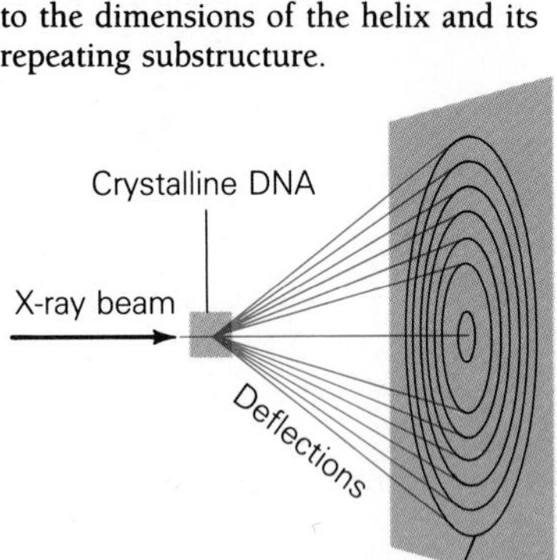

prefix *cyto-* which, of course, means "cell." The least romantic is guanine, first isolated from *guano,* a name which refers to the fecal droppings of bats or seabirds.

Chargaff's Rule Remember that one of the arguments against DNA's being the hereditary molecule was that nothing so simple and *repetitive* could carry much information. But in 1950 Erwin Chargaff showed that DNA from different sources had different base ratios, that is, different frequencies of the four subunits. DNA from *Escherichia coli,* for instance, has about 25% A, 25% T, 25% C, and 25% G, fitting Levene's original theory—but the DNA of humans and other mammals is about 21% C, 21% G, 29% A, and 29% T. As Chargaff looked at the DNA of more and more organisms, he found increasingly different ratios, but he also discovered one general rule: The amounts of A and T in his samples were always equal, and the amounts of G and C were always equal (Table 12.1). Or, in shorthand, A = T and G = C. (This finding became known as Chargaff's Rule.) In addition, one could also state that A + G = T + C = 50%. That is, regardless of the source of DNA, exactly half of the nucleotide bases are purines (adenine and guanine), and exactly half are pyrimidines (thymine and cytosine).

We now know that the reason for these equalities is the specific pairing of nitrogen bases, but to Chargaff they were only intriguing, mysterious observations. Nevertheless, they were key observations that, as we will see, enabled Watson and Crick to work out the structure and function of DNA.

X-Ray Diffraction and More Puzzles X-ray crystallography helps scientists explore the fine structure of crystals. Essentially, the technique involves aiming X-rays at a crystal and noting how the rays are bent (diffracted) by the regular, repeating molecular structure within the crystal. The closer together the regularly repeated structures, the greater the angle through which the X-rays are bent or defracted. The pattern produced on a photographic plate consists of whorls and dots, with those farthest from the center of the plate representing the most closely spaced repeating structure (Figure 12.3). In this way, the relatively simple structure of inorganic crystals has been deduced.

Interestingly, organic chemicals can also be crystallized. In recent years, the three-dimensional structures of many proteins and of some RNA molecules have been worked out. In Watson and Crick's time, however, only preliminary work on a few proteins had been done. People were just starting to look at DNA, among them Maurice Wilkins (who received a Nobel Prize with Watson and Crick) and Rosalind Franklin (who died before her work with Wilkins was fully acknowledged). Their studies revealed a few repeated intramolecular distances, namely, 2.0 nm, 0.34 nm, and 3.4 nm, numbers that showed up again and again in the X-ray image but didn't make complete sense at the time. Wilkins and Franklin also recognized in the patterns a helical molecule with the phosphates on the outside. Franklin even argued that there were probably two strands, not one or three, as had been suggested by others. Although Wilkins and Franklin had a general idea of what the molecule was like, they still didn't know the specifics. It was at this point that Watson and Crick appeared on the scene.

The Watson and Crick Model of DNA

Thus Watson and Crick knew several things when they tackled the DNA problem in the early 1950s. They knew that DNA was a polymer consisting of four different nucleotides, they understood the chemical structure of the nucleotides, and they realized that the nitrogen bases dangled off the sides of the long nucleotide polymers. From Chargaff's data they knew that somehow the number of adenines had to equal the number of thymines, and that the number of guanines had to equal the number of cytosines. They also knew of Wilkins and Franklin's ideas about DNA and the results of their X-ray diffraction studies—the intramolecular measurements 2.0 nm, 0.34 nm, and 3.4 nm, as well as the idea of a helix.

Watson and Crick's experimentation was based on real models made of wire, sheet metal, and nuts and bolts (Figure 12.4). Their models provided a graphic, "hands-on" representation of the emerging molecule. (Sometimes the fingers can grasp what the mind cannot.) Biological intuition, plus biophysical and chemical calculations, along with fitting the pieces this way and that, seemed to indicate that there might be two strands wrapped around one another, with the phosphate-sugar backbone on the outside and the nucleotide bases inside, facing one another. But how were the bases arranged inside?

It was about this time that Wilkins and Franklin's numbers began to make sense. The 2.0 nm measurement represented the total width of the double helix. The 0.34 nm represented distance from one base to the next. That is, if the bases were stacked one on top of the other, like pennies in a roll, each layer would be 0.34 nm thick. The 3.4 nm measurement was a tough one, but it turned out that if each base was set slightly off from the one above like steps in a circular staircase, the double backbone would make one complete twist every 3.4 nm along the axis of the molecule. Further, with a little arithmetic (3.4 nm/0.34 nm = 10), we find that there would be exactly ten nucleotide pairs in each helical turn (Figure 12.5).

FIGURE 12.4
WATSON, CRICK, AND THEIR MODEL OF DNA (1953)
Using wire, bits of metal, and intuition, Watson and Crick put all of the available information together to deduce the structure of DNA.

FIGURE 12.5
THE THREE-DIMENSIONAL STRUCTURE OF DNA
The photos are computer-generated images of DNA. Each dot represents an atom.

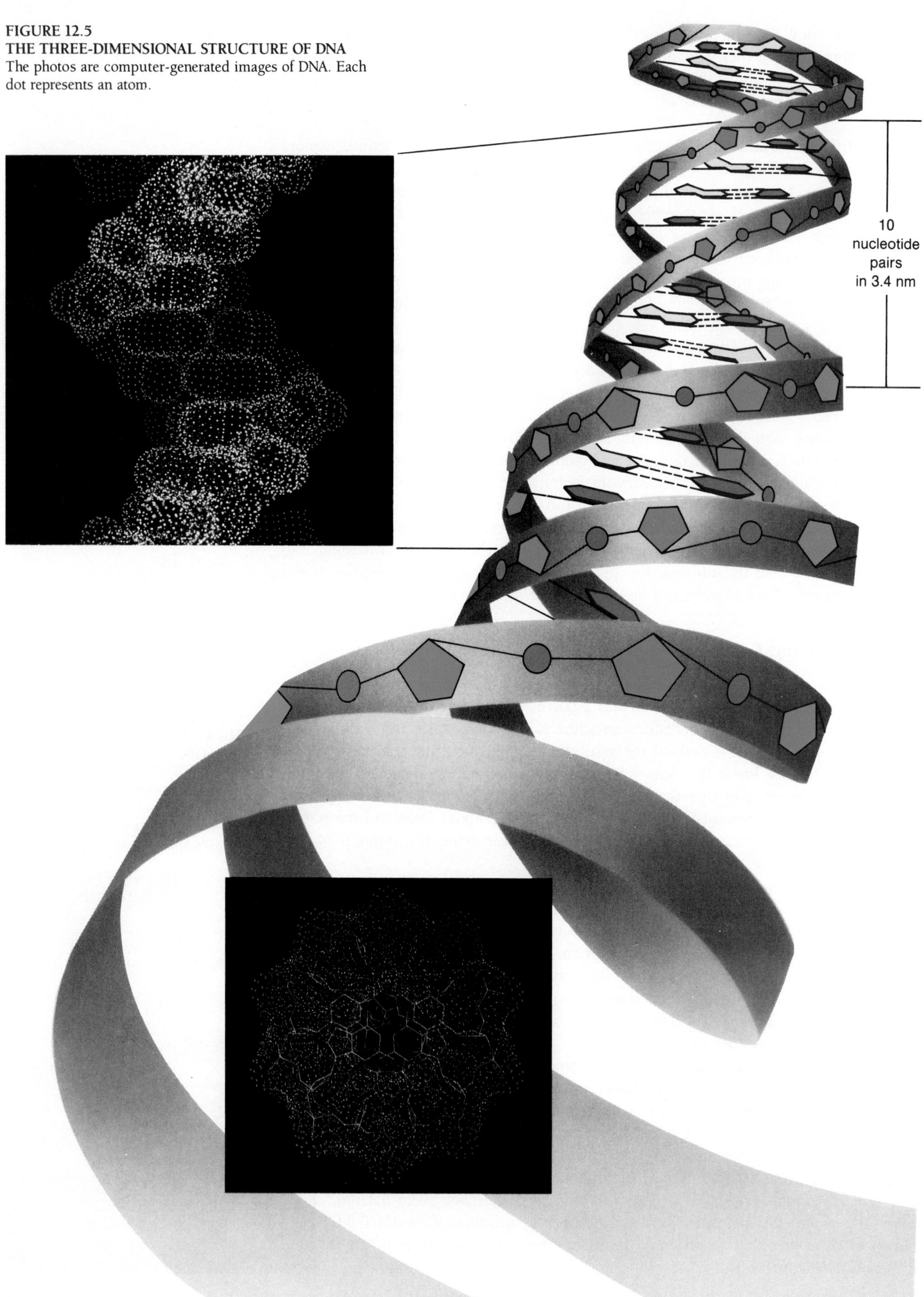

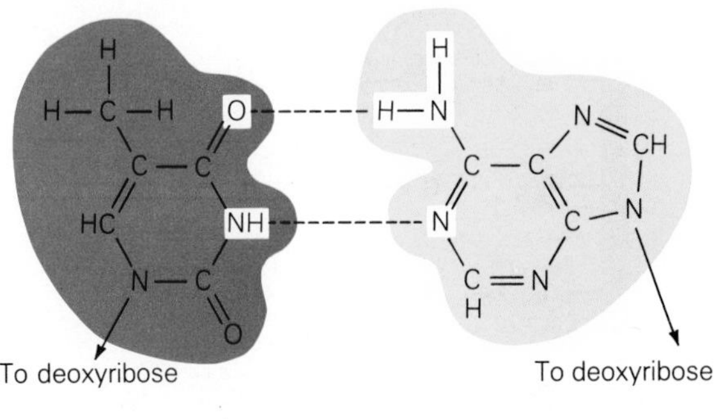

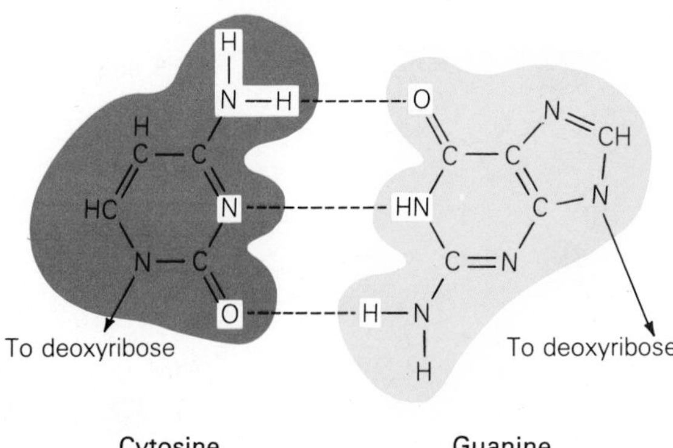

FIGURE 12.6
THE CONCEPT OF BASE PAIRING
Simply put, base pairing in DNA always occurs between A and T and G and C. The pairing of the bases is a result of hydrogen bonding (indicated by the dashed lines) between the bases that span the double helix. The arrangement of atoms to produce the hydrogen bonds is a specific requirement for pairing, and mismatches will not form stable hydrogen bonds.

It was Watson who insightfully grasped the true meaning of Chargaff's strange data about the ratios of A, T, G, and C. In the sheet-metal-and-wire model, two purines would not fit opposite one another within the 2.0 nm confines of the double helix, and two pyrimidines would leave a gap. But one purine and one pyrimidine could fit opposite one another. Thus if a purine were always found across the helix from a pyrimidine, the distance between the sugar phosphate backbones would remain constant. Watson also saw that adenine and thymine would form two hydrogen bonds, while guanine and cytosine would form three hydrogen bonds (see Figure 12.6). This was the only way the DNA molecule could fit together. Thus, the specifics of base pairing represented one of Watson and Crick's major findings. Whatever nucleotides were in one strand, they rigidly fixed the sequence of nucleotides in the other strand. We say that the two strands are **complementary**; they aren't identical, but they fit like a hand in a glove (see Figure 12.7).

The configuration of the two strands of DNA wound around each other is that of the famous double helix. Within the double helix, the two strands of DNA run in opposite directions. They are referred to as antiparallel. That is, if you pictured a DNA molecule vertically on a page, one of the chains would run from top to bottom in its 5′-to-3′ direction, and the other would run from bottom to top in its 5′-to-3′ direction (Figure 12.7).

When you look closely at the space-filling model of DNA (Figure 12.5a) you might note that the spiral of sugar-phosphate groups around the outside of the molecule produces grooves with the bases exposed at the bottom. The one most evident in the model is called the **major groove.** As we will discuss later, the order of bases in this groove is thought to be very important in the control of gene action (Chapter 14).

DNA REPLICATION

Watson and Crick immediately recognized that the structure of DNA, particularly the complementary nature of the paired nucleotides, suggested the core of an elegant mechanism that would ensure the correct replication of the genetic material. "It has not escaped our notice that the specific pairing we have postulated immediately suggests a possible copying mechanism for the genetic material," they said. They conceptualized a simple replication model in which the two strands of the double helix separated, allowing a new complementary strand to be synthesized using each of the old strands for templates. In essence, they were correct. The details, though, are only now becoming clear.

Synthesis means making something, and **replication** means making an exact copy of something. With very few exceptions, DNA synthesis and DNA replication are the same thing, since DNA molecules are made only by copying other DNA molecules. Historically, it was realized very early that one of the properties that the unknown genetic material must have is the ability to replicate when a cell divides. A complete set of genes (within the chromosomes) has to be passed down to each new cell, and any reproducing cell or organism must pass along a full set of genes to descendants.

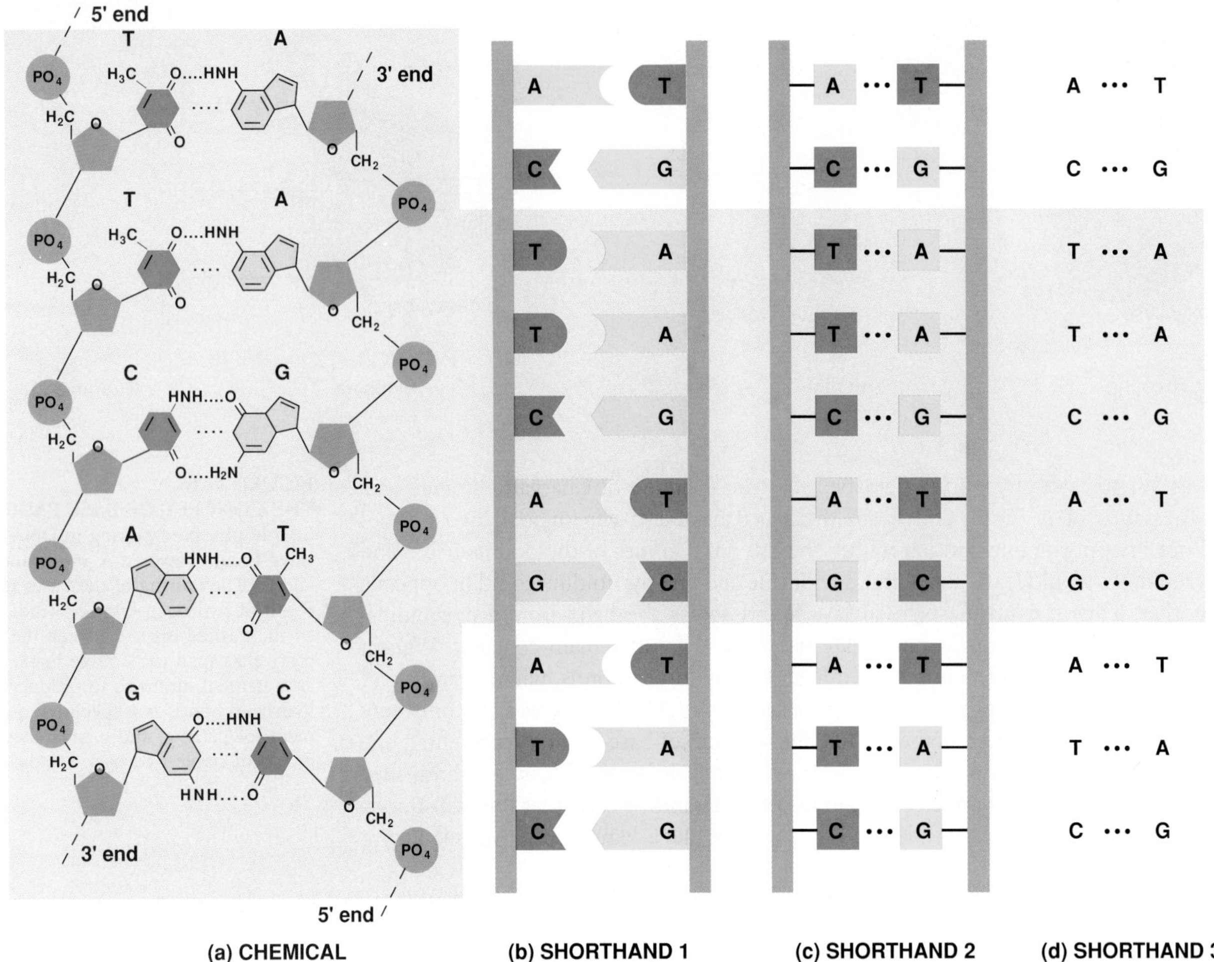

FIGURE 12.7
DIAGRAMMATIC REPRESENTATIONS OF DNA
These are but a few of the many ways that the chemical complexity of the structure of DNA can be reduced to shorthand notations. The most important features required of any shorthand representation are two antiparallel strands separated by nitrogen bases that have specific hydrogen bonds.

The very special way in which nucleotide bases pair in DNA suggests the analogy of a positive and a negative photographic film. The same information is present in both; you can reproduce a positive from a negative and a negative from a positive. Similarly, DNA replication enzymes can produce a "Watson" strand from a "Crick" strand, and a "Crick" strand from a "Watson" strand. In so doing, a new Watson strand and a new Crick strand are produced; thus there are two DNA molecules where there was one before. ("Watson strand" and "Crick strand" are not official terminology, but only a common inside joke that caught on.)

We must confess that all the details of DNA replication aren't known, such as what initiates the process, why it goes out of control in cancer, and why it ceases forever in cells that have reached a terminal state of specialization. But we do know most of the molecular aspects of the process of DNA replication.

The Chemistry of DNA Synthesis

As we said, the concept of DNA replication is quite straightforward. The two complementary strands of DNA are separated from each other, and then those strands are used as templates upon which new strands are constructed. The details of all this, however, are a bit more involved. It all begins with a large number of free nucleoside triphosphates; dATP, dGTP, dTTP, and dCTP (the "d" signifying deoxyribose).

In the synthesis of DNA chains, the high-energy bond between two of the phosphates (recall ATP) is used to form a strong covalent bond between nucleotides. In the process, the two outermost phosphate groups are liberated as pyrophosphate, leaving the innermost phosphate still attached. This remaining phosphate forms a link between two deoxyribose subunits as follows: The free 3′ —OH group of deoxyribose in the first nucleotide reacts with the first 5′ phosphate of the second nucleotide, forming a sugar phosphate sugar linkage. The 3′ —OH group of the second nucleotide is free to interact with the 5′ phosphate group of the third nucleotide, and so on (Figure 12.8). Perhaps the most important thing to remember about this chemistry is that the 5′ phosphate of the incoming nucleotide reacts with the 3′ —OH of the previous nucleotide. The identity of the incoming nucleotide (A, G, T, or C) depends, of course, on the information in the complementary template strand.

Eventually, a long polymer of nucleotides is produced. The covalently bonded backbone of the chain consists of alternating sugars and phosphates linked by shared oxygen atoms. The nitrogen bases are side groups hanging off of the sugars and are not part of

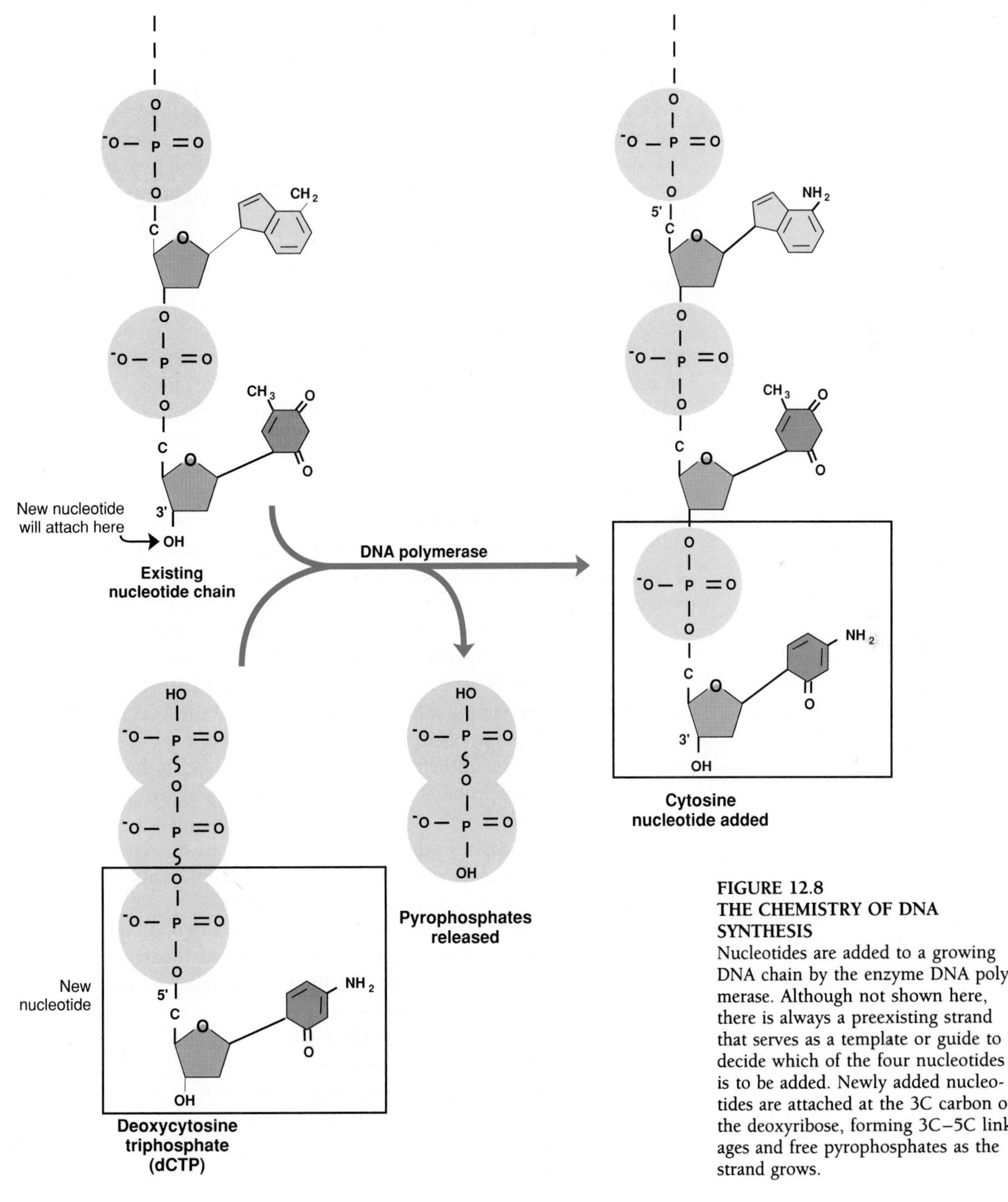

FIGURE 12.8
THE CHEMISTRY OF DNA SYNTHESIS
Nucleotides are added to a growing DNA chain by the enzyme DNA polymerase. Although not shown here, there is always a preexisting strand that serves as a template or guide to decide which of the four nucleotides is to be added. Newly added nucleotides are attached at the 3C carbon of the deoxyribose, forming 3C–5C linkages and free pyrophosphates as the strand grows.

the backbone. The end of the polymer that has the free 5′ phosphate is known as the **5′ end**, and the end with the free 3′ —OH group is called the **3′ end.**

Origins of Replication

The replication process begins somewhere in the middle of the DNA molecule, rather than at one end, at places known as **origins of replication** (Figure 12.9). There, hydrogen bonds are broken, and the paired bases are separated. The helix begins to pull apart—much as two pieces of string wound about each other would separate at the middle. The unwinding is facilitated by an unwinding enzyme called **helicase.** Helicase is a part of the **replication complex,** a group of enzymes and other proteins that take care of the replication process. Two such replication complexes are present at each replication origin, and as unwinding continues they move off in opposite directions, creating two Y-shaped replication forks (Figure 12.9b).

As the strands of the double helix are unwound from each other, an enzyme called gyrase prevents accumulation of twists by making temporary single strand nicks in the DNA. Where the DNA is nicked by the enzyme gyrase, the helix can unwind around the connected strand. After the tension of unwinding is relieved (for a time), the nick is sealed. Replication complexes also contain single strand binding proteins to help keep the template strands apart, and, of course, **DNA polymerase**—the enzyme in DNA synthesis.

DNA Polymerase

It should be clear from our discussion of the chemical basis of DNA synthesis that the main function of the primary enzyme responsible for replication, DNA polymerase, is to add the 5′ phosphate of a new nucleotide to an existing 3′ —OH group. Indeed, for our purposes here, this polymerizing activity is the most important enzymatic activity of several possessed by DNA polymerase. Yet this single polymerizing activity imposes critical limitations on the replication process as a whole.

First, DNA polymerase cannot begin a new daughter strand all by itself; it requires an existing 3′ —OH to add the next nucleotide. So replication actually begins when a different enzyme called **primase** begins to copy the template strand by forming a primer, a short sequence of a few bases of RNA. These few bases of RNA provide the free 3′ —OH for DNA polymerase to begin the polymerization process.

The second limitation is that DNA polymerase can only add new bases to the 3′ —OH end of a growing strand. That is, new DNA strands always grow from their 5′ to their 3′ ends. Yet as we have seen, the two strands of the double helix are antiparallel. So as a replication complex moves off from an origin, it frees up two template strands, and they have opposite polarity. This means that while replication can occur in the 5′ to 3′ direction on one of them, it seemingly must occur in the 3′ to 5′ direction on the other. So how is this problem solved? This question plagued molecular biologists in the early years, and they searched for a form of DNA polymerase that could somehow add nucleotides in the opposite direction. The search was in vain; no such enzyme exists. Eventually, researchers determined that the replication events are quite different on the two template strands.

As the replication complex moves out from the origin along an unwound segment of DNA, synthesis on one strand can proceed continuously from the 5′ to the 3′ direction, to create a *leading* daughter strand (the upper strand in Figure 12.9b). For this reason, synthesis on this part of this strand is referred to as *continuous* replication. But the other parent strand (the lower strand in the figure) is copied *discontinuously*. As the leading strand is being synthesized continuously, unpaired template bases accumulate on the lagging strand until a long enough segment of template has been exposed to allow first primase and then polymerase to begin copying it in its own 5′-to-3′ direction. (You'll note that this copying is in a direction opposite to that of the movement of the replication

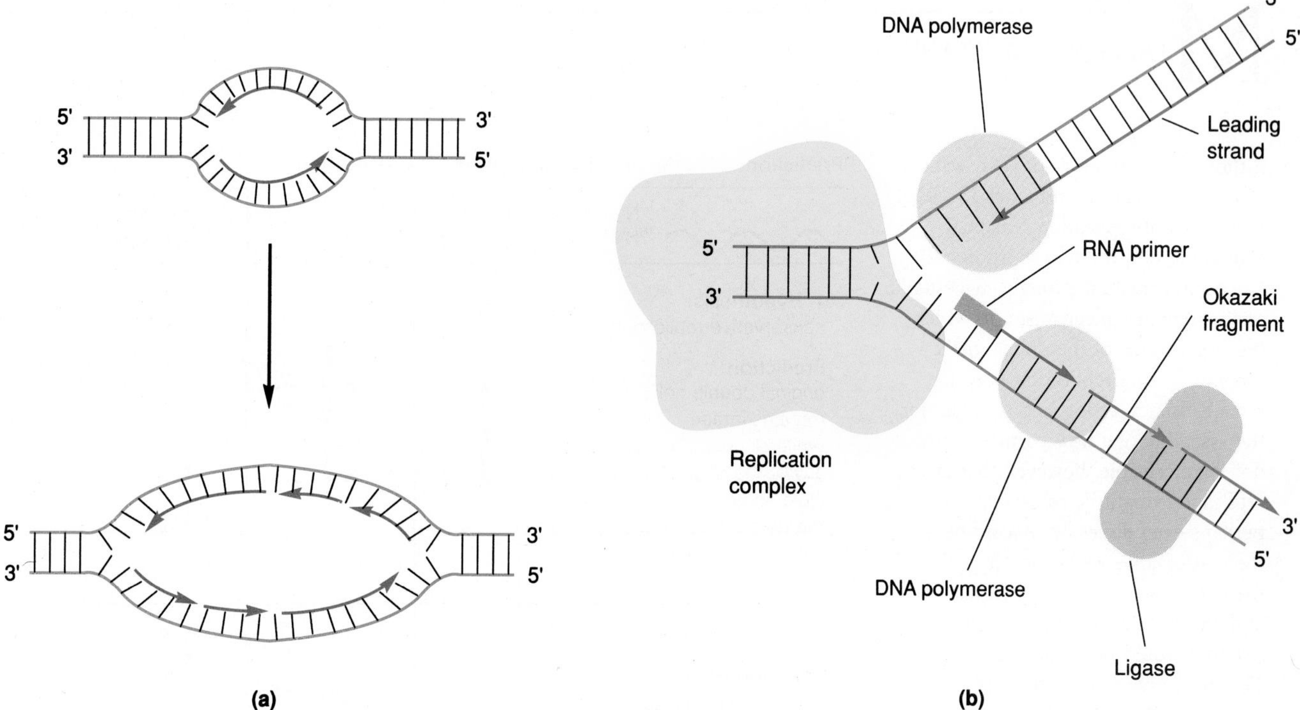

fork.) These short segments of newly assembled DNA are called **Okazaki fragments** (for their discoverer). As replication proceeds and nucleotides are added to the 3′ ends of the Okazaki fragments, they come to meet each other. When DNA polymerase meets the RNA primer from the previous segment, it excises the primers, fills in the gap with DNA, and then leaves the scene. Finally, the ends of the adjacent segments are joined together through the action of the enzyme **ligase.**

In the spite of such complications, we see that in the replication or synthesis of DNA, each individual strand remains intact, and each receives a new fully complementary strand. Such synthesis is termed **semiconservative,** which, as the name implies, means that part is retained and part is new.

There was a time, not long ago, when biologists were doggedly trying to determine the nature of replication; that is, whether it is semiconservative as opposed to either conservative or dispersive (the latter two suggest that the DNA molecule remains completely intact while being copied, or is completely dismantled and copies somehow reassembled). Essay 12.1 reviews one of the truly elegant experiments of this period, the experiment that showed that replication was indeed semiconservative.

FIGURE 12.9
DNA REPLICATION
(a) Localized separation of the two DNA strands at an origin of replication allows the beginning of DNA synthesis. As DNA synthesis proceeds in the 5′ to 3′ direction of the new strand, the "bubble" grows through the movement of two replication forks. However, the limitations of DNA polymerase dictate that only one of the new strands at a fork is continuously synthesized. The other strand can only begin synthesis when the template strand becomes single-stranded. **(b)** A detailed view of a replication fork shows that the discontinuous strand (the bottom one in this case) is synthesized as small fragments, Okazaki fragments, from RNA primers. Ligase then joins the fragments together. The replication complex (including gyrase, helicase, and single-strand binding proteins) leads DNA polymerase through for continuous 5′ to 3′ synthesis of the leading (upper) strand.

DNA Replication in Eukaryotes and Prokaryotes

Replication begins with the two sides of the twisted ladder of DNA being pulled apart at some point along the length of the strand and unwinding, forming a "bubble." Each end of the bubble, where the two single strands of DNA emerge from the double-stranded DNA, is a replication fork. Newly synthesized DNA strands are formed enzymatically on both of the unwound single strands following the Watson-Crick pairing rules. As replication proceeds, further unwinding moves the replication forks in opposite directions, causing the bubble to become larger.

In eukaryotes, each of the numerous chromosomes contains one very long, linear DNA molecule prior to DNA replication. At the initiation of DNA replication, hundreds

ESSAY 12.1

MESELSON AND STAHL

Predictions of competing hypotheses of DNA replication

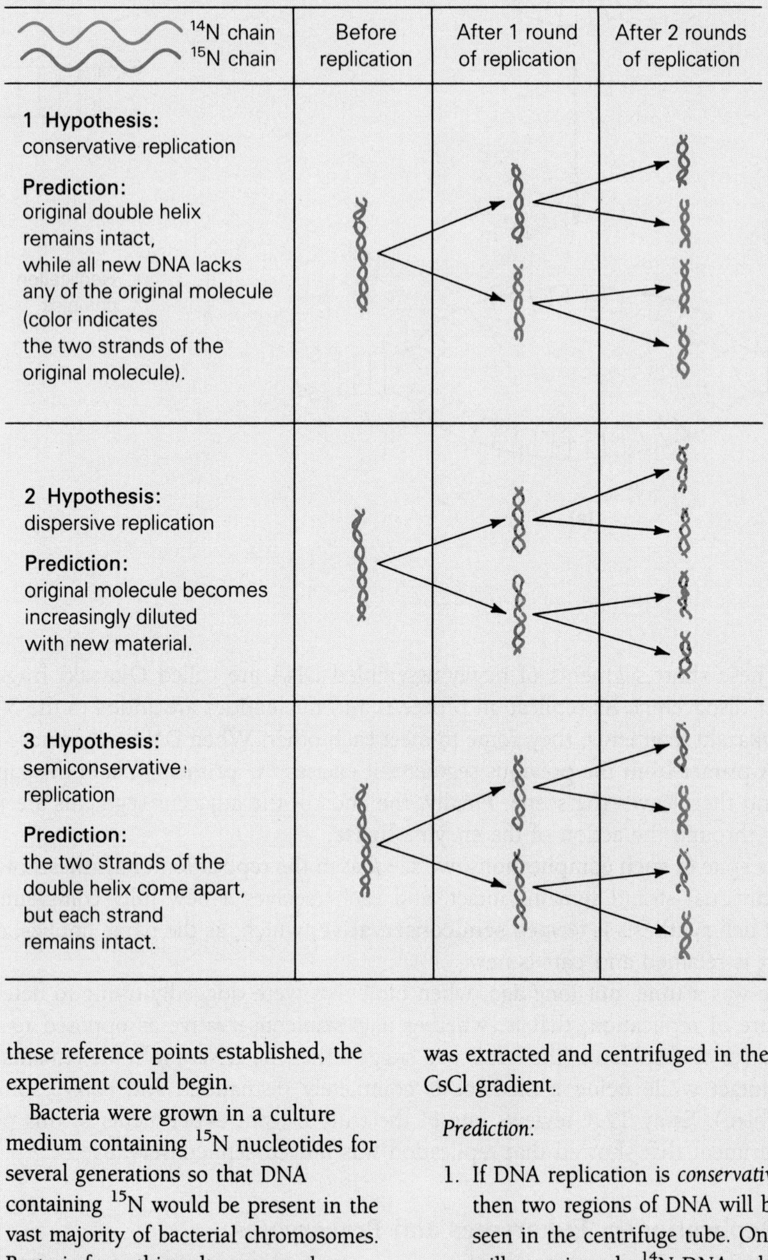

^{14}N chain ^{15}N chain	Before replication	After 1 round of replication	After 2 rounds of replication
1 Hypothesis: conservative replication **Prediction:** original double helix remains intact, while all new DNA lacks any of the original molecule (color indicates the two strands of the original molecule).			
2 Hypothesis: dispersive replication **Prediction:** original molecule becomes increasingly diluted with new material.			
3 Hypothesis: semiconservative replication **Prediction:** the two strands of the double helix come apart, but each strand remains intact.			

Although the structure of DNA was clarified by Watson and Crick in 1953, the details of replication remained something of a mystery. Watson and Crick's information strongly suggested a *semiconservative* mechanism, in which the two strands of the double helix separate during replication but each remains intact, acting as a template for the assembly of a new partner. It was not inconceivable, however, that DNA replication might be *conservative*, or perhaps even *dispersive*. According to the conservative replication hypothesis, the entire molecule would remain intact and another double strand of new material would be replicated alongside. The dispersive replication hypothesis stated that the strand would be dismantled piece by piece and replication would occur along the pieces.

In 1957, Matthew Meselson and Franklin Stahl set out to determine how replication was accomplished. Their procedure utilized two modern tools of biology, isotope-labeled biological chemicals and the ultracentrifuge. Using special techniques for isolating DNA from bacteria, they were able to centrifuge the DNA in a cesium chloride (CsCl) solution, which bands each molecule according to its specific gravity (density). Since bacteria are normally exposed only to nutrients containing common nitrogen 14, a ^{14}N reference line was available. In other words, DNA containing ^{14}N settled in the centrifuge tubes at a certain level, forming what was to be the first reference line. Next, a second reference line was established, this time for bacterial DNA extracted from cells that had been grown for many generations in culture media containing ^{15}N-labeled nucleotides. Since ^{15}N is heavier than ^{14}N, the ^{15}N DNA settled at a somewhat lower position in the centifuge tube. With these reference points established, the experiment could begin.

Bacteria were grown in a culture medium containing ^{15}N nucleotides for several generations so that DNA containing ^{15}N would be present in the vast majority of bacterial chromosomes. Bacteria from this culture were then removed, washed, and resuspended in a culture medium containing only nucleotides with ^{14}N. *One round of replication was permitted.* Cells from this stage were then removed, and the DNA was extracted and centrifuged in the CsCl gradient.

Prediction:

1. If DNA replication is *conservative*, then two regions of DNA will be seen in the centrifuge tube. One will contain only ^{14}N DNA, while the other will contain only ^{15}N DNA.
2. If replication is *dispersive*, all DNA after one round of replication will be halfway between the reference

densities of ^{14}N and ^{15}N DNA.

3. If DNA replication is *semiconservative,* then' only one region of DNA sediment will be found. This region will contain *hybrid* DNA, each molecule containing a ^{14}N strand and a ^{15}N strand. The sedimentation will occur halfway between the two reference lines established earlier.

In other words, one round of replication could not distinguish between the dispersive and semiconservative hypotheses. But other bacteria were allowed to continue into a *second round of replication.* The cells were then removed, and the DNA was isolated and centrifuged in the CsCl gradient.

Prediction:

1. If DNA replication is *conservative,* then the same two sedimentation bands will appear as they did in prediction 1, but the ^{14}N band will contain three times as much DNA as the ^{15}N band.
2. If DNA replication is *dispersive,* then all DNA after two rounds of replication will be uniform; namely, 25% will be ^{15}N and 75% will be ^{14}N. These should appear as a single band appropriately spaced between the two reference points.
3. If DNA replication is *semiconservative,* then two equal sedimentation bands will appear. One will contain *hybrid* DNA and will form a band at the same location as did hybrid DNA in prediction 2. The other will contain ^{14}N DNA only and will settle out at the ^{14}N reference line established earlier.

Results: The results, as diagrammed here, were clear. What Meselson and Stahl saw were the bands shown on the left. Their interpretation is shown on the right—only the semiconservative hypothesis could be supported. Note that the experiment was followed for four rounds of replication, just to be sure.

Results and interpretation

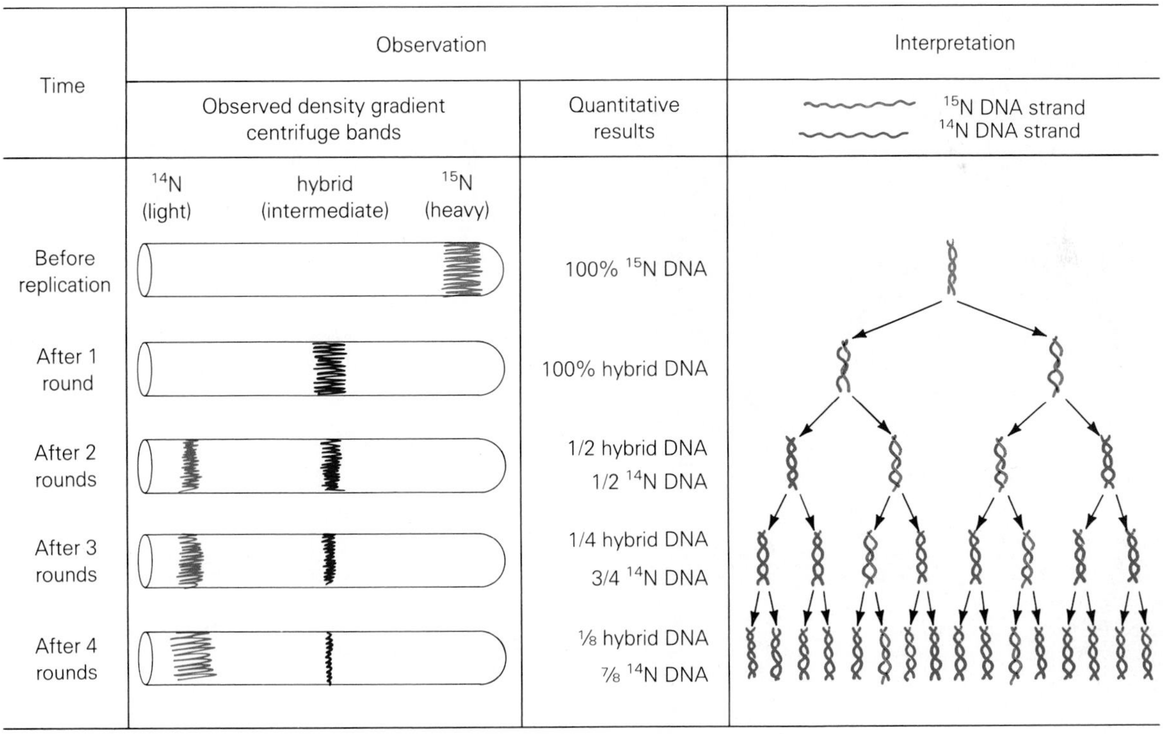

Time	Observation: Observed density gradient centrifuge bands (^{14}N (light), hybrid (intermediate), ^{15}N (heavy))	Observation: Quantitative results	Interpretation (^{15}N DNA strand; ^{14}N DNA strand)
Before replication		100% ^{15}N DNA	
After 1 round		100% hybrid DNA	
After 2 rounds		1/2 hybrid DNA 1/2 ^{14}N DNA	
After 3 rounds		1/4 hybrid DNA 3/4 ^{14}N DNA	
After 4 rounds		⅛ hybrid DNA ⅞ ^{14}N DNA	

FIGURE 12.10
SIMULTANEOUS REPLICATION
The task of replicating all of the very long DNA molecules in a eukaryotic nucleus is immense. Part of the problem is solved by having replication begin in many places, as these several replication bubbles on one *Drosophila* chromosome demonstrate.

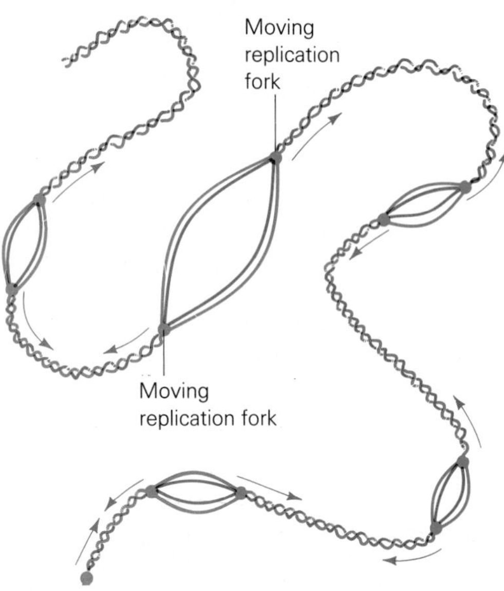

of bubbles are formed along the length of each DNA molecule (Figure 12.10). As the bubbles enlarge, the replication forks of different bubbles run into each other, and the bubbles merge. One replication fork runs off each end of the molecule, and eventually two long, linear DNA molecules exist where there was one before.

Prokaryotes are different in that most require only two replication forks. Multiple replication sites are essential to the eukaryote because its replication complexes work more slowly than do those of prokaryotes. While bacterial replication can proceed at a rate of 1 million base pairs added per minute, eukaryotic rates only range from about 500 to 5000 base pairs per minute. At that rate it would require about one month for a eukaryotic cell to complete replication, but because of multiple replication bubbles, each growing bidirectionally, replication in eukaryotes averages just a few hours. *Drosophila,* the fruit fly, may be a record holder, since the newly fertilized egg forms some 50,000 replication sites, and its four pairs of chromosomes can be replicated in about three minutes.

In prokaryotes, most or all of the DNA is in a single circular molecule that is not nearly as long as a eukaryotic DNA molecule (you have about 1400 times more DNA per cell than an average bacterium). There is no end and no beginning of a circle, of course, but replication in the bacterium *Escherichia coli* always begins at a single, specific initiation point (Figure 12.11). Again, unwinding proteins form a bubble of single-stranded DNA, and the unpaired strands are replicated with new "Watson" and "Crick"

FIGURE 12.11
REPLICATING CIRCULAR DNA
Many prokaryotic chromosomes are circular. As replication proceeds, the two replication forks travel in opposite directions, eventually meeting at the opposite side of the circle to produce the two replicas.

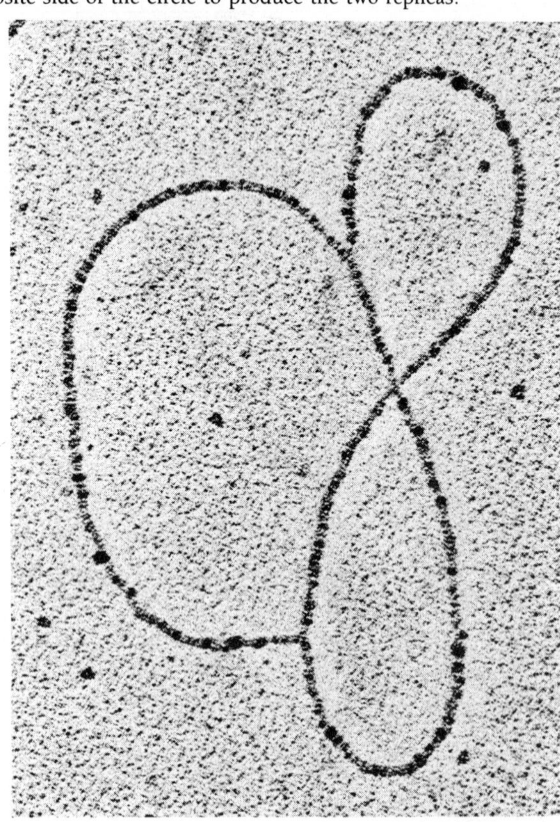

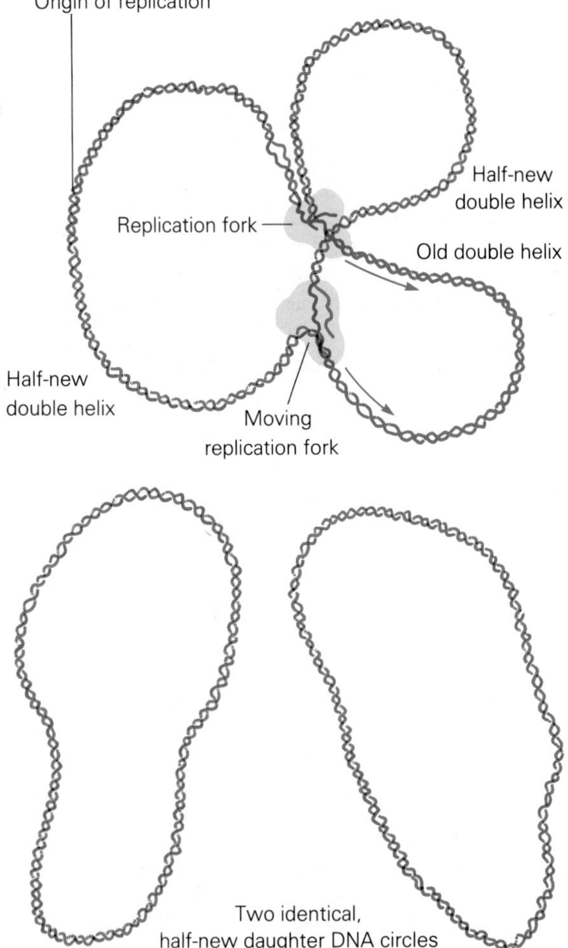

ESSAY 12.2

EUKARYOTIC DNA: A CLOSER LOOK

Earlier, we noted that eukaryotes have about 1400 times more DNA than bacteria. You may not be surprised at such a difference, considering how much more complicated we are than bacteria, but molecular biologists took a different, more searching view of this seemingly simple observation. They wanted to know why.

At least part of the answer comes easily; we have about 60 times more genes than bacteria. But why then, do we have 1400 times more and not 60 times? There must be something essentially different about the eukaryotic and prokaryotic structure. And there is.

It didn't take long for researchers to discover that whereas bacterial chromosomes consist mainly of long stretches of DNA that are unique, eukaryotic chromosomes contain thousands of repetitive base sequences that occur tens, hundreds, thousands, or even millions of times. These repetitive base sequences make up about 30% of human DNA. Let's take a closer look at their distribution, variety, and possible functions.

Satellite DNA

Highly repetitive or satellite DNA makes up about 10% of the total DNA in mouse chromosomes (but its relative concentration varies from species to species). It consists of a relatively small number of simple nucleotide sequences like ATATATATAT, CGCGCGCG, and so on—but each may be present from 100,000 to over 1,000,000 times in the DNA of a single cell. Researchers have discovered that this form of DNA is highly localized in centromeric regions and at the tips of chromosomes. Although no one knows its function for certain, experiments suggest that it may be necessary for chromosome structural stability.

Middle Repetitive DNA

Another form of repetitive DNA found in eukaryotic chromosomes is called middle repetitive DNA, because each individual base sequence is present from 20 to about 100,000 times in a single cell's genome.

Middle repetitive DNA, like highly repetitive DNA, is common in centromeric regions, where it is thought to help maintain their highly coiled structures. But it is also found interspersed between the regions of single copy DNA that represent functional genes. Some middle repetitive DNA may contribute to the control of gene activity. Yet another fraction of a cell's middle repetitive DNA belongs to a class of mobile genetic elements called transposons. Transposon biology is discussed more fully in Chapter 17. But for now you should note that these "jumping genes" are able to insert copies of themselves throughout the genome and are an important source of mutation. As such, they may have served an important role in the evolutionary process. Geneticists have been theorizing for years about what good they are. But whatever the answer, they make up a substantial portion of a cell's genome. One, called Alu, is repeated 30,000 times and makes up three percent of the human genome!

A final example of middle repetitive DNA represents genes that may be repeated from a few to hundreds of times per cell. Such genes usually are ones that code for molecules the cell needs to produce in vast quantities during short times. For instance, in certain active oocytes there are thousands of copies of the genes that produce ribosomal RNA, an essential component of the apparatus that translates genetic messages into polypeptides.

Single Copy DNA

In eukaryotes, the single copy DNA we expect "genes" to be made of is about 70% of the total. An appreciable portion of the DNA is there for structural reasons, some consists of transposons, and some may be important in the control of gene activity. Yet another portion represents genes that are replicated, for various reasons, from a few to hundreds or thousands of times. But still, the single copy fraction is far larger than needed to code for our proteins.

strands in the usual way. The two replication forks—and in prokaryotes there are only two—travel in opposite directions around the circular DNA molecule, eventually meeting each other at a specific termination point halfway around the circle. In Figure 12.11, a smaller *E. coli* chromosome, known as a plasmid, is caught in the act of replicating.

DNA AND GENETIC INFORMATION

We have seen now how DNA meets two of the requirements of the genetic material. It has the capacity to replicate itself precisely. It can contain information by having specific sequences of bases. These properties of DNA are a direct result of its structure. Yet as the story of DNA as the genetic material began to unfold, it was realized that somehow the information held within DNA must be expressed as proteins.

ESSAY 12.3

ONE GENE, ONE ENZYME

The slogan "one gene, one enzyme" was electrifying in its day, some 30 years after Garrod's work. It referred to the Nobel Prize-winning studies of George Beadle and Edward Tatum, who, like Garrod, looked to heredity for an explanation of metabolic pathways. But instead of drawing inferences from human abnormalities or crossing peas, Beadle and Tatum imaginatively chose what was until then a very unusual experimental organism: the fungus *Neurospora,* the same pink mold that may have ruined your bread on occasion. *Neurospora* was to be the first in a series of important microorganisms that would be used in a new field soon to be called molecular biology.

Neurospora crassa (pink mold), *Escherichia coli* (the colon bacterium), *Saccharomyces cerivisiae* (brewer's yeast), and bacteriophages (viruses that infect bacteria) were soon to become the standard tools of latter-day geneticists and molecular biologists. Such microorganisms have two huge advantages for genetic study over mice, flies, peas, humans, and many other organisms: (1) they can be grown cheaply in enormous numbers in a very short time, and (2) unlike higher organisms whose genes occur in pairs, through most of their life cycle these organisms have only one copy of each gene, so many genetic implications can be avoided altogether. Further, the effects of mutation (sudden genetic changes) can be seen immediately, since they aren't hidden by a corresponding normal copy.

Beadle and Tatum produced random mutations by using X-rays to irradiate *Neurospora* spores (dormant, resistant cells important to survival and reproduction; see Chapter 24). They then grew the irradiated spores and screened them for biochemical mutations; that is, they looked for strains that could not grow unless certain simple biochemical compounds were added to the medium (the food on which the fungus was grown). These simple compounds were the metabolites (intermediate products), such as vitamins and amino acids, that are routinely present in biochemical pathways under normal conditions. Their idea was that if a mutant gene was not producing a certain enzyme, then the enzyme's usual product would not be produced, and the biochemical pathway would be brought to a lethal halt. The biochemical pathway could be said to be blocked at a critical step. But adding the missing product of the blocked step would unblock the pathway and allow it to proceed to completion, allowing the fungus to thrive.

Once the nutritional mutants had been identified, the strains could be maintained and used to determine the hereditary basis for enzyme deficiencies. All of the enzyme deficiencies turned out to be the result of simple single gene changes; hence the slogan "one gene, one enzyme," or, in translation, "It requires the action of one gene to produce one enzyme." Garrod's idea had been rediscovered and confirmed: biologists were now more confident that specific genes were responsible for the presence or absence of specific proteins. Actually, the "one gene, one enzyme" slogan has had to be revised since that day. As we will learn in the next chapter, genes actually code for polypeptides. You will recall that many proteins, and the complex enzymes in particular, consist of two or more polypeptides, and these are considerably modified before being incorporated into protein.

In fact, the idea that genetic information was enacted by protein had been around since 1908. In that year, A. E. Garrod, a physician influenced by Mendel's work, published a book called *Inborn Errors of Metabolism.* His subject was *Homo sapiens,* which, at the time, was an unusual experimental organism for genetics research. Garrod was interested in metabolic defects, breakdowns in the complicated biochemical processes of life. He searched for the abnormal products of such defects in the urine, where many metabolic products end up. Of special interest to Garrod was alkaptonuria, a disease in which the urine contains metabolites (products of metabolism) called alkaptones—substances that happen to turn black upon oxidation, and so are easily revealed. Infants with alkaptonuria are usually detected as soon as their diapers start turning black. As the child grows older, black pigments begin to settle in cartilage and other tissues, blackening the ears and even the whites of the eyes. Another more serious effect is a form of arthritis, caused by the accumulation of the metabolite in the cartilage of the joints.

Garrod observed that the disease tended to be found in several brothers or sisters in a single family. By studying family histories, he correctly inferred that alkaptonuria and

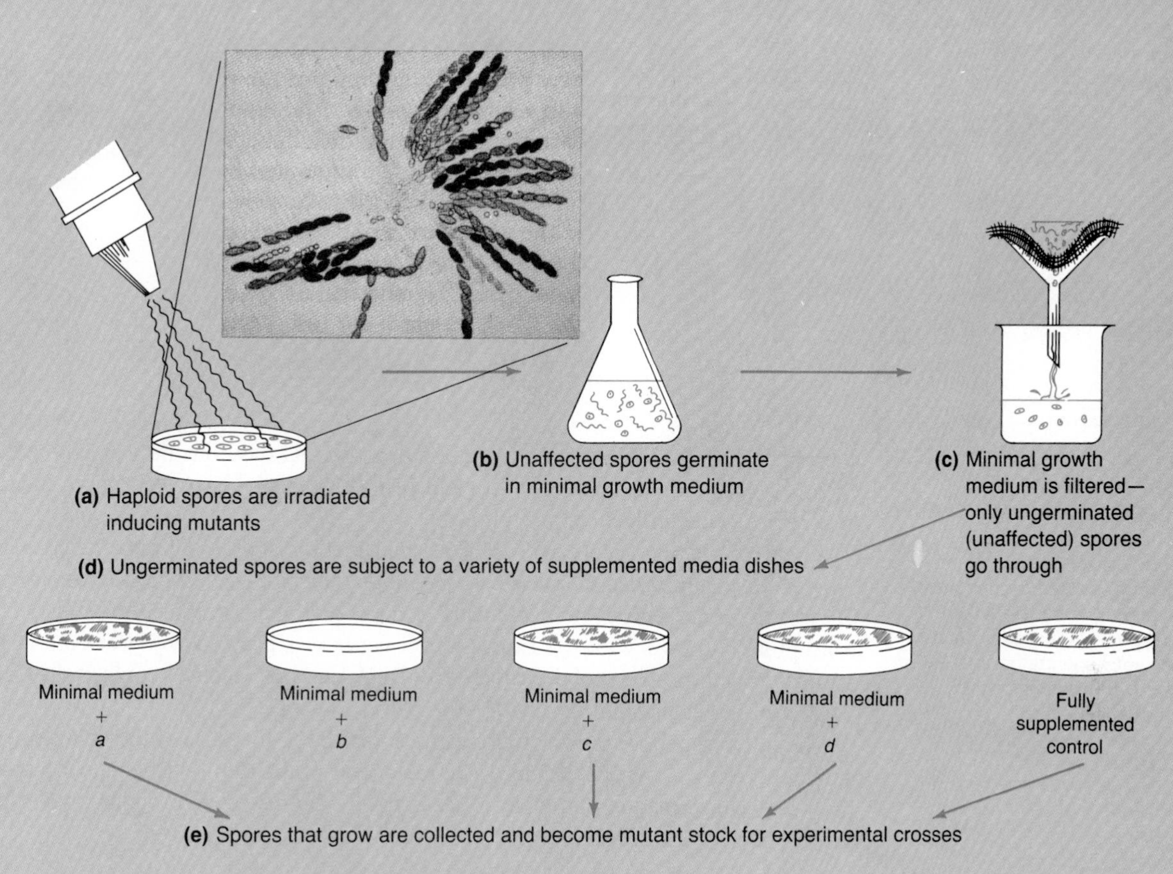
(a) Haploid spores are irradiated inducing mutants
(b) Unaffected spores germinate in minimal growth medium
(c) Minimal growth medium is filtered—only ungerminated (unaffected) spores go through
(d) Ungerminated spores are subject to a variety of supplemented media dishes
Minimal medium + a
Minimal medium + b
Minimal medium + c
Minimal medium + d
Fully supplemented control
(e) Spores that grow are collected and become mutant stock for experimental crosses

certain other inborn errors of metabolism were genetic in origin. The problem, he deduced, is caused by the absence of specific enzymes that are necessary for the long chains of biochemical reactions to occur. If an enzyme for a particular reaction is absent, no reactions can take place past the point where the enzyme is normally a part of the chain, and the substance that the enzyme acts on builds up.

Other inborn errors of metabolism create albinism (a complete or partial lack of melanin pigment in the hair, skin, iris, and sometimes the retina) and phenylketonuria (which also affects hair and skin pigmentation but has a much more severe effect on mental development because of the accumulation of toxic metabolites in the nervous system).

As it turns out, albinism, phenylketonuria, and alkaptonuria are all caused by defects in enzymes that act in the metabolism of the amino acids phenylalanine and tyrosine (Figure 12.12).

Significantly, Garrod discovered that heredity played a role in enzyme activity and that there was a definite connection between heredity and the presence of normal or

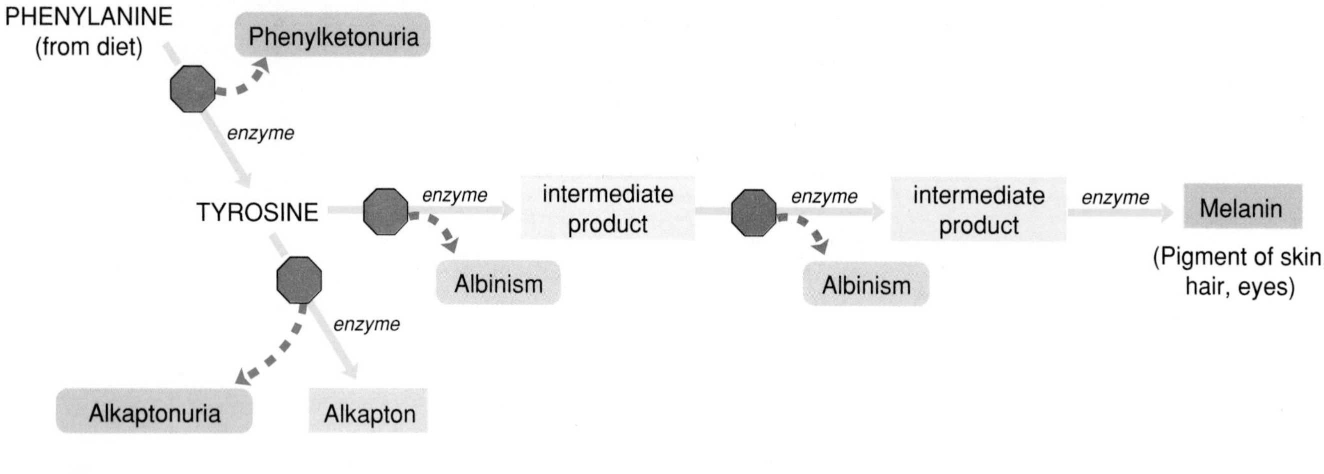

FIGURE 12.12
METABOLIC PATHWAYS OF TYROSINE AND PHENYLALANINE
Garrod's work established the relationship between genes and enzymes. Three conditions caused by the blocking of enzymes are albinism, alkaptonuria, and phenylketonuria. All the blocked enzymes operate on intermediate breakdown products of the amino acid phenylalanine.

abnormal enzymes—quite an accomplishment for his day. But he was ignored. Biochemistry was still an infant science, and geneticists of the time were more interested in how the genes influenced morphology.

So even after the discovery of DNA as the genetic material, the remaining issue was "how does the information stored in the DNA molecules get expressed as proteins?" That issue would lead researchers finally to the core of gene structure and function. The conceptual framework for the expression of genes was provided by Watson and Crick soon after their discovery of the structure of DNA, and has become known as the **Central Dogma** of molecular biology.

The initial aspect of the Central Dogma is that DNA (because of the complementary strands) directs its own replication. The remainder of the Central Dogma concerns how the genetic information stored in DNA is finally expressed as proteins, and it is the subject of the next chapter.

APPLICATION OF IDEAS

In the prehistory of life, nucleic acids might have come before proteins, or proteins before nucleic acids (this is still being argued). How could a purely protein organism store information or reproduce such information for the next generation? Is there some other way around the problem?

KEY IDEAS

HIGHLIGHTS FROM THE DISCOVERY OF DNA AS THE GENETIC MATERIAL

Mendel helped establish that heredity was controlled by "factors," and chromosomes were soon suspected of carrying the factors (genes).

The Early Efforts

Miescher identified DNA in 1869, and in 1914 Feulgen perfected a specific DNA stain (Feulgen stain); however the connection between DNA and heredity was not made for many years.

Transformation

In 1928 Griffith, experimenting with virulence in *Pneumococcus,* determined that nonvirulent strains could be transformed (genetically changed) to virulent strains if the remains of dead virulent bacteria were made available. In 1944, Avery concluded that the transforming material was DNA.

The Experiments of Hershey and Chase

Using a bacteriophage and the bacterium *Escherichia coli,* Hershey and Chase were able to show that only viral DNA entered the host; thus it was DNA that directed the production of new viral particles. This strongly suggested that DNA was the genetic material.

THE STRUCTURE OF DNA

Four different deoxynucleotides, or nucleotides, the structural units of DNA, are assembled into long polymers of DNA strands, or nucleic acids. Prior to assembly, they are in the form of *nucleotide triphosphates* similar to ATP.

DNA Nucleotides

1. Each nucleotide contains the three parts: phosphate, deoxyribose, and a nitrogenous base, in that order.
2. The four bases of DNA, their designations and their triphosphate form are **adenine** (dATP), **guanine** (dGTP), **thymine** (dTTP), and **cytosine** (dCTP).
3. In 1950, Chargaff developed the principle of base-pairing. He determined the relative amounts of A, T, G, and C in a variety of cells, proving that A = T and G = C and that there is exactly as much **purine** in the nucleus as there is **pyrimidine.**
4. Through the use of X-ray crystallography, Wilkins and Franklin determined that DNA was double stranded, probably formed a helix, and had intramolecular measurements of 2.0 nm, 0.34 nm, and 3.4 nm.

The Watson and Crick Model of DNA

1. In 1953, having used critical information from the work of others and by constructing models of their own, Watson and Crick determined the structure of DNA, including its phosphate-sugar backbone, specific (A-T, G-C) base-pairing of purines and pyrimidines, and the meaning of the intramolecular distances. (A-T or T-A), and guanine with cytosine (G-C or C-G).
3. In the double helix, the two polymers run in opposite directions (5′-to-3′ and 3′-to-5′). Many millions of nucleotides may be present.

DNA REPLICATION

1. Replication is the preparation of DNA copies prior to reproduction of the cell or organism.
2. Because of specific base pairing, upon separation of the DNA double strand, each strand can reproduce the other ("Crick" strands can form "Watson" strands, and "Watson" strands "Crick" strands).

The Chemistry of DNA Synthesis

1. When incorporated into DNA, a pyrophosphate is released from each nucleotide triphosphate. Nucleotides are joined by their phosphates and sugars, which form the backbone of the polymer with the nitrogen bases projecting off the side.
2. Synthesis of DNA polymers proceeds from the **5′ end** to the **3′ end.** In its finished form, DNA is a double strand of nucleotides wound into a double helix.

Origins of Replication

1. Replication is carried out by **replication complexes,** which include the unwinding enzyme helicase and the nucleotide adding enzyme DNA polymerase.
2. The helix is unwound, the separated strands form replication forks, and new nucleoside triphosphates are added according to base pairing.
3. The addition of bases to the leading end of a polymer occurs smoothly, one base at a time, but at the lagging end, **Okazaki** fragments must first be assembled in the 5′-to-3′ direction, and then, utilizing the enzyme ligase, they are added in.
4. Because each new polymer is base-paired to an old one, DNA replication is called **semiconservative** (half is conserved).

DNA Replication in Eukaryotes and Prokaryotes

1. Replication sites "bubble out" as they form, and bubbles lengthen as replication proceeds in both directions.
2. In eukaryotic replication, multiple replication forks work simultaneously, forming the many bubbles. In prokaryotes, only two replication forks (one bubble) form along the circular chromosome, but replication in prokaryotes is much faster.

DNA and Genetic Information

1. Garrod identified metabolic disorders such as alkaptonuria by the presence of abnormal metabolites such as alkaptones in the urine. He determined that such conditions were inherited and surmised that they involved abnormalities in the enzymes of metabolic pathways. He correctly associated the abnormal metabolites with abnormal enzymes and such enzymes with abnormal genes.

REVIEW QUESTIONS

1. Briefly summarize Griffith's observations. What was his conclusion, and what would we add to this conclusion today? (273-274)
2. Describe how the use of radioactive tracers by Hershey and Chase led to identifying DNA as the hereditary agent of the phage virus. (275-276)
3. If Chargaff had worked with the bacterium *E. coli* only, how would this have affected the discovery of base-pairing? Describe what he actually found. (278)
4. What did the measurements 2.0 nm, 0.34 nm, and 3.4 nm mean to Wilkins and Franklin? What do they actually represent? (279)
5. List three aspects of DNA structure, previously discovered by others, that led Watson and Crick to construct an accurate model of DNA. Which of these led to Crick's prediction of how replication would work? (279-281)
6. Using a simplified drawing, identify each component of a nucleotide and explain where they are bound together. (277)

7. List the four nucleotides in DNA, and identify whether they are purines or pyrimidines. (277)

new nucleotides attach and in what direction is the chain synthesized? (284-285)

8. Explain how a nucleoside triphosphate is added to a growing polymer of DNA. Specifically, where do the

9. Describe the final structure of DNA. Include its geometric form, what holds it together, and some idea of its length. (280-281)

10. Using the terms *hydrogen bonding, pyrimidine, purine,* and the letters *G, A, C,* and *T,* explain the manner in which nitrogen bases fit together in the completed DNA molecule. (280-281)

11. Explain why base-pairing is so essential to the process of replication. (284)

12. List the three components found in a replication complex, and briefly explain what each does. (286)

13. In what direction *must* replication proceed? Since replication on each DNA strand proceeds in both directions, what problem does this introduce? How is the problem solved? (286)

14. Compare prokaryote and eukaryote replication in terms of replication forks and bubbles. Suggest reasons why the *rate* of replication in prokaryotes may be so much faster than it is in eukaryotes. (288)

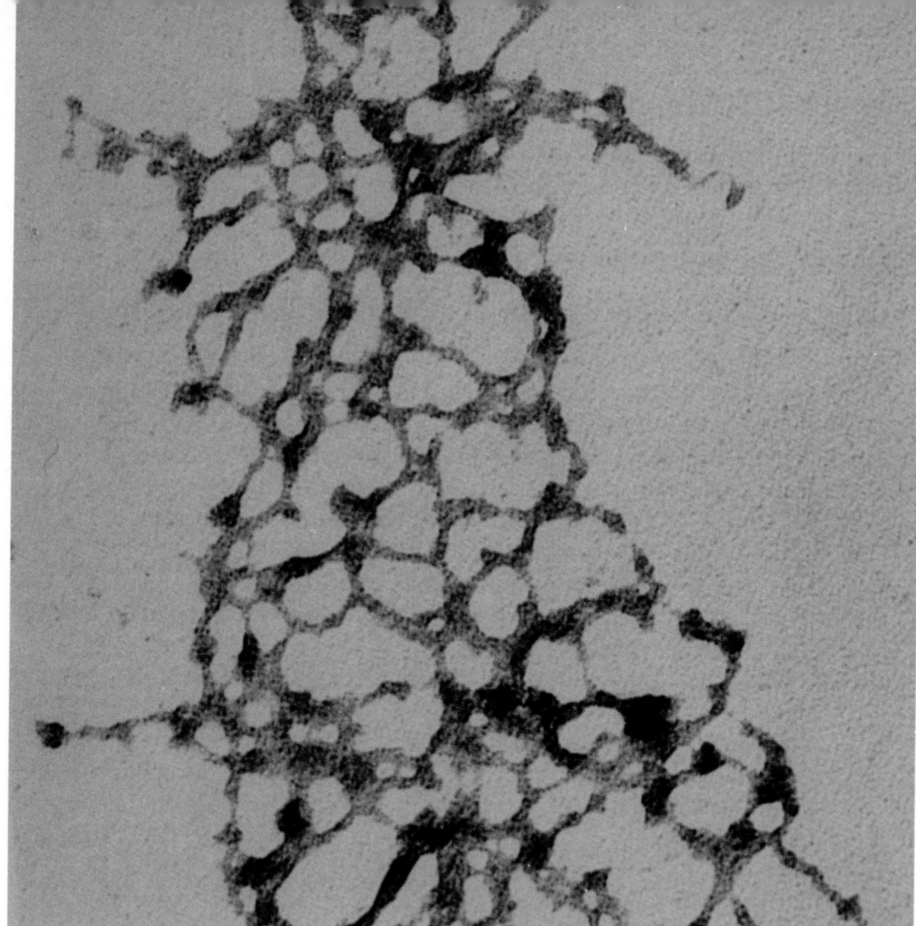

13

GENES IN ACTION

THE STORY OF THE SEARCH FOR THE "STUFF OF HEREDITY" AND THE UNveiling of its structure indeed makes a good tale. But there is more of the story to tell—the part about how genes actually work. Here, we will tell the rest of the story. Particularly, we will see how the information stored in genes dictates the formation of proteins, which by their structure or by their enzymatic function determine the phenotype of the organism.

A full description of DNA as the genetic material requires that two major points be established. The first point is that DNA contains all of the information needed to produce new cells and even entire individuals, and this information is faithfully replicated prior to cell division. We saw this process at work in Chapter 12. In fact, as we left Chapter 12, we briefly mentioned how replication fits into a concept referred to as the Central Dogma (Figure 13.1). (While the use of the term "dogma" is probably improper within any scientific context, its use here is generally accepted.) The second major point involves the rest of the Central Dogma. As shown in Figure 13.1, DNA meets the rest of its requirements as the genetic material as it dictates the business of the cell by transcribing genetic information into RNA, which is in turn translated into polypeptides. The Central Dogma, then, states that genetic information is encoded in DNA, transmitted by DNA replication, transcribed into RNA, and translated into protein.

The discovery that genes are composed of DNA and that their information lies in the linear arrangement of DNA nucleotides was indeed an enormous breakthrough, but it also left many questions unanswered. For example, how exactly are the genetic instructions stated? And how do they determine the order of amino acids in proteins? It didn't take researchers long to determine that DNA is not *directly* involved in making proteins. It couldn't be, because in eukaryotes, DNA remains in the nucleus, while protein assembly occurs in the cytoplasm. So DNA directs protein synthesis by long distance. But how? The answer is by sending its instructions on molecules of RNA.

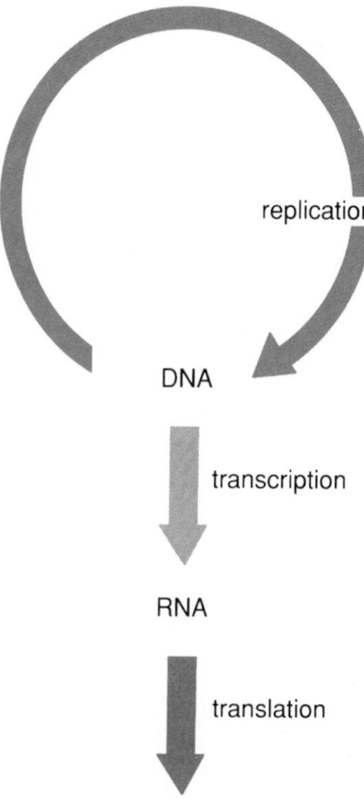

FIGURE 13.1
THE CENTRAL DOGMA
This scheme of information flow is the cornerstone of our understanding of the way in which the hereditary data is stored and replicated as DNA, yet enacted and expressed as protein. The three processes involved are replication, transcription, and translation. We have already discussed replication in Chapter 12. Transcription is the process by which the information stored in DNA is converted to RNA, while translation (which quite logically involves changing languages—from nucleotides to amino acids) is the process that converts the information from RNA into protein.

RNA STRUCTURE AND TRANSCRIPTION

RNA and DNA are very similar molecules. However, their roles in protein synthesis are distinct and dependent upon the differences that *do* exist between them.

Comparing RNA to DNA

Let us begin our discussion of RNA with a brief review of its structure (Figure 13.2).

1. The 5-carbon sugar in RNA is *ribose* instead of *deoxyribose*. All this really means is that RNA has a hydroxyl group attached to the 2′ carbon instead of a hydrogen. (Thus DNA lacks an oxygen there, hence the term *deoxy*.)
2. While both RNA and DNA contain adenine (A), guanine (G), and cytosine (C), the fourth nucleotide base differs in the two molecules. DNA contains thymine (T), whereas RNA contains **uracil** (U), a closely related but slightly different nitrogen base that base-pairs with A the same as thymine.
3. DNA almost always occurs as a double-stranded helix. RNA almost always occurs as a single-stranded molecule, which can have complex, twisted, and folded secondary and tertiary structures.
4. DNA molecules are almost always much longer than RNA molecules—typically, a thousand to a million times longer.
5. DNA is generally more stable than RNA; it is more resistant to spontaneous and enzymatic breakdown, and damage can be repaired because the opposite strand contains the complementary information. RNA is more reactive partly because of the additional reactive —OH side group of ribose, and direct repairs are not possible.
6. Although there is only one type of DNA, there are several classes of RNA, each with its own function.

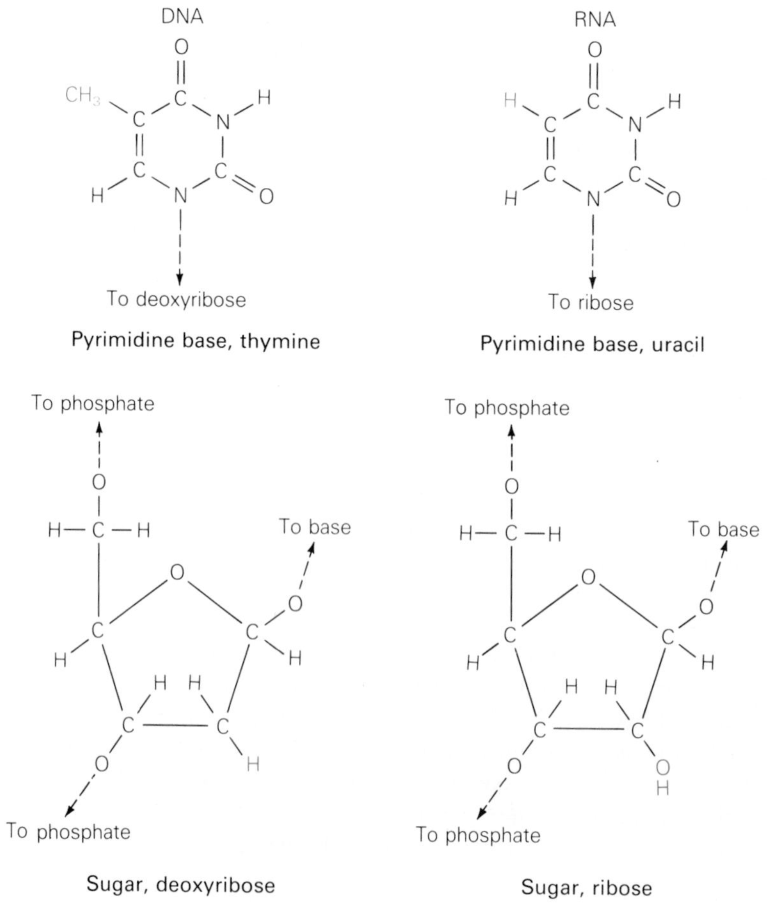

FIGURE 13.2
RNA VS DNA
RNA nucleotides differ from those of DNA in two chemical ways. First, RNA contains the sugar ribose, which contains an —OH group at the 2′ carbon rather than the —H at the 2′ carbon of deoxyribose. Second, in RNA uracil replaces thymine. Uracil and thymine are structurally similar (thymine has a methy group where uracil has a hydrogen), and both base-pair with adenine.

DNA

To deoxyribose

Pyrimidine base, thymine

RNA

To ribose

Pyrimidine base, uracil

To phosphate

To base

To phosphate

Sugar, deoxyribose

To phosphate

To base

To phosphate

Sugar, ribose

Transcription: RNA Synthesis

The process by which the chemical information encoded in DNA is copied into RNA is called transcription. Generally, only one strand of the double-stranded DNA is transcribed. Therefore, the segment of the DNA molecule from which an RNA molecule is transcribed is in a very real sense equivalent to a gene. In the past few chapters, we've been discussing genes as units of information. Now we can begin to tighten our definition and say that a **gene** is a segment of DNA that (through RNA) specifies a protein. Most genes are composed of two general parts: a **coding region** to specify that portion of a gene or DNA molecule that is transcribed into RNA, and a **regulatory region** to regulate transcription of the coding region.

The enzymes involved in transcription are **RNA polymerases,** which happen to be among the largest known enzymes. Eukaryotes contain three different types of RNA polymerases, each of which helps form a specific type of RNA.

RNA polymerase can begin assembling chains of new RNA bases only after it identifies and binds to a specific DNA sequence in a regulatory region known as a **promoter.** A promoter is defined as the DNA within the regulatory region that directs the binding of RNA polymerase and the subsequent transcription of a coding region. (We will return to promoters in the next chapter.)

Following the binding of RNA polymerase to the promoter, the giant enzyme apparently begins to break the hydrogen bonds of the complementary bases, unwinding the DNA. It unwinds about one full turn of the DNA helix, exposing that segment of unpaired DNA bases. It is here that transcription will begin and the base-pairing between DNA and RNA will occur. As a segment is transcribed, the opened helix rewinds while a new segment ahead unwinds (Figure 13.3). Interestingly, as the new RNA bases are brought in and base-paired with the template DNA strand, they temporarily form a DNA:RNA double polymer, as though a kind of replication were occurring. However,

FIGURE 13.3
TRANSCRIPTION: DNA→RNA
In the synthesis of RNA during transcription, the enzyme RNA polymerase moves along the DNA helix, unwinding small portions of the helix as it goes. Once RNA for a particular region has been made, the double helix quickly reforms, thereby displacing the growing single strand of RNA.

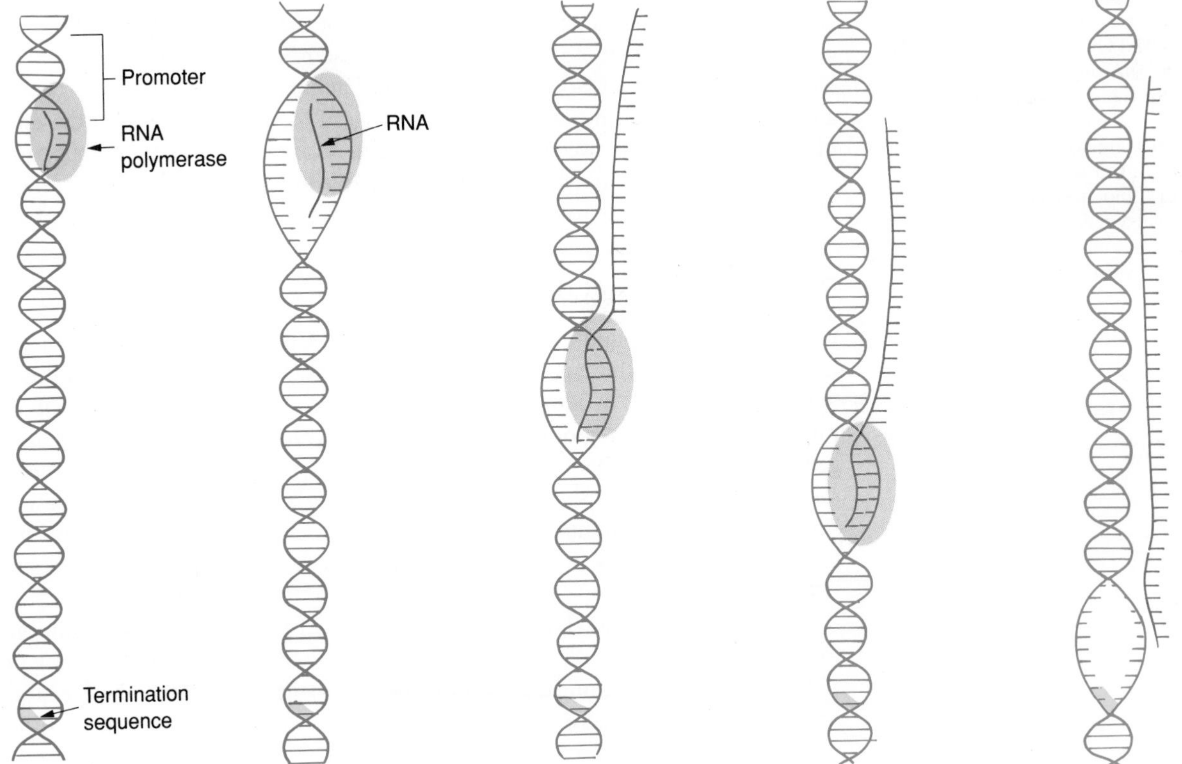

the original DNA helix soon reforms, with the RNA displaced by the nontranscribed DNA strand.

Except for the substitution of uracil for thymine, the base-pairing rules and chemical aspects of RNA synthesis are the same as those of DNA replication, and the behavior of RNA polymerase is similar to that of DNA polymerase. The RNA strand is synthesized in the 5′–3′ direction, using ribonucleoside triphosphates (CTP, GTP, ATP, and UTP). By way of summary, then, for every C in the template DNA strand, RNA polymerase puts in a G ribonucleotide; for every G, a C; for every T, an A; and for every A, a U. When the process is completed, the new RNA has the same order of bases as the appropriate stretch of the nontranscribed strand of DNA, with U substituting for T (see Figure 13.3).

Transcription of a DNA strand continues until the moving polymerase encounters what is called a **termination signal.** Like the promoter, this consists of a special sequence of DNA bases, but it acts to dislodge the growing RNA strand and to release the bound RNA polymerase.

Many RNA molecules can be transcribed from the same gene simultaneously, for as soon as the first few bases of a sequence have been copied and the RNA polymerase has physically moved out of the way, another RNA polymerase can bind to the promoter and initiate transcription (Figure 13.4).

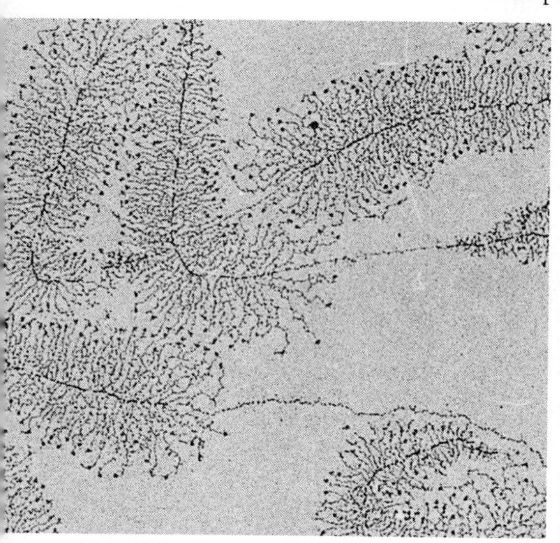

FIGURE 13.4
SIMULTANEOUS TRANSCRIPTION
Once an RNA polymerase molecule has begun transcription, there is no reason why a second one cannot begin transcription before the first is finished. In fact (as seen in this photograph) if much RNA is needed in a short time, a gene can become loaded with a large number of polymerase molecules, all transcribing at the same time.

VARIETIES OF RNA

Remember that the final role of the Central Dogma (Figure 13.1) is the production of proteins from the information that has been stored in DNA. We have stated, in general terms, that RNA plays a key role in that process. Let's now examine the different classes of RNA to see what specific role each of them plays. Once we have described these players, we will put them all on stage to describe the process itself.

There are three kinds of RNA in most cells: **ribosomal RNA (rRNA), messenger RNA (mRNA),** and **transfer RNA (tRNA).** Each is transcribed from DNA, as described above, by one of the three classes of RNA polymerase. Ribosomal RNA and transfer RNA may be thought of as "bit players," but in their supporting roles they function in the expression of virtually every gene. Ribosomal RNA contributes significantly to the structure of ribosomes (the site of protein synthesis, as we will see). Transfer RNA is the key intermediary, responsible for bringing the proper amino acid to be put into the protein. Messenger RNA is the star. It is in the direct line of information flow between DNA and protein. As its name implies, mRNA carries the coded message that will determine the polypeptide to be produced. The messenger RNA for each gene is unique to that gene, so there are many thousands of different mRNAs.

Ribosomal RNA and the Ribosome

Ribosomal RNA is found in ribosomes. While this isn't too surprising, its transcription is quite special. In the eukaryotes, ribosomal RNA, unlike other RNA, is transcribed within the nucleolus. In addition, the DNA involved in the transcription is found within **nucleolar organizer regions,** regions of the DNA that contain multiple copies of the gene responsible for rRNA transcription. Why are so many identical genes needed? It seems that at certain times the demand for rRNA is much greater than at other times. Thus, when large amounts of rRNA are needed, a large number of RNA polymerases can travel along copies of the transcribing genes, spinning off strand after strand of rRNA.

There are three major forms of rRNA that are all transcribed as one unit, one very long primary transcript of the rRNA gene. Following their production, the long primary rRNA transcripts are immediately processed to yield the specific shorter strands of ribosomal RNA needed for ribosome assembly. The three forms are called **18S, 5.8S,** and **28S rRNAs.** A fourth member of the group, **5S rRNA,** is transcribed from a separate gene and prepared outside the nucleolus. ("S" is a sedimentation or density unit used in describing the results of ultracentrifugation and reflects the size and shape of a molecule or particle. Basically, the larger the S value, the larger the particle.) While the various rRNAs will form the skeleton of the ribosome, the remainder will consist of special ribosomal proteins assembled in the cytoplasm. Such proteins enter the nucleus and find their way to the nucleolus, where they join the rRNA. But while ribosomal assembly begins in the nucleolus, it must be completed in the cytoplasm.

Completed ribosomes are made up of two different-sized subunits (Figure 13.5). The smaller subunit fits into the larger in an elaborate interlocking manner, and when assembled will clamp itself over a messenger RNA molecule. The larger—**60S—subunit** contains the 28S, 5.8S, and 5S rRNA, while the smaller —**40S—subunit** contains the 18S rRNA. Proteins make up about half the ribosomal mass, with the smaller subunit containing some 30 different proteins and the larger 45 to 50, all tightly bound to the rRNAs. Prokaryotic ribosomes, by the way, are somewhat smaller than those of eukaryotes, but still consist of a large and small subunit. (It turns out, in fact, that many of the antibiotics used to fight bacterial infections take advantage of the differences between bacterial and eukaryotic ribosomes in order to kill bacteria without harming the eukaryotic host.)

Ribosomes are the only places where proteins are synthesized. The ribosome is where mRNA, tRNA, and the growing polypeptide chain of a protein all work together to create their magic: making proteins. The ribosome itself is not passive in this process. Its surface has specific attachment sites (specially shaped cavities) that will allow tRNAs and mRNA to be in the proper close contact. There is also a site where the enzyme peptidyl transferase works to form peptide bonds between adjacent amino acids (see Figure 13.5). (You may want to briefly review protein structure in Chapter 3.)

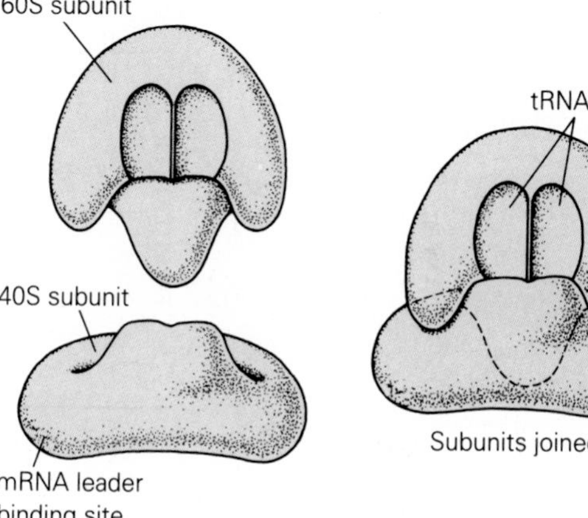

FIGURE 13.5
RIBOSOME
Ribosomes are composed of two different subunits that are physically joined only during the process of translation. The eukaryotic ribosomes consist of a larger 60S subunit and a smaller 40S subunit. The assembled ribosome has two pockets for the binding of tRNAs and a groove between the subunits for the binding of mRNA. Within the ribosome is also a site for the enzyme peptidyl transferase, the enzyme that actually joins amino acids together.

Messenger RNA

As its name suggests, messenger RNA carries the message—in this case, information regarding the sequence of amino acids of the polypeptide to be produced.

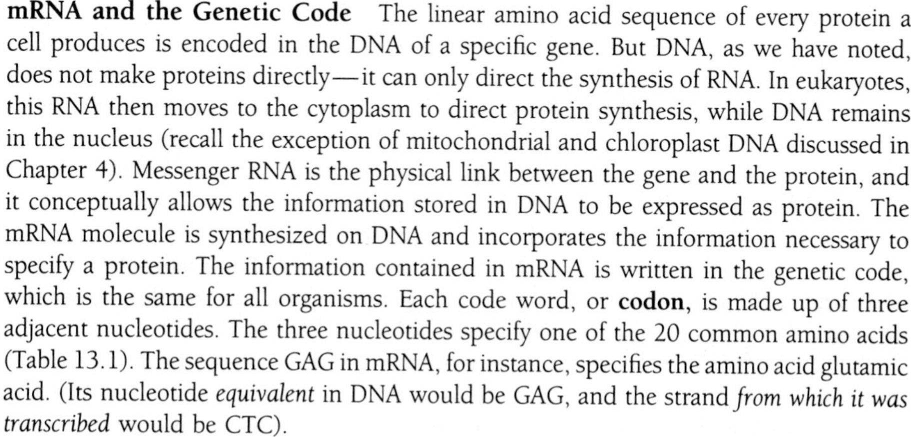

mRNA and the Genetic Code The linear amino acid sequence of every protein a cell produces is encoded in the DNA of a specific gene. But DNA, as we have noted, does not make proteins directly—it can only direct the synthesis of RNA. In eukaryotes, this RNA then moves to the cytoplasm to direct protein synthesis, while DNA remains in the nucleus (recall the exception of mitochondrial and chloroplast DNA discussed in Chapter 4). Messenger RNA is the physical link between the gene and the protein, and it conceptually allows the information stored in DNA to be expressed as protein. The mRNA molecule is synthesized on DNA and incorporates the information necessary to specify a protein. The information contained in mRNA is written in the genetic code, which is the same for all organisms. Each code word, or **codon,** is made up of three adjacent nucleotides. The three nucleotides specify one of the 20 common amino acids (Table 13.1). The sequence GAG in mRNA, for instance, specifies the amino acid glutamic acid. (Its nucleotide *equivalent* in DNA would be GAG, and the strand *from which it was transcribed* would be CTC).

The information in DNA, mRNA, and proteins is referred to as **colinear.** That is, there is a linear relationship between the order of nucleotides in DNA and the order of amino acids in the protein (Figure 13.6). The beginning of the mRNA is always its 5′ end, and the 5′ end of the mRNA always corresponds to the amino terminus of the resulting polypeptide. However, while there is a one-for-one correlation between the DNA and RNA nucleotides, it takes three nucleotides to code for one amino acid.

TABLE 13.1 THE GENETIC CODE

The genetic code can be described in terms of the codons in mRNA. The table is read in the following manner: There are 64 possible codons. The left-hand column contains the first letters of the codons. Across the top are the second letters, and at the right are the third letters. If you wanted to known which amino acid was coded by CAU, you would find C at the left, A at the top, and U at the right. Where the three letters intersect in the table, you will find CAU and the abbreviation His, which stands for the amino acid histidine.[a] Note that the code is *synonymous*, meaning that there is more than one codon for each amino acid (with two exceptions). UAA, UAG, and UGA do not code for amino acids; they signal *stop*, and are known as *terminators*. And one more irregularity needs to be mentioned. AUG has two purposes—it codes for methionine and it also means *start*. Every mRNA begins with AUG, so it is an *initiator*.

FIRST LETTER	SECOND LETTER: U		C		A		G		THIRD LETTER
U	UUU	Phe	UCU	Ser	UAU	Tyr	UGU	Cys	U
	UUC	Phe	UCC	Ser	UAC	Tyr	UGC	Cys	C
	UUA	Leu	UCA	Ser	UAA	*stop*	UGA	*stop*	A
	UUG	Leu	UCG	Ser	UAG	*stop*	UGG	Trp	G
C	CUU	Leu	CCU	Pro	CAU	His	CGU	Arg	U
	CUC	Leu	CCC	Pro	CAC	His	CGC	Arg	C
	CUA	Leu	CCA	Pro	CAA	Gln	CGA	Arg	A
	CUG	Leu	CCG	Pro	CAG	Gln	CGG	Arg	G
A	AUU	Ile	ACU	Thr	AAU	Asn	AGU	Ser	U
	AUC	Ile	ACC	Thr	AAC	Asn	AGC	Ser	C
	AUA	Ile	ACA	Thr	AAA	Lys	AGA	Arg	A
	AUG	Met *start*	ACG	Thr	AAG	Lys	AGG	Arg	G
G	GUU	Val	GCU	Ala	GAU	Asp	GGU	Gly	U
	GUC	Val	GCC	Ala	GAC	Asp	GGC	Gly	C
	GUA	Val	GCA	Ala	GAA	Glu	GGA	Gly	A
	GUG	Val	GCG	Ala	GAG	Glu	GGG	Gly	G

[a]Amino acid abbreviations: alanine, Ala; arginine, Arg; asparagine, Asn; aspartic acid, Asp; cysteine, Cys; glutamic acid, Glu; glutamine, Gln; glycine, Gly; histidine, His; isoleucine, Ile; leucine, Leu; lysine, Lys; methionine, Met; phenylalanine, Phe; proline, Pro; serine, Ser; threonine, Thr; tryptophan, Trp; tyrosine, Tyr; valine, Val.

Since there are four different RNA nucleotides that can occur in any of the three positions of a codon, there are $4 \times 4 \times 4 = 64$ different codons. Three of these, **UAA, UAG,** and **UGA,** are **stop codons** that specify the end of a protein, like the period at the end of a sentence. The remaining 61 codons specify the 20 amino acids. Obviously, there are more types of amino acid-specifying codons than there are types of amino acids, so most amino acids are coded by more than one codon. For instance GGU, GGC, GGA, and GGG all code for one amino acid, glycine. These are called **synonymous codons.**

One codon, **AUG,** is quite special. It can either specify the amino acid methionine in the middle of a protein or serve as an **initiation or start signal.** Ribosomes begin the translation of any mRNA at the first AUG they find in the sequence. Hence, all proteins at least start out with a methionine as their first amino acid.

mRNA Structure There is more to mRNA structure than the sequence of bases that will specify a protein. In both prokaryotic and eukaryotic mRNAs there is a portion of the 5′ end of the mRNA, before the initiation codon, that is not translated into protein. This is called the **5′ leader** (or simply the **leader**). These bases offer the physical space to which ribosomes will bind so that they then can move down the mRNA to the initiation codon and the part of the mRNA that actually encodes the protein. That part of the mRNA that actually codes for the protein can be called the **cistron.** After the termination codons of the cistron, towards the 3′ end of the mRNA, comes another series of bases called the **3′ trailer,** or simply the **trailer.** The entire mRNA, from the beginning of the 5′ leader to the end of the 3′ trailer, is transcribed from the coding region of a gene.

In eukaryotes, the primary mRNA transcripts cannot be translated without posttranscriptional modification or RNA processing (Figure 13.7). First, the mRNA must undergo **capping,** a process in which a special methylated version of triphosphate guanine nucleoside is added to the 5′ end. Capping apparently aids the messenger in later binding and positioning it on the ribosome. The second posttranscriptional modification consists of adding a long series of adenines (called a **poly-A tail)** to the 3′ end of the molecule. Poly-A tails may, in some way, help to transport the mRNA out of the nucleus and determine the number of times an mRNA can be translated before it is degraded.

In addition to the 5′ leader and the 3′ trailer, eukaryotic mRNAs also contain other regions that do not code for protein. These regions are called **intervening sequences** or **introns.** After the initial or primary mRNA has been produced, the introns must be identified and removed, leaving only expressed regions, or **exons,** in the mature mRNA. The removal of introns to produce the mature mRNA is called **splicing** (Figure 13.8).

Splicing is accomplished in the nucleus with the aid of large RNA-protein complexes called **spliceosomes.** Spliceosomes apparently

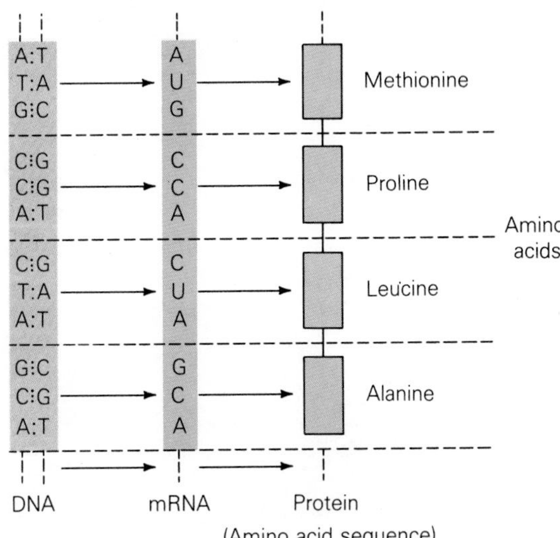

FIGURE 13.6
COLINEARITY BETWEEN DNA, RNA, AND PROTEINS
There is a clear, fundamental relationship among the molecules involved in transcription and translation. They are colinear, in that the nucleotides of DNA and RNA, and the amino acids of the resulting protein are found in the same order, from beginning to end. As you can see, the resulting chain of amino acids is colinear, though not in a one-for-one ratio with nucleotides. We will see that it takes three nucleotides to specify an amino acid.

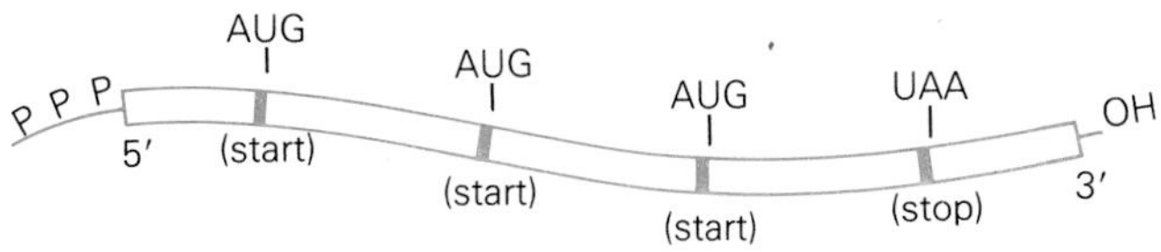

(a) Prokaryote (polycistronic) mRNA

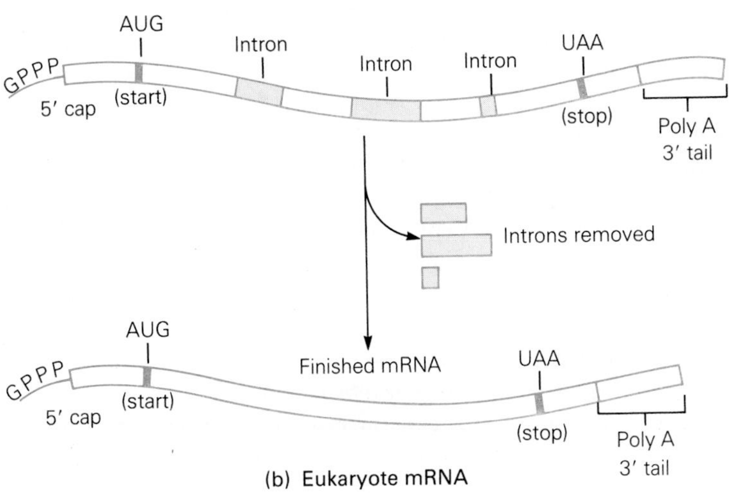

(b) Eukaryote mRNA

FIGURE 13.7
PROKARYOTIC AND EUKARYOTIC mRNA
Messenger RNA in prokaryotes **(a)** differs from eukaryotic mRNA **(b)** in two important ways.

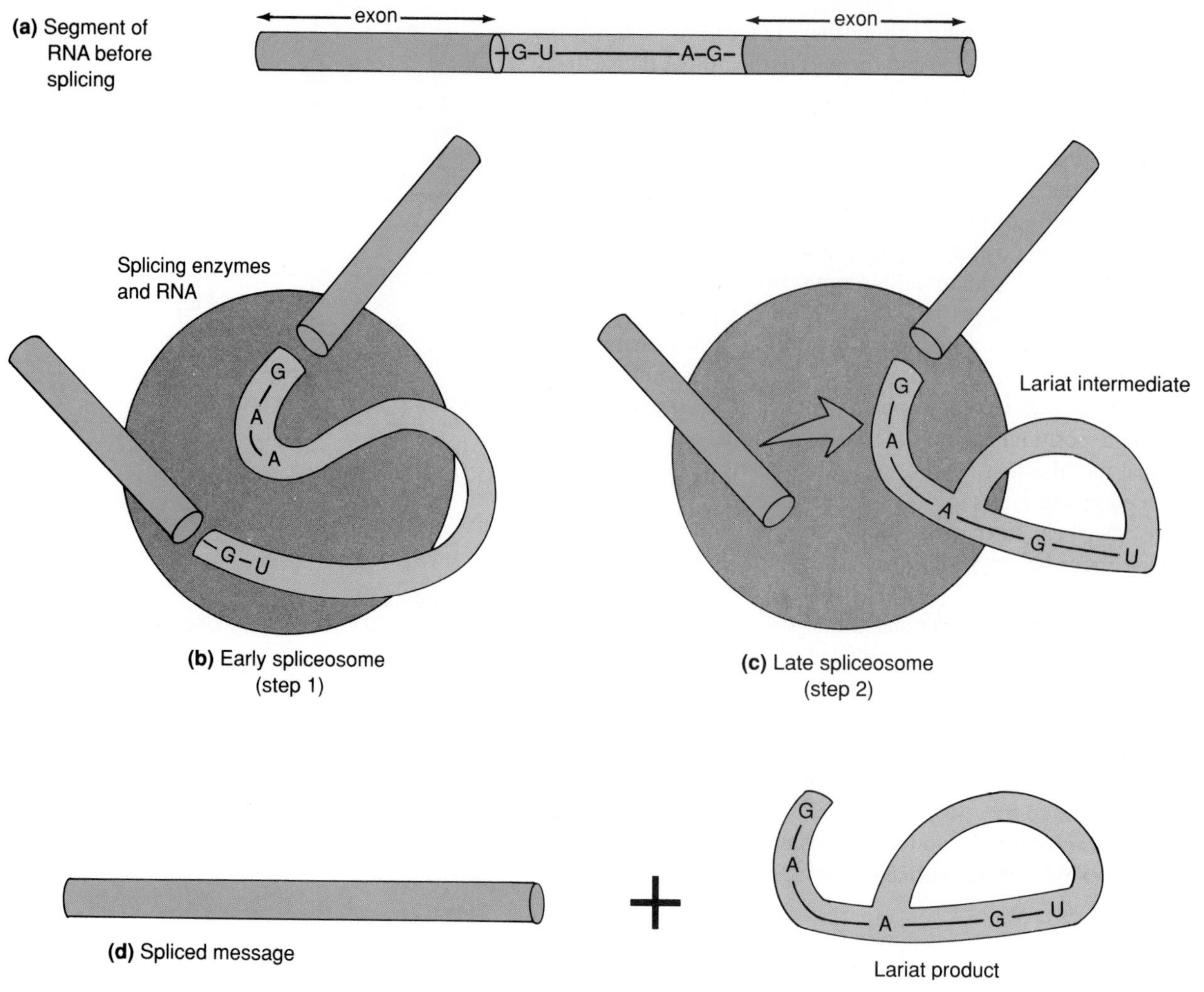

FIGURE 13.8
SPLICING OUT INTRONS
Removal of introns is a complex process. The major points are that an mRNA **(a)** becomes associated with a multiprotein/RNA complex **(b)** that clips out the intron while holding the ends of the exons in close proximity **(c)**. The end result **(d)** is a spliced mRNA and the lariat-shaped remains of the intron.

contain many different enzymes and several different kinds of RNA that help with the splicing process, but one particular component caught the attention of molecular biologists. This is a short piece of RNA that contains a core base sequence that complements short base sequences found at the two ends of nearly every intron. Earlier researchers decided that splicing must involve pairing between the spliceosome RNA and the RNA at the ends of an intron. This would bring the two ends of adjacent exons together so that they could be covalently bonded together end-to-end. The looped intron could, at the same time, be discarded. We now know that splicing is a two-step process and that it involves several different spliceosome RNAs (see Figure 13.8).

Prokaryotic mRNA is quite different from its eukaryotic counterpart. For one thing, it lacks the methylated cap and the poly-A tail. Also, there are no introns in prokaryotic mRNA. Therefore, there is no posttranscriptional processing before the message is ready for translation, and the mRNA is mature as soon as it is made. In fact, translation can even begin on the 5′ end of the mRNA before the 3′ end is finished transcribing. In addition, it is common for prokaryotic mRNA to be **polycistronic** — to contain coding regions (cistrons) for producing more than one polypeptide. Each of these cistrons contains its own initiation and stop codons, so three separate proteins are produced. We will see in Chapter 14 that grouping these cistrons together is a way to ensure equal production of several proteins that are needed at the same time.

Transfer RNA

As we will see, the mRNA carries the coded message to the ribosome, where it is decoded into protein. But, the ribosome itself cannot tell one codon from the next. Deciphering the codons, one at a time, is the job of our last type of RNA: transfer RNA, or tRNA. This small RNA molecule is a critical contributor to the success of translation, as it is responsible for translating the language of nucleotides to the language of amino acids. Let's examine just how the molecule accomplishes this feat.

Each tRNA molecule is a relatively short length of RNA, consisting of about 90 nucleotides. In its primary form, when it is first transcribed from DNA, the tRNA is somewhat longer, but before it becomes active it undergoes some posttranscriptional modification (Figure 13.9). Many primary tRNAs contain introns, which must be excised. In addition, the molecule is "tailored," as special enzymes remove segments from each end. Other enzymes make chemical modifications of some of the bases in special places on the different tRNAs so that the completed molecule contains "exotic" RNA bases in addition to the usual four. These exotic RNA bases may serve in part to help preserve the molecule by retarding enzymatic degradation. Yet another enzyme adds three more nucleotides to the 3′ end of every tRNA, so that all completed tRNAs end with the sequence —CCA.

The mature tRNA molecule is precisely coiled and loops back on itself in a characteristic conformation. The folded tRNA has three loops and a stem, and finally the whole molecule is held in a twisted, L-shaped configuration by hydrogen bonds between its nucleotide bases and between the bases and the free —OH side groups of the ribose

FIGURE 13.9
tRNA STRUCTURE
Transfer RNA molecules are transcribed in the nucleus and are generally about 90 bases long. Some of the bases are enzymatically modified. In its final form, the tRNA molecule has a folded secondary structure **(a)** caused by extensive base-pairing with itself. This folding creates several loops, or unpaired regions, the most important of which is the anticodon loop, which pairs with the codon of the mRNA. The 3′ end of the tRNA is unpaired and always contains the sequence CCA for the attachment of the amino acid. In its tertiary form **(b)** we see that the amino acid attachment site and the anticodon loop are at opposite ends of a molecule that is shaped like a three-dimensional capital *L*, and that the other loops help to form the structure.

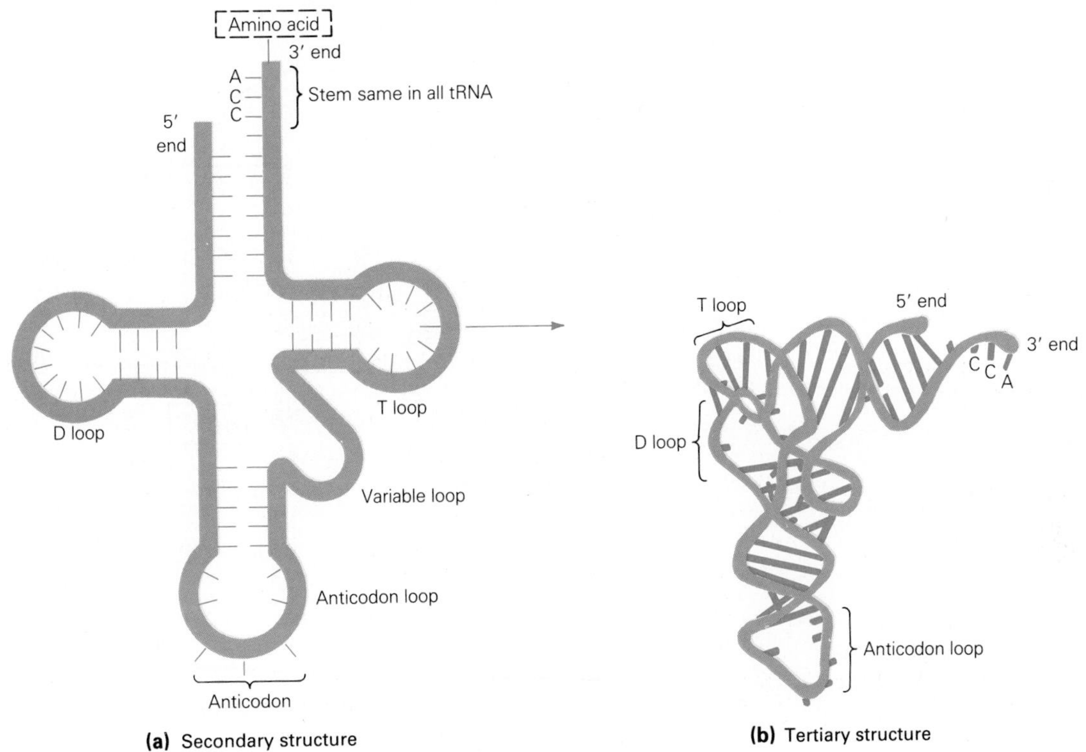

units (Figure 13.9). Such contortions would be quite impossible for DNA because deoxyribose lacks the free —OH group at the 2′ position on the sugar ring.

One of the two key features of a tRNA molecule is that it can be covalently linked to an amino acid. **Charging enzymes** (Figure 13.10) attach amino acids to the —CCA 3′ ends of specific tRNAs. As would be expected, there are different tRNAs for each of the amino acids found in proteins. It follows that there are 20 different enzymes and at least 20 different tRNAs (actually more, because of synonymous codons). The charging enzymes recognize each particular type of tRNA and link it to its own special amino acid with a high-energy covalent bond. Thus, one such enzyme, called an alanine tRNA charging enzyme, binds an alanine tRNA in one of its receptive sites, an alanine amino acid in another site, and (with the aid of ATP) joins alanine to its specific tRNA (Figure 13.10).

The second key feature of the tRNA is the **anticodon.** The anticodon is a series of three bases, physically located on the "opposite" side of the mature tRNA from the amino acid attachment site, that can base-pair with the codons found in mRNA. Each tRNA has its own special anticodon, which recognizes a specific codon found in the genetic code. Do you see how important tRNA is to the process of translation? One end of the molecule recognizes a codon (written in the language of nucleic acids), while the other end carries the specific amino acids (to write in the language of proteins) that correlates with that codon.

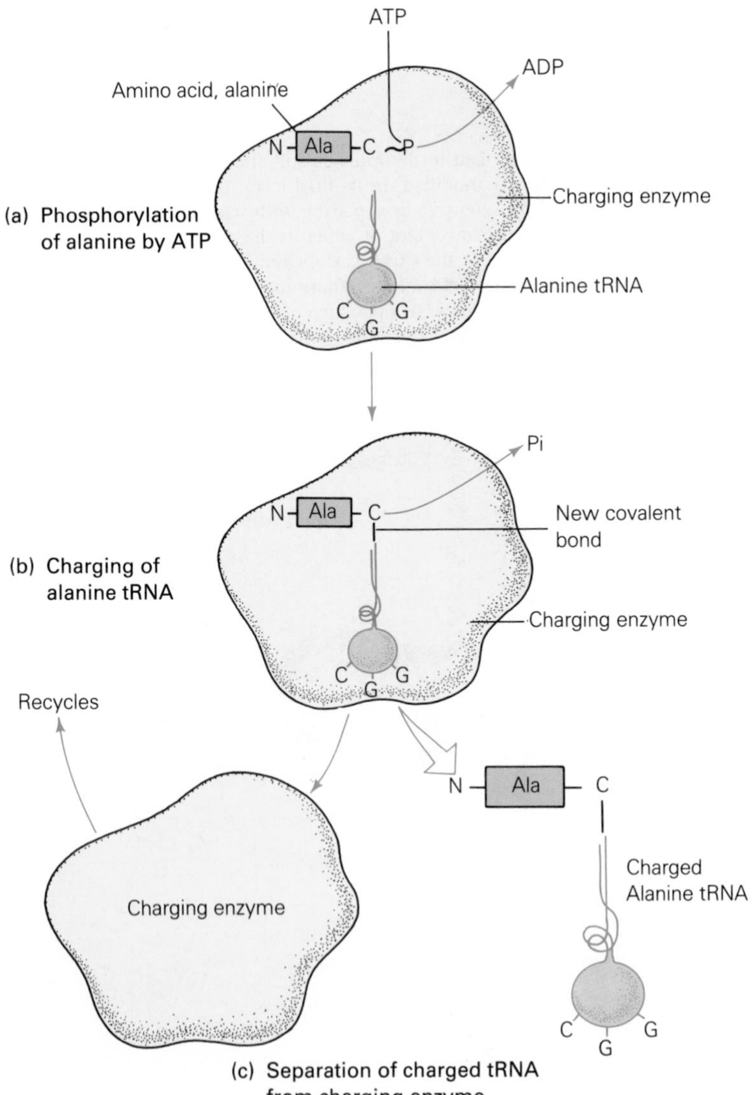

FIGURE 13.10
CHARGING
Charging is the process by which an amino acid is attached to the proper tRNA. In the first step (**a**), the energy-rich phosphate of an ATP is transferred to the amino acid. Then (**b**) the energy in that phosphate bind is used to form a covalent bond between amino acid and its tRNA. (**c**) The charged tRNA is then released from the enzyme.

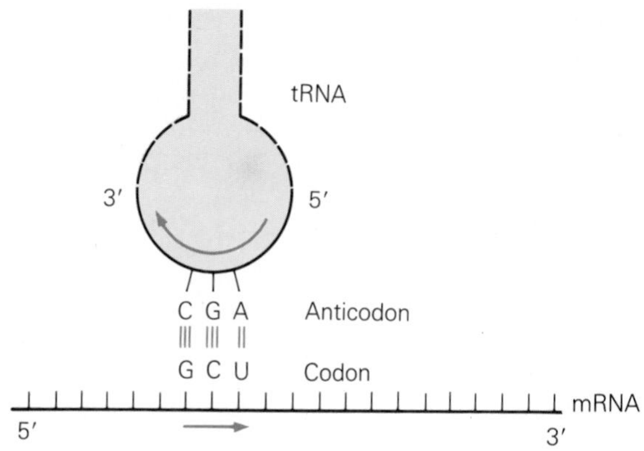

FIGURE 13.11
CODON AND ANTICODON
The codon is the series of three bases on the mRNA that encode a specific amino acid (see Table 13.1). The anticodon of the tRNA is a complementary series of three bases that form specific base pairs with the codon. When the correct tRNA is present, three base pairs form between the codon and anticodon, with the end result being the placement of an amino acid (in this case alanine) in its proper position in a protein.

The "recognition" between codon and anticodon is yet another example of the elegance and power of specific base pairing by hydrogen bonds. In this case, U in the codon pairs with A in the anticodon, C pairs with G, and so on. Thus in the example shown in Figure 13.11, the tRNA anticodon 5′-AGC-3′ pairs with the mRNA codon 5′-GCU-3′, which codes for the amino acid alanine. Remember that in all cases (DNA with DNA, mRNA with DNA, or tRNA with mRNA) the strands run in the opposite directions for base pairing to occur.

TRANSLATION: HOW PROTEINS ARE ASSEMBLED

Having now described all of the players in the show, let's now look at how they all interact to accomplish the process of translation: converting the information in an mRNA into a protein. Translation is divided into three parts: **initiation, elongation, and termination.** All we have said about the RNAs and the genetic code now comes together, and we begin with the ribosome and initiation.

Initiation

Initiation begins when the leading end of the mRNA associates with the small ribosomal subunit. This step is quite precise, and if all goes well, the initiator codon AUG will have aligned itself in the left-hand pocket, or **P site** as it is called (named for the peptidyl enzyme above). The other pocket is named the **A site** (aminoacyl). Thus, the first amino acid to be incorporated into a polypeptide is always methionine. (Or, in prokaryotes, it is a modified methionine called N-formyl methionine.) A special initiator tRNA with a 5′-CAU-3′ anticodon recognizes and pairs with the initiation codon 5′-AUG-3′. It also binds with the small ribosome subunit in an energy-utilizing reaction involving at least three specific initiation proteins. Only after this **initiation complex** (the three initiation proteins, the smaller ribosomal subunit, a charged methionine tRNA, and the mRNA initiator codon) is formed can the large subunit join the complex to form a functional, intact ribosome (Figure 13.12).

The methionine will form the **amino-terminal** (or **N-terminal**) end of the growing polypeptide (see also Figure 13.6 on colinearity). This simply means that the amino group (NH_2) of the first amino acid in a polypeptide will be exposed, since the amino groups of any newly arriving amino acids will all attach to the terminal carboxyl group of the polypeptide being synthesized. Thus, the final amino acid of any polypeptide will have its carboxyl group (—COOH) exposed, making up the **carboxy-terminal** (or **C-terminal**) end of the polypeptide. By convention the N-terminus is written at the left and the C-terminus at the right.

The initiation step is most critical, as it sets the reading frame for the translation process. The genetic code has no commas or any other punctuation to tell the ribosomes which group of three letters to use as codons. The only thing that marks the beginning of the cistron is the initiation codon. After that first AUG, *all* subsequent codons are read in register, three bases at a time.

FIGURE 13.12
INITIATION
(a) The required elements are mRNA, the two ribosomal subunits, and charged methionine tRNA. **(b)** The initial event is base pairing between the mRNA initiator, AUG, and the anticodon UAC, which is found only on methionine tRNA. As the base pairing occurs, the smaller ribosomal subunit joins the RNAs. **(c)** Only then can the larger subunit move in and complete the polypeptide assembling complex.

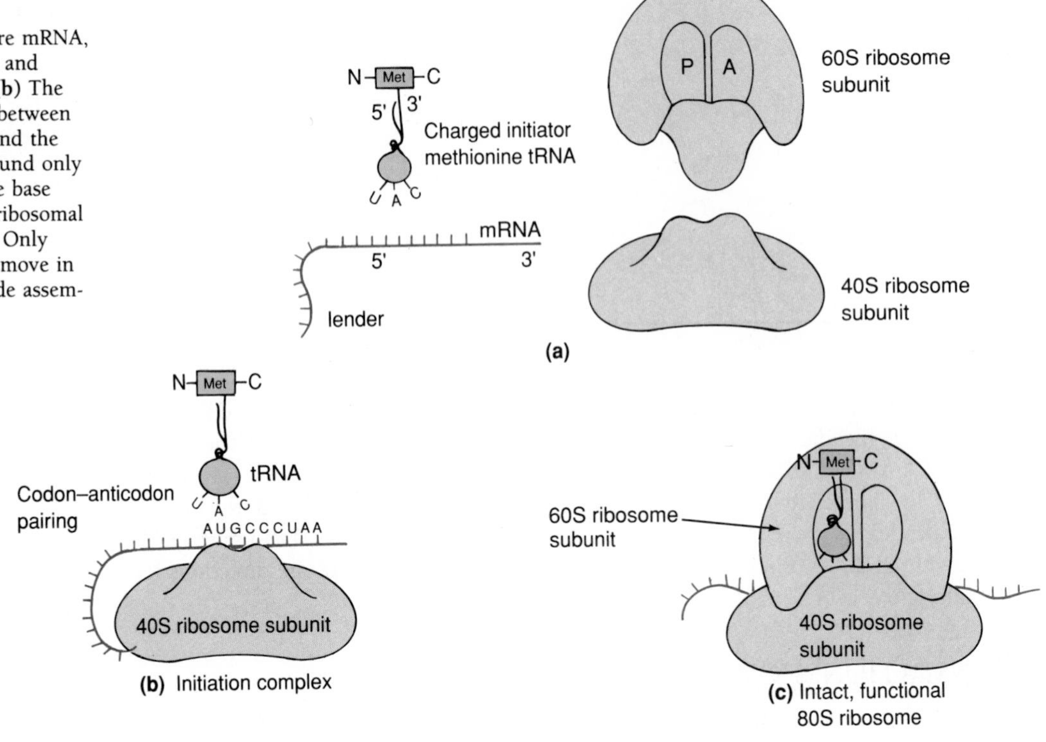

Elongation

That is how a protein starts. The next question is, how does it grow? How are additional amino acids added to the chain?

Elongation is the process by which all the rest of the amino acids in the protein are joined, in order, during translation. It begins immediately with the transfer RNA whose anticodon complements the next codon of the message.

Note in Figure 13.13 that CCC is the next codon, and the tRNA with the GGG anticodon (and proline attached) joins the complex in the ribosome's A site. During the time that the A site was empty, all sorts of small molecules randomly bumped into and out of the empty pocket in the ribosome, including any tRNA molecules that might have happened to be in the neighborhood. Sooner or later a charged protein tRNA wandered in, and it fit so well that it stuck. The good fit resulted from a combination of the shape of the pocket, which fits all charged tRNAs, and the matching of the anticodon with the codon.

FIGURE 13.13
ELONGATION: FORMATION OF PEPTIDE BOND
The initial part of elongation involves the juxtaposition of the two charged tRNAs, then the formation of a peptide bond between the two amino acids. The key here is that after the first amino acid is joined to the second, they remain attached to the second tRNA. The first tRNA is ready to be released.

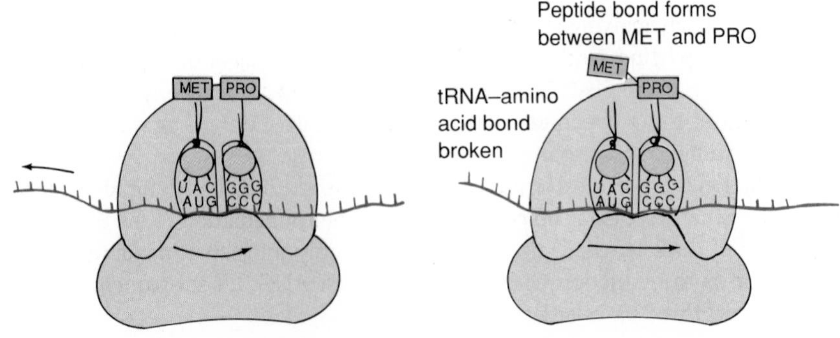

Figure 13.13 shows the two tRNA pockets in the ribosome occupied by the methionine and proline tRNAs. The methionine has been transferred from the stem of its tRNA to the amino group of the proline, as a peptide bond is formed between the carboxyl group of the methionine and the amino group of the proline. The energy for peptide bond formation was provided by the bond that initially held the amino acid to its tRNA (and thus, indirectly, by ATP in the tRNA charging reaction).

Methionine tRNA, now no longer "charged" with its amino acid, loses its affinity for the ribosome, which only binds to charged tRNAs. The methionine tRNA will soon drift out of the pocket, eventually to be recharged with another methionine so that it can participate in translation once again.

In Figure 13.14, the proline tRNA has moved from the pocket of the A site to the pocket of the P site, bringing along with it both the amino acids and the mRNA, which is still bound to its anticodon. This step is called **translocation.** Translocation, you will notice, moves the ribosome three nucleotides to the right along the mRNA. As a consequence, the UUA codon now lies along the floor of the newly empty pocket of the A site. Soon a charged leucine tRNA will randomly bump into place, and the elongation process will proceed. This, then, is why we can compare the ribosome with the head of a tape recorder: the ribosome not only "reads" the mRNA, but moves it along. (It is arbitrary whether we visualize the mRNA moving past the ribosome, or the ribosome traveling down the length of the mRNA.)

The question not often asked about translation is, how does everything know where to go? The scheme just described is indeed straightforward and elegant, but it prompts questions. For example, the movement of the molecules appears to be totally random, and there is every reason to believe that it is. Does this mean, then, that the base pairing of codon and anticodon is just a matter of chance contact? Of course, no combination but the correct one will work, and all others will be rejected. One way of improving the odds is for a great deal of charged tRNA to be around and in motion. Actually, it seems that most chemical events in cells depend on random motion and accidental but predictable collision.

FIGURE 13.14
ELONGATION: TRANSLOCATION
After releasing the first tRNA, the ribosome moves one codon, so that the second tRNA (with its growing chain of amino acids attached) now occupies the pocket of the P site. The A site is now open to receive the charged tRNA for the next codon. The process then cycles from peptide bond formation through translocation as the ribosome moves further down the mRNA.

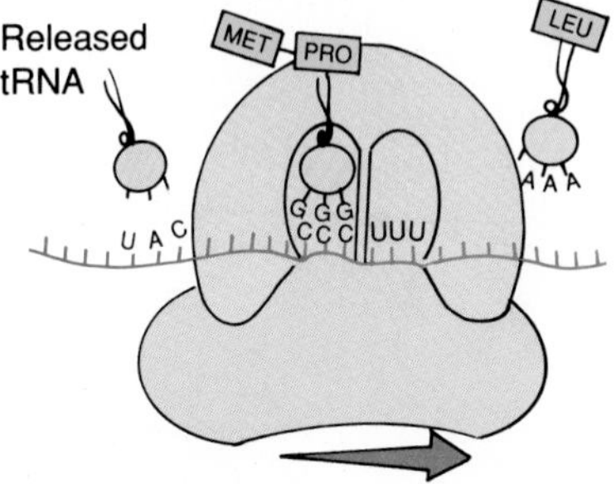

Ribosome moves one codon to the right along the mRNA and is now ready for the next charged tRNA

Termination

Well before the ribosome reaches the end of the mRNA molecule, it runs into a chain terminator, or stop codon, which signals the end of the protein coding information. There are three of these: UAA, UAG, and UGA. Sometimes there are double stops (for example, UAA-UAG), apparently just to be sure that the ribosome gets the idea.

None of the tRNAs have anticodons that are complementary to any of the three stop codons. Instead, there are specific proteins that apparently occupy the A site once a stop codon has been reached. Without a tRNA in the A site, elongation stops. Then, yet another protein factor frees the C-terminal carboxyl group from the last tRNA. Following this, the completed polypeptide is released, and the ribosome falls apart into its two components (Figure 13.15). The entire translation process is reviewed in Figure 13.16.

FIGURE 13.15
TERMINATION
(a) In the first event, the ribosome has reached a terminator codon (UAA). There is no opposing anticodon, since no AUU-bearing tRNA exists. The amino acid just added will be the last. **(b)** The second event is a "derailing" of the ribosome. The details of this process aren't yet clear, but there appear to be special proteins involved that have been named "releasing factors" (this term is deliberately vague in anticipation of new information). Somehow, the presence of UAA triggers a change in the ribosome, and its subunits separate as the final tRNA is released. Our polypeptide moves away for final shaping into secondary, tertiary, and perhaps quaternary form, and the ribosomal subunits go back to "start" for another round of translation.

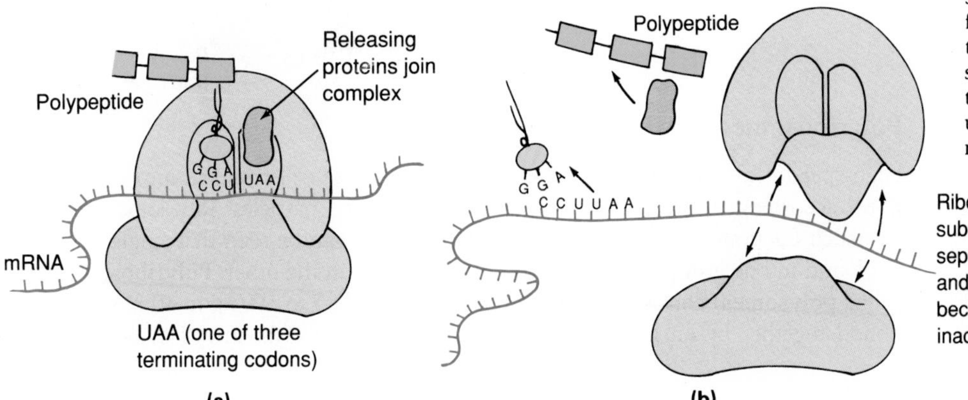

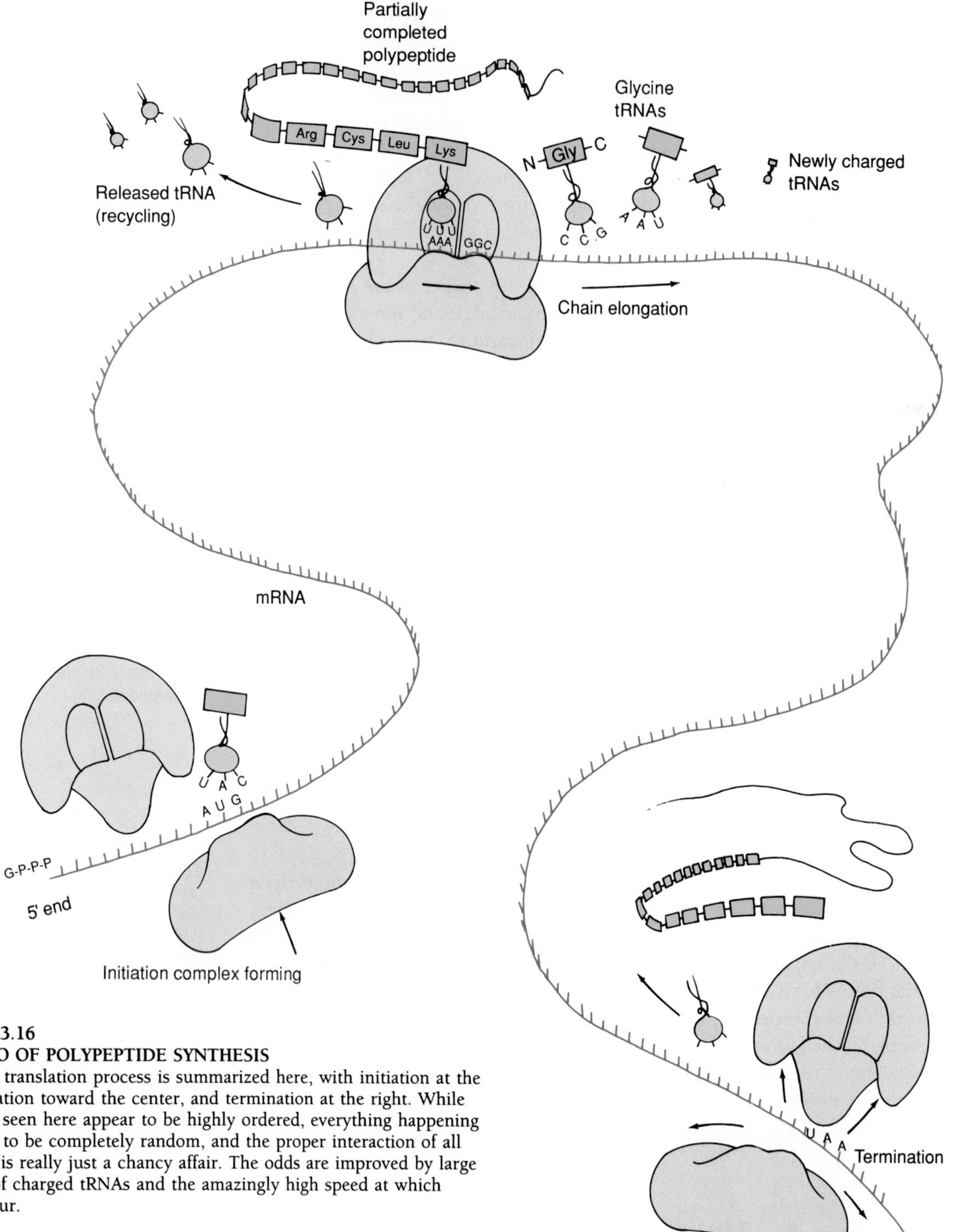

FIGURE 13.16
SCENARIO OF POLYPEPTIDE SYNTHESIS
The entire translation process is summarized here, with initiation at the left, elongation toward the center, and termination at the right. While the events seen here appear to be highly ordered, everything happening is thought to be completely random, and the proper interaction of all molecules is really just a chancy affair. The odds are improved by large numbers of charged tRNAs and the amazingly high speed at which events occur.

Polyribosomes

High-speed centrifugation of crushed cells can separate cell contents into various fractions, according to the size and specific gravity of the particles. Ribosomes appear in two such fractions. Under the electron microscope it can be seen that single ribosomes are found in one group, and *poly*ribosomes are found in the other. **Polyribosomes** (also called **polysomes**) consist of several ribosomes, usually 5 to 10 (up to 40 in some cells), bound together by an mRNA molecule. They look like little strings of beads (Figure

13.17). One might wonder why they are bound together in groups. It seems that different ribosomes are reading the same mRNA molecule and are spaced along it at appropriate intervals—a minimum of 25 nucleotides apart. Each ribosome will travel the whole length of the cistron of the messenger, from the initiation codon to the termination codon; then each will fall apart and drop off. Meanwhile, other ribosomes will assemble themselves at the initiation codon and begin moving along the mRNA.

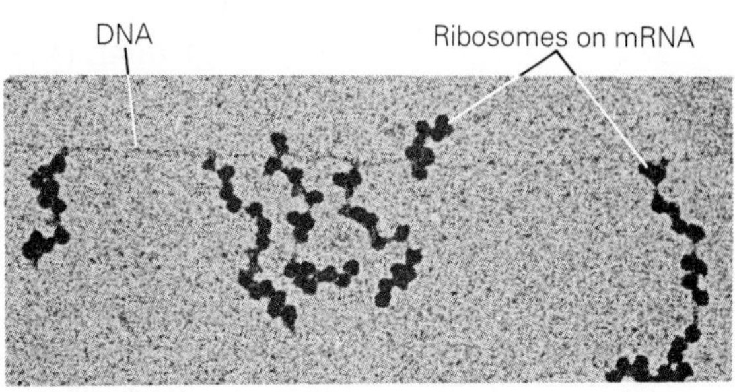

FIGURE 13.17
POLYSOMES IN A PROKARYOTE
Polyribosomes (or polysomes) are shown in this electron micrograph. The amount of protein synthesis is greatly increased when several ribosomes read the mRNA code at the same time. In *Escherichia coli,* polyribosomal translation is so rapid that ribosomes often move along mRNA that is still being transcribed along the chromosome.

The several ribosomes reading the same mRNA are like a group of ancient Talmudic scholars all reading different parts of the same long scroll. While the last scholar to arrive begins with reading Genesis, another may be just finishing Deuteronomy, while others are working on Exodus, Leviticus, and Numbers. When the first scholar is finished, he can take a break, begin a new scroll, or go back to "In the beginning. . . ." Thus, the different ribosomes in a polysome can be producing different copies of the same polypeptide simultaneously, each working on a different portion of the sequence.

Free and Bound Ribosomes

In Chapter 4, we noted that ribosomes occur in two general places, floating free in the cytoplasm or bound to membranes. In higher organisms, bound ribosomes are attached to one side of the membranes of the rough endoplasmic recticulum; in fact, their pebbly appearance gives the rough endoplasmic reticulum its name (see Figure 4.12). You may recall that the endoplasmic reticulum, or ER, is part of the membrane system that specializes in storing and transporting (and modifying) newly synthesized substances within and out of the cell. Bound ribosomes are also seen in bacteria, but prokaryotes lack an ER, and the ribosomes are bound to the inner surface of the plasma membrane itself.

The polypeptides that are produced along the ER (or the membrane of a bacteria) experience a different fate from those produced by the free ribosomes. Whereas polypeptides produced by free ribosomes are simply released into the cytoplasm, those produced on the bound ribosomes are moved across the membrane into the lumen of the rough endoplasmic reticulum (Figure 13.18) (or in the case of bacteria, the external environment).

The most recent thinking by cytologists is that bound ribosomes are no different from free ribosomes, and that their presence on the ER is a function of the polypeptide they happen to be producing, rather than any specialization of their own. In other words, if they are producing polypeptides destined to enter the ER, then they become bound to the ER.

Cytologists have long wondered how the polypeptide gets from the bound ribosome into the lumen of the ER, and new observations seem to provide an answer.

Polypeptides destined for the ER contain at their N-terminus a special short segment of amino acids, called a **signal sequence.** Most of the amino acids of the signal sequence are hydrophobic (water-fearing; see Chapter 3), so they will have a natural affinity for the lipid-rich membrane of the ER. The signal sequence acts to identify a specific receptor site on the ER. Once contact is made, the polypeptide is actively transported, bit by bit while it is translated, through the membrane into the lumen. There the signal sequence, which is of no further use, is clipped off by an enzyme called signal peptidase. The growing polypeptide strand continues to move into the ER as fast as it is synthesized, keeping both the ribosome and the associated mRNA bound tightly to the ER surface.

The evidence for the role of signal peptides and receptor sites is compelling. For example, proteins or polypeptides that are customarily synthesized and used in the free cytoplasm never contain a signal sequence. Further, when polypeptides known to enter the ER for packaging are synthetically produced outside the cell, they always contain the special signal region. These same proteins normally synthesized in the cell and found

FIGURE 13.18
BOUND RIBOSOMES
Recent discoveries presented in this scenario strongly suggest that the ribosomes studding the rough endoplasmic reticulum are at least partially held there by the specific polypeptide they are synthesizing. **(a)** A free-floating ribosome has begun the synthesis of a polypeptide. (Note that in this instance it is the mRNA, not the ribosome, that does the moving.) At its leading (*N*-terminal) end, the polypeptide contains a signal peptide that will bind with a specific surface receptor protein on the membrane of the rough ER. Once bound, the signal peptide is drawn into the rough ER. **(b)** A bound ribosome continues translating its polypeptide, which is continually drawn into the lumen of the rough ER. Within the lumen, an enzyme called *signal peptidase* cleaves the signal peptide from the growing polypeptide, and the remainder can then take on its usual function. Polypeptides accumulate in the rough ER, which will later form vesicles of the Golgi body where final protein-forming modifications will occur. **(c)** Once a polypeptide nears completion and mRNA reaches its stop codons, termination occurs and the ribosomal subunits separate and are freed from the rough ER.

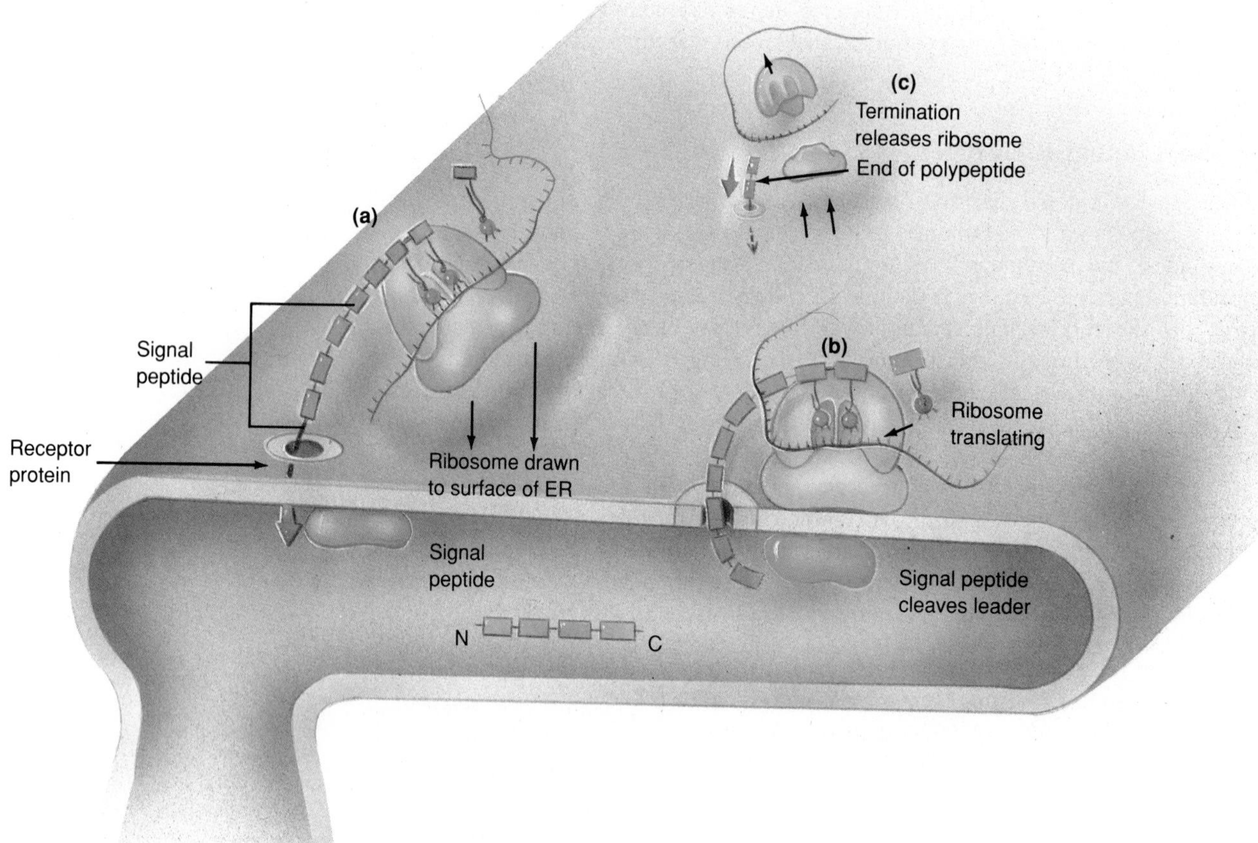

in the ER lack the signal peptide, which, as we mentioned, has been enzymatically removed. Geneticists, always eager to apply their own tools toward the solution of such problems, have isolated mutant bacteria and provided some answers. Certain mutant strains of *E. coli* produce polypeptides with faulty signal peptides, and these polypeptides remain in the cell cytoplasm instead of becoming integrated into the plasma membrane as they normally would. In addition, geneticists have succeeded in hybridizing membranal proteins containing the signal peptide with cytoplasmic proteins, and these hybrids find themselves inserted into the membrane.

We can now see how the process of translation is the mechanism by which the information stored as genes in DNA becomes expressed as proteins. In a very real sense, translation finishes a story that was begun when Mendel first gathered data on the inheritance of traits.

APPLICATION OF IDEAS

It has recently been shown that UGA specifies tryptophan and CUA codes for threonine in the mitochondrial translation systems of yeasts and hamsters. What do these codons usually specify? What does that do to the universal nature of the code? Comment on the significance, if any, of these minor departures from the code.

KEY IDEAS

RNA STRUCTURE AND TRANSCRIPTION

Comparing RNA to DNA

1. RNA differs from DNA in that it contains the sugar ribose, substitutes uracil for thymine, is single-stranded, readily breaks down, and exists in several forms.

Transcription: RNA Synthesis

1. Copying the code from DNA **(transcription)** occurs on the transcribed strand or template, leaving the opposing nontranscribed, idle.
2. In transcription, **RNA polymerase,** the principal enzyme, binds to a DNA **promoter** region, the helix unwinds, and, proceeding in the 5′-to-3′ direction, RNA triphosphate nucleotides are paired to the exposed DNA bases. U is always substituted for T.
3. Transcription ends when RNA polymerase encounters a **termination signal.** Multiple transcription along a DNA strand is common.

VARIETIES OF RNA

1. The three kinds of RNA are **ribosomal (rRNA), messenger (mRNA),** and **transfer (tRNA).** All must undergo modification from the precursors to the final mature form.
2. Ribosomal RNA makes up most of the ribosome, a two-part, interlocking organelle that is the site of protein synthesis. Each ribosome has attachment sites for mRNA and tRNA.
3. In eukaryotes, the long primary mRNA transcript undergoes posttranscriptional modification, where noncoding segments called **intervening sequences,** or **introns,** are removed, leaving only coding segments called exons. Introns are absent in prokaryotes.
4. Eukaryote mRNA contains a **capped** or methylated 5′ end, a **leader,** a **trailer** and a **poly-A tail.** Prokaryotic mRNA lacks the cap and tail and is **polycistronic** (contains several cistrons, polypeptide coding regions).
5. The linear arrangement of bases in mRNA constitutes the genetic code. The amino acids are specified by **codons**—nucleotides in groups of three. There are 64 possible codons, so most amino acids have more than one **(synonymous)** specifying codon. One codon, the initiation codon, specifies both *start* and an amino acid, while three codons specify *stop.*
6. **Charging enzymes** bond tRNAs to their specific amino acids. Each tRNA has a specific amino acid binding site, a ribosomal binding site, and an **anticodon** loop. The anticodon matches a codon on mRNA, assuring that the amino acid carried by a tRNA will be inserted correctly in the polypeptide.
7. The primary tRNA transcripts are first *tailored,* then folded into a cloverleaf secondary shape, and finally folded again into the tertiary "L" shape.

TRANSLATION: HOW PROTEINS ARE ASSEMBLED

Translation includes **initiation, elongation,** and **termination.**

Initiation

1. Initiation requires the smaller ribosomal subunit, an initiation tRNA (with a 5′-CAU-3′ anticodon), and the mRNA initiator codon (5′-AUG-3′), all of which form the ribosomal **initiation complex.**
2. When each component is in place, the larger ribosomal subunit joins the complex, and a second amino acid can be inserted.
3. Polypeptides in eukaryotes all begin with methionine, and each has an **N-terminal** (NH_2) end and a **C-terminal** carboxyl (COO^-) end.

Elongation

1. The elongation of polypeptides occurs through **translocation**—the formation of the peptide bond and the movement of a tRNA from the A to the P site.

2. As a polypeptide grows, a charged tRNA whose anticodon matches the mRNA codon in the A pocket becomes attached. A peptide bond forms between its amino acid and the last one in the polypeptide above, and translocation occurs again.
3. Following translocation, the tRNA in the P site is released and drifts away to recycle.

Termination

1. When the ribosome reaches a chain termination **(stop) codon,** proteins block the pockets, and the final tRNA is released along with the completed polypeptide.

Polyribosomes

1. **Polyribosomes,** or **polysomes,** occur in clusters on one mRNA molecule, where they carry on simultaneous translation.

Free and Bound Ribosomes

1. Bound ribosomes are located along the rough endoplasmic reticulum (and on the plasma membrane of prokaryotes). Their polypeptides enter the ER.
2. Polypeptides destined to enter the ER are tipped by a signal sequence. The growing polypeptide enters the lumen, where its signal sequence is removed. As the polypeptide grows it binds the ribosome to the ER.

REVIEW QUESTIONS

1. List four ways in which RNA differs from DNA. (296)
2. If the sequence of nucleotides in a transcribed strand of DNA was 5′-A-A-G-C-C-T-T-A-G-G-C-A, what would be the sequence of nucleotides in the RNA transcript? (297-298)
3. Briefly describe the process of transcription. Include mention of the primary enzyme and the role of the promotor and termination signal. What, actually, are the last two elements? (297-298)
4. Describe the organization of the ribosome as we now see it. (298-299)
5. Briefly discuss what happens to eukaryotic mRNA during its posttranscriptional modification. What are introns and exons? (300-302)
6. Describe the three regions of the mature eukaryotic mRNA transcript. In what three ways does prokaryotic mRNA differ? (300-302)
7. Following the key points mentioned, describe the organization of the genetic code. (a) What is a codon? (b) How many codons are there? (c) Since only 20 codons are required, what happens to the extras? In what molecules do we find the code written? (299-300)
8. List the steps and the molecules involved in the charging of a tRNA. (304-305)
9. Briefly describe the form taken by tRNA and identify key regions. (302-304)
10. List the different elements of the initiation complex. Explain how they get together to initiate translation. (305)
11. Using a simple drawing, show your understanding of chain elongation by illustrating the key steps involved. (306-307)
12. Briefly explain how chain termination occurs. (307)
13. What are polysomes? How does their presence affect the efficiency of protein synthesis? (308)
14. Are bound ribosomes really part of the ER, as they seem to be? Explain fully. (308)
15. Briefly describe the role of signal sequences. (309)

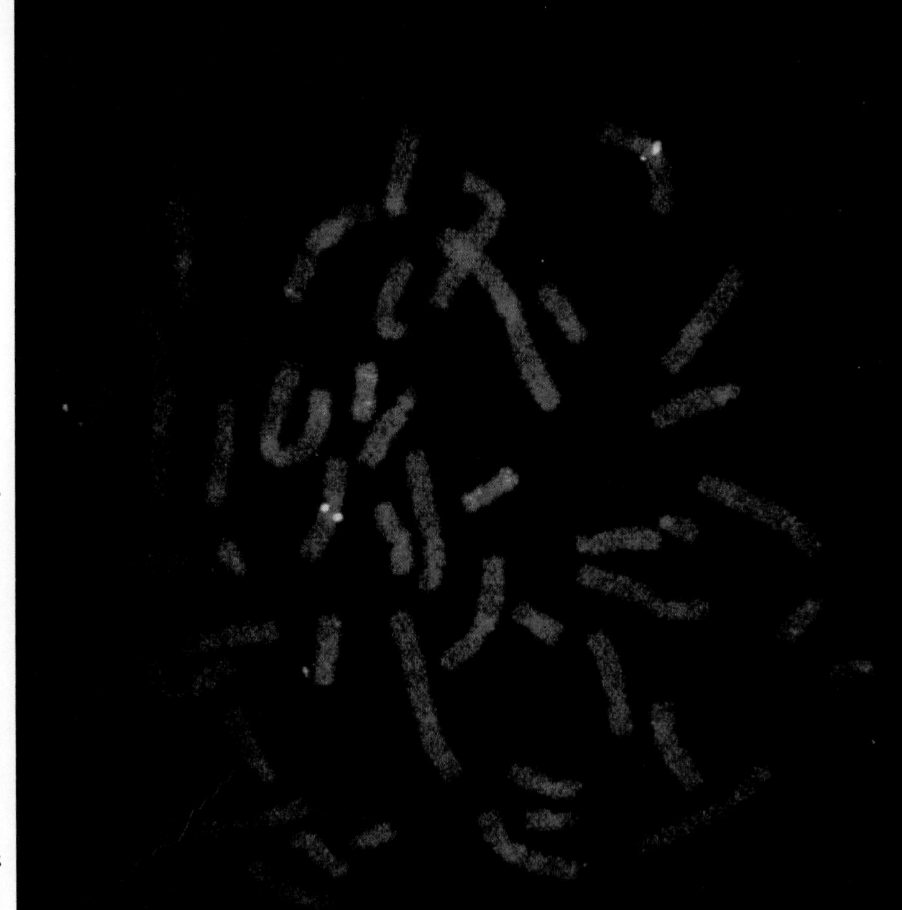

Courtesy, Peter Lichter and David C. Ward, Yale University.

14

GENE REGULATION

MOST ORGANISMS CONTAIN THOUSANDS OF GENES. EVEN THE LOWLY bacterium *E. coli* has some 2500 different genes. But in all species, most of the genes, most of the time, are shut down. Only a small portion are busy transcribing. This, after all, makes sense. Since most cells in an individual contain the same genes, if all those genes were transcribing at once, every cell would make the same products, and every one would be the same. How, then, could a muscle cell be different from a nerve cell? Indeed, although any cell *can* make just about any proteins that any others can make, it makes only certain ones, and so cells, tissues, and organs can become specialized for their different roles in the life of an organism.

The genetic adjustments to those roles, as we will see, are even further refined by various active genes transcribing at various rates. Add to this increasingly complex picture the fact that cells must be able to respond to changes in their environment by adjusting the expression of their genes, and the stage becomes set for us to examine the processes by which genes are selectively and precisely regulated. Indeed, without the precise coordination of genes, cellular metabolism would be more like biochemical chaos than ordered enzymatic pathways.

We have seen in earlier chapters that enzyme products can be regulated through feedback inhibition. (As a product builds up, the process that produced it is slowed.) This kind of pathway regulation is virtually instantaneous and very effective. However, feedback inhibition is somewhat wasteful as a long-term strategy. If the enzymes in a pathway are only needed at certain times, it is much more efficient to synthesize those enzymes only when they are needed, saving the resources and energy that would otherwise be required for constant transcription and translation.

In this chapter we will explore some of the ways cells regulate the expression of their genes. Before we get into it, though, we should be aware of some of the terminology we will be using. As shown in Figure 14.1a, we note that the "gene" is actually composed of two basic parts, a **coding region** that codes for the mRNA and protein product, and

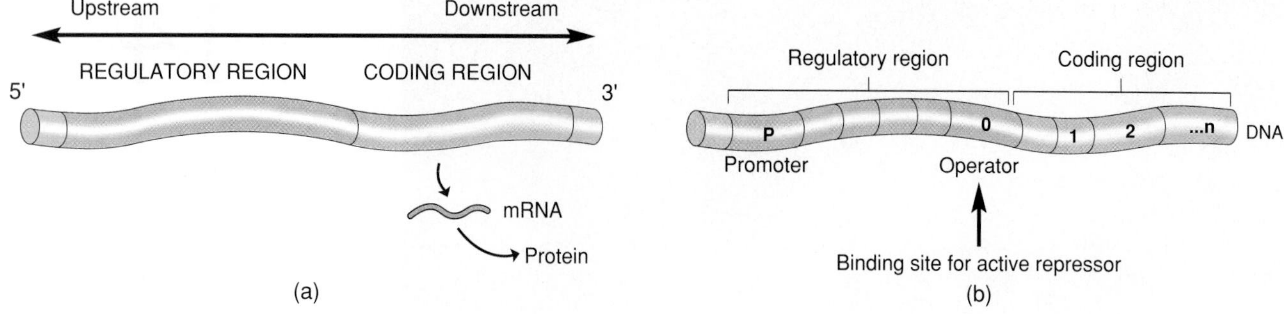

FIGURE 14.1
GENERAL ASPECTS OF GENE STRUCTURE
(a) Most genes consist of two separate parts, a coding region that actually is transcribed into mRNA, and a regulatory region that controls the process of transcription. The 5′ end of the mRNA is the boundary between the regions. Movement toward the 3′ end of the mRNA is in the "downstream" direction, while movement in the opposite direction is "upstream." (b) The key features of an operon are (1) the grouping of several protein coding regions under the control of a single regulatory region, and (2) a regulatory region that consists of a promoter (P) and an operator (O). The operator is the site where a repressor molecule may bind, preventing RNA polymerase from beginning transcription from the promoter.

a **regulatory region** that controls the expression of the coding region. We will use the convention that orients the coding region such that the 5′ end of the resulting mRNA is to the left, and the 3′ (naturally) to the right. In most cases, and in all the cases that we will examine, the regulatory region of the gene is found at the 5′ end of the coding region. The term **promoter** has become accepted as another way to refer to the regulatory region, though it is sometimes used to refer only to that portion of the regulatory region that actually binds the RNA polymerase to start transcription (see Chapter 13). Also, because RNA polymerase transcribes genes from 5′ to 3′, toward the 3′ is called "downstream" and toward the 5′ is "upstream."

We should also be aware that not all genes are highly regulated. After all, some enzymes and proteins are needed at all times, such as those of the glycolytic pathway. So, the genes for these enzymes are continuously transcribed, or **constitutive.** It is important to note also that not all constitutive genes are transcribed at the same rate. For any gene, transcription *rate* is determined by the relative ability of the promoter to bind RNA polymerase.

PROKARYOTIC GENE REGULATION: THE OPERON

Until very recently, bacterial gene systems provided our only insights into the regulation of transcription. Even though we now have some ideas as to how eukaryotic genes are regulated (as we'll see in the next sections), we still have our best comprehension of these processes in bacteria.

There are two recurring themes that characterize coordinated gene regulation in bacteria. The first is that the coding regions that produce the enzymes of the various steps in a single metabolic pathway can be grouped together, occupying adjacent segments of DNA all under the control of a single regulatory region. Since all the coding regions for the pathway are under the control of that one regulatory region, the presence or absence of all the enzymes in the pathway is automatically synchronized. The second underlying theme is that transcription of genes can be prevented by the binding of a **repressor** to a certain area of the regulatory region. A repressor is a protein that binds to DNA to inhibit transcription. If a repressor is bound to the regulatory region, no transcription occurs. If the repressor is not bound to the regulatory region, transcription may occur, and the enzymes encoded by the grouped coding regions are all transcribed at the same time (Figure 14.1b). In fact, all of the coding regions are transcribed into one long mRNA that is referred to as *polycistronic* (see Chapter 13).

In 1961, two scientists at the Pasteur Institute in Paris, Francois Jacob and Jacques Monod, first combined these two themes into a model for gene regulation called the **operon.** An operon, then, is a set of coding regions (usually involved in a single biochemical pathway) clustered together under the control of a single regulatory region, and transcription is regulated by the presence or absence of a repressor protein bound to a segment of the regulatory region that they called the **operator** (hence the name "operon"). All of these basic themes are presented in Figure 14.1.

Let's examine two operons and see just why they are so effective in gene regulation.

Lac: An Inducible Operon

E. coli can grow well with glucose as the only source of carbon, which is a good thing because glucose is so common in the intestine where *E. coli* normally makes its home. However, the bacterium can also utilize another sugar, lactose (milk sugar). Interestingly, the enzymes that act on lactose are usually not made by *E. coli.* Only when lactose becomes the most abundant sugar in the environment does the bacterium even initiate the process that will produce the enzymes that will digest lactose.

It takes three enzymes to allow *E. coli* to digest lactose: a beta-galactosidase to cleave the lactose, a permease to bring the lactose into the cell more efficiently, and a transacetylase whose function is not clearly defined. The coding regions for these three enzymes are linked together under the control of a single regulatory region, and this complex is called the ***lac* operon** (Figure 14.2). The **lac** operon is referred to as an **inducible** operon because it is normally not transcribed until prompted by an **inducer,** in this case lactose.

Linked quite closely to the *lac* operon is a gene called the **i gene,** which makes the *lac* repressor protein. The i gene constitutively transcribes mRNA for the repressor protein at a low rate, resulting in a constant level of about 10 repressor protein molecules per cell. The key characteristic of the repressor molecule that allows it to act as a regulatory protein is that it can bind to either the operator DNA or lactose, but not both at the same time. The binding of a repressor protein to either DNA or lactose is reversible, so that any given moment the repressor is bound most often to whichever of the two it finds in highest concentration.

By itself, the *lac* repressor protein has a strong affinity for the operator sequence of the *lac* promoter. As mentioned before, when the repressor is bound to the operator, no transcription occurs. You might think of this as a simple blocking mechanism, an immovably large repressor boulder sitting on the track of the RNA polymerase train.

Remember, however, that the repressor also has the ability to bind lactose, but it cannot bind both lactose and the operator at the same time. It can bind only one or the other. Thus, if lactose levels rise to high concentrations, the lactose molecules will quickly bind to the free repressor proteins in the cell. Also, when the repressor that is bound to the operator releases (because of the reversibility of the binding), chances are that it will bind to one of the many lactose molecules rather than to the one operator. The binding to lactose is also reversible, such that repressors will also release lactose molecules, but the chances are high that they will bind another lactose before one would again bind the operator. The net effect is that the operator is no longer occupied by the repressor. Thus transcription can proceed from the promoter through to the coding regions—at precisely the time when it is needed most.

The coding region of the *lac* operon is transcribed as a single polycistronic mRNA, which is translated into the three lactose metabolizing enzymes. As the enzymes become available, they begin the process of digesting the lactose. They will continue to be produced as long as there is lactose present in the environment to keep the repressor molecules from binding to the operator.

But by their very nature, the *lac* operon enzymes remove lactose from the environment by digesting it as a source of carbon and energy. Thus if the external supply of lactose diminishes or disappears, these enzymes will soon digest all of the lactose within the cell. Because of the reversibility of the binding between lactose and the repressor, as the concentration of lactose gets low, even the lactose molecules that were bound to repressor molecules can be degraded. This action frees up the repressor molecules to once again bind to the operator and shut down transcription. Thus, when the enzymes are no longer needed (because the lactose is all gone), the operon ceases transcription.

This level of regulation by the repressor is an elegant solution to the seemingly simple question of when to transcribe the *lac* operon. However, the question is not quite as simple as it seems, and so the *E. coli* regulatory system is even more subtle and elegant than we have described so far. Basically, the presence of lactose itself is not reason enough to switch over to lactose metabolism, nor, as it turns out, is it sufficient to induce the *lac* operon. It turns out that it makes energetic sense to utilize glucose, a simpler

FIGURE 14.2
THE *lac* OPERON: THE JACOB-MONOD OPERON MODEL
The production of the three inducible enzymes responsible for the metabolism of lactose is under control of a system known as the lac operon. The operon consists of three principal parts, as shown in the diagram: the regulator gene (*i*), the promoter/operator region, and the structural genes, *z*, *y*, and *a*.
(a) The production of enzymes is shut down. The repression of the structural genes is accomplished by an interaction between regulator gene *i* and the operator. ① The regulator gene transcribes messenger RNA that ② is translated into a repressor protein, which ③ coats the operator, blocking the action of RNA polymerase along the promoter. Thus the transcription of mRNA in the coding region ④ is inhibited.
(b) When lactose enters the cell ⑤ it acts as an *inducer,* ⑥ tying up the repressor protein. Transcription is now allowed, but very little takes place because the RNA polymerase by itself has a poor affinity for the promoter.
(c) However, when cAMP interacts with CAP ⑦ binding of RNA polymerase to the promoter is very efficient and mRNA is transcribed ⑧. The mRNA is translated into the three lactose metabolizing enzymes ⑨. When all of the lactose is digested, newly formed repressor can once again shut down transcription.

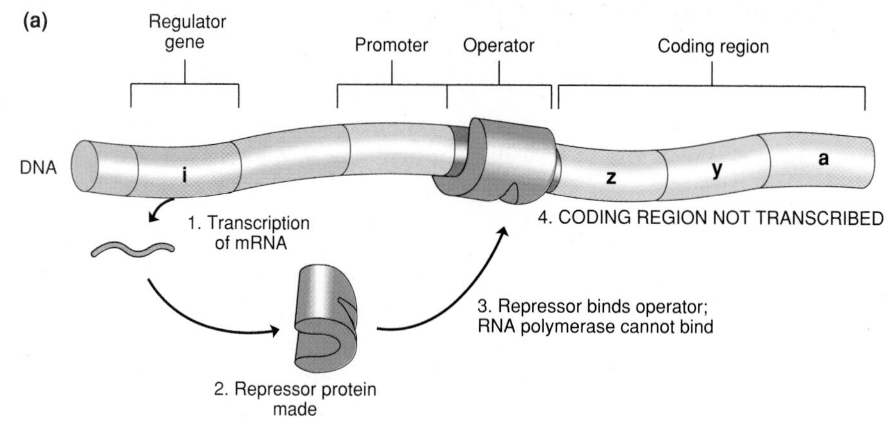
(a)
Regulator gene
Promoter
Operator
Coding region
DNA
i
z
y
a
1. Transcription of mRNA
2. Repressor protein made
3. Repressor binds operator; RNA polymerase cannot bind
4. CODING REGION NOT TRANSCRIBED

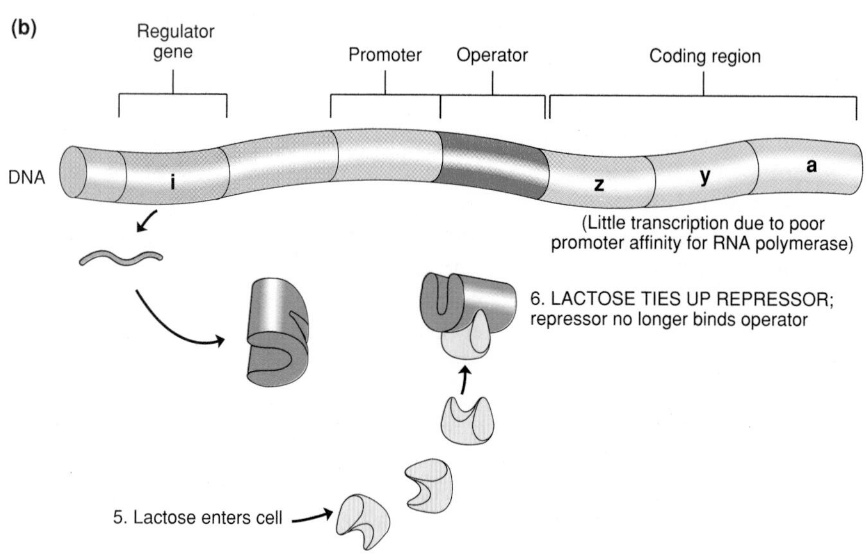
(b)
Regulator gene
Promoter
Operator
Coding region
DNA
i
z
y
a
(Little transcription due to poor promoter affinity for RNA polymerase)
6. LACTOSE TIES UP REPRESSOR; repressor no longer binds operator
5. Lactose enters cell

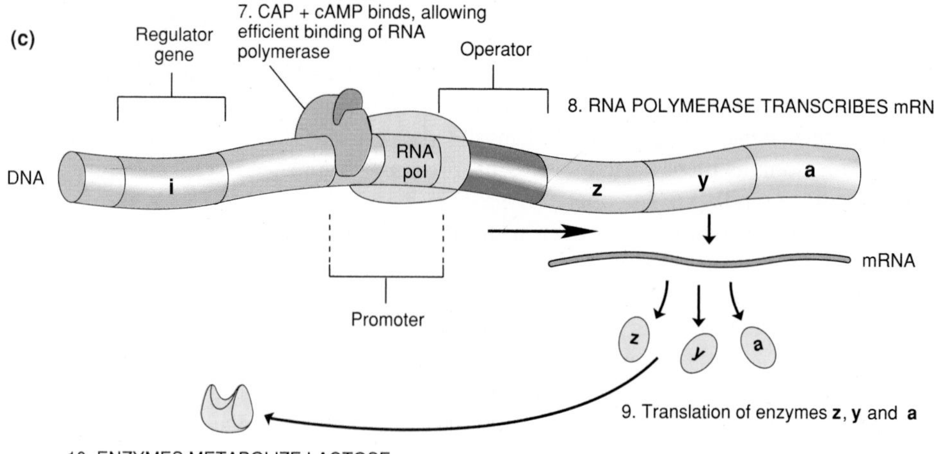
(c)
Regulator gene
7. CAP + cAMP binds, allowing efficient binding of RNA polymerase
Operator
8. RNA POLYMERASE TRANSCRIBES mRNA
DNA
i
RNA pol
z
y
a
mRNA
Promoter
9. Translation of enzymes z, y and a
10. ENZYMES METABOLIZE LACTOSE

sugar, when there is a choice between glucose and lactose. Thus, the bacterium should not waste energy making lactose metabolizing enzymes when there is plenty of glucose present. The *lac* operon, then, should be induced only when there is lactose present *and* no glucose around.

This additional level of control is accomplished by yet another protein-DNA interaction in the promoter sequence. The *lac* promoter by itself binds RNA polymerase quite poorly. Thus, even if the repressor is not bound to the operator, transcription does not readily occur. Therefore, the mere presence of lactose does not induce much transcription. There is, however, an accessory protein that can help the promoter bind RNA polymerase efficiently. This accessory protein is called **CAP, catabolite activator protein.** As opposed to the repressor, CAP is a positive regulatory protein. When it binds to the promoter it greatly stimulates the interaction of RNA polymerase, and thus transcription. But CAP binds to the promoter only when there is no glucose present in the cell. How does CAP know whether there is glucose around? When the level of glucose in the cell falls to a low level, there is an increase in the level of a molecule called **cyclic AMP (cAMP),** a metabolic derivative of ATP. cAMP will bind to CAP, and only then will CAP bind to the *lac* promoter and stimulate transcription. The bottom line is that the *lac* operon will be transcribed only when lactose is present *and* there is no glucose.

Thus we have seen that the *lac* operon is inducible only under conditions that make outstanding energetic sense, and that the induction is accomplished by the interaction of regulatory proteins that have the dual capacity to detect the cellular environment (in this case by binding to lactose or cAMP) and, when appropriate, to bind to DNA. We have seen, already, examples of both positive (CAP) and negative (repressor) control exerted by proteins.

trp: A Repressible Operon

This same kind of metabolic logic that applies to the *lac* operon can work in an opposite manner. We might imagine a situation when the products of an operon are normally needed, and only under unusual circumstances would the operon need to be shut down. A repressor protein can accomplish just this kind of regulation.

E. coli normally must manufacture its own tryptophan (one of the amino acids) because tryptophan is rarely found in its environment. The biosynthetic pathway for the production of tryptophan consists of five enzymatic steps. The coding regions for each of the enzymes in the pathway are located within the polycistronic ***trp* operon** (Figure 14.3). The promoter of the *trp* operon has a relatively high affinity for RNA polymerase and needs no additional activating protein (like CAP) to efficiently transcribe the operon. The operon is normally actively transcribed in order for the cell to produce enough tryptophan for its use. But if tryptophan appears in the environment, it makes sense for the bacterium to utilize the free tryptophan, rather than invest in the enzymatic work to produce it internally. The trick, then, is to inactivate the operon when tryptophan is present.

This is accomplished by a twist of the repressor theme. Much like the *lac* repressor, there is a constitutive gene near the *trp* operon that produces a constant but small amount of *trp* repressor. But unlike the *lac* repressor, the *trp* repressor is, by itself, incapable of binding to the *trp* operator. But when there is excess, free tryptophan around, the repressor acquires the ability to bind to the operator. This occurs because the tryptophan itself acts as a co-repressor, reversibly binding to the *trp* repressor to form an active repressor complex. This complex binds to the operator and shuts down transcription.

Thus, when tryptophan is present in the surrounding environment, the bacterium makes use of that free source of the amino acid. If the tryptophan levels decline, the repressor is less likely to be bound to a tryptophan molecule, the operator site is freed, and transcription can begin again. Once again, a repressor molecule offers the opportunity for a finely tuned regulation—the more tryptophan is available, the less the operon is

FIGURE 14.3
THE *trp* OPERON
(a) The repressible tryptophan operon works in the opposite way of the inducible lactose operon. In this system, cells grown in glucose continually produce the enzyme tryptophan synthase ①, which is essential in producing the amino acid tryptophan. Although the regulator gene produces the repressor protein ②, it cannot bind the operator gene in its present form. Therefore, the system remains turned on. **(b)** If the cells are fed the amino acid tryptophan ③, the system immediately shuts down. Tryptophan joins the repressor protein to form a tryptophan-repressor complex. This complex, in turn, coats the operator, blocking the action of RNA polymerase along the promoter, ④. This shuts down the coding region ⑤ that produces tryptophan synthase. As you can see, this is a conserving process. Why produce the enzyme when its product is abundant?

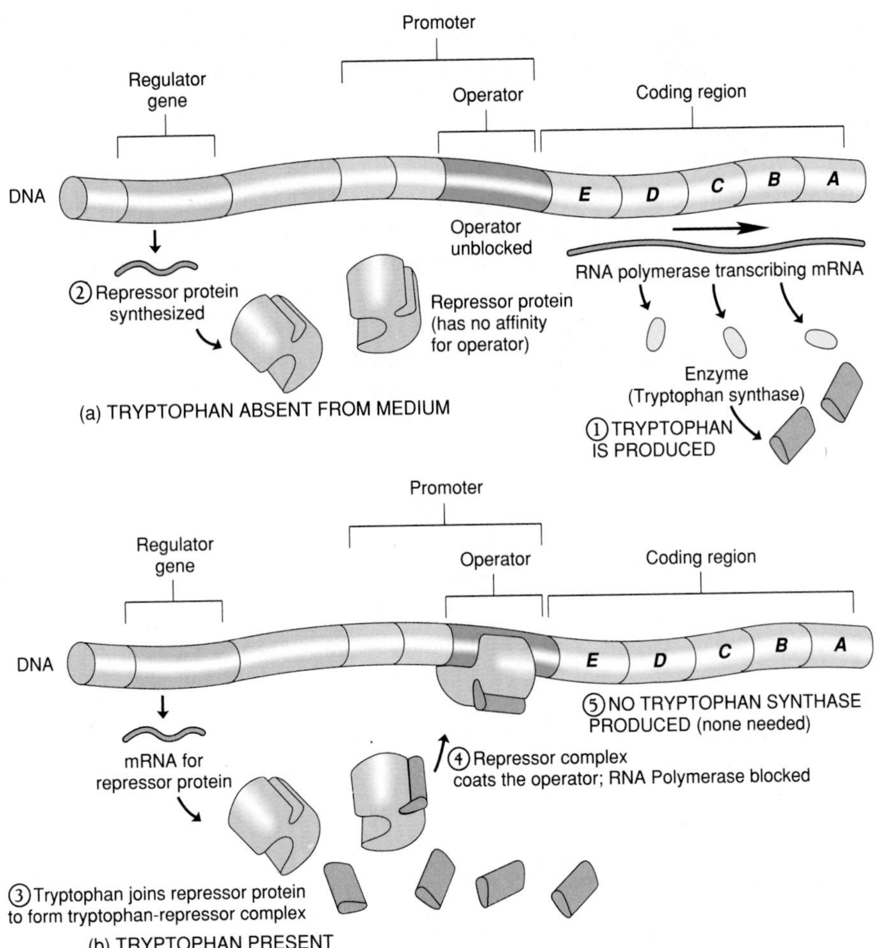

transcribed; the less tryptophan is available, the more the operon is transcribed. All of this carefully balanced bit of biochemical regulation is accomplished by the properties of the *trp* repressor.

The fundamental characteristics of the repressor molecules are the distinguishing features of the two basic types of operons. These features are summarized in Figure 14.4.

We have seen now that bacteria have evolved fairly complex and subtle ways to regulate the expression of certain genes. The most obvious hallmark is the grouping of coding regions to be coordinately regulated into operons. This type of physical arrangement is apparently limited to the prokaryotes. However, the principal features of tran-

Inducible operons	Repressible operons
Repressor alone binds Operator ∴ Operon usually OFF **Inducer** binds Repressor ∴ Repressor does not bind to Operator, allowing operon to turn ON	**Repressor** alone cannot bind to Operator ∴ Operon usually ON **Co-Repressor** binds to Repressor to form Active Repressor complex ∴ Active Repressor complex binds Operator to turn operon OFF

FIGURE 14.4
DISTINGUISHING CHARACTERISTICS OF INDUCIBLE AND REPRESSIBLE OPERONS

scriptional regulation, the negative action of repressor molecules and the positive action of stimulating factors such as CAP, are more generalized phenomena. We will see some examples of the eukaryotic version of these approaches in the next section.

EUKARYOTIC GENE REGULATION

Evolution had a relatively simple problem in the design of bacterial gene regulation. But most eukaryotes consist of a variety of differentiated cell types with different capacities to transcribe different sets of genes. This immediately makes the regulatory problem more complex. In addition, each individual eukaryotic cell can be under the same demands and constraints that we've discussed for prokaryotic cells. There are the metabolic demands of energy acquisition as well as other responses to environmental conditions. But eukaryotic cells face a more complex task, especially if the individual cell is a part of a multicellular organism.

Opportunities for Regulation

We have seen that the bacteria have elegant mechanisms for gene control, but we have also seen that these mechanisms are fairly direct. That is, when the messenger RNA is made, the protein is translated immediately. In fact, the protein translation can begin even before the mRNA is completely assembled.

Eukaryotes cannot do this. The presence of the nuclear membrane necessitates the completion of transcription and the transport of the mRNA to the cytoplasm before translation can occur. But rather than an impediment, these additional processes offer a complete set of new opportunities to control the eventual expression of a gene. After all, we don't see the effects of the gene just because it is transcribed; we see it only after it produces an active protein.

Figure 14.5 outlines the steps in the expression of a eukaryotic gene. Each step represents a possible point at which the process could be selectively blocked—in a real sense, regulated. We will take a look at these possible regulatory levels in the following sections. We will begin with the direct regulation of transcription, a situation where the parallels with the operons will be most apparent. Then we will move on those levels that are seemingly unique to eukaryotes.

Inducible Genes

There are many instances where a gene need be expressed in only certain cell types or under certain environmental conditions. We will discuss two eukaryotic genes that serve as examples of selective gene transcription: genes that are expressed only in response to rises in environmental temperature, and genes that are expressed only in certain developmentally determined cells.

Heat Shock Genes One of the well-studied eukaryotic gene regulatory systems involves a general cellular response to the environmental temperature. Most cells induce, or express, a specific set of genes, the **heat shock protein *(Hsp)*** genes, when they experience a sudden shift to elevated temperatures. While the precise mechanisms are

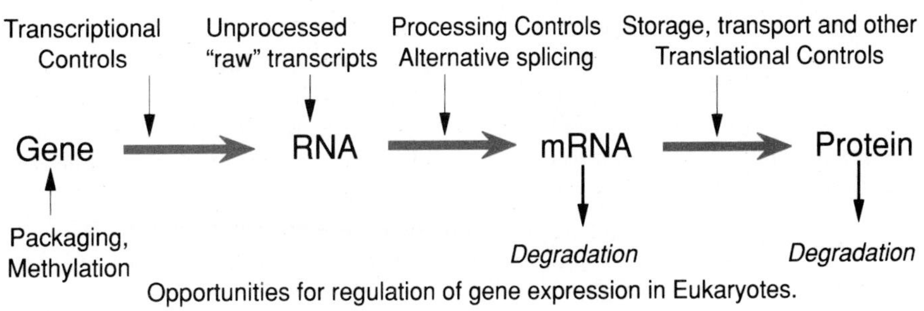

FIGURE 14.5
OPPORTUNITIES FOR GENE REGULATION IN EUKARYOTES
This chart parallels our description of the Central Dogma and points out the several steps in the production of a eukaryotic protein where regulation may occur.

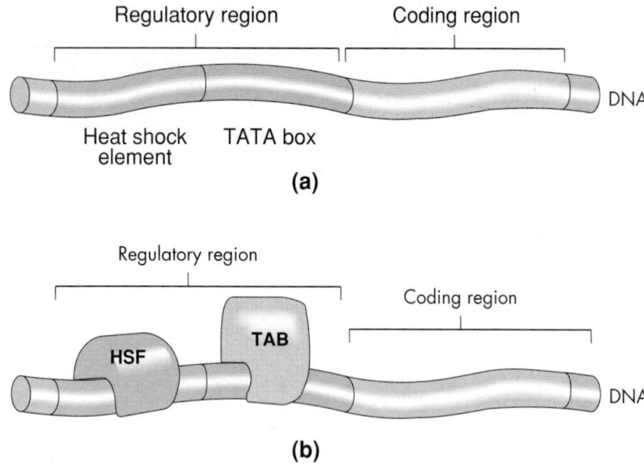

FIGURE 14.6
***Hsp* GENE STRUCTURE**
(a) The regulatory region of a heat shock *(Hsp)* gene consists of two parts, a TATA Box and the heat shock element where the heat shock factor binds to stimulate transcription by RNA polymerase. (b) Transcription occurs only when the heat shock factor is bound to the HSF site *and* when TATA binding (TAB) proteins are bound to the TATA box. The binding of these regulatory proteins stimulates transcription of the coding region by RNA polymerase.

not well understood, the products of these genes provide important protective features that allow the cells to continue living at temperatures that would potentially destroy many biochemical processes. Let's examine the details of the thermal induction of a particular *Hsp* gene from *Drosophila* as an example.

As we might expect, *Hsp* genes are not usually transcribed but are very rapidly induced within minutes of a rise in cellular temperature of a few degrees. This induction of gene activity involves protein interactions with the DNA of the promoter, which (like the prokaryotic situation) lies just upstream of the coding region of the gene (Figure 14.6). There is an area of the promoter, located about 35 bases upstream from the beginning of the mRNA, called the **TATA box,** which serves as the binding site for RNA polymerase. It gets its name from the fact that its sequence consists of TATAAA. The TATA box appears to be a pervasive feature of many, if not all, genes in addition to the heat shock genes. The TATA box apparently serves a fairly general function in all eukaryotic genes, to bind proteins **(TAB proteins)** that in turn help direct the binding of RNA polymerase. While the presence of these TAB proteins is a necessary prerequisite for transcription, they cannot work alone to stimulate RNA polymerase.

In the *Drosophila Hsp* genes, the TAB proteins appear to be bound to the TATA box at all temperatures. This reinforces the idea that their presence alone is not sufficient to accomplish transcription, but that they do act to set the stage for transcription. When the cells experience elevated temperatures, a second protein, the **heat shock factor,** or **HSF,** binds to a specific sequence (called the **heat shock element**) of the promoter DNA, right next to the TAB protein (Figure 14.6). The binding of the HSF, and the presence of the TAB proteins, stimulates RNA polymerase to transcribe the gene. If the cells return to a lower temperature, the HSF will no longer bind to the promoter, and transcription of the *Hsp* gene ceases.

The exact mechanism(s) by which the HSF is selectively able to bind to the promoter is not yet known. It appears that protein modifications, such as phosphorylation, may play a role in specifying the particular binding characteristics of the HSF. In addition, there is more than one *Hsp* gene in *Drosophila,* and each one is activated by the binding of HSF to its promoter. Thus, a whole bank of genes that need to be expressed during heat shock is activated by the HSF. This provides one eukaryotic answer as to how to coordinately regulate a number of separate genes.

The action of the HSF serves as an appropriate example of eukaryotic versions of regulatory proteins responding to environmental signals. However, one of the hallmarks of a eukaryotic organism is the ability to have only certain genes activated in certain cells at specific times.

FIGURE 14.7
STEROID HORMONE REGULATION OF GENE EXPRESSION
In this model, steroid hormones, such as estrogen, bind to cellular receptor proteins. The protein-hormone complex can stimulate transcription if an adapter protein is present and bound to the regulatory region of a gene.

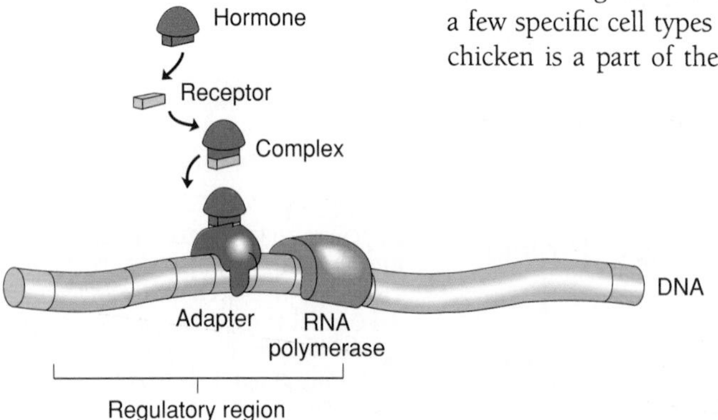

Steroid-Hormone-Induced Genes The steroid hormones, such as estrogen, are produced by specific endocrine glands (see Chapter 37), released to the bloodstream, and carried throughout the body. But each hormone can influence the gene activities of only a few specific cell types called **target cells.** One of the main estrogen target cells in the chicken is a part of the lining of the oviducts. When the oviduct cells are stimulated by estrogen, they begin to make and secrete egg albumin. The presence of estrogen in the environment of a cell in an oviduct stimulates albumin gene transcription. But how? This fairly complicated system of control is shown in Figure 14.7. Estrogen target cells have a special estrogen receptor protein. These are the only cell types that have this protein, which identifies or "tags" them as estrogen-responsive cells.

When an estrogen molecule contacts such a receptor, it forms a complex with it. Then the estrogen-receptor complex apparently migrates to the albumin gene regulatory region.

ESSAY 14.1

HOMEOTIC GENES AND THE HOMEOBOX

As we'll see in Chapter 43, there are many questions that can be asked as an organism develops from the fertilized egg. After just a few cell divisions, we could ask which end will be the head and which the tail? How many segments will be made, and which will carry wings?

It has been known since the early work in *Drosophila* genetics that there are several genes that can drastically alter the answers to these developmental questions. In fact, we now know that there are some eight different genes that control critical developmental steps in fruit fly maturation. These genes are generally referred to as *homeotic* genes (*homeo*-, "same"), since they can result in the replacement of organs by structures similar to those found elsewhere on the fly. For example, the *antennapedia* gene can result in the production of legs on the head, where the antennae should be. Other homeotic genes can control the number of segments in the body, the number of wings, and the head-to-tail orientation of body segments. In short, the eight different homeotic genes are related in that they control key developmental processes, processes that must require the coordinated action of many different genes. The homeotic genes are considered, then, to be a class of regulatory genes, genes that control the expression of other genes.

Researchers are now beginning to understand the way in which these regulatory genes exert their controlling effects. While the protein products of the various homeotic genes are distinctly different, they all share one feature. They are all regulatory proteins that contain a section of approximately 60 amino acids, called the *homeobox*. Computer analysis of the three-dimensional structure of the proteins reveals what they have in common, and why it may be important in their role as regulatory proteins. The homeobox region of the proteins is structurally similar to other proteins known to bind to DNA. In fact, it is the same design that the CAP protein uses to bind the *lac* promoter.

So part of the mystery of how homeotic genes act is solved. The homeotic genes apparently exert their regulatory functions through protein products that directly bind to DNA, probably the promoter DNA of the many genes required to accomplish the elaborate tasks of building organs and body parts.

Once there, it binds to the albumin gene promoter, but only if another DNA-binding protein called the **adapter** is already bound to the promoter. If this adapter is in place on the promoter, the estrogen-receptor complex can bind to stimulate transcription by RNA polymerase.

In summary, the presence of the proper hormone receptor protein gives a cell the ability to be stimulated by the hormone, and the adapter protein identifies the genes that should be activated by the hormone/receptor complex.

Other examples of eukaryotic gene regulation exist, and more are coming to light. In particular, some of the more complex questions of regulated gene expression, such as how regulatory proteins recognize and bind to specific sequences of DNA and how the intricate modifications of development and differentiation occur, are being unravelled (see Essay 14.1).

DNA Packaging and Regulation

We saw in Chapter 9 that eukaryotic DNA is packaged and organized within the nucleus as chromatin, by specific associations with histone proteins. Now we will see that the nature of chromatin structure, the character of the packaging, has a direct impact on gene regulation.

Chromatin Chromosomes, we have seen, are chromatin that is condensed into large, highly visible bodies. At least this is the familiar form they take during cell division. During interphase, when the chromosomes "relax" (or, more properly, become diffused), they form very long, thin threads. We know that it is only the DNA in the diffuse portions of the chromatin net that is transcribing RNA. The condensed portions are inactive. After all, imagine RNA polymerase trying to reach specific DNA sequences that are buried within the highly condensed chromatin network. They just can't get in. It is important to note, however, that chromatin can be condensed to different degrees and that such differences can act as a form of gene regulation.

Recall from Chapter 9 that the primary organizational unit of chromatin is the nucleosome (see Figure 9.2). We might expect that DNA wrapped around a nucleosome would have some difficulty interacting with specific proteins such as the HSF. Indeed, it appears that areas of DNA that are important for regulation (such as promoters) are often kept free of nucleosomes, or that removal of nucleosomes from promoters is necessary for transcription.

DNA in its most highly condensed state becomes quite prominently visible in the light microscope and is called **heterochromatin.** Since it is condensed well beyond the 30 nm stage (see Fig. 9.2), all heterochromatin is considered to be transcriptionally inert. There are two basic types of heterochromatin. One, called **constitutive heterochromatin,** is found near the centromeres of all chromosomes and, as its name implies, never decondenses and probably has no active genes. The other is called **facultative heterochromatin** because its state of condensation can change. Presumably, facultative heterochromatin can contain genes. Even in **euchromatin,** which is much less condensed and contains potentially transcribable genes, the majority of the chromatin is condensed beyond the 30 nm stage. Condensation must be relieved before regulatory proteins and RNA polymerase can bind to promoters and initiate transcription.

There are two striking examples of the relationship between the degree of condensation and gene activity. The first example is the inactivation of one of the X chromosomes in mammalian females to form the transcriptionally inert, highly condensed Barr body (see Chapter 11). The other conspicuous example is gene specific decondensation during the larval stages of *Drosophila* development. In fruit flies, and many other insects, the larval stage is a period of intense gene activity. In Figure 14.8 we see that gene activity

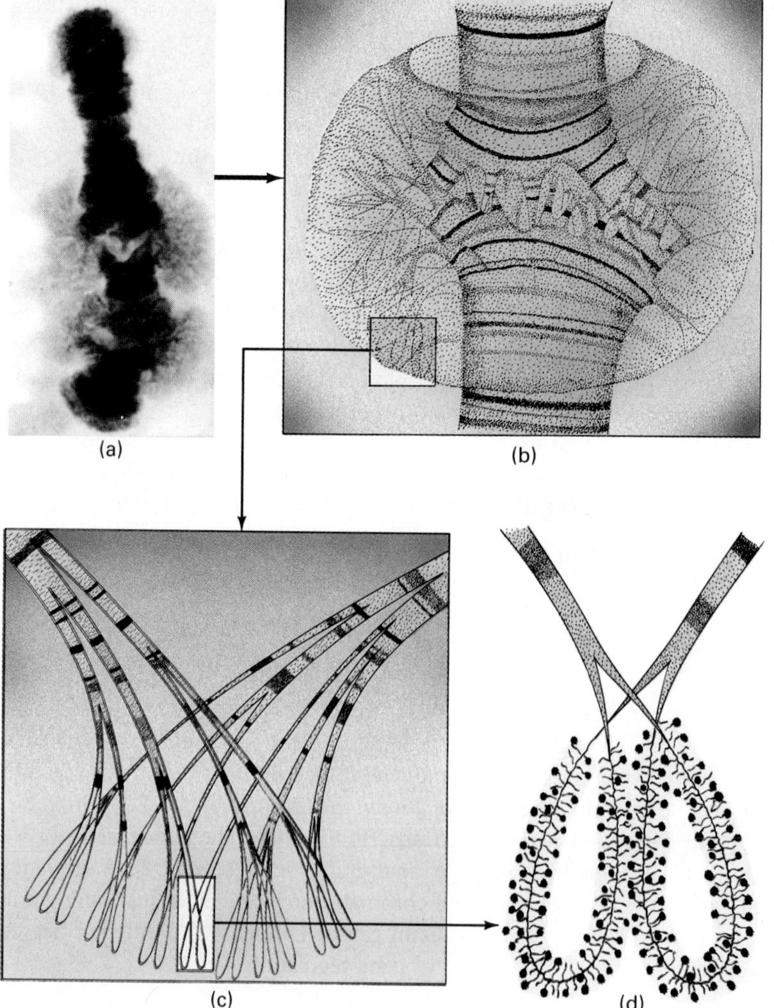

FIGURE 14.8 DECONDENSATION OF CHROMATIN IN THE PUFFS OF *DROSOPHILA* SALIVARY CHROMOSOMES
Puffs are regions of the polytene salivary chromosomes where transcription is known to occur. **(a)** An actual photomicrograph of a puff. **(b-d)** Interpretations of the structure of a puff, showing that several stages of decondensation accompany transcription of this area of the chromosome.

in larval salivary glands is signaled by visible changes in chromatin structure. (Remember, again from Chapter 9, that the chromosomes of the *Drosophila* salivary gland offer a rather unique view. The DNA has been replicated many times, but without segregation into daughter cells. These "endoreplicated" chromosomes consist of thousands of copies of the DNA strands lying next to each other, and are therefore much more visible under the microscope.) The figure shows the regions of transcribing genes as decondensed "puffs" along a *Drosophila* polytene chromosome. Puffs actually consist of loops of DNA that are much less condensed than the DNA in neighboring areas; this means the DNA is accessible to RNA polymerase and regulatory protein.

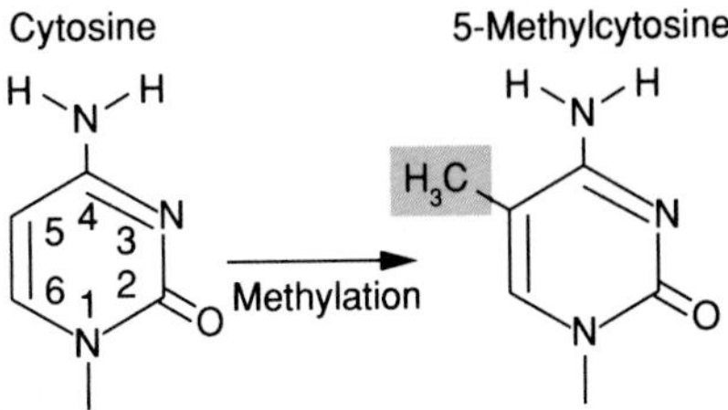

FIGURE 14.9
5-METHYL CYTOSINE

Chromatin that is able to decondense seems to have less H1 histone. Consequently, the nucleosomes should be less stable and less tightly bound together than in chromatin stabilized by H1. It may even be that in this less tightly bound state, nucleosomes can be more easily displaced from a promoter by regulatory proteins. (But apparently nucleosomes need not come completely free of the DNA. Instead, some think that they roll along out of the way like so many yo-yos on a long string.)

Chemical Modification of DNA: Methylation Eukaryotes have a second generalized method for inactivating DNA, involving **methylation,** addition of a methyl group (—CH_3) to certain bases. The primary example is methylation of cytosine nucleotides. Depending on the source of the DNA, up to 5% or more of the cytosine nucleotides within the genome may be modified by the addition of a methyl group to the pyrimidine ring (Figure 14.9). This additional chemical group does not influence the hydrogen bonding characteristics of the cytosine, but it seems to have a rather drastic effect on gene transcription.

Genes that contain a high amount of methylated cytosines are simply not transcribed. There are many examples (though there are also some exceptions) where the methylation state of genes is directly related to whether or not the genes are expressed.

The underlying mechanism relating gene activity and methylation is as yet unknown. One possibility is that methylation plays some role in the packaging of DNA into higher-order, nontranscribable chromatin. Another possibility is that the extra methyl group on a cytosine might well interfere with the binding of specific regulatory proteins.

Posttranscriptional Regulation

As we discussed earlier, the presence of the nuclear membrane offers eukaryotes the ability to separate the two processes of transcription and translation in time and space (see again Figure 14.5). The many steps that accompany this separation of translation and transcription offer several opportunities for regulation.

The first of these opportunities involves the processing of introns from the initial RNA transcript (see also Chapter 13). Since mRNA is not transported to the cytoplasm until the introns are removed, some transcripts could be held in the unprocessed state as a sort of ready reserve. When the gene product is needed, the only thing to be done is to process the message and transport it to the cytoplasm. It is even possible to enlist these intron processing steps in determining the eventual structure of the protein (Figure 14.10). Several examples exist where alternative splicing schemes can be used to generate two different proteins from the same transcript.

The use of alternative selection of exons can be even further expanded—alternative promoters can be used to drive transcription of one gene. Take for example the mouse α-amylase gene shown in Figure 14.10b. It has two promoters, one that is used in the salivary gland and one that is used in the liver. The two different promoters individually control the expression of the gene, at different transcriptional rates, depending upon the organ. The promoters are separated by some distance at the beginning of the gene, with the salivary promoter located upstream of the liver promoter. When transcription occurs from the salivary promoter, the liver promoter is transcribed into RNA. However, the RNA of the liver promoter region is removed from the primary transcript as an intron. Therefore, the coding region of the mRNA is constant, whichever promoter is

(a) Alternative splicing patterns in the rat troponin T gene.

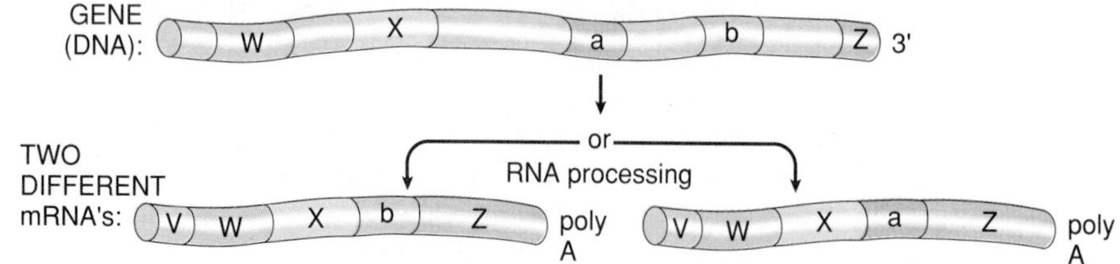

(b) Transcription of the αamylase gene from different promoters in the mouse salivary gland and liver.

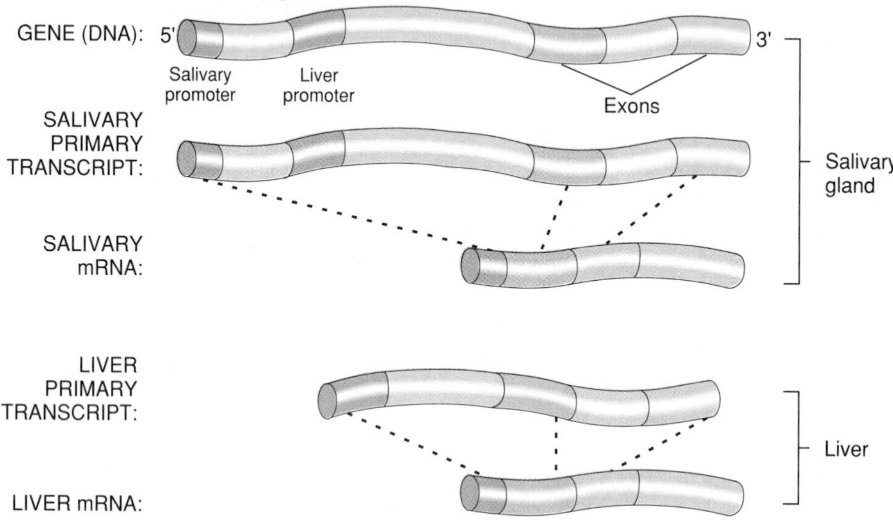

FIGURE 14.10
ALTERNATIVE SPLICING AND THE USE OF ALTERNATE REGULATORY REGIONS
(a) Alternative splicing patterns can result in two different proteins being translated from the same initial gene transcript. In this example, the rat Troponin T gene, variable processing includes either exon α or exon β as part of the final mRNA. (b) The mouse α-amylase gene has two different regulatory regions. The one furthest upstream is active in the salivary gland, while the one closer to the coding region is used in the liver. In the salivary gland, the primary transcript includes the liver regulatory region. But that region is now processed out as part of the first intron. The portions of the mRNA that code for the α-amylase protein are identical. Only the 5′ leader is changed.

utilized for transcription. The final liver and salivary mRNAs differ only in the sequence of their first exon, which consists of 5′ leader sequences.

Another form of posttranscriptional regulation involves the storage of mature mRNAs in the cytoplasm. When the mature mRNA is stored, it can be activated very quickly and with no nuclear involvement. Consider the following example. The rapid, initial DNA replications and cell cleavages that occur shortly after the fertilization of eggs preclude very much transcriptional activity. Therefore, many mRNAs that are translated at these early times are activated from stores of mRNAs. The maternal mRNAs are sequestered as stable protein-RNA complexes, stored until certain conditions, brought on by fertilization make the mRNA available for translation.

Our discussion of posttranscriptional control must finally consider the metabolic lifetime of initial transcripts, mature mRNAs, and the proteins they encode. We can't say why, but we do know that different mRNAs have different lifetimes in the cell. Certainly an mRNA that is around longer will result in more translation and a higher concentration of the protein product. Two genes that produce mRNA at the same rate may differ widely in their apparent gene expression simply because of the different lifetimes of their respective mRNAs. It is also known that different mRNAs can be translated at different rates, and that a given mRNA may be translated at different rates in different cells. So, too, the relative lifetime of the protein product will also play a role. Given equal mRNA levels and translation rates, proteins that degrade faster will appear at a lower concentration in the cell relative to more stable proteins.

Even with these brief discussions, it becomes clear that gene expression is not a simple matter of regulating the amount of transcription that occurs. Indeed, there is a whole host of regulatory possibilities that don't directly involve transcription at all, and it is very likely that the ultimate expression of all eukaryotic genes is subjected to some degree of management at each of these levels.

APPLICATION OF IDEAS

We tend to think of gene regulation and gene expression only as ramifications of selective transcription. At what other "stages" or "levels" might the biology of living cells regulate the phenotype of an individual?

KEY IDEAS

1. Of the thousands of protein coding genes in humans and bacteria, many are active only part of the time, suggesting a controlling mechanism.
2. While prokaryote gene-controlling mechanisms are well understood, the details of eukaryotic transcriptional regulation only now being elucidated.

PROKARYOTIC GENE REGULATION

THE OPERON

1. The **operon** model of prokaryotic gene control was reported by Jacob and Monod in 1961.
2. The **lac** operon in *Escherichia coli* is inducible, its three lactose-metabolizing enzymes (symbolized, *z*, *y*, and *a*) are only produced when lactose is present in the cell.
3. Elements of the operon include the coding regions for producing the enzymes, the **operator** (o), the **promoter** (p) and the repressor gene (i).
4. The lactose operon functions as follows:
   a. Enzyme synthesis repressed. The regulator produces a **repressor** protein that binds with the operator, blocking RNA polymerase so that it cannot interact with the promotor and begin transcription.
   b. Enzyme synthesis induced. Lactose enters the cell and binds the repressor protein, freeing the operator. RNA polymerase interacts with the promoter, transcription begins, and mRNA is translated into the three enzymes. Transcription is efficient, however, only if *CAP* is also bound to the regulatory region.
   c. Enzyme synthesis ends. The lactose-metabolizing enzymes hydrolyze lactose, breaking it down, thus releasing the repressor protein, which once more binds the operator and stops synthesis.
5. The ***trp*** operon controls the production of enzymes that synthesize that amino acid. In this system the enzymes are synthesized continuously, even though a repressor protein is produced. Here, the repressor protein cannot bind the operator unless it first joins a tryptophan molecule.

EUKARYOTIC GENE REGULATION

1. Comparatively little is known about eukaryotic gene control, but it is becoming evident that control systems built upon the principles of regulatory proteins that bind to DNA, much like the operon repressors, also exist in eukaryotes.
2. Eukaryotic heat shock genes appear to require the binding of a TATA box binding protein and a heat shock factor. Steroid-induced genes require the binding of a steroid to a receptor, and the subsequent binding to DNA with an adapter protein.
3. The packaging of DNA into chromatin in eukaryotes offers mechanisms for gene control. DNA condensed beyond the formation of the 30 nm fiber (heterochromatin) is inactive, as is DNA that has been methylated.
4. The various posttranscriptional modifications that must occur in the formation of eukaryotic mRNA offer other points for the control of the expression of genes. In particular, the selective processing of introns can regulate the apparent expression of eukaryotic genes.

REVIEW QUESTIONS

1. How does a repressible operon differ from one that is inducible? Be specific. (315-317)
2. Explain the role of the following in the lactose operon of *E. coli:* (a) promoter, (b) regulator, (c) operator, (d) coding region, (e) repressor protein. (315-317)
3. Describe the proteins bound to a heat shock gene promoter under normal temperature conditions, and after heat shock. (319-320)
4. Describe how the presence of a steroid hormone in the bloodstream activates a gene that should respond to that hormone. (320-321)
5. What is chromatin, and what are its levels of organization? How does packaging relate to gene activity? (321)
6. What is the chemical modification referred to as methylation? What is the effect of methylation on gene activity? (323)
7. Describe how the processing of introns can profoundly influence apparent gene activity. (323)

15

GENES IN VIRUSES AND BACTERIA

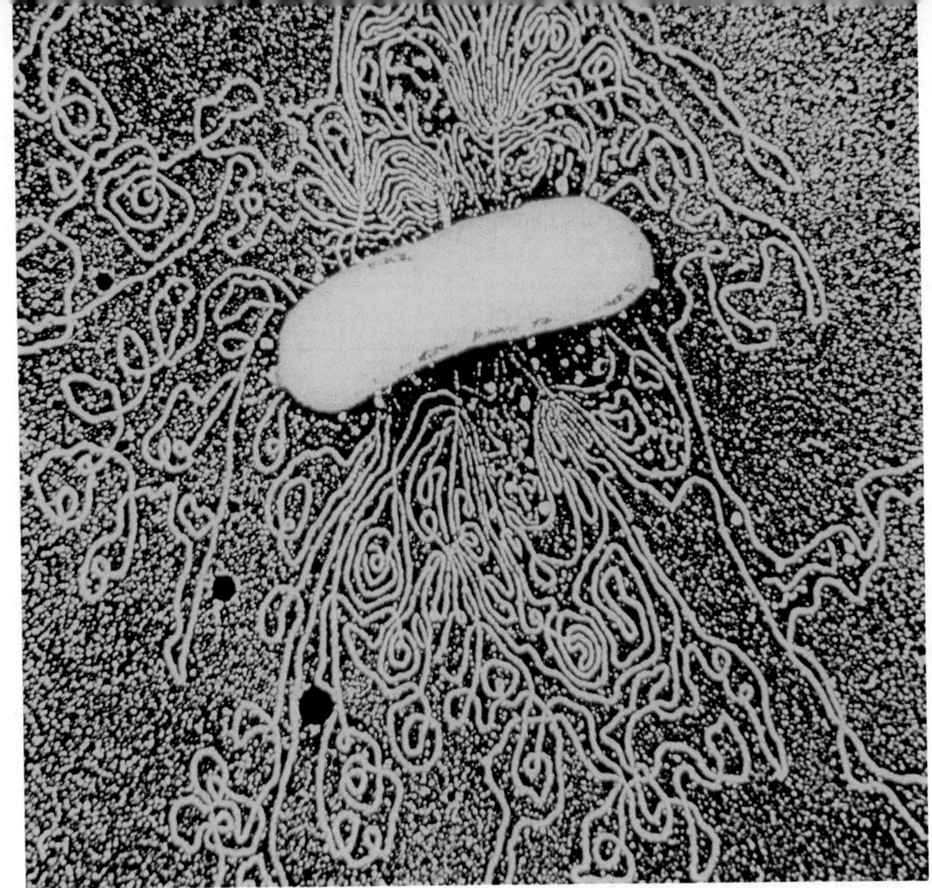

OVER THE YEARS, DIFFERENT ORGANISMS HAVE TENDED TO DOMINATE THE study of genetics. First we had Mendel's true-breeding pea plants, then the hardy, prolific, and amazingly versatile fruit fly. Later, the cutting edge of genetic research focused on the corn plant, *Zea mays,* after which Beadle and Tatum brought the pink mold *Neurospora crassa* into the spotlight. Then, in the late 1940s, a somewhat surprising organism gained the attention of geneticists. They began to focus on *Escherichia coli,* the common colon bacterium, and the viruses that infect it.

Perhaps the emergence of *E. coli* should not have been so surprising. After all, the thrust of molecular biology was to reduce problems to their simplest terms, and bacteria, when all is said and done, are structurally much simpler than fruit flies, corn, or even mold. The viruses that infect bacteria are simpler still. It was thought that the wisest move would be to try first to understand these simpler organisms and then to progress to more complex ones.

E. coli did, indeed, offer a number of advantages as an experimental subject. Not only were they genetically simpler (they are essentially haploid, with only one chromosome), but bacterial cells have little internal structure and are easily broken open so that their cellular machinery can be isolated and analyzed biochemically. More important, bacteria and viruses can be grown in enormous numbers in very short periods of time—*E. coli* populations can double in number in 20 minutes—so experiments can be done quickly. Billions of such organisms can be grown in a test tube, so statistical sampling is never a problem, and even very rare events (such as the occurrence of specific mutations) will occur at a dependable frequency. Furthermore, *E. coli* can get sick; it can be experimentally infected with a virus.

VIRUSES

Viruses, as we will see later, fall into that nether world between life and nonlife. That is, they have many of the traits of living things, yet they can crystallize and develop all the charm of table salt. Nonetheless, certain kinds—specifically bacteriophages—do infect bacteria, and experiments on them have told us a great deal about prokaryotic genetics.

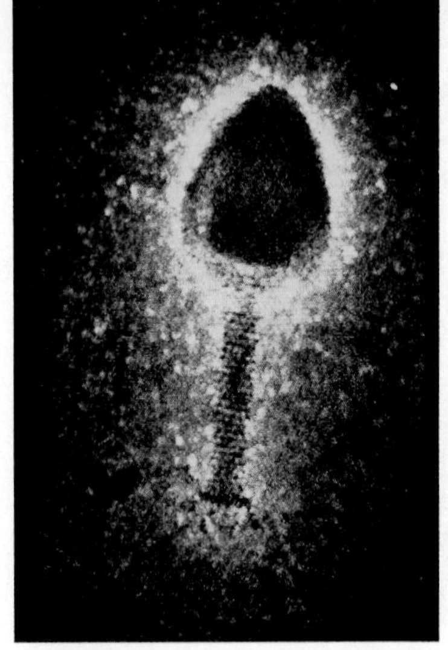

FIGURE 15.1
BACTERIOPHAGE
The T_2 bacteriophage is an excellent example of viral structure. The polyhedral head contains the DNA, and the rest of the bacteriophage is an intricate structure and system designed to deliver the DNA into the host cell.

The Life and Times of the Bacteriophage; The Lytic Cycle

The **bacteriophage**, or phage virus, a parasite of bacteria, has been mentioned with respect to Hershey and Chase's classic experiment, which showed that only the viral DNA entered the bacterial host cell while the empty coat stayed outside (see Chapter 12). Let's look more closely at the bacteriophage and then see what happens when it finds its host. We will consider a typical bacteriophage called T_2, one that commonly invades *E. coli.*

As viruses go, T_2 phage is fairly complex, consisting of a DNA core surrounded by a polyhedral protein head, which ends in a cylindrical, enzyme-laden tail (Figure 15.1). Attached to the tail are a peculiar set of leglike, protein tail fibers, completing a form that is strikingly like one of NASA's moonlanders. Upon "touching down" on its host, the enzymes go to work breaking down the bacterial cell wall, whereupon the tail firmly attaches itself, and the viral DNA is injected into the cell (Figure 15.2a). Once the viral DNA gets inside, it may immediately run into trouble with one of the host's protective mechanisms. However, bacteriophage can also develop mechanisms to overcome those defenses.

As this seesaw evolutionary battle continues, the result is an increasing host-parasite specificity. In other words, every strain of bacteria is subject to infection by only a few bacteriophage strains, and every strain of bacteriophage is able to infect just a few host strains.

If the viral DNA gets past the host's defenses, it may encounter a friendly host protein: RNA polymerase. RNA polymerase isn't able to distinguish between its own DNA and foreign DNA. It reacts to the viral DNA as if it were its own and transcribes mRNA from it, producing viral mRNA. The viral mRNA is then blindly translated by the host's ribosomal machinery into viral proteins.

At this point the virus is in a position to turn the tables on the host. It has now produced new viral enzymes that can readily distinguish between host DNA and viral DNA. These enzymes chop the host DNA into fragments, presumably so that the nucleotides can be recycled into viral DNA. Other viral enzymes make copies of the viral DNA; some even make more viral mRNA to speed up the process. The viral genes also make coat proteins, the leglike "landing gear," and viral penetration enzymes. The final act of the viral proteins is to **lyse** (rupture) the envelope of the host cell, now hardly more than a bag of virus particles. Lysis of an *E. coli* cell may release hundreds of new, infectious particles (Figure 15.2b) that can then move on to infect more host cells, in a continuing cycle called the **lytic cycle.**

FIGURE 15.2
BACTERIOPHAGE IN ACTION
This electron micrograph shows bacteriophage attacking a bacterial cell. The end result is a bacterial cell full of newly formed bacteriophage particles, and eventually cell lysis.

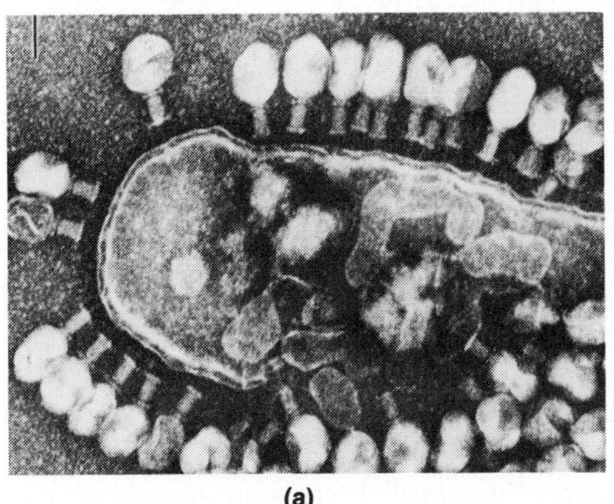

(a)

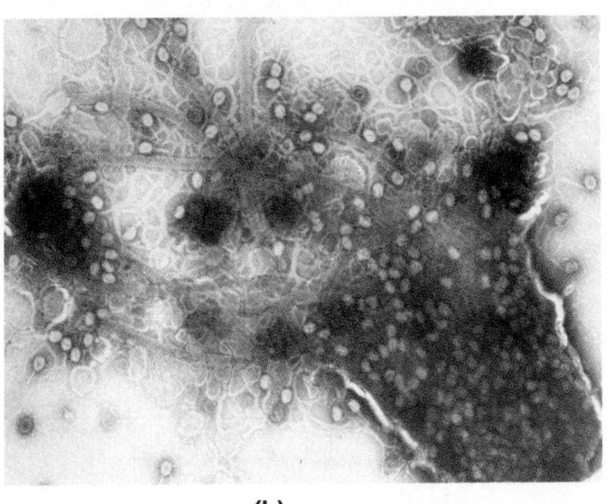

(b)

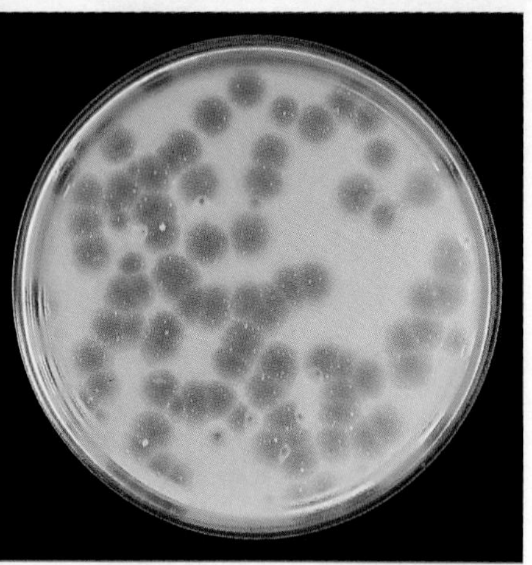

FIGURE 15.3
GROWING BACTERIOPHAGE
Unlike bacteria, bacteriophage and other viruses will not grow on organic media—they require a living host. In this figure, a mixture of bacteriophage has been spread out over a layer (or "lawn") of bacterial cells on the surface of nutrient agar in a Petri dish. The clear circles are places where the bacteria are now absent, an indication that they have been lysed by the bacteriophage, which continues to invade the cells in an ever-widening circle.

The Lysogenic Cycle

Although the life histories of most bacteriophages are based on the lytic cycle, others go through a different sequence, one perhaps more insidious. This one is called the **lysogenic cycle.** Instead of destroying the host chromosome, the phage DNA cuts into and joins the host DNA (a process called *lysogeny*—Figure 22.20). The viral DNA then undergoes replication with the rest of the bacterial chromosome and is passed on to all the bacterium's descendants. While in the chromosome, it does very little except to put out a repressor protein that protects the cell from further attack by other viruses. Then, in some future generation, the viral DNA may break away from the host DNA and begin to make copies of itself. The cell will be lysed, releasing hundreds of new virus particles.

Counting Viruses, or Holes in the Lawn

Viruses are completely parasitic and can only grow in living cells; thus they do not grow by themselves in laboratory medium. However, they can be grown in the laboratory if they are diluted and then spread on agar plates that have been especially prepared with a "lawn"—a continuous surface layer—of susceptible bacteria. Each active virus particle infects just one bacterium, but after lysis its progeny spread to the neighboring bacteria until there is a clear hole in the lawn for each virus that was there to begin with. These holes, called **plaques**, are then counted to determine how many viruses there were (Figure 15.3).

Variations on the Viral Theme

There are many different kinds of viruses, and virtually every kind of cell is subject to viral infection. In spite of such an array of these parasites, most utilize either the lytic or the lysogenic life cycle. There are, however, variations on this theme. The most significant sort of variation involves whether the virus uses DNA or RNA as its primary genetic material, and if it uses RNA, how it handles the replication of that RNA (see Table 22.2).

Some viruses, indeed, use RNA as the nucleic acid of the mature viral particle. The RNA remains single stranded, but even so, it is perfectly capable of carrying the genetic information for the virus. However, some RNA viruses must first convert that information into DNA, which is then transcribed to produce viral protein as well as more copies of the RNA. So RNA makes DNA, which is used to make more RNA (as well as proteins). This process utilizes a very special enzyme, encoded by the viral RNA, known as **reverse transcriptase.** The name of this enzyme tells you what it does. It makes the DNA copy of RNA. Once the first strand complement of the viral RNA is made, the host cell's DNA polymerase can make the second strand. This double-stranded DNA copy of the viral RNA can then be used to direct transcription of more viral RNA. Also, the double-stranded DNA can be incorporated into the host's genome in a process similar to lysogeny. (The AIDS virus is an example.)

There are also RNA viruses that take things a step further and encode an enzyme called **RNA replicase.** This enzyme is capable of making RNA copies from RNA templates, thus eliminating entirely the need for DNA.

BACTERIA

Genetic experiments, such as gene mapping and the making of mutants, can be easily done in bacteria. All it takes is a few experimental refinements designed to recognize mutations in bacteria and allow their manipulation.

Locating Mutant and Recombinant Bacteria

Genetic experiments with bacteria obviously require the ability to identify those with special traits. Among the most easily located "abnormal" bacteria are nutritional mutants—those unable to grow on simple media readily used by wild-type, or unmutated, bacteria. Such growth media are known as **minimal media.** Then various nutrients are added, one at a time. When the nutritional mutant shows signs of growth, the researcher knows that its metabolic machinery was unable to provide that nutrient, and it is labelled accordingly.

Another technique for isolating nutritional mutants involves adding penicillin to the minimal medium. Penicillin is a powerful antibiotic that kills many kinds of actively growing bacteria by interfering with cell-wall synthesis (see Chapter 22). In this procedure, the experimenter puts a large, known quantity of wild-type bacteria into a minimal medium containing penicillin (Figure 15.4). Among the bacteria are a small but unknown quantity of nutritional mutants. The normal cells metabolize the medium, grow, try to divide, and die. But the mutants survive. They remain inactive because they are unable to metabolize and cannot grow and enter cell division. They therefore don't need new cell wall material. So the penicillin doesn't kill the mutant cells; they just sit there, inactive but alive. In this way, the experimenter can substantially increase the frequency of nutritional mutants in the subject population, and these mutants can then be tested on minimal medium.

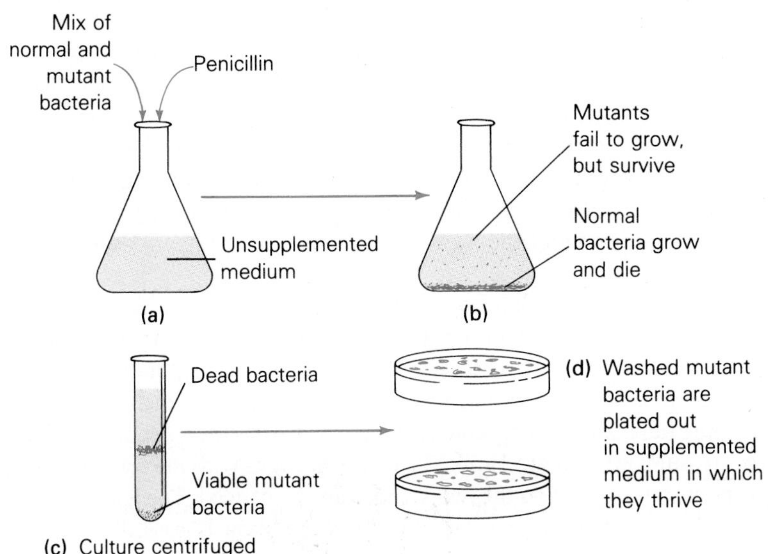

FIGURE 15.4
RECOVERING BACTERIAL MUTANTS
Bacterial mutants can be recovered through a technique involving penicillin. In a mixture of normal and nutritionally mutant cells (a), the normal cells attempt to grow, but die (b). Because the nutritional mutants don't grow, they survive. When the surviving cells (c) are placed in correctly supplemented media, they grow and divide to produce a colony of the mutant for further study (d).

Another ingenious way of recovering mutations is by replica plating, a technique developed by Esther and Joshua Lederberg. As you can see in Figure 15.5, replica plating is a sort of "now you see it, now you don't" process. Here, the bacterial colonies that do not grow after being transferred are the nutritional mutants. But because the colonies have been replica plated, the mutants can be recovered from the original plate. Among other things, the replica plate experiment proved for the first time that nutritional mutants occur spontaneously.

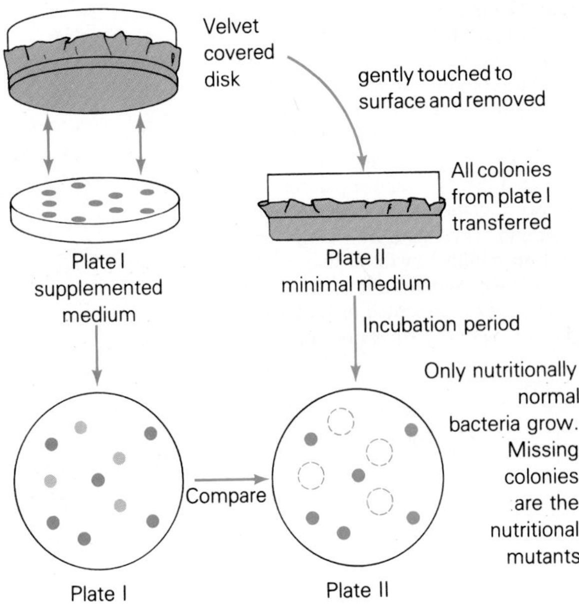

FIGURE 15.5
REPLICA PLATING
Replica plating is a method of locating mutant bacteria. Plate I, containing completely supplemented medium that can support any growth, is coated with a mix of mutant and normal bacteria. The resulting bacterial colonies form a random pattern of dots on the agar. After growth has occurred, a velvet covered disk is used to transfer the pattern of colonies to a minimal medium in Plate II. A comparison of plates I and II soon reveals the mutants, which can be recovered from plate I.

Recombination in Bacteria

Sex in bacteria is so different from sex in even the lowliest eukaryotes that you may not think they really have sex at all. But it's just a matter of semantics. If the broadest possible definition of sex is any mechanism that combines genetic material from two cells into one cell, then bacteria have sex. In fact, they have a fairly varied sexual repertoire.

Different strains of bacteria are able to recombine their DNA in various ways that we will discuss shortly. (Here we broaden our definition of the term *recombination* to include any exchange of DNA, not just that exchange that occurs during meiotic prophase.) However, recombination doesn't involve sex as we know it in eukaryotes. What passes for sex in bacteria is varied, bizarre, and infrequent—so infrequent that the odds of a bacterium having sex are about one in a million. In order to observe such infrequent behavior, it's obviously necessary to take some shortcuts. No one wants to check millions of bachelor bacteria before finding one in the act of mating.

Here is a shortcut. Suppose you have two nutritionally mutant strains of *Escherichia coli,* each of which requires a different amino acid in its nutrient medium. One strain needs tryptophan, and another needs arginine (the first is symbolized trp^-, arg^+; the second, trp^+, arg^-.) (The minus means the nutrient is required, and the plus means that it is not.) Neither can grow on minimal medium. To find that one-in-a-million recombinant, you would mix 50 million or so bacteria of each type, let them mate (or undergo whatever other kind of genetic recombination event they can manage), and transfer them to minimal medium. The trp^-, arg^+ bacteria and the trp^+, arg^- bacteria will both just sit there, unable to grow. Those of the two strains that have managed to recombine to produce trp^+, arg^+, however, will form colonies (Figure 15.6).

Now let's look at some of the ways that bacteria do manage to exchange DNA.

Transformation The first kind of recombination observed in bacteria was **transformation** (see Griffith's experiment, Chapter 12), which involves bacteria taking up fragments of DNA from other bacteria into their cells and incorporating them into their own chromosome. Such recombination takes place with extremely low efficiency. Interestingly, the fragment is incorporated into the bacterial chromosome by what amounts to double crossing over (Figure 15.7). The fragment aligns with the corresponding section of the host's chromosome, and the two sections are exchanged. The leftover host segment is then simply dismantled by enzymes. The new addition is passed along to the cell's descendants when the cell divides.

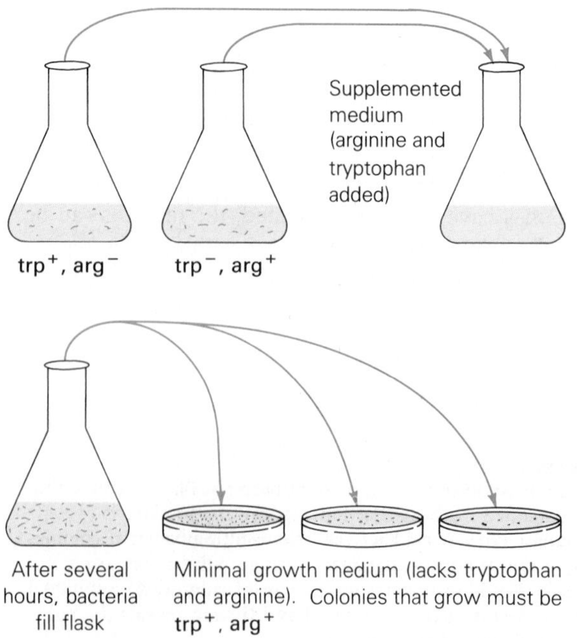

Supplemented medium (arginine and tryptophan added)
trp^+, arg^-
trp^-, arg^+
After several hours, bacteria fill flask
Minimal growth medium (lacks tryptophan and arginine). Colonies that grow must be trp^+, arg^+

FIGURE 15.6
BACTERIAL SEX AND NUTRITIONAL MUTANTS
Biochemical and genetic evidence for sexual reproduction in bacteria was first gathered with the techniques shown in this scheme. Two strains, trp^+, arg^- and trp^-, arg^+, which were selected through replica plating, were grown together in a supplemented medium. Then samples were plated out on minimal medium that lacked tryptophan and arginine. If colonies grew, they represented genetic recombinants with the trp^+, arg^+ genotype.

Transduction **Transduction** involves the transfer of genetic material from one bacterium to another using a bacteriophage as the carrier. What happens is this: Sometimes, when the phage coat protein envelope has already formed and is beginning to fill with DNA, there may still be fragments of the host's DNA floating about. The developing viruses may take up this bacterial DNA in place of or in addition to viral DNA. The developing virus coat doesn't distinguish between them. After lysis, the viral particle with the bacterial DNA attacks a new host, injecting the piece of foreign bacterial DNA. When the new DNA encounters the DNA of the host, recombination can take place. Essentially, then, the protein body of the virus has brought bacterial DNA from one strain into another (Figure 15.8). It has not escaped the attention of biologists that this means of genetic transfer by such omnipresent agents as viruses could have important evolutionary significance in the transfer of genetic information between organisms.

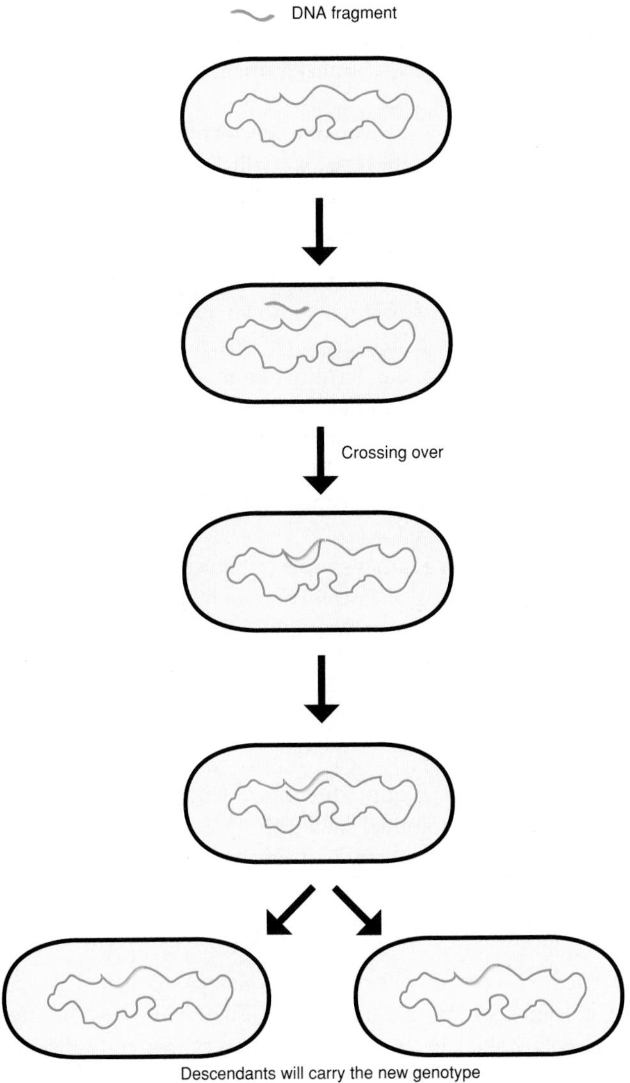

FIGURE 15.7
BACTERIAL TRANSFORMATION
In the process of transformation, a bacterium takes in a fragment of DNA from another bacterium. As the illustration shows, once inside, the fragment aligns itself along the opposing segment of the circular DNA and becomes integrated through double crossing over. The original part of the chromosomal DNA is degraded. All of the bacterium's descendants will then carry the new recombinant chromosome.

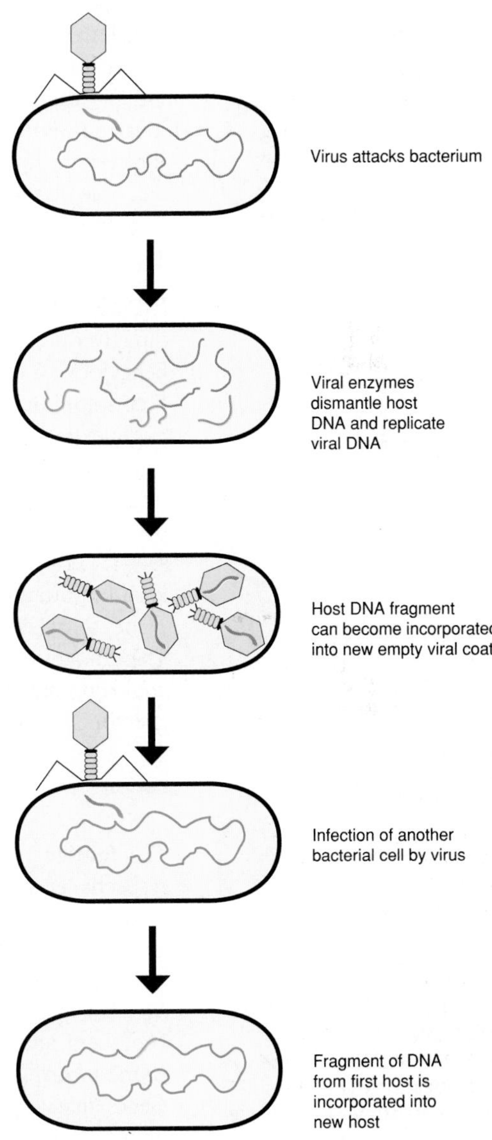

FIGURE 15.8
BACTERIAL TRANSDUCTION
Transduction is a sort of DNA hitchhiking transfer. During phage replication, a segment of the host's fragmented DNA is packaged in the phage head together with the phage DNA. When that virus infects a new host, the segment of bacterial DNA that it carries can be added to the new host's DNA through recombination.

Plasmids and Conjugation Some bacteria contain, in addition to their large circular chromosome, one or more very small circular chromosomes called **plasmids.** Plasmids contain genes and generally replicate in synchrony with the main chromosome and are passed on to progeny of the cell at the time of cell division.

Some plasmids have genes that code for the proteins that make up special structures called **pili** (singular: **pilus**). While most pili are simple proteinaceous spines, some are larger and tubelike. The simpler pili aid the cell in maintaining contact with surfaces. The larger pili are believed to be instrumental in sexual reproduction, drawing cells together and forming a bridge or **conjugation tube**, a hollow tube through which genes can pass. For this reason, the larger pili are often called **sex pili** (Figure 15.9a).

If a plasmid-containing (plasmid-plus) bacterium and a plasmid-free (plasmid-minus) bacterium come close to one another, a sex pilus is formed, and the plasmid-plus bacterium draws up against the plasmid-minus bacterium. The plasmid DNA undergoes an extra round of replication, and as it does so, one of the new copies opens up and is transferred (as a linear DNA molecule) through the sex pilus into the plasmid-minus bacterium (Figure 15.9b). Once inside its new host, the DNA becomes circular again. Other plasmid genes then produce a cell surface substance that prevents further plasmid invasions. Both cells are now plasmid-plus. Plasmid transfer, then, represents a powerful mechanism for genetic recombination in bacteria.

Some plasmids transfer themselves quite easily from one species of bacteria to another. This is unfortunate for us, because some plasmids, as we will learn, carry genes that make bacteria simultaneously resistant to many different antibiotics.

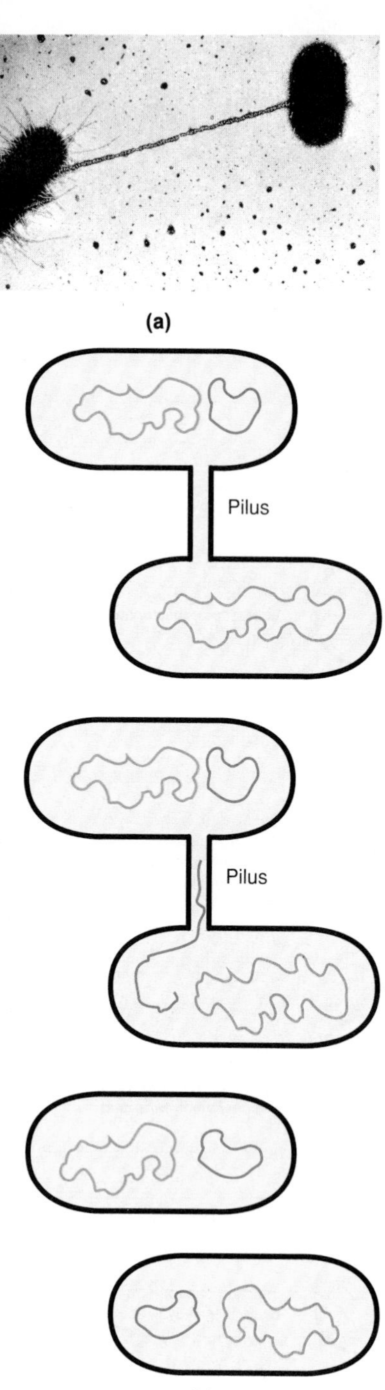

FIGURE 15.9
THE PLASMID AND CONJUGATION
(a) The slender tube between the two bacteria of different strains seen in this electron micrograph is known as the pilus. Its proteins are produced in a plasmid-plus bacterium by genes on the plasmid. **(b)** Once the pilus forms, the plasmid undergoes a round of replication. The new plasmid passes through the pilus into the plasmid-minus bacterium. The arrival of the plasmid changes the recipient into a plasmid-plus bacterium.

***F* Plasmids** Our understanding of another kind of genetic recombination is based directly on the work of Joshua Lederberg and Edward Tatum. They worked out the behavior of a certain kind of plasmid that was originally known as the *F* episome. The *F* episome, or *F* plasmid, got its name from a "fertility plus" strain of bacteria. As you might expect, there was also an opposite strain, the "fertility minus" strain. These bacteria were particularly interesting because genetic material from the positive strain always moved across the sex pilus into the negative strain. Furthermore, after such a union, the negative strain changed, itself becoming positive. But we are getting ahead of ourselves. Let's go back a few years and review a bit of history that shows why this plasmid is so important to modern geneticists.

In 1946, Lederberg and Tatum performed a rather simple but imaginative experiment. They mixed two nutritionally deficient strains of *E. coli* bacteria together to see if they could get any wild-type progeny. The experiment was unusual because, at the time, everyone was convinced that bacteria were confirmed celibates. The two strains were:

Strain A: met^-, bio^-, thr^+, leu^+, thi^+

Strain B: met^+, bio^+, thr^-, leu^-, thi^-

The nutrient symbols refer to methionine, biotin, threonine, leucine, and thiamine, respectively. Neither strain could grow on minimal medium.

It turned out that about one cell in 10 million grew perfectly well on minimal medium and therefore had the following recombinant genotype:

Recombinants: met^+, bio^+, thr^+, leu^+, thi^+

Lederberg and Tatum next tried to work out linkage relationships, as geneticists are prone to do when mapping chromosomes (see Chapter 11). They had a terrible time. Genes that appeared to be closely linked in one experiment were evidently distantly linked in another. It was in this way, years later, and after any number of ingenious (but erroneous) hypotheses were discarded, that biologists concluded that the bacterial chromosome was circular. Imagine the problems anyone would have in trying to work out linkages with circular chromosomes!

Other investigators, including William Hayes, began to employ Lederberg and Tatum's techniques, and as new data came in, a strange and confusing picture began to emerge. For example, it was discovered that bacterial strains could be divided into two groups, which were promptly (perhaps too promptly) dubbed "male" and "female." The names seemed appropriate because males mated only with females, and mixtures of two male strains or two female strains never produced recombinant progeny. These days the "male" strain is simply called F^+, and the "female" strain is called F^-.

What distinguished male from female was that gene transfer was always in one direction, from the male type to the female type. But, surprisingly, the male usually transferred only a few genes at a time. This is obviously a very different situation from that of the truly sexual eukaryotes, whose zygotes receive an equal input of genes from both parents.

A Closer Look at Bacterial Sex How was it discovered that only a few genes from the males are transferred to the female? Let's look at a typical cross between two strains, as it was done in the early days. The fact that a combination of strains with different nutritional requirements can give rise to completely wild-type progeny doesn't tell us much other than that recombination does take place. To get to the root of the matter, it was necessary to look at the transfer of unselected loci—that is, the inheritance of nutritional mutants under conditions in which either the plus or the minus allele could survive. Such unselected genes were called **markers** because they generally were simply used to mark places (loci) on the chromosomes and did not affect whether or not the bacterium would survive. To be sure that recombination is taking place, a medium supplemented with only some of the nutrients is used. Then the recombinant progeny are looked at more closely, with replica plating, on other kinds of partially supplemented agar plates.

For example, suppose the original strains,

Strain A: *met*⁻, *bio*⁻, *thr*⁺, *leu*⁺, *thi*⁺

Strain B: *met*⁺, *bio*⁺, *thr*⁻, *leu*⁻, *thi*⁻

are mixed and plated on a medium supplemented with biotin, leucine, and thiamine but lacking methionine and threonine. All the surviving progeny would have to be met^+, thr^+, indicating that a mating has taken place. But what are their genotypes with regard to *bio, leu,* and *thi,* which are acting as the three genetic markers? There are actually eight different possible genotypes among the recombinant colonies (Table 15.1). To diagnose the genotypes and to count the number of colonies of each type, three replica plates are made. Each replica plate is supplemented with two out of three nutrients. Linkage relationships can be worked out from the relative numbers of colonies of each of the eight types. For instance, if leu^+, thi^+ and leu^-, thi^- combinations were common, while leu^+, thi^- and leu^-, thi^+ were both rare, one could conclude that the leucine locus and the thiamine locus were close together on the chromosome.

TABLE 15.1 EIGHT POSSIBLE GENOTYPES AMONG THE RECOMBINANT COLONIES

	GROWTH ON REPLICA PLATES[a]		
GENOTYPE (ALL met^+, thr^+)	LACKING ONLY BIOTIN	LACKING ONLY LEUCINE	LACKING ONLY THIAMINE
bio^+, leu^+, thi^+	+	+	+
bio^+, leu^+, thi^-	+	+	−
bio^+, leu^-, thi^+	+	−	+
bio^+, leu^-, thi^-	+	−	−
bio^-, leu^+, thi^+	−	+	+
bio^-, leu^+, thi^-	−	+	−
bio^-, leu^-, thi^+	−	−	+
bio^-, leu^-, thi^-	−	−	−

[a] + grows on medium; − does not grow on medium

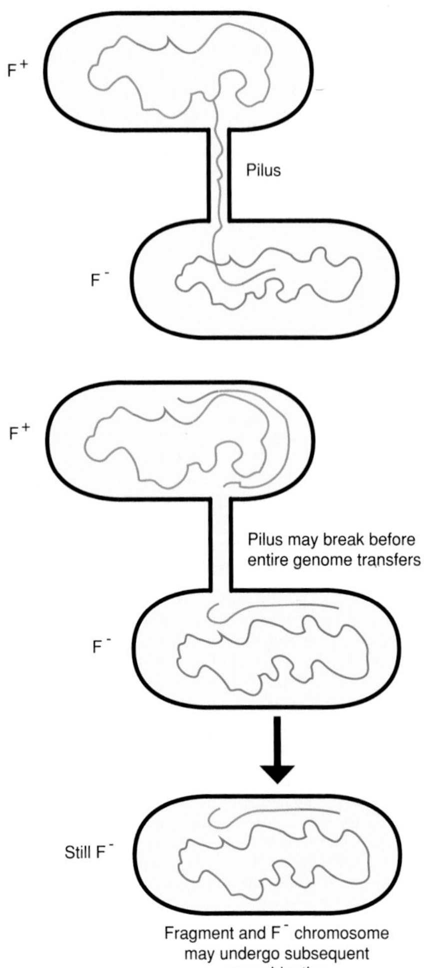

FIGURE 15.10 TRANSFERRING THE INTEGRATED PLASMID
In rare cases the F plasmid joins the entire bacterial chromosome. When this type of F^+ bacterium mates with an F^- type, the entire chromosome replicates, beginning with the integrated plasmid. The replicated plasmid starts through the pilus, dragging the lengthy bacterial chromosome with it. Any portions of the bacterial chromosome that make it through the pilus before the DNA breaks may be incorporated into the recipient by recombination.

The results of experiments such as these also showed that the males contributed relatively little to the recombinant progeny. For instance, if strain A were male and strain B were female, most of the recombinant progeny in the above experiment would be met^+, bio^+, thr^+, leu^-, thi^-—just like their strain B "mothers" for all unselected marker loci.

What recombinant genotype would be the most common if strain A were female and strain B were male? In line with the above reasoning, the most common of the met^+, thr^+ recombinants would be bio^-, leu^+, and thi^+—just like their strain A "mothers." Let's move on now to some very clever related work and find out how some of the riddles seen so far were solved.

The *F* Plasmid as an Agent of Gene Transfer What about the transfer of genes located on the main bacterial chromosome? After all, that's what makes the *F* plasmid so interesting to scientists in the first place. It turns out that while simple transfer is very common, in rare cases the *F* plasmid may become inserted into the main chromosome of its host. The plasmid has no preferred site of insertion and can show up anywhere in the host's circular chromosome. After insertion, the *F* DNA is replicated right along with the host DNA and can be passed on to all the bacterial offspring. So far, this sounds like phage lysogeny. But later, when the plasmid forms a conjugation tube and attempts the transfer to an F^- bacteria, strange things happen. First, the *F* plasmid opens up, as usual. But since it is now part of a larger circle, the entire structure—host chromosome and all—is changed into a linear DNA molecule. Then, when one end of the *F* plasmids goes through the conjugation tube, it starts to pull the entire host chromosome in after it. Because the chromosome is very long, it takes about 89 minutes for the whole thing to get through. In fact, the tube usually breaks apart before transfer is completed. In this case, since the recipient cell usually gets only a piece of the F^+ sequence, it isn't transformed into an F^+ bacterium, but the bacterial DNA that is brought in can recombine, by crossing over, with the DNA of the recipient ("female") bacterium (Figure 15.10).

It wasn't long before Hayes and others were able to isolate several bacterial strains in which the *F* plasmid had been integrated into the host chromosome. Every bacterium in such a strain had the plasmid integrated in exactly the same place. These strains showed a potential for recombination of the main bacterial chromosome that was 1000 times greater than that of the ordinary F^+ strains, because in all the bacteria of these strains each *F* plasmid was integrated, a rare occurrence in other F^+ strains. These were called ***Hfr,*** for **high frequency of recombination.**

Mapping the *E. coli* Chromosome: The Great Kitchen Blender Experiment Progress in molecular biology often follows the development of new technology. In this case, the technology was provided by Fred Waring, a popular band leader of the 1950s. When he was not leading his group, The Pennsylvanians, in song, he was busy inventing the Waring blender. The blender, of course, beats food into a mush, and in so doing it disrupts cells. Molecular biologists needed a good way to separate mating cells, so they tiptoed away with the blenders from their kitchens.

In one of the first blender experiments, Francois Jacob (of the Jacob and Monod team) and Elie Wollman broke apart mating bacteria with such a blender. They let *Hfr* and F^- bacteria mate for precisely measured lengths of time before pushing the button, rudely interrupting the transfer in a primitive form of coitus interruptus.

As we have mentioned, it takes about 89 minutes for the *Hfr* chromosomes to transfer to the F^- bacterium. Thus, if the transfer is interrupted after just a few minutes, only part of the chromosome would be transferred and only a few of the known gene loci would make it across, while most of the others would remain behind. As the periods of mating were extended, more and more loci would be transferred and could be recovered in the progeny, although at lower efficiencies because of spontaneous breakups of the bacterial couples. The last genetic loci come through, but with extremely low efficiency, toward the end of the 89 minutes.

FIGURE 15.11
GENETIC MAPPING OF *E. COLI*
The genetic map of the circular chromosome of *E. coli,* produced by recombinant studies of interrupted *Hfr* matings, is divided into 89 parts, corresponding to the 89 minutes required for a complete chromosome transfer. Each gene locus is identified by code letters that represent the ability or inability to synthesize various substances. Note that many of the codes represent amino acids using the standard abbreviations. The technique used in developing the map includes sequential interruption of gene transfer and, following this, identifying the genes that made it through by plating on different media.

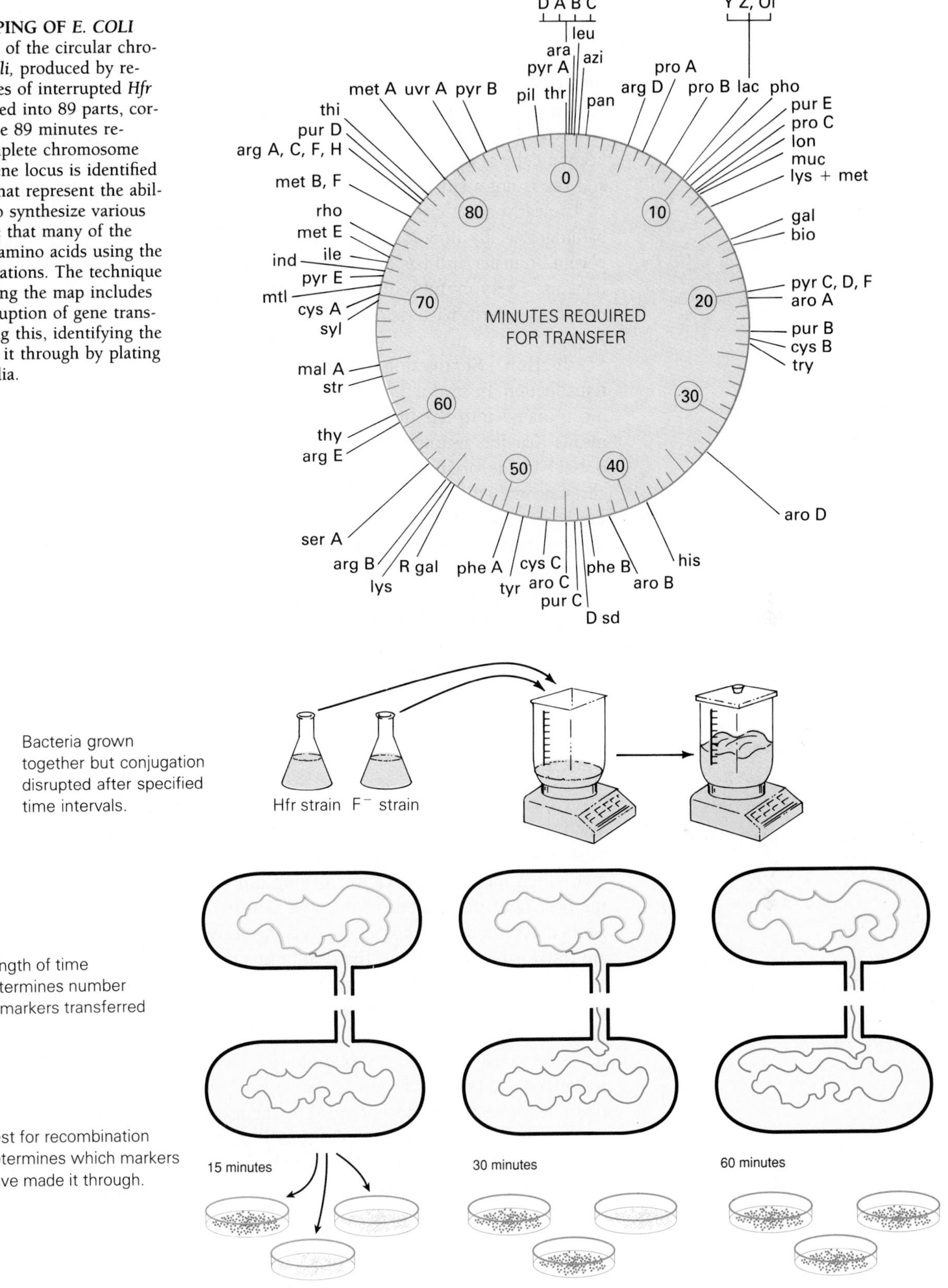

Jacob and Wollman discovered that they could rapidly map *Hfr* bacterial chromosomes with their blender. They found that on a standard *Hfr* chromosome the first locus passed through in about two minutes, and each subsequent marker was given a position appropriate to the time it entered the recipient cell. Figure 15.11 shows the genetic map of *Escherichia coli*. It was derived by noting the time at which the individual loci transferred.

One question that might have occurred to you is whether the donor *Hfr* cell survives the transfer of its chromosome. Recall that plasmid transfer is associated with replication. Apparently, replication occurs *simultaneously* with transfer, with one of the replication fork strands going through the tube and the other staying behind. The donor cell always keeps a copy of the entire *Hfr* chromosome. The answer then is yes, the *Hfr* cell survives, and sex is not a suicidal act for a male bacterium.

The early geneticists forced all sorts of sexual interpretations upon their originally baffling data. By now it seems clear that "bacterial sex" is, in fact, a rare aberration of plasmid transfer and has little to do with sex as we eukaryotes know it. But it was, and remains, a very useful tool in molecular biology. It may also prove to play a significant role in long-term bacterial evolution.

Sexduction **Sexduction** is admittedly an awful term—a combination of sex and transduction. In sexduction, a small bit of the large host chromosome somehow becomes incorporated into the free *F* plasmid. It can then be transferred to any F^- cell. This means that the recipient cell is functionally diploid for those bacterial genes that are present in the plasmid (that is, it has two copies of each such gene, one on its own chromosome and one on the *F* plasmid). In such partially diploid bacteria, gene loci can be heterozygous, and one can even investigate dominance and recessivity relationships in bacteria, where such things don't normally occur. It was through the use of sexduction that Jacob and Monod were able to work out some of the intricacies of the operon in bacteria (see Chapter 14). Also, sexduction clearly suggests a means by which the molecular biologists could manipulate genes. If they could somehow get a plasmid to take up a strand of selected DNA, it might be possible to get bacteria to do some work for them. If this sounds farfetched, just keep it in mind as we go on to Chapter 16.

Plasmid Genes

The *F* plasmid is only one of many kinds of bacterial plasmids that have been discovered. The others are not noted for their ability to transfer the host's chromosomal genes to other bacteria, but they do play roles in bacterial genetics. They have genes of their own, often including genes for making pili and for coating the cell surface to protect the host (and the plasmid) from infection by other plasmids. In addition, some have genes that protect their hosts (and thus themselves) in other ways. For example, plasmids are responsible for some of the deadly toxins produced by disease-causing bacteria.

Some plasmids even carry genes that make their hosts resistant to antibiotics. One plasmid, known as **R6**, endows its host with resistance to six important antibiotics (Figure 15.12). R6 can also be transferred between distantly related species of bacteria. It was previously believed that plasmids were a kind of parasite, and that plasmid infection resembled a disease infection in some ways; we have to modify that now since some plasmids protect their hosts (and thus themselves) against the ravages of antibiotics.

Now that we know about "infectious antibiotic resistance," some mysteries have been cleared up. Time and again physicians have successfully treated bacterial infection with antibiotics, only to find that the same disease turns up again but is no longer cured by the same antibiotic. What has happened is that natural selection has been at work, and our very success at battling microbes has paved the way for the rise of new, more potent strains. Fortunately, the drug industry has more or less kept up with the challenge by continually developing new or altered antibiotics.

Because of the resistance phenomenon, the indiscriminate use of antibiotics is today a controversial issue in medicine. Antibiotics are routinely used in livestock feed to accelerate growth rates and to protect against disease, and those antibiotics can be transferred to meat-eating consumers. (Some livestock breeders have recently agreed to forego the use of antibiotics in this way, partly because Europeans have refused to buy such treated beef.) Fortunately, the use of antibiotics to retard spoilage in foods such as milk is illegal, and these foods are monitored by public health agencies. It is also clear that too many physicians routinely prescribe antibiotics for all manner of minor ailments, in part to guard against secondary infections and malpractice suits, but mainly to satisfy their patients' demand that something dramatic be done to end the sniffles. It is now believed that this indiscriminate overuse of antibiotics has had a positive selective effect on monstrous superresistant bacterial strains.

On the other hand, there is evidence to back up a different view, which is that the widespread use of antibiotics by American and European physicians may have had a net beneficial effect. It has been maintained that such widespread use has managed virtually to eliminate rheumatic heart disease and salmonellosis in those parts of the world where antibiotics are readily available. (Rheumatic heart disease, which is caused by a *Streptococcus* bacterial infection, was until recently a leading cause of death among young and middle-aged adults. Salmonellosis, until recently, was a major debilitating disease. Both diseases are still raging in less developed parts of the world, but both are susceptible to standard antibiotics.) Antibiotics have also greatly reduced the prevalence of syphilis, once the leading cause of insanity. Unfortunately, resistant strains of the syphilis organism are on the loose, and some strains of *Salmonella typhimurium,* the agent for salmonellosis, have picked up the R6 plasmid, with some strains showing resistance to antibiotics.

Plasmids have recently taken on a new significance in molecular biology. As we will see in Chapter 16, many of the most recent advances in genetic engineering and gene cloning have involved plasmids. Their small genomes are readily isolated and manipulated, and they are widely used as vectors (carriers) of spliced genes in DNA experiments.

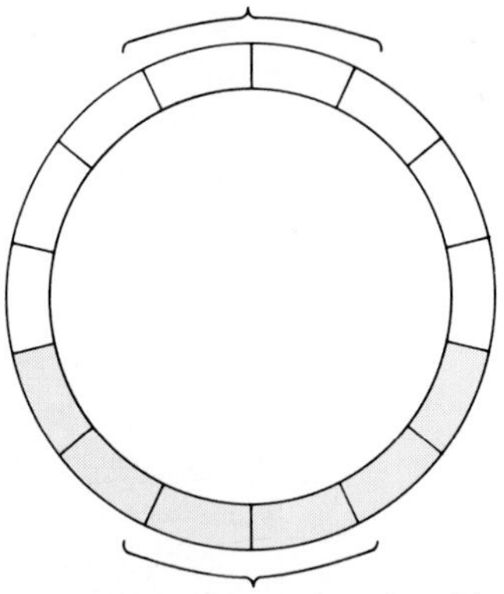

FIGURE 15.12
THE R6 PLASMID
The R6 plasmid is a circle of DNA that contains genes that endow the bacterial cell with resistance to certain antibiotics.

KEY IDEAS

VIRUSES

The Life and Times of the Bacteriophage; The Lytic Cycle

1. Bacteriophage are viruses that inject their DNA into a bacterial host.
2. If the phage succeeds in avoiding the hosts' defenses, its next act will be to use the host's RNA polymerase to transcribe the viral DNA and synthesize viral enzymes.
3. Viral enzymes disrupt host DNA and recycle its nucleotides for use in replicating many copies of viral DNA. In addition, viral DNA is transcribed into RNA to be used in producing viral coat proteins. When viral reproduction is complete, the host cell will be **lysed,** and many infectious viruses released.

The Lysogenic Cycle

An alternative to lysis is lysogeny. The viral DNA is incorporated into the host chromosome, and the viral genes are replicated along with the host's as reproduction occurs. Then, in some future generation, the virus may again enter its lytic cycle.

Counting Viruses, or Holes in the Lawn

Phage activity in a bacterial plate can be recognized as **plaques**—clear areas in an agar plate that otherwise has heavy bacterial growth.

Variations on the Viral Theme

There are viruses whose genome is made of RNA. These viruses replicate their RNA by **reverse transcription** or **replicase.**

BACTERIA

Locating Mutant and Recombinant Bacteria

1. Mutations that affect nutritional requirements of bacteria are readily identified.
2. Replica plating includes transferring colonies from a supplemented to a minimal medium and looking for colonies missing from the pattern.

Recombination in Bacteria

1. Although bacterial sex is infrequent, true sexual reproduction—the combination of DNA from two individuals—can occur through several mechanisms.
2. **Transformation** is the uptake and incorporation of fragments of stray DNA into the bacterial chromosome.
3. In **transduction,** stray DNA fragments carried into the bacterial cell by a phage virus are incorporated into the bacterial chromosome.
4. **Plasmids** are small, supplemental DNA molecules in bacteria; one early group was labeled episomes.

5. One kind of plasmid codes for the synthesis of the **sex pilus** act as **conjugation tubes.**
6. Bacteria with the plasmid are labeled plasmid-plus, others, plasmid-minus. Such plasmids can be transferred to plasmid-minus bacteria, changing them to the plus form.
7. The **F plasmid** represents the first thoroughly investigated sexual activity in bacteria.
8. In 1946, Lederberg and Tatum demonstrated sex in bacteria by crossing two nutritionally mutant strains and obtaining wild-type progeny.
9. Linkage studies from crosses of nutritional mutants revealed that during sexual exchange, only a few genes are transferred at a time, and that transfer occurs in one direction only.
10. Through his experiments, William Hayes added the notion that bacterial sex was catching. Maleness, the ability to transfer genes, could be passed from one individual to another. (It was later determined that it was a sex plasmid that was transferred.)
11. The *F* episome, or plasmid, codes for the sex pili, which permits copies of itself to be transferred. After replication, it opens to a linear form to pass through the sex pilus into the F^- recipient.
12. In rare instances, the *F* DNA will insert into the host's main chromosome (and, as in phage lysogeny, can be replicated and passed along to progeny).
13. Once integrated into the host chromosome, the *F* plasmid can then open to the linear form and the *host's main chromosome* passes through the sex pilus (usually just a portion makes it). Bacteria that did this more commonly were isolated and grown in pure cultures and termed ***Hfr*** **(high-frequency of recombination).**
14. The peculiar chromosome transferring ability of the *Hfr* strain made genetic mapping of the *E. coli* chromosome possible.
15. In **sexduction**, a small portion of main chromosome is incorporated into an *F* plasmid.

Plasmid Genes

1. There is concern with the R6 plasmid since its appearance in a dangerous pathogen could make that bacterium impossible to control. The R6 plasmid may already be responsible for the increases in many antibiotic-resistant strains around today.
2. The R6 risk factor is intensified by the heavy use of antibiotics in the cattle industry and their overprescription by many physicians.
3. The intense use of antibiotics has, however, all but eliminated certain important diseases from developed nations.

REVIEW QUESTIONS

1. Suggest several advantages for using the bacterium *E. coli* in studies of genetics. (326)
2. Briefly outline the general procedure for isolating nutritional mutants of bacteria. (328-329)
3. Suggest a procedure that could be followed to determine whether genetic recombination had occurred in bacteria. (330)
4. Suggest several ways in which sexual reproduction in bacteria is different from sexual reproduction in the eukaryote. (330)
5. Explain how a bacteriophage reproduces itself. (327)
6. Distinguish between lytic and lysogenic cycles in bacteriophage. (327-328)
7. Outline a procedure for visibly detecting the presence of bacteriophage. (328)
8. In what way is transformation similar to transduction? How are they different? (330)
9. Specifically, what is a plasmid? How might a plasmid get from one bacterium to another? (330-332)
10. Using the terms F^+ and F^-, explain how conjugation occurs in *E. coli.* (334)
11. List two things that happen to a "female" or F^- bacterium that has undergone recombination with an F^+ bacterium. (334)
12. Briefly explain how the F^+ plasmid was able to transfer genes from the larger bacterial chromosome. (334)
13. What special capabilities were there in Hayes' *Hfr* strain? (334)
14. Explain the kitchen blender experiment, including the purpose, the timing, the reasoning behind it, and the choice of experimental organism. (334-336)
15. What is an R6 plasmid? In what way can such plasmids be threatening to us? (336-337)

16

GENETIC ENGINEERING: THE FRONTIER

GENETIC ENGINEERING HAS BURST UPON THE PUBLIC CONSCIOUSNESS WITH an impact unusual for such technical matters. In the past, the tremendous attention given by the public to some scientific issues has proven to be transient, finally leaving scientists working quietly in their labs as public interest diminished. Genetic engineering, however, has proven to be a bit different. The public has been fascinated by the idea of working directly on the genetic makeup of living things. This particular scientific brew has continued to ferment, generating its own heat. As new findings rise to the surface and burst upon the scene, they constantly change the atmosphere in which the process works.

Indeed, the promise of genetic engineering is a great one, and the potential benefits of the technology will be continually revised upward as new findings open other avenues for development. We no longer have to squeeze infinitesimal amounts of growth hormone from the pituitaries of cadavers. Thanks to genetic engineering, we can now make it by the vatload. We have the potential to supply normal genes to individuals crippled by inherited maladies, or to diagnose other diseases long before they begin to take their toll. We can alter the amino acid balance in fruits and vegetables, and grow crops that will be less susceptible to chilling frosts. Truly, we are at the beginning of a genetic revolution of tremendous magnitude. No topic in biological science is more worthy of our attention and understanding.

Before going any farther, we might ask, just what is genetic engineering? **Genetic engineering** is the process of altering genetic material by the purposeful manipulation of DNA. It is a subset of the more general kind of endeavor known as biotechnology, the use of biological systems for the production of materials.

DNA TECHNOLOGY

Basically, what genetic engineers do is introduce specific pieces of DNA into cells in such a way that the introduced or "foreign" DNA is replicated in the cell. As the cell containing the newly introduced DNA divides over and over, a **clone** is produced. In a clone, all the cells are identical and, in this case, contain copies of the introduced DNA. Cells containing that new DNA can be grown in any quantity. The foreign DNA is then available in large amounts for study or manipulation. Any product of the genes of introduced DNA can also be made in large quantities as the gene is duplicated again and again, each copy able to make specific proteins. The trick, then, is to attach the DNA segment of interest to a proper carrier DNA molecule, called a **vector,** so that the host cell will treat the combination of vector and foreign DNA (the **recombinant DNA)** as a natural part of its genetic constitution. In the discussion that follows, we'll examine the enzymes used in recombinant DNA research, then show just how these enzymes are used to clone genes. All of the enzymatic tools required to perform the process are natural parts of biological systems, and all of the vectors and DNA transfer technologies are extensions of normal genetic components and processes.

Restriction Enzymes and Sticky Ends

While the basic biology of DNA had been studied for decades, it wasn't until the discovery of restriction enzymes that recombinant DNA became possible. **Restriction enzymes** are enzymes capable of cutting the phosphate backbones of DNA molecules at specific base sequences.

Genetic engineers did not invent restriction enzymes. These enzymes exist naturally in bacteria, where they are a part of a complex defense mechanism against viral infection (see Chapter 15). Specifically, bacteria use restriction enzymes to degrade invading viral DNA. Restriction enzymes are of great use to genetic engineers because, although their function in nature is to cut viral DNA, the enzymes will also cleave those same sequences of nucleotides along any strand of DNA, from any source. The specific nucleotide sequence recognized by each restriction enzyme is called the enzyme's **restriction site.**

Each species of bacterium produces at least one restriction enzyme. By now researchers, by working with many bacterial species, have isolated literally hundreds of different restriction enzymes that recognize and cut hundreds of different restriction sites. Many restriction enzymes (as well as other enzymes used in cloning and molecular biology) are commercially available from molecular biology supply companies (see Figure 16.1). As an example of how one restriction enzyme works, let's take a look at *Eco* RI. (Most restriction enzymes are given names that are derived from the species from which they were isolated. *Eco* RI, for example, was the first, thus I restriction enzyme isolated from *E. coli.*)

When *Eco* RI is added to DNA, it will cut the DNA everywhere it finds its restriction site. In the case of *Eco* RI that restriction site is six bases long:

5′ . . . GAATTC . . 3′
3′ . . CTTAAG . . 5′

. . . . G   AATTC . . .
. . . . CTTAA   G . . .
"sticky ends"

The enzyme cleaves this sequence between the G and A in the staggered manner shown, so that the ends of each resulting **restriction fragment** (cut piece) of DNA contain short, single-strand tails that are called "sticky ends." The term "sticky" comes from the fact that the tails can easily realign with tails from other fragments with the same sticky end. Fragments can be held together, temporarily, by the hydrogen bonding between complementary bases of the sticky ends.

The restriction sites of some other restriction enzymes are presented in Table 16.1. You will note that some enzymes recognize a four-base sequence instead of the six-base sequence recognized by *Eco* RI. In fact, there are also restriction enzymes that recognize five- and eight-base restriction sites.

Table of Contents

FIGURE 16.1
MAIL ORDER MOLECULAR BIOLOGY
This is a reproduction of the table of contents of one of many companies in the business of selling reagents for molecular biology research. Note the large number of different restriction enzymes available. You will also find many of the other enzymes mentioned in this and previous chapters, such as DNA and RNA polymerases and DNA ligase.

Splicing and Cloning DNA Molecules

Any fragment of DNA can be "cloned" by inserting the fragment into a compatible restriction site in a vector DNA molecule. The most commonly used types of vector molecules are bacterial plasmids and viruses that have themselves been engineered to facilitate the production of recombinant DNA. In either case, cloning a fragment is fairly straightforward and is outlined in Figure 16.2.

First, the vector molecule is cleaved by the same restriction enzyme used to generate the fragment of foreign DNA (Figure 16.2a). That way, the sticky ends of the vector will be the same as those of the DNA fragment to be inserted. Then the cleaved vector molecules are mixed with the foreign DNA under conditions that allow the comple-

TABLE 16.1 DNA SEQUENCES RECOGNIZED BY SOME COMMON RESTRICTION ENDONUCLEASES

ORGANISM	ENZYME	TARGET DNA SEQUENCE WITH CLEAVAGE SITES MARKED
Bacillus amyloliquefaciens	*Bam* HI	5′ .... G|GATCC ....3′ 3′ .... CCTAG|G ....5′
Escherichia coli	*Eco* RI	5′ .... G|AATTC ....3′ 3′ .... CTTAA|G ....5′
Haemophilus parainfluenza	*Hind* III	5′ .... A|AGCTT ....3′ 3′ .... TTCGA|A ....5′
Haemophilus parainfluenza	*Hpa* II	5′ .... C|CGG ....3′ 3′ .... GGC|C ....5′
Moroxella bovis	*Mbo* I	5′ .... |GATC ....3′ 3′ .... CTAG| ....5′

mentary bases of the sticky ends to hydrogen bond, or anneal. At this point, the enzyme ligase is added (Figure 16.2b). Ligase covalently joins the sugar-phosphate backbones, making the foreign DNA an integral part of the plasmid.

The newly ligated DNA plasmid is added to bacterial cells that have been treated with calcium to make their cell walls permeable. Thus, some of the treated bacteria will absorb the DNA. (This is essentially an experimental version of the process of transformation described in Chapters 12 and 15.) The introduced DNA will then be replicated each time the bacteria divides.

Since transformation is not a very efficient process, there is a problem at this stage in distinguishing those cells that have taken up plasmid from those that haven't. Fortunately, this is a simple job nowadays. This is because most of the plasmids that are used in cloning experiments contain genes that confer antibiotic resistance to their host cells. So the mixture of cells in a transformation experiment are simply plated on media containing an antibiotic that kills those cells that have not taken up the plasmid. The antibiotic ampicillin is commonly used for this purpose. A typical plasmid is shown in Figure 16.3.

It may have occurred to you that the sticky ends of the vector could simply rejoin without incorporating the new DNA at all. And you are absolutely right; this happens. So how do we separate those bacteria harboring the reconstituted plasmid from those that have the plasmid with the newly inserted DNA?

The solution is quite simple in concept. The vector plasmids are engineered so that the restriction site into which the DNA is to be inserted is in the middle of a gene. When DNA is inserted into the middle of a gene that gene becomes nonfunctional. Therefore, plasmids that have simply rejoined without inserting the new DNA will retain the activity of that gene, while those plasmids that have the extra DNA will now have a nonfunctional gene. We only have to look for loss of that gene activity to know those bacteria that have the recombinant DNA.

In practice, one method of identifying bacteria having plasmids with inserted DNA is to use a cloning restriction site that is in a second antibiotic resistance gene, say for tetracycline. In this case we would then look for those bacteria that could grow on ampicillin plates (because the ampicillin resistance gene is intact), but not on tetracycline plates (because the tetracycline resistance gene is inactivated by the inserted DNA). An alternative procedure is to use a vector plasmid in which the cloning restriction site lies within a gene whose product is capable of turning a colorless indicator chemical dark blue (such as the lacZ gene on pUC19 of Figure 16.3). In this case, those bacterial colonies that turn blue contain only the reconstituted original plasmids. As shown in

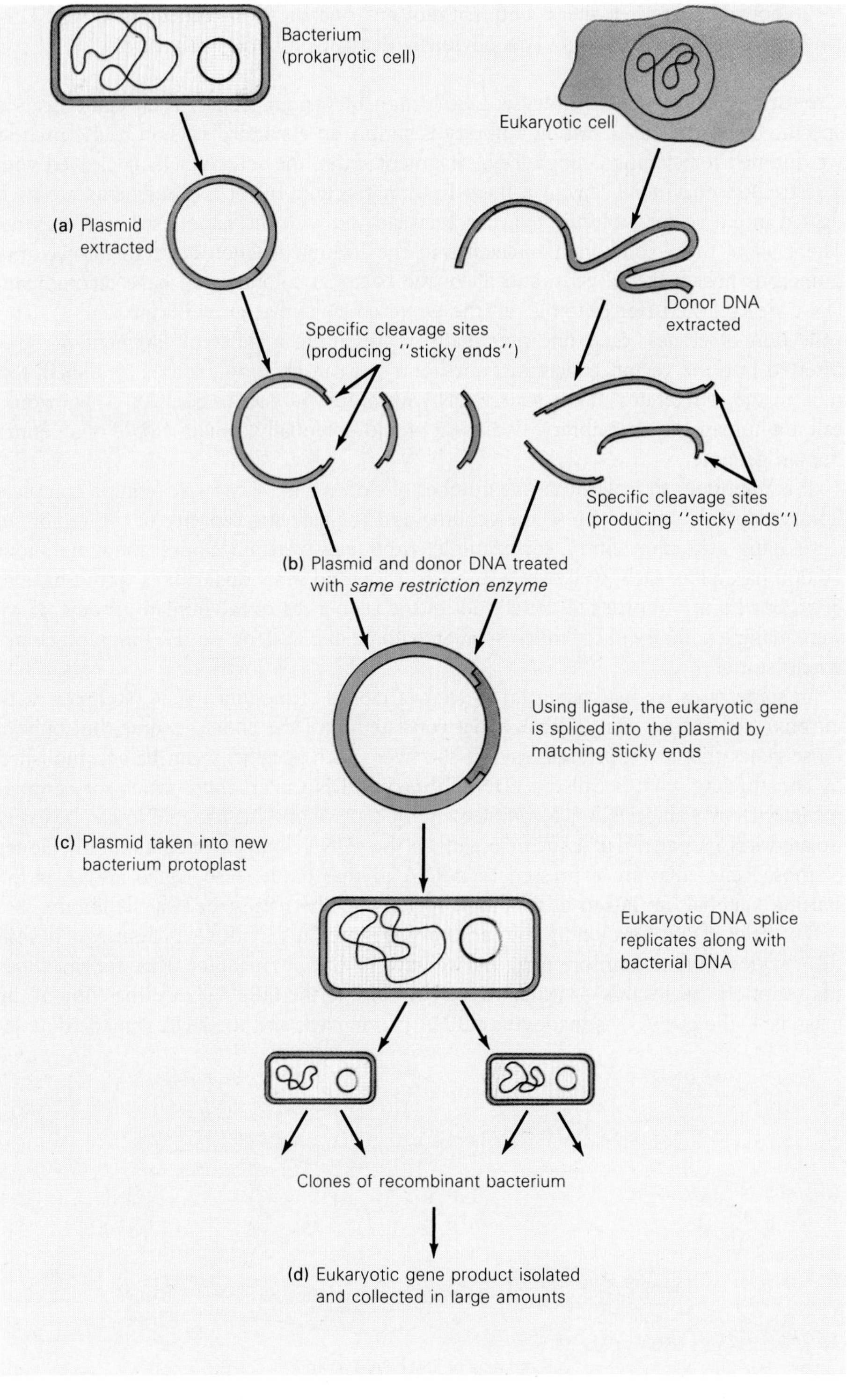

Bacterium
(prokaryotic cell)
Eukaryotic cell
(a) Plasmid extracted
Donor DNA extracted
Specific cleavage sites
(producing "sticky ends")
Specific cleavage sites
(producing "sticky ends")
(b) Plasmid and donor DNA treated with *same restriction enzyme*
Using ligase, the eukaryotic gene is spliced into the plasmid by matching sticky ends
(c) Plasmid taken into new bacterium protoplast
Eukaryotic DNA splice replicates along with bacterial DNA
Clones of recombinant bacterium
(d) Eukaryotic gene product isolated and collected in large amounts

FIGURE 16.2
RECOMBINANT DNA
Donor DNA from nearly any source can be inserted into a plasmid, where it will be replicated and transcribed as part of the bacterial genome. **(a)** Bacterial plasmids and segments of donor DNA are obtained by using the same restriction enzyme. **(b)** The plasmid and donor DNAs are mixed, and the enzyme ligase is then used to fuse the matching sticky ends of the donor DNA to the sticky ends of the plasmid. Next **(c)**, the plasmid is transformed into a bacterial host, where it will be replicated along with the bacterial DNA.

FIGURE 16.3 (BELOW)
pUC 19: A WIDELY USED PLASMID FOR RECOMBINANT DNA
(a) The ampicillin resistance gene in pUC 19 allows only those bacteria harboring this plasmid to grow on media containing the antibiotic ampicillin. The lacZ gene produces an enzyme capable of turning the compound X-GAL from colorless to deep blue. When DNA is inserted into the cloning restriction sites **(b)**, the lacZ gene becomes nonfunctional, and the bacterial colony cannot turn blue.

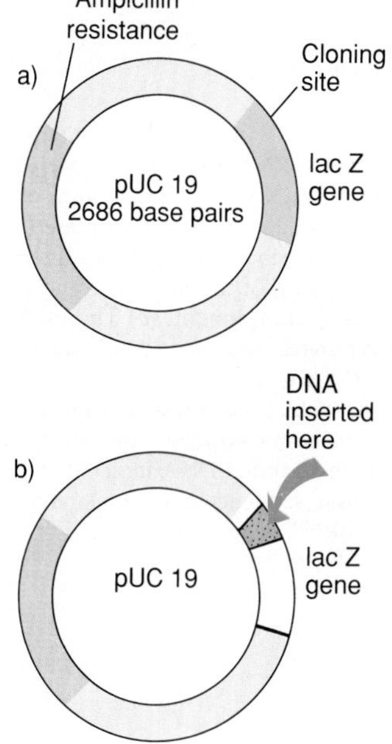

Figure 16.4, those colonies that remain white have the recombinant DNA plasmids. A white colony can be picked off the indicator plate and grown in test tubes, liter bottles or 20-gallon vats, all the while producing millions of copies of the plasmid DNA.

Libraries We can see, then, that it is actually quite simple to produce molecular clones of any piece of DNA. That is, if you already have that piece of DNA. The hard part is getting precisely the section of DNA you want.

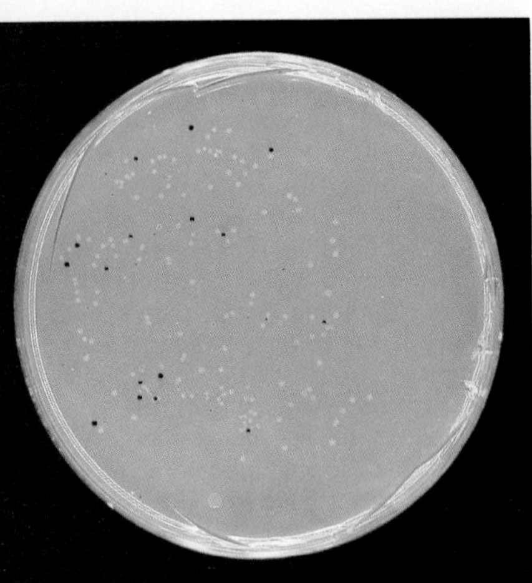

FIGURE 16.4
IDENTIFYING COLONIES WITH RECOMBINANT PLASMIDS
Because this petri plate contains ampicillin, all of the bacterial colonies contain the pUC 19 plasmid. Those with the nonrecombinant plasmid have a functional lacZ gene and are therefore blue. The white colonies contain recombinant plasmids with donor DNA inserted within the lacZ gene.

So how do genetic engineers find that molecular needle in the genetic haystack? They may check it out of a library. This obviously demands a bit of explanation.

Creating a Library A **library** is a collection of various clones. Each clone bears a specific piece of DNA. Creating a library is simply an expanded version of the process we outlined for cloning a single DNA fragment. First, the source DNA is cleaved with a restriction enzyme. Then, in a mass ligation reaction, all of the fragments are each ligated into a vector molecule that has been cleaved with the same restriction enzyme. Then, all of the recombinant molecules in the ligation reaction are transformed into numerous host bacterial cells, and all of the bacterial colonies that have recombinant DNA are collected (for example, all the white colonies that grow on ampicillin). This collection of clones, each one potentially representing a different fragment of DNA inserted into the vector, constitutes our library. It can be simply stored in a small test tube in the refrigerator. If the source DNA were, say, human nuclear DNA, we would call it a human genomic library, because it would potentially contain clones of an entire human genome.

It is important to realize that the number of clones necessary to represent a complete library depends on the size of the genome and the carrying capacity of the vector. In developing a human library, for example, producing plasmid clones averaging about 10,000 base pairs each, well over a hundred thousand individual clones would have to be in our library in order to contain all of the sequences of the human genome. If we were making a library of the much smaller genome of *E. coli,* only a few thousand clones would suffice.

In some cases we may not want to recover clones of the intact gene (complete with introns), or we may want a library that consists not of the entire genome, but only of those genes that are expressed, say, in the liver. Such specificity can be accomplished by constructing what is called a **cDNA library.** (cDNA is an abbreviation for complementary DNA.) The cDNA is a complementary copy of mRNA. The mRNA can be easily isolated from a particular tissue or organ, so the cDNA library will contain only clones of those genes that are expressed as mRNA in that tissue. Also, since mRNA is the starting material, no intron or promoter regions will be represented in the library.

To make cDNA for cloning into a library (Figure 16.5), mRNA is first treated with the enzyme reverse transcriptase. As we saw in the last chapter, this enzyme does transcription "backwards"—it makes a DNA copy of the mRNA. Once the copy of the message—the cDNA—is made, the mRNA is removed, and the DNA is made double-

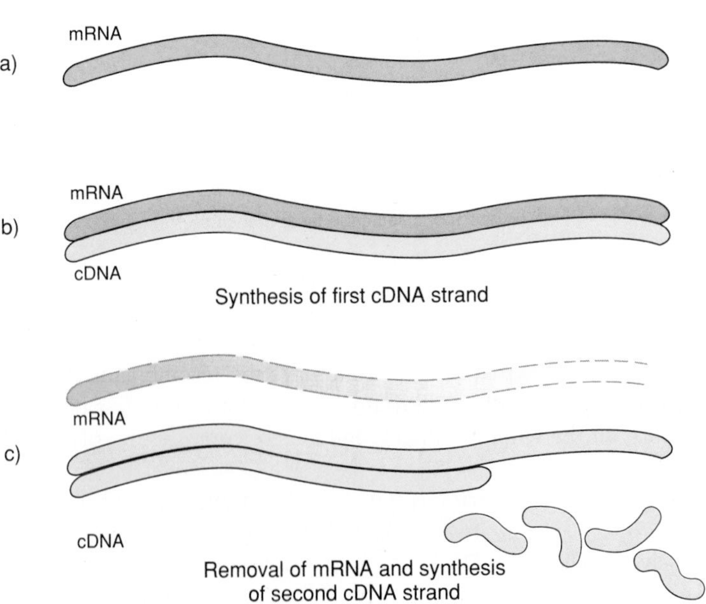

FIGURE 16.5
cDNA
To make cDNA, a complementary copy of mRNA, one needs a preparation of mRNA (**a**). The addition of the enzyme reverse transcriptase (**b**) and deoxyribonucleotides results in the copying of the mRNA strand into its DNA compliment. (**c**) The mRNA is removed, and the second strand of the DNA is made by adding the enzyme DNA polymerase and more deoxyribonucleotides. The result is a double-stranded DNA molecule with the base sequence complementary to the mRNA at the start.

stranded by incubation with DNA polymerase and nucleotides. The double-stranded cDNA is then ligated into a plasmid vector to create the cDNA library.

Remember that the term *library* implies a collection of many different things. So once we have any library, we are faced with the problem of sorting it out. We must be able to go through it to find that one clone of the DNA we are interested in from among the thousands in the library.

Screening a Library Much like the libraries that you are used to sorting through, it is possible to search a recombinant DNA library for the fragment of interest. If you were looking for a volume in a book library, you could look at each book one by one, until you came upon the one you wanted. This would be very tedious and time-consuming. A better approach is to search the whole library at once, using a computer or card catalog. In a similar fashion, it is possible to screen the whole recombinant DNA library at once.

Figure 16.6 outlines the procedure. The first step is to grow the bacteria in our library on petri dishes. Then a sterile filter is placed down on top of the colonies. A portion of each colony sticks to the filter; the rest remains on the plate. The cells on the filter are then lysed by treatment with alkaline solutions. During this process the DNA is released from the cells, is denatured (becomes single-stranded), and binds directly to the filters. The filter now has a spot of DNA at every place where there was a colony on the plate.

The actual search of the library is accomplished by placing the filter in a solution that contains a **DNA hybridization probe**—a single-stranded piece of radioactive DNA that is complementary to the DNA we are after. This solution allows the DNA probe to come into contact with the DNA from the bacterial colonies. For most of the colonies, the DNA probe will not base-pair (i.e. hybridize) with the DNA from the clone. However, in those few cases where the DNA sequences are complementary, the radioactive DNA will hybridize with the cloned DNA and thereby be attached to the filter. After hybridization, the filters are washed to remove the excess, nonhybridized radioactive probe. The filter is then put next to X-ray film, which will get exposed at the areas of radioactivity by a process called **autoradiography.**When developed, the film will have spots where the radioactive probe hybridized to the DNA on the filter, indicating the position of bacterial clones with plasmid DNA homologous to the probe. The X-ray film can be placed under the original petri dish, and the bacterial clone over the spot can be picked and grown up in quantity.

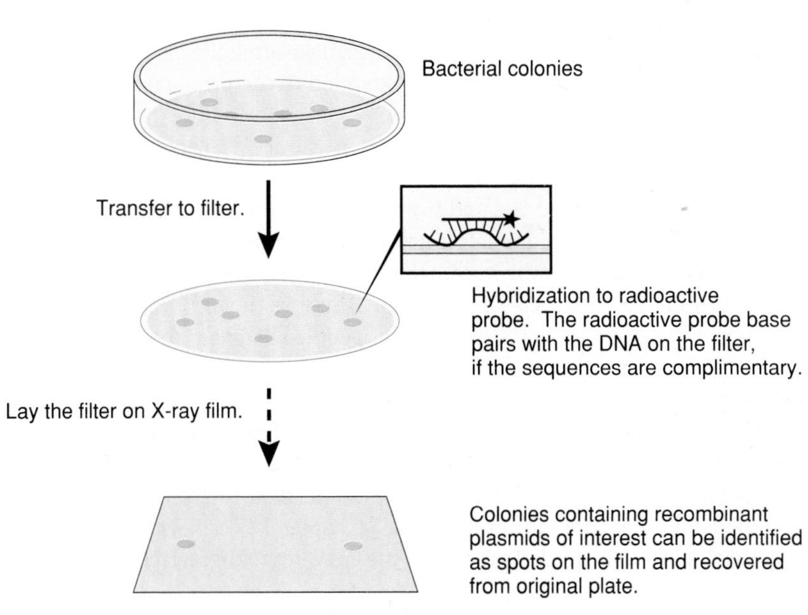

Bacterial colonies
Transfer to filter.
Hybridization to radioactive probe. The radioactive probe base pairs with the DNA on the filter, if the sequences are complimentary.
Lay the filter on X-ray film.
Colonies containing recombinant plasmids of interest can be identified as spots on the film and recovered from original plate.

FIGURE 16.6
SCREENING A RECOMBINANT DNA LIBRARY
When a piece of nitrocellulose filter is placed on the surface of the petri plate, most of the bacterial colony is transferred to the filter, but some of the colony is left behind. Treatment of the filter with alkaline detergents releases the DNA from the bacteria, denatures the DNA to single-stranded form, and bonds the DNA to the filter. A radioactive DNA probe solution is added to the filter. The radioactive probe will bind to DNA from those colonies harboring plasmids that contain sequences complementary to the probe.

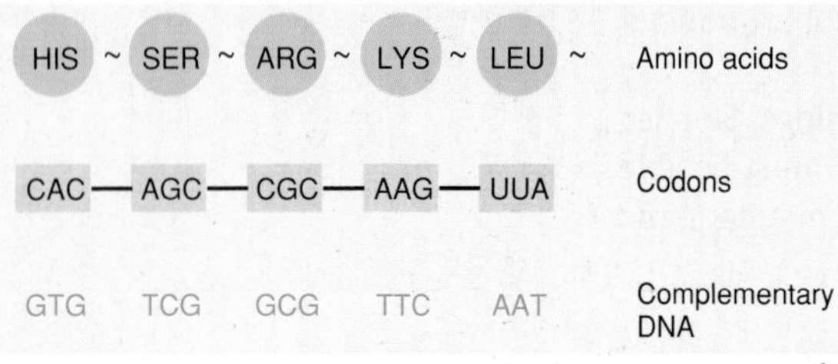

FIGURE 16.7
CONSTRUCTING A PROBE WITH A GENE MACHINE
(a) In some cases, knowledge of the amino acid sequence of a protein can be used to deduce the sequence of bases necessary to construct a hybridization probe for detecting clones of the protein. Ambiguities must sometimes be built into such probes where, for example, the amino acid serine can be coded for by more than one possible codon. **(b)** Gene machines are capable of producing short segments of DNA (up to about 100 bases long) of any base sequence. The technician has only to enter the sequence of bases into the control keyboard.

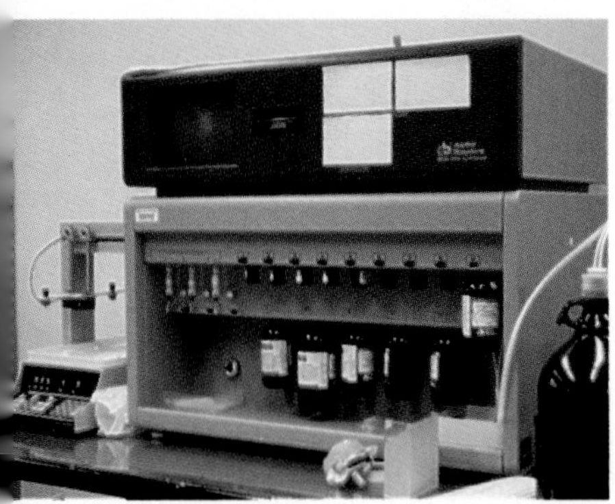

But where do you get the probe? There are several possibilities, and the choice of probes depends on the experiment and the amount of information available.

One possibility is to simply build a probe in a **gene machine** as described in Figure 16.7. A gene machine is an automated chemical synthesis apparatus that will build a sequence with A, T, G, and C in any order specified by the engineer. Therefore, if you know the amino acid sequence of the protein corresponding to the gene you want, you simply use the genetic code to translate the amino acid sequence into the nucleotide sequence, enter the sequence into the keyboard and have the machine build it. (At present the gene machines can produce only a hundred bases or so at a time. Therefore, only part of the necessary sequence for most proteins can be built during any one run of the machine.) Once the DNA is made and recovered from the machine, an enzyme named **kinase** will add a radioactive phosphate group to the 5′ end, and your radioactive probe is ready for use.

You can still recover a bacterial clone containing a particular DNA fragment even if you don't know the sequence of the amino acids. If purified mRNA for the specific gene is available, it can be labeled with kinase and used as a probe. Yet another option is to use the gene from one organism to search a library for clones of that gene in another organism. This is possible in those cases where sequence similarities have been conserved through evolution. There are other more intricate procedures for identifying clones of DNA fragments, and a major amount of time in a cloning laboratory can be spent in the process of determining new and different ways to identify clones.

So we see how DNA libraries are produced and how genes or parts of genes are recovered. We've covered a number of complex ideas to this point, so now let's see how genetic engineers put their skills and techniques to useful work.

APPLICATIONS OF GENETIC ENGINEERING

Now that we have some insights into the process, let's look at some examples of the application of genetic engineering to practical problems.

Manufacture of Biomolecules

Early on in the development of DNA technology, it was realized that the major strength of genetic engineering is the seemingly simple capability to produce large amounts of what was originally a rare piece of DNA. We have seen that any one of the clones from our human DNA library represents a very minor part of the human genome, yet as part of a plasmid we can recover large quantities of that piece of DNA. We can do this simply by growing the bacteria that harbor the plasmid, then isolating the plasmid DNA. Because DNA can code for proteins, that idea can be extended to the production of proteins that are otherwise rare or hard to obtain.

Take for example insulin. Real human insulin was once available only by harvesting human pancreatic tissue from cadavers. Since this source was insufficient to provide our medical needs, insulin was also isolated from pigs and cows and then prepared for use by diabetics. Insulin from other animals is not identical in amino acid sequence to human insulin, but it works in the same way. However, because it *is* subtly different, some people develop allergic reactions to injection of animal insulin. Since the supply of authentic human insulin was very small indeed, people allergic to the pig or cow insulin were in trouble.

Genetic engineers saw this as a good opportunity to apply their technology. A complete cDNA clone for human insulin was recovered from a library, providing all the coding information for true human insulin. In order to have that cDNA information expressed as insulin protein, the cDNA was placed in a plasmid next to a promoter sequence. In fact, the now familiar *lac* operon promoter (see Chapter 14) can serve to transcribe the mRNA for insulin. Now, instead of producing the enzymes to digest lactose, that promoter initiates the production of human insulin within the bacteria. This insulin is human in every detail, each amino acid is the right one, in the right order. Since these bacterial clones can be grown in any quantity desired (Figure 16.8), the amount of authentic human insulin is now essentially unlimited.

In fact, the process of producing insulin, or any other protein, can be streamlined even more by further genetic engineering. Our plasmid can be further modified by adding a short DNA segment between the promoter and the cDNA, so that the resulting protein has something called a signal peptide at its N-terminal end. This signal peptide turns the insulin into a secretory protein (see Chapter 13). As the insulin is synthesized, it is released to the surrounding medium. In the process the signal peptide is removed, thus leaving the medium rich in authentic human insulin. So now we don't even have to harvest the bacteria to recover the insulin. Large growth vats of the engineered bacteria can be supplied with fresh media at one end, while the medium containing the secreted insulin can be drawn off at the other end.

We have seen some of the immediate promise of genetic engineering: the production of insulin, growth hormone, and other useful substances. The immediate benefit of such technology to people is obvious. But the technology holds other promises that, in the long run, may have just as many implications for us. In particular it may improve human life through its influences on agriculture and medicine.

FIGURE 16.8
LARGE SCALE GROWTH OF BACTERIA
Industrial fermenters allow the production of essentially unlimited amounts of bacteria containing recombinant DNA.

Agriculture

For centuries, farmers have been waging war with plant pests and diseases, infertile soils, and the ongoing need to produce more to feed the world's burgeoning population. Genetic engineering offers new hope as it raises the possibility of making plants resistant to insects and bacterial and fungal disease, and able to grow in conditions previously unsuitable. Productivity may also be improved by selectively increasing the size of the edible plant parts, such as storage roots, seeds, or fruits. Furthermore, those foods can be made more nutritious through alterations in their essential amino acid content. (Most plant foods lack one or more of the "essential amino acids," those our bodies cannot manufacture.) So, we are on the threshold of a genetically-engineered agricultural revolution.

In one sense, plant genetic engineers have a real advantage: Several important species (carrots, cabbage, citrus, and potatoes) can be grown from single somatic cells. Potentially, after a plant cell has been modified by the introduction of a novel gene, the researcher can simply develop a clone of the altered cell, and then, using formulas of carefully balanced plant growth hormones and nutrients, produce the engineered plant (see Essay 27.1).

The first step in producing plants (or, as we will see in the next section, animals) with engineered genomes is to put the genes we want to transfer within a vector capable of carrying it into the plant cell. So far, the most reliable vector of plant genes is a plasmid called **Ti**, found in the bacterium *Agrobacterium tumefaciens,* the agent of a plant disease called *crown gall.* During infection of a plant by the *Agrobacterium* bacteria, the Ti plasmid is transferred to the plant's chromosomes. From then on it is a stable part of the plant's genome. In nature, the transfer of the Ti plasmid brings on the symptoms of crown gall disease. However, genetic engineers have now developed Ti plasmids that have had the disease-causing genes removed. So we now have a harmless natural gene vector.

Genes can be spliced into the Ti plasmid by the usual methods, reintroduced into the *Agrobacterium,* which then infects plant cells and transfers the DNA to the plant genome. The problem is, only certain plants (dicotyledons) can be transformed with Ti plasmids. These do not include the important grains such as corn, wheat, rye, and oats, by far the world's most important crops. However, research into extending the range of plants infected by *Agrobacterium* continues with some promise, and advances have re-

FIGURE 16.9 RESTRICTION FRAGMENT DETECTION OF SICKLE CELL ALLELES
(a) Electrophoresis of DNA fragments separates them by size, with the smaller fragments moving further through the gel. (b) A map of the β-globin gene from humans shows that the β^A globin allele region contains three sites for the restriction enzyme Dde I. In the β^S allele that is responsible for the sickle cell trait, the base sequence does not contain the central Dde I site. (c) DNA isolated from homozygous ($\beta^S\beta^S$, $\beta^A\beta^A$) and heterozygous ($\beta^S\beta^A$) for the sickle cell allele, as it would appear after digestion with Dde I (see also Essay 16.1).

(a)

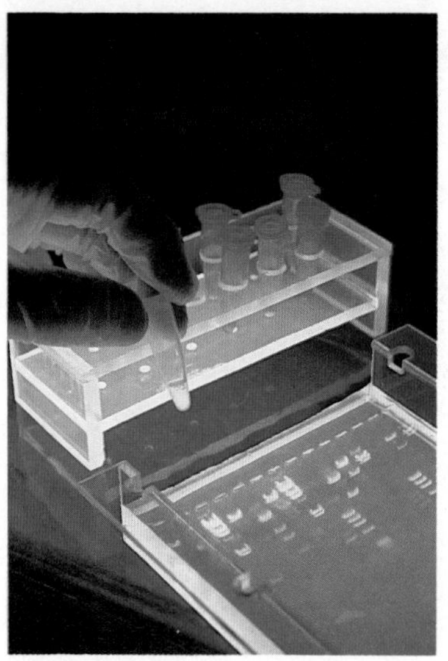

(b)

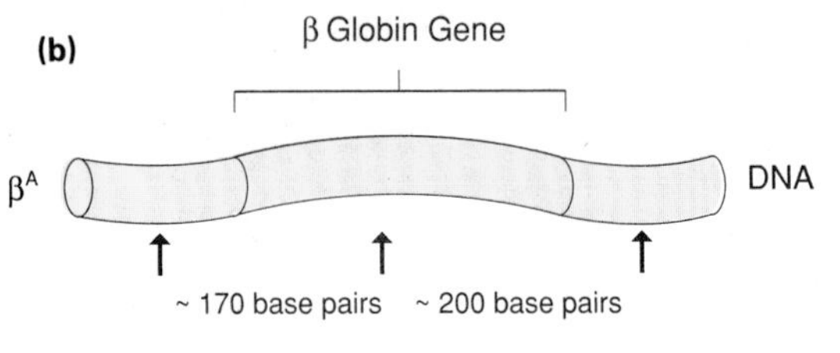

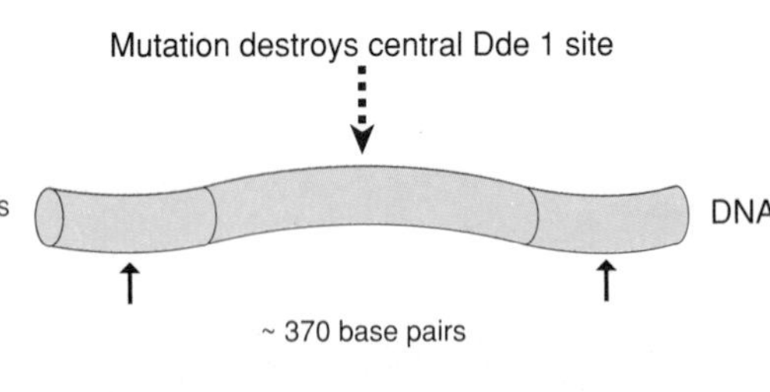

(c)

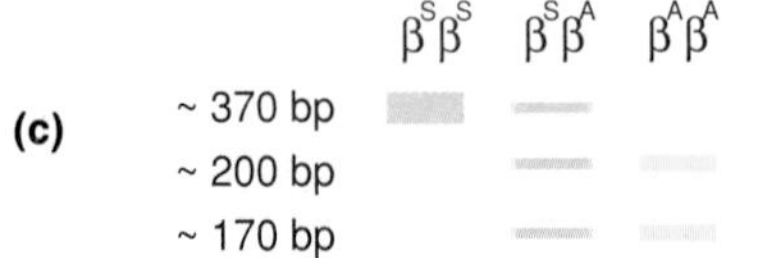

cently been made using miniature needles to microinject DNA directly into plant and animal cells, as well as microprojectile "guns" that shoot DNA-coated metal particles into cells.

Plant genetic engineering is difficult for another reason: most characteristics that need improvement, such as growth rate, relative size of edible parts, and a balance of essential amino acids, are polygenic—controlled by many genes. Furthermore, the identities of such genes are generally unknown. But even if the genes are identified, replacement of 5 or 10 or even 100 genes might be terribly difficult.

Still, there have been some impressive examples of successful genetic engineering in plants. For example, not too long ago it was discovered that plants resistant to certain herbicides possessed special enzymes that protect those plants from the killing action of the herbicide. Genetic engineers stepped in, cloned the enzyme responsible for the resistance, and are now in the process of transferring that gene into valuable crops in hopes of solving major weed control problems. The list of such examples continues to grow, and it probably won't be long until genetically engineered plants are quite common.

Medicine

Equally astounding progress has been made in the application of genetic engineering technology to medically related fields. This list of accomplishments continues to grow and can be summarized under two types of use: diagnosis or treatment.

Diagnosis: Genetic Engineering on the Trail of Killers There are well over 2000 metabolic genetic errors in the human population. Some, such as Tay-Sachs disease and sickle cell anemia, occur with high frequency in certain races or ethnic groups. Others show no racial or ethnic bias, plaguing all groups about equally. Some disorders appear in the young and mark them for life, while others, such as Huntington's disease, have late ages of onset. In the latter, the individual may have reproduced and passed on the genes for the disease long before being aware of his own fate (see Chapter 11). In the past, we could predict the probability of a disease by pedigree analysis and then, often, just wait for it to occur.

Things are changing—rapidly. For example, researchers are now able to detect sickle cell carriers (heterozygotes) by analysis with restriction enzymes. (See Essay 16.1 on Electrophoresis and Hybridization.) It is important to realize that these and other genetic diseases can be traced to changes in the DNA of the afflicted genes. When by coincidence these changes occur in a restriction site, the corresponding restriction enzyme will no longer react there. Thus, in the case of sickle cell anemia, shown in Figure 16.9, there will be one large fragment of DNA containing the β-globin gene, instead of the normal situation where the gene is carried on two smaller ones. In other cases, mutation may create a restriction site, thereby increasing the number of fragments. All of this analysis can be done with DNA collected from blood samples, using electrophoresis and specific DNA probes recovered from gene libraries.

This type of analysis is a powerful diagnostic tool. People who are heterozygous for sickle cell disease or who have Huntington's disease can be identified and counseled, perhaps in time to prevent the tragedy of afflicted children. All this because of the advent of restriction enzymes and the ability to clone DNA.

This ability to determine the molecular genotype of an individual has two more applications that extend the utilization of DNA technology. First, these DNA analyses can be performed on cells recovered during amniocentesis. Thus, fetuses at risk of having an inborn error can be unambiguously identified at a very early stage of development. Second, forensic scientists have applied the DNA technologies to identify criminals—rapists are the prime example. Using DNA that can be isolated from minute quantities of blood or semen, molecular biologists can use restriction enzymes and a variety of DNA probes to create a molecular "fingerprint" of the criminal. These DNA

ESSAY 16.1

ELECTROPHORESIS AND HYBRIDIZATION: POWERFUL TOOLS

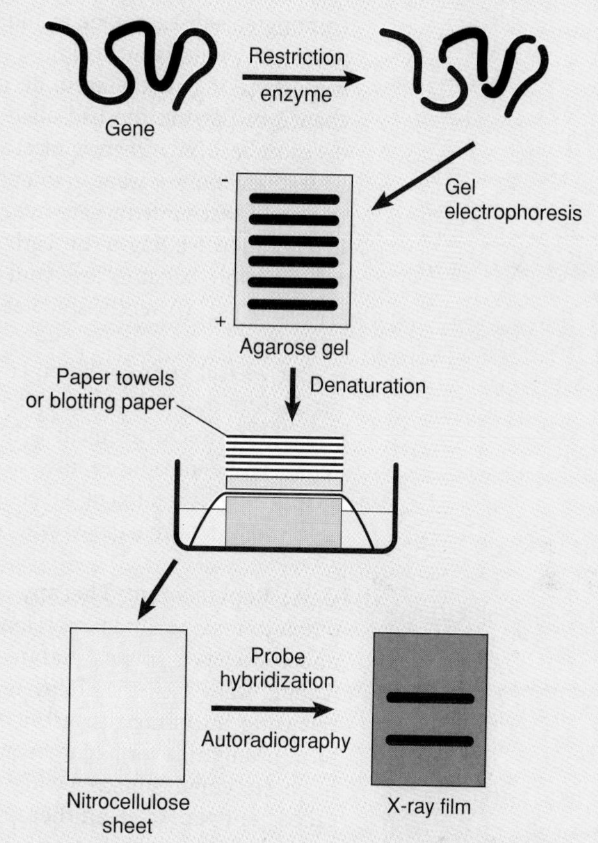

In many situations it may be impossible, or simply not necessary, to clone the gene that we would like to examine. For example, we may wish to characterize a gene from many individuals, without cloning the gene from each person (see, for example, the analysis of globin genes in Figure 16.9). Much information can be derived from some simple restriction digests and an application of the hybridization technology that we described for screening libraries. In this application we use radioactive probes to tell us the size of the restriction fragment(s) carrying the gene.

After digesting the DNA sample with the restriction enzyme, the myriad of fragments produced can be easily sorted by size, using the process called *electrophoresis*. Electrophoresis, as its name implies, uses electrical current to move DNA molecules. (The DNA, because of its backbone containing negatively charged phosphate groups, tends to migrate to the positive electrode.) The electrical field provides the motive force, but a gel matrix is required to achieve separation of DNA fragments. The gel acts like a three-dimensional sieve, and in order to move through the gel the DNA fragments must work their way through, over, and around the gel matrix material. Naturally, the smaller a DNA fragment is, the less its progress is impeded by the gel. The larger a fragment is, the harder it is for it to make it through the matrix. The end result is that the DNA fragments are nicely separated in the body of the gel according to their length.

When genomic DNA is subjected to this analysis, the resulting distribution of fragments is incredibly complex. There are so many fragments produced that it is impossible to distinguish individual bands on the gels. So to detect the size of the DNA fragment bearing the gene of interest, we must use a hybridization probe. The hybridization probe can identify the band on the gel corresponding to the gene (or part of the gene) in the same way that it can point to a specific clone in a library.

After the DNA is separated on the gel, the gel is treated with alkaline solutions to denature the DNA, separating it into single strands. Then the DNA is transferred to a filter membrane by blotting. The filter membrane then contains a molecular image of the separation achieved by the gel, and the image consists of the denatured, single-stranded DNA fragments. Single-stranded, radioactively labeled probe DNA is then added to the filter. After a few hours, the probe DNA will have hybridized to the DNA fragments on the filter that have the complementary base sequence. After washing, the filter can be placed next to X-ray film to detect the positions of DNA fragments that have hybridized. The result is called a *Southern blot,* named for the person who first described the technique.

Thus it is possible to detect and characterize the pattern of restriction sites in or near a gene from any source without going through the hassle of cloning the gene each time.

fingerprints are extraordinarily accurate—it is essentially impossible for two people on the earth (except identical twins) to have the identical DNA fingerprint. DNA fingerprints have already helped put a Florida rapist in jail. It is equally important that DNA fingerprint analysis also has already resulted in the acquittal of a man mistakenly imprisoned for rape.

We can see then that DNA analysis can be a powerful diagnostic implement, potentially able to detect many genetic disorders such as phenylketonuria, sickle cell anemia, and Huntington's disease. One would think that people with a family history of one of these disorders would want to know whether they have the allele. This information would help in making an intelligent decision about having children. (Suppose you had an even chance of passing the bad allele to any children you might have. What would your decision be?) But, whereas most people would want to protect any children they might have, many do not want to know about their own fate. Think about it; would you want to be told that sometime after your 35th birthday you will begin a period of rapid neural deterioration leading to an early death (Huntington's disease), or that by age 45 you will probably begin to lose your memory and soon after most of your mental abilities (Alzheimer's)? (A recent study at Johns Hopkins revealed that 60% of us don't want to know.)

But what if your condition could be treated by providing you with the normal gene product? (Now all this gene-splicing business becomes very relevant!) Would you change your mind? Such treatment is becoming increasingly likely since, as we've seen, some genes can be introduced into bacteria where they can produce their products in vatload quantities. But treatment is becoming much more direct. We are on the threshold of a time when defective genes will be replaced with normally functioning genes.

Gene Replacement Therapy Hypogonadism is a relatively common recessive condition in mice and humans. Homozygotes for the allele responsible, the *hpg* allele, have underdeveloped gonads, and (in mice anyway) seem to have no idea of how to go about mating. They lack the ability to produce a chemical messenger called **gonadotropin-releasing hormone.** Its task is to stimulate the release of pituitary hormones that, in turn, prompt the formation of gametes and sex hormones by the gonads. This condition has been "cured" in mice by injecting eggs with copies of the normal gene—a procedure called **gene replacement therapy.** Because eukaryotic cells often fail to correctly process and translate messages that lack promoters and introns, the starting point for gene replacement therapy is the genomic library, where the gene is still complete. When eggs were taken from females and microinjected with copies of the normal allele (Figure 16.10), about 20% of them incorporated at least one copy of the gene in one of their chromosomes. After fertilization and reimplantation in a surrogate mother, the altered embryos grew into normal fertile mice.

FIGURE 16.10
MICROINJECTION OF DNA INTO A FERTILIZED MOUSE EGG
The injecting needle is at the bottom. The egg itself is held in place by a suction pipette from the top.

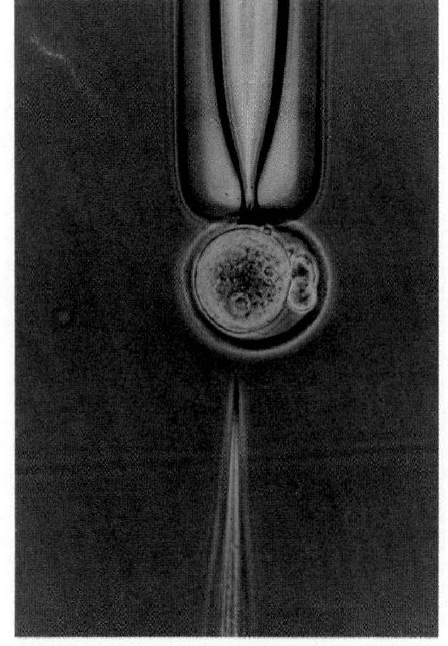

Several other gene replacements have been tried with less success. A few years ago there were glowing reports of genetic engineers curing genetically dwarf mice through gene therapy. Such mice have subnormal levels of growth hormone, so the cure was to microinject eggs with a complete human growth hormone gene (promoter and all). Some grew into healthy mice, and their photos were published around the world. But as it turned out in this case, transcription of the gene accelerated out of control, and the mice grew into sterile giants (meaning simply that we need a more thorough understanding of gene regulation to be better engineers).

Gene replacement therapy is the newest application of genetic engineering, so there are many unknowns. While no one is seriously considering using human eggs (think of the possibilities), replacing defective genes in somatic cells is a much more acceptable idea. One instance is the treatment of immune deficiencies. In such cases the cells will be obtained from bone marrow and cultured in the laboratory. (Bone marrow houses cells that give rise to cells of the immune system; see Chapter 41.) A viral vector will then be used to introduce the normal gene into such cells, and once this succeeds, the cells will be reintroduced into the patient's bone marrow. The results of preliminary experiments with mice indicate that the gene should function properly, correcting the deficiency. Any cells descended from that line will have the corrected genome, so the cure should be permanent for that individual. In this case, since his (or her) reproductive cells weren't treated, the defective gene could still be passed on to children.

GENETIC ENGINEERING: A SOCIAL ISSUE

When the use of plasmids for genetic engineering was introduced by Paul Berg and others in the late 1970s, the possibilities were so dramatic and so bizarre that the new technology immediately spawned a raging controversy. Trouble started when the very people who invented the techniques called for a research moratorium so that possible dangers could be evaluated.

Supporters of the new gene-splicing technology claimed that it held great promise for humanity, and that some of our more pressing problems soon might be solved. They spoke glowingly of cancer cures and of a possible end to all genetic disease. Some even visualized a possible end to hunger with the synthesis of new plants that combine, say, the productivity and photosynthetic efficiency of corn with the nitrogen-fixing ability and protein production of peanuts.

In the minds of genetic engineering proponents, the principal benefit is that scientists like themselves will be able to do more experiments and to learn things faster—and that's nothing to scoff at. There are also more practical benefits, some of which have been realized and some of which are on the horizon. For example, there are thousands of growth-hormone-deficient people who, with the help of human growth hormone, can attain normal height. In the past, growth hormone had to be extracted from the pituitary glands of human cadavers and was extremely expensive. Even though the glands of 50 cadavers were needed for only one dose, a few thousand seriously undersized adolescents were treated with some beneficial results, as well as one man who grew to normal height after having been only four feet tall until the age of 35. Perhaps, with such hormones now easily available, no one will have to be any shorter than he or she wants to be. Does that seem to you like a blessing, or like cavalier tampering with nature?

Alarmed critics of the new techniques saw nothing but disaster ahead. Their fears ranged from the possible release of newly created disease organisms—genetic monsters capable of creating uncontrollable plagues—to new kinds of cancer. Nonsensical scare fiction began to appear. Some critics accused the new biologists of "playing God," of messing around with primal forces, or of seeking a demonic new power. Politicians tried to capitalize on the scare; the city officials of Cambridge, Massachusetts passed an ordinance prohibiting recombinant DNA research within the city limits (which include Harvard University and the Massachusetts Institute of Technology). The National Institutes of Health (NIH) set up rigid guidelines as to what kind of research was to be allowed and what precautions were to be taken.

Can you now see why some people were alarmed? It's possible to put the gene for botulism toxin in *E. coli,* a bacterium that is already adapted for thriving in your gut. It's also possible to clone the entire genome of a cancer-causing virus, or the specific cancer-causing genes. Unlikely, you think? This, too, has been done. That was one of the scare items, but now that it has actually happened, the isolation of cancer-causing genes (oncogenes) has been hailed as the research breakthrough that may lead science to a final understanding of the cancer process and possibly to a dependable cure. Not that anyone can afford to be careless: We mustn't forget that the last two minor epidemics of smallpox in Europe—including the very last death ever caused by this virus—were caused by escaped laboratory strains. The deadly smallpox virus, which only two decades ago infected millions of people, is now extinct except for laboratory cultures.

After a time, the genetic engineering controversy died down somewhat. Interestingly, this happened at about the time gene splicing became big business, with Wall Street taking a lively interest. The city of Cambridge relented, and the NIH guidelines were relaxed. The controversy has not completely died out, however, as scientists continue vigorously to monitor themselves and their colleagues, "aided" by unsolicited assistance from alarmed and vocal amateurs.

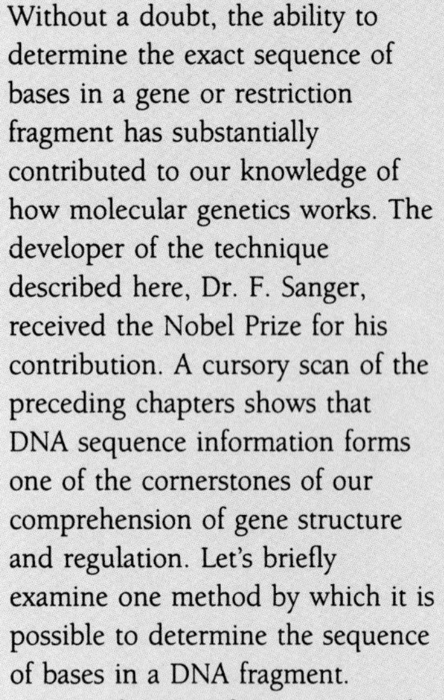

ESSAY 16.2

DNA SEQUENCING

Without a doubt, the ability to determine the exact sequence of bases in a gene or restriction fragment has substantially contributed to our knowledge of how molecular genetics works. The developer of the technique described here, Dr. F. Sanger, received the Nobel Prize for his contribution. A cursory scan of the preceding chapters shows that DNA sequence information forms one of the cornerstones of our comprehension of gene structure and regulation. Let's briefly examine one method by which it is possible to determine the sequence of bases in a DNA fragment.

First, the DNA fragment must be cloned, and for this process a convenient vector is the bacteriophage M13. During infection with M13, the *E. coli* host extrudes thousands of copies of the M13 virus (plus any inserted DNA) into the growth medium, where it can easily be purified. Interestingly, the DNA within the M13 viruses is single stranded. This single-stranded DNA is an excellent material to begin the DNA sequencing protocol.

Sequencing begins by dividing the DNA into four test tubes, one for each of the reactions that will identify the positions of the A, T, G, and C residues in the fragment. In each tube a short piece of DNA, a primer, is allowed to hybridize to the M13 genome right next to the place where the DNA fragment was inserted. Each tube then receives a mixture that contains the following components: trace amounts of radioactive nucleotides

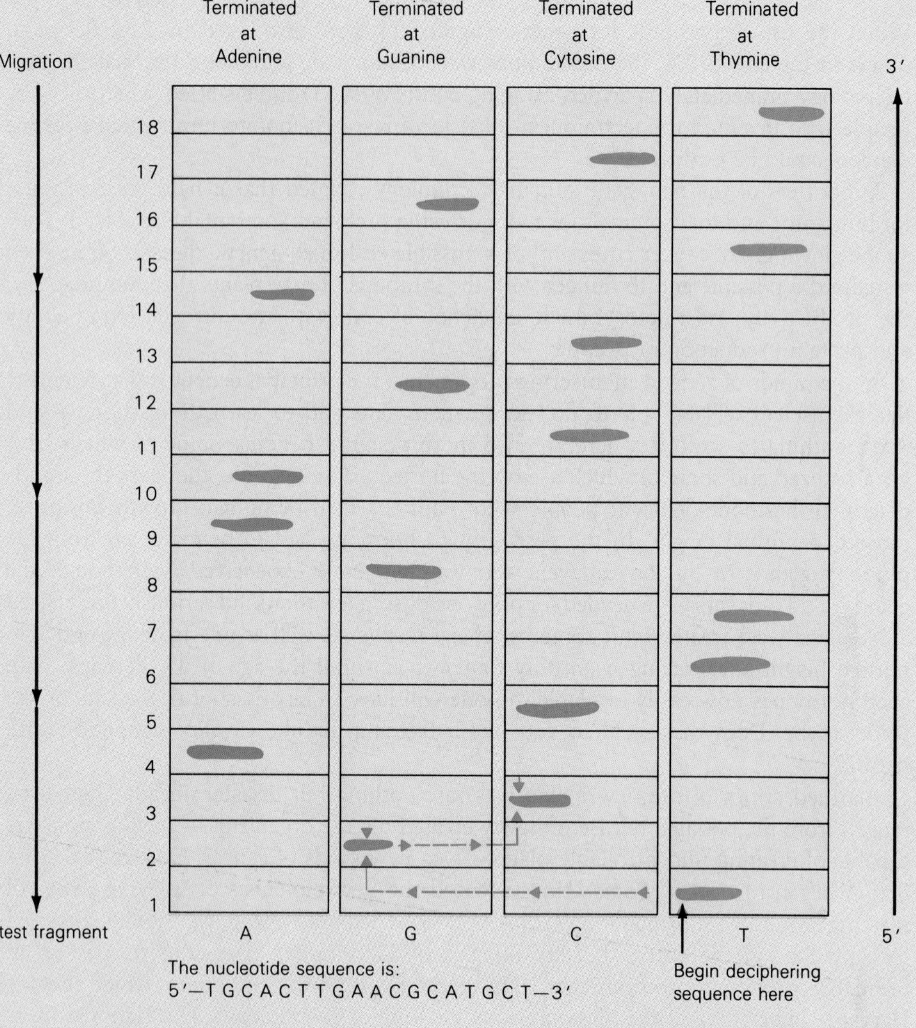

In 1987, there was in fact a problem with a quantity of genetically engineered human growth hormone. Apparently, a considerable amount was stolen, and authorities feared it would appear on the black market. Who would be willing to pay exorbitant prices for growth hormones? Obviously candidates might be the people who inject themselves with steroids, ignoring the apparent risks. In modern society, tallness has many rewards other than athletic prowess. Studies clearly reveal that height is a positive factor in social and business success. How far will people go to succeed in our keenly competitive world? You already know the answer. It is abundantly clear that social issues arising from genetic engineering are just beginning to appear.

(so that we can see the DNA by autoradiography), nonradioactive nucleotides, and the enzyme DNA polymerase. Just as in replication, DNA polymerase will add nucleotides to the primer, forming a complementary strand. The polymerase would completely copy the template, except for the final component in each of the reaction tubes. That final component is a nucleotide analog called a *di*deoxy nucleotide triphosphate. The A tube gets some dideoxy A, the T tube some dideoxy T, and so on. The dideoxy is a very interesting nucleotide mimic. It is capable of being joined to the growing chain, but stops any further progress by the polymerase because the dideoxy triphosphate has no 3′OH to allow the next base to be joined. The growing strand in the A tube then terminates wherever there is a dideoxy A incorporated. Since both the dideoxy nucleotide and the normal nucleotide are present in the same reaction, it is a chance event that a dideoxy will be incorporated and the chain will terminate. While any one chain can terminate in just one place, in the population of molecules being built in the reaction tube all possibilities will appear. Therefore the tube for A will have some molecules that terminate at each of the different As within the sequence.

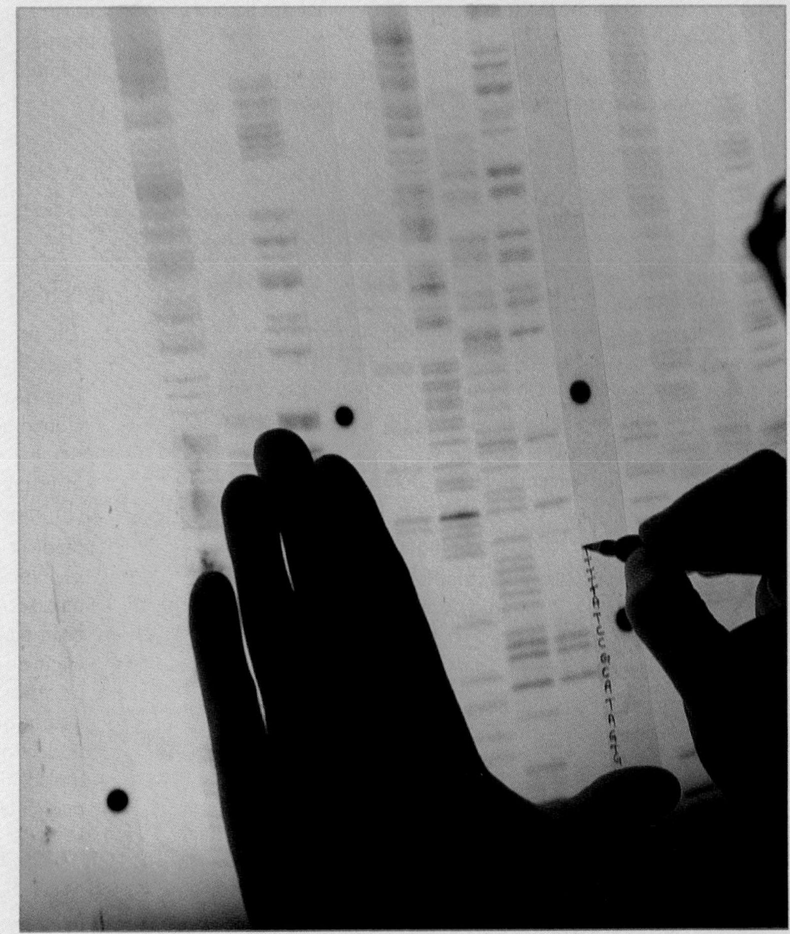
Dan McCoy/Rainbow

To read the sequence, the radioactive DNA from each reaction is loaded onto a gel for electrophoresis. Because these fragments differ in size by only a single base, a very dense matrix of polyacrylamide is used to separate the fragments, and the gels themselves can be up to a yard long. After electrophoresis, X-ray film is placed on the gel, and the individual bands are detected by autoradiography. It is a simple matter to read the sequence by looking for which lane—the A, T, G, or C lane—is the next one to have a band.

We see then, that the potential of genetic engineering is immense. There probably isn't a genetic defect that can't be reversed in some way, a protein that cannot be made in bacterial cultures, or a gene that cannot be replaced. Of course, most genes will not be altered, but it is important to know that we have the technology to do it. The only question is whether technology will outpace our abilities to use such powerful tools wisely and well.

APPLICATION OF IDEAS

1. Consider the abilities of genetic engineers to produce proteins in vast quantities, and also consider the fact that proteins can be altered simply by introducing changes in the DNA sequence. Now, what applications can you see for such products?
2. As science develops DNA probes that are useful for detecting changes in gene sequences, it is becoming possible to define a person's genetic predisposition to certain diseases and conditions. What if a person was denied a job in (for example) the chemical solvent industry because it was determined that he or she was potentially at medical risk by working there? Just how far should genetic screening go?

KEY IDEAS

DNA TECHNOLOGY

1. Gene splicing involves producing chosen DNA sequences (genes), inserting them into plasmids, reintroducing the plasmids to bacterial cells, cloning the cells, and harvesting large amounts of the gene products (such as insulin).
2. **Restriction enzymes** cleave DNA at specific DNA sequences known as **restriction sites.**
3. A major gene splicing technique involves
   a. obtaining plasmids from bacteria;
   b. using restriction enzymes to cleave plasmid DNA, introducing foreign DNA strands to be cloned, and using the enzyme *ligase* to reattach the sticky ends;
   c. preparing recipient cells and permitting the new plasmids to be taken in; and
   d. producing large clones and harvesting the gene product.
4. Gene libraries consist of collections of many individual clones, each one representing a different DNA fragment.
5. A specific gene can be recovered from a library through the use of a **hybridization probe,** a radioactive piece of DNA complementary to the desired gene.
6. The technology is available for determining the amino acid sequence of any protein, then synthetically producing a DNA strand that will code for that sequence. The DNA can then be spliced into a plasmid and the protein produced in quantity.

APPLICATIONS OF GENETIC ENGINEERING

1. The production of protein products from genetically engineered microorganisms is already a real industry.
2. It is possible to genetically engineer plants using cloned genes and the natural plant vector *Agrobacterium tumefaciens. Agrobacterium* is capable of inserting part of its plasmid genome directly into the chromosomes of dicotyledonous plants.
3. Medical science now employs cloned genes and gene segments to diagnose carriers of certain genetic diseases. Gene replacement therapy in mammals has been partially successful, in that it is possible to engineer genes into fertilized eggs and have them be incorporated into the genome. However, all of the ramifications of proper gene regulation are not well understood, and it will be some time before gene replacement therapy will be a viable part of human medicine.

GENETIC ENGINEERING: A SOCIAL ISSUE

1. Genetic engineering remains controversial, although industry now employs many of the technical procedures. There is an element of medical and genetic risk, but much of the controversy centers around the moral and philosophical issues.
2. Even after the self-imposed moratorium on genetic engineering research, scientists continue to monitor and regulate gene research. However, as research proceeds, most regulatory guidelines have continued to ease.

REVIEW QUESTIONS

1. Outline the general steps involved in cloning any particular piece of DNA. (341-342)
2. Define the enzymatic activity and application in genetic engineering of restriction enzymes and ligase. Be specific. (340-342)
3. What is a gene library? How is a gene library screened? (342-345)
4. What is a gene machine? What are its uses in genetic engineering? (346)
5. Describe the process of cloning the gene of a particular protein, and then the engineering of that gene for industrial-level expression of the protein. (346-347)
6. Describe the way in which *Agrobacterium tumefaciens* acts as a natural genetic engineer. (347)
7. What is the basis for the use of gene probes as diagnostic tools in medicine? (348)
8. What is gene replacement therapy, and how might it be accomplished? (350)

17 WHEN DNA CHANGES

IT IS HARD TO IMAGINE THE GENETIC DIVERSITY OF LIFE ON THIS PLANET. Not only are all the species different, but within the species—even within populations and families—each individual varies genetically from the next. It is just this variation that is so important to the process of natural selection, as some genetic combinations prove reproductively more fortuitous than others, and so come to predominate in future generations. Where does all this genetic variation come from? What causes such diversity? Fundamentally it begins with genetic changes called mutations.

We should first note that mutations can have quite different effects, depending on where and when they appear. Mutations in the cell that develops into sperm or eggs may result in seriously ill, malformed (or dead) offspring. If the mutations do not prevent reproduction, the changes will be passed on to the next generation. Mutations in other cells of the body, say those of the skin or liver, are called **somatic** mutations (Figure 17.1) and will not be passed on to the next generation, but can cause cell death or even cancer. Fortunately, many mutations are benign, that is, they neither help nor harm the individual.

So, in general, some mutations must occur in order to maintain variability. But this must be balanced against the deleterious effects of drastic or numerous mutations in an individual.

FIGURE 17.1
SOMATIC MUTATIONS
Mutations that occur in nongerminal cell lineages are called somatic mutations and will not be passed on to the next generation. Here we see an example of a somatic mutation in a human being.

THE STABILITY OF DNA

The stability of DNA is largely due to the way its nucleotides are arranged. On the outside lies only the sugar-phosphate backbone, with all the potentially reactive side groups of the sugar already covalently bonded. The nucleotide bases lie protected inside this backbone, their reactive side groups usually immobilized by hydrogen bonds. The stability of the structure is further enhanced by the geometric tightness of the molecule.

DNA in higher organisms has an additional means of protecting itself from chemical damage. It is tightly wound in nucleosomes (see Chapter 9). The bound histones and the specific configuration of the higher-order chromosome package offers few opportunities for direct access to the DNA.

DNA is, in all organisms, also protected from the permanent effects of damage by another of its inherent characteristics, complementary base pairing (see Chapter 12). Any changes in the bases or the base sequence in one strand has the potential for being corrected as long as the other strand is unaltered. It's like having photographic prints and negatives. If either is damaged, the other can be used to replace it.

In spite of these seemingly formidable protective barriers, changes in DNA occur with an ongoing regularity. Here we will look into several categories of changes, the agents responsible, and the specific way in which they alter DNA. Then we'll see that much of the initial damage is quickly and efficiently reversed. Keep in mind that, although chromosomes have ways of protecting themselves against mutations, mutations provide the raw material for the evolution that has given rise to life itself.

Basically there are three levels of mutational change: (1) **point mutation**, which involves changes in one or a few nucleotides in DNA, and (2) **chromosomal mutation,** which includes rearrangement, loss, or duplication of whole sections of chromosomes. These point mutations and chromosomal mutations are the result of unrepaired or improperly repaired damage to DNA. Finally, there are (3) **transpositions,** or insertions of copies of DNA into new positions in the genome. Before we consider each of these three classes, we'll look into some causes of DNA damage and their molecular effects.

THE NATURE OF DNA DAMAGE AT THE NUCLEOTIDE LEVEL

Any agent that causes a mutation is called a **mutagen.** The chief mutagens are radiation and chemicals. Mutagenic radiation may include electromagnetic waves or subatomic particles. High energy electromagnetic radiations such as gamma and X-rays are extremely powerful mutagens, as are alpha and beta particles. Ultraviolet radiation, such as that emanating from the sun, although far less "powerful" than X- or gamma rays, is a highly significant source of mutation.

The thought of chemical mutagens may conjure up visions of murky, foul-smelling puddles in chemical dumps, one of our latest national disgraces. But the problem is not isolated to such places. To illustrate, a number of common food additives are mutagenic under certain circumstances. Sodium nitrite, a common additive that prevents bacterial spoilage of bacon, hot dogs, and lunch meats, has the potential for conversion into a chemical mutagen.

Whatever the mutagen, its effect is to generate random chemical changes in DNA. Such changes are known as **primary lesions.** Fortunately, for reasons we will get into shortly, most primary lesions are temporary, quickly eliminated from the gene by highly efficient DNA repair processes. Those that are not repaired, or improperly repaired, become mutations.

One of the more common forms of primary lesions is a break in a single strand of the double helix which, like a break in a zipper, interferes physically with its functions (Figure 17.2a). Single-strand breaks are particularly prevalent in DNA that has been exposed to X-rays and other high energy radiations. They interfere with the polymerase enzymes, compromising transcription and replication. Such high energy radiations may also cleave both strands of the DNA molecule, leaving the way open to chromosome-level mutations of the sort we will discuss shortly.

Those of us who expose our skin to ultraviolet radiation from sunlight or tanning sunlamps can expect our DNA to accumulate primary lesions called **pyrimidine dimers.** When DNA is exposed to UV, thymines (and sometimes cytosines) that are adjacent on the same strand can become linked together, side-by-side (Figure 17.2b). Pyrimidine dimers interfere with replication and transcription, in this case in much the same way that sewing together two teeth in a zipper interferes with its operation.

Chemical mutagens may also cause random single- and double-strand breaks in the DNA, but most have more specific biochemical effects. We can distinguish three basic

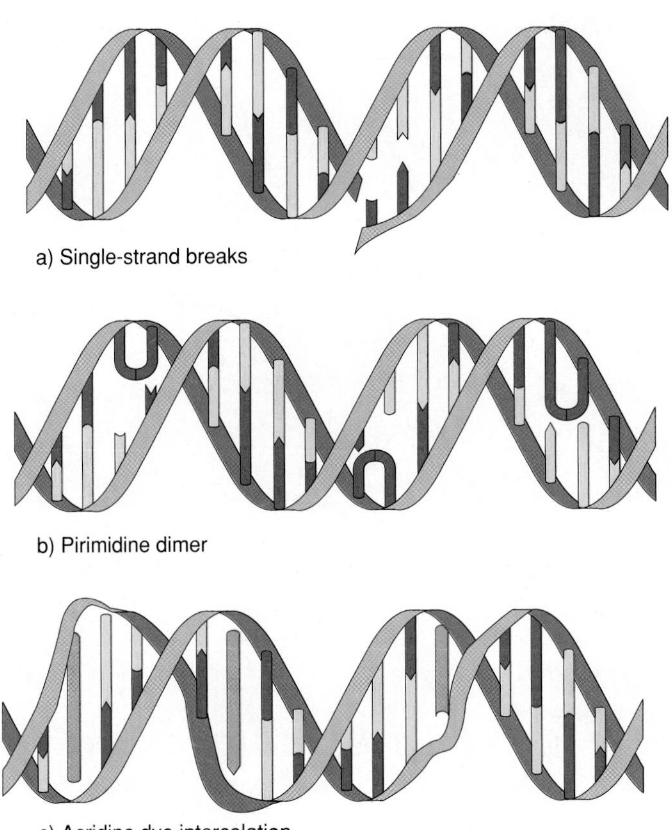

FIGURE 17.2
PRIMARY LESIONS IN DNA
These modifications of DNA, if left unattended, will result in mutations. The major structural modifications are **(a)** breaks in the sugar-phosphate backbone, **(b)** pyrimidine dimers, and **(c)** intercalated molecules.

types of chemical mutagens: **intercalating agents**, **base analogs**, and **DNA modifying agents.** Intercalating agents are usually flat molecules that can squeeze between (intercalate) adjacent base pairs of the DNA helix. For instance, molecules of the dye acridine orange lodge between bases along the length of the double helix (Figure 17.2c). This distorts the molecule, potentially interfering with replication and transcription. During replication, the distortion can result in addition or deletion of single bases in the daughter strands. Base analogs are compounds that are quite similar to the normal nucleotides and are therefore readily incorporated into growing DNA chains. However, many base analogs can exist in two forms, forms that differ in their hydrogen-bonding character and thus cause the wrong bases to be inserted in the next round of replication. DNA modifying agents chemically react with the DNA, adding chemical groups to the nitrogen bases. (An example of such a DNA modifying agent is the deadly, lung-searing mustard gas of World War I fame.) Some DNA modifying agents change the hydrogen-bonding character of bases and cause polymerases to misread bases, while others interfere with transcription and replication in other ways.

We have seen that gamma rays and X-rays can directly affect DNA, but they can also cause damage indirectly. Such radiation is deeply penetrating and energetic enough to dislodge electrons from many of the cell's molecules. This ionizing effect has led to their being labeled **ionizing radiation.** Ionizing radiation creates a variety of highly reactive, electrically-charged molecular groups called **free radicals** (Chapter 20). The free radicals, in turn, create genetic damage by reacting with and altering the nitrogen bases of DNA.

DNA REPAIR: THE CLEANUP CREW

Primary lesions in DNA occur at a fairly regular pace and are not at all unusual. In fact, the sources of primary lesions are a normal part of our daily existence. After all, we daily encounter various levels of mutagenic chemicals and radiation. For that matter,

FIGURE 17.3
EXCISION REPAIR OF A PYRIMIDINE DIMER
(a) The repair enzyme complex detects a pyrimidine dimer. **(b)** The damaged section of DNA, plus about 10 bases on either side of the dimer (about 20 bases in all) is removed. **(c)** A repair polymerase resynthesizes the excised region, using the bases present on the remaining strand as a template. The newly synthesized segment is joined to the original strand by the enzyme ligase.

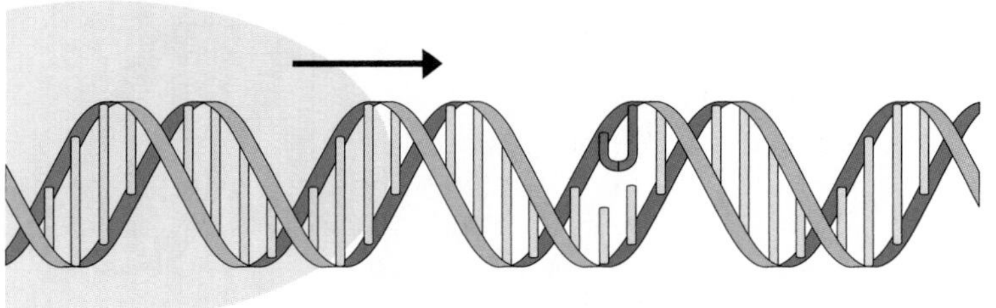

a) Excision repair complex

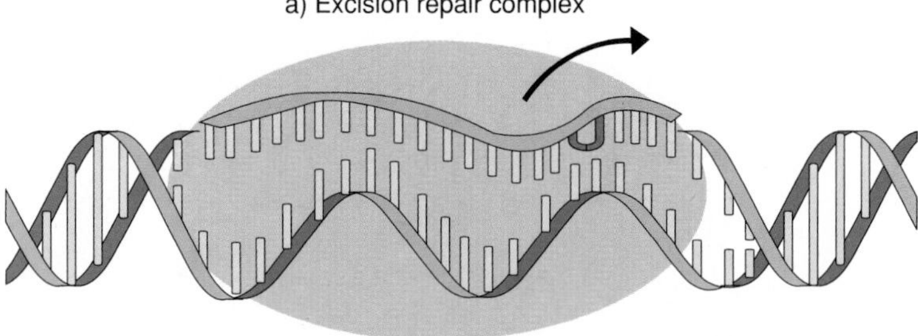

b) Pyrimidine dimer (plus 20 nucleotides) removed

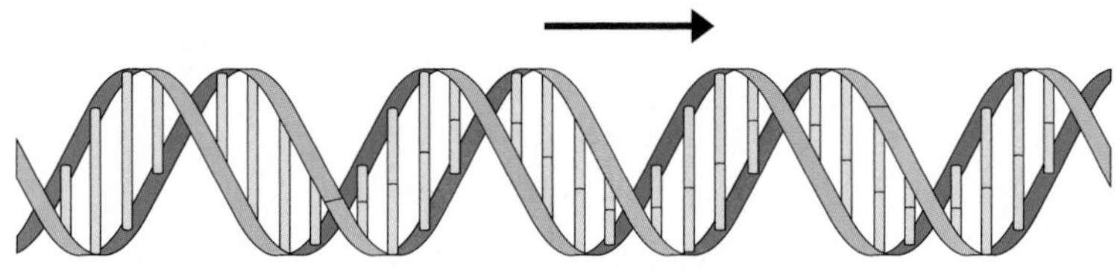

c) Repair polymerase and ligase add relacement strand

many of the processes of life produce potentially harmful compounds. There are many ways in which primary lesions are repaired, but we'll consider two of the best known, each involving the repair of pyrimidine dimers.

The most familiar repair mechanism is called **photoreactivation** repair. It is effective against UV-induced pyrimidine dimers in a wide range of organisms from bacteria to humans. Blue light wavelengths activate a repair enzyme that cleaves the abnormal bonds that form dimers, restoring the proper arrangement of adjacent pyrimidines. If photoreactivation doesn't repair the dimers directly, the primary lesion can still be fixed by a process known as **excision repair** (Figure 17.3). An enzyme moving along the DNA excises the dimer (or, in fact, some other anomalies) along with a few bases on either side. Then DNA polymerase and ligase (Chapter 12) fill in the gap with the proper nucleotides, using the opposite strand for a template.

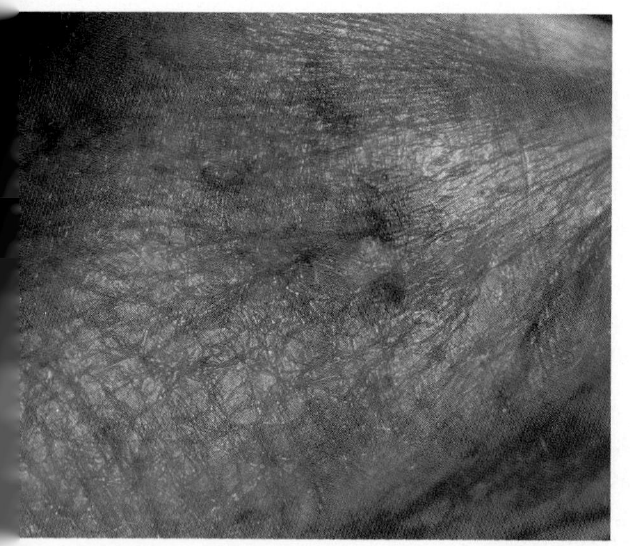

FIGURE 17.4
SUN AND SKIN CANCER
The sun is a potent source of ultraviolet radiation. Extended exposure to its DNA-damaging effects can result in sun-damaged skin and skin cancer.

Repair enzymes are not always perfectly efficient, especially it seems, in humans, as we grow older. Consequently, people who spend too much time in the sun or under tanning lights accumulate skin cell mutations and develop dry, wrinkled, leathery skin—"older" than it should be. Skin cancer, the most prevalent malignancy in this country, is due mostly to sun exposure (Figure 17.4).

There is a rare recessive disorder in humans called **xeroderma pigmentosum** ("dry pigmented skin"). Persons with the disorder have an extremely high rate of skin cancer and must constantly protect themselves from sunlight—even to the extent of becoming nocturnal. The cause of xeroderma pigmentosum is a failure of excision repair enzymes to repair primary lesions, particularly UV pyrimidine dimers.

POINT MUTATIONS

Now that we have some idea of how DNA is altered by physical and chemical means, and how at least some of these damages can be repaired, we can look into the ramifications of mutations. We will begin with point mutations, those small changes that may involve only a single base pair. We will see that even these small alterations can have sweeping effects on the phenotype.

Substitutions

Base substitution mutations are just what they sound like, substitutions or replacements of a single base pair. There are three outcomes of base substitutions, each of which can be illustrated using the first six codons of the human β-globin gene (Figure 17.5a), which produces one of the protein subunits that comprise hemoglobin.

One outcome would have absolutely no effect on the structure of the protein, and thus would be called a **silent mutation.** For example, consider amino acid number 3, leucine, that is encoded by the β-globin sequence presented in Figure 17.5. The codon specifying proline is UUG. If a substitution occurs in the third letter of that codon (Figure 17.5b), leucine can still be put in its proper place in the protein. One need only consult the genetic code table (see Table 13.1) to see that the redundant, synonymous codons of the genetic code will allow the substitution of an A for the third letter for leucine, and that, in fact, most amino acids are specified by more than one codon.

Another outcome would change the structure of the protein but would have no effect on its performance, and thus would be called a **neutral mutation.** An example is the Toguchi allele of β-globin (Figure 17.5c). Here, a substitution occurs in the first letter of the second codon such that the amino acid tyrosine is encoded instead of the normal histidine. (Consult Table 13.1 to be sure you see why.) However, as chance would have it, the substituted amino acid has no discernible effect on the performance of the hemoglobin protein.

FIGURE 17.5
BASE SUBSTITUTIONS AND MUTATIONS IN β-GLOBIN
The β-globin gene offers excellent examples of the results of base substitutions. These examples are taken from real life—they are actual sequences of alleles found in human beings.
(a) The first 21 bases of the coding region of the normal β-globin gene, and the resulting first 7 amino acids.
(b) A silent mutation, where the change of a base pair has no effect whatsoever on the resulting amino acids.

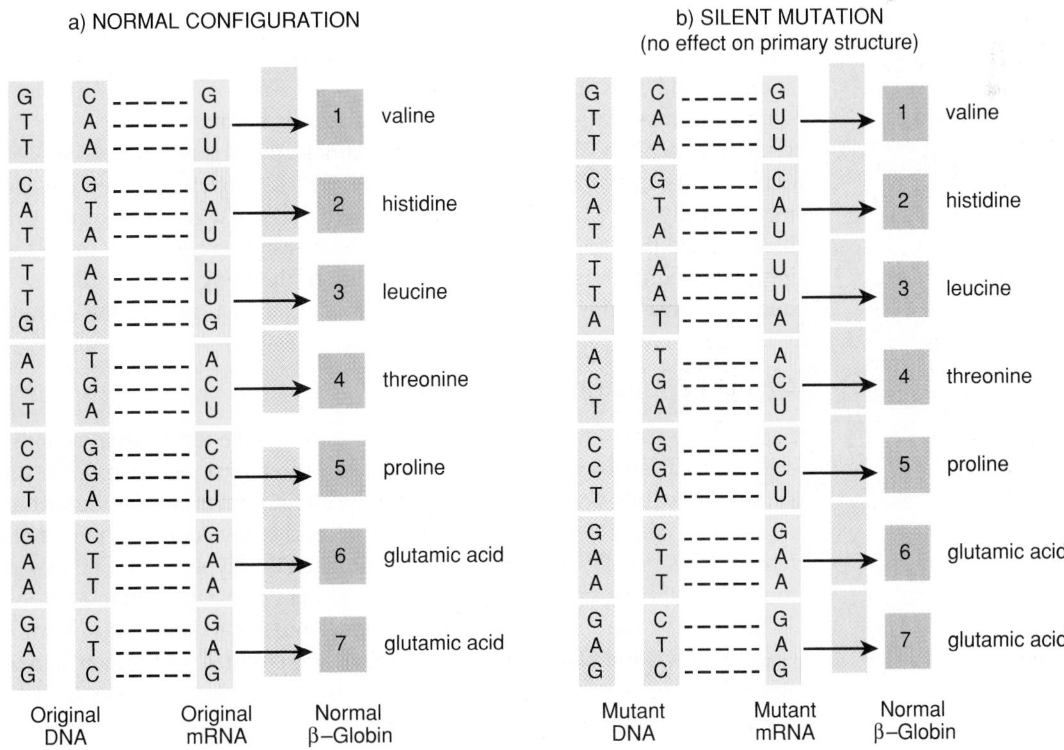

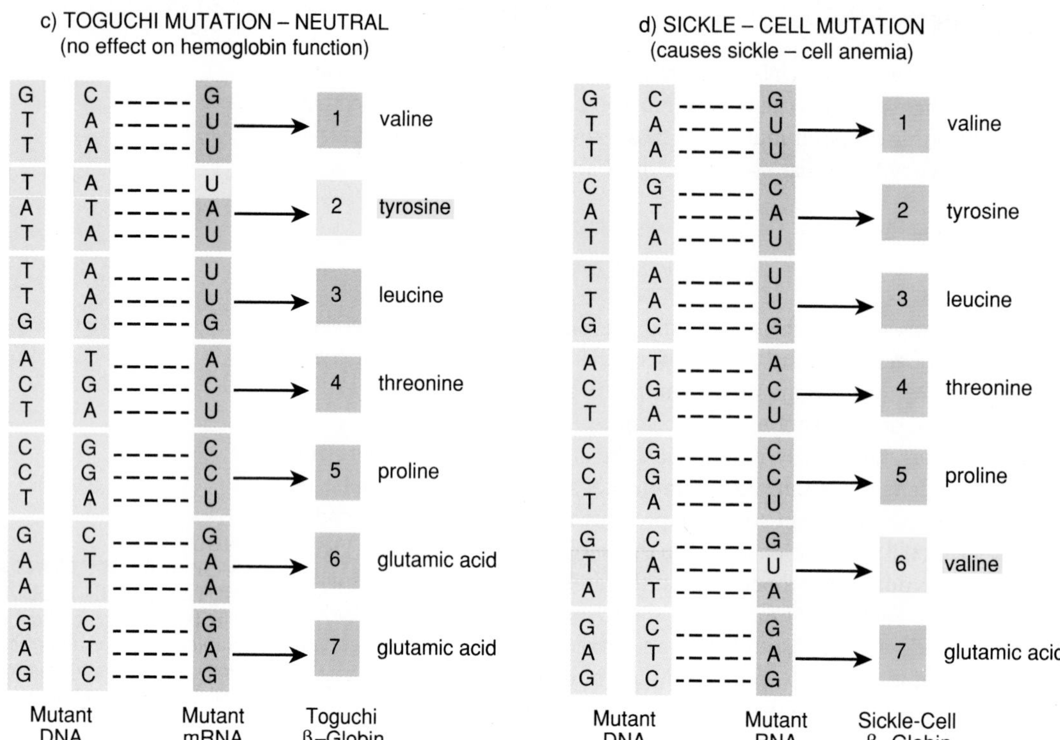

FIGURE 17.5 (cont.)
BASE SUBSTITUTIONS AND MUTATIONS IN β-GLOBIN
(c) A neutral mutation, where the change of a base does alter the sequence of the amino acids by substituting a tyrosine for a histidine. However, this substitution is called "neutral" because it has no physiological effect on the performance of the hemoglobin molecule. **(d)** Unfortunately, the base change that results in the substitution of a valine for the glutamic acid at amino acid position 6 has a drastic effect on the ability of hemoglobin to carry oxygen.

While many point mutations are silent or neutral, many others can have drastic effects, as illustrated by the sickle cell allele of β-globin (Figure 17.5d). In the sickle cell allele, the codon for amino acid number 6 is mutated from GAA (glutamic acid) to GUA (valine). It seems a small change, but it has serious effects on the future structure and performance of hemoglobin. In fact, the hemoglobin of humans bearing this substitution in both of their β-globin alleles (homozygous persons) periodically undergoes changes that distort the red blood cells into peculiar crescent (sickle) shapes. These altered blood cells have a drastic effect on the health of the individual and cause the symptoms of sickle cell anemia. The irregular cells become trapped in the smaller blood vessels, forming blockages that seriously damage the organs being served. Sickling generally occurs any time the red blood cells have a low oxygen content, as they would in metabolically active regions of the body where they must give up oxygen. Unfortunately, blocking the flow to such organs further decreases the oxygen content of the surroundings, amplifying the sickling effect. In addition, the body is further deprived of oxygen due to anemia—the affected red cells have shorter than normal life spans.

Chain-Termination Mutations

Some of the most interesting base substitutions involve termination codons. If the new codon produced by a mutation is one of the three termination codons, the base substitution mutation is called, not surprisingly, a **chain-termination mutation.** Unless the terminating codon is very close to the normal end of the coding region, premature chain terminations usually result in totally nonfunctional polypeptides. The majority of lethal mutations, at least in microorganisms, are of the chain-termination type.

Sometimes, the normal termination codon mutates to some other form, say UAA to GAA. Now, instead of terminating, a glutamic acid is added to the chain, and the ribosome simply continues to add amino acids coded by the sequence of bases on the trailing end of the mRNA. Following the normal stop codon is (probably) a random assemblage of bases, followed by the poly-A tail mentioned in Chapter 13. The ribosomes would continue translation until, by chance, they encountered a stop codon or the end of the mRNA.

Additions, Deletions, and Frame Shift

It will not surprise you to learn that an **addition** involves the insertion of a base into a nucleotide sequence, and a **deletion** involves the subtraction of a nucleotide. The profound changes caused by these simple alterations may be surprising, however. In fact, additions or deletions of base pairs have more serious effects than do most substitutions because when they happen within the expressed sequences of a gene, they change the reading frame of the message (see Chapter 13). Note, for example, in Figure 17.6 where the insertion of a single base pair changes the sense of the message from there on and creates a completely novel peptide chain. Because additions and deletions cause shifts in the reading frame of messages (Chapter 14), they are called **frameshift mutations.** Frameshift mutations are also likely to cause misreading of translation stop words and to create new stop words in the middle of the message. They nearly always result in formation of a grossly abnormal polypeptide.

Point Mutations in Noncoding Regions of DNA

What about point mutations that occur in introns (intervening sequences)? Can they possibly affect gene function? One might expect that most mutations in introns would be silent, having no effect on the resulting protein. Indeed, most have no effect. But, surprisingly enough, it has been shown that mutations within introns may cause dramatic changes in gene function. For example, consider the genetic condition known as thalassemia, a disorder that leads to serious anemia due to very low amounts of hemoglobin. Thalassemia can be caused by any of 30 known point mutations due to substitutions in the globin gene. Several are located within introns, and apparently these affect the efficiency of intron splicing. Reduced splicing efficiency results in reduced hemoglobin formation and hence the disease symptoms.

FIGURE 17.6
THE FRAMESHIFT MUTATION
Addition or deletion of bases usually has a very drastic effect on the resulting protein. For example, the addition of a base in this sequence creates a totally different protein, since each codon is "shifted" in the reading frame. Almost all of the sequence of the mRNA is unchanged, but the resulting protein bears no resemblance to the original protein.

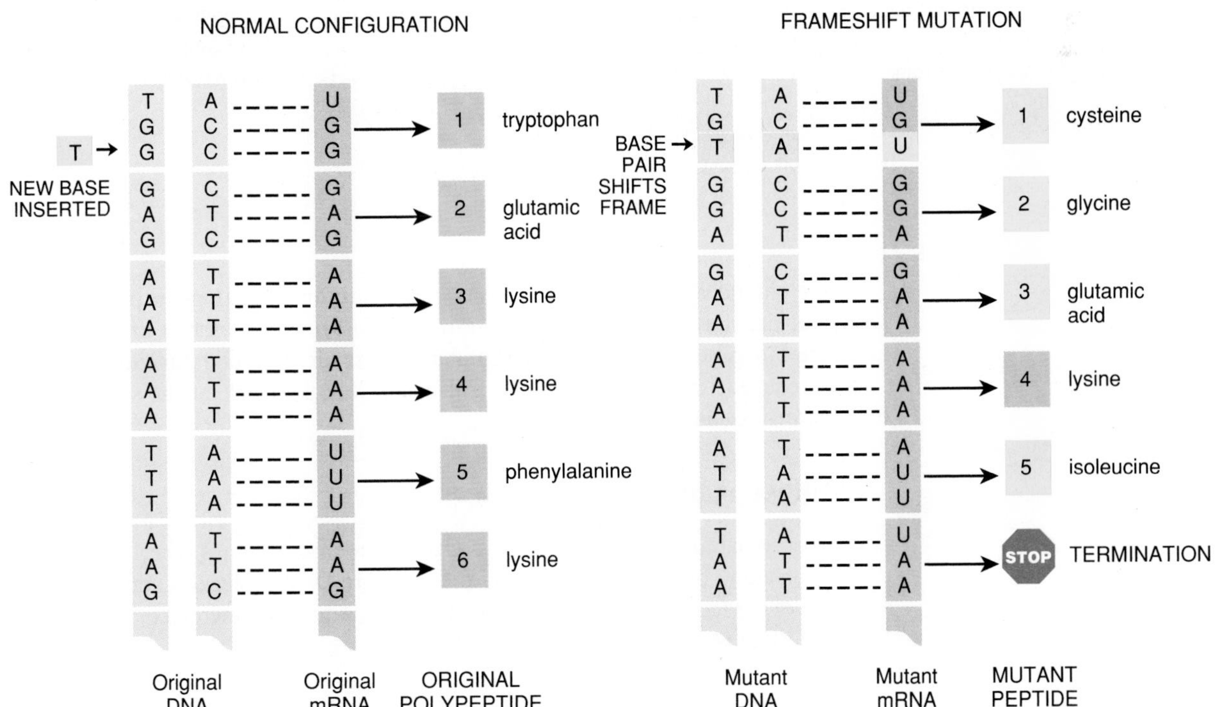

ESSAY 17.1

MUTATION, ONCOGENES, AND HERITABLE CANCERS

For most of this century, cancer research has focused on viruses as possible cancer-causing agents. Although they have now been ruled out as a major cause of human cancer, virus research has helped identify certain human genes that are implicated in at least some cancers.

Researchers discovered that some cancerous cells contain genes that can be incorporated into viruses and then transmitted to normal cells. Once there, they can be associated with cancer. It has also been discovered that virtually those same genes are actually a normal part of the cell's genome. That is, they occur in normal cells. In normal cells, however, the *protooncogenes,* as the cancer-related genes are known, are under tight control. So far about 30 protooncogenes have been identified.

Since the hallmark of cancer is uncontrolled cell proliferation, it is not surprising that most protooncogenes function in normal cell division. Apparently, these genes can be mutated or activated inappropriately, converting them to functional, cancer-causing *oncogenes* and causing the rapid and unrestrained growth of cancer cells, which subsequently invade and displace normal tissue.

How does the virus-transported oncogene affect the normal cells? One possibility is the viruses may also pick up and transmit the DNA sequence that can activate uncontrolled cell proliferation. Or it may be that the viruses transmit the protooncogens into healthy cells without also passing along the DNA control segments, those that restrain the gene and keep it from acting inappropriately (as an oncogene). Thus, while viruses themselves are not actually cancer-causing agents, their control makes them important accessories.

The mechanism of triggering cancer suggested by viral research could also help explain how some chemical carcinogens and ultraviolet rays operate to cause cancer. In fact, we now suspect that several different cancers (particularly leukemias) begin when X-rays, UV light, or certain chemical carcinogens cause chromosome breaks or mutations near or within certain protooncogenes.

Chromosome-level mutations such as inversions, translocations, or deletions may convert protooncogenes to oncogenes by severely affecting their expression. Point mutations can also convert protooncogenes to oncogenes. For example, researchers think that genes called *ras* code for regulatory proteins that normally switch back and forth between active and inactive states, depending on conditions. But one of two different substitution mutations (in either the twelfth or the sixty-first codon) is all that is needed to create the oncogene whose product is locked in the active state. Mutation of *ras* is likely to lead to common forms of lung, pancreas, and bladder cancer. *Ras* is particularly susceptible to mutation by chemicals found in tobacco smoke.

Such findings are promising, but final answers are not around the corner. Before we can say that we truly understand cancer, some hard questions have to be answered. For instance, the notion that cancer begins when a single oncogene is moved from one place to another or mutated is incompatible with the long lapse between exposure to a carcinogen and development of the disease. (Many researchers feel that cancer begins only after two or more oncogenes have been created. This helps explain things, but it must be proven.) It might also be possible that the stage is set for carcinogenesis when a single oncogene is created, but that cancer initiation occurs only when normal changes in our bodies, such as those associated with aging, provide the right conditions for oncogene expression.

We should also mention the effects of substitution mutations in promoter regions. Five of the thirty substitutions that lead to thalassemia are located in the promoter of the globin gene. Apparently these mutations impair promoter function and result in very reduced levels of hemoglobin. All in all, less than half of the substitutions that cause thalassemia are in exons.

CHROMOSOMAL MUTATIONS

Chromosomal mutations involve gross changes in chromosomes. They involve the complete breaking of the double helix in one or more DNA molecules and the subsequent failure of the repair process. Either the molecules remain fragmented or they are rejoined in abnormal ways.

Breakage and Repair

Because any gross chromosome change involves so many nucleotides, one would think that chromosomal breakage would be fatal to the cell, but in fact, it seems to happen frequently, and many times without permanent effects. Usually, the two broken ends simply come back together and are rejoined by DNA repair systems typical of those that correct other kinds of damage. In such breaks in *Drosophila* sperm chromosomes, it has been shown that more than 99.9% are corrected by repair mechanisms.

Deletions, Inversions, and Duplications

Problems arise when chromosome fragments are not properly rejoined. One problem arises because a fragmented chromosome results in one of the two fragments, the **centric** fragment, having a centromere while the other, the **acentric** fragment, does not. This means that during anaphase of mitosis or meiosis (when chromosome separation occurs) the centric fragment moves normally but the acentric fragment tends to lag behind, moving too slowly to end up in the daughter cell where it belongs. (We will come back to one fate of acentric fragments shortly). The daughter cell lacking the lost fragment suffers what is referred to as a **chromosomal deletion** (as opposed to the deletion of single nucleotides that we considered earlier).

While some deletions have no apparent effect, others can cause serious problems. For example, the cri-du-chat ("cat cry") syndrome occurs in individuals that lack a small terminal portion of their number five chromosomes. (Their plaintive voices remind one of the cry of a cat.) While the loss is not fatal, such individuals suffer severe mental and physical defects. In most of the cases studied, the parents' chromosomes were perfectly normal, indicating that most incidents of cri-du-chat emerge from newly mutated gametes.

Other problems can arise when two chromosome breaks occur in the same nucleus at the same time. Two breaks mean four unhealed broken ends. The fusion repair enzymes aren't choosy and will rejoin any two ends, sometimes rightly and sometimes wrongly. When the products of such events end up in gametes, and such gametes succeed in fertilization, the effect on an embryo can be disastrous.

Consider what can happen when both breaks occur in the same chromosome. This means that there will be three fragments, one of them with two unhealed ends. The two end fragments may rejoin and leave out the middle fragment, resulting in a deletion of genetic material. A small deletion affecting only one or a few genes may be passed on to the next generation, but most larger deletions are immediately lethal, and there is no next generation. Also, the middle fragment with two broken ends can form into a ring. If this fragment is large and contains a centromere, it may become an abnormal structure called a **ring chromosome** (Figure 17.7a). There are reasonably healthy people walking around today with ring chromosomes in every one of their cells. Another possible outcome is that the middle fragment may flip over before being rejoined to the two end fragments, resulting in a chromosome with an inversion, or inverted segment (Figure 17.7b).

If the two breakpoints occur in separate but homologous chromosomes, and in different positions on the two homologues, abnormal fusion repair can result in one chromosome being too short and the other being too long. The short chromosome will have a deletion, and the long homologues will have a duplication of genetic material.

Translocations

As a final example of chromosome mutation, consider **translocations.** A translocation results when two nonhomologous chromosomes break, and an incorrect fusion repair results in the attachment of part of one chromosome to another. If both chromosomes break more or less in the middle, such misrepair can result in two reciprocal translocation chromosomes. When the breaks happen to be close to the ends in both chromosomes,

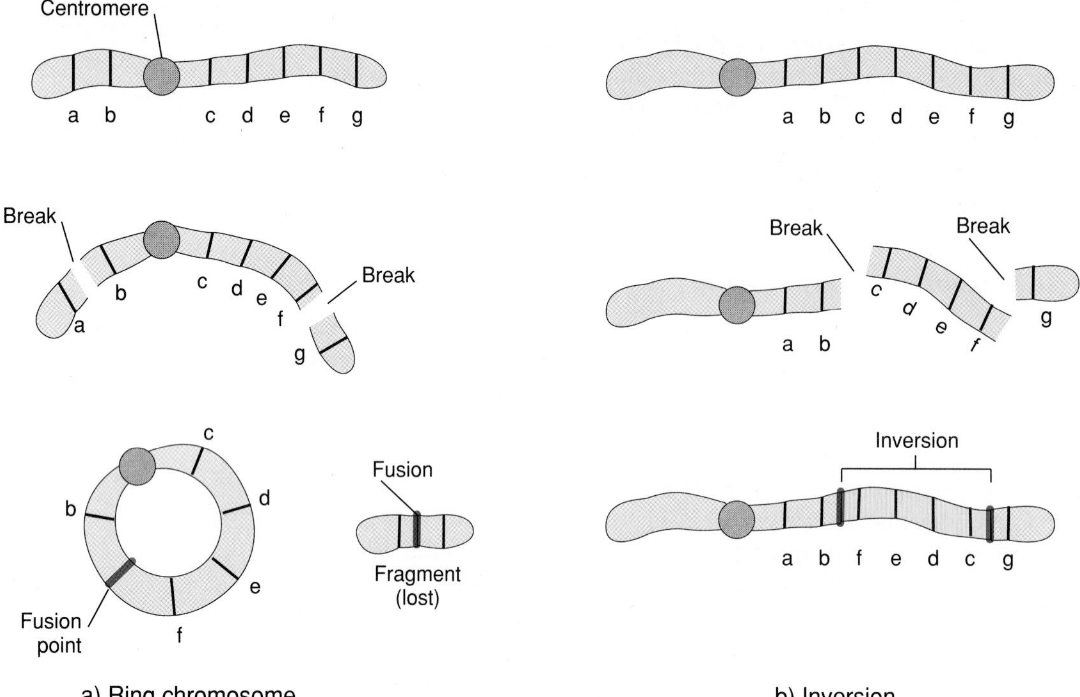

FIGURE 17.7
RING CHROMOSOMES AND INVERSIONS
If both chromosome arms break, the two arms may join to form a ring chromosome and a small fragment that lacks a centromere **(a)**. The small fragment will be lost, but the ring chromosome could be replicated and carried in a fairly normal fashion. If one chromosome arm breaks in two places **(b)**, the internal piece may rejoin in a way that creates a simple inversion. An inversion is just what it sounds like: the piece joins back in the inverted orientation. Since no genetic material is lost, most inversions are not harmful, although they may produce difficulties at synapsis Prophase I of meiosis.

the tiny end fragments may become lost, and the only surviving chromosome will be a chromosome containing most of the genetic material of both chromosomes. The best-known example occurs with chromosomes 14 and 21 and produces what is called **translocation Down's syndrome.** Although the outcome is the same as Down's due to nondisjunction (Chapter 9), translocation Down's differs in that it is inherited. Because the extra part of chromosome 21 is permanently carried along with chromosome 14, as shown in Figure 17.8, Down's syndrome is inherited as if it were linked to chromosome 14.

Translocations and other chromosomal rearrangements, like individual gene mutations, aren't always harmful. Most are, but rare harmless or beneficial ones are the stuff of evolution. Even closely related species as a rule show substantial chromosomal differences that must have originated as mutations. For instance, while we (normal) humans do indeed have 46 chromosomes per cell, our closest relatives, the chimpanzees, gorillas, and orangutans, have 48. A close study of the chromosomes of the four species shows that one of our human chromosomes arose as a fusion translocation mutation in some possibly apelike ancestor.

TRANSPOSITIONS

The third kind of mutation that we will consider is transposition. Transpositions are insertions of copies of DNA segments into new positions in the genome. They are fundamentally different from the kinds of mutations we've already seen. Some of the *results* may be similar, but the cause is not chemical or radiation damage. The cause is the actual movement of pieces of DNA from one place in the chromosome to another.

"Jumping genes" is a fanciful reference to **transposons,** those pieces of DNA capable of moving from place to place in the genome. The story of how these jumping genes were discovered is one of the more compelling tales in modern biology.

It all started some 50 years ago when Barbara McClintock (a shy but tenacious researcher whose demands for experimental equipment included "a decent pair of eyeglasses") wondered about the mutations associated with complex inheritance of color

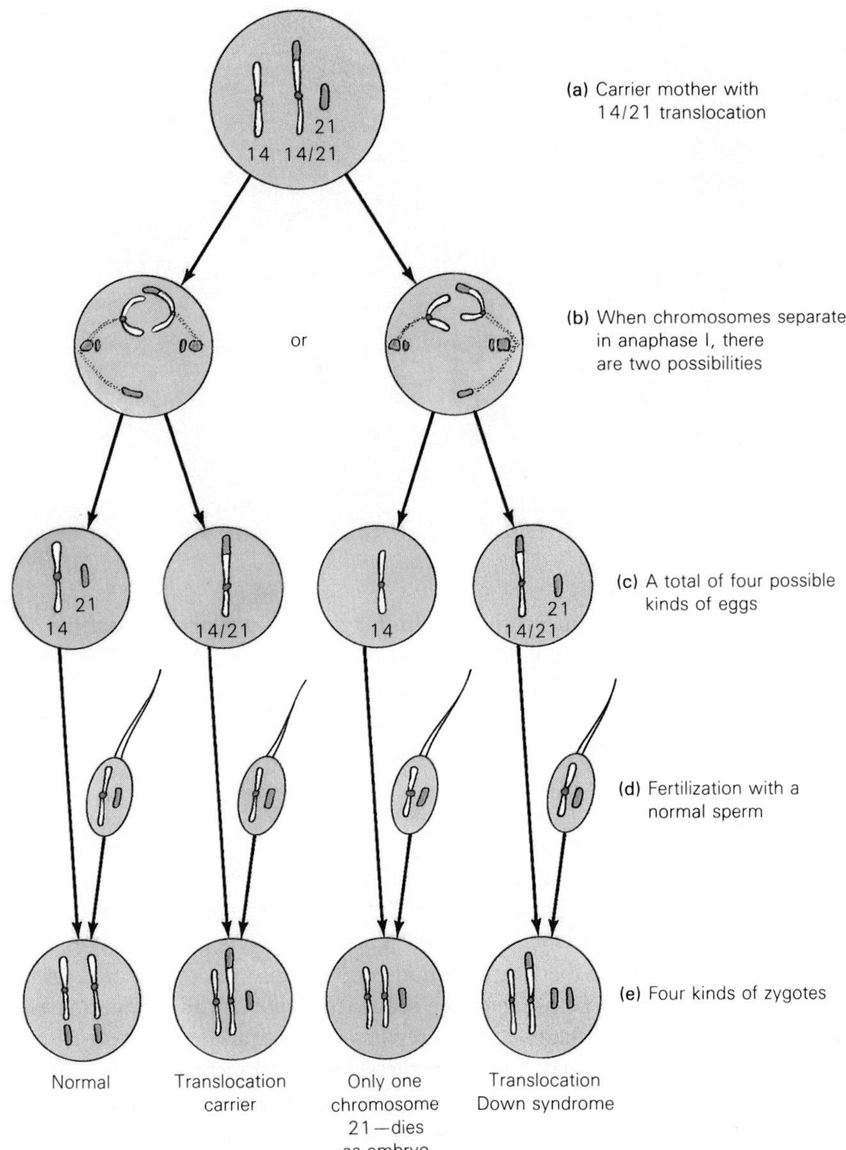

FIGURE 17.8
TRANSLOCATION 14/21

patterns of Indian corn, the corn with multicolored kernels sometimes used as autumn decoration. She concluded that this could only be accounted for if genes moved around from place to place on the chromosomes of corn plants.

McClintock found that some of these mutations were unstable. To illustrate, some strains, that by her calculations, should have had all red kernels had many yellow ones instead. At first she thought the results may have been due to some kind of "back-mutation" to the original yellow condition, but back mutations are relatively rare, and her observations were quite common. Eventually, she conceived of the existence of a mobile genetic element. She called it a "controlling element." Such an element, she proposed, moved around and, upon settling next to a color gene, promptly inactivated it. She also found that the element could move during the development of the kernel, thereby changing the color it produced, and resulting in the variegated coloration such as that seen in Figure 17.9.

McClintock published her first paper on jumping genes in the 1940s. But even though some biologists were interested in her results, most felt that they were supremely unimportant in the overall scheme of things. But she doggedly followed her interests,

FIGURE 17.9
MUTATIONS CAUSED BY TRANSPOSABLE ELEMENTS
Transposable elements can disrupt gene function. Here, what should have been colored kernels lack pigmentation because a transposable element lies in or near the gene for color production. During development of the kernels, sometimes the element leaves the gene, restoring color production in that cell and all its descendants. This results in the colored stripes.

without benefit of funding and without help, for most of her professional life. Unlike Mendel, who suffered the same kind of neglect throughout his lifetime, McClintock was finally vindicated. In the late 1960s and early 1970s, other researchers began to report similar mobile elements in other organisms. In time, jumping genes became the rage. In 1983, Barbara McClintock received the Nobel Prize.

Transposon Structure

But why have these mobile control elements become so important to our understanding of genetics? What are they? What do they do? Transposons (there are hundreds of kinds) consist, basically, of two parts as shown in Figure 17.10. The central part is the gene for **transposase,** an enzyme that enables a transposon to insert a copy of itself elsewhere. On either side of the transposase gene are sequence elements called **inverted repeats** (because the nucleotide sequence of these regions is identical, but in opposite orientations). Simple transposons may consist of these inverted repeats and the transposase gene only. Others, though, are more complex, containing up to 40,000 nucleotide pairs. At the transposon's core may be one to several complete, functional genes in addition to the transposase.

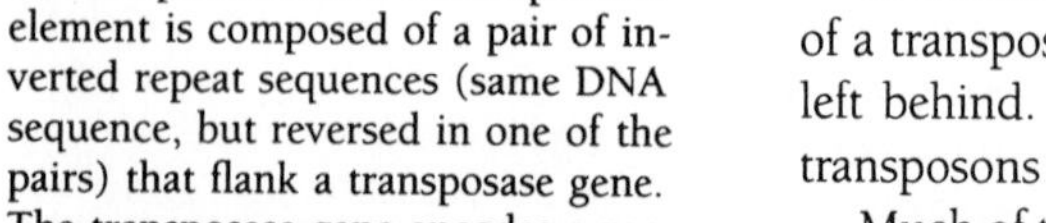

Inverted repeats
transposase gene
Inverted repeats

FIGURE 17.10
A SIMPLE TRANSPOSABLE ELEMENT
The simplest form of a transposable element is composed of a pair of inverted repeat sequences (same DNA sequence, but reversed in one of the pairs) that flank a transposase gene. The transposase gene encodes a protein that is capable of recognizing the inverted repeat sequences of the element, cutting the element out of the surrounding DNA, and reinserting the element at a new location.

Transposons as Mutagenic Agents

How do transposons act as mutagens? Obviously, because they move around. It is easy to see how insertion of DNA elements within a chromosome can disrupt gene function. First, insertion into promoter sequences could completely inactivate transcription. Second, insertion of large elements could physically separate the promoter from its coding region by thousands of bases. And finally, insertion into exons could cause frameshifts or insertions of completely new amino acids.

In addition, the process of transposition can itself be mutagenic. Often the excision of a transposable element is not clean, and small insertions of several base pairs may be left behind. Also, some elements leave behind broken chromosomes, so the action of transposons may occasionally cause chromosomal mutations as well.

Much of the time, transposon movement appears to be random. However, McClintock suggested in her Nobel lecture that, since the transposition rate increases manyfold when cells are placed under life-threatening stress, transposition may be an organism's last-ditch attempt to generate new variability. Thus, under some circumstances, jumping genes may be vital to the continuity of life.

ESSAY 17.2

MUTATION AND THE EVOLUTION OF GENE STRUCTURE

Previously, we estimated the number of genes in the genome of various kinds of organisms, such as *E coli* (about 2000), *Drosophila* (about 5000), and humans (about 120,000). Obviously, gene numbers today vary tremendously from species to species.

But at some time in the distant past there were fewer forms of life with fewer genes each. What is responsible for the changes from this primordial condition, and by what means have they occurred? Intuitively, we might deduce that increases in organism complexity were associated with more enzymes, structural proteins, and regulatory proteins, and therefore more genes. So the question arises, how did the number of genes in evolving populations increase? Most of the answers are based on what we know of (1) transposition and (2) gene duplication. Consider the human hemoglobin gene complex as an example.

The hemoglobin molecules of adult humans are made up of the products of separate genes that code for the α- and the β-globin polypeptide chains. You may recall that in its quaternary level of organization, each hemoglobin molecule consists of two α and two β subunits (see Figure 3.18). But at some time in the past, an ancient version of animal hemoglobin is thought to have consisted only of a pair of α-like subunits. So the initial step in the evolution of hemoglobin, as we know it, must have been gene duplication of a sort that resulted in an extra copy of the α gene. This could occur by the action of translocations or transposons.

Following the creation of two separate α genes (by whatever means), point mutations must have accumulated in the separate genes to change the structures of the two polypeptides slightly, finally creating a pair of molecules that could combine and work together efficiently to carry oxygen in some primordial creature's body fluids.

But the story of the development of hemoglobin doesn't end there. Mammals, you see, have at least four different kinds of hemoglobin; each is used during a different period of development and life. Embryonic hemoglobin appears soon after the embryo is formed. This molecule is well suited to pick up oxygen from the surrounding fluids. Then, somewhat later, following the development of the placenta, embryonic hemoglobin is replaced by fetal hemoglobin, which is well suited to exchange gases with the mother's circulation. Finally, a little before birth, red blood cells carrying one of two forms of adult hemoglobin begin to appear. Adult hemoglobin is especially well-suited for picking up oxygen from the lungs and transporting it to the tissues. Just how these important changes were engineered was a good mystery until it was discovered that each of the mammalian globin genes is actually a tightly linked gene complex. There are four β-globin genes on one chromosome and three α-globin genes on another.

Thus gene duplication, transposition, and evolution have worked to allow an elegant solution to the human body's changing oxygen-carrying demands during development.

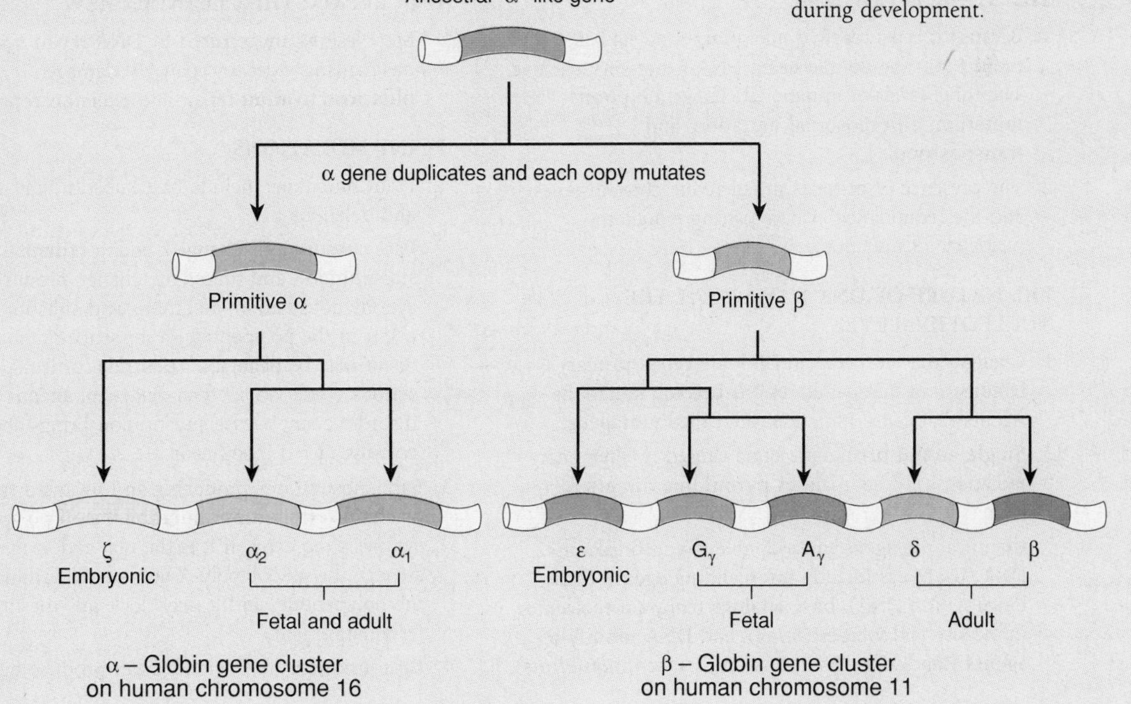

So you see that mutation can occur in a wide variety of ways, with an even wider array of possible effects. Mutations can indeed be harmful, leading to disease, physical abnormalities, and a good deal of suffering. (There are over 3000 human genetic abnormalities, and prevention of these is the major reason for the concern over the rate at which we are adding mutagens to our environment.) But mutations can also be beneficial, in that they provide the variation upon which natural selection can act, and so they have been the basis for the evolution of life as we see it today.

APPLICATION OF IDEAS

1. Recall for a minute the various kinds of mutations and mutagens that have been discussed in this chapter. Then consider the environment that we live in, both the "natural" environment and the one we humans have created. How do your thoughts impact that idea that there are two kinds of mutation, "naturally occurring variation" and "induced mutations"?
2. One of the most frightening things about life in today's world is the tremendous number (70,000-plus) and density of potential mutagens our society is producing. Write a letter to your Congressional representative explaining why you are worried about this problem. (You might even want to send it!)
3. On the other hand, mutagens have, from the very first, been a fact of life. Solar and other forms of radiation have always been with us, as have chemical mutagens (in the form of chemicals that plants produce to protect themselves against being eaten, and even in the form of oxygen, which, as it crosses over cell membranes, generates DNA-damage "free radicals"). Yet, evolution has manifestly failed to derive foolproof means of avoiding DNA damage or repairing it. Why should this be? (Imagine that it most certainly must have been possible and consider the fact that fruit flies are less well-protected against mutation than humans.)

KEY IDEAS

THE STABILITY OF DNA

1. Because it lacks reactive side groups and its bases are tucked into the double helix, DNA is not very reactive. The three levels of mutational change are **point mutation, chromosomal mutation**, and **transposition.**
2. The presence of histones in eukaryote chromosomes and the redundancy of base-pairing reduce the incidence of mutation.

THE NATURE OF DNA DAMAGE AT THE NUCLEOTIDE LEVEL

1. Changes that do occur in DNA are called primary **lesions,** and unrepaired lesions become mutations. Agents that cause mutations are called **mutagens.**
2. **Single strand breaks** are often caused by high energy radiations. UV light causes **pyrimidine dimers** to form.
3. Chemical mutagens cause double-strand breaks in DNA. Examples include **intercalating agents** (that lodge within DNA), **base analogs** (copy-cat molecules that cause faulty base-pairing), and **DNA modifying agents** (that cause replication and transcription errors).

DNA REPAIR: THE CLEANUP CREW

1. Most lesions are repaired by DNA repair systems. Two mechanisms exist to repair UV damage, **photoreactivation** repair and **excision repair.**

POINT MUTATIONS

1. Point mutations include base substitutions, additions, and deletions.
2. Base substitutions change a codon (**silent** substitution), and unless the change produces a synonymous codon, an amino acid substitution will occur in the polypeptide (a missence mutation). The result may be negligible **(neutral),** or it may produce a serious condition such as sickle-cell anemia. In this disorder a single base substitution brings about the collapse of red blood cells.
3. Base substitutions producing an unwanted *stop* codon, or **chain-termination mutations,** stop polypeptide synthesis too early. If it is the *stop* codon that is altered, the mRNA poly-A tail will add amino acids to the polypeptide, and it may "lock up" the ribosome as termination fails.
4. Base insertions and deletions can produce highly

abnormal proteins with lethal effects on the cell. They cause reading-frame shifts, in which all codons from the insertion or deletion onward will be altered.

CHROMOSOMAL MUTATIONS

1. Chromosome breakage and fusion repair is very common, and most repairs occur without problems.
2. When the wrong sticky ends are rejoined, **deletions**, **ring chromosomes**, inversions, and **translocations** can result.
3. In translocations, the greater part of a broken chromosome is fused to a normal chromosome. Generally, minor portions are lost.
4. **Translocation Down's syndrome** is produced by the fusion of part of chromosome number 21 with normal number 14. When a translocation egg is fertilized by a normal sperm, the resulting extra chromosome 21 produces a syndrome identical to trisomy-21. Unlike trisomy-21, a product of nondisjunction, translocation Down's syndrome is inherited.

TRANSPOSITIONS

1. There are naturally occurring genetic elements that can physically move around the genome. These transposable elements or **transposons** were first described by Barbara McClintock, who observed the effects of transposon insertion into the genes of corn.
2. Transposons act as mutagenic agents. When they insert into DNA they can disrupt gene activity, and when they leave a gene they can leave behind base changes in the DNA.

REVIEW QUESTIONS

1. In what ways does the double helix itself help prevent primary lesions or mutations from occurring? How do histones help? (355-356)
2. Describe DNA repair systems. What characteristic of DNA makes such a system feasible? (357-358)
3. How does a point mutation differ from a chromosomal mutation? Which tends to be more serious? (359, 362)
4. Describe an instance in which a base substitution would have no effect on a polypeptide. What, if any, adaptive significance does this suggest? (359)
5. Using an example, explain how a single base substitution could be disastrous. (360)
6. Why would a base addition or deletion tend to be more serious than a base substitution? (361)
7. What effect on a polypeptide would a mutation in the *start* or initiator codon have? Could this be detected? Explain. (general)
8. Describe the events which lead to inversions and translocations. (363-364)
9. What is the structure of a simple transposon? A complex transposon? (366)
10. In what ways may a transposon act as a mutagen? (366)

3
EVOLUTION

18

NATURAL SELECTION AND ADAPTATION

THE QUESTION OF WHETHER SPECIES CHANGE, WHETHER ONE LIFE FORM becomes another life form over generations, did not stem from Charles Darwin. The question is about as old as philosophy itself. In fact, both Plato and his student Aristotle had considered the question and had resolutely declared that species do not change.

THE HISTORY OF AN IDEA

Plato believed in an ideal, a perfect form, from which any variation is an imperfection, and that the perfect form was stable. Furthermore, he believed the ideal to be a form that transcends the real world. (For example, according to his notion, a real organism has no more to do with the ideal form than any given circle has to do with the concept of circularity.) Aristotle (Figure 18.1), Plato's best known protege, developed his own ideas of what life was about by devising a "natural scale" in which organisms are arranged in an ascending order from simple to complex, with man at the very top (just above the Indian elephant). He disagreed with Plato in that he concluded that the essence of an organism was in the organism itself, not in some abstract ideal from which the organism deviated. Both believed, though, in the immutability of species. How, they asked, could you pass on to your offspring something different from what you are?

Biological thinking continued to be firmly embedded in Aristotelian philosophy for some 2000 years. The notion of immutability was furthermore bolstered by Judeo-Christian theology, which described all life as having been created in a few days, with the various species being immutable. The dogma of immutability stemmed largely from the logical extension of the alternative. That is, if life changed through time, and if one species could arise from another, then unless humans were somehow exempt from natural law, they *could* have arisen from some other life form. This, of course, flew in the face of the biblical teaching that man was created in the likeness of God.

By the 1700s it was generally believed that all species, or "kinds" of living things were created in their present form—in other words, that they had not changed in any way over the millennia. This was certainly the view of the Swedish botanist Carl von Linne (1707–1778), who devised a system of classification for all living organisms and, in his fondness for Latin, even called himself Carolus Linnaeus. Many argued that Linnaeus' designations, which involved lumping the species together on the basis of their similarities, were artificial and based on personal whim. Few, however, considered criticizing his assumption of special creation or his concept that each species was fixed in its original form, never to change.

One small departure was suggested by a French contemporary, George-Louis Leclerc De Buffon (1707–1788). In 1753 Buffon proposed that in addition to those animals that had originated in the Creation, there were also lesser families "conceived by Nature and produced by Time." He explained that changes of this kind were the result of imperfections in the Creator's expression of the ideal—a philosophical, rather than a biological, point of view and undoubtedly based on Plato's beliefs.

FIGURE 18.1
ARISTOTLE
The Greek philosopher and naturalist, Aristotle, shared his observations on life with his students in the pleasant garden surroundings of the Lyceum, possibly the first biology classroom.

Other scientists were also beginning to toy with the notion of the heritability of change. In France Jean Baptiste de Lamarck (1744–1829), a protege of Buffon's, boldly suggested that not only had one species given rise to another, but man himself had arisen from other species. A passionate classifier, Lamarck believed that every organism had a relative position on the "scale of nature"—with man, of course, as the highest form of life. He pointed out, however, that the fossil animals found in older layers of rock were simpler forms and contended that they had become higher animals through a gradual progression. In his view there was a "force of life" that caused an organism to generate new structures and organs to meet its biological needs. Such structures then continued to develop through use, with the change in each generation transmitted to succeeding generations through the inheritance of acquired characteristics. Lamarck cited as an example the long neck of the giraffe (Figure 18.2), which he maintained had evolved as each generation of giraffes had stretched their necks in an effort to reach the topmost branches of trees and then transmitted genes for this longer neck to their offspring. Lamarck's notions, called the "law of use and disuse," have become largely discarded.

At the time, Lamarck's arguments did little to persuade his lecture audiences. With the foment of the French Revolution, there was some lively discussion in intellectual

FIGURE 18.2
ACQUIRED TRAITS AND INHERITANCE
A narrow interpretation of one of Lamarck's ideas on evolution suggests that organisms inherit acquired traits, an idea that became known as the "law of use and disuse." It maintains, for example, that the giraffe's long neck evolved through generations of "neckstretching." It's unfortunate that we only remember Lamarck for his error, since he clearly challenged the predominate view of immutability of species and stimulated a lively interest in evolution.

circles, but in society at large the firm conviction that each form of life had arisen through special creation ruled out serious consideration of any other concept.

The intellectual climate into which Darwin was born was far more conservative than that in France. The English had been horrified by the French Revolution, and ideas such as those held by the "French atheists" were either dismissed out of hand or viewed with extreme suspicion. Partly for this reason, the church continued to hold strong sway over the sciences, and biologists in England continued to adhere rigidly to the traditional tenets. This was the philosophical milieu into which Charles Darwin's ideas were thrust.

Enter Charles Darwin

Charles Darwin, as one of the most influential figures in modern history, has by now probably generated as much research about himself as about his work (Figure 18.3). Historians and biologists alike are intrigued with the question, how did he come to his ideas? When was he struck with the concept of natural selection? What triggered his thinking along such lines? Obviously, his visit to the Galapagos islands was pivotal in his thinking. The usual myth has the whole thing turning on his observations of those Galapagos finches. This can hardly be the case, however, because he didn't even learn they *were* finches at all until some years after his return to England. He wasn't even particularly taken with the little birds while he was in the Galapagos Islands. In fact, much of what he reported about them was admitted hearsay by members of the crew. Many of them had taken the opportunity to roam the strange little islands and had offhandedly witnessed things that Darwin hadn't seen. He may have been inspired by his visit to the islands, but many of the details of his theory were pounded out by sheer intellectual labor after his return to England.

It has been suggested from time to time that the subject was "in the air," that the stage was set by circumstance, and that the notion of "descent with modification" (Darwin didn't use the word "evolution" at first) by natural selection was inevitable. Darwin vigorously disagreed that the concept was an inevitable extension of the thinking at the time. He noted that he had often tried to explain his idea to able men but had "signally failed" to make them understand.

DARWIN'S THEORY

Beginning in 1837, Darwin started to keep a journal entitled *Transmutation of Species* and was soon making entries referring to "my theory." Reading that journal today, and the journal of the voyage of the *Beagle*, is like reading a detective story after you already know whodunit—the suspense is in watching the detective sift through clues for the right answer.

It is fascinating to watch his ideas develop in his writings. Keep in mind that Darwin was a country boy. He knew a lot about agricultural practices, and he knew about livestock breeding. Any good farmer knows that a line can be improved by selecting only certain individuals of each generation for breeding (Figure 18.4). (The approach is called artificial selection.) By longstanding folk tradition, the best animals, whether cattle, fowl, or dog, were honored annually at country fairs and chosen for further breeding. The reasoning was simple: Like begets like, offspring tend to resemble their parents, and the "best" parents produce the "best" offspring. (We will shortly consider artificial selection as evidence supporting the theory of natural selection.)

But Darwin wondered, would selection operate without human intervention? And, if so, how? Artificial selection involved the conscious choice of the breeder. If selection occurred in nature, then who was the selector? This line of thought seemed at first to lead right back to a supernatural factor. Without a conscious selector, the inferior individuals were as free to breed as were the most superior, in which case no improvement or change or adaptation would occur.

Did selection have to be conscious, then? Perhaps not. Darwin was impressed with an essay on population that had been written by the Reverend Thomas Malthus three

TABLE 1.1

MILESTONES IN BIOLOGY

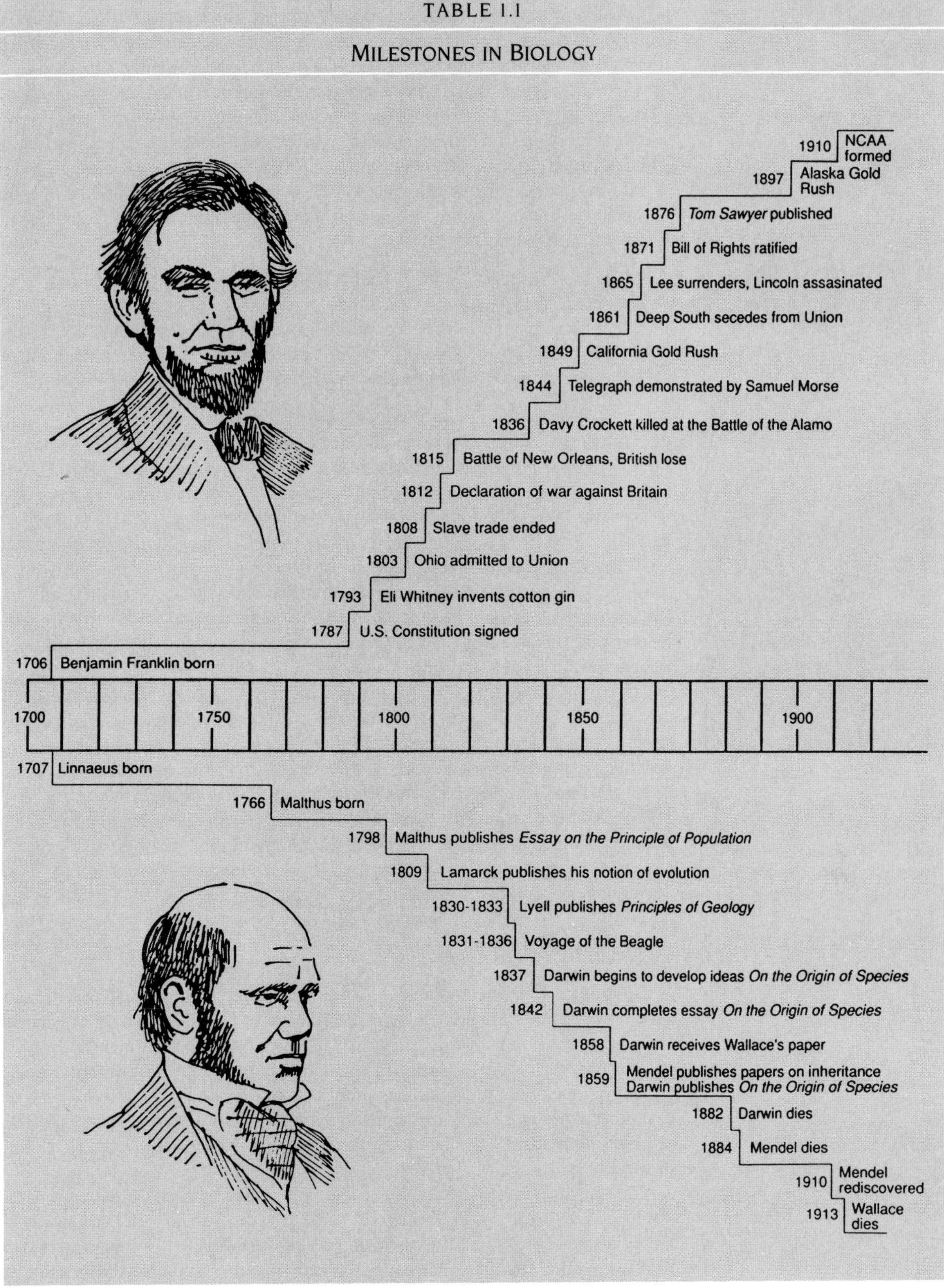

1910 NCAA formed
1897 Alaska Gold Rush
1876 Tom Sawyer published
1871 Bill of Rights ratified
1865 Lee surrenders, Lincoln assasinated
1861 Deep South secedes from Union
1849 California Gold Rush
1844 Telegraph demonstrated by Samuel Morse
1836 Davy Crockett killed at the Battle of the Alamo
1815 Battle of New Orleans, British lose
1812 Declaration of war against Britain
1808 Slave trade ended
1803 Ohio admitted to Union
1793 Eli Whitney invents cotton gin
1787 U.S. Constitution signed
1706 Benjamin Franklin born
1700
1750
1800
1850
1900
1707 Linnaeus born
1766 Malthus born
1798 Malthus publishes Essay on the Principle of Population
1809 Lamarck publishes his notion of evolution
1830-1833 Lyell publishes Principles of Geology
1831-1836 Voyage of the Beagle
1837 Darwin begins to develop ideas On the Origin of Species
1842 Darwin completes essay On the Origin of Species
1858 Darwin receives Wallace's paper
1859 Mendel publishes papers on inheritance
Darwin publishes On the Origin of Species
1882 Darwin dies
1884 Mendel dies
1910 Mendel rediscovered
1913 Wallace dies

FIGURE 18.3
DARWIN'S PLACE IN HISTORY
A time line depicts familiar historical events correlated with advances in evolutionary thought.

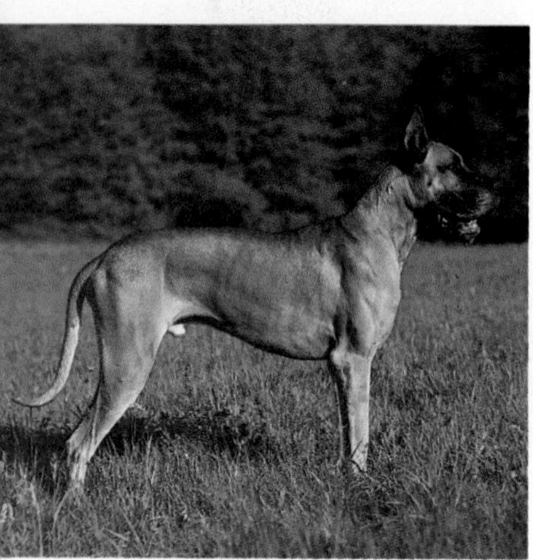

FIGURE 18.4
ARTIFICIAL SELECTION
Generations of selective breeding have produced some amazing variations in *Canis familiaris*, the dog species.

decades earlier. Malthus stressed that all species had enormous reproductive capabilities and that their numbers tended to expand geometrically (2, 4, 8, 16, 32, and so on) unless held in check by starvation or disease. Natural populations, he argued, reached a balance in which all but a few of the young of each generation were forced to perish. *All but a few*. So, mused Darwin, perhaps the inferior individuals were not free to reproduce after all, precisely because they were inferior! The environment, then, could select the individuals that were allowed to breed, and these would be the hardiest. And it all happened through natural means (Figure 18.5).

Darwin's journal shows that by 1838 he had solved the major riddle of evolution. The mechanism of evolution, he said, was **natural selection.** His idea of natural selection has based on four points:

1. There is variation among the individuals of most natural populations.
2. Some of that variation is inherited.
3. Populations tend to produce more offspring than the environment can support.
4. Those individuals whose traits best adapt them to the environment will survive better and leave more offspring than those with less adaptive traits.

A penciled manuscript in 1842 laid out the entire theory of the origin of species through natural selection. Darwin did not publish it, however. He knew that the idea would arouse fierce resistance and that he would have to back up every part of his idea with evidence. He set about preparing to defend his argument.

However, Darwin's robust health mysteriously began to fail. He began to vomit frequently, and he complained of headache, nausea, fatigue, and heart palpitations. In time, because he tired quickly, he virtually stopped seeing anyone, even his valued scientific colleagues. He resigned as secretary of the Geological Society. His father bought him a country house, and his wife became his nursemaid. Strangely enough, he continued to look completely healthy. His mind was as vigorous as ever, but his strength was gone. He continued to work, but only for a few hours a day.

The cause of Darwin's illness remains a mystery. Some medical historians believe it may have been psychosomatic. Others think it may have been the result of his great curiosity as a young man. In August 1835, he deliberately "experienced an attack of the Benchuga, the great black bug of the Pampas . . . ." It turns out that the bloodsucking Benchuga (now known as the barbeiros), *Panstrongylus megistus,* is the carrier of the protozoan *Trypanosoma cruzi,* which causes Chagas' disease. After a latency period of some years, the parasites usually invade the heart and the intestines, causing weakness, intestinal distress, and heart palpitations. New evidence has weakened the Benchuga bug argument, and so his illness remains a puzzle as much to us as it was to him.

Gathering Evidence

In those times when he was able to work, Darwin set about compiling information to back up his theory. He even suggested further tests. Darwin was also an experimental biologist, and his experiments showed that life can spread from mainlands to islands, such as the Galapagos, quite easily. He found that a variety of species can survive extended periods of time in seawater and that airborne organisms can easily move from one place to another, perhaps by being transported by storms. Such evidence was important because much of the foundation of his developing theory depended on the ability of species to disperse.

Working only a few hours a day, Darwin also published lengthy works on the systematics, or natural classification, of barnacles (still a major reference work), on his theory of the origin of coral islands, and on geology. But he continued to develop his theory of natural selection. He showed his early manuscript on natural selection to only one close friend, the botanist Joseph Hooker, who remained doubtful. Twenty years passed, and finally in 1856 Darwin began writing his major work, to be called *Natural Selection*. It was planned as an enormous, six-volume monograph. But in 1858 his work on the manuscript was suddenly interrupted. Unexpectedly, another manuscript, a very

FIGURE 18.5
USE AND DISUSE VS. NATURAL SELECTION
(a) According to Lamarck, constant use of certain organs led to changes in the organs themselves. For example, stretching of the neck, in the case of the giraffe, led to its gradual lengthening. Each generation inherited the newly acquired length from the previous generation until the modern giraffe emerged. **(b)** Darwin maintained that the environment selected the longer-necked giraffes from a population of variants. Thus those with longer necks survived to pass their "long-neck" genes along.

brief one, arrived in the mail from A. R. Wallace, a young naturalist working in Malaya. Wallace asked politely whether Darwin would care to make any comments on the manuscript. The article was a short but well-written statement about evolution and natural selection—Darwin's ideas had been quite independently deduced by someone else! Darwin was mortified. In a letter to his friend Hooker he lamented, "So all my originality, whatever it may amount to, will be smashed . . . Do you not think that this sketch ties my hands? . . . I would far rather burn my whole book, than that he or any other man should think that I have behaved in a paltry spirit." Darwin was about to be scooped; it is not an uncommon experience in science. Would his own work now be thought to have been based on the ideas of another?

Darwin's friends persuaded him to allow his previously unpublished 1842 summary and the text of a long letter describing his ideas to be presented together with Wallace's paper before the Linnean Society of London. Because of Darwin's much more substantial evidence, his paper was given first. The two were presented in July and were published in August 1858. Now Darwin went furiously to work and, putting aside the idea of a huge, definitive monograph (which never was written), he quickly finished an "abstract" of it, the famous *On the Origin of Species,* published in 1859. The first edition sold out on the first day.

The Origin of Species

In the first chapter of *On the Origin of Species,* Darwin sought to establish that animals and plants under domestication are extremely variable, that the variation is heritable, and that the many domestic varieties have arisen, under artificial selection, from wild ancestors. In the second chapter, he drew together what evidence he could find to show that plants and animals in nature were variable too. In the third chapter, he discussed Malthus' idea of the struggle for existence, the idea that the natural reproductive capacities of living things greatly outreach the ability of the environment to support them, so that many organisms perish by starvation or disease without reproducing (Figure 18.6). In the fourth chapter, he introduced natural selection:

FIGURE 18.6
REPRODUCTIVE OUPUT IS SET BY NATURAL SELECTION
The number of offspring produced by most organisms is determined by selective processes. That accounts for the fact that many will die or fail to reproduce for other reasons.

> How will the struggle for existence . . . act in regard to variation? Can the principle of selection, which we have seen is so potent in the hands of man, apply in nature? I think we shall see that it can act most effectively . . . If such (variations) do occur, can we doubt (remembering that many more individuals are born than can possibly survive) that individuals having any advantage, however slight, over others, would have the best chance of surviving and of procreating their own kind? On the other hand, we may feel sure that any variation in the last degree injurious would be rigidly destroyed. This preservation of favourable variations and the rejection of injurious variations, I call Natural Selection.

The New Synthesis

Charles Darwin developed a cornerstone of modern biology in his description of evolution through natural selection. But he had a great deal of trouble convincing his contemporaries of this relationship. Most scientists of his time were quick to accept the fact that evolution occurs, but many of them failed to see just how natural selection was involved. Darwin was unable to convince them, primarily because he had no *mechanism* whereby natural selection could proceed. The problem stemmed from the fact that Darwin didn't even know genes *existed.* So as he set about trying to explain how natural selection might work, he began grasping at straws. He continued to believe that inherited factors were somehow blended in the offspring. Furthermore, with each revision of *On the Origin of Species,* he lapsed deeper into Lamarckian explanations that even his contemporaries knew didn't hold water.

We now know that, not far away in what was then a part of the Austrian-Hungarian empire and is now Czechoslovakia, Gregor Mendel was generating precisely the information that would clear up Darwin's great dilemma. But since Mendel was the only modern geneticist in the world, he had trouble making people see the importance of his work—including Darwin. In fact, a copy of Mendel's paper was found in Darwin's library, completely unmarked.

As we saw in Chapter 10, Mendel's work was rediscovered in the early part of this century, long after the hard-working monk had gone to his reward. The scientific world, newly fascinated with genes and how they are carried along from generation to generation, began to wonder just what genes have to do with natural selection. By the 1930s, the broad principles of the relationships between genes and natural selection had been worked out. Then in the 1940s, Ernst Mayr and others finally produced what they called the *New Synthesis,* the integration of the principles of genetics and evolution.

The New Synthesis explained a great deal and laid the framework for further investigation. That investigation, not unexpectedly, has resulted in the alteration of some of the original statements and has in fact, produced arguments over rates of evolution and just how evolution proceeds (see Chapter 19). Many scientists see such disagreements as no attack at all, but simply as fine-tuning an established principle that will remain, perhaps altered, but alive and well, and a linchpin of modern scientific thinking. With this critical integration of ideas in mind, then, let's see just how evolution does, in fact, proceed by natural selection working on variation, favoring some forms while weeding out others. We can begin by noting some of the many ways that variation can exist in populations.

FIGURE 18.7
VARIATION IN POPULATION
Penguins, like humans, form large populations of sexually reproducing individuals. No two penguins are exactly alike—at least to another penguin. Every king penguin here knows its mate, its offspring, and all its nesting-ground neighbors.

Human beings, like all large populations of sexually reproducing organisms, are extremely variable. No two individuals are genetically alike, with the exception of monozygotic (single-egg) twins.

SOME WAYS LIFE CAN VARY

In any population, individuals are likely to vary from each other as each bears novel traits or combinations of traits. We are familiar with some of the ways our own species can vary (Figure 18.7), but we must keep in mind that variations in other species may not be so obvious.

For example, we humans are particularly keyed to physical variations of the face. We can often distinguish identical twins whom we know well. Dogs, on the other hand, are less sensitive to our facial features, but very attentive to differences in our scent. Apparently our scents are quite distinctive, but those differences go largely unnoticed by us, at least in better social circles.

Animals can vary, not only in their appearance, but in their behavior. Some dogs, for example, may bite you even if they are fully aware of who you are, while others are simply not disposed to bite at all. If you approached a herd of fighting bulls, some will walk toward you while others trot away. (Of course, this difference can be due to mood as well as genetics, but the breeders prefer those that routinely challenge.) There is even behavioral variation among ants of the same caste. Some work efficiently while others run back and forth along the line, accomplishing nothing. Other variations are even less apparent. Researchers have found that we vary biochemically in any number of ways, from our blood proteins to digestive enzymes.

There are two basic points here. The first is that variation can occur in a wide array of traits. The second is that we, with our own special sensibilities, may not be readily aware of the variation. Whether the variation is apparent to us, though, doesn't matter. If one type confers any survival or reproductive advantage, natural selection can act on it.

It is important to note that natural selection operates on variation that is genetically based. In this way, certain genes or combinations of genes can be favored and come to increase in succeeding populations.

Sources of Variation

Variation in a population arises from two primary sources, *mutation* (Chapter 17) and *genetic recombination* (Chapter 11).

Mutations are random heritable changes in the genetic material. Only those mutations that arise in gametes can be passed on to the next generation; those in the somatic (body) cells cannot be. (These, though, can cause cancer.)

There are two kinds of mutations: point mutations and chromosomal mutations. Point mutations, you may recall, affect only a single base in a strand of DNA and are not likely to cause a change in the gene's protein product. In some cases they do cause a change in the product, but even when this happens, the phenotype is still not likely to be affected to such an extent that natural selection can operate on the variant. Not all point mutations are so benign, however. They may change the phenotype markedly, and when this happens, the effect is likely to be deleterious. After all, a fully functioning existing gene cannot be randomly changed without a likelihood of interfering with its functioning.

Chromosomal mutations are massive, spontaneous rearrangements of chromosomes. The rearrangements can involve duplication or deletion of segments of the chromosomes, as well as reversals (inversions) of segments. Chromosomal mutations are much more likely to cause changes in the phenotype, exposing the rearranged gene to natural selection. Chromosomal mutations, like point mutations, are almost always harmful, but by increasing variation in the population they are also the source of new material that can confer benefits, in some cases, so that adaptation through natural selection can proceed. In fact, mutation is the *only* source of new genes in a population.

Researchers have estimated that there are one or two mutations per gamete, but that these are usually rather inconsequential as sources of variation within a population. Most variation, they say, arises as a result of individuals recombining existing genes in unique ways. This recombination occurs with sexual reproduction, as random combinations of genes are brought together as the egg and sperm join, and are then reshuffled by meiosis and crossing over. However, in asexual species, such as bacteria, mutation can be an adequate source of variation. One reason is that, with these species' short generation time, any beneficial mutation can become increasingly common in the population very quickly. In higher plants and in animals, however, most of the variation in a population arises from uniquely recombining existing genes from the gene pool. Once the variation occurs in a population, the question arises, how is it maintained?

Maintaining Genetic Diversity

Successful populations must have ways of maintaining their genetic diversity in the face of natural selection. After all, natural selection tends to move populations toward genetic uniformity by removing unfavorable genotypes, that is, those that are other than optimum. There are several mechanisms that help populations to maintain diversity. We will consider three: balancing selection, heterozygote advantage, and frequency-dependent selection.

Balancing Selection Part of the basis for variability lies in the tendency of recessive traits to "go into hiding" whenever they are rare in the population. They are maintained in the population because, unless they express themselves in the homozygote, natural selection cannot operate on them. There are cases in which an allele continues in a population even though it produces a trait that is selected against. The process that inhibits the loss of variant alleles (even when they are detrimental) is called **balancing selection**, and the phenotypic result is called a **balanced polymorphism** (*poly-*, "many"; *morph*, "form"). Balanced polymorphism is marked by populations with variant individuals where the ratios of those variants do not change noticeably from one generation to the next. Each morph (or variant) may have some particular advantage under certain conditions (Figure 18.8). For example, balanced polymorphisms can occur in "patchy" environments, where a population extends over a number of distinct habitat types. The habitat of the British land snail may vary greatly, some areas being shaded, some lighted, some brown, some green, some mottled, and so forth. Since the snails' shells differ in color and marking, some individuals will be difficult for predatory birds to find under any of these habitat types; thus, the great variation in these snails is maintained.

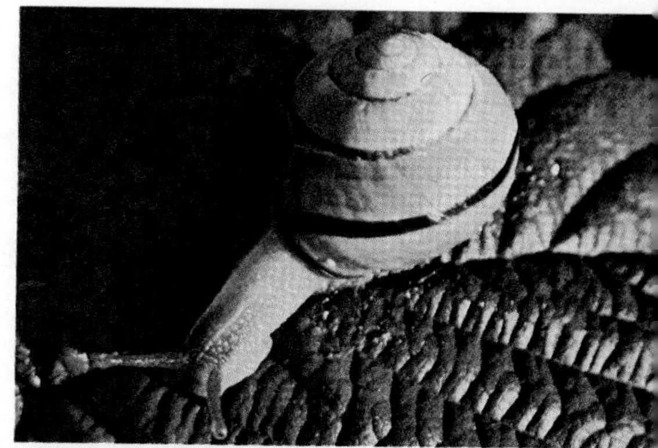

FIGURE 18.8
GENETIC VARIATION IN SNAILS
The land snail, *Capaea nemoralis*, varies markedly in two ways: shell color and banding patterns. The two traits are inherited independently, so each color can occur with each banding pattern. The result is a dramatic variability.

Heterozygote Advantage Genetic diversity can also be maintained by **heterozygote advantage** (also called overdominance), in which the heterozygous genotype has a greater reproductive success than either homozygote. We see an example of heterozygote advantage in the sickle-cell and normal alleles of beta chain hemoglobin in populations where malaria is endemic (see Essay 18.1). Since each of the two homozygous types is inferior to the heterozygote, natural selection keeps both alleles in the population in intermediate frequencies.

Frequency-dependent Selection **Frequency-dependent selection** occurs when the reproductive success of a genotype depends on its frequency in the population. (We will discuss the term *frequency* in more detail in the next chapter. For now just keep in mind that it simply refers to the *ratio* of an allele in a population.) If the success of a genotype is dependent on how frequently it appears in a population, the results may be a stable polymorphism. This happens if a genotype has a net advantage when it is rare and a net disadvantage when it is more common. The problem is, net advantage when rare will result in an increase in allele frequency; when this occurs, the genotype becomes less rare, and its advantage decreases.

For example, some tropical freshwater fish populations are polymorphic for a common gray morph and a relatively rare red morph. The red fish, by their color alone, intimidate the other fish and almost always win out in fish-to-fish competitions. On the other hand, the red morph is easier to see and is more subject to predation by birds. Thus, when the red fish are rare, they have a net benefit because of their powers of intimidation. However, if the red form becomes common, their gray competitors encounter them frequently and become aware that those red fish aren't so tough after all. The gray fish that are harder to intimidate produce more offspring than the more timid gray fish, so a bolder gray morph evolves. Thus, each morph has its advantages, but their relative numbers are important. The red forms are kept at an *equilibrium frequency* at which the benefits of being red just balance the disadvantages (Figure 18.9).

Now that we have seen how variation can arise and how it can be maintained in a population, we can discuss how natural selection operates on that variation.

NATURAL SELECTION

Natural selection has come to be defined as the process by which the environment selects those genotypes that are best adapted to the prevailing conditions. By way of review, natural selection operates only on that variation that is heritable, or genetically based. From among the various genes and combinations of genes in a population, then, some will confer certain advantages to the individual that bears them. If those advantages lead to greater reproductive success (more offspring), then those kinds of genes will come to increase proportionately in the population. Natural selection, then, results in **evolution,** *which is the changing of the ratio of alleles in a population over time.* We will consider, now, two ways that natural selection can result in evolution. The process, we will see, is much more rapid in the first case than in the second.

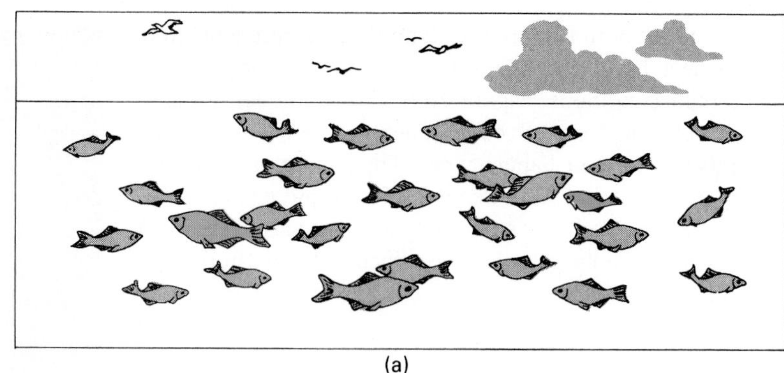

(a)

(b)

FIGURE 18.9 FREQUENCY-DEPENDENT SELECTION
By virtue of their bright color, the red variety of fish **(a)** are more reproductively successful. **(b)** However, as their numbers increase, their bright colors also make them easier for predators to find; thus, the numbers of red fish never overwhelm the gray. The red color also helps in competitive encounters with the gray fish, as long as the red morph does not become too common.

The Peppered Moth: Natural Selection in a Natural Population

Biston betularia is a British moth, commonly called the peppered moth. It occurs in two **morphs**; that is, it may have either of two distinct forms or appearances, as shown in Figure 18.10. One morph is light and mottled (or peppered), and the other morph is black. The British have a long tradition of butterfly and moth collecting, and records on the peppered moth go back two centuries. The black morph, whose color is controlled by a single dominant allele, originally showed up in eighteenth-century collections as a rare, highly prized variant. In the early stages of the industrial revolution (in the 1840s), the black form began to show up in greater frequencies in collections, especially near cities. The black morph continued to become more and more common in industrialized areas, until it greatly outnumbered the light peppered morph. In Manchester, England's industrial center, the dominant black morph achieved a frequency of 98%. Meanwhile, the light peppered morph remained the predominant form in rural areas.

The environment had changed, and the species, through differential mortality and change in allele frequencies, adapted to it. The environmental factor was soot from burning coal. Industrial England, as the nineteenth century proceeded, quietly submitted to its dark cloak of carbon. Bird predation is probably the principal cause of death in these moths. Over the long course of evolutionary time, the moth had achieved a camouflaging coloration that blended well with the light, peppered appearance of lichen-covered tree trunks. But pollution killed the lichens and blackened the trees, making the light peppered morph highly visible and extremely vulnerable to predation. In industrialized areas, the black morph achieved a significant selective advantage because it was less easily spotted by birds.

From a graph of the frequency increase of the black form in the historical data, J.B.S. Haldane (one of the founders of population genetics) calculated its relative fitness in an industrial environment to be twice that of the more conspicuous peppered form. But a

TABLE 18.1 KETTLEWELL'S MARK-AND-RECAPTURE EXPERIMENT WITH MOTHS

	PEPPERED MORPH	BLACK MORPH
DORSET, ENGLAND UNPOLLUTED WOODLAND		
Marked and released	496	473
Recaptured after predation	62	30
Percentage recaptured	1.9%	6.3%
Relative survival	1.00	0.507
BIRMINGHAM, ENGLAND SOOT-BLACKENED WOODLAND		
Marked and released	137	447
Recaptured later	18	123
Percentage recaptured	13.1%	27.5%
Relative survival	0.477	1.00

British naturalist, H.B.D. Kettlewell, performed the crucial experiment. He released known numbers of marked black and light peppered moths in unpolluted woodlands and in polluted, soot-blackened woodlands. In each habitat, after a period of time had elapsed, he recaptured a portion of the released moths. Kettlewell's mark-and-recapture data are seen in Table 18.1.

For the first data set, released in the unpolluted woodland, almost exactly twice as many light forms survived as black forms. That's equivalent to a 100% advantage of the light type in a brief exposure to predation. In the second data set, selection was against the light peppered morph. Almost exactly twice as great a percentage of the favored black type survived.

Incidentally, England has been doing pretty well of late in its battle against air pollution. The woodlands near the cities are once again becoming covered with lichens, and the soot is disappearing. As one might predict, the black morphs of *Biston betularia* are now declining in frequency.

With only one allele for natural selection to work on, evolution can proceed quickly,

FIGURE 18.10
THE PEPPERED MOTH
In this "bird's eye view," the black and peppered morphs of *Biston betularia* are seen on the natural lichen-covered and unnatural soot-covered backgrounds. Their frequencies are drastically affected by predatory birds, and can change quickly, over few generations.

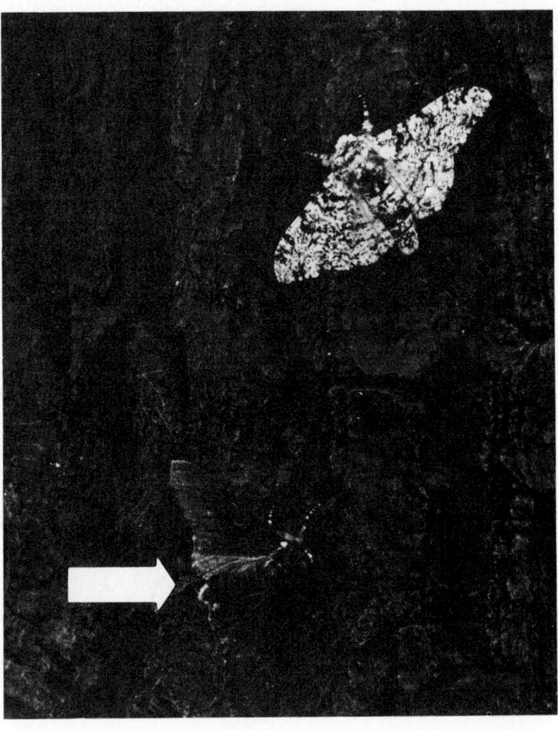

as with the peppered moth. But usually it proceeds much more slowly as natural selection operates on large numbers of alleles. Let's consider, then, how such a process might produce evolutionary changes.

How the Leopard Got Its Spots

Actually, we don't know how the leopard got its spots, but we can consider an imaginary scenario that shows how natural selection might operate on a trait controlled by many alleles and thereby mottle the cat. Assume that somewhere in the leopard's past, its ancestors were a nondescript tawny color, like lions. Then suppose one group (population) of this ancestral cat moved into an area where the ability to climb trees was important. Climbing might be important, say, to an animal that was fond of eggs, or to one that needed to climb to escape larger predators. It might also need to climb trees to cache its kills there, safe from other carnivores that couldn't climb. In any case, climbing became important, so the cats began to spend more of their time in trees. It is easy to see that once the pattern of tree climbing was established, it would become to the cat's advantage to be inconspicuous in the leafy limbs (Figure 18.11).

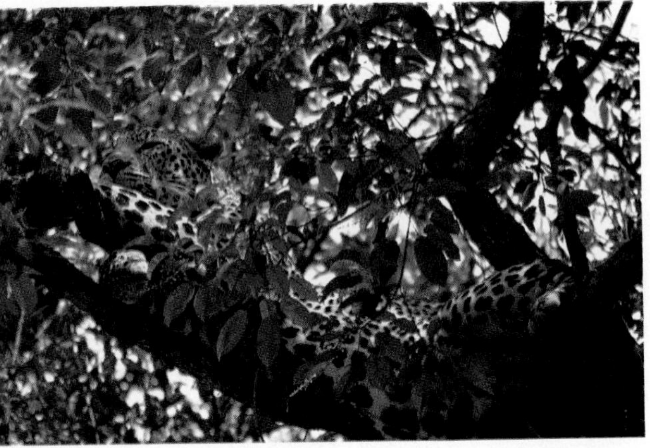

FIGURE 18.11
A LEOPARD AND ITS SPOTS
The mottled fur of the leopard may have developed gradually.

Among the offspring of each generation, some kittens might tend to be a bit more mottled than the others, and these individuals would be less conspicuous among the splotchy, sun-bathed leaves. Since they would be camouflaged from both prey and predators, they would tend to survive more frequently than their tawny siblings and to leave more spotted offspring in the population. In each generation, the more mottled individuals would survive to reproduce more frequently and so, in time, the entire population would be strongly spotted and hard to see. It is important to understand that an adaptive trait does not have to appear full-blown and completely developed before it confers some advantage to its bearer. A slight change can bestow slight benefits and then become magnified in the population through time.

Fitness

It is sometimes said that natural selection favors the fittest individuals. Actually, though, natural selection *determines* the fittest individuals. The subtle distinction arises because of the technical definition of the word *fit*. **Fitness** is a measure of one's relative contribution to the gene pool of the next generation. Thus, it relates directly to one's reproductive success. Natural selection, as we have seen, simply winnows out the genes of those individuals who are relatively unsuccessful reproducers.

Incidentally, the term itself has evolved since Darwin's time. When Darwin spoke of the "fitness" of an individual, or of selection favoring the most "fit," he meant that evolutionary processes favored the organism whose phenotype and behavior were most appropriate to its role, the one that most closely *fit* into its ecological niche or its "place in the economy of nature." Since Darwin, the idea of the "survival of the fittest" has worked its way into our language, and we now may speak of "feeling fit as a fiddle," or of "physical fitness," equating *fitness* with *health* and *vigor*. Darwin was closer to the mark. The *fittest* may be closest to the average, when the average phenotype is one that is already well adapted to the environment.

In some cases (those in which a trait is controlled by only a few, preferably two, alleles), different genotypes can be assigned different values and, therefore, relative fitness can be compared. In such a case the reproductive output of organisms with either of two traits is measured. If one is reproductively more successful, that genotype is given the number 1.00. The reproductive output, then, of the organisms with the other genotype is expressed as a percentage of that figure. For example, if the less successful reproducer leaves only 65 percent of the offspring of its more successful colleague, its relative fitness is 0.65. The difference between these two fitness values (here 0.35) is called the **selection coefficient.** Selection coefficients can generally be regarded as only rough estimates of comparative reproductive success.

EVIDENCE OF EVOLUTION

How good is the evidence supporting the theory of evolution? How well has Darwin's idea held up? Pretty well, it turns out. Let's consider some of the most powerful supporting evidence for the theory, drawing specifically from the fossil record, biogeography, comparative anatomy, biochemistry, and genetic similarity.

The Fossil Record

Darwin is best known as a biologist, but he was also one of the foremost geologists of his time. His interest in rocks and strata and his enthusiasm for digging fossils contributed powerfully to his development of evolutionary theory. In his "geologizing" in South America, he uncovered the fossil remains of an extinct giant armadillo (Figure 18.12). The remains were found deep in ancient rocks, deposited long ago. Above were smaller armadillos of a different sort, alive and scurrying over the stony graves of the giants. The extinct species, Darwin realized, must have been ancestral to the living species.

Many fossils have been discovered since the time of Charles Darwin, and the known record is now immense. Fortunately for an aging Darwin, one of the most amazing records was uncovered during his lifetime. In 1879, Yale University paleontologist O.C. Marsh published a comprehensive study on the evolution of the modern horse, *Equus* (Essay 18.1). These findings lent considerable credibility to Darwin's theory. Thomas Huxley, master debater and Darwin's greatest defender, put these findings to good use in arguing the case of evolution through natural selection.

Unfortunately, few lines of fossil evidence are as complete as that of the horse. This is because of the problems associated with the fossilizing processes, and with reading the record. Fossils generally form from hardened body parts, such as bone, shells and exoskeletons, and woody plant parts, so softer-bodied organisms and softer body parts are largely missing from the record. Paleontologists suggest that two-thirds of the forms that ever lived did not form fossils. Even those species with bones, teeth, or shells—parts that would permit fossilization—had to die in a very special place for those remains to be preserved. Their corpses generally had to lie in mud or silt or be embedded in resin. Because of such problems, our understanding of the variety of life through time is limited.

Even those that did fossilize have often left us scant or confusing evidence. In many cases the fossils were simply destroyed by erosion or pressure, while others were scattered about or altered by moving waters or shifting earth. (The best fossils come from quiet

FIGURE 18.12
THE EXTINCT GIANT ARMADILLO
Fossils such as *Glyptotherium*, the giant armadillo, suggested to Darwin that the usual nineteenth-century explanations of life were not enough.

ESSAY 18.1

THE SAGA OF THE HORSE

Tracing the origin of the horse has taken paleontologists back some 54 million years to the early Eocene epoch of the Cenozoic era (see the geological table, inside front cover). In that distant time lived *Hyracotherium*, a timid, doglike woodland creature that browsed on the soft parts of plants and literally tiptoed around the forest. (Like the dog, it had multiple toes and footpads.) From this decidedly unhorselike creature arose a succession of forms that gradually took on the appearance of a respectable horse. The changes represented a continuing adaptation to a new source of energy, grass. Grasses first appeared in the Eocene epoch and began their gradual spread that culminated in today's vast grasslands.

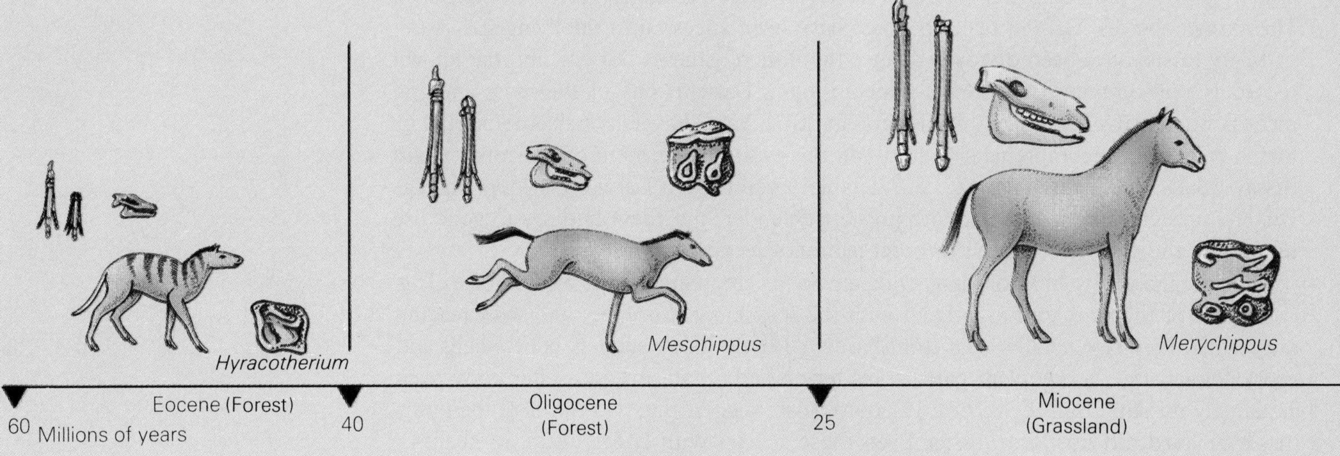

waters in geologically stable areas.) The fossil record has indeed told us a great deal about the life that preceded us, but it cannot tell us the whole story.

The fossil record has provided fascinating indirect evidence of just what kinds of selective pressures might have been operating on ancient populations. For example, in some reptile species, and in some island situations, evolution has tended to produce increasingly larger individuals. According to Steven Stanley of Johns Hopkins University, such a trend could be produced by natural selection operating on species, just as it does on individuals.

The idea is that, as new species arise from existing strains, those that survive the longest and generate the most new species will have the greatest impact on evolutionary trends. Stanley calls the process **species selection.** Notice in Figure 18.13 that evolution in such a case does not proceed linearly, one species giving rise to the next. Instead, each species branches, giving rise to new species even as each continues its own existence.

As new species appeared, they could be larger or smaller than the parent species, but if larger lines continued longer than the smaller lines, the larger species would have more opportunity to produce new species. The result would be a marked trend toward larger size in a number of related species, as we see in Figure 18.14.

Adapting to the grasslands required many changes in the horse's anatomy, not the least of which was in dentition (teeth). Grasses are highly abrasive—feeding requires large incisors for nipping and molars with broad, hardened surfaces for chewing. Of course, as the primitive horses left the forests, predators followed. Because concealment was more difficult in the open spaces, natural selection favored "early warning systems" in the form of long necks and better vision and hearing. It also favored the ability to respond with great speed; thus, we see major changes in the lower legs and feet, particularly the single toe that would become the hoof.

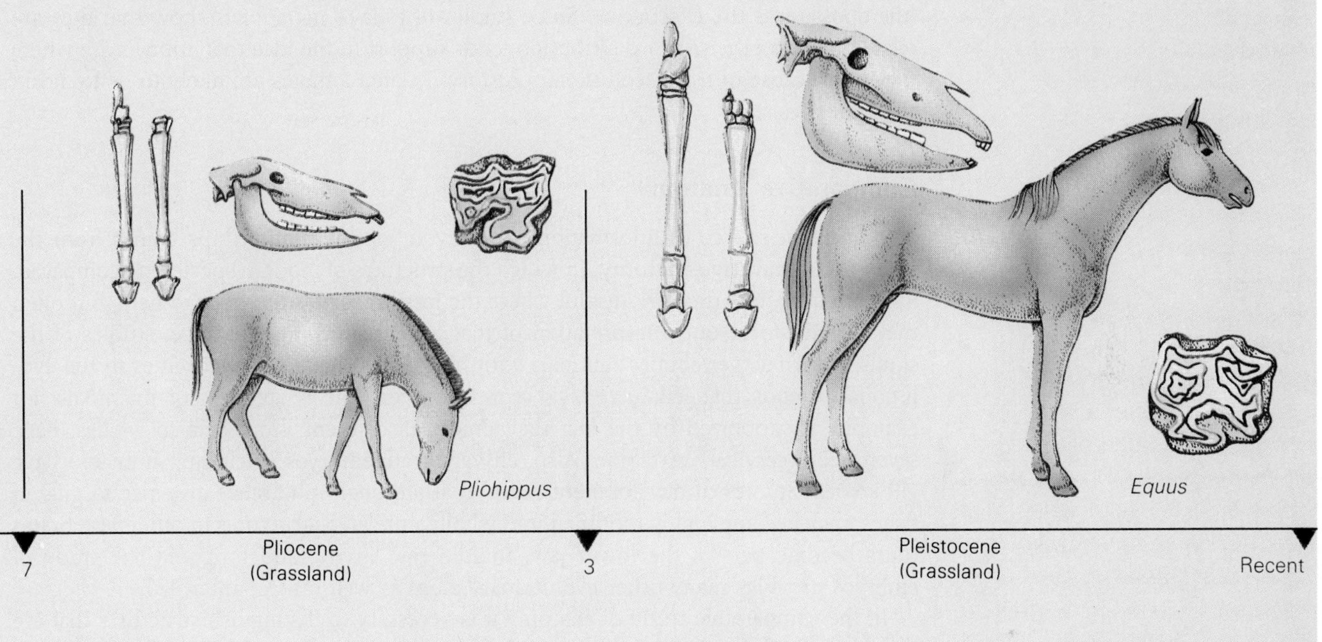

FIGURE 18.13
SPECIES SELECTION
New species arise as branches from older stems. Those that survive longest and give rise to the most branches have the greatest impact on later generations.

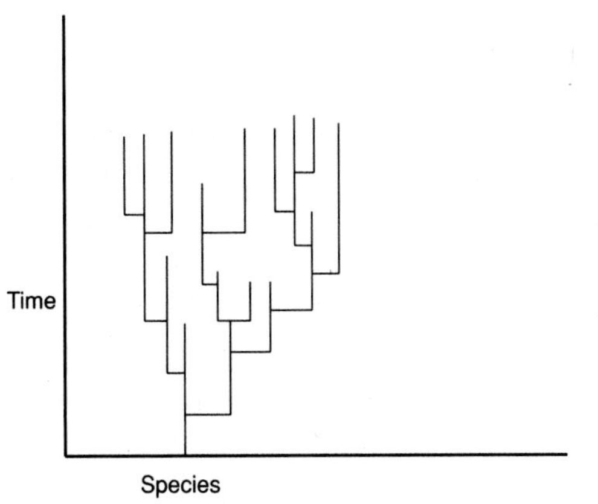

FIGURE 18.14
SPECIES SELECTION FOR LARGENESS
Here, some species selection has resulted in a range of animals generally larger than their evolutionary forebears. Notice that each line gives rise to both larger and smaller lines, but the larger lines survive longer and give rise to more new species.

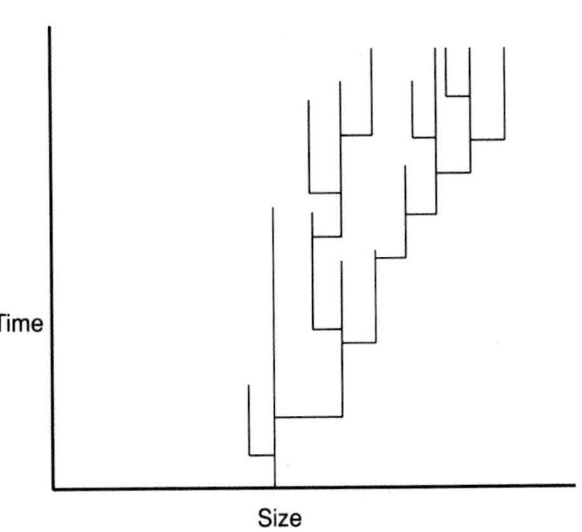

Biogeography

The study of the distribution of living things over the earth—**biogeography**—offers another line of supporting evidence for evolution. Why does each geographically isolated region have its own peculiar assemblage of plants and animals? Recall from Chapter 1 that Darwin noticed that the mara, or Patagonian hare, exists where one would expect to find rabbits, and that it has, in fact, many rabbitlike characteristics. South America, Darwin deduced, doesn't have rabbits because rabbits can't get there. Instead, another kind of animal has evolved, occupying the rabbit's niche.

Darwin also noticed that remote oceanic islands have few mammals except those that travel well over water, such as bats. With the exception of bats and introduced animals, Australia, an isolated island continent, is populated only by primitive marsupial mammals (pouched mammals that lack a true placenta) and by the even more primitive monotremes (duck-billed platypus and spiney anteater). Marsupials are rare over the rest of the earth (the opossum is the exception). Since studies of today's marsupials show that they are related to each other, their distribution lends support to the idea that animals are where they are because of their evolutionary history. Related animals are likely to be found in the same area.

Comparative Anatomy

An important source of information about evolutionary relationships comes from the study of **comparative anatomy,** in which the structures of modern species are compared. Such studies are particularly useful where the fossil record is too poor to help, but often they just reinforce our interpretation of that fossil record. For instance, studies of the skulls of various vertebrates (animals with backbones) have provided clues to the evolutionary trends that led to today's representatives. The relatedness of mammals, for example, is supported by the fact that almost all of them, from bats to whales, have seven neck (cervical) vertebrae. Also, all vertebrate embryos (including humans) have gill arches. Studies of development reveal that although gill arches give rise to gills in fishes, they become highly modified into wholly unrelated structures in other vertebrates (ours become part of the lower jaw, middle ear, larynx, and tongue). The study of embryos provides many other evolutionary clues as well (see Chapter 43).

In the comparative study of anatomy it is necessary to distinguish structures that are **homologous** from those that are **analogous.** Homologous structures are those with common origins in the embryo, indicating common evolutionary origin. Actually, though, they may look quite different and may even have taken on different functions. Analogous structures are those that have similar functions and often appear similar, but form differently in the embryo, so they are presumed to have different evolutionary origins. The wings of insects and birds are often cited as examples of analogous structures. What inferences can you make from a consideration of the forelimbs of the cat, whale, bat, bird, and horse, and the upper arm of a human (Figure 18.15)? A study of the embryos of these vertebrates reveals that, in each case, the limbs emerge from similarly-formed limb buds.

A closer look at the whale's anatomy and that of the python presents another slant to the idea of homology. Both lack hind limbs, but tucked away in the body are useless vestigial bones that are clearly remnants of what was once a functional pelvic (hip) girdle. There are even vestigial hindleg bones. Both animals are apparently descended from four-legged land dwellers. Vestigial organs are looked upon as rare but welcome records of evolutionary events.

Comparative Biochemistry of Amino Acids

You may recall that cytochrome *c* is an iron-containing protein that acts as a carrier in the electron transport chain of cellular respiration. It is found in all vertebrates and in many invertebrates, including insects. The cytochrome *c* molecule, however, varies slightly among the major groups of animals. In fact, a comparison of the amino acid sequence of cytochrome *c* from any mammal or reptile reveals 15 differences along the

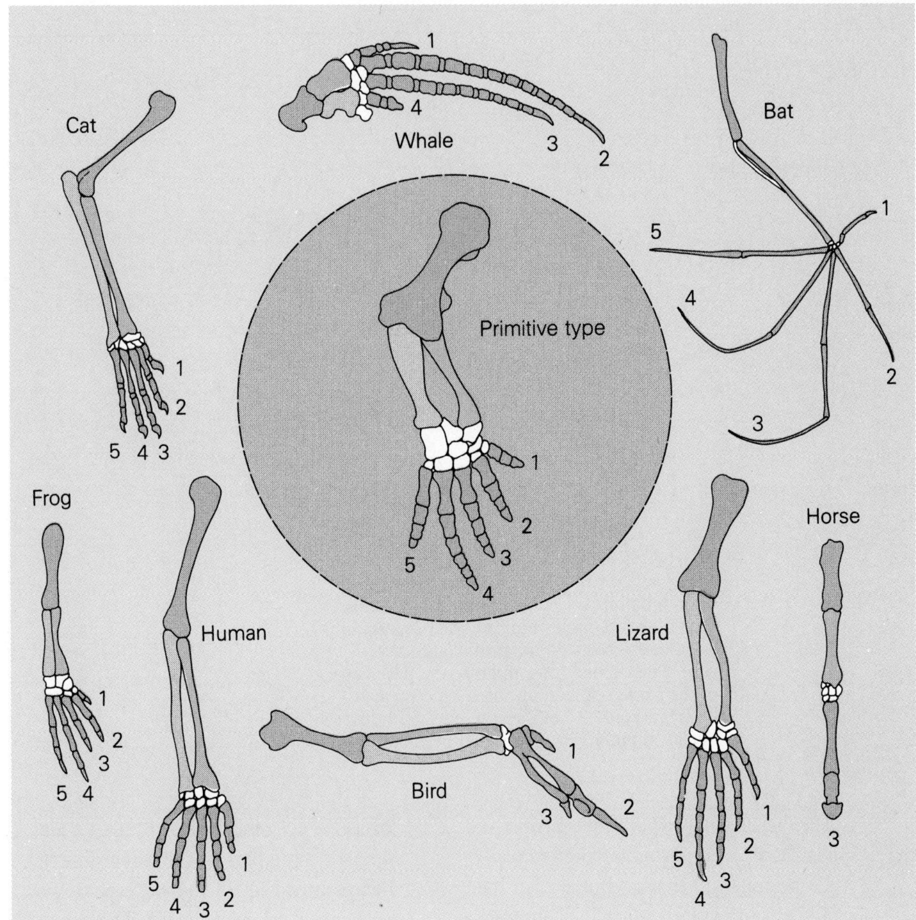

FIGURE 18.15
HOMOLOGOUS STRUCTURES
The forelimbs of several representative vertebrates are compared here with those of the suggested primitive ancestral type. Since they all have the same embryological origin, they are said to be homologous. The most dramatic changes can be seen in the horse, bird, and whale. In these animals, many individual bones have become smaller, and some have even fused. Another interesting modification is seen in the greatly extended finger bones of the bat.

sequence of the molecules. Since mammals and reptiles diverged about 265 million years ago (in the Paleozoic era), the 15 differences in amino acids amount to roughly one mutation per 17 million years (Figure 18.16). When the amino acid sequence of cytochrome *c* is compared for other species, again it works out that there is roughly one change for every 17 million years of evolution. The very slow and steady rate of mutation in amino acid sequences, then, can be used as a kind of molecular clock to help determine just how long ago two lines may have diverged.

FIGURE 18.16
CYTOCHROME *C* PHYLOGENY
This computer-generated phylogenetic tree represents amino acid differences in the polypeptide sequence of respiratory pigment cytochrome *c*, which is found in all aerobic organisms. Each amino acid difference represents at least one DNA mutation that has been "accepted" in evolution. The estimated number of DNA changes (numbers on the tree) determine the evolutionary distance from one organism to another. In general, the tree derived from one short protein is consistent with other trees based on morphology or the fossil record, although the computer seems to think that the turtle is a bird.

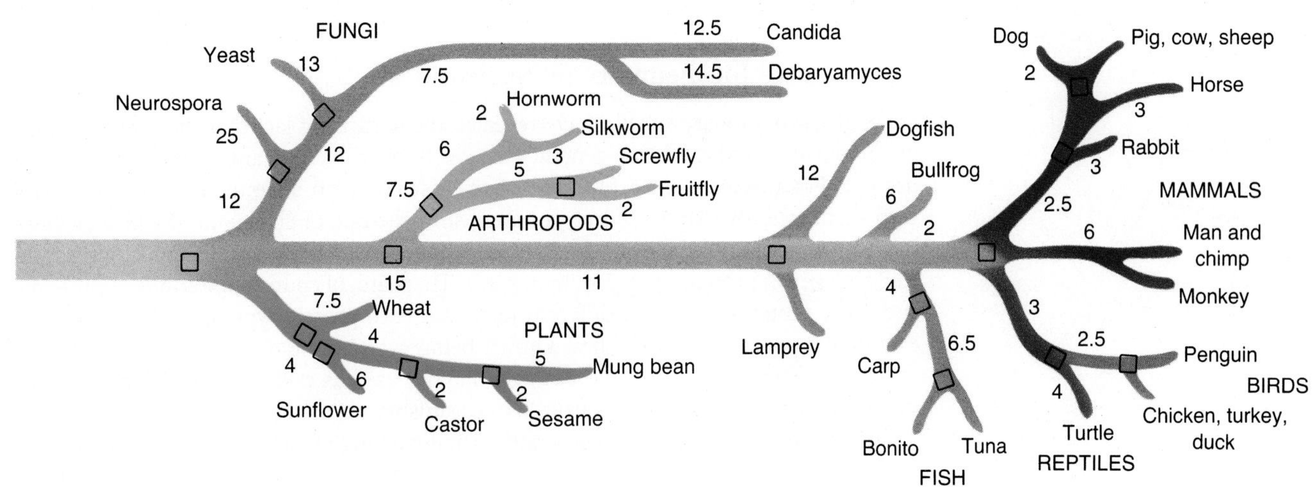

ESSAY 18.2

SELECTION FOR GOING UP OR DOWN

The accompanying graph shows response to selection, in two directions, for *geotaxis* in *Drosophila*. Positive geotaxis is the tendency to move up. Jerry Hirsch invented an ingenious maze that enabled experimenters to select flies for geotaxis (see accompanying illustration). The maze consists of a series of interconnected chambers. As the fly enters each chamber, it finds only two exits: one straight up and one straight down. To get from one end of the maze to the other, the fly must make an up-down choice 14 times. At the end of the maze are 15 collection vials with fly food in them. The flies that end up in the top vial have to have made the "up" choice 14 times in a row; the flies that end up in the bottom vial have to have made the "down" choice 14 times in a row. The flies that go up seven times and down seven wind up in the middle vial (vial 8). The experimenter can simply put a population consisting of hundreds of flies in one end of the maze and collect the high scorers and low scorers from the top and bottom vials at the opposite end of the maze. In a selection experiment, only the high-scoring flies are kept in the "up" line, and only the low-scoring flies are kept in the "down" line.

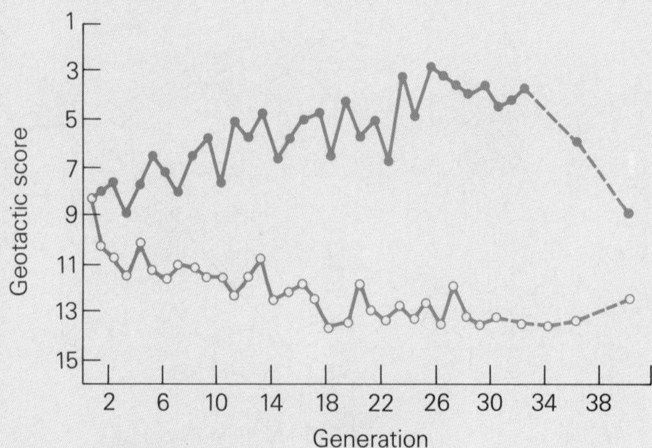

(a) The results of selection for geotaxis in *Drosophila*. Open circles indicate the progress of populations selected for negative geotaxis (down); solid circles indicate the progress of populations selected for positive geotaxes (up). The dashed lines indicate the progress of the same populations after selection had been "relaxed." Note that the up line, in particular, rapidly lost its selected behavior.

In the original experimental runs, the flies showed no particular preference—as a population, that is. But within the original population, some flies showed a very slight preference for going up and some showed an equally slight preference for going down. This behavior is a polygenic trait; there are many gene loci that are variable for alternate up and down alleles in the *behavioral phenotype*. A selection response in both directions continued for 12 to 16 generations. By that time, almost all the flies in the down line always ended up in the extreme down vials (numbers 14 and 15). The average of the up line was around vial number 4, which corresponds to three down choices and 11 up choices.

As usual for selection experiments, the fitness of the extreme up and extreme down lines declined markedly. While it is normal for a fly population to have a mix of both kinds of alleles, it is not normal to accumulate too many

Comparative Biochemistry of Nucleotides

Molecular evolutionary studies have revealed the surprising fact that the rate of change in proteins (an indication of genetic change) is roughly constant in different lines of descent. For instance, since the time of the last common ancestor of a human and a carp—an ancestor that was surely a bony fish—the rate of change in globin molecules has been the same in the carp line of descent and in the human line, about one amino acid change per protein every 7 million years. There are, of course, thousands of proteins, and other proteins change at different rates, but each protein type has its characteristic evolutionary pace (one of the best known being cytochrome *c*).

The nucleotide sequence of DNA also may be used as a kind of clock. In this case, single-stranded DNA molecules from two organisms are mixed together and allowed to "hybridize" (base-pair) along their length. The matching of base pairs along two lengths

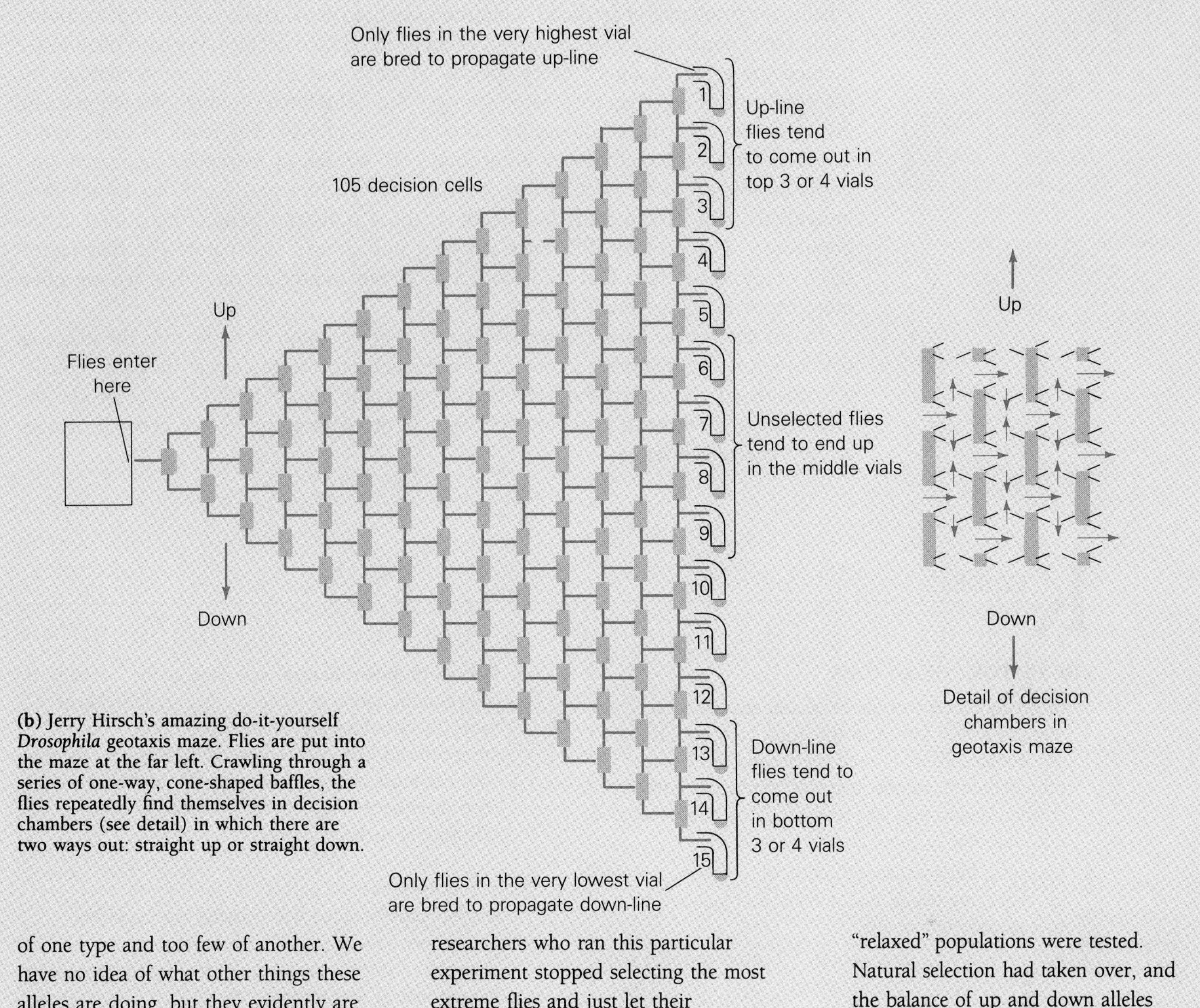
Only flies in the very highest vial
are bred to propagate up-line
Up-line
flies tend
to come out in
top 3 or 4 vials
105 decision cells
Up
Flies enter
here
Unselected flies
tend to end up
in the middle vials
Down
Down-line
flies tend to
come out
in bottom
3 or 4 vials
Only flies in the very lowest vial
are bred to propagate down-line
1
2
3
4
5
6
7
8
9
10
11
12
13
14
15
Up
Down
Detail of decision
chambers in
geotaxis maze

(b) Jerry Hirsch's amazing do-it-yourself *Drosophila* geotaxis maze. Flies are put into the maze at the far left. Crawling through a series of one-way, cone-shaped baffles, the flies repeatedly find themselves in decision chambers (see detail) in which there are two ways out: straight up or straight down.

of one type and too few of another. We have no idea of what other things these alleles are doing, but they evidently are harmful in the wrong balance and combinations. After 20 generations, the researchers who ran this particular experiment stopped selecting the most extreme flies and just let their populations mate and reproduce as they pleased. Every few generations the "relaxed" populations were tested. Natural selection had taken over, and the balance of up and down alleles returned toward normal in both populations.

of DNA is taken as an indicator of genetic similarity and, hence, relatedness. Interestingly, morphologically similar forms may show marked differences in DNA, suggesting that they have taken separate evolutionary directions earlier than appearance alone would suggest.

It may have occurred to you that a slow, gradual accumulation of changes in either amino acid or nucleotide sequence is not compatible with the image of natural selection relentlessly wheedling the gene pool and causing changes that result in adaptation. This is, in fact, an argument used by neutral selectionists (Chapter 19), who maintain that most mutations do not result in adaptation, that most mutations simply accumulate and are neither beneficial nor harmful. Some biologists doubt that this is the case and, furthermore, argue that molecular clocks cannot be used to date the origins of taxa. Most biologists are willing, though, to use molecular clocks to determine the *sequence* of branches in an evolutionary tree. Such information is, therefore, quite useful to certain evolutionary theorists, such as the cladists described in Chapter 20.

Artificial Selection

Finally, the principles of **artificial selection** have been as valuable to us in understanding natural selection as they were to Darwin when he developed the idea. We have relentlessly molded the genes of almost every species we have had an interest in, sometimes for practical reasons (as when we created stronger and faster horses), sometimes whimsically (as we developed hairless, mouselike dogs or wrinkled rats). The result of some of our selection has occasionally been unfortunate, as we see in extremely aggressive and unpredictable pit bulls. Nonetheless, we have proven time and again that by selecting individuals with certain traits for breeding, those traits can become magnified in the population. The primary difference between our choices and nature's is that nature selects only those traits that lead toward successful reproduction, while we are often more capricious. (See Essay 18.2.)

In this chapter we have followed the concept of evolution from the time the idea was developed to the theory's acceptance and integration into biological thought. Finally, we considered the modern evidence that supports the theory. Next we will consider the development of the earth's far-flung species in terms of the accumulation of small changes at the molecular level.

KEY IDEAS

THE HISTORY OF AN IDEA

1. In Aristotle's view of life he placed animals in a "natural scale," with man (the most complex) at the top. He and Plato believed species did not change (immutability), an idea that was compatible with creation dogma and was held firmly by intellectuals until the time of Lamarck and Darwin.
2. Linnaeus, the Swedish naturalist who devised a system for classifying plants and animals, also believed that species were immutable.
3. Buffon held with immutability but proposed that some lesser organisms arose independently of creation and went through change.
4. Lamarck, a strong proponent of evolution, cited the fossil record as evidence of change, including change in humans. As a mechanism of change he proposed that constant use and improvement of a structure led to that structure's improvement, and that such improvements were then inherited. This notion became known as "the law of use and disuse."

Enter Charles Darwin

By Darwin's time the intellectual climate was right for serious consideration of the idea of "descent with modification," or evolution.

DARWIN'S THEORY

1. Darwin's thinking on evolution was partly a product of his knowledge of animal breeding, wherein the traits to be retained are determined by the breeder. He began to consider whether such selection could occur in other ways. The writings of Thomas Malthus, who maintained that populations tended to increase faster than the food supply, provided some insight.
2. Darwin proposed **natural selection** as the mechanism of evolution, citing as evidence that (1) individuals vary, (2) variations are inherited, (3) more offspring are produced than can survive, and (4) those offspring with the most adaptive variations survive and reproduce better than others, thus leaving more offspring of their own type.

Gathering Evidence

1. Darwin experimented with various aspects of his theory, particularly on island colonization and the dispersal of organisms. He published several works on natural history and geology.
2. Darwin's intention was to write an exhaustive review of natural selection and evolution, but because A.R. Wallace had reached similar conclusions, he presented his ideas much sooner and published them in a one-volume book, *On The Origin of Species.*

The Origin of Species

In his book *On the Origin of Species,* Darwin built his case by first discussing genetic variability in domestic plants and animals, then going on to variability in wild species, and then to the subject of overproduction of offspring. These subjects led to his presentation of natural selection.

The New Synthesis

1. The weakest part of Darwin's theory was his inability to explain how genetic variation arose and how it was maintained in populations, both of which were poorly understood in his time. The relationships between genes and natural selection were not worked out until the 1930s.

2. The "New Synthesis" incorporates the principles of genetics into the theory of evolution.

SOME WAYS LIFE CAN VARY

Variation includes behavioral and biochemical traits as well as the common visible traits. Such variation must be genetically based to be significant to evolution.

Sources of Variation

1. The two principal sources of variation are mutation and genetic recombination. For a mutation to be of evolutionary significance, it must be present in the gametes.
2. Until many accumulate, most point mutations (single base changes) are less likely to affect phenotype. Chromosomal mutations are massive and can drastically alter phenotype. Both are usually harmful.
3. Although mutation occurs at a constant rate, most diversity in sexual organisms arises from recombination in existing genes. In haploid, asexual organisms, all mutations affect the phenotype.

Maintaining Genetic Diversity

1. Genetic diversity is maintained in several ways.
2. Recessive alleles are more likely to be retained because their expression is masked by the dominant homologous alleles.
3. Genetic variation is also maintained by **balancing selection,** which occurs in spotty, variable environments where first one allele and then its alternatives are favored. The resulting phenotype is called a **balanced polymorphism.**
4. The frequency of recessives alleles is also maintained through **heterozygote advantage**, wherein individuals heterozygous for a certain gene survive better than the homozygous dominant and the recessive.
5. Diversity is maintained by **frequency-dependent selection,** in which a particular combination of alleles offers a definite advantage, but when the numbers increase they are selected against by other factors.

NATURAL SELECTION

1. Natural selection is defined as selection by the environment for genotypes that best adapt an organism to present conditions. The test of any genotype is its effect on reproductive output.
2. **Evolution** is the change in allele ratios over time.

The Peppered Moth: Natural Selection in a Natural Population

1. A classic case of natural selection is seen in *Biston betularia,* the **peppered moth.** A black variant, or **morph,** is rare in rural areas, but predominant in industrial regions, where it blends in with soot-covered trees. The peppered morph is much more common in the clean forests, where it blends in with lichen-covered trees.
2. Release and recapture studies of both moths in both environments clearly revealed that background coloration and predation were the factors affecting survival. In more recent times, air pollution control has meant a decrease in the black moth in industrial regions.

How the Leopard Got Its Spots

Spots in the leopard's coat, like other such traits, may begin as minor variations that offer modest advantages when accompanying other changes, such as in behavior (tree climbing). Then, through continued selection, such traits are favored more and more, until they become a prominent aspect of the phenotype.

Fitness

Natural selection determines fitness—an individual's contribution to the gene pool relative to that of other individuals. In determining relative fitness, the fittest individual (the one leaving the most offspring) is assigned a value of 1.00, and values less than 1.00 are assigned to those less fit. The difference between the two is the **selection coefficient**, a measure of reproductive success.

EVIDENCE OF EVOLUTION

Natural selection, although modified in detail, remains an essentially valid and respectable theory today.

The Fossil Record

1. The fossil record, such as that of the modern horse, provides important information about evolutionary change. Much of the fossil record is far less complete than that of the horse. Soft-bodied creatures do not fossilize well, and geological conditions must be just right for the formation and maintenance of fossils.
2. The record shows evidence of **species selection,** in which species that exist longer give rise to more lines. Such species are seen as side-branches in the trees.

Biogeography

The distribution of plants and animals provides evidence of past evolutionary activity. Where a range is continuous, although extending over great distances, there are strong similarities among the related inhabitants. Where geographical isolation has occurred, we find greater differences among related species, the degree of difference depending on the time in isolation. In those regions long isolated from others, plants and animals may be quite unique. South American mammals differ considerably from those of North America, but even greater differences are seen between Australian mammals and mammals in other parts of the world. Exceptions include organisms such as birds, which travel over vast distances, or those plants (and their seeds) that can be easily transported by animals, wind, or water.

Comparative Anatomy

Evidence of evolution is seen in anatomical similarities among adult animals, particularly in the vertebrates. **Homologous** organs, those that are of similar embryological origin, indicate evolutionary relatedness. Those that appear similar, or function similarly, but are not of common embryological origin are **analogous** and do not imply close evolutionary relatedness.

Comparative Biochemistry of Amino Acids

Differences in the sequence of amino acids in cytochrome *c*, when analyzed statistically, support older evolutionary schemes. Analysis further indicates that mutations occur at a regular rate over time.

Comparative Biochemistry of Nucleotides

The degree to which DNA nucleotides from different species match up is an indicator of evolutionary relatedness. However, in some instances, species believed to be closely related for other reasons sometimes exhibit large differences in their nucleotide sequences. This suggests to some that many nucleotide changes may be neutral, that is, have no effect on phenotype, thus escaping selection altogether.

Artificial Selection

Artificial selection also provides evidence for natural selection. Great differences from ancestral stock are seen in animals produced in deliberate breeding programs where certain traits are selected for by the breeder.

REVIEW QUESTIONS

1. Discuss immutability or "fixity" of species. Where did this idea originate? Why did it persist so long? (372)
2. What departure from the doctrine of immutability did Buffon propose? Was he safe in making such a proposal? Explain. (373)
3. What was Lamarck's view on the fixity of species? In what important way did his ideas differ from those of Charles Darwin? (373)
4. Darwin's theory is often referred to as "an idea whose time had come." What does this mean in terms of the intellectual climate at the time? (374)
5. Discuss three aspects of Darwin's background that influenced his ideas on evolution. (375-376)
6. List the four principal observations Darwin used to support natural selection. (376)
7. Darwin was able to experiment with certain aspects of evolutionary theory. Describe one such experiment. (376)
8. What prompted Darwin to publish his ideas long before he had originally intended? What did he have to offer that Wallace did not have? (377-378)
9. What was the greatest weakness in Darwin's thesis? Why was this important, and what was he able to do about it? (378)
10. When did the "New Synthesis" begin? What is it about? (378-379)
11. Why must variation be genetically based to be significant? Relate this to a major fault in the evolutionary mechanism proposed by Lamarck. (377)
12. Essentially, how do most mutations affect the individual? Describe the two categories of mutation and compare their effects. (380)
13. What is the source of most genetic diversity in sexual organisms? Why is mutation so significant to asexual organisms? (general)
14. How does a highly varied environment affect genetic diversity? What is the selective response called? (381)
15. Explain heterozygote advantage and provide an example. (381)
16. Using the text example, explain how frequency-dependent selection maintains genetic diversity. (381)
17. What is the ultimate test of any variation?
18. Define evolution in terms of allele ratios. What does this mean? (381)
19. Briefly summarize the "peppered moth" episode. What would you expect if the British had not the output of industrial smoke? (382-383)
20. Explain the meaning of "fitness". How is the selection coefficient determined? (384-385)
21. In what way is the fossil record the most significant evidence of evolution? Why would one expect the record to be incomplete? (385-386)
22. According to the idea of species selection, what should an evolutionary tree look like? (386)
23. In what way is the unique assemblage of animal life in Australia evidence for evolution? What does its uniqueness suggest about the length of time Australia was geographically isolated? (388)
24. With the uniqueness of Australian animals in mind, predict what you might find if you were to compare hooved mammals from northwestern North America with those from Northeastern Asia. Explain your reasoning. (388)
25. Distinguish between homology and analogy and explain what each means in terms of evolutionary relatedness. (388)
26. What is the obvious difference between artificial and natural selection? In what way can the products of artificial selection be used as evidence for natural selection? (392)

19

MICRO-EVOLUTION AND CHANGING ALLELES

THE THEORY OF EVOLUTION IS ONE OF THE MOST PERVASIVE AND explanatory themes in modern biology. As an intellectual fulcrum it has been used to pry loose countless gems from the complex matrix of life. In fact, it is so useful that one wonders how biology could even have been done without a clear understanding of its principles. Of course, as we know, much biology was done without it, but as we also know, much of it was wrong. So as we continue to look at the venerable old idea, we find that some parts of it have weathered, aged, hardened, and cured, while other parts have been changed, and new parts—parts that Darwin could never have imagined—have been added as the concept of evolution has itself evolved.

We begin by reminding ourselves that most evolution proceeds by natural selection acting on genetically based variation. Variation arises simply with the appearance of individuals that differ from other members of the species. Some of these differences will inevitably confer reproductive advantages to some individuals over others. The descendants of those with such advantages can be expected to increase in the population. In a sense, then, nature selects the most successful reproducers, just as a farmer selects the best layers to produce the next generation of chickens. By such means, a population changes through time, and this, we know, is the essential idea of evolution.

ALLELES AND ALLELE FREQUENCIES

It is important to remember that *individuals do not evolve*—at least not in the sense that we will consider evolution here. Instead, *populations evolve*. Such evolution is evidenced by changes in the **gene pool**, which includes all the genes of any **population** at any given time. More precisely, evolution involves changes in allele frequencies. In the last chapter we mentioned that **frequency** referred to the proportion of a certain allele in a population. Now, we can be more precise. Frequency, we are reminded, has nothing to do with how frequently something happens, or how often something occurs in time, such as when we refer, say, to the frequency of tornadoes in the spring. Instead, as we saw, frequency is a *proportion,* namely a proportion of items of a particular kind in a more general class. For example, if there are 10,000 registered voters in town, and 4300 are Republicans, the *frequency* of Republicans in this town is 4300/10,000 = 0.43. Frequencies are always numbers between 0 and 1 because they are always a fraction consisting of a part divided by the whole. Both the numerator and the denominator are counts of individual items, and any individual that appears in the numerator must also appear in the denominator. For instance, the 10,000 registered voters include the 4300 registered Republicans. **Allele frequency**, then, refers to the proportion of a given allele in a population. As allele frequencies change, evolution proceeds.

Now, as the frequency of one allele increases, the frequency of an alternate allele decreases. Such a change can spread—perhaps slowly, perhaps rapidly—through a local population or an entire species, resulting in evolution. Evolution, by the way, is not always the result of natural selection. It can also proceed randomly as this or that gene accumulates by mere chance. In this chapter, though, we will focus on the effects of natural selection on evolution. Specifically, we will first ask how allele frequencies can remain stable, and then how they can change.

THE HARDY-WEINBERG LAW

Godfrey Hardy, an eminent mathematician, had few professional interests in common with Reginald Punnett, the young Mendelian geneticist, but they frequently met for lunch and tea at the faculty club of Cambridge University. One day in 1908, Punnett was telling his colleague about a small problem in genetics. Rumor had it that Gudny Yule, a strong critic of the Mendelians, had said that if the allele for short fingers was dominant and the allele for normal fingers was recessive, then short fingers ought to become more and more common each generation. Within a few generations, Yule thought that no one in Britain should have normal fingers at all. Punnett didn't think this argument was correct, but he couldn't explain why.

Hardy said he thought the problem was simple enough and wrote a few equations on his napkin. He showed that, given any frequency of alleles for normal fingers and alleles for short fingers in a population, the relative numbers of people with normal fingers and people with short fingers ought to stay the same for generation after generation as long as there was no natural selection involved. Today, we say that the population is in **genetic equilibrium** for the gene.

Punnett was excited and wanted to have the idea published (on something besides a napkin) as soon as possible, but Hardy was reluctant. The idea was so simple and obvious, he felt, that he didn't want to have his name associated with it. But Punnett prevailed, the equation was published, and the relationship between genotypes and phenotypes in populations quickly became known as *Hardy's law*. Hardy, who was indeed one of the great mathematical minds of the day, is now known almost solely for this modest contribution. Yule, incidentally, denied ever having said that dominant traits should increase from generation to generation, so this little incident in the history of science was based on a misunderstanding from the outset.

In Germany, within weeks of the publication of Hardy's short paper, the same law was described by the German physician Wilhelm Weinberg. Eventually, the formula became known as the **Hardy-Weinberg law.** (Later, some chose to call it the Castle-Hardy-Weinberg law, in recognition of the belated discovery that an American, William Castle, had published a neglected exposition of the same observation in 1903.) In any case, the Hardy-Weinberg law is the starting point for studying the population genetics of diploid sexual species. The law, restated, is that *both gene and genotype frequencies will remain unchanged—in equilibrium—unless outside forces change those frequencies.*

How the Hardy-Weinberg Law Works

To approach such questions as Britain's "finger-length problem," we must begin by considering how alleles behave in populations. We have been using the term population, and you may have taken it to mean a group of individuals. You were not wrong, but at this point we can give it a more precise meaning. In biology, a population is a group of interbreeding or potentially interbreeding individuals. With this in mind, let's consider the frequencies (or ratios) of alleles that can influence a specific characteristic in a population.

Imagine a population of only two individuals in which the male is homozygous for dominant trait **A** and the female is homozygous for recessive trait **a**. We know that all their F_1 offspring will be heterozygous **(Aa)** for that characteristic. Next, assume that the F_1 individuals mate and produce an F_2 generation. A Punnett square shows us that in the F_2 generation, three out of four individuals will show the dominant trait, and only one will show the recessive trait (Figure 19.1). It might appear, then, that we are on the way to eliminating the recessive allele from the population, since the frequency of the trait has decreased from ½ to ¼. However, if we plot the succeeding generations (F_3, F_4, F_5, and so on), we will find that the proportion of dominant and recessive alleles in the population has not changed at all. In fact, if we think again of the F_1 and F_2 generations, we see that there is a 1:1 ratio of the two alleles even at these stages:

F_1 All Aa (Ratio of A to a is 1:1)
F_2 ¼ AA, ½ Aa, ¼ aa } (Ratio of A to a is 1:1)

A population of two individuals is unrealistic, but it serves to point out how the phenomenon occurs in larger populations. We can show by means of Punnett squares that the frequency of alleles for any characteristics will remain unchanged in a population through any number of generations—unless this frequency is altered by some outside influence such as natural selection. Thus, according to the Hardy-Weinberg law, if there is no natural selection, mutation, or any other force that changes allele frequencies in populations—and if random mating is permitted—the frequencies of each genotype will remain constant through the following generations.

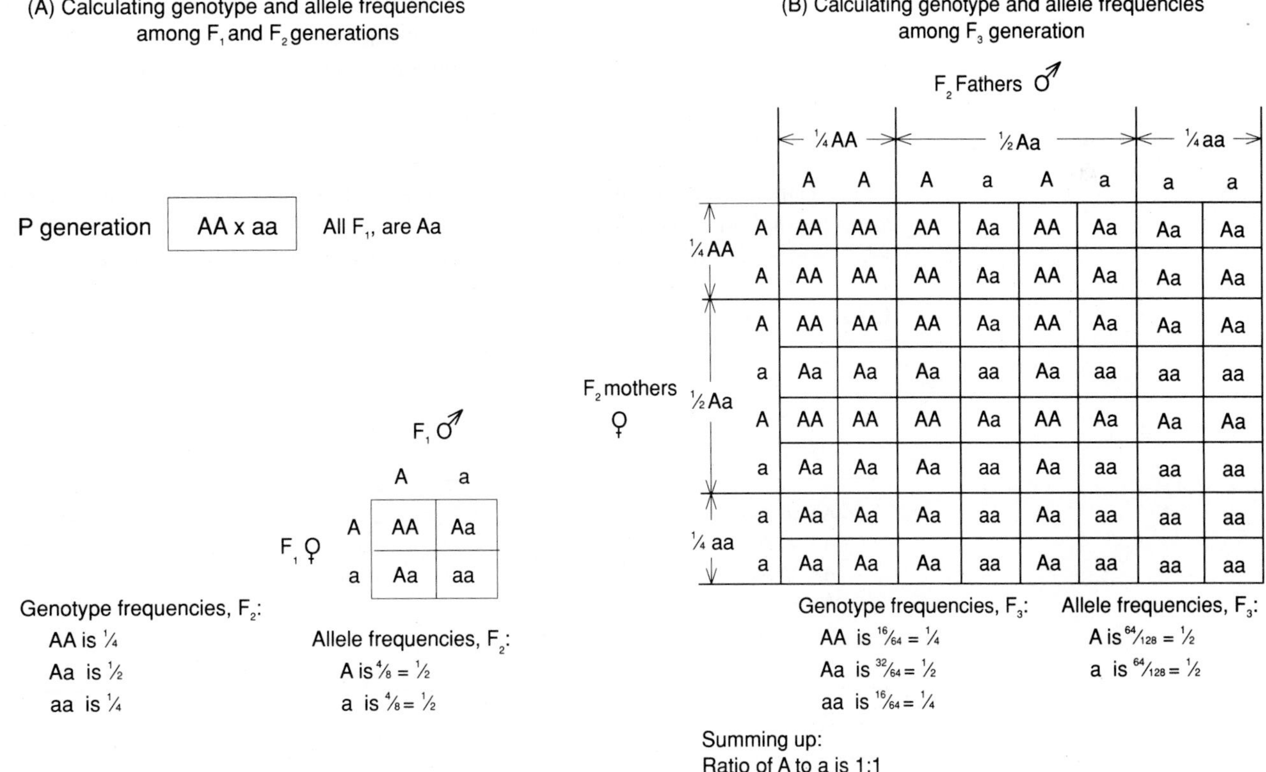

FIGURE 19.1
TRACKING THE ELUSIVE RECESSIVE ALLELE
(a) Following an F_1 cross, a look at phenotypes in the F_2 generation may suggest that the recessive allele is diminishing. However, a count of alleles proves that the frequency is as it was in the *P* and F_1. **(b)** Carrying out all of the possible crosses among F_2 individuals leads to an F_3. A simple summing of the dominant and recessive alleles in the table proves, once again, that the recessive allele does not simply diminish.

Note the frequencies of the combinations **AA**, **Aa**, and **aa** in the F_2 generation in Figure 19.1, which is our first opportunity to see all the possible combinations. We see in the F_2 that one-fourth of the population is **AA**, one-half is **Aa**, and one-fourth is **aa**. In order to see what will happen in the next generation, we can list all the different kinds of matings and how often these should be expected to occur. With three different genotypes, there are three kinds of males and three kinds of females, or nine types of matings altogether. If we combine **reciprocal matings**, such as **AA** × **aa** and **aa** × **AA**, there are still six different types. Each type of mating can be expected to produce certain kinds of offspring in the usual Mendelian ratios.

These different types of matings, together with the resulting proportions of each of the three offspring types, are listed in Table 19.1. For instance, one-fourth of the matings are **AA** × **Aa** (or **Aa** × **AA**). Since this should produce a 50:50 Mendelian ratio of **AA** and **Aa** offspring, among the F_1 a total of one-eighth will be **AA** offspring and one-eighth will be **Aa**.

From Table 19.1 and Figure 19.1, we can see that under the required Hardy-Weinberg conditions, random mating of the F_2 produces an F_3 generation that is ¼ **AA**, ½ **Aa**, and ¼ **aa**, just as in the F_2. The F_4 and F_5 will also have the same genotypic ratios. And that is why recessive traits continue to exist in a population.

An Algebraic Equivalent

In the example above, the two alleles start out at the same frequency—half **A** and half **a**. However, the Hardy-Weinberg distribution also maintains stable genotype frequencies in populations in which the allele frequencies are not the same. To illustrate this algebraically, let p be the allele frequency of allele **A**, while q is the allele frequency of allele **a**. So, $p + q = 1$ (1 = all).

The Hardy-Weinberg distribution says that the expected genotype frequency of **AA** is $p \cdot p$ (or p^2), the expected genotype frequency of **Aa** is $2pq$, and the expected genotype frequency of **aa** is $q \cdot q$ (or q^2). We can show that these frequencies will also be stable. As before, we can list the six (combined) types of matings, the frequencies in which they should occur, and the distribution of offspring of each type of family. The genotype frequencies of the offspring will be p^2, $2pq$, and q^2, just as in the parents' generation. Thus we have the Hardy-Weinberg distribution. Where p represents one allele and q, the other, we have

$$p^2 + 2pq + q^2 = 1$$

The Hardy-Weinberg distribution is easier to understand if we forget about random mating of diploid individuals and just consider the random association of gametes (Figure 19.2), which amounts to the same thing. After all, when an egg and sperm meet—either

TABLE 19.1 EXPECTED F_3 GENERATION

		OFFSPRING EXPECTED		
MATING	FREQUENCY	AA	Aa	aa
AA × AA	1/16	1/16	—	—
AA × Aa	1/4	1/8	1/8	—
AA × aa	1/8	—	1/8	—
Aa × Aa	1/4	1/16	1/8	1/16
Aa × aa	1/4	—	1/8	1/8
aa × aa	1/16	—	—	1/16
Total	1	1/4	1/2	1/4

in the open sea, as with sea urchins, or in the dark confines of a human oviduct—the parents' diploid genotypes no longer really matter. All that matters is the haploid genotypes of the two gametes, and if there is random mating, the probabilities of each gamete will be p and q of being **A** or **a**, respectively.

The Hardy-Weinberg law has very specific implications. For example, if we know the prevalence of a recessive trait such as albinism (the absence of normal melanin pigment) in the population, we can predict, within limits, the probability that any couple in that population will have an albino baby. Here's how it works. Normal skin and eye pigment in humans, **A**, is dominant over the albino condition **a**. The genotype **aa** occurs in about one of every 20,000 people. According to the Hardy-Weinberg equation, the frequency of genotype **aa** is represented by q^2 so that

$$q^2 = 1/20{,}000$$

and the frequency of a single allele **(a)** for this trait is thus

$$q = \text{square root of } 1/20{,}000 = 1/141$$

The frequency of the dominant allele **A** would then be

$$p = 1 - q$$

or

$$p = 1 - 1/141 = 140/141$$

The heterozygous condition **Aa** would, therefore, occur in the population with a frequency of

$$\begin{aligned} 2pq &= 2 \times 140/141 \times 1/141 \\ &= 1/70, \text{ or about } 1.4\% \end{aligned}$$

Since 1.4% of 20,000 is 280, about 280 people in every 20,000 will be carrying a recessive allele for albinism, while, as we have seen, only one will be affected. Hence, in the absence of a family history of this characteristic in either parent, the chance that any couple will have an albino child is very slim.

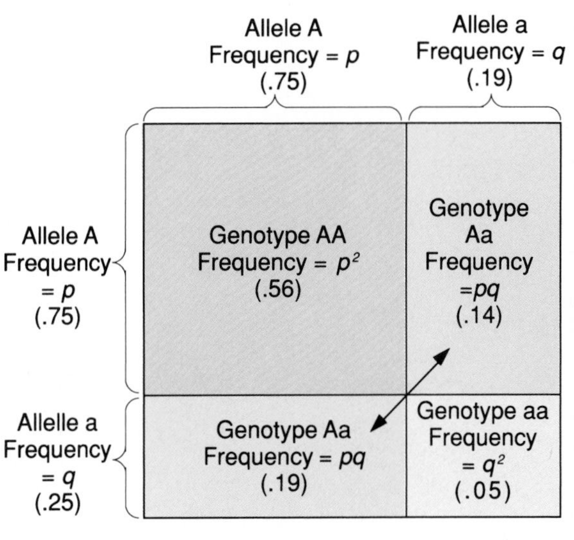

FIGURE 19.2
THE HARDY-WEINBERG EQUILIBRIUM
If there are two alleles, A and **a**, occurring in relative frequencies p and q (the values here are .75 and .25) respectively, and mating is random, a proportion p of the sperm will carry the A allele, and p of the eggs will also carry the A allele. For each zygote formed, the probability that the egg and sperm will both carry A alleles is $p \times p$ or p^2. Similarly, the probability that both of the uniting gametes will carry **a** alleles is q^2. There are two ways that an **Aa** zygote can be formed: A sperm uniting with **a** egg, or **a** sperm uniting with **A** egg. The total probability of one of these two events occurring is $pq + pq$, or $2pq$. (The arrow connects the two boxes that represent the same **Aa** genotype.) The relative frequencies of the three kinds of genotypes in the population will be equal to the individual probabilities of each kind of event: p^2, $2pq$ and q^2 will be the frequencies of genotypes **AA**, **Aa**, and **aa** respectively. (The dimensions of the blocks (colors) represent a situation where the frequency of p is .75, and q is .25.)

The Restrictions of the Hardy-Weinberg Law

It may have occurred to you that such purely mathematical constructions may be a little short on realism, or that they may only operate under very limited conditions. You are correct. In fact, the Hardy-Weinberg law has a number of stringent restrictions:

1. Mating must be completely random.
2. There can be no mutation.
3. There can be no immigration or emigration.
4. The alleles must segregate according to Mendel's first law.
5. The expectations are *exact* only if the population and the sample are infinitely large (and this is never the case).
6. There can be no selection operating on the population.

In a sense, then, the Hardy-Weinberg predictions are most useful when they don't come true, because when they don't, something is happening to the *allele frequencies.* That is, evolution may be occurring. We have listed several conditions in natural populations that will give rise to departures from Hardy-Weinberg expectations.

As we see, two of the factors that can disrupt the Hardy-Weinberg equilibrium are mutation and natural selection. As we found in the last chapter, evolution is fundamentally based on these two events. So we can now consider them in terms of population genetics.

MUTATION: THE INPUT OF NEW INFORMATION

Natural selection is a primary force determining the direction of evolution, but it must have something to work with. That raw material is genetic variation. Of course, the genetic recombination that occurs in meiosis and crossing over continually reshuffles genes to ensure new variation in combinations of existing alleles, but the source of new variations in species is *mutation,* the chance alteration of DNA. In Chapters 17 and 18 we saw how mutations arise through rare, unrepaired changes in base sequences and through chromosome breakage and rejoining. Now we are interested in the fate of mutations as they enter the gene pool.

Balancing Mutation and Selection

In some cases a mutation can be beneficial, but we must keep in mind that this is an extremely rare event. Few mutations will make a gene work better than it did before; the great majority will make the gene work less well or not at all. It's as if you raised the hood on your Porsche and let your neighbor's kid randomly bang the internal workings with a hammer. He may make precisely the adjustment needed to make the car run better, but what are the odds of that? (As a completely irrelevant aside, when the Apollo astronauts were having trouble with a delicate, sophisticated mechanism on their moon rover, Mission Control deliberated 12 hours before coming up with a solution: Try banging on it with a hammer. It worked.) The point of all this is simply that most mutations are harmful. Still, randomly occurring changes provide the variability on which natural selection will ultimately act.

It is also important to realize that, evolutionarily, mutations have the same effect whether they kill an individual or just keep him or her from reproducing. In either case, that person's genes are not passed along; they die with him or her. Thus, while lethal alleles are constantly fed into the gene pool by mutation, they remain rare because they are not passed along. However, if the mutant allele is only partially limiting in its effect, afflicted individuals may reproduce but at a reduced rate, and the mutant form may increase in the population. It will still be held in check, though, by natural selection. Population geneticists tell us that eventually an equilibrium will be attained, so that the number of affected individuals in the population will be directly proportional to the mutation rate, but inversely proportional to the individual's loss of reproductive fitness (Figure 19.3). That is why minor genetic abnormalities are so common: a mutant allele for, say, buck teeth is passed from generation to generation, producing whole families with malocclusion but with otherwise good health. (Essay 19.1, a story about Manx cats, may help in understanding how aberrant genes reach equilibrium.)

Without regard to the severity of the impact on an individual or population, a given gene in any population will mutate at one time or another. In fact, each gene undergoes mutation in a statistically regular and predictable manner, Thus, there is a constant and measurable input of new genetic information into the gene pool. Typically, there is about one new mutation per gene locus per 100,000 gametes. Some of the most severe mutations go unnoticed, however, because no offspring are produced. (It has been estimated that *a third* to *a half* of all human zygotes fail to develop because of dominant mutations. The potential parents are usually quite unaware that anything untoward has happened, even as the lethal gene destroys the developing embryo.)

It is likely that most of the mutations that are transmitted from generation to generation are recessive, since recessive alleles either fail to function at all or function at a reduced level. Many of these recessive mutations are also lethal, but only in the homozygous state—and this is not likely to happen until many generations after they first enter the gene pool.

The easiest mutations to study, of course, are dominant mutations that cause visible changes. An *invisible* change, for example, might be an alteration in some enzyme that had little or no effect on the organism.

FIGURE 19.3
BALANCE BETWEEN MUTATION AND SELECTION
Water in the beaker represents genes in the gene pool. Water enters the beaker at a constant rate (spontaneous mutation). Water flows out of the beaker (natural selection) at a rate that depends on the current level in the beaker. (a) At *mutation equilibrium,* the flow of mutant alleles into the gene pool through spontaneous mutation exactly equals the loss of mutant alleles through natural selection. (b) If the mutation rate is increased, the level of mutant alleles in the gene pool will rise until the outflow again equals the inflow. (c) Similarly, if natural selection against the mutant alleles is reduced, the level of mutant alleles in the gene pool will rise until a new equilibrium is reached.

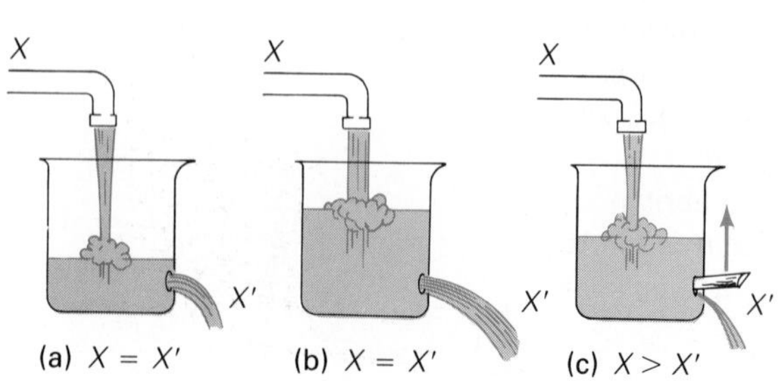

ESSAY 19.1

THE TAIL OF TWO GENOTYPES

To see how natural selection can act in rather unexpected ways, consider what might happen in the relatively simple case of *lethal* alleles. If an individual harbors a harmful allele, death removes the individual and the allele from the gene pool. To illustrate, consider Manx cats (see photo).

Manx cats are peculiar genetic anomalies. They have rather large hind legs and no tails (or very short tails). No one has ever been able to develop a strain of true-breeding Manx cats for the simple reason that the tailless animals are all heterozygotes. Normal cats, with tails, are **TT** homozygotes; Manx cats, without tails, are **Tt** heterozygotes. The homozygous **tt** genotype is an embryonic lethal; that is, it kills the embryo. So already we see a strange thing. The **t** allele is dominant for one trait, and recessive for another. It is dominant for the absence of a tail, and it is recessive for the absence of a kitten. Thus, two Manx **(Tt)** cats, mated, produce ¼ normal cats, ½ Manx cats, and ¼ dead (lost or "absorbed" as early embryos). Actually, then, the only litter you would see from such a cross would be ⅔ **Tt** Manx and ⅓ **TT** alleycat.

Now suppose that someone should populate a remote island with a whole shipload of Manx cats, which would then run wild, yowling and scratching and mating randomly, as cats are wont to do. What would happen? The frequency of the Manx allele, **t**, starts out at $q = 1/2$. In one generation it is reduced to $q = 1/3$. What happens then? With random mating, the third generation of *zygotes* will be $p^2 = 4/9$ **TT**, $2pq = 4/9$**Tt** (Manx), and $q^2 = 1/9$ **tt** homozygous lethal. (See the accompanying text for a detailed explanation of the algebra.)
But when the recessive homozygotes are removed by natural selection, the remaining cats are now half Manx and half alleycat. The frequency of the recessive lethal **t** allele has gone from ½ to ⅓ to ¼ in three generations, and in succeeding generations it will fall further to ⅕, ⅙, and so on. Meanwhile, the proportion of homozygous lethal **tt** zygotes will decrease accordingly (applying the Castle-Hardy-Weinberg formulas): 1/4, 1/9, 1/16, 1/25, 1/36 and so on. Eventually Manx cats will be fairly rare on our hypothetical island. But, significantly, the severe selection against the **t** allele will diminish. The accompanying graph plots the course of the genotypes over 70 generations. Note that the allele frequency of **t** has fallen to 1.37% (1/70) by the end of the 70 generations, and that about 2.8% of the cats will then be Manx. It would take another 70 generations to bring the recessive allele frequency down to 0.7%.

Selection doesn't have to be so severe, of course. As a general rule, the speed of gene change is proportional to the amount of selection against the unfit genotype. For instance, in the above example, suppose that the recessive genotype **tt** wasn't lethal, but merely reduced the individual cat's reproductive ability by one tenth. Then it would take 700 generations, rather than 70, to go from $q = 50\%$ to $q = 1.37\%$.

You might wonder why there are any Manx cats at all. The truth is that people are impressed by anything bizarre in cats and tend to keep the Manx kittens, while disposing of the alleykittens. So in the final analysis, it is human intervention—what can be called *artificial selection*—that keeps the Manx gene going.

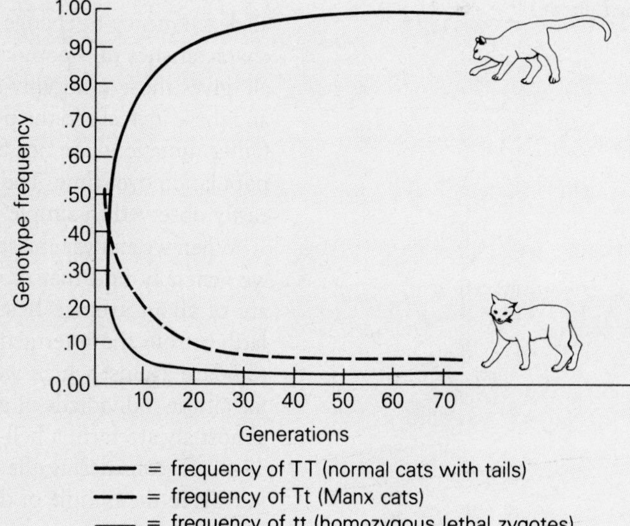

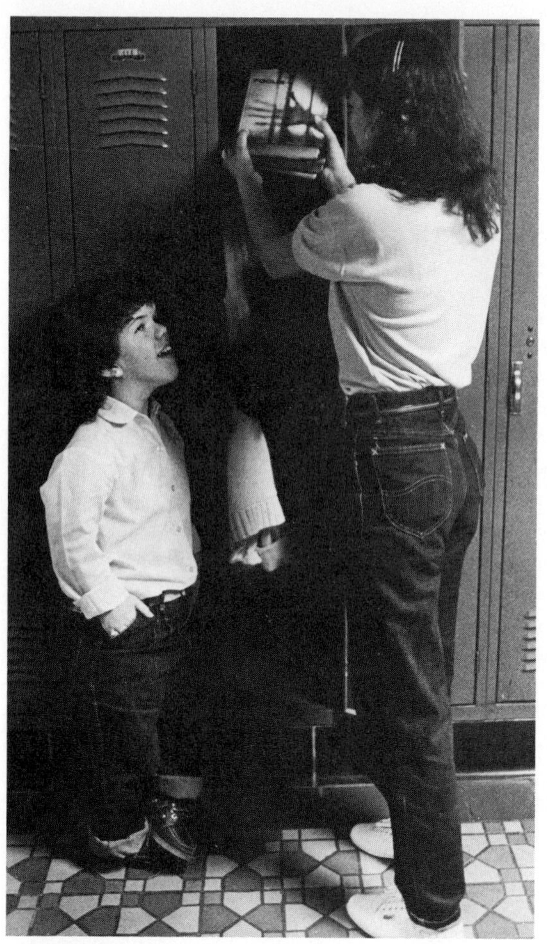

FIGURE 19.4
ACHONDROPLASIA
Since achondroplasia (a form of dwarfism) is dominant in humans, its occurrence in a family with normal parents represents a new mutation in the father or mother. (*Milwaukee Journal* photo.)

A *visible* change, on the other hand, might be dwarfism. A peculiar type of human dwarfism, identified earlier as achondroplasia (see Chapter 11), occurs in one out of about 12,000 births to normal parents (Figure 19.4). Since the condition is readily visible and also dominant, each dwarf born to normal parents represents a newly mutated gene.

With an average mutation rate of about 1 mutation per gene locus per 100,000 gametes, and with an estimated 80,000 genes per human zygote, everyone is likely to be carrying a new gene mutation. Since many of these will accumulate and be passed on to future generations, the total number of people with less than ideal genotypes is very large indeed—in fact it includes all of us if we include the potentially devastating recessive alleles that we all carry. The more minor the genetic defect, in general, the most readily it is passed on, and the more people it will affect. Some of the more common of these less-than-ideal genetic conditions are the familiar nuisances of missing teeth, malocclusion (such as buck teeth), nearsightedness, and deviated nasal septum.

Of course, some genetic conditions are more severe, and to some of the debilitating conditions we have already discussed (such as achondroplasia, hemophilia, albinism, sickle-cell anemia, and phenylketonuria) we can add schizophrenia, manic-depressive syndrome, early-onset diabetes, hereditary deafness, cystic fibrosis, Tay-Sachs disease, and literally thousands of other known genetic disorders. (We must always keep in mind, however, that not all mutation is bad. After all, mutation is one of the reasons you are not an amoeba, or even a random collection of methane, ammonia, and mud).

Once mutation has destabilized the genetic equilibrium of the population, then natural selection will determine whether the change has been beneficial or detrimental. It "evaluates" the change by diminishing it in the gene pool if it interferes with successful reproduction, or by magnifying it in the population if it contributes to reproductive success.

THE DIRECTIONS OF NATURAL SELECTION

The Hardy-Weinberg law tells us that under certain conditions, the allele frequencies in a population remain the same. Of these conditions, there are two that are the most powerful and ubiquitous in nature: mutation and natural selection. Mutation causes the allele frequency to change this way or that, and probably would have little effect on the characteristics of a population if it weren't for natural selection. Natural selection, after all, gives those randomly appearing changes different weights—that is, some it favors, and these increase in the population, causing it to take on a different set of characteristics. Other mutations are not favored, and they and the traits they code for diminish in the population over time. The changes in populations wrought by natural selection are most easily observed in simple systems, and when a single trait is being analyzed.

When we analyze the structure of a population for a single trait, say height in humans, we generally find that most individuals are intermediate for the trait (so most humans are of about average height). There are increasingly fewer individuals as one departs farther from the intermediate (or average) condition. This observation then gives us a standard against which we can determine change. For example, if we plot a single trait for all the individuals of a population graphically, we find that the resulting graph will almost always form a bell-shaped curve—the statistician's **normal distribution** (Figure 19.5). In considering the effects of natural selection, the question is, how do the individuals in the middle of the distribution thrive compared with those on either extreme? Depending on how they fare, we can find three trends: **stabilizing selection, directional selection,** and **disruptive selection** (Figure 19.5b, c, and d). Each of these can be observed to the degree that they vary from the normal distribution of a stabilized population.

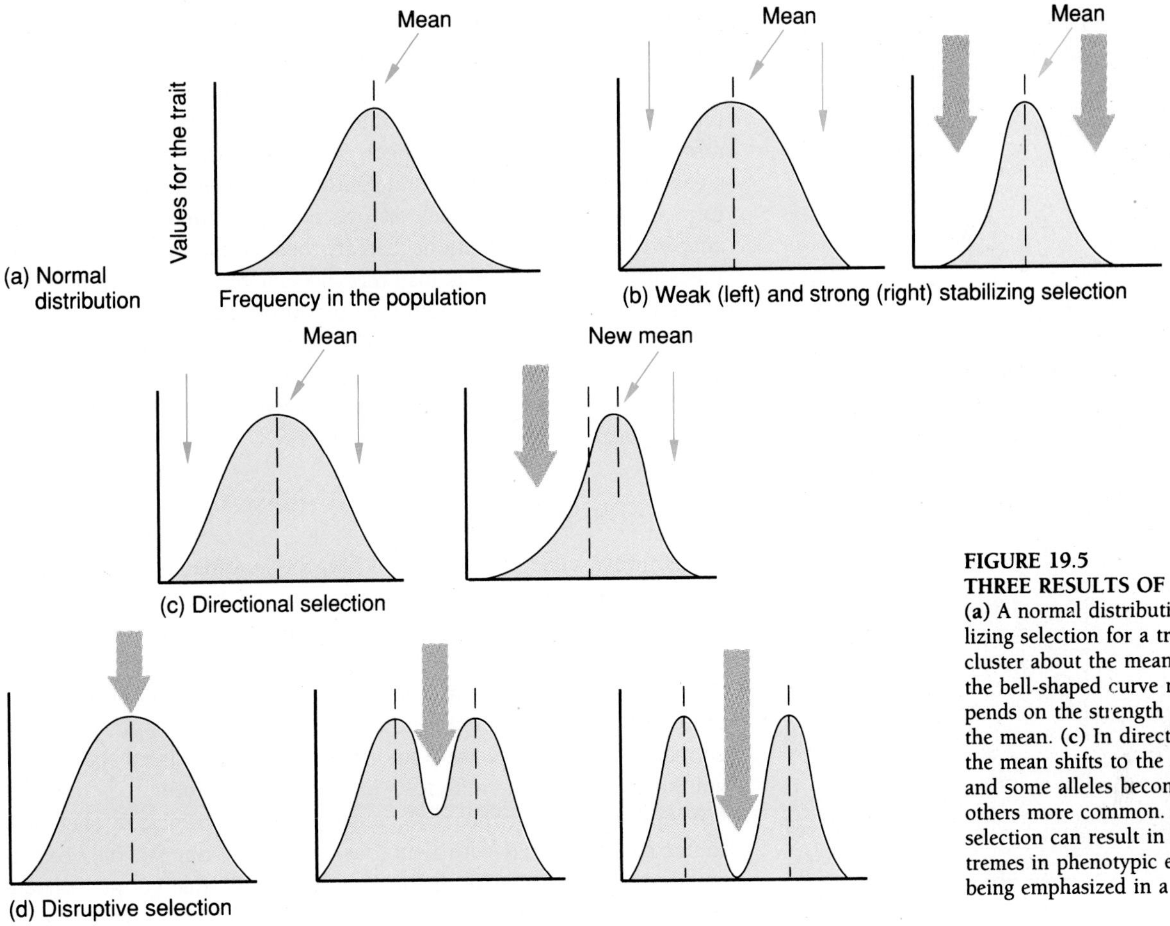

Mean

Values for the trait

(a) Normal distribution

Frequency in the population

Mean

Mean

(b) Weak (left) and strong (right) stabilizing selection

Mean

New mean

(c) Directional selection

(d) Disruptive selection

FIGURE 19.5
THREE RESULTS OF SELECTION
(a) A normal distribution. **(b)** Stabilizing selection for a trait maintains a cluster about the mean. The shape of the bell-shaped curve resulting depends on the strength of selection for the mean. **(c)** In directional selection, the mean shifts to the right or left, and some alleles become scarce and others more common. **(d)** Disruptive selection can result in the two extremes in phenotypic expression being emphasized in a population.

Stabilizing Selection

Stabilizing selection (Figure 19.5b) usually is associated with a population that has become well adapted to its particular surroundings. Any genetic change in the individuals of such a population, therefore, is likely to be harmful. Although genetic variability still exists, selection tends to favor the mean, or average, individual. Because most populations are well adapted to their environments most of the time, stabilizing selection is the most common kind of natural selection.

Perhaps the best-studied example of stabilizing selection is that of birth weight in human babies. (The data are readily available from hospital obstetric wards). If we plot survival rate against birth weight, we find that, not surprisingly, abnormally small babies have relatively low rates of survival. But abnormally large babies also have lower survival rates (Figure 19.6). The highest survival rate is for babies around 3.4 kg (7.4 lb). In this case, the optimal birth weight (as determined by survival rate) is almost exactly the average birth weight. Selection, then, is working against genes for both high and low birth weight. In essence, the average tends to be the best, and "survival of the fittest" becomes "survival of the most average."

FIGURE 19.6
SELECTION FOR BIRTH WEIGHT
Plotting the survival rate of babies of different birth weights indicates that there is an optimal weight of about 2.7 to 3.8 kg (6 to 8 lbs).

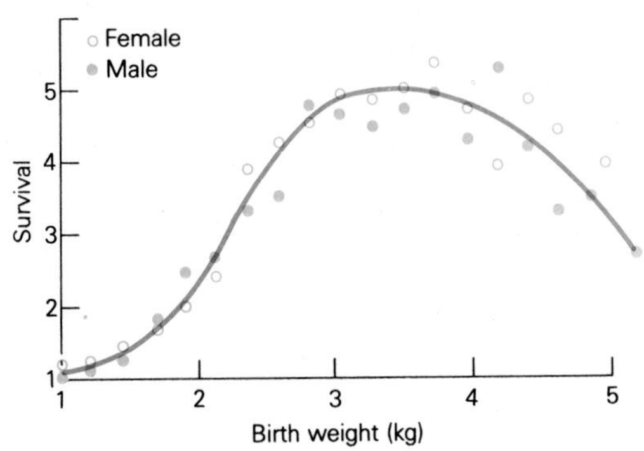

Directional Selection

Directional selection (Figure 19.5c) favors one extreme of the phenotypic range—one end of the curve. This is the kind of selection practiced by dairy

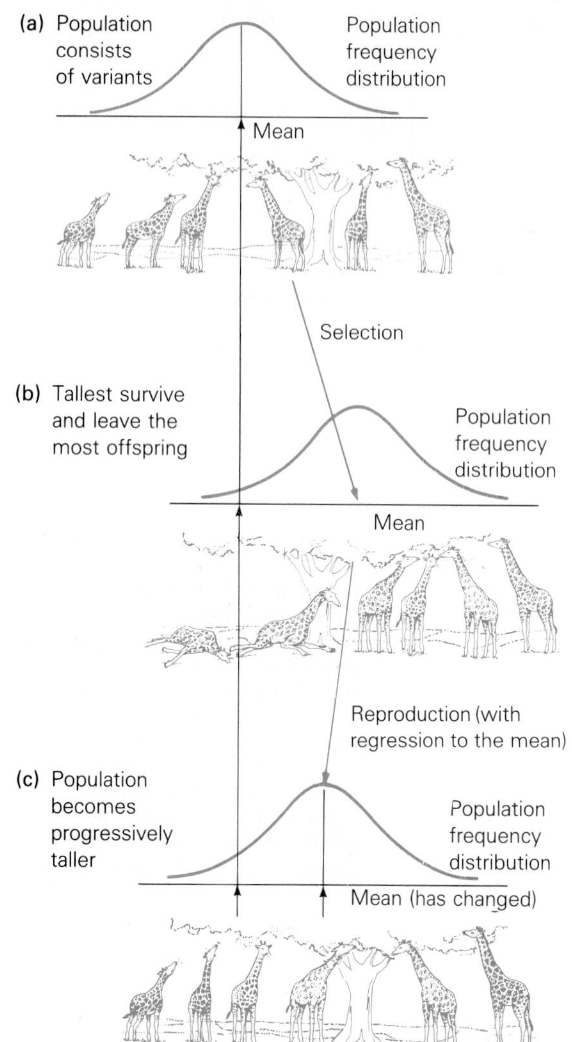

FIGURE 19.7
SELECTION HISTORY OF GIRAFFES
In the evolutionary history of the giraffe, an animal that browses on tall trees, height has been the critical factor. Among the antelope-like ancestors **(a)** height was variable, but as competition for the foliage increased **(b)**, natural selection began to favor taller individuals that could reach the still-untapped resources of the trees. (Natural selection was probably favoring taller trees as well.) As we see, the mean shifted. The taller individual survived, leaving the most offspring. **(c)** The offspring of the survivors tend to resemble their successful parents, although there is some regression to the mean. The average height increases over the course of one generation (exaggerated here). Over many generations, giraffes become taller and taller. (And so, incidentally, do the trees, as only the tallest trees escape defoliation by giraffes.)

breeders who want only the offspring of the cows that give the most milk. In nature, directional selection may be a response to a change in the environment that begins to favor individuals at one extreme. For example, a population may find itself in an unfamiliar territory, a new place offering new challenges. Or the species may suddenly lose a competitor and have new food sources open to it. In such cases, the formerly aberrant individuals at one end of a curve may be better adapted to the new conditions than are those at the center of the curve. Previously unfavored traits may then become the new optimum. Subsequent evolution can be rapid, as the population quickly adapts to its new environmental demands. The peppered moth story, discussed in Chapter 18, is an excellent example of directional selection at work. Also, the early evolution of giraffes represents a classic case of directional selection, as shown in Figure 19.7.

Disruptive Selection

In disruptive selection (Figure 19.5d), the intermediate types are selected against, and those at the extremes are favored. In this way, disruptive selection produces a *bimodal* (or two-humped) distribution curve.

One of the best known examples of disruptive selection is that of the bent-grass plants of Wales. In one area where copper is mined, copper-laden soil is piled up into mounds of waste called spoils. Botanists noticed that the mounds were bare—apparently, they thought, the copper is so toxic that nothing can grow in such places. Then they found spoils covered with bent-grass plants thriving on the copper-laden soil. Further investigation revealed that the copper-tolerant plants were freely exchanging pollen with nearby nontolerant plants. The result was that plants containing genes from both groups were sprouting all over the area, but those with the tolerance trait survived only on the spoils and those without the trait grew everywhere else. Disruptive selection then, was favoring the two extremes.

RANDOM CHANGE: GENETIC DRIFT, POPULATION BOTTLENECKS, AND THE FOUNDER EFFECT

So far, we have been considering the effects of mutation in large populations that were then subjected to the effects of natural selection. Now we will consider different scenarios by which evolution might proceed.

You recall that one of the conditions by which the Hardy-Weinberg equilibrium would be maintained is a large population (infinitely large). This is because equilibrium depends upon the laws of chance. (Recall from Figure 10.6 that if you flip a penny ten times, you *may* get heads nine times—90% of the time—but if you flip it 10,000 times, you are more likely to get heads 50% of the time, as the laws of probability work their way.) This is why a large population size can hamper the speed of evolution. As random mutations arise, moving an allele frequency in one direction, they are likely to be "swamped" or cancelled out by new mutations in the other direction. In some cases, though, they are not cancelled out, and the allele frequency of the population shifts. Any random change in allele frequency in a population not due to natural selection, is called **genetic drift**, and the phenomenon is most easily observed in small populations. In very small populations, in fact, small shifts in allele frequency can have momentous effects on the population. To take an extreme example, suppose in a population of 10,000 individuals, 5%, or 500 individuals, bear a certain allele. If some disaster wipes out 100 of them, the allele is still present in 400 individuals. Now suppose the population size is 100 individuals and 5%, or five individuals, carry the allele. If all five happen to be standing on the wrong side of the mountain some day, a landslide would wipe them out all at once, and the allele would be lost entirely from the population. The larger the population, then, the greater the probability that at least some individuals carrying the allele will survive any catastrophic event.

Random changes in allele frequency can also occur in populations that experience catastrophes, such as flood or famine, that can wipe out most of the individuals in the entire population. As the greatly reduced population recovers, its allele frequencies might be altered through the chance loss of certain alleles. Thus, the allele frequency in the renewed population would depend on just which alleles happened to have been carried by the few survivors. In the jargon of genetics, the population would have gone through a **population bottleneck** (Figure 19.8). A population bottleneck, then, is the result of a population becoming reduced in a brief period of time, resulting in a random change in gene frequencies.

Population bottlenecks can result in sudden increases in the frequency of deleterious alleles. For example, mammologist Lloyd Ingles noted in the 1950s that a species of California deer, the dwarf or tule elk, suffered a drastic reduction. Whereas the herd once ranged throughout California's vast central valleys and mountains, it became restricted to one 1100-acre park in Kern County. The tule elk has recovered, but now suffers an increased frequency of shortened lower jaw (Figure 19.9), a condition that produces grazing difficulties. It is quite possible that the change can be attributed to a bottleneck effect.

Bottlenecks can also occur when a few individuals stray out of their normal species range and establish a successful colony in a new habitat. This situation can produce what is called the **founder effect.** The founder effect occurs when a small part of a population establishes itself in a new area, separate from the original population, and brings with it a different frequency of alleles. In this way, previously rare genes could become common among the descendants of the founder population.

There are many examples of population bottlenecks and founder effects in human history. The numerous Afrikaaners of South Africa all descend from some 30 seventeenth-century European families. As you know, all humans carry their share of recessive mutations, and most of these are harmful, but they are very rare and are seldom expressed in homozygotes. However, the Afrikaaners' genes have gone through a bottleneck of

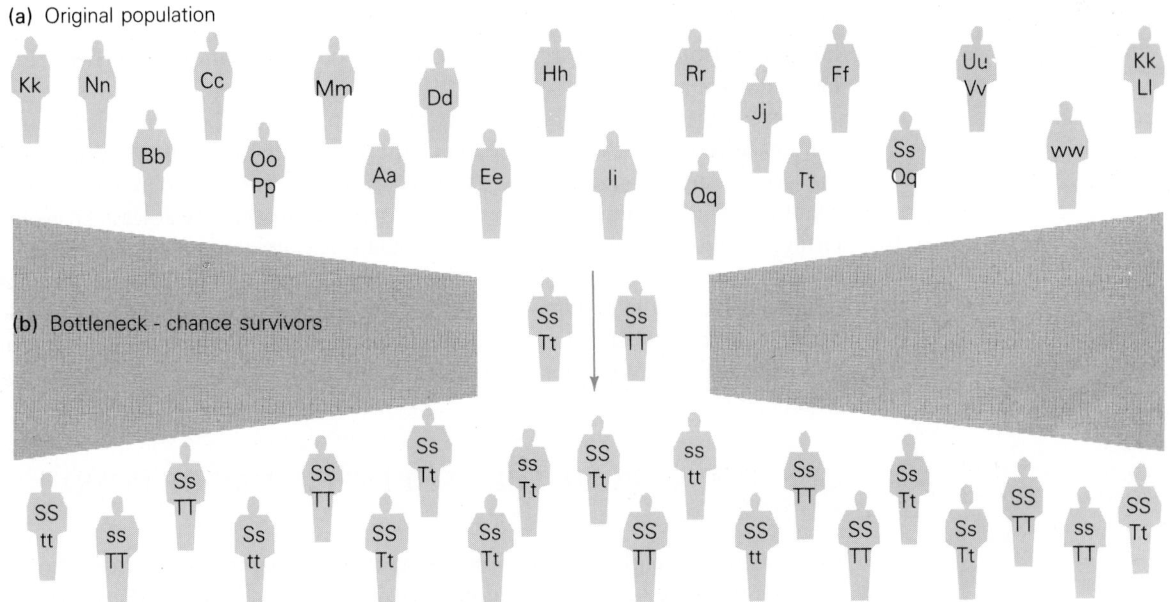

FIGURE 19.8
BOTTLENECK EFFECT
(a) In any large, diploid population, most individuals will be heterozygous carriers for rare recessive alleles at several different gene loci. **(b)** At the time of a population bottleneck, only a relatively few individuals survive. The survivors will carry a random sample of the rare alleles that were present in the formerly large population. **(c)** After the bottleneck, the few survivors will produce large numbers of descendants. Many of these progeny will carry some of the same recessive alleles, which will no longer be rare. Numerous individuals will become homozygous for these alleles. On the other hand, many rare recessive alleles from the original population will not occur at all in the new population.

FIGURE 19.9
POPULATION BOTTLENECK
California tule elk suffered a drastic population decline in the 1950s. Today's survivors have an increased incidence of shortened lower jaw.

only 30 families, so present-day Afrikaaners suffer from a unique set of recurrent recessive genetic diseases that are seldom seen in other populations. For example, a normally rare condition called porphyria variegata is common among the South African settlers of Dutch ancestry. It is a metabolic disorder that is characterized by excess iron porphyrins in the blood (the heme group of blood hemoglobin; see Figure 3.18), red urine, acute sensitivity to light, and eventual liver damage. On the other hand, Afrikaaners are almost completely free of other recessive genetic diseases; the 30 families obviously did not carry those genes to Africa. Similarly, Jews of Eastern European ancestry harbor a different but equally distinctive array of recessive genetic diseases, one of which is Tay-Sachs syndrome, a storage disease discussed earlier (see Chapter 4). This can be taken as circumstantial evidence for one or more population bottlenecks in the history of this group.

Neutralists versus Selectionists

In recent years there has been some controversy over how much natural variation is due to neutral mutations. The argument has produced two camps. On the one hand, there are the **selectionists,** those who attribute most, if not all, evolutionary change to natural selection. The **neutralists** say that much variation on the molecular level is due to selectively neutral mutations that accumulate through genetic drift (see Chapter 17). For example, there are frequently several slightly different allelic forms of any enzyme in a population, differing from each other by only one or two amino acids. Neutralists hold that most of this structural variation has no effect on the function of the enzyme. They think that functionally equivalent alleles just happen by chance mutation and that the allele frequencies drift around meaninglessly. Such random changes, they say, help explain the long periods of rather constant rates of evolution. Take note that the disagreement centers over normally *invisible traits,* not over obvious differences in phenotypes. Neither side maintains that any *visible* phenotypic variation is likely to be meaningless. The neutralist and selectionist hypotheses are restricted to the question of whether normally *invisible* details of molecular structure are subject to natural selection.

It can be shown mathematically that a completely neutral mutation would usually disappear by chance, but that it does have a finite probability of spreading through the population and thus creating a molecular polymorphism (variations in specific proteins). There is also a somewhat smaller chance that such a neutral mutation would drift to fixation (that is, to be carried by all individuals of the species), effortlessly ousting its fully equivalent predecessor. Let's look at some possible examples of genetic drift at the molecular level.

Horse beta globin and human beta globin are exactly the same at 129 out of 146 amino acid positions but are different at the 17 remaining positions. Since the time of the last common ancestor of the two species, which was a little insectivore that lived about 80 million years ago, the beta globin gene has undergone at least 17 evolutionary changes—17 occasions on which a new mutant has replaced an older allele, in one line of descent or the other. Has all this change been adaptive? Are these changes examples of the replacement of an inferior allele by a superior new one through natural selection? Or have some or most of the changes been the result of meaningless neutral mutations and chance alone? We don't know, but the question has prompted vigorous academic arguments.

QUESTIONS ABOUT THE GENETIC FUTURE OF *HOMO SAPIENS*

Now we come to a philosophical and moral question that inevitably arises about genetic variation in future human populations. The question is, if better medical and social care saves the lives of persons with adverse genetic conditions, thus allowing them to reproduce, what is to become of us?

It's not hard to find examples that illustrate the problem. The genetic condition **pyloric stenosis,** an abnormal overgrowth of a stomach valve muscle, was once invariably fatal in infancy. Since the 1920s a simple surgical procedure has saved the lives of nearly all affected infants (in developed countries). But about half the offspring of the saved people are also affected, and also need surgery. In this way, then, the alleles for the condition have actually increased in frequency.

What has happened, in effect, is that a *severe* genetic condition for "certain death in infancy by intestinal obstruction" has been transformed to a relatively *mild* genetic condition for "simple abdominal surgery needed in infancy." Suppose that the remaining risk of the condition, including possible delayed diagnosis and surgical mishap, is such that now about one affected infant in 20 fails to survive. The mutant genes for the condition will continue to increase in the population until, many millennia from now, a new equilibrium between mutation and selection will be reached, resulting in many times the frequency of affected individuals as there was before the operation was developed. If, after all that time, the condition still takes the life of one in 20, the ratio of persons dying from the effects of the gene will have once again reached its 1920 level. In the meantime, many lives will have been saved, and many people with slightly aberrant genotypes will have joined the human population. Is that good or bad?

In fact, this sort of thing has already happened repeatedly; every species adapts to changes in its environment, even such changes as the availability of skilled surgeons. In other cultures, such as those of primitive hunting and gathering tribes, **myopia** (nearsightedness) must be a condition that is often lethal. Presumably, nearsighted aborigines can't find roots and berries efficiently, let alone a zebra. But long before the invention of corrective lenses, the stable social conditions that came with villages and agriculture allowed myopics to survive, possibly even giving them a measure of protection as male myopics stayed closer to the home base and dealt with matters that did not require distance vision. In any case, myopia is much more common among people with a long history of agriculture and urban civilization than among groups that have more recently given up nomadism or the hunting-gathering life. American Indians and American blacks are blessed with much better visual acuity, on the average, than are Americans of European or Oriental ancestry, people long involved in agriculture.

With selection dampened, then, what *is* to become of us? The question simply cannot be answered at present—we will have to wait and see. But we should keep in mind that selection may well be operating on our population in new ways that are not yet apparent. Perhaps, for example, there is an advantage in being able to function well under crowded or regimented conditions. Just what other kinds of traits will be important in this new kind of world we are preparing for ourselves? We don't know, but we can be sure that selection is operating on us, even if we aren't sure how.

APPLICATION OF IDEAS

1. In a herd of wild mustangs, Greg Meddlesome counted 10 palominos, 29 dark (brown or black), and one white. (a) If the genotypes are **Aa**, **AA**, and **aa**, respectively, what are the respective genotype frequencies? (b) What is the allele frequency of **a**? (c) Is the group in approximate Hardy-Weinberg equilibrium? Calculate the HW expectations. (d) Greg further noticed there was only one stallion, which happened by pure chance to be the white **(aa)** horse. The rest constituted a harem of mares—not an unusual situation with groups of wild horses. What will be the approximate allele frequency of **a** in the next generation, assuming that the group remains isolated? (e) Would you expect the next generation to be in HW equilibrium? Explain.
2. Among Americans of European descent, about 70% find weak solutions of the chemical phenylthiourea (also called phenylthiocarbamide, or PTC) bitter and distasteful. The remaining 30% are unable to taste the chemical unless it is extremely concentrated. The difference in ability to taste the chemical is genetically determined by a single gene locus with two alleles. Tasting the chemical is dominant over not tasting, so tasters are designated as **TT** and **Tt**, while nontasters are *tt*. What is the frequency of each allele (**T** and **t**) in the above population? What are the frequencies of the three genotypes **(TT, Tt, tt)?** What is the probability that any taster in the population is homozygous?
3. It is always fascinating to speculate on the future course in the evolution of life, particularly of human

life. Using what you have learned so far, prepare a scenario depicting humans (or what humans will have become) a few million years into the future. Try to base your assumptions on what you know about the directions of selection today. Which of the human attributes (strength, muscular coordination, intelligence, craftiness, and so on) would one expect to persist and perhaps be further emphasized? Which will be lost? Will the pace of evolution have increased over what it was in the past few million years? Why?

KEY IDEAS

ALLELES AND ALLELE FREQUENCIES

1. Evolution occurs in populations, not in individuals. It begins with changes in the **gene pool** (all of the population's genes).
2. Changes in genes occur thorugh mutation, while changes in **allele frequencies** occur through **natural selection** and random events.

THE HARDY-WEINBERG LAW

1. The **Hardy-Weinberg law** predicts the behavior of alleles and allele frequencies in a model population. It explains why in the absence of selective forces, allele frequencies cannot change. Such populations are in genetic equilibrium for those alleles and remain so unless acted upon by forces of change.

How the Hardy-Weinberg Law Works

1. A **population** is defined as a group of interbreeding individuals, and **frequency** refers to a fraction or proportion of one type of item to all of the items in a general class.
2. In the absence of outside influences, allele frequencies remain constant. This can be proven by using an individual cross of **Aa** × **Aa**, and carrying the cross through any number of generations. The frequencies of **A** and **a** remain ½.

An Algebraic Equivalent

1. The algebraic equivalent of a population in terms of the alleles for any gene can be stated as $p^2 + 2pq + q^2 = 1$. In the formula, p^2 equals the frequency of the homozygous dominant genotype; $2pq$, the frequency of the heterozygote genotype; and q^2, that of the recessive genotype.
2. From the HW law, it is possible to predict, within its limits, the probability and outcome of matings in a population.
   a. This requires, first, that the allele frequencies of the gene in question be determined.
   b. Such frequencies can be determined through application of the formula, $p + q = 1$. Here p is the frequency of the recessive allele.
   c. Together their frequencies make up all of those alleles in the population, or, 1.
   d. Allele q is usually detectable simply because it is recessive. First, all of the recessives (qq) in a population are identified and counted. Since the number equals qq or q,^{2} q is the square root of that number.
   e. When q is known, p is readily determined, since $p = 1 - q$.
   f. When both p and q are known, the genotype frequencies can be determined by substituting the numbers for factors in the general HW formula. $p^2 + 2pq + q^2 = 1$.
3. The HW formulas can be used for predicting the probability of any couple in a population producing certain genotypes in their offspring. Albinism is an example:
   a. recessive genotype frequency: $q^2 = 1/20{,}000$
   b. recessive allele frequency: $q = 1/141$
   c. dominant allele frequency: $p = 1 - 1\,1/141 = 140/141$

   Since knowing p and q permits the genotype frequencies to be determined, applying the multiplicative law enables predictions of certain matings to occur.

The Restrictions of the Hardy-Weinberg Law

1. The HW law operates under stringent restrictions in the model population: random mating, no mutation, no immigration or emigration, simple Mendelian segregation of alleles, infinite population size, and no selection.
2. Any departure from the restrictions indicates that evolution is occurring.

MUTATION: THE INPUT OF NEW INFORMATION

Genetic variation is maintained by recombination, but new variations arise through mutation.

Balancing Mutation and Selection

1. Most mutation is harmful. The more harmful the effect, the stronger the selection. Less harmful alleles are likely to be retained in the gene pool for long periods.
2. Mutation occurs in a regular and predictable manner. (One new mutation per gene locus per 100,000 gametes.) All humans harbor certain harmful, recessive mutations.
3. Whereas mutations are generally harmful, they are the source of new genetic variability, and are thus the raw material of evolution.

THE DIRECTIONS OF NATURAL SELECTION

When height in a human population is analyzed and plotted, the data form a bell-shaped curve.

Stabilizing Selection

Stabilizing selection occurs in populations that are well adapted to their surroundings, favoring the intermediate phenotype and selecting against the extremes. Human birth weight is an example.

Directional Selection

When changes occur in the environment or perhaps in a predator or a major food item, selection may begin to favor less common phenotypes, those at one end of the scale. Such directional selection probably occurred in the evolution of the giraffe.

Disruptive Selection

When the environment becomes unstable, selection may shift from the intermediate phenotype to both extremes. When the distribution of phenotypes is plotted graphically, a bimodal curve is often seen.

RANDOM CHANGE: GENETIC DRIFT, POPULATION BOTTLENECKS AND THE FOUNDER EFFECT

Neutralists Versus Selectionists

Selectionists attribute most evolutionary change to natural selection. **Neutralists** believe a considerable amount of change at the *molecular level* occurs randomly, particularly where such changes do not affect the performance of an enzyme or other active molecule. Studies of hemoglobin from horses reveal that many such innocuous changes have occurred.

QUESTIONS ABOUT THE GENETIC FUTURE OF *HOMO SAPIENS*

Medical intervention in human genetic disorders, such as **pyloric stenosis** and **myopia**, circumvents the usual negative effects of natural selection, thus promoting increases in the frequency of such alleles.

REVIEW QUESTIONS

1. Explain the statement, "Individuals do not evolve, populations evolve." (395)
2. What, exactly, constitutes a *gene pool?* (395)
3. Explain what the term *frequency* refers to in population genetics. (395)
4. Carefully define the term *population.* (395)
5. Using the logic of Hardy, explain why, in the absence of selection, the frequency of a recessive gene remains constant. (396-397)
6. Using Punnett squares and the alleles **B** and **b**, prove that the recessive **b** remains constant through three generations.(396-397)
7. Define each of the terms of the HV expression: $p^2 + 2pq + q^2 = 1$. (398)
8. A certain recessive genotype appears in 16% of the individuals in a population. Determine the frequency of its two alleles, **T** and **t**. (399)
9. What is the probability of two heterozygotes mating in the population in number 8? of two heterozygotes mating and producing an offspring that is also heterozygous? (399)
10. The HW distribution applies to model populations. List the six assumptions that must be made to make the application valid. (399)
11. Considering your answer to the last question, how can the HW law be of any use to geneticists? (399)
12. List three ways the HW law can be applied. (399)
13. Explain the relationship between evolution, variation, mutation, and natural selection. (400)
14. What is the usual mutation rate? What are the chances that you carry a new mutation? (400)
15. List five human maladies that can be traced to mutations. (400, 402)
16. Why is it that nearly all mutations are harmful? (400)
17. Draw three graphs, showing the frequency of a certain allele undergoing directional, stabilizing, and disruptive selection. (402-404)
18. Suggest a situation that might tend to encourage disruptive selection. (404)
19. What kinds of environmental changes might encourage directional selection? What other influences might favor this? (403-404)
20. Specifically, how does genetic drift differ from selection? (404)
21. Mention two ways that populations can go through a "bottleneck." How might such occasions affect the allele frequencies of the parent population? (405)
22. What do the neutralists maintain about variation? The selectionists? (406)
23. Suggest three or four instances in which humans in modern society seem to overcome selection against harmful alleles. In a larger sense, do we really thwart this important evolutionary force—that is, in these instances will natural selection simply work in some more subtle manner? (406-407)

20

MACRO-EVOLUTION: THE ORIGIN OF SPECIES

IT IS, IN A SENSE, UNFORTUNATE THAT CHARLES DARWIN WAS DISPOSED to remove himself from most social interaction after he married his cousin and retired to work in the countryside. It can be argued that his chronic illnesses simply did not allow him the luxury of socializing (past a few visits to his brother's soirees in London). But if his writings are any indication, he would have been a marvelous storyteller. Consider this line he wrote in his diary after seeing the Galapagos Islands:

> Both in space and time, we seem to be brought somewhat nearer to that great fact—that mystery of mysteries—the first appearance of new beings on this Earth.

What had elicited such prose from this most observant of men was the life forms he saw on those islands. Even then, he suspected that he was witnessing the *beginnings* of certain forms—the origin of species, as it were. And he knew that in such origins arose the diversity of life.

Here, then, we will consider the sources of that diversity, the ways the different life forms could have appeared. We will search primarily for patterns in the appearance of new forms, and how we have come to think about those patterns. Here, we are dealing with **macroevolution,** the grand changes in living things that generally take many generations, and that occur above the population level. Macroevolution, for example, includes such momentous events as speciation and extinction. We will consider how we catalog and name the life on earth, after we take a look at the sticky old species question.

WHAT IS A SPECIES?

One of the simplest questions in biology is also one of the most confounding: What is a species? You would think that the question would have been decided and dismissed ages ago, but it almost seems that the more we learn about life, the more complex the question becomes. We will first consider a working definition of a species, and then we will see some of the problems associated with the definition. Later in the chapter we will see how new species arise and, finally, how they remain distinct.

Species is Latin for "appearance," and indeed, Carolus Linnaeus, the Swedish founder of modern taxonomy (the science of naming), put organisms into this species or that based on what they looked like. In most cases, in fact, this works quite well, but it doesn't solve all the problems. For example, some species of brittle stars (a spindly starfish) come in a wide range of appearances (perhaps as a way to confuse predators) yet some distinct animal species have remarkably similar appearances, such as the red-bellied and golden-fronted woodpeckers (See Figure 20.1).

The problem of distinguishing different species became a bit easier with the development of population genetics in the early 1900s. Population genetics introduced the notion of interbreeding. A shared gene pool became the fundamental criterion for defining a species. Then in 1942, the great biologist Ernst Mayr proposed that a species be defined as *a group of actually or potentially interbreeding natural populations, each population reproductively isolated from the others.*

"Actually" refers to populations that do, indeed, interbreed. "Potentially" means they *could* interbreed if they were able to reach each other, even though they may never do so. Texas grackles (large, black, raucous birds) are larger than the grackles of Puerto Rico, but their general appearance and behavior suggest that they might well be able to interbreed if their populations were somehow joined. They are placed in separate species for now, based on size differences, but if breeding experiments proved that they could produce generations of healthy offspring, they would be placed in the same species.

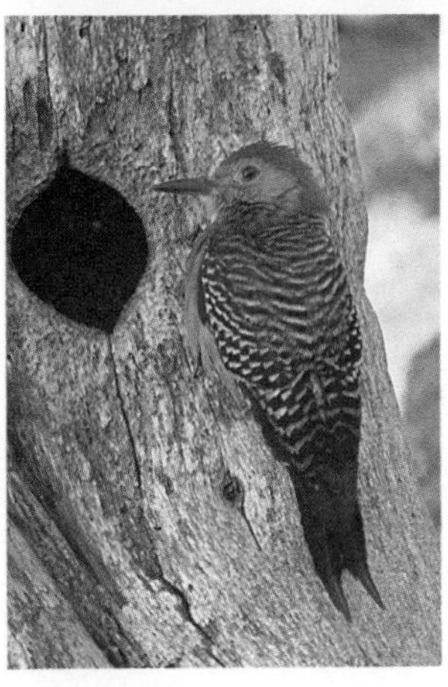

FIGURE 20.1
VARIATIONS
The outward appearance of the organism is not always reliable in identifying its species. Whereas the brittle stars are of the same species, the woodpeckers are not.

Problems with the Species Concept

Even Mayr's carefully crafted definition of the species can run into trouble. For example, one might ask: Is the golden jackal a different species from the timber wolf? That seems like an easy enough question. In the first place, jackals don't look much like wolves. Further, jackals are adapted to hot, dry, open country, and timber wolves are adapted to cold, damp forests. Jackals live in Africa, and timber wolves live in northern Europe, Siberia, and North America. They are apparently two different species.

But it turns out that jackals can mate with domestic dogs, and the offspring are fertile. Furthermore, domestic dogs can and do successfully interbreed with wolves. They can also interbreed with coyotes and dingos. Thus we see that, according to the criterion of "potential interbreeding," timber wolves, jackals, coyotes, and dingos could reasonably be considered one large, highly variable species—because they can all breed with domestic dogs (Figure 20.2).

But the story grows even more complicated. It turns out that there are three quite distinct species of jackals: the golden jackal *(Canis aureus)*, the side-striped jackal *(Canis adjustus)*, and the black-backed jackal *(Canis mesomelas)*. And while they will all breed with dogs, they apparently don't produce hybrids with each other. In the Serengeti of Eastern Africa, the ranges of all three jackals overlap, and there they simply ignore each other. A highly territorial jackal of one species won't even bother one of a different species that wanders through its personal domain. On the other hand, the range of the coyote overlaps slightly with that of the now-rare red wolf *(Canis rufus)* in Texas, and there, wolf-coyote hybrids have been found (Figure 20.3). Still, wolves remain wolves and coyotes remain coyotes, and the two are physically quite distinct animals. Aside from this and the promiscuity of "man's best friend," then, the lack of interbreeding among the eight species of the genus *Canis* appears to be more one of unwillingness rather than inability. Is unwillingness, then, as important a barrier as geographic separation?

To illustrate another kind of complexity with naming species, consider the case of a group of salamanders in California. Their range extends over a long, circular route

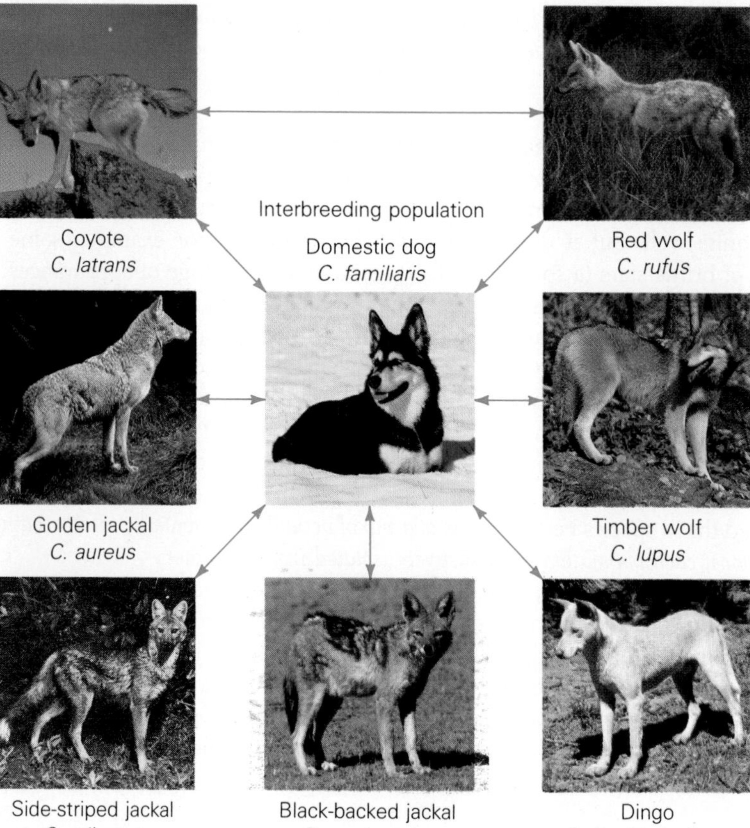

FIGURE 20.2
THE GENUS *CANIS*
Certain problems with the species concept are shown here. Each of the *Canis* species here is considered separate, but each is known to mate readily with the domestic dog. However, they generally do not mate with the others in the group, even if they are given the opportunity. If these matings were to occur, the offspring would presumably be fertile hybrids. So are they all members of one highly variable species? The arrows between recognized groups indicate well-documented interbreeding.

(Figure 20.4). They breed continuously along their range, except at the southerly end, where two variants overlap. There, the variants are dissimilar in appearance, and they treat each other as different species. However, according to our definition, they are not different species. The entire group is considered one species consisting of a number of distinct geographical units called **subspecies.** Two of these subspecies, though, seem to deny their heritage.

Our definition of species also does not cover the case of asexually reproducing species. (How can interbreeding be a factor if they don't breed?) Here, then, we must fall back on physical appearance. We must also resort to physical appearance in determining species among fossils. You can probably see why.

TAXONOMY AND SYSTEMATICS

Humans come in two kinds: those who like asparagus and those who don't. Or those who like country music and those who don't. Or those who like biology and those who don't. Furthermore, those who fall into one group or another are usually given a label that identifies them with their group. (For example, those who like country music go into the group with others of inordinately good taste.) The point is, it's only human to divide the world up, pigeonholing and labelling everything.

Pigeonholing living things has proven difficult, however, because people disagree on how large any "hole" is, and what should go into it. We have already mentioned part of the problem: living things vary. How much can an animal vary from the others before it goes into a different hole (that is, before it gets a different name)? Another problem is that in each pigeonhole are yet smaller holes, and these have holes that are smaller yet. It's not too difficult to place most things in the largest holes. For example, if it has leaves and roots, we can safely place it in the great pigeonhole labeled "plants." If it has ears and it bites, it's definitely an "animal." But not all plants have leaves, and not all

(a)

(b)

animals have ears, so the story already grows complicated. At some point, though, decisions are made about just what collection of traits defines this species or that. The next question is, how to name it.

FIGURE 20.3
HYBRIDS IN THE DOG GENUS
(a) Hybrids sometimes occur between species in nature. In Texas, where the range of the red wolf *(Canis rufus)* overlaps with that of the coyote *(Canis latrans),* natural hybrids are common, such as this 16-month-old male. **(b)** Interbreeding between wolves and domestic dogs is fairly common, even though they are recognized as separate species. The parents of this superb animal were a dog *(Canis familiaris)* and a timber wolf *(Canis lupus).*

How Species Are Named

Each species is given what is called a scientific name. Scientific names follow a **binomial** ("two-name") **system,** wherein the designation includes two terms: **genus** (usually a noun) and **species** (often an adjective but sometimes a person's name). For example, living humans are called *Homo sapiens.* Our genus, or generic name, is *Homo.* Our species, or specific name, is *Homo sapiens.* The scientific name assigned to each species is used only for that species. By convention, the names are italicized. Generic names are often abbreviated, especially after having been written in full; thus, we see our name written as *H. sapiens.*

The two-name system of classification was developed by Karl von Linne (1707–1778), who latinized his own name to Carolus Linnaeus. Linnaeus is responsible both for establishing categories within categories and for naming things with two Latin names. Latin was chosen because it is a dead language, not commonly spoken anywhere outside a few academic halls or in religious ceremonies. Thus, it isn't likely to change much. Also, because Latin is the root of a number of present-day languages, latinized names can transcend many language barriers.

In a monumental undertaking, Linnaeus classified and named many of the earth's creatures. Today, however, the rules for assigning names are rigidly enforced by international commissions. In fact, Charles Darwin sat on the very first such commission.

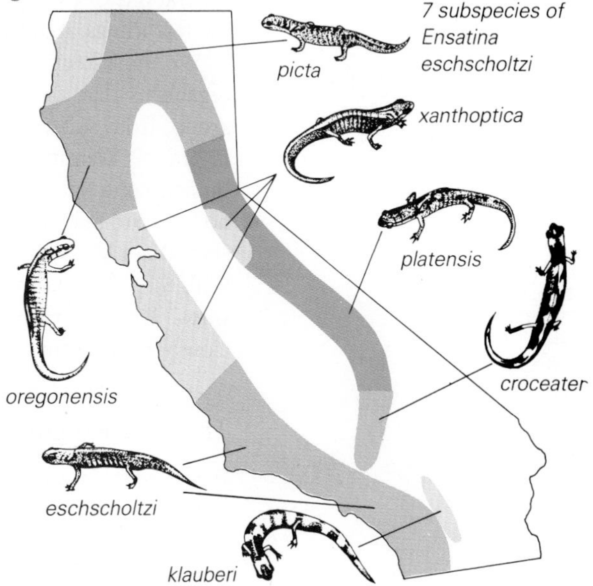

FIGURE 20.4
***ENSATINA* AND THE SPECIES PROBLEM**
Populations of the salamander *Ensatina eschscholtzi* are found in California along the coast and in the inland mountains. Skin color and size vary more or less continuously along the range, which forms a circular shape, as shown on the map. In the southernmost end of the range, two variants overlap, sharing the same territory. While individuals of each type may live just a few yards from each other (see *E.e. eschscholtzi* and *E.e. klauberi*), there is no evidence of interbreeding between types. At the same time, studies made along the range of the salamander indicate that other neighboring variants can and do interbreed. If variants along a range are of the same species but individuals in the extremes of the range do not interbreed, are they two of the same species? Many scientists disagree on this issue; others believe the question to be meaningless, arguing that the species concept itself is invalid.

There are now international nomenclature commissions for zoology, botany, bacteriology, and virology (not to mention enzymology, organic chemistry, and just about any other branch of science in which names are important).

Now let's get back to the idea of "categories within categories" and see how it is important in classifying living things.

Taxonomic Organization: Kingdom to Species

First, we should say that the science of identifying, naming, and classifying organisms is called **taxonomy**, whereas determining evolutionary relationships is the role of **systematics.** Systematists, then, deal with determining which species go into which genus and which genera go into which **families** (the next highest category), and so on. The categories form a hierarchy, which, beginning at the lowest (most exclusive) becomes: *species, genus, family,* ***order,*** *class, phylum,* and *kingdom* (Figure 20.5).

The **kingdom,** then, is the largest, most inclusive, category of life. Most taxonomists believe that life should now be divided into five great kingdoms (although a sixth kingdom is now seriously considered; see Chapter 22). If we were describing political organizations, kingdoms would be the equivalent of nations. They are generally subdivided ito what taxonomists call **phyla** (singular, *phylum*), roughly equivalent to our own states, if we continue the political analogy. Each phylum is itself divided into **classes** (counties?), and so the divisions continue. It is easier at this point to look at the total organization as we make our way down to the species level. The increasingly finer divisions, along with the divisions and subdivisions, can then be likened to boxes within boxes, as seen in Figure 20.6. The increasingly finer divisions, along with a mnemonic to help remember them, are:

TAXONOMIC CATEGORIES	MEMORY AID
Kingdom	"King
Phylum*	Philip
Class	Came
Order	Over
Family	From
Genus	Greece
Species	Singing
Subspecies	Songs"

(The nonsense about King Philip, or some ribald version of it, has been memorized by generations of biology students, some of whom find, after many years, that it is one of the few things about biology they still remember.)

We must keep in mind that, in essence, these are just names applied to groups of organisms that are assumed to be related, to one degree or another. Each taxonomic group, from subspecies to kingdom, is called a **taxon** (plural, *taxa* from *taxis,* "to put in order"). Naming, of course, is necessary, but we must keep in mind that names are human contrivances, created for human purposes, and have no other importance and no separate reality in themselves. At the same time, if a name is going to be useful, it should not be totally arbitrary but should reflect *some* kind of reality. When all is going well, then, taxonomy should reflect systematics. That is, names should reflect relationships. Thus, two species within the same class should share more ancestry (are more closely related) than two species in different classes.

The Five Kingdoms

We must be aware that the definition of life's kingdoms is constantly changing. The prevailing notion, representing an uneasy truce among biologists, is that there are five kingdoms, and that the five kingdoms are fundamentally related to each other as seen in Figure 20.7. The kingdoms are:

*In botanical terms, the phylum is replaced by the *division* (whereupon the mnemonic becomes, "King *David* came over . . . ").

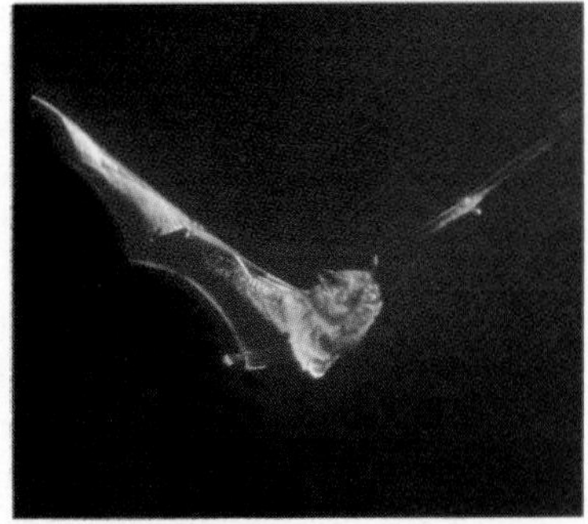

Kingdom:	Animalia	Animalia	Kingdom:	Plantae
Phylum:	Chordata	Chordata	Division:	Anthophyta
Subphylum:	Vertebrata	Vertebrata	Subdivision:	
Class:	Mammalia	Mammalia	Class:	Dicotyledonae
Order:	Cetacea	Chiroptera	Order:	Sapindales
Family:	Balaenopteridae	Craseonycteridae	Family:	Aceraceae
Genus:	*Balaenoptera*	*Craseonycteris*	Genus:	*Acer*
Species:	*B. musculus*	*C. thonglongyai*	Species:	*A. rubrum*

1. **Monera:** The prokaryotes or bacteria. As we will see in Chapter 22, biologists are now in the process of reorganizing the bacteria into two kingdoms, Archaebacteria and Eubacteria.
2. **Protista:** Various, mostly single-celled eukaryotes, including both *protozoa* (nonphotosynthetic protists) and *algae* (photosynthetic protists). Protists may include multicellular forms, but these do not have pronounced tissue differentiation.
3. **Fungi:** Mostly multicellular, parasitic, and scavenging organisms (the molds and mildews, mushrooms and toadstools, yeasts, and elements of the lichens).
4. **Plantae:** Multicellular, photosynthetic organisms ranging from evolutionarily simple mosses to the evolutionarily advanced flowering plants.
5. **Animalia:** Multicellular animals, including sponges (parazoa) and other animals (metazoa).

FIGURE 20.5
TAXONOMIC DESIGNATIONS
The two animals whose classification from kingdom to species are listed here represent the largest and smallest members of a diverse class, the mammalia. The great blue whale *(Balaenoptera musculus)* grows to about 100 feet in length, can weigh over 150 tons, and devours 8 tons of food per day. Kitti's hog-nosed bat *(Craseonycteris thonglongyai)* weighs about 2 grams (0.07 oz), and has a wing span not exceeding 17mm—altogether about the size of a bumblebee. Both these animals have hair, are warm-blooded, suckle their young, and nourish their embryos through a placenta. As it happens, neither species is divided into subspecies. The red maple *(Acer rubrum)* is a flowering plant of considerable size and is rather closely related to those other species that are called maples.

FIGURE 20.6
"BOXES WITHIN BOXES"
The hierarchy of taxonomic groups comprising a taxon may be thought of as "nested boxes." Inside each box are additional boxes, and inside those, more still. The smallest groupings within any box might represent members of the same genus. What might the largest boxes represent?

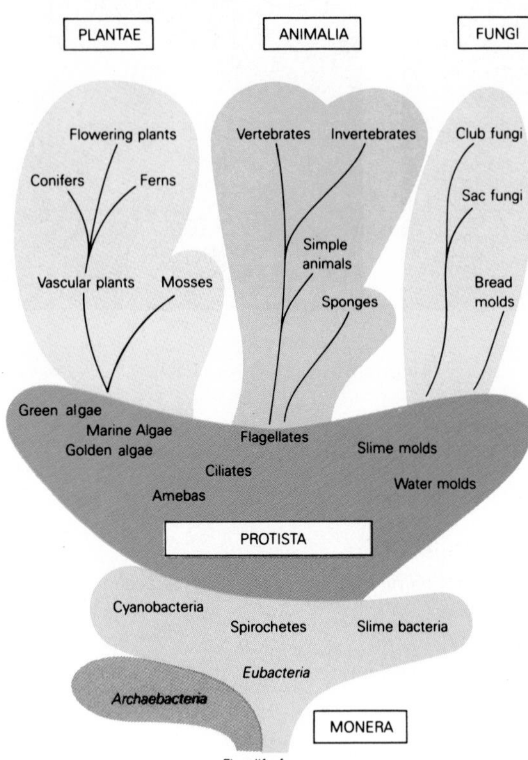

FIGURE 20.7
KINGDOMS OF LIFE
The five kingdoms are arranged in a manner that suggests their phylogenetic relationship to each other. The underlying kingdom, known as Monera, is now being reorganized into two kingdoms, Archaebacteria and Eubacteria, containing two quite diverse groups of prokaryotes. Above this is the large assemblage provisionally called Protista. Some of its members are descended from the same lines that gave rise to plants, fungi, and animals.

One of the problems with this organization, you may have noted, is that there is no mention of the viruses. It might be argued that they should have their own kingdom since they are unlike anything else (Chapter 22), but kingdoms are supposed to reveal evolutionary relationships, and since the evolutionary origins of viruses are unknown, that step can't yet be justified.

Within the five-kingdom scheme we find further problems, particularly but not exclusively in the protists. Kingdoms are supposed to be monophyletic units—that is, all of their members should be traceable to a common evolutionary ancestor. Yet the protozoa and algal protists may be polyphyletic—that is, having possibly evolved from different ancestors. However, they are lumped together provisionally until their evolutionary origins become better known.

METHODS IN TAXONOMY

In the days since Linnaeus, the science of taxonomy, naming living things, has taken a number of turns and has become increasingly sophisticated. Still, however, some of the basic problems remain. Evolutionists have taken to getting at these questions by extending their investigations from particular philosophical stances. Each stance, however, has its limitations, as we will see shortly.

Of course, simply naming species is not the problem. The problem is that the names themselves should reflect evolutionary relationships. Furthermore, when we say species A is more closely related to species B than to species C, we are implying that A and B share a more recent common ancestor than either does with C. So, immediately, we see that taxonomy (naming) is intricately associated with systematics (relationships).

As we think about taxonomy and systematics, we should carefully avoid the assumption that primitive species are giving way to advanced species. There is no such thing as a primitive species. After all, a slime-mold's ancestry is just as long as our own. Thus, we do not speak of primitive and advanced species. However, we may speak of *primitive* and *derived traits*. Primitive traits are those that arose relatively early in evolutionary history and have been inherited with little change from remote ancestors. Derived (advanced) traits are those that have undergone recent change. Most animals are "mosaics" of primitive and derived characters. For example, the platypus is heavily endowed with primitive traits, such as egg-laying and the presence of a cloaca. But its mouth is very recently derived and highly specialized. Humans, on the other hand, have a number of derived traits, such as bipedalism and a simple aortic arch, but our jaw is remarkably primitive. Such differences in the rates of evolution of various structures have presented evolutionists with great problems of naming and classifying the various life forms.

Another problem in developing a unified theory of evolution is that taxonomists must make judgement calls. Such decisions often reflect a very personal philosophy, since they are simply a matter of choice—of how one chooses to look at the problem. Some taxonomists, for example, will look at two groups and be impressed by their similarities. Others will be impressed by their differences. The former scientists (the "lumpers") will tend to place them in the same category. The latter scientists (the "splitters") will place them in different categories. For example, some taxonomists put tigers and house cats in the genus *Felis,* while others, impressed with the size differences, segregate the big cats into the genus *Panthera.*

These, then, are some of the problems confronting taxonomists. Nevertheless, the challenges have been accepted, and these researchers have forged ahead, trying to make some sense of the vast array of life, past and present. We will now consider three approaches to the questions—the *classical,* the *phenetic,* and the *cladistic.*

The Classical Approach

Historically, classifying organisms has been based on homologies (similarity due to common descent of a structure, that is, descent from a common ancestor). The first steps in discovering homologies generally involve comparative anatomy and are based simply on what the organism looks like. If it looks like a bear, it's going to be treated by classical taxonomists as a bear—at least at first. As they go further, the creature's ancestry will be considered. Paleontologists (those who study fossils) will be consulted in a search for the creature's predecessors. If the predecessors were not bearlike, the animal's classification will be reconsidered. For example, it was once thought that bears were distinct carnivores, not closely related to any other living carnivore. But, more recently, fossil evidence indicates that bears are, in fact, rather closely related to dogs. Thus, bears have been repositioned on the evolutionary tree in light of the newer findings.

Classical taxonomists also consider the embryological development of the animal in question. Its embryology will be compared to that of other species. Similarities in embryological development are considered evidence of homology and indicate a fundamental relatedness. Some of the results of embryological comparisons can be unexpected. You may be surprised to find, for example, that the echinoderms (such as sea stars) are placed near our own phylum (chordata) in evolutionary trees (see Figure 30.4). Obviously, this is not based on any outward resemblance between us and them, but rather on very fundamental embryological considerations (see Chapters 30 and 31).

Classical taxonomists, then, rely on a "constellation of characters," an approach that involves comparative anatomy, paleontology, and embryology (always with an eye toward discovering homology). However, proponents of the next two approaches reject the classical approach as being entirely too subjective. Their approaches are, they say, far more coldly objective. Interestingly, their approaches are based on almost exactly opposite premises.

The Phenetic Approach

Phenetics is not based on assumptions about homology. Furthermore, the approach does not presume anything about relatedness. Since any statement regarding relatedness is regarded as conjectural, pheneticists let numbers speak for themselves. Essentially, this process examines a great many characters (say about 100), using a numerical scoring system as shown in Figure 20.8a. Furthermore, no assumption is made about the evolutionary significance or derivation of any character, or about any other factor regarding its development. The idea is that homologous structures will be distinguished from analogous structures by sheer numerical weight of the first over the second. Thus, when enough data are available, the numerical pheneticists argue, computers can score the organisms according to the number of characters they share, thereby revealing the relationships of organisms.

Critics reply that phenetics looks for relatedness based on phenotype alone and that such outward appearances can mask a host of internal or genetic differences. Eastern and western meadowlarks (Figure 20.8b), for example, would be unlikely to be distinguished by such methods. Furthermore, the critics note that the phenetic approach is too readily confounded by **convergent evolution.** Plants such as the ocotillo of North America might be confused, for a time, with the unrelated allauidia of Madagascar, based on appearances alone (Figure 20.8c). There are probably few strict pheneticists today, but their quantitative techniques have had a strong impact on other approaches to taxonomy.

Cladistics

Cladistics (*clados,* "branch") classifies organisms according to *when* they branched from common ancestors. It does not take into account *how much* they diverged from the ancestral group.

FIGURE 20.8
SPECIES DESIGNATION THROUGH PHENETICS
Problem: to determine the relationships among eight species: A, B, C, D, E, F, G, H. (**a**) Each species is paired by computer with the species with which it has the most in common, according to a large number of observable traits. The number 1.0 indicates complete agreement, 0.5 indicates agreement in half the characters. The level of agreement is shown by the horizontal lines joining the groups. (**b**) Here, each pair is matched with the pair it most closely resembles in the range of characters being considered. (**c**) Each grouping of pairs is then compared to other such groupings and placed with the groups with which they have the most phenotypic traits in common. In this way, descendency is determined by the sheer number of observable traits. Any analogous structures that are included will, according to the theory, be swamped by the greater number of homologous structures.

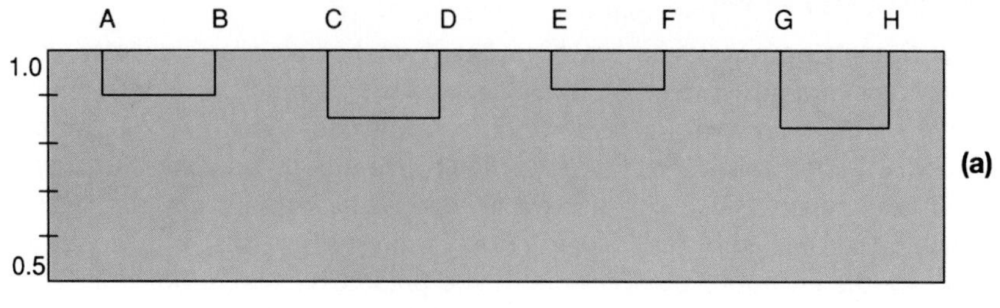

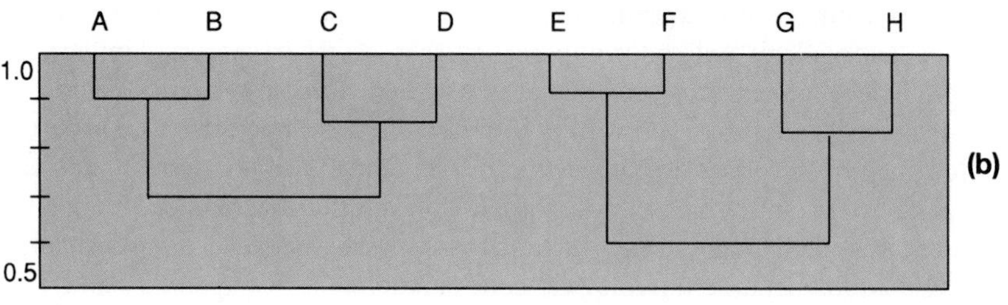

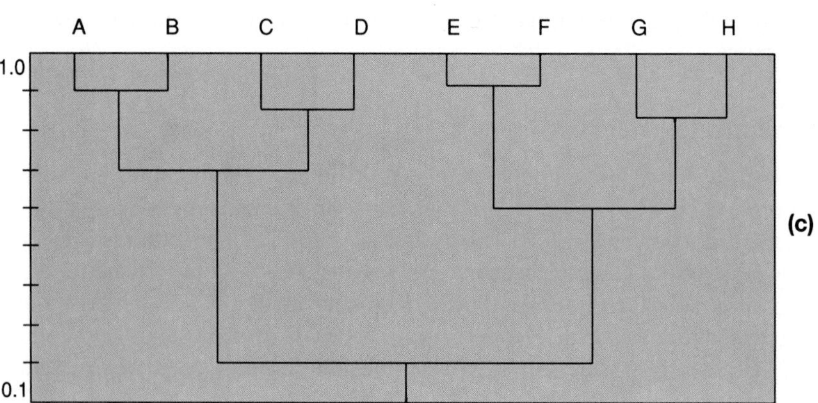

The **cladogram**, the evolutionary tree produced by cladistics, is based on particular kinds of homologous traits called shared derived characters. These derived chracters must have appeared *after* a branch diverged from the ancestral stock. Those traits that are common to all species on a cladogram are called shared primitive characters. For example, all of the groups shown in Figure 20.9 have a nerve cord. This trait (a shared primitive character) was present in their last common ancestor, and so it is useless in helping to diferentiate one group from another. The branch points in the cladogram are, we see, based on shared derived characters. These are new, or novel, structures that are homologous in all the groups that appeared earlier.

The cladists, those who espouse cladistics, have caused quite a stir in the scientific world, in large part because of some of the evolutionary trees produced as cladograms. The problem is, the relationships indicated in some cladograms fly in the face of our usual assumptions. For example, according to the phenetics view, crocodiles are more closely related to lizards than they are to birds. This view, of course, would be intuitively supported. However, the cladists, using shared derived characters, place crocodiles and birds closer together than crocodiles and lizards (Figure 20.10). Their reasoning is that the same line that diverged from ancestral lizards gave rise to both crocodiles and birds, but that birds evolved far more rapidly and added so many novel traits along the way that they now appear vastly different from their close relatives, the crocodilians. Their arguments have been so persuasive that this is the view now held by most systematists.

We see, then, that the business of grouping living things with their nearest relatives is not an easy matter. Furthermore, tracing lines of descendancy is more difficult still. Nonetheless, such efforts must continue, since our understanding of evolutionary processes are strongly based on just such information. We will now take a close look at some of these processes.

THE MECHANISMS OF SPECIATION

One of the most interesting and challenging questions in biology is, how does **speciation** (the formation of new species) come about? What factors encourage this development? A critical factor, it turns out, involves geography. We will first see how species can arise in geographically separated populations and then in overlapping populations.

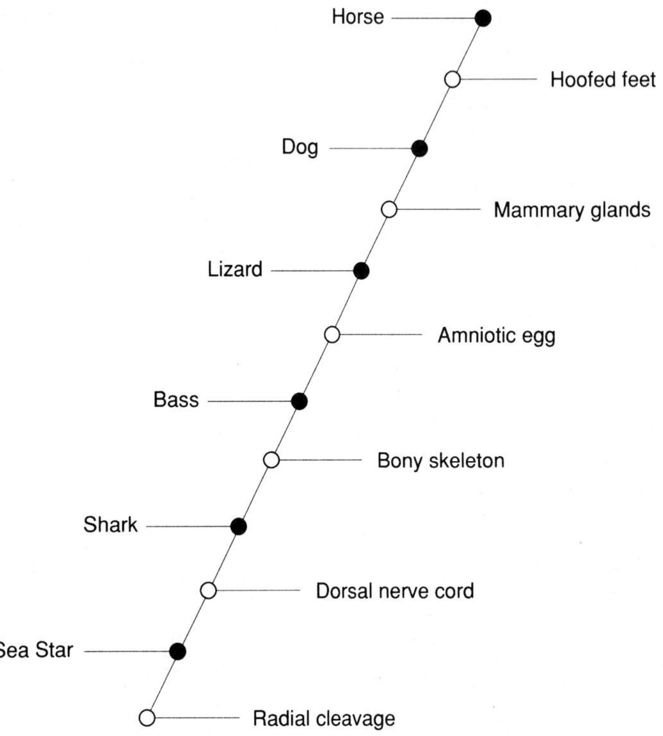

FIGURE 20.9
THE CLADOGRAM: SPECIES DESIGNATION THROUGH CLADISTICS
Each branch indicates the presence of an evolutionary novelty that sets it apart from what went before. All of the animals started with a nerve cord (circular in the sea star). Thus, they are all set apart from organisms with diffuse nervous systems (such as cuidarians). They all have dorsal nerve cords, except the sea star, and three have amniotic eggs. If some horses did not have mammary glands or if some lizards had hooves, this cladogram would be abandoned in favor of another arrangement.

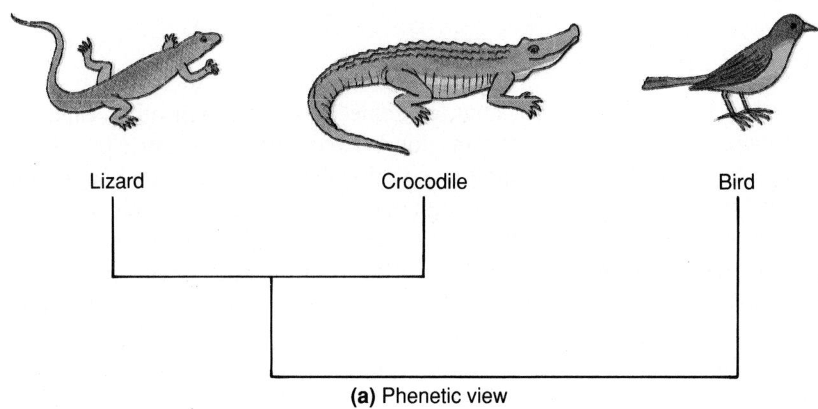

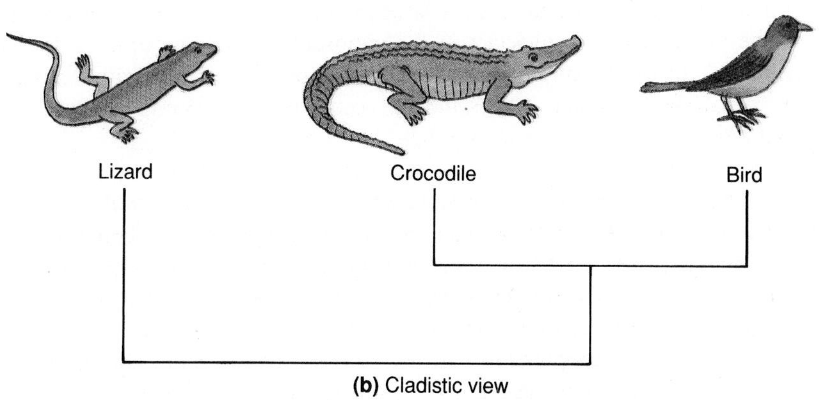

FIGURE 20.10
TWO VIEWS OF EVOLUTIONARY HISTORY
(a) In the phenetic view, lizards and crocodiles share more similarities and homologous structures with each other than either does with birds. It follows that they are more closely related to each other than either is to birds. (b) The cladistic view ignores homology and considers only when the groups diverged from each other. This tree is based on the belief that the last common ancestor of crocodiles and lizards predated the last common ancestor of crocodiles and birds.

Allopatry and Speciation

Allopatry refers to populations occupying different geographic areas. Such separation, we now know, is important in most kinds of speciation events. The formation of new species after the geographic separation of once continuous populations is known as **allopatric speciation.** The idea is that, when a population is split, each subgroup takes its own distinct evolutionary route until finally the two subgroups have diverged so much that interbreeding is no longer possible, even if they should rejoin. It is at this point that they are regarded as separate species.

There are two primary ways that populations of a species might become isolated. One way is that a group of individuals (or a seed, or even one inseminated female), might find itself in a new but hospitable place. Individuals can reach such places in a number of ways. Ocean islands, for instance, are occasionally populated by the descendants of unwilling, drenched, and thoroughly disgusted passengers on driftwood logs. Species may also reach islands by air, as when birds and flying insects are blown to some island by powerful winds. The finches Darwin observed on the Galapagos Islands may have arrived there in just such a way (Essay 20.2).

The second primary way that populations can be separated is by the slow process of geological change dividing a previously continuous group. This can happen on a local scale, as when the Grand Canyon gradually divided a population of squirrels, or it can occur on a grander scale. Historically, the most dramatic separation of populations has been due to **continental drift,** the separation and moving apart of immense land masses (Essay 20.1). It is believed that there have been at least two such episodes, the first in Precambrian times and the second during the Mesozoic era, about 230 million years ago. The final stage of the latter episode began some 65 million years ago when Africa and South America parted company. The unique array of species on the various continents, and particularly those on Australia and South America, has been the result of millions of years of allopatric speciation after the stage was set by continental drift.

Parapatry and Speciation

Parapatry refers to populations that abut, or come up against each other. Since this means that there is a strong line of demarcation between them, parapatric populations are probably rare in nature for the simple reason that living organisms do not generally respect "lines." If the two populations are of the same species, there is likely to be some gene flow between them, thereby keeping them from taking separate evolutionary directions. Thus, **parapatric speciation,** the formation of distinct species in parapatric populations, is probably rare in nature. If it occurs it is likely to be found in strongly heterogeneous ("patchy") environments where populations occupying adjacent patches are subject to strong selection, and where the abutting populations, or their genes, do not readily disperse.

Sympatry and Speciation

Sympatry refers to populations occupying the same geographical area. Whereas allopatric speciation—the formation of new species in geographically isolated groups—is by far the most common type of speciation, **sympatric speciation**—speciation within a population occupying a single habitat—does indeed occur. The best examples are found in plants.

Sympatric Speciation in Plants Sympatric speciation is quite common in some plants. In particular, it occurs through two main processes, hybridization and **polyploidy.** Among the flowering plants, in particular, there are many examples of new species arising by the hybridization of existing species. We have mentioned that in animals hybrid offspring, such as mules, are usually sterile. The mule's sterility can be traced to gametogenesis, where during prophase I of meiosis, the horse and donkey chromosomes are so different that they fail to synapse (see Chapter 9). (Pairing up is essential for both

ESSAY 20.1

CONTINENTS ADRIFT

Have you ever noticed that the bulge of eastern South America would fit rather nicely against the western coast of Africa? So has almost every fourth-grader who ever played with a globe. But after the initial observation no one usually pursues the matter further.

No one, that is, except people like Alfred Wegener. Wegener was captivated by the shape of the continents surrounding the Atlantic Ocean, and he too noted the particularly good fit between South America and Africa. In a work first published in 1912, he was able to coordinate this jigsaw puzzle analysis with other geological and climatological data to propose the idea of continental drift. He proposed that about 200 million years ago, all of the continents were joined together into one enormous land mass, which he called *Pangaea*. In the ensuing millennia, according to Wegener's hypothesis, Pangaea broke apart, and the fragments began to drift northward (by today's compass orientation) to their present location.

Wegener was not treated well in his lifetime. His geologist contemporaries attacked his naivete as well as his supporting data, and the theory was pretty much discarded until about 1960. About that time, a new generation of geologists revived the idea and found new data to support it. The most useful data has been the determination of magnetism in ancient lava flows. When a lava flow cools, it permanently fossilizes a small sample of the earth's magnetic field, recording for future geologists both its north-south orientation and its latitude. Detailed maps of the positions of the continents through the ages can then be made. It now seems that Wegener's (and all those fourth-graders') insight was absolutely right. Not only did continental drift occur as Wegener hypothesized, but it continues to occur today.

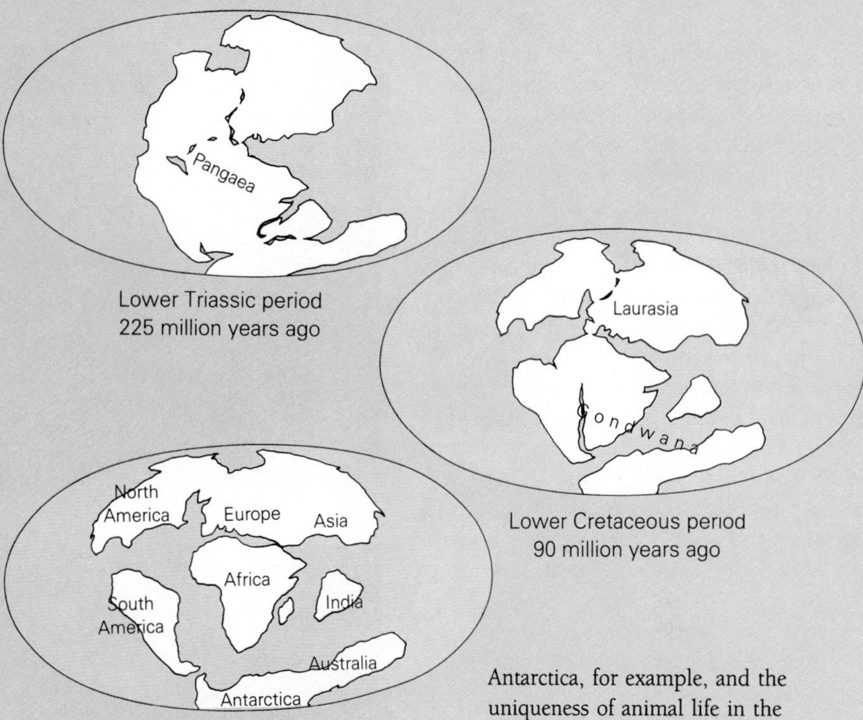

Lower Triassic period
225 million years ago

Lower Cretaceous period
90 million years ago

Paleocene epoch 65 million years ago
END OF MESOZOIC ERA

Geologists have long maintained that the earth's surface is a restless crust, constantly changing—sinking and rising through incredible unrelenting forces below. These constant changes are now known to involve large distinct segments of the crust known as *plates*. At the edges of these immense masses, new ridges are constantly built up and, in response, some edges are ground down. Where continents or pieces of continents are slammed together, mountain chains have formed. When ridges are built up in the ocean floor, the oceans expand. For example, astoundingly precise satellite studies reveal that the Atlantic Ocean is growing 5 cm wider each year.

In addition to its fascinating geological implications, the theory of continental drift is vital to our understanding of the distribution of life on the planet today. It helps explain the presence of fossil tropical species in Antarctica, for example, and the uniqueness of animal life in the Australian continent and South America.

As the composite maps indicate, the disruption of Pangaea began some 230 million years ago in the Paleozoic era. By the Mesozoic era, the Eurasian land mass, now named *Laurasia*, had moved away to form the northernmost continent. *Gondwanaland*, the mass that included India and the southern continents, had just begun to divide. Finally, during the late Mesozoic, South America and Africa completed their separation. For a time Australia and South America remained connected through Antarctica, which was more temperate then. During this time the marsupials spread over all three continents, but were driven to near-extinction by placental mammals in all but Australia. Both the North and South Atlantic Oceans would continue to widen considerably up into the Cenozoic, a trend that is continuing today. So we see that whereas the bumper sticker "Reunite Gondwanaland" has a trendy ring to it, it's an unlikely proposition.

ESSAY 20.2

THE GALAPAGOS ISLANDS, HOME OF DARWIN'S FINCHES

The variety of finches of the Galapagos Islands (see Chapter 1 and Figure A) is a result of allopatric speciation. Compared to the giant tortoises, strange flightless cormorants, and impish sea iguanas living there, Darwin's 13 species of finches are not particularly interesting—that is, not until the saga of their evolution is revealed.

The finches are all 10 to 20 cm long, and both sexes are drab-colored browns and grays. Six are ground species, each feeding on different, appropriately-sized seeds or cactus, while seven are tree finches. In each species the beak has become modified for its specific diet. One of the strangest tree-dwellers is the woodpecker finch. Lacking the long, piercing tongue of the woodpecker, it uses barbed cactus spines to pry insect grubs out of cracks and crevices in the trees.

On the *Beagle's* historic visit to the Galapagos Islands, Darwin collected everything he could find or catch, including the ordinary little brown finches. He took no special interest in them, but a London bird taxonomist later examined the specimens and noted that the inhabitants of different islands, though very similar to each other, were clearly different species. Ordinarily, birds living in one locality tend to be very different from one another. Whereas these birds were all clearly finches, many of them were seen doing things that finches don't ordinarily do.

Why have these little birds become so important to our understanding of speciation? Darwin concluded (and his conclusions are backed by years of careful research by others since) that the different species of birds were all descended from the same stock. Long ago, about 10,000 years ago by current estimates, the volcanic islands were colonized by South American finches that probably were blown out to sea by

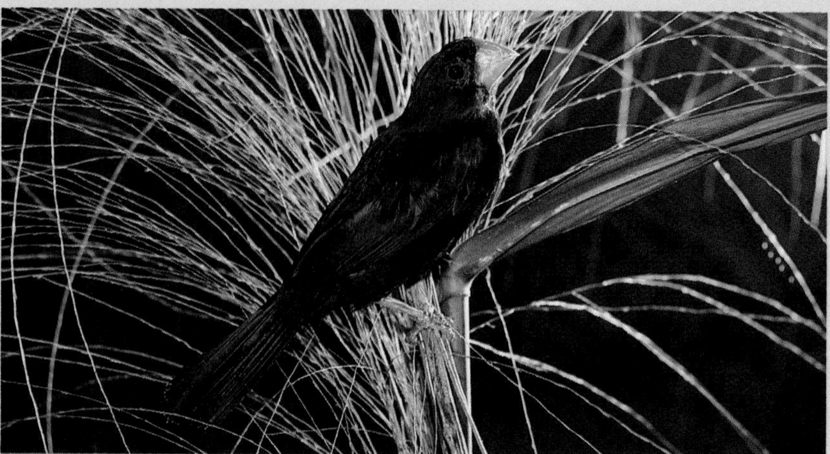

a storm. Apparently, conditions on the islands were favorable, and the "castaways" flourished. Their descendants eventually populated all the islands by occasional island hopping. However, the island hopping was rare enough to ensure the virtual isolation of each population. What followed then is referred to as adaptive radiation—the branching of an evolutionary line through the invasion of new environment, accompanied by adaptive evolutionary changes.

According to our scenario, the little birds had the islands to themselves, as far as they were concerned. There was a great variety of food. They were already well adapted for foraging for small seeds on the ground, but there were other plentiful untapped food resources—food not ordinarily eaten by finches.

Soon enough the expanding populations were depleting the available supply of small seeds. Thus, natural selection began to favor birds that could also cope with larger seeds and with other food resources. In time, the bird's bill sizes began to change as each population began to adjust itself more closely to the different kinds of food found on each island. Natural selection was working its way. Eventually, the isolated birds of the different islands differed genetically to the extent that any island hoppers would find themselves to be reproductively incompatible with the residents of other islands. Clearly, speciation, the formation of new species, had occurred.

As differences in lifestyles and specialization became magnified among speciating populations, competition would have been reduced. If the genetic differences led to differences in resource utilization, competition might become so low that two emerging species eventually could coexist on the same island. Their coexistence would also set the stage for further change. For example, with two species of finches trying to survive on one island, natural selection would favor the individuals in each population that were as different as possible from those in the other population, thereby further reducing competition between them. The tendency for differences in competing species to become exaggerated as each specializes in different directions is called character displacement.

After thousands of years of finches occupying the Galapagos Islands and separating, changing, specializing, and rejoining, the different populations today are totally unable to interbreed. This means that several of the species can exist side-by-side on every island. Each species uses the resources of the island in its own unique manner, in some cases filling niches that are occupied by other kinds of birds on the mainland.

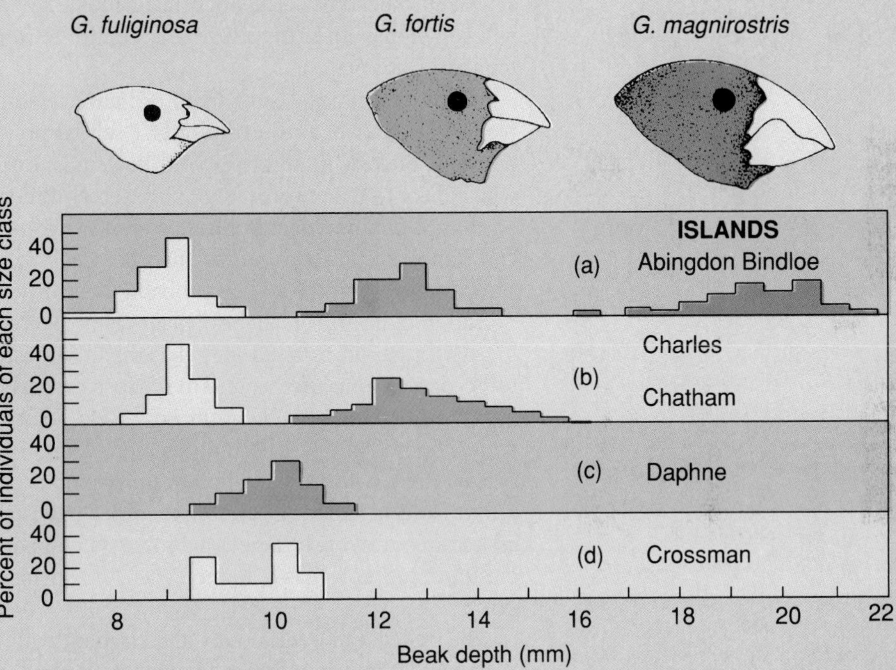

crossing over and for the proper alignment and separation of homologous chromosomes in the first division of meiosis.) Such a failure results in abnormal, random chromosome separations in anaphase, and eventually, in abnormal gametes.

Why don't hybridizing plants have the same problem? First, plant species that appear to be very different physically may be genetically similar enough that hybrids between them can undergo normal or nearly normal meiosis. (Where the range of such species overlap, there may be extensive mixing and partial hybridization, thus forming highly variable **hybrid swarms**—irises often exist in hybrid swarms.) Fertile hybrids, of course, allow the exchange of genetic material between species and stretch the traditional working definition of species. However, such hybridization between "good" (established and recognized) plant species is common and, surprisingly enough, doesn't seem to result in the breakdown of either species or their merging into a single species. This could be because hybrid swarms occasionally find their own niche and become new, distinct species.

New species can also result in **polyploidy**, in which there is an increase in the number of chromosomes, usually caused by the doubling of chromosome sets. This happens spontaneously from time to time in the mitotic divisions of the growing plant. The chromosomes double normally, in preparation for cell division, but for some reason the cell fails to divide. Thus, the abnormal cell and all of its progeny will have four complete sets of chromosomes. The abnormal cell is now called a **tetraploid** cell, and its progeny will sometimes form the tissue of an entire tetraploid branch of a plant, complete with tetraploid flowers.

When tetraploidy occurs in an ordinary diploid plant, the resulting tissues (or whole plants) are called **autotetraploids** ("self-tetraploid" or, more precisely, "self-four-genomes"). Meiosis in autotetraploid flowers is, once again, abnormal. The chromosomes will still try to pair two-by-two, but each chromosome will now have *three* homologues. So clumps of homologous chromosomes form, and meiotic segregation becomes jumbled and imprecise. Perhaps a few balanced seeds will be produced, but autotetraploids are comparatively infertile. They probably have an extremely limited role in evolution.

Tetraploidization in hybrid plants is a different matter. As we just saw, the reason most hybrids fail in meiosis is that the chromosome sets from the parental species fail to "recognize" one another and to synapse properly. Because the hybrid plant cell nucleus has two different haploid chromosome sets, some chromosomes simply can't be matched with a homologue (a matching pair member). But when spontaneous tetraploidization occurs, there will suddenly be two different but complete *diploid* chromosome sets. Such a tetraploid is called an **allotetraploid** ("other-tetraploid"). When flowers form in the allotetraploid hybrid, there is no longer a compatibility problem in meiosis, since every chromosome now has a homologue, and meiosis can then proceed normally (Figure 20.11).

The pollen and ovules of an allotetraploid will be diploid, but the plant can pollinate itself to reconstitute new, entirely allotetraploid seeds and new, entirely allotetraploid, and entirely fertile plants. But the allotetraploid plants are fertile only with themselves and with each other; if they are crossed to either parental species with the haploid pollen and ovules, they produce only *triploid* seeds, which are infertile. This reproductive isolation means that the new self-fertile allotetraploid plant constitutes, in fact, *an instant species*. The new species will soon fail, of course, unless it happens to occupy a new niche or unless it can outcompete another plant species—perhaps one of its parents—in its own territory.

Interestingly, polyploidy can occur repeatedly. Thus, we find hexaploids (six genomes, or three doubled copies of three different parental genomes), octaploids, and so on (the term *polyploid* covers any level of chromosome duplication higher than diploid). All of this may seem freakish, but in fact it is not even a rare process in flowering plants. Many plant species, both wild and domestic, are polyploids. Wheat, for example, is actually an allohexaploid, a genetic combination of the chromosome complements of three entirely different Middle Eastern species of wild grasses. *Triticale* (Figure 20.12) is a human-engineered application of the tendency of wheat and rye grasses to form

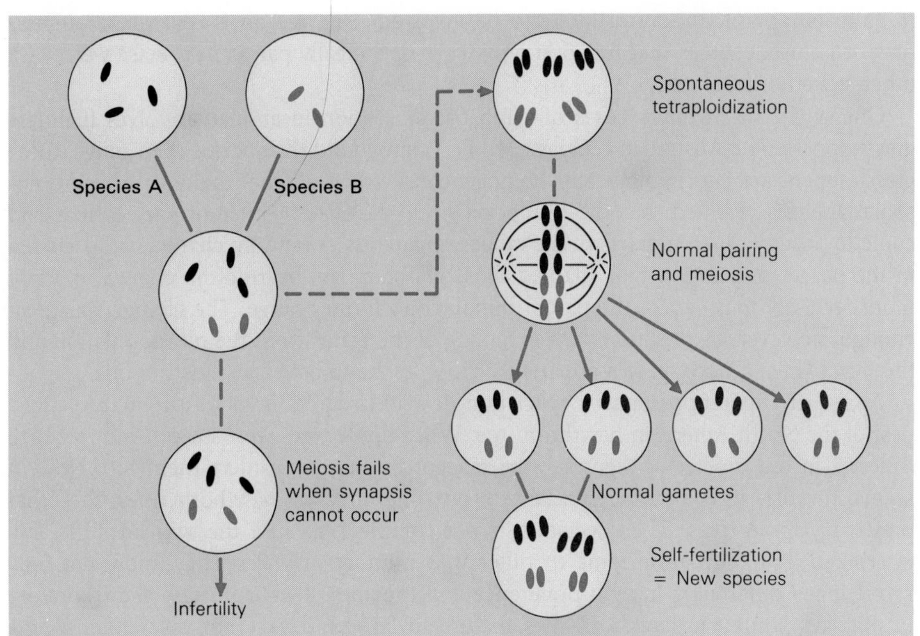

FIGURE 20.11
HYBRID STERILITY
While a plant hybrid may readily be produced in nature, it is often sterile. Sterility arises when, during meiosis, unmatched chromosomes fail to find their homologues. Here, species A is successfully pollinated by species B, producing a hybrid. Since there is no match in the chromosome complement, the hybrid cannot carry out meiosis and is therefore sterile. However, should a spontaneous doubling of chromosomes occur, as it sometimes does, normal pairing-up and meiosis occur, and normal gametes are produced. Following self-fertilization, a new viable species is produced.

allotetraploids. In the final analysis, it turns out that polyploidy is an excellent mechanism for increasing plant diversity through rapid speciation.

For reasons not fully understood, polyploid species may have much greater tolerance for harsh climatic conditions. In one study, for instance, only 25% of the plants sampled in lush tropical regions were polyploids, while 85% of all the flowering plant species in the raw environs of northern Greenland were polyploid.

Thus, hybridization followed by polyploidization creates instant, sympatric species of flowering plants, ready to be tested by the forces of natural selection. This versatility helps explain, hypothetically at least, how flowering plants arose rather abruptly in evolution and very quickly radiated out over the landscape to create the incredible diversity of plant species that dominate our world.

Sympatric Speciation in Animals Sympatric speciation in animals is probably much rarer than it is in plants. In fact, for years theorists have argued over whether sympatric speciation occurs at all in animals. However, a few rather clear-cut examples have been found. The processes in animals, however, are quite unlike those in plants. For example,

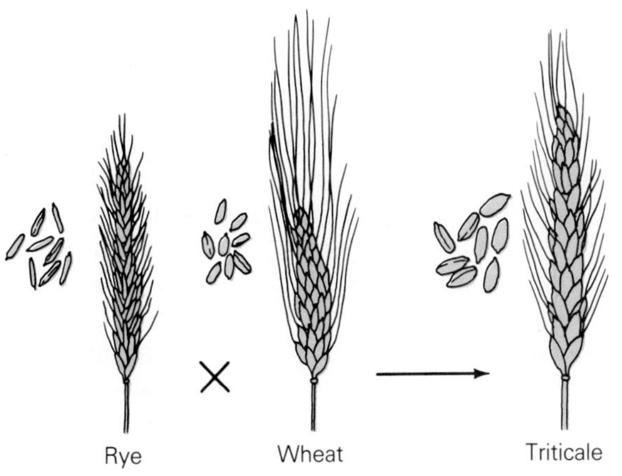

FIGURE 20.12
MIRACLE GRAIN
Triticale is a hybrid grain produced by crossing wheat and rye. (The name is a combination of *Triticum* and *Secale,* the genus names of the two.) Actually, the hybrid formed is sterile because of the usual chromosome incompatibility. But agricultural scientists solve the problem by treating the hybrid with *colchicine,* a mitosis inhibitor that causes a doubling of chromosomes—synthetic tetraploidy. Following this treatment, the treated plant will carry out a normal meiosis and produce viable gametes and a viable F_2, F_3, and so on. *Triticale* combines the vigor of rye with the high grain yield of wheat. Further, the genes of the rye add the amino acid lysine to the usual high quality of wheat protein, resulting in a more nutritious product.

in animals, polyploidy is not likely to be involved. One reason is that many animals have sex chromosomes that function only when specifically paired in precise ways with other sex chromosomes.

One of the best known cases of sympatric speciation in animals involves flightless grasshoppers. The Australian biologist M.J.D. White found that, across their range, these grasshoppers are morphologically homogeneous—they all look alike—but chromosomal analysis revealed that the group is comprised of two different species that had come to occupy different parts of the range. Apparently, a random change had occurred in the parent population that enabled those grasshoppers bearing the change to adapt more precisely to a part of the parent population's former range. The change was great enough, however, to preclude further mixing of the genes from the old population and the new. Thus, a new species apparently arose in the midst of an existing one.

A case of apparent sympatric speciation now in progress involves the maggot fly, a pest of the North American hawthorn tree. When apple trees were introduced to North America in the nineteenth century, the maggot fly began to infect them too. Now, it seems, the flies have become specialized, one line infecting hawthorn trees, the other preferring apple trees. The speciation is not complete because the two lines can still interbreed, but they have come to differ in a number of ways. They show not only pronounced differences in fruit preference, but the apple flies mature in the laboratory in 40 days while the hawthorn flies mature in 54–61 days. They also differ in the frequency of several enzyme-coding genes.

REPRODUCTIVE ISOLATING MECHANISMS

As species form, what keeps them from rejoining in hybridization, intermingling genes and cancelling out their special traits? (Of course, if they did rejoin and interbreed freely, they were not separate species to begin with.) We find that there are a number of such **reproductive isolating mechanisms**, means of enabling species to maintain their integrity and avoid hybridizing. These mechanisms are essentially of two types: prezygotic barriers and postzygotic barriers.

Prezygotic Barriers

Prezygotic means "before the zygote," and so **prezygotic barriers** are those mechanisms that ensure reproductive isolation by acting before the zygote is formed. Here we will consider five such barriers.

FIGURE 20.13
A TIGLON
Tigers and lions have been successfully crossed in captivity, although the offspring are probably sterile. One product, the tiglon, had a siberian tiger father and lion mother. (The reciprocal would be a liger.) Differences in geography and other considerations preclude such hybridization from occurring in the wild.

1. **Geographical Barriers:** Species obviously cannot interbreed if they can't reach each other, and geographical barriers can keep potential breeders apart. The lions of Africa and the tigers of Asia do not meet and therefore do not interbreed. Interestingly, though, when they are allowed to mingle in zoos, they do interbreed, producing fertile *tiglons* (when the tiger is the father) and *ligers* (when the lion is the father) (Figure 20.13).
2. **Ecological Barrier:** When two similar species share the same habitat, they may occupy different parts of it and therefore rarely come into contact (too rarely for mingling of the gene pools). For example, until the last century, the ranges of tigers and lions did overlap in India, yet no hybrids were discovered. One reason may have been that lions tended to live in open grassland (just as they do in Africa today), whereas tigers hunted in the deep shadows of the Asian forests.
3. **Behavioral Barriers:** Animals that might otherwise attempt to interbreed may not be drawn to each other because of differences in their behavior. In some cases, those critical differences may not be related to reproduction. For example, when the lions and

tigers of Asia overlapped, the species may have rarely encountered each other because, not only were they attracted to different habitats, but they had quite different social tendencies. The lions mingled easily in family groups known as prides, while tigers tend carefully to avoid others of their kind, except when breeding. (Although recent evidence suggests they may be more sociable than we once believed.)

In other cases, the behavioral barriers may be directly related to reproduction. Birdsong is a reproductive advertising device that attracts potential mates. However, it does not attract those species not cued to respond to the message. Such reproductive advertising thus brings individuals of a species together without attracting other species. The ranges of the strikingly similar eastern and western meadowlark (Figure 20.14), for example, overlap, yet they are not attracted to each other, apparently in part because of quite distinctive songs.

Behavioral barriers may also involve temporal (time) isolation. Brown and rainbow trout may live in the same waters in the Rockies, but they cannot normally interbreed because the browns mate in the spring and the rainbows in the fall.

4. **Mechanical Barriers:** Some species cannot interbreed simply because their genitalia don't fit. Notably, insects have quite distinct reproductive parts. The male structure may be a tortuous device with hooks, barbs, and protuberances that would seem to discourage any but the most precise of fits. (This barrier in insects may have been overemphasized since, in some species, the female reproductive parts are not so specialized.) Interestingly, some flowers bar hybridization through marked specialization in floral parts; they are compatible only with specific pollinators that do not visit other species of plants.
5. **Gametic Barriers:** Even if individuals of two species should mate successfully, fertilization might not take place because of incompatibility between eggs and sperm. In animals with internal fertilization, the reproductive tract of the female is often a hostile environment for the sperm of another species. The problem is routine, of course, where species such as sea urchins release eggs and sperm into the water. Hybridization, here, may be blocked by the presence of species-specific receptors on the surfaces of gametes. In plants, the pollen tube may not be encouraged to begin growth on the wrong stigma, or if it does, it may soon be stopped by an incompatibility with the female.

FIGURE 20.14
MEADOWLARK SPECIES
Eastern and western meadowlarks are quite similar and have overlapping ranges, yet they do not interbreed.

If all of these prezygotic barriers should be crossed, a hybrid zygote will form. Then an array of postzygotic barriers will come into play. It seems that the world does not welcome those of confused lineage.

Postzygotic Barriers

Postzygotic means "after the zygote", and so **postzygotic barriers** are those mechanism that ensure reproductive isolation by acting after the zygote is formed. Here we will consider four such barriers.

1. **Hybrid failure:** The first barrier discouraging successful hybridization after the sperm and egg have joined is the chance that, because of genetic incompatibility, the genetic material cannot fuse in fertilization and begin the intimate intermingling that sets the stage for mitosis.
2. **Hybrid inviability:** If the genetic material should fuse in fertilization, genetic incompatibility can lead to such severe mismatch problems at meiosis that, after a number of mitotic divisions, the zygote dies.
3. **Hybrid sterility:** In some cases, a hybrid will be produced, forming a vigorous healthy organism, but one that is unable to reproduce successfully because abnormal meiosis in the hybrid produces abnormal gametes. Mules are hybrids between horses and donkeys (Figure 20.15). They are indeed robust and powerful beasts but because of problems within their gamete, they cannot reproduce.

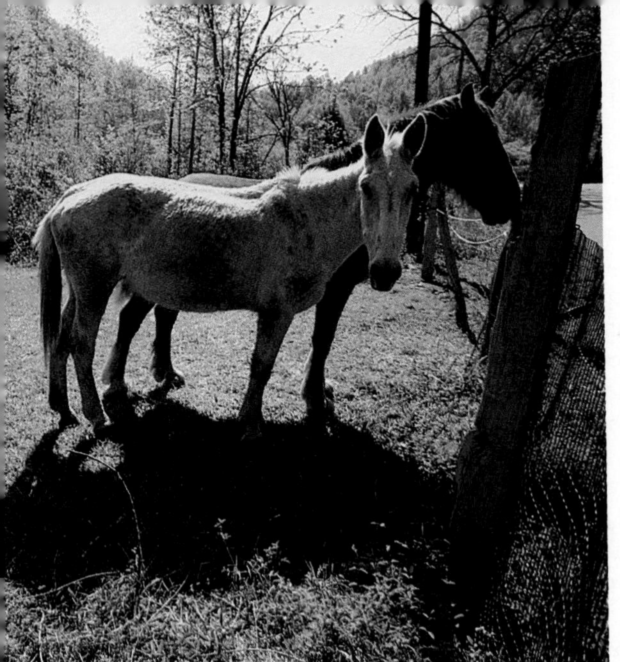

FIGURE 20.15
THE MULE
Some modern-day philosopher described the mule as a sad creature, one that had no past and no future—both genetically accurate observations. Although prized for their stamina and strength, mules are dead-ended because of sterility.

4. **Hybrid breakdown:** In hybrid breakdown, the F_1 hybrids are vigorous and fertile and reproduce easily. The problem comes with a weak and defective F_2 generation that cannot produce viable offspring.

PATTERNS OF EVOLUTION

Now that we have seen how the far-flung species on the planet can appear and how they can be maintained as distinct entities, let's step back and take a broader look at just how this diverse array can be changed and channeled by the forces of natural selection. We will see first how species come to be increasingly different from each other, then how they become more similar, and finally, how they sometimes determine each other's course of evolution.

Divergent Evolution

We used the term **adaptive radiation** to describe the manner in which speciation occurred in the Galapagos finches. The finches spread into new ecological niches, and in the process of adapting they became more different from one another. Put another way, they diverged from each other. The process is called **divergent evolution** and it means that species tend to become increasingly different from each other over time. The idea of divergence has strong ecological overtones, since it often occurs when a species finds some new and different way of using the environmental resources, thus establishing a new ecological niche or, in Darwin's phrase, a new "place in the polity of nature." The opening of new niches is greatly encouraged by natural selection, since it relieves (for a time) the unrelenting competition for resources.

Convergent Evolution

While one may be struck with trends towards divergence in evolution, in some cases different species may grow more alike. The process, called **convergent evolution,** is due to different species undergoing similar adaptations to the same kind of environment. Such convergence occurs on all levels, from biochemical to morphological. Thus, organisms may grow more alike in enzyme systems or photosensitive pigments, as well as in coloration or body shape.

Darwin was struck by the evidence of convergent evolution as he traveled. As we saw earlier, he recorded in his journal that the South American mara, or Patagonian hare, *Dolichotis patagonum,* was quite similar in both appearance and behavior to the European rabbit (see Figure 1.3). Close examination of the mara, however, revealed that it was a rodent, quite distinct from the rabbit, a lagomorph.

Convergence is dramatically illustrated in comparisons between placental mammals and the distantly related, geographically isolated marsupial (pouched) mammals of Australia. Marsupials have established niches very similar to those of many placental mammals of other continents. Through long periods of isolation, selection, and adaptation, unrelated species—placental and marsupial—have often taken on a striking resemblance to one another. Thus in Australia we have the rabbit bandicoot, the marsupial mouse, the marsupial mole, the flying phalanger, the Tasmanian wolf, and the banded anteater (Figure 20.16). All have a placental counterpart on other continents.

Coevolution

Quite often the direction of evolution in one species is strongly influenced by what is happening in the evolution of another species, particularly if the two species are somehow dependent on each other. The reciprocal influence of two species in which they determine each other's evolutionary direction is called **coevolution.** The most obvious examples of coevolution are to be found in predator-prey relationships. As natural selection im-

Placental hare

placental wolverine

placental flying squirrel

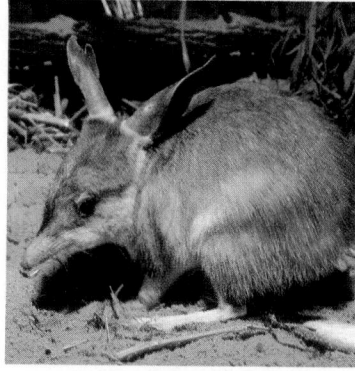

marsupial rabbit bandicoot

Tasmanian devil

marsupial sugar glider

FIGURE 20.16
CONVERGENT EVOLUTION
Convergent evolution is well supported by selected comparisons of certain North American placental mammals with marsupial (pouched) counterparts in Australia. Each pair has made similar adaptations to the environment and shows striking similarities in body structure.

proves the predator's skill, selection also favors greater evasiveness in the prey. As the prey develops better defenses, the predator must develop better means of detection or pursuit. The prey population responds once again, and the "evolutionary chase" goes on.

We see similar adaptations between parasite and host species. Parasites, of course, can kill, but sometimes an animal or plant may harbor parasites without obvious ill effect. The parasite has thus evolved, not only ways of invading the host, but also of not doing too much damage once there. (After all, it is dependent on the host.) The host, for its part, has evolved ways of tolerating the successful parasite. Thus, each species adapts to the other as they "fine tune" their relationship.

In some cases, coevolution has been responsible for great specificity in reproductive modes. For example, the flowering plants and their insect pollinators may come to be intricately adapted to each other through coevolution. In fact, some flowering plant species are pollinated by only one insect species. The yucca is entirely dependent on the yucca moth for pollination, and the yucca moth in turn is entirely dependent on the yucca flower, where it lays its eggs and where its larvae grow on a diet of yucca seeds. The yucca moth must ensure pollination, since only properly pollinated flowers will produce the seeds the moth larvae need.

An interesting example of herbivore-plant coevolution is that between passion-flower vines and a butterfly *(Heliconius)* whose larvae specialize in eatings its leaves (Figure 20.17). This plant has succeeded in manufacturing poisons that prevent most other insects from devouring its leaves and young shoots, but the butterflies have evolved the necessary biochemistry to detoxify the poisons. The female lays bright yellow eggs on the young leaves and shoots. This bright color is a warning to other female butterflies that it would be best for them to find their own plants, and the other butterflies have evolved the response of avoiding laying eggs where there are eggs already, thereby cutting down the competition their own larvae will face.

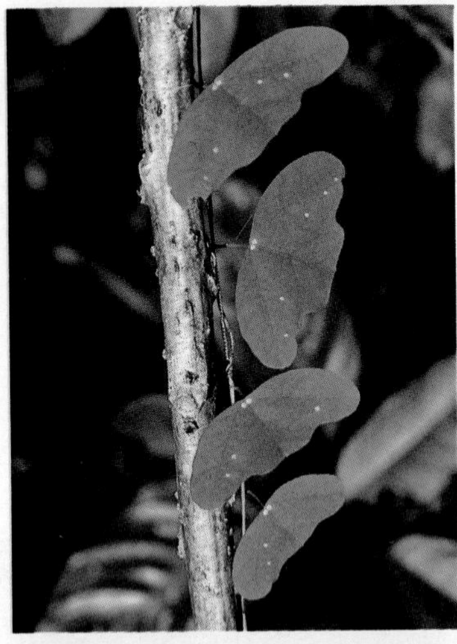

FIGURE 20.17
COEVOLUTION
The passion flower vine (*Passiflora*) continually evolves new strategies to resist the parasitic activity of the *Heliconius* larvae. Displaced nectar glands (yellow spots) resembling the butterfly's eggs discourage "additional" egg laying. (In this case the mimicry has not worked. The brighter spot is the egg of *Helioconius* amid the less colorful spots produced by the plant.) Variation in leaf shape may also mislead egg-laying females searching for a specific shape and size.

We see, then, that natural selection molds populations in a variety of ways, that its influences can stem from a variety of sources and can proceed in a number of directions. One of the precepts of Darwinian evolution is that these changes occur gradually, that they are the result of a relentless accumulation of small changes. However, this idea has recently been challenged.

SMALL STEPS OR GREAT LEAPS?

Charles Darwin's assumption that species change slowly through time because of the accumulation of small changes, is now called **gradualism** (Figure 20.18a). Darwin was fully aware that there were certain problems with this concept, particularly as evidenced in the fossil record. He noticed that many species seemed to go unchanged for great periods of time while other forms seemed to change drastically over geologically brief periods. He was convinced that the problem lay in the "incompleteness" of the fossil record. After all, he reasoned, we are likely to find only pitifully small remnants of the vast array of life that once existed. Further, the remnants themselves are often incomplete and so severely treated by time that it can be hard to judge what the living thing might have looked like, or how rapidly its kind had changed.

As time went on, there was increasing evidence in the fossil record that species go unchanged, sometimes for millions of years, and then suddenly and dramatically undergo marked changes—so marked, in fact, that they produce new species. Niles Eldridge of the American Museum of Natural History and Stephen Jay Gould of Harvard University proposed that these abrupt and great changes in the fossil record were not due to its incompleteness, but because of the fact that evolution is not a slow, gradual process. They claimed, quite simply, that evolution generally proceeds in fits and starts, that species, once formed, go unchanged for great periods, and that when they do change, they are likely to change markedly. This concept is called **punctuated equilibrium** (Figure 20.18b). The "suddenness," they stress, must be viewed in geological time, perhaps tens of thousands of years. This may seem like a long time, but keep in mind that species last, on the average, several million years on the planet, so even tens of thousands of years is but a small part of that time.

The concept of evolution, we see, encompasses a range of complex and interactive ideas. Since its inception, it has also provided us with a host of questions, enduring puzzles that continue to resist our intellectual attacks. Someday we may know the answers to all our vexing questions about how and why life changes and how the various forms are related. And yet, perhaps not. It is possible that some of the events of our biological past will never be unraveled, and that these same questions will remain forever as unweathered monuments to human limitation. Thus, whenever our species passes into extinction, we may carry with us many of the same questions that we ask today.

EXTINCTION AND EVOLUTION

Just as speciation marks the birth of species, extinction heralds the death of species. Extinctions have always been a part of the evolving drama of life on the planet, from those first failed droplets to the much publicized modern species that have dwindled into oblivion. These latter cases are usually isolated instances and not part of some sweeping change in which many species die out at once. In addition to these isolated extinctions, there are rather constant, ongoing and low-level "background" extinctions that have always marked life on earth.

The history of life has also been marked by more dramatic levels of extinction, those involving a great number and variety of species. Specifically, some scientists estimate that there have been five great extinctions since life appeared. These were so important because they were critical to the evolution of the remaining life. The first occurred at the end of the Cambrian period, during the Paleozoic era (about 500 million years ago). The next great extinction occurred at the end of the Ordovician period (about 425 million years ago), the next at the end of the Devonian period (some 345 million years ago).

Then came the fourth great extinction. This one was perhaps the most sweeping, marking the demise of a great many species at the close of the Permian period (some 225 million years ago). It is estimated that more than 90 percent of marine life went extinct at that time. The fifth, and so far the last, major extinction occurred at the end of the Mesozoic Era (some 65 million years ago).

In 1983, D. Raup, J. Sepkowski, and D. Daveys reported evidence of great die-offs every 26 to 28 million years throughout life's tenure on earth. Their conclusions have been heavily debated, even while a number of theories were developed to explain the occurrences. Some believe the fatal events to be celestial (see the Alvarez hypothesis, Chapter 32) while other believe that natural changes on earth were sufficient to cause the die-offs.

However frequently and for whatever causes the great extinctions occurred, they certainly took a great toll on existing life and opened new and unchallenged evolutionary directions for the survivors. Now, however, the question of extinction has taken on a new and more personal significance.

Extinction and Us

There is disconcerting evidence that we are apparently in the midst of another, truly devastating mass extinction, and this time there is little question of the cause. As Pogo once said, "We have met the enemy, and he is us." Indeed, the extinction rate is presently estimated by a number of experts to be between a thousand and several thousand species per year. If we continue on our present course of environmental destruction, by the year 2000 the extinction rate could be about 100 species per day. Within the next several decades, we could lose a quarter to a third of all species now alive. This rate of loss is unprecedented on the planet. Furthermore, the problem is qualitative, as well as quantitative. Not only are we losing *more* species, but we're also losing different *kinds* of species. The earlier mass extinction involved only certain groups of species, such as the cycads and the dinosaurs. The other species were left more or less intact. Presently, though, species are dying out across the board. That is, the current extinction affects all the major categories of species. Of particular note, this time the terrestrial plants are involved. In the past, such plants provided resources that the surviving animals could use to launch their comeback. With the plants also devastated, any comeback (marked by a period of rapid expansion and speciation) by animals will be greatly slowed.

The present great extinction will also prove to be devastating to the resurgence of species for another reason. This time we are killing the systems that are particularly rich in life, including tropical forests, coral reefs, salt water marshes, river systems, and estuaries. In the past these systems have provided genetic reservoirs from which new species could spring and replenish the diversity of life on the planet. In effect, we are drying up the wellspring of future speciation.

In the past, extinctions have been due to two major processes, environmental forces (such as climatic change) and competition. For the first time, a single species (our own) has had the opportunity to cause mass extinctions at both levels. Much of the habitat

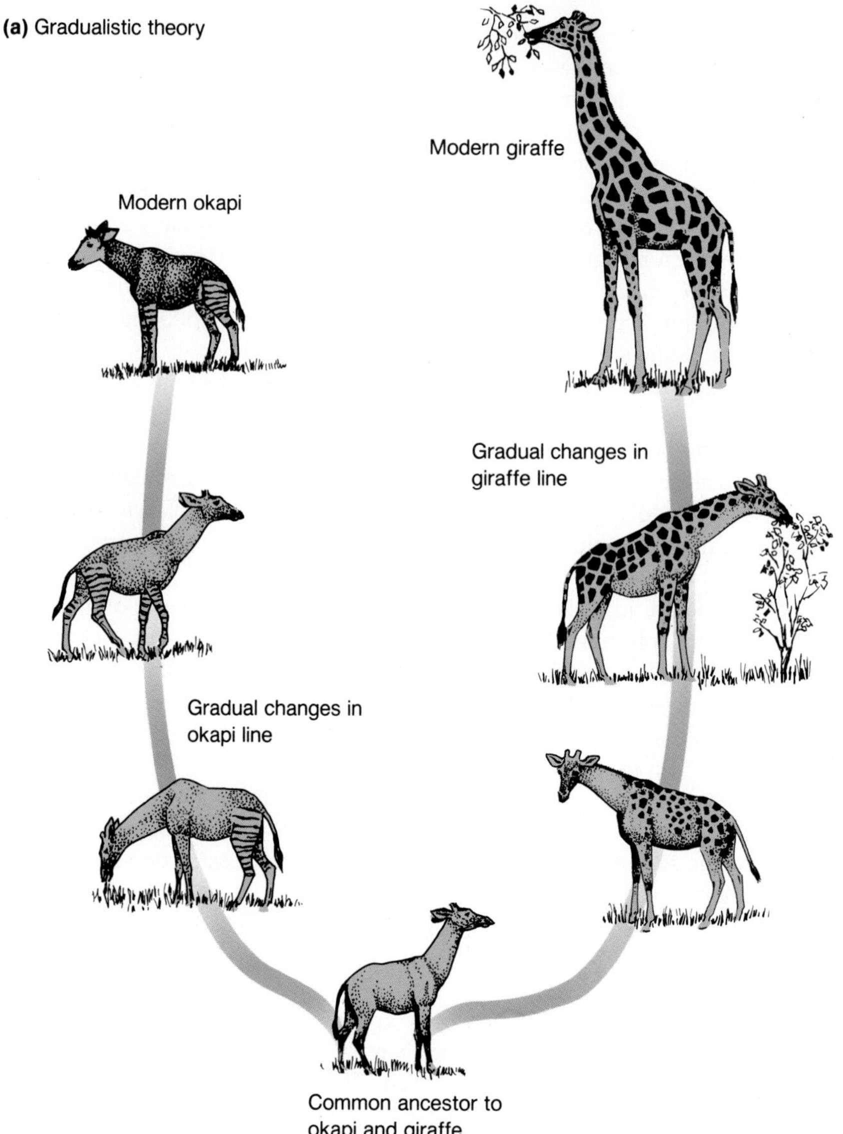

FIGURE 20.18
GRADUALISM VS. PUNCTUATED EQUILIBRIUM
The okapi and giraffe are believed to have descended from a common ancester. **(a)** According to the traditional, gradualistic model, an early division produced two lines, each of which underwent gradual evolutional transitions. The accumulation of countless minute changes leads to the modern representatives. **(b)** The punctuated model proposes that the giraffe line witnessed long periods with little change, interrupted by widely separated episodes of speciation, each resulting in the sudden rise of new species. The most recent episode of speciation produced the modern giraffe. The okapi line branched early from the ancestral line and, following its initial changes, remained relatively unchanged over time.

destruction has been due to humans needing the land where other species lived, forcing them into extinction as they failed in their competition with us. And now we find ourselves in the remarkable position of being able to alter environmental forces. As the Amazon basin is destroyed, the trees are no longer available to cycle water back to the atmosphere, causing many experts to predict sweeping changes in the weather.

We are continuing to interact in new ways with the environment. As we continue to release chlorofluorocarbons into the atmosphere, the ozone holes grow larger and the delicate veil of life on earth becomes increasingly bathed in destructive radiation. And we are learning that the oceans have become perilous places for many forms of life because we continue to use the great waters as dumps for dangerous or unknown chemicals.

There are those who say that our course, by now, is irreversible, that the damage is done and now we must sit back and prepare to reap what we have sown. Others, though, argue that there is still time to alter our course, to salvage much of what is left and to protect it from further damage. The danger is in accepting the first alternative if the second is really the case.

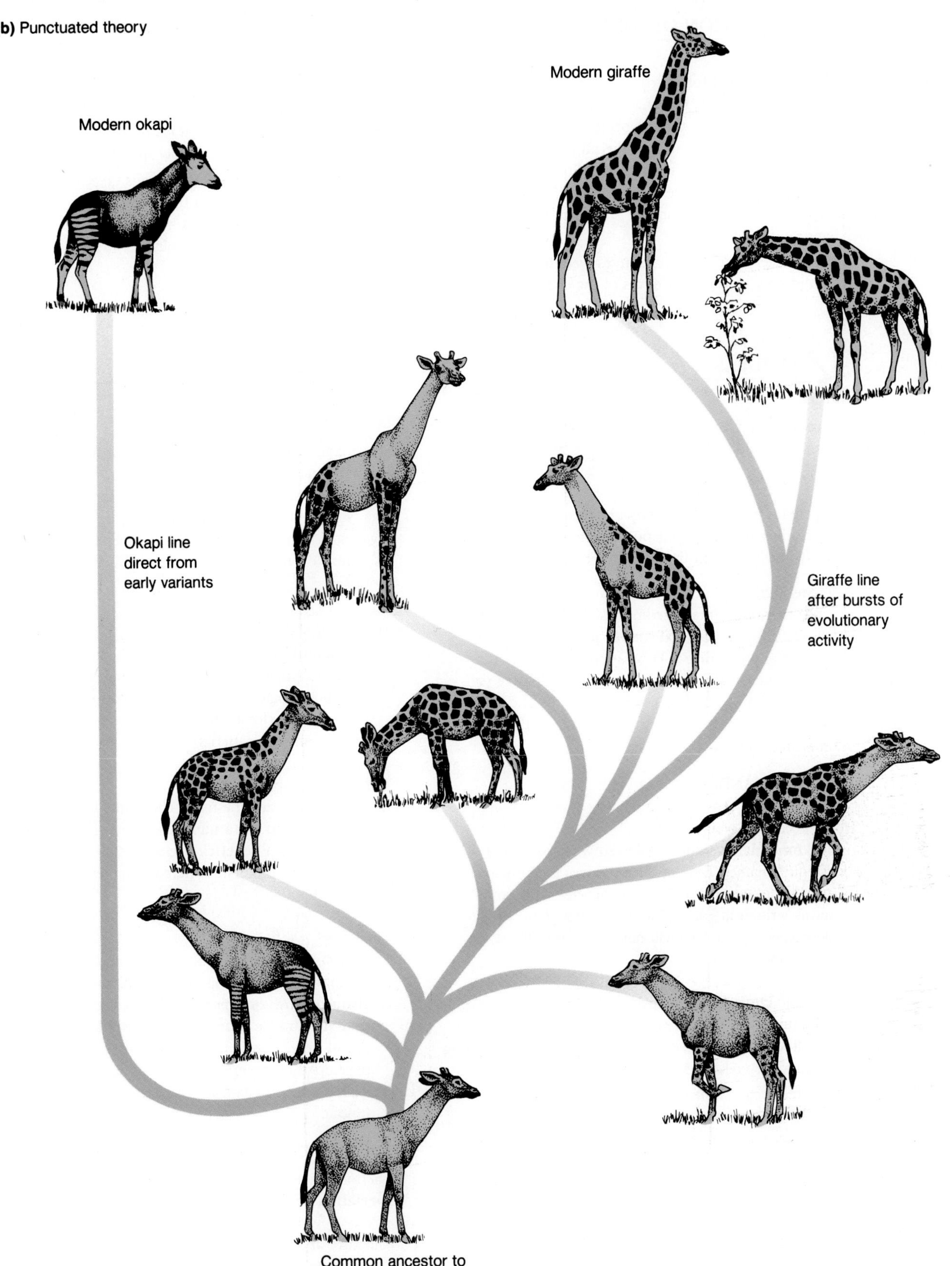
(b) Punctuated theory
Modern okapi
Modern giraffe
Okapi line
direct from
early variants
Giraffe line
after bursts of
evolutionary
activity
Common ancestor to
okapi and giraffe

APPLICATION OF IDEAS

1. The fossil record for nearly all forms of multicellular life on earth begins rather abruptly shortly before the Cambrian period of the Paleozoic era. This raises numerous questions and, in the minds of some, adds fuel to the creationist's arguments. Forgetting "special creation" for the moment, offer two different hypothetical explanations for the seemingly abrupt appearance of such life.
2. Until recently, phylogenetic trees were hypotheses that could only be tested in terms of the characteristics contrived by the tree's originator. This is no longer true. Describe two important discoveries of this century that offer other ways of testing these hypotheses and explain how such testing is done.
3. In assigning organisms to various taxa (or in determining which taxa they have previously been assigned), a dichotomous key is commonly used. Find an example of a dichotomous key in the literature and explain how one is used.

KEY IDEAS

Macroevolution involves the rise of new species and the extinction of old species.

WHAT IS A SPECIES?

A **species** is a group of actually or potentially interbreeding natural populations, each group reproductively isolated from the others.

Problems With the Species Concept

The criterion of interbreeding in defining species is not always sufficient. Several species in the dog family successfully interbreed. Populations of a salamander species successfully interbreed all along their very long, U-shaped range, but when individuals occupying the two ends of the range are brought together, interbreeding fails.

TAXONOMY AND SYSTEMATICS

How Species are Named

The **binomial system** is followed when assigning scientific names. Each species receives a **genus** and species designation (e.g., *Homo sapiens*) with both terms written in Latin. The binomial system was developed by the Swedish naturalist Linnaeus (Carl von Linne).

Taxonomic Organization: Kingdom to Species

1. **Taxonomy** involves identifying, naming, and classifying organisms; **systematics** involves establishing evolutionary relationships.
2. The hierarchy of taxa into which organisms are assigned—in the descending order—is: **kingdom, phylum, class, order, family, genus, species**, and **subspecies.**

The Five Kingdoms

The five kingdoms include **Monera** (prokaryotes, including Archaebacteria and Eubacteria), **Protista** (primarily single-celled and colonial eukaryotes, some multicellular), **Fungi** (mostly multicellular, parasitic, and scavenging eukaryotes), **Plantae** (multicellular, photosynthetic, eukaryotes), and **Animalia** (multicellular, heterotrophic eukaryotes).

METHODS IN TAXONOMY

Systematics involves establishing relationships based on common ancestry. The terms "primitive" and "advanced" pertain to traits rather than to species. In establishing taxa, some systematists ("lumpers") tend to organize many organisms into large groups, while other taxonomists ("splitters") favor the establishment of many smaller groups.

The Classical Approach

In establishing evolutionary relationships through traditional methods, systematists concern themselves with homologies (structures of similar origin), paleontology (fossil record), and embryology (developmental patterns).

The Phenetic Approach

Pheneticists examine many characters in a group of organisms and use a numerical scoring system to draw comparisons. **Phenetics** stresses visible phenotype, which raises problems where convergent evolution is encountered.

Cladistics

Cladists organize their taxonomic trees, or **cladograms,** according to primitive and derived characters. All organisms in the cladogram exhibit shared primitive characters. (All chordates have a dorsal nerve cord and a notochord.) Branches are then developed by grouping organisms according to the shared derived characters each group thus being unique.

THE MECHANISMS OF SPECIATION

The formation of new species from old is called **speciation.** It occurs in several ways.

Allopatry and Speciation

Allopatric speciation is the rise of new species following geographical isolation. Isolation may occur through mistakes in navigation or other accidents. In a longer time frame, isolation can be produced by barriers formed through geological events such as erosion, earthquake, volcanism, **continental drift,** and others.

Parapatry and Speciation

Parapatric speciation occurs between abutting populations of a species between which gene flow occurs but is diminished. The flow is, in turn, overpowered by different selective factors operating on the two populations, so speciation occurs. Since the usual tendency is for gene flow to dilute any such changes, parapatric speciation is rare.

Sympatry and Speciation

1. **Sympatric speciation** occurs within a continuous population.
2. In plants, sympatric speciation occurs through hybridization and **polyploidy.** Hybridization readily occurs through genetically similar species, sometimes with **hybrid swarms** occurring along overlapping ranges.
3. Whereas hybrids may be sterile because of chromosome incompatibility at meiosis, this can be overcome through polyploidy—a spontaneous chromosome increase by doubling, tripling, etc., of chromosomes. Whereas **autotetraploidy** (doubling of chromosomes in non-hybrids) produces abnormalities in meiosis, **allotetraploidy** in hybrids results in compatible meiotic chromosomes and fertile offspring. New species thus arise. *Triticale* is an allohexaploid. Polyploids seem to tolerate harsh climates well, and 85% of Greenland's flowering plants are polyploids.
4. Sympatric speciation in animals is rare, but examples include the Australian flightless grasshopper and the maggot fly. One line of maggot fly infects apple trees, another, hawthorn trees, and although they can still interbreed, several significant differences have come about.

REPRODUCTIVE ISOLATING MECHANISMS

Several barriers prevent diverging species from rejoining.

Prezygotic Barriers

Prezygotic barriers (before the zygote) include **geographical** (distance), **ecological** (different habitat preferences), **behavioral** (different social tendencies such as mating behaviors), **mechanical** (incompatible genitalia and pollination structures), and **gametic** (sperm barriers or incompatible pollination mechanisms).

Postzygotic Barriers

Postzygotic barriers include **hybrid failure** (failure of sperm and egg nuclei to fuse), **hybrid inviability** (problems with development), **hybrid sterility** (failure in meiosis and thus abnormal gametes), and **hybrid breakdown** (segregation in F_2 yields weakened or noviable offspring).

PATTERNS OF EVOLUTION

Divergent Evolution

Divergent evolution involves newly formed species becoming progressively different.

Convergent Evolution

In **convergent evolution**, different species living in similar environments take on chemical and physical similarities. There are many examples between placental and marsupial mammals.

Coevolution

Coevolution involves reciprocal influences between two species. It is commonly seen in close relationships such as those of predator and prey, parasite and host, and herbivore and plant.

SMALL STEPS OR GREAT LEAPS?

The notion of evolution occurring through the gradual accumulation of small changes is called **gradualism.** An opposing view, called **punctuated equilibrium,** maintains that evolution occurs in intermittent spurts of activity following long periods of inactivity. Some gaps in the incomplete fossil record may be attributed to punctuated equilibrium.

EXTINCTION AND EVOLUTION

Evolution is marked by ongoing low-level extinctions and by dramatic mass extinctions. The fossil record reveals one that occurred 500 million and another, 225 million years ago. The last great extinction occurred some 65 million years ago.

Extinction and Us

Evidence today suggests we are in another period of mass extinction, this one related to environmental destruction by humans.

REVIEW QUESTIONS

1. Applying the interbreeding provision, define the term "species." (411)
2. Describe two situations where the interbreeding provision of the species definition fails. (411)
3. Distinguish between the work of the taxonomist and systematist. (412)
4. Explain how the bionomial system is used in naming species. Why is the Latin language used? (413)
5. List the five levels in the taxonomic hierarchy and classify humans in these levels. (414)
6. Name the five kingdoms and list examples of organisms in each. (415)
7. Why is it erroneous to characterize a species as primitive or advanced? (416)
8. Compare the way "lumpers" and "splitters" look upon the grouping of organisms. (416)

9. List three main sources of information used by the traditional systematist when establishing taxonomic relationships. (416)
10. Explain how pheneticists establish phylogenetic relationships. What problems does convergent evolution bring to them? (417)
11. How does the cladist make use of shared primitive and shared derived characters in establishing phylogenetic trees? (417-418)
12. What did the cladists find that was surprising about the evolutionary relationship between birds and crocodiles? (418-419)
13. Using the Galapagos finches as an example, explain how allopatric speciation occurs. (420)
14. Suggest several ways geographical barriers might arise. (420)
15. What is parapatry? Why is it rare? (420)
16. What is sympatric speciation? What organisms would one best look to for examples? (420, 424)
17. What is the usual effect of hybridization in animals? Why does this happen? (420)
18. Compare autotetraploidy and allotetraploidy in plants. Which provides a means for meiosis and gamete formation to succeed? (424)
19. In what kinds of environments do polyploids seem to survive best? (425)
20. Give an example of an economically important polyploid species and state its parentage. (425)
21. List examples of sympatric speciation in animals. (426)
22. What are reproductive isolating mechanisms? How are they significant to speciation? (426)
23. Briefly discuss three prezygotic barriers. (426-427)
24. Briefly discuss three postzygotic barriers. (427-428)
25. Using the Galapagos finches as an example, explain divergent evolution. (428)
26. What factors characterize convergent evolution? Provide three examples of convergence between Australian marsupials and placental mammals. (428)
27. List three kinds of relationships in which coevolution might occur. (428-429)
28. How would you characterize a successful parasite? (429)
29. Characterize the way in which evolution occurs according to both gradualism and punctuated equilibrium. (430)
30. Discuss two ways of interpreting the incompleteness of the fossil record. (430)
31. Summarize the history of mass extinctions. Which was the most recent? (430-433)
32. List two dangerous, man-made environmental changes that are going on right now. (433)

21

ORIGIN OF LIFE

WE HAVE NOW HAD A LOOK AT SOME OF THE PRINCIPLES GOVERNING EVOlution, and we have explored its classical triad of forces: mutation, variation, and natural selection. There is indeed a substantial body of information on such processes, yet many of the most fundamental questions remain. The ultimate question, of course, is how did life begin? The question is as old as humanity itself. The ancients throughout the world were absorbed by this mystery, and their conclusions have lingered in our consciousness, often forming the basic premises of many philosophies and religions.

Undoubtedly, many of the ancient cultures had great faith in their ideas. But faith in an idea, while it may serve as some sort of motivating force, is only a starting point. The faith must stand ready to be tested, and the burden of proof falls on the faithful.

But how does one investigate an improbable event that may have happened only once, several billion years ago? It is clear that we can never *prove* how life came to be on this planet. We can, however, examine any number of seemingly plausible notions of how life *might* have arisen. All scenarios or speculations about the origin of life must involve suppositions. These suppositions can lead to predictions, and it is the predictions that can be tested. Many such speculations have been trotted out, and many of the predictions based on their suppositions have been tested. Most of the proposed schemes of how life might have originated couldn't possibly have worked; they have depended on assumptions about nature that have proven to be untrue. But through the process of elimination, we have narrowed the range of possibilities and have highlighted the most promising avenues for further examination. At this time, scientists generally agree that the preponderance of evidence indicates that *life successfully arose only once* in the earth's history and that the conditions at that time were unique, probably never to occur again.

Note: Geological time is charted inside the front cover.

A HYPOTHESIS ON THE ORIGIN OF LIFE

Although Darwin speculated that life might have arisen in a warm, phosphate-rich pond, the first serious proposals concerning the spontaneous origin of life (those that were based on sound biochemical and geological information) began to appear some 60 years ago. Such schemes were presented by J.B.S. Haldane, a Scottish biochemist, and by A.P. Oparin, his Russian counterpart. They proposed that soon after the earth's formation, when conditions were quite different from those existing today, a period of spontaneous chemical synthesis began in the warm ancient seas. During this era, amino acids, sugars, and nucleotide bases—the structural subunits of some of life's macromolecules—formed spontaneously from the hydrogen-rich molecules of ammonia, methane, and water. Such spontaneous synthesis, explained Haldane, was only possible because there was little oxygen in the atmosphere, an important condition since oxygen will react spontaneously with many organic substances. The energy for this synthesis was, at this time, readily available in the form of lightning, ultraviolet light, heat, and higher energy radiations. Since there were no organisms to degrade the spontaneously formed organic molecules, they accumulated until the sea became a "hot, thin soup" (of, perhaps, about the consistency and nutritive quality of chicken bouillon) (Figure 21.1).

FIGURE 21.1
SCENARIO OF EARLIEST EARTH CONDITIONS
Torrential rains formed the young seas while volcanoes released gases into the developing atmosphere. Heat and electrical discharge provided energy for chemical synthesis.

Both Haldane and Oparin suggested that polymerization—the joining of the structural subunits into macromolecules—was greatly encouraged by the ever-increasing concentrations of precursor molecules. Amino acids joined to form the first polypeptides and eventually the first complete proteins. Certain active versions of these proteins became the earliest enzymes. The rate of chemical interaction must have been very slight at first, but there was plenty of time. The early enzymes may not have been particularly effective, but in time they were able to catalyze a number of kinds of activity, including that

which would have produced more proteins like themselves. Collections of these new catalysts then became enclosed by simple membranelike, water-resistant protein or lipid shells. All the while, these active little bodies would have perpetuated themselves by making use of the energy-rich nutrients of the ancient sea.

At first these conglomerates divided simply because they became too large and fell apart. But eventually the crude membrane-surrounded droplets, or **coacervates** as Oparin called them, took in new types of molecules, *autocatalytic* molecules that in some way induced the formation of molecules somewhat like themselves. These would have been the precursors of genes. With such systems the coacervate can be called a **protocell** (prototype cell). Keep in mind that in Haldane and Oparin's time, not much was known of DNA or of its crucial genetic role. These first protocells, with their crude genetic mechanism, persisted and increased in numbers. The presence of this reproducing molecule would have increased the odds that these protocells could have continued their existence in the primordial sea. The scene was now set for the inexorable forces of natural selection. The survival of the fittest had begun. According to Haldane and Oparin, from these early and persistent successes, the forces of natural selection molded and winnowed the tiny, membranous droplets and eventually produced the first simple cellular life.

The **Haldane-Oparin hypothesis** relies heavily on a number of assumptions about the physical conditions of the early earth, as well as the occurrence of largely random events. Nonetheless, although some of the assumptions are now known to be incorrect and the original hypothesis has had to be modified, the overall hypothesis is yet to be disproven.

Haldane's and Oparin's hypothetical scheme remained a neglected intellectual curiosity for more than a quarter of a century. But in 1952, Nobel laureate Harold Urey (the discoverer of deuterium, a heavy isotope of hydrogen) and Stanley Miller, his graduate student, began testing some of the assumptions in earnest.

The Miller-Urey Experiment

The crucial proposition of the Haldane-Oparin hypothesis was that the precursors of the molecules of life would form spontaneously in the primitive atmosphere. Miller and Urey created a laboratory apparatus at the University of Chicago that attempted to simulate what were then believed to be the conditions of the primitive earth (Figure 21.2). They introduced a small amount of water and a mixture of gases including

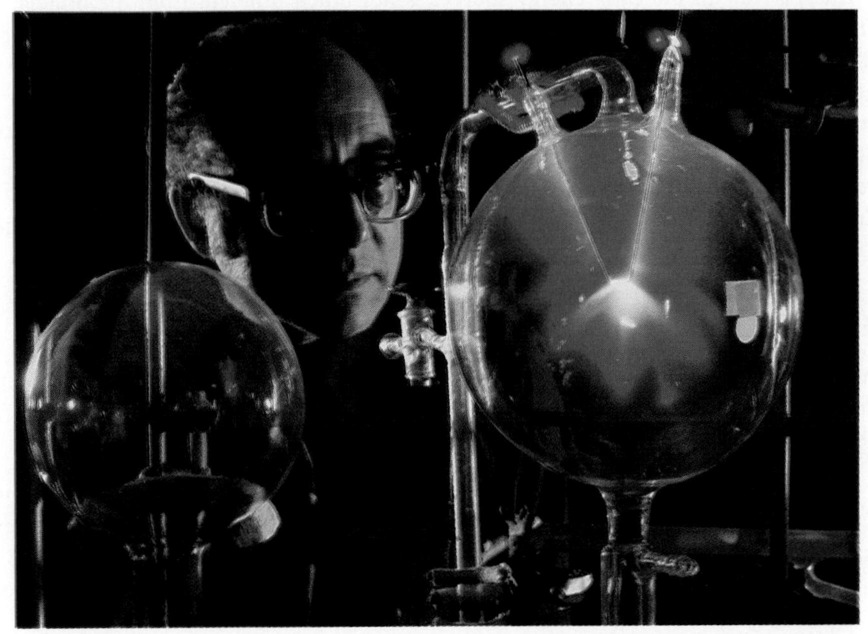

FIGURE 21.2
THE CLASSIC MILLER-UREY EXPERIMENT
Heated gases of the theoretical primitive atmosphere were subjected to electrical discharges in a sealed, sterile environment. Residues were collected in the lower chamber and periodically analyzed. Results indicated that some of the simple monomers of life could be produced spontaneously under the test conditions.

methane, ammonia, water vapor, and hydrogen—but no free oxygen—into the apparatus. As an energy source, they produced repeated electrical discharges (lightning?) through the atmosphere of the upper flask. After a week, they analyzed the sediments that collected in the lower flask. Among the various molecules they found aldehydes, carboxylic acids, and, most interestingly, amino acids, all of which are commonly found in living cells.

Although these small molecules were a far cry from living things, the fact that they were produced at all provoked a lively revival of interest in the Haldane-Oparin hypothesis. But before we discuss the work that followed Miller and Urey's breakthrough, let's review what scientists today believe the early earth was like.

THE EARLY EARTH

The best estimates suggest that the earth took form about 4.6 billion years ago, along with the sun and the other planets of our solar system. Prior to this, the precursor of the solar system was a vast, flattened cloud of gases, dust, and other debris (Figure 21.3). Recent theories maintain that the cloud was cold, but that as the sun and planets coalesced, a great deal of heat was generated from the press of gravitational forces, supplemented with heat from radioactivity. When it took form, the earth's crust was a molten semiliquid. Some 600 to 800 million years were to pass before it solidified. (Figure 21.4 is a timetable of earth history.)

When the crust finally cooled to below the boiling point of water, torrential rains began to fall, initiating the formation of the oceans. The atmosphere, of volcanic origin, consisted largely of water vapor, methane, carbon monoxide, carbon dioxide, ammonia, hydrogen, and hydrogen sulfide. This is close to the list proposed by Oparin and Haldane years ago, but now we know more.

Recently, atmospheric scientists, applying new data (some derived from NASA probes of the solar system), have carried out computer simulations of the primitive atmosphere. They revealed that the hydrogen-rich precursors of the Haldane-Oparin atmosphere would have been rapidly broken down by ultraviolet radiation, and that most free hydrogen would have been lost to outer space. So the list of conditions and chemical constituents has been revised. The constituents of the early atmosphere are now thought to have been water vapor (H_2O), carbon dioxide (CO_2), carbon monoxide (CO), molecular nitrogen (N_2), and possibly some hydrogen (H_2). Happily, recent experiments, similar to those of Miller and Urey but using the revised list of gases, produced even greater yields of small organic molecules.

Today's view of the early energy sources is not much different from the original—scientists still maintain that energy abounded in the primitive atmosphere. For instance, ultraviolet light was plentiful; the ozone layer, which today screens out much of this energy, had yet to form. (Ozone is O_3, a molecular variant of oxygen.) Other energy sources included lightning, heat, and what geologists call shock energy (from earthquakes and tremors). So, the scientific consensus today is that the major requirements of chemical evolution—reactive gases and sources of high energy—were in abundance. In fact, many scientists view the spontaneous formation of monomers as an inevitability rather than a mere possibility.

THE HYPOTHESIS TODAY

Scientists today continue to test the hypothesis of the spontaneous generation of life. In particular, five aspects of the hypothesis must receive attention if the whole is to be taken seriously:

1. Verification and clarification of the physical conditions of the primitive earth must continue. Scientists must continue with their efforts to understand the conditions of the primitive earth. Without such understanding, any hypothesis becomes a house of cards, ready to tumble if its foundations are disturbed (status: very good and getting better).

FIGURE 21.3
THE YOUNG SOLAR SYSTEM
The hypothetical events in the formation of the solar system from a flattened cloud of cold gases: *(top)* the flattened dust cloud, *(center)* sun and planets forming, and *(bottom)* the completed solar system.

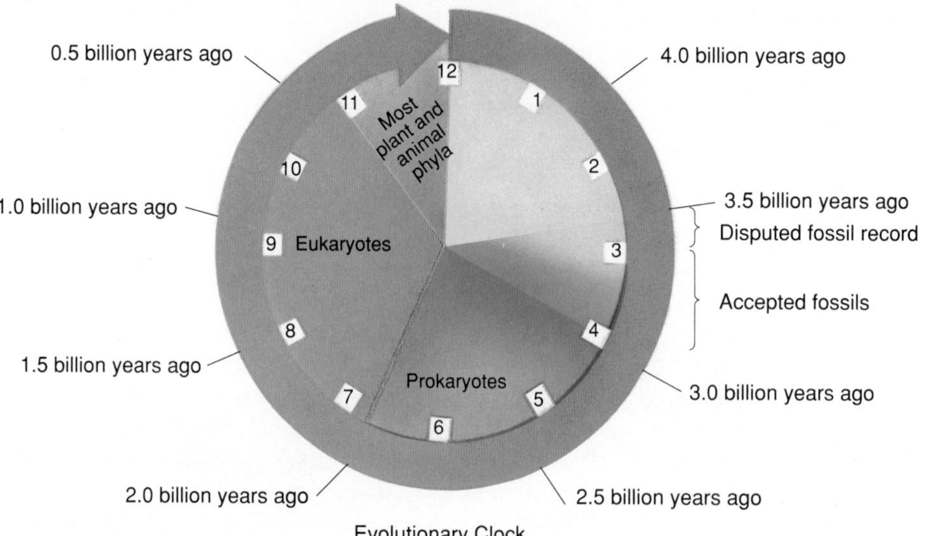

Evolutionary Clock

12 Midnight –	Earth forms	11:00 AM –	First vertebrates
2:00 AM –	Crust forms	11:30 AM –	Age of dinosaurs
3:00 AM –	First undisputed life	11:50 AM –	Age of mammals
3:00 AM –	Prokaryotes	11:59:00 –	First hominids
6:45 AM –	First eukaryotes	11:59:40 –	First humans
10:45 AM –	Primitive animal phyla evolving	11:59:59 –	All of human history
10:54 AM –	First terrestrial plants	12:00 Noon –	Present

FIGURE 21.4
AN EARTH CALENDAR
Here, the earth's history is shrunk into a period of 12 hours—from midnight to noon. The events are in chronological order, and the spacing indicates elapsed time.

2. It must be shown that essential monomers, such as amino acids, nitrogen bases, and simple sugars, can be produced under primitive conditions, in the absence of enzymes or other biological activity (status: well established).
3. It must be shown that the familiar polymers of life—proteins, nucleic acids, and so forth—can form spontaneously from monomers (status: not well established).
4. The spontaneous formation of active, well-defined, cell-like bodies with isolating membranes or borders must be verified (status: fairly good).
5. It must be shown to be at least possible that all of this will result in the production of simple, self-replicating systems, with repositories of genetic information and the ability to maintain metabolic processes from generation to generation (status: not yet established).

The Monomers

The early work of Miller and Urey gave rise to a host of similar experiments. The atmospheric constituents and the energy sources have been varied, especially as our knowledge of primitive conditions has been refined. The results have been rewarding. The list of laboratory-synthesized monomers now includes all of the nucleotide bases of DNA and RNA, along with the essential sugars, all of the amino acids, and most essential vitamins. (As an aside, it turns out that adenine, the most widely occurring nucleotide base in today's life forms, has been the easiest to synthesize.) Thus, research supports the idea that many of the monomers associated with life were produced through spontaneous generation on the early earth.

The Polymers: Macromolecules of Life

It is one thing to produce such monomers and quite another to induce them to join, forming polymers, without the assistance of enzymes. It is here that the original hot, thin soup hypothesis is weakest. All biological polymerizations involve *dehydration linkages* between the monomers—that is, removal of water to produce the linkage (see

Chapter 3). Researchers agree that such reactions would *not* have been energetically favorable in the primitive sea, or indeed in any aqueous medium, because of the mass action law. In water, biological polymers slowly dissociate back into monomers, and heat just accelerates the process. Without the catalyzing effects of enzymes, spontaneous polymerization is possible only when the concentration of monomers in water is high.

With such problems, then, how could polymerization have been achieved? There have been many suggestions. For example, biologist Carl R. Woese, now with the University of Illinois, has proposed that life began, not in the sea, but in the hot, extremely dense atmosphere of the *very* early earth.

Quite a different view is proposed by Sidney Fox of the University of Miami. Fox has demonstrated that polymerization of amino acids occurs readily under hot, drying conditions such as might be found along the edges of volcanoes or even on the hot beaches of ancient seas. Pools of organic precursors, rich in amino acids, could have been concentrated by evaporation and heated to allow the spontaneous formation of polypeptides. Fox has succeeded in producing polymers of 200 or more amino acids under hot, drying conditions. Fox calls aggregations of these spontaneously generated polymers **thermal proteinoids.** We'll return to Fox's work shortly.

One of the greatest problems in reconstructing the development of life lies in explaining how the nucleic acids, DNA and RNA, might have formed. Investigators have even stacked the deck in trying to form such molecules. They have boiled and dried concentrated energy-primed nucleotide triphosphates in the presence of single-stranded templates of DNA, producing an energetically favorable direction of reaction. However, without the appropriate enzymes, no second strand of DNA is produced. The units do not link to form polymers. Linkages can be forced, but they occur in the wrong places. To date, there is no evidence that nucleic acid polymers can be produced spontaneously; hence, some scientists are now theorizing that the first synthesis of nucleic acids might have come long after the actual origin of life.

Self-Replicating Systems One of the questions that bothers some investigators is which macromolecules formed first: nucleic acids or proteins? The problem is that contemporary organisms use nucleic acids—DNA, mRNA, tRNA, and rRNA—to synthesize the polypeptides incorporated into enzymes, yet at the same time, enzymes appear to be necessary to synthesize nucleic acids. It's the ultimate chicken-and-egg problem. Those who favor proteins as primitive macromolecules point out the relative ease with which peptide bonds are produced and the great generality of protein's enzymatic activity. Those who favor nucleic acids dwell on the information-carrying feature of the genetic material and the fact that some enzymelike characteristics are associated with certain RNA molecules. They suggest that rRNA and tRNA are the remnants of a once-large class of nucleic acid "enzymes" or enzymelike molecules. It's possible that the two systems originated together. We'll return for another look at RNA shortly.

FIGURE 21.5
THE COACERVATE
When proteins or other polymers are introduced into water, they tend to cluster together into distinct droplets called coacervates. The coacervate surrounds itself with a boundary layer that is selective in admitting kinds of molecules. When coacervates reach a critical size and mass, they divide spontaneously, a process characteristic of cells.

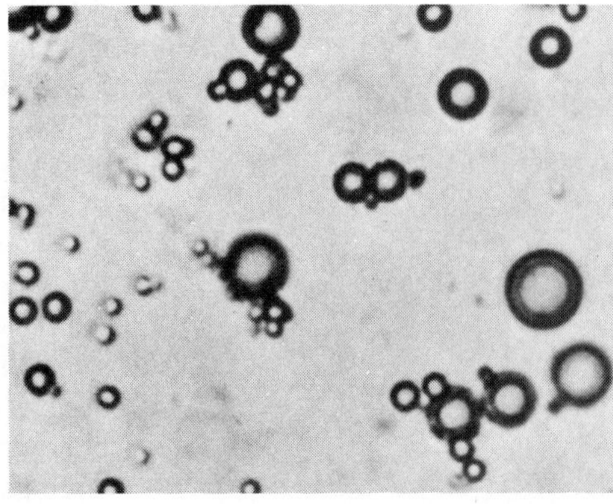

MODEL PROTOCELLS
Oparin and the Coacervate

After his proposal on the spontaneous generation of life was made, Oparin spent some 50 years trying to test his assumptions. He developed a unique experimental approach, using those colloidal droplets called coacervates as crude models of cells (Figure 21.5). Coacervates are tiny spheroid bodies that form when certain macromolecules are introduced to water under carefully controlled ionic and pH conditions. Each droplet includes a cluster of the large molecules and a surrounding membranelike shell of water molecules. An interesting property of coacervates is that they "grow." A coacervate continually takes in selected substances from its surroundings, thus increasing in size. Then, at some critical mass, the coacervate divides (shears in two). Then its growth resumes, only to divide again. In this way its numbers increase. In some instances, the droplets become even more selective through the formation of a more complex membrane beneath the watery shell. So we see that coacervates have some of the properties of life—the selective intake of materials, growth, division, and increased numbers.

Oparin and his followers produced coacervates in a great variety of ways, adding selected substances and enzymes and observing some of the chemical activity associated with cell metabolism. The activities could be made quite specific by controlling the medium and the substances to be added. But coacervates are not living entities, and their more significant feats depend upon their being supplied with enzymes that have previously been extracted from living things. Nevertheless, coacervates have helped researchers to envision a critical stage in the origin of life.

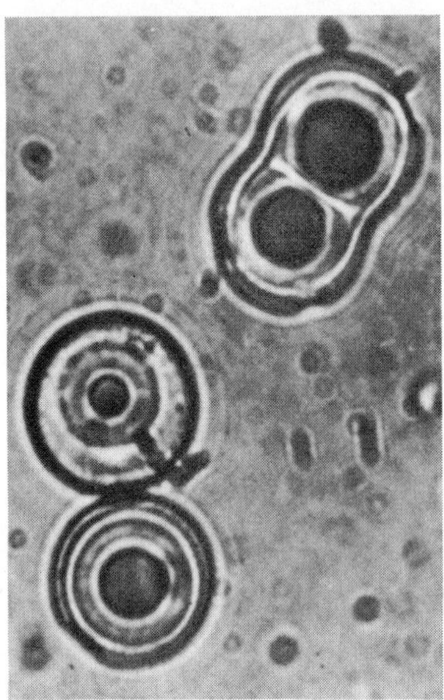

FIGURE 21.6
PROTEINOID ORGANIZATION
Proteinoids are polypeptides that polymerize spontaneously from evaporating concentrations of amino acids. Proteinoid microspheres could have formed in small, hot pools of spontaneously formed monomers located in regions of intense volcanic activity.

Fox and Thermal Proteinoids

Sidney Fox's thermal proteinoids have also given us strong indications of how cells may have arisen. When placed in water, the proteinoids cluster into bodies similar to coacervates. Fox calls these bodies **proteinoid microspheres** (Figure 21.6). Such spheres automatically form two-layered membranes, isolating themselves from their water surroundings, again in the manner of coacervates. Further, the spheres take up molecules from the surrounding environment, behaving as though their surrounding membrane were selective. They grow, fuse with other microspheres, produce tiny buds from their surface, and under certain conditions divide their mass. Fox is still experimenting with these tiny bodies, but unlike Oparin's, Fox's approach is conservative. His experiments center on autocatalytic activity; thus enzyme systems derived from cells are not used.

A Bit of Clay

Some researchers have turned to clay particles as an answer to some of the problems of spontaneous polymerization, and even as forerunners to living cells. Why clay? Under experimental conditions organic monomers adhere to clay particles, forming concentrations on the particle surfaces. Because clay particles carry electrical charges, they interact with monomers and in some instances catalyze their polymerization. Amino acids have formed proteinoids on the surface of clay particles. Interestingly, the amino acids favored by clay particles (and as a matter of fact, the sugars that also bind to clay) are of the same chemical structure as those found in life. The energy needed to polymerize amino acids into proteinoids could have come from the environment, as Fox has proven—heat will do it. It could also have come from some primitive, energy-yielding versions of ATP. Recall that ATP is actually a triphosphate RNA nucleoside, and nucleotides of RNA and DNA can form spontaneously.

Speculations on the Models

Oparin's and Fox's work suggests how a wide variety of organized droplets might have arisen, each group specific in its content and chemical activity. Extending such observations, we can surmise that in the primitive earth, there may have been competition among the droplets for the spontaneously forming, energy-rich monomers in the thin, hot soup. In any case, as time went on, such events must have become less and less random as less efficient droplets failed, their constituents to be used by their more stable associates. Of course, stability through generations of droplets could only be achieved by the droplets having some means of faithfully duplicating themselves.

One might imagine a system in which the molecules inside a droplet might act as a template (mold) that could organize smaller molecules to form a replica of the template molecule (Figure 21.7) prior to division. In this way the droplet could remain functionally stable through generations of its kind (roughly the way genetic mechanisms work today). Of course, this is a giant step in our scenario of how cells came to be. There is little indication that simple molecules can behave in such a way.

Of course, not everyone is convinced that such active droplets, or protocells as they

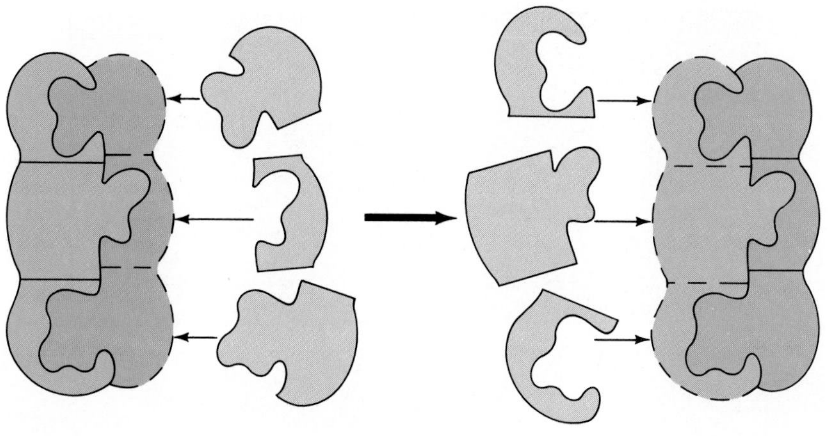

FIGURE 21.7
SELF-REPLICATING MECHANISMS
The figure at the left serves as a template from which the figure at the right can be formed. Such self-perpetuating mechanisms were a necessary step in the evolution of early life.

became known, represent a vital step in the development of life. Some biologists are convinced that self-perpetuating genetic systems arose first, probably in the form of DNA or RNA. Such nucleic acids have the self-duplicating or "replicating" quality needed. Supposedly, such molecules would have then become surrounded by an active droplet and membrane. The intriguing thing about this notion is that there are such models around today in the form of viruses. The simplest of these consist only of a nucleic acid core (the genes) and a protective protein coat. But there are also difficulties with this hypothesis: Every virus known today is parasitic, unable to carry on vital metabolic and reproductive activity outside a living host cell (see Chapter 23). Then, as we learned in Chapter 3, unlike amino acids, which readily polymerize into polypeptides, nucleic acids will not form from nucleotides in the absence of enzymes.

THE EARLIEST CELLS

By taking the giant step from metabolically active colloidal aggregates to self-reproducing **protocells** (leaving huge gaps for future theorists to deal with), we can apply some informed speculation to many questions about early cellular life: What were the earliest cells like? What were their energy sources? How do we get from the earliest stages to Awramik's 3.5 billion-year-old cyanobacteria, a complex photosynthetic prokaryote?

RNA: Gene and Enzyme?

A growing number of researchers are becoming convinced that the first bona fide genes were encoded in RNA rather than DNA. RNA, like DNA, has information-storing capabilities, yet it is much simpler. And, whereas DNA nucleotides do not polymerize spontaneously under primitive conditions (Figure 21.8a), researcher Leslie E. Orgel of the Salk Institute has succeeded in producing RNA polymers by concentrating RNA nucleotides in a saline environment. Such polymers, Orgel observes, can also replicate, forming copies of themselves through RNA base-pairing. The process, admittedly slow, nonetheless happens without the intervention of replicating enzymes. In other words, we now know that RNA can make RNA (Figure 21.8b). So a major obstacle in the understanding of molecular evolution may have been overcome. This replication process, in addition, is much simpler than the complex, multi-enzyme process of DNA replication (see Chapter 12), which probably arose much later in the history of life.

Researchers have also noted other examples of catalytic properties in RNA. In the processing of transfer RNA by *E. coli*, an enzyme containing a small molecule of RNA is used. When the RNA is removed and used alone, it succeeds in cutting and splicing the raw tRNA transcripts. Further, recent studies of intron excision in the protist *Tetrahymena* reveal that the raw transcripts of ribosomal RNA remove their own introns, also without the aid of an enzyme. Speculating on the significance of this, James E.

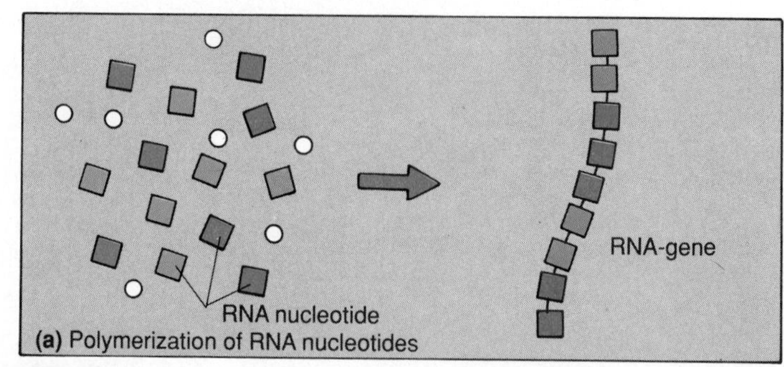

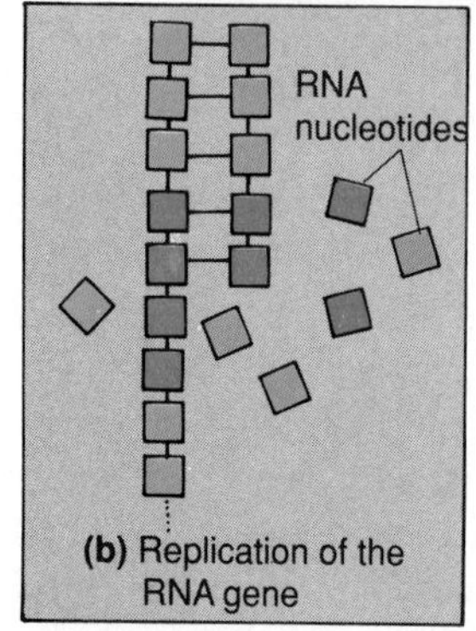

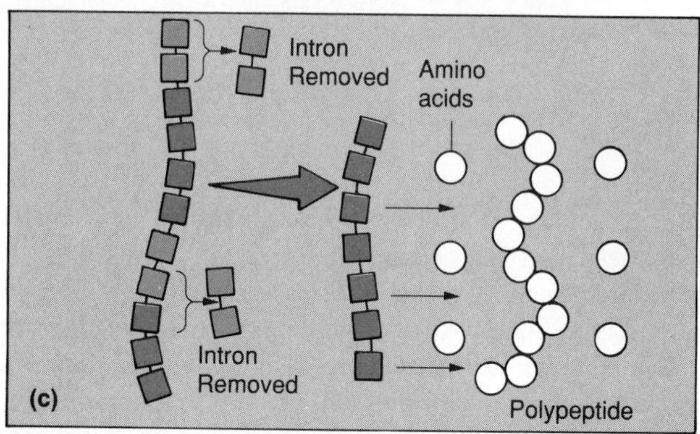

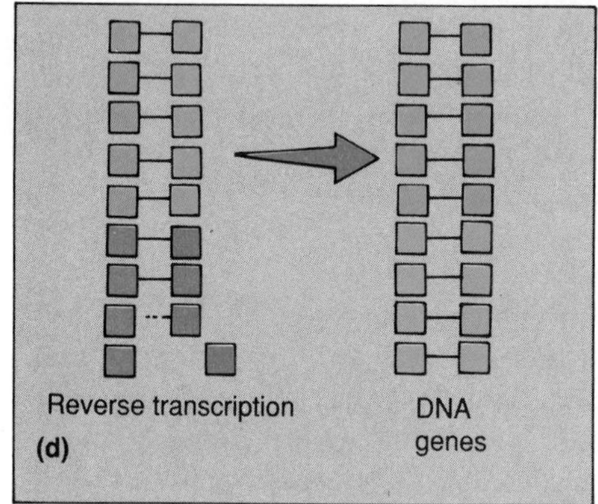

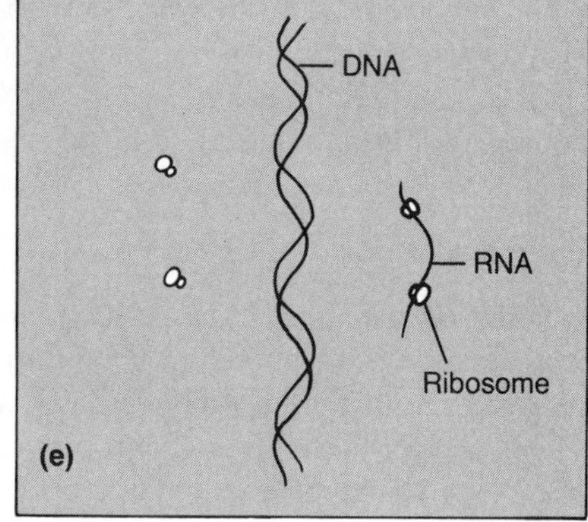

FIGURE 21.8
THE FIRST GENETIC SYSTEM
(**a**) The first genes may have formed from the spontaneous polymerization of RNA nucleotides. (**b**) An early capability of the crude RNA gene would have been replication, which can occur in RNA without the presence of protein enzymes. (**c**) Such RNA would most likely have contained introns, which were clipped out before translation of the coding into a polypeptide. (**d**) The presence of reverse transcriptase, one of the first enzymes, would have made the formation of DNA possible. (**e**) The earliest cell, or progenote, may well have contained DNA, still containing the introns collected randomly in earlier events.

Darnell, Jr. suggests that the early RNA genes, more or less randomly organized at first, must have contained various introns (intervening sequences) that had to be clipped out and the exons (expressed sequences) spliced together (Figure 21.8c). Darnell is thus suggesting that introns arose very early in the history of life.

The first DNA, suggest the researchers, may well have been produced through reverse transcription (Figure 21.8d), a biochemical trick well known in single-stranded RNA viruses (see Chapter 14). For a time biologists thought that reverse transcriptase, the enzyme responsible for making DNA from an RNA template, was unique to certain viruses. But now similar enzymes are known to be present in all eukaryotic cells, a finding that suggests that the enzyme is indeed ancient. But, why did DNA arise at all? DNA, it turns out, offers advantages over RNA. Included are greater stability, more efficient information storage, and certainly a built-in system redundancy (its double strand) that aids in the correction of replication errors and in the deletion of mutations.

The conversion of RNA to DNA early in the history of life explains even the presence of introns in eukaryotic cells today. They have been present since the first genes formed. The prokaryotes, which are intron-free, they suggest, have since streamlined their protein-synthesizing machinery, eliminating the introns and thus becoming far more efficient protein synthesizers. This viewpoint represents a turnabout in how today's bacteria are to be perceived. We ordinarily think of them as primitive, but here, even this time-honored notion is being questioned.

The Progenote, Ancestor of All Cells

Along this line of informed speculation, researchers are suggesting that the very first cell, which biologist Carl Woese calls the **progenote**, had undergone the transformation to DNA, complete with its introns. It also had ribosomes for translating the genetic message into polypeptides (Figure 21.8e). The progenote may have given rise to three separate lines of life, two of which produced the prokaryotes—the archaebacteria and eubacteria—and a third that gave rise to the eukaryotes (Figure 21.9). This new thinking, as we will see in the next chapter, is a clear departure from early concepts wherein eukaryotes are thought to have arisen from prokaryotes.

The Early Metabolic Cell

There is considerable disagreement as to the metabolic characteristics of the earliest cells. Some hold that the first life forms were **photoautotrophic** or **chemoautotrophic.** Both are **autotrophs**, deriving their energy from light or inorganic chemicals, respectively, and living independently of other organisms. Others propose that the earliest cells were primitive versions of today's *Clostridium,* a soil bacterium that is rapidly killed by oxygen. These life forms probably relied heavily on the comparatively simple anaerobic respiration (see Chapter 8). (Remember that there was no available oxygen or oxidative respiration.)

The original energy supply for the early heterotrophic cells may well have come from the spontaneously produced monomers that were still available in the ocean. But we can surmise that expanding populations of the new, living cells soon began to use up the available resources, and the resulting competition led natural selection to favor those cells able to exploit new energy sources or to exploit old ones more efficiently. Cells could prey on one another, but this simply redistributed the limited and dwindling supply of organic nutrients. The next major advance was to break out of this limited food chain altogether. The cells that could begin to do this, even somewhat inefficiently, would begin a line that would replace others. Thus we have the rise of autotrophy.

The Early Autotrophic Cells

The primitive photoautotrophs today obtain their hydrogen from dissolved hydrogen sulfide and their energy from sunlight. It's a good bet that the earliest successful autotrophs also used light energy in the simplest of photosystems to pry the hydrogen away. But the number of places in the world that provide both hydrogen sulfide and abundant

FIGURE 21.9
PROGENOTES: ANCESTOR TO ALL LIFE
The hypothetical progenote, the first organism with an organized DNA-based genome, is presumed to be the ancestor to all other life.

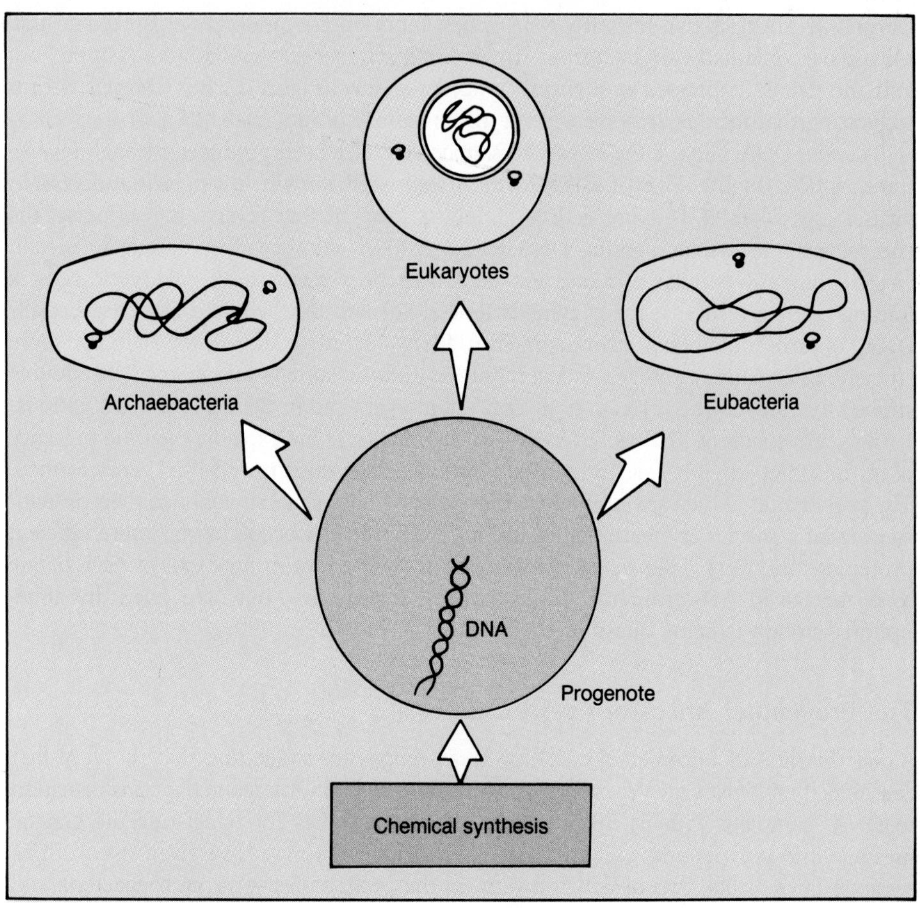

sunlight is severely limited. At some point, some ancestral cyanobacterium began to obtain its photosynthetic hydrogen supply from an energetically less favorable but far more abundant source: water. This step required a complex photosytem, as we discussed in detail in Chapter 7. But the accomplishment was a success, and it was to change the earth forever.

Where water is the source of hydrogen in photosynthesis, the waste product is molecular oxygen. As the early cyanobacteria flourished, exploited new niches, and multiplied, the amount of oxygen they released became significant, "poisoning" the water for their anaerobic competitors. At first, the regions of oxygen poisoning would have been local, just a thin later of oxygenated water in the sea or in a shallow pond, but this gradually changed.

Although many oxygen-sensitive organisms undoubtedly became extinct, being literally driven into the mud, new forms less sensitive to the poisonous gas were to emerge through mutation and natural selection. At first, they probably only developed ways of detoxifying oxygen. Later, however, the corrosive power of oxygen was actually utilized, put to work in extracting the energy from organic foodstuffs. Thus, oxidative aerobic respiration came into being.

It took about 2 billion years, but eventually photosynthesis changed the strongly reducing atmosphere of the early earth to the oxygen-rich, strongly oxidizing atmosphere of today. The early earth contained many **oxygen sinks** that absorbed oxygen as fast as it could be produced. For example, the elemental iron, elemental sulfur, and abundant iron sulfide of the early earth's crust took in enormous amounts of oxygen as they were transformed into iron oxide and various sulfates. Only when these oxygen sinks were finally saturated could free oxygen exist in the air. By the time oxygen became a significant atmospheric gas, organisms of a new kind—the eukaryotes—had already made their presence felt.

New modes of nutrition arose, too. The burgeoning cyanobacteria themselves represented an abundant new source of food for any heterotroph able to engulf and assimilate their stored energy. Thus the world saw the emergence of the first herbivores. Those anaerobes unable to cope with oxygen and eat other organisms were soon relegated to the backwaters of life's great progression. They could exist only if hidden away from poisonous oxygen in pockets of the earth's crust, in nutrient-rich muds and in deep recesses of stagnant waters. And it is in such places that they remain to this day.

In our imaginative scenario, we have seen two metabolic forms of life emerging: the photosynthetic, oxygen-producing phototroph and the aerobically respiring, oxygen-utilizing heterotroph (Figure 21.10). Now that we have arrived at the time of the first fossil evidence of life on earth, we can leave this hypothetical world and begin to consider the known world. We leave many questions unanswered, and we must put away those explanations based on conjecture and on an imperfect knowledge of the primitive earth. But this is the way of science; those questions will not lie untouched on some intellectual shelf. They will be repeatedly brought out, dusted off, tested again—and perhaps altered by the weathering effect of new data.

We will move on now to learn more about the prokaryotes, the organisms believed to be the most direct descendants of the earliest forms of life on earth. We have visited this remarkable group before, specifically, in our comparisons of cells and our discussion of prokaryotes as experimental subjects. In the next chapter, we will consider the prokaryotes as our first subjects in our consideration of the diversity of life.

FIGURE 21.10
THE EARLIEST CELLS
A history of the evolution of cells from the simple anaerobic heterotroph, sopping up spontaneously formed nutrients, to the more sophisticated aerobic, photosynthetic cells that preceded known life.

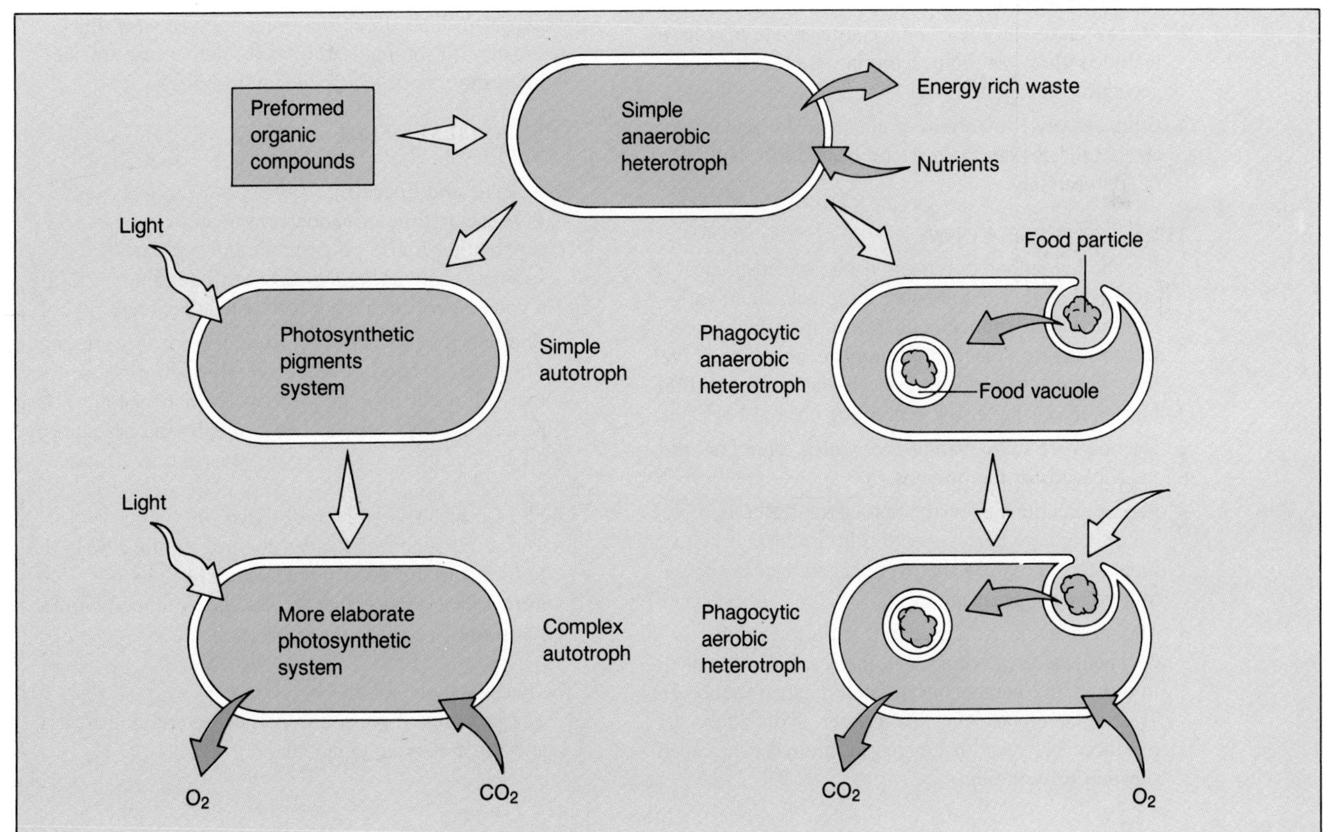

KEY IDEAS

Compelling evidence suggests that life arose spontaneously on the earth, at a time when unique conditions were present.

A HYPOTHESIS ON THE ORIGIN OF LIFE

Haldane and Oparin were the first to speculate on the spontaneous origin of life. The **Haldane-Oparin hypothesis** proposed that the precursors of life's molecules formed from inorganic sources, and these underwent polymerization to form macromolecules. Such macromolecules included primitive versions of enzymes. Droplets, or **coacervates**, formed and were enclosed by protein or lipid shells. A next hypothetical step required the formation of genelike **autocatalytic** molecules that could assure the faithful reproduction of such enzymes.

The Miller-Urey Experiment

Miller and Urey tested the Haldane-Oparin hypothesis with a device that simulated the primitive environment. Using a mix of gases and applying an electrical discharge, they succeeded in synthesizing amino acids, aldehydes, and carboxylic acids, all known to be monomers of cells.

THE EARLY EARTH

1. New knowledge of ultraviolet light has changed ideas on the primitive earth atmosphere. The atmosphere in which life arose was more likely to have consisted of warter, carbon dioxide, carbon monoxide, molecular nitrogen, and some free hydrogen.
2. Energy sources in the primitive atmosphere may have included ultraviolet light, lightning, heat, and geological shock energy.
3. Estimates place the origin of life some 3.8 billion years ago; fossils of cyanobacteria are believed to be 3.5 billion years old.

THE HYPOTHESIS TODAY

1. For the hypothesis to remain viable, scientists must be confident about the following: proposals about early conditions, the spontaneous formation of monomers of life and their polymerization into the polymers of life, the spontaneous formation of cell-like bodies, and the formation of simple, self-replicating chemical systems.
2. Scientists are still debating over which came first, the nucleic acids or the proteins.
3. Recent experiments on the revised atmospheric conditions have been successful in producing the usual monomers and a few that were not formed in the Miller-Urey experiment.
4. The mass action law suggests that polymerization is not likely to have occurred in the sea, but more likely in heated and highly concentrated pools of monomers. Treating amino acids in this manner, Sydney Fox produced polymers that aggregate into what he called **thermal proteinoids.**
5. Efforts to polymerize functional nucleic acids from nucleotide triphosphates, even when using single-stranded DNA for templates, fail in the absence of polymerizing enzymes.
6. Whether enzymes preceded nucleic acids or vice versa is unresolved, and strong cases can be made for either.

MODEL PROTOCELLS

Oparin and the Coacervate

Oparin experimented with coacervates by inducing metabolic activity via enzymes he supplied. Under contrived conditions some of their behavior simulated life.

Fox and Thermal Proteinoids

In water, Fox's proteinoids formed **proteinoid microspheres,** which had some of the characteristics of coacervates. Fox concentrates on natural systems.

A Bit of Clay

Because of its physical characteristics, clay forms an active nucleus for some polymerizing reactions. This suggests a possible role in the spontaneous generation of life.

Speculations on the Models

In an imaginative scenario, active droplets may have entered into competitive activity. In time, successful droplets with self-replicating systems established persistent lines. In an opposing point of view, life began with self-replicating molecules. Viruses make interesting models of such simple systems.

THE EARLIEST CELLS

RNA: Gene and Enzyme?

1. RNA has both information-storing and catalytic capabilities. It also polymerizes and replicates spontaneously under the early earth conditions, so it is a good candidate for the first genetic system.
2. The first RNA genomes probably had introns present. They were excised by RNA itself through processes that still occur in some protists. The conversion of the genome to DNA was made possible by the presence of an ancient form of the enzyme reverse transcriptase.

The Progenote, Ancestor to All Cells

1. The earliest life form, called the **progenote,** is believed to have had a DNA genome, which retained the introns that arose earlier. It was also equipped with ribosomes.

The Early Metabolic Cell

1. The progenote gave rise to Archaebacteria, Eubacteria, and the eukaryote kingdoms.

The Early Autotrophic Cells

1. There is disagreement on the metabolic characteristics of the early *protocells*. Some suggest they were chemotrophic or phototropic, while others claim they were simple anaerobic heterotrophs. The continued autocatalytic synthesis of monomers may have provided carbon compounds for early heterotrophs.
2. Keen competition for energy-rich monomers may have encouraged variants that could extract hydrogen from inorganic sources, using light energy in simple photosystems. The utilization of water as a hydrogen source probably came later, since it requires complex photosystems.
3. When the use of water in photosynthesis occurred, oxygen joined the gases of the atmosphere for the first time.
4. The buildup of significant amounts of oxygen in the atmosphere required an enormous time period because of the presence of **oxygen sinks**—elemental substances such as sulfur and iron that readily combined with the gas.
5. As cell populations grew, new modes of heterotrophic nutrition arose. Food chains emerged as cells began developing cells through phagocytosis.

REVIEW QUESTIONS

1. Briefly describe Haldane and Oparin's hypothesis on the origin of life. (438-439)
2. According to most biologists, the spontaneous origin of life occurred only once in the earth's history. What is the basis for this restricted statement? Why not several times? (438)
3. What steps are involved in going from inorganic matter to the coacervate level? What, according to Haldane and Oparin, were the energy sources? (439)
4. What is the status of the original Haldane-Oparin hypothesis today? (439)
5. Describe the experimental apparatus and materials used by Miller and Urey in testing the spontaneous origin hypothesis. What did Miller and Urey actually establish? (439-440)
6. In what specific ways has our understanding of the primitive atmosphere changed since the days of the Miller-Urey experiment? What effect on the original hypothesis has subsequent experimentation produced? (440)
7. List five requirements that must be met if the "spontaneous origin of life" hypothesis is to continue to hold up. With which of the five requirements are scientists making satisfactory progress? (440, 442)
8. What problems, if any, do researchers find with attempts at spontaneously synthesizing nucleic acids? Did nucleic acids necessarily precede enzymes? Explain. (442-443)
9. Briefly describe the polymerization experiments of Sydney Fox. Why didn't he use a water solution such as might have been found in the early oceans? (444)
10. Describe Oparin's experimental work with coacervates. Did he actually produce protocells? Has he proven the spontaneous origin hypothesis? Explain. (443-444)
11. List several reasons why RNA is a logical choice as the first genome. (445,447)
12. What is a progenote? To what cell lines did it give rise? (447)
13. Why might the earliest phototrophs have used hydrogen sulfide as a hydrogen source rather than water? In what way did the later shift to water "change the earth forever"? (447-448)
14. According to the latest thinking, it took 2 billion years for significant oxygen to accumulate in the atmosphere. Why was such a long period of time required? (448)

4

DIVERSITY AND FUNCTION: MICROORGANISMS AND FUNGI

22

PROKARYOTES AND VIRUSES

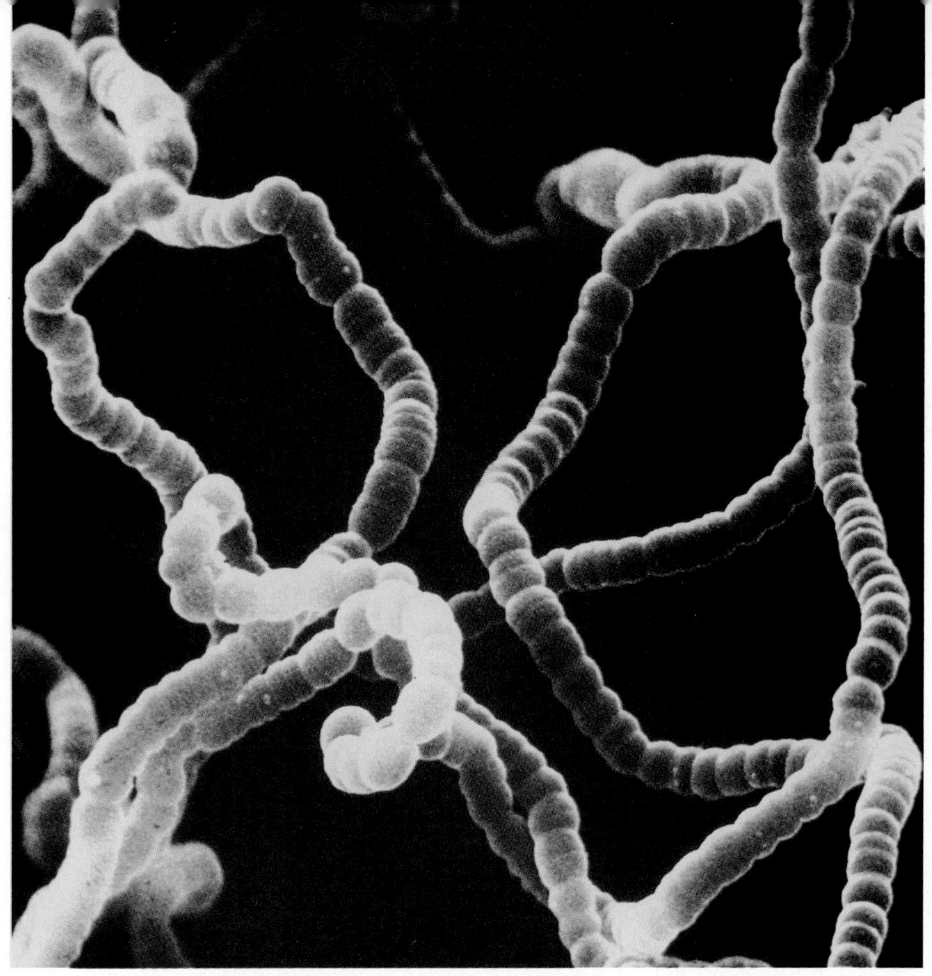

We like to think that we are the dominant life form on the planet (although, occasionally, someone says something about cockroaches or rats). However, the question of dominance doesn't make much sense. Do we measure it in terms of power? Then is a tiger dominant over a human? Under what conditions? Don't both we and the tiger depend intimately on the delicate plants around us? The question of dominance makes even less sense when we consider which life form has lived on the planet for the longest time. In this context we can lay little claim to dominance because we are indeed newcomers. In fact, the group with the longest history are the seemingly insignificant prokaryotes, the bacteria.

Bacteria are in the kingdom **Monera** and are called prokaryotic. The term, **prokaryote** (*pro*, before; *karyon*, kernel) refers to the fact that bacteria have no organized nucleus.

ORIGIN OF PROKARYOTES

Bacteria are indeed ancient, their fossils having been found in deposits 3.5 billion years old. (According to very recent thinking, the first eukaryotes probably arose 2 to 2.5 billion years ago.) The best evidence of prokaryote antiquity is seen in strange columnlike deposits known as stromatolites (Figure 22.1). These are highly laminated deposits of sedimentary rock, first seen in coastal regions. Each layer was produced by dense populations of the earliest cyanobacteria. The cells were easily fossilized after having continually taken in calcium carbonate and other minerals from the surrounding water. Today's cyanobacteria, formerly known as "cyanophytes" and "blue-green algae," are advanced photosynthetic prokaryotes.

Some authorities doubted that these strange columns actually represented fossilized life forms until it was found that there are still active stromatolite-forming cyanobacteria at work along the shores of Shark Bay on Australia's west coast. Similar mineral-incorporating formations are also seen in Yellowstone National Park in the United States.

Later in the chapter we will look into some new and striking ideas on prokaryotic evolution and classification. They are based on recent studies that use the powerful new analytical techniques of molecular biology. In fact, these techniques show so much promise in resolving taxonomic (naming) and systematic (relationships) problems that microbiologists are keeping bacterial taxonomy on hold for now. When enough information is available, Kingdom Monera will probably undergo a complete taxonomic revision. We'll see that this shakeup has already begun and that the first important changes are lofty ones, at the kingdom level itself. It seems that the vast realm of prokaryotes is now being divided into two new kingdoms: **Eubacteria** and **Archaebacteria**, but we'll come back to this after we focus on the bacteria themselves.

BIOLOGY OF PROKARYOTES

Everyone knows a little about bacteria. They are the organisms most people call "germs," those generally thought of as harmful. They are the organisms people go to so much trouble and expense to avoid. In spite of endless gargling, spraying, and bathing, the fact remains that we are literally awash in a sea of the tiny creatures. We may also be vaguely aware that some are actually helpful to humans, although most of us hesitate to admit that the tangy flavors of some of our favorite foods are caused by bacterial secretions. Bacteria are a fascinating group in many ways, as you may agree after you have seen what they are and how they make a living.

Cell Structure

One of the most striking characteristics of the prokaryotic cell is its small size. Most are about one-tenth the size of typical eukaryotic cells. Further, their internal structure is much simpler (Figure 22.2). As we mentioned above, there are no membrane-surrounded organelles, so the bacterial cytoplasm is nearly devoid of recognizable structure. (People are rarely impressed by their first look at bacteria through a light microscope.)

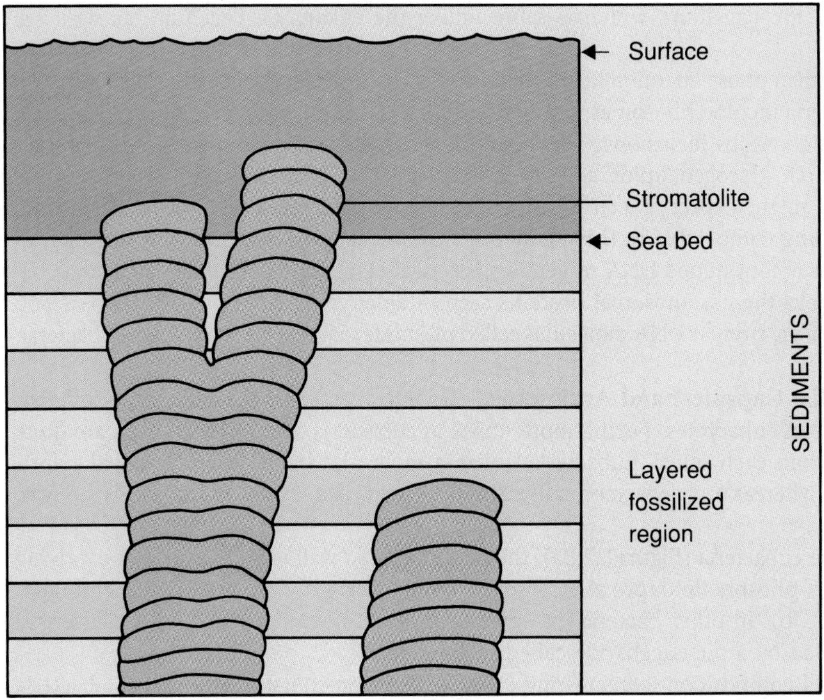

FIGURE 22.1
THE OLDEST FOSSILS
Fossilized stromatolites are thought to represent the most ancient evidence of life on the earth. They arose on rocky ledges in the tidepools of shallow Precambrian seas. As new populations arose at the surface, their predecessors underwent fossilization below, leaving a permanent record of past struggles.

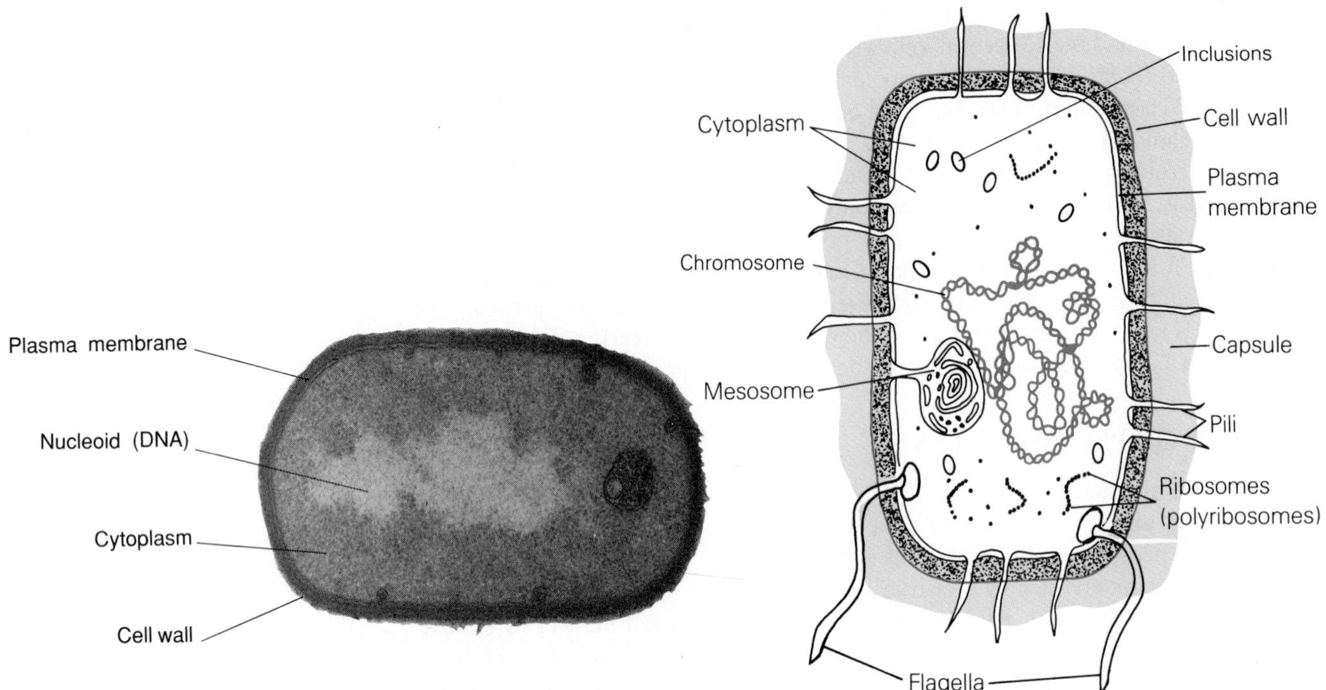

FIGURE 22.2
THE PROKARYOTIC CELL
The typical prokaryotic cell lacks the membrane-surrounded organelles of the eukaryote, including the organized nucleus. The cell wall is often surrounded by a slimy sheath. Free ribosomes and polyribosomes are common, as are fibrous protein projections known as pili. Membranous mesosomes are often seen in dividing cells. Flagella are corkscrew-shaped, tubular, rotating structures.

Cytoplasmic Structures Surrounding the cytoplasm is a plasma membrane that, in eubacteria, is similar in many respects to the eukaryotic version. It contains the usual bilayer of straight-chain phospholipids with numerous proteins interspersed. In addition, the membrane may contain assemblages of respiratory and phosphorylating enzymes, analagous to those of the eukaryotic mitochondrion. In some photosynthetic bacteria the plasma membrane becomes greatly extended and folded internally, forming a **thylakoid**, complete with photosynthetic pigments. Another infolding forms the membranous **mesosome**, which apparently functions during cell division. It may aid in chromosome separation and in the formation of a new cross-wall between daughter cells.

Bacterial cells often contain a number of **pili**, slender extensions composed of the protein **pilin.** Pili aid the parasitic bacterium in adhering to the host cell (the agent of gonorrhea uses its pili in this manner). Other pili, called **sex pili**, aid in the transfer of genes during rare episodes of sexual reproduction. Another prominent structure in some bacteria is the flagellum, which is quite unlike the eukaryotic flagellum, as we'll see shortly.

Other than those mentioned, the remaining discernable cytoplasmic structures in most bacteria include ribosomes, often occurring as chain-like polyribosomes (see Chapter 13), and various inclusions. The latter are generally aggregations of stored materials such as starch, glycogen, lipids, and in a few bacteria, phosphorus and sulfur compounds.

As you might expect, bacteria do not form a nuclear envelope, and in other ways lack anything comparable to the highly organized eukaryotic nucleus. The single chromosome is a continuous DNA molecule (effectively circular), described as "naked" because it lacks the chromosomal proteins seen in eukaryotic chromosomes. And as you know, smaller, circular DNA molecules called plasmids may also be found within bacteria.

Cell Walls, Capsules, and Antibiotics Bacterial cell walls are chemically different from those of eukaryotes. Furthermore, those in eubacteria and archaebacteria are quite different from each other. Eubacteria utilize a molecular building block called **peptidoglycan** whereas archaebacteria make use of proteinaceous and other, poorly known, polymers.

In some eubacteria (Figure 22.3a), the peptidoglycan wall is quite thin, but overlying it is a thin phospholipid/protein complex, often referred to as an "outer membrane" (Figure 22.3b). In other bacteria, however, the peptidoglycan layer is quite dense and is penetrated by a polysaccharide called **techoic acid.**

Cell wall composition is interesting for several reasons. First, many of those bacteria

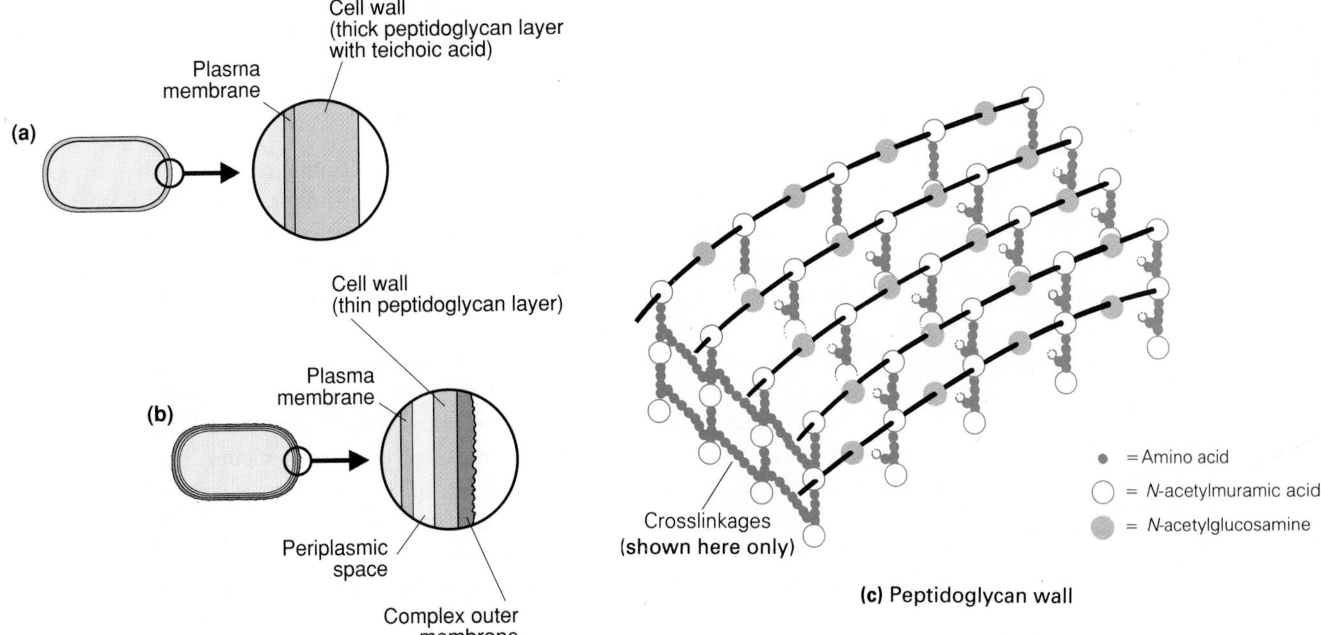

FIGURE 22.3
THE EUBACTERIAL CELL WALL
While eubacterial cell walls contain peptidoglycan, there are two basic differences in their organization. In one group **(a)** the peptidoglycan is dense and is penetrated by techoic acid. In the second group **(b)** the peptidoglycan is thin and is covered by a complex outer membrane. These differences are reflected in how diagnostic stains are taken in and how the cell reacts to antibiotics. The structural units of peptidoglycan **(c)** are two amino sugars, *N*-acetylglucosamine and *N*-acetylmuramic acid. To form the cell wall, the subunits are interconnected by peptide bridges, forming dense rows of the tough conglomerate.

with the thicker wall are highly susceptible to penicillin. In its presence, peptidoglycan-synthesizing enzymes in newly-divided cells are inhibited, and cell walls cannot be formed. Without the tough, resistant wall, the bacterial cell, rich in solutes, cannot oppose or resist the natural influx of water, and the membrane soon bursts. Thus the bacteria's fate is the inverse of our own, since their death means that we survive a troubling or perhaps life-threatening infection.

Second, bacteria with the thicker cell wall readily absorb and retain crystal violet, a deep purple cellular dye used in what microbiologists call the **Gram stain.** Such bacteria are designated **Gram positive** (Figure 22.4).

Those bacteria with the thinner walls do not retain the crystal violet of the Gram stain, and are designated **Gram negative.** Further, most Gram negative bacteria are not susceptible to penicillin.

As you can see, the Gram reaction has an obvious medical application. Many pathogenic bacteria identified as Gram positive can be quickly controlled with penicillin or penicillin-like antibiotics. But if the physician knows that the invading bacterium is Gram negative, some time is saved because alternative antibiotics are prescribed. Tetracycline and streptomycin are effective against many Gram negative bacteria. Rather than halting cell wall synthesis, they kill bacteria by inhibiting protein synthesis. The Gram reaction has also been used by microbiologists for years as an important criterion in bacterial classification.

In addition to the cell wall, many bacteria are surrounded by a slimy layer or **capsule** consisting of complex carbohydrates or polypeptides. Slimy coats aid bacteria in adhering to surfaces that provide nutrients, an important ability for both parasitic and free-living types. In parasitic types, the capsule also helps the bacterial cell avoid being engulfed by the host's phagocytic white blood cells. On the other hand, certain capsule carbohydrates are antigenic, invoking a reaction in which antibodies attach to and clump invading cells. This makes the engulfing by the phagocytes much easier.

Endospores Many bacteria become transformed into highly resistant, thick-walled **endospores** (Figure 22.5) when unfavorable conditions such as food shortages arise. Endospores are dehydrated cells, in which all metabolic activity has ceased; the cellular components are held in a state of dormancy. Some endospores are so resistant that they can survive boiling, as long as it is not prolonged. One report claims that spores 150,000 years old were able to germinate. Germination is the return to the metabolically active

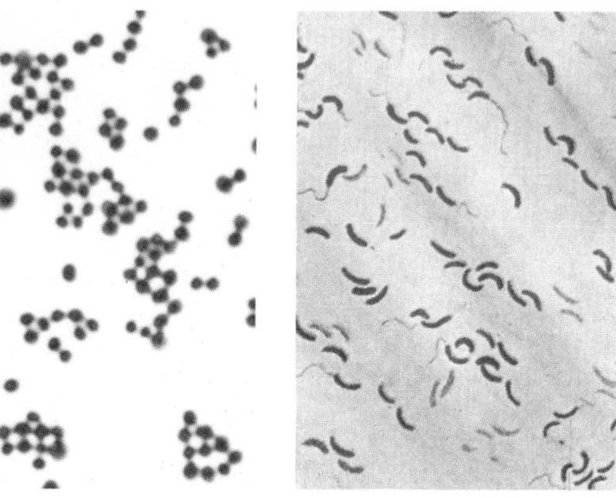

FIGURE 22.4
GRAM REACTIONS
The bacteria on the left are Gram positive (purple), while those on the right are Gram negative (red). The reaction to Gram staining is an important diagnostic tool, particularly in clinical use. The red coloring within Gram negative bacteria is a "counter-stain" that is retained, making them easier to spot.

state that occurs when conditions become more optimal. The ability of bacteria to form spores is another taxonomic consideration.

Cell Form and Arrangement

In addition to spore formation and diagnostic staining reactions such as the Gram stain, bacteria are identified and classified according to a number of other characteristics. Among these is cell shape and arrangement. The most common shapes of bacterial cells are rodlike, spherical, spiral, comma-shaped and filamentous. Arrangements vary from individual cells to pairs, octets, chains, and clusters (Figure 22.6).

The rodlike cells, or **bacilli** (their more technical name, singular: **bacillus**), occur as single cells and chains, with some chained cells enclosed in sheaths. (It is in the bacilli where endospore formation is most common.) The spherical cells, or **cocci** (singular: **coccus**), have the most varied arrangements, occurring singly, paired (**diplococcus**), in beadlike chains (**streptococcus**), grapelike clusters (**staphylococcus**), and sometimes in groups of eight (**sarcina**). Spiral-shaped cells fall into three categories according to motility. One group actually consists of many cells that form helical filaments, moving through a gliding action. **Spirochetes,** a second group, have flexible walls, moving through the action of axial filaments beneath an outer membrane. The third spirally-shaped group, called **spirilla,** have rigid cell walls and move through the action of flagella. The **vibrio**, or comma-shaped bacteria, are actually a curved version of the rod shape, but the difference can be important to clinical microbiologists. A particularly nasty Gram negative vibrio produces an exotoxin (released poison) that is the cause of cholera.

Mycoplasmas: The Smallest Cells Before we leave our considerations of the prokaryotic cell itself, let's consider what is perhaps the oddest bacterial group of all, the **mycoplasmas.** Mycoplasmas are among the smallest living things, with the cells of some species being less than 0.16 μm in diameter, far smaller than most other bacteria. In fact, they are smaller than some viruses. Oddly, they lack a rigid cell wall and therefore have no definite shape. Almost all are parasites of animals including humans, but some mycoplasmas have recently been found in plant cells. They are completely resistant to penicillin, presumably because penicillin normally kills by interfering with the growth of bacterial cell walls. They sometimes form colonies resembling certain fungi. Reproduction is by binary fission. One form is responsible for a relatively mild form of human pneumonia (Figure 22.7).

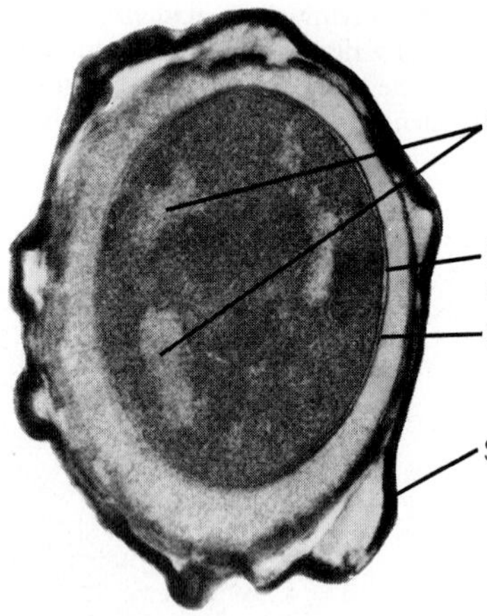

Chromosome
Plasma membrane
Cell wall
Spore coat

FIGURE 22.5
THE ENDOSPORE
Many bacteria form tough-shelled resistant endospores. Within each endospore is a chromosome and the dehydrated cytoplasm. Under more ideal conditions the endospore will take in water, and the cell will resume activity.

Bacterial Movement

Many bacteria make use of the flagellum for movement, but the bacterial flagellum is structurally and functionally quite different from the version seen in eukaryotes. Recall that the eukaryotic flagellum moves with whiplike undulations, based on sliding microtubular filaments. The microtubules are composed of tubulin microspheres, and the whole apparatus is surrounded by an extension of the cell membrane (see Figure 4.23). In contrast, the bacterial flagellum is a stiff, tubular, rodlike structure, permanently bent into a gently curved corkscrew shape and composed of the crystalline protein **flagellin.** The bacterial flagellum protrudes from its anchorage in the cell wall, and since it rotates on its axis, it pulls the cell through the water (like a ship's propeller) (Figure 22.8).

A less familiar form of movement in bacteria is gliding. The so-called **gliding bacteria** are thought to move by exuding a slippery mucilage and then moving over it. Alternatively, some researchers believe the mucus is secreted in a concentrated form, and as it takes up water, it expands, pushing the organism along.

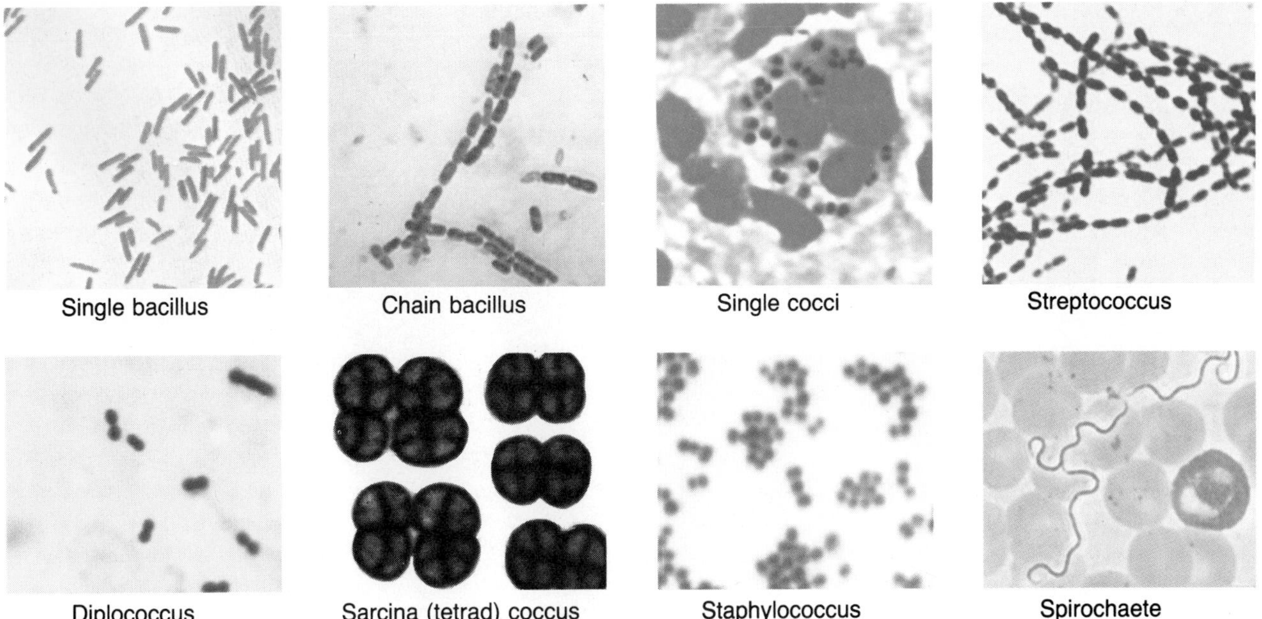

FIGURE 22.6
EUBACTERIAL FORM AND ARRANGEMENT
The three principal forms in bacteria are rod shaped, spherical, and spiral or corkscrew. Arrangements vary considerably, and in the spherical or coccus form are seen doubles, chains, and clusters.

Gliding bacteria, or **myxobacteria** as they are classified, are among the strangest prokaryotes in other ways as well. Under drying conditions they will mass together, soon forming an upright stalk that bears a **fruiting body** at its tip. Spores form within this body, and when they become scattered, those that land in favorable circumstances will germinate and form a new vegetative cell. In this respect, the myxobacteria are very much like some fungi.

Bacterial Reproduction

By far, the most common mode of reproduction in bacteria is asexual. It occurs through a primitive process called **fission:** simple cell division without mitosis or meiosis. Of course, fission is preceded by DNA replication, and if all goes well during the division process, each daughter cell gets one replica of the bacterial chromosome. What ensures the equitable division of replicas isn't well understood, but the observations seem clear

FIGURE 22.7
THE SMALLEST CELLS
Mycoplasmas, the smallest cells, have only those structures essential to life. Note the absence of a cell wall, unusual for a prokaryote.

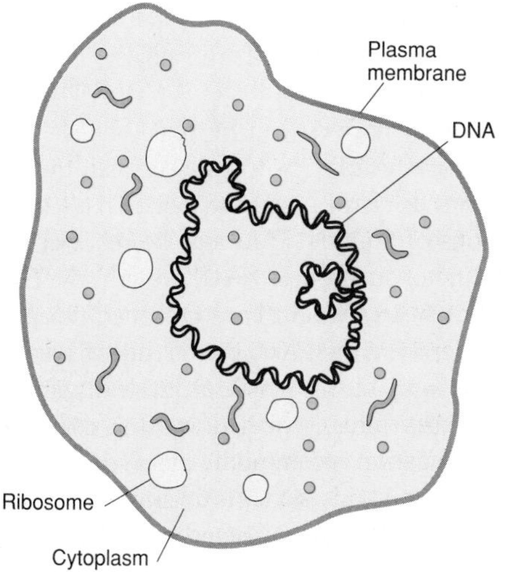

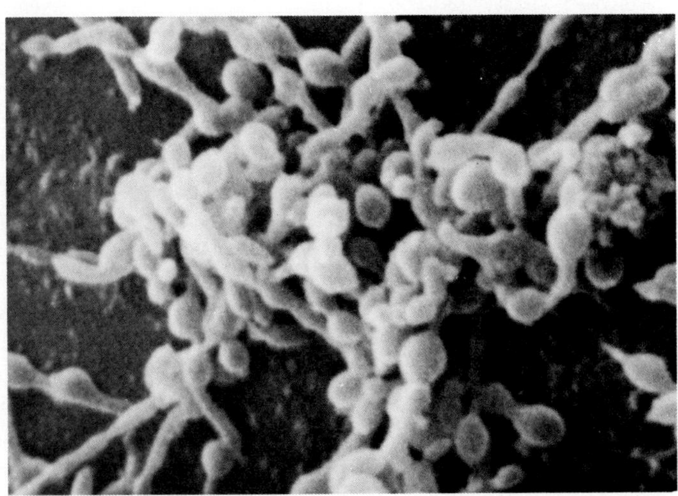

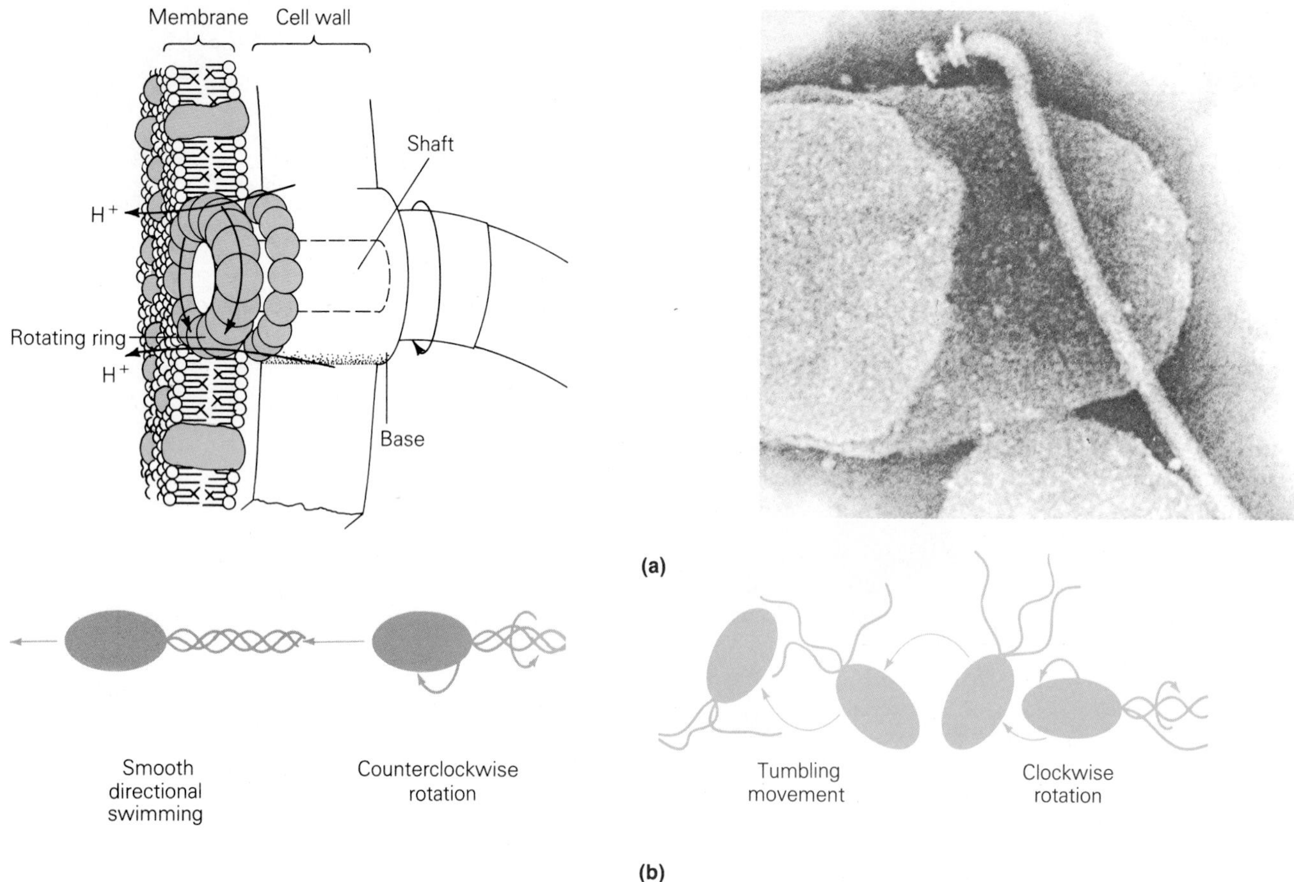

FIGURE 22.8
THE BACTERIAL FLAGELLUM
(a) The prokaryotic flagellum is a tubular, rotating structure permanently bent into a helical configuration and composed of the protein flagellin. High magnification by the electron microscope reveals that the flagellum is anchored to a hooklike shaft, which penetrates the cell wall and is anchored in two ring-shaped bases in the plasma membrane. **(b)** When groups of bacterial flagella rotate counterclockwise, they join to produce a smooth synchronous movement that propels the cell along in a definite direction. But when they spin clockwise, the individual flagellum straightens somewhat, and their spinning sends the cell into a tumbling movement that is more random in direction.

enough. Following chromosome replication, the replicas, which are attached to the mesosomes, move apart, the membrane extends inward, and finally, a new cell wall is laid down between the two daughters (Figure 22.9).

The rate of growth and cell division in bacteria can be phenomenal. For instance, under ideal conditions, our own colon bacterium, *Escherichia coli,* can double every 20 minutes. Potentially, 72 generations could form in just one day. (That's 4.7×10^{21} or 4.7 sextillion bacteria), literally millions of pounds. (You may be relieved to know this has never happened. Within a few hours of such unrestricted growth, the colony would have run out of space and food, and its wastes would have poisoned its environment.)

On rare occasions, as we saw earlier (Chapter 15), prokaryotes undergo a simple kind of sexual process called **conjugation**, during which a one-way transfer of genes may occur. This requires the aid of a sex pilus, which draws the cells together and permits the transfer to occur. While conjugation is commonly observed in highly selected laboratory strains, not much is known about this sexual exchange under natural conditions. (Other mechanisms of gene transfer among bacteria were also discussed in Chapter 15.)

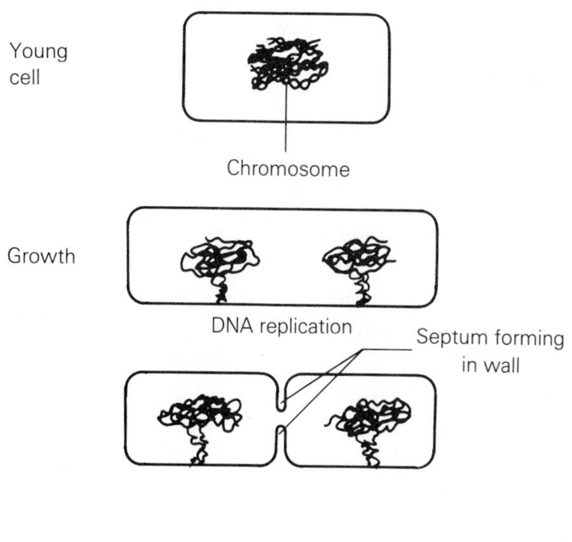

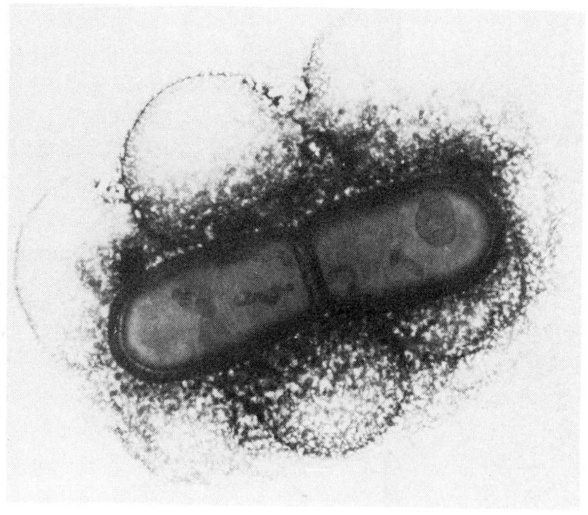

FIGURE 22.9
ASEXUAL REPRODUCTION IN BACTERIA
Cell division in bacteria occurs through fission. Note the attachment of chromosome replicas to the cell membranes, the membrane's inward growth, and the synthesis of a new wall.

Metabolic Activities of Bacteria

Many bacteria are **heterotrophs** ("other feeding"); thus, their nutrients must come from other organisms. Among these are the **decomposers**, bacteria that bring about the decay of organic matter. Other heterotrophs are **parasites**, often referred to as **pathogens.** They cause many of the diseases of plants and animals, including those of humans. Several groups of bacteria are **autotrophs** ("self-feeding"), either as **photoautotrophs** or **chemoautotrophs.** The photoautotrophs synthesize their essential molecules from simple inorganic molecules, using the energy of light to drive the reactions. The chemoautotrophs also utilize inorganic molecules from the earth's crust, but their energy is derived from the chemical reactions they bring about, rather than from light.

Bacteria also vary in other aspects of their metabolism, in the use of oxygen, for example. Whereas some live as **aerobes** (oxygen-users), others are **anaerobes** (those that do not utilize oxygen). The **obligate aerobes** must have oxygen, but the **obligate anaerobes** are poisoned by it. Others function either way, with or without oxygen (like the eukaryotic yeasts: see Chapter 8) and are therefore called **facultative anaerobes.**

A Word About Bacterial Energetics Heterotrophic bacteria, at least the aerobic (oxygen-using) types, carry on cell respiration in a manner surprisingly similar to what we saw happening in the mitochondrion (see Chapter 8). Oxidative respiration, occurring in the cell (as it does in the mitochondrion) utilizes high-energy electrons from foods oxidized in the cytoplasm. The electrons are sent through electron transport systems in the membrane, powering numerous proton pumps that eject the protons from the cell. As a result, the protons establish a steep chemiosmotic gradient between the immediate surroundings and cytoplasm, representing substantial potential energy for the cell. The membrane also contains the familiar phosphorylating enzymes (Chapters 7 and 8), and as the protons pass down their gradient through channels leading into these bodies, ATP is synthesized (Figure 22.10a).

The proton gradient also powers a number of active transport mechanisms, one of which brings nutrients into the cell. Further, the unique, spinning bacterial flagellum receives its propulsive force from an influx of protons.

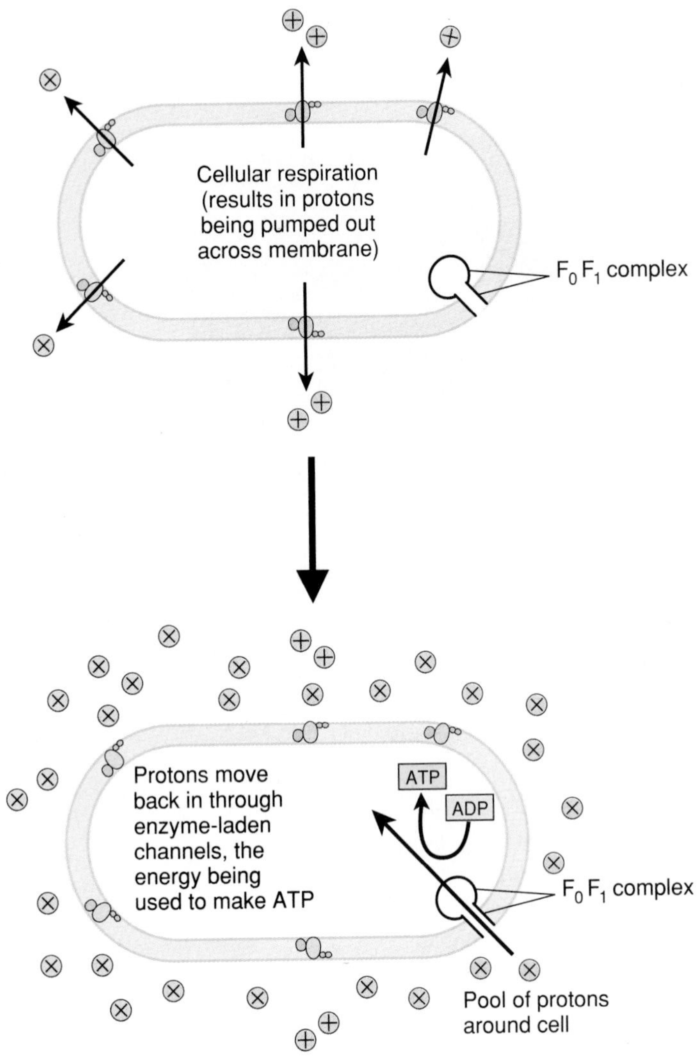

FIGURE 22.10
CHEMIOSMOSIS IN BACTERIA
Aerobic bacteria make use of a steep chemiosmotic gradient to form ATP. Protons pumped from the cell return, passing through phosphorylating enzymes.

Bacterial Heterotrophs

Bacterial heterotrophs are so diverse that there is hardly an organic molecule that some bacterium cannot use. Bacterial nutrients include those making up the remains of dead organisms, which brings us to the decomposers.

Decomposers Most heterotrophic bacteria gain their energy through decomposition or decay. They secrete enzymes that cause the breakdown of organic matter in dead organisms and their wastes, a practice that has earned them an alternative name, **saprobe** (*sapros* from Greek, meaning "rotten" or "putrid"). However offensive it might seem, this activity is ecologically crucial and is undoubtedly the most important of all bacterial activities. Decomposition releases such key ions as nitrates, phosphates, and sulfates, which then become available to other organisms (see Chapter 47). Their availability is particularly important to the growth and metabolism of both eukaryotic and prokaryotic photoautotrophs, organisms that bring solar energy into the living realm. Sometimes we lose our objectivity about decomposers. If they get to our food before we do, rendering it unpalatable, we may find it difficult to appreciate their role in the "big picture."

Bacterial species become quite specialized in their role as decomposers. Each has its own nutritional requirements, and quite often the energy-rich waste products of one group provide the nutrients for another group. Thus, a succession of different bacterial decomposers arises in a food source. (If you leave a hunk of beef on your desk for several days, you may notice that its fragrance will vary over time as different groups of decomposers take over.) Such a progression is apparent in the **nitrogen cycle** (see Figure 47.7), where the conversion of dead organisms to useful nitrates or ammonia requires the sequential activities of several specialists. As a final note on the indispensability of this group, we need only remind ourselves that without the decomposers, many essential minerals would be tied up for long periods, and the countryside would be littered with corpses.

Inhibiting Decomposers We go to a lot of trouble to keep bacterial decomposers from getting to our food and other goods. In general, the optimal conditions for decomposers include an appropriate food supply (ours does nicely), warmth, moisture, and for the aerobes, adequate oxygen. Importantly, an abundance of oxygen will retard the growth of anaerobes, some of which are serious pathogens as well as food-spoilers. Such information naturally suggests ways to inhibit the growth of the decomposers. Among the most common means of inhibiting bacterial growth are cooling, freezing, drying, salting, and sugaring. Whereas cooling slows all metabolic activity, the other methods listed essentially remove water from the food or set up osmotic conditions that are unfavorable to bacterial cells. Of course, we can begin by killing the bacterial cells that already exist on the foodstuffs. This is done through sterilization, using intense heat, gases such as ethylene oxide, or irradiation. In canning, the food is heat-sterilized and sealed in airtight containers, thus preventing new bacteria from reaching or growing in the food. Pasteurization, a gentler process involving a brief and limited application of heat, is usually

FIGURE 22.11
MICROBIAL DELIGHTS
Parmesan, Cheddar, and Swiss cheese owe their distinctive flavors and other characteristics to specific bacterial agents. The holes in Swiss cheese are naturally produced by trapped gases from the same propionic acid bacteria that are responsible for the tangy flavor.

reserved for milk and other fragile foods. It is intended to kill only certain disease-causing bacteria. Many decomposers, especially spore-formers, survive pasteurization, and even an unopened container of milk soon spoils. Chemical preservation, common in the food industry (just read the labels), includes the use of small amounts of sorbic acid, sodium nitrate, calcium propionate, or other chemicals, to retard bacterial growth.

Economically Useful Decomposers Before we continue our long litany of accusations against bacteria, we are compelled to note that they can be quite helpful. For example, some are responsible for the flavors of your favorite foods. We owe the tartness of pickles and sauerkraut to the metabolic wastes of lactic acid bacteria. Still other bacteria provide flavor to yogurt, buttermilk, sour cream, and many cheeses, such as Parmesan, Cheddar and Swiss (Figure 22.11).

Bacteria are also used commercially to generate enzymes, such as amylase (the starch-splitter used in brewing) and others used in laundry detergents and pharmaceuticals. Additional harvested bacterial products include amino acids, hormones, vitamins, and antibiotics. Many of the antibiotics used today are produced by highly selected strains from the genus *Streptomyces,* members of a group of soil bacteria called actinomycetes.

Let's not forget that bacteria are of primary importance to genetic engineering, which holds so much promise for making rare products available in large amounts (see Chapter 16). On a more personal level, the vast populations of *E. coli,* wriggling in your bowel this very minute, help make vitamin A available to you from what would have otherwise been wasted. And finally, the normal bacterial flora that live harmlessly on and in the body, by their very presence, compete with would-be pathogens, thus restricting their success.

Pathogens of Animals Now let's return to our complaints against bacteria. We can first note the obvious: bacteria can cause disease in animals, including ourselves. Pathogens occur in each of the three bacterial forms we reviewed earlier. For instance, among the bacilli alone we find the agents of such dread diseases as Hanson's disease (leprosy), typhus, black plague, diptheria, and tuberculosis.

Over the years, medical science has responded to the threat of pathogens through an intensive campaign, and we have been notably successful. In spite of such progress, we are still threatened by some bacteria. Even as you read, thousands of infants are now dying from a disease called **salmonellosis.** You won't see much about this on the evening news because it's not a problem of medically sophisticated nations. But in Third World countries, a lack of sanitary facilities such as flush toilets and proper sewage disposal, and the absence of clean drinking water (things we take for granted) condemn millions of infants to death each year. Salmonellosis is a debilitating disease at best: its symptoms include fever, ulcerated intestines, and devastating vomiting and diarrhea. The agent can be any of several species from the genus *Salmonella,* which are motile, Gram negative bacilli. *Salmonella* is commonly water-borne, finding its way into water supplies when they are contaminated by human wastes that contain the pathogen. *Salmonella* species are also responsible for one of the most common types of food poisoning and for the highly lethal typhoid fever.

Another problem that has resisted conquest by modern technology is the extremely dangerous type of food poisoning called **botulism.** When the highly resistant endospores of another bacillus, the soil anaerobe *Clostridium botulinum,* are introduced into foods during the canning process, these spores germinate. The resulting bacteria thrive in the nutrient-rich, airless environment. When such food is eaten, the deadly bacterial secretions form one of the most powerful poisons known. (It has been said that a single teaspoonful has the potential to kill off the entire human species!) With the decline of home canning, botulism is now rare, although the threat is always present, even in commercially prepared foods. Fortunately, the food-canning and distribution industries and governmental agencies are on constant alert.

Clostridium tetani, a close relative of *C. botulinum,* thrives in oxygen-free pockets of the soil, where it plays an important role in decomposing organic material. Unfortunately it can also thrive in deep, oxygen-poor wounds, from which its toxins spread to the nervous system producing **tetanus,** an excruciatingly painful condition with spasmodic muscle contractions that leave the body rigid and arched for prolonged periods (which led to the old name, "lockjaw").

C. perfringens is the agent of **gas gangrene,** literally a rotting of the flesh surrounding an infected wound. Untreated, gas gangrene is invariably fatal.

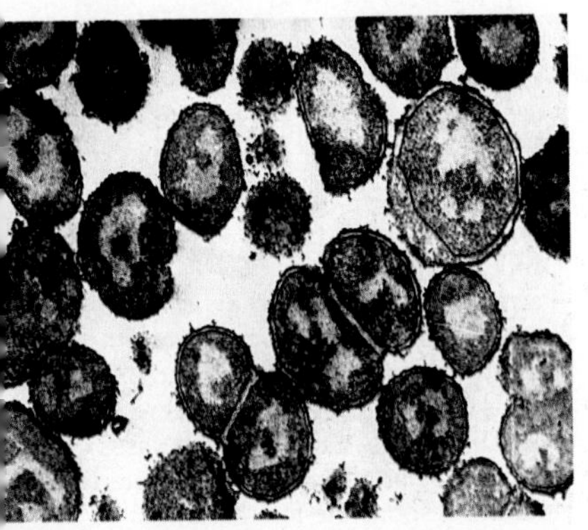

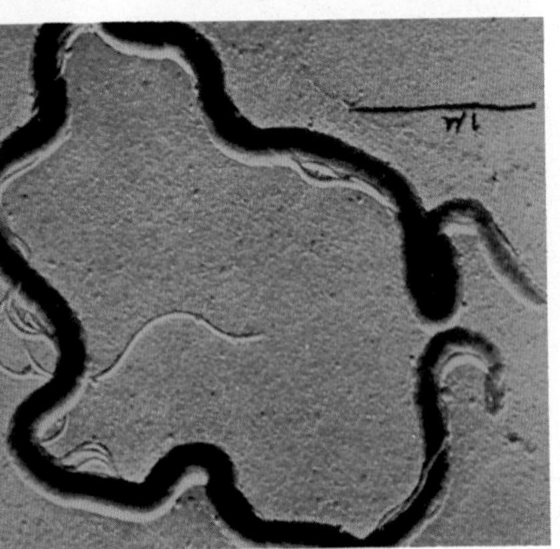

FIGURE 22.12
AGENTS OF SEXUALLY TRANSMITTED DISEASE
Two common agents of sexually transmitted disease are *Neisseria gonorrhoeae* (top) and *Treponema pallidum* (bottom), the eubacteria of gonorrhea and syphilis, respectively. Gonorrheal bacteria are diplococci, occurring in pairs with a capsule. The agent of syphilis is a spirochete.

Coccoid pathogens, such as certain staphylococci, are commonly involved in minor skin infections, boils, and pimples, but under some conditions they can create enormous, dangerous infections. One notorious group crops up occasionally in hospitals and creates troublesome, dangerous, and resistant infections, especially among the newborn. *Streptococcus pyogenes* causes strep throat and other important diseases. Actually, the body's own reaction to the streptococcus is the real threat. Its attempts to fight the microbes results in destruction of body tissues, leading to the diseases scarlet fever, rheumatic heart disease, and rheumatic nephritis, which are still significant causes of death, especially in undeveloped regions.

Less common agents of disease are the **rickettsias,** extremely small obligate parasites that grow only within cells. They cause Rocky Mountain spotted fever and are introduced into humans through the bite of a wood tick.

Sexually Transmitted Diseases *Neisseria gonorrhoeae* is a socially objectionable diplococcus that causes the sexually transmitted disease **gonorrhea** (Figure 22.12). Signs of gonorrhea are readily apparent in men (painful urination along with a pus discharge) but often go undetected in women since the infection is more internalized. Should the infection spread into the pelvic region it can cause sterility. Since gonorrhea can be transmitted during birth and there is a risk of its causing blindness in the baby, it is routine to treat the eyes of newborns with drops of silver nitrate. For a time, gonorrhea was well controlled by penicillin therapy, but evolution permits bacteria to adapt to the challenges of their changing environment, and now we are forced to deal with new, antibiotic-resistant strains.

The most notorious of the spirally shaped bacteria is *Treponema pallidum,* the screw-shaped spirochete of **syphilis**, another common sexually transmitted disease (Figure 22.12b). An early sign of syphilis is usually the appearance of a "chancre," a small, ulcerated sore on the genitals or mouth. However, it soon disappears, and an unwary person might believe that all is well. It is not. In its more progressive stages, syphilis has been called the "great pretender" since its widespread effects mimic many other diseases. If it is allowed to go untreated, the spirochetes will eventually enter the nervous system, permanently damaging brain tissue and causing blindness, insanity, and death. Sadly, the spirochete is also known to cross the placenta, infecting the embryo, and producing serious birth defects in the baby.

Another group of pathogens, the **chlamydias,** cause a disease of the same name. It has some of the symptoms of gonorrhea and may be the most prevalent of the sexually transmitted diseases.

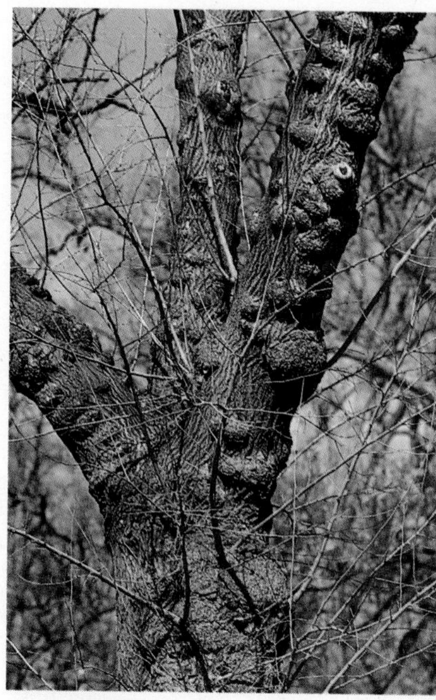

FIGURE 22.13
BACTERIAL DISEASE IN PLANTS
Crown gall is a tumorous growth brought on by the insertion of DNA from the bacillus *Agrobacterium tumefaciens* into some of the plant's chromosomes.

Pathogens of Plants Occasionally, we hear of some dire threat to crops by some microorganism. Usually the newsmakers are fungal or viral, but bacteria are also quite capable of creating agricultural havoc. Included in the list of bacterial diseases of plants are such earthy-sounding names as **blights, wilts, galls,** and **rot.** The effects range from simple discolorations of fruit and leaves to certain death of the afflicted plant. Virtually all of the bacterial plant pathogens are bacilli, and most are flagellated.

The presence of blight is heralded by dead and dying tissue on flowers, leaves, and stems. Fire blight, a particularly devastating infection caused by *Erwinia amylovora,* a motile bacillus and a relative of *Escherichia coli,* quickly kills young pear and apple trees. If pear trees fail to grow in your locale, the reason might well be the presence of *E. amylovora.*

Wilts are attributed to members of the genus *Corynebacterium,* a group that also includes the agent of human diphtheria. Wilt, characterized by the drooping of leaves and stems, suggests what is happening inside the plant. The bacterium invades the vascular system, hindering the movement of water, minerals, and food in the plant. Wilts may threaten such significant crops as alfalfa, beans, squash, and watermelons, each invaded by a different species of *Corynebacterium.*

Gall is a form of plant tumor. Crown gall, among the most common, is initiated by *Agrobacterium tumefaciens.* While not generally a fatal disease, it is quite disfiguring (Figure 22.13). Such tumors have an interesting origin. Plant pathologists have known for some time that the afflicted cells become genetically transformed when bits of bacterial DNA are incorporated into their chromosomes. The cells then divide out of control, a typical cancerous response, and the galls emerge. This conclusion took a bit of creative sleuthing before the connection to *A. tumefaciens* was made, since by the time the tumors have actually formed, the invader is nowhere to be found. You may recall our mentioning *A. tumefaciens* in connection with genetic engineering in Chapter 16.

Bacterial Autotrophs

Photoautotrophs The best known of the photoautotrophic or photosynthetic bacteria are the cyanobacteria, a widely distributed group containing some 2500 species. They live in aquatic environments including oceans, ponds, lakes, tidal flats, moist soil, swimming pools, and even around leaky faucets. Some flourish in hot springs where temperatures reach 75° C (167° F). They can often be recognized by their blue-green color, but they also may be black, purple, brown, or red. Cyanobacteria abound in the marine environment where, through their photosynthetic activity, they release significant amounts of oxygen and contribute to the organic matter on which marine life depends.

Cyanobacteria occur as single cells, as colonies of cells, and even in a simple multicellular state (Figure 22.14). Many cyanobacteria occur in long chains of cells organized into filaments, which may or may not be branched. In other instances, cyanobacteria

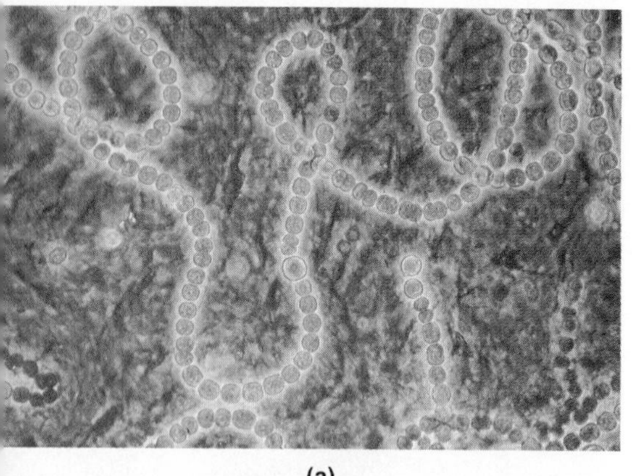
(a)

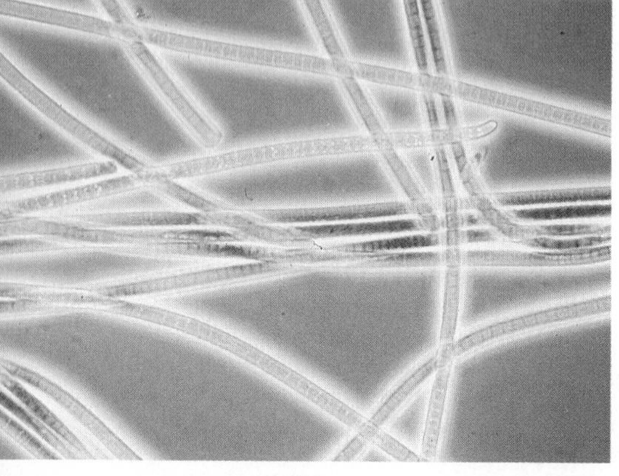
(b)

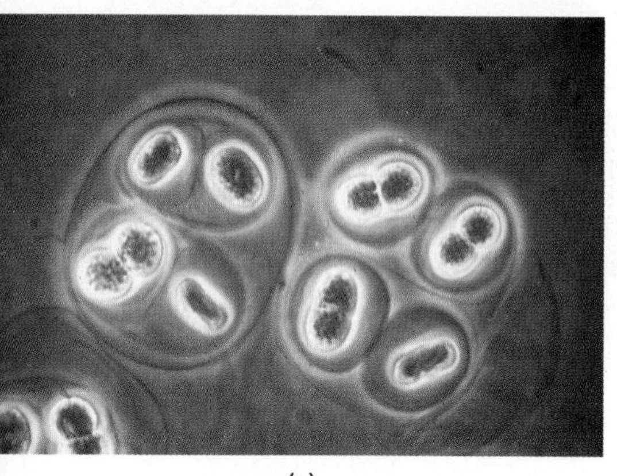
(c)

FIGURE 22.14
THREE CYANOBACTERIA
A few of the diverse cyanobacteria are the filamentous, beadlike *Nostoc* (a), the slowly undulating, filamentous *Oscillatoria* (b), and the spherical *Gloeocapsa* (c) enclosed in its gelatinous wall. All are inhabitants of fresh water. The larger, thick-walled cells of *Nostoc* are known as *heterocysts,* cells that specialize in nitrogen fixation.

form gelatinous masses of no definite shape. The cyanobacteria can move: filamentous forms such as the common *Oscillatoria* rotate in a screwlike manner, while the gelatinous forms glide along in a mucuslike slime they produce. As far as is known, reproduction in the cyanobacteria is through fission only.

The cells of cyanobacteria reveal a considerable level of complexity (Figure 22.15). As we mentioned earlier, their chlorophyll is integrated into thylakoids, extensions of the plasma membrane. Actually, the entire photosynthetic cell is comparable to a eukaryotic chloroplast. Photosynthesis in the cyanobacteria is nearly identical, biochemically, to that of algae and plants. Like the algae and plants, their photosynthetic pigments include chlorophyll *a* and the accessory pigment beta carotene, although they lack chlorophyll *b*. In addition, cyanobacteria produce red and blue pigments called **phycobilins,** which are important in the capture of light energy and contribute to the characteristic color of these bacteria. The glucose produced through their photosynthesis is stored in their own form of starch, which is similar to animal glycogen. These characteristics make cyanobacteria likely predecessors of modern plants, or at least of modern chloroplasts.

A number of cyanobacteria produce specialized, nitrogen-fixing cells called **heterocysts.** Their role is to incorporate atmospheric nitrogen into a form useful for producing amino acids and other nitrogen-containing molecules. Interestingly, the formation of nitrogen-fixing heterocysts is inhibited in many species when alternate sources of nitrogen—ammonia or nitrates—are added to their medium.

Halobacteria (*halo,* "salt") form a photoautotrophic group of the archaebacteria. Their name refers to the salty environment in which they thrive. In the simplest of photosystems, the halobacteria capture light energy with the purple pigment, **bacteriorhodopsin,** and use that energy to pump protons out of the cell. This creates the usual steep chemiosmotic gradient, which the bacterium then uses to generate ATP (see Chapter 6). ATP is formed as the protons are readmitted, passing down the chemiosmotic gradient through phosphorylating enzymes at special sites in the membrane. (Interestingly, a form of rhodopsin is a visual pigment in the vertebrate eye—the "visual purple" that reacts with light, triggering the neural impulses associated with vision.) Because halobacteria do not use any form of chlorophyll during photosynthesis, it is assumed that they did not contribute to the evolution of modern plants.

Chemoautotrophs Chemoautotrophs obtain their energy from the oxidation (removal of electrons) of simple inorganic compounds in the earth's crust. Much of this energy is then used in molecule building, and as with many photoautotrophs, the usual starting molecule is carbon dioxide.

One group of chemoautotrophs obtains its energy through the oxidation of sulfur or sulfur compounds. Among these are sewage and swamp bacteria that commonly utilize the hydrogen from hydrogen sulfide, storing the remaining sulfur as granules in their cells. One sulfur-using group produces sulfuric acid, which ionizes in soil water, producing sulfate and releasing hydrogen ions, creating a highly acidic soil. Other chemoautotrophs get their energy by oxidizing iron and manganese where they occur in a reduced state, while yet others make use of hydrogen gas. (For a preview of how one sulfur-utilizing bacterium provides a source of energy for the strange organisms of the great oceanic rifts, see Essay 47.2.)

As nitrogen-fixers, certain soil bacteria as well as aquatic cyanobacteria also qualify as chemoautotrophs. They reduce atmospheric nitrogen (N_2) to ammonia for their own use. Nitrogen-fixing soil bacteria carry on a mutualistic existence within the roots of certain flowering plants, where an exchange of surplus ammonium ions for photosynthetic products cements the partnership. But of more global significance, surpluses are also released into the enviroment, where they become available to other plants. Essay 22.1 takes us further into the nitrogen-fixing process itself.

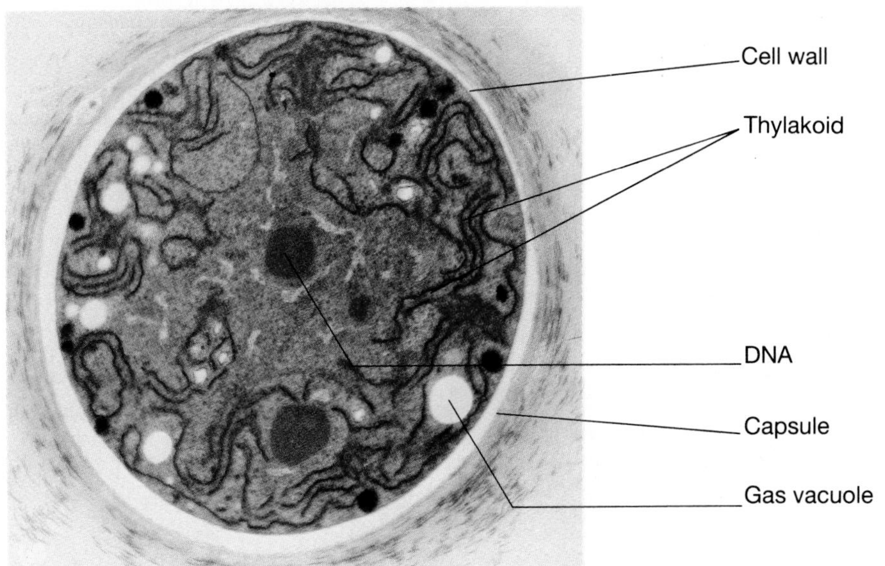

FIGURE 22.15
CYANOBACTERIAL STRUCTURE
Cyanobacteria are unusual prokaryotes in that they have many membranous, chlorophyll-containing thylakoids, extensions of the plasma membrane.

EUBACTERIA, ARCHAEBACTERIA, AND A NEW PHYLOGENY

As recently as 1975, biologists were quite comfortable with the organization of prokaryotes into Kingdom Monera, which contained two major groups, the bacteria and the cyanobacteria. But as you are well aware, opinions about how things should be grouped can change rapidly in biology. Intensive research pioneered by Carl R. Woese, who used new molecular techniques such as RNA nucleotide sequencing for establishing phylogenetic relationships, has changed the way we classify microorganisms. A proposed phylogeny based on this work is shown in Figure 22.16.

Woese introduced the idea that the prokaryotes actually include two distinct, unrelated groups—*Archaebacteria* and *Eubacteria* (the "first, or ancient, bacteria" and "true bacteria," respectively). They are now appearing as separate kingdoms in a number of schemes. While both groups are definitely prokaryotic, they are different in enough ways to indicate separate origins from progenotes, now believed to represent the earliest forms of life. The progenotes are also believed to have produced the first eukaryotes (see Chapter 21). The two kinds of bacteria look much alike when viewed through the light microscope, but the electron microscope and chemical analysis reveal basic differences in both structure and chemistry (Table 22.1). The life style of the archaebacteria can be truly fascinating.

TABLE 22.1 DIFFERENCES BETWEEN ARCHAEBACTERIA AND EUBACTERIA

STRUCTURE	ARCHAEBACTERIA	EUBACTERIA
Cell wall	Variety of substances, often proteinaceous	Peptidoglycans
Plasma membrane lipids	Modified branched fatty acids	Straight-chain fatty acids
Ribosomes	30S, 50S subunits greater similarity to eukaryotic than to eubacterial	30S, 50S subunits, unlike archaebacterial and eukaryotic
Flagella	Unknown	Tubular, rotating (protein flagellin)
Photosynthetic pigments	Bacteriorhodopsin	Bacteriochlorophyll and chlorophyll *a*

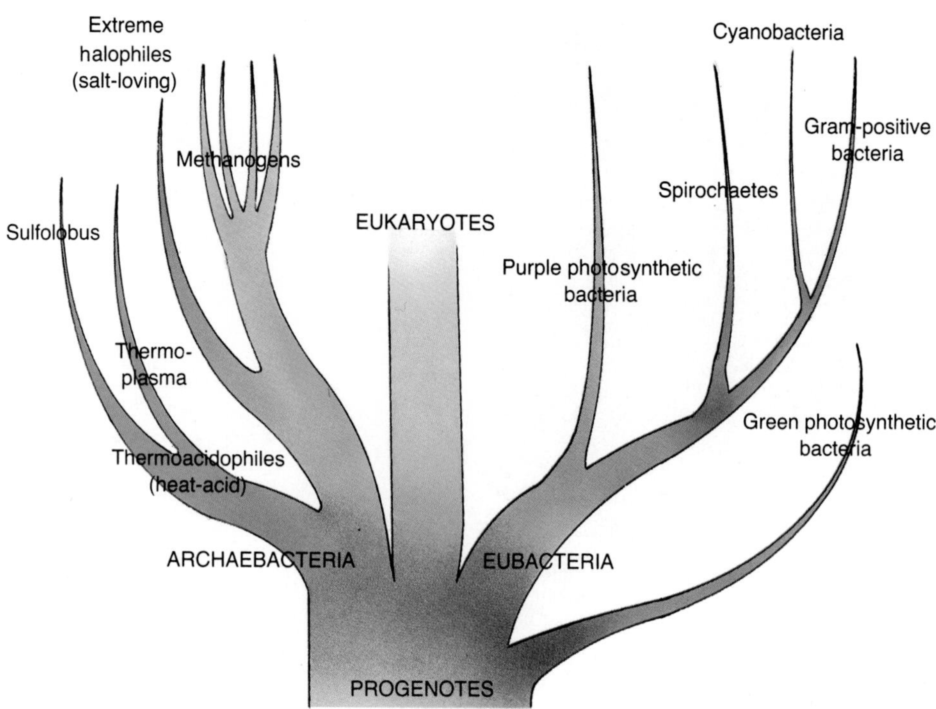

FIGURE 22.16
PHYLOGENY OF PROKARYOTES
Recent phylogenetic studies recognize two prokaryotic kingdoms, Archaebacteria and Eubacteria. The distinction is based on striking biochemical differences. The ancestors of both, and of the eukaryotes, were the progenotes, the first life forms to emerge from a long period of chemical synthesis.

Archaebacterial Life

Archaebacteria are not easy to grow in the laboratory and therefore are not as familiar to bacteriologists as the eubacteria. When it first became apparent that prokaryotes tend to fall into two distinct groups, it was assumed that the archaebacteria were rare, "primitive," and possibly relics of the earliest form of bacterial life. They were also considered to be quite bizarre in that the most familiar archaebacteria lived in strange and improbable habitats, such as near-boiling hot springs and very acidic or salty ponds (harsh conditions not unlike those of the primitive earth). But the largest group of archaebacteria, the **methanogens** (methane generators), are found in habitats where carbon dioxide and hydrogen are readily available, but where there is little or no oxygen. We find them in anaerobic marshes, in sewage treatment plants, in the mucky, anaerobic sea and lake bottoms (such as the Black Sea), and, as we have seen, in the oxygen-deficient bowels of animals, including humans. The familiar *Escherichia coli,* an aerobic eubacterium, is better known than our archaebacterial gut inhabitants only because it is so much easier to grow in laboratory cultures. There are no hard and fast rules about oxygen utilization by prokaryotes, but some archaebacteria are obligate anaerobes—that is, they cannot survive in the presence of oxygen. Others have the ability to use oxygen in their metabolism.

Methanogens use hydrogen gas to reduce carbon dioxide, producing methane gas (CH_4, or marsh gas, as it was first known) and water. The reaction (which requires a battery of enzymes) is:

$$4H_2 + CO_2 \rightarrow CH_4 + 2H_2O$$

Incidentally, sewage treatment plants can help conserve energy by using the methane gas they produce as a fuel for electrical generators.

Other archaebacterial types include the aptly named extreme halophiles ("salt lovers") discussed earlier, extreme thermophiles ("heat lovers"), and the thermoacidophiles ("heat and acid lovers"), names suggesting rather drastic living conditions. The halophiles thrive in Great Salt Lake, the Dead Sea, and in salt-evaporation facilities, where, as we mentioned, they use bacteriorhodopsin to harness the sun's energy in their unique version of photosynthesis.

ESSAY 22.1

THE NITROGEN FIXATION PROCESS

The chemical process of nitrogen fixation can be quickly summarized by the simple formula:

$N_2 + 3H_2 \rightarrow 2NH_3$ (ammonia)

$NH_3 + H_20 \rightarrow$
NH_4^+ (ammonium ion) + OH^-

But such simplicity is truly deceptive. The atoms of nitrogen gas (N_2) are held tenaciously together by three extremely resistant covalent bonds, and so nitrogen does not readily enter into chemical reactions. For instance, consider the widely used Haber industrial process of nitrogen reduction, in which most synthetic fertilizer is produced. This process uses iron as a catalyst, and requires heat exceeding 500° C and pressure exceeding 300 atmospheres. All of this is needed just to pry the nitrogen molecule apart so that its fragments can be reduced by hydrogen. So how do the tiny nitrogen-fixing bacteria overcome these seemingly impossible requirements?

Bacteria fix nitrogen through the use of a two-part enzyme complex, the iron-containing protein **reductase** acting as a reducing agent, and a highly specific enzyme known simply as **nitrogenase**. The reactions converting nitrogen gas to ammonium ions occur in two main stages, summarized in the illustration. Here, we see that electrons from the respiratory chain along with energy from ATP, provides the reducing power.[1] Reductase then passes its electron to nitrogenase, and this potent enzyme then reduces atmospheric nitrogen, forming ammonium ions that provide nitrogen for the manufacture of amino acids.

How expensive is the process in terms of ATP? The nitrogen-fixing bacteria in the roots of the common garden pea use fully *one-fifth of the ATP produced by the plant.* Each ammonia molecule (or ammonium ion) produced requires 6 ATPs. There is an interesting epilog to this story. It has to do with genetic engineering and the nitrogenase enzyme. It is clear that the industrial reduction of nitrogen is an energy-costly process, but what are the alternatives in a hungry world?

Why not turn crop plants into nitrogen fixers? That is, why not use the recombinant DNA technology to provide corn, wheat, and rice with their own nitrogenase coding genes? Geneticists have been trying to do just that. The nitrogenase gene, or **nif** as it is called, present in the nitrogen-fixing bacterium *Klebsiella pneumoniae,* has been excised and spliced into an *E. coli* plasmid. The engineered plasmid was then reintroduced into colonies of the common colon bacterium, and some nitrogen-fixing activity has been detected. The next step, a formidable one and one that has not been done, is to get the plasmid into a plant. There are serious problems to solve. One is that the enzyme is readily inactivated by oxygen, so its surroundings must be kept anaerobic. (Evolution has provided the pea plant with a form of hemoglobin in its root nodules. As the hemoglobin binds the oxygen, an anaerobic environment is maintained.) Another drawback is the heavy ATP requirement mentioned above. Such a demand might not be easy for corn or wheat to fulfill. Work today is stalled around these problems.

[1]The source of electrons in photosynthetic N-fixers is the light-activated electron transport system, and that in heterotrophs is the respiratory electron transport system.

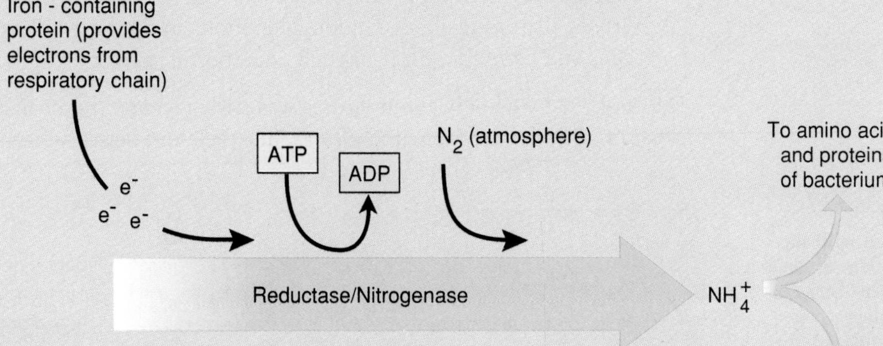

Following its reduction by NADH, the activated iron-containing protein (RH) joins with the nitrogen-fixing enzyme, nitrogenase. The complex then reacts with molecular nitrogen, forming ammonium ions.

The thermophiles live under incredibly harsh conditions, some of them thriving at 90° C (194° F, near boiling). In the hot springs of the deep-sea Galapagos rift, bacteria survive at even higher temperatures. (Recent studies of *Pyrodictium,* a rift bacterium, reveal that it thrives at 360° C, but the great pressure apparently helps in preventing enzyme inactivation. The thermoacidophiles live under even harsher conditions, some thriving in strong acids heated to near boiling. Surprisingly, the internal pH of such archaebacteria remains close to neutral.

THE VIRUSES

As we muster our intellectual forces and try to come to grips with the nature of life, we are inexorably led to the realm of the **virus.** Viruses are minute, biologically active particles made up of a nucleic acid core, a covering of protein, and sometimes an enzyme or two. While they can indeed be biologically active in cells, as anyone with a simple cold can tell you, some can also become crystallized, like common table salt, and remain inactive for a seemingly indefinite period.

Even those scientists who place them among living organisms are uneasy about where they fit in the organizational scheme of things. Adding to the biologist's dilemma, there are many apparently unrelated viruses that share only a few features, such as their extremely small size and highly limited mode of life. Viruses range in size from 20 to 300 nm, a range that extends from the size of large molecules to that of the smallest bacterial cells (Figure 22.17).

All things considered, it is patently clear that viruses aren't cells. They lack just about everything cells have, except the genetic instructions—the coded nucleic acids (DNA or RNA)—needed to ensure their own perpetuation. Because of this, all viruses are obligate parasites. They can neither carry on metabolic activity nor can they reproduce unless they are in a host cell.

Of course, our interest in viruses is not entirely academic. After all, we humans are susceptible to several hundred viral diseases, not to mention those that infect our crops and domestic animals. The list of human maladies caused by viruses includes not only the common cold, but smallpox, polio, German measles, chicken pox, mumps, and the many forms of influenza that periodically sweep through human populations. More recently, viruses known as oncogenic viruses, long known to be related to cancers in laboratory animals, have been implicated in certain human cancers. And now humans are confronted by the most confounding invader, the virus of AIDS.

What do viruses do to cells? Let's have a brief preview. Viruses damage their hosts in two principal ways, with short and long-term effects:

1. They physically disrupt host cells, often killing them as they reproduce and make their escape. If enough cells are killed, the host will certainly die.
2. Viruses also bring about random alterations in the host cell's genome. As we have seen, one common effect of such alteration is cancer.

Most of what we know about viruses was fairly recently discovered—a product, more or less, of 20th-century research efforts. But the story actually begins long before that.

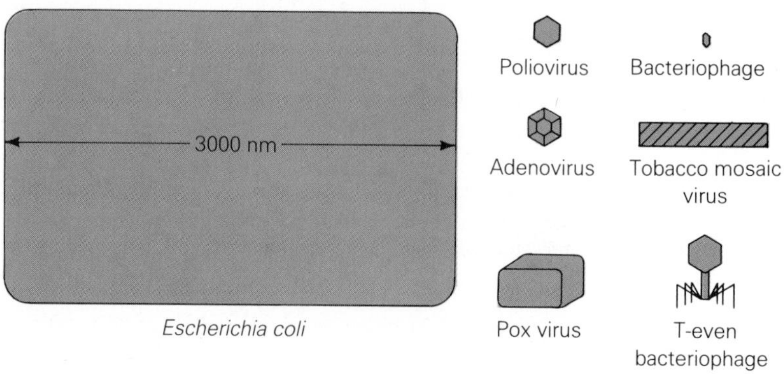

FIGURE 22.17
VIRAL DIMENSIONS
Comparing several viruses with the colon bacterium *Escherichia coli* gives us some idea of the size of things in the viral realm. The smallest virus shown, a polio virus, is only 30 nm in diameter, compared to the 500 nm diameter of the bacterium. On the same scale, one of our own red blood cells would stand about shoulder height, and a human egg cell would be about equal to the width of a large swimming pool.

(a)

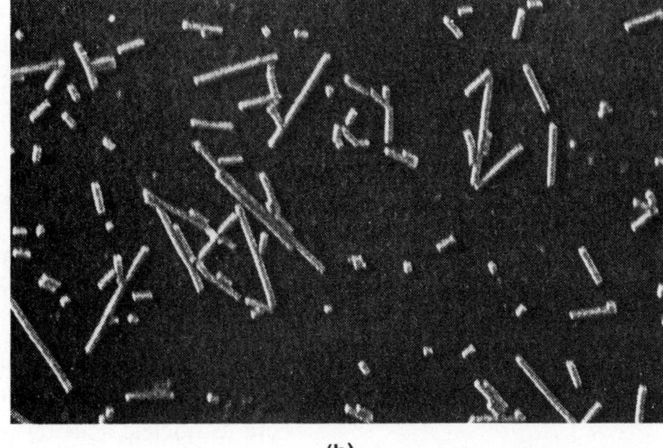

(b)

FIGURE 22.18
TOBACCO MOSAIC VIRUS
(a) The leaves of an infected tobacco plant are mottled in color and wrinkled in texture. (b) In this electron microscope view the crystallized virus takes on a cylindrical or rod-shaped appearance.

The Discovery of Viruses

The discovery of viruses came at a time when the newly emerging field of bacteriology was making its first great gains. By the late 1800s, Louis Pasteur, Robert Koch, and others had convinced the world that bacteria were the agents of some diseases. But other diseases apparently were caused by something else. The methods that usually proved so successful in finding the bacterial villains simply didn't work for these diseases. Oddly enough, the stymied bacteriologists, although unable to find the culprits, were able to produce vaccines that were effective against some of them. For example, vaccines for the prevention of smallpox and rabies in humans and hoof-and-mouth disease in cattle were developed long before the viral agents of these devastating diseases were found. A number of causative factors were proposed, some rather imaginative. As an example, the term *virus* itself means "poison," and for ages, people attributed viral diseases such as yellow fever and smallpox to poisons carried by the "deadly night air."

By 1892, studies of a contagious disease in tobacco, soon dubbed the **tobacco mosaic virus**, revealed several important clues to the nature of viruses. Dimitri Iwanowski, the Russian biologist, extracted juices from infected plants, strained the liquid through an extremely fine filter (the type used to remove bacteria from growth media too delicate to sterilize by heating), and using the filtrate, succeeded in spreading the disease to healthy plants. From his work, Iwanowski was able to establish three facts about the disease agent:

1. It was smaller than a bacterium.
2. It could be transferred from plant to plant; that is, it was contagious.
3. It somehow increased in number in the host plant.

By the turn of the century, the filtration technique had been broadly applied, and the list of these filterable viruses, as they were once known, grew substantially. But such progress was not without its problems. Despite persistent efforts, the early bacteriologists were unable to grow the filterable agents in the usual bacterial culture media. This meant that the techniques so useful in identifying bacterial disease agents could not be fully applied to the tiny viral agents. Further, they still could not tell whether the filterable agent was a living entity or just a chemical of some kind. The puzzle persisted.

Then, in 1935, the American microbiologist Wendell Stanley trained his light microscope on a drop of filtrate from infected tobacco plants and became the first person to see a virus. Actually, what he saw was the crystalline form—the long, slender crystals of the dormant tobacco mosaic virus (Figure 22.18). (In that same year, the large, rod-shaped virus particle itself was first observed through the newly invented electron microscope.) Stanley also showed that while the tobacco mosaic virus appeared in many ways to be nonliving and could not be grown by any usual bacterial culture methods, it could somehow reproduce if it was present in healthy tobacco plants. Stanley's procedure was straightforward. He introduced a small drop of dilute viral filtrate into a healthy plant. The plant would then develop the disease, and from its tissues he could recover a much greater quantity of the viral substance. The search was on to learn more about these peculiar little entities. Since then, particularly in the past few decades, we have learned a great deal about the biology of viruses.

The Biology of Viruses

One of the most important discoveries about viruses is that they are all parasites. The reason they cannot be grown on a bacterial growth medium is that they lack the enzymes and metabolic pathways commonly found in living cells; therefore they must invade cells and use the metabolic machinery of their hosts to complete their life cycles.

Each virus particle, or **virion,** as the individual particles are known, consists of a core of nucleic acid (either DNA or RNA), a protein coat called a **capsid,** and usually one or more cell-penetrating enzymes. A few contain their own DNA or RNA polymerase.

Whereas an inactive virion contains an array of genes, it cannot carry out even the simplest requirements for living without a living host. So viruses must be considered the most specialized of parasites, lacking all structure except that which facilitates the invasion of a host and the production of more viruses.

Viral Shapes Most viruses are either helical or polyhedral (many-sided) in shape, but a few are cubic, and some are brick-shaped. The shape of a virion depends ultimately on the arrangement of protein subunits, or **capsomeres,** in the capsid. The helical viruses, for example, consist of a spiral of nucleic acid surrounded by a capsid whose capsomeres follow the nucleic acid spiral. Included among them is the tobacco mosaic virus, the first virus to be described. The polyhedral viruses may sometimes appear spherical, but a close look at some reveals that the spheres are actually polyhedrons, many-sided forms. Included in the polyhedral viruses are the familiar "even numbered" or T-even bacteriophages (discussed in Chapters 12 and 15), with a polyhedral head and a cylindrical tail. Helical and polyhedral viruses may be enclosed in a complex covering or envelope (Figure 22.19). Among the polyhedral viruses that are surrounded by an envelope are members of the disreputable *Herpes simplex* group. Among the cubic or brick-shaped viruses is the agent of smallpox, which, thanks to the efforts of the World Health Organization, is now believed to be extinct outside the laboratory. (The last reported case of smallpox occurred as a laboratory accident in 1977). The retrovirus of AIDS, incidentally, is actually spherical and has a surrounding envelope.

Biochemical Differences In addition to these variations in form and covering, viruses can also differ in the kind of nucleic acid core they contain. For example, in some viruses the genetic material is standard double-stranded DNA, just as it is in all metabolizing organisms. But in other viruses, we may find (1) single-stranded DNA, (2) double-stranded RNA, (3) a single large molecule of single-stranded RNA, or (4) several small strands of RNA. The nucleic acid can be circular or linear, and the single-stranded nucleic acid can be either the transcribed strand or the nontranscribed strand. Many of these differences are listed in Table 22.2.

Viruses and the Host Cell

How Bacteriophages Invade **Bacteriophages**, you'll recall (see Chapter 15), infect bacteria. They begin their attack on cells by attaching to the cell surface, in some instances through the use of special enzymes. The site of penetration is often quite specific, requiring the biochemical recognition of certain host cell surface molecules by viral coat proteins prior to attachment. In the T-even bacteriophages, the tail releases **phage lysozyme,** an enzyme that digests its way through the bacterial cell wall. The capsid or coat remains outside, and the nucleic acid core is injected into the host cell.

Once inside a bacterial cell, what happens next depends upon the specific virus and its capabilities. There are generally two possible outcomes. Some phages, such as the

FIGURE 22.19
VIRAL DIVERSITY
(a) In the naked tobacco mosaic virus, the capsomeres (protein units) follow the spiral form taken by the nucleic acid core. **(b)** The influenza virus is surrounded by a dense envelope, complete with a number of spiked projections, important to cell penetration. **(c)** The adenovirus takes the form of a polyhedron. The tiny spheres are capsomeres. **(d)** Herpes simplex is a polyhedral virus surrounded by a spiked envelope.
(e) One of the most complex of the viruses is the T-even bacteriophage, with its "moon-lander" polyhedral head and complicated cell-puncturing tail.

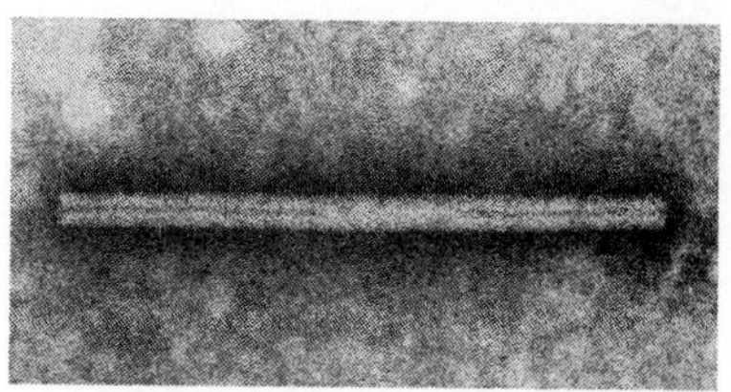
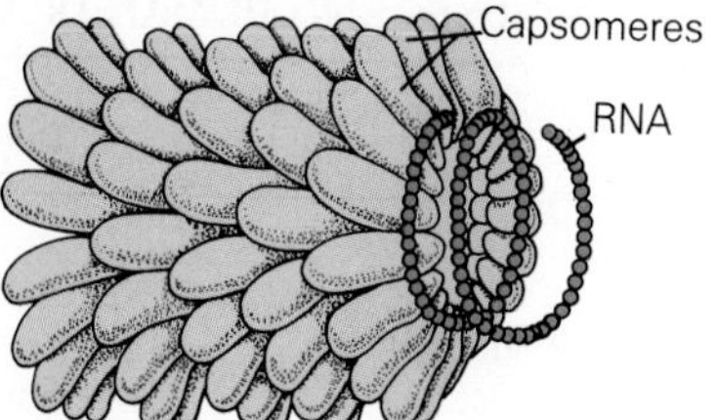

(a) **Naked spiral**
(tobacco mosaic virus)

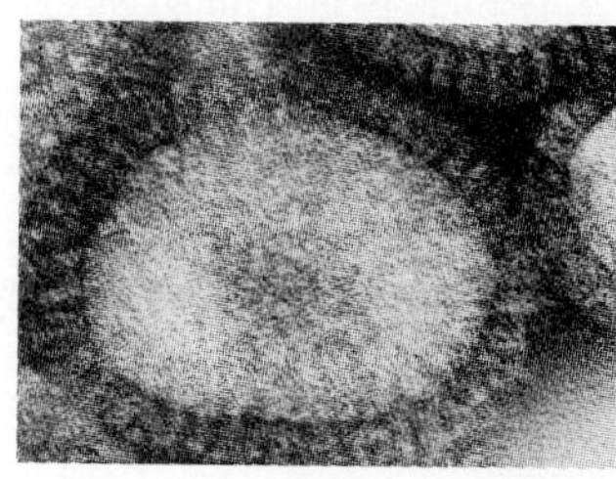
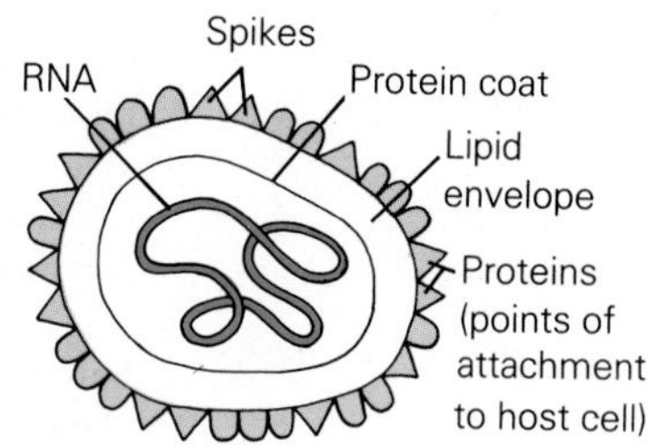

(b) **Enveloped spiral**

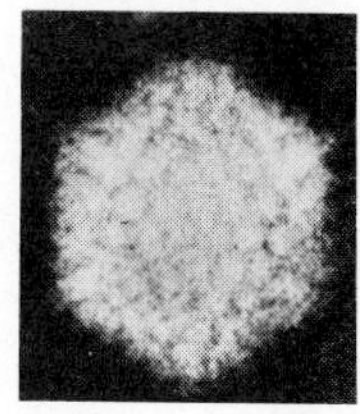
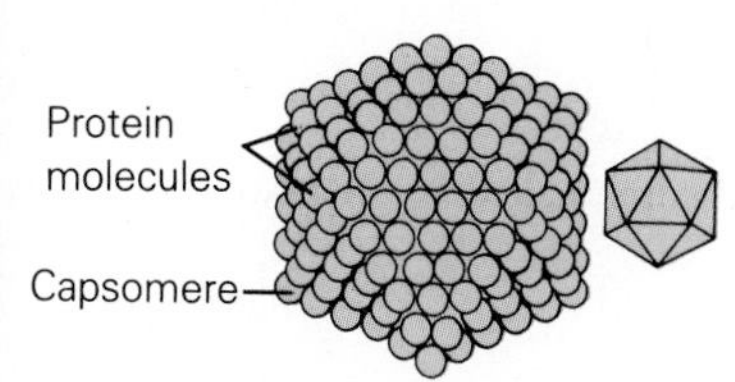

(c) **Naked polyhedron**

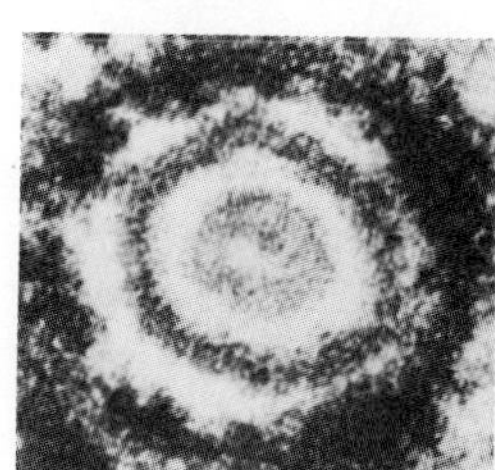
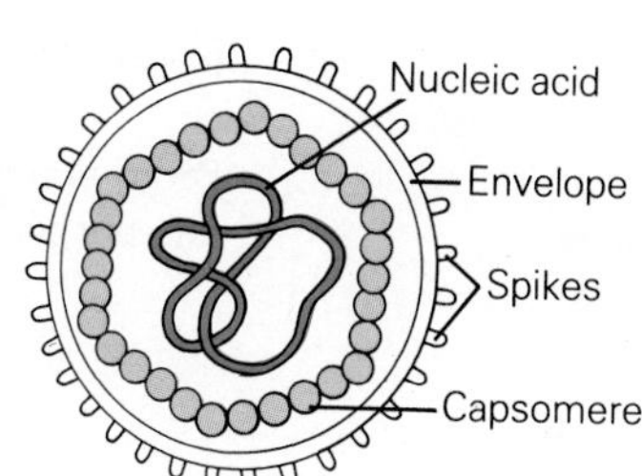

(d) **Enveloped polyhedron**

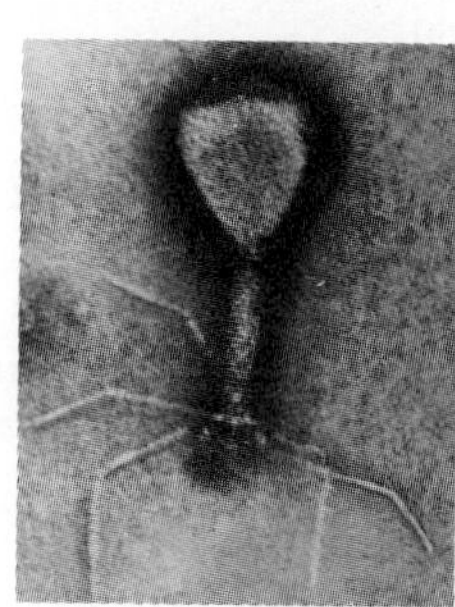
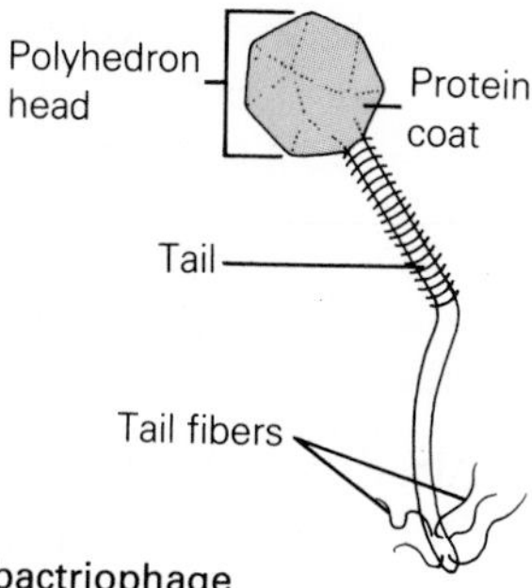

(e) **T-even bactriophage**

TABLE 22.2 VIRUSES AND HUMAN DISEASES

VIRUSES	DISEASES
DNA VIRUSES	
Poxviruses (brick-shaped)	
Vaccinia type	Smallpox
Paravaccinia	Nodules on hands
Herpesviruses (icosahedrol, enveloped)	
Group A	
Herpes simplex 1	Fever blisters, respiratory infections, encephalitis
Herpes simplex 2	Genital infections, cervical cancer (?)
B virus	Encephalitis
Group B	
Varicella-zoster	Chicken pox
Cytomegalovirus	Jaundice, liver/spleen damage, brain damage
Ungrouped	
Burkitt's lymphoma	Burkitt's lymphoma, Hodgkin's disease, infectious mononucleosis
Adenoviruses (icosahedral, naked)	
Humanadenoviruses	Respiratory infections
Papovaviruses (ichosahedral, naked)	
Human papilloma	Warts
RNA VIRUSES	
Picornaviruses (icosahedral, naked)	
Enteroviruses	
Poliovirus	Poliomyelitis
Coxsackie A	Muscle and nerve damage, common cold, meningitis
Coxsackie B	Meningitis, paralysis
ECHO	Paralysis, diarrhea, meningitis
Rhinoviruses	Common cold
Reoviruses	Pathogenicity unknown
Arboviruses (helical, enveloped)	
Dengue	Fever, rash
California encephalitis	Encephalitis
Myxoviruses (helical, enveloped)	
Influenza viruses	
Type A_0, A_1, A_2	
Type B_0, B_1, B_2	
Type C	
Paramyxoviruses	
Sendai virus	Common cold
Mumps virus	Mumps
Pseudomyxoviruses	
Measles	Measles, pneumonia, common cold
RNA Tumor Viruses	Breast cancer(?)
Retroviruses	
HTLV-I, HTLV-II	Leukemia, lymphoma (still in question)
HIV	AIDS (acquired immune deficiency syndrome)
VIRUSES OF UNCERTAIN CLASSIFICATION	
Rubella virus	German measles
Hepatitis Viruses	
Type A	Short incubation hepatitis (infectious)
Type B	Long incubation hepatitis (serum)
Type nonA-nonB or C	Hepatitis

HTLV, human T-cell lymphotropic virus; HIV, human immunodeficiency virus.

T-evens, are capable of a lytic relationship only, in which the virus immediately reproduces and then lyses or destroys the cell. They are aptly named **lytic phages.** Others such as the **phage lambda** of *E. coli,* also lyse cells, but may first enter a quiescent, **lysogenic state.** Phages with this capability are called **temperate phages.**

Following invasion by a lytic phage (see Figure 22.20a), the invading phage disassembles the bacterial host's DNA and uses it for its own replication. It then uses the host transcription and translation machinery for producing many new viral coats, tails and tail fibers (landing gears), and viral enzymes. The amino acids to be assembled into these viral proteins are, of course, also "appropriated" from the host (like your freeloading Uncle Charlie, who, having eaten all of your groceries, borrows your overcoat as you show him out the door). Finally, the host bacterium is lysed, literally bursting, and the new infectious virions are released.

When the bacteriophage lambda of *E. coli,* a temperate phage, invades a cell, it may immediately enter its lytic cycle, or it may delay this for a time (Figure 22.20b). In the latter case, a lysogenic relationship ensues, during which the phage genome is integrated into the host chromosome. Integration of phage DNA occurs in places along the bacterial chromosome where base sequences form a base-pairing match with those of the phage DNA. Insertion usually occurs at very specific gene loci. Afterward, the phage, now called a **prophage,** has its DNA replicated right along with the host DNA, and when fission occurs, the replicas are distributed to the newly forming daughter cells. This can easily go on for generation after generation of cells. As a result, the prophage may find its numbers greatly increased, as part of the genome of an entire bacterial colony.

The quiescent state of a prophage may persist for many generations. Such an inactive state is maintained by a repressor protein, synthesized by the bacterium, according to mRNA coding transcribed from one of the hitchhiking viral genes. Should the repressor gene itself be inactivated, however, or should DNA replication in the bacterium itself become inhibited, the remaining prophage genes will become active, undergoing replication and transcription. New viral particles will emerge, and the cell will be lysed. Prophage activation can be readily induced in the laboratory by exposing an infected bacterial culture to physical agents, including certain chemicals and ultraviolet light.

Invasion by Plant Viruses Viruses are, of course, serious parasites of plants as well as of bacteria and animals. In most instances plant viruses are spread by insects feeding from one plant to another, or by insect pollinators carrying infected pollen from plant to plant. Viral infections are particularly dangerous to crops, especially potatoes, sugar beets, wheat, soybeans, and tobacco. Infected plants are easily recognized by the presence of mottled and wrinkled leaves, loss of color, tumors, dying tissue, and stunted growth. The tobacco mosaic virus does so well in the domestic tobacco plant that the minute viral particles can account for one tenth of an infected plant's dry weight!

Invasion by Animal Viruses The manner in which animal viruses penetrate their hosts is generally similar to the action of T-even phages. But, as you might expect, the differences between prokaryotic and eukaryotic cells will have some influence on the way viral infection occurs. The capsids of animal viruses, like those of the T-even phages, contain specific sites that have complementary target sites on the host plasma membrane. In the adenoviruses (the viruses of respiratory infections), active sites are at the corners of the polyhedron, while in the enveloped myxoviruses (measles, mumps, and some flu viruses) they are on "spikes" on the envelope. The AIDS virus makes use of a glycoprotein in its initial attachment.

There are no injecting mechanisms in animal viruses, so they must enter the cell in a different manner (Figure 22.21). In those viruses lacking an envelope, the entire virion enters the host. In the enveloped viruses, the envelope itself sometimes fuses with the membrane at a complementary site, and the viral nucleic acid enters there. In other instances the host cell obliges the virus by engulfing it through phagocytosis, enclosing the virus in a food vacuole. The host's lysosomal enzymes are then released into the

FIGURE 22.20
LYSIS AND LYSOGENY
(a) In the lytic state the phage produces many new viruses, which are released as the cell ruptures. These may go on to infect other cells. **(b)** In the lysogenic state the viral genome is incorporated into the bacterial chromosome. It may remain there for many generations, replicating along with the host's DNA, but under certain conditions a lytic state **(c)** may begin.

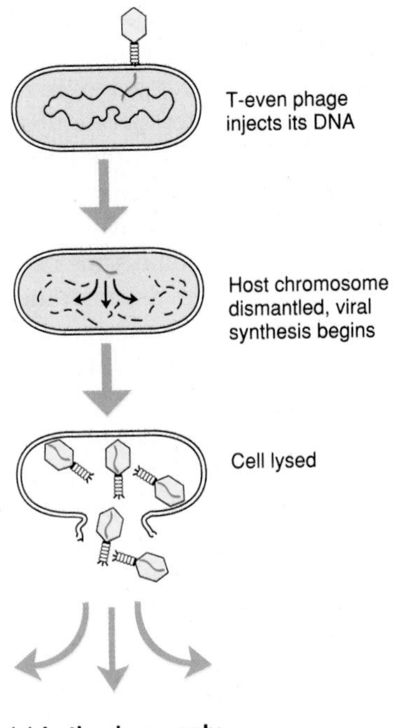
T-even phage injects its DNA
Host chromosome dismantled, viral synthesis begins
Cell lysed
(a) Lytic phase only

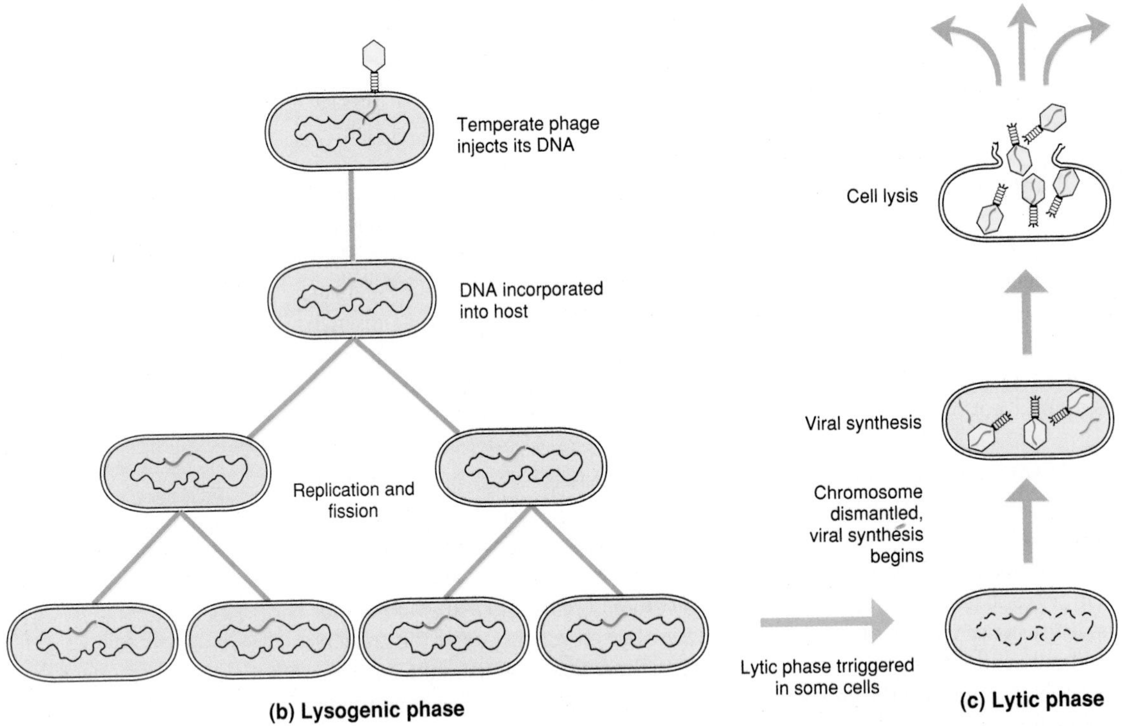
Temperate phage injects its DNA
DNA incorporated into host
Replication and fission
(b) Lysogenic phase
Lytic phase trriggered in some cells
Chromosome dismantled, viral synthesis begins
Viral synthesis
Cell lysis
(c) Lytic phase

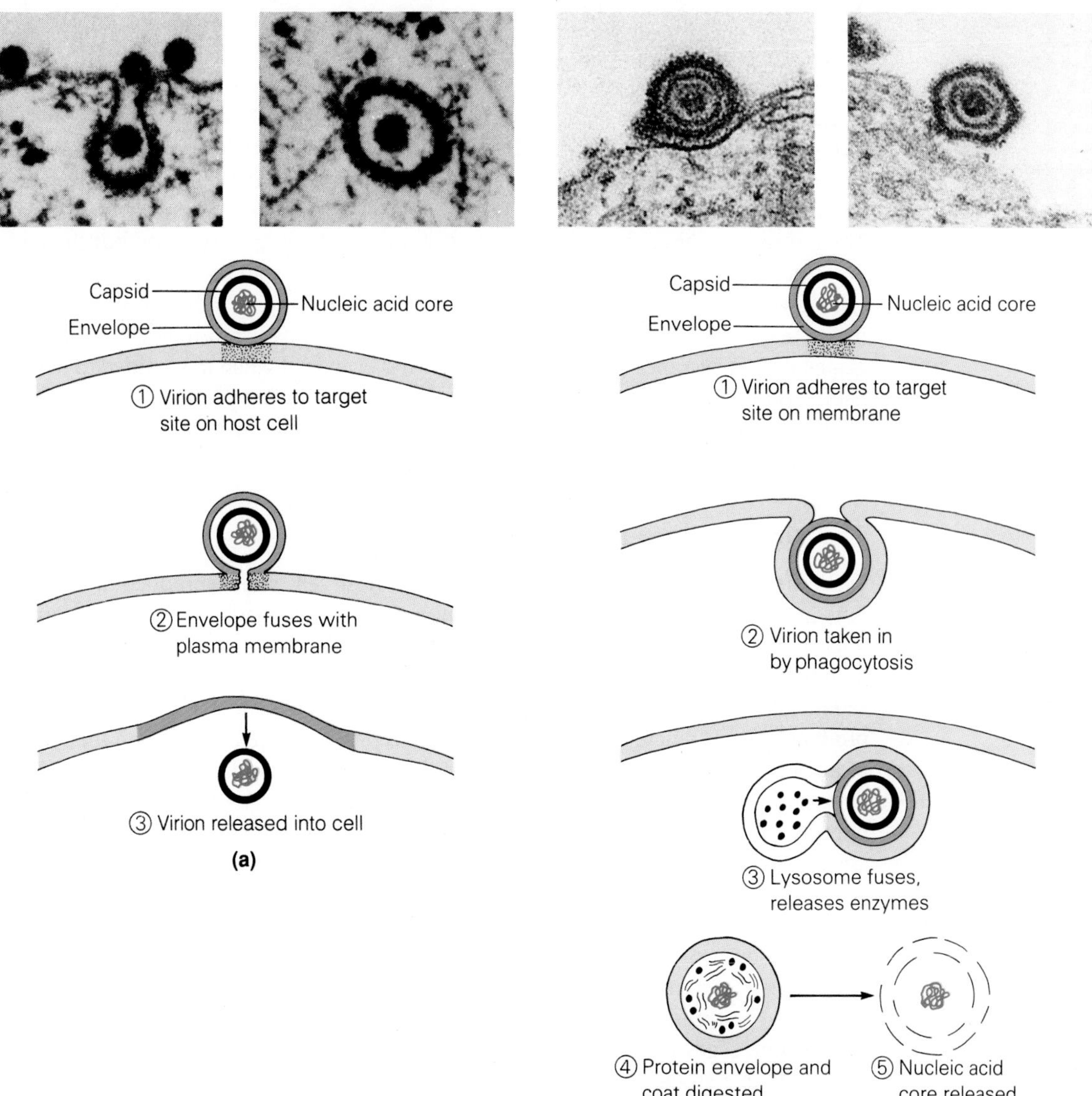

Capsid
Envelope
Nucleic acid core
① Virion adheres to target site on host cell
② Envelope fuses with plasma membrane
③ Virion released into cell
(a)
Capsid
Envelope
Nucleic acid core
① Virion adheres to target site on membrane
② Virion taken in by phagocytosis
③ Lysosome fuses, releases enzymes
④ Protein envelope and coat digested
⑤ Nucleic acid core released
(b)

FIGURE 22.21
HOW ANIMAL VIRUSES ATTACK CELLS
(a) To penetrate its host cell, the herpes simplex virus first adheres to special sites on the cell. Then its envelope fuses with the plasma membrane and the nucleic acid core enters the cell. **(b)** In other instances, animal viruses are taken into the host cell by active phagocytosis. Once inside, the protein envelope and coat may be digested away by host lysosomal enzymes and the nucleic acid core released.

vacuole and attack the virion just as they would a food particle. However, the enzymes are only able to hydrolzye the protein capsid. This releases the active, infectious core of the virus, and the invasion is complete.

Herpes Simplex By now, we've all heard about herpes. Herpes is caused by viruses from a group designated **herpes simplex.** The two best known are **herpes simplex I** and **herpes simplex II.** A familiar characteristic of the two is their cyclic shifts between the innocous dormant phase (analogous in this way to the lysogenic state) and the lytic phase, or, as it is usually phrased, from a state of remission to the active state.

Herpes simplex I, usually a simple annoyance, brings on the familiar "cold sores" or "fever blisters," generally about the lips and mouth. In its dormant state, herpes simplex I hides away in the spinal nerves, and we are unaware of its presence—it produces no outward symptoms. But, for some of us, a day in the summer sun, or some stress or illness, can trigger the lytic cycle. The virus migrates into tissues of the lips and enters into a frenzy of reproduction and cell lysis. The product of this is the familiar, often painful, fever blister.

Herpes simplex II, which causes a sexually transmitted disease known as **genital herpes,** has similar habits. No one knows for sure what triggers the active lytic state, although sometimes, it seems to be associated with physical or mental stress. When active, the virus attacks cells in the mucous membranes, generally those of the mouth and genitals, and can be passed from one to the other. In the lytic state, viral reproduction is quite rapid, numerous cells are lysed, and disfiguring and painful blisters erupt. The fluids released as the blisters break are laden with viral particles, so that at this time a herpes sufferer is highly contagious and can readily infect a sex partner. Afterwards, the virus quietly subsides, retreating into the spinal nerves, and a dormant phase begins. The remission is thought to be the result of increased antibody activity, but the respite is just temporary, and new outbreaks are likely to occur soon.

At the moment, genital herpes is permanent and incurable, although there are treatments that reduce the length and severity of the symptoms. Unfortunately, like syphilis, genital herpes becomes visciously dangerous and even fatal when transmitted to newborn babies, though it does not generally cross the placenta during pregnancy. Where herpes is known to be present in a pregnant woman, the physician may elect to deliver the baby by Caesarian section rather than risk exposure through normal vaginal delivery.

The Viral Genome and Its Replication

Once the viral DNA or RNA is released into a host cell, viral reproduction can begin. But, just how replication and capsid synthesis occur depends on the organization of nucleic acids in the virus. In the double-stranded DNA viruses, the first viral genes to be transcribed code for the replication enzyme DNA polymerase, and viral DNA replication follows. Next, transcription and translation occur, and the protein capsids are produced. In the single-stranded RNA viruses, replication often begins with the synthesis of a special, double-stranded RNA, which becomes the template for the replication of more viral single-stranded RNA. In **retroviruses,** the single-stranded RNA becomes opposed by a single strand of matching DNA nucleotides. Next, the RNA nucleotides are dismantled and replaced by more DNA nucleotides, and the viral genome becomes a typical DNA molecule. Recall that it is the special viral enzyme, **reverse transcriptase,** that makes this reversed form of transcription (RNA to DNA) possible (see Chapter 16). In the poliovirus, the single-stranded RNA of its genome doubles as messenger RNA, attaching to host ribosomes and producing both replicating and protein-synthesizing enzymes. The translation product, it turns out, is an enormous polypeptide that is then cleaved into segments, each making up a viral enzyme or capsid protein.

The double-stranded RNA viruses include the poorly understood **reoviruses.** Their genome is divided into several double-stranded RNA segments. Each is replicated much like DNA, using enzymes contributed by the virus itself. In the case of the enveloped viruses, such as the myxoviruses (influenza, measles, and mumps), the new viral genomes and coats are assembled as usual, but the surrounding viral envelope, we find, is literally borrowed from the host's plasma membrane. "Borrowed" is accurate, in a sense, since the next cell to be invaded will have the same envelope added to its membrane.

HIV: Portrait of a Killer

Our list of human retroviruses includes three recently discovered groups: HTLV-I, HTLV-II, and HIV. The first two may be responsible for two rare forms of leukemia, but the third, by far the best known, is the agent responsible for AIDS. These three viruses are notorious for their devastating effects on the human immune system (also see Chapter 41). More specifically, **HIV (human immunodeficiency virus)** shows a special preference for certain white cells, the lymphocytes, cells capable of what is called the immune response. Most commonly it is one subset of the lymphocytes, referred to as **helper T-cells,** that the AIDS virus attacks. Helper T-cells play a pivotal role in initiating and organizing the immune system's usual response to invading disease organisms and to the rise of cancers. When the T-cells fail, the vital immune response is feeble at best, and other invading parasites soon have their way.

This is all spelled out in the name **AIDS**, which by the way, is not the name of a disease; it is the acronym for **acquired immune deficiency syndrome**, a set of symptoms brought on as the result of a crippled immune system. Among the symptoms is the appearance of rare diseases, such as pneumocystic pneumonia and Kaposi's sarcoma, previously known only as "textbook diseases" by most physicians. Normally, these diseases are readily opposed and stopped by the immune system, but both are common to AIDS victims, whose immune systems have been crippled by the virus.

While most of what we know about the deadly HIV involves the T-lymphocytes and immune system, it is now almost certain that the virus has a direct effect on the brain. As brain cells are destroyed, an irreversible mental deterioration begins.

The Human Immunodeficiency Virus Like the other human retroviruses, HIV is spherical (Figure 22.22). Its surrounding envelope is formed from many spherical glycoprotein units, each penetrating the bilayer of a lipid membrane below. (The membrane represents a bit of plasma membrane borrowed from the last host.) Within the envelope is a core of protein surrounding the viral genome, a number of single-stranded RNA molecules, and the enzyme reverse transcriptase.

Typical of animal viruses, HIV enters the host cell through a receptor-mediated process. Certain glycoproteins on the virus's surface form a match with recognition sites on the helper T-lymphocyte's plasma membrane. Once a match is made, the two membranes fuse, and the naked protein core containing the RNA viral genome is released into the cytoplasm. It is there that the retrovirus begins its deadly work (Figure 22.23).

Following a successful invasion, the retrovirus undergoes reverse transcription as described earlier, and the single-stranded viral RNA is transformed into double-stranded viral DNA, now carrying the viral genome. The DNA is then incorporated into the host cell's genome, somewhere in its chromosomes. There it remains, quite inactive but replicating along with the host chromosome from one generation of cells to the next.

Fortunately, the dormant state is the most common status of an AIDS virus. No one knows how many humans carry the virus this way, but medical researchers are desparately seeking the answer so that they may better understand the severity of the present epidemic (see Chapter 41).

The AIDS virus may remain inactive for a long time, often years (some say up to ten), before it begins to reproduce. What seems to set off its devastating effect is the normal activation of HIV-carrying T-cells. As they begin their response to some routine infection, the implanted viral DNA enters into a frenzy of transcriptional activity, pro-

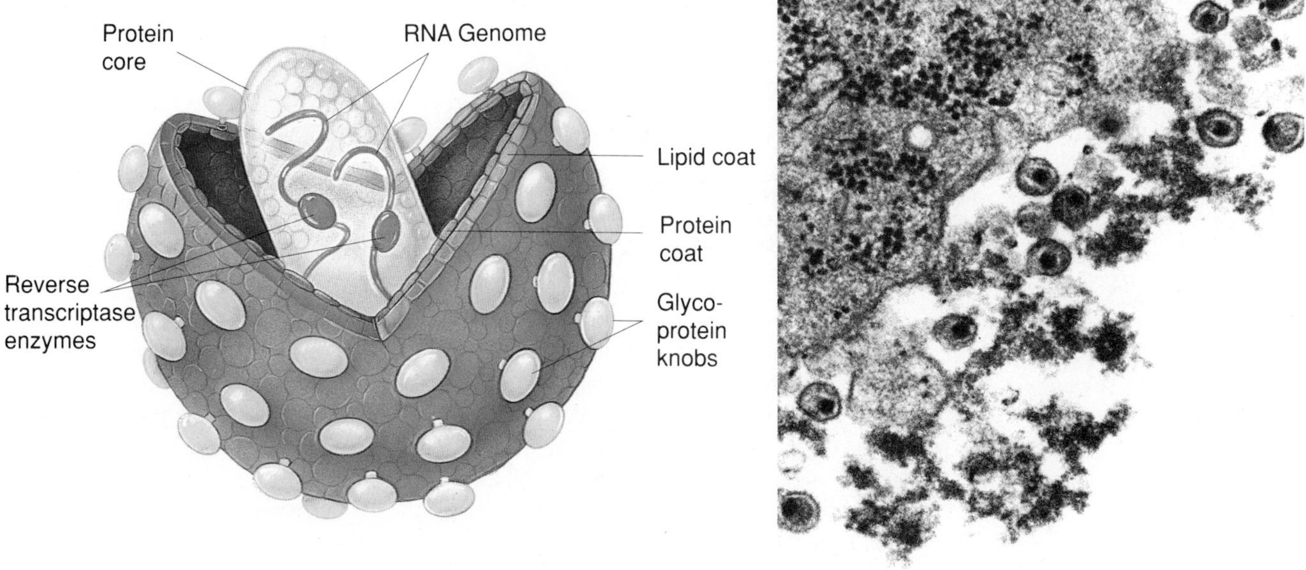

FIGURE 22.22
HIV: THE RETROVIRUS OF AIDS
The AIDS virus is an enveloped sphere with a complex lipid and protein coat. At the heart of each virion are two copies of its single-stranded RNA chromosome, each with the all-important enzyme reverse transcriptase. HIV's destructive potential is not apparent from its featureless appearance through the electron microscope (photo).

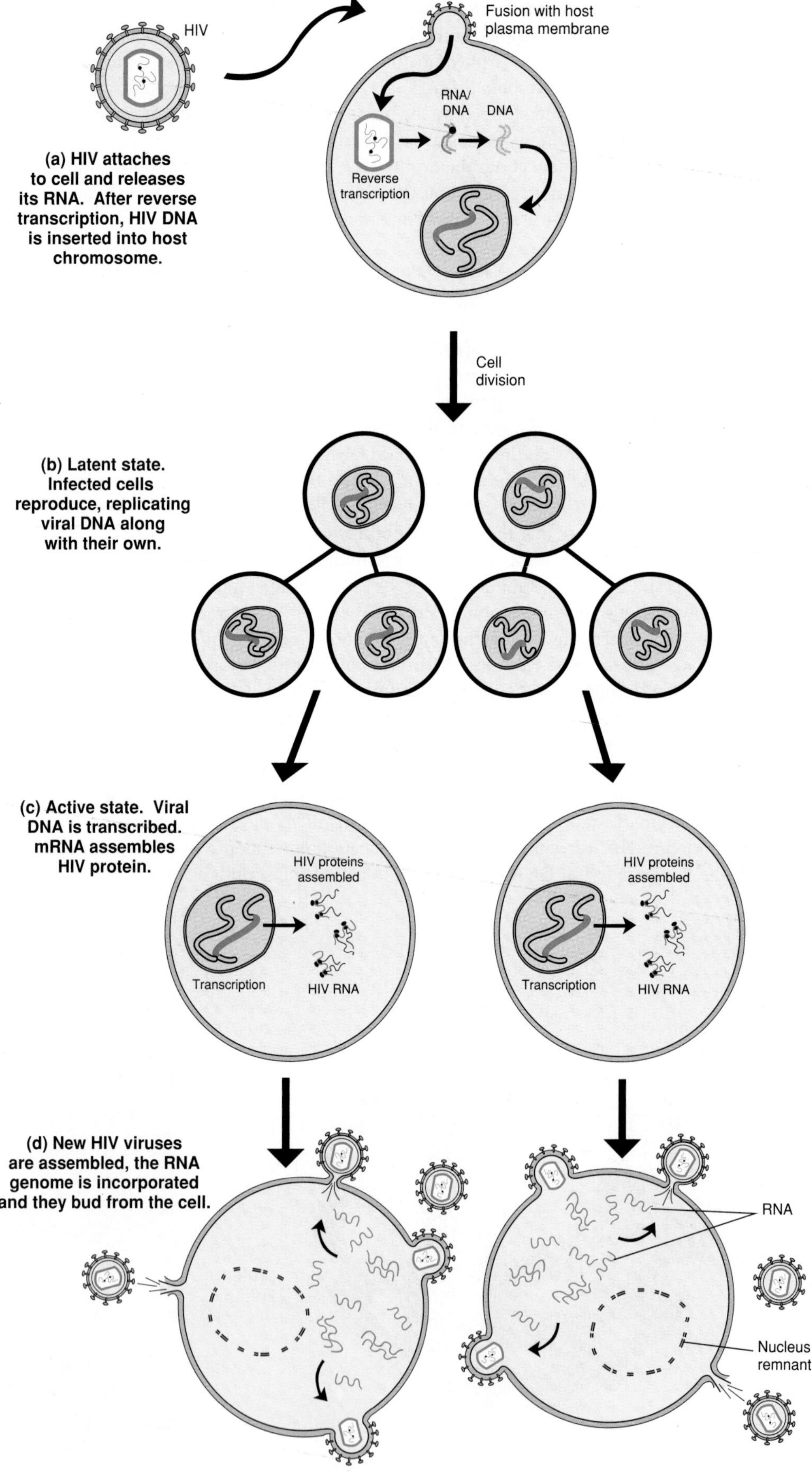
HIV
Fusion with host plasma membrane
RNA/ DNA
DNA
Reverse transcription
(a) HIV attaches to cell and releases its RNA. After reverse transcription, HIV DNA is inserted into host chromosome.
Cell division
(b) Latent state. Infected cells reproduce, replicating viral DNA along with their own.
(c) Active state. Viral DNA is transcribed. mRNA assembles HIV protein.
HIV proteins assembled
Transcription
HIV RNA
HIV proteins assembled
Transcription
HIV RNA
(d) New HIV viruses are assembled, the RNA genome is incorporated and they bud from the cell.
RNA
Nucleus remnant

FIGURE 22.23
LIFE CYCLE OF A KILLER
(**a**) The deadly AIDS virus fuses with the host cell membrane, releasing its single-stranded RNA inside. The viral RNA then uses reverse transcriptase to make a complementary strand of DNA. The viral DNA is then inserted into a host chromosome. (**b**) The insertion may remain intact through many rounds of cell division by the host cell, or, alternatively, it may become activated, entering a period of intense transcription (**c**) and producing many single-stranded RNA transcripts. The necessary viral proteins are then produced, and upon assembly, (**d**) each new viral particle is budded from the surface of the doomed cell.

ducing copy after copy of the single-stranded RNA that makes up the HIV genome. This RNA contains the coding needed to make new viral envelope proteins, protein coats, and more reverse transcriptase. As usual, these activities are carried out at the expense of the host cell's supply of amino acids and energy, also making use of its ribosomes, transfer RNAs, and whatever else is needed.

Newly transcribed RNA along with raw proteins migrate to the cell's plasma membrane, where they become integrated right into the membrane's lipid bilayer. The membrane is then used as both a viral assembly site and as the newly forming envelope itself. As the raw transcripts are modified, some form new envelope proteins, whereas others become core proteins or the critical viral enzymes. As each virion is assembled, two RNA transcripts are incorporated into the core. The fully assembled virions then bud from the cell surface, breaking away as new, infective HIV, ready to invade more lymphocytes, or perhaps even new hosts, should the occasion arise.

The formation and budding of viral particles can be so intense that soon the plasma membrane is virtually riddled with leaking holes. As you might expect, this kills the cell, effectively eliminating a key participant in the immune reaction and thus leaving the victim open to the infections that will soon end his or her life.

Like any successful parasite, HIV must have a means of infecting a second host; otherwise the host's death will bring about its own demise. HIV's two primary avenues of transmission are blood and semen. Like the bacterial agents of syphilis and gonorrhea, the virus of AIDS is considered to be sexually transmitted. As we will find, however, there are other modes of transmission that have little to do with sex (see Chapter 41).

The Origin of AIDS Where did the AIDS virus come from? Prior to 1981, virtually no one had heard of AIDS, and then, almost overnight, it was headline news. In spite of rumors to the contrary, AIDS did not originate in the homosexual communities, nor did it arise in Haiti. The best guess is that HIV had its origin in equatorial Africa, where AIDS is a serious problem today. Interestingly, AIDS in Africa affects equal numbers of men and women.

The AIDS virus most likely evolved from an innocuous ancestor called SIV (simian immunodeficiency virus), which occurs in the African green monkey. Epidemiologists have even uncovered several intermediate retroviruses that seem to bridge the gap between the green monkey virus, which infects humans but is not pathogenic, and the deadly HIV. Somehow the monkey virus found its way into humans, and then, through mutation, gave rise to the deadly AIDS virus. Some researchers believe this happened over 40 years ago.

Epidemiologists have searched for HIV antibodies in blood stored in blood banks prior to 1970, and all tests have been negative except those from stored blood originating in a small region in equatorial Africa. There, the results have been positive in blood samples taken as far back as the 1950s. One possibility is that HIV had spread across central Africa by the 1970s, leaving that continent and reaching Haiti, whose population has its roots in equatorial Africa, in the late 70s. From Haiti, once a favorite vacation site for gay American men, it appears to have spread into the Americas and Europe. As you know, AIDS has now reached pandemic proportions; that is, AIDS has now become a global epidemic.

This brief account of one of modern society's most urgent unsolved medical problems has undoubtedly left you with many questions. How does the virus single out specific kinds of cells? How do lymphocytes normally perform their functions? What are some of the difficulties in overcoming AIDS? We'll return to these questions as we look into the functioning of the immune system in Chapter 41.

APPLICATION OF IDEAS

1. There are strong arguments against the liberal use of antibiotics. Numerous kinds of bacteria live on and inside the body, often in an innocuous manner. Many are heterotrophs while others are marginal parasites, and still others are a constant threat as virulent parasites. What, if anything, does the frequent use of antibiotics have to do with these normal populations? Why not kill all the bacteria possible?
2. Despite the great care taken in the commercial canning of foods, occasional cans of spoiled food appear in supermarkets. A common form of spoilage can be identified by a general bulging of the can. What causes the bulging? Suggest ways in which bacteria might survive the canning process. What kind of bacteria might live in a sealed environment?
3. By gathering together pertinent information about bacteria, develop a table or chart that could be used in identifying individual groups. Use such characteristics as form, arrangement, nutritional requirements, and other requirements of a biochemical nature.
4. You are in remote back-country in your spanking new 46-foot self-contained motor home when you become mired in the mud. Adding to your problems is a hole punched in the fuel tank—you've lost your gas, and your electrical generator is useless. It will be at least a week before help arrives. You've plenty of fresh fruits, vegetables, milk, and meats, but they are quickly warming up in the fridge. Fortunately, you read about bacterial growth in your introductory biology course. Describe the steps you will take to keep your food from spoiling. Include at least three methods.

KEY IDEAS

ORIGIN OF PROKARYOTES

1. Among the earliest evidences of life are 3 to 3.5 billion-year-old columnlike **stromatolites**, laminated deposits produced by dense bacterial populations.
2. Recent revisions of prokaryote taxonomy have divided the older kingdom **Monera** into two groups (kingdoms), **Archaebacteria** and **Eubacteria.**

BIOLOGY OF PROKARYOTES

Cell Structure

1. The eubacterial plasma membrane is associated with respiratory and phosphorylating enzymes. Membranal infoldings form photosynthetic **thylakoids** and **mesosomes,** the latter participating in fission. Cellular extensions called **pili** aid in attachment and DNA transfer. **Ribosomes** and storage inclusions also occur. The genome is a single, protein-free chromosome, although some genes are in plasmids.
2. Cell walls are of **peptidoglycan** in eubacteria and protein in archaebacteria. In Gram positive cells, the peptidoglycan layer is thick and contains **techoic acid,** but in Gram negative types, it is thin and has a complex, overlying outer membrane. Gram positive bacteria are often susceptible to penicillin, whereas Gram negatives are usually not. Penicillin causes problems in cell wall synthesis. A slimy capsule may occur outside the bacterial cell wall.
3. Some bacteria form **endospores,** highly resistant, thick-walled dehydrated survival structures.

Cell Form and Arrangement

Bacteria assume the following forms: rodlike **(bacillus),** spherical **(coccus),** corkscrew **(gliding-helical, spirochete,** and **spirillum),** and commalike **(vibrio).** Bacteria may be arranged as singles, doublets **(diplo-),** octets **(sarcina),** chains **(strepto-),** or clusters **(staphylo-).**

Bacterial Movement

The bacterial flagellum is a coiled S-shape, is solid and rotating, and is made up of **flagellin.** Movement in **gliding bacteria** occurs through the use of a layer of mucus.

Bacterial Reproduction

Asexual reproduction occurs through **fission,** which includes DNA replication, replica separation, inward extension of the membrane, and daughter cell separation. A life cycle may occur in only 20 minutes. Sexual reproduction is rare. One form, **conjugation,** includes a one-way transfer of DNA, aided by the **sex pilus.**

Metabolic Activities of Bacteria

1. **Heterotrophs** depend on other organisms for nutrients. Heterotrophs include **decomposers** and **parasites** or **pathogens. Photoautotrophs** and **chemoautotrophs** synthesize food from inorganic molecules. **Obligate aerobes** require oxygen: **obligate anaerobes** are poisoned by it, and **facultative anaerobes** tolerate it.
2. Aerobic bacteria utilize electron transport systems to build a chemiosmotic proton gradient, which is then utilized in generating ATP. The gradient also powers flagellar movement and active transport. Anaeorbes utilize substrate-gained ATP to power proton transport.

Bacterial Heterotrophs

1. Because they break down dead organisms, decomposers **(saprobes)** are ecologically significant recyclers of

essential elements such as sulfur and nitrogen. Decomposition can be inhibited by altering growth requirements. Methods include cooling, drying, high heat, and using chemical agents. Decomposers are used in the production of some foods and other useful products. Harmless symbiotic bacteria of the body help resist pathogen invasion.
2. Animal pathogens of importance today include the agents of the intestinal disease **salmonellosis** and agents of **botulism, tetanus, gas gangrene,** and strep throat. Important sexually transmitted bacterial diseases include **gonorrhea, syphilis,** and **chlamydia.**
3. Plant pathogens cause discoloring blights, deadly wilts, and disfiguring galls.

Bacterial Autotrophs

1. Cyanobacteria live under wide-ranging, often harsh, conditions. They occur as single cells, in filaments, and in masses. **Phycobilins** and **chlorophyll *a*** harness light energy, and **heterocysts** fix nitrogen. Halobacteria create proton gradients using light energy captured by **bacteriorhodopsin.**
2. Chemoautographs obtain energy through the oxidation of simple inorganic compounds such as those of sulfur or iron. Nitrogen fixers (cyanobacteria and certain soil bacteria) reduce atmospheric nitrogen, forming useful nitrogen compounds that enrich the earth's waters and soils. N-fixers include cyanobacteria and plant symbiotes.

EUBACTERIA, ARCHAEBACTERIA, AND A NEW PHYLOGENY

Differences between eubacteria and archaebacteria include cell wall content, membrane lipids, ribosomal RNA, and photosynthetic pigments.

Archaebacterial Life

1. Archaebacteria commonly live in hot, acidic, and salty environments. Most are **obligate anaerobes.**
2. Examples of archaebacteria include methanogens—anaerobic methane generators, **extreme halophiles, extreme thermophiles,** and **thermoacidophiles,** bacteria that live in dense salt concentrations, very hot water, and hot, acidic conditions, respectively.

THE VIRUSES

Taxonomic relationships within the **viruses** and between viruses and other life are unknown. They range in size from that of large molecules to the smallest bacteria. Viruses are responsible for many human, animal, and plant diseases.

The Discovery of Viruses

1. Pasteur and Koch had developed vaccines against viral diseases long before anyone knew what viruses were. This is reflected by the name *virus* since it means "poison."
2. Iwanowski experimented with the **tobacco mosaic virus** in the late 1800s, transferring a virulent filtrate from plant to plant. By 1900, viruses were regularly isolated by filtration, and they became known as filterable viruses.
3. W. M. Stanley identified tobacco mosaic virus crystals in 1935, the same year they were first seen in the newly developed electron microscope. The fact that the virulent material increased while in the plant supported the idea that it was alive.

The Biology of Viruses

1. Viruses are obligate parasites—they cannot carry out life processes outside a host cell.
2. Each virus particle or **virion** contains a nucleic acid core and a protein coat or **capsid.**
3. Viruses are helical (tobacco mosaic virus), polyhedral **(herpes simplex),** or cubic and brick-shaped (smallpox). Shape depends on the arrangement of coat proteins or **capsomeres.** Some have surrounding envelopes (HIV, the AIDS virus).
4. Some viruses have the standard double-stranded DNA, others double-stranded RNA or single-stranded RNA.

Viruses and the Host Cell

1. **Bacteriophages** utilize **phage lysozyme** to penetrate bacteria. **Lytic** phages disrupt the host chromosome, generate many new virions, and lyse the cell. **Temperate phages** may enter a lytic state or a **lysogenic** state; in the latter the viral genome joins the host chromosome as a **prophage.** Lytic activity is repressed, but the prophage replicates with the host DNA and may infect a large clone population. When activated, it enters the lytic state.
2. Typically, plant viruses are transmitted by insects. Many crops are susceptible.
3. Animal viruses enter cells at specific membrane sites. The membrane of an enveloped virus fuses with the host membrane, permitting the protein-nucleic acid core to enter. Unenveloped viruses are taken in through phagocytosis, the protein envelope is digested, and the nucleic acid core released.
4. **Herpes simplex I** (fever blisters of lip) and **herpes simplex II** (genital herpes) remain inactive in nerve cells, but migrate to mucous membranes where they cause cell lysis. Lysis results in the runny blisters in genital herpes, which are contagious. Genital herpes, when transmitted during birth, can be lethal to the newborn.

The Viral Genome and Its Replication

Double-stranded DNA viruses replicate and transcribe in the usual manner. **Retroviruses** use **reverse transcriptase** to convert their single-stranded RNA genome to double-stranded DNA, which is then inserted into the host chromosome. When activated, the insert transcribes single-stranded RNA, the viral genome, which directs viral protein synthesis and virion assembly. Double-stranded RNA viruses replicate and transcribe like DNA.

HIV: Portrait of a Killer

1. **HIV (human immunodeficiency virus)** infects **helper**

T-cells, a type of lymphocyte, thus crippling the immune system and bringing on the fatal condition called **AIDS (acquired immune deficiency syndrome).** Death is the result of other, often rare, diseases.

2. The AIDS retrovirus is spherical and enveloped, and its genome is single-stranded RNA. It enters the host cell at receptor sites matching those of its glycoprotein envelope. Following a transformation to double-stranded DNA, it inserts into a host chromosome, where it may remain dormant for as long as 10 years. When activated, it transcribes the viral genome, the host cell is lysed, and infectious virions are released. Transmission is through blood and semen.
3. HIV may have evolved from SIV (simian immunodeficiency virus), supposedly originating in the African green monkey some time after 1950.

REVIEW QUESTIONS

1. Describe the stromatolite and explain its formation. What evidence best supports the notion that living cells produce stomatolites? (454-455)
2. How far back in time have the prokaryotes been traced? What present day group represents the first known prokaryote? (454)
3. Prepare a simple line drawing of a bacterium, adding and labeling: cell wall, outer membrane, capsule, plasma membrane, flagellum, chromosome, thylakoid, and mesosome. Suggest a function for the last four structures. (455-456)
4. Describe an endospore. Of what adaptive value is the endospore to the bacterium? (457-458)
5. List several ways in which the prokaryotic flagellum differs from that of the eukaryote. (458)
6. Name the two categories of response to the Gram stain and explain this reason in terms of the cell wall. What does the Gram reaction have to do with antibiotics? (457)
7. Using simple drawings, illustrate the four shapes of bacterial cells. Give their technical names. (458)
8. Name and describe the process associated with asexual reproduction in bacteria. (459-460)
9. Distinguish between anaerobe and aerobe, between obligate anaerobe and obligate aerobe. Which of the latter two would most likely employ an electron transport system in its respiratory activity? Explain. (461)
10. Suggest what you think is the most ecologically significant activity of bacteria and defend your choice. (general)
11. List four important bacterial diseases, name the pathogen, and briefly describe the disease symptoms. (464-465)
12. Name the photosynthetic pigments of the eubacteria and archaeobacteria. Which is most like those of the eukaryotes? (466)
13. What is nitrogen fixation? Why is it ecologically important? (466, Essay 22.1)
14. List four ways in which eubacterial and archaebacterial cells differ. (467-468)
15. What, precisely, are viruses? How does this explain why viruses require a living host? (470)
16. What was Iwanowski's contribution to our understanding of viruses? (471)
17. How did Stanley show that viruses could reproduce? (471-472)
18. List the three general forms taken by viruses. What determines the form? (472)
19. List four variations in the nucleic acids occurring in viruses. (472)
20. Briefly summarize the events in the lytic and lysogenic states of the phage virus. (472-475)
21. Review two ways in which animal viruses enter the host cell. (475, 477)
22. Compare replication in double-stranded RNA viruses with that in retroviruses. (478)
23. Describe the behavior of the herpes simplex virus in each of its forms, I and II. At what point is herpes simplex II contagious? (477-478)
24. Prepare a drawing of HIV virion. Label envelope, glycoprotein, protein capsid, single-stranded RNA, and reverse transcriptase. (479-481)
25. What specific cells does HIV tend to attack? What, in its active phase, does it do to these cells, and why is this so disastrous? What condition does this bring on? (479-480)

23

THE PROTISTS

IN DIMENSIONS THAT FALL BETWEEN THOSE OF BACTERIA AND THE SMALLest animals and plants, there exists a fascinating realm of life so diverse in its makeup that we are at a loss to find enough common ground for a satisfying description. A few drops of pond water, especially one including nutrient-rich bottom sediments, seen under the microscope can quickly confirm the existence of the microscopic eukaryotes that make up much of Kingdom Protista. While protists are chiefly creatures of the earth's waters, places where their numbers may become astronomical, some have found ways to exist in the soil and on the trunks of trees, while others hide out in the bodies of larger eukaryotes. But not all protists are microscopic. In fact, as we'll see, the protist kingdom includes several groups that are primarily multicellular and one with a few giants, albeit relatively simple.

It may seem peculiar, but the first person to report having seen protists was a seventeenth-century Dutch cloth merchant who had an insatiable curiosity and a penchant for grinding lenses. His name was Antony van Leeuwenhoek. Leeuwenhoek built his own microscopes, and although they contained only one lens and were very unrefined by today's standards (Figure 23.1), he is credited with some amazing discoveries. For instance, in one of his many letters to London's Royal Society, Leeuwenhoek carefully described certain minute microorganisms that could only have been bacteria. Their existence was not verified by others until a century later. If you've had occasion to try focusing a modern compound (multilens) light microscope on tiny pond creatures or on living bacteria you can better appreciate his observation. In addition to his many reports on protists (which Leeuwenhoek called "cavorting beasties"), this skilled observer is credited with original observations of mite parasites in fleas, fish red blood cells, human sperm, aphid reproduction (which he correctly interpreted as parthenogenetic—"without mating"), muscle structure, and the venom-injecting structures of spiders.

As you'll see, our familiarity with the protists has increased enormously since Leeuwenhoek's day. In fact, today we know of more than 100,000 living species, and some 35,000 are known only by their fossils. We'll look at the diverse characteristics of protists shortly, but first let's briefly consider their evolutionary origin, while making some hypothetical connections between them and the prokaryotes.

ORIGIN OF THE PROTISTS

Asking where and how the protists arose is also asking about the origins of eukaryotes themselves, because we are reasonably certain that today's protists evolved from the earliest eukaryotes. The early eukaryotes, in turn, apparently emerged from one or more prokaryote lines. Evolutionists estimate that eukaryotic life arose as far back as 2.5 billion years ago, some 1 billion years after the first prokaryotes (see Chapter 22). The oldest eukaryotic fossils, debatably, are of red and green algae, unearthed in the Bitter Springs limestone deposits of central Australia. Since these fossils are fully formed eukaryotic cells, we are left with significant gaps to fill, gaps that cannot be clarified by the fossil record. Nevertheless, there are a number of ideas on how the eukaryotes arose, one of which is receiving a lot of favorable attention today.

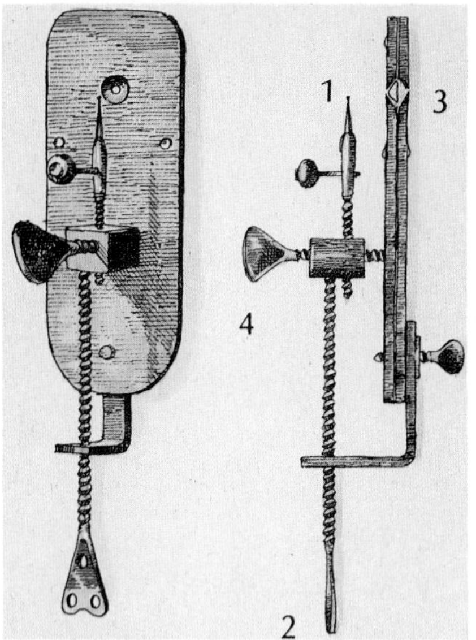

FIGURE 23.1
ANTONY VAN LEEUWENHOEK
The elegantly coiffured gent is Antony van Leeuwenhoek. The device is a microscope of his own design. The specimen was placed on the apparatus at (1), brought into position vertically by turning the lower screw (2), and moved toward or away from the lens (3) by turning the shorter screw (4).

The Serial Endosymbiosis Hypothesis

The **serial endosymbiosis hypothesis** is an ingenious and widely accepted explanation of the origin of the eukaroytic cell. It maintains that the eukaryote line arose when prokaryotic cells with various specific capabilities were incorporated as **endosymbionts** into other cells. (*Symbiosis* simply refers to a close relationship between two species.) The invaders lived somewhat independently at first, but soon an interdependence or **mutualism** was established. It was through these incorporations, goes the hypothesis, that the newly emerging eukaryote came into possession of mitochondria, chloroplasts, and a microtubular system as seen in today's eukaryotic flagella, cilia, and centrioles.

The hypothesis is not universally accepted among biologists, but its credibility has been vigorously advanced by Lynn Margulis of Boston University, who maintains that the primitive eukaryotic cell was derived by at least three separate events that involved the union of four separate prokaryotic lines (Figure 23.2).

Line A, which she calls the **protoeukaryote**, had evolved the ability to move its plasma membrane, and thus was able to engulf particles and form digestive vacuoles and other internal membranous structures. It, then, was the first predator. It was capable only of anaerobic respiration (glycolysis), but it may have had multiple chromosomes and a nuclear membrane. There is no such organism alive today.

Line B was an *aerobic* bacterium, not too dissimiliar to *E. coli,* which was engulfed by the line A cells. Eventually, a **mutualistic symbiosis** (mutually beneficial coexistence) developed so that engulfed line B cells were not digested, but instead began to help break down other foodstuffs ingested by the larger cell. In time, the two became completely dependent on one another. Finally, line B cells lost the ability to live outside their hosts, and, Margulis hypothesizes, their descendants exist today as mitochondria. Over time, much of the mitochondrial genome was lost, but even today, mitochondria retain a complete functioning set of tRNAs, bacterialike ribosomes, and a bacterialike circular chromosome of nearly naked DNA. That, we are told, was the first symbiotic event.

The second symbiotic event, she suggests, perhaps involves a bit more faith. More to the point, it generated the most controversy and is patently rejected by many biologists. Here, a new organism enters the picture: *line C.* According to Margulis, line C resembled a modern prokaryote, the spirochete, in that it was long, thin, and highly motile. Margulis assumed that it contained microtubules in a 9 + 2 arrangement. The line C organism first attached to the outer surface of the line AB complex to become the first flagellum. It also introduced the protein *tubulin,* which was to give rise to cilia, flagella, basal bodies, centrioles, and spindle fibers. Acquisition of cilia and flagella gave further mobility to the evolving eukaryotic cell. With the latter, the stage was set for mitosis and meiosis.

The third symbiotic event was the acquisition of cells of *line D.* Margulis proposes that line D cells were simply ancient cyanobacteria. It was from these that the emerging eukaryotes gained chloroplasts. Note that in none of these steps did it pay the host to

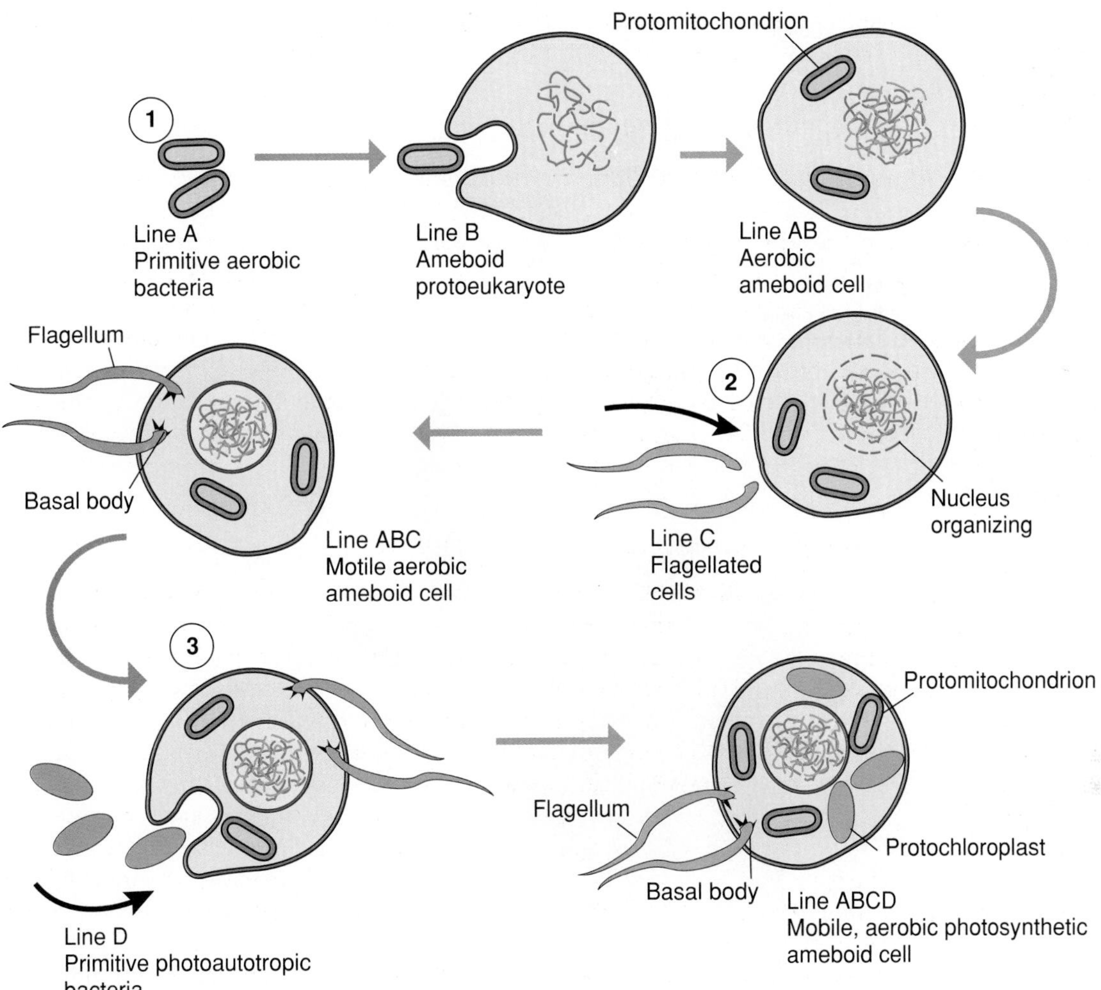

FIGURE 23.2
THE SERIAL ENDOSYMBIOSIS HYPOTHESIS
According to the serial endosymbiosis hypothesis, evolution of eukaryotes began when aerobic bacteria (line A) became established as endosymbionts in a larger phagocytic, anaerobic, *proto*-eukaryote (line B). Next is the most controversial aspect of the hypothesis. With its new "mitochondria" in place, the novel cell line (AB) was next invaded by flagellated bacteria (line C), adding greater motility to the aerobic cell (line ABC). Finally, the new line took in simple bacterial photosynthesizers (line D), and a new line (ABCD) of versatile eukaryotes arose, leading eventually to some of today's photosynthetic protists. Sometime early in the process, other internal membrane-surrounded organelles such as the eukaryotic nucleus also arose.

be too voracious. The eukaryotic cell able to engulf photosynthetic bacteria without digesting them acquired a reliable source of energy and nutrients. Part of the evidence for this last step is that chloroplasts retain their own prokaryotic type of ribosomes and circular DNA, and even have their own tRNAs. Margulis argues that there are hundreds of known cases in which modern host organisms have taken photosynthetic symbionts into their cytoplasm.

The phylogenetic tree in Figure 23.3 includes essentials from the serial endosymbiosis hypothesis. (Also see Essay 4.2.)

CHARACTERISTICS OF THE PROTISTS

Although many of the protists are clearly single-celled organisms, others form colonies of cells wherein members become specialized for tasks such as locomotion, feeding, and reproduction. Such specialization may improve the colony's efficiency, but it is limited. Each cell can generally get along on its own if circumstances require it. Yet other species, as we mentioned, are definitely multicellular, in which case the cells become specialized for different functions and eventually become totally dependent upon each other. Some protists are even complex enough to form simple specialized tissues and organs.

Protists practice all common modes of feeding. Some are active predators, capturing prey through phagocytosis. Others more quietly absorb simple organic nutrients across the body wall from their surroundings, while still others are active, damaging parasites, obtaining their nourishment within the cells of larger eukaryotic hosts. Many protists are photosynthetic, trapping sunlight energy in pigments organized into photosystems in well-formed eukaryotic chloroplasts.

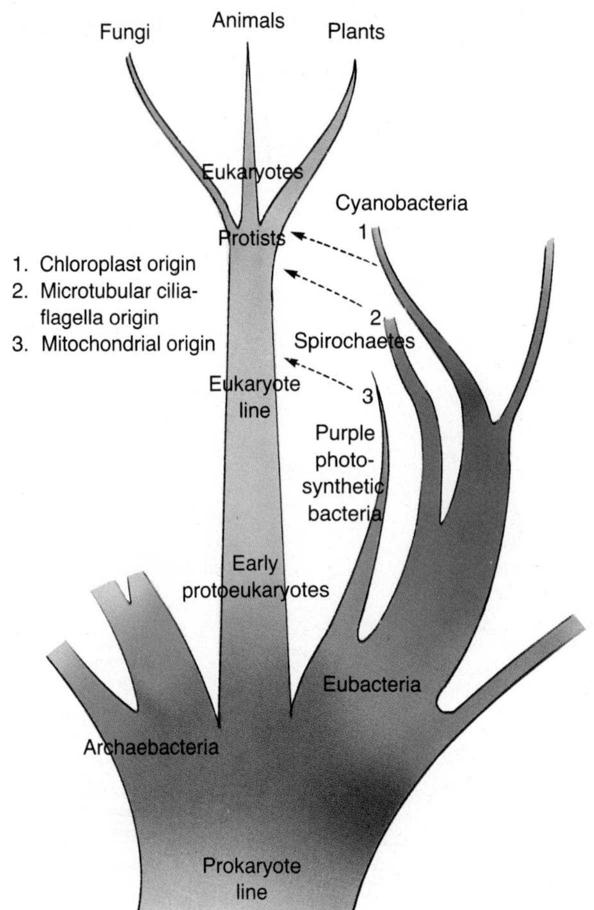

FIGURE 23.3
EUKARYOTIC ORIGINS
According to the serial endosymbiosis hypothesis, the eukaryotic line was produced from several independent endosymbiotic events.

Movement, where it occurs, is chiefly through microtubular cilia and flagella, although ameboid groups move through flowing cytoplasmic extensions called **pseudopods** ("false feet").

Asexual reproduction in protists is common and is a means of rapidily increasing population size. Sexual reproduction is less common and has never been observed in a considerable number of protists. Asexual reproduction in the simpler protists occurs through binary fission, but unlike binary fission in prokaryotes, it involves mitosis. A variant of fission called **fragmentation** is often observed in parasite life cycles. In fragmentation, repeated mitoses occur without cell division, but eventually the cytoplasm divides, and each nucleus ends up in a tiny cell of its own.

Sexual exchanges vary greatly. Some species simply fuse entire cells, while others conjugate and exchange haploid nuclei. Still others follow the familiar pattern of meiosis, gamete formation, and fertilization. In a peculiar variant of sexual reproduction, called **autogamy**, meiosis occurs, but its products are recombined within the same organism. This is followed by cell division. Although this precludes new genetic input, it does produce genetic variation in the daughter cells. In other variations on sexuality, species that have gametes may form **isogametes**, where the gametes are morhpologically identical, or they may form **heterogametes** with large stationary eggs and small motile sperm. Variations on each of these themes in protists seem endless, as we will see. But while sexual activity varies enormously, we can boil things down to arrive at the three typical kinds of life cycles found in eukaryotes.

Life Cycles in Protists and Other Eukaryotes

Although life cycles vary greatly among eukaryotic organisms, there are three basic or fundamental patterns we can discern (Figure 23.4). These are the **zygotic**, **gametic**, and **sporic** cycles.

The Zygotic Cycle The zygotic cycle is considered primitive, probably having evolved much earlier than the others. We find it to be the reproductive pattern in some of the simpler algae, some protozoa, and in most fungi. In these organisms nearly all of the individual's life is spent in the haploid state. At some point, certain haploid cells (isogametes are common) or their nuclei fuse, bringing on the diploid state. The diploid individual may enter an extended dormant period, or, more commonly, it will immediately enter meiosis, restoring the haploid state.

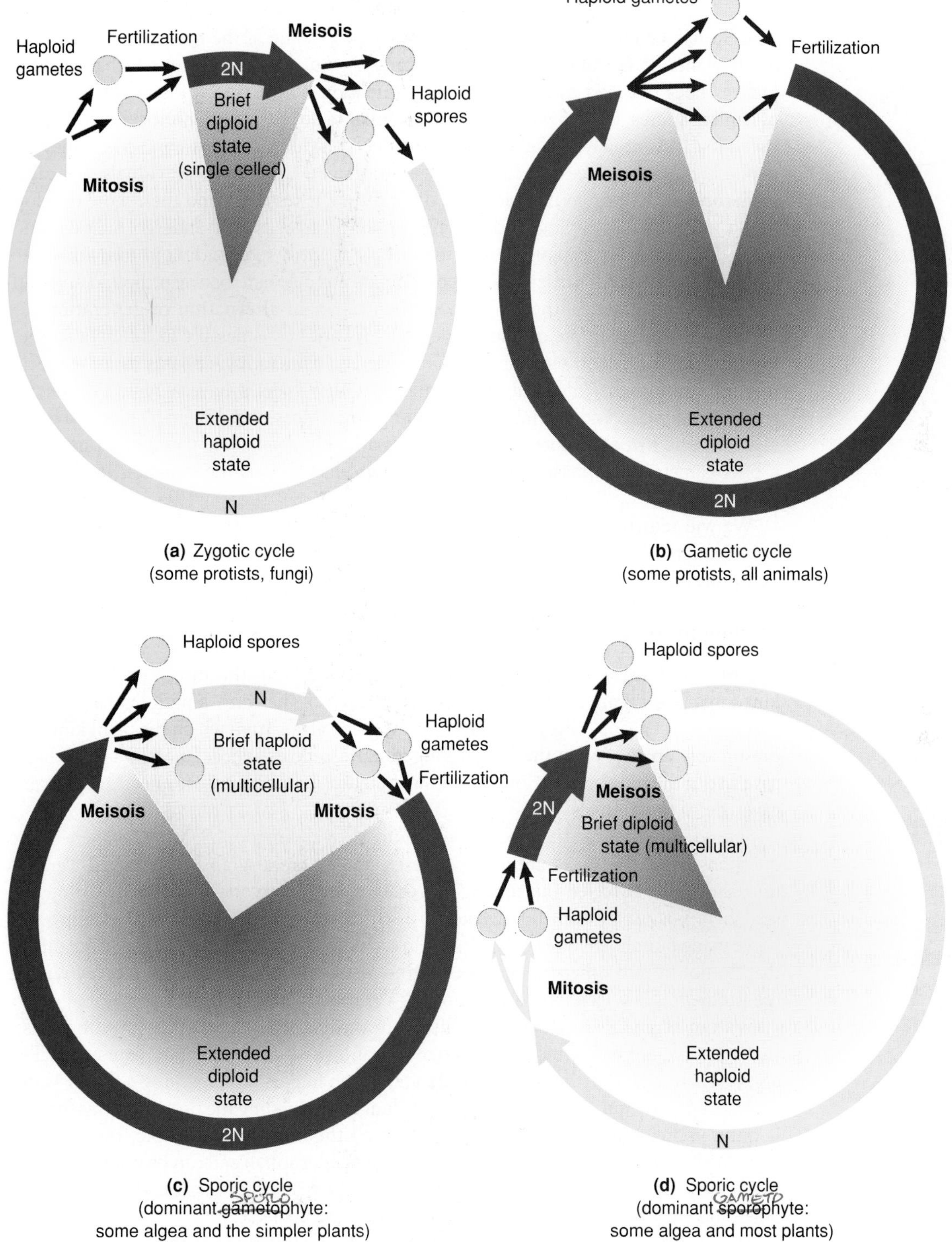

(a) Zygotic cycle (some protists, fungi)

(b) Gametic cycle (some protists, all animals)

(c) Sporic cycle (dominant gametophyte: some algea and the simpler plants)

(d) Sporic cycle (dominant sporophyte: some algea and most plants)

FIGURE 23.4
THREE LIFE CYCLES
Most of the earth's eukaryotes fit into one of the three principal kinds of life cycles: (a) zygotic, (b) gametic, and (c,d) sporic.

The Gametic Cycle The gametic cycle occurs in some protists and in all animals, including humans. It is quite direct and uncomplicated. In the gametic cycle, the cells are diploid except for the gametes—the sperm and egg cells (heterogametes) or their counterparts (isogametes)—which are haploid. In animals, gametes are produced as certain diploid cells undergo meiosis. The diploid state is restored as gametes unite in fertilization. While this is a conceptually comfortable cycle, many of the world's creatures have evolved quite different systems. The next theme, in fact, is pretty much the opposite of the animal life cycle.

The Sporic Cycle The sporic life cycle is seen in some of the multicellular algae and in all plants. It is quite varied in its expression but has one common characteristic. Meiosis does not lead directly to gamete formation but instead leads to the formation of haploid spores, each of which has the potential to develop into a multicellular, haploid individual. Eventually, cells in this individual produce gametes, but since they are haploid, mitosis suffices. Because they form gametes, haploid individuals are called **gametophytes.** Upon fertilization, the diploid state is restored, and the zygote develops into a multicellular, diploid individual. Certain of its cells will undergo meiosis, once again, forming more haploid spores. Because they form spores, diploid individuals are called **sporophytes.** Further, since sporic organisms alternate between diploid and haploid multicellular states, the life cycle is often called an **alternation of generations.**

There is one final note on the sporic cycle. It varies considerably in different species. For instance, we'll soon see that the sporophyte and gametophyte phases in some species of algae are virtually *identical* in appearance. Other species have a highly *dominating gametophyte* phase and a brief, simple sporophyte. (Mosses and their relatives have this characteristic.) In still other species, there is a highly *dominating sporophyte* phase with a brief gametophyte phase consisting of a few cells. (This is the situation in nearly all plants.)

We will return frequently to the three life cycles as appropriate in this and the next two chapters. You may find it useful to return to this discussion and to Figure 23.5 as you proceed.

Codosiga botrytis

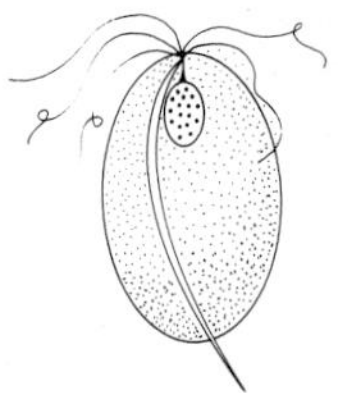

Trichomonas vaginalis

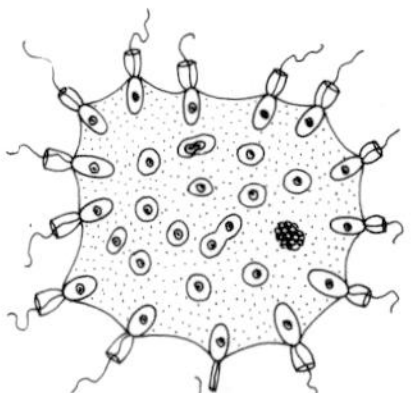

Proterospongia

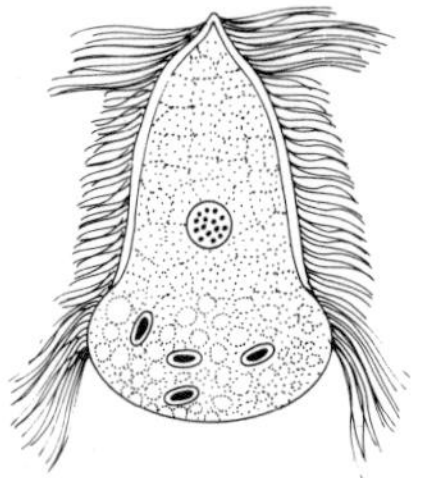

Trichonympha campanula

FIGURE 23.5
FLAGELLATES
Examples of the subphylum Mastigophora include *Codosiga botrytis* a branching colonial, *Trichomonas vaginalis,* whose species designation is quite descriptive, the colonial *Proterospongia* which may share ancestors with the sponge, and the multiflagellated *Trichonympha campanula*, which lives as a symbiont in the termite gut.

Phylogeny of Today's Protists

The protists form a broad polyphyletic group underlying the three other eukaryotic kingdoms (see Figure 23.3). As this indicates, each of the three kingdoms has its roots in some protist group. One group includes the green algae, from which the plant kingdom presumably arose. Below the fungal kingdom are protists whose ancestors most likely gave rise to the fungi, and below the animal kingdom are those whose ancestors probably gave rise to the animals.

Should the protists then be divided into several kingdoms? Maybe. There is no question that the protists are a disparate group—a literal grab bag of types. Perhaps they should be parceled out among the other three eukaryotic kingdoms? As you can see, there is always room for debate in discussions of phylogenetic relationships, but we've made our choice. We will continue to include in Kingdom Protista those organisms that do not fit into other kingdoms until the new tools and methods of biology resolve the problem. Now let's look at the protists themselves in more detail.

We can begin by dividing the protists into three branches. By tradition, the protists that resemble animals are called **protozoans**, and those that resemble plants are called **algae.** There are also the **slime molds** and other funguslike creatures such as **water molds** that were, until recently, classified as fungi. Subdividing the protists in this manner is quite useful, but it doesn't begin to cover all the cases. For example, there are the puzzling little *Euglena*. Like the algae, they contain chlorophyll and can be photosynthetic. But if they are grown in the dark so that they can't photosynthesize, their green area fades and they become heterotrophic, like true protozoans.

The task of sorting all this out must fall to the evolutionists of the future, who will rely on the modern tools of molecular biology. For now, let's keep in mind that our categorizing is tenuous and subject to change should new information become available.

THE PROTOZOANS

A favored taxonomic scheme divides the protozoan protists into seven phyla, three of which are significant to us. Included are phylum **Sarcomastigophora**, the flagellates and amebas; phylum **Apicomplexa**, the immotile spore-formers; and phylum **Ciliophora**, mostly ciliated forms. As you can see, locomotion plays an important role in the classification of protozoans.

Sarcomastigophora: The Flagellates and Amebas

Flagellates The flagellated protozoans occupy a subphylum called **Mastiogophora**, formerly a phylum (Figure 23.5). Most members propel themselves by flagella, which occur singly, in pairs, or occasionally in greater numbers. The flagella have the 9 + 2 microtubule structure typical of eukaryotes. Movement through undulation of the flagella permits the flagellate protozoan to move efficiently in any direction (see Figure 4.21).

Flagellated protozoa feed in a variety of ways. They may hunt and capture prey or absorb nutrients through their body covering. Reproduction is primarily by asexual means, with mitosis followed by binary fission. In the flagellates, binary fission occurs along a longitudinal plane. Little is known about sexual exchange among the protozoan flagellates.

Each of our three phyla of protozoa has its parasitic forms. One of the most notorious flagellates, *Trypanosoma gambiense,* is the causative agent of **African sleeping sickness** (Figure 23.6). The **trypanosome** is carried by the infamous tsetse fly, whose bite injects the parasite into the mammalian victim. Control of the disease is very difficult, since nearly all large mammals in tropical Africa harbor the parasite.

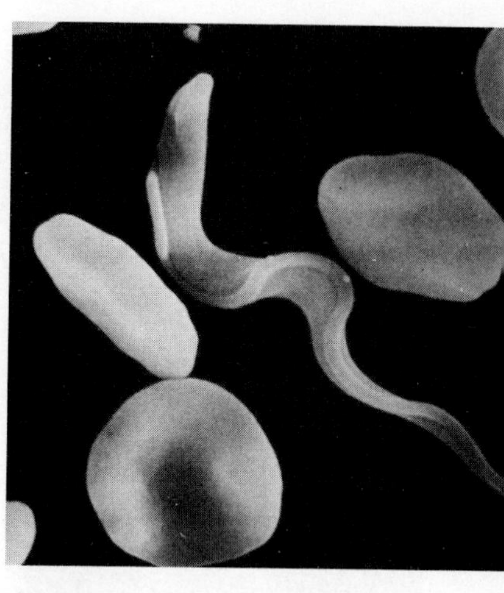

FIGURE 23.6
TRYPANOSOMA GAMBIENSE
The flagellum of *T. gambiense,* attached along the body by a thin, membranous flap, produces an undulating motion as the parasite makes its way through the host's blood cells. When humans are infected, the trypanosome reproduces in the blood and lymph glands and eventually enters the cerebrospinal fluid, where it brings on the symptoms of African sleeping sickness *(trypanosomiasis).* As the parasites invade the central nervous system, tremors, headache, apathy, and convulsions commence, leading to coma and death.

The Amebas Imagine a toughened, determined, well-trained army on the move into some primitive land. Then imagine that army brought to a dead halt, as soldiers drop their weapons, clutch at their bellies, and dart for the nearest cover. It has happened. Armies have literally been stopped by a tiny protozoan called *Entamoeba histolytica,* the agent of **amebic dysentery.** The ameba is transmitted in food or drinking water contaminated by infested human feces. That, of course, sounds highly unlikely in a medically sophisticated country such as our own, but keep in mind that hepatitis and salmonellosis, common enough problems in the United States, are often transmitted the same way. Amebic dysentery though, is complicated by the fact that it can be transmitted by people who carry the disease but show no symptoms.

Entamoeba histolytica is just one example from subphylum **Sarcodina.** The free-living "naked amebas," or sarcodines, as members of this group are known (Figure 23.7), are remembered particularly for their physical plasticity. They have no definite form and constantly change their shape. They do this by sending out pseudopods—temporary extensions of the cell—in any direction as they move and feed.

Ameboid movement is explained in terms of changing cytoplasmic state: an alternation between a firmer, jellylike "gel" state and a highly liquid and flowing "sol" state. The gel state is found along the perimeter of the ameba, forming what is called **ectoplasm.** The thinner sol state or **endoplasm** is found within. Pseudopods appear where a clear, softer, **hyalin cap** forms somewhere along the body. The endoplasm then begins a continuous flow into the softer cap, whereupon it turns outward and stiffens into a gel state, thereby forming a tube. Next, the tube becomes anchored to the substratum, permitting its continuing formation to move the ameba in a forward direction. Of course, if this continued, the ameba would soon run out of sol. But the flow keeps going because, at the other end, the gel is converted to sol once again (see Figure 23.7). (Certain white blood cells are creeping through the tissues of your body in just such a way this very moment!)

Sarcodines typically feed through phagocytosis. Food particles or prey are surrounded by converging pseudopods, or they may simply adhere to the cell surface; either way they become trapped in a food vacuole formed from the ameba's plasma membrane. Digestion occurs within food vacuoles, and the products are subsequently absorbed into

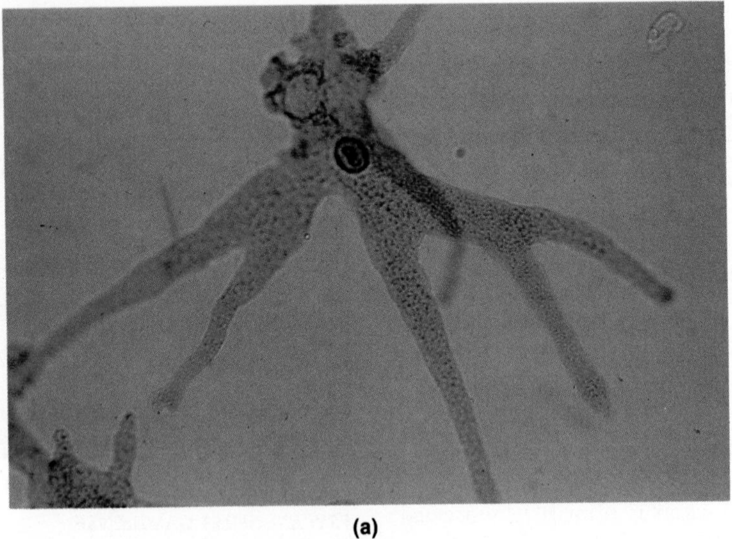

(a)

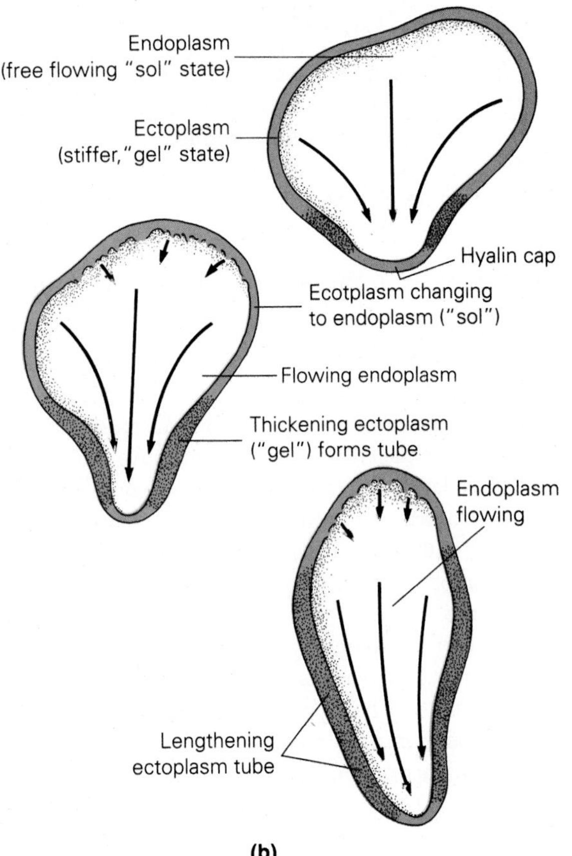

(b)

FIGURE 23.7
NAKED AMEBA AND AMEBOID MOVEMENT
(a) The extensions of the body seen in *Amoeba proteus* are pseudopods or "false feet," tubes of flowing cytoplasm that occur as the ameba moves. **(b)** Movement in amebas begins as the more liquid inner cytoplasm, known as *endoplasm,* flows towards the leading edge. As it flows to the sides, it is converted to a stiffer, gel-like *ectoplasm,* which forms a tubelike border, lending shape to the enlarging pseudopod. Note that at the trailing edge, the stiffer ectoplasm is converted back to the free flowing endoplasm, so that the protist can sort of "keep up with itself."

the cytoplasm. Undigested residues are expelled by exocytosis. Recall (see Chapter 5) that in this process, the vacuolar and plasma membranes fuse, and the vacuole's contents are released to the outside. As you see, the plasma membrane has its lost portion returned.

Reproduction in the amebas is chiefly through mitosis and binary fission, although some sexual reproduction also occurs. Where it involves the naked amebas, whole individuals undergo meiosis, and their products simply fuse together.

Some amebas are surrounded by hardened, shell-like tests. The **heliozoans,** for example, produce a hardened capsule of silicon dioxide (glass) from which extend long, slender **axopodia.** The axopodia are actually needlelike pseudopods upon which the cytoplasm creeps back and forth, gathering food through phagocytosis. The axis of each axopodium includes a large number of microtubules in a spiral arrangement. The assembly and disassembly of microtubules permits the axopod to be extended or withdrawn (Figure 23.8). In their sexual reproduction, some helizoans withdraw into cystlike states and undergo meiosis. Surprisingly, polar bodies are formed and discarded, leaving one viable cell per event. Haploid individuals then fuse.

Another group, the **radiolarians,** also produce coverings of silicon dioxide (Figure 23.9). They have been on earth a long time, and their siliceous corpses on the ageless ocean floor form deposits known as the radiolarian ooze. The **foraminiferans** also form skeletons, elaborate shells of calcium carbonate. The shell is pitted with numerous pores through which the pseudopods move in and out as the organism feeds. Foraminiferans also contribute to the dense ooze of the ocean floor. In fact, some land areas that were once under the sea are still composed of dense deposits of foraminiferan shells. The white cliffs of Dover, England, are a good example. When foraminiferans reproduce sexually, they undergo meiosis and form flagellated, haploid sexual cells. Such cells fuse, restoring the diploid state, whereupon they pass through many rounds of mitosis and cell division.

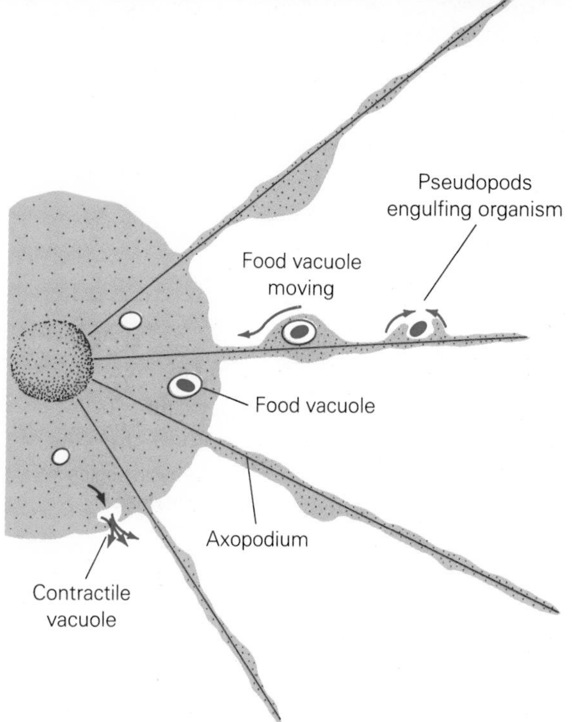

FIGURE 23.8
A HELIOZOAN
Actinosphaerium has a spherical body covered with long, fine axopodia, made up of numerous microtubules and surrounded by streaming cytoplasm. Following phagocytosis, waves of cytoplasmic movement carry trapped food particles into the main body. Food vacuoles in the cytoplasm are the sites of digestion. Contractile vacuoles aid in maintaining a suitable water balance.

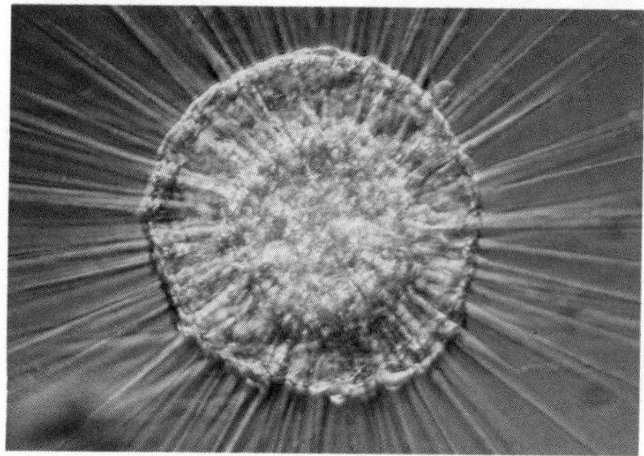

Apicomplexa: The Sporozoans

From one point of view, the **sporozoans** may be among the most important of all protozoa to humans. That's because members of one genus can cause **malaria**, a disease that has been called humanity's greatest curse. (Some people think humanity's greatest curse is a pop quiz!) Sporozoans have three major characteristics: (1) many form spores or spore-like stages; (2) they commonly live as parasites; and (3) in their mature stages, they have no special means of locomotion. The spore, in this group, is defined as a cell enclosed in a protective casing and capable of infecting a host. Sporozoans may have incredibly complex life cycles, with both sexual and asexual phases, and the two phases may occur in different hosts. Because malaria has been so important in human history, we'll focus on its agent as an example of a sporozoan with a complex life cycle.

Malaria is spread from person to person by the female *Anopheles* mosquito, a tiny insect harboring the sporozoan parasite, *Plasmodium vivax* (the most common of several *Plasmodium* species that infect humans). This decidedly dangerous mosquito is alive and well throughout the world. One of the most effective mosquito control agents has been the pesticide DDT, but because of its deleterious effects on animals, particularly birds and mammals, its use is highly controversial. Besides, today we are confronted by DDT-resistant mosquitoes.

During its life cycle (see Figure 23.10), *P. vivax* reproduces in both the mosquito and the vertebrate host. The cycle begins when a mosquito injects its anticoagulant into a human that it has just pierced with its tubular mouthparts. This injection carries with it a number of **sporozoites**, an ameboid stage in the parasite's life cycle that matures in the mosquito. The sporozoites make their way through the bloodstream and enter cells of the liver, where they undergo a quiet asexual phase involving fragmentation. First, the sporozoite nucleus enters mitosis again and again, and then the cytoplasm divides, separating each nucleus into its own cell. At this point the host still shows no reaction. However, the parasite is reproducing rapidly and changing to the **merozoite** phase. Eventually, large numbers of merozoites enter the bloodstream. Their task is to invade the red blood cells, where they reproduce simultaneously, rupturing the blood cells and releasing toxins (poisons) throughout the host's body. The toxins bring on the familiar fever and chills of malaria. The cycle repeats itself every 48 hours, 72 hours, or longer periods, depending on the species of *Plasmodium* involved.

Eventually, some merozoites follow a new developmental path, transforming into **gamonts** (gamete-forming cells), and *Plasmodium* enters its sexual phase. Gamonts cannot

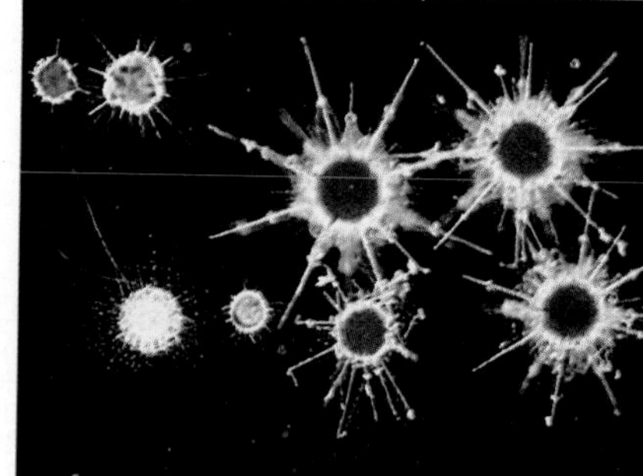

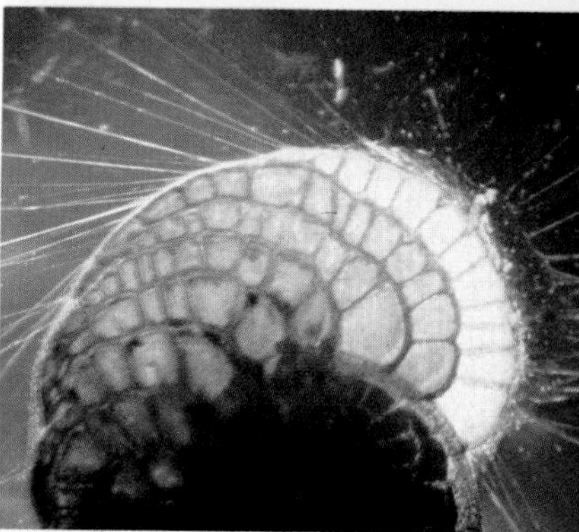

FIGURE 23.9
TWO RADIOLARIANS
The radiolarians (above) are typically spherical, with highly sculptured skeletons. The foraminiferans (below) tend to form spiral shapes reminiscent of tiny snails.

(Gamonts injested by mosquito.)

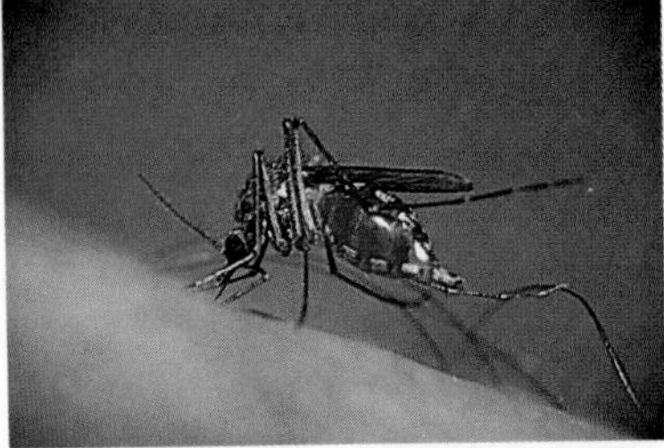

Anopheles mosquito

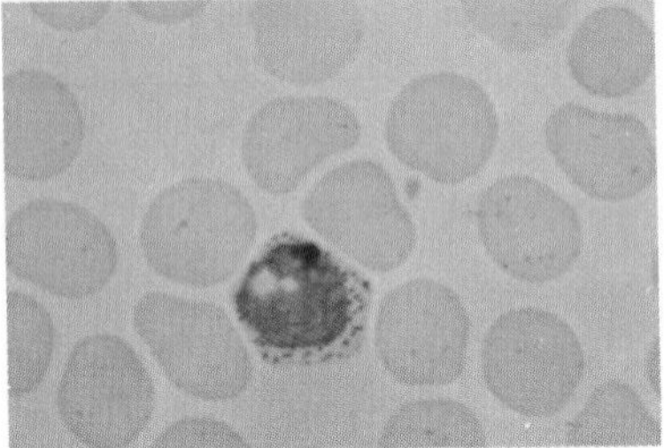

Plasmodium vivax

FIGURE 23.10
***PLASMODIUM* LIFE CYCLE**
Plasmodium life cycle and the symptoms of malaria. **(a)** Injection of sporozoites into host. **(b)** Asexual reproduction and formation of merozoites in the host liver. **(c)** Infection of red cells, cyclic reproduction, and fever. **(d)** Formation of gamonts in host blood cells. **(e)** Reinfection of mosquito and sexual reproduction. Note the repeated rhythm of merozoite increase and fever, followed by chills.

mature in the human host, but must be taken in by a female *Anopheles* mosquito as it feeds on an infected human. (This is why isolating malaria victims is a sound medical practice.) In the mosquito's gut the gamonts mature into gametes, male and female, which join in fertilization. The zygotes develop in the walls of the intestine, where each divides continuously, and large numbers of sporozoites form. Following their migration into the mosquito's salivary gland, the sporozoites are ready to start a new infection cycle.

Ciliophora: Ciliates

The **ciliates** are a highly diverse group of protozoans, and some species undoubtedly represent the most complex single cells on earth. In size alone they range from about 10 micrometers to 3000 micrometers (approximately the same relative difference between shrews and blue whales, see Figure 23.11).

Ciliates are identified by the cilia that appear at some stage in their life cycle, and that are used in movements, or feeding, or both. The cilia occur in rows, either longitudinal or spiral, or they can occur in fused tufts. Each cilium can perform a precise rowing motion (see Chapter 4), but for efficiency in both locomotion and feeding, a large number of cilia must perform in a coordinated manner, moving in sequence much like rows of wheat bending before gusts of wind. This coordination is made possible by an elaborate cytoskeletal network of nervelike fibers that connect one ciliary basal body with the next. In *Paramecium,* well-known ciliates, some of the cilia are fused into a kind of membrane that lines the **cytostome,** a funnel-like feeding structure. In other genera, such as *Euplotes,* the cilia may be arranged in tufts, which serve as many "legs," enabling members of this genus to scamble over the bottom sediment or paddle along through the water above.

The body covering of the ciliate is often a tough but elastic **pellicle,** a thin, translucent envelope of secreted material outside the plasma membrane. Because it is elastic, the

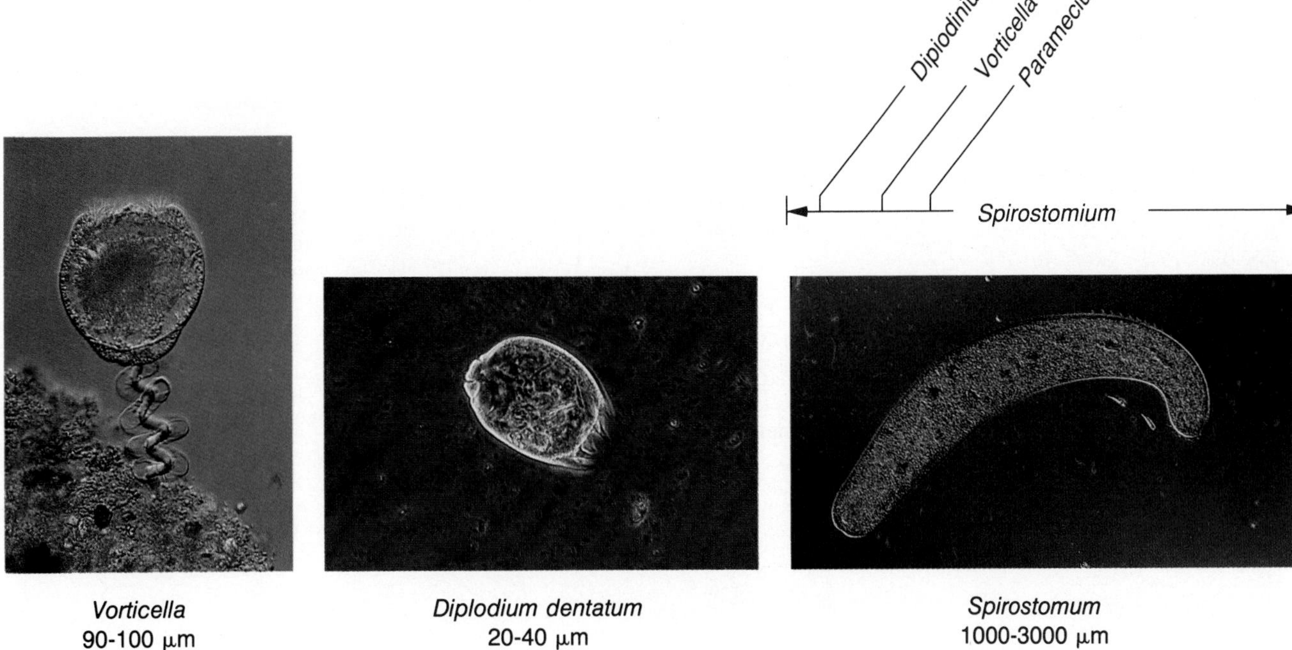

FIGURE 23.11
A SAMPLE OF CILIATE DIVERSITY
The ciliates are enormously varied in both size and structure. One of the largest, *Spirostomum,* is 3000 μm (or 3 mm) in length—easily visible to the unaided eye. *Diplodinium,* one of the smallest ciliates, is 1/300 the size of *Spirostomum. Vorticella,* of intermediate size, has a long, contractile stalk.

ciliates can bend and wriggle and contort and manage to get past or through all sorts of obstructions.

Alternating with the ciliary basal bodies are slender structures called **trichocysts.** Trichocysts can be forcefully discharged from the body surface en masse when the *Paramecium* is disturbed. Each discharged body is a long, threadlike cylinder with a barblike head. Some ciliates have toxic trichocysts that are used in capturing prey or in defense, while others use the trichocyst to moor themselves in place while feeding.

Ciliates, like many freshwater protozoans, also have **contractile vacuoles** with which they actively pump out excess water (see Figure 23.12a). This is an important means of maintaining proper osmotic conditions, since the cells often live in a hypoosmotic (watery) environment. Both the filling and emptying cycles appear to operate by ATP-powered contractions of cytoplasmic microfilaments.

Most ciliates feed in an animal-like manner, taking in bulk food such as bacteria or other protists, rather than by simple absorption. Some are seemingly ravenous predators; one species, *Didinium,* although relatively small, can engulf ciliates considerably larger than itself. *Paramecium* feeds less conspicuously, principally on bacteria, which it sweeps along its **oral groove** leading to its cytostome and on into **food vacuoles** that form through phagocytosis. The site of food vacuole formation, as one might expect, is a thin region of the body wall, containing only the plasma membrane. Once filled, the food vacuole breaks away and moves through the cytoplasm in a well-defined path. As digestion occurs, the nutrients diffuse out of the vacuole into the cytoplasm along the way. The undigested residues are then expelled from the body through exocytosis. If the cytostome is the protozoan's mouth, then the **cytopyge** is its anus. It appears periodically—always in the same place—when solid wastes are ready to be expelled (Figure 23-12b).

Reproduction in ciliates is extremely complex and rather fascinating. *Paramecium caudatum,* a laboratory favorite, has a complex sex life. Sexual reproduction involves conjugation, which in these protozoans is a temporary fusion of two cells, followed by meiosis and the exchange of haploid nuclei (Figure 23.13). Whereas there are no recognizable sexes in *Paramecium,* there are "mating types"—individuals that are different in less obvious ways. Different mating types must be present for conjugation and sexual exchange to occur. There are other twists to the story as well.

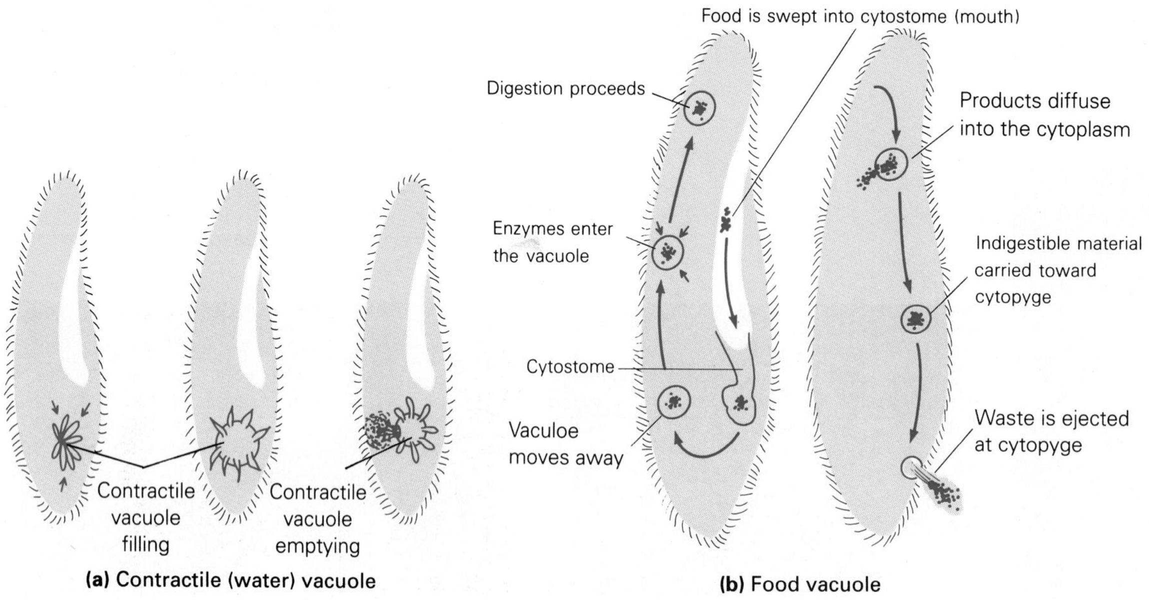
Contractile
vacuole
filling
Contractile
vacuole
emptying
Food is swept into cytostome (mouth)
Digestion proceeds
Enzymes enter
the vacuole
Cytostome
Vaculoe
moves away
Products diffuse
into the cytoplasm
Indigestible material
carried toward
cytopyge
Waste is ejected
at cytopyge

(a) Contractile (water) vacuole

(b) Food vacuole

FIGURE 23.12
VACUOLES IN A CILIATE
Ciliates such as *Paramecium* use vacuoles for the digestion of food and for maintaining water balance. **(a)** Water vacuoles occur at specific sites. They balloon out when filling and contract suddenly as they empty their contents to the outside. **(b)** Food particles taken into the cytostome (gullet) enter food vacuoles for digestion.

In *Paramecium,* the genetic material is separated into a germ line (involved in reproduction) and a somatic line (involved in the maintenance of the individual). The germ-line DNA is confined to the diploid **micronucleus.** The somatic DNA, from which all RNA is transcribed, occurs in the polyploid **macronucleus** (having multiple copies of each chromosome). In any case, during conjugation, meiosis in both partners produces several haploid micronuclei (Figure 23.13c). One from each partner then passes through a cytoplasmic bridge between the two. In each individual, the incoming micronucleus fuses with a resident micronucleus. The result is a new diploid combination of genetic material in each of the conjugants.

After conjugation, the macronucleus disintegrates while the new diploid micronucleus undergoes mitosis. One of the two daughter nuclei enters the germ line, where it remains inactive until the next round of mitosis or meiosis. The other undergoes multiple replications, eventually forming the new polyploid macronucleus.

Asexual reproduction in *Paramecium* is much simpler. The macronucleus simply pinches in two, and each half is retained. Then the micronucleus undergoes replication and mitosis in the usual manner. The cytoplasm divides across the center of the cylindrical cell as each new half regenerates the parts it is missing (Figure 23.14). As we have come to expect, asexual reproduction accounts for rapid population increases.

MYXOMYCOTA AND ACRASIOMYCOTA : SLIME MOLDS

The poetic name *slime molds* is perhaps unfortunate, since by most schemes, these admittedly slimy creatures are not molds. Some biologists include these with the fungi, but for reasons that will soon become obvious, we will not. Actually, there are two distinct groups, the **acellular slime molds (Myxomycota)** and the **cellular slime molds (Acrasiomycota),** and they may not be closely related to each other. Most slime molds join the decomposer bacteria in their ecologically vital activity of recycling mineral nutrients.

One of the best-known acellular slime molds, *Physarum polycephalum,* has all of the characteristics of a huge ameba except that it is multinucleate, its many nuclei the product of numerous mitotic events without cell division. *P. polycephalum* may commonly be found creeping over the moist underside of rotting tree trunks. As it moves along, it feeds by phagocytizing bits of organic matter. The mass shows some sensitivity and avoids obstacles and dry areas (Figure 23.15).

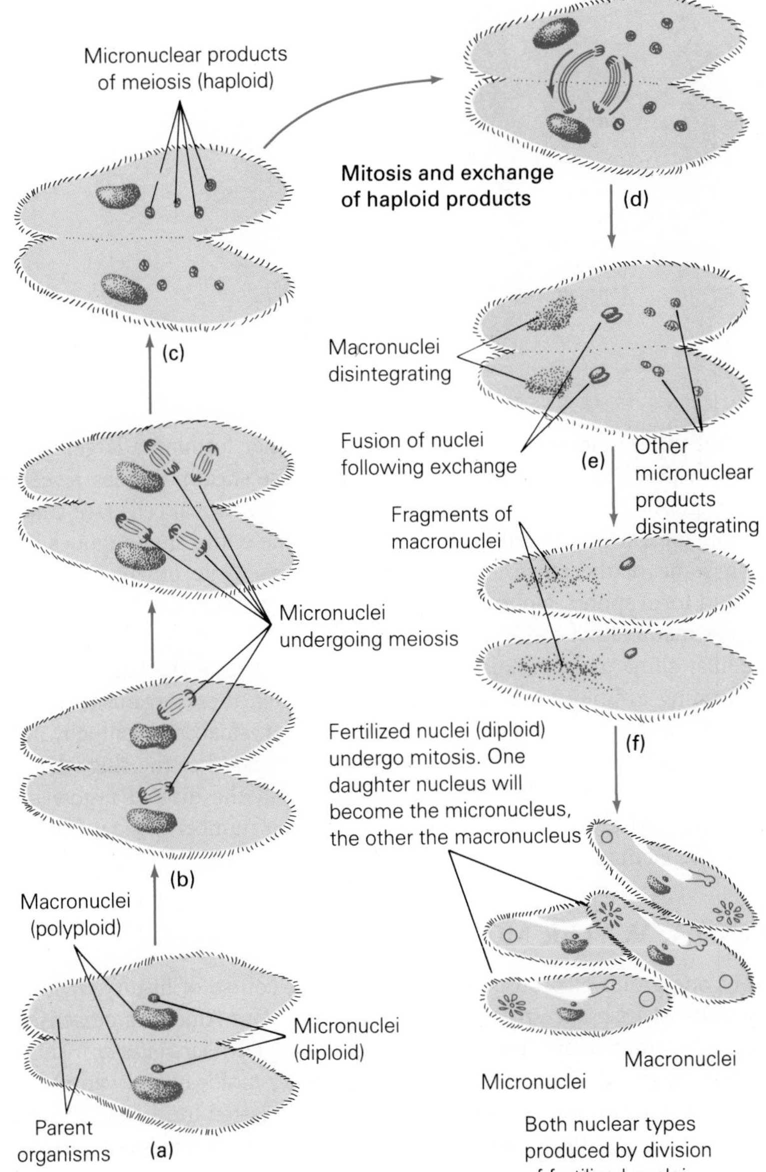

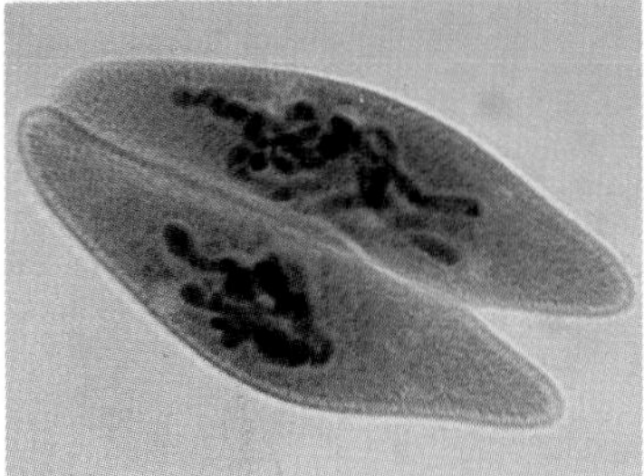

FIGURE 23.13
SEXUAL EXCHANGE IN *PARAMECIUM*
During conjugation, micronuclei undergo meiosis, and an exchange of the haploid products occur between cells. As a result, after fusion the diploid micronucleus in each cell contains a new combination of genes from the two strains. Afterwards, a new macronucleus is produced through mitosis. *Paramecium* later enters into asexual reproduction, increasing its numbers through mitosis and cell division.

At some point in the creeping ameboid stage of its life cycle, *P. polycephalum* will seek out a drier habitat (or perhaps its moist habitat will dry out). It is only then that it shows the characteristics of a fungus. Like many fungi, the drying mass produces slender vertical props—**sporangiophores**—topped by spore-forming **sporangia.** Each sporangium undergoes a number of meiotic divisions, producing numerous haploid spores. The spores emerge to be carried aloft by air currents. If a spore lands in a suitable place, it begins to divide and produce either an ameboid **myxameba** or a flagellated **swarm cell.** If conditions aren't quite suitable, either the myxameba or swarm cell can suspend activity, forming a dormant cyst from which it will emerge later. Otherwise the myxameba may divide a number of times, producing many ameboid descendants. Finally, either the myxamebas or the swarm cells can pair off and fuse, producing the diploid state once more. From there, either can form a new multinucleate ameboid stage and again repeat the life cycle.

The cellular slime molds differ fundamentally from the acellular group in that the small, individual amebas retain their individuality, even when swarming, an act for which the cellular slime molds are famous and one that ends in spore formation. Swarming usually begins when the small amebas run short of the bacteria they feed

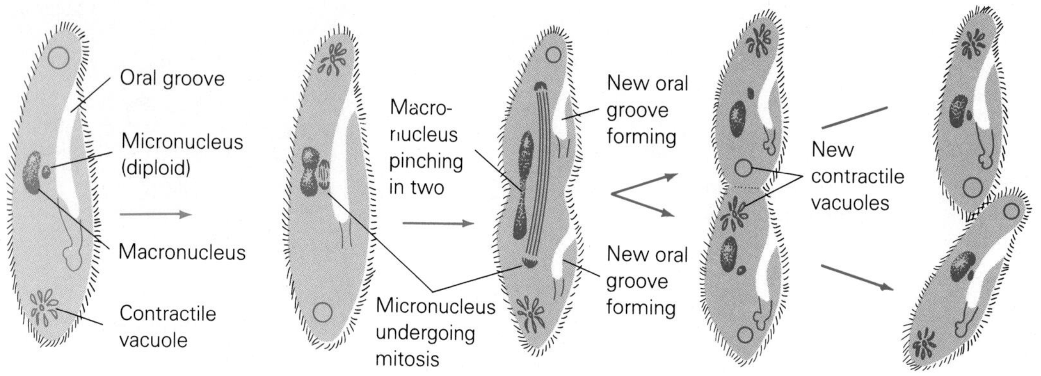

FIGURE 23.14
ASEXUAL REPRODUCTION IN *PARAMECIUM*
As the macronucleus pinches in half, the micronucleus undergoes mitosis. The cell divides transversely, and new organelles begin forming.

upon. At that time, numerous individuals mass together forming a large, many-celled "slug" stage, which becomes surrounded by a cellulose sheath. Studies reveal that the release of cyclic AMP, a common chemical messenger in animal cells (see Chapter 37), is responsible for drawing the individuals together. The entire slug migrates for a time, whereupon some of the amebas form erect processes while others move into these processes and form spores. Upon dispersal, the spores may germinate into new amebas, but the other members of the slug simply die off.

The cellular slime mold's sexual cycle begins when two of the haploid amebas fuse together, forming a diploid zygote (the only time when these organisms are diploid). The cell then proceeds to engulf a number of other amebas, eventually forming a **macrocyst** ("large" cyst—but actually quite small compared to the slug) that becomes surrounded by a cellulose wall. Within the macrocyst, the diploid zygote undergoes meiosis, the haploid products divide mitotically, and a number of new amebas escape and resume the feeding stage.

OOMYCOTA: WATER MOLDS

Members of one group of protists helped change the course of history, at least for the Irish. If you are of Irish ancestry, you just might owe your American citizenship to the activity of a highly destructive species of **water mold,** *Phytophthora infestans.* This parasite, known as "late blight," kills potato plants. (Not all water molds live in water.) Late blight was the cause of the famous Irish potato famine, which lasted from 1843 through 1847. Many Irish starved, and those who could left their homeland, emigrating to the United States.

Members of phylum Oomycota are named partly for the large eggs of some species and partly for their filamentous body (*oo-,* "egg"; *-mycetes,* "threadlike"). Because of the threadlike or filamentous growth form in many, the water molds were once classified with the fungi. They were considered primitive because of their flagellated spore stage, their cellulose cell walls, and their simple vegetative stage. By contrast, fungi use chitin to form cell walls and have nonmotile spores. Therefore, water molds have been placed with the protists. Actually, such organisms are quite interesting to biologists, since they tend to span or bridge taxonomic groups, indicating how one may have given rise to another in the dim past.

Some water molds actually live in water, where many species invade injured or diseased plants or animals. Other members, the terrestrial species of the oomycetes, are dangerously parasitic. This group includes **downy mildews** and **blights,** both of which can be seen on the undersides of infected leaves. You can expect them on your house or garden plants if you tend to overwater or undercultivate. Many crops, such as potatoes, beans, melons, sugar beets, and grains, are especially susceptible to infection. *Phytophthora,* the villain of the Irish potato famine mentioned above, grows on moist leaves by penetrating the numerous stomata (leaf pores) with its many filaments. These natural openings on leaves permit the mold to penetrate the photosynthesizing cells, absorbing

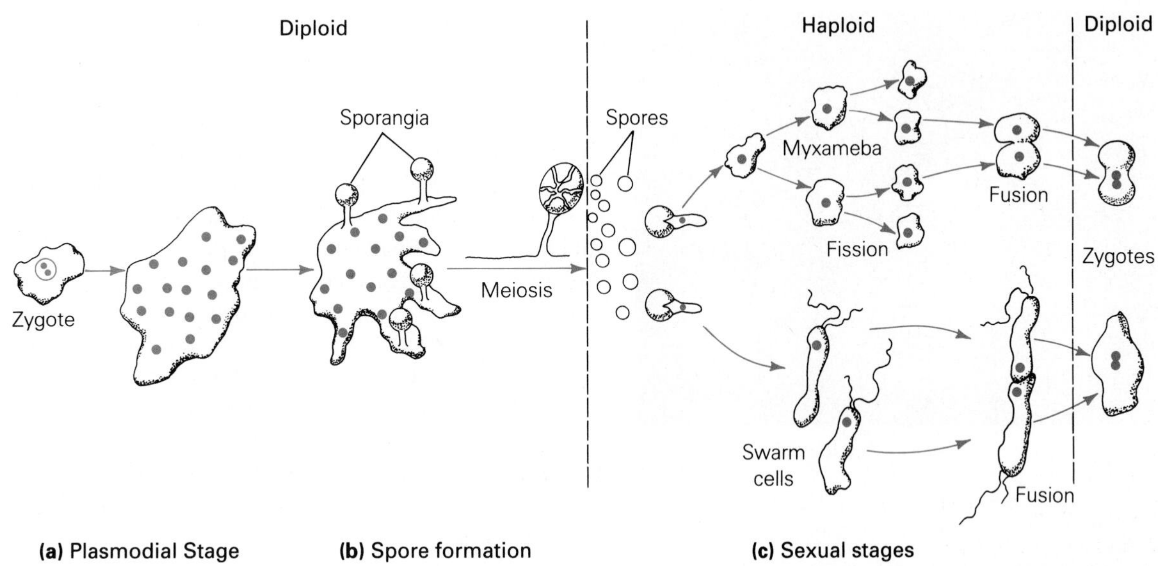

TABLE 23.1 CHARACTERISTICS OF THE HETEROTROPHIC PROTISTS

PHYLUM	USUAL CHARACTERISTICS
PROTOZOANS	Protists with animal-like movement and feeding characteristics
Sarcomastigophora	(a) Flagellates. Movement by flagella; feeding by absorption and predation; asexual reproduction: binary fission; sexual reproduction: unknown (b) Amebas. Naked and with supporting glassy or calcareous skeleton; ameboid movement; feeding by phagocytosis; asexual reproduction: binary fission; sexual reproduction: meiosis and fusion of cells; amebic dysentery; vast marine calcareous deposits
Apicomplexa	Sporozoans. Nonmotile adults; feeding by absorption; many are parasitic; many form spores; malaria agent *Plasmodium vivax* carries on asexual reproduction via binary fission and fragmentation in mammalian or bird host and sexual reproduction via fusion of gametes in mosquito.
Ciliophora	Ciliates. Complex organelles; movement and feeding involve cilia; asexual reproduction: binary fission; sexual reproduction: conjugation and exchange of haploid micronuclei; some parasites
SLIME AND WATER MOLDS	Protists with some fungus-like spore formation
Acrasiomycota	Cellular slime molds. Small ameboid cells feed by phagocytosis; asexual reproduction: swarm, forming large cellular slug which produces spores in fungal fashion; sexual reproduction: fusion of amebas, formation of macrocysts and new amebas
Myxomycota	Acellular slime molds. Multinucleate ameboid mass, feeding by phagocytosis; asexual reproduction: mass forms spores in fungal fashion; sexual reproduction: fusion of amebas or swarm cells.
Oomycota	Water molds. Fungal-like filamentous growth, but with cellulose walls and flagellated stages; asexual reproduction: flagellated spores, sexual reproduction: with large egg, small nonmotile sperm, parasites of plants and animals; cause of late blight

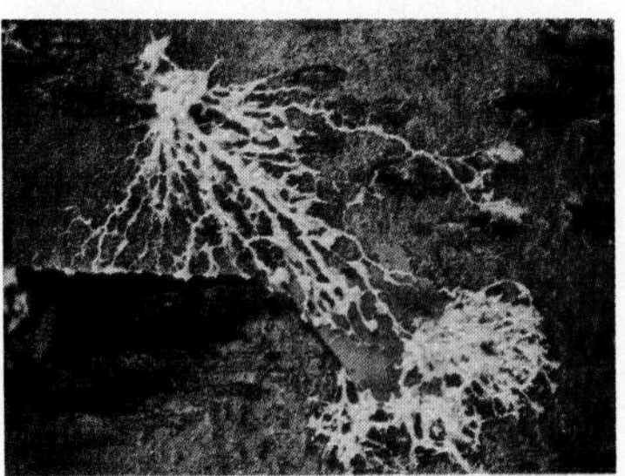

FIGURE 23.15
LIFE CYCLE OF AN ACELLULAR SLIME MOLD
(a) The acellular (ameboid) stage is a creeping, feeding phase in which the cytoplasm enlarges, and numerous rounds of mitosis occur without cell division (photo). **(b)** This is followed by meiosis and spore formation (photo). **(c)** Spores enter a sexual phase in which either myxameba or flagellated swarm cells form. They fuse, forming diploid cells, which may then enter a new ameboid stage.

(a) Infected leaves

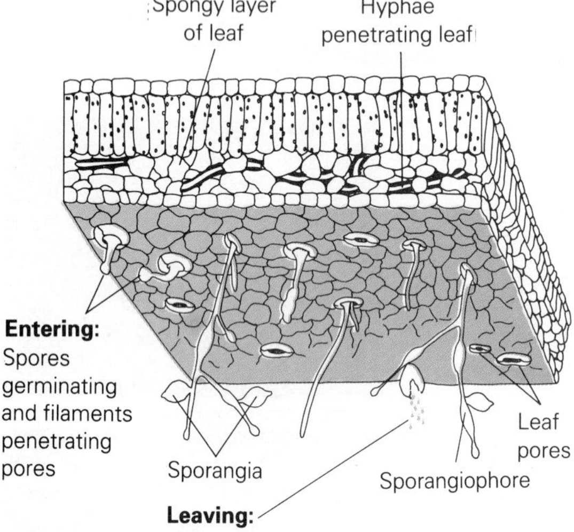

(b) Asexual reproduction

FIGURE 23.16
PLANT PARASITES
(a) When a downy mildew or late blight infection occurs in a plant, the fuzzy filamentous growth can be seen on the underside of the leaves. **(b)** *Phytophthora infestans,* the agent of late blight, sends its filaments throughout the spongy tissue of the leaf, where it absorbs the plant's sugars. Reproductive filaments then emerge through the leaf pores, form sporangia at their tips, and release spores that can then carry the infection to other plants.

the nutrients as they are produced. Eventually, an asexual reproductive stage begins, and the filaments grow back out through the stomata, producing spore-forming structures at their tips (Figure 23.16).

In their asexual cycle, the water molds produce sporangia. From these, flagellated spores emerge and begin to swim about. If they encounter a food source, growth occurs, and a new colony develops (Figure 23.17a).

Sexual reproduction in water molds (Figure 23.17b) begins with the emergence of unusually thick filaments that produce egg cells in spherical **oogonia** (sing. **oogonium**). Fertilization is accomplished in an unusual way. Filaments growing near an oogonium send fingerlike branches over the spherical body. Within these branches are the **antheridia**, wherein sperm nuclei arise. The branches form fertilization tubes that penetrate the oogonia, permitting the sperm nuclei to reach the eggs. Some of the events are similar to fertilization in higher plants; however, the similarity may be a simple coincidence—a case of convergent evolution.

THE ALGAE

The algal protists, or algae, are photosynthetic and plantlike. They are extremely widespread, living in all aquatic habitats and a few terrestrial ones. Significantly, algal protists make up much of the **phytoplankton** (small, floating, aquatic photosynthetic organisms). Phytoplankton are of immense ecological importance because they form the energy base of many marine and freshwater food chains. Along with the **seaweeds** and **kelp** of the marine environment, the phytoplankton account for a substantial part of the earth's total photosynthetic yield. In addition, many algae live as photosynthetic symbionts in the cells of corals and other animals, and some join fungi in forming the lichens.

The algae are a diverse group, ranging from some of the simplest eukaryotes to the more complex. We have divided the algae into the small, unicellular forms (a few of them colonial) and the larger multicellular types, but size is no indication of their importance. After all, the unicellular algae make up the ecologically vital phytoplankton. On the other hand, the multicellular algae can form huge undersea forests that cannot fail to impress. The largest, the kelps, have evolved highly specialized structures, and as we will see, can have complex life cycles, some quite plantlike and others curiously like our own.

Pyrrophyta: The Dinoflagellates

Visitors to tropical waters (and some northern waters as well) may be surprised and delighted as they are rowed back to their anchored ship in a dugout canoe after a rousing night ashore. Each time the paddle slides into the water, the water seems to explode with tiny iridescent lights. Even the wake of the canoe is aglow. Objects tossed overboard leave a shimmering trail as they disappear into the briny depths. It seems like magic, but it is the magic of **dinoflagellates** ("whirling flagella"). The name "Pyrrophyta" means "fire plants."

Some of the dinoflagellates are also responsible for another dramatic and far more dangerous phenomenon: the dreaded **red tide.** When conditions are right, such as with sudden increases in mineral nutrients, certain dinoflagellates multiply to incredible densities called "blooms." The species with reddish pigments may also contain a potent nerve poison, and the results of their periodic blooms spell death for enormous numbers of fish. The water turns a rusty or blood color, and fish by the thousands go belly up (Figure 23-18). Clams, oysters, and mussels are unaffected by the dinoflagellates, but they can accumulate the poisons, making their flesh toxic to humans. The illness, called "mussel poisoning" (after the bivalve mollusc that commonly harbors the agent), can be quite serious and even fatal. Since this is primarily a summertime problem, people who

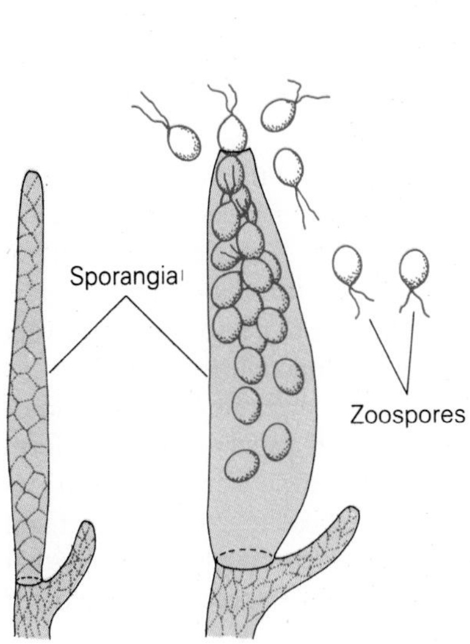

(a) Asexual reproduction

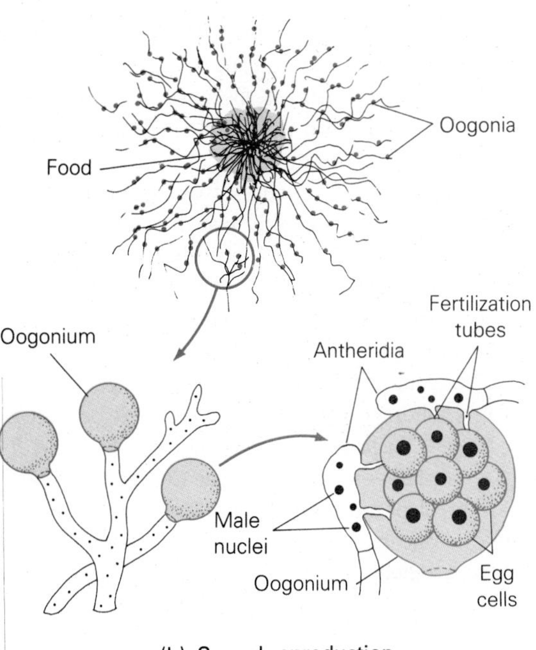

(b) Sexual reproduction

FIGURE 23.17
ASEXUAL AND SEXUAL REPRODUCTION IN WATER MOLDS
(a) During asexual reproduction in *Saprolegnia,* sporangia produce numerous flagellated zoospores, which then germinate, form filaments, and repeat the asexual cycle. **(b)** In their sexual cycle, some water molds develop complex gamete-forming structures.

collect their own seafood generally avoid clams, oysters and mussels in months without an "r." (This means May, June, July, and August—but don't bet your life on it!)

Several dinoflagellates are seen in Figure 23.19. Many have two flagella, which lie in prominent grooves, their movement giving the cell a spinning motion. The photosynthetic species have chloroplasts, which contain the pigments chlorophyll *a* and *c* along with carotenoids. Their cell walls are stiff cellulose plates whose arrangements gives some the appearance of armored helmets.

Although many dinoflagellates are free-living, others are important as photosynthetic symbionts within larger organisms. Their numbers in coral, for instance, can reach astronomical proportions (some 30,000 per square mm). Through photosynthesis, they provide the coral animals with much of the energy expended in the formation of the great coral reefs. Heterotrophic dinoflagellates also occur, some feeding on other organisms and others living as parasites.

The dinoflagellate nucleus is unique, with large, permanently condensed chromosomes containing far less protein than is seen in those of other eukaryotes. They are permanently attached to the nuclear envelope. Further, the nuclear envelope is not dismantled during mitosis and appears to play a role in chromosome separation (as does the mesosome in the prokaryotes—see Chapter 22). Mitosis occurs without the elegant interplay of spindle elements within the nucleoplasm. Instead, bundles of microtubules simply form a cytoplasmic channel between grouped chromosomes.

The chief method of reproduction in dinoflagellates is asexual, using binary fission. Sexual unions commonly occur through the fusion of two entire haploid cells, producing a brief diploid state. After this, meiosis occurs and the haploid state resumes.

FIGURE 23.18
RED TIDE
The dinoflagellates can multiply to incredible densities with seasonal increases in mineral nutrients. These densities, called blooms, are known as "red tide."

Euglenophyta: The Euglenoids

The best known **euglenoid**, *Euglena,* was mentioned earlier as an example of why classification can be so difficult. (For instance, *Euglena* has a triple nuclear membrane, which is another reason why it remains in its own distinct group.) But this protist is also interesting for other reasons.

The name **Euglenophyta** comes from *eu-,* "true", *glena-,* "eye"; and *-phyton,* "plant"—the plant suffix referring to their former theoretical association with the plant kingdom. The visible "eye-spot" of euglenoids (Figure 23.20) is really not an eye at all, but a shield of red pigment lying next to a light-sensitive area called a **photoreceptor.** Because of the shield's arrangement, euglenoids can detect light and move toward it. When the sensitive area is not shielded, the organism simply begins to move. The result is that it continually approaches the light—an adaptation for photosynthesis. Actually, only about a third of euglenoid species are photosynthetic. The others presumably lost their chloroplasts over time.

Euglenoids are unicellular, flagellated, and asexual. *Euglena* is the most widespread genus and can serve as an example. Like all euglenoids, *Euglena* lacks rigid cell walls and has a flexible body, but tends to maintain a flasklike shape. A single long flagellum arises from a mouthlike depression. There is actually no evidence that *Euglena* ingests solid food particles, but other genera of euglenoids are known to do so. A second very short, rudimentary flagellum is present in the gullet, but its function is not known.

Euglena contains chlorophyll *a* and *b* and carotenoids, incorporated into a chloroplast—all quite similar to what we find in plant cells. It stores its starches in the form of **paramylon,** which is quite different from plant and other algal starches. So far, no one has observed sexual activity in *Euglena,* but asexual reproduction, as you might expect, occurs by mitosis and binary fission. Euglenoids can reproduce so rapidly that they may impart a green color to a pond, except for one ruddy species that colors it red.

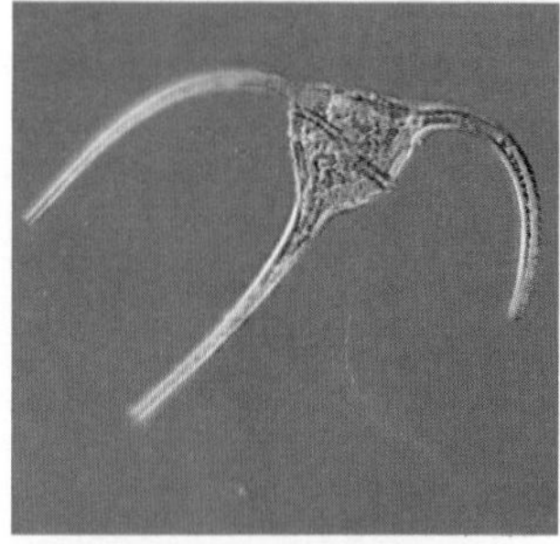

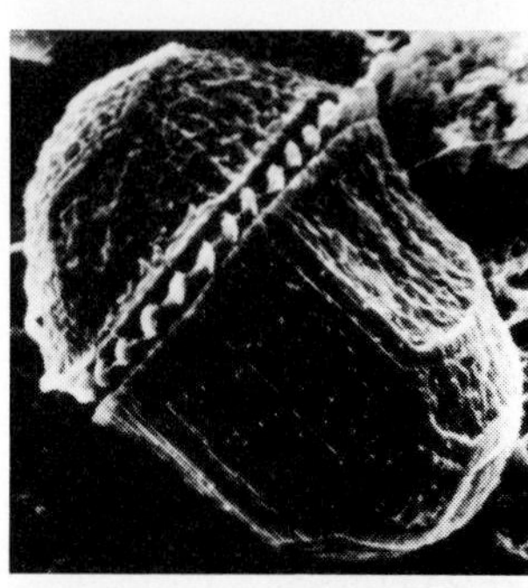

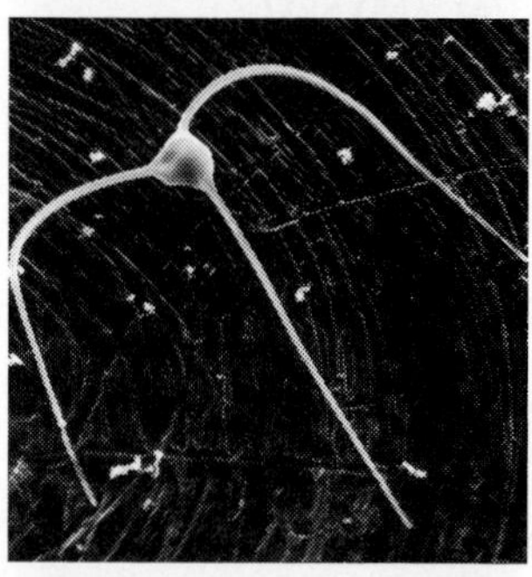

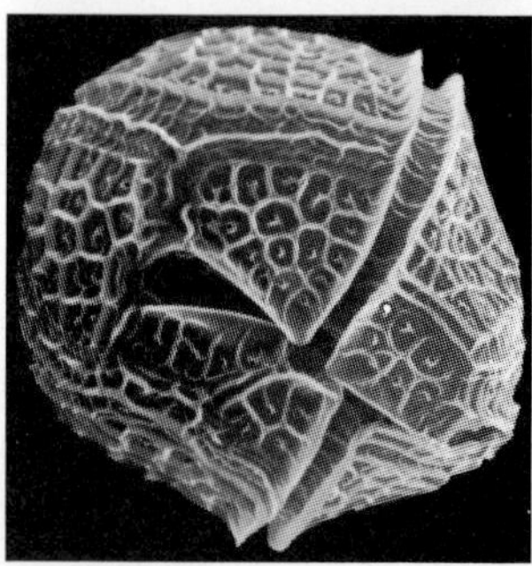

FIGURE 23.19
COMMON DINOFLAGELLATES
There is considerable variation in size and shape. *Gymnodinium* (top) and *Gonyaulax* (bottom) are red tide species.

Chrysophyta: Yellow-Green Algae, Golden-Brown Algae, and Diatoms

Both the **yellow-green algae** and the **golden-brown algae** make up part of the phytoplankton of marine and fresh waters, forming a vital part of the food chains. The names are derived from the color of their light-absorbing pigments. All members of **Chrysophyta** ("golden plants") contain high concentrations of carotenoids, of which the most abundant is called **fucoxanthin.** Fucoxanthin absorbs much of the light, transferring it to molecules of chlorophyll *a*. The chrysophytes store their carbohydrates in the form of **chrysolaminarin.** Chrysophytes are a diverse group. Some surround themselves with cell walls that are essentially glass (silicon dioxide) boxes. Some chrysophytes are flagellated, others are not; most are solitary, but a few genera produce filamentous colonies; some are multinucleate, and some are truly multicellular.

Although they are photosynthetic, some ameboid species of golden-brown algae can ingest solid food. Many yellow-green algae can be blown about by the wind and are often found growing on tree trunks, rocks, or soil.

The Diatoms **Diatoms,** common photosynthetic marine and freshwater chrysophytes, live in glass boxes. The boxes are indeed very fragile and peculiar structures consisting of an inner box and an outer lid, both made of the colorless silicon dioxide mentioned earlier (Figure 23.21).

When they reproduce asexually, diatoms undergo mitosis and divide within their shell, whereupon the two old halves of the shell each become the outer lid of a daughter cell as new inner half-boxes are secreted. Because glass boxes don't stretch very well, one of the daughter cells, the one that inherited the inside half of the parent cell, must be smaller than the other. This presents the diatom with a problem. After all, it wouldn't do to grow smaller with every generation. So, what's the solution?

Eventually the smallest diatom undergoes meiosis in its box, producing haploid gametes. In many species (chiefly those in fresh waters), the gametes are ameboid and are not differentiated. In others (mainly marine), a small flagellated sperm and a larger egg are formed. In either case, the haploid gamete sheds its box and becomes the free and naked sexual form. Gametes meet and fuse to produce the diploid zygote, called an **auxospore** (*auxo,* "increasing") because it increases in size for a while before reaching the adult size. The auxospore then matures, secretes a new glass covering, and the cycle continues (Figure 23.22).

The walls of marine diatoms are highly ornamented and beautiful to the human eye.

The intricate designs are produced by the arrangements of tiny holes through which gas and water are exchanged. Diatoms can be radially or bilaterally symmetrical.

Ocean floor sediments consist largely of diatom shells. Deposits of ancient diatom shells form **diatomaceous earth,** which is used in toothpaste, swimming pool filters, and insulating material. The minute glassy particles have an abrasive quality which is useful in polishing. The thickest known deposits of diatomaceous earth (about 1 km in depth) occur off the coast of Lompoc, California.

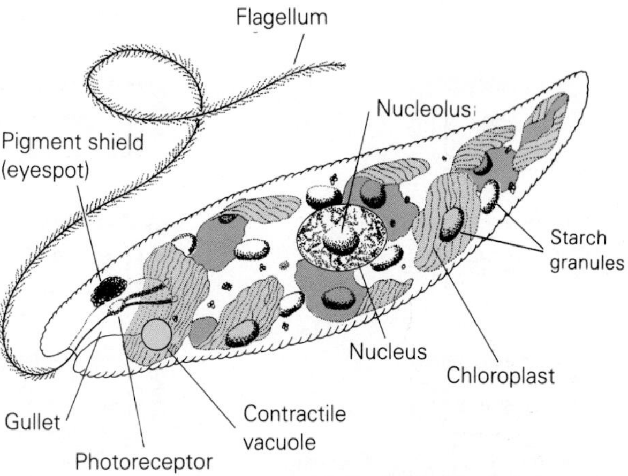

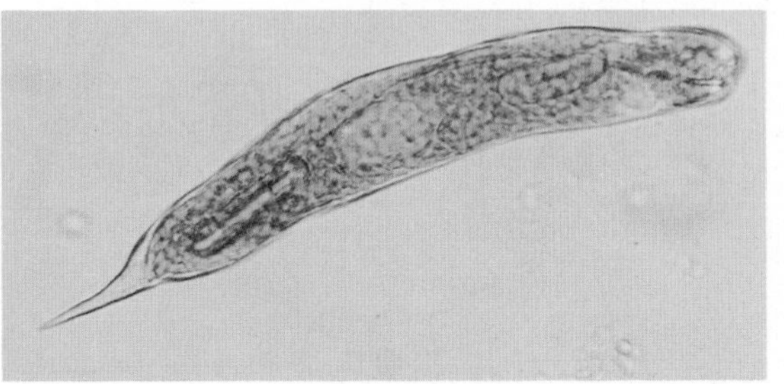

FIGURE 23.20
***EUGLENA GRACILIS,* A WELL-KNOWN EUGLENOID**
Note the large, well-organized chloroplasts, specialized starch (paramylon) bodies, and photoreceptor.

Multicellular Algae

A true multicellular condition is common among the red, brown, and green algae, although single-celled species are seen in the latter. Nevertheless, these groups contain the largest protists. As we'll see, their size range is enormous, from minute, single-celled green algae to truly gigantic ocean kelps.

Rhodophyta: The Red Algae **Rhodophyta** (*rhodo-,* "red"; *-phyta,* "plants") includes about 300 species, all called **red algae.** All contain the pigments **phycocyanin** (meaning "algal blue-green") and **phycoerythrin** (meaning "algal red"), as well as chlorophyll *a.*

Flagella and cilia are not found in the red algae. Even the sperm (called *spermatia*) lack flagella and must float passively to receptive female cells.

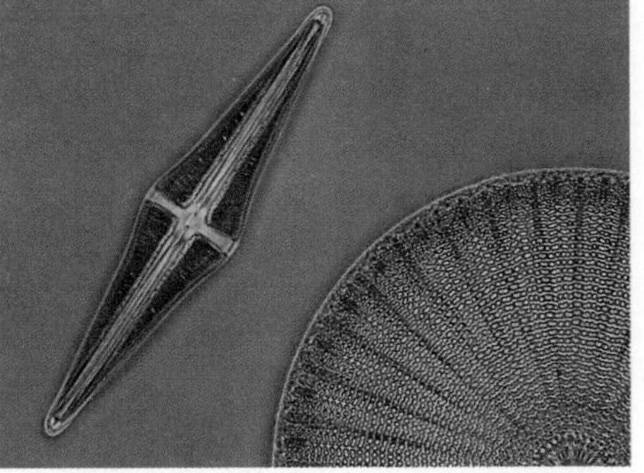

FIGURE 23.21
SCULPTURED GLASS
Diatoms are well known for their diverse, beautifully sculptured, glassy skeletons. Some are etched so finely that they are used to test the quality of light microscope lens systems.

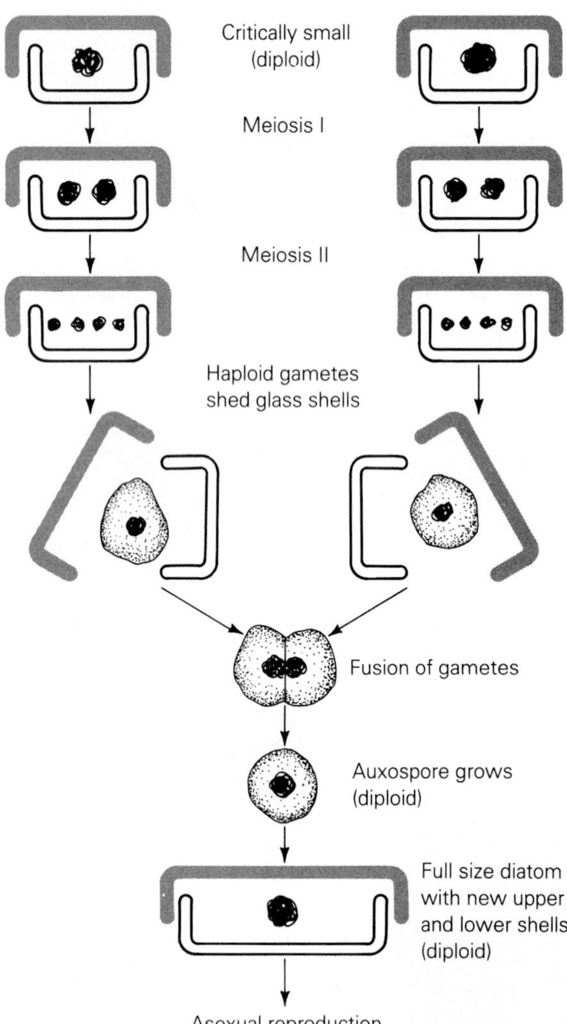

FIGURE 23.22
SEX IN A DIATOM
When continued cell division results in cells that become critically small, meiosis occurs and gametes form. In this species of diatom, one ameboid gamete fuses with an identical gamete from another diatom, forming a diploid auxospore. When the auxospore grows and matures, it produces a new shell, large enough to go through many asexual divisions before the sexual process must be repeated.

All but a few red algae are aquatic, and nearly all of them are marine (Figure 23.23). You have undoubtedly seen red algae if you've ever looked into a tide pool. In fact, red algae comprise a large portion of what we commonly call *seaweed.* You might not recognize them as red algae, since they are often green or black, and some are blue or violet, but those growing well below low tide are often red. Actually, color can't be used to identify species. Individuals of the same species often are found to be of different colors at different depths, in what is believed to be an adaptive response to the changing intensity of light available for photosynthesis.

Marine red algae grow primarily on rocky coasts, where they attach firmly to the seabed by specialized structures known as **holdfasts.** Holdfasts are simply anchors, not roots. After all, algae grow in water and have no need of a root system for the extraction of water and minerals.

Red algae store their foods in the form of a starch called **floridean starch** and produce a number of other polysaccharides. One of these, **agar-agar,** is used by Indonesians to thicken soup and by biologists as a jellylike medium on which to grow bacteria. The bacteria cannot metabolize the agar but use only the nutrients that have been added to it. Although red algal polysaccharides cannot be metabolized by most microorganisms or by anything else, we may eat them anyway. A species of red alga known as *Irish moss* is harvested in enormous quantities for its polysaccharide **carrageenan,** which gives a fake richness to chocolate-flavored dairy drinks and fast-food milkshakes.

As we examine the multicellular algae, particularly those in which there is a pronounced alternation between the sporophyte and gametophyte phases, we find that this phenomenon can be accented in two ways. In the first, known as **isomorphic alternation,** the sporophyte and gametophyte generations are identical in appearance, differing only in chromosome number. In the second, **heteromorphic alternation,** one of the two phases is quite dominant in the life cycle. It is usually larger, longer lived, and nutritionally independent. The other phase may be greatly reduced, often to just a few cells, and is often nutritionally dependent on the first.

The common red algae *Polysiphonia* follows a sporic life cycle of alternating sporophyte and gametophyte phases, as do many of the seaweeds and kelps (Figure 23.24). In this genus, the sexual haploid and the asexual diploid individuals are both multicellular, with branching growth forms that closely resemble one another—what we just described as isomorphic alternation of generations. The phases can be distinguished only by inspecting the reproductive structures under the microscope (or by chromosome counts). Neither generation is dominant—that is, neither is decidedly larger or more common than the other.

Cells within the diploid sporophyte of *Polysiphonia* undergo meiosis, producing haploid spores that are released into the surrounding sea. Successful spores then develop into male or female gametophytes, which as expected produce gametes. The red algal version of sperm, nonmotile cells called **spermatia** (singular, **spermatium**), arise in **spermatangia** formed by male gametophytes. Egg cells form in **carpogonia,** the female gametophyte's counterpart to spermatangia. Each egg cell develops a lengthy, hairlike extension called a **trichogyne,** which protrudes from the carpogonium, and to which drifting spermatia become attached. Following attachment, the spermatial nucleus penetrates the trichogyne, soon reaching and fusing with the egg nucleus. Thus, fertilization occurs within the confines of the female gametophyte.

The zygote gives rise to an intermediate diploid structure, a **carposporophyte,** which remains in the gametophyte, producing and releasing numerous **carpospores.** The carpospores, in turn, develop into independent sporophytes, and a new cycle begins.

Some species of red algae are heteromorphic, with either the sporophyte or the gametophyte generation clearly dominant. In some, the gametophyte generation is reduced to a single-celled stage, and eggs and sperm develop directly by meiosis, just as in animals. Many life cycles have yet to be worked out. There aren't too many hard and fast rules about alternating phases, even in a group of 300 species.

The fossil history of the red algae is not very complete, but it seems likely that they shared the earth with the green algae (as well as with the far older cyanobacteria) as far back as the Cambrian period at the beginning of the Paleozoic era, some half billion years ago. (See the geological timetable inside the front cover.)

FIGURE 23.23
RED ALGAE
Most red algae are small seaweeds, measuring from a few centimeters to perhaps a meter in length. Most species are marine and are found in warm waters, usually attached to rocks by their *holdfasts*. Some of the branched bodies form widened, flat blades, but most are frilly and delicate.

Phaeophyta: The Brown Algae The **Phaeophyta** includes about 1000 named species. The **brown algae** are distinguished by their characteristic brown pigment, fucoxanthin. Brown algae at least have the decency to be brown—all of them. All of them also store carbohydrates in the form of **laminarin** and **mannitol**, and they have characteristic structural polysaccharides as well (notably **algin**, an important constituent of commercial ice cream and frozen custards). Unlike the red algae, the brown algae have flagellated sperm, and strangely enough, some female reproductive cells are flagellated as well.

FIGURE 23.24
THE LIFE HISTORY OF A RED ALGA
Polysiphonia's sporophyte is diploid, produced by the growth of a carpospore. The sporophyte generation ends with meiosis and the production of haploid spores, which form the gametophyte. The male gametophyte produces nonmotile sperm cells, which, when released, adhere to the trichogyne protruding from the egg. Fertilization starts a new diploid sporophyte generation within the pouchlike carposporangium. Many diploid carpospores emerge.

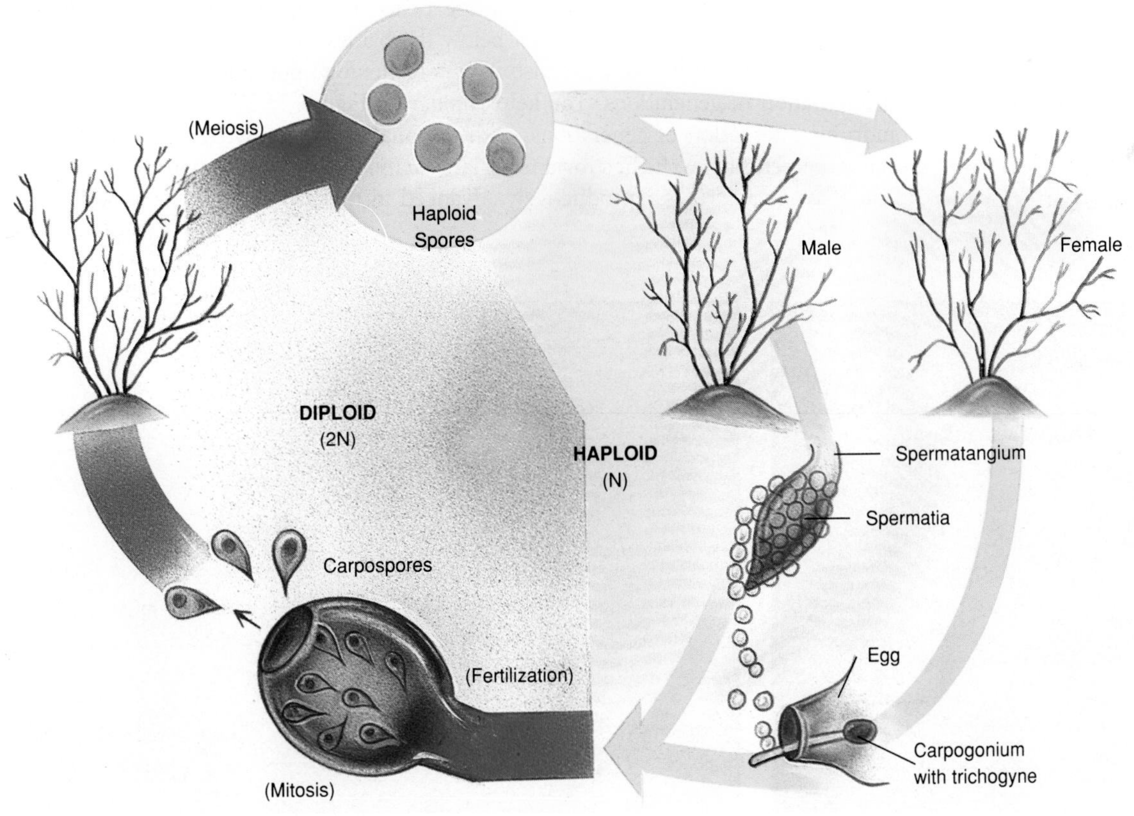

FIGURE 23.25
THE BROWN ALGA *SARGASSUM*
Sargassum, a floating seaweed, forms branching stipes and flat, leaflike blades. The round objects, called bladders, act as floats, keeping the seaweed near the surface where light is plentiful.

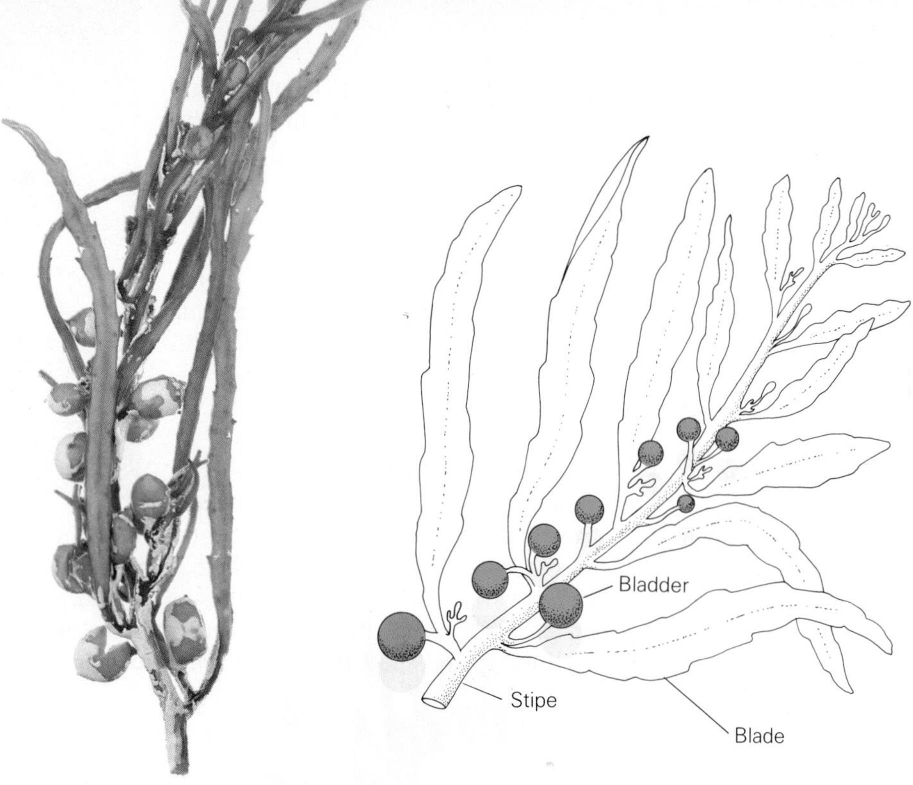

The brown algae live only in the ocean, particularly in cold coastal waters. An exception of sorts is the genus *Sargassum,* which is found in great masses in the warm water that flows through the middle of the Atlantic Ocean, an area called the Sargasso Sea (Figure 23.25).

Brown algae come in all sizes, from microscopic deep-water filaments to the famous giant kelps living in dense marine forests in the shallow waters offshore (Figure 23.26). Some of the kelps are enormous; *Macrocystis* and *Nereocystis* are record holders, some exceeding 60 meters (nearly 200 ft) in length. *Macrocystis* is regularly harvested along the southern California coast by kelp-cutting boats. Its regrowth is rapid, thanks to **meristems** actively dividing tissues just below the blades.

The life histories of brown algae also vary considerably, but most are sporic, with a clear alternation of generations. The kelp *Laminaria* (Figure 23.27) has a strongly heteromorphic alternation of generations, with dominating sporophytes and separate microscopic gametophytes. Its heterogametes, large stationary eggs and small motile sperm, indicate that *Laminaria* is reproductively advanced in this aspect of its life as well.

FIGURE 23.26
THE GIANT KELPS
The giant kelp *Macrocystis* (left) thrives in cold oceans. Seen here are its highly branching stipes and gas-filled bladders at the base of each blade. Some kelps reach enormous sizes, forming undersea forests (right).

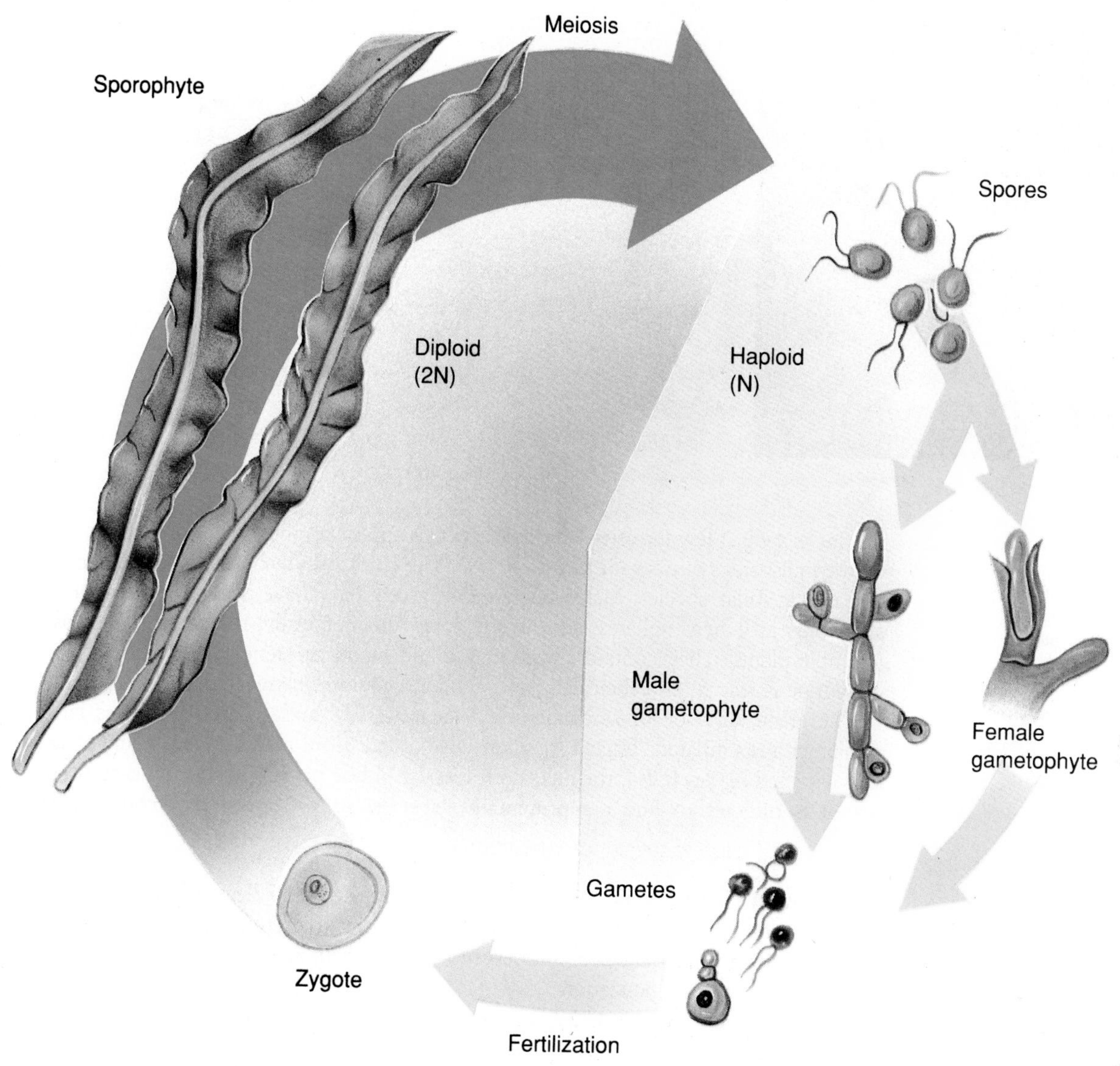

FIGURE 23.27
LIFE CYCLE OF *LAMINARIA*
The life history of *Laminaria,* a brown alga, is sporic. It is characterized by a very dominant sporophyte generation and a brief gametophyte generation. The biflagellated spores develop into male and female gametophytes. Flagellated sperm cells are released from the antheridium, swim to the oogonium, and fertilize the egg cells. The diploid zygote will then divide rapidly and develop into the mature sporophyte.

For an alga, the rockweed, *Fucus,* is full of surprises. There is no alternation of generations, and its life cycle is gametic (as is our own). That is, the products of meiosis are gametes, not spores. The diploid adults form hollow **conceptacles** at the tips of their highly branched blades (Figure 23.28). Cells in the conceptacles give rise to antheridia and oogonia, male and female gamete-forming structures respectively, and it is within these that meiosis occurs. Following fertilization, the zygote develops into the adult rockweed.

The kelps appear to be the most structurally complex brown algae. For example, *Macrocystis,* the giant kelp, has highly specialized tissues and organs (see Figure 23.26). These include the convoluted holdfast; **stipes,** which are stemlike structures supporting the **blades;** blades themselves, which resemble flattened leaves; and spherical, hollow **bladders** (or floats), which keep the photosynthetic cells of the kelp near the surface. The brown alga *Postelsia* even has specialized conducting tissue that closely resembles that of some plants. But in general, algae do not require elaborate conducting systems. Everything each cell needs is available in the surrounding seawater.

Chlorophyta: The Green Algae There are about 7000 named species of **green algae** and whereas most of them are freshwater forms, there are a respectable number of marine

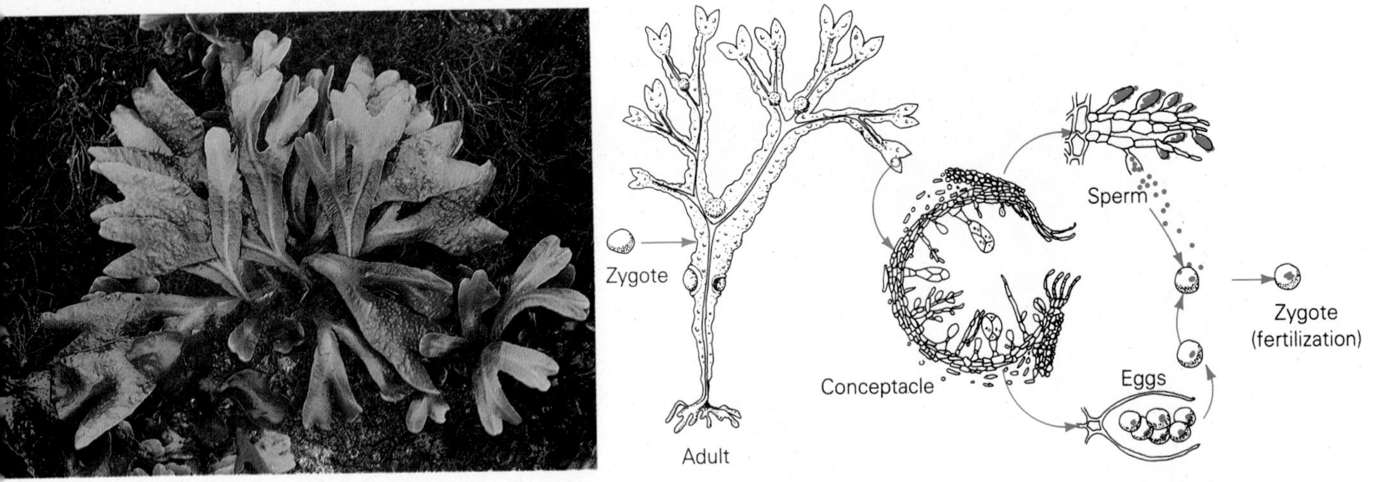

FIGURE 23.28
LIFE CYCLE OF *FUCUS*
Fucus, a brown alga, is surprisingly animal-like in its life cycle. Within the diploid adults, specialized cells undergo meiosis, forming large non-motile eggs and small, motile sperm. Fertilization ushers in the diploid adult stage once again.

species as well. A few terrestrial/airborne species appear on the surface of melting snow, on the moist sides of trees, or free in the soil. The **chlorophytes** include both unicellular and multicellular species. Many single-celled green algae have become photosynthetic symbionts in lichens, ciliates, and invertebrates. Biochemically, the green algae closely resemble plants. They contain chlorophyll *a* and *b* and carotenes. A chief storage carbohydrate is starch, and their cell walls contain cellulose, hemicellulose, and pectin.

The green algae occur as single-celled flagellated or unflagellated forms, as chains or filaments, as inflated "fingers" (*Codium,* also called "dead man's fingers"), and as delicate flattened blades (*Ulva,* the delicate *sea lettuce* of tide pools). The group is so diverse that it is difficult to find a representative species, so we'll arbitrarily choose a few examples.

Single-Celled Green Algae The most primitive of the green algae are the *single-celled* and *colony-forming* types. As far as anyone can figure, this group, although related to the multicellular green algae, is not descended from them and thus simply has retained the cellular level of organization.

Chlamydomonas (Figure 23.29) is a favorite organism of many biologists. It is easily grown, and its genetics and physiology have been studied in detail. The cells have two flagella, a single light-sensitive **red eyespot,** and one cup-shaped chloroplast. Like *Euglena, Chlamydomonas* is positively phototactic (it moves toward light), and if kept in a transparent container of water, individuals will congregate on the side closest to the light. *Chlamydomonas* is usually seen in a haploid chromosome state, wherein it reproduces asexually by mitosis and cell division and builds huge clone populations. It will not enter into a sexual cycle unless conditions are right.

The right conditions include the presence of opposite mating types (the plus and minus strains) as well as some form of environmental stress, such as absence of nitrogen compounds. When both mating types are present, the haploid *Chlamydomonas* produces isogametes mitotically. (Recall that isogametes are identical in appearance.) The two isogametes fuse by butting, head on, and entering a wildly spinning "nuptial dance" as their nuclei slowly fuse. Following the union, the zygote, now representing the diploid state, forms a tough, resistant **zygospore.** The zygospore may remain dormant for a considerable time or, if conditions are right, it may immediately enter into meiosis, producing four haploid **meiospores**. This restores the haploid state, and the individuals mature into the familiar haploid population. As we see, then, *Chlamydomonas* follows the more primitive zygotic life cycle, in which the haploid state dominates.

Colonial Green Algae Another remarkable green alga is *Volvox* (Figure 23.30). Although it is not related to any truly multicellular form, its history tantalizingly suggests what the earliest beginnings of multicellularity *might* have been like. *Volvox* is composed of a spherical colony of virtually identical cells, each one, in fact, very similar to an

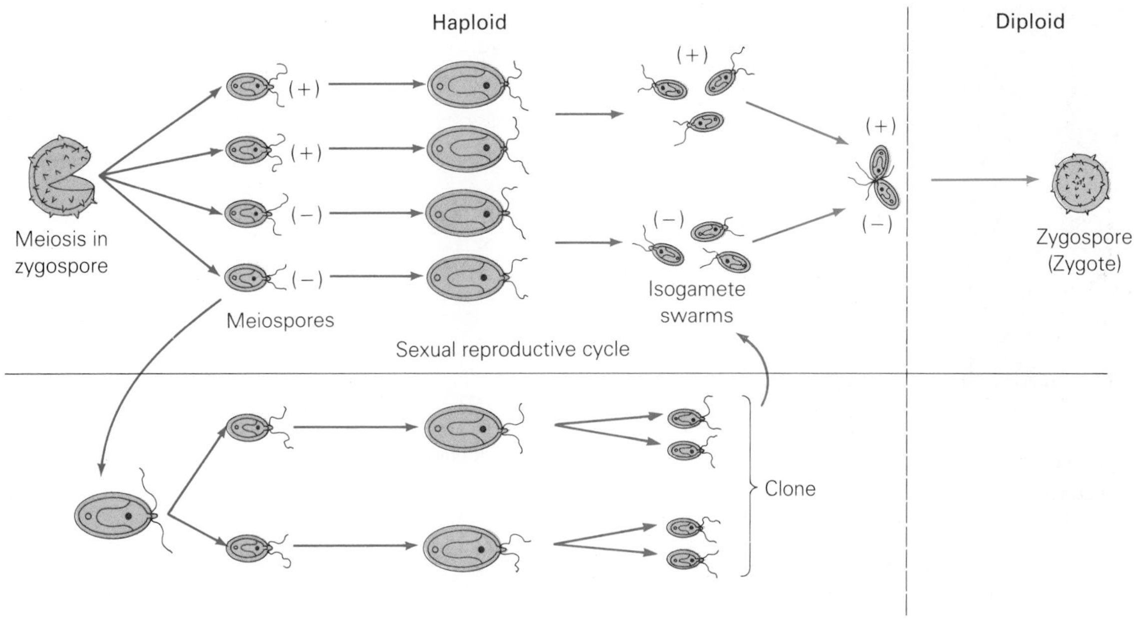

FIGURE 23.29
ZYGOTIC LIFE CYCLE OF *CHLAMYDOMONAS*
The diploid zygospore is a brief interlude in an otherwise lengthy haploid history. Zygospores enter meiosis and produce flagellated meiospores. These can enter either a sexual phase, producing many isogametes, or an asexual phase, producing enormous cloned populations through repeated mitotic divisions. Sexual reproduction occurs when different (+ and −) mating strains meet. Flagellated isogametes fuse head-to-head, producing diploid zygospores once again.

individual *Chlamydomonas*. So, in a sense, *Volvox* is a group of individuals in a sphere behaving as an organism. Or is the *sphere* really the individual? The seeming simplicity of this sort of question is deceptive. Biologists have been trying to answer it for years.

Volvox is interesting not only because it suggests a way multicellularity could have arisen, but also because it shows a kind of rudimentary differentiation within the colony (Figure 23.30). The sphere swims in an organized manner, with the flagellated cells in front pulling and those in the rear pushing. In the sexual phase some cells become specialized for reproduction, producing a few large eggs or many smaller sperm. In asexual reproduction, pockets of the sphere depress inward, and new spheres of cells pinch off and come to lie inside the parental sphere. The new, young spheres are inverted, however, with their flagella directed inward. The daughter *Volvox* colony has to turn itself inside-out within the parent structure. Eventually it is released through a hole in the parent colony.

Siphonous Green Algae Green algae of another group, the **siphonous algae**, have taken off in a different evolutionary direction. Old botany texts once called it "an evolutionary dead end," but this is probably unfair because, after all, they're still around—and doing quite well in fact. Algae of this group aren't exactly multicellular, but they're not exactly unicellular either. Although the nuclei undergo mitosis, the cytoplasm doesn't divide, and cell walls are not laid down between the newly separated nuclei. The result is a multinucleate mass of cytoplasm (or **coenocyte**) within an enveloping plasma membrane and cell wall. The group includes *Acetabularia,* a weird little alga that looks like a toadstool or parasol. Its single cell stands about 5 to 9 cm tall. Some *Acetabularia* are multinucleate coenocytes, but in other species, the multiple nuclei are not dispersed through the cytoplasm; instead, they clump into a single compound nucleus in the foot, so that the structure of the whole organism is very like that of a single cell, but an extraordinarily large one (Figure 23.31).

Multicellular Green Algae *Ulva* and *Ulothrix* are among the more structurally complex green algae. *Ulva,* familiar to tidepool enthusiasts as "sea lettuce," is broad and leaflike, although only two cells in thickness. Its leafy form is the product of cell division in two dimensions. On the other hand, cell division in *Ulothrix* and other filamentous algae occurs in one dimension only. The life cycle of the two species is also quite different. *Ulva* exhibits a clear-cut alternation of generations, although, like the red alga, *Polysi-*

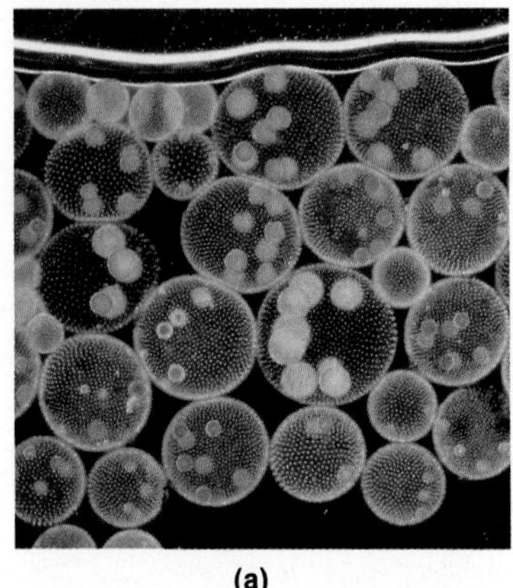

(a)

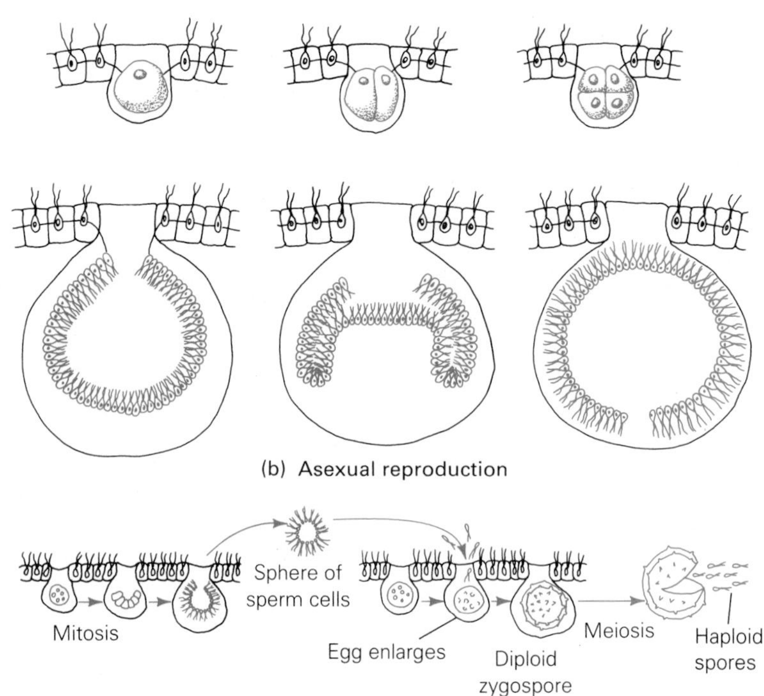

FIGURE 23.30
VOLVOX, A COLONIAL ALGA
(a) *Volvox,* an alga, is a spherical colony of tiny, interconnected, flagellated cells. The dense spherical bodies seen within the colony are concentrations of cells that were produced asexually. **(b)** During asexual reproduction, individuals of the sphere increase in size, divide, grow, and divide again, but eventually they are all reduced down to the original size. The result at first is a miniature of the original colony, except that the flagella are all pointed inward. Next the new colony undergoes inversion—it actually turns inside out—and a more typical flagellated colony emerges, soon to escape and live on its own. **(c)** In sexual reproduction, flagellated cells of the sphere differentiate into sperm- and egg-producing structures. With fertilization, a zygospore is produced. *Volvox* colonies are zygotic, so the zygospores enter meiosis, and the long haploid state is restored.

phonia, it is isomorphic. That is, except for chromosome numbers, the gametophytes and sporophytes are similar. The flagellated spores and gametes produced respectively, by sporophyte and gametophyte individuals, are also quite similar.

There is no alternation of generations in *Ulothrix*. This species follows a zygotic life cycle, so like *Chlamydomonas,* it is essentially haploid. Isogametes formed through mitosis in specialized cells fuse, giving rise to zygospores. Meiosis occurs immediately, restoring the haploid state (Figure 23.32).

The themes of alternation of generations and of the tendency of one phase to become dominant continue in the evolution of the land plants. Although there is no fossil evidence strongly supporting the thesis that the ancestors of the land plants were closely related to present-day multicellular green algae, most authorities agree that this is the case. We will soon see that some of the filamentous green algae probably share ancestors with the plants.

(a)

(b)

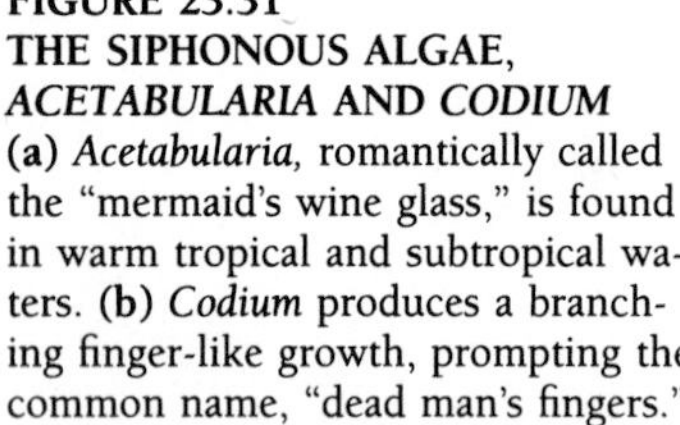

FIGURE 23.31
THE SIPHONOUS ALGAE, *ACETABULARIA* AND *CODIUM*
(a) *Acetabularia,* romantically called the "mermaid's wine glass," is found in warm tropical and subtropical waters. **(b)** *Codium* produces a branching finger-like growth, prompting the common name, "dead man's fingers." It is often found in tidepools, growing on rocks and on the shells of mollusks.

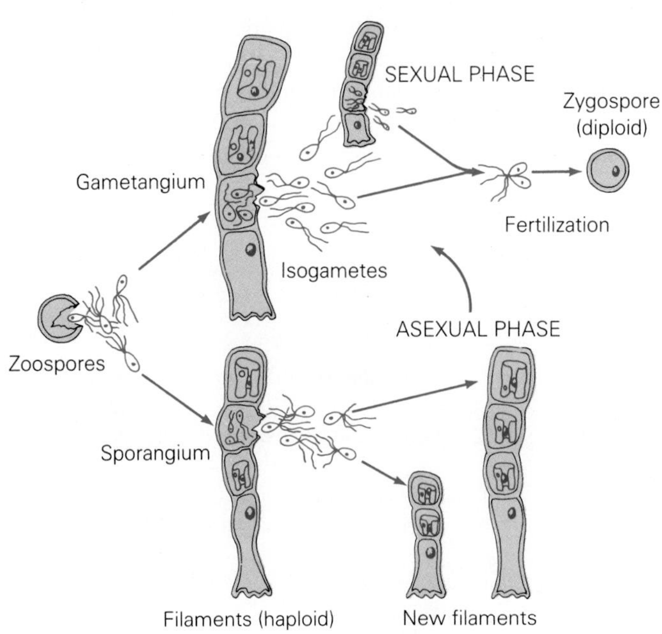

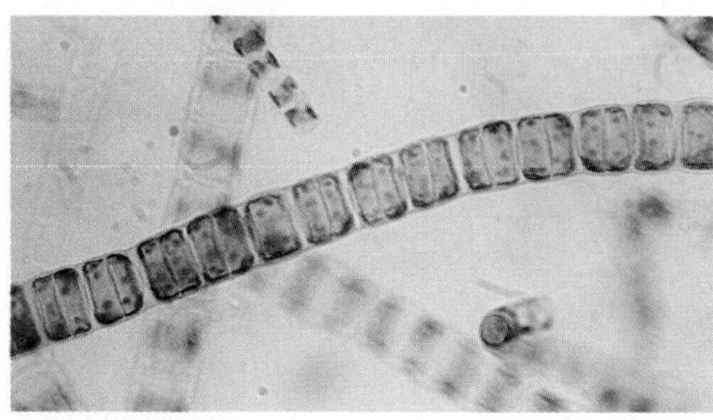

FIGURE 23.32
***ULOTHRIX*, A FILAMENTOUS ALGA**
Like *Chlamydomonas, Ulothrix* has a zygotic life cycle. The diploid zygospore enters meiosis producing numerous zoospores. In a lengthy asexual phase, the zoospores produce filaments that give rise to motile, tetraflagellate spores, which then grow into more filaments. In the sexual phase, some filaments develop gametangia, which produce biflagellate isogametes. Different mating types fuse to produce diploid zygospores.

TABLE 23.2 CHARACTERISTICS OF THE AUTOTROPHIC PROTISTS

PHYLUM/DIVISION*	USUAL CHARACTERISTICS
Pyrrophyta	Dinoflagellates. Single-celled; flagella in many; Chl *a* and *c*; peridinin (a carotenoid); cellulose walls; storage carbohydrate: starch; asexual reproduction: binary fission; sexual reproduction: cell fusion; red tide organisms
Euglenophyta	Euglenoids. Single-celled; flagella; Chl *a* and *b*; carotenoids; no cell wall; storage carbohydrate: paramylon; asexual reproduction: fission; sexual reproduction: unknown; *Euglena gracilis*
Chrysophyta	Yellow-green and golden-brown algae and diatoms. Mostly single-celled; some with flagella; Chl *a* and *c*; fucoxanthin (a carotenoid); cellulose or silicon dioxide walls; storage carbohydrate: chrysolaminarin; asexual reproduction: binary fission, with diminishing size in diatoms; sexual reproduction: in diatoms, fusion of gametes; important to marine food chains; form diatomaceous earth
Rhodophyta	Red algae. Multicellular: nonmotile; Chl *a* and *d*; phycocyanin and phycoerythrin; cellulose walls; storage carbohydrates: floridean starch, agar, carrageenan; asexual reproduction: spores; sexual reproduction: sporic cycle, alternation of generations, both isomorphic and heteromorphic species; seaweeds
Phaeophyta	Brown algae. Multicellular, some very large; motile gametes; Chl *a*; fucoxanthin; cellulose walls with algin as structural polysaccharide; storage carbohydrates: laminarin and mannitol; asexual reproduction: some with flagellated spores; sexual reproduction: sporic cycle (a few with gametic cycles), heteromorphic alternation of generations; seaweeds, kelps
Chlorophyta	Green algae. Single-celled, colonial, or multicellular; some with flagella; Chl *a* and *b*; carotenoids; cellulose walls with hemicellulose and pectin; storage carbohydrate: starch; asexual reproduction: flagellated spores; sexual reproduction: zygotic cycle in some, sporic cycle in others with isomorphic alternation of generations; pond scums, sea lettuce

*The group designation "division" is traditionally used by botanists where others use the term "phylum."

The green algae were probably the dominant form of aquatic life in the Cambrian period and throughout the Ordovician and Silurian, too (405 to 570 million years ago). Most authorities believe that the first terrestrial plants descended from the filamentous green algae, possibly during the Lower Devonian period (some 400 million years ago). The brown and red algae remained in the sea, specializing in other directions.

algae Protists that are plantlike; single-celled and multicellular, with photosynthetic pigments, firm cell walls in most, and complex storage polysaccharides.

ameba (also amoeba) Those protozoans of phylum *Sarcomastigophora* that move or feed through ameboid movement.

alternation of generations In the sporic life cycle, the alternation between the diploid sporophyte and haploid gametophyte phases.

ciliate Protozoans of phylum *Ciliophora;* those that move and feed through the use of cilia.

flagellate Those protozoans of phylum *Sarcomastigophora* that move or feed through the use of one or more flagella.

gametic cycle Life cycle with a dominating diploid state, a haploid state limited to gametes with fertilization restoring the diploid state (some protists, all animals).

gametophyte In the sporic life cycle,the haploid phase during which gametes are formed through mitosis.

protozoa Animallike protists that are chiefly single-celled. Locomotion occurs through amoeboid movement or through the use of flagella or cilia. All are heterotrophic, living as predators, parasites, and scavengers.

sporic cycle Life cycle with generations alternating between a multicellular diploid sporophyte and a multicellular haploid gametophyte. Meiosis in the sporophyte yields spores that form the gametophyte, which in turn, produces gametes whose union restores the sporophyte. Either generation may dominate or they may be equal, and they may occur in separate individuals or within one (some algae, all plants).

sporophyte In the sporic life cycle, the diploid phase during which spores are formed through meiosis.

sporozoan Protozoans of phylum *Apicomplexa;* nonmotile as adults and usually parasitic.

zygotic cycle A life cycle with a dominating haploid state, in which gametes form through mitosis and in which the brief, limited diploid state is followed by meiosis and a return to the haploid condition (some protists, all fungi).

APPLICATION OF IDEAS

1. The use of DDT has been greatly curtailed for important ecological reasons, yet it has proven in the past to be the greatest malaria deterrent known in parts of the world where other measures have been ineffective. Many organizations would like to see this form of mosquito control resumed. Discuss the issue of human health versus preservation of the environment. (Read *The Silent Spring,* by Rachel Carson, for some insight into the DDT controversy.)
2. Euglenoids, and in particular *Euglena gracilis,* have been classified as plants by botanists, animals by zoologists, and protists by others. Review the characteristics of *Euglena* and discuss the basis for the taxonomic disagreement. In what ways does the protist designation help? What does such a problem reveal about the categories contrived by taxonomists?
3. One of the problems of controlling African sleeping sickness has been the movement of nomadic tribes and their cattle in and out of endemic areas. Your job as an official attempting to control the spread of the disease is to explain to the tribes why they can no longer move about freely and why some of their cattle have to be disposed of for health reasons. How would you explain the life cycle of the trypanosome and the spread of sleeping sickness to these primitive people?
4. Some protists, particularly very large ciliates, seem to represent exceptions to the cell theory, and the term *unicellular* or *single-celled* seems inappropriate. Using one of the larger ciliates as an example, make a case for using the term *acellular*. What might substitute for the term "organelles" in an acellular organism?

KEY IDEAS

ORIGIN OF THE PROTISTS

The first eukaryotes arose some 2.5 billion years ago. The oldest eukaryotic fossils are of red and green algae, but they were probably not the first eukaryotes.

The Serial Endosymbiosis Hypothesis

1. The **serial endosymbiosis hypothesis** proposes that eukaryote arose from a number of prokaryotic cell lines that became **endosymbionts** through invasions and incorporations. Such events explain the presence of such organelles as mitochondria, chloroplasts, basal bodies, flagella, cilia, and centrioles.
2. According to the endosymbiosis hypotheses,
   a. *Line A* **protoeukaryotes** were phagocytic predators capable of anaerobic respiration only
   b. *Line B,* an aerobic bacterium, was taken into line A, and was the forerunner of the mitochondrion.
   c. *Line C* cells, when incorporated, brought in the 9 + 2 microtubular flagellum containing the protein *tubulin,* giving rise to centrioles, basal bodies, and cilia and flagella.
   d. The final incorporation involved photosynthetic cells of *line D,* possibly a cyanobacterium. This incorporation established the chloroplast.
3. Evidence supporting the serial endosymbiosis hypothesis comes from several sources. The mitochondrion and chloroplast have critical similarities to bacteria. Included are ribosomal RNAs, cytochrome *c,* and amino acid sequences. The link to a flagellated bacterium is weak, although flagellate symbionts are known to exist in certain protists.

CHARACTERISTICS OF THE PROTISTS

Protista is a grab bag kingdom; its members include single-celled, colonial, and multicellular types. Feeding includes all common modes: predation, grazing, decomposition, parasitism, and photosynthesis. Movement is through cilia, flagella, and **pseudopodia.** Asexual reproduction is mainly through mitosis followed by binary fission. **Fragmentation** involves repeated mitoses before cell division finally occurs. Sexual reproduction can occur via conjugation, **autogamy** (meiosis and recombination within the individual), and gametogenesis with either **isogametes** (identical in form) or **heterogametes** (egg and sperm).

Life Cycles in Protists and Other Eukaryotes

1. The zygotic cycle of many protists and all fungi includes a lengthy haploid phase, followed by sexual union, with or without gametes, and meiosis with the resumption of the haploid state.
2. The gametic cycle of animals and some protists includes a lengthy diploid phase, followed by meiosis, gametogenesis (usually), sexual union, and resumption of the diploid state.
3. The sporic cycle of plants and many algae includes a diploid, or **sporophyte,** phase, ending with meiosis, which leads to the formation of haploid spores (not gametes). This is followed by a multicellular haploid or **gametophyte** phase, gametogenesis (usually), sexual union, and restoration of the diploid phase. The haploid and diploid phases vary greatly in length among species.

Phylogeny of Today's Protists

The polyphyletic Kingdom Protista includes **protozoans** (protists resembling animals), **algae** (protists resembling plants), and **slime molds** and **water molds** (protists resembling fungi). There are many protists that do not fit neatly into these groupings.

PROTOZOANS

Sarcomastigophora: Flagellates and Amebas

1. Most flagellates have the 9 + 2 flagellum. They may capture food or simply absorb it through the body wall.
2. **Trypanosoma gambiense,** the agent of **African sleeping sickness,** is one of several important flagellate parasites.
3. *Entamoeba histolytica,* an ameboid protist, is the waterborne agent of amebic dysentery.
4. Most ameboid protists move about by the formation of pseudopods (false feet), temporary extensions of the cell into which the body flows. They feed mostly through phagocytosis.
5. Heliozoans have retractible microtubular **axopodia,** spinelike extensions covered by a moving ameboid cytoplasm used in feeding.
6. **Radiolarians** and **foraminiferans** have hardened skeletons, of silicon dioxide and calcium carbonate, respectively. Both contribute to ocean floor deposits, including the radiolarian ooze.
7. Asexual reproduction in sarcodines is through mitosis and cell division. Sexual reproduction is not often seen but usually involves meiosis and cell fusion. Foraminiferans produce flagellated haploid cells that fuse. In heliozoans, three-fourths of the gametes are discarded as polar bodies during meiosis, and fertilization is through simple fusion.

Apicomplexa: The Sporozoans

1. **Sporozoans** are mainly spore-forming, nonmotile parasites.
2. Medically, the most important sporozoan is *Plasmodium vivax,* an agent of **malaria.** It is carried by the female *Anopheles* mosquito. The most effective control agent is the highly controversial insecticide, DDT.
3. The life cycle of *P. vivax* includes sexual stages in the mosquito and asexual stages in the host. The mosquito injects **sporozoites** into the host, and these enter the host's liver, where many **merozoites** form. When freed, they invade the red blood cells and release toxins that bring on the alternating fever and chills of malaria. Some enter a sexual stage, forming **gamonts,** which when taken back into the mosquito form sperm

and eggs. The zygote divides asexually to form many sporozoites, which are then injected into the next host.

Ciliophora: Ciliates

1. **Ciliates** have an enormous size range. Cilia are commonly used in movement and in feeding. The ciliate covering, or **pellicle,** is made up of a tough, secreted material. Some ciliates release threadlike **trichocysts** for capturing prey and for mooring. Water balance is maintained by **contractile vacuoles** that collect and force water out of the cell. Food taken in through the **oral groove** and the **cytostome** is commonly digested within **food vacuoles,** and wastes are expelled through the **cytopyge.**
2. Germ-line DNA (for replication and mitosis) in *Paramecium* is maintained in one or more **micronuclei,** which somatic-line DNA (for transcription) is maintained in the **macronucleus**.
3. Sexual reproduction **(conjugation)** involves meiosis, cell fusion, and the exchange of haploid micronuclei. There are no gametes or sexes, but different mating types occur. Asexual reproduction occurs through mitosis and cell division.

MYXOMYCOTA AND ACRASIOMYCOTA

1. The **acellular slime mold,** *Physarum polycephalum,* has a giant multinucleate, creeping and feeding ameboid stage that ends with a spore-forming stage. Haploid spores are formed through meiosis in **sporangia,** which develop atop erect **sporangiophores.** Such spores germinate into an ameboid **myxameba** or flagellated **swarm cell.** Either can form a dormant cyst, and either can fuse with others, becoming diploid and forming another feeding stage.
2. **Cellular slime molds** consist of small separate amebas that swarm together into a spore-forming slug stage, typically when the food supply runs short. Sexual reproduction may occur via fusion of haploid amebas followed by **macrocyst** formation and subsequent meiosis, forming new amebas.

OOMYCOTA: WATER MOLDS

1. A parasitic **oomycete,** *Phytophthora infestans,* also called **late blight,** created the Irish potato famine by infecting the potato plant.
2. Oomycetes receive their name partly from the large egg cell produced in an **oogonium.** The **water molds** are funguslike in appearance, but they produce flagellated spores and have cellulose cell walls.
3. Many aquatic water molds are parasitic. Terrestrial species include the parasitic crop-destroying **downy mildews** and **blights**. *Phytophthora* invades the leaf through the stomata and returns through these pores to produce spores.
4. In asexual reproduction, aquatic water molds produce flagellated spores. In sexual reproduction, spherical oogonia produce egg cells. Fingerlike **antheridia** invade the oogonia, permitting sperm nuclei and egg nuclei to meet.

THE ALGAE

Algae—photosynthetic protists—range from single-celled to complex multicellular forms. Most are aquatic, living in all of the earth's waters, and in a few instances on land. Included is the vast, floating, microscopic marine **phytoplankton** and the marine **seaweeds** and **kelps.**

Pyrrophyta: The Dinoflagellates

1. Dinoflagellates produce the famous **red tides,** which bring about the death of fish and other marine life.
2. Dinoflagellates are mainly flagellated, single-celled photosynthesizers. Some are free-living, and others live in the cells of other organisms as symbionts. They have a primitive form of cell division in which the single nuclear membrane does not break down. Reproduction is chiefly asexual, through binary fission, but sexual unions may occur.

Euglenophyta: Euglenoids

Euglena, a single-celled representative euglenoid, is a flagellated, very flexible swimmer that orients to light through a light-sensitive **photoreceptor.** Its chloroplast contains chlorophyll *a* and *b* and carotenoids. Its starches are stored as **paramylon.** It reproduces asexually by mitosis and transverse cell division—no sexual reproduction has been observed.

Chrysophyta: Yellow-Green Algae, Golden-Brown Algae, and Diatoms

1. **Chrysophytes** inhabit fresh and salt water, and owe their color to concentrated light absorbing carotenoids (mainly **fucoxanthin**). Carbohydrate storage is in the form of **chrysolaminarin.**
2. Some chrysophytes have glassy coverings, and some are flagellated, while others produce colonies, a few are multicellular, and some are even terrestrial.
3. Because of their glassy, boxlike coverings, **diatoms** must solve the problem of getting progressively smaller as they undergo mitosis and cell division. Their solution is sexual reproduction. Many produce ameboid gametes that abandon the covering and fuse, forming **auxospores.** Following this, a full-size covering is produced, and the cycle repeats. The boxes fall to the seabed to form **diatomaceous earth.**

The Multicellular Algae

1. Most **red algae (Rhodophyta)** are marine seaweeds that grow on rocky coasts, using **holdfasts** to fasten themselves in place. They produce **floridean starch** and other polysaccharides, including **agar-agar** and **carageenan.**
2. In alternating generations, the sporophytes and gametophytes may be **isomorphic** (identical) or **heteromorphic** (different). The alternation of generations of *Polysiphonia* is isomorphic, but in others it is heteromorphic. Meiosis in the sporophyte results in the production of haploid spores and the start of the gametophyte phase. The gametophyte produces nonmotile haploid gametes that fuse in

fertilization, restoring the sporophyte phase. There are no motile cells in the red algae.

3. Fossils of red algae date back to the Cambrian period.
4. **Brown algae (Phaeophyta)** contain the pigment fucoxanthin and store their carbohydrates as **laminarin** and **mannitol.** They have flagellated sperm and, sometimes, flagellated eggs.
5. The Sargasso Sea contains great masses of the floating brown alga *Sargassum.*
6. *Fucus* is a reproductively advanced alga. There is no alternation of generations and the life cycle is gametic, like our own.
7. Brown algae include the giant kelps, *Nereocystis* and *Macrocystis.* Their specialized bodies include anchoring holdfasts, stemlike **stipes,** and large, leaflike **blades.**
8. **Green algae (Chlorophyta)** occur as single cells, coenocytic strands, filaments of cells, and flattened multicellular blades. They contain chlorophyll *a* and *b,* and produce cell walls of cellulose and hemicellulose in a pectin matrix—all plant characteristics.
9. *Chlamydomonas,* a representative single-celled green alga, is motile with two flagella and has a **red eyespot.** It exists primarily in the haploid state, where it reproduces through mitosis. Its life cycle is zygotic. When opposite mating types meet, **isogametes** form, fertilization occurs, and a brief diploid **zygospore** forms. Zygospores undergo meiosis, producing **meiospores.**
10. *Volvox* forms a spherical colony of small flagellated cells. Some specialize in movement, some as eggs and sperms, while others form small asexual spheres.
11. **Siphonous green algae** are **coenocytic** (multinucleate). They undergo repeated mitosis and become quite large, but little cell division occurs.
12. **Multicellular green algae** are the most complex green algal forms. *Ulva,* a multicellular marine form, has a distinct isomorphic alternation of generations. *Ulothrix* follows a zygotic life cycle.
13. Filamentous green algae are important, since they are believed to represent the ancestral group from which plants arose.

REVIEW QUESTIONS

1. Discuss how the protists form an underlying or ancestral group for the other kingdoms. (486)
2. Briefly explain what problem the serial endosymbiosis hypothesis addresses, and list the eukaryotic organelles concerned. (486-487)
3. Beginning with the protoeukaryotic cells in line A, list the hypothetical events leading from the prokaryotic to the eukaryotic cell. (486-487)
4. For which eukaryotic organelles is the serial endosymbiosis hypothesis most satisfactorily supported? Briefly summarize the supporting evidence. (486-487)
5. List five modes of nutrition utilized by the protists. Is there any you know of that they do not utilize? (Consider Chapter 22.) (487)
6. Use a simple circular diagram to characterize the zygotic life cycle. List two groups of organisms in which this occurs. (489)
7. Humans follow the gametic life cycle. Briefly state what this means in terms of human life. (490)
8. Briefly discuss the one overriding characteristic of sporic life cycles. To which groups of organisms does this cycle pertain? (490)
9. Review the characteristics of Mastigophora. (491)
10. Briefly review the life history of the agent of African sleeping sickness. Why has it been so hard to control its spread? (491)
11. Describe the best-known sarcodine parasite, its effect on the human host, and the symptoms it produces. (491)
12. Explain the formation and functioning of a pseudopod. What are its two functions? (491-492)
13. Explain what axopodia are, where they are found, and how they are used. (492)
14. In what ways are the skeletal parts of radiolarians different from those of foraminiferans? What evidence suggests that they were incredibly numerous in former times? (492)
15. Describe an example of sexual reproduction in the sarcodines. (492)
16. List three characteristics of the sporozoans. (493)
17. Briefly summarize the life cycle of *Plasmodium vivax* and, where appropriate, relate events to the symptoms of malaria. (493-494)
18. Describe the organization of the cilia in *Paramecium* and in *Euplotes.* What are their two principal uses? (494)
19. Describe the trichocysts and name their functions. (495)
20. Summarize the steps involved in feeding in *Paramecium.* (495)
21. Discuss the peculiar organization of DNA in *Paramecium.* (496)
22. Summarize the events involved in sexual reproduction in *Paramecium.* (496)
23. In what way is *Physarum* a typical protist? How does it get from a spore state to a multinucleate state again? (496-497)
24. Describe the manner in which late blight infects potato plants. (498)

25. Explain how sexual reproduction takes place in the aquatic water molds. Why is this considered to be an advanced condition? (500)
26. Using a simplified diagram, explain how alternation of generations between sporophyte and gametophyte might appear in an isomorphic species. What event always starts a gametophyte phase? A sporophyte phase? (490)
27. Explain in detail how mitosis and meiosis differ between a dinoflagellate and *Paramecium*. (500-501)
28. List several animallike and several plantlike characteristics of *Euglena*. (501-502)
29. Describe the problem diatoms have with asexual reproduction, and explain how it is solved. (502)
30. Summarize the events in the alternation of gametophytes and sporophytes in *Polysiphonia*. (504)
31. List three polysaccharides of the red algae and name some uses for these products. (504)
32. List the characteristic pigments and polysaccharides of the brown algae. (505)
33. Briefly characterize the life history of the brown alga *Fucus*. Why is its sexual reproduction considered advanced? (507)
34. List several specialized structures of *Macrocystis* and describe their functions. (507)
35. List several important characteristics of the chlorophytes. (508)
36. Summarize the life history of *Chlamydomonas*. Is this protist primitive or advanced? Why? (508)
37. Using *Volvox* as an example, explain the colonial form of life. How does this differ from multicellularity? (508-509)
38. Describe the organization of the siphonous algae. Why aren't they seriously considered as ancestral to plants? (509)
39. Describe the life history of *Ulva*, the sea lettuce, characterizing its alternating phases. (509-510)
40. Of the several groups of chlorophytes, which seems to be most closely related to the plants? Why? (546)

24

THE FUNGI

YOU MAY HAVE BEEN WALKING IN A DAMP WOODLAND ONE DAY AND suddenly have come upon a large mushroom on the forest floor. It can be an intriguing experience, especially if you don't know much about mushrooms; their size, shape, and even coloration may seem rather outlandish (Figure 24.1). They have a distinct sense of mystery about them, possibly because they are so often associated with dark, wet forests—pensive places. If the area is remote, the day is overcast, and you are alone, it is not hard to secretly believe in the "little people," if only for a moment.

If you have experienced a moment like this, you may not be pleased to hear that the mushroom is a fungus. A fungus! The word *fungus* is simply *not* associated with beauty and mystery. A fungus, to many people, is bad; it grows on our food, lurks on shower floors, and at times invades our very bodies, where it can cause stubborn, often embarrassing, and sometimes fatal infections. You won't be any happier to learn that, in addition to feeding off the living as parasites, many of these creatures make a living as **saprobes**—feeding off the dead. Nonetheless, mushrooms, along with baker's yeast and a number of other friendly organisms, are fungi.

WHAT ARE THE FUNGI?

All fungi are heterotrophs. Most are multicellular with relatively simple bodies, although they often have highly elaborate reproductive structures (for example, the familiar mushroom.) The fungal body, or **mycelium** as it is known, consists of extensive, spreading, threadlike filaments called **hyphae** (Figure 24.2). The hyphae may be numerous slender cells joined end to end, or a single hypha may be one long, tube-like growth lacking complete cross walls, surrounding a continuous cytoplasm. In many hyphae the cytoplasm is multinucleate, or coenocytic, wherein mitosis occurs without cytoplasmic division. You may recall this organization in the siphonous green algae *Codium* and *Acetabularia* (see Chapter 23).

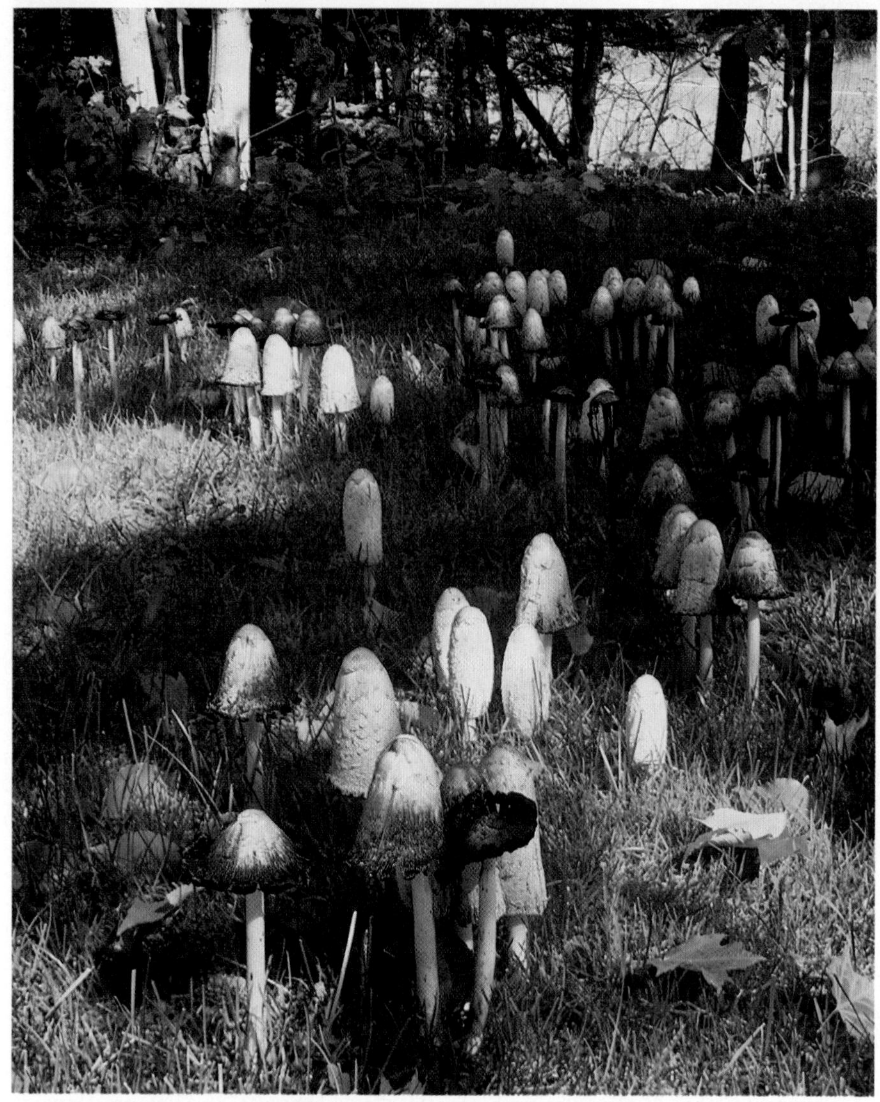

FIGURE 24.1
THE FUNGAL WORLD
The cluster of mushrooms probably originated when spores from some distant mushroom were carried by air currents to this ideal location, where dead vegetation, moisture, and warmth made growth possible. Beneath the surface is a vast fungal growth that spread from a central point.

Mycelial growth rates can be phenomenal; a single mushroom can produce up to a kilometer of new hyphae in only one day. Such extensive growth must be supported by an efficient nutrient supply system. In fungi, cytoplasmic streaming distributes food and synthesized materials throughout the mycelium. Fungal growth is so extensive in fertile soil that it has an important effect on soil texture itself, greatly increasing its water-holding characteristics.

Feeding in the Fungi

Fungi require an organic food source much the same as other heterotrophs, although some are able to use simple organic substances while others require food sources as complex as our own. Like bacteria, fungi feed by **absorption.** The hyphae secrete digestive enzymes into their surroundings, which happen to be their foods. The products of digestion diffuse into the cell or are transported there through active (energy-requiring) processes. Parasitic fungi produce special hyphae called **haustoria,** which penetrate cells, making it possible for the fungus to absorb nutrients directly from the host's cytoplasm.

Fungi share with bacteria the role of decomposers, breaking down the corpses of dead organisms. Thus, like bacterial decomposers, fungi are critical to the

FIGURE 24.2
FUNGAL GROWTH
Most fungi grow by producing filamentous hyphae, fine tubes of cytoplasm in a chitinous wall, that penetrate the substrate to obtain food. The entire mass of hyphae is known as the mycelium. When conditions favor such growth, the sporangiophores rise up and produce sporangia at their tips, eventually releasing numerous haploid spores into the environment.

recycling of carbon and important mineral nutrients (for example, nitrogen and phosphorous) that would otherwise be tied up in corpses. People who study such things estimate that there is about one ton of bacterial and fungal decomposers present in the upper eight inches of an acre of fertile soil.

Of course, decomposition has another side as well. Fungal heterotrophs do not distinguish between the fallen trees or dead squirrel and what humans consider valuable. They attack anything that is organic, including our food, clothing, stored goods, books, and even our houses.

We have learned to put some heterotrophic fungi to work commercially. Various species are used in the manufacture of cheeses, antibiotics, linen, bread, wine, and beer. A recent addition to the list of pharmaceutical products, **cyclosporine,** is the product of a soil fungus. Cyclosporine is widely used today in organ transplantation, since it can relatively safely suppress the body's natural immune response, which otherwise leads to transplant rejection. Earlier immunosuppressant drugs caused a variety of unacceptable side effects.

The list of parasitic activities among the fungi is long and infamous and ranges from rose mildew to athlete's foot. The fungal parasites of plants can be particularly devastating, so much so that our wellbeing worldwide virtually depends upon our ability to control the parasites of food crops, and especially those of wheat, rye, corn, rice, and other grains.

Other plants, however, depend on a mutualistic (mutually beneficial) fungal association in their roots, forming what are called **mycorrhizae** (singular, **mycorrhiza**; "fungus-roots"). Here, the widespread fungal hyphae absorb minerals, supplying them to the plant. In turn, the fungus takes up nutrients produced by the plant (also see Chapter 28).

Reproduction in Fungi

Asexual reproduction in fungi usually involves the formation of spores. Fungi are prolific spore–formers, with each individual capable of forming millions of the minute bodies. They are formed by mitosis and cell division throughout most of the life cycle, each

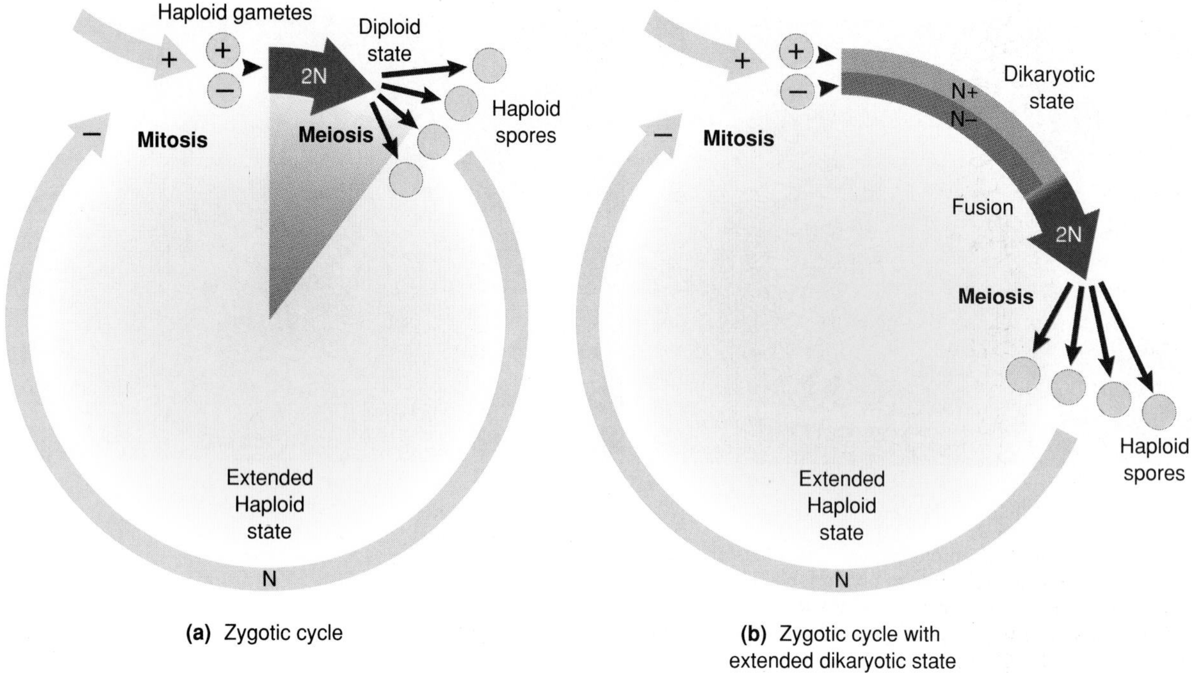

FIGURE 24.3
ZYGOTIC LIFE CYCLE
(a) Most of the fungal life cycle is spent in a haploid state, with only the briefest diploid pause following fertilization. Spores are produced mitotically throughout the extended haploid state, but they are also formed through meiosis in the diploid individual. **(b)** In many fungi, even after the hyphae of different strains fuse, the two haploid nuclei, although in the same cell, do not fuse but remain separate, resulting in a dikaryotic state. A considerable time may go by before fertilization is completed.

spore consisting of a haploid nucleus, a cytoplasm greatly reduced by dehydration, and a protective covering or spore case. Since the fungal mycelium is fixed in place, the spore provides a means for the organism to establish itself in new food sources, often at some distance away. Aiding in spore distribution are air currents and water. Further, spores can often survive long periods where growth conditions are not favorable (dryness, heat, intense sunlight, etc.), so they also represent a mechanism for survival.

Sexual reproduction also occurs. Fungi, you may recall, follow what is clearly a zygotic life cycle (Figure 24.3a). Recall (see Chapter 23) that zygotic organisms remain haploid through most of their life cycle, reaching a diploid state only when fertilization occurs. We also noted that the diploid state is notoriously brief, followed immediately by meiosis (which leads to spores) and the return to the haploid state. Such an extensive haploid state is, as we've seen, also common in the protists.

Sex in fungi occurs through conjugation, the fusion of individuals, which, in zygotic organisms, results in a new cell containing haploid nuclei from the two mating strains. Ordinarily we would conclude by mentioning fusion of the two haploid nuclei, but fungi are exceptional. They undergo a peculiar, delayed fertilization, at least peculiar when compared to fertilization in other kingdoms.

In many fungi, the fusion of the two haploid nuclei is delayed while an extensive new growth occurs and reproductive structures emerge. Cells bearing two such nuclei are said to be in a **dikaryotic** ("two-nucleate") state, what we might call the "N + N" condition. Throughout all of this, the two nuclei, chastely residing side by side, go through round after round of mitosis as the mycelium grows. When fertilization finally does occur, each cell, in typical fungal fashion, enters immediately into meiosis, and the haploid state resumes (Figure 24.3b).

Fungal Relationships

Fungi were once included in the plant kingdom, mainly because of their lack of motility and the presence of cell walls. But as researchers unveiled more and more differences between fungi and plants, the two were finally placed in different kingdoms. The differences are indeed numerous. For instance, whereas plant cell walls are essentially cellulose, fungal cell walls are composed of chitin. Chitin is a nitrogenous carbohydrate that also occurs in the exoskeletons of arthropods (such as insects and crustaceans). Further, the fungi carry on a primitive form of mitosis and meiosis seen elsewhere only in certain protists (Chapter 23). The nuclear envelope in many species is not dismantled during mitosis, but remains intact, constricting between the clusters of daughter chromosomes. In some fungi, the envelope is only partially broken down or is dismantled late in the process. Curiously, the mitotic spindles are formed within the nucleus. Further, although the usual microtubular spindles form, centrioles are lacking, just as they are in plants. This is not all that surprising, however, since centrioles are characteristic of organisms that have cilia and flagella in at least some stage, and these are absent in the fungi, at least among those organisms we are including in Kingdom Fungi. Finally, it was recently discovered that, compared to those of other eukaryotes, the fungal chromosomes have very little histone protein bound to them. Most of these traits indicate why fungi are no longer considered to be plants.

There are a number of theories regarding the origin of fungi, but most biologists now believe they sprang from the colorless, heterotrophic, flagellated protists. Of course, this further separates the fungi from the plant line, which probably evolved from a green algal ancestor. Whether the fungi are a monophyletic or a polyphyletic group is not completely resolved, but we will later review the arguments for both cases.

KINGDOM FUNGI

The 100,000 named species of fungi occupy four[1] divisions[2]: **Zygomycota** (conjugating molds, including bread molds), **Ascomycota** (the sac fungi, including brewer's and baker's yeasts), **Basidiomycota** (the club fungi, including mushrooms), and **Deuteromycota** (also called **Fungi Imperfecti**). We will refer to them by somewhat more familiar names: **zygomycetes, ascomycetes, basidiomyctes**, and **fungi imperfecti.**

Zygomycota: Conjugating Molds

The zygomycetes are sometimes called the "lower fungi" because their hyphae lack cross walls, or at least regularly spaced, complete cross walls, and thus the cytoplasm is multinucleate. The name "zygomycetes" comes from the fact that newly formed zygotes enter a spore stage called a **zygospore**. Zygospores are also common along algae such as *Chlamydomonas,* where we find the zygotic life cycle as well. Most members of this division are saprobic. One of the zygomycetes, *Pilobolus* (the "spitting mold"), has a bizarre way of disseminating its spores, as we see in Essay 24.1.

***Rhizopus*, A Bread Mold** The common bread mold, *Rhizopus stolonifer,* turns up in everyone's refrigerator at one time or another (Figure 24.4). But don't necessarily look for it on bread because *Rhizopus* grows vigorously on many foods—and besides, most of our bread is so full of preservatives that fungal growth is inhibited. Spores that land on any suitable medium will first absorb water and then germinate. Hyphae sprout in

[1]Some authors include water molds and slime molds in Kingdom Fungi but we have elected to follow those who place them in Kingdom Protista.

[2]The designation "division" is traditionally used by botanists, where others use the term "phylum."

(a)

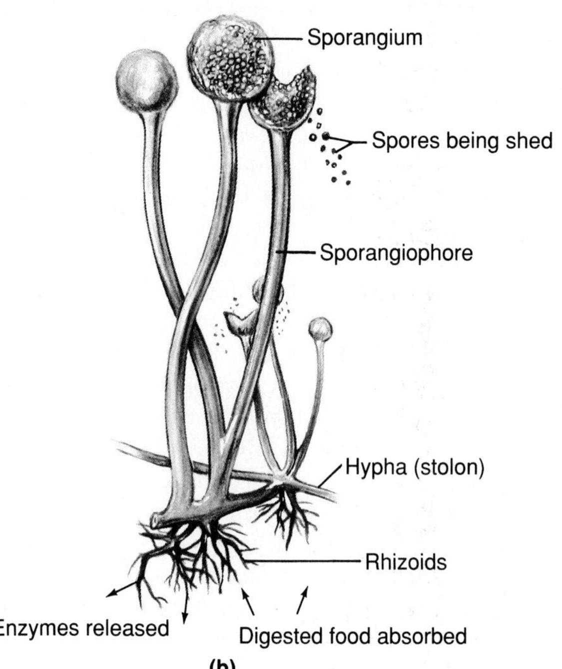

(b)

FIGURE 24.4
***RHIZOPUS*, A BLACK BREAD MOLD**
(a) The mycelium of *Rhizopus* grows in a tangled mat throughout the medium and is not easily visible, but its sporangia, which are black when mature, form the dots on the large mass seen in the photograph. **(b)** *Rhizopus* secretes enzymes from the hyphae into the food, where digestion occurs extracellularly. The simplified products of digestion are then absorbed into the cells.

all directions, branching time and again until the mycelial mass has fully penetrated the food. The favored conditions for the growth of *Rhizopus* and many other fungi—in addition to a supply of proper nutrients—are moisture and warmth, although they can do rather well in one's refrigerator. When the bread mold has grown for a time, it develops long horizontal hyphae known as **stolons**, which venture along the surface of the food, sending rootlike growths called **rhizoids** back down into the food. Much of this is invisible to the unaided eye, but another part of the mold's growth is quite prominent.

You may have noticed a tangled fuzzy growth topped by tiny black dots just before you began spreading the peanut butter on the last slice of bread in the house. The fuzzy growths are aerial hyphae, whereas the black dots are **sporangia.** The latter appear at the tips of vertical hyphae called **sporangiophores.** In addition to suggesting a late night trip to the market, their presence is a sure sign that *Rhizopus* has begun asexual reproduction. Numerous haploid spores arise through mitosis on each sporangium. Each spore contains a haploid nucleus, a dehydrated and metabolically inactive, or dormant, cytoplasm, and a surrounding protective coat. When released, the spores will be lofted by the air, and the rest is up to chance. If the proper conditions are encountered, a spore will germinate; that is, it will become metabolically active, taking in water, breaking through the spore case, and starting a new mycelium. Since many of the spores fall right back near the sporangium that bore them, the food may become covered with *Rhizopus* overnight.

Sexual Reproduction *Rhizopus* also has a fairly simple sex life (Figure 24.5). It begins with **conjugation**, the fusing of two cells and the subsequent union of nuclei. There are no true males and females, so we refer to compatible mating types as *plus* (+) and *minus* (−). When different mating types are grown in the same medium, specialized club-shaped hyphae are produced. The hyphae of plus and minus mating types can fuse to form closed chambers known as **gametangia**, which contain many haploid nuclei. The gametangia then fuse, and their adjoining walls break down, permitting haploid nuclei from each strain to meet and fuse. The common chamber, now containing a number of diploid nuclei, then forms a tough resistant wall, becoming a zygospore. The diploid period may be short, or the tough, resistant zygospore may remain dormant for a considerable period. Under favorable conditions meiosis soon occurs within the zygospore, and many haploid nuclei—four times the number of diploid nuclei—are formed.

Next, the zygospore germinates, and a hypha emerges, immediately producing a sporangium. The haploid nuclei produced earlier are then incorporated into spores to be released. As the spores germinate, each new organism will have a rearrangement of genes because of genetic recombination during meiosis. Keep in mind this general pattern: sexual reproduction always requires the presence of plus and minus (+ and −) mating types and is often associated with times of environmental stress, long periods of dormancy, and eventually, the exploitation of newly favorable environments.

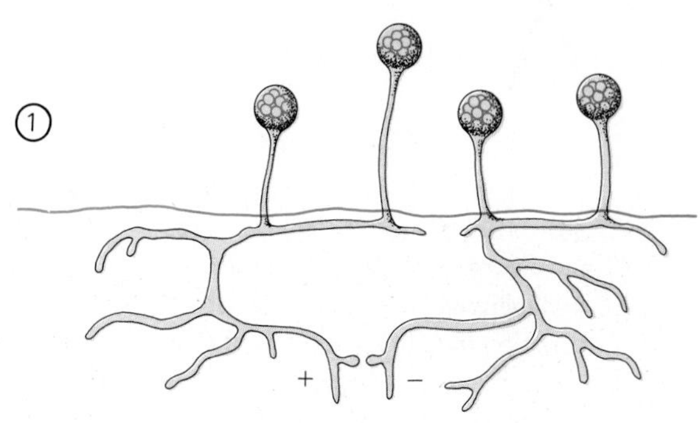

Ascomycota: Sac Fungi

Division Ascomycota and division Basidiomycota are often referred to as *higher fungi*. This elevation in status is primarily due to their septate condition—that is, cell divisions are routinely accompanied by the formation of crosswalls, or **septa.** Generally, in haploid hyphae, each resulting cell compartment has one nucleus, but the septa have perforations through which cytoplasm and nuclei can pass. Thus, in spite of the crosswalls, some hyphal cells may become multinucleate.

The ascomycetes, or **sac fungi** as they are also known, include some familiar members such as the edible morels and truffles, the powdery mildews (not to be confused with "downy mildew" mentioned in the last chapter), and the blue and green molds of citrus fruit. Some yeasts are also placed in this group. The sac fungi include the infamous chestnut blight, *Endothia parasitica,* an Asian mildew that brought about the extinction of vast forests of the graceful American chestnut. A sac fungus with rather ghastly implications for our own species is one known as *Claviceps purpurea,* a parasite of rye and other grasses.

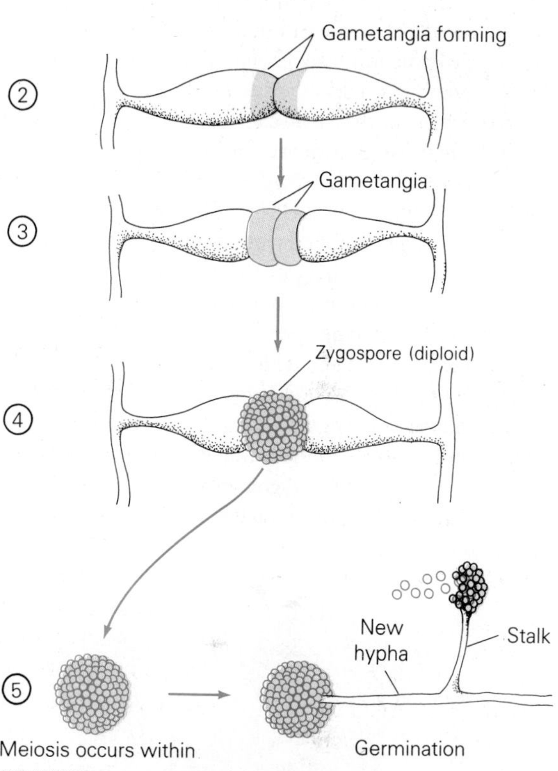

FIGURE 24.5
SEXUAL REPRODUCTION IN *RHIZOPUS*
Rhizopus enters into a sexual phase if plus (+) and minus (−) mating types are present. Neighboring hyphae form clublike gametangia, which fuse and mature into tough, thick-walled zygospores. Following meiosis and germination, hyphae will emerge. Asexual reproduction occurs when sporangia produce haploid spores.

Claviceps produces a plant disease called **ergot,** which in rye is characterized by the formation of a compact, black mycelium. Bread that has been made from "ergoted" flour contains certain alkaloids with the peculiar ability to constrict blood vessels in the body extremities. People who continue to eat bread made from such flour suffer from a condition known medically as **ergotism,** but historically as *Saint Anthony's fire.* The old name reflects the common symptoms of ergotism: burning sensations in the hands and feet accompanied by hallucinations. The restricted blood flow to these parts can bring on gangrene, often requiring amputation of the affected part. Ergot poisoning can even prove fatal, since respiratory and heart failure may occur. Such horrors were common in past centuries but are now rare. (Some historians believe that the peculiar behavior of the "witches of Salem" and the "voices" heard by Joan of Arc were actually products of ergot poisoning.)

Reproduction in Sac Fungi During asexual reproduction, the sac fungi characteristically produce spores known as **conidia** (sing. **conidium**). Conidia are produced in long chains at the ends of numerous specialized hyphae known as **conidiophores** (Figure 24.6).

Sexual reproduction among the sac fungi is quite diverse. However, there are some underlying similarities among the various types. The name *sac fungi* originates from the production of **ascospores** in a saclike container called the **ascus** (plural, **asci**). A group of asci are usually found in a fruiting structure known as an **ascocarp,** which occurs in a variety of shapes. We'll consider the sexual process in the sac fungus, *Peziza* (Figure 24.7).

Today, derivatives of ergot are used clinically in tranquilizers, for treating migrane headache, for bringing on abortion, and for controlling difficult cases of uterine bleeding after childbirth.

ESSAY 24.1

A FUNGUS THAT SPITS

The talent to survive on this planet takes many forms, but rarely does it involve spitting. However, *Pilobolus*, a zygomycete, depends on the ability to expectorate in an accurate and timely manner. It turns out that spitting is its primary way of ensuring that its offspring—actually its spores—will find their way into the same fortunate circumstance that previously led to the success of the spitter. In this case, the fortunate circumstance involves a manure pile. Let's see what is going on here.

After gorging itself on delicacies in the dung of an herbivore, *Pilobolus* shifts from feeding to asexual reproduction. It produces a typical upright sporangiophore, tipped by a sporangium filled with hardy, resistant spores. But below the sporangium lies a swelling, bulbous structure, the *subsporangial swelling*, not seen in other zygomycetes (see photo). Further, this fungus doesn't wait for wind or water to carry its spores away to shift for themselves. Under the right conditions, its subsporangial swelling, water-filled to the bursting point, will literally explode, ejecting the sporangium for a distance of two meters or more. If fortune smiles on the sporangium, it will land in and adhere to a blade of grass. If fortune smiles twice, the grass will be eaten by an herbivore, and the spores will survive the rigors of the digestive tract to emerge with the animal's feces, guaranteed of a food supply and a good prospect for future success. The adaptive value of this aiming system seems obvious. It is far more advantageous to send the spores upward and outward than it would be to shoot them down into the exhausted food reserves.

It turns out that sporangium ejection is not a random process but is closely oriented to the direction from which it receives light. In aligning itself, the stalklike sporangiophore bends first in one direction and then in another, but eventually aims the sporangium into the light. At this point, an explosion in the subsporangial swelling is triggered, and off goes the sporangium toward the light.

Such a description leaves us with many questions, particularly about the light-seeking growth of the sporangiophore. And precisely what triggers the rupturing of the subsporangial swelling? Researchers have suggested a lens effect whereby light is focused directly on a photoreceptor that controls the direction of growth in the sporangiophore. When the sporangium is aimed directly at the light, the photoreceptor receives maximum stimulation. However, this is all speculation, and researchers have yet to put their finger on the precise mechanisms involved.

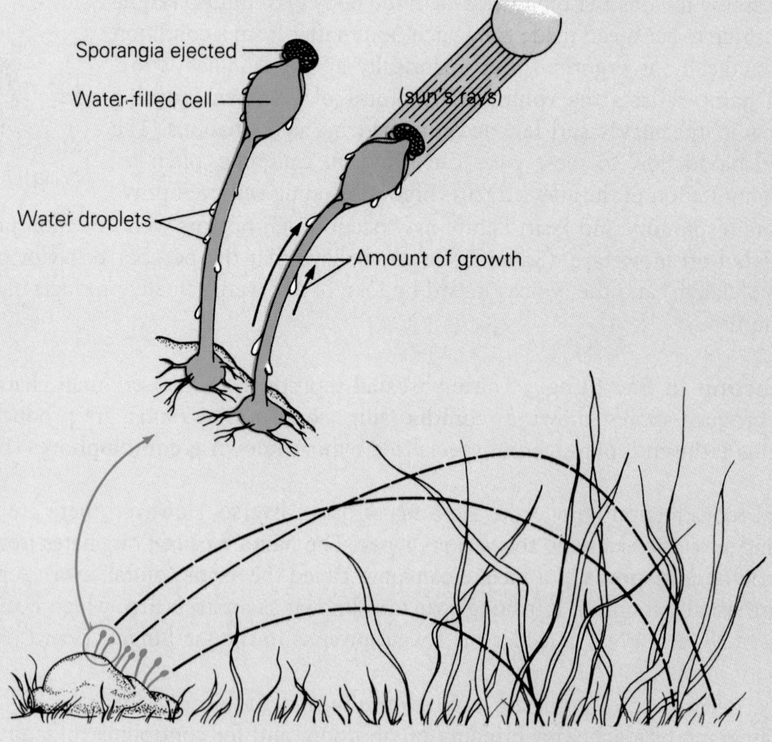

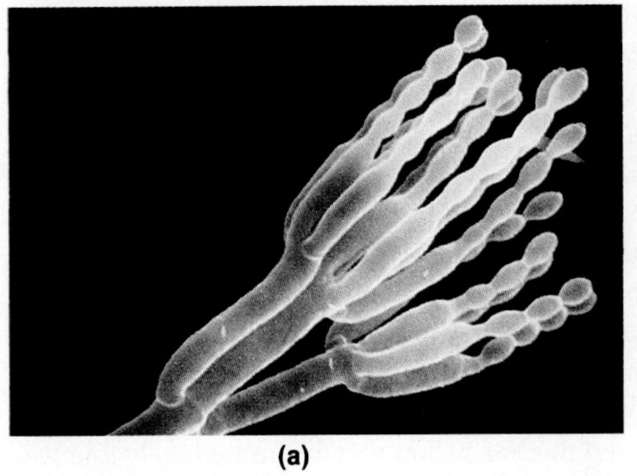

(a)

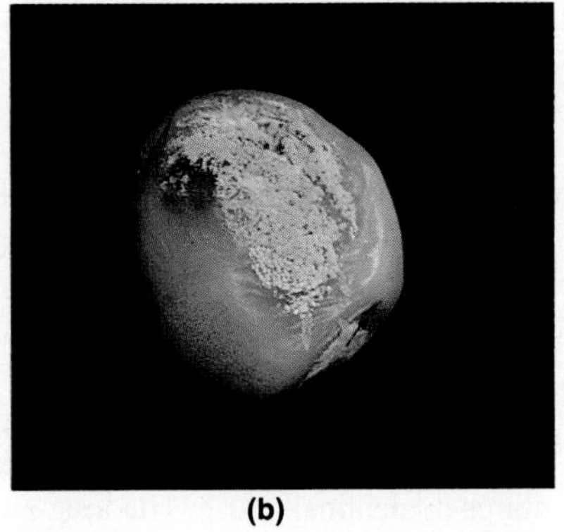

(b)

FIGURE 24.6
SPORE FORMATION IN SAC FUNGI
(a) In each conidiophore, the terminal cell rounds out and matures into a spore called a conidium, followed by the one below and so on, until a long chain of such haploid spores forms. They will then break away and become airborne, some landing on new food sources. **(b)** Fruit spoiled by a sac fungus.

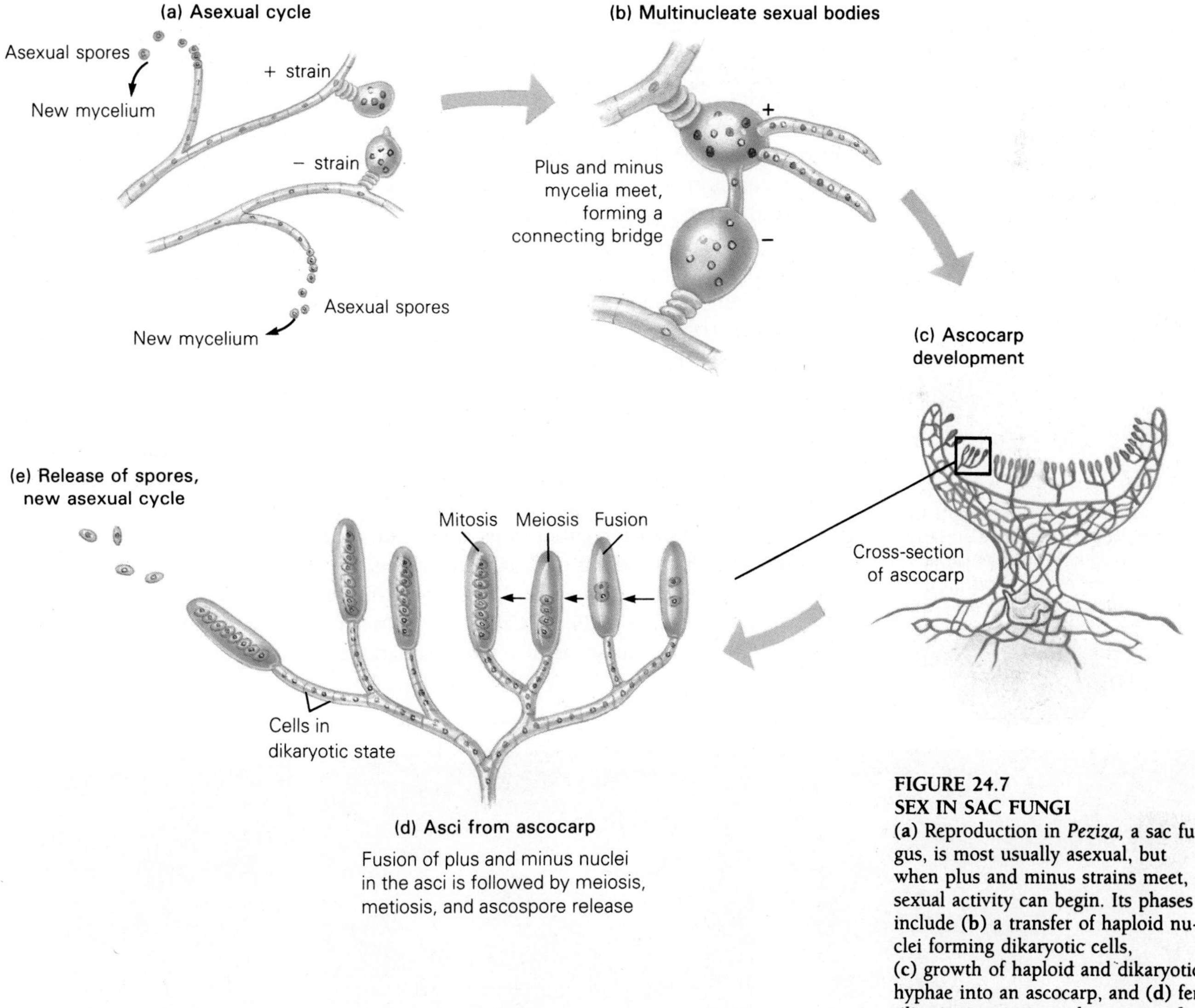

(a) Asexual cycle

Asexual spores

New mycelium

+ strain

− strain

Asexual spores

New mycelium

(b) Multinucleate sexual bodies

Plus and minus mycelia meet, forming a connecting bridge

+

−

(c) Ascocarp development

Cross-section of ascocarp

(e) Release of spores, new asexual cycle

Mitosis Meiosis Fusion

Cells in dikaryotic state

(d) Asci from ascocarp

Fusion of plus and minus nuclei in the asci is followed by meiosis, metiosis, and ascospore release

FIGURE 24.7
SEX IN SAC FUNGI
(a) Reproduction in *Peziza*, a sac fungus, is most usually asexual, but when plus and minus strains meet, sexual activity can begin. Its phases include **(b)** a transfer of haploid nuclei forming dikaryotic cells,
(c) growth of haploid and dikaryotic hyphae into an ascocarp, and **(d)** fertilization, meiosis, and ascospore formation, followed by **(e)** release of ascospores.

When the hyphae of opposing mating types of the sac fungus *Peziza* come into contact, each produces large multinucleate swellings. One, the **ascogonium**—the fungal version of a female sexual structure—will contain many haploid nuclei that remain in place. The other, the **antheridium**—an equivalent male structure—will also contain numerous haploid nuclei. Soon a bridgelike conjugation tube, the **trichogyne**, will form between the two bodies, and haploid nuclei from the antheridium will cross to enter the ascogonium. So far, this sounds pretty typical of sexual reproduction, but at this point the typical aspects end. This is because the two newly associated haploid nuclei fail to fuse, instead entering the dikaryotic state mentioned earlier. The dikaryotic cells, along with the older "sterile" hyphae, cooperatively produce the ascocarp, a chamber that becomes lined with young dikaryotic asci (sacs). It is within the asci that plus and minus nuclei finally fuse. The long-awaited nuclear fusion may be immediately followed by meiosis, or it may be further delayed. Nevertheless, when meiosis occurs, the four haploid daughter cells undergo one round of mitosis each, forming eight haploid ascospores. When the ascospores are fully mature, the ascus breaks open, and the spores disperse. Unlike the asexual spores, which are genetically identical, each pair of ascospores will contain a unique genome, the product of genetic recombination as it occurs during meiosis. Each ascospore is capable of producing a new mycelium, which will remain haploid until another sexual encounter once again ushers in the brief diploid phase.

The Lichens Another well-known group of sac fungi lives with certain algae or cyanobacteria. Together, they form **lichens.** (It should be mentioned that some of the other fungi are also capable of this relationship.) It is generally believed that the alga provides photosynthetic products (sugars) to the fungus in exchange for moisture, housing, and anchoring. This symbiotic association is usually considered to be mutualistic—beneficial to both parties. But in reality, one benefits more than the other because the algal or cyanobacterial partner can get along without the fungus. In fact, laboratory studies show that it does much better on its own. The fungus, however, thrives only if supplied with the complex nutrients it normally gets from its partner. For this reason, some authorities consider this often-touted example of pure mutualism to actually be parasitism. (If they are correct, we have just lost another long-cherished notion.)

Both the fungi and the algae that comprise lichens reproduce asexually through spore production. They are also capable of growing from fragmen[illegible] may wash away and settle in new habitats. Often, the algal spores that are release[illegible] associated with short lengths of fungal hyphae, and so, if conditions are right when they settle, they are set to go into business.

About 17,000 lichen combinations are known. They are distributed from the Arctic and Antarctic to the deserts and high mountains, some living on bare rock, others on trees, and still others on bare soil. Lichens are arranged into three groups, according to their appearance (Figure 24.8). **Crustose** (encrusting) species form dry, limited, but colorful colonies on sun-scorched rocky surfaces where they act as hardy pioneers,

(a)

(b)

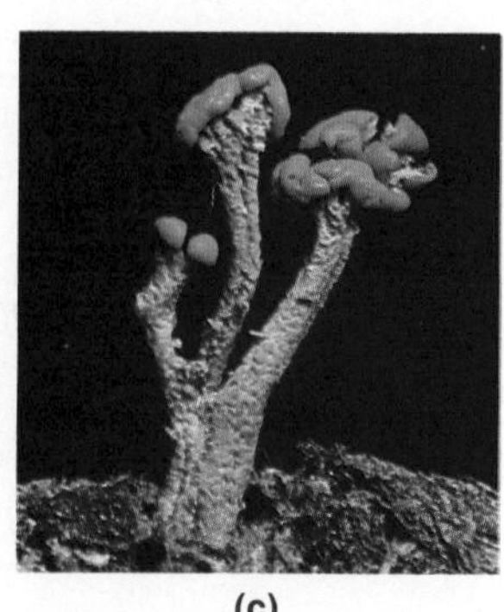

(c)

FIGURE 24.8
LICHENS
(a) A crustose lichen forms a crusty colorful mat over bare rock surfaces. **(b)** Foliose lichens, such as *Parmotrema tinctorus,* often appear on tree surfaces, whereas **(c)** *Cladonia cristotelli,* a fruticose lichen, lives on the soil surface.

hastening the weathering process and beginning the long soil-building process. Where cyanobacteria make up the photosynthetic component, their nitrogen-fixing activities further enrich the soil. **Foliose** ("leafy") lichens lead a more aerial existance, growing in trees where some resemble Spanish moss (a flowering plant). **Fruticose** ("shrubby") lichens live on the soil surfaces, often forming tangled, matlike growths. One fruticose lichen, **reindeer moss**, forms great expanses of ground cover in some Arctic regions. As the name suggests, it is a food source of reindeer.

Lichens are known to take up substances rapidly from the surrounding air and from rainfall. This is generally a very useful adaptation, except that in recent times these substances have included industrial contaminants such as sulfur dioxide, heavy metals, and radioactive materials. As such substances are accumulated, these highly sensitive organisms are quickly destroyed (the photosynthetic partner dying first). For this reason lichens have become useful "indicators" of air pollution problems. Observing their general condition and determining their chemical content is now a routine part of environmental studies.

Yeasts Sac fungi also include **yeasts.** A yeast is a single-celled fungus; the cells may be round or oval and may occasionally be found in short, branched chains. Yeasts grow by **budding,** a mitotic process with unequal cytoplasmic division. The buds (the parts receiving the lesser amount of cytoplasm) enlarge and finally separate from the parent cell. A complete organism grows from these tiny, dispersing bodies.

As you may now be aware, even the simpler fungi may not lead very simple lives. To drive the point home, consider the sex lives of the yeasts. Common bakers' yeast can be diploid or haploid, and either form can undergo asexual reproduction by budding (Figure 24.9). The diploid form can also undergo meiosis. The four haploid products of meiosis become separate ascospores inside the original cell wall of the parent, which thus becomes the saclike ascus. (That, by the way, is why these yeasts are classed as ascomyctes or "sac fungi.")

Two of the four ascospores will be the *alpha* mating type, and two will be the *a* mating type. They can undergo immediate fusion to regain the diploid state, or sexual fusion between alpha and *a* cells can occur much later, as long as the two types are

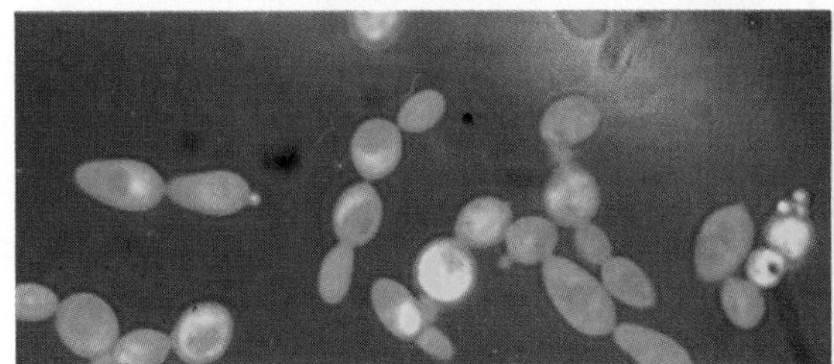

FIGURE 24.9
YEAST LIFE CYCLES
(a) In the diploid state, yeast cells reproduce asexually by budding (photo), or may undergo meiosis **(b)**, producing four haploid ascospores within an ascus. Two will be designated *alpha,* and two will be *a.* **(c)** Such haploid ascospores emerge, budding for a time, so that large haploid populations are produced. **(d)** *Alpha* and *a* individuals fuse, restoring the diploid state.

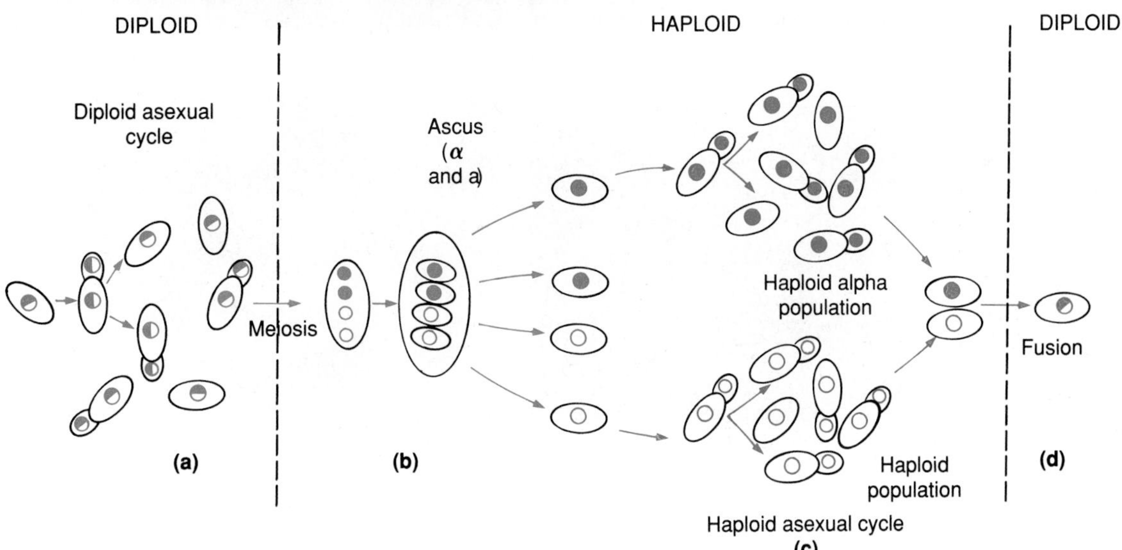

present. Generally, diploid strains do not undergo meiosis when conditions are good; meiosis usually is induced by food shortages and drying.

Like other eukaryotes, yeasts have mitochondria, and when oxygen is available, they can oxidize sugar completely to carbon dioxide and water. But mitochondria cannot function under anaerobic conditions, and in the absence of oxygen, yeasts derive their energy from anaerobic glycolysis. When food and oxygen are both abundant, the yeast may still utilize the anaerobic process.

As you may recall, under anaerobic conditions, glycolysis in yeast is followed by the fermentive process in which ethyl alcohol and carbon dioxide are the end products (see Chapter 8). Both products are immensely important to humans and have been throughout history. The carbon dioxide is what makes bread rise and gives some beverages their bubbles. The ethyl alcohol is what brewers and vintners are after when they make beer and wine. As far as the yeasts are concerned, they continue to ferment until their own alcoholic waste poisons them. This occurs when the concentration of alcohol reaches about 13%, the percentage of alcohol in unfortified wines. It is possible to get a higher concentration of alcohol from fermentation products (up to 16% from selected yeast strains) of course, but distillation is required to further increase the alcohol level of beverages.

Basidiomycota: Club Fungi

Basidiomycetes, or **club fungi**, are significant to us in a number of ways. Although the group includes such familiar varieties as mushrooms, toadstools, and shelf fungi, it also includes some devastating parasites (Figure 24.10). Among these are wheat rust and corn smut. Considering the worldwide dependence on wheat and corn, it is easy to appreciate the concern over these parasites. In fact, the development of rust-resistant strains of wheat is one of the great success stories of applied genetics. Saprobic club fungi are also partly responsible for the decomposition of dead trees and litter on forest floors.

FIGURE 24.10
SEVERAL CLUB FUNGI
Basidiomycetes can be large and colorful, as is **(a)** the edible mushroom, *Lepiota*. **(b)** Another mushroom, *Amanita phalloides*, looks harmless enough, but has been dubbed the "death cap." Such mushrooms are commonly called toadstools. **(c)** Shelf fungi are often seen perched on the sides of living trees, where they cause wood rot. **(d)** Corn smut, a devastating crop parasite, can quickly destroy an entire crop.

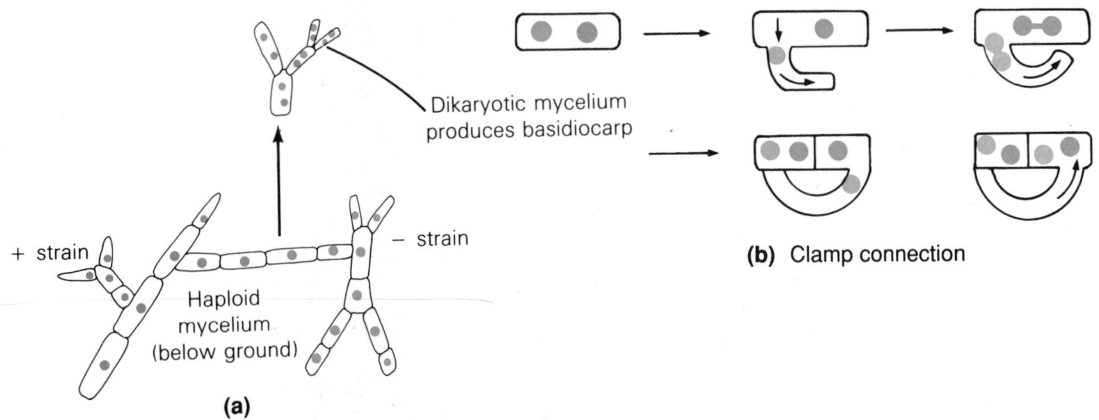

FIGURE 24.11
THE SUBTERRANEAN MUSHROOM
(a) Most of the mushroom's growth is subterranean, where a mycelial mass digests organic matter and grows. Plus and minus hyphae join, producing cells of a dikaryotic state. They continue to grow and feed for a time. **(b)** During cell division in a dikaryotic cell, a cytoplasmic bridge called a clamp connection forms between daughter cells. Its role is to ensure the proper distribution of newly formed plus and minus nuclei in daughter cells.

Like the sac fungi, the club fungi are septate (or partially so). They also reproduce sexually and produce complex sexual structures. The group can be divided roughly into two groups that differ both nutritionally and structurally. We will use the saprobic mushroom to represent one group and the parasitic wheat rust to represent the other.

Growth and Sexual Reproduction in the Mushroom The familiar mushroom with its thick **stalk**, topped by a dome-like or flattened **cap**, is really the end of the story. Long before the cap appears, a dense, penetrating subterranean growth occurs. Its hyphal mass actively digests and absorbs nutrients from decaying organic matter in the soil. In some forms, the mycelium may also form the symbiotic, mycorrhizal relationship with the roots of plants, mentioned earlier.

For sexual reproduction to begin, hyphae from plus and minus strains must meet (Figure 24.11a). Following conjugation, a dikaryotic state ensues, and the N + N cells may persist in the soil for some time, entering mitosis and cell division as the mycelium grows and feeds. Not surprisingly, this requires many mitotic cell divisions, but it also introduces a logistics problem: if the dikaryotic state is to be retained, then a replicated plus and minus nucleus must end up in each pair of daughter cells. How does mitosis provide for this? It doesn't. In the basidiomycetes, the **clamp connection** ensures that this happens. Let's look into this (Figure 24.11b).

Clamp connections are a sort of cellular detour, wherein the cell undergoes a branching growth that later rejoins the main cell. One of the nuclei enters the branch, and one remains behind. Then mitosis occurs in each nucleus, and the replicas separate. The original nuclei end up with replicas at opposite ends of the cell. Finally, a cell wall is laid down between the two newly separated pairs of nuclei. As a result, two cells emerge, each with a plus and minus nucleus, and the dikaryotic state is retained.

At some point, the dikaryotic hyphae go on to form the dense, fleshy reproductive structure, the **basidiocarp**, which appears above ground (Figure 24.12a). Thus, the familiar mushroom is a sexual structure, one in which nuclear fusion will finally occur. Typically, the site of nuclear fusion is to be found on the underside of the cap, in minute **basidia** (sing. **basidium**), clublike structures arranged in rows along the thin **gills** that radiate out from the stalk (Figure 24.12b). (The basidium is, of course, the source of the taxonomic name, Basidiomycota.) Fusion, or fertilization, is followed by meiosis and spore formation. Each haploid nucleus becomes incorporated into a **basidiospore**, and enormous numbers of such basidio spores may be released from each mushroom.

We can't leave the subject without a word about picking your own wild mushrooms for the table. Early in 1988, the news carried a story about two people who used an illustrated mushroom picker's guide to gather wild mushrooms. One of the two was a veteran mushroom picker. Their carefully gathered harvest was taken home, sautéed in butter and garlic, and promptly eaten.

It turned out that while their selection closely resembled those labelled "edible" in the guide, they had actually collected the highly toxic toadstool aptly named "the death

FIGURE 24.12
LIFE CYCLE OF THE MUSHROOM
(a) After sufficient growth has occurred, the basidiocarp emerges above ground, and the cap expands and opens, exposing numerous gills on its underside. **(b)** Each gill contains tiny, clublike basidia, where fusion of plus and minus nuclei finally occurs. **(c)** Each fused nucleus then undergoes meiosis, producing four haploid basidiospores.

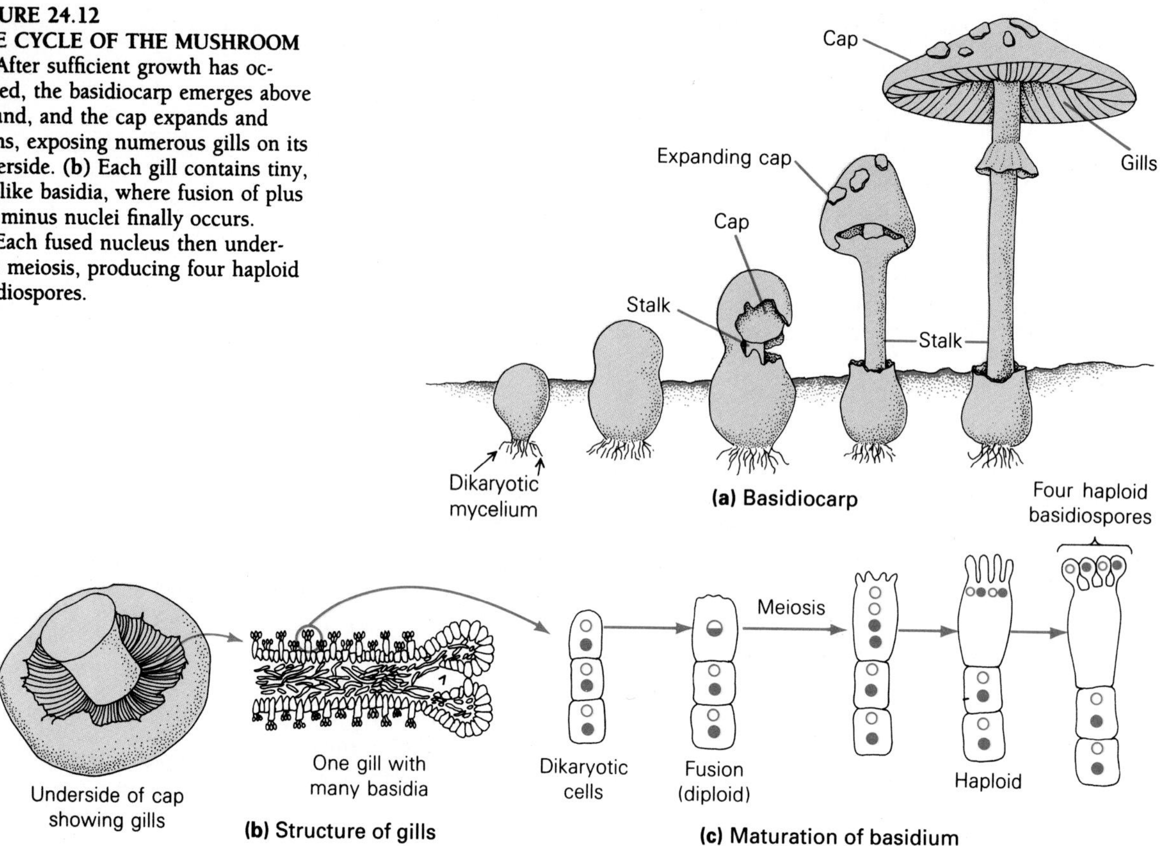

cap." The two gourmets became critically ill, hovering near death for a time. (About half of such cases end in death.) But both survived, their lives spared when suitable donors were found and their badly damaged livers replaced. The best bet, it seems, is to pick mushrooms only from the source expert mycologists use—the corner supermarket.

Wheat Rusts Now let's consider wheat rust, *Puccinia graminis,* a species with a truly complex life cycle, most of which is spent in the dikaryotic state, and which involves two different hosts and several kinds of spores. One host, of course, is wheat, but the parasite also must infect the common barberry plant, *Berberis vulgaris.*

While in the wheat, the rust fungus produces dikaryotic spores of two types. One—a red spore, or **uredospore**—simply infects other wheat plants, while the other—a black spore, or **teliospore**—is the rust's investment in future generations. Teliospores survive the winter season and germinate in moist soil in the spring. Until then the black spores remain in a dormant state. As such they are not active and cannot infect the wheat or the barberry. Then, just prior to teliospore germination, each diploid nucleus undergoes meiosis in the usual fashion, producing what become four basidiospores. Upon their release, the basidiospores infect the wild barberry.

It is only in the barberry leaf that the sexual phase occurs. As usual in fungi, plus and minus strains must be present. When they meet, a union occurs, with plus and minus nuclei remaining separated, thus restoring the dikaryotic state. Finally, another round of spore formation occurs, this time with dikaryotic **aeciospores** emerging. It is the aeciospore that can infect the next wheat crop.

Wheat infected by *Puccinia graminis* is invariably weakened or killed. After the wheat rust's life cycle was discovered, it became common practice to break the infection cycle by burning contaminated fields and by systematically destroying barberry bushes, the intermediate host.

TABLE 24.1 CHARACTERISTICS OF THE FUNGI

DIVISION	USUAL CHARACTERISTICS
Zygomycota	Conjugating molds. Hyphae without complete cross walls; asexual reproduction: upright sporangiophore and spores; sexual reproduction: conjugation, fertilization, and formation of zygospores; black and pink bread molds
Ascomycota	Sac fungi. Hyphae with cross walls; asexual reproduction: conidiophores with chains of conidia; sexual reproduction: conjugation, then lengthy dikaryotic state, formation of ascocarp, fertilization, and growth of ascospore; morels, powdery mildews, blue and green molds, *Claviceps purpurea* (cause of ergot)
	Lichens. Mutual association of sac fungi and algae or cyanobacteria
Basidiomycota	Club fungi. Hyphae with cross walls; asexual reproduction by spores; sexual reproduction: conjugation, then lengthy dikaryotic state, formation of basidiocarp, fertilization, and growth of basidiospores; mushrooms, shelf fungi, bracket fungi, wheat rust, corn smut
Deuteromycota	Fungi imperfecti. Mixed group whose sexual stages are unknown; asexual reproduction: spore formation; penicillin mold, parasites of athlete's foot, thrush, yeast infection

Deuteromycota: Fungi Imperfecti

A fungal group with the name "fungi imperfecti" will need explaining. *Imperfecti* is a botantical term referring to lack of sexual reproduction, and the approximately 24,000 named species of fungi imperfecti have no known sexual phase. Actually, the fungi imperfecti still require a lot of taxonomic work. Undoubtedly, many will eventually be placed with the sac and club fungi. But a vital part of fungal classification is based on sexual stages, and if these phases have not yet been observed, the fungus cannot be further classified.

The fungi imperfecti include a number of human parasites, including *Trichophyton mentagrophytes,* the agent of athlete's foot. Then there is the yeast-like *Candida albicans,* which causes a mouth infection called "thrush" and troublesome vaginal infections. Such problems commonly follow prolonged episodes of antibiotic therapy, generally for some unrelated infection. During such treatment, many bacteria and fungi, harmless symbiotic residents of the body, are destroyed. Without their presence, there is little competition, and the antibiotic-resistant *C. albicans* becomes a successful invader.

But on a more positive note, we also find some harmless and useful species. The genus *Penicillium* is a source of the important antibiotic penicillin. From the same genus, consider *Penicillium roquefortii* and *Penicillium camembertii* as examples whose names give them away as agents used in the flavoring of famous cheeses. Then there is *Aspergillus oryzae,* which, along with lactic acid bacteria, is responsible for the special flavor of soy sauce and, in part, for the fermenting of saki. *A. oryzae* is also important in the fermentation and nutritional enrichment of livestock feeds.

Some species are predators that actually catch certain tiny roundworms (nematodes) that live in the soil (Figure 24.13). They develop ringlike structures that constrict like an inflated noose when a hapless roundworm inadvertently passes through. Then fungal haustoria enter the prey and digestion begins.

FIGURE 24.13
A PREDATORY FUNGUS
Arthrobotrys, a microscopic fungus, bears many snares (open loops). These swell rapidly upon touch, trapping anything inside. Gotcha! The nematode worm becomes ensnared by a rapidly closing noose. Following the capture, haustoria from the fungus will penetrate the worm's body and digestion and absorption will begin.

ANOTHER LOOK AT FUNGAL ANCESTORS

Tracing the origin of fungi from ancestral stock is at best a conjectural exercise. There are many possibilities, but none, so far, is based on the fossil record or an elaborate biochemical analysis of amino acids, nucleic acids, or other compounds. The principal guesses have included the following:

1. Fungi arose monophyletically from colorless, heterotrophic, flagellated protists. The connection is represented by the slime molds, which, evolving from amebas, developed aerial spore formation as an adaptation to land. Fungi then developed from slime molds. (This is currently the favored hypothesis.)

2. Fungi are monophyletic, derived from green algal stock. The derivation was accompanied by a loss of the photosynthetic pigments of algal forerunners, and the retention of cell walls.

3. Fungi arose from ancient red algae. Like the red algae, the fungi lack centrioles and motility.

4. Fungi are a polyphyletic group, some having been derived from algal ancestry and others from flagellated protists. Aerial spore formation is an example of convergent evolution, since it represents a very successful adaptation to life on land.

Fossil fungi are dated tentatively back to the Precambrian era, some 900 million years ago, but these aren't very reliable finds. Fossils from the Ordovician period, between 450 and 500 million years ago, have been clearly identified as fungi, and mycorrhizial associations have been preserved in Silurian deposits dating back over 400 million years. (See the geological timetable inside the front cover.)

As far as taxonomic association goes, we can presently do little more than compare structures and life cycles, look for possibly favorable comparisons, and arrange the phylogenetic tree accordingly. Recent cytochrome *c* sequence studies indicate that cellular slime molds are closer to the protist *Euglena* than to the true fungus *Neurospora,* so in spite of some of the decidedly fungal characteristics of the slime molds, they have been reassigned to the protists. There are many, many questions about fungal origins, but until more fossil or biochemical evidence is available, the questions will remain unanswered.

MULTICELLULARITY

We have seen that some of the "simple" protists and fungi are not so simple after all. Their life cycles are complex and elusive; they appear in all sorts of forms and colors, and they are rather complex in structure. Some are not only multicellular, but contain a variety of cell types. Each cell type plays its own role in the life of the organism, and such specialization enables the protist and fungus to exist under quite precise and demanding conditions.

Let's consider for a moment the biological implications of multicellularity and cell specialization. Why are so many successful species composed of many cells of different types? The answers may provide an important insight to the way living things have adapted to such a complex planet.

The Origin of Multicellularity

We can begin by defining multicellularity. *A multicellular organism is one that is composed of a number of cells that cooperatively carry out the functions of life.* This definition is meant to distinguish the multicellular organisms from aggregations and colonies of single-celled organisms, although the distinction is not always clear. For our purposes, multicellular organisms are made up of cells that cannot survive independently under natural conditions.

We assume that multicellularity began when aggregations of cells became interdependent. This would have happened when different cells became adapted for different roles. For example, some cells might have become specialized for producing or obtaining food, others for transport, and still others for moving the aggregate to a more favorable place. We might imagine that, finally, some cells came to specialize in reproduction.

As you know, the organisms of the earth are divided into taxonomic groups according to shared characteristics. Two of these groups, the bacteria and the protists, are commonly referred to as unicellular. This description should exclude these taxa from membership in the multicellular club. However, the taxonomy is not that clear-cut. Some protists are clearly multicellular. Further, some of the fungi are unicellular, although this is thought to be a derived condition.

Fungi help bridge the gap between unicellular-colonial and multicellular organisms. While yeasts are unicellular, the reproductive structures of sac and club fungi may be quite complex, some species of the latter containing specialized cells for stalk, cap, gills, and covering. The noose of the predatory fungus *Anthrobotrys* is one exceptional example of fungal cell specialization in nonreproductive tissue.

But keep in mind that for the most part there is little specialization in the fungal mycelium. Cell specialization is almost entirely reserved for the reproductive structures. It is the plants and animals that have made the transition to somatic cell specialization—to the tissue, organ, and organ system levels of organization. We can't let this transition slip by unnoticed.

It seems obvious that when the necessary labors of life are doled out to specialized cells in a colony, then each will become better at performing its particular task, and the result will be greater efficiency. It follows, then, that multicellular organisms must somehow be better off than the single-celled organisms. But this is not necessarily so. Single-celled organisms have survived very well through evolutionary history without being displaced by multicellular life. We have emphasized that some of the protists, such as the ciliates, are highly complex and nicely adapted creatures. Their lineage has certainly continued without the drastic adaptive changes that have marked ours. And they have become successful without following the multicellular trend. If single cells are so successful, then, why did multicellular life develop?

The answer has already been stated: in a word, it's *specialization*. This term has two connotations, however. First, cells become specialized and interdependent so that the success of one type depends on the success of others. In addition, the relative numbers of each type, the absolute numbers of the cells (the size of the multicellular body), and the direction of their specialization have permitted the organism to exist in various kinds of environments. For example, some multicellular bodies may have disproportionate numbers of contractile (muscle) cells so that they can survive under environmental conditions that require strength. However, they may not have many cells that are specialized to detect soundwaves, so in environments where hearing is important, they must yield before other organisms that perhaps have fewer muscles but better hearing. Thus, cellular specialization means that organisms may come to occupy different niches and may branch out, specialize, and diversify, creating the staggering array of living things we see around us.

KEY IDEAS

Fungi get a lot of bad press because they are either parasites, invaders of living organisms, or **saprobes**, feeders of dead materials such as our foods and organic goods.

WHAT ARE THE FUNGI?

1. Fungi are multicellular, nonmotile heterotrophs, lacking tissue organization except in their reproductive structures. Some are coenocytic, others cellular; cell walls are formed from chitin.
2. The mycelium is often made up of threadlike tubular cells called **hyphae**. Hyphae may be cellular **(septate)** or coenocytic **(nonseptate)**. The mycelium is essentially a feeding structure that gives rise to reproductive structures.

Feeding in the Fungi

1. Fungi feed through **absorption**, secreting enzymes outside and absorbing the digested food. Parasitic fungi feed via penetrating hyphae called **haustoria.**
2. While fungi are commonly parasites, many are reducers, joining soil bacteria in breaking down nutrients and recycling essential mineral ions. One group forms a mutual symbiosis with plant roots called **mycorrhizae**, in which they assist the roots in mineral and water absorption. We use fungi as food, in food production, and for their pharmaceutical products.

Reproduction in Fungi

1. Asexual reproduction often occurs through mitotic spore production in **sporangia.** Spores are easily dispersed and highly resistant.
2. Each fungus is haploid except for a brief diploid state following fertilization. An exception is the **dikaryotic** state, in which following sexual reproduction, the two haploid nuclei in each cell remain separated during the growth of elaborate sexual structures. Eventually they fuse, but the diploid cell then enters meiosis, producing haploid spores.

Fungal Relationships

Fungal characteristics such as chitinous cell walls, primitive mitosis and meiosis (within the nuclear envelope), an absence of centrioles and thus of motility, and chromosomes with little protein, make the kingdom unique and separate from others. It is proposed that fungi evolved from heterotrophic, flagellated protist ancestors.

KINGDOM FUNGI

Kingdom Fungi contains divisions **Zygomycota, Ascomycota, Basidiomycota,** and **Deuteromycota.**

Zygomycota: Conjugating Molds

1. Zygomycetes are coenocytic, and most are saprobic.
2. The ideal growth conditions for *Rhizopus stolonifer* (bread mold), a common contaminant of foods, are sufficient nutrients, moisture, and warmth. Its **stolons** (horizontal hyphae) send **rhizoids** (penetrating hyphae) into food. **Sporangiophores** (upright hyphae) are tipped by sporangia in which mitotic spore formation occurs.
3. Sexual reproduction occurs through a form of **conjugation** involving plus and minus strains. Haploid nuclei gather in **gametangia,** which then join, permitting the nuclei to fuse. Several diploid nuclei form a thick-walled **zygospore,** which may become dormant.

Ascomycota: Sac Fungi

1. Ascomycetes, a group of higher fungi, is septate (cellular), and its members include powdery mildews, citrus molds, and certain species of yeasts.
2. *Claviceps purpurea* produces **ergot**—a compact, black mycelium—in the rye plant. When ergoted rye flour is eaten, the symptoms of **ergotism**—restricted blood flow to extremities, burning sensations, hallucinations, and gangrene—may develop. Derivatives of ergot are used in tranquilizers, to control migraines, and to slow bleeding after childbirth.
3. In asexual reproduction, **sac fungi** form **conidiophores** on which spores called **conidia** emerge through mitosis.
4. Sexual reproduction in *Peziza* requires plus and minus strains. One produces an **ascogonium,** the other an **antheridium,** and these are joined by a tubular **trichogyne,** through which nuclei migrate. The plus and minus nuclei join in cells, but enter a dikaryotic state. Plus and minus parent hyphae along with the new dikaryotic hyphae cooperatively produce an **ascocarp,** in which the dikaryons develop saclike **asci.** Fusion of plus and minus nuclei occurs in the asci, followed immediately by meiosis and the formation of haploid **ascospores.** They then germinate to form a new mycelium.
5. The **lichens** include some 17,000 specific associations, each of which is an intimate mutualistic symbiosis, made up of an alga or cyanobacterium and a fungus. The photoautograph supplies organic nutrients, while the heterotroph reportedly provides moisture and anchorage. Recent studies suggest that only the fungus benefits. **Crustose** lichens live on rocks, **foliose** lichens live on trees, and **fruticose** lichens live on the soil. All are important soil builders, and one type, **reindeer moss,** forms the energy base of certain Arctic food chains. In asexual reproduction, spores are produced by each member.
6. **Yeasts** are single-celled sac fungi that increase through **budding** (mitosis and highly unequal division). A colony can be diploid or haploid, and the diploids may undergo meiosis, producing ascospores within the old cell. The meiotic spores can immediately fuse, restoring the diploid state, or this may happen after many mitotic divisions and budding events. Some yeasts are important to us in producing bread, wine, beer, and liquor, while others are parasites, causing serious infections.

Basidiomycota: Club Fungi

1. **Club fungi** include saprobic mushrooms and shelf fungi, along with important parasites such as wheat rust and corn smut. Most of the mushroom mycelium is below ground, while above is the **basidiocarp,** a fruiting body that produces the **basidiospores.**
2. Following the subterranean union of plus and minus mushroom hyphae, a dikaryotic mycelium forms and may continue growing for a time. Cytoplasmic bridges called **clamp connections** between newly forming daughter cells ensure proper distribution of plus and minus daughter nuclei. The dikaryotic mycelium emerges from the ground to produce the basidiocarp. Its cap contains **gills** lined with **basidia** in which fusion of plus and minus nuclei occurs, followed by meiosis and basidiospore formation.
3. **The wheat rust,** *Puccinia graminis,* requires two hosts, the wheat and the barberry plant. While in wheat, it produces dikaryotic red **uredospores,** which infect other wheat, and dikaryotic black **teliospores,** which winter over. In the spring, the winter dormancy ends, and the diploid nucleus of each telispore undergoes meiosis. The emerging plus and minus haploid basidiospores then infect the barberry. Sexual reproduction between the plus and minus hyphae follows. More dikaryotic cells emerge and enter another round of spore formation. The dikaryotic **aeciospores** produced then infect the new wheat crop.

Deuteromycota: Fungi Imperfecti

Fungi imperfecti are unclassified fungi with no known sexual phase. Many are probably sac and club fungi.

Members include the athlete's foot fungus and several **Penicillium** species used in flavoring foods. Another species captures and feeds upon soil roundworms.

ANOTHER LOOK AT FUNGAL ANCESTORS

Fungi probably arose from colorless, heterotrophic, flagellated protists. Other possibilities include a green algal origin, a red algal origin, and polyphyletic derivation from algae and protists. The fossil record of fungi, which is very tentative, can be traced back 900 million years (to the Precambrian).

MULTICELLULARITY

The Origin of Multicellularity

1. Multicellular organisms are composed of many cells cooperatively carrying out the life functions. They may have originated from casual aggregations that became interdependent. The gap between unicellularity and multicellularity is bridged by the fungi (and some of the algae).
2. While the ability of cells and tissues to specialize is vital in multicellular organisms, this does not mean that unicellular organisms are inefficient.

REVIEW QUESTIONS

1. List four characteristics shared by most fungi. (517)
2. What is unusual about the cell walls of most fungi? How might this be used as an argument against including fungi in the plant kingdom? (521)
3. Distinguish between the terms *coenocytic, multinucleate,* and *nonseptate.* To which group or groups of fungi do these apply? (517)
4. Explain how a mycorrhizal association serves as an example of a mutualistic symbiosis. (518)
5. Distinguish between the terms *mycelium, hypha,* and *haustoria.* (517-518)
6. Describe the absorptive feeding mode. How does this differ from the manner in which digestion occurs in animals? (518)
7. What is the function of a spore? How is this function adaptive to the fungus? (518)
8. In what way does timing in the union of fungal haploid nuclei (fertilization) seem unusual when compared to plants and animals? (519)
9. Among the protists, which is the best guess for a fungal ancestor? What protist groups may have sprung from this ancestor as well? (521)
10. From what reproductive characteristic is the name *zygomycete* derived? (521)
11. Describe asexual reproduction in the common bread mold, *Rhizopus stolonifer.* (521-522)
12. List the steps taken during conjugation in *Rhizopus.* (522-523)
13. What is the "sac" of the sac fungi? (523)
14. What is ergoted wheat? Ergotism? (523)
15. List three ways in which the activities of *Claviceps purpurea* have benefitted humans. (523)
16. How does the development of conidia differ from the development of the asexual spores in bread molds? (523)
17. Trace the events in sexual reproduction in the sac fungus, using such terms as *ascogonium, dikaryon, trichogyne, sterile hyphae, plus and minus strains, antheridium, asci,* and *ascospore.* (523-526)
18. What are lichens? What is the most recent thinking about the symbiotic relationship in lichen associations? (526-527)
19. List two ecologically significant activities of lichens. (527)
20. Describe the specific activity in yeasts that places them in the sac fungi. What are the two fates of haploid ascospores? (527-528)
21. List two economically important uses for yeast, and tell which products of yeast metabolism are important to each. (528)
22. Trace the events leading to the production of the mushroom basidiocarp. What reproductive event happens in the basidiocarp? (528-529)
23. In what way(s) is sexual activity in the mushroom similar to that in the sac fungi? (529-530)
24. In which host does the sexual phase of the wheat rust life cycle occur? Trace the events. (530)
25. At what point in the life cycle of *Puccinia graminis* can the terms *diploid* and *haploid* be applied? Explain the events surrounding this. What nuclear term applies the rest of the time? (530)
26. Why would one combine burning the wheat fields with removal of barberry plants as a rust-preventing measure? Why not just remove the barberry plants? Would burning have to be repeated? (530)
27. What is the basis for assigning fungi to the fungi imperfecti? List three familiar examples. (531)
28. List three possible explanations of fungal origins. (531-532)

DIVERSITY AND FUNCTION: PLANTS

25

PLANT EVOLUTION AND DIVERSITY

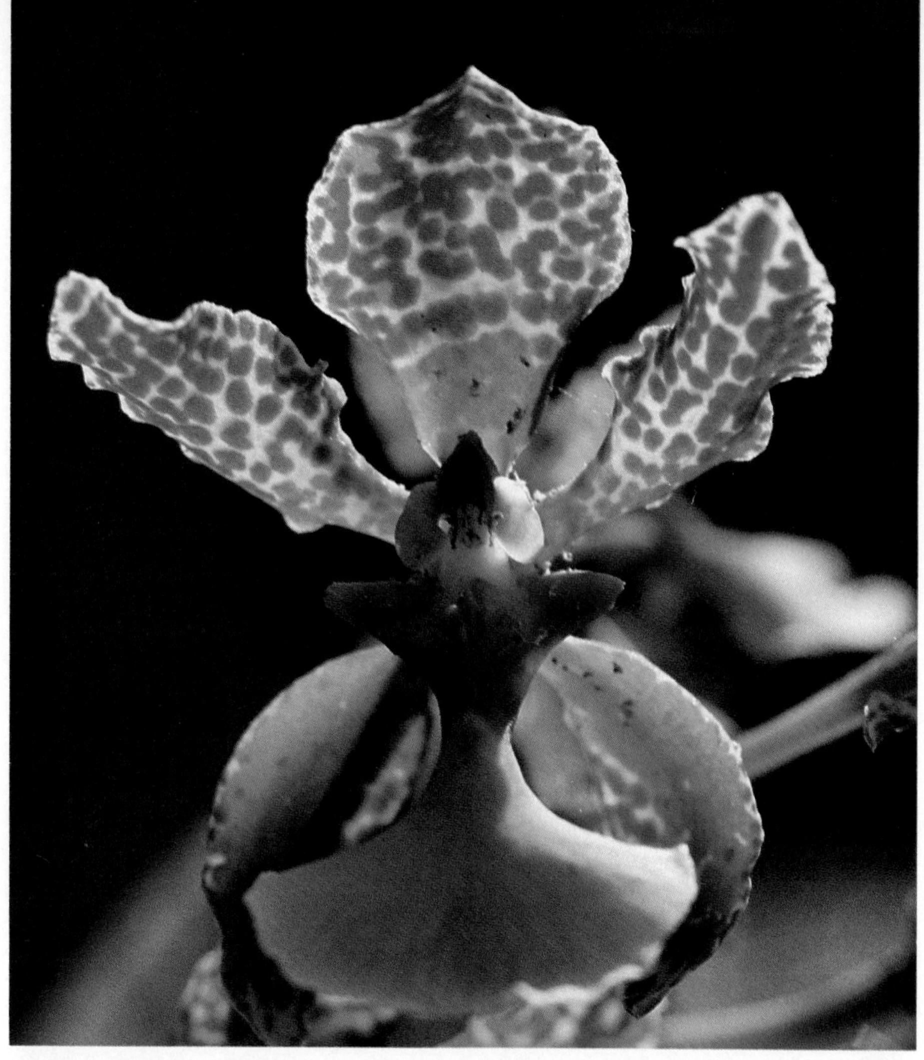

WITH OUR ARRIVAL IN THE PLANT KINGDOM WE COME TO THOSE ORGANISMS that are familiar to everyone. These are the life forms that grace our lives by lending soft hues to our surroundings, by providing our shelter, and by warming our hearths. But perhaps most important, plants are our ultimate source of oxygen and food. What, then, are plants? The question seems simple, but it is not. We can safely assume that an elm is a plant, but reputable scientists have argued over whether slime molds, fungi, and even bacteria belong in this group as well. However, the five-kingdom scheme we have chosen to follow (see Figure 20.6) will help us handle some of the problems and inconsistencies more easily.

WHAT ARE PLANTS?

Formally, plants are multicellular, photosynthetic organisms with specialized cells and tissues. Their photosynthetic pigments include chlorophylls *a* and *b* and carotenoids, and their primary storage polysaccharide is amylose starch. The most common structural polysaccharide is cellulose, which is used in cell wall construction and is often joined there by strengthening agents such as pectins and lignins. All plants exhibit a life cycle involving an alternation between diploid sporophyte and haploid gametophyte phases. Plants also produce *multicellular* embryos, which are housed within *multicellular* gametophyte tissues, a fairly clear departure from algae.

You may have noticed that many of these characteristics are shared with one or another of the algal protists (see Chapter 23). So, the list must be considered collectively if it is to define the plant kingdom. With this in mind, let's review alternation of generations in plants and then consider some trends that characterize plant evolution.

Alternation of Generations in Plants

The plant life cycle is described as **sporic** (Chapter 23). There are alternating diploid sporophyte and haploid gametophyte states in what is traditionally called an alternation of generations. In such alternating states, meiosis in the diploid sporophyte yields single-celled, haploid spores, which have the potential to develop into the multicellular gametophytes. Then, certain cells in the gametophyte form gametes: sperm and egg. When sperm and egg meet in fertilization, a new diploid sporophyte generation begins. This is all quite different from the gametic animal life cycle. In animals, meiosis always leads directly to gamete formation, and there is nothing remotely equivalent to the multicellular, haploid gametophyte of plants.

The sporic life cycle in plants reveals two very different trends (Figure 25.1). In the life cycle of the so-called **nonvascular plants,**[1] the bryophytes (mosses, liverworts, and hornworts), the gametophyte is clearly dominant, and the sporophyte is reduced. In some instances the sporophyte is even difficult to see without magnification. In all other plants, the opposite is true; the sporophyte is highly dominant, accounting for most of the life cycle. The gametophyte is highly reduced and may occur either as a tiny separate and independent plant or as a relatively small number of haploid cells, surrounded by sporophyte tissue. The latter, we might point out, is clearly the trend in the most advanced, more recently evolved, plants.

In summary, should you nap in a mossy glade, the gametophyte provides your velvety cushions. But if your preference is a grassy slope under a graceful elm, you can thank the sporophyte for both your cushion and your shade.

Megaspores and Microspores Many **vascular plants** have followed another evolutionary trend over time, a departure from **homospory**—producing one form of spore only. The trend has been toward the **heterosporous** (*hetero-*, "different") condition, where two distinct types of spores are formed. These are **microspores** ("small spores") and **megaspores** ("large spores"). Microspores give rise to male gametophytes, or **microgametophytes.** In seed plants the microgametophyte is represented by the minute pollen grain. Megaspores develop into **megagametophytes,** which occur in the ovules, located in the ovary of the flower. The gametophyte gives rise to the egg cell, which when fertilized, and in conjunction with other tissues of the ovule, differentiates into the seed. We will look further into the gametophytes of seed plants in the next chapter.

With these evolutionary trends in mind let's move along to the nonvascular plants, where dominating gametophytes and homospory prevail.

DIVISION BRYOPHYTA: NONVASCULAR PLANTS

Bryophyte Characteristics

The **bryophytes—mosses, liverworts,** and **hornworts**—are multicellular plants with simple tissue organization. However, the tissues are differentiated enough to divide the tasks of life efficiently. Although most are terrestrial, the bryophytes retain a reproductive feature common to the aquatic algae: a swimming sperm that must have water to reach the egg. But that water need only be a few drops of dew or splashing raindrops.

In the bryophyte life cycle (see Figure 25.1a), the gametophyte is easily seen (the mossy cushion), long-lived, and thus quite dominant. The sporophyte, on the other hand, is usually far less conspicuous, shorter-lived, and in many instances nutritionally dependent upon the gametophyte.

Bryophytes, as we have seen, lack the well-developed vascular tissue seen in vascular plants. Vascular tissue contains tough fibers and thick, hardened cells, both of which lend great strength to the vascular plants. This permits such plants to grow quite tall.

[1]While bryophytes are designated nonvascular (without vascular tissue), some conducting elements similar in function to those of vascular plants (those with vascular tissue) are seen. Further, some vascular plants have only very primitive conducting tissues.

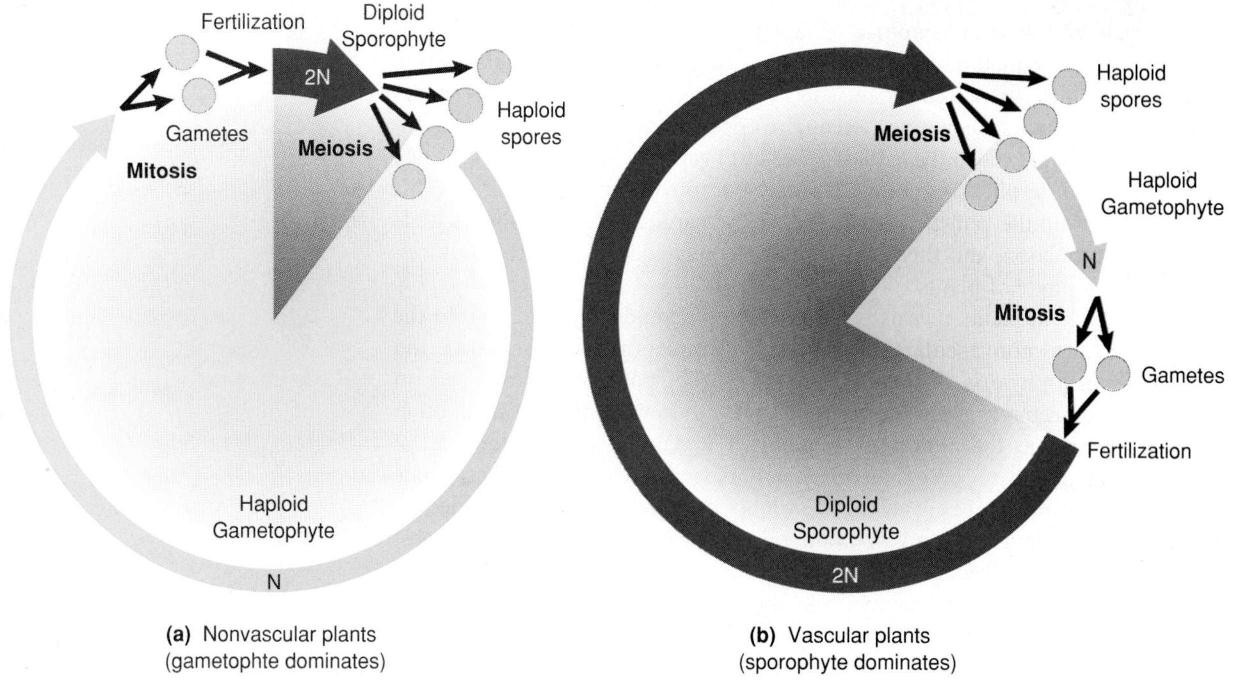

FIGURE 25.1
GENERALIZED PLANT LIFE CYCLE
(a) The sporic life cycle in nonvascular plants has a highly dominant gametophyte phase and a brief, reduced sporophyte. **(b)** In the vast majority of vascular plants, the emphasis is reversed, and the sporophyte is the dominant phase.

Because bryophytes lack such supporting tissues, there are no tall mosses, liverworts, and hornworts. Instead they cling close to the earth or hang from the trunks and branches of vascular plants. The bryophyte body (actually the gametophyte in most of our discussion) contains leaf-like, stem-like, and root-like structures. The terminology is deliberate, since true leaves, stems, and roots, by definition, have vascular tissues. Analagous structures in bryophytes carry out some of the same functions. For example, threadlike rhizoids anchor the plants to the soil, but unlike true plant roots, most are not involved in the absorption of water.

In spite of their simplicity, bryophytes have been quite persistent and certainly successful—surprisingly so, considering that they originated some 350 million years ago (see Table 25.1). They filled ecological niches that were seemingly ignored by the vascular plants. While vascular plants underwent the evolutionary changes associated with large size, and adapted to drier and drier climes, many bryophytes became specialists in inhabiting moist places, perhaps less suited to many vascular plants. Bryophytes weren't entirely restricted to moist habitats, however, since many have successfully adapted to hot dry desert conditions, high windswept rocky mountain outcroppings, and the frigid polar regions. Further, along with lichens, these simple plants contribute to the dense tundra ground cover of the Arctic. It seems the only conditions bryophytes cannot tolerate are those produced by humans—they are notably absent in areas of severe air pollution.

There are as many as 16,500 named species of bryophytes, two thirds of them mosses. The others, primarily liverworts, are less common. Chances are that you can't recall ever having seen a liverwort, but we are about to change this.

Class Hepaticae: Liverworts

The name liverwort dates from the ninth century, when the plant was popularly used for the treatment of liver ailments. (The liverwort's lobelike growth was thought to resemble the liver. "Wort" means herb.) There are some 6000 species.

Liverworts produce a rather simple gametophyte, which can be either "leafy" or **thalloid,** the latter tending to be flat and resembling (for lack of a better analogy) branching, green cornflakes. Just about everyone's favorite liverwort is *Marchantia,* a thalloid liverwort. Its dark green gametophyte is a branching, ribbonlike growth that remains flattened against the earth (Figure 25.2a). Each branch is notched at its tip, and within the notches are a number of **apical** (growing) cells.

Also present on the leafy gametophyte are pore-like openings composed of several cell layers. Like the pore-like **stomata** of vascular plants, these openings permit an exchange of gases, including the inward diffusion of carbon dioxide so essential to the photosynthetic cells within. The pores also respond to changes in surrounding humidity, essentially narrowing when it is low and dilating when the humidity is high. They thereby help the plant avoid excessive water loss. Simple single-celled, threadlike rhizoids and multicellular scales anchor the plant. In liverworts, the rhizoids also absorb water from the soil.

Asexual reproduction in *Marchantia* occurs in two ways: **fragmentation** and **gemmae** (sing. **gemma**) production. In fragmentation, small, isolated portions of the plant body grow into a complete gametophyte. Gemmae are small multicellular bodies that form in concave structures called **gemma cups** (Figure 25.2b). Like a gametophyte fragment, a dislodged gemma is capable of forming an entire body.

TABLE 25.1 GEOLOGICAL HISTORY OF PLANTS

ERA	PERIOD(S) (MILLIONS OF YEARS AGO)	CONDITIONS	PLANT HISTORY	
Cenozoic	Quarternary Tertiary	Glaciation, mountain building, cooling	Extensive grasslands	AGE OF ANGIOSPERMS
	65			
Mesozoic	Cretaceous	Rocky mountains Extensive lowlands	Flowering plants	AGE OF GYMNOSPERMS
	135			
	Jurassic	Lowlands, inland seas	Conifers Cycads	
	197			
	Triassic	Mountains, drying		
	225		Most recent continental drift begins	
Paleozoic	Permian	Glaciers Inland seas dry up	Ginkgos Earliest conifers	
	280			
	Carboniferous		Ferns, horsetails whisk-ferns, club mosses	GREAT PALEOZOIC FORESTS
	345	Mountain building		
	Devonian		Earliest byrophytes	AGE OF SEEDLESS PLANTS
	405		Lycophytes	
	Silurian	Extensive shallow seas, mild climate		
	425		Earliest vascular plant fossils (rhyniophytes)	
	Ordovician			PLANTS INVADE LAND
	500			
	Cambrian		First plant fossils	PLANT LIFE BEGINS
	570			
Precambrian			(Algae)	AGE OF ALGAE

(a)

(b)

FIGURE 25.2
THE LIVERWORT
(a) *Marchantia,* a liverwort, has a spreading, ribbonlike gametophyte. (b) Gemma cups containing multicellular gemmae are seen.

Marchantia has complex sexual structures. (Figure 25.2). Sexual reproduction begins when the gametophyte produces two kinds of **gametangia:** egg-forming **archegonia** and sperm-forming **antheridia.** Such gametangia occur within peculiar stalked structures, **archegoniophores** and **antheridiophores,** which resemble miniature palms and parosols, respectively. Archegonia and antheridia may occur on the same individual, or they may occur on separate individuals, depending upon the species. In the plant world, the two conditions are termed **monoecious** ("one house"), with male *and* female parts on the same individual; and **dioecious** ("two houses"), with one individual having male *or* female parts. When the gametangia mature, swimming sperm are dislodged by raindrops, and as the drops splatter, the sperm are literally splashed against the archegonia. Upon entering an archegonium, a sperm will fertilize the egg within, and with this union, a new diploid sporophyte phase in the life cycle begins (Figure 25.3).

The minute liverwort sporophytes remain small and attached to the gametophyte, eventually developing **sporangia** structures in which meiosis occurs and haploid spores are formed. Liverworts, like other bryophytes, are homosporous. Thus, the spores are all similar. This ends the brief diploid sporophyte phase of the life cycle.

Spore dispersal is assisted by the presence of odd, spiral cells called **elaters,** cells with spiral thickenings. The elaters respond to minute changes in humidity by coiling and uncoiling, and this twisting movement dislodges and frees the spores. Each spore, under suitable conditions, can produce a new gametophyte.

Class Anthocerotae: Hornworts

The hornworts are a relatively minor group, with only 100 or so named species. Although the gametophyte somewhat resembles that of the liverworts, its lobed growth forms a rosette rather than a ribbon. Further, in *Anthoceros,* a representative genus, each cell contains one large chloroplast along with a starch-storing pyrenoid. These are very primitive traits, characteristic of some green algae. Antheridia and archegonia are produced directly in the gametophyte. Fertilized egg cells develop into unusual diploid sporophytes—long, slender, hornlike growths (from which the name "hornwort" was derived) that form slender sporangia. When the sporangia split, spores are released.

Two other aspects of the sporophyte's growth are unusual. First, it can grow continuously, producing spores over a long period of time. This is made possible by an active **meristematic region,** an area of simple, undifferentiated, actively dividing cells at its base. Second, under exceptional conditions it can survive on its own should the supporting gametophyte die. The hornwort sporophyte has a water-resistant coating or cuticle and exchanges gases with the air via structures called stomata, (pores formed by **guard cells,** that open and close the pore through shape changes) both of which are characteristic of more complex plants. Like the young moss sporphyte, it is photosynthetic. Following meiosis in the sporophyte, the haploid spores are carried away to begin a new gametophyte generation. Some hornworts form a symbiotic relationship with nitrogen-fixing cyanobacteria, hosting them in their tissue spaces. This association provides the hornwort with a rich source of nitrogen compounds.

Class Musci: Mosses

The *mosses* are the most numerous and common of the bryophytes, with about 9500 named species (second only to the flowering plants). They grow best in moist conditions but are found in a variety of habitats. Mosses even thrive in deserts , some having made a simple adaptation to the drying conditions. They do not resist dehydration as do most other land plants. When their habitat becomes dry, these mosses simply dry out too, and, as you would expect, metabolic activity slows dramatically. The tinder-dry moss is not dead, however. With the next rainfall or heavy dew, water diffuses into the dry tissues, and the moss literally "comes to life," seemingly resuming where it left off.

Interestingly, a certain terrestrial animal can do the same thing. One group of *tardigrades* (obscure, microscopic animals that somewhat resemble arthropods), are also capable of being desiccated (sometimes for years) and of then rehydrating successfully. It is no coincidence that the tardigrade species are found in an appropriate locale—living and browsing among the mosses.

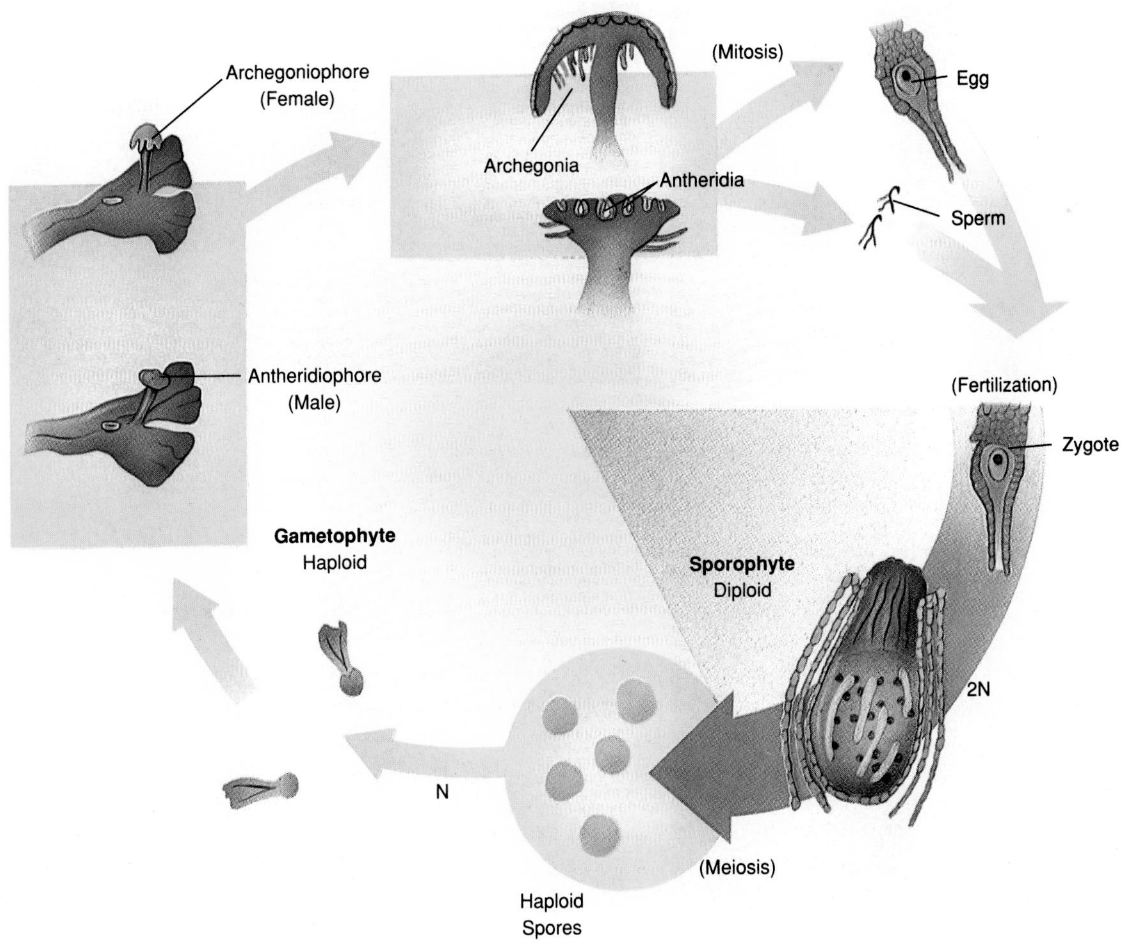

FIGURE 25.3
LIFE CYCLE OF A LIVERWORT
The sporophyte phase in *Marchantia* enters meiosis, yielding haploid spores that begin the gametophyte phase. The haploid gametophyte soon produces archegonia on the underside of umbrellalike archegoniophores and antheridia in the upper surface of antheridiophores. Fertilization starts a new sporophyte phase.

Mosses are of considerable ecological importance, particularly in Arctic regions where, along with lichens, they provide much of the ground cover. As with lichens, they are also well known as pioneer plants, species responsible for colonizing previously uninhabited regions and producing the first soils. Thus, we often find mosses growing on bare rocky outcroppings. Gardeners commonly use mosses for decorative purposes, and *Sphagnum* (peat moss) is used as a mulching or bedding material for lawns and gardens. In Ireland and a few other places, *Sphagnum* decays, forming peat, which is dried and used as heating and cooking fuel (and is also burned to provide the smoky flavor of Scotch whisky).

The Moss Life Cycle We begin the moss life history with the haploid spore and the gametophyte phase (Figure 25.4a). Once a haploid spore has established itself in a suitable place, it will absorb water and sprout into a young threadlike moss gametophyte

FIGURE 25.4
TYPICAL MOSSES
(a) Gametophytes of the common moss *Polytrichum juniperinum* form a soft carpet over the soil. **(b)** Sporophytes are represented by sporangia raised up on setae (note explosive release of spores). **(c)** Each vase-like archegonium of *Mnium* gives rise to an egg cell and later houses the young sporophyte. **(d)** The small cells packed in the antheridium of *Mnium* are developing sperm.

(a)

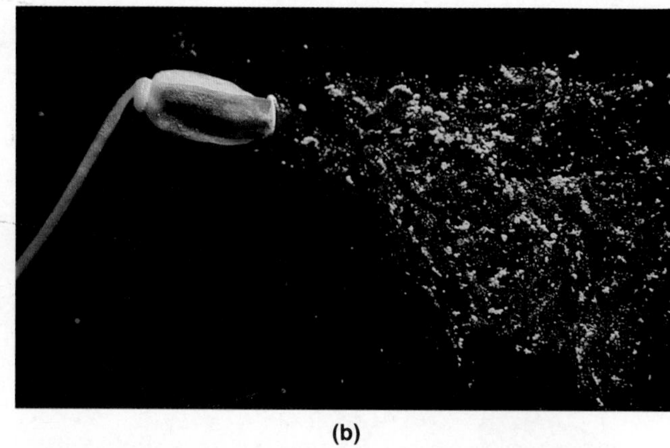

(b)

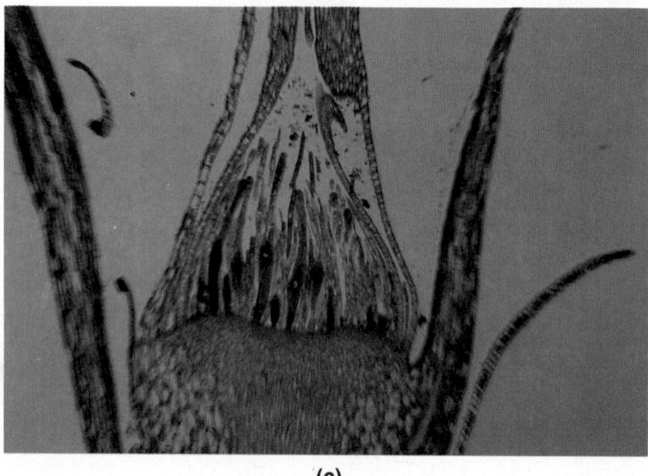

(c)

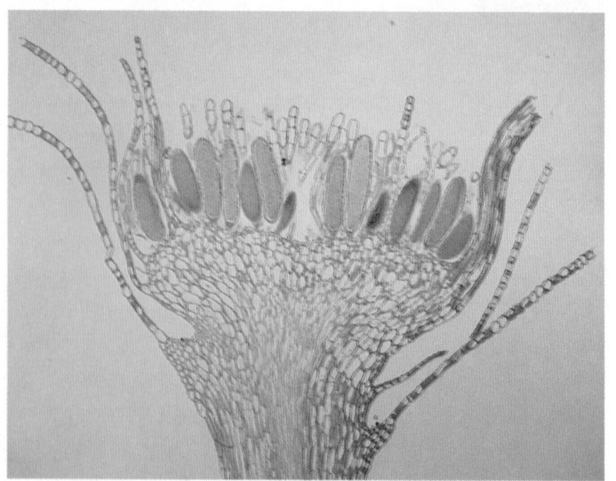

(d)

stage called a **protonema**. The protonema is very similar to certain filamentous green algae. If conditions are favorable, the protonema will bud in several places, finally producing the mature leafy gametophyte.

The Gametophyte The moss gametophyte is anchored in the soil or to a rocky substrate, by rhizoids, elongated threads of cells that emerge from the base of the leafy gametophyte. The moss rhizoid, however, does little to bring in water and minerals. The "stem" is a complex cylinder of many cell layers surrounded by a photosynthetic **epidermis** (covering layer). The epidermis surrounds a region of **cortical cells,** which in turn contains a **central cylinder** of thick-walled, toughened cells. Many mosses contain cells that resemble the conducting tissues of vascular plants. The moss equivalent of a leaf is a flattened, scale-like growth, only one cell in thickness and lacking in supporting and conducting elements. In some species, a water-proofing layer of waxy *cutin* covers the leafy scales.

At maturity, the moss gametophyte enters into sexual reproduction. First, gametangia (Figure 25.4 c and d) form at the top of each plant. Some species are monoecious, so that both archegonia and antheridia occur on the same individual. Others are dioecious, in which case the two types of gametangia occur on separate plants. Both archegonia and antheridia are surrounded by a protecting layer of nonreproductive cells called **sterile jacket cells.** These cells are an important terrestrial adaptation, providing protection for the embryo while in the archegonium.

If water is available, sperm emerge from the mature antheridium, enter a neighboring archegonium, and fertilize the egg. As with the liverworts, the water often takes the form of raindrops, which literally splash the sperm from the antheridia to the archegonia. Some researchers maintain that the sperm are directed into the archegonium by chemical attractants.

The Sporophyte Fertilization ushers in the diploid sporophyte generation. The typical moss sporophyte (Figure 25.4b) consists of a broad **foot** firmly anchored in the archegonium and a slender stalk called a **seta,** which has a sporangium at its tip. The seta is exceptional for a bryophyte structure, in that it contains *vascular tissues.* **Hydroids** and **leptoids** conduct water and food, respectively.

Moss sporophytes are generally quite small, but in some species the stalk may become as long as 20 cm (about 8 in.). At first, the sporophyte is photosynthetic, but ultimately it becomes nutritionally dependent upon the gametophyte. Like vascular plant sporophytes, some moss sporophytes have a water-resistant cuticle and many stomata.

Within the sporangium, cells undergo meiosis, producing numerous haploid spores. In some species, when the capsule is mature and spore-filled, the **operculum** (cap) falls away, revealing the **peristome,** a ring of tooth-like barriers. Under moist conditions, the teeth remain interlocked, but when conditions are dry, the teeth bend outward, and the spores escape. The likelihood of any one spore landing in a suitable place is not very good, but the low odds are countered by the great number of spores each capsule produces (up to 50 million). Since spore formation involves meiosis, it brings in the gametophyte generation once again. The moss life cycle is reviewed in Figure 25.5.

Bryophytes and the Evolutionary Origins of Vascular Plants

How do the bryophytes fit into the evolution of plants? We can report with confidence that no one today seriously believes that the bryophytes were ancestors of the vascular plants. But here the consensus ends. Some botanists maintain that the mosses, liverworts, and hornworts evolved independently of the vascular plants from their own green algal ancestor. They cite a number of similarities of bryophytes to certain green algae, including early growth stages in mosses that resemble filamentous green algae, almost identical pigments (including chlorophylls *a* and *b* and xanthophylls) as well as the pyrenoids, starch-storage bodies curiously present in the photosynthetic cells of hornworts and some primitive green algae.

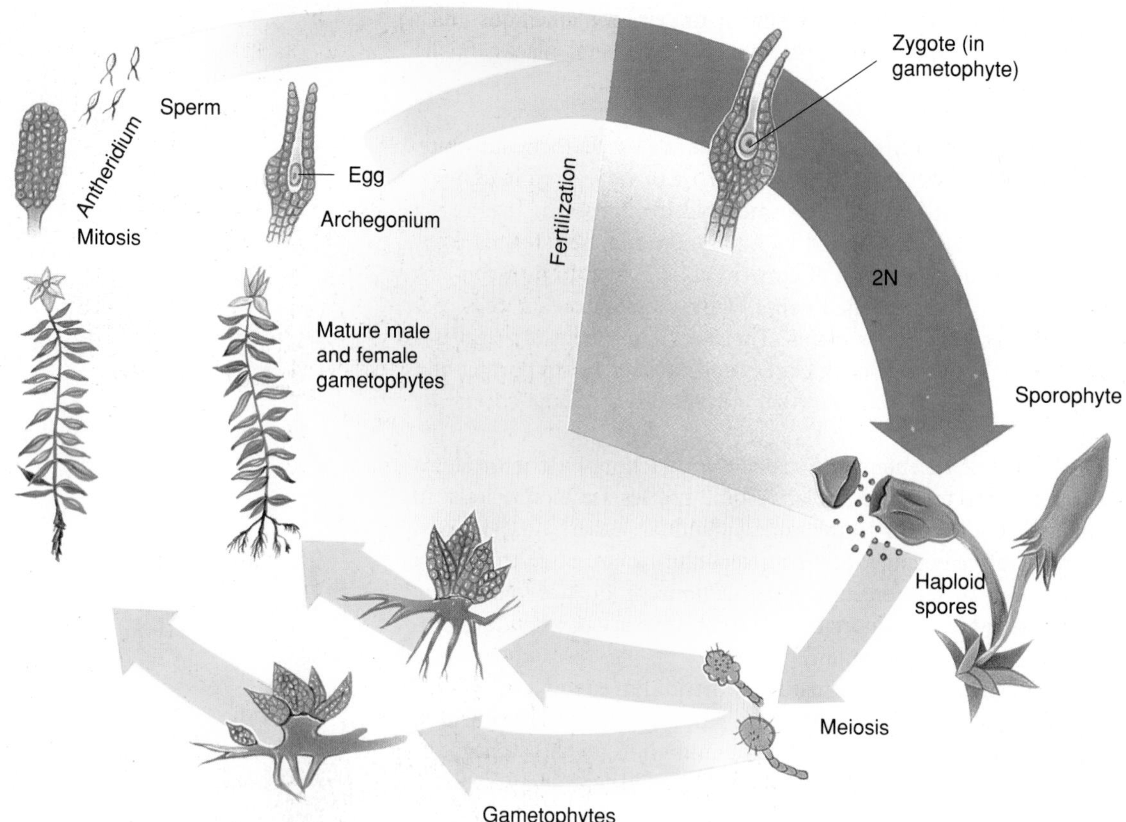

FIGURE 25.5
ALTERNATION OF SPOROPHYTE AND GAMETOPHYTE PHASES IN A MOSS
The moss sporophyte, a stalked sporangium, produces haploid spores that give rise to the gametophyte. The dominating leafy gametophyte produces archegonia and antheridia in which egg and sperm form. Fertilization starts a new sporophyte, which emerges directly from the gametophyte.

There are other theories, however. In particular, some botanists maintain that the bryophytes and vascular plants share a common green algal ancestor. They propose that the bryophytes diverged from the early vascular plant line about 50 million years after it had begun. The bryophyte's lack of vascular tissue, they suggest, is a product of evolutionary simplification—a loss of whatever vascular development had occurred in their immediate ancestors. Further, they add that the unique characteristics of bryophytes are the product of evolutionary specialization, so the bryophytes may not be as primitive as some people think. In fact, they may be relative newcomers. Recognizable fossils of mosses can only be traced back to the Devonian period, about 350 million years ago, while those of vascular plants can be traced back to the Silurian, over 400 million years.

One of the best guesses is that the bryophytes and vascular plants share a common ancestor, and that it resembled today's green alga *Coleochaete* (Figure 25.6). One line of evidence is the manner in which mitosis and meiosis occur. As in the plants, the nuclear envelope in *Coleochaete* is dismantled during cell division, and following this, formation of the cell plate between daughter cells occurs through the coalescing of Golgi vesicles (see Chapter 9, Figure 9.8).

THE VASCULAR PLANTS

Bryophytes are interesting, but if someone were to ask you to think of a plant, you would probably think of a vascular plant, such as an oak, rose, or pine, and not a hornwort. Most people simply are far more familiar with vascular plants than with the others.

The vascular plants are often referred to by the more technical term **tracheophyte** (*tracheo-*, "tubes"; *-phyton*, "plants"), which refers to the tubelike vascular elements of the **xylem** and **phloem**, which transport water and foods, respectively. Our comments will be general here, since these tissues are discussed in detail in Chapter 27. The most important point we should make is that vascularity was a smashing evolutionary success, providing an answer to some of the terrestrial problems of evolving plants. First, vascular tissue, including the woody, hardened xylem and tough accompanying supporting fibers, allowed plants to grow to immense sizes. Apparently, large size provided a novel way of adapting to the terrestrial environment—one not available to the bryophytes. Second, and closely related, once the problem of desiccation was managed, the successful transition to terrestrial life also depended on getting water up from the soil to the photosynthetic tissues. Thus, in the presence of increasing competition among rapidly emerging species, natural selection often favored root systems that could penetrate deeply and spread widely in the soil, stems that could hold the foliage higher, and in some instances, leaves that grew longer and broader.

The origin of vascular plants is lost in antiquity, but the earliest fossils are to be found in the 400 million-year-old rocks of the Silurian period of the Paleozoic era. Among the earliest vascular plants is an extinct group called the **rhyniophytes.** They made their debut along with the first vertebrates—those clumsy, unimpressive armored fishes with jawless, sucking mouths. (But even these creatures were probably not very impressed by the seemingly insignificant rhyniophytes.)

Consider *Rhynia,* whose fossils are well known. It grew in foul marshes, probing the air with simple leafless stems arising from **rhizomes**, a form of stem that grows horizontally (Figure 25.7). The peculiar growth of the fingerlike stems was important for two reasons. First, their branching and rebranching provided a vascular framework for the later evolution of the flattened leaf blade. But equally significant, the primitive stems contained a system of tiny tubelike, nonliving, water-filled elements we call **tracheids**, a part of the water-conducting xylem common to today's vascular plants. With the tracheids came the promise of new horizons for, as we have seen, xylem not only conducts water, but its tough, fibrous elements provide vital support in many plants. Xylem forms the wood of woody stems, providing the strength that permitted vascular plants to grow tall, altering the very appearance of the earth.

Today, vascular plants inhabit nearly every part of the planet that protrudes above water, and a few that do not. As they radiated out over the landscape, they adapted to their environments in different ways and thus produced a wide variety of forms. Most developed true (vascular) leaves, stems, and roots (and some evolved seeds). A finishing touch was the development of mechanisms for fertilization that did not require the flagellated sperm or a watery medium for sperm travel. You may recall that the more recently evolved plants, and the fungi, lack centrioles and the 9 + 2 microtubular flagella that characterized the swimming sperm of their ancestors.

Vascular plants have been classified in a number of ways, including one in which the tracheophytes are grouped under division[1] *Tracheophyta*. But, according to most taxonomists, there are nine divisions of vascular plants, four of which release spores that develop into inconspicuous—but independent—gametophytes. The great 18th century taxonomist, Linnaeus, was so confounded by the apparent lack of familiar reproductive structures in these plants that he named them **cryptogams**, which quaintly translates as "hidden marriage." The rest of the plants, with their highly visible cones and flowers, came to be known as **phanerogams**, or "apparent marriage."

Of the nine divisions, six probably will be unfamiliar to everyone except perhaps a few knowledgeable gardeners. Note that the vascular plants are further divided into those that do not produce seeds and those that do. Be aware that the names will differ from one source to another.

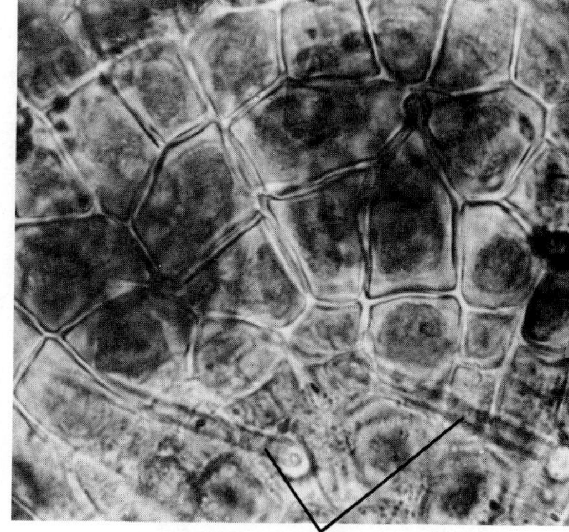

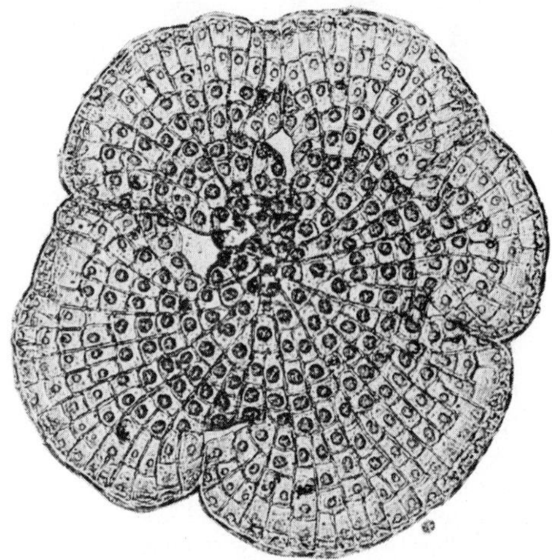

FIGURE 25.6
AN ALGAL ANCESTOR
Green algae similar to *Coleochaete* are candidates for the ancestor of both bryophytes and the vascular plants. These algae are quite small, about 0.5 mm in diameter. Although it is still around and doing well in the shallows of lakes and ponds. *Coleochaete* is thought to be quite primitive, exhibiting a number of characteristics that were present in its ancestors.

[1]The taxonomic term *division* is used by botanists instead of phylum.

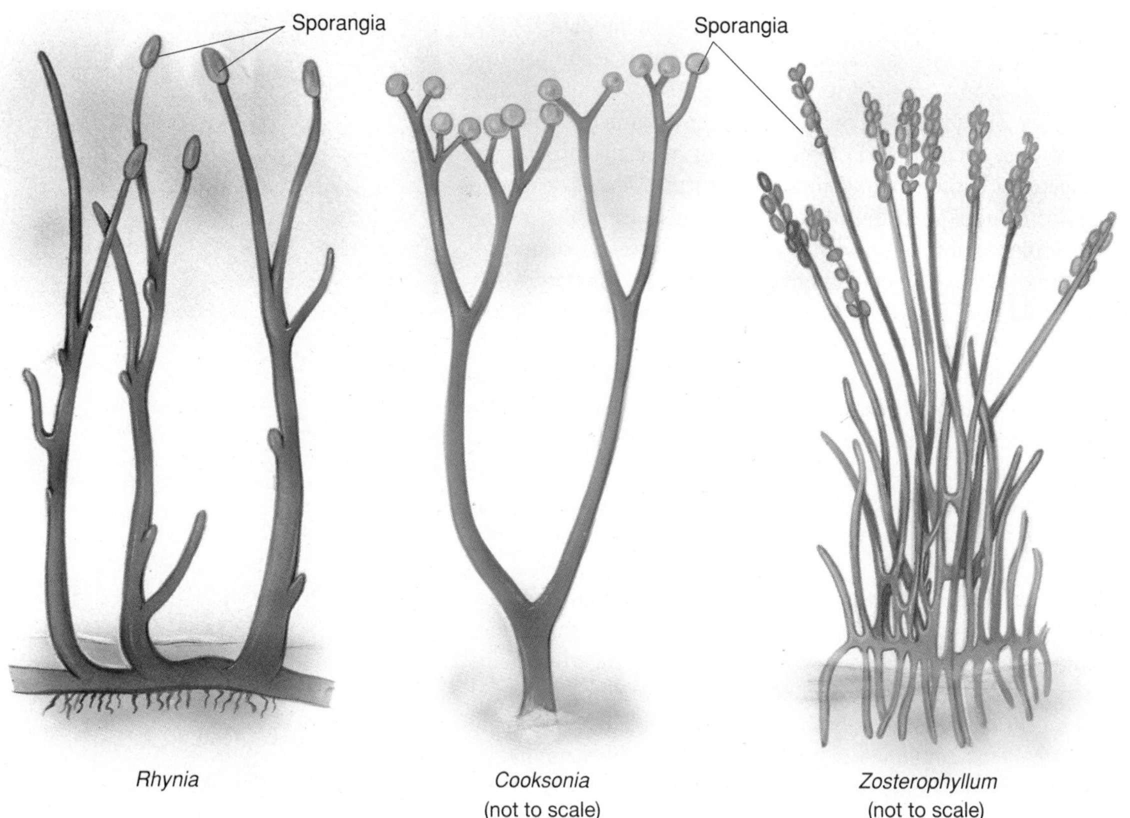

FIGURE 25.7
THE EARLIEST VASCULAR PLANTS
The early vascular plants were quite small. Some, like *Cooksonia* and *Zosterophyllum,* were just a few centimeters tall (*Rhynia* was nearly 50 cm.). Although they lacked leaves and had simple, branching stems, tipped by primitive sporangia, the presence of vascular tissue signalled the start of an enduring trend. Vascular plants make up the vast majority of plant species today.

The nine divisions are:

I. **Seedless plants:** Cryptogams (tiny independent gametophytes)
   Division Psilophyta ("naked plants"): whisk ferns
   Division Lycophyta ("spiderlike plants"): ground pine or club mosses
   Division Sphenophyta ("wedge plants"): horsetails
   Division Pterophyta ("winged plants"): ferns

II. **Seed plants:** Phanerogams (tiny gametophyte incorporated into sporophyte)
   A. **Gymnosperms** (naked seeds)
      Division Cycadophyta: cycads
      Division Ginkgophyta (maidenhair tree): ginkgo
      Division Gnetophyta: gnetophytes
      Division Coniferophyta ("cone-bearers"): pines and other conifers
   B. **Angiosperms** (seeds with fruit)
      Division Anthophyta ("blossom or flower plant"): the flowering plants

(Also see Table 25.2.) The first three divisions were once an important part of the earth's greenery, but their heyday is over, and they are now relatively minor groups. However, we will consider them anyway because they are interesting and because some of their characteristics may represent significant stages in the evolution of the seed plants. Of an estimated 258,000 named species of vascular plants, only about 1000 species comprise the first three groups, and of these, all but 19 are found in division Lycophyta. Thus, in considering these divisions, we will be looking at a very few survivors of another time, before the ferns and then the seed plants became the dominant forms of plant life on earth. Of the second group—the seed plants—the first four divisions are commonly grouped into the **gymnosperms,** whereas the fifth group, the flowering plants (Division Anthophyta), are usually called the **angiosperms.** Figure 25.8 summarizes the evolutionary history of the vascular plants.

Division Psilophyta: Whisk Ferns

Psilophytes, members of Division Psilophyta, are found in the fossil record of 375

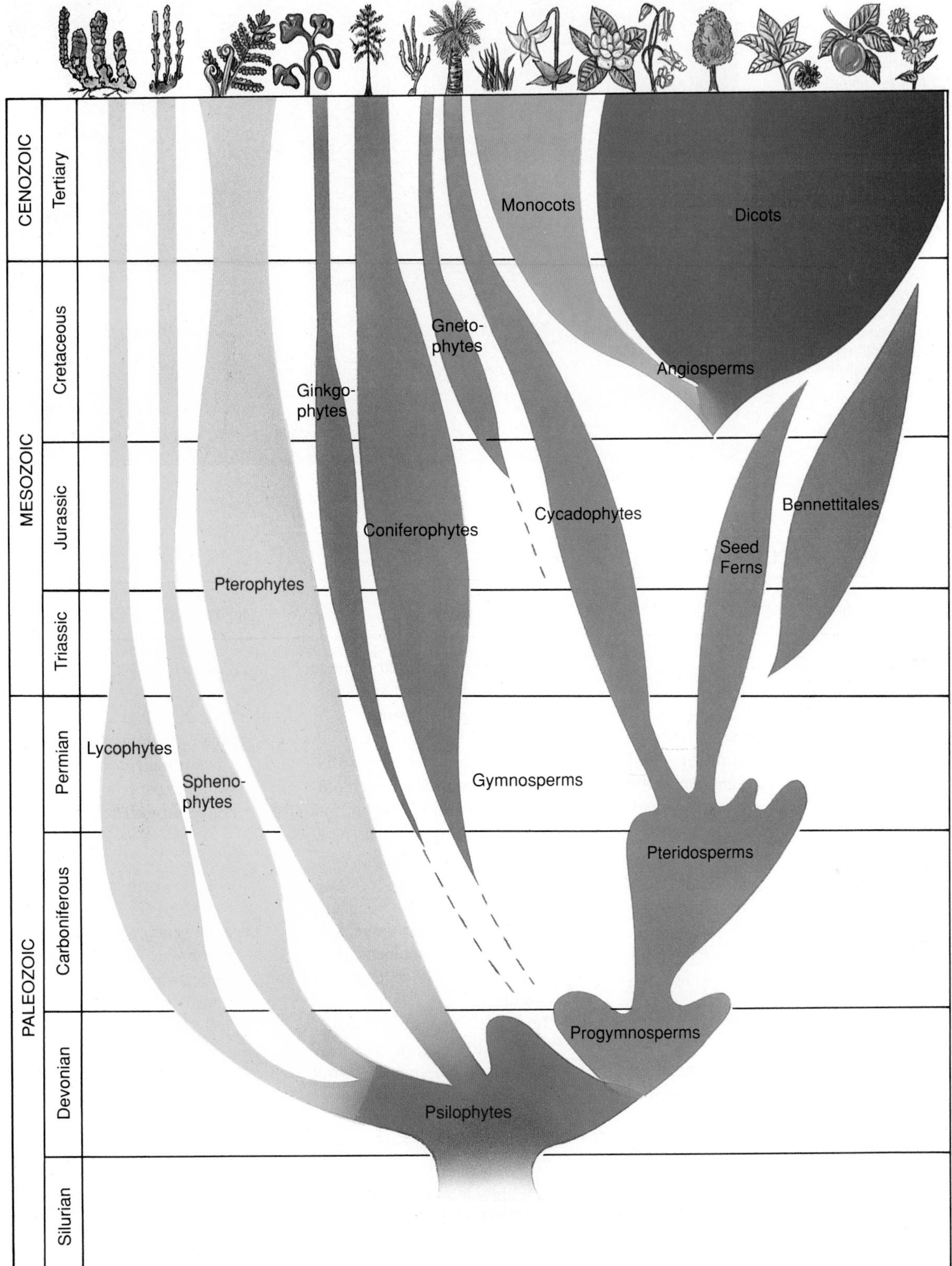

FIGURE 25.8
VASCULAR PLANT PHYLOGENY
One of several possible phylogenetic trees of vascular plants. The width of each line of descent (color) indicates in a comparative way how successful each group was through time. The dashed line indicates uncertain origins and affiliations. The lines of extinct groups stop short of present times.

FIGURE 25.9
PALEOZOIC FOREST
Paleozoic times. The earliest forests became established in the Devonian period of the Paleozoic era. By the Carboniferous period, huge forests of lycophytes, psilophytes, sphenophytes, and primitive seed ferns flourished. The luxurious growth seen here is now represented on earth by fossil fuel deposits. (Note the primitive animals.)

① Eusthenopteron
② Eogyrinus
③ Diplovertebron
④ Meganeuron
⑤ Eryops
⑥ Seymouria
⑦ Limnoscelis
⑧ Varanosaurus
Ⓐ Eospermatopteris
Ⓑ Calamites
Ⓒ Lepidodendron
Ⓓ Sigillaria
Ⓔ Cordaites

TABLE 25.2 SUMMARY OF THE PLANT KINGDOM

KINGDOM PLANTAE: DIVISIONS	EXAMPLES	CHARACTERISTICS	LIFE CYCLE
THE NONVASCULAR PLANTS		Little or no organized tissue for conducting water and food (xylem and phloem)	
Division Bryophyta (15,600 species)	Mosses, liverworts, hornworts, *Polytrichum, Riccia, Anthoceros, Marchantia*	Terrestrial and freshwater aquatic; multicellular with tissue specialization; chlorophylls *a* and *b*; many adaptations to land	Separate generations; gametophyte usually dominant; swimming sperm; egg and zygote protected
THE SEEDLESS VASCULAR PLANTS: **Cryptogams**			
		Organized vascular tissue; limited root development; many primitive traits; chlorophylls *a* and *b*	Separate generations; dominant sporophyte; minute gametophyte; swimming sperm
Division Psilophyta (4 species)	Whisk ferns, *Psilotum*	Terrestrial; tropical; primitive vascular tissue; no true leaves	
Division Lycophyta (about 1000 species)	Ground pine or club moss, *Lycopodium, Selaginella*	Terrestrial; widespread; well-developed vascular tissue in some; true leaves	
Division Sphenophyta (15 species)	Horsetails, *Equisetum*	Terrestrial; widespread; underground stems (rhizomes); simple, hollow aerial stems; well-developed vascular tissue	
Division Pterophyta (11,000 species)	Ferns: *Sphaeropteris, Platycerium, Osmunda*	Terrestrial; moist soils to dry; underground stems; highly developed variable leaf; well-developed vascular tissue	
THE SEED-PRODUCING VASCULAR PLANTS: **Phanerogams**			
		Seeds; organized vascular tissues; extensive root, stem, leaf development; chlorophyll *a* and *b*.	All with dominant sporophyte, which retains the gametophyte; pollen producers

million years ago, making them a remarkably persistent group of plants. The psilophytes, along with the lycophytes and sphenophytes, formed the vast, swampy Paleozoic forests (Figure 25.9). The psilophytes may be representative of an early stage in the evolution of vascular plants, but we can't be sure because there are fossils of other, more advanced plants that date even further back. The division is represented today by two genera and only four species, all found mainly in the tropics. Sporophytes of *Psilotum nudum,* the "whisk fern," (Figure 25.10), produce hairy rhizoids similar to those of the bryophytes, but no true roots. The stems are fairly well developed, with simple, primitive vascular tissue. The leafless stems are photosynthetic and bear simple, primitive sporangia above tiny scalelike appendages. The sporophyte phase in the life cycle of the psilophytes is quite dominant. The gametophyte, a tiny, separate, underground plant, produces both antheridia and archegonia. Following fertilization, the sporophyte emerges from the gametophyte in a manner reminiscent of the mosses. However, it soon becomes independent, and the tiny gametophyte withers and dies.

Psilotum ranges along our southernmost states and in Hawaii and Puerto Rico, where subtropical and tropical conditions prevail. Its sister genus, *Tmesipteris,* lives chiefly as an **epiphyte** (a plant that grows on larger plants, but not parasitically). It occurs only in Australia, New Caledonia, New Zealand, and some southern Pacific islands.

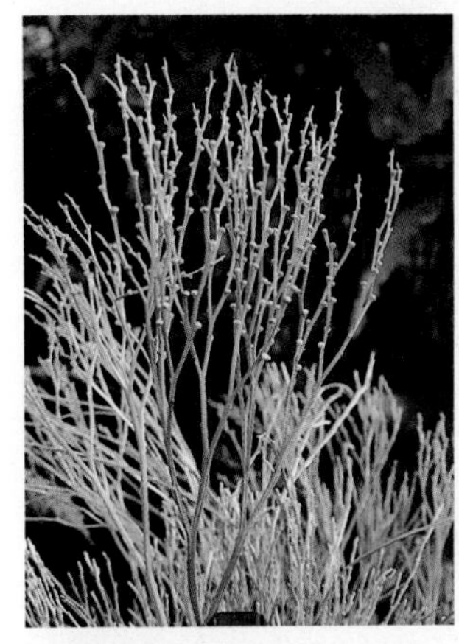

FIGURE 25.10
A LIVING FOSSIL
Psilotum, a vascular plant with many primitive traits, produces an erect, branching sporophyte with simple vascular tissue. Spores are produced in clustered sporangia.

KINGDOM PLANTAE: DIVISIONS	EXAMPLES	CHARACTERISTICS	LIFE CYCLE
Gymnosperms		Seeds not enclosed by fruit; water conduction by tracheids only; terrestrial	
Division Ginkgophyta (1 species)	Maindenhair tree, *Ginkgo biloba*	Native of China, widely cultivated; large tree; fan-shaped leaf	Swimming sperm (utilizes plant fluids)
Division Cycadophyta (100 species)	Cycads: *Zamia, Cycas*	Widespread but mainly tropical; thick, partially subterranean stem; palmlike foliage; very large pollen and seed cones	Swimming sperm (utilizes plant fluids)
Division Gnetophyta (71 species)	*Gnetum, Welwitschia, Ephedra*	Warm-temperate regions, deserts; angiosperm features: bladelike leaf, two types of water-conducting tissue (tracheids and vessels), flowerlike pollen cones	Nonmotile sperm
Division Coniferophyta (550 species)	Pines, redwoods, firs, junipers, larch, cypress, hemlock	Widespread in temperate and subarctic areas; needle leaves; common; evergreen; multiple cotyledons	Nonmotile sperm; wind-pollinated
Angiosperms		Flowers; seeds with fruit surrounding	Nonmotile sperm; wind, water, and animal pollinated
Division Anthophyta			
Class Dicotyledonae (about 170,000 species)	Dicots: magnolia, cabbage, tobacco, cotton, willow, apple, oak, maple, bean	Worldwide; widely cultivated; net-veined leaf; floral parts in 4s and 5s or their multiples; two cotyledons; primary and secondary growth	
Class Monocotyledonae (about 65,000 species)	Monocots: iris, orchids, grains (grasses)	Worldwide; widely cultivated; parallel-veined leaf; floral parts in 3s or its multiples; one cotyledon; most have primary growth only	

FIGURE 25.11
CLUB MOSSES
Club mosses (lycophytes) are not mosses at all, but vascular plants with true roots, stems, and leaves. Note the microphylls (primitive leaves) that emerge from the stem epidermis. In this species, sporophylls are clustered into strobili or cones at the tips of leafy upright stems.

Division Lycophyta: Club Mosses

The most familiar **lycophytes** are those in the genus *Lycopodium* (Figure 25.11), whose members range widely from the Arctic to the tropics. Many tropical species are epiphytes. You may have seen *Lycopodium* sporophytes without knowing it, since they are often mistaken for pine seedlings. In some parts of the United States they are known as **ground pine** or **club moss.** They are green year-round and quite conspicuous in the winter when other ground cover has died back. It was once considered quaint to gather them to make a Christmas wreath, but the practice has actually endangered this interesting little plant and its disturbance is now prohibited by law in many places.

Lycopodium sporophytes are dominant, and they have vascular roots, stems, and leaves. In the adult plant, the roots are of the **adventitious** type; that is, they emerge from the stem rather than from a primary root. The vascular system of *Lycopodium* is well developed, extending from the adventitious roots through stem and leaves. The leaves, however, are small and arranged in a whorl (a spiral) around the stem. They are called **microphylls** (small leaves) and are believed to have evolved as simple outgrowths from the stem epidermis (outermost covering tissue). They contain a single vein of vascular tissue. In contrast, the leaves of higher vascular plants are called **megaphylls** (large leaves). They contain many veins and owe their evolutionary origin to the fusion of branching stems (such as those seen in *Rhynia*). Some lycophyte species produce horizontal runner stems, which eventually form a mat over the forest floor, with occasional roots and vertical stems springing from the runners.

In lycophytes, spores are formed in sporangia associated with **sporophylls** (since they are really specialized microphylls). The sporophylls are arranged along the stem in some species, but in others they cluster together, forming **strobili** (singular, **strobilus,** also called a cone) at the ends of the branches. When the haploid spores are released, they germinate and grow into tiny gametophytes that produce both antheridia and archegonia. Fertilization occurs when there is sufficient water over the gametophyte for sperm to swim to the archegonia. Following fertilization and development, the diploid sporophyte emerges from the gametophyte, becomes independent, and a new cycle begins.

Selaginella **A Reproductively Advanced Lycophyte** Most of the 700 species of the lycophyte genus, *Selaginella,* are tropical, favoring moist regions, although a few inhabit deserts. One desert species, the "resurrection plant" of Mexico, Texas, and New Mexico, loses its green color, curls up, and "dies" (becomes dormant) during droughts, only to "return to life" at the first sufficient rainfall.

Selaginella is a genus of considerable evolutionary interest because it is heterosporous—it produces two types of spores: microspores and megaspores. The plants we have considered so far, including *Lycopodium,* are **homosporous;** only one kind of spore is produced. It follows then, that gametophytes are also monoecious. Heterosporous plants are dioecious: microspores give rise to gametophytes that produce antheridia, and megaspores give rise to separate gametophytes that produce archegonia.

Selaginella produces its microspores and megaspores in **microsporangia** and **megasporangia,** structures located within its strobili. Following the release of the microspores, their development leads to the formation of a **microgametophyte,** which at maturity contains a sperm-producing antheridium. Each megaspore, it follows, forms a **megagametophyte,** which will in turn produce a number of archegonia, each of which bears one egg. When sufficient water is available, the antheridia burst, releasing motile sperm that swim into the necklike opening of each archegonium and fertilize the egg cell within. Until photosynthesis can begin, growth and development of the young sporophyte embryo is supported by nutrients from cells in the surrounding megagametophyte (Figure 25.12).

We've invested some time on the life cycle of *Selaginella* because its life cycle is surprisingly like that of seed plants. We'll come to these shortly, so you will hear more about microspores and megaspores, and microgametophytes and megagametophytes.

TABLE 25.3 COMPARISON OF HOMOSPOROUS AND HETEROSPOROUS SPECIES

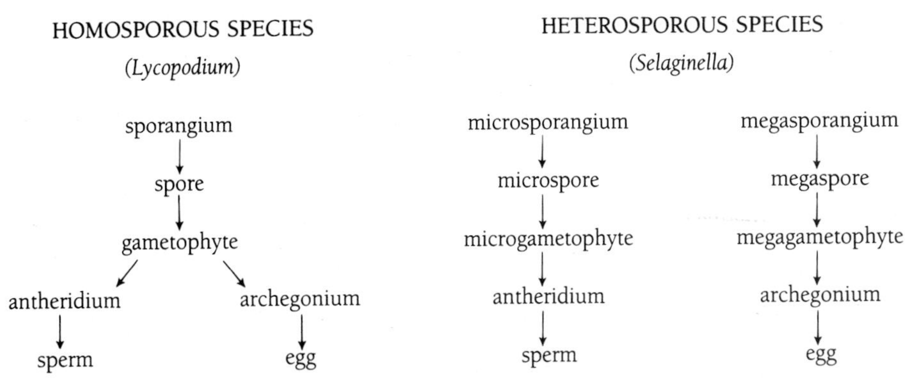

HOMOSPOROUS SPECIES (*Lycopodium*)	HETEROSPOROUS SPECIES (*Selaginella*)	
sporangium	microsporangium	megasporangium
↓ spore	↓ microspore	↓ megaspore
↓ gametophyte	↓ microgametophyte	↓ megagametophyte
↙ antheridium ↘ archegonium	↓ antheridium	↓ archegonium
↓ sperm ↓ egg	↓ sperm	↓ egg

The lycophytes were not always the humble, ground-hugging plants we see today. One order, the Lepidodendrales, produced great treelike plants over 50 meters tall and 2 meters in diameter. In fact, 300 to 400 million years ago, the lycophytes were the dominant plants of the Devonian and Carboniferous forests (see Figure 25.9). But change is inexorable, and with the drying climate of the Permian period, the primitive giants died, possibly replaced by the newly evolving gymnosperms, but leaving only a few remnants of a once prominent group. Their ghosts now haunt us—the corpses of these great plants became partly decomposed to form vast coal and oil reserves, so-called fossil fuels that influence modern global politics as we squabble over the remains.

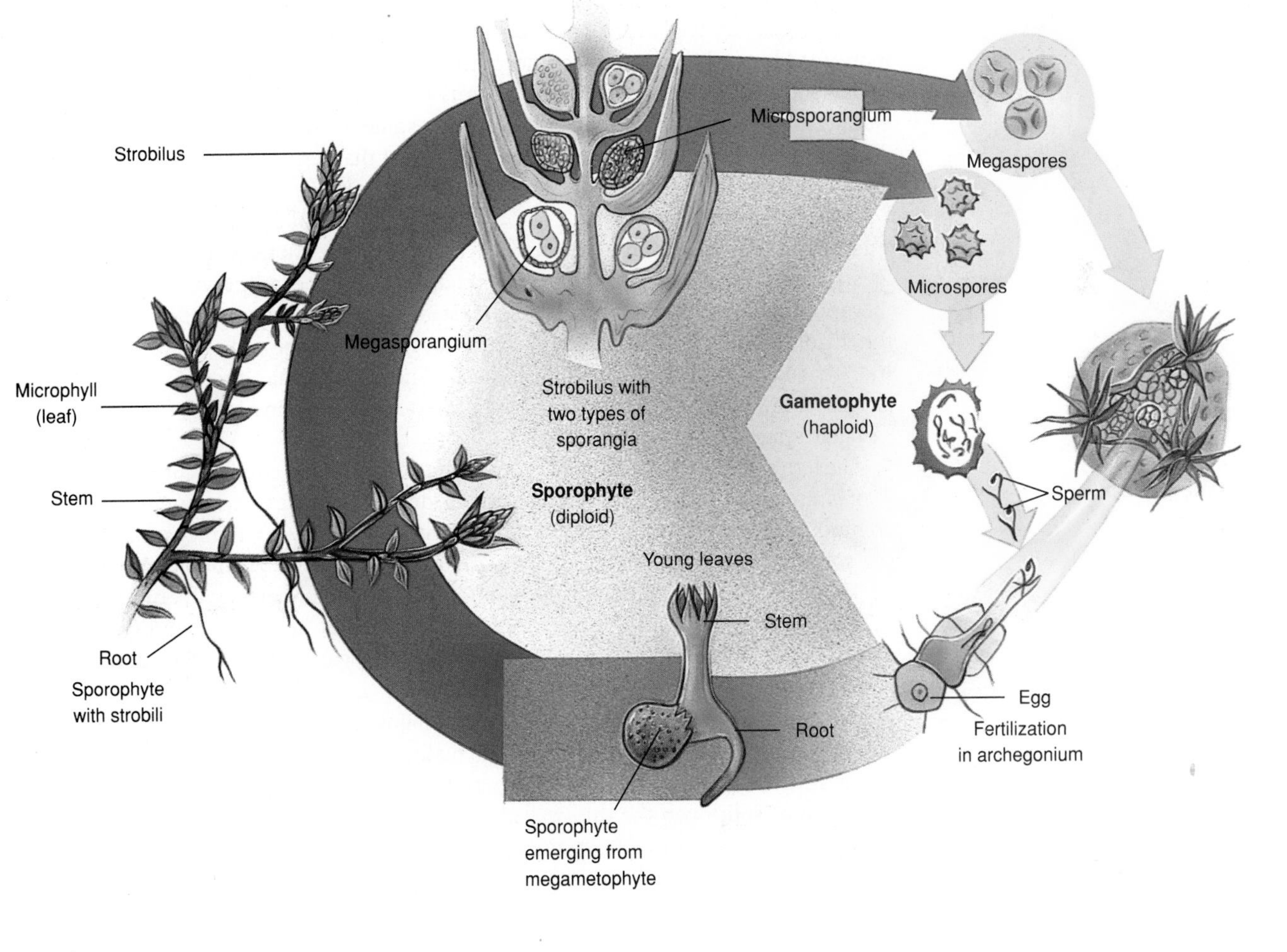

FIGURE 25.12
A HETEROSPOROUS LYCOPHYTE
Selaginella, a heterosporous lycophyte, produces strobili that bear microsporangia and megasporangia. Microspores and megaspores subsequently form separate microgametophytes and megagametophytes. Fertilization occurs when a sperm escapes from the antheridium and swims to the archegonium. The new sporophyte emerges from the megagametophyte.

Division Sphenophyta: Horsetails

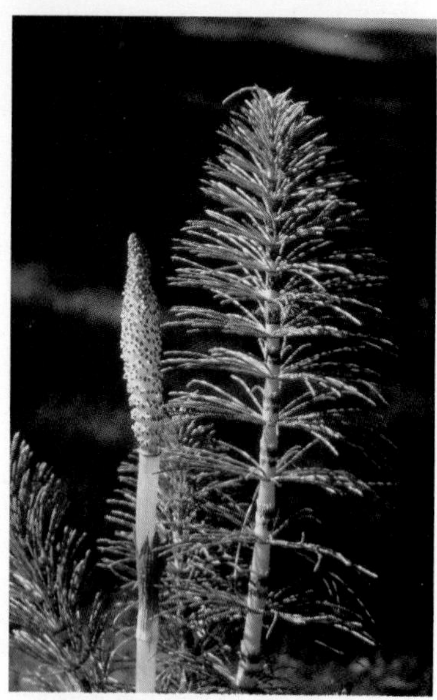

FIGURE 25.13
THE HORSETAIL
Some species of horsetail *(Equisetum)* produce two types of shoots. One is vegetative, with photosynthetic branches, whereas the other, a "fertile shoot," is nonphotosynthetic, unbranched, and bears a spore-producing strobilus at its tip.

The **sphenophytes**, like the lycophytes, had their day in the late Paleozoic, when they contributed significantly to the lush Carboniferous forests. The few surviving species are distributed worldwide, often favoring moist habitats, such as along the banks of rivers and streams. Today there is only one genus, *Equisetum* (*equi-*, "horse"; *setum,* "bristle"), with 15 or so species. Most species grow to less than a meter in height. One tropical giant, *E. giganteum,* is reported to be 6 m tall, but its scrawny stems are less than 3 cm (1.25 in) in diameter. It is commonly supported by neighboring trees.

Sphenophytes are often called "scouring rushes," a name reflecting their "pre-Brillo" use as pot cleaners. They are well suited for this because their cell walls contain glassy silica, a good abrasive.

Some species produce two kinds of shoots. Vegetative shoots bear scalelike microphylls and whorls of short, lateral branches. Reproductive shoots produce spores in prominent strobili and are unbranched (Figure 25.13). In other species vegetative branches bear strobili at their tips. Sphenophytes are homosporous. The spores develop into pinhead-sized, independent gametophytes that bear both types of gametangia.

Although the gametophytes have the potential to produce both antheridia and archegonia, some produce antheridia only, thus preventing self-fertilization. Once again, flagellated sperm swim from antheridia to archegonia, and with the advent of fertilization a new sporophyte generation begins.

Division Pterophyta: Ferns

Pterophytes—ferns—are undoubtedly among the most enchanting of plants. The spring forest, dripping with cool rain, is accented by delicate ferns rising with tiny bowed heads from the damp floor. Later, the plants will lend an exotic touch to the woods as they stand full grown, their leaves splayed, as if placed there as decoration.

Ferns have survived in great numbers since Paleozoic times, apparently having adapted more readily to changing environments than have the other primitive vascular plants. Today some 11,000 named species of ferns are widely distributed over the earth, including both tropical and temperate regions; some even live in arid climates (Figure 25.14). Further, a glance at Figure 25.8 will reveal that ferns have enjoyed considerable success since their inception in Devonian times.

The fern sporophyte (Figure 25.15) typically consists of a thick rhizome containing a number of tiny roots. Most ferns have mycorrhizal associations (see Chapter 24), which greatly enhance their absorption of water and minerals. The large divided leaves are true megaphylls, having evolved from branch systems. The leaves emerge above ground from the rhizomes as **fiddleheads,** each of which uncoils into a large and often frilly compound leaf (that is, a leaf subdivided into leaflets). The leaves come in all sorts of sizes and shapes. A few, such as the tree ferns, have tall stems supporting a leafy rosette. The fern vascular system is well-developed.

The fern's sporophyte is decidedly dominant and produces copious numbers of spores. Spores are usually produced on the underside of the leaves in sporangia, which are commonly clustered into structures called **sori** (singular, *sorus*). In their younger stages, the sori are sometimes hidden under a scalelike cover called an **indusium** (see Figure 25.15 c, d). You have undoubtedly seen the sori on the underside of ferns. They often occur as rows of brown dots and may be mistaken for an insect or fungal invasion. Their sudden appearance has sent many an alarmed gardener scurrying to the nearest plant nursery for pesticides.

The sporangia may be quite complex, containing elaborate mechanisms for spore release. In some species, spore release—actually, forceful ejection—occurs when rapid changes in the thick-walled cells of the **annulus** (a row of cells along one wall of the sporangium) split open the thin-walled **lip cells.**

Following their release, the spores are widely distributed by air currents. Those spores that settle into moist, well-protected, surfaces will germinate. Nearly all ferns are homo-

FIGURE 25.14
DIVERSITY IN FERNS
A small sample of the diversity in ferns includes **(a)** the tree fern *(Sphaeropteris)*, **(b)** the maidenhair fern *(Adiantum)*, **(c)** the floating fern *(Salvinia)*, and **(d)** the staghorn fern *(Platycerium)*. Ferns are a diverse group, perhaps because they were not seriously diminished by geological changes as were the other primitive vascular plants.

(a)

(b)

(c)

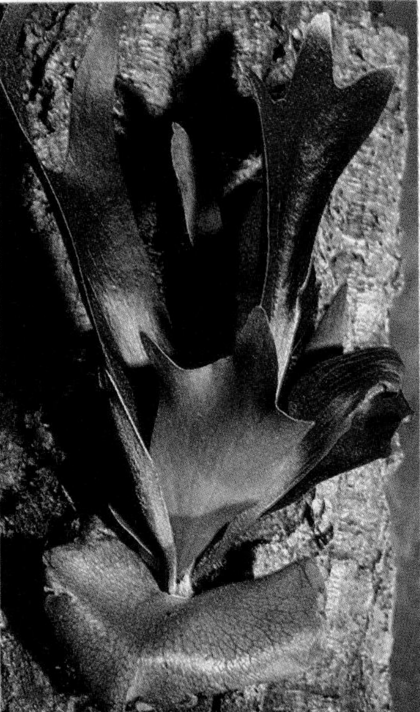

(d)

sporous, and as expected, the gametophytes produce both antheridia and archegonia. (Interestingly, a number of extinct species were heterosporous—more of the seemingly unpredictable course of evolution to ponder.)

The delicate fern gametophyte, typically flattened and heart-shaped, is known as a **prothallus** (Figure 25.16). Although short-lived, it is photosynthetic and independent, anchored to the soil by minute rhizoids. When water is available, the sperm escape from the antheridia and swim to the archegonia, where fertilization occurs. The diploid embryo, housed and protected for a brief time by the prothallus, develops into the familiar leafy sporophyte. Figure 25.17 summarizes the life cycle of a fern.

One might wonder whether ferns can self-fertilize by using such a system. After all, it usually isn't very far from an antheridium to the nearest archegonium on the wet surface of a single fern gametophyte. Actually, self-fertilization is not uncommon, but cross-fertilization is more adaptive from a genetic point of view. Accordingly, many ferns have evolved mechanisms for inhibiting self-fertilization. The first prothallus to mature produces a hormone known as **antheridogen** (similar to the hormone gibberellic acid, discussed in Chapter 29). The hormone *suppresses* male (antheridial) development in the gametophyte in which it is produced but *stimulates* male development in neighboring gametophytes. Thus, the hormone's effect is to produce separate male and female prothalli, promoting cross-fertilization. In other species, sperm and eggs from the same gametophyte are physiologically incompatible, and only fertilizations involving different gametophytes are successful.

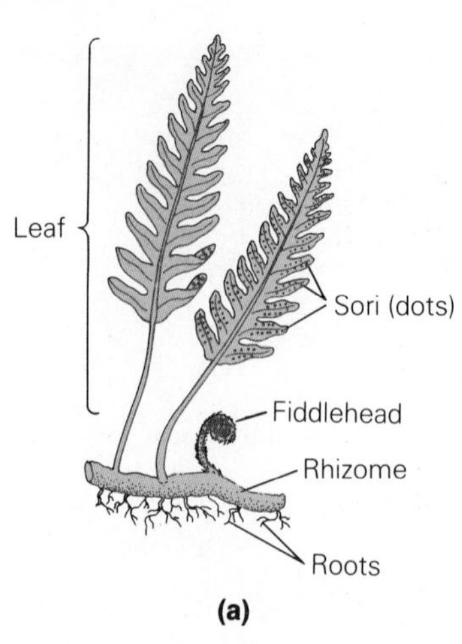

(a)

(b) Sori

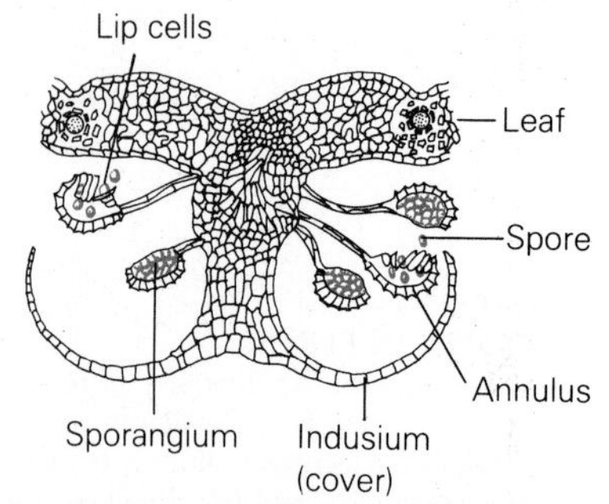

(c) Cross-section of Sorus

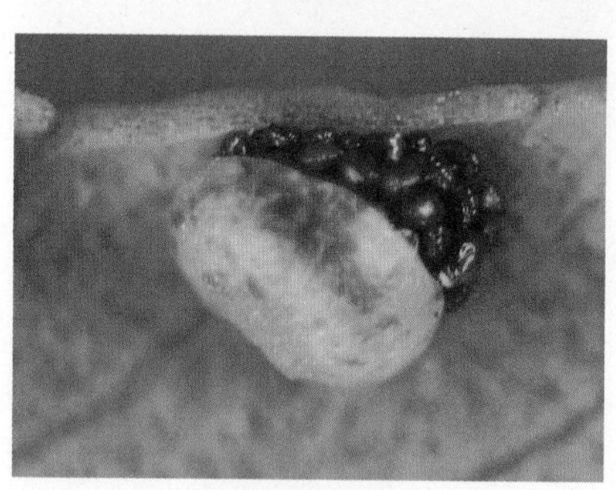
(d) Sorus

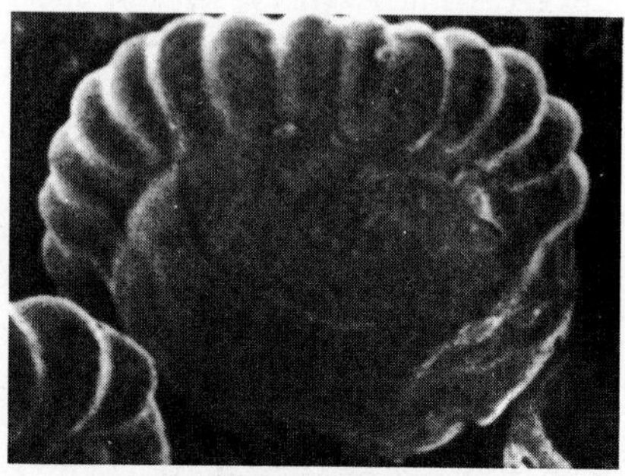
(e) Sporangium

FIGURE 25.15
FERN SPOROPHYTE
(a) Anatomy of a fern sporophyte. Below the soil, an extensive rhizome (underground stem) sends its leaves above ground. Leaves emerge from the rhizome as highly curled *fiddleheads*, each of which unwinds into its final complex, divided form. The rhizome also produces hairlike roots. (b) In many species, spore-forming structures called sporangia are borne in sori on the underside of the leaf. (c) Spores are produced meiotically in the sporangia and are catapulted out when lip cells opposite the annulus split. (d-e) Photos of one sorus and one sporangium.

THE SEED PLANTS

The primitive vascular plants and their less conspicuous contemporaries, the earliest seed plants, persisted through the Devonian, Carboniferous, and Permian periods. Toward the close of the Permian period, at the end of the Paleozoic era (about 225 million years ago), the plant life of the earth experienced a strange and dramatic change. It may have been the same sort of drastic change that brought on the extinction of the dinousaurs some 100 million years later. Powerful geological movements produced a completely different landscape and climate. The earth had been a rather smooth globe, and its surface had permitted countless warm and shallow lowland seas to cover the planet. Rather quickly (in geological terms), things changed. Soggy lowlands were uplifted to form vast mountain ranges. The monotonous warm climate gave way to a general cooling, even as the new uplands dried. Thus, the primitive, marsh-loving giants of the late Paleozoic era perished, leaving only a few scattered remnants of the once-prominent groups.

With the demise of the ancient forests, the competitive edge in the plant world passed to the inconspicuous gymnosperms, seed plants that were better prepared to survive on this new kind of earth. They continued to evolve, becoming larger and better adapted to the land, and soon appeared everywhere over the Mesozoic landscape.

Among the primitive gymnosperms were the ancestors of today's conifers. The secret of their newfound success was their ability to draw water from deep below the dry surface of the earth, transfer it in efficient vascular systems, and conserve it in water-resistant stems and leaves. They also developed a different mode of reproduction, with the female gametophyte fully enclosed within the sporophytic tissues. In primitive vascular plants, as we have seen, the gametophyte is separate, often small and ground-hugging, and altogether seeming quite vulnerable. It may well have been the new drier and harsher conditions that brought about the loss of many gametophytes from the

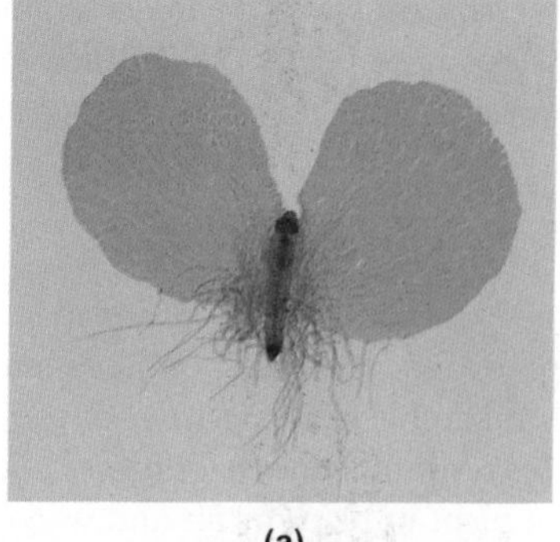

(a)

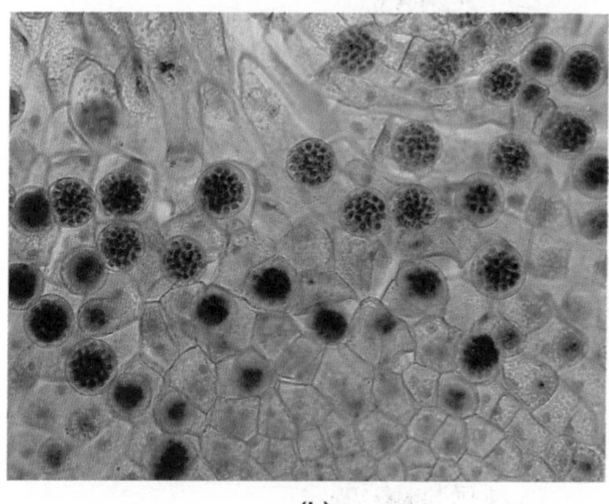

(b)

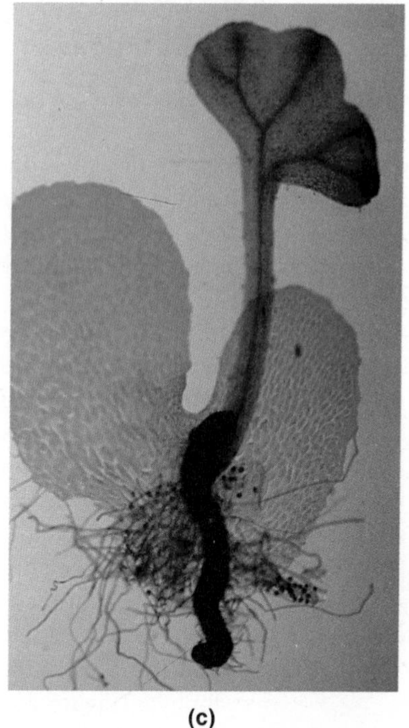

(c)

FIGURE 25.16
THE FERN PROTHALLUS
(a) The flattened, heart-shaped fern prothallus is a tiny, independent plant, specializing in gametogenesis and housing the young sporophyte. **(b)** Sperm arise in antheridia and eggs in archegonia. **(c)** The young sporophyte arises from a zygote within an archegonium.

FIGURE 25.17
ALTERNATION OF SPOROPHYTE AND GAMETOPHYTE PHASES IN A FERN
The young sporophyte begins its growth from a zygote housed in the gametophyte archegonium. The mature sporophyte produces haploid spores meiotically, and upon germination, the spores develop into a heart-shaped gametophyte, the prothallus, which bears antheridia and archegonia. Fertilization ushers in the sporophyte generation.

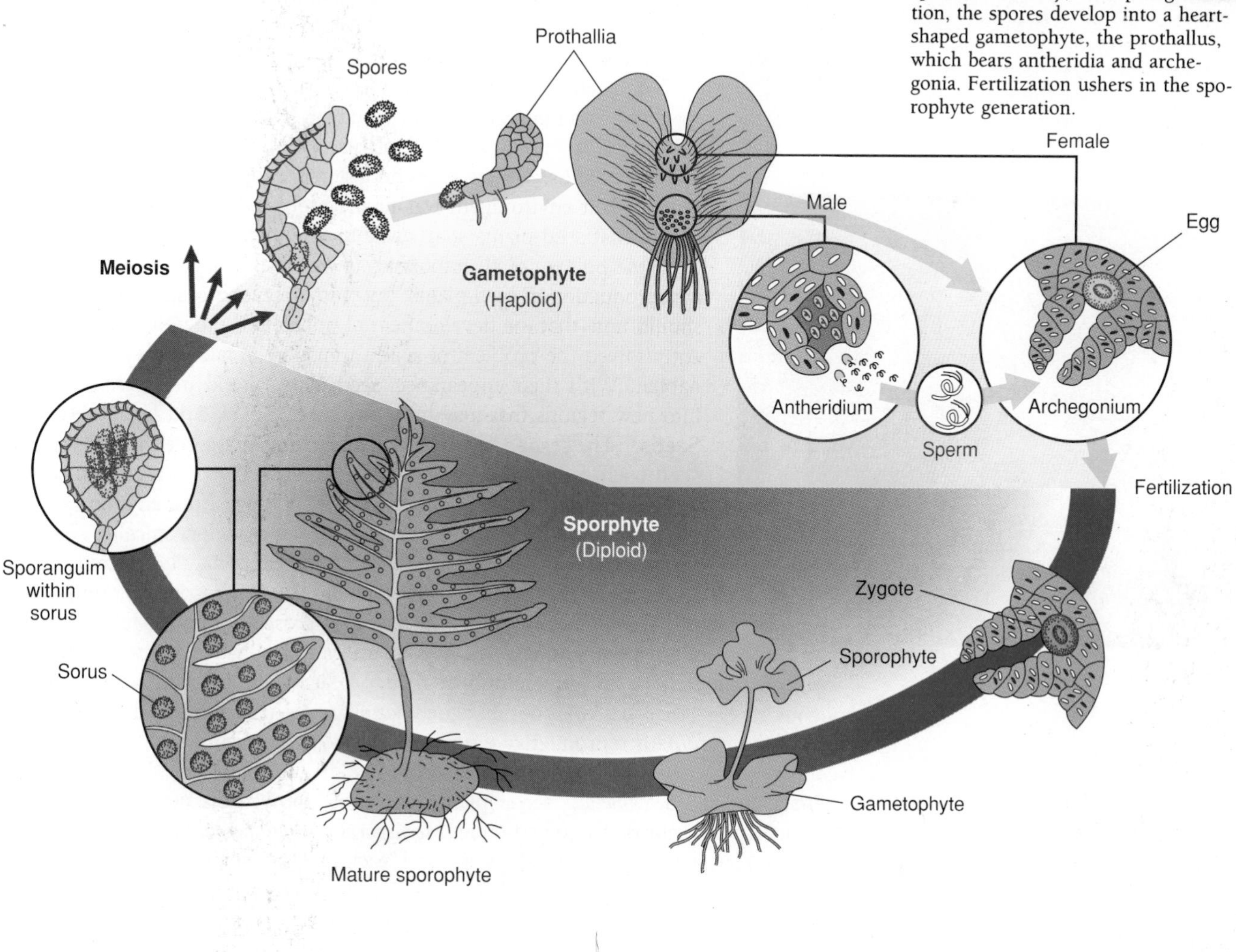

Prothallia
Spores
Female
Male
Egg
Meiosis
Gametophyte
(Haploid)
Antheridium
Sperm
Archegonium
Fertilization
Sporanguim
within
sorus
Sporphyte
(Diploid)
Zygote
Sorus
Sporophyte
Gametophyte
Mature sporophyte

earth. The changing earth also presented certain problems for the emerging seed plants, such as how to get the sperm from one tall plant to the egg in another, or simply, how to get the two together when sufficient moisture was not normally available.

Reproductive Adaptations in the Emerging Seed Plants

In large part, seed plants succeeded because they are heterosporous. (Here is where we again encounter microspores and megaspores and the corresponding gametophytes.) In the seed plant, microspores are formed through meiosis, as usual, but these microspores undergo mitosis and develop into minute, resistant, **pollen grains.** Pollen, as you are certainly aware, can be easily carried from plant to plant by air and water currents, and in some instances, by animals, notably insects.

Pollen Included in each mature pollen grain are a **generative cell** and a **tube cell.** (The presence here of more than one cell, incidentally, helps us distinguish pollen grains from the simple spores in the plant groups discussed earlier.) The generative cell gives rise to two sperm cells, and the tube cell directs the growth of the **pollen tube,** another device unique to seed plants. Pollen tubes are fine tubular growths that emerge from germinating pollen after **pollination** has occurred. (Pollination is the deposition of pollen by wind, water, or animals on a receptive cone or flower.) Pollen tubes literally digest their way to the female gametophyte, thereby permitting the sperm to reach the egg *without the necessity of external water.* The pollen tube and the cells within make up the seed plant's microgametophyte (Figure 25.18).

Pollen grains fossilize well, and their fossil records help reconstruct the nature of the flora of bygone periods. In fact, our limited knowledge of flowering plant origins is based heavily on the analysis of fossil pollen from the soil and rocks by scientists called **palynologists.**

With such an efficient and protected delivery system, motility in the sperm was no longer important. Today in fact, swimming, motile sperm are absent in nearly all seed plants. (We'll look at exceptions in a moment.) Thus, we see that the last reproductive link to the ancient watery plant environment was finally broken. The venture was so successful that seed plants soon came to dominate the new environment.

These points are all important as we set the stage for a closer look at reproduction in seed plants, but from an evolutionary viewpoint, we should note that the development of pollen and the pollen tube apparently solved the problem of sperm transport in tall plants and in a dry habitat. With their appearance, seed plants could more easily spread into new regions that may have once been hostile and forbidding.

Seeds The second reproductive innovation in the seed plants was the **seed** itself, a structure that consisted of an embryo, a region of stored food, and hardened seed coats—quite a change from anything so far. Or is it? If we consider how a seed develops embryonically, some of the mysteries about its appearance on the earth may be cleared up.

As we've indicated, all seed plants are heterosporous; thus, they produce a megasporangium and a microsporangium. Meiosis in the megasporangium yields haploid megaspores, which in turn form the megagametophyte. The megagametophyte then produces archegonia, each of which contains an egg cell. Does this sound vaguely familiar so far? It should, since these are some of the important reproductive events in *Selaginella,* our reproductively advanced lycophyte. But, how do we get to a seed from this?

First, unlike *Selaginella,* seed plants do not form resistant walls around their megaspores, nor do they shed them. *Megaspores are retained in the sporophyte body,* where the megagametophyte develops. Second, each megasporangium contains a fleshy

FIGURE 25.18
POLLEN
(a) The cells within the male gametophyte include a tube cell and a generative cell, the latter of which will divide to form two sperm cells. **(b)** The pollen grain (which houses the immature male gametophyte) seen in the scanning electron micrograph is that of the northern white cedar.

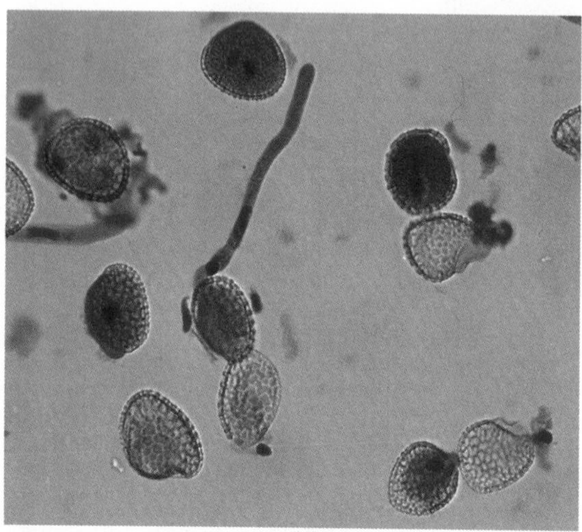

(a)

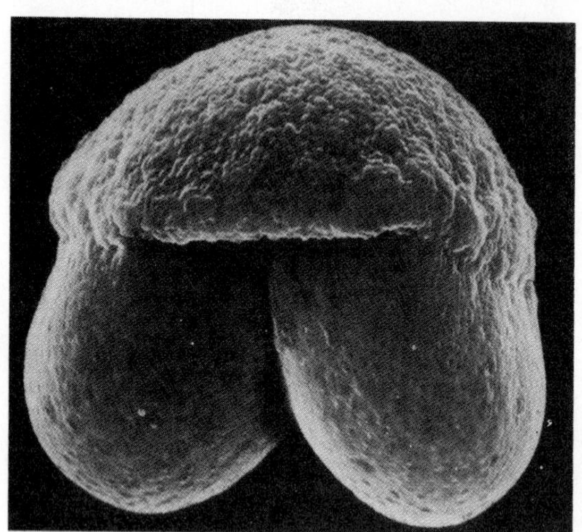

(b)

surrounding layer of sporophyte tissue, the **nucellus**, which is in turn surrounded by more sporophyte layers known as **integuments.** Except for a minute opening called the **micropyle**, the integuments completely enclose the megasporangium, which will go on to form the megagametophyte, which in turn, gives rise to the egg. The integuments, nucellus, and the megasporangium within make up the **ovule.** How does this become a seed?

After fertilization, the embryo develops, along with a quantity of food (which will later support growth in the seedling). Finally, the embryo enters a dormant period, and the surrounding integuments develop into hardened, protective seed coats (Figure 25.19). So, as you see, there are some fundamental similarities between reproduction in the seed plants and their forerunners, at least early in the reproductive process. Actually, the evolution of ovules and seeds represents an extension of a fundamental plan that was laid down much earlier in the primitive vascular plants.

Where do seeds form? The seeds of gymnosperms form in cones or cone-like structures, that in the lycophytes and other cryptogams, are restricted to spore-formation alone. In flowering plants (angiosperms), the seeds form within a floral structure called the ovary. The ovary later forms or contributes to the fruit surrounding the seeds.

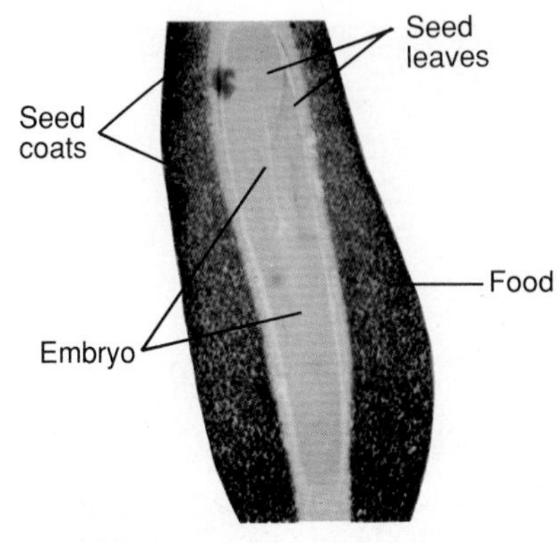

FIGURE 25.19
THE SEED
The pine seed contains a dormant embryo, a quantity of stored food, and hardened protective seed coats.

The Gymnosperms

The name *gymnosperm* literally means "naked seed," and it refers to plants that have seeds without fruit. The term is not an official taxonomic grouping, although in past use it has designated both a division and class. Today, however, the gymnosperms are considered to encompass four divisions, the Ginkgophyta, Cycadophyta, Gnetophyta, and Coniferophyta. The first three are not very widespread today, but the fourth division includes the familiar conifers, the common "evergreens" that cover our mountains, adorn our homes, and form a vast belt across many northerly regions of the world. We will return to the conifers after a brief look at the other gymnosperm divisions.

The Ginkgo The one ginkgophyte species that still exists is often referred to as a "living fossil." *Ginkgo biloba,* the **maidenhair tree** (Figure 25.20) was once known to Europeans only from fossils and was thought to be extinct. But the living *Ginkgo* was subsequently found on the grounds of Oriental temples, and finally, in 1946, growing wild in China. They are now commonly cultivated as decorative plants throughout the world. Today it seems strange that great parts of the earth were covered by ginkgos in the early Mesozoic.

FIGURE 25.20
***GINKGO BILOBA;* ONE OF A KIND**
The leaves of the *Ginkgo* are fan-shaped, divided slightly into two lobes. Its seeds have malodorous, fleshy coverings. *Ginkgos* are diocecious, either male or female.

From its appearance you might think that *G.biloba* is an angiosperm—it even sheds its leaves in autumn, after they turn from green to lovely hues of yellow. However, its fruitless seeds reveal its membership in the gymnosperm club. Reproductively, the *Ginkgo* retains a primitive reproductive trait—the swimming sperm. Male *Ginkgo* trees (the species is dioecious) produce airborne pollen grains that germinate near the ovules in female trees. The microgametophyte then grows toward the egg cell. On reaching the vicinity of the egg cell, it releases two motile, multiflagellated, swimming sperm cells. Fluids for the very short distance the sperm must travel are produced by the receptive sporophyte itself. Following fertilization, the female *Ginkgo* produces seeds in the manner of a typical gymnosperm.

FIGURE 25.21
A CYCAD
Except for the prominent pollen and ovulate cones, many cycads look like a cross between a palm and a pineapple. These interesting gymnosperms grow wild in the tropical and subtropical regions of most continents and are garden favorites wherever they can be grown.

The Cycads Plants resembling modern **cycads** had their day during the late Triassic period of the Mesozoic, some 200 million years ago. We know from fossils that they were among the most common plants in those ancient forests. They may even have been an important part of the diet of giant reptiles of that era. Today, 100 or so remaining species are found mainly in tropical regions (Figure 25.21). You may have seen cycads in museums and parks, and you may have mistaken them for palms or ferns.

In the cycads, pollen is produced in conelike strobili. Sexes are separate, and pollen is carried by the wind to the female cones. The windborne pollen produces sperm cells that, like those of the *Ginkgo,* are flagellated (our last look at swimming sperm in plants).

The Gnetophytes The ancestors of the 70 or so named species of gnetophytes alive today may have been related to the ancestors of today's flowering plants. We can't be sure because information on the evolutionary origin of flowering plants is woefully scarce. *Ephedra,* widespread in North American desert regions, is a highly branched shrub. Another gnetophyte, *Gnetum,* reproduces as a true gymnosperm but has bladelike leaves resembling those of the cherry tree. *Welwitschia,* produces pollen cones that resemble flowers. Within the stems of gnetophytes are water-conducting **xylem vessels.** While such vessels are common in the angiosperms and gnetophytes, other gymnosperms produce only the more primitive, water-conducting tracheids.

Welwitschia, by the way, may be the world's most bizarre plant (Figure 25.22). It produces only two leaves, but they grow continuously from their bases, splitting and resplitting, spilling their twisted, tentaclelike growth over the ground, as if the object of some relentless torture. *Welwitschia* lives in the very dry deserts of coastal southwestern Africa, drawing its water almost entirely from fog that spreads inland nightly.

The Conifers The **conifers** (division Coniferophyta) include nine families containing only about 550 named species, not a large number in spite of their great populations

FIGURE 25.22
A DESERT GNETOPHYTE
The splitting, twisting, and tortured foliage seen here belongs to *Welwitschia mirabilis,* a desert gnetophyte. The pink structures are pollen cones.

in the coniferous forests of the world. The most common and best known are in the pine family, Pinaceae. In addition to the familiar pines, this family includes firs, spruces, hemlocks, Douglas firs, junipers, and larches. The conifers, with a few exceptions, are the evergreens, the cone bearers whose corpses decorate your house at Christmas. Most species are found in the cold northerly climates of the earth (Figure 25.23). The conifers form a vast, worldwide belt of forests called the **taiga** (see Chapter 46). Cold weather species are also found in the higher altitudes of high mountains further south. But conifers are not restricted to the colder climates. For example, there are great pine forests over much of the southeastern United States, and one of the tallest plants known, the redwood, *Sequoia sempervirens,* thrives in the milder, foggy climate of coastal California and Oregon.

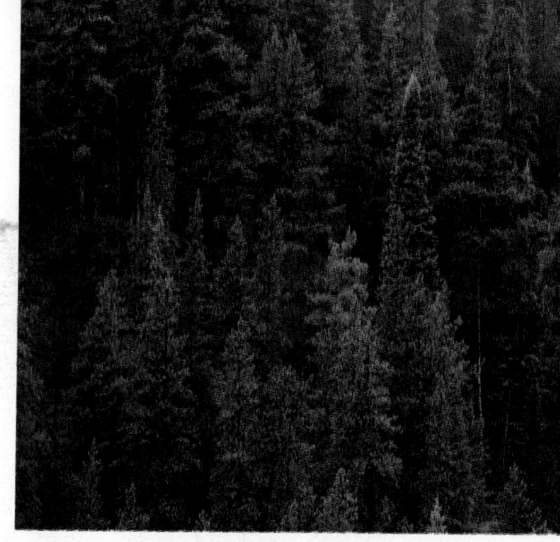

FIGURE 25.23
CONIFEROUS FORESTS
Coniferous forests are found chiefly in the cold regions of the earth. They form a continuous belt across North America, northern Europe, and the northern Soviet Union. Some types of conifers also thrive in more temperate regions.

Typically, conifers produce their needlelike or scalelike leaves seasonally, but unlike other seasonal plants, the shedding of old dead leaves is gradual and continuous so that the trees remain green ("evergreen") all year round. The unique leaf form is an adaptation to arid conditions brought on by both limited precipitation and extreme cold (where water is tied up as ice and snow). Because of the needlelike or scalelike form, conifer leaves have little surface area from which water can be lost. In addition, each leaf is surrounded by a thick waterproof cuticle, with the stomata (leaf pores) deeply recessed.

Reproduction in a Conifer Conifers typically bear separate male **pollen cones** (pollen-bearing) and female **ovulate cones** (ovule-bearing). So while the tree is monoecious, a cone is either male or female. Not surprisingly, each species has distinctive cones. Let's look at some of the complex reproductive events in the pine, a typical conifer.

As you can see in Figure 25.24, events begin in certain sporophyte cells tucked away in the scales of the pine cone. **Microspore mother cells** in pollen cones undergo meiosis, giving rise to numerous microspores. Within the hardening walls of each microspore, the haploid cell undergoes mitosis to form a number of cells, among them a generative cell and a tube cell. The pollen grain, with its cells, is the immature male gametophyte. It is at this time that pollen is carried by wind from the pollen cone to the ovulate cone and pollination begins. Let's see what has been happening in the female counterpart.

The female counterpart will be found in the ovulate cone. Each of the familiar scales of the ovulate cone bears two ovules—each of which you will recall, consists of a nucellus, surrounding integuments, and, at first, a **megaspore mother cell.** It is from the megaspore mother cell, of course, that megaspores are formed. Keep in mind that so far, all of these structures are actually part of the pine sporophyte—meiosis has yet to occur.

Upon meiosis, the megaspore mother cell gives rise to four haploid megaspores (beginning the pine female gametophyte generation), three of which disintegrate. The remaining megaspore produces the female gametophyte. This requires many rounds of mitotic cell division, followed by differentiation in which two or three egg-bearing archegonia are formed. All is then ready for fertilization.

Curiously, pollination and pollen tube growth commences long before meiosis occurs in the megaspore mother cell. Pines take their time with all of this, and it will be some 15 months from the time pollen arrives at the ovulate cone to the moment of fertilization. Prior to pollination, the scales of young ovulate cones are obligingly parted, and drops of sticky fluid are exuded. The fluids trap the windblown pollen near the micropyle of each ovule. Soon, the fluids shrink inward, drawing trapped pollen grains through the micropyle to the nucellus. The scales then close, and pollination is complete. Upon contacting the nucellus, the pollen grains form their pollen tubes, which slowly digest their way towards the developing female gametophyte. Prior to the pollen tube's formation (or afterwards; it varies), the generative cell divides, yielding two sperm cells. The tube cell and one of the sperm will disintegrate, but the other sperm will fertilize the egg nucleus. The act is repeated in archegonia throughout each ovule, and a number of zygotes begin to form embryos. But, as it turns out, only one per ovule survives.

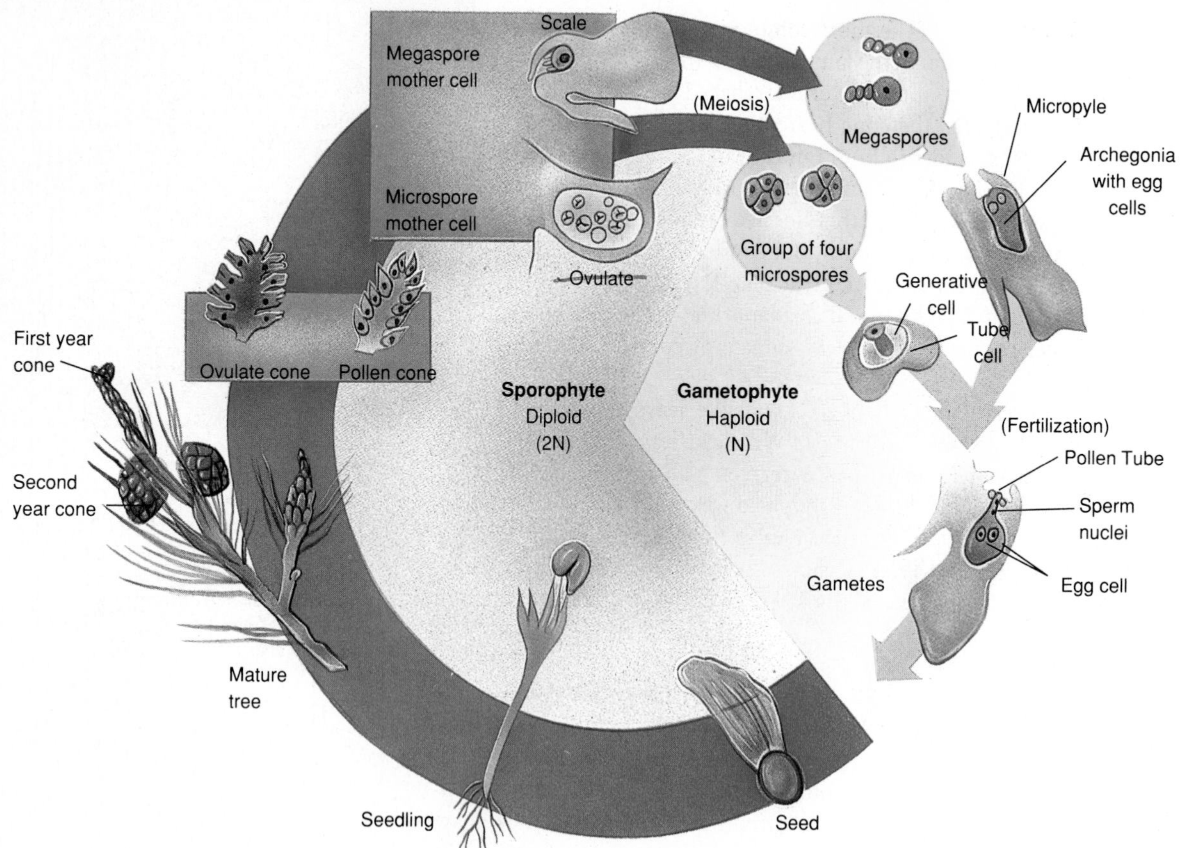

FIGURE 25.24
LIFE CYCLE OF A CONIFER
The gametophytes of conifers are produced and maintained within sporophyte tissues. Microspore mother cells within each scale of the pollen cone undergo meiosis, each forming four haploid microspores. The microspores develop into pollen grains, and later the generative nucleus divides, yielding two sperm cells. Each female cone scale has two ovules on its surface. A megaspore mother cell in each ovule undergoes meiosis, and one of the products becomes the female gametophyte. Pollen grains eventually produce pollen tubes, releasing sperm in the vicinity of archegonia, each of which contains an egg. Fertilization occurs when a sperm unites with an egg. Seed development follows.

The zygote gives rise to the embryo, which, following a period of growth and differentiation, becomes dormant. At this time the embryo is made up of a **hypocotyl** or embryonic root, two **apical meristems** (regions of simple, undifferentiated cells), and eight fingerlike **cotyledons** (also called **seed leaves**). The embryo is still surrounded by gametophyte (haploid) tissue which will nourish the seedling for a time (see Figure 25.19). The entire ovule participates in the formation of the seed, so in addition to the embryo, each pine seed consists of a hardened seed coat derived from the old sporophyte integuments and remnants of the nucellus. By the time seed development has been completed, a second year will have passed—as we said, pines take their time.

THE END OF AN ERA

The lives of the gymnosperms and the great dinosaurs were once strangely intertwined. For eons, these great beasts foraged through immense Mesozoic forests of gymnosperms, but then, for some reason, they both began to die off. The dinosaurs had marched to extinction by the end of the Mesozoic, but some gymnosperms, we know, managed to survive. With the arrival of the Cenozoic era about 65 million years ago, a new kind of plant gained dominance, the upstart flowering plants—the angiosperms, as we will call them. With their greater adaptability to fluctuating conditions, the angiosperms fanned out over much of the planet in those distant days. Although the colder north temperate and mountainous regions remained the domain of the hardy gymnosperms, the angiosperms abounded throughout the tropical and temperate regions.

The most recent chapter in the saga of the interaction between gymnosperms and angiosperms is an interesting one: the gymnosperms seem to be on the comeback trail. In the last few million years, extensive northern spruce and pine forests have displaced vast tracts of hardwood angiosperms.

Angiosperms: The Rise of the Flowering Plants

The astounding diversity of flowering plants today—some 235,000 species—stands in stark relief against the sameness of the conifers. Such diversity is further dramatized by the speed with which it seems to have come about. Whereas biologists are certain that angiosperms evolved from one of several extinct gymnosperm ancestors, there are no clear fossil links between the two. Further, most of the angiosperm fossil record dates back only about 80 million years, near the close of Mesozoic, but by 75 million years ago, many of today's angiosperm families had appeared. So it seems that flowering plants had literally burst upon the scene, diverging into every conceivable habitat and changing the surface of the earth.

We are left with a host of questions, some still unanswered. When did the earliest flowering plants actually appear, and where are their fossils? Were the first angiosperms inconspicuously distributed among dominating Mesozoic gymnosperms? (This was exactly the case with the Mesozoic mammals, who were vastly overshadowed by the dinosaurs.) And especially, what conditions would have promoted the sudden explosion of flowering plants?

The apparent paucity of early angiosperm fossils suggests that the first flowering plants populated drier upland regions of the earth. The lowlands were ideal locations for formation of fossils, and many fossilized gymnosperms are found in the sedimentary rocks originating in such regions. The upland locales were too dry for much fossilization, so the angiosperm record is scant. Fortunately, pollen is readily fossilized, so in spite of the extremely spotty record of other plant parts, we know from their pollen that flowering plants diverged from their gymnosperm ancestors some 125 million years ago in the Cretaceous period of the Mesozoic. The enormous diversity seen today probably came about much later, towards the end of Mesozoic (see Figure 25.8).

The sudden emergence of flowering plants has been linked, hypothetically, to geological changes at the end of the Mesozoic and drastic climate changes during the early Cenozoic (a startling drop in the average temperature of the earth of perhaps 20°C). The changes in climate were brought about by three events. The first was a period of mountain building. The second was a series of events that geologists call **plate tectonics** or **continental drift** (see Essay 20.1). According to evidence from studies of the ocean floor, the major land masses of the earth began shifting at about the start of the Mesozoic era, some 200 million years ago. Most of the world's land mass broke apart, and the pieces drifted northward, so that by the Cenozoic the continents had roughly assumed their present positions. Apparently, the angiosperms were versatile enough to have survived both these changes.

In the third event, the end of the Mesozoic era may have been marked by a major catastrophe—the collision of a gigantic asteroid with the earth. The dust it raised, according to what has become known as the **Alvarez theory** (after paleontologist Walter Alvarez and his physicist father, Louis Alvarez), blocked the sun, causing massive weather changes and large-scale extinctions. Such drastic changes on the earth's surface would have caused the extinction of many living things and provided countless opportunities for surviving organisms to expand and diversify. As fantastic as the asteroid theory may seem, it has received a great deal of attention and is now provisionally accepted by a number of prominent scientists. (Atmospheric scientists are quite concerned with a similar scenario, called *nuclear winter,* which many believe would follow nuclear war.) We will take a closer look at the Alvarez hypothesis in Essay 32.1.

Were climatic and geological events the only factors responsible for the rise of angiosperms? Perhaps not. Perhaps the great diversity seen in flowering plants suggests another factor. It has been suggested that flowering plants have a greater variability and can undergo more rapid speciation than can many other living things. There is some compelling evidence that this is true. For example, the ability of angiosperms to form fertile hybrids is well known. Hybrid flowering plants can even form "instant species" through polyploidy, a doubling, tripling, etc. of chromosome number brought on by irregularities in meiosis or mitosis. As we saw in Chapter 20, hybrids are usually infertile due to the failure of the parental chromosomes to synapse during meiosis. However, the problem is solved in plants where a spontaneous chromosome doubling has occurred. Meiosis goes on without a hitch, and the fertile hybrids give rise to new lines that are quite distinct from the parental lines. Thus we have an "instant species."

A critical factor in the rise of the flowering plants could well have been an evolving partnership with the insects. Insects became pollen carriers as they were drawn to flowers by the lure of sweet nectar and nutritious pollen. The mutual adaptation presents many classic cases of coevolution. Wind pollination is quite effective in grasslands and in coniferous forests, where there are great numbers of the same species. But wind is certainly an inefficient vehicle in a mixed forest, where the next individual of one's species may be quite distant. Not only is insect pollination more efficient than wind pollination under these conditions, but it is much more selective; once a bee finds nectar in one flower, it will return to that kind of flower. Such specificity on the part of insects makes it easier for new species to reproduce and for many different species to coexist. We will return to this point in the next chapter.

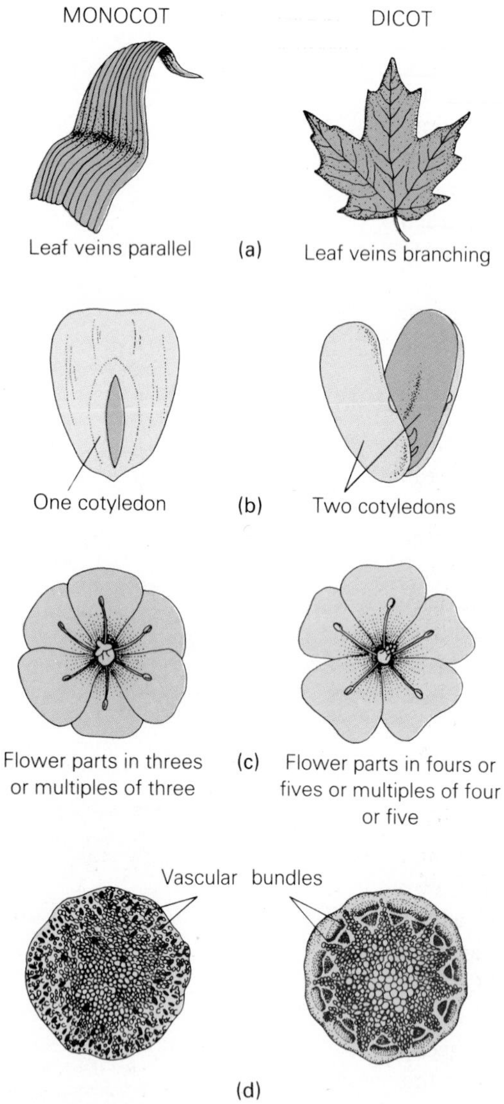

MONOCOT DICOT

Leaf veins parallel (a) Leaf veins branching

One cotyledon (b) Two cotyledons

Flower parts in threes or multiples of three (c) Flower parts in fours or fives or multiples of four or five

Vascular bundles

(d)

FIGURE 25.25
MONOCOTS AND DICOTS
Monocots and dicots differ in four obvious ways (a) in their leaf venation, (b) seed cotyledons, (c) flower parts, and (d) vascular systems.

The Angiosperms Today

In a sense, the angiosperms have inherited the earth—for the time being. Today they indeed constitute the vast majority of plant species (estimated at 235,000 or so). In spite of such diversity, Division Anthophyta, the flowering plants, contains only two living classes—**Dicotyledonae** (the **dicots**) and **Monocotyledonae** (the **monocots**).

The dicots are clearly in the majority, with 170,000 species. Briefly, the dicots form about 10 families (botanists differ on the number), whose membership includes such familiar examples as the magnolia, oak, maple, buttercup, cabbage, hibiscus, potato, willow, mint, rose, apple, melon, bean, locust, carrot, and dandelion. The 65,000 species of monocots make up four families, represented by such familiar plants as the lily, date palm, rye grass, wheat, corn, sugar cane, iris, and orchid. The two classes represent distinct evolutionary lines, the monocots presumably having emerged from the dicots. The two differ in many ways, some of the more obvious of which are seen in Figure 25.25.

We won't dwell on the biology of angiosperms here, although we have included a typical life cycle in Figure 25.26. Their biology is the subject of the next four chapters. Here we'll consider only certain aspects of flowering plant evolution.

Origin and Phylogenetic Relationships There are a number of taxonomic schemes that attempt to show evolutionary relationships among the angiosperms. There is some agreement that the family Magnoliaceae (the magnolias) in the order Ranales, is a living representative of the ancestral group. The fossil record isn't much help in resolving the issue since early in the strata of the Cretaceous period and in greater numbers in the early Cenozoic. Most of our presumed plant relationships, therefore, are based on the study of today's plants, particularly the comparison of their flowers. As an example, the buttercup is suggested as an ancestral type because of its primitive flower structure, as we will see.

When botanists use the floral structure to determine taxonomic relationships, they must first decide what is primitive (original equipment) and what is advanced (new)—and that presents problems.

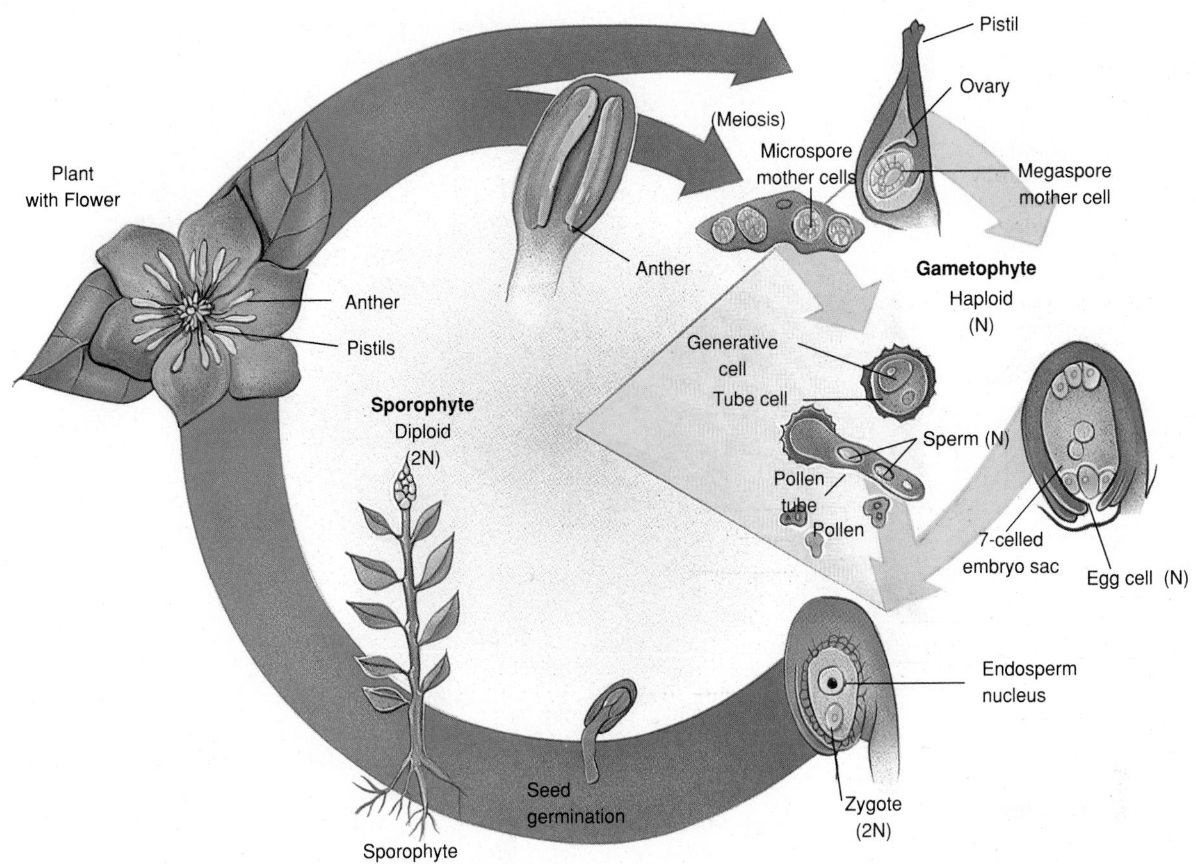

FIGURE 25.26
LIFE HISTORY OF AN ANGIOSPERM
The angiosperm life cycle begins with fertilization of the egg and development of the embryo and seed. The seed germinates and grows into a mature plant, which produces the flower. The gametophyte will develop in the ovules and anthers of the flower. Following meiosis in the female reproductive structure, a megaspore goes through three rounds of mitosis, and the haploid female gametophyte (megagametophyte) generatioan commonly emerges as a seven-celled (eight-nucleate) embryo sac. Following meiosis in the anthers (the male counterpart to the ovary), each resulting haploid microspore enters a round of mitosis, producing a pollen grain. It contains a tube cell and a generative cell.

Following pollination, the pollen germinates, and a pollen tube grows toward the ovules. The generative nucleus divides through mitosis, producing two sperm. Upon reaching the embryo sac, one sperm fertilizes the egg cell, and the other the nearby binucleate cell.

The concepts of *primitive* and *advanced* are always troublesome. We must keep in mind that every living organism has exactly as long an evolutionary history as every other living organism—some 3 to 3.5 billion years—even though some groups may have changed rapidly while others appear to have been marking time. So perhaps it is best not to label any living organism, or even any current taxon, as either primitive or advanced. Instead, we may consider *individual features* within a group to be primitive (present in the original founders of the group) or advanced (evolved subsequently by only some members of the group and different from the earlier condition). If only specific features are considered as primitive or advanced, many semantic and philosophical problems disappear. We might, then, keep in mind that every living organism has a mix of primitive and advanced features.

In keeping with this reminder, botanists have worked out what they consider to be "primitive" floral features. Any departure from these types represents an advancement or divergence from the ancestral type. Let's consider the buttercup, *Ramunculus macranthus,* to illustrate the primitive condition (Figure 25.27):

1. Floral parts are arranged in spirals.
2. **Carpels** (ovule-containing parts of the ovary) are always superior to (that is, above) the **receptacle** (the enlarged end of floral stem) and other floral parts.
3. Carpels and **stamens** (pollen-producing parts) are numerous and not fused.
4. **Petals** are separate or completely divided, never fused.
5. The flower has **radial symmetry**, not **bilateral symmetry** (disk-like rather than with right and left sides).
6. The flower is **complete** and **perfect** (has all parts, including both accessory and reproductive).

FIGURE 25.27
A PRIMITIVE FLOWER
(a) The buttercup flower *(Ranunculus macranthus)* is considered to be primitive. Its floral parts (sepals, petals, stamens, and carpels) are numerous and unfused. In addition, it is radially symmetrical (spherical or disklike; see glossary). **(b)** The orchid *(Cypripedium calceolus)* is considered advanced, principally because of its fused parts and its bilateral symmetry (right and left-sided; see glossary).

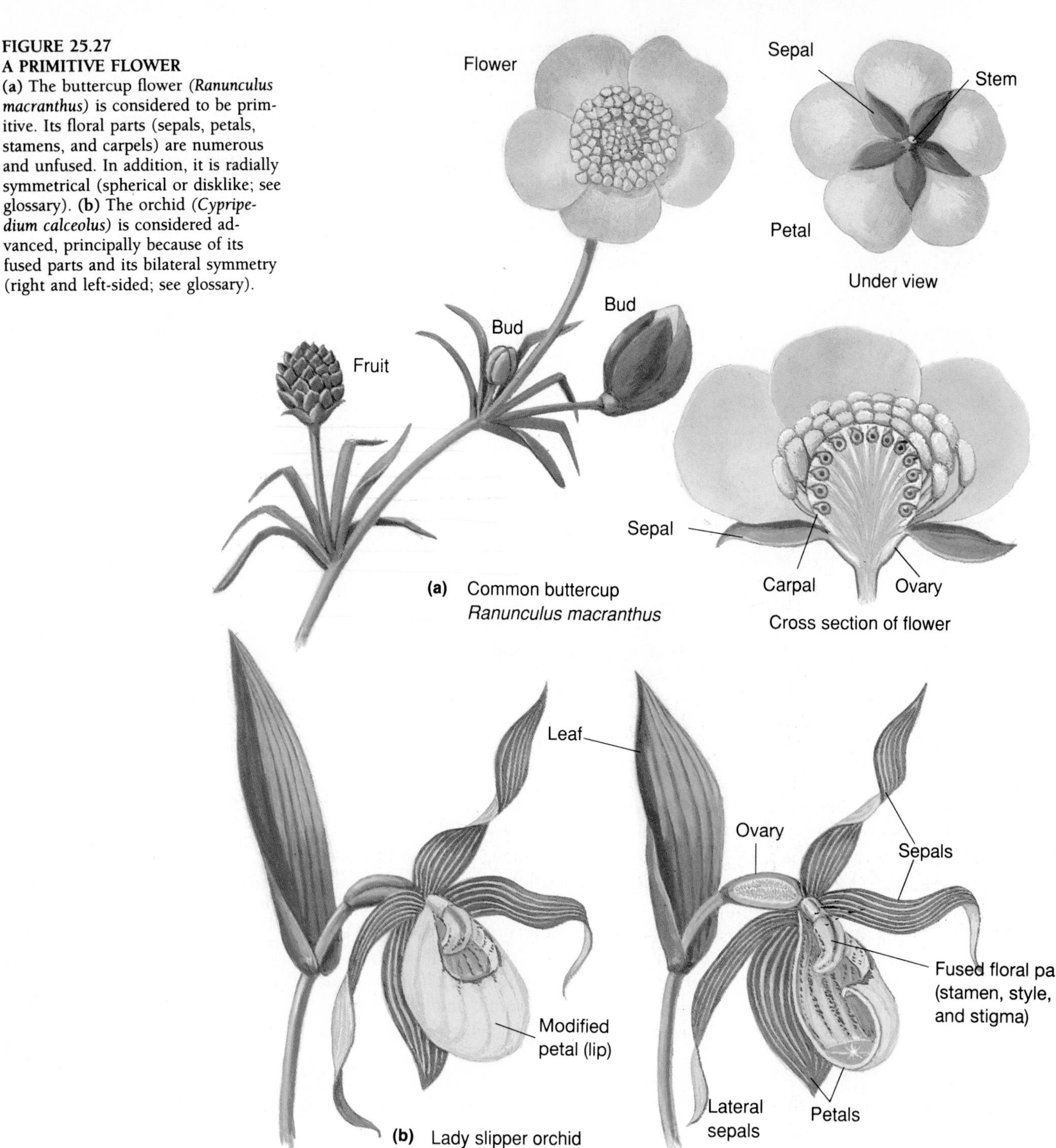

Among the greatest evolutionary trends has been the remarkable development of plants. They have silently but relentlessly winnowed their way into virtually every available nook and cranny, constantly changing and adapting with such a pervasive influence that they have also markedly affected a wide range of other forms of life. We have seen that in terms of sheer numbers of species, the most successful of the plants today are those that produce flowers. Because flowering plants are so inextricably linked to other forms of life, we will next focus specifically on this fascinating group.

angiosperm Any seed plant whose seeds are surrounded by the mature ovary, or fruit (rose, oak, bean, corn, maple, geranium).

antheridium In bryophytes and cryptogams, gametophyte tissue that produces and releases the sperm (not generally applied to seed plants).

archegonium In plants, gametophyte tissue that produces and houses the egg, and for a time, supports the embryo (not applied to flowering plants).

dioecious ("two houses") The condition in which sperm and eggs are produced in separate individuals.

generative cell A cell in the immature male gametophyte (pollen grain) of seed plants that divide to form two sperm.

gymnosperm Any seed plant whose seeds are not surrounded by an ovary or fruit (conifer, ginkgo, cycad, gnetophyte).

heterosporous Pertaining to plants that produce microspores and megaspores, which give rise, respectively, to sperm-producing and egg-producing gametophytes (includes all seed plants).

homosporous Pertaining to plants that produce one type of spore only, thus their gametophytes may produce both sperm and eggs (mosses and ferns).

megagametophyte The female gametophyte, the product of a megaspore and a structure in which an egg forms.

megaspore In heterosporous plants, spores that form the female gametophyte (megagametophyte), which, in turn produces the egg.

megaspore mother cell In the seed cone and in the ovary of a flower, a diploid cell destined to undergo meiosis and form four haploid megaspores, one of which gives rise to the female gametophyte.

microgametophyte Male gametophyte, the product of a microspore and a structure in which sperm form.

microspore In heterosporous plants, spores that form the male gametophyte (microgametophyte), which produces sperm.

microspore mother cell In the pollen cone and in the anther of a flower, a diploid cell destined to undergo meiosis and form haploid microspores, cells that form male gametophytes.

monoecious ("one house") The condition in which sperm and eggs are produced in the same individual (also hermaphroditic).

ovule In the sporophyte, tissue containing the megaspore mother cell, and a surrounding nucellus and integuments. It later contains the egg cell, and, following fertilization and development, forms the seed.

phloem Vascular tissue specialized for the transport of sap containing sugars and other substances.

pollen grain The microspore of a seed plant, haploid cells that comprise or give rise to the male gametophyte (microgametophyte).

APPLICATION OF IDEAS

1. Present an argument for dividing the plant kingdom into two kingdoms. Why would one consider this in the first place? What would the kingdoms contain, and from what ancestral types would they have emerged?
2. Describe the conditions (climatic, topographic, and so on) that would have to prevail for the bryophytes to return to the prominence they once held. Which animal groups might persist under these conditions?
3. The fossil record clearly demonstrates the presence and even prominence of fernlike but seed-producing plants in the Devonian forests. The significance of these plants to the evolution of today's seed producers is unknown. Present two competing hypotheses that might account for seed evolution having occurred twice. Support the more plausible of the two.
4. The involvement of animals in plant reproduction may have been a key factor in the rise and spread of angiosperms. Discuss this proposition and explore two aspects of angiosperm reproduction that involve animals today. How might these have influenced animal evolution?

KEY IDEAS

WHAT ARE PLANTS?

Key plant characteristics include:

1. Structure: multicellular with specialized cells and tissues
2. Pigments: chlorophylls *a* and *b* and carotenoids
3. Polysaccharides: amylose starch and cellulose
4. Life cycle: **sporic**—alternating haploid and diploid phases
5. Reproduction: multicellular embryos housed in multicellular gametophyte tissues.

Alternation of Generations in Plants

1. In the sporic life cycle, sporophyte cells undergo meiosis, producing spores that form haploid multicellular gametophytes in which gametes arise. Gametophytes dominate in the **nonvascular plants,** and sporophytes dominate in the **vascular plants.**
2. Whereas more primitive plants are **homosporous,** advanced plants are **heterosporous.** The latter form **microspores** (which develop into **microgametophytes**), and **megaspores** (which develop into **megagametophytes**).

DIVISION BRYOPHYTA: NONVASCULAR PLANTS

Bryophytes include **mosses, liverworts,** and **hornworts.** Although they are terrestrial, water is required for reproduction. An absence of conducting and supporting vascular tissue limits the size of bryophytes. Typically, they thrive in moist lowlands, but many have adapted to rocky places, deserts, and tundra.

Bryophyte Characteristics

The gametophyte phase dominates, Leaf-, stem-, and root-like structures provide most functions served by those structures in vascular plants. Threadlike rhizoids provide anchorage.

Class Hepaticae: Liverworts

Marchantia, a **thalloid** liverwort, produces a ribbonlike, sectioned, and branching gametophyte. Porelike epidermal openings permit gas exchange and water conservation, and **rhizoids** and scales absorb water. Asexual reproduction is through fragmentation and **gemmae** production. **Gametangia (archegonia** and **antheridia)** form on palmlike **archegoniophores,** and parasol-like **antheridiophores.** After fertilization, certain cells in the sporophyte undergo meiosis, forming spores that are expelled by **elaters.** Spores develop into gametophytes.

Class Anthocerotae: Hornworts

The gametophyte of the hornwort *Anthoceros* is a low rosette form. Each cell contains one large chloroplast and a pyrenoid (similar to green algae). The sporophyte is hornlike and undergoes continuous growth and spore formation, its growth emerging from a **meristematic** base. It may survive independently. Hornwort sporophytes form a water-resistant cuticle with pore-like stomata.

Class Musci: Mosses

1. Mosses make up the great majority of bryophytes. They grow in many habitats including deserts, where, through dormancy, they withstand long droughts. Mosses are important as pioneer plants they are active soil builders. Peat moss *(Sphagnum)* is used for mulching and forms peat, an important fuel.
2. The moss life cycle includes a haploid spore, a threadlike **protonema** stage, and the leafy gametophyte. The moss gametophyle is anchored to the ground by rhizoids. A photosynthetic "stem" **epidermis** surrounds a cylinder of **cortical cells,** which contain thick-walled cells of a supporting **central cylinder** (which conduct water to some extent). The "leaves" are a flattened array of photosynthesizing cells, which in some species secrete a waterproofing **cutin.**
3. The gametophyte produces antheridia and archegonia, which are surrounded by protective **sterile jacket cells.** Rain splashes sperm to the archegonia, where fertilization occurs. The nutritionally independent, diploid sporophyte—consisting of a **foot, seta,** and **sporangium,** extends from the gametophyte. The sporophyte is unusual in that it contains a cuticle and numerous **stomata.** Spores are released as the caplike **operculum** breaks away. A new cycle starts as haploid spores are produced through meiosis in the sporangium

Bryophytes and the Evolutionary Origins of Vascular Plants

There are several theories of bryophyte origin. One suggests a different ancestor from the plants. Another more recent one is that bryophytes branched off the vascular line and through evolutionary simplification lost their vascular tissue and took on other bryophyte characteristics. The fossil record supports this view. The common ancestor to both groups might have resembled the green alga, *Coleochaete.*

THE VASCULAR PLANTS

1. Vascular plants, often called **tracheophytes,** have tubelike vascular tissues, the **xylem** and **phloem,** a highly dominant sporophyte, and a greatly reduced gametophyte.
2. The presence of efficient, water-conducting vascular tissue made it possible for vascular plants to grow very large, to draw water from deep in the soil, and to transport it up to the leaves.
3. The **rhyniophytes** of the Silurian, among the earliest vascular plants, had leafless, upright shoots arising from **rhizomes** (horizontal stems). These simple stems contained tough, supporting xylem tissue with water-conducting **tracheids.**
4. Evolutionary developments in the rapidly spreading vascular plants included true roots, stems, and leaves. In some, evolution produced nonmotile sperm, new mechanisms for fertilization, and an embryo protected within a **seed.**

5. Vascular plant divisions include the seedless plants—**Psilophyta, Lycophyta, Sphenophyta,** and **Pterophyta** (ferns), and the seed plants—**Coniferophyta** (conifers), **Cycadophyta, Ginkgophyta, Gnetophyta (gymnosperms),** and **Anthophyta,** (**angiosperms** or flowering plants).

Division Psilophyta: Whisk Ferns
The dominant sporophyte of **psilophytes,** or whisk ferns, has anchoring rhizoids but no true roots, photosynthetic true stems, and primitive sporangia. Psilophytes are homosporous, their spores producing a tiny separate gametophyte, which in turn produces sperm and eggs in antheridia and archegonia. Upon fertilization the young sporophyte becomes independent. Psilophyte fossils date back 375 million years, but their phylogeny is not well known.

Division Lycophyta: Club Mosses
1. The dominant sporophyte of **lycophytes** (ground pines or club mosses) has true and **adventitious** roots (arising from the stem), true stems, and true leaves. The leaves are **microphylls** (derived from the epidermis, as opposed to larger **megaphylls,** derived from branching stems). Spores form in many **sporophylls** or in a single, terminal **strobilus.** Separate gametophytes produce the gametes, and after fertilization the young sporophyte becomes independent.
2. Spore formation is either primitive and homosporous (spore of one type) or advanced and heterosporous (spores of different types). The latter includes **microspores** (small) and **megaspores** (large). The first produces **microgametophytes,** which give rise to antheridia (which produce sperm gametes), the second, **megagametophytes,** which give rise to archegonia, each forming an egg.
3. During the Devonian and Carboniferous periods, lycophytes formed forests of large trees, the remains of which form fossil fuels.

Division Sphenophyta: Horsetails
One surviving genus of Sphenophyta, *Equisetum,* forms slender photosynthetic sporophytes with branched or unbranched stems with whorls of microphylls. Homosporous spore formation occurs in a strobilus. The gametophyte is tiny and separate. **Sphenophytes** contributed to the lush Carboniferous forests.

Division Pterophyta: Ferns
1. Ferns were among the dominant species of the Paleozoic forests.
2. Ferns are widespread and numerous. The typical dominant sporophyte consists of thick rhizomes (underground stems), which give rise to **fiddleheads** that uncoil to form large, highly divided leaves (megaphylls). The roots often form mycorrhizol associations.
3. Homosporous spores are produced in sporangia, often clustered into **sori** and sometimes covered by an **indusium.** Elaborate spore-ejecting mechanisms involve the **annulus** and **lip cells** of sporangia.
4. When a haploid spore germinates, it forms a photosynthetic gametophyte—the **prothallus,** which forms archegonia and antheridia. In some, the hormone **antheridogen** helps prevent self-fertilization. Following fertilization, the young independent sporophyte emerges.

THE SEED PLANTS

The demise of the ancient forests and the rise of the seed plants may have been brought about by vast geological changes—upheavals that raised the land and subjected it to drying and cooling trends. The gymnosperms were efficient at gathering and conserving water; they incorporated the gametophyte into their protective tissues.

Reproductive Adaptations in the Emerging Seed Plants
1. Pollen provides the means for the sperm to reach the egg in the cones and flowers of seed plants. Seed plants produce microspores, which develop into **pollen grains,** containing **generative** and **tube cells.** Following pollination, pollen forms **pollen tubes,** which provide a passage that the sperm follow into the female gametophyte. This eliminates the need for outside water. Swimming sperm do not occur in the more recently evolved seed plants.
2. In seed plants, megaspores are retained and give rise to megagametophytes within the sporophyte. This occurs within the **ovule,** a sporophyte tissue that consists of a **nucellus, integuments,** and **megaspore mother cell.** Following meiosis in the megaspore mother cell, one haploid megaspore develops into the female gametophyte, which gives rise to an egg cell. After fertilization, the ovule develops into a seed, which includes an embryo, stored food, and surrounding seed coats.

The Gymnosperms
1. **Gymnosperms,** now an informal term, include ginkgos, **cycads,** gnetophytes, and **conifers.**
2. **Ginkgophytes** are composed of a single, broad-leaved species, *Ginkgo biloba,* the maidenhair tree. It is wind-pollinated, produces seeds in cones, and retains the swimming sperm, for which it provides a fluid environment.
3. The **cycads** are palmlike plants with separate sexes. Pollen is produced in large conelike strobili. Flagellated sperm persist. Cycads were prominent in the Triassic, some 200 million years ago.
4. Gnetophytes reproduce as do gymnosperms, though some have angiosperm characteristics. One species has broad leaves, one produces flowerlike cones, and another has advanced water-conducting elements called **xylem vessels.**

5. Conifers form great forests in the northern hemisphere and in mountain regions. Pines are the most common tree. The leaves, typically needlelike or scalelike, are replaced gradually. The leaf from is an adaptation to dryness. Conifers bear separate male (pollen-producing) and female (ovule-producing) cones.
6. After meiosis and microspore production in the male or **pollen cone,** mitosis yields a microgametophyte containing a **generative cell** and a **tube cell.** The microgametophyte becomes enclosed in a tough covering, forming a **pollen grain.** Before fertilization, the generative cell divides to form two nonmotile sperm. The tube cell guides the development of a **pollen tube,** which penetrates the female gametophyte, and one sperm fertilizes the egg.
7. In the female pine cone, the megagametophyte develops in the **ovule**—a megasporangium surrounded by **integuments** and a **nucellus.** After meiosis, a single surviving meagaspore produces the megagametophyte, which develops archegonia, each with one egg. Sperm enter through a **micropyle,** fertilizing the egg. The one successful embryo from each ovule consists of a **hypocotyl,** two meristems, eight **cotyledons,** and a quantity nutrient stored in gametophyte tissues.

THE END OF AN ERA

The gymnosperms and the dinosaurs fell from prominence together at the close of Mesozoic. The period is marked by the rise and divergence of the angiosperms, but in recent times the conifers appear to be making a comeback.

Angiosperms:
The Rise of the Flowering Plants

1. Flowering plants are far more diverse than conifers. Flowering plants probably originated in the Mesozoic, but until the Cenozoic they were apparently restricted to drier uplands, where fossils rarely formed.
2. The sudden rise of flowering plants is attributed to a rapidly changing climate accompanied by mountain-building and **plate tectonics,** or **continental drift.** The latter began some 200 million years ago and represented a northerly shift of the continents.
3. According to the **Alvarez theory,** a third great geological event, the collision of a gigantic asteroid with the earth, darkening the atmosphere with dust and creating massive weather changes, brought about many extinctions.
4. The replacement of gymnosperms by angiosperms may have also been due to the angiosperms' greater variability, adaptiveness, and high rate of speciation. Additional factors were efficient insect pollination and the coevolution of pollinators and plants.

The Angiosperms Today
The 235,000 angiosperm species form two major classes, **Monocotyledonae (monocots)** and **Dicotyledonae (dicots).** They differ in seed structure (number of cotyledons), flower anatomy, and vascular tissue distribution in stem and leaf.

The ancestral angiosperm is represented by the family Magnoliaceae and specifically by the magnolia. The fossil record is scant and of little help, so plant phylogeny is based mainly on comparative features of living angiosperms.

The primitive floral condition, as seen in the buttercup, includes spiral flower parts, **carpels** above the **receptacle** and other parts, numerous and free carpels and **stamens,** separate **petals** (not fused), **radial symmetry** rather than **bilateral, perfect** and **complete** flowers (has all parts, including male and female).

The great diversity of flowering plants may be attributable in part to the continuing adaptation (coevolution) of the plant and its animal pollinator. In the ongoing competition for pollinators and the increasing efficiency of pollination, natural selection may have favored variants

REVIEW QUESTIONS

1. List five significant characteristics of plants. Which if any is (are) unique to plants only? (538-539)
2. Summarize two evolutionary trends involving life cycles and spore formation in more recently evolved vascular plants. (539)
3. In what way are the bryophytes "tied" to the ancestral watery environment? (539)
4. What, in addition to conduction, does vascular tissue do? What two limitations does its absence impose on bryophytes? (539-540)
5. List three reasons why today's bryophytes might be considered successful. (540)
6. The current theory of bryophyte origins suggests that they shared common ancestors with the vascular plants. How, according to this, does one account for the simplicity of bryophytes and their lack of vascular tissue? (545-546)
7. What evidence recommends the green alga *Choleochaete* as a common ancestor to plants? (546)
8. What are "true" roots, stems, and leaves? What characteristic of bryophytes permits them to get along fine without these? (540)
9. Characterize the life cycle of *Marchantia.* How does this compare with other plants? (541-542)
10. Describe the gametophyte of *Marchantia.* What purpose do the gemmae serve? In what structures do the liverworts produce sperm and egg? What kind of help do sperm get in reaching the eggs? (542)
11. How does the hornwort sporophyte differ nutritionally from the liverwort sporophyte? How

does its growth differ from that of other bryophyte sporophytes? (542-543)

12. Describe a desert adaptation of mosses. (543)
13. List two ecologically important activities of mosses. (544)
14. Summarize the life cycle of a moss, beginning with the germination of a haploid spore. (544-545)
15. Explain how the presence of vascular tissue has influenced the distribution and the size of vascular plants. (546-547)
16. Describe the body form of the rhyniophytes. Why are they biologically important? (547)
17. List four divisions of vascular plants that do not produce seeds. In what two ways is reproduction in these groups primitive? (547-548)
18. Briefly describe the body form of the whisk fern, *Psilotum nudum*. Which of its features would one consider primitive? (548-549)
19. Describe the sporophyte of *Lycopodium*. How do its *microphylls* differ from *megaphylls?* (552)
20. In what way is spore formation in *Selaginella* advanced over spore formation in *Lycopodium?* How does the location of the gametophyte phase differ? Which of the two species appears to be the most similar to seed plants? (552-553)
21. Describe the body form of a horsetail. Is its reproductive cycle advanced or primitive? Explain. (554)
22. Characterize the distribution of ferns. What does this indicate about their evolutionary success? (554)
23. Describe the body form of a fern, using the terms *rhizome, fiddlehead,* and *divided megaphylls.* (554)
24. Using the terms *sorus, indusium, sporangia, annulus,* and *lip cells,* describe the structures responsible for spore formation and dissemination in a fern. (554)
25. Briefly summarize the life cycle of a fern, starting with the haploid spore. (554-555)
26. Summarize the geological changes that may have prompted the end of the ancient Paleozoic forests. How might the small independent gametophyte have added to the survival problems of the early vascular plants? What plants replaced the lycophytes, ferns, and psilophytes? (556)
27. Explain how the nonmotile sperm of an advanced seed plant gets from the pollen to the egg cell. (558)
28. What constitutes a seed? How might the evolution of seeds have given the gymnosperm an advantage over earlier vascular plants? (558-559)
29. Which of the gymnosperm divisions is most prominent today? In what regions are most of these to be found? (561)
30. In what ways are the ginkgo and cycad reproductively primitive? (559-560)
31. List three characteristics of gnetophytes that make this group related to the angiosperms. (560)
32. What specific adaptation do the needle- and scalelike leaves of the conifer represent? In what other ways are conifer leaves adapted to this situation? (561)
33. Outline the following aspects of pine reproduction: pollen formation, egg formation, pollination. (561)
34. When did the angiosperms rise to prominence? Where were their forebears living during the time gymnosperms were prominent? (562-563)
35. Briefly describe three factors that might have produced vast changes in life at the close of the Mesozoic. (563)
36. Name the two groups of angiosperms and list several characteristic differences between them. (564)
37. In what way do the evolutionary terms "primitive" and "advanced" present usage problems? Are individual species entirely primitive or entirely advanced? How might one avoid this semantics problem? (564-565)
38. List four flower characteristics that would be considered primitive. Suggest a specific flower to which they apply. (565)
39. Explain how the ongoing coevolution of insect pollinator and plant might increase diversity. Describe an extreme case of specialization between flower and insect. (564, 575-576)

26

FLOWERING PLANT REPRODUCTION

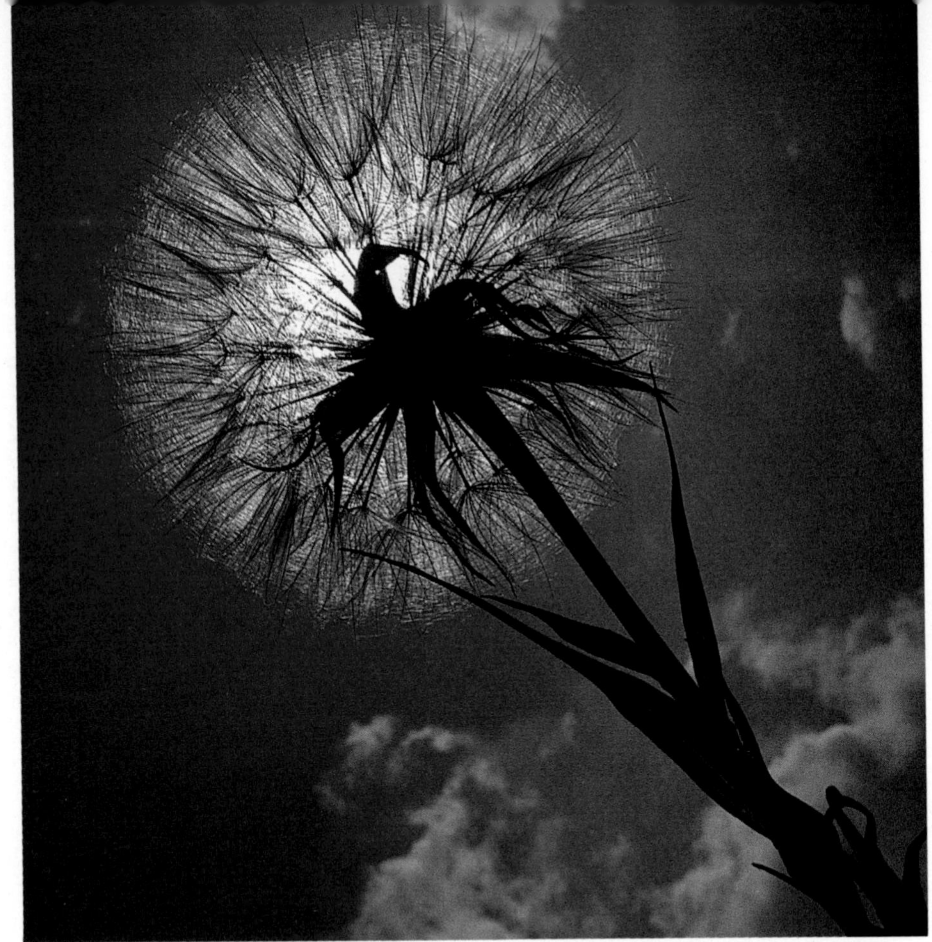

AS PLANTS BEGAN TO EXPAND OVER THE EARTH'S OFTEN PARCHING SURFACE, they were met by a host of unexplored opportunities. At the same time, they encountered perhaps as many real threats to their very existence. Among those threats was the problem of reproducing in the dry air. As we will see here, that problem was largely solved by flowers.

In the flowering plants—the angiosperms—the role of sexual reproduction falls on the flower itself. The flower is, to many of us, the most notable part of an angiosperm, and we extol its beauty and fragrance. We must keep in mind, however, that they are not simply decorations; they have a job to do. Furthermore, we should keep in mind that after fertilization, the flower's petals drop off, and part of the flower enlarges to become the seed-bearing *fruit*. The seed, we know, is the flowering plant's investment in the future. So let's begin our consideration of sexual reproduction in angiosperms with the flower.

SEXUAL REPRODUCTION

The Flower

A typical flower is composed of as many as four **whorls** of parts around the **receptacle** or **base** (Figure 26.1). The term **whorl** means simply that the floral parts are repeated in a circle. Actually, each member of a whorl is a modified leaf, but some are much more modified than others. The outermost whorl consists of the **sepals**, which are usually green and leaflike. They surround and protect the flower bud before it opens and are commonly photosynthetic. The whorl of sepals is collectively known as the **calyx** (meaning "cup").

Just within the calyx lie the **petals**; these are the second whorl of floral parts. They are often large and colorful, but most are still somewhat leaflike. A whorl of petals is known as the **corolla** (meaning "garland").

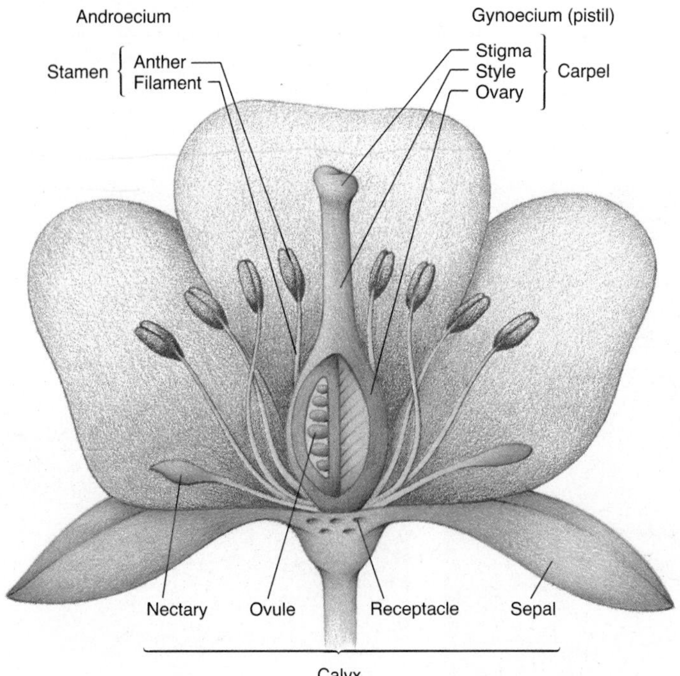

FIGURE 26.1
A GENERALIZED FLOWER
Flowers generally consist of four major parts, occurring in whorls and supported on a receptacle. They are the calyx (sepals), corolla (petals), stamens (filaments and anthers), and carpel or carpels (also pistil, made up of ovary, style, and stigma).

The third whorl of floral parts are the **stamens** which collectively make up the **androecium** ("house of man"). The stamens, highly modified leaves, can be considered as male floral parts. Each stamen usually includes a slender stalk, the **filament,** which leads to an enlarged bilobed structure, the **anther.** The two lobes contain four microsporangia, or **pollen sacs,** as they are also known. As we've seen, cells in the microsporangia produce **microspores,** which in turn develop into **pollen grains** (see Chapter 25).

The fourth region of the flower is the **gynoecium** ("house of woman"), which includes the **carpel** or **pistil,** names that are often used interchangeably. (We will use the first.) The carpel, which often takes the shape of a bowling pin, can be considered as the female floral part. Like stamens, carpels are also highly modified leaves. There are many variations in number and arrangement, depending on the species. In some species, there are numerous carpels, arranged in a whorl of separate individuals. In others, a number of carpels may be partially or completely fused into a single unit. There may, of course, be just one carpel per flower. Each carpel consists of three parts: the **ovary,** the **style,** and the **stigma.** The ovary produces the ovules, which contain the megasporangia. Extending from the ovary is the style, a stalk that supports and elevates the stigma, an enlarged structure at its tip. The stigma is often hairy or sticky and specializes in receiving pollen.

In summary, the flower consists of a calyx (whorl of sepals), corolla (whorl of petals), **androecium** (whorl of stamens), and **gynoecium** (carpels). The last two whorls are the most intimately involved in reproduction. Sepals and petals are called **accessory floral parts,** since they play no direct role in sexual reproduction. However, the petals can be quite important to the process, particularly in insect-pollinated species.

Variation in Flowers

It seems that natural selection has molded, warped, changed, hidden, and amplified flower parts in endless ways (Essay 26.1). For example, there tend to be clear differences between monocots and dicots (Class Monocotyledonae and Dicotyledonae; see Chapter 25). Most dicot floral parts are arranged in fours and fives or multiples of these numbers, whereas monocot flowers often have parts arranged in threes and multiples of three. The most primitive floral types have well-defined parts; the sepals, petals, carpels, and

ESSAY 26.1

DIVERSITY IN FLOWERS

Daffodils (*l*) and lilies (*r*) are monocots. Note the number of floral parts (6). Both are radially symmetrical and simple in design.

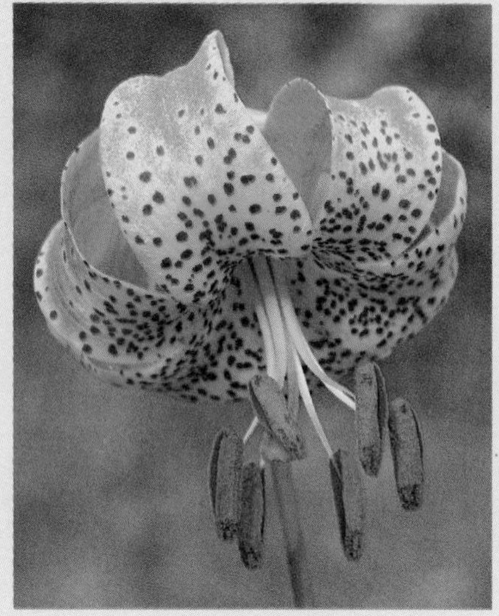

The San Diego hibiscus (*l*) and the red pimpernel (*r*) are both dicots, have floral parts in fives, are radial symmetrical, and follow simple designs.

The snapdragon (*l*), a dicot, and lady slipper (*r*), a monocot are bilaterally symmetrical (have right and left sides), and very complex flowers. Both represent a clear departure from the more primitive, simple, radial flowers.

Another dicot, the sunflower, a composite flower (actually a whole bouquet), occurs as a broad, disk-shaped, head. Whereas most of its members are tiny disk flowers, large ray flowers form the periphery.

stamens occur in a **radially symmetrical** pattern—that is, circular and disklike. A line drawn through the center of the disk produces equal halves. Such flowers act as platforms upon which a pollinator may land.

Highly evolved flowers, in contrast, are often **bilaterally symmetrical** or **irregular**; they have right and left halves that are essentially mirror images (as is true of humans and most other animals). A few highly advanced flowers seem to have lost their symmetry altogether (see Figure 25.27b). Bilateral flowers often fit the pollinator's body like a lock-and-key mechanism, and they tend to attract very specific animal pollinators.

Other variations include flowers with missing parts. A flower with one or more basic parts absent is called **incomplete,** as opposed to the **complete** flower, which has all of the usual parts. If either (or both) of the sexual parts, the carpels or stamens, is absent, the flower is referred to as **imperfect.** If they are both present, the flower is **perfect.** Obviously, flowers that are imperfect must also be incomplete, and flowers that are complete must be perfect. Cornflowers are imperfect and are either **carpellate flowers** (female) or **staminate flowers** (male), but both occur on the same plant, which tells us the corn plant is actually monoecious. Date palms are dioecious, so carpellate and staminate flowers grow on separate trees, and to provide for pollination a date-grower usually plants one staminate tree per 10 or so carpellate trees.

In some instances flowers are grouped together into clusters, or **inflorescences,** which take various forms such as the heads of daisies, marigolds, and sunflowers. Thus a single sunflower is really a bouquet of flowers. **Composite flowers** are inflorescences composed of tiny, individual flowers, each of which produces a single seed-bearing fruit. In the daisy, marigold, and sunflower heads, the outer rows of **ray flowers** are usually carpellate, or sometimes sexless flowers, their large petals serving to attract pollinators to the central **disk flowers.**

Flower Structure and Pollination

We all know that flowers can be beautiful, elegant, and fragrant. Some can also produce a sweet fluid called **nectar.** Color, fragrance, and nectar all have a role in flowers that depend on animals such as insects, bats, and birds for transferring pollen from one flower to another. The color and fragrance attracts pollinators to the flowers with the promise of sweet, energy-rich nectar (usually secreted from **nectaries** near the base of the flower). We, too, are attracted to flowers because of their color and fragrance. It is interesting that we share the insect's appreciation for flowers, because insects and people sense quite different things according to their different sensory abilities. For example, some insects see ultraviolet light, so they may find purple patterns in flowers that to us appear white. And since most insects can't see red at all, a beautiful red rose appears to them as black. As certain flowers evolved, it must have become adaptive for them to increasingly specialize according to the sensory abilities of various kinds of pollinators. Flower structure, appearance, and scent were modified to attract and assist specific kinds of pollinators, whether beetles, flies, moths, butterflies, bats, or others (Table 26.1).

Animal Pollinators and Flower Specializations From an evolutionary viewpoint, then, insects and other animal pollinators may "specialize." That is, they may be attracted to only one kind of flower. A pollinator that has been rewarded with nectar even once may become "keyed" to that particular type of flower, ignoring all others, and will transfer pollen only between flowers of the same species, or even of the same strain. Natural selection may thus tend to favor plants with the most distinctive and easily identified flowers. A new strain with a slightly modified flower can therefore become genetically isolated if it is not recognized by the insect that pollinates the parental strain. In this way, the flower might find itself on a new evolutionary pathway without ever having undergone the usual geographical isolation from its parents as speciation occurred (see Chapter 20). Such evolutionary separation would lead to specialization, of course, as each strain came to adapt more precisely to the most efficient pollinator available.

Actually, there are many examples of insect specialization that have led to angiosperm diversity. You may recall from Chapter 20 our discussion of coevolution between the

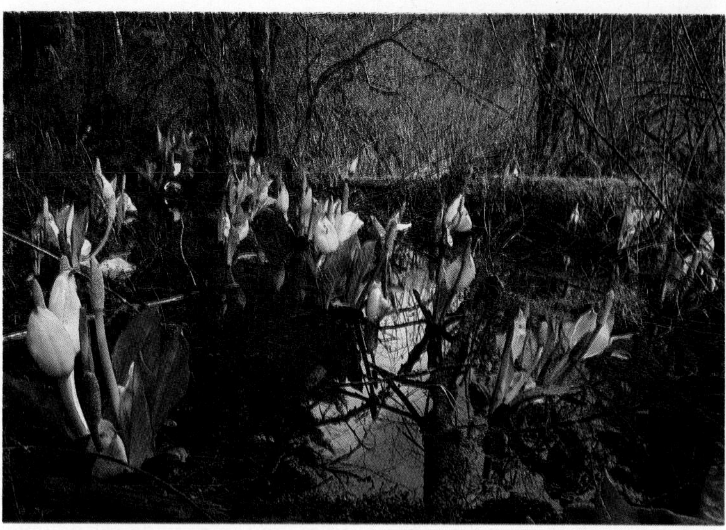

FIGURE 26.2
THE YUCCA AND YUCCA MOTH
The *Yucca* flower and yucca moth depend on each other for survival. The plant serves as a source of food for the moth's larvae, and the moth is the plant's chief pollinating agent.

FIGURE 26.3
A FOUL-SMELLING FLOWER
Lysichiton americanum (western skunk cabbage) gives off a pungent "skunk" odor. It is pollinated by flying beetles.

passion vine and a butterfly that lays its eggs in the vine's leaves (see Figure 20.22). In another instance, the *Yucca*, a desert plant, requires a specific insect, the yucca moth, to accomplish pollination, but the interaction goes far beyond that (Figure 26.2). When in bloom, the *Yucca* sends up a long growth that bears large clusters of white flowers at its end. As the yucca moth pays its nightly visits to yucca flowers, it gathers pollen and rolls it into small balls. At successive stops, it pierces the ovary using its sharp ovipositor (egg-burying structure) and injects a number of eggs among the young ovules. It then places a ball of sticky pollen into the opening (perhaps a form of "sealing behavior," since many insects entomb their eggs). Upon hatching, the larvae eat their way through the carpels and reproductive tissue, but they emerge before they destroy too much. The remaining undamaged ovules form seeds and drop to the ground about the same time the moth larvae become pupae. Both moth and plant have evolved to a state where they are entirely dependent on each other. The moth cannot survive without the plant, and the plant is not naturally pollinated by any other means.

Another example of specialization is seen in the familiar "skunk cabbage" *(Symplocarpus)*. This marsh plant produces a flesh-colored structure, which bears flowers within. Its pungent, unpleasant odor, like that of rotting flesh, will have you searching through a nearby marsh, looking for an unfortunate cat. Honeybees avoid this flower, but flies swarm over it. Flies, in fact, lay their eggs in the flower and in so doing pollinate it. Another example of a foul-smelling flower is seen in Figure 26.3. Some large, drably colored, nectar-rich, night-opening flowers are pollinated by bats, which are attracted to the strong fruitlike or fermenting odor. The coevolving relationship has gone even further. In one species, the pollen has a much higher than usual protein content, which, as it turns out, is an important dietary supplement to the bat pollinator of that species. Two examples of adaptations to insect pollinators involve **nectar guides** and mimicry, the latter of which is probably the "ultimate effort" made by any flowering plant (Figure 26.4).

This is not to say that all flowering plants are such specialists. Plants that are generalists take a different adaptive route and depend on a variety of animal species. They thus develop traits that are attractive to a number of pollinators. For example, many insect pollinators are attracted to flowers with distinctly separated petals as well as aromatic secretions. The adaptations for attracting a variety of pollinators are, of course, tempered by other factors such as environmental influences. It wouldn't do to have big, floppy flowers on dry, windy slopes.

Wind Pollination There is also a very large group of flowering plants that are wind-pollinated. Among the wind-pollinated monocots are grasses such as bluegrass, wheat, rye, and corn. Wind-pollinated dicots include most species of temperate-zone trees, for

TABLE 26.1 FLOWERS AND ANIMAL POLLINATORS

ANIMAL	VISUAL CUES	CHEMICAL CUES
Beetles	Not significant; flowers dull colored or white	Strong odors—fruity, spicy, or foul
Bees	Bright colors; yellow or blue (ultraviolet perception); highly divided floral parts	Strong fragrance
Flies	Large flowers; dull, flesh-colored	Musky to rotting odors
Day moths and butterflies	Bright colors—reds, oranges, yellows, blues	Strong fragrance
Birds	Bright colors—reds and yellows	Copious, sugary nectar; little odor
Bats	Color not significant (night flyers)	Copious nectar; fruity, fermenting odors

example oak, birch, poplar, and alder. Of course, the gymnosperms described in the last chapter are largely wind-pollinated.

It may seem strange that grasses produce flowers, but remember, they are angiosperms. Grasses have stamens and carpels, but they are plain, small, and inconspicuous (Figure 26.5). There would be little adaptive value in such energy-costly amenities as colorful petals and sweet nectaries in wind-pollinated flowers. Instead, these flowers produce

(a)

(b)

(c)

(d)

(e)

FIGURE 26.4
INSECT POLLINATION DEVICES
Some plants evolved elaborate devices for ensuring pollination. *Wedelia* has nectar guides, pigment lines, that are easily visible to insects and guide them into the nectaries. In some cases, the guides are not visible to us **(a)**. but show up under ultraviolet light, which is visible to bees **(b)**. **(c)** One orchid, *Oncidium,* which has specialized anthers with very adhesive surfaces, mimics male bees, and transfer of the pollinium (pollen sac) occurs when an aggressive, territorial male bee is deceived into picking a fight with it. **(d)** Note the pollinium on the bee hindleg. When a bee enters the orchid, its head or body brushes the adhesive projections, which become firmly attached. As the bee backs out, these break away. The pollinia are then deposited in the next orchid visited by the bee, and new pollen sacs may be picked up. **(e)** Flowers of the orchid *Ophrys speculum* mimic female wasps. Male wasps try to copulate with them but manage only to pollinate the orchid.

Wild oat

Bog cotton

FIGURE 26.5
WIND-POLLINATED FLOWERS
Wind-pollinated flowers lack color and fragrance. Also absent are sepals and petals. They often have feathery, plumelike, pollen-catching stigmas, stamens usually in threes, and pollen that does not tend to stick together as it does in insect-pollinated species. Wind pollination is efficient where many plants of the same species are clustered.

copious amounts of very light, easily dispersed pollen and elaborate stigmas for trapping it. Even so, wind pollination seems to be inefficient, and it's true that much of the pollen is lost (some of it up the nose of hay fever sufferers). Recent studies involving the use of wind tunnels with gynnosperm cones and grass flowers, however, indicate that wind pollination may be more efficient than we think. Cones and grass inflorescences, acting like snow fences, tend to slow and trap wind-borne pollen.

Sexual Activity in Flowers

Earlier we noted that a few of the primitive vascular plants (for example, *Selaginella;* see Chapter 25), and all seed plants, are heterosporous. That is, they produce two kinds of spores, megaspores and microspores. We also saw that as plants evolved, the gametophyte phase of the life cycle became more and more reduced and dependent on the sporophyte. In the flowering plants, the female gametophyte and at least the younger phase of the male gametophyte are tucked away in the ovaries and anthers of the flower. With these points in mind, let's look into the process of sporogenesis, the formation of spores, and their development into the gametophyte in flowering plants.

Events in the Ovary Within the soft tissues of the flower's ovary are found the young **ovules.** The ovules originate from a surrounding tissue known as the **placenta** and for a time remain attached to it by a stalklike **funiculus.** Each ovule consists of a megasporangium, or **nucellus,** as it is known in flowering plants, and two **integuments**—skinlike protective coverings that will much later form the seed coat. At the exposed end of the integument is a small opening called the **micropyle.** But most importantly, each ovule contains one large cell called the **megaspore mother cell,** and it is this cell which undergoes **megasporogenesis,** which we will now describe.

The megaspore mother cell, like its male counterpart in the stamen, the **microspore mother cell,** will undergo meiosis, producing haploid spores. Typical of plants, these meiotic products are not gametes, but instead they will enter mitosis to produce the gametophyte, which, in turn, will produce the gametes (Figure 26.6).

Meiosis in the megaspore mother cell involves the usual two divisions, so at the end of the process there are four haploid spores—the megaspores. Usually, three of the megaspores simply disintegrate. The surviving meagaspore enlarges considerably before its haploid nucleus undergoes three successive mitotic divisions, giving rise to the gametophyte or, more accurately, the **megagametophyte** (the female gametophyte generation of the flowering plant). If our arithmetic is correct, after three mitotic divisions, the megagametophyte will contain eight identical haploid nuclei, all in one greatly enlarged cell (one divides into two, two into four, and four into eight). With the mitotic events completed, the megagametophyte reaches an eight-nucleate stage.

But there's more. The next events may vary according to the species, but we will consider what commonly happens. In most cases the eight nuclei are equally distributed to the two ends of the developing megagametophyte, after which two of the nuclei, one from each end, migrate to the center of the cell. Finally, new cell walls are laid down, and the cytoplasm is divided into seven separate, unequal cells. The result of all this is that one cell, often the largest one, ends up in the center with two nuclei (see Figure 26.6j). The binucleate cell will later give rise to the **endosperm** (a food storage tissue in the seed), but for now it is known as the **central cell** (and sometimes as the **endosperm mother cell**). Its nuclei are called **polar nuclei.** Of the remaining six cells, the one nearest the micropyle (site of sperm entry), will probably become the **egg cell.** The completed megagametophyte is now called an **embryo sac** (a welcome relief from the six-syllable term). (We should point out that once again, the flowering plants are diverse,

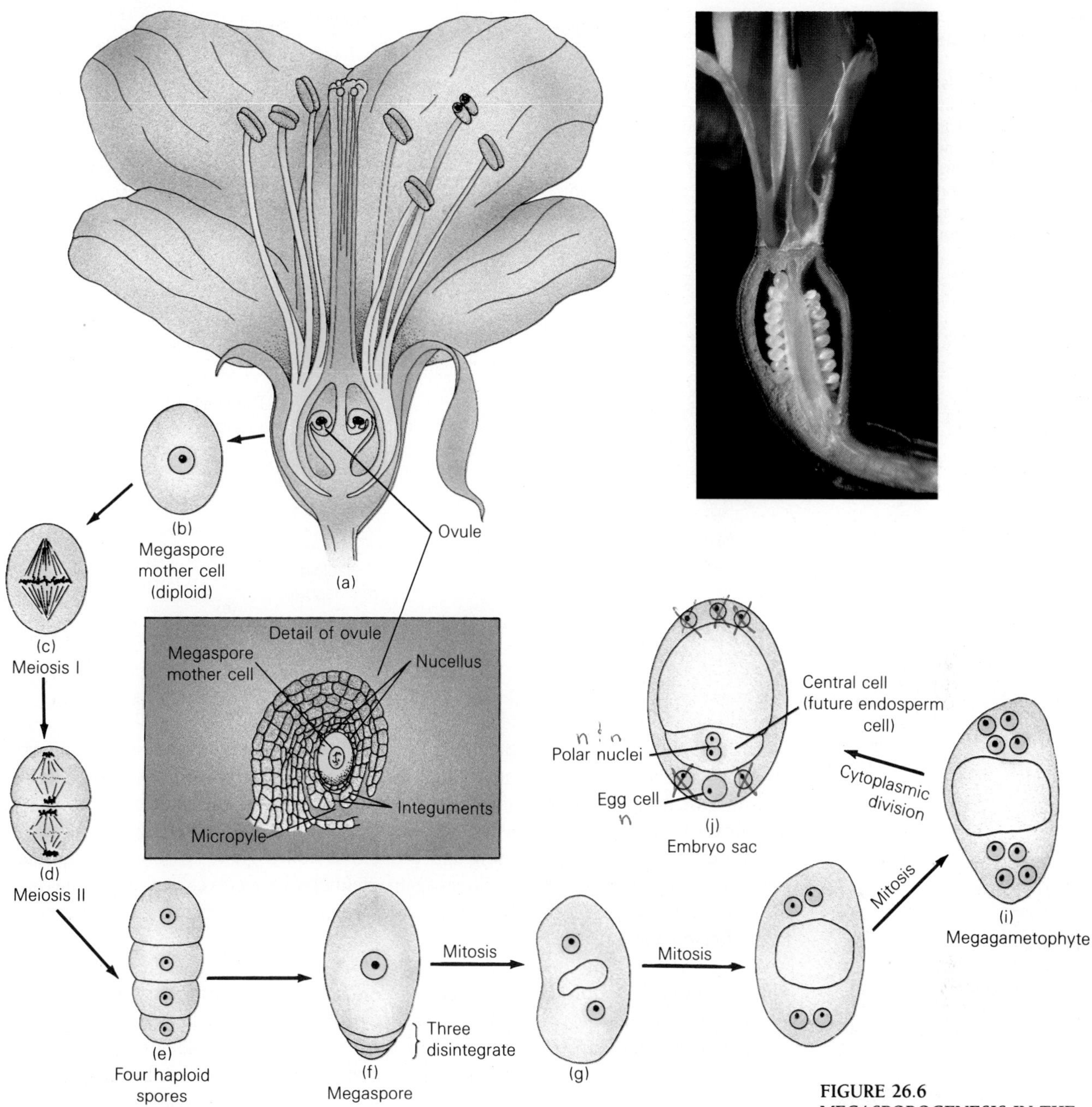

FIGURE 26.6
MEGASPOROGENESIS IN THE OVARY
The female gametophyte develops within the ovules **(a)**. Each ovule contains a megaspore mother cell that undergoes meiosis, producing four haploid megaspores **(b-e)**. Three disintegrate, the fourth enlarges and undergoes mitosis three times **(f-i)**, resulting in eight nuclei. With cell wall formation, the embryo sac, a seven-celled structure, emerges **(j)**. The large central cell becomes the endosperm mother cell. One of the smaller cells becomes the egg cell. The embryo sac is the female gametophyte.

and whereas the eight-nucleated embryo sac is commonplace, it is not universal.) There will be no more nuclear divisions in the embryo sac until after fertilization. At that time, both the egg cell and the central cell will participate in a unique and impressive event known as **double fertilization.**

Events in the Anther While the embryo sac has been undergoing its development, similar but less complex events have been occurring in the anthers, the sites of **microsporogenesis** (Figure 26.7). Typically, each anther contains four pollen sacs—chambers that contain numerous diploid microspore mother cells, each of which will enter meiosis. By the end of meiosis, each microspore mother cell will have produced four microspores. The microspores will then begin their development into male gametophytes (microgametophytes), which will then produce sperm cells.

FIGURE 26.7
MICROSPOROGENESIS IN THE ANTHER
Microspore mother cells within the pollen sacs **(a)** undergo meiosis **(b)**, forming four microspores. Each microspore then undergoes one round of mitosis, and a generative cell and tube cell emerge. **(c)** Note that the generative cell is actually suspended in the tube cell cytoplasm. With the development of a tough wall, a pollen grain is formed.

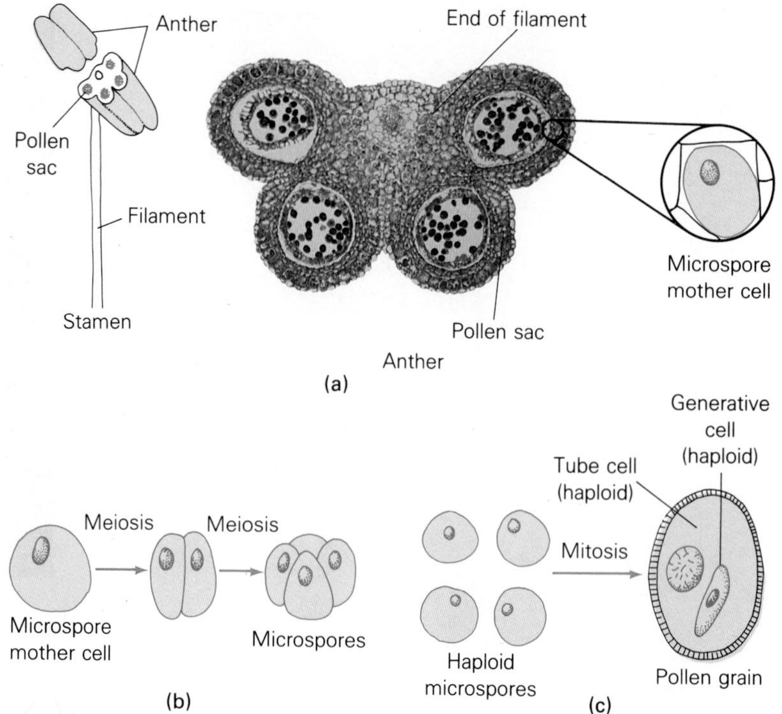

The haploid nucleus within each microspore will undergo mitosis, and when the two resulting daughter cells differentiate, they become a **generative cell** and a **tube cell**, both of which remain within the microspore. With mitosis, the male gametophyte's (microgametophyte's) formation is underway. Meanwhile, each emerging microgametophyte develops a resistant coating and becomes a pollen grain. In some species, the generative cell will soon undergo mitosis again, yielding two sperm cells, thus completing the male gametophyte's formation while in the pollen grain. In others, this event occurs later, after pollination.

FIGURE 26.8
POLLEN TUBES
Under proper conditions, pollen germinates, giving rise to pollen tubes, which convey the sperm cells into the ovule.

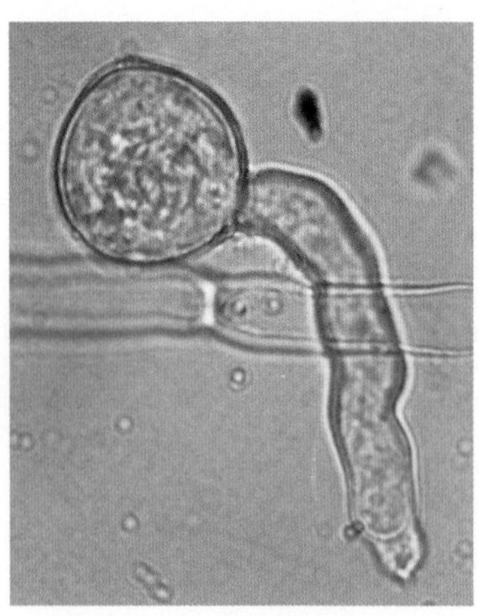

Pollination and Double Fertilization Technically, **pollination** occurs when pollen is deposited on a receptive stigma. The source of the pollen may be the male parts of the same flower, other flowers on the same plant, or another individual. We are aware of the dangers of inbreeding, and since self-fertilization is obviously inbreeding of the severest sort, it is not surprising to learn that some plants have ways of avoiding self-pollination, such as the position of their anthers, the timing of microspore and megaspore production, or even physiological incompatibilities. Nonetheless, strangely enough, self-pollination is a regular event in some species.

When pollen germinates, a **pollen tube** forms, soon lengthening and progressing through the style towards the ovary (Figure 26.8). The tube cell makes this growth possible through the production of proteolytic enzymes that literally digest the tissue ahead of the lengthening tube. In those species where mitosis in the generative cell is delayed, it is about this time that it occurs, and the two long-awaited sperm cells form. This final episode of mitosis completes development of the male gametophyte.

Finally, the tube penetrates the ovule at the tiny, porelike micropyle. The two sperm cells then enter the embryo sac, where the unique double fertilization mentioned earlier occurs (Figure 26.9). One sperm fertilizes the egg cell, and a diploid zygote is formed. This event ushers in the next sporophyte generation in the life cycle of flowering plants.

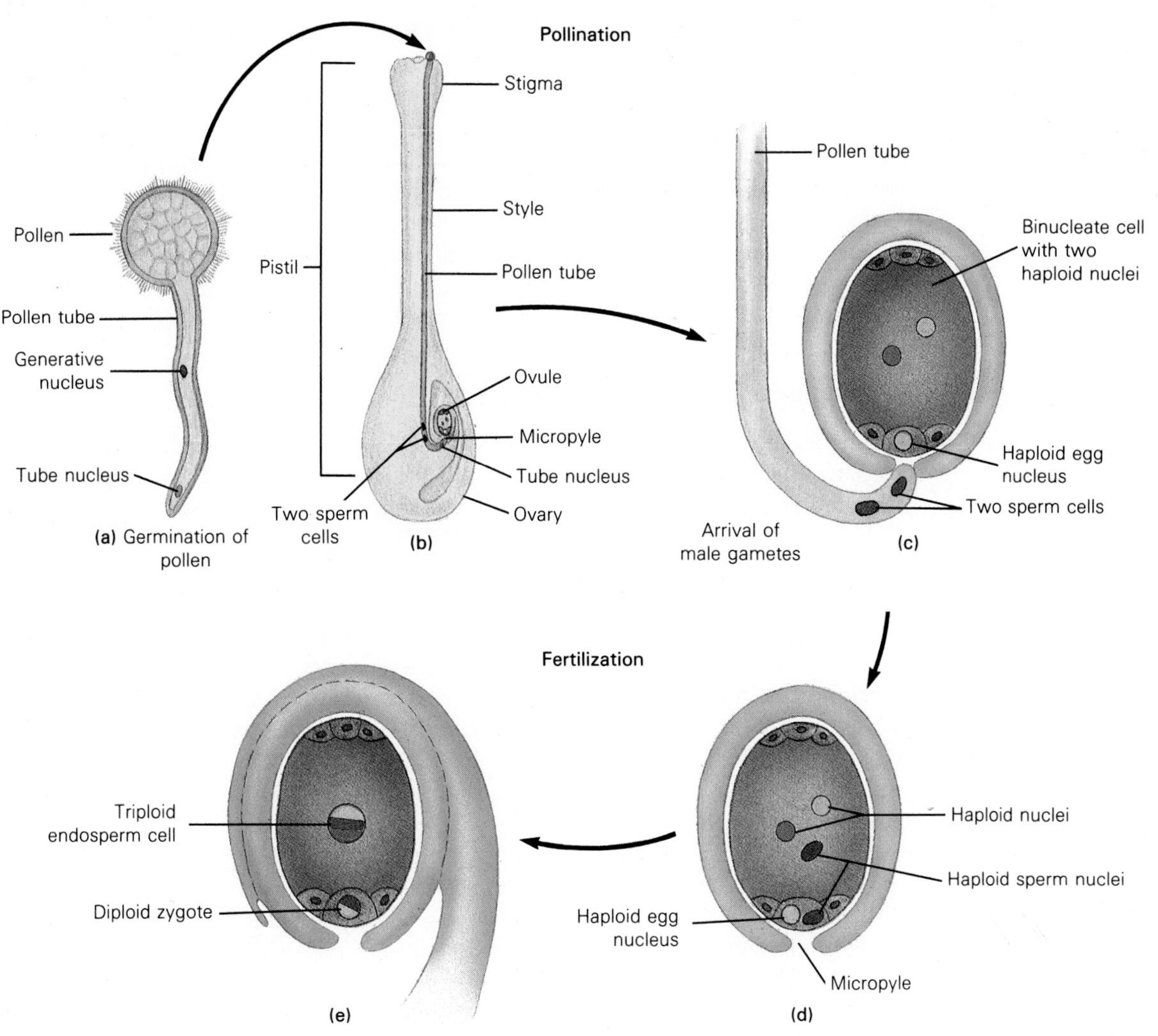

FIGURE 26.9 POLLINATION AND FERTILIZATION
(a) Pollination is followed by pollen tube growth. (b) As the pollen tube penetrates the style and ovary, the generative cell may undergo mitosis at this time, producing two sperm cells. The tube then penetrates the ovule (c) and the two sperm enter, one fertilizing the egg, and the other the central cell (d and e) (note the triploid endosperm).

The fertilized egg, or zygote, will develop into the plant embryo. The second sperm penetrates the central cell, where its nucleus joins the two polar nuclei already present, producing a triploid **primary endosperm nucleus.** (While a triploid primary endosperm nucleus is common in many species, it may in others be diploid, or even pentaploid, depending on the number of polar nuclei.) The endosperm provides food for early development after seed germination occurs. Double fertilization, as seen here, represents one of the basic differences between angiosperms and gymnosperms. You may recall from the last chapter that only one sperm is functional in each gymnosperm microgametophyte.

After fertilization, the flower begins to change, and its beauty may fade. Commonly, those parts that do not sustain the seeds begin to wither and fall off. Then the ovary itself will swell, sometimes to enormous proportions. This begins the development of **fruit,** another structure characteristic of angiosperms. In everyday usage, "fruit" usually refers to a sweet and juicy structure such as the familiar apple, grape, or banana. Botanically, however, the fruit is the ripened or mature ovary. Whereas it does commonly consist of a sweet fleshy structure, it may also include any other mature ovary, such as the familiar string bean, cereal grain, pumpkin, or tomato. It even includes certain hardened parts of coconut, walnut, and the so-called sunflower "seed." For a more complete look at fruits and their development, see Essay 26.2.

As an aside, the United States Supreme Court at one time was required to decide whether the tomato was a fruit or a vegetable. It seems that Congress had passed a 10% tariff on vegetables, but there was no tariff on fruit. The nine learned judges, wise men all, decided that the tomato was not sweet enough to be a fruit. It was clearly a vegetable, and the tariff was imposed. (Subsequently, other august bodies have tried to do science. It might interest math majors to know that one legendary state legislature proposed a law to solve the awkwardness of *pi:* they rounded it off!)

Development of the Embryo and Seed

After fertilization, the new diploid zygote, the primary endosperm nucleus, and all the associated maternal tissues begin mitosis and the differentiation that will produce the plant **embryo** and its life support systems. The finished product—the seed—will consist of (1) an embryo, (2) some kind of food supply, and (3) hardened, protective **seed coats** or their equivalent.

Seed development varies widely among flowering plants, so we will follow the process in selected representatives from the dicots and monocots.

Development in the Bean: A Dicot The most familiar dicot seeds are probably foods such as peas, beans, peanuts, and sunflower seeds—those eaten without much alteration (except, perhaps, for getting rid of the thin, papery peanut seed coat—the thing that gets stuck in your throat and ruins the movie for you). The embryo, a miniature sporophyte sporting two large cotyledons, takes up most of the dicot seed. Consider the peanut again. The shell in this case is the fruit wall. Inside are two seeds (usually), each of which can be easily separated into two parts, the cotyledons. If you look closely, you can see the embryo, a tiny, cylindrical object still attached to one of its cotyledons. With this elementary lesson in mind, we'll look into the events that lead up to seed formation in the bean. You will want to refer to Figure 26.10 as we proceed.

After fertilization, the ovule contains an embryo sac with a diploid zygote, a primary endosperm nucleus, and several layers of surrounding and supporting diploid cells of the maternal tissue (Figure 26.10a). For convenience, what follows can be divided into three events: developments in the endosperm, growth of the cotyledons, and growth of the rest of the embryo, or **embryo axis**, as it is called.

Within the primary endosperm, the nuclei divide continuously until a multinucleate mass surrounds the zygote (see Figures 26.10b and c). (This stage is carried to an extreme in the coconut, a monocot, since "coconut milk" is nothing but a mass of white, fluid endosperm cytoplasm with free nuclei floating in it.) Ultimately, in most dicot seeds, cell walls are laid down around the endosperm nuclei.

The zygote itself does not usually begin to develop into an embryo until the endosperm has grown to a considerable size. Then the zygote beings a series of cell divisions that will eventually produce the embryo. The first few divisions produce a vertical row of cells. Then, at the inner end of the row, cell divisions speed up and, more important, begin to occur in many planes, producing a three-dimensional ball of cells (see Figure 26.10d). Unlike the cells in animal embryos, which lack walls and grow between, around, and over each other as the embryo takes form, plant embryo cells must do this chiefly through changes in division planes. At any rate, it is this ball of cells that becomes the embryo. A **suspensor** is formed by cells at the other end of the row, which divide much more slowly. One becomes a large **basal cell**, swollen by the uptake of water. The suspensor anchors the embryo in place and aids in the transfer of food to the rapidly growing embryo.

At this point, we see a fundamental difference between the monocot and dicot

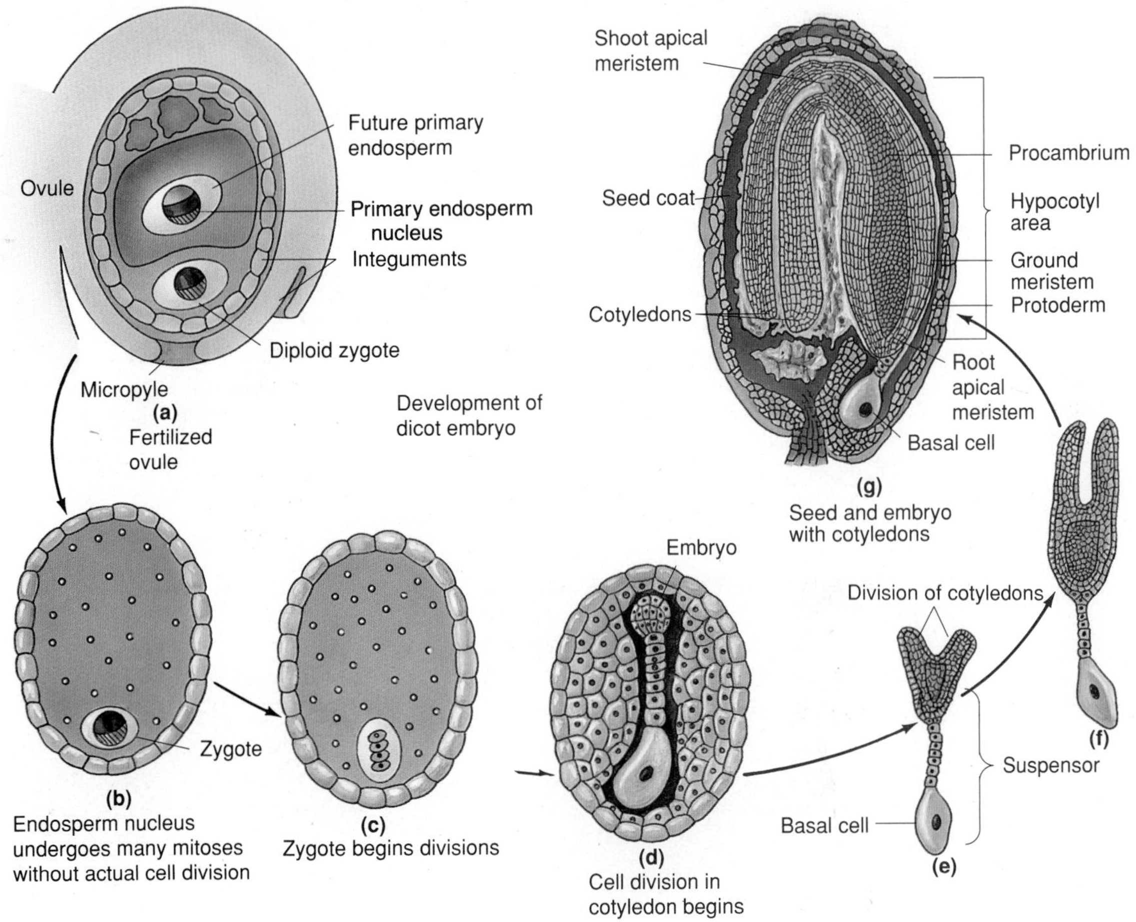

FIGURE 26.10
SEED AND EMBRYO DEVELOPMENT IN A DICOT.
After fertilization, the embryo sac (**a**) contains a diploid cell, the zygote, and a triploid primary endosperm nucleus. Numerous mitotic events occur in the endosperm (**b**), and the embryo undergoes its first cell divisions (**c**). Further cell divisions occur in the embryo, and cell walls are laid down in the endosperm (**d**). A suspensor forms below anchoring itself in the parental tissue and the two cotyledons emerge (**e, f**). As the seed matures, enlarging cotyledons have made a U-turn, and signs of differentiation are seen in the remainder of the embryo (**g**). Much of the embryo is hypocotyl, an embryonic root or radicle that contains several primary tissues (procambium, protoderm, and ground meristem). A shoot apical meristem has formed just above the cotyledons.

embryos. In the rapidly growing dicots, the cells of the embryo form a two-pronged, heart-shaped structure (see Figures 26.10e and f). In the monocots, equally rapid cell division produces a simple, elongated structure. The two lobes of the heart will produce the two cotyledons of dicot plants; the single growth will form the single cotyledon of monocots.

In beans and many other dicots, the cotyledons continue to grow, absorbing all or nearly all of the endosperm and filling most of the space of the growing embryo sac. As they grow, the planes of division change once more, and the result is that the cotyledons form a U-turn, growing back toward the base of the seed (see Figure 26.10g). We should add that not all dicot cotyledons are so fleshy. Those of the castor bean, for instance, are thin and more leaflike, surrounded by a large amount of endosperm.

As the cotyledons form, the remainder of the embryo begins to take shape. At its upper portion, where the cotyledons divide, a naked dome of tissue, the **shoot apical meristem,** forms. It will give rise to the future shoot and leaves. **Meristematic tissue** in plants is special in that it is made up of simple, undifferentiated cells. Meristem is a source that the embryo draws upon and that the adult plant also uses throughout its life as a source of cells for the ongoing formation of new tissues.

The shoot apical meristem, and anything it produces in the embryo, is called the **epicotyl** ("above the cotyledon[s]"). In some embryos the epicotyl is no more than simple meristem, but in others, the bean embryo included, the epicotyl includes a **plumule,** tiny embryonic leaves. In the emerging seedling, the plumule is regarded as the first bud (Figure 26.11b).

ESSAY 26.2

FLOWERS TO FRUITS, OR A QUINCE IS A POME

A fruit is a ripened ovary that sometimes exists in association with certain floral parts. There are three basic types of fruits: simple, aggregate, and multiple, depending on the number of ovaries in the flower or the number of flowers in the fruiting structure.

Simple fruits may be derived from a single ovary or, more commonly, from the compound ovary of a single flower. They can be divided into two groups according to their consistency at maturity: simple fleshy fruits and simple dry fruits.

Simple fleshy fruits include the **berry,** **pome,** and **drupe.** The berry has one or several united fleshy carpels, each with many seeds. Thus the tomato **(a)** is a berry, and each of the seed-filled cavities is derived from a carpel. Watermelons, cucumbers, and grapefruits are also berries (but, oddly enough, blackberries, raspberries, and strawberries technically are not berries). Pome **(b)** means "apple," and the group includes apples, pears, and quinces. In the pome, only the inner chambers (roughly, the "core") are derived from the ovary, and most of the flesh comes from accessory tissue. (Most of the flesh comes from the receptacle, which grows

(a) *(left)* Young, simple fleshy fruit (berry) of the tomato with only sepals remaining. The flower is seen at right.

(right) Mature fleshy berry of the tomato. A cut at right angles to its axis reveals five fused carpels, each containing the seed-bearing, fan-shaped parts of the ovary.

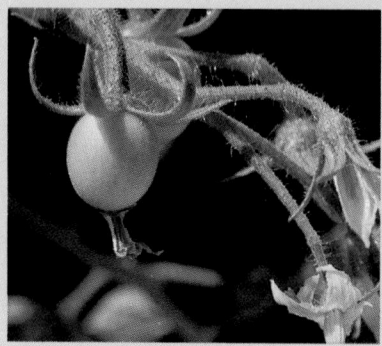

(b) *(left)* The organization of a pome becomes apparent in the young fruit, as the receptacle forms the *floral tube* surrounding the ovary.

(below) In the mature fruit, most of the floral parts have withered away, but the floral tube has greatly enlarged, producing the sweet, edible portion of the fruit that contains the ovary (core). In the cross section we see the remnants of the flower, including the united carpels.

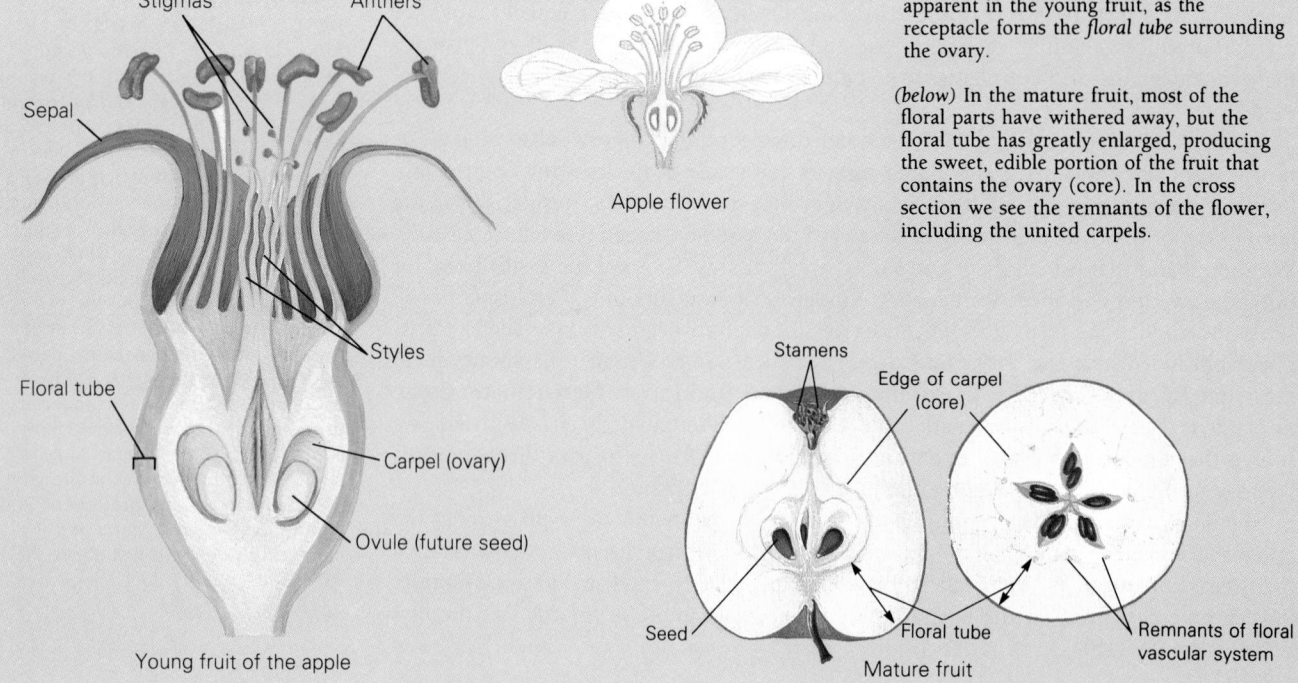

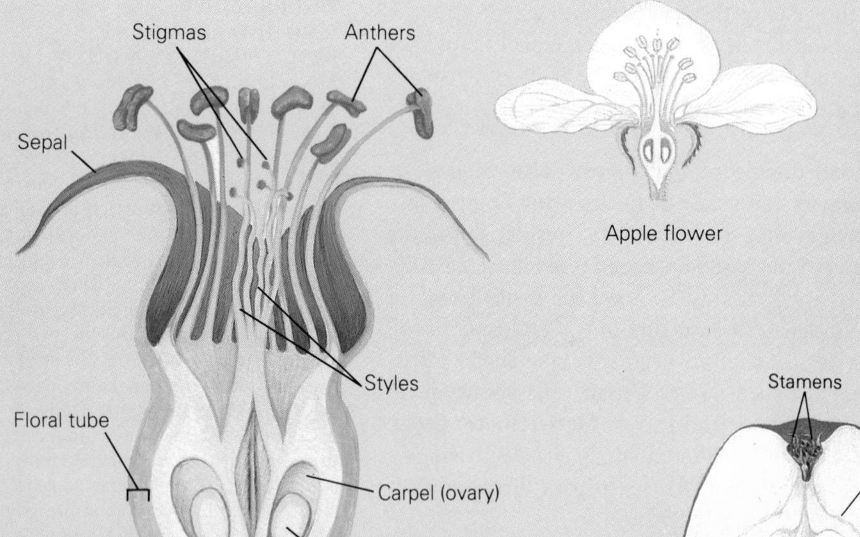

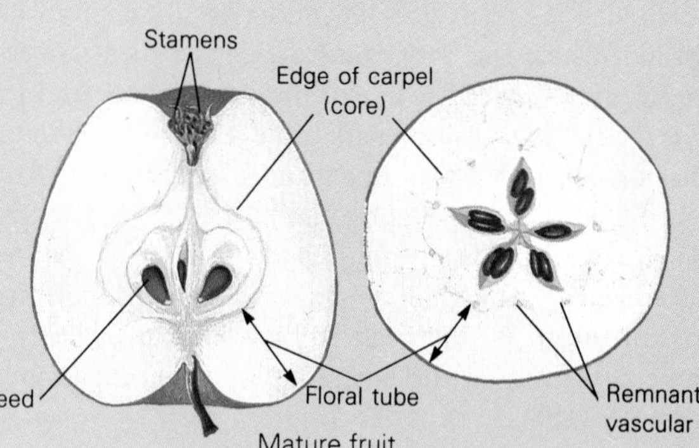

up over the ovary.) A drupe—what a wonderful word—is also derived from one to several carpels, but only a single seed in each develops to maturity. The ripened ovary consists of an outer fleshy part and a hard, inner stone, containing the single seed. Peaches and cherries are drupes, as are coconuts.

There are many kinds of simple dry fruits, but they are neatly categorized as follows: (1) those with many seeds, which split open and release their seeds, and (2) those with few seeds, which do not split open or release seeds. The first group is called **dehiscent (c)**, from the verb *dehisce,* to split or to open, and includes poppies, peas, beans, milkweed, snapdragons, and mustard. Dehiscent fruits split at maturity. The second group is called **indehiscent (d)** (nonsplitting). Its members include sunflowers, dandelions, maples, ash, corn, and wheat. Indehiscent fruits do not split at maturity.

Aggregate fruits (e) are derived from numerous separate carpels of a single flower. Blackberries, raspberries, and strawberries are aggregate fruits, Aggregate fruits consist of many simple fruits clumped together on a common base.

Multiple fruits (f) are formed from the single ovaries of many flowers joined together, as seen in the mulberry, fig, and pineapple. The pineapple starts out as a cluster of separate flowers on a single stalk, but as the ovaries enlarge, they coalesce to form the giant multiple fruit. (The commercial variety, the kind we most commonly see, is a seedless hybrid.)

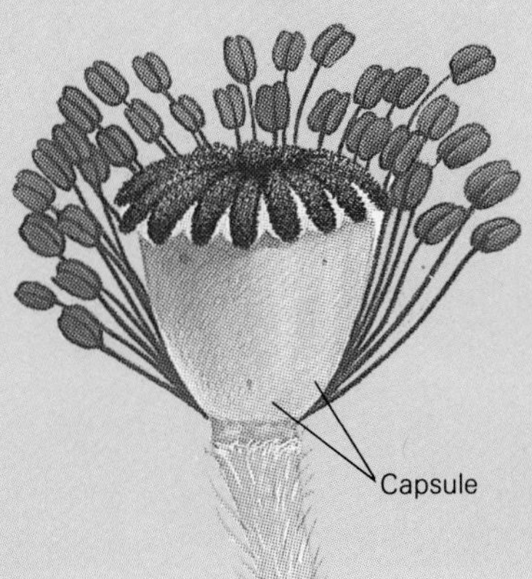

Young fruit

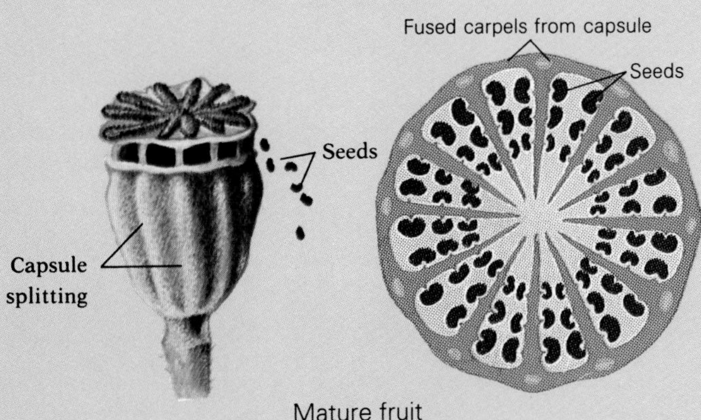

Mature fruit

Flower of poppy

(c) *(center)* The dry dehiscent fruit takes the form of a capsule surrounding the maturing seeds.

(bottom) The capsule, free of other floral parts, becomes a seed-dropping machine as it splits. In cross section, the united carpels are visible in this simple fruit. Each carpel contains numerous seeds in rows along its length.

FLOWERS TO FRUITS, OR A QUINCE IS A POME—cont'd

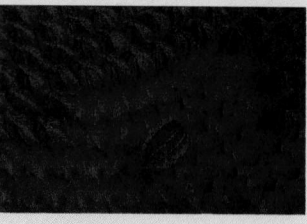

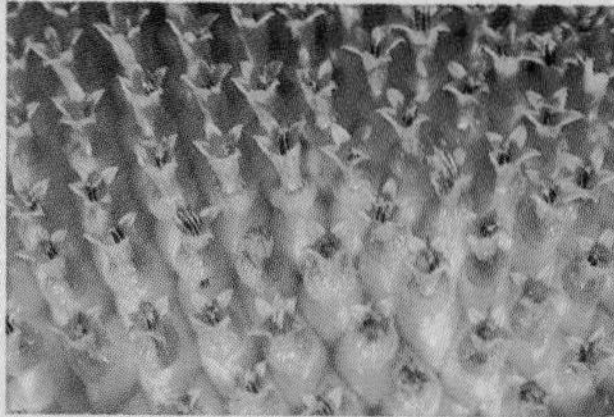

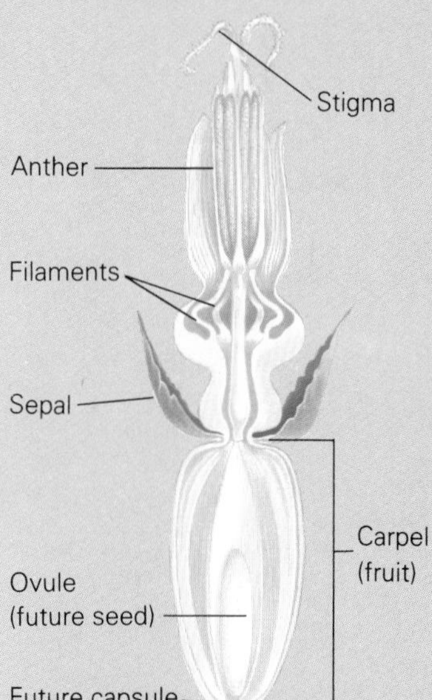

(d) The sunflower, a composite form, consists of many tiny individual disk flowers, making up the *head*. Each flower is simple, consisting of one carpel that will hold a single seed. As the dry indehiscent fruit matures, it will become surrounded by the familiar hardened "shell." Technically, the fruit is called an achene.

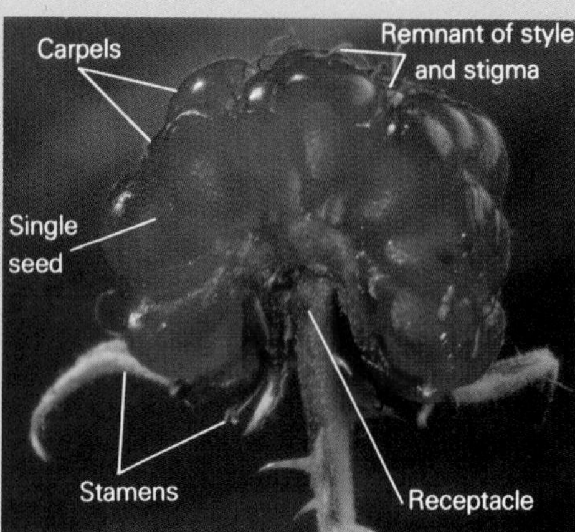

(e) Maturing aggregate fruits of the blackberry. Each of the small spheres, the carpels of this aggregate, is actually a simple fleshy fruit (drupes, in this case) containing a hard seed.

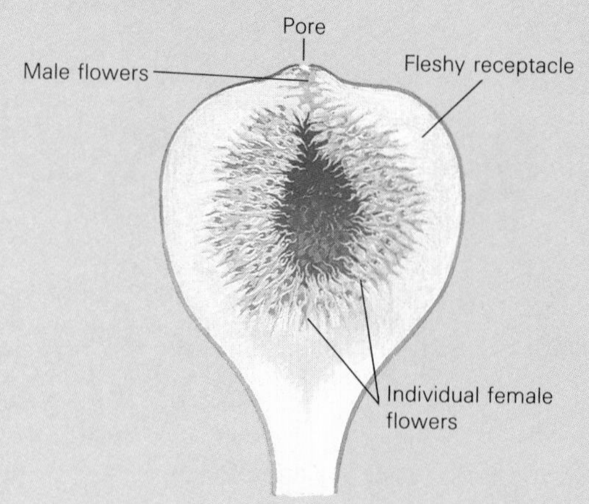

(f) In the common fig, functional male and female flowers both develop within vase-like receptacles, but on different trees. The male receptacle remains tiny (and inedible), producing staminate flowers within its porelike opening. However, it also contains nonfunctional female flowers that serve only to attract the female fig wasp, who lays her eggs within. When the young hatch, the males fertilize the females and then usually die, but the females crawl out, becoming dusted with pollen as they leave. Later, as they prepare to lay their eggs, some wil inadvertently enter female flowers. Since the carpellate flowers arc not suitable, the wasps soon leave, but not before pollinating all of the flowers within. The flowers then mature into a simple achene, forming a swollen, fleshy mass characteristic of the edible fig.

That region of the embryo axis below the cotyledons is logically called the **hypocotyl** ("below the cotyledon[s]"). At the lower portion of the hypocotyl is another region of meristematic tissue, the **root apical meristem,** which may be covered by a cap of cells. This meristem will provide an ongoing supply of cells as new root tissue is formed in the seedling and adult plant. In some embryos, root development begins early and is clearly visible. The embryonic root is called a **radicle.**

At this period in the development of many plant embryos, differentiation of the plant's three basic tissue systems has also begun (see Figure 26.10g). At the bean embryo's outer surface is the **protoderm,** a cell layer that will later differentiate into epidermis (outer skin). Toward the center of the embryonic axis is a line of cells known as **procambium.** These cells will later form the primary phloem and xylem and other cells of the vascular system. The undifferentiated tissue between the protoderm and procambium is called **ground meristem.** It will later contribute to all other types of cells and tissues in the growing plant. We will return to the roles of these tissues, and those of the epicotyl and hypocotyl, when germination and early growth of the dicot are introduced in Chapter 27.

Development in Corn: A Monocot As we've seen, development in monocots and dicots becomes decidedly different as the cotyledon emerges. The single cotyledon of corn and other monocots takes on a cylindrical shape, rather than a heart shape, and becomes quite large. In the onion, another monocot, the cotyledon is comparatively enormous, curving into a spiral and taking up much of the seed space.

The corn embryo axis undergoes a considerable degree of differentiation early in its formation, with the epicotyl soon taking the form of a well-developed, multilayered plumule, and the hypocotyl giving rise to a dominating, well-formed, radicle. It seems that, in corn and other grains, leaf and root development get a head start (Figure 26.11a).

Further distinguishing the corn embryo from the bean is the presence of two protective sheaths, the **coleoptile,** which surrounds the plumule, and the **coleorhiza,** which surrounds the young root. You've seen the fine, green, coleoptile-covered shoots if you have watched the emergence of corn seedlings in your garden or grass seedlings in a newly seeded lawn. The plumule or young leaf soon breaks through the coleoptile and begins to unfurl into the sunlight. The corn root is very fast growing and likewise soon breaks through the coleorhiza. But we are getting ahead of our story.

The cotyledon in corn and a number of other monocots is called the **scutellum** (see Figure 26.11a). Unlike the bean cotyledons, which function in food storage, digestion, and absorption, the scutellum of corn carries out digestion and absorption only. The food lies adjacent to, but outside, the scutellum. Food storage in corn and some other monocots (and some dicots) occurs primarily as undifferentiated starchy endosperm, which takes up much of the space in the kernel.

The large corn endosperm is surrounded by an outer, protein-filled layer of cells, called the **aleurone.** The covering of the corn grain or kernel itself (that part that always get stuck between your front teeth on first dates) is the **pericarp,** which is actually part of the ovary. The usual integuments, those that become the seed coats of the bean, fuse with the pericarp and cannot be distinguished.

A similar organization is seen in wheat. What you have heard referred to as "wheat germ" is the wheat embryo, including its scutellum. The wheat germ, along with the pericarp and aleurone, are commonly removed before milling, and only the starchy endosperm becomes flour. Wheat germ is highly nutritious, but it also spoils easily, which is one reason why it is removed. The pericarp and aleurone become what are called "bran." Bran is primarily cellulose, which cannot be digested, but as we are now aware, it is an excellent bulk food. Because people have recently become more aware of the need for good nutrition and cellulose bulk, both wheat germ and bran are now frequently added back into foods. They are also used in making animal feed. (Considering the contents of grains, it is clear why they have become so basic to the nutrition of the world's hungry billions.)

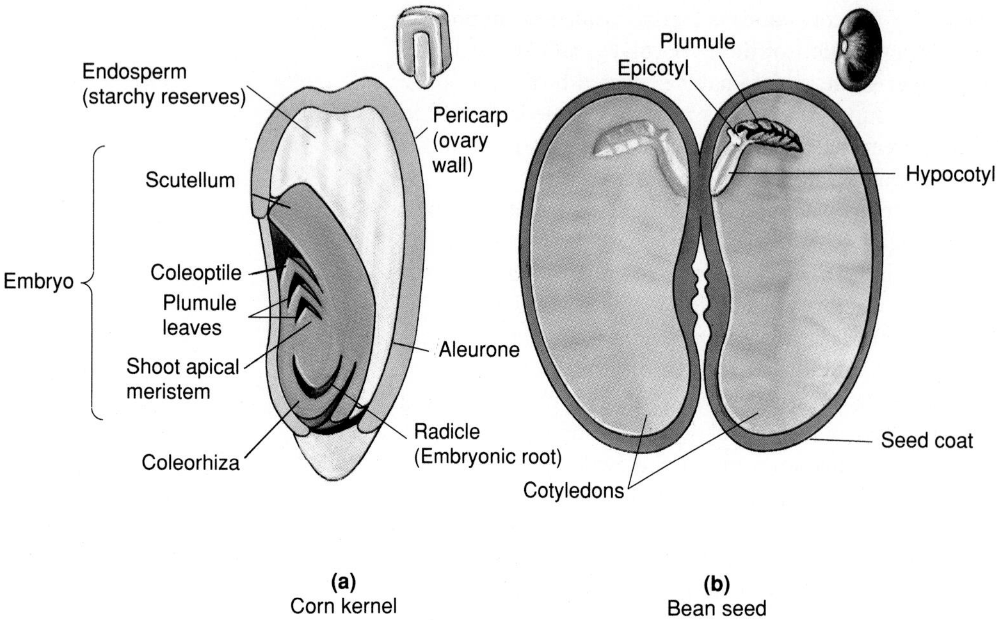

FIGURE 26.11
MONOCOT AND DICOT SEEDS
(a) In the corn kernel (a fruit), food reserves occur in the form of an unorganized starchy endosperm. During germination and early growth, the lengthy scutellum will act as a food-absorbing structure, taking in digested nutrients from the endosperm. **(b)** In the bean seed, food is stored in cells of the paired cotyledons, which occupy much of the seed. Upon germination, the bean cotyledons, which emerge from the soil, develop chlorophyll and for a time carry on photosynthesis.

You might have noticed that we have avoided using the term "corn seed" or "wheat seed." This is because, technically, a kernel or grain is not a seed, but is actually a complete "fruit." Fruit, by definition, is a mature ovary surrounding one or more seeds. In the carpellate (female) corn flower, each kernel is a complete ovary. (The fine hairlike "corn silk" extending from each ovary is the *style* of the corn flower. The staminate [male] flower, we might add, is the "tassel" atop the plant.) The mature kernel, as we mentioned above, is surrounded by a pericarp—the former wall of the ovary—and because of this, it qualifies as a fruit. Where is the corn seed? It is everything within the pericarp. A corn kernel is really the equivalent of the bean pod, which is the bean fruit. Actually, the corn fruit is more similar to a peach or plum—simple, dicot fruits that enclose one seed (see Essay 26.2).

Seed Dormancy

Following development, most seeds and grains enter a dormant period. They lose most of their water and enter a state of greatly reduced metabolism that lasts until germination occurs. They do not lie unprotected, however, because their outer coverings (seed coats and pericarps) have hardened. While in their dehydrated, hardened state, seeds and grains resist the growth of bacteria and fungi, and they travel well.

The dormant period varies greatly, and although the seeds of the majority of species are viable for just a few years (some much less), those of some species remain viable for much longer. Lotus seeds found in peat deposits near Tokyo germinated after 2000 years of dormancy. The record, though, is held by a delicate flower of the Yukon, *Lupinus arcticus.* Some of its seeds grew into fine seedlings after having lain dormant in frozen soil for over 10,000 years.

Seed Dispersal

In order for seeds to succeed, they must begin their growth under suitable conditions—generally away from the competition of parents and other seeds. Thus, seed dispersal is a particularly critical factor in the life cycles of plants, and they have intricate means for accomplishing this end.

Annual plants, those that live for one growing season, have a brief time to produce and disperse their seeds. Many live in disturbed habitats, and their seeds must be well scattered to assure a few of success in the transitory conditions to which they have adapted. Perennials, plants that live on for many seasons, particularly the large trees, have a different kind of problem. Undispersed seeds that germinate beneath their parent's boughs must compete with it for water, mineral resources, and, especially, sunlight.

But, back to the questions of dispersal. How do seeds bearing the young embryos get around? The primary natural seed carriers are water, wind, and animals (Figure 26.12). Perhaps the best-known example of a water-borne seed is the coconut. Coconut palms have become established on even the most remote and tiny islands of the Pacific and are found throughout the tropical and subtropical regions of the earth. Their seeds simply fall from the trees along beaches. When captured by the tides, they float about, and some eventually wash up on hospitable shores. There they germinate and establish themselves. Wind dissemination is common in the maples and elms, which produce winged seed-bearing fruits. As they fall from trees they gain speed until at a certain rate of descent they begin to fly sideways, spiraling to new frontiers away from their parents. Dandelion fruits are carried in the wind on fine plumes derived from the calyx. An odd adaptation to wind dispersal is the tumbleweed: the whole plant forms a large and very light ball that dries up, breaks off at its base, and rolls across the prairie, scattering seeds. The tumbling plants inspired songwriters to put such antics to music.

Seeds can be carried by animals in two ways—on the outside and on the inside. Those that are carried on the outside often have spiny, barbed fruit (the geranium and bur clover are familiar examples) or barbed or sticky seed coats. What dog owner hasn't had to remove cockleburs or foxtails from the animal's ears or paws?

Animals transport seeds on the inside, usually by eating the digestible fruit that surrounds indigestible seeds. When eaten, the seeds are carried for a while in the animal's digestive tract to be deposited, intact and ready to germinate, in the animal's excrement. This is what bright-colored, fleshy, sweet-tasting fruits and berries are all about—at least from a plant's point of view. Many such fruits also contain powerful laxatives to help the process along. In some cases, the seed itself is attractive and nutritious to the dispersing animal. (Birds at feeders are attracted to the seeds.) Since these seeds are small, a plant can produce many of them, and, although they can be digested, a number will escape the animal's digestive processes to germinate later in a nitrate- and phosphate-rich dropping. Some animals also store seeds for later use, and many an oak has grown from an acorn that some absentminded squirrel has buried and forgotten.

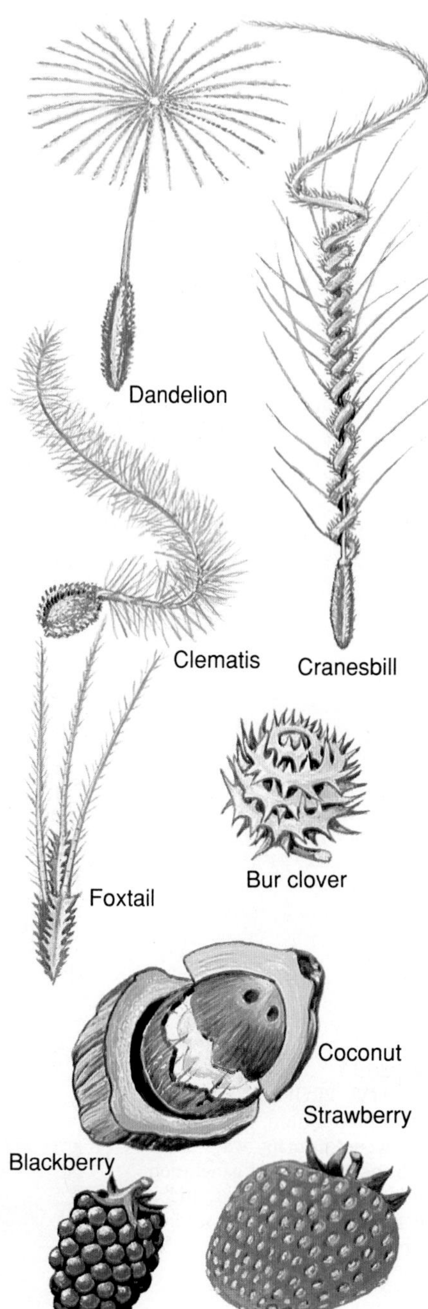

FIGURE 26.12
DIVERSITY IN SEED DISPERSAL
Seeds and seed-bearing fruits are dispersed by wind, water, or animals. Plumes and wings help in lofting the seed-bearing structures of *Clematis* and dandelion into breezes that will carry the offspring well away from their parents. The cranesbill, foxtail, and bur clover have adaptations for clinging to the fur of animals. The seeds within blackberry and strawberry fruits are carried in an animal's digestive tract for a time, but are eventually deposited in the feces. The coconut seed, well protected by its tough fruit (husk) is uniquely adapted for drifting in the sea.

ASEXUAL REPRODUCTION

Asexual reproduction in flowering plants is commonplace. The most common form it takes is **vegetative propagation,** the development of new individuals from a fragment of the parent plant. Of course, vegetative propagation results in individuals that are genetically identical to the parent plant. Agriculturists have long realized certain advantages in the use of vegetative propagation over seeds, where practical. Vegetative propagation ensures that the desirable qualities of certain crop plants will be maintained from one crop to the next with little effort on the grower's part. Retaining those qualities through seed production can be complex and costly, often requiring carefully managed genetic breeding programs for each generation of seeds to be used. We'll consider some of the specifics.

Vegetative Progagation at Work

It is generally known that many plants can be propagated from **cuttings.** People sneak into other people's yards, snip a twig, and tiptoe into the shadows, their heads full of images of the splendid plant that will soon attract neighbors to their own yards. Sometimes it works and sometimes it doesn't, because not all plants can generate roots from a planted stem. In nature, however, broken branches may fall to the ground and take root, a fact we often overlook. Willows, for example, can reproduce this way. When people attempt to propagate plants from cuttings, they usually keep them in water until roots appear and then plant them. Sometimes, they add the plant hormone **auxin** to the water in which the plant is kept. Root growth is generally preceded by the appearance of **callus** tissue, which is an undifferentiated mass of thin-walled cells produced at the site of injury.

New plant growth can also emerge from what are known as **adventitious buds.** These commonly formed on severely damaged stems or on cut off stems. They are also found on the roots of some plants such as the silver leaf poplar and black locust. In these cases, hidden buds along underground roots begin to sprout, sending young stems upward and suffusing the ground beneath with tiny roots. They are often referred to, rather ungallantly, as *suckers.*

In a few instances, shoots can emerge vegetatively by **leaf generation,** such as in *Bryophyllum* (Figure 26.13a). If a bryophyllum leaf falls into a moist environment, it may produce a new plant. Home gardeners often propagate bryophyllum by floating the leaves on water until young shoots appear.

Some plants send **runners** or **stolons** along the ground. These are wispy stems that snake away from the parent plant until they send down roots at some distance from their origin and eventually develop into independent plants (Figure 26.13b). The runner stems have intermittent nodes, where adventitious roots and shoots are produced. Hardly anyone grows strawberries from seeds, since it is such a simple matter to pinch off a runner and transplant the shoots arising from it.

Tubers are also stems, but they are typically thick and are found underground. The domestic potato is our best example (Figure 26.13c). The potato doesn't look much like

(a)

(b)

(c)

FIGURE 26.13
VEGETATIVE PROPAGATION
Vegetative propagation, an asexual process, occurs in many ways, some of which include **(a)** leaf generation, **(b)** the formation of runners, and **(c)** the sprouting of stems from buds in tubers.

a stem, but careful examination reveals typical stem features. The most significant is the eye, which is actually a bud, fully capable of producing stem and adventitious root growth. In fact, a new potato crop can be produced by planting pieces of potato that include at least one eye.

Crab grass and bermuda grass can reproduce vegetatively through underground stems known as **rhizomes.** These stems produce nodes at various distances from each other. Each node sends roots down and stems up. In addition to their insidious method of spreading, these grasses are C4 plants with improved photosynthetic efficiency, so that propagating bermuda grass is easy. A cylindrical core cutter is used to obtain "plugs" of the grass, which can then be set into new ground.

Apomixis: Seeds Without Sex

Sexual reproduction in higher plants is always through seeds, but interestingly, not all seeds are produced sexually. In some instances they are produced asexually—that is, without the union of sperm and egg nuclei. This form of asexual reproduction in plants is known as **apomixis**, and the phenomenon can occur in several ways. Usually, the megaspore mother cell fails to carry out meiosis; thus, the diploid condition is retained, and the embryo begins development without the intervention of the male gamete. Of course, the new embryo has exactly the same chromosomes and genetic characteristics as the single parent sporophyte on which it is borne. Thus, all of the descendants of an apomict line are essentially clones. Examples of apomict species include Kentucky bluegrass, hawthorn, and blackberry.

It is interesting that while completely asexual plant species are very common, none of these species appears to have been around very long in evolutionary terms. Unless the environment remains exceptionally stable, none of them can be expected to hold out in the evolutionary long run. Most apomicts, however, can be expected to survive because they can also reproduce sexually. It's as though these species are "hedging their bets," going the asexual route when it is advantageous but always maintaining its main evolutionary sexual line.

Natural Selection and Apomixis Evolutionary theorists have an explanation for the appearance of this peculiar phenomenon. Natural selection seems to proceed on the profound and unbiased principle that whatever works, works. And sometimes asexual reproduction is extremely adaptive to a plant. For example, a particularly successful genotype can be saved from the vagaries of genetic recombination that occur during meiosis and when unrelated haploid cells are brought together in fertilization. 100 percent of an apomict's genes are passed along to all its offspring, and this favored genotype can quickly spread through a habitat. But, one might ask, under what conditions could a lack of variability be adaptive? Isn't variability in a population the key to success?

Plants living under drastic conditions, such as those in Arctic regions, cling to existence through the most precarious and fragile adaptations. Any disruption of its precise genotypes could spell disaster for a population. Further, in such rigorous environments, plants may be so far apart as to make pollination difficult, so asexual reproduction through apomixis is highly adaptive.

Apomixis may also speed up the speciation process by increasing the chances of successful hybridization. Recall from our earlier discussion that hybrids are frequently sterile because of the failure of chromosomes to synapse at prophase I of meiosis. There are no homologues available, and gametogenesis fails. In our earlier example, we pointed out how spontaneous chromosome doubling provides one solution. Apomixis is an even simpler solution, because meiosis is not necessary. Thus, apomict hybrids undergo seed development without a hitch. This has happened time and again with the highly variable Kentucky bluegrass.

The problem with apomixis should be apparent. When the environment changes, as it inevitably will, the apomictic population may not contain the variability needed to ensure even minimal survival. Those plants that can shift to a sexual phase, however, can shuffle genes, producing new combinations and leaving more variable offspring. These kinds of plants will have an improved chance of success when the environment changes.

Since sexual organisms have more variability built into their genetic systems, they have a long term advantage and will once again take over the habitat. Then, as the environment stabilizes, the newly succeeding variants may become apomictic. We can see, then, how apomixis could have arisen, but judging from the fact that strict apomicts don't have long evolutionary histories, it seems that sex is the safer bet in the long run.

anther In flowers, a male reproductive structure containing microspore mother cells—those that undergo meiosis giving rise to microspores.
carpel The entire female reproductive structure of the flower; includes the ovary, style, and stigma. Also called a pistil.
coleoptile A protective, sheathlike covering over the plumule in some monocot embryos.
cotyledon Also "seed leaf." Paired food storage structures in the dicot seed, and a single digestive structure (the scutellum) in the monocot grain.
embryo sac The female gametophyte (megagametophyte), often containing seven haploid cells, including an egg cell and a binucleated central cell.
endosperm A food storage region in the angiosperm seed.
epicotyl That region of the embryo located above the union of the cotyledons.
fruit The mature or ripened ovary (or ovaries) surrounding the seeds of flowering plants.
hypocotyl That region of the embryo located between the union of the cotyledons and the embryonic root tip or radicle.
ovary In flowering plants, the female reproductive structure that contains the ovules.
pollination In flowering plants, the depositing of pollen on the stigma.
radicle An embryonic root in wich differentiation has already begun.
staminate flower A flower containing stamens but lacking carpels.
seed The fully mature ovule consisting of an embryo, stored food, and surrounding seed coats.
sporangium In plants, a sporophyte structure specialized for meiosis and spore formation.
tube cell A cell in the male gametophyte (pollen) that directs the formation of a pollen tube.
vascular tissue Tissues (xylem and phloem) specialized for transport.
xylem Vascular tissue specialized for the transport of water and dissolved minerals and for providing woody structural support.

APPLICATION OF IDEAS

Characteristically, the great diversity in angiosperm floral parts can partly be explained through the dependence of plants on animals for pollen transfer. However, one large group has become adapted for pollination by wind. Which group is this? Would you expect much diversity in flowers of this group? In what other ways might wind-pollinated flowers differ from those pollinated by insects? Compare the specific structures in a typical wind-pollinated angiosperm flower with that of an insect-pollinated flower.

KEY IDEAS

SEXUAL REPRODUCTION

The Flower

1. The flower consists of modified leaves arranged into **whorls**, which include:
   a. **Accessory parts**: the **receptacle**, the **calyx (sepals)**, and **corolla (petals)**.

b. Sexual parts: the **androecium** (**stamens** or microsporophylls, consisting of **filaments** topped by **anthers**), and the **gynoecium** or carpel[s], each of which contains an **ovary**, a necklike **style**, and a **stigma**.

2. Floral structure varies with the pollinating agent. Attractants include odor, **nectar**, (within **nectaries**), pollen, color, and structural pattern.

Variation in Flowers

1. Floral parts in monocots occur in threes and its multiples, while dicot floral parts occur in fours or fives or their multiples. More primitive flowers are **radially symmetrical**, while advanced types are often **bilaterally symmetrical** or **irregular**.
2. **Complete flowers** have all four basic parts, but **incomplete flowers** may have one or more missing. **Perfect flowers** have both carpels and stamens, but **imperfect flowers** are **carpellate** (female—lacking stamens) or **staminate** (male—lacking pistils).
3. Some flowers are clusters, or **inflorescences**, and if they contain individual flowers, each with a single fruit, they are **composite flowers**. Composite flowers have many central **disk flowers** and modified, surrounding **ray flowers**.

Flower Structure and Pollination

1. Through coevolution, many flowers and insects reach a specialized relationship that assures the plant of pollination and the insect of a food supply. *Yucca* requires the yucca moth, which lays its eggs in the flower's ovary. Their relationship has become obligate. The smell of skunk cabbage attracts the fly as a pollinator. Fermenting odors attract bats. Some flowers are generalists; they are pollinated by a variety of insects.
2. Wind pollination occurs in a large number of flowering plants, including dicots and monocots. Wind-pollinated flowers are inconspicuous, lacking colors and odors. Floral parts are modified for capturing pollen. Recent studies suggest that the shape of wind-pollinated flowers and cones makes wind pollination more efficient than was formerly thought.

Sexual Activity in Flowers

1. Flowering plants are heterosporous, producing megaspores and microspores, with the gametophyte restricted to cells in the ovules and anthers.
2. Ovules originate in **placental** tissue to which they are connected by a **funiculus**. An ovule is a megasporangium or **nucellus**, containing a **megaspore mother cell** and accompanied by skinlike **integuments**. An opening, the **micropyle**, occurs at one end.
3. In the ovary, megaspore mother cells undergo meiosis, producing four haploid products. Three disintegrate, but the fourth divides mitotically three times, producing an eight-nucleate **megagametophyte**. Upon cell division, six small, single-nucleated cells emerge, one of which is the **egg cell**. A seventh, large **central cell** (or **endosperm mother cell**), containing two **polar nuclei**, will later give rise to the **endosperm**. The completed megagametophyte is called an **embryo sac**.
4. **Microsporogenesis** begins in the **anthers**, where each **microspore mother cell** undergoes meiosis, producing four haploid **microspores**. The nucleus of each divides mitotically, producing a **generative cell** and a **tube cell**, thus beginning the formation of the male gametophyte or microgametophyte, which becomes a **pollen grain**.
5. **Pollination** is the transfer of pollen onto a stigma. Many plants have structural or physiological adaptations for avoiding self-pollination.
6. Germinating pollen produces a **pollen tube**, which digests its way through the style to the ovary. At the time of pollen tube growth (or earlier) the generative cell undergoes mitosis, producing two sperm, one of which fertilizes the egg and the other, the central cell. In a **double fertilization**, the first fertilization gives rise to the diploid zygote, and the second, the triploid **primary endosperm nucleus**.
7. Following fertilization, the ovary matures into the seed-bearing **fruit**.

Development of the Embryo and Seed

1. Mitosis and differentiation in the primary endosperm and zygote will produce the **embryo**, which along with a food supply and **seed coat**, make up the seed.
2. Developmental activity often begins with the endosperm, which undergoes repeated mitosis, becoming multinucleated. Then the embryo begins its mitotic divisions. The early embryo is a ball of cells anchored by the large **basal cell** of the **suspensor**.
3. In the dicot, the embryo takes on a heart shape as two cotyledons take form. In monocots, the cotyledon is single. As the bean cotyledons grow they make a turn and enlarge greatly, enclosing the **embryo axis** and comprising much of the seed.
4. Continued growth in the embryo produces a **shoot apical meristem** at its tip (made up of simple, uncommitted **meristem tissue**), which in some species is surrounded by a **plumule**. Any growth above the origin of the cotyledons is called **epicotyl**. Most of the embryo axis in the bean is **hypocotyl**. At its base is the **root apical meristem**, which produces the **radicle**. Primary tissues include the outer **protoderm**, central **procambium**, and **ground meristem**.
5. In monocots such as corn, each kernel develops from an ovary, so it is technically a fruit. The fruit includes a starchy endosperm, an embryo, and a digestive structure, the **scutellum** (cotyledon). The endosperm forms an **aleurone**, which is surrounded by the **pericarp**. The corn radicle and plumule are protected by a **coleorhiza** and **coleoptile**, respectively. The embryo and pericarp are removed prior to the milling of flour.

Seed Dormancy

Many seeds enter a dormant period in which they harden

through dehydration. Some dormant seeds have remained viable for thousands of years. In their dormancy, seeds have little metabolic activity.

Seed Dispersal

1. Successful dispersal of seeds reduces competition and improves the chances of survival.
2. Seeds are dispersed by water, wind, and animals. Wind-dispersed seeds may have winglike structures or plumes. Some seeds are eaten by animals and are distributed in the animals' wastes.

ASEXUAL REPRODUCTION

The chief means of asexual reproduction is **vegetative propagation,** development of a new individual from some fragment or portion of another plant. It is agriculturally useful, since the absence of genetic recombination ensures the precise continuation of desirable qualities.

Vegetative Propagation At Work

Applications of vegetative propagation include raising new plants through the use of stem **cuttings, adventitious buds,** and **leaf generation.** Some plants send out **runners** or **stolons** that put down roots periodically, establishing new growths. Potatoes and other **tubers** are underground stems that produce growth at buds ("eyes"). Underground stems called **rhizomes** periodically produce new growths along their lengths.

Apomixis: Seeds Without Sex

Apomicts are plants that can produce embryos and seeds without fertilization. Most can also reproduce sexually. Theorists suggest that **apomixis** is an adaptation to very harsh conditions, those in which any genetic variation would most likely be disadvantageous to the plant. It may also be adaptive where sparsely distributed populations have difficulty in achieving pollination. Apomixis may speed up speciation, since otherwise infertile hybrids avoid meiosis altogether, thereby avoiding the problem caused by unmatched homologues. In the long run, sexual systems, those that produce variability, are probably more adaptive.

REVIEW QUESTIONS

1. List the four major regions of the flower. Which are accessory, and which are the "male" and the "female" parts? (572-573)
2. What are equivalent terms for microsporogenesis and megasporogenesis? (578-579)
3. What colors and odors would you expect in flowers that are pollinated by the following: bats, carrion flies, bees? (575-576)
4. What is an imperfect flower? Can an imperfect flower be complete? Explain. (575)
5. From the viewpoint of natural selection, explain why grasses lack colorful, fragrant flowers. (576-578)
6. Using flower symmetry and numbers of floral parts as a basis, distinguish between primitive and advanced plants and between monocotyledons and dicotyledons. (564-565)
7. Could a carpellate flower be complete and perfect? Explain. Give an example of a plant that produces separate carpellate and staminate flowers. (575)
8. Using the example of *Yucca* or the passion flower, explain how plants and insects specialize through coevolution. (575-576)
9. How do the flowers that attract beetles differ from those that attract bees? Suggest reasons for the difference. (576)
10. Suggest some advantages and disadvantages that wind pollination might have. (576-578)
11. Describe events in a megaspore mother cell that lead to the production of the eight-nucleate megagametophytes. (578)
12. List the parts of an embryo sac. Which participate in fertilization? (578-579)
13. Trace the events that carry a microspore mother cell through microsporogenesis, ending with a microgametophyte. What is another more common name for the latter? (579-580)
14. Beginning with pollination, describe fertilization in the flowering plant. (580)
15. Give a technical definition of the term *fruit.* Describe the fruit of corn and string beans. (581-582)
16. What are the three parts of a seed? (582)
17. Trace the development of the dicot embryo. Include a description of the suspensor, the heart-shaped embryo, and the cotyledons. What are the three primary tissues? (582-583, 587)
18. Using a simple drawing of an embryo, locate the epicotyl, a hypocotyl, radicle, and plumule. From what groups of cells does each arise? (583, 587)
19. What are two basic differences between the monocot and dicot embryo and seed? (582, 587)
20. List three agents of seed dispersal. Why is dispersal an important problem for plants? (589)
21. Of what strategic value to the plant is a sweet, fleshy, and nutritious fruit? (589)
22. Define and provide three examples of vegetative propagation. Why is it valuable to agriculture? (589)
23. How do runners differ from rhizomes? Give an example of a plant utilizing each. (590)
24. What is apomixis? Explain how it might overcome the problem of sterility in hybrids. What effect might this have on evolutionary rates? (591)
25. Suggest two situations, not involving hybrids, in which apomixis might be advantageous to plants. (591-592)

27

PLANT GROWTH AND STRUCTURE

MIGHTY OAKS, INDEED, FROM TINY ACORNS GROW. WE SEE THE ACORNS, AND we see the oaks, and we know their relationship. However, there is a fascinating story to be told about how all this comes about. It involves profound and highly organized changes as the acorn is stirred into an awakening and as it begins to draw sustenance and energy from its surroundings. It builds tissue and stores energy, constantly growing and reorganizing until it takes its place among the other forest stalwarts. By now, of course, it is producing its own acorns, which will soon play their own role in the ongoing saga.

Considering the importance of plants to all of us in our everyday life, the story of growth and development is well worth knowing. It is these processes, among flowering plants, that we will consider now.

GERMINATION AND GROWTH OF THE SEEDLING

Germination is the beginning of growth from a seed. The seed, essentially a dormant stage, is now aroused as it begins a renewed period of metabolic activity. This renewal is triggered, however, by a rather precise set of conditions.

Germination Conditions

Most plants have the same requirements for germination: an adequate supply of water, favorable temperatures, and the availability of oxygen. Metabolism resumes when water is absorbed into the dry interior of the seed and its enzymes are mobilized.

While all plants have these requirements, many also require special conditions to get germination going. The seed coats of some species are especially tough and resistant.

Some seeds require a period of freezing temperatures to split their coats, while some require the heat of fire. A few require the abrasive action of sand particles carried by flowing water, and still others require the hydrolyzing action of animal digestive enzymes. On the island of Mauritius there are 11 huge Tambalacoque trees, a very rare species. All of the trees are about 300 years old; there are no younger trees. Every year they put out a crop of huge seeds with very thick seed coats. But the seeds never germinate. It seems that normal germination requires that the seed be passed through the crop of another native of the island, a bird called *Raphus cucullatus,* better known as the dodo. But, thanks to us, the dodo has been extinct for about 300 years, so the seeds just lie around waiting for the bird that will never come. (You may be pleased to learn, however, that recently, the naturalist who worked out this curious relationship managed to get several of the seeds to germinate by passing them through a turkey.)

Most plants, as evidenced by their numbers, do not have such unusual requirements for germination. Monocots and dicots generally move quite easily from seed to seedling. The apparent ease, though, masks a plethora of important changes going on within the plant.

The Seedling

In a number of monocots such as the grains, those with large unorganized endosperms, the uptake of water is followed by the mobilization of the starchy reserves stored therein. This is an indirect process involving hormones known as gibberellins produced by the embryo (see Chapter 29). The target of the gibberellins is the aleurone cell layer of the endosperm, just below the seed coats. The aleurone cells respond to the hormone by producing alpha-amylase, a starch-digesting enzyme that hydrolyzes the starch to mobile, soluble sugars that the embryo can put to use.

Corn As germination occurs in corn, a monocot, the embryonic root or radicle emerges first, breaking through the coleorhiza and pushing down into the soil (Figure 27.1a). Next, the shoot, surrounded by the protective coleoptile, elongates rapidly, pushing upward through the soil. Soon after, the first leaf breaks through the coleoptile and emerges into the sunlight. The rapidly growing young root below is soon accompanied by more roots. Oddly, in corn and some other monocots, these newcomers will emerge not from the primary root, but from the stem region above.

Onion The emergence of the onion seedling, another monocot, is quite different. Its lengthy cotyledon elongates rapidly, emerging from the seed in a U-shape. At its free lower end, a fast-growing root develops. The U-shaped cotyledon, acting as a "bumper," grows toward the surface. After the cotyledon emerges from the soil it straightens, drawing what is left of the seed out of the soil and holding it aloft. The first leaf emerges far below the seed remnant, from the base of the cotyledon. Eventually the cotyledon, an embryonic structure, withers, and the adult plant takes form (Figure 27.1b).

Bean Dicots vary considerably in their early growth, but the early visible signs of germination include the swelling of the seed and the splitting of the restrictive seed coats. As germination in the bean proceeds, the young root emerges and quickly penetrates the soil, forming many lateral roots and anchoring itself securely. The hypocotyl then elongates rapidly (Figure 27.1c). (The intake of water into cells is an important factor in elongation, since it is the resulting force exerted against the young cell wall that lengthens the cell.) As the hypocotyl grows it forms a sharp curve—the **hypocotyl hook**—which acts as a protective bumper for the delicate parts below. In its upward growth, the hypocotyl draws the cotyledons along behind it. Upon breaking through the surface the hypocotyl straightens, the cotyledons open up, and the young plumule is exposed to the light. At this point the epicotyl begins its elongation, lifting the expanding leaves up to the light.

When the bean seedling breaks the surface, its chloroplasts mature, turning the plant a deep green. The embryonic leaves are exposed to energy-laden light as their growth

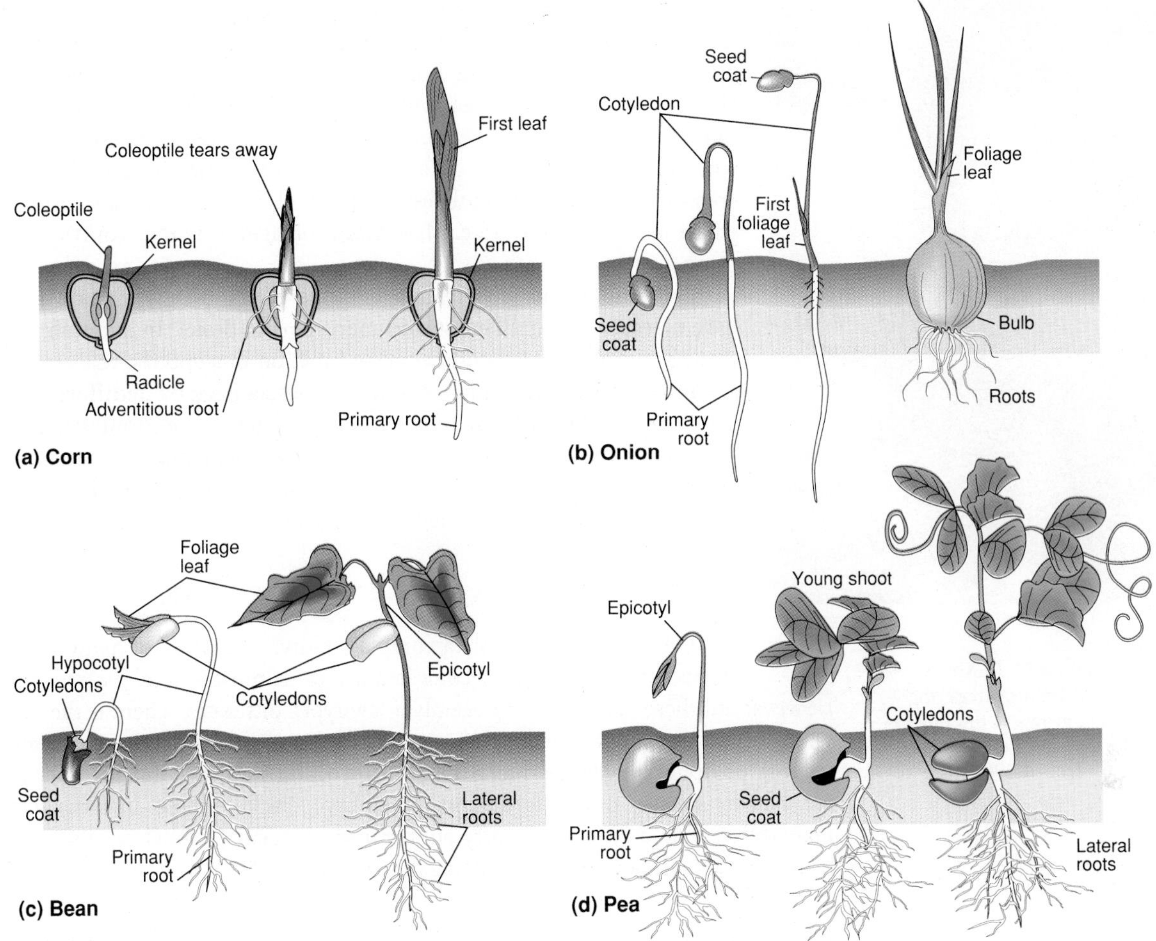

FIGURE 27.1
GERMINATION AND GROWTH
(a) As corn germinates, the radicle breaks through the coleorhiza to form the primary root. Next, the coleoptile, with the plumule enclosed, pushes up through the soil, whereupon the plumule breaks through its sheath. The primary root is soon obliterated by numerous adventitious roots, which emerge directly from the stem. **(b)** In the onion seedling, the hypocotyl curves downward with the young root emerging. The cotyledon breaks through the soil, raising the seed coat aloft. Note the position of the first leaf. **(c)** In the bean, the rapidly elongating hypocotyl forms a curved "bumper," grows upward, and draws the cotyledons behind it. Upon emerging from the soil, the hypocotyl straightens, raising the cotyledons and plumule. Just below the plumule, the epicotyl raises the young foliage even higher. Meanwhile, the young primary root has produced many lateral roots as the root system matures. **(d)** As the pea germinates, the epicotyl breaks through the soil, and its plumule expands. The hypocotyl curves downward, forming the primary root.

continues. At this time, the food reserves in the cotyledons diminish, and they wither and fall away, leaving telltale scars on the young shoot. The young bean seedling is then fully on its own.

Pea Although the cotyledons of the sprouting bean are drawn from the soil by the emerging hypocotyl, the garden pea does things differently. In the emerging pea seedling, the cotyledons remain behind in the soil. This time it is a fast-growing epicotyl that forms the U-shaped "bumper," soon emerging from the soil. Upon breaking the surface, its plumule is held aloft, and the leaves unfold and increase. Meanwhile, the pea hypocotyl has turned downward from the seed, and the young primary root has emerged. Lateral roots soon break through the walls of the primary root, and the root system increases (Figure 27.1d).

Primary and Secondary Growth

Seed germination and the early growth of the shoot are examples of **primary growth.** Primary growth goes on throughout the life of a plant. It is responsible for continued growth in the length of shoots and roots, and for the production of leaves, flowers, and new branches. In contrast, **secondary growth** accounts for growth in girth or thickness. While all of the conifers are capable of secondary growth, within flowering plants secondary growth is seen mainly in dicots. However, not all dicots undergo extensive secondary growth. In annuals, for instance, it may be absent or quite restricted. They germinate, grow, develop, and reproduce in a single season. As you might expect, then, extensive secondary growth in flowering plants occurs chiefly in the perennials, such as long-lived shrubs and trees.

Perennial plants are also unique among living things in that they exhibit what is

FIGURE 27.2
ONE OF THE OLDEST PLANTS
One of the world's oldest trees, the bristlecone pine, grows at the timberline of California's White Mountains. One specimen reached 4900 years of age.

called **open growth,** or **indeterminate life span.** This means that barring fire, injury, or disease, theoretically they can live forever. There are the usual arguments over which plant holds the record, and for years the documented champion was a 4900-year-old conifer, a bristlecone pine from the White Mountains of California (Figure 27.2). Today, we're not so sure. Some cottonwoods may have reached an age of 8000 years, and the age of certain desert creosote bushes was recently estimated at 11,700 years. In any case, we can be sure that somewhere on earth grows a plant that is the world's oldest living thing.

In a sense, plants with open growth never completely mature. They always have a reserve of undifferentiated meristematic tissue. In addition, many cells that have differentiated and begun to function in a specific role are capable, in a manner of speaking, of having their clocks set back by **dedifferentiation** (see Essay 27.1). They thereby return to an immature state, from which they have the potential to mature or differentiate all over again, perhaps along new routes. We will return to this point, but let's prepare for a closer look at primary and secondary growth by learning something about plant tissues.

TISSUE ORGANIZATION AND THE PLANT

The plant body is certainly a complex entity, composed of many types of cells organized into seemingly countless layers within the roots, stems, and leaves. However, all these are composed of only a few types of tissues. There is the perpetually young apical meristem, which produces three types of primary tissue or **primary meristems: protoderm, ground meristem,** and **procambium.** We noted earlier that these are the first specific tissues formed in the embryo. Each of these, in turn, gives rise to more differentiated tissues that carry out the many specialized functions that characterize plant life. This organization is illustrated in Figure 27.3.

Protoderm and the Epidermis

Mature protoderm contributes to the outermost layers of cells, or **epidermis.** In the leaf and young primary stem, epidermal cells are commonly flattened and irregular in shape. They are covered by a waxy **cuticle** of **cutin,** which provides waterproofing and prevents water loss from the more delicate tissues within. In addition to the simple epidermal cells, pairs of intricate **guard cells** surround pores, called **stomata** (singular, **stoma**), which permit the exchange of gases between photosynthetic cells and the atmosphere. Epidermal cells just above the root tip have the vital job of absorbing water. They produce numerous, lengthy **root hairs** that greatly increase the absorbing surface area. Where secondary growth occurs, the epidermis is replaced by a multilayered tissue called **periderm.** The periderm is made waterproof by the addition of **suberin,** another waxy substance.

Ground Meristem

Ground meristem differentiates into three important tissues: **parenchyma, collenchyma,** and **sclerenchyma.** Parenchyma is widely distributed throughout the plant. It is an often loosely arranged tissue consisting of large, thin-walled, irregularly shaped cells. Parenchyma takes in a considerable amount of water, producing the turgid condition important to leaves and the shoots of young plants. It is also a food storage tissue, often packed with starches. Leaf parenchyma (also called *chlorenchyma*) contains most of the chloroplasts and is the most important site of photosynthesis. Parenchyma tissue is capable of cell division and differentiation into other tissue and is responsible for the growth of new tissues in wound healing.

Collenchyma and sclerenchyma tissues have some roles in common, but structurally they are quite different. Collenchyma cells remain alive at maturity. The cells commonly form a cylinder of supporting tissue just within the epidermis of young stems and in

FIGURE 27.3
PRIMARY MERISTEMS AND THEIR TISSUE DERIVATIVES
During primary growth, primary meristems give rise to:

Protoderm
Protoderm differentiates into covering tissues, including the epidermis of roots, stems, and leaves. More specialized examples include guard cells, leaf hairs, and root hairs.

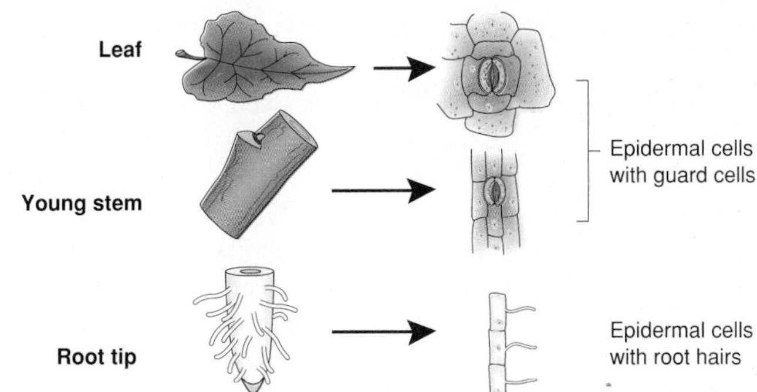

Ground meristem
Ground meristem differentiates into three basic tissue types:

Parenchyma is widely distributed in the stem and root and makes up the photosynthetic tissues of the leaf. The cells are large and thin-walled, often involved in storage.

Collenchyma is primarily involved in support. Its thick-walled cells form tough but flexible cylinder of tissue below the epidermis, within vascular tissue, and in the supporting portions of the leaf.

Sclerenchyma, in its *fiber* form, strengthens shoots. In its *sclereid* form, it provides hardness for seed coverings and shells.

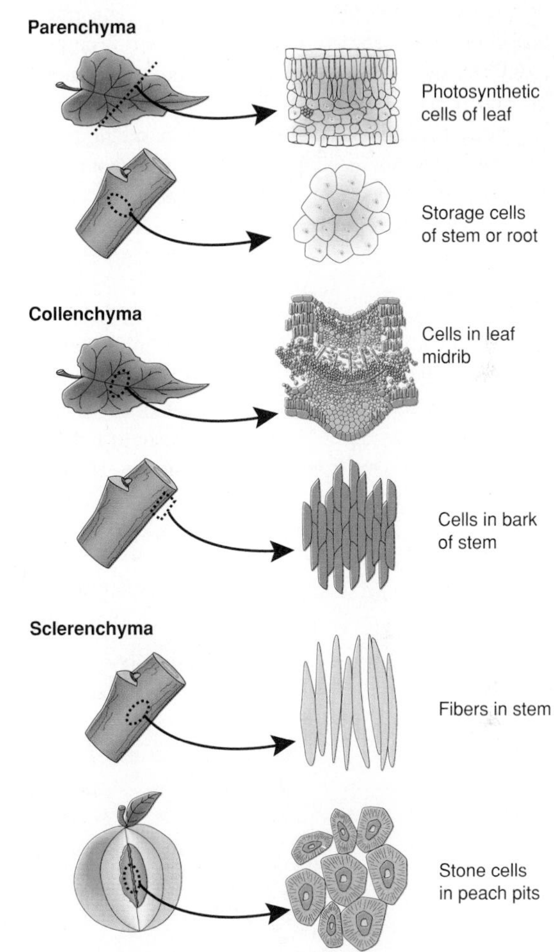

Procambium
Procambium differentiates into xylem and phloem, the conducting (vascular) tissue of roots, shoots, and leaves. Xylem specializes in water and mineral transport; phloem, in food transport.

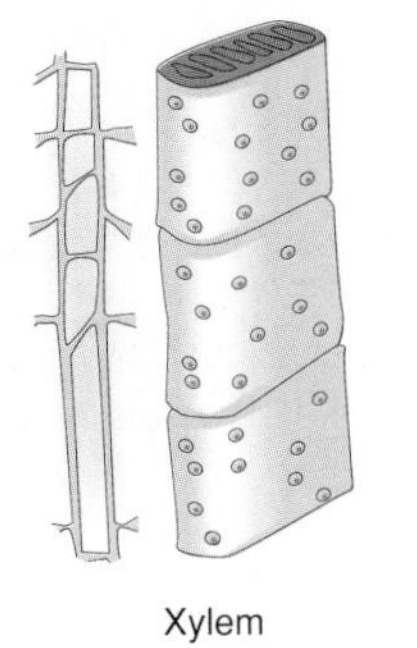

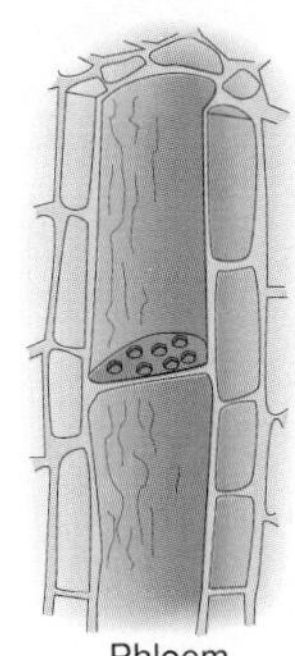

ESSAY 27.1

THE PROBLEM OF DIFFERENTIATION

Scientists have indeed uncovered detail after detail about plant development until it seems that we surely must have the big problems solved. But one haunting question prowls the attics of our minds: "How do plant cells differentiate?" This, of course, is a most fundamental question, so we might wonder, what do we really know? The answer is essentially the same for plants and animals: we haven't yet developed entirely adequate explanations. Perhaps none exist, but biologists are making progress on a number of fronts.

Botanists in particular have provided valuable basic information. This may well be because in some ways plants make ideal subjects for studying cell growth and proliferation under controlled conditions. For one thing, some of them can regenerate from bits and pieces. Many plants can regenerate roots from stem cuttings, stems from bits of root, and even entire plants from leaves. This knowledge has been invaluable in agriculture through the ages, but for biologists it suggests clues to the puzzle of differentiation. It supports the idea, for example, that plant tissues are *totipotent*—that is, their cell nuclei retain the capability needed to produce the entire organism from which they come. If this is true, then, can individual plant *cells,* acting alone, duplicate the regenerative feat we see in cuttings? This question has been partly answered through studies using tissue culture techniques.

Plant researchers turned to tissue culture methods in the 1930s and 1940s. A pioneer in these efforts, Johannes van Overbeek, discovered that a suitable medium was coconut milk. Besides important nutrients, coconut milk contains some critical hormones needed for plant growth. Van Overbeek succeeded in growing individual cells that he had separated out from young carrot embryos **(a)**. He cultured them in his coconut milk and later planted them in soil. Lo! They produced normal adult carrot plants. The results indicated that cells in the embryo, at least, were totipotent.

It wasn't until the late 1950s, however, that mature tissue was first cultured in a similar manner by F. C. Steward at Cornell University. Mature carrot cells were successfully removed and grown in a nutrient medium. The mature cells grew into rootlike structures that, when planted, produced entire carrot plants **(b)**. More recently, there has been excellent progress in culturing redwood trees, orchids, and many other plants. Plant physiologists have also succeeded in producing clones of potato varieties from cells in the leaf. In so doing, they have provided agriculture with a potentially economical way of growing potatoes, and biology with yet another "experimental organism." The potato clone promises to lend itself to investigations into the unyielding mysteries of development.

Extending this work, researchers at Kansas State University obtained experimental material from leaf cells that were first converted to protoplasts **(c)**. Enzymes were used to remove cell walls and intercellular materials, leaving the naked cells behind. The new protoplasts then massed together,

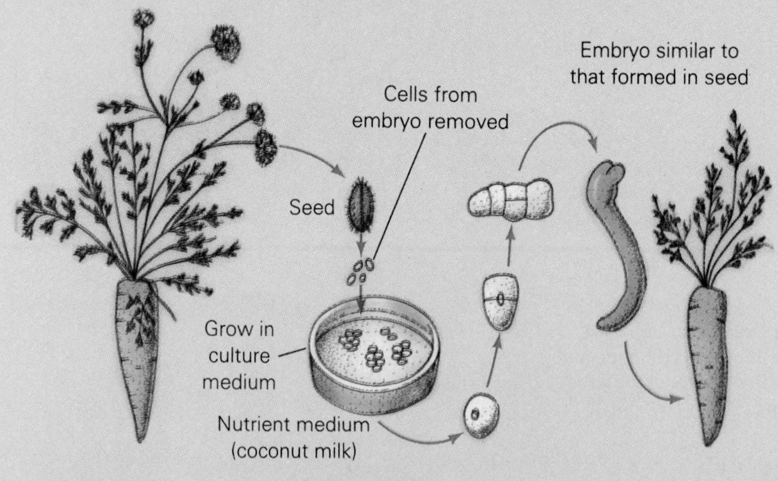

entered cell division, and produced new cell walls. The resulting tissue was then transferred sequentially through several types of culture media. Each medium contained certain mixes of synthetic hormones, each of which had its own effect on differentiation. The first mix produced a large undifferentiated callus. The second encouraged shoot growth and differentiation, and in response to a third mix, the tissue produced roots. Newly differentiated plants were then transferred to regular soil beds, where they were grown to maturity. This procedure proves once again that mature plant cells lose none of their original genetic potency.

Though the work is in its early stages, the potential benefits are clear. Through cloning, agriculturists could produce highly selected, unvarying crop plants (particularly in producing virus-free strains). In keeping with the new era of genetics and molecular biology, potato protoplasts can be used in gene splicing and recombinant DNA studies. The Kansas group has succeeded in fusing nuclei in potato and tomato protoplasts, producing a hybrid that would not readily occur in nature or through sexual, genetic crosses. Such readily manipulated experimental organisms hold great potential for the ongoing study of development.

A more immediate goal in such efforts is to confer the disease resistance of one species on another less resistant one. For example, the tomato plant used in the fusion described above carries with it genes that will enable the potato to resist the water mold that causes *late blight* (see Chapter 23). Not only could this eliminate a constant and serious threat to an important food supply, but it also could eliminate the widespread use of certain dangerous chemicals—the fungicides now used to control the blight.

Let's digress here and consider the seemingly facetious question, "Who needs variable potatoes?" The answer is, we do. In fact, we would do well to begin developing "gene banks" to preserve the variation in the original phenotypes. Once we drive these older varieties into extinction through selective breeding for our favorite traits, there will be no going back. Biologists are becoming alarmed at what might happen if our highly selected food strains should be decimated by some new fungal or bacterial mutant or perhaps some environmental variable that we can't control. We will need those original strains just to start again.

We see, then, that many cells retain the capability necessary to produce the entire plant. Second, when removed from their cell associations, cells can regress to an undifferentiated state from which they can then proceed through normal embryonic development. How does this answer the question about differentiation? It tells us both that genetic information is not lost as development proceeds, and that it is not irreversibly repressed.

Such findings are important from the standpoint of pure research, but they may gain in importance as we seek to provide for an increasingly hungry world.

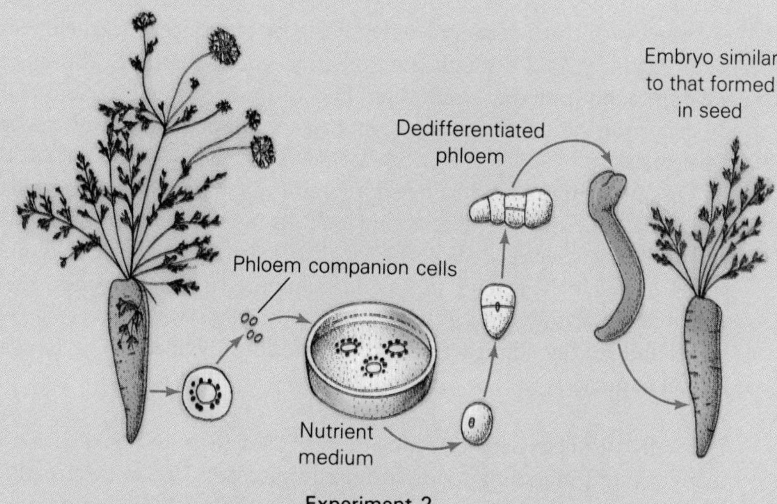

Experiment 2

the supporting structures of leaves. Those "strings" that considerate hosts remove from celery before it is served are collenchyma. Collenchyma cell walls do not become hardened by lignin, but they are often quite thick. Their thickness and flexibility make collenchyma useful supporting material in tissues whose cells are still increasing in length.

Sclerenchyma is often non-living, having lost its nucleus and cytoplasm. Its cell walls are usually thick and hardened by lignin, making them especially useful as supporting elements in tissues that have finished their growth. Sclerenchyma occurs in two forms, **fibers** and **sclereids.** Because of their great strength, the fibers are one of the most important structural components in vascular plants. These same characteristics make the fibers from jute and hemp useful when manufactured into products such as string, rope, and sacking. Sclereids, shorter branching cells, are common in seed coats and nut shells, and you've probably come across sclereids called **stone cells** in the gritty parts of pear flesh.

Procambium and the Vascular Tissues

Procambium, the third primary meristem, has only one role (but consider its implications): to produce vascular tissue, the **primary xylem** and **primary phloem**. Later, it will contribute to **vascular cambium,** a tissue important to secondary growth. Since we now have some idea of the basic functions of these tissues, we can now focus more tightly on their development and organization. We will return to vascular cambium shortly.

Xylem As we know, water and minerals move upward from the root to the stem and leaves through the xylem. Xylem in angiosperms includes **tracheids** and **vessel elements** along with nonconducting fibers and parenchyma. Vessel elements, (cells) when attached end to end form the familiar **vessels**. (You may recall that vessels are absent in vascular plants of more ancient lineage, thus vessels are considered to be advanced evolutionary structures.) Both tracheids and vessels are of little use while alive, but after they die, their cytoplasm disappears, and the elongated cell walls form the extensive, lengthy water-conducting structures.

Tracheids are long and slender with tapered ends (Figure 27.4a). They do not lie precisely end to end; their ends form staggered, overlapping connections. One might expect the end walls to be absent where the elements fuse together. The end walls do persist, but they contain very thin, porous regions called **pits** or **pit pairs,** which consist of primary cell wall material only. The pits of adjacent tracheids are aligned so that water readily passes through from one tracheid to the next. The side walls of tracheids are commonly pitted as well; thus fluid can move laterally. Note in Figure 27.4b that pit pairs can be either simple or bordered, although those of tracheids are generally bordered.

Vessel elements tend to be shorter and much wider than the cells of tracheids (see Figure 27.4c). Tapering, if it occurs at all, is much less noticeable in the end walls. The end walls themselves may be heavily perforated or even entirely absent. As a result, vessel elements lying end to end form uninterrupted pipelines, the vessels, through which water readily moves. Like tracheids, the side walls of vessel elements are heavily pitted and readily admit water. The xylem fibers, which can be long and tough, have a role in support.

Phloem While xylem conducts water, phloem is responsible for the movement of many solutes, including foods (primarily sucrose). Unlike xylem, phloem must remain alive to carry out its function. This is a complex tissue that is composed of **sieve elements, companion cells, and phloem parenchyma** (Figure 27.5). (Also included are **phloem fibers,** elements that provide structural support.) Sieve elements occur as **sieve cells** in the gymnosperms and as **sieve tube members** in the flowering plants. Sieve tube members, like the vessel elements of xylem, represent an evolutionary advancement and we will confine discussion to these.

FIGURE 27.4
XYLEM TISSUE
Xylem includes conducting elements known as tracheids and vessel elements. **(a)** Tracheids are long, slender elements with pitted side and end walls. **(b)** Pit pairs are thin-walled regions consisting of only the two adjacent primary cell walls. Pit pairs may be simple or bordered. **(c)** Vessel elements have pit pairs in their side walls, but their end walls are completely perforated. Vessel elements are usually much larger and less tapered than tracheids.

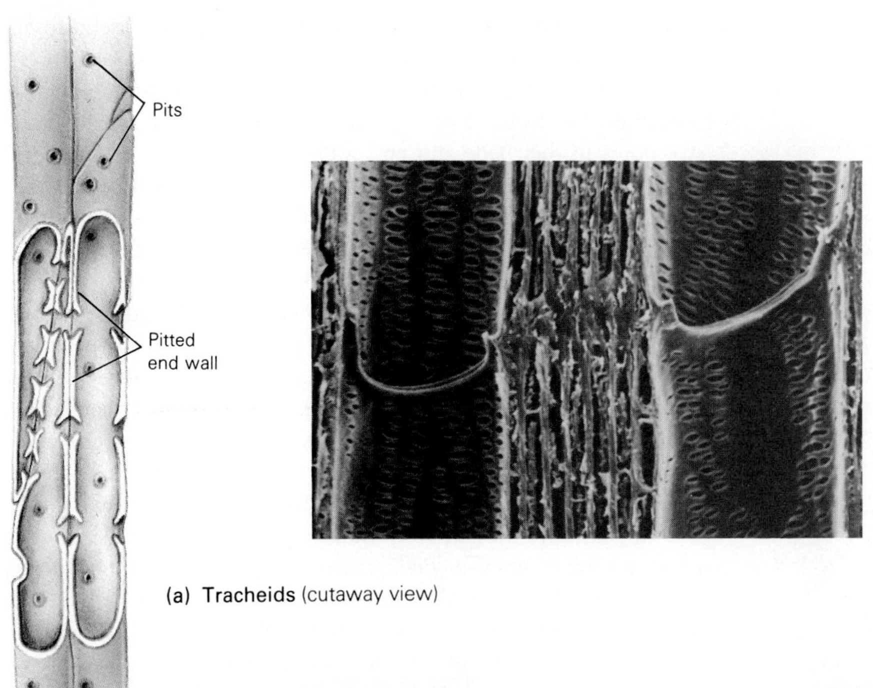

(a) **Tracheids** (cutaway view)

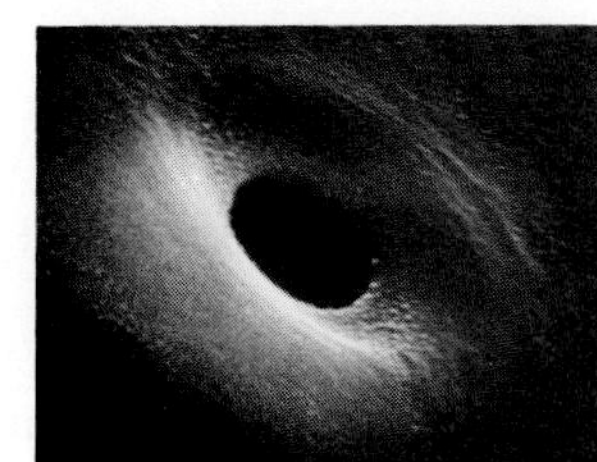

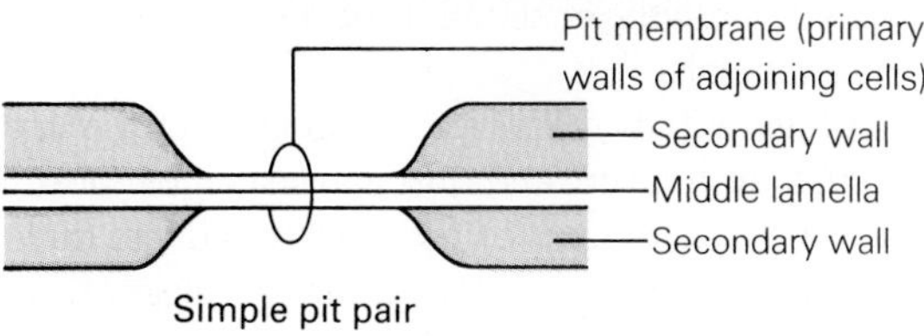

Simple pit pair

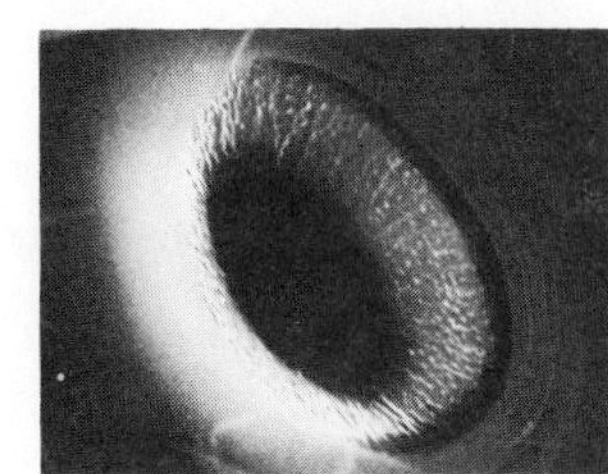

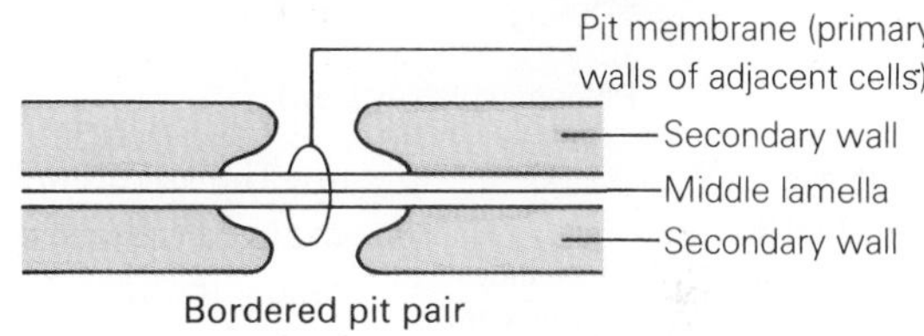

Bordered pit pair

(b)

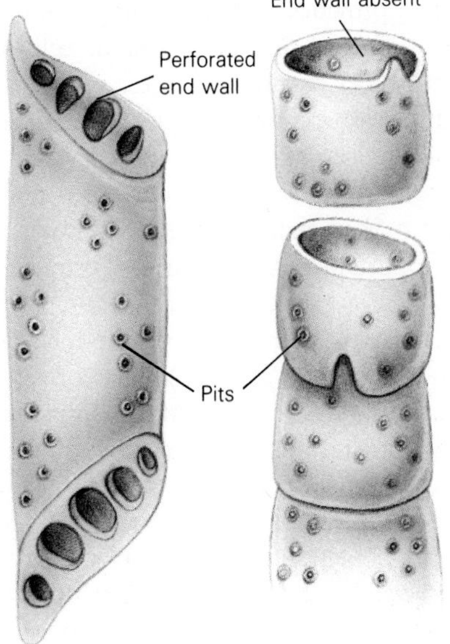

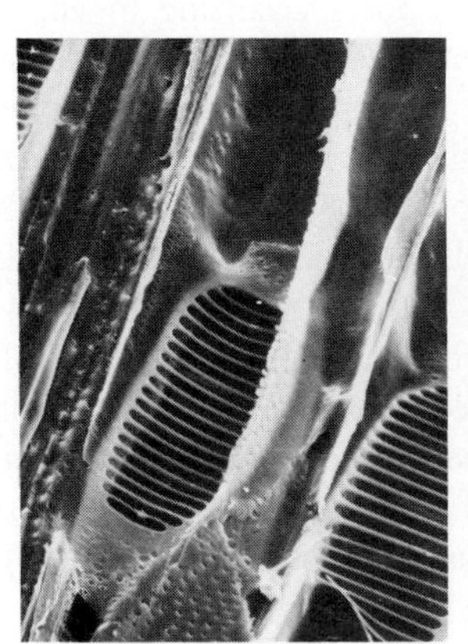

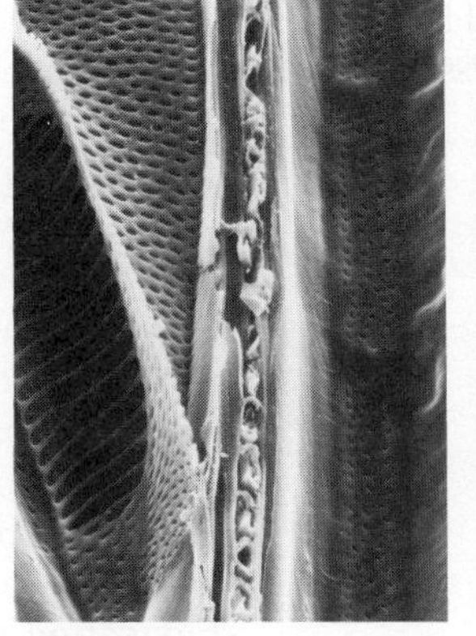

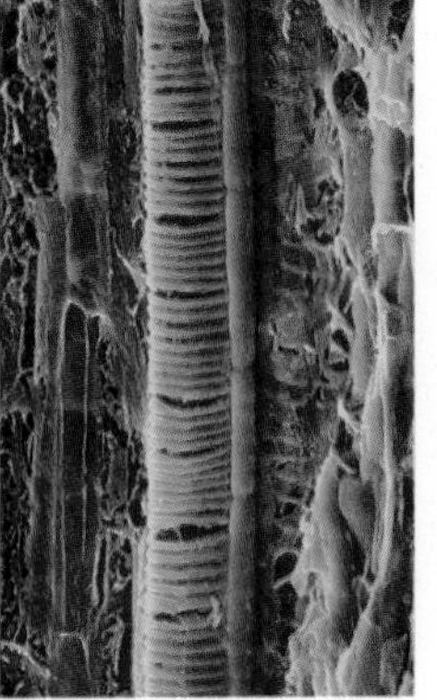

(c) **Vessels**

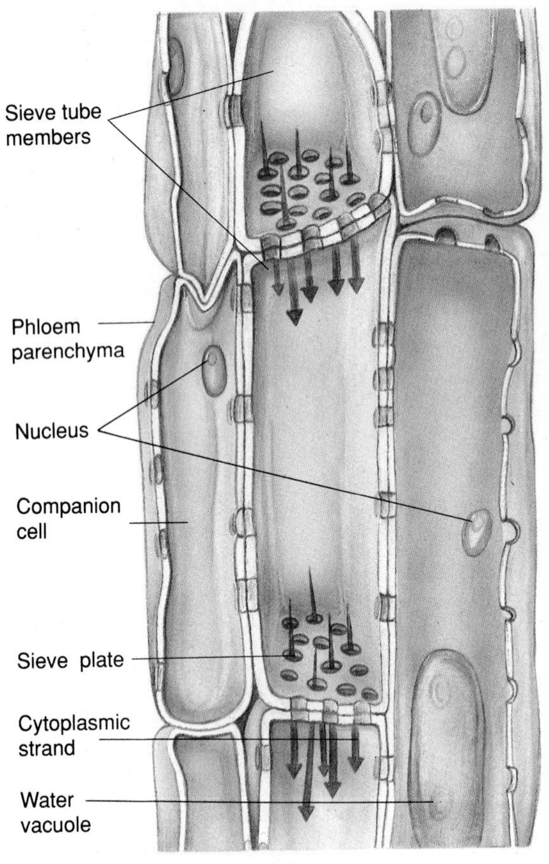

FIGURE 27.5
PHLOEM TISSUE
In flowering plants, phloem includes sieve tube members bordered by companion cells and phloem parenchyma cells. Sieve tube members conduct the phloem stream, while the other two provide physiological support and lateral transport. The end walls of sieve tube members contain sieve plates whose pores are penetrated by strands of cytoplasm from the adjacent cells.

Sieve tube members (individual cells), when arranged end to end, form the **sieve tubes.** (Just as vessel elements form vessels.) It is in the sieve tubes that foods are conducted from one location in the plant to another. The term, "sieve" is appropriate since sieve tube members have pores in their side and end walls. The largest pores are seen in **sieve plates**, which form the end walls between sieve tube members. Thin streams of cytoplasm extending through the enlarged pores of sieve plates conduct substances from one member to the next. Sieve plates, we should note, are absent in the more primitive sieve cells.

Companion cells are aptly named, since they lie adjacent to the sieve tube members and also function in conduction. The exact relationship between the two isn't entirely clear, but we know that while the companion cell contains a nucleus, the sieve tube members do not. Presumably, the companion cell takes care of any needs related to protein synthesis for both types of cells. There is also considerable evidence that the companion cells function importantly in lateral transport. Specifically, they actively transport sugars into and out of the sieve tube members—a vital part of the overall food transporting process, and one we will look into in the next chapter. The third element of the phloem triad, the simpler phloem parenchyma cell, is involved chiefly in storage but also actively transports foods into and out of the sieve tube members.

PRIMARY GROWTH IN THE ROOT

The Root Tip

Primary growth—growth in length—in the plant root takes place at the **root tip.** The root tip is indeed the business end of the root, for its tasks include growth, penetration of the soil, and the absorption of water and minerals.

A look at Figure 27.6 will reveal the organization of a typical root tip. The apical meristem is located at the very end, just above a region of larger, loosely arranged cells called the **root cap.** The root tip is commonly divided into three functional zones, the first of which includes the meristem and forms the region of cell division (note the many nuclei involved in mitosis). Above this lies the region of elongation and further back, the region of maturation. We can now consider each in greater detail.

In the apical meristem, the region of cell division provides a constant supply of cells for the root cap below and for the zone of elongation above. In addition, some cells are held in reserve, not specializing in either direction, to ensure that these provisions continue. The root cap acts as a protective bumper, shielding the delicate meristem behind. As the root tip penetrates the abrasive soil, the cap cells are sloughed away and must continually be replaced. Further, the rupture of the cap cells releases a slimy substance, a polysaccharide that acts as a lubricant to the lengthening root tip. Root cap cells, as we will see in Chapter 29, may also be important to the downward growth of roots.

While new cells added below the apical meristem replace those lost by the root cap, new cells added above will differentiate into the three basic primary meristems: protoderm, ground meristem, and procambium. But first they will play an important role in the zone of elongation. It is within this zone that the cells elongate, providing the "downward push" we associate with root growth.

Elongation in the Root Tip Elongation requires changes in cell walls. In the young cells above the root apical meristem, the primary wall is still thin and pliable, but even

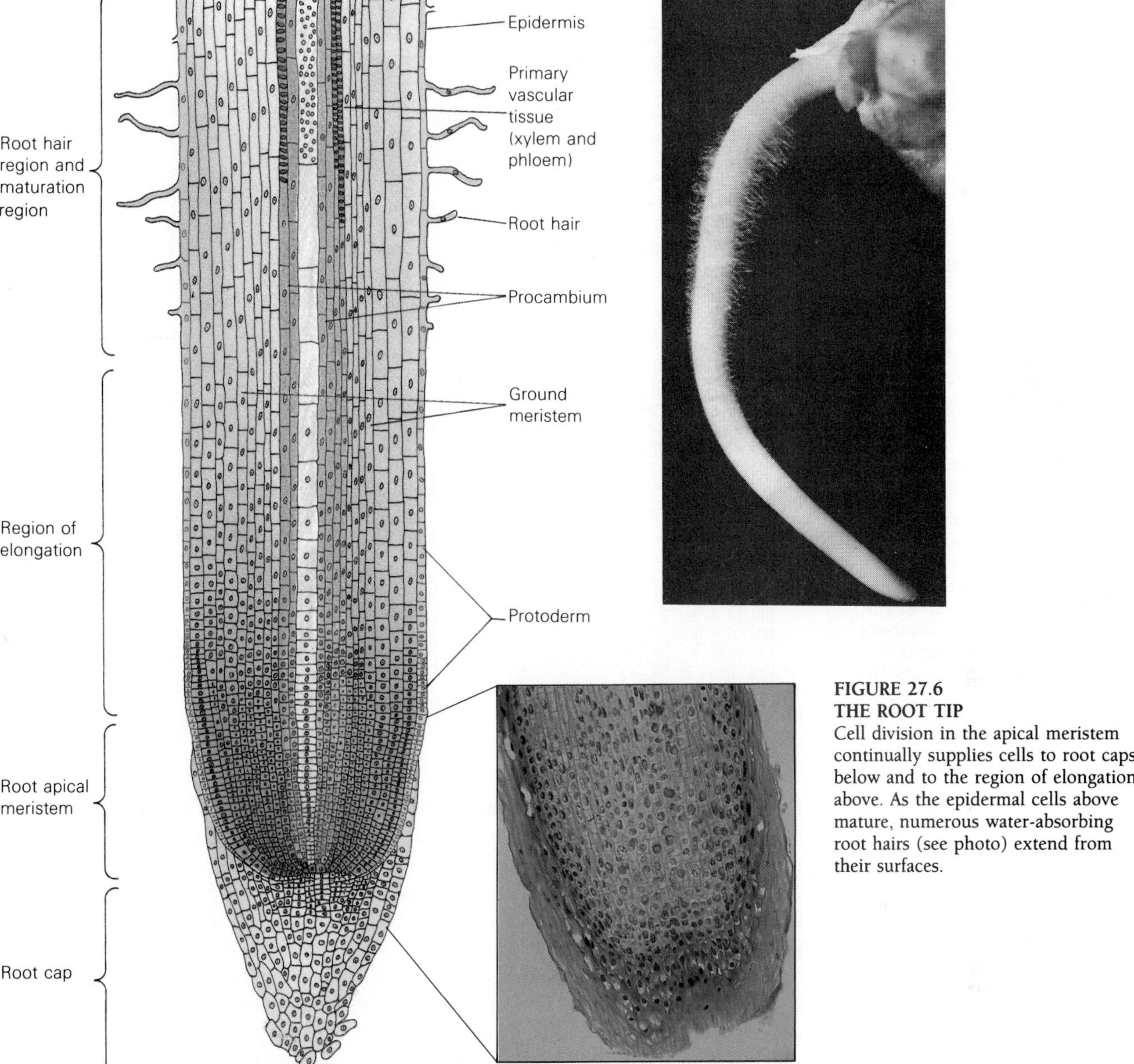

FIGURE 27.6
THE ROOT TIP
Cell division in the apical meristem continually supplies cells to root caps below and to the region of elongation above. As the epidermal cells above mature, numerous water-absorbing root hairs (see photo) extend from their surfaces.

so, the fibers must be further loosened for elongation to occur. Physiologists now believe that this is done by acidification—the addition of acid. The acidic condition either activates certain enzymes that break cellulose bonds, or it directly weakens the cellulose fibrils by disrupting hydrogen bonds between cellulose polymers (see Chapter 3). The source of the acid is hydrogen ions (H^+) within the cell cytoplasm.

Under the influence of the hormone auxin, hydrogen-ion (proton) pumps in the plasma membrane flood the cellulose wall with the ion. After fiber loosening has been accomplished, the ongoing uptake of water and subsequent increase in turgor pressure, usually restrained by a hardened wall, causes the cell to elongate. As a result, the first layers of cellulose fibrils circling the cell change their orientation, first becoming diagonal and then longitudinal. Any new fibrils laid down during this process will be at right angles to the original fibers (Figure 27.7). Once the cell wall is completed and hardened, no further elongation is possible.

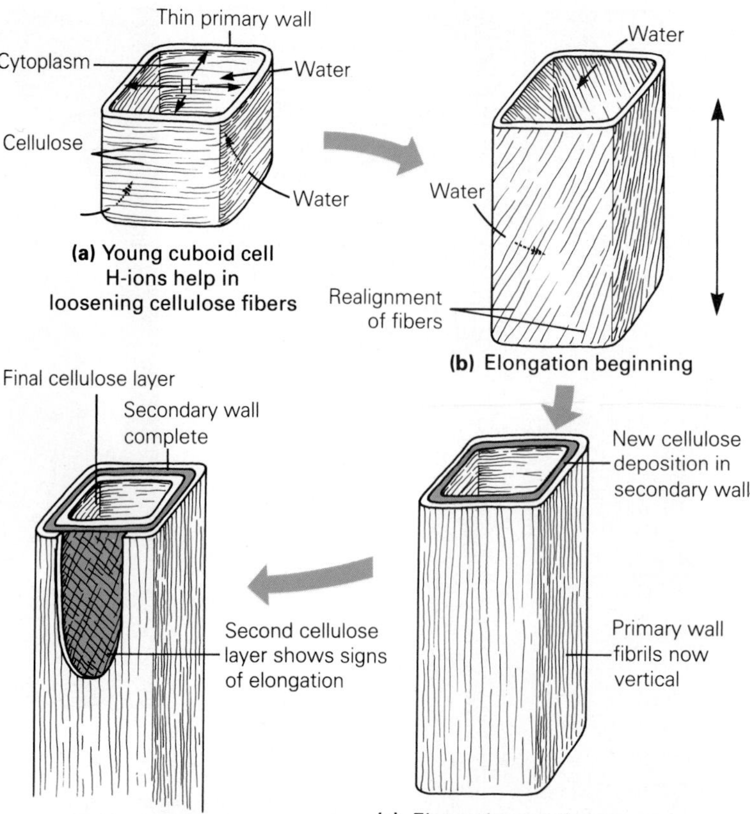

FIGURE 27.7
MECHANICS OF CELL ELONGATION
Cell elongation is a highly precise process. **(a)** Acidification of the wall loosens the cellulose fibers, and **(b)** the uptake of water creates sufficient turgor pressure to cause cell lengthening. **(c,d)** As elongation continues and new secondary walls are laid down, cellulose microfibril patterns change.

Differentiation in the Young Root Once their elongation has been completed, the older cells, which now form the region of maturation, are left behind the growing root tip to mature and begin differentiation. The outer tissues, the protoderm, will form the epidermis, and as we saw in Figure 27.6, many of these young covering cells will produce lengthy root hairs. Ground meristem just within the protoderm forms the **cortex**, a region of thin-walled parenchyma.

Toward the central region of the root, some of the ground meristem will mature into a cylindrical ring of cells known as the **endodermis** (inner skin). The endodermis performs an active role in the vital uptake of water by the plant. Layers of impregnated waterproof material, strategically located in four of the six side walls of the endodermal cells, are called **casparian strips.** Because of their presence, all incoming water is detoured from the usual cell wall route into the endodermal cell, where it must pass through the cytoplasm before moving on (see Figure 28.6). Just within the endodermis lies the **stele.** It comprises the developing vascular system. The stele deserves special attention.

The Stele The stele is surrounded by a cylinder of cells called the **pericycle**, which lies just inside the endodermis (Figure 27.8). The pericycle conducts water and minerals inward to the vascular tissue, but in dicots it has another function as well. It has the capacity to produce **lateral roots** (also called **branch roots**). The young lateral roots emerging from the pericycle closely resemble primary root tips in their organization. Their growth has a brutal aspect, since they must digest and push their way through the endodermis and cortex, crushing many cells and finally bursting through the epidermis of the primary root (Figure 27.9). Eventually, each branch root will develop its own vascular tissue, which will become joined to that of the primary root.

In some dicots, the organization of the vascular tissue within the stele looks (in cross-section) like a four-armed star (see Figure 27.8). The broad arms contain the primary

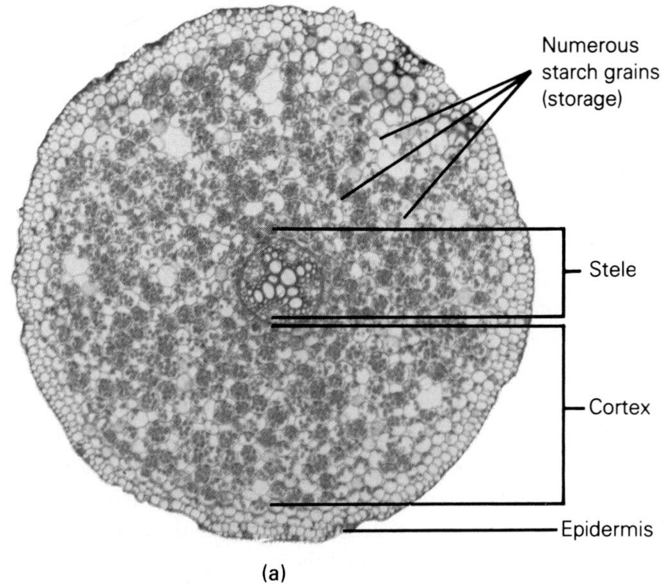

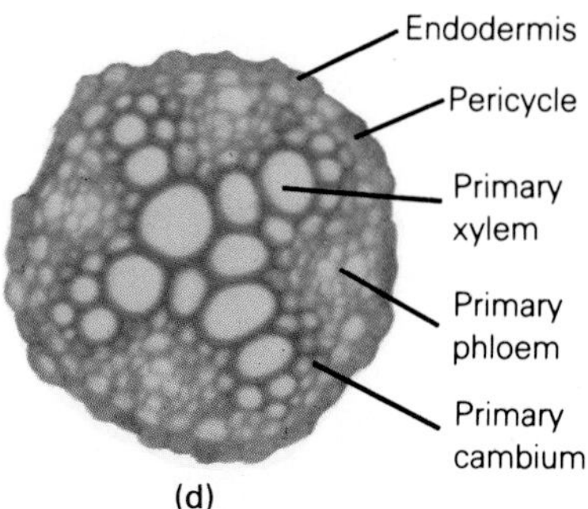

FIGURE 27.8
THE MATURE DICOT PRIMARY ROOT
(a) A cross section through the mature dicot root reveals three primary regions: the epidermis, cortex, and stele. **(b)** In a magnified view of the stele, its vascular tissue is readily visible.

xylem, which is made up of large, thick-walled, water-conducting cells. The primary phloem, with its conspicuous sieve tubes (food-conducting structures), is found between the arms of the star. In plants capable of secondary growth, undifferentiated procambium produces a strip of vascular cambium between the xylem arms and phloem regions. Vascular cambium will later produce the secondary xylem and secondary phloem.

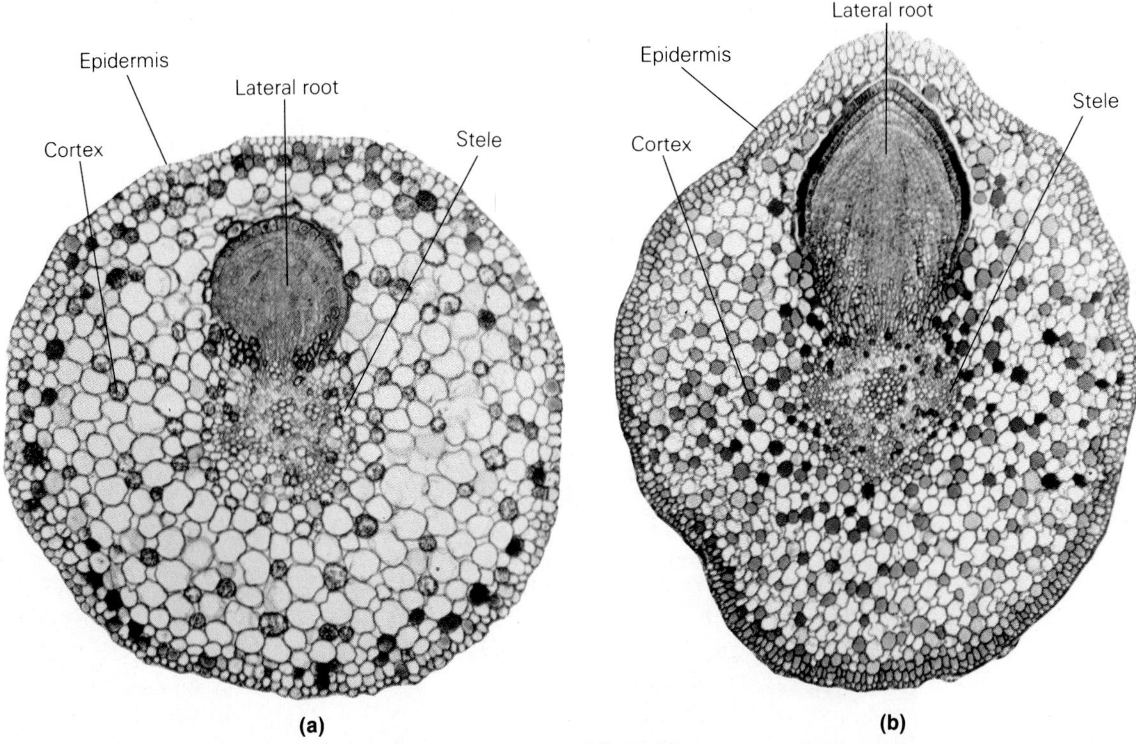

FIGURE 27.9
GROWTH OF LATERAL ROOTS IN *SALIX* (WILLOW)
(a) Lateral roots originate in the pericycle of mature primary roots. Cells there undergo repeated divisions, producing what at first appears to be a formless mass. **(b)** As the emerging root takes form, it elongates, making its way through the cortex and breaking through the epidermis. During this period, an apical meristem develops at the tip, producing a root cap ahead and typical primary tissues in its wake.

Root Systems

Root systems differ between dicots and monocots. While both begin by producing a fast-growing main or primary root, the primary root in monocots does not usually persist as it does in dicots. For instance, in corn, the primary root is replaced early in life by **adventitious roots,** which emerge from the stem. These in turn produce branches that originate from the root pericycle in the usual manner of lateral roots. You may have seen the adventitious roots in corn, since they are often exposed. From their origin in the stem's base, they curve downward into the soil. Adventitious roots arising from above ground are called **aerial roots.** Where they act in supporting the plant (as they do in corn), they become **prop roots.** Aerial adventitious roots are also seen in dicots, for example in English ivy, tropical red mangroves, and banyan trees.

Roots can generally be categorized as either **tap roots** or **diffuse roots** (Figure 27.10), although some plants have combinations of the two. Tap roots are characterized by a large, main root from which lesser branches emerge. Examples are the carrot and sweet potato. Diffuse roots are comprised of a number of roots of roughly the same size. An example is the fibrous, matted roots of grasses. The root density of grasses can be remarkable and in fact, influenced the development of the American West as settlers were constantly frustrated in trying to plow the grassy plains, their efforts earning them the title "sodbuster."

The extent to which roots grow and the power of their growth is legendary. Everyone has seen sidewalks and streets literally lifted up by a meandering tree root. The direction and extent of root growth depends not only upon individual species, but upon physical factors as well. Among these are soil moisture and composition, and certainly soil temperature. The most active water absorption goes on in what are called "feeder roots," usually concentrated in the top meter of soil.

Whereas the roots of some trees, such as beeches, spruces and poplars, are relatively shallow, those of others, such as oaks and pines, grow quite deep. Even the adventitious roots of corn penetrate the soil to a surprising degree, venturing down one and one-half meters. Alfalfa roots reach even further—six meters into the soil. For now, the record holders are desert species: an Egyptian acacia tree's roots reached over 30 meters (almost 100 ft.), and an Arizona mesquite (just a scrubby brush), uncovered in an open-pit mine, sent its roots down an astonishing 53 meters (174 ft.)!

PRIMARY GROWTH IN THE STEM

The organization of the apical meristem of the shoot is somewhat more complex than that of the root (Figure 27.11). As we mentioned earlier, a small mound of tiny, actively dividing apical meristem is found at the very tip of the shoot. As cell division occurs, daughter cells at the lower side of the meristem undergo extensive elongation, carrying

(a)

(b)

FIGURE 27.10
TWO KINDS OF ROOT SYSTEMS
(a) Diffuse root systems are common in monocots such as the bluegrass seen here. This vast, spreading system consists of adventitious and lateral roots. (b) In tap root systems, the primary root maintains dominance, leading the way and producing a profusion of lateral roots.

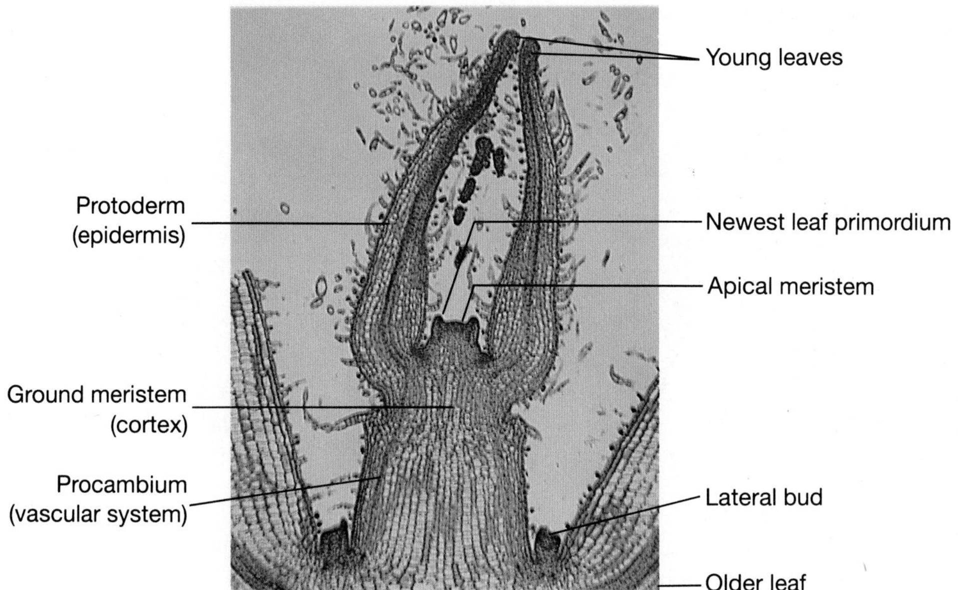

FIGURE 27.11
THE DICOT SHOOT TIP
A dicot shoot tip includes the shoot apical meristem, a number of leaf primordia, and three primary tissues. The primary tissues form the epidermis, vascular system, and cortex of the maturing stem.

the meristem upward or outward. Below this region of elongation, cells that have completed the process undergo differentiation, but their pattern of development is quite different from that of the root tip. In the center, thin-walled parenchyma cells will form the cortex. Just outside of cortex, strands of procambium differentiate into the primary xylem and phloem. At the close of primary growth in dicots, the bundles of vascular tissue will have become arranged in a ring around the perimeter of the cortex. Each bundle (actually a cylinder) contains a strand of procambium, with xylem on its inner side and phloem on its outer. The vascular bundles are surrounded by cortex, which extends to the young differentiating epidermis. In many young stems, the cortical cells just below the epidermis contain chloroplasts and carry on photosynthesis.

A region called **pith** is seen in older portions of the stem. Pith is chiefly parenchymal in origin and can be distinguished from the surrounding cortex by its loose organization, including many intercellular spaces.

The distribution of vascular tissue in the monocot shoot, as you may recall from Chapter 25, is fundamentally different from that in the dicot shoot. While the vascular bundles commonly form a rather neat ring within the dicot stem, in monocots they typically are scattered throughout the stem. Figure 27.12 compares the arrangement in corn and sunflower plants.

As the shoot grows, it leaves behind it differentiating tissue as we described above, but it also leaves behind patches of meristem that give rise to **leaf primordia** and **branch primordia.** The primordia are patches of embryonic tissue that give rise to new tissues. General regions of potential leaf production or leaf attachment are known as **nodes.** The stem region between successive nodes is, logically enough, called an **internode.**

Nodes and internodes are most easily seen in the growing tips of deciduous woody plants—seasonally growing plants that shed their leaves in the autumn, such as oak, maple, or hickory trees (Figure 27.13). When a leaf falls from such a plant, it leaves a scar that remains visible for a time on the young branch. Above each **leaf scar** is a tiny **lateral** or **axillary bud,** which in the younger region may produce leafy stems in the spring. The age of a young woody stem can be determined by counting the number of groupings of terminal **bud scale scars** below the **terminal bud** at the stem tip.

Lateral buds may be inhibited from sprouting into lateral branches by the action of the hormone, *auxin,* which originates in the apical meristem. Under such inhibition, the uppermost growing tip grows fastest, with a decreasing gradient of inhibition forming below. Thus a tree may take on a triangular shape, or at least will have somewhat stepped arrangement in its branches. But under certain conditions, depending on the position of the bud relative to the apex and other factors, the lateral branch bud may be released from its dormancy and sprout into a lateral branch. Or, if a principal branch is lost, a

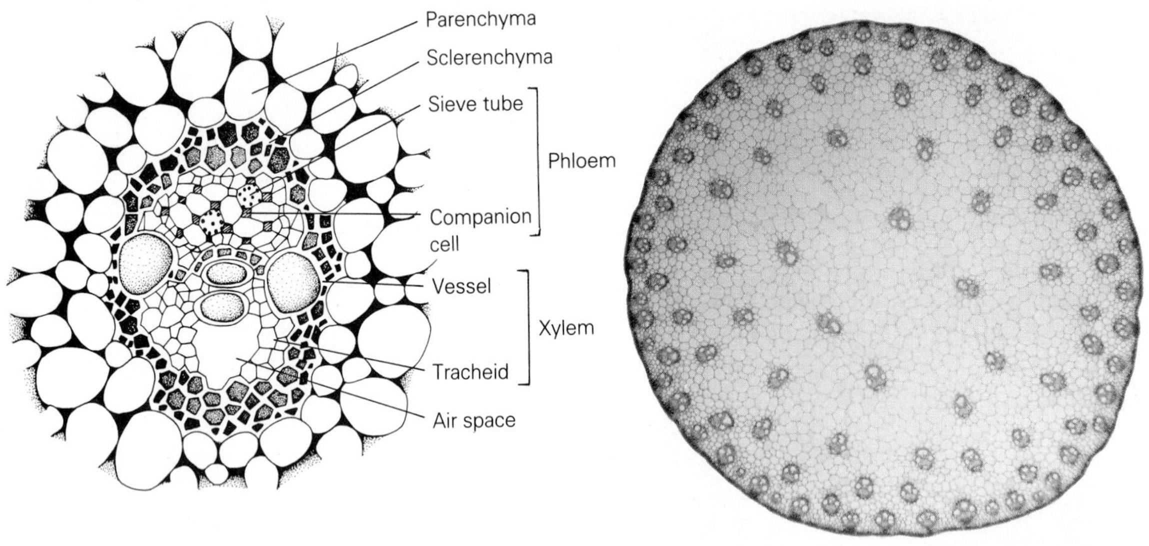

(a) **Monocot** (corn) vascular bundle

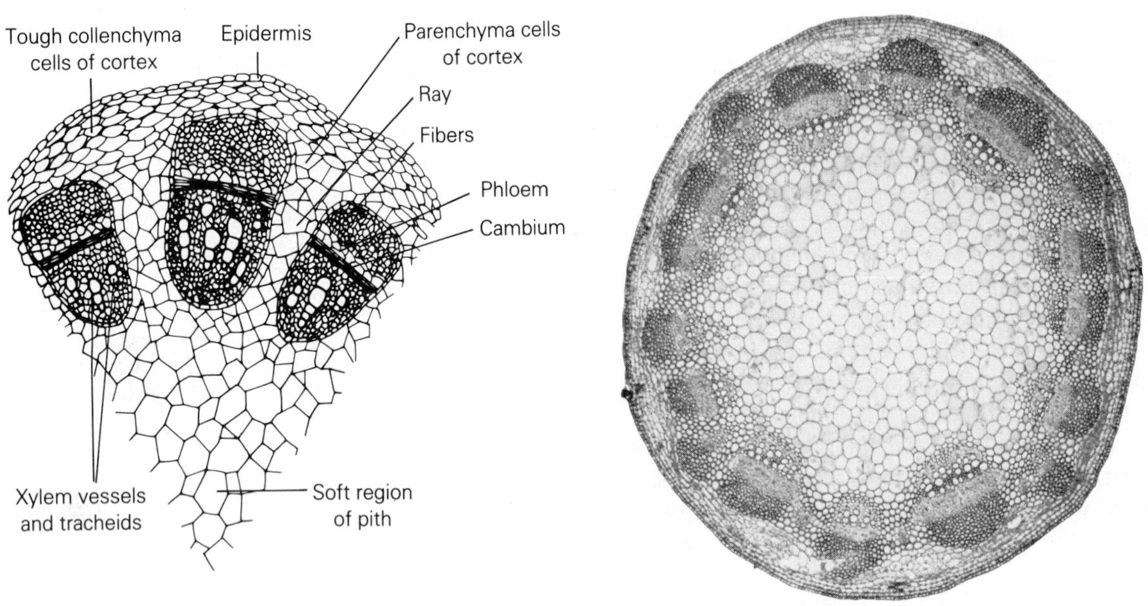

(b) **Dicot** (sunflower) vascular bundles

FIGURE 27.12
MONOCOT AND DICOT STEMS
(a) A cross section through a corn stem reveals a scattered distribution of vascular bundles. Each bundle (see detail) contains xylem and phloem surrounded by tough sclerenchyma cells (fibers). **(b)** The vascular bundles of the sunflower, a dicot, occur as a cylinder organized around the stem's perimeter. The primary phloem and primary xylem are separated by a region of procambium. Most of the stem's interior is soft pith.

nearby dormant apical meristem, suddenly released from its auxin-inhibited state, will begin a growth surge, thus replacing the branch.

It should be noted that grasses do not follow the path of growth that we have described here, perhaps, it is thought, because grasses evolved under strong pressure from animal grazers. If grasses grew from their tips, it would be difficult for them to recover from grazing. But new growth in grasses continually emerges from a meristematic reserve of tissue near the roots. Thus, the loss through grazing does not interfere with growth.

LEAVES

In its development, each leaf requires contributions from the basic tissue formed by the three primary meristems. Protoderm contributes to the highly specialized epidermis, which, you will recall, has the conflicting tasks of slowing water loss yet admitting air. The light-trapping photosynthetic cells within the leaf are essentially parenchyma, arising from ground meristem. Also originating from ground meristem are the many collenchyma cells that form important supporting elements. And finally, because of the activity of the procambium, the leaf contains an extensive vascular system—xylem and phloem elements—which brings in water and minerals and takes away newly produced sugars.

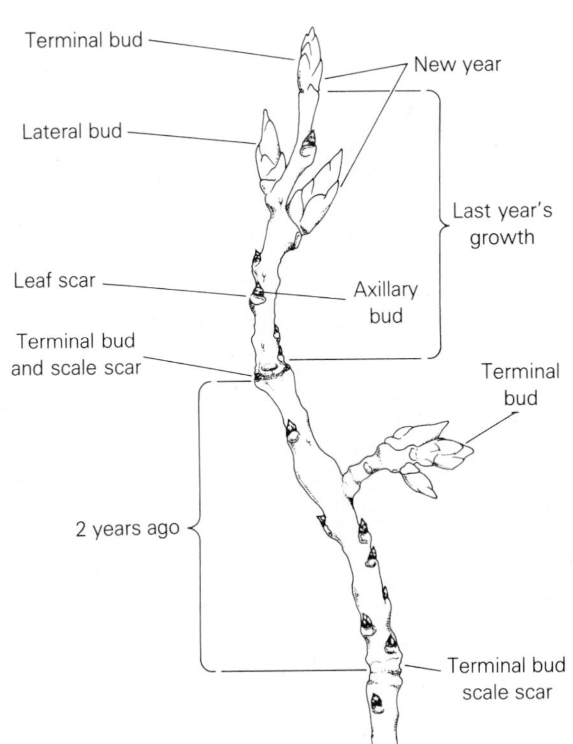

FIGURE 27.13
SHOOT ANATOMY
This dormant deciduous woody stem tip reveals two years of growth history. The dormant terminal bud will emerge as the next season's growing tip. The last season's growth can be determined by measuring the distance between the terminal bud and the first group of terminal bud scale scars.

Anatomy of a Leaf

The typical dicot leaf (Figure 27.14) is attached to the stem by a stalklike **petiole,** which extends to the flattened **leaf blade.** The vascular system of the stem passes into the leaf through the petiole and into the blade along a large central vein called a **midrib.** Typically in dicots, the central midrib has a number of major branches or veins that penetrate the blade on either side, branching and rebranching so that no cell in the blade is very far from an extension of the vascular system. The smaller veins are surrounded by specialized cells that form what is called the **bundle sheath,** and all materials passing in or out of the veins must pass through the bundle sheath cells. (You may recall the special role of the bundle sheath cells in C4 plants discussed in Chapter 7.)

Looking at the remaining tissue organization of the dicot leaf (Figure 27.15), we see that the **upper epidermis** (that side most exposed to the sun) consists of large, simple, flattened cells whose outer walls are typically shiny—coated with a waxy layer of cutin. These cells lack chloroplasts and are very transparent. Within the leaf are several layers of photosynthetic parenchyma, or **leaf mesophyll,** the tightly packed, vertically arranged, **palisade parenchyma,** and below this, the very loosely arranged **spongy parenchyma.** The latter is aptly named since the loose arrangement contains numerous spaces filled with water or very moist air. The spaces provided are vital avenues for carbon dioxide and oxygen diffusion.

Carbon dioxide enters and water vapor exits the leaf through numerous porelike stomata. Each stoma is a tiny pore surrounded by two guard cells, which, through changes in shape, can open and close the pores (see Chapter 28). Technically, each pair of guard cells and the pore they form are called a stomatal apparatus, but we will usually just call it a stoma. The stomata are found primarily in the **lower epidermis**—the underside of the leaf. Lower (and often upper) epidermal cells also often produce lengthy **leaf hairs,** which give the surface a fuzzy appearance. Leaf hairs tend to impede the movement of air over the pore-marked surface, cutting down on evaporation in windy weather. In addition, some hairs are sharp and hooked and can repel voracious insect larvae. Some hairs are glandular, containing chemical toxins, or sticky, trapping substances. Some even contain "appetite suppressants," that inhibit insect feeding behavior.

FIGURE 27.14
DICOT LEAVES
Dicot leaves are typically net-veined, with a major, central vein giving rise to smaller veins at either side. The smaller veins subdivide to form finer and finer veins.

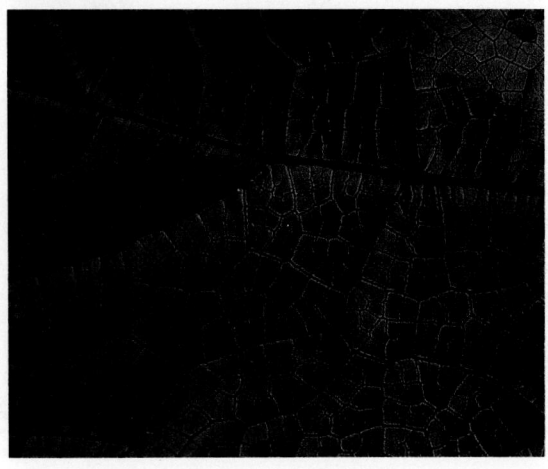

That's the dicot leaf. The organization of the monocot leaf (see Figure 27.16) is quite different. Monocots, such as grasses, have no petioles. The leaf consists of two parts; the leaf sheath, which surrounds the stem, and the blade, which extends outward. The sheath is extremely tough, as you can attest if you have ever tried to tear off a corn or bamboo leaf. In addition, the veins do not branch from a central midrib but emerge in parallel rows, interconnected by many small veins.

SECONDARY GROWTH IN THE STEM

While primary growth provides the growing plant with its basic tissues and accounts for growth in length, secondary growth produces growth in girth. Since secondary growth in roots and stems is similar in many ways, we will confine our discussion to the stem. Secondary growth

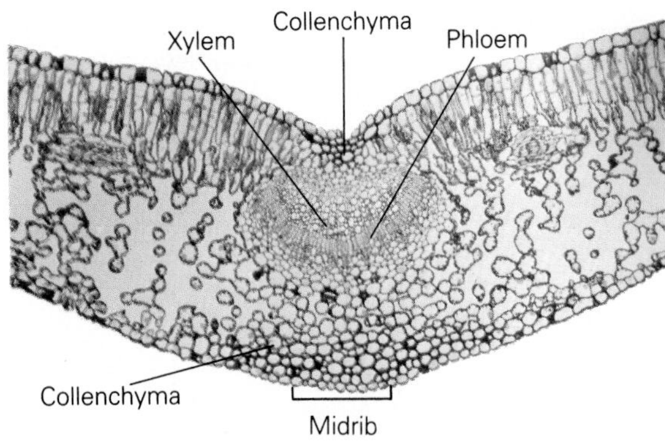

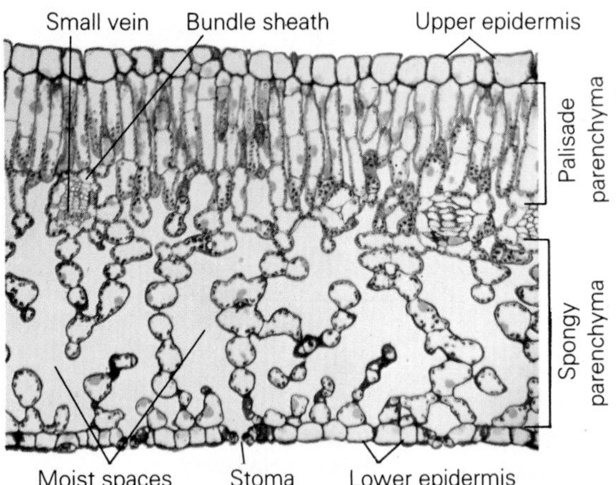

FIGURE 27.15
DICOT LEAF ANATOMY
The tightly packed layer of longitudinal cells in the dicot leaf is the palisade parenchyma. Below this is the loosely arranged spongy parenchyma, with its many spaces containing water and moist air. The lower epidermis is interrupted by a number of pore-like stomata, formed by paired guard cells. Note the presence of vascular and supporting tissue in the midrib.

arises from regions of undifferentiated cambium (meristemetic tissues) that become active. As we have mentioned, secondary growth among angiosperms is characteristic of dicots and is most notable in woody shrubs and trees.

Transition to Secondary Growth

As we have seen, primary growth in the dicot stem[1] ends with the vascular bundles becoming arranged into a ring or cylinder. A tiny region of procambium remains between patches of xylem and phloem elements within the bundles. Remaining between the bundles are areas of cortex consisting of versatile parenchyma cells, as we see in the highlights of secondary growth shown in Figure 27.17.

As the procambium between the xylem and phloem becomes active, rapid cell division ensues, and the expanding tissue links up with the parenchyma cells between the vascular bundles. This forms a cylinder of active tissue, the **vascular** or **secondary cambium.** As the cells in the ring of newly formed vascular cambium divide, those inside will differentiate into new xylem, and those outside into new phloem (just as the procambium did in primary growth). It is more accurate to now refer to these tissues as **secondary xylem** and **secondary phloem.** Vascular cambium also contributes cells to the rays—spokelike lines of cells. These are called **vascular rays,** or simply **rays.** They consist of strands of parenchyma produced by the vascular cambium. Vascular rays store materials and provide a means of lateral transport and help accommodate the stress created by rapidly expanding cylinders of xylem as the tree grows in girth. Without the rays, such stress could readily split the trunk of a tree.

As growth continues, the vascular cambium forms a ring of rapidly expanding tissue, with secondary xylem forming inside the ring and secondary phloem forming outside. As you might expect, pressure resulting from the rapid production and growth of new cells in internal regions of the stem can cause mechanical problems. As new phloem and xylem emerge, their outwardly expanding front reaches delicate older phloem and parenchyma tissues, crushing them and eventually obliterating this primary organization. Soon, only the tougher primary xylem remains in the pithy center to mark the earlier stages. The newly emerging xylem enlarges and matures, and as it grows it pushes the vascular cambium outward. Likewise, the new phloem pushes outward,

[1]Some monocots (e.g., palms) also have secondary growth, but it is not derived from vascular cambium.

FIGURE 27.16
MONOCOT LEAF VENATION
Monocot leaves are parallel-veined, with a number of similar-sized veins running alongside each other. These are connected by a fine network of smaller veins (see inset).

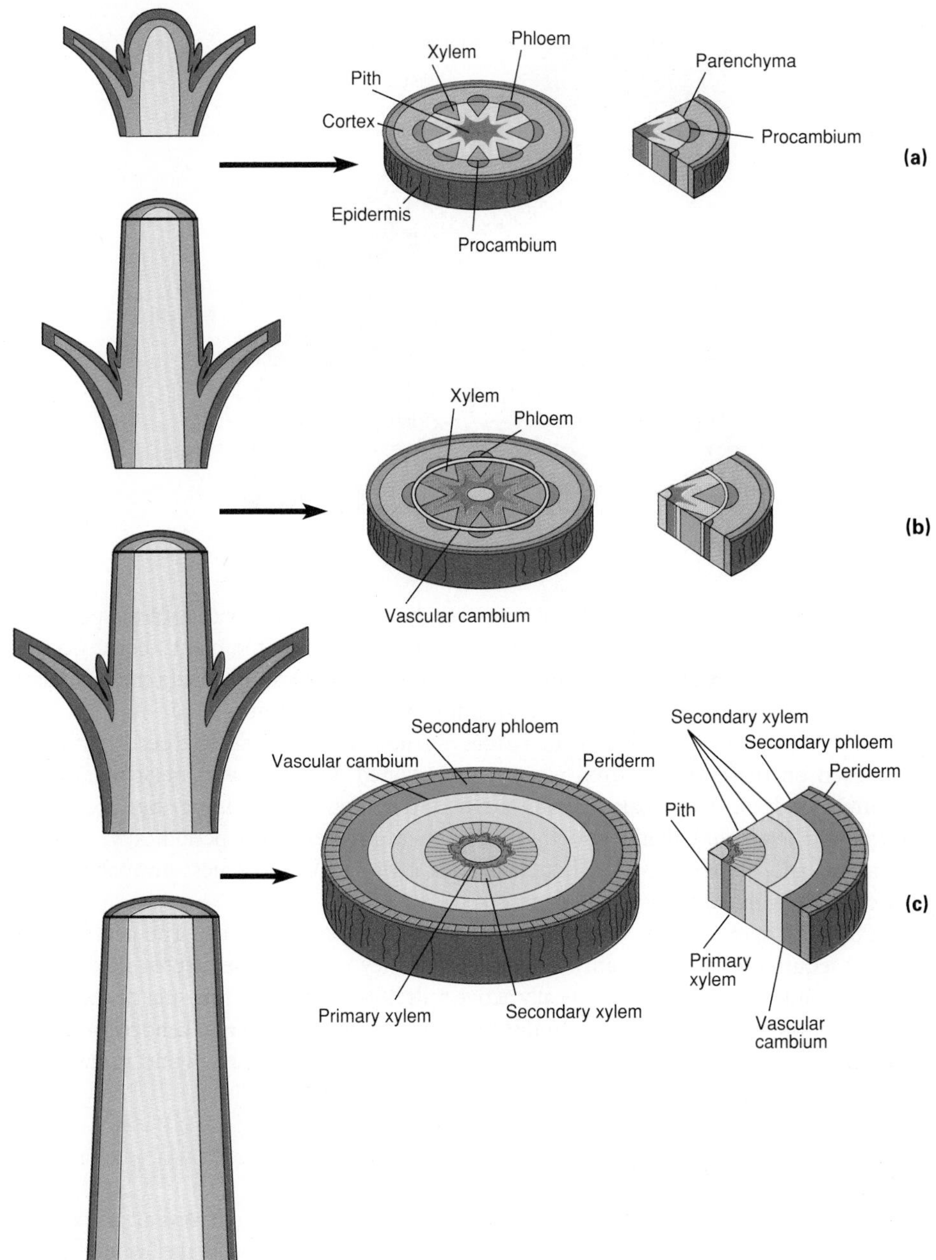

FIGURE 27.17
FROM PRIMARY TO SECONDARY GROWTH
Primary growth begins at the stem tip and is completed further down **(a)**. At its completion, secondary growth begins, and a solid cylinder of vascular cambium emerges **(b)**. As secondary growth proceeds, primary tissues are crowded out, and eventually annual rings of secondary xylem appear **(c)**.

crushing the fragile parenchyma tissue of the cortex. Eventually, the ring of vascular cambium and phloem comes to rest near the perimeter of the stem. Thus, we see that secondary growth also has its brutal aspects.

Growth of the Periderm

Secondary growth in the vascular tissues within the stem is accompanied by changes in the shoot epidermis, a primary tissue which is replaced by the periderm. Cells in the neighboring cortex take on a new role, producing the **cork cambium** (Figure 27.18). The cork cambium, like the vascular cambium, undergoes continuous cell division, contributing layer after layer of cells that expand outward, rupturing and replacing the old epidermis. This new tissue constitutes the shoot's **periderm**, whose function is to

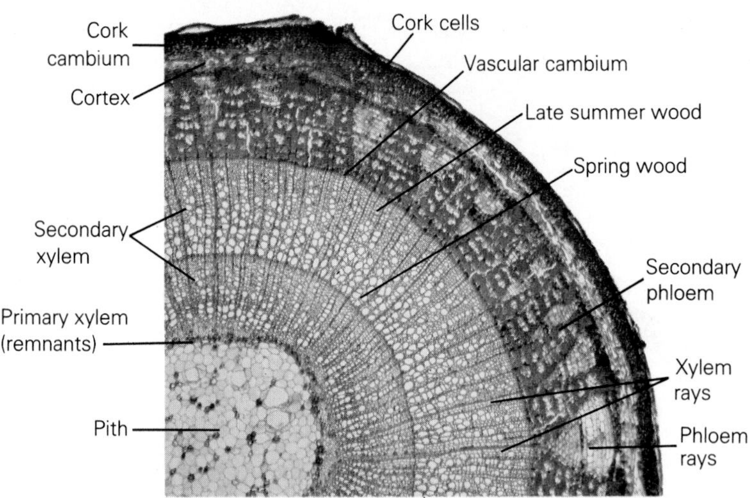

FIGURE 27.18
OLDER WOODY STEM
Several different tissues are visible in the stained cross section of a two-year-old woody stem. The outer ring of cork and cork cambium is followed within by a region of cortex, then by a ring of triangularly arranged phloem and phloem rays. Just within this is a thin line of dark cells, the vascular cambium, and on its inner side, two wide, distinct bands of xylem, the annual rings.

provide a tough, water-resistant covering over the stem. The outer cells of the periderm will be impregnated with waterproofing suberin and become the plant's **cork.** Cork is incorrectly called **bark**, but actually, *bark* refers to all stem tissue outside the secondary xylem, including the vascular cambium, phloem, and periderm.

For the record, the cork in wine bottles is true cork. But it doesn't occur naturally, not even on the cork oak from which it is obtained. Cork growers must remove the natural periderm of the oak, causing the tree to respond to the injury by "obligingly" forming a new, smoother regrowth. This is then stripped away periodically, cut into cylinders and placed into wine bottles in such a way that they almost invariably break or crumble when you try to get them out.

The cork layer is highly water-resistant, but it is not entirely sealed. The living tissues below require a constant supply of oxygen, and they must release carbon dioxide, just as we would expect from metabolically active cells. It turns out that this vital gas exchange occurs through minute openings in the periderm known as **lenticels.** Lenticels are most numerous over metabolically active stem regions, and they are also found in the outer skin of pears, apples, and some other fruit.

The Older Woody Stem

Older regions of the woody stem, those that have gone through several seasons of secondary growth, are mostly composed of nonliving xylem tissue—the "wood." Some of this is **heartwood,** a region, commonly darker in color, that no longer conducts water and minerals. (Heartwood is responsible for the beauty of natural wood products such as fine furniture and paneling.) The most recent layers of xylem, the **sapwood,** are those that conduct water and minerals.

The vascular cambium, phloem, and periderm—all lying outside the xylem—form a rather thin cylinder of living material around the perimeter of the stem. In fact, while many large trees can survive a burned out interior or woody region, removing a strip of bark all the way around the trunk (called girdling) will kill even the most robust forest tree. (In a dispute between environmentalists and loggers recently over the fate of a particularly fine redwood, someone solved the problem by simply girdling the forest giant with a chain saw.) Why is girdling so effective? Consider what the loss of phloem might mean to the roots of a tree.

Annual Rings Cross sections cut through the stems of temperate zone trees, trees that undergo seasonal growth, reveal very prominant annual rings (see Figure 27.18).

Annual rings form because trees grow rapidly in spring and summer, tapering off in autumn and stopping growth in winter. The earliest layers of xylem cells are very large, while those produced later in the season tend to be smaller. There is normally less water available and less time for cell enlargement as summer passes and autumn draws near. The tree's annual rings can be counted to determine the age of the tree. Annual rings can be compared over the years, telling us something about climatic change, particularly in very old trees such as the sequoia and bristlecone pine, or in fossilized tree trunks. The poor growing seasons result in smaller cells and thus thinner annual rings. More recently, researchers have begun to correlate ring thickness with another factor, increasing levels of air pollutants such as ozone and acid rain.

annual ring (also growth ring) Alternating dark and light rings visible in cross sections of stems in seasonally growing woody plants. A complete ring, a dark band and its associated light band, represents a season's or year's growth.
bark All tissues outside the vascular cambium of a woody plant (e.g. phloem, cortex, periderm).
cork The outer, water-proofing layer of stem cells, orginating in the cork cambium.
germination The resumption of metabolic activity and growth in the dehydrated and dormant seed, preceeded by the uptake of water.
leaf mesophyll Photosynthetic tissues between the upper and lower leaf epidermis.
phloem Food-conducting vascular tissue; includes sieve elements (sieve cells and sieve tube members), companion cells, phloem parenchyma, and phloem fibers.
primary growth Essentially, growth in length, provided through cell division and elongation in primary meristems (shoot and root apical maristem).
secondary growth Essentially, growth in girth, provided through cell division in secondary tissue, specifically in vascular cambium and cork cambium.
secondary phloem In secondary growth, phloem originating in vascular cambium.
secondary xylem In secondary growth, xylem originating in vasculalr cambium.
stele In the vascular plant root, a central cylinder of vascular and supporting tissue lying within the endodermis.
stomata A minute pore in the leaf or stem epidermis, bordered by guard cells. (The plural in stoma. Both stomata and guard cells are called the stomatal apparatus).
tracheid Tubular water and mineral conducting structures composed of individual tracheid elements connected end to end, each with angular, pitted end walls (all vascular plants).
vascular cambium Undifferentiated tissue that gives rise to secondary xylem and secondary pholem; provides for secondary growth.
vessel Tubular water and mineral conducting structures composed of vessel elements connected end to end, each with end walls that are fully perforated or entirely absent (in agiosperms and gnetophytes).
xylem Water and mineral conducting vascular tissue; includes vessel elements, tracheids, fibers, and xylem parenchyma. The wood of woody plants.

APPLICATION OF IDEAS

1. An important characteristic of many plants is indeterminate growth. What does this mean, and how does the organization of plants provide for such growth? Do any animals share this characteristic? Is there anything comparable in the other kingdoms? (Consider a clone of protists.)
2. A young boy carved his initials about eye-level in a small tree. Returning as a man, he noted that both he and the tree had grown and that the initials were still there. In what position were the man's eyes when he read the initials—looking up, down, or straight ahead? Explain your answer with a brief but technical discussion of growth in the woody stem.
3. Dendrochronology involves the study of growth rings in wood stems—both intact and fossilized. Of what do growth rings consist, and what kinds of information might a dendrochronologist gain by their study in plant fossils? How might this information be useful?

KEY IDEAS

GERMINATION AND GROWTH OF THE SEEDLING

Germination Conditions

For many seeds, **germination** (the emergence from dormancy) simply requires the uptake of water, suitable temperatures, and the availability of oxygen, but others require special conditions such as exposure to freezing temperatures, fire, abrasion, or exposure to animal digestive enzymes.

The Seedling

1. Following the uptake of water by monocot grains, hormones called **gibberellins** stimulate the aleurone to release enzymes that hydrolyze the endosperm starches into sugars.
2. In the corn seedling, the radicle emerges from the coleorhiza and grows downward, absorbing water and anchoring the seedling. The coleoptile-enclosed shoot grows upward, breaks free, and unfurls the first leaf.
3. The U-shaped onion cotyledon breaks through the soil first, drawing behind it the seed remnant. The first leaf emerges far below the seed remnant, and an elongating hypocotyl gives rise to the root below ground.
4. Seedling growth in the bean, a dicot, begins with the emergence of the hypocotyl, which forms a bumperlike **hypocotyl hook.** Its growth raises the seed out of the soil, whereupon the epicotyl rapidly elongates and spreads the plumule out to face the sunlight.
5. The pea epicotyl emerges from the soil, but the seed remains behind. The lengthening hypocotyl forms the young root.

Primary and Secondary Growth

1. **Primary growth** includes growth in length in the shoot and root, along with the production of leaves, flowers, branches, and **branch roots. Secondary growth,** seen primarily in dicots (and in conifers), provides for growth in thickness.
2. Perennial plants have **open growth,** or **indeterminate life span** (unlimited growth); thus they have no maximum life span. Continuous growth is made possible by reserve of undifferentiated meristematic tissue, but mature plant cells can also undergo **dedifferentiation** and then differentiate into some other cell type.

TISSUE ORGANIZATION IN THE PLANT

Apical meristem produces three kinds of **primary meristems: protoderm, ground meristem,** and **procambium.** Each gives rise to specialized tissue systems.

Protoderm and the Epidermis

Protoderm produces the leaf **epidermis; guard cells,** which surround the pores, forming the **stomata;** root epidermis with **root hairs;** and the complex **periderm.** Waterproofing materials include **suberin** (in the periderm) and **cutin** of the leaf or stem **cuticle.**

Ground Meristem

Ground meristem differentiates into the very common thin-walled **parenchyma** of leaves, stems, and roots; the thick-walled, supporting **collenchyma fibers** of stems; and the **sclerenchyma,** including **sclereids,** which occur as **stone cells** in pears.

Procambium and the Vascular Tissues

1. Xylem includes water-conducting **tracheids** and **vessel elements** and surrounding xylem fibers. The tracheids are long and slender and have pitted end and side walls. (**Pits** are thin regions containing primary wall only.) **Vessel elements**—the cellular units of **vessels**—are wider and shorter, with open-ended walls and pitted side walls.
2. Phloem consists of food-conducting **sieve elements, phloem parenchyma, phloem fibers,** and **companion cells.** Sieve elements include primitive **sieve cells** and advanced **sieve tube members,** the latter of which have **sieve plates.** When assembled, sieve tube members form **sieve tubes** with a continuous cytoplasm. Since **sieve tube members** lack nuclei, nuclear functions are carried out by companion cells, which are also involved in the active transport of nutrients into sieve tubes. Phloem parenchyma specializes in storage and active transport.

PRIMARY GROWTH IN THE ROOT

The Root Tip

1. The **root tip** contains a mitotically active root apical meristem, which gives rise to replacement cells for the **root cap** region ahead and primary tissues behind.
2. Behind the meristem, cells stimulated by the hormone auxin elongate, pushing the root tip through the soil. Under the influence of auxin, hydrogen ions (acid) from the cytoplasm enter the cell wall, beginning a softening process. Turgor pressure produces the lengthening effect.
3. In the primary tissues, protoderm cells produce the young epidermis, which forms numerous root hairs, while those of the ground meristem begin formation of the root **cortex** and **endodermis.** Within the stele, procambium forms the xylem and phloem of the **stele,** a cylinder of conducting and supportive tissue.
4. The stele includes an outer **pericycle** and regions of primary xylem and primary phloem. In plants capable of secondary growth, procambium forms regions of **vascular cambium,** which later gives rise to **secondary xylem** and **secondary phloem.** The pericycle gives rise to branch roots.

Root Systems

Monocot root systems often include **adventitious roots** that emerge from the stem. **Tap root systems** have a prominent main root and many branching or secondary roots, while all of the roots in **diffuse root systems** are similar in size. Feeder roots near the soil surface absorb most of the required water and mineral ions.

PRIMARY GROWTH IN THE STEM

1. A dome-shaped shoot apical meristem gives rise to cells that elongate, producing shoot growth. When the young primary tissues differentiate, they form a thin epidermis along the outside and islands of vascular tissue that surround an inner region of thin-walled parenchyma called the **pith.**
2. Meristem also gives rise to leaves, **leaf primordia**, and **branch primordia** at **nodes.** The expanses of stem between nodes are called **internodes. Lateral branch buds** or **axillary buds** give rise to branches unless suppressed by hormones. **Terminal buds** form **terminal bud scale scars** at the end of each growth season.
3. Grasses emerge continuously from a **basal meristem.**

LEAVES

While protoderm contributes to the complex leaf epidermis, ground meristem differentiates into photosynthetic parenchyma **(leaf mesophyll)**, and procambium gives rise to the leaf vascular tissues.

Anatomy of A Leaf

1. The structures of leaves include a supporting **petiole**, the flattened **leaf blade**, and a central vein called a **midrib.** Smaller veins are surrounded by **bundle sheath** cells, through which materials must pass.
2. A cross section through the leaf reveals a simple **upper epidermis**, then **palisade parenchyma**, followed by **spongy parenchyma** with air spaces, and then the more complex **lower epidermis.** The lower epidermis includes cells with extensive **leaf hairs** and the porelike stomata.
3. While veins in dicot leaves tend to be branching (net-veined), those in monocots generally run parallel (parallel-veined).

SECONDARY GROWTH IN THE STEM

Transition to Secondary Growth

During secondary growth, secondary xylem and phloem arise from vascular cambium **(secondary cambium).** The growth of newly produced **secondary xylem** and **secondary phloem** expands outward, forming a continuous ring and crushing older growth in its path.

Growth of the Periderm

In secondary growth, the epidermis is replaced by a periderm which include the **cork cambium** and **cork.** The cork cells are suberized, but gases can be exchanged through the **lenticels.**

The Older Woody Stem

1. Older woody stems are composed mainly of xylem or "wood" (nonconducting heartwood and conducting sapwood). Living tissues are restricted to a thin outer layer called **bark.**
2. **Annual rings** in the wood represent seasonal growth patterns. Spokelike patterns—the **vascular rays**, or **rays**—aid in transport and help the stem expand.

REVIEW QUESTIONS

1. Compare the early growth of the corn seedling with that of the bean. (596)
2. Distinguish between primary and secondary growth. List two groups of plants in which secondary growth is possible. (597)
3. How do plants make provision for indeterminate growth? (598)
4. In general, what does the protoderm provide? List three of its tissue derivatives. (598)
5. List three tissues derived from ground meristem, and mention a function of each. (598, 602)
6. Compare the structure of tracheids and vessel elements. Which would seem better able to transport water? Describe the structure that makes provision for lateral transport among tracheids and vessel elements. (602)
7. List the four kinds of cells in phloem tissue and state a function for each. (602, 604)
8. What is the function of the sieve tube? List two provisions that make transport possible from one sieve tube member to the next. (604)
9. How is the sieve tube member able to function without a nucleus? (604)
10. Prepare a diagrammatic drawing of a root tip, labeling the following: root cap, root hairs, epidermal cells, apical meristem, stele, cortex, regions of elongation and maturation, procambium ground meristem, and protoderm. (604)
11. Draw a cross section of a mature region in a root labeling the stele, epidermis, endodermis, pericycle, cortex, primary xylem, vascular cambium, and primary phloem. (606-607)
12. Contrast the tap root system with the diffuse root system. In what plants would one most likely find the latter? (608)
13. Compare the arrangement of the vascular system in the young shoot with that of the root. (608-609)

28

PLANT TRANSPORT MECHANISMS

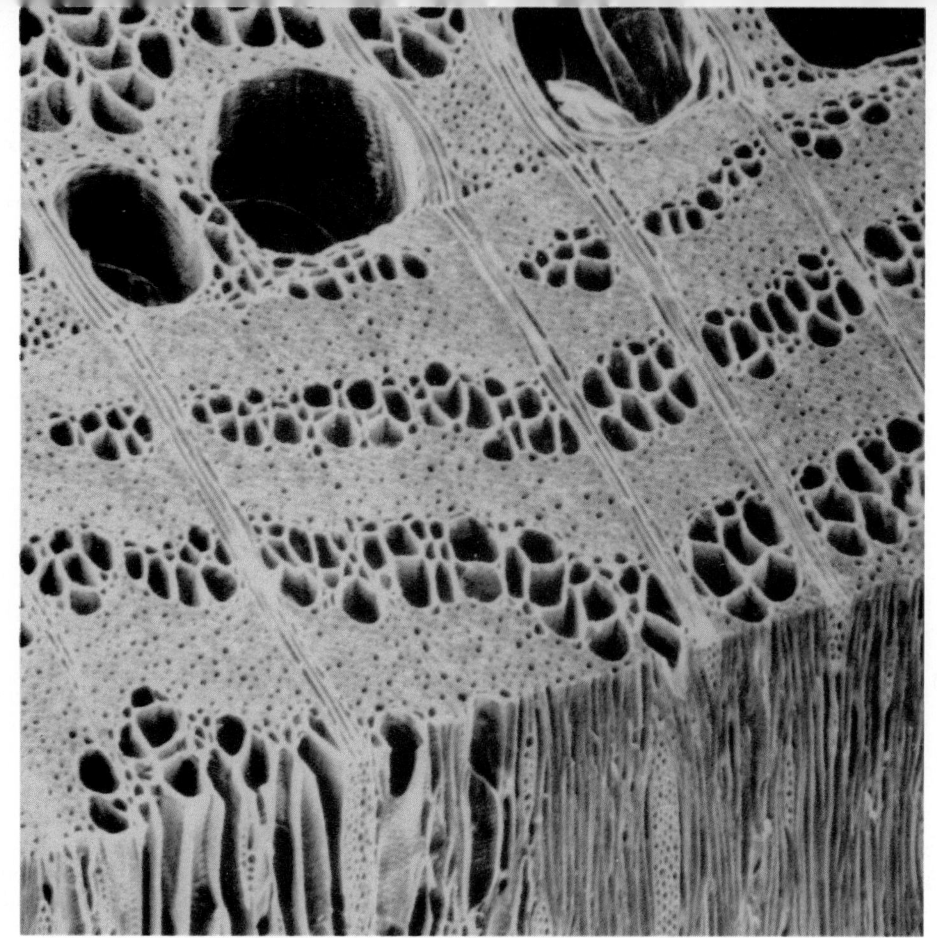

AS LIVING THINGS GO WE CAN COUNT OURSELVES AMONG THE GIANTS. THE average size of organisms on the planet is closer to that of a mouse. However, there are many living things far larger than we are, and the greatest of these are plants—towering, invincible plants, great trees such as redwoods that have resisted every threat except our need for lawn furniture. Why have so many of the plants become so large? One reason is that there are decided advantages to being tall, especially if one is a forest plant, since here there is intense competition for sunlight. The lofty giants are able to expose their leaves to unobstructed light. Also, the vast root system typically produced by large plants allows them good access to water and minerals and provides firm anchorage. Furthermore, tall trees can disperse their seeds and fruit farther from home than can their shorter contemporaries. There are, then, a number of advantages to being a forest giant.

Largeness, however, has its price. The scaly leaves of a giant redwood (Figure 28.1) constantly lose water to the atmosphere. That water must be replaced by a root system in the soil far below. On the other hand, those roots must be fed by carbohydrates manufactured in the distant leaves. The plant must somehow bring water and minerals from the soil up to its leafy canopy and food from the leaves down to the roots. Thus, large plants have had to pay for their large size by developing extensive vascular systems for the transport of water and nutrients.

The cost of large size does not end there. The high forest canopy must be supported, sometimes against strong winds and heavy rains or snow. Therefore, much of the plant tissue must be devoted to physical support. The forest giants, and other higher plants as well, solve the problems of both support and water transport with the same tissue—the woody xylem. Xylem is well adapted for its functions. Wood is marvelously strong supporting material, and through it water can be raised great distances above the earth. This ability to lift water has long mystified engineers since it is done with *little apparent direct expenditure of plant energy*. It is important to point out that there are no known active transport mechanisms that directly move individual water molecules.

The other part of the vascular system is the phloem, which specializes in carrying food molecules. Food molecules can be moved by known processes that involve energy expenditures, but there are many aspects to this movement that also remain a mystery. For example, food molecules do not haphazardly pass from the leaves downward, but are somehow routed according to the specific requirements of the different tissues below. Furthermore, the distribution can shift as conditions change. Food can even be moved upward from storage regions in the root; in spring, trees and bushes that have gone through the winter with bare, leafless branches routinely move food molecules up from their roots to the newly forming leaves and twigs. Indeed, there are a number of riddles regarding the movement of nutrients through the phloem. We will come back to these later, but first let's look at the role of xylem.

FIGURE 28.1
A KING-SIZE TRANSPORT PROBLEM
This towering California coast redwood *(Sequoia sempervirens)* is one of the world's tallest trees (111 m or 364 ft). If we think about its problem of water transport—from root to leaf—our awe and appreciation of the coast redwood can only be enhanced.

THE MOVEMENT OF WATER AND MINERALS

Water from the soil enters the plant through countless thin-walled root hairs, crossing the root cortex mainly along the porous cell walls between cells, but some passing through the cytoplasm as well. The water then passes through the endodermis and moves into the **stele** (the root vascular cylinder that contains both xylem and phloem; see Chapter 27). Upon entering the stele, water passes into the xylem, the vessels and tracheids of the root vascular system. It then continues through the xylem of the stem and on to the foliage. There, it passes out of the xylem, moving through numerous cells to eventually enter the air spaces in the leaf's spongy mesophyll. It is in the moist air spaces that evaporation occurs, and water, in its gaseous state, diffuses out of the plant through the leaf stomata (Figure 28.2), except for a relatively meager but vital amount that is used in photosynthesis. Water loss from the leaf through evaporation and diffusion is called **transpiration.**

A great deal of water is shifted from the soil to the atmosphere through transpiration. A single potato plant, for instance, can lose about 95 liters of water in one growing season, and in the same amount of time 206 liters of water can evaporate from the leaves of a single corn plant (see Table 28.1). In Chapter 47 we will find that the transpiration from a forest has immense ecological and meteorological significance, affecting both organisms and climate. (We are learning this too late in the Amazon rain basin, where vast changes in rainfall patterns are in store as the forests are ravaged.)

How is so much water moved through a plant? And more to the point, *why* is so much water moved? Answering the second question is relatively simple. Plants require water to carry on photosynthesis and to provide the turgor pressure that holds leaves and young stems erect. The first question is a little tougher. The problem of water moving up through, say, a tall tree, boils down to the two basic explanations. Either something is *pushing* the water up, or something is *pulling* it up. The first explanation that often comes to mind is that somehow the root provides the force needed to push water up into the leaf. After all, water must enter the root first.

Root Pressure: Is the Root a Kind of Pump?

If you've ever celebrated into the wee hours of the morning, you may have gotten your knees wet as you made your way home across your lawn. At the time, you may have thought the moisture was dew. Perhaps it was, but often the presence of water drops on the grass is the result of its having been pushed up by the roots through the plant's tiny vascular system. The blades of grass have special openings near their tips, where excess water can escape. Such loss of water in its *liquid* phase is known as **guttation,** not to be confused with transpiration, which involves the evaporation of water. Guttation occurs mainly at night, when evaporation is reduced and water excesses occur. The water has been forced up to the tips of the grass blades by **root pressure** from below.

Root pressure, a water-moving force developed in the root through osmosis, occurs when ions are actively secreted into the stele from nearby living cells. The presence of

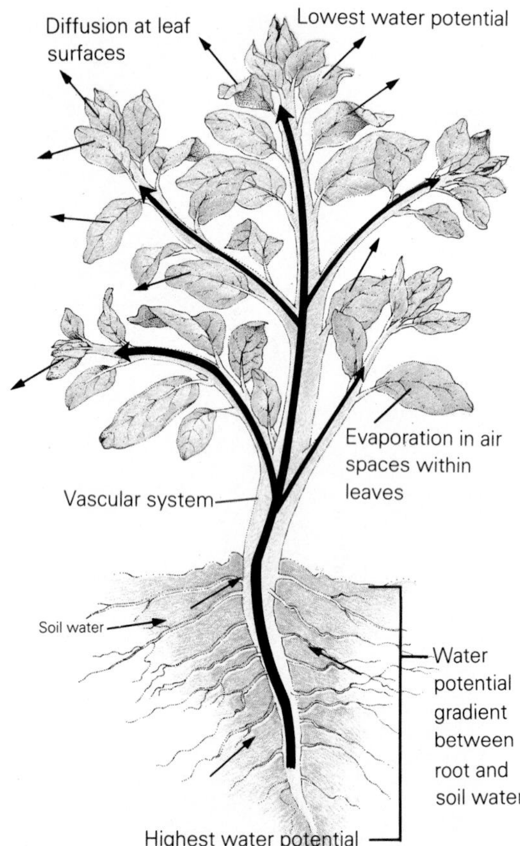

FIGURE 28.2
PLANT WATER POTENTIAL GRADIENT
The movement of water and minerals through a plant follows an established water potential gradient from soil water to air surrounding the leaf.

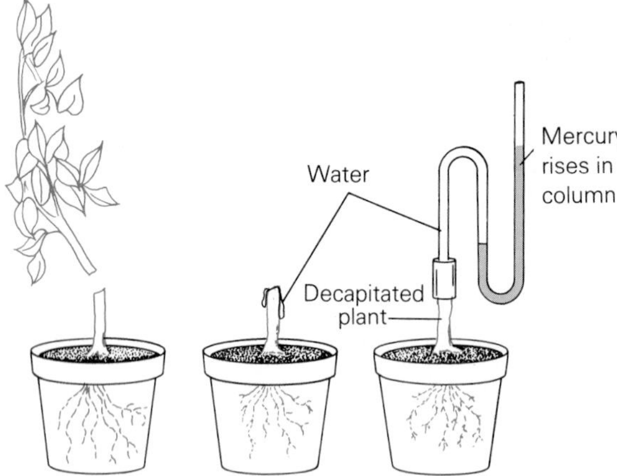

FIGURE 28.3
DETECTING ROOT PRESSURE
Root pressure can be measured by substituting a manometer for the stem and foliage of a potted plant. Mercury in the S-shaped tube rises as water is exuded from the vascular system of the stem.

ions in the xylem fluids sets up a water potential deficit (see Chapter 5), and water moves by osmosis into the stele. The endodermis, we know, has the ability to actively transport ions in one direction, resisting their diffusion back into the cortex. If the root is a pump, then the pressure the pump generates can be traced back to an energy-requiring mechanism, the active transport of ions. It creates an osmotic condition favoring the passive inward movement of water.

Root pressure can be demonstrated by removing the stem from a small plant and attaching a mercury manometer (a device for measuing fluid or air pressure) in its place (Figure 28.3). A rise in the manometer's column of mercury is used as a measure of pressure. With this apparatus it has been possible to demonstrate that roots can generate pressures of about 3 to 5 atmospheres (3 to 5 times greater than that of the atmosphere). So the immediate question is: Is this sufficient to push water to the top of a tall tree? The answer is no. The weight of a 100-meter column of water is at least two times too great to be supported by a root pressure of even 5 atmospheres. That is, over 10 atmospheres is the minimal requirement. The weight of such a column ought to push water right out of the vascular cylinder, back into the earth. Yet trees continue to stand there with water constantly moving upward through their enormously tall stems and out the lofty foliage.

One more observation should dispel the notion of root pressure as a total explanation of upward water transport. During periods of peak water loss, the pressure inside the trees is less than the pressure outside. If the roots were generating great pressure, cutting into the xylem would produce a *spurt of water*. Actually, however, cutting produces a sucking action as air is drawn *in*. In summary, then, root pressure apparently plays a role in water transport, but it cannot be considered to be a major force, at least not in tall trees.

Most plant physiologists now believe that root pressure, when it occurs, is only an indirect effect of the active transport of inorganic nutrients, such as nitrates, potassium, and phosphates, into the roots. So we find that roots can provide only a modest push, certainly not enough to explain water transport through tall plants. Therefore, we must consider the alternate explanation, the possibility of a pulling force. Is it possible to pull water? This is an unusual prospect, and one for which we will have to lay out some background.

We might first review the characteristics of water and its movement. In a truly literal sense, the vascular plant has evolved "around" such characteristics. The special peculiarities of root hairs, the root endodermis, the xylem vessels and tracheids, the leaf mesophyll, and the stomata are all adapted to take advantage of the special properties of water.

Water Potential and the Vascular Plant

Forces that bring about the movement of water in plants, regardless of their origin, do so by affecting water potential.[1] As you may recall from Chapter 5, such forces are varied and include concentration gradients, solute content, temperature, pressure, and matrix characteristics (the matrix being the surrounding materials). While the forces are varied, they all affect water potential by increasing or decreasing the free energy state of water. We'll see some of these forces at work in a moment, but for now let's just consider a working definition of water potential.

[1]Water potential values are determined by comparing the water under consideration with pure water at atmospheric pressure and at the observed temperature.

TABLE 28.1 TRANSPIRATION RATES IN LITERS PER DAY FOR SELECTED PLANTS

	LITERS PER DAY
Cactus	1/50
Tomato	1
Sunflower	5
Ragweed	6
Apple	19
Coconut palm	75
Date palm	450

Where a water potential gradient exists in a plant (that is, where a higher free energy state of water exists), there will be a tendency for water to move, whether through diffusion or bulk flow, down that gradient. (Unlike diffusion, where molecular movement is random, molecules involved in bulk flow move in one direction only.) As a result there will be a drop in the free energy of the original system.

In terms of living systems, a greater water potential in one region (a cell or intercellular region) relative to the water potential in another will result in the movement of water towards the region with lower water potential. Perhaps the most familiar way in which water potential in a cell may be altered is through changes in its solute concentration.

Solutes, recall, are ions or molecules in solution. Consider two cells with the same water potential. If the first cell has its solute concentration increased (perhaps by active transport), then that cell's water potential would be lower than the unchanged cell alongside. Water from the second cell would pass into the first. On the other hand, if the first cell had its solute concentration decreased, its water potential compared to that of the second cell would be greater, and the reverse movement of water would occur. Again, water moves from regions of higher water potential to regions of lower potential.

In relating water potential to the functioning plant, we can point to a well-substantiated observation: Water moves from the soil into the root and from the root through the stem to the leaves, and finally out of the leaves to the surrounding air. Thus, we can infer that the water potential is high in the soil, somewhat lower in the plant, and lowest in the surrounding atmosphere. When, for any reason, this gradient is disrupted, the plant is in trouble. This can happen, for example, when solutes in the soil water (such as nitrates and phosphates in fertilizers) are too concentrated around the plant. This would decrease the water potential in the soil water outside the root. Thus, water would tend to move out of the root. In response, the leaves would rapidly lose water, and with the subsequent loss of leaf cell turgor, they would collapse or wilt. Fortunately, the loss of turgor also occurs in guard cells, which would thereby change shape, closing the stomata and slowing transpiration. If the normal water potential gradient is not disrupted too long, the leaf changes would be reversed.

So far we have only described the movement of water along a water potential gradient. We have not seen how the gradient is established and how it is maintained in the plant. But most important, we have yet to see how such a gradient can be translated into the substantial *pull* needed to move water to the foliage of tall trees.

The Leaf, Transpiration, and the Pulling of Water

Plants have to be porous; they cannot be waterproof. If photosynthesis is to occur, carbon dioxide must be able to diffuse from the surrounding air into the leaves. Any such avenue permitting the inward diffusion of gases must likewise permit the escape of gases, and such gases include water vapor. So the loss of water by photosynthesizing plants is unavoidable. Let's review this common avenue of gas exchange.

The tissue organization of the typical dicot leaf, as you will recall (see Chapter 27), includes an upper epidermis, a dense palisade parenchyma just within, and a very loosely arranged spongy parenchyma below this. Both groups of parenchymal cells are photosynthetic. Finally, there is the lower epidermis. Numerous stomata occur in the lower epidermis and often the upper epidermis as well (Figure 28.4).

The open moist air spaces between parenchymal cells are avenues of diffusion, both for incoming carbon dioxide and for outgoing water vapor. The rate of water loss depends primarily on two factors: temperature at the leaf surface and air spaces, and relative humidity in the air outside the leaf. Each of these factors profoundly affects the rate of transpiration from the leaf. As many a dismayed gardener has found, the rate of water loss in plants on hot, dry days far exceeds such losses on other days.

Evaporation from the leaf surface immediately affects water potential in tissues just within the leaf, setting in motion the movement of water throughout the plant. Let's look at the details.

We can begin by noting that water evaporating from the leaf spaces is replaced by water from the cells surrounding the spaces (see Figure 28.4). This leaves these cells with a reduced water potential. The result is that water will move into these cells from others even closer to the source. The source, of course, is the water-filled xylem of the leaf veins. Thus, the continued evaporation of water establishes a steep water potential gradient between the vascular system and the spongy region of the leaf, and water tends to move rapidly down its gradient. In addition to its movement along porous cell walls, water movement through the leaf cells is facilitated by the presence of numerous **plasmodesmata**—minute cytoplasmic connections passing through pores in adjacent cell walls and connecting the cytoplasm of one cell to that of another (see Chapter 5).

Transpiration at the leaf surface translates into a powerful water potential gradient within the leaf, which produces a major pulling force. Measurements reveal that this force, in water-depleted leaf cells, create pressures up to 12 atmospheres—enough to lift a xylem-sized column of water 130 m high! This force is therefore quite sufficient to raise water from the soil to the tops of the tallest trees.

We have seen that the movement of water from cell to cell in the leaf is always along a gradient from higher water potential to lower water potential, and that it is a *passive* process, not a direct result of biochemical work done by the plant. The process is powered by the water potential gradient set up by transpiration, the evaporation of water from the air spaces of the leaf. The **free energy of evaporation**, as it is called, is an indirect form of solar energy, since the sun affects the wind, humidity, and temperature, all of which influence the rate of transpiration. These forces operate with other important physical influences in the xylem to produce the impressive movement of water.

FIGURE 28.4
WATER POTENTIAL IN THE LEAF
Evaporation from the leaf sets up a water potential gradient between the outside air and the leaf's air spaces. The gradient is transmitted into the photosynthetic cells and on to the water-filled xylem in the leaf vein.

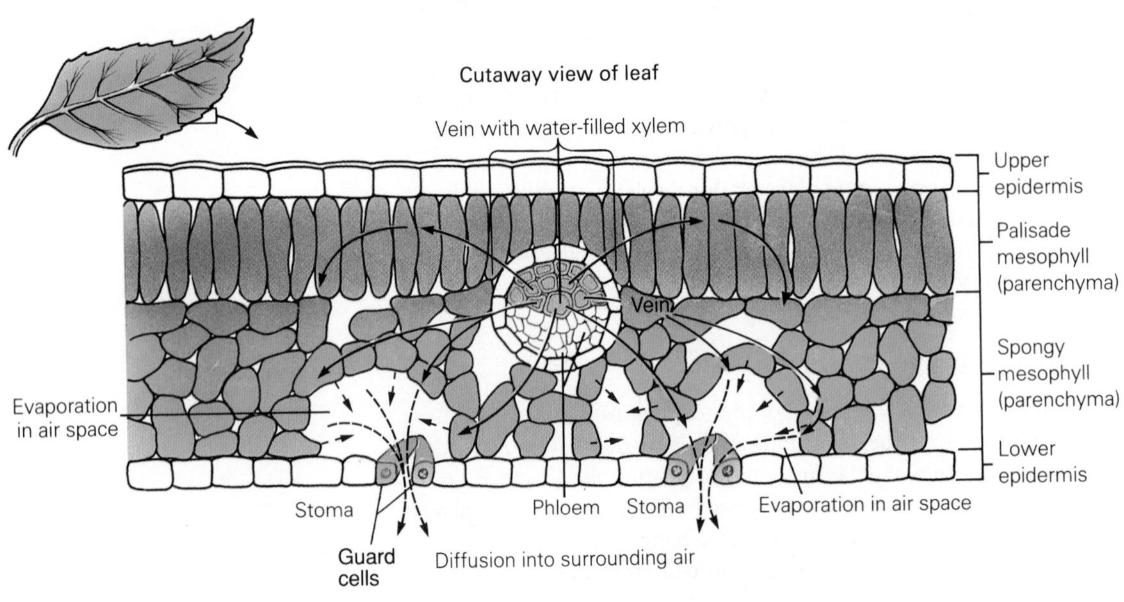

Water Movement in the Xylem Through the TACT Mechanism

Four important forces influence the movement of water in the xylem elements: **transpiration, adhesion, cohesion,** and **tension,** which we will call the **TACT forces.** We have already discussed transpiration as a major force in creating the water potential gradient, so let's briefly look at the roles of the others. The second force, adhesion, refers to the attraction of water molecules to materials such as glass or the cellulose in plant cell walls. Cohesion is the attraction of certain molecules to each other. (We considered both in detail in our discussion of water in Chapter 2. You may recall that it is the polarity of water that attracts the molecules to each other.) Tension, the fourth force, refers to the stress produced by exerting a pull on a narrow column of water—namely, the pull on water in the vessels and tracheids of xylem during transpiration.

The adhesion of water to cellulose cell walls may be a major water-moving force throughout the plant since it produces water's "creeping" movement from cell to cell and through the cellulose-lined vessels and tracheids of the xylem. Evidence suggests that it is adhesion that causes water to leave the leaf mesophyll cells and enter the surrounding air spaces where evaporation occurs. Accordingly, some authorities claim that it is these adhesive forces that set in motion the "pull" of transpiration-pull. To support their claim, they point to the immensity of the combined surface area of cells surrounding the air spaces of a leaf. Is the force of adhesion enough to lift long columns of water to the leaves of the tallest trees? It may be, but there is much more to the process.

Cohesion—the attractive force between the molecules of a substance—is particularly evident in water. The positive (hydrogen) ends of one molecule are attracted to the negative (oxygen) ends of the next, forming countless, shifting hydrogen bonds. It is cohesion that gives water its physical viscosity ("liquid-stickiness"). When water is present in a cylinder of narrow diameter such as that of the vessel or tracheid, its cohesive forces are greatly magnified. Such a column has significant tensile strength—the ability to be pulled without breaking. For example, steel wire has more tensile strength than cotton string, which has more tensile strength than cooked spaghetti. Water would seem to have even less tensile strength than cooked spaghetti, but the tensile strength of a column of a fluid depends on all sorts of complex factors, and prominent among these is its diameter. By some calculations, the tensile strength of a column of water in a tiny xylem vessel approaches that of a steel wire of the same diameter!

Tension, as we mentioned, is the stress placed on an object by a pulling force. As you might expect, when a plant is heavily transpiring, the tension on the water columns within the many xylem vessels and tracheids can be enormous. In fact, the pull has a measurable squeezing effect on the trunks of trees; their trunk diameters actually decrease during intense transpiration! (Recall our mention of the faint sucking observed when the xylem is punctured.) Essay 28.1 explains how such delicate measurements are made. The important point is that as tension increases, the cohesive strength of water in the xylem elements, and their adhesion to the xylem walls, prevent breakage. The pull on the column is transferred downward, creating a decrease in water potential in the root stele below. In response, water moves into the stele, permitting the upward movement to continue.

Thus, water movement in the plant is caused by the combined effect of four cooperating "TACT" forces. In review, we've seen that transpiration produces a water potential deficit that through a chain reaction produces a water potential gradient, which along with adhesive forces creates a powerful pull on minute water columns in the leaf xylem. Because of incredibly strong cohesive forces characteristic of water in minute columns, the columns remain intact, and water is literally lifted through the entire vascular system. This all works very well, but as you might expect, it can only continue if water intake by the root can keep up with the movement of water through the stem and out of the leaf.

Water Transport in the Root

We have already seen that osmotic conditions in the root can provide a certain amount of push in the transport of water. But let's emphasize the point that during intense transpiration, root pressure is totally overwhelmed by transpiration pull and is no longer

ESSAY 28.1

TESTING THE TACT THEORY

Parts of the transpiration, adhesion, cohesion, and tension (TACT) theory can be readily tested. To demonstrate cohesion and tension, simple techniques similar to those used to demonstrate root pressure can be applied. (a) The stem of a plant is cut (under water so as not to break the water columns) and connected to a water-filled glass tube. (If dye is added to the water, it will soon appear in the leaves and flowers; florists sometimes use this trick to dye flowers odd and unnatural colors.) If the lower end of the water-filled glass tube is put into a dish of mercury, the column of mercury rises as the water evaporates above, indicating a strong pull. However, the analogy fails in some ways, since the glass tubing is far greater in diameter than are vessels and tracheids.

But how can it be demonstrated that it is evaporation and not some other force that is involved in transpiration? A clever mechanical model seems to support the evaporation principle. (b) The setup is the same as before, but this time the top of the glass tube is attached to a porous clay *potometer,* instead of the crown of a plant. There is no air in the potometer, only water. The wetted microscopic pores of the clay potometer permit water to literally creep out to the surface of the clay cylinder through the forces of adhesion and cohesion (much like what is believed to go on in the spongy tissue and air spaces of the leaf). Moisture evaporates from the damp outer surface, and the column of water rises, creating a partial vacuum in the tubing that is filled by the rising column of mercury. The rise is much faster if a fan is used to blow air around the potometer, just as transpiration would be more rapid on a dry, windy day.

A very clever experiment lends strong support to the TACT theory. D. T. MacDougal, a plant physiologist, wondered about the enormous tension that must be placed on water columns in the immense xylem of a tree if the transpiration pull idea were correct. He reasoned that such a tension might even affect the stem diameter, perhaps causing an actual decrease during peak activity. (c) MacDougal tested this idea by devising an instrument that could measure minute changes in the stem. The instrument, a *dendrometer* (later called a *dendrograph* when a recording drum was added), did in fact detect the expected changes. Not only did the stem decrease in diameter, but it did so in regular day-night cycles. It is common knowledge that plants transpire much less at night—in the absence of sunlight—when evaporation rates are low, and MacDougal's measurements coincided with this very nicely. How might increased temperature accompanied by decreased humidity have affected MacDougal's dendrograph tracings?

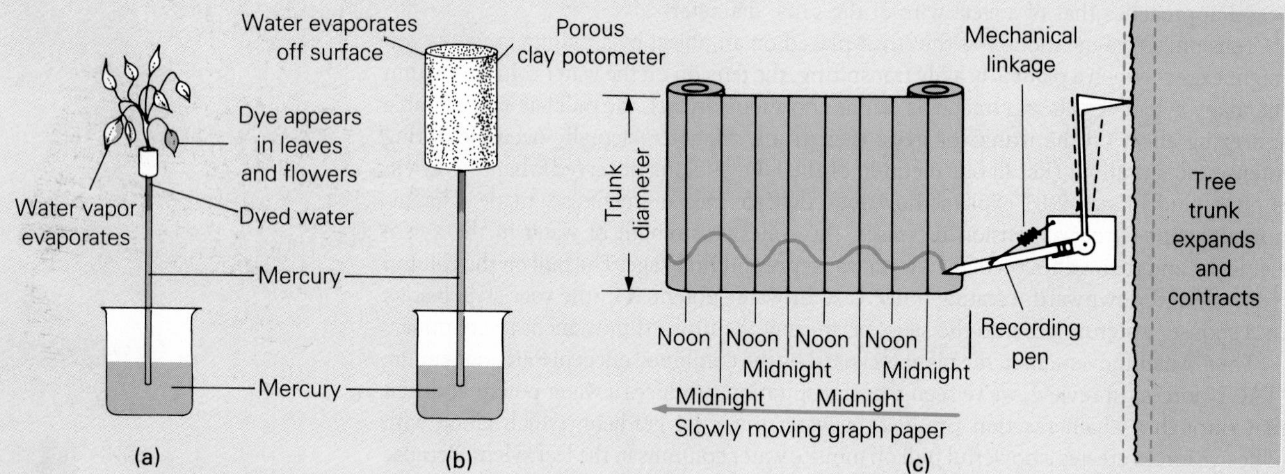

Figure 23.1

(a) The pulling force of transpiration can be demonstrated. As the foliage transpires, dye appears in the leaves and water is drawn into the tube with a force that lifts the mercury up the tube. (b) Evaporation from a clay potometer is analagous to evaporation from the leaf surface. (c) The tracing shows changes in a tree's diameter over several days. The peaks and valleys occurring in 12 hour intervals coincide with periods of maximum and minimum transpiration. Plants transpire most at midday and least at night.

considered by biologists to be a major force in the movement of large quantities of water. Root pressure as a force in water transport is probably most important when transpiration is slow or when water potential in the soil water is low. At those times, a proper osmotic gradient in the root would be essential to the continued movement of water into the root and stele. It may also be important to ground-hugging plants.

Within the root, the ongoing pull occuring in the xylem creates a water potential gradient in the other cells of the stele that is transmitted across the cortex to the extensive root hairs of the epidermal cells (Figure 28.5). So, as long as soil water is plentiful, water will passively enter the epidermal cells and pass across the root to the xylem. In its transit through the cortex, water passes mainly along the highly porous parenchyma cell walls, along what is called the apoplastic route. Some water passes along the symplastic route, through the cytoplasm. In the cortical cells this movement is facilitated by the presence of many plasmodesmata, cytoplasmic connections between cells. Upon reaching the endodermis, however, all water bound for the stele must follow the symplastic route—through the endodermal cytoplasm. Let's see why.

The endodermal cell walls contain a waxy region—the **casparian strips**—arranged in such a way as to direct the water flow across the plasma membrane and through the cell cytoplasm rather than along its cell walls (Figure 28.6). The cytoplasm, of course, is a living and highly regulated substance, so the passage of water may not be a simple process there. Because of the casparian strips, it is possible that the endodermis may exercise both a physical and osmotic directional control over the movement of water, although the importance of this during intense transpiration isn't clear. It is possible that one function of the endodermis is to help maintain an inward water potential gradient by actively transporting mineral solutes into the stele. This is important in preventing water loss when transpiration slows or when water potential in the soil drops. In a real sense, the casparian strips waterproof the stele. Of course, the endodermis also transports ions into the stele, where they can be carried to the leaves for metabolic use.

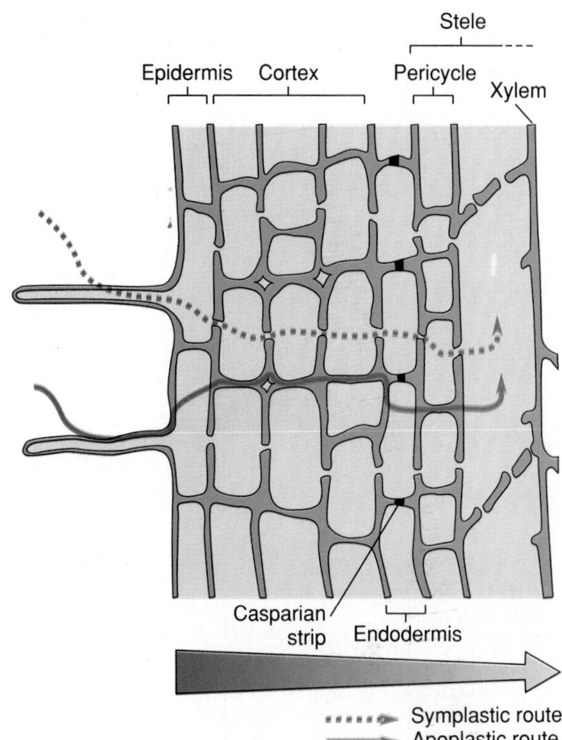

FIGURE 28.5
THE MOVEMENT OF WATER IN THE ROOT
While some water passes through the cytoplasmic (symplastic) route in root cells, most moves along the porous cell walls (apoplastic route). To enter the stele, water must first pass through the endodermal cytoplasm. From there it resumes its transit along cell walls to the xylem.

Before leaving the subject, we should note that the high water potential responsible for water's inward movement around the root is frequently lost. This usually happens towards the end of an active day—when the root may literally "run dry." But roots are not passive organs; they are capable of rapid growth, particularly at night. This growth increases when new water resources are encountered and slows where dry soil is encountered. So, root growth itself is also an important factor in the transport of water.

Guard Cells and Water Transport

When plants made the evolutionary transition to land, one problem of immediate importance was the danger of drying. The evolution of a watertight cuticle was a key adaptive solution to this problem, but in view of the terrestrial plant's dependence on atmospheric carbon dioxide, it was just half a solution. The second half involved a means of permitting carbon dioxide to enter the leaf, but in a controlled manner, since, as we've seen, the inward diffusion of carbon dioxide is unavoidably accompanied by the outward diffusion of water. The adaptive solution in this case was the versatile **stoma** (plural, **stomata**). The stoma is a pore, formed by the pairing of **guard cells** that are distributed in the epidermis (Figure 28.7). The whole assemblage makes up the **stomatal apparatus.** We will usually refer to the apparatus simply as the stoma.

Guard cells can increase and decrease the size of the stoma, freely admitting an exchange of gases at one time but restricting such an exchange at others. In this way the stomatal apparatus helps curtail the excessive loss of water from the plant interior.

FIGURE 28.6
THE CASPARIAN STRIPS
(a) The partial suberization of endoderm provides a barrier that blocks the usual apoplastic route of water. (b) Thus, water is directed into the endodermal cytoplasm where its movement is more controlled.

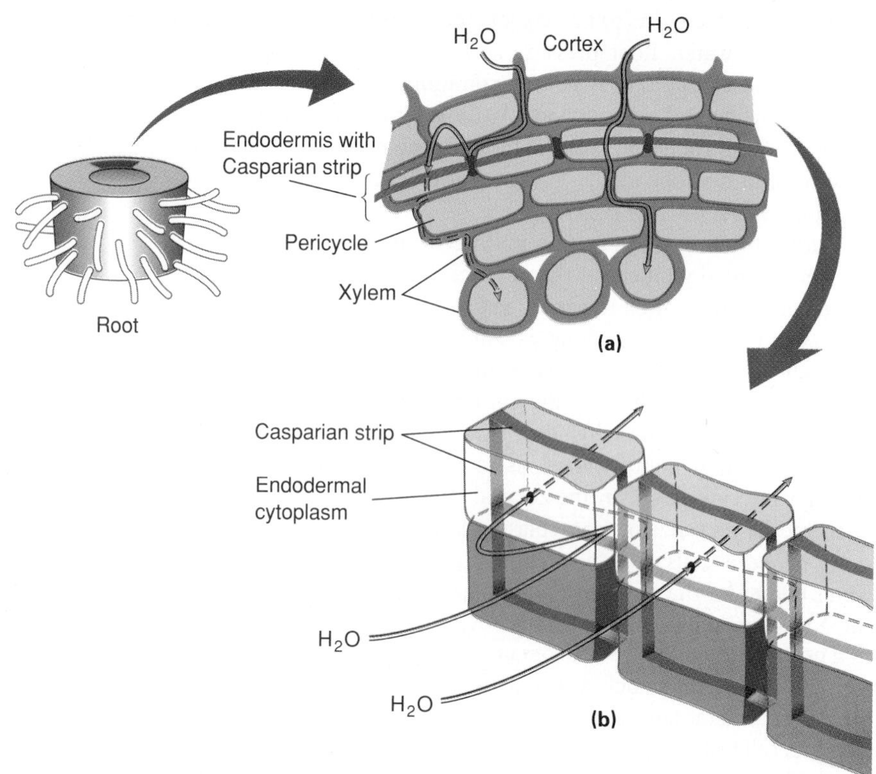

FIGURE 28.7
STOMATA
Stomata, pores formed by paired guard cells, occur in the leaf epidermis, although they also appear on green, photosynthetic stems. They are often recessed, as seen in the scanning EM view.

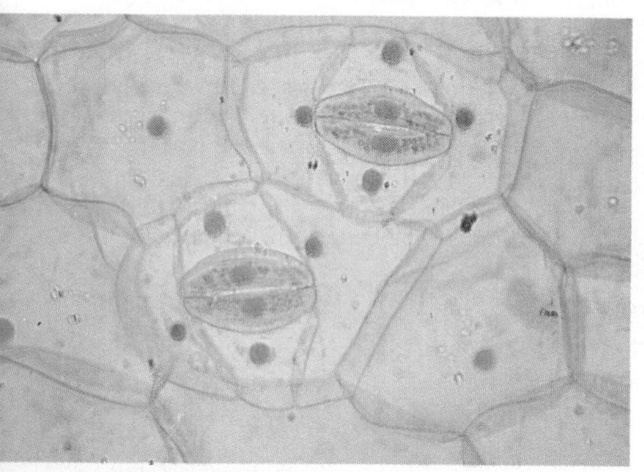

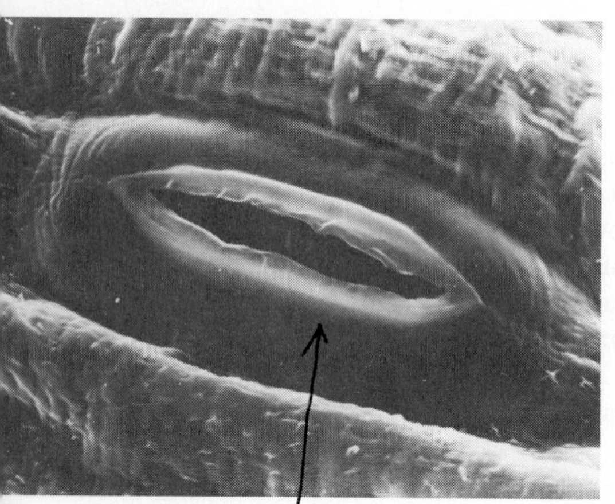

Stomata are located in the leaf epidermis, usually concentrated in the lower epidermis, but not restricted to that location. They also occur on the photosynthetic, green stems of herbaceous plants such as cacti and on some green fruit. Whereas the average density is about $10{,}000/cm^2$, some leaves contain as many as 100,000 in the same area.

The stomatal opening is regulated by turgor in the guard cells (Figure 28.8). Put simply, a pair of guard cells pressed together reduce the stoma size, and the same pair, when bent apart, increase it. At first glance, it might appear that an increase in turgor should press the guard cells together, closing the stomatal opening. But this doesn't happen, for a good reason. The inner or facing walls of the paired guard cells are fused at both ends. Therefore, in spite of turgor increases, the inner facing walls must remain the same overall length. Further, the orientation of cellulose microfibrils in the walls of the guard cells is radial. As turgor increases, this orientation permits the outer margin of the turgid guard cells to lengthen, but keeps them from bulging out or getting thicker. What happens if such cells get longer along their outer margins, but remain the same length on their inner margins? They bend. It is this bending that increases the stomatal opening. When turgor is lost, the cells straighten, resuming their previous shape, and the stomatal opening decreases. So when guard cells are flaccid, short, and straight, the stomata are closed; but when they are turgid, long, and bent, the stomata are open.

In general, the turgor increases during daylight and decreases at night. In this manner, gases are allowed to pass through the stomata when the plant is engaged in photosynthetic activity, and water loss is restricted when it is not.

Turgor Changing in the Stomatal Apparatus How does turgor change in guard cells? As you might expect, water enters and leaves guard cells in response to changes in solute concentrations, which affect water potential. When solutes increase in guard cells, water potential lessens, and water enters from surrounding cells. When solutes decrease, water potential increases, and water leaves the guard cells. What then causes the changes in solute concentration? The answer is not clear, but several independent factors may be involved. For a time

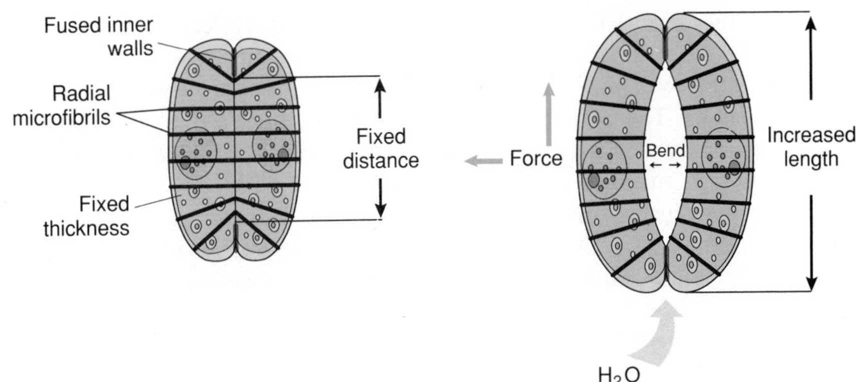

FIGURE 28.8
THE STOMATAL MECHANISM
The radial orientation of microfibrils and the strategic fusion of facing inner walls in the paired guard cells translate increasing turgor into cell bending, which enlarges the stoma.

the presence of chloroplasts in the guard cells, which are usually lacking in other epidermal cells, seemed to provide a major clue. The reasoning was that the chloroplasts carry on photosynthesis early in the day, and the resulting carbohydrate molecules increase the solute concentration in the guard cell, thus reducing the water potential there. Water passing down the newly established gradient enters the guard cell, increasing its turgor and opening the stoma. Then, when the sun goes down, the processes slow down, sugar is metabolized, and the gradient is reversed. Water leaves the guard cell, turgor decreases, and the stomata close. It's a rather neat explanation and it fits our expectations for day/night behavior of guard cells. But it's wrong.

As usual, some restless soul came along and carried out one experiment too many. The sugar-osmosis hypothesis didn't hold water, as it were. It turns out that the changes in turgor occur much too rapidly to be explained by the synthesis of sugar solutes in photosynthesis.

In the late 1960s, physiologists in the United States found that plants in direct light show an increase in the active transport of potassium ions into the guard cells. (Actually, they made use of an observation first reported in Japan in 1943, but we weren't holding very many conventions with Japanese scientists that year.) It turns out that only the blue part of the light spectrum is effective in potassium transport because the process is triggered by a blue-sensitive pigment. The entrance of potassium increases the solute concentration of the cell, speeding the uptake of water and resulting in stomatal opening (Figure 28.9).

Unfortunately, the **potassium ion transport mechanism**, which is now well-substantiated, has not answered all the questions. The link between blue light and the guard cell bending may be direct, with the pigment itself powering potassium transport; or it may be indirect, with the light reactions of photosynthesis providing a coded signal for potassium transport. The latter explanation is preferred at the moment, although the relationship between light reaction activity and potassium pumping is not known.

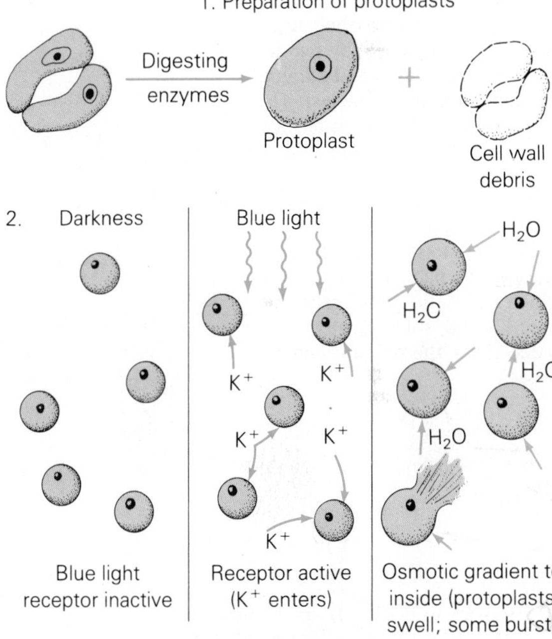

FIGURE 28.9
POTASSIUM TRANSPORT HYPOTHESIS
Studies of potassium transport in guard cells indicate that a light-mediated process brings about changes in turgor. When protoplasts (cells with walls removed) are subjected to light, particularly certain blue wavelengths, they rapidly take in potassium ions (K^+). The uptake of K^+ is followed immediately by water, with cell size increasing by as much as 50%.

Furthermore, a severe water loss by the plant apparently alters the potassium transport mechanism. If water loss is severe, plants undergo what is called "water stress," and in some species a hormone known as abscisic acid (or ABA) is released by nearby cells. In the presence of ABA, potassium transport is reversed, and the ion leaves the guard cells, which rapidly lose turgor, and the stomata close. Plant physiologists have also studied simpler variables affecting the activity of guard cells, such as carbon dioxide concentrations, temperature, and light.

Carbon dioxide has a positive, reinforcing effect on the light-mediated transport of potassium ions, but in an indirect way. Experiments with corn reveal that the stomata open widely when internal carbon dioxide levels are experimentally *decreased,* even when this is done in the dark. This indicates that as CO_2 in the leaf cells diminishes, K^+ transport into the guard cell increases. The increase of this solute, as we've seen, results in the further entrance of water into the guard cell, and subsequently, an increase in the stomatal opening. The CO_2 factor makes good sense, since a diminishing level of the gas is associated with intense photosynthetic activity, and it is at this time that having wide open stomata is adaptive, since it makes more CO_2 available.

Considering the variables involved, physiologists have come to the conclusion that the stomatal apparatus has multiple controls. Chief among the factors influencing turgor are light, K^+, CO_2, H_2O, and ABA. Figure 28.10 presents a model system incorporating these elements.

Finally, there are the **CAM plants**, a group that seemingly ignores all of our hard-won data.

CAM Photosynthesis, or How to Hold Your Breath All Day In some plants the opening and closing cycle of the stomata is reversed. The stomata are closed all day and open all night. Included are many cacti and other succulents (fleshy-bodied plants common to the desert) and members of the family Crassulaceae—the "stone-crops." This strange adaptation helps prevent intolerable water loss in the excessively arid desert environment in which the plants live. It also means that access to atmospheric carbon dioxide is cut off during photosynthesis, when it is needed. Yet the plants carry out the light-independent or sugar-synthesizing reactions of photosynthesis in a manner similar to other plants. How do they handle their carbon dioxide requirements?

The plants use what is called **crassulacean acid metabolism,** or **CAM.** Carbon dioxide is admitted at night and temporarily incorporated into certain organic acids. Then, in the daylight, the chemical reactions are reversed, and the carbon dioxide is released within the cell. From there it is used in the production of glucose according to the usual biochemical pathway—the Calvin cycle of the light-independent reactions (see Chapter 7). The energy and hydrogen sources, ATP and NADPH, are provided, as usual, from the light reactions. All this chemical activity involves the use of precious ATP reserves, but that's a price they pay for living in a less competitive environment.

FOOD TRANSPORT IN PLANTS

In 1671, just six years after he described the cells in cork, Robert Hooke and a colleague, Robert Brotherton, "girdled" a tree. That is, they removed a ring of bark around the trunk of the tree, including the soft, moist layer beneath the dead outer cork. The tree

FIGURE 28.10
FACTORS INFLUENCING GUARD CELLS
(a) Potassium ion transport into the guard cell is increased by two factors: the presence of light and a low carbon dioxide level in the leaf. The ion's presence leads to water uptake by the guard cell. As long as water loss is not exceptional, the guard cells will remain turgid, keeping the stoma open. **(b)** When water loss is excessive, water stress ensues. Leaf cells produce the plant hormone ABA, which causes the guard cells to release K^+ and, subsequently, to lose water. Thus, the stoma become closed.

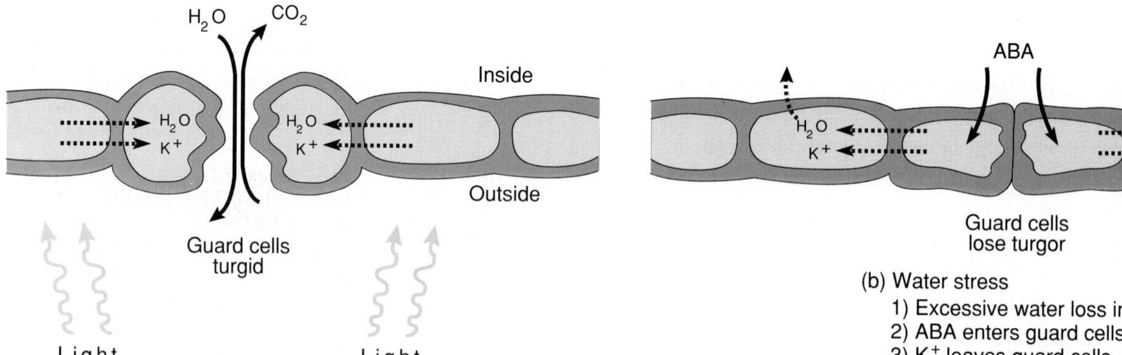

(a) Normal functioning
1) Light reaches light receptor
2) K^+ transported into guard cells (reinforced by low CO_2 levels in active leaf)
3) Water diffuses into guard cells
4) Stoma opens

(b) Water stress
1) Excessive water loss in plant
2) ABA enters guard cells
3) K^+ leaves guard cells, followed by water
4) Guard cells lose turgor, stoma closes

eventually died, but they observed that it continued to grow for some time, putting out new leaves and branches. They also noted something odd. The trunk of the tree increased in diameter above the ring *but not below it.* The roots did not grow, and in fact began to shrivel. Hooke and Brotherton concluded that the material a plant receives from the roots is transported downward in the bark, while the material it receives from the roots is transported upward in the wood. Hooke and Brotherton's conclusions have remained unaltered for well over 300 years.

Actually, sugar and other nutrients move through the phloem in either direction, down toward the roots when the leaves are photosynthesizing and up toward the branches at other times, especially in spring when new leaf buds are growing. It is now clear that some form of active transport is involved in the movement of sugar and other nutrients through the phloem. We will look at the role of active transport after making a few general observations about food transport in the phloem.

Mechanisms of Phloem Transport

We learned earlier that in the movement of water from roots to foliage through the xylem, plants make use of the energy of evaporation, water potential gradients, and those molecular peculiarities called adhesion and cohesion. The TACT forces, we found, provide what is essentially a pulling force that creates great tension within the tracheids and vessels. Only the thickness and strength of the cellulose xylem walls prevent these conducting elements from collapsing under this great force. This situation is quite different in the phloem, though. The content of the sieve tubes, the **phloem sap**, is instead, *pushed* along. The pushing force, as we will see, is provided by *hydrostatic pressure,* which can be considerable in the sieve tubes.

A Little Help from the Aphids Aphids take advantage of the high hydrostatic pressure of phloem sap by using their long, hollow mouthparts to drill tiny holes in individual sieve tubes, essentially creating little artesian wells. The tiny insects then relax and let the nutrient-laden sap flow into their bodies. In fact, the sap often flows right through their bodies and accumulates in drops at the other end, where it is euphemistically called "honeydew"—a delicacy among ants. (If you have parked your car under an aphid-infested tree, you may already know about honeydew.) Plant physiologists sometimes take advantage of the aphids' drilling technique—a feat they have technical difficult achieving—in studying phloem sap. The procedure, developed in the 1950s by two insect physiologists, is to let the aphids drill into the phloem and then anesthetize them with a gentle stream of carbon dioxide. Next a sharp razor blade is used to cut the imbedded mouth parts away from the head, and the resulting liquid flow, a veritable "artesian well," becomes a source of pure sap for analysis (Figure 28.11). The amount of sap that can be captured in this manner is considerable, up to a milliliter of fluid per hour, the flow lasting up to about four days. Actually, the two researchers were really interested in studying aphid nutrition, but they immediately became aware of the applications possible to their plant physiology colleagues. Such methods have been continually in use since that time.

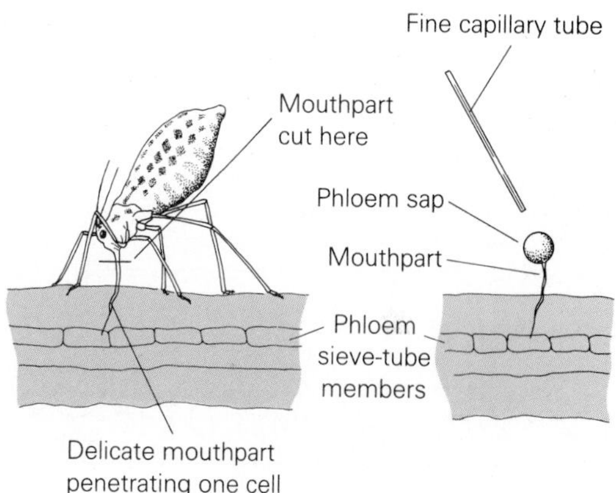

FIGURE 28.11
APHIDS AND PHLOEM ANALYSIS
Aphids use their fine, tubelike mouthparts to pierce the phloem and intercept the sugar-laden phloem stream, which is under pressure. Taking advantage of this veritable "artesian well," plant physiologists simply anesthetize the busy aphid and cut it away from its mouth parts. Fine capillary tubes are then touched to the oozing sap, and a sample is available for analysis.

It turns out that sap is actually a rather dense fluid, especially when the plant is photosynthetically active. Nonetheless, it moves rather rapidly through the phloem stream, up to one meter per hour. Sap is mainly water and dissolved solutes, with sucrose making up about 90% of the latter. Other sugars, along with nutrients, hormones, and amino acids, are also transported in the phloem. The movement of such substances in the plant is referred to as **translocation.**

Flow from Source to Sink As we mentioned, phloem sap flows in either direction along the vascular system, according to the plant's needs. It first flows into developing leaf primordia, and then, when the leaf matures and begins photosynthesizing, it moves out of the leaf and on to the roots, the maturing fruit and growing seeds, the apical meristem, or very often into storage regions of the stem and root. We call these places where food is destined for use or storage **sinks** (a term borrowed from engineering, as in "heat sink"). The place where moving food originates, whether the site of photosynthesis or storage, is called the **source.** The flow of sap, therefore, proceeds from *source to sink.* Active transport is essential in both regions.

The active transport of nutrients through the phloem is often against the concentration gradient. Sugars in the leaf cells, for instance, are often at a much lower concentration than they are in the nearby phloem elements into which they are being loaded. In turn, such sugars may be unloaded from the phloem into food storage cells with even higher concentrations of sugar. Moving materials against their concentration gradient takes energy, provided, as usual, by ATP. Ordinarily sucrose, the principal sugar transported in the phloem stream, is converted to starch for storage and hydrolyzed into sucrose once again for transport in the phloem. Because of their size, the large starch molecules are more apt to stay put than are the smaller more mobile sucrose molecules.

The Pressure Flow Hypothesis We have seen that active transport is used in loading nutrients into the sieve tubes at the source and unloading them at the sink, but what accounts for the flow of such materials in between? How do they pass through the sieve tubes? Many hypotheses have been proposed to account for phloem sap flow. Interestingly, the most favored one—the **pressure flow hypothesis**—is far from the newest. Proposed back in 1927 by Ernst Munch, a German physiologist, it is based on differences in water potential between the phloem and xylem—differences that are created by the active transport of nutrients.

The idea is straightforward: the active transport of sugars into the phloem stream greatly decreases the water potential there in comparison to the high water potential in the nearby water-filled xylem elements. Water responds by moving out of the xylem, down the gradient, into the phloem. The inward movement of water raises the hydrostatic pressure within the phloem sieve tubes. This pressure forces along the phloem sap with its load of nutrients and water. Then, at the target tissues—the nutrient sink—sugars are moved out of the stream by active transport. This loss of solutes increases the water potential within the phloem, and the water escapes from the phloem stream, usually finding its way back into the adjacent xylem (Figure 28.12).

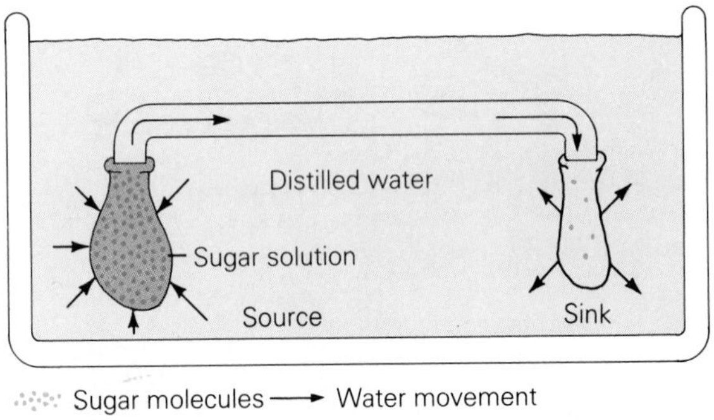

Thus, water apparently circulates in plants, moving from xylem to phloem at the source, and moving from phloem back to xylem at the sink. The sieve tube elements themselves may play a passive role in transport, with the companion cells and phloem parenchyma providing the energy for active transport and somehow determining whether nutrients are to be loaded or unloaded. A cellular scheme of nutrient transport is seen in Figure 28.13.

FIGURE 28.12
PRESSURE FLOW MODEL
In this model, water follows its potential gradient into the dense sugar solution in the sack at the left (the "source"). The resulting hydrostatic pressure creates a flow through the tubing to the water-filled sack on the right (the sink). This model simply illustrates the flow; the net movement of water ceases when the sugar reaches equilibrium. (Note that the sacks are permeable to water but not sugar.)

A half-century-old idea that explains such a vital aspect of plant biology as food transport reminds us that our own generation has no monopoly on creativity. But, venerable though it is, the pressure flow hypothesis is not without its problems and critics. Scientists would still like to know, for instance, how the phloem sap is isolated from the phloem cytoplasm. How does it move through the sieve pores without disturbing the cytoplasmic extensions there? And what determines whether nutrients are to be loaded or unloaded? In other words, what controls translocation? Obviously, science still needs bright and inquiring minds.

GAS TRANSPORT IN PLANTS

Plants continually exchange gases with the environment while carrying out both photosynthesis and respiration. But in general, plants don't have specific systems to deal with gases; they rely primarily on stomatal exchanges. Few plants have developed ways to

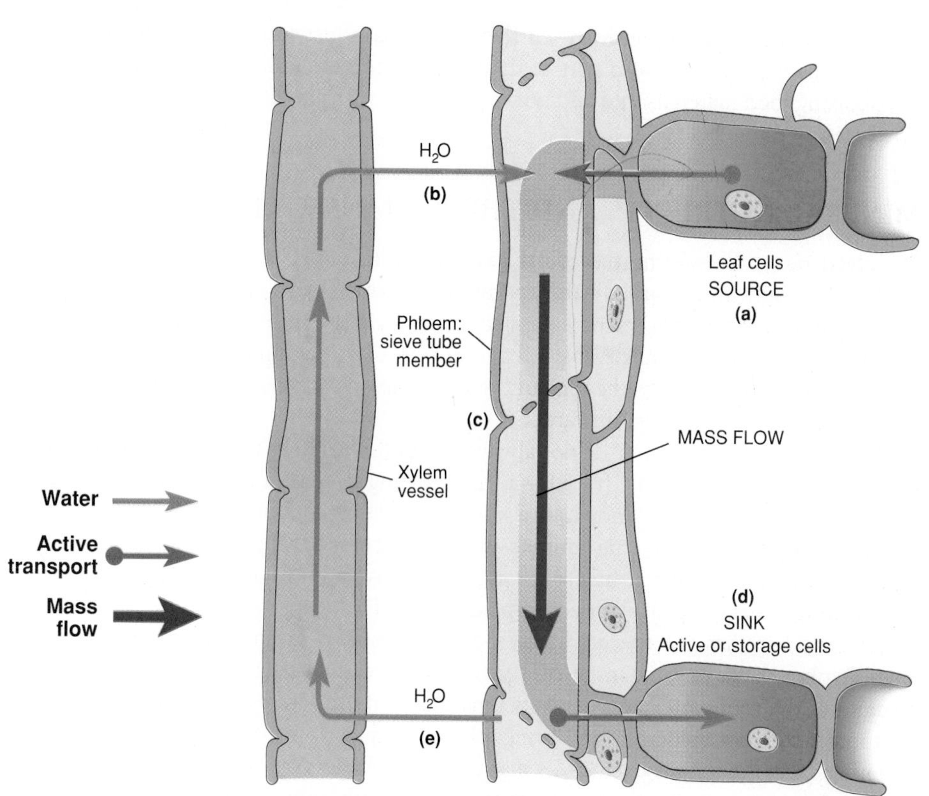

FIGURE 28.13
THE PRESSURE FLOW HYPOTHESIS
The movement of sugars in the phloem begins at the source, where **(a)** sugars are loaded (actively transported) into the sieve tube. Loading of the phloem sets up a water potential gradient that facilitates the movement of water into the dense phloem sap from the neighboring xylem **(b)**. As hydrostatic pressure in the phloem sieve tube increases, pressure flow begins **(c)**, and the sap moves through the phloem. Meanwhile, at the sink **(d)**, incoming sugars are actively transported out of the phloem. The loss of solute produces a high water potential in the phloem, and water passes out **(e)**, returning eventually to the xylem.

move gases around in their own tissues. In monocots, the vascular bundles commonly contain an air channel along with the xylem and phloem and supporting fibers. So at least some plants have made provision for the passage of air. Otherwise, gas is usually transported by diffusion where thin-walled cells, such as those of the root tip epidermis, those just within the lenticels (pores along the woody stem), and the spongy parenchyma bordering the air spaces within the leaves, contact the air or water. If oxygen and carbon dioxide are to be transported in the xylem, however, they must be dissolved, since air bubbles destroy cohesion and tension in these columns. Should this happen, water will move into adjacent columns through pits in the walls of the tracheid or vessel (see Figure 27.4).

Root hairs are good gas exchangers when soil conditions permit. Since they arise near areas of intense metabolic activity—growth, maturation, and active transport—their oxygen requirement is great. The exchange of gases takes place across the cells of the root tip and the root hair surfaces. The exchange is easy here because the thin-walled cells have no resistant cuticle, and the tremendous surface area of root hairs provides a large interface for gas exchange.

Where roots are submerged in water, gas exchange and oxygen availability are adversely affected. For this reason, the roots of plants that are grown in tanks of water must be aerated just as one would do for aquarium fishes. This is also why you can kill your potted plants by overwatering them: without gaseous oxygen in the soil, they simply drown. Some plants, however, have adapted to flooded, oxygen-poor soils. These include marsh and swamp plants, such as red maple and swamp alder, and domesticated plants such as rice. It turns out that many of these plants have large, hollow air channels in their stems (not to be confused with the much smaller channels associated with the vascular bundles of monocots). The **pneumatophores** ("air roots") of the black mangrove tree *(Avicennia germinans)* extend above the surface of the water, permitting air to diffuse into the loosely arranged tissue and then down into the roots (Figure 28.14).

Lenticels, common in woody stems, provide an avenue of gas exchange for the very active tissues in the bark. Recall that both the vascular cambium and the cork cambium

FIGURE 28.14
GAS EXCHANGE IN THE BLACK MANGROVE
The black mangrove often grows in water that is low in oxygen content. It has adapted to its environment through the evolution of pneumatophores, porous root extensions that reach above the surface, where air penetrates their spongy tissues.

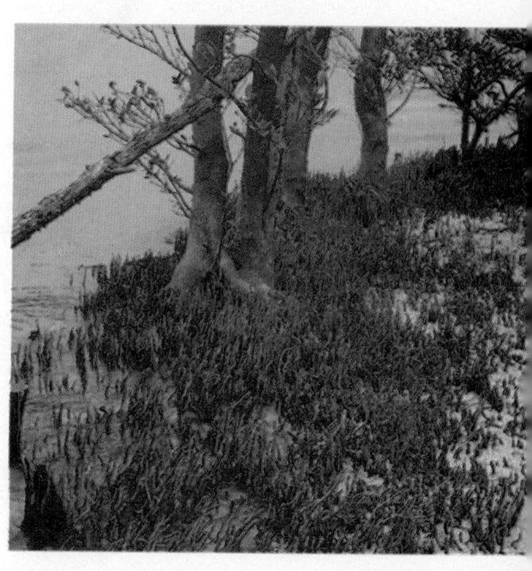

are intensely active during growth. Further, as we have just seen, the phloem elements and their associated cells, also part of the bark, are metabolically active, using ATP for transporting food materials into and out of the phloem stream. Thus, an adequate supply of oxygen is essential.

MINERAL NUTRIENTS AND THEIR TRANSPORT

Dissolved minerals entering the root hairs and epidermal cells follow a route (Figure 28.15) that takes them through the cytoplasm of parenchyma cells in the cortex. Apparently most of their movement is passive until they reach the endodermis, whereupon active transport and a considerable investment of ATP is required. Active transport in the endodermal cells passes the ions to the pericycle, which then secretes them into the water columns for transport upward.

The uptake of mineral ions is not always straightforward. In most plant families, the plant gets a little help from a friend—a friendly fungus, that is. An association of plant root and fungal mycelium known as a *mycorrhiza* ("fungus root," Figure 28.16) is one of those rare instances of true mutualism: both members benefit. However, the relationship may not be clear at first. After all, plants have a lot to offer a fungus, but what does a fungus have that a plant could possibly need?

It turns out that the mycorrhizae efficiently absorb and concentrate certain ions, notably phosphate. In some instances the fungal mycelium with its mineral load actually penetrates the root cortex and deposits ions there. In other cases, the mycelium simply surrounds the root epidermis and brings the plant into close association with the ions. In either case, the fungus extends its many mycelial fingers out into the soil, acting like a second root system for the plant. In transporting phosphates directly into the plant or concentrating them near the epidermis, the fungus makes this valuable ion more

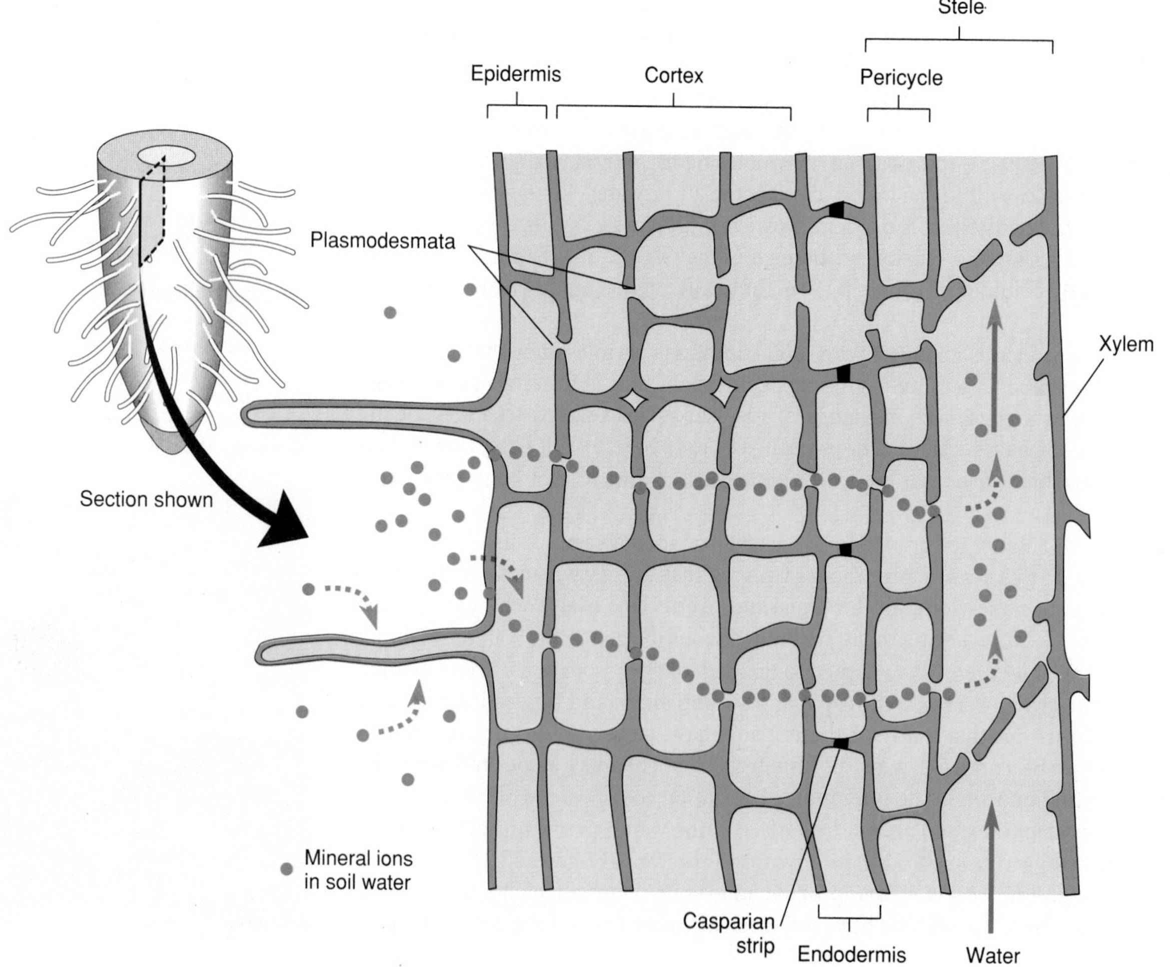

FIGURE 28.15
PASSAGE OF IONS THROUGH THE ROOT
Unlike most movement of water in the root, ions tend to follow the symplastic pathway. Movement from cell to cell is facilitated by numerous plasmodesmata.

available to the plant. The fungus, for its part, absorbs complex carbon compounds such as sugars and amino acids—plant products that are initially produced through photosynthesis. Of the many plant species studied for mycorrhizal relationships, this fungal—plant-root association has turned up in about 90%.

Agricultural scientists have long been interested in the mineral requirements of plants. While it is technically easy for researchers to establish a long list of chemical constituents of plants, the challenge is to determine which are truely essential nutrients, and which are coincidental with no role in plant nutrition. One approach to the question has involved deprivation experiments. Plants are grown in aerated distilled water of rigidly controlled purity. The only mineral ions present are those added by the experimenter. Once the minimal nutrients needed for growth are known, researchers can determine the particular effect of any mineral simply by witholding it from the solution (Figure 28.17).

From such experiments botanists have accumulated a list of important mineral nutrients essential to plant growth (Table 28.2). Notice that the table is divided into **macronutrients** and **micronutrients** (trace elements). These terms simply reflect the relative quantities required. For example, ions such as chloride (Cl^-) are required in such minute quantities that they are very difficult to detect, yet photosynthesis cannot proceed without them. Trace elements often cause problems for plant scientists in trying to control experiments—if you can't find it, you can't always keep it out.

FIGURE 28.16
A STUDY OF THE FUNGUS-ROOT ASSOCIATION
In this experiment, all of the seedlings were first grown in a nutrient solution with all essential materials provided. The group on the left were then transplanted directly into prairie soil. The group on the right were planted first in forest soil where mycorrhizal associations are common, and then, after such associations had formed, transplanted to prairie soil.

Many of the mineral nutrients of plants are made available through cycles that are often quite intricate. These are usually referred to as **biogeochemical cycles**, some of which are described in Chapter 47. Key elements such as nitrogen, sulfur, phosphorus, and calcium, as well as carbon, oxygen, and hydrogen, are constantly recycled between the living and nonliving realms of the earth. When these cycles fail, the result can be infertile soil and limited plant growth.

FIGURE 28.17
IRON DEFICIENCY
The iron-deprived tomato plants *(right)* clearly reveal problems of iron deficiency through stunted growth and foliage lacking the deep-green color of the control plants.

(a)

(b)

(c)

FIGURE 28.18
INSECTIVOROUS PLANTS
(a) The bladderwort *(Utricularia)* is a water plant that forms many bladder-like chambers, each equipped with a trigger-activated trapdoor. When the trap is tripped, the door opens inward, the prey is sucked in by inrushing water, and the door shuts behind it. **(b)** The pitcher plant *(Sarracenia)* has evolved a one-way passage into its trap. Insects venturing into the vaselike structure find their return blocked by downward-pointing hairs. **(c)** The sun-dew *(Drosera)* makes use of a thick, sticky fluid that traps the insect.

Perhaps the strangest way for a plant to fulfill its nutrient requirements is that seen in **insectivorous plants,** those that feed on insects. Insect capture by plants is believed to be an adaptation to life in nitrogen-poor soils such as those of swamps and bogs. There, many of the essential nutrients, including nitrogen compounds, are leached away by water or consumed by anaerobic bacteria. (Oxygen is in short supply as well.) Unlike carnivorous animals, however, such plants do not use their prey as an energy source; they can only make use of the nitrogen.

Examples of insectivorous plants include the bladderwort *(Utricularia),* the Venus flytrap *(Dionaea),* the pitcher plant *(Sarracenia),* and sundew *(Drosera),* some of which are seen in Figure 28.18. Both the bladderwort and the Venus flytrap (discussed in

TABLE 28.2 ELEMENTS AND NUTRIENTS ESSENTIAL TO PLANT GROWTH

ELEMENT	HOW TAKEN IN	EXAMPLES OF USE
MACRONUTRIENTS		
Calcium	Calcium ion (Ca^{2+})	Cell wall, plasma membrane, coenzyme activity
Carbon	Carbon dioxide (CO_2)	Proteins, lipids, carbohydrates
Hydrogen	Soil water (H_2O)	Proteins, lipids, carbohydrates
Magnesium	Magnesium ion (Mg^{2+})	Chlorophyll molecule
Nitrogen	Nitrate ion (NO_3^-), Ammonium ion (NH_4^+)	Amino acids, purines, pyrimidines, protein
Oxygen	Atmospheric oxygen (O_2)	Cell respiration
Phosphorus	Phosphate ion ($H_2PO_4^{2-}$)	Nucleic acids, phospholipids, ATP
Potassium	Potassium ion (K^+)	Plasma membrane, enzyme activity, guard cell mechanism
Silicon	Silicate ion ($HSiO_3^-$)	Cell walls
Sulfur	Sulfate ion (SO_4^{2-})	Proteins, coenzyme A
MICRONUTRIENTS		
Boron	Borate ion (BO_3^-), Tetraborate ion ($B_4O_7^{2-}$)	Cell elongation, carbohydrate translocation
Chlorine	Chloride ion (Cl^-)	Accumulates as HCl in chemiosmotic photosynthesis
Copper	Cupric ion (Cu^{2+})	Enzyme activity
Iron	Ferrous ion (Fe^{2+}), Ferric ion (Fe^{3+})	Enzyme activity, chlorophyll synthesis
Manganese	Manganese ion (Mn^{2+})	Enzyme activity
Molybdenum*	Molybdenum ion (Mo^{3+})	Enzyme activity, including N-fixation
Zinc	Zinc ion (Zn^{2+})	Hormone activity

* Indirectly important in the nitrogen cycle

Chapter 29) actively capture prey. Interestingly, if essential nitrogen compounds are provided to these plants, they go right on trapping insects. (But then, just to further prove that variation and adaptation are endless in life, there are some insectivorous plants that stop growing traps when sufficient nitrogen is available.)

We see, then, that as life became more complex, that complexity often involved an increase in size. With this growth came a need for transporting the molecular necessities of life from one part of the organism to another. This need has been met in plants in a variety of ways that apparently employ simple, basic law of physics. But in some cases these laws don't seem so simple after all, and the search for the precise mechanisms of transport in plants goes on.

APPLICATION OF IDEAS

The statement is made in this chapter that vascular plants have literally "evolved around the characteristics of water." Discuss the full meaning of this statement. Begin by explaining exactly how vascular plants have taken advantage of water's peculiarities. Then rephrase the statement in such a way as to introduce the terms *natural selection* and *adaptation*. Complete your discussion by considering how this evolutionary direction made great increases in size possible, and how this result has been advantageous to evolving plants.

KEY IDEAS

1. A major problem in plant physiology has been to explain how water passes from the roots to the foliage of tall trees with little direct expenditure of plant energy.
2. The vascular system of plants has the dual function of transporting water through its **xylem** and foods through its **phloem.** In the latter, the plant uses active transport and expends considerable energy.

THE MOVEMENT OF WATER AND MINERALS

1. Water travels from the soil to the air around the leaf via root hairs, root cortex, **stele,** root xylem, stem, leaf, leaf mesophyll, air spaces, **stomata,** and outside.
2. The evaporation and subsequent diffusion of water from the leaf is called **transpiration.** In many plants, the quantity transpired daily can be measured in liters.

Root Pressure: Is the Root a Kind of Pump?

1. **Root pressure,** a force originating from osmotic conditions in the root, is responsible for **guttation,** the forcing of water in its liquid phase from the tips of grass blades.
2. Root pressure is produced when ions are actively transported into the stele, setting up a water potential gradient.
3. Root pressure is not a major force in the movement of water to the tops of tall plants. When the xylem of a tree is tapped during active transpiration, a sucking action rather than a spurt occurs, indicating a negative rather than a positive root pressure.

Water Potential and the Vascular Plant

1. Water always moves from regions of greater water potential to regions of lesser water potential. Water potential is the inverse of the solute concentration in cells.
2. The normal water potential gradient of a plant is from high potential in soil water to low potential in air surrounding the leaf.
3. A substantial loss of water potential in the soil leads to a loss of water from the root and wilting in the foliage.

The Leaf, Transpiration, and the Pulling of Water

1. Water loss is unavoidable if CO_2 is to be available for photosynthesis. Plants make use of the continued water loss to initiate the mechanism that raises water from the roots.
2. Transpiration creates a steep water potential gradient through the cells of the leaf, and water enters from the leaf xylem. The cell-to-cell movement is facilitated by numerous plasmodesmata.
3. The water potential gradient in the leaf produces a pulling force that creates pressures of up to 12 atmospheres, enough to raise water through the tallest trees.
4. The movement of water through the stem and leaves is provided by the **free energy of evaporation.** Indirectly, this energy comes from the sun.

Water Movement in the Xylem Through the TACT Mechanism

1. **Transpiration** and **adhesion**, the first two **TACT forces**, set up the pulling force that raises water through the xylem.
2. Adhesion, the attraction of one type of molecule to another (such as water to those of surrounding surfaces), may be a major force, since the adhesion of water to the cellulose walls of leaf cells may initiate the pulling action. It is also a factor in cell-to-cell movement.
3. **Cohesion**, the mutual attraction of similar molecules, is critical in the tiny water columns of the xylem. It provides the tensile strength needed to hold them together when under tension.
4. **Tension** is the stress placed on the water columns by the pulling forces above. During intense transpiration, tension can be sufficient to cause a narrowing of tree trunks.
5. These combined forces provide and maintain the pull needed to raise water through the plant.

Water Transport in the Root

1. During intense transpiration, root pressure is overwhelmed, and the pull from above is simply transmitted across the root. Root pressure may be significant when transpiration slows or when water potential outside the root is low.
2. Water entering the root follows mainly the **apoplastic** (cell wall) **route.** Because water is directed around the **casparian strips** and through the endodermal cytoplasm–to a **symplastic** route–the endodermis may somehow influence water movement. It does transport ions into the stele. Root growth is usually vital to the ongoing uptake of water.

Guard Cells and Water Transport

1. Paired **guard cells** form stomata in the leaf epidermis. Because of the radial arrangement of cellulose microfibrils in their cell walls, guard cells bend when turgid, increasing the stomatal opening. A loss of turgor closes the **stoma.**
2. Stomata are generally open in daylight and closed in darkness.
3. Turgor changes in guard cells are brought about by shifting solute concentrations. In the **potassium ion transport mechanism,** light activates transport of the ions into the guard cell. This decreases their water potential, water enters, and the turgid cells bend apart.
4. A decreased CO_2 concentration also increases K^+ transport. This usually occurs during intense photosynthetic activity—a time when more CO_2 is needed.
5. Severe wilting causes the release of ABA, which causes guard cells to lose K^+ and then water. This brings about closing of the stomata, and the plant is protected.

FOOD TRANSPORT IN PLANTS

In 1671, Hooke and Brotherton determined that substances produced in leaves were transported downward, while those collected by the roots were transported up.

Mechanisms of Phloem Transport

1. While water is moved by a pulling force, a push in the form of hydrostatic pressure is responsible for food transport.
2. Aphid mouthparts are used in sampling phloem sap, a dense solution containing sucrose, other nutrients, and hormones. The movement of foods in the plant is called **translocation.**
3. Foods are commonly moved from **source** to **sink.** They are actively transported into and out of the phloem stream.
4. The favored explanation of phloem transport is the **pressure flow hypothesis.** Sugars are actively transported into the phloem stream at the source. This increase in solute decreases the water potential in the phloem elements, and water enters, increasing hydrostatic pressure. The increased pressure forces the dense phloem sap along to the sink, where the solutes are actively transported out. The loss of solute raises the water potential of the phloem, and water reenters the xylem.

GAS TRANSPORT IN PLANTS

1. Plants must carry on gas exchange in their metabolically active tissues. Commonly, in woody plants, oxygen enters and carbon dioxide leaves the stem via the porous lenticels. In some monocots the passage of gases occurs through open air channels.
2. In the metabolically active root, gases are readily exchanged across the extensive root hairs in the root tip epidermis.
3. Marsh plants may have a problem with providing oxygen to the roots. Some have hollow, air-conducting channels in their stems. Black mangrove trees have **pneumatophores** that extend above the water surface, permitting gases to diffuse into the roots.

MINERAL NUTRIENTS AND THEIR TRANSPORT

1. The transport of mineral ions through root cells is primarily active.
2. Many plants are assisted in mineral uptake by a fungus-root association called a mycorrhiza. The fungus absorbs and concentrates ions near the root hairs or within the root, and in turn absorbs useful compounds from the plant.
3. Mineral nutrient requirements are commonly determined through deprivation experiments. A mineral is withdrawn and the plant carefully studied for deficiency symptoms.
4. Mineral nutrients are commonly provided through **biogeochemical cyles,** where the ions cycle between organisms and the physical environment. Nutrients are divided into **macronutrients** and **micronutrients.**
5. **Insectivorous plants** obtain their nitrogen from the digestion of insects.

REVIEW QUESTIONS

1. Name the two different tissues of the vascular system and summarize their roles in transport. (619-620)
2. In what major way does the mechanism of food transport differ from the mechanism of water transport? (619-620)
3. Starting with the epidermis, trace the movement of water through the plant, naming the important tissues through which it passes. (620)
4. What is transpiration? How could transpiration possibly affect the weather? (620)
5. Explain how root pressure is created. In what way is root pressure visible in grasses? (620-621)
6. Explain why root pressure cannot be responsible for the rise of water to the tops of tall trees. What function does root pressure more likely fulfill? (621)
7. What is the rule about water potential and the direction in which water will flow? What effect do solutes have on water potential gradients? (621-622)
8. Explain transpiration again, this time by describing activity within the air spaces of a leaf. What effect does evaporation have on cells bordering the leaf spaces? On cells further in? (623)
9. Describe the water potential gradient of the leaf, beginning with the air outside a stoma and ending with the water-filled xylem. (624)
10. It has been said that the energy for transpiration is unlimited. Explain what this means. (624)
11. Name and briefly describe the TACT forces. (624)
12. Assuming that transpiration and adhesion create the necessary pulling force, what is the role of cohesion in water transport? In what way is the diameter of xylem vessels and tracheids related? (624)
13. What two kinds of evidence do physiologists find that suggest the presence of tension in the transpiring tree? (624-625)
14. Under what two conditions might root pressure be significant to the transport of water? (624-626)
15. Why is it reasonable to suspect that the endodermis has a role in the control of water uptake by plants? (626)
16. What is the usual cycle of stomatal opening and closing in the plant? (627)
17. Describe the shape of a pair of guard cells in the turgid condition. What factor apparently influences turgor? (627-628)
18. Explain the potassium transport mechanism of guard cell function. (628)
19. Describe how CO_2 levels and severe water loss affect guard cell turgor. (628)
20. Describe the peculiar stomatal cycling of the CAM plants. How can they do this and still produce carbohydrates in the daytime? (629)
21. Compare the pressures in the phloem stream with that of the xylem stream. (630)
22. Explain how physiologists make use of the aphid in studying phloem sap. Summarize their findings on the general makeup of phloem sap. (630)
23. Briefly describe how events at the source produce a flow in the phloem. Is this a push or a pull? (631)
24. Explain what happens to the nutrients and to the hydrostatic pressure when the phloem stream reaches the sink. (630-631)
25. What function do lenticels, pneumatophores and root hairs have in common? (632)
26. Explain how the formation of a mycorrhizal association is beneficial to the plant. How is it beneficial to the fungus? (633-634)

29

PLANT REGULATION AND RESPONSE

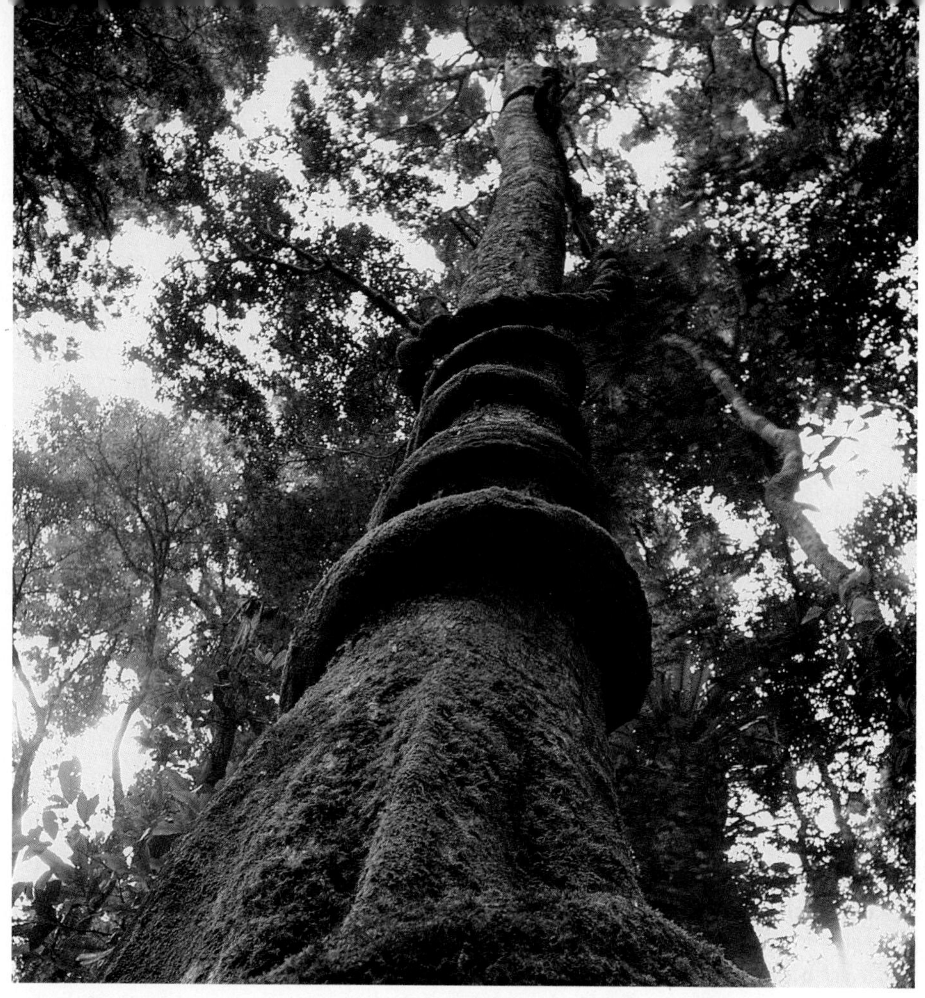

CHARLES DARWIN'S INVESTIGATIONS INTO EVOLUTION LED HIM INTO A WIDE array of scientific fields, from shell fish taxonomy to behavior. In fact, his publication *The Expression of the Emotions in Man and the Animals* (1872) was to become the foundation of animal psychology. But Darwin realized that the area was so complex that even the most basic questions presented overwhelming challenges, so he sought to reduce behavior to its simplest terms. And what, he thought, could be simpler than studying the behavior of plants? Thus Darwin became the first "plant psychologist" and, some would say, the last.

Of course, one is tempted to ask: Do plants actually *behave?* The answer is, they do—if behavior is considered to be a response to a stimulus.

Darwin began his studies with the most interesting of all plant behavior, that of the insect eaters such as the Venus flytrap. In his 1875 publication *Insectivorous Plants* he described in exquisite detail the triggering of the plant and its response, but he was at a loss to explain just *how* such rapid movement was accomplished. So he changed the focus of his research; he looked into the slower and seemingly simpler behavior of climbing plants.

The tips of climbing plants, he found, move about in circles while they grow, "as if seeking something to twine around." If no object is encountered, the tip eventually grows straight upward. But if a twig or a beanpole is encountered, the slowly growing tip begins to spiral around it and to tighten up. In this way, vines and climbing plants can reach great heights without producing thick woody supporting structures such as those found in trees (Figure 29.1).

Darwin also found that on a hot day a new shoot of a hop plant can revolve once in 2 hours, 8 minutes. After 27 such revolutions without encountering anything, the tip of Darwin's plant was describing a circle of 19 inches in diameter and, by looking closely, Darwin found the movement was discernible to the human eye.

Darwin described many variations of this basic pattern and (not suprisingly) interpreted them in terms of natural selection. He showed that the underlying mechanism was the elongation of first one and then another part of the stem below the growing shoot, causing it to bend first one way and then another (*Climbing Plants,* September 1875). But what brought about this elongation?

LIGHT AND THE GROWTH RESPONSE

Darwin, now 71 years old, worked on the problem with his son, Francis (Figure 29.2). Their experimental organism was canary grass. Like other grasses, in its early stages of growth, it produces a tubular sheath, the **coleoptile.** As the shoot emerges from the soil, the primary (first) leaf remains within the protective coleoptile for a brief time. It is this young stage that researchers have found so interesting. Darwin and his son found that when the coleoptile is illuminated from one side, bending occurs toward the light. The bending, they noted, does not actually occur at the tip, but rather in the elongating part of the coleoptile, well below the tip. Was light acting directly on the bending part? Apparently not. Darwin illuminated only the growing tip of the coleoptile, and it bent as before. With a small piece of foil, he covered the tip and exposed the rest of the plant to light. Nothing happened; it didn't curve. Darwin then knew that something was happening in the growing tip that caused certain areas of the shoot beneath to elongate. In *The Power of Movement in Plants* (1880), Darwin ascribed this fundamental discovery to "some matter in the upper part which is acted upon by light, and which transmits its effect to the lower part. . . . When seedlings are freely exposed to a lateral light some influence is transmitted from the upper to the lower part, causing the latter to bend." Thus, in an indirect way, Charles Darwin and his son were the first to propose the existence of a plant hormone.

Early in this century, a Dane, Peter Boysen-Jensen, and a Hungarian, A. Paal, extended Darwin's observations. Boysen-Jensen decapitated a coleoptile of an oat seedling a few millimeters from the tip, put a tiny block of gelatin (a porous material) on the stump, and replaced the tip, which he then illuminated from one side. The coloeptile bent as before, showing that Darwin's "influence" was something that could move through gelatin. To verify this, Boysen-Jensen placed tiny slivers of impermeable mica between the coleoptiles and the stumps. There was no bending. When he inserted mica into slits cut partway through coleoptiles and then illuminated them from different sides, he found that the active substance did not move down the illuminated side, but only down the side away from the light (Figure 29.3). Thus, the stimulus that caused the normal cell elongation and bending was present on the darkened side of the coleoptile only.

In his efforts, Paal decapitated oat coleoptiles and then restored the tips on the growing stubs, but off to one side or the other. The shoot always bent away from that side, even in the dark. This clearly suggested that there was a substance emanating from the tip, and that this substance stimulated the elongation of cells just below. The next logical step was to isolate this material and see precisely what it could do.

The Isolation of Auxin

By 1926, Fritz Went, a Dutch scientist (later working in the United States), had succeeding in doing just that. Went's technique consisted of decapitating oat seedlings and placing the tips on agar blocks as seen in Figure 29.4. (Agar is also porous.) He then took tiny squares of the agar and placed them on the side of the decapitated seedlings. Went kept his subjects in a darkened environment throughout the experiment, so that

FIGURE 29.1
CLIMBING PLANTS
Climbing plants are adapted to rapid upward growth using other plants or objects for support.

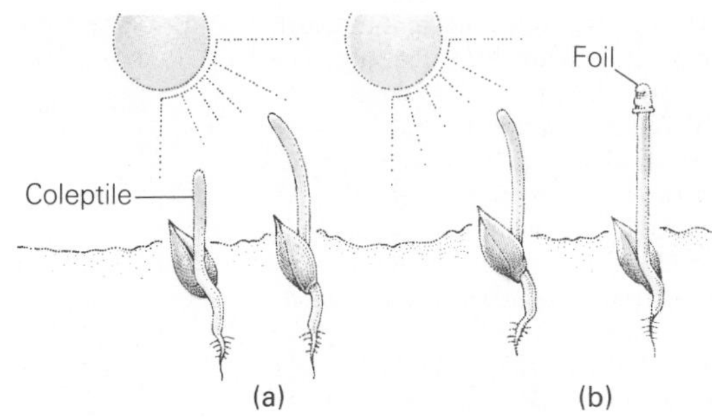

FIGURE 29.2
DARWIN'S PLANT TROPISM EXPERIMENT
Charles Darwin and his son Francis were the first to report studies of the response of plants to light. They subjected canary grass seedlings to light from one side, noting the bending growth **(a)**. To determine where the photosensitive region was, they used thin metal caps to cover various parts **(b)**. Their results indicated that something from the tip was influencing bending in the shoot region below.

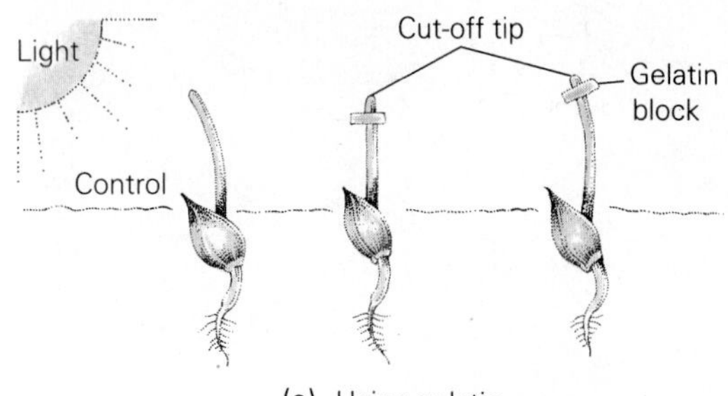

(a) Using gelatin

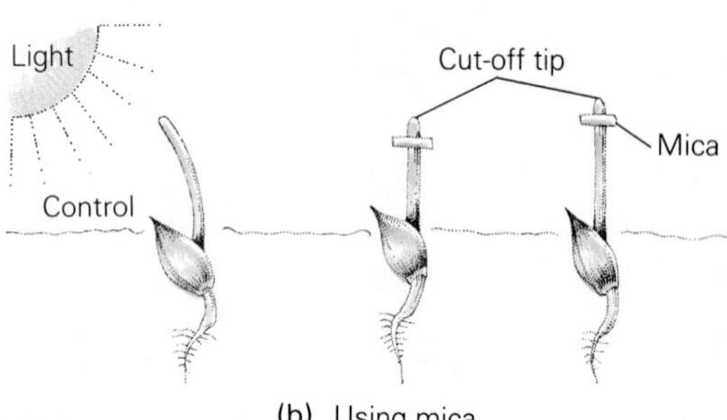

(b) Using mica

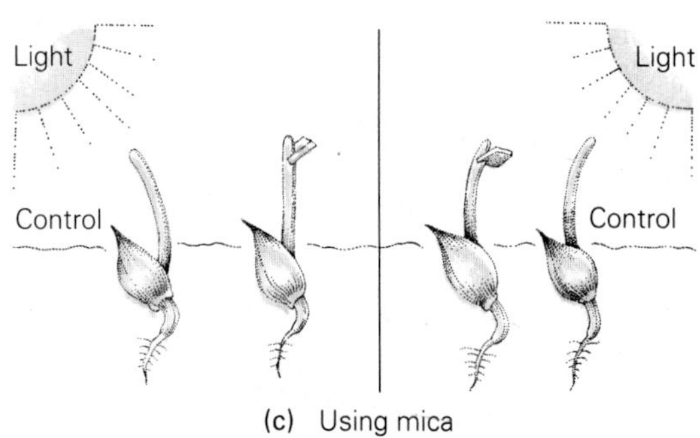

(c) Using mica

FIGURE 29.3
THE BOYSEN-JENSEN EXPERIMENT
Boysen-Jensen elaborated upon Darwin's experiments, using gelatin slabs and slivers of mica. (**a**) Cut off growing tips were placed on tiny blocks of gelatin balanced on the stem. The plants grew toward the light as usual, suggesting that a diffusible substance had passed through the gelatin. (**b**) When impermeable mica was used instead of gelatin, no response to light was seen. (**c**) When mica slivers were inserted part way into the stems on opposite sides of plants, the response depended upon which side was lighted. This indicated that the agent passed down the side *away from the light* (or was perhaps inactivated by light).

any bending response could be attributed to substances in the agar block alone. In other words, he permitted ony one variable at a time in his experiment. Typically, bending occurred within an hour after the blocks were applied, proving that an active, collectible agent was the cause of bending. Although he did not chemically characterize the active substance, Went named it **auxin** (from the Greek word *auxein*, "to increase"). Later, when more of its chemistry became known, it would be called **indoleacetic acid**, or **IAA**.

Any biochemical isolation requires a **bioassay system**, a way of measuring whether a substance is present, and in what concentrations. An assay system for the coleoptile growth substance was soon worked out with oat seedlings, using Went's basic methods. Hormone concentrations are calculated by the degree of bending seen in the oat seedlings. The greater the angle, the more hormone present.

Such bioassay methods were worked out for other hormones as well. In its time, this procedure represented the most reliable technique biologists had for determining the presence and quantity of hormones in plant tissues or extracts. Hormones often occur in extremely minute quantities. Newer, more direct methods are now used for these purposes.

Auxin: Its Structure and Roles

Unknown to Boysen-Jensen, Paal, or Went, the active substance of auxin had been isolated from fermentations back in 1885 by a pair of biochemists, E. and H. Salkowski. However, the Salkowskis hadn't a clue about its biological significance. The substance was not successfully isolated again until 1934, when it was finally purified from, of all things, human urine. We now know that the mysterious coleoptile growth substance is actually a widespread and fundamental plant growth hormone. The structure of auxin is very similar to the amino acid tryptophan; in fact, tryptophan is its primary building block.

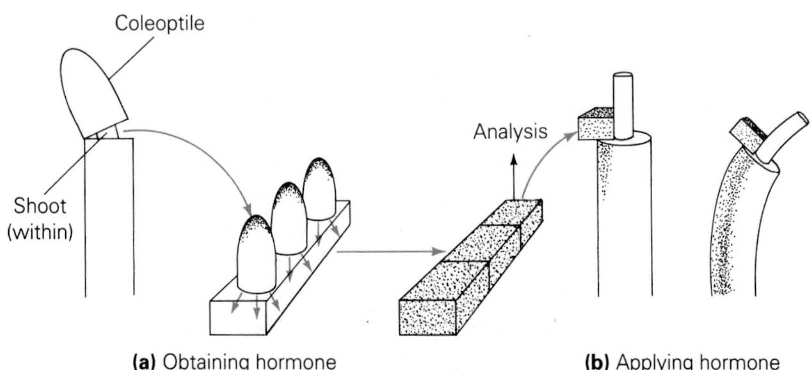

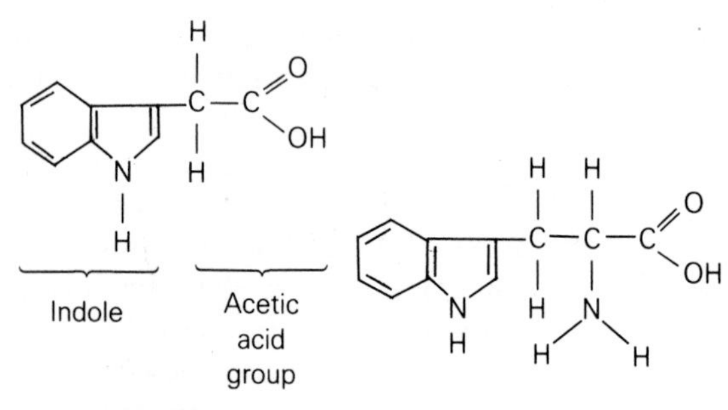

FIGURE 29.4
WENT'S AUXIN ASSAY
(**a**) Fritz Went was able to collect the growth-stimulating substance in an agar gel. He then cut the agar into tiny squares, which he placed on the coleoptile. Their presence brought about bending of the shoot. (**b**) Went's assay procedure determines the presence and quantity of the hormone—the quantity being directly proportional to the amount of bending.

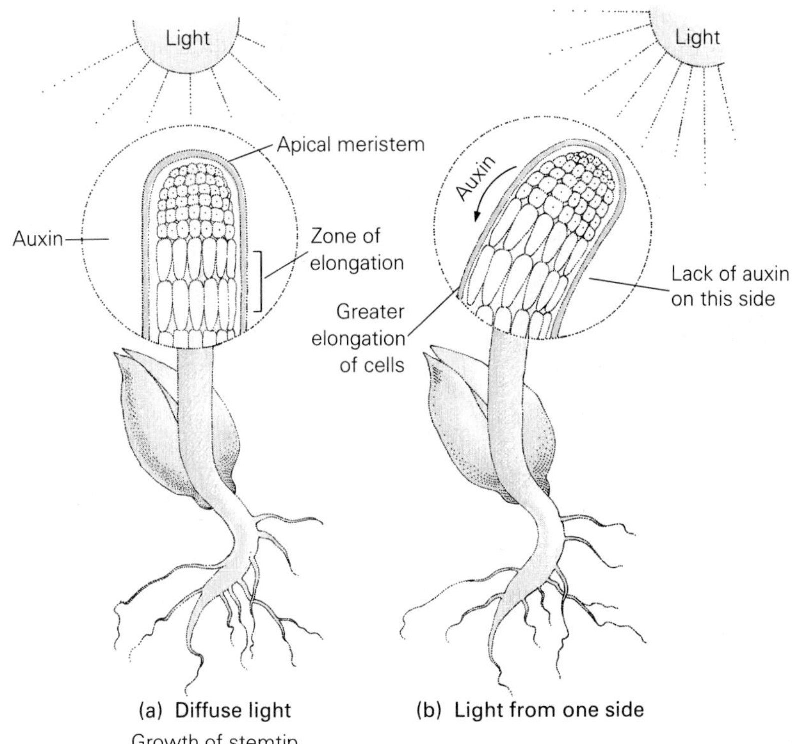

FIGURE 29.5
THE ACTION OF AUXIN
Auxin is known to promote cell elongation in the growing shoot. (**a**) When light is multidirectional, elongation proceeds equally about the stem. (**b**) When light is directed to one side, the greatest elongation occurs on the opposite, unlighted side.

Auxin doesn't cause cells to proliferate by mitosis, but rather promotes cell enlargement through elongation (Figure 29.5). Cell elongation, a primary plant growth mechanism, occurs during the growth of stems after the tiny, tightly packed daughter cells form through mitosis in the apical meristem (see Chapter 27). In auxin's presence, the cell wall is infiltrated by hydrogen ions pumped from the cell. The resulting acidity may then activate enzymes that break certain bonds in the cellulose fibers of the wall. This loosening effect permits formerly restrained turgor pressure in the cell to bring about elongation (see Figure 27.7).

Auxin is also involved in seasonal leaf **abscission**, or leaf fall, but in a negative way. As auxin diminishes, many valuable materials within the leaves are mobilized and with-

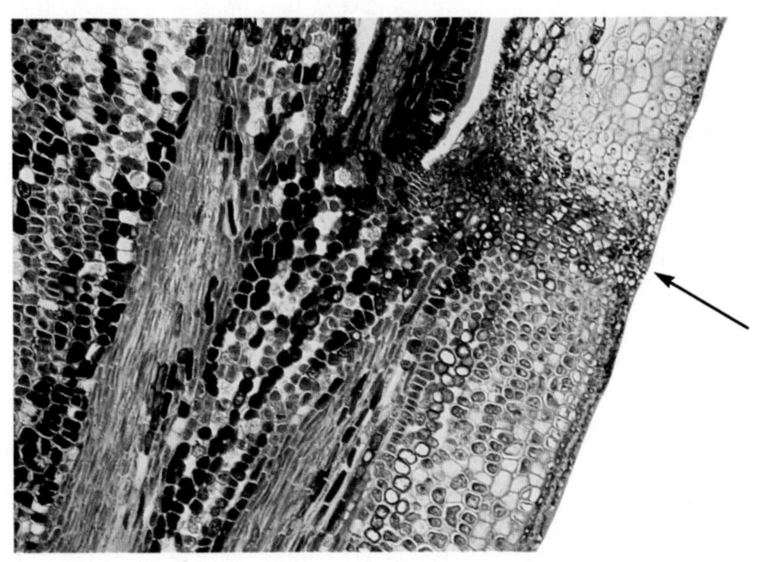

FIGURE 29.6
THE LEAF ABSCISSION LAYER
Before a leaf falls, the cells of the abscission zone (arrows) die and harden. Eventually, the leaf petiole will break away and fall. The dead abscission zone forms a sealing scar.

drawn into the plant. Next, a second hormone, ethylene (which is discussed further along), triggers the release of cellulases—cellulose-splitting enzymes—that break down cell walls in the **abscission zone,** a region of cells at the base of the leaf petiole (Figure 29.6). As this happens, new cells with a highly suberized content may form along the inward side of the abscission zone of each leaf, effectively sealing off the plant's vascular system. Then, thin-walled parenchyma cells along the abscission zone enlarge, weakening the remaining connection. Finally the leaves break away (and the raking begins).

The events preceding leaf fall are important for several reasons. As we've seen, the plant recovers valuable materials from the leaves which can then be recycled for further use. Then, by sealing off the vascular system, the plant avoids the risk of excessive water loss. Such sealing also closes off openings, that, like an open wound in our own skin, permits the invasion of parasites.

Auxin has many other roles in the growth and responses of plants. Its influence on cell elongation is quite central to growth in the root tip, which also occurs through elongation (see Chapter 27). There is further evidence of auxin's importance to root differentiation. When auxin (or one of its analogs) is applied to stem cuttings, it greatly accelerates the growth of adventitious roots (those emerging from the stem). As we will see shortly, auxin also influences growth in negative ways. Its inhibitive effects are responsible for the common triangular shape taken by the crown of some trees. We will be considering these and other of auxin's influences as we proceed.

OTHER PLANT GROWTH HORMONES

A number of other plant growth hormones have now been identified, and some of their stories are quite fascinating. We will consider the best known of these.

Gibberellins

The **gibberellins,** a large family of molecules, received their name from the source of their discovery, *Gibberella fujikuroi,* a fungus. The fungus produces a condition the Japanese called "foolish seedling disease." The symptoms in rice plants include abnormal flowers and strangely elongated, weak stems that tend to break as the grains form. At one time the fungus threatened rice harvests in Japan.

Between 1926 and 1935, Japanese botanists had already isolated and purified the active substances produced by the fungus, but it wasn't until the 1950s that other nations took an interest in these strange molecules. Today, we are continuing to study the natural production of gibberellins to learn more about how they operate in conjunction with auxins to control cell elongation. Some of the results are finding their way into agricultural use. Gibberellins are used to promote stem growth in sugar cane and celery, to increase starch digestion during the malting process in brewing, and to bring about a loosening effect on grape clusters, thus improving the effectiveness of fungicide spraying.

FIGURE 29.7
GIBBERELLINS AND STEM GROWTH
Gibberellins have a dramatic effect on stems. The control plants at the left and the experimental plants at the right were treated identically, except that gibberellins were applied to the group at the right.

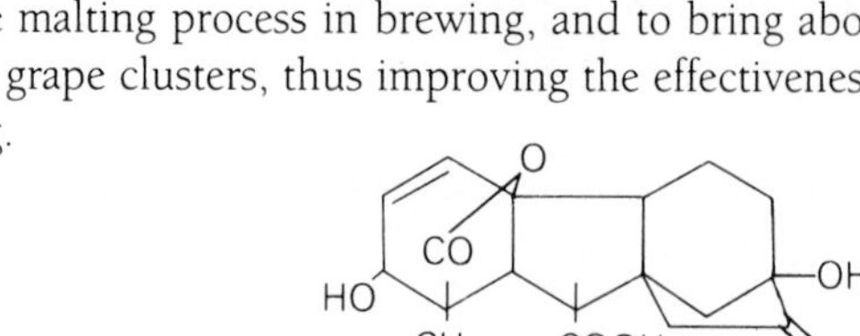

A gibberellin

Gibberellins are formed in young leaves around the growing tip, and possibly in the roots of some plants. (However, we don't know what role they play in root activity.) The power of gibberellins in stem elongation has been most dramatically illustrated in experiments with genetic dwarfs. Dwarf corn, for instance, can be induced to grow to normal height after the application of gibberellins. This indicates that a hormonal failure causes the plants' shortness and, except for this abnormality, dwarf corn has the potential to grow tall. Incidentally, the degree of growth in a dwarf depends on the quantity of gibberellins applied. Thus the botanist has an excellent method of gibberellin bioassay (Figure 29.7).

Gibberellins also have another role. They act as chemical messengers to stimulate the synthesis of an enzyme called **alpha amylase** and other hydrolytic enzymes in grains such as barley and corn (see Chapter 27). As these grains germinate, the embryo secretes gibberellins, which move to the cell layer that surrounds the starchy endosperm, the aleurone. The cells of the aleurone, apparently stimulated by the hormonal messenger, begin to produce alpha amylase, which then breaks down starch and makes sugars available to the growing plant (Figure 29.8). There may be many steps or only a few between the arrival of the hormone messenger and the transcription activity required for producing the enzyme. As you may recall from past discussions (see Chapter 14), the specific hormonal gene-activating mechanisms are still under investigation.

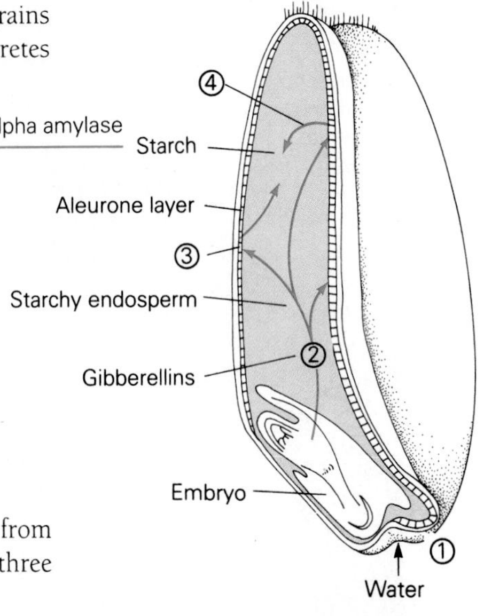

FIGURE 29.8
GIBBERELLINS AND SEED GERMINATION
Gibberellins bring about the conversion of starch to sugar, as seen here. (1) Water is absorbed by the seed. (2) The embryo secretes gibberellins. (3) The hormone reaches the aleurone cell layer, stimulating these cells to produce alpha amylase. (4) Alpha amylase, released into the starchy endosperm, hydrolyzes the starch, forming sugars, which are then taken in by the embryo.

Cytokinins

Most of what we know about **cytokinins** springs from work begun in the 1950s, when it was found that plant growth could be influenced by something from corn kernels. In 1964, the first of the molecules, **zeatin**, was described; since then, three others have been identified.

CH_2OH
|
$C—CH_3$
‖
CH
|
CH_2
|
NH
|
(purine ring: N, N, N, N—H)

Zeatin

Prior to the discovery of zeatin, research scientists had found that in order to grow cultures of plant tissues, they had to add coconut milk to the medium. What was so special about coconut milk and corn kernels? They began to isolate chemicals found in both sources, and the mysterious substance turned out to be a group of hormones they called cytokinins.

The first thing biologists found out about cytokinins was that they stimulated cell division in plants, although they had to be coupled with other plant hormones to do this. Using tissue taken from tobacco plants, researchers mixed **kinetin** (one of the cytokinins) and auxins in different ratios and then grew tobacco pith cells in the various media. The interaction of the two hormones is quite complex. The growing pith at first produced only an undifferentiated **callus,** the same tissue that forms around wounds. Then, when hormones were administered, it turned out that auxin encouraged root growth, while kinetin encouraged shoot growth—but only when the opposing hormone was in low concentration. Further, when very low concentrations of both hormones were present, little callus growth and no differentiation occurred. At high concentrations of both, the callus grew but no differentiation occurred (Figure 29.9). Thus cytokinins also play a role in plant differentiation.

The studies indicated that the undifferentiated and ordinary cells of the pith, the

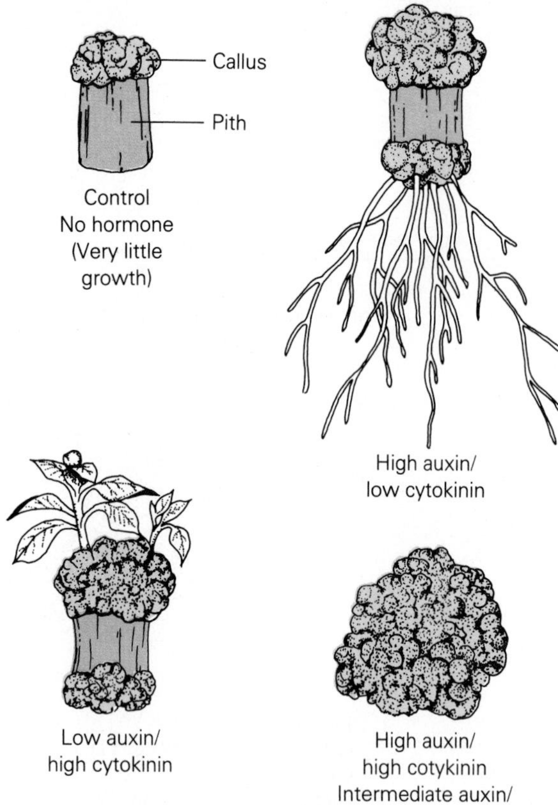

FIGURE 29.9
CYTOKININS AND DIFFERENTIATION
In this experiment, the pith is treated with varying combinations of auxin and cytokin, each of which produces some variation in growth and differentiation.

parenchyma cells, also contain all the genetic information necessary to develop into other kinds of plant cells, producing a variety of tissues and organs. Even differentiated tissue, grown on the proper medium, would sprout into complete, normal plants. The potential value of such studies staggers the imagination—whole individuals grown from a piece of tissue! This capability in cells is referred to as **totipotency**—a phenomenon that first involves **dedifferentiation**, the return to a simple undifferentiated state, and then differentiation, usually along a different path from that of the original tissue. The concept of totipotency suggests that plant cells, no matter how specialized, never lose the genetic capability of the fertilized egg cell. More on this subject is found in Essay 27.1. Later we will find that this may not be true of animals (see Chapter 43).

Another role of the cytokinins is associated with **senescence**, or aging, in plants, a process in which chlorophyll breaks down and other leaf pigments fill the leaf cells. For example, senescence occurs in leaves prior to their seasonal fall from the branches of deciduous trees. Leaves also age if they are picked while in the prime of life. Then, of course, they die. When picked leaves are treated with cytokinins, however, aging is retarded. The chlorophyll does not break down (so the leaves stay green), protein synthesis continues, and carbohydrates do not break down. Synthetic cytokinins have been applied to harvested vegetable crops such as celery, broccoli, and other leafy foods in order to extend their storage life.

We have a few ideas about how cytokinins do their work. It is known, for example, that some of the tRNA molecules contain cytokinins as a functional part of their structure, so it is possible that some cytokinins may facilitate protein synthesis. Frankly, however, plant physiologists are groping at this point, the precise answer to how cytokinins work constantly eluding them.

Ethylene

Ethylene is a relatively simple compound when compared to other plant hormones:

$$\begin{matrix} H & & H \\ & \diagdown \; \diagup & \\ & C{=}C & \\ & \diagup \; \diagdown & \\ H & & H \end{matrix}$$

Ethylene

Ethylene is a gas—one you can smell around ripening fruit. In fact, it controls the ripening process. It is synthesized by altering the amino acid methionine, commonly found in cells. Since it is a gas, it readily diffuses out of plants; thus the concentration of ethylene in plant tissues depends on its rate of production and escape.

In addition to its role in initiating the ripening process, ethylene is believed to play an important part in the emergence of seedlings from the soil. As some seeds sprout, the upper portion forms a sharp curve, sheltering the young fragile leaves underneath (Figure 29.10) as the tough, thickened curve of the shoot plows its way up through the soil. Somehow, the presence of ethylene prevents the plant from straightening and keeps the fragile leaves from unfolding to the sky until the shoot is free of the ground. Once this happens, the light ethylene gas readily escapes, lowering its concentration, permitting the stem to straighten and the leaves to expand.

The concentration of ethylene in plants is important, as is the case with all plant hormones. When the concentration rises to an abnormal level, the result can be disaster. For example, ethylene in the form of air pollution can cause defoliation and eventually

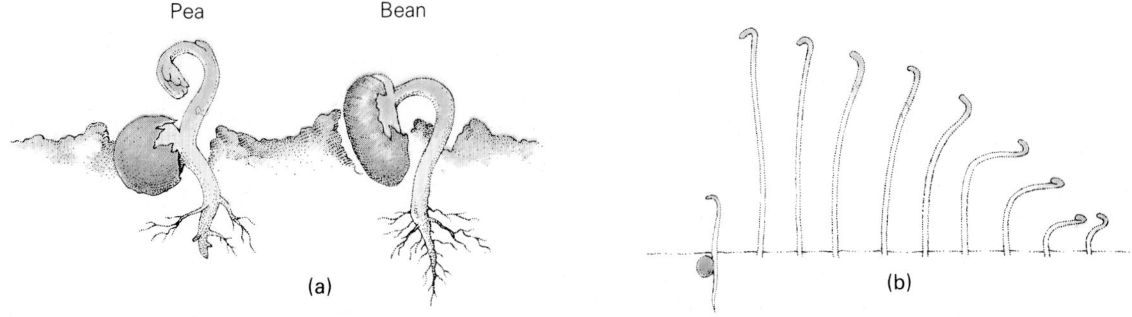

FIGURE 29.10
TESTING ETHYLENE'S ROLE IN HYPOCOTYL CURVATURE
(a) Many dicot seedlings form a curved, protective "bumper"—the hypocotyl hook—as they push their way through the soil. The curve is maintained by a surrounding concentration of ethylene. When the seedling breaks through the soil, the curve straightens. **(b)** The effects of ethylene on curvature in pea seedlings are demonstrated by this experiment, in which seedlings on the right have increasing ethylene concentrations.

death in plants. As we've seen ethylene promotes the production of the cellulases that normally increase just before regular abscission occurs.

Ripening fruits produce ethylene naturally, but ethylene gas is also a petroleum product—the same one that is polymerized to make polyethylene plastic. The synthetic gas has long been used to initiate the ripening of fruits in transit to the world's markets. The banana industry in particular owes much of its success to this simple chemical. It is now possible to ship unripened fruit to world markets in prime condition without worrying about untimely ripening. The release of ethylene over green bananas in warehouses or in ship cargo holds will produce ripe, yellow bananas at just the time they reach the consumer.

Ethylene may also help determine sex in certain flowers. In plants that bear separate male and female flowers, treatment with gibberellins encourage male flowers to develop, while treatment with ethylene encourages female flowers. The artificial stimulation of female flowers in the fields can be quite profitable if one is marketing the fruit.

Abscisic Acid

Our last example of a plant hormone, one we have already mentioned, is **abscisic acid** (ABA). Its molecular structure consists of a carbon ring with a short carbon chain containing a carboxyl (acid) group:

Abscisic acid

It is generally believed that abscisic acid induces winter dormancy by suppressing mRNA production, which would inhibit the growth-promoting hormones, auxin and gibberellin. Dormancy is especially important to deciduous plants of the temperate regions. Dormancy, we might say, is the deciduous plant's adaptation to both cold temperatures and water shortages (the second brought on by the liquid waters being locked up as ice and made unavailable).

A final role ascribed to ABA involves the stoma. Recall that under periods of water stress (see Chapter 28), ABA enters guard cells where it brings about the outward transport of potassium ions. The outward movement of water follows, and the flaccid guard cells shrink, closing the leaf's stomatal openings. The plant is thus protected from further water loss.

Applications of Plant Hormones

Investigation of plant hormones has led to the development of increasingly precise analytical techniques, such as the hormone assay. The improved techniques have enabled us to regulate and understand (sometimes in that order) a wide range of hormones. Such knowledge has been extremely useful to people in a variety of ways. Not only can we produce the specific plants we want, but we can also destroy them—including some weeds, and at one time, the trees of Vietnam's forests (which hid the activities of the Viet Cong and North Vietnamese soldiers).

Since the discovery of auxins, chemists have developed an entire family of analogous compounds, some of which have even greater growth-promoting effects than the original auxin. One such compound is 2,4-dichlorophenoxyacetic acid, known mercifully as 2,4-D. While 2,4-D promotes growth at very *low* concentrations, it kills plants at higher concentrations. This compound is a very common and potent agent of weed control. Dicots, including the broad-leaved plants we often think of as "weeds" are much more sensitive to the herbicide than monocots (wheats, oats, and so on). This means that you can kill the clover in your lawn with 2,4-D while sparing the bluegrass and bermuda grass. Of major ecological importance, 2,4-D offers the advantage of rapid biodegradability. (It rapidly ceases biological activity when exposed to the elements and organisms of the soil.)

Weed control agents are not uncontroversial, however. For example, there have been serious questions about the effects of a chemical called 2,4,5-T (2,4,5-trichlorophenoxyacetic acid), closely related to 2,4-D but used in controlling the growth of trees. It reportedly has caused birth defects in mammals—and we are mammals. For this reason, its broad use for valid purposes such as fire prevention has to be measured carefully against other risks.

The problem is, we create a weed problem by clearing the land, building roads, and otherwise disrupting the natural habitat. Then we solve the problem by applying unnatural controls, which in turn may create additional, often unexpected problems. Actually, the real risk, as we now see it, does not arise so much from the control agents themselves as from a lack of quality control in the manufacturing process. In the past, 2,4,5-T is known to have been contaminated by dioxin, the infamous and frightening environmental pollutant common to illegal (and legal) chemical dumps. We now know that the 2,4,5-T used in the defoliant **agent orange** of Vietnam fame was also heavily contaminated with dioxin. The potential consequence of the frequent exposure of American military personnel (and who knows how many unfortunate Vietnamese citizens) to agent orange during Vietnam conflict is still a volatile issue. Veterans' groups have claimed that such exposure has caused them a number of drastic health problems and increased birth defects in their children.

PLANT GROWTH RESPONSES AND MOVEMENTS

As we now know, auxin and other plant hormones are responsible for many of the specialized growth responses and movements that so intrigued Darwin. Some movements are based on other mechanisms, but whatever the basis, plant movement can be highly visible and often quite fascinating. Some of the visible responses are called **tropisms** ("to turn"). As the derivation implies, tropisms involve bending, an unequal growth response generally toward or away from a directional stimulus. For example, growth influenced by directional light, such as the auxin response discussed earlier, is **phototropism.** There are others. Unequal growth influenced by gravity is called **gravitropism** (formerly "geotropism"). And some plants exhibit uneven growth in response to touch, a response called **thigmotropism** (*thigma*, "touch").

Other movements, not generally related to growth, include the **nastic responses.**

Nastic responses (*nast,* "to press down"), involve the movement of leaves (sometimes flowers). They are generally produced by changes in turgor pressure in certain cells in the base of the leaf petiole. As we'll see, some of these movements occur with startling suddenness, often through **thigmonasty**, a nastic response to touch. The best known example of thigmonasty is the sudden closing of the toothed leaves of the insectivorous plant, the Venus flytrap.

FIGURE 29.11
APICAL DOMINANCE IN CONIFERS
The leading growing tip releases auxin that retards growth in the lateral branches below. The older branches, however, have had longer to grow, even at a reduced rate, so the result is the familiar triangular appearance in plants of this group.

Phototropism

Most researchers believe that during the phototropic response, light stimulates something in the growing tip and causes the auxin to move laterally in the apical meristem. The process appears to be controlled by a photoreceptor, probably a yellow pigment. Although this initial photoreceptor has not yet been identified, it is known to be most responsive to blue light. The entire growing tip produces auxin in quantity, but as it diffuses down the shoot it crosses over to the unlighted side, where cells respond by elongating and thereby, bending the tip toward the light. Such unequal growth produces the typical bending of the shoot (see Figure 29.5).

As with many hormones, auxin is short-lived. Instead of accumulating in the stem, it is inactivated by specific enzymes in the lower parts of the plant. The inactivation process tightly regulates the level of auxin within the plant. As we've seen, excess auxin can actually inhibit growth.

Auxin and Apical Dominance The familiar triangular profile of conifers (Figure 29.11) is the result of a phenomenon known as **apical dominance**, the inhibition of the growth of other branches by hormones from the uppermost tip of the plant. It turns out that auxin, produced in the dominant growing tip, inhibits growth in lateral branches below and even stops the development of any new branches. Thus, the plant's energies are devoted to upward growth. If a plant is "topped"—its dominant tip removed—the inhibitory effects of apical dominance are removed. The highest remaining lateral branches then begin to grow upward, seeming to vie with each other for dominance. But what is happening is that auxin suppression has not yet been established in the lateral branch tips. Ultimately, all but one of the lateral branches will be suppressed, and that one—the new suppressor—will become the main trunk of the tree. Apical dominance enables plants to survive extensive damage from browsing animals, windstorms, and inept gardeners.

Apical dominance is very dramatic in conifers but less pronounced in broadleaved trees. These trees may have a single trunk up to a certain height and then branch out into several large limbs, with no one limb truly dominant. Some form of hormonal inhibition dominance may occur within each of these major limbs (Figure 29.12a).

(a)

(b)

FIGURE 29.12
APICAL DOMINANCE IN BROAD-LEAVED PLANTS
Broad-leaved plants often lack a single dominant tip (**a**) and produce a spreading growth. (**b**) Very dense decorative hedges are produced through constant pruning. With repeated loss of growing tips, each stem continues to produce new lateral branches.

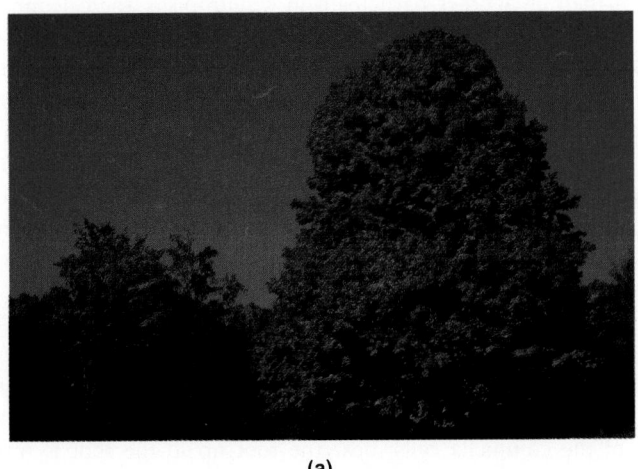

Continued removal of growing tips, as is done in pruning, increases the number of lateral shoots. This results in a decorative bushy plant growth (Figure 29.12b). Pruning can also stimulate a response in fruit trees, whereby the plant's resources are put into production of fruit rather than wood.

Roots and Gravitropism

When a seedling is placed on its side, the shoot tip bends upward, and the root tip bends downward. A simple enough observation, but there are two distinct mechanisms at work here. The shoot turns upward in response to processes we have already discussed, but the root turns downward for quite different physiological reasons. Essentially, we will now see that the root turns downward in response to gravity. The response is gravitropism (formerly "geotropism," Figure 29.13a).

Any explanation of gravitropism, until recently, was awash in a sea of contradictory hypotheses based on equivocal lines of evidence. For example, we know that auxin promotes elongation in shoot tips (Figure 29.5) but, for some reason, it seems to *inhibit* elongation in root tips. Physiologists were faced with the question, How can a hormone stimulate and inhibit the same response?

The key to such opposing responses, once again is a matter of concentration. That is, low concentrations of auxin stimulate elongation, whereas high concentrations inhibit it. Before this peculiarity became known, however, scientists cast far and wide for other explanations. In the long run, the long and confounding search paid off, since it brought to light several aspects of gravitropism that have finally helped to solve a complicated puzzle.

In seedlings lying horizontally on their sides, auxin accumulates in greater abundance in the lowermost cells and inhibits growth there. The uppermost cells, though, grow normally, and so the root tip bends downward. (When auxin is experimentally applied to cells on one side of the root apical meristem, those cells fail to elongate, and bending occurs, Figure 29.13b.) This helps explain why the root bends toward the earth, but how does the auxin end up on the lower side of a horizontal root?

In seedlings, when the root is in a vertical position, auxin moves down through the core of the root tip, eventually reaching the root cap. There, some of it simply diffuses out in the soil, but the rest moves back up the root tip, through cells of the root's perimeter (Figure 29.13c). It is important to note that the return flow in the vertical root is evenly distributed around the root and that the quantity is insufficient to inhibit elongation.

An important change takes place when the seedling is turned horizontally. Now, not only does most of the auxin travel along the lower side of the root, but its concentration there is great enough to inhibit elongation. This explains a lot about root behavior, but some questions remain. For example, why does the auxin accumulate on the lower side of a horizontal plant? Auxin is too light for gravity to draw it down through plant tissues.

It turns out that there is another participant in gravitropism, the calcium ion. Like auxin, calcium moves in a root placed horizontally. Specifically, it collects along the lower side of the root cap. But what effect does calcium have on root-bending? When experimenters added calcium-blocking agents to roots, the gravitropic response failed to occur. When calcium was restored, bending soon resumed. In fact, the roots of young seedlings can be made to turn and twist at the whim of the experimenter—into loops and spirals—simply through the continued application of calcium to one side or the other. (So botanists have a source of amusement comparable to physicians who like to tap our knees with mallets.) But what causes the calcium ions to begin moving, and for that matter, what does calcium have to do with auxin?

To get at the answer, researchers turned their attention to the root cap. They now believe that the trigger of the gravitropic mechanism is located in the root cap's centermost cluster of cells, the **columella.** Each of its cells contains a number of heavy, starch-storing bodies called **amyloplasts.** When the root tip is vertical, the amyloplasts gather along the bottom of the columella cells, near the root tip. If the root is turned to a horizontal position, they descend again, to what were the sides of the cells (Figure

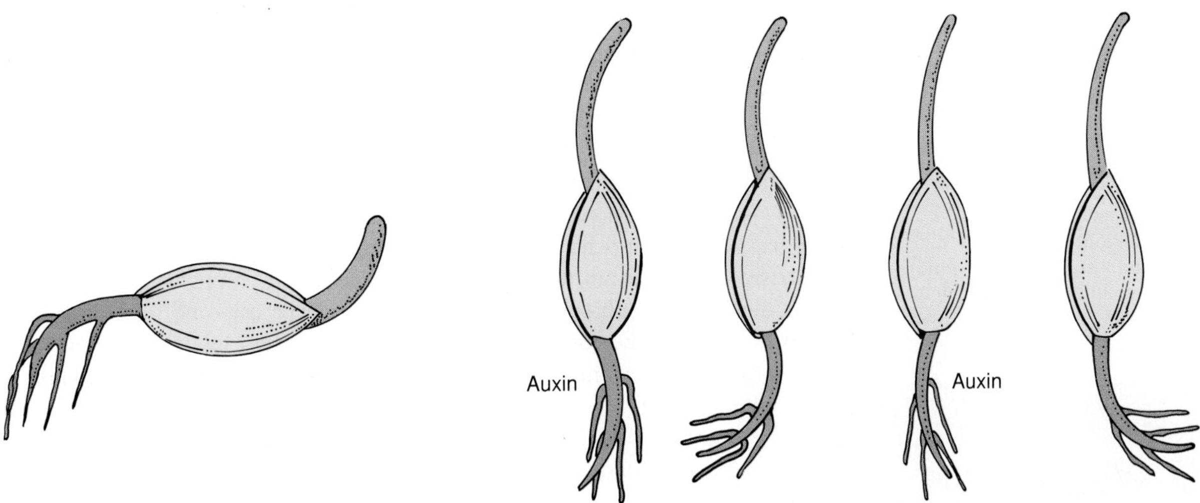

(a) Gravitropism in the seedling root

(b) High auxin concentration inhibits elongation

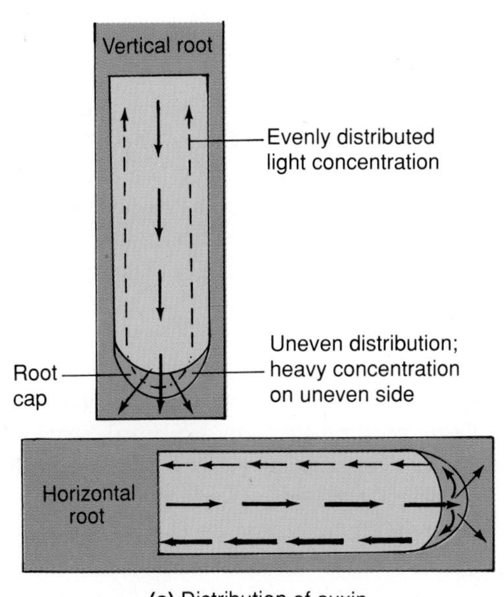

(c) Distribution of auxin

FIGURE 29.13
GRAVITROPISM IN ROOTS
(a) In the gravitropic response, roots curve toward gravity. This response is attributed to concentrations of auxin at the "gravity side" of the root, where it inhibits cell elongation. (b) This role of auxin has been experimentally verified. (c) Auxin's natural distribution is quite different in the vertical and horizontal root tip, the latter permitting an inhibiting quantity to gather "towards gravity." In the hypothetical cellular mechanism, falling amyloplasts in the horizontal root cause the release of calcium ions, which activate calmodulin. Calmodulin, in turn, activates enzymes that start auxin and calcium pumps, and the two leave the columella cells. Auxin accumulates on the lower side of the root, where it inhibits cell elongation.

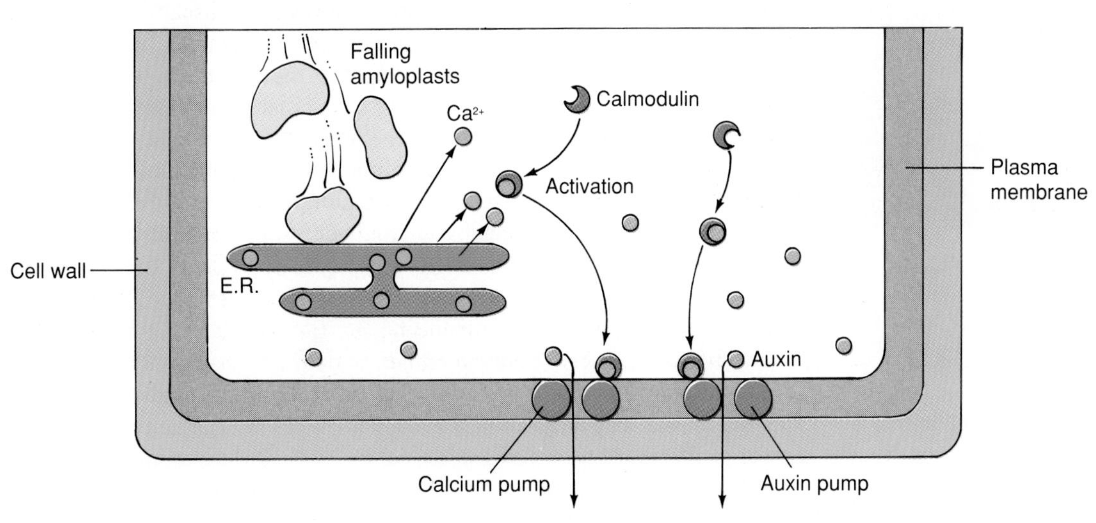

(d) Cellular mechanism

29.13d). Is the new location of the amyloplasts related to auxin and calcium ion migration? At this point the experimental observations aren't complete, but researchers are not hesitant to guess.

They have hypothesized that when the dense amyloplasts fall to the lower part of the root, they come to rest against the lowermost portions of the endoplasmic reticulum, where they bring about a release of calcium ions that are stored within the ER (Figure 29.13e). The calcium, now free in the cytoplasm, sets in motion a cascade of enzymatic reactions, starting with the activation of calmodulin, a well-known enzyme activator. The activated enzymes, in turn, start calcium and auxin pumps working in the nearby plasma membrane. Both substances leave the columella cells and migrate to the lower margin of the cap, whereupon the auxin begins its journey along the lower side of the root—its accumulation there leading subsequently to the inhibition of cell elongation.

When researchers apply calmodulin inhibitors to root caps, calcium and auxin migration do not occur, and the gravitropic response is strongly retarded. However, the role of amyloplasts, while intriguing, remains quite tentative. The connection between calcium ions and calmodulin is better established. In any case, the pieces of this puzzle seem to be coming together as one more complex mechanism in the picture of life emerges.

FIGURE 29.14
THIGMOTROPISMS
Upon touching a surface, grape tendrils begin a coiling growth, which soon fastens the plant securely to the object.

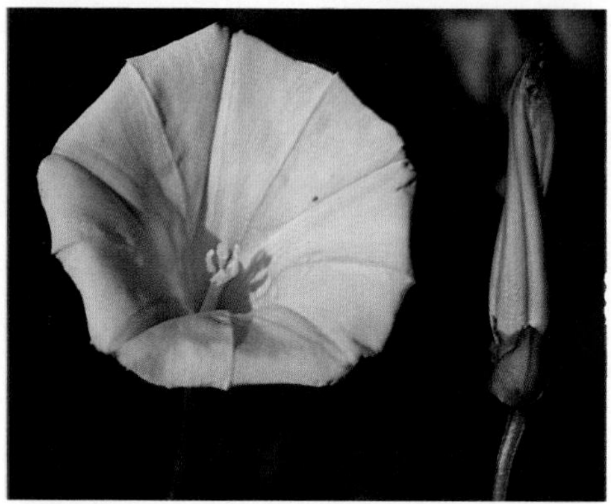

FIGURE 29.15
SLEEP MOVEMENTS
Nastic responses in *Albizzia julibrissin* include sleep movements. In daylight hours, the leaves or leaflets are fully exposed to sunlight, but at night they turn or fold into a vertical position.

Thigmotropism

The slender, coiling growths at the tips of young grape and pea plants are known as **tendrils.** Tendrils are modified leaves or stems that are highly sensitive to touch. This response is a thigmotropism. When tendrils encounter a solid object, their tips respond by growing toward it, perhaps coiling around the object, thereby providing a firm anchor, which in turn permits further growth (Figure 29.14). Compared to most plant responses, the coiling can be quite rapid, with a complete turn forming in less than an hour (so it moves faster than the minute hand of a clock).

As you might expect, coiling is produced by unequal cell elongation, with the greatest cell lengthening occurring opposite the touching surface. While the actual mechanism hasn't been satisfactorily explained, some research suggests that it is light-dependent. Experiments show that pea tendrils maintained in the dark do not coil when touched. Once exposed to light, however, they resume their normal thigmotropic response.

Nastic Responses

Unlike tropic responses, nastic responses are usually nondirectional and reversible. Included are the so-called "sleep-movements," wherein leaves or leaflets that are splayed out to the sunlight in the daytime, fold or turn into a vertical position at night (Figure 29.15). Like other nastic responses, this one involves groups of cells at the base of the leaf petiole or the base of leaflets that form a special structure called a **pulvinus** (plural, **pulvini**). Its components, known as **motor cells**, specialize in the pumping of potassium ions (K^+) from one place to another, thereby altering the water potential in such cells. As we've seen again and again: where ions go, water quickly follows, and the resultant changes in turgor bring about the raising and lowering of petioles that characterize sleep movement. (Recall that a K^+ transport mechanism also regulates the turgor in guard cells.)

The redwood sorrel *(Oxalis oregana)*, a delicate, shade-tolerant plant of redwood forests, thrives along the forest floor. It does well in this semidark habitat only because it is highly efficient at absorbing what light there is. The price for this adaptation is exacting: the redwood sorrel cannot survive well in bright sunlight. But as the sun's position changes, the plant's shady habitat is subject to intermittent rays of bright sunlight that penetrate small openings in the forest canopy. A second adaptation helps the sorrel avoid potential damage. In response, its leaves slowly fold downward, rising again only when the sun has moved on.

FIGURE 29.16
SOLAR TRACKING
Sunflowers turn to follow the sun throughout the day.

Solar Tracking As another example of the nastic response, consider the familiar sunflower and a common experience. Driving west past a field of sunflowers on a sunny morning can be a memorable experience, for two reasons. The flowers are strikingly beautiful, and they are all facing in your direction (Figure 29.16). Returning the same afternoon, though, can be disconcerting. The flowers, still quite beautiful, greet you once more, but they face the opposite direction! You haven't entered the "twilight zone," what you are witnessing is called **solar tracking** (also **heliotropism**). While many plants gradually *grow* towards the sun (a *phototropic* response), the faster, reversible response of sunflowers (along with cowpeas, soybeans, cotton, and others), once again, relies on a changing turgor. (As you see, the term heliotropism is really a misnomer; growth is not involved.)

Negative solar tracking responses are also known. The leaves of some drought-resistant plants regularly fold or expose only their edges when struck by bright sunlight. This keeps the surface temperature down, an important water-saving adaptation.

Thigmonastic Response The ability of some plants to respond abruptly when touched, a thigmonastic response, has fascinated people for ages. The mechanism, again involving changes in turgor in pulvini, is similar in principle to sleep movement. Our two examples are *Mimosa pudica,* the "sensitive plant," and *Dionaea muscipula,* the famous Venus flytrap.

Touching or pressing any of the many fine leaflets of *M. pudica* sends its leaves into an almost spasmlike reaction. The leaflets fold, and the leaf petioles suddenly droop (Figure 29.17a and b). Some theorists suggest that this response may discourage browsing animals, while others maintain that it simply helps the plant avoid excessive water loss when hot, dry winds blow. The second hypothesis is supported by the finding that *M. pudica* is heat-sensitive, responding to heat as it does to touch.

The thigmonastic response in *M. pudica* involves turgor changes in the motor cells of its pulvini. It is preceded by the transport of K^+ out of cells that lose turgor and into cells that gain turgor. This is typical of nastic responses.

Interestingly, when just one leaflet in *Mimosa pudica* is stimulated, say by a burning match, the responses progresses throughout the plant, from one petiole to the next. The original stimulus apparently initiates a measurable electrical disturbance, an actual current flow that passes through the plant, primarily through parenchyma cells in the vascular system. Interestingly, the electrical characteristics of this current flow are quite similar to those of the animal nerve impulse. When it reaches a pulvinus, the electrical disturbance sets up the familiar postassium-related shift in turgor, and the leaflet or leaf responds. The current is regenerated in each pulvinus and thus passes on to the next, and so on throughout the plant.

The trap of the Venus flytrap, *Dionaea muscipula,* (Figure 29.17b), actually a modified leaf, lies open when at rest. Each half of the trap has three tiny, hairlike triggers that spring the trap when brushed by a hapless insect. When triggered, the toothed leaf, now tightly closed, presses the insect against digestive glands on the inner surface. The rest is history as the plant fills its nitrogen requirements.

We should mention that the trap isn't easily fooled. The triggers have a built-in code; two hairs must be touched in succession, or one hair must be touched twice before the trap goes into action. Generally, this means that a falling twig or inert object brushing the trap will have no effect. A wandering insect, on the other hand, is more apt to stumble across the code.

The rapid closing of the Venus flytrap, while involving current flow and sudden changes in turgor, actually has a growth aspect, since it involves cell wall elongation. Recall that cell elongation in the root tip (see Chapter 27) began with the active transport of hydrogen ions (acid) into the wall, an act that brings about the loosening of the cellulose fibers. Turgor pressure then causes the cells to lengthen. The process occurs quite rapidly in the Venus flytrap, where it is concentrated in the cells forming the trap's

FIGURE 29.17
TOUCH RESPONSES IN PLANTS
(a) *Mimosa,* the touch-sensitive plant, visibly "cringes" when touched along its leaflets. **(b)** *Dionaea,* the Venus flytrap, springs its trap on a hapless insect if its triggers are touched in a certain way.

two outer walls only; those making up the inner walls remain unchanged. Opening of the trap occurs through the gradual elongation of the inner cell walls, requiring about 10 hours. It isn't surprising that the sudden and massive pumping of hydrogen ions into so many cell walls is quite energy-costly: researchers note that the ATP reserves in such tissues drop nearly 30% with each closing of the trap.

LIGHT AND FLOWERING

The response of organisms to changing lengths of day and night is known as **photoperiodism,** a phenomenon important to most forms of life. The response of many birds to lengthening days, for instance, is the formation of their spring plumage and an increase in the size of their gonads as the breeding season approaches. As the season progresses, one species of flowering plant after another begins to blossom, each in perfect synchrony. This highly coordinated photoperiodic response may mean greater reproductive efficiency since it results in the presence of enough flowers of each kind to guarantee pollination, and thus, fertilization. Such synchrony may also coincide with the life cycle of the insect or other animals responsible for pollinating that plant species.

Throughout much of the Northern Hemisphere, the early-flowering crocus is the famous herald of spring. As the days lengthen, species after species bursts into bloom in a rather predictable sequence. As summer gives way to fall, and days again begin to grow shorter, still more species will unfurl their colors. Hardly any other good news in this life is so dependable.

But one wonders, how does the crocus "know" that spring has arrived? And how does the ragweed "know" that fall is here? Or, to regain some scientific objectivity, what physiological processes stimulate the tissues of the plant to shift to flower production?

The questions are not simple ones. Biologists would like to have a better idea of how to both stimulate and inhibit the reproductive process. If the process could be controlled, florists could have their product ready just in time for Mother's Day, and even more important, farmers could regulate the production of fruit and develop a more steady market. As a matter of fact, botanists are learning to control these very things, and they are getting better at it with the passing years.

Still, the answers are not all in by any means. For example, flowering seems to be under hormonal control, but no specific flowering hormone has ever been discovered. Auxin is at least indirectly involved, but in a negative way. The concentration of auxin drops in plants that are about to go into flower; furthermore, the application of auxin can sometimes be used to prevent or delay flowering.

Photoperiodicity

As to how a plant knows what the season is, we do have some answers. The critical factor for many species is the *length of the night,* rather than the length of day, as was once believed. Of course, the length of night varies with location and with the season, but plants somehow are able to "count the hours." When the period of darkness reaches a certain length, the plant responds in a predictable manner, as though it "knows" April, June, or January has arrived.

The response of plants to changes in the length of night can be roughly organized into three categories. Plants that begin the flowering process before the summer solstice, June 21, are called **long-day plants** because they flower only when the lengthening days reach some set length (set for that species). Examples include henbane, lettuce, and some varieties of wheat, potatoes, and spinach. Plants that do not flower until after the days shorten to some critical length, generally after the summer solstice, are called **short-day plants.** Included are strawberries, poinsettias, chrysanthemums, and primroses (Figure 29.18).

The third category of flowering plants are the **day-neutrals,** which appear to be indifferent to cycles of light and darkness. They may flower continuously or respond to other stimuli. Examples include gorse and dandelions. The terminology is a little misleading because, as we said, day length is not really the critical factor. It is actually the length of night that matters. We'll show you the evidence shortly.

The Dark Clock and Flowering How was it determined that it was the duration of *darkness* that triggers flowering, and not the duration of *daylight?* Actually, the demonstration is rather simple and can be done in laboratories, where the light and dark periods can be regulated. Under such conditions it was found that the period of light can be changed without causing any flowering changes, but if the length of darkness is

(a)

(b)

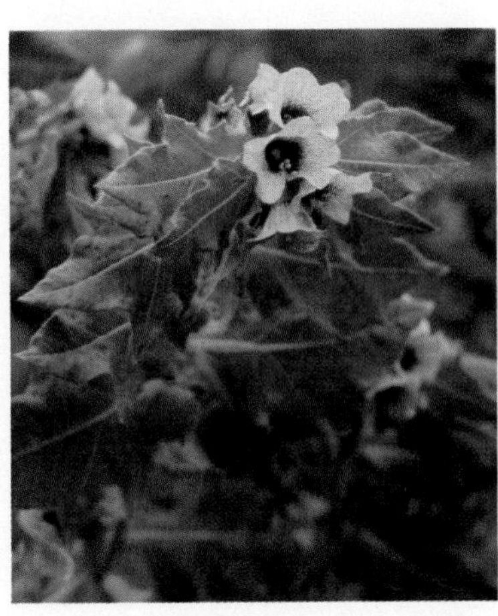
(c)

FIGURE 29.18
PHOTOPERIODICITY
(a) The chrysanthemum is a short-day plant that produces flowers in autumn. **(b)** Poinsettias are also short-day plants and will bloom when the night is longer than 13 hours. **(c)** For its flowering, the henbane requires a long day (short night), usually in excess of 12 hours.

tampered with, the plant responds. Quite dramatically, a single, relatively brief exposure to intense light in the middle of the night can "trick" a long-day (short-night) plant into responding as if the short nights of spring had arrived. Such a plant can then be made to bloom in any season. The same treatment, if done every night, can prevent a short-day (long-night) plant from flowering at its normal late summer or autumn time, since it will react as if the nights were still too short for such activity.

This bit of academic tinkering was immediately put to practical use by chrysanthemum growers. For many years they had extended the chrysanthemum season into early winter, artificially lengthening the days by keeping bright lights turned on for several hours after sunset. But now they found they needed only to turn on the lights for a few minutes each night in order to get the same results. Similarly, but much more important economically, sugar cane growers can stall flowering in their fields by turning floodlights on at night. Sugar-laden sap that might otherwise be used to produce commercially useless flowers can then continue to accumulate in the stems.

The cocklebur has proved to be an ideal organism for studying the initiation of flowering. It is so hardy that researchers have been able to alter it drastically and still count on its survival. The cocklebur is a short-day (long-night) plant, and will put out homely little green flowers if exposed just once to a period of darkness longer than 8 ½ hours.

Armed with that knowledge, plant physiologists began to toy with the plant. They found that varying the wavelengths of the light had curious results (Figure 29.19). The most effective light for establishing the photoperiod was found to be in the red or orange-red region. Far-red light, with its longer wavelength, had the opposite effect; that is a flash of far-red light actually reversed the effect of either white or red light if it immediately followed either of them. However, if the far-red flash *preceded* the red flash, or was *delayed* for more than 35 minutes after the red flash, the red flash had its full effect. Curious indeed.

The midnight flash experiments gave rise to the hypothesis that there was some receptive pigment involved. The receptive pigment could not be chlorophyll (which is abundant in the photosensitive leaves) because it responds to different wavelengths from those produced by the floodlights. The scientists suggested that perhaps there is another pigment that occurs in two forms: P_r, the red-absorbing form, and P_{fr}, the far-red-absorbing form. (P_r absorbs maximally at 600 nm, whereas P_{fr} absorbs maximally at 730 nm.) Absorption of light of the appropriate wavelength would change this hypothetical pigment from one form to another. During the day, the predominance of red light over far-red would convert all the pigment from the P_r form into the P_{fr} form, but at night there would be a spontaneous reversion of the pigment back to the P_r form. A midnight flash of red light would immediately convert the pigment once again into the P_{fr} form, but in this state it could again be reversed to P_r by absorbing far-red light.

Light-induced change:

Far-red light $P_{fr} \rightarrow P_r$

Red Light $P_r \rightarrow P_{fr}$

Spontaneous dark reaction: $P_{fr} \rightarrow P_r$

There was joy in the streets when the hypothetical pigment was found. It was named **phytochrome**, and it turned out to be a large membrane-bound protein complex. It now seems that phytochrome is involved in a number of other light-induced phenomena, such as the turning of leaves toward light and the rapid orientation of chloroplasts as they move broadside to the light source.

Phytochrome indeed occurs in the two forms hypothetized, but apparently it also goes through many other conformational changes, that are not at all simple or well understood. At one time it was thought that the conversion of one form of phytochrome to another was a slow process that might "time" the duration of the night, but in fact these conversions are relatively rapid. There is clearly some kind of "dark clock" involved, but the phytochrome itself is apparently only the detector of light and the trigger that

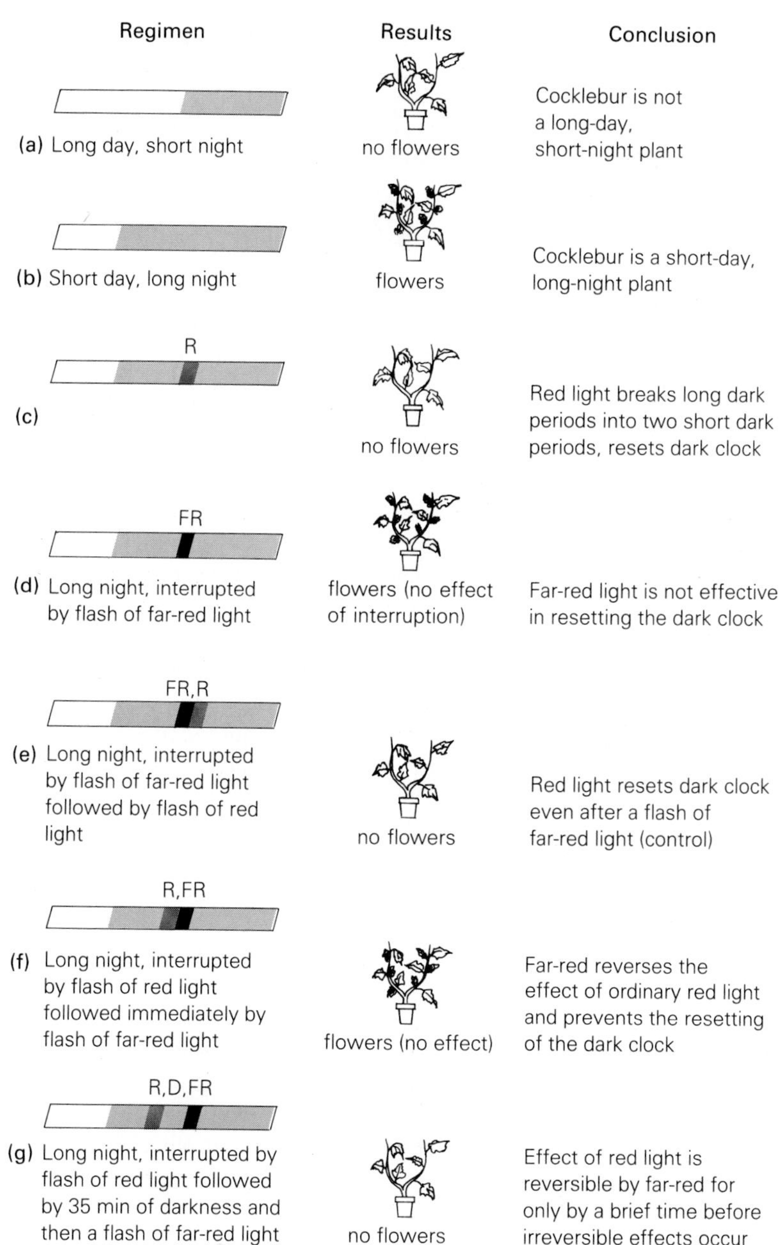

Regimen
Results
Conclusion
(a) Long day, short night
no flowers
Cocklebur is not a long-day, short-night plant
(b) Short day, long night
flowers
Cocklebur is a short-day, long-night plant
R
(c)
no flowers
Red light breaks long dark periods into two short dark periods, resets dark clock
FR
(d) Long night, interrupted by flash of far-red light
flowers (no effect of interruption)
Far-red light is not effective in resetting the dark clock
FR,R
(e) Long night, interrupted by flash of far-red light followed by flash of red light
no flowers
Red light resets dark clock even after a flash of far-red light (control)
R,FR
(f) Long night, interrupted by flash of red light followed immediately by flash of far-red light
flowers (no effect)
Far-red reverses the effect of ordinary red light and prevents the resetting of the dark clock
R,D,FR
(g) Long night, interrupted by flash of red light followed by 35 min of darkness and then a flash of far-red light
no flowers
Effect of red light is reversible by far-red for only by a brief time before irreversible effects occur

FIGURE 29.19
PHOTOPERIODS IN THE COCKLEBUR
(a,b) The cocklebur, a short-day plant, will flower when the nights become long. Its requirement of a specific dark period to initiate flowering is demonstrated at **(c)**, where the period is interrupted by a flash of red light. **(d)** A similar flash of far-red light has no effect. **(e)** Preceding the red flash by a flash of far-red light apparently has no effect on the inhibiting effect of red. **(f)** But when the flashes are reversed, with far-red following red, the effect of red is reversed, and the flowering occurs. **(g)** The timing of these flashes is critical, since a lapse of 35 minutes or more between red and far-red flashes is apparently enough time for the effect of red light to reset the "dark clock," inhibiting flowering.

sets the dark clock. According to this idea, a flash of red light in the middle of the night resets the dark clock back to zero, just as it is set to zero at the end of a normal day.

But just what *is* the dark clock, and how does it work? Again, we don't know. It is presumably some kind of chemical reaction, but unlike most chemical reactions it is almost completely insensitive to temperature differences within normal ranges. (This would be appropriate for any light-activated molecule.) It is a most peculiar phenomenon indeed.

Transmitting the Stimulus The dark clock and its photoreceptor, phytochrome, are located in the leaves. The signal that actually induces flower development has to be delivered from the leaves to other parts of the plant. In the resilient cocklebur, an isolated, amputated leaf can be subjected to an appropriate flower-inducing photoperiod regimen and subsequently grafted back onto a plant, causing the plant to flower. The signal is evidently hormonal, but, as we said, no flower-inducing hormone has been found. In some experiments using plants other than the cocklebur, it appears that there are *repressing* hormones produced by the leaves at all times *except* when the dark clock indicates that the appropriate time for flowering has arrived.

FIGURE 29.20
THE LEAF, A PHOTORECEPTOR
This dramatic experiment lends strong support to the idea that the photoreceptor is in the leaf and that a transmitted agent is involved. Only one leaf in this series of six grafted cocklebur plants is exposed to the proper photoperiod, yet all six plants produce flowers.

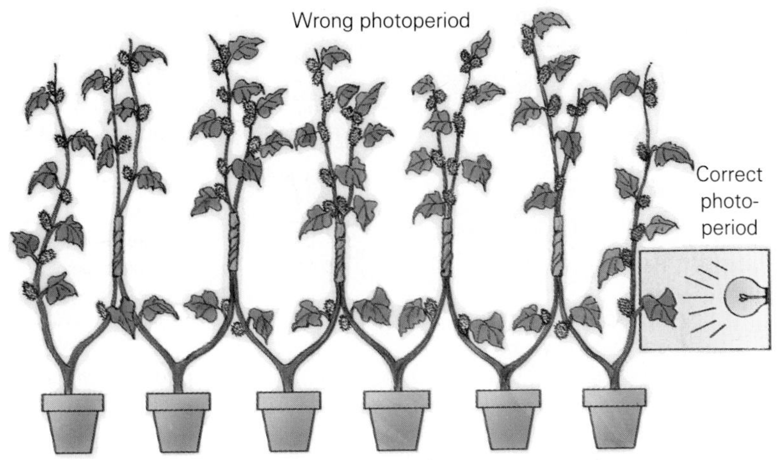

The hormonelike transmission of the photoperiod effect has been demonstrated most convincingly by an experiment in which six cocklebur plants were grafted together in a row (Figure 29.20). One leaf of the plant at one end of the line was enclosed in a box and given the appropriate photoperiod treatment, after which *all six plants* flowered, one after the other, right down the line. In other grafting experiments, the cambia of the two plants were separated by a sheet of paper, but the message got through anyway, just as we would expect if the signal were hormonal.

The hypothetical flower-inducing hormone has somewhat presumptuously been named **florigen.** However, because it has proven to be so elusive, some plant physiologists believe that there is no special flower hormone at all but that the message is conveyed by particular levels and combinations of auxin, gibberellins, and other known plant growth hormones.

We are aware that, paradoxically, auxin has an inhibitory effect on flowering. Since growing tips of shoots produce auxin in quantity, flowering can sometimes be induced or increased by simply cutting off the shoot tips. Rose growers have discovered this, and the clicking of pruning shears can be heard over the countryside at the beginning of the floral season.

Pineapples and Auxin Pineapples are different: in pineapples, auxin helps induce flowering. Pineapples are also tropical plants, living where there is little seasonal change in day length. Not surprisingly, they are indifferent to photoperiods and normally bloom and bear fruit on irregular schedules of their own, all year long. In the past, pineapple pickers had to roam the fields daily, looking for ripe pineapples—an expensive and time-consuming process. Now, growers synchronize the flowering of whole fields by the application of artificial auxins. The plants dutifully flower and ripen all at once, and can be efficiently harvested by machines. That's why pineapples are cheaper than they used to be. (But do you suppose we're paying hidden costs by spraying artificial auxins into the environment?)

APPLICATION OF IDEAS

1. Formerly, plant physiologists attributed the success of plant roots in reaching water to *hydrotropism,* a water-initiated growth response. While the idea has now been discarded, root growth is known to be most dense in moist areas of soil. Suggest an alternative hypothesis to explain this observation. How do roots happen to find water? How might your hypothesis be tested?
2. Many angiosperms respond to photoperiods in their flowering activities. What might be some advantages to complex flowering systems that relate to day or night length? If photoperiodism is adaptive to plants, how can the success of the day-neutral plants be explained?

KEY IDEAS

Charles Darwin was among the first to experiment with plant growth mechanisms.

LIGHT AND THE GROWTH RESPONSE

1. Working with canary grass, Darwin found that the young **coleoptile**-covered shoot would respond to light by bending toward it. By experimenting, he found that the agent responsible was formed in the tip and passed downward.
2. Boysen-Jensen decapitated oat seedlings, mounted tiny blocks of gelatin, a very permeable substance, on the stump, and then placed the tips on the gelatin. On others, he used a tiny chip of impermeable mica to separate the tip from the stump. The response toward light occurred only in the group where gelatin was used, indicating something from the tip had to pass downward for the bending response to occur.
3. Paal decapitated oat seedlings and used the tip alone for experiments. He placed the tip to one side of the stump, and the bending response always occurred on that side. He concluded that some influence from the tip caused cell elongation.

The Isolation of Auxin

1. Fritz Went placed oat seedling tips on agar for a time and then used tiny agar blocks in his experiments. Arranging the agar blocks on the seedling stumps in various ways, and keeping his seedlings in darkness, he observed the usual bending, indicating that the active substance had diffused into the agar.
2. Went named the active substance **auxin** (later identified as **indoleacetic acid** or **IAA**). He also worked out a **bioassay** system.

Auxin: Its Structure and Roles

1. Auxin's primary action is to stimulate cell elongation. The mechanism may involve lowering the cell wall pH and thus activating microfibril-loosening enzymes.
2. Auxin is involved in **abscission** (leaf fall). Prior to leaf fall, changes in cells at the base of the petiole produce an **abscission zone.** Then, as auxin diminishes, materials in the leaf are reclaimed by the plant, cell walls in the abscission zone break down.

OTHER PLANT GROWTH HORMONES

Gibberellins

Gibberellins are known to cause dramatic stem elongation. They restore normal growth to genetic dwarf corn. When released by the embryo of grains, they stimulate the aleurone to produce starch-digesting **alpha amylase.**

Cytokinins

1. **Cytokinins** such as **zeatin** and *kinetin* stimulate cell division. Together, auxin and kinetin influence differentiation in the **callus** formed from tobacco pith cells. The relative proportions and concentrations of the two hormones determine whether the growth will form shoots, roots, both, or neither.
2. The capability of plant cells to **dedifferentiate** from a specialized state and then redifferentiate into a new tissue type is called **totipotency.**
3. Cytokinins retard **senescence**, or aging, in plants. These hormones may interact with tRNA, facilitating protein synthesis.

Ethylene

Ethylene is a gas derived from the amino acid methionine. It hastens fruit ripening and helps retain the hypocotyl hook in some emerging seedlings.

Abscisic Acid

Abscisic acid inhibits auxin and gibberellins, bringing on plant dormancy in preparation for winter. It also participates in the stomatal water-stress response.

Applications of Plant Hormones

Many plant hormones or their analogs are now produced synthetically. 2,4-D is a **biodegradable** dicot weed control agent. 2,4,5-T, a defoliant, is used in controlling tree growth.

PLANT GROWTH RESPONSES AND MOVEMENTS

Plant growth responses, called **tropisms**, include **phototropism**, **gravitropism**, and **thigmotropism**, are directional. **Nastic responses,** movements based on reversible turgor changes, are generally non-directional.

Phototropism

1. In phototropism, a blue-sensitive photoreceptor detects light direction, and, through processes still unknown, the migration of auxin is directed to the unlighted side of the shoot, where elongation then occurs.
2. Auxin plays an inhibitory role in **apical dominance,** whereby one growing shoot elongates and the rest are inhibited to some degree. Loss of the dominant tip releases inhibition, and another branch becomes dominant.

Roots and Gravitropism

1. The gravity-oriented bending of roots, gravitropism, occurs when unequal distributions of auxin occur in the root tip. In the vertical root, auxin is equally distributed, cell elongation occurs equally, and no bending occurs. When a seedling root is placed horizontally, auxin concentrates on the lowermost side, inhibiting elongation, whereas the lighter concentration on the upper side permits elongation, and thus the root bends toward gravity.
2. Calcium affects the distribution of auxin in the root tip. It is stored in the endoplasmic reticulum (ER) of cells in the **columella** of the root cap and released when **amyloplasts** press against the ER. In the vertical root tip, the resting amyloplasts cause the release of calcium ions into the cytoplasm below. The ions

activate calmodulin, which in turn activates enzymes that affect the pumping of calcium ions and auxin, which are evenly distributed. In the horizontal root tip, the amyloplasts cause the release of calcium ions near the side wall, activating pumps there and thereby pumping calcium ions and auxin into the lowermost cells, which are inhibited.

Thigmotropism

Touch-induced growth responses occur in **tendrils**, which respond by uneven cell elongation opposite the touching part. This produces a coiling growth about solid objects.

Nastic Responses

1. Nastic responses include the folding and turning of leaves and flowers. They are brought about by turgor changes in **motor cells** of **pulvini**, caused by the transport of potassium ions from one part of the petiole base to another. Included are sleep movements, the daily unfolding and folding or turning of leaves.
2. The continuous daily turning of leaves or flowers towards or away from the sun is called **solar tracking.**
3. Touch-initiated nastic responses, as seen in *Mimosa pudica* and *Dionaea muscipula,* are rapid. In the first, touch or heat causes a current flow that activates potassium pumps in the pulvini. Subsequent turgor changes bring about the collapse of leaf petioles and leaflets. In *D. muscipula,* the trap—a modified leaf—is triggered by the movement of hairs, whereupon standard cell elongation in the outermost cells causes the trap to close.

LIGHT AND FLOWERING

Photoperiodism is any response to changing lengths of night or day. Flowering in plants is often photoperiodic.

Photoperiodicity

1. In their response to photoperiods, plants fall into three categories: **long-day, short-day,** and **day-neutral.**
2. Plants respond to the length of night rather than the length of day. A brief exposure to light during the night will induce flowering in long-day plants and stall flowering in short-day plants.
3. The most effective light for interrupting flowering is in the red region of the spectrum, while far-red reverses the red effect if it follows within 35 mintues. Two interconvertible variations of a light receptor, P_r and P_{fr}, have been proposed. The receptor has been identified as **phytochrome,** a large, membrane-bound protein, but its operation as a "dark clock" isn't understood.
4. Experiments with the cocklebur strongly suggest a hormonal mechanism that may work through the repression of flowering at all times except during the proper photoperiod.

REVIEW QUESTIONS

1. Describe Charles Darwin's experiments. What conclusion did he reach? (639-640)
2. What information did Boysen-Jensen and Paal add to what Darwin had already determined? (640)
3. Describe the technique used by Went, and explain how his assay procedure worked. (640-641)
4. Describe the specific action of auxin on plant cells and the related enzymatic mechanism. (642)
5. List the main steps involved in leaf abscission. What role does auxin play? (643)
6. Briefly describe how gibberellins were discovered. What effect do they have on dwarf corn? (643-644)
7. Explain the role of gibberellins in the germination and growth of grain embryos. (644)
8. Explain the concept of totipotency. How do agriculturalists make use of this capability? (644-645)
9. What effect do cytokinins have on senescence? How might cytokinins interact with auxins in this role? (645)
10. How might abscisic acid help temperate zone plants survive? (646)
11. Name two auxin analogs and describe how they are used. (647)
12. Using the idea of a photoreceptor and the pattern of auxin diffusion, present an explanation of phototropism. (648)
13. Briefly explain how pruning affects apical dominance in stems. (649)
14. Suggest two basic differences between tropic and nastic responses and provide two examples of each. (637)
15. Explain how the distribution of calcium and auxin in vertically and horizontally arranged seedling roots differs, and how this affects root tip growth. (649-650)
16. What role do amyloplasts and calmodulin play in gravitropism? (649-650)
17. Review an example of the nastic response and explain the role of pulvini and motor cells. (651-652)
18. Summarize the events involved in the thigmonastic response in *Mimosa pudica.* How does this response differ from sleep movement? Explain the sudden closing of the Venus flytrap. Why might this be considered to be a tropism? (651-653)
19. List and define the three photoperiodic categories of plants. (653-654)
20. Briefly explain how plant physiologists determined that the period of darkness was the significant factor in the photoperiodic flowering responses. (654-655)
21. Describe the experimental observations that led to the conclusion that the flowering photoreceptor had two forms, P_r and P_{fr}. Write a simple equation showing the effects of red and far-red light. (655)
22. Explain how a repression system might work to instigate flowering. (657)
23. What is the status of the flowering hormone hypothesis today? (657)

6

DIVERSITY AND FUNCTION: ANIMALS

30

INVERTEBRATES I

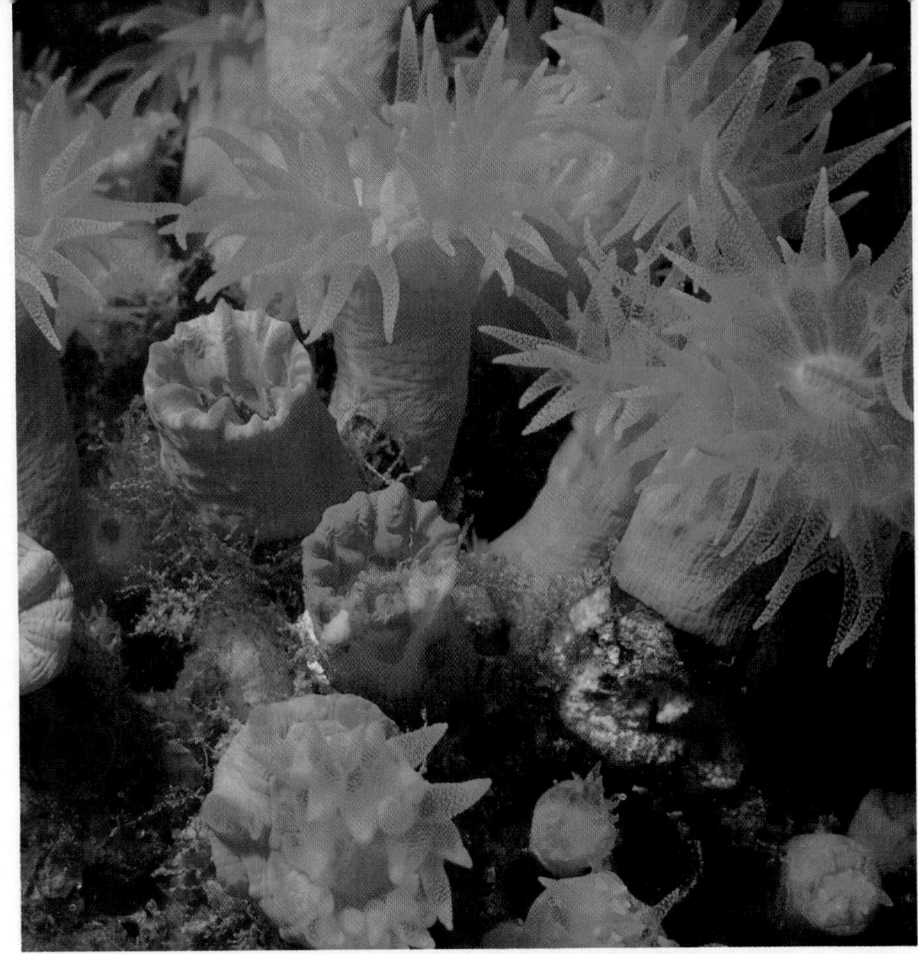

IF YOU WERE DRIVING ACROSS THE COUNTRYSIDE OF KENYA AND PASSED A troop of baboons, would it be safe to stop, get out, and walk among them, or are they dangerous? Would you be afraid of wolves if you walked across Isle Royale in Michigan? Why don't people use watchcats instead of watchdogs? Can a horse be trusted to guard the pasture? Is it safe to pick up a centipede? Obviously, our fellow animals on the planet are quite diverse. Yet through all of them there runs the common thread of origin. We will now consider the diversity of animal life that shares this place. We will focus not only on the bewildering, fascinating, and even perplexing array of animal life, but also on those great underlying themes that bind us all together.

WHAT IS AN ANIMAL?

If someone asked you what an animal is, you could probably come up with a fair definition. But you would probably be wrong just the same. The reason is that animals are so varied that exceptions can be found to almost every rule. So what kinds of rules can we make? Some would simply say that animals move around and have mouths and eat things. That's not a bad start. If we add "and are multicellular," we've pretty much covered the animal kingdom. But some animals would virtually shout, "Not me!"

Because of the difficulties of finding commonality among such diversity, any description of the group must be both broad and precise. Generally, then, we can say that animals have the following traits:

1. *Multicellularity*. Animals are made up of numerous, complex eukaryotic cells.
2. *Organization*. Most animals have well-organized organ systems that carry out the various life functions (such as digestion, reproduction, gas exchange, and so on).
3. *Movement*. Animals are able to move by the contraction of organized muscle or contractile elements.

4. *Support.* Cells and tissues are bound together by the complex animal protein collagen, which, in vertebrates, is the most common protein.
5. *Nutrition.* Animals require complex nutrients, including carbohydrates, fats, proteins, and vitamins.
6. *Reproduction.* Animals are essentially diploid, with gametic life cycles. Meiosis leads directly to gamete formation, characterized by large nonmotile eggs and minute, motile sperm. Many species reproduce asexually.

As we said, these are general statements. There are some very conspicuous exceptions to these rules.

Most of us are probably familiar with at least a few representatives of the most common phyla. Interestingly, even members of the same phyla may be quite different, so although we may be somewhat familiar with two kinds of animals, we may not be aware that they are related. For example, humans are chordates (phylum **Chordata**), yet we share the phylum with creatures with odd names like "sea squirts" and "salps." (When you're looking at a salp it can take all your concentration to believe it's a phyletic relative.) Other commonly encountered phyla are represented by insects (phylum **Arthropoda**), earthworms **(Annelida)**, and snails **(Mollusca)**. You probably know that your dog may have roundworms **(Nematoda)**, and if you frequently swim in the sea, you may have felt the sting of a jellyfish **(Cnidaria)**. Coastal dwellers also know that tidepools abound with sea stars **(Echinodermata)**, sponges **(Porifera)**, and flatworms **(Platyhelminthes)**.

We have just named the nine major animal phyla, but there are just as many rarer ones (although we will not consider them here). Each phylum has certain characteristics that set it apart from others, and each has a considerable range of diversity within it. All are quite fascinating, and some may include species that might be regarded as weird.

ANIMAL ORIGINS

The questions "Who are we?" and "Where did we come from?" have long interested scientists, as well as philosophers, mystics, theologians, primitives, and drunks. From the scientific view, it seems certain that today's plants, animals, and fungi sprang from ancient protists. The consensus among biologists is that the animal kingdom has its roots in the ancestors of two different protist lines. One of these, a group of flagellate protozoans (see Chapter 23), produced the subkingdom **Parazoa**, consisting of one phylum, Porifera, the sponges. The other gave rise to the other subkingdom, **Metazoa**. It includes the rest of the earth's animals, all probably derived from a single metazoan line, which in turn was derived from a protozoan of some sort. No one knows exactly what sort, but one leading hypothesis is that the unicellular ancestor of the metazoans was a flagellated protozoan, although one different from the parazoan ancestor. Flagella in today's animals are found in sperm.

Unfortunately, there is no fossil evidence that bears on the issue at all. In fact, the oldest eukaryotic fossils are similar to some of today's metazoa. These animals, which lived some 680 million years ago, were not very different from modern jellyfish. This leads us to the question, "How did the first multicellular animal come into being?" Let's see if we can marshal our informed ignorance to develop a scenario that might put together some of these ideas.

The basic question is, how does one go from a single-celled state to a multicellular one? According to one theory, through the aggregation of single-celled forms. As single-celled organisms joined, they would have formed groups that would have become increasingly coordinated and interdependent. The interdependence, according to the scenario, would have grown stronger as the various cells began to specialize, taking on different functions. Some, for example would have specialized in movement, others in feeding, and still others in reproduction. With such specialization, the coordination of the cells would have become ever more diverse and efficient. Then, as the cells reached a state of total interdependence, they would achieve true multicellularity.

Most biologists agree that the earliest developments toward multicellularity in animals must have occurred in the ancient seas, among marine protozoans (Figure 30.1). Perhaps

FIGURE 30.1
MARINE FLATWORM
Thin-bodied marine flatworms are seen gliding over rocky surfaces of the sea bed. Their simple organization suggests what some of the first metazoan animals might have looked like.

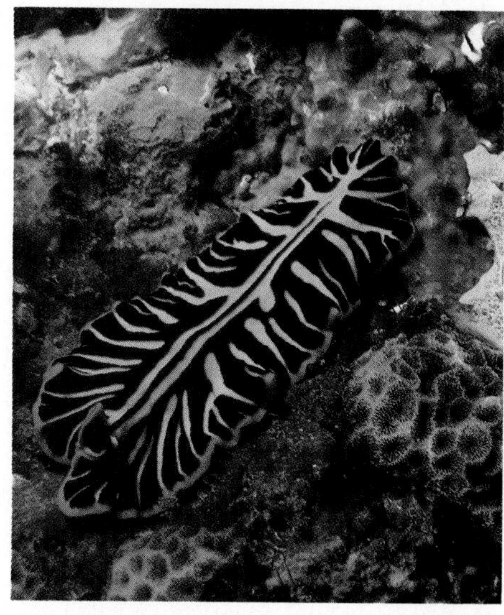

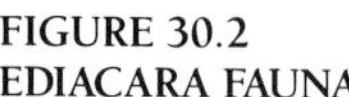

FIGURE 30.2
EDIACARA FAUNA
The Ediacara fossils from the late Precambrian seas are dominated by the fossilized remains of thin-bodied cnidarians. Among these are (**a**) the jellyfish—not unlike those seen today—and (**b**) stalked, featherlike corals, nearly identical to today's "sea pens." The bottom dwellers include several species of annelids (**c**), segmented worms that are identifiable by the lines crossing the body. Arthropods (**d**) are also represented, as are a few shelled mollusks (**e**) and some fossils from phyla that are now extinct (**f**). The egglike mass (**g**) is believed to be algae.

the simplest crawling or stationary forms were followed by floating or swimming versions, or maybe animals anchored by long stalks—any adaptation that could enable the organism to begin to exploit new food resources. Figure 30.1 shows an existing animal that is thought to be similar to some of the earliest animal life. (See also Chapter 24 for a discussion of the evolution and advantages of multicellularity.)

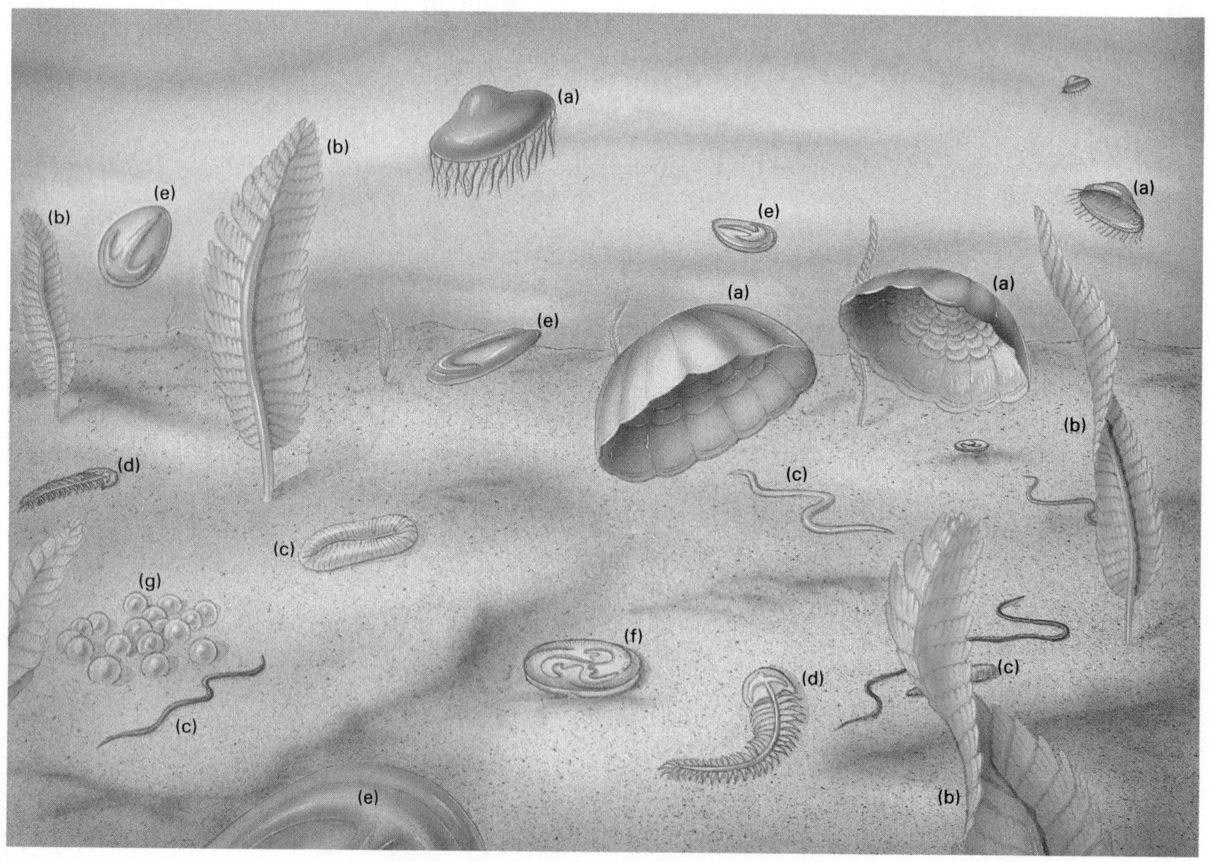

The Early Fossil Record

Most authorities agree that animal life extends back no further than one billion years. The earliest indisputable animal fossils, the **Ediacara fauna** are found in late Precambrian deposits from southern Australia. The deposits are 600–700 million years old, and marked by scattered remains of numerous fossil jellyfish, soft-bodied corals, and worms (Figure 30.2). (See Table 30.1 for a review of geological time.) Still older rocks in the area bear what seems to be fossilized tracks and burrows that might have been made by marine worms.

An even richer fossil deposit, estimated to be about 530-550 million years old, occurs in the **Burgess Shale Formation** of Western Canada. Its remarkable fossils, many of which are of soft-bodied animals, include representatives from *all* of today's major animal phyla. Many of the animals were apparently more complex than those of the Ediacara fauna (Figure 30.3).

The great diversity of animal life seen in the late Precambrian appears to have arisen very suddenly, at least in geological terms. The most compelling explanation for this sudden burst in speciation and evolutionary divergence is a widespread and massive animal extinction known to have occurred some 650 million years ago. Such extinctions have occurred again and again in the history of animal life, and each is followed by new bursts of speciation among the survivors. Recall that such sudden episodes of intense evolutionary activity are explained through the theory of punctuated equilibrium. (See Chapter 20 and Essay 32.1.)

TABLE 30.1 GEOLOGIC TIMETABLE

ERAS (YEARS SINCE START)	PERIODS	TIME PERIOD IN MILLIONS OF YEARS
Cenozoic	Quaternary	
	Holocene (present)	last 10,000 years
	Pleistocene	.01–2
	Tertiary	
	Pliocene	2–6
	Miocene	6–23
	Oligocene	23–35
	Eocene	35–54
(65,000,000)	Paleocene	54–65
	Cretaceous-Paleocene discontinuity	
Mesozoic	Cretaceous	65–135
	Jurassic	135–197
(225,000,000)	Triassic	197–225
Paleozoic	Permian	225–280
	Carboniferous	280–345
	Devonian	345–405
	Silurian	405–425
	Ordovician	425–500
(570,000,000)	Cambrian	500–570
Precambrian	Most invertebrate phyla present	570–4500
(Origin of earth, 4.5 billion years)		

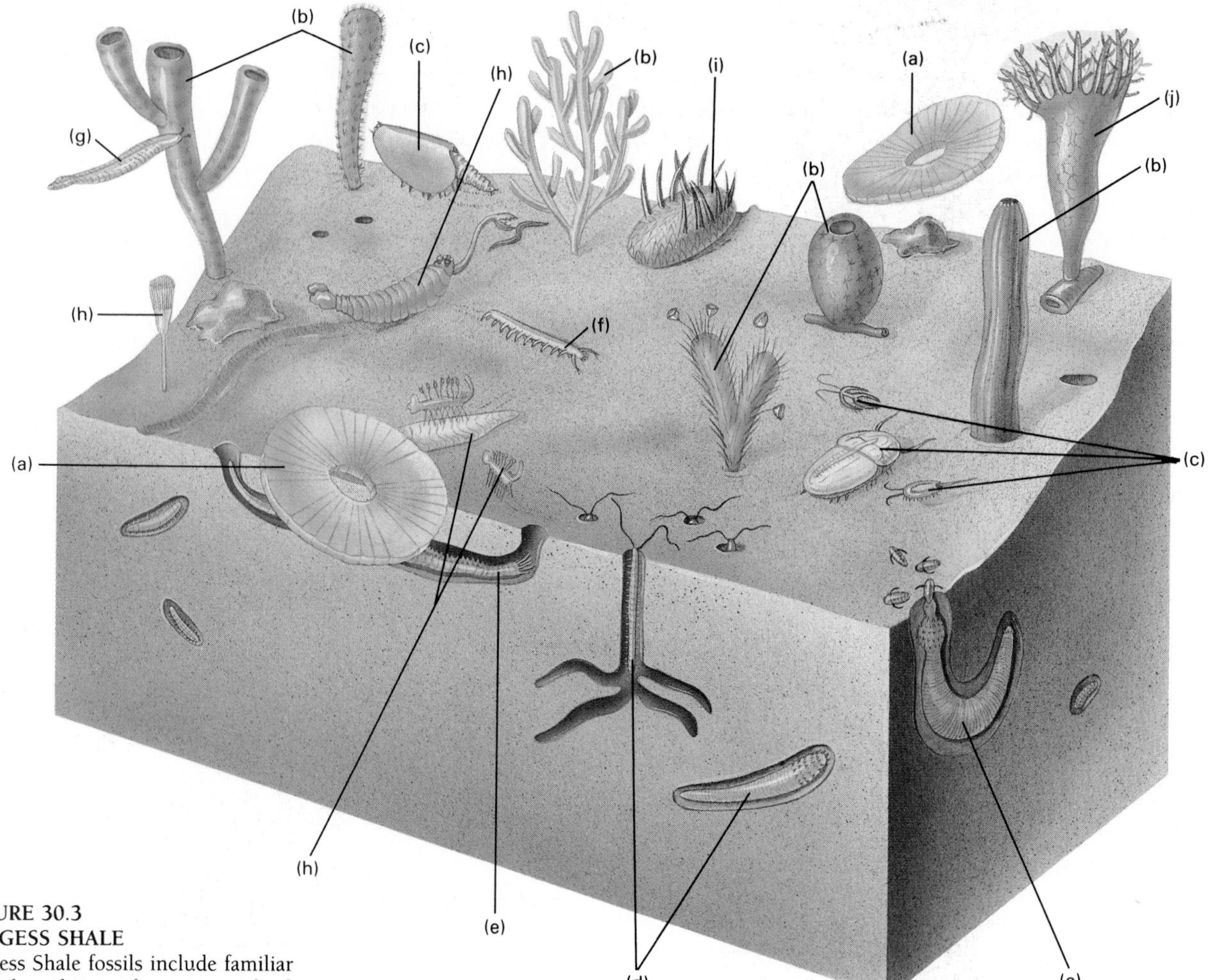

FIGURE 30.3
BURGESS SHALE
Burgess Shale fossils include familiar animals and some that are completely unknown to zoologists. The "pineapple slices" (**a**) are cnidarians, an animal group not nearly as common here as in the earlier Ediacara fauna. The large stalked animals (**b**) are primarily filter-feeding sponges, while the many-legged creatures scurrying across the ocean floor (**c**) are arthropods. Tube-dwelling annelids (**d**) are also seen. The large wormlike creatures in the U-shaped burrows (**e**) are called priapulids. The ancient marine onychophoran (**f**) closely resembles the modern terrestrial onychophorans. The Burgess Shale fossils include primitive chordates closely resembling today's *Branchiostoma* (**g**). Some species are unknown today (**h**), but the peculiar, spiny creature (**i**) is a mollusk, and the stalked animal with the feathery tentacles (**j**) is an echinoderm.

TABLE 30.2 CHARACTERISTICS OF THE MAJOR ANIMAL PHYLA

PHYLUM	LEVEL OF ORGANIZATION	BODY SYMMETRY	CLEAVAGE PATTERN	LARVAL TYPE	DIGESTIVE TRACT	BODY CAVITY	SEGMENTATION
SUBKINGDOM PARAZOA							
Porifera	Cellular	Radial	NA	Flagellated	None	Primitive spongocoel	No
SUBKINGDOM METAZOA							
Cnidaria	Tissue–Organ	Radial	NA	Ciliated planula	Gastrovascular cavity	None	No
PROTOSTOMES							
Platyhelminthes	Organ system	Bilateral	NA	Similar to trochophore	Gastrovascular cavity	None (acoelomate)	No
Nematoda			Spiral cleavage	None	Complete gut	Pseudocoelom	
Mollusca				Trochophore		Coelom (schizocoel) (Greatly reduced in mollusks)	Yes (greatly reduced in mollusks)
Annelida							
Arthropoda				Nauplius in some crustaceans			
DEUTEROSTOMES							
Echinodermata	Organ system	Pentaradial	Radial cleavage	Dipleurula or similar	Complete gut	Coelom (enterocoel)	No
Hemichordata		Bilateral					
Chordata				Unique where occurring			Yes

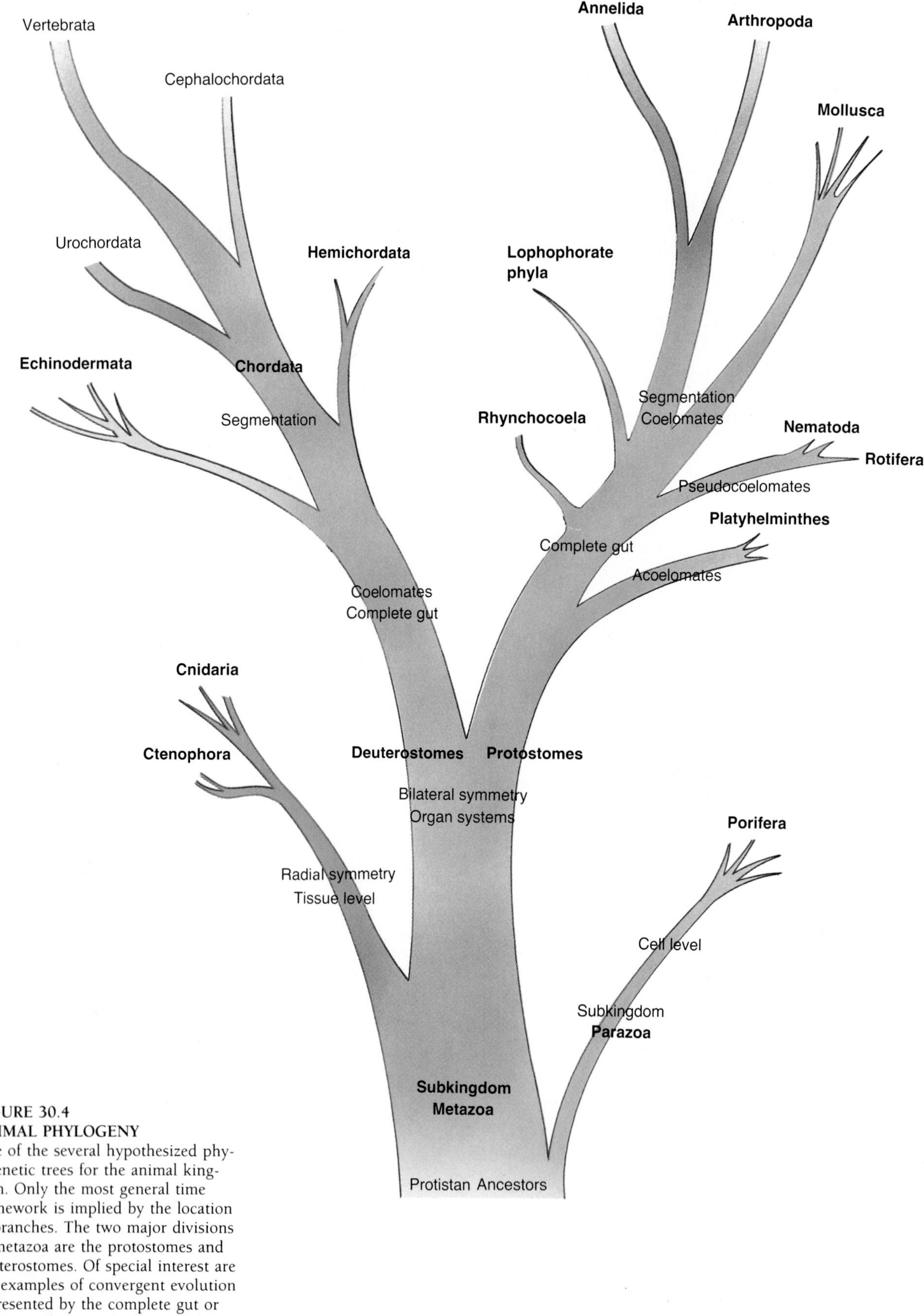

FIGURE 30.4
ANIMAL PHYLOGENY
One of the several hypothesized phylogenetic trees for the animal kingdom. Only the most general time framework is implied by the location of branches. The two major divisions of metazoa are the protostomes and deuterostomes. Of special interest are the examples of convergent evolution represented by the complete gut or coelom and segmentation in both the deuterostomes and protostomes.

Phylogeny of the Animal Kingdom

The evolutionary history and relationships among animal phyla can be represented by a phylogenetic tree (see Figure 20.7). The tree shown in Figure 30.4 is only one of several possibilities, but it does indicate what are believed to be some of the major milestones of animal evolution. As with all such trees, it uses several sources, including the fossil record, comparative biochemistry and anatomy, physiology, and embryology.

Note the tree's major branchings. Each branch marks significant evolutionary changes that gave rise to the various major animal phyla. Note in particular that the largest, most fundamental divergence is between what are called **deuterostomes** and **protostomes**, designations that involve most of the animal phyla. Our own phylum, Chordata, is located on the left branch along with echinoderms and hemichordates. The right branch contains several other large phyla, the annelids, arthropods, and mollusks. Strange as it seems, the profound differences between protostomes and deuterostomes exist largely because of the timing of the appearance of the mouth in the developing embryo, the way the early embryo undergoes cleavage and other even more subtle distinctions.

The terms "protostome" and "deuterostome" literally mean "mouth first" and "mouth second." In protostomes the mouth opening forms earlier in development than it does in deuterostomes. We will look further into the distinctions between protostomes and deuterostomes in the next chapter (see Essay 31.1 and Table 30.2, p.665).

LEVELS OF ORGANIZATION AND PHYLUM PORIFERA

In considering the various phyla, we will note landmark evolutionary changes as we move from one group to the next. Specifically, we will be looking at five conditions:

1. level of organization
2. body symmetry
3. number of embryonic germ layers
4. presence or absence of body cavities
5. presence or absence of a complete gut

To begin, let's note that anatomically, most animals have several **levels of organization**: cell, tissue, organ, and organ system (Figure 30.5). The cell constitutes the fundamental level that still has the properties of living systems. The second level, the **tissue**, consists of a group of specialized cells with a common function. The lining of the intestine, for instance, is a tissue that specializes in food absorption. Tissues, in turn, are organized into **organs**, such as the intestine itself, which perform all or part of a major function. The intestine contains absorbing tissue along with muscle, nerve, and secretory tissues. Organs generally interact with other organs to form the **organ system** or, simply, **system.** The digestive system, including the mouth, esophagus, stomach, intestine, and so on, is an organ system, as are the respiratory and reproductive systems.

Sponges (phylum **Porifera**) exhibit only the cellular level of organization, making them the simplest animals. Essentially, a sponge consists of only a few specialized but loosely organized types of cells with little coordination among the cells. In fact, sponges can be completely separated into individual cells in a laboratory dish, and their cells will move back together and form new sponges.

Earlier, we noted that sponges probably evolved from a line of flagellated protists, but the time in which this occurred is not known. Fossil sponges, as we've seen, were present in the Ediacara fauna from Precambrian times. By the Paleozoic era, calcareous (chalky) sponges were common, and the siliceous (glassy) sponges became widespread by the mid-Paleozoic.

The Biology of Sponges

Most sponges live in clear, shallow ocean waters around the world, although a few live at great depths and some live in fresh water. The adults are sessile (fixed in place), living quietly attached to the bottom, where they exhibit no apparent movement. While sponges are often small, some are more than two meters across.

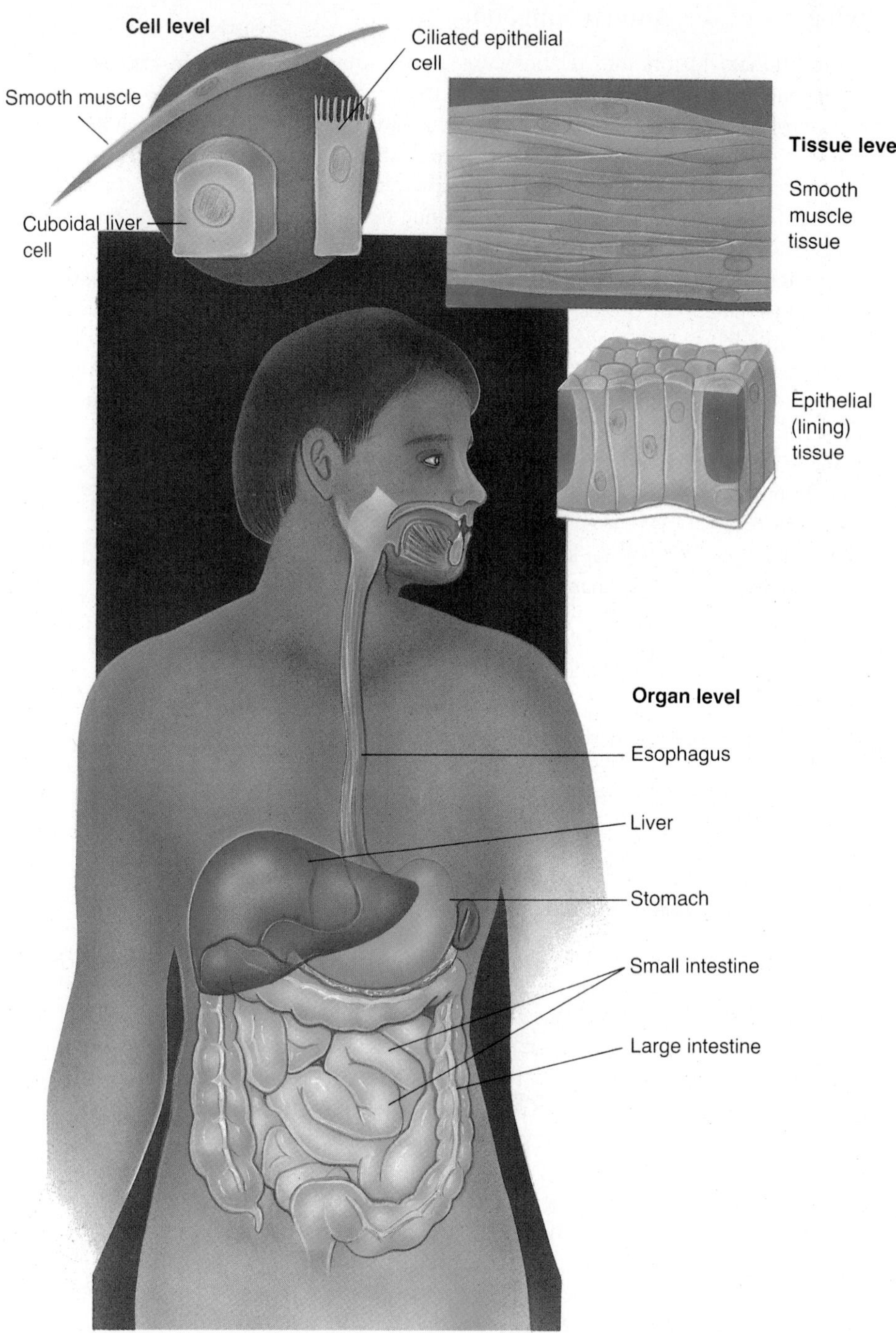

FIGURE 30.5
LEVELS OF ORGANIZATION
Most animals have several levels of organization: cell, tissue, organ, and organ system.

In the simpler sponges, the body is vaselike, with a large central cavity or **spongocoel**, and an opening, the **osculum**, at the top (Figure 30.6). The outer wall of the thin body is riddled with porelike cells called **porocytes**, which admit water into the central cavity. A jellylike, unstructured region, the **mesohyl**, separates the outer cell layer from the inner and houses wandering **amebocytes.** The inner wall is made up of numerous flagellated collar cells, the **choanocytes**, which line the central cavity and there create water currents that draw sea water in through the porocytes. The incoming water brings with it fresh supplies of oxygen and food. As filter-feeders, sponges feed on microscopic organisms and bits of organic matter that become trapped in mucous secretions of the choanocytes. These trapped particles pass down the collar of the choanocytes to the base of the cell below, where they are phagocytized and digested in food vacuoles. The water currents that brought in the particles then pass out through the osculum, laden with carbon dioxide and other cellular wastes.

In more complex sponges, the body wall is relatively dense and the body cavity highly branched. The pores open into canal systems that frequently widen into chambers lined with collar cells before leading into a reduced central cavity.

Sponges maintain their body shape by **spicules** or fibers that are scattered through the body wall and that act as skeletal elements. The spicules, produced by amebocytes, differ in each of the three classes of sponges. In calcareous sponges the spicules contain calcium carbonate, while those of the beautiful glassy sponges consist of silicon dioxide (glass). In a third group, the **proteinaceous sponges** or "bath sponges," the skeletal material is **spongin**, a fibrous protein. (Most cleaning sponges sold today are synthetic.)

Sponges are usually **hermaphroditic**, with each individual producing both sperm and eggs. Sperm release (often in clouds) is aided by the currents from the collar cells. The egg cells are often located just beneath the collar cells, where the sperm can easily penetrate. A fertilized egg escapes as a highly flagellated, swimming larva, which after a brief period settles to the bottom and begins to sit out its adult life. Some sponges can also reproduce asexually, releasing **gemmules**, clusters of cells surrounded by a resistant wall. The gemmules of freshwater sponges may remain dormant through hard times, such as cold or dry spells, before becoming active and producing a new sponge body.

As adults, the sponges' ability to respond to the environment is extremely limited and is similar, in fact, to responses by colonial flagellates. There are no nerve cells although a rudimentary means of cell to cell communication may exist. Restriction to the cellular level of organization means that functions are carried out by individual cells or aggregates of cells, with little of the coordination of activity seen in the tissue level.

FIGURE 30.6
SPONGE ANATOMY
In the simple, vaselike sponges, the body wall consists of two organized cell layers surrounding a central cavity, the spongocoel. Between lies the gelatinous region in which are embedded supporting calcareous spicules. Flagellated choanocytes (See inset) sweep in water that contains microorganisms and bits of organic debris, both of which are used as food. Amebocytes produce spicules that form skeleton elements (inset).

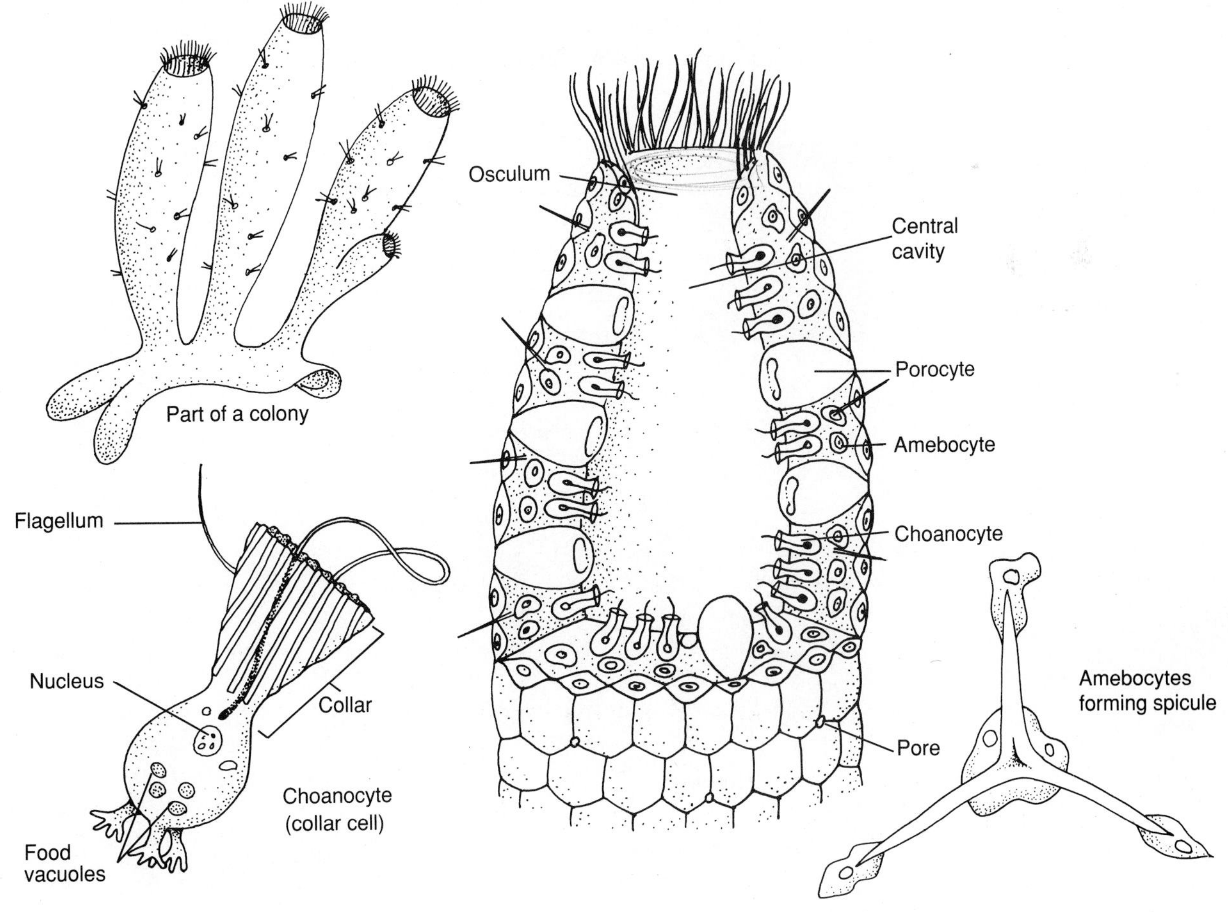

TISSUES AND THE RADIAL PLAN: PHYLA CNIDARIA AND CTENOPHORA

The body plans of metazoan animals reveal two basic kinds of symmetry, **bilateral** and **radial** (Figure 30.7). Both are conceptually simple, but the differences have far-reaching evolutionary implications. The bilateral body plan has only *one* plane of symmetry, which basically divides the body into right and left sides. We humans are a good example of a bilateral animal. The radially symmetrical form is quite different. The body is essentially disklike or cylindrical (sometimes spherical), perhaps with radiating arms. Theoretically, there are unlimited planes of symmetry. In other words, any plane that passes through the center body from top to bottom will divide the body into roughly equal halves.

Radial symmetry in today's animals exists in only a few phyla, most markedly in the cnidarians and ctenophores. However, adult echinoderms such as sea stars, sea urchins, and sand dollars have an odd, **pentaradial** or five-part radial symmetry. There is no agreement about how the radial body form arose, but many zoologists theorize that the direct ancestor to all metazoans, including the radial ones, was probably a bilateral inhabitant of the muddy ocean bottom (where bilateral symmetry works well). It may have resembled the **planula larva**, an early developmental stage of cnidarians that we will look at shortly. The theorists go on to suggest that a radial offshoot from these simple animals did well as a swimming form, and that this line gave rise to the radiates.

Phylum Cnidaria

Phylum Cnidaria (pronounced "nidaria") includes hydrozoans, jellyfish, corals, and anemones. Most cnidarians are organized along the tissue level, but some have what could be called organs. For example, some have tentacles, and tentacles are composed of several tissues (including skin, nerve cells, and contractile fibers), which seems to qualify them as organs. In most species of cnidarians the tentacles are armed with stinging cells called **cnidocytes** ("nettle cells"). When triggered, the cnidocytes release harpoonlike stinging structures called **nematocysts** ("thread-pouch").

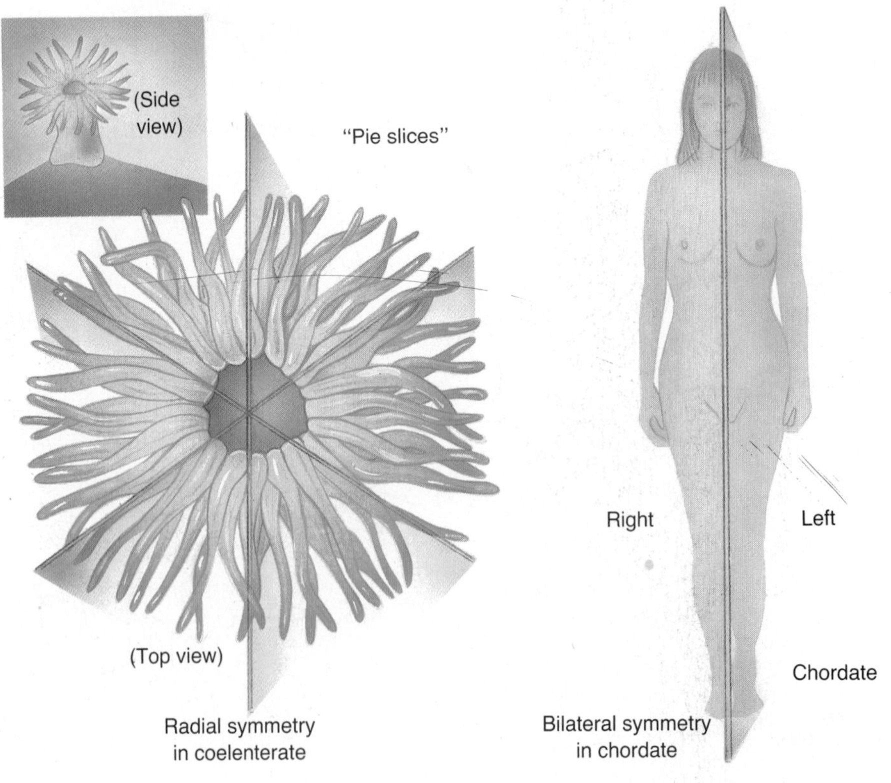

FIGURE 30.7
BODY SYMMETRY
In radial symmetry the body is essentially spherical, disk-shaped, or cylindrical. Any plane drawn through the center produces roughly equal left and right halves. Although three such planes are shown, there are infinitely many. Bilateral symmetry, on the other hand, means that the body can be equally divided by one plane only. This division produces mirror-image left and right halves that contain similar structures.

Cnidarians are thin-walled animals; their bodies are composed essentially of an outer and inner layer of cells. Sandwiched between these layers is a cellular, jellylike matter—the **mesoglea.** Wandering ameboid cells, nerve processes, and contractile fibers are found in the mesoglea. The outer cells (or epidermis) are mainly protective, while the inner cells (or gastrodermis) form a saclike gut called the **gastrovascular cavity** because it serves both digestive and circulatory functions. A simple opening serves as both mouth and anus—or, to put it another way, cnidarians have a mouth but no anus and must spit out the undigested remnants of whatever they swallow.

As with other radial animals, there is little centralization of the nervous system. In fact, it takes the form of a **nerve net,** a diffuse network of nerve cells and their extensions, covering the entire animal. In addition, the neurons or nerve cells of cnidarians are quite unusual in that nerve impulses can travel in either direction from one cell to the next. In bilateral animals, most neural conduction is one way. The cnidarian nerve cells are organized to receive impulses from sensory cells (cells that receive external stimuli) and bring about responses in the contractile fibers and cnidocytes.

Cnidarians may take either of two forms: the **polyp** or the **medusa.** Polyps are usually sedentary, attached to rocks, wharf pilings, and other objects, whereas medusas (jellyfish) can swim. In some cnidarian species, both polyp and medusa stages exist in an unusual alternation of generations. Unlike alternation of generations in plants, however, both polyp and medusa forms are diploid.

Class Hydrozoa In **hydrozoans** the polyp state usually dominates, and although some solitary species occur, most form dense colonies. They reproduce asexually by **budding,** that is, by outgrowths in the form of miniature polyps. Some polyps are specialized for capturing and digesting food, which is then moved by ciliary action throughout the continuous hollow body of the colony. Others specialize in sexual reproduction, many producing medusae that form sperm and eggs that are released to join in the open water.

The life cycle of one well-known colonial hydrozoan, *Obelia,* is shown in Figure 30.8. Note the stage called a planula, a ciliated, swimming larva. In a sessile animal, the swimming larval stage makes it possible for the species to disseminate in the environment.

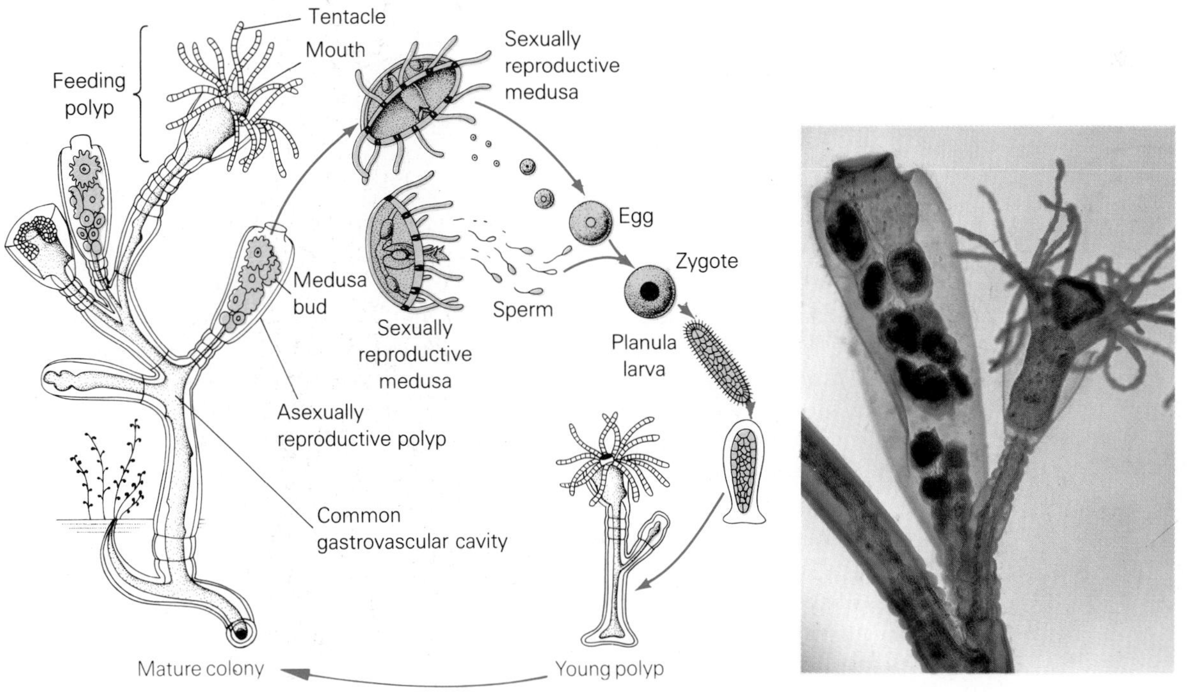

FIGURE 30.8
LIFE CYCLE OF A HYDROZOAN
In its life cycle, the hydrozoan *Obelia* develops a number of polyps that specialize in feeding or reproduction. The reproductive polyps form swimming medusae, which produce eggs and sperm. Following fertilization, the zygote develops into a ciliated planula larva. The planula becomes an adult hydrozoan polyp, and the cycle repeats.

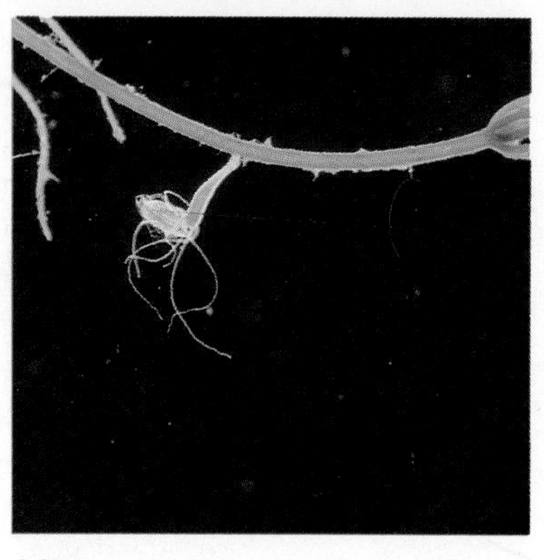

(a)

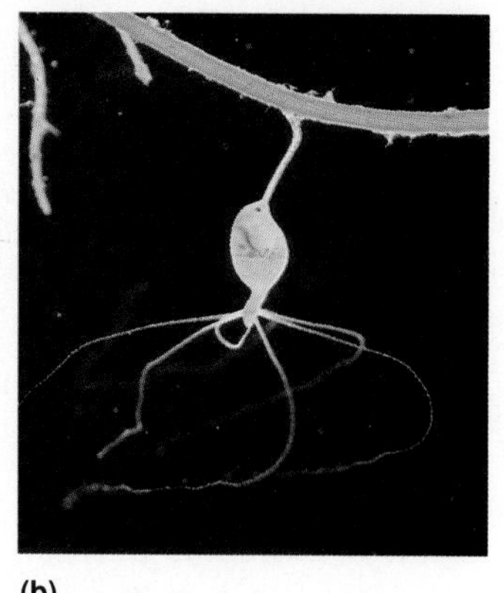

(b)

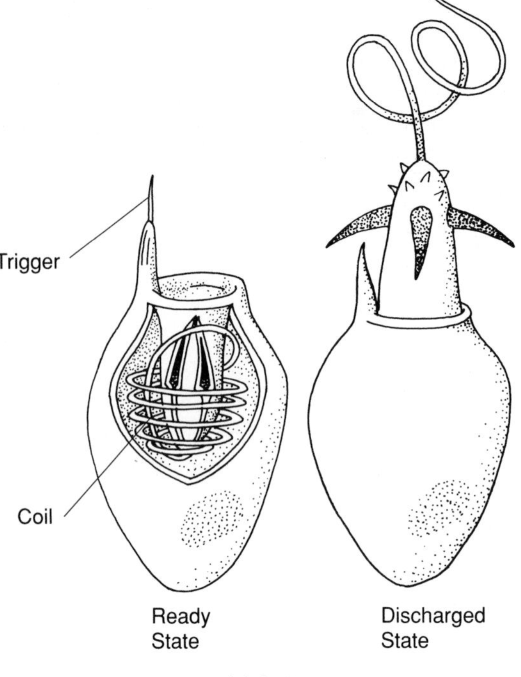

(c) Cnidocytes

FIGURE 30.9
A FRESH WATER HYDROZOAN
(a) The hydra stuns its prey with stinging cnidocytes. (b) Using its tentacles, it draws the subdued prey into its gastrovascular cavity, where digestion occurs. (c) A closeup of the cnidocytes shows both the "armed-and-ready" state and the discharged state. A simple touch of the trigger brings about an explosive release of the threadlike nematocyst.

One of the most familiar hydrozoan groups consists of the tiny, freshwater species commonly called **hydras** (Figure 30.9). Hydras are solitary and quite small, their slender, vaselike bodies rarely exceeding 3 centimeters in length (about 1.2 inches). Further, there is no medusa stage and no planula larva. The hydra polyp produces asexually by budding new polyps. As in so many other minute freshwater forms, sexual reproduction occurs only under harsh conditions, and the zygote is protected by a toughened, resistant case. When conditions improve, the encased zygote develops into a tiny polyp. Most hydra are brown in color, but *Chlorohydra* is bright green due to the presence of an intracellular photosynthetic symbiont, a green alga upon which it depends for nutrition.

Other hydrozoans are quite unlike these. For example, the large Portuguese man-of-war (*Physalia*) looks like a jellyfish, but is actually a floating hydrozoan colony. Each colony includes a gas-filled float and feeding and reproductive polyps with long trailing and stinging tentacles. A fish, perhaps seeking food or refuge in the dense "foliage" of the hydrozoan's tentacles, is at once paralyzed by the venom of thousands of ejected nematocysts as it brushes past the graceful, often brightly colored lures. The fish becomes totally ensnared and is then subjected to digestive enzymes from the feeding polyps.

Class Scyphozoa In the scyphozoans, or jellyfish, the dominant stage is the swimming medusa, which in this class is called a **scyphomedusa** (Figure 30.10a). Polyps are present in most species, but they are highly reduced. In effect, the jellyfish life cycle is opposite that of the hydrozoan.

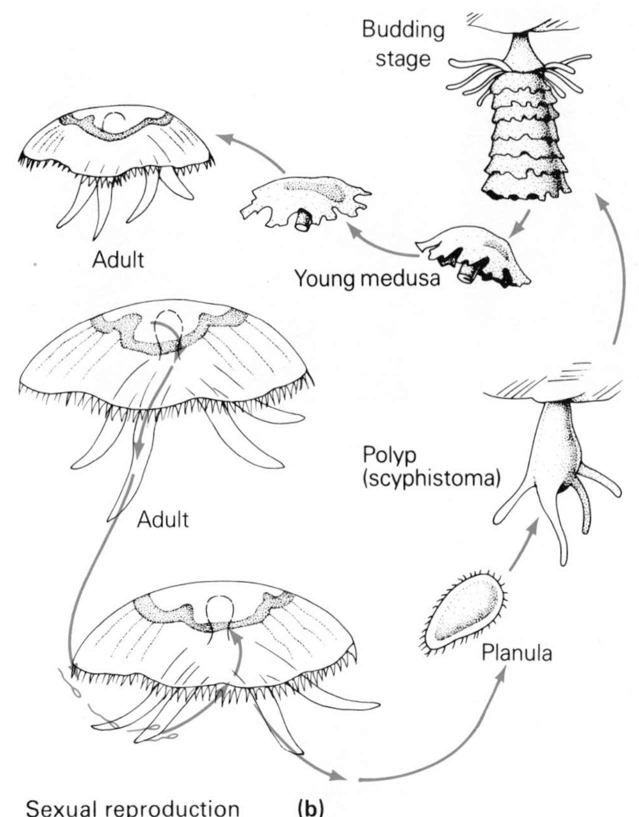

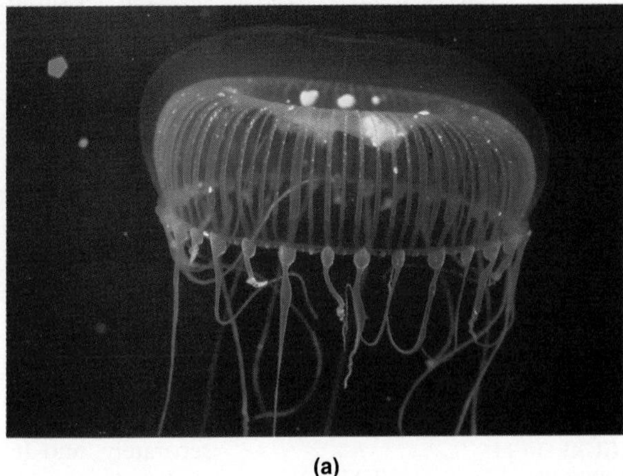

FIGURE 30.10
LIFE CYCLE OF A SCYPHOZOAN
(a) Jellyfish are free-swimming animals with a highly reduced polyp stage. (b) Sexes are separate, and after fertilization a ciliated planula larva forms. It attaches as a polyp to the sea bed. The polyp undergoes anasexual budding process that yields many young medusae. Each will mature into an adult jellyfish.

Most jellyfish have a bell-shaped body with the mouth centrally located on the underside, surrounded by tentacles. For the most part, they are drifters, but they can swim by contracting muscle fibers around the bell. While most are modest in size, some, such as *Cyanea,* attain large proportions—up to 2 meters in diameter with tentacles 70 meters long. These enormous jellyfish are indeed dramatic sights when drifting in clear, blue waters, far out at sea.

Sexes are separate in nearly all jellyfish, and fertilization is usually external. In the common Atlantic jellyfish, *Aurelia* (Figure 30.10b), gametes pass through the mouth opening into the water. The eggs become lodged in brood pouches in the arms, where fertilization occurs. A swimming planula larva develops and soon settles to the ocean floor, generally becoming fixed to a rock. There it undergoes a transition into a **scyphistoma,** a tiny polyp that at first looks very much like a hydra. In *Aurelia,* ongoing divisions across the entire scyphistoma produces young medusae, one after another in assembly-line fashion. As the outermost medusa matures, it breaks away and is soon transformed into an adult jellyfish.

Class Anthozoa Anthozoans—the corals and anemones—exist only as polyps (Figure 30.11). The most familiar anthozoans in temperate waters are the heavy-bodied sea anemones. Anemones abound in rocky tidepools and in shallower coastal waters. Corals are colonial anthozoans whose polyps secrete surrounding walls of limestone (calcium carbonate). The lovely coral formations sold in curio shops are actually their dead skeletons. The cumulative effects of coral secretions (along with significant contributions by certain mollusks, tube worms, and corralline algae) over millions of years have produced coral atolls (islands), as well as the barrier reefs that surround many islands in the Pacific. Although all corals can feed with tentacles, most tropical corals play host to photosynthetic, symbiotic algae that provide most of their nutrients.

It was mentioned earlier that polyps can reproduce asexually by simply budding new polyps. In their sexual reproduction, female and male polyps produce eggs or sperm

(a)

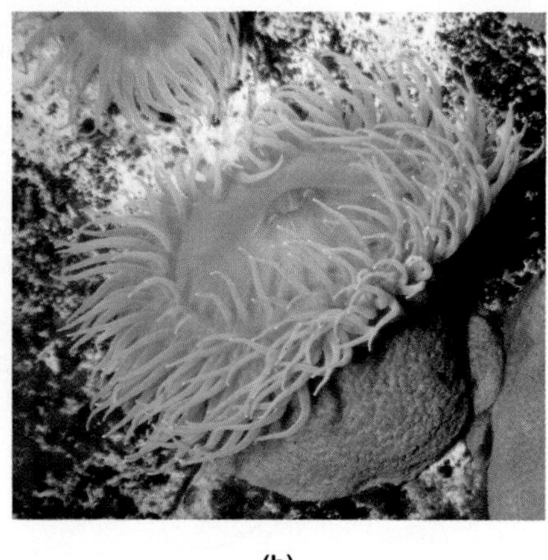

(b)

FIGURE 30.11
REPRESENTATIVE ANTHOZOANS
(a) Many corals are colonial polyps living in extensive limestone coverings that they secrete. The vase-like body is usually extended, but when disturbed it quickly retracts. **(b)** Anemones are often more solitary and are usually attached by their bases to the substrate.

separately, and fertilization occurs in the water. The zygotes develop into swimming planula larvae that form more anthozoans.

Phylum Ctenophora: Comb Jellies

Ctenophorans (Figure 30.12), commonly called "comb jellies," appear to be close relatives of cnidarians, but their similarities may be a case of convergent evolution. Like the cnidarians, comb jellies are radially symmetrical marine animals. Also, the body consists of two cell layers separated by a jellylike mesoglea, with the innermost layer surrounding a gastrovascular cavity. Unlike cnidarians, comb jellies have eight rows of **"combs"** consisting of fused cilia. The coordinated beating of these combs moves the animals along. And whereas cnidarians have tentacles armed with stinging cells, the tentacles of comb jellies are armed with "glue cells." They snare the prey rather than sting it. Their chief food is plankton, minute drifting marine organisms. Ctenophorans have another distinguishing, even enchanting trait. They are bioluminescent; that is, they glow. One species, *Cestus veneris* or "Venus' girdle," has a glowing, ribbonlike body nearly a meter long.

BILATERALITY, MESODERM, AND PHYLA PLATYHELMINTHES AND RHYNCHOCOELA

In the animal kingdom, the radially symmetrical species are the minority. Most animals are bilaterally symmetrical (see Figure 30.7). With the evolution of bilateral symmetry, a significant trend in body organization known as **cephalization** arose. Animals developed a head or leading end and a tail or trailing end. At first the head may have been simply a concentration of muscles, an adaptation for burrowing in the soft sea beds of ancient oceans, which is where the oldest fossils of bilateral animals are found. But leading ends soon became equipped with sensory structures for the detection of food, light, sound, and other stimuli. With senses concentrated in the leading end, an animal could more quickly perceive what sort of environment it was moving into. It wouldn't do to go around backing into new environments. Such sensory structures would have required neural support and integration; thus, clusters of nerve cells were concentrated at the head end, and the evolution of the brain was underway.

The Three-Layered Embryo

The cnidarians and ctenophorans are much simpler than the other metazoans. Their simplicity is related to their having only two fundamental types of embryonic tissues, or **germ layers**, as they are called. These are the **ectoderm** ("outer skin") and **endoderm** ("inner skin"). All the other metazoans have a third germ layer called the **mesoderm** ("middle skin"). Those animals whose embryos have two germ layers are called **diploblasts** ("double germ"), whereas those with three embryonic germ layers are the **triploblasts** ("triple germ").

Each of the germ layers contributes to specific portions of the developing body, laying down the basic framework from which that region develops (see Figure 30.16 and Figure 43.10). For example, whereas the nervous system is essentially ectodermal in its origin, inner linings such as those of secretory glands, the gut, the blood vessels, and those of the respiratory and reproductive systems are endodermal in origin. The mesoderm contributes to muscle, and in vertebrates, to the internal bony skeleton and the blood. Actually, once the basic framework is laid down, each organ eventually receives tissues derived from the other germ layers.

FIGURE 30.12
CTENOPHORANS
Delicate comb jellies are propelled through the water by rows of cilia making up the combs. The combs are in rows on vertical ridges around the body wall, and the central pouchlike object is the gastrovascular cavity. This is the comb jelly *Pleurobrachia,* about 2 cm in diameter.

Phylum Platyhelminthes: The Flatworms

The phylum **Platyhelminthes** contains the flatworms, including one class that is free-living and two that are parasitic. Flatworms have bilateral body symmetry, so unlike radiate animals, they have an anterior (head) and posterior (tail) end. They have also clearly arrived at the organ-system level of development, although some of the organ systems are quite simple. Tissue of mesodermal origin in these triploblastic animals, include well formed muscle and associated parenchyma cells.

Flatworms lack special respiratory and circulatory structures. Because of their flattened bodies, most cells are close to the body surface, making simple cell-to-cell diffusion of gases sufficient to accommodate metabolic needs. In the parasitic species, the nervous and digestive systems are greatly reduced or virtually absent. (Do you see why this might be the case? Remember, tapeworms usually live in the gut of a host animal.)

Unlike other triploblastic animals, flatworms have solid bodies. Although a gastrovascular cavity is present in many, there are no internal cavities, those that occur between the gut and body wall of most other triploblasts. For this reason flatworms are called **acoelomates** ("without a cavity").

Class Turbellaria: Free-Living Flatworms Although many of the **turbellarians**—the free-living flatworms— are marine, the most familiar of the group is the very common freshwater species, *Dugesia,* better known as **planarians** (Figure 30.13a). While many species are quite small, a few millimeters in length, their well-developed organ systems make them attractive as laboratory subjects. Thus, planarians have been the subject of anatomical, developmental, and behavioral studies for many years.

There is some evidence of centralization in the nervous system. It is concentrated at the anterior end, where large **ganglia** (clustered nerve cell bodies) and eyespots (light receptors) are located. Two ventral nerves run down the length of the body, sending out branches from a number of smaller ganglia in a ladderlike arrangement (Figure 30.13b).

Planarians are predators and scavengers, feeding through a protrusible, muscular **pharynx,** while pinning the unlucky victim in a slimy mucous layer secreted below. Food enters the highly branched gastrovascular cavity, where digestion begins (Figure 30.13c), but bits of food are also phagocytized by cells lining the cavity, and digestion is completed in food vacuoles within the cells. Thus digestion is both extracellular and intracellular.

In freshwater flatworms, **osmoregulation**—the maintenance of water and ion balances in the body—is accomplished by the action of the **protonephridia,** also known as the **flame cell system.** In many species, the protonephridium consists of two highly branched systems of tubules with a number of blind sacs containing the **flame bulbs.** Each bulb is made up of a group of ciliated **flame cells** (the waving, or "flickering" of the cilia, as seen under the microscope, inspired their name). Their primary function is getting rid of excess water that continually enters the body because of a natural osmotic gradient.

Planarians swim in a curious sort of crawl. The smaller species secrete a layer of mucus underneath them and then beat their way through it with the cilia that abound

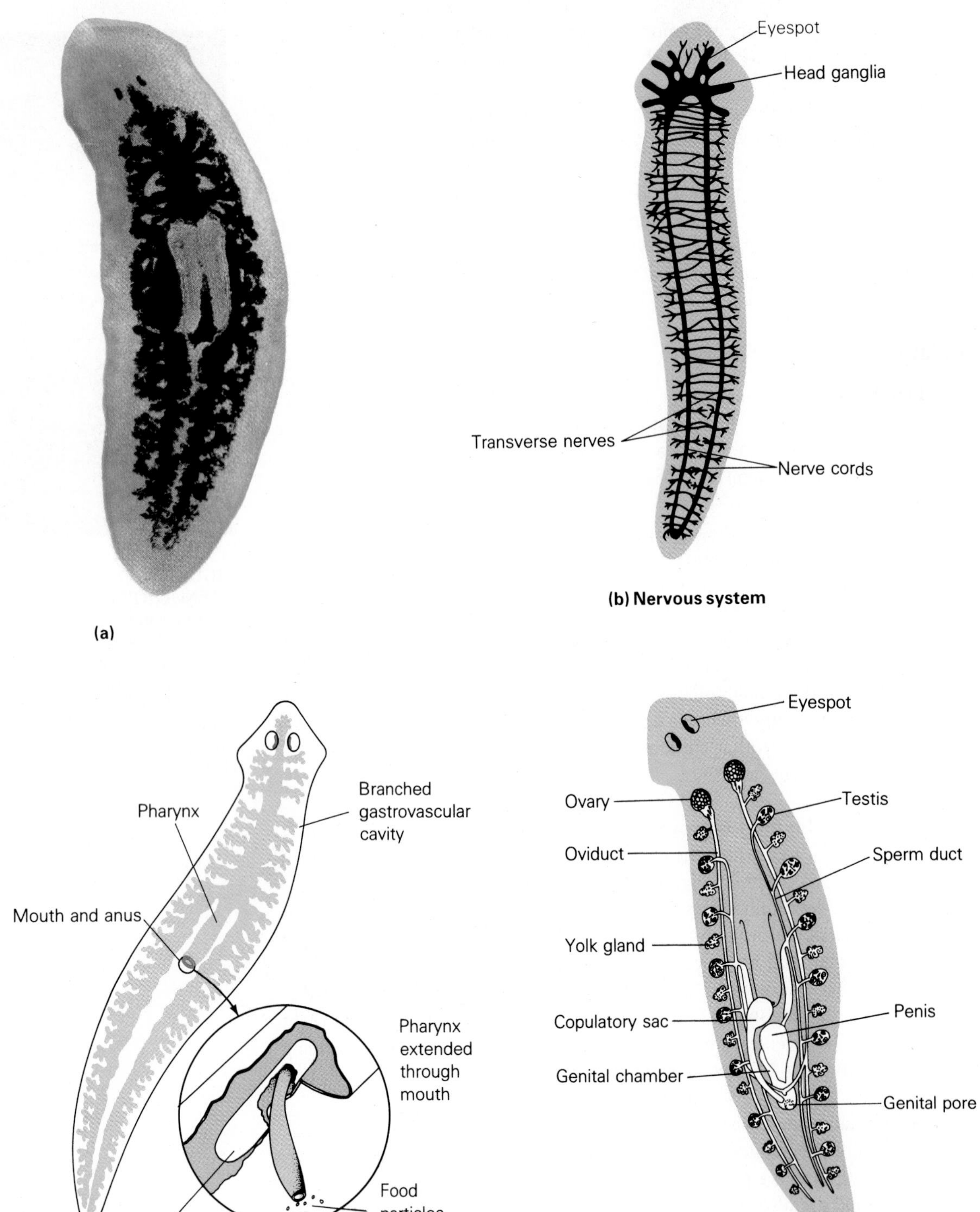

FIGURE 30.13
PLANARIAN ANATOMY
Dugesia (**a**), a free-living planarian flatworm, is only a few millimeters long, yet it has several organ systems that permit the animal to feed, reproduce, and respond to external stimuli. *Dugesia* is about 15 mm long. (**b**) Light-sensitive eyespots help the flatworm respond to light, and a ladderlike nerve network coordinates movement of its muscles. (**c**) Digestion occurs in a highly branched gastrovascular cavity, which begins with a complex, muscular pharynx. (**d**) The reproductive system includes both male and female organs.

on their ventral body surface. Larger species rely on an undulating movement of the flat body.

Planarians reproduce both asexually and sexually. In asexual reproduction, the individual undergoes fission—simply pinching in half across the body. Each half then regenerates the missing parts. The regenerative power of planarians is legendary; almost any portion of the body can be restored following its removal. Planarians are hermaphroditic; each individual has both male and female sexual structures. They include well-defined ovaries, testes, and copulatory organs, the latter permitting internal fertilization (Figure 30.13d). Yet planarians do not usually self-fertilize. During sexual reproduction, two individuals copulate with the simultaneous exchange of sperm. Although planarians develop directly into miniature adults, many marine turbellarians produce a ciliated larval stage.

Class Trematoda: The Flukes The **trematodes**, or **flukes**, are parasites. The flukes are somewhat similar to the free-living flatworms, although the flukes are not ciliated and their gastrovascular cavity occurs in two parts that are tubular rather than branched. Flukes are equipped with large suckers, an anterior "oral" sucker and a ventral sucker, both of which are used to attach to their host's tissues. Flukes invade many kinds of animals and make their home in various parts of the host, including the gills, lungs, liver, and intestines. One species has even found its niche in the urinary bladder of frogs. One human liver fluke *(Chlonorchis sinensis),* common to Asia, has an enormously complex life cycle involving three very different hosts—human, snail, and fish. Since sexual reproduction occurs in the human, we are the **primary host.** The snail and fish are **intermediate hosts.** However, each host is essential to the parasite's life cycle. Note in Figure 30.14 the ongoing change from **miracidia** to **sporocysts, redia**, and **cercaria,** each a specialized state in one bizarre and demanding life cycle. Since each stage reproduces asexually, one fertilized egg can lead to a large number of adult parasites. As with other such parasites, the human liver fluke relies on great numbers of offspring to increase the likelihood of survival.

An important fluke in the Nile Valley of Egypt today is the human blood fluke, *Schistosoma mansoni,* which has recently become widespread in the area. The extensive irrigation canals fed by the Aswan Dam have allowed aquatic snails to flourish. The snails serve as an intermediate host to the fluke, which causes the debilitating disease **schistosomiasis.** Generally, it weakens the host to the point that other diseases cause early death. While in the cercaria stage, the parasite enters the body directly through the skin when people wade in the water. It is estimated that well over 60% of the population in the Nile delta is now infested with the blood fluke. The disease is also common in some other tropical regions.

Class Cestoda: Tapeworms Members of class **Cestoda**, the **tapeworms**, are quite different from other flatworms. These parasites have no digestive cavity and must absorb digested food of the host directly across the body wall. To this end, the outermost external membrane is highly folded, greatly increasing the surface area. Virtually all vertebrates harbor tapeworms, generally with little harm to the host animal. Humans are host to seven different species—something to think about as you look around the room.

Typically, the sexually mature tapeworm, consists of an extremely small, rounded, anterior **scolex** that is equipped with suckers and hooks for attachment to the host's intestinal wall (Figure 30.15a). Extending from the scolex is a lengthy chain of flattened blocks called **proglottids**, produced continually through budding. The proglottid is a complete reproductive unit, each containing ovaries, testes, and male and female copulatory structures (Figure 30.15b). Whereas cross-fertilization with other individuals is the general rule, self-fertilization among proglottids and even within a proglottid is known to happen. Fertilized proglottids enlarge, and finally the mature proglottids, each containing thousands of fertilized eggs, break away from the chain and pass out the

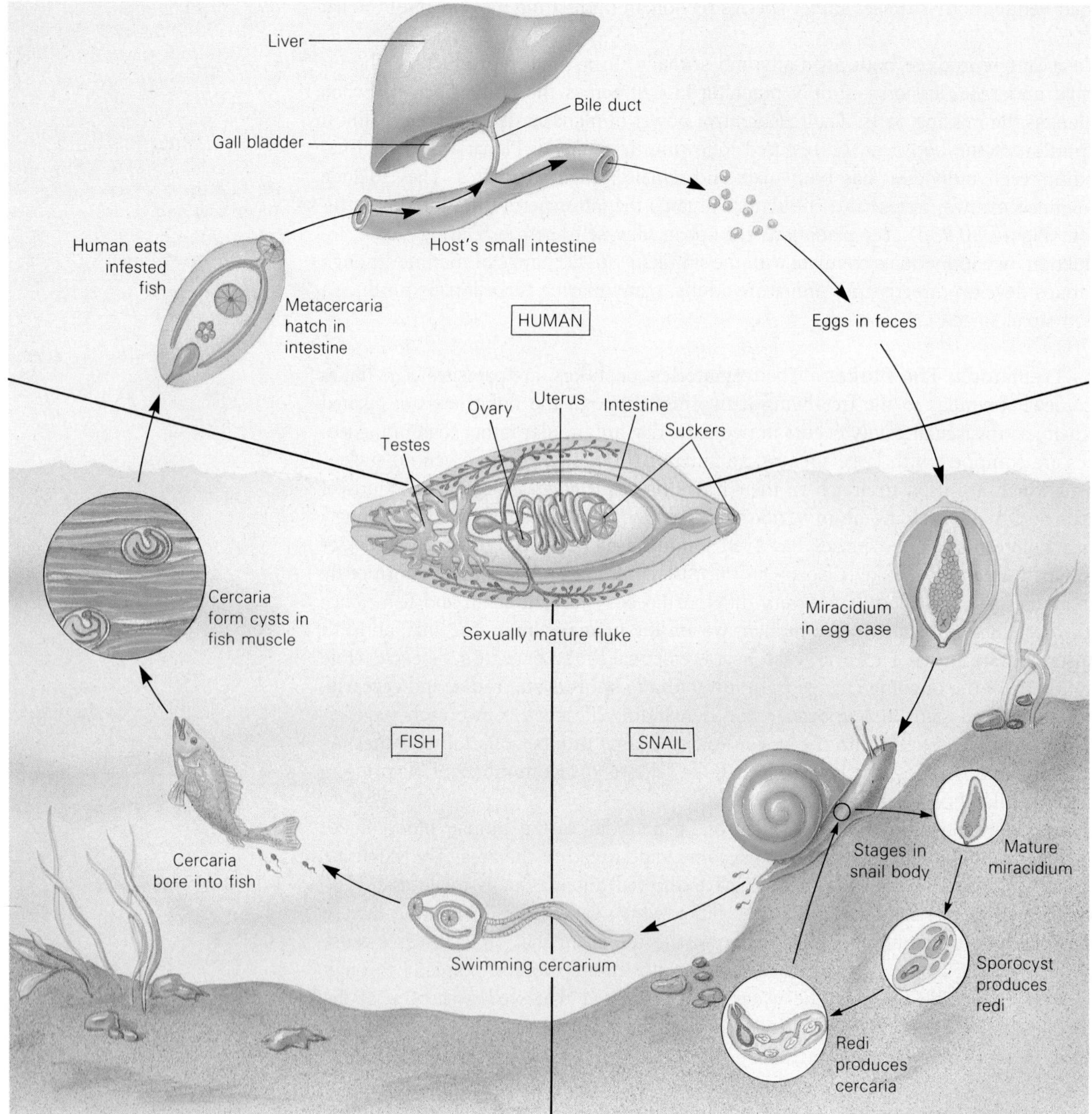

FIGURE 30.14
HUMAN LIVER FLUKE
The human liver fluke, *Chlonorchis sinensis*, infects the liver, where it can bring on cirrhosis and death. A sexually mature fluke is seen at the center. Its life cycle requires human, snail, and fish hosts. Sexual reproduction in the human host gives rise to egg cases containing miracidia that leave the host in the feces. The miracidia then infect snails that feed on the feces. Sporocysts form next, and these asexually produce numerous redia, which in turn produce many swimming cercariae. The latter escape from the snail, attach to a fish, and bore into its muscles, where they again form cysts. When a human eats the fish raw or partially cooked, as is often the custom in Asia, the young flukes emerge and make their way to the liver, where the cycle repeats.

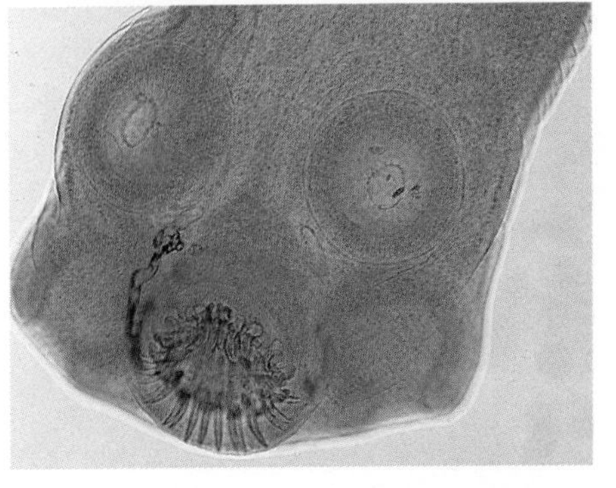

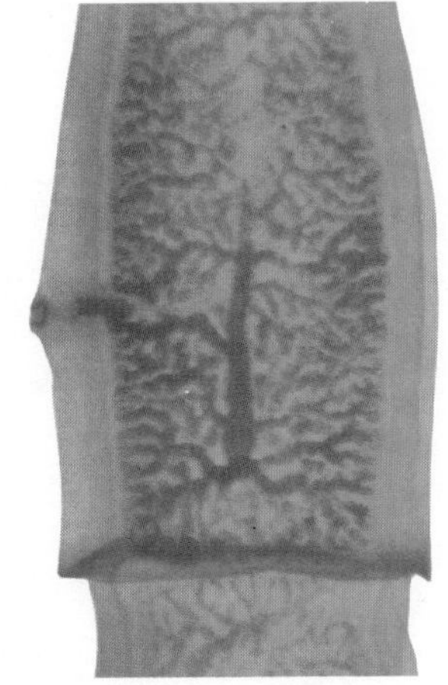

FIGURE 30.15
TAPEWORM ANATOMY
(a) The tapeworm scolex is equipped with a ring of hooks and several suckers for grasping the host's intestinal wall. (b) The remainder of the body, consisting of multiple reproductive units known as proglottids, buds continuously from the scolex. Each proglottid contains well-defined ovaries and testes, along with related ducts and glands.

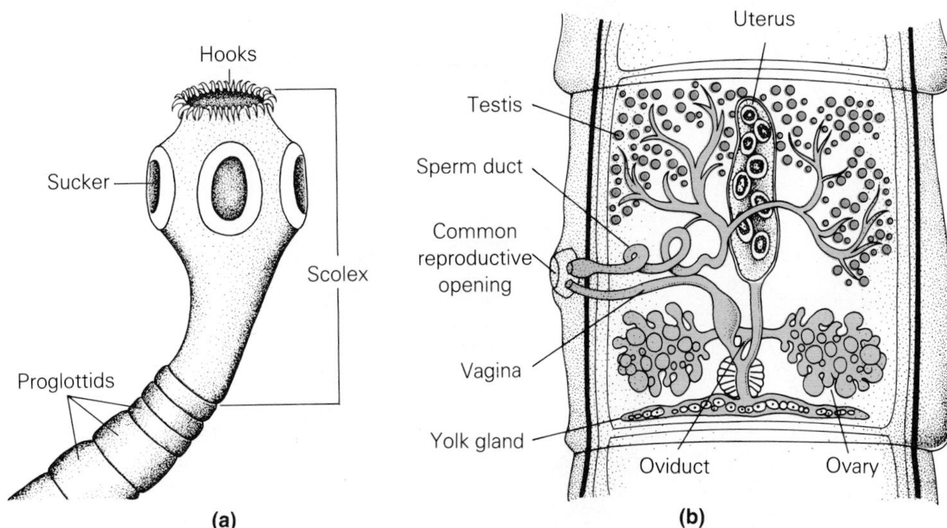

host's body in the feces. The fertilized eggs must be ingested by the next host in order for the life cycle to continue.

In addition to the primary host, where sexual development and reproduction occur, most tapeworms have an intermediate host. It is usually a different species, typically an animal that is eaten by the primary host. For example, pigs, cattle, and fish serve as intermediate hosts to some human tapeworms. In the pork tapeworm the fertilized eggs, passing out of the human host, are ingested by the pig as it feeds. (Obviously this link is easily disrupted where modern sanitation is available, but remember that much of the world lacks such amenities.) Upon entering the pig's intestine, the egg develops into a tunneling creature that bores its way through the intermediate host's tissues, generally ending up in the muscles. There it forms an encapsulating, protective cyst around itself and develops into a stage known as a **bladderworm**. The bladderworm becomes dormant, remaining so until the pork is eaten uncooked or undercooked, whereupon it emerges as a young tapeworm, migrates, attaches to the host intestine, and repeats its life cycle. For the trendy folks, sushi (uncooked fish) is now implicated in the transmission of such parasites.

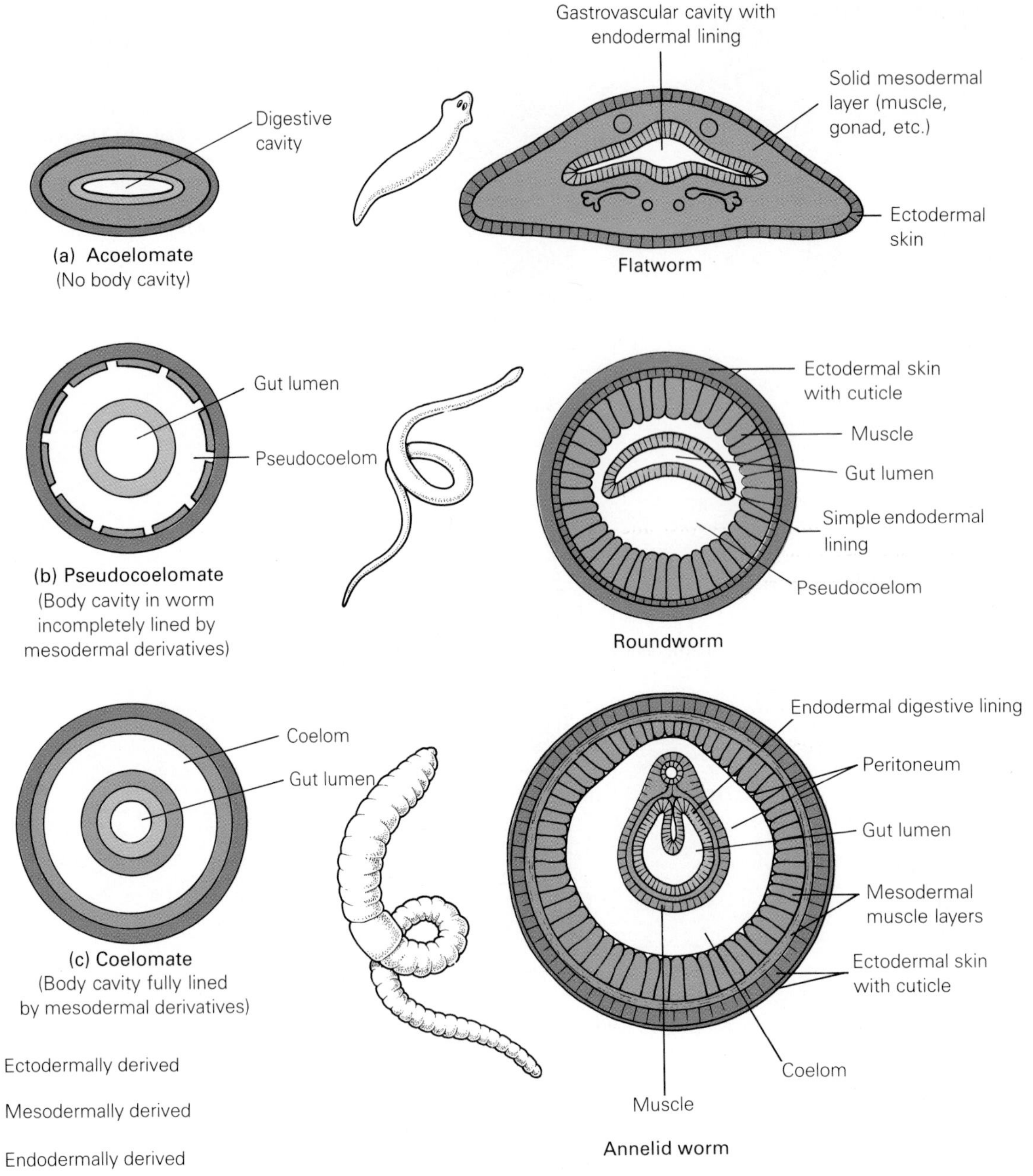

FIGURE 30.16
BODY CAVITIES
The coelom is a mesodermally derived and lined body cavity between the gut and the body wall. **(a)** Flat worms lack a body cavity and are designated acoelomates. **(b)** Round worms are pseudocoelomates. Although their body cavities are extensive, they are not fully lined by tissue of mesodermal origin. **(c)** In coelomates, the coelom is fully lined by mesodermally derived tissue.

Phylum Rhynchocoela: Ribbon Worms

You quite likely have never seen a ribbon worm (phylum **Rhynchocoela**). Although they aren't really rare, they usually don't make headlines either. These, we are happy to say, do not burrow into your soft tissues: most are free-living marine predators. Ribbon worms have some characteristics in common with the flatworms and some with more complex animals. For example, they lack internal body cavities (they're acoelomates), and their body is ciliated. Further, they make use of flame cells in osmoregulation.

But unlike flatworms, ribbon worms have a complete one-way, tubelike gut, with a mouth at one end and anus at the other. This is a landmark development that we emphasize further along. Ribbon worms feed through the use of a **proboscis**, a long muscular tube that is shot out through a pore. The proboscis coils about the prey, often an annelid worm, and its bladelike stylet stabs the prey repeatedly before the proboscis draws it to the mouth for swallowing.

BODY CAVITIES, A ONE-WAY GUT, AND PHYLA NEMATODA AND ROTIFERA

As animal life on our planet continued to evolve and specialize, ever adapting to particular habitats and life styles, yet new landmark evolutionary changes appeared. We have already mentioned differences in symmetry and the importance of a third germ layer, the mesoderm. Now we will see how the development of a body cavity was associated with the formation of a one-way gut and how this arrangement permitted yet other novel developments on the animal scene. With the development of a one-way gut we find a digestive tract with a mouth, intestine, and anus. The resulting body plan resembles a "tube-within-a-tube," since the gut lies within the body wall. (Keep in mind that "gut" is not an indelicate term in biology.)

The body cavity—the space between the tubes—is the **coelom.** Animals with a coelom are known as **coelomates,** while those lacking such cavaties are acoelomates, as we saw in the flatworms and ribbon worms. Not all hollow spaces within the body qualify as coeloms. A true coelom is lined entirely by tissue derived from embryonic mesoderm. This lining is called the **peritoneum.** In the **nematodes** and **rotifers,** our next two phyla, we encounter a body cavity, but in this instance it is lined partly by mesodermally derived tissue and partly by tissue derived from ectoderm. A body cavity incompletely lined with mesoderm is called a **pseudocoelom** ("false coelom"). Figure 30.16 compares the three conditions. Coelomate animals are the subject of the next chapter.

Phylum Nematoda

The nematodes or roundworms are best known for their parasitic members, some of which are truly horrendous. Simply reciting the life cycle of some of them can cause nightmares. But let's stay calm and objective. Roundworms are slender, cylindrical, and usually tapered at both ends. They are bilaterally symmetrical, although from outside appearances, they lack a distinctive head. Most are surrounded by a strong and flexible cuticle. Since there is no rigid skeleton, the turgid, fluid-filled body cavity—a pseudocoelom—acts as a **hydrostatic skeleton,** helping the worm maintain its shape and acting as a resistant base for muscle action. Nematodes move by flexing longitudinal muscles first on one side and then on the other, producing a wriggling or undulating motion. This wouldn't get one very far in the open water, but many nematodes live in the soil, and the wriggling action helps them thread their way between the solid particles. Nematodes come in a range of sizes, but they are usually easy to recognize since they all look like smooth, glistening worms (Figure 30.17a).

Nematodes thrive in nearly every conceivable moist and aquatic habitat, from the soil of flower gardens to the world's oceans. Most of the estimated half million species of nematodes are free-living. Winnowing their way along, they feed on protozoans,

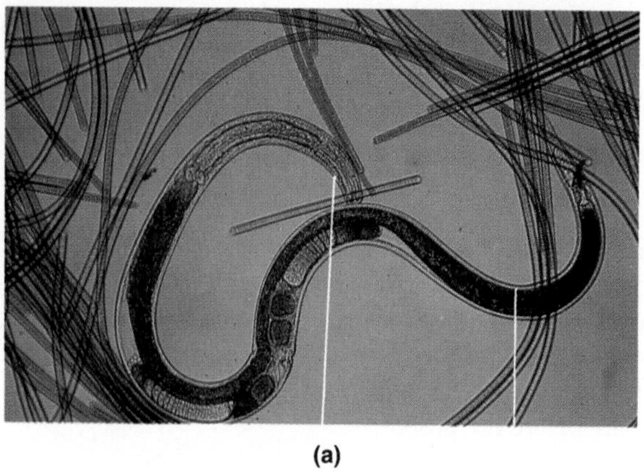
(a)

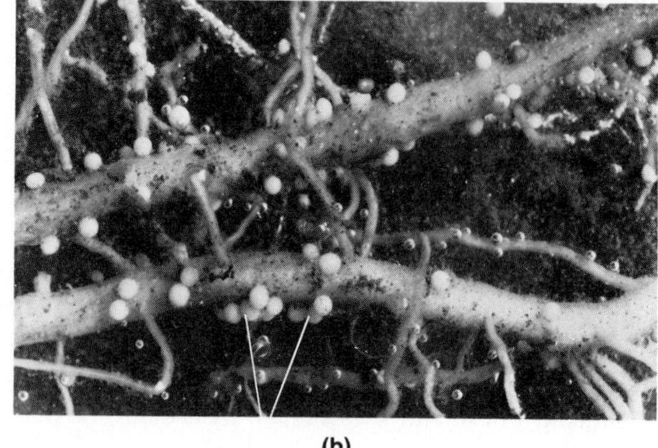
(b)

FIGURE 30.17
SOIL NEMATODES
(a) Nematodes abound in most soil. **(b)** Although most are scavengers and predators, many nematodes infect plants, making them important agricultural pests. The parasites may form cysts on the plant's root.

rotifers, small earthworms, and each other. Some capture their prey with a paralyzing saliva. Others make use of a tiny bladelike device that impales the hapless prey while the sucking lips and pumping pharynx drain its juices. Because of their astounding number, free-living nematodes are an essential link in the earth's ecological organization.

Many nematodes are very important parasites of plants (Figure 30.17b). Plant nematodes may have a devastating effect on agriculture and require constant control at great expense. In fact, sometimes the farmer must shift to more resistant crops in order to make the contaminated land usable again.

Parasitic roundworms probably infest all vertebrate species. In fact, we harbor at least 50 rather harmless species and 10 more dangerous ones. Two species that are particularly dangerous to humans are *Ascaris lumbricoides* and *Trichinella spiralis.*

Ascaris lumbricoides, a giant of its phylum, inhabits the bodies of both hogs and humans (Figure 30.18). Sexes are separate. Females often exceed 20 to 35 cm (8 to 14 inches) in length. The males, shorter and narrower, are easy to identify by their hooked posterior extremity. At the tip of the hook is located a pair of bristlelike spicules, which males use to hold the female's genital pore (reproductive opening) open during copulation.

Ascaris reproduces in the host's intestine, and its resistant egg cases pass out with feces. The tough egg cases containing developing embryos may survive in the soil for many years. As with other internal parasites, the odds against an embryo finding the proper host are very great, so *Ascaris* counters this by producing incredible numbers of offspring. A female may contain about 30,000,000 eggs, which can be released at a rate of 200,000 a day! There is no intermediate host, so the eggs are passed from one person or hog to the next.

The larvae hatch in the small intestine and begin a strange odyssey through the wall of the gut, to the bloodstream, which they follow to the heart and lungs, up and out of the breathing passages, to the pharynx, where they are swallowed, finally coming to rest again in the intestine. By the time they reach the intestine they may be 2 or 3 millimeters long. But once there, they grow to their full proportions. *Ascaris* survives in its host by neutralizing trypsin, a key enzyme in protein digestion. Without the enzyme, heavily parasitized children may suffer dietary protein deficiency resulting in impaired growth and mental capacity.

FIGURE 30.18
ASCARIS LUMBRICOIDES
Male ascarid worms are smaller than females and have a curved posterior containing several bristles that are used in mating.

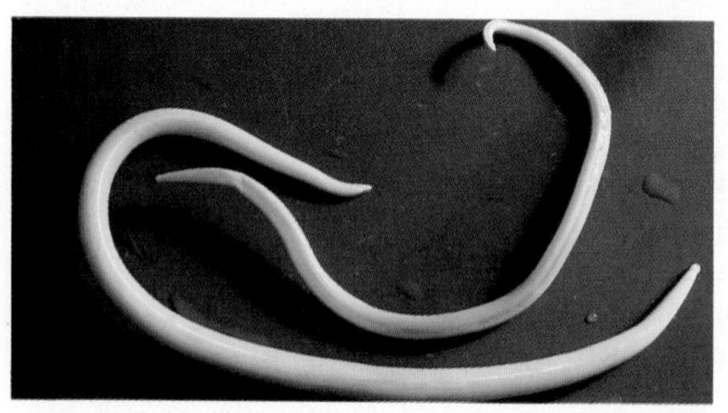

The round worm *Trichinella spiralis* is a common inhabitant of pigs, common house (brown and black) rats, and humans. It is passed from one animal to the next when contaminated flesh is eaten. The encysted larvae escape in the intestine and bore into its lining, where they mature and reproduce. Each female can produce about 1500 young. The young then begin to migrate through the lymph channels, eventually finding their way into the voluntary muscles, tissue rich in blood and oxygen. There the worms mature and surround themselves with a protective cyst (Figure 30.19). The cycle ends there unless the hog is eaten by another hog or some other carnivore, such as ourselves. If we eat infested pork or pork products without first killing the worms, they may travel through our own lymph channels, to encyst in our own muscles. It has been estimated that an ounce of heavily infected sausage can contain 100,000 or more encysted larvae, which could produce at least 100 million venturesome offspring in their new host. Pork or pork products can be made safe by thorough cooking (58° C, 137° F) or deep freezing (-23° C, -10° F) for 20 days. (Well-cooked pork is alway gray—never pink.) The symptoms of this infestation, called **trichinosis,** include muscle pain, fever, blood disorders, edema, and gastrointestinal disturbances. There is no way of killing the parasites, but most people survive the invasion unless other medical complications set in.

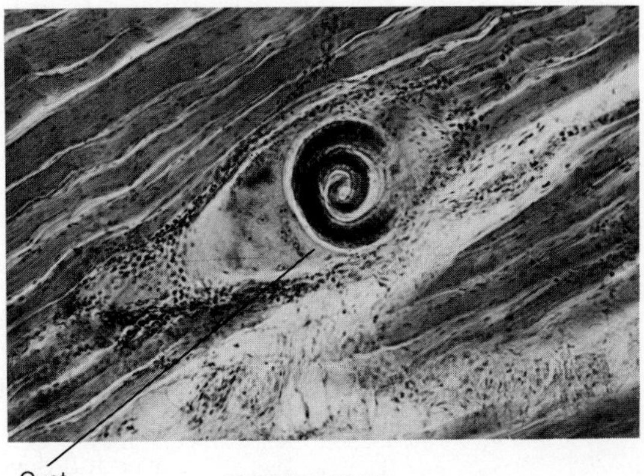

Cyst

FIGURE 30.19
SOURCE OF TRICHINOSIS
Trichinella spiralis is known to encyst in pork. If such meat is eaten without sufficient cooking, trichinosis may result. The worms will leave their cysts and venture through the body, forming new cysts, generally in the diaphragm, ribs, tongue, eye muscles, and larynx.

Phylum Rotifera

The rotifers, we can report, are not parasitic. In fact, their elaborate feeding system is unlike that of all other pseudocoelomates. They feed by sweeping microscopic organisms into their mouths through the action of double rings of cilia around their head area. Food is swept into a grinding gullet, or **gizzard**, which prepares it for digestion. The digestive system contains other specialized organs, including a stomach, two digestive glands, intestine, and anus (Figure 30.20). Rotifers are equipped with protonephridia (flame cell systems) like those of the flatworms. However, in flatworms, the protonephridia regulate only water and ion balance (osmoregulation), while in the rotifer, the protonephridia form a true excretory system, solving the problems of both nitrogen waste disposal and osmoregulation. Waving cilia direct fluids along a system of ducts to a **urinary bladder**, which subsequently empties through an opening to the outside. As you see, the digestive and excretory systems are quite advanced in such otherwise simple animals.

Female rotifers are much larger than the males. Males are also scarce, and, in some species, they have never been found. One reason may be that the male digestive system is so rudimentary and inadequate that, apparently, they die soon after mating. The short interlude of the male's existence would seem to threaten the survival of these animals

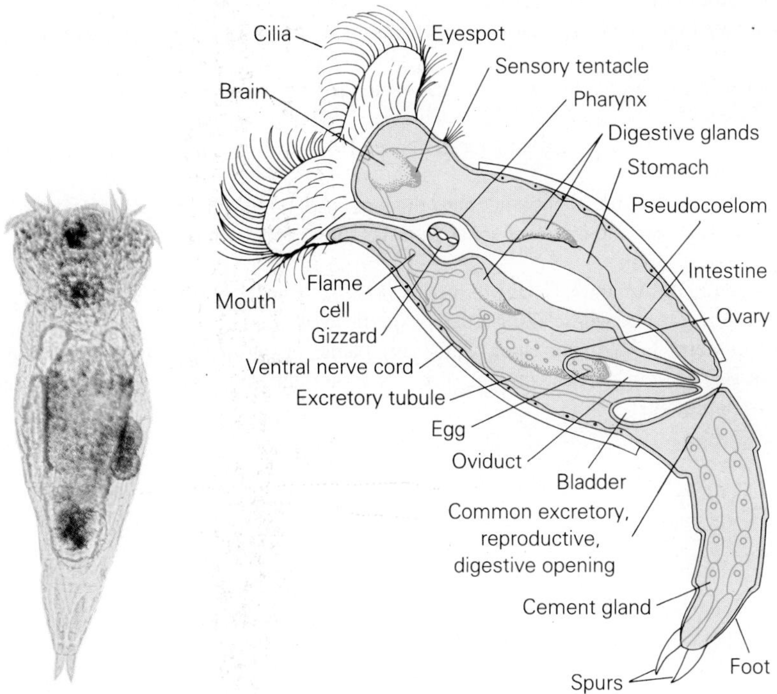

FIGURE 30.20
ROTIFER ANATOMY
Rotifers, extremely common freshwater animals, are about the same size as many protists (most less than one mm in length) but are much more complex. Note the well-defined "brain" and the complex digestive and reproductive systems. Cement glands at the base of the foot permit the animal to temporarily stop its whirling motion and anchor itself to the bottom.

were it not for a surprising ability of the females to compensate. The females, it turns out, are of two types: those that mate with males and those that don't. Before you begin to extrapolate, we hasten to explain.

In rotifers that reproduce sexually females may first go through several generations of reproduction without fertilization, an asexual process called **parthenogenesis.** The eggs do not enter meiosis, so they remain diploid. These begin to divide, rapidly developing into more females. Rotifer populations increase quickly in this fashion, some doubling every two days. Then, when some (unknown) environmental cue is received, the females switch to meiosis, and haploid eggs are formed. If unfertilized, these eggs develop into males. Upon maturity, the haploid males produce sperm that fertilize any haploid eggs they encounter. Such zygotes, interestingly, become surrounded by thickened, resistant walls, and can remain dormant for prolonged periods, some wintering over and developing only when warm weather returns. These develop into females. So, in the world of rotifers, males are reduced to the simplest of roles. Their contribution is an occasional input of genes, adding some genetic variability to an otherwise all-female population.

Here, then, we have reviewed a number of kinds of animals with a variety of evolutionary advances. In particular we have seen the development of bilaterality and a one-way gut. Next, we will see species in which this gut passes through a mesoderm-lined cavity. This may seem an insignificant development at first, but we will quickly find that it is not.

acoelomate Those solid-bodied animals that form neither a pseudocoelom nor a coelom (e.g. flatworms).

bilateral symmetry Body organization with one plane of symmetry, essentially a midline dividing the body into right and left sides (most coelomate phyla).

cephalization In most animals, the condition in which a head or leading end can be distinguished from its tail or trailing end, and generally where neural structures are concentrated.

coelomate Animals that produce a coelom: a body cavity between the body wall and gut whose lining, the peritoneum, is derived from embryonic mesoderm.

ectoderm In the young embryo, germinal tissue destined to form neural and covering tissues.

endoderm In the young embryo, germinal tissue destined to form lining tissues such as those of the lungs, gut, blood vessels, and various glands.

gastrovascular cavity In simpler animals, an extensive body cavity that serves in digestion, food distribution, and gas exchange.

intermeidate host In the life cycles of some parasites, a host in which only asexual reproduction occurs.

mesoderm In the young embryo, germinal tissue destined to form peritoneum, muscle, and the skeleton in enchinoderms and vertebrates.

metazoa One of two animal subkingdoms, this one made up of all animal phyla except porifera, the sponges.

parazoa One of two animal subkingdoms, this one made up of phylum porifera, the sponges.

primary host In the life cycles of some parasites, the host in which the parasites undergoes sexual reproduction.

pseudocoelomate An animal that produces a pseudocoelom, which is a body cavity between the wall and gut, and whose lining is derived partly from mesoderm and partly from ectoderm.

radial symmetry Body organization with many planes of symmetry (radii). Where the body is disk-shaped or cylindrical any "pie slice" has equivalent structures to any other (cnidarians and ctenophorans).

APPLICATION OF IDEAS

1. Interestingly, one group of animals and one group of plants represent an aside from the main line of animal and plant evolution. Both emerged from protist ancestors, although very different ones, and each is doing well today but in restricted environments. Describe these two groups and how they relate to the main lines. In what way does each differ from main line organisms?
2. Explain the importance of animal embryology to the development of evolutionary theories and phylogenetic trees. Give examples. List some specific taxonomic problems that might arise if embryos were ignored.
3. Radially symmetrical animals are all aquatic. Suggest reasons why this type of symmetry is not found in the terrestrial environment. If it were, how might such an animal function?

KEY IDEAS

WHAT IS AN ANIMAL?

Animals are multicellular, eukaryotic heterotrophs, usually organized at the organ-system level with specialized organs, tissues, and cells. Most exhibit active movement through muscle action, and are heterotrophic. Animals produce a cellular matrix of collagen. They reproduce sexually, producing large, stationary eggs and small, flagellated sperm.

ANIMAL ORIGINS

1. The two animal subkingdoms, **Parazoa** and **Metazoa,** probably arose from different flagellate protist ancestors.
2. The transition from single-celled to multicellular life probably began with casual aggregations in which cells became interdependent, then specialized as they reached the colonial level of organization.

The Early Fossil Record

1. Reliable animal fossils date back to the Precambrian Era some 580 to 680 million years ago, although fossil worm tubes and tracks may be as old as 700 million years.
2. The **Ediacara fauna** fossils included jellyfish and worms; the slightly newer **Burgess Shale formation** includes fossils from most of today's phyla.

Phylogeny of the Animal Kingdom

The phylogenetic tree representing animal evolution reveals the separate origins of metazoans and parazoans. An early major split produced the **protostomes** ("mouth first") and **deuterostomes** ("mouth second"), the latter of which include the vertebrates.

LEVELS OF ORGANIZATION AND PHYLUM PORIFERA

Within organisms, the cell represents the fundamental level of organization, followed by the **tissue** level, the **organ**, and the **organ system.** The sponges of phylum **Porifera** represent the cellular level.

The Biology of Sponges

1. Sponges are all aquatic, and mostly marine. Their body consists of a few kinds of cells, including **porocytes, choanocytes**, and **amebocytes**, the latter occurring in a gelatinous **mesohyl.**
2. The amebocytes secrete skeletal elements—**spicules**—which in **calcareous sponges** are of calcium carbonate, in the **glassy sponges**, silicon dioxide, and in the **proteinaceous sponges**, the protein **spongin.**
3. Sponges are filter-feeders that draw water in through the body wall and phagocytize tiny food particles in the body cavity or **spongocoel**, before sending it out through the **osculum.**
4. As **hermaphrodites**, sponges produce sperm and eggs, and upon fertilization, a ciliated larva forms. In asexual reproduction, **gemmules**, clusters of cells, break away to form a new body.
5. Sponges lack nerve cells, so all responses are carried out by individual cells.

TISSUES AND THE RADIAL PLAN: PHYLA CNIDARIA AND CTENOPHORA

Most animals are **bilateral**—that is, the body can only be divided symmetrically by forming left and right halves. A few are **radial**, meaning that *any plane* passing through will divide the body into right and left halves. Echinoderms have a five-part **pentaradial** symmetry.

Phylum Cnidaria

1. Phylum **Cnidaria** is organized primarily on the tissue level with some organ development. They are thin-walled, saclike animals with tentacles and stinging cells—**cnidocytes**—that release **nematocysts.** The body wall contains an inner **mesoglea** that contains nerve cell processes, contractile fibers, and ameboid cells. The central **gastrovascular cavity** carries on digestion and respiration. Responses are facilitated by a widespread **nerve net** of nerve cells and processes.
2. Cnidarians include two basic forms, the stationary, attached **polyp** and the swimming **medusa** (jellyfish).
3. **Hydrozoans** are mostly marine, forming a stationary feeding and asexually reproducing colonial polyp. An asexually produced swimming medusa carries out sexual reproduction, and after fertilization a swimming **planula larva** forms a new polyp colony. Hydra, a freshwater cnidarian, is solitary and reproduces

asexually by budding and sexually by sperm and egg. It has no medusa or swimming larvel stage. *Physalia*, the Portuguese man-of-war, is a floating hydrozoan colony.

4. **Scyphozoans** (jellyfish) drift in the ocean, capturing prey with stinging tentacles. Following fertilization a swimming planula larva forms a fixed **scyphistoma** (polyp), producing many young jellyfish asexually.
5. The **Anthozoans** (corals and anemones) are strictly polyps. Following fertilization, the usual planula larva forms, but it develops directly into a new polyp. Most coral polyps secrete limestone coverings, forming coral beds and reefs.

Phylum Ctenophora: Comb Jellies

Ctenopharans the comb jellies, are organizationally similar to cnidarians but for the fused cilia or **combs** and tentacles bearing glue cells.

BILATERALITY: MESODERM AND PHYLA PLATYHELMINTHES AND RHYNCHOCOELA

Part of the bilateral trend included **cephalization**, an emphasis on a head or leading end, bearing feeding and sensory structures.

The Three-Layered Embryo

Metazoans have two embryonic **germ layers**, **endoderm** and **ectoderm**, from which all body tissues are derived. A third layer, the **mesoderm**, first seen in the flatworms, makes many new tissues possible, including organized muscle, blood, and internal skeleton (endoskeleton).

Phylum Platyhelminthes: The Flatworms

1. Phylum **Platyhelminthes**, the flatworms, includes free-living and parasitic members, many with the organ-system level of organization.
2. The free-living **turbellarians** include *Dugesia* **(planaria)**, a complex species with the following characteristics:
   a. an extensive, branched gastrovascular cavity;
   b. a ladderlike nervous system with large anterior **ganglia** and light-sensitive structures;
   c. **osmoregulation** (water management) through **protonephridia**—the **flame cell system**—in which excess water is forced through collecting tubules to the outside by ciliated **flame cells** within **flame bulbs;**
   d. movement by cilia, which push against mucus released by the body;
   e. feeding through a complex protrusible **pharynx**, with digestion beginning in the gastrovascular cavity and being completed within the lining cells;
   f. asexual reproduction through fission and sexual reproduction involving copulation and complex sexual organs. Planarians are hermaphroditic.
3. The **flukes** (class trematoda) are all parasitic, using a sucker mouth to feed on the host. Important human parasites include the human liver fluke and human blood fluke, the latter of which causes **schistosomiasis.** The life cycle of some may occur in several hosts.
4. The **tapeworms** (class **Cestoda**) are found in most vertebrates. They attach to the host intestine via a small hooked and suckered **scolex.** While in the **primary host**, the tapeworm produces numerous **proglottids** in which the sex organs and eggs and sperm develop. Ripened proglottids filled with fertilized eggs pass out with the host's feces. When eggs are taken into an **intermediate host**, they rupture, and a burrowing form enters muscle tissue, where it encysts as a **bladderworm**, remaining dormant until the muscle is eaten by the primary host.

Phylum Rhynchocoela: Ribbon Worms

The ribbon worms are a manor phylum but have evolutionary importance in that they have characteristics of both flatworms and roundworms.

BODY CAVITIES, A ONE-WAY GUT AND PHYLA NEMATODA AND ROTIFERA

Coelomate animals have a mesodermally lined body cavity—the **coelom**—and a tubelike, one-way gut, complete with mouth and anus. The gut contains a muscular wall. In roundworms, the cavity is a **pseudocoelom** or false coelom. It is not mesodermally lined, and the gut lacks muscle. **Acoelomates** lack such body cavities.

Phylum Nematoda

1. The **Nematodes**, or roundworms, are slender, cylindrical pseudocoelomates with little cephalization. They are soft-bodied but are made firm by fluid pressure forming a **hydrostatic skeleton.** Most are free-living predators, but others are important parasites of plants and animals.
2. The largest nematode, *Ascaris lumbricoides,* infests humans and hogs, where males and females sexually reproduce. The fertilized eggs pass out with the host's feces. When swallowed, the eggs hatch in the intestine, make a circuit through the lungs and back down to the intestine, where the life cycle repeats.
3. *Trichinella spiralis,* or the trichina worm, reproduces sexually in the hog intestine, and fertilized females bore into the muscle and reproduce asexually, forming great numbers. When the muscle is eaten by a human or another hog, the cycle repeats, producing a disease called **trichinosis.**

Phylum Rotifera is made up of tiny but complex, free-living aquatic pseudocoelomates. Their digestive system includes several specialized organs, such as a grinding **gizzard.** A simple flame cell system carries out osmoregulation and excretion, and a **urinary bladder** stores liquid excretory wastes. Reproduction commonly include **Phylum Rotifera parthenogenesis**—development of eggs without fertilization.

REVIEW QUESTIONS

1. List several important characteristics of animals. List two that are exclusive to animals. (660–661)
2. List four systems found in most animals and suggest the primary function of each. (General)
3. Cite a basic difference between Parazoa and Metazoa. (661)
4. What evidence suggests that the animals evolved from flagellate ancestors? (661)
5. Approximately when does the *undisputed* animal fossil record begin? What animal groups were present then? Were these the first multicellular animals? Explain. (663)
6. What do the Burgess Shale fossils tell us about the state of animal evolution at the end of the Precambrian? (663–664)
7. List the levels of organization found in organisms and define each. (667)
8. Prepare a drawing of a simple sponge and label the osculum, spongocoel, and porocytes. Using arrows, show the path of water. (668–669)
9. List four specific kinds of sponge cells and explain their functions. (668–669)
10. Discuss two reasons why the sponge is relegated to the cellular level of organization. (668–669)
11. Describe the manner in which sexual reproduction occurs in sponges. (669)
12. Define and give examples of bilateral and radial symmetry in animals. (670)
13. Describe the typical cnidarian body plan and include the terms *mesoglea, gastrovascular cavity, tentacles, cnidocytes, nematocysts,* and *nerve net.* (670–671)
14. What is the difference between a polyp and a medusa? In which cnidarian class is each of these prominent? (671–674)
15. Describe the peculiar way in which hydrozoans carry out sexual reproduction. How is this similar to alternation of generations in plants? What is the fundamental difference? (671)
16. Describe the life cycle of a typical scyphozoan cnidarian. (672–673)
17. In what two significant ways does the anthozoan life cycle differ from that of most hydrozoans and scyphozoans? (673–674)
18. List two differences between ctenophorans and cnidarians. (674)
19. With the advent of bilateral symmetry, what new ways of exploiting the environment may have been made available? What evidence is there that bilateral symmetry may be as old as 700 million years? (674)
20. List the three embryonic germ layers. What structures does the flatworm owe to the presence of a third layer? (674–675)
21. Describe the osmoregulatory organ system in the turbellarian and explain how it functions. (675)
22. Compare the digestive systems of planaria and hydra, listing primitive and advanced characteristics where applicable. (672, 675)
23. Briefly review the life cycle of the human liver fluke, *Chlonorchis sinensis.* How might such a complex cycle both help and hinder our control of the parasite? (678)
24. Describe the body form of a typical tapeworm. In what ways is it less complex than other flatworms? Is the lack of structure an advanced or primitive condition. (677–678)
25. Briefly review the life cycle of a two-host tapeworm. (679)
26. Using simple drawings, carefully distinguish between the terms *coelomate, acoelomate,* and *pseudocoelomate.* (681)
27. Describe the hydrostatic skeleton of the roundworm and explain how it helps in movement. (681)
28. In what habitat are most nematode species located? Why are they ecologically important? (681–682)
29. Describe the life cycle of *Ascaris.* (682)
30. At what stage would *Trichinella spiralis* produce the symptoms of trichinosis? Describe these and explain briefly how such a problem can be avoided. (682–683)
31. For their minute size, rotifers are amazingly complex. Using the digestive system as an example, support this comment. (683)
32. Describe the usual method of reproduction in a rotifer population. Under what conditions do males become important? (683–684)

31

INVERTEBRATES II

AS ANIMAL LIFE EVOLVED ON THE EARTH, NEW TRAITS CONTINUALLY appeared, most of which were weeded out by the relentless winnowing of natural selection. Occasionally, though, some lines developed a new feature that gave them an immediate and sweeping advantage over other forms. One of these highly adaptive structural innovations was the **coelom**, the mesoderm-lined body cavity between the gut and the body wall. Those species possessing such a wonderful cavity are called **coelomates**, and because of the simple fact that they have a coelom, they have been able to succeed under a variety of environmental circumstances and to radiate out over the earth in numerous adaptive directions.

The invertebrate coelomates are sometimes referred to as "advanced invertebrates," because they generally have a number of other advanced features, traits that evolved later in the history of life.

THE COELOM AND COELOMATES

The coelom is lined with a tough, glistening, mesodermally derived **peritoneum** from which arise broad, membranous folds, the **mesenteries.** The mesenteries hold the organs within the coelom in place while allowing a certain freedom of movement. In species with circulatory systems, blood vessels run through the mesenteries, carrying blood to and from the intestine.

In many species, the coelom is fluid-filled and lined with beating cilia that keep the fluids in motion. Such movement can be an efficient means of transporting digested foods and metabolic wastes. The oldest function of the coelom, one involving both primitive and modern-day worms, was probably locomotion. In such soft-bodied animals, the coelomic fluids form a hydrostatic skeleton, as muscular pressure against the fluids gives the body a certain rigidity and provides a resistant base for the action used in burrowing. Fossilized worm burrows, you may recall (Chapter 30), are the oldest evidence of animal life on earth.

The principal phyla of invertebrate coelomates we will consider here are the mollusks, annelids, arthropods, and echinoderms. The first three phyla are protostomes (as are the flatworms, nematodes, and rotifers), whereas the echinoderms are deuterostomes (along with hemichordates and chordates). We will also look at the **lophophorates**, a peculiar group that is considered protostomate by some zoologists and deuterostomate by others.

While both protostomes and deuterostomes may form a coelom, it is believed to have evolved independently in the two groups and, in fact, the coelom actually forms differently in the embryos of the two groups. (The presence of a coelom in both animal groups is thus a major example of convergent evolution.)

The Significance of the Protostome/Deuterostome Lines

At this point let's again consider the importance of the evolutionary divergence of the protostome and deuterostome lines (see Figure 30.4). With relatively few exceptions, the protostome line was to emphasize great diversity, small bodies, extremely rapid reproductive cycles, and great population size, all characteristics that more than compensate for limited learning ability.

The early deuterostomes weren't much different—if anything, they were less intellectually gifted and less adventurous. They were also small, simple creatures with extremely limited nervous systems. Yet somehow there was amazing potential in this kind of animal. The deuterostomes were to produce not only the echinoderms, such as the witless sea star, but the chordates, which would in turn produce the largest and brainiest creatures of all. However, hundreds of millions of years separated the early deuterostomes from the vertebrates.

The first embryonic opening—the **blastopore**—forms the anus in deuterostome embryos and the mouth in protostome embryos. The two groups also differ in the origin of the skeleton. In the protostomes, the exoskeletons of the arthropods, the shells of the mollusks, and the bristles of the annelids consist of nonliving, noncellular materials secreted by the ectoderm-derived epithelium. In contrast, most deuterostomes produce **endoskeletons**—stiff, internal skeletons that are formed by mesodermal cells. In most vertebrates the endoskeleton is bone, a hardened matrix containing numerous living bone cells.

The important question for evolutionists is, how could both protostomes and deuterostomes have arisen on the earth? One would think that the first animal species to have developed such a major advantage as the tube-within-a-tube body plan would have taken over the world. Yet these two kinds of coelomates, protostomes and deuterostomes, coexisted. Both these evolutionary lines underwent massive adaptive radiations in the Ediacara period of the late Precambrian, and the phyla that descended from each are today in competition with one another. However, ecological and evolutionary theory suggests that the earliest protostomes and deuterostomes must somehow have failed to fall into competition with one another at the time they first became established. There are at least two ways this might have happened.

The first hypothesis is that the early protostomes and early deuterostomes were protected from direct competition because they were *ecologically* isolated. Whereas the earliest protostomes were burrowing worms, the primitive deuterostomes were upright suspension-feeders. So while the annelid-arthropod-mollusk ancestors were crawling about in the muck, the echinoderm-chordate ancestors—our own forebears—were sitting on stalked rear ends straining seawater. Who could have predicted which of the two lines would eventually produce eagles and astronauts?

The second hypothesis is that the earliest protostomes and deuterostomes were protected from direct competition because they were *geographically* isolated. You will recall from our discussions of continental drift (Essay 20.1) that the present-day land masses were produced by the breakup and drift of the supercontinent, Pangaea, which began about 230 million years ago. Studies of magnetic lines and forces in ancient lava beds indicate that continental drift also occurred at an even earlier time. Current thinking is that Pangaea itself was formed through the union of two great land masses or supercontinents. The northern supercontinent, **Proto-Laurasia**, consisted of what is now North America, Siberia, and China; the southern supercontinent, **Proto-Gondwana**, included everything else. But what has this to do with protostomes and deuterostomes? Simply that some early fossil evidence suggests that the protostomes began in the shallow shelf waters of the southern supercontinent and that the deuterostomes radiated out from the shallow waters of the northern supercontinent. Thus, they were firmly established, diverse, and occupying different niches before the shifting land masses brought them the pleasure of each other's company.

(a)

(b)

(c)

FIGURE 31.1
LOPHOPHORATES
The lophophorates include three phyla that all have a structure known as the lophophore. (a) One of the phyla, Ectoprocta, forms either crusty colonies (on rocks) or seaweedlike fronds. (b) Brachiopods are often mistaken for clams, but the shell halves are arranged dorsoventrally rather than laterally. (c) Phoronids feed by extending their lophophores from their burrows like spirally coiled fans.

THE PROTOSTOME LINE

We will begin our look at protostome coelomates with the lophophorates. What comes to mind when you think of a lophophorate? Probably not much. The lophophorates are comprised of three rather obscure phyla. Some of these phyla, though, include some rather familiar forms of life.

The Lophophorate Phyla

The three lophophorate phyla are **Ectoprocta**, **Brachiopoda**, and **Phoronida** (Figure 31.1). Ectoprocts, a group containing some 4,000 species, are known more familiarly as bryozoans. These animals are often seen along rocky coasts at low tide. They form branched, crusty structures that look something like seaweed but are actually colonies of individuals formed by asexual budding. Each tiny individual bears tentacles and feeds in the fashion of a coral, but there the similarity ends. The tentacles are attached to a curved ridge called the **lophophore**, and each tentacle is covered with a ciliated lining. Some feathery ectoprocts are dried and sprayed with green paint, and sold in grocery stores as "living air ferns—requiring no care."

Brachiopods are shelled lophophorates that superficially resemble a clam or scallop. However, the two valves or shells hinge so as to form a top and bottom, rather than two sides (dorsoventral rather than lateral). The shell encloses a coiled ridge of ciliated tentacles that comprise the lophophore.

The phoronids are small wormlike creatures that live in tubes that penetrate the oxygen-starved sediments of estuaries and bays. They also bear ciliated lophophores. Phoronids, brachiopods, and ectoprocts all produce free-swimming larvae, typical of marine invertebrates that are fixed in place as adults. Such larvae make up part of the marine plankton.

While today's lophophorates are only minor phyla in terms of their species numbers, they were once more important. In fact, they are well represented as the oldest fossils. Both brachiopods and ectoprocts are found in Cambrian deposits over 500 million years old. Some rock strata are composed almost entirely of their fossilized shells. The brachiopods have dwindled from about 3000 named species in Ordovician fossil beds to about 200 named species alive today.

Phylum Mollusca

The earth is burgeoning with **mollusks.** In fact, we already know of about 100,000 species, which, in terms of sheer numbers, puts them behind only the nematodes and arthropods. Some mollusks are minute, living in tiny inconspicuous shells, while the giant North Atlantic squid that swims confidently through the cold ocean waters reaches over 18 meters (59 feet) in length. Of the seven classes, we will consider only the four larger classes as representatives of the phylum (Figure 31.2).

The term *mollusk,* derived from Latin, means "soft-bodied," though for an invertebrate, this isn't very descriptive. Aside from being soft-bodied all mollusks are distinguished by a muscular **foot** that contains sensory and motor systems and may be used for swimming, creeping, digging, holding on, or capturing prey. Some have external **shells** produced by secretions of a fleshy covering known as the **mantle,** which is also present in nonshelled members of the phylum. The **mantle cavity** (the space enclosed by the

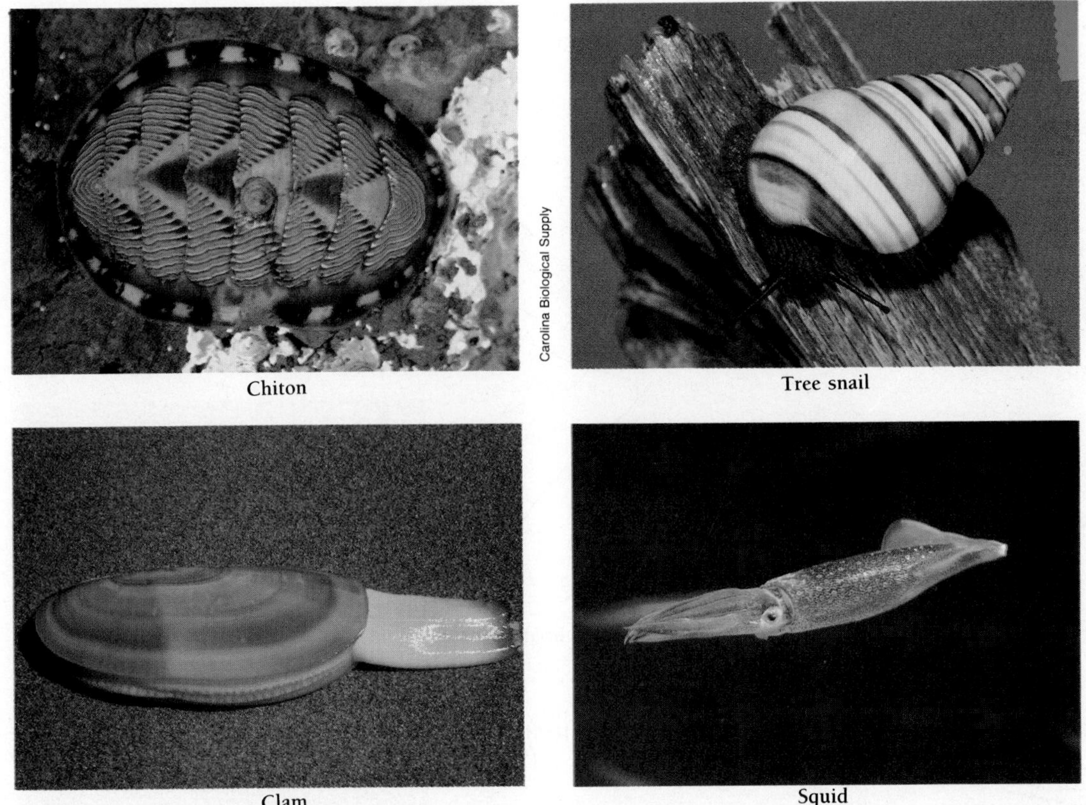

Chiton

Tree snail

Clam

Squid

mantle) houses feathery respiratory **gills** in aquatic mollusks. In terrestrial mollusks, the lining of the mantle cavity is highly vascular and serves as a respiratory membrane across which oxygen and carbon dioxide can pass (Figure 31.3).

In mollusks, the coelom is conspicuous in embryos but is very much reduced in adults, often present only as an open region surrounding the heart, with some remnants in the nephridia and gonads. Except for the cephalopods (squids and octopuses), mollusks have an **open circulatory system.** Unlike **closed circulatory systems,** those in which the blood remains within vessels, the vessels in open systems end some distance from the heart, the blood then flowing into spongelike sinuses and cavities before returning to the heart.

The digestive system is, of course, tubelike, with both a mouth and anus. Mollusks (except for bivalves and tuskshells) have a rasping tonguelike structure called the **radula**

FIGURE 31.2
DIVERSITY IN MOLLUSKS
Examples from each of the four major classes provide a scant sample of molluskan diversity since some one hundred thousand species are known.

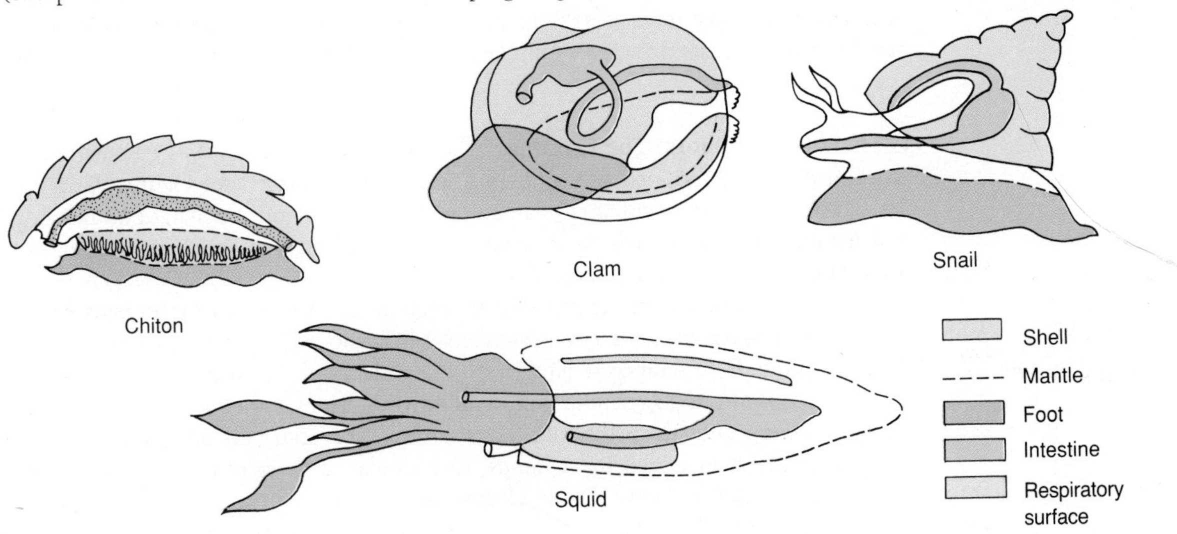

FIGURE 31.3
THE MOLLUSK BODY PLAN
All mollusks share a basic body plan, which includes a fleshy foot and mantle, and commonly, a shell or its remnants. Members of the phylum are classified according to variations in the basic structures.

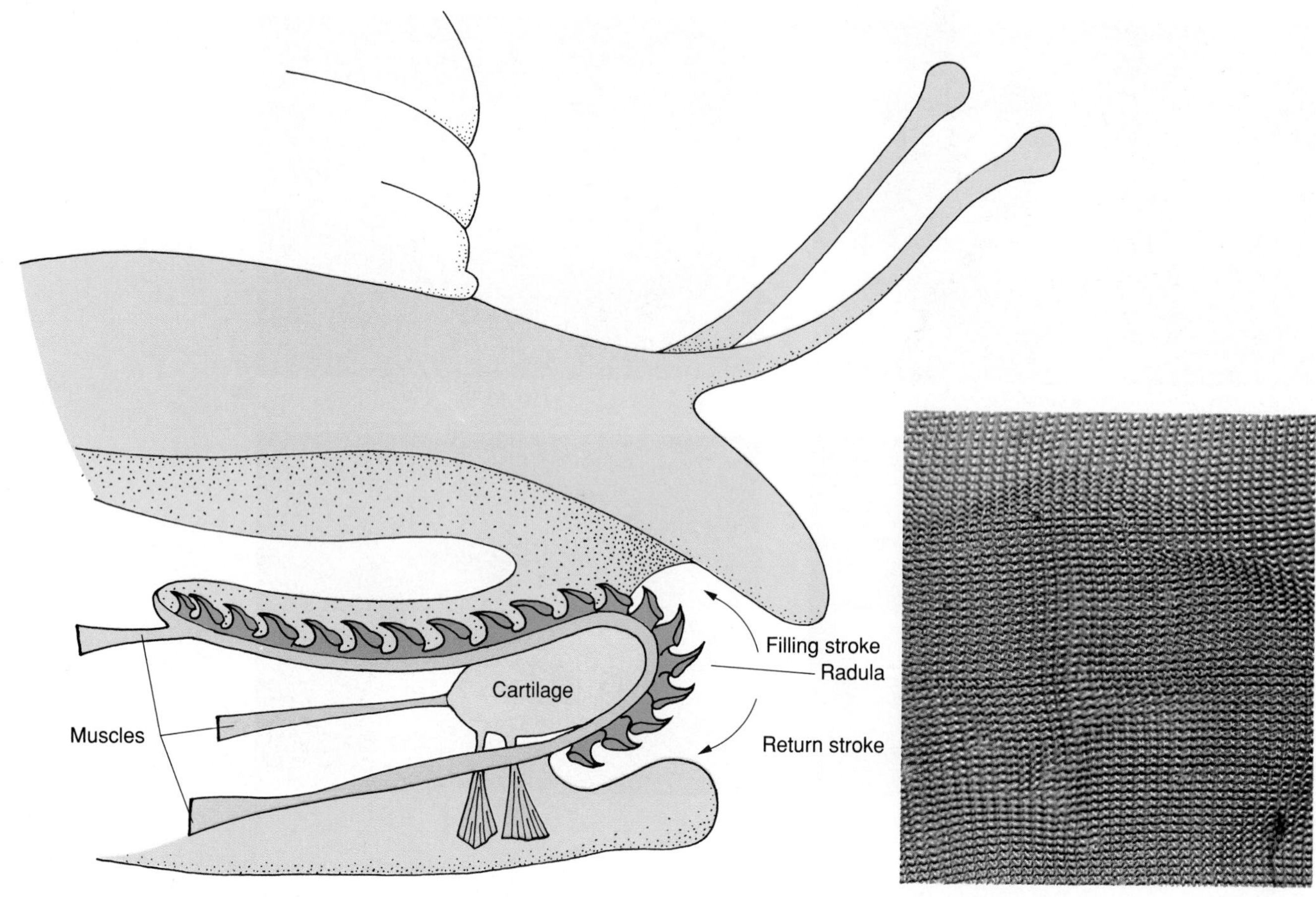

FIGURE 31.4
THE RADULA
The radula is a ribbonlike membrane containing a row of teeth, all pointing backward. It lies over a bed of cartilage and slides back and forth as it is pulled first by one set of muscles, then by another. The rasping action tears and shreds food. Photo shows a surface view of the radula.

with which they file their food into small particles (Figure 31.4). Mollusks make use of **nephridia** to carry on osmoregulation and excretion. We will look more closely at the nephridium in the next section.

The evolutionary relationship of mollusks to other protostome coelomates is uncertain, mainly due to their apparent lack of segmentation and their highly reduced coelom. Both segmentation and a well-formed coelom are key characteristics of annelids and arthropods. Yet, like marine annelids, some mollusks produce a **trochophore larva,** a ciliated, pear-shaped swimming form (see Essay 31.1). In addition, studies of certain fossil mollusks indicate that the coelom may at one time have been much more prominent than it is now. Since there is no convincing evidence of segmentation in the fossils, it may be that mollusks diverged from coelomate ancestors before segmentation became established.

Class Polyplacophora: The Chitons The **chiton** is a mollusk with a flattened body, an elongated foot, and a shell composed of eight simple plates (see Figures 31.2a and 3a). The gut, beginning with a radula-equipped mouth, extends the length of the body. The mantle forms a skirtlike arrangement just above the foot and houses numerous feathery gills.

Chitons are slow-moving creatures that creep along over the rocky bottoms of tide pools through wavelike contractions of muscles in the foot. They graze as they go, using their sharp radula to scrape the algae from the surfaces of the rocks. When disturbed, their defense is to put powerful foot muscles to work gripping the rock, thereby making themselves very difficult to dislodge. Alternatively, they can curl up, presenting only their tough plates to a predator. Chitons, very similar to those of today, appear in the fossil record in 500-million-year-old Ordovician deposits.

ESSAY 31.1

EMBRYOS AND EVOLUTION: PROTOSOME AND DEUTEROSTOME CHARACTERISTICS

Embryos often provide interesting clues to evolutionary history, sometimes filling in the many gaps of the fossil record. We can, for instance, usually distinguish between two great animal groups—the protostomes and deuterostomes—on the basis of several distinctive embryological differences. The two groups express these differences almost from the start of life.

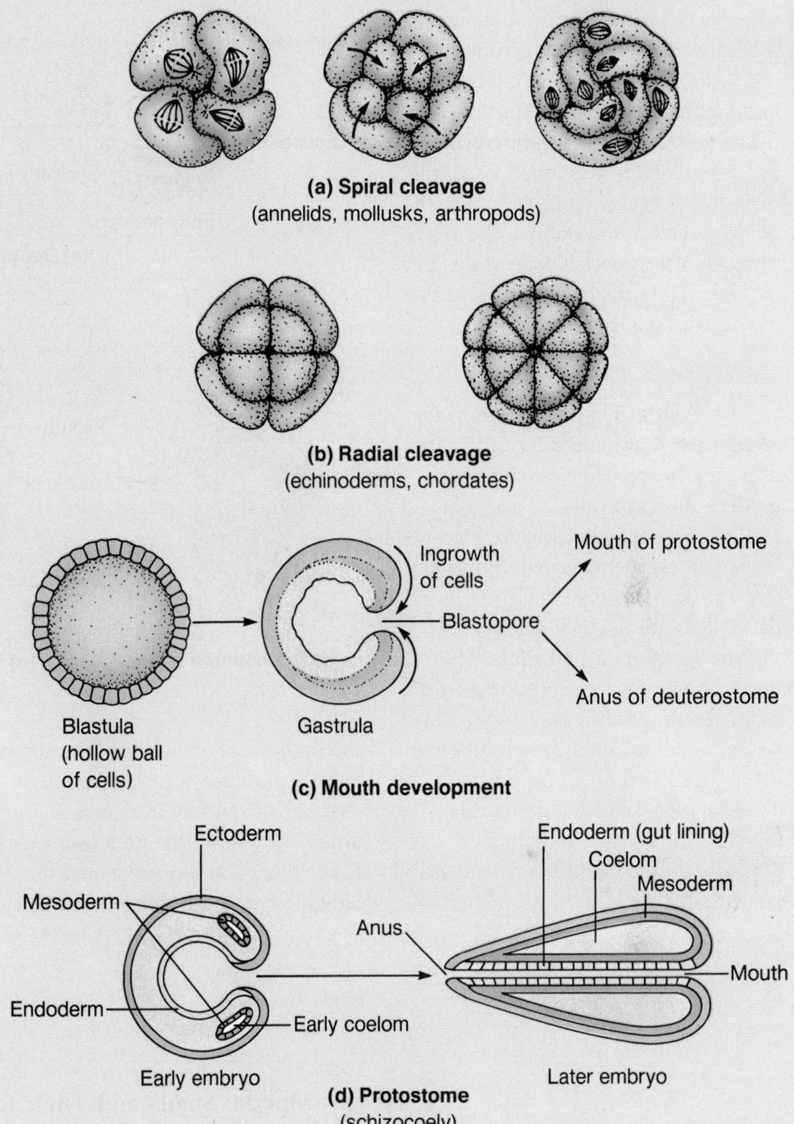

(a) Spiral cleavage (annelids, mollusks, arthropods)

(b) Radial cleavage (echinoderms, chordates)

(c) Mouth development

(d) Protostome (schizocoely)

CLEAVAGE PATTERNS

Shortly after fertilization, the zygote begins the first of many cell divisions or cleavages that will provide the immense bank of cells from which the embryo will emerge. The zygote first divides into two cells, then into four, then eight, sixteen, and so forth. If the animal is protostome, by about the third round of divisions the upper cells come to lie on the dividing line of the cell layer below, rather than directly over them **(a)**. Thus we have **spiral cleavage.** With the deuterostomes, the cell divisions occur in radial fashion, called **radial cleavage.** Radial cleavage places each new cell neatly atop another **(b)** so that the cleavage lines are nicely aligned.

MOUTH DEVELOPMENT

A second important distinction explains the terms *protostome* and *deuterostome* ("first the mouth" and "second the mouth," respectively). During the normal course of events in an embryo's development, the cleavages produce a ball of cells called a blastula. Then, certain cells begin to sink in, or invaginate, producing a **gastrula** stage. The invagination itself is called a blastopore **(c)**. In the protostomes the blastopore often marks the origin of the embryonic mouth, but in the deuterostomes, this region marks the site of the anus. The deuterostome mouth develops from a secondary opening called the **stomadeum.**

SCHIZOCOELS VERSUS ENTEROCOELS

A third distinction is seen in the manner in which the coelom develops following the ingrowth of cells mentioned earlier. Although protostomes and deuterostomes are both coelomates, they are believed to have evolved independently, and the presence of a coelom in each represents convergent evolution. Thus, as you might expect, there are differences in how the coelom forms. In the protostome, the **schizocoelous** process is seen. The term "schizocoelous," not surprisingly, refers to a "splitting"—in this instance the coelom forming as a split in the mesoderm, which itself has

ESSAY 31.1

EMBRYOS AND EVOLUTION: PROTOSTOME AND DEUTEROSTOME CHARACTERISTICS, (cont'd)

formed in patches near the blastopore **(d)**. The split produces two hollow patches of mesoderm that expand by cell division, filling the old cavity and forming the mesodermally lined coelom. In the deuterostomes, the **enterocoelus** process occurs. In this instance **(e)**, the mesoderm forms two pouches along the hollow center of the embryo, and cell division in the pouches spreads the mesoderm into the hollow, forming the mesodermally lined coelom.

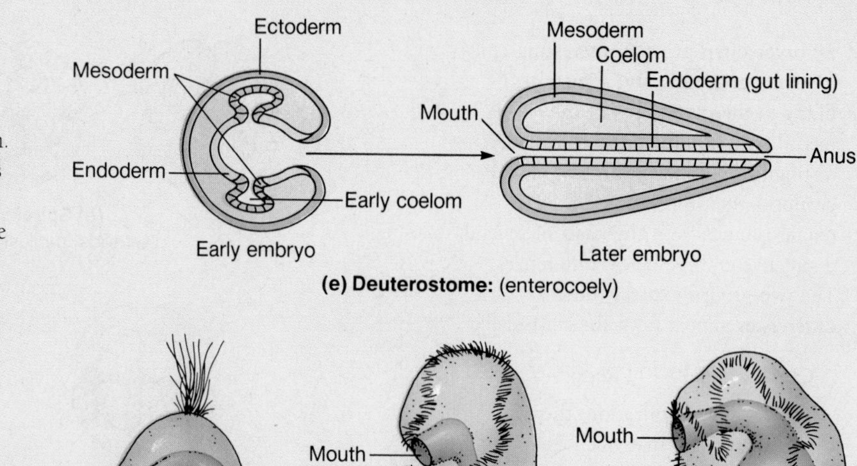

(e) Deuterostome: (enterocoely)

SWIMMING LARVAE

A final distinction can be seen in the appearance of swimming larvae in those protostomes and deuterostomes that produce this embryological stage. Mollusks and annelids produce a pear-shaped larva generally referred to as the trochophore **(f)**. While the larvae in these phyla are certainly not identical, they bear important similarities. The deuterostomes, when represented by the echinoderms, produce several types of larvae, which, according to evolutionary theory, are based on an hypothetical ancestral type called the **dipleurula larva.** In the sea star, the specific dipleurula is the bipinnaria larva **(g)**, while in the brittle stars, sea urchins, and sand dollars, the larval derivative is called a **pluteus.** In each echinoderm class, the larvae vary from a basic hypothetical form that no longer exists. The hypothetical dipleurula and the early bipinnaria are also quite similar to the larval form of the acorn worm, another deuterostome. Note the similarity in body plan and arrangement of the gut and ciliated bands between the bipinnaria and the **tornaria larva** **(h)** of the acorn worm.

(f) Trochophore larva

(g) Bipinnaria larva

(h) Tornaria larva

Class Gastropoda: Snails and Their Relatives Gastropods glide along on a large, extendable, muscular foot much as the chitons do. The foot lies just below most of the digestive organs, thus giving rise to the poetic name "gastropod" or "stomach foot." Some, such as snails and slugs, have successfully adapted to land life, while others glide over the seabed, their beautifully sculptured shells accenting the ocean floor. Whereas marine gastropods generally possess well-formed feathery gills, gas exchange in terrestrial and freshwater snails and slugs is provided by a lung consisting of a heavily vascularized region within the mantle cavity. Interestingly, freshwater snails, and a few marine species, have lungs instead of gills and must come to the surface to breathe. It is theorized that, like the marine reptiles and mammals, such species arose from terrestrial lines, returning to the water after lungs had evolved.

The basic organization of the snail is unusual in that during development, its body undergoes a process called **torsion** (twisting or rotation) brought about by uneven muscle growth. This causes all sorts of internal changes, and no one is quite sure what the advantages might be. As you can see in Figure 31.5, the body undergoes a 180° twist, bringing the mantle cavity forward and the anus around to a position just above the

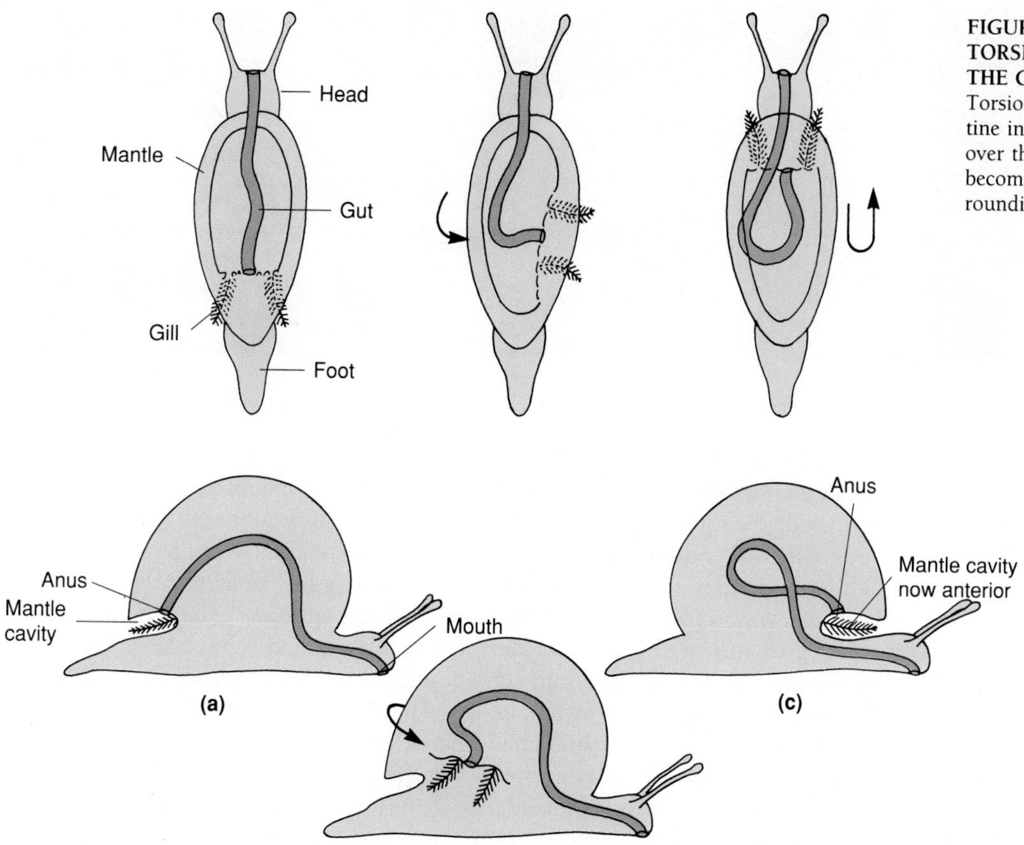

FIGURE 31.5
TORSION AND ITS EFFECTS ON THE GASTROPOD
Torsion in gastropods puts the intestine into a loop, ending with the anus over the head. The mantle cavity becomes displaced and ends up surrounding the head.

head (certainly not the most esthetic arrangement possible). The overall configuration provides considerable space for retraction of the entire head-foot into the shell, which may be the chief advantage—protection of the delicate gastropod head. Attached to the foot is a tough lid, or **operculum**, composed of a horny material. Once the operculum is in place across the opening into the shell, the snail is safe from many predators. Interestingly, the coiled shell itself evolved independently of torsion, perhaps even preceding it.

There are a lot of gastropods around, somewhere in the neighborhood of 35,000 species. They vary from those with elaborate twisted shells to shell-less creatures such as the slug, nudibranch, and sea hare (Figure 31.6).

Class Bivalvia (Pelycypoda): The Bivalves The bivalve (two-valved) mollusks differ from gastropods primarily in that the shell is organized into two hinged **valves.** The head and sensory appendages are greatly reduced, although the foot is fleshy and highly extensible—used by clams for burrowing. The hinged shell is drawn closed by two powerful muscle groups.

Bivalves lack the radula and are chiefly filter-feeders, using mucus to trap small particles of food from the water. Currents of water, bearing algae and other minute organisms, are drawn into the mantle cavity through an **incurrent siphon** by the action of cilia that line the large gills. The food particles are then trapped in heavy mucus secretions, which are moved by cilia along food grooves to the **labial palps.** The labial palps then move the food into the mouth. The gill surface also serves in the exchange of carbon dioxide for oxygen, with water passing on its way to the **excurrent siphon.** The circulatory system of bivalves is typical of mollusks. The heart is located in its coelomic cavity. As you can see in Figure 31.7, the intestine also passes through the cavity; in fact, the heart is actually wrapped around the gut.

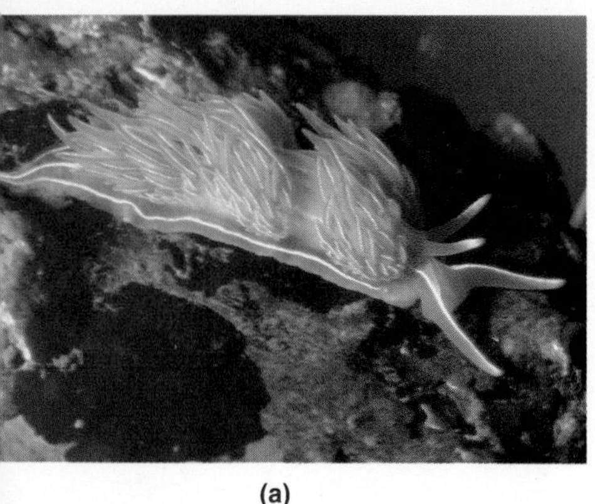
(a)

(b)

(c)

FIGURE 31.6
REPRESENTATIVE GASTROPODS
There are 35,000 species of gastropod mollusks. These include (**a**) the beautiful nudibranch, (**b**) the cone snail, seen here eating a paralyzed fish, and (**c**) the keyhole limpet.

Class Cephalopoda: The Squid and Octopus The **cephalopods**, "head-foot" mollusks, include the squid, octopuses, cuttlefish, and nautiluses. All but the nautiluses lack the external shell. Squid and cuttlefish have greatly reduced internalized shells. In cuttlefish, the spongy shell ("cuttlebone") is used mainly for flotation.

In many ways, cephalopods are highly specialized creatures. They are active and voracious predators, feeding on fast-moving invertebrates and vertebrates. Their circulatory system is closed, an advanced condition that provides the rapid circulatory efficiency needed by large fast-moving animals. They have a centralized, pumping heart but in addition there are two **branchial hearts,** chambers that provide an additional boost to deoxygenated blood as it moves into the gills. Water is moved over the gills by the pumping of the muscular mantle, in contrast to the ciliary gill currents found in other marine mollusks. In cephalopods, the head is greatly enlarged and is covered by the mantle itself. The thick, muscular molluscan foot is highly modified into tentacles, some complete with suckers. The mouth is armed with a horny beak used in ripping and tearing prey.

Because of their size and active behavior, the cephalopods tend to attract the attention of predators. Their defense is their speed and agility. Octopuses and squid move by forcefully ejecting water through their siphons in a jetlike action. The siphon can be turned in any direction. They may also quickly change colors or release a cloudy "ink" to foil predators. Octopuses often hide in burrows and can squeeze their large bodies through and into tiny cracks and crevices. Like other cephalopods they have highly developed sensory receptors, including eyes that are very similar to those of vertebrates even though they evolved independently (Figure 31.8). The octopuses also have a large, well-developed brain and are surprisingly intelligent invertebrates.

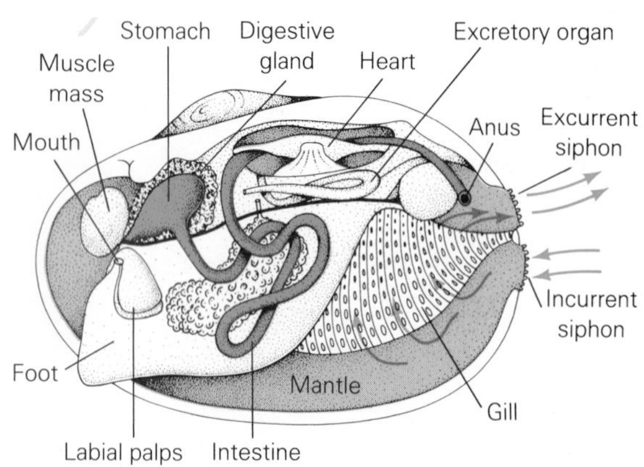

FIGURE 31.7
ANATOMY OF A CLAM
In this diagram, one valve, one pair of gills, and part of the body wall have been cut away. The clam's head is highly reduced. The foot makes up much of the body. Clams bring in a steady flow of water through the incurrent siphon, pass it through pores in the gills, and then send it back out through the excurrent siphon. Rows of cilia sweep any particles forward, and labial palps detect whether an object is edible. The mouth opens into a short gullet that leads into a stomach pouch. Food then moves into a lengthy intestine, which actually passes through the head and on to the anus.

The Segmented Body Plan

Along with the coelom and a newly specialized digestive system, a third evolutionary milestone appeared in most of the coelomates: the **segmented body plan** or **metamerism.** Here, both the body and the coelom are divided by transverse septa (cross walls) into a sequence of units, or **metameres.** In the simplest forms of segmentation the units are more or less repetitious. Such structural repetition is most obvious in certain annelids and in arthropods such as millipedes and centipedes. A more complex version of the theme is also found in other coelomate animals. In vertebrates, for example, we find a form of segmentation in the repeated vertebrae, ribs, spinal nerves, and trunk muscles. However, vertebrate segmentation is believed to have evolved independently of that of invertebrates. Nonetheless, the theme prevails in coelomate animals, thereby attesting to its evolutionary importance. As we will see, many adaptive variations have sprung from this basic body plan.

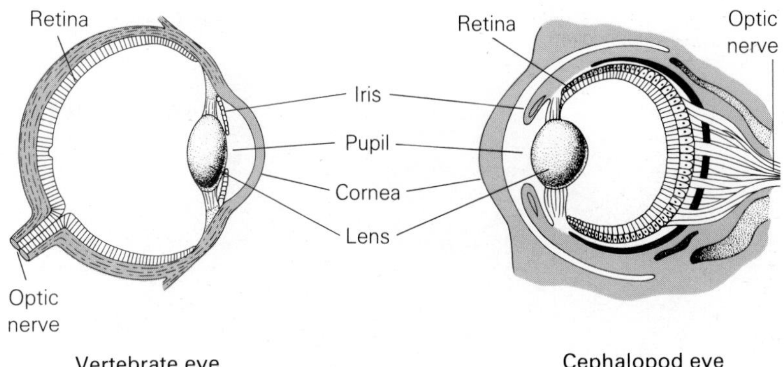

FIGURE 31.8
CEPHALOPOD VISION
The eyes of cephalopods (such as the octopus and squid) are remarkably similar to those of vertebrates.

Phylum Annelida

The extreme segmentation in the annelids is believed to be an adaptation for burrowing. The trait developed hand in hand with the fluid-filled coelom, which produces the hydrostatic pressure found in both earthworms and roundworms. The hydrostatic skeleton was an early alternative to the hardened skeleton that arose later.

The annelids include three classes: **Oligochaeta** (earthworms), **Hirudinea** (leeches), and **Polychaeta** (marine worms).

Class Oligochaeta: Earthworms Oligochaetes include the terrestrial earthworms, freshwater worms, and a few marine forms. Oligochaetes have little differentiation in the head region, lacking eyes and other elaborate sensory structures. Of course, when one uses the head for digging, such elaborations might create a few problems. The most obvious characteristic of the earthworm body, whether viewed externally or internally, is segmentation. Internally, each segment contains elements of the circulatory, digestive, excretory, and nervous systems (Figure 31.9). Their circulatory system, like our own, is closed. The blood is virtually always enclosed in blood vessels, not allowed to percolate freely through tissue spaces as is common in many invertebrates. The earthworm's circulatory system includes five pairs of **aortic arches** (hearts), which are essentially pulsating vessels, along with arteries, veins, and capillaries. The blood of many annelids contains hemoglobin, an oxygen-carrying protein, but it is not bound within blood cells, as is our own hemoglobin. Instead, it is dissolved in the circulating fluid.

The excretory system is also highly developed in the terrestrial worms. Nearly all segments have paired **nephridia.** A nephridium is considered to be quite advanced over a protonephridium (see Chapter 30) because it not only takes up fluids, ions, and nitrogen wastes, but it recovers most of the water and valuable ions, sending them back into the blood. (As a terrestrial animal, the earthworm must be a water conserver.) Thus, the nephridia perform functions similar to the vertebrate kidney, maintaining water and ion balances and removing nitrogen-containing metabolic wastes from the coelomic fluids.

The earthworm is hermaphroditic, with complex male and female reproductive organs present in each individual, but is not self-fertilizing. When earthworms copulate, they exchange sperm reciprocally, and each stores the sperm in its **sperm receptables** until the proper time for fertilization. The **clitellum,** the smooth, whitish cylinder of external tissue on earthworms, secretes a mucous cocoon that will slide forward along the body, receiving eggs and sperm from special reproductive pores. Fertilization occurs within the cocoon, which is then shrugged off. It will house the embryos while they develop. There is no larval stage, and the young hatchlings resemble the adults.

The digestive system is complete, following the tube-within-a-tube plan. The tube is subdivided in the first 20 or so segments into swallowing, storing, and grinding regions. The intestine, a digestive and absorptive structure, continues to the anus. A prominent

FIGURE 31.9
EARTHWORM BODY PLAN
Note the prominent segmental body plan, the smooth, glandular clitellum, and the circular and longitudinal muscle groups in the body wall of the earthworm. Transport is carried out by a closed circulatory system that includes five paired "hearts" (aortic arches) and an extensive system of blood vessels. The gut, suspended in the coelom, contains several specialized regions. The nervous system consists of a pair of enlarged ganglia above the pharynx and a lengthy ventral nerve that gives rise to ganglia in each segment. Nearly all segments contain paired nephridia, complex tubular structures that eliminate nitrogen wastes and regulate body water.

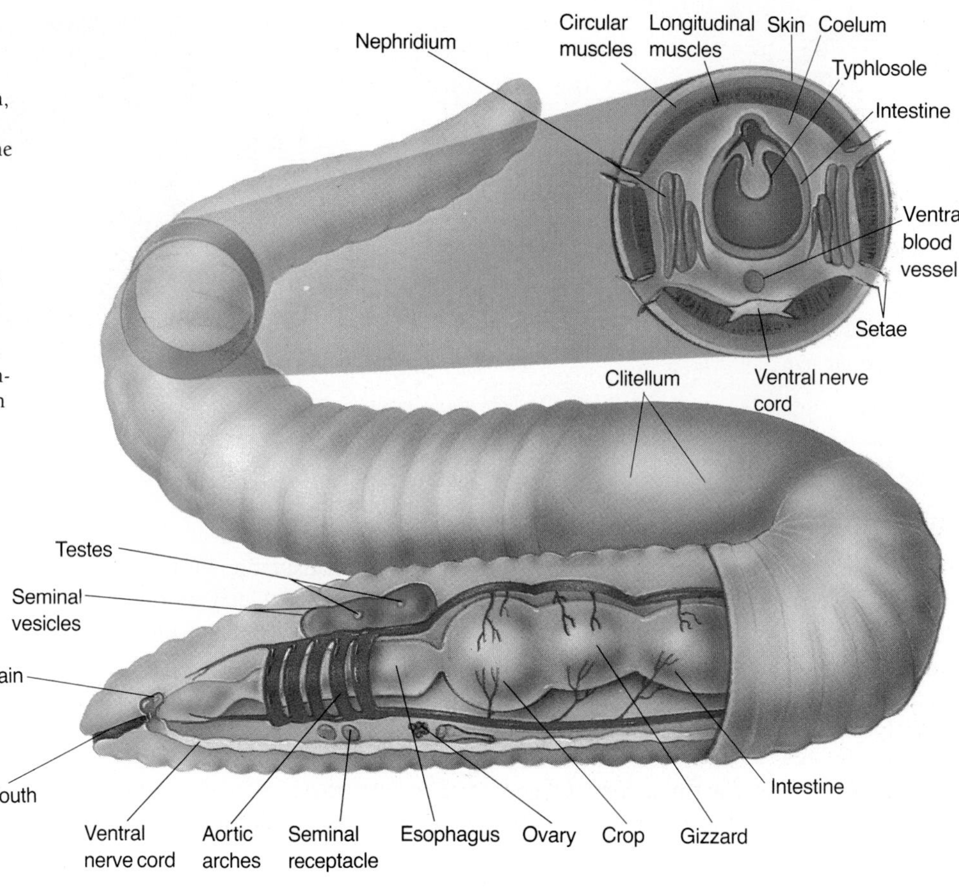

fold in the intestinal wall, the **typhlosole**, increases the digestive surface (see Figure 31.9). Earthworms feed on organic matter in the soil.

The earthworm's nervous system includes a brain and a fused pair of solid **ventral nerve cords.** (Invertebrate nerve cords are all solid; chordates have hollow ones.) The brain consists of a ring of nerve tissue around the pharynx, just behind the mouth. This nerve ring includes enlarged paired ganglia above and below the pharynx, with those below giving rise to the ventral nerve cords. The nerve cords, arranged closely parallel to each other, extend along the body to the most posterior segment. Paired ganglia form in each segment, with branches that innervate the surrounding tissues.

Burrowing and other movements are accomplished by two layers of body wall muscles—circular and longitudinal—which extend and shorten the body, respectively. Acting as a hydrostatic skeleton, the turgid, fluid-filled body provides a flexible but resistant base for muscle action while maintaining the earthworm's shape. During burrowing, anchorage is provided by the **setae** (singular, **seta**), chitinous bristles found on most segments, which can be inserted into the burrow to hold some segments fast while other parts of the body are extended or contracted (see Chapter 33).

FIGURE 31.10
THE LEECH
Leeches are armed with sharp, piercing mouthparts and suckers for holding onto their hosts and drawing blood. After feeding, they drop off and fast for long periods.

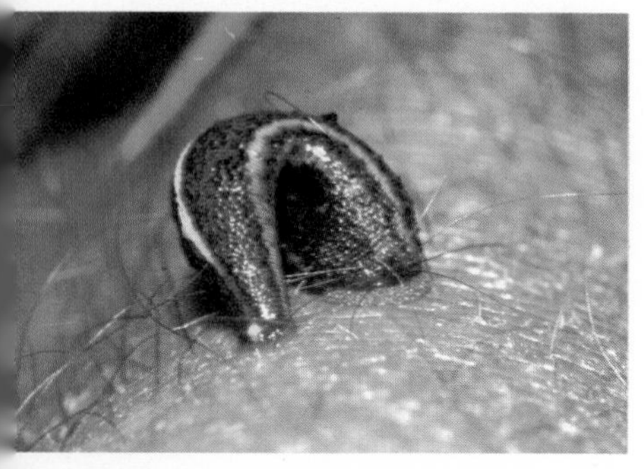

Class Hirudinea: The Leeches **Leeches** live primarily in fresh water, although some species are found in marine and moist terrestrial habitats. Although commonly thought of as parasitic, there are many predatory and scavenging species. The parasites are **ectoparasites**—that is, they attach themselves to the skin of their hosts, which include humans, and draw blood.

The segmented body of the parasitic leech is usually flattened, with suckers at the anterior and posterior ends (Figure 31.10). Suction in the anterior sucker is applied by a muscular pharynx, and some species have horny teeth that cut through the skin of the host organism. When they have made an incision, they secrete an anticoagulant called **hirudin** into the wound. Leeches were once sold in pharmacies as a popular remedy for the swelling and discoloration of "black eyes," and in earlier times to "bleed" patients as a common treatment for illness.

The body of the leech is formed of modified segments. The segmentation is apparent in the nervous, reproductive, and excretory systems, but the coelom itself is not divided. Like the earthworm, leeches are hermaphroditic but not self-fertilizing. They are also similar to earthworms in their copulation, egg laying, and development.

Class Polychaeta: Segmented Marine Worms Polychaete worms generally have well-formed head regions with eyes, specialized sensory structures, and numerous fleshy extensions called **parapodia.** Although clearly segmented, some polychaete worms have specialized body regions, unlike earthworms and leeches. Some are active swimmers and burrowers, while others are sedentary, remaining in burrows or tubes and exposing only their anterior parts for feeding and gas exchange. A number are particle feeders, trapping minute marine plankton and bits of detritus (loose material) in cilia or mucus covering numerous tentacles. Others like the "clamworm," *Nereis virens* (Figure 31.11) live a different life.

Nereis makes its home in sandy and muddy tidal flats. Although it tends to burrow reclusively in the sediments, it can be roused to swim when startled or when ready for mating. Externally, the most striking features of the clamworm are its powerful, retractable jaws, the numerous pairs of fleshy parapodia along its segments, and the sensory appendages at its anterior end. The parapodia, used both in movement and in gas exchange, are muscular extensions of the body wall that usually bear setae.

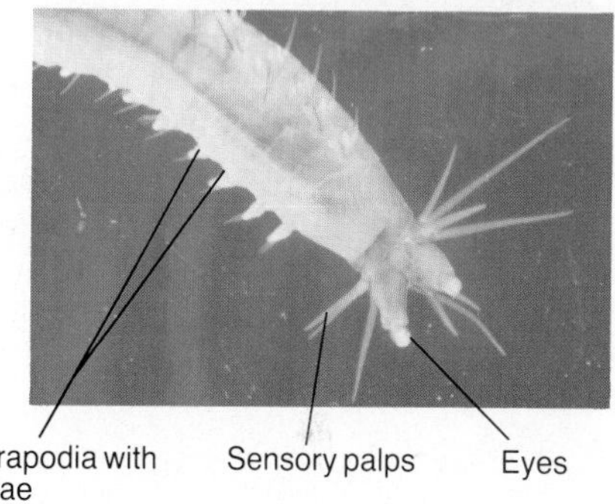

FIGURE 31.11
NEREIS, A MARINE ANNELID
The clamworm is an aquatic relative of the earthworm but is highly specialized for its marine existence. It has a well-differentiated head with eyes, sensory projections, and a set of retractable grasping jaws.

Unlike what we've seen in earthworms and leeches, sexes are usually separate in marine polychaetes. Gametes are produced seasonally by the coelomic lining. Some species simply release their gametes into the surrounding water, but in others the females swell with eggs and then literally burst. In one particular exotic species, the Samoan palolo worm *(Eunice viridis),* a large number of posterior segments become filled with mature gametes. Then during a night in late October, when the moon is in its last quarter, the ripened, specialized regions of countless worms break away simultaneously and swim to the surface (guided there by light receptors in each segment). At about dawn, with the surface literally crawling with the odd creatures, each bursts, and sperm and egg find each other in an orgy of fertilization. The fertilized eggs then form swimming **trochophore** larvae that become part of the marine zooplankton. The trochophore is typical of many other marine invertebrates, including mollusks. It gets its name from the spinning motion (Greek; *trochos,* "wheel") of its ciliated, pear-shaped body (see Essay 31.1).

Among the polychaetes are a number of tube-dwelling species. Some construct tubes of sand particles, cemented with mucus or calcium secretions. They are easy to see in their burrows because their feathery plumes (which are actually feeding and respiratory devices) constantly pop in and out. These colorful ciliated tentacles have inspired common names such as "fan worm," "feather worm," and "peacock worm."

Recall that burrowing worms, very likely annelids, may have a truly ancient evolutionary history. Fossilized burrows from Precambrian strata have been dated at about 700 million years, representing the oldest signs of animal life (see Chapter 30). The earliest tracks are random and dispersed, but later ones appear aggregated, possibly signifying changes in the social behavior of the worms.

Phylum Arthropoda

There are a lot of **arthropods.** You can get the idea by imagining Noah trying to load his ark with animals. If, by chasing, swatting, and kicking, he could have gotten one pair aboard per minute, he would have spent 18 months, night and day, simply loading his quota of crickets, crabs, lice, flies, centipedes, aphids, wasps, weevils, dragonflies, lobsters, ticks, and the like into the vessel.

To date, some 900,000 species of arthropods have been described, and some authorities estimate that about 1 million more await identification. We must conclude that as a phylum, **Arthropoda** is the most diverse on earth.

So what are these pervasive creatures? The name *arthropod* tells us that they have "jointed feet," actually jointed legs. They are also segmented, but these segments are not simply repeating units as in earthworms; they may instead be highly specialized for different tasks. Arthropods have an **exoskeleton** (*exo-*, "outside") made of chitinous material, often hardened with calcium salts. It serves as a protective covering and for the attachment of muscles used in movement. Joints are formed at thin, flexible regions of the exoskeleton. As arthropods grow they must **molt**, that is, periodically shed their protective coverings and form new exoskeletons. Most arthropods shed or molt several times during development, and some continue the process throughout their adult life.

Arthropods have successfully invaded most of the earth, often developing amazingly narrow specializations. (Consider the adult mayfly, a delicate creature that emerges without mouthparts with which to feed, and which must mate and leave offspring in the few precious hours of life allowed it.)

Arthropods have varying diets, living as omnivores, herbivores, carnivores, scavengers, ectoparasites, and endoparasites. They live on, under, and above the surface of land and water, from deep ocean trenches to the highest mountain peaks. In may cases, because of their highly specialized niches, numerous species can exist side-by-side without seriously competing with each other.

Arthropod success is due to other important factors as well. Their reproductive capacity can be phenomenal. In addition, many produce larvae that have an entirely different diet from the adults, thus expanding the niche and avoiding competition between parent and offspring. Because of the large brood size, particularly in insects, the opportunity for genetic diversity is great. This, along with a very short generation time, helps the insect overcome all sorts of environmental deterrents. (For example, DDT-resistant mosquitoes have evolved in only a few generations.)

The phylum Arthropoda is divided into three subphyla: **Trilobita, Chelicerata,** and **Mandibulata** (Figure 31.12).

Subphylum Trilobita: The Trilobites There are no living **trilobites.** Trilobites originated in Precambrian times and extended into the Cambrian and Ordovician, where they peaked, undergoing extinction some 300 million years ago during the Carboniferous period. In their peak they dominated ancient Paleozoic seabeds, burrowing or shovelling their way through bottom mud and sand. To date, over 4000 fossil species have been described, most of which are rather small, 5 to 7.5 cm (2–3 in.), but some species got to be nearly 70 cm (about 28 in.) in length. The taxonomic position of trilobites is unsettled. Some zoologists maintain that they were ancestral to the crustaceans, but others consider them a separate line.

Trilobites have a characteristic three-lobed body that is divided by two burrows running its length. They are easily recognized by their shieldlike shape and highly segmented body with its numerous similar paired appendages (Figure 31.13), which were apparently used in movement and respiration.

Subphylum Chelicerata Subphylum Chelicerata includes horseshoe crabs, daddy longlegs, spiders, ticks, mites, and scorpions. They all lack jaws and have six pairs of appendages, including four pairs of legs. The first two pairs of appendages are sensory palps and **chelicerae** (fangs). None of the chelicarates have antennae.

Class Arachnida The **arachnids** (see Figure 31.12) don't win much affection with their habit of sucking the juices of other organisms. Spiders, equipped with hollow fangs and venom, are all carnivores. Typically, when a spider bites its prey, it injects venom, which weakens the unfortunate animal. Next it injects digestive enzymes into its host's body that liquify the tissues there. Then it sucks the fluid out.

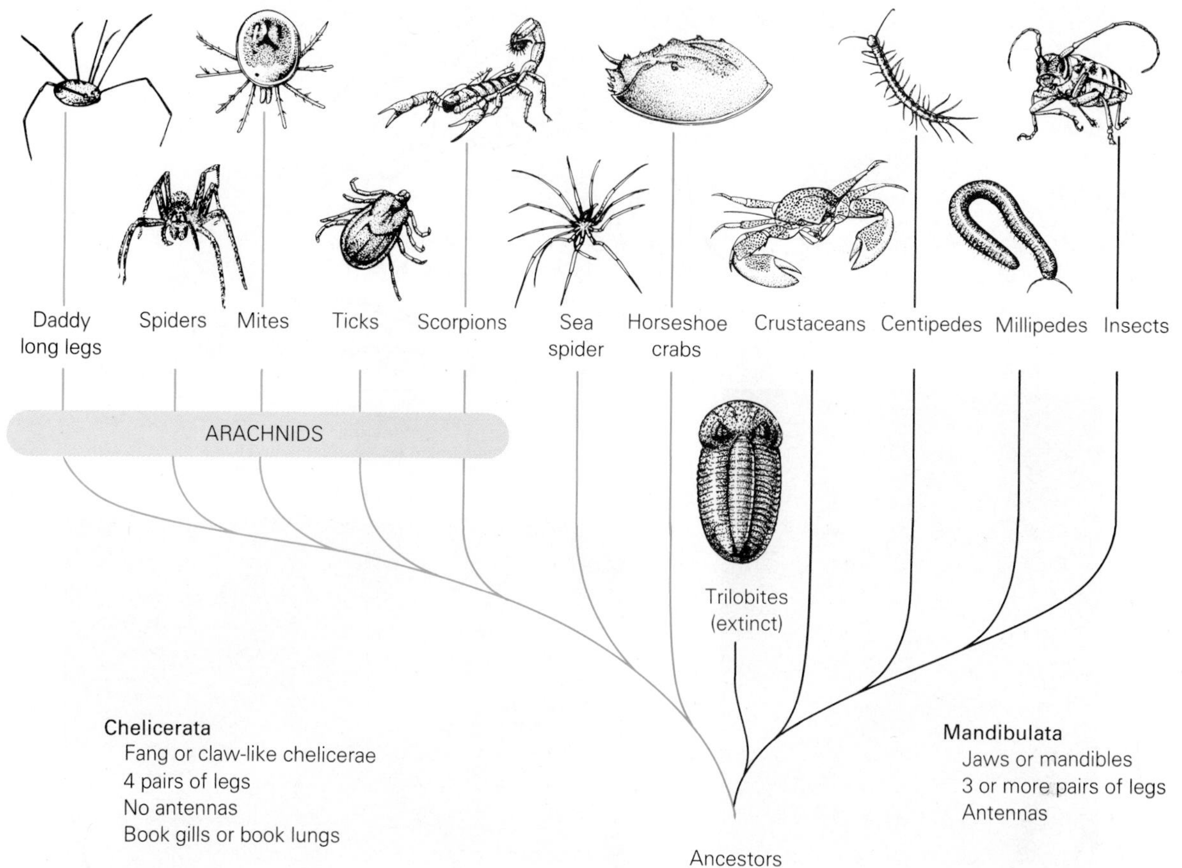

FIGURE 31.12
PHYLOGENY OF ARTHROPODS
The phylum Arthropoda can be organized into two contemporary subphyla on the basis of mouthparts and appendages. The chelicerates, shown in the left branch, lack jaws and antennae but produce clawlike chelicerae (which in the spiders are venomous fangs). In addition, they have four pairs of walking legs. The mandibulates, branching off to the right, have jaws and three or more pairs of walking legs. Most have antennae. Trilobites form an extinct subphylum.

FIGURE 31.13
TRILOBITES
Although now extinct, trilobites were commonplace through the early Paleozoic era.

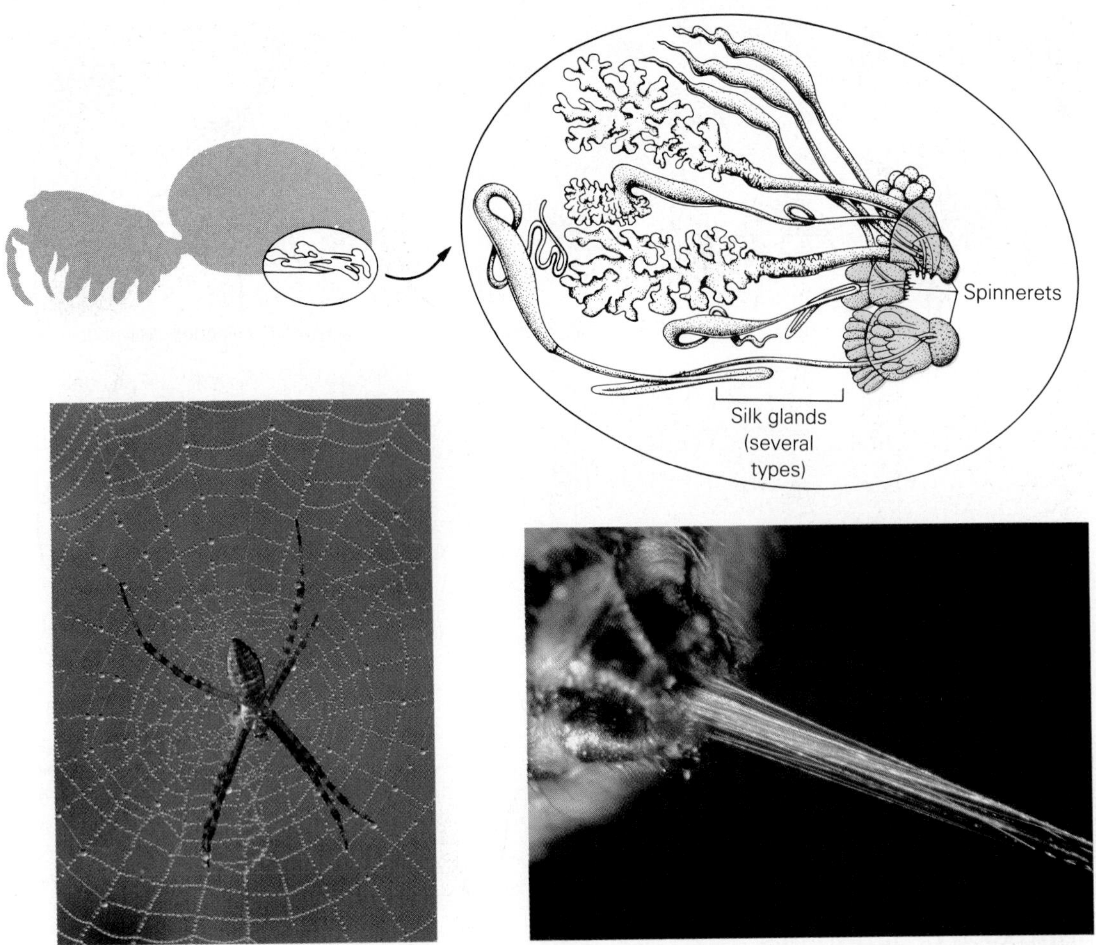

FIGURE 31.14
SILK GLANDS OF THE SPIDER
The orb spider, a master web-maker, can produce several kinds of silk. The silk glands, as shown here, are located in the abdomen. They open into the spinnerets which emit threads of different diameters. The web in the photograph belongs to an orb spider. The spider can move about quickly on the web, but it must know where the sticky parts are or it will become ensnared in its own trap.

Spiders show little evidence of external segmentation. They follow a two-part body plan which includes a cephalothorax (fused head and thorax) and abdomen. Spiders also differ from other arthropods in that they respire through a structure known as the **book lung.** Book lungs are internal sacs that open to the outside through slits. The sacs are lined with leafy folds, similar to the pages of a book. Blood passing through the thin "pages" exchanges gases with air in the sac.

Most spiders are equipped with silk-producing glands connected by ducts to external devices called **spinnerets** (Figure 31.14) Both the **web** and the **cocoon** of the spider are produced by this system. The silk of spiders (and insects) is a protein known as **fibroin**, composed of the amino acids glycine, alanine, and tyrosine. Spiders as a group are known to produce seven different kinds of silk, each with its own function. Webs have both sticky parts with which to ensnare prey and safe, dry parts along which the spider can run.

Subphylum Mandibulata Members of the subphylum Mandibulata have **mandibles** (jaws), as well as antennae and various numbers of paired appendages, including three or more pairs of walking legs. There are four major classes and two rather minor ones. The major classes are **Crustacea** (including marine and freshwater crabs, lobsters, copepods, barnacles, and sowbugs), **Chilopoda** (centipedes), **Diplopoda** (millipedes), and **Insecta** (flies, grasshoppers, bees, and so forth). Figure 31.15 illustrates the wide variety within the subphylum. The crustaceans are not closely related to the other mandibulates, and the tendency in recent years has been to elevate them to a subphylum of their own.

Class Crustacea **Crustaceans** live in both marine and fresh water. Their size ranges from microscopic ostracods and copepods to giant crabs from the western North Pacific.

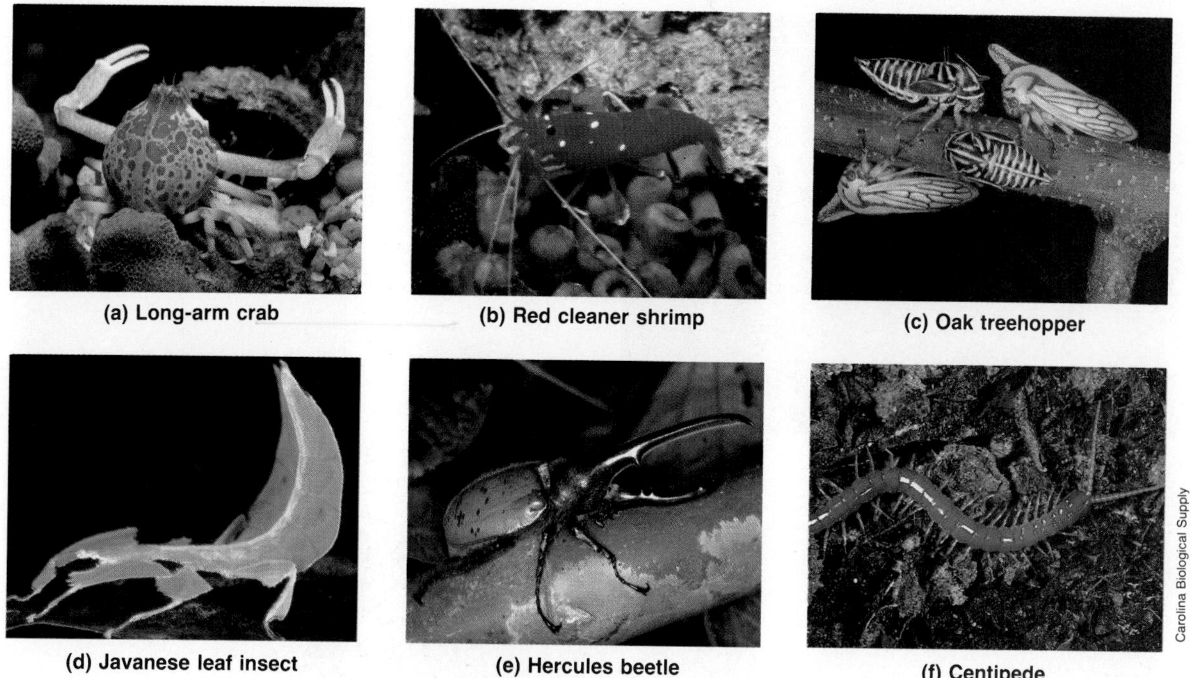

(a) Long-arm crab (b) Red cleaner shrimp (c) Oak treehopper (d) Javanese leaf insect (e) Hercules beetle (f) Centipede

FIGURE 31.15
MANDIBULATES
These mandibulate species can give you some idea about the diversity of appearance and habitat. There are crustaceans such as **(a)** the long-arm crab, and **(b)** the red cleaner shrimp. **(c)** The oak treehopper is an attractive insect common to deciduous forests. **(d)** The exotic Javanese leaf insect contrasts with the Hercules beetle **(e)**, whose long jaws serve to attract females. The centipede **(f)** captures prey with venomous jaws.

Species in this diverse group, such as the familiar pillbugs (or sowbugs) clearly demonstrate their relationship to the primitive annelids, while others (like the crabs) have become much more advanced, departing dramatically from the ancient plan. The segments of the most primitive crustaceans bear paired appendages, each somewhat like the next. In fact, they are reminiscent of the polychaete worms in this respect.

Crustaceans are gill breathers, with gill cavities partly covered by folds of the body wall. The active movement of certain appendages creates water currents that pass over the feathery gills (see Chapter 40). Because they are encased in a semirigid, secreted exoskeleton, to allow for growth crustaceans frequently molt, shedding their exoskeleton in a hormonally regulated manner (see Chapter 37).

Among their highly developed sensory structures is a **compound eye,** often borne on a moveable or retractable stalk. The eye, similar to that of insects, is composed of a large number of visual units called **ommatidia.** Each has a **corneal lens** and crystalline cone lens below. A light-sensitive transluscent cylinder below each set of lenses responds to incoming light (Figure 31.16). The compound construction and stalked position enable the animal to obtain a 180° view of its surroundings.

Many of the marine forms of the diverse crustaceans are of great ecological importance. For example much of the ocean's minute floating life, the plankton, consists of microscopic crustaceans. They are so numerous that one order, Euphausiacea, commonly known as "krill," is the main diet of the largest whales, as well as an important part of the ocean's longer food chain.

Classes Chilopoda and Diplopoda Centipedes (Chilopoda) are sometimes confused wth millipedes (Diplopoda). But centipedes have flattened bodies, while millipede bodies tend to be cylindrical; and centipedes have only one pair of appendages per segment (except for the first), while most segments of millipedes have two pairs. And then, if it stings you, it's a centipede. Millipedes are harmless herbivores, while centipedes are predators and scavengers. The centipede's first pair of legs are modified into perforated claws with which they inject poison into prey. The sting of common temperate zone centipedes is painful but probably no more life-threatening to humans than a wasp sting. However, the sting of larger tropical species can put an adult in bed with a fever for several days.

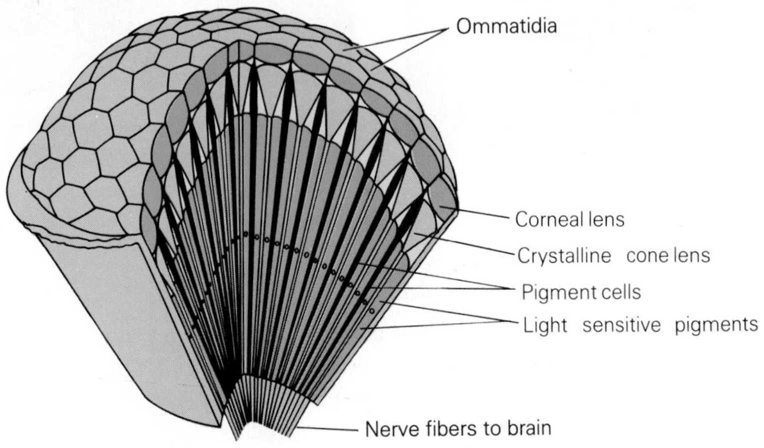

FIGURE 31.16
THE COMPOUND EYE
Many individual light-receiving ommatidia make up the compound eye. Their arrangement provides a very wide visual field, over 180° in the stalked crayfish eye. Each ommatidium has two lenses, an outer corneal lens and a crystalline cone lens below. The two focus light into a translucent cylinder that contains light-sensitive pigments. Slender pigment cells surrounding the unit keep light from passing into adjacent ommatidia—as long as the light is bright. Each active ommatidium, then, contributes to the final image. In dim light, the pigment cells withdraw, and light from several lens systems finds its way to one cone. Whereas the image is not as acute this way, it does provide enough light stimulus for the eye to respond.

Class Insecta The majority of arthropods and, indeed, the majority of animals, are **insects.** Except in the sea where the crustaceans hold sway, insects rule the earth in terms of numbers and kinds. We are continuously engaged in conflict with them and, even today, we often lose. Insects are of such importance ecologically and economically that we literally could not have reached our present population size and prominence among the world's creatures without understanding something about them.

The segmented body of insects has undergone considerable modification from the ancestral form, primarily through the fusion and alteration of segments. The modification has been so sweeping, in fact, that only in the abdominal region and in the embryo is it possible to see a clearly segmented body. The exoskeleton, like that of many other arthropods is basically of chitin, but it retains a flexibility not seen in the crustaceans.

The bodies of insects have three major regions: the **head,** the **thorax,** and the abdomen (Figure 31.17). The thorax gives rise to three pairs of legs and, in many species, wings. The legs are modified in different species for running, jumping, catching, holding, or simply resting. The appendages on the abdomen have been modified to form the genitalia

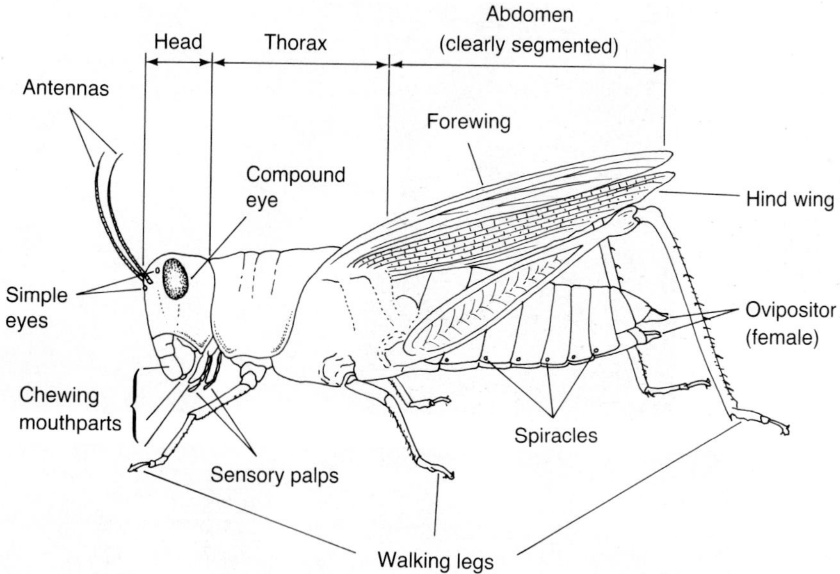

FIGURE 31.17
INSECT BODY PLAN
The insect body consists of three main parts: head, thorax, and abdomen. Segmentation is most apparent in the abdomen. The thorax and head are fused. The grasshopper has a pair of compound eyes, two or three simple eyes, and a pair of sensory antennae. Its mouthparts include the jaws, lips, and sensory palps. The thorax gives rise to three pairs of walking legs and two pairs of wings, While the abdomen contains 12 hinged segments. Paired spiracles open into each segment, admitting air. The last segments bear the reproductive structures and the anal opening.

and anal structures. In many species, females have an **ovipositor**, a hollow appendage that is used to dig into the soil or bore holes in plants, in which they lay their eggs.

During development, most insect species (about 88%) pass through a **complete metamorphosis.** That is, their life cycles include four stages: **egg**, **larva**, **pupa**, and **adult** (Figure 31.18a). Insect larvae, known variously as caterpillars, grubs, and maggots, are wormlike and often have chewing mouthparts. The larvae are usually voracious eaters, and it is in this stage when those identified as agricultural pests do most of their damage. As the larval stage continues, the wings are formed within the body. Eventually, the larvae form a surrounding case or cocoon and enter the pupal stage. The pupa may become dormant, spending the winter in this state, but it is in this stage that development is completed. A full-grown adult emerges from the pupal case.

Other insects, such as grasshoppers, mayflies, and mantids, undergo an **incomplete metamorphosis,** going from the egg to a **nymph** (or in aquatic insects, the **naiad**) to adult. Nymphs tend to resemble the adult, but lack wings. The wings develop externally as the nymphs gradually increase in size.

In a relatively small number of insects, chiefly the wingless types such as silverfish and springtails, development is direct. They emerge from the egg as juveniles—miniatures of the adult—and begin their growth to full adult size.

Insects breathe by admitting air into a complex, tubular, **tracheal system** through external openings known as **spiracles.** The air passes into highly branched **tracheae**, which subdivide into smaller **tracheoles**, finally reaching individual cells where gas exchange occurs. Some trachaea end in balloonlike air sacs.

The major sense organs and mouthparts are in the head, where you would expect them to be. Consider, for example, the head of the grasshopper. It has one pair of sensory antennae, a pair of compound eyes, and three simple eyes, or **ocelli.** Its mouthparts consist of a single **labrum** (upper lip) and a **labium** (lower lip) bearing **sensory**

FIGURE 31.18
INSECT METAMORPHOSIS
In the moth's complete metamorphosis (four upper photos), an egg develops into a larva, a feeding stage. Following a period of intense growth and development, the larva is transformed into a pupa, and development is completed. The complete adult emerges from the pupa. In incomplete metamorphosis as seen in the grasshopper, growing juveniles resemble adults.

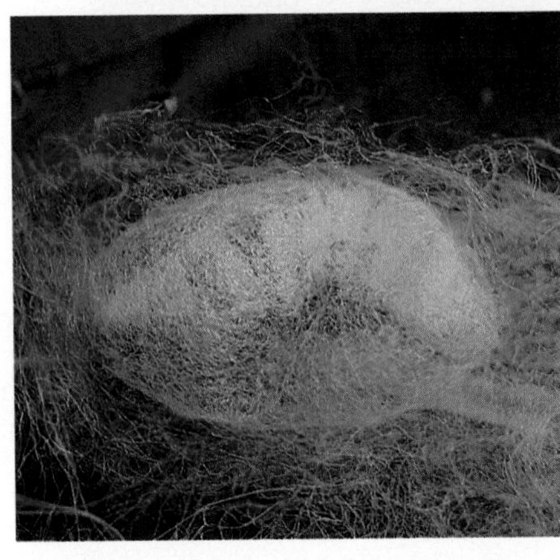

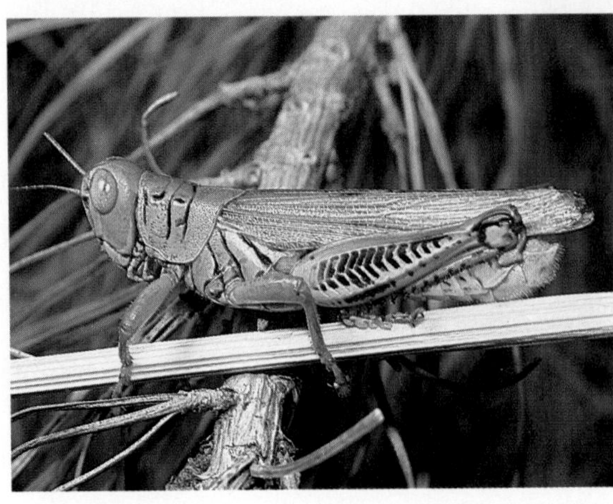

palps, and there are two laterally movable mandibles (jaws) that do most of the chewing. These are assisted by a pair of **maxillae** also bearing sensory palps. The entire apparatus is magnificently adapted for biting and chewing leaves. The sensory palps detect texture and flavor, the jaws rip off and chew bits of vegetation, and the lips hold the food in place during chewing.

Mouthparts in other insects are often highly specialized to fit their specific feeding niche. For example, the maxillae of butterflies are modified into a long, flexible siphoning device (neatly coiled when not in use). On the other hand, the cicada has a piercing and sucking mouth, while the housefly has a large, complex, tonguelike apparatus used for licking (Figure 31.19).

Returning to our grasshopper, let's examine its locomotive structures. First, although some grasshoppers are great jumpers, they are also flying insects. They have two pairs of wings; one pair actually propels the insect, and the other serves as a protective wing cover. The desert locust, a close relative of the grasshopper, can fly over 200 miles in a day. We will look into the details of insect flight in Chapter 33.

The legs of insects are also highly specialized for specific modes of life. The grasshopper, for example, has two pairs of legs of similar size for walking and grasping, while its third pair is greatly enlarged. The combination of extremely powerful muscles in the third pair and the light exoskeleton produces a great mechanical advantage, enabling the grasshopper to jump great distances from a standing start. The praying mantis has

FIGURE 31.19
INSECT MOUTH PARTS
(a) The grasshopper's chewing mouth is nicely adapted for eating foliage. Two sets of palps contain taste receptors, while biting is done with the paired mandibles and maxillae. The labium and labrum help by holding the food. **(b)** Butterflies use their lengthy mouthparts in siphoning up plant juices. The long, extendable paired maxillae are held together like the sides of Ziploc bags, forming a long drinking tube. When not in use, the tube is kept tightly coiled by flexible rods. **(c)** The housefly has its own feeding strategy. It salivates on its food, stirring it with the large spongelike labium. Its labrum, part of a tubelike device, sucks up the partially digested mess.

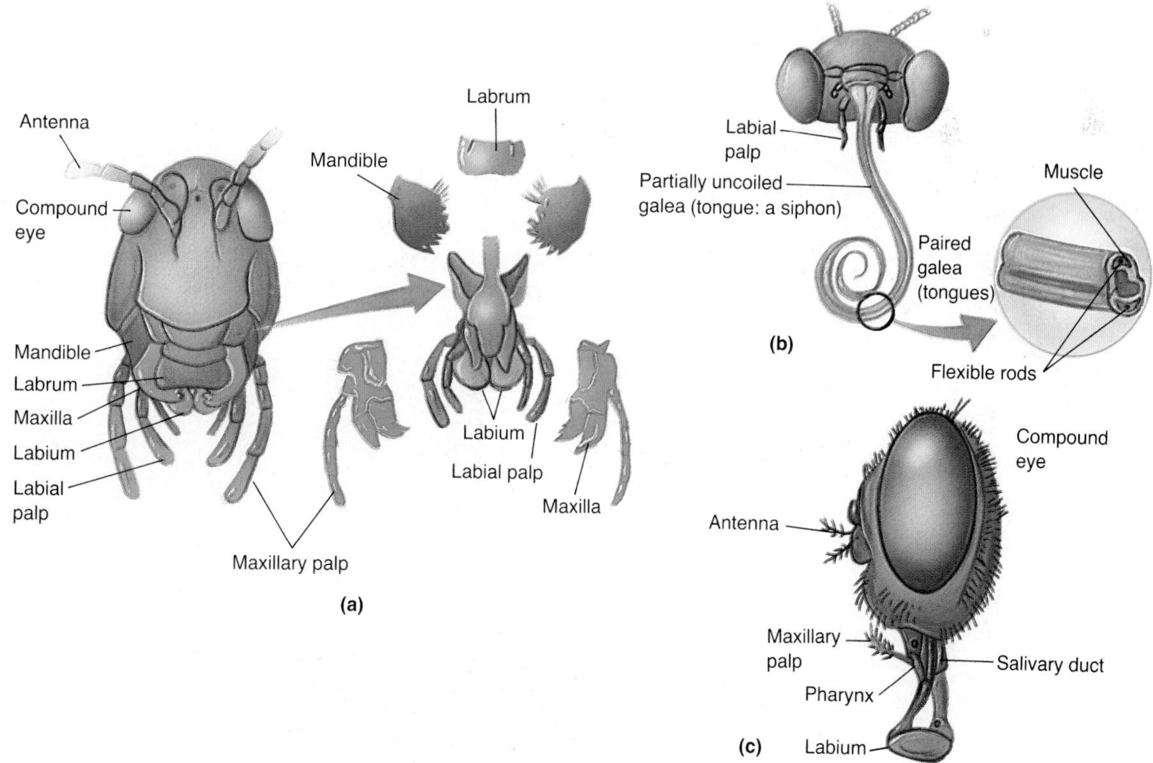

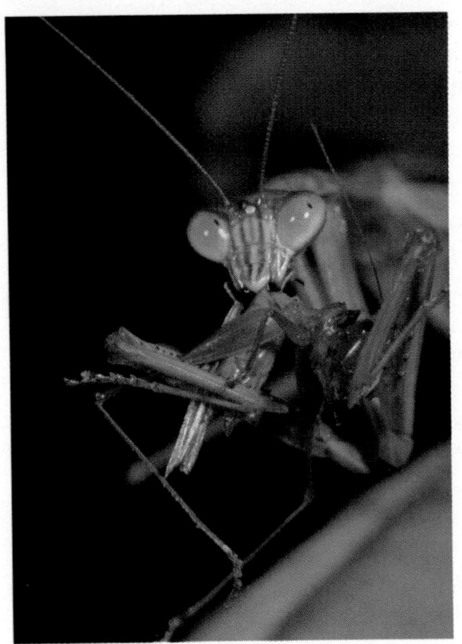

FIGURE 31.20
PRAYING MANTIS
The praying mantis is a voracious predator. This mantis has torn its insect prey open with its powerful barbed forelegs and is feeding with its strong jaws.

very strong, spiked forelegs that snatch and hold prey, which is then ripped apart by powerful jaws (Figure 31.20). Honeybees have walking legs modified for pollen collection (Figure 31.21). The first pair is equipped with a pollen brush for collecting pollen from the body and a circular indentation for cleaning the antennae. (Pollen gathering is a messy business.) The middle pair of legs are less specialized, having a spur that is used in removing wax (used in hive building) from abdominal glands. The posterior legs contain pollen compactors, for packing the pollen, and hollow grooves, or pollen baskets, for carrying it. Pollen is a major source of bee food. And, finally, consider the sensitive legs of roaches, which detect delicate air currents set up by anything that might be after them, such as an irate apartment dweller.

Of course, these are only a few examples of insect diversity. We will discuss additional aspects of their biology elsewhere, but Table 31.1 should underline the immense variation in this group.

Arthropods have been around a very long time. Fossils of marine forms are found in Cambrian strata and possibly in the Precambrian. By the Ordovician period, some members had left the sea and had begun to explore the terrestrial environment. This escape from their marine enemies was only temporary, though; by the Devonian, the first amphibians had struggled ashore and have been eating insects ever since. But the arthropods did not fall easily before the hungry vertebrates. They ran, flew, and crawled away, invading every part of the globe and establishing all sorts of new niches.

Phylum Onychophora

Phylum **Onychophora**, which includes only about 65 to 70 named species, is so minor that it could well be ignored if it weren't for the fact that some researchers consider it an evolutionary link between Annelida and Arthropoda, or perhaps representative ancestors of both. The onychophorans are believed to represent an ancient animal that arose from the primitive arthropod line very shortly after the arthropods diverged from the early annelids. Onychophorans are seen among the fossils of the famed Burgess Shale fauna (see Figure 30.3). After leaving the arthropod line they made the difficult transition to land, where they assumed the rather restricted existence we see today (Figure 31.22).

The best-known onychophoran is the peculiar *Peripatus,* which lives among the damp leaves of the tropics in Africa, Southeast Asia, New Zealand, Australia, and the West Indies. *Peripatus* has a thin, soft cuticle covering a muscular body wall. It has two muscle layers that run in circular and longitudinal directions. The animal is highly segmented, revealing its annelid heritage, but the segmentation is largely confined to its appendages and internal structure; its body from the outside is fairly smooth. However, it also has several features characteristic of arthropods. Tiny claws or small pincers arm its soft appendages, and its head even has antennae. It breathes through a series of spiracles

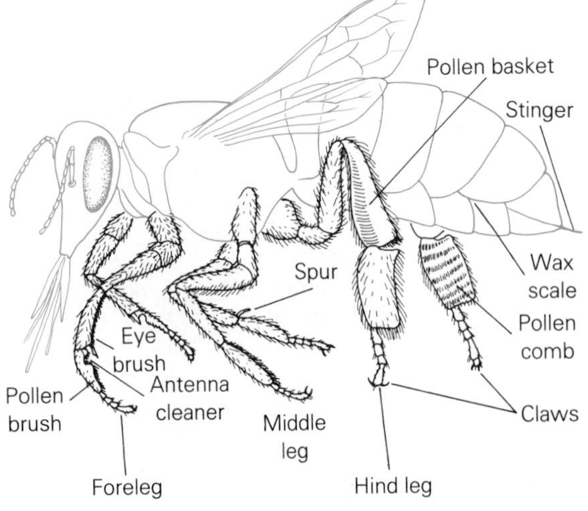

FIGURE 31.21
HONEYBEE LEGS
Honeybee legs are specialized for the various tasks associated with pollen gathering and hive building.

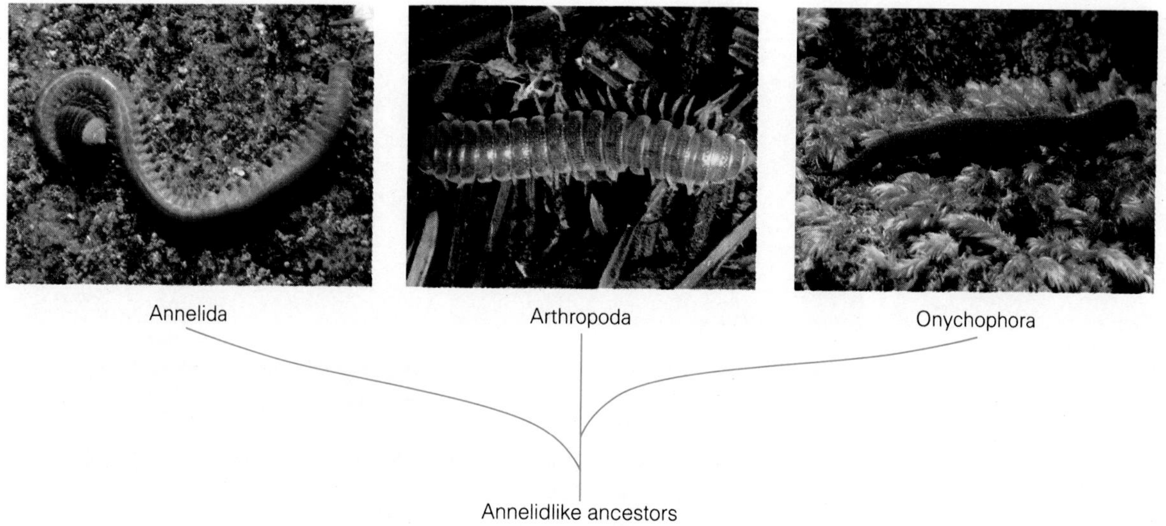

FIGURE 31.22
AN EVOLUTIONARY LINK
The phylum Onychophora has many characteristics of both the annelids and the arthropods. The thin-walled, segmented onychophoran body and the serial duplication of internal structures is reminiscent of the annelids. On the other hand, onychophorans have jointed legs tipped with claws and an open circulatory system, both of which are arthropod characteristics.

along its body, which open into tiny pits connected to a number of tracheae. This arrangement permits air to reach the internal organs easily, but there is no way of controlling the size of the opening, so *Peripatus* avoids water loss by living in moist habitats.

We have reviewed, now, the major groups in the prostostome line, those animals in which the mouth appears before the anus. Now we will review the invertebrate deuterostomes. As you will see, we have considered only one invertebrate phylum in this group.

THE DEUTEROSTOME LINE

The deuterostome line is composed principally of the echinoderms, such as sea stars, and the chordates. Since we will consider the chordates in the next chapter, here we will focus only on the echinoderms.

Phylum Echinodermata

Your first look at **echinoderms** ("spiny-skinned" animals) may lead you to believe that you are on the wrong road to the vertebrates. Sea stars, brittle stars, sea urchins, sea cucumbers, and sand dollars (Figure 31.23) appear unlikely to be relatives of amphibians, fishes, mammals, and birds, not to mention humans. A comparison of the anatomy of echinoderms and vertebrates won't instill much confidence in you either. To see the relationship between the echinoderms and the vertebrates, we must look at their earliest embryonic development, when the first cleavages occur, and later at the developing mouth. It is there that the echinoderms are unveiled as deuterostomes (see Essay 31.1). Their embryos indicate that they are related, not directly to the vertebrates, but to their soft-bodied cousins, the hemichordates, which in turn are related to the chordate phylum. We will review this relatedness in the next chapter.

Echinoderms are unusual in many respects. Their spiny, crusted covering is actually an endoskeleton, even though it seems to be on the outside. Echinoderms are covered by an epidermis, beneath which lie the calcareous plates that are secreted by the dermis, which is of mesodermal origin. (Recall that the shells of mollusks and the exoskeletons of arthropods are both secreted by epidermal cells that stem from ectoderm.)

The body of adult echinoderms is essentially a pentaradial (five-part radial) construction, a scheme found in no other phylum. The larvae are bilateral, so the pentaradial condition turns out to be only a peculiar secondary state. Echinoderms show no obvious segmentation, indicating that they are not closely related to the annelid line.

Broad arm star

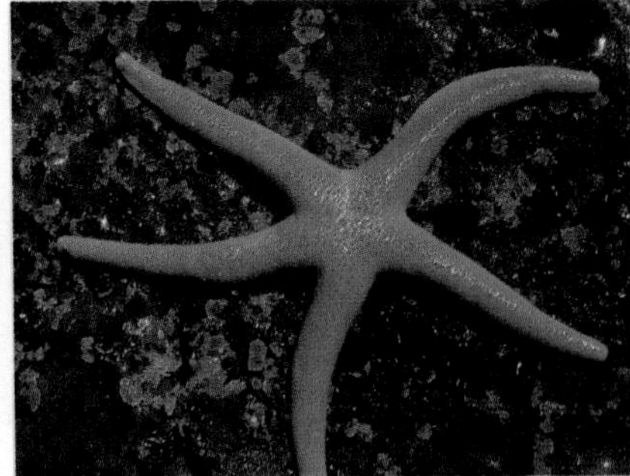

Sea star

Brittle star

Sea urchin

Feather star (crinoid)

FIGURE 31.23
REPRESENTATIVE ECHINODERMS
One of the most striking features of the echinoderms is their five-fold radial symmetry, most easily seen in the sea stars and the brittle stars. These marine animals have a spiny skin, with the longest spines found in the sea urchins.

One of the most unusual features of the echinoderm is the **water vascular system,** which is composed of a series of water-filled canals, which by hydraulic pressure help extend their numerous muscular **tube feet.** In many echinoderms, each tube foot is equipped with a terminal sucker that is used in grasping. We will come back to this unique characteristic shortly.

Echinoderms are exclusively marine and brackish water animals. Many species are found in the shallow waters of the continental shelf, particularly just below the reaches of the lowest tides. However, one group, the brittle stars, abound in the deepest trenches of the ocean.

The phylum consists of five classes: **Crinoidea** (sea lilies), **Holothuroidea** (sea cucumbers), **Echinoidea** (sea urchins and sand dollars), **Asteroidea** (sea stars), and **Ophiuroidea** (serpent stars and brittle stars). Each class differs considerably from the others, with perhaps the greatest departure being the soft-bodied, wormlike sea cucumbers, which have only scattered elements of an endoskeleton.

FIGURE 31.24
REGENERATION IN SEA STARS
Most sea star species can regenerate an entire new body from any portion of the central disk. (One species, *Linckia,* can even regenerate a new body, complete with disk and arms, from very small pieces of any arm.)

Class Asteroidea: The Sea Stars As a representative of the phylum, we will concentrate on one class of echinoderms—the sea stars (see Figure 31.23a). Sea stars (or starfish, if you insist) are well known for their voracious appetite when it comes to gourmet foods such as oysters and clams. Obviously, they are the sworn enemy of oystermen, although these same oystermen may have inadvertently helped the spread of the sea stars. At one time, when they caught a sea star, they chopped it apart and vengefully kicked the pieces overboard. But they were unfamiliar with the regenerative powers of the sea star. The central disk merely grows new arms, and a single arm, with a bit of central disk attached, can form a new animal (Figure 31.24).

Since sea stars are slow-moving predators, their prey must be even slower-moving, or better yet, immobile. A favorite meal is the bivalve mollusk. The sea star's ability to get past a bivalve's primary defense is legendary. As you know, oysters, clams, and mussels rely on their closed valves for protection against predators. In attacking the

TABLE 31.1 MAJOR INSECT ORDERS[a]

ORDER (NUMBER OF NAMED SPECIES WITHIN ORDER)	EXAMPLES	CHARACTERISTICS
COLEOPTERA (300,000)	Japanese beetles Stag beetles	Two pairs of wings, horny and membranous Heavy, armored exoskeleton Biting and chewing mouthparts Complete metamorphosis Herbivores, predatory carnivores, scavengers Serious agricultural pests
LEPIDOPTERA (140,000)	Moths Butterflies	Two pairs of wings Hairy bodies Long coiled tongue for siphoning Complete metamorphosis Serious agricultural pests
HYMENOPTERA (90,000)	Bees Wasps Ants	Two pairs of membranous wings Head mobile Well-developed eyes Chewing and sucking mouthparts Stinging Complete metamorphosis Many are social Important as pollinators and predators of other insects
DIPTERA (80,000)	Flies Mosquitoes	One pair of wings and halteres (balance organs) Sucking, piercing, lapping mouthparts Complete metamorphosis Medically important as carriers of dangerous diseases. (malaria, yellow fever, sleeping sickness, encephalitis, filiariasis)
HEMIPTERA (40,000)	True bugs: Bed bugs Assassin bugs	Two pairs of wings, horny and membranous Piercing mouthparts Many agricultural pests Incomplete metamorphosis
HOMOPTERA (32,000)	Aphids Scale insects Cicadas Plant lice	Two pairs of wings, horny and membranous Piercing, sucking mouthparts Agricultural pests Incomplete metamorphosis
ORTHOPTERA (30,000)	Roaches Grasshoppers Mantises Crickets	Two pairs of wings, horny and membranous Biting and chewing mouthparts Incomplete metamorphosis Herbivores (except mantids)
ODONATA (5000)	Dragonflies Damselflies	Two pairs of wings Biting mouth food basket (legs and spines) Incomplete metamorphosis Predator

[a]The number of named species in each of the thirteen orders listed ranges from a very modest 1000 in order Dermaptera to 300,000 in order Coleoptera. In fact, by conservative estimate, it is accurate to say that one in four animals roaming the earth is a beetle. Most of the remainder of the animals are butterflies, moths, flies, mosquitoes, bees, ants, and wasps (and their near relatives).

TABLE 31.1 (continued)

ORDER (NUMBER OF NAMED SPECIES WITHIN ORDER)	EXAMPLES	CHARACTERISTICS
NEUROPTERA (4000)	Ant lions Dobson flies Lacewings	Two pairs of membranous wings Biting mouthparts Complete metamorphosis Silk cocoon
ANOPTURA (2400)	Lice	Wingless Sucking or biting mouthparts Small with flattened body, reduced eyes Clawlike appendages (for clinging to skin) Highly host-specific (in humans, head, body, and pubic lice are of different species) Medically important (carry typhus and cause relapsing fever) Incomplete metamorphosis
ISOPTERA (2000)	Termites	Two pairs of wings, but some stages are wingless Chewing mouthparts Social, division of labor for reproduction, work, defense Incomplete metamorphosis
SIPHONAPTERA (1200)	Fleas	Small, wingless, laterally compressed Piercing and sucking mouthparts Jumping legs Complete metamorphosis Medically important (carry bubonic plague, typhus)
EPHEMEROPTERA (1000)	Mayflies	Two pairs of wings Vestigial mouthparts in adults (do not feed) Few days of adult life (reproduce and die) Incomplete metamorphosis
DERMAPTERA (1000)	Earwig	Two pairs of wings, leathery and membranous Biting mouthparts Large pincerlike appendages in males Incomplete metamorphosis

shellfish, the sea star wraps its arms around it, using its tube feet to hold the prey in position gape side up. It then everts its delicate, thin-walled stomach, and slips it into minute crevices between the valves. (One-tenth of a millimeter is all some sea stars need.) The openings may be further increased by the pulling action of the sea star's tube feet. The stomach secretes powerful digestive enzymes into the prey's body, and soon the strong adductor (valve-closing) muscles weaken and fail. That's it for the shellfish.

Tube feet are located in double rows in the grooved underside (the oral side, since the sea star's mouth is on the bottom) of each arm (Figure 31.25). Each tube foot is attached to a short **lateral canal**, which in turn connects with a pipelike **radial canal** that extends along the arm. The radial canals from each arm all connect to the hollow **ring canal** in the central disk. Changing pressures in the water vascular system may be brought to equilibrium by a vertical tube, the **stone canal**, which has a sieved opening called a **madreporite** (mother pore) on the upper side of the sea star.

Each tube foot contains longitudinal muscles that can contract, shortening the delicate structure. Above each foot is a rounded sac, an **ampulla,** which is similar to a squeeze-bulb. When muscles contract a water-filled ampulla, water is forced into the foot, extending it hydraulically. The sucker end of the foot attaches to a surface, and the longitudinal muscles then contract, shortening the tube and exerting a pull. Thus, the tube feet, contracting in a highly coordinated fashion, can attach to an oyster or draw the animal over the rocky bottom of its habitat.

The sea star, like all echinoderms, lacks a centralized nervous system. The greatest concentration of nerves is in a ring around the mouth, with branches extending into the arms. Sensory receptors are limited in the sea star to one or more sensory tentacles and a light-sensitive region at the tip of each arm.

Echinoderms are either male or female, but it is hard to tell which is which. Sperm or eggs, produced in the well-defined testes or ovaries found in each arm, are shed into the water through tiny openings in the arms. Fertilization takes place in the water. As described in Essay 31.1, the sea star, a deuterostome, undergoes radial cleavage in its early embryological development (as opposed to the spiral cleavage of protostomes). The embryo differentiates into a free-swimming **bipinnaria larva** that is distinctly different from the trochophore larvae of other invertebrates but similar to that of other deuterostomes. Thus, whereas adult sea stars don't resemble vertebrates at all, echinoderm embryology suggests that our ancestors were probably rather closely related.

Fossils of the feather-like **eocrinoids,** the earliest echinoderms, are found in Cambrian deposits along with fossil protostomes. They were stalked animals somewhat different from their living descendants the **crinoids** (sea lilies and feather stars). Other classes of echinoderms appear first in the Ordovician, the next oldest period.

With the echinoderms, then, we have set the stage for our entrance into a more familiar world, that of backboned animals and their considerably less familiar chordate relatives.

FIGURE 31.25
WATER VASCULAR SYSTEM
The water vascular system of echinoderms consists of a sievelike madreporite, a number of canals, and numerous tube feet. Water is used to fill the ampullae, which in turn are used to extend the tube feet. Each tube foot ends in a sucker tip. When several of the tube feet contact a surface, they can contract by muscular action. Working in a series, the tube feet pull the sea star along the sea bottom. They also assist in holding prey.

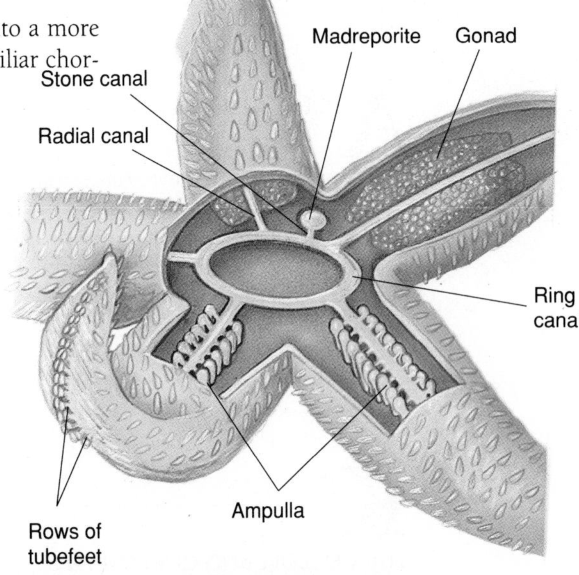

Tube feet extended

Arms circling prey

closed circulatory system A circulatory system in which the blood remains within closed vessels and exchanges occur across the vessel walls (see open circulatory systems).

coelom In coelomate animals, a body cavity lined by mesodermally derived peritoneum and located between the gut and body wall.

complete metamorphosis In insects, a life cycle that includes the egg, larva, pupa, and adult stages.

deuterostome Those animal phyla whose embryos undergo late mouth formation, have radial cleavage, and where larvae occur are of the dipleurula type (e.g. echinoderms and chordates).

mantle In mollusks, a generally fleshy structure that covers the gills and secretes the shell in some, forms a hood over the head in others, and often encloses a cavity where gas exchange goes on.

open circulatory system A circulatory system in which blood is pumped through vessels that empty into open sinuses and cavities in the body and limbs before the blood returns to the heart.

protostome Those animal phyla whose embryos undergo early mouth formation, spiral cleavage, and where larvae occur, are of the trochophore type (e.g. mollusks, annelids, and arthropods).

segmented body plan The organization in which the body is divided serially into transverse segments or metameres and the structures within the segments more-or-less repeated (e.g. annelids and arthropods).

ventral nerve cord In protostomes, the major nerve trunk running from the brain to the body and located along the ventral body wall.

water vascular system A unique system of movement in echinoderms that involves numerous, grasping tube feet that are extended by hydrostatic pressure generated by muscular ("squeeze-bulb") ampullae.

APPLICATION OF IDEAS

1. A popular post-Darwinian concept of evolution placed humans at the pinnacle of a long, gradual progression of forms, all evolving toward the human line. (One such scheme placed nineteenth-century English gentlemen at the very top. Guess where this idea originated.) Other animals were represented as failures or cases of arrested progress. Present a strong argument against this point of view. Does evolution really lead anywhere? Have any of today's animals "given rise" to any other of today's animals? Are humans and other species still evolving?
2. Using the mollusks as examples, explain the evolutionary terms *primitive, specialized, generalized,* and *advanced.* Why are such terms commonly inappropriate when applied to the entire organism?

KEY IDEAS

THE COELOM AND COELOMATES

Coelomates are animals with a permanent, mesodermally lined body cavity and muscular gut. The **peritoneum** gives rise to supporting **mesenteries.** The fluid-filled coelom forms a hydrostatic skeleton, which is particularly useful for burrowing. The fluids provide for other functions as well. The **coelom** is present in most protostomes and all deuterostomes, but its development is different in the two groups.

The Significance of the Protostomel Deuterostome Line

1. Protostomes tend to occur with great diversity and have small bodies, rapid life cycles, and limited learning. Although some deuterostomes are similar to protostomes, an opposite trend began with the chordates. Further, in deuterostomes, the exoskeleton is replaced by a mesodermally derived **endoskeleton.**
2. The divergence of protostomes and deuterostomes occurred in the late Precambrian. Two hypotheses seek to explain how such divergence occurred:
   a. In one, *ecological* isolation occurred as one group specialized in burrowing for food and the other formed stalks and filtered food from the sea.
   b. A second hypothesis proposes that isolation was *geographical.* Prior to the formation of Pangaea, there existed **Proto-Laurasia** and **Proto-Gondwana,** two supercontinents in which, respectively, the deuterostomes and protostomes arose.

THE PROTOSTOME LINE

The Lophophorate Phyla

Lophophorates have characteristics of both protostomes and deuterostomes. The three phyla include **Ectoprocta** (bryozoans), **Brachiopoda**, and **Phoronida**. The **lophophore** is a group of ciliated tentacles attached to a curved ridge. Brachiopod and ectoproct fossils occur in Cambrian strata.

Phylum Mollusca

1. **Mollusks** are among the most numerous invertebrates, ranging in size from microscopic **bivalves** to the giant squid. All have a variation of a **foot**, **mantle**, **mantle cavity**, **shell**, and in most aquatic forms, a **gill**. The coelom is highly reduced, an **open circulatory system** is present in most, and many feed with a rasping **radula**. Nephridia are present.
2. The evolutionary relatedness of mollusks to other protostomes is in question, but they do produce a trochophore larva, and some fossil mollusks reveal segmentation and a well-developed coelom.
3. **Chitons** have a primitive body plan, with traces of segmentation in the eight shell plates and rows of simple gills.
4. Most **gastropods** ("stomach-feet") are snails or snaillike, although some lack shells. While marine snails have gills, freshwater and terrestrial snails use vascularized mantle-cavity for lungs. Many snails undergo **torsion** (a 180° twist) as they develop. A horny **operculum** helps protect the retracted body.
5. The bivalves, clamlike mollusks with paired, hinged shells, often have a powerful, retractable digging foot. Clams are filter-feeders, using cilia to draw water into the body through an **incurrent siphon**, whereupon food is removed and then taken in by **labial palps**, gases are exchanged in the gills, and water leaves by an **excurrent siphon**.
6. Cephalopods have reduced and internalized shells, well-developed brains, image-forming eyes, closed circulatory systems, muscular mantles, and a foot highly modified into grasping tentacles. The octopus and squid move rapidly by jetting water from the mantle.

The Segmented Body Plan

In the **segmented body plan**, the body is composed of repeated segments or **metameres**. Segmentation is also seen in vertebrates.

Phylum Annelida

1. In annelids, extreme segmentation and a hydrostatic skeleton may have evolved as an adaptation to burrowing.
2. In the earthworm, an oligochaete, each segment contains elements of most organ systems. Earthworms have a **closed circulatory system** with five pairs of pumping **aortic arches** and blood that contains hemoglobin. Excretory and osmoregulatory structures called **nephridia** are greatly advanced over the protonephridium. Earthworms are hermaphroditic, each with **sperm receptacles**. Fertilized eggs are released into a slime cocoon, secreted by the **clitellum**. The gut includes specialized areas for swallowing, storing, grinding, digesting, and absorbing. The nervous system includes a brain (enlarged ganglia) and a **ventral nerve cord** with branches in each segment. The body wall contains circularly and longitudinally arranged muscles.
3. **Hirudinea**, the **leeches**, are **ectoparasites**. They attach themselves via toothed suckers to warm-blooded hosts. A secretion, **hirudin**, prevents clotting. Although segmentation is modified, its systems are similar to those of the earthworm.
4. Polychaetes are marine annelids. *Nereis*, the clamworm, has prominent jaws and fleshy, vascular, **parapodia** that function in swimming and gas exchange. In marine worms, sexes are separate, and following fertilization, a swimming **trochophore** larva forms, similar to that of many other marine invertebrates. Many polychaetes are tube-dwellers, and the fossils of this group are the oldest traces of eukaryote life.

Phylum Arthropoda

1. Nearly 1 million **arthropod** species are known. Common characteristics include segmented body, jointed legs, ectodermally secreted **exoskeleton**, open circulatory system, and ventral nerve cord. They live in all major habitats and produce great numbers of genetically variable young, which often live on a different diet from adults.
2. The trilobites, a major arthropod group, have been extinct for 300 million years. They have a segmented, three-lobed, shieldlike body, with numerous paired appendages.
3. Chelicerates are arthropods that lack antennae and jaws, and have six pairs of appendages. They have venom-injecting fangs known as **chelicerae**. The largest group are **arachnids**, the spiders, all of which are venomous carnivores. They exchange gases in book lungs, produce **fibroin** silk for **webs** and **cocoons** with silk glands and **spinnerets**.
4. Mandibulates, arthropods with jaws, have antennae and three or more pairs of walking legs.
5. **Crustaceans** are mainly aquatic, both marine and freshwater. Their body plans range from the highly segmented primitive condition to advanced states with little visible segmentation. They are mostly gill breathers, with a chitinous exoskeleton made rigid with calcium salts. They have **compound eyes**, consisting of multiple **ommatidia**. Ecologically important crustaceans include the floating marine and freshwater plankton that are important to the aquatic food chains.
6. Chilopods and diplopods are represented respectively by the venomous, predatory centipede and the herbivorous millipede.
7. The majority of animal species are **insects**. Their characteristics include a flexible chitinous exoskeleton, segmentation chiefly in the abdomen,

three pairs of legs, one pair of antennas, and sometimes wings.

8. Other insect features include a three-part body with **head, thorax,** and **abdomen,** greatly modified legs, and complex genitalia with a specialized **ovipositor.**
9. Insects with **complete metamorphosis** pass through **egg, larval, pupal** and **adult** phases. Where **incomplete metamorphosis** occurs, eggs develop into **nymphs** or **naiads,** from which adults emerge. A few species have direct development, from egg to juveniles, which simply increase in size to become adults.
10. Gases are exchanged through a **tracheal system** composed of branched **tracheae** and **tracheoles** that sometimes end in air sacs. **Spiracles** control the external openings.
11. Insect sensory structures include compound eyes and simple **ocelli.** Mouthparts include the **labrum** and **labium** (lips), chewing **mandibles, maxillae,** and **sensory palps.** Insect mouthparts vary greatly.
12. Insect leg specializations include modifications for jumping, grasping, pollen collecting, and detecting intruders by movement of air.
13. Arthropod fossils extend back to Precambrian times.

Phylum Onychophora

The onychophorans, typified by *Peripatus,* have the bidirectional body wall muscles and pronounced internal segmentation of the annelid, yet, like arthropods, they also have clawed appendages and antennae, and exchange gases through a tracheal system complete with spiracles. They may have diverged from ancient arthropods shortly after that group diverged from annelids.

THE DEUTEROSTOME LINE

Phylum Echinodermata

1. The phylogenetic link between **echinoderms** and vertebrates is seen in the embryo, where similarities in cleavage planes and mouth development are seen.
2. The spiny skin of the echinoderm is secreted by mesodermally derived tissue and is covered by an epidermis. Although adult echinoderms are pentaradial, their larval forms are bilateral. The **water vascular system** with its **tube feet** is an exclusive feature.
3. Five echinoderm classes include **Crinoidea** (sealilies), **Holothuroidea** (sea cucumbers), **Echinoidea** (sea urchins), **Asteroidea** (sea stars), and **Ophiuroidea** (serpent stars).
4. Asteroidea (sea stars), predators of bivalves, use their tube feet and water vascular system in moving about and in feeding. Their thin-walled stomach is everted through gaps in the valves to digest the prey. Tube feet occur in double rows within grooves in each arm. Each connects to a **lateral canal,** then to a **radial canal,** which in turn connects to the **ring canal.** The ring canal gives rise to a **stone canal** that opens to the outside through a sieved **madreporite.** Tube feet are extended by water pressure from an **ampulla,** whereupon the sucker end fastens to an object and minute muscle fibers shorten the foot.
5. The simple nervous system consists of a ring and branches into each arm.
6. Separate individuals produce eggs and sperm, and fertilization is external. The swimming **bipinnaria larva** is unlike the protostome trochophore.
7. The earliest echinoderms were stalked Cambrian **eocrinoids.**

REVIEW QUESTIONS

1. List several features of the coelomate body and gut. (688)
2. Briefly summarize an ecological hypothesis and a geographical hypothesis that explain how the protostome and deuterostome evolutionary lines became established without conflict on the early earth. (689–690)
3. What is the leading characteristic of the lophophorate? List the three phyla and name a member of each. (690)
4. What evidence links the mollusks with the annelids and arthropods? (692)
5. Describe the chiton body plan (shell mantle, foot, and gut). Why might biologists consider it primitive? (692)
6. Explain what torsion is and how it affects the final body plan of the gastropod mollusk. (694–695)
7. Describe functional specializations in the foot, mantle, and gills of the clam. (695)
8. Describe the anatomical and functional specializations seen in the foot and mantle of the cephalophod mollusk. (696)
9. List several ways in which octopuses and squid are well adapted to their roles as predators. (696)
10. Briefly explain metamerism and give an example of a highly metameric animal. (696)
11. Explain how a hydrostatic skeleton is used in movement. (698)
12. Describe the earthworm's circulatory system and blood. Would this be considered primitive or advanced? Explain. (697)
13. Use a simple drawing to explain the functioning of the earthworm's nephridia. (697)

14. What role does the earthworm's clitellum play in reproduction? (697)
15. List examples of specialization in the earthworm's gut. (697–698)
16. What evidence of cephalization is found in the earthworm? (698)
17. Explain how the earthworm makes use of its muscles and setae in burrowing. (699)
18. Describe the feeding specializations that occur in the leech. (698–699)
19. List three anatomical differences between the clamworm and the earthworm. (699)
20. In what phyla is the trochophore larva found? What is the phylogenetic significance of this? (694, 699)
21. List the four molluskan structures that vary greatly from class to class. (690–691)
22. Briefly describe the characteristics of arthropod appendages, skeleton, nervous system, and size. (700)
23. Suggest how the following may be significant to the success of arthropods: rate or reproduction, potential variability, life cycle with distinctly different stages, small size. (700)
24. List three characteristics of the chelicerates that are not present in the mandibulates. (700)
25. Briefly explain how spiders produce and design their webs. (702)
26. List the four major classes of mandibulates and cite an example from each. (702)
27. How does the crustacean exoskeleton differ from that of the insect or arachnid? (703)
28. Describe the structure of the compound eye and explain what it actually enables arthropods to see. (703)
29. List two vital ecological roles played by the marine crustaceans. (703)
30. What are some ways in which chilopods differ from diplopods? (703)
31. Name the three main body parts of insects. In which of these is the evidence of segmentation most obvious? (704–705)
32. List three major characteristics of insects. (704)
33. Compare the structures of gas exchange in spiders and terrestrial insects. (702, 705)
34. Describe the mouthparts of the grasshopper. For what are they specialized? (705, 707)
35. Describe three specializations in the legs of insects. (707–708)
36. List both the arthropod and annelid characteristics of the onychophorans. Where do these strange animals seem to fit in invertebrate phylogeny? (708–709)
37. List two embryological characteristics of echinoderms that suggest their evolutionary relationship to chordates. (709)
38. How does the spiny covering of the echinoderm qualify as an endoskeleton? (709)
39. List the elements of the water vascular system and briefly explain how the tube feet extend, fasten on, and shorten. What are two roles of this unique system? (710, 712–713)

32 THE CHORDATES

THE HAIRY-NOSED WOMBAT CANNOT RUN VERY FAST. NEITHER CAN ITS cousin, the naked-nosed wombat. But coyotes can, and cheetahs can outrun coyotes. None of these animals jump from the tops of trees, but flying squirrels and birds do. And a bird may go up as well as down. Both birds and chimpanzees build nests in trees, but only one gives milk, and neither can hold its breath as long as a turtle. Most fish don't have to come up for air at all. Fish, however, do have certain things in common with turtles. In fact, all these animals have enough traits in common that they are placed in the same phylum: **Chordata.** In addition they are all in the same subphylum: **Vertebrata;** that is, they all have backbones.

The chordates are indeed a fascinating and highly diverse group of animals that, while seemingly quite different from each other, have certain fundamental traits in common. It may be a little hard to see, at first, how a species as grand and noble as our own could be related to something that sits like a leathery little pouch on a wharf piling or one that wriggles through the sandy ocean bottom. But, as we will see, the backboned animals do share a number of characteristics with such creatures, the nonvertebrate chordates. Furthermore, all chordates share certain traits with another group, even more distantly related, that bears mentioning—the hemichordates.

PHYLUM HEMICHORDATA

The **hemichordates** ("half-chordates") are represented by the acorn worm, a burrowing marine animal with a conical (acornlike) **proboscis** or nose, and just behind it a **collar** (Figure 32.1a). Note the rows of **gill slits** in the wall of the pharynx, behind the collar. These slits reveal the worm's affinity to the chordates, since chordates have gill slits at some time during their development—even if only as embryos. In the acorn worm, the apparatus acts as a kind of gill and helps in feeding as well.

While the gill slits link the hemichordates with the chordates, another seemingly insignificant trait links the acorn worms to the echinoderms—the close resemblance of their larvae (Figure 32.1b). In addition, the embryos of echinoderms, hemichordates, and chordates undergo radial cleavage (see Chapter 31). It seems that in the distant past, the ancestors of echinoderms, chordates, and hemichordates were rather closely related. This theoretical relationship is explained in Figure 32.2.

PHYLUM CHORDATA

The chordates are customarily divided into three subphyla: **Urochorodata, Cephalochordata,** and **Vertebrata.** In addition to the gill slits just mentioned, chordates have **dorsal, hollow** (tubular) **nerve cords,** a peculiar cartilaginous structure called the **notochord,** and a **postanal tail.** In primitive chordates, the gill slits become sieves for feeding, and in the fishes, the **gill arches,** supporting structures between the slits, will later support the gills. In other vertebrates gill slits appear only in the embryo.

Our own spinal cord is a good example of a dorsal, hollow nerve cord. (Recall that the invertebrates in the last chapter had *paired ventral solid* nerve cords.) We will learn in Chapter 43 how the dorsal nerve cord is formed during embryological development.

Chordates also have, at least at some point in their development, a notochord. The notochord is a flexible, turgid rod running along the back that serves as kind of skeletal support. It consists of large cells, apparently under considerable hydrostatic pressure, encased in a tight covering of connective tissue. In nearly all living chordates the notochord exists only in the embryo or larva. In vertebrates it begins to dissipate as the backbone appears, early in embryonic development.

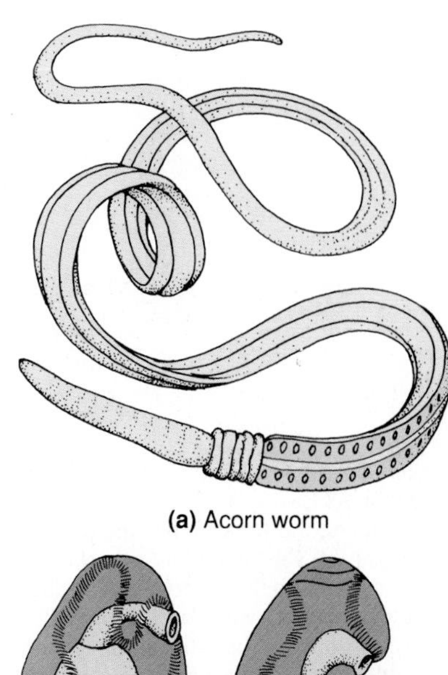

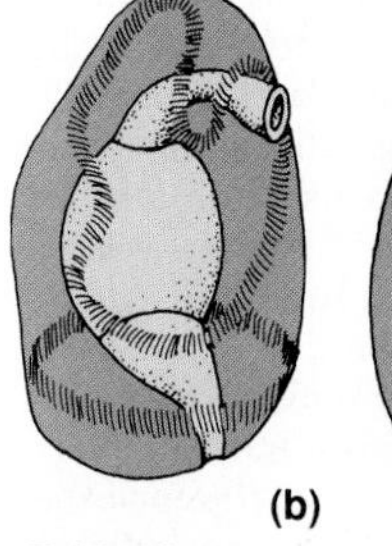
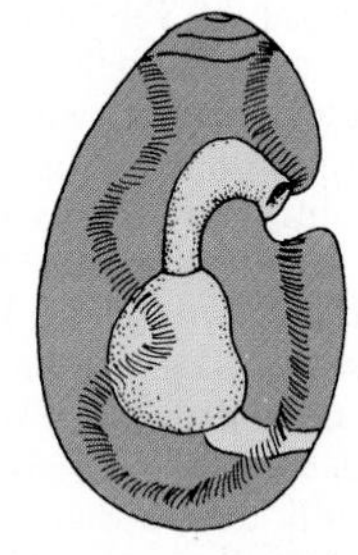

FIGURE 32.1
A HEMICHORDATE
(a) The acorn worm has a large number of pharyngeal gill slits, a chordate trait. It burrows in mud flats, using its proboscis somewhat in the manner of an earthworm. Water and organic debris enter the mouth. The nutrients in the debris are sorted and digested by the gut, while the water passes out through the gill slits. **(b)** Evidence of the hemichordates' affiliation with echinoderms is seen in their larvae, which are similar in the two groups.

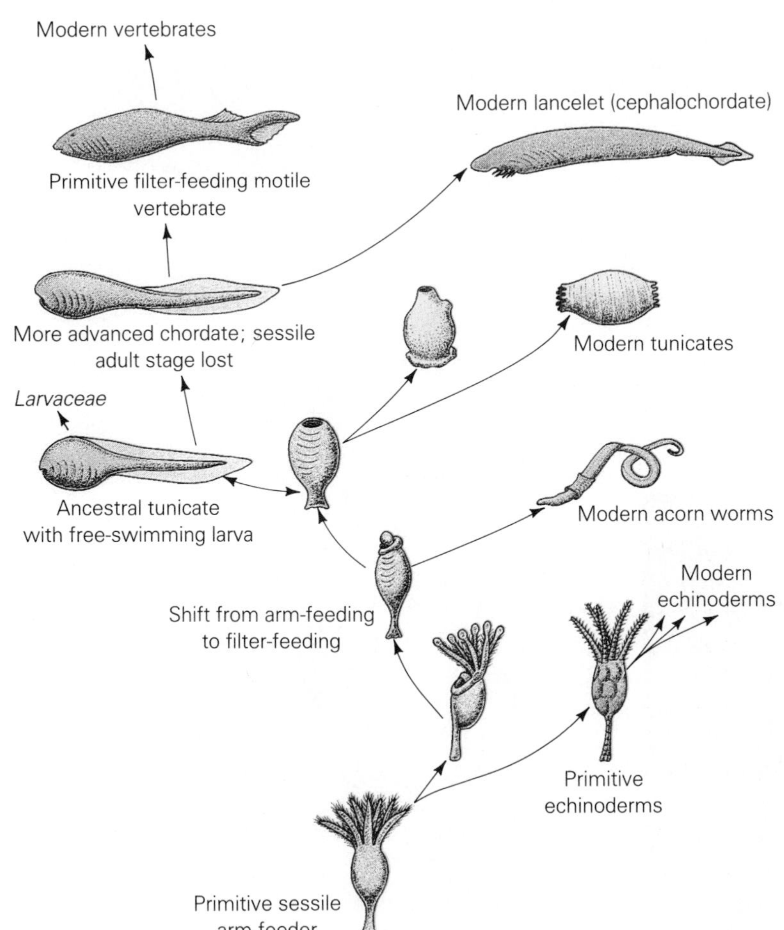

FIGURE 32.2
DEUTEROSTOME EVOLUTION
According to one theory, echinoderms, hemichordates, primitive chordates, and vertebrates arose from a sessile, deuterostome. These simple animals produced a swimming larva. From these early creatures arose the echinoderms and hemichordates, both of which still retain the free-swimming larva. Following these divergences, the deuterostome line divided once more. One branch produced the urochordate line (tunicates and salps); the other arose from certain larvae that had undergone neoteny. They had somehow retained their juvenile body form yet reached sexual maturity. From this evolutionary experiment arose the cephalochordates and the vertebrates.

The final defining characteristic of chordates, the postanal tail, is present in at least some stage of development in all chordates. Whereas adults of one chordate group, the **tunicates,** generally do not have such a tail, it is found in their larva, just as it is in human embryos.

Subphylum Urochordata: Tunicates and Salps

Tunicates, also called **ascidians** or simply "sea squirts," are the most common urochordates. The name *tunicate* refers to the transparent, tough covering, or **tunic,** on the outside of the soft, saclike body. The urochordates also include the bizarre, transparent, free-swimming **salps** along with a minor but highly significant group, class **Larvacea.**

Tunicates are revealed as chordates by their tadpolelike larvae, which bear a notochord, a dorsal, hollow nerve cord (which enlarges anteriorly), gill slits, and postanal tail (Figure 32.3). In its transition to the adult form, the tunicate loses its obvious chordate characteristics. Its tail, notochord, and nerve cord are absorbed into the body, leaving only the enlarging **gill sac** (or gill basket) as a clue to its chordate relationship. Tunicates are filter-feeders, using their gill clefts as strainers. The adult circulatory system is open, consisting of little more than a bizarre heart that pumps blood first one way, then the other. To further confuse the biologist, the tunic contains cellulose. Cellulose, a common cell wall material in plants, is rarely found in animals.

Though the tunicate larva changes to a saclike adult form, the larvacean retains the juvenile tadpole body plan, with its gill slits, tail, and notochord. Thus, except for sexual maturity, it never reaches what is the adult stage seen in other urochordates. The retention of the juvenile form in reproductively mature animals is called **neoteny;** to evolutionary

FIGURE 32.3
TUNICATES
As a larva, the tunicate has the pharyngeal gill slits, a lengthy notochord, and a dorsal, hollow nerve cord—all hallmarks of the chordates. The body form of the adult is rather simple and has no specialized sensory devices, although it maintains the pharyngeal gill slits throughout adult life. Currents of water are produced by cilia in the gill structure drawing food particles into the digestive tract. Respiration occurs as water exits through the body wall.

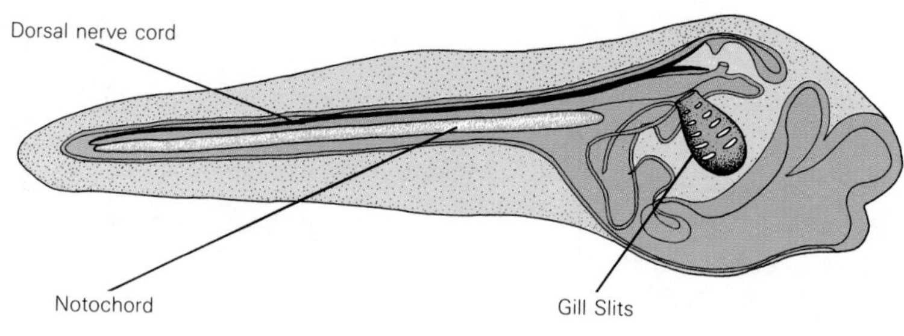

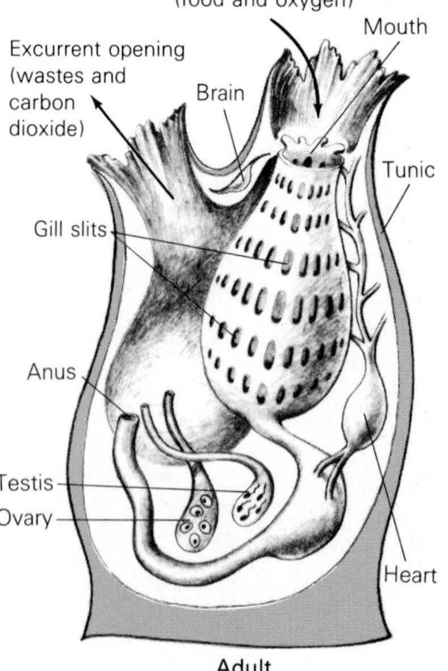

theorists, this odd phenomenon suggests how the main chordate line might have originated (see Figure 32.2).

According to the theory, all cephalochordates and vertebrates can trace their roots to the fortunate evolutionary accident that led to the arrival of these perpetual juveniles. From this primitive line arose species that at first were awkward, slow, jawless creatures, sucking up nutrients from the mucky sea bottom sediments. But in 500 million years, evolution has time to exert its powerful effects, and from these humble beginnings arose a species that would someday change the very face of the earth. We will come to this species shortly, but first let's consider a subphylum whose members more closely resemble their Precambrian ancestors.

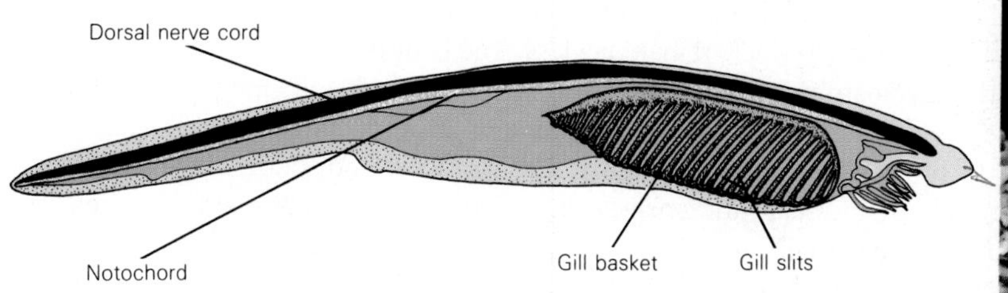

FIGURE 32.4
THE LANCELET
In its body plan, *Branchiostoma* is clearly similar to the basic vertebrate pattern. The notochord runs the length of the body and functions as an anchor for the paired muscle segments (myostomes). There are also a dorsal, hollow nerve cord and pharyngeal gill slits.

Subphylum Cephalochordata: Lancelets

Cephalochordate fossils, similar to today's species, have been found in the Burgess Shale of Precambrian origin, so they are a truly ancient group (see Figure 30.3). The cephalochordates, or **lancelets**, as they are commonly known, are nonetheless members of our own phylum. Whereas they look like fish (Figure 32.4), the notochord is prominent in adults. In fact, the notochord protrudes beyond the brain. (This is how the creature got its name.) The relationship between cephalochordates and vertebrates is best seen by comparing the lancelet to the larval lamprey, a primitive vertebrate (Figure 32.5). The adult cephalochordates retain the pharyngeal gill slits and develop a dorsal, hollow nerve cord. Another prominent feature is the body musculature, composed of repeating units called **myotomes** (an example of segmentation, a trait found in most fish and all vertebrate embryos), which aid in both swimming and burrowing. Lancelets are filter-feeders. Beating cilia draw water into a complex mouth whose tentacles separate out food particles to be taken in for digestion. Water taken into the mouth passes through the gills where oxygen and carbon dioxide are exchanged, enters an outer chamber, the **atrium**, and exits through an opening near the tail.

SUBPHYLUM VERTEBRATA: ANIMALS WITH BACKBONES

Most **vertebrates** have several traits in common in addition to the standard chordate characters. They include a vertebral column (backbone), cranial brain development, a ventrally placed heart and dorsal aorta, gills or lungs for gas exchange, a maximum of two pairs of limbs, one pair of eyes, paired kidneys, and separate sexes.

There are seven classes of living vertebrates. The first three are fishes, while the rest includes the familiar amphibians, reptiles, birds, and mammals. We will also consider one extinct class of fishes.

Carolina Biological Supply

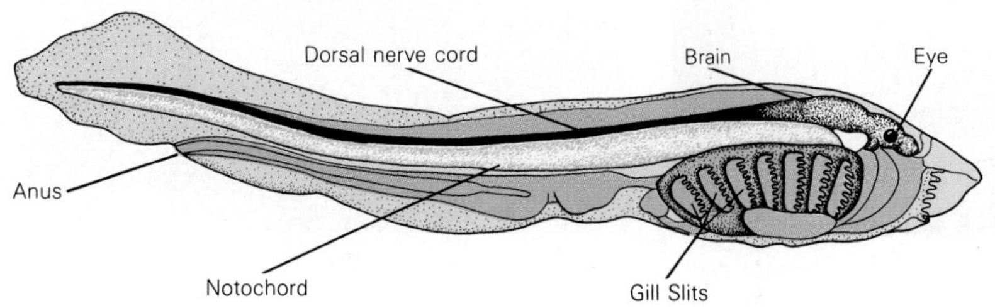

FIGURE 32.5
AN AMMOCOETE LARVA
The ammocoete larva of the lamprey bears many similarities to the lancelet (and to the tunicate larva).

Class Agnatha: Jawless Fishes

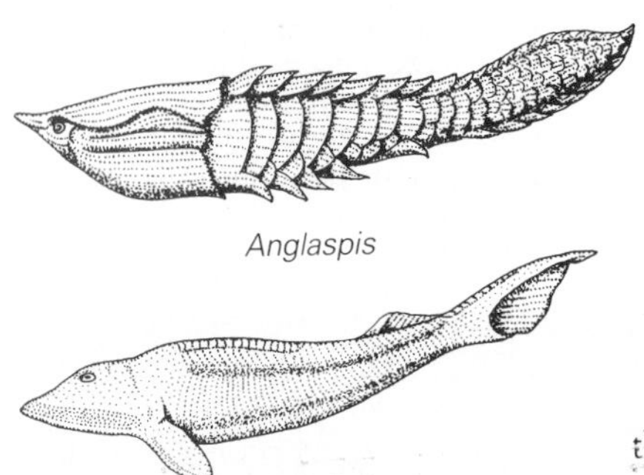

FIGURE 32.6
THE EARLIEST FISHES
The ostracoderms were jawless fishes of the Ordovician period. *Anglaspis* was more typically fishlike and was probably more adapted to swimming than bottom dwelling. *Hemicyclaspis* was probably a bottom dweller. Its heavy, armored head was quite likely used to plow through the mud as it searched for worms and mollusks, which it sucked up into its mouth.

The oldest well-defined vertebrate fossils on earth are a group of **agnathans** called the **ostracoderms.** These were heavy, armored, jawless fishes (Figure 32.6). Their fossils date back half a billion years, occurring in the sediments of the Ordovician, Silurian, and Devonian periods (see Table 30.1). Their fin structure, which does not include the two pairs of fins seen in later fishes, tells us that they were sluggish swimmers. They probably depended on their armor and bottom-dwelling habits for protection against the large invertebrate predators of the day. Without jaws, the ostracoderms were probably not predacious but more likely strained their food from the water.

Today's agnathans, boneless remnants of the armored fishes (which died out at the end of the Devonian period), remain jawless. The lampreys and hagfish are members of the subclass **Cyclostomata** ("rounded mouths"). Their bodies are long and cylindrical, with simple median fins adapted for wriggling along the ocean bottom. They lack the paired fins of other fishes, their skeleton is cartilaginous, and, oddly, their notochord persists throughout life. Parasitic species of lamprey feed with the aid of a rounded sucker mouth armed with horny spikes and a rasping tongue (Figure 32.7). The hagfishes, or slime hags, as some species are called, lack the sucker mouth and feed by a rasping device, boring into the bodies of dead or dying animals. By some standards they aren't a particularly admirable group.

The parasitic lamprey life cycle includes a filter-feeding larva, the **ammocoete**, which resembles the adult lancelet (see Figure 32.5). During the very long larval period, the animal inhabits fresh water, but the adult form of most species normally lives at sea until it is time to spawn. The sea lamprey has shown its physiological adaptability by surviving in the fresh waters of the Great Lakes, where it now completes its life cycle and is a very successful parasite.

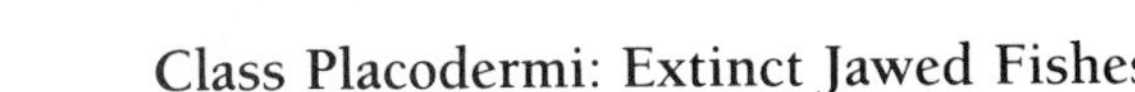

Class Placodermi: Extinct Jawed Fishes

The **placoderms** are believed to have descended from one line of ostracoderms, but the evidence is poor. They originated back in the Silurian period, their numbers swelling during the Devonian and dwindling out some 150 million years later. Clearly, they must be considered a successful group in terms of their tenure on earth.

Placoderms were armored, with hinged jaws. There is little doubt that the evolution of moveable, biting jaws was one of the most significant events in vertebrate history, quickly permitting the establishment of many new predatory feeding niches and the broadening of others. (It also gave vertebrates a way to make life miserable for each other.)

The placoderm jaw is known to have derived from the gill arches (the bony structure that supports the gills). This, of course, required a considerable modification in both

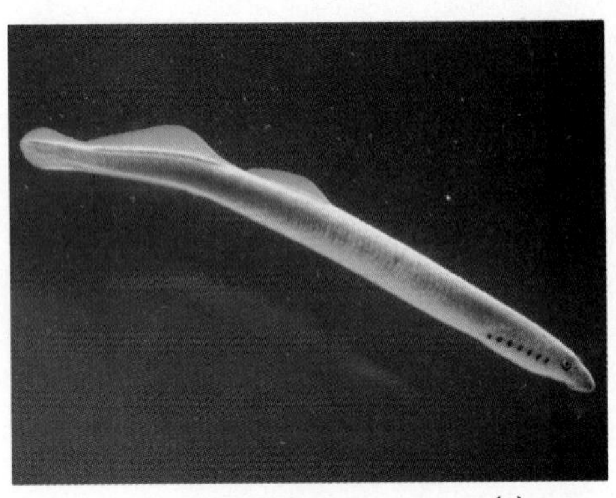
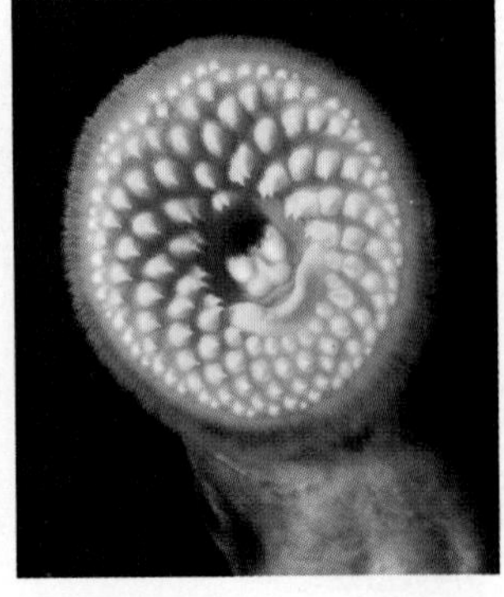

(a)

(b)

FIGURE 32.7
THE CLYCLOSTOMES
Today's jawless fishes, lampreys and hagfish, lack a bony skeleton. A notochord persists in the adult along with the rudimentary cartilaginous skeleton. **(a)** The parasitic lamprey uses rasping and sucking mouthparts in feeding. The hagfish **(b)** is a bottom scavenger known to attack weakened or injured fish.

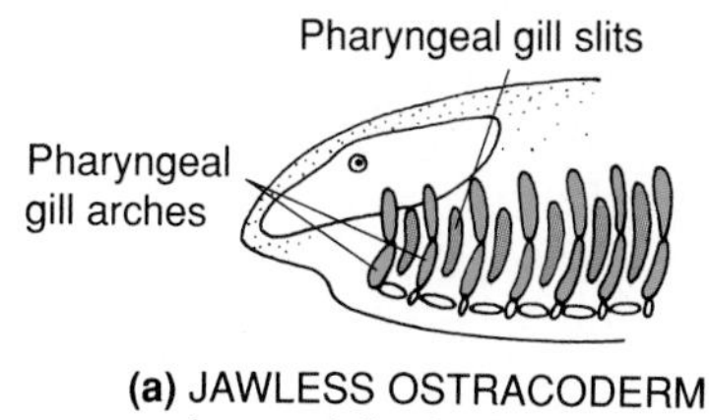

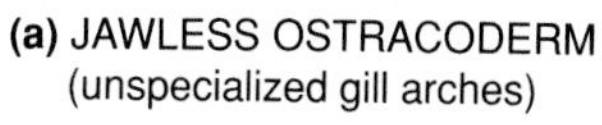
(a) JAWLESS OSTRACODERM
(unspecialized gill arches)

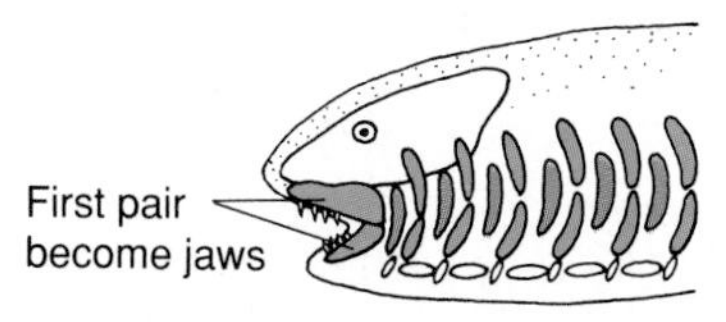

(b) PRIMITIVE JAW OF PLACODERM
(1st pair specialized)

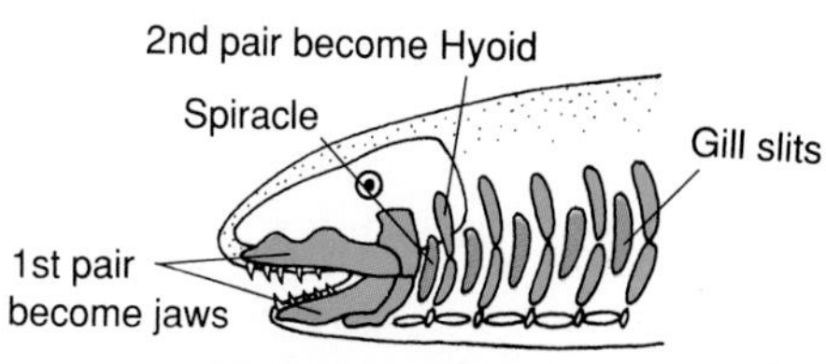

(c) JAW OF SHARK
(1st and 2nd pair specialized)

FIGURE 32.8
EVOLUTION OF JAWS
The vertebrate jaw is believed to have evolved from the primitive gill arches. **(a)** In ostracoderms, the gill arches are all similar and unspecialized. **(b)** In the placoderms, the first gill arches have been modified into the primitive upper and lower jaws, accompanied by related muscle development. **(c)** In a modern shark, the first pair of gill arches are greatly modified into strong, hinged upper and lower jaws. The second pair have become the hyoid bones. One gill slit has become the spiracle—a small, dorsal opening.

the position and strength of the gill arches, as well as the pharynx. Today, the upper and lower jaws of most fish form a hingelike structure that is only loosely attached to the cranium (Figure 32.8). The hinge improved jaw mobility and enabled the primitive vertebrates to become predators, a role formerly dominated by invertebrates. One of the more fascinating species was the gigantic *Dunkleosteus* (Figure 32.9), about the size of a modern gray whale and possibly one of the most fearsome predators that ever lived.

From the placoderm line arose the two large classes of jawed fishes we see today—the cartilaginous fishes (class Chondrichthyes) and the bony fishes (class Osteichthyes).

Class Chondrichthyes: Sharks and Rays

Class **Chondrichthyes** includes the cartilaginous fishes, a fascinating group of predators and scavengers made up mainly of sharks and rays. Many of these modern predators have replaced the protective armor and the heavy skeleton of their ancestors with a tough skin, light frame, and great speed. Their origin remains a puzzle, but their fossils first appear in the early Devonian period. By the Mesozoic era, such fossils dwindled as the numbers of bony fishes burgeoned. But beginning in the Jurassic period, they again increased gradually, and since then they have held their own.

The cartilaginous skeleton of sharks and rays was once taken as evidence of their primitive nature. Evolutionists more recently have established that the cartilaginous skeleton of sharks is a *derived* condition—that is, the ancestors of cartilaginous fishes had bony skeletons. (Recently, scientists have found exciting new evidence of residual bony tissue in the spinal column of sharks.) Some biologists believe the evolution of a lightweight cartilaginous skeleton is an adaptation to deep-water life. In support, they note that sharks and rays lack the hollow, buoyant swim bladder found in bony fishes, although some buoyancy is provided by a large store of lightweight lipids.

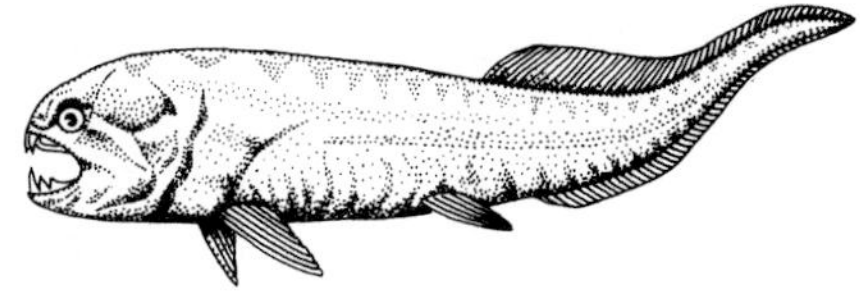
Dunkleosteus

FIGURE 32.9
AN EARLY JAWED FISH
Note the teeth and hinged jaws of *Dunkleosteus,* a 10-meter-long predator. The placoderms were armored, some with massive plates surrounding the head and other portions of the body.

In addition to their cartilaginous skeleton, sharks and rays are unique in several other respects. For example, their body and tail shape is unlike that of most bony fishes (Figure 32.10). The shark's tail is asymmetrical, more like that of the extinct placoderms than those of modern bony fishes. The body, particularly at the anterior, is dorsoventrally flattened, rather than laterally as in typical bony fishes.

(a)

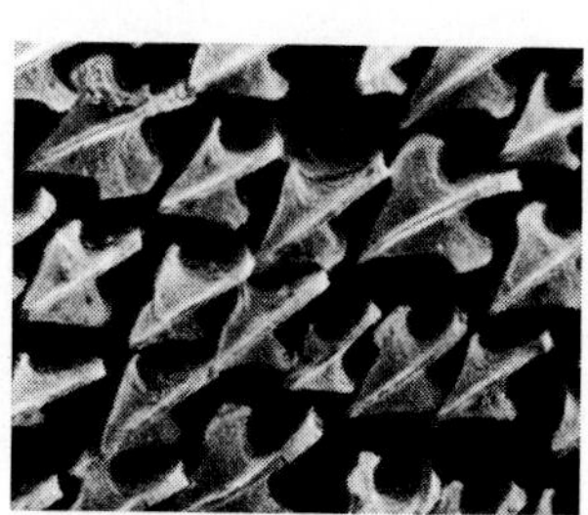
(b)

FIGURE 32.10
SHARK CHARACTERISTICS
(a) The body shape is an excellent lesson in streamlining. Sensory structures include an extensive lateral line organ that detects nearby movement, along with keen olfactory and visual receptors. **(b)** The skin is tough and flexible, consisting of placoid scales, each a miniature toothlike structure.

The skin of sharks and rays is very rough, embedded with miniature "teeth" known as **placoid scales** (see Figure 32.10). Each is anchored into the dermis by a basal plate. Embryologically, these scales form the same way as teeth, and in fact, the shark's teeth are simply larger versions of the same structure. The teeth themselves develop continuously throughout life, with newly formed teeth migrating forward from rows inside the mouth to replace older or broken teeth. The older teeth fall out and sink to the bottom, often washing up on beaches. The rays and several kinds of sharks have heavy, flattened "paved" teeth, useful for feeding on shellfish. The largest sharks, the whale sharks and basking sharks, are filter-feeders that eat small zooplankton they strain from the sea.

A curious feature in the shark's short digestive system is the **spiral valve** (also seen in lampreys and some bony fishes), which consists of a spiraling flap of absorptive tissue that extends into the gut and greatly increases its surface area (see Chapter 38). Sharks don't chew, and the large chunks of food they swallow require a considerable time for digestion. The spiral valve provides a means for slowing the movement of food, permitting the powerful digestive enzymes to do their work. The digestive system ends in a **cloaca,** a posterior chamber that also receives ducts from the reproductive and excretory systems and ends with the **vent.** Thus the cloaca not only receives feces and urine, but is also a sperm receptable and a birth canal. The cloaca is found in all vertebrate classes, but exists only in the embryo in most mammals (see Chapter 43).

Sharks have keen senses and locate prey by smell, by sight, and by certain patterns of vibrations (distress movements) that are picked up by the **lateral line organ.** The lateral line organ, characteristic of all fish, is formed of canals running the length of the body, containing groups of sensory cells that respond to water movements. Other sensory canals on the head are sensitive to bioelectrical disturbances or weak electrical fields created by nearby animals, although little is known of the receptor mechanism.

Contrary to the concern of some scuba divers, the toothsome, gaping grin on the mouth of an approaching shark is not necessarily anticipatory. It is generally accepted that its open mouth ensures a continuous flow of oxygen-laden water over the gills and out through the gill slits.

Fertilization in sharks and rays is internal, but embryo development varies from simple, primitive egg-laying in some species to a more advanced regime of protecting the independent egg and embryo in the uterus until hatching in others. A few shark species are quite reproductively advanced, retaining the egg and embryo in the uterus and providing nourishment, removing wastes, and exchanging gases, just as mammals do. In some shark species, although embryos are produced in a steady assembly line, only the oldest survive. They feed by devouring their younger brothers and sisters in the uterus as fast as they appear.

Class Osteichthyes: The Bony Fishes

The name **Osteichthyes** means "bony fishes" (*os,* "bone"; *ichthyes,* "fishes"). Like many modern animal groups, the bony fishes have expanded into just about every conceivable aquatic niche. They are a very diverse group (Figure 32.11) with well over 20,000 named species. It should be made clear that bony fishes are not so closely related to sharks as a first glance might indicate. Not surprisingly, the constraints of an aquatic life on a swimming vertebrate tend to limit body and fin shape, so there are general similarities. But there are also a number of important differences between the two groups.

To begin, the fins of many bony fish are much more refined, greatly promoting maneuverability. Sharks are great swimmers, but they can't stop short or make the quick darting turns so characteristic of bony fish. The body surface of the bony fish is also distinctly different from that of sharks. The bony fish's scaly skin has numerous mucous glands that produce the familiar slimy covering that can reduce drag, or water friction, by nearly 70 percent.

Bony fish have a **swim bladder**, a hydrostatic organ that improves balance and permits the animal to remain stationary at varying depths. In its most advanced form, the swim

Bay blenny
Butterfly fish
Blue tang
Psychedelic fish
Angler fish
Lionfish

FIGURE 32.11
CLASS OSTEICHTHYES
Bony fish in a great diversity of shapes. Most of the structural differences are in fin and body shape, but many of the rules are seemingly broken by such odd species as the lionfish, and even more dramatically by the creatures of the abysses such as the angler fish.

bladder has two specialized regions, a **gas gland,** which secretes gas into the bladder, and a **reabsorptive area,** which removes gas. In a more primitive state, the swim bladder wall is not so specialized in the release and uptake of gases, but is connected by a duct to the pharynx, as in the air-gulping lungfishes. In what is believed to be its most ancient state—in the primitive gar pike, for instance—the bladder has the spongy nature of a lung, suggesting that today's swim bladder may once have been a breathing device.

Many bony fish actively draw water into the mouth and pump it over the gills. The gill structures include five rows of gills on each side, located in a common **gill chamber** and each protected by a bony **operculum,** a moveable external flap. Each gill consists of a supporting gill arch with toothed **gill rakers** (that keep food out of the gills) and rows of **gill filaments.** Gas exchange occurs in the gill filaments, which are extremely thin-walled with a rich supply of fine blood vessels just beneath the surface. Respiration is closely supported by the circulatory system, which includes a two-chambered heart. In its circuit through the body, blood is pumped from the heart by the single, muscular **ventricle** (larger pumping chamber) directly to the gills, where it exchanges carbon dioxide for oxygen before being circulated to the body and returned to the heart (see Chapters 39 and 40).

Reproduction is as varied in bony fishes as it is in the cartilaginous fishes. Most are egg-layers that employ external fertilization. Some species release enormous numbers of eggs freely into the environment, while others release far fewer eggs into sheltered places or nests they have built. Others do not lay eggs but maintain the embryo within the female body and bear relatively developed young.

Bony fish are believed to have evolved from some kind of air-breathing, freshwater ancestor that had already developed lungs. From this distant stock, the bony fishes split into three groups: **Actinopterygii** (ray-finned fishes—those with thin, flattened fins with supportive spines—from which most of today's bony fishes evolved. **Dipneustei**

FIGURE 32.12
A FAMILY TREE OF FISHES
Included are three living classes (Agnatha, Chondrichthyes, and Osteichthyes) and one extinct class (Placodermi). The osteichthyes, or bony fish, are divided into three groups: ray-finned fishes, lungfishes, and lobe-finned fishes. Most living fish belong to one of the three orders of ray-finned fish. Three living genera of lungfish remain, but there is only one species of lobe-finned fish—the coelacanth *Latimeria*. The lobe-finned fishes gave rise to the terrestrial vertebrates.

(lungfishes), and **Crossopterygii** (lobe-finned fishes, those with flattened fins attached to fleshy lobes, Figure 32.12). While ray-finned fishes make up the vast majority of species, only three genera today represent the lungfish. They live in swamps in Australia, Africa, and South America (Figure 32.13). These strange fishes have highly vascularized lungs that connect to the pharynx, which unlike that of any other living fish, is connected by openings to the nostrils. Lungfish survive in highly stagnated water by filling their lungs with oxygen gulped from the surface. Some species can even survive periods of pond drying by lying dormant in the mud.

Lobe-finned species were always few in number, but they have had enormous evolutionary influence. They gave rise to the amphibians—the first animal land invaders—

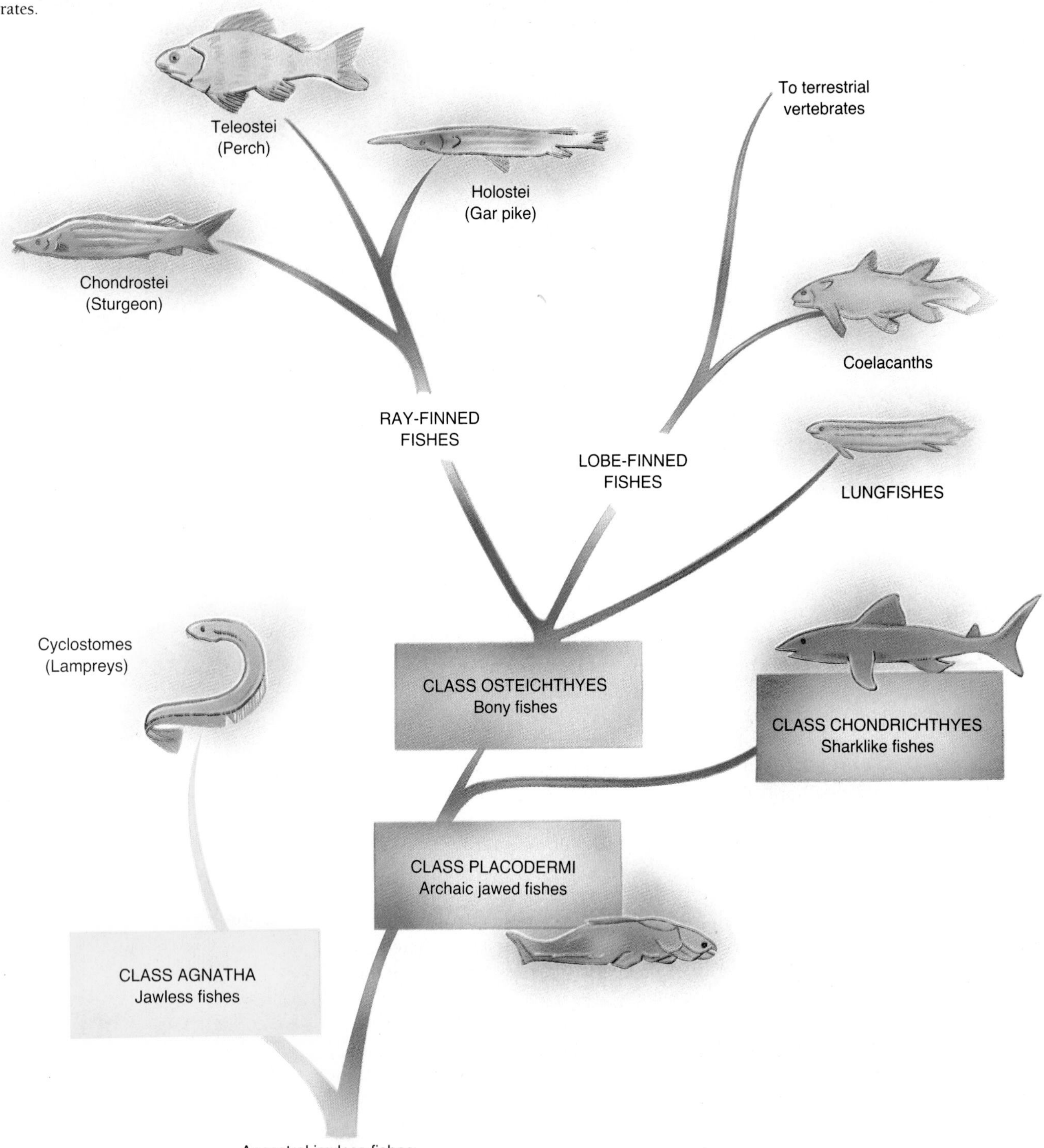

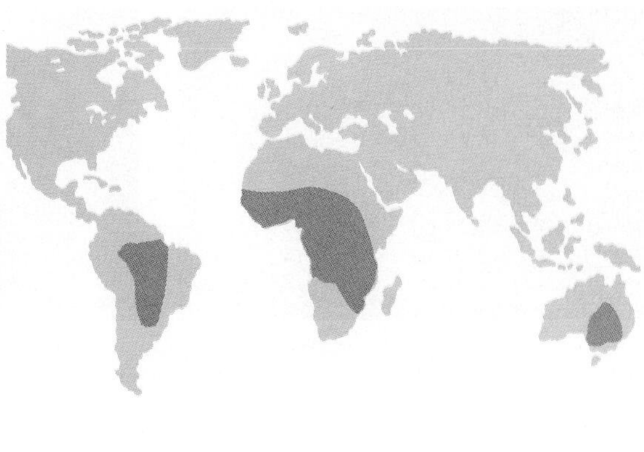

FIGURE 32.13
LUNGFISH DISTRIBUTION
The presence of lungfish on three widely separated continents is thought to be the result of continental drift. The lungfishes are presently found in tropical regions where seasonal droughts are common.

and from these arose the reptiles, and from these, the birds and mammals. Today, Crossopterygii includes just one species—the rare, deep-sea **coelacanth** *(Latimeria chalumnae),* the survivor of a once rather large and diverse group that was long believed to be extinct (Figure 32.14). Latimerians themselves are not ancestral to anything alive today and do not even belong to the specific crossopterygian group from which land vertebrates evolved. *Latimeria* is a distant relative indeed, but it is nonetheless more closely related to us than is any other living fish.

Class Amphibia: The Amphibians

Amphibians today are represented by three orders: **Urodela** (the salamanders), **Anura** (frogs and toads), and **Apoda** (the wormlike caecilians). Nearly all amphibians reproduce and develop in aquatic habitats, though some are well adapted to drier environments and a few even live in deserts. Both modern amphibians and modern reptiles are believed to have evolved from members of one ancient terrestrial group, the **labryinthodonts** (Figure 32.15), whose fossils are found in Carboniferous rocks.

FIGURE 32.14
A LIVING "FOSSIL"
The coelacanth, *Latimeria,* was found off the East African coast. Since their discovery in 1939, a number of *Latimeria* have been caught and subjected to intense study. Note the fleshy lobed fins. Interestingly, the adults retain the notochord throughout life and have vestigial, fat-filled lungs.

Salamander

Bullfrog

Caecilian

Reptile line

Primitive amphibians

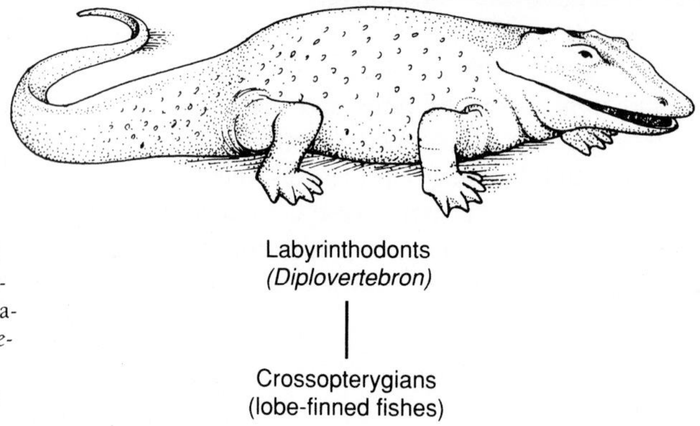

Labyrinthodonts
(Diplovertebron)

Crossopterygians
(lobe-finned fishes)

FIGURE 32.15
THE AMPHIBIAN LINE
The amphibian line is believed to have arisen from certain lobe-finned fishes. These are then somehow connected to the labyrinthodonts. The labyrinthodont shown here *(Diplovertebron)* may have been the primitive amphibian from which both reptiles and amphibians emerged.

Note the differences in the skeletons of a lobe-finned fish and a fossil amphibian in Figure 32.16. Among the skeletal modifications necessary for terrestrial life is increased development and size in the vertebral columns and the shoulder and hip girdles. In the evolution of other terrestrial vertebrates, particularly those to whom speed was important, the legs gradually came to be located alongside or under the body and projected downward rather than sideways. Locomotion came to depend on the muscles of the limbs and limb girdles, rather than on lateral undulations of the entire body (Figure 32.17).

Amphibians have three-chambered hearts and go the fish one better by having a blood circuit through the lungs that is separated from the rest of the body. This arrangement permits oxygenated blood to be returned to the heart for a second pump before entering the body circuit. The separation is imperfect, however, and some mixing of oxygenated and deoxygenated blood occurs in the single ventricle (see Chapter 39).

Amphibians have moist, highly vascularized skin, which, along with the lungs, is an important organ of respiratory exchange. (Amphibian lungs have little surface area since they are essentially vascularized sacs and not spongy like those of other vertebrates.) A moist skin is essential for gas exchange, but if gases can cross the skin, so can water. Thus, the amphibian is always at risk of dehydration, and accordingly, many live a semiaquatic life. There are important exceptions, though. The desert spadefoot toad, for example, survives its desert habitat by spending long dry spells in a fully dormant state, buried in the soil.

Like the desert spadefoot, most other toads and frogs fertilize the eggs externally, in water, but fertilization in caecilians and most salamanders is internal. Male salamanders release their sperm in small

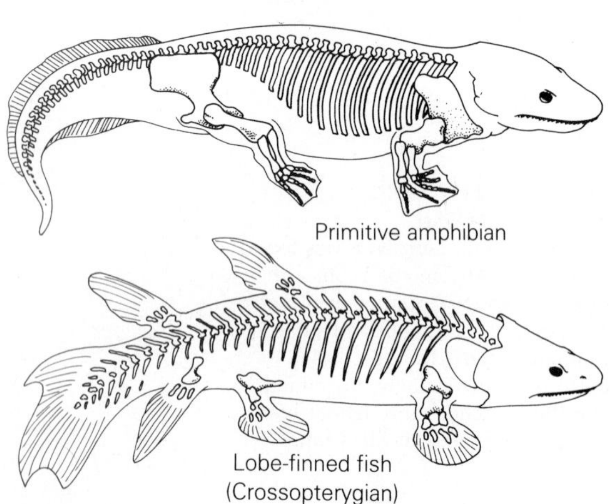

FIGURE 32.16
FISH TO LAND VERTEBRATES
The legs of amphibians evolved from the bony, muscular fins of their crossopterygian ancestors. The lobe-finned fish probably began adapting to land by crawling from pond to pond or possibly by climbing out of the water to avoid aquatic predators.

gelatinous packets called **spermatophores.** The females crawl over the packets and squat, picking them up in their cloacas, where fertilization occurs before the eggs are laid. Caecilians actually copulate, the male introducing sperm directly into the female's cloaca.

Some amphibians have managed to escape a water-bound existence. *Pipa pipa,* the Surinam toad of South America, carries the fertilized eggs in moist pouches on its back, assuring the young a relatively safe developmental period. In *Rhindoerma,* a small Chilean frog, the male scoops up the eggs in his mouth, where the embryos then complete their development. But in spite of such evolutionary adaptations to drier environments, amphibians remain a semiaquatic group. The first vertebrates to live a truly terrestrial life were to be reptiles.

Class Reptilia: The Reptiles

Eventually, backboned animals conceived, lived, and died entirely on the dry earth—and these were the scaly creatures called **reptiles.** The new land dwellers continually adapted to the dry environment and, in so doing, created a wide variety of specializations. Changes in their sexual reproduction and development were especially significant. Reptiles retained the cloaca, but males developed a penis for efficient copulation and internal fertilization, a prerequisite for full-time land dwelling.

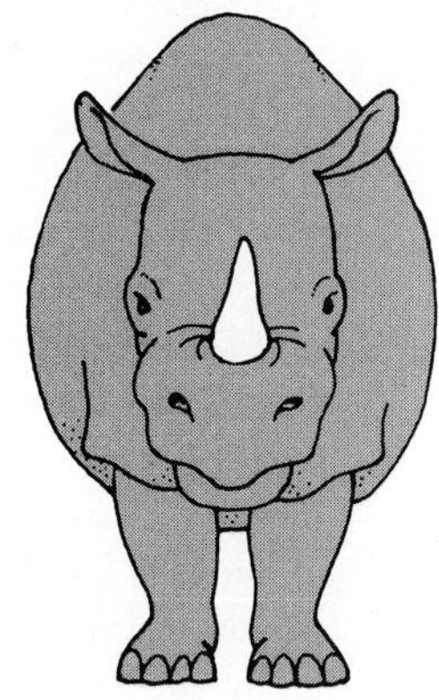

FIGURE 32.17
VERTEBRATE LIMBS AND POSTURE
The adaptation to terrestrial life is seen as a transition in the positioning of limbs in the tetrapods (four-legged vertebrates). **(a)** Many amphibians have their thin, lightly muscled legs angled or splayed out to the side. Thus the body weight is borne on flexed joints. **(b)** Reptile legs are alongside the body, but they still angle out to the side. **(c)** Mammals have their bodies raised above the ground with the limbs alongside or below the body.

A highly significant change was the evolution of the **amniotic egg**, in which the embryo is surrounded by a fluid-filled, membranous sac known as an **amnion** (characteristic of reptiles, birds, and mammals). The amniotic egg of the reptile is a porous, leathery case, containing a food supply, vascularized membranes that transport food and oxygen into the embryo and remove wastes, and sufficient fluid to cushion it from injury and protect it from drying. These delicate, extensive **extraembryonic membranes**, as they are known, are produced by the embryo and grow along with it (Figure 32.18).

Adult reptiles avoid water loss in a number of ways. Like their embryos and like the birds and terrestrial insects, they produce uric acid, a crystalline nitrogen waste that doesn't require much water for elimination. Water loss is minimized by a dry skin covered with protective scales and few mucus-secreting glands. The dry, scaly skin lacks elasticity and must be shed periodically as the animal grows.

Reptile lungs, unlike the simple saclike lungs of amphibians, are composed of many smaller compartments and thus have a spongy texture and a greatly increased surface area. Like that of the amphibian, the heart in most reptiles is three-chambered, but a partial septum divides the ventricles. In crocodiles and alligators, the ventricular septum is complete, forming a four-chambered heart. Reptiles are **ectothermic** (commonly known as "cold-blooded"). Ectotherms do not have efficient physiological mechanisms, as do **endotherms** ("warm-blooded" animals), for increasing or decreasing body heat as outside temperature changes. However, ectotherms often make use of behavioral responses to either take in or avoid environmental heat. Thus lizards and turtles warm themselves by basking in the sun and escape the sun when there is a risk of overheating (see Chapter 36).

Most of the reptiles are carnivorous, except for a few herbivorous lizards, tortoises, and turtles. Because their metabolism is slower, reptiles require considerably less food than do birds and mammals of the same size. Reptiles locate whatever food they do eat by a variety of means, including vision, heat detection, olfaction (smell), and hearing, and some use venom to subdue prey.

Reptile History The earth doesn't harbor as many reptiles now as it once did. In fact, the Mesozoic era, which lasted over 160 million years, is known as the "Age of the Reptiles." However, the reptiles actually originated much earlier, branching off from the amphibian line in the Carboniferous period of the Paleozoic era (see Table 30.1). They peaked during the Jurassic and dwindled down to the comparatively few species of today.

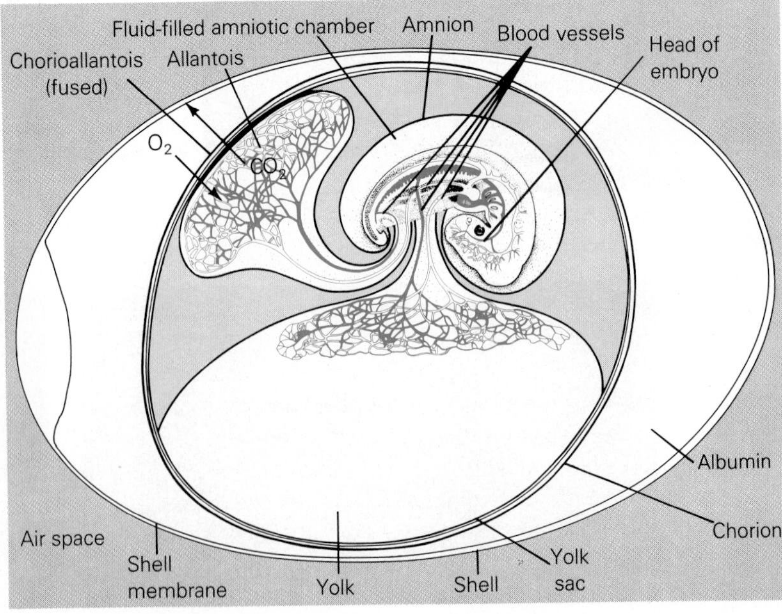

FIGURE 32.18
THE LAND EGG
The amniotic egg of birds and reptiles represents a transition to terrestrial life. The egg provides conditions similar in some ways to the aquatic environment. The amniotic cavity formed by the amnion is fluid-filled, protecting the embryo. The yolk sac grows out over the food mass of the yolk, traversed by blood vessels that bring the vital food supply into the embryo. The allantois, an extension of the urinary bladder, collects the embryo's nitrogen wastes. The vascularized chorion assists the embryo in its exchange of respiratory gases by providing the necessary surface area. Finally, the egg case itself, often leathery in the reptile and calcified in the bird, protects the contents while permitting gas to be exchanged with the surroundings.

FIGURE 32.19
FLYING REPTILES
The pterosaurs represented an experiment in flight. Some had a wingspan of 7 to 10 m, yet the wing was merely membranous skin attached to an enormously elongated digit (finger). There is no evidence of large wing muscles, indicating that the pterosaurs probably did more gliding than flying. Most pterosaur fossils are in marine strata, suggesting that they may have soared down from seaside cliffs in a pursuit of fish.

From the earliest reptiles, called the **cotylosaurs** or "stem-reptiles," there emerged a fascinating array of species. The land, for a time, was theirs. There were both giants and dwarfs. They climbed, hunted, hid, lurked, and threatened. As their numbers increased, competition and predation probably led some to return to the water from which their ancestors came. Others protected themselves with armor or sheer speed, striding around on their long hind legs. At least two lines of flying reptiles developed, one of which led to modern birds. The other, a much earlier and immediately successful group, were the **pterosaurs**, the first flying vertebrates (Figure 32.19). They were not large as ancient reptiles go, but some reached a meter in length. Pterosaurs were persistent, surviving from the Jurassic period to the end of the Cretaceous, but like so many reptile lines, they became extinct and thus are not ancestral to anything alive today. Some recently discovered fossils contained a surprise: *Pterosaurs, at least the smaller ones, had fur.* Thus, like mammals, they may have made inroads into **endothermy**—producing body heat to provide a more constant body temperature.

By the start of the Cretaceous period, some 135 million years ago, large was "in." This was the age of the "ruling reptiles," the great dramatic beasts known as **dinosaurs** (Figure 32.20). Even the pterosaurs produced a giant—*Pteranodon,* the size of a small airplane. The capstone of these great beasts was *Tyrannosaurus* ("tyrant lizard"), probably the largest terrestrial carnivore ever. But even larger dinosaurs stalked the earth. Great lumbering herds of plant-eating beasts devastated the taller trees, stripping them of their foliage. One called *Diplodocus* holds the size record at 30 meters long and weighed in at 30 metric tons.

The reptiles were hugely successful, as witnessed by their prominence on earth for so many millions of years. Why did they die off? Their demise came with the closing events of the Cretaceous period, and no one is sure what brought it about. We do know, however, that the delicate webs of life are highly vulnerable to change, and the loss of one member affects all. Perhaps a change in the climate started the process of their extinction. We know, for example, that their habitat was cooling at about this time. Some scientists blame the emergence of the early mammals, which might have preyed on the eggs of the giant reptiles. An exciting and well-substantiated theory has to do with the vast effects brought about by a giant asteroid that collided with the earth at the close of the Mesozoic. This idea, the Alvarez hypothesis, is discussed in Essay 32.1, p. 734. Another recent theory suggests that the sun has a companion star, named *Nemesis,* that, in cycles of about 26 million years, causes a rain of comets that destroys most life on earth. The geologic record roughly suggests such a cycle but the star has not been found.

FIGURE 32.20
RULING REPTILES
The so-called "ruling reptiles" of the Mesozoic includes some of the largest animals ever to rove the land.

1 Araucarites
2 Ramphorhyncus
3 Allosaurus
4 Schizoneura
5 Matonidium
6 Archaeopteryx
7 Brontosaurus
8 Ginkgo
9 Pteranodon
10 Quercus
11 Cornus
12 Pandanus
13 Anatosaurus
14 Sassafras
15 Ankylosaurus
16 Palmetto
17 Sabalites
18 Salix
19 Struthiomimus
20 Magnolia
21 Triceratops
22 Stegosaurus
23 Tyrannosaurus
24 Neocalamites

Jurassic Period

(gymnosperms and ferns)

Modern reptiles probably descended from four separate branches of the stem-reptiles and today form four orders (Figure 32.21, p. 735). The turtles and sidenecked turtles form one line of descent; the alligators and crocodiles another; the lone species, (*Sphenodon punctatum* or tuatara), a "living fossil," constitutes a third order; and a fourth includes the lizards and snakes. Snakes are believed to have evolved away from one branch of the lizard line about the start of the Jurassic, midway through the reptile era. The four lines survived the rigors of the Cenozoic era and are represented today by 6000 to 7000 species.

Class Aves: The Birds

The fossil ancestors of modern birds can be traced back to newly discovered *Protoavis*, from the late Triassic period, about 225 million years ago. It had feathers and presumably could fly. The best known fossils are of *Archaeopteryx* (Figure 32.22, p. 736), whose skeleton is, interestingly, almost indistinguishable from that of the thecodonts, the reptilian stock that produced crocodiles.

Modifications for flight are found throughout the modern bird's body (Figure 32.23, p. 736). The skeleton is light and strong. Many bones are hollow, containing extensive air cavities crisscrossed with bracings for strength. The reptilian jaw has been drastically lightened, and teeth have been replaced by a light horny beak. The neck is long and flexible, and the bones of the trunk (pelvis, backbone, and rib cage) have become fused into a semirigid unit. The breastbone (sternum) is greatly enlarged and possesses a large keel to which the large flight (breast) muscles attach in flying species. The tail is greatly reduced and, with the exception of a few individual bones, has become fused. The feet are often specialized, for example, for perching or grasping.

Cretaceous Period

(flowering plants and gymnosperms)

Feathers, another important modification for flight, are believed to have evolved from reptilian scales. Feathers are hollow and rodlike, which makes them unusually strong for their weight. Part of their strength results from their interlocking barbs (branches). Softer, noninterlocking down feathers provide insulation in all young and many adult birds.

Birds, like mammals, are endothermic, which permits adaptation to a range of climates. This innovation is quite costly in colder climates, however, since considerable food reserves must be oxidized just to generate body heat. Not only does endothermy require a greater intake of food, but it also requires a supply of oxygen for cell respiration. To meet the greater oxygen demand, both birds and mammals have highly efficient four-chambered hearts, which provide two fully separated circuits: the pulmonary circuit for oxygenating the blood and the systemic circuit for gas exchange in the rest of the body. In the bird lung, we find an extremely extensive, spongy respiratory surface, utilizing unique, one-way passages rather than the tiny, blind-ending and balloonlike alveoli (air sacs) in the lungs of most other air-breathing vertebrates. Birds do have blind air sacs, but they are large and outside the lung, serving primarily as temporary air reservoirs. We will take another look at the bird lung in Chapter 40.

Water conservation is important to most birds since the diet is often dry and drinking opportunities are limited. Most birds lack a urinary bladder. The excretory waste, uric acid, is simply added to the digestive wastes in the cloaca and most of the accompanying urinary water is reabsorbed. The water-conserving regime and lack of a bladder help keep the bird's weight down. And many female birds have only one ovary, perhaps further trimming for flight.

Birds have continued the reptilian tradition of producing large, resistant, amniotic eggs, with basically the same supporting extraembryonic membranes and food supply

ESSAY 32.1

EXIT THE GREAT REPTILES

The demise of the great reptiles has long been a puzzle. Hundreds of hypotheses have been considered. Some people once believed that dinosaurs were simply outcompeted by brainier mammals, or that the little rascals ate the dinosaur eggs. Others say that drastic climate changes marked the close of the Mesozoic or that a companion star to the sun causes comets to rain down every 26 million years. The problem is, none of these ideas could account for the truly dramatic suddenness of the mass extinction in which the dinosaurs disappeared.

Perhaps the answers can be found in the rocks. There are places on the earth where the 65-million-year-old rock stratum that marks the boundary between the Mesozoic and the Cenozoic can be found. In the waters off Gubbio, Italy, the marine deposits are particularly clear. What can they tell us? Below the boundary, layer after layer of Mesozoic carbonate rocks show a clear assemblage of Mesozoic plankton skeletons, dominated by very large foraminiferans (protists; see Chapter 23). But above the Mesozoic rocks, in the strata that were deposited in the Cenozoic, the fossil life forms are dramatically different. In layer after layer of Cenozoic rocks, we find fewer, much smaller species. Even more interesting, at the boundary between the two layers of plankton deposits is a single layer of clay about a centimeter thick.

In 1980 the paleontologist Walter Alvarez and his Nobel Prize-winning physicist father, Luis Alvarez, reported that this boundary layer has about 50 times more of the elements iridium and platinum than would be expected. Iridium is extremely rare in the earth's crust, but is a common metal in meteorites and asteroids. The Alvarez hypothesis, as it has come to be known, is that an asteroid collided with the earth 65 million years ago, and that's what ended the Mesozoic. The idea is not so far-fetched as it seems; astronomers calculate that enough asteroids pass through the earth's orbit that we can expect one to hit our planet every 100 million years or so. From the amount of iridium they found in that thin layer, they calculated that the asteroid was about 10 kilometers in diameter. A rock six miles across is enormous; it could knock you down. But this one may have done more than that. Computer simulations tell us that the dust thrown up by the impact brought the world into total darkness—much darker than the darkest night—for about three months. The effect of such darkness would be enormous. Since the sun's rays could not reach the earth, photosynthesis would stop, and plants would begin to wither and die. Many of the earth's animals would have been blinded, and certainly the weather would have changed. Tropical species would have felt subfreezing temperatures for the first time. Further research has corroborated the Alvarez hypothesis, and the iridium-rich Mesozoic-Cenozoic boundary layer has been found in many places.

Few organisms were equipped to survive such a catastrophe. Seed plants and marine plankton that formed resistant cysts were seemingly unaffected. But animal species changed significantly; for example, no animal species weighing more than 26 kg (57 lb) survived into the Cenozoic. Most of the earth's great dinosaurs, then, would have been killed by the earth's great darkness.

The demise of the reptiles was obviously not quite complete, since we have a considerable number with us today. As you watch turtles going about their business, keep in mind that they are descendants of survivors of the asteroid impact. But other species are descended from those Mesozoic reptiles too. One had great prospects in store. That species would build highways, develop fascinating ways to annihilate fellow members, vote twice, and devise elaborate justifications for all its behavior. But that was still a long way off.

(see Figure 32.18). As a rule, birds produce fewer eggs than do reptiles, a reflection of the parental care so characteristic of nearly all bird species. (With such care, a greater proportion of eggs are likely to hatch.)

There are about 8600 named species of birds, divided into about 27 orders. In spite of such extensive speciation, however, all birds have rather similar general structures. Apparently, the demands of flight have superimposed a specific set of design requirements. Birds differ most strikingly in the development of their beaks and feet, both of which are often related to their feeding behavior (Figure 32.24, p. 737).

Class Mammalia: The Mammals

Mammals are hairy animals with glands that give milk. In fact, it is the **mammary gland**, or mamma, for which the class—and one of our parents—is named. The hair or fur

FIGURE 32.21
SURVIVORS OF THE REPTILE AGE
Today's reptiles are believed to have evolved through four separate lines. The snakes and lizards descended from a line of smaller reptiles. The lizardlike tuatara of New Zealand is the lone remnant of another group. Right out of the central line of dinosaurs emerged the crocodilians, forming their own line. The side-necked turles have a long separate history of their own. Both kinds of turtles, like the snakes and lizards, sidestepped the giants and emerged from ancient stock early in the Mesozoic. Many extinct groups are also shown here. The widths of the colored lines roughly indicate the relative numbers of species at any point in time; dashed lines indicate our best guesses in the absence of fossil data.

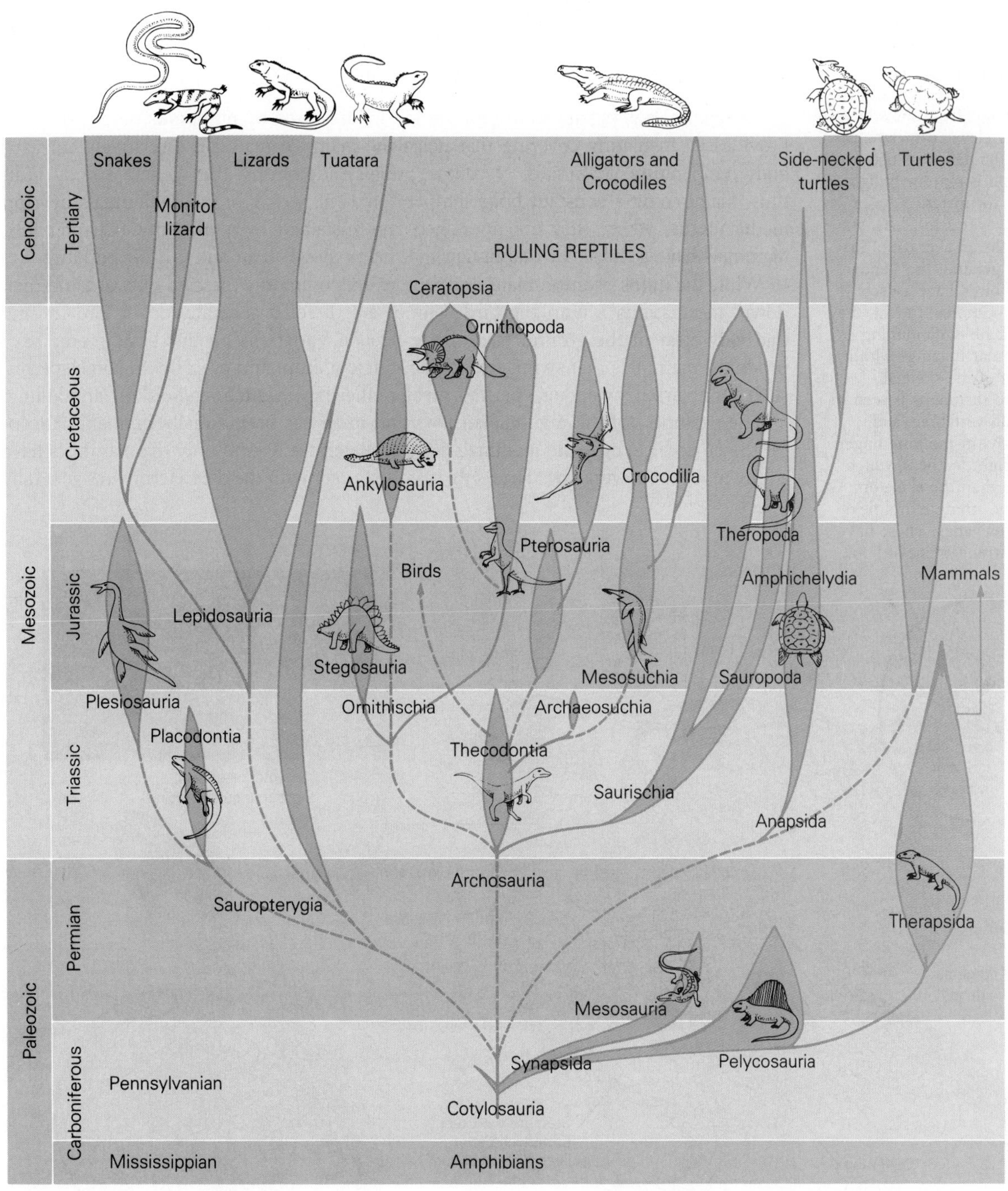

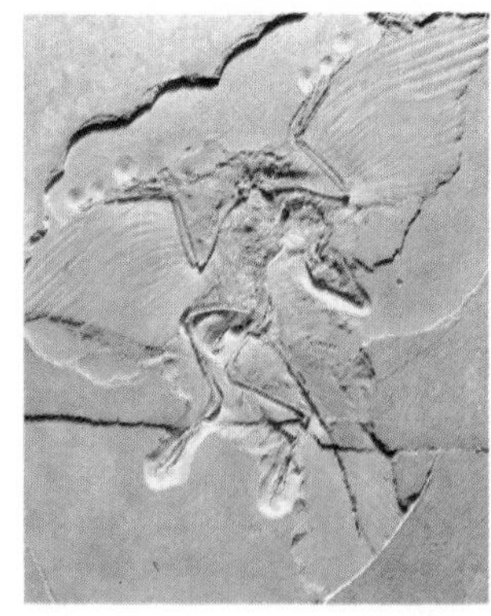

FIGURE 32.22
ARCHAEOPTERYX
Although birdlike in some ways, *Archaeopteryx* is not ancestral to today's birds, Note the clawed fingers in the reconstruction. Were it not for the feathers, *Archaeopteryx* would probably have been identified as just one more small dinosaur.

of the mammals is produced in follicles and is lubricated by numerous oil glands. Hair provides an insulating covering that helps the mammal maintain a relatively constant body temperature. Mammals, of course, are all endotherms, they generate body heat, thus, maintaining a constant body temperature. The young of most mammals develop in the uterus, where they are nourished and sustained by a **placenta** (Chapter 43). Mammals have a muscular diaphragm that helps move air in and out of the lungs.

While the limbs of amphibians and reptiles are commonly directed outward and then down, the mammal's (and the bird's) limbs are directed generally downward, raising the body clear of the ground (see Figure 32.17). Variations on this basic theme have produced mammals that swim, run, climb, burrow, leap, and even fly. In some species, such a limb arrangement also permits greater efficiency in catching, holding, and killing.

The evolution of the mammalian jaws and teeth has been equally dramatic. Except for the crocodiles, only the mammals have socketed teeth. And only the mammals have a fully mobile, chewing jaw. The teeth, of course, vary with the diet. Herbivores generally

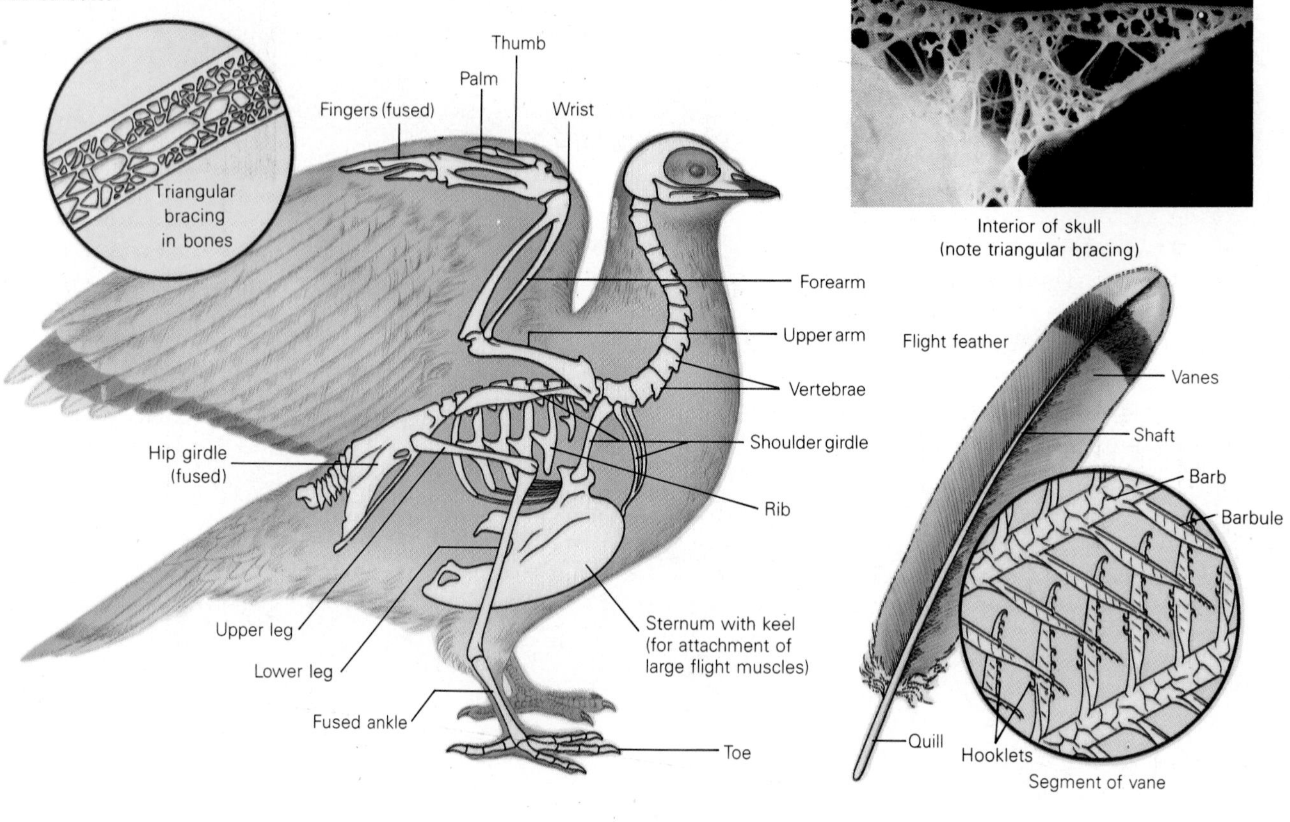

FIGURE 32.23
DESIGN FOR FLIGHT
Modifications for flight are seen in nearly every aspect of the bird's anatomy and physiology, from the streamlined form to the elevated metabolic rate. In spite of the demands placed on it, the skeleton is extremely light. The frigate bird, for instance, has a wingspan of just over 2 m (7 ft), yet its skeleton weighs an average of just 113 g (4 oz). In general, the slender, hollow bones of birds have a deceivingly delicate appearance; in fact, they are strong and flexible, containing numerous triangular bracings within (see photo). Part of the skeletal strength is due to fusion, as is seen in the hip girdle, tail vertebrae, and, most spectacularly, in the long finger (wing) bones. Flight feathers, which may weigh more than the skeleton, owe their extreme strength and flexibility to numerous vanes. These have an interlocking arrangement of hooklike barbules.

Triangular bracing in bones
Thumb
Palm
Fingers (fused)
Wrist
Forearm
Upper arm
Vertebrae
Shoulder girdle
Rib
Hip girdle (fused)
Sternum with keel (for attachment of large flight muscles)
Upper leg
Lower leg
Fused ankle
Toe
Interior of skull (note triangular bracing)
Flight feather
Vanes
Shaft
Barb
Barbule
Quill
Hooklets
Segment of vane

(a)

(b)

(c)

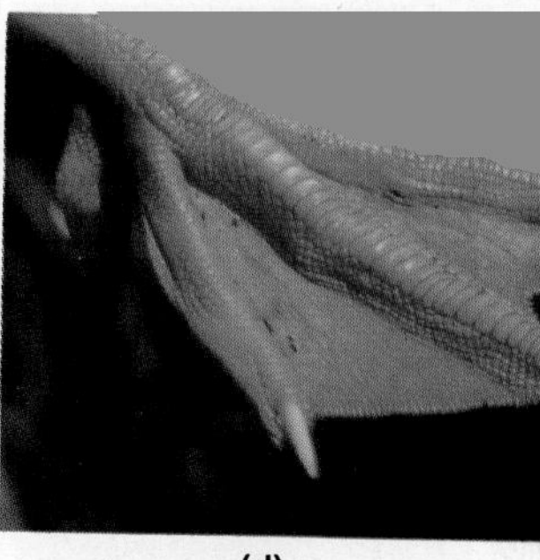
(d)

(e)

FIGURE 32.24
BIRD DIVERSITY
These foot variations permit birds to explore many food sources and habitats.
(a) The flicker's hooked hind toes easily cling to the bark of trees.
(b) The hawk uses its sharply curved talons to capture prey.
(c) and (d) The lobed feet of the coot and webbed feet of the duck are well suited to their aquatic habitats.
(e) The emu, a flightless bird, relies on its speed afoot to escape enemies.

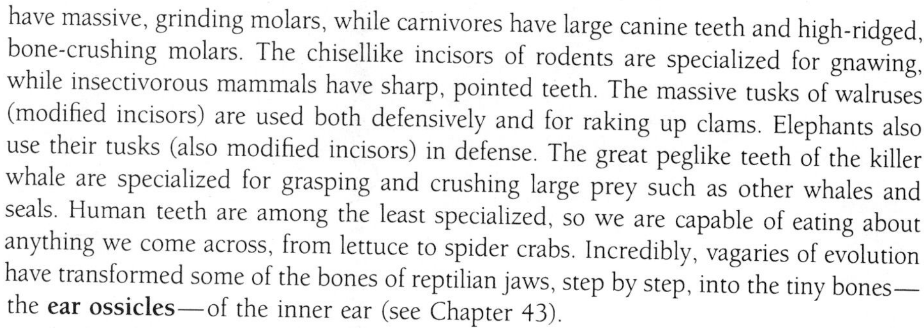

have massive, grinding molars, while carnivores have large canine teeth and high-ridged, bone-crushing molars. The chisellike incisors of rodents are specialized for gnawing, while insectivorous mammals have sharp, pointed teeth. The massive tusks of walruses (modified incisors) are used both defensively and for raking up clams. Elephants also use their tusks (also modified incisors) in defense. The great peglike teeth of the killer whale are specialized for grasping and crushing large prey such as other whales and seals. Human teeth are among the least specialized, so we are capable of eating about anything we come across, from lettuce to spider crabs. Incredibly, vagaries of evolution have transformed some of the bones of reptilian jaws, step by step, into the tiny bones—the **ear ossicles**—of the inner ear (see Chapter 43).

The Mammalian Brain Above all, the hallmark of a modern mammal is its relatively large and versatile brain. Today's mammals are smarter than other vertebrates; that is, they (1) rely less on genetically programmed instinct and (2) adjust more readily to their environment, basing more of their behavior on individual experience and learning. Their remarkable characteristics are reflected in the structure of their brains. Not only has the cerebrum (the part associated with learning and conscious thought) become larger, but new parts have been added, such as the **corpus callosum**, which integrates the mental activities of the left and right halves of the brain (see Figures 35.5 and 35.9).

Keep in mind, however, that increased learning capacity is simply one evolutionary solution to the challenge of survival. For example, insects have a comparatively limited ability to learn, yet they are our chief competitors for the earth's resources. Their alternative path of evolution obviously works well for them.

Mammal Origins The mammals had their reptilian origin somewhere in the Permian period, branching off very early from a large group that preceded the ruling reptiles, known as the **therapsids**, sometimes known as the "mammallike reptiles." The transition into the full mammalian condition was gradual, with no sharp dividing line seen in the fossil record. Some early mammals are said to have been doglike in appearance, but this is not very accurate, and you could probably clear out a bar if you walked in with one on a leash (Figure 32.25).

The therapsids were flesh eaters that developed the forelegs and musculature necessary for running. The angle of their elbows was quite different from that of the first terrestrial reptiles, whose elbows angled out

FIGURE 32.25
THE EARLIEST MAMMALS
The mammals arose from the therapsids, an early reptile line. The earliest therapsids retained many reptilian traits, including scales, but by the Triassic period (197 to 225 million years ago), features such as the limbs and stance had already changed. Raising the entire body from the ground was an adaptation for running and leaping, but it introduced new problems in balance and coordination—problems that required simultaneous adaptive changes in the brain.

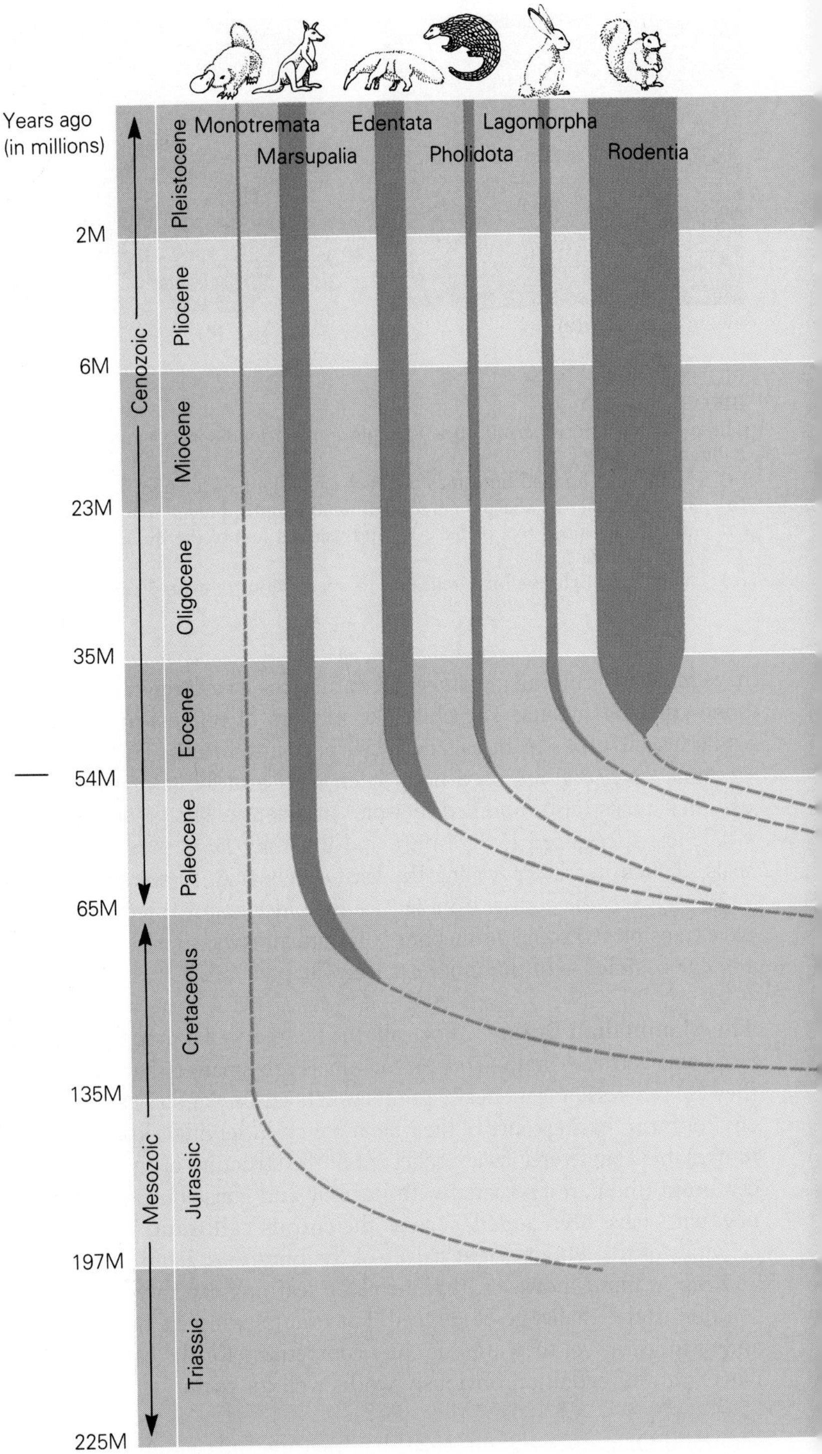

FIGURE 32.26
THREE TYPES OF MAMMALS
Three lines of mammals probably evolved from the ancestral therapsids. Monotremes (duck-billed platypus, top) are hairy egg layers. Their young feed by simply lapping milk off the mother's fur. The marsupials (kangaroo, middle) give birth to immature young, which climb to the pouch where they fasten to a nipple and remain there as they complete their development. The placental mammal (horse, bottom) completes its embryonic development in the uterus, supported by membranes that form the placenta. It also nurses for a time after birth.

to the side. Other indications of their mammalian association include tooth structure as well as skull and jaw features.

Survival in the developing mammals may have been enhanced by the ability to carry embryos within the body and to bear live young, rather than deserting the eggs to predators or having to defend a stationary nest. Another advancement was the production of milk, a very convenient food that provided the young with high-protein, high-calorie nutrition during the critical stages of growth. In turn, the young would have tended to remain with at least one parent in order to be fed, and this continued association would have afforded the opportunity for the offspring to learn from the parent. This opportunity would have reinforced selection toward braininess.

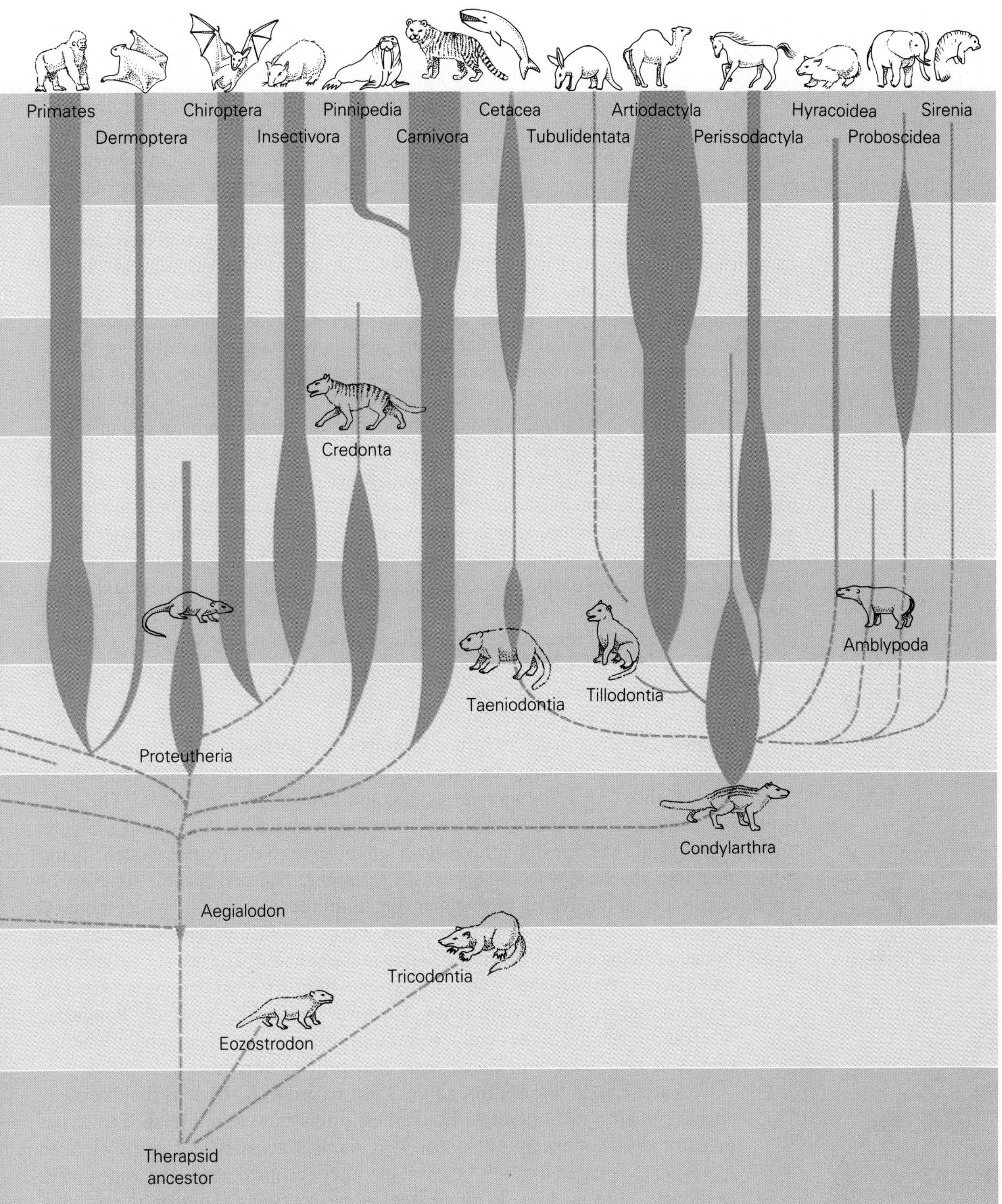

FIGURE 32.27
THE DIVERGENCE OF MAMMALS
Today's mammals occupy 18 or 19 orders and are dispersed into innumerable niches over the earth. Mammalian roots reach back to the early Mesozoic era but do not appear to branch out significantly until the Cenozoic, an era of great climatic changes that saw the end of the ruling reptiles. Numbers of species are indicated by line thickness. Dashed lines represent hypothetical relationships.

The dawning Cenozoic era ushered in the "Age of Mammals." By the end of the Mesozoic era, three types of mammals had diverged from the original line: **monotremes** (order Monotremata: the egg-laying mammals); **marsupials** (order Marsupalia: the pouched mammals); and **placental mammals** (mammals that form a true placenta). Figure 32.26 shows an example of each.

The monotremes never became widespread, and today there are just two species: the duckbilled platypus and the spiny anteater (also called the echidna). The marsupials are much more similar to the placental mammals, differing primarily in some aspects of development. Whereas in the placental mammals, the placenta provides for the needs of the embryo throughout its development in the uterus, marsupials produce only a "pseudoplacenta," a modification of the yolk sac. It supports the young for a brief time in the uterus before a very premature birth occurs. The newborn marsupial then migrates to the pouch, where development is completed.

With the exception of the ubiquitous opossum and a few ratlike pouched mammals in South America, marsupials are absent in all continents except Australia. Interestingly, marsupials evolved in the New World and were once prominent in both North and South America. Then, at the close of Mesozoic, their displacement began as placental mammals from Asia made their way into North America across existing land bridges. This displacement soon spread to South America, and by some 50 million years ago, most marsupials were restricted to then-subtropical Antarctica and Australia, which were still joined by a land bridge (see Essay 20.1). The modern opossum *(Didelphis)* reentered North America only 3 million years ago when the Isthmus of Panama formed. Later, Antarctica became inhospitable to marsupials, and they began to die out there. But by that time, Australia had separated from Antarctica, carrying its thriving marsupials with it. Apparently the placentals did not keep up, for with the exception of bats, the only placental mammals in Australia today are those introduced by European colonists.

The three modern mammalian subclasses radiated from an explosion of species in the early Cenozoic era. By the Eocene period, which began 54 million years ago, the main lines of mammalian evolution had already become established. The Cenozoic era was a disruptive time indeed, marked by drastically fluctuating climatic patterns and intense selection. Many Cenozoic mammals didn't survive the severe stress; however, those that did underwent rapid speciation and divergence, filling many new and highly specialized niches. Today's 4500 species are organized into 18 orders (or 19, depending on whether you consider the pinnipeds—walruses, seals, and sea lions—to be carnivores or in their own order). Each of the orders represents a major line that appeared in the early Cenozoic (Figure 32.27, p. 739).

The Primates Modern primates, order **Primates,** are divided into two major suborders, the first consisting of the more primitive tree shrews, tarsiers, lemurs, and lorises (Prosimii), and the second, the monkeys, apes, and humans (Anthropoidea). The latter is further subdivided into the New World monkeys (Ceboidea), Old World monkeys (Cercopithecoidea), and apes plus humans (Hominoidea). New World, western hemisphere monkeys are those with the prehensile (grasping) tails and nostrils set apart by a wide nasal septum, typified by the familiar "organ-grinder monkey" or spider monkey. The Old World, Eurasian and African monkeys are more likely to have short tails and doglike snouts with the nostrils close together, as in the baboons. The great apes (gibbons, orangutans, chimpanzees, and gorillas) and humans, members of superfamily Hominoidea, are called **hominoids.** The great apes belong to family **Pongidae,** whereas humans are in the family **Hominidae.** All members of our family, whether extinct or extant, are thus called (note the spelling) **hominids.**

FIGURE 32.28
THE FIRST PRIMATE
Plesiadapis, a rodentlike tree mammal from the Paleocene epoch, is believed to be ancestral to the primate line. *Plesiadapis* was rather small, about the size of a housecat—not very impressive, but it had a bright future.

Primates can be traced through the fossil record as far back as the Paleocene epoch, some 65 million years. The earliest primates were unlike modern types in many ways. For example, they had long snouts and claws, and actually looked more like rodents. In fact, it is believed that they occupied nearly the same niches as modern rodents, some living in trees, some scurrying about on the ground, and some in burrows feeding on grubs. Our best-known example is *Plesiadapis* (Figure 32.28). The divergence of the various primate groups is represented in Figure 32.29. Note how recently humans have diverged from common ancestry with the gorillas and chimpanzees.

How do primates differ from other mammals? This would seem to be an easy question, and it is unless one attempts to become too specific. Primates are often described as generalized mammals, meaning that they have many unspecialized features, some even quite primitive.

The limbs and body of *most* primates are well adapted for arboreal (tree-dwelling) life. For example, all have prehensile (grasping) hands divided into clawless (nailed) fingers. Most have long arms and some have prehensile tails.

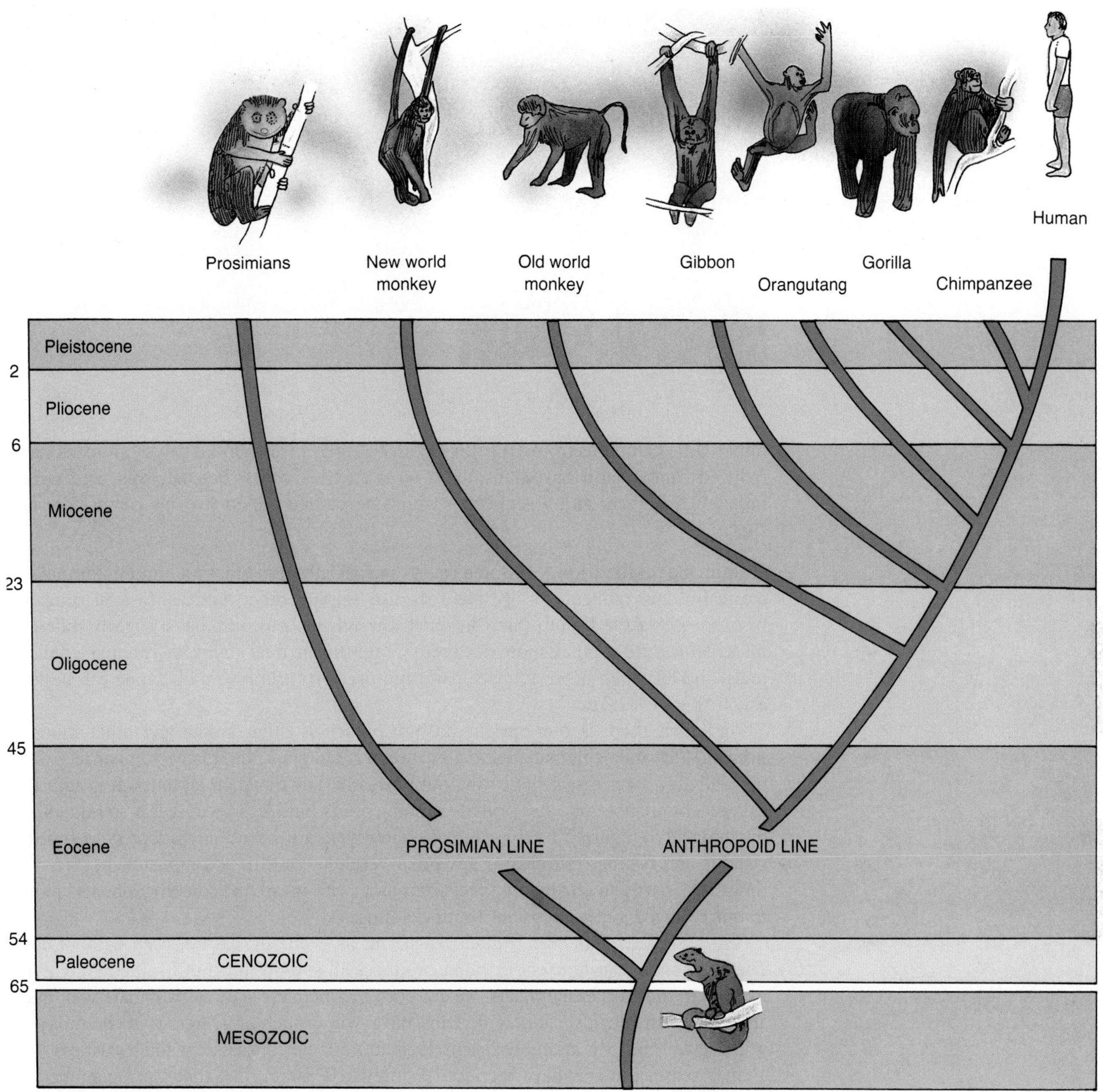

Primates have also developed binocular, stereoscopic vision with eyes located in a forward position. If you are going to swing on limbs, you need good eye-hand coordination—and the ability to judge distances (Figure 32.30).

Of course, all of these anatomical specializations require a substantial degree of brain development, with strong emphasis on centers dealing with coordination and vision. But more significantly, the primate brain is highly adapted for learning. This ability is especially developed in chimpanzees and gorillas (and the capuchin, a South American monkey), but humans are without parallel in intelligence and relative brain size.

Primates tend to be omnivores, eating all sorts of food, so their teeth are relatively unspecialized. However, with the exception of humans, primates, especially males, have

FIGURE 32.29
PRIMATE HISTORY
This tentative phylogeny of the primates places *Plesiadapis* at the base. The earliest branch produced the prosimian line, followed by the New World monkeys. Old World monkeys diverged some 10 million years later, subsequently followed by the great apes. The most recent divergence occurred between humans and chimpanzees.

FIGURE 32.30
BRACHIATION
Good brachiation (swinging from limb to limb) requires prehensile hands, long arms, and keen hand-eye coordination. **(a)** The gibbon is one of the most talented brachiators. Note the lengthy arms and the four-fingered grasp. **(b)** Both the hands (outside) and feet (center) of the orangutan are well suited for grasping and swinging. The thumb is positioned lower than our own and out of the way.

(a)

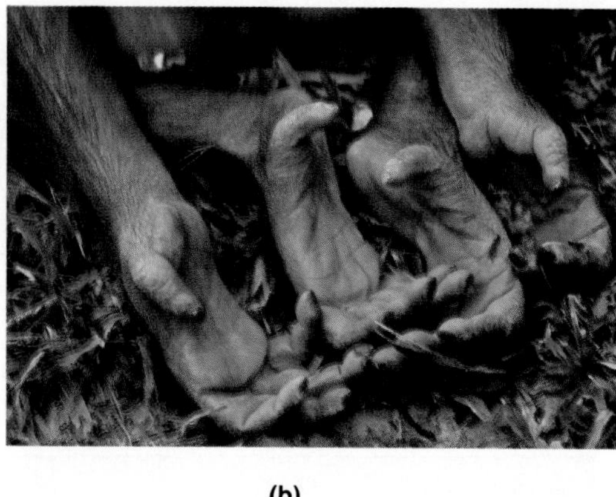

(b)

rather large canine teeth, used primarily in fighting, threat, and defense. Humans have reduced canines and have substituted tools such as rocks, baseball bats, and cruise missiles. An officer and a gentleman would never bite the enemy; his canines are too short.

Human Specialization In some cases the specializations that are unique to humans are really only refinements of traits that also exist in other animals. Several primates have an opposable thumb that can touch the fingers, but ours has a slightly different musculature and is much more dexterous. With this manual ability we can make precise tools, and although other animals make and use very primitive tools, none can build a watch (Figure 32.31a).

And then there is our upright posture, which is quite unlike any other modern primate's. Savanna chimpanzees stand to see over tall grass, and many chimpanzees walk bipedally (on two legs) a short way, but normally they move on all fours. It is amusing to see a chimpanzee or other primate running on its hind legs because it is so humanlike but ungainly (Figure 32.31b). Humans, however, are beautifully adapted to bipedal walking and running, surpassing any other vertebrate in this accomplishment. We owe this ability partly to our highly specialized foot, with its arched construction, and partly to our enlarged *gluteus maximus* (buttock) muscles.

We pride ourselves on our intelligence, but its evolution has not been without cost. The large skull that houses our brain produces all sorts of birth difficulties. And, even more than most other mammals, we are born in a helpless state. A baby hare will lunge and hiss at an intruder, and some baby birds will crouch and "freeze" at the mother's warning call. But we are more defenseless, and our newborns seem totally witless.

In a sense, humans have sacrificed a "prewired" brain, one that is programmed to react in certain ways under given situations, for one with greater potential. That is, we rely less on genetically based instincts and more on raw intelligence than do other species. Certainly this makes it possible for us to cope with a more variable and unpredictable environment. But we must undergo long periods of learning and therefore have evolved extended parental care. The fact that we have adapted to our world through intelligence, learning, and nurturing (rather than, say, having thousands of children and leaving their survival to chance, as some organisms do) has undoubtedly altered our social patterns to an enormous degree.

Aside from our magnificently overgrown brain, the principal and unique specialization of the human species is our capacity for language. Language is not easily separated from intelligence because much of our brain is directly related to this function. Despite highly publicized and controversial studies, there is no evidence whatsoever that dolphins or apes can tell each other about their experiences in any but the most basic terms. The capacity for creating abstract symbols to communicate concrete ideas, which is what language is, appears to be absent in even the most highly trained animals.

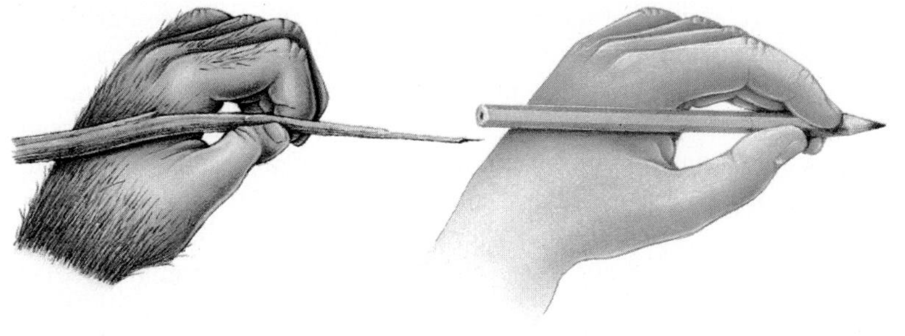

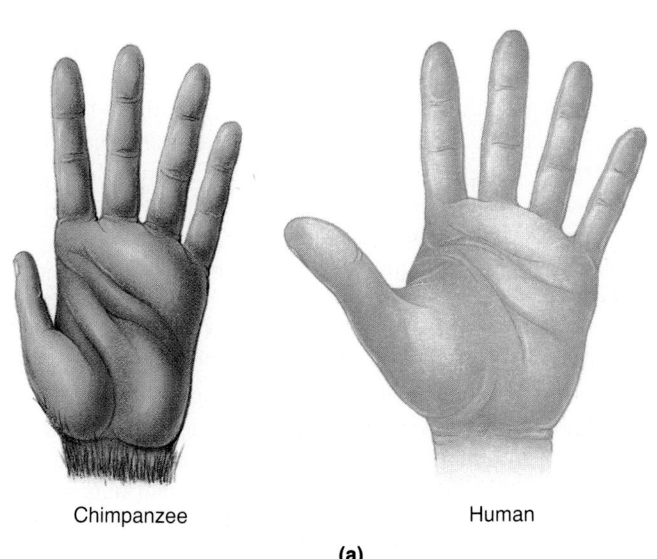

(a)

(b)

FIGURE 32.31
SMALL BUT IMPORTANT DIFFERENCES
(a) The hands of humans and chimpanzees appear generally similar, but important differences include the length and musculature of the thumb. The thumb on a chimp's hand has an "out of the way" location. This is important to a brachiator but makes the handling of tools far less precise. **(b)** While chimpanzees can assume a bipedal posture, neither hip nor leg structure supports this very well.

With language, evolution entered a new dimension. Humans are capable of *cultural* evolution in addition to genetic evolution. We can tell our children the truths (and lies) that we learned from our parents. Factual information can be transmitted via word of mouth over tens of generations; the historically accurate account of the siege of Troy, handed down by illiterate singing bards for hundreds of years before being committed to paper, is but one example. Language also allows for politics: even unlettered tribes manage social organization far above the level that could ever be achieved by gibbering apes. Written language, of course, has extended the scope and power of cultural evolution by yet another order of magnitude.

HUMAN EVOLUTIONARY HISTORY

As is usual when humans try to assess themselves, their genes, or their heritage, the air becomes full of indignation and accusations. The issue has been a volatile one since Darwin's day, when the very idea that humans had an evolutionary ancestry at all was offensive to many. Even Alfred Wallace, the codiscoverer of the principle of natural selection, couldn't bring himself to believe that human beings had originated in the same way other species had. Darwin himself firmly believed that humans and apes had evolved from a common ancestor, but there was no fossil evidence for this contention. Opponents of the idea of human evolution made much of the absence of a "missing link" and continued to make much of it long after many hominid fossils had been discovered.

In the early 1960s, biochemists Allan Wilson and Vincent Sarich began comparing primate proteins and DNA sequences. Their data clearly indicated that humans were more closely related to the chimpanzees and gorillas of Africa than to the orangutans of Asia. They were aware that molecules in these substances change due to mutation at a steady and dependable rate and thus provide us with a kind of "molecular clock." This clock suggested that humans and African apes diverged from a common ancestor not more than 6 million years ago—a startling notion at the time. Anthropologists had routinely assumed a much more ancient divergence and were mostly hostile to the new ideas. Another furious, often boring scientific feud was underway. However, other data, both biochemical and paleontological, backed up the biochemists rather than the traditional anthropologists, causing us to adjust our notions of human lineage. Let's review the current theory.

The Australopithecines

About 2 to 3 million years ago, a number of humanlike forms lived in relatively dry open grasslands of Africa. Their fossils have been given a great number of contradictory names, but when the dust settles, five early species can be recognized: *Australopithecus robustus, Australopithecus africanus, Australopithecus boisei, Australopithecus afarensis,* and *Homo habilis.* It is not conclusively established that any of these forms was directly ancestral to *Homo sapiens,* but it is clear that the first three *Australopithecus* species were widespread and successful, and persisted almost unchanged until at least 1 million years ago, long after the genus *Homo* was well established. Still, our own ancestors could have evolved from early offshoots of the *Australopithecus* line.

The australopithecines (members of the *Australopithecus* line) were rather small-boned, light-bodied creatures, about 1 to 1.5 m (3.5 to 5 ft) tall. They had humanlike teeth and jaws with small incisors and canines, and they walked upright (Figure 32.32). They apparently hunted baboons, gazelles, hares, birds, and giraffes. Their weapons were clubs made from the long bones of their prey. (Near some *Australopithecus* bones, anthropologist Raymond Dart found a fossil baboon skull with a fossil antelope leg bone jammed into it.) The australopithecine **cranial capacity** (a measure of brain size) ranged from 450 to 650 cc, as compared with 1200 to 1500 cc for modern adult humans.

The first specimen of *A. afarensis,* the oldest known hominid (dubbed "Lucy" after a Beatles song from the 1960s), was found in the northern Ethiopian desert region in 1974 by Donald Johanson. The fossil remains, estimated to be over 3 million years old, consisted of little more than half of the skeleton of an upright-walking female. The following year, Johanson's luck blossomed. He unearthed the remains of 13 more "Lucy" types, all in one area. After several years of puzzling over these slightly built creatures, Johanson reached a decidedly unorthodox yet insightful conclusion. He proposed that *A. afarensis* was indeed more primitive than other australopithecines, placing it at the base of the hominid tree. The idea, hotly contested at first, is now well accepted by many researchers.

FIGURE 32.32
EARLY HOMINIDS
The australopithecines were small but muscular creatures. Their heads were more apelike than human in appearance, with a low cranial profile and little or no chin. The jaws were large and forward-thrusting.

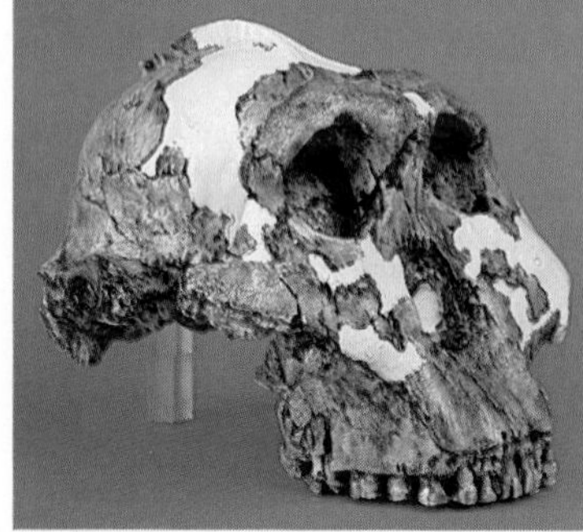

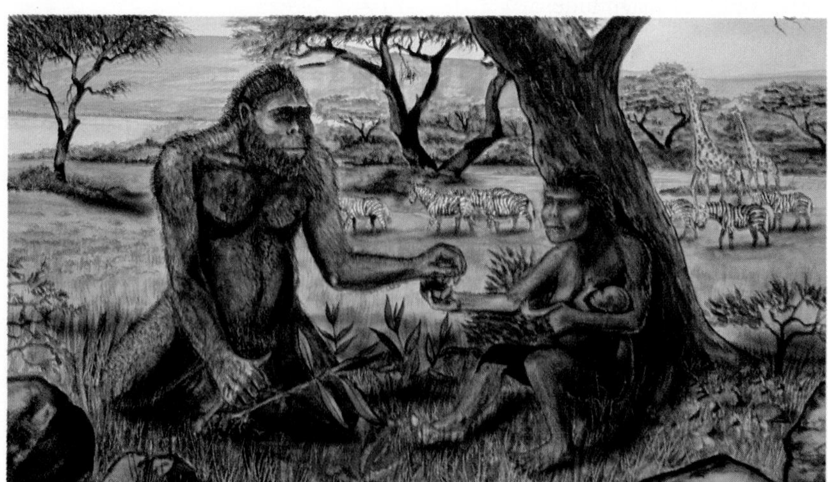

Carolina Biological Supply

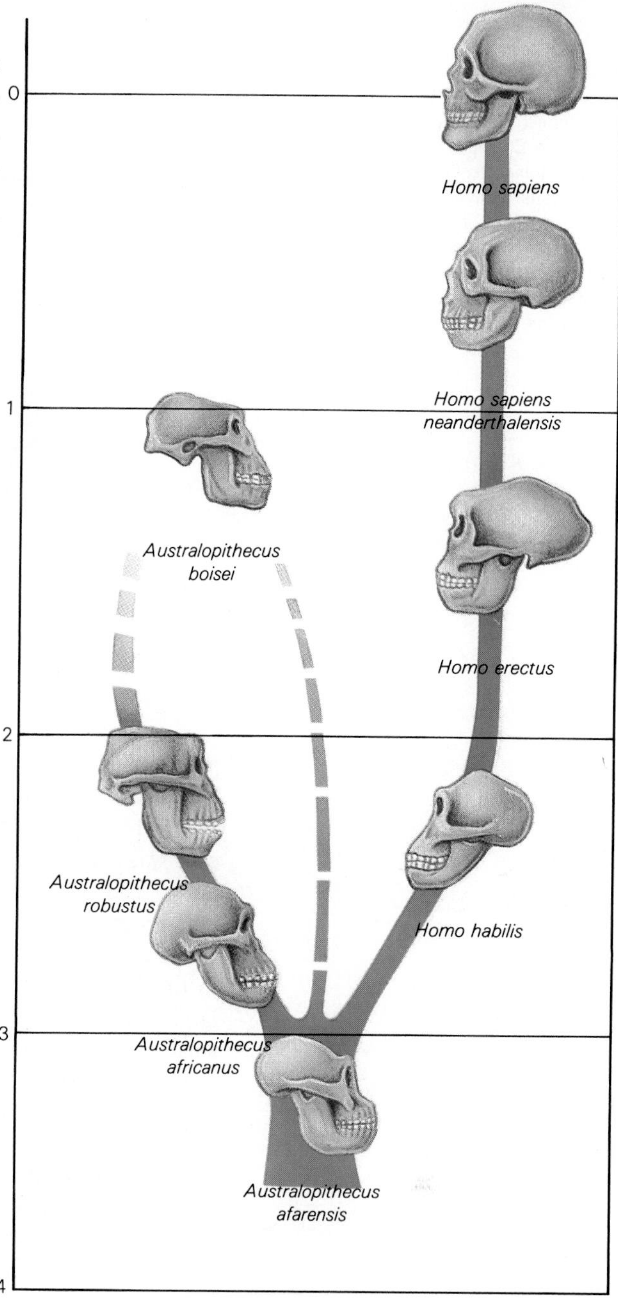

FIGURE 32.33
HOMINID HISTORY
The known hominid history spans a period of nearly 4 million years. In a scheme based partly on the conclusions of Donald Johanson, a line represented by *Australopithecus afarensis* produced a branch containing other australopithecines and another branch leading to *Homo*.

A number of physical anthropologists go along with the scheme seen in Figure 32.33. According to this scheme *A. afarensis* produced two or three branches, one of which led to the genus *Homo*, which includes modern humans. The other line or lines produced *A. africanus, A. robustus,* and *A. boisei.*

A. boisei and *A. robustus,* as the latter's name implies, were larger boned—though not taller—than *A. africanus,* the oldest of the three. Most strikingly, the two newer arrivals had much larger jaws and teeth with greatly expanded cheek bones to accommodate the massive jaw muscles. These features reach their extreme in *A. boisei.* From a traditional view this progression may seem to be backward—toward an apelike condition—but other physical evidence and dating are to the contrary. The shape of the teeth and the curve in the jaws are decidedly human, not apelike. The tooth wear, incidentally, suggests that much of the diet of these australopithecines consisted of plant food.

The Human Line

Homo habilis is the name given by famed anthropologist Louis Leakey to certain fossils from the Olduvai Gorge of East Africa. As the name indicates, these fossils seem to be close to the modern human form, with an average cranial capacity of about 656 cc. Critics have argued that the new find is just a variant of *Australopithecus* and that Leakey was unjustified in trying to place his fossils within the genus *Homo*.

More recently, Leakey's widow Mary and his son Richard found a hominid skull of unusual interest at Koobi Fora. At 1.6 million years old, it is older than some *Australopithecus* fossils, but it appears to be much closer to the human line than is *Australopithecus*. For one thing its cranial capacity—over 800 cc—is greater than those of either *Australopithecus* or the earlier disputed *Homo habilis* finds. The Leakeys, apparently fed up with the sterile arguments that have raged over the naming of hominid fossils, simply identified their unique find by its arbitrary field identification: fossil skull 1470.

Homo erectus The fossil beds of the eastern shore of Lake Turkana have more recently yielded fossils that are closer to the modern human. The new species has been called *Homo erectus* and is considered to be an extinct member of our own genus. They are most interesting in that they appear to be of the same species as some of the first hominid fossils ever found—those once called "Java Ape Man" and "Peking Man" and designated *Pithecanthropus*. The earliest fossil *Homo erectus* skulls are known to be more than 1.5 million years old. But the Lake Turkana fossils were only about half a million years old, and some *Homo erectus* fossils may be even less than 200,000 years old. In other words, *Homo erectus* flourished, relatively unchanged, for well over a million years. During its first 300,000 years it coexisted with other, more primitive hominid species, including the australopithecines and possibly *Homo habilis*. In at least the last 100,000 to 200,000 years of its existence, *Homo erectus* overlapped with yet another upstart species, *Homo sapiens*. Fossils of *Homo erectus* have been found in China, Europe, and southern Africa, and of course, East Africa (Figure 32.34). The evidence left by *Homo erectus* provides fascinating grist for the mills of our imagination.

Homo erectus had a small brain, heavy brows, and strong jaws and teeth. But from the neck down, they apparently were very much like us. In addition to their fossilized bones, *Homo erectus* left behind crude stone tools. The tools, consisting of hand axes and scrapers, are designated **Lower Paleolithic** ("Old Stone Age"). There is evidence that these Lower Old Stone Age people built shelters, hunted small game, and gathered

FIGURE 32.34
HOMINID DISTRIBUTION
The richest australopithecine finds are located in Ethiopia, Tanzania, and Kenya. *Homo erectus* was far-ranging, with fossils recovered in eastern and southern Africa, Europe, China, and Indonesia. Neanderthal fossils have been unearthed throughout much of Europe, the Mideast, North Africa, and China.

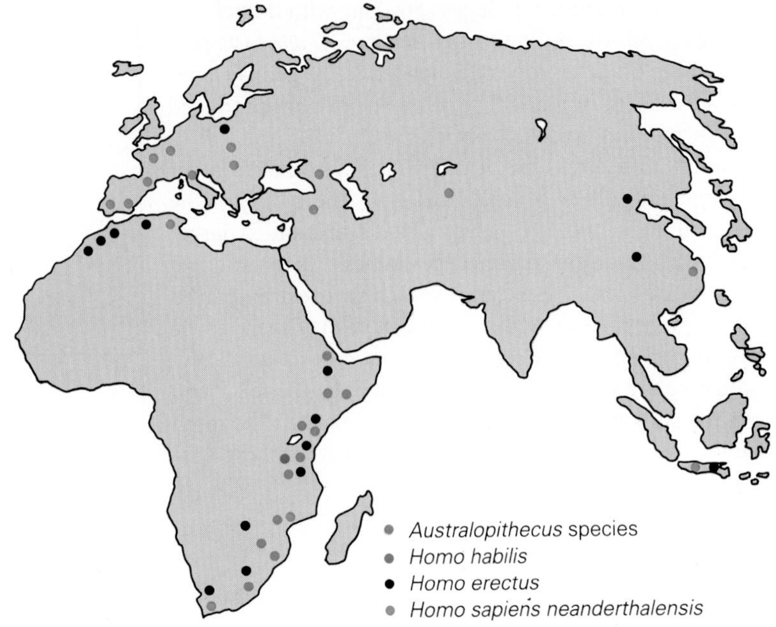

plant foods. The era of *Homo erectus* apparently ended just after the first glaciers of the Pleistocene receded.

Recent findings at a well-known site at Zhoukoudian, China, reveal that one large cave was continuously inhabited by *H. erectus* for over 200,000 years, and finally abandoned 230,000 years ago. During this incredibly long period of continuous residence, both the anatomy and the technology of *H. erectus* changed significantly. Skulls unearthed in the oldest debris of the cave revealed a cranial capacity of about 915 cc, while those in the most recent layers had reached 1140 cc. The stone tools had progressed from large, crude, hastily fashioned choppers and scrapers in the oldest, deepest layers, to smaller, much more refined tools in the newer surface layers. The cave inhabitants used fire from the start, hunted both large and small game, and ate a variety of nuts, fruits, seeds, and other plant matter.

Neanderthals and Us Even while Darwin was studying and writing at his country estate, workers in a steep gorge in the Valley of Neander (in German, *Neanderthal*) were pounding at something that turned out to be a skeleton, inadvertently smashing it to bits but leaving enough for researchers to see evidence of a new and different kind of human. (A similar skull had been unearthed at Gibraltar a few years earlier but had not created much of a stir.)

Scientists are rarely at a loss for words, and an explanation was immediately forthcoming. A professor Mayer of Bonn examined the heavy-browed skull and proclaimed that the skull and bone fragments belonged to a Mongolian cossack chasing Napoleon's retreating troops through Prussia in 1814; an advanced case of rickets had caused him great pain and his furrowed brow had produced the great ridges; he was so distraught he had crawled into a cave to rest, but alas, had died there. This scenario has been rejected in favor of the idea that the bones were those of a member of an early form of human that became extinct—the **Neanderthal Man.**

Members of *Homo sapiens neanderthalensis* thrived, or survived, until around 40,000 years ago. They left a rich record of their fossil remains and some indication of their tools and culture. Their geographic range was large, including Europe, Asia, and Africa (see Figure 32.34).

Neanderthals had large brow ridges and sloping foreheads but were not as radically different from modern humans as was once believed. Their chins did not protrude as far as those of *Homo sapiens*, but their necks and bodies—when they didn't have rickets—

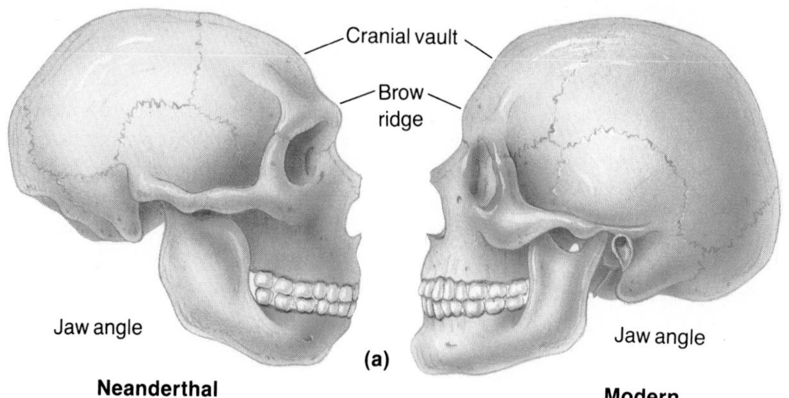

FIGURE 32.35
FACE-TO-FACE CONFRONTATION
A modern human and Neanderthal reveal many general similarities and some striking differences. **(a)** Notable in Neanderthal is the slope of the cranial vault (forehead), the large brow ridges, and receding lower jaw. The jawbone itself has less of an angle where it curves upward to form its articulation with the skull. **(b)** The facial reconstruction produces a decidedly modern human image.

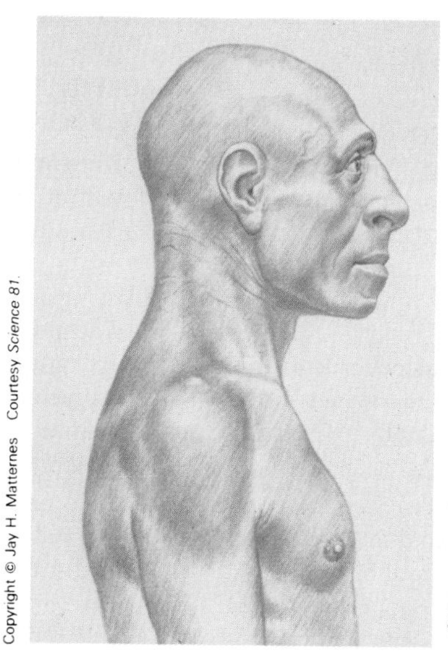

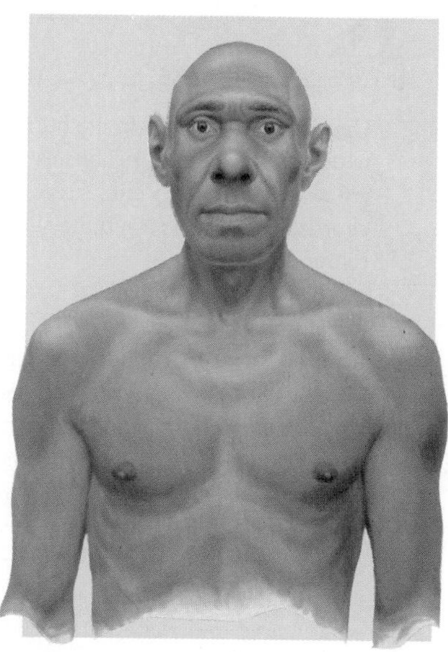

(b) Neanderthal

were like ours and indeed like those of *Homo erectus* (Figure 32.35). There is no way to know whether they had a spoken language. However, we do know, from fossil pollens found in their remains, that Neanderthals sometimes covered their dead with flowers before burial. Whether or not that's sophisticated, it certainly is human—touchingly so.

Neanderthal-type fossils are not found in rocks younger than about 40,000 years. Why did the Neanderthals disappear? One school has proposed that the classic Neanderthal line and the line of modern humans diverged as long ago as 250,000 years, perhaps when early *Homo sapiens* was separated geographically by the great glaciers of the Pleistocene. About 40,000 years ago, the scarce *Homo sapiens sapiens*—the modern human—suddenly burgeoned, and Neanderthal became extinct, possibly by being wiped out by *Homo sapiens sapiens,* a subspecies famous for its eagerness to go to war.

A second theory minimizes the differences and states that *Homo sapiens* in the Neanderthal age was one highly variable species that included both the classic Neanderthal types and individuals much like ourselves. This theory places the Neanderthals squarely in the mainstream of human evolution. Modern humans, the theory maintains, simply evolved from somewhat archaic Neanderthal ancestors. Therefore the Neanderthals didn't die out, but merely changed. Those who believe this interpretation suggest that earlier anthropologists simply put too much emphasis on some extremes of normal variation. Much of the remainder of human evolution is called *history* and is within the domain of historians and archaeologists.

APPLICATION OF IDEAS

1. Compare a bony fish (a vertebrate) with the lancelet (a cephalochordate). In the comparison try to determine which characteristics in each represent primitive or advanced states.
2. The Cenozoic era is often referred to as the Age of Mammals. Did the mammals evolve in the Cenozoic? If not, why the label? When did mammals first appear, and what did some of the earliest representatives look like?
3. Suggest three anatomical differences that clearly distinguish humans from other primates. Are the human characteristics abrupt departures from general primate characteristics, or are they differences in degree? Cite several examples. Comment on how some of these differences may have been responsible for the uniqueness of human accomplishments.

KEY IDEAS

PHYLUM HEMICHORDATA

Hemichordates include the burrowing acorn worm, named for its conical **proboscis** protruding from a **collar**. While its **gill slits** suggest its relatedness to chordates, its larvae are quite similar to those of echinoderms.

PHYLUM CHORDATA

1. Chordates—the **Urochordata, Cephalochordata**, and **Vertebrata**—have gill slits, **dorsal, hollow nerve cords**, a **notochord**, and a **postanal tail.**
2. In primitive chordates, the gill slits are used as sieves, while in fishes, the **gill arches** between them support the gills. The dorsal, hollow nerve cord in chordates contrasts to the ventral, solid nerve cord of many invertebrates. The notochord, a firm supporting rod, is seen mainly in the chordate embryo. The postanal (after the anus) tail is present in most adult chordates and in the embryos of the rest.

Subphylum Urochordata: Tunicates and Salps

1. **Tunicates (ascidians** or sea squirts) are the best-known urochordates, but the subphylum includes **salps** and **larvaceans.**
2. The name *tunicate* comes from the saclike body covering, or **tunic.** While the swimming, tadpolelike, larval tunicate has all of the chordate traits, the only one remaining in the stationary adult is the **gill sac.**
3. Larvaceans retain the larval form even when sexually mature, a condition called **neoteny.**

Subphylum Cephalochordata: Lancelets

1. Cephalochordates **(lancelets)** are fishlike in shape and exhibit all of the chordate characteristics. Segmentation is revealed in the **myotomes**, repeated muscle units in the body wall.
2. Lancelets use cilia for filter-feeding and have a complex mouth surrounded by tentacles. Gases are exchanged when water, leaving the mouth, passes over the gills and out of an opening in the surrounding **atrium.**

SUBPHYLUM VERTEBRATA: ANIMALS WITH BACKBONES

Vertebrates have a vertebral column, cranial brain, gills or lungs, no more than two pairs of limbs, one pair of eyes, kidneys, and separate sexes.

Class Agnatha: Jawless Fishes

1. The oldest vertebrate fossils are **agnathans,** half-billion-year-old armored jawless fishes called **ostracoderms.** Living remnants of the class are the eel-like jawless and boneless **cyclostomes**—the lampreys and hagfish.
2. Parasitic lampreys feed via a rasping sucker mouth. Normally, the adults live at sea but reproduce in rivers and streams, where a long larval stage develops.

Class Placodermi: Extinct Jawed Fishes

The extinct **placoderms,** armored fishes with hinged jaws, arose during the Silurian, giving rise to the cartilaginous and bony fishes.

Class Chondrichthyes: Sharks and Rays

1. Chondrichthyans have cartilaginous skeletons, an asymmetrical tail, a dorsoventrally flattened body, and unique, **placoid scales**—miniature but true teeth.
2. The digestive, reproductive, and excretory systems of sharks and rays open into a common **cloaca,** which leads to the external **vent.** Digestive surface in the gut is increased by a **spiral valve.**
3. Sharks have a highly developed **lateral line organ,** which senses water movement. As they swim, water passes over the gills where gases are exchanged.
4. Sharks utilize internal fertilization, and while some shark species are egg-layers, others retain the embryos throughout development. A few species provide the embryo with nutrients and remove metabolic wastes.

Class Osteichthyes: The Bony Fishes

1. The diverse bony fishes of class **Osteichthyes** are often highly maneuverable, with a mucus-covered body surface that helps eliminate drag. Most have an

adjustable **swim bladder** for maintaining buoyancy at various depths. Its gases are generated in a **gas gland** and removed by a **reabsorptive area.**

2. Gas exchange occurs in thin-walled **gill filaments,** supported by gill arches with toothed **gill rakers,** located in two **gill chambers,** each protected by an **operculum.** In most bony fishes water is pumped over the gills. Blood flowing through fine vessels in the gills receives oxygen, which is then distributed to the body. Respiration is supported by circulation in a two-chambered heart.
3. Most bony fishes are egg-layers, and external fertilization is common. A few retain the eggs during development.
4. The bony fishes arose from an air-breathing ancestor to form the subclasses **Actinopterygii** (ray-fins), **Dipneustei** (lungfishes), and **Crossopterygii** (lobe-fins). The vast majority are ray-finned fishes, while only three species of lungfish persist. They breathe by gulping air and forcing it into the swim bladder. Only one species of lobe-finned fish, the **coelacanth** *(Latimeria chalumnae)* is known. The lobe-finned line is ancestral to the terrestrial vertebrates.

Class Amphibia: The Amphibians

1. Amphibians today occur in three orders: **Urodela** (salamanders), **Anura** (frogs and toads), and **Apoda** (ceacilians). Amphibians and reptiles arose in the Carboniferous period.
2. Most amphibians have lungs, but to some extent also use their moist, vascularized skin for gas exchange. Gas exchange is also facilitated by the presence of a new circuit through the lungs, made possible by a third heart chamber.
3. Reproduction in amphibians ranges from simple external fertilization and development in water to internal fertilization, by actual mating or by the transfer of sperm packets **(spermatophores).**

Class Reptilia: The Reptiles

1. Specific adaptations of **reptiles** for dry terrestrial life include internal fertilization and development in the independent, self-supporting, **amniotic egg** named for the amnion, a membrane surrounding the embryo. It contains a food supply and supporting **extraembryonic membranes** developed by the embryo. There are a dry, scaly skin; complex, spongy, protected lungs; and, because reptiles are **ectothermic,** a behavioral repertoire for avoiding the effects of extreme temperatures.
2. Reptiles arose from **cotylosaurs** ("stem-reptiles") to become the prominent vertebrates during the Mesozoic (the "Age of Reptiles"). The highly diverse group included **pterosaurs,** large flying reptiles, some of which may have been endothermic (warm-blooded).

Class Aves: The Birds

1. The earliest known bird, *Archaeopteryx,* appeared in the early Jurassic. Reptilian traits include scales on the legs and the general reptilian skeletal framework.
2. Flight modifications in birds include forelimbs modified into wing bones, a light, strong skeleton with fused units and reduced tail, a light beak (no teeth), enlarged breastbone with a keel, perching and grasping feet, and scales modified into feathers used for flight surfaces and insulation.
3. Birds are endothermic, a condition supported by efficient circulatory and respiratory systems. The heart is four-chambered, and the pulmonary (lung) and systemic (body) circuits are completely separated. The lungs are unusual among vertebrates in that they provide for a one-way flow of air. Numerous air sacs are present.

Class Mammalia: The Mammals

1. Mammals, named for the **mammary glands** (milk glands), produce hair and are endothermic. Nearly all young develop in the uterus supported by the **placenta.** Breathing is assisted by a muscular diaphragm. The jaws are quite mobile and contain socketed teeth that vary widely with feeding and dietary habits and are thus important to classification.
2. The mammalian brain, with its enlarged cerebrum and newly evolved **corpus callosum,** is large and versatile and provides for more learning than that of other vertebrates.
3. Mammals originated from the **therapsids,** flesh-eaters whose limbs were modified for running on all fours.
4. By the Cenozoic, the Age of Mammals, three groups had diverged: monotremes (egg-layers), marsupials (pouched mammals), and placentals. The first is minor today with only a few species, while the second, the marsupials, are concentrated in Australia. Placental mammals, those with a true placenta, make up the majority.
5. Order **Primates** includes the more primitive Prosimii and the Anthropoidea. The latter is subdivided into New World monkeys, Old World monkeys, and apes and humans. Apes and humans make up the superfamily **Hominoidea,** or the **hominoids.** The human family, **Hominidae,** including extinct forms, are the **hominids.**
6. Primates are generalized mammals whose characteristics include long forelimbs (arms) and prehensile hands with fingernails rather than claws. Some have prehensile tails. Vision is binocular and stereoscopic, and the brain is highly adapted to learning. Most are omnivores.
7. Human specializations include an opposable thumb that is more versatile than that of other primates, a naturally upright posture, arched foot, and bipedal gait. Most striking is human intelligence, the capacity for abstract thought, learning, and a facility for language. These are coupled with a very long and dependent developmental period. Humans experience cultural as well as physical evolution.

HUMAN EVOLUTIONARY HISTORY

Theories of human origins have a history of controversy. Newer data from the study of hominoid protein amino acid sequences suggest that humans and apes diverged no more than 6 million years ago.

The Australopithecenes

1. The known fossil hominid line begins with the **Australopithecines,** *Australopithecus afarensis, A. robustus, A. Africanus,* and *A. boisei.* The last three lived up to about 1 million years ago.
2. The australopithecenes were about four feet tall, had humanlike teeth, walked upright, and hunted with crude bone weapons. Their cranial capacity was somewhat larger than the gorilla's. *A. robustus and A. boisei* had a larger frame and larger jaws and teeth than the others.
3. Anthropologist Donald Johanson maintains that *A. afarensis* is the most primitive of the genus, suggesting that it is ancestral to both the australopithecenes and humans.

The Human Line

1. The Leakey discovery, *Homo habilis,* is tentatively placed near the start of the human line. Somewhat closer to the human line is skull 1470, another Leakey find.
2. Fossils of *Homo erectus* are found throughout much of the Old World, persisting for as long as 1.3 million years and overlapping *H. sapiens.* The body was like a modern human's, but the brain was smaller (about 915-1140 cc) and the brows and jaws larger. Their tools are classified as **Lower Paleolithic** (Lower Old Stone Age).
3. The earliest *H. sapiens* were the **Neanderthals** *(Homo sapiens neanderthalensis),* who ranged across the Old World as recently as 40,000 years ago. Their bodies were like those of modern humans, and their heads were similar except for a receding lower jaw and a sloping crown. All fossils since the Neanderthal's demise are *Homo sapiens sapiens* (modern humans). The fate of Neanderthal is unknown, but they may have simply become *H. sapiens.*

REVIEW QUESTIONS

1. What characteristics of the acorn worm suggest that it is related to both the echinoderms and the chordates? (718–719)
2. List and briefly describe the four leading chordate characteristics. Which appears in the embryos of humans but not in the adults? (719)
3. Considering what tunicates look like, what is the strongest evidence that they are really chordates? (720–721)
4. Using an example from the urochordates, explain the concept of neoteny. How does this concept help theorists bridge the gap between the earliest chordates and the vertebrates? (720–721)
5. Compare lancelets, juvenile lampreys, and larvaceae. What does the comparison suggest? (721)
6. List five important characteristics of vertebrates. (721)
7. List several characteristics of the ostracoderms. When were they prominent? (722)
8. Name the two survivors of the Agnatha. Describe their mode of feeding. (722)
9. Why are the placoderms so significant in the history of vertebrates? Briefly explain how their major evolutionary innovation came about. (722–723)
10. According to recent thinking, what is the origin of the cartilaginous skeleton? How might it be adaptive to the shark or ray? (723)
11. Describe the digestive system of the shark, beginning with the teeth and ending with the vent. Include the spiral valve. (724)
12. Characterize the wide range of reproductive modes in the shark. Which is most advanced? (724)
13. What is the function of the modern swim bladder? How does the fish compensate for various depths? (724–725)
14. Compare the gill structure of bony fishes and sharks and explain how gills are ventilated in each. (724–725)
15. Name and briefly describe the three groups of bony fishes and rate them in terms of their numbers and importance today. (725–726)
16. In what way were the lobe-finned fishes significant to vertebrate evolution? (726)
17. Name the three groups of living amphibians and list common examples of each. (727)
18. Beginning with the salamanders, compare body stance and limb position in the amphibian, reptile, and mammal vertebrates. (728)
19. What separate circuit does the three-chambered amphibian heart provide that was not present in fishes? Would you expect pulmonary branches of the circulatory system to enter the skin? Why? (728)
20. List two examples of adaptations that permit amphibians to reproduce out of water. (729)
21. List four specific ways in which reptiles have adapted to the terrestrial environment. One should involve the embryo. (729–730)
22. Trace the history of reptiles. When did they originate and in what geological period were they most prominent? Briefly, how is the demise of the ruling reptiles explained? (729–730)
23. List five specific ways in which the bird's body is modified for flight. (732–733)

24. Describe the four-chambered heart and complex respiratory system of birds and explain why both might have been important to the evolution of flight. (733)
25. List three characteristics that are unique to the mammals. (734, 736)
26. Describe five significant variations in the jaws and teeth of mammals, and explain how they relate to food-getting or defense. (736–737)
27. Suggest ways in which the mammalian brain and behavior differs from that of other vertebrates. (737)
28. How might nursing the young have affected learning in mammals, and how might this have continued to influence the evolution of the brain? (General)
29. List the three subclasses of mammals and describe reproductive differences among them. (739–740)
30. List five ways in which the primates are different from other mammals. Why is it difficult to be specific? (740–741)
31. Trace the evolutionary history of the primates, mentioning the original type, and listing the main branches as they occurred. (740–741)
32. List four significant ways in which our anatomy is different from all other primates. (742)
33. Name the australopithecene species, briefly describe them, and suggest how they fit into the hominid evolutionary line. Where does *Homo habilis* fit in? Is it properly named? (744)
34. Describe *H. erectus*, its range, its technology, and degree of success. What happened to *H. erectus?* (745–746)
35. Compare the physical features of Neanderthal to modern humans. (746–747)
36. Discuss two hypothetical fates of Neanderthal. (747)

33

SUPPORT AND MOVEMENT

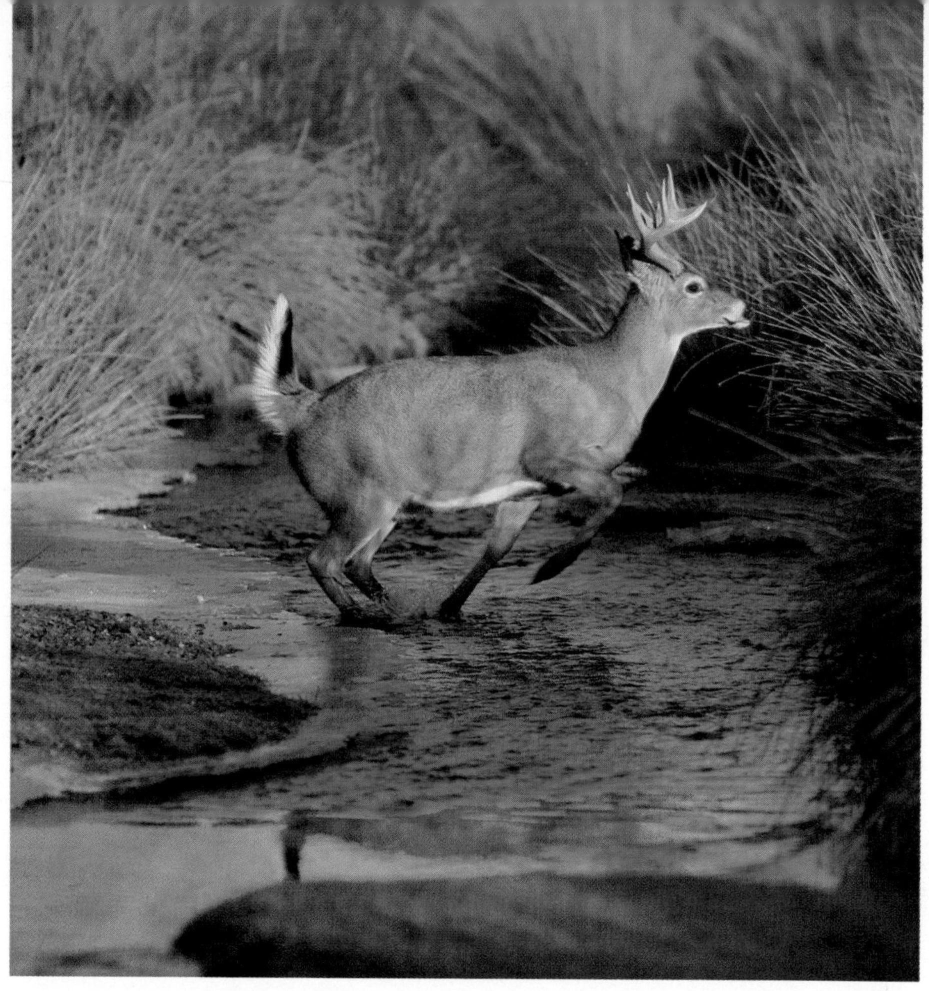

COMPARED TO OTHER CELESTIAL BODIES, THE EARTH IS A RATHER UNSPECtacular place. Here, for example, we lack those fascinating black holes through which astronomers tell us time and space can be stretched. Even our temperatures are moderate compared to those of other planets. And then consider how tranquil the earth's surface is compared to the great storms of Jupiter and the scorching surface temperatures of Venus. Perhaps, in fact, it is precisely because the earth is so moderate and unspectacular that life has formed here. That life, though, has had to face some problems. After all, the earth is not entirely a benign place.

One challenge arises because the earth is not only large, it differs greatly from one place to another. It also changes with time. This means that some areas are more suitable for life than others, and areas that are quite hospitable at one time may be unsuitable (even deadly) at another time. Thus, the success of many animals depends partly on their ability to move from one place to another. Some travel easily, such as the plover, a bird whose migratory distance between its wintering grounds and breeding grounds would take it around the world each year. Other animals, such as the lowly chitons, creep slowly over the wave-washed rocks, scraping off morsels of slimy food as they go (Figure 33.1). Nonetheless, movement is as important to one as it is to the other.

Hand in hand with movement goes the problem of support. This problem arises because massive celestial bodies tend to pull objects toward their surfaces. Not only do apples fall, but animals find it hard to jump. Gravity exerts a continuous tug on our bodies, seeking to flatten us against our great globe. But we resist with muscles that contract and defy that pull, and with skeletons that support our bodies and give our muscles something to pull against. Natural selection has produced in animals an interesting array of systems that serve in support and movement. We will begin our survey of these systems with the invertebrates.

INVERTEBRATE SUPPORT AND MOVEMENT

Many animals have easily recognizable types of skeletons, but some—the soft-bodied animals—seem to have no visible means of support. However, soft-bodied animals are often supported by a **hydrostatic skeleton.** As unlikely as this may sound, such skeletons efficiently maintain form and provide a firm base for muscle movement.

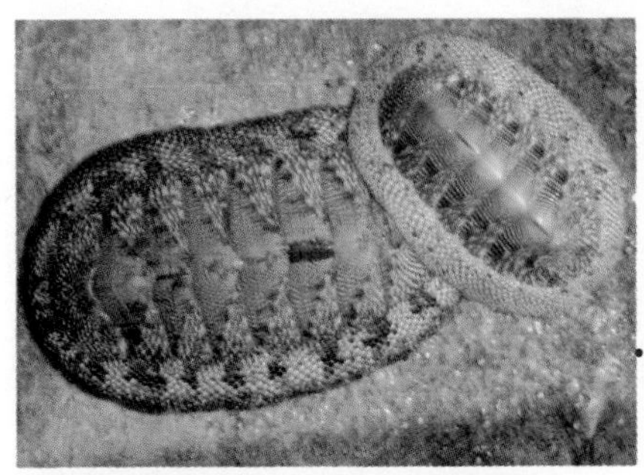

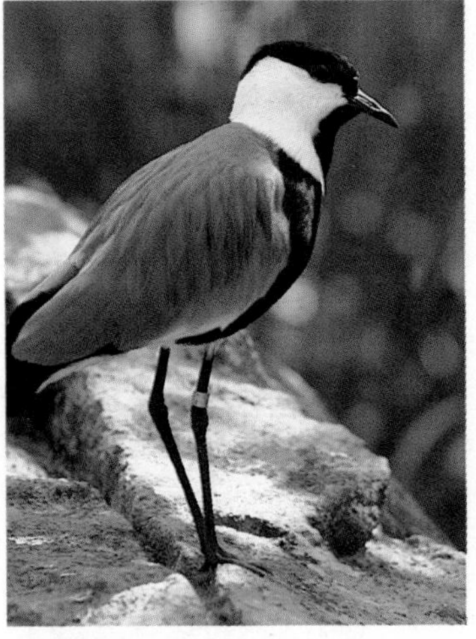

FIGURE 33.1
TWO WORLD TRAVELERS
Both the chiton and the plover use their skeletal and muscular structures in movement—but with quite different results. The chiton's travels, as it creeps along in its rocky marine world, are measured in centimeters. The plover, in contrast is a world traveler capable of migratory and navigational feats that stagger the imagination.

Hydrostatic Skeletons

The hydrostatic skeleton, as its name implies, involves fluids (*hydro,* "water"; *stasis,* "maintenance"). Many soft-bodied animals make use of the pressure within fluid-filled spaces for maintaining body form and for support and movement. We find examples in cnidarians, nematodes, mollusks, and annelids. How can fluids provide for these important functions? Let's consider the simplest example, the hydrostatic skeleton at work in a free-living soil nematode.

Nematodes Recall from Chapter 30 that nematodes are slender, hairlike worms that move through moist soil particles with a whiplike, thrashing motion, the body bending first one way and then the other. In water, where solid particles are absent, their movements are not too effective. Also recall that nematodes have an extensive, fluid-filled body cavity, the pseudocoelom. The body wall surrounding the pseudocoelom contains **muscle fibers** that run longitudinally (along the body), and these are encircled by a tough, elastic, **cuticle** (Figure 33.2). The worm's firmness, and thus its body shape, is maintained through pressure exerted by the cuticle against the noncompressible fluids within. The hydrostatic skeleton is reinforced in these worms by a network of collagen fibers that crisscross the pseudocoelom.

Movement in nematodes occurs as muscle fibers along one side of the turgid body contract, bending the body toward that side and stretching the muscles of the opposite

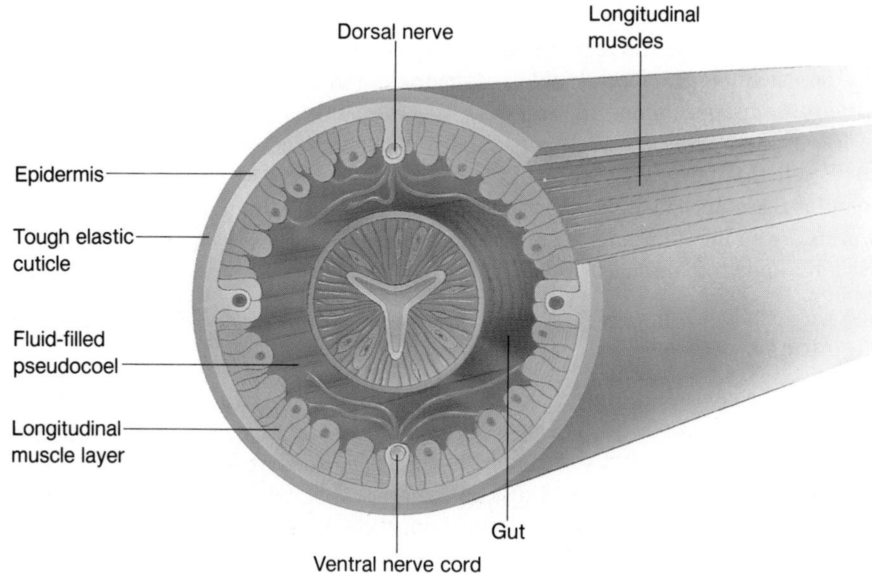

FIGURE 33.2
THE HYDROSTATIC SKELETON
A turgid, fluid-filled pseudocoelom provides the resistant base against which the nematode's longitudinally arranged muscles work. Contraction flexes the body, and the rebounding of the stretched cuticle on the extended side returns it to its former shape. Through coordinated contractions, the undulating movement of the roundworm is produced.

side. Upon relaxation of the muscle, the cuticle rebounds, and the displaced fluids return the body to its original shape. Then, opposing muscles contract, and the body bends the other way. So we see that the trapped fluids and the muscles work in opposition along the body to complete this simple motion. The fluids resist the bending and tend to return the body to its original shape.

Mollusks When digging, the clam makes use of the hydrostatic principle. It pumps blood into its fleshy foot, extending it and forcing it into the soft seabed. There the foot swells, acting as an anchor. Next, powerful retractor muscles draw the heavy body toward the firmly anchored foot. Then, as the hydrostatic pressure is again applied, the foot narrows and is extended again for another cycle of probing, anchoring, and pulling. The clam, of course, wears its skeleton on the outside in the form of its shell, so these hydrostatic properties are involved in movement only.

Annelids The earthworm has a fluid-filled coelom and a muscular body wall, so its hydrostatic skeleton is fundamentally similar to that of the nematode. However, its organization in the earthworm is much more complex.

First, we should keep in mind that earthworms are segmented internally as well as externally. Thus, fluids in each segment are isolated or partitioned off from those in the next. Second, whereas nematodes have only longitudinal muscles, earthworms have two layers: an inner longitudinal layer and an outer circular layer (Figure 33.3). Contraction of the longitudinal layer brings about a shortening and thickening of the body, while contraction of the circular layer extends and narrows the body. (How might it make turning movements?) Where the contraction of two groups of muscles in any animal produces an opposing effect such as this, the muscles are termed **antagonists.** In spite of the name, however, such muscle groups work in a highly coordinated and cooperative manner.

Each of the earthworm's segments has its own bundles of longitudinal and circular muscles and is capable of independent movement. Herein lies the key to the worm's ability to burrow. It begins to burrow by contracting circular muscles in its first segments, and relaxing the longitudinal ones, which extends the forward segments of its body into the soil. Next, stiff bristle-like **setae** are extended outward into the soil, firmly anchoring the first segments. Then, longitudinal muscles in these segments contract as the circular muscles relax, shortening the region and drawing the rest of the body forward (Figure 33.3c). The extending-anchoring-pulling sequence continues with other segments, and the worm soon disappears into the soil.

The use of the hydrostatic skeleton for burrowing, although very efficient, is quite ancient. Recall that burrowing worm tracks represent the earliest evidence of animal life. Some of these ancient trails show marks like those made by the setae of modern earthworms. From these findings it seems that the hydrostatic skeleton was most likely the first kind known to animals.

Exoskeletons

We now leave the concept of watery skeletons to consider those that are more familiar: hard skeletons. These can occur as **exoskeletons**, outside the soft parts, and as **endoskeletons,** inside the soft parts. Exoskeletons and endoskeletons are further distinguished by their embryological development. Exoskeletons are formed by secretions of epidermal cells, derived in the embryo from ectoderm. Endoskeletons in vertebrates are formed essentially by cartilage and bone cells. Such cells originate for the most part in embryonic mesoderm. No matter how the skeleton originates, however, its functions are similar: support, movement and protection.

Hard skeletons function in movement by providing a firm base for the attachment of muscles that move limbs and other parts. As you would expect, such muscles are arranged in opposing pairs referred to earlier as antagonists. Since antagonists create

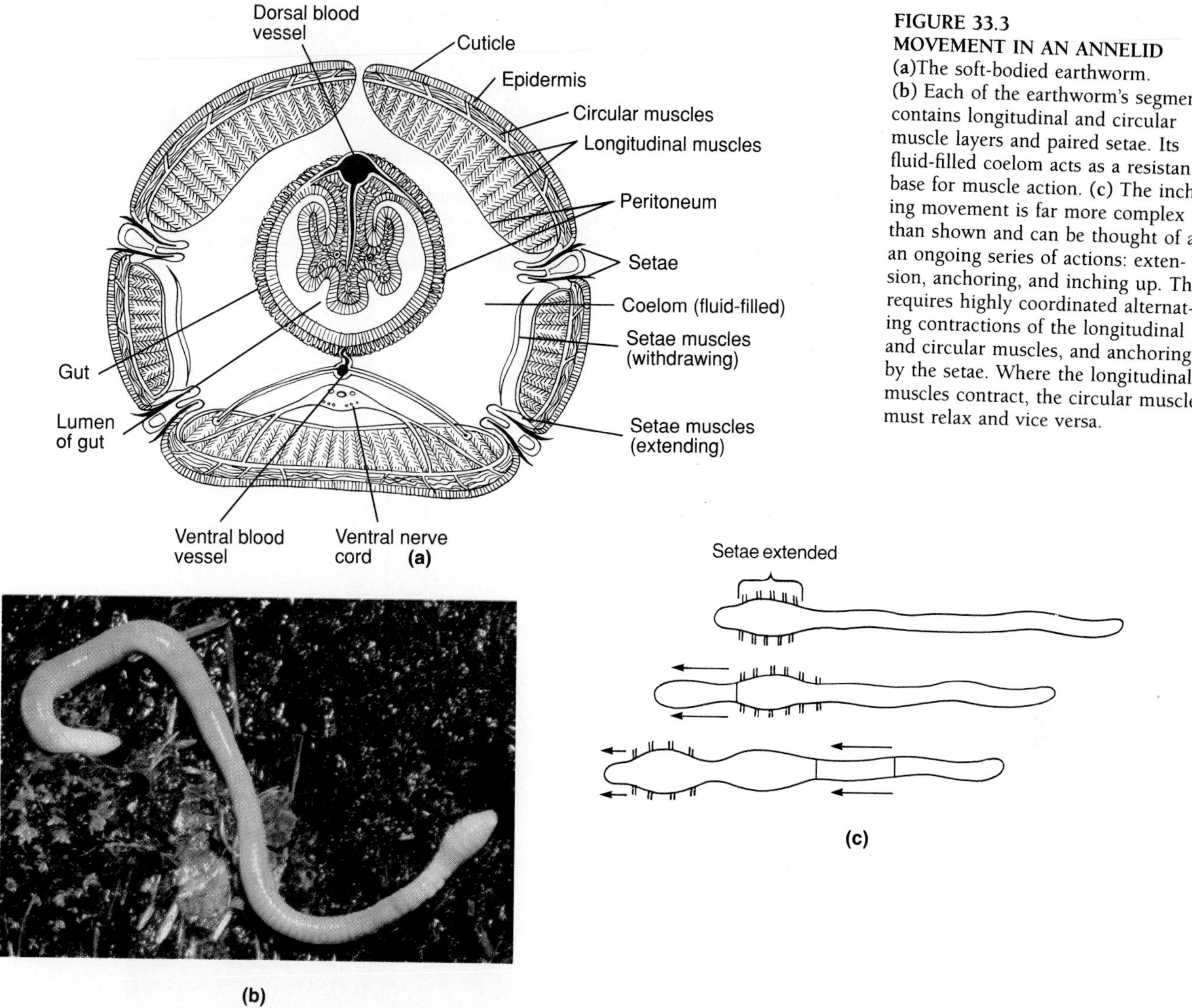

FIGURE 33.3
MOVEMENT IN AN ANNELID
(a) The soft-bodied earthworm. **(b)** Each of the earthworm's segments contains longitudinal and circular muscle layers and paired setae. Its fluid-filled coelom acts as a resistant base for muscle action. **(c)** The inching movement is far more complex than shown and can be thought of as an ongoing series of actions: extension, anchoring, and inching up. This requires highly coordinated alternating contractions of the longitudinal and circular muscles, and anchoring by the setae. Where the longitudinal muscles contract, the circular muscles must relax and vice versa.

opposing movements, one muscle group extends a limb or part, and the other draws it back. Opposing muscle groups are essential, since animal muscles do their work through contraction only. (Muscles pull; they do not push.)

The protective function of an exoskeleton is quite obvious. It acts as a hardened shield around the body. Consider, for example, the protection offered a clam by its shell. Or consider the hardened exoskeleton of the insect.

Arthropods Hardened, jointed exoskeletons are characteristic of the phylum Arthropoda, the gigantic phylum that includes insects, crustaceans, spiders, and other joint-legged types. In addition to providing a firm muscle base and protection for the animal, the arthropod exoskeleton, or cuticle as it is more specifically known, helps retard water loss in terrestrial species. You may recall that its principal constituent is chitin, a complex carbohydrate. In crustaceans (lobsters, shrimp, crayfish, crabs, and so on), the chitin is hardened with calcium carbonate. The cuticle at the joints remains thin and flexible, but tough like leather.

As we see in Figure 33.4, one muscle group flexes a limb, whereas another (antagonistically) extends that limb. Similarly, muscles that bring the chewing mouthparts together oppose those that open the mouth parts.

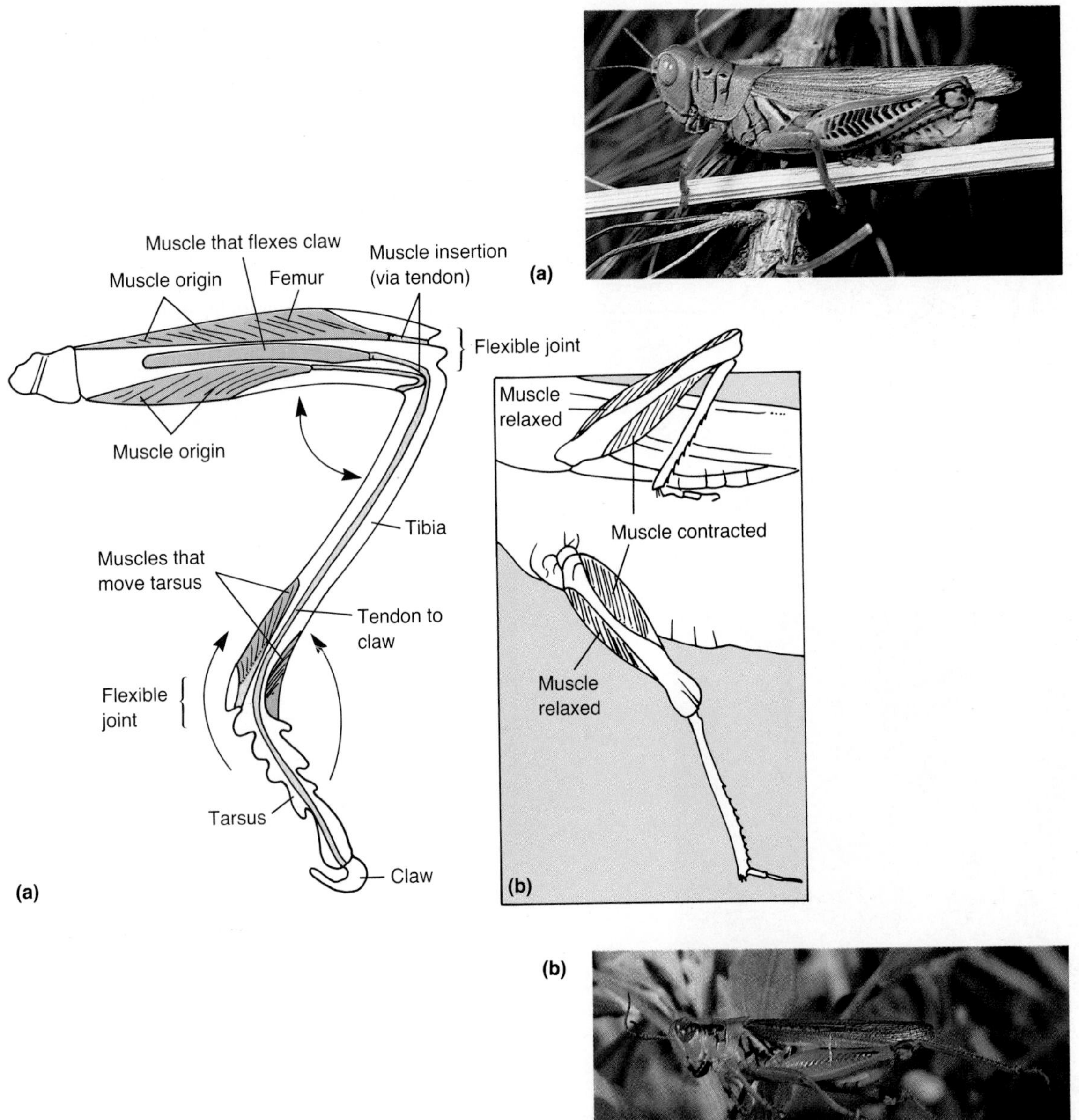

FIGURE 33.4
COCK-AND-RELEASE MECHANISM
The jumping legs of the grasshopper contain powerful muscle groups responsible for its impressive, sudden leaps. The muscles originate within the femur and insert, via tendons, on the tibia below. **(a)** As the muscle on the underside of the femur contracts, it brings the tibia close to the femur, and the insect assumes a crouched position (inset). **(b)** Rapid contraction by the upper muscle extends the tibia with startling suddenness, raising the insect's body as a jump begins. A third muscle in the femur moves the tarsal claw via a very long tendon, whereas a fourth pair of muscles in the tibia flexes and extends the tarsus. The tarsus and claw are used in slower climbing movements.

Each muscle has an **origin** and an **insertion.** The origin is the end of the muscle that is attached to a more stationary base, whereas its insertion is toward the opposite end (the end attached to the more moveable part of the skeleton). In the arthropod limb, muscles originate in an inner segment of the exoskeleton, which may house all of the muscle. The insertion is in the next segment out—the segment to be moved. The connection between the end of the muscle and its point of insertion occurs via a long tendon that passes through the joint and makes a firm attachment to the exoskeleton.

It might seem that exoskeletons would be cumbersome, limiting the movements of the animal (like the solid armor used by knights of the Middle Ages). However, just the opposite is true. Insects, we know, are quite agile and, in fact, are capable of feats of great strength. We've all been impressed by ants dragging burdens that are heavier than they are. And you are all too aware that a flea can jump from the floor all the way up into your pants leg (usually making their presence known just as you make your way into heavy traffic). So, imagine how high a horse-sized grasshopper could jump! The answer is, it couldn't—it would hardly be able to move.

The reason is that arthropod exoskeletons are essentially just hollow, thin-walled, jointed tubes with complex muscle attachments inside. Such tubes are quite strong and light—as long as they remain small. For a substantial increase in size to occur without a loss in strength, the exoskeleton would have to undergo a corresponding increase in thickness.Thin tubes of large diameter would simply buckle when placed under stress. So at some critical size, the sheer weight of an exoskeleton would curtail most movement, and the "armored horse" would just lie there, feebly kicking at the air.

The remarkable leaping ability of some insects can be partly explained by what is called a "cock-and-release" mechanism. It includes a buildup of tension in an appendage and the sudden release of that tension. This mechanism permits the impressive jumps and sudden changes in direction that may have frustrated your last efforts at catching grasshoppers (see Figure 33.4).

Insect Wings and their Muscles. The movement of thin, chitinous, sheetlike insect wings is somewhat unusual. There are no flight muscles in the wing itself, but there are two pairs of powerful muscles in the insect **thorax.** Some species have **direct flight muscles** attached to the wings *directly,* through a complex lever arrangement. In others, those with **indirect flight muscles,** the movement is *indirect* and depends on distorting the cuticle of the thorax, which in turn moves the wing. Both kinds of movement are explained in Figure 33.5.

Small insects can beat their wings at incredible rates. Midges ("no-see-ums") have been recorded at over 1000 wing beats per second—far faster than nerve impulses can be generated. How are the wings activated? Apparently, the muscles themselves are capable of maintaining the cycle between neural impulses, but beyond that supposition, researchers are stalled.

Mollusk Shells Mollusks, such as snails, clams, and nautiluses, make use of both hydrostatic skeletons and exoskeletons. The exoskeletons are, of course, the shells, a hardened covering secreted by the fleshy mantle. The shells of snails and bivalves, such as clams, are important to their protection. In the cuttlefish, a cephalopod mollusk, the shell is internalized and highly specialized for a novel function (Figure 33.6). It has become a flotation device. The **cuttlebone,** as it is called, has many hollow spaces supported by strong calcium carbonate walls. The cuttlefish changes its buoyant state by constantly altering the contents of the cuttlebone's spaces between fluids and nitrogen gas. The resulting buoyancy permits it to remain at a desired depth with little swimming effort. The same principle is applied in modern submarines, where the ballast tanks serve as flotation chambers.

Although the shells of mollusks, like arthropod exoskeletons, are secreted by epidermal tissue, unlike arthropods, mollusks don't shed their exoskeletons as their bodies grow. Instead, their shells grow by ever-widening increments at the edges. Such growth

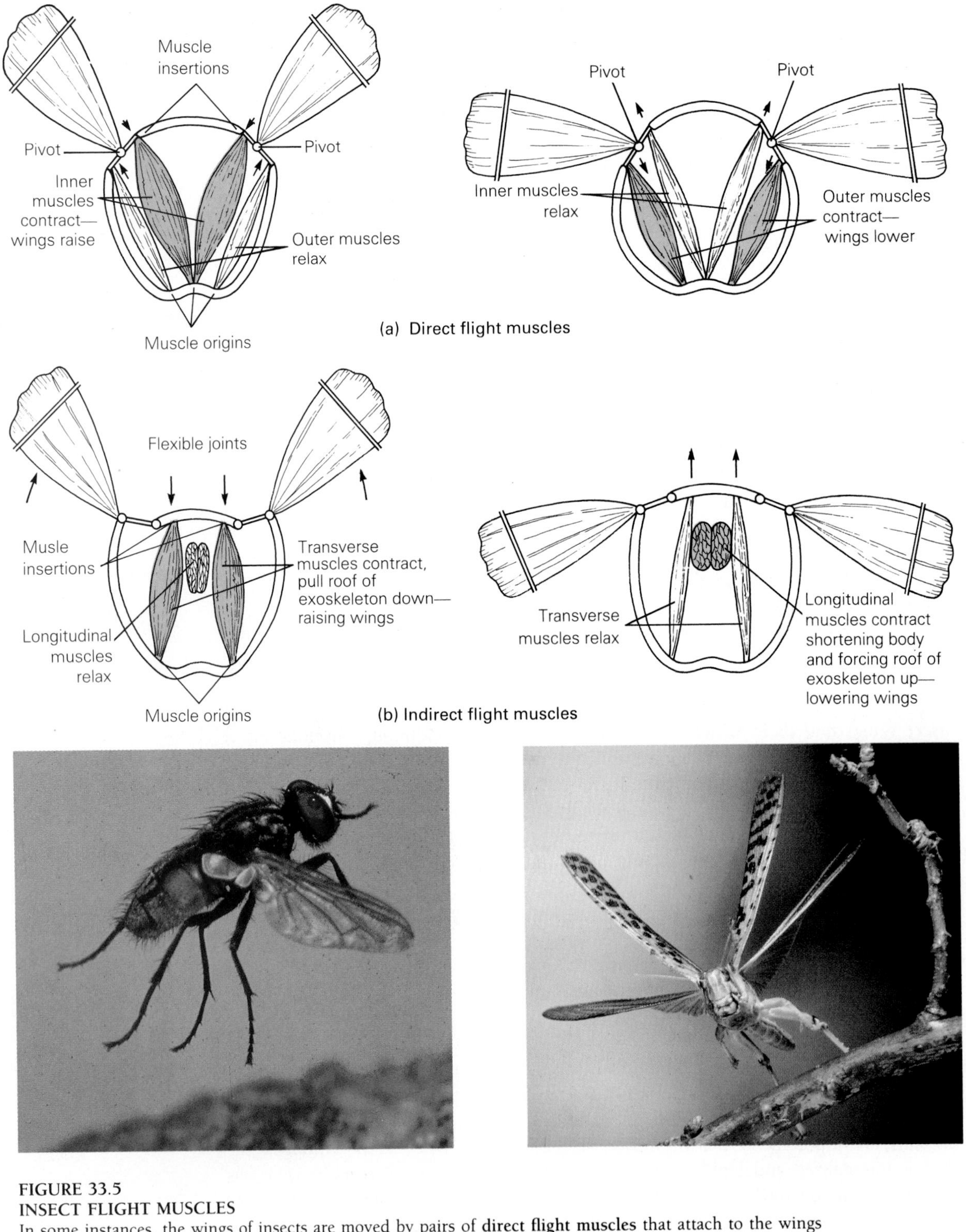

Muscle insertions
Pivot
Pivot
Inner muscles contract—wings raise
Outer muscles relax
Muscle origins
Pivot
Pivot
Inner muscles relax
Outer muscles contract—wings lower

(a) Direct flight muscles

Flexible joints
Musle insertions
Transverse muscles contract, pull roof of exoskeleton down—raising wings
Longitudinal muscles relax
Muscle origins
Transverse muscles relax
Longitudinal muscles contract shortening body and forcing roof of exoskeleton up—lowering wings

(b) Indirect flight muscles

FIGURE 33.5
INSECT FLIGHT MUSCLES
In some instances, the wings of insects are moved by pairs of **direct flight muscles** that attach to the wings themselves via a complex "seesaw" pivot (**a**). One set of muscles raises the wing, while the other lowers it. In other insects, **indirect flight muscles** move the wings by acting on the thorax (**b**). Vertically arranged muscle pairs contract, pulling the roof of the thorax down. As you can see by their attachment, this raises the wings. Downward movement is provided by longitudinal muscles, whose contraction pops the thoracic roof back up. The actual "lift" in insect wing action is provided by the constant changing of wing angles, which describes a figure 8 in each complete cycle. The wings tilt downward on the downbeat, scooping air over the leading edge. During the upbeat, the wings tilt upward, spilling the air over the trailing edge. So far, no one has been able to explain how hovering insects do their stunts.

follows some interesting patterns. In the most primitive mollusks, the shells are simple, straight "dunce cap" cones with new shell added evenly around the edge. In others the conical shape remains but is modified by simple coiling, that is, coiling in a single plane. In the chambered *Nautilus,* for example, new shell growth occurs fastest on one side, so the simple spiralling coil forms. This works fine in a swimming aquatic animal where the weight is supported by water, but a chambered *Nautilus* living on land, or remaining on the seabed, would tend to fall over on its side. Most snails and other gastropods have evolved shells with a complex "leaning coil," where the angle of the coil as well as the size of the shell opening undergoes constant change. Such growth ensures a more equalized distribution of shell and body weight (Figure 33.7). Further, the shell of the land snail is far thinner and lighter than that of its marine counterpart.

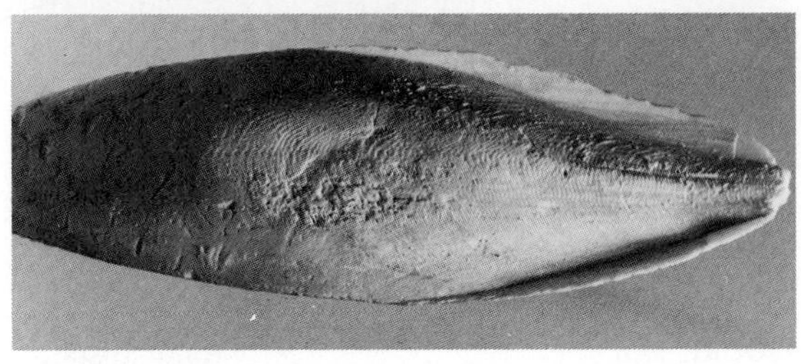

FIGURE 33.6
NEW ROLE FOR A SHELL
The cuttlebone is an internalized exoskeleton of the cuttlefish. This highly porous flotation device is located just behind the head and along the inside of the dorsal surface.

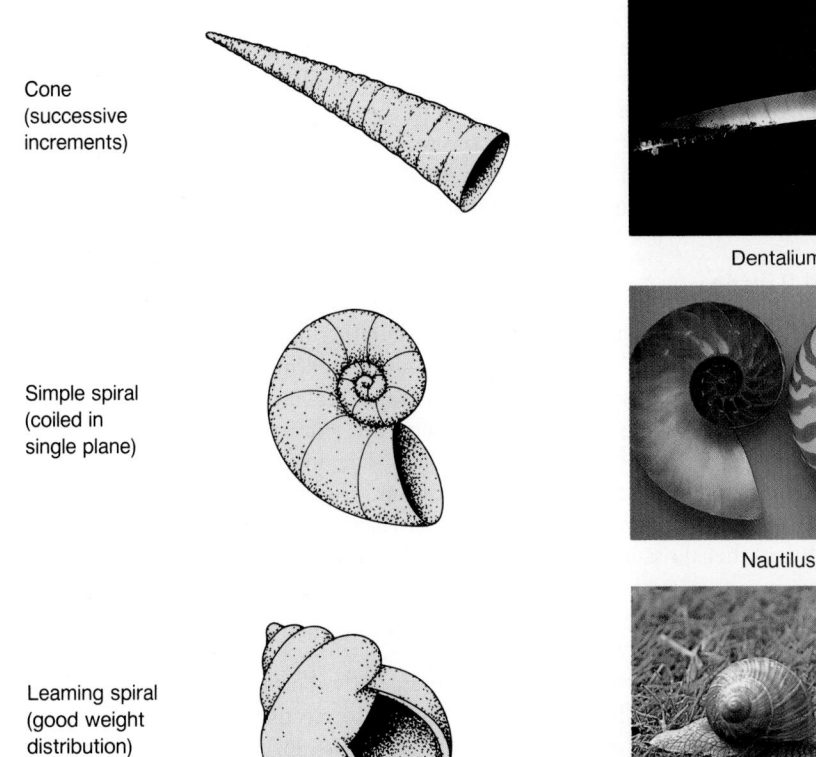

Invertebrate Endoskeletons

Endoskeletons lie within the body and are initially deposited by embryonic mesoderm. We are most familiar with this sort of skeleton because ours is an endoskeleton. But another phylum shares this distinction with us.

Echinoderms The spiny-skinned echinoderms, such as sea stars, also have endoskeletons, just as do the vertebrates. This may not be too surprising if you remember that both groups are deuterostomes,

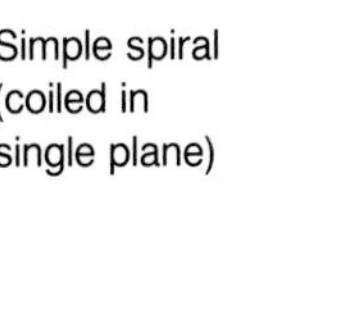

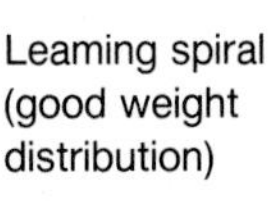

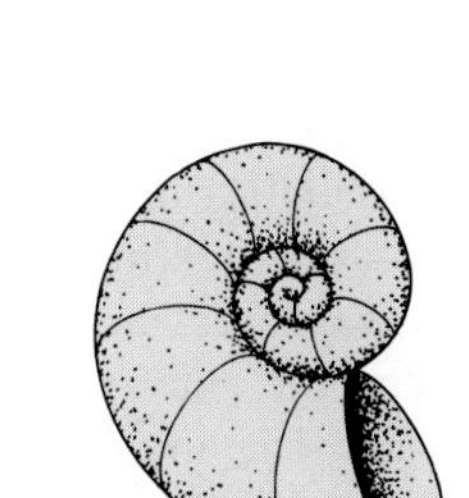

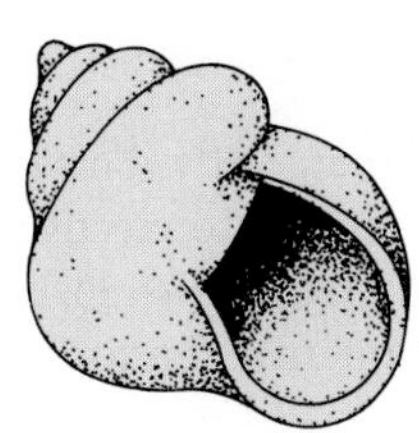

FIGURE 33.7
SHELL GROWTH
Mollusk shells grow continuously through the addition of shell material, secreted by the mantle. The most primitive shells are simple cones and cones spiralled in a single plane. The shell of the land snail forms a far more complex leaning spiral, which provides a more even distribution of shell weight.

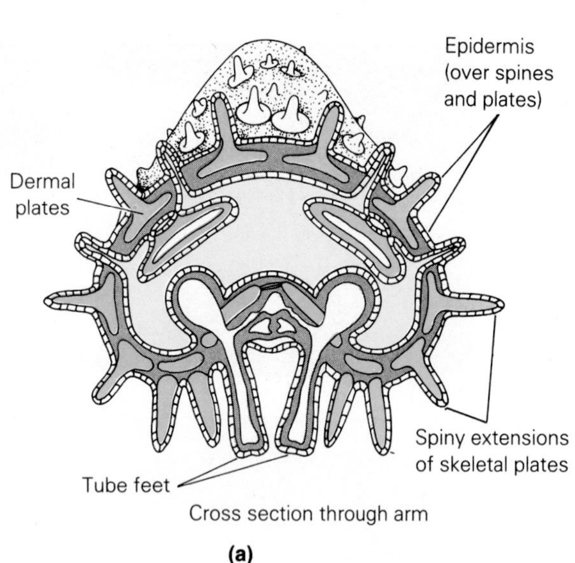

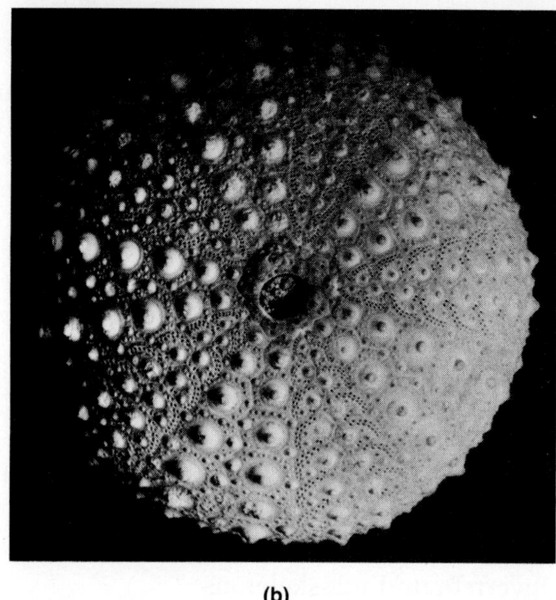

FIGURE 33.8
SEA STAR ENDOSKELETON
(a) In the sea star, the skeleton consists of numerous plates of calcium carbonate. Some plates produce blunt spines. Note the thin epidermis overlying the endoskeleton. **(b)** The empty sea star endoskeleton reveals the plates very clearly.

quite distant from the protostomes with their exoskeletons. Under the echinoderm epidermis, the thick dermis (of mesodermal origin) secretes a series of plates that fit closely against each other. The plates themselves are composed of many tightly interlocked little spined bodies, formed chiefly of calcium carbonate, a common material in marine animals. As the skeletal plates form, they erupt over the entire upper surface of the sea star, creating short, blunt, immovable spines (Figure 33.8). The spines of sea urchins, on the other hand, are very long, often very sharp, and movable.

VERTEBRATE TISSUES

Before we get into vertebrate supporting structures, we will look briefly at some kinds of tissues. **Tissue,** you will recall, is an aggregation of similarly specialized cells. There are just four basic tissue types that contribute to the many organs and organ systems of all metazoans: **epithelium, connective tissue, muscle,** and **nerve.** Muscle tissue is explored in detail later in this chapter, and nerve tissue is the subject of chapters to come, so here we will concentrate on the first two.

Epithelial Tissue

Epithelium, or **epithelial tissue,** forms most surface linings or coverings of the organism, both interior and exterior. Thus the **epidermis**—the outermost layer of the skin—is epithelial. Epithelial tissue also lines the mouth, the nasal cavities, the respiratory system, the coelom, the tubes of the reproductive system, the gut, and the body cavities and forms the interior, or **endothelium,** of blood vessels. The lining tissues of glands contain a thickened epithelium called **glandular epithelium.** The cells of the glandular epithelium both line the ducts (openings) and secrete whatever that gland secretes. As you might expect, epithelial tissues differ greatly in different parts of the body (Figure 33.9).

Connective Tissue

Connective tissue is found throughout the vertebrate body. This is probably a good thing since its role is essentially to hold the body together. Connective tissue consists of cells that produce a noncellular **connective tissue matrix** (Figure 33.10). Examples of familiar connective tissues include bone, cartilage, **ligaments, tendons,** and even blood. (The matrix of blood is called plasma.)

The principal and most abundant substance in most connective tissue matrices is **collagen,** the fibrous protein that accounts for at least a third of the total body protein

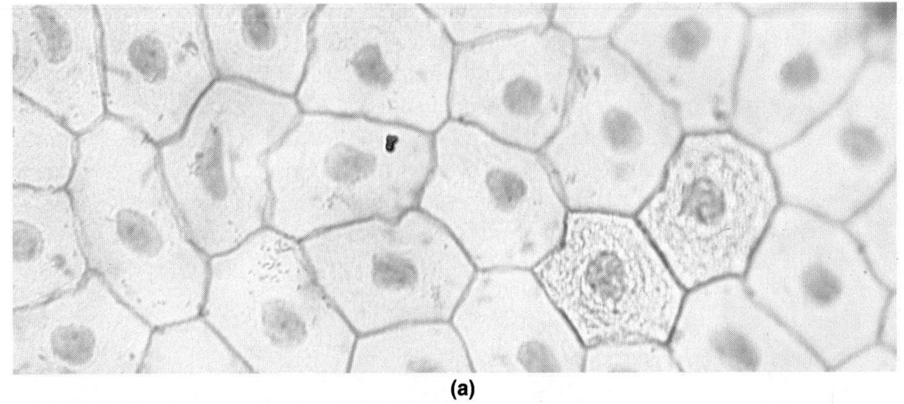

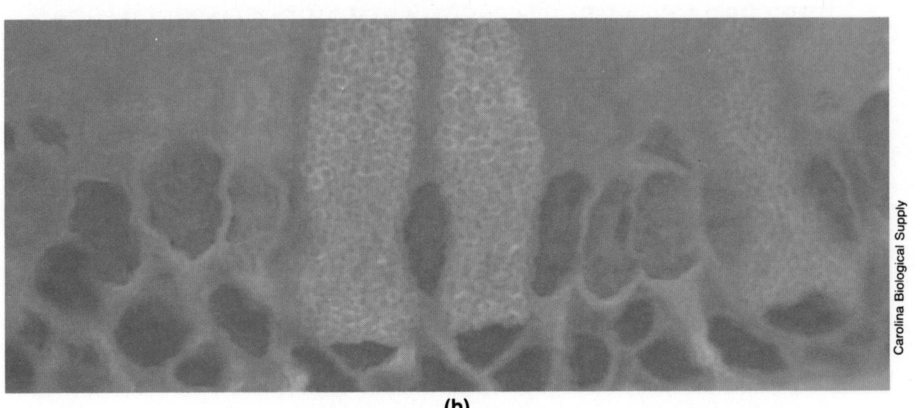

FIGURE 33.9
EPITHELIAL TISSUE
(a) Simple linings are formed by flattened squamous cells, which may form one or more layers. (b) The respiratory linings—nasal passages, trachea, and bronchi—contain columnar cells, commonly including glandular (or secretory) and ciliated cells.

in the larger vertebrates. Collagen is the "glue" of the animal body and forms extremely tough structures, such as the cornea of the eye, the tendons and ligaments, and the all-important disks that cushion the spine. Collagen is even the principal nonmineral component of bone.

Collagen is produced and secreted by cells known as **fibroblasts.** As the protein is secreted, it takes the form of tough, lengthy fibers that form the connective tissue matrix (Figure 33.11). The fibroblasts are vitally involved with wound healing and the mending of broken bones. In wound healing, a number of different kinds of connective tissue cells cooperate, reforming normal structures and, when necessary, forming tough collagenous **scar tissue.**

FIGURE 33.10
CONNECTIVE TISSUE
(a) In loose connective tissue, the cellular elements lie scattered in a loosely arranged matrix of dense collagen and fine elastic fibers. The blue objects are nuclei. (b) Dense connective tissue contains much more of the fibrous matrix arranged in parallel rows, with the living cells crowded among individual collagen fibers. The blue objects are nuclei.

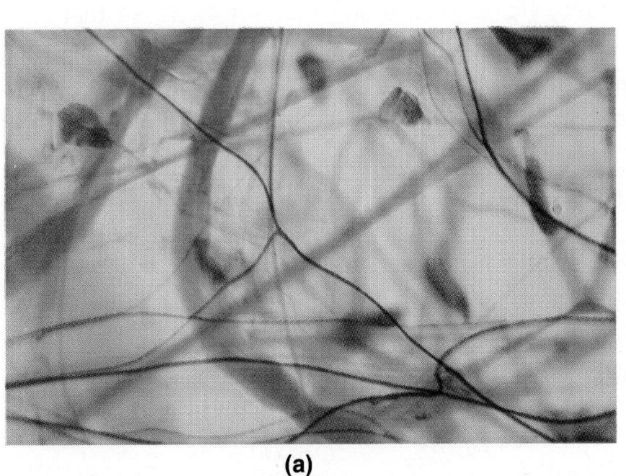

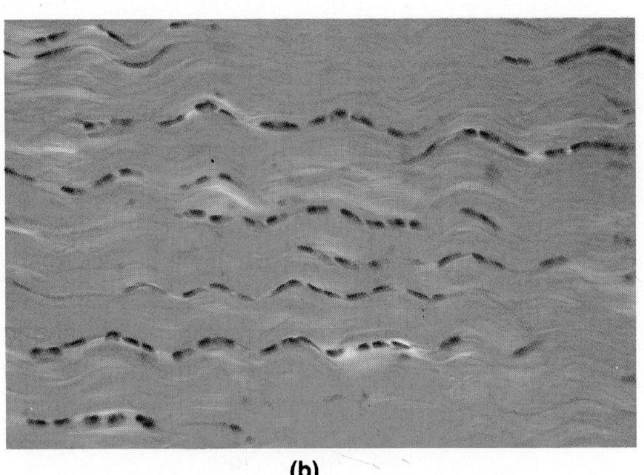

(a)

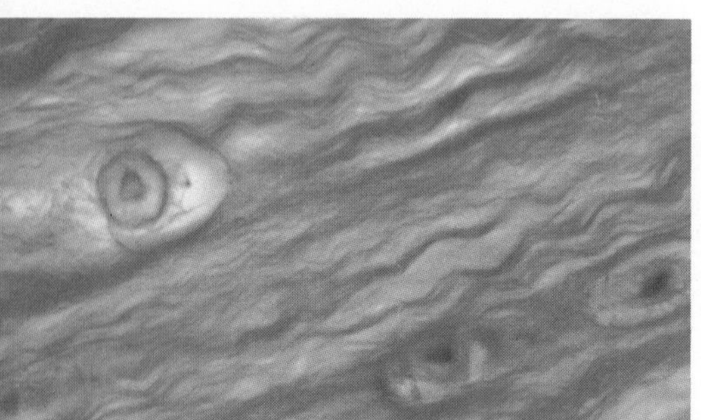

(b)

FIGURE 33.11
COLLAGEN
(a) Collagen fibers are the primary material of tendons—strong cordlike structures attached to bones.
(b) Individual collagen fibers in tendons form a wavy pattern, with distinct cross-banding of dense protein.

In the mending of bones, fibroblasts form a thin sheet of connective tissue around bones and temporarily fill the fracture site with tough fibers of collagen. Eventually, the new bone forms and hardens, permanently fusing the break.

Elastin, another important protein of connective tissue, is a major component of elastic fibers. Such fibers have "memory"—they can stretch to several times their length and snap back to their original size. Elastin is particularly important to arteries, notably the larger ones whose flexibility helps maintain blood pressure. Interestingly, elastin is common in the necks of grazing mammals—those that spend so much time with their heads near the ground. Can you see the role of elastin here?

With this overview of selected vertebrate tissues, we can now take a closer look at vertebrate skeletons.

VERTEBRATE SKELETONS: HARDENED CONNECTIVE TISSUE

As vertebrates, when we think of supporting structures, we normally don't think of cuticles, shells, or spiny plates. We think of bone. Bones are the light, hardened structures that hold us erect (keeping us from lying about, dotting the landscape in great heaps of protoplasm) and shield our more delicate parts from damaging environmental objects. So, we might well ask, what is bone, and what gives it its toughness?

Except for the cartilaginous fishes (see Chapter 32), vertebrates have skeletons composed primarily of bone (although some cartilage usually persists, as we shall see). **Cartilage** makes up the skeleton of the jawless fishes (lampreys and hagfishes) and the sharks and rays. It has a clean, almost glassy consistency, and is made up of **chondrocytes** (cartilage cells) embedded in a surrounding gel-like protein matrix that is heavily invested with collagen fibers. In the vertebrate embryo, cartilage is the forerunner of bone, but in the mature vertebrate it persists only in a few places such as the ears and nose, and in the joints.

The bony skeleton is important in at least five ways: (1) it supports the body and (2) serves as an attachment for muscles. (3) Certain parts of the skeleton are protective. In particular, the skull and rib cage protect the delicate structures within; the skull is nature's version of the motorcycle helmet. Further, some bones (4) produce red blood cells in their interior spaces, and (5) all bones act as a natural reservoir for the body's calcium.

The Structure of Bone

Typically, the long bones of the limbs are surrounded by a thin membrane of connective tissue called the **periosteum.** Within, we find two kinds of bone, **spongy** and **compact** (Figure 33.12a). Spongy bone is well named since it is porous and spongelike, made up of a weblike structure of hard bone (laid down along lines of stress so that this seemingly fragile bone is very strong). The spaces in the web are filled with soft tissue. Spongy bone is found at the enlargements at each end of long bones. Some bones, such as the ribs, sternum, vertebrae, and hip bones, contain **red marrow,** the site of red blood cell production.

Compact bone makes up the shaft—the cylindrical, lengthy portion of the long bone—and is thick and dense. Its central cavity contains **yellow marrow,** a dense fatty material. Much of the red marrow in young vertebrates is replaced by yellow marrow in the adult. Despite its stonelike appearance, compact bone is definitely a living tissue. It contains numerous metabolically active bone cells, as well as nerve cells and blood vessels. Most of the bony mass, however, consists of hardened calcium phosphate in a collagen matrix. The source of compact bone's great strength is seen best in its microscopic organization.

Under the microscope, thin sections of compact bone reveal intricate, repeated units of structure called **Haversian systems,** or **osteons.** Each Haversian system consists of

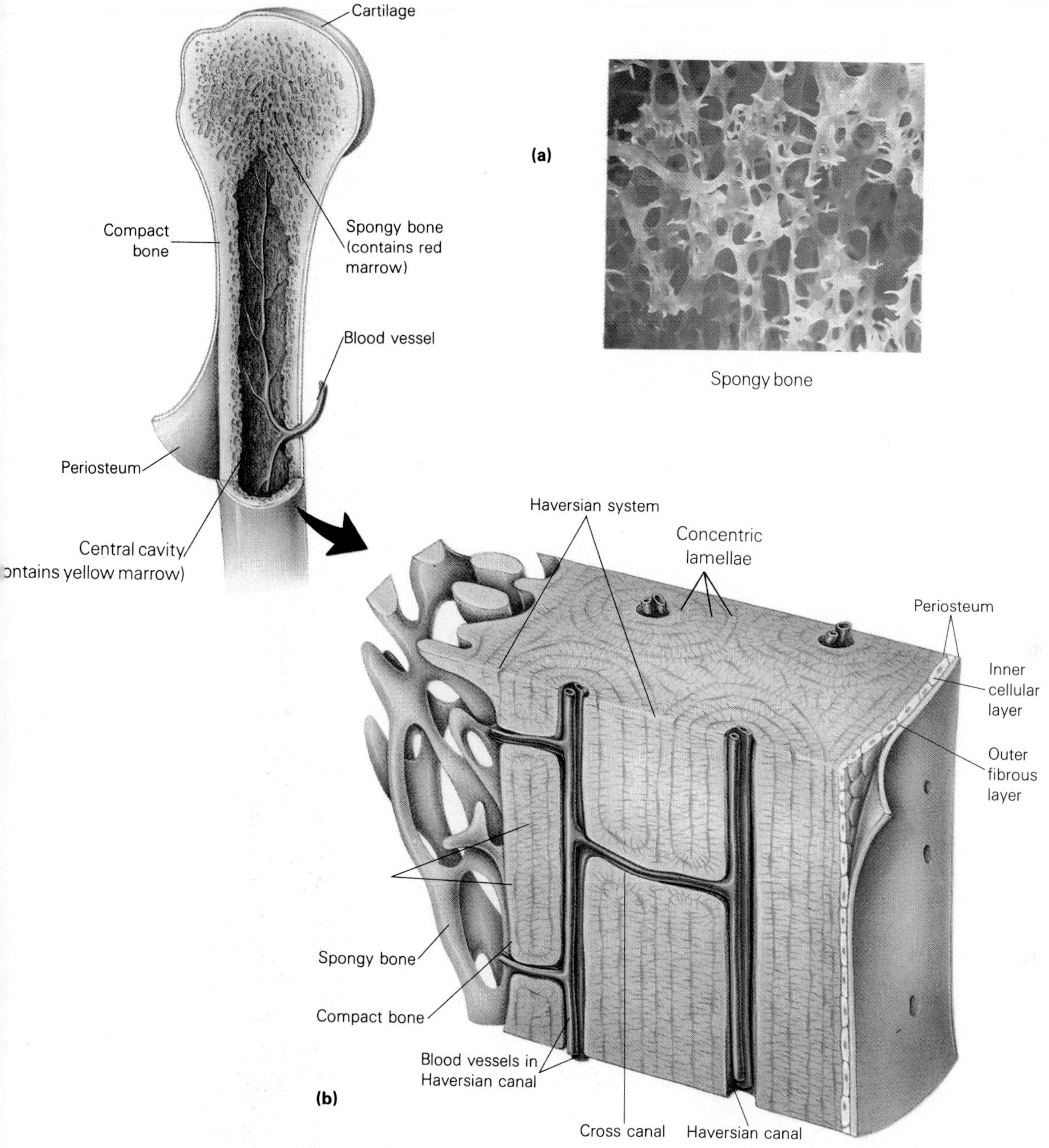

Cartilage

Compact bone

Spongy bone (contains red marrow)

Blood vessel

Periosteum

Central cavity (contains yellow marrow)

(a)

Spongy bone

Haversian system

Concentric lamellae

Periosteum

Inner cellular layer

Outer fibrous layer

Spongy bone

Compact bone

Blood vessels in Haversian canal

(b)

Cross canal

Haversian canal

Haversian system of compact bone

FIGURE 33.12
STRUCTURE OF BONE
(a) The sectioned long bone reveals a hard shaft of compact bone and a central cavity of yellow marrow. Spongy bone is seen near the joints. **(b)** Numerous Haversian canals, responsible for carrying blood vessels and nerves, are surrounded by hardened concentric rings, or lamaellae, containing bone cells (osteocytes). Such cells are entombed in lacunae, tiny cavities that communicate with each other via minute crevices called canaliculi. Such structures are grouped into Haversian systems.

a **central canal** that contains blood vessels and nerves, and a surrounding region of concentric rings or **lamellae**, consisting of calcified bone (Figure 33.12b). The rings are, in effect, tubes within tubes—a laminated tubular construction that imparts great strength and resiliency to the bone.

Within tiny cavities or **lacunae** (singular, **lacuna**) we find the bone cells, or **osteocytes.** Each lacuna is interconnected with others via minute canals, aptly named **canaliculi**, which pass throughout the hardened bone. The osteocytes touch each other via long projections through the canaliculi. Some of these extensions reach the central canal, permitting the cells to carry on an exchange of materials with the circulatory system. The osteocytes are important in calcium deposition and withdrawal, processes that are controlled by hormones (see Chapter 37).

Organization of the Vertebrate Skeleton

At this point we can take a closer look at the organization of the bones of the vertebrate skeleton. We should begin by noting that anatomists divide the vertebrate skeleton into two regions: the **axial skeleton** and the **appendicular skeleton.** The axial component includes the **cranium** (skull), the **vertebral column**, the **rib cage** and **sternum** (breastbone). The appendicular component includes the limbs (always two or fewer pairs) and the two **girdles, pelvic** (hip) and **pectoral** (shoulder). Much of the appendicular skeleton, by the way, is absent in most snakes, which is why they are not broad shouldered. Before getting into the details of all this, let's pause to consider a few joints.

The Joints The various bones of the body meet at **joints.** Each joint allows only a certain kind and degree of movement, and some, such as the **sutures** between the bones of the skull, allow no movement at all (try flexing your skull). The more movable joints, though—for example, the knee, elbow, or hip—are both complex and elegant. The articulating (contacting) surfaces are covered with a thin layer of exceedingly smooth cartilage. The bones are held together by short bands of white fibers of collagen called **ligaments** (familiar because they often suffer sprains—or tears). Where the ligaments form an enclosing capsule they produce a **synovial joint.** The entire joint is constantly lubricated by fluid, secreted inside the capsule. The presence of smooth cartilage and lubricating fluid is highly advantageous. It helps keep the joint from wearing out by reducing friction and allows for quieter movement. As you can imagine, a hungry lion creaking and squeaking around in the grass is likely to come up short.

The movable joints are generally named according to the type and degree of their movement. Thus there are **gliding, pivotal, ball-and-socket,** and **hinge** joints (Figure 33.13).

The Axial Skeleton The axial skeleton includes the **skull** and jaw, structures that have traditionally yielded a great deal of information about the evolution of vertebrates. In addition, the axial skeleton includes the vertebral column and the rib cage.

The Skull Skulls from the vertebrate classes make a rather nice comparative series. In the fishes the cranium consists of a large number of loosely connected, individual plates. The jaws are simple and form a strictly one-directional hinge, which means they can bite but there is little of the lateral movement that makes efficient grinding possible. In fish, both the top and bottom jaws move, a feature also seen in birds but one that does not occur in mammals. Our upper jaw, the **maxilla,** is fused to the skull.

As we compare the skulls of other vertebrates in the order in which we believe they progressed through their evolution, the trend is toward fusion of parts, resulting in fewer bones (Figure 33.14, p. 768). (For an account of an interesting evolutionary problem with the vertebrate skull, see Essay 33.1.) Amphibians and reptiles have more mobility in the jaws than do fishes, but like birds, they swallow food whole. In fact, the jaws of snakes unhinge to swallow prey that is often larger in diameter than the snakes themselves.

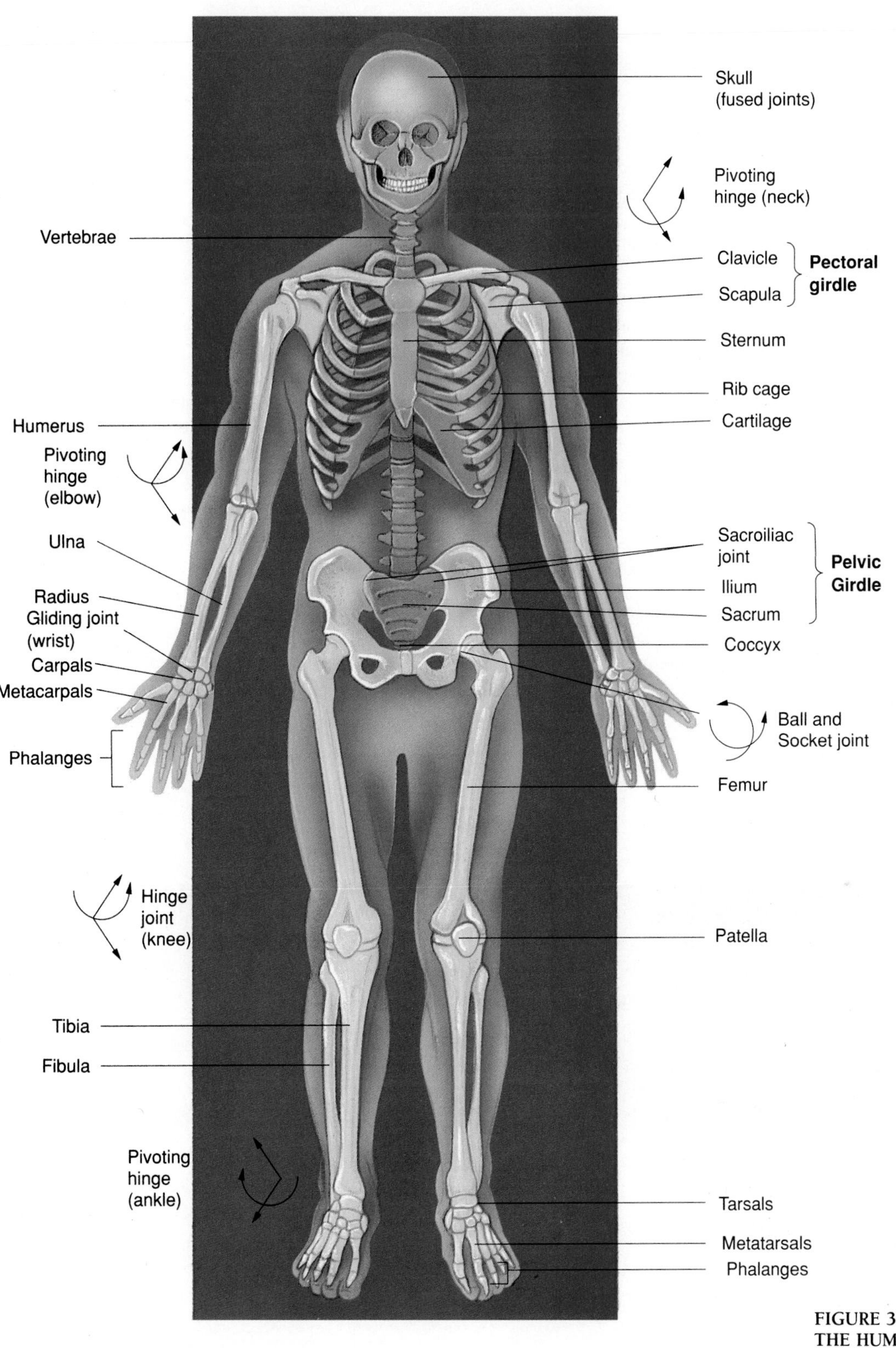

FIGURE 33.13
THE HUMAN ENDOSKELETON
The axial skeleton includes the skull, vertebral column, rib cage, and sternum. The appendicular skeleton includes the pectoral and pelvic girdles and the appendages. Four types of moveable joints are seen in the appendicular skeleton.

ESSAY 33.1

EVOLUTION OF THE MAMMALIAN SKULL

The history of the mammalian skull provides an unusual illustration of evolution at work. We could call this example "How to repair a mistake, or you can't hardly get here from there." Originally, most of the bones of the skull weren't part of the axial skeleton at all. The axial skeleton consisted of the vertebrae, the ribs, the gill arches (later, jaws), and a tiny **braincase**, suitable for a tiny brain. Most of the bones of the head were **dermal bone**, protective plates of bone developed in the lower layers of the skin, as we see in fish today.

When fish became terrestrial, the dermal bone plates fused and hardened over time to become the skull of the ancient amphibians and early reptiles. There were holes for the eyes and a simple, membranous, partly open braincase deep inside. There is one problem with this design: the jaw muscles are inside the skull, between the skull and the open, membranous braincase **(a)**. One result is that when the jaw muscles contract and therefore bulge, there is no place for the bulge to go except to squeeze the brain. It would be much better to have the jaw muscles on the outside of the skull.

Evolution, as Darwin argued tirelessly, tends to proceed by small increments. Thus, it would seem that there could be no way to get the heavy jaw muscles from the inside of the skull to the outside without leaving a string of maladapted monsters halfway between.

But we now know that, in the evolutionary descent of reptiles, membranous openings gave the bulging jaw muscles someplace to go, relieving pressure inside the skull. Meanwhile, flangelike projections from the dermal bones of the roof of the skull developed between the jaw muscles and the brain, further protecting it. At the same time, similar processes grew up from the floor of the braincase **(b)**.

Eventually, the protective flanges, or side processes, that developed over the brain fused, creating a new and more solid encasement for this delicate— and enlarging—organ. And the openings in the dermal skull grew larger and larger **(c,d)**, until eventually all that was left of the original skull outside the jaw muscle was the thin, fragile, **zygomatic arch** (cheekbone). In mammals, the zygomatic arch served as the origin of a new muscle, the **masseter**, which allows side-to-side chewing. It took many tens of millions of years to accomplish this by evolution and natural selection, with each step along the way a slight improvement over the one before. But in the end, the jaw muscles were outside the skull where, logic seems to tell us, they belonged in the first place.

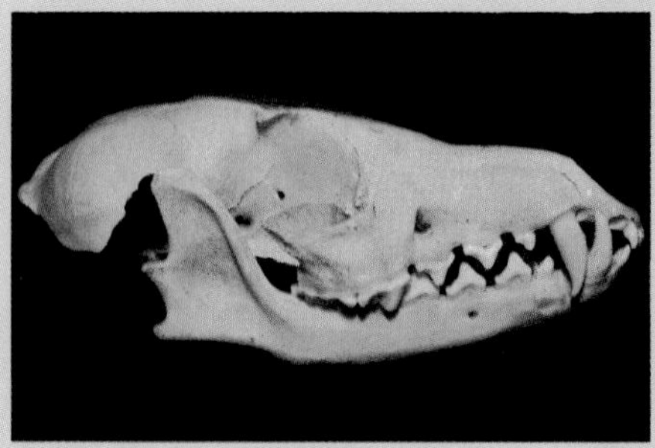

Dog

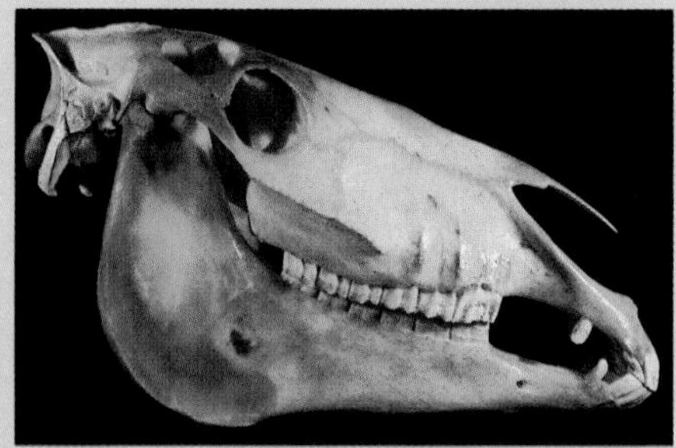

Horse

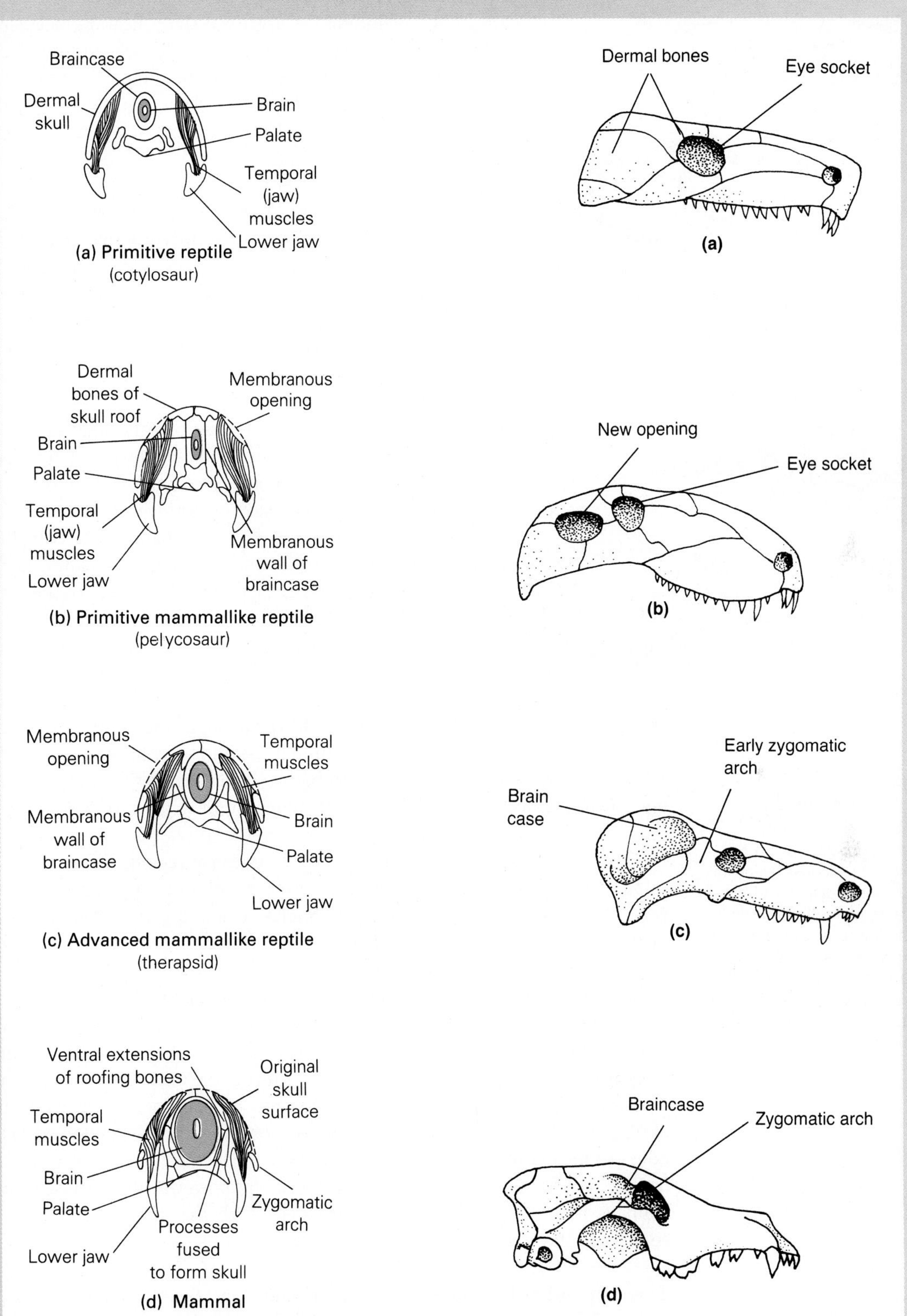
Braincase
Dermal skull
Brain
Palate
Temporal (jaw) muscles
Lower jaw
(a) Primitive reptile
(cotylosaur)
Dermal bones
Eye socket
(a)
Dermal bones of skull roof
Membranous opening
Brain
Palate
Temporal (jaw) muscles
Lower jaw
Membranous wall of braincase
(b) Primitive mammallike reptile
(pelycosaur)
New opening
Eye socket
(b)
Membranous opening
Temporal muscles
Membranous wall of braincase
Brain
Palate
Lower jaw
(c) Advanced mammallike reptile
(therapsid)
Early zygomatic arch
Brain case
(c)
Ventral extensions of roofing bones
Original skull surface
Temporal muscles
Brain
Palate
Zygomatic arch
Processes fused to form skull
Lower jaw
(d) Mammal
Braincase
Zygomatic arch
(d)

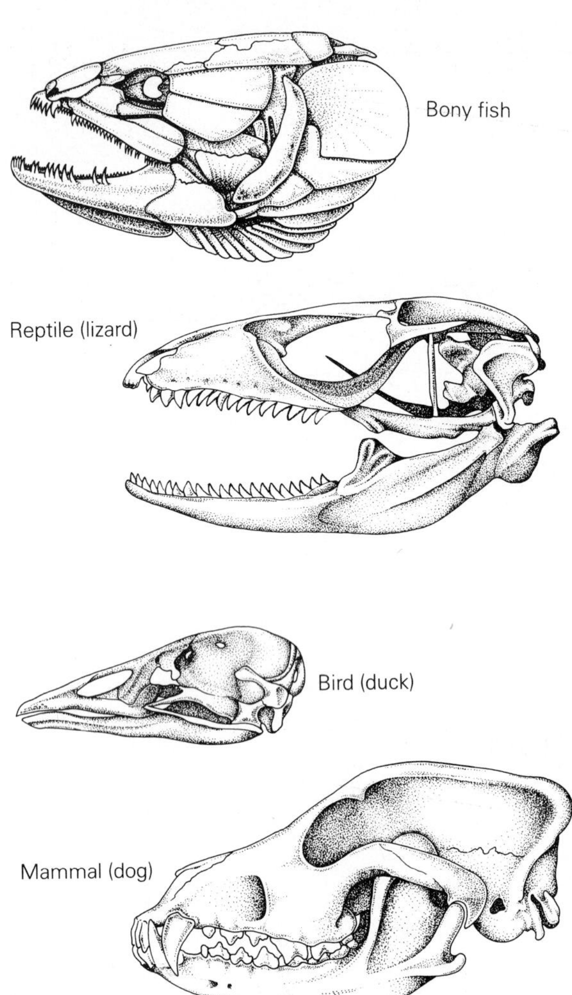

FIGURE 33.14
VERTEBRATE SKULL SERIES
In the evolution of the vertebrate skull, the trend has been toward the fusion of bones. We see less fusion in the skulls and jaws in fishes and reptiles, and more rigid skulls in birds and mammals, due to fusion. Note that the mammalian jaw, while comparatively simple, is very moveable. Its great mobility permits chewing—a seemingly common act, but actually unusual among vertebrates.

They are able to do this because the lower jaw is loosely attached to the skull and can be pulled away from it.

In vertebrates the skull not only protects the brain but also houses a complex group of receptors. As you know, many of the senses are "up front," including the organs of sight, hearing, balance, smell, and taste. The sensory structures are actually isolated from the brain itself, but the major nerve pathways enter the braincase via openings known as **foramina** (singular, **foramen**). The largest of these, the **foramen magnum,** is at the base of the skull and serves as an opening for the spinal cord.

The Vertebral Column Throughout the evolution of vertebrates, the backbone has carried out its supporting role in a number of ways. For example, in the vast majority of terrestrial animals that walk on all fours, the backbone has often had to support the great weight of the belly. On the other hand, aquatic species are buoyed by water, and their reduced weight places far less stress on the vertebral column. Thus, the backbone or vertebral column can be quite varied and specialized.

The backbone consists of a series of bones that form a flexible axis for the body. Each bone is individually known as a **vertebra,** and their number varies greatly from one class of animal to another. The reptiles have the record, with as many as 400 (most of them very similar to each other). The birds, on the other hand, have comparatively few, partly because the posterior ones are fused. All mammals, from giraffes to shrews to whales, have exactly seven vertebrae in the neck, although the number in the remainder of the column varies.

Humans have 33 vertebrae, although there may be one more or one less, depending on how many make up the **coccyx** or "tailbone." All vertebrae, other than those that have fused, are separated by **intervertebral disks** of **fibrocartilage.** Fibrocartilage is cartilage containing numerous reinforcing strands of collagen. Each disk has a softer, compressible center that provides cushioning and helps make lateral movement possible. Intervertebral disks take a lot of wear, and painful ruptures sometimes occur.

The human vertebral column is organized into five regions, as seen in Figure 33.15. In addition to its supporting role, the vertebral column protects the spinal cord. The cord passes through what is called the **neural canal,** a series of openings or foramina in the individual vertebrae. Vertebrae in the thoracic or chest region also provide attachments for ribs.

Besides the skull and the vertebral column, the axial skeleton includes the ribs and the sternum. The rib cages of many species have become highly modified, producing some interesting variations. For example, frogs and toads have very little rib development. They also lack a diaphragm and breathe by moving the muscles in the mouth and throat. Most land vertebrates make use of the rib cage in breathing. Another unusual variation of rib structure is found in the turtle, whose ribs and vertebrae are joined to the platelike bones of the **carapace** (upper shell). In birds, the sternum gives rise to the **keel,** which has a large, flat vertical projection to which the powerful flight muscles are attached. Humans have 12 pairs of ribs. All articulate with the thoracic vertebrae, but not all connect to the sternum (see Figure 33.13).

The Appendicular Skeleton The appendicular skeleton includes the two girdles (rings) of bone and bones of the limbs (arm and leg bones).

The Pectoral and Pelvic Girdles Humans (and birds) are unusual among vertebrates in that they are bipedal (walk on two feet). Therefore, our pectoral girdles, greatly

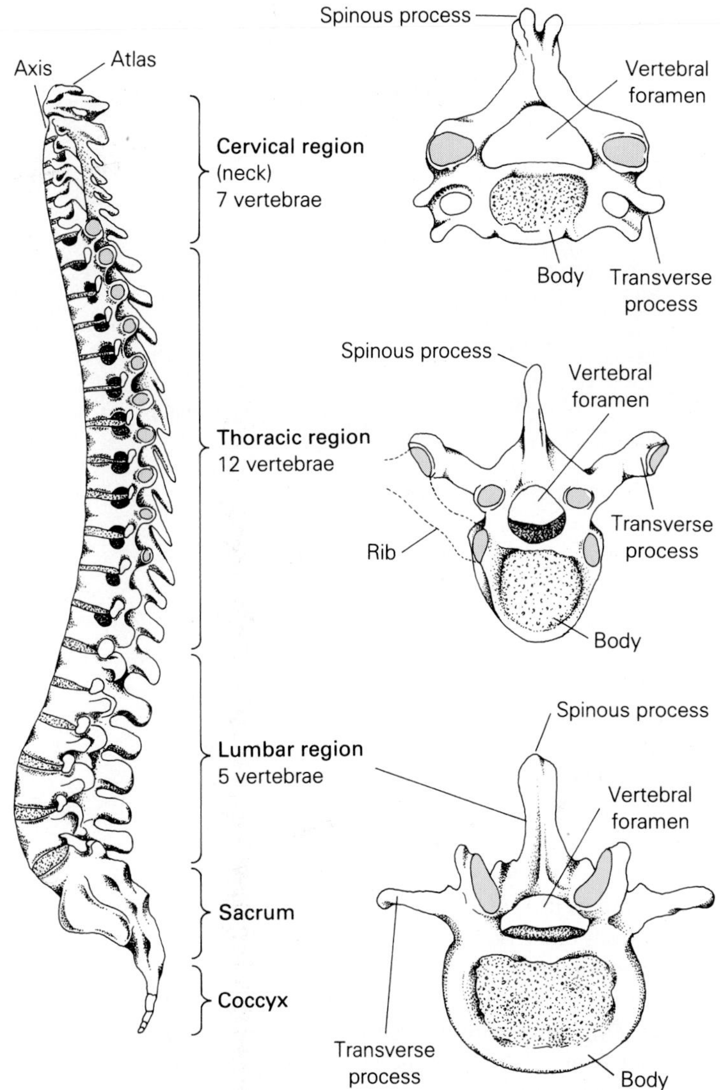

FIGURE 33.15
THE HUMAN VERTEBRAL COLUMN
The vertebral column, in profile, takes on a gradual S-shape. It is divided into the cervical, thoracic, and lumbar regions, plus the fused sacrum and coccyx. Vertebrae in the thoracic region contain special indentations (facets) for the attachment of the ribs. Note the more massive lumbar vertebrae.

relieved of their weight-supporting burden, are considerably reduced. The human pelvic girdles are also different even from the other primates, which are not well adapted to an upright gait. As you can see in Figure 33.16, our upright posture is associated with a distribution of weight that is unusual among primates. The upper torso rests entirely on the **sacroiliac joint** (the joint between the **sacrum** and the **ilium**), with some help from certain back muscles that distribute the weight to the ilium. The bones and joints of the human pelvic girdle are seen in Figure 33.13.

The pectoral girdle consists of two **scapulae** (shoulder blades) and two **clavicles** (collar bones). The shoulder blade is a triangular, flattened bone whose outer margin is thicker and forms part of the shoulder joint, a ball-and-socket arrangement (see Figure 33.13).

Vertebrate Limbs As we have mentioned, vertebrates have no more than two pairs of legs. Those with fewer than two are rather rare and include snakes, whales, dugongs, and kiwis. We see in the flying vertebrates limbs that have been greatly modified for locomotion. In the two living groups, birds and bats, the forelimbs and especially the **digits** (fingers) have been greatly modified for flight (see Figure 32.23).

With the exception of the thumb and the arch of the foot, human appendages are rather generalized. In Chapter 32 we described in detail our famous opposable thumbs,

FIGURE 33.16
COMPARISON OF GORILLA AND HUMAN SKELETONS
Of all the primates, only humans are truly bipedal. Compare the pectoral and pelvic girdles, position of skull, relative length of arms and legs, and the shape of the foot in the human and the gorilla. Note the more horizontal position of the pelvis in the gorilla compared to its vertical position in the human. The pectoral girdles are more generally similar, with greater mass in the scapula of the gorilla, as might be expected in a heavy, brachiating primate.

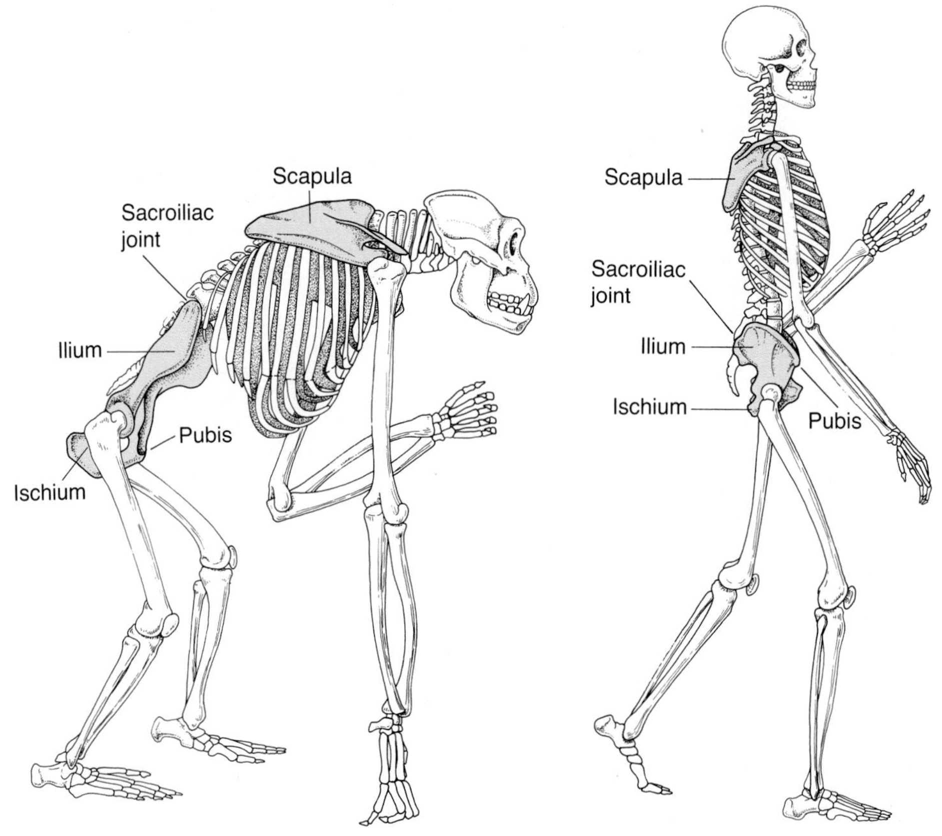

our arched foot, and well-developed gluteus maximus (buttock muscles), each of which distinguishes us from other primates.

The mammalian forelimbs, such as our arms, are made up of three long bones: the large **humerus** of the upper arm and the **radius** and **ulna** of the lower arm. The elbow is a complex, pivoting hinge joint, while the wrist forms a flexible hinge. (The bones of the hands and wrists are seen in Figure 33.13.)

The human leg, like the arm, consists of a single upper bone, or **femur** and two lower bones, the **tibia** and **fibula**, as shown in Figure 33.13. Most vertebrates have this general plan. The upper and lower bones meet at the knee joint, but only the tibia articulates with the femur. The fibula articulates with the tibia just below the knee. A fourth bone, the chestnut-shaped **patella**, or kneecap, is embedded in a tendon.

The human ankle and foot bones are also seen in Figures 33.13 and 33.16. The famous arch of the human foot—not seen in other primates—is not formed by the bone structure itself, but by the bindings of its ligaments.

VERTEBRATE MUSCLE: ITS ORGANIZATION AND MOVEMENT

Vertebrate muscles can be divided into three types according to their microscopic structure, characteristics of contraction, and means of control. The three types are **smooth muscle, cardiac muscle**, and **skeletal muscle.** Whereas each is composed of muscle cells, such cells have become so specialized that they are generally referred to as muscle fibers.

Smooth Muscle

Smooth muscle is also called **involuntary muscle** and **visceral muscle**, terms that remind us that we have little conscious control over this muscle and that it can be found in the viscera or gut. In addition to this location, smooth muscle also occurs in the walls of blood vessels, at the base of each hair, in the iris of the eye, in the uterus, and other places. You may have noticed that the hair on the back of your neck rises when you look a politician in the eye, or that you can't stop your stomach from growling when meeting your date's parents, or that you can't call back a blush once it has begun. These are all due to involuntary actions of smooth muscle. Smooth muscle even influences sex, since it is the muscular control of blood entering and leaving the genitals that accounts for changes occurring during sexual arousal. Smooth muscle is under the control of the **autonomic nervous system,** a branch of the nervous system that controls many activities below the conscious level (see Chapter 35). The autonomic nervous system can both inhibit and trigger smooth muscle contraction.

Under the microscope, smooth muscle tissue is, logically enough, smooth in appearance, at least in comparison to the other types of muscle (Figure 33.17a). The individual cells or fibers are spindle-shaped and tapered at their ends, and each has a single nucleus. Smooth muscle tends to occur in sheets rather than in the dense bundles seen in skeletal and cardiac muscle. In the digestive tract, the sheets occur in a tubelike form, making up one of the several layers that comprise the gut. Smooth muscle contains the same contractile proteins as other muscle types, but its contractile proteins do not have the same highly ordered organization.

Cardiac Muscle

Cardiac muscle is heart muscle. Its remarkable strength and endurance are lengendary, and rightfully so. After all, there is little room for error in its functioning. Whereas it is unique in some ways, cardiac muscle has some things in common with the other muscle types. Like smooth muscle, cardiac muscle fibers contain a single, centralized nucleus. Also like smooth muscle, cardiac muscle is largely under autonomic control. But like fibers of skeletal muscle, cardiac muscle fibers are cylindrical and **striated** (striped) as seen in Figure 33.17b. The stripes reflect the highly ordered arrangement of contractile proteins.

Unlike both smooth and skeletal muscle, cardiac muscle fibers are branched and present a woven appearance. For this reason, heart muscles contract in a twisting motion that, when magnified throughout the heart, results in a "wringing" or "squeezing" action in the chambers with each beat. Where individual cardiac muscle fibers join, their membranes are highly folded, forming interlocking junctions named **intercalated disks.**

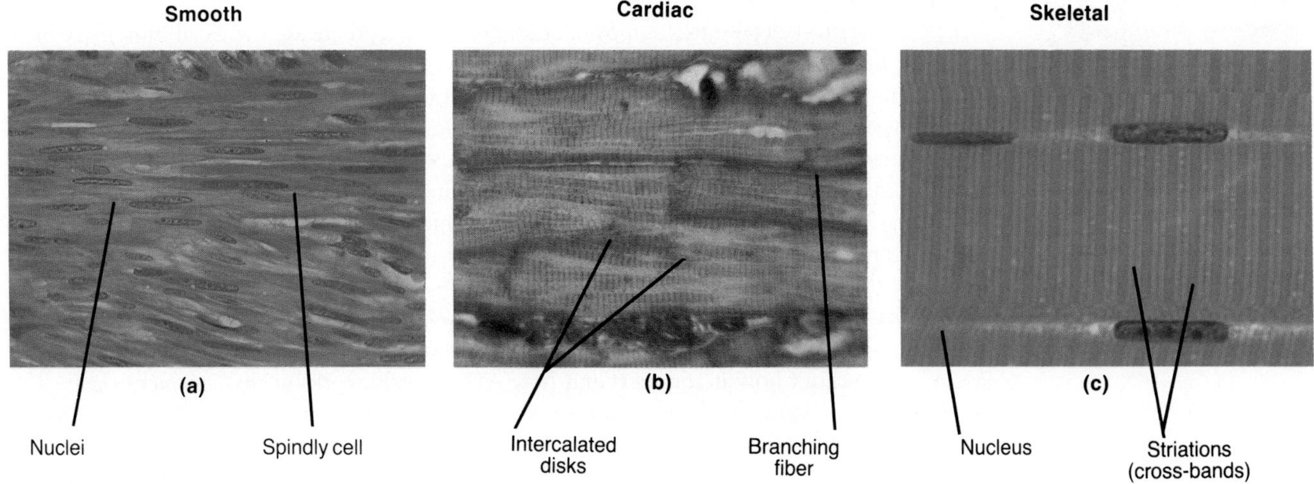

FIGURE 33.17
THREE TYPES OF MUSCLE TISSUE
(a) Smooth muscle consists of long, spindly cells, each containing a single nucleus. **(b)** Cardiac muscle branches and rebranches and is interrupted by intercalated disks. Like skeletal muscle, it is striated. **(c)** Skeletal muscle is heavily striated and multinucleate.

These disks are extremely "leaky," that is, they permit an easy flow of electrical currents between cells and thus throughout the tissue. Finally, cardiac muscle will contract without outside stimulation; contraction is intrinsic—it comes from within. Contraction will even occur in isolated cells, those that have been teased apart and kept alive in a tissue culture medium. We will look further into the unique properties of cardiac tissue in Chapter 39. Again, the most impressive feature of cardiac muscle is its tireless, rhythmic beating.

Skeletal Muscle

Skeletal muscle is also known as **voluntary muscle** and **striated muscle.** It is controlled by **motor neurons** that are part of the **somatic** (voluntary) **nervous system.** As for its three names, it is *skeletal* because these are the muscles that move the skeleton; it is *striated* because of its molecular organization; and *voluntary* because it is *possible* to move the muscles at will. Skeletal muscle can contract much more rapidly and forcefully than the other types, but it cannot sustain such contraction without tiring.

Skeletal muscle cells are striated, unbranching, cylindrical, and multinucleate, with the nuclei lying just beneath the cell surface (see Figure 33.17c). The fibers may also be quite long, sometimes several centimeters in length. Each fiber is packed with precisely arranged contractile proteins. We will return to the details of this arrangement shortly.

Gross Anatomy of Skeletal Muscle Skeletal muscle fibers are bound into bundles known as **fascicles,** which are surrounded by a tough sheath of connective tissue. The fascicles in turn are bound together to form the muscle, and it is the fascicles that give meat its stringy appearance. Blood vessels run throughout the fascicular bundle, supplying oxygen and nutrients and carrying off wastes. Nerves also penetrate the bundle, their branches dividing ever more finely until they reach each fiber. These nerves will carry the messages that cause the fibers to contract.

Skeletal muscles are enclosed in a tough casing of connective tissue, the **fascia,** which is continuous with the inner connective tissue sheaths. (*Fascia* is Latin for a "band" or "bandage." It may help to remember that *fascism* is a rather binding form of government.) At the ends of the muscle, the fascia coalesces into denser collagenous tissue, forming the cordlike **tendons,** which may be broad and short or thin and long. Tendons, in turn, integrate with the periosteum covering the bones, thus ensuring the continuity of force from muscle to bone. Earlier we noted that each muscle has an origin (fixed base) and an insertion (moveable end). Figure 33.18 shows the arrangement of major muscle groups.

Not all skeletal muscles move bones. For instance, there are the facial muscles, which in humans are quite significant since we rely so heavily on facial expressions in communicating. Tongue muscles also have complex origins and insertions. Thus, the tongue has considerable dexterity, which is essential for eating, speaking, and cleaning the teeth in expensive restaurants. Some muscles form **sphincters,** rings that surround passages within the gut and at the anus and mouth. Other muscles may form into flattened sheets that join with broad, thin tendons called **aponeuroses.** A familiar example is the sheet of abdominal muscles that hold your paunch in.

The Ultrastructure of Skeletal Muscle Now let's take a look at the details of muscle ultrastructure and contraction. We will look into the finer structural levels of muscle and find out how it contracts and how ATP actually works at the contractile site.

Muscle structure is best understood by mentally dissecting it down through its levels of organization, beginning with the gross tissue itself (Figure 33.19a). We have already discussed the organization of muscle down to the fiber level. We saw that the muscle itself is subdivided into bundles, each containing numerous fibers. Describing the individual muscle fibers in detail (Figure 33.19b) brings us into the ultrastructural level of organization, where most of our knowledge comes from electron microscope studies. We can review a few aspects about muscle fiber first and then look at its contractile structure.

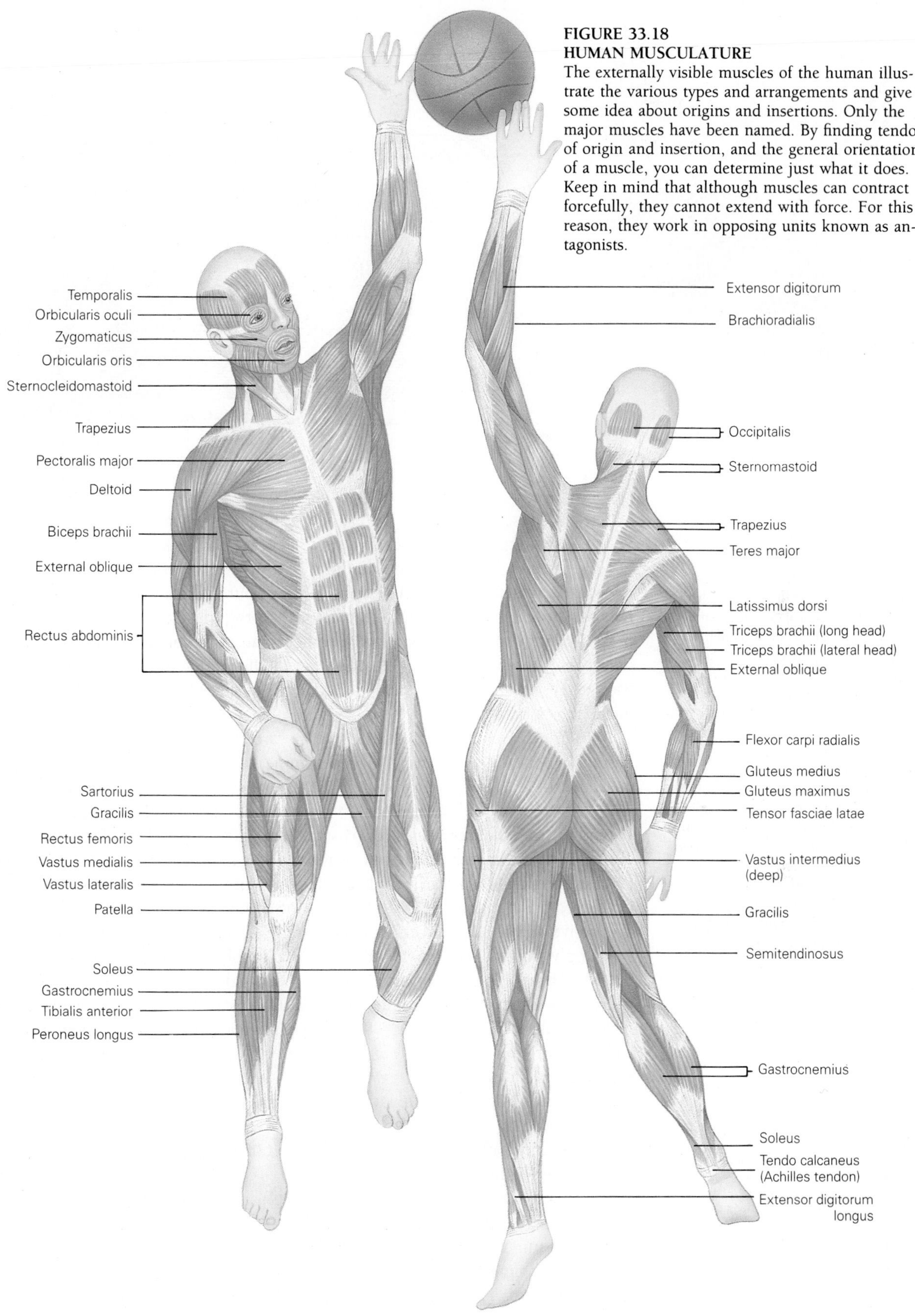

FIGURE 33.18
HUMAN MUSCULATURE
The externally visible muscles of the human illustrate the various types and arrangements and give us some idea about origins and insertions. Only the major muscles have been named. By finding tendons of origin and insertion, and the general orientation of a muscle, you can determine just what it does. Keep in mind that although muscles can contract forcefully, they cannot extend with force. For this reason, they work in opposing units known as antagonists.

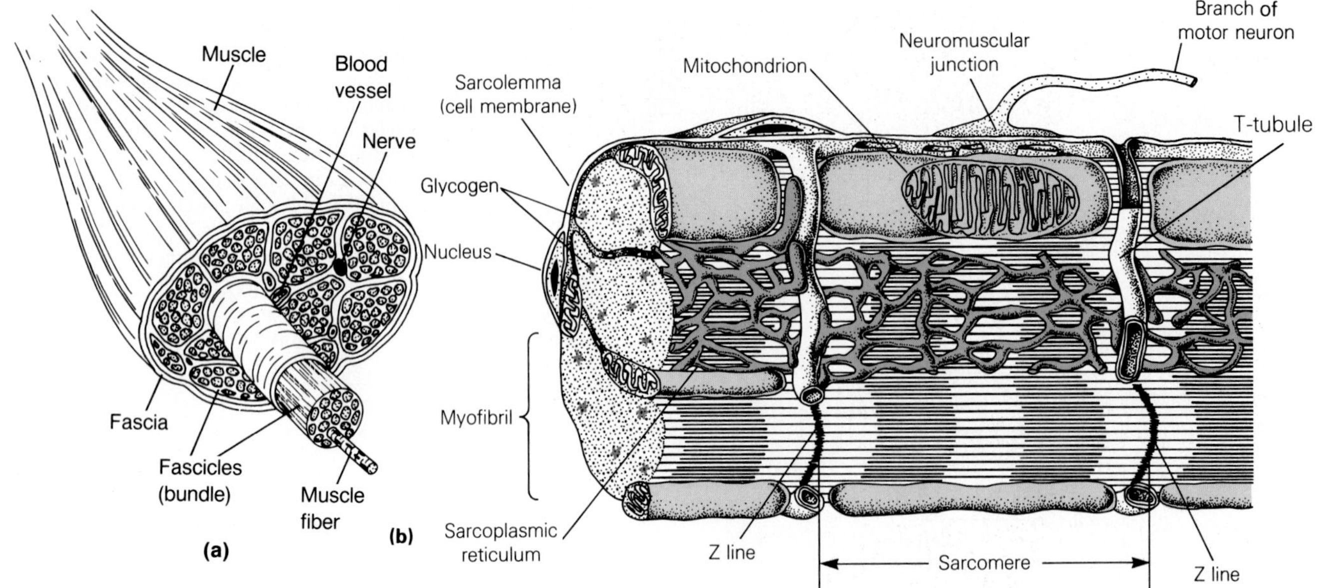

FIGURE 33.19 SKELETAL MUSCLE ORGANIZATION
(a) Whole muscles are surrounded by a fascia, which is continuous with the tendons of origin (the anchor) and insertion (the part to be moved). The muscle's thicker midregion is commonly called the belly. Within the muscle are fascicles (bundles), each fascicle consisting of many fibers bound together in a connective tissue matrix. **(b)** Each muscle fiber is surrounded by a plasma membrane called a sarcolemma, which contains numerous T-shaped extensions called T-tubules. Within the sarcolemma are nuclei, the membranous sarcoplasmic reticulum, and mitochondria. Each muscle fiber contains many myofibrils, which form the contractile units. The myofibrils reveal the banding typical of skeletal muscle.

Skeletal muscle fibers range from a few millimeters in length to several centimeters—enormously long as cells go. Each muscle fiber (or cell) is surrounded by its equivalent of a plasma membrane, the **sarcolemma.** The sarcolemma receives the endings of motor neurons at **neuromuscular junctions.** Motor neurons relay impulses from the central nervous system (the brain and spinal cord) to the muscles. These impulses stimulate muscle contraction, as we will see. The neuromuscular junctions are the nerve-muscle interfaces where motor impulses are received. Distinct neuromuscular junctions, by the way, are absent in smooth muscle; there, the junctions are marked only by slight swellings in the motor neurons.

Within the muscle fiber, just below the sarcolemma, are found a number of nuclei and mitochondria and many glycogen granules—just what one would expect for such an active tissue. Lying just beneath the sarcolemma is the **sarcoplasmic reticulum,** which, like the endoplasmic reticulum of other cells, is a membranous, hollow structure. Extending from the sarcolemma or plasma membrane are numbers of rod-shaped tubules, also hollow and membranous but somewhat larger than the tubes making up the sarcoplasmic reticulum. These are called **transverse tubules**, or **T-tubules.** (They are poorly developed in smooth muscle.) So far, we have named only supporting cellular structures. The active contractile elements of the muscle fiber are actually cylindrical **myofibrils** (or simply **fibrils**). The myofibrils form the visibly striped pattern as seen in Figure 33.19b. Their presence suggested the name *striated muscle.*

Isolating one myofibril permits us to obtain a clear view of these visible striations of skeletal muscle (Figure 33.20). To understand the pattern, we must first realize that each myofibril contains numerous rod-shaped filaments, or **myofilaments**, of the protein **myosin**, and an even greater number of thinner filaments of the protein **actin.** When viewed in cross section, it is common to observe each thick filament surrounded by six thinner filaments.

The two kinds of filaments are also arranged in a highly specific longitudinal manner, as we see in Figure 33.20.

We can now look at the arrangement of myofilaments from the surface of the myofibril, beginning with the very prominent **Z lines.** The region between the Z lines is the contractile unit, also known as the **sarcomere.** Moving inward from either Z line, we come to a very light zone called the **I band.** It consists of actin only, with no overlapping of myosin. Further inward is a broad, dark **A band**, with a smaller and lighter central strip, the **H zone.** The two darker parts of the A band consist of overlapping actin and myosin myofilaments. The lighter central H zone represents myosin alone unless the muscle is contracted whereupon the H zone disappears.

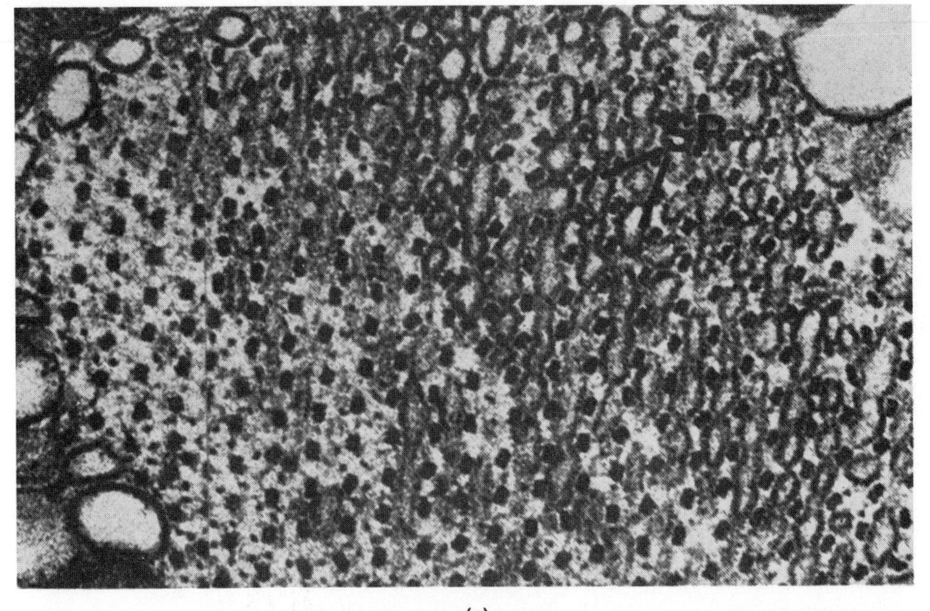

(a)

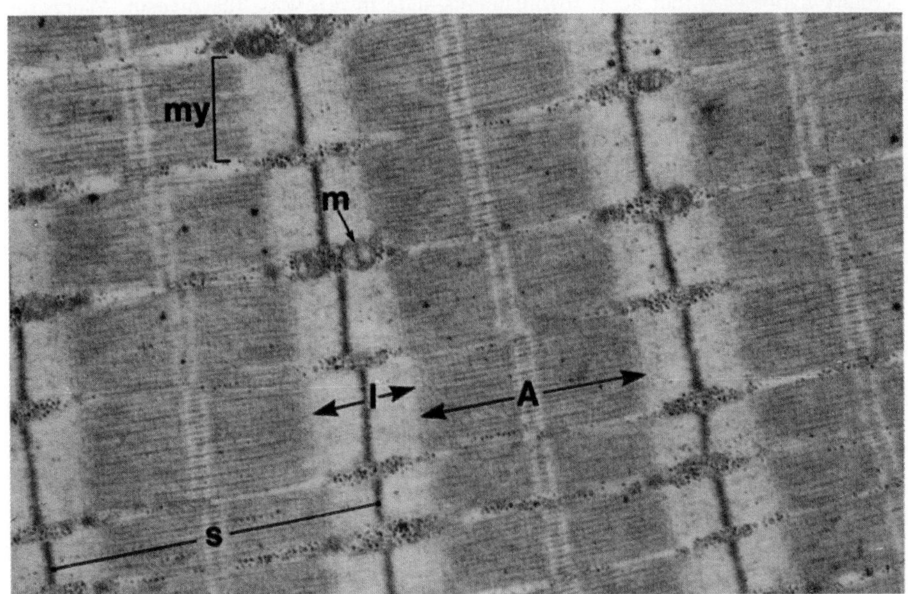

(b)

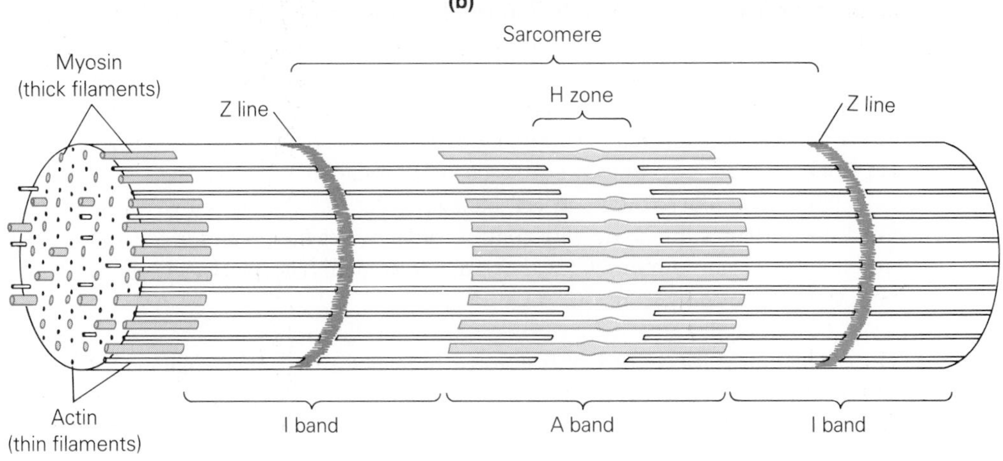

(c)

FIGURE 33.20
THE MYOFIBRIL LEVEL
(**a**) When viewed from cross section, the arrangement of the thin actin myofilaments around each thick myosin myofilament is visible. (**b**) The organization of a contractile unit is clearly seen in the lateral view through the electron microscope. (**c**) In the reconstruction, each sarcomere in this myofibril is bordered by two Z lines. Just inside each Z line are the lighter I bands, containing actin only. Between the I bands is the darker A band, containing both actin and myosin. Its lighter center is the H zone, with myosin only. The end view, taken through the darkest part of an A band, shows the arrangement of actin myofilaments around each myosin.

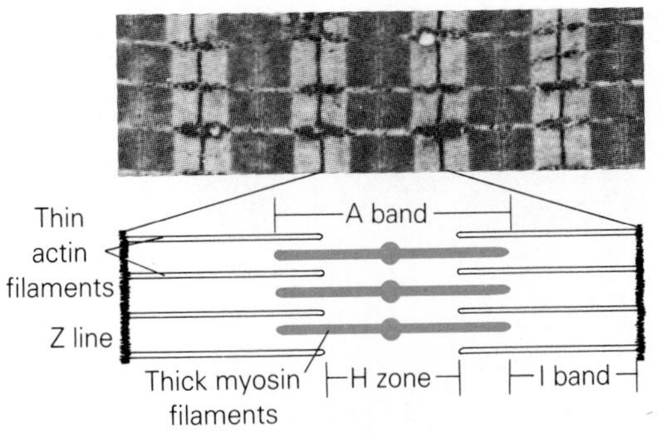

(a) Relaxed

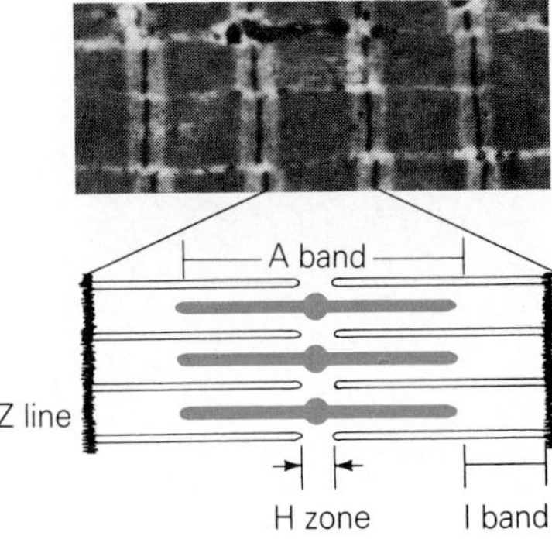

(b) Contracted

FIGURE 33.21
RELAXED AND CONTRACTED MUSCLE
Electron micrographs of muscle show what happens at contraction. **(a)** The muscle in a relaxed state: note the distance between Z lines, the width of the I bands, and the density of the A band. **(b)** Now observe the muscle in a contracted state: the lines have been included as guides to reveal changes in Z line distance and I band width. Note these changes and the increase in density in the A band. Has the A band width changed? What has happened in the H zone? How would these changes affect an entire muscle?

Contraction at the Ultrastructural Level When the muscle fiber is stimulated, the Z lines move closer together and the sarcomere is shortened. This changes the banding, as you can see in Figure 33.21, darkening the region where the H zone was. What actually happens is that actin myofilaments slide inward past the myosin. When the muscle is fully contracted, the light I bands are greatly reduced because the actin and myosin are then almost completely overlapping. Notice that the width of the A band remains constant—it is exactly equal to the length of a myosin filament. It's important to keep in mind that the filaments themselves don't change in length, although sometimes the actin appears to pile up in the center of the sarcomere.

All this leaves us with a host of questions. What causes actin filaments to slide? What is actin, anyway? What is the relationship between its movement and ATP? Only since the development of electron microscopes have biologists been able to answer some of these questions. For example, at very high magnification, researchers have found numerous tiny bridges extending from the myosin filaments to the surrounding actin filaments (Figure 33.22). The bridges are globular extensions of myosin, called **myosin heads.** During muscle contraction the myosin heads apparently reach out at an angle, attach to the nearest actin filament, forming bridges, and then they straighten, pulling on the actin and causing it to slide along. The sequence is repeated again and again, the myosin heads leaning out along the length of the actin, attaching, straightening, releasing, and then leaning out to get a new grip further along. The pulling action is not unlike the hand-over-hand movements a line of sailors might use to raise the mainsail.

The myosin heads contain ATPases, enzymes that can hydrolyze ATP by cleaving its outer phosphate group. Recall that this yields ADP, P_i, and some free energy. In this instance, the energy is used in the formation and movement of myosin bridges.

As you can see in Figure 33.22c, the pulling action occurs on each side of the H zone, which explains why the Z lines draw together as each contractile unit shortens. When the action is repeated throughout a muscle, the entire muscle shortens.

We've seen how the action of numerous myosin heads within the contractile units can bring on muscle contraction. This bit of information, we might add, represents a major finding by biochemists and cell biologists. But the myosin bridges are only half of the story. The rest has to do with control. What causes the bridges to form, and what prevents them from forming when they are not needed? The answers are found at the molecular level of muscle action.

The Molecular Aspects of Contraction In their search for controlling mechanisms, muscle physiologists turned their attention to isolated actin and myosin. Applying chemical analysis and electron microscopy (this time using magnifications of up to half a

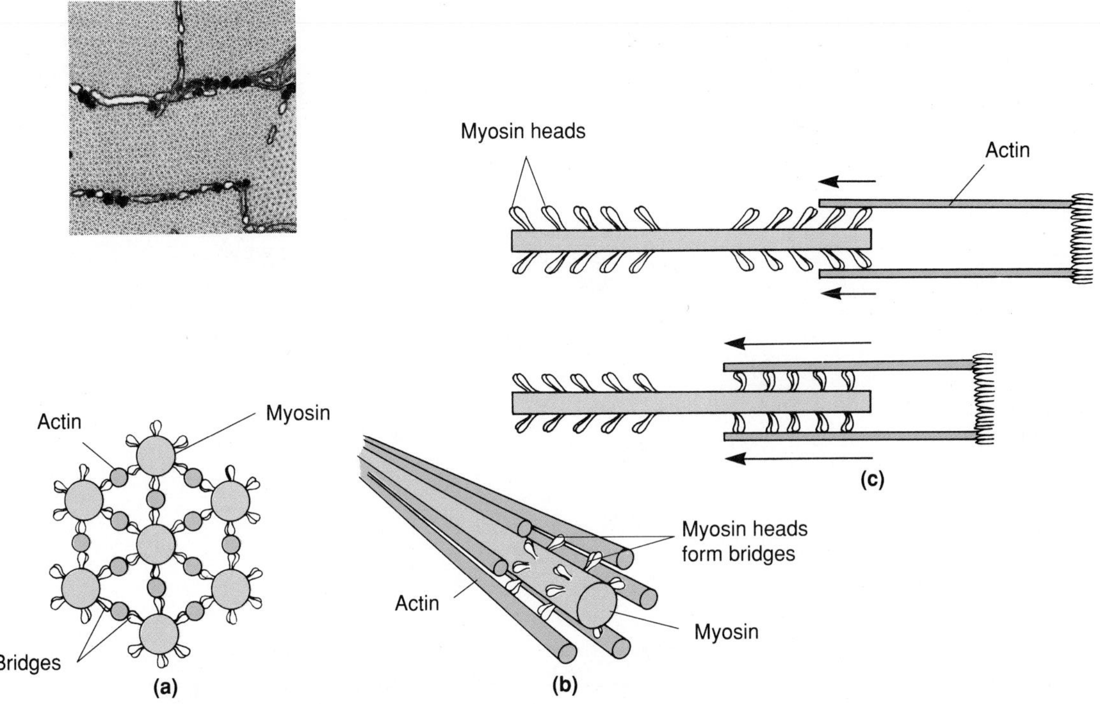

FIGURE 33.22
MYOSIN BRIDGES
(a) The electron micrograph of a cross section through the myofibril and the accompanying drawing show the bridges between myosin and the surrounding actin. **(b)** Notice the relationships of the myofilaments and their bridges in the three-dimensional drawing. **(c)** During contraction, these bridges actively pull the actin fibers inward, thus shortening the contractile unit.

million diameters), they found the molecular detail needed to develop the necessary models of what occurs in living organisms.

What they found was that each myosin filament consists of numerous spirally wound, rodlike proteins. Where individual protein molecules end, they turn outward from the filament, forming the club-shaped myosin heads (Figure 33.23a). The heads occur in clusters at specific distances along the filament. The thin actin filaments are more complex, each consisting of three protein components: actin, **tropomyosin**, and **troponin** (Figure 33.23b). The largest component is the actin itself, which consists of two long strands arranged in a gradual helical twist. Each strand contains numerous globular polypeptide subunits. The protein tropomyosin is also long, double-stranded, and helical, but is much thinner than actin. It lies along the grooves formed by the actin helix. The third component, the globular protein troponin, occurs in regularly spaced clusters, bound to the actin strands.

As we've seen, actin does the actual moving during contraction. It provides binding sites for the myosin bridges, and it is this binding that triggers ATPase activity in the myosin heads. Tropomyosin and troponin control muscle contraction because they block the actin-myosin binding sites—thus inhibiting contraction (Figure 33.24a). The sites will stay blocked, in fact, until the appearance of the calcium ion (Ca^{2+}). When Ca^{2+} is present in the contractile unit, it binds to troponin, which somehow alters tropomyosin so that it can no longer block the binding sites (Figure 33.24b). Actin's myosin-binding sites are then exposed, and the myosin heads attach and become active. Upon the release of energy from ATP, the bridges draw the actin filaments inward, and the muscle contracts. It turns out that each complete action by a myosin head requires two ATPs, the first to straighten the myosin head, the second to break its attachment so that it may reattach to a new binding site.

The next question is, what controls the calcium? In the resting muscle, calcium ions are stored in the sarcoplasmic reticulum, continually transported there by ATP-powered

FIGURE 33.23
ACTIN AND MYOSIN
(a) The thicker myosin myofilament consists of two long, slender proteins spirally wound into the filament. At the end of each rod there emerges a thick, club-shaped tip—the myosin head. **(b)** The thinner actin myofilaments contain three kinds of protein. The major protein is globular actin. The individual actin spheres are arranged in a long, slender, helical filament. Tropomyosin, a short filamentous protein, follows the curve of the helix for a distance, ending with a molecular of troponin. The tropomyosin-troponin-actin complex of the actin filament controls the formation of myosin bridges.

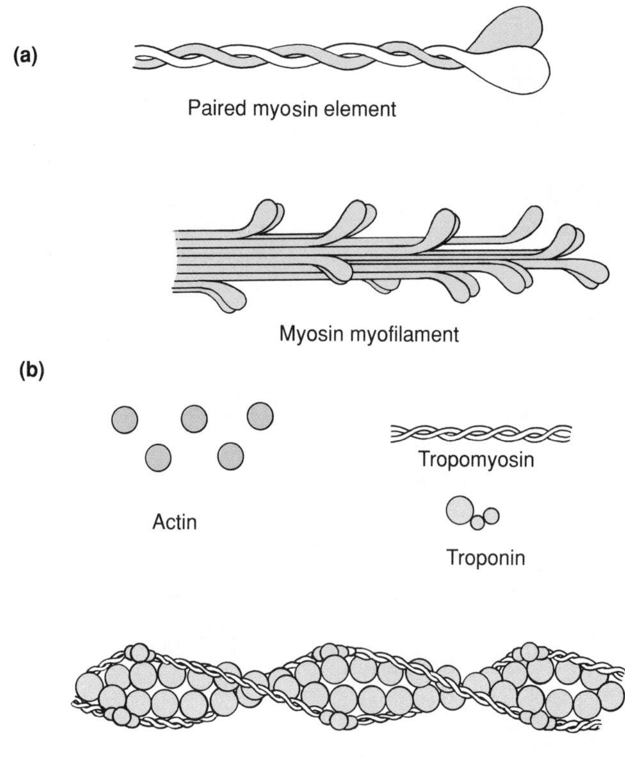

membranal pumps. The pumps thus keep the contractile units clear of calcium until muscle contraction is needed.

We already know that neural impulses bring on muscle contraction, so all that remains is to fill in one more blank. When a neural impulse from a motor neuron reaches the neuromuscular junction on the sarcolemma, it sets in motion an electrical disturbance. The disturbance spreads over the sarcolemma and then along the T-tubules to the membranous sarcoplasmic reticulum—within which the calcium ions are stored. The disturbance causes calcium ion channels in the sarcoplasmic reticulum to open, and as the ions suddenly leak out, they diffuse rapidly down their gradient into the surrounding contractile units. The muscle then contracts. All of this occurs so rapidly that all of the contractile units in each muscle fiber can contract almost simultaneously.

The muscle will contract as long as calcium ions are present, and calcium remains as long as the neural impulses continue. When the impulses cease, the events of contraction reverse themselves. The calcium ion pumps quickly clear the ions from the contractile units (Figure 33.24c), the tropomyosin/troponin complexes block the myosin-binding sites of actin, and the muscle relaxes.

Summing Up We've covered a lot of ground here, so let's briefly review the major points, starting with the neural impulse.

1. Neural impulses create electrical disturbances that cause the release of calcium ions from the sarcoplasmic reticulum.
2. Calcium ions enter the contractile units, where they alter the troponin/tropomyosin inhibitors, exposing actin's myosin-binding sites.
3. Myosin heads bind to actin, energy from ATP is released, the heads change shape, and the actin myofilaments are drawn along.

● Ca^{2+} (calcium ions)

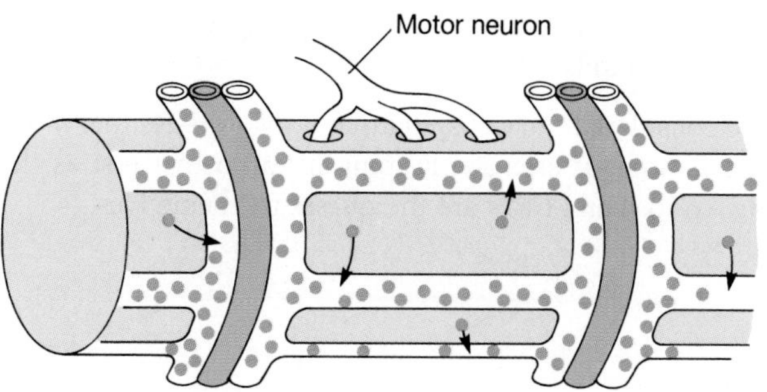

(a) Resting state Ca^{2+} pumped into sarcoplasmic reticulum.

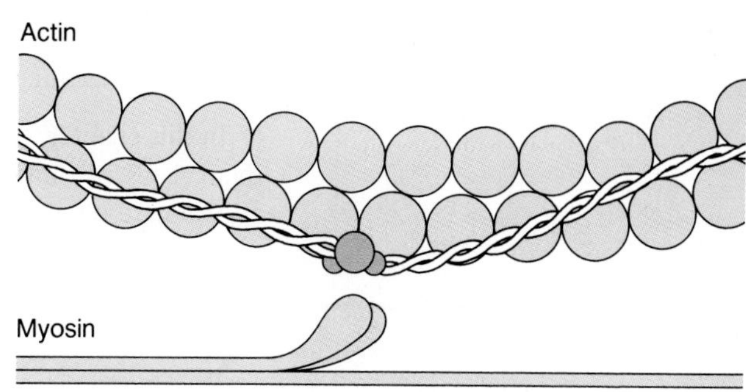

Bridge inhibition (myosin attachment site blocked)

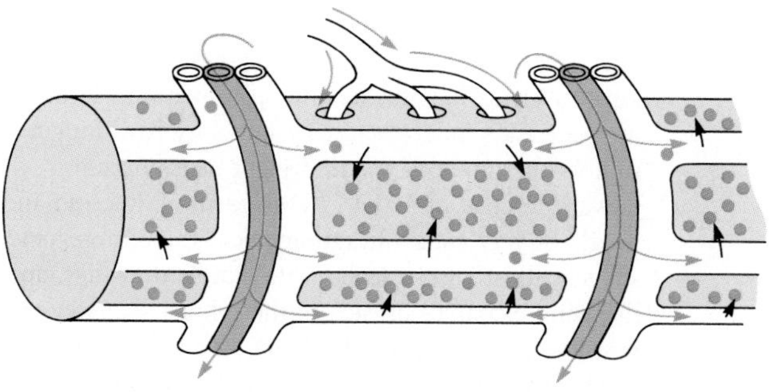

(b) Motor impulse occurs Impulse reaches muscle and spreads to surface into T-tubules, then to sarcoplasmic reticulum. Ca^{2+} leaves sarcoplasmic reticulum and enters contractile units.

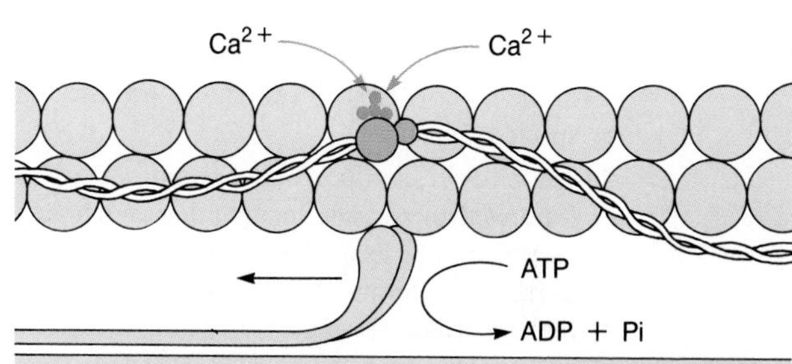

Inhibition removed (myosin attachment site exposed)

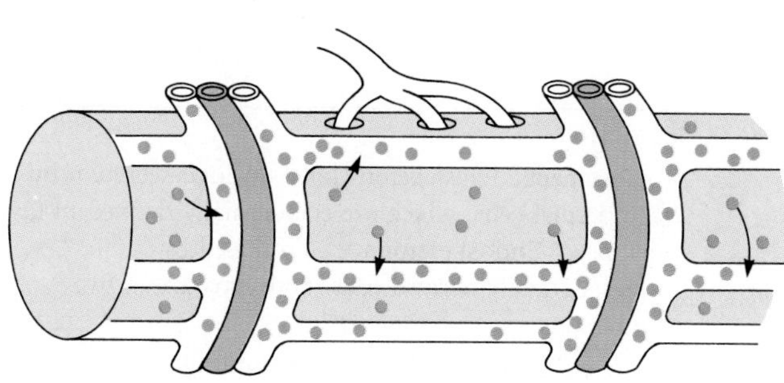

(c) Muscle recovers When impulse ceases, the calcium-ion pumps again transport Ca^{2+} into the sarcoplasmic reticulum.

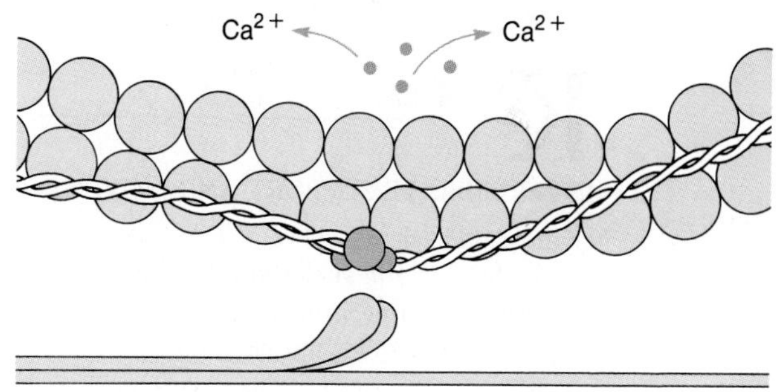

Inhibition returns (myosin attachment site again blocked)

FIGURE 33.24
CALCIUM AND CONTROL
(a) In the muscle's resting state, calcium ions are pumped into and stored in the sarcoplasmic reticulum, and the muscle is polarized. In this state, actin-myosin binding (bridge formation) is inhibited. (b) When an impulse from a motor neuron reaches the muscle surface, a depolarizing wave is created, passing down the T-tubules to the sarcoplasmic reticulum. Ca^{2+} floods the contractile units, the bridges form, and ATP in the bridges is cleaved to ADP and P_i in the usual energy-releasing action. Each bridge shortens, doing its part in moving the actin filament inward. Spent bridges take in a second ATP and are released, reattaching further down the actin filament to tug again. (c) When neural stimulation ceases, Ca^{2+} is removed by ion pumps. Troponin and tropomyosin again block the bridge attachment sites, and the muscle returns to its resting state.

4. A second ATP provides the energy for restoring the myosin heads so that they may reattach and continue pulling.
5. When the neural impulses cease, the calcium ions are pumped back into the sarcoplasmic reticulum, and the inhibited (relaxed) state is restored.

In this chapter, then, we have considered skeletal systems and the muscles that move them. Let's continue now with our look at animals, keeping in mind that just as the answers are often so fascinating, so in many cases are the questions themselves.

APPLICATION OF IDEAS

1. The evolution of skeletons has occurred in two distinctly different directions in the protostomes and deuterostomes. How do skeletons in the two groups differ? What causes the drastic terrestrial size restriction in one of these trends? Is there a restriction in the other? Are such restrictions present in aquatic species? Explain.
2. The bird skeleton probably has more specialized features than are seen in any other vertebrate. Describe several specific examples. Also cite examples of the use of triangular forms. What is the value of this? How do humans make use of such a form in architecture?
3. Describe how muscle studies have involved the merging of different scientific disciplines. For example, the visual aspects of contraction have now been interpreted on the biochemical level. Explain where one leaves off and the other begins in the contractile process.
4. Levers are classified according to the position of the fulcrum, the force applied, and the weight moved. *First, second,* and *third class levers* are described respectively as follows: weight and downward force at opposite ends, fulcrum near weight (pry bar lifting a stone); fulcrum at very end, weight near fulcrum, upward force at other end (wheelbarrow); fulcrum and weight at very ends, lifting force near the center (hand shovel). Illustrate the levers with simple drawings, and then find examples of each in the human skeleton and muscles. Determine which, if any, offers the best mechanical advantage (the least amount of force needed to move a part the greatest distance).

KEY IDEAS

INVERTEBRATE SUPPORT AND MOVEMENT

Hydrostatic Skeleton

1. The **hydrostatic skeleton** firms the animal body through fluid pressure against a resistant wall. It also provides a base for muscle action.
2. Roundworms are fluid-filled, with muscles directed longitudinally, permitting lashing movements only.
3. Earthworms have complex muscle tissue arranged in two directions around their fluid-filled coelom. The body is extended by the squeezing action of circular muscles and shortened by contraction of longitudinal muscles. The combined action of muscles and anchoring **setae** makes the earthworm an excellent burrower.
4. Certain mollusks make use of the hydrostatic property for making digging movements with the muscular foot.

Exoskeletons

Exoskeletons surround the body, providing a base for muscle attachment and levers to be used in body movement. **Antagonists** provide for opposing movements. Exoskeletons form through secretion by epidermal cells, which are ectodermally derived in the embryo. **Endoskeletons** are produced by mesodermally derived cells. Both have protective functions.

1. The arthropod exoskeleton—the **cuticle**–provides a muscle base and leverlike parts, prevents water loss, and protects soft body parts. The cuticle is leathery and flexible at the joints. Muscles that move the body have fixed origins and moveable insertions. Muscles occur as antagonists, groups that have opposing movements. Because of weight problems, the exoskeleton is useful only to small animals. The agility of insects is partly explained through muscle "cock-and-release" mechanisms.
2. Wing movement in flying insect species is either *direct* or *indirect.* In the first, opposing **direct flight muscles** simply move the wings up and down, while the second, **indirect flight muscles** are vertically and longitudinally arranged and move the **thorax** up and down, and the wings respond through a complex

hinge. A tilted angle provides lift. Wingbeat rates can exceed neural impulses because of a resonating mechanism in the muscles.

3. Mollusk shells protect the body, but parts such as the fleshy foot and eyestalks are often moved through hydrostatic mechanisms. Air is withdrawn from spaces in the spongy cuttlebone, a shell remnant in cuttlefish, making it an effective flotation device.
4. In mollusks, external shells are continuously secreted by the mantle and take several forms.

Invertebrate Endoskeletons

The dermis of echinoderms secretes calcareous plates and spines that originate below the epidermis.

VERTEBRATE TISSUES

1. The vertebrate body consists of four tissue types: **epithelial, connective, nerve,** and **muscle.**

Epithelial Tissue

1. **Epithelium** or **epithelial tissue** covers the body as **epidermis** and lines internal organs as well. Blood vessels have an **endothelium,** while glands have a **glandular epithelium.**

Connective Tissue

Connective tissue contains cells in a noncellular **connective tissue matrix** as seen in bone, cartilage, **ligaments, tendons,** and blood. **Collagen** is a common binding or tough matrix material. Cells called **fibroblasts** secrete collagen and are active in bone mending. **Elastin** fibers provide elasticity to artery walls.

VERTEBRATE SKELETONS-HARDENED CONNECTIVE TISSUE

The bony endoskeleton provides support, a muscle base, protection, a source of blood cells, and acts as a calcium reservoir.

The Structure of Bone

1. Bone is surrounded by the sheathlike **periosteum.** Within are regions of weblike **spongy bone,** containing **red marrow** where red cells form, and regions of **compact bone,** containing fatty **yellow marrow.** Bone consists of hard calcium phosphate in a collagen matrix containing numerous bone cells, blood vessels, and nerves.
2. Compact bone is organized into **Haversian systems** or **osteons,** each with a **central canal** and hard, concentric, laminated **lamellae. Osteocytes** (bone cells) reside in hollow **lacunae,** interconnected by weblike **canaliculi.**

Organization of the Vertebrate Skeleton

1. Vertebrate skeletons consist of two divisions. The **axial** division contains the **cranium** (skull), **vertebral column** (backbones), and ribs. The **appendicular** division contains the two **girdles—pelvic** (hip bones) and **pectoral** (shoulder bones and limbs).
2. Immovable joints include the fused **sutures** in the skull. Movable joints have articulating surfaces of cartilage, and the bones are held in place by **ligaments** that often form the capsules of **synovial joints.** Lubrication is provided by fluid secretions.
3. Movable **joints** include **gliding** (wrist bones; slight back and forth), **pivotal** (head and neck; rotation in one plane), **ball-and-socket** (shoulder; free rotation), and **hinge** (knee; hingelike extension and flexion in one plane).
4. The skull or cranium in modern fishes includes many separate bones with little fusion. Fusion increases in the amphibians, reptiles, and birds, and reaches its peak in mammals. Jaw mobility is also greatest in the mammals, although the **maxilla** (upper jaw) is fused to the skull. Sensory receptors located on the cranium communicate with the brain through **foramina.** The **foramen magnum** at the base of the skull admits the spinal cord.
5. The number of **vertebrae** in the **vertebral column** varies among vertebrate classes. Reptiles have the most, while in birds they may have fused. Mammals all have seven neck vertebrae.
6. The 33 human vertebrae include four in the **coccyx.** The free vertebrae are separated by compressible, soft-centered **intervertebral disks** of **fibrocartilage.** The vertebral column houses and protects the **spinal cord,** which passes through the **neural canal.**
7. Each half of the human pelvic girdle contains three bones. On each side, the three bones form the hip socket. The two ilia and the **sacrum** form the **sacroiliac joint** in the back. The pectoral girdle includes the two **scapulae** and **clavicles,** a loose assemblage that forms the shoulder joint.
8. Human forelimbs are very generalized, except for subtle thumb modifications. The **humerus** (upper arm) joins the ball-and-socket shoulder joint, while the **radius** and **ulna** (lower arm) form a complex pivoting hinge joint at both ends.
9. The human leg contains the large **femur** (upper leg) and the **tibia** and **fibula** (lower leg). The hinged knee joint, formed by the femur, tibia, and **patella** (knee cap), depends on a capsule of many tough ligaments for its strength.

VERTEBRATE MUSCLE: ITS ORGANIZATION AND MOVEMENT

Smooth Muscle

Smooth muscle (also **involuntary** or **visceral**), is slow moving and rhythmic, occurring in the gut, blood vessels, hair roots, iris, uterus, and other places under the unconscious control of the **autonomic nervous system.** Individual cells occur in sheets, are long and tapered, and have a single nucleus.

Cardiac Muscle

Cardiac muscle (heart) is involuntary. Its fibers are cylindrical and branching, with interlocking borders called **intercalated disks,** important to contraction. It has a rhythmic and tireless contractile characteristic in which isolated cells contract spontaneously.

Skeletal Muscle

1. Skeletal muscle (also **voluntary** or **striated**), is controlled at the conscious level through the **somatic nervous system.** Cells, called **fibers,** are striated and multinucleated.
2. Muscle fibers form bundles called **fascicles,** and a number of fascicles bound by connective tissue form muscle bundles. Muscles contain a rich supply of blood vessels and nerves. A sheathlike **fascia** surrounds the muscle, forming the **tendons** at its ends. Tendons connect muscle to bone, forming **insertions** on the part to be moved and **origins** on the stationary part.
3. Muscles also move the face and the tongue, and others form ringlike **sphincters** in the gut. Sheetlike muscles in the abdomen originate and insert via flattened tendons called **aponeuroses.**
4. Muscle fibers are surrounded by the **sarcolemma,** which receives **motor neurons** at **neuromuscular junctions.** Each fiber contains several nuclei, many mitochondria, and an extensive **sarcoplasmic reticulum.** Hollow **transverse tubules,** or **T-tubules,** form boundaries for each contractile unit. The pattern of contractile **myofibrils,** or **fibrils,** produces the striations.
5. Myofibrils contain **myofilaments,** or **filaments,** made up of two kinds of protein: **myosin** and **actin.** Each myosin filament is surrounded by six actins. The striations are seen in the longitudinal view. Prominent **Z lines** mark the **sarcomere,** or contractile unit. Within the Z lines lies a light **I band** (actin alone), and further in lies a darker **A band** (actin and myosin), ending in a lighter central **H zone** (myosin alone).
6. Upon contraction, the Z lines move closer together, the I bands are reduced, and the H zone darkens. However, the A band remains constant, indicating that it is the actin that moves, while the myosin is stationary. The patterns change because minute **myosin** heads form bridges with the actin filament, pulling it inward.
7. The energy for bridge movement is provided by ATP, which is hydrolyzed within the myosin heads which contain ATPases.
8. Myosin heads emerge at regular intervals from spirally wound protein filaments. Actin filaments contain the proteins actin, **tropomyosin,** and **troponin.** The latter two lie on the thicker actin proteins. Tropomyosin and troponin control contraction by inhibiting myosin attachment. The presence of calcium alters their arrangement, making myosin bridge attachment possible. One ATP provides for bridge straightening, and a second provides for reattachment.
9. In contraction, a neural impulse reaches the muscle fibers and travels along the T-tubules to the sarcoplasmic reticulum, which responds by suddenly releasing calcium ions into the contractile unit. The ions alter the inhibition state, and the bridges connect and straighten sequentially along the actin filaments, pulling them inward as contraction continues.
10. Restoration of the resting state begins when neural stimulation stops and the calcium ions are actively transported back into the sarcoplasmic reticulum. The action occurs simultaneously throughout the fibers served by a motor neuron.

REVIEW QUESTIONS

1. Explain how fluids in the nematode body provide the resistant base needed for coordinated muscle movement. Describe the arrangement of muscle fibers and the movement they provide. (753)
2. Describe the muscle arrangement of the earthworm, and explain how it makes use of this in burrowing. Of what specific use is hydrostatic pressure in the coelom? (754)
3. Use a chart to compare the location, manner of formation, embryological origin, and functions of exoskeletons and endoskeletons. (754–755)
4. Using an example, explain the antagonistic arrangement of muscles in the arthropod. Why are antagonists essential? (755, 757)
5. What size restrictions does the arthropod exoskeleton impose? Why is this so? (757)
6. Compare the direct and indirect arrangement of flight muscles in the insect. (757)
7. Explain how the molluskan shell grows. To what geometric forms is such growth restricted? (759)
8. Describe the endoskeleton of the echinoderm. Considering that much of it protrudes through the epidermis, what qualifies it as an endoskeleton? (760)
9. What is the general role of epithelial tissue? List four different kinds and mention where they are found. (760)
10. List five human structures that consist primarily of connective tissues. (760–761)
11. What are two roles of collagen? How is it produced? (761)
12. List five important functions of the vertebrate skeleton. (762)

13. Compare spongy and compact bone. In which are the red and yellow marrows found? (762)
14. With the aid of a simple diagram, illustrate an osteon, labeling lamellae, canaliculi, lacunae, osteocytes, and the central canal. Which is (are) nonliving? (762–763)
15. List four types of joints and, using examples from the human body, describe the movement possible in each. (764)
16. Describe a typical synovial joint, including the role of ligaments and cartilages. (764)
17. Describe the organization of the human vertebral column, naming the main regions and listing characteristics of vertebrae in each. Also describe the structure of the disks separating the vertebrae. (768)
18. List the bones that make up the pelvic girdle and name the three joints they form. (768–769)
19. List the bones that make up the pelvic girdle and name the three joints they form. (769)
19. List the bones of the pectoral girdle. In what way is the human pectoral girdle different from those of most mammals? (769–770)
20. Briefly discuss a unique human characteristic in the hands and feet. (760–770)
21. List one unique characteristic each for skeletal, cardiac, and smooth muscle tissue. (770–772)
22. List two ways in which cardiac and smooth muscle are similar, and one way in which cardiac and skeletal muscle are similar. (770–772)
23. How do the roles of ligaments and tendons differ? (764, 772)
24. Distinguish between tendons of origin and tendons of insertion. Do tendons always insert or originate on bones? Explain. (772)
25. Starting with the complete muscle, write the following terms in a logical descending order: Complete muscle, myofilament, troponin, fiber, actin, myofibril, tropomyosin, sarcomere, myosin, fascicle. (772–774)
26. Using a simple drawing, explain the organization of a contractile unit. Begin with two Z lines, and fill in between. Explain the composition of each part of the pattern. (774)
27. Describe the arrangement of actin and myosin in a contractile unit, and in so doing explain how the actin moves during contraction. (774–776)
28. Describe the protein components of myosin and actin. (776–777)
29. What keeps the myosin bridges from spontaneously attaching to actin? What happens to ATP when attachment does occur? (777)
30. How does the absence of CA^{2+} affect the contractile unit? Where are these ions when the muscle is at rest? (777–778)
31. Beginning with an impulse moving along a motor neuron, describe the sequence of events leading to contraction. (780)

34

NEURAL CONTROL I: THE NEURON

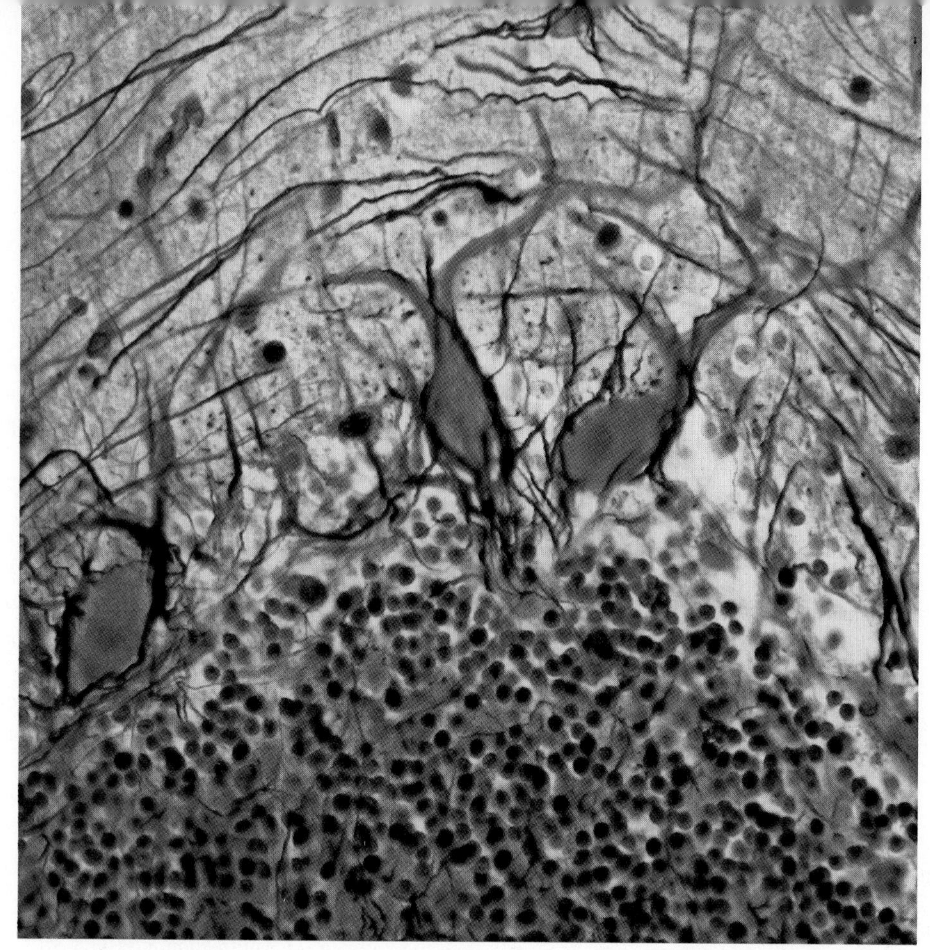

OUR PLANET IS A LIVELY PLACE, AND ONE OF MIXED BLESSINGS. IT PROVIDES us not only with opportunities, but dangers. In fact, that is what much of life is about: taking advantage of the opportunities and avoiding the dangers. This means, of course, that living things must be able to assess the nature of the environment accurately and to respond to it appropriately. Deer, rabbits, and grasshoppers are able to sense the presence of young, tender plants. As the plants quietly submit to the ravages of the plant eater, they may be avenged by some sharp-tooth predator peering from behind a rock. The predator is able to detect and evaluate signals emanating from the prey as surely as the plant eaters can find grass. Both kinds of animals share the same environment, but their sensory abilities are specialized to react to different aspects of it. Here we will see not only how signals are detected but also how they move through the body, activating a pattern of responses that help the animal succeed in a complex and changing world.

Essentially, what we will see is a network of highly specialized cells that are devoted to the task of transmitting information. These are the nerve cells, or **neurons.** Their function is to generate neural signals and conduct them from one part of the body to another. The signals are electrical in nature, brought about by the shifting of charged particles (ions) across the neural membrane. The signals themselves are simple enough; the complexity of the nervous system in animals such as vertebrates is due to the elaborate circuitry with its incredibly intricate arrangements and interconnections, involving literally billions of neurons.

Functioning neural networks accomplish three things. They *detect* events in their surroundings and send a signal to coordinating centers. These centers, which consist of other neurons, *integrate* the information and formulate a response. Additional neurons carry this decisive message to some body part, where a *response* is initiated. As you can see, the three key terms are "detection," "integration," and "response." The environmental cue that started all of this is referred to as a **stimulus.** Figure 34.1 illustrates the three aspects at work in a real-life situation.

FIGURE 34.1
ANIMAL RESPONSES
The ability to respond rapidly to incoming environmental cues is an outstanding characteristic of animals. Specialized receptors detect events in their surroundings, and this "raw information" is directed to the animal's brain for processing, integration, and response.

THE NEURON
Cellular Structure

Neurons come in many sizes and shapes, but they are characterized by having a **cell body** from which a number of processes (lengthy structures) extend (Figure 34.2). These processes can be extremely long, since some neurons must carry messages to a distant part of the body (such as those that reach from your foot to your spinal cord).

The cell body contains the nucleus and, generally, most of the cell's cytoplasm. The cytoplasm contains the typical organelles, such as ribosomes, an endoplasmic reticulum, and numerous secretory bodies. The cell body can produce **neurotransmitters**—chemical messengers that ordinarily must be present for a **neural impulse** to be relayed from one neuron to another. Neurotransmitters are transported within the slender neural processes along a highly organized cytoskeletal framework. Eventually they are released at specific sites.

The processes that extend from the cell body are of two principal types: **dendrites** and **axons.** The highly branched dendrite ("little tree") is the receiving end of the neuron. It receives stimuli from its surroundings, often other neurons, and conducts this information in the form of an electrical signal toward the cell body and axon. Actually, the cell body need not receive all its signals from the dendrite; it can be stimulated directly.

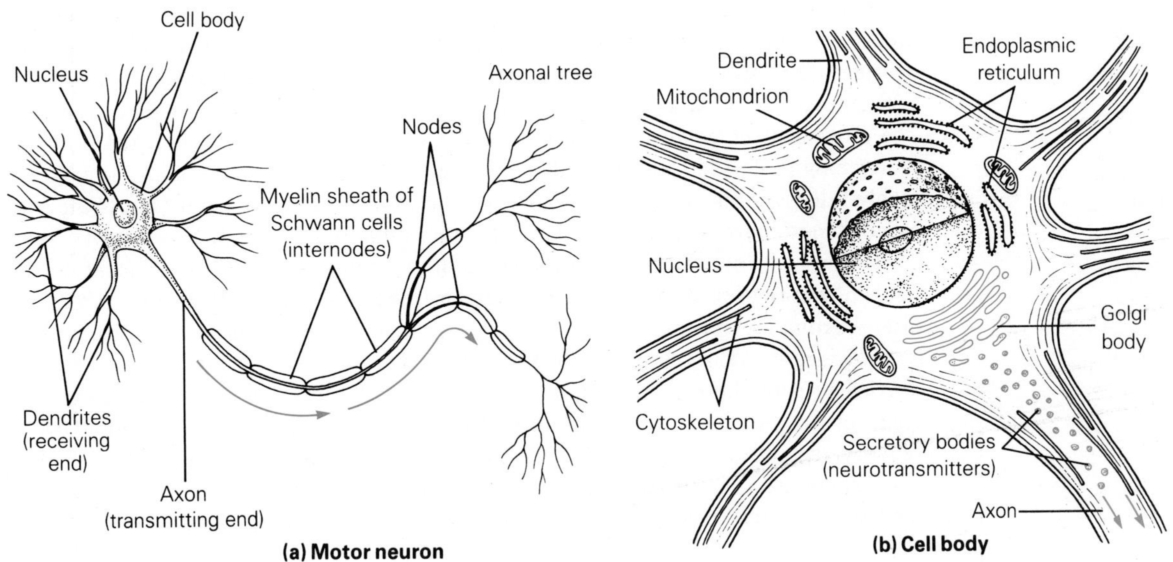

FIGURE 34.2
THE NERVE CELL
(a) Each vertebrate neuron has three principal parts: receiving processes called dendrites, a cell body, and a sending portion called an axon. **(b)** The cell body produces neurotransmitters, which are carried by vesicles along the axonal cytoskeleton to the axon's branched endings.

Axons are lengthy processes that are specialized for transmitting signals over long distances in the animal body. (Axons are absent in some of the smaller neurons of the brain, those that transmit their signals over very short distances.) Axons function by converting signals from the dendrite or cell body into a special kind of signal, or **neural impulse,** also called an **action potential.** Action potentials usually originate in the **hillock,** a region at the base of the axon (see Figure 34.2), but they can also originate anywhere along the axon.

The axon transmits neural impulses *away from* the cell body. An axon may communicate with other neurons, relaying its signal along to them, or it may directly stimulate an **effector.** An effector is any organ capable of a desired response, such as movement or secretion. Effectors can be any of the muscles, including those of skeleton, heart, blood vessels, and intestine. An effector can also be one of the many glands, such as a salivary gland, which upon stimulation pours its secretions into the mouth.

A neuron often has many dendrites, but it usually has only one axon. The single axon, however, may branch at any point along its length. An axon commonly divides and redivides at its tip, forming a terminal **axon tree** (also called an arborization), each branch of which forms a knoblike ending. It is from these endings that the axon releases a neurotransmitter that either triggers an impulse in the next neuron or activates an effector. Where the tips of the axons innervate a muscle fiber, they spread to form **neuromuscular junctions** (see Figure 33.24).

The axon seen in Figure 34.2 is surrounded (jelly-roll fashion) by a **myelin sheath,** a flattened sheath of fatty material typical of that found in many, though not all, types of vertebrate neurons. Like any lipid, myelin has great electrical resistance and acts as an insulator. But the sheath is interrupted at frequent intervals by the **nodes** (also called nodes of Ranvier), where the axon is in direct contact with the surrounding intercellular fluid. The nodes are small spaces between the end of one wrapping cell, or **internode** as the wrapped regions are called, and the beginning of the next. The fatty myelin sheath is composed of 50 to 100 layers of the flattened and rolled plasma membrane of a specialized axon-encasing cell. Outside the central nervous system, myelin sheaths are formed from **Schwann cells** (Figure 34.3), while within the central nervous system (that is, in the brain and spinal cord) they are formed from somewhat similar cells called **oligodendrocytes.** The myelin sheath and the nodes have much to do with the speed of neural impulses, as we will see shortly.

Cells of the Nervous System

The nervous system consists of two kinds of specialized cells, the neurons we have already mentioned and **neuroglia.** The neuroglia, or **glial cells,** comprise some 90% of the cellular component of the vertebrate brain. They are believed to perform supporting functions. Glial cells include the myelin sheath–forming Schwann cells and oligodendrocytes just discussed. In addition, they provide a structural framework for the fragile neurons. They also carry out helpful metabolic functions. Neurobiologists believe that another function of the glial cells is clearing potassium ions from the intercellular spaces. The movement of potassium ions in and out of neurons is an essential part of their functioning. One kind of glial cell, the **astrocyte,** appears to be responsible for the highly selective transport of materials from the capillaries to the brain (across the so-called "blood-brain barrier" that keeps the brain's internal environment constant).

Neurons fall into three categories according to their general functions in the three fundamental phases of neural activity. **Sensory neurons** are responsible for *detection,* **interneurons** are responsible for *integration,* and **motor neurons** are responsible for *response*. See Figure 34.4 for a view of representative shapes.

As their name implies, sensory neurons (also called **afferent neurons**) sense what is going on in the animal's surroundings. They usually receive such information from **sensory receptors,** highly specialized structures that are responsive to gravity, touch, light, chemicals, heat, and other environmental stimuli. Sensory neurons carry information concerning both the external and internal environment; thus we are also able

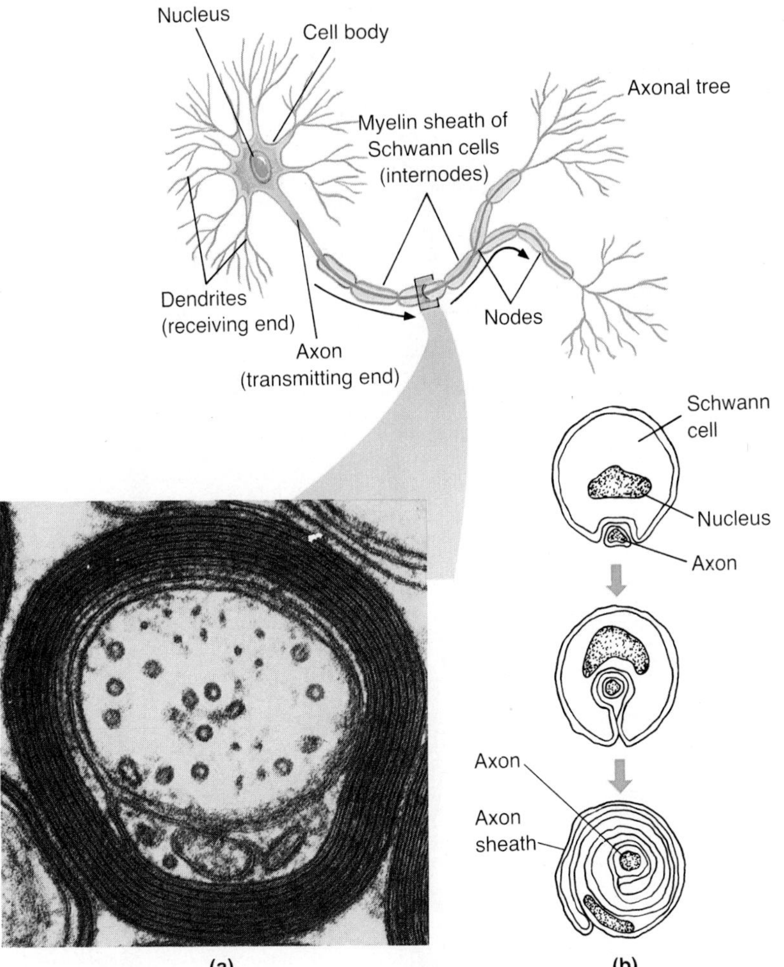

FIGURE 34.3
A MYELINATED NEURON
(a) The electron micrograph is of a cross section through a myelinated axon. The encircling membranes are formed by Schwann cells. **(b)** During development, a sheath-forming cell begins its enveloping action by simply surrounding part of an axon. Then one end of the plasma membrane grows under the other, advancing along in a "burrowing" action until it has wrapped itself about the axon several times.

to receive "situation reports" from our internal organs. Sensory neurons from the skin have lengthy processes called **peripheral axons** that originate at the sensory receptor. A peripheral axon conducts signals to the cell body, and from there they are relayed to a **central axon.**

Interneurons are nerve cells that communicate only with other nerve cells. In the vertebrate nervous system, interneurons make up much of the neural component of the spinal cord and brain. Initially, interneurons receive input from sensory neurons, usually relaying their signals to batteries of other interneurons. The complex circuitry of interneurons in the vertebrate spinal cord and brain is largely responsible for the integration of stimuli and coordination of responses. Interneurons, of course, eventually direct impulses to motor neurons.

Motor neurons (also, **efferent neurons**) receive signals from interneurons and transmit signals called **motor impulses** to effectors, such as muscles and glands, which then respond. In the last chapter we learned about the complex response of skeletal muscle to motor impulses arriving at neuromuscular junctions on the surface of muscle fibers.

Nerves The axons of many neurons often travel together throughout the body. In such parallel arrangements, they form larger structures called **nerves** (Figure 34.5). A nerve can be likened to a telephone cable carrying many individual lines, each insulated from the others. The large glistening nerves are surrounded by their own coverings of touch connective tissues. As you are probably aware, damaging or severing a major nerve can produce, in the organs served, such devastating effects as paralysis or loss of sensitivity, or both.

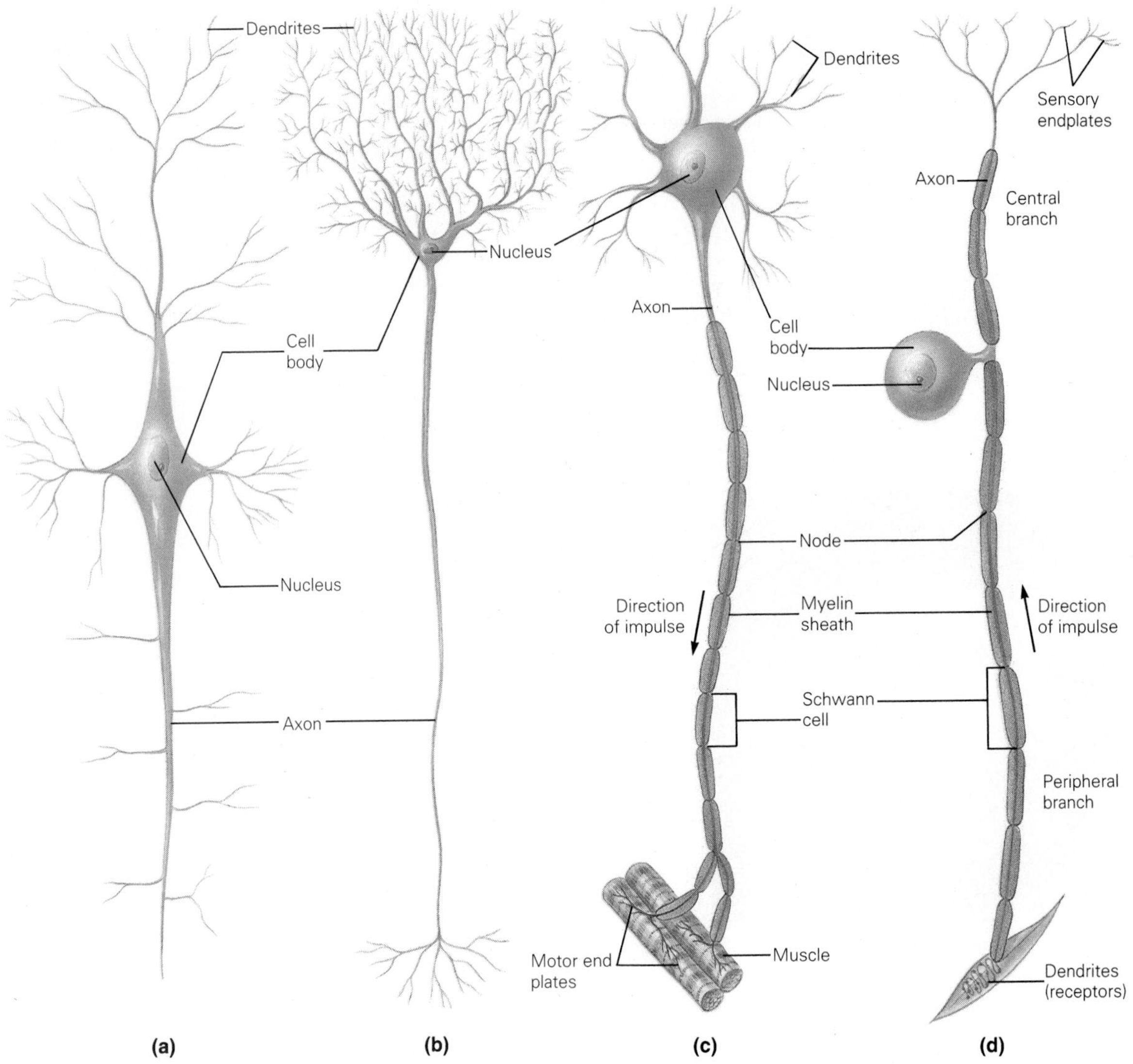

Dendrites
Cell body
Nucleus
Axon
(a)

Nucleus
(b)

Dendrites
Axon
Cell body
Node
Direction of impulse
Myelin sheath
Schwann cell
Motor end plates
Muscle
(c)

Sensory endplates
Axon
Central branch
Cell body
Nucleus
Direction of impulse
Peripheral branch
Dendrites (receptors)
(d)

FIGURE 34.4
VARIATION IN NEURONS
Four neurons found in human beings show the diversity of these cells: **(a)** and **(b)** show clear differences in cell bodies, axons, and dendrites, **(c)** is a motor neuron with axons that run from the nervous system to the effector (in this case, a muscle), **(d)** is a sensory neuron that runs from the receptor to the spine. Note that the sensory neuron has no true dendrites. The nodes in the myelin sheath are where one Schwann cell ends and another begins. A single nerve cell may be nine feet long, such as those that run from the base of a giraffe's spine down its hind leg.

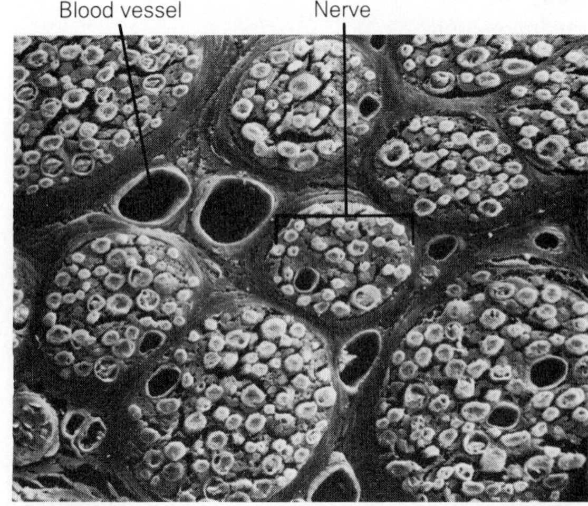

FIGURE 34.5
A NERVE "CABLE"
Several nerves, as seen in cross section through the scanning electron microscope, contain numerous neural processes, possibly both axons and dendrites. Each contains its own insulating sheath, and the entire ensemble is surrounded by dense connective tissue that binds it into its cablelike structure. Blood vessels penetrate the nerve, providing the exchanges needed to maintain the neurons.

ns, action potentials (mentioned earlier) that travel along gnals that travel along dendrites and cell bodies. Passive its in that they are affected by resistance and diminish s, graded signals are quite sufficient for short distance dendrites and cell body (and among dense clusters of use they diminish over distance, they are unsuitable for Sending signals a long distance requires the action po- f the axon. We will confine much of our discussion in tial and find out how the nervous system has become transmission.

cal in nature, does not behave as an electrical current an electrical current diminishes with distance, an action in its strength, seemingly regenerating itself as it travels. on potentials may travel, consider what happens when ow.) Impulses originating in your brain are relayed to l. The motor impulses follow the axon out of the cord, euromuscular junctions in the big toe muscles. There eat for a single cell.

ve the shifting movements of ions across neural plasma nvolves highly selective and controlled ion movement, satile membrane channels. Both active transport and are involved. To understand precisely how an axon transmits a signal so efficiently along its length, we must look more closely at the special nature of its membrane as well as the ions involved. We can divide the events surrounding the action potential into three stages:

1. Establishing a polarized, resting state in the neuron
2. Generating an action potential and propagating it along the axon
3. Restoring the polarized, resting state

The Resting State: A Matter of Ion Distribution Actually, the term "resting state" is not very appropriate. One thing that living neurons never do is "rest." Instead, we should note that the "resting" neuron is in a *polarized state,* that is, a condition in which *the electrical charges outside the axon's plasma membrane are different from those inside.* Specifically, resting axons are more negative inside than outside. The maintenance of this charge difference involves the positioning of certain ions, with their electrical charges, on one side of the membrane or the other. When the axon is polarized in this way, it is in a precarious state indeed. If disturbed, the ions will rush to equilibrium. This tendency is the basis for the action potential. But before we get to how the impulse travels, let's look at the conditions behind the polarized resting state.

The ions most important to neural activity are sodium (Na^+) and potassium (K^+), along with much larger, negatively charged proteins. Whereas the small sodium and potassium ions are quite mobile, the proteins, because of their large size, cannot cross the plasma membrane and so must remain within the axon. In the neuron's resting state, the sodium ions just outside the membrane outnumber those inside by about ten to one, resulting in a steep sodium ion gradient. Potassium ions have an opposite gradient, with many more present on the inside of the axon than on the outside.

One reason for this distribution is that the axon's plasma membrane contains many **sodium/potassium ion exchange pumps.** Such pumps move three sodium ions out of the axon and two potassium ions in for each pump cycle (Figure 34.6a). (The stepwise operation of the pump may be reviewed in Figure 5.10.) From what we have said about the polarized resting state so far, you can see that the exchange pump is of considerable importance in neuron function.

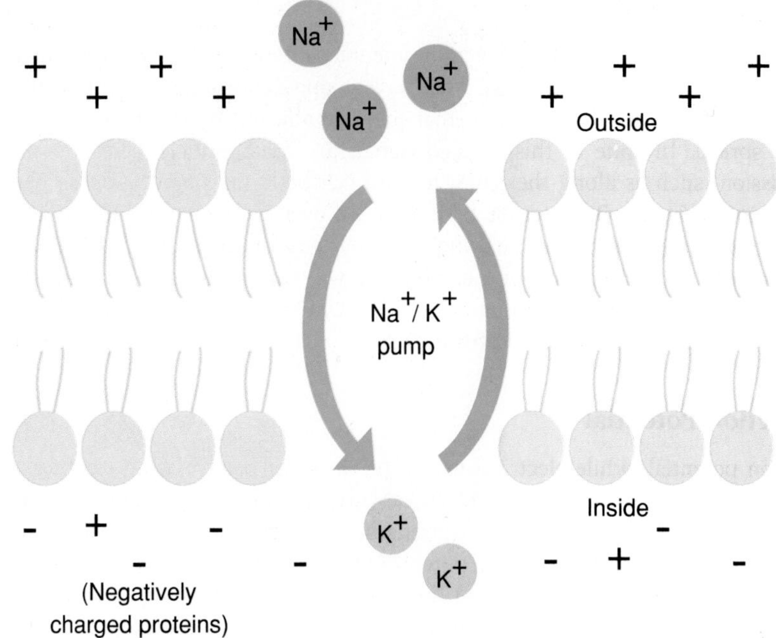

FIGURE 34.6
AXONAL ION EXCHANGE PUMPS AND ION DISTRIBUTION
ATP-powered Na^+/K^+ ion exchange pumps transport Na^+ out of the neuron and K^+ into the neuron in a 3-to-2 ratio. Both ions form steep diffusion gradients across the membrane. Because of immobile negatively charged proteins within the membrane, the axon is more negative inside than out.

Other factors that maintain the two ion gradients have to do with membrane permeability and electrical charges. When at rest, the axon's plasma membrane is not very permeable to Na^+, so as these ions are pumped out, most remain there. This explains the steepness of the Na^+ gradient. Conversely, the axon's membrane is much more permeable (leaky) to K^+, and as these ions are pumped in, many diffuse right back out across the membrane to the cell exterior. Because the membrane is leaky to K^+, one might expect K^+ to reach equilibrium (to equalize in concentrations on both sides of the membrane). This fails to happen because of the fundamental behavior of charged particles: Like charges repel and unlike charges attract. Thus, the many positively charged sodium ions already outside repel the positively charged potassium ions and the many negatively charged proteins on the inside of the axon attract the potassium ions. Thus potassium remains in a greater concentration inside the axon than outside (Figure 34.6b).

The Resting Potential Now that we've considered how the ion distribution produces the polarized state in an axon, let's look at the axon's electrical characteristics. Again, the net electrical charge inside a resting neuron, despite the presence of K^+, is negative relative to the outside, where Na^+ ions abound. The polarized state of the resting neuron can be verified experimentally by placing a tiny electrode on the surface of the axon and inserting another electrode into the cytoplasm within (Figure 34.7). Typically, the charge difference across the membrane results in an electrical voltage of about -70 mV (millivolts: 1/1000 of a volt). This voltage is called the **resting potential.**[1] The minus sign indicates the negatively charged state of the axon's interior relative to the outside surface. The electrical and chemical characteristics of the polarized resting neuron, as

[1]A resting potential of -70 mV represents the equilibrium potential for both ions. The equilibrium potential is a measure of the precise electrical force required to stop the *net* diffusion of an ion across the membrane. (Note that the use of the term "equilibrium" does not at all imply that the concentration of ions is in equilibrium.)
The equilibrium potential for K^+, the voltage that will just oppose its net diffusion across the membrane, falls between -70 and -100 mV. The equilibrium potential for Na^+ falls between $+50$ to $+65$ mV. Because K^+ diffuses far more readily than Na^+ in the resting axon, the resting potential is much closer to the equilibrium of K^+. Should the permeability of the membrane to Na^+ become large, as happens when an action potential begins, the membrane potential would change drastically.

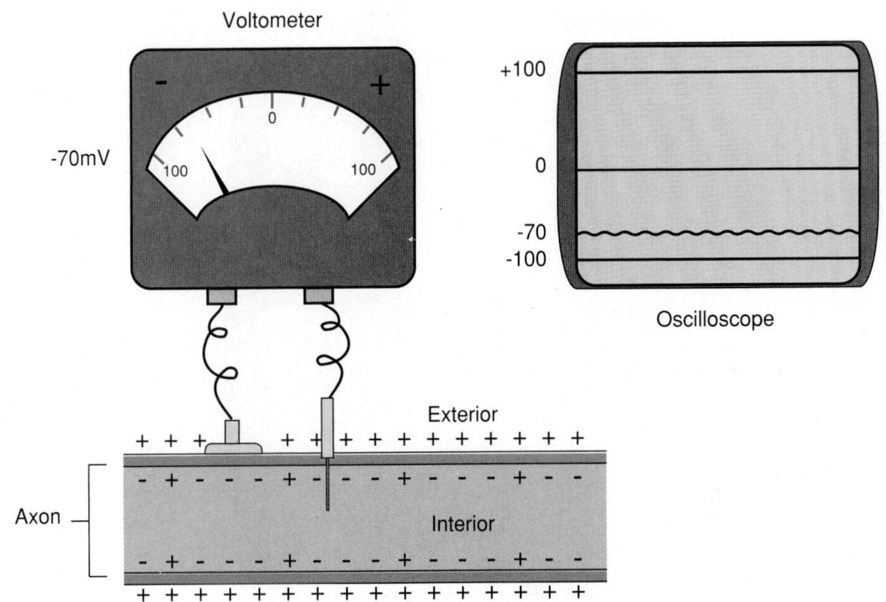

FIGURE 34.7
RESTING POTENTIAL
In its resting state a neuron is polarized. Its inside surface, just within the plasma membrane, is negative relative to the outside. The resting potential for many neurons is about −70 mV.

you may have noticed, are not unlike those of the fully charged chemiosmotic system in chloroplasts and mitochondria (see Chapter 6). At that time we described the condition as an "electrochemical gradient," a fitting term since it takes into account both the chemical and the charge gradient. Considering all of this so far, the term "resting neuron" seems even less fitting than before. Consider that the "set" mouse trap you gingerly slide under the refrigerator is also "at rest."

The electrochemical gradient of the polarized neuron represents a significant amount of potential energy. To tap this energy, some of the system must be allowed to "run down," to move toward equilibrium. And this brings us back again to the action potential.

Generating Action Potentials The axon, we've said, is able to generate and conduct action potentials. An action potential is a self-propagating wave of *depolarization* that moves along the axon to its end. When a neuron is stimulated, say on a dendrite or on the cell body, a passive electrical signal spreads to the axonal hillock. There, the action potential arises and sweeps along the axon to its terminal branches. The terminal branches then can activate another neuron or an effector.

An action potential begins with a sudden increase in the permeability of the axon's plasma membrane to sodium, and these ions, no longer held in check, rush inward. At that point, the precarious balance of charges is suddenly lost, and the membrane potential shifts rapidly from −70 mV to about +30 mV. Thus that area of the membrane becomes depolarized. (Note that the shift in membrane potential from −70 mV to +30 mV represents a 100 mV difference.)

Action potentials at first affect only a small part of the membrane. But once they begin, they form an ongoing depolarizing wave that sweeps along the axon. The wave is quite dramatic when seen on an oscilloscope (Figure 34.8). Like an ocean wave, each action potential is followed by a trough. In the neuron, the "trough" represents *repolarization*—the restoration of the resting state. An action potential at any given point along the axon is extremely short-lived, the period from depolarization to repolarization requiring about two milliseconds (two thousandths of a second).

Special Characteristics of Action Potentials An important characteristic of the action potential is called the "all-or-none" principle. Axons have a **threshold voltage** requirement (see Figure 34.8) that must be attained if action potentials are to begin. For many

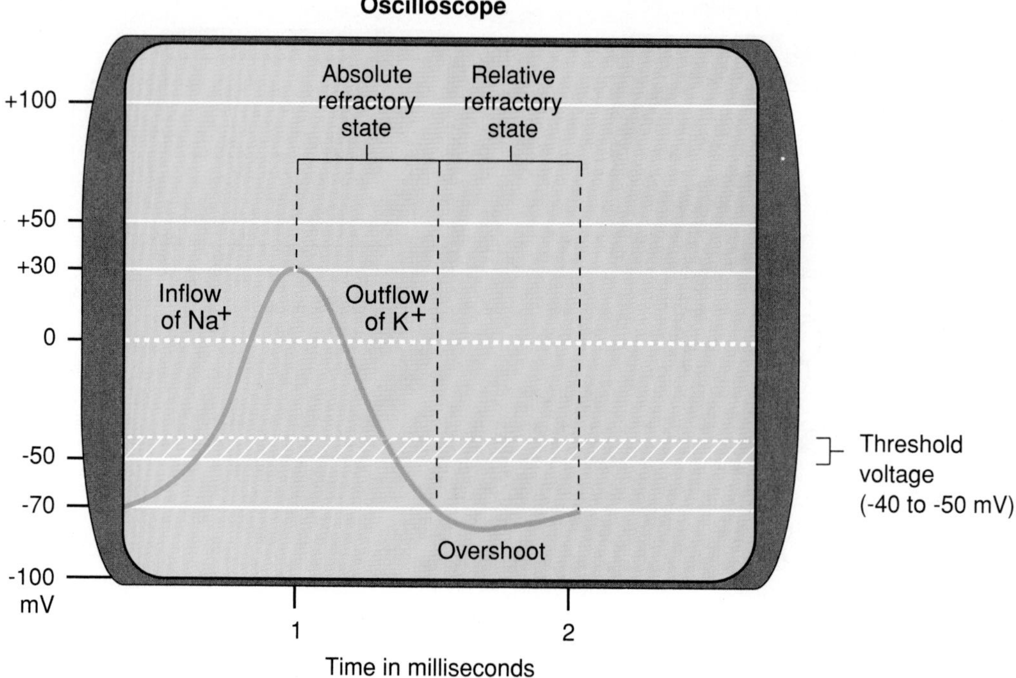

FIGURE 34.8
THE ELECTRICAL CHARACTERISTICS OF AN ACTION POTENTIAL
The oscilloscope tracing reveals the electrical characteristics of an action potential. Because of a sodium ion influx, the membrane becomes depolarized, the membrane potential rising from -70 mV to $+30$mV. Repolarization, involving an outpouring of potassium ions, occurs rapidly. Until the resting potential is attained, the neuron is in a refractory state. The small dip below the resting potential, a slight overshoot, is the relative refractory state.

neurons, the threshold requirement is a depolarization to about -40 mV to -50 mV. Anything less, and the action potential will not be sustained; it will simply fade out. Once the threshold requirement is reached, however, the action potential will proceed unfalteringly along the entire length of the axon.

Another characteristic of action potentials is that, once generated, their magnitude and their speed cannot be altered by increases or decreases in the stimulus (the event that triggered the sudden increase in Na^+ permeability). Action potentials remain constant from one end of the axon to the other. Thus, in a given neuron, all action potentials look about the same on the oscilloscope, and they all move at the same speed. However, one factor that is affected by stimulus intensity is the *rate* at which action potentials arise. The brain interprets a greater frequency of action potentials as being due to a stronger stimulus. Stepping on a thumb tack generates more action potentials than stepping on a pea (your own verbal response to the two stimuli may also reflect the difference).

Reestablishing the Resting Potential The rapid influx of Na^+ in an action potential is cut short as the plasma membrane once again becomes impermeable to that ion. (We'll see how shortly.) This renewed impermeability marks the start of repolarization—restoring the resting potential. However, the essential ion of recovery is not sodium, but potassium. In repolarization, potassium ions rapidly diffuse out of the neuron, continuing down their gradient until their numbers roughly balance the number of sodium ions that entered. Why potassium? Recall that in the resting neuron, outward potassium diffusion was curtailed by the attractive force of negative proteins within the neuron. However, with sodium now competing for these negative charges, potassium is more free to diffuse to the outside. Thus, it is potassium that restores the membrane potential to -70 mV.

So now there is excess K^+ outside the neuron and excess Na^+ inside, but the Na^+/K^+ ion exchange pumps quickly restore the former ion distribution. As it turns out, the restoration isn't immediately critical. Whereas a 100 mV shift may seem like a lot, in actuality such a change involves relatively few ions. Neurobiologists estimate that only one sodium ion in ten million crosses the membrane. Further, they have found that even when the exchange pumps are experimentally inhibited with chemical agents, a neuron may generate thousands of action potentials before the changing distribution of ions begins to show an effect.

The time of repolarization, when potassium ions are leaving the axon and the membrane potential is again moving towards -70 mV, is called the **refractory period.** Figure 34.8 shows two parts of the refractory period. In the **absolute refractory period,** new action potentials cannot be generated at any point along the axon where recovery is occurring, regardless of the intensity of a stimulus. In the **relative refractory period,** which is characterized by a dip or "overshoot" below the resting potential, new action potentials can be generated, but only by a very strong stimulus. As you see, then, neurons have a maximum rate of firing that is limited by the period required for recovery.

Ion Channels and Gates

One may wonder, at this point, just how a membrane can suddenly change in permeability to certain ions. Until a few years ago, we knew very little about such changes, but now the discovery of certain fascinating structures in the axon's plasma membrane is shedding new light on the behavior of sodium and potassium during an action potential.

The neuron, it turns out, contains large numbers of special protein-lined **ion channels** all along its axon, each capable of admitting a specific kind of ion. For example, there are channels that admit ions of sodium, potassium, calcium, or chlorine. The channels are not simple pores. They have moveable regions in their protein constituents, the **ion gates,** that can open and close the channels. The ion channels of neurons are of two types: those whose gates respond to electrical disturbances, and those whose gates respond to specific chemicals. We'll look here into the electrically activated ion gates.

The resting potential of the axon, its depolarization, and its repolarization all involve the operation of **sodium ion gates** and **potassium ion gates,** both of which are voltage-sensitive. Each sodium ion channel actually has two kinds of gates, an outer **activation gate** and an inner **inactivation gate.** Although the Na^+ activation and inactivation gates are both voltage-sensitive, they respond to different voltages. Figure 34.9 shows a model of the Na^+ and K^+ gates and illustrates their operation.

The Sodium Gates Recall that the resting axon's plasma membrane is impermeable to the sodium ion. This means, then, that at least one of the two sodium gates must be closed at this time. It turns out that it is the sodium activation gates that are closed; the inactivation gates remain open during the resting state (Figure 34.9a). When an electrical signal of sufficient strength reaches the axon, the nearest sodium activation gates respond by opening. Sodium ions thus rush into the axon, starting the depolarization that characterizes the action potential (Figure 34.9b). This depolarization generates still more electrical signals, and sodium activation gates a little further along the axon open in response. The sweeping movement of an action potential is made possible by a chain reaction of gate openings. As we've seen, repolarization follows along behind the wave. Logically, then, the ion gates are also involved in the recovery process.

As the membrane potential reaches $+30$ mV, the sodium inactivation gates close, blocking the further entry of Na^+ (Figure 34.9c). The inactivation gates will remain closed until the surrounding region of the neuron is repolarized. It is important that the two kinds of gates respond to different voltages. Were both sodium gates to open at the same voltage, and remain that way, the neuron would be locked in a depolarized state, and the resting voltage could not be restored. The repolarization, you will recall, is brought on by the movement of potassium, so we'll consider its gated channels next.

The Potassium Gates Potassium gates also respond to the initial electrical disturbance, but the K^+ gates respond much more slowly—opening fully only when the $+30$ mV membrane potential is reached, or just about the time that the sodium inactivation

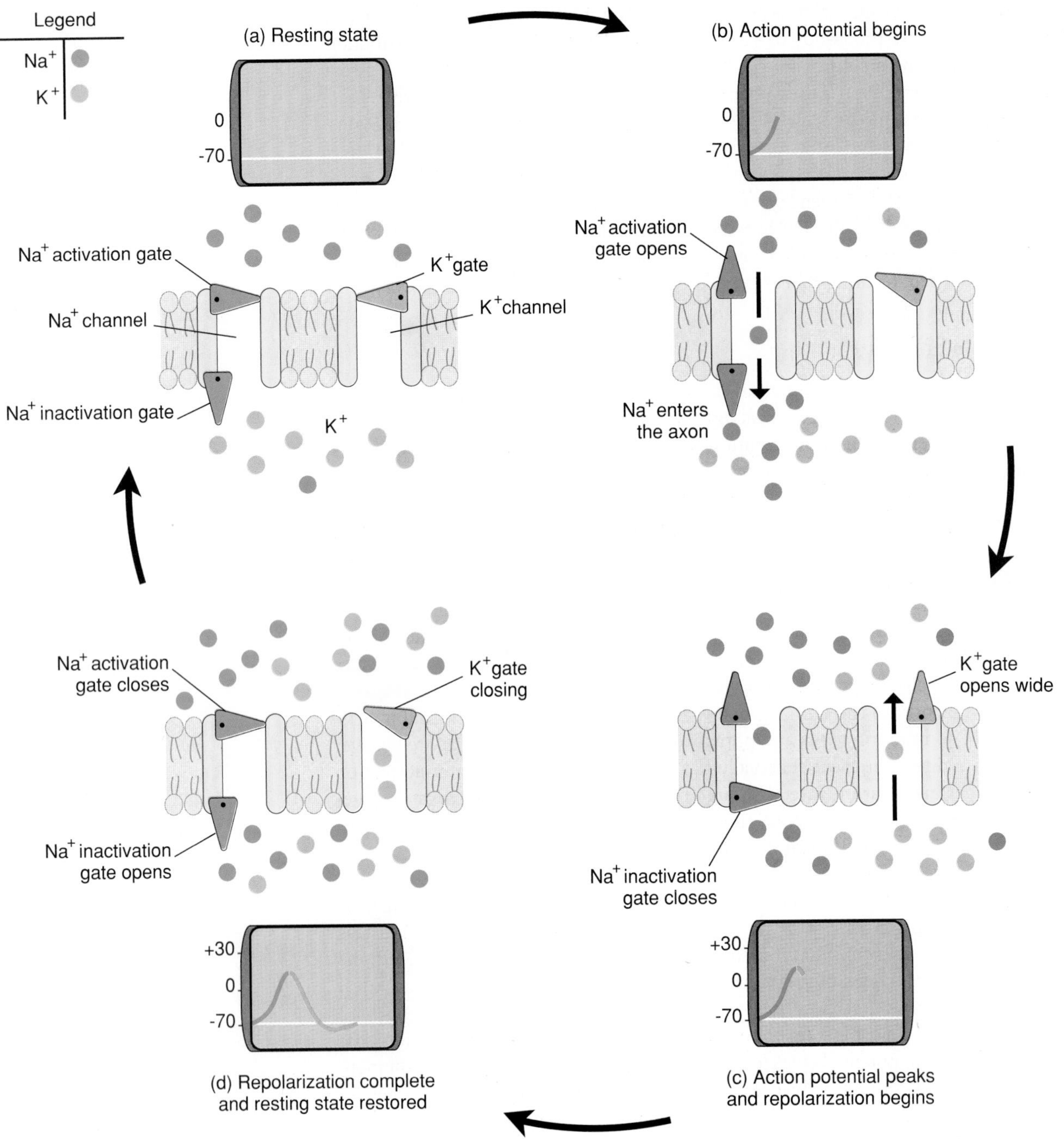

FIGURE 34.9
GATED SODIUM AND POTASSIUM CHANNELS

Gated, voltage-sensitive sodium and potassium channels bring about changes in plasma membrane permeability to these ions. **(a)** In the resting state, Na^+ activation gates are closed, and inactivation gates are open. Potassium gates are also closed. **(b)** When a neuron is sufficiently stimulated, Na^+ activation gates in the immediate region open, and Na^+ diffuses in. **(c)** When the membrane potential in that locale reaches +30 mV, the Na^+ inactivation gates close, stopping the Na^+ influx, and recovery begins. The K^+ gates react slowly to the neuron's initial stimulus, opening wide as the membrane potential reaches +30 mV. Potassium ions are then free to diffuse down their gradient, out of the neuron. **(d)** The exit of K^+ brings the membrane potential back down to about −70 mV, the resting potential. Once repolarization is complete, the K^+ gates and the Na^+ activation gates close, and the Na^+ inactivation gates open.

gates close. With the sudden exit of potassium ions, that region of the axon is repolarized, and the resting potential of -70 mV is restored (Figure 34.9c,d). At that voltage, all three gates respond. The sodium activation gates and potassium gates close (the latter barring the further escape of K^+), and the sodium inactivation gates open (Figure 34.9d). Thus, the resting state is fully restored.

The closed position of the sodium inactivation gates during the recovery period (from $+30$ mV to -70 mV) explains why the refractory period occurs. With the inactivation gate closed, there is no way for a second inrush of sodium ions to occur. Thus, no new action potentials arise until the resting membrane potential is regained and the sodium inactivation gates open.

Let's summarize what we know about the action potential so far.

1. During the resting state, the axon is polarized (-70 mV), with many more negative charges inside than outside. The ion balance is such that Na^+ is in greater concentration outside, while K^+ is in greater concentration inside.
2. When a neuron is stimulated, the resulting electrical signal reaches the axon and causes sodium activation gates to open. Sodium ions enter, depolarizing the region (to $+30$ mV), and this new disturbance opens Na^+ activation gates further away. Thus, the action potential sweeps along the neuron (Figure 34.10).
3. Repolarization begins at $+30$ mV, with the closing of Na^+ inactivation gates and the complete opening of the K^+ gates. This transition marks the beginning of the refractory period.
4. The outpouring of K^+ restores the resting potential (-70 mV). The K^+ gates and Na^+ activation gates close, and the Na^+ inactivation gates open.

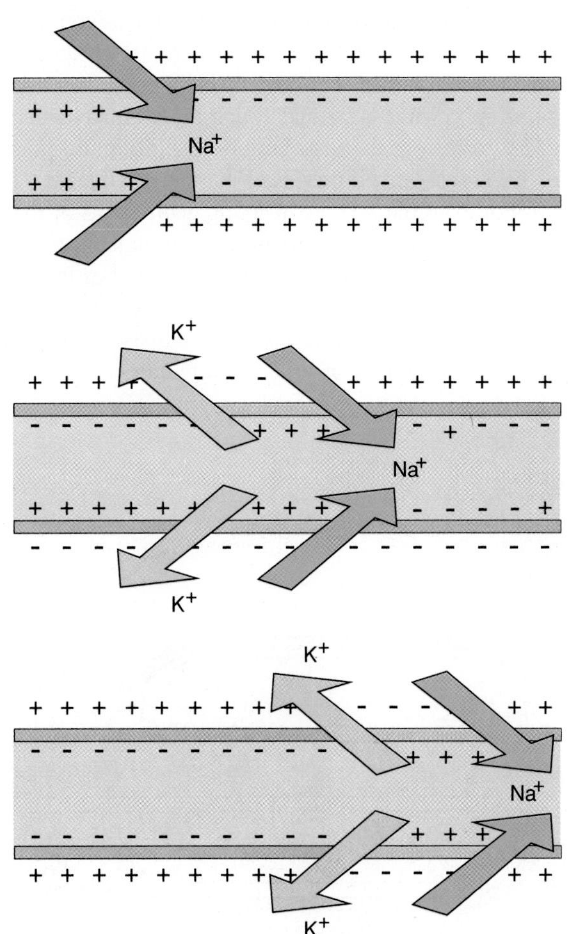

FIGURE 34.10
THE CHEMICAL DYNAMICS OF AN ACTION POTENTIAL
The ongoing scenario illustrates how the displacement of sodium and potassium ions leads to a moving action potential.

Myelin and Impulse Velocity

As we've seen, axons in vertebrates are commonly surrounded by myelin sheaths—the membranes of Schwann cells or oligodendrocytes. These act both as insulators and as a mechanism for speeding up impulses. In those human neurons that are myelinated, for example, impulses are conducted at a speed of up to 100 meters per second, while the speed of conduction in our nonmyelinated neurons and in those of invertebrates is much slower. But how do myelin sheaths function? We can see how the fatty sheaths would insulate neurons, but how do they increase impulse velocity? And for that matter, how can an action potential be generated in an axon that is insulated from the surrounding, sodium-rich extracellular fluid?

The answer to both questions lies in the arrangement of the myelin sheath. We must admit now that the smooth-flowing, wavelike nerve impulse that we described in such detail only pertains to nonmyelinated fibers. In the myelinated neurons, the action potentials occur only at the nodes, those gaps in the myelin found at regular intervals along the axon. Further, repolarization in myelinated axons is different than in nonmyelinated axons. Myelinated axons of vertebrates lack gated potassium channels and thus must rely entirely on **potassium leak channels** for recovery.

The increased speed of the impulse in myelinated fibers occurs because the neural impulse virtually "jumps" from one node to the next all along the axon. Neurobiologists call this **saltatory propagation** (*salto,* "jump") (Figure 34.11). The generation of an action potential in one node produces a minute current that spreads quickly throughout the internode. When this current reaches voltage-sensitive sodium activation gates in the next node, those gates open, a new action potential is generated, and the impulse continues down the axon.

Like a skillfully thrown rock skipping along a lake surface, saltatory propagation permits a greatly increased velocity of transmission. And, as you can imagine, this method of propagation is more energy-efficient, since there are so few ions to be pumped in and out later. In one comparison between myelinated and nonmyelinated axons conducting at the same speed, it was estimated that the nonmyelinated axon required 5000 times as much ATP to restore the distribution of sodium and potassium ions.

The speed of impulses can be increased in non-myelinated axons if they are very large in diameter. For example, the fast-reacting giant axons of the squid that permit it to "jet-propel" itself away from danger are nearly a millimeter in diameter, whereas most others are only 1/100 to 1/1000 that size. But even the fast impulse conduction of the squid's giant axons, at some 30 meters per second, is slow compared to conduction in the myelinated axons of vertebrates. Giant axons are also found in cockroaches, earthworms, crayfishes, and a number of other invertebrates, where, as in the squid, they

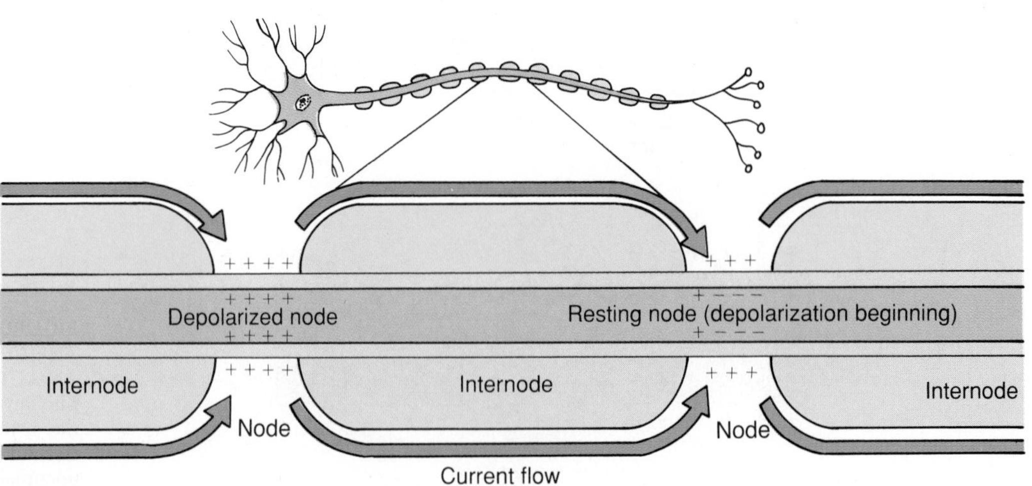

FIGURE 34.11
SALTATORY PROPAGATION
Action potentials occur only at the nodes of myelinated axons—jumping from one node to the next. The internodal "leaps," called saltatory propagation, are possible because electrical currents generated in one node during depolarization reach voltage-sensitive sodium activation gates in the next node, starting a new action potential there. Thus the impulse seems to jump from node to node.

promote rapid escape movements. Thus in the evolution of nervous systems we see that two alternatives, giant axons and myelinated axons, were both favored by natural selection.

COMMUNICATION AMONG NEURONS

Neurons communicate with each other at special sites called **synapses.** At a synapse, one neuron triggers an impulse in another neuron (or it might trigger an effector, such as a muscle). Typically, the axon of one neuron will form a synapse with a dendrite or cell body of a second neuron. Although a neuron typically has but one axon, that axon's highly branched ending may provide hundreds to thousands of synaptic connections with other neurons. There are two distinct kinds of synapses: electrical and chemical. Chemical synapses are far more common, but both have unique advantages.

Electrical Synapses

Electrical synapses occur at gap junctions between neurons. You may recall from Chapter 5 that gap junctions are places where adjacent plasma membranes press together and minute pores form. The pores make the cytoplasm of one cell continuous with that of another. Such an intimate relationship between neurons permits current to pass readily from an active neuron to a resting neuron, thus prompting an action potential in the second.

The main advantage of the electrical synapse is speed. Chemical synapses require several steps for action potentials to be generated in a second neuron, and this requires more time than the current flow associated with electrical synapses. Electrical synapses have been found in several invertebrates and are associated with giant axons and escape movements. In the crayfish, for instance, electrical synapses occur between a giant nerve fiber from the brain and a large motor axon that activates the abdominal muscles. In response to nerve stimulus, the muscles produce a powerful thrust in the paddlelike tail that propels the crayfish backwards, away from danger. Electrical synapses also occur in vertebrates. In fishes, for example, electrical synapses activate the sudden flip of the tail and make possible the quick starts and turns used in the feeding chase by both predatory and prey.

Chemical Synapses

In **chemical synapses,** the neurons do not actually make physical contact, and action potentials do not simply pass from one neuron to the next. Action potentials must be generated anew in receiving neurons. In chemical synapses, neurons are separated by a minute, 20 nm space known as the **synaptic cleft** (Figure 34.12a). Communication occurs through the action of chemicals called neurotransmitters. Neurotransmitters released by the first or **presynaptic** neuron diffuse across the synaptic cleft, where they activate specialized receptors on the second or **postsynaptic** neuron. Receptor activation triggers a new action potential.

Chemical synapses can be quite complex. Axons branch near their ends, forming many terminals that end in minute swellings called **axonal knobs.** Many axonal knobs may be involved in the communication between just two neurons. In fact, some neurons have hundreds to thousands of axonal knobs, all synapsing with a single target neuron.

While slower than electrical synapses, chemical synapses offer much more versatility in their responses. This is made possible by a number of variables, such as the variety of neurotransmitters and a range of potential receptors. We'll consider the common neurotransmitter **acetylcholine** as we look into what actually goes on at a chemical synapse.

Action at the Synapse Within each axonal knob are a large number of vesicles that store the chemical neurotransmitter, in this case acetylcholine. Its release involves ions

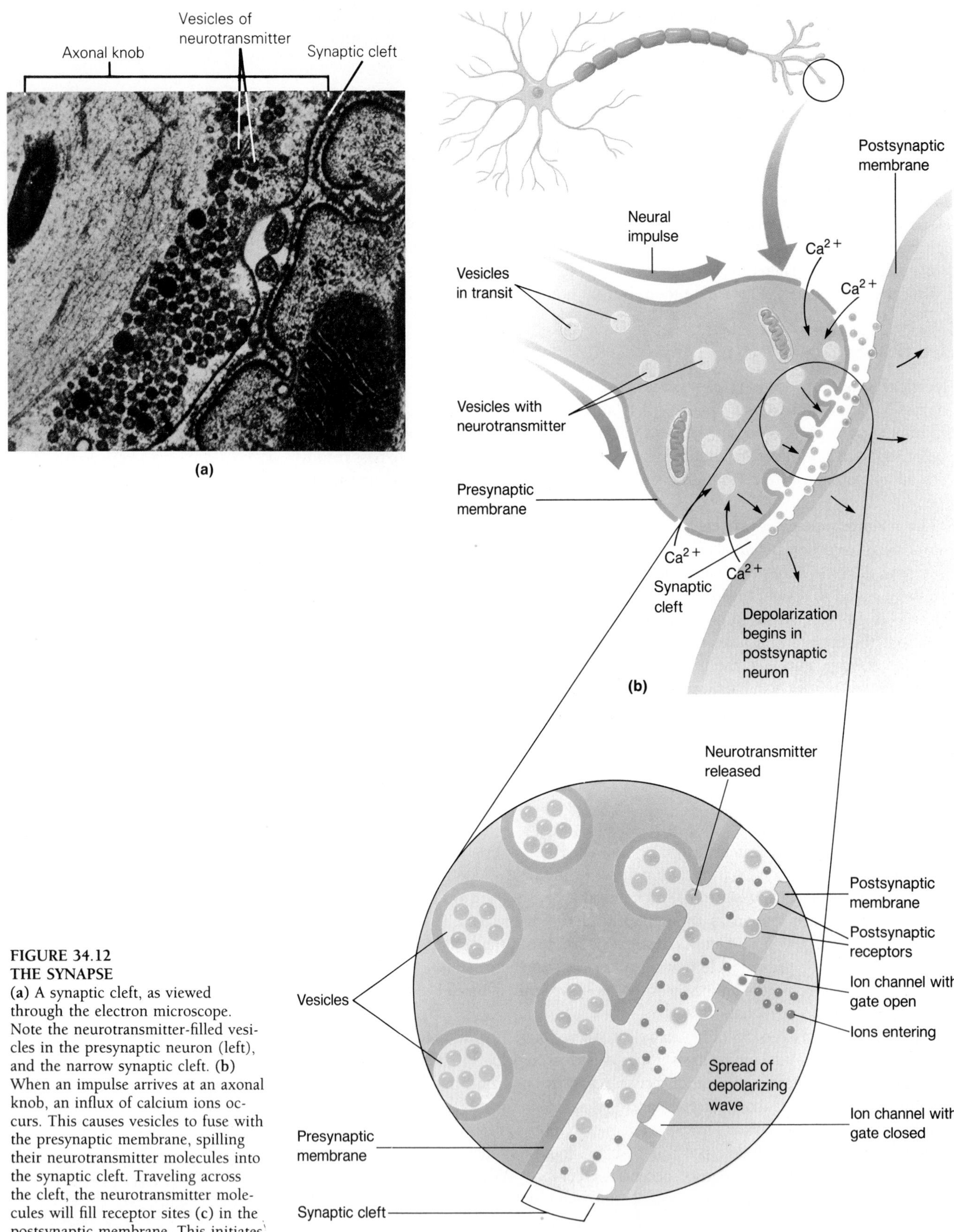

FIGURE 34.12
THE SYNAPSE
(a) A synaptic cleft, as viewed through the electron microscope. Note the neurotransmitter-filled vesicles in the presynaptic neuron (left), and the narrow synaptic cleft. (b) When an impulse arrives at an axonal knob, an influx of calcium ions occurs. This causes vesicles to fuse with the presynaptic membrane, spilling their neurotransmitter molecules into the synaptic cleft. Traveling across the cleft, the neurotransmitter molecules will fill receptor sites (c) in the postsynaptic membrane. This initiates a new depolarizing event that sweeps along the receiving neuron.

of calcium (Ca^{2+}) that are present in extracellular fluids outside the knob. When an action potential reaches the axonal knob, voltage-sensitive gates of calcium ion channels in its membrane open, and calcium ions diffuse inward. In their presence, the transmitter-laden vesicles fuse with the **presynaptic membrane**, where they rupture, releasing numerous acetylcholine molecules into the synaptic cleft. The molecules then diffuse across the cleft and bind to specialized receptors on the **postsynaptic membrane**—the membrane of the second neuron (Figure 34.12b).

The acetylcholine receptor sites of the postsynaptic membrane are *chemically* gated ion channels. That is, their gates respond to chemicals rather than to electrical currents. The presence of the neurotransmitter causes the gates to open, thereby admitting sodium ions (sometimes potassium) into the second neuron. When a sufficient number of such channels has opened in the postsynaptic membrane, the current generated by the shifting ions will spread to the axon's hillock. If the resulting voltage surpasses the required threshold, an action potential will begin. Whereas one such event would probably have little effect, the presence of many synapses on a single receiving neuron helps ensure this reaction.

By its very organization, the synapse acts as a one-way valve between most neurons. The transmission is one way because the postsynaptic membrane has no neurotransmitters to release, and the presynaptic membrane has no receptor sites. One-way transmission is an important means of coordinating the work of the nervous system. Coordination is accomplished in other ways as well, such as through inhibitory synapses.

Inhibitory Synapses So far we have been discussing **excitatory synapses**, those that "excite" the receiving neuron, causing action potentials there. Another kind of synapse does not trigger action potentials, but *inhibits* them, making them less likely to occur. These are called **inhibitory synapses.** One way such inhibition occurs is through **hyperpolarization**, in which the net negative charge inside the neuron becomes considerably increased. This may happen when specialized receptors respond to a neurotransmitter by opening chemically gated chloride ion channels. Once opened, they admit the negative chloride ion (Cl^-) into the neuron's interior, which increases its polarized state. Consequently, the hyperpolarized neuron requires considerably more than the usual threshold voltage for action potentials to start.

Inhibition can also occur if the neurotransmitter causes chemically gated potassium channels to open—channels that permit the escape of potassium ions. Their escape counteracts the uptake of sodium ions, brought about by nearby excitatory synapses, effectively cancelling out depolarization. At first this stage may seem counterproductive, but actually it has a critical role.

Typically, a neuron or an effector will possess both inhibitory and excitatory synapses coming from other neurons. Whether or not the second neuron fires or the effector reacts depends ultimately on the net effect of both types of synapses (Figure 34.13). At

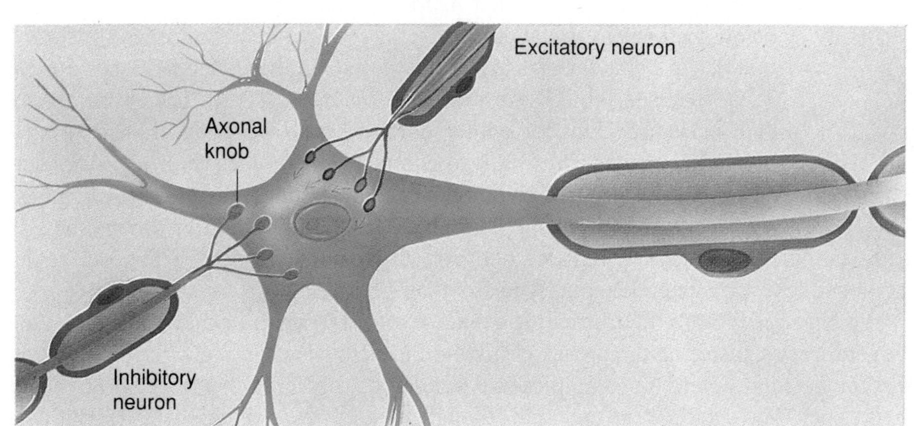

FIGURE 34.13
EXCITATORY AND INHIBITORY SYNAPSES
Combinations of excitatory and inhibitory synapses produce closely regulated activity in receiving neurons. Whether or not action potentials are generated depends upon the sum of the two effects.

times, inhibition can be vital. It is inhibition, for example that permits sleep. Certain inhibitory neurons, by suppressing action potentials, can screen out routine incoming stimuli—background noises or random thought patterns—whatever might disturb your sleep. But then, a rustling sound under the bed might stimulate enough excitatory neurons to overwhelm the inhibitory neurons. In the next chapter you'll learn that the cerebellum, the brain structure responsible for coordinating voluntary movement, does much of its work through inhibition.

Recovery at the Synapse It is essential that a neurotransmitter not be allowed to linger in the synaptic cleft. Should this happen, the receiving neuron would continue to be excited or inhibited, and the coordination of neural activity would be lost. Some neurotransmitters are simply recycled back to the presynaptic membrane for uptake and storage. Others are enzymatically broken down. For example, acetylcholine is dismantled into choline and acetyl groups by **acetylcholinesterase,** an enzyme that is often bound to collagen fibers within the synaptic cleft. The choline group is then taken up by the presynaptic membrane and recycled.

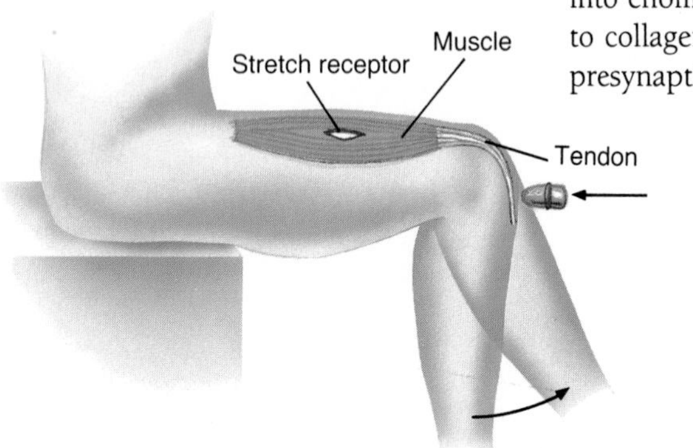

The removal of a neurotransmitter from the synaptic cleft is so rapid that its presence is quite fleeting, lasting less than a millisecond. Interestingly, nerve poisons such as pesticides of the organophosphate class do their deadly work by blocking acetylcholinesterase activity. How might this affect other functions in the animal?

The Reflex Arc: The Simplest Model of Neural Activity

The simplest behavioral response occurs through the **reflex arc,** a neuronal circuit in which impulses from a sensory neuron may be relayed directly to a motor neuron. Such a reaction, then, involves as few as two neurons, but in most cases, the impulse is simultaneously transferred to a number of interneurons. Because of its simplicity and speed, the reflex arc can be quite important to adaptive behavior. For example, should you step on a hot coal at the beach, your reaction may involve a sensory neuron that perceives the stimulus, and a motor neuron that brings about the graceless "high-stepping" move that entertains your friends and gets you off the coal. By the time your brain is aware of the problem, having been alerted via additional impulses shunted there by interneurons, your reflexive behavior has prevented a more serious burn.

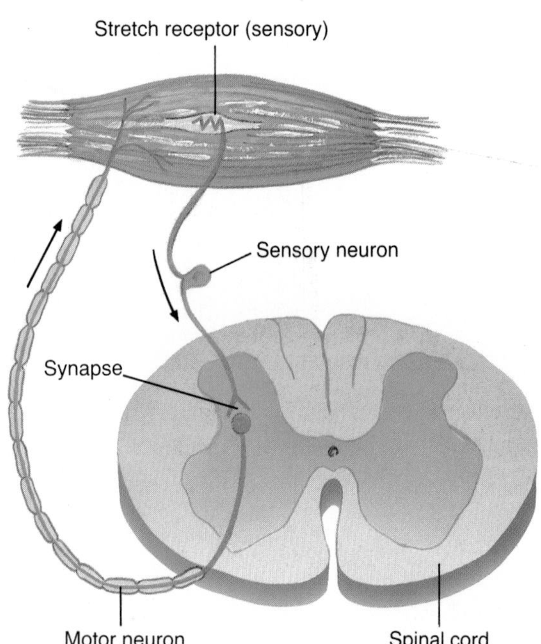

FIGURE 34.14
A REFLEX ARC
In the knee-jerk reflex, striking the tendon causes the muscle above to extend slightly. Stretch receptors sense this change and carry impulses to the spinal cord, where they synapse with motor neurons. Impulses from the motor neurons bring about contraction in the muscle without intervention by the brain.

The **patellar response,** known also as the "knee jerk reflex" is also simple, directly involving just two neurons (Figure 34.14). Your physician taps you just below the knee cap with a small rubber hammer, and in response, your lower leg kicks out slightly. This may seem to be for the physician's amusement, but he or she is actually trying to rule out certain neurological disorders. What happens is that the blow causes the tendon below the knee to stretch slightly. This activates stretch receptors (neurons that detect slight changes in skeletal muscle) in the large quadriceps muscle above the knee, which flash a message to the spinal cord. There, the sensory neuron synapses directly with a motor neuron that sends an impulse back to the quadriceps, causing it to contract. That's when your leg kicks out. The reflexive response is sudden and involuntary, and the brain is only indirectly involved (as a passive and surprised observer). The reflex can be important at other times, as when the knee inadvertently buckles, activating the stretch receptor, which causes the leg to suddenly straighten, keeping you from falling.

We have seen, then, the fascinating events involved in neural behavior. By the time we get through analyzing the processes involved in such events, we may have the feeling that the nervous system is so complex and ponderous that it sets the stage for error and inefficiency. We should keep in mind, though, that all these things happen routinely and with dazzling swiftness and accuracy. The evolution of neural mechanisms, such as we've just seen, indeed underscores the remarkable efficacy of natural selection.

We'll move now to other aspects of neural function, looking particularly at the organization of neurons and neuroglia into complex animal nervous systems, including that of the human. We will also explore the fascinating sensory organs, those structures with which organisms perceive their surroundings—perceptions that elicit important responses.

action potential In the axon, a self-regenerating, depolarizing wave, that sweeps along an axon to the axonal knobs.

interneuron A neuron of the spinal cord or brain, one that receives neural signals from other neurons, aids in the integration of information, and transmits signals to motor neurons.

motor neuron A neuron that receives signals from the central nervous system and transmits them to an effector such as a muscle or gland.

neural impulse A fast-moving depolarizing wave in a neuron, including both graded and action potentials, that sweeps along a neuron.

neuroglia Massive numbers of supporting glial cells interspersed with neurons and making up the central nervous system.

neuron A nerve cell. A cell specialized in the transmission of neural signals.

neurotransmitter A chemical substance, produced and secreted by a neuron, and intended to activate another neuron or an effector.

reflex arc The simplest complete neural reaction, involving a sensory neuron that carries impulses to the spinal cord where it forms a synapse with a motor neuron whose impulses stimulate a response.

resting state The polarized state in a neuron during which the region outside the neural membrane is more positive than the neuron's inside. A neuron prepared for a neural impulse.

sensory neuron A neuron specialized for receiving input from sensory receptors and transmitting neural impulses to interneurons for integration and response.

synapse A region where two interacting neurons come together, separated by a narrow synaptic cleft.

APPLICATION OF IDEAS

1. There are a number of types of neural cells in the vertebrate body. Some of those are myelinated, using rapid saltatory propagation, and some are unmyelinated and conduct impulses more slowly. Using what you know of the principle of natural selection and evolution, how might myelinated fibers have evolved? Remember, there are many kinds of cells in the body (some with high fat content), many of them lying in close association but with unrelated roles.
2. A recent finding has been the discovery of the roles of ion channels and gates. Explain the roles of these structures in terms of depolarization and repolarization and graphically relate their behavior to the waves of an action potential shown on an oscilloscope.

KEY IDEAS

THE NEURON

Cellular Structure

1. **Neurons** consist of a **cell body** with the usual cellular organelles and receiving and sending processes called **dendrites** and **axons**, respectively. One neuron activates a second, or an **effector**, through chemical substances called **neurotransmitters** that are produced in the cell body and released at knobby ends of the **axon tree**. Neurotransmitters that activate muscles are released at **neuromuscular junctions**. Dendrites are receiving processes that conduct signals to the cell body. Axons conduct **neural impulses** over long distances. Such impulses arise at the axonal **hillock**.

2. Many vertebrate neurons are wrapped in **myelin sheaths**, produced by **Schwann cells** outside the brain and spinal cord and **oligodendrocytes** within. The sheaths or **internode** regions, contain minute gaps called **nodes.**

Cells of the Nervous System

1. Neurons include **sensory neurons** (also **afferent neurons**), **interneurons**, and **motor neurons** (also **efferent neurons**). The first receive stimuli such as light and heat and communicate with the second, which integrate the response and transmit it to the third, which produce the response. **Neuroglia** (or **glial cells**), especially **astrocytes**, are extremely common in the brain. They supply structural support, regulate potassium, and act as a selective barrier to materials moving into and out of the brain's neurons.
2. **Nerves**, composed of numerous axons and dendrites, form cablelike structures surrounded by connective tissue.

NEURAL SIGNALS

Signals conducted along the dendrites and cell body are passive, diminishing as they spread. Neural impulses or **action potentials** do not diminish and are thus suitable for longer distances.

The Action Potential

1. Action potentials do not diminish with distance. They occur in axons where controlled ion movement, active transport and diffusion, are involved. Their three aspects are: (a) polarization (resting state), (b) propagation **(AP)**, and (c) repolarization.
2. In the resting state, the axon is polarized; the interior, just within the membrane, is more negative than the exterior, just outside. The charges involved come from negatively charged proteins and from positively charged sodium and potassium ions. **Sodium/potassium ion exchange pumps** transport Na^+ out of the axon and K^+ in. Although the membrane is more permeable to K^+ (because of leak channels), its attraction to the negatively charged proteins within keeps potassium from reaching equilibrium.
3. The charge difference across the membrane of the resting neuron is -70 mV.
4. Neural impulses begin at the axonal hillock, where a sudden change in sodium permeability brings on an inrush of the ion and depolarization. The membrane potential changes from -70 mV to $+30$ mV.
5. The action potential proceeds as a fast-moving wave, following immediately by repolarization.
6. Action potentials require a specific **threshold voltage** for their generation. Once generated their magnitude and speed are constant, although the rate at which they are generated depends upon the strength of a stimulus.
7. Repolarization is brought on by an efflux of K^+, which continues until the -70 mV, the **resting potential**, is attained.
8. The period of recovery, between $+30$ mV and $+70$ mV, is the **refractory period**, a time when a second action potential cannot occur.

Ion Channels and Gates

1. The movement of sodium and potassium ions during a neural impulse is controlled by the behavior of voltage-sensitive **ion gates** that control **ion channels.** The gates operate in response to small current flows in the neuron.
2. In the resting channel, sodium **inactivation gates** are open, whereas sodium **activation gates** and **potassium ion gates** are closed.
3. At the start of an action potential the sodium activation gates open, and sodium ions rapidly diffuse in, causing depolarization. At the peak of the action potential, the sodium inactivation gates close, and the potassium gates open. Potassium ions diffuse out, and the resting potential is restored.
4. When the resting potential is reached, the potassium ion gates and the sodium activation gates close. The sodium inactivation gates open.

Myelin and Impulse Velocity

1. Myelinated neurons conduct impulses much faster than nonmyelinated neurons, since depolarization occurs only at the nodes.
2. Action potentials at the node create enough current flow to activate the sodium gates in the next node, so the impulse jumps from node to node in what is called **saltatory propagation.** Electrical disturbances pass almost instantaneously through and around internodes.
3. Since myelinated axons lack potassium channels, recovery involves **potassium leak channels.**
4. Faster propagation in invertebrates such as squid, earthworms, and cockroaches, occurs through giant axons.

COMMUNICATION AMONG NEURONS

Neurons communicate at **synapses.** An axon may form thousands of such connections with its effector.

Electrical Synapses

Electrical synapses occur at gap junctions between neurons. A moving action potential simply sets off an action potential in the second cell. Such synapses are fast and simple and are associated with escape movements. Such movements may involve giant axons in invertebrates or myelinated axons in vertebrates.

Chemical Synapses

1. **Chemical synapses** involve the secretion of chemicals called neurotransmitters, which bring on action potentials in a second neuron. Action potentials themselves do not cross the **synaptic cleft**, a minute space between neurons.
2. Impulses reaching the **axonal knobs** trigger the opening of calcium ion channels. Calcium entering the

knob brings on the fusion of vesicles with the **presynaptic membrane** and the release of the neurotransmitter.

3. The neurotransmitter, **acetylcholine**, activates chemically gated K^+ or Na^+ channels in the **postsynaptic membrane**, and the influx of ions creates the current necessary to produce an action potential.
4. **Inhibitory synapses** counteract **excitatory synapses** by hyperpolarizing the second neuron or by opening potassium channels, thus negating action potentials.
5. The cooperative action of excitatory and inhibitory synapses are an important part of neural coordination and regulation.
6. Activity at the synapse is stopped by the re-uptake of neurotransmitter by the presynaptic membrane or by the enzymatic breakdown of neurotransmitter.

The Reflex Arc: The Simplest Model of Neural Activity

1. **Reflex arcs** may involve as few as two neurons, a sensory neuron and a motor neuron. In the **patellar response**, a sensory neuron detects muscle lengthening and flashes an impulse to the spinal cord, where it synapses with a motor neuron whose impulses cause the muscle to shorten.
2. Reflexive reactions can be fast and simple and are an important part of escape reactions.

REVIEW QUESTIONS

1. List the three things neurons do. (784)
2. Name the three functional parts of a neuron and explain what each does. (785)
3. List the three types of neurons and state their general functions. (786–787)
4. List three probable functions of neuroglia. (786)
5. Describe the structure of a nerve. (787)
6. Why is the action potential more suitable for long distance communication than the passive, graded potential? (789)
7. Describe the arrangement of sodium and potassium ions in the resting neuron. What two factors account for this distribution? (789–790)
8. Describe the arrangement of charges in the resting neuron. State the resting potential and explain what this means. (790–791)
9. What happens to sodium at the start of an action potential? How does this affect the membrane potential? (791)
10. In what two ways is a neural impulse somewhat like an ocean wave? (791)
11. Explain the "all-or-none" principle. How might this be adaptive to an animal? (791–792)
12. What is the significance of the rate at which neural impulses might be occurring along an axon? (792)
13. Describe the events that bring about repolarization. What is the refractory period? (792–793)
14. What is the role of the sodium/potassium ion exchange pump in the neural impulse? Is its work between impulses urgent? (793)
15. Describe the position of the various ion gates in the resting neuron. (793)
16. What happens to the sodium and potassium gates during an action potential? How about during repolarization? (793–795)
17. Suggest a logical reason for the presence of two sodium gates rather than just one. (794–795)
18. Explain how a neural impulse proceeds through a myelinated axon. How does this affect speed and energy cost? (796)
19. What provisions are there for fast neural transmission in invertebrates? How are these adaptive to the squid, earthworm, and cockroach? (796–797)
20. Describe the electrical synapse. What are its advantages over the chemical synapse? (797)
21. Using a simple drawing, describe the organization of a chemical synapse, labeling the axonal knob, postsynaptic membrane, synaptic cleft, neurotransmitter, vesicle, receptor sites, calcium gates, and presynaptic membrane. (797)
22. Briefly describe the activity at the synapse, starting with the arrival of a neural impulse at a synaptic knob and ending with the opening of gates in the next neuron. How do these gates differ from the sodium and potassium gates farther along? (797–798)
23. What keeps the neurotransmitter from continually activating a neighboring neuron? What special situation exists for acetylcholine? (800)
24. Suggest two ways in which a neurotransmitter can inhibit the generation of an action potential in the next neuron. (799–800)
25. Explain the organization of a reflex arc. Of what adaptive value are such simple arrangements? (800)

35

NEURAL CONTROL II: NERVOUS SYSTEMS

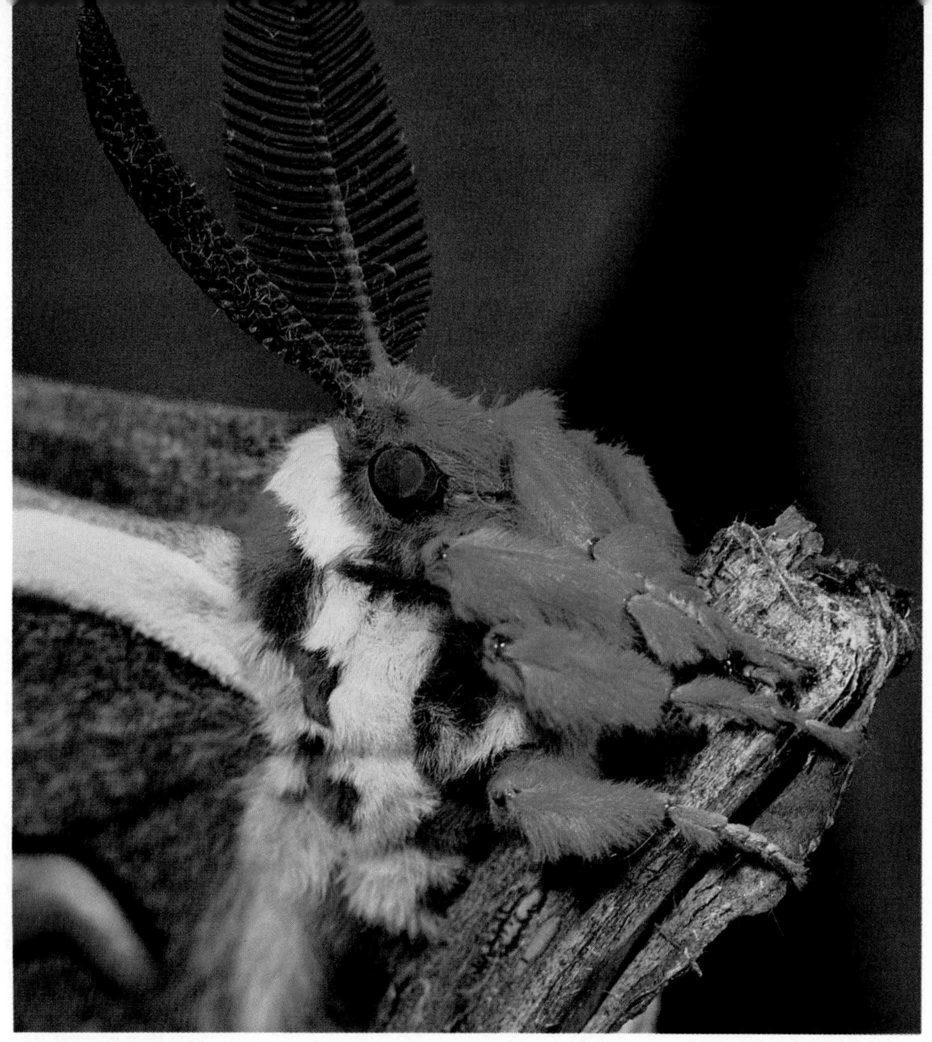

WE HUMANS PRIDE OURSELVES, RIGHTLY OR NOT, ON OUR INTELLIGENCE, and we are aware that the seat of that intelligence resides in that great gray structure we call the brain. We are also aware that the brain and the long stem that descends from it, the **spinal cord** are very complex associations of the sorts of neural cells that we have just discussed. So here, then, we will focus on the structure and activities of both the brain and the spinal cord, which together compose what is called the **central nervous system.** We will also take up the **peripheral nervous system,** including a consideration of the special senses—the brain's "window" to the world around it.

Before we launch into a discussion of brains and intellect, we might first note that many kinds of animals seem to do just fine without either. So perhaps we should begin by placing all this in some kind of evolutionary perspective. The brain, after all, is the result of an overdevelopment of one end of the nervous system. Note that in talking about ends, we must be talking about animals with bilateral body plans. Sea stars don't have ends. Neither do sea anemones. Being radially symmetrical, these animals are as likely to move off in one direction as another. Worms, on the other hand, have ends, and they tend to move along in the direction of their body axis. In the course of evolutionary history, it seems that it would have become advantageous for animals with elongated bodies, as they move about, to lead with the same end. Logic and evolutionary expediency thus seem to dictate that this leading end would be the site of a concentration of receptors that could tell the animal the nature of the environment into which it was moving before its whole body was exposed to that environment. It wouldn't do to back into an unfamiliar place. So **cephalization,** the development of an anterior end, set the stage for the formation of the brain by localizing neurons in the leading end of a bilaterally symmetrical body. The brain was probably a subsequent specialization of a highly sensory and reactive area. With increasing cephalization animals came to rely less on "hard-wired," genetically determined behavior, and more on the benefits of learning through experience.

Let's now consider a few of the specializations in a selected sequence of animals, moving from those that don't have brains to those that do. We will arrange these species from simple to complex with the tacit assumption that we may be tracing, albeit roughly, the evolutionary development of the central nervous system. The assumption is a common one, but keep in mind that each representative animal has had its own long evolutionary history and does not necessarily represent some sort of evolutionary stepping stone. We are only looking for suggestions that might help lead us through the unchronicled history of the brain.

INVERTEBRATE NERVOUS SYSTEMS

The nervous systems of invertebrates are quite diverse, and, as we mentioned knowing about them may shed some light on how the brain evolved. Here, we will note the increasing organization of the nervous system in certain invertebrates. We will begin with an animal with no brain at all, just a diffuse, netlike nervous system. Then we will see an animal with neurons aggregated into what might be considered a rudimentary brain. In other groups we will see neural columns extending from such brains. Then in yet other animals, these are joined in the course of development to form a single, ventral nerve cord with specialized areas along its length.

Simple Nerve Nets, Ladders and Rings

The simplest metazoan neural organization, the **nerve net**, appears in the radially symmetrical animals of phylum Cnidaria, such as the hydra (Figure 35.1a, see also Chapter

FIGURE 35.1
NERVE NETS, LADDERS, AND VENTRAL NERVE CORDS
(a) In the hydra's the netlike arrangement of nerve fibers, neurons are scattered over its body, with the greatest number near its mouth and tentacles. **(b)** In planarians, many neurons accumulate in the head region, forming ganglia. Two nerves emerge from the brain, forming a ladderlike arrangement. **(c)** In the sea star, the nervous system consists of a central ring with major nerves radiating into each arm. There is little cephalization. **(d)** The earthworm has a complex, segmented nervous system with large paired ganglia above and below the esophagus. A ventral, solid nerve cord produces paired ganglia in each segment.

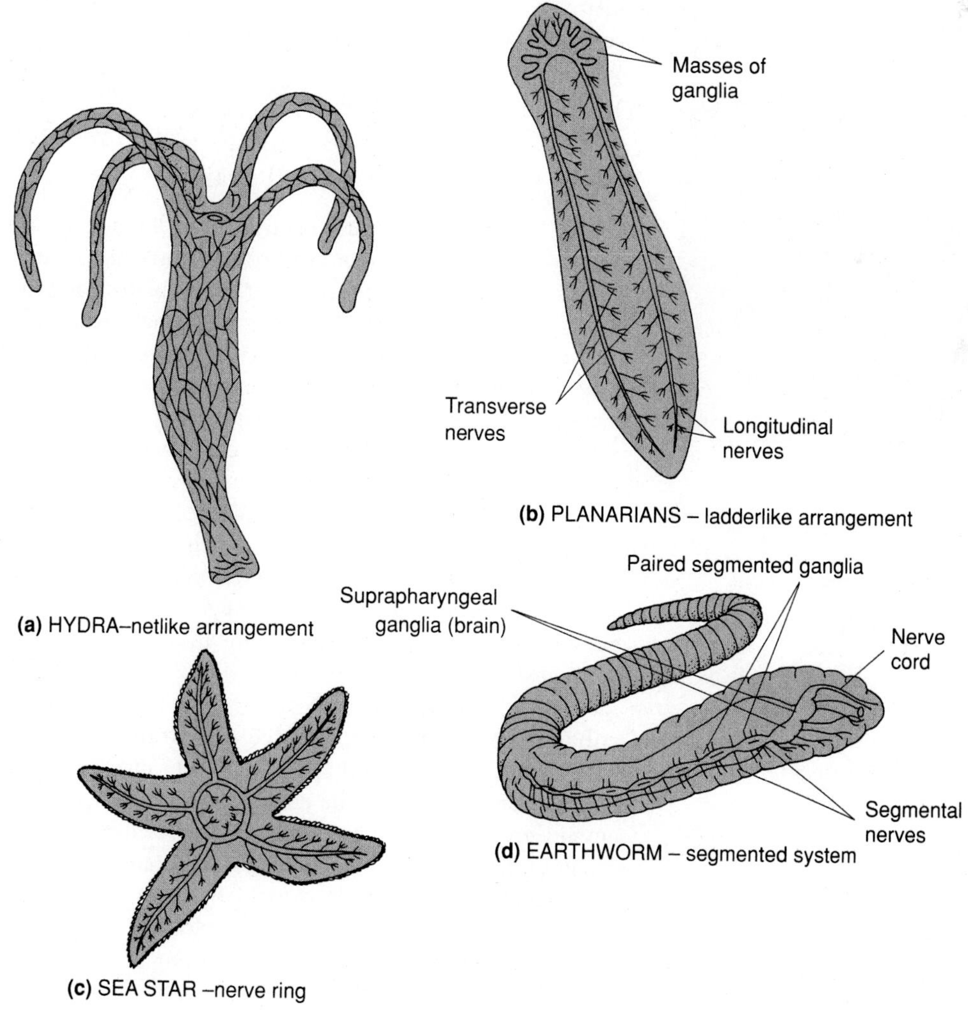

30). Theirs is a diffuse nervous system in which there is no clustering of nerve cells or their bodies into ganglia, as seen in more centralized systems. The neurons are, however, more concentrated about the tentacles and mouth. The arrangement of axons in the cnidarian gives the appearance of a net, but often the crisscrossing neurons do not actually touch. Thus an impulse travelling along any axon would not stimulate another axon because they were touching. Presumably, such separation would be more versatile than a true net since it would permit pathways to become established. This in turn would allow for certain tissues or organs to react independently of others.

The most primitive flatworms have a netlike nervous system similar to that of the cnidarian hydra, but in others the system is somewhat modified. For example, moderately primitive flatworms may have up to five nerve cords. In many planarians the nervous system resembles a ladder, consisting of only two parallel nerve cords with runglike cross connections and a concentration of nerve cells in the head region (Figure 35.1b). The anterior end of planarians has both a higher rate of cellular activity and greater sensitivity than the rest of the body. In addition, it is in the planarians that we first see large, anterior **ganglionic masses.** Ganglia are clusters of neurons or their cell bodies, often accompanied by supporting cells and surrounded by a sheath. Flatworms also boast cellular specializations in their nerve cells, having motor neurons, sensory neurons, and interneurons.

Echinoderms and Nerve Rings Echinoderms, among the oddest of all animals because of their unique pentaradial (penta, five) body symmetry and water vascular system, have an extremely limited nervous system. Early in their evolution they departed from other deuterostomes and proceeded along an avenue of narrow specialization. Apparently, echinoderms have no need for a centralized brain or even a rudimentary one. The greatest concentration of neurons in the seastar is the nerve ring around the mouth. This ring gives rise to branches that enter each arm, where they parallel the water vascular system (Figure 35.1c).

Massed Ganglia to Organized Brains

The annelids, we know, are segmented worms. In the earthworm, large ganglia surround the pharynx in the head of the worm (a common arrangement in annelids, arthropods, and other higher invertebrates), and each segment along the length of the worm has a pair of fused ganglia emanating from the nerve cord (see Figure 35.1d). Within the ventral nerve cord lie several **giant axons,** whose quick conduction of impulses permits the rapid burrowing and escape behavior of a startled worm (see Chapter 34).

What about mollusks? If you have ever owned a pet clam, you may have noticed that it really wasn't much company. Clams and other bivalves are notoriously short on personality, perhaps because they are not very intelligent. Like many more sedentary mollusks, their nervous system is quite simple. The nervous system of the chiton takes on a ladder form roughly similar to that of a flatworm, except that is is much more complex. The greatest concentration of nerves forms a nerve ring around the mouth and a dense ladderlike arrangement along the foot.

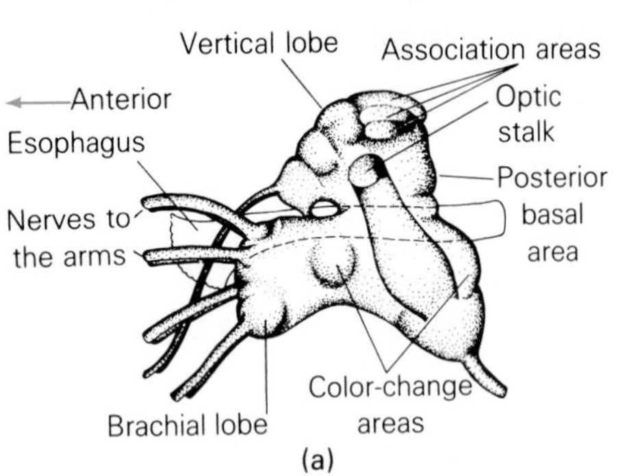

FIGURE 35.2
THE CEPHALOPOD BRAIN
The octopus has a well-developed brain with specialized regions related to vision, rapid movement, and intricate muscle control.

Cephalopods and the Appearance of a Brain Surprisingly enough, another mollusk is rather bright and has been called "smarter" than some vertebrates. We're referring to the octopus, a remarkably active predator. It, like many invertebrates, has a ganglionic mass of neural tissue around its esophagus, but the mass is so differentiated that it may rightfully be called a brain. For example, a distinct and identifiable part of that complex serves as a respiratory center; another part controls the animal's rapid color changes; other parts deal with spatial associations and are hence important in learning; and other parts control eye movement (Figure 35.2a). The octopus, in fact, has been an important research animal in the study of learning and memory.

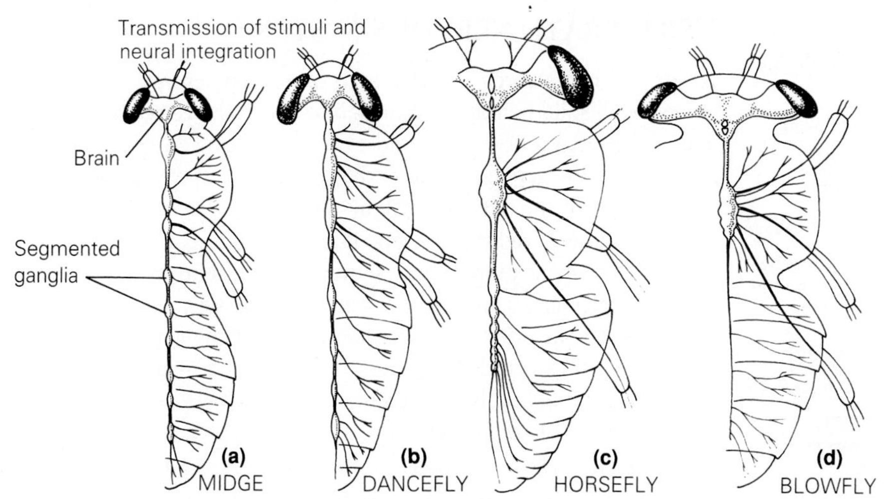

FIGURE 35.3
SPECIALIZATION IN INSECTS
Evolutionary trends in many insects include consolidation of the segmental nervous system into a more centralized one. Although there are larger ganglia in the head and thorax of the midge (**a**) more primitive segmental ganglia remain in the abdominal region. The dancefly (**b**) shows some increase in consolidation in the ganglia, while this is even more apparent in the horsefly (**c**). Finally, segmental fusion of ganglia in the blowfly (**d**) is complete.

The squid, another cephalopod mollusk, is noted among physiologists for its giant axons, which form a complex over the mantle and are important to the "jet-propelled" escape movements for which the squid is noted. Much of what we know about neural propagation has come from studies of the giant axons of squid.

Arthropods and a Segmented Nervous System Arthropods are interesting from an evolutionary viewpoint because not only are their bodies visibly segmented and specialized (Chapter 31), but within, the nervous system has also specialized, evolving different ganglia along its length to serve the different parts of the body. In many groups the ganglia have massed to form a more centralized arrangement, with a great reduction in the segmental ganglia. Compare the brain and segmental ganglia in the midge and blowfly in Figure 35.3. In the crayfish and other crustaceans capable of rapid darting movements, such movement is instigated by giant axons, as we saw in the annelids and mollusks.

The insect brain is composed of a large ganglionic mass above the esophagus and a smaller one below (Figure 35.4). Each of these brain parts has very specific roles, with much of the larger mass devoted to integrating information from the eyes, antennae, and other sensory structures so abundant on the insect body. The ganglionic mass below the esophagus controls the mouthparts and salivary glands. Interestingly, in arthropods that have just the simple eye—one that crudely registers only light intensities—the visual centers take up about 3% of the brain. But in highly visual species, such as the housefly, the visual centers may occupy as much as 80% of the brain.

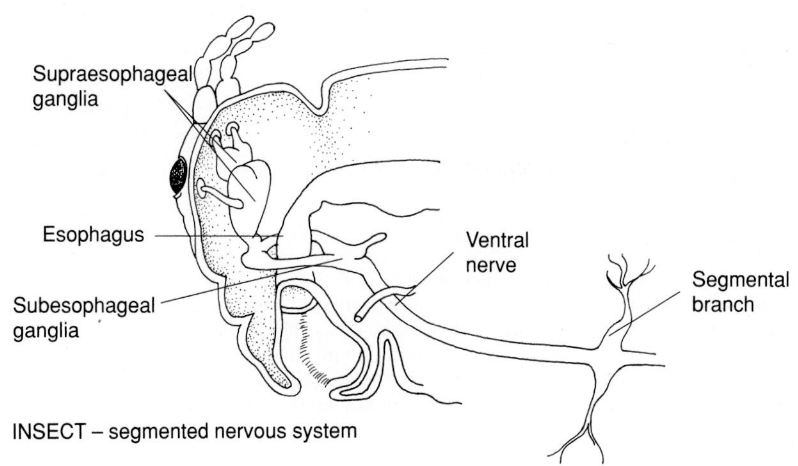

FIGURE 35.4
THE INSECT BRAIN
The insect nervous system includes ganglia above and below the pharynx, a ventral, solid nerve cord, and ganglia serving most segments.

VERTEBRATE NERVOUS SYSTEMS

So far, the nerve cords we have seen have been ventral. Furthermore, these are all solid neural structures. Now, with the vertebrates, we find a dorsal hollow nerve cord and—at its bulbous anterior end—the large, complex, ganglionic mass that is the brain. We can begin with an overview of the nervous systems in various classes of vertebrates.

Among the vertebrates, there is a pronounced tendency toward increased cephalization. That is, the more recently evolved classes generally have larger brains in relation to body size. The living classes of vertebrates include the jawless lampreys and hagfish, the sharks and rays, bony fishes, amphibians, reptiles, birds, and mammals. Of these, the most recently evolved, and hence "higher" classes, are the birds and mammals. On the other hand, our lowly cephalochordate relative, the lancelet, evolved much earlier. It has a hollow spinal cord but no centralized brain at all, a trait it presumably shares with our own Precambrian ancestors. All vertebrate brains have three parts: the **forebrain, midbrain,** and **hindbrain.** The forebrain includes the prominent **cerebrum** and certain regions below. In some vertebrates the forebrain is small, the other two brain regions being disproportionately larger. The fish brain, for example, is dominated by the midbrain, but there is also a relatively large hindbrain. In many vertebrates, the midbrain is important in analyzing visual and chemical (olfactory) stimuli, but in humans it serves primarily as a communicating bridge between the forebrain and hindbrain. Compare the brains of the vertebrates in Figure 35.5. You can see that in mammals, the midbrain and hindbrain have been overgrown by the forebrain's cerebrum. So, it is in the mammals that we find the most complex cerebral development. In addition, the wrinkled outer region of the cerebrum—the **cortex**—is far more prominent in the mammals. For an in-depth look at the mammalian nervous system, we will turn again to that brainiest of all creatures.

The Human Nervous System

The human nervous system, like that of other vertebrates, is divided anatomically into two major parts: the central nervous system (CNS) and the peripheral nervous system (PNS) as shown in Figure 35.6. The CNS contains the brain and spinal cord, while the PNS includes all neural structures outside these two, including a number of major nerve trunks and the sensory structures. The PNS is, in turn, divided into the **somatic nervous system** and the **autonomic nervous system.** The somatic system has **motor** and **sensory divisions,** that is, it is composed of motor neurons that move muscles and activate other effectors, and sensory neurons that transmit impulses from the sensory organs to the brain or spinal cord. The autonomic system comprises two divisions, the **sympathetic division** and the **parasympathetic division.** It is heavily involved in homeostasis; thus, it coordinates many of the regulatory activities of the body, doing so chiefly below

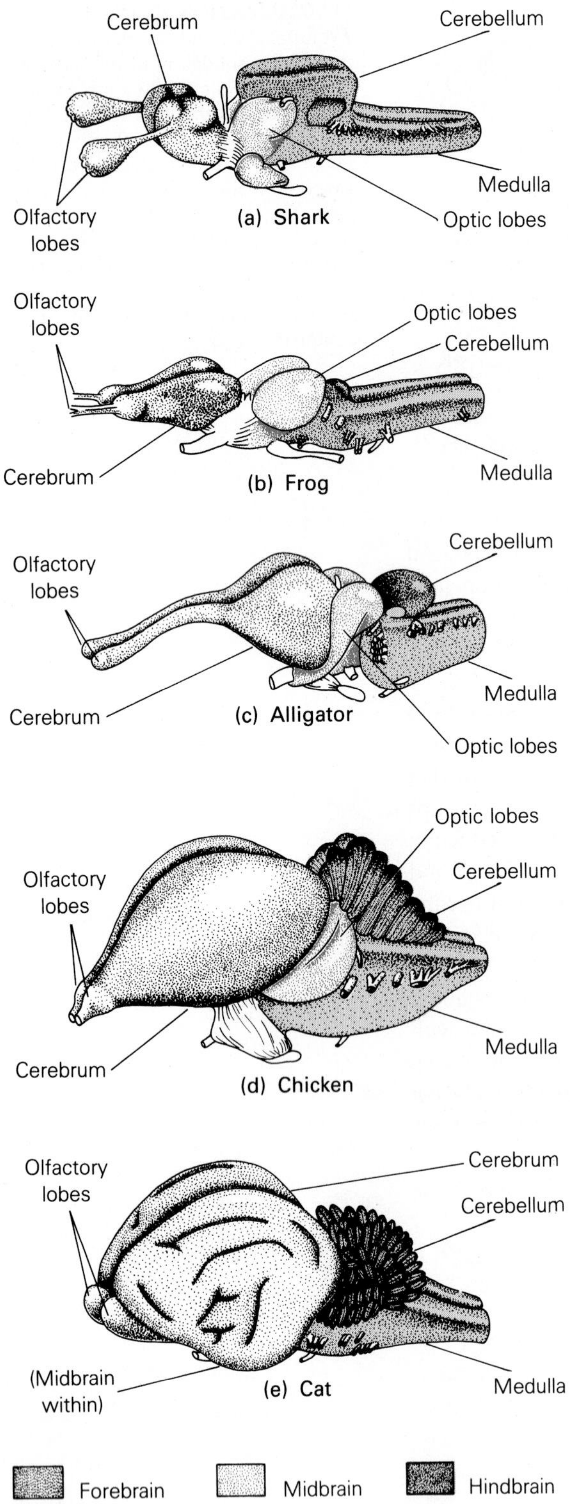

FIGURE 35.5
THE VERTEBRATE BRAIN
Comparing the anatomical structures of the brain in five classes of vertebrates (cartilaginous fish, amphibian, reptile, bird, and mammal) reveals general evolutionary trends and specific trends in specialization. **(a)** Note the relatively large olfactory lobes in the shark's brain. It clearly reflects the importance of chemical detection to this predator. **(b)** The frog feeds by visual means, as does the chicken. Note the relative size of their optic lobes. The trend toward increasing dominance of the cerebrum in vertebrate evolution is also apparent, beginning with the alligator **(c)** and becoming more pronounced in the bird **(d)**. **(e)** The trend is greatest in the mammal, with increased convolutions in the cortex. Convolutions or foldings are a way of increasing cerebral size without greatly increasing cranial size.

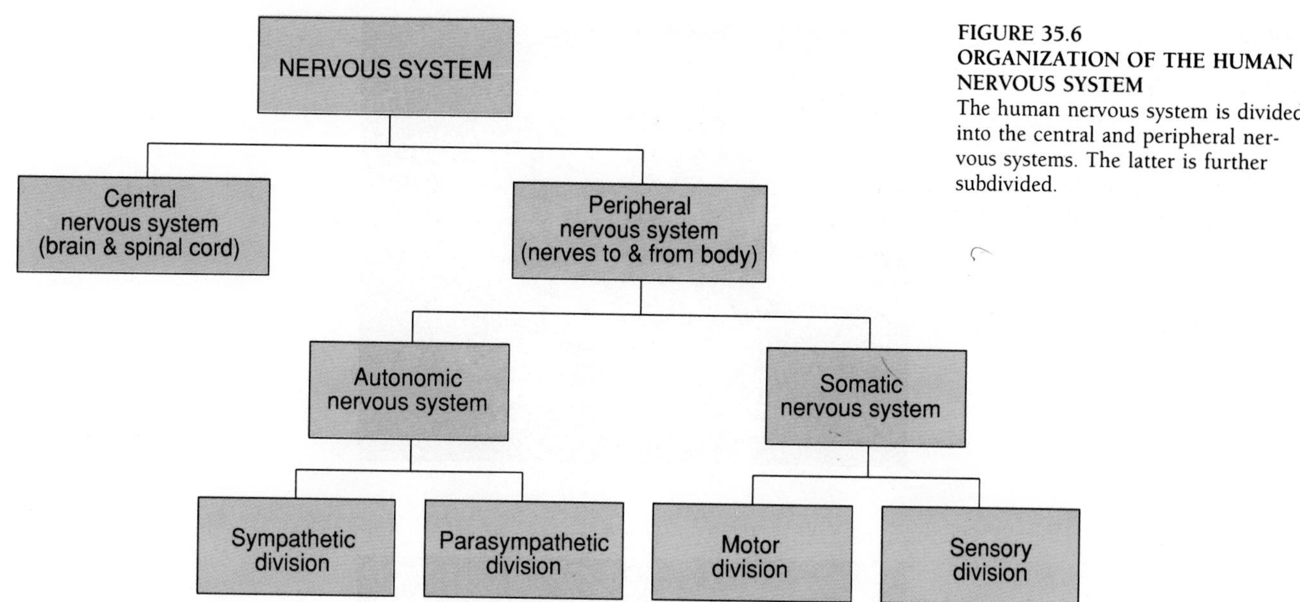

FIGURE 35.6
ORGANIZATION OF THE HUMAN NERVOUS SYSTEM
The human nervous system is divided into the central and peripheral nervous systems. The latter is further subdivided.

the conscious level. The autonomic system receives the axons of many motor neurons that originate in the brain and **spinal cord** those that relay motor impulses to the internal organs.

The Human Brain

The human brain is an absolutely fascinating structure given to self-congratulation. (In these words we find one complimenting itself.) In fact, an enlarged brain and an enlarged gluteus maximus are two of the most prominent traits of our species. One might wonder what the planet would look like had the human brain not motivated us to action and had we spent more time resting dolefully on our glutei maximi. The human brain has devoted a great deal of time to thinking about itself, but to this day many of its processes remain unknown. Furthermore, the things we discover are often hard to believe. For example, there is some evidence that every word you have ever uttered or heard in your entire life may be filed away in your brain, even though you will go to your grave having recalled hardly any of that information. (How nice it would be to be able to retrieve the exact words of our last conversation with a loved one.)

The human brain (Figure 35.7) weighs about 1.4 kilograms (three pounds), has a volume of about 1300 to 1500 cubic centimeters, and contains some 100 billion or more neurons and at least 10 times that many supporting **glial cells.** Each neuron may form many synapses, and the number of alternative pathways of impulses, along with the coordination required for an appropriate response, produce a veritable harmonic neural symphony. The brain and the spinal cord contains **gray matter** (cell bodies) and **white matter** (myelinated axons). The gray matter of the brain lies on the outside (the reverse of what we will find in the spinal cord). ("Gray matter" is what people like to accuse one another of lacking.)

The delicate brain with its gelatinlike consistency requires an effective means of support and protection. In addition to the surrounding bony skull, it is enclosed by three tough, protective membranes called **meninges.** The spaces within the meninges, and any open spaces within the brain itself, are filled with pressurized, shock absorbing **cerebrospinal fluid.** Finally, the billions of fragile neurons themselves are embedded in an even vaster matrix of glial cells, which make up much of the mass of the brain.

The circulation of blood through the brain is quite interesting. Metabolically, the brain is highly active and makes great demands on the circulatory system. For example, it receives about 15% of the blood output from the heart (when the body is at rest). It

FIGURE 35.7
THE HUMAN BRAIN
(a) The cerebrum, with its two hemispheres and highly convoluted surface, dominates the human brain. The convoluted walnutlike cerebellum lies on either side of the medulla oblongata, which itself gives way to the spinal cord. (b) This view of the brain further reveals the cerebrum's dominance. Note the open spaces, which in life are filled with cerebrospinal fluid. The smooth regions are chiefly the axons of myelinated neurons.

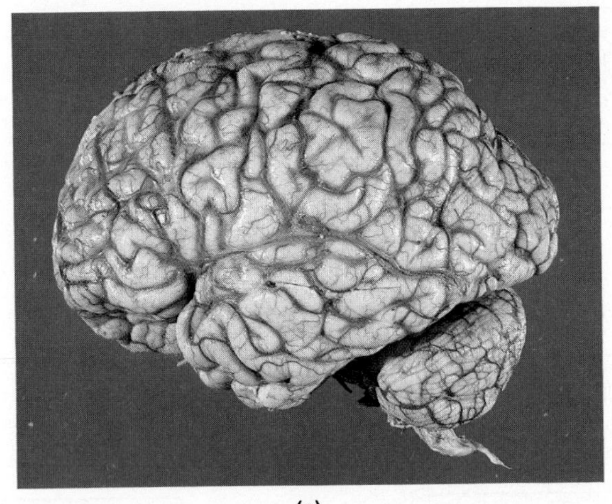

(a)

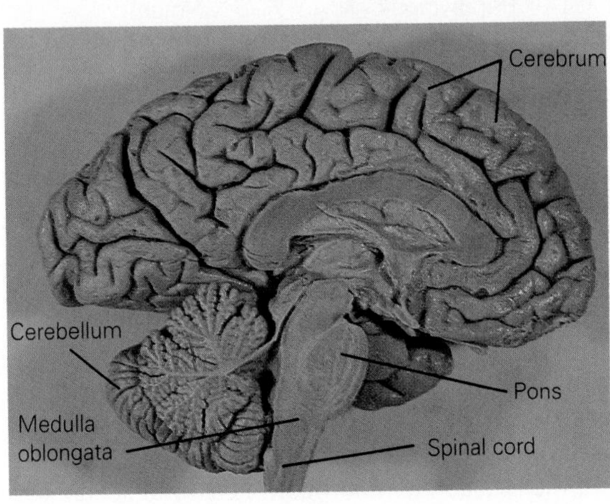

(b)

has a substantial oxygen demand and quickly malfunctions when oxygen-deprived. Glucose supplies most of the brain's energy, and, in fact, the brain utilizes about 75% of the body's intake of that sugar.

The arteries carrying blood into the brain have a special arrangement that helps ensure a sufficient supply of blood to this vital organ. They form a circle, the **circle of Willis,** at the base of the brain, there giving rise to smaller arteries that serve various parts of the brain. Each branch serves more than one part, however. Thus, should an artery branching from the circle become blocked, the brain region served by that artery will receive blood from another branch of the circle, thus minimizing damage.

The brain is also largely protected against toxic substances normally carried in the blood. To enter the brain tissue, substances carried by the blood must first cross the **blood-brain barrier,** a group of mechanisms that restrict the passage of materials from the blood into the brain. Part of this barrier is provided by the capillaries themselves, since the endothelial cells that make up the brain's capillaries have tightly joined and overlapping cell membranes. (In other regions of the body the union of capillary cells is looser, providing a porous surface through which materials can readily pass.) In addition, materials that do leave the blood must first pass into highly selective **astrocytes,** glial cells that attach to the capillaries via footlike processes (Figure 35.8). The presence of the effective blood-brain barrier, incidentally, makes it difficult for researchers to carry out experiments testing the effects of new drugs on the brain. Most such chemicals simply do not reach the brain tissue.

You may be surprised to learn that our three-pound brain isn't the largest on earth; whales and porpoises have larger brains. But intelligence (since we invented the term) seems to be related to the *relative size* of the brain to the total body, and here we humans lead the pack. (Besides, one might say, whales and porpoises don't have opposable thumbs, so what would they do with great intelligence anyway?)

We have mentioned that in all vertebrates, the brain consists of three regions, elegantly named the forebrain, midbrain, and hindbrain. We will take a close look at these now, beginning with the forebrain (Figure 35.9).

The forebrain, the largest and most dominant part of the human brain, is responsible for conscious thought, reasoning, memory, language, sensory reception and decoding, and conscious movement of the skeletal muscles. In addition to the cerebrum, the forebrain includes the **thalamus, reticular system, hypothalamus,** and the **limbic system.** As we'll see, some of the so-called "systems" are overlapping. We will focus first on the cerebrum, the human brain's "thinking center."

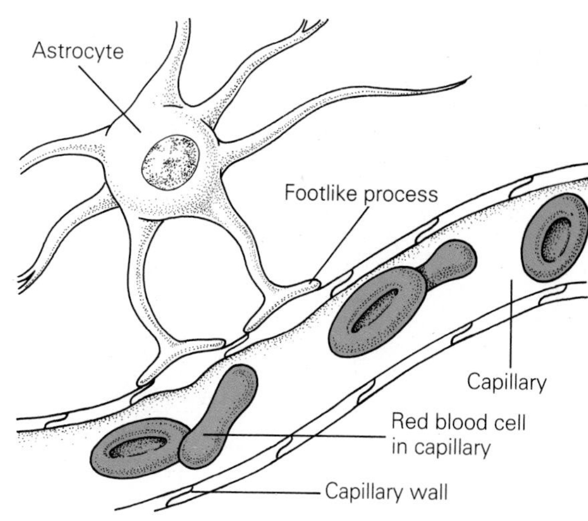

FIGURE 35.8
THE BLOOD-BRAIN BARRIER
Materials entering and leaving the brain tissue must pass across what is called the "blood-brain barrier." Note the tightly joined cells of the capillary wall and the footlike attachments of the astrocyte. Most materials are believed to pass through the astrocytes prior to entering the brain tissue.

The Cerebrum To most people, the word *brain* conjures up an image of two large, deeply convoluted gray hemispheres. Those hemispheres actually make up only part of the brain, the cerebrum—the largest and most prominent part of the human forebrain.

The left and right halves of the cerebrum are the **cerebral hemispheres**, and the outer layer of gray cells is the **cerebral cortex.** The cerebral cortex consists of a thin but extremely dense layer of nerve cell bodies (about 15 billion in all) and their dendrites, overlying the more solid white matter below. The white region consists of myelinated nerve fibers, reaching in all directions throughout the brain (see Figure 35.9b). These fibers bring information to the cortex for processing and carry the integrated messages from the cortex to other parts of the brain. All vertebrates have a cerebrum, but it differs

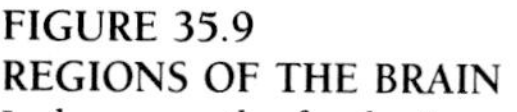

FIGURE 35.9
REGIONS OF THE BRAIN
In humans, the forebrain—the cerebrum and the structures it encloses—make up most of the brain. The midbrain contains connecting pathways between the forebrain and hindbrain. The hindbrain includes the pons, medulla oblongata, and cerebellum.

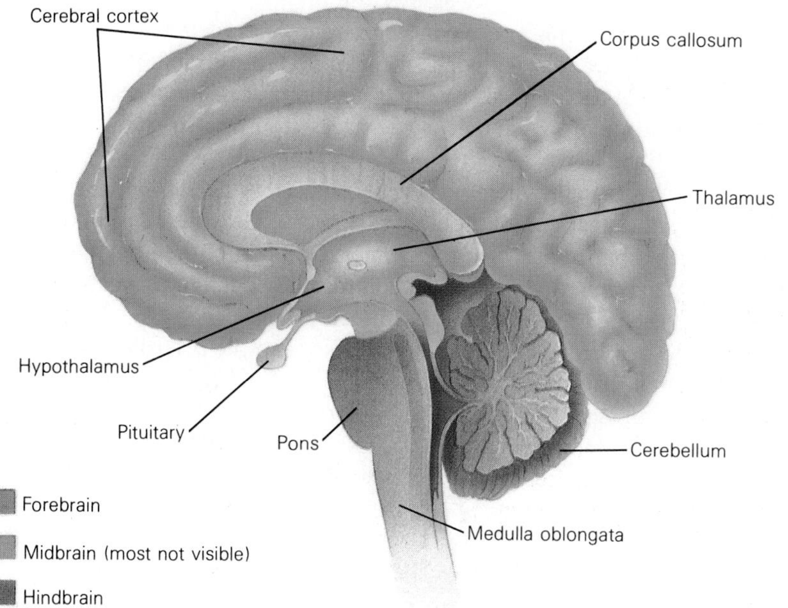

markedly from class to class, especially in the degree of development of the cortex. In some kinds of vertebrates the cerebrum may serve essentially to refine behavior that could be performed to some degree without it. These kinds of animals are not generally regarded as deep thinkers, but rely more on genetically programmed behavior emanating from the "old brain"—that is, the noncerebral brain, the part that evolved first. In more advanced animals, the cerebrum takes on greater importance and, as in the case of the visual and auditory centers, often takes over functions that were once the responsibility of other, older parts of the brain.

For example, if the cerebrum of a frog is removed, the frog will show relatively little change in behavior. If it is turned upside down, it will right itself; if it is touched with an irritant, it will scratch; it will even catch a fly. Also, sexual behavior in frogs can occur without the use of the brain—but it is probably best not to extrapolate from that. A rat is more dependent on its cerebrum. A rat that is surgically deprived of its cerebrum can visually distinguish only light and dark, although it seems to move normally. A cat with its cerebrum removed can meow and purr, swallow, and move to avoid pain, but its movements are sluggish and robotlike. Dogs treated this way are more helpless and just stand around, eventually starving unless food is thrust into their mouths. A monkey whose cerebrum has been removed is severely paralyzed and can barely distinguish light and dark. The result of massive cerebral damage in humans is total blindness and almost complete paralysis. Although such persons can breathe, and swallow, they soon die.

It seems that, from an evolutionary standpoint, more and more of the functions of the lower brain are transferred to the cerebrum in the more "cerebrated," or intelligent species. Generally, the degree to which the cerebrum has taken over neural control is reflected in its size. Thus more "advanced" animals have relatively larger cerebrums (see Figure 35.5). However, size is not the only indicator of the cerebrum's complexity. Convolutions, for example, increase the surface area of the cortex without enlarging the braincase. The deep convolutions seen in the human brain are lacking in the brain of the rat (but note that rats, with their smooth and tiny brains, are notoriously hard to foil). The highly touted and undoubtedly intelligent dolphin has a highly convoluted cerebrum, but with fewer layers than the human cerebrum.

The Lobes of the Cerebrum In humans, each cerebral hemisphere is divided into four lobes, occipital, temporal, frontal, and parietal (Figure 35.10). At the posterior is the **occipital lobe.** It contains a region that receives raw, visual sensory input from the optic nerve and begins the analysis of that input. If the occipital lobe is injured, black "holes" appear in the part of the visual field that is registered in that area.

The **temporal lobes** are at either side of the brain, under the temples. Each lobe roughly resembles the thumb of a boxing glove and is bordered anteriorly by a deep groove, the **lateral fissure.** The temporal lobe helps to process input from senses relating to hearing and smell. This lobe also helps with the processing of visual information, possibly constructing more comprehensive images from cruder information received from the occipital lobe.

The **frontal lobe** is right where you would expect to find it—at the front of the cerebrum. It underlies that part of the skull that people hit with the palm of their hand when they suddenly remember what they forgot. One part of the frontal lobe regulates precise voluntary movement. Another part controls the bodily movements that produce speech and is considered to be part of the speech center.

The area at the very front of the frontal lobe is called the **prefrontal area.** Whereas it was once believed that this area was the seat of the intellect, it is now apparent that its principal function is sorting out sensory information. In other words, it places information and stimuli into their proper context. The gentle touch of a mate and the sight of a hand protruding from the bathtub drain might both serve as stimuli, but they would be sorted differently by the prefrontal area.

The **parietal lobe** lies directly behind the frontal lobe, and the two are separated by a deep cleft called the **central fissure.** The parietal lobe contains the sensory areas for the skin receptors and the cortical areas that detect body position. Even if you can't see your feet right now, you probably have some idea of where they are, thanks to receptors

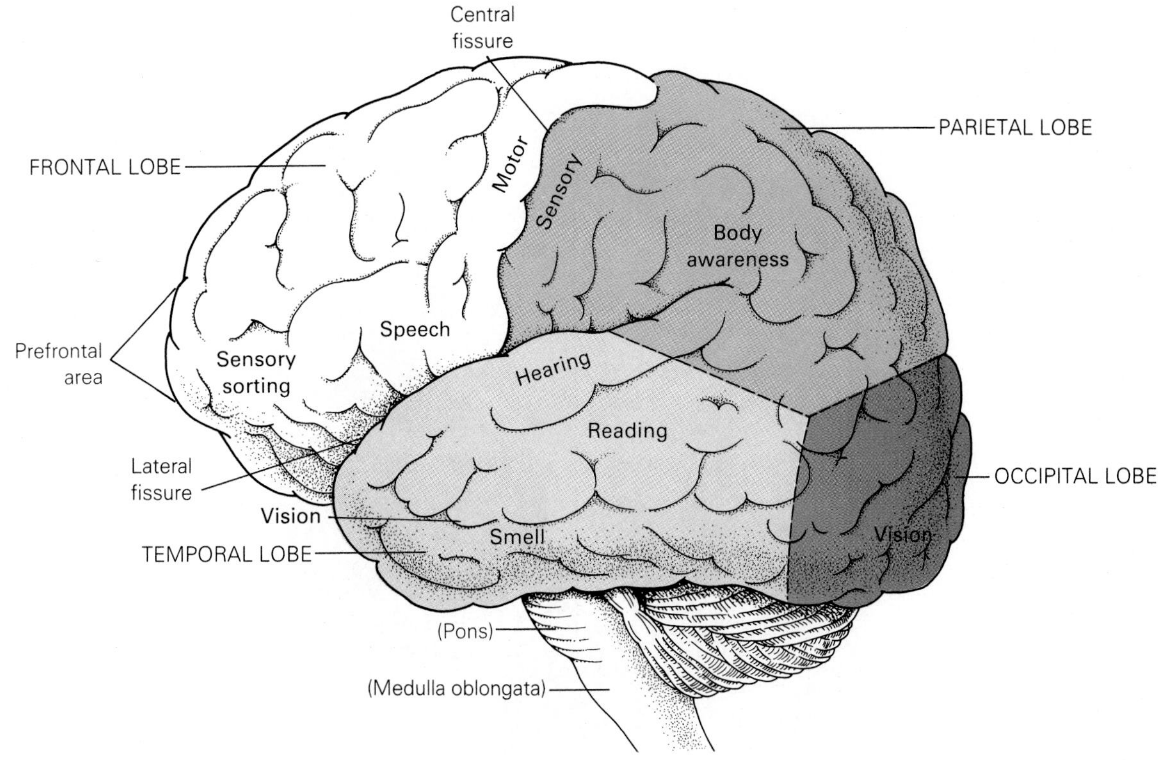

FIGURE 35.10
CEREBRAL LOBES
The human cerebrum is divided into four prominent lobes: frontal, temporal, parietal, and occipital. Each is somewhat specialized in its functions, and all represent the so-called higher centers of the brain.

in your muscles and tendons that innervate centers in the parietal lobe (and the **cerebellum** as well). Damage to the parietal lobe can cause numbness and a sense that one's body is wildly distorted. In addition, the victim may be unable to perceive the spatial relationships of surrounding objects.

Sensory and Motor Regions of the Central Cortex By stimulating various parts of the cortex with electrodes, investigators have located two large regions that are specialized for certain motor or sensory activities. The sensory and motor areas are located in specific **gyri** (singular, **gyrus**) prominent raised folds between fissures (see Figure 35.10). The motor area lies just anterior to the prominent central fissure. It originates the motor impulses that bring about voluntary muscle movement. Just posterior to the same fissure lies a second gyrus, one that receives sensory input from the skin (touch, pressure, etc.) and taste receptors.

Mapping the two gyri reveals that each is subdivided according to the region of the body served. The map is reconstructed in Figure 35.11. The distortion of the organs has been done, not out of an appreciation for the macabre, but to indicate the relative amount of motor and sensory cortical regions devoted to each body part shown. For example, much of the sensory gyrus area is organized for the reception of touch input from just the tongue, face, fingers, and genitals. But whereas much of the motor gyrus is devoted to movement of the tongue, face, and fingers, little is involved with the genitals. We can conclude from this that even though the genitals are very sensitive to touch, we have little conscious control over them. You may have already discovered this.

Two Brains Though the two cerebral hemispheres are roughly equal in size and potential, they are not completely identical in form and are far from identical in abilities. For example, the control of certain learned patterns takes place primarily in only one hemisphere. Other evidences of a difference between the hemispheres is seen in handedness.

FIGURE 35.11
SENSORY AND MOTOR CORTEX
The sensory and motor areas of the brain are located on the opposite sides of the central fissure, the prominent groove between the frontal and parietal lobes. In the caricature, the size of the body parts reflects the relative number of sensory or motor neurons serving that part.

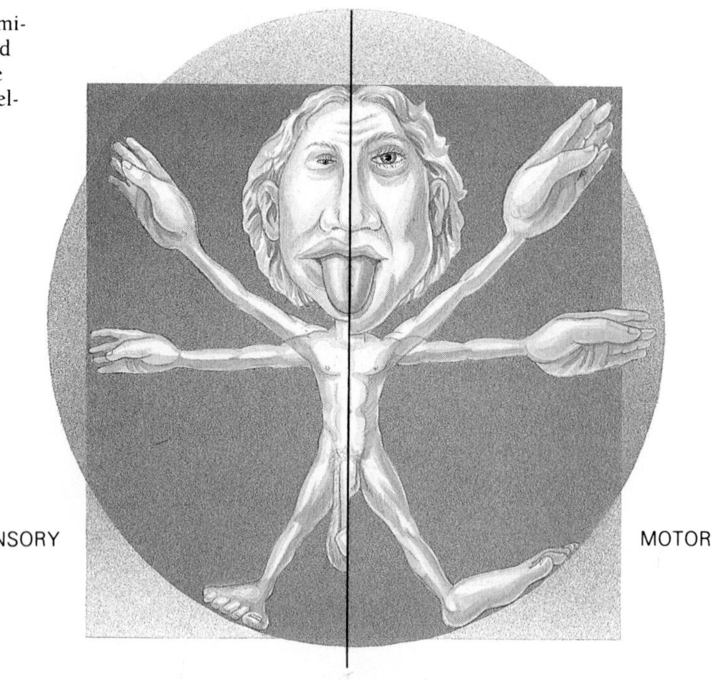

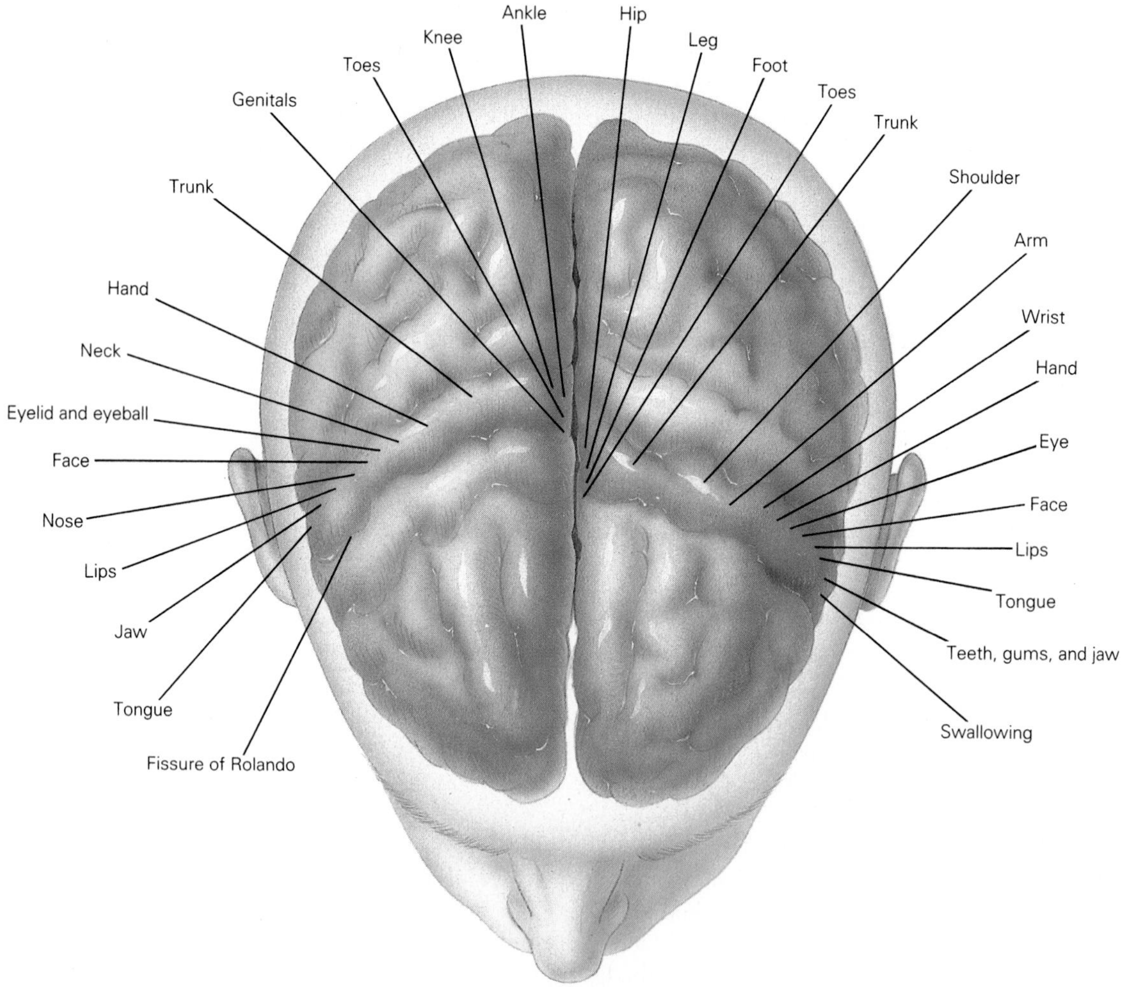

Genitals
Toes
Knee
Ankle
Trunk
Hand
Neck
Eyelid and eyeball
Face
Nose
Lips
Jaw
Tongue
Fissure of Rolando

Hip
Leg
Foot
Toes
Trunk
Shoulder
Arm
Wrist
Hand
Eye
Face
Lips
Tongue
Teeth, gums, and jaw
Swallowing

It is interesting that other species, such as rats and parrots, also show right- and left-handedness, although not in the proportions seen in humans. About 89% of humans are right-handed, but about half of rats and parrots are right-handed. Nerve tracts (bundles of axons or dendrites) cross from one side of the body to the other side of the brain (see Essay 35.1), so the right side of the body is controlled by the left side of the brain (and vice versa). Thus in right-handed people, the left half of the brain is dominant. It is also slightly larger than the right half. There are many mysteries to handedness. For example, no one can account for the unusual prevalance of southpaws among professional tennis players and artists.

There seems to be two kinds of left-handedness. In the more common type, the brain's left half is dominant (just as in right-handed people). These lefties are the ones who write by crooking their left hand around so as to write "upside down." In the less common type of left-handedness, the *right* half of the brain is dominant. It contains the functional speech center and controls the left hand, which is held at a more conventional writing angle. A few people have both traits, a sort of double negative: their right hands are controlled by dominant right hemispheres, so that they write with their right hands—which are held upside down!

Other functions, such as speech, perception and different aspects of IQ test performance are also more likely to be located in one hemisphere than the other, and in some cases, the distribution of centers is related to handedness. About 95 percent of right-handers and 70 percent of left-handers have the speech center in the left hemisphere. (About 15 percent of people have bilateral speech centers.) The left hemisphere also seems to be the seat of analytical thinking, while spatial perception is right-hemisphere function. Left-hemisphere brain damage can result in **aphasia**—the inability to speak or understand language—while right-hemisphere brain damage may result in the inability to draw the simplest picture or diagram (but does not affect the ability to write letters and numbers, which is a left-hemisphere function).

In spite of these specializations, the right and left hemispheres operate as an integrated functional unit. The primary route of communication between the right and left cerebral hemispheres is the **corpus callosum** (see Figure 35.9). It seems that if one side of the brain learns something—for instance, by feeling an object with just one hand—the information will be transferred to the other hemisphere. However, if the corpus callosum has been severed, the left brain literally doesn't know what the left hand is doing (Essay 35.1).

It should be pointed out that a certain degree of compensation is possible between the two halves of the brain. For example, if the dominant hand is injured, it is possible to learn to use the other hand with almost equal facility. If one hemisphere of a young child's brain is severely injured, the other side will eventually take over some or all of its functions. This ability to shift, like so many other abilities, declines with age.

Language Centers in the Cortex The human brain has highly developed areas devoted to the processing of language. We are just now beginning to understand how the centers interact, and much of what we are learning comes from studies of stroke victims. When a stroke occurs, part of the brain suffers an oxygen and nutrient loss, causing the death of brain tissue. Among the more typical results of stroke damage is aphasia, the loss or impairment of speech, especially when the damaged area is in the left side of the cortex. From stroke studies, neurophysiologists conclude that three regions of the left hemisphere are important to speech—**Broca's area**, **Wernicke's area**, and the **angular gyrus** (Figure 35.12). Of course, other regions of the cortex are also important to language, since linguistic articulation also involves hearing, seeing, even writing—not to mention thinking (although some people seem to have no trouble speaking without thinking much at all).

When we hear words, like all sounds, they are initially received by the auditory receptors, but unless the incoming neural information is processed in Wernicke's area its meaning will not be deciphered. The neural source of our spoken words, however,

ESSAY 35.1

THE GREAT SPLIT-BRAIN EXPERIMENTS

Some years ago, Roger W. Sperry and Robert E. Meyers experimentally separated the two hemispheres of the brains of cats by severing both the corpus callosum and the optic chiasma. This meant that the right eye was now connected only to the right half of the cerebral cortex and the left eye to the left cerebral cortex. Intensive testing then showed that the animals behaved as if they had two separate brains. A cat could be trained to perform a task by using one eye (the other covered) and when presented with the same task viewed by the other eye, the animal would respond as if no learning had occurred.

Later, Sperry and Michael Gazzaniga severed only the corpus callosum in human epilepsy patients as treament for their condition. With their optic chiasmas intact they were tested in an apparatus that allowed a particular image to fall entirely on only one half of the retina (see figure). This meant an image could be effectively transmitted to only the right or the left hemisphere of the brain. The results were intriguing: the patients reported "seeing" only what was transmitted to the left hemisphere. If a word such as "heart" were shown but divided by a partition so that each eye could see only part of the word ("he" with the left eye and "art" with the right), then "he" fell on the right half of the brain, "art" on the left, and the subjects reported seeing "art."

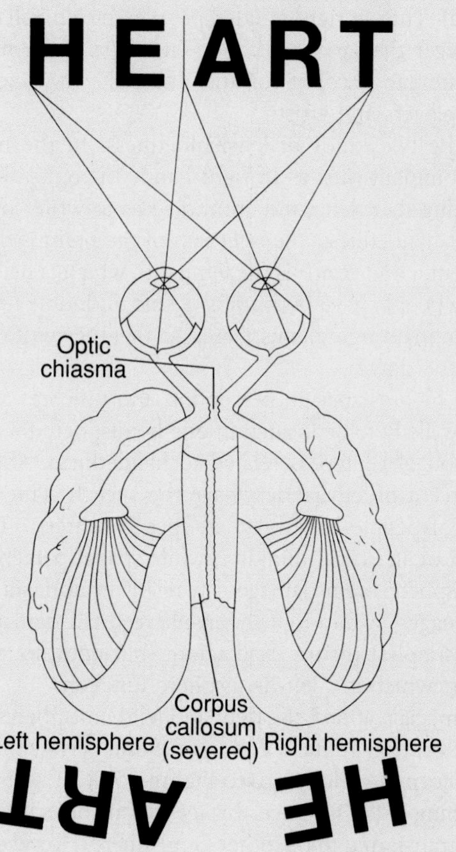

If the corpus callosum is cut, the halves of the brain are capable of functioning as separate and roughly equivalent units, although each has its own special qualities. If the optic chiasma is also cut, images seen by one eye can't be transferred to the opposite cerebral hemisphere. When the right eye is covered, a person can learn tasks and make discriminations using only the left eye. But if the left eye is then covered, the person must relearn the same tasks.

It was quickly discovered that the results were not due to visual phenomena alone. The patients could only tell researchers what fell on the left halves of their brains because that is where the centers of speech are located. On the other hand, if they were allowed to express themselves nonverbally, they could immediately describe what the right half of the brain "saw." Because the right side of the cortex controls the left hand, and vice versa, a person can first feel an object, say with the left hand, and then pick out that object with the same hand from a collection hidden behind a screen, simply by feel. However, because the information is traveling to the right half of the brain and that half cannot communicate with the left half, the person cannot name the object. Interestingly, the person will be able to pick out the object faster by using the left hand than the right because the right cortex is far superior at dealing with spatial relationships. Further experimentation has revealed that the human brain is, in a sense, effectively divided into two halves—two brains, as it were. The left half indeed tends to deal more with rational, verbal, and logical information, while the right is more intuitive. However, the information has fallen into the hands of pop psychologists, whom we now find breathlessly explaining that the problems of the world are due to too much dependence on the cold and logical left hemisphere and that we should be training ourselves to depend on the soulful and holistic right side. The secret, some say, is encompassed in the teachings of Eastern religions. Those who make such claims, however, seem to neglect the fact that the talents of each half of the brain are mutually dependent. They work together, and have historically, to produce that peculiar combination of behaviors that are so distinctively human.

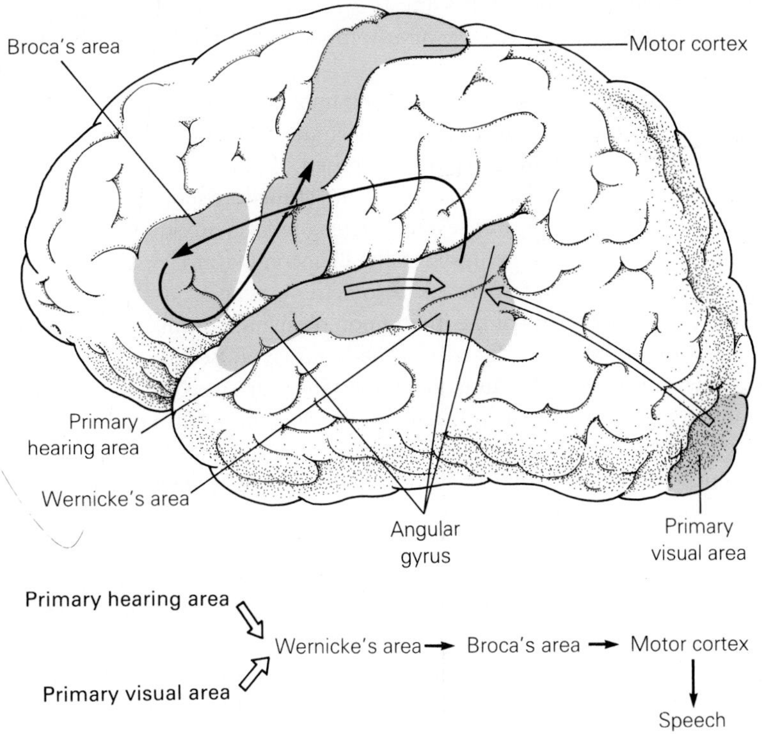

FIGURE 35.12
LANGUAGE PROCESSING
Using language requires the cooperation of neural centers in two cortical regions, known as Broca's area and Wernicke's area, along with visual, hearing, and motor centers. The principal pathways used in repeating what one has just heard, and in reading aloud, are indicated by the arrows.

is generated in Wernicke's area and transmitted to Broca's area along a special pathway. Broca's area then activates the motor cortex, providing information needed to properly activate the vocal cords, mouth, lips, tongue, and jaw, so that the proper sounds may be uttered.

Reading aloud involves the third region, the angular gyrus, as well as the visual and auditory areas of the cortex. Neural messages from the eyes reach the visual field of the cortex for processing, and this processed visual information is transferred to the angular gyrus, where the symbols are translated into words. In a manner of speaking, the angular gyrus "hears" the words we have just seen, by associating the visual message with auditory patterns in Wernicke's area. (Can you "hear" the words you see?) Speaking the words then requires the transfer of neural patterns to Broca's area and on to the motor cortex as usual. Poor readers mouth the words they read, but most of us learn to turn off our motor cortex unless we are deliberately trying to read aloud.

Damage to Broca's area results in badly impaired grammatical structure in both spoken and written sentences. Verbs and pronouns are commonly absent, although, with some effort, the sentences can be understood. Lesions in Wernicke's area result in an inability to decipher incoming messages properly. Speech, in this case, will include the verbs and pronouns and recognizable phrases, but these are strung together in a nonsensical manner. Neurologists conclude, then, that Wernicke's area produces the basic format of the spoken sentence, but Broca's area refines the structure and coordinates its actual vocalization.

The Thalamus The thalamus is located at the base of the forebrain. It is a paired structure, with one half on either side of a central fluid-filled cavity called the **third ventricle** (see Figure 35.9). The thalamus has been rather poetically called the brain's "great relay station." It consists of densely packed clusters of cell bodies called **nuclei,** through which most sensory input to the cerebrum must pass.

The thalamus integrates the sensory information that constantly bombards the body, subconsciously sorting it out and channeling the various signals to the appropriate parts of the cerebrum. The thalamus also receives signals from the cerebrum and sends the appropriate information to specific parts of the cerebellum. If a boy is to walk a fence successfully, the cerebrum must consciously focus on the problem. However, the required balance is actually maintained by the cerebellum. It is the cerebellum, then, that helps him make the delicate adjustments that keep him from falling off the fence in front of his friends.

Finally, the thalamus has the special task of relaying impulses to the cortex that help maintain consciousness. In this last function, the thalamus relies on a portion of its tracts known as the **reticular system**, a region composed of clusters of neurons that run throughout the thalamus and as far as the lower part of the forebrain. The reticular system is still somewhat of a mystery, but several interesting facts are known about it. For example, it "bugs" your brain, tapping virtually all incoming and outgoing communications. It seems that the reticular network is something of an arousal system that serves to activate the appropriate parts of the brain upon receiving a stimulus. The more messages it intercepts, the more a specific part of the brain is aroused.

The reticular system also seems to function importantly in sleep. You may have noticed that it is much easier to fall asleep when you are lying on a soft bed in a dark, quiet room than on the floor of a noisy bar. Under the quieter conditions, there are fewer incoming stimuli; as a result, the reticular system receives fewer messages, and the brain is allowed to relax and initiate the processes associated with sleep.

The Hypothalamus The hypothalamus lies, as the name implies, below the thalamus (see Figure 35.9). Actually, it makes up the floor of the brain's centrally located third ventricle. It is densely packed with cells that help regulate the body's internal environment and certain general aspects of behavior. In its principle function, the hypothalamus receives sensory input from many internal organs via the thalamus and makes use of the input in coordinating such internal conditions as heart rate, appetite, water balance, blood pressure, and body temperature. It also influences such basic drives as hunger, thirst, sex, and rage. Electrical stimulation of various centers in the hypothalamus can cause a cat to act hungry, angry, cold, hot, benign, or horny. In humans it is known that a tumor pressing against a "rage center" can cause the person to behave violently and even murderously.

Another major function of the hypothalamus is the coordination of the nervous system with the endocrine system (see Chapter 37). In fact, the hypothalamus prompts much of the hormone secretion that goes on in the pituitary, a major endocrine gland.

The Limbic System The hypothalamus and the thalamus, along with certain pathways in the cortex, are functionally part of what is called the limbic system (Figure 35.13). The limbic system links the forebrain and midbrain and contains a number of nuclei that control certain aspects of muscle tone, such as the positioning of an arm so that the fingers are in the best position to type or play the piano. Since it includes the hypothalamus, the limbic system encompasses hypothalamic functions and so influences emotions such as fear, rage, sexual arousal, aggressiveness, and motivation. For example, the **amygdala**, lying within the limbic system, can produce rage if it is stimulated and docility if it is removed. Finally, the amygdala, along with the **hippocampus**, a region of the temporal lobe, figures importantly in memory storage and recall. Memory processing will be discussed in Chapter 44. Other regions of the limbic system include the **cingulate gyrus** and the **fornix.**

The Midbrain and Hindbrain Essentially, the midbrain joins the hindbrain and forebrain, by numerous connecting tracts between the two (see Figures 35.5 and 35.9). Certain parts of the midbrain receive sensory input from the eyes and ears. All auditory input of vertebrates is processed here before being sent to the forebrain. In most vertebrates, visual input is first processed in the midbrain, but in humans and other mammals the visual information is sent directly to the forebrain. And while the midbrain is involved in complex behavior in fishes and amphibians, many of these same functions are assumed by the forebrain in reptiles, birds, and mammals.

The hindbrain consists of the **medulla oblongata**, the **cerebellum**, and the **pons** and is continuous with the spinal cord. It contains the lower, more posterior brain

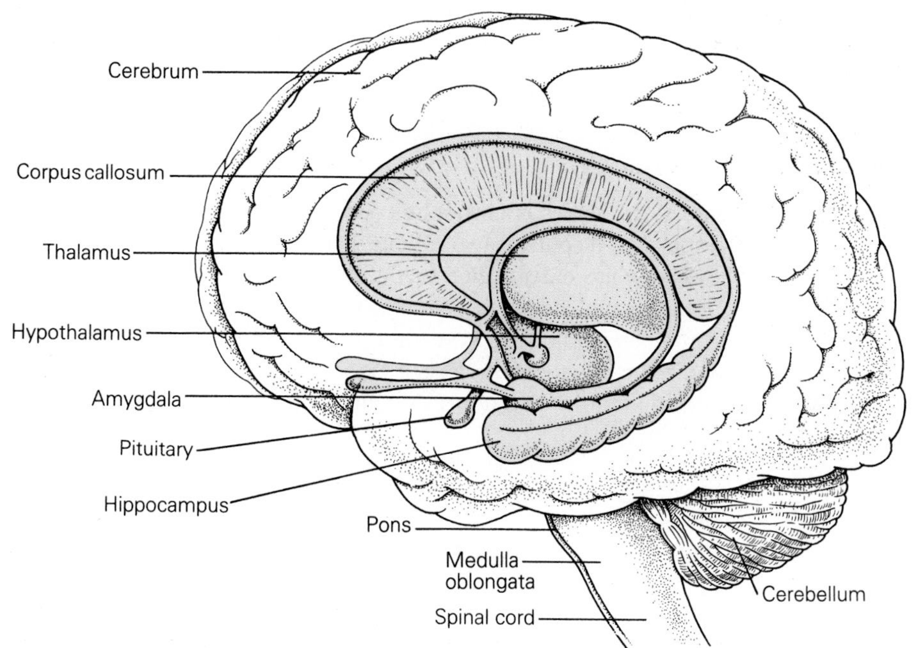

FIGURE 35.13
THE LIMBIC SYSTEM
The major components of the limbic system include the amygdala, hippocampus, thalamus, hypothalamus, fornix and cingulate gyrus.

centers (see Figure 35.13). As a rough generality, the unconscious, involuntary, and mechanical body processes are directed by these lower regions.

The Medulla Oblongata The lowest part of the brain, the medulla oblongata, or simply the medulla, contains centers that control breathing rate, heart rate, and blood pressure. In addition, all communication between the spinal cord and the brain must pass through the medulla.

The Cerebellum The cerebellum is a small, bulbous, paired structure with about the general appearance of the two halves of an enlarged walnut. It lies above the medulla and toward the back of the head. The function of the cerebellum is the coordination of movement. A particularly skilled gymnast can leave us quite impressed with his or her performing skill. Interestingly, those smoothly executed movements are not entirely voluntary. Much of what we see is made possible by the cerebellum, which operates below the level of consciousness.

The cerebellum coordinates all voluntary movement in the limbs and body and aids in the maintenance of posture and balance. It does this chiefly through inhibition, limiting the force with which a muscle contracts and the distance a limb travels, and generally dealing with several muscle groups at the same time. To coordinate this activity, the cerebellum requires constant sensory input informing it about the degree of muscle contraction and the position of the limbs and body. Such input comes from the eyes, balance organs, and from the muscles themselves. If it is to intercede in voluntary movement, the cerebellum must also sense *impending* movement; that is, it must know of such movement ahead of time. Voluntary movement itself is directed by conscious centers in the cortex, which send impulses along motor pathways through the brain and spinal cord and out to the muscles. Such impulses are also shunted directly into the cerebellum by branches from the same motor pathways. So the cerebellum literally "bugs" the cortex, "listening in" on its activities. Again, we are not consciously aware of the cerebellum's vital coordinating functions.

The Pons The pons (Latin for "bridge") is the part of the hindbrain just above the medulla oblongata. It contains ascending and descending neural tracts that run between the brain and spinal cord and between the cerebrum and cerebellum, linking the functions of the cerebellum with the more conscious centers of the forebrain. The pons also receives tracts to and from certain **cranial nerves**, the large nerves that emerge directly from the brain to form connections in both the somatic and autonomic nervous systems. The pons also aids the medulla in regulating breathing.

The Spinal Cord

The spinal cord (Figure 35.14) serves as a major link between the brain and the peripheral nervous system. The spinal cord contains two regions that, in cross section, appear white and gray to the eye. The white outer region of the cord is white because it contains immense numbers of myelinated axons. They run in very specific pathways or tracts to and from the brain, some extending far from their neural cell bodies. The butterfly-shaped inner region of the cord is gray because it is made up chiefly of the neural cell bodies of nearly countless numbers of interneurons and motor neurons, along with vast numbers of glial cells.

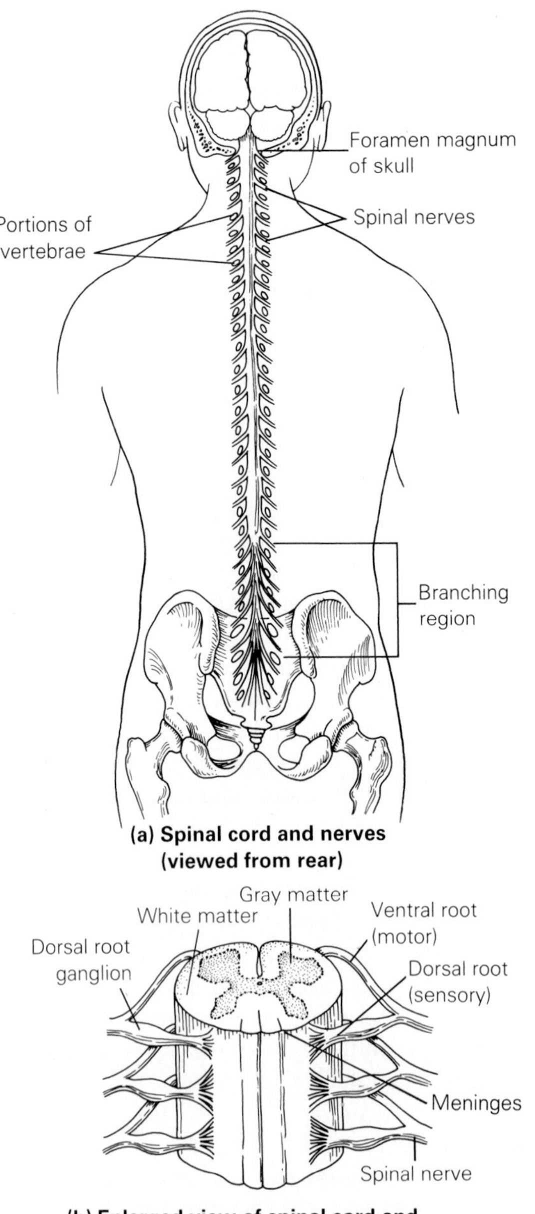

FIGURE 35.14
THE SPINAL CORD
(**a**) The spinal cord extends from the base of the brain into the lumbar region of the spine, where it branches into many descending nerves. The spinal nerves are major branches that emerge from between vertebrae. (**b**) A cross section of the spinal cord shows two distinct regions. The gray, double-winged region consists primarily of cell bodies and nonmyelinated neural fibers, while the outer white region is largely composed of myelinated axons that form the major spinal tracts. The emerging spinal nerves contain motor and sensory neurons. (**c**) Cell bodies of motor neurons lie in the spinal cord, while their axons emerge through the ventral root. Cell bodies of sensory neurons reside in dorsal root ganglia, and their peripheral axons extend from these ganglia to all parts of the body. Their central axons pass through the dorsal root of the spinal cord and synapse with interneurons.

The spinal cord begins as a narrow continuation of the brain, emerging from the foramen magnum ("big hole") at the base of the skull. It lies sheltered within the vertebral canal, a continuous channel formed by the neural arches of the vertebrae. The spinal cord, like the brain, is surrounded by a tough, three-layered sheath, the meninges, which contains the cushioning cerebrospinal fluid.

Paired spinal nerves emerge from the great cord through the spaces that lie between adjacent vertebral arches. Each of these spinal nerves is formed from two roots in the cord, a dorsal (toward the back) **sensory root** and a ventral (toward the belly) **motor root** (see Figure 35.14). The cell bodies of the motor neurons, like the cell bodies of interneurons, lie in the gray matter of the cord, with their lengthy axons passing out through the spinal nerves and onward to the voluntary muscle. On the other hand, the cell bodies of sensory neurons are clumped in large **dorsal root ganglia** just outside the cord.

The spinal cord has a certain degree of autonomy in that it can carry out synaptic reflexes between sensory and motor neurons without consulting the brain (see Figure 34.14). It is good to remove your hand from a hot stove as soon as possible, and letting the spinal cord control the situation reflexively helps preclude wasting time with cerebral debate. The spinal nerves carry sensory messages such as those telling you that your hand is on that hot stove, and the voluntary motor messages that activate the appropriate muscles to get your hand off. Their neurons also form a large part of the autonomic nervous system—the constellation of nerves and nerve cells that is involved in the involuntary activities of your internal organs.

Chemicals in the Brain

We know that the brain contains billions of neurons and that these neurons interact in a delicate and coordinated manner, integrating and shunting information from one place to another. This interaction includes both excitation and inhibition of adjacent neurons through very specific action in the trillions of synapses.

Whereas the peripheral nervous system uses only a few neurotransmitters—primarily acetylcholine and **norepinephrine**—the brain has at least 50, with more being discovered all the time. This number probably shouldn't be too surprising considering the specificity that is necessary in orchestrating the brain's vast network of cells.

The brain's neurotransmitters may be simple **monoamines**, such as norepinephrine, **dopamine**, **histamine**, and **serotonin.** Each of the above monoamines is a modified amino acid, but in some instances unmodified amino acids, such as glycine and glutamate, act as neurotransmitters.

The **neuropeptides** are a more recently discovered class of neurotransmitters. As their name suggests, they are short chains of amino acids (from two to about 40).

Among the more interesting brain neuropeptides are the **enkephalins** and **endorphins.** These modify our perception of pain and also have an elevating effect on mood. Anything acutely painful, such as running a marathon race or shooting oneself in the foot will stimulate the release of enkephalins. The "runner's high," a slight euphoria that may appear after about a 10-mile run, is also attributed to the release of enkephalins (although some runners simply may be ecstatic at covering that distance without dying). These peptides are also called the **opioid neurotransmitters** because morphine and other opiates will also bind to the neural enkephalin receptors and mimic their action.

Our rapidly expanding knowledge of the neurotransmitters is also shedding some light on the action of certain other drugs that affect the central nervous system (and vice versa). Amphetamines, for example, are believed to imitate norepinephrine, a neurotransmitter that stimulates the brain and accelerates blood pressure and heart rate. Cocaine has a different, more drastic effect. It amplifies the action of neurotransmitters such as serotonin, norepinephrine, and dopamine by blocking the enzymes and reuptake mechanisms that normally clear them from the synaptic cleft after they have acted. Drugs such as LSD, mescaline, and psilocybin mimic the role of other natural neurotransmitters. They produce their hallucinogenic effects by artificially stimulating action at the synapses.

Electrical Activity in the Brain

Each active neuron in the brain creates impulses that have electrical characteristics, although about a million neurons must fire simultaneously in order for this energy to be detected from outside the body. The instrument used in such detection is called an **electroencephalograph**, and the record obtained, an **electroencephalogram** (EEG). Electrodes leading to the device are fastened at various places on the head, and the feeble currents they pick up (primarily from the cortex) are amplified and recorded.

The electroencephalograph is not very useful in determining what, specifically, is going on in the brain, but it is useful in detecting abnormal electrical discharges and other gross changes in brain activity, such as the differences between normality and epilepsy, and even between wakefulness and sleep. Figure 35.15 compares the EEG of a normal person with that of a person with epilepsy. As the recording shows, epileptics are subject to sudden, random bursts of electrical activity. In a few instances, this abnormality can be traced to apparent physical defects, such as scars or lesions in the brain tissue, but most forms of epilepsy are simply not understood.

Sleep Electrical activity in the brain does not cease with sleep. Quite the contrary. EEGs reveal that sleep is accompanied by a considerable amount of brain activity. Sleep has several distinct phases (Figure 35.16). By far the most interesting of these is called paradoxical sleep, or **REM sleep** (rapid eye movement sleep). During REM sleep the other skeletal muscles are in a highly relaxed state, but the eyes dart about beneath the eyelids (thus, the paradox). The EEG recording at this time is more similar to that produced by wakefulness. If a person is awakened at this time, he or she will report vivid dreams (that will ordinarily be forgotten if the person is allowed to waken normally). REM sleep appears to be essential, although no one really knows why, just as no one really knows why dreaming might be useful. Experimental studies reveal that people deprived of REM sleep wake up tired, and in subsequent sleep periods they will extend the REM period (contrary to myth, you can make up for lost sleep). If the deprivation is continued, anxiety sets in and concentration becomes difficult. Eventually, personality disorders arise.

Psychologists and biologists generally agree that, for whatever reason, sleep appears to be highly essential and is somehow restoring. Theorists suspect that the sleep period, and REM sleep in particular, is a period of information sorting and storage and may be essential to long-term memory.

The Peripheral Nervous System

The peripheral nervous system is a vast network of nerves that spreads from the central nervous system to all parts of the body. It contains motor neurons and sensory neurons, running side-by-side within the same nerves. The peripheral nervous system is further divided into the somatic nervous system and the autonomic nervous sytem (see Figure 35.6). There are motor neurons in both systems, but only the somatic system has sensory neurons. The somatic nervous system carries the impulses that our conscious minds are most aware of: sensations that inform us of events in the world around us and the commands to our voluntary muscles that enable us to do something about those perceptions. The autonomic nervous system is more concerned with our internal workings and with our less conscious, less voluntary reactions.

The autonomic nervous system itself has two parts, the sympathetic division and the parasympathetic division. While the somatic, sympathetic, and parasympathetic entities have their own neurons, they often share the major peripheral nerves from the brain and spinal cord. Thus, a cranial nerve may carry elements of both the autonomic system and the somatic system. The vagus nerve, for instance, carries sensory and motor neurons of the pharynx, yet it also contains many motor neurons of the parasympathetic division of the autonomic nervous system.

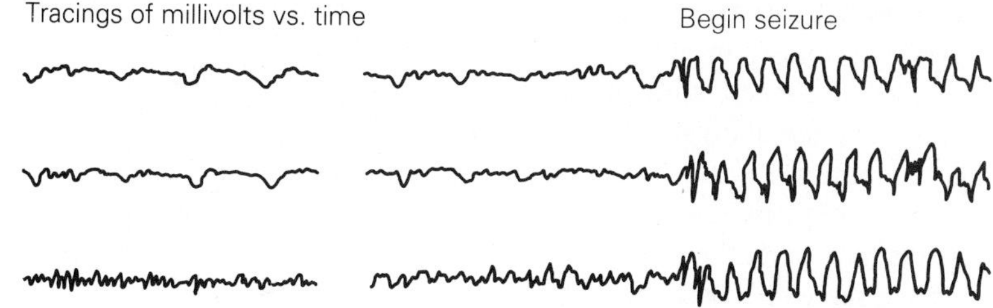

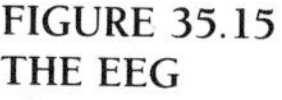

FIGURE 35.15
THE EEG
The normal electrical activity of the brain is seen at the left, where several EEG tracings from different electrode locations have been made. In the epileptic, normal electrical activity in the brain may occur between seizures, but when the episodes do occur, the electrical disturbances are quite obvious.

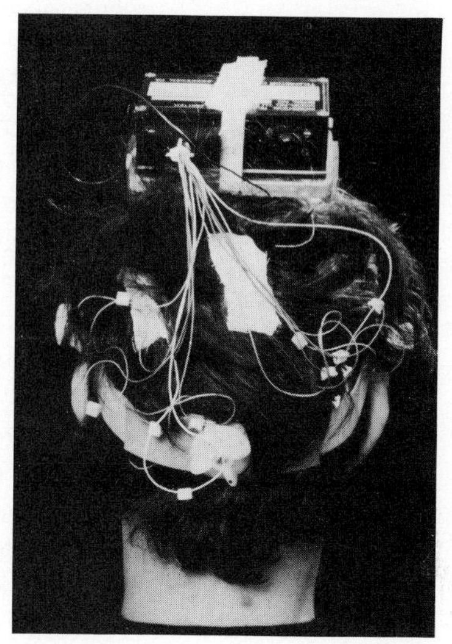

Awake
Stage 1
Stage 2
Stage 3
Stage 4
REM

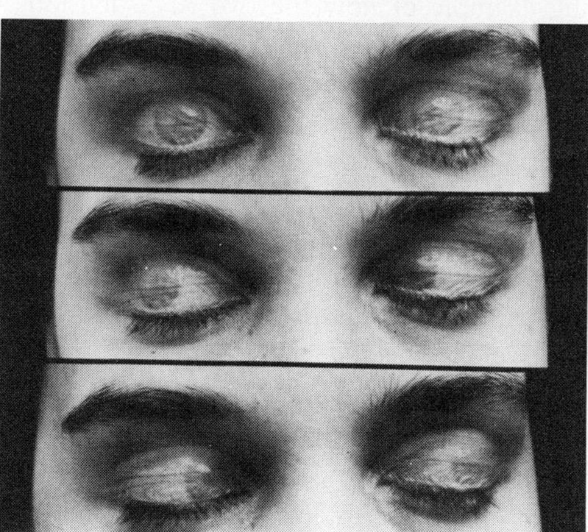

(a)

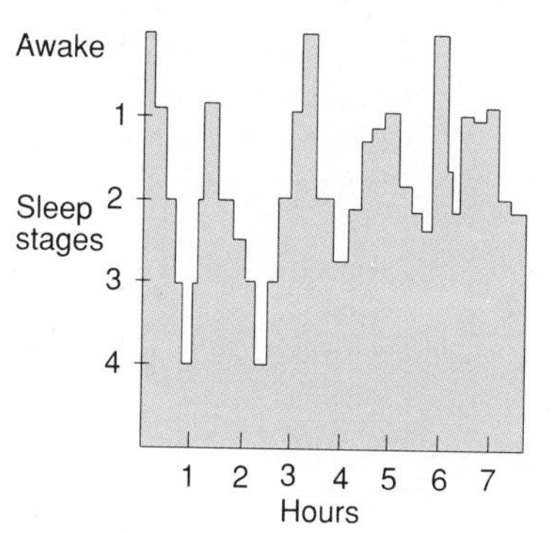

(b)

FIGURE 35.16
THE EEG IN SLEEP
(a) By studying the electrical activity of the brain and episodic eye movement, researchers have identified four sleep stages: (1) drowsiness, (2) light sleep, (3) intermediate sleep, and (4) deep sleep. Each has its own pattern of electrical activity as seen in the EEG. Typically, sleep is interrupted by a burst of electrical activity whose patterns resemble wakefulness. Since they are associated with rapid eye movement, they are called REM periods or REM sleep. **(b)** The second graph was derived from a number of night-long recordings. It shows that several periods of REM sleep (dark areas) occur in a typical night.

The Autonomic Nervous System The autonomic nervous system (ANS) is essentially a *motor* system, carrying impulses from the brain and spinal cord to the organs it serves. In doing this, it works in concert with the central nervous system in regulating the activity of the viscera, the internal organs. These include not only the prominent contents of the thoracic and abdominal cavities (the heart, lungs, digestive organs, kidneys, and bladder), but also the smooth muscle of the blood vessels, the iris, nasal lining, sweat glands, salivary glands, and even the tiny **arrector pili muscles** that cause our hairs to stand on end when we see who's been elected.

Working under the direction of the central nervous system, especially the medulla and hypothalamus, the ANS promotes **homeostasis**. That is, it coordinates the adjustments needed to maintain an overall stability or constancy in the face of changing conditions. For such a system to function efficiently requires a considerable amount of sensory feedback from the organs served. This is carried out by sensory nerves of the somatic nervous system, which, by keeping the brain informed, also play an important role in the process of homeostasis.

The two divisions of the autonomic nervous system, the sympathetic and parasympathetic, have opposite effects and operate in a highly coordinated manner to produce an overall adaptive effect. One familiar example is seen in heart rate. The human heart, without outside influence, contracts about 70 to 80 times per minute but will speed up when stimulated by a sympathetic nerve called the cardioaccelerator and slow down on a signal from a parasympathetic nerve (the vagus; see Figure 35.17). How can neural impulses from two different nerves have opposing effects on heart rate? The answer has to do with the specific neurotransmitters released by those neurons at effectors. In many species, most, but not all, of the sympathetic neurons secrete the neurotransmitter norepinephrine at their target organs, while neurons of the parasympathetic system secrete acetylcholine (as do motor neurons that move skeletal muscle). Norepinephrine accelerates heart rate, while acetylcholine slows it down.

An interesting example of how the two ANS divisions effect such changes in the viscera and muscles is seen in the "fight or flight" response we will discuss in Chapter 37. In an emergency, the response is instigated by the sympathetic nervous system. After the emergency has passed, the parasympathetic division takes over, and conditions return to normal. The autonomic nervous system has a variety of effects on other organs as well (Table 35.1).

As you might expect, the divisions of the autonomic nervous system differ anatomically as well as physiologically. As seen in Figure 35.17, most of the nerves and nerve tracts that make up the parasympathetic division originate in the brain and toward the lower

TABLE 35.1 RESPONSES TO SYMPATHETIC AND PARASYMPATHETIC NERVES

ORGAN	SYMPATHETIC EFFECT	PARASYMPATHETIC EFFECT
Pupil of eye	Dilates	Constricts
Heart	Accelerates	Slows
Intestine	Decreases movement	Increases movement
Salivary glands	Decreases secretion	Increases secretion
Stomach glands	Decreases secretion	Increases secretion
Lungs	Dilates air passages	Increases secretion
Blood vessels		
Respiratory system	Dilates	Constricts
Heart	Dilates	Constricts
Skin	Constricts	Dilates
Stomach	Constricts	Dilates
Metabolism	Increases	Decreases

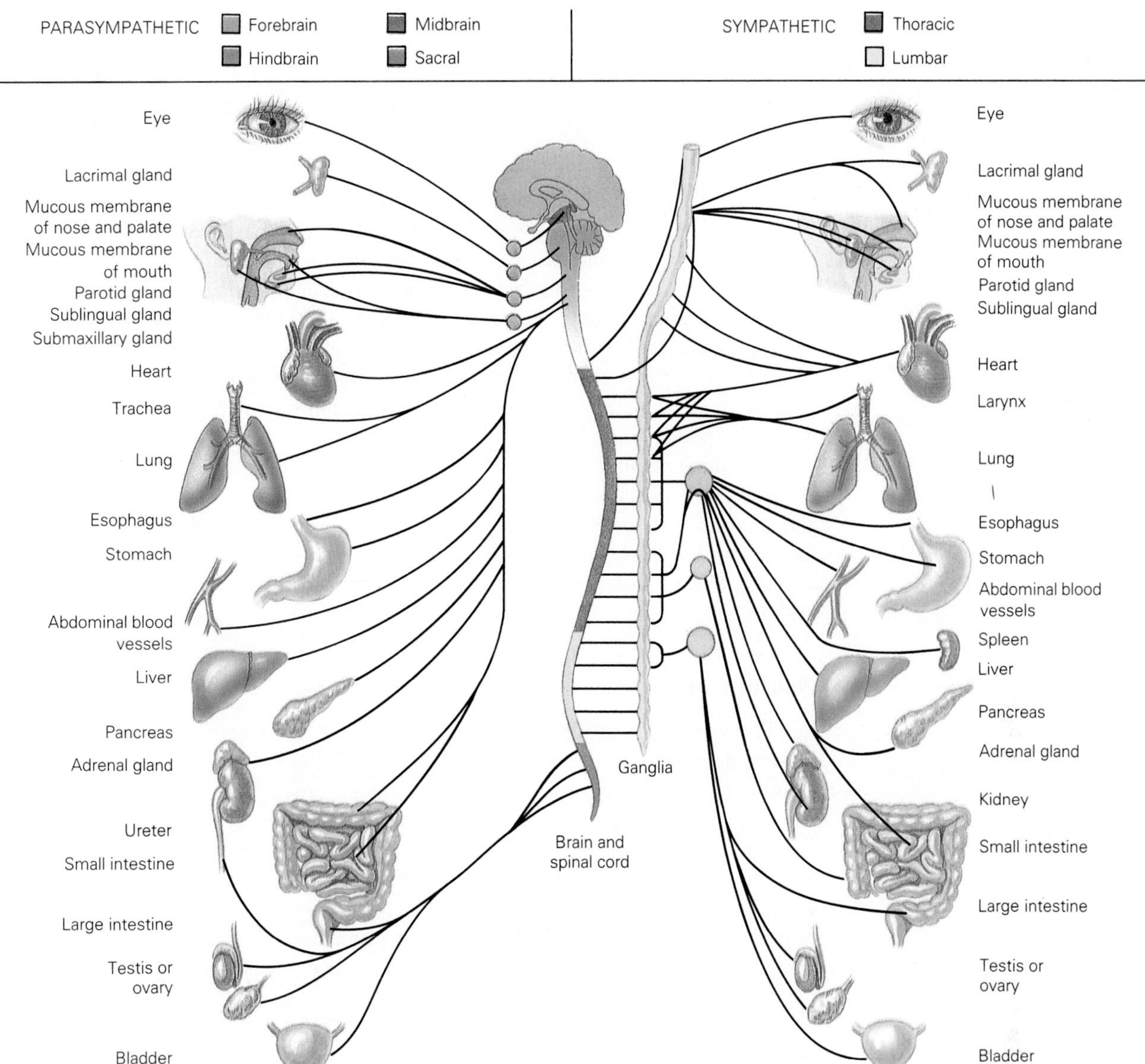

region of the spinal cord. Some go directly to their target organs; others end in external ganglia, where they relay impulses to nerves continuing toward the target organ. The sympathetic neurons all originate in the gray matter of the spinal cord. Furthermore, all of the sympathetic nerves leaving the cord pass through or synapse within rows of **sympathetic ganglia** just outside the cord. Some sympathetic nerves synapse once again in ganglia at various locations in the body. One of these, the **celiac plexus**, better known as the **solar plexus**, is often the target of a blow from a boxing or martial arts opponent, bringing about a temporary paralysis of the diaphragm and a momentary adjustment of attitude.

FIGURE 35.17
AUTONOMIC NERVOUS SYSTEM
The organization of the autonomic nervous system, showing the various organs it serves. The autonomic system is motor in function, so its nerves transmit impulses from the brain and spinal cord to the organs served. Its principal role is regulatory—keeping the internal organs operating in a carefully coordinated manner in response to constantly shifting conditions and needs.

THE SENSES

Animals require a constant input of information from their surroundings, input that includes what our own brain interprets as light, sound, heat, odors, or tastes. Such information is an animal's link to the outside world, and short- and long-term responses are based on it.

Sensory input is first intercepted by sensory receptors. These can be quite simple, with little more than a specialized dendritic region, as with free neural endings that lie scattered beneath the surface of the skin. Some sensory receptors are incredibly complex, as is the vertebrate eye, where a great many individual receptors along with other structures are organized into **sensory organs.**

Sensory receptors have several characteristics in common. They are **transducers**—that is, they convert the varied stimuli they receive into electrical signals. Receptors, like neurons, undergo depolarization, but instead of action potentials they generate what are called **receptor potentials.** Receptor potentials, in turn, stimulate action potentials in closely associated sensory neurons. Unlike the "all-or-none" ("go" or "don't go") action potential, receptor potentials vary in intensity. Increases in receptor potential intensity become translated into a higher frequency of action potentials in the sensory neuron, where the rate of firing has information value.

If you think about it, such "grading" of stimuli can be quite useful, since it is through versatile input that we make very fine distinctions between bright and dim light, soft and loud sounds, delicate aromas and eau de goat. We'll briefly consider how such distinctions are made, an interesting problem, for as we've seen, neurons only transmit action potentials along their axons, and all action potentials are the same. The answer lies in neural coding.

Neural Codes

Action potentials, instigated by sensory receptors and transmitted in sensory neurons, must be deciphered if the animal is to get a clear reading of its surroundings. The sorting, identifying, interpreting, and integrating is done primarily in the brain. As we've seen, one way the brain can decode incoming impulses is through their *frequency*. A mild stimulus, such as light touch, might produce just a few impulses per second and might even be ignored by the brain. A much stronger stimulus, such as hitting one's finger with a hammer, produces a virtual barrage of impulses in a very short time and registers as pain. The intensity of a stimulus can also be perceived according to the *number* of neurons being stimulated. The hammer stimulates many sensory neurons at once, making the stimulus difficult to ignore (you can test this for yourself).

Decoding, however, involves more than just distinguishing between degrees. It involves making distinctions in "kind" as well. Distinguishing specific kinds of incoming information—touch, light, taste, sound, temperature, pain—depends on how the sensory neurons and the integrating interneurons are connected, or "wired." Partly through learning and experience, regions of the brain specialize in interpreting sensory information, and each sensory neuron connects via specific tracts to its corresponding brain region. As we've seen, some centers of the brain deal only with sound, for example, and when they are stimulated, "hearing" results. Interestingly, stimulating such centers directly with electrodes will likewise be interpreted as sound (or memories of sound). Finer levels of distinction and integration also exist; thus most of us can distinguish a reasonably accomplished tenor from an alley cat hitting the same note.

Tactile Reception

Tactile receptors, or mechanoreceptors, are extremely sensitive, fast-firing neurons that respond to any force that deforms or alters the shape of their plasma membrane. In some cases, nonliving bristles or hairlike structures protruding from the surface of the plasma membrane touch objects in the environment before the organism's body actually reaches the objects. Such structures act as "early warning systems" that, when activated, permit the animal to make appropriate, timely responses before danger can increase or a potential food morsel can evade capture.

Touch in Invertebrates A number of arthropods, from caterpillars to spiders, are fuzzy. One adaptive advantage of their covering is related to an increased sensitivity to touch, since the bristles extending from the body are connected to tactile receptors.

Tactile sensitivity may be important for a variety of reasons. For example, web-building spiders have hairy legs that respond to vibrations set up by prey caught in the web (Figure 35.18). Cockroaches have tiny hairs protruding from their abdomen that, when stimulated by very light air currents, send them scurrying for safety. Many aquatic insects have bristles on their heads that are sensitive to water currents. R. S. Wilcox has accumulated a fascinating body of evidence that indicates how certain aquatic insects use water vibration in communicating with each other. Touch sensitivity, in general, is common among a wide range of invertebrates and is usually employed in finding food, mating, and avoiding predators.

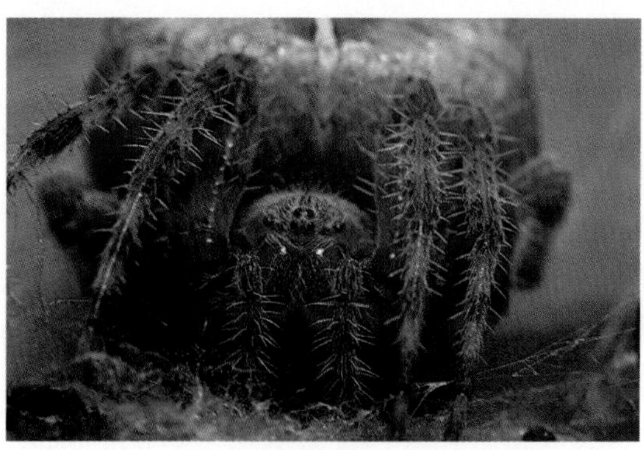

FIGURE 35.18
TOUCH RECEPTORS IN AN INVERTEBRATE
Touch receptors in spiders are often associated with hairs. When a hair is bent or moved, it activates a sensory neuron that transmits its impulse to an associated ganglion for integration. Some sensory hairs are fine enough to be moved by light air currents.

Touch in Vertebrates Vertebrates have two kinds of tactile receptors. **Distance receptors,** such as the lateral line organs of fishes (Figure 35.19), are sensitive to water motion. And then there are **contact receptors.** In humans and other mammals, these are of two types: pressure and touch. The pressure receptors are located deep in the

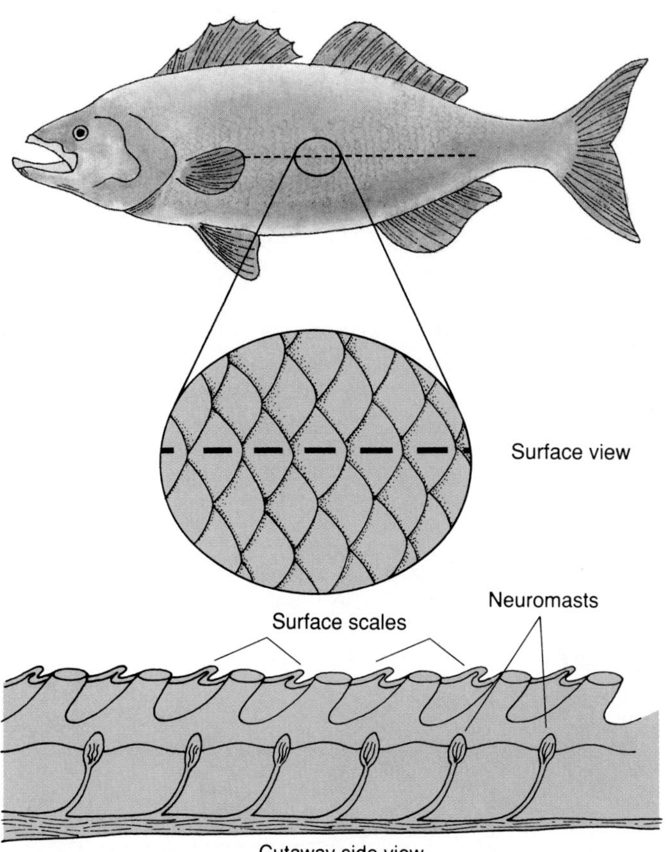

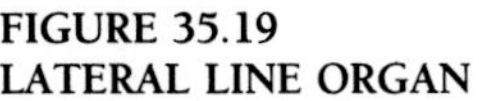

FIGURE 35.19
LATERAL LINE ORGAN
The lateral line organs of bony fishes and sharks can detect water movement. Such disturbances are picked up by receptors known as neuromasts, clusters of sensory hairs embedded in a gelatinous mass. Movement of water in the canal bends the neuromast, moving the hairs, which stimulates associated neurons to fire and sends impulses to the brain. In photo, the row of dark dashes is the lateral line.

skin and seem to consist of encapsulated nerve endings called **Pacinian corpuscles.** Light touch, on the other hand, is believed to be registered in **Meissner's corpuscles,** which lie near the surface of the skin (Figure 35.20). Among primates, sensitivity to touch is greatest around the lips, nipples, eyes, and fingertips. In addition to touch receptors, the skin abounds with simple, unspecialized, free nerve endings that register pain and thermoreceptors that respond to heat. A special kind of mechanoreceptor found in the arteries, the **baroreceptor,** is activated by blood pressure within the vessels.

In humans, as in most mammals, touch sensitivity is particularly great around hairy areas. As you may have noted, this includes the hairline around the face, and the genitals. The "whiskers" of many animals are especially sensitive to touch. While the scant body hairs of humans are of limited value in keeping us warm, they are good sense organs: try moving a body hair without feeling anything.

Until 1982 it was generally believed that hairs, in their sensory function, were simply dead, mechanical levers that when touched would jostle the sensory nerves surrounding their roots. Then another theory was developed based on information that the hair protein keratin is so highly structured as to be essentially crystalline. In fact, it was found that each hair comprises a single **piezoelectric crystal.** A piezoelectric crystal discharges electricity when it is deformed. (Cheaper phonograph pickups use piezoelectric crystals to translate needle movements into electric currents.) Researchers found that like other transducers, a perfectly dead hair generates a small electrical discharge when it is bent, and that nerve endings respond to this electricity.

Incidentally, the outer layer of skin is also made primarily of keratin, and it too has a piezoelectric effect. Bending or depressing the epidermis generates detectable electric discharges that are picked up by the touch-sensitive free nerve endings.

Thermoreception

Thermoreception is important in a wide range of animals. Many of the processes of life are possible only within a specific and often narrow range of temperatures, and animals must be able to sense the temperature of their environment in order to place themselves under the most nearly optimal conditions.

Heat Receptors in Invertebrates While thermoreception may be important to many invertebrates, little is known about its mechanisms. Heat detection is particularly important to the ectoparasites of birds and mammals: leeches, fleas, mosquitoes, ticks, and lice—parasites that must find a warm-blooded host. Their thermoreceptors are generally located on the antennae, legs, or mouthparts.

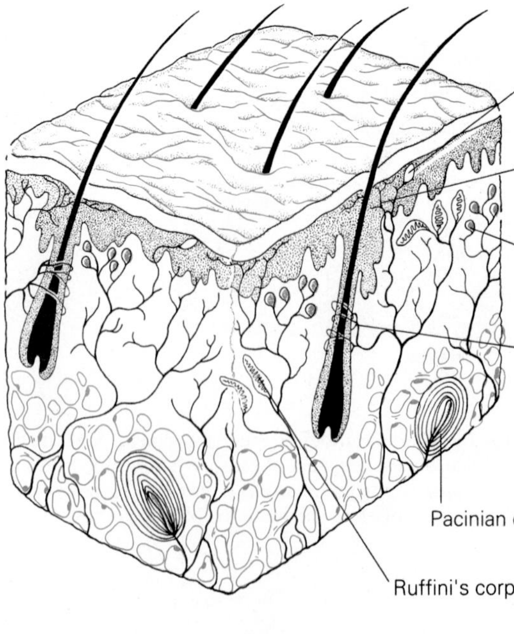

Meissner's corpuscle (light touch)
Free nerve endings (touch, pain)
Krause's end bulbs (cold)
Free nerve endings with hair shaft (touch)
Pacinian corpuscle (heavy pressure)
Ruffini's corpuscles

FIGURE 35.20
SKIN RECEPTORS
The skin contains a variety of sensory structures, each specialized for detecting certain stimuli. Interestingly, the hair shaft itself is a sensory structure. A slight bending of the hair causes it to discharge a very small electrical impulse, which is picked up by neurons that surround the hair root.

Heat and Cold Receptors in Vertebrates Many vertebrates have specific heat receptors. The pits of a pit viper, such as a rattlesnake, are a pair of indentations between the eyes and nostrils. These are loaded with heat (infrared) receptors that tell the snake when it is facing a living thing that is generating metabolic heat (Figure 35.21).

There is some disagreement over whether humans and other mammals have specific receptors for detecting heat and cold, or whether free nerve endings register these stimuli. It has even been suggesteed that a single neuron can register both heat and cold, simply by firing in different patterns. Some physiologists believe that heat receptors are located deeper in the skin than cold receptors. Further, whether the brain registers "warm" or "cold" (within limits) depends on the immediate previous experience of the receptors. If the skin has been warm and touches something less warm, the object will seem cold, but if the skin has been cold and touches something cool, the second sensation will be exaggerated warmth. This can be demonstrated by switching your hands from ice water to cold water.

Other physiologists ascribe thermoreception to the free nerve endings and to specialized skin receptors called **Ruffini's corpuscles** and **Krause's corpuscles** (or **Krause's end bulbs**; see Figure 35.20). The latter are far more numerous and are stimulated by a greater temperature range than are the former.

FIGURE 35.21
HEAT SENSORS IN THE PIT VIPER
Heat sensors, used for detecting prey by a pit viper, are located in depressions near the eyes. Each pit consists of an outer chamber that ends in a thin membrane covering an inner chamber below. By moving its head back and forth, the pit viper is able to use incoming thermal cues from the two pits to zero in on its warm-bodied prey.

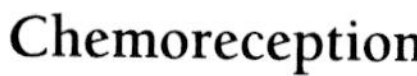

Chemoreception

Chemoreception is the ability to perceive specific molecules. These molecules are often important cues to the presence of specific entities in the environment. Nearly all animals and a great number of protists exhibit chemoreception. It is essential to many animals in finding food, locating a mate, and avoiding danger.

Chemical Receptors in Invertebrates. Planarians locate food by following chemical gradients in their aquatic surroundings. Their simple **chemoreceptors** are located in pits on their bodies, over which they move water with their beating cilia. Certain insects have taste and smell receptors of astounding sensitivity. Among insects, chemoreceptors may be found in the body surface, mouthparts, antennae, forelegs, and, in some cases, the ovipositor (since many insects lay their eggs only on certain plants).

Among the incredible stories of insect olfactory (sense of smell) abilities we find the tale of certain moths. The moth smells with its antennae, and adult males have enormous antennae with thousands of sensory hairs (Figure 35.22). About 70% of the adult male receptors respond to only one complex molecule called **bombykol**, a sex attractant released by females. As the molecules drift downwind finally to touch the antennae of a wandering male, they enter the tiny pores of a "hair," or **sensillum**, which houses his olfactory receptors. The molecules then dissolve in the fluid that moistens the receptor and interact with its membrane. Upon perception of the molecule, the male becomes excited, reorients his flight, and heads upwind, a behavior that sooner or later should lead him to the waiting siren.

Chemical Receptors in Vertebrates. Vertebrates detect chemicals in a number of ways, employing three major means: general receptors and two types of specialized receptors, **gustatory** (taste) and **olfactory receptors.** For example, many aquatic vertebrates have generalized chemical receptors scattered over the body surface. Olfaction and gustation in vertebrates are usually accomplished by moving chemical-laden water or air into a canal or sac that contains the chemical receptors.

In land vertebrates, the keenest olfactory senses are found among the mammals, especially the carnivores and rodents. Good olfaction is essential to carnivores; wolves and lions, for example, often locate their prey by smell. It is less clear why rodents have evolved such strong olfactory ability, since their plant food sources are not usually widely

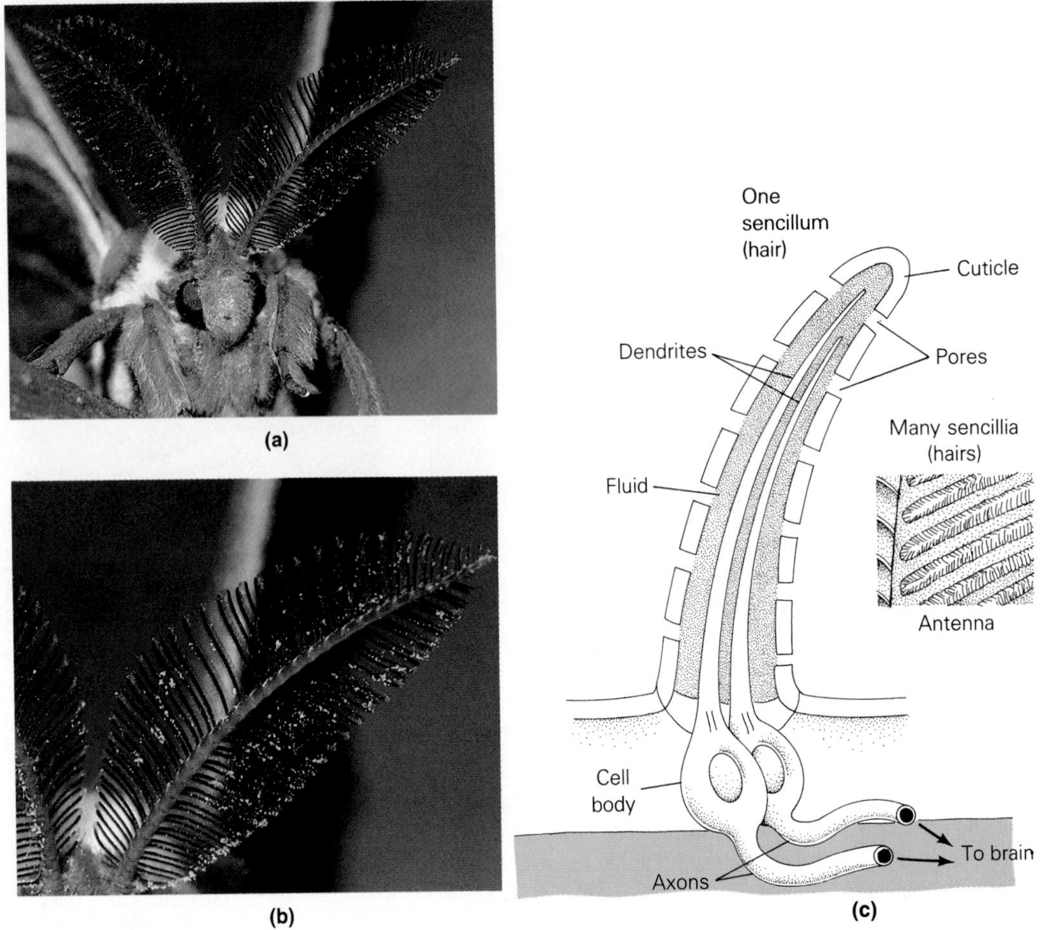

FIGURE 35.22
CHEMORECEPTORS IN THE MOTH
(a) The antennae of the male moth are remarkably large, complex, and sensitive. (b) Each large bristle of an antenna has numerous hairlike sensilla extending outward. (c) Within each sensillum is a fluid-filled cavity containing sensory neurons. Airborne molecules stimulate dendritic endings, producing impulses that are sent to the brain.

dispersed and certainly do not run away. However, their sense of smell may be associated less with food-finding than with their complex social organization. Rats, for instance, may use their sense of smell to discriminate among families (or tribes) and the individuals within them.

Nearly all vertebrates, including most mammals (but not primates), have olfactory receptors called Jacobson's organs in the roofs of their mouths. In amphibians, Jacobson's organ simply gives the animal a better notion of what it's eating, but snakes and lizards actually smell with the structure. As they flick their forked tongues in and out, the moist tips capture scent molecules that are delivered into the twin openings of their Jacobson's organ inside the mouth.

Compared to many other mammals, olfaction in primates is rather poor. For example, chimpanzees smell about like you do (no offense). The olfactory neurons of primates, like those of other mammals, are located in nasal epithelium, specialized as **olfactory epithelium**, where molecules in the air encounter the moist surface produced there (Figure 35.23). Scattered among the epithelial cells are specialized olfactory neurons that synapse with neurons of the **olfactory bulb** of the brain, an extension located just above the bony roof of the nasal cavity.

In all land-dwelling vertebrates, taste receptors, or **taste buds**, are confined to the mouth area. Humans experience four basic tastes: sweet, sour, salty, and bitter. Humans are omnivorous; that is, we'll eat almost anything, plant or animal. So it has undoubtedly been important to survival that through our long evolutionary history, we have become able to distinguish among a variety of tastes and smells. To illustrate, many alkaloid poisons taste bitter, and unless we deliberately train ourselves otherwise, we tend instinctively to avoid them. Sweet, on the other hand, is the taste of carbohydrates. Many kinds of fruit taste sweet, especially when they are ripe and have their highest nutritional value. Sour is not one of our favorite tastes, being the taste of acid and of unripened fruit that will benefit us more if we wait until it ripens. Gustation, then, is not only a matter of taste, but of survival.

Proprioception

Proprioception is the ability to determine the position of the body, or the position of one part of the body relative to another. It is made possible by sensory receptors in the muscles, joints, and tendons.

Proprioceptors in Invertebrates. It is particularly important for animals with many body parts, such as arthropods, to be able to coordinate those parts—to know what each is doing. Sensors in arthropods may be located peripherally, for example, at the base of "hairs," or deep within the muscles. Cockroaches, which are running animals, have proprioceptive hairs on the side of their legs. These respond to flexion of the knees, so they are aware of whether a leg is extended or flexed. Other arthropods—the crabs and lobsters— have proprioceptors located within the muscles themselves.

Proprioceptors in Vertebrates. Some proprioceptors in vertebrates, as well as in some invertebrates, take the form of stretch receptors (mentioned earlier in our discussion of the reflex arc). These respond to the stretching of skeletal muscle and tendons that is produced as limbs are extended or flexed. You may not be able to see your foot under the table, but you could probably point to it because your proprioceptors tell you where it is. Proprioception is well developed in mammals such as the tree-dwelling primates, who must move with incredible agility through their forest canopy.

Auditory Reception

Audition, or hearing, is similar in some respects to lateral line reception in fishes, since in both a distant stimulus is transmitted to the receptors via a medium (air or water). In fact, evolutionary theory holds that the structure of balance and hearing in the vertebrates evolved from increasingly complex lateral line organs.

Hearing in Invertebrates. Most invertebrates don't have specialized sound receptors, but many are sensitive to vibrations in the air, water, or soil in which they live. A notable exception among the invertebrates are the insects, a group that boasts several kinds of

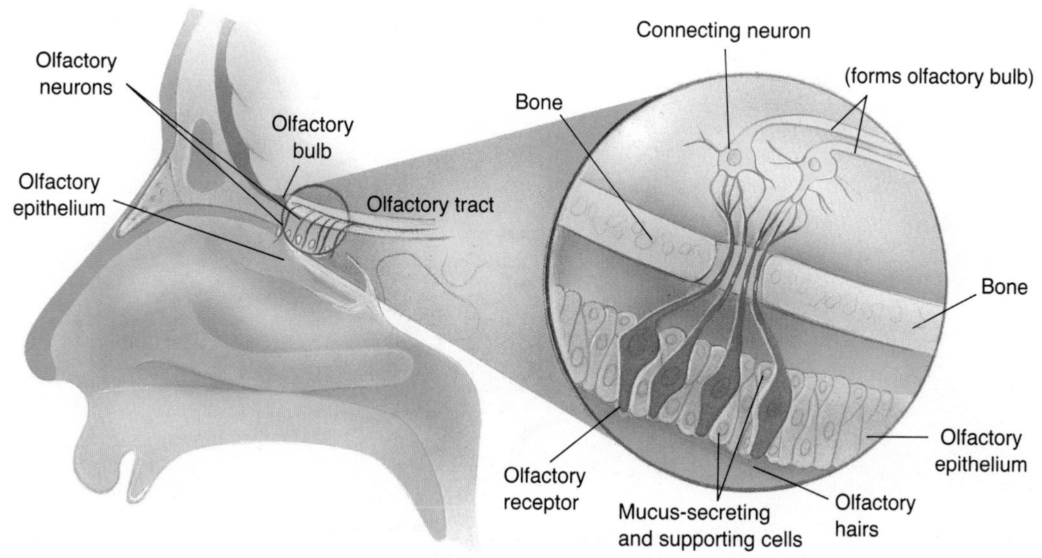

FIGURE 35.23
OLFACTORY RECEPTORS IN HUMANS
The olfactory receptors in the human nose connect to the olfactory bulb of the brain. These receptors are able to distinguish a wider variety of stimuli than are the taste buds. The olfactory neurons are part of the nasal epithelium dispersed with other cells. Each neuron has numerous olfactory "hairs," actually modified cilia, that protrude from the epithelium.

sound receptors. For example, most species have sensory hairs that respond to low-frequency vibrations of air, but others have a specialized organ, located in a leg, that is sensitive to movements of whatever the insect is standing on. Yet others, including grasshoppers and crickets, have **tympanal organs** that respond to high-frequency vibrations, much like the human eardrum. For a special case of insect hearing and its adaptiveness, see Essay 35.2.

Hearing in Vertebrates. The sound receptors in most species of vertebrates are located in the **inner ear**; however, vibrations reach the inner ear in a variety of ways. In many fishes, sound vibrations in water are conducted directly to the inner ear by vibrating water, but in others (such as the minnows, catfishes, and suckers), a set of tiny bones—the **Weberian ossicles**—connect the swim bladder to the inner ear. Sound vibrates the air-filled bladder, which moves the bones and stimulates the receptors in the inner ear. The fishes that have such an apparatus can detect a much wider frequency of sound than those that lack it.

Land vertebrates have an **auditory canal** and one to three **middle ear bones** that transmit vibrations of the **tympanum** or **tympanic membrane** (eardrum) to the inner ear. The tympanic vibrations usually stimulate auditory receptors, which then carry the impulses to the brain.

In humans and other mammals, the auditory apparatus consists of the **external, middle,** and inner **ear** (Figure 35.24). The external ear includes the **pinna** ("ear"), the auditory canal, and the tympanum. Most mammals are able to move the pinna so as to

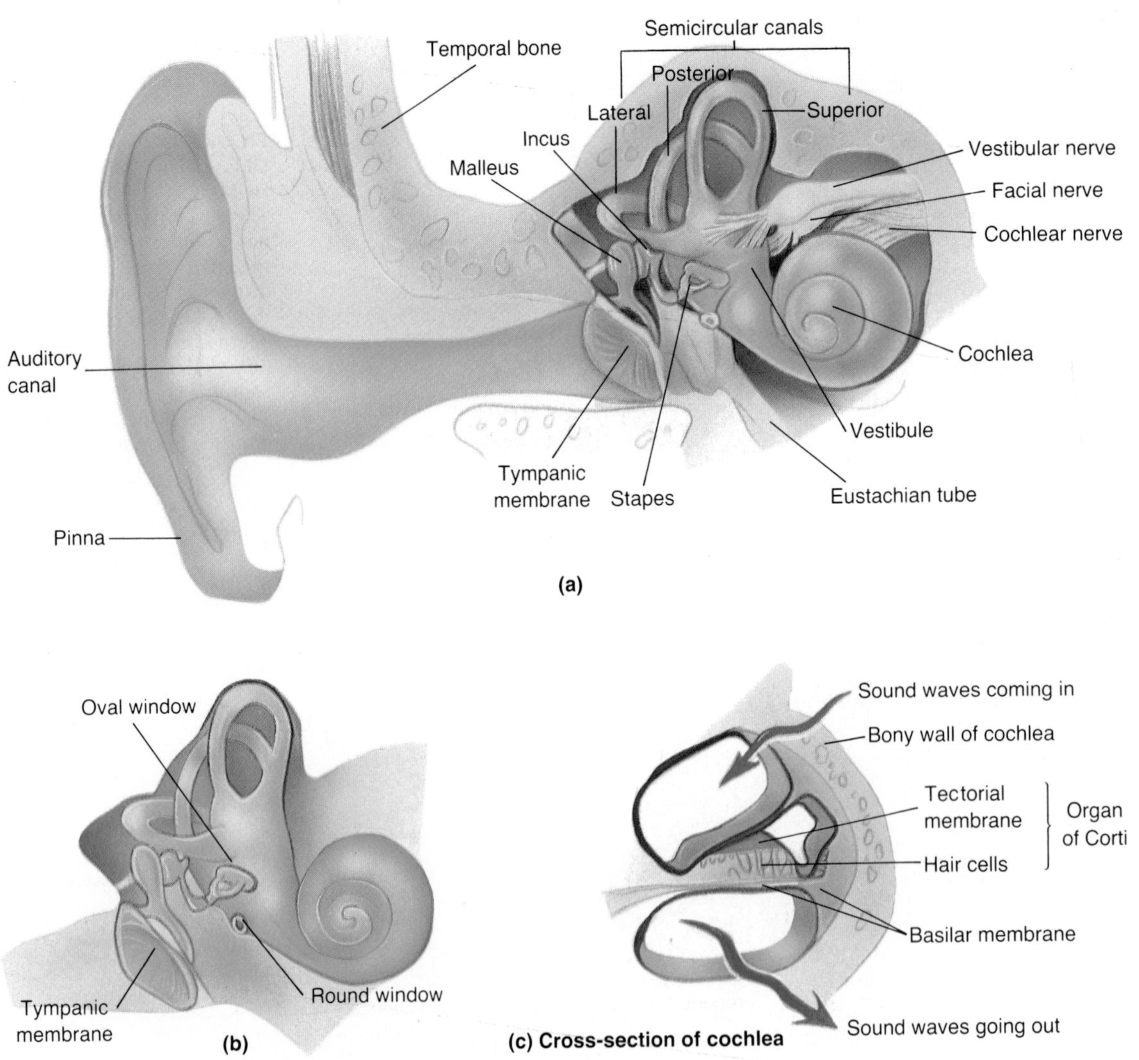

FIGURE 35.24
STRUCTURES OF HEARING IN HUMANS
(a) The structures of hearing in humans. **(b)** Sound waves vibrate the tympanic membrane, setting in motion three tiny leverlike bones: the malleus, incus, and stapes. The stapes, attached to the oval window, sets fluids in motion within the snail-shaped cochlea. The cochlea is divided by the basilar membrane. **(c)** Sensory hair cells of the basilar membrane are embedded in the gelatinous tectorial membrane. The two membranes and the hair cells are called the organ of Corti. The sound impulses pass inward over one surface, exciting the hair cells of the organ of Corti. Different regions of the basilar membrane are sensitive to different sound frequencies.

maximize sound input and locate its source. (Humans have largely lost this ability, and those who can move their ears are often in great demand for social events.)

The three bones of the middle ear are the **malleus, incus,** and **stapes** (which translate to *mallet, anvil,* and *stirrup,* respectively). Acting as a jointed lever, they transfer vibrations of the eardrum to a thin membrane in the **cochlea** of the inner ear.

The inner ear of mammals consists of the cochlea and the **vestibular apparatus.** The cochlea is a lengthy fluid-filled tube, doubled back on itself and then coiled in the manner of a snail shell. It is hard to visualize how the structure works, but if we can imagine it to be straight—as it is, in fact, in birds—we see what is actually a U-shaped tube containing the sensory neurons involved in hearing (see Figures 35.26 b and c). One end of the U-tube holds the **oval window,** to which the stapes is affixed, while the other end contains the highly flexible **round window.**

As sound waves strike the eardrum, the resulting vibrations are transferred by the three middle ear bones to the oval window, which, in response, vibrates rapidly. This creates a wave movement in the fluid of the outer tube and, at its far end, a compensating bulging of the round window (so the round window serves to dissipate the sound energy). As fluid pulsates within the tube, it activates what is called the **organ of Corti.** As you see in Figure 35.24c, the organ of Corti contains a **basilar membrane,** which bears sensory neurons called **hair cells** (modified cilia). The hairlike tips of these cells are embedded in the gelatinous **tectorial membrane.** As the basilar membrane moves, the sensory hairs are bent, creating receptor potentials, which in turn activate neurons leading to the cochlear nerve. This nerve leads to the midbrain, where the neurons synapse with those that carry impulses to be processed in the hearing centers of the cortex.

Sounds, we know, vary in intensity and pitch. The perception of intensity, or loudness, seems to depend on the *number* of auditory neurons that fire, as well as the *frequency* of their firing. Difference in pitch (highness and lowness) depend on which auditory neurons are stimulated. The basilar membrane is narrow at the broad base of the cochlea and wide at its apex end. Hair cells at its thin, more rigid beginning respond better to higher frequencies, which are interpreted by the brain as higher pitched sounds. Hair cells in the wider, more flexible apex of the snail-shaped chamber respond better to lower frequencies, those that translate into lower-pitched sounds. As we have described, sound waves travel up one side of the U-shaped tube and down the other. Each particular sound frequency traveling up one side of the basilar membrane will be exactly in phase with the same frequency traveling down the other side of the membrane only in one region. The in-phase resonance (or reinforced vibration) at that region vibrates the basilar membrane, producing the sensation of pitch.

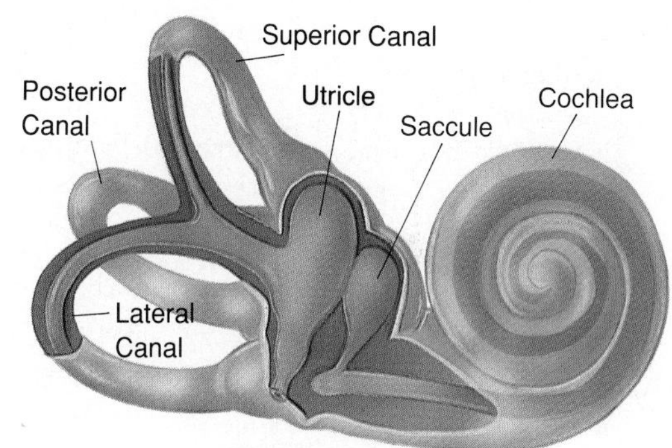

FIGURE 35.25
VESTIBULAR APPARATUS
The labyrinth of the inner ear contains sensory receptors that detect changes in body position and body movement. The utricle and probably the saccule specialize in detecting and reporting head position. Hair cells embedded in a gelatinous matrix are disturbed by moving calcium carbonate crystals as body position changes. The three fluid-filled canals detect motion. Each contains a number of hair cells embedded in a gelatinous dome. When the head is moved, fluids press against the dome, bending it and thereby activating the hair cells.

Gravity and Movement

Gravity sensors are important in determining body position and maintaining balance, two obviously related functions. Most invertebrates detect the body position through organs called **statocysts.** In the crayfish, the statocyst is a chamber lined with sensory hairs upon which lie a number of fine sand grains. As the body changes position, the sand grains move against the hairs, setting off impulses. The interpretation of these impulses enables the crayfish to determine its position with respect to gravity. Researchers have verified this mechanism by rearing crayfish in tanks in which iron filings are present, but no sand grains. Some of the iron filings then end up in the statocysts. When a magnet is placed over the statocysts, the iron filings are lifted off the sensory hairs and the crayfish, sensing that it is "upside down," flips over onto its back.

In humans and other mammals, body movements, position, and balance are detected in the inner ear by the vestibular apparatus (Figure 35.25). It is composed of the **semicircular canals,** and **saccule,** and the **utricle.** These three structures are closely associated with those of hearing. Each semicircular canal lies in a different plane, at right angles to each of the other two. This arrangement permits the sensing of move-

ESSAY 35.2

PREDATOR STRATEGIES AND PREY DEFENSES

The study of insect hearing has provided us with some good stories about scientific detective work, and one of the best ones is about the coadaptation of moth and bat hearing. It seems that noctuid (night flying) moths are a favorite prey of certain bats. These bats fly swiftly and can turn on a dime to capture their prey on the wing. The bats locate the prey by emitting high-frequency sounds that bounce back from any structure in the environment—the principle of sonar. The echos provide the bat with information on the flying insect, and the bat simply intercepts the hapless moth. In fact, the term "echolocation," familar in submarine jargon, was first coined to describe the use of sonar by bats.

But just as bats have evolved ways of catching moths, noctuid moths have developed ways of avoiding bats. When they hear the cry of a bat, they take evasive measures. As you might suspect when dealing with bats, this is easier said than done. It turns out, though, that the moth does it with a simple hearing apparatus.

Noctuid moths, like many insects have paired *tympana,* one on either side of the thorax (the inner midsection). Kenneth Roeder found that each tympanum has only two receptors. One, called the *A1 cell,* is sensitive to low-intensity sounds. The other, the *A2 cell,* responds only to loud sounds. Surprisingly, neither kind of receptor is very good at distinguishing frequencies (high versus low notes)—a sound of 20,000 hertz (cycles per second) elicits the same neural action potential as one of 40,000 hertz, a much higher sound.

As sound becomes louder, however, the A1 cell fires more frequently and with a shorter lag time after the stimulus. The A1 cell also shows a greater firing frequency in response to pulses of sound than to continuous sounds. And it just so happens that bats emit pulses of sound.

In a sense, the moth has beaten the bat at its own game. Its very sensitive A1 neuron is able to detect bat sounds long before the bat is aware of the moth. The moth can not only detect the distance of the bat, but it can tell whether the bat is coming nearer, as the sound of an approaching bat grows louder.

In addition, the moth is able to detect the direction of the bat. The mechanism is simple. If the bat is on the left side **(a)**, the left thoracic receptors of the moth will be exposed to the sounds while the receptors on the right will be shielded (see illustration). Therefore, the left receptor fires sooner or more frequently than the right if the bat is on the left. If the bat is directly behind **(b)**, both neurons will fire simultaneously. Thus, the moth can determine the distance and direction of the bat. But what about its altitude?

If the bat is above the moth **(c)**, the bat sounds will be deflected by the upward beat of the moth's wings. If the bat is beneath the moth, however, the wing beats will have no effect on the pattern of neural firing. The moth, then, decodes the incoming data, probably in its thoracic ganglion (from which the auditory neurons emerge) so that it pretty well has the bat pinpointed.

But what does it do with this information? If the bat is some distance away, the moth simply turns and flies in the opposite direction, thus decreasing the likelihood of ever being detected. The moth probably turns until the A1 cell firing from each ear is equalized. When the bat changes direction, so does the moth.

Bats fly faster than moths, though, and if a bat should draw to within 2.5 m (8 ft) of the moth, the moth's number is up—at least if it tries to outrun the bat. So it doesn't. If the bat and moth are on a collision course, that is, if the moth is about to be caught, the sounds of the onrushing bat will become very loud. At this point, the A2 fiber begins to fire—the signal of imminent danger. These messages are relayed to the moth's brain, which then apparently shuts off the thoracic ganglion that had been coordinating the antidetection behavior. Now the jig is up and the moth changes tactics. Its wings begin to beat in peculiar, irregular patterns or not at all. The insect itself probably has no way of knowing where it is going as it begins a series of unpredictable loops, rolls, and dives. But is also very difficult for the bat to plot a course to intercept the moth. If all goes well, the erratic course will take the moth safely to the ground, where the echoes of the earth will mask its own echoes.

The noctuid moth's evolutionary response to the hunting behavior of the bat serves as a beautiful example of the adaptive response of one organism to another. It also shows clearly that the sensory apparatus of any animal is not likely to respond to elements that are irrelevant to its well-being. It is not important for moths to be able to distinguish frequencies of sound, but it is important that they are sensitive to differences in sound volume. Anyone who tried to train a moth to respond to different sound frequencies could only conclude that moths are untrainable.

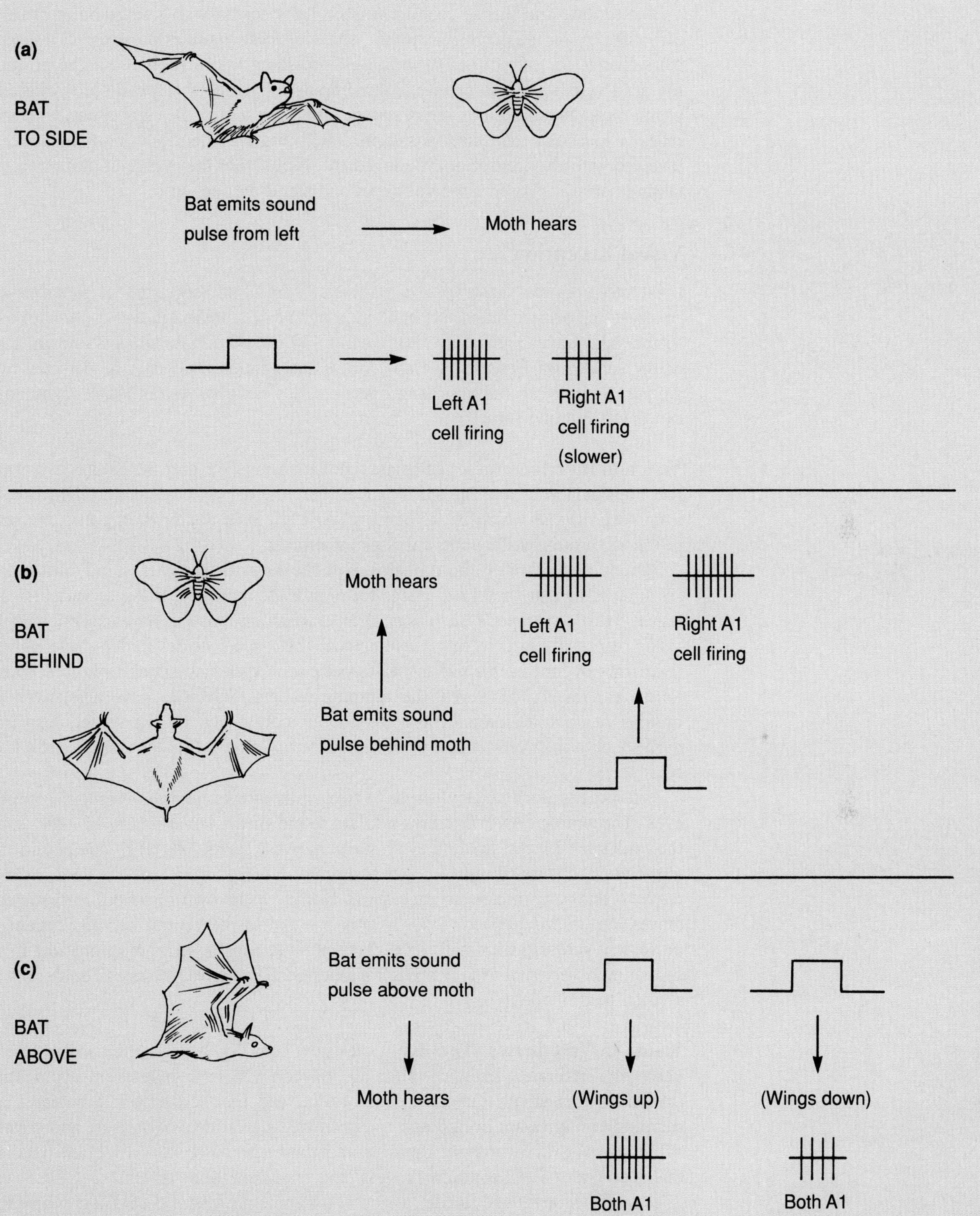
(a)
BAT
TO SIDE
Bat emits sound
pulse from left
Moth hears
Left A1
cell firing
Right A1
cell firing
(slower)
(b)
BAT
BEHIND
Moth hears
Left A1
cell firing
Right A1
cell firing
Bat emits sound
pulse behind moth
(c)
BAT
ABOVE
Bat emits sound
pulse above moth
Moth hears
(Wings up)
(Wings down)
Both A1
cells firing
Both A1
cells firing
(slower)

ment—acceleration or deceleration—in any direction. Each canal is filled with fluid, and its movement jostles sensory hairs that extend into the canals. As the fluids move, they bend the hairs, creating receptor potentials, which in turn activate sensory neurons that send impulses to the brain.

The saccule and utricle contain sensory hairs coated with fine granules of calcium carbonate. As in the crayfish statocyst, shifts in these granules pressing on the sensory hairs change the rate of neural impulses, providing information about the position of the head with respect to gravity. Some impulses travel to the spinal cord, where body position can be adjusted by reflex action; others are sent to the cerebellum, where other reflexive muscular coordination is orchestrated; and yet others move on to higher centers involved with the control of eye movement. Input from the eyes is important in maintaining balance. (Try to close your eyes and stand on one leg.)

Visual Reception

Light receptors are sensitive to a particular part of the spectrum of electromagnetic energy—the part we call visible light, although some animals can detect ultraviolet light. Visible light wavelengths range from about 430 to 750 nm. As far as is known, considerably shorter wavelengths, such as X-rays and gamma rays, can't be detected by any animal. Neither can the very long ones, such as radio waves, although many animals can detect infrared (heat).

Vision in Invertebrates. The planarian flatworm (see Figure 30.13) has two eyespots that are shaded on opposite sides. It can therefore tell which direction light is coming from according to which eye is being stimulated. Such ability is, of course, important to these bottom-dwelling and dark-seeing animals.

Among invertebrates, the cephalopod mollusks—the octopus, squid, and others—are unique in that they have image-forming eyes, quite like those of vertebrates (see Figure 31.8). Arthropods, such as spiders, crayfish, and insects, have exceptionally good vision, especially in detecting movement in their visual fields. Spiders have eight eyes (two rows of four each), but in most species the eyes lack enough photoreceptors to form clear images. However, the jumping spiders *(Salticidae)* have a relatively large number of photoreceptors. These are undoubtedly beneficial; the spiders leap from a distance onto their prey, and it wouldn't do for a weak-eyed spider to leap onto a hungry bird.

Crustaceans, such as crayfish, have two kinds of eyes: **simple eyes** and **compound eyes.** The simple eyes lack lenses and are found in the larval stages. In many species, the adult's sensitive compound eye rests on movable stalks. The term "compound" refers to the numerous visual units known as **ommatidia** (Figure 35.26). The ommatidia can't move to follow an image, so each stares blankly in its own direction until something moves into its visual field. Then it fires a signal to the central nervous system. Any movement across the animal's field of vision stimulates a series of ommatidia in turn, so that even very slight movements are detected. The convex surface of such eyes gives a visual field of about 180°.

Vision in Vertebrates: The Eye Let's now consider vision in humans as representative of vertebrates. Humans, after all, rely on vision to help them avoid danger, coordinate movement, maintain equilibrium, read the faces of other humans, avoid stepping in unpleasant things, and to experience a great deal of pleasure and creativity. While much is known about the eyes and their operation, how the brain handles all this is an enormously complex process, and many questions remain.

We will begin with a brief discussion of the anatomical aspects of vision, where things tend to be straightforward. The human eyeball (Figure 35.27) is roughly spherical, but protruding somewhat in front. It is surrounded by a tough, white outer layer, the collagenous **sclera,** except for the anterior bulge, which is covered by the transparent **cornea.** The sclera maintains the shape of the eyeball and serves as an attachment for

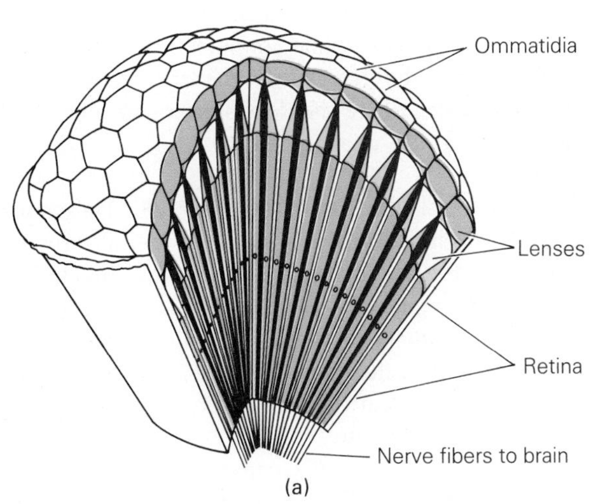

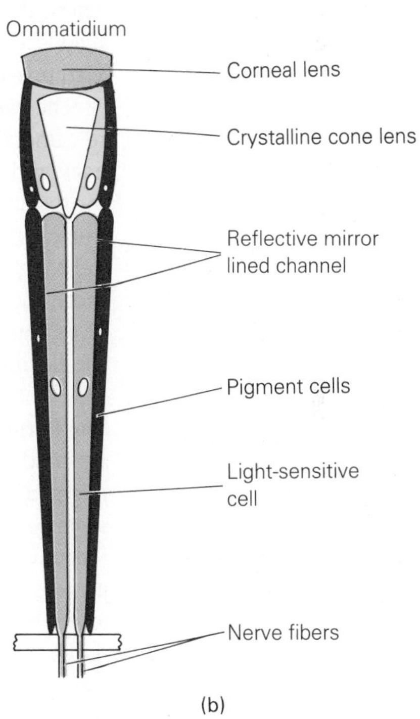

FIGURE 35.26
THE COMPOUND EYE
(a) Compound eyes are common among insects and other arthropods. Each eye consists of many units known as ommatidia. **(b)** Each unit has a compound lens (a thickened corneal lens and a crystalline cone lens), a translucent cylinder of light-sensitive cells below, and a surrounding layer of versatile pigment cells. The latter restrict the movement of light from one ommatidium to the next, at least in bright light. But in dim light, the pigment cells withdraw, and light from several ommatidia merges, thereby providing enough stimulus to activate a few of the units. In the latter case, the image is believed to be far less clear but is useful in detecting the movement of objects in the visual field.

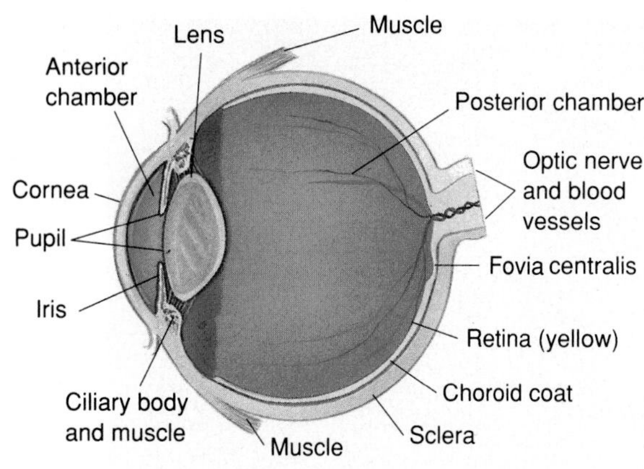

FIGURE 35.27
THE HUMAN EYE
The human eye is multilayered, with two major fluid-filled chambers separated by the lens. The amount of light entering the eye through the pupil is regulated by the iris, which is able to change the diameter of the pupil. The light is focused on the retina by changes in the shape of the pliable lens. The retina contains light-sensitive cells (the rods and cone).

the voluntary muscles that move the eyes. Because of its curvature, the cornea aids in focusing light. Interestingly, the cornea has no blood supply and must receive O_2 by diffusion from outside the eye. This is why contact lenses must be porous if they are to be left in place over a long time. Incidentally, the absence of a blood supply means that immune cells are also absent, and thus corneal transplants are not easily rejected by the body's immune defenses.

Just within the sclera is the highly vascularized **choroid.** At its anterior portion the choroid supports the **iris** of the eye, a thin diaphragm of smooth muscle. At the center of the iris lies the **pupil**, an opening whose size the iris regulates. The ringlike iris contains pigments we recognize as eye color. Behind the iris is the dense, transparent **lens,** held in place by another ring of muscle, and then a large space filled with transparent gel. Finally, there is an innermost layer of tissue at the back of the eye—the **retina**, a complex region of light-sensitive and supporting neurons. The neural fibers form the **optic nerve**, a common pathway out of the eye. Let's now see how each of these structures works.

Light entering the eye passes through the transparent cornea, and then through the **anterior chamber**, filled with a fluid called the **aqueous humor**, and through the pupil of the iris. The iris adjusts the pupil size according to the brightness of entering light—like the light-controlling diaphragm of a microscope or camera but far more delicate and responsive. It does this through two very fine muscle layers, one circular and the other longitudinal. The first constricts the pupil while the second dilates it, and each is under the control of a different part of the autonomic nervous system. From the pupil, light traverses the lens and then passes through the clear gels and fluids—called the **vitreous humor**—of the **posterior chamber** to fall finally upon the retina.

The eye is often compared to a camera—the lens to the camera lens and the retina to camera film. However, the camera lens adjusts to close or distant objects by moving back and forth (interestingly, sharks focus the same way), but our lenses accommodate for distance by changing shape. This is done by a ring of precisely coordinated **ciliary muscles** and their delicate supporting ligaments, which attach directly to the lens. Relaxing the ciliary muscles flattens the lens, while contraction forces it to assume a more rounded form. As we grow older, the flexibility of the lens decreases, and many of us must wear reading glasses in order to see the details of nearby objects. Actually, the flexibility of the lens peaks early in life, at about age 10.

The Retina. The retina consists of four layers of cells (Figure 35.28). The deepest layer, attached to the inner surface of the choroid coat, is pigmented. It absorbs light that might otherwise reflect about inside the eye and create visual problems. In many vertebrates, including dogs and cats, the pigmented layer is modified at night, revealing a reflective layer that increases their sensitivity to low levels of light. Overlying the pigmented tissue are the **rods** and **cones**, the actual light receptors. You might expect them to be in the direct path of incoming light, but they are covered by two more layers of rather transparent neurons. When the rods and cones are stimulated by light, they activate the overlying **bipolar cells**, which in turn synapse with the layer of **ganglion cells** just above. Ganglion cells gather from all parts of the retina to form the **optic nerve tract**, which carries impulses to the brain.

The slender rod cells far outnumber the cones, by about 18 to 1 (125 million rods to 7 million cones in each eye). Impulses from the rods are decoded by the brain into black, white, and gray images; different colors (wavelengths of light) are not distinguished. However, the rods make up for this limitation by being highly responsive to dim light, far more so than cones. In fact, the rods provide most of our night vision.

It may come as no surprise to learn that cones are cone-shaped. Unlike the rods, each cone responds optimally to only its specific portion of the visible spectrum. Vision is most acute in the **fovea centralis**, or simply **fovea**, a slight depression directly in the path of light focusing on the retina. The fovea in mammals contains a rich concentration

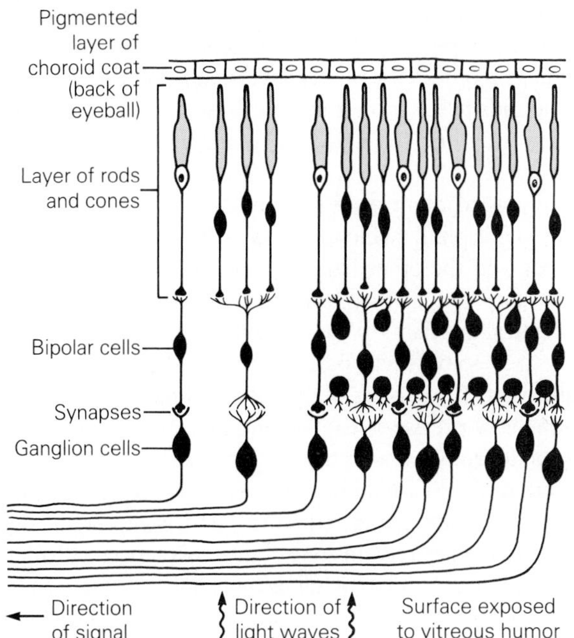

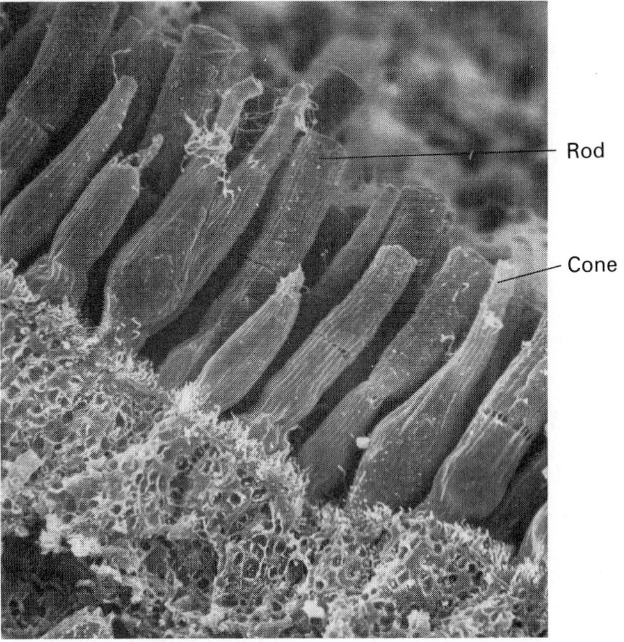

FIGURE 35.28
THE RETINA
The retina contains two types of photoreceptors: rod cells and cone cells. Overlying the retina are the bipolar and ganglion cells, the latter forming the optic nerve.

of cone cells with few other retinal cells. For this reason, cones are capable of much finer discrimination of detail in bright light than are rods, although they hardly respond at all to dim light (which is why we can see light but not much color at night). Since cones are not very functional at night, it is easier to see a distant object, such as a faint star, if we look to the side of that object, avoiding the cone-rich fovea.

According to the most prevalent theory of color vision, there are three types of cones: red-sensitive, green-sensitive, and blue-sensitive. We can see so many variations of color because the sensitivity ranges of these receptors overlap. The color sensitivities of the cones depend on what sorts of visual pigments (light absorbers) they contain—or, more specifically, on variations in the proteins that join the visual pigments.

The actual light-detecting regions in rods are the stacked disks that form the upper part of the cells. (Each rod may have up to a thousand of such disks.) The disks contain **rhodopsin,** a photoactive pigment that breaks down in the presence of light into two colorless products: the protein **opsin** and a derivative of vitamin A called **retinal.** The chemical reaction itself sets events in motion that create neural signals in the bipolar and ganglionic cells above (see Figure 34.28).

The rods work in a peculiar manner—they are only active in the absence of light. At that time their gated sodium channels are open, and they are in an ongoing state of depolarization. In this state they secrete an inhibitory neurotransmitter where they synapse with the bipolar cells above, thus inhibiting the bipolar cells from firing. When light is absorbed and rhodopsin is broken down, the sodium gates close, and the rods become hyperpolarized. The secretion of neurotransmitter slows considerably, and passive or graded receptor potentials begin in the bipolar cells. The bipolar cells synapse with ganglion cells, and any resulting action potentials pass through the optic nerve to the brain.

Neurobiologists have yet to determine how such a negative system of operation might be adaptive, but they marvel over the acute sensitivity possible in rod cells. Apparently rod cells are excellent amplifiers. A single photon can produce a detectable electrical signal in the retina, and the human brain can actually "see" a cluster of five photons.

Opsin and retinal eventually enter into a chemical pathway in which they are recombined into rhodopsin. Apparently, dietary vitamin A must be supplied continually to keep the pathway going in the right direction. If the body is deprived of vitamin A, severe night blindness can result. If a knock on the door brings you from a very brightly lit room into near total darkness, you may experience a temporary form of night blindness. Under intense light, most of the rhodopsin in the rods bleaches to the opsin/retinal form, and a certain period of time is required for the reconversion to rhodopsin to catch up. While you are waiting for the rods to function, the cones—which aren't very useful at night—at least help you peer dimly into the darkness to see whatever brought you to the door in the first place.

While humans have reasonably good night vision, many day-flying birds lack rods altogether and are almost totally night blind, which explains why so many birds come to roost at twilight. Owls and bats, on the other hand, avoid daylight and are quite at home in the dark. It is not surprising, therefore, to learn that their eyes have only rods.

We see, then, that it is the chemical, mechanical, and electrical interplay of specialized neural cells that enables us to sense some small part of the world around us and to respond appropriately to our situation. The success of the animal kingdom on this planet has been, in large part, due to the very precise sensitivity and reactivity of its members. We have seen that this highly organized and adaptive responsiveness is based on the peculiar irritability of a constellation of cells called neurons. We know something about the kinds of neurons that exist, their supporting structures, and even certain details about how neurons work. Yet neural biology remains one of the most resistant challenges in all of science. Nonetheless, researchers continue to tell us more about just how animals are able to respond so precisely to both their internal and external environments in a very complex world.

APPLICATION OF IDEAS

1. Use the terms *primitive* and *advanced* to describe the major parts of the human brain. Explain the criteria on which you based your decisions.
2. The central nervous system, particularly the conscious center, exercises what might be called "noisy control," as they receive, integrate, and respond to a barrage of sensory messages. The autonomic system, on the other hand, can be referred to as a "quiet" system, since its activity is generally below the conscious level. Elaborate on these ideas by considering how the two systems respond to problems in osmoregulation and in thermoregulation (in the latter include problems of both heating and cooling).
3. Compare the functioning of the image-producing human eye with that of the compound eye of arthropods. We know what *we* see, but what might the arthropod actually see? What are some of the advantages of each kind of eye?
4. Compare the degrees of development in special senses in various vertebrates. Explain how an emphasis on certain of these can adapt the animal to its specific environment.

KEY IDEAS

INVERTEBRATE NERVOUS SYSTEMS

Simple Nerve Nets and Ladders

1. Cnidarians show no evidence of **cephalization.** They are radially symmetrical and contain a **nerve net** whose cells concentrate at the mouth and tentacles.
2. The flatworm nervous system resembles a ladder with two **ganglionic masses** (clusters of nerve cell bodies) at the anterior end and the ladder arrangement proceeding posteriorally.
3. Echinoderms lack any evidence of cephalization, their nerves clustered into a ring with branches extending into each arm.

Massed Ganglia to Organized Brains

1. The annelids have large ganglia above and below the esophagus and a ventral nerve cord with segmental ganglia. **Giant axons** in the cord support fast escape movements.
2. While the bivalve nervous system is fairly simple, that of the cephalopod is complex and features a functionally differentiated brain and complex behavior. Giant axons are found in the squid mantle.
3. Arthropod specializations include reduction in the number of segments and ganglia and much greater cephalization than is found in the annelids. The insect brain includes an enlarged ganglionic mass above the esophagus, which, in those with **compound eyes,** includes greatly enlarged visual integrating centers.

VERTEBRATE NERVOUS SYSTEMS

1. Vertebrate nervous systems include a pronounced trend toward cephalization and a dorsal, hollow nerve cord.
2. The vertebrate brain follows a three-part plan of **forebrain, midbrain,** and **hindbrain.** The **cerebrum** and surrounding **cortex** of the forebrain reach their greatest development in mammals.

THE HUMAN NERVOUS SYSTEM

The human nervous system is divided into the **central nervous system** (CNS)—the brain and **spinal cord**—and the **peripheral nervous system** (PNS)—the **somatic** and **autonomic systems.** The somatic system is largely composed of sensory receptors and sensory neurons and the voluntary motor nerves that move skeletal muscle, while the autonomic system is motor and involuntary, dedicated mainly to homeostasis.

The Human Brain

1. The human brain weighs about 1.4 kg, has a volume of 1300 to 1500 cc, contains some 100 billion neurons and 10 times that number of **glial cells.** Cell bodies form its gray outer regions, while myelinated fibers make up its white inner mass. **Cerebrospinal fluid** and the **meninges** cushion and protect the brain respectively.
2. The brain has a high oxygen and glucose demand. One safeguard against deficiencies is found in the arterial organization, the **circle of Willis**, which gives rise to arteries serving the brain. Should blockage occur, an ongoing blood supply to most of the brain is assured. Toxic substances carried by the blood cannot enter brain cells because of the **blood-brain barrier**—tightly junctured capillary cells and selective **astrocytes** that communciate with them.
3. The forebrain is involved in conscious thought, sensory reception, voluntary movement, and other voluntary acts.
4. The cerebrum, the largest region of the forebrain, is divided into left and right **cerebral hemispheres,** which are covered by the wrinkled **cerebral cortex.** The cortex contains some 15 billion cell bodies and dendrites. The white matter below includes the myelinated fibers.
5. The importance of the cerebrum to common voluntary acts is minimal in frogs but increasingly important in rats, cats, dogs, monkeys, and humans.
6. The cerebrum is made up of a number of lobes. The **occipital lobe** receives and analyzes visual information, while the **temporal lobe** processes auditory input and some visual information. The **frontal lobe** regulates voluntary movement and speech. The **prefrontal area** of the frontal lobe sorts sensory input, putting it into proper context. The **parietal lobe** processes sensory information, including body position.
7. The processing of sensory input and the instigation of voluntary muscle action occurs in specific regions of the cortex.
8. The cerebral hemispheres are functionally distinct, and many learned patterns take place in just one hemisphere. The left hemisphere predominates in right-handed persons and in the more common type of left-handedness, but in the rare form of left-handedness, dominance is in the right. Speech is primarily a left hemisphere function, as is analytical thought. Damage to this hemisphere can provide a loss of language abilities.
9. Connections between hemispheres occur through the **corpus callosum.**
10. Knowledge of language centers comes mainly from studies of stroke victims. Three left hemisphere regions, **Broca's area, Wernicke's area,** and the **angular gyrus,** are involved.
11. When heard, words are processed by Wernicke's area. Words to be spoken begin in this area but are transmitted to Broca's area, which activates the motor regions involved in voice.
12. Reading aloud involves all three areas. Visual images of the words go to the angular gyrus, which associates the words with auditory patterns in Wernicke's area. Wernicke's area produces the basic sentence structure and coordinates vocalization.
13. The **thalamus,** which is a relay structure, connects various parts of the brain and includes the **reticular system,** which taps incoming and outgoing communications. It also acts as an alarm system and suppresses irrelevant stimuli, thus permitting sleep.
14. The **hypothalamus,** which monitors many functions, acts as a homeostatic regulator (heart rate, blood pressure, body temperature, thirst, hunger, sex drive). It also stimulates hormonal activity in the pituitary and is subject to negative feedback.
15. The **limbic system,** containing a number of specific **nuclei,** includes the hypothalamus, thalamus, and some cortical pathways. It links the fore- and midbrain and is involved in emotion (for example, when areas of the **amygdala** are stimulated, we may experience rage).
16. The midbrain forms connections between the hind- and forebrain and receives sensory input from auditory and visual receptors.
17. The hindbrain consists of the **medulla oblongata, cerebellum** and **pons.** The first controls breathing and heart rate and contains many pathways, including some from **cranial nerves**.
18. By integrating sensory input from the eyes, balance organs, and skeletal muscles, and by tapping output from the motor cortex, the **cerebellum** coordinates and refines voluntary muscles movement. It does this at the unconscious level, chiefly through delicately applied neural inhibition.
19. The **pons** contain tracts traveling between the forebrain and cord and to and from the cerebellum.

The Spinal Cord

1. The spinal cord includes pathways between the brain and much of the peripheral nervous sytem. The

white regions are myelinated neurons, while the gray are nerve cell bodies. Many reflexive acts occur at the cord level.

2. The cord emerges from the brain through the **foramen magnum** and lies within the **vertebral canal.** It is bathed in the cerebrospinal fluid and surrounded by the meninges, three layers of supporting connective tissue.
3. Motor neurons arise from cell bodies in the cord to pass out through the ventral **motor root** of the spinal nerves. Sensory neurons enter the cord via the dorsal **sensory root** from rows of ganglia outside, where their cell bodies cluster.

Chemicals in the Brain

1. The brain has at least 50 different neurotransmitters. The **monoamines,** modified amino acids, include **norepinephrine, dopamine, histamine,** and **serotonin. Neuropeptides,** short chains of amino acids, include **enkephalins** and **endorphins,** important in pain and mental mood.
2. Other neuropeptides include the **enkephalins** and **endorphins,** mood elevators and pain modifiers. They are called **opioids** because their effects can be mimicked by opiates. Amphetamines cause the retention of neurotransmitters, while cocaine blocks their normal enzymatic degradation, and LSD, mescaline, and psilocybin mimic neurotransmitters.

Electrical Activity in the Brain

1. Electrical activity in the brain can be detected with an **electroencephalograph** and recorded as an **electroencephalogram,** or EEG. Such recordings are useful for detecting gross abnormal patterns.
2. EEGs reveal that a considerable amount of electrical activity accompanies sleep. Electrical activity during **paradoxical sleep,** or **REM sleep** (rapid eye movement), a time of dreaming, is similar to that of wakefulness.

The Peripheral Nervous System

1. The peripheral system includes the somatic and autonomic systems. The autonomic system is made up of the **sympathetic** and **parasympathetic divisions.**
2. The autonomic nervous system ANS is essentially homeostatic in function. Its motor neurons carry impulses from the brain and spinal cord to the viscera, blood vessels, irises, secretory glands, and other involuntary structures, bringing about fine adjustments as necessary.
3. The sympathetic and parasympathetic divisions have opposing functions, with the former generally increasing an action and the latter decreasing it. For example, the sympathetic **cardioaccelerator** nerve releases neurotransmitters that increase heart rate, while the parasympathetic vagus nerve releases those that slow the heart rate.
4. In "fight or flight" events, the sympathetic division increases heart and breathing rate and blood pressure and sends more blood to the muscles and brain. Afterward, these actions are reversed by the parasympathetic division.
5. Most parasympathetic nerves originate in the brain and a few in the lower spinal cord. Some go to external ganglia, where they synapse with neurons that innervate the target organ. Sympathetic nerves all emerge from the cord, and all pass through, or synapse within, the **sympathetic ganglia** along the cord. Secondary ganglia also occur, including those of the **celiac plexus** or **solar plexus.**

THE SENSES

Sensory receptors are highly specialized for the stimuli they receive, but all act as **transducers,** converting external stimuli first into **receptor potentials** and then into action potentials. Reception potentials are graded, increasing from threshold level to an intense level. (In this way a strong stimulus can increase the rate of action potential propagation, thus providing greater information about the stimulus.)

Neural Codes

While all action potentials are the same, the number and speed of impulses can act as a code in the central nervous system. Interpretation of sensory information depends on neural organization in the brain, along with experience, memory, and learning.

Tactile Reception

1. Tactile (touch) or mechanoreceptors respond to deforming force. In invertebrates, they are often associated with sensory hairs. They detect moving solid objects and air and water currents.
2. In vertebrates mechanoreceptors include the **distance receptors** (such as the lateral line organ of fishes) and **contact receptors** that specialize in touch. In humans, **Pacinian corpuscles** are activated by pressure, while **Meissner's corpuscles** respond to light touch. **Baroreceptors** in the arteries respond to changes in blood pressure. Body hairs and the outer dead skin layer act as **piezoelectric crystals,** discharging current when deformed.

Thermoreception

1. **Thermoreception** is the detection of changes in temperature. Heat detection in invertebrates is especially important to ectoparasites of the warm-bodied mammals and birds.
2. In vertebrates, heat sensors include the pit of pit vipers, and **Ruffini's corpuscles** and **Krause's end bulbs** in mammals. Thermoreception in humans is poorly understood and, up to a point, the perception of heat and cold may depend on the last experience of the receptors involved.

Chemoreception

1. **Chemoreception** is widespread in animals. Insects have chemoreceptors on many body parts, but the **sensillum** of certain moths is the most acute known and can be activated by a single molecule of **bombykol,** the sex attractant.

2. Vertebrate chemoreceptors include general receptors and **gustatory** and **olfactory receptors** (taste and smell).
3. In rodents, a keen olfactory sense aids in social interactions. Reptiles sample chemicals with their forked tongue, which is inserted into the paired **Jacobson's organ** in the mouth, where receptors are located. In mammals, an **olfactory epithelium** contains sensory neurons that respond to chemicals, synapsing with neurons in the **olfactory bulb** above.
4. In fish, taste receptors may be scattered over the body surface, while in terrestrial vertebrates they are in the mouth. In humans, taste receptors are located in **taste buds** on the tongue, where there is some specialization for sweet, sour, salty, and bitter.

Proprioception

Proprioception—sensing the position of the body or its parts—in invertebrates occurs through sensors in the body hairs and muscles. In vertebrates, proprioceptors include the stretch receptors involved in the reflex arc.

Auditory Reception

1. **Audition** in insects includes the specialized **tympanal organ,** a drumlike device.
2. Some fishes have an **inner ear** containing tiny bones, **Weberian ossicles** that move in response to sound, stimulating action potentials in nearby sensors. In land vertebrates, the hearing organs include an **auditory canal** and one to three **middle ear bones** that transmit sound from the **tympanum** *(tympanic membrane)* to the inner ear.
3. The human hearing organ includes the **external ear (pinna,** auditory canal and tympanum). The middle ear bones include the **malleus, incus,** and **stapes,** which connect the tympanum with the **cochlea.** The inner ear includes the snail-shaped cochlea and the **vestibular apparatus.**
4. The cochlea is a fluid-filled, U-shaped tube with the **oval window** (stapes attachment) on one end and the **round window** at the other. Sound sets up motion in the fluids, which activates the **organ of Corti**—a **basilar membrane** containing **hair cells** embedded in the **tectorial membrane.** Bending of the hair cells creates generator potentials, which in turn start action potentials in neurons from the cochlear nerve. Differences in intensity and pitch are produced by activity at various parts of the basilar membrane.

Gravity and Movement

1. Some invertebrates have **statocysts,** chambers containing grains that move against sensory hairs when changes in body position occur.
2. In humans and other mammals, movement, position, and balance are detected by the vestibular apparatus. Three fluid-filled **semicircular canals,** lying in three planes, detect acceleration and deceleration as the fluids move against sensory hairs. Fine granules in the **saccule** and **utricle** shift with movement that activates sensory hairs, which inform the brain about position.

Visual Reception

1. Visual receptors in arthropods include **simple eyes** and **compound eyes,** the first of which lack lenses. Compound eyes contain immovable units called **ommatidia** that register images.
2. The vertebrate eyeball consists of tough, outer **sclera,** with a forward, transparent **cornea** and more internal **choroid.** The choroid supports the **iris,** which surrounds the **pupil.** The **lens** separates two fluid-filled chambers. The light-sensitive **retina** within the eyeball communicates with the **optic nerve.**
3. Light entering the eye passes through the **anterior chamber (aqueous humor),** lens, and **posterior chamber (vitreous humor),** to the retina. Focusing is done through changes in lens shape brought about by the **ciliary muscles.** The amount of light entering is altered by the iris, which determines the size of the pupil.
4. The retina consists of four cell layers, including an innermost pigmented layer, a layer of **rods** and **cones** (the sensory receptors), a layer of **bipolar cells,** and a final layer of **ganglion cells.** The bipolar cells and ganglion cells are neurons.
5. Rods respond to all wavelengths of the visible spectrum and function well at night, while each cone responds to either red, green, or blue light wavelengths and has little night function. Cones are most concentrated in the **fovea.**
6. Upon exposure to light, **rhodopsin** in rods bleaches to **opsin** and **retinal,** and the chemical change activates the synapses with the neurons above, where action potentials begin. The two products recycle, using vitamin A to produce rhodopsin. Since much of the rhodopsin is in the opsin/retinal form in bright light, temporary night blindness occurs when the surroundings are suddenly darkened.

REVIEW QUESTIONS

1. Describe the arrangement and concentration of neurons in a cnidanian. What is such an arrangement called? (805–806)
2. What evidence of bilateral symmetry can you find in the flatworm's nervous system? Is there any cephalization? Explain. (806)
3. Name two invertebrates with giant axons. What role do such axons play? (796, 806)
4. What characteristics of the cephalopod central nervous system qualify the usual invertebrate ganglionic mass as a full-fledged brain? What capabilities does this provide to the cephalopod? (806–807)
5. Name the three parts of the vertebrate brain and describe the general evolutionary trends seen in each part. (808)

6. In addition to neurons, what other cells are common in the brain? What is the approximate ratio of these cells to neurons?
7. What is the function of the cerebrospinal fluid? Where is it found in the brain?
8. What is the blood-brain barrier? How are materials actually transferred from capillaries to the brain cells?
9. In which vertebrates is the midbrain prominent? What is its significance to humans?
10. Compare the importance of the cerebrum in the rat, frog, and primate. What general trend in the localizing of functions does your answer suggest?
11. State a function that is localized in each of the following cerebral lobes: prefrontal, occipital, temporal, and frontal.
12. Describe the organization of sensory and motor areas about the central sulcus. Compare the amount of sensory and motor control area devoted to the hand, genitals, leg, tongue, and toes.
13. What is the general rule about the cerebral hemispheres and control of right and left sides of the body? How does this affect right- and left-handedness?
14. What evidence is there that the left hemisphere contains the speech centers?
15. Explain the manner in which the three speech areas of the cortex interact when one reads aloud. In general, what are the roles of the Wernicke and Broca areas?
16. What are the main functions of the thalamus and its reticular system?
17. List five processes that are regulated by the hypothalamus. With what other system does it closely interact?
18. What structures, in addition to the hypothalamus and thalamus, make up the limbic system? Name some activities of these.
19. Using the terms *cerebral hemisphere, cerebral cortex, convolutions,* and *gray and white matter,* describe the organization of the cerebrum.
20. List the three regions of the hindbrain and state their general functions.
21. Describe the organization of the spinal cord, including its covering membranes, its gray and white regions, and the manner in which spinal nerves enter and leave.
22. Explain, theoretically, how it came to be that there are about 50 neurotransmitters in the central nervous system and only a few in the peripheral nervous system.
23. What is the role of enkephalins and endorphins? Why are they called opioid neurotransmitters?
24. Compare the action of amphetamines, cocaine, and LSD in the brain.
25. During which sleep episode is brain activity close to that of wakefulness? Why is this period of sleep so important?
26. What characteristic clearly distinguishes between the somatic and autonomic nervous systems?
27. Describe the organization of the autonomic nervous system, naming its divisions and explaining their anatomical arrangement.
28. Compare the role of the autonomic divisions in the "fight or flight" response.
29. List three important characteristics of sensory receptors.
30. In what major way does receptor potential differ from an action potential? How is this difference adaptive?
31. List the two types of mechanoreceptors found in the human skin and state the specialization of each.
32. What is a piezoelectric crystal? What tactile structures qualify as such structures?
33. In what specific instance might sensitive thermoreceptors be highly adaptive to invertebrates?
34. What are the specializations found in Ruffini's and in Krause's corpuscles?
35. Describe the structure of the moth antennae, stressing the organization of an individual sensillum. Comment on their sensitivity. What is the adaptive value of this elaborate sensory arrangement?
36. Explain how Jacobson's organ functions in the reptile.
37. Describe the mammalian olfactory epithelium and its connections to the brain.
38. Discuss ways in which the human and primate taste specializations may be adaptive.
39. List the parts of the human external, middle, and inner ear, and include a description of the cochlear organization.
40. Describe the events of hearing, beginning with air waves buffeting the tympanum and ending with an action potential in the auditory nerve.
41. Explain how a crayfish senses its body position. In what way is this similar to the operation of the saccule and utricle of humans?
42. List the structures or regions of the eye through which light passes on its way to the retina. Where appropriate, explain how each structure functions.
43. Describe the organization of the retina. In which region is color vision most acute? Why is this?
44. In what way is the retina specialized for receiving red, blue, and green wavelengths? Other colors?
45. Briefly summarize the chemical events occurring in the rod cells. When is rod vision most essential? Why?

36

THERMOREGULATION, OSMOREGULATION, AND EXCRETION

ABOUT 200 YEARS AGO, CHARLES BLAGDEN, THEN THE SECRETARY OF THE Royal Society of London, proved that he was one of the most persuasive people on earth. He managed to talk two friends into joining him in a small room in which the temperature had been raised to 126°C (260°F). Being aware that water boils at 100°C, they naturally were a bit reluctant. But Blagden prevailed, and the men, taking along a small dog and a steak, entered the room. They emerged 45 minutes later and were surprised to find that they were not only alive, but in good shape. The dog was fine, too. But the steak was cooked!

Thus, Blagden demonstrated the remarkable abilities of living organisms to adjust to severe environmental conditions by regulating their internal processes. These regulatory abilities should be expected, though, as the delicate processes of life demand extremely constant and specific conditions.

Let's consider another example. A young woman had thought she was in good shape living in the city, but now she is finding it pretty tough going while backpacking high in the mountains. Like the dog in the hot room, she too is panting, but for a different reason. The level parts of the terrain are easy enough, but when climbing, she has to stop repeatedly just to catch her breath. Her heart races, she feels light-headed and sick to her stomach, and she has a sharp pain in her left side. After each rest, though, her heavy breathing subsides, she feels more comfortable, and she is able to resume climbing. Her companion, who has been in the mountains all summer and is having no problem, tells her that she'll get used to the altitude in a week or two. Her companion is right. In fact, by the next morning she is already feeling a bit better, and in a couple of weeks she will hardly notice the altitude at all.

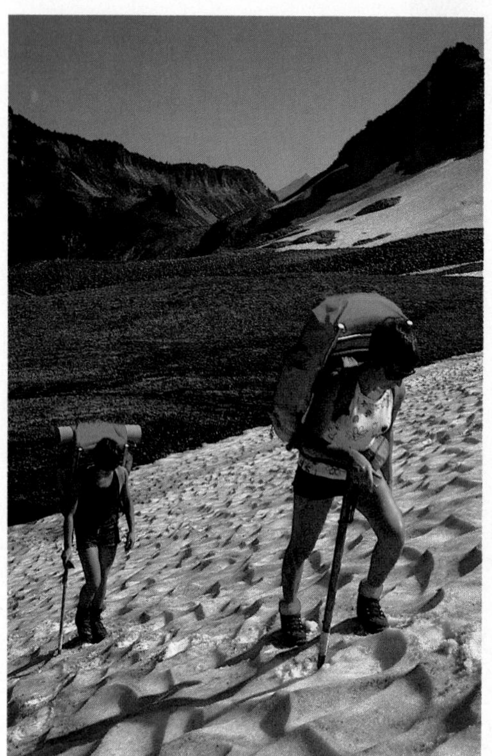

FIGURE 36.1
HOMEOSTATIC MECHANISMS AT WORK
As the backpacker climbs higher onto the mountain, her body undergoes many physiological changes that make it possible for her to meet new demands.

HOMEOSTASIS

Both Blagden and the climber have demonstrated what are called homeostatic responses. **Homeostasis** ("same status") is the maintenance of the physiological conditions required to support the life of the organism. In a sense, homeostatic mechanisms keep physiological conditions within a certain range conducive to life.

Our definition is necessarily broad since homeostasis pervades all aspects of life and functions in any number of ways on a variety of organ systems. In fact, every organ system functions, in some way, through homeostatic mechanisms, including the nervous system, respiratory system, circulatory system, and immune system. Even behavior has homeostatic significance. Homeostasis, then, is one of the pervasive themes in biology.

We should also keep in mind that different homeostatic responses can operate at the same time. For example, in Blagden's experiment, the dog probably sought to leave the room—a behavioral response. Then he probably gave up and lay there panting, as physiological changes began to occur. Regarding the novice mountain climber, her pause by the trailside to "catch her breath" was a response to falling levels of oxygen in her blood. Her behavior was changing even as certain physiological responses were kicking in (Figure 36.1). Neither she nor the dog had much control over the panting; it's an almost completely automatic adjustment to the stimuli. The sharp pain in the climber's side is an indication of another homeostatic mechanism. Oxygen levels are low at high altitudes. So with exertion increasing the oxygen demand, the spleen must release more blood cells, making them available to carry whatever oxygen there is. Her pain, then, is caused by the spontaneous muscular contraction of her spleen. The nausea is due to the shunting of blood from her gut to her oxygen-demanding brain and muscle tissue. On a different time scale, continued exposure to high altitudes stimulates the climber to produce more red blood cells. How, then, is the body able to respond so sensitively to changes in the environment? The responses are usually automatic, brought on through what is called feedback, which can be either negative or positive. **Feedback** occurs when the result of some process returns to affect the subsequent course of that process.

Negative and Positive Feedback Loops

The **negative feedback loop** refers to a situation in which a stimulus produces a reaction that ultimately reduces the stimulus. Thus, the negative feedback loop is inherently stabilizing, maintaining the organism in a steady state. Negative feedback loops tend to be delicate in nature, their operation surging and fading intermittently as an optimal condition is reached (Figure 36.2). The thermostat in a house also works along the negative feedback principle.

Let's consider a homey analogy. A baby cries when it is hungry; the parent feeds it; the baby is no longer hungry and stops crying. From the baby's point of view, the stimulus was hunger, and its response was crying, which ultimately stopped the hunger. From the parent's point of view, the stimulus was the crying and the response was feeding the baby, which stopped the crying.

Sometimes, the loop fails in one way or another. The baby may cry not because it is hungry, but because it is in pain from diaper rash. If the parent tries to feed the baby, it cries harder, so the parent tries harder to feed it, so it cries even harder, and so on. This reinforcing cycle constitutes a **positive feedback loop**, in which the stimulus evokes a response that further increases the stimulus.

Positive feedback loops are uncommon in nature and are usually associated with illness. High blood pressure, for instance, can damage arteries and arterioles, and the damaged vessel walls can become infused with lipid materials, scar tissue, and cellular growth. This in turn restricts the size of the vessel opening, further increasing blood pressure and arterial disease. So high blood pressure can actually raise the blood pressure further. As another example, we all know of people who are depressed because they are overweight, and so they eat because they are depressed. However, positive feedback or reinforcement can be a normal part of some processes. A familiar example to most of us is the heightening of sensory pleasure that peaks with orgasm, although even here, a strong negative feedback aspect soon follows the peak.

In this chapter, we will consider two important examples of homeostasis. One is **thermoregulation**, the regulation of body temperature. The other is **osmoregulation**, which involves ways in which animals control their internal osmotic conditions.

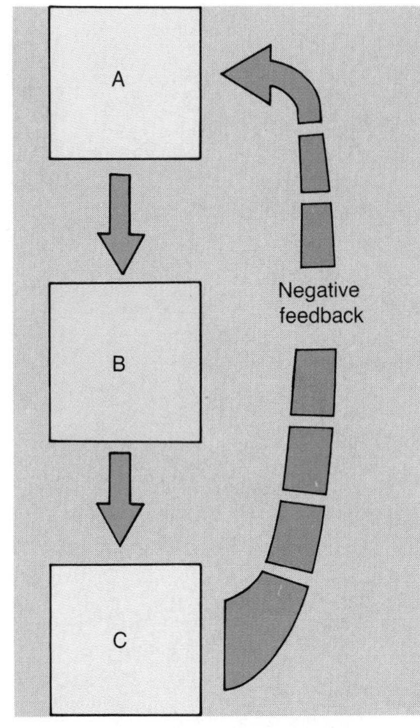

FIGURE 36.2
A NEGATIVE FEEDBACK LOOP
In the negative feedback system shown here, as A forms B which forms C, the accumulation of C suppresses the activity of A. As A diminishes, so does C, resulting in A increasing again.

THERMOREGULATION

Thermoregulation is the ability of an organism to maintain its body temperature either at a constant level or within an acceptable range. Some animals have very little of this ability, yet others are highly specialized for it.

Why Thermoregulate?

Animals must thermoregulate if they are to remain active beyond certain environmental temperatures. Under extreme conditions, low temperatures can slow metabolic processes, and since such metabolic processes generate body heat, a further cooling comes about. In this case the stage would be set for a dangerously irreversible positive feedback loop.

Frigid temperatures can also freeze the water in living cells, tying up that water and causing the remaining cellular material to become too concentrated. Ice crystals can also disrupt delicate membranes and destroy tissue.

Thermoregulation also involves control of excessive body heat. In fact, overheating often causes more problems than does overcooling. For example, high temperatures can accelerate rates of biochemical reactions to unacceptable levels. High temperatures can denature enzymes and other critical proteins, bringing metabolic activity to a halt and severely altering cell structure. Such damage is often irreversible, whereas damage done by low temperatures, within certain limits, is more likely to be temporary. Animals, of course, have ways of avoiding excessive body temperature increases and of ridding the body of excessive heat, although such strategies are often more limited than those involving adaptation to the cold.

As we will see, there are many adaptive solutions to the problem of maintaining optimal body temperatures. Some of these solutions are metabolic and others are behavioral. Some solutions simply involve retaining what heat there is.

Categories of Thermoregulation

Traditionally, those animal species that thermoregulate most (those retaining a rather constant internal temperature) are called **homeotherms.** Those that do not actively thermoregulate (those whose body temperatures tend to be near that of their surroundings) are called **poikilotherms.** The lay terms for the two conditions are "warmblooded" and "coldblooded," respectively. At one time the term *homeotherm* was reserved for birds and mammals, and all other animals were thought to be poikilotherms. But as this idea was tested, it soon became clear that things weren't quite so simple. Although birds and mammals are certainly warmblooded, ongoing studies have resulted in a growing list of other animals that retain surprisingly constant body temperatures. Such studies have also unveiled a third category of thermoregulation. Those in this category, the **heterotherms,** thermoregulate part of the time and allow their temperatures to track that of the environment the rest of the time.

There are two major processes that animals use to warm their bodies: endothermy and ectothermy. **Endothermy** involves the generation of heat through internal metabolic processes. The production of heat through metabolic means is called **thermiogenesis**. You may recall that metabolic energy transfers are not 100% efficient and that at each step, some energy is lost as heat. That heat, when generated in sufficient amounts, can then be used in thermoregulation.

In contrast, **ectothermy** involves the utilization of external sources of heat. Ectothermic animals, then, absorb heat from the environment. A frequent method ectothermic animals use to gain heat is to expose themselves deliberately to the sun. Heat can be gained in other ways as well, as when a fish living in a hot spring moves closer to the vent or on a cold day when a flea migrates to a dog's warmer body parts. We can say, then, that in endothermy, heat comes from the inside, whereas in ectothermy it comes from the outside.

Endothermic Regulation

Endothermy offers a certain stability in a varying world. If an animal maintains a constant temperature, the chemical reactions associated with metabolism can go on at a more constant rate, not being slowed drastically by cold or greatly accelerated by heat. Thus, the animal can remain active at a wide range of environmental temperatures. The advent of endothermy has been critical in evolutionary history, since it provided a new way for animals to cope with new environments, those that varied from blazing deserts to the frigid Arctic.

Keep in mind that when we talk of temperature constancy, we're referring to a *range* of internal temperatures. After all, even among the strictest homeotherms, body temperatures can vary—both from time to time and place to place within the body. We find that in ourselves for example, our temperatures vary throughout the day and according to how much metabolic energy we're expending. We also find that our skin and extremities are cooler than our internal organs. (Thus the advent of the rectal thermometer.)

FIGURE 36.3
HIBERNATING ANIMALS
The dormouse isn't dead; it's in a state of hibernation. During hibernation the usual body thermoregulatory controls are overridden, body core temperatures fall as much as 30° C, and metabolism drastically slows.

The greatest constancy for any homeotherm is in **core temperature**, or temperature deep within the body. Even core temperature varies widely among endotherms. Consider those that hibernate (such as bats) and those that enter metabolic stupor (such as bears). In both groups the core temperature sometimes drops drastically (Figure 36.3). Inactivity during cold seasons is adaptive in that it reduces the need for energy when food is not abundant, or at least not in the abundance needed by endotherms.

In fact, a principal disadvantage of endothermy is that it requires a great deal of energy. Since the source of energy is food, endotherms must feed frequently. But there are other disadvantages to endothermy. With few exceptions, birds and mammals cannot tolerate much change in core body temperature. (Variations of two or three degrees in our own internal temperatures, for instance, send us running to our physicians.) Endothermy also places body size restrictions on animals, primarily limiting how small they can be. Below a certain body size, heat production cannot keep up with heat loss. The traditional explanation is that smaller animals have a proportionately greater surface area per unit of volume than do larger animals, so their heat loss is greater. (Interestingly, endothermy does not seem to limit how large a species can get.) Actually, the relationship between heat loss and body size is now being challenged and there may be more to the phenomena than simple geometry.

Endothermy in Cold-Adapted Moths Owlet moths (family Noctuidae, Figure 36.4a) are cold-adapted temperate zone species and part-time endotherms. We referred to this condition earlier as heterothermy. As with many other flying insects, their bodies cool to surrounding temperatures when they are at rest. But when they prepare for flight the moths shift to endothermy. Specifically, a moth's thoracic temperature must reach about 30°C before the flight muscles will function properly. (Recall from Chapter 33 that the insect flight muscles lie within the thorax.) Thermiogenesis occurs through shivering in the flight muscles. Shivering is caused by the simultaneous contraction of opposing muscles, an act that requires a lot of work but produces little movement (like isometric exercise). Most of the metabolic energy is thus converted to heat, which produces the temperatures required for flight.

It is important to the owlet moth that the heat generated through shivering be retained in the thorax, the location of the flight muscles. Heat is retained in the thorax in three ways. First, the thorax is covered with modified scales that have a "fluffy" appearance and provide an effective insulation against heat loss to the outside. Second, a cluster of air sacs just posterior to the thorax slows heat escape to the abdomen. Lastly and of the most significance, these moths have a **countercurrent heat exchanger**, a circulatory arrangement found in many cold-adapted animals, and one we will see again and again. In such arrangements, warm blood leaving the core region of the animal body passes near cooler blood returning from the outer regions. Since the fluids are moving in opposite directions, generally in closely adjacent vessels, heat is shunted from the warmer to the cooler blood along the entire flow. By the time the core blood reaches the outer regions it has cooled down close to environmental temperature, its heat having been recycled to incoming blood. Such heat exchanges are possible anytime fluids of different temperatures pass close by in opposite directions. In the owlet moth, as you see in Figure 36.4b, the heat transfer occurs primarily in two exchangers, one in the thorax and another in the abdomen. Looking at the abdominal heat exchanger, we see that warm blood is carried toward the abdomen through an open blood sinus. (Recall from Chapter 31 that insects have open circulatory systems.) Blood in the sinus gives off heat to cool blood returning to the thorax through a blood vessel. Thus the heat remains in the thorax.

Many moths are not cold-adapted and do not have such an exchanger. In these species, heat from the muscles is lost to the abdomen and is then lost from the body. But this brings up an important point about the directions of adaptation and evolution. It turns out that under warmer conditions the cold-adapted moths can rapidly overheat, falling into a metabolic stupor until they cool down. Under the same conditions, other moths are able to lose heat easily and to function normally. In this way the two kinds of insects are able to utilize the same environment at different times and under different conditions, which may reduce the level of competition between them.

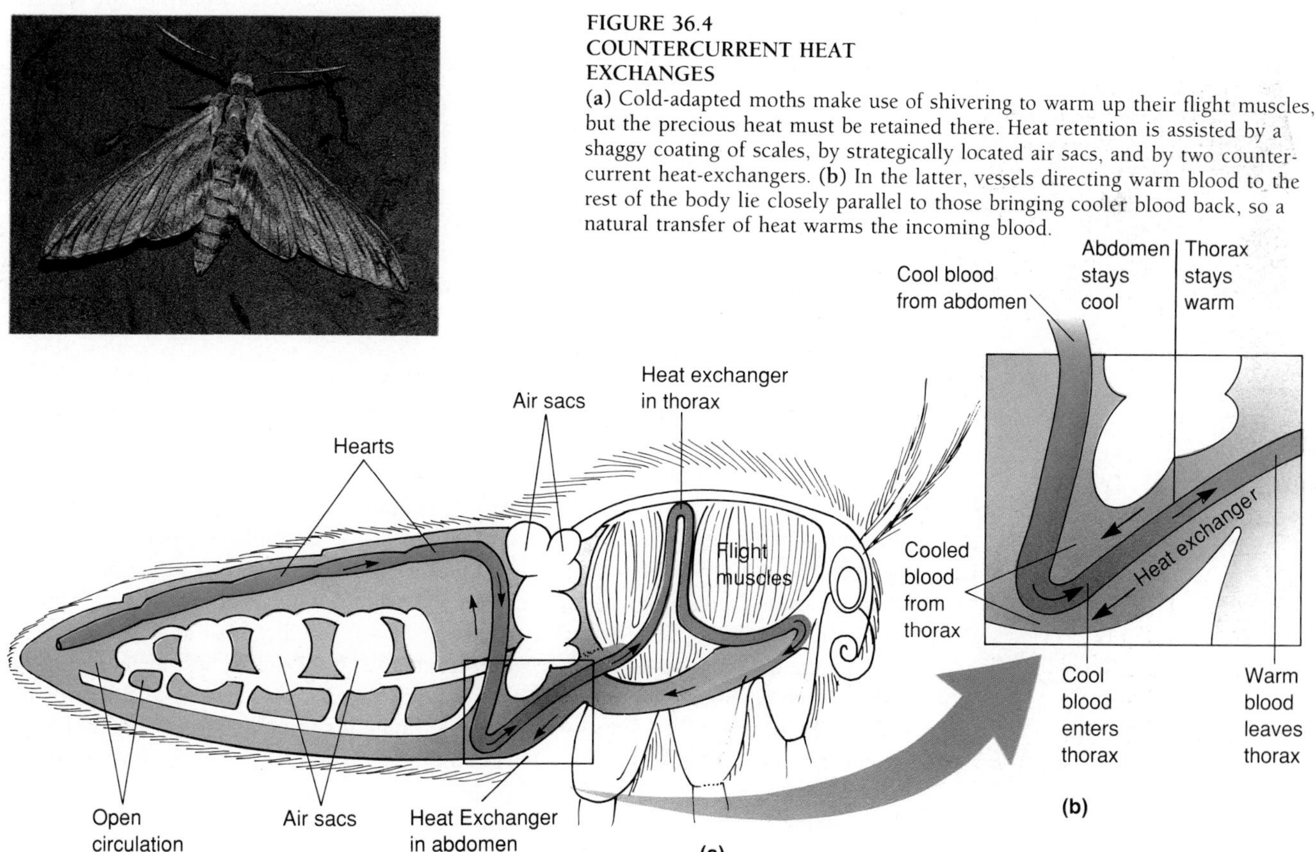

FIGURE 36.4
COUNTERCURRENT HEAT EXCHANGES
(a) Cold-adapted moths make use of shivering to warm up their flight muscles, but the precious heat must be retained there. Heat retention is assisted by a shaggy coating of scales, by strategically located air sacs, and by two countercurrent heat-exchangers. **(b)** In the latter, vessels directing warm blood to the rest of the body lie closely parallel to those bringing cooler blood back, so a natural transfer of heat warms the incoming blood.

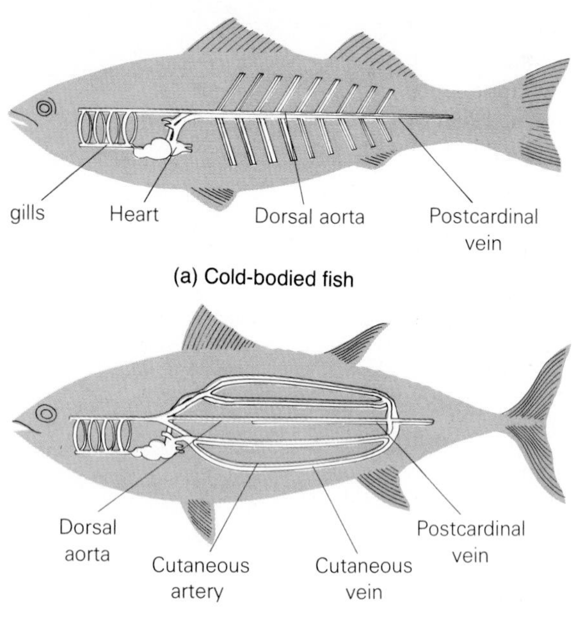

FIGURE 36.5
CIRCULATION IN A THERMOREGULATING FISH
(a) In most fishes the arrangement of blood vessels is such that core heat is quickly distributed to the rest of the body. **(b)** In the bluefin tuna, the retention of heat in the core is aided by a countercurrent heat exchange among major vessels. Note parallel arrangement of vessels carrying warmed blood out of the core region with those bringing cooler blood in from the extremities.

Now let's consider a fascinating example of a countercurrent heat exchanger at work in a vertebrate.

Endothermy in the Bluefin Tuna: A Cold-Adapted Fish Most open sea fishes have little thermoregulatory ability. Those that do thermoregulate rely almost exclusively on endothermy, the metabolic generation of body heat. (Obviously, sunbathing after the fashion of ectotherms would be a problem, and there are few concentrated heat sources beneath the surface.) The best known endothermic fish is the bluefin tuna, a fast-moving, cold-water predator. The bluefin is, in fact, among the fastest swimmers in the sea. Furthermore, its swimming muscles are able to function actively even in the coldest waters of its range. It is able to function so well in cold water because it is a true endothermic homeotherm, a warm-blooded animal.

Compare in Figure 36.5 the circulatory systems of non-regulating and thermoregulating, or endothermic species of fish found in similar habitats. Notice that in the first, the major blood vessels give rise to branches that serve the propulsion muscles and other portions of the body. Such an arrangement can do little to retain body heat in a central place. On the other hand, the major arteries and veins of the homeothermic bluefin tuna tend to run the length of the fish, parallel to each other with blood flowing in opposite directions, thus forming a countercurrent heat exchanger.

Much of the bluefin's body heat is generated by its large, dark, swimming muscles, which make up most of the body core. The body core is some 10° warmer than the skin, which remains at about the temperature of surrounding sea water (Figure 36.6a). As intense muscle activity generates heat, this heat is efficiently retained in the core region by use of a countercurrent heat exchanger. As in the cold-adapted moths, this involves warm core blood passing cooler peripheral blood as it leaves the core region.

Notice in Figure 36.6b the presence of a second, much more complex heat exchanger in the **rete mirabile** ("wonderful net"). The dark muscle making up much of the tuna's body core, and the warmest part of its body, is served by a vast network of arteries and veins, all arranged in a parallel fashion. They make up a dense, highly efficient countercurrent heat exchanger that keeps the dark swimming muscles quite warm.

Thermoregulation in Birds and Mammals Mammals and birds have developed endothermy into a fine art. They have evolved very effective ways of generating and conserving heat. In fact, they can develop and retain so much heat that they must have efficient cooling mechanisms as well. We will first see how body size is important to heat generation.

Body Size and Heat Generation An important aspect of thermoregulation in animals is **metabolic rate,** the total body energy produced per unit time. In birds and mammals, metabolic rate, like heat loss, is generally related to body size. Specifically, the smaller the body size, the greater the metabolic rate (Figure 36.7). The greater the metabolic rate, the more heat produced. The hummingbird, for example, supports its intense metabolic activity through a large sugar (nectar) intake, so high in fact that at night when it can no longer feed, it survives only by falling into a metabolic slump. Thus it may be regarded as heterothermic. Among the mammals, the small shrews (about the size of large cockroaches—which are on their menu) have the highest metabolic rates, whereas elephants have the lowest. Although such shrews have about one millionth the mass of the elephant, they consume about 100 times as much oxygen per gram of body weight. For a carnivore, their food consumption is also monumental, their daily intake being about 75 percent of their body weight.

Physiologists are now beginning to challenge the traditional explanation of the link between metabolic rate and heat loss. It has long been held that small animals, with

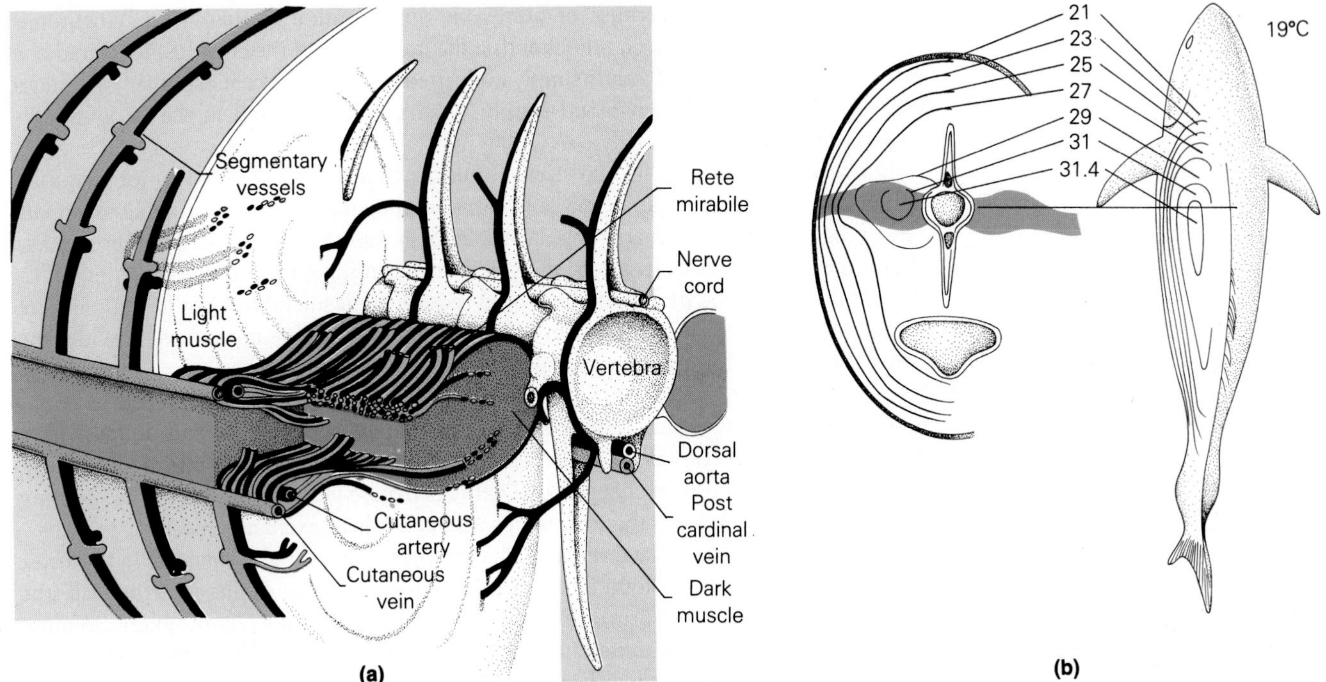

FIGURE 36.6
THERMOREGULATION IN THE BLUEFIN TUNA
(a) The bluefin tuna also has a countercurrent heat exchanger in the mass of parallel veins and arteries making up the rete mirabile. It retains heat in the core region around the dark swimming muscles. (b) The bluefin tuna's body core is much warmer than its extremities and its cold water surroundings. Heat concentrates in the dark swimming muscles.

their proportionately larger surface areas, must maintain high metabolic rates just to keep up with heat loss. If this were so, then one would expect small *ectotherms* to have cooler bodies than large ectotherms. This isn't the case; both tend to be in equilibrium with their surroundings. Certainly, more is involved in heat loss than simply body mass and surface area. We will continue to watch these developments with interest.

Thermiogenesis Birds and mammals generate and retain heat in a variety of ways, both actively and passively. We will begin with the active generation of heat through thermiogenesis. Thermiogenesis occurs through three principal mechanisms, all involving the conversion of chemical energy into heat energy. The first is by oxidative respiration, a process that produces heat during ATP synthesis. The second is the generation of heat through shivering, which, as we have seen, involves the rapid contraction of opposing muscle groups. The third way is by the utilization of brown fat (also see Chapter 8).

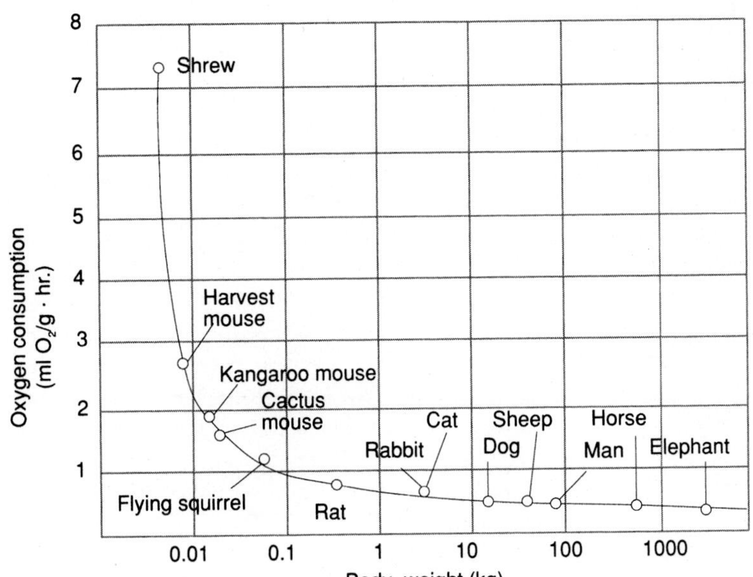

FIGURE 36.7
BODY SIZE AND METABOLIC RATES
The metabolic rate in selected mammals, as determined through oxygen consumption, indicates that this rate is greater in the smaller animals than in the larger.

Brown fat is brown because of numerous mitochondria (actually the cytochrome oxidase they contain). In such mitochondria the free energy of the chemiosmotic gradient isn't used to generate ATP, but is simply converted to heat. Unlike most other fat storage tissue, brown fat has a rich blood supply, so the heat generated in the fat is rapidly dispersed throughout the body.

Brown fat thermiogenesis occurs in a number of animals. We find it, for example, in human infants, in bats, and especially in hibernating mammals. Hibernating mammals eat prodigiously before hibernating, laying in great reserves of fat, including brown fat. When they enter that deep sleep, their body temperatures drop by as much as 30°C. Regaining normal body temperature would require a considerable amount of time were it not for brown fat. The metabolism of brown fat, triggered by hormones, provides the body heat needed for a quick arousal.

Slowing Heat Loss Once heat is produced, it must be retained, and animals have developed a number of ways to slow heat loss. One way they do this, one we have seen, is by countercurrent heat exchangers that help retain core heat. Furthermore, the shunting of warm blood away from the extremities assists the heat exchange mechanism.

Humans, for example, have very effective countercurrent heat exchangers in the arms. When the body is warm, blood returns from the hands through veins near the skin (the ones you can see in your arm), thus permitting heat to escape. As the body becomes chilled, though, blood is shunted into deeper veins that run parallel to the arteries, thus enabling the body to conserve heat through a countercurrent heat exchange between the opposing vessels (Figure 36.8). Shunting blood deeper into the body to conserve heat is a costly defense. Cooler limb muscles do not work as efficiently, and the extremities, deprived of warming blood, can become numb and eventually frostbitten. In frostbite, tissue is damaged by freezing. Thus a person can actually lose fingers, toes, and ears before the core temperature reaches a danger level. The internal organs receive first priority in the battle for survival.

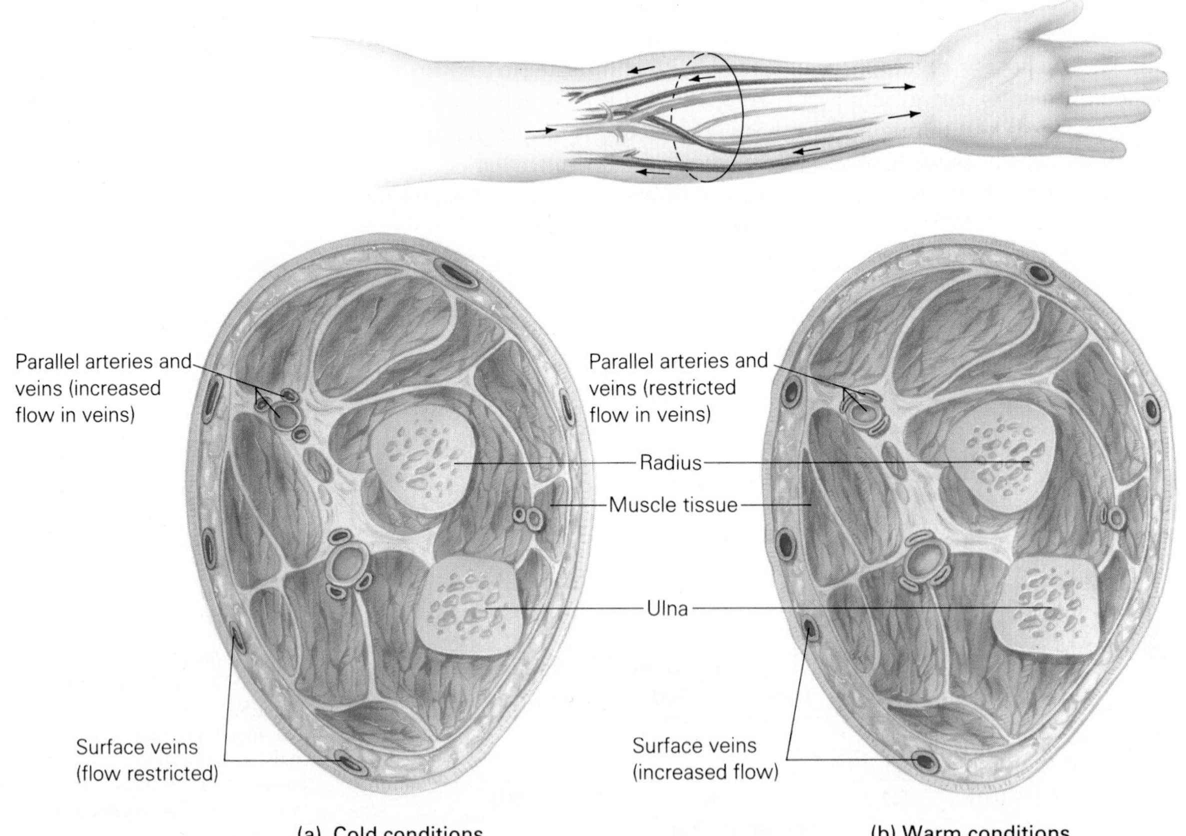

FIGURE 36.8
THE COUNTERCURRENT HEAT EXCHANGER IN HUMANS
(a) The large arteries deep in the arm closely parallel some of the larger veins. There, they form a countercurrent heat exchanger, thus helping to retain core heat. **(b)** Large veins also occur just below the skin. When heat must be released, the return flow is directed to those surface veins.

FIGURE 36.9
BLUBBER
A dense layer of fat forms in effective insulating layers in arctic and antarctic mammals.

Watching wading birds in the dead of winter can send a shiver through the most stalwart of souls. The birds casually wade in water that is near freezing (in the case of sea water, perhaps below freezing). However, they still maintain their normal body temperature. In fact, even their legs stay warm, due to countercurrent heat exchangers in the feet. A high level of unsaturated fats in cell membranes of tissues in the feet is an additional adaptation to cold. Such fats stay liquid when at very low temperatures.

Mammals also utilize fat for protection from extreme cold. Marine mammals such as whales, seals, and walruses are protected from heat loss by thick layers of **blubber**, a specific type of fatty tissue (Figure 36.9). Blubber, an excellent insulator, is not simply passive but plays a versatile role in thermoregulation. Unlike most fatty tissues but like brown fat, blubber has a rich blood supply. In frigid polar waters the vessels close down, and the blubber is virtually blood-free. In that state, it resists nearly all heat loss. But when a whale encounters warmer waters, or when a sea lion basks on a sunny shore, reduction of the core temperature becomes essential. Blood vessels in the blubber then open and blood enters, heat escapes, and overheating is avoided.

Most mammals rely on body hair, and all birds rely on feathers for insulation. Both can be "fluffed," or made to stand on end, thereby producing tiny, heat-trapping air spaces. Perhaps you've seen chickadees perched at your bird feeder on a cold winter day, their feathers fluffed, looking like tennis balls with beaks (Figure 36.10).

FIGURE 36.10
FEATHER FLUFFING
Birds increase the insulating effects of feathers by fluffing them, thereby creating minute heat-retaining air spaces around the skin. Fluffing increases with drops in temperature.

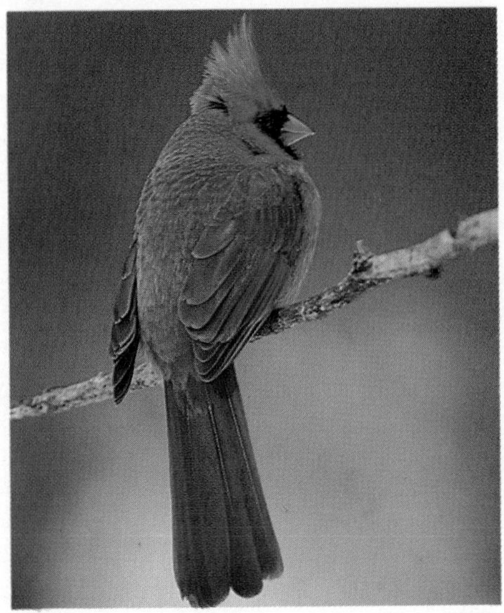

Many mammals have dense, soft **underhair** that acts as insulation, and long, coarser **guard hair** over it that protects against moisture and abrasion and provides for coloration. In aquatic mammals, the layer of guard hair is virtually waterproof, and thus they can swim with impunity in frigid water (Figure 36.11). The combination of hair and blubber insulating the Arctic harp seal holds in virtually *all* of the animal's body heat. Canadian researchers were recently thwarted in their attempts to use airborne infrared tracking devices to follow the seals' movements. The animals simply do not give off the body heat needed to provide infrared images. Furthermore, when the researchers switched to ultraviolet detectors, they could find the seals all right, but they appeared as negative images only—that is, they did not reflect ultraviolet energy from their coats as expected. This was a startling situation. It turned out that, whereas the surrounding snow reflected ultraviolet light, the seals absorbed it. Further studies

FIGURE 36.11
COLD-ADAPTED MAMMALS
Mammals of the Arctic are well adapted for normal activity in common subzero air temperatures. Their well-insulated bodies keep in most of the heat that is generated metabolically.

revealed that the individual hairs on many arctic mammals, especially those of the polar bear, act as excellent solar collectors. Polar bear hairs have fibrous cores that, upon absorbing ultraviolet light, act as tiny fiber-optic cables, conducting the energy to the black, light-absorbing skin below. Engineers are now trying to apply the same principle to the manufacture of solar collectors.

When it comes to using body hair for insulation, humans are exceptions. Your cousin Walter aside, humans don't have enough hair on their bodies for effective insulation. Anthropologists tell us that humans evolved in hot climates where the furry coat of our ancestors was traded for the ability to sweat profusely, thereby cooling the body—something few other mammals can do as well. Thus, we own much of our success in colder climates to behavioral, not physiological strategies.

We'll turn now to another aspect of thermoregulation in endotherms, the problem of overheating.

Encouraging Heat Escape The most important avenues for heat escape in birds and mammals are the skin and the respiratory passages. Heat escapes from such surfaces in three major ways: radiation, conduction, and convection.

Heat, we know, radiates into cooler surroundings. Radiation is important in dissipating heat from the peripheral circulation, such as through the skin. In addition, heat escapes through conduction (in which heat passes by contact to a cooler solid, such as a tile wall). In convection, heat is carried away by the cooler air molecules. Convection figures prominently in cooling through evaporation.

Evaporative cooling is a vital part of thermoregulation in both birds and mammals. The evaporation of water from the surface of an animal is accompanied by a transfer of heat to the surroundings and a subsequent cooling of the body surface. To encourage such cooling, animals may wet themselves, some making use of standing water, others spreading saliva or even urine on their warm bodies. Most mammals have sweat glands in the skin that become active as the body warms, wetting and cooling the skin. Sweat glands are particularly important in humans, horses, and certain apes. (A human in a hot desert may lose as much as 1.5 liters of water per hour through sweating. At this rate, and without replacement, one would be near death by a single day's end.)

Mammals and birds also carry out evaporative cooling through the respiratory passages, generally exhausting most of the heat through the open mouth. Birds actively pant, but in addition, they vibrate the thin-walled floor of the mouth, accelerating evaporative cooling there through what is called **gular flutter.** As dogs pant, they inhale through the nose and exhale through the open mouth, their wet tongues extended to increase the area of evaporative cooling. The use of evaporative cooling, of course, depends on the availability of body fluids, which must be maintained at certain critical levels in all animals. An excessive loss of water can quickly bring about dehydration.

Behavioral Adaptations in Endotherms Endotherms not only regulate their temperatures physiologically, but behaviorally as well. Behavioral regulation is rather straightforward. It simply involves doing those things that result in cooling or heating the body so that the core temperature remains within certain limits. In endotherms, behavioral regulation of temperature usually occurs when the animal perceives that the internal mechanisms are not keeping up. For example, evaporative cooling slows with increasing humidity. Under hot, humid conditions, then, animals must find other ways to cool their bodies. So they may resort to behavioral strategies. At midday, birds may stop singing and rest quietly in the trees, while some may bathe in standing water. Dogs like to dig holes in newly planted lawns, using the cool, moist, freshly dug soil to take up heat from their warm bellies. Desert rodents, unable to sacrifice much body water for cooling their bodies, retreat to more humid burrows until nightfall. One species of mammal simply turns on the air conditioner.

Cessation of activity is important in avoiding overheating. Some time back, one of your authors (Wallace) suffered heat stroke in a marathon. He reports:

> The temperature was 75°F, and the humidity was 80%. Hardly ideal, but all seemed well until the 24th mile. My time was about 6½ minutes per mile when, suddenly, I felt strange—remote, dizzy, and spacy. I slowed down, thinking I would quickly recoup. But I didn't recoup. My goal for the 26 miles was anything under 2 hours and 50 minutes, and with 2 miles to go my time was 2 hours and 36 minutes. I tried to mentally subtract 36 from 50 to see how fast I would have to run the final 2 miles, but any mental calculations were impossible. I began to realize I was in trouble, and when I felt my arm I found it was dry and chilled. A wave of dismay swept over me because I knew continued stress could cause kidney or brain damage, but I wanted to finish, so I slowed down even more. Then I began to almost fall asleep again and again while running. I managed to stay awake and trotted in at 2:55. Recovery, with water and mineral-laden drinks, was rapid. Strangely enough, a blanket was needed to fend off what felt like a chill wind. Physiologists later told me that I had possibly altered my thermoregulatory mechanisms so that I had lost too much heat, and that the sleepiness could have been the same sort described by people who are freezing.

The point is, the thermoregulatory mechanisms are in delicate balance, and if shifted off balance, the results may not be merely drastic, but unexpected.

Internal Thermoregulation An organism can change in many ways to meet the challenge of variable environmental temperatures. The question then is: How are these responses regulated and integrated? In particular, how are they handled in humans? Humans, after all, maintain an average internal temperature of about 37°C (98.6°F) within an amazingly narrow range. Surely some form of sophisticated internal control is required since we have so many independent ways to regulate heat.

The brain constantly monitors the temperature of blood flowing through it. The responses initiated by the brain override all other controls; hence, the brain can be thought of as the real thermostat of the body. The presence of a thermoregulatory center in the brain is not surprising. The brain is highly sensitive to temperature changes and easily malfunctions when its temperature falls outside a narrow range of tolerance. (In fact, the brain is cooled by blood vessels that also pass through the nasal area where evaporative cooling occurs.)

The heat-monitoring part of the brain is the hypothalamus. It indirectly receives temperature information originating in sensory receptors in the skin, and, more directly, through its own sensors, which detect blood temperature in its own vessels.

Actually, the hypothalamus behaves very much like an ordinary home thermostat, but with far more precision. Physiologists have been able to detect neural activity in the hypothalamus when blood temperature varies by as little as 0.01°C.

The hypothalamus, in its constant monitoring of blood temperature, can elicit a variety of responses (Figure 36.12). For example, through its influence of the autonomic (sympathetic and parasympathetic) nervous system, it can increase the rate of heat production by stimulating the adrenal gland (a hormone-secreting body), causing it to release the hormone epinephrine into the bloodstream. Epinephrine speeds up the conversion of glycogen to glucose in the liver. The release of glucose into general circulation and its subsequently increased availability to cells enables cells to increase their respiratory activity, which creates heat. Epinephrine also stimulates thermiogenesis in brown fat. In addition, the hypothalamus can prompt the autonomic nervous system to reduce blood flow in the skin.

The hypothalamus also stimulates the pituitary gland to release thyroid-stimulating hormone, which in turn causes the thyroid gland to increase the output of its hormone,

FIGURE 36.12
THE CENTER FOR THERMOREGULATION
The hypothalamus receives thermal information from thermoreceptors in the skin and by direct sensing of blood arriving from the core of the body. Its response to changing temperatures is carried out in two ways: through activation of the autonomic nervous system and of the endocrine system. Negative feedback occurs through the continued monitoring of blood temperature.

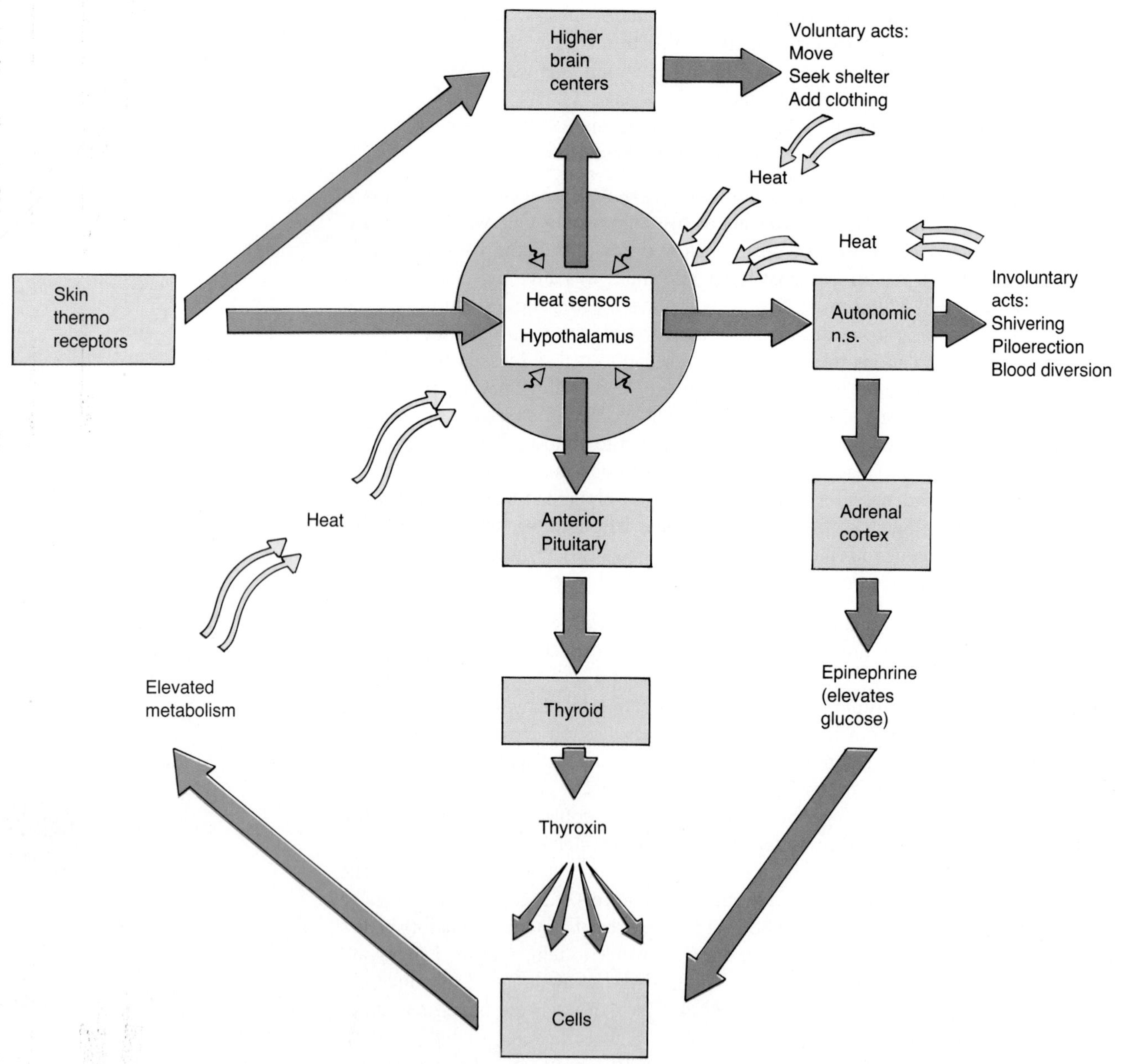

thyroxin (see Chapter 37). The effect of thyroxin is to increase cellular respiratory activity and thereby produce body heat. Increases in body heat are subsequently detected by the hypothalamus, which then eases off on the pituitary, slowing heat output—another example of a negative feedback loop.

Thus, information arriving from peripheral and internal sensors results in a considerable variety of hypothalamic actions. As a homeostatic control structure, the hypothalamus keeps the organism in a finely tuned and responsive state, maintaining body temperature at an optimal level.

Ectothermic Regulation

The chief advantage of ectothermy is its low metabolic cost. It's like living in a house that's not heated or air-conditioned. If it's cold outside, the house is cold; if it's hot, the house is hot. But your energy bills are low. Because they do not require extra food just for keeping body temperature up, ectothermic animals need far less food than endotherms, and many can fast for long periods. There are, however, marked disadvantages to ectothermy. Since ectotherms rely mainly on outside sources of heat, their activities may be curtailed at certain times, such as at night and in the colder seasons. We will begin our look at ectotherms with problems related to low temperatures.

Some Cold Adaptations What does freezing do? Interestingly, animals are not irreversibly affected by extreme cold unless ice crystals form within cells. There, as we've seen, ice crystals disrupt membranes and cellular organelles and cause the death of the tissue made up of those cells. The formation of ice crystals outside of the cells, in the intercellular spaces, is not as serious. In fact, such freezing can be part of a protective mechanism in cold-adapted species. Researchers report that in certain beetles and midge larvae, the freezing of extracellular fluids is encouraged by the presence of particles that act as ice nuclei. How does this protect the cells? The cells themselves are protected because as water is locked up as ice in the extracellular spaces, the solutes left behind produce an osmotic effect. That is, their presence encourages the osmotic movement of water out of nearby cells, which then become dehydrated. So, even though ice forms in the extracellular spaces, the dehydrated cells avoid the problem, surviving nicely until warmer temperatures return.

A number of ectotherms resist freezing through the presence of specific antifreeze substances in their body fluids. The parasitic wasp, *Brachon cephi,* for example, increases its body's glycerol content, which lowers the freezing point of its body fluids to about -17°C. Certain insect larvae do even better, resisting ice formation at temperatures as low as -47°C. Another form of antifreeze, a glycoprotein, protects the Antarctic ice fish, *Trematomus,* from freezing (Figure 36.13).

FIGURE 36.13
FISH WITH ANTIFREEZE
Concentrations of glycoproteins in the body of the Antarctic ice fish *Trematomus nicolai* prevent the formation of ice crystals. Ocean temperatures below the ice hover between -1° and -2°C throughout much of the year.

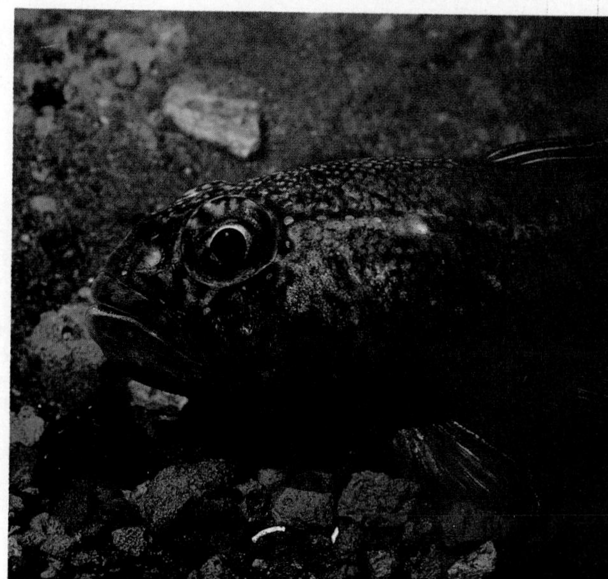

Behavioral Adaptations in Ectotherms Since the body temperature of ectotherms depends on the temperature of their surroundings, behavioral thermoregulation is very important to this group. They must go where the ambient temperature suits their needs, and once there, they must do the things that cause the loss or gain of appropriate amounts of heat. Among insects, for example, the desert locust crawls into the sunlight at dawn and orients its body sideways to the sunlight so that the greatest possible surface area is presented to the warming rays. Honeybees concentrate heat by huddling together in cold weather and, in hot weather, they promote evaporative cooling of the hive by bringing in water and fanning it with their wings.

Fishes that are not endotherms have little opportunity for thermoregulation. As we've seen, behavioral strategies are not generally available, especially in the open sea where environmental temperatures vary little from place to place. Fishes in shallower fresh waters, on the other hand, can move from warmer, sun-bathed shallows to cooler, deeper waters and back, as circumstances demand.

FIGURE 36.14
THERMOREGULATION THROUGH BASKING
Early in the day, the lizard's problem is to warm its muscles in the sun but not expose its sluggish body to predators. It often begins by exposing its head, warming it first before exposing its entire body. At midday it may linger in shadows or dig into the sand to avoid overheating. Then in the afternoon, it may again expose its entire body to the waning sun.

It turns out that the degree to which reptiles can regulate their body temperature depends on where they live. Reptiles from the tropics, for example, can't regulate their body temperature nearly as well as those from temperate regions, where weather changes more drastically. In fact, historically, the success of reptiles in invading temperate zones from the tropics has depended heavily on their ability to thermoregulate behaviorally. The strategies of temperate zone lizards are apparently very successful, since these reptiles maintain remarkably constant body temperatures as they move about.

Reptiles often absorb heat by basking in the sun, and the temperate desert lizards have developed the technique into an art. They avoid excessive heat gain by avoiding the heat of midday and by simply facing the sun's direction, thus exposing as little surface as possible. Heat is absorbed maximally by presenting more of the body to the sun (Figure 36.14).

The horned lizard (or "horny toad," as some insist) is also a behavioral thermoregulator, but it supports behavioral efforts through a cellular mechanism. It alternately reflects or absorbs sunlight energy through changes in its skin color, thus cooling or warming itself as conditions warrant (Figure 36.15). Skin color changes are accomplished through the movement of pigments in skin cells. When the pigment granules are drawn together, the skin lightens, and more light energy is reflected. When the granules become dispersed the skin color darkens, and more solar energy is absorbed. Thus through pigment migration, the reptile considerably improves its basking efficiency.

Questions about the extent of thermoregulation in reptiles also extend to species that are no longer around. Since endothermy is not taxonomically restricted, there is an ongoing controversy over whether dinosaurs were endotherms or ectotherms, and the current scarcity of dinosaurs only adds to the problem. It has been suggested, in any case, that the strange, highly vascularized backbone plates of the stegosaurs were excellent

(a) Low temperature

(b) High temperature

FIGURE 36.15
SKIN COLOR AND SUNLIGHT
The horned lizard (like many reptiles and amphibians and some fishes) is able to change its color through pigment migration—an automatic response. **(a)** When exposed to cooler temperatures in experimental situations, the lizard's color darkens. **(b)** At higher temperatures, it lightens considerably.

heat radiators and absorbers (Figure 32.20). Other dinosaurs (for example, *Dimetrodon*) had high dorsal fin structures that probably served both to capture solar heat and to radiate unwanted heat, depending on the animal's orientation to the sun and surrounding temperatures.

There are many other examples of heat regulation in the so called "coldblooded" animals—those that were once believed to be nonregulators. The point here, in addition to introducing the subject and pouncing on some popular old ideas, is to illustrate the continuous diversity of form and function in living things.

OSMOREGULATION AND EXCRETION

Now let's look at another problem of regulation in animals—the problem of how animals regulate the water and ion concentrations in their cells. Most cells are about two-thirds water, but in some cases, "about" won't suffice. In certain cells, the amount of water is critical, as is the relative abundance of various ions in the cell fluids. The process of maintaining the proper water and ion balance is called **osmoregulation.** And for many animals, this necessarily leads us into the methods by which excesses of water or ions are removed from the body by the processes of **excretion.** Excretion refers to the removal of metabolic wastes—the byproducts of cellular reactions—from the body.

Producing Nitrogen Wastes

Amino acids, you recall, are the constituents of proteins. Typically, proteins must be broken down into individual amino acids during digestion so that they can be absorbed across the gut wall and into the bloodstream. Those same amino acids can be used to build new proteins in the cells; they may also be converted to fatty acids or carbohydrates for storage, or used as fuel in respiration. Some of these changes produce leftover fragments of nitrogen, and accumulations of these nitrogen fragments can be extremely poisonous.

During **deamination** (Figure 36.16), the amine group, $—NH_2$, is removed from an amino acid as NH_3, or ammonia. In the presence of protons from water, it readily forms ammonium ions (NH_4^+). (Note in the figure the role of the cellular coenzyme NAD here—another example of the multiple use of molecules.) Ammonia is highly toxic, but

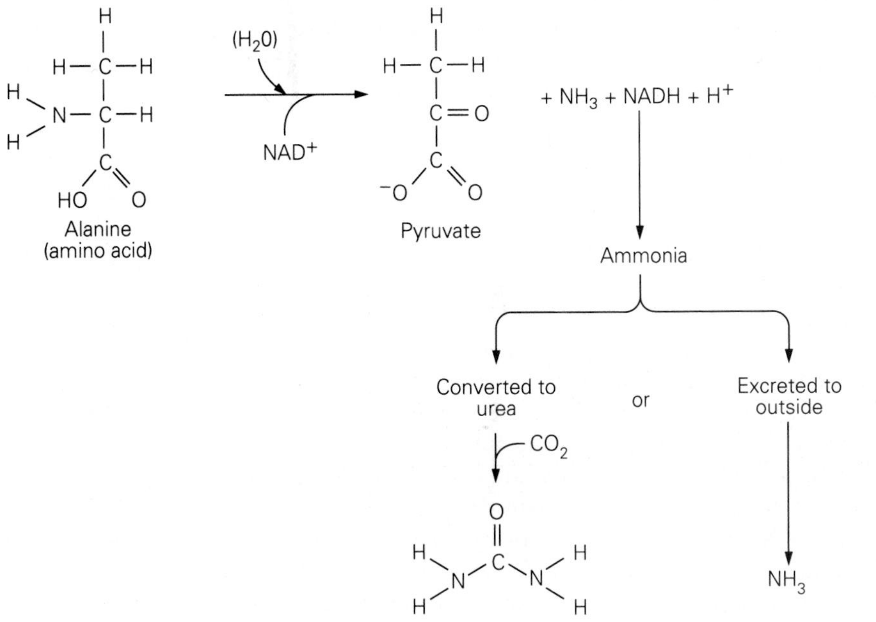

FIGURE 36.16
NITROGEN WASTES
In the metabolic breakdown of alanine, pyruvate is formed, along with the waste product, ammonia. Pyruvate is then free to enter one of several pathways (such as the citric acid cycle). Ammonia may be excreted directly, or it may be converted to urea.

many organisms can safely handle it if they live in fresh water and are small enough to exchange materials easily with their environment. However, in many terrestrial animals, a group that generally must conserve water, the volume needed to dilute ammonia to a safe concentration may not be available. In such cases, the ammonia must be converted to something more manageable.

In many animals, ammonia is converted to **uric acid** or to **urea.** Uric acid is a semisolid, insoluble waste. It is the primary nitrogen waste of insects, reptiles, and birds, and is usually produced as a dense paste. Urea is highly soluble and is the primary nitrogen waste of earthworms, many fishes and amphibians, and all mammals. Urea is dissolved in water and released as urine. Frogs are interesting in that their aquatic young, the tadpoles, excrete their nitrogen as ammonia. Only when they metamorphose into the partially terrestrial, water-conserving adults do they switch to urea production. A certain salamander is even more versatile; it produces urea while on land, but when it reenters the water to breed, it quickly switches to the less costly route of excreting ammonia. So, the precise way that nitrogen wastes are handled is also dictated by the need to conserve water. We will say more about this shortly.

The Osmotic Environment

Each kind of environment presents its own osmoregulatory problems for animals. Over the eons, as the species attempted to explore new environments, they faced a multitude of water-regulation problems. The results today are varied, as we will find as we review how natural selection has resulted in widely differing adaptations to animals that live in salt water, fresh water, and on land.

The Marine Environment Animals of the sea live immersed in a complex solution of ions, including sodium, potassium, calcium, magnesium, chloride, and others. When we think of salt water we usually think of the 3.5% sodium chloride content of that solution, but we can't neglect the others. In their totality, such solutes create osmoregulatory problems for the ocean's animal inhabitants. To survive in the marine environment, organisms must adapt to the troublesome osmotic conditions. The surrounding salt water is hyperosmotic to their bodies. It contains a greater concentration of sodium, chloride, and other ions than do their body fluids. Conversely, their body fluids are hypoosmotic to the sea, that is, they contain a higher concentration of water than do the surroundings. So we see two concentration gradients occurring and thus two tendencies: The ions tend to diffuse in, and the water tends to diffuse out. If unabated, either can be life-threatening.

To survive this double threat, marine animals have evolved a variety of adaptive strategies, most of which place the animal in one of two broad categories: they are either **osmoconformers** or **osmoregulators.**

Osmoconformers As the term *osmoconformer* implies, the solute concentration within the body simply conforms, changing passively, as it were, to that of its surroundings. Thus it simply matches the solute concentration of, or is *isoosmotic* to, the seawater. Many marine animals are osmoconformers, as this method requires little energy expenditure.

As an example, the limpet *Acmaea limatula* is a resident of the intertidal zone (Figure 36.17a). As such, it commonly experiences changes in salinity as the tides change or as runoff from the land dilutes its surroundings. Rather than fighting to maintain any particular internal salinity, the limpet simply conforms to these salinity changes. Remarkably, its body fluids can remain isoosmotic in a salinity range of 1.5 to 5%. For some reason, these changes have little effect on the animal. A similar change in our body solutes would produce instant death.

While most vertebrates are osmoregulators, sharks and rays are conformers, or at least partly so. Their adaptation to the hyperosmotic marine environment is to retain sufficient urea in the blood and tissue fluids to create an isoosmotic condition with the

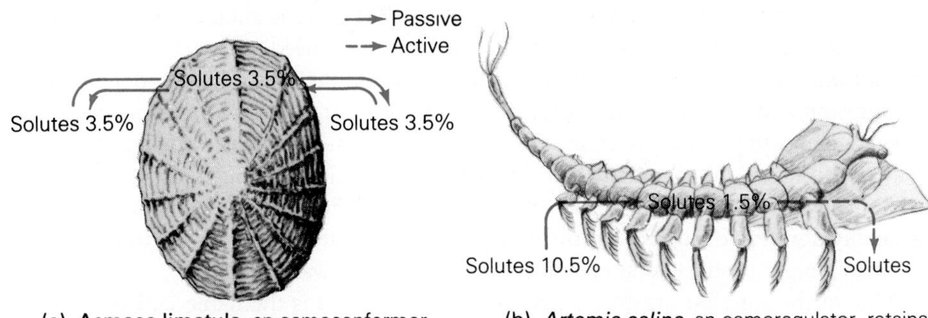

(a) **Acmaea limatula,** an osmoconformer, remains isotonic to changing salinity by allowing free diffusion of ions in and out

(b) ***Artemia salina,*** an osmoregulator, retains a stable solute balance by actively transporting Na^+ across gill membranes

FIGURE 36.17
SOLUTIONS TO THE SALINITY PROBLEM
(a) Osmoconformers simply permit body salinity to vary with the surroundings, whereas osmoregulators **(b)** maintain a specific body salinity through active transport.

seawater outside. The nitrogen waste, then, is put to use. If this seems peculiar, keep in mind that as far as osmosis is concerned, it doesn't make much difference what the solutes are as long as the total concentrations of *water* inside and outside the organism are equal. When they are unequal, there will be a net movement of water down its gradient, in or out, until an equilibrium is reached. Incidentally, sharks and rays help avoid the problem of excessive salt intake by actively transporting it back out. The site of transport is a special gland opening into the rectum.

Osmoregulators Osmoregulators maintain a relatively constant internal solute concentration despite changes in salinity around them. Osmoregulation, though, involves work—active transport. Many of the marine arthropods are osmoregulators. One of the best-known regulators is the crustacean *Artemia salina* (see Figure 36.17b), which lives in salt ponds. It can maintain constant solute concentrations in its body when environmental salinity varies from less than 1% up to 30%! Only a highly specialized and efficient osmoregulator could survive under such varying conditions. The osmoregulatory organ of *Artemia salina* is the gill, which has salt glands that actively transport salt out of the blood. Interestingly, similar structures are used by marine fishes.

Most marine vertebrates are osmoregulators. The bony fishes have the same problem as the invertebrates: they lose water and gain salt. They make up for water losses by drinking seawater, but only enough to replace what water they lose. A terrestrial mammal (such as a human) that drank seawater would subsequently require huge quantities of fresh water to remove the salt from its body. But because of the special salt-screening cells in its gills, the fish has no such problem. With salt actively transported from the body, seawater becomes perfectly suitable for drinking. The other ions are either excreted by the kidneys or passed through the gut unabsorbed.

Marine birds and reptiles also take in seawater with their food. The solute concentrations of their bodies, however, are about the same as those of terrestrial species. So what do they do with all those salts? Unlike their terrestrial cousins, they do not rely on the kidney; instead, they actively excrete salt through special glands located near the eye that drain through a duct into the nose. Their kidneys, in fact, are not very efficient salt removers. Studies of gulls reveal that whereas the salt concentration of the excretory waste is well below 1%, the fluid exuded by their salt glands is about 5% salt.

Unlike marine fishes and birds, marine mammals have highly efficient kidneys and can actively transport ions into their urinary collecting ducts. For them, then, the kidney is the salt-excreting gland. However, most marine mammals avoid drinking seawater and rely on relatively low-solute fluid from the body fluids of the fish they eat.

The Freshwater Environment Just as salt water presents osmoregulatory problems, fresh water, too, has its share of hazards. Fresh water is hypoosmotic to its inhabitants, which tend to take in the fresh water through osmosis. If the organism isn't able to keep excess water out, its cells may swell and rupture. But there is another problem. Fresh water doesn't provide its denizens with the ions they need. In particular, sodium, potassium, and magnesium are in short supply. Thus, the ions tend to obey the laws of diffusion and leak out of the organisms into the environment. The organisms must expend energy to pump these errant molecules back in by active transport. Such transport may occur across the membranes of the skin, gills, or kidney tubules.

A freshwater existence also has certain advantages. For example, the animals there generally don't have much of a problem with handling nitrogenous wastes. In fact, many of them simply produce ammonia and flush it out with water.

One of the simplest organized osmoregulatory systems is the **flame cell system**, or **protonephridium**, of the planarians, freshwater flatworms. This system is chiefly an osmoregulatory device and possibly an excretory system as well (Figure 36.18). (Interestingly, protonephridia are absent in marine flatworms, which do not have the problem of excess water.) The flame cell system consists of numerous minute blind sacs leading into a complex network of tubules that empty to the outside via excretory pores. Water from intercellular fluids is transported into the flame cells by pinocytosis ("cell drinking"). Each flame cell contains numerous cilia facing the tubular lumen of its **flame bulb**, and the cilia set up currents that carry the fluids through the tubules to the pores.

The kidneys of freshwater fishes, as you might expect, are not adapted for conserving water, and they produce copious amounts of highly dilute urine. Their kidneys are, however, extremely efficient at retaining salts, which would otherwise be quickly depleted. As we will see in the next section, in the filtering action of the kidneys, the first filtrate that leaves the kidneys contains many valuable materials that must be recovered before the urine is finally formed. But despite its efficiency, the kidney cannot prevent some loss of salts. To counter this loss, the freshwater fish must transport salt into its body against its gradient, using active transport. Interestingly, once again the active transport of salt takes place in the gill, but in the opposite direction of what we saw in marine fishes.

The excretory system of amphibians is also rather highly developed. In frogs, the tadpole produces and excretes its nitrogen waste in the form of ammonia, much the way a fish does. This is possible because of its watery habitat and because most kinds of tadpoles exist on vegetarian diets light on proteins and heavy on carbohydrates. As mentioned earlier, adult frogs shift to water-conserving urea as the principal nitrogen waste, but even so, most frogs drink and excrete huge amounts of water—up to 30%

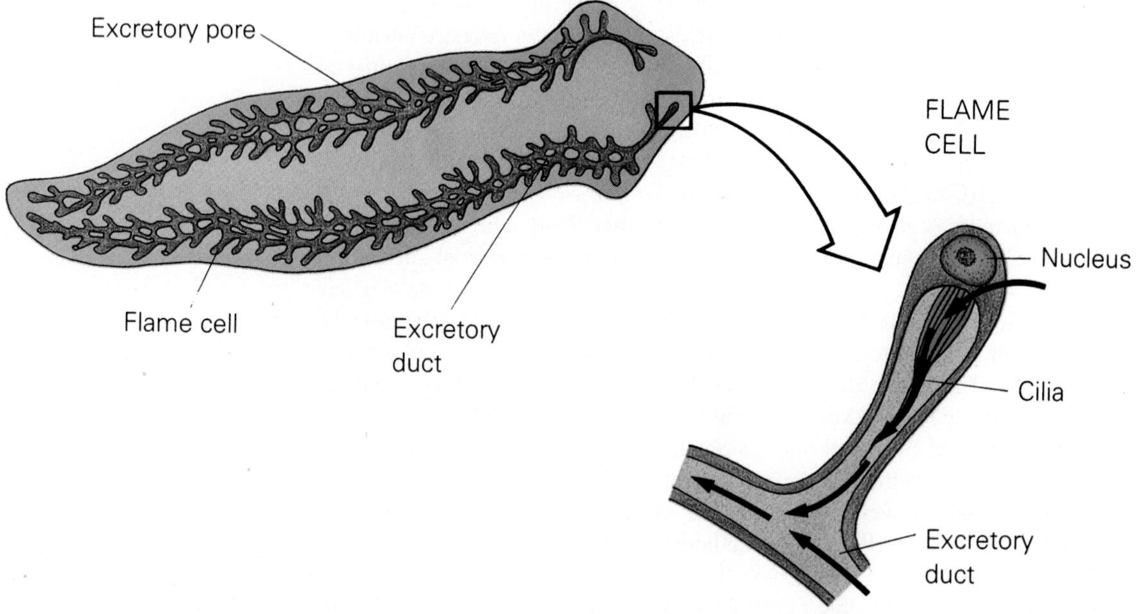

FIGURE 36.18
FLAME CELLS
Freshwater planarians have an extensive osmoregulatory system composed of protonephridia—ciliated flame cells that expel fluids into tubules leading out of the body.

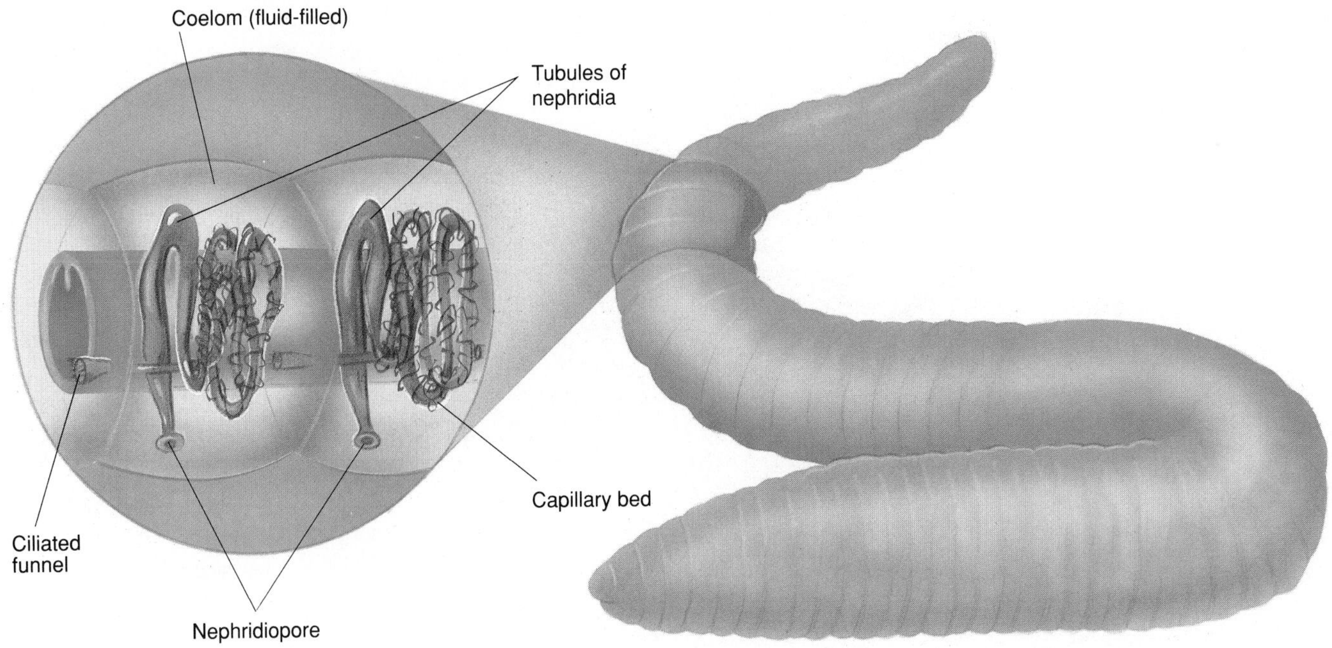

FIGURE 36.19
NEPHRIDIA
Paired nephridia occur in most segments of the earthworm. Their function is to excrete wastes and excess water. As the coelomic fluids pass through a nephridium, water, ions, and other valuable materials are returned to the blood.

of their body weight daily. Like fishes, amphibians tend to lose body salts in the urine. Also like fishes, they actively transport salts from their water surroundings back into the body, but the site of the active transport in amphibians is the skin.

The Terrestrial Environment In one important way, the terrestrial environment presents a problem similar to that of the oceans: how to avoid water loss. Terrestrial animals solve their water-retention problem through high water intake, highly efficient water-conserving excretory systems, watertight skin, and behavioral patterns that help deal with the problem of dessication. We will first consider a few land-dwelling invertebrates.

Invertebrates of the Terrestrial Environment Though the earthworm is technically terrestrial, it lives in a perpetually moist environment. Nonetheless the earthworm is a water conserver. The fluids in each segment are constantly filtered and recycled by rather complex structures known as **nephridia.** (In some ways, the nephridium is similar to the nephron, the filtering structure of the vertebrate kidney, which we will consider shortly.) As you can see in Figure 36.19, each paired nephridium cleans the fluid in the segment just ahead of it. Any fluid moving through the ciliated funnel must pass a rich supply of capillaries in the tubule wall, where materials can be exchanged between the fluids and the blood. Water, minerals, and other essential materials are reabsorbed into the blood, while dissolved wastes pass to the outside through an opening known as the **nephridiopore.** Here, then, osmoregulation and excretion are accomplished by the same organ, just as in the vertebrates. But before leaving the invertebrates, let's consider some specializations in that enormous group called arthropods.

The arthropods of dry land have developed some very distinct ways of solving their osmoregulatory and excretory problems while reducing water loss. Their adaptations involve a water-resistant, waxy cuticle, a well-protected respiratory surface, and the elimination of nitrogen wastes in the form of semisolid uric acid.

The excretory system in insects consists of many blind, hollow, tubular structures known as **Malpighian tubules.** They emerge at about where the mid- and hindgut join (Figure 36.20). The tubules extend into the coelomic fluids and here absorb body wastes. Inside the tubules, nitrogen wastes from the coelomic fluids are converted to insoluble uric acid crystals, which pass through the lumen of the Malpighian tubules directly into the gut. Once in the gut, the uric acid joins the digestive wastes, passing to the outside at defecation. This method of excretion conserves water in two ways. First, the nitrogen

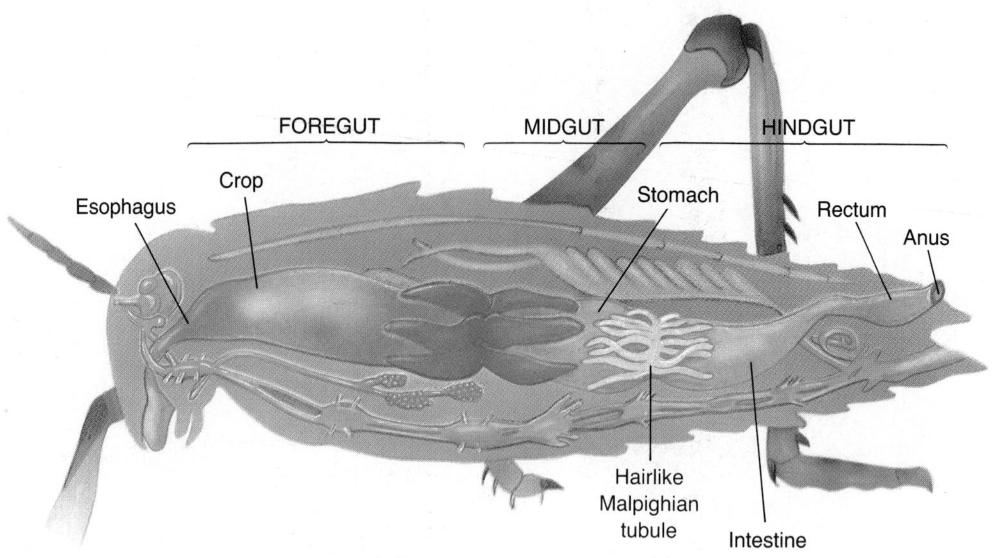

FIGURE 36.20
MALPIGHIAN TUBULES
In insects, Malpighian tubules collect nitrogen wastes from the coelomic fluids (blood), before emptying into the gut. Water is reabsorbed from the fecal wastes as they form.

waste is a solid that doesn't require water for dilution. Second, fluids from the Malpighian tubules can be reabsorbed by cells in the lining of the hindgut. Thus the insects produce a dense, fairly dry digestive and excretory waste. The disadvantage of the uric acid route is the metabolic expense, requiring energy and a complex pathway.

Vertebrates of the Terrestrial Environment Vertebrates that roam the land must constantly replace water that is lost during breathing and during the removal of nitrogen wastes; thus their movements are restricted by the availability of water. Aside from an essentially waterproof skin, their most important means of conserving water is a specialized kidney. Two vertebrate groups, the reptiles and the birds, conserve water through the production of uric acid, which can be eliminated in a semisolid state. Uric acid production is also an adaptation to development with an egg, since it can be safely isolated from the embryo. (Imagine the problem in the bird or reptile egg if ammonia collected.) The primary nitrogen waste in mammals, however, is urea (with some uric acid produced through the breakdown of purines), which requires a constant supply of water for its elimination. This is why a water-conserving kidney is so important in mammals. The human kidney serves as an example of this efficient organ.

THE HUMAN EXCRETORY SYSTEM

Whereas humans in developed nations are very wasteful of water, their kidneys are not. (Consider that in the home, Americans generally use about three gallons of highly purified water to flush away a few ounces of urine.) Like other mammals, humans have centralized, complex, and efficient kidneys, whose primary roles are the excretion of the nitrogen waste urea and the conservation and control of the body's water and mineral content. Let's first concentrate on the structure of the kidneys themselves and then discuss how they work and how their function is controlled.

Anatomy of the Human Excretory System

The human **excretory system** (Figure 36.21) includes the **kidneys, ureters, urinary bladder, urethra,** and the blood vessels of the **renal circuit** (see Chapter 39). The renal circuit includes the paired **renal arteries**, along with a great number of special branches, and the paired **renal veins.** Blood entering the renal arteries comes directly from the aorta—the body's largest artery; thus it is under considerable pressure. High pressure is essential to the kidney's operation, and should blood pressure fall drastically, the kidney's vital functions will begin to fail.

Figure 36.22 illustrates the major features of the human kidney. Its anatomical regions include the **cortex**, the **medulla**, and the **renal pelvis.** The dense cortex consists of the filtering units of the kidney—the **nephrons** and their related blood vessels. Each nephron consists of a spherical capsule and a long tubule. The tubules form long loops that venture down into the medulla and return to the cortex. These in turn are connected to numerous **collecting ducts** that join to form the dense, fan-shaped **pyramids.** The pyramids narrow down, leading into funnel-shaped **calyces** (singular, **calyx**) that empty into the renal pelvis.

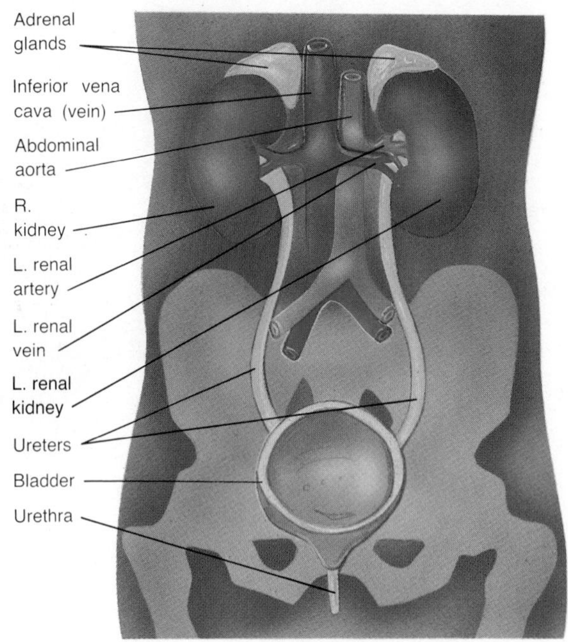

FIGURE 36.21
THE HUMAN EXCRETORY SYSTEM
The human excretory system includes paired kidneys, their blood vessels, the ureters, urinary bladder, and urethra.

The Urine-Forming Pathway Briefly, the urine-forming pathway begins when the nephrons receive materials that have been filtered out of the blood, and from this filtrate the urine forms. Once formed, the urine passes into the collecting ducts to empty into the renal pelvis. This is the first place where pools of urine collect. Urine passes from there to the bladder via the ureters. From the bladder, urine is voided from the body through the urethra.

Microanatomy of the Nephron

The nephrons, which number about 1 million per kidney, are the functional or filtering units. Their structure reflects their function, as is so often the case in life (see Figure 36.22c). The nephron begins with **Bowman's capsule**, a hollow bulb or cup surrounding a ball or tuft of capillaries known as a **glomerulus.** Extending from the capsule is a lengthy tubule with a peculiar hairpin loop in its midsection. We will return to the loop, but first let's take a more detailed look at the glomerulus and the other blood vessels of the nephron.

The glomerular capillaries arise from the **afferent** (incoming) **arterioles**—tiny branches of the renal artery. As the blood finally emerges from the glomerulus, it enters the **efferent** (outgoing) **arteriole.** The efferent arteriole branches into a second capillary network, the **peritubular capillaries,** which form a fine network over the entire nephron. The peritubular capillaries coalesce to form a venule, which is subsequently joined by venules from the other nephrons. These merge to form the renal vein. This peculiar routing of the blood vessels into two capillary beds is essential to glomerular filtration and tubular reabsorption, major processes in the nephron.

Bowman's capsule leads into a lengthy tubule that makes up most of the nephron. The first region, the **proximal convoluted tubule**, winds a meandering route and then forms the long hairpin bend called the **loop of Henle.** This loop is found in all water-conserving kidneys and is extremely prominent in the kidneys of desert-dwelling mammals for reasons we hope to make clear. After forming the loop of Henle, the tubule again begins to twist and contort into another convoluted section, this time called the **distal convoluted tubule.** It then joins a collecting duct, which also receives tubules from a number of other nephrons.

Note in Figure 36.22d that the afferent arteriole forms a passing connection with the distal convoluted tubule. This is the **juxtaglomerular complex**, an association that is vital to the kidney's role in sodium reabsorption.

In summary, the elements of the nephron proper, in the order of urine flow, are (1) Bowman's capsule, surrounding the glomerulus, (2) the proximal convoluted tubule, (3) the descending limb of the loop of Henle, (4) the ascending limb of the loop of Henle, and (5) the distal convoluted tubule, which leads to a collecting duct.

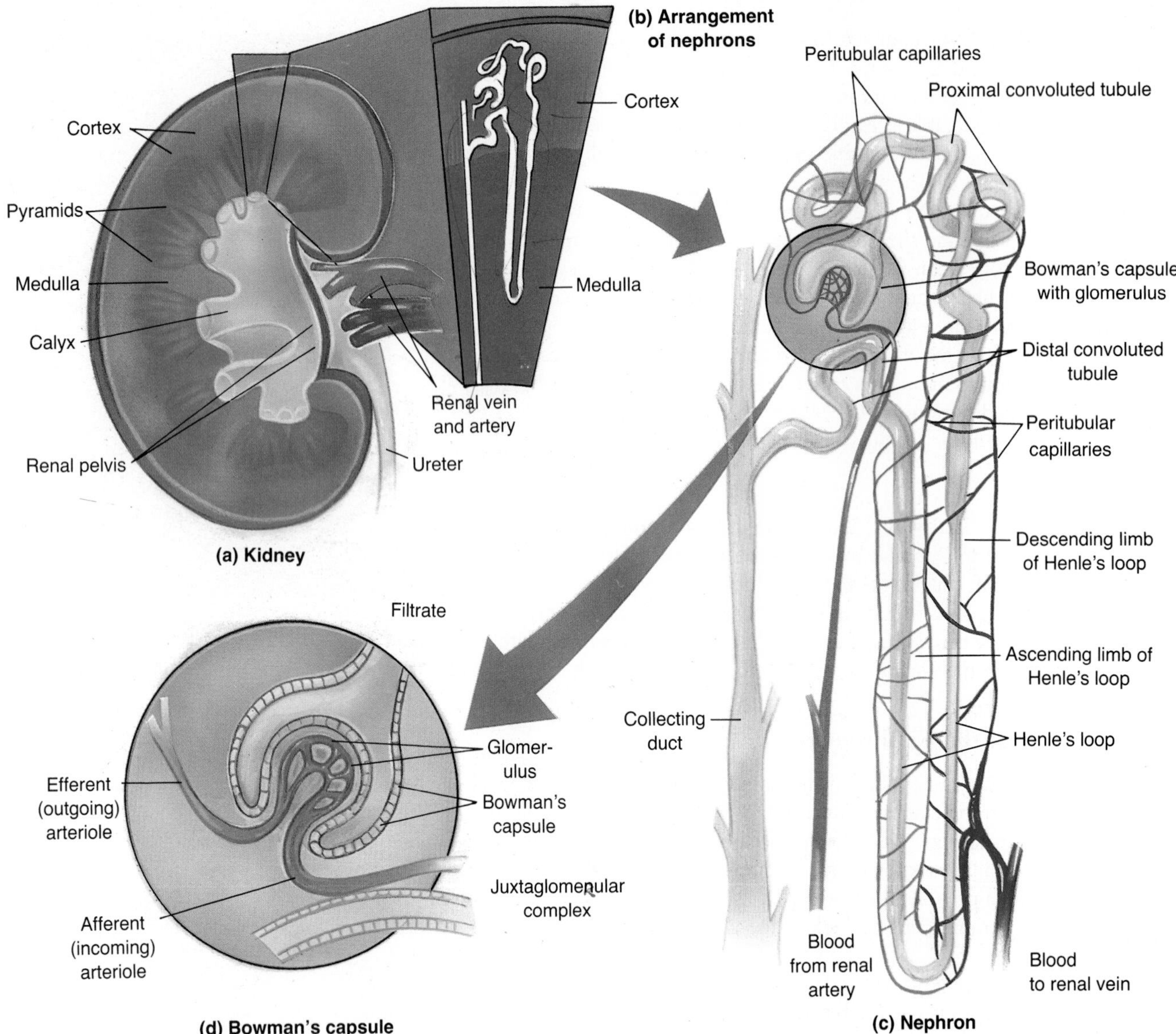

FIGURE 36.22
ANATOMY OF THE KIDNEY
(a) The kidney contains an outer cortex, an inner medulla that contains the pyramids, several large funnellike calyces, and a final collecting region known as the renal pelvis. (b) Note the relationship between the cortex and medulla of the kidney and the loop of Henle. (c) The functional units of the kidney are the nephrons. Each includes the following: Bowman's capsule, proximal convoluted tubule, loop of Henle, and distal convoluted tubule. Each nephron is joined to a nearby collecting duct. Blood enters the nephron at the Bowman's capsule, where an afferent arteriole has branched into a mass of smaller vessels that comprise the glomerulus. Emerging from the glomerulus is an efferent arteriole that immediately branches to form the extensive peritubular capillary network over the entire nephron. (d) The juxtaglomerular region of the nephron is essential to the control of sodium reclamation.

The Work of the Nephron

The critical nature of the nephron's role is underscored by its use of several transport processes. For example, we will encounter force filtration, active transport, endocytosis, exocytosis, diffusion, and osmosis. We will also see a special countercurrent exchange mechanism at work, one that is, in principle, quite similar to the countercurrent heat exchangers discussed earlier. These forces are summarized in Table 36.1 and Figure 36.23. To see how all of this comes together in excretion, let's begin at Bowman's capsule with force filtration.

TABLE 36.1 THE NEPHRON AT WORK

REGION	PROCESS	SUBSTANCES IN TRANSIT
Bowman's capsule	Force filtration from blood	Water, ions, glucose, urea, amino acids
Proximal convoluted tubule	Active transport out	Glucose, amino acids, Na^+
	Diffusion out	Cl^-, water
Descending loop of Henle	Diffusion out	Water
	Diffusion in	Na^+ and Cl^-
	Countercurrent concentration of salt and urea in kidney medulla	
Ascending loop of Henle	Active transport out	Cl^-
	Diffusion out	Na^+, K^+
	Diffusion in	Urea
Distal convoluted tubule	Active transport out	Na^+
	Diffusion out	Cl^- and water
	Secretion into tubule	H^+, NH_3, K^+ (varies with aldosterone)
Collecting duct	Diffusion out of duct	Water (varies with ADH), K^+, urea (joins salt in kidney medulla)
	Secretion into duct	NH_3, H^+, K^+ (K^+ varies with aldosterone)

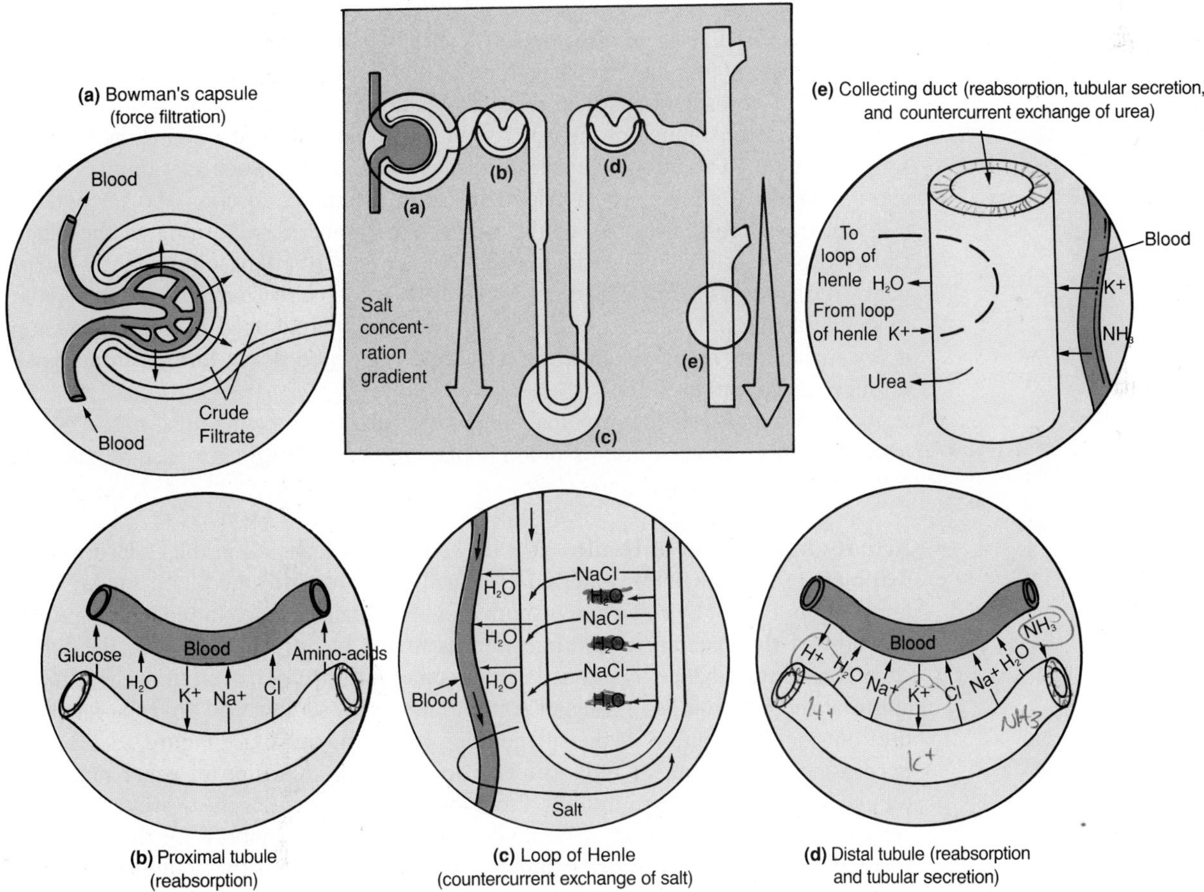

FIGURE 36.23
TRANSPORT PROCESSES IN THE NEPHRON
Each region of the nephron makes use of its unique anatomy and various transport processes to carry out its function. As a result wastes, excess water and salts end up in the urine. Through selected recovery mechanisms, valuable blood constituents are recovered, and the body's water and ion balances are maintained. Note the steep solute gradient produced around the loop of Henle, a product of the countercurrent flow of salt and urea.

Bowman's Capsule Recall that afferent arterioles emerging from the renal artery enter Bowman's capsule, where they form a mazelike tuft of capillaries called the glomerulus (Figure 36.23a). Arterial blood is pushed into the glomerulus with considerable force and is thus under great hydrostatic pressure. About one fifth of its volume is forced through the capillary walls (the process being enhanced by the presence of small pores in the capillary walls) and into the cavity of Bowman's capsule. This filtering process is called **force filtration,** a truly descriptive term.

Force filtration is nonselective, and only the larger molecules of the blood escape its effects. Substances force-filtered from the blood include most ions and smaller molecules, the latter including glucose, amino acids, and urea. Of the blood plasma, only the large protein constituent (molecular weight above 30,000) escapes force filtration. Since many of the substances in this filtrate are extremely valuable, this may seem to be an extreme measure for getting rid of urea and a small amount of excess water and salt. It is hard to account for the evolution of such a gross process, except to remind ourselves that in nature, what works, works.

We might point out that while the crude filtrate entering the glomerulus is considerable, not much of this ends up as urine. For instance, while the adult kidneys on an average day (depending on fluid intake) filter about 180 liters of fluid from the blood, usually no more than 1.2 liters of urine are formed. Thus more than 99% of the water filtered out of the blood (and large quantities of salt) is returned to the blood before urine leaves the collecting ducts.

Reabsorption in the Proximal Convoluted Tubule The filtrate, in its primary, highly dilute state, enters the nephron, passing into the proximal convoluted tubule (Figure 36.23b), and some rather viscid, concentrated, hyperosmotic blood is carried off by the efferent arteriole that leads from the glomerulus. This hyperosmotic condition results in a steep osmotic gradient that encourages the reentry of water from the dilute primary urine; thus, much of the water from the proximal tubule reenters the blood by simple osmosis. In addition, valuable substances in the filtrate, such as sodium, glucose, and amino acids, are actively transported out of the tubule to enter the surrounding cells and then cross the walls of the peritubular capillaries, returning to the blood. Chloride ions passively follow sodium ions. The active transport of sodium ions is carried out by the familiar sodium-potassium ion exchange pump (see Chapter 5). So in addition to sodium being pumped out of the filtrate, potassium is pumped in. Since potassium is a valuable ion, much of it will be reclaimed before the newly formed urine passes through the collecting ducts.

As the filtrate leaves the proximal convoluted tubule it contains urea, miscellaneous toxic substances (small amounts of ammonia, creatine, and uric acid), some salt, and, still, much of the original water.

Activity in the Loop of Henle It is in the loop of Henle and in the collecting ducts that most of the remaining water and salt will be returned to the blood. As far as we know, water cannot be actively transported by membranal pumps in biological systems, yet most of the water in the filtrate returns to the blood. This is where the special countercurrent exchange mentioned earlier comes into play. The countercurrent mechanism of the nephron is a means of concentrating salt (sodium and chloride ions) and urea outside the nephron in the surrounding kidney medulla (see Figure 36.23c). The hyperosmotic environment thus formed encourages the diffusion of water out of the tubule into these salty surroundings. From there, the water enters the nearby peritubular capillaries, where the blood is in an even greater hyperosmotic state than the salty surroundings the water had just entered. The important point is that while water cannot be actively transported, the active transport of salt through a countercurrent mechanism provides an osmotic environment that brings about the passive movement of water in the necessary direction. How, then, is the countercurrent mechanism set up?

Anatomically, the descending limb of the loop of Henle, the ascending limb of the loop, and the collecting tubules form a curious S-shape, with all three portions lying more or less side-by-side (see Figure 36.22c and 36.23c). Bowman's capsules and the convoluted tubules lie in the cortex (outer part) of the kidney, while the loops and the collecting tubules extend into the medulla (inner part). The tissue fluids of the medulla are relatively salty (hyperosmotic), so as the dilute urine flows down the descending limb of the loop, it loses water and gains salt, both by passive diffusion. At the bottom of the loop, the urine, still carrying with it the unwanted wastes, has lost much of its water, but it has picked up a lot of salt, and salt must be conserved.

In the ascending limb of the loop of Henle—the portion headed back toward the cortex—salt is actively transported from the nephron into the surrounding tissue (see Figure 36.23c). (Actually, only chloride (Cl^-) is pumped out, but sodium (Na^+) and potassium (K^+) ions follow the negative chloride ions.) This active transport both enables the blood to recapture the salt and maintains a high salt concentration in the kidney medulla, thus encouraging the movement of water out of the nephron. The water and ions then enter the nearby peritubular capillaries.

Reabsorption in the Distal Convoluted Tubule and Collecting Ducts In the distal convoluted tubule, as in the proximal tubule, sodium ions are actively transported out, followed by chloride ions, adding still more solute to the surroundings (Figure 36.23d). As before, water follows the ion gradient. Following the loss of salt and water, filtrate in the tubule tends to become isoosmotic with blood in the nearby peritubular capillaries, so there is no further tendency for water to move in or out. But the newly formed dilute urine next flows through the collecting ducts (Figure 36.23e) and has to traverse the salty medulla one more time. As it again enters regions with an increasing outward osmotic gradient, the urine may lose yet more water. We say "may" since, as we will see, a special hormonally regulated mechanism is at work in the collecting ducts. In addition, urea itself diffuses out of the collecting duct and into the medulla, further increasing the hyperosmotic state therein. Urea also diffuses back into the hairpin loop of Henle at its bottom region, thus setting up its own countercurrent mechanism with this substance.

Tubular Secretion While reabsorption, involving both active and passive transport, is responsible for reclaiming valuable materials and getting them back into the peritubular capillaries, another process, **tubular secretion,** also occurs (Figure 36.23b, d, and e). But unlike reabsorption, tubular secretion transports substances *out of* the peritubular capillaries and *into* the collecting duct and distal convoluted tubule. Not all of the blood's water (with its low molecular weight solutes) is forced into Bowman's capsule during filtration. Certain molecules and ions left in the blood after it leaves Bowman's capsule, such as ammonia, hydrogen ions, potassium ions, organic acids, and creatine, are actively transported from the capillaries to the nephron. Tubular secretion, then, is a last mechanism for getting rid of additional wastes. Table 36.1 reviews the structures of the nephron, the forces at work in each structure, the contents of the nephron at key points, and the contents of the peritubular capillaries.

Control of Nephron Function

The hypothalamus, in addition to its many other homeostatic functions, measures the osmotic pressure, or water content, of the blood passing through its capillaries. If the blood is hypoosmotic (has excess water), the kidney can begin to release more water as urine. If the blood registers as hyperosmotic (meaning the body is beginning to run short of water), the kidney is put on a water-rationing regime. In such cases, neurons of the hypothalamus secrete **antidiuretic hormone**, or **ADH**, into capillaries of the posterior lobe of the pituitary, and the hormone is then released into the blood. Its targets are the epithelial cells of the collecting ducts (Figure 36.24a). As a clue to ADH's function, consider that a **diuretic** is an agent that increases urine flow; thus an **antidiuretic** does the opposite.

ADH increases the permeability of these cells to water. Thus, more water leaves the

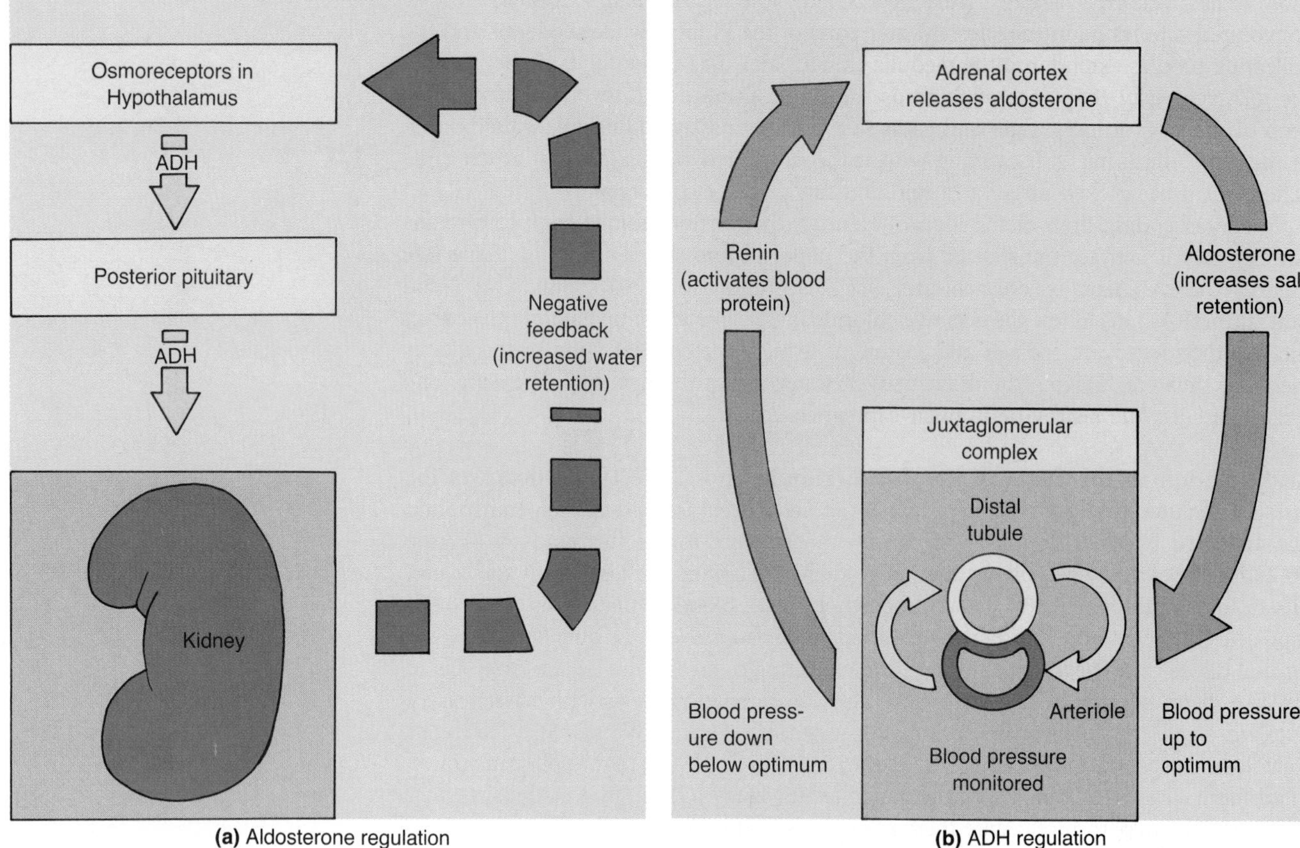

(a) Aldosterone regulation

(b) ADH regulation

FIGURE 36.24
HORMONAL CONTROL OF OSMOREGULATION
(a) Water reabsorption is influenced in part by the hypothalamus via the posterior pituitary and (b) in part by fine blood pressure adjustments made through an interplay between the juxtaglomerular complex and adrenal cortex.

collecting duct to reenter the blood. When the osmotic consistency of the blood is restored to a normal range, the stimulation of the hypothalamus slows and ADH secretion falls off, producing an increase in urine volume. The stimulation of the hypothalamus also creates the sensation of thirst. As the individual drinks, the blood becomes diluted, and again, the hypothalamus stops sending out antidiuretic hormone and the kidney increases its output of fluid. Eventually, another sensation is produced, this time from the stretch receptors of the bladder. And a new behavior is initiated.

Another hormonal mechanism operates to control the retention of sodium chloride. In this case the hormone is the steroid **aldosterone** (see Chapter 37), that is released by the cortex of the adrenal gland. The link between sodium chloride retention and aldosterone release is indirect and not entirely understood.

Cells of the juxtaglomerular complex of the nephron's distal tubule (see Figure 36.22d) monitor minute changes in blood pressure in the nearby afferent vessel (the one carrying blood into the glomerulus). If blood pressure is below optimum, the complex stimulates the release of the enzyme **renin** into blood passing through the afferent vessel. Renin has no effect on the nephron, but it does activate a blood protein that, on reaching the adrenal cortex, stimulates aldosterone secretion. Aldosterone's target cells are in the nephron's distal convoluted tubule and the collecting ducts. The target cells respond by speeding up their transport of sodium ions out of the tubule and back to the blood. As expected, water follows, and the additional water increases blood pressure enough to form a negative feedback loop back to the juxtaglomerular complex.

This mechanism has two important aspects. First, it ensures that the blood pressure in the kidney itself is great enough to maintain an efficient force filtration phase. Second, the aldosterone-related increase in sodium transport also means an increase in the excretion of potassium ions, mentioned earlier. The active transport mechanism involved is identical to the familiar sodium-potassium ion exchange pump of blood and nerve cells. We will learn in the next chapter about another equally complex mechanism of blood pressure control, this one involving the heart itself and a hormone called ANF (atrial natriuretic factor). ANF is released by cells of the atria in response to changes in pressure exerted against the atrial walls.

We should add that the excretory system is also involved in other homeostatic mechanisms—for instance, in the control of blood acidity. The pH of our blood can change, depending on what we eat, but it is generally kept within fairly narrow limits. Since the acidity of urine can vary tremendously, from pH 4 to 9, excess hydrogen and hydroxide ions in the blood, like other toxic substances, are eliminated by secretion through the kidneys. Aldosterone, by the way, also speeds the movement of hydrogen ions into the urine.

In this chapter we have seen only a few of the many delicate, interacting, and highly coordinated mechanisms that keep the body's internal environment within the extremely precise limits critical to life. Remember, we live in what is essentially a disruptive environment. To remain organized in the face of potential disruption requires an ongoing, uphill battle against entropy-increasing forces. That battle is best fought under optimal physiological conditions, and the body's homeostatic mechanisms help ensure those conditions.

countercurrent heat exchanger A circulatory arrangement in which warm blood leaving the body core passes cooler blood returning from the extremities in closely paralleled vessels, thus providing the opportunity for heat to pass into the cooler blood all along the vessels.

ectotherm An animal that regulates its body temperature through external sources of heat, generally through basking and other behavioral strategies.

endotherm An animal that regulates its body temperature by varying thermiogenesis and employing cooling mechanisms.

excretion The removal from the body of metabolic wastes such as ammonia, urea, and others.

excretory system The body system responsible for the elimination of metabolic wastes such as ammonia and urea.

heterotherm An animal that regulates its body temperature part of the time only.

homeostasis The complex process whereby a steady physiological state is maintained in the face of constantly changing conditions.

homeotherm An animal that regulates its body temperature around some set point ("warm-blooded").

negative feedback loop An automated, circularized physiological control mechanism in which a stimulus initiates an action, and that action, directly or indirectly, *reduces* the stimulus.

osmoregulation The regulation of ions and water resulting in a state of constancy for those materials in the body.

poikilotherm An animal that cannot regulate its body temperature ("cold-blooded").

positive feedback loop A circularized, automated control mechanism in which a stimulus initiates an action, and that action, directly or indirectly, *increases* the stimulus.

thermiogenesis The generation of body heat through an acceleration of the metabolic rate, for example by increasing the rate of cell respiration.

thermoregulation The maintainance of body temperature at a constant level through behavioral or physiological means.

APPLICATION OF IDEAS

1. Many of the animal body regulatory functions are automated, occurring with little, if any, conscious intervention. What are some adaptive advantages to this kind of control? What is the alternative, and how might it affect the general efficiency of an animal's metabolic activities?
2. Homeostatic mechanisms involve far more than thermoregulation and osmoregulation. Describe examples of homeostasis in other systems, such as digestive, respiratory, and circulatory, and in the action of skeletal muscle.
3. Once the problem of cross-matching tissue types is solved, transplanting a kidney is a comparatively simple and straightforward procedure. Knowing what you do about kidney function, explain this statement. Exactly what is required for a kidney to function, and why are its needs so simple?
4. Is thermoregulation in an endotherm more likely to fail under conditions of extreme heat or extreme cold? Explain fully.

KEY IDEAS

HOMEOSTASIS

1. **Homeostasis** is defined as the tendency for physiological systems to maintain internal stability through the coordinated response of its parts to anything tending to disturb such stability.
2. All of the body's organ systems contribute to the homeostatic state.
3. Homeostasis can operate through short term, immediate responses or through long term, cyclic responses.

Negative and Positive Feedback Loops

1. Many homeostatic mechanisms are regulated by **negative feedback loops,** whereby a stimulus creates a response that in turn alters or removes the stimulus, thus lessening or stopping the response.
2. **Positive feedback loops** also occur, but they are often a sign of physiological trouble. In this case a stimulus creates a response that in turn intensifies the stimulus, thus intensifying the response, and so on.
3. Two homeostatic mechanisms in animals are **thermoregulation** and **osmoregulation.** The latter is often involved with another function, **excretion,** which occurs in the **excretory system.**

THERMOREGULATION

Thermoregulation is the ability of an animal to increase or decrease its body temperature.

Why Thermoregulate?

1. Through thermoregulation, animals maintain metabolic activity at low temperatures, avoid ice formation in cells, and avoid the denaturing effects of high temperatures.
2. Animals prevent excessive heating and cooling through behavioral and physiological strategies.

Categories of Thermoregulation

1. Whereas the **core body temperature** of **homeotherms** is relatively constant, that of **poikilotherms** varies with external temperatures. Probably few pure poikilotherms exist. **Heterotherms** are part-time thermoregulators.
2. Endotherms maintain a constant body temperature through **thermiogenesis**—generating metabolic heat.
3. Ectotherms utilize external heat, making use of behavioral strategies for both heating and cooling the body.

Endothermic Regulation

1. The body temperature of endotherms varies in its, the core temperature being more constant. Some endotherms experience falling core temperatures at hibernation.
2. Endotherms release more heat energy and thus require greater food intake. They have a low toleration for core temperature changes.
3. **Endothermy** places limitations on smallness, since small animals have proportionally larger surfaces from which heat can escape.
4. Cold-adapted noctuid moths are heterotherms, remaining at ambient temperatures until flight is required. They warm the thoracic flight muscles through shivering thermiogenesis (rapid but nonproductive muscle contraction). Heat escape from the thorax is prevented by insulation and through **countercurrent heat exchangers.** The flow of warm blood from the thorax is opposed by cooler blood entering; thus, heat passes to the cooler blood. The arrangement produces overheating problems in warm periods.
5. The bluefin tuna retains a warm core temperature around its swimming muscles through the use of two countercurrent heat exchangers, one involving the general circulation and the other, a massive exchanger called the **rete mirabile.** Through its retention of heat, the tuna remains active and fast-moving in very cold water.
6. Birds and mammals maintain a constant core temperature, with strategies for both warming and cooling the body.
7. In general, the **metabolic rate** increases as body size decreases, but size alone is not thought to be the determining factor (the cause and effect relationship is complex). Some small mammals and birds enter a metabolic slump when feeding is interrupted.
8. Thermiogenesis in birds and mammals includes oxidative respiration, shivering, and brown fat metabolism. In the latter, oxidative respiration is hormonally induced, but instead of ATP being generated, the energy from oxidized fuels is converted to heat. Brown fat thermiogenesis is important to thermoregulation in human infants and, in some mammals, to recovery from low hibernation temperatures.
9. Birds and mammals resist heat loss by shunting blood away from the skin and extremities and through the use of countercurrent heat exchangers. The latter occur in our own arms and legs, wherein returning blood has two alternative routes: along the skin (for heat loss) or alongside deep arteries (for heat retention through the countercurrent exchange).
10. Arteries and veins in the legs of wading birds form countercurrent exchangers that retain body heat in core regions.
11. Many mammals, particularly marine species, make use of thick layers of **blubber** to provide insulation from frigid surroundings. Whereas the restriction of blood from the blubber aids in retaining core heat, shunting blood into the blubber aids in body cooling.
12. Hair and feathers provide insulation that can be improved by fluffing. Arctic seals are adapted to cold through the presence of blubber, insulating **underhair**, and highly waterproof **guard hair.**

Virtually no heat escapes. Some hairs contain fibers that conduct ultraviolet light to the skin, where it converts to heat.

13. Shunting blood toward the skin aids in cooling the body through radiation, conduction, and convection. Evaporative cooling, involving the use of external water or water from sweat glands is an efficient means of speeding heat loss. Panting also speeds heat loss. Birds make use of both panting and **gular flutter.** Fluid intake must balance fluid loss.
14. Behavioral strategies become most important when physiological mechanisms of thermoregulation have reached their limits. High humidity thwarts evaporative cooling mechanisms, and cessation of activity is then essential.
15. The body also regulates metabolically—increasing or decreasing heat output by varying the rate of cell respiration.
16. Temperature changes are detected by sensory neurons, and reactions may include increased heat production, shivering, and/or behavioral responses.
17. Heat measurements are also made by the hypothalamus of the brain. In response it may
    a. stimulate the adrenal medulla to release the hormone epinephrine, which prompts the liver to release glucose for increased cell respiration and heat output;
    b. prompt the pituitary to release thyroid-stimulating hormone, which in turn causes the thyroid gland to release the metabolism-elevating hormone, thyroxin.
18. The hypothalamal thermostat may measure calcium levels rather than heat. Suffusing the brain with calcium ions for cryogenic surgery will cause a reduction in metabolic activity and a cooling of the body.

Ectothermic Regulation

1. **Ectothermy** has the advantage of low energy (food) costs. A clear disadvantage is the failure of behavioral strategies when outside energy is not available.
2. The chief danger from cold is the formation of ice crystals within cells. Some ectotherms utilize mechanisms that accelerate the freezing of extracellular fluids, thereby creating a hyperosmotic extracellular environment. Water diffuses out of the cells, and dehydration prevents cell freezing. Some cold-adapted fishes have antifreeze substances in their body fluids.
3. Animals actively seek out areas where temperatures are optimal. Reptiles, generally considered ectotherms, use basking behaviors, in which they expose their bodies maximally to warm up and minimally to cool down. Color changes through pigment migration in some species help in either the absorption or reflection of sunlight. Physiological mechanisms of thermoregulation in reptiles are suspected.

OSMOREGULATION AND EXCRETION

Osmoregulation is the ability of animals to regulate ions and water in the body. Osmoregulation is closely related to excretion, the removal of metabolic wastes.

Producing Nitrogen Wastes

1. During **deamination,** the amine groups are removed from amino acids as toxic ammonia, NH_3, which is excreted as is by some aquatic animals.
2. In insects, reptiles, and birds, the primary excretory waste is **uric acid,** while for many other invertebrates and the fishes, amphibians, and mammals, the primary excretory waste is **urea.**

The Osmotic Environment

1. The marine environment is a hyperosmotic medium in which organisms tend to lose water and gain ions. The tissues of **osmoconformers** conform to the surroundings, becoming isoosmotic. **Osmoregulators** have mechanisms for actively removing ions.
    a. Osmoconformers include the limpet, *Acamea,* along with sharks and rays and the coelocanth. The latter three maintain their isoosmotic condition by retaining urea in the blood and body fluids. Sharks and rays also osmoregulate by actively pumping salts out of their bodies.
    b. Osmoregulators such as the crustacean *Artemia* and the marine bony fishes actively secrete salts out through the gill. Such secretion permits the bony fish to drink seawater. The nitrogen waste of bony marine fishes is ammonia, which is excreted across the gill. Marine birds and reptiles secrete salts from glands located near the eye, while marine mammals use the kidney for this purpose.
2. The freshwater environment is hypotonic, so water tends to enter the body through osmosis. Conversely, ions tend to diffuse out of the body. Since water conservation is not necessary, nitrogen wastes can be readily flushed out with water.
    a. Some freshwater invertebrates use the **flame cell system** or **protonephridium.** Excess fluids are transported into the **flame bulb** by pinocytosis, and the waving cilia push the fluids through the tubules and out through pores in the body wall.
    b. The kidney of the freshwater fish recovers some ions, but some must also be actively transported in through gill structures.
    c. The tadpole excretes ammonia, but the adult frog shifts to urea excretion. Amphibians actively transport ions in through the skin.
3. Osmoregulatory problems in the terrestrial environment primarily involve conserving water.
    a. The earthworm's many **nephridia** remove nitrogen wastes and excess salts, releasing them through the **nephridiopore,** but reclaim water and other essential materials. Arthropods have a watertight cuticle and often rely on metabolic water. Their **Malpighian tubules** collect nitrogen wastes from coelomic fluids and produce uric acid, which is excreted into the gut.

The hindgut reabsorbs essential water, producing a semidry waste.

b. All terrestrial vertebrates have specialized, water-reabsorbing kidneys. While reptiles and birds and their embryos produce uric acid, which can be excreted in a semidry state, mammals produce urea, which requires water for its excretion.

THE HUMAN EXCRETORY SYSTEM

Anatomy of the Human Excretory System

1. The human **excretory system** consists of the **kidneys, ureters, urinary bladder, urethra,** and the **renal circuit (renal arteries** and **renal veins).**
2. The kidney includes an outer **cortex,** middle **medulla,** and the **nephrons.** The nephrons include a capsule and a looping tubule that joins others to form the **collecting ducts,** making up the **pyramids.** The pyramids empty into the **calyces,** which lead into the **renal pelvis.**
3. The nephrons form urine, which passes from the collecting ducts to the renal pelvis. The renal pelvis empties into the ureters, which conduct urine to the urinary bladder, and the urethra voids the urine from the body.

Microanatomy of the Nephron

1. The nephron begins with **Bowman's capsule,** which surrounds the **glomerulus,** a ball of capillaries arising from an **afferent arteriole** of the renal artery. Leaving the glomerulus is an **efferent arteriole,** which forms the **peritubular capillaries,** where reabsorption takes place. These spread over the nephron to later form a venule that joins others to make up the renal vein.
2. Bowman's capsule leads to the **proximal convoluted tubule,** the **loop of Henle,** and the **distal convoluted tubule,** which joins a collecting duct. The afferent arteriole also connects with the distal convoluted tubule, forming the **juxtaglomerular complex.**

The Work of The Nephron

1. Each part of the nephron functions as follows:
   a. *Bowman's capsule.* **Force filtration** in Bowman's capsule causes much of the water and ions and smaller molecules to leave the blood and enter the proximal convoluted tubule.
2. *The proximal convoluted tubule.* The peritubular capillaries contain blood in a hyperosmotic state, so much of the water filtrate reenters the blood by osmosis. Active transport also returns sodium (chloride follows passively), glucose, and amino acids to the blood.
3. *The loop of Henle.* The ascending loop actively transports chloride ions (sodium ions follow passively) into the surrounding area, recycling salt and creating a hyperosmotic state in the kidney medulla. The hyperosmotic state is further increased by urea, which diffuses out of the collecting ducts.
4. *The distal convoluted tubule.* The active secretion of sodium ions occurs with chloride ions and water passively following. Potassium ions enter the tubule.
5. *Collecting ducts.* Water leaves the collecting ducts in response to **antidiuretic hormone (ADH),** which is secreted by the posterior pituitary in response to osmotic conditions in the blood (actually detected by the hypothalamus).
6. **Tubular secretion** forces ammonia, hydrogen ions, potassium ions, organic acids, and creatine into the tubule.

Control of Nephron Function

Nephron control is hormonal, with water reabsorption controlled by ADH from the posterior pituitary and sodium chloride reabsorption controlled by **aldosterone** from the adrenal medulla. Sodium chloride transport is monitored by the juxtaglomerular complex. The arteriolar cells secrete **renin,** which stimulates the adrenal cortex to secrete aldosterone. Aldosterone increases the reabsorption of sodium chloride and the excretion of potassium.

REVIEW QUESTIONS

1. Define the term *homeostasis* and list several homeostatic mechanisms (outside of thermoregulation and osmoregulation). (846)
2. Using the household thermostat as an example, explain how a negative feedback loop works. (846)
3. State an example of positive feedback. In biological systems, what do most instances of positive feedback indicate? (846–847)
4. Distinguish between homeothermy, poikilothermy, and heterothermy. Why is the term poikilothermy troublesome? (847)
5. How do heat sources differ in endotherms and ectotherms? (847–848)
6. Why is it more accurate to refer to "core temperature" than body temperature? (848)
7. State an advantage and a disadvantage to endothermy and ectothermy. (847–848)
8. Describe thermiogenesis in the cold-adapted noctuid moth. (848–849)
9. Using a simple drawing of paired pipes, one carrying cold fluid and the other warm fluid, explain the principle of the countercurrent heat exchanger. (849)
10. Describe the arrangement of two countercurrent heat exchangers in the bluefin tuna. (836)

11. State the relationship between body size and metabolic rate in mammals. Name a mammal at each extreme. What is the principal disadvantage of a very high metabolic rate? (850–851)
12. Explain how thermiogenesis through brown fat metabolism works. Why is this source of heat important to hibernating mammals? (852)
13. Cite an example of a countercurrent heat exchanger in humans and explain how it works in retaining core heat. (852)
14. List an advantage and a disadvantage in the shunting of blood away from the body extremities during extreme cold. (852)
15. Describe the two sources of insulation used by the Arctic harp seal. What evidence can you offer about its effectiveness? (853–854)
16. Discuss ways in which the skin functions in cooling the body. Include the role of water. (854–855)
17. Under what conditions do behavioral cooling strategies in mammals and birds become more important than those that are physiological? List several of these strategies. (854–855)
18. Where in the human is body temperature controlled? What does this region actually measure? (855–856)
19. What is the role of the thyroid gland in maintaining a constant internal body temperature? (856–857)
20. Specifically, what danger does freezing produce in animals? Explain two ways in which the tissues of some ectotherms cope with this danger. (857)
21. Explain how a temperate-zone lizard might increase and decrease its absorption of the sun's energy. (858)
22. Define osmoregulation. What substances are involved? (859)
23. List the three common nitrogen wastes. Through what chemical activity are such wastes produced? What kind of diet would tend to produce the most nitrogen waste? (859–860)
24. What are two osmoregulatory problems of animals living in the marine environment? (860–861)
25. Describe two ways in which marine animals respond to their hyperosmotic surroundings, and cite examples of each. (860–861)
26. Compare the means of osmoregulation in the following: marine birds, reptiles, bony fishes, and mammals. (860–861)
27. What osmoregulatory problems confront the freshwater animal? (861–862)
28. Describe the protonephridia of planarians. What aspect of osmoregulation is this system concerned with? (862)
29. How is it that freshwater fishes can safely form and excrete the nitrogen waste ammonia? (862)
30. Compare the method and form of nitrogen waste excretion in the earthworm and insect. Which, if either, is better adapted to dry terrestrial habitats? (863–864)
31. Why must most terrestrial mammals drink copious amounts of water on a regular basis? (864)
32. Describe the gross anatomy of the human excretory system, listing five significant parts. Trace the flow of urine through these parts. (864–865)
33. What microstructures make up most of the cortex, the medulla, and the pyramids of the kidney? (865)
34. List the five parts of a nephron in their functional order, and carefully explain the arrangement of related blood vessels. (865–867)
35. List the forces at work in each part of the nephron. (Table 36.1)
36. Returning to microanatomy, describe what happens to each of the following in each part of the nephron: urea, water, sodium ions, chloride ions, amino acids, glucose. (Table 36.1, Figure 36.23)
37. In general, what is the function of the loop of Henle? What is the significance of its hairpin shape? What does the collecting duct contribute? (868–869)
38. What are the targets of antidiuretic hormone? How do the target cells respond? How is antidiuretic hormone regulated? (869–870)
39. Describe the process of tubular secretion. What are the materials involved, and in which direction are they secreted? (869)
40. Explain how the secretion of aldosterone is regulated. (870)

37

HORMONAL CONTROL

LIVING THINGS RESPOND. IF THEY DIDN'T, THEY WOULDN'T EXIST, PLAIN AND simple. Responses take many forms, some sudden and fleeting, others slower and longer lasting. Some responses are even permanent. Whatever the time frame, though, many of life's responses are quite complex and tightly coordinated. If you poke a sleeping dog with a stick and scream, he may leap straight up into the air. The response may seem simple enough, but unseen within his body, chemicals and neural changes interact in immediate and complex ways to help him adjust to this new situation. In the longer term, other neural changes may be associated with his changing his attitude towards you. We discussed in Chapters 34 and 35 how neural mechanisms help us respond to stimuli from both the external and internal environment. Here we will look at chemical mechanisms, focusing primarily on hormones. We will also see how neural and chemical events interact to produce adaptive responses.

THE CHEMICAL MESSENGERS

Many biologists have chosen their profession simply because they are fascinated by the complexities of life and want to help unravel the mysterious weave that makes up life's fabric. The tighter the weave (the more complex the problem), the greater the challenge. And among the greatest of biological challenges has been to try to understand how cells, tissues, and in fact the whole organism come to respond so precisely when quietly touched by molecules of a particular configuration (Figure 37.1). In many cases, the molecules act as "messengers," sent perhaps from far away in the body.

Until recently, all chemical messengers were known simply as **hormones.** They were defined as chemical substances produced in one part of the body and transported in the blood to other parts of the body, where they produced a specific effect. While attractive in its simplicity, this definition now must be broadened. Today we know that many chemical messengers don't travel very far at all. Some affect nearby cells and tissues, and others affect only the cell in which they are produced. Now endocrinologists speak of **endocrine hormones** (chemical messengers that act in areas other than where they are produced), **paracrine hormones** (those that act in adjacent cells), and **autocrine hormones** (those that act in the cell in which they are produced). In nearly all instances in which we use the term "hormone" here, we will be referring to endocrines, and most of our attention will be directed to this group. Endocrine hormones are produced by **endocrine glands,** the *ductless* glands that send their secretions into the blood. In contrast, *ducted* glands such as sweat and mucus glands are termed **exocrine glands.** With our terminology sorted out now, we will begin our look into hormonal control by contrasting it with neural control.

Neural and Hormonal Control Compared

Neural and hormonal control generally work in close concert and often have overlapping functions, particularly in homeostasis. Interestingly, they have similar chemical characteristics. For example, some vertebrate hormones are similar and some are even identical to certain neurotransmitters. (Recall that neurons activate each other and their effectors through the secretion of chemical neurotransmitters, see Chapter 34.)

A primary difference between neural and hormonal control involves the time frame in which they work. Neural activity is generally instantaneous—neural impulses arise and subside with startling swiftness. Chemical control is generally much slower, and its effects are longer lasting. Further, whereas the action of neurotransmitters is restricted to the synaptic cleft (which they cross to activate the next neuron or target organ), endocrine hormones are usually released into the bloodstream or body fluids for distribution to other cells.

The similarities between neural and hormonal control have strong evolutionary implications. According to some biologists, neural control preceded hormonal control, the latter evolving as non-neural cells took on the task of synthesizing neurotransmitters. Other biologists believe that hormonal control preceded neural control. They point out that true nervous systems require at least a tissue level of organization, but organisms below the tissue level rely heavily on purely chemical control. The evolutionary sequence has not been completely resolved.

While neural and hormonal control may be perceived as distinct, one particular fact makes their close relationship quite clear: some hormones are secreted by neurons. Such secretions, fittingly called **neurosecretions** (or *neurohormones*), are produced in neurons and travel through axons to their terminals, where, like neurotransmitters, they are

FIGURE 37.1
HORMONES AT WORK
Hormones play many roles in the lives of animals, most not visible to the eye, but some startlingly so.

released. But unlike neurotransmitters, which are released into synaptic clefts, neurosecretions are released into the blood. Neurosecretions are quite common in invertebrates, and although less common in vertebrates, there are important vertebrate examples. For example, we saw in the last chapter that the human hormone ADH, actually a neurosecretion, is critical to water recovery in the kidney's collecting ducts.

Molecular Structure of Hormonal Messengers

Most hormones fall into either of two general categories: peptides and lipids (Figure 37.2). The peptide category ranges from parathyroid hormone and insulin, which contain 84 and 51 amino acids respectively, to lightweights such as thyroxin, with two amino acids, and epinephrine, with just one. In the lipid category are the familiar four-ringed steroids such as the sex hormones and a group of relative "newcomers" to biologists, the long-tailed lipids known as **prostaglandins.** Prostaglandins tend to fall into the autocrine and paracrine hormone category, since their effects remain local.

Characteristics of Hormonal Control

A number of hormones are likely to be circulating through the body at any given time. Although present in very low concentrations, hormones are extremely powerful molecules, able to cause sweeping changes. So obviously, they must be under some form of closely coordinated control if the body is to avoid complete chaos. Much of this coordination is afforded by the nature of hormonal control itself.

To begin, although hormones may circulate widely in the animal, they elicit responses only in specific **target cells.** Hormones encounter a variety of cells in their travels and in such a system it is up to the cell to identify the messenger. The identification is accomplished by the presence of very specific receptor sites on or in the target cells, sites that are able to bind only with certain hormonal molecules.

The precision of hormonal control is also made possible by the brevity of the chemical's existence. Hormones remain active in many cases for less than an hour and are usually enzymatically degraded after they have stimulated the target cell. The short hormonal lifetime is quite essential; after all, it wouldn't do to have every hormonal response repeated endlessly. Further, the rapid degradation of hormones enables the body to promote both rapid, short-term effects, such as the familiar urgent responses to danger, and longer term effects, such as those involving growth and development. The short-term effect is accomplished through a burst of hormone release, whereas the long-term effect occurs through a slow but steady release. As usual, there are exceptions; not all hormones are so short-lived. In testing a synthetic version of the hormone MSH (melanocyte-stimulating hormone), researchers found that it remained bound to receptor sites on the skin cells for months, continuously prompting the cells to produce dark melanin pigment. (The effects might be thought of as a sort of "hormonal suntan.")

Cells may secrete hormones in response to a number of cues. Hormone release may be stimulated by neurons, by hormones from other cells, or by conditions in the extracellular environment. In the latter case, recall that cells of the juxtaglomerular complex of the nephron respond to minute changes in blood pressure, sending chemical messages to the adrenal gland, which responds by secreting the hormone aldosterone.

Commonly, hormone secretion is regulated through negative feedback (see Chapter 36). When a certain level of hormone concentration is reached, or when the effects it causes in its target cells reach a given peak, the gland cells releasing the hormone are inhibited, and hormone release slows. Then, when the hormone level falls sufficiently, gland inhibition lessens and hormone secretion again increases.

Identifying Hormones

Since hormones often produce noticeable anatomical, physiological, and behavioral changes, investigators were able to develop very straightforward procedures for hormone

PEPTIDE HORMONES

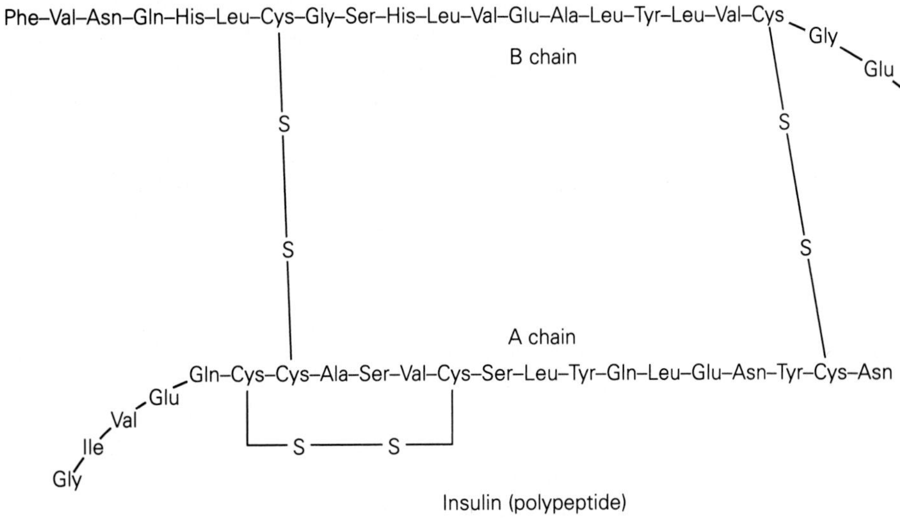

Insulin (polypeptide)

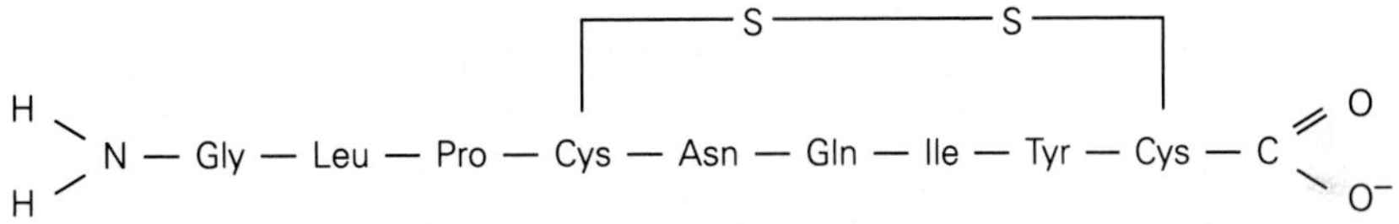

Oxytocin (polypeptide)

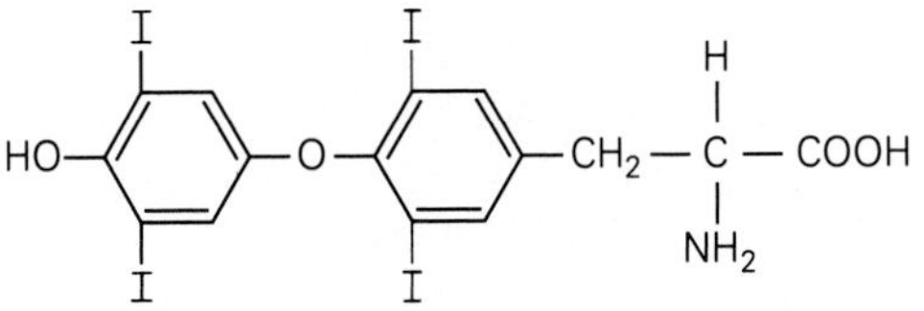

Thyroxin (dipeptide)

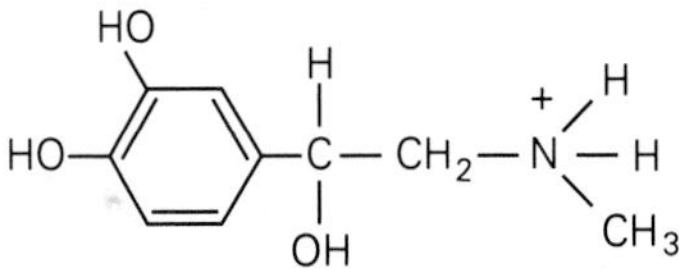

Epinephrine (amino acid)

LIPID HORMONES

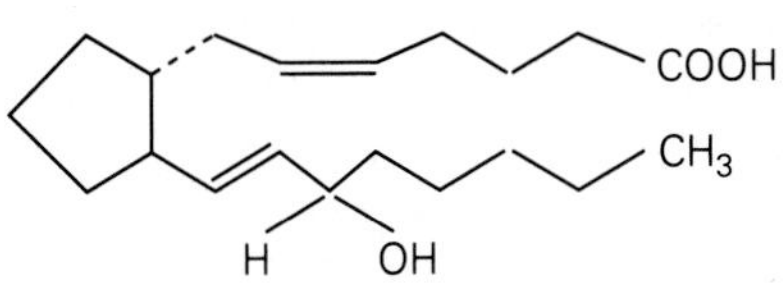

Prostaglandin E_2 (lipid)

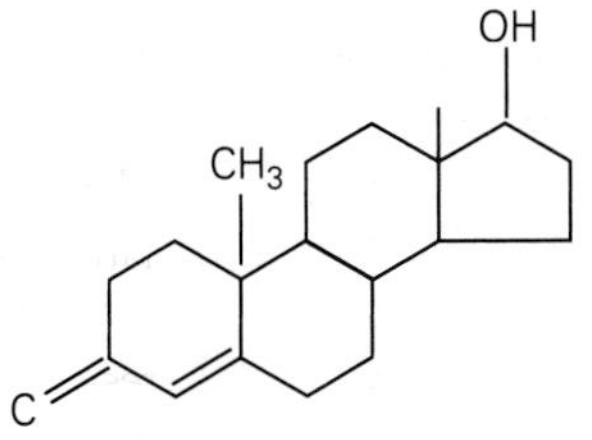

Testosterone (steroids)

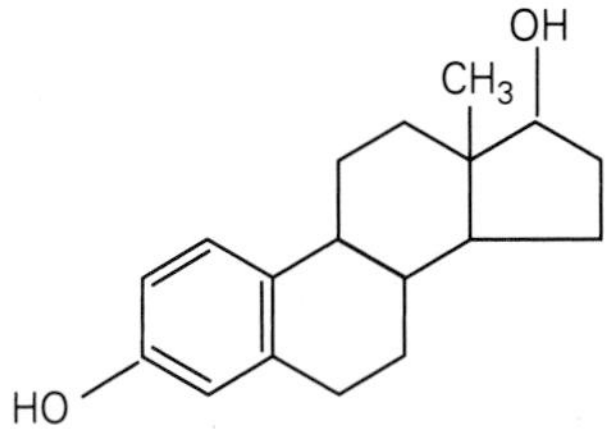

Estradiol (steroids)

FIGURE 37.2
CHEMICAL STRUCTURE OF HORMONES
Two classes of hormones are peptides and lipids. Some peptides are very simple, while others are large enough to qualify as proteins. Lipids include steroids and prostaglandins.

identification. In traditional procedures, they began by making an educated guess about whether a certain tissue might be producing the hormone responsible for some condition. The tissue was then removed (where practical), and the investigators watched for any symptoms caused by a deficiency of the hormone in question. (In the case of insulin, surgical removal of the pancreas in dogs resulted in diabetes.) If a symptom appeared, extracts from the tissue were then introduced into the animal. If the extract alleviated the symptoms, the presence of an active substance in that tissue was confirmed. The next steps involved chemically isolating and identifying the active substance, and then more tests. Such procedures and the isolated hormones they yielded became the basis for replacement therapy, the treatment of hormonal deficiencies in humans by injections of purified isolates.

Such procedures were quite successful in the early days of endocrinology, but there were drawbacks. For example, there are organs that cannot be removed without killing the experimental organism, and whereas death is an easily recognized symptom, no one has yet been able to bring about its alleviation. So today other procedures are also employed, some of which were unavailable not long ago. For example, instead of using an extract of a tissue that may contain other material as well as the hormone, a highly purified form produced through recombinant DNA techniques (see Chapter 16). It is then administered in cases in which its deficiency is suspected. If it alleviates the deficiency symptom, it is revealed as the hormone in question.

Other contemporary techniques involve radioactive tracers. Target cells can readily be identified if hormones containing radioactive elements are introduced in a test animal. It is also possible today to locate infinitesimal amounts of hormones through the use of a procedure called **radioimmunoassay.** The incredible sensitivity of such tests has been compared to that of detecting the presence of a half cube of sugar in a very large lake.

Radioimmunoassay combines the use of radioactive tracers and the immune reaction between antibodies and antigens. Briefly, antibodies against the hormone are prepared, generally by injecting the subject hormone into a "reservoir" animal. The animal's immune system responds by forming highly specific antibodies against the hormone, a typical response against any antigen (the invading chemical). The antibodies are then extracted from the animal's blood plasma and used as a testing agent for the presence of matching antigens. Thus, any time the newly acquired antibody comes in contact with the hormone that stimulated its production, molecules of the two bind together, forming an easily detectable, insoluble precipitate. In this way the presence of the hormone in any sample is known. By varying the procedure, it is also possible to quantify the results, that is, to determine the precise amount of hormone/antigen present in a sample.

Quantification is done by introducing a carefully measured amount of a radioactive form of the same antigen. The radioactive form of antigen then competes with the nonradioactive antigen for binding sites on the antibody, and consequently some of the precipitate is radioactive and some is not. Following this, the excess unbound radioactive antigen is washed away, so only bound antigen remains. The amount of hormone/antigen in the test sample is then calculated from the ratio between radioactive and nonradioactive bound antigen/antibody. Figure 37.3 illustrates the quantifying procedure. Finally, we should point out that there are many new refinements and variations on the basic procedures of immunoassay, and detecting the presence of hormones has become an art of great precision.

CHEMICAL MESSENGERS AND THE TARGET CELL

Now that we've considered the nature and chemical makeup of hormones, their control, and their identification, we'll see what happens at the target cell. Just how does it respond to the hormone? The target cell's response depends on whether it has cell surface receptors or cytoplasmic receptors. By this, we see that hormones generally work in one of two ways.

The peptide hormones attach to the surface receptor of a target cell and trigger a cascade of events within the cytoplasm through intermediaries called **second messen-**

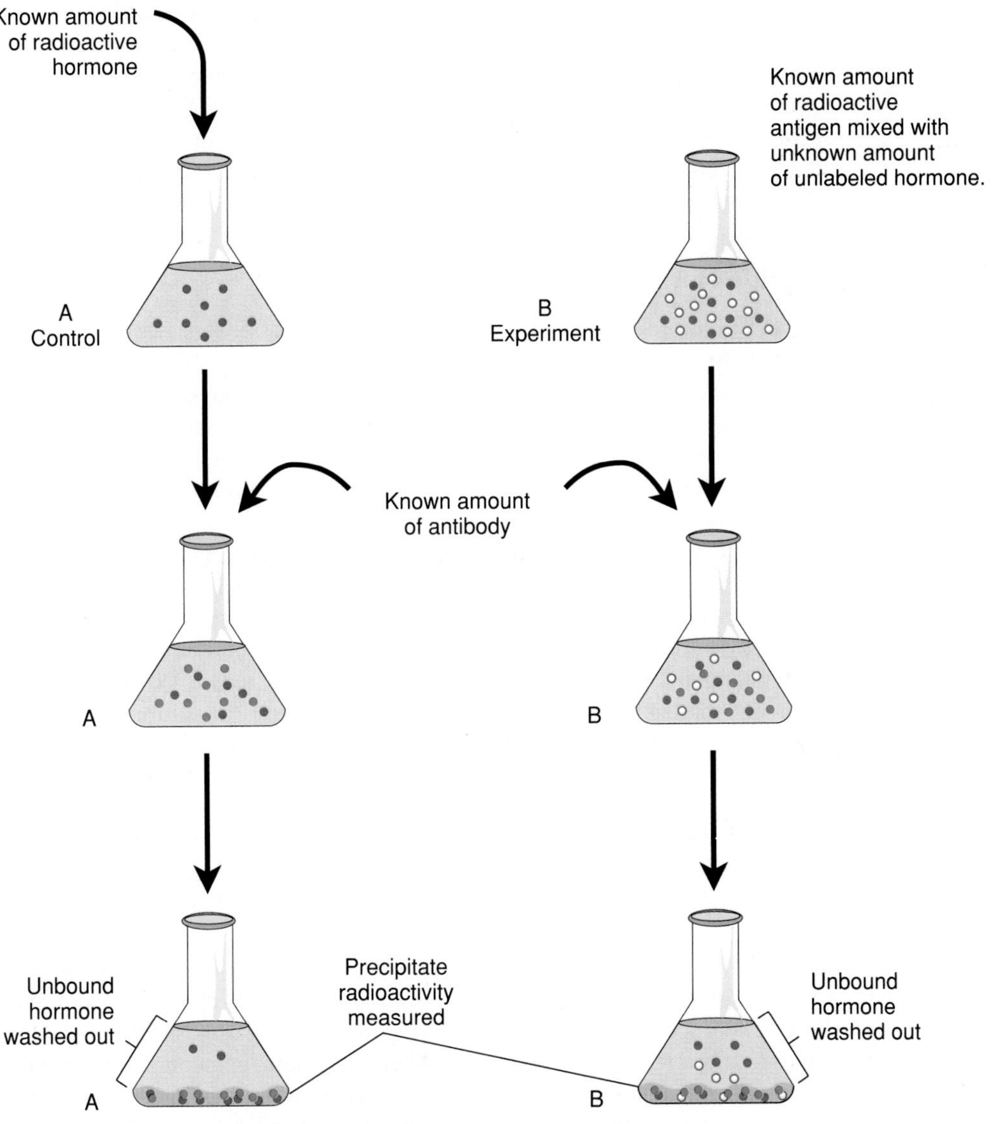

FIGURE 37.3
RADIOIMMUNOASSAY PROCEDURE
(a) In the control sample, a known amount of radioactive hormone (antigen) is combined with a known amount of antibody. Unbound antigen is washed away, and the radioactivity of the bound antibody-antigen precipitate is measured. (b) The experimental sample has the same contents except that an undetermined amount of unlabeled (nonradioactive) hormone is mixed in with the initial step. From this procedure, the hormone concentration in the experimental sample can be determined.

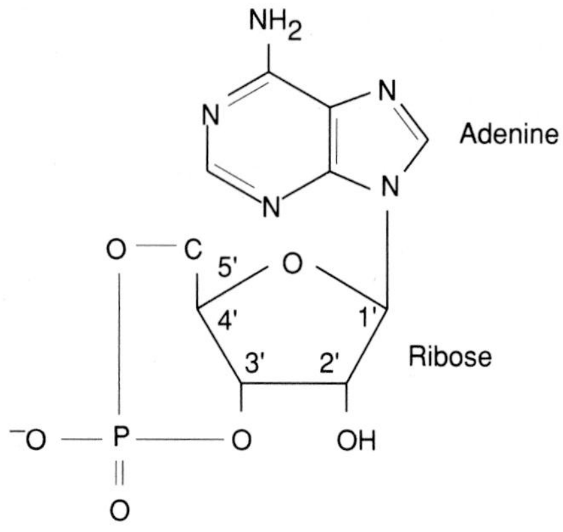

gers. During this cascade, inactive enzymes in the cytoplasm are activated, and the hormonal reaction begins. Other hormones, in particular the steroid hormones, actually penetrate the nucleus of the target and alter the expression of selected genes there, often triggering the production of new enzymes. So the peptide hormones activate existing enzymes, and the steroid hormones trigger the production of new enzymes. We'll begin with the peptide hormones and surface receptors.

Cyclic AMP

FIGURE 37.4
CYCLIC AMP, A COMMON SECOND MESSENGER
Cyclic AMP is a modified version of AMP (adenosine monophosphate), the dephosphorylated version of ATP. Note the bonds linking the single phosphate back to ribose.

Peptide Hormones and Second Messengers

If we're going to concern ourselves with second messengers, we must assume there are first messengers. Although they are rarely called by that term, the peptide hormones that are released from endocrine glands are themselves the first messengers. These travel to the target cell, attach to specific matching receptors on the cell surface, and trigger a series of events within the cell. As this scenario is multiplied for numerous cells within a tissue, that tissue reacts, as does the organ to which the tissue contributes.

In 1971 the Nobel Prize was awarded to E. W. Sutherland of Vanderbilt University for the discovery of a second messenger. This is a molecule called **3′,5′ cyclic adenosine monophosphate** (Figure 37.4), usually shortened to "cyclic AMP" when spoken, and to **cAMP** when written. We now know that there are other second messengers, such as **3′,5′ cyclic guanosine monophosphate**, called **cGMP**, and **inositol triphosphate** or **IP_3**. It turns out that some second messengers are widespread in nature; cAMP is the same molecule that causes the legendary aggregation of solitary cellular slime molds (see Chapter 23).

cAMP, Epinephrine and the Liver Cell One of the effects of the hormone epinephrine is a sudden increase in the blood glucose level, which is accomplished through the enzymatic breakdown of glycogen reserves in the liver. The chain of events leading to the elevated glucose level is now rather well understood.

Epinephrine, released from the adrenal medulla, is carried by the blood to its target cells, such as those of the liver (Figure 37.5). Like all target cells of peptide hormones, liver cells have very specific hormone receptor sites along their plasma membranes. Once fixed to its specific receptor in a liver cell plasma membrane, epinephrine activates an enzyme known as **adenylate cyclase.** This enzyme is part of the membrane system itself, but it remains inactive until the epinephrine is in place. But when adenylate cyclase is activated, it immediately converts cytoplasmic ATP to cAMP, which activates critical enzymes and eventually leads to the breakdown of glycogen.

Specifically, cAMP activates an enzyme called protein kinase. (Kinases have the special property of being able to activate other enzymes.) Next, the protein kinase activates a second enzyme, phosphorylase kinase, which in turn activates glycogen phosphorylase *a.* The latter then cleaves each glycogen molecule into many glucose units, actually glucose-1-phosphate and glucose-6-phosphate. The phosphate is then removed, and simple glucose passes through the plasma membrane of the liver cell and enters the blood. (Interestingly, protein kinase simultaneously inactivates a second liver enzyme, glycogen synthetase, which is responsible for the opposing reaction, linking glucose subunits into glycogen when glucose is in excess.)

An important aspect of the second messenger reaction is called a **cascade.** The initial reaction—the binding of a hormone to its receptor site—involves relatively few molecules. But with each following step, many more molecules are involved—thus, a few molecules of epinephrine bound to the plasma membrane activate many cAMP molecules, which in turn activate a great number of enzyme molecules. Each enzyme, acting over and over and with lightning speed, as enzymes do, produces immense numbers of

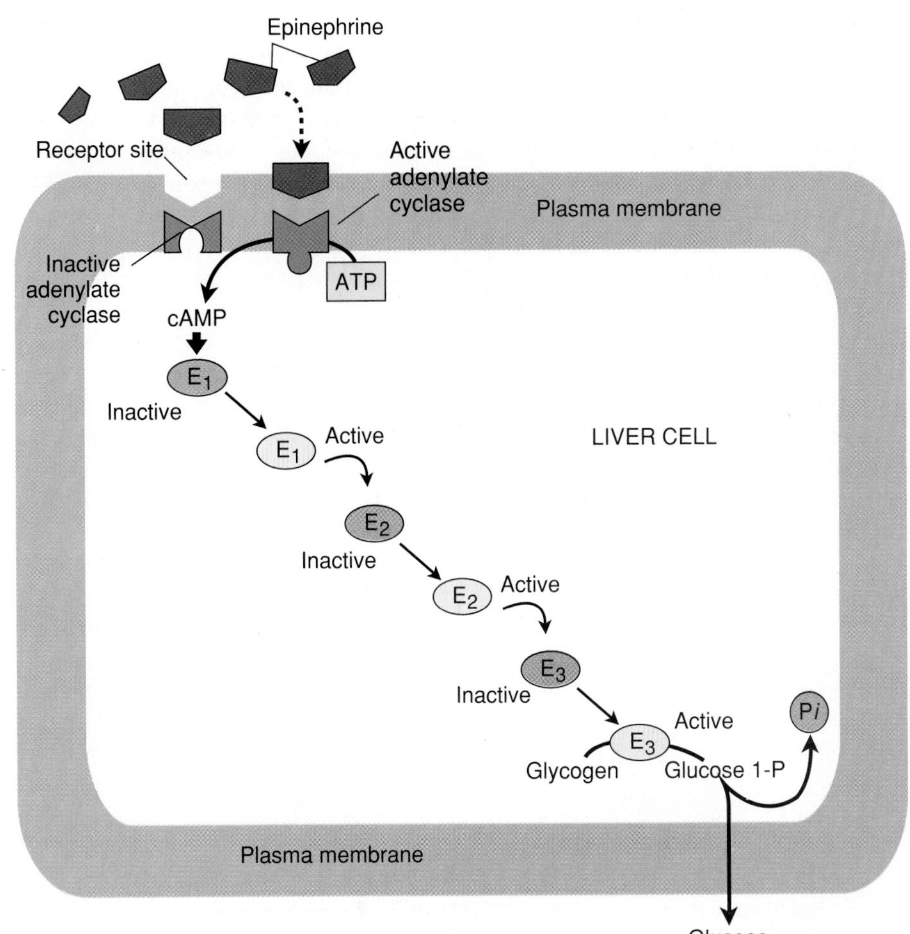

FIGURE 37.5
ACTION OF CYCLIC AMP
When epinephrine binds to its plasma membrane receptor on a liver cell, it sets in motion a sequence of reactions that leads to the breakdown of glycogen and the transport of glucose out of the cell.

product molecules. One estimate is that through the cascading effect, a single molecule of epinephrine can account for the release of 100 million molecules of glucose.

Another important characteristic of the second messenger operation is rapid degradation. The cyclic bonds of cAMP and cGMP are readily cleaved, producing AMP and GMP. These two are recycled into the usual respiratory pathways, and ATP and GTP are regenerated. Further, when the second messenger is deactivated, the protein kinases return to an inactive state. In this way, the entire system shuts down, remaining inactive until another barrage of epinephrine reaches the target cells.

It is now known that, except for thyroxin and insulin, all peptide hormones work through second messengers. This raises an important question. We know that cAMP is involved in the action of several hormones and that each of these hormones has a different effect. But how can one kind of agent bring on different effects? The answer lies in the ways that cells differ in their cytoplasmic machinery, that is, in the enzyme systems awaiting activation. For instance, whereas the response to cAMP in liver cells may be the breakdown of glycogen and liberation of glucose, the same second messenger activated in follicle cells of the ovary will bring on the release of a sex hormone.

Dual Second-Messenger Operation In many tissues, two different hormones can elicit opposing effects. Each has its own kind of receptor sites, and each stimulates the formation of its own second messenger. Such a dual control system works in both smooth and cardiac muscle.

In smooth muscle, that of the gut and blood vessel walls for example, one hormone activates adenylate cyclase, which in turn catalyzes the conversion of ATP to cAMP, just as we described earlier. This brings on muscle relaxation. A second hormone activates the membrane enzyme guanylate cyclase, which in turn catalyzes the conversion of GTP to the second messenger, cGMP. cGMP's effect is to cause smooth muscle to contract (Figure 37.6). The hormone in greater concentration (filling more receptor sites) has an inhibiting effect on the lesser hormone's ability to stimulate its second messenger. Thus, as cAMP increases, cGMP decreases, and vice versa.

In our discussions of the autonomic nervous system (see Chapter 35), we saw such a dual system at work at the neural level, another example of the close relationship between neural and chemical control. Parasympathetic nerves serving the heart secrete the neurotransmitter acetylcholine, which slows heart rate. Opposing this are sympathetic nerves whose secretion, epinephrine, speeds heart rate. It is now known that the second messenger activated by acetylcholine is cGMP, whereas the one activated by epinephrine, as in the liver cell, is cAMP.

Steroid Hormones and Gene Control

The other great class of chemical messengers, the lipid hormones, includes the steroids. As we've seen, this group does not employ second messengers, nor does it involve

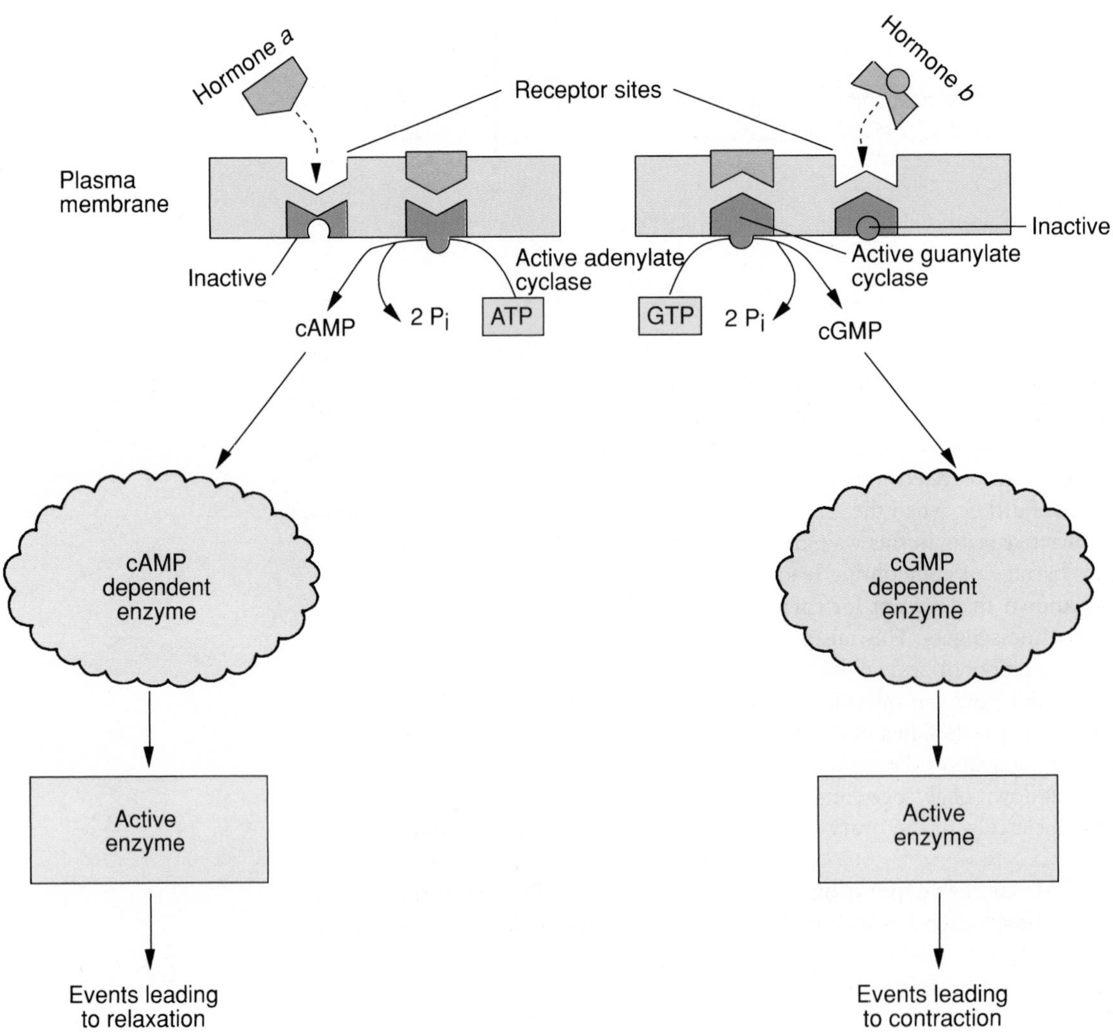

FIGURE 37.6
MULTIPLE SECOND MESSENGERS
In this model of a double second-messenger system, a smooth muscle cell relaxes or contracts in response to two different hormones. Each has its own receptor, membrane enzyme, and second messenger.

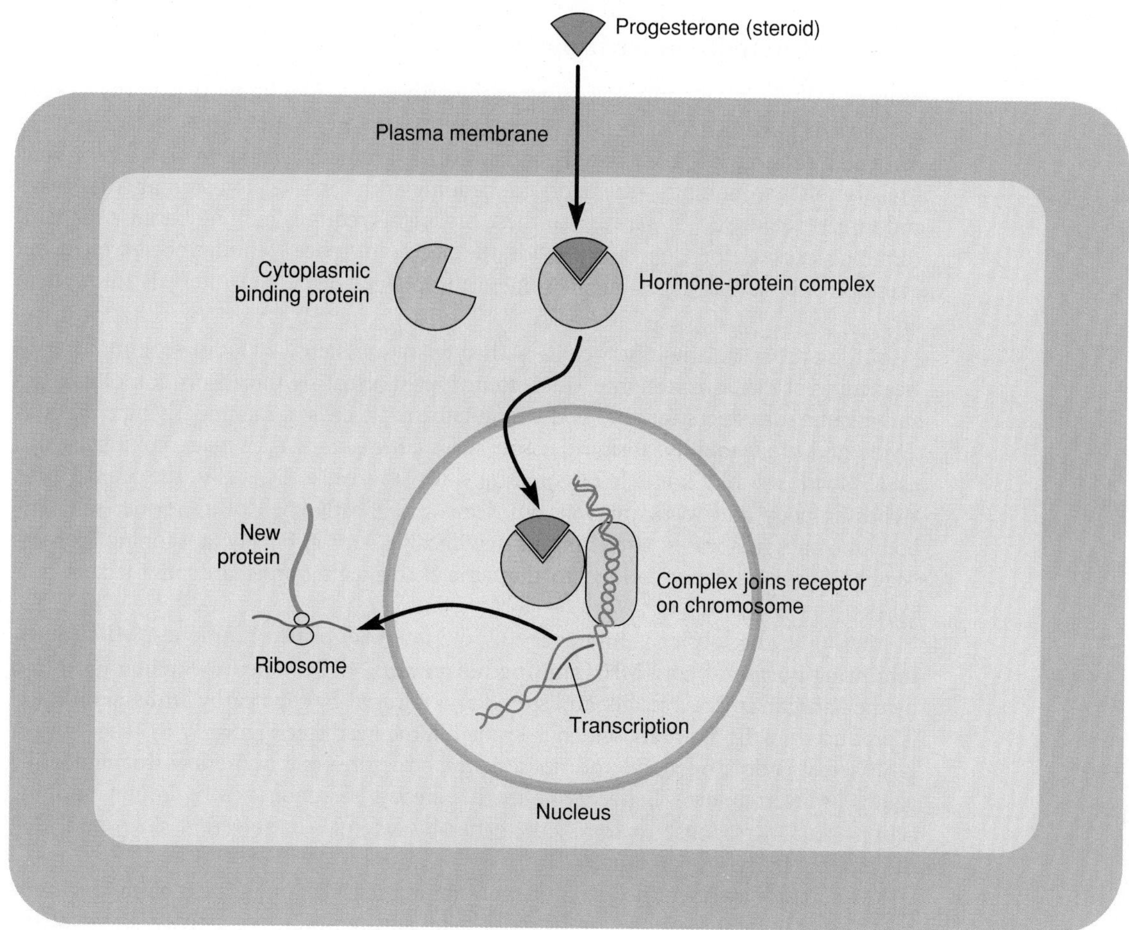

enzymes already present in the cytoplasm. Instead, steroid hormones (as well as thyroid hormone) penetrate the cell membrane, enter the nucleus, and direct the action of genes there.

Upon entering the cytoplasm of a target cell, the steroid hormone becomes tightly bound to a **cytoplasmic binding protein.** The newly formed protein-hormone complex then moves into the nucleus, where it joins with chromosomal receptors, yet other proteins, this time part of the chromatin complex. The complex of hormone, cytoplasmic binding protein, and chromosomal receptor activates certain genes, bringing on their transcription. Subsequently, polypeptides are assembled that often go on to form enzymes (Figure 37.7).

FIGURE 37.7
STEROIDS IN THE CELL
Progesterone, a lipid-soluble steroid, readily passes through the plasma membrane and joins a receptor called a cytoplasmic binding protein. Together they penetrate the nucleus, where the hormone-protein complex joins a chromosomal receptor which then activates a DNA segment. RNA is transcribed, and a protein is produced along ribosomes.

The system was first worked out by Bert W. O'Malley and his co-workers, then at Vanderbilt University. It was known that estrogen and progesterone treatments could cause the oviducts of female baby chicks to produce egg-white albumin and other proteins. These researchers labeled steroid hormones with radioactive tags and followed them through the sequence described above. At each step, the various intermediate compounds were isolated and identified. And at last they unravelled the story of how steroid hormones activate genes.

Now that we've seen the primary ways that hormones function at the cellular level, we can consider, in both invertebrates and vertebrates, some specific hormones, their sources, and their effects on target cells.

INVERTEBRATE HORMONES

Probably all invertebrates rely on some form of hormonal regulation. In many cases the chemical messengers are neurosecretions, those secreted by neurons rather than by specialized gland cells. Because human fortunes are so intimately tied to those of the arthropods, we know more about chemical messengers in the jointed-legged creatures than we do about those in the other invertebrates.

Hormonal Activity in Arthropods

Examples of chemical control abound in the arthropods. We know that hormones are involved in growth, reproduction, pigmentation, osmoregulation, and metabolism. In particular, researchers have worked out many of the hormonal actions involved in **ecdysis** ("to strip off") or molting—the periodic shedding of the exoskeleton seen in crustaceans and insects (Figure 37.8). Growth in these organisms requires molting because the hard confining exoskeleton does not provide the necessary space. Molting occurs most frequently in the larval stage, but in some arthropods, it occurs at intervals throughout life.

In the crustacean, molting begins with preparatory steps, including a thinning and weakening of the hard, tough exoskeleton brought about by the absorption of calcium and organic substances into the body. Simultaneously, cells in the underlying epidermis divide and grow rapidly, forming a soft, new cuticle. It is then that molting begins, usually with the old exoskeleton splitting down the back. Typically, the animal then arches its body and works its way out, leaving a ghostly shell of its former self. The body then absorbs water, stretching its new flexible cuticle. Finally, a "tanning" process begins. Calcium salts are added to the cuticle, thickening and hardening it as a new exoskeleton takes form.

Molting in the lobster is influenced by the interplay of two hormones, **MIH (molt-inhibiting hormone)** and **MH (molting hormone).** MIH is a neurosecretion produced by the **X-organ** in the eyestalk and stored and released by the nearby **sinus gland.** MH is produced by the **Y-organ,** which is located in the head (see Figure 37.8). The Y-organ is an actual endocrine gland and not a neural structure (one of the few true endocrine glands in invertebrates). Between molts, the ongoing secretion of MIH inhibits molting activity, but in response to some cue, generally external, its secretion slows and MH, the molting hormone, dominates.

Researchers have identified a number of external cues that influence molting. Included are light, temperature, injury, nutritional state, and reproductive activity. Some of the cues tend to be specific to certain groups of arthropods. For instance, some burrowing crabs will not molt if kept in constant light, whereas crayfish will not molt if kept in continuous darkness. Some environmental conditions influence molting in all groups. For example, low temperatures retard molting in all arthropods, as does starvation.

Hormones and Development in Insects Now we come to development in those remarkable arthropods called insects. Insect development has received a great deal of scientific attention, partly because some people would like to bring much of it to a screeching halt. It is difficult to overemphasize the importance of this research area,

FIGURE 37.8
MOLTING HORMONES
The molting hormone, MH, is secreted by the Y-organ, whereas molt-inhibiting hormone, MIH, is secreted by the X-organ.

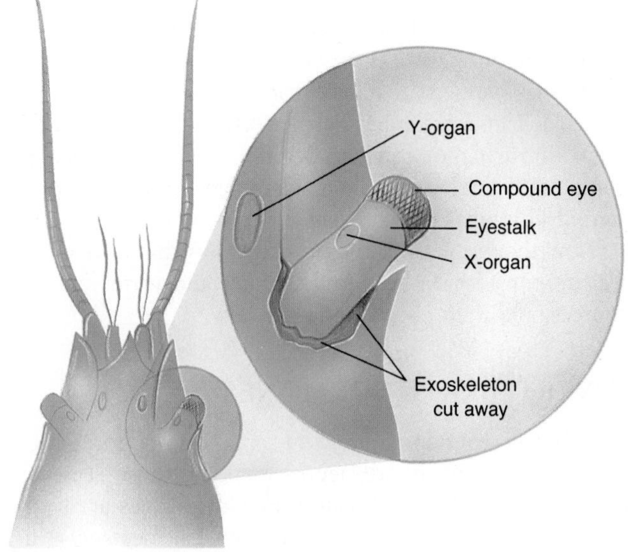

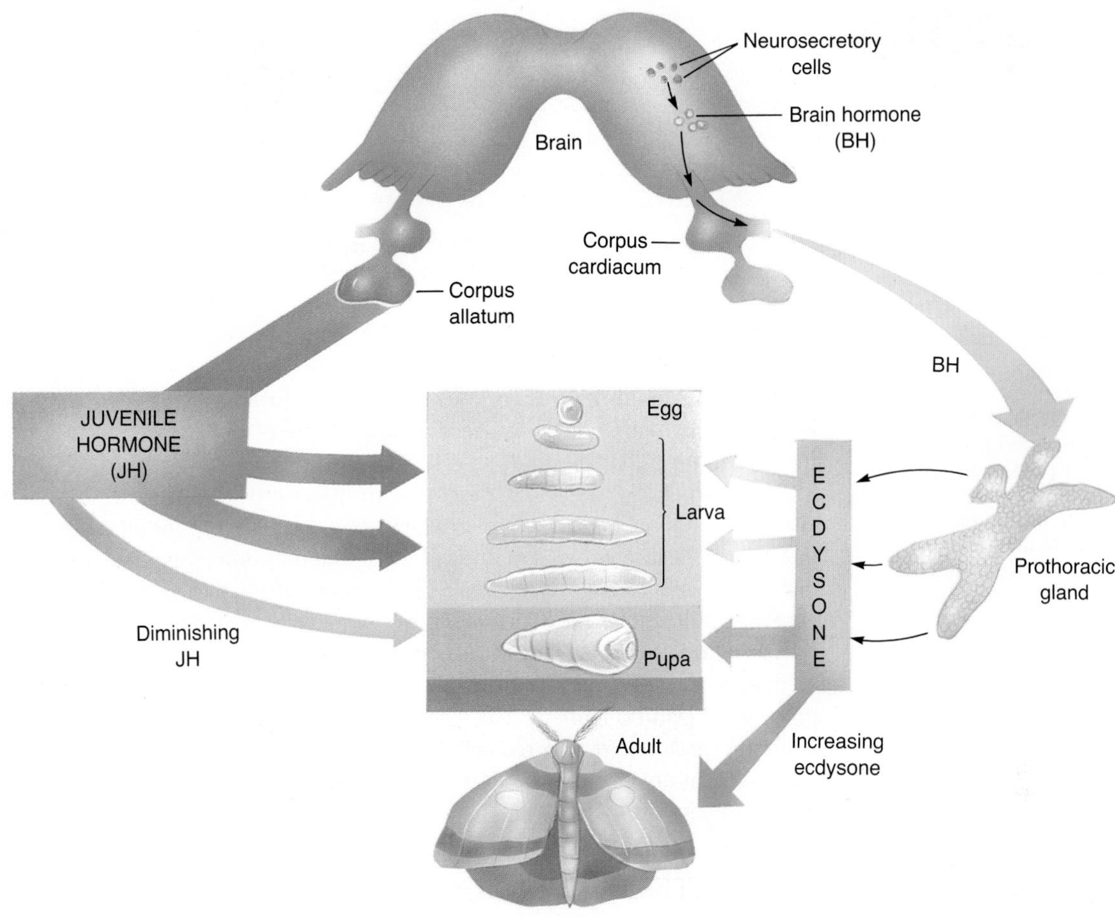

FIGURE 37.9
HORMONES OF METAMORPHOSIS
When enough juvenile hormone is present, it dominates metamorphosis, permitting larval growth, but inhibiting pupal formation. As its concentration wanes, ecdysone dominates, a pupa forms, and the adult soon emerges. Pupation time varies greatly.

since our ability to control insects—probably our leading ecological competitors—depends on an exact knowledge of all aspects of their lives.

Much of what we know about their development has come from a few in-depth studies such as those on the American silkworm, *Hyalophora cecropia*. We can use it to show a developmental program called a complete metamorphosis. In a complete metamorphosis, the insect passes through all developmental stages: egg, larva, pupa, and adult (see Chapter 31). The structural changes from one stage to another are strikingly dramatic. The silkworm larva goes through several molts before entering pupation. It spends a winter in the pupal stage, emerging from the pupa case in the soft light of spring. It has carefully conserved its stored energy through the winter by existing at a very low metabolic rate in **diapause,** a common developmental state in insects.

Metamorphosis in all insects is controlled through a complex interplay of two hormones (Figure 37.9). The first, **ecdysone,** is secreted by the **prothoracic gland,** which is in turn aroused by a neurosecretion called **brain hormone** or **BH.** BH, not surprisingly, is secreted by neurons of the brain. Ecdysone has two roles: it supports molting and instigates the changes that bring on new developmental stages. Specifically, it brings the larval stage to an end by prompting the emergence of the pupal stage. Later, it triggers the appearance of the adult stage. (Interestingly, ecdysone is quite similar chemically to estradiol, a vertebrate female sex hormone.) The second hormone, **juvenile hormone (JH),** is produced by neurons in paired brain extensions known as **corpora allata** (singular, **corpus allatum**). JH counters the effects of ecdysone. Whereas it does not

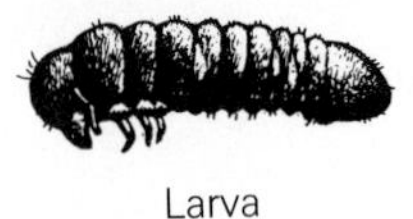

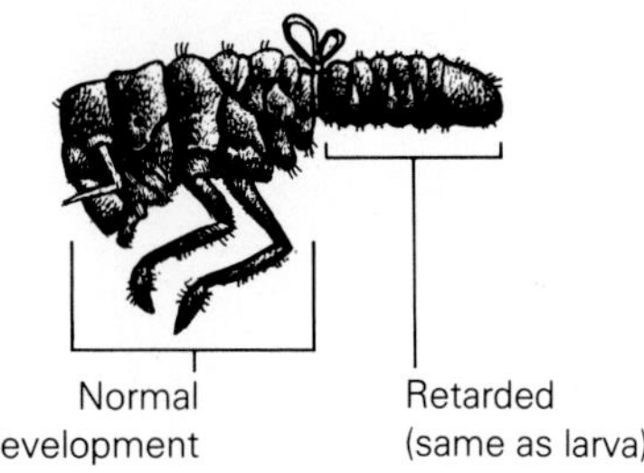

FIGURE 37.10
RESTRICTION OF ECDYSONE
If ecdysone is restricted from reaching the posterior parts of an insect larva, those parts will remain in the larval stage while the head and thorax undergo normal adult development. If the body is tied off just behind the prothoracic gland, the head area will mature normally, but the rear area, not reached by ecdysone, will molt unchanged.

interfere with molting *per se,* it does maintain the juvenile state. That is, as long as the level of JH is high enough, it overrides the effects of BH and ecdysone, and the final transition into pupa and adult stages cannot occur.

Some rather ingenious experiments with insect growth hormones have led to some fascinating findings. For example, when several larval corpora allata are transplanted into larvae, the insects begin molting as usual, but they don't pupate—they keep growing into larger and larger larvae! Since the corpora allata secrete juvenile hormone, the experiment illustrates the overriding effect of juvenile hormone over brain hormone and ecdysone, when it is present in sufficient levels.

In another series of experiments, threads were tied around the larvae just behind the prothoracic gland (Figure 37.10). The part of the insect anterior to the ligature pupated, while the posterior region remained in its larval state. Apparently, the difference was due to the failure of ecdysone to reach the hindpart of the insect, leaving it in a juvenile condition when it molted.

THE VERTEBRATE ENDOCRINE SYSTEM

As the vertebrates evolved, both the nervous system and hormonal system became increasingly important as a means of maintaining homeostasis. The hormones work in a rather similar fashion in many of the vertebrates, so, for the most part, we will again draw on that representative mammal with the generalized teeth and the untalented toes.

We should begin by noting that anatomically, the human endocrine system consists of a number of distinct glands as well as less organized tissues in various parts of the body (Figure 37.11 and Table 37.1). These are highly interactive, and their coordination is critical to the normal development and functioning of the animal.

The Pituitary

The first of the vertebrate endocrine structures we will consider, using humans as our main example, is the **hypophysis,** also called the **pituitary gland.** It is a bilobed structure about the size of the tip of your little finger, located at the base of brain, where it is cradled in a saddlelike pocket of bone rising from the floor of the skull (see Figure 37.11). If you point one finger directly between your eyes and stick another into your ear, you will not only gain the attention of other people on the bus, but the lines will intersect at about the location of the pituitary. Considering its small size, it may seem surprising that the pituitary is one of the busiest of the endocrine glands, releasing at least 10 hormones.

Actually, the pituitary has two distinct lobes or structures, the **adenohypophysis,** a true gland, and the **neurohypophysis,** which is a neural structure. Although the first manufactures hormones, the second simply receives neurosecretions carried there by the axons of neurons originating elsewhere. In humans and other primates the adenohypophysis and neurohypophysis are referred to as the **anterior pituitary** and **posterior pituitary** respectively, terms we will be using most of the time.

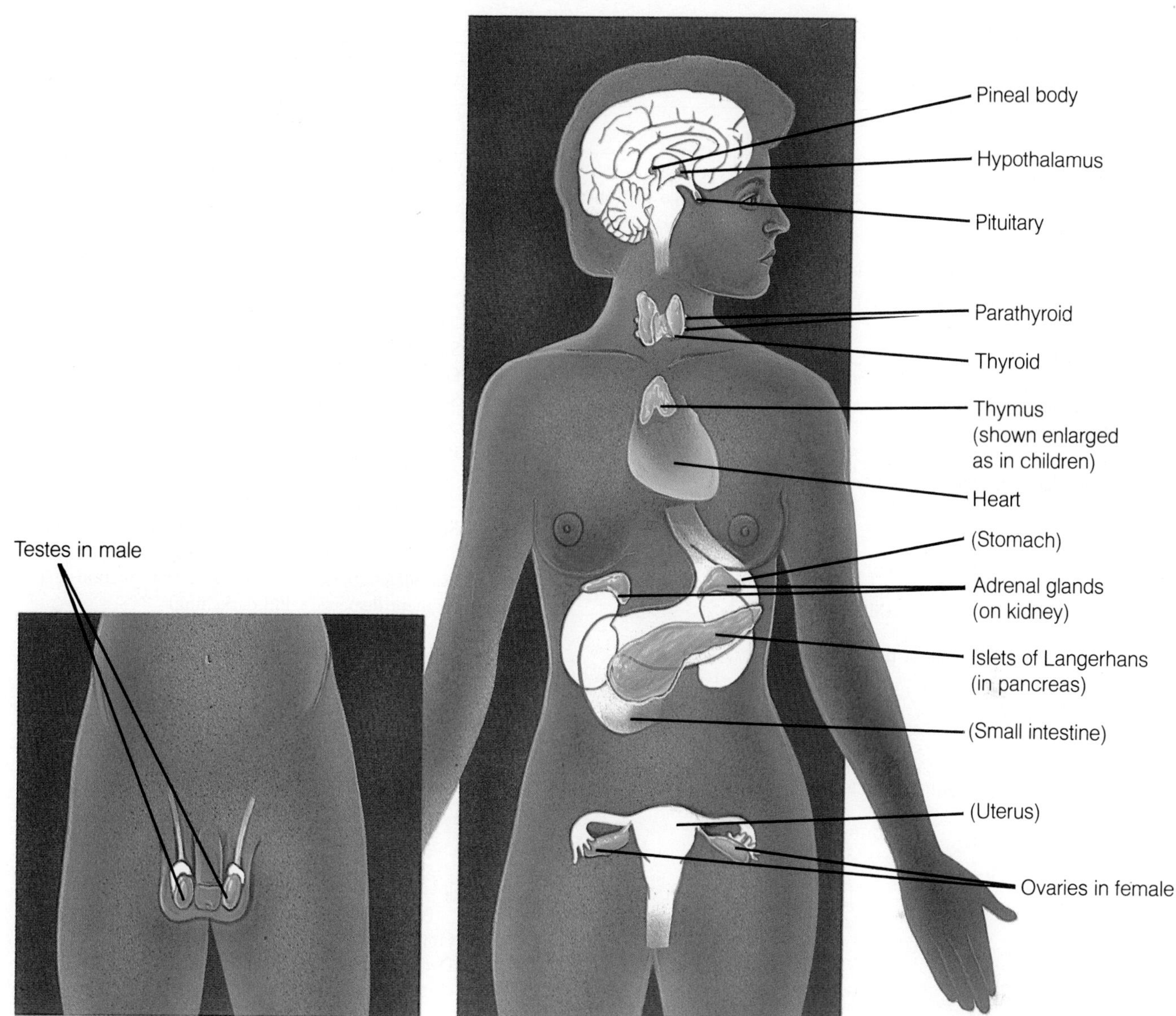

FIGURE 37.11
THE HUMAN ENDOCRINE SYSTEM
The hormone-secreting structures in humans consist of several distinct ductless glands and a number of less organized tissues in the stomach, small intestine, and placenta.

In addition to their functional differences, the two regions are also embryologically distinct: The adenohypophysis—or anterior pituitary—develops from embryonic ectoderm from the embryonic mouth, whereas the neurohypophysis—or posterior pituitary—forms as an outgrowth of the developing forebrain, so it is indeed a neural structure.

The pituitary interacts intimately with the hypothalamus, a part of the forebrain mentioned several times in earlier chapters. It is located just above the stalk on which the pituitary is suspended (see Figure 37.11).

Control by the Hypothalamus In the previous chapter, we saw that the hypothalamus plays a key role in maintaining body temperature and in osmoregulation, two critical homeostatic functions. In its relationship with the pituitary we find even more homeostatic functions. Specifically, the hypothalamus determines *what* hormones the two pituitary lobes will release and *when* they will be released. The hypothalamus actually interacts with the pituitary in two ways. Its relationship to the anterior pituitary is hormonal, but its relationship to the posterior pituitary is strictly neural. A close look at the anatomy will help with this distinction.

TABLE 37.1 ENDOCRINE SYSTEM

STRUCTURE AND SECRETION	TARGET	ACTION
HYPOTHALAMUS		
Releasing and inhibiting hormones	Adenohypophysis	Stimulate and inhibit hormone release
*Oxytocin Vasopressin (ADH)	Also considered neurohypophyseal hormones	
ADENOHYPOPHYSIS (anterior pituitary)		
Adrenocorticotropic hormone (ACTH)	Adrenal cortex	Secretes steroid hormones
	Fat storage regions	Fatty acids released into blood
Growth hormone (GH)	General (no specific organs)	Stimulates growth; amino acid transport Growth
Thyroid-stimulating hormone (TSH), also called thyrotropin	Thyroid	Secretes T_3 and T_4
Prolactin	Breasts	Promotes milk production
Follicle-stimulating hormone (FSH)	Ovary and testis	Stimulates growth of follicle and estrogen production in females, spermatogenesis in males
Luteinizing hormone (LH)	Mature ovarian follicle	Stimulates ovulation, conversion of follicle to corpus luteum, and production of progesterone
	Interstitial cells of testis	Stimulates sperm and testosterone production
Melanocyte-stimulating hormone (MSH)	Melanocytes	Pigment dispersal (skin darkening)
NEUROHYPOPHYSIS (posterior pituitary)		
*Oxytocin	Breasts	Stimulates release of milk
	Uterus	Contraction of smooth muscle in childbirth and orgasm
	Seminal vesicles	Ejaculation
*Vasopressin, also called antidiuretic hormone (ADH)	Kidney	Increases water uptake in kidney (decreasing urine volume)
THYROID		
Thyroxin (T_4)	General (no specific organs)	Increases oxidation of carbohydrates; stimulates (with GH) growth and brain development
Triiodothyronine (T_3)		Storage form of T_4
Calcitonin	Kidney, bone	Decreases blood calcium level; increases excretion of calcium by kidney, inhibits release from bones
PARATHYROID		
Parathyroid hormone (PTH)	Intestine, kidney, bone	Increases blood calcium level; decreases excretion of calcium by kidney and speeds absorption by intestine and release from bones
HEART		
Atrionatriuretic Factor (ANF)	Kidneys	Increases sodium and water excretion

*The hypothalamic hormones, oxytocin and vasopressin are carried by axons into the neurohypophysis.

The Anterior Pituitary Connection As we see in Figure 37.12a, the hypothalamus is anatomically linked to the anterior pituitary by a special circulatory arrangement. Capillaries in the hypothalamus merge to form a short vein that passes directly into the anterior pituitary. Circulatory arrangements of this kind, two capillary beds connected by a vein, are called **portal circuits.** Neurons of the hypothalamus release neurosecretions into capillaries nearby. The secretions are then carried by the blood through the portal

TABLE 37.1 ENDOCRINE SYSTEM—*continued*

STRUCTURE AND SECRETION	TARGET	ACTION
ISLETS OF LANGERHANS		
Alpha cells: glucagon	Liver	Stimulates liver to convert glycogen to glucose; elevates glucose level in blood
Beta cells: insulin	Plasma membranes	Facilitates transport of glucose into cells; lowers glucose level in blood
ADRENAL CORTEX		
Mineralocorticoids—aldosterone	Kidneys	Increased recovery of sodium and excretion of potassium and hydrogen ions; uptake of chloride ion and water
Glucocorticoids—cortisol, corticosterone	General (no specific organs)	Increases glucose synthesis through protein and fat metabolism; reduces inflammation
Sex steroids	General (many regions and organs)	Promotes secondary sex characteristics
ADRENAL MEDULLA		
Epinephrine Norepinephrine	General (many regions and organs)	Increases heart rate and blood pressure; directs blood to muscles and brain; "fight or flight mechanism"
OVARIES		
Estrogen Progesterone (ovarian source replaced by placenta during pregnancy)	General (many regions and organs) Uterus (lining)	Development of secondary sex characteristics; bone growth; sex drive (with androgens); regulates cyclic development of endometrium in menstruation; maintenance of uterus during pregnancy
TESTES		
Testosterone	General (many regions and organs)	Differentiation of male sex organs; development of secondary sex characteristics; bone growth; sex drive
PINEAL BODY		
Melantonin	Melanocytes, other targets uncertain in humans—perhaps hypothalamus and pituitary	Pigment aggregation (blanching of skin); may have some influence over hypothalamus or pituitary in cyclic activity
THYMUS		
Thymosin	Lymphocytes	Stimulates development of B-lymphocytes
NONSPECIFIC ORIGIN		
Prostaglandins	General (many regions and organs)	Presence in semen stimulates contraction in female genital tract, aiding sperm movement; stimulates ovulation in chickens, perhaps humans; aids birth through uterine contraction and cervical relaxation; affects blood clotting

circuit, directly to the anterior pituitary. There they make contact with the glandular endocrine cells, where they regulate the release of pituitary hormones.

There are some nine hypothalamic secretions that control the anterior pituitary, each having a specific effect. Among these are **releasing hormones** and **inhibiting hormones**, secretions that encourage or inhibit, respectively, the release of hormones from that lobe. For example, **growth hormone releasing hormone (GHRH)** prompts the release of

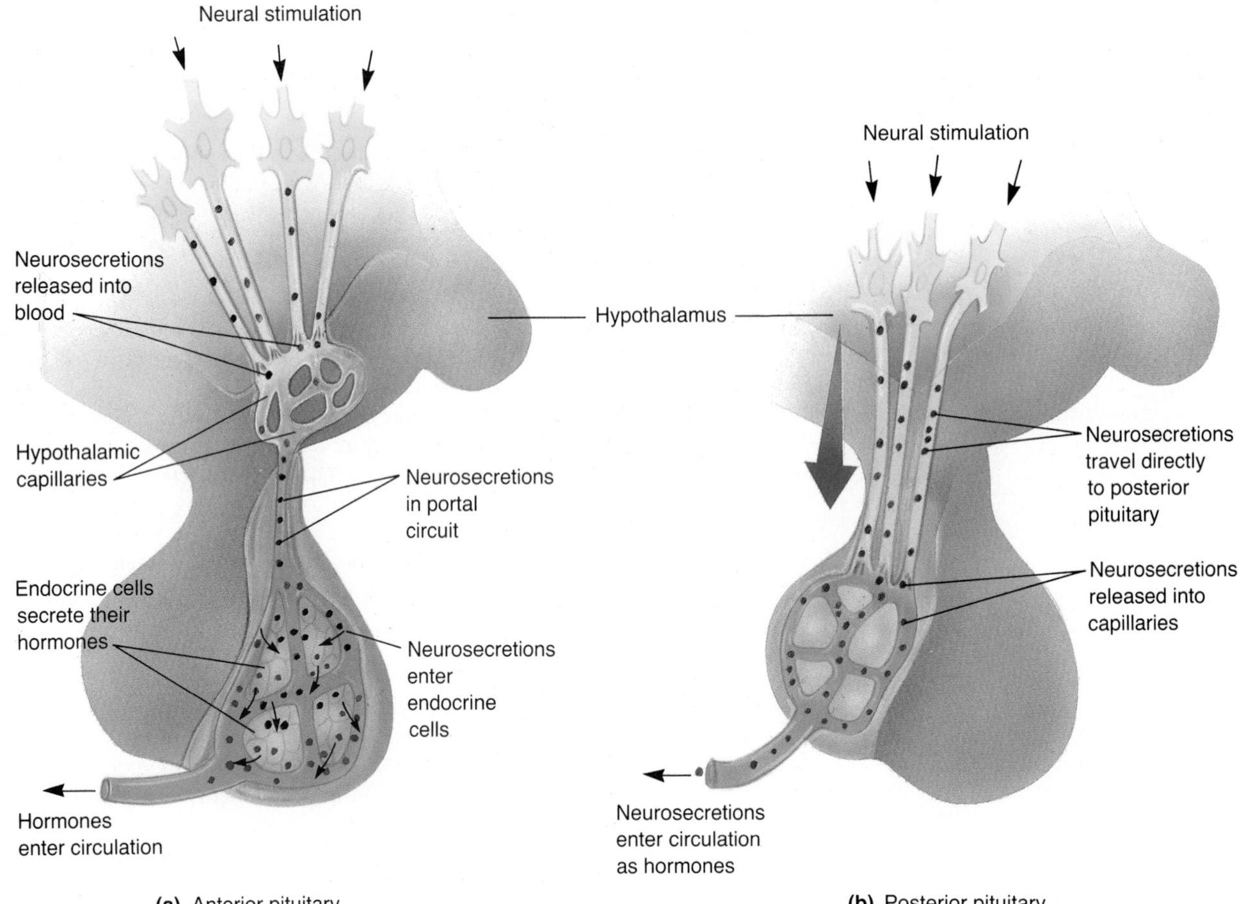

FIGURE 37.12
THE PITUITARY AND HYPOTHALAMUS
(a) The hypothalamus stimulates hormone secretion in the anterior pituitary through its neurosecretions (releasing and inhibiting hormones). After they are released, the neurosecretions pass through the portal circuit to the anterior pituitary, entering cells there and stimulating hormone release. **(b)** The posterior pituitary receives its hormones from axons emerging from cell bodies in the hypothalamus; thus, it acts only as a reservoir.

growth hormone (GH), whereas **growth hormone release inhibiting hormone** does just what you would expect—it inhibits the release of GH. A list of releasing and inhibiting hormones is given in Table 37.2. Note from the table that the secretion of three hormones of the anterior pituitary (GH, prolactin, and MSH) is controlled through releasing and release-inhibiting hormones, whereas the others are controlled through negative feedback.

Releasing and inhibiting hormones are released into the blood in incredibly small amounts. Were such amounts to be released into the main circulation they would be too diluted to be effective. But because of the short and direct portal circuit, these key messengers move unerringly to their target cells in the anterior pituitary.

The Posterior Pituitary Connection The relationship between the hypothalamus and the posterior pituitary, as we see in Figure 37.12b, is even simpler and more direct. As we've indicated, this lobe is not glandular; it merely receives and releases neurosecretions formed in the hypothalamus. Axons originating in nerve cells in the hypothalamus extend through the pituitary stalk, entering the posterior pituitary and making contact with its capillaries. The neurosecretions are then released into the capillaries for distribution to the body.

Thus, we see that the pituitary is subservient to the hypothalamus. So, does this mean that the hypothalamus is autonomous? Not at all. The hypothalamus is controlled both by a number of negative feedback loops involving hormones and an ongoing barrage of neural information from the central nervous system. Again, the relationship between hypothalamus and pituitary illustrates the intimate association of the nervous and endocrine systems in animals.

TABLE 37.2 HYPOTHALAMIC RELEASING AND INHIBITING HORMONES

HORMONE	RESPONSE BY ANTERIOR PITUITARY	
	NEGATIVE FEEDBACK CONTROL	
CRH	Corticotropin releasing hormone	Release of ACTH
TRH	Thyrotropin releasing hormone	Release of thyrotropin (TSH)
GnRH	Gonadotropin releasing hormone	Release of FSH and LH
	CONTROL THROUGH RELEASING AND RELEASE INHIBITING HORMONES	
GHRH	Growth hormone releasing hormone	Release of GH
GHRIH	Growth hormone release inhibiting hormone	Inhibition of GH release
PRH	Prolactin releasing hormone	Release of prolactin
PRIH	Prolactin release inhibiting hormone	Inhibition of prolactin release
MSHRH	Melanocyte-stimulating hormone releasing hormone	Release of MSH
MSHRIH	Melanocyte-stimulating hormone release inhibiting hormone	Inhibition of MSH release

Hormones of the Anterior Pituitary We will begin our look at the anterior pituitary's hormones with those that are controlled by negative feedback: ACTH, TSH, FSH, and LH (see Tables 37.1 and 37.2). **ACTH,** or **adrenocorticotropic hormone** is targeted for the **adrenal cortex,** the outer region of each **adrenal gland** (see Figure 37.11). When stimulated by ACTH, the adrenal cortex secretes a number of steroid hormones that have various effects. As the hormone levels rise, negative feedback begins. Since hormones enter the general circulation, they reach all parts of the body, including the hypothalamus and the anterior pituitary. Both are sensitive to rising levels of the hormones and at critical levels become inhibited, and as a result the release of ACTH is slowed (Figure 37.13).

ACTH also has another, more general, role. It influences the metabolism of fats, regulating their release for redistribution in the body.

Thyroid-stimulating hormone, or **TSH** (also known as **thyrotropin**) is responsible for stimulating the **thyroid,** another endocrine gland, to release the hormones **thyroxin** and **triiodothyronine.** These hormones regulate the metabolic rate of the body, and we will have a closer look at their functions when we consider the thyroid gland. Thyroxin and triiodothyronine are widely targeted, so many tissues respond.

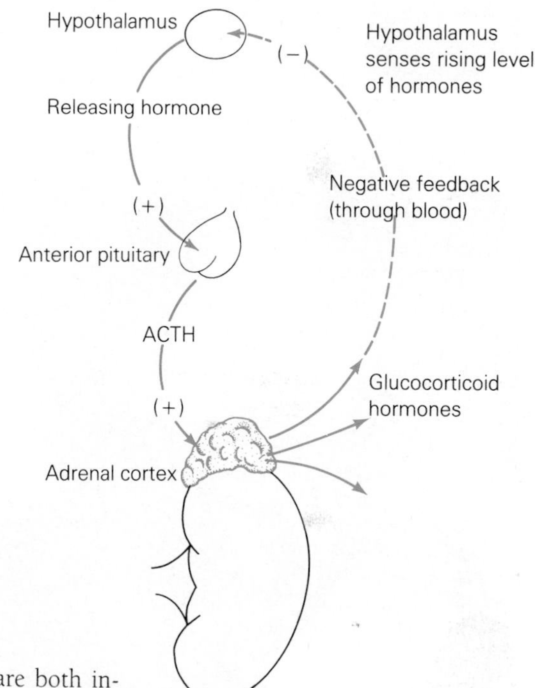

FIGURE 37.13
ACTH AND THE NEGATIVE FEEDBACK LOOP
In negative feedback loops, the product of an action has a suppressing effect on that action. Here the hypothalamus (1) has stimulated the pituitary (2) to release ACTH. In response, the target organ—the adrenal cortex—secretes steroid hormones (3). The rise of steroids in the blood acts as a negative feedback loop. (4) When sufficient quantities are present, they inhibit the hypothalamus, and the system slows or shuts down.

Follicle-stimulating hormone (FSH) and **luteinizing hormone (LH)** are both involved in stimulating the gonads to produce sperm or eggs and sex hormones. Since FSH and LH affect the gonads they are known as **gonadotropins.** In human females, the two hormones play a regulating role in the monthly menstrual cycle. We will return to the specifics of FSH and LH action in Chapter 42.

The remaining anterior pituitary hormones are controlled more directly by the hypothalamus through both releasing and inhibiting hormones. Releasing hormones, we've seen, stimulate hormone secretion, whereas inhibiting hormones slow such secretion.

GH or **growth hormone** (also called **somatotropin**) has very broad influences, activating many tissues. It stimulates growth by promoting the uptake of amino acids into cells, thereby accelerating protein synthesis. In addition, GH stimulates the release of fats from fatty tissues, in effect shifting the body more towards the metabolism of fats as an energy source. GH also stimulates the breakdown of liver glycogen into glucose (as does epinephrine, one of the adrenal hormones).

Proper levels of GH in the body are essential to normal growth. This becomes

(a)

(b)

FIGURE 37.14
GROWTH ABNORMALITIES
(a) When GH secretion is insufficient, the result can be growth retardation and the development of a pituitary dwarf. Oversecretion during adolescence can have the opposite effect, producing a pituitary giant. **(b)** Sudden increases in GH after maturity and when bone lengthening has ended produce growth in certain body parts, such as the face and hands. Here we see the result, a condition known as acromegaly.

distressingly obvious when something goes wrong. Excessive secretion (such as might occur with a pituitary tumor) during the early years can produce **giantism (pituitary giants).** Many people with this condition range from seven to nine feet tall. Conversely, in young people, an underactive pituitary with lower than normal GH levels produces **pituitary dwarfs,** often called **midgets** (Figure 37.14a). There are several thousand pituitary dwarfs in the United States today. Until recently, treatment for growth deficiencies was difficult and expensive. Stimulating growth in even one pituitary dwarf required a daily injection of all the GH that could be isolated from several human bodies. But since GH is a protein, new sources are now available through recombinant DNA technology (see Chapter 14).

Should GH levels rise suddenly in an adult human, the resulting condition is known as **acromegaly.** In this abnormality, growth that has stopped on time mysteriously resumes, but it is restricted to areas where cartilage persists, such as the hands, feet, jaw, nose, and some internal organs (Figure 37.14b).

Prolactin promotes milk production in mammals, including humans. Toward the end of pregnancy, the blood levels of prolactin increase to 28 times that in nonpregnant females. But milk is still not produced, because it is inhibited by high estrogen and progesterone levels associated with pregnancy. These levels fall off dramatically at birth and milk flow then begins. The actual release or ejection of milk is in part under the influence of oxytocin from the posterior pituitary. Thus while milk *production* is influenced by prolactin, milk *ejection* is brought on by oxytocin.

Prolactin is common throughout the vertebrate classes, where its effects differ depending on the group. For example, in freshwater fishes it aids osmoregulation; in many mammals it stimulates milk production; and in birds it functions in fat metabolism. In pigeons and a few other birds prolactin stimulates the production of "crop milk." This milky fluid is produced in the pigeon's crop, a storage organ, and regurgitated to the young. Interestingly, crop milk is very similar in its protein and lipid constituents to the milk of some mammals. Through its use, the adult birds are freed from the need to find specific food for the young and can go on with the familiar, opportunistic feeding for which pigeons are famous (that is, eating whatever is available).

Because prolactin is found in so many vertebrate groups, it is believed to have appeared early in vertebrate evolution. Theorists suggest that its diverse roles in the vertebrate groups arose through evolutionary changes not in the hormone, but in its target cells.

In humans, **melanocyte-stimulating hormone,** or **MSH,** is secreted by the anterior pituitary, but little is known about its function in our bodies. In other vertebrates, however, MSH influences the migration of pigment granules in cells called **melanocytes.** When MSH binds to cell surface receptors, melanin granules disperse throughout the melanocytes, bringing about a darkening of the skin. As the MSH concentration diminishes, the granules begin to cluster about the nucleus, and the skin lightens in color (Figure 37.15). This reversal may be brought about by another hormone, melatonin from the pineal body.

Hormones of the Posterior Pituitary As we've seen, the "hormones" of the posterior pituitary, or neurohypophysis, are actually neurosecretions. They are produced by neurons of the hypothalamus and delivered to the posterior pituitary via axons of those neurons. About ten different vertebrate neurosecretions are distributed in this way. They form a chemical family known as **neuropeptides,** each composed of nine amino acids and differing slightly from each other in their amino acid composition. Most vertebrate

FIGURE 37.15
MSH AND SKIN PIGMENTATION
Coloration in the two frogs depends upon the distribution of melanin pigments that occur in versatile melanocytes. Lighter coloration occurs when melanin granules form tight clusters. In the presence of MSH, the granules spread out, distributing melanin throughout each skin cell and bringing on the darkened appearance.

groups produce two or three different neuropeptides, although four are found in some mammals.

The role of neuropeptides differs in the various vertebrate classes, but there are two general categories. One group is uterotonic, that is, promotes contractions of the uterus (and oviducts). In some species, hormones from this group prompt the release of milk. The best known of the milk releasers is the neuropeptide hormone **oxytocin.** The second group goes by the general name **vasopressin** and consists of hormones that help the body maintain blood pressure.

In women, oxytocin is released by, among other things, the stimulation of nipples and by sexual intercourse, and causes uterine contractions during orgasm. The uterine activity, known as "tenting," may help ensure fertilization by actively drawing semen into the uterus (see Chapter 42). And we've seen that oxytocin functions with prolactin to stimulate milk release.

Oxytocin plays an important role in bringing about the contractions of the uterine muscles during labor and delivery (see Chapter 43). But once the baby is born, suckling stimulates oxytocin production, which helps bring the distended uterus back down to normal size. Synthetic oxytocin, known clinically as "pitocin," is usually administered to stimulate uterine contractions if labor is unusually prolonged or the contractions too feeble. Although the role of oxytocin in men hasn't been fully clarified, it is thought to stimulate the rhythmic smooth muscle contractions associated with ejaculation. When injected into male sheep and cattle, oxytocin increases semen output.

In humans, the posterior pituitary receives oxytocin and vasopressin from hypothalamic neurons. Human vasopressin, known more familiarly as ADH (antidiuretic hormone), helps adjust blood pressure levels by bringing about an increase in the recovery of water in the kidney, prior to the final formation of urine. ADH is secreted in response to impulses from hypothalamic pressure sensors that constantly monitor the osmotic condition of blood passing through (see Figure 36.24).

Evolutionary Relationships of the Neurohypophyseal Hormones The chemical similarity of the hypophyseal neuropeptides has important evolutionary implications. Biologists believe the 10 or so variants can be traced back to a primitive "ancestral" molecule, and that chemical variations in today's neuropeptides represent a number of gene mutations that have occurred over vertebrate history. Oxytocin, or one of its variants,

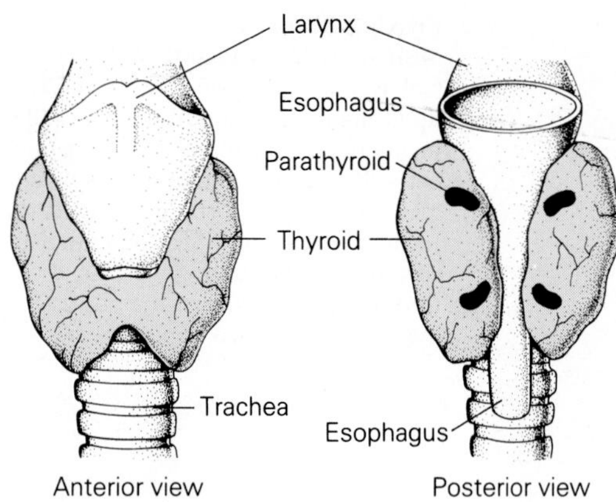

FIGURE 37.16
HUMAN THYROID AND PARATHYROIDS
The thyroid gland is a bilobed structure nearly surrounding the trachea at a point just below the larynx. Its hormones include thyroxin, triiodothyronine, and calcitonin. Note the position of the parathyroid glands on the posterior side.

is believed to be limited to placental mammals, in which its two principal functions are stimulation of milk ejection and uterine contraction. Considering the evolutionary history of vertebrates (placental mammals being the last to appear), it seems likely that oxytocin is the most recently evolved neurohypophyseal secretion.

Whereas oxytocin is absent in the pouched and egg-laying mammals and members of the other vertebrate classes, some version of vasopression occurs in all. (Interestingly, the human fetus has a more primitive version of vasopressin that is not produced in the adult.) The most ancient version is probably **arginine vasotocin**, or **AVT.** It is present throughout the vertebrates and is the only neurosecretion of its kind occurring in the jawless fishes (class Agnatha), whose ancestors were the first known vertebrates.

The Thyroid

In humans, the thyroid gland is shaped somewhat like a bow tie and is located in an appropriate place for one, slightly below the larynx ("Adam's apple," Figures 37.11 and 37.16).

The thyroid produces two nearly identical hormones, thyroxin and triiodothyronine, both of which are made up of two covalently bonded molecules of the amino acid tyrosine, with the addition of iodine. Whereas thyroxin, also called $\mathbf{T_4}$, contains four atoms of iodine (see Figure 37.2), triiodothyronine, or T_3, contains only three. This caused some puzzlement among endocrinologists until recently, when it was determined that T_3 is actually inactive and stable, even at the target cell, and is really a storage form of T_4. T_4 is both active and unstable at the target cell, where it does its job and quickly breaks down, just as a hormone should.

Thyroxin increases the metabolic rate, specifically by accelerating activity in biochemical pathways where carbohydrates are oxidized. Precisely how it does this is not known, but unlike most peptide hormones, thyroxin passes through the cell membrane, becoming active in the cytoplasm. Since it is known to enter the mitochondrion, it may express its effect there.

Overactive and underactive thyroid glands produce **hyperthyroidism** and **hypothyroidism**, respectively. The symptoms of hyperthyroidism are nervousness, hyperactivity, insomnia, and weight loss. Skinny, active people are often accused of being hyperthyroid, but in fact most skinny, active people are perfectly normal—they're just skinny and active. In Graves' disease, a hyperactive thyroid produces, in addition to the usual symptoms, a characteristic bulging of the eyes brought about by fluid accumulation in the tissues behind them (Figure 37.17a).

A **goiter** is unsightly (see Figure 37.17b), but it is actually a normal response to low levels of iodine in the diet. The prominent, bulging overgrowth of the thyroid gland increases the efficiency of iodine utilization and retention so that more nearly normal levels of thyroxin can be maintained. Before the introduction of iodized salt early in this century, such goiters were common where foods grown in iodine-depleted soil could not provide the necessary supply. Seafood has always been an excellent source of dietary iodine, but today an adequate supply is available in iodized salt sold in grocery stores.

Hypothyroidism (underactive thyroid) causes a general slowing of the metabolic rate, bringing on a condition in adults called **myxedema.** Such individuals are usually overweight, physically and mentally sluggish, and have a puffy, bloated appearance due to fluid and protein accumulation beneath the skin.

In some cases, genetic abnormalities can cause a lack of thyroid hormone from birth. If left untreated, **cretinism,** a condition marked by extreme mental and physical retardation, may develop. At present, if the sluggish thyroid condition is diagnosed early enough, cretinism can be treated effectively by the administration of a daily dose of thyroid hormone.

Thyroid control occurs through a negative feedback loop. When TRH (thyrotropin releasing hormone) from the hypothalamus reaches the anterior pituitary, the latter

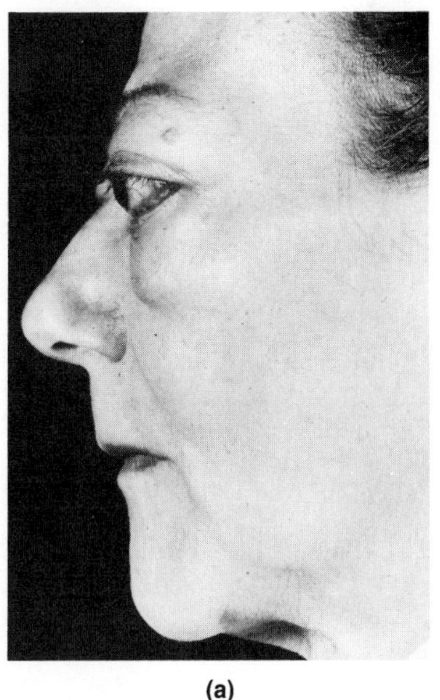
(a)

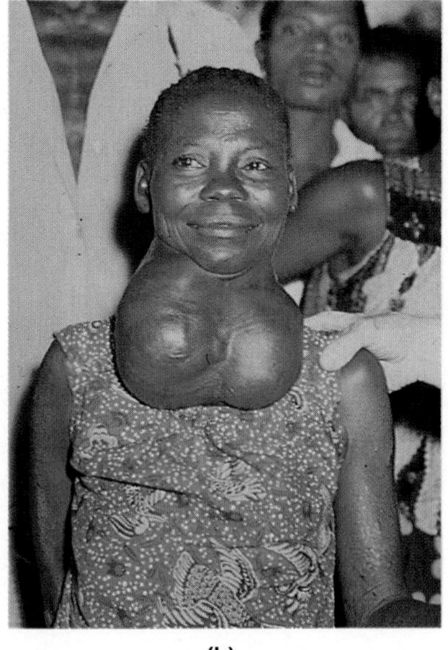
(b)

FIGURE 37.17 THYROID-RELATED ABNORMALITIES
A malfunctioning thyroid can cause a number of important clinical problems. A hyperactive thyroid may produce rapid metabolic rate, general nervousness, and a failure to gain weight irrespective of diet. In Graves' disease (**a**), the symptoms also include bulging eyes. (**b**) Iodine deficiency can lead to the formation of goiter, an enlargement of the thyroid.

responds by secreting TSH (thyroid-stimulating hormone or thyrotropin). TSH stimulates the thyroid to secrete T_3 and T_4, the latter of which elevates the rate of aerobic respiration, as we've seen. But rising levels of T_3 and T_4 also begin to inhibit the hypothalamus and anterior pituitary. Then TRH and TSH secretions subside, and the thyroid, no longer stimulated, slows its own hormone secretion (Figure 37.18).

The third thyroid hormone, **calcitonin**, together with secretions from the parathyroid, regulates calcium ion (Ca^{2+}) levels in the body.

The Parathyroid Glands

The **parathyroid glands** are pea-sized bodies embedded in the tissue of the thyroid, two or three in each wing of the "bow-tie" (see Figure 37.16). You will not be startled to learn that they produce **parathyroid hormone (PTH).** PTH is a giant among peptide hormones, its 84 amino acids making it larger than insulin.

PTH, working in concert with calcitonin from the thyroid, regulates calcium ion levels in the blood. Abnormally low levels of blood calcium ion disrupt calcium-dependent processes such as blood clotting, membrane permeability, enzyme action, and muscle function. In the latter case, calcium ion deficiencies can bring on muscle spasms and convulsions. Abnormally high levels of calcium lead to serious deterioration of bone, which, considering the high calcium content of bone, seems paradoxical.

The two calcium-regulating hormones have opposite effects—PTH increases the blood calcium level, and calcitonin decreases it (Figure 37.19). When calcium levels fall below optimal, the parathyroid responds by releasing PTH, but when calcium levels rise above optimal, the thyroid increases its release of calcitonin.

The cells most directly involved in calcium management, and the targets of the two hormones, are those of the kidneys and bones. (In the case of PTH we can add the lining cells of the small intestine.) Cells in the nephrons of the kidney respond to PTH by decreasing calcium excretion, and to calcitonin by increasing its excretion. In bone, the body's main calcium reservoir, PTH stimulates activity in bone-dissolving cells called **osteoclasts.** When they are active, calcium is released into the blood. Calcitonin, on the other hand, stimulates **osteoblasts,** the bone-forming cells. They withdraw calcium from the blood and deposit it in bone. Finally, cells lining the small intestine absorb calcium in response to PTH, but in its absence they permit the calcium to pass unabsorbed through the digestive tract. Vitamin D, we might add, aids PTH in the uptake of calcium.

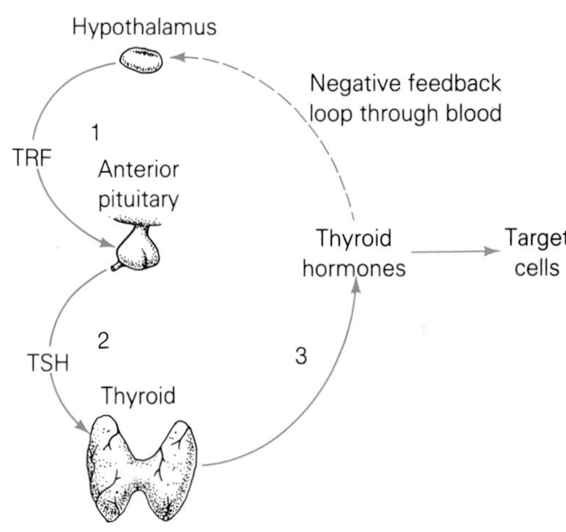

FIGURE 37.18 NEGATIVE FEEDBACK AND METABOLIC RATE
When thyroid hormonal level in the blood is low, the hypothalamus responds by stimulating the anterior pituitary (**1**) via a specific releasing hormone, TRH. In response, the pituitary secretes TSH (**2**) which in turn stimulates the thyroid to secrete its hormones (**3**). The body responds with an elevated metabolic rate. While this is happening, the hypothalamus and pituitary sense the rise in the level of thyroid hormone in the blood (**4**) and at some precise threshold they are inhibited.

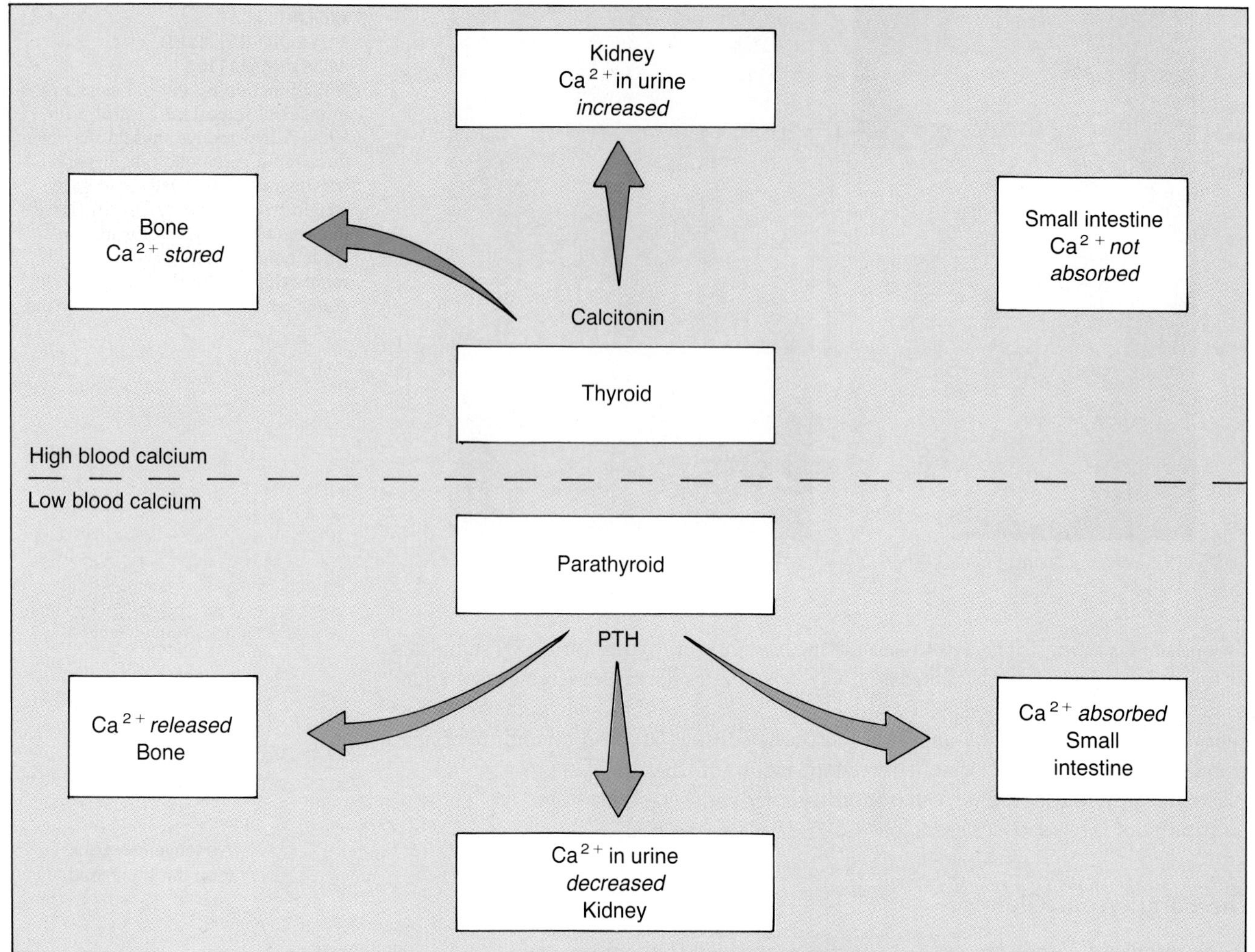

FIGURE 37.19
CALCIUM REGULATION
Parathyroid hormone (PTH) and calcitonin cooperatively regulate body calcium. PTH secretion increases when blood calcium drops. It brings that level up by withdrawing calcium from the bones, increasing intestinal uptake of calcium, and slowing the excretion of calcium by the kidneys. Calcitonin has the opposite effect. When blood calcium is high, it decreases the rate at which calcium leaves the bone, and increases excretion of calcium through the kidneys.

This, of course, is why mothers buy vitamin D-fortified milk and pester their children to drink it; mothers know about vitamin D's influence on PTH.

The Pancreas: Islets of Langerhans

The pancreas is essentially an exocrine gland, responsible for the secretion of a battery of digestive enzymes. But within the pancreas lie clusters of endocrine cells known as islet cells, or more formally, **islets of Langerhans** (Figure 37.20). Within the islets, cells designated "alpha" cells secrete the hormone **glucagon**, and cells designated "beta" secrete **insulin.**

Glucagon is a polypeptide consisting of a single chain of 29 amino acids. Its principal role is to stimulate target cells in the liver to break down glycogen into glucose. Glucagon is released when blood glucose levels fall below optimum.

Insulin has the opposite effect on blood glucose—that is, it decreases blood glucose levels. The two islet hormones complement each other in maintaining the delicate balance of glucose in the blood (Figure 37.21).

Insulin starts off as a protein that contains 81 amino acids. It is inactive in this form and must be altered by enzymes before it can do its job. The enzymes remove a number of the amino acids from the middle of the sequence, producing active insulin composed of two short chains with disulfide cross-linkages (see Figure 37.2). Insulin's best-known role is to help move glucose across plasma membranes, but it is also thought to encourage glycogen synthesis in the liver. Along with GH, insulin promotes the uptake of amino acids and their incorporation into protein, while at the same time inhibiting the conversion of amino acids into glucose.

In its role in promoting glucose uptake across cell membranes, insulin is broadly targeted: Most cells have insulin receptor sites. The mechanism by which insulin works is not well-understood. This may seem surprising considering the many years of clinical studies of this common hormone. We do know that when insulin finds its binding site in the membrane-bound receptor, the plasma membrane becomes much more permeable to glucose. Upon entry into the cell, glucose is usually converted to glucose-6-phosphate, a step that prevents it from diffusing out of the cell. Any failure in glucose regulation, whether through an under- or overabundance of insulin, or through deficiencies or defects in its cell receptors, creates havoc with the body. Such failures produce the wide-ranging symptoms of sugar diabetes, or **diabetes mellitus**, the most common form (see Essay 37.1).

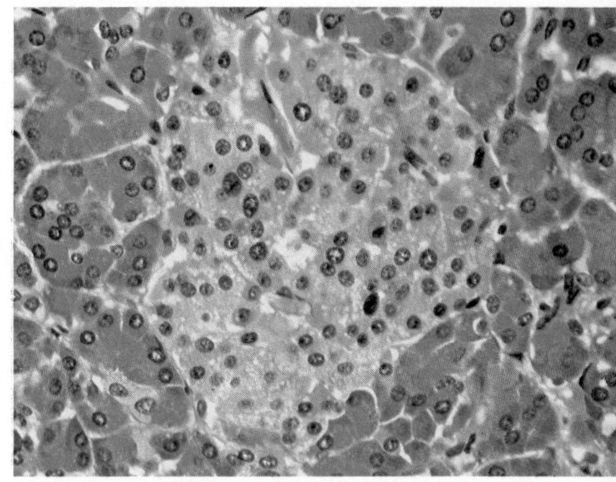

FIGURE 37.20
ISLETS OF LANGERHANS
The islets of Langerhans are endocrine tissues clustered in the pancreas (an exocrine gland). Note the differences in the islet cells and the surrounding pancreatic cells. Cells in the islets secrete insulin or glucagon.

The Adrenal Glands

Hormones of the Adrenal Cortex In humans, the paired adrenal glands are situated atop each kidney (*ad,* "upon"; *renal,* "kidney"; Figure 37.22), whereas in other vertebrates they may simply be near the kidney. The **corticosteroids**, steroid hormones of the adrenal cortex, include three groups, the **mineralocorticoids**, **glucocorticoids**, and **sex steroids.** The most familiar mineralocorticoid is aldosterone, which, as we saw in Chapter 36, affects blood pressure by increasing the reabsorption of sodium by the kidney, which subsequently results in the recovery of water. Since ADH also promotes water recovery in the kidney we see that the two hormones have overlapping functions. Aldosterone also decreases the reabsorption of potassium and hydrogen ions. Any change in hydrogen ion concentration, of course, alters the delicately balanced blood pH. For some newer discoveries on the management of blood pressure, see Essay 37.2.

Cortisol, a common glucocorticoid, promotes the conversion of certain amino acids to glucose, increases blood glucose levels, and speeds the mobilization and utilization of fatty acids. Cortisol also has anti-inflammatory and antiallergenic qualities. Hydrocortisone, a pharmaceutical preparation of cortisol, is available as an over-the-counter drug commonly used to reduce inflammation.

The last of the corticosteroids are sex steroids. These closely resemble the estrogens

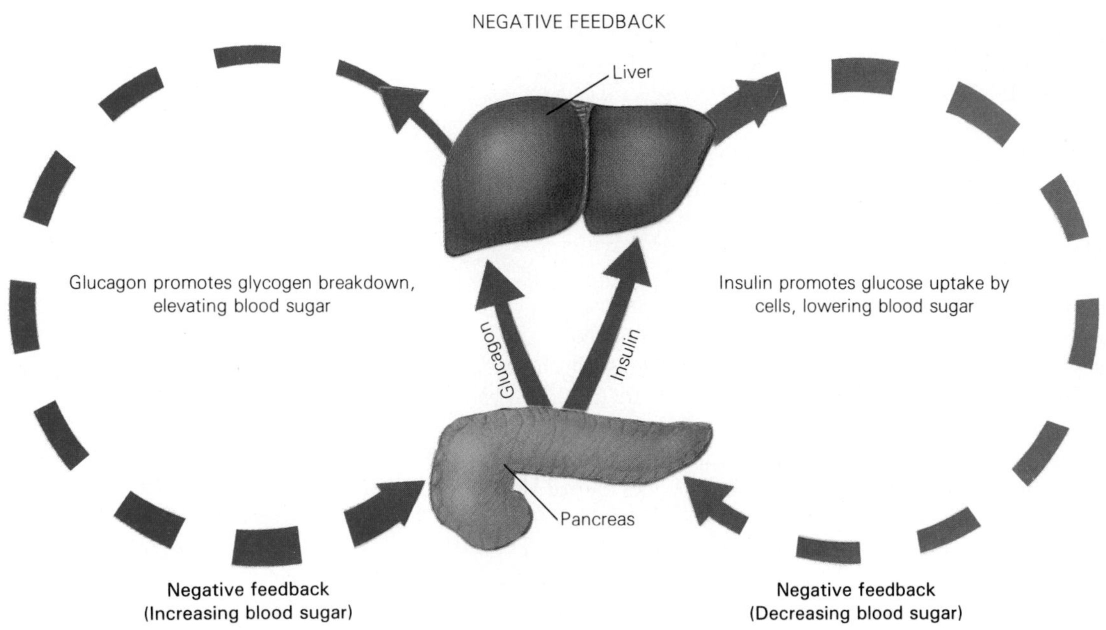

FIGURE 37.21
CONTROL OF BLOOD SUGAR
Low blood sugar and high blood sugar bring on the release of glucagon and insulin, respectively. While they have opposing actions, the controlling factors are the same. In both instances it is the actual correction of the condition that initiates a negative feedback loop that then inhibits the hormonal release.

ESSAY 37.1

DIABETES MELLITUS AND OTHER BLOOD SUGAR DISORDERS

Deficiencies in insulin activity produce **hyperglycemia** (high blood sugar), a condition that characterizes **sugar diabetes** or **diabetes mellitus.** In some diabetics, the ultimate cause of the disease is an actual deficiency of insulin, which can be caused by insufficient insulin production or by increased levels of **insulinase**, an enzyme that destroys insulin. However, most adult diabetics have normal levels of insulin but a deficiency of receptor sites on the membranes of the target cells. In some cases, the target cells have enough receptor sites but simply fail to respond properly.

The symptoms of diabetes include glucose in the urine, greatly increased urine volume, dehydration, constant thirst, excessive weight loss, exhaustion, fatty liver, ulcerated skin, local infections, blurred vision, and many other problems. Some symptoms are the result of glucose starvation of the cells—though glucose may be present, an insulin deficiency prevents it from crossing the plasma membranes. In addition, the untreated diabetic has a remarkable halitosis. The exhaled breath contains ketones, especially acetone (one of the strong-smelling chemicals in some fingernail polish removers).

Mild cases of some forms of diabetes in adults can be controlled by a well-regimented diet in which carbohydrates and fats are carefully regulated. More advanced cases can often be controlled by insulin injections, as long as the problem does not involve insufficient cell surface receptors.

Diabetes in young people (those under 20), called juvenile-onset diabetes, accounts for about 10% of all cases of the disease. It is quite unlike adult-onset diabetes, in that its cause is singular, centering specifically about a degeneration of the beta cells of the pancreas. This, of course, leads to a chronically high blood sugar, since not enough insulin is produced to bring about glucose uptake by cells. With glucose unavailable, the body mobilizes its fat reserves, which brings on the accumulation of ketones and other organic metabolites in the blood. This lowers the blood pH, sometimes to a dangerous or even fatal level.

Juvenile-onset diabetes may not have a genetic basis, or at least not the strong genetic basis we find in adult-onset diabetes. Clues come from the study of diabetes in identical twins, where any genetic factor should be clear-cut. If the dominating factor is the gene, then when one twin becomes diabetic we can expect diabetes to appear soon in the other. This is exactly what happened in twins over 50 years of age; when one became diabetic, the disease soon afflicted the other. Further, most cases of adult-onset diabetes involved causes other than insulin deficiency. But in younger diabetics, in only about half of the cases did the second twin become diabetic. Further, the younger group of diabetics were clearly insulin-deficient. Juvenile-onset diabetes, then, is different from the adult-onset disease and seems to be strongly triggered by non-genetic factors. Ongoing studies suggest several possibilities, including autoimmune disease (where the immune system attacks the beta cells) and viral disease, where certain viruses preferentially attack beta cells.

Cells may also be glucose-starved because of **hypoglycemia**, or low blood glucose. A pancreatic tumor may be at the root of this problem. As the pancreatic tissues grow out of control, too much insulin enters the blood. Any blood glucose is cleared away rapidly and sequestered as intracellular glycogen. The brain cells, which depend on glucose as a fuel, are the first to be affected. Low blood sugar in the brain results in dizziness, tremors, violent temper, and sometimes blurred vision and fainting.

In the absence of pancreatic tumors, some people can have milder (but sometimes serious) types of chronic hypoglycemia. In such cases, one of our much-praised negative feedback mechanisms goes haywire. An increase of glood glucose brings about an increase in insulin secretion, as you might suspect. The insulin in turn facilitates the conversion of blood glucose into tissue glycogen, bringing the glucose level back down. Unfortunately, there is a time lag, and the pancreatic beta cells may overproduce insulin, causing the blood glucose level to fall sharply.

ESSAY 37.2

THE HEART AS AN ENDOCRINE STRUCTURE

The heart pumps blood. That's its job, and we've known about it since William Harvey's 1628 essay on circulation. (Actually Harvey may have been scooped in 1553 by a Spaniard, Miguel Servet, who Latinized his name as *Servetus.* His novel ideas received little attention except from the members of the Inquisition, who condemned Servetus for heresay and had him burned at the stake. (His book joined him in the fire.)

Now, though, we've learned the heart has another job: it helps regulate the excretion of both sodium and water (see Chapter 6), and it does it by secreting a hormone called atrial natriuretic factor, or ANF.

The presence in the body of a substance such as ANF has long been suspected, simply because the known regulatory process couldn't account for all known changes in water and sodium retention. In 1935, John Peters of Yale University said there must be some other mechanism, in or near the heart, to monitor the volume of fluid passing through the heart and to regulate that volume. Indeed, it was found that water and sodium excretion in the kidneys were directly related to the degree which the atria were distended when filled with blood. Seemingly, a third factor (in the atria) was operating in conjunction with the secretion of aldosterone, whose role in regulating blood pressure and blood volume was known.

In 1956, very dense granules were discovered in the cells of guinea pig atria, and were later found to be present in the atria of all animals. In 1974, Marc Cantin and Jacques Genest of the University of Montreal found that the granules were similar to storage granules in the cells of certain endocrine glands. Key experiments then revealed the granules to be sites of hormone concentration. For example, when a rat atrium was homogenized and injected into another rat, the recipient showed a brief, but massive surge in water and sodium excretion.

We now know that ANF acts in complex ways in various centers throughout the human body, including the hypothalamus, the pituitary, the adrenals, and the kidneys. Essentially it regulates the volume of blood passing through the heart by influencing the excretion of water from the kidneys. ANF probably works by causing the glomeruli to become more permeable so that more water and sodium can be filtered from the blood. ANF is part of a complicated negative feedback loop that operates as the ANF inhibits the production of angiotensin II, a powerful constrictor of smooth muscle, such as that around the arterioles that pass into the kidney. Angiotensin II also prompts the adrenal gland to release aldosterone which stimulates both the kidney and the posterior pituitary to inhibit water loss.

Interestingly, people with congestive heart failure usually have high blood pressure, high water retention and high levels of sodium in the blood. However, their ANF release is also very high. It seems that in this instance, the kidneys simply do not respond to ANF.

and testosterone secreted by the gonads (as we will see shortly). The male corticosteroid hormone, referred to as **adrenal androgen**, is secreted in far greater amounts than are the female hormones. Adrenal sex steroids influence sexual behavior, particularly the female libido, and some aspects of sexual development. Excessive adrenal androgen secretion, seen more in older women, is known to produce masculinizing effects such as the growth of coarse, dark facial hair.

Hormones of the Adrenal Medulla The adrenal medulla produces two fascinating modified amino acid hormones: **epinephrine** and **norepinephrine** (commonly called *adrenalin* and *noradrenalin,* old trade names originally coined by a pharmaceutical company). The two differ slightly in that epinephrine has one more methyl ($—CH_3$) group than does norepinephrine. Their actions in the body are nearly identical, but epinephrine's effects are widespread, while those of norepinephrine are more limited. (See Chapter 35 for a review of their work as neurotransmitters in the autonomic nervous system.)

Epinephrine and norepinephrine cause a wide variety of dramatic changes in the body when an emergency arises. Their release, usually a reaction to danger, fright, or anger, brings on what is called the "fight or flight" response. The two hormones affect

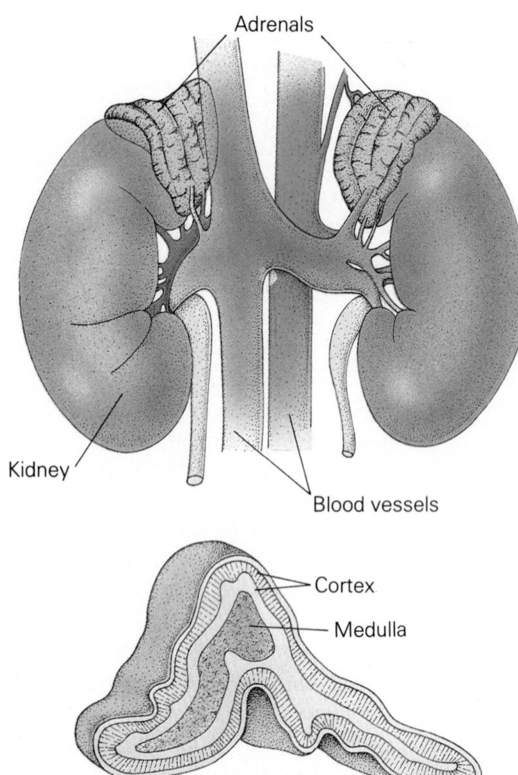

FIGURE 37.22
ADRENAL GLANDS
The adrenal glands are roughly triangular structures situated on the top of the kidneys. The gland is divided into two secretory parts, an outer cortex and an inner medulla. Each produces its own battery of hormones.

circulation by accelerating heart rate, thus increasing cardiac output, and by shunting blood into the skeletal muscles and away from other organs such as those of digestion and reproduction. This is accomplished by selected vasoconstriction and vasodilation, which closes and opens capillary beds, respectively. Blood flow in the capillaries near the body surfaces is usually restricted (causing the skin to pale and reducing the extent of bleeding). In addition, blood pressure increases as glucose is suddenly released into the bloodstream from the liver. We have all heard stories of people exhibiting superhuman feats of strength at these times, and you may have had such an experience yourself.

Adrenal Hormones and Physiological Stress Stress is a topic of increasing concern to a host of people, from physiologists to psychologists, sociologists, and apologists. If you're sitting there trying to get through a book of this size and finals are coming up Thursday, you have some idea of the concept. As you haul the book around, you may gain a different appreciation of the term. Stress, then can arise on many levels. Furthermore, it is not an exclusively human malady; stress can occur in any vertebrate. What does stress do and what causes it?

When a human or other vertebrate is injured or subjected to irritating stimuli, the body responds in an effort to meet the problem and counteract the effects of the stimulus. Both psychological and physiological stress are associated with sudden increases in the levels of certain adrenal hormones. Their surge begins as higher brain centers prompt the hypothalamus to increase its secretion of corticotropin releasing hormone. The anterior pituitary responds by increasing its ACTH output, which stimulates the adrenal cortex to increase considerably its output of cortisol, and to a lesser extent aldosterone. This results in elevated blood glucose, suppression of the immune and allergic responses (an effect of cortisol), and elevated blood pressure through increased water and sodium reabsorption (an effect of aldosterone). Whereas some of these reactions have straightforward survival value, others do not. One must wonder, what possible use would it be for the immune system to be suppressed following severe injury? It may be that the suppression merely keeps the immune response under control so that it can function over the long term; but we just don't know.

Stress also increases the output of hormones from the adrenal medulla, including epinephrine and norepinephrine, which further elevate blood sugar, heart rate, and blood pressure, and in general prepare the body for some form of exertion. Vasopressin (ADH) secretion by the posterior pituitary is another result of stress, so water retention and blood pressure are further increased. Increased blood pressure, of course, would become important following an injury that resulted in massive bleeding. Falling blood pressure is a principal cause of physiological shock and kidney failure. (Of course, increased blood pressure would also temporarily increase the loss of blood.)

From a biological point of view, stress responses are adaptive—*if they are transient.* A situation arises, the body responds, and the response is adaptive. But ongoing or long term stress, such as humans might experience in a fast-paced, problem-ridden, and highly competitive life style, is apparently different. The prolongation of such dramatic physiological responses can bring on permanent damage to the mind and body. When rats are stressed by being artificially crowded together, they develop a range of aberrant behaviors, from asexuality to cannibalism and infanticide. Monkeys forced to make decisions amid unpleasantness develop ulcers. What sort of experiments might we be currently running, unaware, on our own species?

The Gonads: Ovaries and Testes

The endocrine function of the ovary and testis is discussed in Chapter 42, so let's just briefly note some of the major points here. First, the hormones of the ovary—**estrogens** and **progesterone**—and the principal hormone of the testis—**testosterone**—are all steroids. Estrogen, by the way, includes a whole family of molecules, among them **estradiol**, **estriol**, and **estrone.** Estradiol is the major estrogen secreted by the ovary. For a look at the significance of sex hormones in embryonic development, see Chapter 43.

In addition to their reproductive function, sex hormones have an important role in skeletal development. Their sudden increase in the blood at the onset of puberty stimulates the lengthening of the long bones in humans, and causes a marked spurt of growth. (Interestingly, it also causes a simultaneous spurt in mental growth as measured by test performance). The onset of puberty is variable: some of us are precocious and some are definitely late bloomers. Toward the end of puberty, the sex hormones have the opposite effect. They promote the final fusion of growth regions in the long bones, after which no further increase in height is normally possible. Thus late-maturing adolescents, those in which bone fusion is delayed, may grow to be unusually tall. Low levels of estrogen secretion may delay puberty and produce tall girls. The same relationship, by the way, holds for testosterone and boys. Remember that tough, well-muscled kid who matured way ahead of everyone else—and then quit growing, so that by high school he was a "short guy" and quit picking on you after your own hormones finally got going?

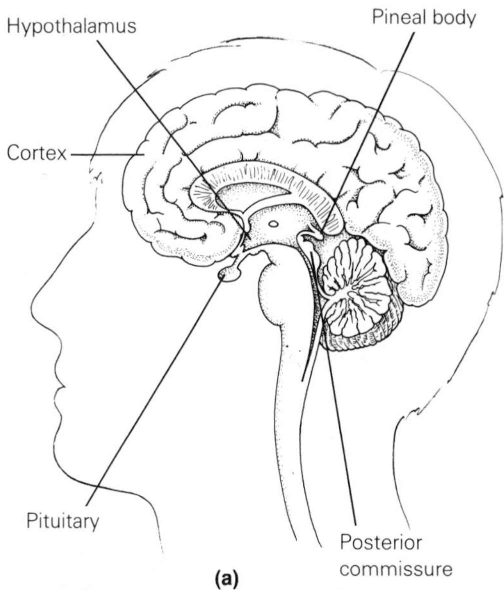

FIGURE 37.23
THE PINEAL BODY
(a) In humans, the pineal body is located in the roof of the third ventricle. (b) In reptiles, it occurs as the "third eye," beneath a scaly lens.

The Thymus

The **thymus**, located just behind the sternum in adult mammals (see Figure 37.11), is a rounded, spongy organ. Its only known function, a vitally important one, is providing cells for the immune system, particularly a class of cells called lymphocytes. Such cells originate in the red bone marrow and apparently migrate to the thymus before birth. Some of the lymphocytes venture out into other places, such as the spleen and lymph nodes. Soon after birth, the thymus begins to secrete the hormone **thymosin**, which is targeted to certain members of the resident lymphocyte population. Their response is to mature into B-cells, lymphocytes that later specialize in forming antibodies. The human thymus changes with time. It is most prominent in children, peaking in its growth at puberty, whereupon it begins to atrophy, becoming almost unrecognizable in older people.

The Pineal Body

In humans, the **pineal body** is a pea-sized gland, located deep in the brain, just above the brain stem and protruding from the roof of the third ventricle (Figure 37.23a). Some biologists have serious reservations about whether the pineal body has an endocrine function, but others are convinced that it does. The first strong evidence of an endocrine function was the discovery that when a pineal extract from cattle is injected into frogs, a general blanching of the skin occurs. Scientists were intrigued with the notion that cattle extracts could cause this effect in frogs. Because of its pigment-related effects, the extract was called **melatonin.** Further observations revealed that such color changes were the result of a clustering together of melanin pigments—just the opposite of the MSH effects mentioned earlier. Recall that MSH brings on pigment dispersal and a darkening of the skin.

Melatonin is also thought to play a role in **circadian rhythms** (*circa,* "about; *dia,* "day"; see Chapter 45). Circadian rhythms are behavioral and physiological changes that take place on a daily basis. It is known that many circadian rhythms in vertebrates are regulated by melatonin. These rhythms include daily cycles of sleep and activity, and they even affect receptivity to medicines or other hormones. If the pineal body is removed in birds, the cyclic behavior ceases; if it is implanted in a non-cycling bird lacking the body, cycling activity resumes. Researchers have found that when even a few pineal cells are grown in a culture dish in darkened surroundings, they secrete melatonin, roughly on a cyclic 24-hour schedule, just as they do in the living organism.

Melatonin may also be a gonadal inhibitor in mammals. In rats exposed to long periods of light, melatonin secretion is slowed, and the rat enters reproductive readiness. In continuous light, the rat remains in constant readiness. The connection between light and the pineal is indirect, with nerve pathways relaying their impulses several times between the retina of the eye and the pineal body. In birds, the light receptor is not the eye at all but some unknown region of the brain. This receptor has been located in frogs and lizards and is called the "third eye," or "median eye." Removal of the median eye results in an increase in sexual activity in these vertebrates.

People have continued to speculate about the pineal, particularly about its evolutionary history. The answer is a bit startling. It turns out that it may be the remains of some ancient and vestigial second pair of eyes.

Here is a case in which an understanding of evolutionary origins was a real help in working out the physiology and function of an organ. The fossil skulls of the earliest ostracoderms (jawless fishes) have holes for three eyes. The third eye is in the middle, right on top of the skull. Only a remnant of the eye remains in modern fish (and humans as well), but some amphibians and reptiles have retained a rudimentary but recognizable eye. In the New Zealand reptile *Sphenodon punctatus,* called the tuatara, the small median or third eye even has a lens and retina. Some very common American lizards also have a tiny, degenerate third eye on the top of their heads, complete with light receptors and an eyehole in the skull covered by a special translucent scale (Figure 37.23b). This third eye can't form an image, but it is sensitive to light. It is, in fact, the light receptor by which the lizard sets its biological clock, behaving one way when days are long and another way when short winter days draw on.

Physiologists are notably vague in discussing the role of the pineal body in humans. It is known to secrete melatonin, and the quantity released follows a day-night cycle, rising in the night and falling in the daytime. Some physiologists suggest a gonad-suppressing effect in humans, but this has not been substantiated.

Prostaglandins

The **prostaglandins** make up the most recently discovered class of vertebrate hormones. In 1982 the Nobel Prize for this discovery was awarded to Sune K. Bergstrom and Bengt Samuelsson of Sweden and John R. Vane of England. Some of the prostaglandins are among the most potent biological materials known. Perhaps the reason for their late discovery is that they are not produced by specialized organs. Prostaglandins of some sort are produced by most kinds of tissues.

The release of prostaglandins can be brought about by other hormones or by almost any irritation of the tissues, including mechanical agitation. Not surprisingly, some prostaglandins are involved in inflammatory responses and in the sensation of pain. Aspirin may be effective against pain, inflammation, and fever because it inhibits the synthesis of prostaglandins.

Prostaglandins have been found to have other actions. For instance, one type causes uterine contractions. In the 1970s, Samuelsson found that blood platelets, which are involved in blood clotting reactions, produce a prostaglandin called **thromboxane** that causes platelets to stick together and the walls of arteries to squeeze shut. But Vane discovered another prostaglandin, **prostacyclin**, that has exactly the opposite effects—it prevents clots, keeping blood fluid, and prevents arteries from closing. Since prostacyclin is produced by the cells that line the blood vessels, it seems that prostaglandins may be important in maintaining normal circulation.

We mentioned that aspirin inhibits prostaglandin synthesis. It turns out that thromboxane synthesis is inhibited by very low levels of aspirin, while such levels have little effect on prostacyclin. So is aspirin good or bad for us? It can be both. First, there is evidence that some heart attacks, arterial disease, and stroke are caused by an imbalance of thromboxane, where spontaneous clotting (embolus formation), and arterial constriction bring on the familiar symptoms. Seemingly, one answer is to lower thromboxane levels, and aspirin can do this. In fact, recent medical evidence indicates that aspirin in low doses (one tablet every other day) is beneficial. It is known that habitual aspirin users have far lower rates of arterial disease and heart attacks than do nonusers. On the down side, aspirin in high doses, such as those followed by arthritis, can cause serious stomach bleeding, so caution is advised.

We see, then, that animals have developed highly complex and specific ways of chemically coordinating their activities and responses. We should add that our discussion has not been exhaustive. There are still other chemical messengers. Some of these we discuss elsewhere, such as the digestive hormones and hormones secreted by the mammalian placenta. Nevertheless, we should now be able to appreciate the important regulatory, responsive, and homeostatic roles of chemical signals in the daily lives of animals.

APPLICATION OF IDEAS

1. As two nineteenth-century physiologists attempted to determine the function of the pancreas, they observed flies swarming around the urine of dogs from which the pancreas had been removed. This observation was an initial step in our understanding of the role of insulin in sugar metabolism. What might the next steps have been in isolating insulin, and what rigorous rules should the investigators have followed in establishing its function?
2. Assuming that insect hormones could be synthesized in sufficient quantity, suggest a specific hormone that might be applied to insect control. Explain your choice, how it might work, and reasons why this method of insect control might be more ecologically suitable than the use of chemical insecticides.
3. While the anterior pituitary and thyroid glands are known to control growth, malfunctions in any of the endocrine glands can also affect growth. Considering each of the endocrine glands, explain why this is so.
4. The controlling functions of the endocrine system often overlap those of the nervous system. Cite examples of this overlap and comment on the adaptive significance, if any, of this.

KEY IDEAS

THE CHEMICAL MESSENGERS

Hormones form three categories: **endocrine** hormones (acting at some distance, **paracrine** hormones (acting in cells nearby), and **autocrine** hormones (acting within the cell of origin). Endocrine glands are ductless; they secrete hormones into the blood.

Neural and Hormonal Control Compared

1. Although both neural and hormonal control involve chemicals, in neural control the chemicals are secreted into the synaptic cleft. In chemical control, they are generally secreted into the blood. Neural control is transient and fast working; chemical control is slower and longer lasting.
2. Whereas neurons generally secrete neurotransmitters, in some instances their secretions, or **neurosecretions,** are released into the blood.

Molecular Structure of Hormonal Messengers

Chemically, most hormones are peptides or lipids. The smallest peptide hormones have one amino acid, whereas the largest has 84. The lipid group includes steroids and prostaglandins.

Characteristics of Hormonal Control

1. Each kind of hormone has specific **target cells** that bear matching receptor sites, either on the surface or in the cytoplasm.
2. Hormones are shortlived. Most are chemically degraded in less than one hour. Short- and long-term effects are determined by the rate of secretion.
3. Physical and chemical cues prompt hormonal secretion.
4. Many hormones are controlled by negative feedback loops involving the hormone or a product of its action.

Identifying Hormones

1. An older method for positively determining a hormone's presence and action involves removing the source tissue or gland, observing subsequent deficiency symptoms, introducing extracts of the excised tissue, and observing alleviation of the symptoms.

2. Among many more recent techniques are the use of isotope-labeled hormones (to establish target cells) and **radioimmunoassay.** The latter requires the generation of antibodies sensitive to the suspected hormone and then using the antibody to determine its presence. The quantity of the hormone in question can also be determined.

CHEMICAL MESSENGERS AND THE TARGET CELL

Most peptide hormones interact with membrane receptors on the target cell, thereby activating second messengers. Lipid hormones act directly at the gene level.

Peptide Hormones and Second Messengers

1. **Second messengers** include **cyclic AMP, cyclic GMP, and IP_3.**
2. When epinephrine binds to its receptor site on a liver cell, it activates the enzyme **adenylate cyclase**, which converts ATP to cAMP. cAMP activates enzymes that break glycogen down into glucose, which diffuses out of the cell.
3. The action of second messengers brings on a **cascading** effect, with more and more molecules involved as the steps proceed. Second messengers are short-lived, their rapid deactivation making control possible. Upon their deactivation, the related enzyme system also becomes inactive.
4. The response of a cell to cAMP depends ultimately on which of many possible enzyme systems is present in that particular cell.
5. Some cells have dual second-messenger systems, each activated by its own hormone. Their actions are opposite; thus, in smooth muscle the formation of cAMP causes relaxation, and the formation of cGMP brings on contraction.

Steroid Hormones and Gene Control

Steroids pass through the plasma membrane and join **cytoplasmic binding proteins.** The new complex enters the nucleus and binds to a **chromosomal receptor.** The new association then activates a gene, and protein synthesis ensues. Typically, the protein is an enzyme whose appearance enables the cell to carry out the required hormonal response.

INVERTEBRATE HORMONES

The source of hormones in many invertebrates is the nervous system.

Hormonal Activity in Arthropods

1. In molting, or **ecdysis**, skeletal materials ae absorbed, the remainder of the cuticle is secreted and hardened.
2. In crayfish, the Y-organ secretes MH, the molting hormone, while the nearby X-organ secretes molt-inhibiting hormone (MIH). Light is an essential part of regulating ecydysis.
3. Many insects pass through a complete metamorphosis, including egg, larva, pupa, and adult states. The juvenile state is maintained by juvenile hormone (JH). Final molting and the end of the larval state occur when brain hormone from the corpus cardiacum stimulates the prothoracic gland to secrete **ecdysone**, which brings on the pupal and adult stages.
4. Transplanted larval corpora allata cause an extension of the larval state. If the transport of hormones is stopped, parts of the body mature while other parts remain in a larval state.

THE VERTEBRATE ENDOCRINE SYSTEM

The vertebrate endocrine system includes a number of distinct endocrine glands and less-defined tissues.

The Pituitary

1. The hypophysis, or **pituitary gland**, has two functionally and embryologically distinct regions, the adenohypophysis, or **anterior pituitary**, and the neurohypophysis, or **posterior pituitary.**
2. The **hypothalamus** controls secretion by the pituitary. It interacts with the anterior pituitary through neurosecretions that are released into blood. The blood carries the neurosecretions directly to the anterior pituitary via a special blood **portal circuit.**
3. Hypothalamic neurosecretions include a number of specific releasing hormones, each of which stimulates the anterior pituitary to release one of its seven corresponding hormones. The release of some anterior pituitary hormones is inhibited by **negative feedback** alone, but several are controlled by additional neurosecretions called inhibiting hormones.
4. The two hormones of the posterior pituitary are not synthesized there but are received as neurosecretions from axons originating in the hypothalamus.
5. The secretion of **ACTH, TSH, FSH,** and **LH** is inhibited by negative feedback only. ACTH stimulates the adrenal cortex to release its steroid hormones and helps regulate the metabolism of fats. TSH stimulates the thyroid to release its two hormones, thyroxin and triiodothyronine. FSH and LH stimulate the gonads to produce sex hormones and gametes.
6. The secretion of **GH, prolactin,** and **MSH** is inhibited by inhibiting hormones. GH promotes growth through increased uptake of amino acids, stimulates the release of fats, and prompts the conversion of liver glycogen to glucose. Oversecretion of GH brings on **giantism** and **acromegaly**, and undersecretion causes dwarfism. **Prolactin** promotes milk production, but milk release requires oxytocin. Prolactin is common through vertebrates. It promotes osmoregulation in fishes and crop milk synthesis in some birds. MSH stimulates the dispersal of pigment melanocytes, bringing about darkening of the skin.
7. Vertebrates produce some 10 neurohypophyseal, known as **neuropeptides**, which have two general functions. One group is **uterotonic,** causing contractions in the uterus and oviducts—the best known member is **oxytocin.** The second group, the **vasopressins** (ADH in humans), helps regulate blood pressure.

8. In humans, oxytocin is involved in smooth muscle contractions associated with sexual arousal, orgasm, ejaculation, and the birth process. It also brings on milk release.
9. The oldest vertebrate neuropeptide is thought to be one called **arginine vasotocin,** which is present in all classes, including the jawless fishes. Oxytocin is more recent, occurring only in placental mammals.

The Thyroid

1. Two of the three thyroid hormones, **thyroxin** (T_3) and **triiodothyronine** (T_4) differ only in their iodine content. In amphibians thyroid hormone stimulates metamorphosis, while in humans it influences carbohydrate metabolism.
2. **Hyperthyroidism** produces weight loss and nervousness, while a shortage of iodine produces a **goiter,** a visibly enlarged thyroid gland. **Hypothyroidism** causes **myxedema**—sluggish behavior, body puffiness, and fluid retention. In children, hypothyroidism leads to cretinism, severe mental and physical retardation.
3. **Calcitonin,** the third thyroid hormone, works with the **parathyroid glands** to regulate calcium.

The Parathyroid Glands

1. **Parathormone (parathyroid hormone** or PTH), a polypeptide, cooperates with calcitonin in regulating the distribution of body calcium. Parathormone increases calcium levels in the blood by increasing intestinal absorption and kidney reabsorption, and by stimulating the bone-decalcifying activity of **osteoclasts.** Calcitonin performs roughly the opposite actions, including the stimulation of bone-calcifying activity by **osteoblasts.**
2. Low levels of PTH produce blood calcium deficiencies resulting in severe muscle spasms and loss of muscle control. High PTH levels cause severe bone calcium loss (**osteoporosis**) and irregularities in muscle contraction.

The Pancreas: Islets of Langerhans

1. The **islets of Langerhans** are clusters of insulin-secreting **beta cells,** and glucagon-secreting alpha cells.
2. Glucagon stimulates the liver to convert glycogen to glucose, which elevates the blood glucose level. Insulin stimulates cells to take up glucose, thus lowering blood levels.
3. Deficient insulin levels cause high blood sugar, commonly called **diabetes mellitus.**

Adrenal Glands

1. The **adrenal glands** have two layers—the outer cortex and inner medulla. The adrenal cortex secretes mineralocorticoids (aldosterone), glucocorticoids (cortisol), and sex steroids.

a. Aldosterone increases potassium excretion by the kidney and slows sodium excretion. Aldosterone hypoactivity causes excessive salt and water loss and potassium retention, causing a potentially fatal acidosis.

b. Cortisol increases blood levels of glucose, amino acids, and fatty acids and is an anti-inflammatory agent.

c. Adrenal steroid sex hormones include **adrenal adrogens.** The sex steroids influence libido, and development.

2. Adrenal medullary hormones include **epinephrine** and **norepinephrine** (adrenalin and noradrenalin). Their general effect is to prepare the body for emergencies (for example, increasing heart and breathing rate, elevating blood sugar, and shunting blood away from areas such as the digestive tract).
3. The adrenal gland plays a role in physiological stress, responding by secreting aldosterone, cortisol, epinephrine, norepinephrine, and vasopressin. Elevations in these hormones aid in raising blood pressure, suppressing inflammation, and increasing "fight or flight" responses, all intended to aid the body in meeting emergencies. Prolonged stress responses, common to some human life styles, produce physical and psychological damage.

The Gonads: Ovaries and Testes

The ovaries produce two kinds of steroids, **estrogens** (**estradiol, estriol,** and **estrone**) and **progesterone,** while the testes produce the steroid hormone **testosterone.** Sex hormones influence reproduction but also influence body development and bone growth.

The Thymus

The thymus secretes **thymosin,** which stimulates lymphocyte development.

The Pineal Body

Pineal extracts containing melatonin cause pigment withdrawal in frog melanocytes, thus blanching the skin. Melatonin may also influence circadian rhythms—affecting the alternating periods of sleep and activity. Light may influence melatonin action, via the optic nerves in some groups, the "third eye" in frogs and lizards. The third eye may only affect the biological clock.

Prostaglandins

Prostaglandins are produced almost universally in the body. Some are involved in inflammation, others cause uterine contractions; one (**thromboxane**) aids blood clotting, while another (**prostacyclin**), slows clotting. Aspirin inhibits thromboxane synthesis, and new studies on aspirin suggest that this may be related to reduced incidences of heart and arterial disorders.

REVIEW QUESTIONS

1. How do endocrines, paracrines, and autocrines differ? (877)
2. List a similarity and a difference between neural and chemical control. (877–878)
3. Name and generally describe the two chemical classes of hormones. (878)
4. List two characteristics of hormonal control that help promote coordination and control. (878)
5. List the four steps in the traditional procedure for determining whether an observed response or activity is under hormonal influence. (878)
6. List the steps that an endocrinologist might follow when identifying a new hormone. (878–879)
7. Outline the events that follow the binding of epinephrine to its cell membrane receptor. Which substance represents the second messenger? How does the system shut down? (882)
8. How is it possible for several different hormones to promote different responses, yet make use of the same second messenger? (883)
9. Summarize the operation of a dual second-messenger system. (883,885)
10. Outline the events that follow the entry of a steroid hormone into the cell. (884–885)
11. How, in terms of promoting enzyme activity, does the effect of a steroid hormone differ from that of a peptide hormone? (884–885)
12. Describe the interplay between the X-organ and Y-organ of the crayfish. (886)
13. Name and briefly describe the four stages of molting. (886)
14. List the secretions of the insect corpus cardiacum, corpus allatum, and prothoracic gland. Which promotes the final larval molt and emergence of the pupa and adult? (887)
15. What happens when ecdysone is present in the insect head and thorax but is prevented from reaching the larval abdomen? (888)
16. Compare the functional relationship of the anterior and posterior pituitary lobes with the hypothalamus. What does this tell you about the posterior lobe as a gland? (888–889)
17. List a releasing hormone and an inhibiting hormone, explain what they do. (891–893)
18. Review the negative feedback loop established between the hypothalamus, pituitary, and adrenal cortex. (893)
19. List the targets of the anterior pituitary hormones ACTH, TSH, GH, and prolactin. (893–894)
20. What are the effects of too much GH in the child? In the adult? (893–894)
21. Under what conditions do prolactin levels rise in the blood? (894)
22. What is the relationship between oxytocin and prolactin? (895)
23. What is the true source of oxytocin and antidiuretic hormone? What is the target and action of the latter? (891–892)
24. How does the chemical structure of triiodothyronine differ from that of thyronine? Compare their actions. (896)
25. Compare the symptoms of an overactive thyroid with those of an underactive thyroid. Under what conditions does cretinism occur? (896)
26. Compare the calcium-regulating action of calcitonin with that of PTH. What are the symptoms of an overabundance of the latter? (897–898)
27. Name the types of islet cells and their secretions. (898)
28. Briefly discuss the interaction of glucagon and insulin in the precise regulation of blood sugar. (898)
29. What are some possible causes of hypoglycemia? What symptoms accompany the abnormality? (Essay 37.1)
30. Name the two regions of the adrenal gland and list their hormones. How do the chemical structures of the two groups of hormones differ? (899)
31. Describe the action of the hormone cortisol. Which of its actions are also produced by the drug hydrocortisone? (899)
32. Other than their reproductive functions, what roles do the sex hormones have? (902–903)
33. What are the two known actions of melatonin, the pineal secretion? (903–904)
34. Where are the prostaglandins produced? List two actions of prostaglandins in the circulatory system. What might a low level of prostacyclin cause in the blood? (904–905)

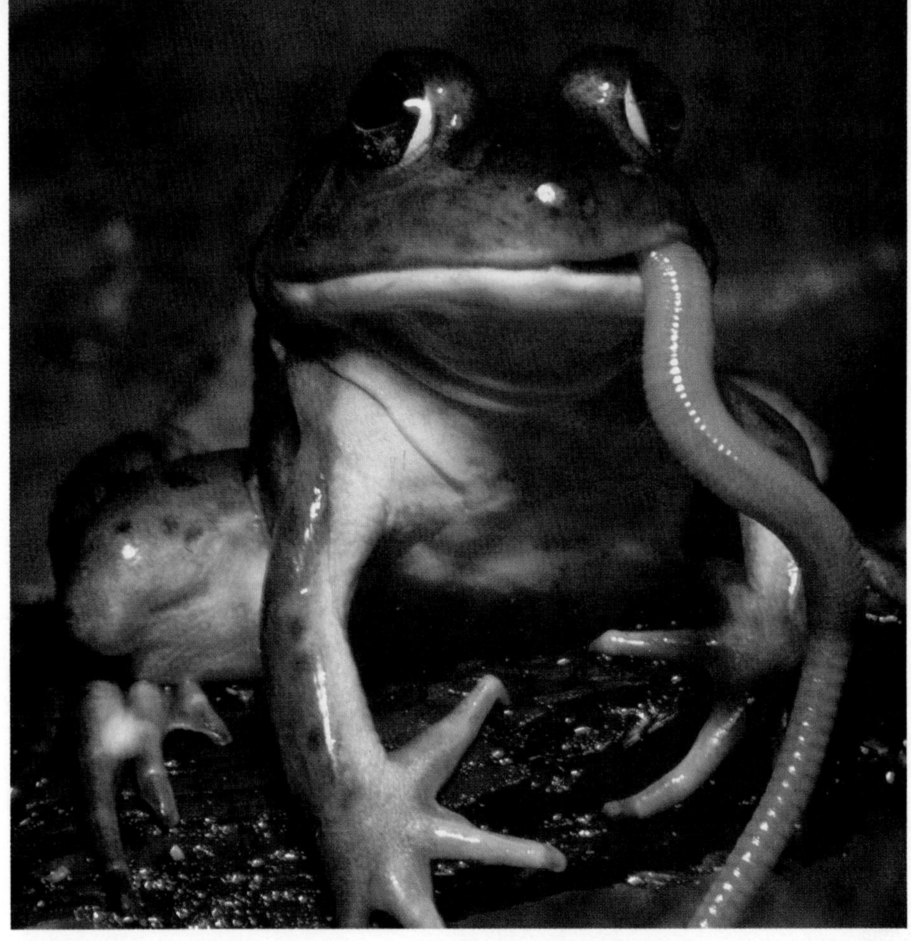

38

DIGESTION AND NUTRITION

IT'S ALWAYS A LITTLE RISKY TO GENERALIZE ABOUT EVOLUTIONARY TRENDS since evolutionary change runs in many directions at very different rates. Some traits change rapidly, some change slowly. If a species happens to harbor many rapidly changing traits, then the species itself changes rapidly. By the same token, a species with a great proportion of slowly changing traits will likely be very similar to its distant ancestors. Even within the same kind of animal, different organs may evolve at different rates and show different levels of specialization. And in some cases, species with many advanced traits can also harbor some very primitive traits. For example, we humans consider ourselves advanced, but we have primitive mouths. The specialized bill of the seemingly primitive platypus, on the other hand, is considered one of the most advanced mouth structures among vertebrates. But the gut! We humans can be proud of our gut (though we rarely boast about it). We can be proud because we have a very highly evolved gut—not as highly evolved as that of the cow with its four-chambered stomach, but still very efficient and rather complicated. It is only fitting, then, to learn more about the specializations of such a nice gut, but first we'll look at those of other animals and see how they have done.

DIGESTIVE SYSTEMS

Digestion refers to the mechanical and chemical breakdown of complex nutrients (foods) into smaller molecular components so that they may be absorbed. Once absorbed, the products of digestion are used as an energy source and as raw materials for synthesizing the molecules required for life. Animals go about getting food in a great variety of ways, and, in fact, much of the pageantry of life is centered over acquiring something to digest and in avoiding being so acquired (Figure 38.1).

Nutrition is an encompassing term that refers to any process involved in nourishment or being nourished. Here we will consider not only nutrients, but their sources, digestion, chemical characteristics, and roles in metabolism.

FIGURE 38.1
FOOD-GETTING IN ANIMALS
Animals spend much of their time in the search for and capture of food. Countless variations occur in this vital activity.

Some organisms do not actually ingest food but use **extracellular digestion.** Fungi and bacteria simply secrete digestive enzymes into the surrounding nutrients and break the food materials down into simpler subunits that then move into the cell. On the other hand, many eukaryotes are capable of taking food particles into their cells, where they are digested **intracellularly** in **digestive vacuoles.** You can readily see digestive vacuoles in large protozoans such as *Paramecium* and *Amoeba* (see Figure 23.13).

We will now briefly survey digestion and digestive systems in a few representative animals, beginning with the invertebrates. Then we'll move on to the vertebrates, concentrating on humans.

Saclike Systems in Invertebrates

Food-getting in the sponges—animals that lack organ systems—is the responsibility of the flagellated **collar cells,** or **choanocytes** (Figure 38.2a, and see Figure 30.6). These active cells line the body canals, and their beating causes seawater to swirl and eddy throughout the body of the sponge. The water carries in tiny food particles, which are trapped by a layer of mucus on each collar, passing down to enter the cell below. From there the food is then moved to other cells, where it becomes enclosed in digestive vacuoles. Molecules of digested food are then transported throughout all the cells of the sponge. It is believed that peculiar, wandering cells called **amebocytes** aid in the distribution processes by carrying food from one place to another. So the sponge, a simple multicellular animal, carries on intracellular digestion without a well-defined digestive system.

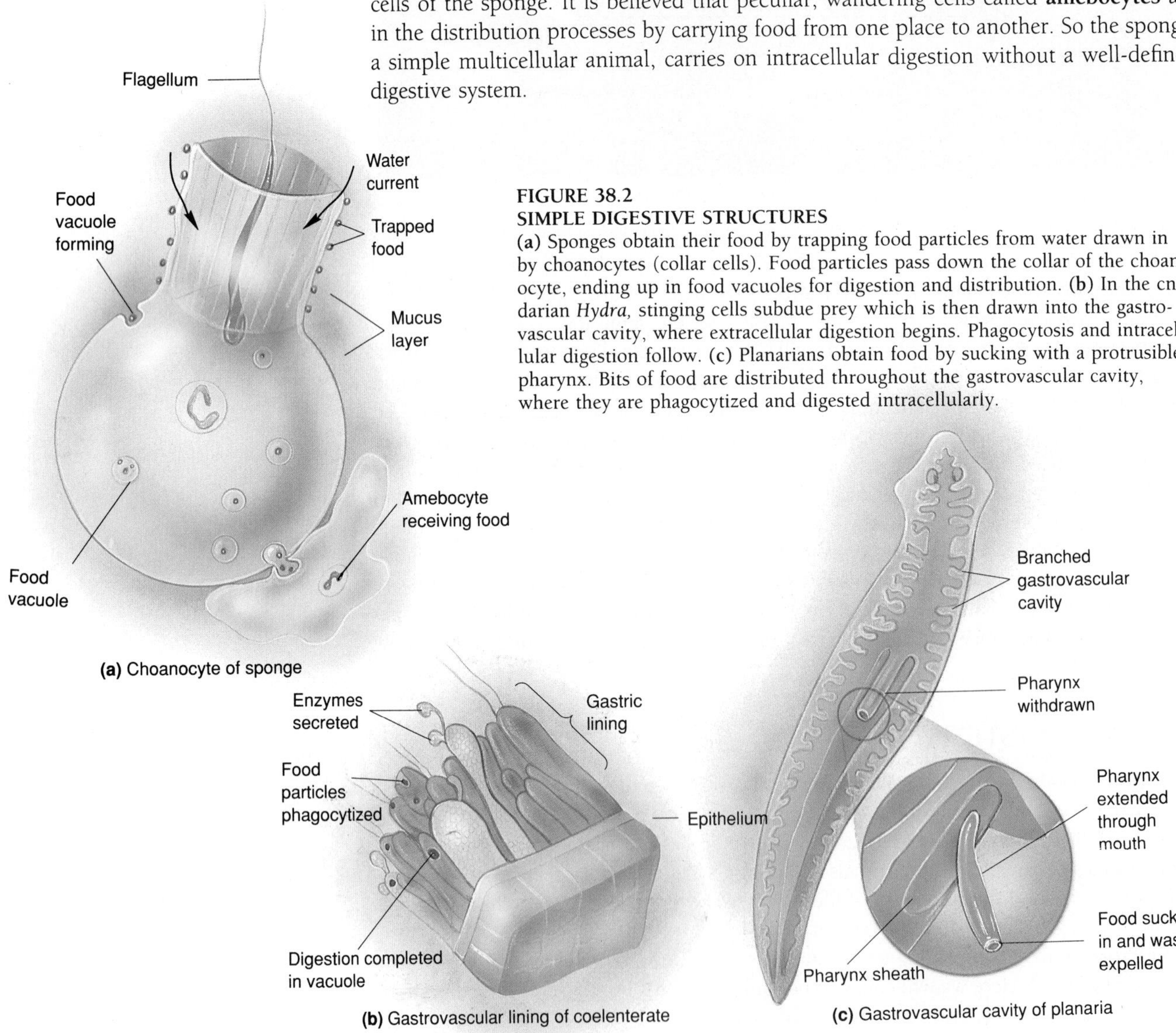

Flagellum
Water current
Food vacuole forming
Trapped food
Mucus layer
Amebocyte receiving food
Food vacuole

(a) Choanocyte of sponge

Enzymes secreted
Gastric lining
Food particles phagocytized
Epithelium
Digestion completed in vacuole

(b) Gastrovascular lining of coelenterate

Branched gastrovascular cavity
Pharynx withdrawn
Pharynx extended through mouth
Food sucked in and wastes expelled
Pharynx sheath

(c) Gastrovascular cavity of planaria

FIGURE 38.2
SIMPLE DIGESTIVE STRUCTURES
(a) Sponges obtain their food by trapping food particles from water drawn in by choanocytes (collar cells). Food particles pass down the collar of the choanocyte, ending up in food vacuoles for digestion and distribution. **(b)** In the cnidarian *Hydra,* stinging cells subdue prey which is then drawn into the gastrovascular cavity, where extracellular digestion begins. Phagocytosis and intracellular digestion follow. **(c)** Planarians obtain food by sucking with a protrusible pharynx. Bits of food are distributed throughout the gastrovascular cavity, where they are phagocytized and digested intracellularly.

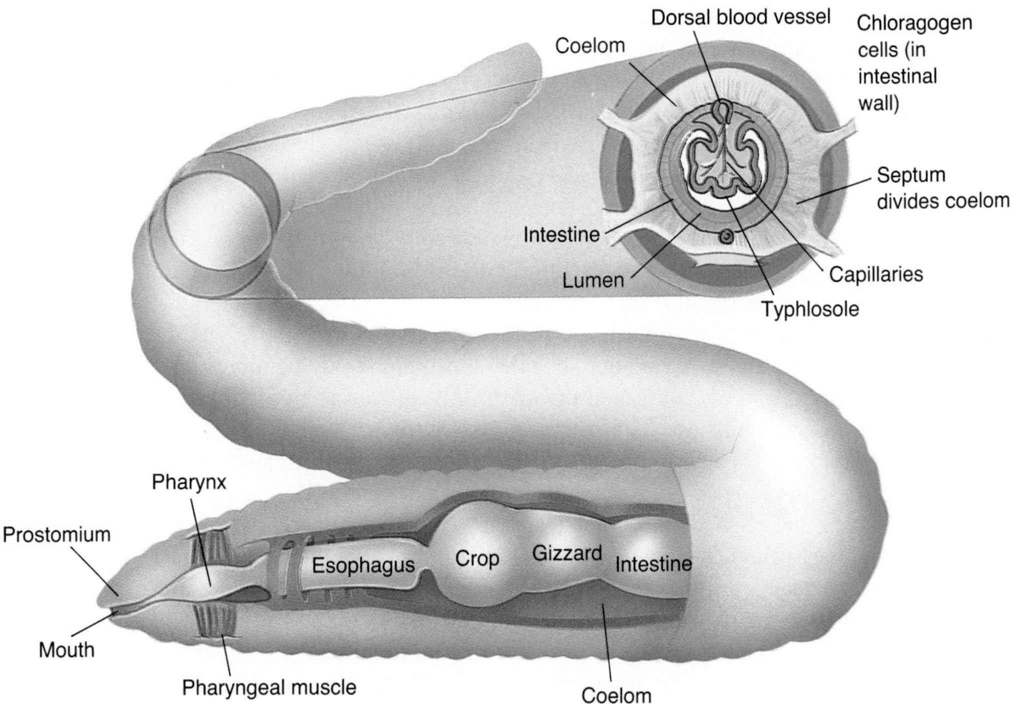

FIGURE 38.3
SPECIALIZATIONS IN THE EARTHWORM GUT
The complex earthworm digestive system is "complete," forming a tube-within-a-tube plan. Food moves onward through a tube from mouth to anus. The muscular gut has several specialized regions, including the pharynx, crop, gizzard, and lengthy intestine. The absorbing surface is greatly increased by an intestinal fold known as the typhlosole.

The cnidarians (Figure 38.2b) and flatworms (Figure 38.2c) have somewhat more complex digestive structures, including a prominent saclike gut or **gastrovascular cavity** with a complex lining of enzyme-secreting, phagocytic (food-engulfing), and ciliated cells. Digestion begins extracellularly when enzymes are released into the gastrovascular cavity, but after the food is partially broken down, digestion is intracellular. Particles are engulfed by phagocytic cells lining the gut, and digestion is completed. (For a review of phagocytosis, see Chapter 5.) In both cnidarians and flatworms, the presence of a blind gastrovascular cavity means that indigestible material must eventually be ejected through the organism's mouth. It is not surprising that such an arrangement is considered primitive (see also Figures 30.9 and 30.13c).

The Tube-within-a-Tube Digestive Plan

Beyond the flatworms and cnidarians, nearly all other animals have the tube-within-a-tube body plan and use extracellular digestion. (Recall from Chapters 30 and 31 that this plan evolved with the pseudocoelomates and coelomates.) Yet some intracellular digestion persists. The so-called "liver" of many invertebrates, such as mollusks and crustaceans, actually consists of intestinal side passages lined with phagocytic cells, and some food is digested within these cells after having been broken down into small particles.

As an example of digestive systems in invertebrates, we can consider two quite different groups: earthworms and insects. As you can see in Figure 38.3, the gut in earthworms is well defined and associated with several specialized organs. As the earthworm burrows through the soil, it feeds on decaying organic matter. Its mouth is a simple opening, followed by a very muscular **pharynx** that swallows the organic matter and forces it along. A short **esophagus** directs the mass into the thin-walled **crop**, which is a temporary storage organ. The ingested material is next moved into the thick-walled, muscular **gizzard.** The gizzard is filled with bits of gravel and, like the gizzard in seed-eating birds, is used to grind the food.

From the gizzard the food moves into the **intestine,** where it is digested by enzymes secreted by cells in the intestinal lining. The digestion and uptake of food occur along the entire length of the intestine. The **typhlosole,** a large, foldlike extension of the gut wall, speeds up absorption by greatly increasing the surface area (see Figure 38.3). From the absorptive surface, the food molecules pass into the circulatory system for distribution throughout the body. Although the earthworm has nothing to compare with the vertebrate liver, groups of epithelial cells known as **chloragogen** cells perform some of the same functions, including the conversion of glucose to glycogen, the **deamination** (removal of nitrogen) of amino acids, and the formation of urea. Worms get their color partly from the accumulation of some of these products. In addition (strangely enough), roaming phagocytes pick up whole bits of soil, which pass through the earthworm's intestine and then move to the skin, with the resulting earthy coloration presenting a major visual problem to the early bird.

It is difficult to generalize about insect feeding structures. In fact, the group serves as a lesson in digestive variation, one that illustrates the tremendous range of adaptations to the varied energy sources of the earth. It is difficult to think of anything that at least some insects don't eat. This versatility has been particularly troublesome to humans since the dawn of agriculture; the availability of food in large patches has made it possible for some insect species to increase in number dramatically.

Biting and chewing mouthparts are very common in insects, particularly in predators and foliage eaters (see Figure 31.19). They can be clearly seen in the grasshopper, which has, essentially, a five-part mouth. The grasshopper tastes its food with its **palps**, pairs of sensory structures attached to the mouthparts. Biting and shearing are done with the large saw-toothed **mandibles**, while the smaller **maxillae** deftly handle the food and do some finer grinding. The upper and lower lips, the **labrum** and **labium**, assist by holding food so that the shearing and grinding can occur efficiently. Saliva from well-formed salivary glands in grasshoppers helps with the grinding and begins digestion.

The grasshopper's digestive system, like that of many insects, consists of three parts, the **foregut**, **midgut**, and **hindgut** (Figure 38.4). The foregut is divided into a pharynx, esophagus, and crop. The crop functions primarily in food storage. In some insects this region has a roughened surface that is used to further grind the food. The midgut is a digestive structure where both digestion and absorption occur.

Surrounding the junction of the fore- and midgut are 12 **gastric ceca** (singular, **cecum)**—pouchlike accessory structures that secrete digestive enzymes into the midgut. The hindgut is mainly a water-absorbing structure that receives both digested food and cellular nitrogenous wastes. As digestive residues pass through, the water is removed and conserved, and the feces leave in a semidry condition, an important adaptation to the dry terrestrial environment.

The digestive system of the housefly is slightly different; it carries on some digestion outside the gut. You should be aware that the housefly walking across your chocolate cake is doing two things. First the fly is tasting the icing with its feet—a place where many insects have taste receptors. Second, when it bends its head down, it is disgorging

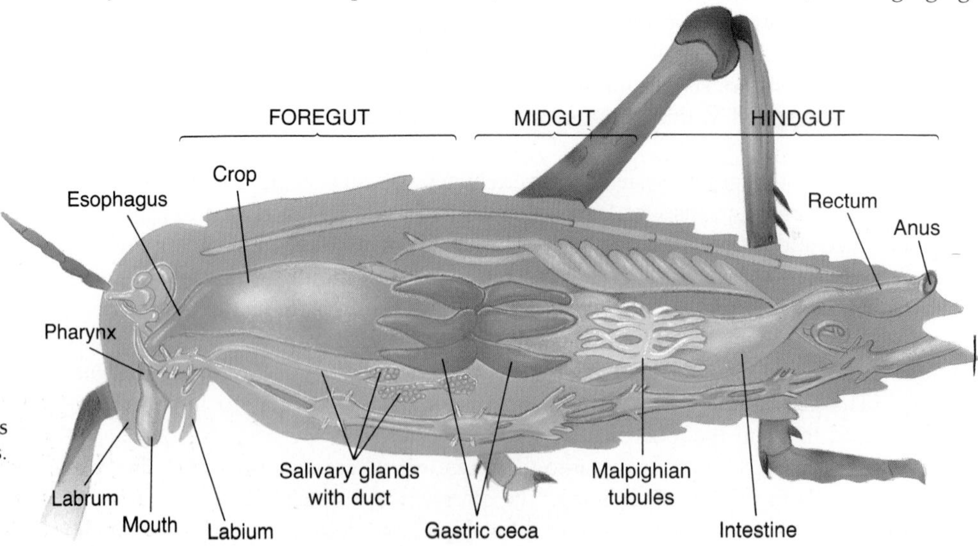

FIGURE 38.4
INSECT DIGESTIVE TRACT
The grasshopper's digestive system is similar to that of many other insects. The digestive tract typically consists of three main divisions, the fore-, mid-, and hindgut.

its enzymes onto the surface and stirring it all around with its hairy labium. But don't worry—it will suck up any mess it makes with its hollow, tubelike mouthparts. The rest is yours.

Plant eaters, by the way, frequently have piercing and sucking mouthparts. Recall the aphids that suck up the phloem sap (see Chapter 28). There are also fruit-eating moths that tear the fruit first, then suck up the exposed fluids. Of course, piercing and sucking mouthparts are found among the blood-sucking and predatory species as well. When the ant lion sinks its huge curved mandibles and maxillae into the ant's vulnerable abdomen, it doesn't chew. It uses channels formed between the paired mouthparts to suck out the ant's body fluids.

Feeding and Digestive Structures in Vertebrates

The jaws, teeth, and beaks of vertebrates provide us with clues to what they eat. For example, if you find a large skull with conical teeth, you've probably found a fish eater. If the skull is very large, though, the conical teeth may have once graced a killer whale who used them to crush the bodies of large prey such as seals. On the other hand, if the skull is very small and the teeth more pointed, the animal probably ate insects. You can rest assured, however, that the creature with the pointed teeth was not a grazer. Let's find out about feeding and digestive specializations in one of the sharper-toothed types from the briny depths.

Sharks and Bony Fishes. Sharks are certainly carnivores, despite the fact that they swallow practically anything. One of your authors, while on an expedition, helped dissect a 12-foot tiger shark and found a Polaroid negative that had been thrown overboard six days earlier and 200 miles away! Finding a picture of oneself in a shark's stomach can lessen the pleasures of seafaring life.

Two aspects of the shark's food-getting and digestive apparatus are especially noteworthy. First, its teeth occur in rows, most of which are around the periphery of the mouth (Figure 38.5a). As the teeth grow, the rows migrate outward so that the outermost ones fall out (and sink to the ocean floor, thereby providing tourists with necklaces). The jaw itself is a hinge joint, moving in one plane only. Instead of chewing, the shark uses the sideways head–thrashing and tearing action typical of shark attacks.

The second unusual aspect of the shark's digestive system is the **spiral valve** in its relatively short intestine. The valve (Figure 38.5b) resembles a circular stairway. Apparently, its function is to slow the passage of food through the shark's short intestine to give digestive enzymes time to act. It also provides additional absorptive surface.

Bony fish are extremely diverse, with over 20,000 species described, several of which will be neglected here. Like the sharks, the jaws of the bony fishes can only move in one plane and are used for biting

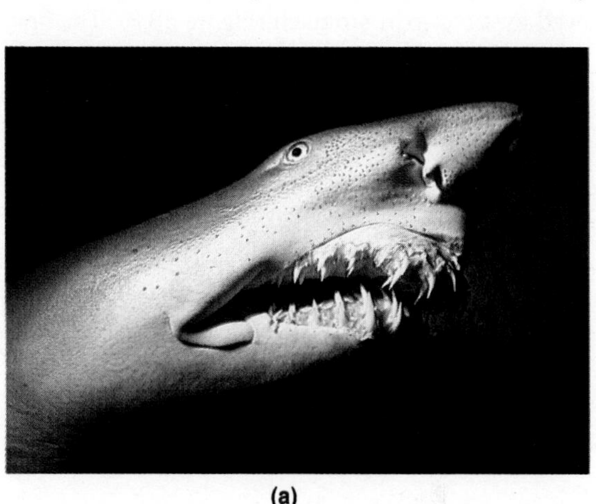
(a)

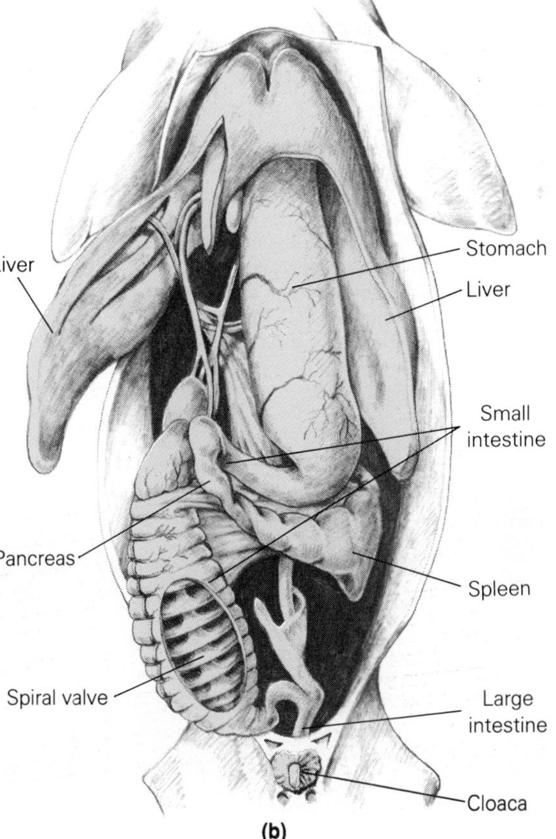

(b)

FIGURE 38.5
AN "EATING MACHINE"
(a) The fearsome mouth of the shark is lined with rows of teeth. The outer rows are the oldest and are often broken or lost as the shark feeds. However, new teeth soon move forward and outward to replace them. **(b)** While the shark gut is short and abrupt, its absorbing surface is greatly expanded by the presence of the spiral valve. Spiral valves are also found in some of the more primitive bony fishes.

and sucking only. The teeth are numerous, and in many species they grow from the roof of the mouth as well as from the upper and lower jaws.

The herbivorous fishes have interesting mouthparts. For one thing, they lack tearing teeth. They have a row of fine teeth borne on a horny ridge. Some patches of teeth on the floor of the mouth, where they grind bits of plant material against leathery patches on the roof. Other herbivorous fishes filter minute particles from the water with highly branched gill rakers through which water passes as it enters the gill chambers. Herbivorous fishes usually have very long, highly coiled intestines, typical in animals that extract nourishment from cellulose-enclosed plant cells.

Amphibians, Reptiles, and Birds. The amphibians have a digestive system similar to that of bony fishes but with some important differences. For example, whereas fish have rather immobile tongues, the tongue in amphibians is often highly movable. In fact, frogs use their tongue to capture insects. The tongue is attached to the front of the mouth, permitting the sticky organ to be flicked far out with considerable speed and accuracy. The prey is then crushed against a peculiar patch of teeth on the roof of the mouth and swallowed whole—amphibians can't chew either.

The reptiles are a rather diverse group and boast a wide variety of techniques in detecting and catching prey. For example, some reptiles have many teeth, while in others, the teeth are fewer in number but more specialized. Crocodiles and alligators are unusual in that their teeth are seated in sockets in the bone, much like our own. Evolution has provided some venomous snakes, including the North American pit vipers, with retractable grooved or hollow **fangs** with intricate injecting structures. Other venomous snakes—cobras, kraits, and mambas, for instance—have permanently erect fangs and must "chew" the venom into their prey. All snakes, by the way, are carnivorous and have undergone a unique modification of the jaw for swallowing whole prey (Figure 38.6).

Snakes and lizards find food with the aid of well-developed eyes and a specialized olfactory (odor sensing) organ at the roof of the mouth. This structure, called **Jacobson's organ,** detects molecules (odors) from the air, which are brought to it by the moist, flicking tongue. The pit vipers orient to warmblooded prey through the use of acutely sensitive heat-detecting receptors in the facial pits (see Figure 35.21).

The dentition of birds is as scarce as hen's teeth since all of today's birds have bills rather than teeth. The bill or beak is a bony structure with a tough, continuously growing covering of keratin, a horny protein also found in fingernails and hair. While the basic structure of the bill is similar in all birds, there is considerable variation in its shape and size. Each type strongly reflects the feeding habits of the bird (Figure 38.7).

In some instances, feeding habits are reflected in the feet as well. Typically, raptors, or birds of prey, have long, curved talons for grasping prey. Ground foraging species such as grouse and pheasants have heavy, strong feet for scratching the soil, while the feet of the ostrich and emu can be deadly weapons.

The digestive system of seed-eating birds includes a large crop for storing and moistening food, followed by a two-part **stomach** (Figure 38.8). The first portion—the very

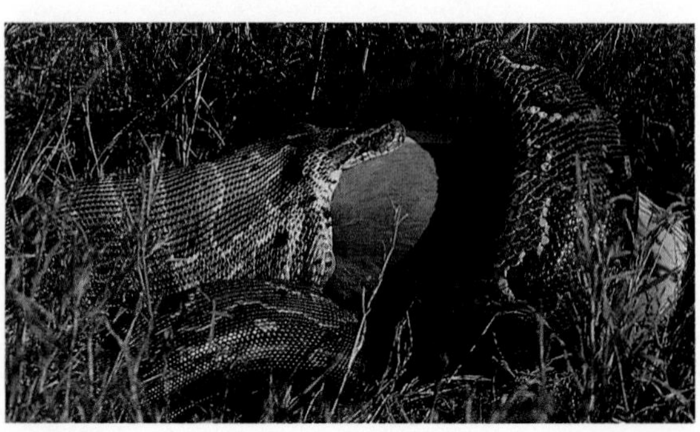

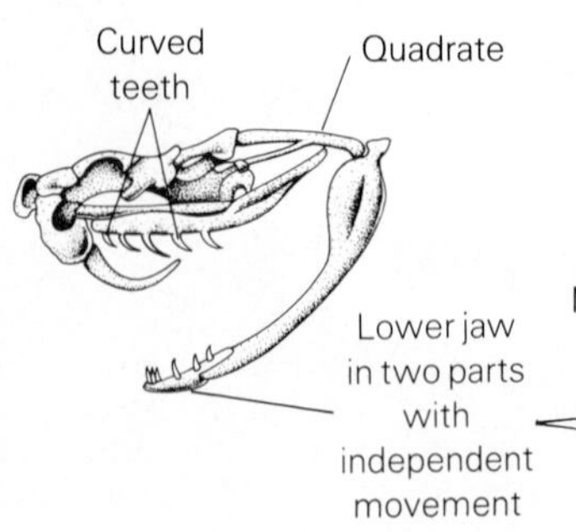

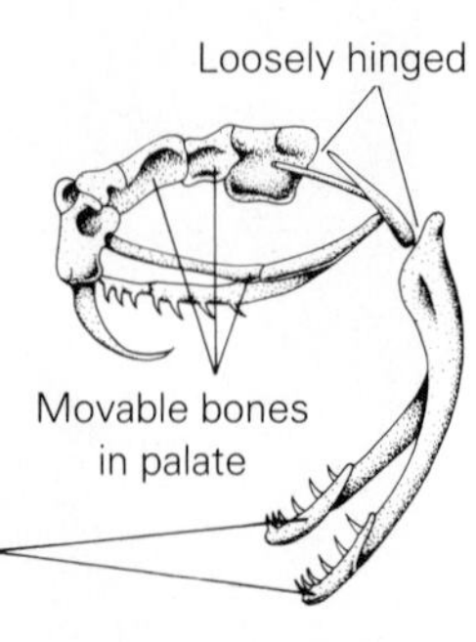

FIGURE 38.6
FEEDING IN THE SNAKE
Snakes cannot tear or chew their food. Captured prey, sometimes quite large, is swallowed whole. This is possible because the lower jaw is loosely attached to the skull, and the bones of the palate are moveable, as are the left and right halves of the lower jaw. Stretching of the esophagus and stomach and the absence of a sternum help in moving the prey through the digestive tract.

White Ibis

Bald Eagle

Sword-billed Hummingbird

Oyster Catcher

Small Ground Finch

FIGURE 38.7
BIRD FOOD-GETTING SPECIALIZATIONS
The bills of birds provide clues to their feeding habits.

glandular **proventriculus**—secretes gastric juices into the coarse food. It then passes into a very muscular **ventriculus** (gizzard), where it is pulverized with the aid of a hardened lining and abrasive sand grains the bird routinely swallows.

Foraging and Digestive Structures in Mammals. Mammalian teeth grow in sockets in the jaws, a trait that distinguishes them from almost all other vertebrates. Generally, the mammal begins life with temporary teeth, or milk teeth, which are later replaced by permanent teeth. The basic structure of each permanent tooth is the same in all mammals (Figure 38.9).

While the teeth of most fishes and reptiles show little specialization, mammals have four types of teeth, each usually with its own specialized function. The front teeth, or **incisors,** highly developed in the rodents and rabbits, are chisel-shaped for gnawing and cutting. The second group, the **canines,** are used for capturing and killing prey, tearing food, and defense. They are very prominent in carnivores and completely absent in some of the herbivores. (Imagine feeding a carrot to a horse with fangs.) The last two groups, the **premolars** and **molars,** are specialized for grinding. They are large and flat in the herbivores and are used to break down the cellulose walls and fibers of plants. Some carnivores, such as dogs, have large, sharp-edged premolars specialized for shearing and crushing bones.

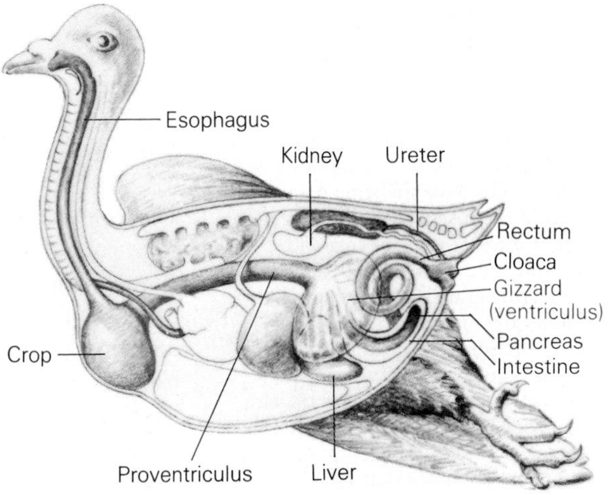

FIGURE 38.8
BIRD DIGESTIVE SYSTEM
Herbivorous birds have crops and gizzards, organs specialized for handling plant foods. In addition, they have a special stomach region called the proventriculus, which saturates food with digestive enzymes before it enters the ventriculus, or gizzard.

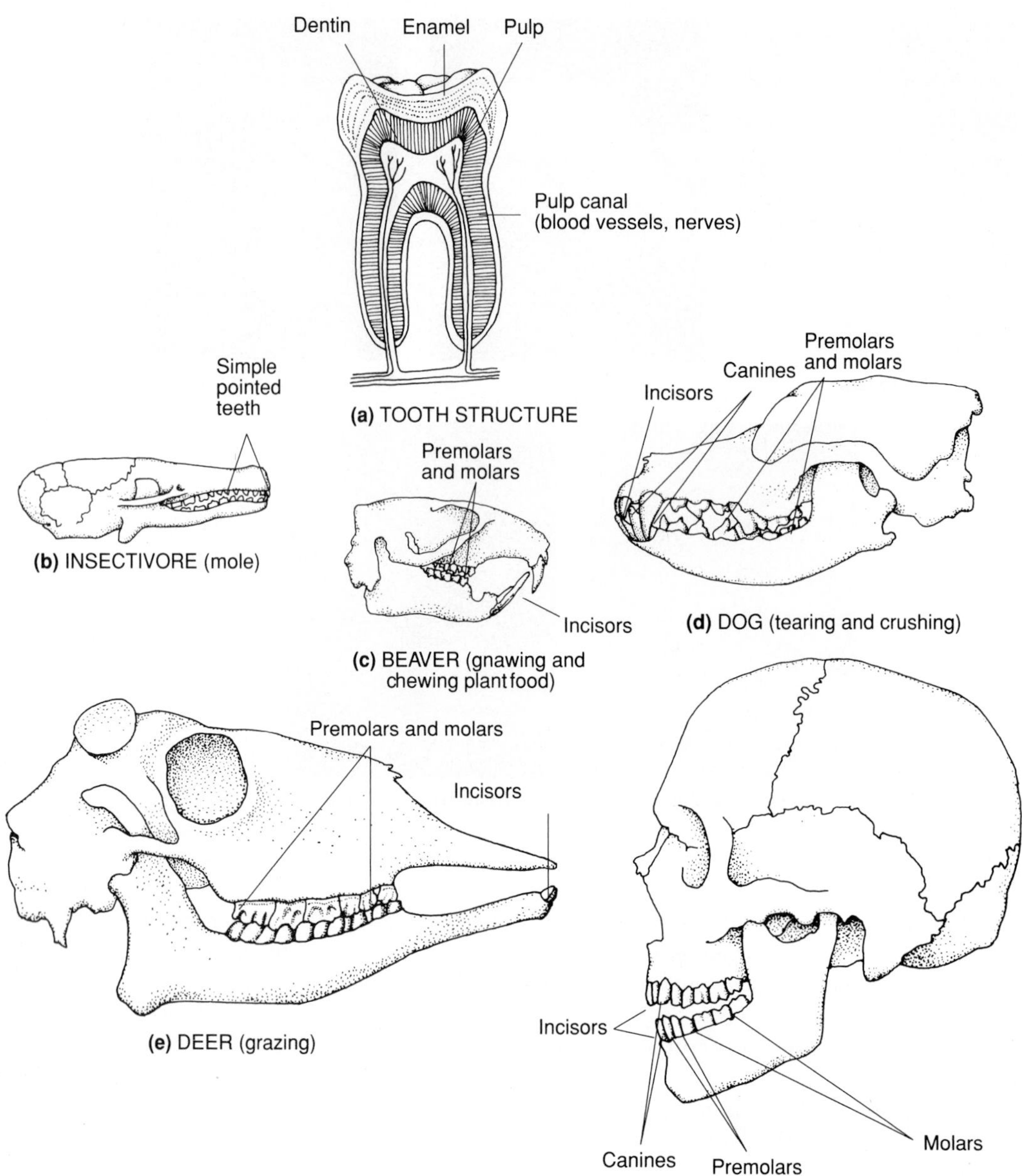

FIGURE 38.9
DENTITION IN MAMMALS
Nearly all mammals share the same basic tooth structure (**a**): a hardened layer of enamel surrounding a softer dentin region. Within the dentin is a pulp cavity, penetrated by blood vessels and sensory nerve endings. The arrangement and individual shapes vary considerably, however, according to the diet to which the species has adapted. (**b**) Note the simple pointed teeth of the insect-capturing insectivore. Compare the gnawing and grinding teeth of a rodent (**c**) with the tearing and crushing teeth of a canine (**d**). Also note the nipping lower incisors of the ruminant (**e**) and the absence of upper incisors—a horny covering meshes with the lower incisors for nipping foliage. Human dentition (**f**) is quite uniform and relatively unspecialized.

Dentition in humans is rather generalized, lacking the prominent, specialized, exaggerated features often found in other mammals. Because of the lack of specialization, some zoologists regard human dentition as primitive (see Figure 38.9).

The Ruminants: Grazing Mammals Plant eating mammals, particularly foliage eaters, have solved the difficulties associated with cellulose digestion. You may recall that although cellulose is a potentially rich source of energy, breaking the beta linkages between the glucose subunits requires enzymes that few animals possess. Cellulose-digesting enzymes are present principally in microorganisms such as fungi, protists, and bacteria. Those mammals that carry out cellulose digestion have evolved special relationships with organisms with such capabilities.

Cattle, horses, and other herbivores begin the process with heavy, flat teeth that are specialized for grinding. The digestive tracts of cattle and other grazers, or **ruminants**

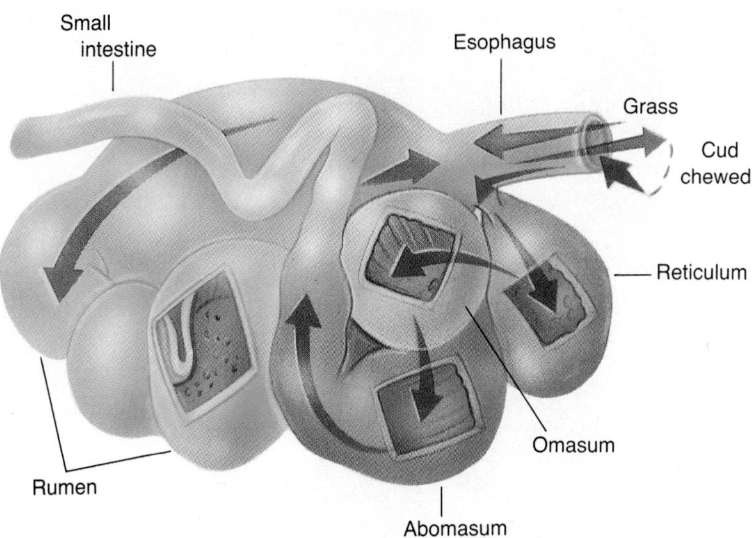

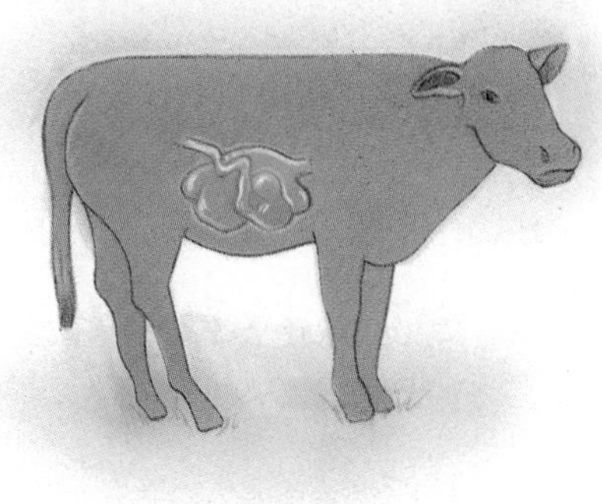

FIGURE 38.10
THE FOUR-PART RUMINANT STOMACH
Digestion in the cow begins in the large rumen, where vast numbers of protozoans and bacteria begin the breakdown of cellulose. The products are regurgitated as the cud, rechewed, and swallowed, this time entering the reticulum (follow the colored arrow). The partially digested mass then enters the omasum and finally the abomasum (the true stomach), where digestion of the microorganisms themselves begins. Digestion is completed in the small intestine.

(including deer, giraffes, antelope, and buffalo), have four-chambered "stomachs" including the **rumen** from which the name *ruminant* is derived, and the **reticulum**, **omasum**, and **abomasum**. Cattle don't produce enzymes that can digest cellulose but instead, like termites, harbor certain protozoans and bacteria in their rumen. These organisms *can* break down cellulose, forming nutritious products that, along with the microorganisms themselves, can begin digestion in the abomasum, the actual stomach, in the usual manner. The unusual aspects of digestion in the four-part stomach, including "regurgitation" and "cud chewing," are described in Figure 38.10.

THE DIGESTIVE SYSTEM OF HUMANS

The human digestive system (Figure 38.11) is rather representative of the mammalian system in many respects. We will now follow the enchanting path of food through the human tract, discussing the specialized structures along the way. We will then take a close look at the chemistry of digestion and some aspects of nutrition.

The Oral Cavity and Esophagus

We have already discussed human dentition, so now let's concentrate on some of our other feeding structures, such as the lips and tongue. In all mammals, the lips have a rather essential role in eating. If you don't know what that role is, try eating a meal without closing your lips (but do it somewhere else). Fortunately, even those crass souls who chew with their mouths open must close them to swallow. So, the lips hold food in and seal the mouth to permit swallowing.

The tongue is also important to eating. In addition to its role in moving food into position for chewing, swallowing, and sorting out fish bones, it constantly monitors the texture and nature of foods. This is a valuable chore, since it warns us about peculiarities in time to prevent our swallowing something we shouldn't. Further, the tongue *tastes,* that is, it can distinguish certain chemicals by specialized chemoreceptors clustered in sense organs called **taste buds.** These receptors distinguish four basic tastes: salty, sour, sweet, and bitter (Figure 38.12). In addition to informing us of flavor, the stimulation of taste buds enhances the flow of **saliva.**

There are three pairs of salivary glands, the **parotids, submandibulars,** and **sublinguals** (see Figure 38.11, upper inset). Each salivary gland empties its secretions into the mouth. These secretions, collectively, form the saliva, which consists mainly of water, ions, lubricating mucus, and the starch-splitting enzyme **amylase.** Salivary amylase, however, is probably less important in digestion than it is in oral hygiene. It helps break

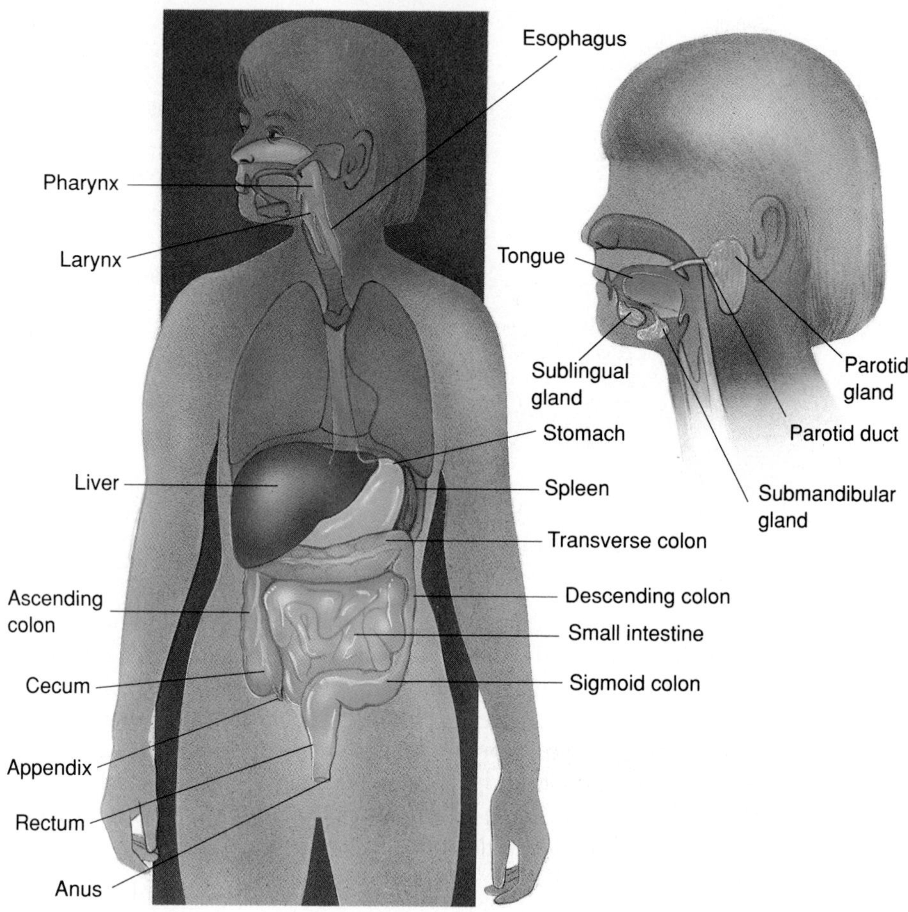

FIGURE 38.11
THE HUMAN DIGESTIVE SYSTEM
The three salivary glands (inset) open into the oral or mouth cavity. The pharynx and esophagus are primarily simple, muscular passages secreting only lubricating mucus. The stomach is quite complex, secreting hydrochloric acid and protein-digesting enzymes. The small intestine is both secretory and absorptive, its villi and microvilli greatly increasing its absorbing surface. It also receives the secretions of the liver and pancreas. The colon, or large intestine, is chiefly involved in water absorption and the compacting of digestive wastes. It also provides a suitable environment for enormous populations of bacteria, some of whose waste products are useful to the host.

down starchy food particles caught between the teeth. Saliva is mainly important for moistening and lubricating food.

The pharynx, the throat region in the rear of the oral cavity, forms a common passageway with the nasal cavity. Just below the base of the tongue the pharynx divides, forming the anterior **larynx** (Adam's apple) and the posterior **laryngopharynx.** During swallowing, food is pressed by the tongue against the **soft palate**, closing the nasal passageway; the larynx is then raised, bending the **epiglottis** over the **glottis** (laryngeal opening), thus closing the air passageway to the lungs and directing food into the esophagus (Figure 38.13). Should this action fail, food will enter the larynx, producing violent spasms of coughing, or even choking (Essay 38.1).

The **esophagus** (the food tube between the pharynx and stomach) has no digestive function but simply moistens food and moves it to the stomach. Its upper portion contains skeletal (voluntary) muscle, while the rest has smooth (involuntary) muscle. Thus, swallowing begins as a conscious act but soon becomes automatic.

The lower esophagus, by the way, is very similar in its tissue organization to the rest of the digestive tract. As you can see in Figure 38.14a, there are three principal layers of tissue plus a fibrous outer coat, or **serosa.** The inner lining, the **mucosa**, contains a mucus-secreting epithelium (covering), overlying a region of connective tissue. The second tissue layer, the **submucosa**, contains more connective tissue with blood and lymphatic vessels, nerves, and glands. The third layer, the **muscularis**, is a region of smooth muscle with an inner circular layer and an outer longitudinal layer.

The muscles of the esophagus move food along by wavelike **peristalsis.** Food being swallowed is pressed into a **bolus** (clump) by the tongue and pharynx. As the bolus approaches the involuntary muscles of the esophagus, the circular layer just ahead of it relaxes and the muscles behind the bolus contract, pushing food along (Figure 38.14b).

FIGURE 38.12
THE TONGUE
Taste buds on the tongue surface tend to specialize in detecting one of the four flavors. The scanning electron microscope study reveals flattened, columnlike projections that contain the taste buds.

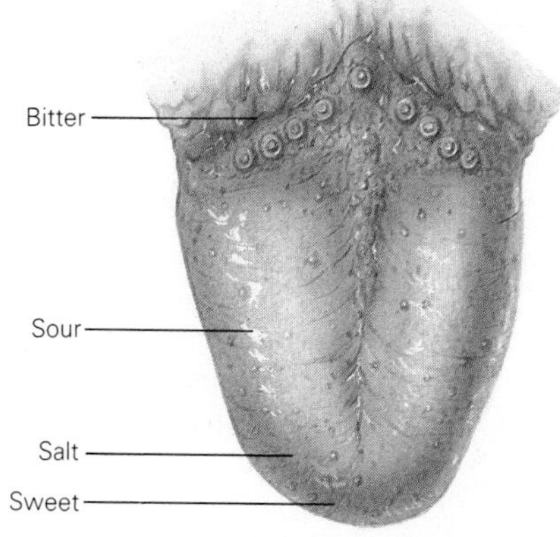

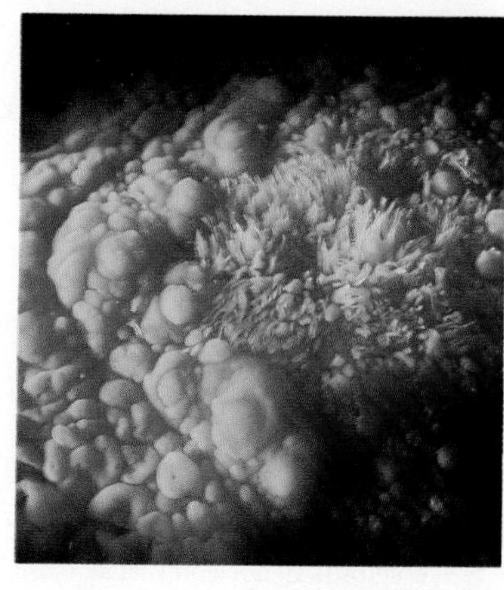

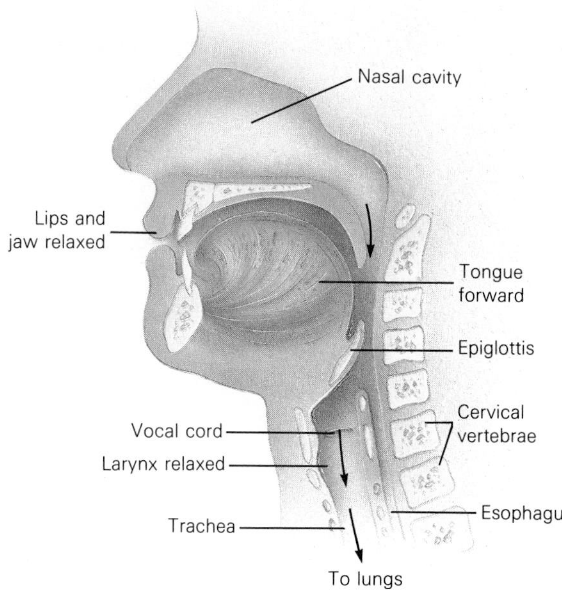

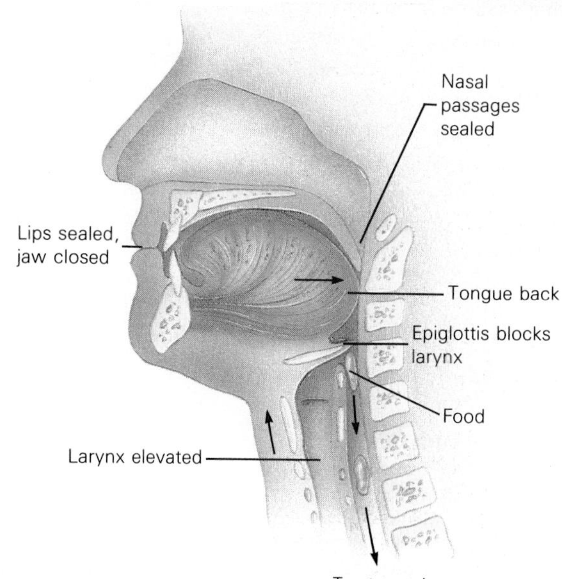

FIGURE 38.13
SWALLOWING
Swallowing begins as a food mass is pressed upwards by the tongue against the soft palate and back toward the pharynx. As the involuntary stages begin, the soft palate elevates, closing off the nasal cavity. The larynx elevates, bending the epiglottis over the glottis and sealing off the breathing passage. Food then enters the laryngopharynx, which forces it into the esophagus.

The muscles are coordinated by nerve endings that are activated by the presence of food. Because peristalsis is under local (intrinsic), hormonal, and autonomic control, it is an involuntary action, and unless you have an embarrassingly noisy intestine you are probably not aware of the process.

The Stomach

The human stomach is structurally and functionally similar to that of most other vertebrates. It temporarily stores food and begins its digestion. In addition, its acids and enzymes kill many of the microorganisms we swallow. The stomach can be closed off at either end by two muscle sphincters (Figure 38.15a). When contracted, these circular groups of muscles permit the stomach to churn and liquefy food without forcing it back into the esophagus or into the intestine before it is ready. You may have noticed that the upper ring of muscle, the **cardiac sphincter**, sometimes fails when the stomach is overfull or filled with gas, and allows acidic fluids to enter the esophagus, creating "heartburn." The overproduction of these acids is also related to emotional stress.

The muscle layers of the stomach are more complex than those of the esophagus. Here, we encounter a third, diagonal layer of muscle just inside the circular layers. This one produces a twisting action that accompanies the usual wringing (by circular muscles) and shortening (by the longitudinal muscles). Also, peristalsis in the stomach is nondirectional, and moves food back and forth in a sequence of contortions until it reaches the well-churned state required by the intestine.

ESSAY 38.1

THE HEIMLICH MANEUVER

Human evolution, unfortunately, has resulted in the openings of the trachea and esophagus being closer together than is the case in many of the species. The result is a marked propensity for food to "go down wrong"—that is, for an occasional food particle to move into the air passages. This happens when the epiglottis is not completely closed during swallowing. In this case the glottis spasmodically contracts and causes choking. In some cases the food is simply coughed up, but in extreme cases, the victim is completely unable to breathe. Some eight Americans die this way each day.

However, a rather simple action can save many of these lives. It is called the Heimlich Maneuver, and it works as follows: (1) Stand behind the victim. (2) Wrap your arms around his waist. (3) Make a fist with one hand, knuckles directed upward and inward against the victim. (4) Place the knuckle between the rib cage and the navel. (5) Cup the other hand over the fist. (6) Quickly press inward and upward against the victim's abdomen. (7) Repeat if necessary.

If the victim is lying on his back, kneel, place your knees on either side of his hips, and with the heels of your hands (one on top of the other) on the abdomen as above, press upward with a quick thrust. (Repeat if necessary).

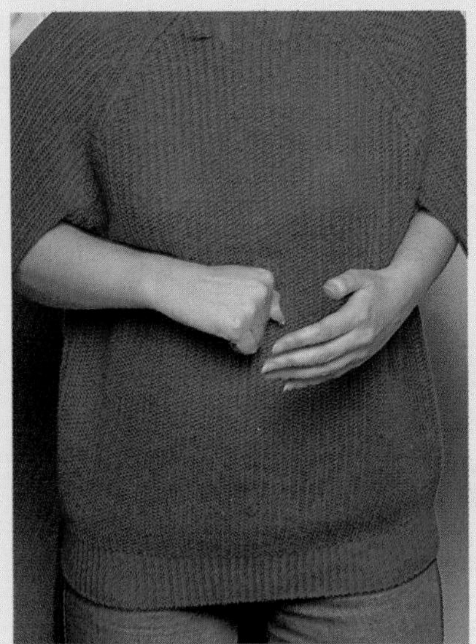

FIGURE 38.14
ORGANIZATION OF THE GUT
(a) The basic structure of the tube includes three tissue layers and the outer coat or serosa. **(b)** Food moves onward through peristalsis. Circular muscle layers ahead of the food relax while those behind, contract.

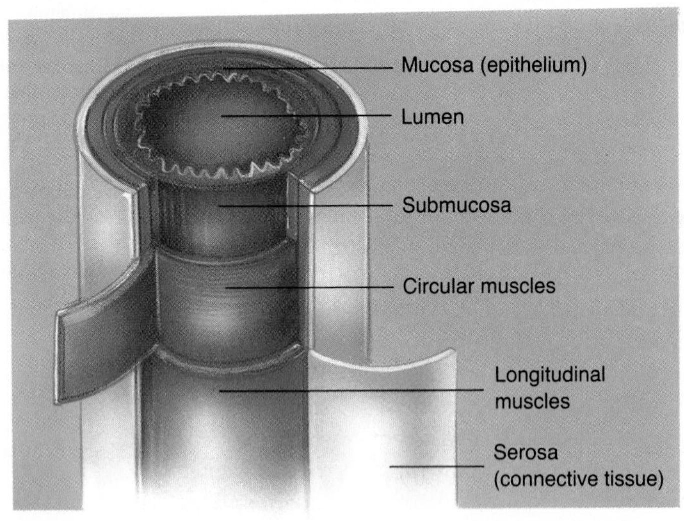

(a) Structure of esophagus

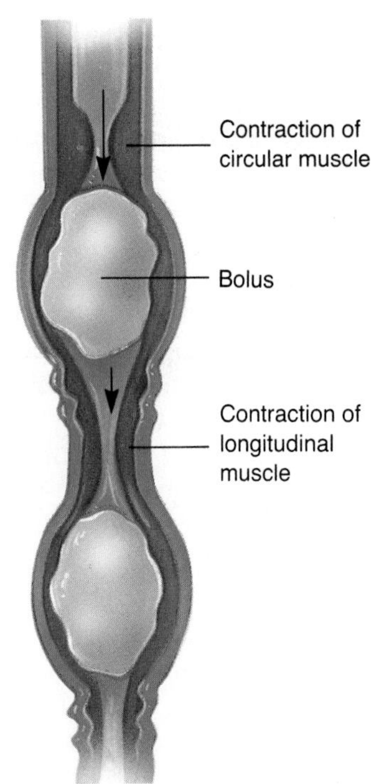

(b) Peristalsic

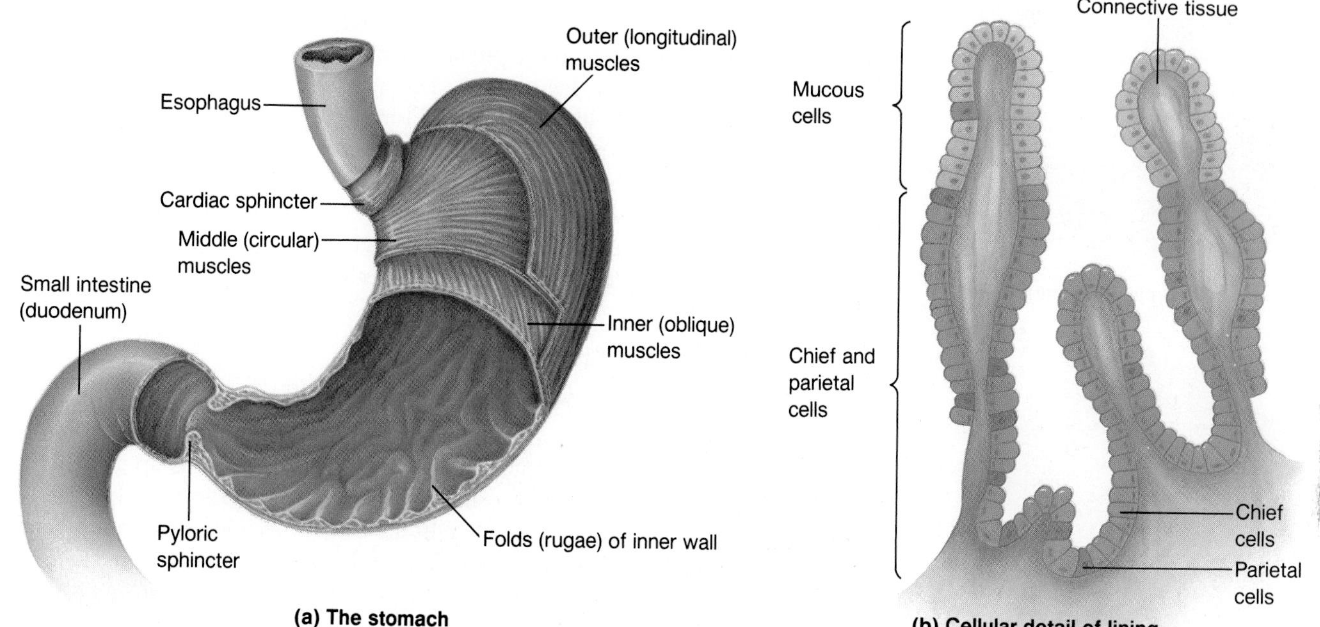

FIGURE 38.15
STOMACH
(a) The stomach is a J-shaped pouch with a sphincter at either end. Its muscular layers run in three directions and can produce powerful writhing and wringing motions. (b) The complex lining bears mucus-secreting cells at the tips of long issue columns. In the deep crevices between the columns, pepsinogen-secreting chief cells and acid-secreting parietal cells release their products.

The stomach lining (mucosa) contains many long tubular glands that secrete the gastric juices. The glands contain **chief cells**, which secrete the protein **pepsinogen**, and **parietal cells**, which secrete hydrochloric acid (HCl) (see Figure 38.15b). The acid activates the pepsinogen, forming the digestive enzyme **pepsin**, which initiates the digestion of protein. Other glands secrete water, mucus, and small quantities of **gastric lipase**, a fat-splitting enzyme. The presence of HCl produces a very low pH of 1.6 to 2.4 in the stomach fluids. The acid not only kills many bacteria and other microorganisms taken in with food, but also opens the folds and turns of globular proteins, better exposing their peptide bonds to pepsin. But the acidity of the stomach and the potency of its enzymes could (and sometimes do) endanger the lining itself. One reason we don't digest our stomachs is that a layer of **mucin**, an insoluble mucoprotein, forms a coating over the stomach lining.

The Small Intestine

Once the food reaches a liquefied state, now referred to as **chyme**, the **pyloric sphincter**, another ring of muscle, relaxes a bit, allowing a small amount of liquefied food to move into the **small intestine**. The small intestine (see Figure 38.11) is about 6 m (20 ft) long in humans, consisting of three regions: the **duodenum**, **jejunum**, and **ileum.** The small intestine has two roles: digestion and absorption. Most of the digestion and all of the absorption of foods occurs in the small intestine. Much of the absorption involves active transport, but passive transport is also involved. In fact, about 90% of the water present in foods diffuses into the small intestine.

The small intestine is uniquely adapted for absorption, having an enormous surface area (estimated at about 700 m^2, or the floor area of four or five three-bedroom houses). Its inner surface has the appearance of a badly wrinkled bath towel (Figure 38.16a). Closer examination of the surface reveals that there are many folds, each covered with tiny projections called **villi** (Figure 38.16b). Each villus is in turn covered with columnar epithelial cells bristling with **microvilli**, fingerlike projections of the plasma membrane (Figure 38.16c).

Within each villus is a **capillary bed** and a **lacteal** (a lymphatic vessel). The villi are capable of rather vigorous movement, like millions of wiggling fingers. They help in mixing food and enzymes in the gut and assist in absorption by creating pressures that tend to concentrate digested food in the many recesses of the lining.

Until recently, it was believed that digestive enzymes were free in the intestinal lumen. But careful studies of the epithelial cells have led physiologists to believe that most of the enzymes are actually bound to the plasma membranes, forming a frilly surface that is part of the **glycocalyx** (see Figure 5.2). This precise positioning of the enzymes presumably helps prevent the lining of the small intestine from being digested.

The cellular lining of the small intestine is thought to be very temporary and always in a state of replacement. This isn't surprising if you consider the amount of frictional wear that it suffers. Abrasion scuffs away an estimated 17 million cells each day. However, new cells are continuously produced at the base of each villus, and these migrate upward in an orderly fashion, replacing those worn away at the tip.

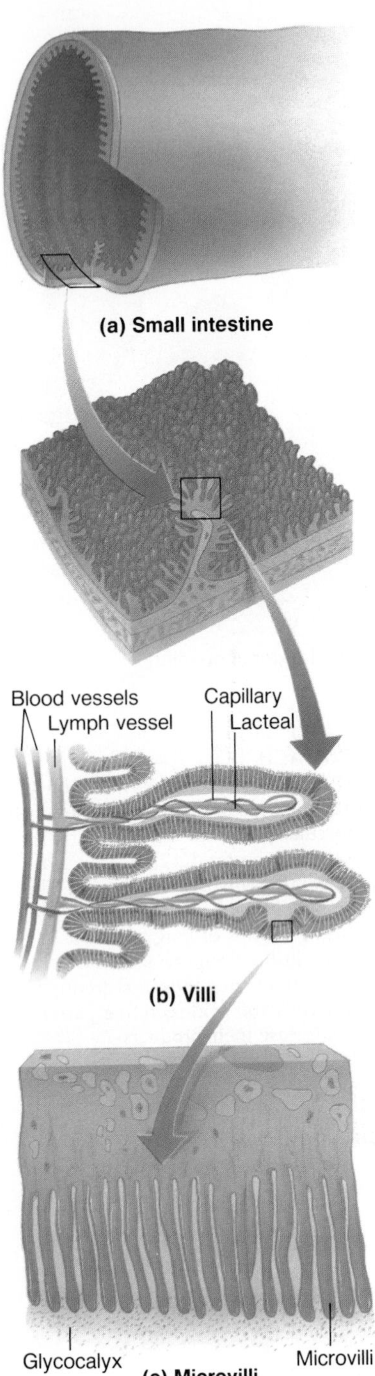

FIGURE 38.16
THE SMALL INTESTINE'S LINING
(a) The inner surface or mucosa of the small intestine contains a number of folds, each of which is covered by the fingerlike intestinal villi. (b) Each villus consists of a column of cells containing a capillary network and a lacteal (a blind ending of a lymph duct), into which digested foods move. (c) The surface cells of the villus have their membranes folded into numerous microvilli. At the surface, membrane-bound enzymes form the hazy glycocalyx.

Accessory Organs: The Liver and Pancreas

The duodenum of the small intestine receives the secretions of two accessory organs, the **liver** and the **pancreas.** The liver, an organ of many functions, aids in digestion by secreting a slightly alkaline fat emulsifier known as **bile** that breaks fats up into minute globules. Bile is a complex substance containing cholesterol, bile salts and pigments, water, and modified amino acids. The bile salts are actually steroids and are important in dissolving fats. The bile pigments are products of hemoglobin destruction, since the liver (along with the spleen) is the red blood cell graveyard. These breakdown products of hemoglobin become part of the digestive wastes, providing the characteristic color to feces.

After being produced in the liver, the bile is stored in the **gall bladder** (Figure 38.17). The release of bile is brought about by a hormone whose name defies pronunciation, **cholecystekinin-pancreozymin**, which causes the gall bladder to contract (and also stimulates the release of pancreatic enzymes). Bile reaching the duodenum via the **bile duct** is joined by highly alkaline fluids from the merging **pancreatic duct**, just before it enters the intestine.

In humans, the pancreas is a long glandular organ lying nestled in the first turn of the small intestine (see Figure 38.17). One of its products, sodium bicarbonate, neutralizes the acid accompanying partially digested food from the stomach, protecting the small intestine and raising the pH, thereby enhancing the activities of the next series of enzymes. The pancreas secretes an entire battery of digestive enzymes that are involved in the breakdown of fats, carbohydrates, protein, and nucleic acids. The pancreatic enzymes, together with those of the small intestinal lining, carry out most of the digestive process.

The Large Intestine

The small intestine joins the **large intestine**, also called the **colon** or **bowel**, on the right side of the abdominal cavity. The large intestine consists of the **cecum, ascending colon, transverse colon, descending colon**, and **sigmoid** (S-shaped) **colon** (see Figure 38.11). The last portions of the digestive tract are the **rectum**, the **anal canal**, and the **anus**—where the story ends.

Below the point where the two parts of the intestine join, the large intestine forms a blind pouch, the cecum. Protruding from the cecum is the **appendix**—a hollow, fingerlike extension. At the union of the small and large intestines is the **ileocecal valve**, a one-way valve that prevents a backflow of the food residues into the small intestine.

The primary function of the colon is to absorb water and minerals into the blood and to prepare the feces to leave the digestive tract. The colon takes up about 10–20% of the water in the liquid digestive residues that enter. The mucosa of the colon is glandular and secretes mucus, which is used in the production of feces and lubricates the lower portions of the colon. Very little, if any, digestive activity takes place in the colon itself, although a rapid digestion of food residues is carried on by teeming populations of microorganisms. *Escherichia coli,* along with a number of methane-producing bacteria (see Chapter 22), are prominent inhabitants. Strange as it may seem, a sizable

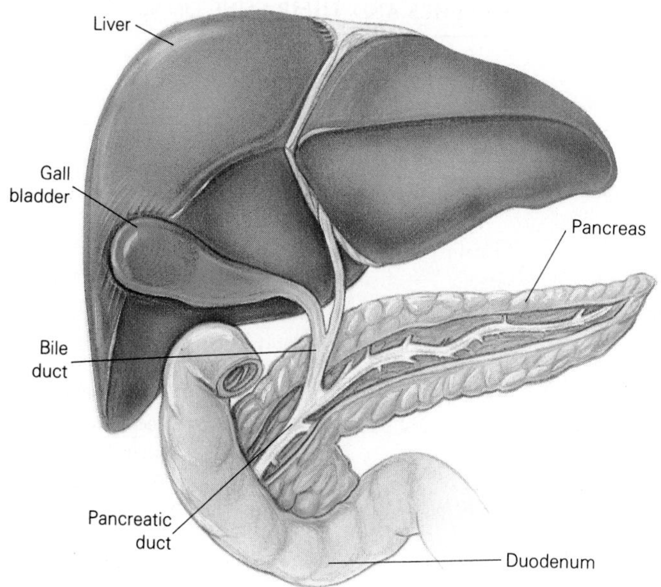

FIGURE 38.17
ACCESSORY DIGESTIVE ORGANS
The liver and pancreas are accessory digestive organs. The ducts from both the gall bladder and the pancreas meet to form a common duct that empties into the duodenum of the small intestine.

amount of fecal material, perhaps a third by dry weight, is composed of bacteria. These intestinal bacteria play an important and rather unexpected role in nutrition. A flourishing bacterial colony can help prevent deficiencies in such vitamins as vitamin K, biotin, and folic acid.

The final region of the digestive system is the rectum. Its structure is basically the same as that of the rest of the colon, except for the presence of folds in its walls. These are the **rectal valves**, which help support the weight of feces so that gravity doesn't become the enemy of propriety. The anus is the external opening. The movement of wastes through the anal sphincter is usually, thankfully, under voluntary control.

THE CHEMISTRY OF DIGESTION

You may recall from our previous discussions that proteins, carbohydrates, and fats are often very large, complex molecules composed of repeating molecular subunits (see Chapter 3). Except for certain triglycerides, these macromolecules can't be absorbed through the lining of the gut. So in the digestive process, enzymes attack the linkages of the molecular subunits, breaking them down into their components.

The bonds are broken through **hydrolysis**, the process in which water is added enzymatically to the linkages, disrupting them and yielding the molecular subunits. Following digestion, the simplified foods cross the gut lining, but while water and some ions move by osmosis and diffusion, respectively, most foods must be actively transported through membranal carrier mechanisms across the intestinal villi into the blood or lymphatic fluids within.

During digestion, carbohydrates are hydrolyzed into simple sugars, fats into fatty acids and glycerol, proteins into their various amino acids, and nucleic acids into free nucleotides. The digestive enzymes and their functions are listed in Table 38.1.

Carbohydrate Digestion

All starch digestion begins in the mouth, where the enzyme salivary amylase breaks some linkages, producing the disaccharide (two glucose units) maltose and many longer polysaccharide fragments. Starch digestion is then temporarily stalled by the acidity of the stomach, but resumes in full swing in the small intestine. What happens next depends on the starch.

TABLE 38.1 DIGESTIVE ENZYMES AND THEIR FUNCTIONS

SOURCES AND ENZYMES	SUBSTRATE	PRODUCT
Salivary glands		
Salivary amylase	Starch	Maltose and starch fragments
Stomach lining		
Pepsin	Protein	Peptides
Rennin	Casein	Insoluble curd
Gastric lipase	Triglyceride	Fatty acids + glycerol
Pancreas		
Trypsin	Peptide linkage	Shorter peptides
Chymotrypsin	Peptide linkage	Shorter peptides
Carboxypeptidase	*C*-terminal bond	Free amino acids
Ribonuclease	RNA	Nucleotides
Deoxyribonuclease	DNA	Deoxynucleotides
Alpha glucosidase	1-6 linkages of amylopectin	Starch fragments
Pancreatic amylase	Starch	Maltose
Pancreatic lipase	Triglyceride	Fatty acids, glycerol
Intestinal lining		
Aminopeptidase	*N*-terminal bond	Free amino acids
Dipeptidase	Dipeptide	Free amino acids
Nuclease	Nucleotide	5-carbon sugar + nitrogen base
Maltase	Maltose	Two glucose units
Sucrase	Sucrose	Glucose + fructose
Lactase	Lactose	Glucose + galactose

Fragments of amylose, a simple, straight chain polysaccharide, are all broken down by amylase from the pancreas into maltose, whereupon the intestinal enzyme **maltase** completes the job by splitting the double sugars into glucose units. But amylopectin, you may recall (see Chapter 3), is a branched polysaccharide. In addition to the usual straight chain with its 1–4 linkages, it has branches that are formed by 1–6 linkages. These linkages are unaffected by pancreatic amylase and must be sheared by another enzyme, **alpha glucosidase.** Following this, digestion is completed by pancreatic amylase and maltase in the usual manner (Figure 38.18).

Sucrose (table sugar) and lactose (milk sugar)— both disaccharides—are also split into simple sugars in the small intestine. The enzyme **sucrase** splits sucrose into glucose and fructose, whereas **lactase** splits lactose into glucose and galactose.

The role of lactase in milk digestion has an interesting aside. Whereas the enzyme is in plentiful supply in the small intestines of all normal human infants, its production diminishes after age two. A large percentage of adults of Northern European origin and adults from a few African dairying tribes continue producing lactase, so milk digestion is no problem for them. (A dietary mainstay of the Masai tribes of East Africa is a concoction of cow's blood and milk.) But some of the adults mentioned above and most Black and Asian adults stop producing lactase altogether. When they consume milk or any food containing lactose, they cannot digest it, and the symptoms of lactose intolerance soon arise. The presence of undigested lactose in the colon brings on diarrhea, and lactose fermentation by colon bacteria produces painful gas buildup. Eating cheese causes less of a problem because its lactose has already been digested by microörganisms used in its manufacture.

Glucose and fructose are actively transported into the capillaries of the villi in the small intestine, and from there they are carried by the blood-stream to the liver for storage. The active transport of glucose is carried out by the same ATP-powered carrier that transports sodium ions into the intestinal cells.

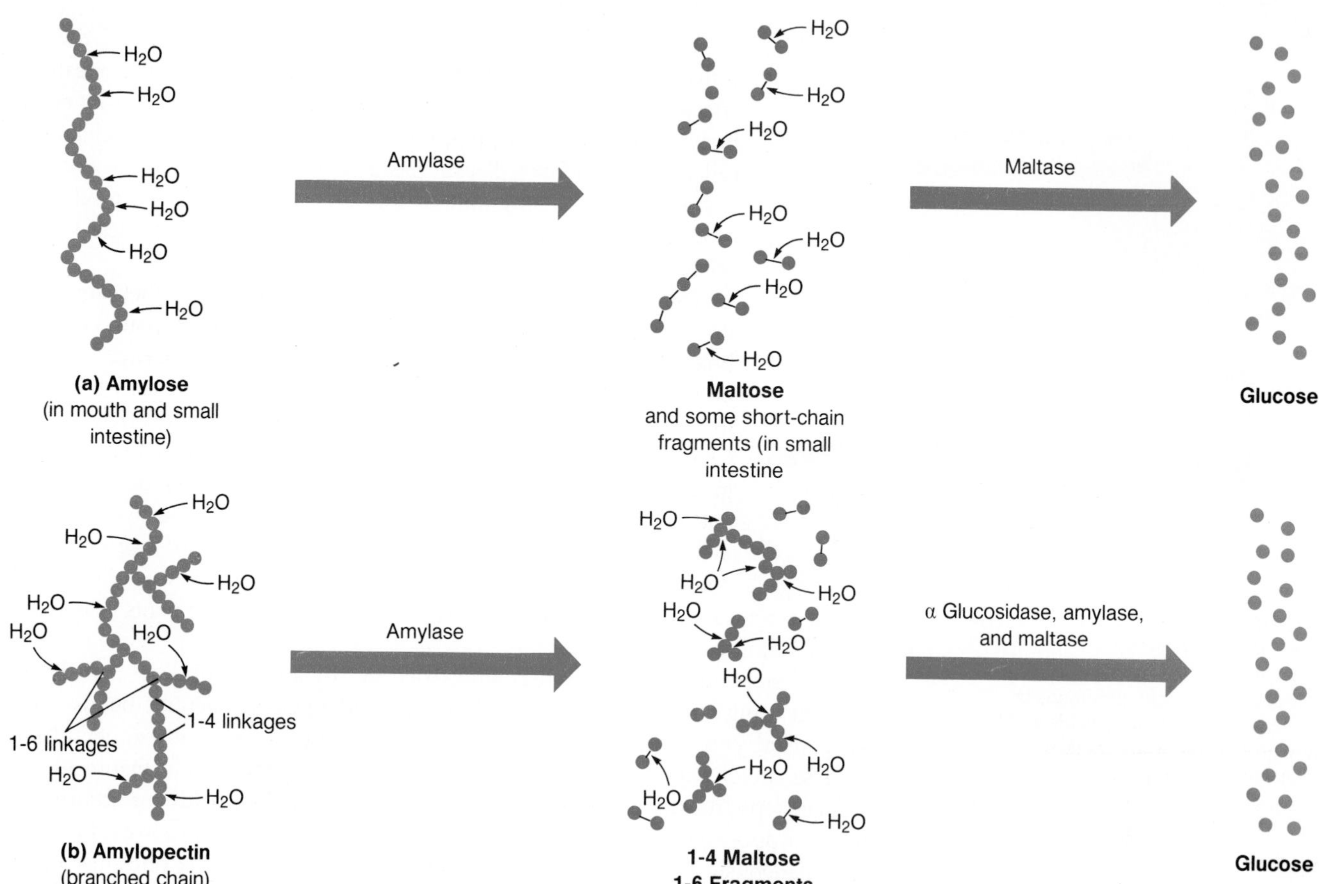

FIGURE 38.18
STARCH DIGESTION
(a) Amylose, a straight chain starch, is readily digested by amylase, a starch splitter that breaks 1–4 linkages. Maltose, the product, is broken down by maltase. (b) Amylopectin digestion is more complex because of the highly branched chain. Amylase cannot attack the 1–6 linkages; thus its products include both 1–4 linked maltose and short, branched chains containing both 1–4 and 1–6 linkages. All of the products are hydrolyzed to glucose by enzymes of the small intestine, but alpha glucosidase is required for the hydrolysis of the 1–6 linkages.

Fat Digestion

With some minor exceptions, fat digestion occurs in the small intestine. Since the digestive environment there is quite watery, we might expect fats and oils to cluster in masses, rejecting their surroundings and making it difficult for hydrolyzing enzymes to do their work. This would be true were it not for bile salts. When fats enter the small intestine, the gall bladder contracts, and bile passes through the bile duct and joins the fats (see Figure 38.17). Bile is a fat emulsifier, which means that it physically breaks the fats down into tiny droplets, dispersing them in the watery surroundings. Bile can do this because it has detergent qualities. Bile salts are steroids that have both fat-soluble and water-soluble regions, so they literally form bridges between the fat and water molecules. How does this help fat digestion? The conversion of large fatty masses into small droplets exposes more of the fat molecules to the fat-splitting enzymes.

Fats are digested by the enzyme pancreatic **lipase.** Lipase breaks down neutral fats (triglycerides) into fatty acids, glycerol, monoglycerides, and diglycerides (Figure 38.19), which simply diffuse across the lipid-soluble core of the plasma membrane of intestinal cells. Once inside a lining cell, these products are reassembled into triglycerides that, along with absorbed cholesterol, then gather into **chylomicrons,** minute bodies enclosed by a thin protein envelope. In this form, they pass across the cell and enter the lacteals of the villi, from which they are transported via lymph vessels to the circulatory system.

Protein Digestion

Proteins are among the largest and most complex molecules known. It is not surprising, then, that their digestion is also complex, requiring a number of different enzymes that form a family called **proteases.**

The first step in protein digestion, as we have mentioned, occurs in the stomach, where an enzyme precursor, pepsinogen, is activated by hydrochloric acid to form the

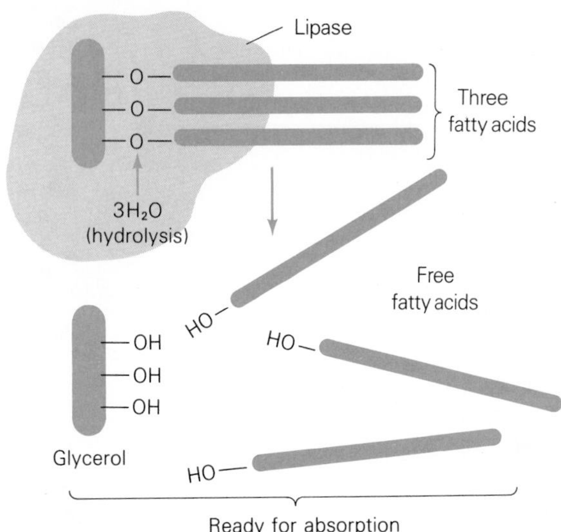

FIGURE 38.19
FAT DIGESTION
Fats are digested by the enzyme lipase, which attacks the bonds between each of the fatty acids and the glycerol. Three water molecules are consumed in the process.

active enzyme, pepsin. Pepsin is an **endopeptidase**, which means it attacks peptide bonds within the protein rather than near its ends. Enzymes that break peptide bonds at the end of a peptide chain are called **exopeptidases.** Pepsin has a very limited action, attacking the peptide bonds of only four of the 20 amino acids. As a result, the products of pepsin's action are usually still lengthy peptides (Figure 38.20).

The peptides pass into the small intestine, where they are acted upon by three pancreatic enzymes: **trypsin, chymotrypsin,** and **carboxypeptidase.** Carboxypeptidase, an exopeptidase, cleaves peptide bonds at the carboxy-terminal end of the peptide, one after the other in a nibbling action, thus freeing individual amino acids. (Recall from Chapter 3 that one end of a protein chain contains a free carboxyl group, and the other end contains a free amino acid group.) Trypsin and chymotrypsin, both endopeptidases, are very selective. Each attacks the peptide bonds on the carboxyl side of just five of the 20 amino acids (Figure 38.20b). The products of action by these three pancreatic enzymes include single amino acids and small peptide fragments of perhaps two to ten amino acids.

The final steps in protein digestion are carried out by aminopeptidases and dipeptidases originating in the intestinal lining. Aminopeptidase cleaves peptide bonds at the amino-terminal end of peptides (thus like carboxypeptidase, it is an exopeptidase). Dipeptidases attack peptide bonds in dipeptides, yielding two amino acids (Figure 38.20). These final steps free the remaining amino acids, which are then ready for absorption.

The absorption of amino acids, like that of glucose, involves ATP-powered carriers that simultaneously transport sodium ions and amino acids into the intestinal lining cells. Once inside the lining cells, amino acids diffuse across the inner walls into capillaries of the villi. They are then carried to the liver, where they are removed from the blood.

In the cells of the liver, amino acids may follow various metabolic pathways, depending on the body's needs. Some are used by the active liver cells to form proteins required there, since the liver has a very rapid protein utilization rate. Other amino acids are deaminated (their amino groups are removed; see Chapter 36) and converted to acetyl-CoA, which is sent into the citric acid cycle where it is used as a source of energy (see Chapter 8). Some amino acids simply converted to other amino acids or used in the formation of the nitrogen bases of nucleotides. Still others are distributed to various cells of the body for their use in protein synthesis. A large number of amino acids are assembled into serum proteins (the proteins of blood plasma), which are released into the bloodstream. We'll look further into the serum proteins in Chapter 39.

The Activation of Proteases It might have occurred to you that the presence of so many protein-cleaving enzymes could present a danger to cells in their surroundings. After all, cells are largely protein, and proteolytic enzymes cannot really distinguish one protein from the next. (Recall that the danger posed by hydrochloric acid to the lining of the stomach and small intestine is averted by the stomach's mucin coating and the small intestine's sodium bicarbonate.)

One safeguard against damage to the digestive lining by proteases is the fact that many of these enzymes first appear in an inactive form. This is particularly important to the pancreas, where many enzymes are manufactured and stored. The pancreatic enzymes trypsin, chymotrypsin, and carboxypeptidase remain in the inactive form of trypsinogen, chymotrypsinogen, and procarboxypeptidase until they enter the small intestine. There, trypsinogen is activated by the intestinal enzyme **enterokinase.** (Recall that *kinases* activate enzymes.) Activated trypsin then modifies the other two enzymes, and they too become active. For this reason the proteases are usually harmless to the pancreas.

But occasionally the safeguards break down, for instance when the pancreatic duct is blocked and the enzyme concentration builds. Then the proteases may become active

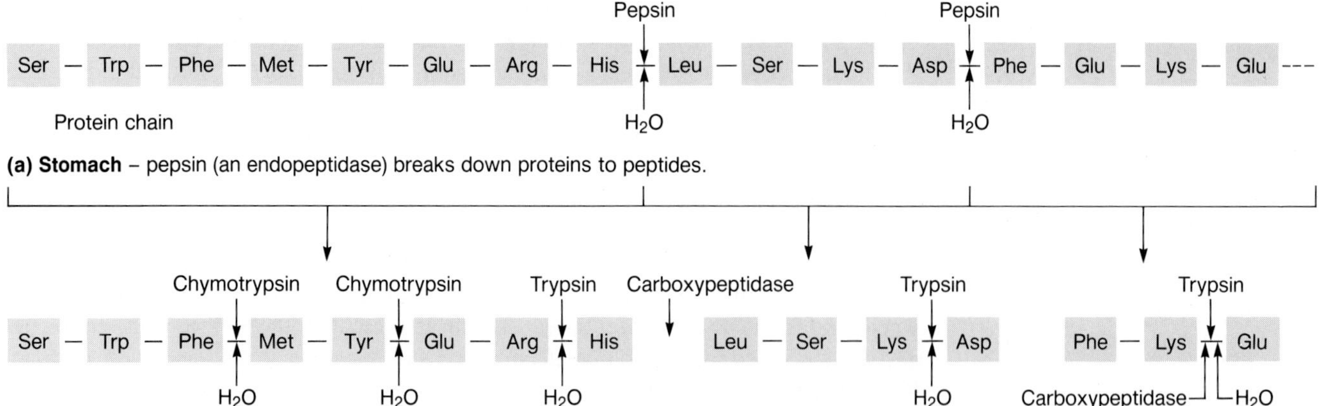

(a) Stomach – pepsin (an endopeptidase) breaks down proteins to peptides.

(b) Small intestine – pancreatic endopeptidase breaks down peptides to smaller peptides. Carboxypeptidase (an exopeptidase) frees the carboxy-terminal amino acids.

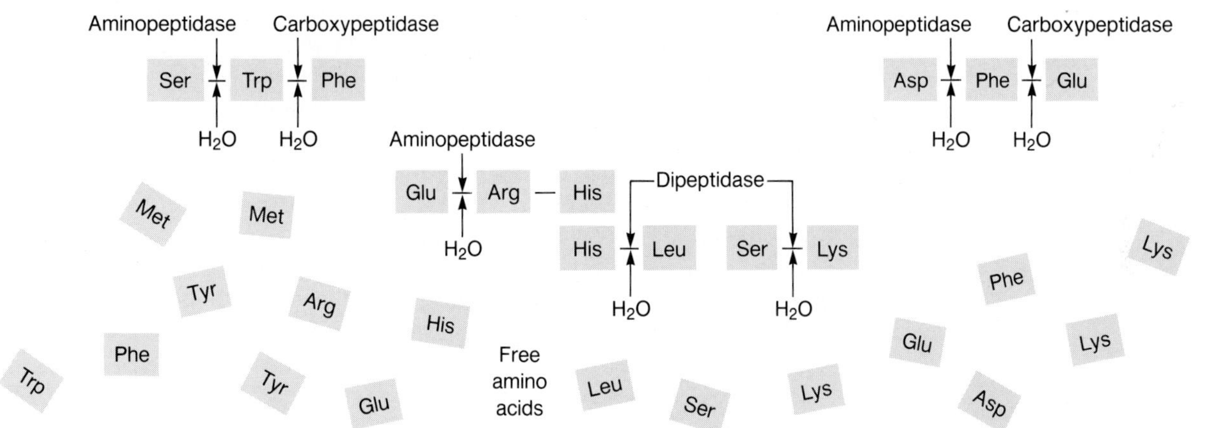

(c) Small intestine – intestinal enzymes (amino and dipeptidases) break down peptides to amino acids. Pancreatic enzyme carboxypeptidase continues to work. Pancreatic enzyme carboxypeptidase.

FIGURE 38.20
PROTEIN DIGESTION
(a) The digestion of protein begins with the action of pepsin. (b) The peptide fragments are then attacked in the small intestine by the pancreatic peptidases. (c) Peptide fragments of varying lengths are then hydrolyzed by peptidases from the small intestine itself. A residue of single amino acids emerges, and protein digestion is complete.

while in the pancreas. This brings on a condition called acute pancreatitis, a potentially fatal disease in which the enzymes attack and destroy the pancreatic tissue itself. Thankfully, pancreatitis is rare—for a special reason. The pancreas produces a trypsin inhibitor, a protein that blocks trypsin's action. And since trypsin activates the other proteases, this is a particularly effective safeguard.

Nucleic Acid Digestion

Since someone is bound to ask, we'd better mention nucleic acid digestion. Almost everything we eat contains nucleic acids. The enzymes that hydrolyze nucleic acids are called **nucleases,** often referred to as RNAases and DNAases. The nucleases break the nucleic acids down into individual nucleotides (see Chapter 12) or into small chains of three or four nucleotides. The enzymes are divided into two classes on the basis of where they attack the giant molecules. **Exonucleases** cleave off the end nucleotides, while **endonucleases** attack bonds within the molecule. Incidentally, the nucleases, and the proteases as well, are important research tools in the study of nucleic acid and protein biochemistry.

Integration and Control of the Digestive Process

The release of digestive enzymes and related substances must be very precisely timed. It turns out that such timing is under at least three types of control—mechanical, neural, and hormonal—with some interaction among the three types. For example, saliva flow can be stimulated mechanically by chewing a tasteless substance like paraffin. Yet, even the thought of eating chocolate cake can evoke salivation. Further along the digestive tract, gastric secretions can also be stimulated by our simply sensing the presence of food. A neural message is sent along a branch of the **vagus nerve** from the brain to the stomach lining (Figure 38.21).

The presence of food in the stomach mechanically evokes gastric secretions by stimulating sensory neurons in the stomach wall itself. Impulses reach the brain, and again the brain sends down the word via the vagus nerve to resume the flow of gastric secretions.

A third mechanism is hormonal and independent of the nervous system. For example, when food, particularly concentrated protein, is present in the stomach, the hormone **gastrin** is released into the bloodstream by cells of the stomach wall itself. Once in the blood, the hormone circulates freely but is ignored by all cells except highly specific target cells, which have plasma membrane or cytoplasmic receptors for that specific hormone (see Chapter 37). Once the hormone has been received and recognized, the reaction is very precise. The target cells, in this case the nearby gastric glands, respond to gastrin by releasing gastric juices. The gastrin mechanism, like many that are hormonally regulated, works through a negative feedback principle. The release of gastric juices, which include HCI, lowers the stomach pH. When it reaches pH2, gastric secretion slows greatly. Thus, the product of the hormone's action—the HCI—has a negative effect, shutting down the hormone's release. The four major hormones involved in the integration and control of digestion are shown in Table 38.2.

SOME ESSENTIALS OF NUTRITION

The subject of nutrition is of increasing interest to the general public. Unfortunately, interest in food doesn't necessarily imply knowledge about nutrition. Many food faddists

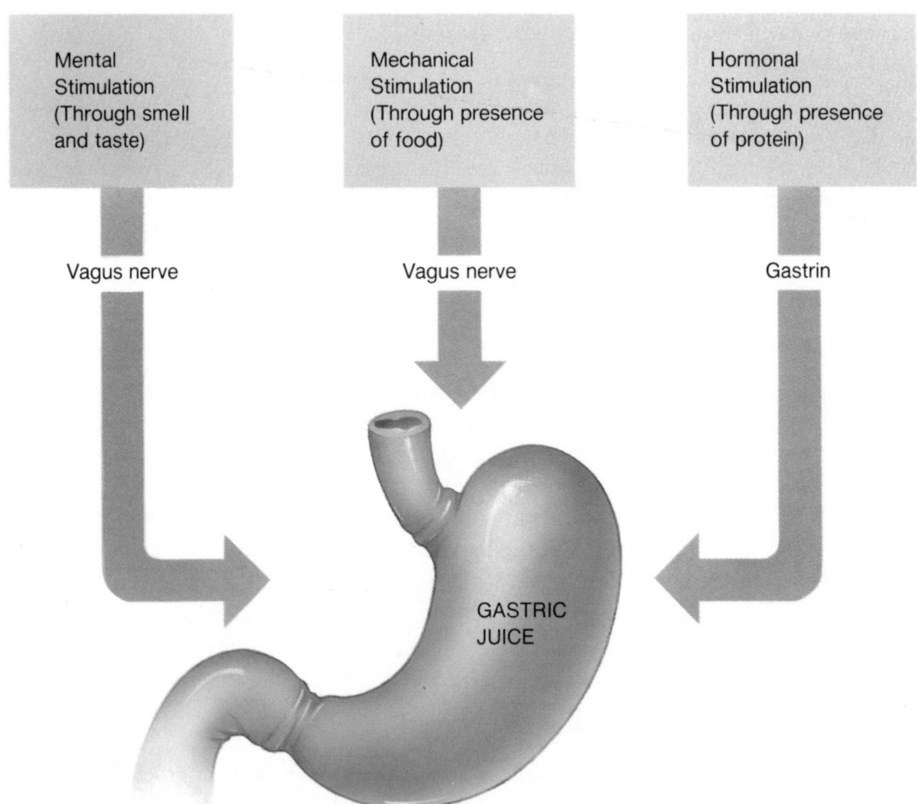

FIGURE 38.21
CONTROL OF GASTRIC SECRETIONS
The flow of gastric juices can be stimulated in several ways. Just the thought, the smell or the taste of food can activate the autonomic nervous system with the vagus nerve stimulating gastric juice release. More directly, the presence of bulk food in the stomach can stimulate release. The presence of protein in the stomach stimulates gastrin (a hormone) release into the blood. Its targets, the glandular cells of the stomach, respond by releasing gastric juices.

TABLE 38.2 DIGESTIVE HORMONES

HORMONE	SOURCE	STIMULATION	FUNCTION
Gastrin	Stomach	Protein in stomach Vagus nerve	Stimulates release of gastric juice
Enterogastrone	Small intestine	Acid state in intestine Fats in intestine	Inhibits gastric juice release and slows stomach contractions
Secretin	Small intestine	Acid state in intestine Peptides in intestine	Stimulates release of sodium bicarbonate
Cholecystokinin-Pancreozymin	Small intestine	Food entering small intestine	Stimulates secretion of pancreatic enzymes and contraction of gall bladder

are almost ignorant of the subject. Some do not seem to realize that "organically grown" food is no better for you than food grown with chemical fertilizers; that rose hip vitamin C is no different from the commercial vitamin C produced by bacteria; that large doses of vitamins are not healthful and may be harmful; that dietary protein in excess of body needs is simply deaminated and converted to glucose or fatty acids in the liver—and fat on the hips. Actually, most Americans eat far more protein than they need, and tests with long-distance runners show that athletes have more stamina on a high-carbohydrate diet than on a high-protein diet. On the other hand, some health faddists prescribe diets so low in protein as to produce virtually the only remaining cases of protein-deficiency disease in America. All this is interesting but confusing, so let's have a closer look at diet and see if the scientific findings have any relevance for us.

Carbohydrates

Carbohydrates are common sources of energy. As we saw in Chapter 8, they are used in generating ATP, but they are used in other ways as well. For instance, once glucose makes its way from the gut into the circulatory system, it is immediately removed and stored as glycogen by the liver. From there it is meted out according to need. Thus, the blood glucose level remains rather constant, with a temporary 10 to 20% increase immediately after a high-carbohydrate meal. Glycogen is also stored in the muscles, where it is broken down into glucose and used as an energy source. Long-distance runners build additional glycogen reserves in muscles by continually depleting their reserves in long training runs and letting them build back up at ever higher levels. For those who have no intention of becoming involved in heavy exercise, it is important to know that excess glucose is readily converted into body fat.

Fats

Fats in the diet are not all bad. The polyunsaturated fats remain an essential nutrient to humans as well as to other animals, although the quantity needed is open to debate. Fats are excellent concentrated energy sources with, gram for gram, twice the energy value of carbohydrates. In addition, they contain the essential fat-soluble vitamins A, D, E, and K. Unsaturated fats are necessary constituents of all plasma membranes, yet vertebrates and most other animals cannot synthesize all the types needed. Some, the essential fatty acids, must be provided in food. Enlightened dieters, then, include at least some fats in the diet. However, fats can be synthesized from carbohydrate and protein excesses, so cutting down on these nutrients is also required if one wishes to lose fat from the body. For this reason, high-carbohydrate or high-protein weight-reducing diets may defeat their own purpose if the intake is not carefully measured.

You are probably aware of the persistent controversy over saturated fats and cholesterol, and may faithfully avoid these in favor of the unsaturated fats heavily touted in

ESSAY 38.2

CHOLESTEROL AND CONTROVERSY

Cholesterol is a vital constituent of plasma membranes. Our own liver can make it readily. Nevertheless, the amount of cholesterol circulating in the blood is greatly influenced by dietary intake. Not only is cholesterol in food taken up directly, but the amount of cholesterol synthesized and its concentration in the blood can be dramatically decreased by reducing either the amount of cholesterol or the amount of saturated triglycerides in the diet, and especially by switching to unsaturated fats.

How did cholesterol get its unsavory reputation? The plaques that plug arteries in chronic arteriosclerosis are heavily infiltrated with cholesterol. The fact that persons with abnormally high levels of circulating cholesterol have abnormally high rates of arterial disease and heart disease suggests a cause-and-effect relationship. Thus, it seems perfectly reasonable to conclude that anyone will benefit from reducing cholesterol intake and replacing saturated (animal and hydrogenated vegetable) fats with unsaturated (nonhydrogenated vegetable) fats. Some doctors still give such advice, but new information is suggesting other routes to good health.

In a recent study, a large group of middle-aged men was divided into an experimental population and a control population. The experimental population was given a low-cholesterol, high-unsaturated diet. Sure enough, their blood cholesterol levels went down, and over the years, their rates of arterial disease and heart disease were lower than those of the control population. Case proven? Not quite. While the death rate due to heart attacks was lower in the experimental group, the overall death rate of the experimental group was significantly *higher* than that of the control group. It turns out that the experimental group had a higher cancer death rate.

What happened? One possibility is that the higher levels of cholesterol in the control group gave them healthier plasma membranes that protected them from cancer and other disease, or perhaps gave them a better-balanced steroid metabolism.

Another, even more likely, possibility is that the higher levels of unsaturated fats in the experimental group were actively toxic. Unsaturated fats spontaneously form free radicals, highly reactive molecules that contain unpaired electrons. Free radicals tend to react with and degrade other molecules in a random manner. Interestingly, ionizing radiation damages cells in a similar manner. It creates free radicals that attack DNA and other cellular constituents. In any case, the study indicates what can happen to those who radically change their diets in response to even a seemingly well-founded scientific opinion. A generation or two of Americans have avoided cholesterol like the plague and, overall, may have gained nothing from their efforts.

advertisements for margarine and cooking oils. Essay 38.2 reveals a startling twist to some prevalent ideas on the subject.

Protein

Since our bodies don't store protein well, we should include it in our daily diet. The amino acids of protein are used in several ways: they are the building blocks of our own protein, some kinds are used to form the nitrogen bases of the nucleic acids DNA and RNA, they can be oxidized for energy, and, as mentioned, they can be converted to fats and carbohydrates.

It is interesting that humans (along with all other animals, protozoans as well) can convert some amino acids to others. In fact, we can synthesize 12 of the 20 amino acids we require for maintenance and growth. The remaining eight must be supplied, intact, in the diet. These are referred to as the **essential amino acids,*** a phrase that delights advertising agencies, although it is something of a misnomer, since all 20 are biologically essential at some level.

Dietary proteins vary in "quality." High-quality proteins contain more kinds of amino acids than do low-quality proteins. Protein synthesis in your own cells cannot proceed unless all of the constituent amino acids are present, so the usefulness of a protein is

*The essential amino acids are: leucine, isoleucine, lysine, methionine, phenylalanine, threonine, tryptophan, and valine. See Figure 3.16.

limited to the relative concentration of its scarcest essential amino acid. In general, animal proteins are of higher quality than plant proteins. Although plants do store proteins in their seeds for the development of new seedlings, these proteins are usually of comparatively low quality for human consumption. Plant storage proteins are typically deficient in tryptophan, methionine, or lysine, and lysine deficiency in particular can become a problem for vegetarians. Nevertheless, a carefully managed vegetable diet can be quite sufficient for human needs, if the right combinations of foods are consumed. A combination of peas and beans, for example, provides the essential amino acids, but because of protein turnover, these foods or other combinations with a similar balance of amino acids must become a regular part of a healthy vegetarian diet.

Vitamins

In addition to essential fatty acids and essential amino acids, a variety of substances, known as vitamins, are also essential in the diets of animals. Most of the vitamins have been clearly identified by biochemists so that we know both their molecular structure and their precise function (Table 38.3). A few are only vaguely understood, however. All we can say about them is that if you don't have them, you will develop something-or-other, and it will probably be bad. Shortages of niacin (vitamin B_6 and riboflavin, for

TABLE 38.3 VITAMINS

VITAMIN	SOURCE	FUNCTION	DAILY REQUIREMENT	RESULT OF DEFICIENCY
A, retinol	Fruits, vegetables, liver, dairy products	Synthesis of visual pigments	1500–5000 IU[a] 3 mg	Night blindness, crustiness about eyes
B_1, thiamine	Liver, peanuts, grains, yeast	Respiratory coenzyme	1–1.5 mg	Loss of appetite, beriberi, inflammation of nerves
B_2, riboflavin	Dairy products, liver, eggs, spinach	Oxidative chains in cell respiration	1.3–1.7 mg	Lesions in corners of mouth, skin disorders
Niacin, nicotinic acid	Meat, fowl, yeast, liver	Part of NAD and FAD cell respiration	12–20 mg	Skin problems, diarrhea, gum disease, mental disorders
Folic acid	Vegetables, eggs, liver, grains	Synthesis of blood cells	0.4 mg	Anemia, low white blood count, slow growth
B_6	Liver, grains, dairy products	Active transport	1.4–2.0 mg	Slow growth, skin problems, anemia
Pantothenic acid	Liver, eggs, yeast	Part of coenzyme A of cell respiration	Unknown	Reproductive problems, adrenal insufficiency
B_{12}	Liver, meat, dairy products, eggs	Red blood cell production	5–6 μg	Pernicious anemia
Biotin	Liver, yeast, intestinal bacteria	In coenzymes	Unknown	Skin problems, loss of hair and coordination
Choline	Most foods	Fat, carbohydrate, protein metabolism	Unknown	Fatty liver, kidney failure, metabolic disorders
C, ascorbic acid	Citrus fruits, tomatoes, potatoes	Connective tissues and matrix antioxidant	40–60 mg	Scurvy, poor bone growth, slows healing (colds?)
D	Fortified milk, seafoods, fish oils, sunshine	Absorption of calcium	400 IU[a]	Rickets
E	Meat, dairy products, whole wheat	Antioxidant	Unknown	Infertility, kidney problems
K	Intestinal bacteria	Blood-clotting factors	Unknown	Blood-clotting problems

[a]IU = International units mg = milligram (1/1000) μg = (1/1,000,000)

example, can lead to serious illness because the coenzymes NAD and FAD, necessary for glycolysis and cell respiration, are derived from these vitamins. Vitamins C and E have a somewhat more general function: both are antioxidants that remove spontaneous free radicals that would otherwise cause damage through random oxidation reactions.

Vitamins can be categorized as fat-soluble or water-soluble. The two behave quite differently in the body. For example, water-soluble vitamins function as coenzymes; fat-soluble vitamins do not. Whereas water-soluble vitamins function in most animals, the fat-soluble function only in vertebrates. Finally, excess water-soluble vitamins are excreted, but the fat-soluble ones are stored in the body fat. This is why it is dangerous to overdose on fat-soluble vitamins—they remain in the body's fat and can be released as that fat is metabolized.

Mineral Requirements of Humans

In addition to the organic nutrients mentioned, animals require a variety of inorganic ions that are generally referred to as **minerals** (Table 38.4). Some of those needed in susbstantial amounts are calcium (a major constituent of bone and of many cellular processes), magnesium (necessary for many enzyme activities), and iron (a constituent of hemoglobin and of the cytochromes of cell respiration). Sodium, potassium, and chloride are also needed in fairly substantial quantities; they are involved in ion balance, and sodium and potassium are also involved in nerve cell conduction.

Trace Elements As we mentioned with plants, elements that are necessary for life but in only very small amounts are called **trace elements.** Iodine is perhaps the best-known trace element, since an iodine deficiency produces **goiter**, a highly visible overgrowth of the thyroid gland (see Figure 37.18b). Iodine is essential in producing thyroxin, the thyroid hormone, and the overgrowth is the thyroid gland's odd way of compensating

TABLE 38.4 MINERAL NUTRIENTS

MINERAL	FOOD SOURCE	FUNCTION	DAILY REQUIREMENT	RESULT OF DEFICIENCY
Calcium	Dairy foods, eggs	Growth of bones and teeth, blood clotting, muscle contraction, nerve action	800 mg	Tetany, rickets, loss of bone minerals and muscle coordination
Cobalt	Common in foods, water	Vitamin B_{12}	1 mg	Anemia
Copper	Common in foods	Production of hemoglobin, enzyme action	2 mg	Anemia
Fluorine	Most water supplies	Prevents bacterial tooth decay	Unknown	Tooth decay, bone weakness
Iodine	Seafood, iodized salt	Thyroid hormone	0.15 mg	Hypothyroidism
Iron	Meat, eggs, nuts, raisins	Hemoglobin (oxygen transport)	10 mg (men) 18 mg (women)	Anemia, skin problems
Magnesium	Green vegetables	Enzyme function	350 mg	Dilated blood vessels, irregular heartbeat, loss of muscle coordination
Manganese	Liver, kidneys	Enzyme function	Unknown	Loss of fertility, menstrual irregularities
Phosphorus	Dairy foods, eggs, meat	Growth of bones and teeth, ATP nucleotides	800 mg	Loss of bone minerals, metabolic disorders
Potassium	Most foods	Nerve and muscle activity	2–4 g	Muscle and nerve disorders
Sodium	Most foods, salt	pH balance, nerve and muscle activity, body fluid balance	0.5 g	Weakness, muscle cramps, diarrhea, dehydration
Zinc	Common in foods	Enzyme action	15 mg	Slow sexual development, loss of appetite, retarded growth

for the iodine shortage. Many other minerals are constituents of coenzymes (for example, cobalt in vitamin B_{12}), are otherwise necessary for enzyme function, or are involved in synthetic processes (for example, zinc is needed for insulin synthesis). The functions of many trace elements are unknown—for instance, it is not known why small amounts of fluoride retard cavities in teeth.

Among the most bizarre recently discovered mineral requirements are arsenic, silicon, and selenium. Arsenic is an extremely deadly poison, yet laboratory animals have died of arsenic deficiency! Silicon is one of the most abundant elements (found in sand and rocks). Yet, silicon deficiencies may also be widespread. The mineral is needed as a cross-linking agent in the elastic walls of major arteries. Finally, selenium is known to be necessary for the functioning of at least one enzyme. Human selenium deficiency disease is common in some parts of China. (In other parts of China the population suffers from selenium poisoning.)

So, what's a body to do? We've seen that low-cholesterol and high-cholesterol diets are both harmful, that the balance between saturated and unsaturated fats mustn't be tipped too far in either direction, that too much protein is harmful but too little is worse. Vegetarianism can be dangerous, but so can meat. A tiny amount of arsenic is deadly, but an even tinier amount may help keep you alive; you can get sick from too much vitamin A or D, or from not enough. Most people consume too much sodium, but everyone needs some. Too little selenium is bad, but so is too much. How can one make intelligent decisions in the face of conflicting information?

There is no simple answer. We must resort to generalities and clichés, such as "moderation in all things" or "variety is the spice of life" or "don't eat so much." No one ever suffered from eating sensible amounts of fresh fruits and vegetables. Get enough roughage. Avoid faddism. Avoid fats, but don't be a fanatic about it (unless you are overweight, in which case a little antifat fanaticism may be in order). Part of the problem is, we live in a food-laden environment. Most of us can choose from a range of foodstuffs, some traditional, others recently and chemically contrived. With the opportunity for choice, however, comes the responsibility of being informed. And since much of what we know has just been learned recently, the responsibility remains a continuing one.

APPLICATION OF IDEAS

1. Select five distinct variations in mammalian dentition (including its absence) and relate these variations specifically to the food consumed by each animal. Suggest how such variation might arise.
2. The feeding habits of animals are highly varied. Omnivores commonly feed on a wide variety of foods, while more specialized species, whether herbivore or carnivore, may feed mainly on one specific item with little deviation. Discuss examples of each and suggest short- and long-term evolutionary advantages and disadvantages in either direction.
3. Careful studies of herbivorous animals reveal that they rarely produce the enzyme cellulase, yet a considerable amount of their diet consists of cellulose. Using examples, explain how mammals extract nutrients from cellulose. Considering that a large part of the carbohydrate produced on earth is in the form of cellulose, suggest reasons why most animals haven't evolved the ability to synthesize the enzyme. What advantage would such a capability offer humans?

KEY IDEAS

DIGESTIVE SYSTEMS

1. **Digestion** is the mechanical and chemical breakdown of nutrients into absorbable parts, while **nutrition** refers to nourishment and the characteristics of essential nutrients.
2. **Extracellular digestion** refers to digestion outside of cells as seen in bacteria and fungi. **Intra-cellular digestion** occurs in cells, generally in **digestive vacuoles.**

Saclike Systems in Invertebrates

1. Sponges sort food particles out of seawater brought in by **choanocytes (collar cells).** Particles caught in the mucus-lined collar cells are phagocytized by the cells below for intracellular digestion and transported about by **amebocytes.**
2. Cnidarians and flatworms begin digestion extracellularly in the **gastrovascular cavity**, with food phagocytized and digestion completed intracellularly.

The Tube-within-a-Tube Digestive Plan

1. Higher invertebrates have the tube-within-a-tube body plan, and digestion is mainly extracellular. Intracellular digestion occurs in the invertebrate liver, which is lined by phagocytic cells.
2. The earthworm gut contains specialized regions. Food is swallowed by the **pharynx** and passes through the **esophagus** to the **crop** for storage. Grinding occurs in the **gizzard**, and digestion and absorption occur in the long **intestine.** The **typhlosole**, a deep fold in the intestinal wall, increases surface area. **Chloragogen** cells convert glucose to glycogen and **deaminate** amino acids to be used as fuels.
3. The grasshopper's chewing mouthparts include sensory **palps**, shearing **mandibles** and **maxillae**, the liplike **labrum** and **labium**, and salivary glands that secrete saliva. The digestive system includes a **foregut** consisting of a pharynx, esophagus, and crop for swallowing and grinding; a **midgut** for digestion and absorption; **gastric ceca** for enzyme secretion; and a water-absorbing **hindgut.** Insect mouthparts are also adapted for piercing and sucking.

Feeding and Digestive Structures in Vertebrates

1. Vertebrate jaw and tooth structure varies with feeding specializations. Carnivorous sharks have continuously growing rows of teeth. In the gut, a winding flap, the **spiral valve,** increases the surface area for digestion and absorption. Teeth in bony fishes vary from numerous sharp teeth to patches on the roof of the mouth. Herbivorous fishes generally have lengthy coiled intestines suitable for the time-consuming cellulose digestion process.
2. Some amphibians have a protruding tongue for capturing insects. The jaw, like that of the reptile, is not suitable for chewing, so prey is swallowed whole.
3. Crocodiles and alligators have socketed teeth, while some venomous snakes have hollow or grooved retractible **fangs** for injecting venom. Snakes and lizards capture air molecules with the tongue and analyze them with the olfactory **Jacobson's organ.** Pit vipers use their heat sensitive pits to orient themselves to their prey.
4. The beak (and feet) of birds are often specialized for their source of food. Digestive specializations in seed-eaters include a storage crop and a two-part **stomach** composed of the glandular, enzyme-secreting **proventriculus** and the gravel-filled and muscular **ventriculus**, or gizzard.
5. Mammalian teeth are socketed and include temporary, or milk teeth, and permanent teeth. Four types of teeth include **incisors, canines, premolars,** and **molars.** The shape and size of each group commonly relates to diet.
6. **Ruminants** are grazing mammals whose four-part stomach includes a **rumen**, where microorganisms digest cellulose. The products and microorganisms are themselves digested by the **reticulum**, **omasum**, and **abomasum** (true stomach).

THE DIGESTIVE SYSTEM OF HUMANS

The Oral Cavity and Esophagus

1. The lips and tongue assist in eating and swallowing. **Taste buds** located on the tongue detect salt, sour, sweet, and bitter flavors. **Saliva,** a watery solution of ions, mucus, and **amylase**, a starch-digesting enzyme, are produced in the **parotid**, **submandibular**, and **sublingual** salivary glands.
2. The pharynx divides to form the **larynx** and **laryngopharynx.** Some of the structures involved in swallowing are the tongue, esophagus, **soft palate, epiglottis, glottis,** and **larynx.** The larynx is raised during swallowing, preventing food from entering the air passage.
3. The esophagus, a food-conducting tube, is made up of an innermost, secretory **mucosa**, a vascular and glandular **submucosa,** and a circular and longitudinal layer of smooth muscle, the **muscularis**, all of which are surrounded by a fibrous **serosa.** Wavelike contractions called **peristalsis**, coordinated by the autonomic nervous system, move the food **bolus** to the stomach.

The Stomach

1. The stomach stores and mixes food, and its acidity helps destroy microorganisms. The **cardiac** and **pyloric sphincters** close off the stomach during its churning peristalsis. A third, oblique muscle layer aids the churning by producing a twisting action.
2. The glandular lining contains **pepsinogen**-secreting **chief cells** and **parietal cells**, whose **hydrochloric acid** secretions activate pepsinogen to the active form, **pepsin.** Other enzymes include **gastric lipase** and milk-digesting **rennin.**

The Small Intestine

1. The **small intestine** consists of the **duodenum, jejunum**, and **ileum.** Its role is digestion and absorption. In addition to what is provided by length, the surface area of the intestine is increased by folding, by projections called **villi**, and by cellular projections called **microvilli.** Each villus contains a **capillary bed** and a **lacteal**, and is capable of movement.
2. The small intestine surface contains bound enzymes forming the **glycocalyx.** New cells continually replace those lost by abrasion.

Accessory Organs: The Liver and Pancreas

1. The **liver** secretes **bile**, an alkaline secretion that emulsifies fats. Its contents include hemoglobin breakdown pigments (bilirubin and biliverdin).

2. Bile is stored in the **gall bladder** and secreted through the **bile duct**, which joins the **pancreatic duct** before entering the intestine. Bile secretion is a response to the hormone **cholecystekinin-pancreozymin**, which is released when fats are present in the gut.
3. The glandular **pancreas** secretes sodium bicarbonate, which neutralizes stomach acids in the intestine. It also secretes a variety of enzymes that act on carbohydrates, fats, proteins, and nucleic acids.

The Large Intestine

1. The **large intestine**, or **colon**, begins with the **ileocecal valve**, which opens into a pouchlike cecum (to which the **appendix** is attached). Included are the **ascending colon**, **transverse colon**, **descending colon**, and **sigmoid colon.** Further along are the **rectum**, **anal canal**, and **anus**.
2. The colon absorbs water, concentrates the feces, and provides a suitable environment for bacteria that secrete useful vitamin K, biotin, and folic acid. Within the rectum, **rectal valves** help support the feces, and the anus controls defecation.

THE CHEMISTRY OF DIGESTION

1. The bonds connecting subunits of complex foods are broken through **hydrolysis,** releasing the simple subunits that can be absorbed or actively transported into cells lining the intestine.

Carbohydrate Digestion

1. Starches are hydrolyzed into maltose by salivary amylase in the mouth and by pancreatic amylase in the small intestine. Amylopectin requires the enzyme **alpha glucosidase. Maltase** hydrolyzes maltose into glucose, which is then actively transported by a sodium carrier into the blood.
2. **Lactase** hydrolyzes milk sugar, but adults in some races fail to produce the enzyme. Upon taking in milk they experience the symptoms of lactose intolerance.
3. Glucose is transported to the liver, where it is converted to glycogen and gradually released again as glucose.

Fat Digestion

1. Following their emulsification by bile, fats are hydrolyzed by **lipase** into fatty acids, glycerol, monoglycerides, and diglycerides. These diffuse into the lining cells, reform as triglycerides, and join cholesterol to form **chylomicrons,** which pass into the lacteals.

Protein Digestion

1. Protein digestion begins in the stomach, where pepsin, an **endopeptidase**, hydrolyzes inner peptide bonds, forming shorter peptides.
2. The pancreatic **proteases, trypsin, chymotrypsin** (both endopeptidases), and **carboxypeptidase** (an **exopeptidase)** are released into the small intestine. The first two produce peptides, whereas the last frees *C*-terminal amino acids.
3. The intestinal protease **aminopeptidase** (an exopeptidase) frees *N*-terminal amino acids. **Dipeptidase** frees amino acids from dipeptides.
4. Amino acids enter the blood for transport to the liver, where some are deaminated and used for energy, while others form serum proteins for transport to needy cells.
5. Pancreatic proteases are safely stored in the pancreas in an inactive form. Trypsin, inactivated by a special pancreatic protein, is activated by **enterokinase** in the small intestine, whereupon it activates the chymotrypsin and carboxypeptidase.

Nucleic Acid Digestion

1. **Nucleases,** nucleic acid hydrolyzing enzymes, include **exonucleases** that work on the ends and **endonucleases** that work within the chain.

Integration and Control of the Digestive Process

1. Digestive enzymes are released through mechanical (physical presence of food), neural (thoughts and detection of food), and hormonal mechanisms. Neural messages travel to the stomach via the **vagus nerve** when the gut lining senses the chemical presence of food. The cells react by releasing a hormone such as **gastrin** into the blood. When hormones reach their specific target cells or tissues, they stimulate them to release enzymes into the gut.
2. Hormonal systems generally work through negative feedback, in which a stimulus evokes the release of a hormone, which removes the stimulus and thereby stops its own release.

SOME ESSENTIALS OF NUTRITION

Carbohydrates

1. Carbohydrate is a vital energy source, but in excess it is converted to body fat.

Fats

1. Fats are excellent energy-providing foods and sources of fat-soluble vitamins. Unsaturated fats are needed for plasma membranes. Although unsaturated fats must come from the diet, most foods can be converted to saturated fats.

Protein

1. Protein must be provided daily because of its constant turnover. Excess amino acids are deaminated and used for energy or converted to fats and carbohydrate.
2. Eight of the amino acids—**essential amino acids**— must be provided by the diet, while the remaining 12 are interconvertible by the body. High-quality dietary protein contains sufficient amounts and kinds of amino acids to provide all that is needed.

Vitamins

1. Fat-soluble vitamins are generally essential in forming coenzymes (NAD and FAD), and water-soluble vitamins as **antioxidants** (removing free radicals) and other uses. The former are harmful in overdoses.

Mineral Requirements of Humans

1. Major essential **minerals**, inorganic ions, include calcium, magnesium, iron, sodium, potassium, and chloride, used in structure, enzyme action, respiration, and osmotic regulation.
2. Trace elements include iodine, cobalt, and silicon. A shortage of iodine can cause **goiter**, an overgrowth of the thyroid.

REVIEW QUESTIONS

1. Distinguish between extracellular and intracellular digestion and provide an example of each from the invertebrates. (910–911)
2. Explain how the sponge takes in nutrients, carries on digestion, and distributes the products. (910)
3. Describe digestion in the gastrovascular cavity of a cnidarian. List three digestive functions served by the cavity. (911)
4. List the specialized regions of the earthworm digestive tract and give a function of each. Include absorbing structures. (911–912)
5. List the three parts of the insect gut and explain what each does. (912)
6. Describe two anatomical feeding specializations in the shark and explain how they operate. (913)
7. List four anatomical feeding specializations in the amphibians and reptiles. (914–915)
8. Explain the functions of the bird crop, proventriculus, and ventriculus. (914–915)
9. List the four types of mammalian teeth and describe specializations in each. (915)
10. List the parts of the four-chambered stomach of cattle and trace the path followed by food. (916–917)
11. Since cows cannot digest cellulose, how do they obtain needed nutrients from grasses? (917)
12. Draw a simple human taste map and label the specialized areas. (917)
13. List the three salivary glands and explain how saliva functions in digestion. (917–918)
14. Explain how peristalsis works and how it is controlled. (918–919)
15. Describe the muscularis of the stomach and its action in moving food. (919)
16. Describe the stomach lining and state the specific functions of the parietal and chief cells. Name a secretion that protects the stomach lining. (921)
17. Describe the four levels of structure that provide the small intestine with its enormous surface area. (921)
18. Draw a simple villus. Label its parts, including any specialized cell surface features. (921)
19. What is the glycocalyx? How is this an improvement over the simple release of enzymes into the lumen? (922)
20. What is the digestive function of the liver? How does the release of its secretion take care of two problems at once? (922)
21. List two general functions of the pancreatic secretions. (922)
22. Briefly discuss three functions of the colon. (922–923)
23. Name and describe the common chemical mechanism of digestion. What is the opposite process called? (923)
24. List the steps in the digestion of starch. What is the end product? (923–924)
25. Explain the peculiar manner in which fatty acids are absorbed. (925)
26. What is an endopeptidase? Name three of these and state where they are produced. (925–926)
27. Where are the exopeptidases produced? What steps in protein digestion do they accomplish? (926)
28. List three different mechanisms controlling enzyme release. (927–928)
29. Explain how gastrin is released and what stops its action. (928)
30. List several essential reasons for including saturated and unsaturated fats in the diet. (929)
31. What determines the quality of dietary protein? What vegetable foods contain high-quality proteins? (930–931)
32. List two specific ways in which vitamins are used by the body. (931–932)
33. List four major and two trace mineral requirements and state their uses. (932–933)

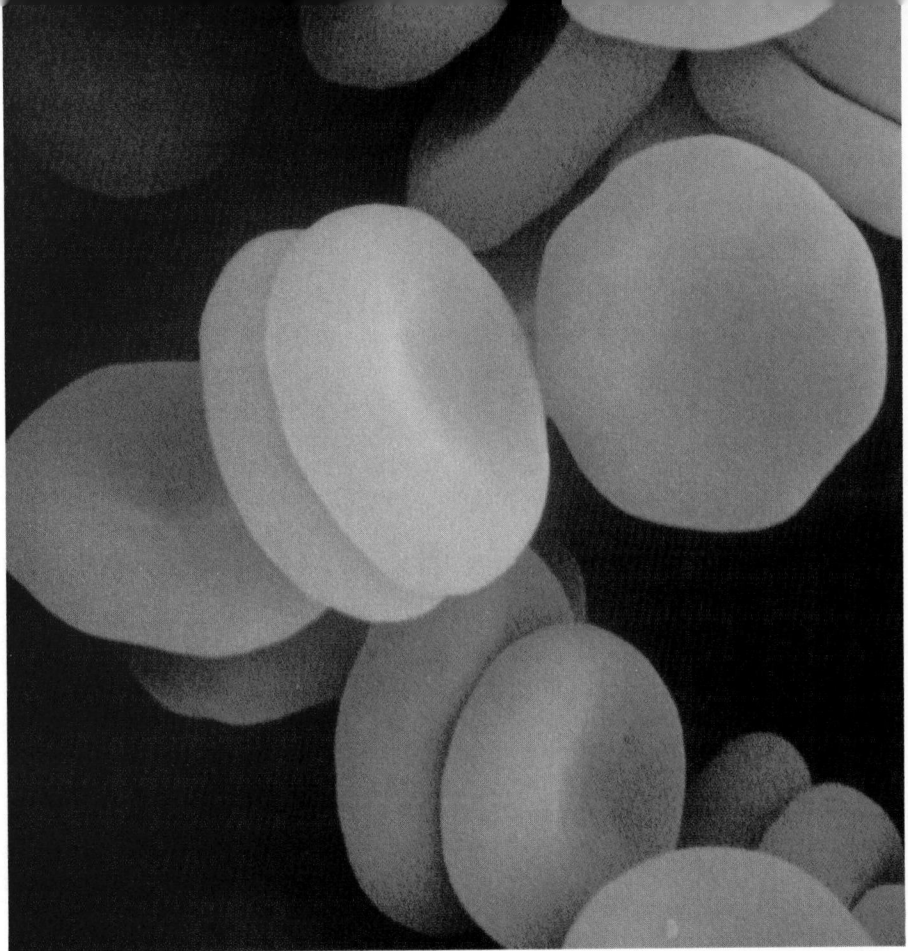

39

CIRCULATION

WATCHING THE EARLY DEVELOPMENT OF A VERTEBRATE EMBRYO IS A fascinating and often reaffirming experience, even to crusty old biologists who have spent their adult lives considering the various manifestations of life. It might seem that the feeling of wonder would eventually diminish, but somehow it doesn't. The sight of the quick pulses of a tiny, unformed heart signalling a new life is perpetually intriguing. The heart is, in a sense, the very symbol of animal life.

In its earliest days, the vertebrate heart gives little indication of how complex it will become since at that time it is only a simple, twitching tube. Curiously, the beating begins hours before there is any blood to pump. In fact, the circulatory system is among the first systems to take form. Why should this be?

Actually, if we consider that the embryo needs an inflow of nutrients and oxygen and the prompt removal of metabolic wastes, we should expect the circulatory system to develop early. But the circulatory system—that is, the heart and the vessels it serves—is important throughout the life of the animal. In this chapter, we'll see why. We can begin by reminding ourselves that the primary mission of the circulatory system is transport: transport of oxygen, carbon dioxide, nutrients, water, ions, hormones, antibodies, and metabolic wastes. The circulatory system, along with the lymphatic system, which we will discuss shortly, constantly shifts water, ions, and proteins about, thus helping to regulate osmotic conditions in the body. Osmotic conditions are somewhat complex since there are actually three fluid regions involved: the fluid of blood itself, the interstitial fluid (freer, watery fluids between and among cells), and the intracellular fluid (fluid within the cell). Since capillaries are usually surrounded by interstitial fluids, all substances passing from the blood to the cells (and vice versa) must first pass through this watery matrix. We will learn more about the nature of interstitial fluids after we look into the structure and work of the circulatory system.

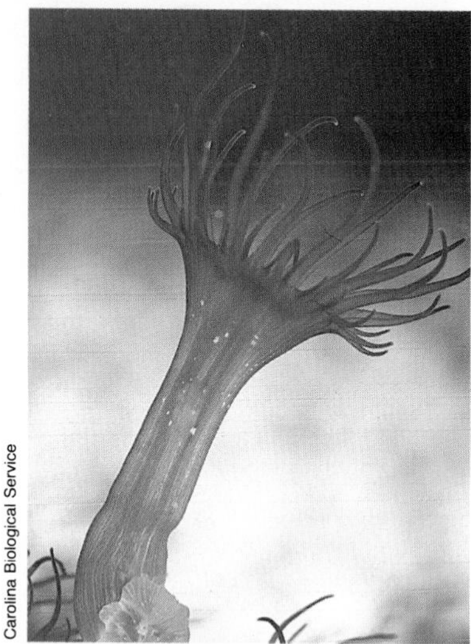

FIGURE 39.1
THE DIRECT EXCHANGE
Thin-bodied aquatic creatures such as the sea anemone and Portuguese man-of-war have no need for elaborate circulatory systems. Cell-to-cell transfers of materials and a direct exchange between body cells and the environment suffice.

ADAPTATIONS IN ANIMALS WITHOUT CIRCULATORY SYSTEMS

As living things evolved, they became diverse, often more complex, and in many cases, larger. With greater size and complexity came new problems. Some of these problems involved how to better service the various tissues of the body, that is, how to transport food and oxygen to them and to carry away their metabolic wastes.

Among the smallest forms of eukaryotic life, the protists (see Chapter 23), there are few, if any, transport problems. Their small size results in a high surface-area-to-volume ratio; therefore, a relatively expansive body covering is in direct contact with the environment. In such organisms, the body surface alone easily accommodates the necessary exchanges of gases, nutrients, and wastes. Cytoplasmic movement within the cell, such as seen in the flowing cytoplasm of an ameba (see Figure 23.8), helps in distributing molecules within the body.

The sponges (see Figure 30.6) are multicellular animals that also rely on their cell surfaces for exchanging materials with the environment. Although some sponges reach considerable size, the direct exchange method still works for them because they have a thin body wall and no cell is far from the surrounding water. Recall that the sponge's choanocytes (collar cells) constantly pump water through its body, bringing in food and oxygen and washing away metabolic wastes (see Chapters 30 and 38). The sponges do, however, make use of ameboid cells in the distribution of foods, providing an assist to cell-to-cell transport.

Cnidarians, the hydrozoans, jellyfish, and corals, also lack an organized circulatory system. As is true of sponges, cnidarians are very thin-walled and hollow; thus, most of their cells carry out exchanges directly with the watery surroundings (Figure 39.1). The extensive gastrovascular cavity provides a means of food distribution. Even the flatworms, with their much denser bodies, rely heavily on cell-to-cell transport and direct exchanges with their environment, aided by an extensive gastrovascular cavity.

CIRCULATORY SYSTEMS IN INVERTEBRATES

The body mass of the larger, more complex animal is too great for its requirements to be met by direct environmental exchanges and simple cell-to-cell transport. Such exchanges occur far too slowly to provide a sufficient supply of oxygen and food and to remove carbon dioxide and nitrogenous metabolic wastes before they become toxic. The evolution of larger animals thus included provisions for rapid transport in the form of circulating fluids contained within circulatory systems. There are essentially two types of circulatory systems: open and closed (Figure 39.2).

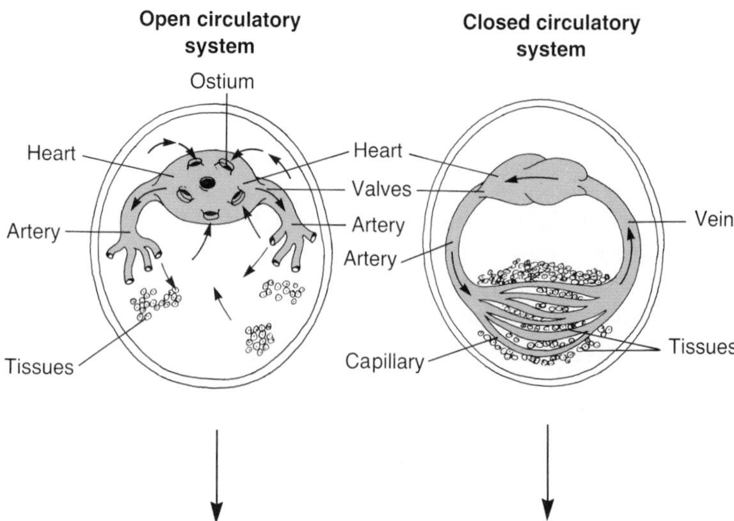

FIGURE 39.2
OPEN VERSUS CLOSED CIRCULATORY SYSTEMS
In an open circulatory system, blood leaves the arteries and percolates through tissues, finally reentering the heart through special openings called ostia. In the closed system, blood is always contained in endothelium-lined vessels, and transfer of materials to and from tissues must take place across the walls of those vessels. Note that the valves in both types of heart are arranged in a way that ensures the forward movement of blood.

Open and Closed Circulatory Systems

In **closed circulatory systems**, the blood remains within the confines of vessels throughout its entire circulation. In **open circulatory systems** blood does not remain in vessels. We'll use the open circulatory system of the arthropod as our example, but keep in mind that arthropods vary immensely due to the sheer size of the phylum.

In many arthropods, the blood is pumped by the heart (or hearts—there are often several) into an artery, a vessel that carries blood away from the heart. But the artery soon ends, and the blood enters channels leading to sinuses or cavities sometimes called **hemocoels** (blood cavities). Because these cavities are lined with **endothelial** cells (flattened epithelial lining cells), they can actually be regarded simply as expanded forms of blood vessels. As the blood percolates through such hemocoels, its fluids are forced around and through the body's various tissues. (Thus, there is no clear separation of blood and interstitial fluids as in closed systems.) As the blood continues to move, it eventually reaches a cavity surrounding the heart. From there the blood is drawn into the heart through openings called **ostia** (singular, **ostium**) and pumped out again for another circuit. If you know about boats or flooded cellars, the blood-collecting work of the arthropod heart may remind you of a bilge pump, a submergible pump that draws water from its surroundings and directs it into hoses.

It is difficult to generalize about the many species of arthropods, but we can say that probably all of them have open circulatory systems. If you compare the circulatory systems of the crayfish and the grasshopper in Figure 39.3, you will see that the system is more complex in the gill-breathing crayfish (also see Figure 40.5). Its blood is involved in respiration as well as transport, whereas in the insect, the circulatory and respiratory systems are independently structured. The hearts of the two arthropods are also quite different. Note that the grasshopper heart is no more than a long dorsal blood vessel with muscular thickenings along its posterior region. In the crayfish, the heart is compact, more central, and more complex.

Open circulatory systems are also found in certain mollusks as well as in other kinds of invertebrates. Some invertebrates, however, have closed circulatory systems. As you may recall, the annelids have a remarkably well-developed closed system with distinct vessels, tubular hearts, and hemoglobin-rich blood. In addition, they have a fluid-filled coelom that assists in the distribution of materials. Note in Figure 39.4 the major vessels and the five pairs of hearts, or **aortic arches.** Valves in the aortic arches ensure a one-way flow of blood. Closed systems are also found in cephalopod mollusks, such as the squid and octopus.

Blood flows much more rapidly in closed systems than in open ones; thus, it would seem to have a marked adaptive advantage in active species that need a ready supply of oxygen to meet their greater metabolic needs. Yet insects, among the most active of all animals (as you know if you have ever tried to catch a fly), have an open system. They are allowed this evolutionary indulgence for a very simple reason. They do not

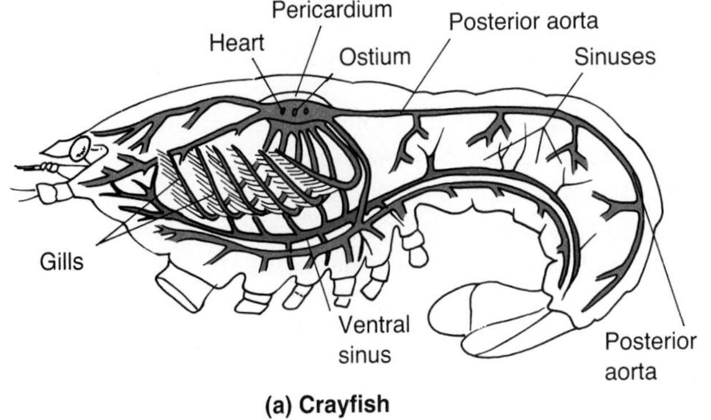

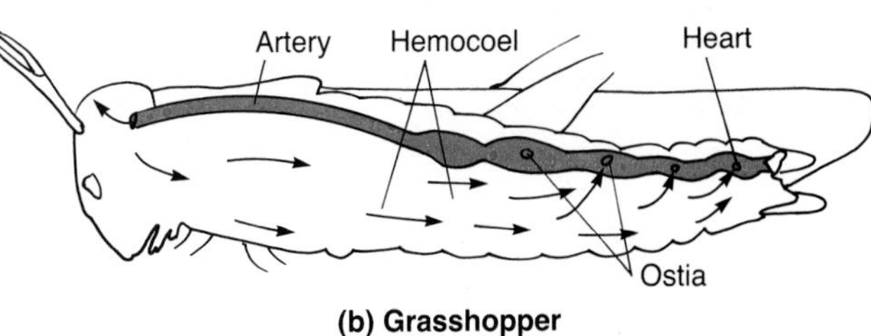

FIGURE 39.3
THE ARTHROPOD OPEN CIRCULATORY SYSTEM
(a) The system of vessels is more extensive in the crayfish, with the blood serving a respiratory function as it passes through the gills. **(b)** The grasshopper has one dorsal vessel that contains several pumping regions or "hearts." Blood is collected by the hearts and pumped forward, first to sinuses in the head and then through the body sinuses, eventually returning to the dorsal vessel.

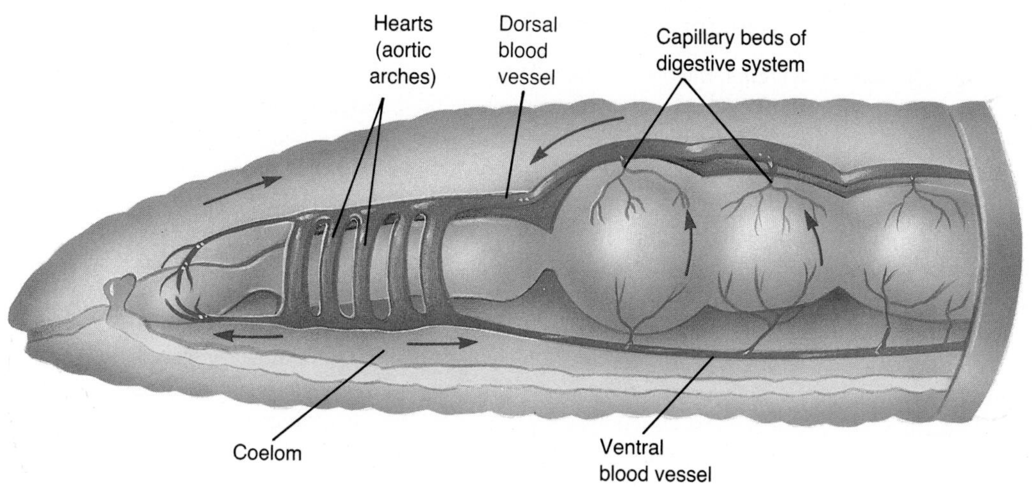

FIGURE 39.4
THE ANNELID CLOSED CIRCULATORY SYSTEM
In the earthworm's closed circulatory system, blood is pumped along by five pairs of aortic arches. Blood moves toward the body and head via the large ventral blood vessel, where, after passing through extensive capillaries, it returns to the aortic arches via the dorsal vessel.

rely much on their blood to carry oxygen. Instead, oxygen is transported through a system of open tubes that penetrate every part of the body, as described in Chapter 40.

BLOOD TRANSPORT IN THE VERTEBRATES

All vertebrates have a closed circulatory system (if we ignore the fact that blood percolates through the liver) and a centrally located heart. The most complex of these is the four-chambered heart of crocodiles, mammals, and birds. But let's first consider the simpler two- and three-chambered arrangements in fish, amphibians, and reptiles.

Fishes and the Two-Chambered Heart

Fish are the most simple-hearted of all vertebrates. Their heart essentially consists of two pumping chambers, a single **atrium** (also called an **auricle**) and a single **ventricle.** The atrium is a generally thin-walled structure that receives blood and transfers it to the ventricle. The ventricle is a larger, thick-walled chamber that sends blood into arteries with considerable force. In fish, incoming blood is first received by an enlarged vein, the **sinus venosus,** prior to entering the atrium, and blood leaving the ventricle first enters an enlarged artery, the **conus arteriosus.** These enlarged chamberlike vessels are also found in the three-chambered hearts of amphibians and reptiles, but in the four-chambered hearts of birds and mammals, they are inconspicuously incorporated into the heart.

Notice in Figure 39.5 that blood leaving the fish's heart passes through the **ventral aorta** going directly into the blood vessels and capillaries of the gills. From the gills, freshly oxygenated blood flows through the body and from there returns to the heart. Blood returning to the heart moves sluggishly, since much of it has passed through major capillary beds both in the gills and in the body tissues, where frictional resistance is great.

Some of the blood has passed through yet other capillary beds, further decreasing its pressure. In one instance, blood that has already passed through the gills later enters capillary beds in the intestine, where it picks up the products of digestion. The capillaries then rejoin, forming the **hepatic portal vein,** which goes to the liver. Here it divides once more into capillaries. It then enters a large vessel, the hepatic vein, for its return to the heart.

In summary, the fish's blood flows from heart to gills via the ventral aorta, through the gill arches, from the gills to the tissues via the dorsal aorta, and from the tissues to the heart. Again, during each cycle, all the blood flows through the gills and continues

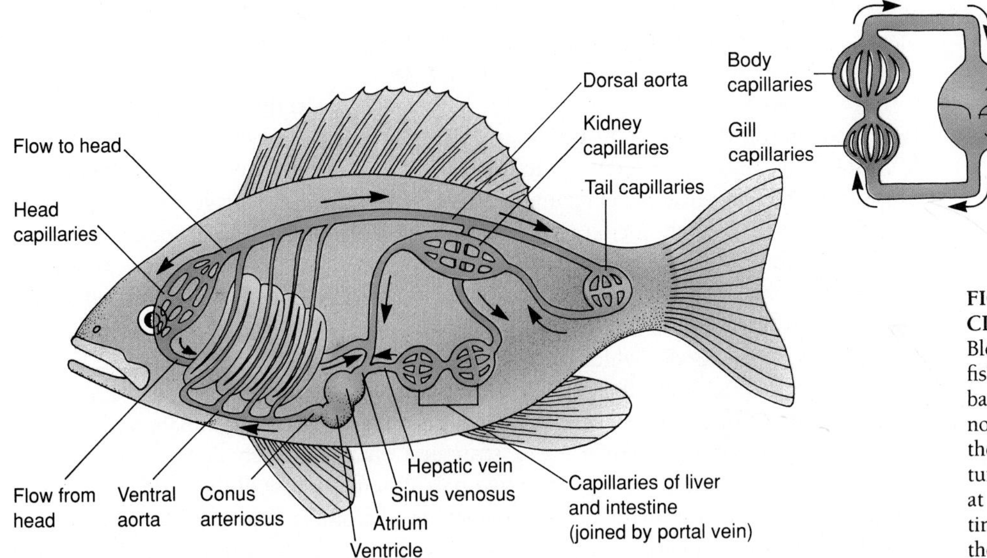

FIGURE 39.5
CIRCULATION IN THE FISH
Blood follows a direct circuit in the fish, from heart to gills to body and back to the heart. Note that there is no separate circuit for oxygenating the blood. Also note that blood returning to the heart has been through at least two capillary beds and sometimes three. (Follow the flow from the heart to the ventral aorta to the gills, dorsal aorta, tail, kidney, and back to the heart.) Three capillary beds include those of the gill, kidney, and liver.

on through the other body tissues. Oxygen is picked up in the gills and released in the tissues prior to the blood's return to the heart.

AMPHIBIANS, REPTILES, AND THE THREE-CHAMBERED HEART

In the evolution of animals that, however temporarily, could survive on land, a new emphasis on lungs and a deemphasis on gills meant that the simple one-cycle circulatory system of the fish was no longer enough. As oxygen demands increased, greater blood pressure and a new way of oxygenating the blood were in order. New circuits in the system arose, with the single atrium becoming divided into two in the amphibians and reptiles. Also, some of the blood pumped from the heart was directed through a new **pulmonary circuit** to the lungs, which became separate from the **systemic circuit** to the other body tissues. Today, we find the three-chambered heart in all amphibians and nearly all reptiles.

As an example we can consider the frog heart. We see in Figure 39.6a the large sinus

FIGURE 39.6
THE THREE-CHAMBERED AMPHIBIAN HEART
The frog's three-chambered heart is seen from both the dorsal (**a**) and the ventral (**b**) views. Note the two atria and single ventricle. An intricate system of flaps and partial valves helps reduce the mixing of oxygenated and deoxygenated blood in the single ventricle. (**c**) Note the presence of a separate circuit for oxygenating the blood (compare with Figure 39.5).

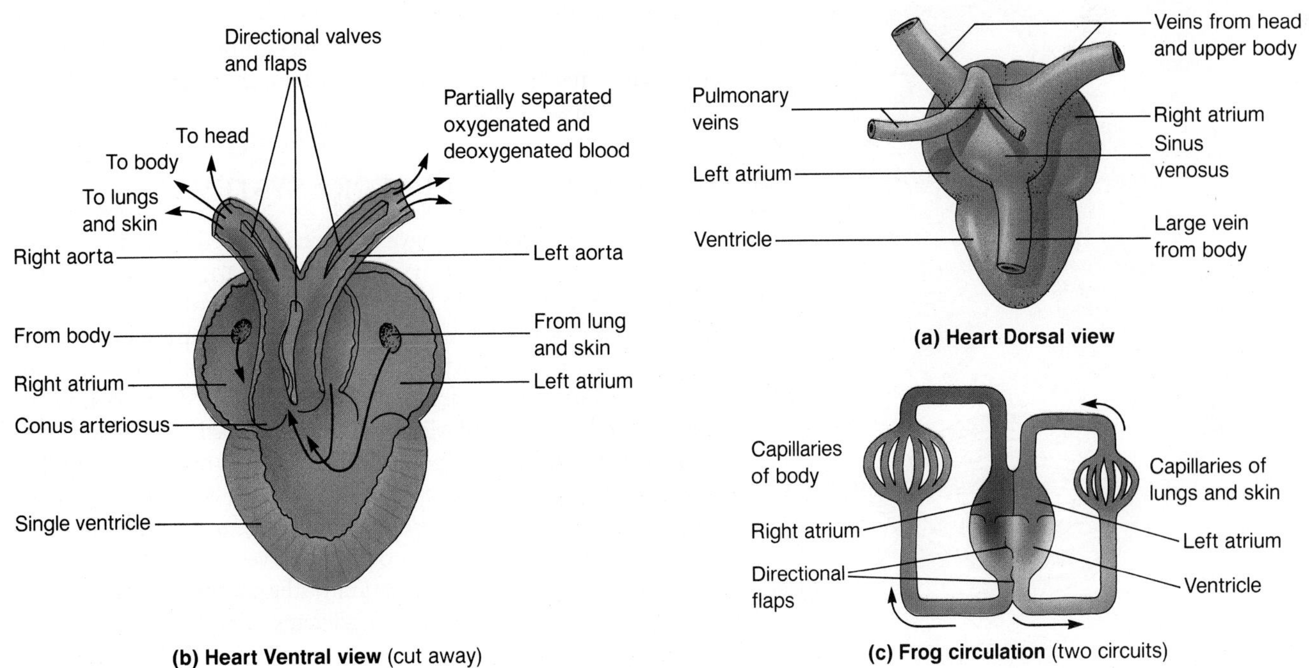

venosus (also present in fish), but here we see *two atria.* The third receiving chamber makes possible the two-circuit system mentioned above. Let's follow the flow of blood in Figure 39.6b and c to see how the system works.

Deoxygenated blood returning from the body enters the **right atrium** and from there is pumped into the single ventricle. Nearly simultaneously, oxygenated blood returning from the lungs and skin enters the **left atrium** and is also pumped into the ventricle. This system has produced a vexing question for biologists. Is the oxygenated blood from the lungs mixed with the deoxygenated blood from the body? The evidence indicates that it does mix, but only partially. In the amphibian heart, total mixing is prevented by flaps and partial valves that separate the oxygenated and deoxygenated blood somewhat and direct it into the proper arteries. Reptiles also have a partial **septum,** or partial partition, in the ventricle, which helps keep the two kinds of blood separated. Even a hint of such mixing would be physiologically unacceptable for birds and mammals—endothermic creatures with much greater oxygen demands than amphibians and reptiles.

In summary, we note that a major structural change is seen in the circulation of amphibians and reptiles. We now see a two-part system because of the added pulmonary circuit. As a result, the extra boost provided by the single ventricle after oxygenated blood returns to the heart from the lungs keeps the blood pressure strong as the blood is sent to the tissues.

Birds, Crocodiles, Mammals, and the Four-Chambered Heart

The evolution of the four-chambered heart was not a sudden thing. It occurred by a succession of tiny steps. We can see indications of some of these intermediate steps among living amphibians and reptiles. First there was the amphibian three-chambered heart with its directing valves. Then a partial septum evolved in the single ventricle, further dividing oxygenated and deoxygenated blood. Then, in the crocodiles, the ventricular division became complete, and birds and mammals were to share this evolutionary innovation.

The four-chambered heart of birds, crocodiles, and mammals includes a right atrium and ventricle and a left atrium and ventricle. The conus arteriosus and sinus venosus are absent in the mammals and birds. In the four-chambered heart, some of the blood is oxygenated in the lungs, and some is deoxygenated in the body tissues in each complete circuit. The crucial difference between this and fish circulation is that a complete circuit in birds and mammals involves two trips through the heart, once through the left side and once through the right side. In fact, it is convenient to think of the four-chambered heart as two separate pumps (Figure 39.7).

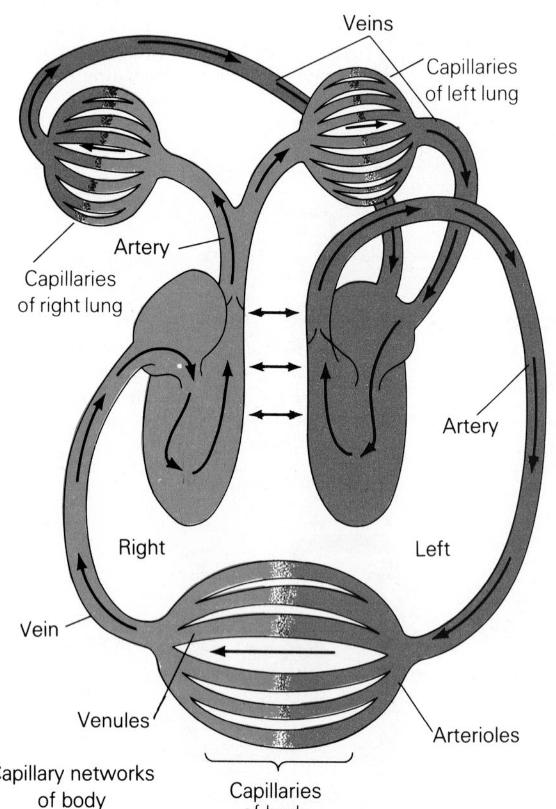

FIGURE 39.7
THE FOUR-CHAMBERED HEART: TWO PUMPS IN ONE
Mammals, birds, and crocodiles have four-chambered hearts. The four-chambered heart essentially allows two separate blood circuits. The right side receives deoxygenated blood from the body and pumps it through the lungs in the pulmonary circuit. The left side receives blood from the lungs and pumps it through the body in a separate circuit.

THE HUMAN CIRCULATORY SYSTEM

Humans, like all mammals, have a four-chambered heart and a closed circulatory system. The system consists of the heart, arteries, capillaries, and veins. The arteries are muscular vessels that always carry blood away from the heart and, with the exception of the **pulmonary arteries,** always carry oxygenated blood; hence, arterial blood is usually bright red. Pulmonary arteries direct deoxygenated blood from the heart to the lungs. **Capillaries** are thin-walled vessels that arise from **arterioles** (smaller arteries), forming highly branched **capillary beds.** All exchanges between the blood and the cells of the body occur through the capillaries. **Venules** (small veins) receive the blood from capillaries and flow into ever-larger veins; these direct blood toward the heart. With the exception of the **pulmonary veins** (those leading from the lungs to the heart), all veins carry deoxygenated blood (Figure 39.8).

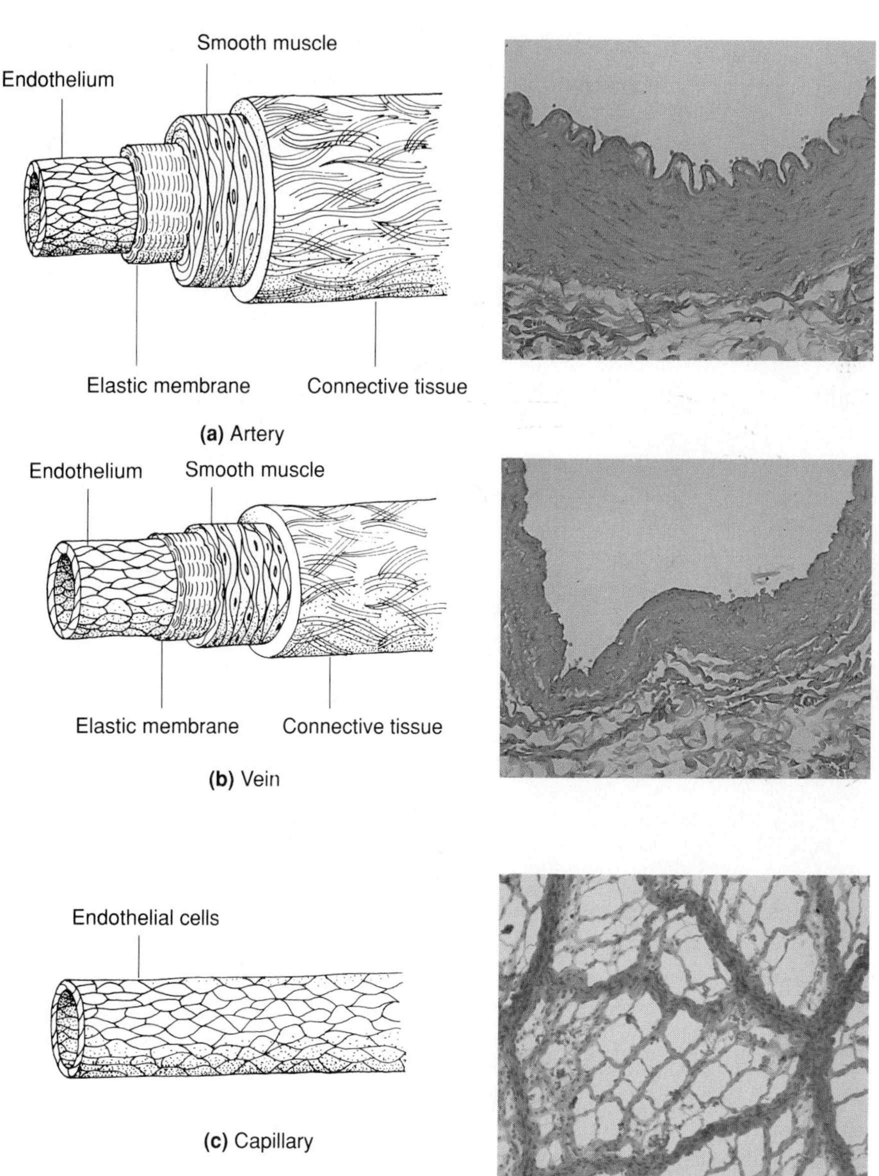

FIGURE 39.8
THE ARTERY, CAPILLARY, AND VEIN
Arteries (a) and veins (b) contain four tissue layers: a tough outer connective tissue, smooth muscle, an elastic membrane, and a lining endothelium. The outer three layers are thicker in arteries, in keeping with their higher blood pressure and need for elasticity. The capillaries (c) are very thin-walled, actually just one cell in thickness, facilitating exchanges between the blood and body cells.

Arteries tend to be round and thick-walled, and are invested with heavy layers of smooth muscle and elastic connective tissue. Constriction or dilation of arteries and arterioles controls the blood flow in specific tissues. Veins, on the other hand, tend to be thin-walled and flattened, with much larger openings in cross section. They also generally lie nearer the surface, so most of the vessels you see under the skin are veins.

Circulation Through the Heart

The Right Side Circulation through the four-chambered heart begins much like that in the three-chambered heart. Deoxygenated blood arrives at the right atrium and, simultaneously, in an oxygenated condition at the left atrium. Blood arriving at the right atrium is carried by two major veins, the **superior** and **inferior vena cavae** (Figure 39.9). (They are named "superior" and "inferior" because in humans one is higher than the other—a result of our upright posture.) From the right atrium the blood flows into a chamber below it, the muscular **right ventricle.** About 80% of the blood entering the right ventricle gets there by flowing into the expanding chamber; the rest is pumped

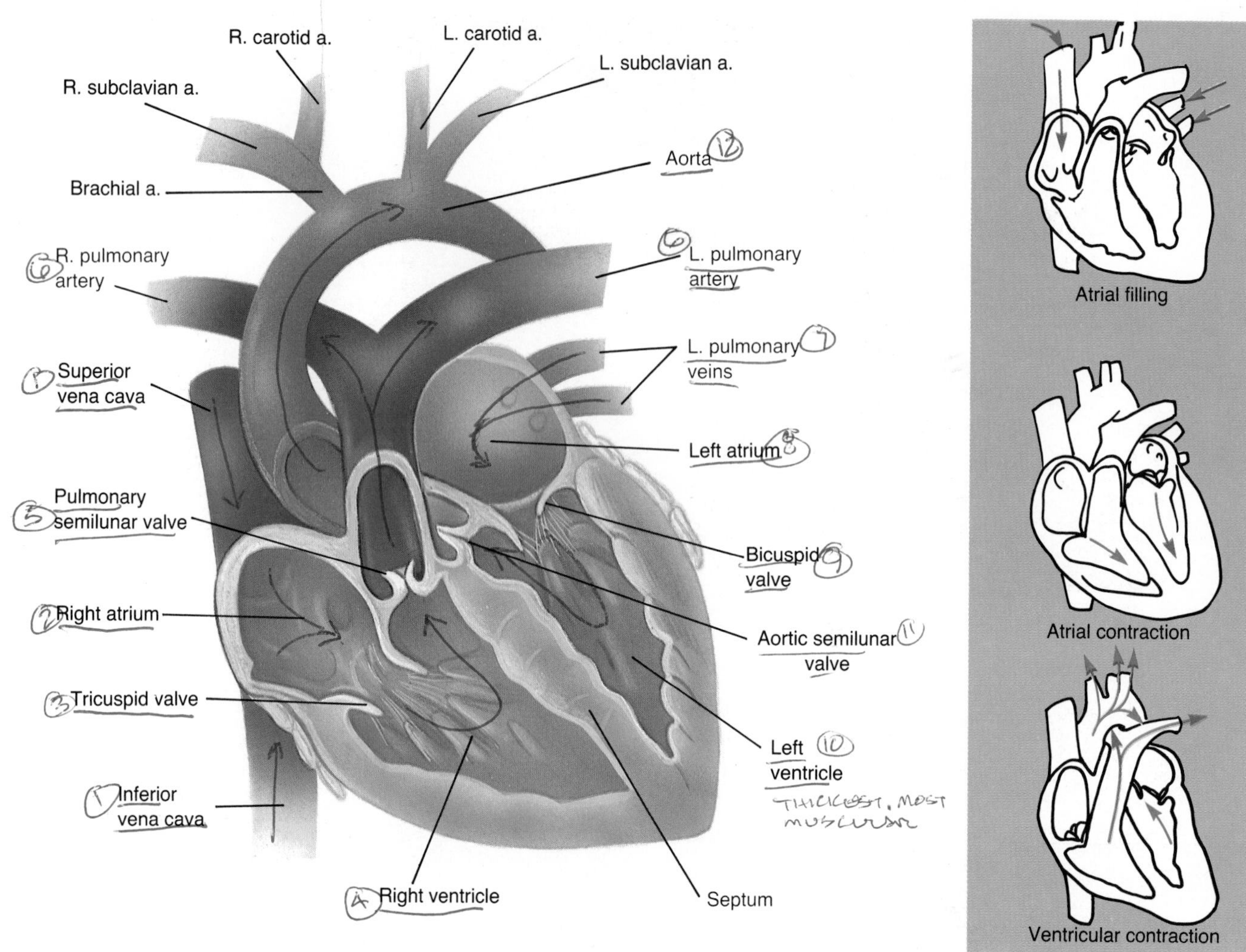

FIGURE 39.9
CIRCULATION THROUGH THE HEART
(a) The route of blood from the body is: vena cavae to right atrium to right ventricle and on to the pulmonary arteries leading to the lungs. Following oxygenation, the route is: pulmonary veins to left atrium to left ventricle and on to the aorta for distribution to the body. Backflow is prevented by strategically located valves. **(b)** Note the sequence of contraction and the squeezing action of the chambers.

by contraction of the right atrium. When the right ventricle contracts, it forces blood into the pulmonary arteries and on to the lungs for gas exchange.

One-way valves between the chambers, and between the chambers and the arteries leaving the heart (Figures 39.9 and 39.10), ensure that the blood flows in the right direction. The valve between the right atrium and ventricle, the **tricuspid valve**, checks any backward flow created by the ventricular contraction. When the tricuspid is closed, it resembles a three-part parachute; the "shroud lines" are a series of **chordae tendineae**, tendinous cords that prevent the three flaps from collapsing backward like an umbrella in a strong wind. As blood passes into the muscular pulmonary artery, its backflow is prevented by the **pulmonary semilunar valve**, a three-part curved flap that closes when the artery is full and expanded. The name *semilunar* refers to the half-moon shape of each of the three flaps.

The Left Side Blood from the right and left branches of the pulmonary arteries finally wends its way into the capillaries surrounding the alveoli of the lungs, where gases are exchanged. The capillaries then rejoin to form the pulmonary veins, which return oxygen-rich blood to the left atrium. When the left atrium contracts, blood passes through the double-flapped **bicuspid**, or **mitral valve** into the **left ventricle.** (The name *mitral valve* comes from the valve's shape, similar to a bishop's hat, or mitre.) This valve has two flaps, as opposed to the tricuspid of the right side, but otherwise has the same "parachute-like" appearance complete with its own chordae tendinae.

The walls of the left ventricle are much thicker than those of the right, which probably

FIGURE 39.10
VALVES OF THE HEART
The larger tricuspid and bicuspid valves between the atria and ventricles have thin-walled flaps restrained by strong, cordlike, chordae tendineae that extend from papillary muscles. The aortic and pulmonary, semilunar valves have small but thick curved flaps.

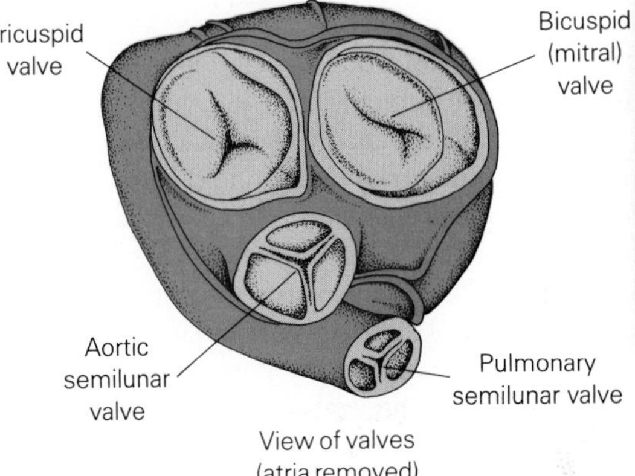

View of valves (atria removed)

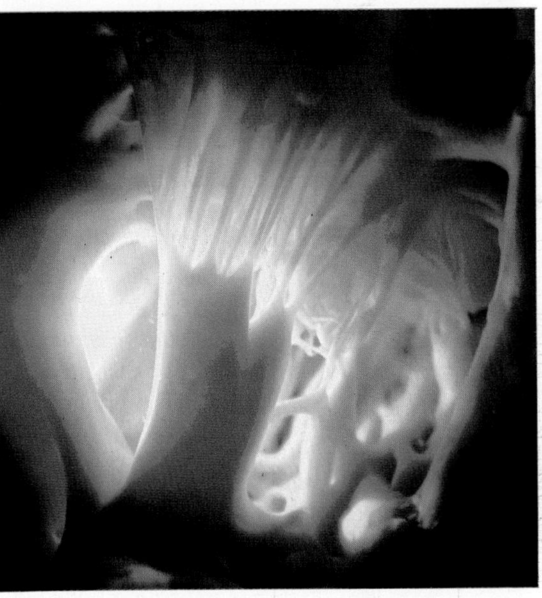

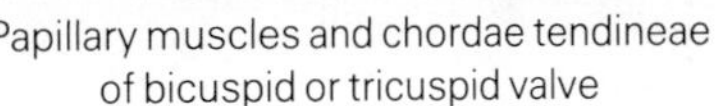

Papillary muscles and chordae tendineae of bicuspid or tricuspid valve

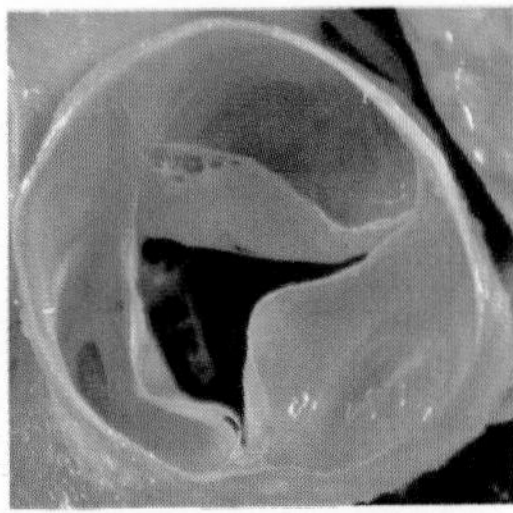

Semilunar heart valve open

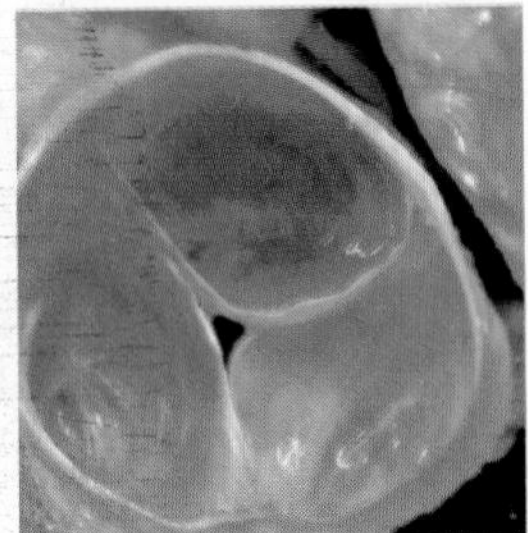

Semilunar heart valve closed

gave rise to the myth that the heart is on the left side of the chest. (Actually, it is in the center, but tilted so when you pledge allegiance, your hand is actually over your left lung.) The left ventricle is large and muscular enough to force the oxygenated blood throughout all the tissues of the body and into the veins that eventually return it to the right side of the heart—an enormous task. Contraction of the left ventricle sends blood through the **aortic semilunar valve** and on into that great artery, the **aorta** (see Figure 39.9). The aorta makes a U-turn to the left, giving rise to branches that serve the head, arms and digestive organs as it turns and passes down through the trunk. Below, it splits into the **iliac arteries**, which branch into the legs.

Control of the Heart

Some might say that the subject of the human heart has been overworked. After all, not only have volumes been devoted to describing its relentless and tireless activity, but we have eulogized and venerated it as the center of love and emotion. In reality, though, it needs no romanticizing, since no discussion can be cold and clinical enough to drain it of its wonder. In Chapter 33 we mentioned that the heart muscle is unique because of its unusual contractile tissue. This was an extreme understatement. There are a number of physiological differences between heart muscle and skeletal muscle, including the branching fibers, interlocking membranes (intercalated disks), and tireless contractions of cardiac muscle cells. One of their most interesting features is that the contraction itself and its rhythmicity are inherent in the cells themselves. The heart, in other words, is capable of beating without outside regulation. This is dramatically illustrated in transplanted hearts, which sometimes go on beating for many years with no nerve connections whatsoever.

Extrinsic and Intrinsic Control In no sense, however, is the normal heart functionally isolated, beating merrily away, independent of the rest of the body. The body's constantly changing demands wouldn't allow that. The heart is, in fact, greatly affected by a number of conditions, even including the emotions. We will find that the heart rate is under both **extrinsic** (from without) and **intrinsic** (from within) control.

Extrinsically, the heart can be fine-tuned by the autonomic nervous system. The nerves that serve the heart originate in the spinal cord and in the medulla of the brain.

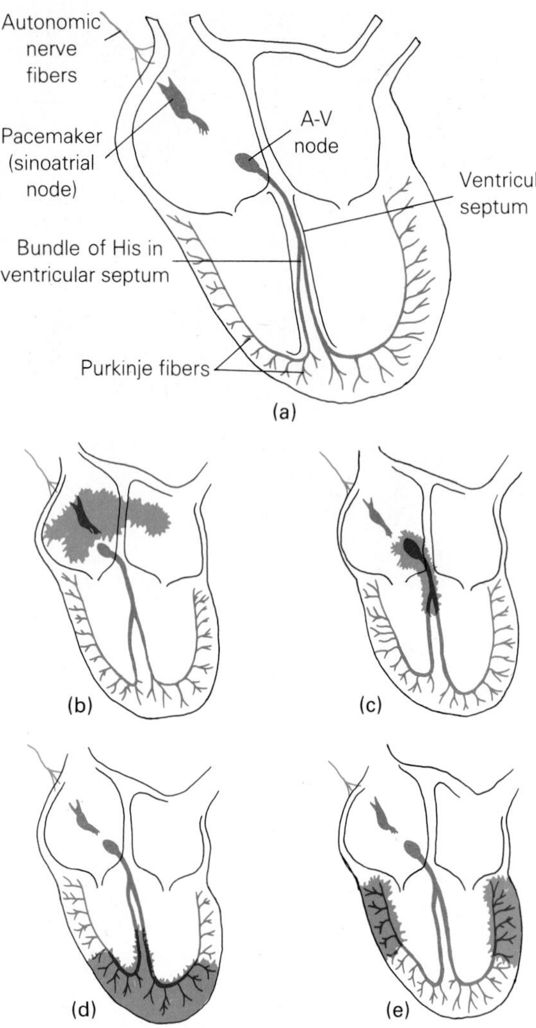

FIGURE 39.11
CONTROL OF HEARTBEAT
(a) Control involves the SA and AV nodes and nerve fibers that extend into the ventricle. (b) In each heart cycle, impulses from the SA node initiate atrial contraction and then **(c)** activate the AV node. There, impulses are regenerated and sent down the bundle of His, through the septum **(d and e)** to Purkinje fibers below. Ventricular contraction begins at the bottom and moves upward, producing a wringing action.

One group, the sympathetic nerves, accelerates the heart rate. The second, the parasympathetic nerves, act to slow the heartbeat. (We considered these nerves in Chapter 35.) The two kinds of nerves permit tight regulation of this vital organ so that it operates in concert with the rest of the body. The heart may be further influenced by hormones from the endocrine system. Epinephrine (adrenalin) from the adrenal gland has the same effect on heart rate as does the accelerator nerve. Also recall that the heart senses blood pressure with its atria, and accordingly, secretes a hormone that helps in its regulation (see Essay 37.2).

One may wonder, how does the heart direct itself if it is influenced by all these factors? It directs itself through intrinsic control.

The origin of the heartbeat is in the modified muscle of the right atrium, at a region known as the **sinoatrial (SA) node** (Figure 39.11). More commonly called the **pacemaker**, this is also the region influenced by the autonomic nerve fibers. The pacemaker transmits impulses across the atrial walls (causing the two atria to contract simultaneously) and on to a second node—the **atrioventricular (AV) node.** There is enough delay in the transmission of the impulse for the atria to complete their contraction before the ventricles begin theirs. The AV node initiates contraction of the two ventricles. Impulses from the AV node pass through a strand of specialized muscle in the ventricular septum known as the **bundle of His.** (The name is the same in both sexes, and anyway, it's pronounced *hiss.*) The bundle branches into right and left halves that travel to the pointed base of the heart, where they branch into **Purkinje fibers** that initiate contraction there. The ventricular muscle fibers are arranged in a spiralling fashion so that their contraction produces a twisting, wringing motion that squeezes the blood out.

The Working Heart

There are two major heart sounds, usually described as "lub" and "dup". The first is that of the sudden closing of the tricuspid and bicuspid valves as they shudder under the tremendous force of the contracting ventricles. The second sharp sound occurs as the aortic and pulmonary semilunar valves are snapped shut by arterial backflow and pressure when the ventricles relax. Then there is a brief pause. Then (one hopes), another *lub dup*. The period of ventricular contraction is known as **systole,** while the longer period during which the chambers of the heart fill up once again, is **diastole.**

It has been calculated that the resting heart rate is generally 72 contractions per minute, and that each contraction forces about 80 ml (2.7 oz) of blood into the aorta. The amount of blood passing through the heart with each heartbeat is the **stroke volume,** comparable to engine displacement. Cardiac output, the amount pumped each minute, is about 5 to 6 liters (11 to 13 pints). During activity, cardiac output can be increased up to 30 to 35 liters per minute—a remarkable reserve power! In case you have sensed an arithmetic problem here, we hasten to explain that your heart doesn't have to increase its beat rate by six to get a six-fold increase in cardiac output—if you are in good condition. During vigorous activity an athlete's heart rate climbs rapidly, but so does the stroke volume. A more sedentary person, on the other hand, must maintain cardiac output primarily through increased heartbeat, with less increase in stroke volume. This is one reason an unconditioned person tires quickly and recovers slowly.

Blood Pressure

Blood pressure is the force of blood against the arterial walls. It is this force that moves blood from the heart through the body. The aorta and major arteries receive the full impact of the powerful surges of blood from the left ventricle. A great many body functions depend on a rather constant pressure in the circulatory system, so the arterial walls have adapted accordingly. When the heart contracts, blood enters the arteries faster than it can leave, so the sudden swell of blood expands the elastic walls of arteries. As the blood moves onward through the arterial tree, the expanded arterial walls contract through their elasticity, squeezing the rapidly decreasing blood volume. The speed of this recoil allows some lowering of blood pressure but prevents a drastic pressure drop. For example, there is a pressure drop in diastole of about 35%, or, as this would be usually expressed, about 40 to 60 mm of mercury (mm Hg). A typical systolic pressure in a young, healthy person is 120 mm Hg; a typical corresponding diastolic pressure is 80 mm Hg. Your doctor would call this set of blood pressure measurements "120 over 80." You've probably had your blood pressure taken with a **sphygmomanometer** (Figure 39.12). The pressure measurement in systole, the time of ventricular contraction, reflects the sudden force exerted against the arterial walls. In diastole, the time between the heart's contractions, pressure is maintained by the elastic arterial wall returning to its original shape.

(a)

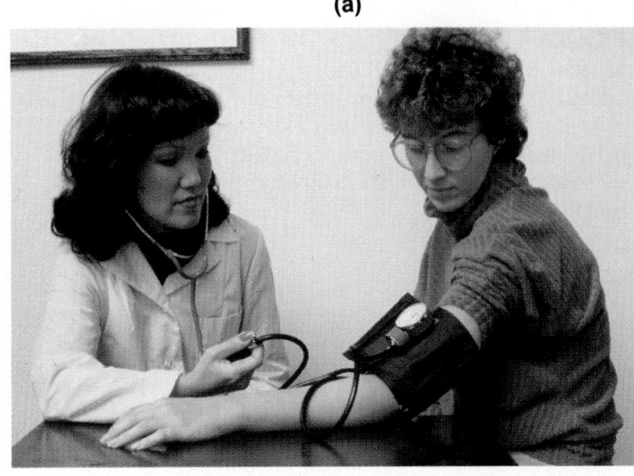

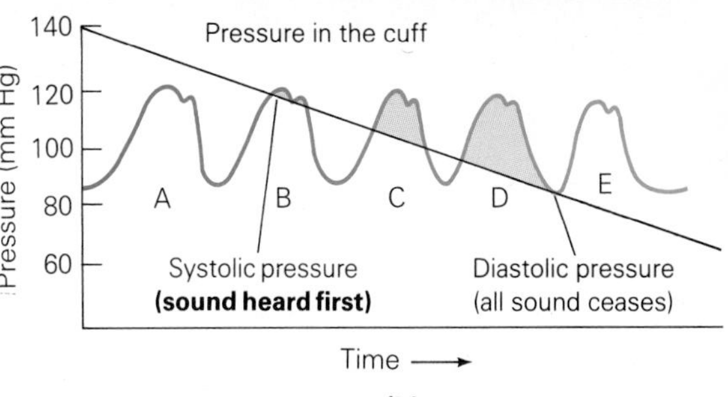

(b)

FIGURE 39.12
MEASURING BLOOD PRESSURE
(a) Blood pressure is commonly measured in the brachial artery with a device known as a sphygmomanometer. As the cuff is inflated, the brachial artery is collapsed, and no sound is heard. Pressure in the cuff is slowly reduced, and when a beating sound begins, the systolic pressure is noted. When the sound can no longer be detected, the diastolic pressure is noted. **(b)** In the graph, the peaks represent systolic pressure and the valleys diastolic pressure. The shaded area is the pressure range at which the arterial sounds are heard. The diagonal line represents pressure decreasing in the sphygmomanometer.

A common problem in older persons is **arteriosclerosis,** or "hardening of the arteries". Arteriosclerosis has several causes, including calcium deposits in the arterial connective tissue, general loss of arterial elasticity, and especially **atherosclerosis**—arterial lesions or thickenings (plaques) characterized by localized growth of smooth muscle cells and deposits of lipids, especially cholesterol (see Figure 3.14). Any of these can result in an effective reduction of the diameter of the vessels which, in turn, raises blood pressure. (You can visualize how this would happen by imagining that you're trying to water a rose bush, but your garden hose won't reach. If you reduce the opening of the hose by putting your thumb over it, the water squirts farther because you have increased its pressure.) As vessels lose their elasticity, they may create a serious burden on the heart as resistance to flow increases.

Vasoconstriction and Vasodilation The smooth muscle lining the walls of arterioles plays a vital role in maintaining and directing the flow of blood into capillary beds. Opening of arterioles, called **vasodilation,** reduces blood pressure, and their constriction, called **vasoconstriction,** increases blood pressure. In the arterioles, these two mechanisms function in a delicate and balanced interplay that keeps the overall blood pressure remarkably constant in spite of great surges of pressure in the arteries feeding arterioles. Vasodilation and vasoconstriction are under the control of the autonomic nervous system and partly under the control of hormones. In addition, oxygen, carbon dioxide, hydrogen ions, and a variety of drugs can bring about changes in the diameter of arterioles.

Circuits in the Human Circulatory System

Though we are not about to discuss all the pathways of the human circulatory system here, we would like to consider a few interesting circuits. A **circuit** can be loosely defined as a distinct and major pathway of blood supply and return, generally where some function is performed. Figure 39.13 shows the four major circuits: pulmonary, hepatic portal, renal, and systemic (general body circuit which includes muscles, skin glands, and so on). In addition to these, there is the vital **cardiac circuit,** the circulatory system of the heart itself. The pulmonary circuit, already described, is a good example of a relatively simple circuit.

Hepatic Portal Circuit In the **hepatic portal circuit,** arteries branching from the abdominal aorta spread across the intestinal membranes and enter the digestive organs.

FIGURE 39.13
MAJOR BLOOD CIRCUITS
Each circuit carries out exchanges of nutrients, waste products, and oxygen between the blood and body cells, but most have other, more specialized functions.

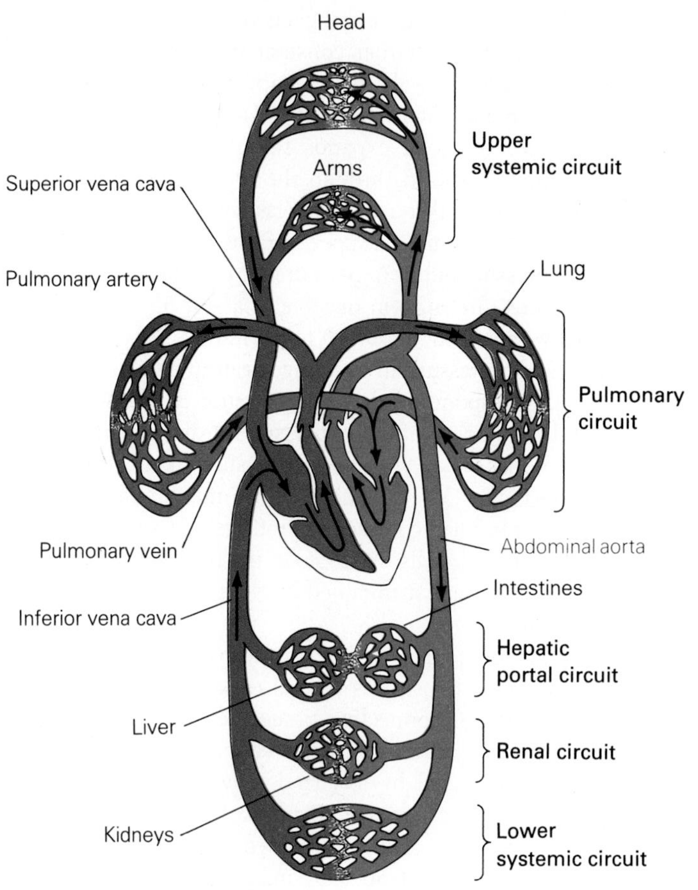

Those entering the small intestine form capillaries within the intestinal villi. Digested foods are collected in the capillaries, which then merge together and, along with vessels from the rest of the intestine, form the **hepatic portal vein** leading to the liver. (Portal veins lie between two capillary beds.) Upon entering the liver, the portal vein divides to form a second capillary bed. Actually, the liver contains not only capillaries, but many branched, epithelium-lined **sinusoids** (minute cavities) through which the blood passes. (As was mentioned, it could be argued that this makes the human circulatory system a partially open one.) From there, the blood, its cargo of nutrients now reduced and cleansed by the liver of toxic materials, returns directly through the inferior vena cava to the heart.

Renal Circuit As it passes through the abdominal cavity, the aorta sends right and left branches, the **renal arteries,** into the kidneys. There, in one of the most complex filtering systems imaginable, the nitrogenous waste urea, excess water, miscellaneous metabolic byproducts, and some salts are removed from the blood (see Chapter 36). Of equal significance, water and salt levels are carefully adjusted, a vital process in the maintenance of the precise osmotic conditions required by the body. Blood returns from the kidneys via the **renal veins** to the inferior vena cava, and then to the heart.

Cardiac Circuit The cardiac circuit includes the arteries, capillaries, and veins of the heart muscle. The heart muscle receives first crack at oxygenated blood leaving the left ventricle, since the first branches of the aorta are the right and left **coronary arteries.** They emerge just beyond the aortic semilunar valve, coursing separately to the right and left as they follow a furrow around the outer surface of the heart, finally joining toward the posterior side (Figure 39.14).

As the arteries branch and rebranch, they continually rejoin each other through what

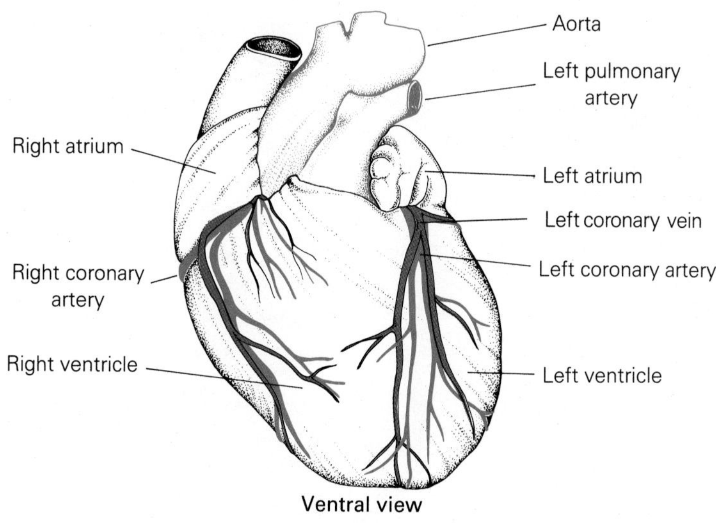

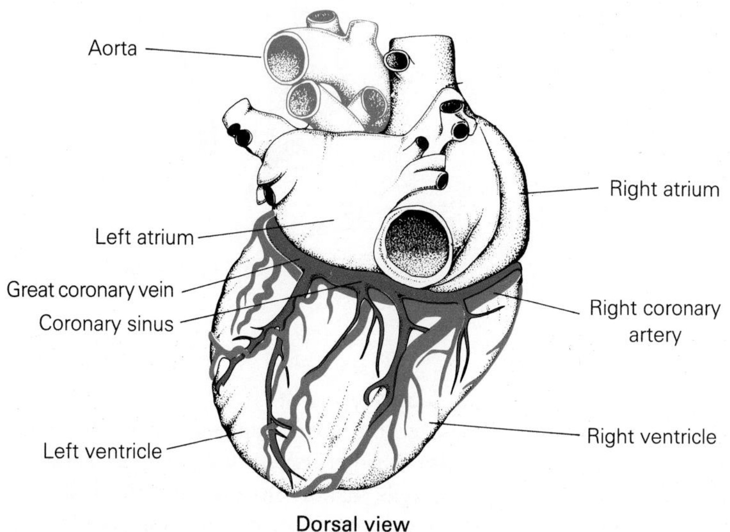

FIGURE 39.14
THE CARDIAC CIRCUIT
The heart muscle has its own vessels, the coronary arteries and veins. The coronary arteries branch directly from the aorta, sending their branches deep into the heart muscle. The coronary veins emerge from heart muscle to form the coronary sinus, which empties into the right atrium.

anatomists call **anastomoses.** The resulting weblike network with its accompanying capillaries is quite significant, since it helps ensure that no sizable part of the heart muscle will be without a blood supply should a blockage, such as a **coronary thrombosis** (a blood clot), occur in some vessel.

Following its transit through the capillary networks of the heart, blood enters the **coronary veins,** whose routes parallel those of the arteries. The veins finally join each other to form the **coronary sinus,** a receptacle that opens into the right atrium.

Systemic Circuit The systemic circuit includes the rest of the body. Here we refer to capillary beds in muscles, glands, bones, the brain, and so on. In these places CO_2 and O_2 are exchanged, required substances for cell maintenance are delivered, wastes are removed, and so on. Capillaries are so profusely distributed that no cell lies far from the blood supply. The walls of the capillaries are usually composed of only one layer of flattened epithelial cells. They are fascinating little vessels and deserve a closer look.

The Role of the Capillaries

As we learned earlier, the capillaries are, in a sense, the functional units of the circulatory system—all exchanges take place in these vessels. One might even say that the role of

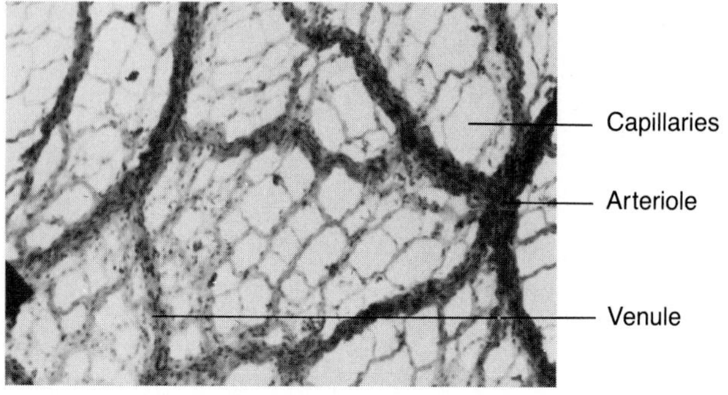

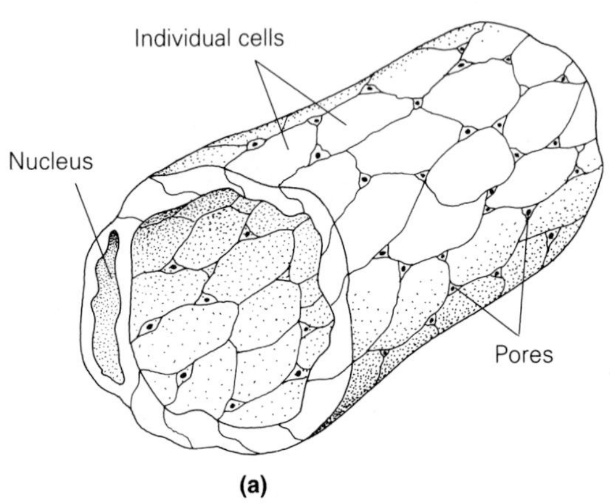

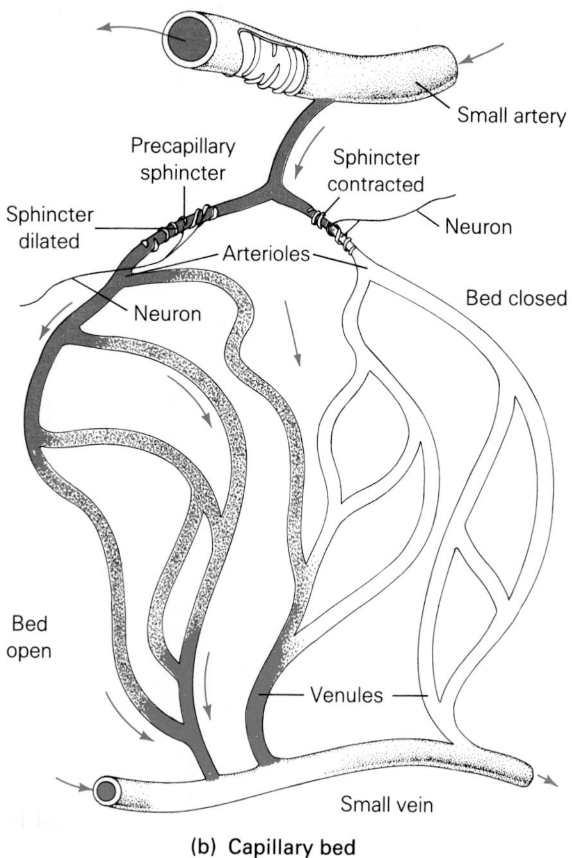

(b) Capillary bed

FIGURE 39.15
CONTROL OF CAPILLARY FLOW
(a) Capillaries are composed principally of endothelial cells. Because their walls are only one cell thick, materials pass rapidly through them. **(b)** Capillary beds arise where arterioles branch. The beds can be partially shut down in various regions by action of precapillary sphincters. In this instance, the left side is open while the right side is shut down.

every other part of the system is simply to assist them. Their walls are made up of interlocking cells in single thickness, so the surroundings are never more than one cell away from the blood (Figure 39.15a). Further, except for brain capillaries (the "blood-brain barrier;" see Chapter 35), the cells making up the capillary wall fit inexactly in places, leaving minute crevices or pores through which small molecules can pass. Finally, electron micrographs of capillaries show that their flat, thin cells engage in extensive pinocytosis (Figure 39.16). Recall from Chapter 5 that in pinocytosis ("cell drinking"), the membrane becomes actively involved in surrounding and taking in dense solutes.

As we pointed out in our earlier discussion of vasodilation and vasoconstriction, one of the more interesting capabilities of the circulatory system is its ability to open and close selected capillary beds. Such changes are made possible by the presence of smooth muscle **precapillary sphincters** located in the tiny arterioles just about where they divide into the capillaries (Figure 35.15b). Thus, for example, in an emergency situation the vasoconstriction of such sphincters in the digestive area will partially close down circulation in the system so that more blood can be sent to the muscles and brain.

FIGURE 39.16
PINOCYTOSIS IN THE CAPILLARY
Capillaries are far more than minute tubes. The flattened cells comprising their walls accommodate simple diffusion, but their versatile plasma membranes carry on a considerable amount of active transport. Note the numerous pinocytic vesicles in the capillary cells.

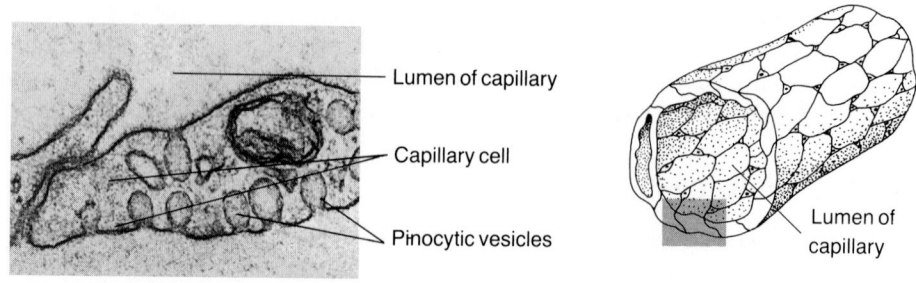

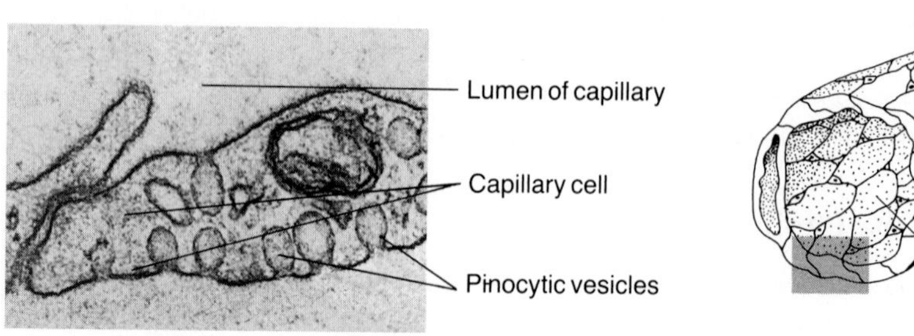

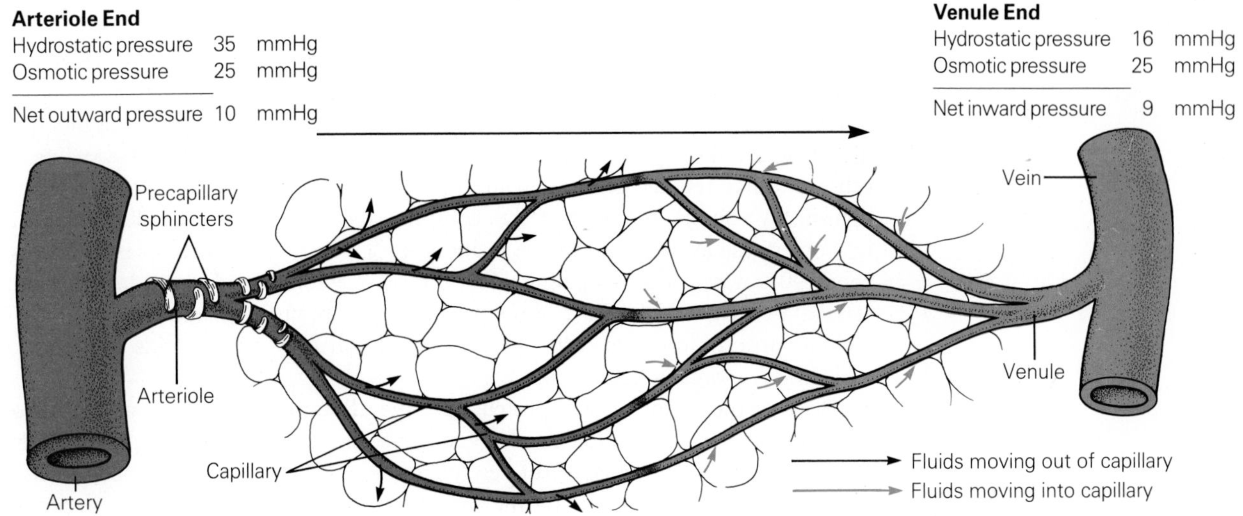

FIGURE 39.17
FORCES IN THE CAPILLARY BED
The movement of water and solutes in and out of the circulatory system occurs in the capillary beds. This movement, in turn, is controlled by pressure relationships in the bed. The two forces at work are hydrostatic pressure and osmotic pressure, and both operate inside and outside the capillary.

As blood passes through a capillary bed, the transport of water and various molecules to and from the capillary depends on several opposing events. (Measurements of the forces involved are summarized in Figure 39.17.) Blood entering a capillary bed will lose nutrients, ions, water, and oxygen. Blood leaving a capillary bed will have picked up ions, water, carbon dioxide, and other metabolic wastes. As you can see, water and ions leave the capillaries and enter the interstitial fluids at one end of the bed, only to return at the other. (The factors involved in the exchange of oxygen for carbon dioxide in capillaries are more complex; see Chapter 40). Diffusion gradients can account for some of this peculiar behavior, but hydrostatic pressure is also important. Hydrostatic pressure is great at the arteriolar end of a capillary bed, and many substances are simply forced through the thin capillary walls in a filtration process. But things change at the venule end. Hydrostatic pressure is reduced, and since water has been forced out of the capillary, the protein-dense plasma left behind becomes a hypertonic medium, and a concentration gradient arises. Thus, most of the water simply returns to the blood through osmosis, and the ions diffuse down their gradient as well. Any excess outside the capillary will eventually be returned to the blood by the lymphatic system.

The Veins

Capillaries join, forming venules, which in turn coalesce to form veins. Because of frictional resistance in capillary beds, blood pressure is lowest in the veins, where it can measure as low as 5 mm Hg. Thus, although blood volume leaving the arteries is equal to the volume entering the veins (except for water loss to the lymphatic system), the force of its movement is greatly depleted. Consider our earlier analogy: if you block the water leaving a hose with your thumb, the pressure in the hose (arteries) is considerable, but as the spray (capillaries) strikes the surface of the ground and the water trickles down the gutter (veins), the pressure is dissipated.

Because of the greatly reduced blood pressure in veins, the blood needs help in getting back to the heart. Many of the veins have one-way, flaplike valves that allow blood to move in one direction—toward the heart. The walls of the veins, though much thinner than arterial walls, do contain some smooth muscles that can contract and help push the blood along (see Figure 39.8). But venous flow receives assistance in other ways as well. In the simple movement of the limbs, for instance, the muscles squeeze and massage the veins, moving blood along. The one-way valves prevent backflow. In the chest area, breathing movements squeeze the walls of the vessels, sending the blood toward the heart.

The Blood

By definition, blood is a tissue. That is, it is composed of several types of cells in more or less resident (unchanging) status. More precisely, it is a connective tissue (see Chapter 33) with **plasma** as its matrix (ground substance). Gentle centrifugation of blood easily divides it into three parts: plasma on top; a thin, clear band of **leukocytes** (white blood cells) and **platelets** in the middle; and the heavy **erythrocytes** (red blood cells) on the bottom. By volume, the erythrocytes comprise about 45% of the whole blood in men. In women, the average percentage is about 43%. The straw-colored plasma consists of about 90% water. The remaining 10%, the plasma solids, includes a long list of substances; by weight, about 70% of the plasma solids are proteins, and the remaining 30% includes urea, amino acids, carbohydrates (mostly glucose), organic acids, fats, hormones, and various inorganic ions.

Three of the major plasma proteins are the **albumins**, **globulins**, and **fibrinogen.** The albumins are large proteins that bind miscellaneous impurities and toxins in the blood and aid in the transport of certain hormones, fatty acids, and metal ions. They are also involved in maintaining the osmotic conditions of the blood, so important to capillary functions. The globulins include the **antibodies**, or **immunoglobulins**, which are important in the immune response, and certain proteins involved in the transport of lipids and fat-soluble vitamins. Fibrinogen is important in the blood clotting process.

The Red Blood Cells Red blood cells are quite small (about 6 to 8 μm in diameter), are biconcave and disklike in shape (Figure 39.18), and, when mature, lack nuclei and therefore cannot undergo further division. Normally, there are about 5 million red cells per cubic millimeter of blood. Red and white blood cells are produced continuously in the **red bone marrow.** In adult humans, red cell formation occurs almost exclusively in the axial skeleton. Red cells have about a four-month life expectancy. When they age, some rupture, spilling their hemoglobin into the blood. Most aging or damaged red cells are phagocytized by certain white cells in the spleen and liver.

The White Blood Cells Unlike red cells, white cells (leukocytes) do not discard their nuclei as they mature. The white blood cells are active in the immune system, which will be discussed in detail in Chapter 41. For now it is important to know that the **neutrophils**, whose numbers make up the majority of circulating leukocytes, are important phagocytic cells. They aggregate at infection sites, engulfing invading microorganisms. **Basophils** are involved in the inflammatory response. **Eosinophils** are also involved in inflammatory responses, but in addition, they help destroy larger parasites. **Lymphocytes** are the backbone of the immune system. In lymphatic tissue they fight off invaders and confer immunity. The **monocytes,** once activated at infection sites, develop into **macrophages** which phagocytize foreign cells and cellular debris. One group of cells derived from lymphocytes, the **natural killer (NK)** cells, are unspecialized leukocytes that destroy diseased cells.

FIGURE 39-18
RED BLOOD CELLS
Mature red blood cells take the form of biconcave disks.

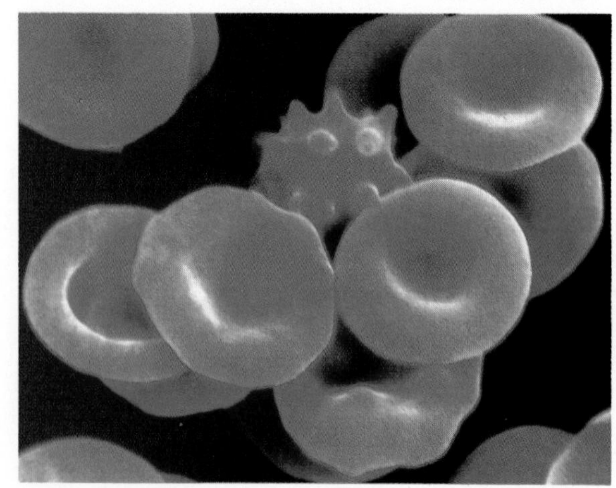

Platelets Blood platelets (or **thrombocytes**) are tiny, numerous structures. Unlike erythrocytes, which are whole cells that have since lost their nuclei, platelets are only fragments of cells. The small, disk-shaped vesicles are formed directly from **platelet mother cells (megakaryocytes),** which also form in red bone marrow. When these large cells mature, they fragment without mitosis—a most peculiar process. Platelets play an important role in blood clotting.

Figure 39.19 illustrates the cellular origin of the blood cells and platelets. Curiously, each of the formed elements of the blood—the cells and platelets—has the same origin, the **hemocytoblast**, or simply **stem cell.**

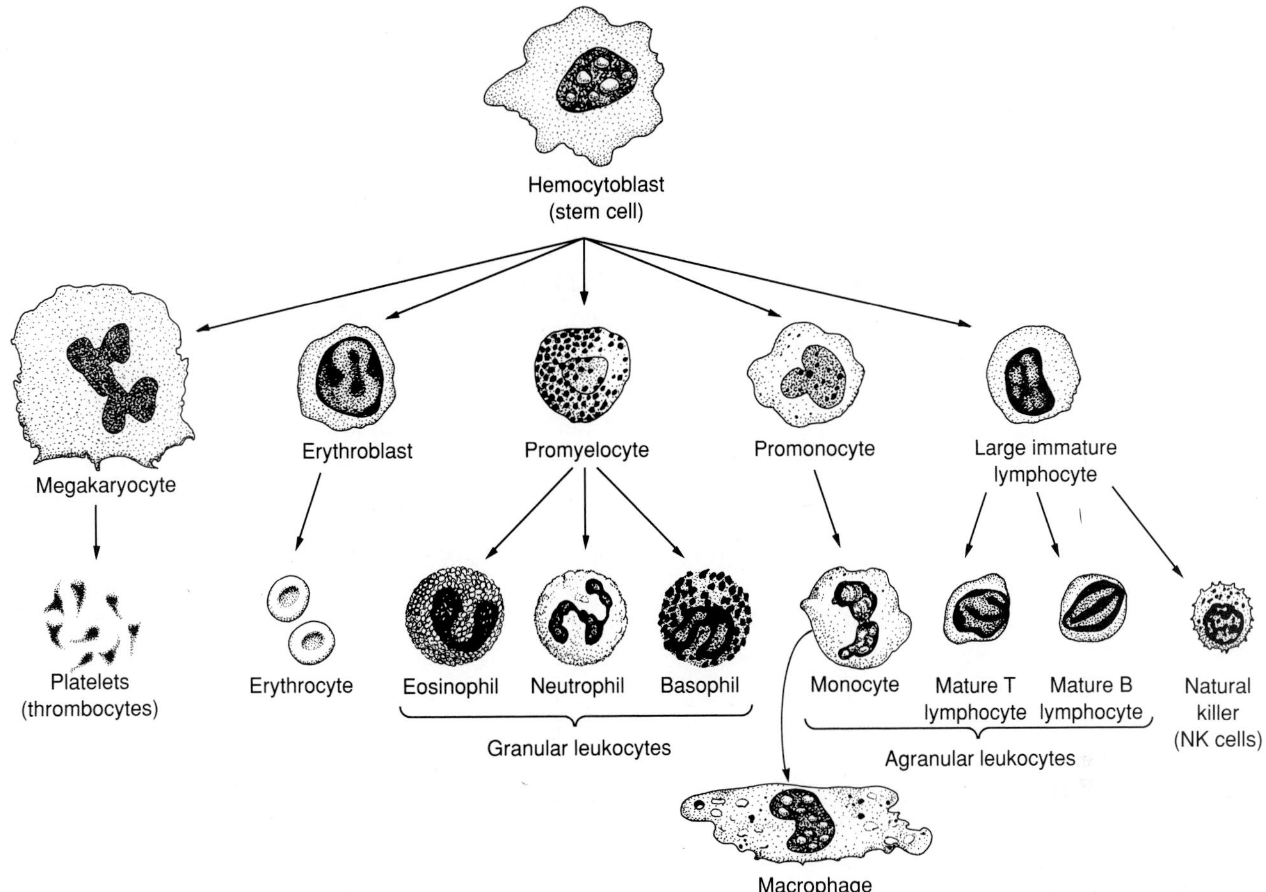

FIGURE 39.19
DEVELOPMENT OF BLOOD CELLS
Stem cells in red bone marrow give rise to each of the formed elements (erythrocytes, leukocytes, and platelets) of the blood.

Clotting Blood clotting is essential to any organism that relies on circulating fluids. The world is full of objects that can pierce our bodies and cause our fluids to leak out, and clotting minimizes such leakage.

Blood clotting is an extremely complex and only partially understood process. At least 15 substances are involved, some of which are part of an intricate arrangement that prevents accidental clotting. Two proteins are basic to the process: **prothrombin,** an inactive clotting protein present in the plasma, and fibrinogen, one of the major plasma proteins. Platelets and damaged cells also play a role. Generally, the clotting process proceeds as follows:

1. A vessel is damaged.
2. Platelets attach at the wound site, form lengthy extensions, and adhere to collagen fibers, forming what is essentially a "plug."
3. The platelets rupture, releasing (a) vasoconstrictors that cause nearby vessels to constrict, reducing blood loss, and (b) **thromboplastins** (enzymes).
4. In the presence of thromboplastins and calcium ions, prothrombin in the plasma becomes **thrombin,** a specific endopeptidase.
5. Thrombin breaks apart the large fibrinogen molecules of the plasma, the smaller pieces forming a fibrous, sticky protein called **fibrin.**
6. Fibrin fibers, along with damaged platelets, red cells, and white cells, form a network that solidifies, becoming a clot that stops the bleeding.
7. The clot contracts, pulling the wound together, further preventing bleeding and encouraging healing.

The absence of any of the many factors participating in clotting, or their failure to perform their function, is quite serious. Earlier we discussed the genetics of one such condition, hemophilia A (see Chapter 11). In this recessive condition, factor VIII (antihemophilic factor), is absent. In hemophilia B, the so-called Christmas disease, the absent substance is factor IX (plasma thromboplastin component). Other reasons for failure of the blood-clotting mechanism include an insufficiency of calcium ions. (Citric and oxalic acids readily take up calcium and are added to blood samples to prevent clotting.) An insufficiency of vitamin K is an important cause of slow clotting. This vitamin is required by the liver for the synthesis of prothrombin.

THE LYMPHATIC SYSTEM

The lymphatic system has four essential roles: (1) it maintains fluid and electrolyte (ion) balances in the body; (2) it transports certain fatty acids from the intestinal villi to the blood; (3) it assists with the work of the immune system; and (4) it provides a route by which interstitial fluids can return to the circulatory system. In its first role, the lymphatic system drains tissue spaces and cavities of fluid and ions that have not been recovered by the capillaries. Such fluids are returned to the bloodstream through ducts near the heart.

The lymphatic system (Figure 39.20) consists of **lymph vessels** and clusters of **lymph nodes** (often called "lymph glands"). Lymph vessels include countless tiny blind endings called **lymph capillaries** (which include the lacteals of the intestinal villi; see Chapter 38). Lymph capillaries collect fluids, solutes, and foreign materials from tissue spaces,

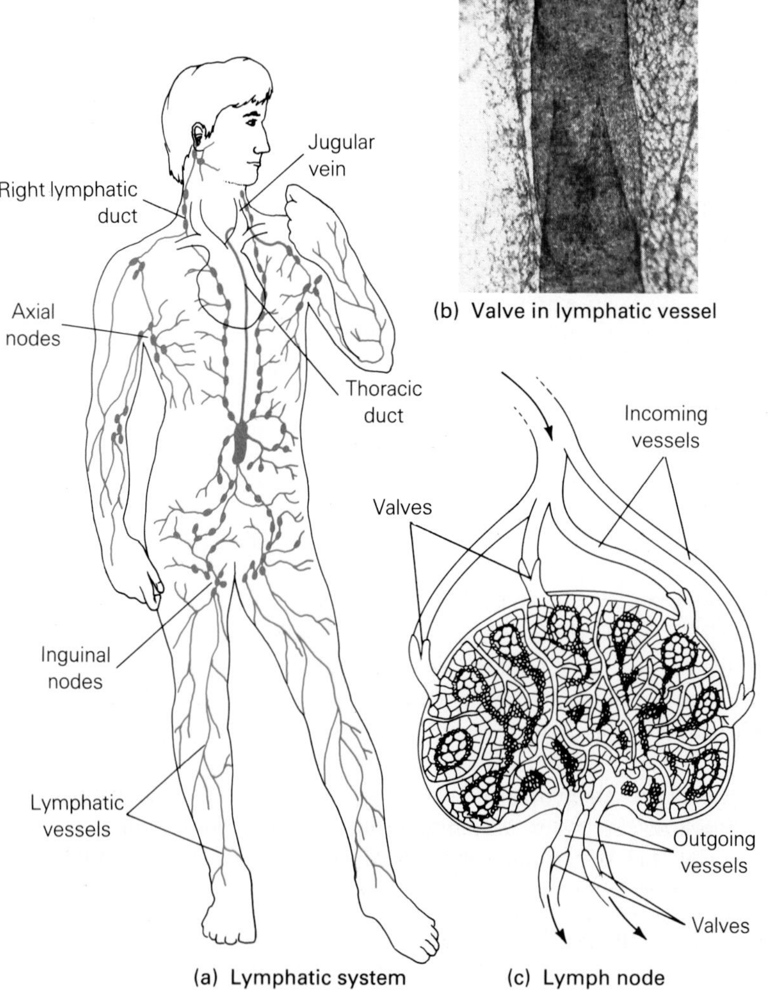

FIGURE 39.20
THE LYMPHATIC SYSTEM
(a) The lymphatic system consists of a number of lymph vessels and nodes. The lymph vessels eventually empty into large veins near the heart. **(b)** Fluids are directed by a number of one-way valves, similar to those seen in veins. **(c)** Lymph nodes are located generally throughout the body but cluster in several regions, including the groin, abdomen, armpits, neck, and head.

emptying into **lymphatic collecting ducts**, which join to form the **lymphatics**, major vessels that carry the lymph to the bloodstream. The watery lymph is pushed along by the squeezing action of muscle and changes in pressure within the thoracic cavity. Like the veins, lymph vessels have one way check valves that help keep the lymph from backing up so that the flow continues toward the heart. Along the way, much of the lymph filters through lymph nodes.

The lymph nodes are located throughout the body, but their greatest concentrations are in the head, neck, armpits, abdomen, and groin (see Figure 39.20). Each node is a compartmentalized mass of tissue that harbors multitudes of lymphocytes—primary cellular agents of the immune system. Foreign materials, bacteria, and viral particles carried into the vessels are swept into the nodes, where they are attacked by resident white cells. Activated lymphocytes cause the nodes to enlarge, so swollen lymph nodes are a telltale sign of infection.

Occasionally, the ducts and vessels in the lymphatic system are turned into deadly avenues for the spread of cancer. While cancer cells entering the lymph nodes are commonly attacked and killed by highly specialized lymphocytes, often some survive and continue their rapid cell division as they are carried by the lymphatic stream throughout the body.

APPLICATION OF IDEAS

1. Tracing the embryological development of the four-chambered heart in humans reveals stages when the heart appears as a simple tube, as two-chambered, as essentially three-chambered, and finally, four-chambered. What does this suggest about the genetic framework upon which the human heart is organized? Could your hypothesis be tested? In what way does the fate of the pharyngeal arches (see Chapter 43) in humans help support your hypothesis?
2. The gravest danger in coronary embolism (blockage) immediately follows its onset, a time when tissue death is occurring. This is true even when only a small amount of heart tissue is destroyed. Why is such loss significant to the normal heart function? What characteristic of the cardiac arteries helps minimize such damage?

KEY IDEAS

In addition to blood, other body fluids include **intracellular fluids** (those inside cells) and **interstitial fluids** (those between cells).

ADAPTATIONS IN ANIMALS WITHOUT CIRCULATORY SYSTEMS

Simpler, thin-walled, aquatic animals, lack circulatory systems, utilizing cell-by-cell transport and direct exchanges with their environment. Included are sponges, cnidarians, and flatworms.

CIRCULATORY SYSTEMS IN INVERTEBRATES

Open and Closed Circulatory Systems

1. In **open circulatory systems**, blood is pumped into vessels but leaves them to percolate through spaces called **hemocoels.** Blood returns to the heart to be drawn up for another circuit. In **closed circulatory systems**, the blood elements remain within vessels.
2. The gill-breathing crustacean uses its circulatory system to transport oxygen and carbon dioxide, and accordingly the system is more complex than in the insect where it does not have a respiratory function.
3. Annelids have closed circulatory systems, distinct vessels, and five pairs of tubular **aortic arches** (hearts) with one-way valves.

BLOOD TRANSPORT IN VERTEBRATES

Fishes and the Two-Chambered Heart

1. The fish heart consists of one **atrium (auricle)** and one **ventricle.** Blood entering the atrium is first collected in a large vein, the **sinus venosus**, and blood leaving the ventricle enters an enlarged artery, the **conus arteriosus.**
2. Blood leaving the fish heart enters the **ventral aorta**, which directs it to the gills, where gas exchange occurs. The blood then passes through the head and body. All blood returning to the heart has traveled through at least two capillary beds. Most blood is returned to the heart via the common cardinal veins.

Amphibians, Reptiles, and the Three-Chambered Heart

1. Terrestrial vertebrates have a second atrium and a **pulmonary circuit**, which carries blood to the lungs and back. In both amphibians and reptiles, some mixing of oxygenated and deoxygenated blood occurs in the single ventricle.
2. In the frog's circulation, deoxygenated blood from the body enters the sinus venosus, then the **right atrium** and single ventricle. At the same time, oxygenated blood from the lungs and skin enters the **left atrium** and then the ventricle. Partial separation is provided by flaps and partial valves in the heart and the conus arteriosus, through which blood passes on its way back to the lungs, skin, and body.
3. Reptiles have only a partial **septum** within the single ventricle, so mixing of oxygenated and deoxygenated blood occurs in these animals.

Birds, Crocodiles, Mammals and the Four-Chambered Heart

The four chambered heart of crocodiles, birds, and mammals includes a right and left atrium and a right and left ventricle, but the conus arteriosus and sinus venosus are absent. The pulmonary circuit is completely separated.

THE HUMAN CIRCULATORY SYSTEM

The human circulatory system consists of the four-chambered heart, arteries, **capillaries** and veins. The muscular arteries carry blood away from the heart, and except for the **pulmonary arteries**, this blood is oxygenated. All exchanges occur in the thin-walled capillaries, following which blood returns to the heart through the veins. While arteries tend to be thick-walled and muscular, veins tend to be larger, to have thinner walls, and to contain less smooth muscle.

Circulation through the heart

1. Deoxygenated blood from the body enters the right atrium from the **superior vena cava** and **inferior vena cava.** From the right atrium, blood enters the muscular **right ventricle**, which pumps it to the lungs for gas exchange. Backflow into the right atrium is prevented by the **tricuspid valve,** a one-way valve whose thin flaps are held in place by **chordae tendineae.** Backflow from the pulmonary artery to the right ventricle is prevented by the **pulmonary semilunar valve.** Deoxygenated blood from the pulmonary artery enters the capillaries of the lung, where its gases are exchanged.
2. Oxygenated blood returns from the lungs via the **pulmonary veins,** enters the left atrium, and moves on to the thick-muscled **left ventricle**, which pumps it into the **aorta.** Backflow into the left atrium is prevented by the **bicuspid valve (mitral valve)**, while backflow into the left ventricle is prevented by the **aortic semilunar valve.**

Control of the Heart

1. While heart muscle has an inherent contractile nature, control of its rate and effort is both **extrinsic** (external) and **intrinsic** (internal). Extrinsic control is through the autonomic nervous system, which accelerates heartbeat via sympathetic nerves and slows it via parasympathetic nerves. Epinephrine from the adrenal glands also accelerates the heart.
2. Intrinsic control originates in the **sinoatrial (SA) node**, or **pacemaker**, which sends contractile impulses across the atrial walls, causing contraction there. The impulse reaches the **atrioventricular (AV) node** and is relayed to the **bundle of His** in the ventricular septum. The bundle's two branches pass to the base of the ventricles and up their outer walls, giving rise to many branched **Purkinje fibers.**

The Working Heart

1. The *lub* of the *lub dup* heart sound is that of the tricuspid and bicuspid valves shutting, while the *dup* is that of the shutting of the aortic and pulmonary valves. The period of ventricular contraction is **systole** while the period between contractions is **diastole.**
2. The amount of blood per contraction is the **stroke volume**, while the **cardiac output** is the rate output per minute.

Blood Pressure

1. Elasticity in the major arteries maintains **blood pressure** during diastole. Clinically, blood pressure is measured with a **sphygmomanometer** and is recorded as systolic over diastolic pressure (for example, 120/80 mm Hg).
2. Problems in maintaining pressure arise in **arteriosclerosis** (loss of elasticity) and **atherosclerosis** (plaque formation), arterial wall diseases that increase resistance, burdening the heart.
3. Arterioles are capable of **vasodilation** (opening) and **vasoconstriction** (closing), shunting blood where need is greatest.

Circuits in the Human Circulatory System

Circuits are circulatory pathways where some special function is performed by the blood.

They include the following:

a. **pulmonary circuit** (gas exchange);
b. **hepatic portal circuit** (food carried from gut to liver for storage and distribution via the **hepatic portal vein**, which forms **sinusoids** in the liver);
c. **renal circuit** (**renal arteries** and **veins** bring blood to and from the kidneys where wastes, excess water, and other substances are removed from the blood);
d. **cardiac circuit** (blood supply to heart muscle, includes **coronary arteries** and **coronary veins.** Arteries from netlike **anastomoses**, which help limit the effect of **coronary thrombosis.** The coronary veins empty into the right atrium via the **coronary sinus**);
e. **systemic circuit** (a catch-all for the rest of the body).

The Role of the Capillaries

1. All transport functions are carried out by the capillaries, which are composed of a single thickness of interlocking cells. Blood is directed into or away from capillary beds by smooth muscle **precapillary sphincters** in arterioles.

2. The functioning of capillaries depends on diffusion gradients, hydrostatic pressure, and active transport (commonly, pinocytosis).
   a. Hydrostatic pressure and to a lesser amount, diffusion gradients, account for losses of water, ions, and nutrients at the arteriolar end of a capillary bed.
   b. Steep osmotic and diffusion gradients bring water and ions back into the capillaries at the venule end.
   c. The exchange of oxygen and carbon dioxide follows the diffusion gradient, and water not reclaimed is recycled by the lymphatic system.

The Veins

Blood pressure is lowest in the veins, but although force is reduced, volume in the veins must nearly equal that of the arteries. The onward movement of venous blood is assisted by one-way valves, muscular squeezing, and breathing movements.

The Blood

1. Blood (a connective tissue) consists of cells (mainly **erythrocytes**, but also **leukocytes** and **platelets**) and **plasma** (a watery matrix with proteins, nutrients, ions, hormones, and wastes).
2. Plasma proteins include **albumins**, which aid in transport; **globulins**, which include **antibodies** or **immunoglobulins**, and **fibrinogen**, which functions in blood clotting.
3. Red blood cells are constantly replaced by production in the **red bone marrow.** Aging cells are phagocytized by **macrophages** in the spleen and liver.
4. White blood cells include **neutrophils**, **basophils**, **eosinophils**, **lymphocytes**, **monocytes**, and **macrophages**, each of which plays a role in the immune responses, which include recognition, phagocytosis, inflammation, and antibody formation.
5. Platelets, or thrombocytes, are cellular fragments formed from platelet mother cells (megakaryocytes).
6. All blood cells originate from one cell type, the **hemocytoblast** or **stem cell.**
7. Clotting is quite complex, with some 15 steps. Following damage to a vessel, platelets gather at the wound, form a collagen plug, and release vasoconstrictors and enzymes called **thromboplastins.** The latter converts **prothrombin** to active **thrombin.** Thrombin cleaves fibrinogen, forming **fibrin,** which becomes the clot.

THE LYMPHATIC SYSTEM

1. The lymphatic system redistributes body fluids and ions, transports lipids, and cooperates with the immune system.
2. Lymphatic structures include **lymph vessels** and **lymph nodes.** The vessels begin as **lymph capillaries,** which lead to **collecting ducts** and then the larger lymphatics. Fluids move through the squeezing action of muscles and by breathing movements.
3. Lymph nodes are the sites where disease organisms and cancer cells are destroyed by lymphocytes and other cells of the immune system. Cancer commonly spreads via the lymphatic system.

REVIEW QUESTIONS

1. List five substances commonly carried in the blood. (937)
2. Explain how open and closed circulatory systems differ and list an example of each from the invertebrates. (939)
3. In what way is the circulatory system of a crustacean more complex than that of an insect? (939)
4. Trace the flow of blood through the earthworm's circulatory system. (939)
5. Draw a simplified scheme of the fish circulatory system and explain the major circuits. What is a portal circuit? (940–941)
6. Name a separate blood circuit in the amphibian and reptile not found in the fish. Is this separation perfect? Explain. (941–942)
7. Beginning with the sinus venosus, trace the flow of blood through the amphibian heart, naming each major structure through which it passes. (942)
8. In what groups of animals has the heart progressed to four chambers? Prepare a simple diagrammatic drawing of the four-chambered heart, naming the chambers and (with arrows) illustrating the flow of blood. (942)
9. Compare the structure of arteries, veins, and capillaries. Which is directly involved in exchanges with the tissues? (942–943)
10. Trace the flow of blood from the vena cavae to the lungs, naming chambers, valves, and vessels along the way. (943–944)
11. Trace the flow of blood from the pulmonary veins to the aorta, naming chambers, valves, and vessels along the way. (944–945)
12. Starting with the SA node, describe the pathway followed by a contractile impulse as it passes through the heart. (945–946)
13. Describe the dual control of heart rate produced by the autonomic nervous system. (946)
14. Distinguish between stroke volume and cardiac output, and compare changes in these factors in a well-conditioned person and a poorly conditioned person during vigorous exercise. (946)
15. Explain how blood pressure is maintained during diastole. (946–947)
16. Describe the renal circuit and hepatic portal circuit and explain the special function of each. (947–948)
17. Describe the vessels of the cardiac circuit. What special arrangement occurs in the smaller arteries, and how is this adaptive? (948–949)

18. Discuss the cellular construction of the capillaries and relate this to their function. (949–950)
19. List the forces at work at either end of a capillary bed and explain how these affect the materials in the blood. (951)
20. What happens to blood pressure when blood reaches the veins? In view of pressure changes, explain how blood is moved back to the heart. (951)
21. List six components of blood plasma. Which is the most common? (952)
22. Discuss the red blood cell, including its size, shape, life expectancy, and what happens to it when it ages. (952)
23. List five types of white blood cells and briefly describe their functions. (952)
24. Summarize the clotting process, mentioning the role of platelets, thromboplastins, prothrombin, and fibrin. (953)
25. Describe the composition of the lymphatic system. What are its three main functions? (954–955)

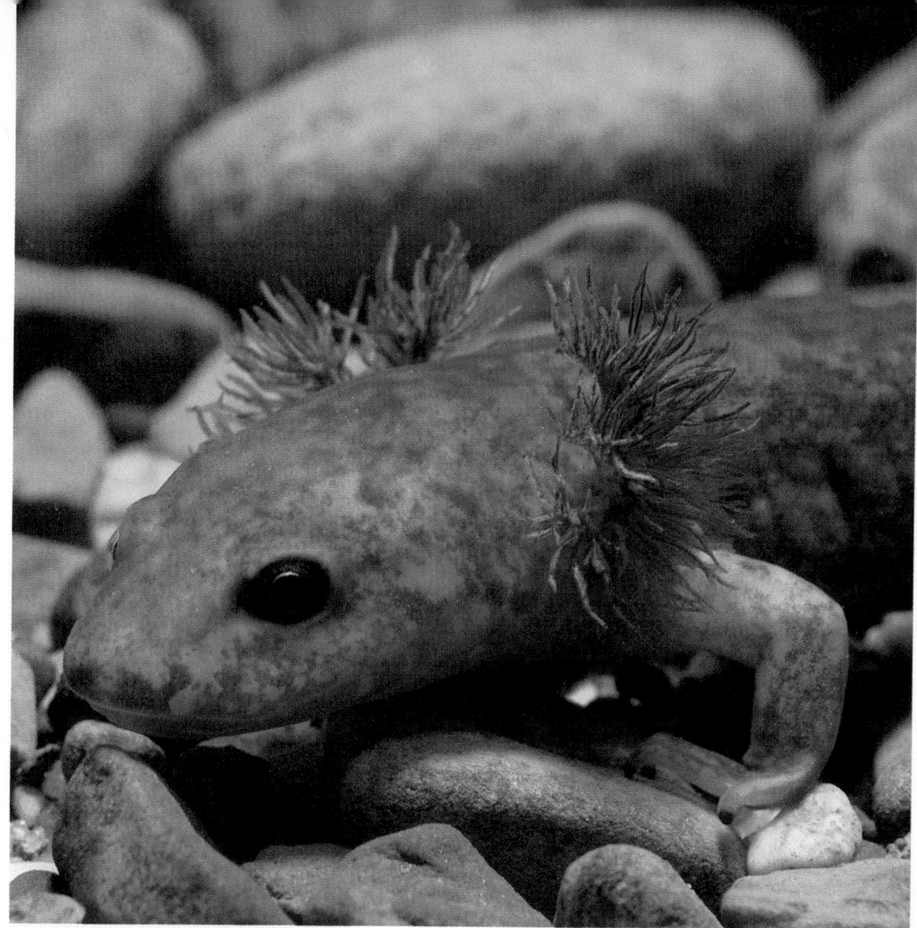

40

RESPIRATION

THIS MAY NOT BE THE BEST OF ALL POSSIBLE WORLDS, BUT IT'S THE ONLY one we've got, and it does have a way of presenting us with an array of offerings. In this world, in a very real sense, the processes of life are involved with overcoming various problems imposed by our environment and with taking advantage of the opportunities that we find before us.

For example, one such opportunity has been afforded by oxygen, but it did not always fall under the category of an opportunity; it was once a dire threat to life. In fact, much of the oxygen present today may have arisen as a form of atmospheric pollution—a toxic product of early photosynthetic life (see Chapter 21). It took many millions of years for living things to adapt to the increasing levels of atmospheric oxygen. At first it must have been a dangerous game indeed, since oxygen can play havoc with the usual chemical reactions of life unless its effect is somehow neutralized. At the time oxygen first appeared, there was little in the atmosphere to absorb ultraviolet radiation, so atmospheric oxygen was quickly converted to ozone (O_3), a highly poisonous oxidizing agent. (Ozone still arises spontaneously in the upper atmosphere, but see Essay 49.2 for a new twist in the ozone story.) The advent of oxygen, then, transformed the surface of an already unpredictable earth into an even more dangerous place.

Today, however, ozone helps protect us from dangerous ultraviolet radiation from the sun. Further, most organisms can actually use molecular oxygen (O_2). Interestingly, perhaps because of common evolutionary descent, these species tend to use oxygen in the same general way—as an acceptor of spent electrons during cell respiration. (Recall that energy-depleted electrons leave the respiratory electron transport systems of the mitochondrion to join oxygen and eventually form water, a metabolic waste product; see Chapter 8.)

Many species have even developed very elaborate systems to distribute oxygen to the cells. In many animals, including humans, the **respiratory system** exchanges carbon dioxide for oxygen, while the circulatory system carries these gases, as well as nutrients, wastes, and hormones, throughout the body.

FIGURE 40.1
ANIMAL RESPIRATORY INTERFACES
The simplest gas exchange interface occurs in skin breathers **(a)** such as the sea anemone, where simple diffusion across a thin body wall is sufficient. Many aquatic animals utilize the external gill **(b)**, as seen in *Necturus,* the mud puppy. The fish gill **(c)** is internalized (at least it is enclosed in a covered gill chamber). Water and blood, moving oppositely, provide an efficient countercurrent exchange. **(d)** Terrestrial insects make use of an extensive internalized tracheal system where minute thin-walled passages provide for gas exchange throughout the body. **(e)** The internalized lung of terrestrial vertebrates provides a protected, efficient gas exchange surface. **(f)** In birds, the lung reaches its most specialized form, with a one-way flow and cross-current exchange.

In this chapter, we'll look at the respiratory systems of a variety of animals, focusing on both their unity and diversity. We will find that solutions to problems of gas exchange cut across all taxonomic lines, and that similarities occur not only because of evolutionary relatedness but because of adaptations to similar environments. In other words, we will be looking at examples of both convergent and divergent evolution of the respiratory system.

GAS EXCHANGE SURFACES

Let's begin by emphasizing that the exchange of gases requires structures with two basic characteristics. First, they must have a permeable surface area, with a thin-walled membrane of sufficient size to meet the animal's requirements. Second, this surface area must be moist, since gases can't normally cross dry membranes (such as our skin, which contains an outer layer of dry, dead cells). These requirements have been met through evolution in an interesting variety of ways. Most familiar are the spongy lungs of air-breathing vertebrates and the frilly gills of fishes and numerous aquatic invertebrates. But there are other ways (Figure 40.1).

The problem of maintaining moistness in the terrestrial environment, in many species, is solved by the secretion of copious amounts of mucus by cells of the exchange membrane. We will refer to the exchange membrane as the **respiratory interface** as we proceed since it is the respiratory structure that is exposed to the environment. The respiratory interface can vary from lungs and gills to body surfaces. Our survey

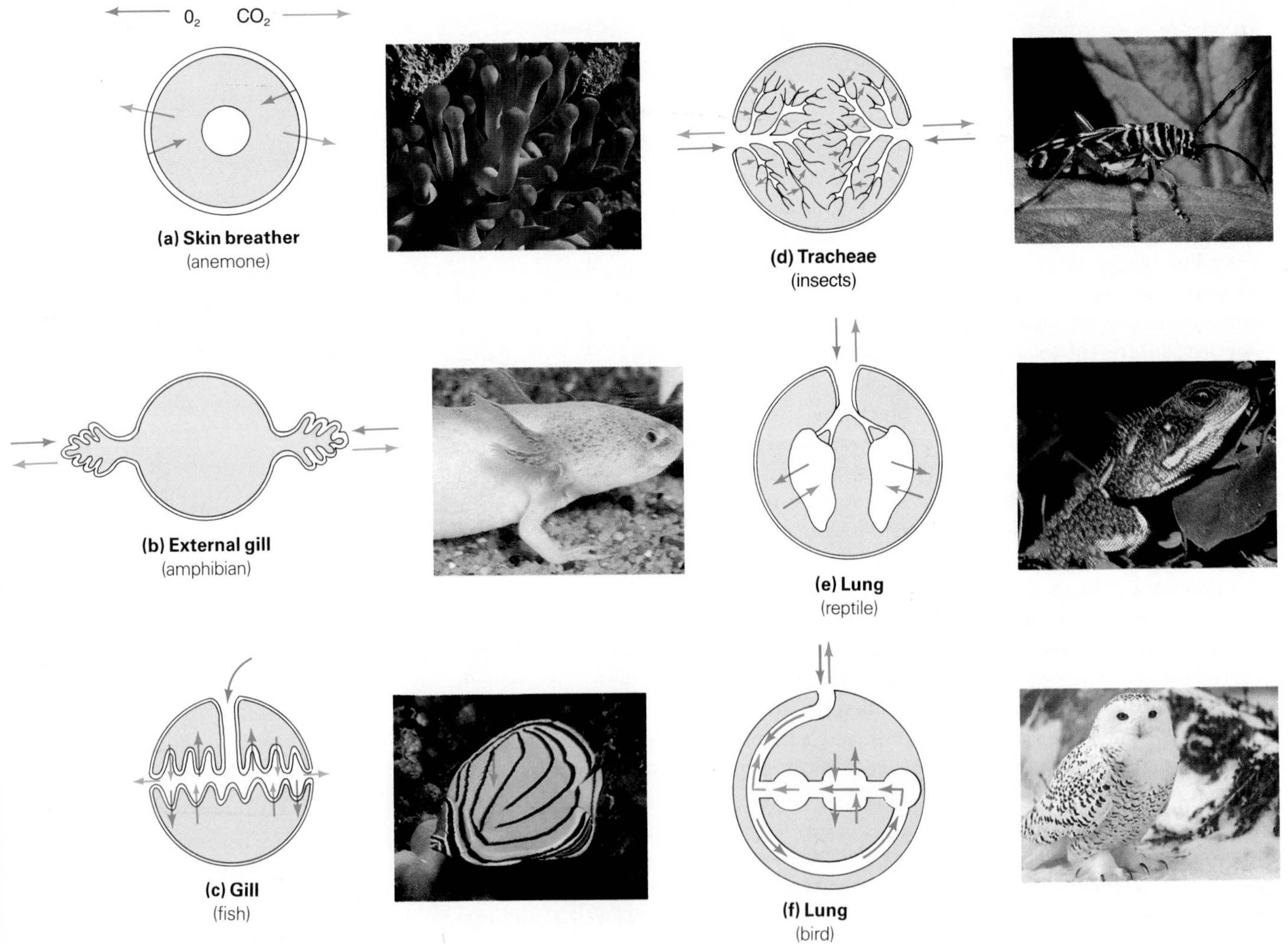

will begin with some of the animals that use their body surfaces for respiration. Then we will consider specific structures that are more complex.

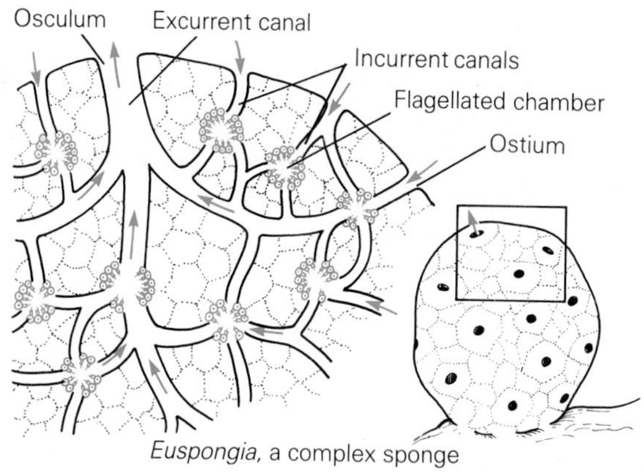

FIGURE 40.2
INCREASED SURFACE AREA IN THE BATH SPONGE
In *Euspongia,* the respiratory surface is increased substantially by an extensive array of interconnecting canals. The canals lead into chambers lined with flagellated collar cells that create strong water currents.

The Simple Body Interface

Use of the body surface or skin as a respiratory interface occurs in a number of unrelated groups. It is an apparently simple solution to the problem of gas exchange but one that has severe restrictions as well. You may recall the general structure of some of the simpler sponges discussed earlier. Their thin-walled, vaselike bodies permit a simple exchange of gases with both the surrounding sea water and a current of sea water carried through the hollow body by flagellated collar cells. However, we should remind ourselves that not all sponges are small, and the more complex sponges may reach an impressive size indeed. How do gases penetrate their mass? While the body surface is still the exchange interface, mass is no problem since the body, rather than being vaselike, is riddled with small canals. These canals are lined with collar cells that create strong currents of oxygen-laden water that reaches all parts of the sponge body before being expelled through the osculum (Figure 40.2).

Similarly, although some species of cnidarians (such as jellyfish) attain considerable size, they require no specialized respiratory structures. Their saclike bodies consist essentially of two cell layers that are virtually always in contact with the external environment. Their outer body wall and extensive gastrovascular cavity provide an adequate exchange interface (see Figures 30.8 and 30.10).

Though flatworms are considerably more dense than cnidarians, they are, nevertheless, flat, and this shape yields a large surface area for its mass. Flatworms also have an extensive highly branched gastrovascular cavity (see Figure 30.13). Because of these characteristics, much of its body mass is exposed to its watery environment, and no cells are far from it. It can, therefore, use cell-by-cell diffusion as a means of exchanging gases with the environment. Of course, as long as it relies on diffusion alone to transport gases, it can never grow very large—its volume would quickly outgrow its surface area. So if you ever see an 8-foot planarian gliding toward you in your favorite swimming hole, ignore it. It doesn't exist. Some marine flatworms grow to as much as 10 cm wide and 60 cm long, but they are only a few millimeters thick.

In the spiny echinoderms, an expansion of the respiratory interface has been accomplished through the **dermal brancheae**, outpocketings of the coelomic wall that protrude through pores in the endoskeleton. Fluids are moved in and out by ciliated action in the coelom (Figure 40.3).

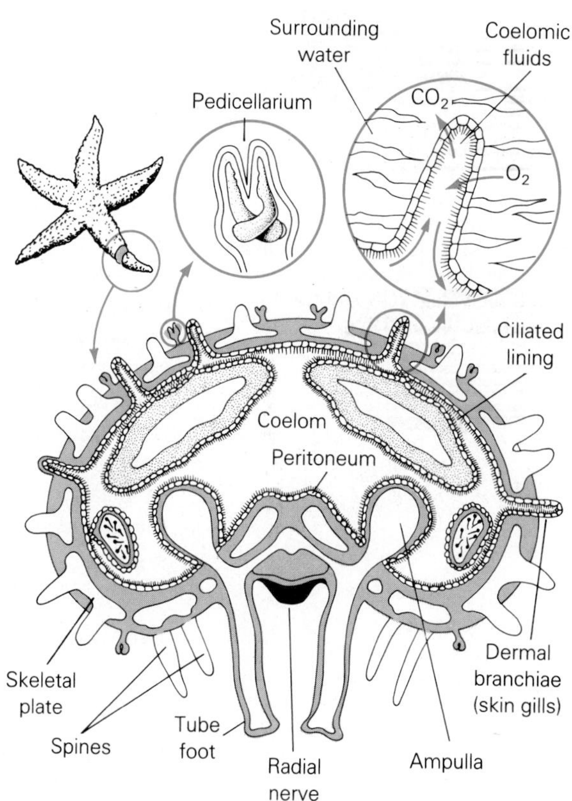

FIGURE 40.3
GAS EXCHANGE IN THE SEASTAR
The dermal brancheae of the sea star are formed from extensions of the skin that protrude through the skeletal plates. The pincerlike pedicellaria work to keep the sea star's surface clear of encrusting organisms. Gas exchange also occurs across the tube feet.

Complex Skin Breathers A few terrestrial animals, such as the earthworm and small lungless salamanders, manage to use their moist skin as a respiratory surface (the salamander also uses its highly vascular pharynx). But these complex animals are very special cases. For example, the earthworm's habitat is really only semiterrestrial since it restricts itself to moist soil. The lungless salamanders are also limited to damp places. They spend their days underground (or under logs), venturing out mainly at night when risk of desiccation is lessened. The skin of earthworms and lungless salamanders, by the way, is kept moist by the secretion of a slimy layer of mucus.

Although such skin breathers have no special structures with which to exchange gas with their environments, their great mass prevents them from relying on the cell-by-cell diffusion of gases. The skin breathing is enhanced by efficient circulatory systems that readily transport gas once it passes into their bodies. Both the earthworm and the

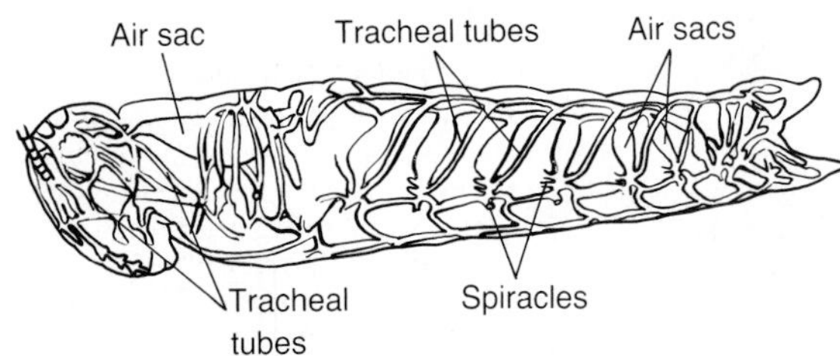

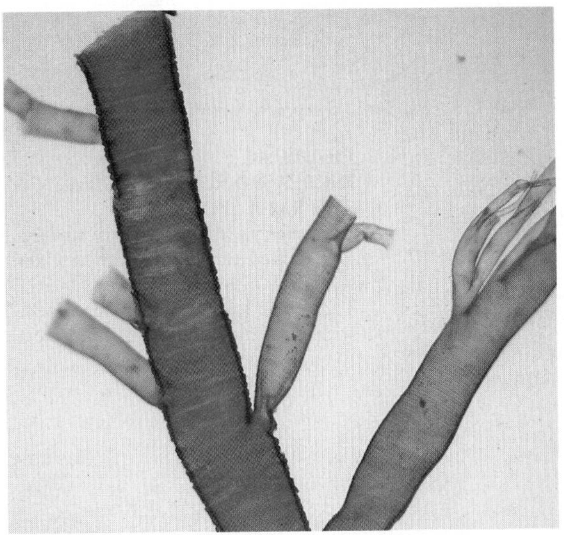

FIGURE 40.4
THE INSECT TRACHEAL SYSTEM
The insect respiratory system is an elaborate network of thin-walled tracheal tubes, some of which end in air sacs. The finest branches carry oxygen to individual cells and tissues throughout the insect.

salamander have closed circulatory systems (see Chapter 39) containing hemoglobin, the oxygen-carrying protein pigment. The salamander's hemoglobin is contained within special blood cells, as is our own, but the earthworm's hemoglobin is dissolved in the blood. Even so, the worm's blood is a far more efficient oxygen carrier than is simple plasma. Thus skin breathing works for some complex terrestrial organisms—if they restrict themselves to a moist habitat and if they have an efficient circulatory system. The earthworm and the lungless salamander, then, share an interesting combination of behavior, anatomy, and physiology that permits their simple skin respiratory systems to work.

Expanding the Interface: Tracheae

Most terrestrial animals have developed more specialized breathing surfaces. These usually involve specialized infoldings of the gut or body surface. By internalizing their respiratory surfaces in this way, land dwellers reduce the problem of excessive water loss. Furthermore, these internal pouches are usually highly folded and convoluted, providing an increased area for gas exchange. The lungs are the most obvious example of such an arrangement.

Among the variations of this internal arrangement are those of the arthropods, a group that does most things differently. Insects, for example, have a **tracheal system.** The insect body is riddled with a series of tiny, highly branched tubules called **tracheae** and **tracheoles** (Figure 40.4). These tubes open to the outside via tiny valvelike openings known as **spiracles,** commonly situated along the insect's sides. In grasshoppers and some other larger insects, air is pumped in and out of the spiracles by a bellows action of the abdomen, the spiracles opening and closing in synchrony with the pumping action. In grasshoppers, bees, and others, the branching tracheae terminate in elastic **air sacs** that expand and contract and aid in the exchange processes. In other species of insects, air simply diffuses through the tracheal system. Oxygen and carbon dioxide are exchanged through the walls of very fine tracheoles, which permeate the body and carry oxygen to the immediate vicinity of every active tissue. The finer branches of the tracheae are filled with fluid through which the respiratory gases can diffuse.

In aquatic insects, the tracheae may terminate in **tracheal gills**—thin-walled extensions of the exoskeletons that contain finely-branched tracheal networks. Although insect circulatory systems are efficient, they appear to serve primarily to distribute food molecules and metabolic products; the respiratory functions are thus left to the tracheal system.

Complex Interfaces: Gills

Aquatic mollusks, arthropods, and chordates are often highly active animals, and their increased respiratory demands are met with a more complex system. Here we find the **gill,** which, like the lung, is a combination respiratory-circulatory exchange surface. We might point out here that the presence of similarly constructed gills in widely divergent animal groups does not indicate common ancestry, but rather the rigorous restrictions for gas exchange imposed by the environment.

Gills are thin-walled, finely divided, and feathery structures with extensive capillary beds through which blood flows, carrying carbon dioxide from the body to be exchanged for oxygen from the surrounding water. Capillaries, you will recall, are the smallest blood vessels, with walls only one cell thick, across which gases can freely pass.

Gills in Mollusks and Arthropods Most mollusks rely on gills. In bivalves (such as the clam) the gill serves in both food filtering and respiration, as was mentioned in

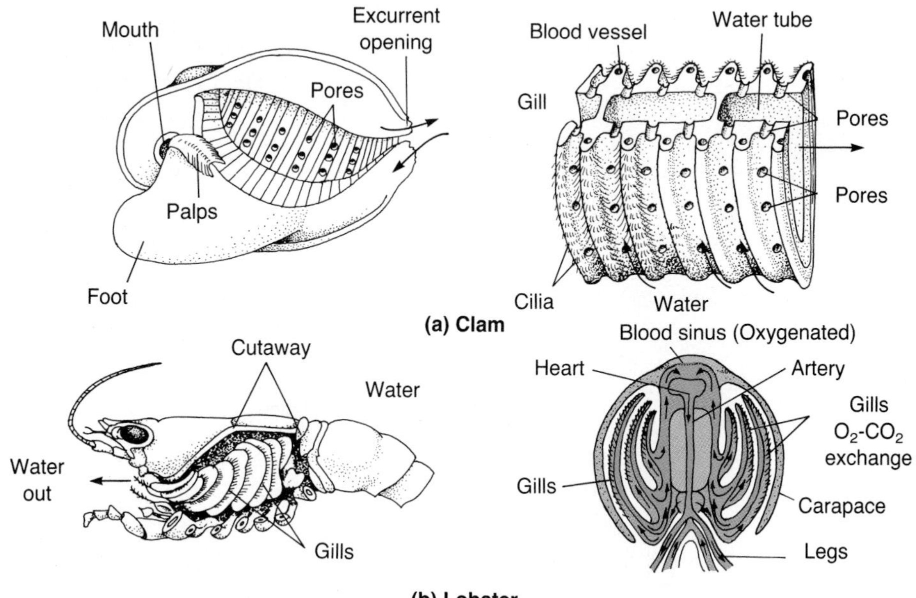

FIGURE 40.5
INVERTEBRATE GILLS
(a) Cilia on the gill surface of the clam generate a stream of water that flows over the gills, where gases are exchanged and food particles are sorted out. **(b)** The lobster gill is ventilated by a forward-moving stream of water generated by tiny appendages called bailers. The lobster's circulatory system pumps blood through capillaries in the gills.

Chapter 31. The clam gill (Figure 40.5a) consists of sheetlike folds that protrude into the mantle cavity. Water is drawn by ciliary action into the incurrent siphon. From there, it passes through microscopic pores in the gills, entering channels that direct it out through the excurrent siphon. As water passes through the channels, oxygen and carbon dioxide are exchanged through tiny blood vessels that permeate the gill partitions. Oxygenated blood is then returned directly to the heart for distribution to the body. Many mollusks and arthropods lack both red blood cells and hemoglobin, but they do have respiratory pigments. These species make use of a copper-containing protein called **hemocyanin** for transporting oxygen. Hemocyanin is also found in many crustaceans, spiders, and scorpions. In hemocyanin, oxygen is bound to copper rather than iron as in hemoglobin, and on oxygenation the pigment turns from colorless to blue.

The arthropod gills are varied, but those of the lobster, a crustacean, will serve as a good example. The lobster gills are feathery extensions of the body wall, attached partly to the bases of the legs (Figure 40.5b). They are covered by a **carapace,** which is open at both ends. Water is drawn under the carapace, across the gills toward the head by the action of a paddlelike appendage near the jaws known as the **bailer.** Blood from a hemocoel (an open blood passage) in the thorax floor passes through channels in the gills, exchanging gases as it goes. Upon leaving this vital circuit, it returns to the heart to be pumped through the body again. As we saw earlier, such a circulatory system—with blood leaving the vessels and percolating through open spaces—is called an open circulatory system. Further, the blood of crustaceans contains hemocyanin, and since this pigment readily combines with oxygen, it is ideal for creatures such as crayfish that live in stagnant, oxygen-poor water.

Gills in Fishes Gills in fishes arise from a number of supporting **gill arches** that contain rows of feathery **gill filaments.** Each gill filament contains a series of platelike **lamellae,** each of which houses a network of capillaries (Figure 40.6). Carbon dioxide-laden blood from the body enters the gill filaments in thin-walled capillaries emerging from **afferent** (incoming) **vessels** in the gill arch. The capillaries branch and rebranch in each lamella, forming the network in which the exchange of carbon dioxide for oxygen occurs. Oxygen-rich blood leaving the gill filament enters an **efferent** (outgoing) **vessel** in the gill arch and, upon leaving the gill, joins with the dorsal aorta to be distributed throughout the fish's body.

Two factors facilitate the exchange of gases within the minute capillary networks of the lamellae. First, since the walls of both the capillaries and the filaments are extremely thin, there is little distance between the blood and water passing over the gills. Second,

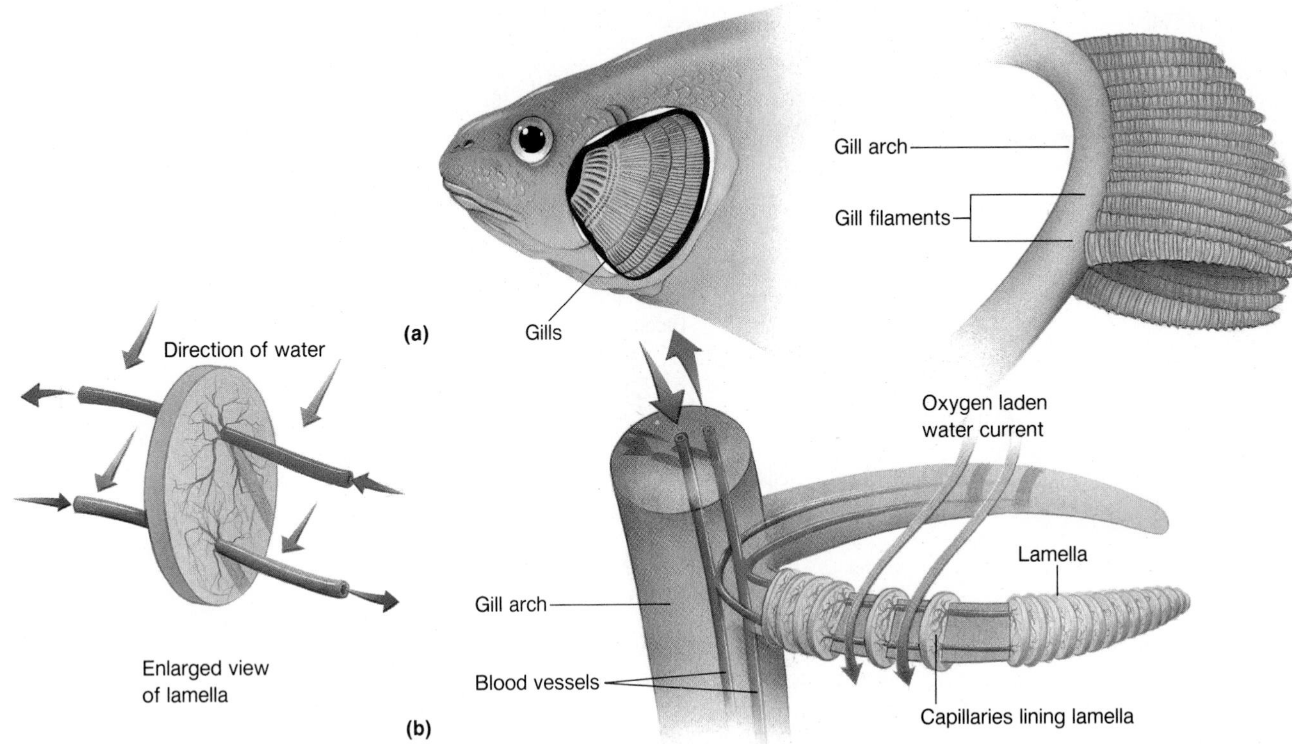

FIGURE 40.6
THE FISH GILL
(a) The fish gill includes a number of gill arches, which support the gill filaments. **(b)** Surface area of the filaments is increased by platelike lamellae that contain a rich supply of capillaries. The opposing movement of blood and water across the lamellae sets up an efficient countercurrent exchange of gases.

the flow of blood in each lamella opposes the flow of surrounding water, setting up a highly efficient **countercurrent exchange.** In the gill, blood passing over the exchange network is constantly confronted by a fresh supply of water; thus, as long as blood is passing across the lamellae, there is an inward diffusion gradient for oxygen and an outward diffusion gradient for carbon dioxide. Such efficiency is especially important to active animals in aquatic environments where getting enough oxygen is difficult.

Even if one has gills, underwater breathing presents problems. Because of oxygen's low solubility, its concentration in water is rather low. The solubility of oxygen in water is inversely proportional to temperature, so in general, cold water contains more oxygen than warm water. But even the coldest saturated water contains less than 1% oxygen by volume. This is paltry compared to the oxygen content of air—21% by volume. Further, the diffusion rate of oxygen through water is vastly slower than in air.

Water's low oxygen content and high viscosity mean that active aquatic animals must devote considerable time and energy just to moving large volumes of water over their respiratory surfaces.

The faster-swimming bony fishes solve their respiratory problems much as do the sharks—they keep moving. Others rely on a pumping action. They keep a continuous flow of water moving over the gills by the subtle action of pharyngeal muscles. To "breathe," the fish closes its opercula (gill covers), opens its mouth, and by expanding its mouth and gill chambers, draws water in. The fish then closes its mouth, contracts the oral cavity, and opens its opercula to let the water flow out across the gills. That's why motionless fish look as though they are gulping.

Complex Interfaces: Lungs

Evolution of the Vertebrate Lung The vertebrate **lung** has had a curious evolutionary history. Early in vertebrate history, perhaps as early as the Precambrian, there lived the ancestor of modern bony fishes, a freshwater inhabitant that obtained oxygen by gulping air. A pocket of highly vascularized tissue became specialized in the pharynx of this ancestral fish as a place to hold its oxygen-laden air between swallows. Eventually, the air pocket grew and branched, and evolved into the vertebrate lungs.

Some of the descendants of that early ancestral fish species retained the lungs as accessory organs, although they developed gills as well. One such group is the lungfishes (see Figure 32.13), a peculiar assemblage of freshwater fishes that are uniquely adapted for survival in ponds that tend to dry up. A different group, the lobe-finned fishes, was later to leave its freshwater habitat and become the ancestor of the terrestrial vertebrates. For this group, the possession of lungs was a vital **preadaptation**—an adaptation to one way of life that, by coincidence, enabled the organism to survive in another.

Those ancient lobe-finned fish then achieved a second evolutionary breakthrough—literally. The nasal cavities in most fish are blind pouches used only for smelling the water. But the ancestor of terrestrial vertebrates evolved a pair of **internal nares**—openings between the nasal cavity and the mouth. It's not clear that they actually did use their internal nares for breathing. (Lungfish today are unusual among bony fishes, in that they have internal nares but also gulp air through their mouths.) Still, this feature proved to be yet another preadaptation that helped greatly when certain descendants invaded the land.

To this day, the internal nares of amphibians and most reptiles open directly into the mouth cavity, so these animals can breathe with their mouths closed. In the evolution of mammals, a bony ledge, the **palate**, finally separated the mouth cavity from the nasal cavity, displacing the internal nares back to the region of the pharynx. Thus we can chew and breathe at the same time, as can crocodiles.

Most bony fish have kept the primitive lung structure, but its function has changed. In these species, the lung became a flotation device—the swim bladder—and the gills regained their status as the organism's sole gas-exchange organ. (In an entirely different role, the swim bladder also aids in sound reception.)

The Vertebrate Lung The lung in vertebrates is an inpocketing, branching tube (basically an extension of the gut) that in some species ends in a multitude of tiny air sacs **(alveoli)**, where blood and air are separated by a thin, moist membrane. Air is usually pumped in and out through a bellows mechanism. By muscular control of its breathing apparatus and by controlling air that passes through its mouth and **pharynx**, the vertebrate can control the amount of air that reaches its lungs. Of course, moisture loss is unavoidable; the exhaled air is laden with water vapor.

The evolutionary pathway from an aquatic to a terrestrial respiratory system is perhaps suggested in today's amphibians. Nearly all amphibians have lungs (the lungless salamanders we mentioned apparently once had lungs, but lost them along the way), and nearly all exchange gases in two ways: through rather simple, baglike lungs and through their moist, highly vascularized skin. The ability to breathe through the skin enables many amphibians to "hibernate" in mud through cold or dry seasons, a time when a restricted oxygen supply requires them to remain inactive. (It's probably hard to be very active while encased in mud anyway.) Not only are the lungs simple, but even breathing movements in amphibians are quite primitive. Since there is no diaphragm or well-developed rib cage, air is forced in and out of the lungs by the raising and lowering of the floor of the mouth. This coarse pumping action is aided by the presence of "check valves" in the nostrils and the **glottis**, a slit-like opening valve that forms the opening of the trachea (Figure 40.7).

Since the dry reptilian skin is impervious to air, reptiles, including the many aquatic species, are strictly lung breathers. Perhaps *strictly* is not the right word. The inevitable exception this time is found in some water-breathing turtles that use a large, heavily vascularized cloaca as a supplemental respiratory surface—a kind of water lung. (You may have trouble picturing this if you remember that a cloaca is the common posterior chamber through which digestive and urinary wastes and eggs pass from the body; see Chapter 32.) The lungs of

FIGURE 40.7
BREATHING MOVEMENTS IN THE FROG
As the frog breathes, air is forced in and out of its lungs by the bellows action of its throat and body wall. Note the four-stroke system and the role of check valves in the nostrils and trachea. (How might holding a frog's mouth open affect its breathing?)

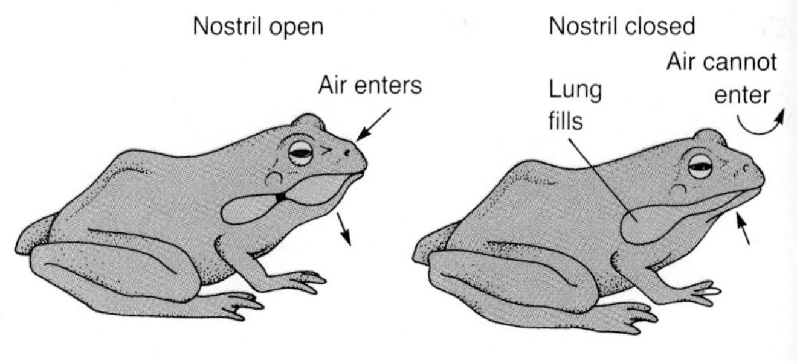

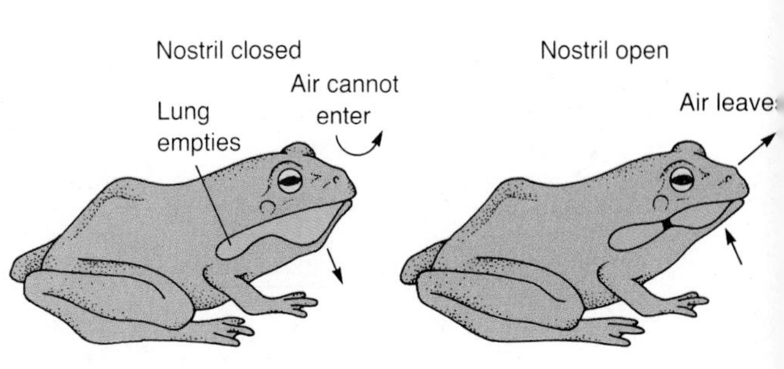

reptiles are essentially saclike, but some are more complex than those of amphibians. For example, some species, such as the monitor lizards, have more subdivided air passages. Breathing movements involve the muscles around the entire body cavity since there is no diaphragm and no separate abdominal and thoracic (chest) cavities.

The Unique Bird Respiratory System The respiratory system of the bird is unlike that of any other vertebrate (Figure 40.8). The lung, like that of the reptile and mammal, is penetrated by many air passages and is thus quite spongy. But most of the similarity ends there. The passages in the bird lung do not end in alveoli (clusters of blind microscopic sacs), but occur as numerous open-ended tubes called **parabronchi** (Figure 40.8b). There is no in-out movement of air as is the case in a human lung. Instead, air enters the bird lung from the posterior end, flows through the many parabronchi, and exits from the lung's anterior end—*a one-way flow* (Figure 40.8c and d). Further, because of the arrangement of capillaries around the parabronchi, the blood passes at right angles to the air flow, forming what is called a **crosscurrent exchange** (see Figure 40.8b). A look at the entire bird respiratory system will help explain how all of this provides for an efficient exchange.

Air enters the respiratory system much the same as in other vertebrates, through the mouth and nostrils, and then into a tubular structure called a **bronchus** (in other vertebrates, the *trachea*). However, the air does not enter the lung directly at this point, but passes into a number of large air sacs (see Figure 40.8a). These air sacs are another unique feature of the bird respiratory system. There are three pairs of anterior air sacs and two pairs of posterior sacs (anterior and posterior to the lung). In addition, the air sacs form extensive branches, some passing into the major long bones. But these extensive air sacs have little to do with actual gas exchange. They function as air reservoirs and as a bellows that force air in and out of the lung. As you can see in Figure 40.8c and

FIGURE 40.8
THE UNIQUE BIRD LUNG
(a) The bird lung lies between anterior and posterior air sacs. We can follow the respiratory sequences in a highly simplified diagram. **(b)** On inspiration, fresh air fills the posterior sacs, and air from the lungs is drawn into the anterior sacs. **(c)** On expiration, air from the posterior sacs is forced into the lungs, and air from the anterior sacs leaves the body. Thus, there is a one-way flow through the lungs. **(d)** The enlarged view of a portion of the lung shows the cylinderlike parabronchi. Gas exchange occurs in a crosscurrent flow with capillaries that pass around the cylinders.

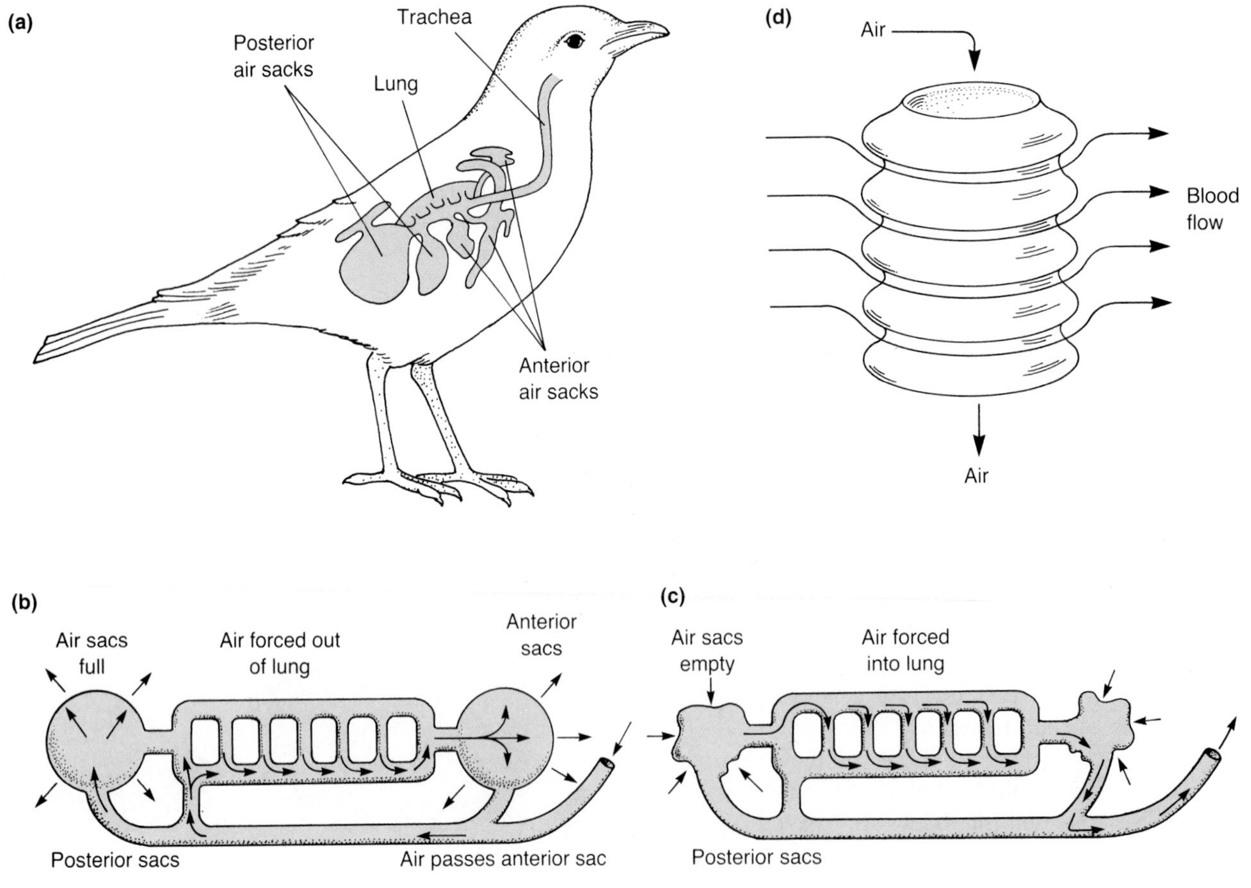

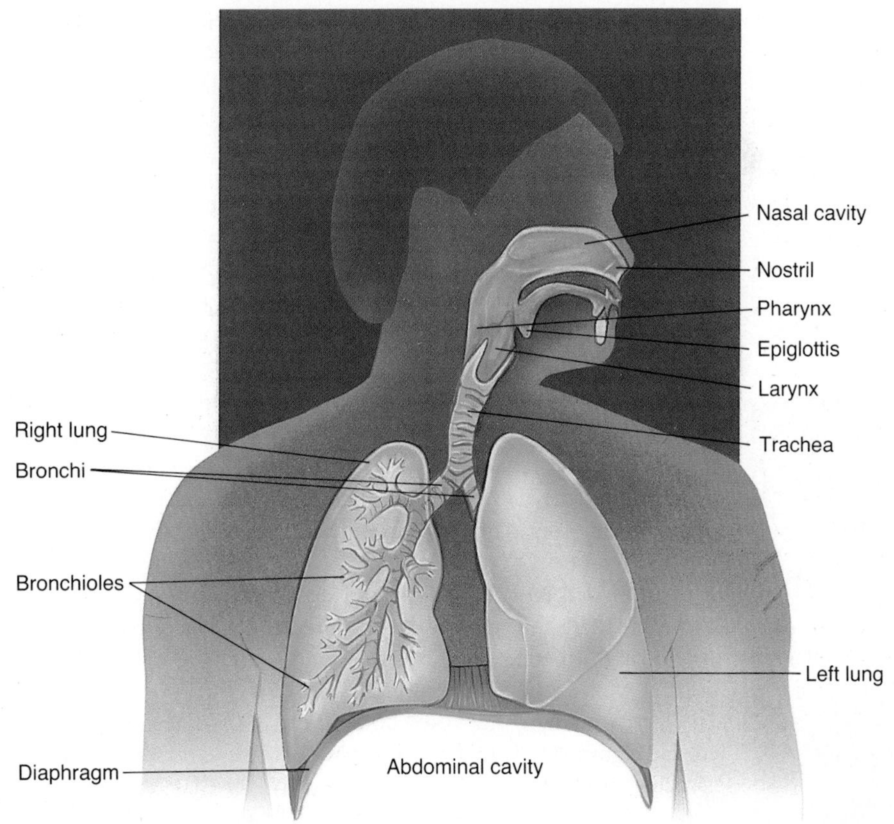

FIGURE 40.9
THE HUMAN RESPIRATORY SYSTEM
The respiratory system includes extensive air passages, where air is warmed and filtered. The passages branch and rebranch throughout the lungs.

d, the sacs inflate on inspiration, as the chest cavity of the bird expands (providing the force), and deflate on expiration, when the chest cavity contracts. Inflation fills the posterior sacs with fresh air and the anterior sacs with stale air (from the lung). Deflation forces fresh air into the lung and stale air into the bronchus and out of the body. Unlike the mammalian lung, which inflates and deflates during breathing, the bird lung changes very little during the process.

The important points are that air flows in a one-way path through the lung—from the posterior air sacs, through the lung, to the anterior air sacs, and this one-way flow provides for a most efficient exchange since it crosses the flow of blood through the lung capillaries. Such an arrangement required dramatic evolutionary changes from the simpler reptilian lung. How were such vast changes adaptive?

The primary advantage of the bird lung is that it adapts the bird to flight at high altitudes where oxygen levels are low. Maintaining rigorous activity at higher altitudes requires the most efficient gas exchange possible, and this, apparently, is where the unique bird respiratory system is most significant. High altitude flying is particularly important to birds that migrate, and while many birds migrate at an altitude of only 1200 to 1500 m, many fly higher. Pilots have reported birds flying at 6000 m (19,685 ft), and radar tracking has indicated that they can fly at 7000 m (about 23,000 ft). Mammals, including the flying bats, subjected to activity at this altitude would have great difficulty functioning at all and would quickly fall into a metabolic stupor. As you might expect, bats are low altitude migrators.

We now turn to a more detailed look at a respiratory system in the mammals, concentrating on humans.

THE HUMAN RESPIRATORY SYSTEM

The respiratory system in humans (Figure 40.9) is fairly typical of those of other mammals. The major structures include the mouth and nasal passageways, the pharynx,

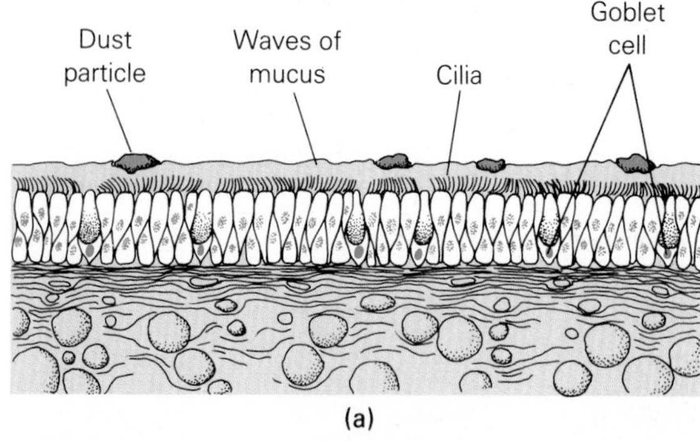

FIGURE 40.10
THE CILIATED RESPIRATORY LINING
(**a**) Much of the surface of the human respiratory passages contains goblet and ciliated cells. (**b**) The scanning electron microscope reveals a dense covering of cilia.

larynx, trachea, bronchi, and lungs. Air first passes through the nasal passages (unless one is a chronic mouth breather). In addition to their role in directing the air flow, these passages are important in filtering, warming, and moistening the air prior to its entering the lungs. The nostril hairs act as an initial filter, trapping dust particles. A nose full of hair may not be esthetically pleasing, but your lungs undoubtedly appreciate it.

In all mammals, the nasal cavity is separated from the mouth by the **hard palate,** a bony shelf in the roof of the mouth ending in a softer muscular region, the **soft palate.** The nasal cavity is lined with mucous membrane, which includes mucus glands, mucus-secreting **goblet cells,** and cells bearing cilia (Figure 40.10). These glands and cells produce a film of mucus, which is constantly swept toward the throat by the ciliary action. In addition, the lining of the nose contains many dense capillary beds that warm the air before it enters the lungs. This countercurrent flow of air and capillary blood is very effective; even on very cold days, the air entering the lungs is warmed almost to body temperature.

The inhaled air moves from the nasal passages into the pharynx and from there through the larynx into the trachea—unless you are swallowing, whereupon the laryngeal opening is closed. You can't breathe while you swallow. (Of course, immediately everyone tries.) The larynx contains the **voice box** and vocal mechanism—the **vocal cords** (Figure 40.11). Below the larynx begins the trachea, a tube that contains many C-shaped rings of stiff cartilage that hold the airway open. As the trachea enters the chest, it branches into right and left **primary bronchi,** which branch again and again into the **bronchioles** that form the **respiratory tree.**

The trachea and some of the other bronchial structures have a secreting and sweeping lining similar to that in the nasal passages, so the air is cleaned of dust and debris once again. (The air going out is probably cleaner than the air coming in, so in a sense, humans help clean the earth's air.) Tiny particles of inhaled dust become trapped in the mucus, which is continually being moved up into the throat and swallowed. This capability is lost to heavy smokers, whose respiratory linings are subject to drastic change as the years and cigarette packs go by (Figure 40.12).

The bronchioles, the tiniest branches of the respiratory tree, end in grapelike clusters of air spaces known as alveoli (Figure 40.13). Each cluster is enclosed in a dense capillary

FIGURE 40.11
THE LARYNX
(**a**) The larynx consists principally of cartilage and muscle. Toward the front is a protruding cartilage commonly called the "Adams apple." (**b**) The larynx houses the vocal cords. As we breathe, the vocal cords form a triangle; but as we speak, the cords tighten to form a slitted opening that is varied in breadth to produce different pitches.

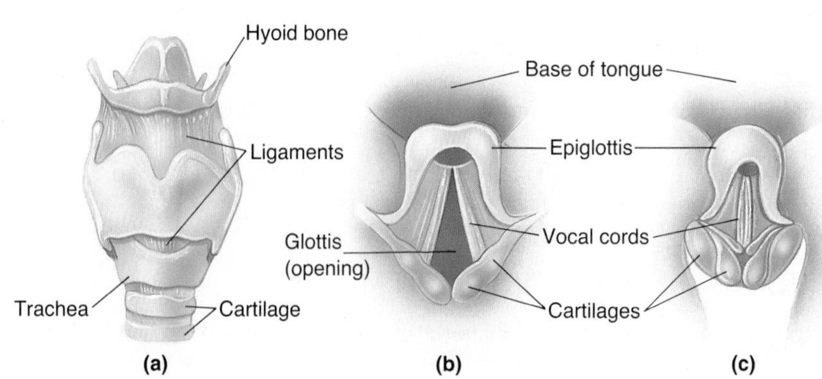

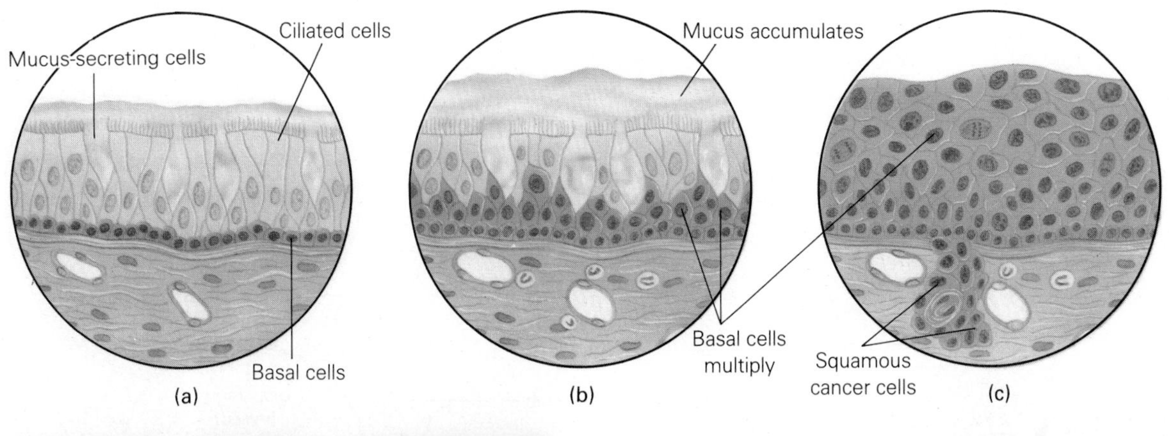

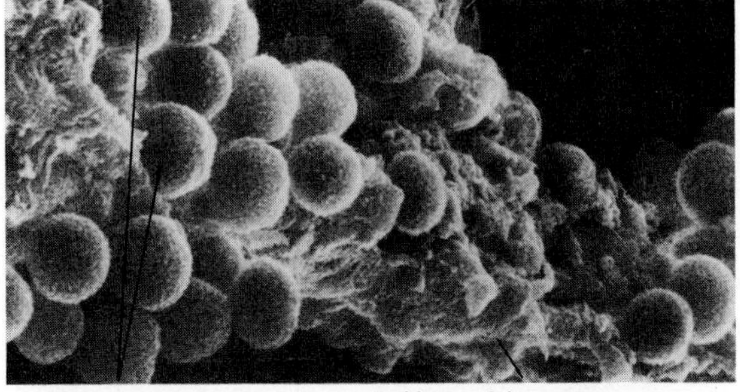

FIGURE 40.12
A BRIEF HISTORY OF LUNG CANCER
The normal ciliated epithelium (**a**) of the respiratory passages includes columnlike, ciliated, and mucus-secreting goblet cells. In the smoker's respiratory lining (**b**), the cilia become partially paralyzed, and the mucus accumulates on the irritated lining. Where an early cancerous state exists, the lowermost basal cells divide more rapidly and begin to displace normal columnar cells. As the cancer progresses (**c**), most of the normal columnar cells are replaced by simple cancer cells that form a spreading tumor. In advanced cases, clusters of cancer cells may be carried away in the lymphatic system, spreading to other parts of the body. The contrast between normal and cancerous cells is seen in the SEM of human lung tissue.

bed, and here the atmosphere and blood are only a membrane apart. The clusters, of course, provide an enormous surface area. In fact, the total surface area of the approximately 300 million alveoli of the human lungs has been estimated at nearly 70 m^2 (750 ft^2), or about the area of a tennis court.

When viewed intact, the lungs appear lobed and roughly triangular in shape, with a broad base. Two airtight, moist baglike membranes, the **pleurae**, enclose the lungs. The inner pleura is attached firmly to the spongy surface of the lung. The outer pleura forms the tough lining of the **pleural cavity**, which is the space between the lungs and chest wall. (*Pleurisy* is an inflammation of the pleura.)

The pleural cavity is bounded at its lower portion by that dome-shaped muscular shelf, the **diaphragm**. The diaphragm, a mammalian characteristic, separates the body cavity into its thoracic and abdominal portions.

The Breathing Movements

Breathing, or **ventilation**, in humans involves both the diaphragm and muscles of the rib cage, the **intercostal muscles** (Figure 40.14). In the relaxed condition, the diaphragm rises into a dome shape and protrudes into the pleural cavity. Inspiration is accomplished by contracting (lowering and flattening) the diaphragm and contracting the external intercostals, which raises the rib cage. This in turn increases the volume of the airtight pleural cavity, creating a partial vacuum. It is the partial vacuum that causes the lungs to fill with air. The weight of the atmosphere itself forces air down the trachea to inflate the lungs. Were it not for the lungs, the air would simply replace the vacuum in the thoracic cavity—the lungs just happen to be in the way.

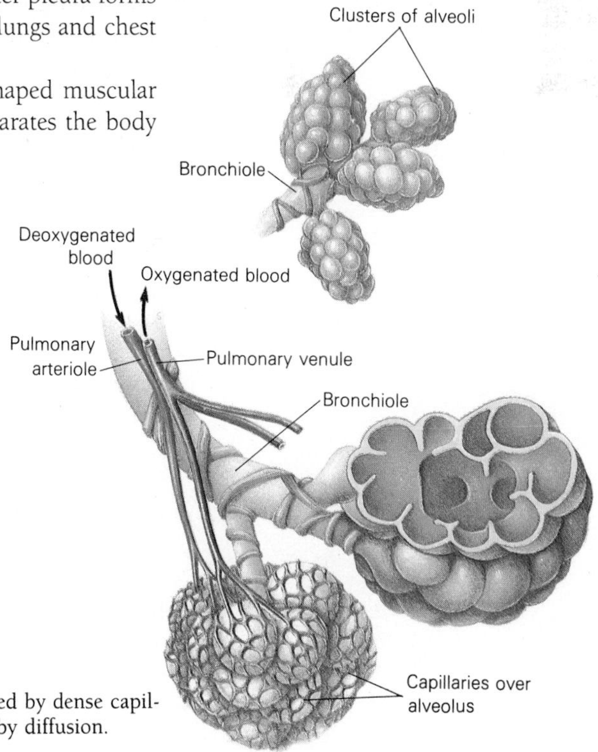

FIGURE 40.13
CIRCULATION IN THE ALVEOLI
Alveoli are blind, thin-walled sacs whose surfaces are covered by dense capillary beds. The thin walls permit a ready exchange of gases by diffusion.

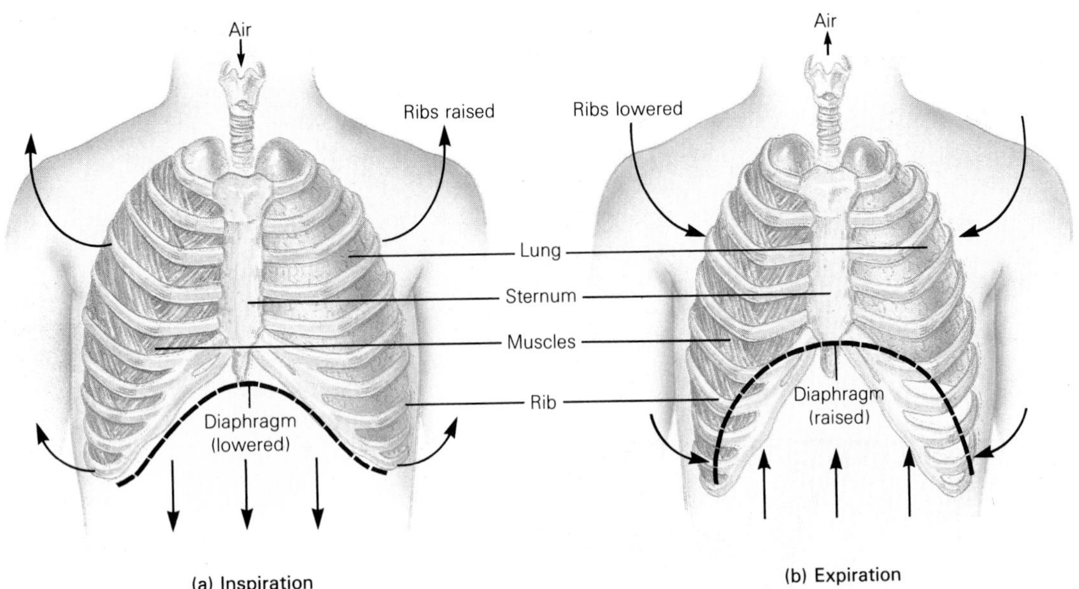

FIGURE 40.14
BREATHING MOVEMENTS
(a) Inspiration occurs when contraction of the intercostal rib muscles elevates the rib cage and contraction of the diaphragm causes it to flatten somewhat. Both actions expand the chest cavity, thereby decreasing air pressure within. As a result, air rushes down the trachea and inflates the lungs. **(b)** In expiration, the intercostal (rib) muscles and diaphragm relax, restoring the previous air pressure to the chest cavity. This compresses the lungs, forcing air out into the trachea.

Expiration, when passive (not forced), is produced by the relaxation of the rib cage and diaphragm, actions that decrease the pleural cavity volume. The elastic lungs resume their former state, and air is forced out through the respiratory passages. Forced expiration involves the contraction of internal intercostal and abdominal muscles, which further exhaust the lungs.

While we are at rest, about half a liter of air is moved in and out with each breath we take. But when we fill our lungs to their greatest capacity, about four liters are moved in and out. This maximal measurement, the **vital capacity**, depends mainly on body size, and varies considerably among individuals and between the sexes. In a well-trained male athlete, the vital capacity can exceed six liters. But no matter how hard one tries to expel all of the air in the lungs, about 1.5 liters always remains. This **residual air** is highly significant since, as we will see, it contains carbon dioxide that is essential in maintaining an adequate respiratory rate. Even in the superefficient bird lung, the main bronchus (trachea), by virtue of its great length, helps retain a critical amount of carbon dioxide in the respiratory system.

The Exchange of Gases

Partial Pressure To understand some of the basic aspects of the transport of oxygen and carbon dioxide, it is helpful to know some things about the behavior and characteristics of gases in general. As you know, our atmosphere is composed of a mixture of gases—primarily nitrogen and oxygen, with a much smaller amount of carbon dioxide. Together, the gases of the atmosphere exert a pressure—the total atmospheric pressure. The pressure exerted by any one gas in the mixture is logically called its **partial pressure**, designated as P_g (the letter g represents any specific gas). At sea level, and under what chemists call **standard conditions**, the total atmospheric pressure is known to be 760 mm Hg (mercury). For instance, air is about 21% O_2; therefore, the partial pressure of O_2 at sea level would be 21% of 760, or 160 mm Hg (P_{O_2} = 160). Carbon dioxide gas accounts for only about 0.04% of the atmosphere; thus it exerts a partial pressure of 0.3 mm Hg (P_{CO_2} = 0.3) (.04% of 760). Most of the remaining total atmospheric pressure is due to nitrogen gas. Since atmospheric pressure decreases with altitude, so do the partial pressures of the atmospheric gases. On a mountaintop, the air is still 21% O_2 and 0.04% CO_2, but since the total atmospheric pressure is much less, so, therefore, are the partial pressures of the two gases. (At 14,000 feet, the total atmospheric pressure may be only 450 mm Hg, while the P_{O_2} would be 95 and the P_{CO_2} only 0.18.) For this reason, we must breathe faster and deeper up there.

The concept of partial pressure is important to our understanding of gas exchange because diffusion of dissolved gases always occurs from regions of higher partial pressure to regions of lower partial pressure. For instance, suppose we placed an open container of a fluid that was rich in CO_2 and poor in O_2 (as measured by the partial pressures of these two gases) inside a closed container with a mixture of gases that was rich in O_2 and poor in CO_2. In due time the fluid and the gas would equilibrate, so that the partial pressure of both gases would be the same in both the fluid and the air. The actual amount, in contrast to the immediate availability, of O_2 and CO_2 in each part of the system would depend on the solubility of each gas in the fluid, which in turn would depend on such factors as temperature and pH. Partial pressure differences are important in setting up diffusion gradients, but as we will see, evolution has added a few twists to the simple exchange of gases.

Exchange in the Alveoli We breathe in air rich in O_2 and breathe out air rich in CO_2. The exchange takes place on the moist inner surfaces of the alveoli, primarily through simple diffusion (Figure 40.15). The blood that enters the lung from the heart has previously been routed through the body tissues, where mitochondrial respiration has depleted the oxygen supply, creating a very low partial pressure of oxygen (P_{O_2} = about 40). At the same time, metabolic activity has increased the partial pressure of CO_2 in the body tissues. The blood has tended to equilibrate with the body tissues. Thus, the blood now carries more carbon dioxide. A typical partial pressure for carbon dioxide in blood entering the alveoli would be about 45.

So, compared to partial pressures in the atmosphere, blood entering the lungs has a low partial pressure of O_2 and a high partial pressure of CO_2. Conversely, in the alveoli, the partial pressure of oxygen is high—about 100 mm Hg—and the partial pressure of carbon dioxide is about 40 mm Hg. (In case the greater than expected partial pressure of carbon dioxide bothers you, recall that alveolar air is not completely replaced during

FIGURE 40.15
PARTIAL PRESSURES AND GAS EXCHANGE
(a) Compare the partial pressures of oxygen and carbon dioxide in the blood entering and leaving the alveolar capillary bed, and note the partial pressures in the air space within the alveolus. The differences tell us that the two gases have moved down their concentration gradients as blood flows through the capillary bed.
(b) An enlarged view of a capillary segment shows the concentration gradient in the form of color variation, and gives you an idea of the relative size of the red blood cells and capillaries.

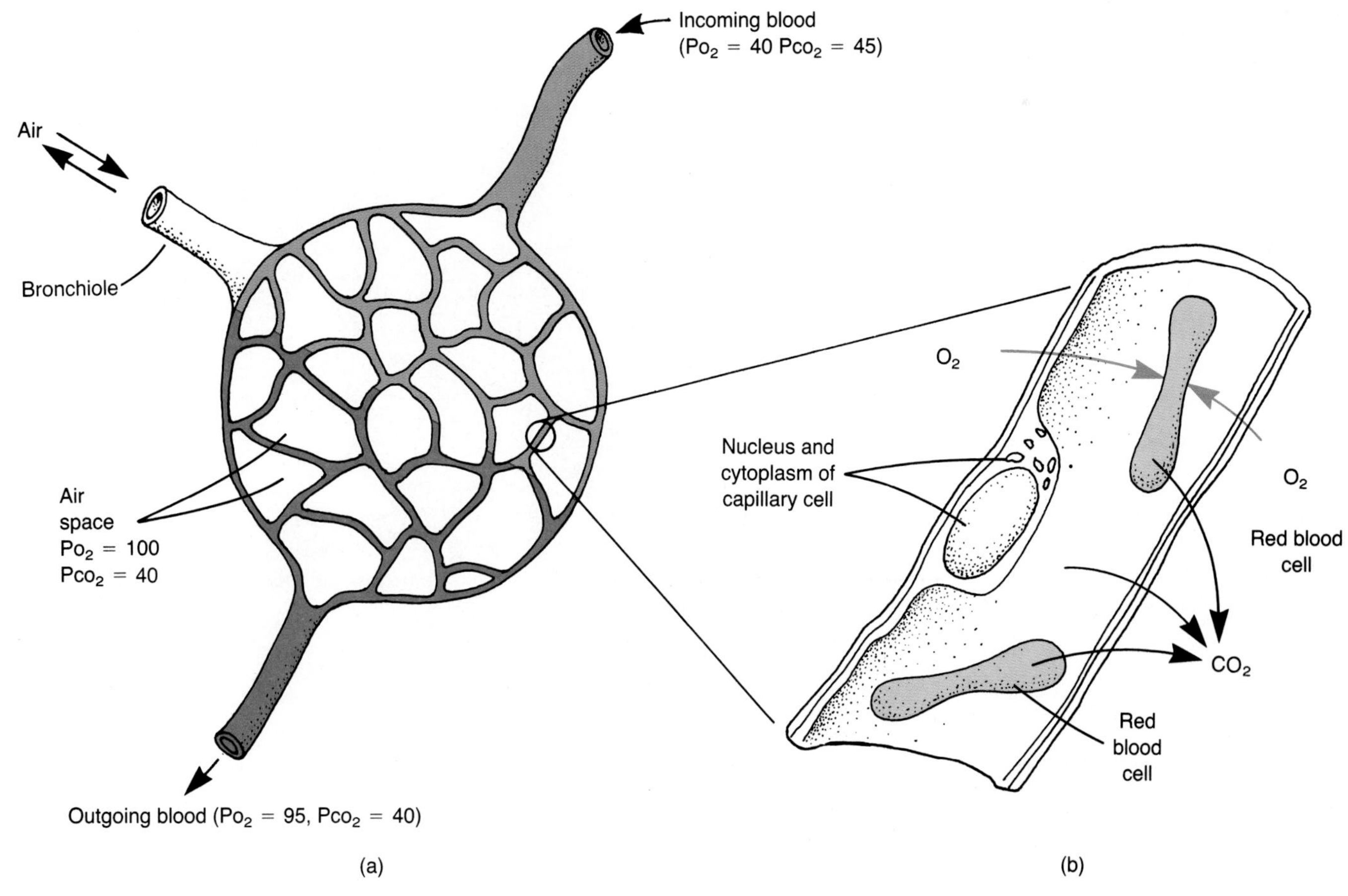

inhalation, but is a mixture of old and new air.) In the brief time that the blood and air are in near contact on opposite sides of the thin alveolar membrane, nearly complete equilibrium takes place. That is, the partial pressures of O_2 and CO_2 in the blood and air become almost equal as molecules of O_2 diffuse in and molecules of CO_2 diffuse out. Then the equilibrated air is exhaled, and fresh air is inhaled while the blood flows continuously through the lung. Blood leaving the lung is relatively high in oxygen, with a partial pressure of about 95 mm Hg. The corresponding partial pressure of carbon dioxide is about 40 mm Hg.

The partial pressure of oxygen in metabolically active tissue shifts dramatically as oxygen is consumed. There its partial pressure may fall to 25. The partial pressure of carbon dioxide, as you would expect, increases, perhaps reaching 46.

Thus, the respiratory gases diffuse down their pressure or concentration gradient, and the organism takes advantage of this "free" means of exchange to keep its oxygen and carbon dioxide at optimal levels. For some organisms, this is about all there is to respiration, but for many, including the vertebrates, there is much more. Passive diffusion is quite important, but it is only a part of the total gas exchange process. Other aspects of the process are directed by complex biochemical mechanisms. Let's consider the process in the human respiratory system.

Oxygen Transport

Oxygen is transported in the red blood cells which are literally packed with hemoglobin. Because of hemoglobin's great affinity for oxygen, it can carry 60 times as much as can an equal weight of water alone. Thus, if we didn't have a specific oxygen carrier, our circulation would have to be 60 times as fast, or our activity would have to be greatly curtailed.

So what precisely is this magic molecule called hemoglobin and how does it work? First, it consists of four polypeptide chains and four heme groups (Figure 40.16), one heme group for each polypeptide chain. Each heme group contains an iron atom that can bind with one molecule of oxygen (O_2). (Thus, each hemoglobin molecule can hold

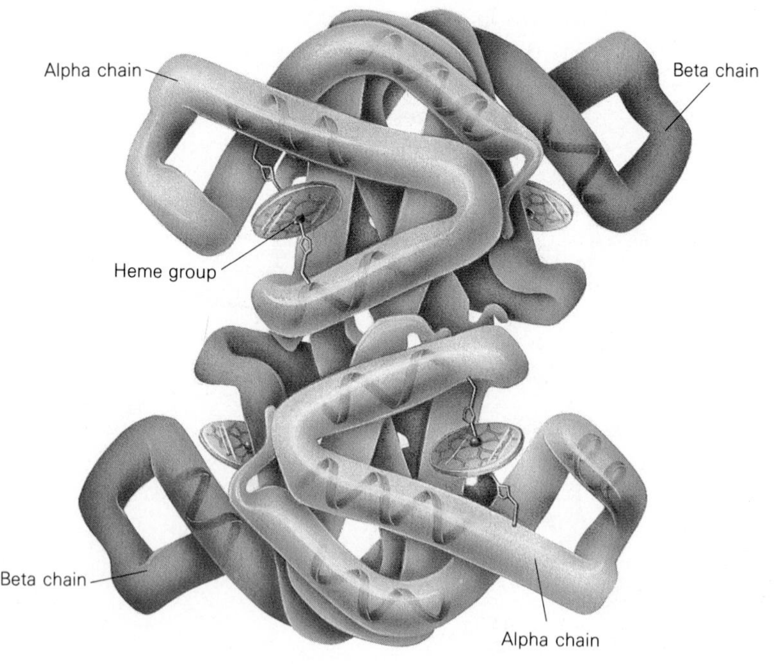

FIGURE 40.16
HEMOGLOBIN
Each of the four polypeptides of hemoglobin bears a heme group, and each heme can bind to one oxygen molecule (O_2).

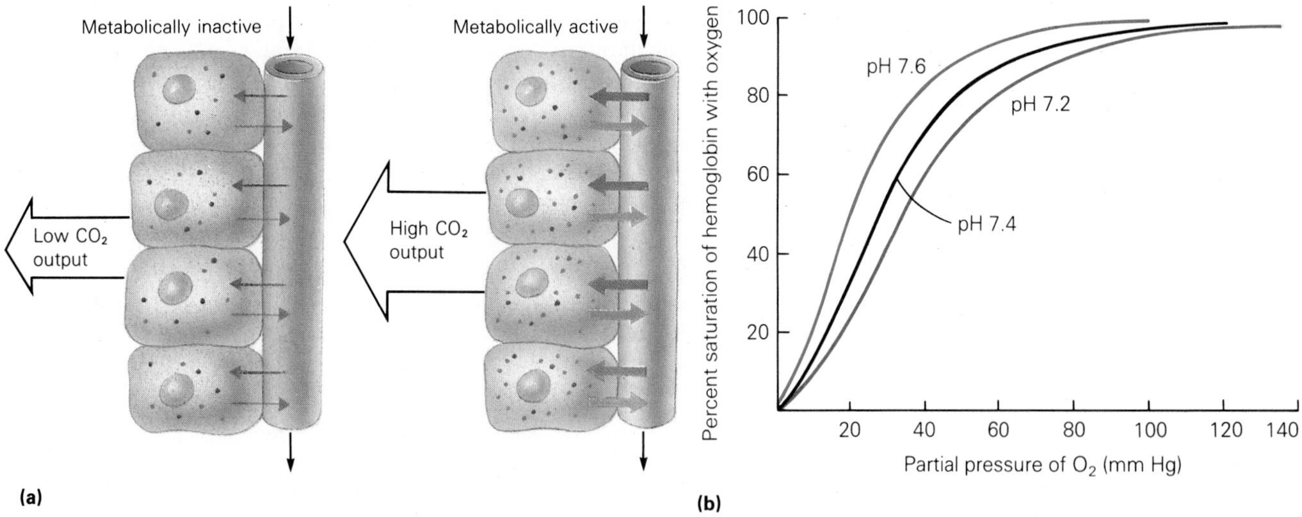

FIGURE 40.17
THE BOHR EFFECT
In the Bohr effect, hemoglobin surrenders its oxygen load more readily in the presence of increasing amounts of carbon dioxide. **(a)** This chemical behavior means that metabolically inactive cells will receive less oxygen than metabolically active ones, regardless of the oxygen concentration gradient. **(b)** The Bohr effect is readily seen in graphs where the percent saturation of hemoglobin with oxygen is plotted against oxygen partial pressure at varying pH levels. As blood CO_2 increases, pH decreases (becomes more acidic), and oxygen dissociates more readily from hemoglobin.

eight oxygen atoms when saturated.) The association and dissociation of O_2 and hemoglobin (Hb) is usually simplified as:

$$Hb + O_2 \rightleftharpoons Hb \cdot O_2$$

The left side of the formula shows **deoxyhemoglobin** and the right side, **oxyhemoglobin.**

One key to the efficiency of our respiratory system is the behavior of hemoglobin itself. Hemoglobin is highly specialized to associate and dissociate with oxygen under certain conditions. Whereas the affinity of an ordinary passive fluid (such as water) for O_2 is simply dictated by partial pressure, hemoglobin has the remarkable ability to "change its mind" about how much oxygen it will accept. To put it a bit more scientifically, hemoglobin can change its affinity for oxygen depending on such factors as pH, the partial pressure of CO_2, and temperature. A shifting affinity for oxygen allows the hemoglobin to give up much of its oxygen in its journey through the body and to become quickly saturated again in the lungs.

One more point about oxygenation of the blood: At sea level, and under normal conditions, blood passing over the alveoli is nearly saturated with O_2. That is, most of the hemoglobin molecules pick up their full allotment of four O_2 molecules. Increasing the O_2 concentration in the lungs has very little effect on the amount that crosses into the blood. There is a considerable safety margin in this relationship because it also works the other way: reducing the oxygen concentration doesn't easily change the amount of oxygen reaching the blood either. At half the partial pressure of O_2 at sea level, the blood will still be 80% saturated. This is why humans can live comfortably at varying altitudes, including high on mountains where O_2 partial pressure is low.

Furthermore, at high altitudes, additional red blood cells are produced—a long-term adaptation. The rate of red cell production is controlled through a feedback system originating in the kidney. Certain cells there detect the oxygen level in passing blood, and when it is not up to a certain standard, the hormone **erythropoietin** is released. Its targets are blood-forming elements in red bone marrow, which respond by producing more red blood cells. Added red blood cells mean an increase in blood oxygen, and the kidney cells, no longer stimulated, slow the erythropoietin release.

Carbon Dioxide Transport

Now let's look at a related phenomenon: carbon dioxide transport. It turns out that where partial pressure of CO_2 is high, hemoglobin has a lower affinity for O_2. When the CO_2 level is low, hemoglobin holds its O_2 more tightly. This peculiar influence by carbon dioxide, known as the **Bohr effect** (Figure 40.17), clearly has adaptive significance. The presence of carbon dioxide indicates that oxidative metabolism is occurring, increasing the oxygen demand. The most metabolically active regions produce the greatest quantities of CO_2, which in turn hastens the release of O_2 from the passing bloodstream. Conversely, oxygen-rich blood passing less metabolically active tissue will release less of its oxygen.

The transport of carbon dioxide in the blood is far more complex than that of oxygen. About 8% of the CO_2 to be transported simply goes into solution in the water of blood plasma. The remainder enters the red cells, where it follows one of two paths. Some combine in a loose association directly with the hemoglobin. Instead of combining with the heme groups as O_2 does, CO_2 reacts with amino side groups in other parts of the protein. The combination of carbon dioxide and hemoglobin is called **carbaminohemoglobin:**

$$\text{Hb} + CO_2 \rightleftharpoons \text{Carbaminohemoglobin}$$

The remaining carbon dioxide reacts with water in the red cells, forming carbonic acid (H_2CO_3). Carbonic acid in turn dissociates into hydrogen ions (H^+) and bicarbonate ions (HCO_3^-). This reaction can also occur in the plasma, but only very slowly. Red cells, however, contain the enzyme **carbonic anhydrase**, which not only speeds up the formation of carbonic acid but can also rapidly convert carbonic acid back to carbon dioxide and water. This is quite important, since carbon dioxide must be reformed quickly if it is to leave the body during the transit of the blood through the alveoli of the lungs.

When carbonic acid dissociates in the red cells, the hydrogen ions are buffered (pH changes resisted) by the protein hemoglobin itself. The bicarbonate ions diffuse out into the plasma, where they are balanced by sodium ions, forming sodium bicarbonate ($NaHCO_3$, or more precisely, the ionic forms, $Na^+ + HCO_3^-$). In addition to providing a means of transporting carbon dioxide, sodium bicarbonate forms an important part of the body's **acid-base buffering system**—that is, it helps neutralize any acids or bases that might form in the blood, keeping the pH constant—near neutral. (The sodium and bicarbonate in salt form, by the way, is identical to commercial baking soda and to the main ingredient in many stomach acid neutralizers.) The reactions so far can be summarized as

$$1)\ \text{Hb} + CO_2 \rightleftharpoons \text{Carbaminohemoglobin}$$

$$2)\ CO_2 + H_2O \underset{\text{carbonic anhydrase}}{\rightleftharpoons} H^+ + HCO_3^-$$

$$3)\ Na^+ + HCO_3^- \rightleftharpoons NaHCO_3$$

As we see, the reactions are all reversible, and in each, it is the quantity of carbon dioxide present that dictates the direction of the reaction (typical of enzymatic reactions; see Chapter 6).

So, in active tissues, where carbon dioxide levels are high, the direction of the reactions is toward carbaminohemoglobin and toward the formation of hydrogen and carbonate ions and sodium bicarbonate. But in the alveolar capillaries, any free carbon dioxide diffuses out of the blood, so its concentration is low. This prompts a speedy cascade of reversing chemical events: (1) The carbaminohemoglobin releases its carbon dioxide; (2) the bicarbonate of sodium bicarbonate in the plasma reenters the red cells; (3) it joins hydrogen ions to form carbonic acid; and (4) with a boost from carbonic anhydrase there is the rapid conversion back to carbon dioxide and water. The carbon dioxide then diffuses out of the capillary and into the alveolus for expiration out of the body.

The Control of Respiration

The perpetual, incessant changes that are the hallmark of living things are usually only minor adjustments—fine tuning—in response to shifting conditions within and without. For example, the need for O_2 and the production of CO_2 are always changing in every part of the body—now a little more here, now a little less there. Considering all such adjustments, however, the body's task is enormous, and its ability to respond is admirable, even dazzling. How can it handle such a complex chore as regulating oxygen and carbon dioxide levels in literally millions of places at once?

First, we should be aware that both the respiratory and circulatory systems must respond in a coordinated way to changing oxygen levels. After all, increasing the breathing rate would be of little value unless one also increased the rate of blood flow. Second, it is important to know that respiratory control is exceedingly complex, involving many types of sensors and a variety of complex interactions. We will explore some of this complexity with a look at how breathing rate is controlled.

Neural Control We can vary the rate and depth of our breathing, but only up to a point. If your little brother holds his breath to get his way, don't worry. He may begin to lose his rosy complexion, but the ruse won't work. No matter how hard he tries, as the CO_2 level in his blood rises, his autonomic nervous system will take over and he will be forced to breathe.

As you see, breathing is under both voluntary and involuntary neural control. In addition, a number of chemical sensors in the body monitor the blood's carbon dioxide and oxygen levels and its hydrogen ion content (pH), sending their information to the brain. Thus respiratory control has a chemical as well as a neural component. Both are only partially understood.

It is known that breathing movements are coordinated by voluntary centers in the cortex (those that permit your little brother to make the threats about holding his breath) and by involuntary centers in the pons and medulla (those that defeat his strategy). While the anatomy of the involuntary centers is far from clear, there appear to be **inspiratory centers** and **expiratory centers**, both located in the medulla. Activity by the expiratory center is restricted to periods of strenuous breathing.

During quiet breathing, the inspiratory center is self-excited, creating impulses on its own. Its impulses pass through the spinal cord to the diaphragm and intercostal (rib) muscles, which contract and bring on inspiration. After about two seconds of activity, the inspiratory center spontaneously rests for three seconds when expiration (a passive process) occurs.

During periods of more strenuous activity, the breathing rate increases sharply. Whereas the inspiratory center still brings on inspiration, now at a faster rate, other factors come into play. For instance, stretch receptors in the lungs—activated by prolonged inspiration—fire inhibiting signals back to the inspiratory center, thus permitting expiration and preventing overinflation. In addition, the expiratory center now comes into play. It sends its messages to different muscles, some in the chest and some in the abdomen, whose contraction adds force to passive expiration, thus increasing the expulsion of air from the lungs.

Chemical Control Chemical control is based on input from several chemoreceptors, two in major arteries and one in the brain (Figure 40.18). The arterial **carotid bodies** and **aortic bodies** are aroused when blood CO_2 levels rise and when the hydrogen ion level (acidity) increases. They also sense decreases in oxygen, but, surprisingly, to a much lesser extent. More sensitive chemoreceptors in the fluid-filled spaces of the medulla oblongata in the brain monitor hydrogen ion levels in the cerebrospinal fluid. When neurons in any of these regions become active, their impulses are relayed to the respiratory centers in the pons and medulla, and the rate and depth of breathing increase. Such breathing, of course, decreases the level of CO_2 and hydrogen ions in the blood and increases the level of oxygen. The chemoreceptors, sensing these changes, slow their output of neural impulses (another example of negative feedback at work). Thus, a balance is set that retains the blood in an optimum chemical state. Interestingly, during vigorous exercise this straightforward chemical regulatory mechanism is overridden by new input from the cerebrum and input from other kinds of sensors, possibly proprioceptors, located in the joints and muscles.

You may have noticed that our discussions of gas exchange have often led to discussions of the circulatory and nervous systems. It is quite difficult, if not impossible,

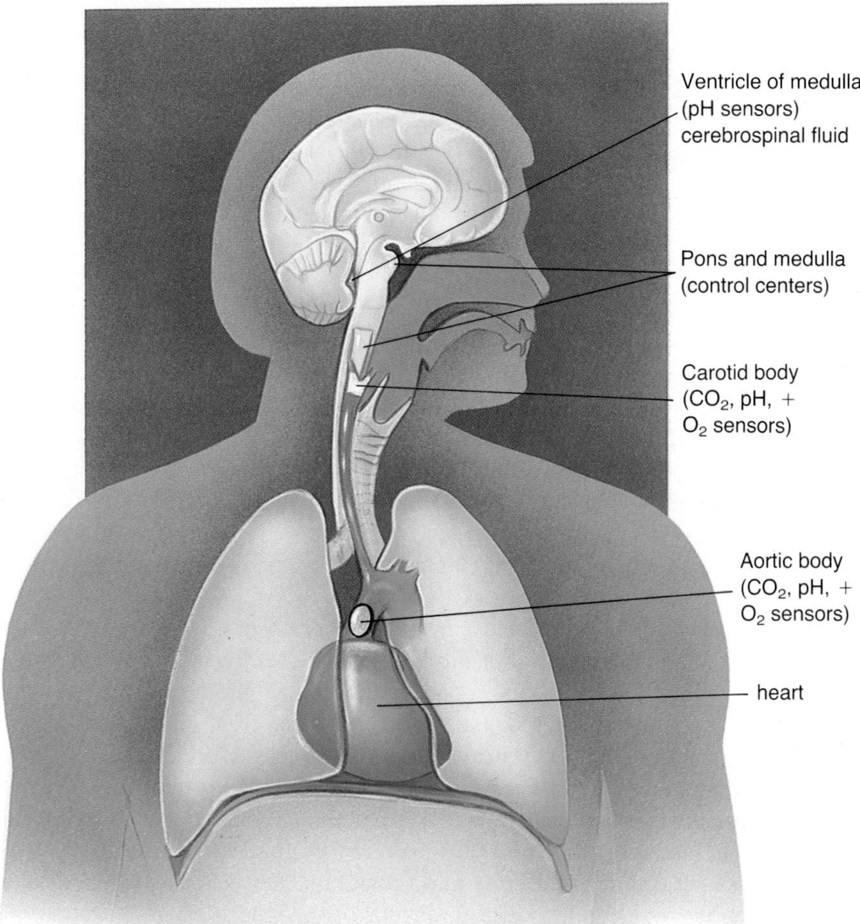

FIGURE 40.18
BREATHING CONTROL CENTERS
Chemoreceptors in the aorta and carotid arteries monitor carbon dioxide, hydrogen ion (pH), and oxygen levels, communicating their information to the respiratory control centers in the medulla and pons. Additional receptors monitor the hydrogen ion content (pH) of cerebrospinal fluid in the medulla's spaces.

to discuss one system to the exclusion of all others. In large part this is because animal systems function in an integrated and interdependent manner, and no system functions alone. As we now focus specifically on the immune system, we will again weave in our findings from other discussions as we continue to try to understand the basic fabric of this thing called life.

APPLICATION OF IDEAS

1. All aquatic mammals, birds, and reptiles are air breathers. No matter how well they have adapted to the water in other ways, they must come to the surface to breathe. What does this clearly indicate about their ancestors? Given enough time, is it probable that these animals would evolve a gill-like device? Explain your conclusion. What might some disadvantages be if gills were present in an endotherm?

KEY IDEAS

Since oxygen can disrupt many life processes and reactions, organisms have had to adapt to the rise of oxygen as an atmospheric gas. Its presence as ozone in the upper atmosphere protects organisms from ultraviolet radiation. Oxygen is essential in aerobic life as an electron acceptor in the cell respiratory process. Oxygen and carbon dioxide are exchanged in the **respiratory system** and transported in the circulatory system.

GAS EXCHANGE SURFACES

Gas exchange in animals requires extensive, thin, moist membranes, such as those seen in the gill and lung. Such membranes are called the **respiratory interface.**

The Simple Body Interface

1. In complex sponges, gas exchange across the body interface works well because of the extensive canal system. It also works in cnidarians because of the extremely thin (two-cell layered) body wall. Denser flatworms increase the respiratory interface through the highly branched gastrovascular cavity, but this severely restricts their size. The surface of the echinoderm is increased by the presence of **dermal brancheae.**
2. Earthworms remain in moist soil most of the time, and the salamander is nocturnal, so they can safely exchange gases through the skin interface. Further, both have an efficient closed circulatory system and blood containing hemoglobin.

Expanding the Interface: Tracheae

Most terrestrial animals have internalized respiratory interfaces that help resist desiccation. Insects have a **tracheal system,** which includes body openings called **spiracles,** highly branched tubes called **tracheae** and **tracheoles,** and, in some, thin-walled **air sacs.** Gas exchange occurs in the fine branches, which are often fluid filled. In aquatic insects, the tracheae terminate in external **tracheal gills.**

Complex Interfaces: Gills

1. The **gill** is similar in invertebrates and vertebrates, with the respiratory interface intimately associated with the circulatory system. Feathery structures contain extensive capillary beds where blood and surrounding water are separated by a single membrane.
2. Through the use of cilia, clams draw water in through an incurrent siphon. Water crosses the vascular gill exchange surface and exits through an excurrent siphon. Oxygen transport is aided by the copper-containing protein **hemocyanin.**
3. The lobster gills lie in chambers below a protective **carapace.** Water is drawn across the gills by an appendage called a **bailer.** The circulatory system is open in most of the body, but blood vessels occur in the gills.
4. The fish gill includes supporting **gill arches** with rows of **gill filaments,** each containing many capillary beds in platelike **lamellae.** Blood enters the filament from **afferent vessels,** crosses the lamellar capillaries and exits the filament to **efferent vessels.** During its transit, CO_2 is exchanged for O_2 across the thin walls. Water flow across the lamellae opposes blood flow, setting up an efficient **countercurrent exchange.**
5. Water holds less oxygen than air, and diffusion is slower; thus aquatic animals must expend energy to move a large volume of water across the gills (or to move themselves through water).

Complex Interfaces: Lungs

1. The earliest fishes gulped air at the surface, using a vascularized **pharynx** for exchange. This system, an example of **preadaptation,** underwent modification into the **lung** in the first terrestrial vertebrates. In addition, **internal nares,** openings between the nasal cavity and mouth, evolved in the early terrestrial animals. In mammals, the **palate** came to separate the pharynx into mouth and nasal cavities. In fishes, the primitive lung evolved into the swim bladder, a flotation device.
2. The vertebrate lung is a highly branched inpocketing that, in many species, contain numerous air sacs, intimately associated with capillaries, and inflated and deflated by a bellows action.
3. The amphibian lung is a simple paired saclike affair, inflated and deflated by a pumping action in the muscular mouth floor. Check valves in the nostrils and **glottis** help coordinate the filling and emptying cycles. Amphibians also use the skin as an exchange surface.
4. Reptile lungs are spongy and complex and, aside from the vascular cloaca of aquatic turtles, represent the only respiratory interface. Breathing occurs through a bellowslike action of the entire body wall.
5. In birds, air passes *through* the lung rather than in and out. Air moving through the trachea bypasses the lung to enter posterior air sacs and from there enters the lung. The passage of air through the lung is crosscurrent to the flow of blood through lung capillaries. The bird respiratory system represents a specific adaptation to flight at higher altitudes, where oxygen is limited.

THE HUMAN RESPIRATORY SYSTEM

1. Major structures of the human respiratory system include the mouth, nasal passages, pharynx, **larynx, trachea, bronchi, bronchioles, alveoli** and lungs. The palate separates the mouth from the nasal passages, which moisten, warm, and filter air entering the lungs. **Goblet cells** secrete dust-trapping mucus in the nasal passages, and cilia move the mucus film toward the pharynx for swallowing. Warming of incoming air is aided by a countercurrent blood flow.
2. Incoming air moves through the larynx, into the trachea and **primary bronchi,** through branching bronchioles, and finally into the alveoli. The larynx contains the **voice box** and **vocal cords.**
3. Much of the **respiratory tree** contains a mucus-secreting and ciliated epithelium that traps dust particles and sweeps them upward, out of the system.
4. The lungs lie in the **pleural cavity,** enclosed by double membranes, the **pleurae.** The dome-shaped **diaphragm** forms the lower part of the cavity.

The Breathing Movements

1. In inspiration, **intercostal muscles** contract, elevating the rib cage, and the diaphragm flattens, creating a partial vacuum in the pleural cavity. The lungs

passively expand in response to the inward movement of air. In exhalation the rib cage drops, the diaphragm resumes its relaxed dome shape, and the elastic lungs resume their former shape, forcing air out.

2. While the minimal (resting) exchange of air is about 0.5 liters, the maximum, or **vital capacity**, is about 4.0 to 6.0 liters. Some **residual air** always remains in the lungs.

The Exchange of Gases

1. While all gases of the atmosphere contribute to total atmospheric pressure, each gas exerts a **partial pressure**—P_g. Under **standard conditions** (total pressure 760 mm Hg) P_{O_2} = 160 mm Hg, and P_{CO_2} = 0.3 mm Hg. P_g decreases with altitude.
2. The diffusion of a gas follows its partial pressure gradient.
   a. In blood arriving at the alveolar capillaries, P_{O_2} = 40 mm Hg, while P_{CO_2} = 45 mm Hg.
   b. In alveoli air, the P_{O_2} = 100 mm Hg and P_{CO_2} = 40 mm Hg. Thus CO_2 leaves the blood and O_2 enters the blood.
   c. In blood leaving the alveoli, P_{O_2} = 95 mm Hg and P_{CO_2} = 40 mm Hg.
   d. In very metabolically active tissues, these values can change to 25 and 46 mm Hg respectively.

Oxygen Transport

1. Hemoglobin can carry 60 times as much oxygen as water. Its four polypeptide chains contain four heme groups, each of which can reversibly associate with one O_2. Thus deoxyhemoglobin forms **oxyhemoglobin.**
2. The association and subsequent dissociation of hemoglobin and oxygen is complex, depending on pH, P_{CO_2} and temperature.
3. Hemoglobin has a built-in safety factor. At sea level the blood leaving the lungs is nearly saturated with oxygen, and increasing the P_{O_2} has little effect. But at half the sea-level P_{O_2} (high altitude), the blood will be 80% saturated. In addition, when a consistently low P_{O_2} occurs in the blood, the hormone **erythropoietin** is released, stimulating new red cell formation, a slower, long-term adjustment.

Carbon Dioxide Transport

1. Because of the **Bohr effect**, the affinity of hemoglobin for oxygen is inversely proportional to the partial pressure of carbon dioxide. Thus, in blood passing metabolically active tissues with a high CO_2 concentration, substantially more O_2 will dissociate.
2. In the transport of CO_2,
   a. about 8% of the CO_2 goes into solution in plasma;
   b. the remainder enters the red cells. Some associates with amino acids in hemoglobin, forming **carbaminohemoglobin,** and the rest forms carbonic acid with water in the red cell;
   c. the speed of dissociation reactions depends on the action of the enzyme **carbonic anhydrase;**
   d. within the red cell, carbonic acid dissociates into bicarbonate and hydrogen ions (acid);
   e. some hydrogen ions are buffered by hemoglobin while the carbonate ions are buffered by hemoglobin while the carbonate ions diffuse out to the plasma, where they are buffered by sodium ions, forming sodium bicarbonate—a part of the body's **acid-base buffering system;** and
   f. all reactions are reversible, according to the mass action law.
3. The reactions of carbon dioxide reverse in the lungs, where the P_{CO_2} in the alveolus is low. The restoration of ions into CO_2 is greatly speeded by the action of carbonic anhydrase.

The Control of Respiration

Breathing involves neural and chemical factors. In quiet breathing the neural inspiratory center in the medulla instigates rhythmic inhaling. Exhaling is passive. With strenuous activity, this action speeds up. Stretch receptors in the lung also act, preventing overinflation by inhibiting the inspiratory center. The neural expiratory center also begins to act, producing more forceful expiration. Breathing rate is influenced chemically by input from several chemical sensors that become active when carbon dioxide and (to a lesser extent) oxygen levels rise and pH falls (acidity increases).

REVIEW QUESTIONS

1. Explain why the circulatory and respiratory systems in complex animals are closely integrated. (959)
2. What are two basic characteristics required of a gas exchange structure? What additional restriction is imposed by the terrestrial environment? (960)
3. List three invertebrates that use the simple body interface for gas exchange, and explain how each satisfies the basic requirements. (961)
4. Explain in specific terms how the flatworm's respiratory requirements restrict its size. (961)
5. The surface area of the earthworm and lungless salamander alone are not sufficient to provide an adequate gas exchange for their dense bodies. What else is required? Explain in detail. (961–962)
6. Describe the insect tracheal system. What is its relationship to the circulatory system? (962)
7. Describe the manner in which the clam gill functions. Name its respiratory pigment. (962–963)
8. Describe the structure of the fish gill using the following terms: *gill arch, gill filament, lamellae, afferent* and *efferent blood vessels,* and *capillary beds.* (963–964)

9. The passage of blood and water in the fish gill is said to be a countercurrent flow. What does this mean, and how does it affect gas exchange efficiency? (964)
10. Discuss the problem of availability of oxygen in the aquatic environment and describe solutions in the bony fishes. (964)
11. Summarize the main events in the evolution of the vertebrate lung. (965)
12. Compare the amphibian and reptilian lung. How does the amphibian supplement its lung surface area, and why haven't the reptiles taken advantage of this? (965–966)
13. Describe the four-stroke breathing mechanism in the frog. (965)
14. List several ways in which the bird lung differs from that of the reptile or mammal. (966–967)
15. Describe the movement of air through the bird respiratory system, beginning with air entering the trachea. (966–967)
16. What special requirements of birds does its extremely efficient respiratory system satisfy? (967)
17. List the structures through which air passes in the human respiratory system, beginning with the nostrils and ending with the alveoli. (967–968)
18. Describe the structure of the nasal passages and explain their role in respiration. (968)
19. Describe the structure of the alveoli, including their structural relationship to the circulatory system. (968–969)
20. Summarize the movements of the rib cage, diaphragm, and lungs in inspiration and expiration. What actually causes the lungs to fill? To empty? (969–970)
21. Carefully define partial pressure and state the partial pressures of oxygen, carbon dioxide, and nitrogen at sea level. How does this change at higher altitudes? (970–971)
22. What factors affect the movement of gases down a partial pressure gradient? (971)
23. State the partial pressures of oxygen and carbon dioxide in blood *entering* the alveoli, and the partial pressures of the two gases in the alveolar air. In what direction do the two gases move as a result? (971–972)
24. Write a formula representing the reactions between hemoglobin and oxygen. What four factors influence the rate and direction of the reactions? (973)
25. Describe the response of the body to a prolonged reduction in P_{O_2} in the arteries. (973)
26. Describe the Bohr effect and explain how it is adaptive. (973–974)
27. What happens to carbon dioxide when it enters the blood? What is the role of carbonic anhydrase? (974)
28. What is the function of sodium bicarbonate formation in the blood plasma? (974)
29. Explain how mass action affects the blood carbon dioxide chemistry in the lung. (974)
30. What two factors in the blood seem to have the most influence on respiratory activity? (974–975)
31. List three major regions where neural sensors measure conditions in the blood. (975)
32. Explain how the respiratory control center responds to increased pH, increased P_{CO_2}, and decreased pH. (975)

41

THE IMMUNE SYSTEM

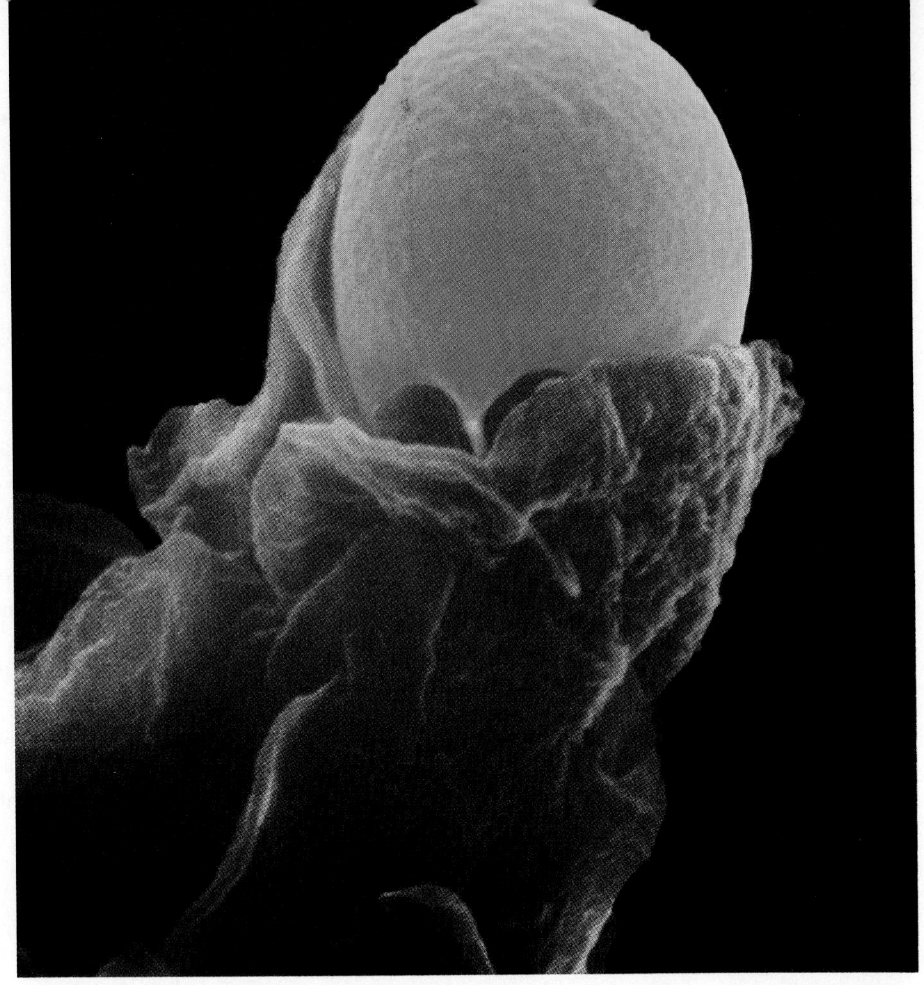

IN EVERY LIVING CELL THERE IS A VERITABLE storehouse of energy-rich nutrients. In this exploitative world, we can't expect such a valuable nutrient source to escape the attention of certain other living things, with their own energy needs. And so we find various life forms invading others, each trying to avail itself of what the other holds. The invasion can be by large forms, as when a tiger avails itself of the energy stored in an antelope. Or it can be by smaller forms, such as when a virus penetrates a single cell. Here we will focus on the various ways living things defend themselves against the smaller invaders. Among them are viruses, bacteria, protozoans, fungi, and a number of tiny animals. Such invaders can do harm in a bewildering number of ways. Some viruses rob organisms of vital cellular components, literally wrecking the cell when they reproduce. Other viruses destroy gene functions by inserting their DNA randomly into the host's chromosomes. Bacteria, protists, and fungi secrete enzymes that break down cell membranes and cell cytoplasm. Further, their toxic secretions often bring about a painful death for some hapless animal, or at best, leave the weakened host open to still other invaders. It seems a wonder that susceptible organisms survive at all, but most would-be hosts are not defenseless against such invasions. In fact, they are protected by a most remarkable **immune system.** The immune system consists primarily of a widespread array of tissues and cells along with certain organs whose primary function is to protect the body against damage from invading organisms and the harmful substances some produce.

Much of what is known about the immune system pertains to humans and a few other mammals, although some research has focused on other animal groups. Across the spectrum of animal life, certain common defense processes emerge. For example, in all cases the body defends itself in two major ways. First, it keeps most invaders *out* quite effectively (your cough and stuffy nose aside), and second, once invaded the body mounts a two-tiered counterattack. We should keep in mind, also, that the body does not recognize "invaders;" it recognizes "non-self." That is, it attacks anything that it deems foreign. We will look first at how the body keeps invaders out (Figure 41.1).

THE FIRST LINE OF DEFENSE

The first line of defense in our own bodies and those of many other animals is the **integument**, or body covering. In humans, the skin and mucous membranes cover body surfaces, presenting an effective barrier against most invasion. The skin has two interesting defenses. In a sense we might say that it is tough and poisonous. It is tough because it contains a protein called keratin that resists the potent enzymes of would-be invaders such as bacteria. Further, the fatty acids of the skin are toxic to certain bacteria, as are secretions from sweat and oil glands. Thus, as long as the skin is intact and unbroken, it is an effective first-line defense.

The body passages also have effective defenses. Remember that the vertebrate body is essentially a "tube within a tube," so that technically, anything in the gut, respiratory, and reproductive passages is still outside the body, and in the case of potentially dangerous organisms, must be kept there. So we find that mucus secretions of the respiratory system entrap countless microorganisms, holding them until they can be swept away by ciliary motion or are dealt with by other defense mechanisms. Most organisms that enter the stomach after being swallowed die in its strongly acidic environment. Further,

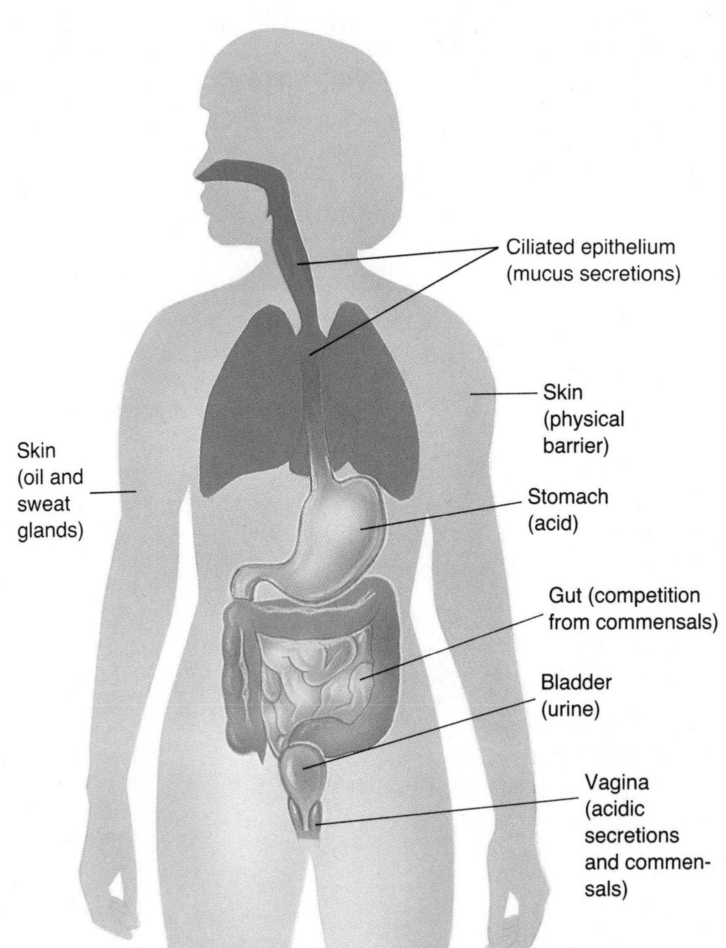

FIGURE 41.1
FIRST-LINE DEFENSES
To gain entry to the body, invaders must first penetrate the external and internal body coverings, both of which provide physical and chemical barriers.

anti-bacterial enzymes called **lysozymes**, found in tears and other body secretions, quickly dispatch some bacteria by cleaving chemical bonds in the cell wall. The urinary passages, another potential avenue of invasion, are naturally protected from bacteria by the flushing action of urine. Finally, we get a little help from our friends. Over the long eons of our evolution, a variety of bacterial, fungal, and even animal species have adapted to living harmlessly on certain of our surfaces. Whereas they are no threat to us, they do present a strong competitive barrier to parasites. In the large intestine, massive colonies of friendly colon bacteria, such as *E. coli,* simply use up most of the available nutrients, keeping invaders in check by essentially starving them out. In the vagina, populations of harmless bacteria feed on the glycogen secretions and secrete protective lactic acid that renders the vagina inhospitable to many other organisms. Such defenses, by the way, are often lost after prolonged antibiotic therapy that kills the useful bacteria. With the competition down, invaders may then establish themselves. So an antibiotic may solve one problem, only to create others.

SECONDARY DEFENSES: NONSPECIFIC AND SPECIFIC

Whereas natural selection has provided us with primary or first-line defenses, consisting of physical and chemical barriers to infection, it has been at work on invaders as well. After all, with the invaders' very energy source continually at risk, we can expect selection for successful variants to be quite strong. As a result, evolution continues to provide pathogens with highly effective countermeasures that permit them to penetrate our bodies, even across previously effective barriers. When the first lines of defense are penetrated, the second-line, or secondary, defenses come into play.

Secondary defenses are of two types, traditionally referred to as **nonspecific** and **specific defenses** (Figure 41.2). Nonspecific defenses tend to be direct and immediate. They involve both chemical or molecular agents and certain types of white blood cells, all of which are in a fully prepared state, needing only to be triggered into action. Specific defenses, however, are complex and require time for their preparation. They also involve chemical agents and cells, but both must be brought to an aroused and ready state. So, we see that both nonspecific and specific defenses have chemical and cellular aspects. Let's begin with the chemical defenses.

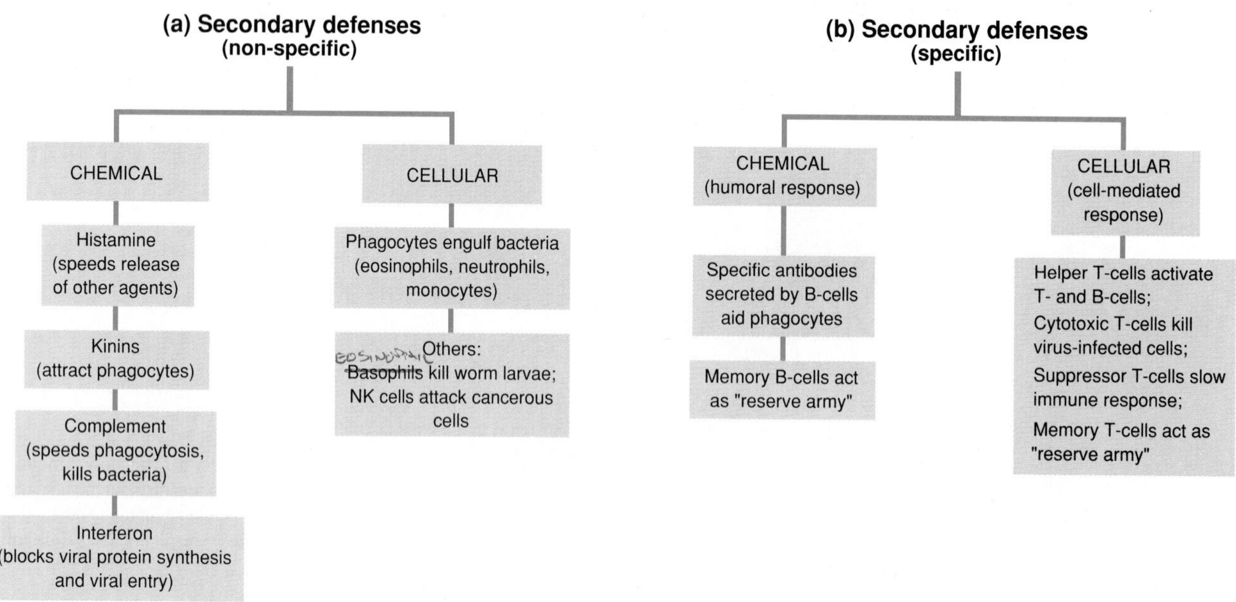

FIGURE 41.2
SECONDARY DEFENSES
Secondary defenses include nonspecific and specific responses to invasion, both of which have chemical and cellular aspects. The nonspecific defenses are in a constant ready state, whereas the specific ones require time for the immune reaction to be mounted.

Nonspecific Chemical Defenses

In the **nonspecific chemical defenses,** the invaded or injured cells themselves release chemicals that help destroy or impede the progress of the invader. There are four major kinds of such chemical defenses.

Histamine **Histamine** is familiar to us all because it brings on the redness and swelling associated with inflammation and infection. Histamine speeds the release of defensive cells and chemicals from the blood at the troubled site. In injuries, this would include blood-clotting agents. Histamine does this first through vasodilation, prompting the opening of arterioles at the site, thereby increasing blood flow. Next, histamine increases the permeability of capillaries in the area, permitting the rapid escape of other defensive agents. (It might have occurred to you that "**antihistamines**" reduce these reactions, especially in the nasal passages of hayfever sufferers and of those with that charming affliction called the common cold.)

Kinins Injured cells also release **kinins,** polypeptides that, like histamine, increase circulation and capillary permeability, thus adding to the inflammatory response. In addition, kinins attract certain phagocytic white cells to the injury or infection site. Kinins also affect nerve endings, causing pain and tenderness and perhaps discouraging further injury as we begin to favor the part until healing occurs.

Complement **Complement** is actually a battery of about 20 plasma proteins that act *directly,* by attacking invading bacteria, and *indirectly* by cooperating with other agents of the immune system. In their direct action the proteins work sequentially, each step magnifying the effects of the step before in a cascading manner. The complement coats the surface of a bacterium. The coating makes the cell far more susceptible to the action of phagocytes, because phagocytes have specific complement recognition sites on their membranes (Figure 41.3). These sites, known as **C3b receptors,** bind to the complement that has attached to the invader, thereby locking phagocyte and bacterial cells together in a process called **opsonization**. Such recognition sites also play major roles in other work of the immune system.

Even more dramatically, complement may destroy invaders directly by assuming ringlike formations that penetrate the lipid bilayer of the bacterial membrane, forming pores that allow water in and literally bursting the bacterial cell (Figure 41.3b).

Interferon **Interferon** stimulates cells of the body to resist attack by viruses. Researchers were put on the track of this mysterious substance (there are actually three types) when they noticed that when the body was under attack by one virus, it generally wasn't susceptible to other kinds of viruses. They eventually discovered that a cell under

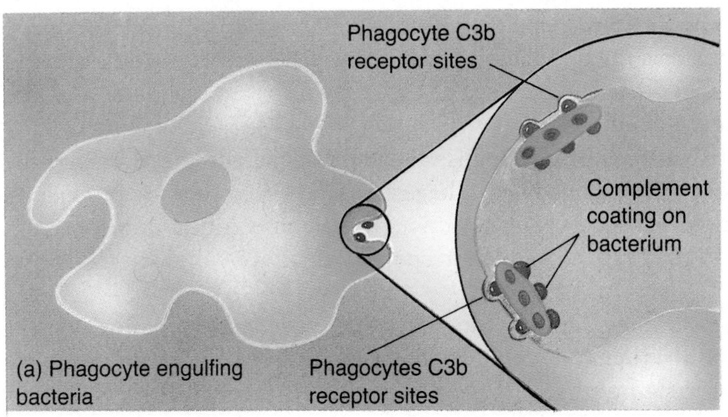

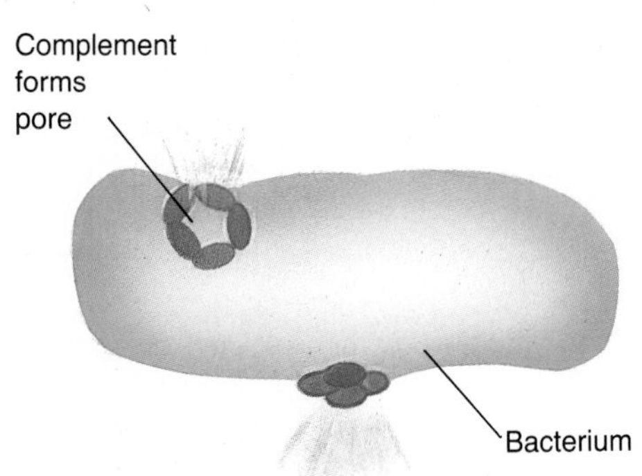

FIGURE 41.3
CHEMICAL DEFENSE: COMPLEMENT
When activated, the battery of proteins known as complement attacks invaders in two ways. **(a)** Complement binds to bacterial surfaces and then to receptor sites on phagocytes. This stimulates the phagocyte to engulf the invader. **(b)** Complement can kill some bacterial cells directly through a perforating effect that creates leaks in the cell.

attack releases a substance that attaches to surface receptors on other cells, making it more difficult for viruses to succeed.

This substance, interferon, apparently works by blocking the synthesis of viral coat protein and by stimulating both the inflammatory and immune responses. It can now be made in large quantities by recombinant DNA techniques. Researchers once saw this increased availability as a steppingstone in the fight against cancer, since interferon does block protein synthesis and hence cell growth. That promise has not yet been realized except in a few specific cases, such as in one rare type of leukemia.

Nonspecific Cellular Defenses

In **nonspecific cellular defenses**, the **leukocytes** (white blood cells) are mobilized to attack the invading agent. Leukocytes, like erythrocytes (red blood cells), are produced continually by stem cells in regions of red bone marrow (see Figure 39.19). Five kinds of leukocytes are involved in these immune responses. Three of these are **phagocytes**; that is, they engulf and digest invaders. They include **eosinophils**, **neutrophils**, and **monocytes** (Figure 41.4). A fourth leukocyte, the **basophil**, is important to the inflammatory response, and the fifth, the **lymphocyte**, includes one type that is involved in nonspecific defenses and several whose roles come under the category of specific cellular defenses.

The Phagocytes First we will consider the remarkable role of the phagocytes—the eosinophils, neutrophils, and monocytes. Phagocytes are commonly associated with connective tissues that help form linings such as those of the liver, kidney, lung, capillaries, spleen, and lymph nodes. Phagocytes are also present in the blood, and some are residents of the brain.

Phagocytes are drawn to the site of an infection by chemicals such as kinins and complement that are released from damaged cells, although in some instances the attracting substances are produced by invading bacteria. Within an hour of the release of such chemicals, great numbers of phagocytes squeeze through the capillary walls, enter the afflicted area, and begin their work engulfing bacterial cells and cellular debris. As usual, materials taken in through phagocytosis end up in vacuoles. Such vacuoles are quickly joined by lysosomes carrying powerful digestive enzymes, and the invader is dispatched. Many phagocytes perish in the battle, and their remains contribute to the familiar substance known as **pus.**

Eosinophils are weakly phagocytic cells that tend to congregate at sites where chemical defense agents have clumped invaders together. Once there, they devour the whole complex. Eosinophils also attack and kill parasitic worms, particularly the more vulnerable, soft-bodied larvae. They bind to the surfaces of the larvae, where they deposit numerous toxic granules they carry. Chemical substances in the granules then kill the larvae by disrupting their body walls.

Neutrophils are the expendable, frontline "foot soldiers." They are generally the first to congregate at an invasion. They are numerous and vulnerable and place great pressure on the invaders by their sheer numbers. Perhaps 100 billion neutrophils are produced in the bone marrow each day, but these survive only a few days and suffer great casualties with every invasion. Like other leukocytes, they are drawn toward chemicals that are emitted by cells at the troubled site.

The monocytes arrive after the neutrophils, and once at the site they undergo remarkable changes, growing into huge **macrophages** (Figure 41.5). They are prodigious eaters, taking in invading cells and cellular debris. They are also critical to the specific defenses, particularly to the primary immune response, which we will discuss shortly.

Natural Killer Cells Most lymphocytes function additionally in the specific defenses, but one type, the **natural killer (NK) cell,** plays an exclusively nonspecific role. NK cells originate as subpopulations of certain lymphocytes (see Figure 39.19), but little is known about their development. We do know that they roam the body, contacting cells

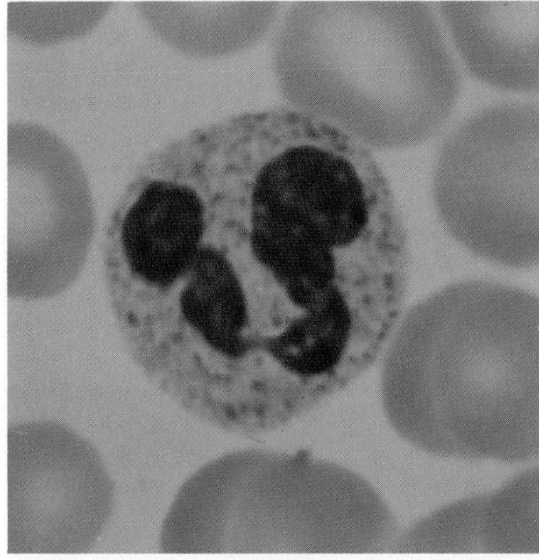
Neutrophil

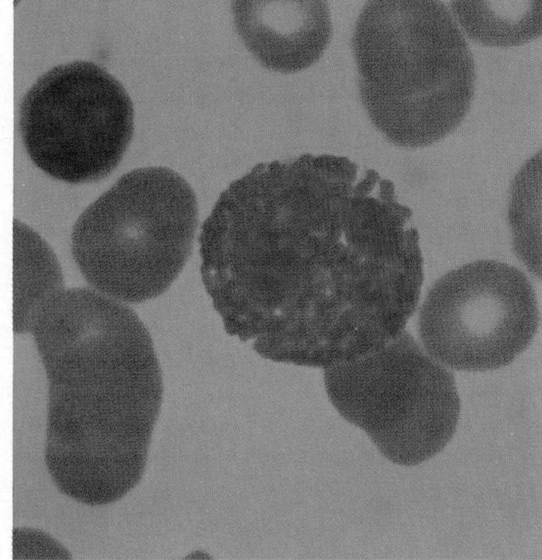
Eosinophil

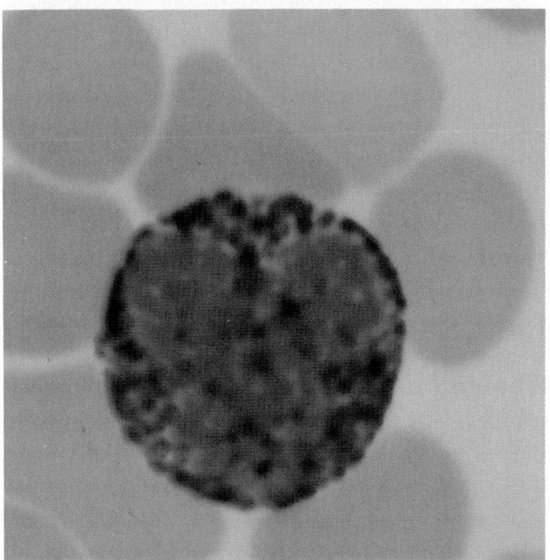
Basophil

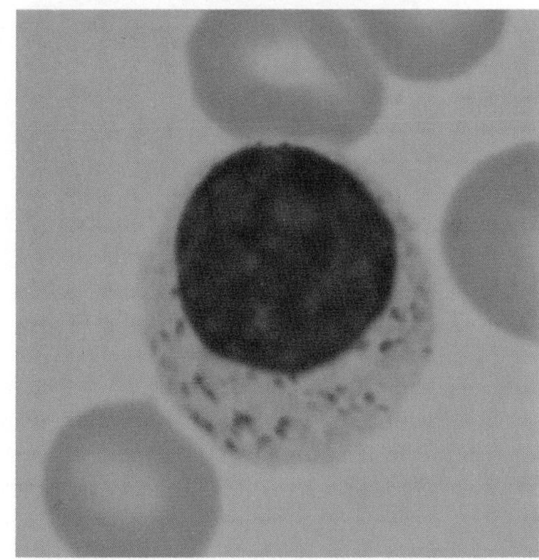
Lymphocyte

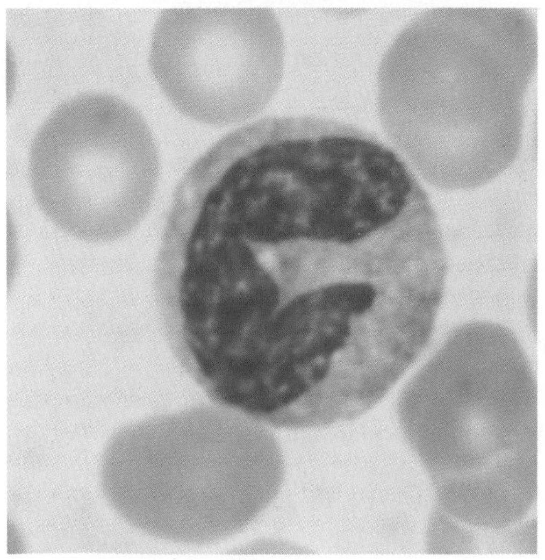
Monocyte

FIGURE 41.4
THE LEUKOCYTES
Stem cells in red bone marrow give rise to five types of leukocytes, each with a specialized task in defending the body against invasion.

FIGURE 41.5
MACROPHAGES AND THEIR FUNCTION
Macrophages are phagocytic leukocytes that actively engulf invading organisms and cellular debris. The largest phagocytes, the macrophages, devour many invading bacteria and cellular debris.

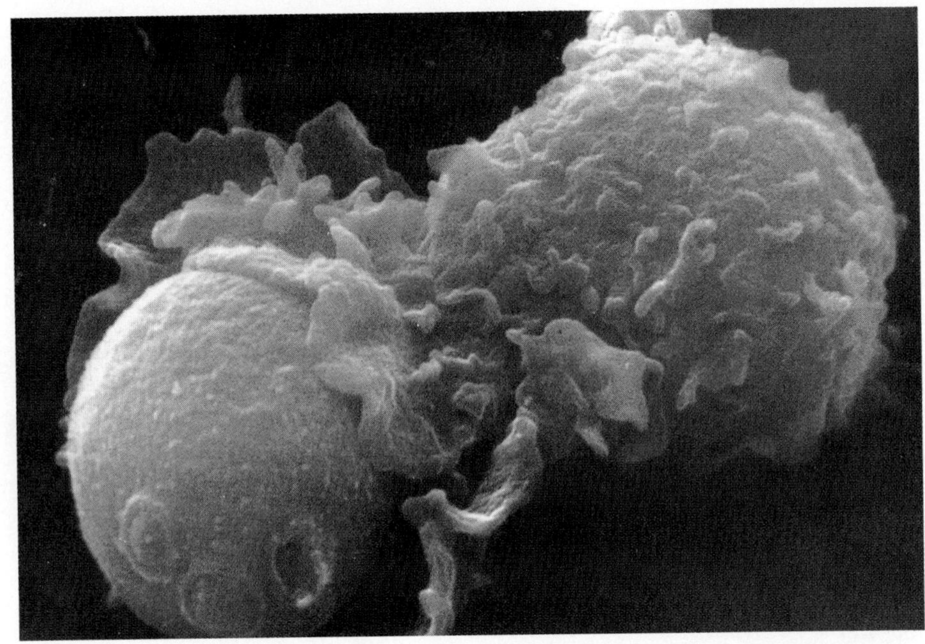

of all types. If they encounter cancerous cells or cells harboring viruses, such cells are immediately attacked and killed. NK cells attack only the body's own diseased cells. In fact, it is suspected that the failure of these first-line defenders, particularly later in life, is a key to the development and spread of cancer.

SPECIFIC CELLULAR AND CHEMICAL DEFENSES

Specific defenses are those that generally come to mind when we think of our own immunity. They form the **immune responses.** Such defenses first require exposure to the invading agent, followed by the programming of specific responses. All of this requires time, so specific defenses can be slow to start. When you received those infamous immunizing "shots" just before starting school, or when going abroad to certain regions of the world, or into military service, your body was being encouraged to produce just such defenses. Specific defenses include those that confer long-lasting protection against disease.

We have long been aware of such defenses, both historically and individually. In the fifth century, the Greeks wrote that people who had recovered from the plague never suffered from it again. School children are aware that once they've had chicken pox or mumps, that's usually it; they don't catch them twice. Such resistance is due to very precise cellular processes that are triggered by the first exposure—processes that enable the body to "learn" the nature of the invader and to defend itself subsequently against those specific traits. This kind of immunity begins with the action of lymphocytes.

Lymphoid Tissues and Lymphocytes

Lymphocytes are white blood cells that arise in the red bone marrow early in fetal development. Regions where lymphocytes congregate are known collectively as **lymphoid tissues.** Functionally, there are two kinds of lymphoid tissues, primary and secondary (Figure 41.6). **Primary lymphoid tissue** exists mainly in the thymus and red bone marrow. (The thymus is a bilobed organ located underneath the breast bone; see Chapter 37.) **Secondary lymphoid tissue** includes the lymph nodes, tonsils, adenoids, spleen, and Peyer's patches (clustered tissues on the small intestine's wall). Lymphocyte development and maturation occur in the primary lymphoid tissues. In the secondary lymphoid tissues, the various mature lymphocytes begin their critical roles in defense.

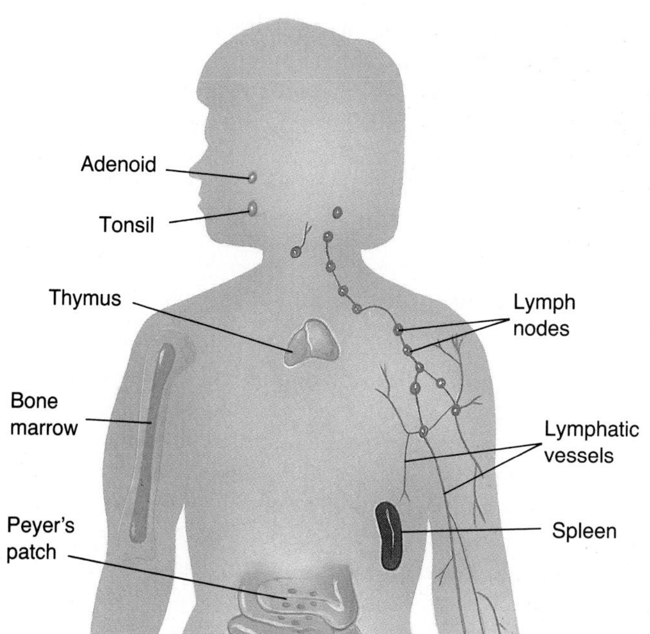

FIGURE 41.6
LYMPHOID TISSUES OF HUMANS
Lymphocytes arise in primary lymphoid tissues (thymus and red bone marrow), where they go through a maturation process. Later, as virgin B- and T-cells, they migrate to secondary lymphoid tissues, where they await selective activation. The activated cells then migrate out to the infection sites.

B- and T-Lymphocytes

Two kinds of lymphocytes are involved in the immune response: B-cells and T-cells. In humans and other mammals, **T-cells** differentiate in the thymus, the organ for which they are named. **B-cells** originate in the red bone marrow and in the fetal liver. (The "B" in the name is actually derived from the *bursa of Fabricius,* a structure in the chicken.) In the inactive state, B- and T-cells are nearly impossible to distinguish visually, but they can be chemically identified on the basis of certain unique cell surface proteins. Once activated, the B-cells form an unmistakably extensive rough endoplasmic reticulum, while the T-cells contain large concentrations of free ribosomes (Figure 41.7).

Humoral and Cell-Mediated Responses The B- and T-cells have quite different roles in the immune response. B-cells do two things: most make and secrete antibodies (discussed next), and a few act as "memory cells." The latter are among the cells that confer lasting immunity, thus helping the body respond quickly to future encounters

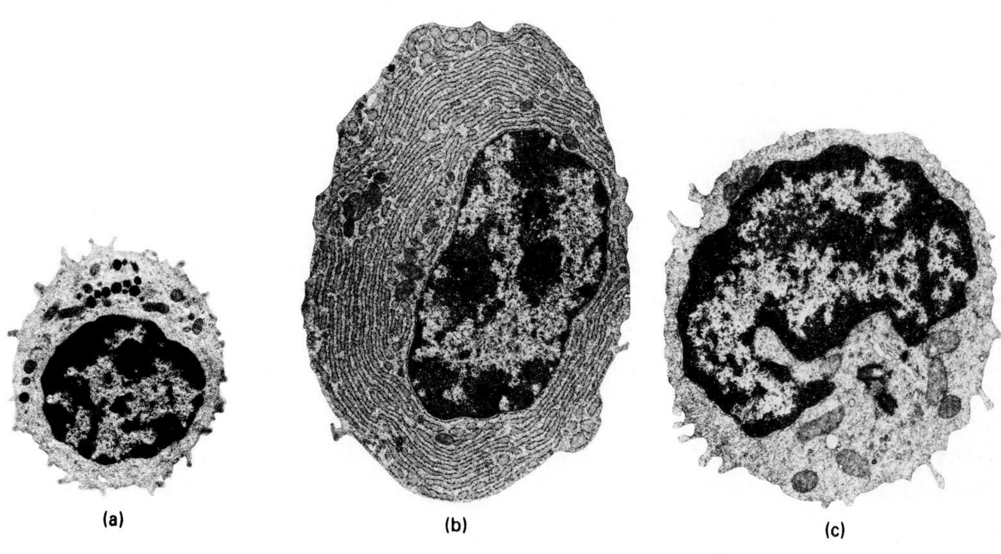

FIGURE 41.7
B-CELLS AND T-CELLS
(a) Immature lymphocyte. (b) Mature, activated B-cells contain an extensive rough endoplasmic reticulum (RER) where polypeptides are assembled into antibodies. (c) Mature, activated T-cells are larger than their immature predecessors, but little other differentiation is visible.

with the same offender. The secretion of antibodies, a specific chemical defense, is called the **humoral response** (*humor,* "body fluid").

Some T-cells specialize in the **cell-mediated response.** This includes a direct attack against diseased body cells, particularly those that have been invaded by a virus and those that are cancerous. Other T-cells activate B-cells, and still others moderate and suppress the immune response. In addition, some T-cells from each of these specialized groups become memory cells, which, like memory-B cells, confer lasting immunity.

Antibodies and the Humoral Response

As we've said, B-cells produce antibodies. **Antibodies** (also called **immunoglobulins**) are special proteins secreted by B-cells in response to the presence of foreign substances called antigens. **Antigens** (also **immunogens**) may be proteins, nucleic acids, or carbohydrates, but are generally restricted to molecules with a molecular weight over 5000. They can take the form of unattached molecules (those not part of an organism), or they can be cell surface molecules on invading organisms such as bacteria and viruses. They also appear on the surface of cancer-ridden cells. Antigens contain localized regions (chemical groups) called **antigenic determinants,** and it is these regions, specifically, that trigger a response by a matching antibody. There may be several antigenic determinants on an antigen, and each of these may prompt the formation of a different antibody.

Thus, antibodies identify and bind to antigens or antigen-bearing invaders, and in various ways, help destroy them. Since antibodies are pivotal to the immune response, we will consider them more closely.

Antibody Structure Antibodies are made up of four polypeptides, two long or **heavy chains,** and two short or **light chains.** The chains are bonded together by disulfide linkages, which hold them into a Y-shaped configuration as seen in Figure 41.8. The branching ends of each chain contain special **variable regions,** representing that part of the antibody that binds to its matching antigen when the two interact. Each kind of antibody differs in the amino acid sequence of the variable region. That name is appropriate since there is nearly unlimited variation possible in this part of each polypeptide chain. (Humans, for example, produce literally millions of different variable regions.) Below the variable region of each chain is the **constant region,** so named because this portion is the same in each antibody of a certain class.

Mammalian antibodies fall into five classes according to the nature of their constant regions. That is, there are just five variations in the sequence of amino acids making up the stems of all antibodies. The five classes are designated IgG, IgA, IgM, IgD, and IgE ("Ig" stands for immunoglobulin.) Each class has a different general function in the

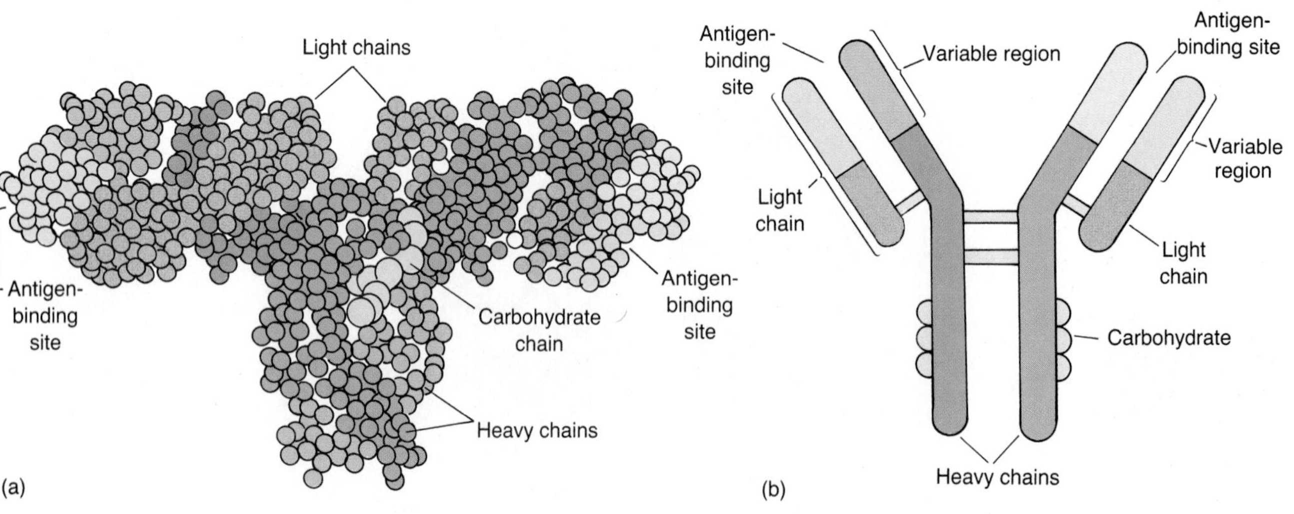

FIGURE 41.8
THE ANTIBODY
(a) A space-filling model of an antibody. The small colored spheres represent individual amino acids; the large white spheres are carbohydrates. Note that the model, a monomer, takes roughly the form of a Y.
(b) Model of an antibody. Antibodies consist of two heavy and two light chain polypeptides. Each polypeptide, in turn, contains a constant region and a variable region. Whereas variable regions become bound to antigens, the heavy chain stem may bind to a matching site on a roving phagocyte or on phagocytic cells lining the liver passages, spleen, and lymph nodes.

humoral response. For example, IgG helps lyse foreign cells, whereas IgA interferes with the ability of an invading cell to attach to a host cell. (The antibody classes are further discussed in Figure 41.9.)

Some classes of antibodies contain both bound and free-floating forms. For example, free-floating IgM antibodies occur as a large five-"Y" complex that moves unattached in the bloodstream. The bound version of IgM, a single "Y", lies embedded in the plasma membrane of the B-cell as do IgD antibodies. Bound antibodies play an important role in antigen recognition by B-cells.

Antibody/Antigen Interactions Antibodies interact with antigens in three basic ways, with some variation. First, antibodies (those with multiple binding sites) bind to several antigens, holding them together in clumps (Figure 41.10) that can be easily located and devoured by phagocytes. Second, certain interactions between antibody and antigen trigger the complement system (mentioned earlier), and complement produces holes in invading cells, thus admitting water that causes the cell to burst. A third interaction involves opsonization, a special situation that deserves closer attention.

FIGURE 41.9
CLASSES OF ANTIBODIES
Human antibodies are grouped into five classes according to the chemical characteristics of their constant regions.

IgG Makes up 70-75% of total human antibody. Monomer structure; common in the blood in late primary infections and throughout secondary infections. Triggers complement activity. Crosses placenta, conferring some immunity on fetus.

IgA Makes up 15-20% of total human antibody. Occurs in dimer and monomer form. Common in human milk (constituent of colostrum), respiratory mucus, saliva, tears, and walls of intestine. Located on cell surfaces where it immobilizes bacteria until phagocytosed.

IgM About 10% of total human antibody. Occurs as a pentamer (5-part) in blood to which it is restricted by its size. A monomer form occurs as a receptor in B-cell membranes where it plays a key role in the arousal of virgin B-cells.

IgD About 1% of total human antibody. Occurs as a monomer on the surface of B-cells where it may function in concert with IgG, as a receptor and activator in the primary immune response.

IgE Occurs in trace amounts as a monomer. Associates with mast cells (from connective tissue) in allergic responses, triggering their release of histamine which is associated with tissue inflammation (as in hay fever).

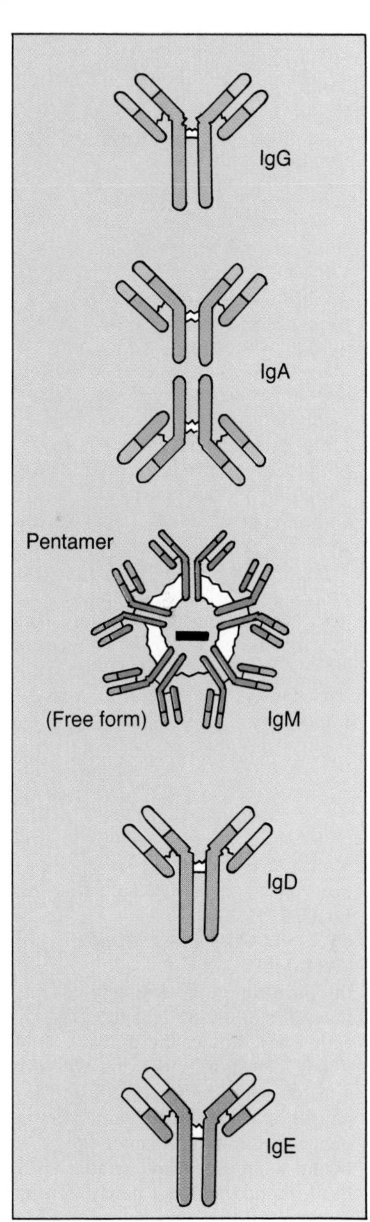

FIGURE 41.10
AN EXAMPLE OF ANTIBODY-ANTIGEN INTERACTIONS
In some instances the binding of antibodies to cell surface antigens brings about the formation of clusters that are readily engulfed by phagocytes.

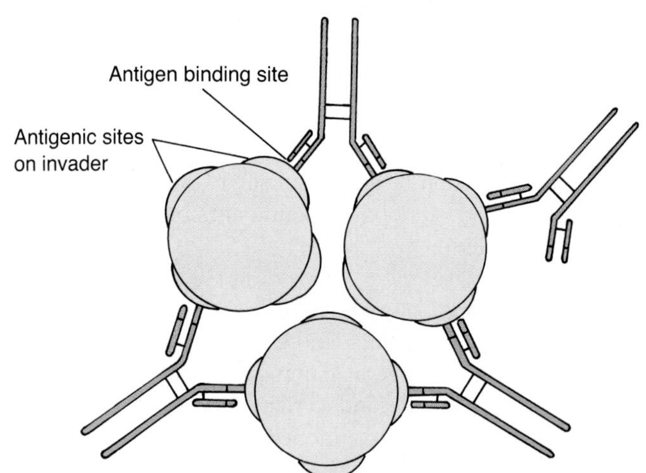

Recall that opsonization involving complement provides a means for a phagocyte to bind to an invading cell. A similar binding is also made possible by antibodies (Figure 41.11). Each invader has many antigenic sites, and each of these has numerous antigenic determinants, so as a result many antibodies will bind to a bacterium, literally coating it. Further, with their variable regions bound to the invader's cell surface, the stemlike constant regions of the antibodies protrude outward. Each stem has what is called an **Fc region**, and these regions have matching **Fc receptors** on the phagocytes. So, while the antibodies are bound to antigen at one end, they also bind to the phagocyte at the other. As more binding occurs between the two, the phagocyte literally rolls over the invader and engulfs it.

Humoral responses, reactions between antibodies and antigens, are a vital part of the immune system's work. We've seen that antibodies are quite specific in their structure, and their great diversity makes it possible for them to react to an enormous number of different antigens. But this just scratches the surface of the immune response. The role of B-cells in the humoral response depends heavily upon two factors: (1) their ability to identify antigens, and (2) the coordinating efforts of other lymphocytes, namely the T-cells. Without the intervention of T-cells, the response of B-cells would be severely curtailed. Nowhere is this more dramatically illustrated than in the devastating work of HIV, the AIDS virus, which preferentially destroys certain T-cells. In addition to the role of B-cell coordinator, T-cells are responsible for the complex cell-mediated response. Both tasks involve the ability of lymphocytes to recognize both invaders and each other, so we'll move now to the cell recognition process and how it develops.

Lymphocytes and Cell Recognition

Let's begin by reminding ourselves of a critical distinction between phagocytes and B- and T-lymphocytes. Phagocytes are nonspecific; they do not discriminate in the same

FIGURE 41.11
OPSONIZATION: A DOUBLE "WHAMMY"
In opsonization, an antibody-coated invader is a "set-up" for the phagocytic army. Any of the protruding antibody stems, or Fc regions, will form a match with an Fc receptor on the phagocyte. The binding stimulates the phagocyte, and as site after site is bound to antibody, the invader finds itself trapped within a newly formed vacuole. All that remains is for a lysosome to fuse and release its potent load of proteolytic enzymes, and the invader's time is up.

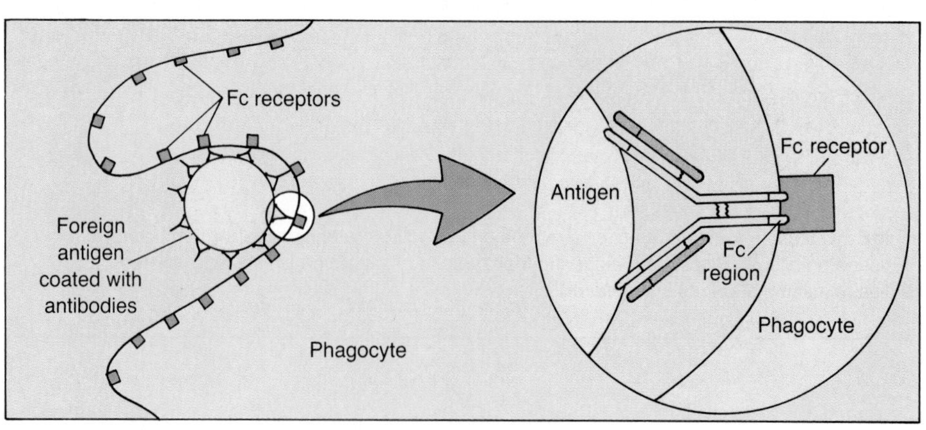

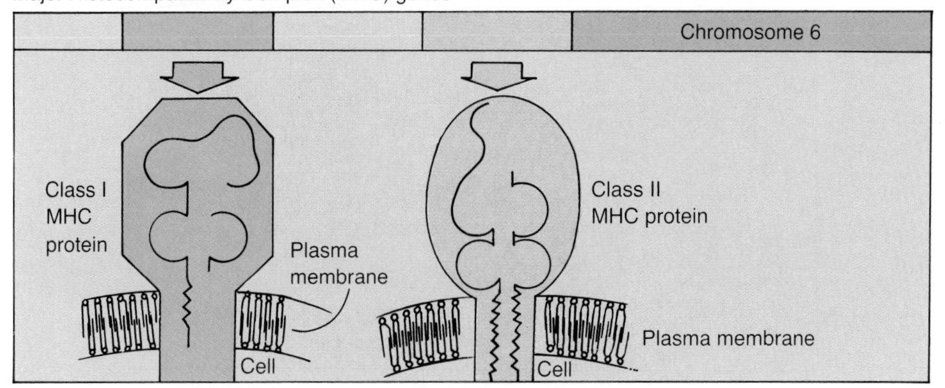

FIGURE 41.12
MHC GENES AT WORK
MHC genes code for three classes of proteins, two of which become bound to the plasma membrane. Each individual's MHC proteins are unique. Seen here is part of the human MHC gene complex and two of its products, Class I MHC proteins and Class II MHC proteins.

way lymphocytes do. They attack all invaders with equal vigor. Lymphocytes, on the other hand, are very specific in their recognition of invaders. In lymphocyte development, each is programmed to recognize only one specific antigen. And no matter which antigen turns up, an antibody soon interacts with it. The response of lymphocytes to literally millions of different antigens (even those manufactured and never found in nature) has presented an array of fascinating puzzles for biologists. How can B-cells make antibodies for so many different antigens? How do they actually recognize the antigens? Further, how can they tell the antigen-bearing cells from the normal cells of the body? We can now begin to answer such questions, starting with the last—recognizing self.

We will digress here for a close look at T- and B-cell recognition systems and how they develop before going on to how these cells function in immune responses. Knowing about this will help you understand how the specific responses come about. Bear in mind as we go that much of our understanding is still tentative and hypothetical, and certainly subject to revision.

The MHC and Recognition of Self Unless you have an identical twin, you are chemically unique. Each of us bears proteins on our cell surfaces that are slightly different. You may already realize this in view of what you know about organ transplants and the associated problem of transplant rejection. Rejection occurs because the immune system identifies cells of the transplant as foreign. (Much of what we know about the immune response comes from organ transplantation studies.)

As you might expect, the basis for these cell surface differences is genetic. On the sixth chromosome of humans are several hundred genes that code for proteins each individual will bear on his or her cell surfaces. The genes that code for these proteins make up what is called the **major histocompatibility complex**, or **MHC.** The MHC has been found in all mammals studied so far.

The human MHC (also called the human leukocyte-associated antigen gene cluster, or HLA) codes for three groups of proteins, designated as Classes I, II, and III. Class III forms complement, the group of chemical defense proteins mentioned earlier. Classes I and II, which interest us here, are the aforementioned cell surface proteins, which are called **MHC proteins** (sometimes referred to as "MHC antigens"). The two protein classes have distinct configurations (Figure 41.12), and each group plays a very specific role in the complex lymphocyte interactions that make up the immune response.

During our fetal development each of our cells becomes "labelled" by Class I and/or Class II proteins coded by the MHC. These MHC proteins will act as a kind of badge that identifies the cells as "self," and consequentially healthy cells will be largely ignored by the body's own immune system, particularly the roving T-cells. Those very badges, though, can come perilously close to being an invitation to attack. This is because T-cells have protein recognition sites called **T-cell receptors** on their plasma membranes. These sites fit the MHC proteins like a lock and key (much in the fashion of enzymes and substrates). The reason the T-cell doesn't attack friendly cells, or self, is that its receptor is actually a *dual site*. Part of a T-cell receptor indeed recognizes self, that is, a

FIGURE 41.13
THE DUAL RECEPTOR
T-cells make use of their dual receptor to identify friendly cells bearing antigen (diseased body cells). Part of the receptor will match one of the many MHC Class I or Class II proteins, and part will match an antigen.

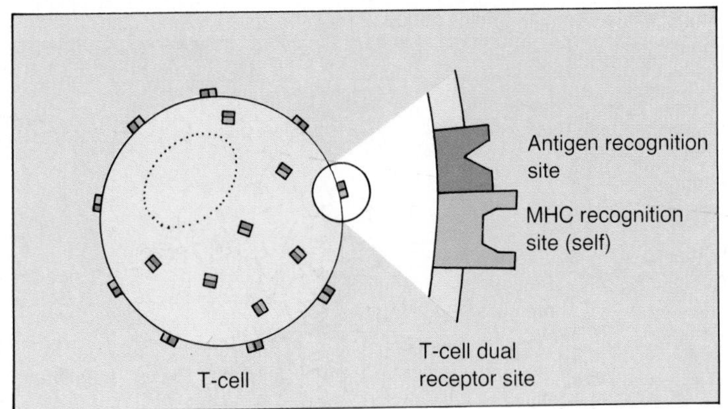

Class I or a Class II surface MHC protein, but the adjacent part recognizes an antigen (Figure 41.13). It is only when both self and an antigen are recognized that the T-cell goes on the attack. The T-cell receptor, indeed, has remarkable abilities of discrimination. How do such abilities arise?

The Dual Receptor and MHC Restriction A prevailing theory proposes that as T-cell populations mature in the thymus, the many and various T-cells assemble slightly different receptors on their surfaces. Some will match the body's Class I or Class II MHC proteins, but most will not. Those T-cells that do not form a match simply die. As a result, each individual's developing T-cells come to recognize friendly body cells. Because their receptors match either a Class I or a Class II MHC protein, the T-cells are now referred to as **MHC restricted.** Further, those T-cells whose receptors fit only Class I MHC proteins are now designated **cytotoxic T-cells,** whereas those whose receptors match with Class II MHC proteins become **helper T-cells.**

We must add one more point about MHC proteins. Whereas all of our nucleated body cells bear Class I MHC proteins (and can thus be identified by cytotoxic T-cells), only macrophages, B-cells, T-cells, and a few others bear Class II MHC proteins. The logic of this will become clear as we proceed into the primary immune response.

Having seen how T-cells recognize self, how do they come to recognize invaders (foreign antigens)? Immunologists are still struggling with this question, but one theory is that each T-cell receptor undergoes a transformation into the dual site mentioned earlier. According to this model, each of the developing T-cell's receptors is at first made up of two polypeptides, a gamma unit and a beta unit. The receptor recognizes Class I or Class II MHC proteins as described earlier. In the next step, the gamma unit is replaced by a newly synthesized polypeptide, the alpha unit. The beta unit still recognizes the body's own cell surface MHC proteins, but the alpha unit recognizes only a specific antigen, one the T-cell has yet to meet. Among the immense T-cell population there is an incredible array of alpha polypeptides making up such recognition sites. This is how T-cells are able to recognize so many different antigens. How the body provides for this diversity brings us back to the gene once more, and we will come back to that problem later in the chapter. The growing T-cell army is now ready to fulfill its role in immune defense: interacting with other cells of the immune system and with the body's diseased, antigen-bearing cells.

Those T-lymphocytes that have completed the development of dual receptors (Figure 41.14) are called **virgin T-cells,** a name that indicates something is yet to come. They will retain their "chaste" state until they encounter a matching antigen. Actually, most never do, but it's comforting to know that such a large reserve army is always available should a need arise.

The B-Cell Receptor B-cells also have surface receptors that interact with antigen. These receptors, as we noted earlier, are actually bound antibodies from the IgM and IgD classes (Figure 41.15), and each virgin B-cell carries some 10,000 of such antibodies

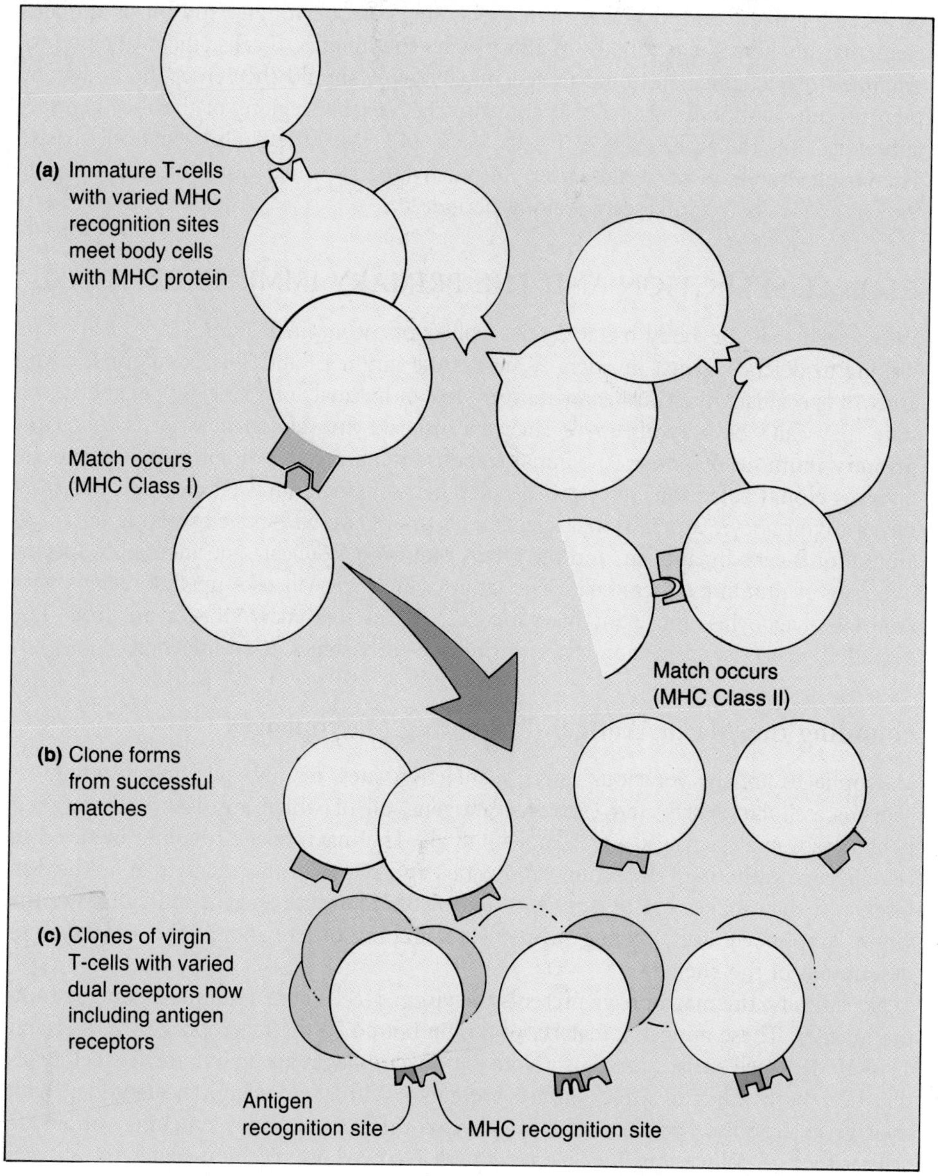

(a) Immature T-cells with varied MHC recognition sites meet body cells with MHC protein

Match occurs (MHC Class I)

Match occurs (MHC Class II)

(b) Clone forms from successful matches

(c) Clones of virgin T-cells with varied dual receptors now including antigen receptors

Antigen recognition site

MHC recognition site

FIGURE 41.14
BUILDING THE T-CELL'S DUAL RECEPTOR
(a) MHC protein receptors at first occur in great variety in the vast immature T-cell populations of the thymus. However, only those populations whose receptors find matches with one of the body's own MHC proteins survive. **(b)** The survivors give rise to clones, whose receptors **(c)** undergo further modification into the all-important dual receptor, part of which recognizes an MHC proteins, or "self," and part of which recognizes a specific antigen it has yet to encounter.

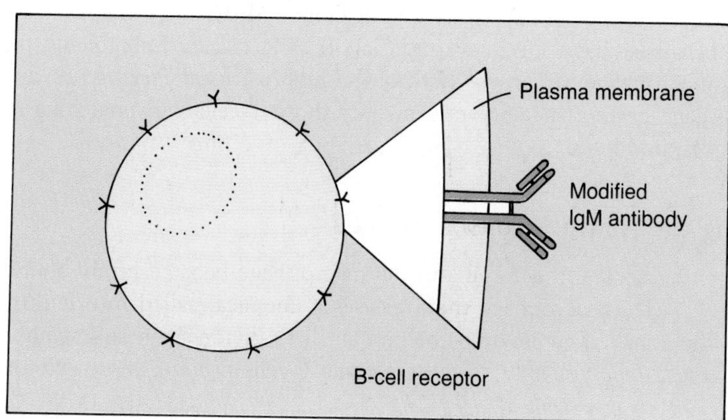

FIGURE 41.15
B-CELL RECEPTORS
Receptor sites on B-cells are actually bound antibodies, their stems anchored in the lipid bilayer of the plasma membrane and their two antigen recognition sites protruding outward.

on its cell surface. Both occur as monomers (single Y's), with an extra bit of antibody stem that anchors the antibody to the plasma membrane. B-cells, then, are preprogrammed to recognize and bind to specific antigens, should they enter the body, and then produce antibody against that antigen. The variable regions of the B-cell surface antibodies, like the alpha units of T-cells, occur in a vast variety, also genetically based. The **virgin B-cells** (a name they keep until activated by a matching antigen) now join the virgin T-cells in a quiescent waiting period.

CLONAL SELECTION AND THE PRIMARY IMMUNE RESPONSE

We've seen that the body has a diverse collection of virgin B- and T-cells ready and waiting to defend against invaders. Whereas the virgin B- and T-cells comprise a vast army of specialists, there are comparatively few individuals of a single type, and in this state they can't be very effective. They are aroused into action in what is called the **primary immune response** (the initial response to an invader or antigen). The process involves **clonal selection**, the rapid development of single lines (clones) selected from the vast lymphocyte army. Clonal selection occurs in two ways, one involving the direct arousal of B-cells by antigen, and the other, requiring a delicate and incredibly specific interplay of immune system cells. The latter, which we will take up first, begins with aroused phagocytes, the giant macrophages. Recall that macrophages are front line defenders that arise from monocytes during the early stages of an infection.

Sounding the Alarm: Antigen-Presenting Macrophages

Macrophages are the voracious eaters at infection sites, readily engulfing invading organisms, cellular debris, and even free antigens, all of which are dealt with by their potent lysosomal enzymes—well, not quite all. The macrophage becomes involved in "intelligence gathering," collecting various bits and pieces of the enemy that are held in reserve. It then makes use of these bits and pieces to arouse specific individuals in the virgin lymphocyte army, giving them their marching orders and, at the same time, a description of the enemy.

Specifically, the macrophage presents the virgin T-cells with a sample of antigen from the invader. These antigenic materials become bound to the macrophage's Class I and Class II MHC cell surface proteins. (Note that macrophages are among the few cell types that bear both kinds of MHC surface proteins.) With their membranes studded with their grisly trophies (Figure 41.16) they move through the body touching various T-cells as they go. When they finally encounter a T-cell whose receptors match the antigen and its associated MHC protein, the primary immune response begins. The presence of **antigen-presenting macrophages** is vital because virgin T-cells do not recognize free antigen.

Some antigen-presenting macrophages will form a match with T-cell populations that have Class I MHC recognition sites (the cytotoxic T-cells), and other macrophages will match T-cell populations with Class II MHC recognition sites (the helper T-cells). These T-cells are "selected" by the macrophage to be cloned; thus, the process is named "clonal selection." Because it has both Class I and Class II MHC cell surface proteins, the antigen-presenting macrophage can arouse both of the highly critical T-cell types, and because the macrophage bears a specific antigen, only those T-cells with matching antigen receptors will respond.

Spreading the Alarm: Aroused T-Cells

When matching occurs, the T-cell and the macrophage become firmly bound together (Figure 41.17). The macrophage then releases a chemical called **interleukin I**, which stimulates the attached cytotoxic or helper T-cell to divide again and again, soon producing a large clone. Not only do the cytotoxic T-cells produce more cytotoxic T-cells

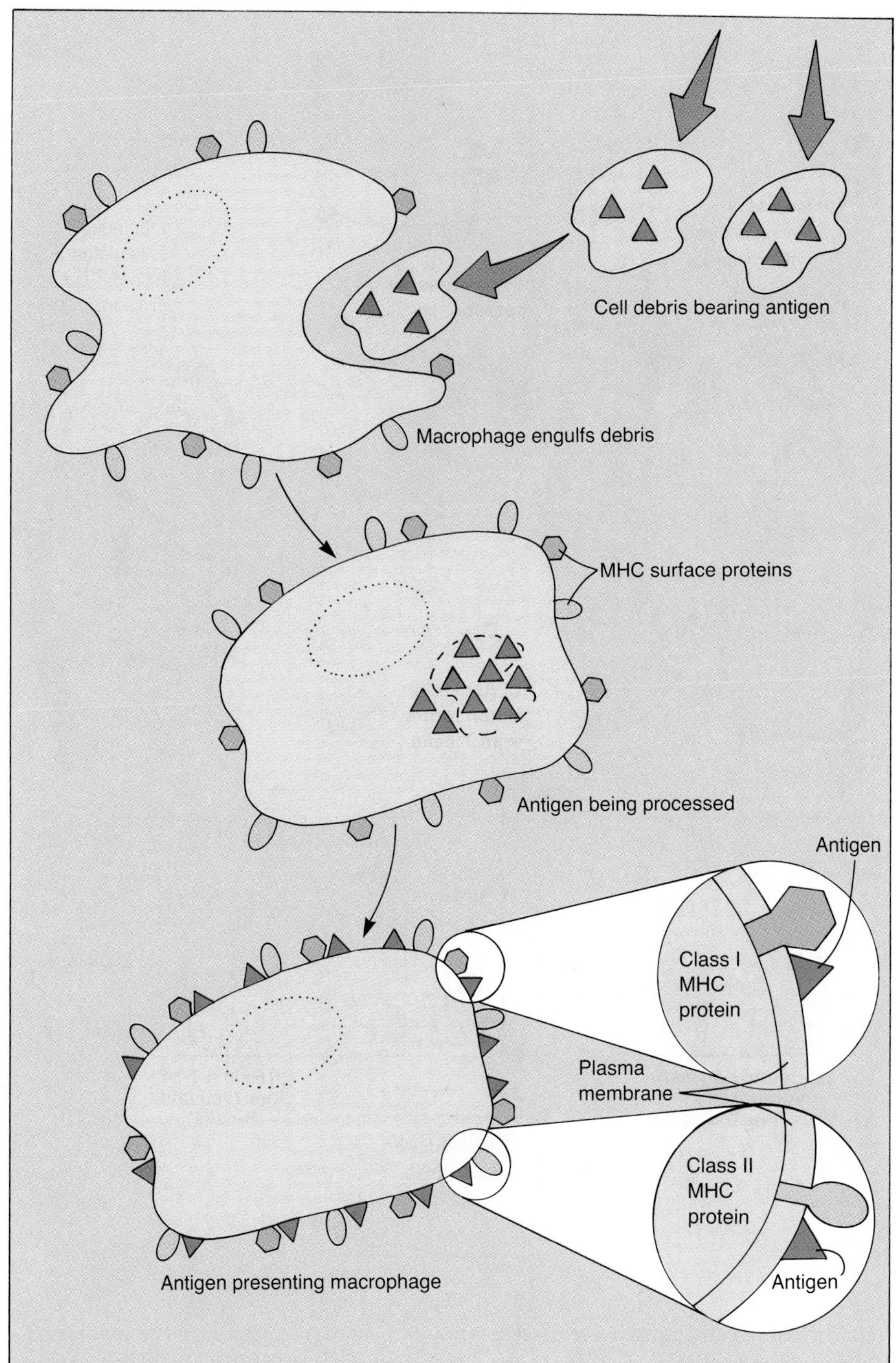

FIGURE 41.16
FORMING AN ANTIGEN-PRESENTING MACROPHAGE
Macrophages incorporate either free antigen or antigen from partially digested invaders into their cell surfaces, where the antigens join Class I and Class II MHC proteins. This produces a dual recognition site that will complement the dual receptor of some specific virgin T-cells.

FIGURE 41.17
CLONAL SELECTION: T-CELL ACTIVATION
(a) From the incredibly diverse virgin T-cell populations, an antigen-presenting macrophage finds a cytotoxic T-cell and a helper T-cell whose receptors match its own. (b) Upon binding with a T-cell, the macrophage secretes interleukin I, which activates the cytotoxic and helper T-cells. The two then produce enormous clones (including some memory cells), each with recognition sites that are identical to the original activated cell. (c) The cytotoxic T-cells then seek out and destroy infected cells while (d) the helper T-cells arouse selected virgin B-cells.

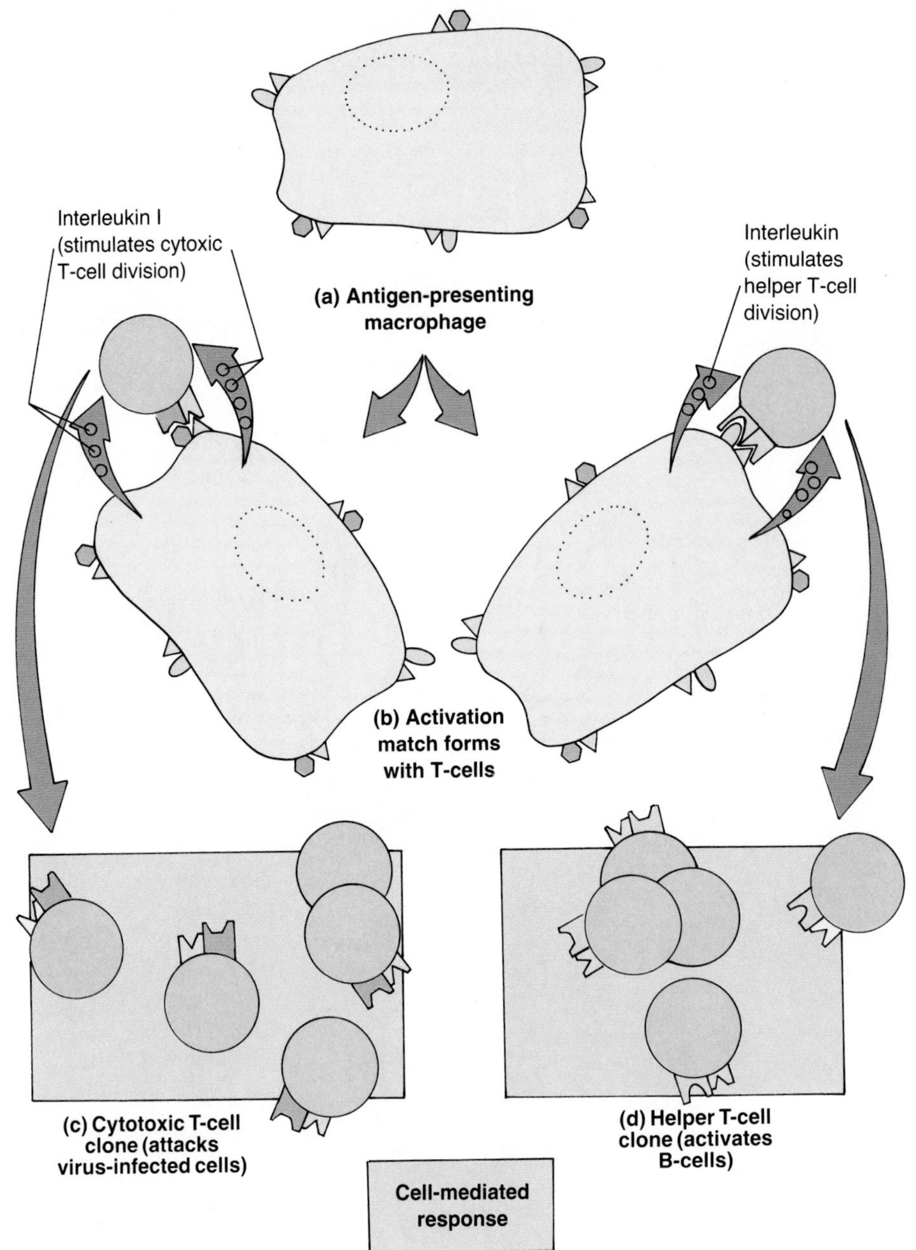

(and helper T-cells more helper T-cells), but each kind also produces the **memory T-cells** mentioned earlier. The key to success against the invader is twofold: first there is the incredibly precise clonal selection process, and second is the million- to billionfold yield of active lymphocytes. The vast numbers are produced through an amplification process; as the numbers increase, so does the stimulus for further cell division in the selected clones.

Cytotoxic T-Cells Kill Infected Cells The two types of aroused T-cells have quite different roles. The cytotoxic T-cell, as we've seen, specializes in the cell-mediated response, that is, in killing infected body cells. Using its dual cell surface receptor, it begins approaching the body's normal cells in a manner we can't help but call "frisking." If the cell bears no antigen, the cytotoxic T-cell goes on its way (much of the movement random). But if a cell harbors a virus, the viral antigens will usually show up on the infected cell's surface. When the dual match (with both MHC protein and antigen) is made, the cell will be killed immediately by lysis, often before the virus has replicated. That is, the cytotoxic T-cell produces holes in the afflicted cell's plasma membrane (much in the manner of complement), an act that permits water to enter and burst the cell. (NK cells also use this procedure when attacking cancer-ridden cells.) The cytotoxic T-cell then resumes its patrol, leaving the mess to be cleaned up by phagocytes.

Cytotoxic T-cells also attack and destroy cancer cells. This is possible because cancer cells produce a cell-surface antigen that will form a match with just the right cytotoxic T-cell. But while such matches do often occur, some cancer cells readily escape the attention of the lymphocytes, breaking out of their primary site and going on a rampage, spreading through the body, where they give rise to many secondary tumors. Then the patient hears the chilling word, *metastasis!* Through intensive studies of metastatic tumor cells in mice, Israeli immunologists have concluded that such cancer cells avoid cytotoxic T-cells because of changes that occur in their MHC cell surface proteins. Such changes mean that while some individuals in the vast army of virgin cytotoxic T-cells have receptors that match the cancer antigen, none have receptors for the metastatic cancer cell's MHC proteins. Without the dual match, the T-cell cannot act.

Helper T-Cells Activate the B-Cells In addition to stimulating growth of the T-cell clones, helper T-cells also activate selected B-cells, whose role, you will recall, is the secretion of antibodies. Antibodies are effective against free antigen, antigen in viral coats, and antigen on the surface of cells such as pathogenic bacteria and fungi. Thus, the B-cell's role is broader than that of T-cells. But before an effective B-cell attack can take place, the proper virgin B-cells must be identified and large clones formed.

The arousal of B-cells is different from that of T-cells; it begins when the surface antibody of a virgin B-cell encounters a free-floating, matching antigen (Figure 41.18). The matching antigen quickly binds to the B-cell's surface antibody, whereupon it joins a Class II MHC protein, forming a receptor much in the way receptors form in macrophages. The now-activated B-cell then goes on its own prowl. Its new dual receptor can form a match with the cell surface receptor on a specific macrophage-aroused helper T-cell, one that has been activated by the same antigen. When such contact is made, binding occurs, and the bound helper T-cell then secretes interleukin I as did the antigen-presenting macrophage mentioned earlier. This prompts the B-cell to enter cell division, forming its own enormous clone. As in the cloning of T-cells, the growth of B-cell numbers is a self-amplifying process as each increase in number prompts still further increases.

B-Cells: Several Offensive Strategies

As the B-cell clone grows, **plasma cells** and **memory B-cells** emerge. The plasma cells are short-lived, lasting only a few days, but during this brief period they begin the humoral response—the production and secretion of copious amounts of antibody. It is estimated that each B-cell can secrete up to 2,000 antibody molecules per second during its short but intensely active life.

Again, because of the incredibly precise clonal selection process, the B-cell clone secretes antibody that is extremely specific to the antigen that aroused it in the first place. (Actually, since antigens commonly have a number of antigenic determinants, many specific B-cells will be aroused, and the overall response will be much broader than we have indicated.)

There is another aspect to the versatility of activated B-cells. Although the all-important surface antibodies are restricted to the IgM and IgD immunoglobulin classes, plasma cells are capable of producing antibodies of the other classes as well (IgG, IgA, and IgE, but primarily IgG and IgA). Whereas the specific variable region of the cell surface IgM or IgD antibody that reacted with antigen earlier is retained, the plasma cells can generate any of the several kinds of constant regions. (Recall that the antibody classes differ in their constant regions.) By producing several classes of antibodies, but each reacting to

FIGURE 41.18
B-CELL ACTIVATION
The activation of virgin B-cells requires interaction with helper T-cells. **(a)** B-cells first capture free antigen that matches their surface receptor antibody and **(b)** incorporate the antigen with one of their Class II MHC proteins. **(c)** A matching helper T-cell then binds to the B-cell, whereupon the helper secretes interleukin I. **(d)** The B-cells proliferate, forming large clones of plasma cells that produce and secrete antibody and memory B-cells that remain inactive until subsequent infections occur.

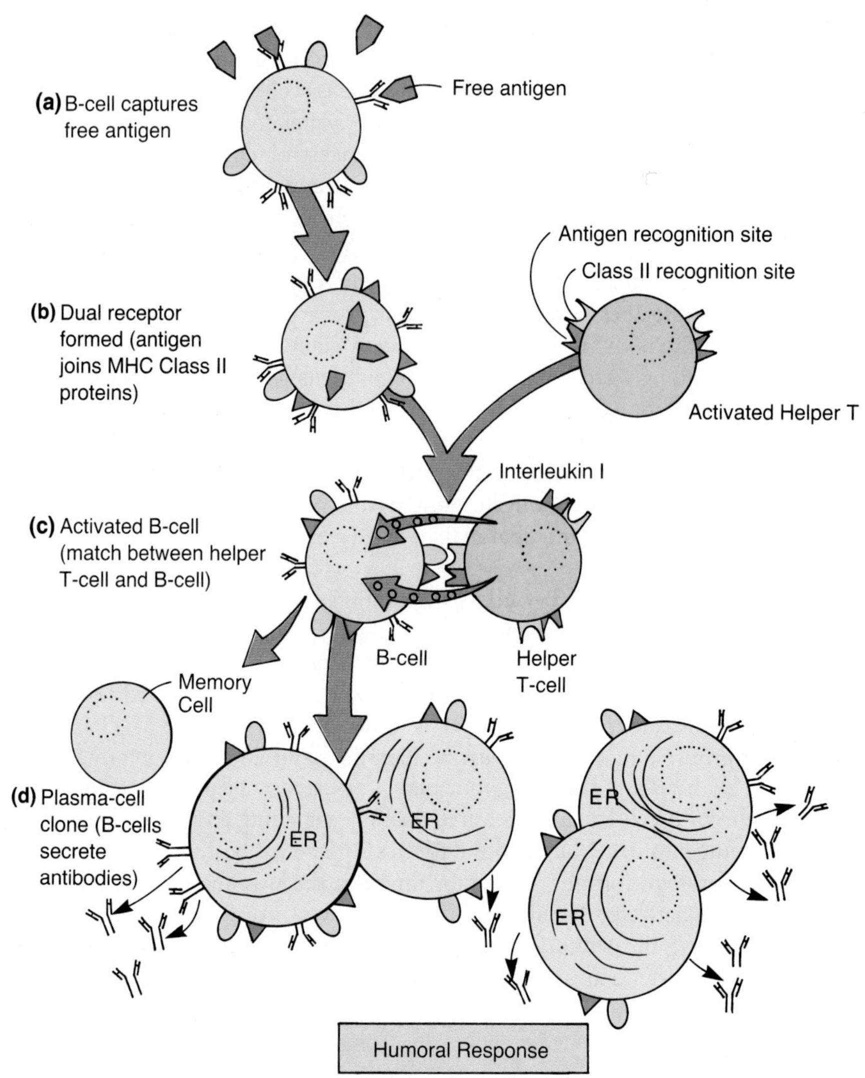

the same antigen (or antigenic determinant), the antibody attack against invaders can be even more varied. (You may want to review the specializations of antibody classes in Figure 41.9.)

As antibody circulates in the blood and body fluids, it will sooner or later encounter and bind to matching free or bound antigen. Masses of antigen, clumped together by antibodies, are then readily engulfed and destroyed by phagocytes. In addition, the interaction of antibody with antigen activates complement, which as we've seen has its own way of dispatching invading cells. We also saw earlier that antibody bound to invaders makes the efficient opsonization process possible (see Figure 41.11). So we see that B-cells provide a many-pronged offense against invaders.

Antigen-arousal of Virgin B-Cells Immunologists are still arguing about its significance, but it is quite possible that virgin B-cells can become activated without the assistance of helper T-cells. Such activation occurs when B-cells contact a select group of antigens. Included among the antigens are certain lipopolysaccharides, the protein flagellin, the polysaccharide dextran, and certain D-amino acid chains, all of which are associated with bacteria. These antigens, it turns out, bind to recognition sites other than the usual cell surface antibodies. Once activated, such selected B-cells would, of course, form large clones of antibody-secreting plasma cells. Although there are many unanswered questions about this means of B-cell arousal, including the nature of the recognition sites, it would seem that such activation is an efficient means of combatting bacterial intruders.

Suppressor T-Cells: The Battle is Won

We have seen how the immune system is activated and how a highly specific but massive response is generated. But how is the immune response stopped once the danger is past? A third kind of T-cell, the **suppressor T-cell,** modulates the immune response, keeping it from running out of control, and eventually bringing it to a halt. We haven't mentioned this one until now because researchers do not yet know how suppressor T-cells arise, and they are not certain of how they work. They are thought to inhibit any remaining and unaroused virgin T-cells or B-cells that are capable of selection and activation. Since the lifetime of activated T-cells and plasma B-cells is limited, blocking the activation of reserves will automatically limit the number of active lymphocytes and eventually halt the immune response. One key to suppressor T-cell activity may be the amount of antigen circulating. Suppressor T-cells may be sparked into action by diminishing antigen, a sign of victory over the invader.

Summing Up the Primary Immune Response

The body has an immensely diverse army of virgin B- and T-cells, capable of identifying and responding to just about any antigen. When an invasion occurs, macrophages combine captured antigen with their Class I and Class II MHC surface proteins, thereby becoming antigen-presenting cells. They form matches with specific cytotoxic T-cells and helper T-cells, binding to them and secreting interleukin I, which stimulates clone formation. The cytotoxic T-cells perform the cell-mediated response, attacking and killing infected body cells. B-cells find matching antigen and form receptors with their Class II MHC protein that will match receptors on helper T-cells. Binding results in the release of interleukin I and the growth of vast antibody-secreting B-cell clones that then carry out the humoral response. The antibodies clump antigen, activate complement, and support opsonization. Suppressor T-cells stop the primary immune response. But now that we are well, what if the same invader should again break through our defenses?

VIGILANT MEMORY CELLS AND THE SECONDARY IMMUNE RESPONSE

The first time we fall ill to a new invader, our reaction to that invader may be quite slow. This is because it takes time for macrophages and lymphocytes to mobilize their response. But given that time and if all goes well, the offenders will be dealt with and we will recover. We are left weaker but "wiser." That is, our immune system is wiser. (*We* may go right on doing whatever it was that got us sick.) This new wisdom is found in the memory B- and T-cells, which are now capable of what is called the **secondary immune response.**

Memory cells have the ability to recognize the specific antigen that earlier aroused their sister cells. While the other activated B- and T-cell clones live short, busy lives, memory cells live on and on, perhaps for decades. This is important, for should the same invader show up a second time, it is immediately recognized, and the body's response is much quicker than before. The selected memory B- and T-cells undergo round after round of cell division, and soon a massive new army of active lymphocytes arises to repel the second invasion. We may not even be aware of the renewed struggle, for symptoms of illness are often absent or quite mild. So, thanks to our memory cells, we suffer many diseases just once.

Active and Passive Immunity

When you got your first vaccinations your body was provided with shortcuts to the secondary immune response. You skipped most of the primary immune response because you developed **artificial active immunity.** The vaccine you received contained weakened or killed disease agents that, while harmless to you, still maintained their antigenic properties and could thus promote the primary immune response with little of the misery and risk of the actual disease. During this shortened response, banks of memory cells were produced as usual, but the aroused lymphocyte army, having found itself with little real work to do, quickly retired. But had you later been confronted with the real disease agent, your immune system would have gone right into the streamlined secondary immune response, and the invader would never have known what hit it.

Where vaccines aren't available or where the disease has already begun, alternatives are possible. One is the injection of an **antiserum** containing the specific antibodies against the antigenic agent. Such antibodies are routinely obtained from animals (and humans) exposed to the disease agent under carefully controlled laboratory conditions. When injected, the antibodies go about the task of immobilizing the invader's antigens. Since the immune system is not activated, this is known as **passive immunity.** Until recently this was the only treatment available against the deadly rabies virus. Antibodies are short lived, and therefore the protective effects are temporary.

We've still left a few questions unanswered, as you may have noticed. For instance, how can so many different types of antibodies arise? That is, how does the body provide for the incredible diversity needed in virgin lymphocytes? After all, there is a seemingly endless number of different antigens around.

THE ROAD TO LYMPHOCYTE DIVERSITY

All mammals and birds, and perhaps all vertebrates, can produce antibodies against any large molecule. If you inject a rabbit with crocodile hemoglobin, the rabbit will make hundreds of different specific antibodies against crocodile hemoglobin. How does the immune system do it? Can rabbits be *preprogrammed* to resist crocodile proteins? Are there enough antibody-coding genes to go around to enable rabbits to resist crocodile molecules? More significantly, are there enough genes to code for antibodies against millions upon millions of potential antigens?

The answer is complex, but let's start by emphasizing that there are not nearly enough genes in the human genome to provide for such diversity. To make all these antibodies, certain genes must be modified through **gene rearrangement.** The genes that code for the variable regions of antibodies occur in about 300 DNA segments located throughout the chromosomes. These segments are rearranged during the immune system's development. That is, the genes coding for the antibody variable regions are actually broken apart and recombined in a seemingly infinite number of ways (estimated at about 18 billion). You may recall that we've said that all the cells of the body have the same genetic information. Obviously, a correction is in order. You can now see that each clone of differentiated B-cells had its DNA rearranged so that it has certain tailormade genes that other cells don't have.

We see, then, that during B-cell maturation, the DNA of the chromosome itself is permanently rearranged to create new variable region genes (Figure 41.19). Since there are many ways to break and rearrange the chromosome, any one of a vast array of possible antibody genes can be created in a given B-cell.

Antibody diversity also can be produced in other ways. One way is by rearranging the gene products, the proteins themselves. In addition, somatic mutations can produce single DNA base substitutions (see Chapter 17), causing slight variations in the antibody (known to occur in the genes coding for IgA and IgG classes).

Now what about the T-cells? How do they produce surface receptors so varied that they can recognize virtually any antigen? It turns out that much less is known about the genetics of the alpha polypeptide component of the T-cell's dual recognition system. However, immunologists strongly suspect that such diversity must be produced in a manner similar to that responsible for the diversity of antibodies, that is, essentially through rearrangements of DNA segments within the genes responsible.

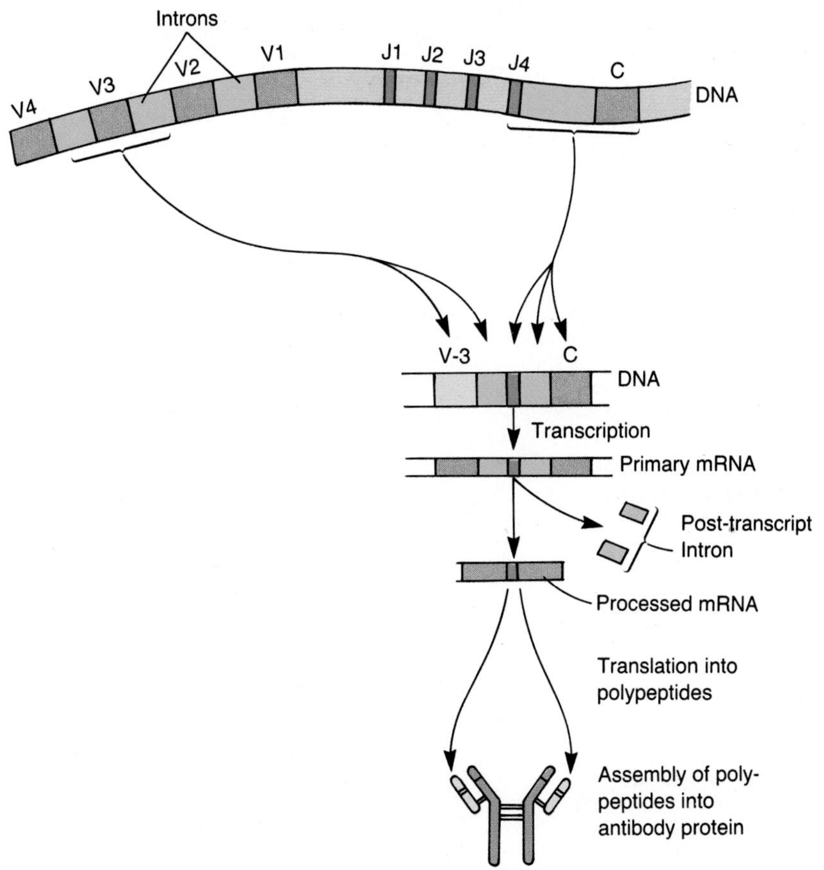

FIGURE 41.19
ANTIBODY DIVERSITY AND ASSEMBLY
Antibody diversity is traced primarily to the manner in which variable region genes are selected and rearranged in each maturing B-cell. The reorganization of a light chain gene and its transcription is seen here. As variable genes are removed, they are assembled along with joining genes and constant region genes to form a special antibody genome, unique to that cell. All that remains is protein synthesis and the formation of light chains.

MONOCLONAL ANTIBODIES

Antibodies have long been used to detect the presence of specific disease antigens and other molecules of research interest. We came across such a use in Chapter 37 (radioimmunoassay). In theory, antibodies can be used to locate any molecule. The chief difficulty is in obtaining a pure sample of the specific antibody. The traditional technique of producing antibodies is to introduce a small amount of known antigen into a laboratory animal, and after a time draw blood and isolate the antibodies. Unfortunately, the introduced antigen usually contains many different antigenic determinants, so a single antigen will activate many different B-cell lines, and thus a number of different antibodies will form. What was clearly needed was a way to sort out B-cells according to their specific antibody production.

In 1976, researchers found a way of doing just that—and began producing what became known as **monoclonal antibodies.** First, an animal is injected with an antigen, and after a time B-cells are obtained from the animal's spleen, an organ of intense B-cell activity. Such B-cells will include some that are producing the desired antibody, but in order to isolate such cells, the numbers of B-cells has to be increased. But B-cells do not do well in tissue culture, so the cells are first fused with **myeloma cells,** those derived from a tumor. (Tumor cells are quite hardy in cell culture, and a line can be maintained indefinitely.) The fused cells, called **hybridomas,** are then cultured, their numbers greatly increased. Then they can be sorted according to the antibody produced, the desired lines are cultured, and monoclonal antibody is collected (Figure 41.20).

There are many exciting uses for monoclonal antibodies. They can be made radioactive or fluorescent and thus can be used to locate readily the antigen that brought about their formation in the first place. For example, the molecule of interest (antigen) could

FIGURE 41.20
MONOCLONAL ANTIBODIES
The production of monoclonal antibodies requires several steps. (**a**) An animal is injected with an antigen containing several antigenic determinants, whereupon its B-cells begin producing specific antibodies. (**b**) The spleen is removed and activated B-cells grown in culture. Myeloma cells and a fusing agent are added, and long-lived hybridomas form. Large numbers of hybridomas are produced, and (**c**) the cells are separated according to the antibody they produce. Large clones are maintained and pure monoclonal antibodies harvested.

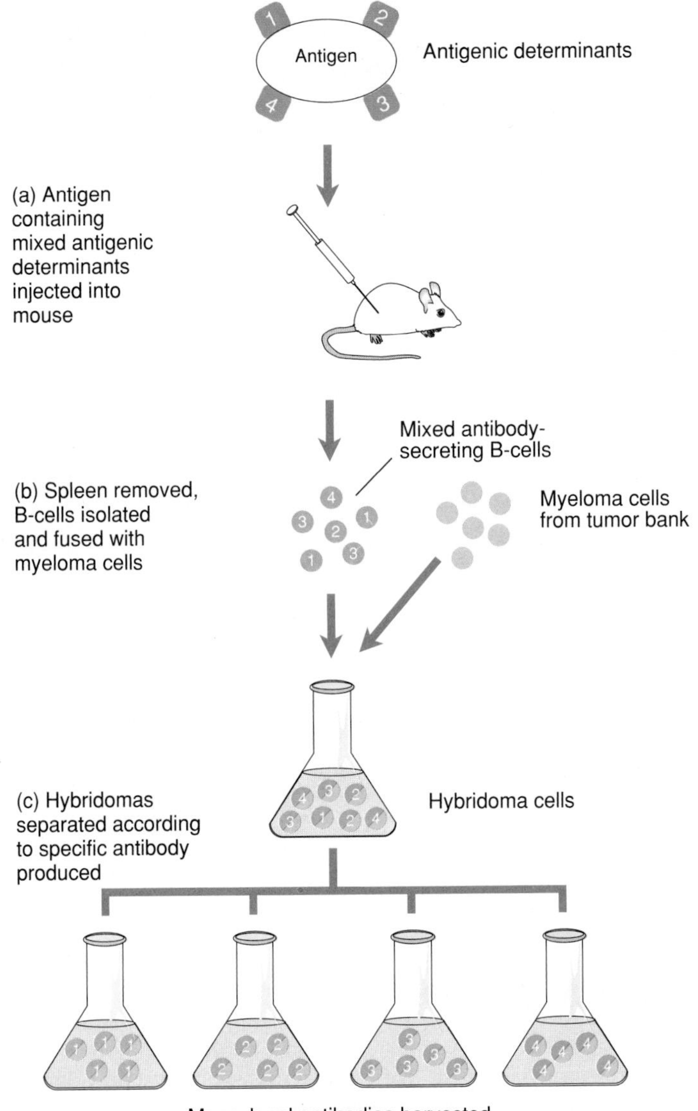
Antigen
Antigenic determinants
(a) Antigen containing mixed antigenic determinants injected into mouse
Mixed antibody-secreting B-cells
(b) Spleen removed, B-cells isolated and fused with myeloma cells
Myeloma cells from tumor bank
(c) Hybridomas separated according to specific antibody produced
Hybridoma cells
Monoclonal antibodies harvested

be a hormone, or even its cell receptor site; either could be identified. Monoclonal antibody can even be used in pregnancy testing by detecting the presence of the hormone HCG (human chorionic gonadotropin; see Chapter 43), produced by the newly formed placenta. Recent applications also include the clinical diagnosis of viral and bacterial diseases. For example, antigens associated with a sexually transmitted disease can be detected in a matter of minutes, whereas the diagnosis used to require days for culturing pathogens sampled from the infection site. Monoclonal antibodies can be used as drug delivery systems, that is, as a means for getting medication to the right place in the body. In the treatment of cancer, for example, physicians are experimenting with the concept of linking potent but toxic cancer-treatment drugs to monoclonal antibodies that will bind to cancer cell surface antigens. The hope is that the drug will go directly to its target and do its job. Thus, dosages of these harsh drugs can be greatly reduced and the difficult side effects eliminated. It seems the potential applications for monoclonal antibodies are limited only by the imagination of researchers.

WHEN THE IMMUNE SYSTEM GOES WRONG

We've marvelled so much over endless antibody diversity and the incredible specificity and preciseness of the immune system that we may have left an impression of perfection at work. This is hardly the case. Things do go wrong with the immune system, and when they do, the results are usually disastrous. We've all heard of the unfortunate boy whose immune system never really kicked in. He had to live within a sterile, room-sized plastic bubble while physicians tried desperately to goad the system into working. More commonly, we are confronted with a previously functional immune system gone awry—either on a rampage against itself, or in a suppressed state, as in AIDS. We will look further into these two immune problems.

Autoimmunity: Attack Against Self

Considering the enormous complexity of the immune system, it should be no surprise that it sometimes makes mistakes. A line of lymphocytes may begin reacting against self, that is, against one of the organism's own proteins or tissues as though it were a foreign invader. Such a reaction is called **autoimmunity**, or **autoimmune disease.** Among the many known or suspected autoimmune diseases are arthritis, nephritis, rheumatic fever, systemic lupus erythematosus, various hormone disorders, certain forms of diabetes, and possibly schizophrenia.

One way autoimmune disease arises is as an aftermath of certain infections. Some disease organisms are particularly insidious in that their surface antigens are sufficiently like our own that the invaders are recognized as self, which lets them slip by our defenses. Eventually, however, they will be detected, and the immune system will come up with antibodies that will knock out these impostors. The problem doesn't end there, though. Unfortunately, the antibodies created may cross-react with our own tissues, causing a severe autoimmune reaction. The *Streptococcus* bacteria that cause strep throat are notorious for this. An infection in the throat leads to the formation of antibodies that can attack tissue elsewhere, notably in the kidneys or heart valves, with serious and sometimes fatal results.

AIDS: The Crippled Immune System

In 1981 the medical community became aware of a problem that had quietly begun to take hold some years earlier. A growing number of young men were coming down with what were formerly "textbook diseases," those so rare that most doctors had only dim, medical school memories of them. These men, living chiefly in coastal cities of the U.S., were developing Kaposi's sarcoma, a rare cancer that weakens blood vessels and announces its presence through purple marks on the skin. Some had pneumocystic pneumonia, an ailment brought about by an invading protist that the body should have resisted easily. Others had peculiar fungal infections of the mouth that caused a fuzzy coating on the tongue that could not be scraped off.

Workers at the Centers for Disease Control in Atlanta determined that the sudden rise of such exotic diseases was attributable to a suppressed immune system, and in 1982 they named the condition **AIDS, acquired immune deficiency syndrome.** They quickly recognized a pattern; almost all of the cases involved young homosexual men. However, the condition soon showed up in other groups, including bisexual men, intravenous drug users, people who had received blood transfusions, the sexual partners of these people, and the children of infected mothers. To most epidemiologists, this information meant that an infectious agent was responsible for the syndrome and that the agent more than likely could be sexually transmitted. Actually, other examples of virally-induced immune system suppression had been known for some time, so the researchers were not entirely in the dark.

In 1984, the infectious agent of AIDS was discovered independently by French and American workers. The culprit was determined to be a retrovirus now called **HIV** for **human immunodeficiency virus**, although for a time the preferred name in the United States was HTLV-III, for human T-cell lymphotrophic virus. The biology of HIV is discussed in Chapter 22.

Although HIV can affect several kinds of cells, it preferentially attacks helper T-cells. As HIV infection progresses, the number of these pivotal lymphocytes dwindles. Their decline is followed by a decrease in B-cell activity, and without the antibodies they secrete, the body becomes a "sitting duck" for rare diseases and other opportunistic infections. The effects are so disabling that most individuals with active AIDS die within three years.

No one yet knows exactly how the virus affects helper T-cells, but some researchers suggest that it renders their receptor sites inoperative. Others believe that HIV brings about an autoimmune response in which lymphocytes begin attacking each other, thus leaving the victim defenseless. Whatever its mode of action, HIV is an insidious invader, attacking the very group of cells that organize the body's resistance.

In Chapter 22 we discussed the mode by which HIV gains entrance to cells and its destructive activities therein, so here we can be brief. After an AIDS virus attaches to its target cell, it invades by penetrating the plasma membrane (see Figure 22.23). Once within the cell, the virus sheds its protective protein coat and releases two copies of single-stranded RNA (its genome), along with the enzyme reverse transcriptase. This versatile enzyme converts the RNA to double-stranded DNA, which is then inserted into one of the host's chromosomes. There the viral genome may remain inactive, simply replicating along with the host chromosome, perhaps for years. Or, the new viral DNA may immediately become active, transcribing new viral protein, replicating viral genes, and assembing many new viral particles. The cell soon ruptures, and the new infective agents are released to infect still more cells or to be passed on to a new host should the occasion arise.

In its attack, the AIDS virus first binds to a cell surface protein called **CD4**, located on the plasma membrane of helper T-cells. Certain glycoproteins on the viral coat form a perfect match with the CD4 molecule. The CD4 protein is also present on other immune cells, including macrophages, about 5% of the B-cells, and on certain cells outside the immune system. Included among the latter are selected duodenal and colon cells, certain skin cells, and the brain's macrophages and glial cells (the latter make up much of the brain's mass). In fact, HIV invasions of the brain have now been well-documented. The susceptible cells of the body have one thing in common—they all bear the CD4 protein.

The "Achilles heel" of HIV may well be its use of the CD4 protein as a binding site. Workers now believe that one effective therapy might be to flood the patient with monoclonal antibodies that can bind with CD4 and thereby block the entry of the AIDS virus into body cells. Another possible therapy may be to introduce large amounts of

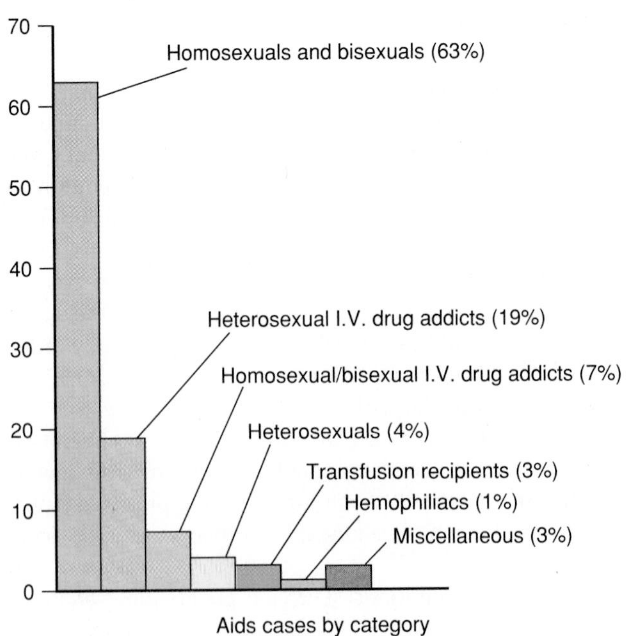

FIGURE 41.21
AIDS STATISTICS
By far, the greatest number of AIDS cases to date in the U.S. have occurred in homosexuals and bisexuals.

CD4, which would then bind to the viral glycoproteins themselves, thereby preventing them from binding to target cells. Such procedures are now being attempted, but so far with very limited success, and immunologists guess that we are still years away from an AIDS solution. Further, HIV is known to mutate frequently, introducing some knotty problems for medical research.

In the meantime, the only effective treatment for AIDS in the U.S. is the antiviral drug **AZT** (azidothymidine). AZT does not cure AIDS; it only prolongs the lives of certain victims. AZT acts by blocking the enzyme reverse transcriptase, the one responsible for converting viral RNA to DNA. During reverse transcription, the enzyme incorporates AZT instead of the usual thymine nucleotide into the growing DNA chain. This immediately blocks the addition of the next DNA nucleotide, so reverse transcription stops and the viral life cycle cannot proceed. Unfortunately, AZT's potential effectiveness turns out to be temporary, and it has serious side effects in some people. Now, researchers fear that new mutant strains of HIV have emerged that are somehow resistant to the effects of AZT. Nevertheless, encouraged by such partial successes, researchers are working intensely at developing drugs with similar properties.

During all of this, the epidemic goes on. Between 1981 and mid-1988, as many as 1.5 million cases of HIV infection are estimated to have occurred in the United States. Also in this time period, a total of 66,464 cases of active AIDS had been reported, and of these, 37,535 of the victims had died. By comparison, in 1952, the worst year of the polio epidemic, 21,000 cases of paralytic polio were reported. The AIDS problem also shows many signs of intensifying. We may be looking at hundreds of thousands of new AIDS cases by the early 1990s. On a worldwide scale, the numbers are frightening. World Health Organization officials estimate 5 million people have been infected by the AIDS virus; some 250,000 active AIDS cases have already occurred; and 180,000 have died. Another million new active AIDS cases are likely to occur in the next five years. (For a look at the situation in Africa, where HIV arose, see Chapter 22.)

U.S. Public Health Service authorities now believe that the virus is present in one out of 30 men between 20 and 50 years of age. While most U.S. AIDS victims are men, by mid-1987 some 2,000 American women and several hundred children had the disease. (In Africa, AIDS is distributed equally between the sexes.) In 1988 (Figure 41.21), the highest risk group continued to be male homosexuals and bisexuals (about 63% of present cases), followed by heterosexual intravenous drug users (19%), homosexual or bisexual intravenous drug users (7%), heterosexuals (4%), clinical blood recipients (3%), hemophiliacs (1%), and cases of undetermined origin (3%). Very recently, epidemiologists have noted a sharp rise in the incidence of AIDS in black heterosexuals, all attributed to illicit intravenous drug use.

The Progression of AIDS It is important to distinguish between the presence of HIV in a person and the onset of active AIDS. For each 100 people identified as having been infected (that is, their blood tests positive for antibodies against HIV), only one of them on the average has full-blown, active AIDS. These active cases are the people who are usually hospitalized for Kaposi's sarcoma, pneumocystic pneumonia, and other rare diseases. A second group, about 10%, will have earlier symptoms, which include weight loss, swollen lymph nodes, prolonged colds and other viral infections, fatigue, diarrhea, and persistent fever (any or all of which could be from other causes). The remainder, some 89%, will harbor the virus but have no outward symptoms. The outcome for this latter group is uncertain, but we know that active AIDS can crop up in people infected as far back as 10 years, maybe longer.

With more and more data becoming available, immunologists are now finding a pattern in the response of the immune system to HIV. Following the progress of the virus over a 10-year period, researchers note that it multiplies rapidly during the first year, but then, as the immune system responds, the virus diminishes drastically. The immune system remains more or less in control of the infection for another five or so years, after which its effectiveness undergoes a decline, reaching a low in the ninth and

FIGURE 41.22
AN INVITATION TO AIDS
The injection of street drugs, often involving shared and contaminated syringes or batches, is an invitation to AIDS.

tenth years. The decline period correlates closely with an ongoing rapid depletion of helper T-cells, those lymphocytes responsible for activating and coordinating the immune response. The decline in helper T's is accompanied by a new peak in the number of AIDS viruses present in the body. It is at this time that the victim begins to suffer the fatal secondary infections.

The Transmission of AIDS All but about 6% of HIV infections occur via semen and blood. HIV may also be spread in other ways, but the frequency of such spread is low and the means of transmission are not clear. HIV has been found in several other body fluids, including tears, sweat, and mucus. Very possibly, it may occur in mother's milk.

Sexual transmission of HIV infection almost always involves viruses from infected semen finding their way into the blood. AIDS is common in homosexual men because many have engaged in anal intercourse. The rectal wall is fragile, and intercourse invariably causes small abrasions through which the viruses enter the bloodstream. Vaginal intercourse with an infected man is less risky for women because the vagina is not normally abraded during the sexual act (although any minute abrasion, whatever its source, is all that is required). Transmission from heterosexual women to heterosexual men is even less likely, since relatively few women harbor the virus (at least in the United States), and normal intercourse does not usually involve access to the man's bloodstream.

The risks accompanying sexual contact can be reduced through the use of condoms, since they capture the semen and prevent the escape of the virus. But it is important to know that condoms must be of high quality; a surprising number of condoms break during use. As in their contraceptive use, condoms must be fitted prior to intercourse. Animal membrane condoms are of little value, since they are readily penetrated by the AIDS virus.

Intravenous drug users represent the fastest growing risk group in the U.S. today. AIDS in this group is spread largely through the use of contaminated hypodermic needles used to inject "street drugs" (Figure 41.22). Many of these drug users are among the least informed people in our society, and so they often continue sharing needles, oblivious to the risk. Some use liquid laundry bleach to sterilize shared needles, and while this precaution is certainly better than none, it is not entirely effective. Some, using their own personal needles, inject themselves with the AIDS virus picked up from contaminated drug batches shared with others.

The risk of contracting the virus among the heterosexual, nonaddicted public is also increasing. There are two main sources of contagion connecting high-risk and low-risk groups. One is through individuals who move in and out of the drug scene, otherwise maintaining an aura of respectability. The other is the bisexual, the male whose sex partners alternate between high-risk homosexual men and low-risk heterosexual women.

The clinical blood supply is another source of infection, albeit a much less frequent one. Whereas screening for HIV is always performed, it is not entirely effective. Some people, anticipating surgery, are now storing their own blood in advance; others bring their own donors, usually family members, to the hospital just in case.

Saddest of all, a growing number of AIDS victims are infants and children. While no one is sure how infants become infected, it is possible that HIV crosses the placenta. Alternatively, it may be passed to infants through breast feeding. What is known is that HIV antibodies, generated by an infected mother, do cross the placental barrier and can be detected in the infant for up to a year. During this period there is no way of determining whether or not the baby has been infected, since the mother's and infant's antibody are indistinguishable.

It is reassuring to know that the risk of HIV transmission in ways other than those mentioned is nearly zero. Studies of households where one or more infected persons' lives have revealed no nonsexual transmission of the disease agent. Other studies indicate that the possibility of insect transmission of HIV is highly unlikely.

IMMUNITY IN OTHER ANIMALS

Apparently some immune responses occur in all animal phyla. In similar organisms, such responses tend to be nonspecific and may only involve ameboid phagocytic cells. You may recall that ameboid cells are present in sponges (see Chapter 30). In many invertebrates, certain proteins first coat an invading microorganism, making capture easier for the phagocytes.

Graft rejection is common in sponges and other invertebrates, just as it is in humans. Such rejection clearly suggests that these animals make use of cell surface molecules to identify self and to reject non-self. Most invertebrate groups also demonstrate a degree of immunological memory. When grafting experiments are repeated, using grafts from the same source, the speed of rejection is accelerated with each succeeding experiment. When grafts are performed repeatedly from different sources, rejection of each new graft takes about as long as first-time grafts.

As one might expect, the most complex immune responses in invertebrates are those of coelomates. Some annelids, for instance, have nearly all of the immune capabilities of vertebrates, including cell responses similar to those of vertebrate B-cells and T-cells. They also exhibit a definite immune memory. Arthropods exhibit a B-cell-like activity, but immunologists are still not certain whether anything similar to vertebrate T-cell activity occurs. The echinoderms and tunicates (both more closely related to vertebrates than to most other animals) have well-developed immune capabilities and even have cells closely resembling vertebrate lymphocytes.

Vertebrates boast the most advanced immune system. They all are capable of producing true antibodies, and they all have B- and T-cells and lymphoid tissue. In fact, immune capabilities are fairly uniform throughout the vertebrates, although lymphoid tissues have become increasingly specialized in the most recently evolved classes—particularly the birds and mammals. In the mammals, we find the greatest specialization in T- and B-cell functions, MHC proteins, and antibodies. Among vertebrates, the jawless fishes (lampreys and hagfish) have the least amount of lymphoid tissue. The jawed fishes, however, have distinct lymphoid structures, particularly in the thymus, spleen, and gut.

It seems, then, that the more recently evolved animals have the most elaborate and specialized immune systems. It is not clear how this came to be, but certainly the ongoing seesaw episodes of coevolution between parasite and host played a key role.

antibody (also immunoglobulin) A protein consisting of two light polypeptide chains, covalently linked to two heavy polypeptide chains, the molecule containing a variable region that matches and binds to a specific antigenic determinate (part of an antigen).

antigen (also immunogen) Any substance capable of prompting the specific immune defenses including humoral and cell-mediated responses.

antigenic determinant A more precise term than antigen, meaning one of several regions on an antigen to which specific, matching antibodies will bind.

cell-mediated response The activation and response of T-cells to invaders, including the activation of B-cells by helper T-cells, the destruction of infected body cells by cytotoxic T-cells, and the regulation of the response by suppressor T-cells.

clonal selection In the primary immune response, the selection process in which a specific antigen or antigenic determinant activates only those lymphocytes with matching membrane recognition sites, such recognition leading to proliferation of that specific line and thus the start of the specific defense.

humoral response The manufacture and secretion of specific antibodies by plasma cells (activated B-cells) in response to the presence of antigen and activation by helper T-cells.

lymphocyte A class of leukocytes including B-cells, T-cells, and NK (natural killer) cells.

MHC (major histocompatibility complex) A gene grouping on the sixth human chromosome that codes for the assembly of complement and cell surface proteins, the latter designated MHC I and MHC II proteins. Such proteins are unique to each individual.

MHC restriction A condition of T-cells in which the receptor sites in each group and its clones recognize only a specific Class I or Class II MHC protein. Cytotoxic T-cells are restricted to Class I MHC proteins, whereas helper T-cells are restricted to Class II MHC proteins.

nonspecific defense The instant arousal of chemical and cellular agents in response to tissue damage and/or invasion by disease organisms or the presence of foreign substances.

phagocyte In the immune system, any cell capable of engulfing and digesting cellular debris, foreign substances, or invading organisms, often with the aid of complement and antibody.

primary immune response The time-consuming arousal of specific defenses by a *first-time* invader, beginning with antigen-presenting cells and continuing with the selection and proliferation of specific T- and B-cell lines.

secondary immune response The rapid arousal of specific defenses by memory T-cell and B-cell lines that, having identified a previous invader, produce immense clones of defending cells.

specific defense Time-consuming immune defenses involving antigen and/or antigen presenting cells, and the arousal of highly selected T-cell and B-cell populations that carry out humoral and cell-mediated responses.

KEY IDEAS

THE FIRST LINE OF DEFENSE

1. The **immune system** provides defenses against foreign substances and invading organisms.
2. The intact skin, its keratin layer and sweat and oil glands, is a first defense. Mucus secretions, stomach acids, body fluids with **lysozymes**, urine flow, and commensal and mutual populations of microorganisms also make up the first-line defense.

SECONDARY DEFENSES: NONSPECIFIC AND SPECIFIC

Secondary defenses include **nonspecific** and **specific** chemical and cellular agents.

Nonspecific Chemical Defenses

1. **Histamines** produce inflammatory responses aided by **kinins**, both of which increase vessel permeability.
2. **Complement**, a number of serum proteins, forms defenses that include **opsonization** by making a link between bacteria and **phagocytes**, and killing bacteria directly by disrupting the plasma membrane.
3. **Interferon** synthesis is prompted by the presence of double-stranded viral RNA. It stops viral protein synthesis, slows division in cancerous cells, and promotes viral resistance in nearby cells.

Nonspecific Cellular Defenses

Leukocytes include **eosinophils**, **neutrophils**, **monocytes**, **basophils**, and **lymphocytes**. The first four provide general defenses, while the lymphocytes provide specific defenses.

1. Eosinophils utilize toxic granules to kill parasitic worm larvae.

2. Neutrophils and monocytes follow chemical gradients to infection sites. They have cell surface receptors that bind to invading organisms, complement, and antibodies. Monocytes grow into **macrophages** at infection sites. Together with neutrophils, they engulf foreign matter, cellular debris, and infected cells.
3. **Natural killer cells** develop from lymphocytes, later roving the body and attacking virus-infected and cancerous cells.

SPECIFIC CELLULAR AND CHEMICAL DEFENSES

Specific defenses, carried out by lymphocytes, require complex recognition systems and molecular learning.

Lymphoid Tissues and Lymphocytes

Lymphocytes arise and mature in **primary lymphoid tissue** (red bone marrow and thymus). They reside as virgin cells in **secondary lymphoid tissues** (lymph nodes, tonsils, adenoids, spleen, small intestine), where they are activated.

B- and T-Lymphocytes

1. **T-cells** differentiate in the thymus and **B-cells** in the red bone marrow. The two differ in surface receptor proteins and in the extensive endoplasmic reticulum of B-cells.
2. B-cells respond to invaders with the **humoral response,** the formation and secretion of antibodies. Some T-cells carry out the **cell-mediated response,** the activation of specialized cells that attack infected body cells. Other T-cells activate B-cells.

Antibodies and the Humoral Response

1. **Antigens** are molecules that elicit specific immune responses by lymphocytes. Such molecules include foreign proteins, nucleic acids, or carbohydrates. **Antibodies** bind noncovalently to **antigenic determinants** located on antigens.
2. The basic Y-shaped antibody is constructed of four polypeptide chains. The two forks contain **light** and **heavy chains,** ending in a **variable region.** The remainder of the Y, including the stem, is made up of **constant regions.**
3. Variable regions bind to antigenic determinants, whereas the **Fc region** of the heavy chain constant region binds to a phagocyte **Fc region.**
4. Whereas antibodies differ from each other in their variable regions, the five classes of antibodies differ in the heavy chain regions forming the stem. Each class has a special function.
5. Free floating antibodies circulate, whereas bound IgM and IgD antibodies make up antigen receptors in B-cells.
6. Antibodies bind antigens together, forming clusters. Some interactions stimulate complement release by cells. Others cause opsonization after the antigen becomes coated by antibody. Such antibodies link invading cells to engulfing phagocytes.

Lymphocytes and Cell Recognition

1. Lymphocytes take on their specificity during their development through the formation of specific cell surface receptor sites.
2. Each cell has several hundred genes that form the **MHC complex.** It codes for three classes of proteins. Classes I and II become cell surface MHC proteins, which are unique in each individual. They are essential to cell recognition, permitting lymphocytes to identify invaders, normal and infected body cells, and each other.
3. As T-cells mature, each produces membrane receptor proteins that have the potential to match any number of potential **MHC proteins.** During development, only cell lines with receptors that match the body's actual MHC proteins survive. All nucleated body cells display Class I MHC proteins, but only certain immune system cells have the Class II type. They next have their receptor sites modified into dual receptors. The second receptor is a protein that will match with an as yet unconfronted antigen. As MHC-restricted cells, subpopulations of **virgin T-cells** become specialists in binding to virus and cancer-infected body cells, macrophages, and other lymphocytes.
4. During B-cell maturation, each incorporates thousands of IgM and IgD antibodies in the plasma membrane. These have the potential to form a match with an as yet unconfronted antigen.

CLONAL SELECTION AND THE PRIMARY IMMUNE RESPONSE

Clonal selection is the process wherein an antigen is used to activate **virgin T- and B-cells** with matching receptor sites. Among the millions of virgin lymphocytes, comparatively few clones will have such matches. A key participant is the antigen-presenting cell.

Sounding the Alarm: Antigen-Presenting Macrophages

Macrophages incorporate free or partially digested antigen into the Class I and II MHC proteins, forming a dual site that will match those of virgin **cytotoxic** and **helper T-cells.**

Spreading the Alarm: Aroused T-Cells

1. Antigen-presenting cells, when bound to T-cells, secrete **interleukin I,** a messenger that stimulates cell proliferation. Clones of T-cells arise.
2. Cytotoxic T-cells use their dual recognition sites to identify, attack, and kill infected cells bearing the antigen that originally brought about their arousal. Other T-cells become **memory cells.**
3. Helper T-cells stimulate other T-cells and activate virgin B-cells.
4. B-cells find a match between their cell surface antibodies and free antigen from the invader, incorporating the antigen into one of their Class II MHC proteins. This new receptor will match that of an aroused helper T-cell. When bound together, the T-cell releases interleukin I, and B-cell proliferation occurs.

B-Cells: Several Offensive Strategies

1. **Plasma B-cells** produce and secrete copious amounts of a specific antibody that matches free antigen and antigen on invaders, preparing them for phagocytosis and activating complement release. B-memory cell clones also form.
2. A few bacterial antigens activate virgin B-cells without T-cell intervention.

Suppressor T-Cells: The Battle Is Won

Suppressor T-cells coordinate the immune response, slowing it as antigen levels fall. They may also bind to receptors of unactivated lymphocytes.

VIGILANT MEMORY CELLS AND THE SECONDARY IMMUNE RESPONSE

1. Memory cells of each aroused clone are long-lived, capable of responding at once to new intrusions by the same invader that brought on their initial activation. This brings on the **secondary immune response.**
2. Memory cells, when stimulated by new exposures to their antigen, immediately proliferate, bringing on a far more rapid response than occurred with initial exposure.

Active and Passive Immunity

1. Vaccines containing weakened or killed organisms promote **artificial active immunity** by bringing on a much reduced version of the primary immune response. The clones of memory cells provide the immunity.
2. Vaccines composed of specific antibody against a disease provide **passive immunity.** No memory clones form, so the effect is temporary.

THE ROAD TO LYMPHOCYTE DIVERSITY

In the diverse B-cell populations there are individuals with receptors for any potential antigen. Such diversity is genetically based, but it is beyond the original capacity of the genome. Unlimited antibody diversity is possible because of **gene rearrangement** involving variable region and joining genes. There is also versatility in the final arrangement of variable gene products. Somatic mutation accounts for additional rearrangements of amino acid sequences in the polypeptides. The genes that determine T-cell antigen receptors are also thought to go through gene rearrangement, but little is known about such genes.

MONOCLONAL ANTIBODIES

Clones of B-cells that produce a single antibody are produced through genetic engineering techniques.

Monoclonal antibodies have many clinical applications. Procedures for obtaining such clones include inducing antibody formation in animals and obtaining B-cells, which are then fused with **myeloma cells** to form **hybridomas.** These are separated and screened, and the monoclonal cells are cultured.

WHEN THE IMMUNE SYSTEM GOES WRONG

Autoimmunity: Attack Against Self

Autoimmunity, self-destruction by immune cells, arises when antigens bear chemical characteristics similar to our own MHC surface proteins. The antibodies produced by B-cells interact with such proteins affecting our own cells.

AIDS: The Crippled Immune System

1. **AIDS (acquired immunodeficiency syndrome)** was recognized in the early 1980s when the incidence of rare diseases rose suddenly. AIDS is a condition in which the immune system is suppressed. The agent is **HIV, human immunodeficiency virus,** a retrovirus that devastates helper T-cells and others. It invades cells by binding to a cell surface protein, **CD4.** Most American AIDS victims are homosexuals, bisexuals, intravenous street drug users, blood transfusion recipients, and babies born to AIDS victims.
2. Upon entry into host cells, the single-stranded RNA genome of HIV is converted to double-stranded DNA and inserted into the cell's genome. There, it may remain inactive or begin at once to reproduce, thereby crippling the immune system.
3. AIDS is spread chiefly through semen and blood. Any surface cut or abrasion that contacts infected semen or blood is sufficient for transmission of the virus. Sexual intercourse and the use of unsterilized hypodermic needles are primary ways in which infection is spread.

IMMUNITY IN OTHER ANIMALS

Defenses in simpler invertebrate animals usually involve nonspecific cellular agents such as phagocytes. Grafting experiments with sponges indicate they have some immunological memory. Coelomates reveal the greatest development of immune systems among invertebrates. Some have specific cellular responses with lymphocyte-like cells. All vertebrates have B- and T-cells and lymphoid tissue. Birds and mammals have humoral and cell-mediated responses similar to those of humans.

REVIEW QUESTIONS

1. Summarize the general functions of the immune system and name the kinds of invaders. (980)
2. List five specific agents or conditions of the body that prevent the invasion of parasites. (Figure 41.4)
3. List four different nonspecific chemical immune defense agents and briefly summarize the work of each. (983–984)
4. List three types of phagocytes and explain how they deal with invaders. (984)

5. Which of the leukocytes specializes in parasitic worm infestations? Describe its performance. (984)
6. What is the specialization of natural killer (NK) cells? What is their probable origin? (984–985)
7. List two primary and four secondary lymphoid tissues. In general, what happens in each? (986)
8. Distinguish between the humoral and cell-mediated specific immune responses. Which leukocytes are responsible for each one? (987–988)
9. What determines whether a substance is an antigen? What, specifically, are antigenic determinants? (988)
10. Prepare a simple drawing of an antibody, labeling the following: heavy chain, light chain, disulfide bond, hinge, variable region, constant region, and Fc region. (988–989)
11. Where are bound antibodies usually found? Free antibodies? (989)
12. List three variations in the antibody/antigen reaction. Use simple drawings to illustrate these. (989–990)
13. What two things about cells are the receptor sites of T-cells able to recognize? Why are both essential? (991–992)
14. What are the MHC proteins? Why are they also called MHC antigens? (991–992)
15. Briefly summarize how T-cell receptors form their dual recognition proteins and what each recognizes. (991–992)
16. What distinguishes a cytotoxic T-cell's dual receptor from that of a helper T-cell? (992)
17. What constitutes the B-cell's cell surface receptor? What is its function? (992, 994)
18. The primary immune response occurs through the clonal selection process. What is clonal selection, and what problem does it solve? (994)
19. Summarize the way in which macrophages become antigen-presenting cells. Why is this a pivotal role in the primary immune response? (994)
20. What determines whether an antigen-presenting macrophage binds to a virgin cytotoxic T-cell or a virgin helper T-cell? What is the first thing that happens in either case when such binding occurs? (994)
21. What is the specific task of the activated cytotoxic T-cell? How do its receptor sites aid in this? (996–997)
22. Explain how cytotoxic T-cells kill their target cells. (996–997)
23. What are two functions of helper T-cells? (997)
24. Describe the three major events in the activation of B-cells. (997–998)
25. Briefly discuss the chain of events that connects antigen, specific B-cell, antibody, and invading organism. (997)
26. How do memory T- and B-cells differ from active T- and B-cells? In what way do they make the secondary immune response happen more quickly and efficiently than the primary response? (999)
27. Distinguish between active and passive immunity; natural and artificial immunity. (999–1000)
28. A simple count of genes in the human genome tells us that there are nowhere near enough to account for the immense antibody diversity of B-cells and the equally immense antigen recognition proteins of T-cells. Describe the central process cells use to provide for this diversity. (1000)
29. What are monoclonal antibodies? Why is having cultures of cells that produce one and only one antibody useful? (1001–1002)
30. What is the agent of AIDS, and what is its specific effect on cells? (1003–1004)
31. Which cells does the HIV favor? What surface molecule does it identify, and how might this present a possible therapeutic avenue? (1004–1005)
32. List four groups of Americans that are considered at high risk for contracting AIDS (highest first) and explain why each is susceptible. (1005)
33. What defensive cells seem to be present in all animal phyla? (1007)
34. Summarize the evidence for some kind of immunological memory in sponges. (1007)
35. Which groups of invertebrates possess the most advanced immune systems? List immune elements found in these groups. (1007)
36. What immune elements do all vertebrates have in common? (1007)

42

REPRODUCTION

WE MAY BE VERY PROUD OF OURSELVES AND THINK WE FIT INTO OUR WORLD very well. However, this is true only of our present world—we cannot assume that we would fit into some future world with equal ease. If when we reproduce, we were to make replicas of ourselves, with each of our offspring bearing 100% of our genes, they would presumably do well as long as the world remained as it is now. But they might be in a lot of trouble if it should change. In fact, our little replicas might not fit into a different kind of world well at all and might die out.

But if we mix our genes randomly with someone else's, those genes having been shuttled around through meiosis, we will produce a motley crew of offspring—all of them different. Then should new conditions arise, at least some of our offspring would be likely to survive to propel our genes into yet other generations. This is the essence of the prevailing explanation for the evolution of sex—that it promotes variation, the variation upon which natural selection can act. Let's look into the various ways in which animals mix their genes and a few ways in which they do not.

MODES OF REPRODUCTION: ADAPTIVE THEMES

Animals have developed countless ways of getting their gametes together. Through this vast array of mechanisms and behaviors, however, there are common themes and trends that can be placed in a certain order, such as from simple to complex. The judicious use of such ordering can perhaps give us a clearer idea of just what the various mating patterns are, and perhaps some insight into how they arose. If we are going to start with the simplest case, however, we will have to set the subject of sexual reproduction aside for the moment and look into its alternative. We've seen it before in the Kingdom Protista, where many member species appear to rely entirely on **asexual reproduction**.

Reproduction without Sex

In the reproductively primitive protozoans, the animal-like protists, sex is quite often not a part of reproduction at all. For example, an ameba may simply replicate its genetic material and divide it equitably through mitosis, following which its single cell divides into two separate cells (as we saw in Chapter 23). This process is called **fission.** We have also seen that more complex animals can reproduce by **budding**, as in the cnidarians, whereupon certain body cells simply produce a new body, usually branching from the parent animal (Figure 42.1). Other animals may reproduce asexually through **fragmentation,** breaking into two or more parts, the cells of each then regenerating into a new individual. This occurs frequently in flatworms and bryozoans. And we have seen that sea stars can reproduce through the **regeneration** of parts (see Figure 31.23). Some sponges produce **gemmules,** collections of viable body cells surrounded by a newly formed body wall and including a skeleton of spicules. Each gemmule may winter over and later develop into a new sponge.

Other animals can reproduce asexually through **parthenogenesis**, that is, the development of an embryo from an unfertilized egg. Parthenogenesis is common among insects such as bees, wasps, ants, and beetles (Figure 42.2a). Some species routinely alternate between sexual reproduction and parthenogenetic asexual reproduction, such as we see in the freshwater crustacean, *Daphnia,* an important part of food chains, especially in northern lakes. Parthenogenesis seems to begin when populations are small and growth conditions optimal, that is, when the living is easy. At such times, *Daphnia* eggs begin development without fertilization, and only female offspring are produced (the development of males is suppressed somehow). The tiny developing females can be clearly seen in the mother's brood pouch (Figure 42.2b). But in harder times, after the populations increase and food shortages dwindle, or when the water temperature falls, parthenogenesis leads to the development of male offspring, and typical sexual reproduction resumes.

The cellular factors involved in parthenogenesis are poorly understood, but certainly complex. In some species, for example, the eggs do not enter meiosis at all, but in others meiosis is completed. In some, meiosis occurs, but the diploid number is restored as meiotic nuclei fuse back together, or when the haploid egg cell fuses with a haploid polar body. Our understanding of the selection for sex in parthenogenesis is not even this clear. Most unfertilized daphnid eggs emerge as females, but unfertilized eggs in honeybees form males (drones).

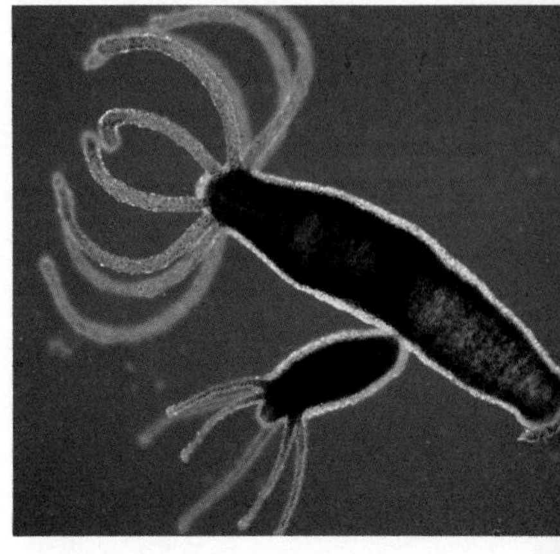

FIGURE 42.1
ASEXUAL REPRODUCTION
The cnidarian *Hydra* reproduces simply and directly through budding. Since only one individual is involved, the new miniature will be genetically identical to its parent.

(a)

(b)

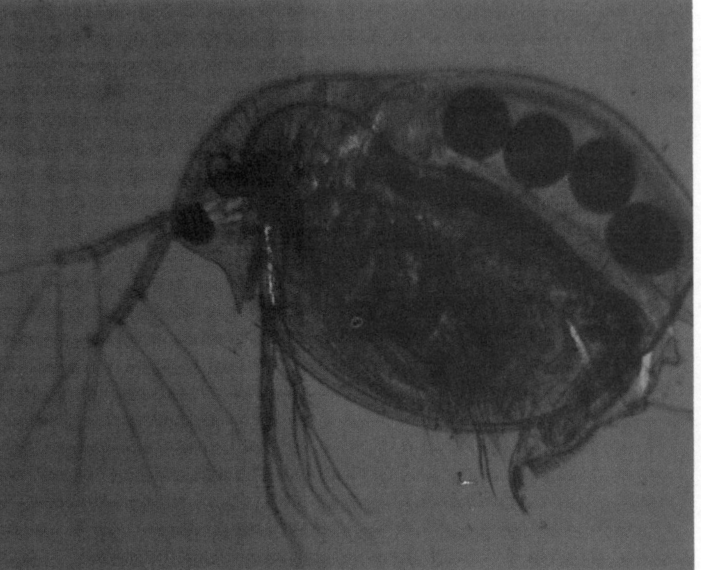

Carolina Biological Supply

FIGURE 42.2
PARTHENOGENESIS
(a) The honeybee colony contains many workers, all females and all formed from fertilized eggs. The males, or "drones," as they are known, develop parthenogenetically—from unfertilized eggs. **(b)** Parthenogenesis in *Daphnia* (the water flea) is the rule rather than the exception. During easier times, the population grows rapidly through this process. Note the presence of partly developed embryos in the brood pouch.

Reproduction with Sex

Virtually all animals reproduce sexually as a matter of course. So let's have a look at them and see how they have become increasingly efficient at (1) getting eggs and sperm together and (2) protecting the developing embryo. Sexual reproduction has seemingly endless variations. One of the basic variations involves just how many sexes are involved, one or two? Another involves how and where the gametes get together; some animals release their gametes for **external fertilization**, while others have the means for carrying out **internal fertilization.**

One Sex or Two? In the lost-and-found column of a newspaper, people who try to describe lost pets usually begin with "male" or "female." That's because they've usually lost a dog or a cat and not a land snail. Land snails are both. When both sexes occur in one individual, the condition is called **monoecious** ("one house"). When referring specifically to animals, the term **hermaphroditic** is also used. When animals are of one sex or another, like horses, dogs, canaries, and us, the condition is described as **dioecious** ("two houses"). Both monoecious and dioecious animals, however, reproduce sexually, with a number of variations on the theme. The **dioecious** condition is considered evolutionarily more advanced than the monoecious condition. This is based on the proposition that there was no sexual differentiation in the earliest sexually reproducing species (see Chapter 23).

External Fertilization External fertilization, in its simplest form, is probably the most primitive and certainly the least cost-efficient form of sexual reproduction. Cost efficiency refers to the numbers of gametes needed to ensure the simple replacement of the individual. The eggs and sperm are released outside the body, where fertilization and development of the embryo occurs. The practice is almost exclusively aquatic—external fertilization on land would present the certain risk of dessication. External fertilization is common to both invertebrates and vertebrates.

Some species—many echinoderms are examples (Figure 42.3)—release their gametes into the water without taking great measures to ensure fertilization. They may, however, increase the odds of successful fertilization by coordinating the release of gametes. Sea stars, for example, often live in dense groups along the rocky sea bed or on wharf pilings. Different individuals in such dense populations, responding to similar cues, release their gametes simultaneously. That is, the release of eggs by one individual prompts others nearby to release great clouds of sperm, and the presence of sperm induces still others to release eggs. Thus there is an amplifying effect. Whereas this does improve the chances of successful fertilization, some sea stars routinely shed as many as 2.5 million eggs in just one spawning. Since the world isn't overrun with sea stars, one would have to assume that in the long run, such numbers are needed merely to maintain a stable population size.

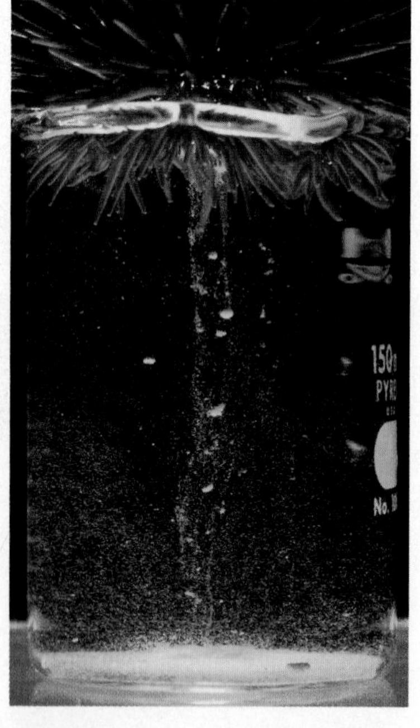

FIGURE 42.3
EXTERNAL FERTILIZATION
Echinoderms shed their gametes into the sea, where fertilization and development take place. The presence of sperm or egg from one individual stimulates others to release the opposing gamete.

Other species also provide a measure of precision in gamete release, both in time and space. For example, the California grunion (Figure 42.4) is a fish that seasonally spawns three or four days after a new or full moon—with the highest tides of the month. Shortly after the tides begin to recede, the females, hotly pursued by the males, ride the incoming waves, thrash their way up onto the wet beach, wiggle their tails into the sand, and deposit their eggs. The males writhe around the females, releasing their sperm. Then, if all goes well, the fish catch the next wave out. (In southern California all does not always go so well; a favorite sport there is "grunion hunting," all very legal as long as only the hands are used.) The grunion zygotes are left to develop, protected in the sand, for a month, until the next high tide stimulates hatching and washes the emerging young out to sea. This brings us to the next evolutionary advancement—the protection of the young.

Internal Fertilization: The Adaptive Advantages In the course of the evolution of sexual reproductive patterns, we assume that the first animals released their gametes willy-nilly into the environment. Then we saw, among some of the more reproductively primitive of today's animals, some attempt at increasing the efficiency of external fertilization by at least releasing the gametes at the same time. The next step was to release them at the same time and place. All these improvements on external fertilization, though, still leave a high degree of risk for the gametes. Many are wasted, and natural selection is, by its very nature, a process that tends to minimize waste.

The next step, then, is to reduce the waste by developing ways to put the sperm directly into the female reproductive tract, where the eggs are. (Think of the time and energy that goes into gamete production and the fact that most are unlikely to enter into fertilization.) Getting the sperm into the partner's reproductive tract has at least three advantages. One, it reduces the number of eggs necessary; two, it vastly increases the probability that all the eggs will be fertilized; and three, it gives the developing eggs a measure of protection—as the mother looks after herself, she protects the embryos within her body, at least for a time.

There is an ecological advantage to internal fertilization as well. With the embryos safely protected in the mother's moist internal environs, sexual reproduction can succeed on dry land. This step, taken hundreds of millions of years ago, opened up a host of new evolutionary opportunities for animals. But let's be careful; the more complex modes of reproduction are not reserved for the more recently evolved animals, nor have they all occurred on land.

FIGURE 42.4
IMPROVING FERTILIZATION EFFICIENCY
Late at night in certain parts of the year, southern California beaches come alive with the small, flapping bodies of the California grunion. Upon reaching the upper reaches of the beach, each female, accompanied by males, digs into the sand and releases her eggs. The eggs are immediately covered by sperm released by the stimulated males. By the next high tide, the young will be ready to hatch as the waves again sweep high onto the beach.

INVERTEBRATE REPRODUCTIVE PATTERNS

It may seem that internal fertilization holds no surprises. After all, there can be only so many ways that this can occur, and we're already somewhat familiar with one of them. What might be surprising, however, is that internal fertilization occurs in the simplest of all animals, the sponges. Most sponges are hermaphroditic, having both egg- and sperm-producing gonads. Sponges often release their sperm in great clouds (Figure 42.5), which promotes the release of similar clouds from nearby sponges. The sperm are carried into neighboring sponges by the inwardly directed currents all sponges generate. They are taken up by collar cells, from which they migrate into the body wall, where they fertilize the eggs.

The most common means of internal fertilization involves **copulation**, the introduction of sperm into the female reproductive tract, usually (but not always) through the use of an intromittent organ, a **penis** or its equivalent. Copulation occurs in a great number of invertebrate groups, many of which are hermaphroditic.

One hermaphroditic species, the land snail *Helix* (Figure 42.6) has a particularly interesting way of stimulating copulation. As two reproductively ready

FIGURE 42.5
SPERM RELEASE IN SPONGES
Sponges are known to release their sperm in great clouds. Some sperm will be drawn into nearby sponges, where internal fertilization will occur. The risks of sperm loss are more than compensated for by their numbers.

FIGURE 42.6
COPULATION IN SNAILS
Hermaphroditic land snails copulate and mutually exchange sperm, using a well-formed penis. Impaling each other with a small dart appears to be an important prelude.

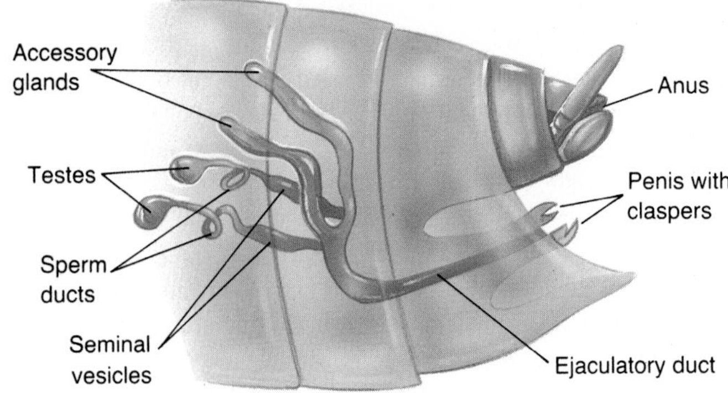

FIGURE 42.7
THE MALE INSECT
The reproductive system of the male insect is complex and well-adapted for copulation and internal fertilization.

snails draw toward each other, they very gingerly begin to touch. After a time they extend their bodies from their shells and gently intertwine. Then, in a bizarre fashion, each suddenly thrusts a tiny, sculptured dart into the body of the other. Both recoil, at first, but each is now strongly stimulated to proceed with copulation. A white spot appears over an eye stalk of each snail; the emerging penis. The *Helix* penis is relatively enormous. When extended, it is almost as long as the snail. Each snail probes the other's body for the reproductive opening, inserts its penis, releases its sperm, and moves away. The sperm are stored in a special receptacle until the eggs are laid.

Male insects are quite distinct from females, and in most species copulation and internal fertilization are the rule. A typical male insect reproductive system (Figure 42.7) is made up of paired testes within which sperm develop, complex paired sperm ducts to conduct the sperm, seminal vesicles for sperm storage, and accessory glands that produce a seminal fluid. The seminal fluid transports the sperm, which are packaged in packets called **spermatophores.** The insect penis, its intromittent organ, is extended outward and inserted into the female's **vagina** (the corresponding female copulatory organ) during mating, whereupon the spermatophores are ejected. The male insect may use elaborate clasping organs for holding the female in position while copulating. Interestingly, the penis is usually barbed, hooked, or braced in such a bizarre fashion that it can only fit into a vagina of closely reciprocal structure. This usually means a female of its own species. Students of evolutionary theory note that this lock-and-key arrangement has evolved to prevent inappropriate matings and unwanted hybridization between similar species. Remember that animal hybrids are usually sterile (unlike many plant hybrids—see Chapter 20). Considering the vast numbers of insect species, it follows that many are quite similar (at least to us), so any mechanism that helps a species avoid wasting its reproductive efforts would be adaptive.

The female insect's reproductive system (Figure 42.8) includes a **spermatheca** ("seed case") in which sperm are stored following copulation. Eggs are produced in paired **ovaries,** and when egg-laying occurs they pass through the **oviducts** to the vagina for release. Fertilization occurs as the eggs move into the vagina. Most insects mate just once, whereupon the females store a lifetime supply of sperm in the spermatheca, enough to fertilize all the eggs she will ever produce. Two or more accessory glands, also leading into the vagina, secrete sticky materials over the eggs, providing them with a means for attaching to some external surface.

Spiders are all carnivores, and they are all venomous. Many species are solitary animals. And a venomous, carnivorous hermit must not be taken lightly. In particular, the habits of spiders add up to problems in mating, especially for the small, weak males who must initiate the process. In some species, they timorously approach the females at mating time, waving special appendages as though they were signal flags. The waving seems to mesmerize the female, at least long enough for the male to mate with her. The orb spider males signal their presence in an even more cautious manner. The male climbs on the female's intricate web and "strums" out a signal that inhibits her feeding behavior. Males of other species even prepare a gift—usually a paralyzed insect, neatly wrapped

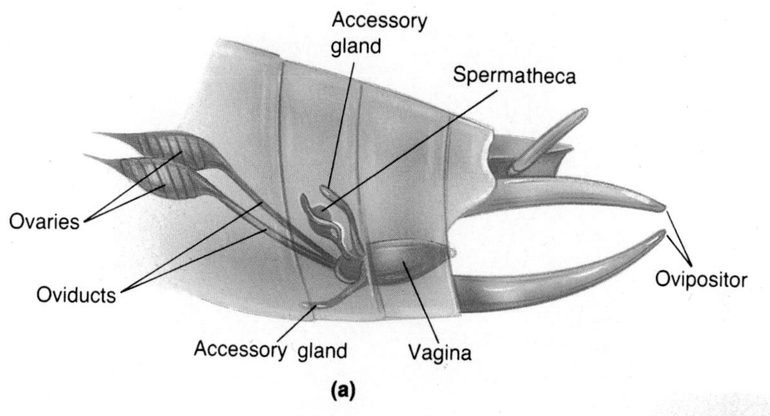

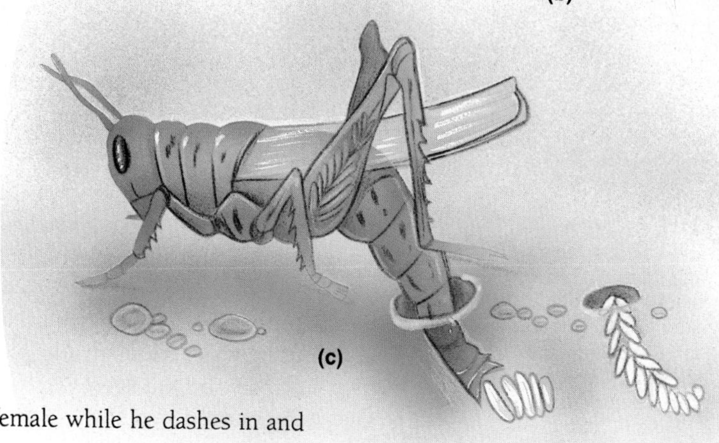

FIGURE 42.8
THE FEMALE INSECT
(a) The female anatomy includes the vagina, oviducts, and ovary, plus a seminal receptacle, the spermatheca. The young generally develop externally within tough egg cases. **(b)** The sticky eggs are often placed on the undersides of foliage, but some females **(c)** have specialized ovipositors that are used to dig or drill recesses in which the eggs are entombed.

in silk—anything to distract or occupy the cannibalistic female while he dashes in and accomplishes his task.

After all this trouble, most species of spiders don't actually copulate. Instead, the male may deposit his sperm on the ground and then scoop it up in a modified appendage. He then either draws near the female or climbs on her back and rakes his sperm-laden appendage over the plates covering her vagina. Then he clears out. He doesn't always make it, but in some species the chances are increased if he pinches the female hard as he leaves. She will then immediately tuck into a defensive ball, and before she can recover, he's gone.

SEXUAL REPRODUCTION IN VERTEBRATES

Now that we are moving into the realm of animals with backbones, it might seem that from here on we would be dealing with evolutionarily advanced traits, such as separate sexes, internal fertilization, and the care and protection of offspring. In many instances, however, this is not the case.

External Fertilization in Vertebrates

Certain ocean fishes, such as cod and plaice, shed their gametes into the sea and then do nothing to protect the young. They do manage to congregate in the same area, though, before discharging their eggs and sperm. The inefficiency of this method, as we've seen, means that great numbers of gametes must be released. (Six million eggs were counted in one female cod. The male's sperm are still being counted.)

Salmon (Figure 42.9) also engage in external fertilization, but they increase the odds of reproductive success by migrating upriver from the sea, over arduous routes, to the very freshwater stream where they were hatched. (It is now well established that they rely on chemical cues, imprinted early in life, for their unerring navigation.) Once in the headwaters, the females release their eggs into simple, pebbly nests, and the males immediately shed their sperm over the eggs.

(a)

(b)

FIGURE 42.9
THE SALMON
(a) The upriver travels of the migrating salmon often take them over obstacles that only the strongest may survive. (b) Intimate spawning behavior and sheltered breeding grounds in the headwaters help increase the chances for success.

As an example of a fish employing highly complex reproductive behaviors, consider the European three-spined stickleback (see Figure 44.12). As the breeding season draws on, hormonal changes begin to occur in the drab little male, and he becomes a splendid fellow indeed, replete with a bright red belly and blue back. The hormones have also triggered changes in his behavior, and he begins to build an elaborate, tunnel-like nest, using water plants that he glues together with body secretions. During his nest-building activity, he will aggressively drive all other males from the vicinity. Then, using precise, elaborate movements—a zigzag dance, body quivering, and some anxious prodding and nudging, the male guides the female into the grassy tunnel. There, she succumbs to his coaxing and releases her eggs. This done, she promptly leaves, having nothing further to do with him or their offspring. The excited male immediately enters the nest and releases his sperm over the eggs. Although the female has lost interest, the male stays and guards the nest. He fans oxygen-laden water over the eggs, and after the young hatch he carefully protects them until they are ready to venture out on their own.

So we see that whereas external fertilization in fishes, when left to chance alone, is fraught with risk, it can also be rather carefully done. Each gamete, and each offspring, can be given the sort of attention that maximizes its chances for survival.

Among amphibians, most frogs and toads increase the odds of reproductive success by taking the mating behavior a step further. The male clasps the female from behind (Figure 42.10), and as she releases the eggs, he sheds sperm directly over them. Their reproductive openings are very close to each other during this process, and it doesn't take much imagination to see that the next step toward increased efficiency in fertilization and protection for the young would be for the male to deposit the sperm directly into the female's reproductive tract. This brings us to copulation in the vertebrates.

FIGURE 42.10
EXTERNAL FERTILIZATION IN THE FROG
Mating behavior in the frog improves the chances of each egg being fertilized. The male stimulates the female to release her eggs and almost simultaneously releases his sperm over the egg mass.

Internal Fertilization in Vertebrates

Internal fertilization in many vertebrates, the cartilaginous fishes (sharks and rays), amphibians, reptiles, and birds, involves the **cloaca**, an organ that is absent in nearly all adult mammals (and most bony fishes). (It is present for a brief period in the mammalian embryo.) Recall that the cloaca is a chamber that receives digestive wastes, excretory wastes, and gametes, opening to the outside via the vent. During copulation, the cloaca is used to receive sperm from the male partner.

Male sharks, as you may recall, have claspers, simple copulatory structures, (modified pelvic fins) that are used to introduce sperm into the female cloaca (Figure 42.11) (see Chapter 32). Of the amphibians, one species of frog, the tailed frog, uses its tail in copulation. Actually, of the amphibians, salamanders are best known for internal fertilization, yet most do not actually copulate. The male releases packets of sperm, which the female places in her cloaca. As an exception, the limbless salamander male has an actual copulatory organ.

Male turtles and crocodiles have a penis complete with erectile tissue, similar in some ways to that of the mammal. Within the cloaca are two spongy ridges and an enlargement called a **glans.** During sexual excitement, the spongy tissue fills with blood, and the ridges close together forming a sperm-conducting groove. The swollen glans, which protrudes from the body at this time, is inserted into the female cloaca, where the sperm are deposited. Oddly, male snakes and lizards have two penises, or **hemipenes,** which can be everted (turned inside out like the fingers of a glove). This accommodates a "side-by-side" mating position so only one penis is used at a time.

Male birds, with the exception of a few species, including ostriches and some ducks, lack a penis, and their mating techniques can best be described as somewhat disorganized—at least from a human perspective. Insemination is accomplished by the male pushing his cloaca against the female's and quickly ejaculating during what has been poetically called "a cloacal kiss." He usually does this first by treading on her tail, causing her to crouch and raise her rear. His tail then cups under hers—cloacal contact is made—and in a flurry of wing-flapping, the deed is done (Figure 42.12). Incidentally, the swifts, which are known as great fliers, are even better fliers than we thought—they mate in mid-air.

Birds are well known for the care they give their offspring. Typically, only a few eggs are produced each season, and then one or both parents lavish them with warm attention. In species of ground-foraging fowl, quail, and in most water birds, the young are born in what is called a **precocial state:** the eyes are functional, feathers are present, and they can immediately walk about and feed themselves. In others, particularly the song birds such as robins and meadow larks, the young may hatch in a completely helpless, semideveloped **altricial** state, doddering weakly about and demanding feeding and protection for many days. The reproductive success of these birds depends, again, on the successful rearing of just a few well-protected offspring.

FIGURE 42.11
REPRODUCTION IN SHARKS
Male sharks can be readily distinguished by the presence of paired, fingerlike claspers near the cloaca. These are erectile structures used in copulation.

Mammals Mammals copulate, their eggs are fertilized internally, and with very few exceptions they produce a **placenta** for the metabolic support of the embryo. Through the placenta the developing placental mammal receives its oxygen and nourishment and voids its wastes and carbon dioxide. Let's turn to the exceptions first, and then look into the details of reproduction in a more typical mammal, one that produces a true placenta.

All mammals, one would think, carry on internal fertilization and development—and one would be mostly right. However, life has again provided us with variations on a theme, and there are exceptions. Consider monotremes, the egg-laying mammals, and, to a lesser extent, the marsupials or pouched mammals.

Monotremes are represented by the spiny anteater (echidna) and the duckbilled platypus (see Figure 32.26). Like birds, these mammals have cloacas, and also like birds, copulation simply involves males and females pressing their cloacas together. The male duckbill's bulblike penis, part of the cloacal wall, then introduces sperm into the female tract. Whereas fertilization is internal in monotremes, the young develop in a large, birdlike egg. Whereas the echidna female tucks the egg away in a temporary groove in the abdomen, the female duckbill hides her eggs in a lengthy burrow, where they are periodically incubated. Monotreme females produce milk, but they do not have nipples. Echidna young suck milk right through pores in the mother's abdominal skin. The female duckbill secretes her milk from numerous pores, and the hatchlings simply lick it from her fur (Figure 42.13).

Reproduction in the marsupials (pouched mammals), certainly a more typically mammalian group (Figure 42.14) than the monotremes, also has some surprises. For example, the "erect" kangaroo penis points downward, while the opposum's penis is a forked

FIGURE 42.12
COPULATION IN BIRDS
Copulation in birds occurs in a fleeting moment when the male and female cloacas are brought together and sperm is transferred. This seemingly risky technique works surprisingly well.

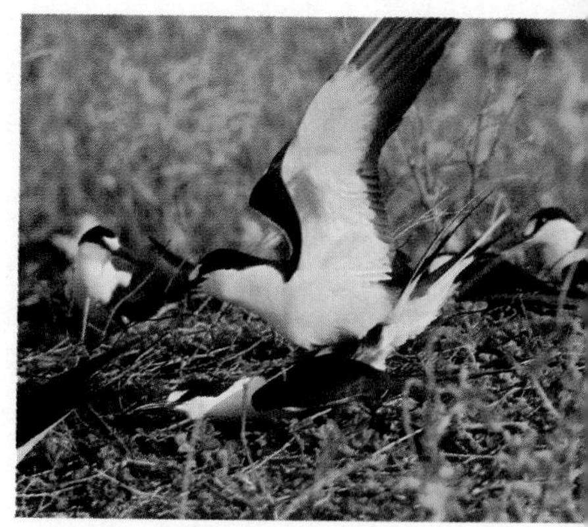

FIGURE 42.13
EGG-LAYING MAMMALS
Like other mammals, monotremes have hair, fertilize internally, and produce milk. However, like reptiles, they lay eggs. The existing monotremes include **(a)** the spiny anteater or echidna (two species) and **(b)** the duckbilled platypus (one species).

structure with grooves through which the semen runs. Each fork correspond to the female's divided vagina. The birth process in marsupials is peculiar in that, by placental standards, it is quite premature. Each largely unformed and glistening offspring, looking like a tiny slug, must leave the security of the uterus in an immature state and make its way along the mother's abdomen to the pouch. There it finds a nipple, to which it remains firmly attached until its development is complete.

Placental mammals copulate, and the embryo completes its development in the secure and protected environs of the uterus. Since humans are in many ways typical of placental mammals, we will look into our own reproductive biology right after this.

REPRODUCTION AND THE SURVIVAL PRINCIPAL

In addition to just wanting to tell about animals, we have included many diverse, often bizarre, examples of animal reproduction to bring up an important principle. Whether we consider gametes or offspring, or whether millions of young simply left on their own, or one or two are carefully guided through maturity, *in the long run, the average number of individuals reaching sexual maturity is one for each reproducing adult.* This has proven sufficient for the perpetuation of the species. So, the various modes of reproduction are merely alternatives in the expenditure of reproductive energy. And natural selection seems to have favored many directions. If gametes are simply to be released in the environment, then an animal's reproductive energy goes into the production of enormous numbers, and a game of chance ensues. Thus, the loss of a few offspring means very little. On the other hand, if safe development and care of offspring is the direction evolution has taken, then out of necessity relatively few gametes are produced, and the reproductive energy is spent in ways that tend to ensure the outcome. The loss of a few offspring in this instance, however, may have severe and tragic consequences. We will explore this interesting idea further in Chapter 48.

HUMAN REPRODUCTION

Humans are reproductively unique in some ways, and we will deal with some of our unusual behavior since it is biologically interesting and essential to our understanding of our own sexuality. We'll begin with the anatomy of males and females, and then go on to some of the physiological and behavioral effects of hormones.

The Male Reproductive System

External Anatomy The genitalia of the human male is rather typical of mammals in general (Figure 42.15a). The external organs include the **penis** and the **testes.** Shortly before birth, the testes descend from the abdomen, where they developed, moving over the public bone and down into the saclike **scrotum.** If the testes fail to descend, sterility results, since the scrotum is cooler than body temperature and sperm will not develop properly at the internal body temperature.

The testes are paired oval bodies that produce both sperm and sex hormones. Much of the testis consists of the highly coiled **seminiferous tubules** that contain a dense **germinal epithelium,** which gives rise to cells that undergo meiosis, forming sperm. Outside the seminiferous tubules are found clusters of **interstitial cells** in a connective tissue matrix—endocrine tissue that secretes testosterone, the male sex hormone. Following their development in the seminiferous tubules, the immature sperm are swept into a network of tubules known as the **rete testis,** which leads to a number of ducts that finally enter the **epididymis** (Figure 42.15b). The epididymis, a sperm storage structure, is a highly coiled tubule that lies perched above and extending down one side of each testis. Each entire testis is surrounded by a tough sheath and contains a rich supply of sensory neurons (which as most males will attest include a seemingly inordinate number of pain receptors).

FIGURE 42.14
POUCHED MAMMALS
In marsupials, the pouched mammals, the young begin development in the uterus but are born early, finishing their development in the pouch.

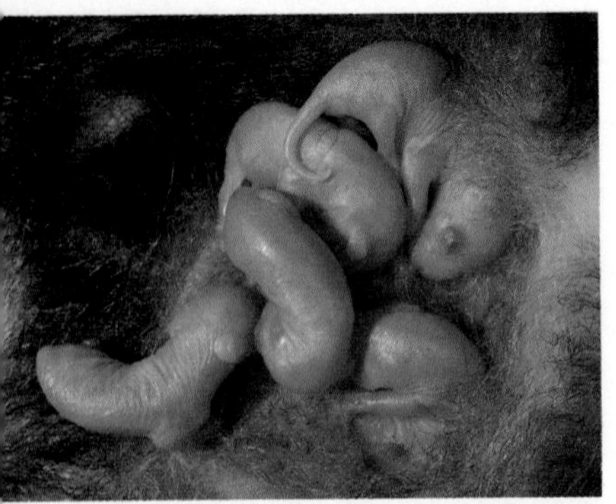

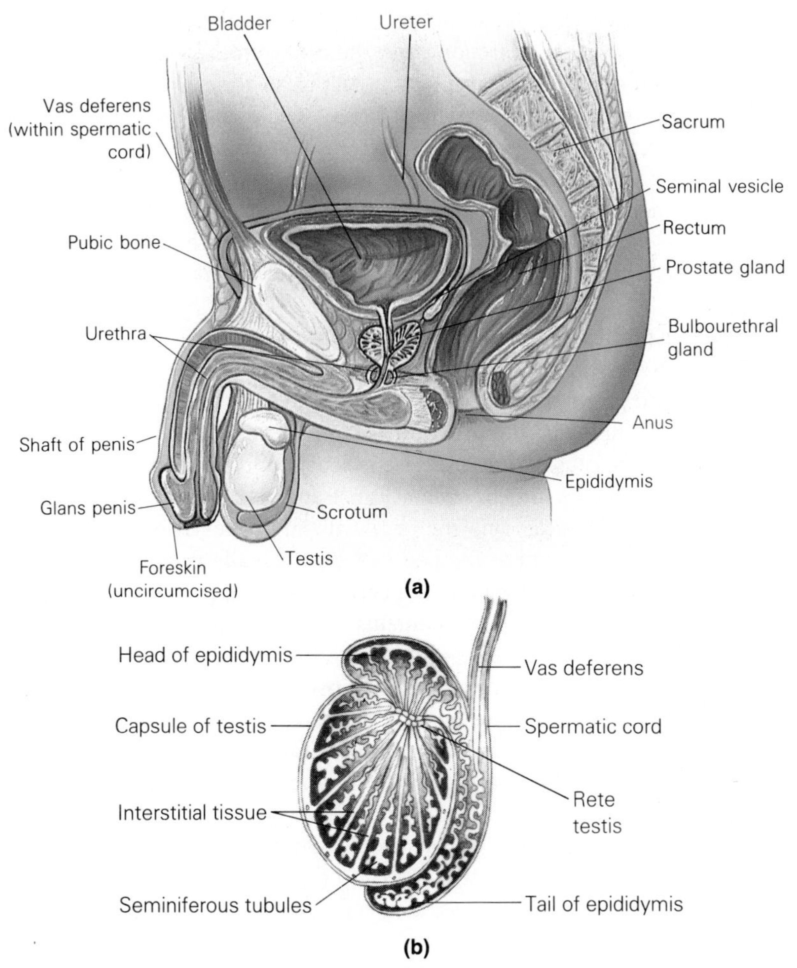

FIGURE 42.15
THE HUMAN MALE
(a) The male genitalia include the penis and testes. **(b)** Coiled seminiferous tubules make up much of the testis.

Internal Anatomy: The Sperm Route Let's follow the route taken by sperm as they leave the testis during ejaculation. Leaving the epididymis, sperm enter the **vas deferens,** which will carry them to the urethra, through which they will exit the body (see Figure 42.15). Along the way, the sperm will receive secretions from three different sources.

The vas deferens passes through its surrounding **spermatic cord** along with the blood vessels and nerves of the testes. Each spermatic cord—one from each testis—extends upward, over the bones of the pubic arch, and then into the body cavity, over the path taken by the testes in their previous descent into the scrotum. The paired vas deferens continue to the underside of the bladder, where they unite, and receive the duct of the paired **seminal vesicles,** the first accessory glands. Unlike this structure in some invertebrates, vertebrate seminal vesicles are not used to store sperm. Instead, they produce most of the volume of the sperm-bearing **semen.** The fluids added from the seminal vesicles contain nutritive and buffering materials. The misleading name "seminal vesicle" is a legacy from early anatomists who named the structure before its function was understood.

After being joined by fluids from the seminal vesicles, the newly formed, sperm-bearing semen passes from the vas deferens to the **urethra.** Surrounding this junction is the fleshy tissue of the second accessory, the **prostate gland.** From the prostate, the semen receives an alkaline secretion that raises its pH and produces its characteristic viscosity, color, and odor. Raising the pH of the semen activates the immobile sperm and, following ejaculation, protects them by neutralizing the normally acidic environment of the vagina. The late activation—just before ejaculation—is vital since each sperm has limited energy resources and, once in the female genital tract, has a lengthy journey to complete.

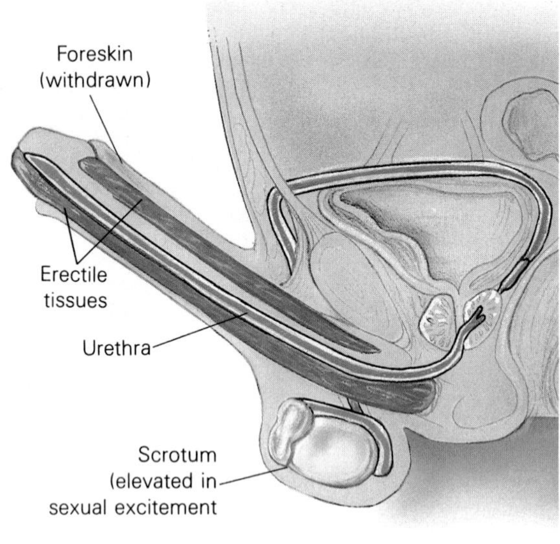

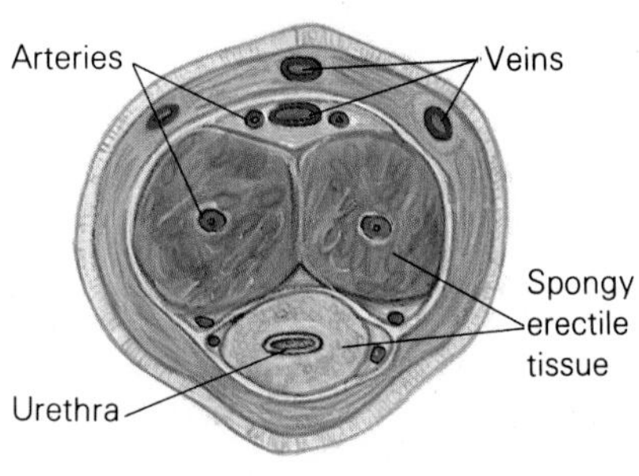

FIGURE 42.16
ERECTILE TISSUE
The three regions of erectile tissue in the penis retain blood during erection. This enlarges, curves, and firms the penis in preparation for copulation.

Following the addition of fluids from the prostate, the semen receives a final secretion from the **bulbourethral glands**, or **Cowper's glands.** While their function isn't clear, they do add slippery mucous secretions to the semen that may help wash residual urine from the urethra, and since these secretions also precede ejaculation they probably have some lubricating value. The semen in its final state follows the route of the urethra through the penis, from which it is forcefully ejected. As you may have surmised, the urethra has the dual role of conducting both semen and urine—at different times.

Erectile Tissue In mammals, the penis consists of very specialized tissue with an erectile capability. The penis must be more or less erect and turgid before it can be introduced into the female vagina. In horses and humans, the **erection** occurs when spaces in the penis fill with blood, causing it to increase in length and girth and to become turgid. In some mammals, such as cattle and sheep, erection may be brought about by extending the penis, already stiffened by cartilage or bone, from deep within the body where it usually rests.

In humans the penis is an erectile shaft tipped by an enlarged region, the glans. The penis contains three cylinders of very spongy tissue (Figure 42.16). Two of these are the paired **corpora cavernosa**, and the third is the **corpus spongiosum**, which extends into the glans. During sexual excitement, the blood flow into the spongy tissue increases, and venous flow back to the body is retarded. As the penis fills with blood, it becomes firm and erect. A complete or partial loss of erection almost invariably follows ejaculation, the normal result of a more restricted arterial blood flow and subsequent drainage of the spongy tissue. We will return to this subject in Essay 42.1, where the human sexual response is taken up.

Sperm and Their Production Spermatogenesis (sperm production) occurs in the walls of the seminiferous tubules (Figure 42.17). The diploid cells in the outermost region of the tubule called **spermatogonia**, continually divide by mitosis. About half of the mitotic daughter cells begin the series of meiotic divisions that will produce the haploid sperm. The uncommitted cells continue to provide a source of new cells for the meiotic process.

When one of the cells enters the meiotic process it becomes known as a **primary spermatocyte.** Then meiosis I separates homologous chromosomes into daughter cells known as **secondary spermatocytes.** Meiosis II follows, with the production of four **spermatids**, each with the haploid chromosome number (see Chapter 9). Each spermatid will then undergo the complete reorganization that is necessary to form mature **spermatozoa.** The spermatids lose most of their cytoplasm at this time and each receives metabolic support from surrounding "nurse" cells called **Sertoli cells.**

In the mature spermatozoa, the chromatin will be condensed into a minute **sperm head.** The cytoplasm itself will differentiate into a midpiece and tail, and excess cytoplasm will be sloughed off. The midpiece contains a peculiar spiral-shaped mitochondrion and a centriole. The tail is a very long, tapered flagellum, which propels the sperm along. At the tip of the sperm head is the **acrosome**, a caplike structure that arises from an enzyme-laden lysosome and overlays the compact nucleus. We'll see in the next chapter how the enzymes of the acrosome function in egg penetration and fertilization. Figure 42.18 compares the sperm from several animals.

The Female Reproductive System

External Anatomy The most prominent part of the external genitalia, or **vulva**, of women is the **mons pubis**, also called the **mons veneris** ("mountain of love,") a hair-

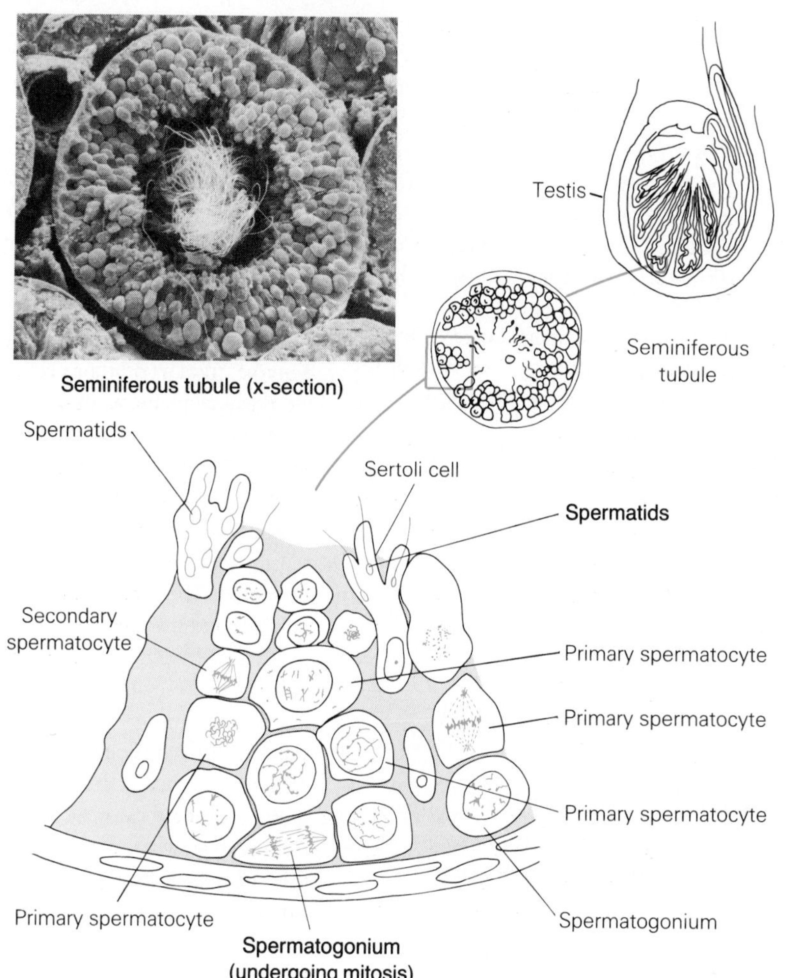

FIGURE 42.17
SPERMATOGENESIS
Spermatogonia, diploid cells lining the seminiferous tubules, undergo mitosis throughout the life of most males, thereby making cells continually available for spermatogenesis. Following the usual stages of meiosis, the resulting haploid spermatids differentiate into spermatozoa within Sertoli cells. Note the numerous sperm flagella in the SEM (left).

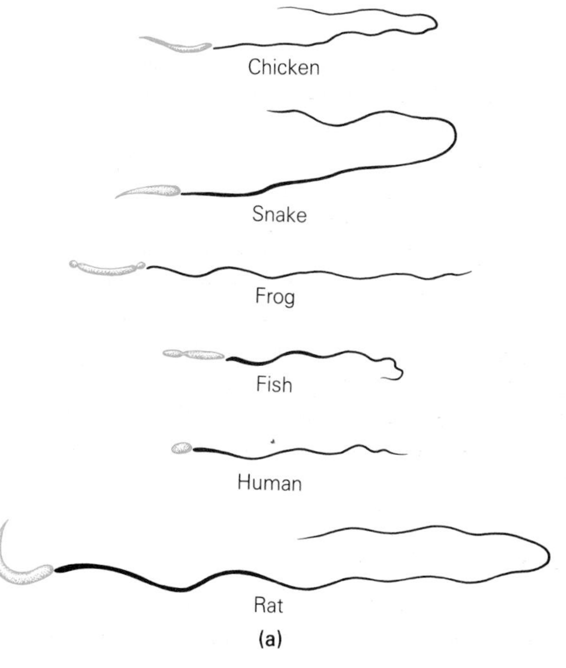

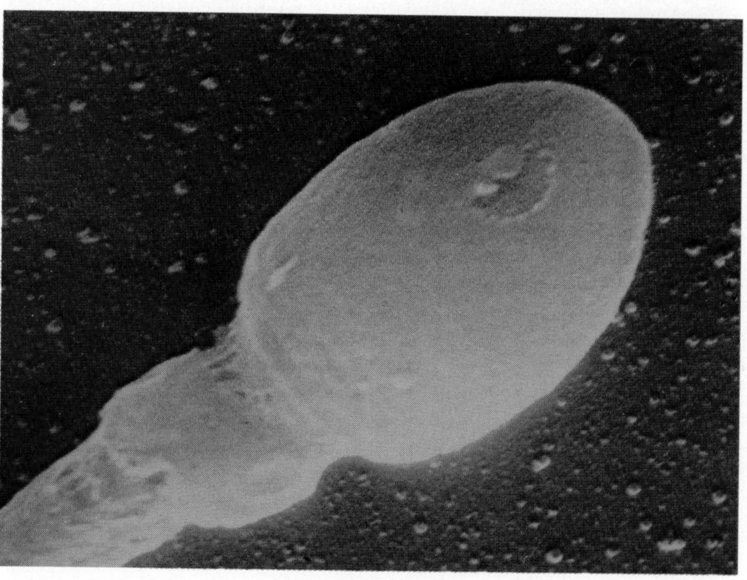

FIGURE 42.18
VARIATIONS IN SPERMATOZOA
(a) Spermatozoa from various vertebrates are shown to scale. While their shapes vary, each contains the same elements. The sperm head contains the highly condensed haploid chromosome complement. The remainder is made up of a midpiece and a flagellum **(b)** SEM of human spermatozoa.

ESSAY 42.1

THE HUMAN SEXUAL RESPONSE

Much of what we know about human sexuality today has grown out of the pioneering efforts of William H. Masters and Virginia E. Johnson, of the Reproductive Biology Research Foundation in St. Louis. Using human volunteers, Masters and Johnson developed observing and measuring techniques for studying many of the physiological aspects of human copulation that were heretofore only known anecdotally, if at all. Not insignificantly, they also laid to rest many erroneous ideas on the subject.

In their analysis, Masters and Johnson divided the sex act, perhaps somewhat arbitrarily, into four phases: **arousal, plateau, orgasm**, and **resolution**, in that order. In their usual context, the four phases refer to actual copulation, but they can of course occur in other ways. We had better add also that not all the events listed occur every time, thus keeping in mind that when it comes to humans, there are variations on variations. Further, the phases are generally continuous, not clearly separated, and the time involved may take anywhere from a few minutes to many minutes.

In women, the preliminary acts of sexual arousal, such as kissing and caressing (variations here are endless) bring on increased heart rate, faster breathing, and a rise in blood pressure. This is accompanied by more outward manifestations such as firming of the nipples, spontaneous muscle contractions, and often what is known as the **sex flush.** This amounts to a reddening of the skin, particularly about the genitals, face, breasts, and abdomen, but not restricted to those areas. Changes occur in the glans clitoridis (analogous to erection in the male), general moistening of the vaginal walls, a marked elevation and parting of the labia majora, and a size increase in the labia minora, such increases brought about by the accumulation of blood in these organs. These responses are physiological preparations for reception of the penis; however, such changes do not necessarily signal psychological readiness of the woman.

In the male, arousal is characterized by somewhat similar events in circulation and respiration. Genital changes include erection, accompanied by a general contraction and elevation of the scrotum, and a moistening of the glans penis. Such events can occur notoriously fast in a male, often preceding correlating events in the female. But for the sake of this idealized discussion, let's assume arousal is simultaneous and copulation has begun.

As women enter the phase known as plateau, sensation and movement heightens and vigorous pelvic thrusting is common. The swollen clitoris withdraws into its hood rather suddenly at this time, and is now exquisitely sensitive to touch. The labial swelling increases, as does swelling in the lower portion of the vagina, especially the opening. The uterus may elevate, tilting backward somewhat. As intensity grows, the genitals reach the **orgasmic platform**—preparatory physiological and anatomical changes leading to orgasm. Physical activity may peak now, with intense thrusting of the pelvis, although there is a great deal of variation here. Characteristically, the facial muscles relax, producing a slack appearance and almost an absence of expression.

In men, the intensity of the arousal phase has also increased steadily. The glans reaches its greatest size, and the penis achieves its greatest curvature and rigidity. The testes increase in size and further elevate in the scrotum. Pelvic thrusting increases, and secretions from the bulbourethral glands increase just prior to orgasm.

The term **climax,** often used to describe orgasm, is fitting. The events in the previous phase, as their intensity increases, now come to a climax, and orgasm begins. In both men and women, the orgasm is a matter of complex muscular contractions, usually accompanied by intensely pleasurable sensations. These contractions are reflexive and involuntary, occurring in a steady series.

In women they emanate from the muscles and thickened tissue about the vaginal opening—joined to a varying extent by activity in the vagina and the uterus. During these contractions, **tenting** may occur: the cervix is drawn upward, increasing the diameter of the surrounding inner vagina. Some authorities claim that tenting creates a suction that causes semen to pool around the cervix, and subsequent contractions in the uterus are thought somehow to draw the semen in, although certainly orgasm is not a prerequisite to fertilization. Uterine and cervical contractions are caused at least in part by the hormone oxytocin, which is released from the pituitary at this time.

The involuntary contractions of orgasm occur in 0.8 second intervals, spreading through the lower pelvis. In an intense orgasm, many of the other muscles of the body may begin to contract spasmodically. The actual origin of an orgasm is a subject of controversy at this time, but the preponderance of opinion is that clitoral stimulation triggers this phase. Stimulation preceding and during

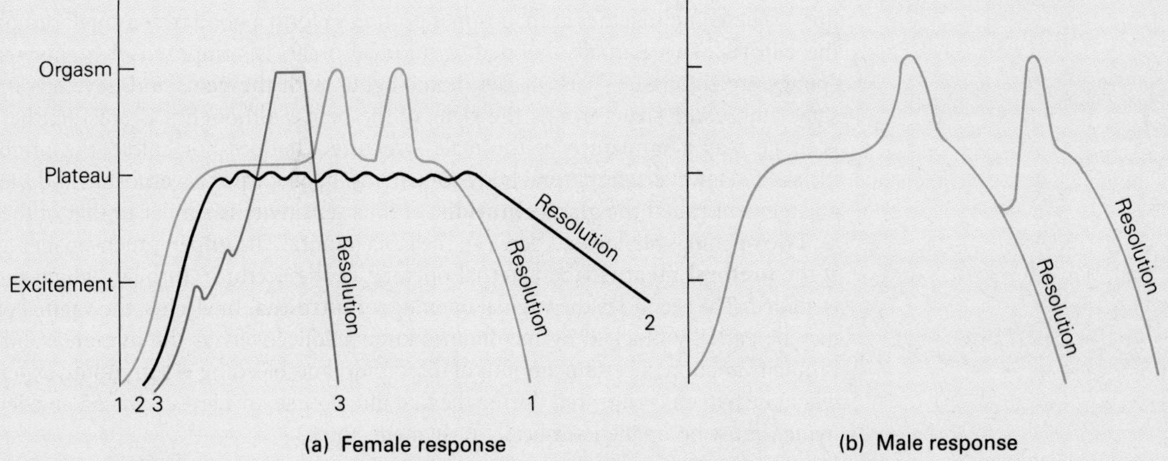

(a) In the highly variable female orgasm, we see three types. In 1, it occurs in several peaks of intense response. The number varies considerably in such multiple orgasms. In 2, there are a greater number of multiple events that are far less intense—just above the plateau level. In 3, the female orgasm is similar to that in males, with one intense period that lasts somewhat longer than the male's. Note the difference in resolution time in the three types of female orgasm. **(b)** Compared to this, the male orgasm is much simpler and more stereotyped, although as the graph indicates, orgasm can occur again in males after a period of resolution. The recovery time between orgasms in males increases with age.

orgasm is probably indirect, involving the clitoral hood, which is apparently moved rhythmically against the clitoris by the thrusting action of the penis and by shoving of the two pelvises. Thrusting also tugs against the labia minora, which, as you can see in Figure 42.19, is continuous with the hood.

The duration of the orgasm in women is highly variable. Several common reactions are shown in the accompanying graph. Orgasms may occur with one intense peak or several, or the peaks may be far more numerous but less intense. The ability of women to experience several orgasmic peaks in a single sexual act is sharply contrasted to the usual single orgasmic peak to which men are limited.

In men, ejaculation is a part of orgasm; however, orgasm can be divided into two distinct parts. The first is **emission.** Emission occurs with the rhythmic, peristaltic contractions in the vas deferens, prostate, and seminal vesicles. As a result, the elements of semen, (spermatozoa and fluids) collect at the base of the urethra prior to expulsion. The second part, **ejaculation,** a spinal reflex action triggered by the presence of semen in the urethra, will immediately follow. This forceful expulsion of semen occurs as powerful striated muscles at the base of the penis contract, also in intervals of about 0.8 sec. The semen is forced through the urethra and out of the penis in several spurts, the first usually bearing more semen than the rest. Ejaculatory contractions usually continue for a short time (several seconds). Involuntary contractions of many muscles of the lower pelvis are usually involved also. On the border between voluntary and involuntary movements are the intense, even violent pelvic thrustings and muscular contractions involving the whole body that commonly accompany this phase of the sex act.

Following orgasm, both males and females retreat from the state of sexual excitement into what is known as **resolution,** a phase that has also been referred to as **afterglow.** Men usually lose erection rapidly and enter a **refractory** period in which rearousal can be quite difficult—at least for some minutes and usually longer. The refractory period varies, typically increasing with age. Resolution in women is somewhat different, since immediate arousal is often possible. In any case, the quiet, relaxed, and often tender resolution phase may have an intense pleasure of its own.

covered mound of fatty tissue overlying the pubic arch (Figure 42.19a). Below the mons pubis lie the larger outer folds, or **labia majora** ("major lips"), of the vulva. In an unaroused state, these folds cover a number of rather sensitive structures. Just inside the labia majora are the less prominent, thinner, inner folds, or **labia minora** ("minor lips"). These join together at their upper portion to form a hood over a small prominence, the **clitoris.** This partially hooded, cylindrical organ is composed of highly sensitive, spongy erectile tissue. It is, in fact, homologous with the penis and develops from the same embryonic structures as the glans of the penis. Although it is erectile, the clitoris is in no way a miniature of the male structure. (It does not enfold the urethra, for instance.) However, the clitoral body contains a similar corpus cavernosum and a terminal enlargement called the **glans clitoridis,** and its sensitivity is similar to that of the penis.

Two openings are located between the labia minora. The upper, much smaller opening is the **urethral meatus** (the external opening of the urethra), through which the urine is voided. The second is the vaginal opening, or **introitus.** In virgins, the vaginal opening may be partially blocked by membranes known collectively as the **hymen.** While there is much variation, a certain amount of discomfort and bleeding is commonly experienced when the hymen is ruptured during the first intercourse. In a few instances, an overgrown hymen must be opened surgically (a hymenectomy).

Internal Anatomy The vagina is a distendable, muscular tube about three inches long in its relaxed state (Figure 42.19b). It is the female copulatory organ, receiving the erect penis and semen during copulation, and it is also the birth passageway. The vagina is marvelously adapted for such diverse functions. Its highly folded walls consist of an inner mucous membrane (a stratified squamous epithelium; see Chapter 33), a middle muscular layer, and an outer layer of fibrous connective tissue. The inner layer presents

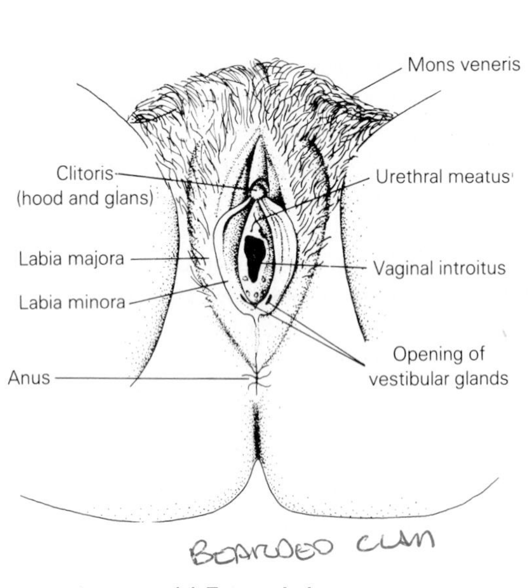

(a) External view

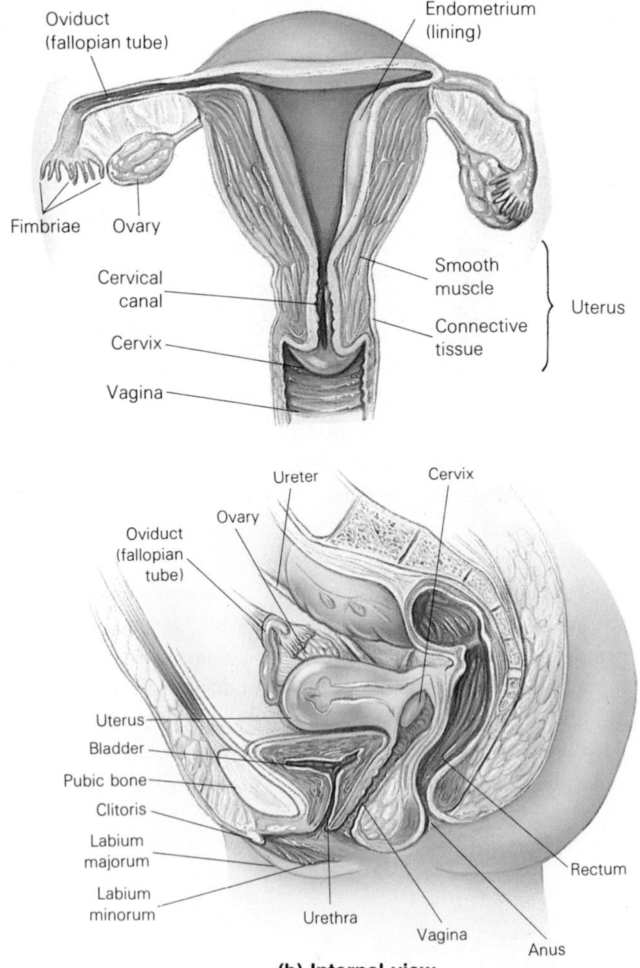

(b) Internal view

FIGURE 42.19
THE HUMAN FEMALE
(a) The female external genitalia include the labia majora and minora, the clitoris, and the vaginal introitus. (b) Internally, the vagina terminates at the cervix, which opens into the uterus. The oviducts (fallopian tubes) branch from the uterus and terminate at fingerlike fimbriae. Although the oviducts receive ova from the paired ovaries nearby, there is no direct connection.

a moistened, yet firm, stimulatory surface for copulation. During sexual arousal the vagina releases lubricating mucous secretions from glands in its inner recesses and through the **vestibular glands**, or **Bartholin's glands**, whose ducts open into the lower margin of the vaginal opening. They correspond somewhat to the bulbourethral glands in the male. The vaginal muscle and connective tissue layers are capable of great extension during passage of the infant at birth, accommodating a head about four inches in diameter. What is more remarkable is the rapid return to the original state from this strained expansion.

Other internal organs of the female reproductive system include the **uterus**, the oviducts, or **fallopian tubes**, and the ovaries, along with their supporting structures. As you can see in Figure 42.19b, the uterus begins at the upper end of the vagina as the **cervix.** Its opening, the **cervical canal**, is surrounded by a ring of dense tissue. The uterus is a hollow, pear-shaped organ that tilts its larger end forward in a slightly folded manner (if you can imagine a folded pear). Most mammals have paired uteri; the single uterus of humans and other primates (and bats as well) is an evolutionary specialization for species that customarily give birth to only one offspring at a time.

The walls of the uterus consist of three specialized layers. The inner layer is the versatile **mucosa.** It produces the soft, highly vascularized **endometrium** that will receive and support a zygote if fertilization occurs. (We will look at the cyclic production of the endometrium a little later in this chapter.) Below the mucosa lies the middle layer, a region of smooth muscle whose fibers run in several directions. The uterus has an enormous capability to expand, as it must when accommodating a fetus. The outermost layer is connective tissue.

The oviducts emerge from each side of the upper end of the uterus and extend for a few inches outward and then downward. They terminate in a cluster of fingerlike processes known as the **fimbriae.** Curiously, there is no direct connection between the oviduct and the ovary; eggs are simply released into the moist surroundings and are drawn into the oviducts by active cilia on the fimbriae.

Ovary and Oocyte Development The ovaries are oval in shape and about 2.5 cm (about one inch) in length. The outermost region of each ovary consists of a germinal epithelium. The germinal epithelium contains the **oocytes**, cells that began meiosis during fetal development. Technically, they are not yet haploid egg cells, although they are commonly referred to as eggs or **ova** (singular, **ovum**). (Recall from Chapter 9 that in human oogenesis, meiosis begins during fetal life, but stops during prophase I, the stage in which those cells remain until sexual maturity is reached.) Figure 42.20 reviews the events in oocyte development, through ovulation.

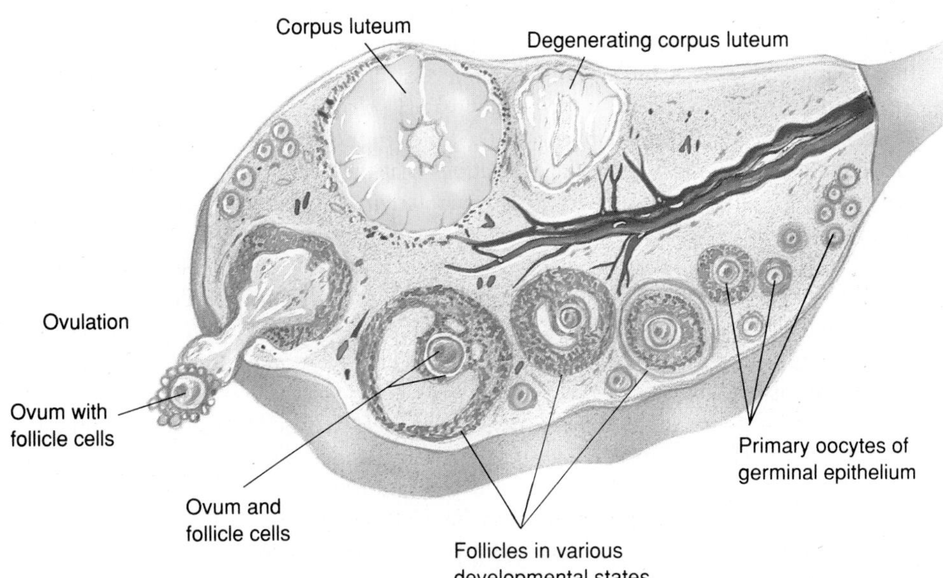

FIGURE 42.20
HISTORY OF AN OVUM
Each month, in the ovarian cycle, an oocyte becomes surrounded by a cluster of supporting cells—the follicle. The follicle enlarges, becoming fluid-filled, with a number of cells directly surrounding the oocyte itself. At mid-cycle, the oocyte completes meiosis I and is ejected from the ovary at ovulation. The vacated follicle will form a hormone-secreting body—the corpus luteum.

Gamete production represents an important difference between males and females. Females are born with all the oocytes they are ever going to have. The male, however, continues to produce new sperm cells throughout his reproductive lifetime—literally trillions of them. Nevertheless, the number of oocytes in the ovaries at birth—some 2 million—is astounding, especially considering how few can ever mature. By puberty this number has shrunk considerably, to approximately 200,000 oocytes in each ovary. Of course, nowhere near that many will mature. In her reproductive lifetime, each human female has about 450 opportunities to become pregnant (about once every four weeks for about 35 years).

Hormonal Control of Human Reproduction

The reproductive activities of humans, as with all higher animals, are under the influence of hormones. The primary sites of human reproductive hormonal activity are the pituitary gland, the gonads, and the placenta. These centers of hormone production are, in turn, under the influence of the hypothalamus. The role of the hypothalamus was described in detail in Chapter 37, but a review here will be helpful. Essentially, the hypothalamus does two things: it initiates the release of certain hormones, and it reads the blood level of other hormones. In this manner it can act as a precise regulator whose operation is subject to delicate feedback mechanisms.

When prompted by gonadotropic releasing hormones from the hypothalamus, the anterior pituitary secretes hormones known as **gonadotropins** (see Chapter 37). As their name implies, gonadotropins stimulate growth or activity in the gonads (both the ovaries and the testes). In addition, they are indirectly responsible for the surge of growth during puberty and help direct the development of those familiar secondary sex characteristics. More directly, the gonadotropins stimulate the production of the sex hormones and initiate the development of sperm and egg cells.

Two specific gonadotropins of the pituitary are *FSH (follicle-stimulating hormone)* and *LH (luteinizing hormone),* both of which are polypeptides. Their targets are the ovaries in women and the testes in men. (LH in men is sometimes called ICSH, for interstitial cell-stimulating hormone, named for its targets, the interstitial cells of the testes.) We will begin with hormonal action in males, where, as you may already have noticed, things are considerably simpler.

Male Hormonal Action FSH is continually secreted throughout the life of the male. Its specific targets are the seminiferous tubules of the testes, where it stimulates sperm production. In addition to promoting sperm production, FSH teams up with LH to prompt the testes to produce **testosterone**, the primary male sex hormone. Recently, it became known that FSH secretion in both males and females is inhibited by another hormone, one dubbed **inhibin**, which in males is produced in the seminiferous tubules. Inhibin is thought to stabilize testosterone levels through a delicate negative feedback loop to the anterior pituitary, where it slows FSH secretion.

Testosterone is produced throughout the testes in the interstitial cells that lie outside the seminiferous tubules. There are indications that the production of testosterone may rise and fall with increased and decreased sexual activity—or even with social interactions with the opposite sex. Testosterone levels commonly decline in all-male crews of ships at sea, for instance, as evidenced by lower rates of beard growth. (More recently, researchers claim that the mere anticipation of sex can produce an elevation in body testosterone.) But even if the male's social life follows the pattern of feast and famine, his testosterone levels will be far more constant than the corresponding ovarian hormones in a female, as we shall see.

Testosterone is vital to the development of the characteristics of "maleness." Although the appearance of the embryonic gonads is genetic, their subsequent development requires testosterone (see Chapter 43). If it is absent during the eighth or ninth month of development, the testes fail to descend into the scrotum. Later on, testosterone is essential for those perturbing events associated with puberty, such as voice changes, the growth of body hair, bone and muscle development, and the enlargement of the testes and penis. Abnormalities resulting from low testosterone levels in the developing male are readily corrected by administering the hormone. Finally, the sex drive itself appears to be greatly influenced by the presence of testosterone.

High levels of testosterone inhibit the release of LH. The critical testosterone level is apparently detected by the hypothalamus, which then slows its stimulation of the pituitary. The pituitary, in response, slows its release of LH, thus lowering the production of testosterone by the interstitial cells. So, as you can see, a feedback mechanism regulates testosterone levels. Actually, things aren't quite that simple. For example, we mentioned earlier that sexual and social activity can influence hormonal levels. Apparently, the frequency of ejaculation and/or sexual arousal has some influence on the hypothalamus, or perhaps directly on the pituitary, or the testes—or all three.

Now let's see if we can simplify some of this. Under the influence of releasing hormone from the hypothalamus, the pituitary secretes the gonadotropins FSH and LH. These hormones stimulate sperm production and initiate the production and release of testosterone by the testes. Levels of testosterone are sensed by the hypothalamus, which responds by adjusting the release of gonadotropins. The result is a fairly steady level of hormone production. As you will see, all this is quite straightforward when compared to what happens in the human female.

Female Hormonal Action The onset of puberty in girls most often occurs between 9 and 12 years of age. The beginning of fertility follows in two to three years. Both events are initiated by a rise in the pituitary gonadotropin FSH, which acts on the ovaries. The ovaries respond to FSH by producing estrogens. In turn, estrogens influence the growth of the breasts and nipples, broadening of the hips, and in general, the development of the usual adult female contours. Less conspicuously, estrogen influences the growth of the uterus, the vaginal lining, the labia, and the clitoris. The appearance of pubic and axillary (underarm) hair, another signal of puberty, is initiated by the combined effects of estrogen and ACTH (adrenocorticotropic hormone from the pituitary) on the adrenal glands (see Chapter 37). In response to these hormones, the adrenals produce androgens (essentially, male hormones), which stimulate the growth of the coarser hair of the genitals and underarms.

Fertility in the female is marked by the start of the **ovarian**, or **menstrual cycle**, in which the gonadotropins and ovarian hormones rise and fall with some regularity and oocytes mature. Other female vertebrates may experience somewhat analogous **estrous cycles** with a regular frequency or perhaps only once or twice each year. The remainder of the time they are sexually unresponsive and infertile. The human female is considered to be unusual among mammals in always being potentially responsive.

The Menstrual Cycle On the average, a mature ovum is released from one of the two ovaries about every 28 days. Coinciding with this event are intense preparatory activities in the uterus. Within a week after ovulation, its temporary lining, the endometrium, must be ready to support the young embryo. These events are closely correlated through the action of pituitary and ovarian hormones, which are themselves subject to timely negative feedback events in the hypothalamus.

In discussing the menstrual cycle, it is common to place the essential event of **ovulation** (the release of an ovum) at midcycle, or the 14th day, and begin counting on the first day of menstrual flow. (Of course, such schedules assume a regular, 28-day cycle.) The accompanying Figure 42.21 will help in understanding how the events are organized.

Days 1 to 14: Proliferation The first half of a menstrual cycle is known as the **proliferative** or **preovulatory phase.** During the first three or four days (sometimes longer), blood and tissues of the endometrium are shed in what is called **menstruation** (the "period"). Since we are talking about a cycle, we'll come back to menstruation again, but first let's see what the hormones are doing.

During the proliferative phase, releasing hormones from the hypothalamus stimulate the pituitary to increase its output of FSH and LH. Their targets are the ovary's **primary follicles**, the arrested oocytes, and the cells clustered about them. At first, the two hormones stimulate several primary follicles to grow, although, usually just one reaches maturity, migrating towards the ovary's surface as it does. It is at this time that meiosis resumes in the oocyte. As a follicle grows, it forms a fluid-filled cavity around the oocyte and becomes known as a **Graafian follicle.** By the midcycle, the Graafian follicle will protrude like a blister from the ovary's surface (see Figure 42.20).

During this time, the growing follicle secretes estrogens, which steadily increase in the bloodstream (estrogens are a family of steroids, the most important being estradiol and estrone.) Their principal target, the lining of the uterus, responds by producing the new endometrium (see Figure 42.21). Cells of the lining undergo rapid cell division, forming dense layers over the entire inner surface. The cell layers are soon penetrated by a profusion of blood vessels from the uterus. The result is a soft, highly vascularized lining, exactly what is required to support the early human embryo. The cervix also responds to rising estrogen levels, secreting a thin alkaline mucus, thus changing the usually acidic vaginal environment to one that is more hospitable to sperm cells.

As you can see, the reproductive hormones tightly coordinate the events in the ovary with those in the uterus. One result is that at the time of ovulation the uterus is nearing a fully receptive state. Should an egg be fertilized, it will reach the uterus in a few days and there find itself in a highly receptive environment.

Day 14: Ovulation On the fourteenth day of the cycle, there is a rapid rise in LH, often referred to as the "LH surge." The mechanism behind this is not clear, but rising levels of estrogens are believed to form a *positive* feedback loop, one that leads the anterior pituitary to increase its output of LH. The target of LH is the enlarged fluid-filled Graafian follicle, which responds by secreting still more estrogens, and as the positive loop increases, a final surge in LH secretion occurs, as mentioned above (see Figure 42.21).

The oocyte within the Graafian follicle, now having completed meiosis I, lies within a cluster of surrounding cells. Then as the fluid content of the follicle increases still further, the follicle bursts, and **ovulation** occurs. The oocyte and its surrounding cells, the **corona radiata** ("radiating crown"), are released and begin the journey to the oviduct. The event is followed by a slight, temporary elevation in body temperature (about 1°F.). This temperature elevation is an important indicator for women trying to establish the "fertile period," either in the interest of conception or contraception.

Days 15 through 28: Secretion Following ovulation, the **secretory**, or **postovulatory phase** of the cycle begins. The vacated ovarian follicle, under the continued influence of LH and FSH, forms a new structure called the **corpus luteum** (*corpus,* body; *luteum,* yellow). The corpus luteum continues its secretion of estrogens and, in addition, secretes the hormone **progesterone.** The target of both the estrogens and the progesterone is the endometrium, where progesterone brings on the secretory phase, prompting blood vessel growth and proliferation and glycogen accumulation.

Thus, progesterone instigates the final preparation of the uterus for the reception of a zygote. And should fertilization occur, progesterone will help maintain the endometrium throughout the ensuing pregnancy. Progesterone also stimulates a temporary thickening and an increase in secretion in the vaginal lining and the growth of milk ducts in the breasts. Finally, the appearance of progesterone is believed to bring on the slight temperature rise following ovulation.

Progesterone levels rise in the blood for about six days following ovulation. During this time, the endometrium of the uterus will have reached its greatest development. Its now loosely packed cell layers become fluid-filled sinuses, penetrated with glandular tissue that produces other secretions. If fertilization has occurred, a tiny embryo, now just a ball of cells about six days old, will have formed. Through the action of its own enzymes, it will become implanted deep in this receptive tissue, literally digesting its

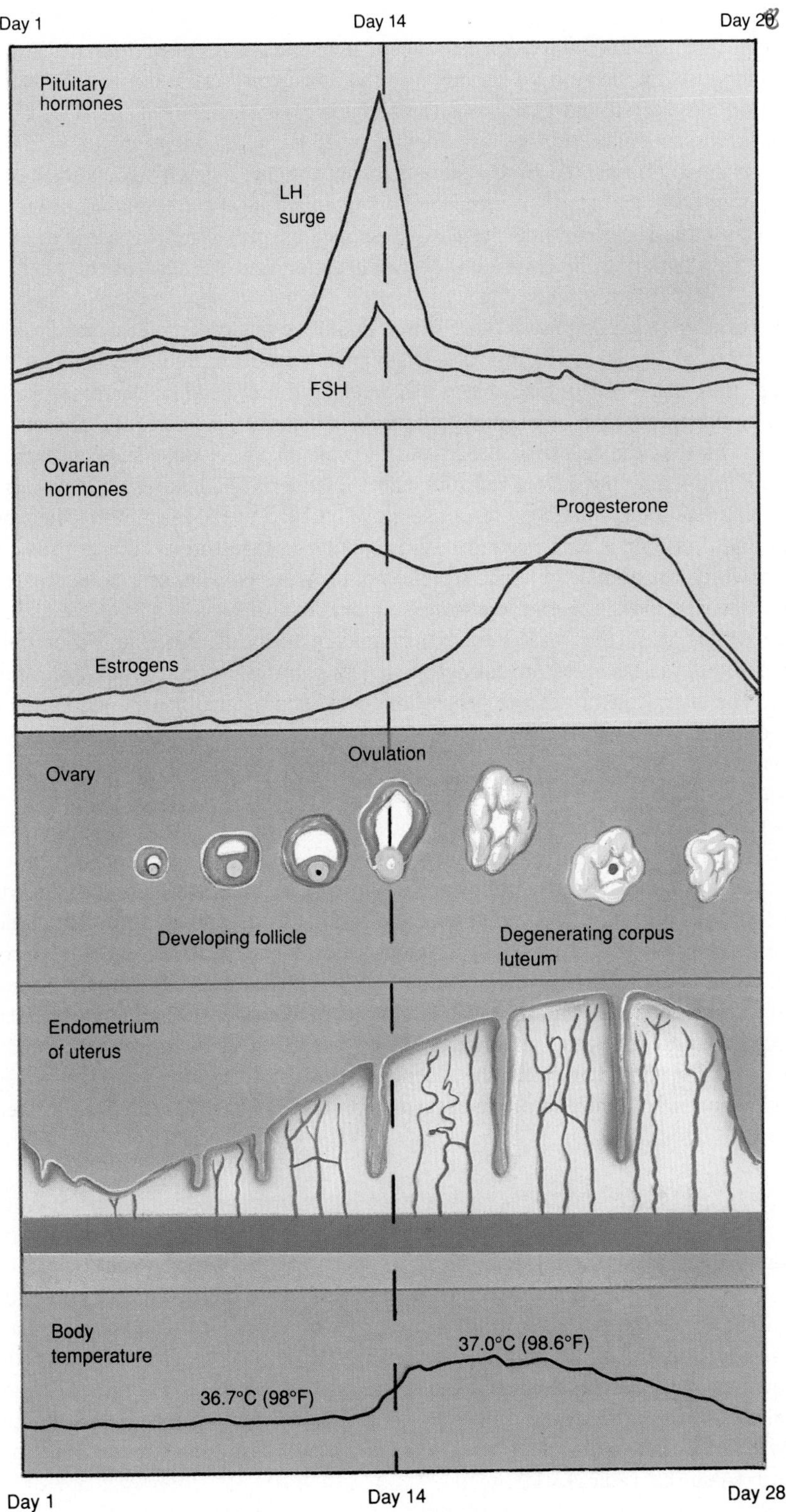

FIGURE 42.21
HORMONAL LEVELS AND THE CYCLE
The rise and fall of various pituitary and ovarian hormones and their effects on the follicle, endometrium, and body temperature are charted through the 28-day menstrual cycle.

way deep into the fluid-filled spaces, where it continues its development. The slight bleeding caused by the embryo settling into the uterus may cause the woman to believe her period has started and that she isn't pregnant.

The uterus remains receptive for about 10 to 12 days following ovulation. Unless an embryo is snugly implanted by that time and putting out its own signals to the corpus luteum, that body will start to degenerate. Its production of estrogen and progesterone will decrease, and the endometrium will lose its receptive state. In a few days it will slough away, signaling the beginning of menstruation and the start of the next 28-day cycle.

Why does the corpus luteum fail? While definitive answers aren't available, there is some consensus on the following. Unless pregnancy has occurred, the signals from the pituitary start to fall off during the fourth week of the cycle. The decrease in LH and FSH may occur because of negative feedback, primarily that of progesterone. Figure 42.21 shows that the levels of progesterone peak on about day 21. The high blood hormone levels may produce a negative effect on the hypothalamus. It slows its stimulation of the pituitary which, in response, slows its LH and FSH secretion. The resulting drop causes its own chain reaction. Without the gonadotropins, the corpus luteum recedes, which causes a drop in the secretion of progesterone and estrogens. And finally, without their support, the endometrium simply breaks down.

However, if pregnancy has occurred, things are very different. The extraembryonic tissue of the fetus becomes an endocrine gland, secreting **human chorionic gonadotropin.** The latter constitutes the progesterone-secreting signal mentioned earlier—the signal directed to the corpus luteum. So the aging corpus luteum continues to secrete hormones that keep the uterine lining hospitable for the demanding little embryo. (The presence of HCG in the urine, incidentally, is the basis for many pregnancy tests.) Later—about the third month of pregnancy—the HCG will diminish as the placenta begins secreting its own progesterone and estrogen.

Let us now summarize the important events of the menstrual cycle. It involves the production of a mature ovum along with the simultaneous preparation of the uterus to receive it. The first half of the cycle is under the influence of FSH, LH, and estrogens, which, respectively, cause follicle and uterine growth. Ovulation is brought about by a surge in LH at midcycle. The second half is principally under the influence of FSH, LH, progesterone, and estrogen (see Figure 42.21). The effect is the continued preparation of the uterus for implantation. In the absence of pregnancy, rising levels of estrogen and progesterone towards the end of the cycle may produce a negative feedback loop, ending the cycle.

Contraception

Conception, at least in the more developed nations, has increasingly involved decision-making processes. The general acceptance of birth control has produced a great deal of technological interest. As a result, there are many options in **contraception** today. We will consider some of these for two reasons. First, it is obviously relevant to people of reproductive age. Second, there is a biological principle at work behind each method. We will organize our discussion of birth control under four major headings, although you will notice that there are instances when the groups overlap or are used in combination. (Also see Table 42.1.)

Natural Methods Some "natural methods" may seem somewhat unnatural in their application. We call them this simply because they don't require chemical, mechanical, or surgical intervention.

Coitus interruptus, withdrawing the penis just before ejaculation, is a surprisingly common method of contraception, particularly among novices, the poor, and the unprepared. But it is risky. There are the problems of premature release of small amounts of semen as the Cowper's gland releases fluids during sexual excitement. And then, couples may engage in intercourse more than once, ignorant of the presence of residual

TABLE 42.1 METHODS OF BIRTH CONTROL

METHOD	APPLICATION	FAILURE RATE	DRAWBACKS
Natural methods			
Coitus interruptus (withdrawal)	Penis withdrawn just before ejaculation	About 30%	Mental stress for both partners; great self-control required; sperm can leak prior to ejaculation
Douche	Vagina flushed with water or chemical solution (vinegar common) after intercourse	About 40%	Slightly better than no precautions—in other words, worthless; sperm can enter the cervix within 1–2 seconds after ejaculation
Rhythm	Intercourse avoided during carefully determined fertile period	Under ideal conditions about 20%	Requires great motivation; fails when cycles are irregular
Mechanical intervention			
Condom	Rubber sheath worn over penis; blocks sperm entry	7%-30%, depending on quality and strict usage; safer with spermicide; withdrawal immediately after ejaculation is essential	Requires strong motivation since it is interruptive and some sensation is lost for men; can break; cost fairly high
Diaphragm with spermicide	Rubber dome; fits over cervix; blocks sperm; spermicide kills sperm	2% with spermicide and completely regimented use	Must be fitted by physician; strong motivation required; somewhat messy
Intrauterine devices (IUDs)	Plastic and/or metal devices inserted into uterus; may contain slow-release progestin (synthetic progesterone); prevents implantation	1%-5%	Must be inserted and monitored by physician; temporary discomfort in some; some expulsion; complications if pregnancy occurs; recommended only after at least 1 child
Chemical intervention			
Oral contraceptives ("the Pill")	Combination or sequential estrogen and progestins, or progestin alone (minipill); prevents ovulation	1%-5%	Costly; must be prescribed and monitored by physician; temporary side effects; greater risk to heavy smokers
Spermicide alone (foams, jellies, creams, or suppositories)	Placed in vagina with applicator before each intercourse; kills sperm	About 10% when properly used	Generally messy and short-lived (newer foams somewhat better); interruptive
Sponge	Polyurethane sponge saturated with spermicide placed in vagina before intercourse	Believed to be same as diaphragm (2%), but data not yet available; no fitting necessary	Interruptive; reliability not proven
Surgical intervention			
Sterilization			
Vasectomy	Incisions through scrotum, section of vas deferens removed and remainder tied off; sperm cannot leave testes	Virtually zero, now 45% reversible	Some psychological effects occur
Tubal ligation	Incision through abdominal wall, oviducts cut and tied or just tied; sperm cannot reach egg	Nearly zero	Hospital stay may be required; costly; simply tying is somewhat reversible; newer microtechniques produce less surgical risk

NOTE: Experimental innovations still being tested or developed include Silastic implants for slow release of progestin (1-6 years); long-lasting injections of progestin (3 mos.), now used outside the U.S.; vaccine that brings on menstruation if implantation occurs; morning-after prostaglandin treatment; and male hormonal sperm suppressors.

sperm cells in the male urethra. But most significantly, there is the psychological frustration of a man's having to do exactly what he is least inclined to do (and some men are less dependable than others). Even when he succeeds, withdrawal may deprive his partner of her own orgasm. Thus coitus interruptus is considered to rank "poor to fair" in terms of dependability (Table 42.1).

Douching after intercourse was a fairly common method of contraception in past years. It involves flushing the vagina with plain water or one of various household or commercial solutions. Diluted vinegar was the most common douche because its acid properties tend to kill sperm. We mention this crude method only because of its great failure rate (about 40%) and to discourage its use. (Failure rate refers to percentages of women who become pregnant per year of dependence on that method.) Even the most hastened douching can fail; sperm cells can pass through the cervix and into the safe confines of the uterus only moments after ejaculation.

The **rhythm method** is perhaps the most biologically interesting way of practicing birth control. The principle itself is simple: there must be no intercourse during the period in which conception is likely (including that with coitus interruptus, a risky practice). Theoretically, this includes four days each month, two days on either side of ovulation. But since no one is sure just how long sperm survive in the female genital tract, an additional day on each side is usually recommended. Traditionally, the rhythm technique first required careful daily monitoring of body temperature for a few months to detect the small rise in temperature accompanying ovulation, and through this the establishment of an accurate calendar of the cycle. A new technique in determining the time of ovulation has been helpful. At the time of ovulation, the mucus coating the cervix reaches a certain consistency and quantity that can be ascertained by self-examination. The test, however, requires some prior instruction by a qualified clinician. Under the best of conditions, the rhythm method fails 20% of the time.

Mechanical Devices Mechanical devices work by blocking sperm entry or by inhibiting implantation (Figure 42.22). We include the **condom, diaphragm,** and **intrauterine device (IUD).** Condoms are very thin rubber or animal-membrane sheaths designed to fit over the penis. They are probably the most widely used means of contraception, although the data indicates a failure rate of 7–10%. The lower figure indicates faithful, proper, and timely use of a quality product. (Among other things, correct use involves avoiding leakage by withdrawing the penis *immediately* after ejaculation.) The use of condoms has become more popular recently, but not for contraceptive reasons. They

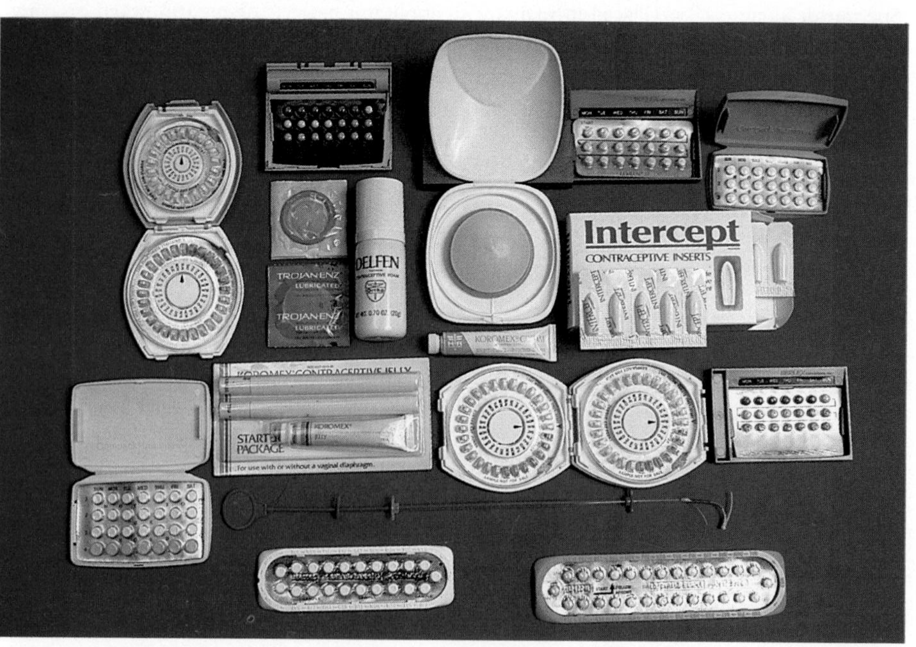

FIGURE 42.22
CONCEPTION BARRIERS
Sperm barriers include the condom and diaphragm. The IUD prevents conception, but its exact effect is not known.

are recommended for minimizing the spread of AIDS, especially where casual sex is involved or where new sexual relationships are being formed—at least until both partners are sure AIDS is not present. By the way, the animal-membrane condoms, and certain inferior brands of rubber condoms, are not particularly effective. The AIDS virus can penetrate the animal skin type, and inferior condoms are known to break easily.

The diaphragm is a rubber dome with a thick rim that is worn over the cervix during intercourse. Generally, the inner side is coated with a spermicidal jelly or cream before it is inserted. Data on the diaphragm's reliability are inconsistent, with failure rates varying from 2% to 20%. The lower figures represent faithful use with a spermicide.

The intrauterine device or IUD has been a widely used, very effective contraceptive device. It is also the least understood. Although first developed in the early 1930s, interest in the IUD didn't really grow in the United States until the late 1960s, when the very promising reports of a 30-year study were published. Since that time, a variety of forms of the IUD have been used all over the world. It appears that the IUD interferes with implantation of the embryo. Whatever its action, its failure rate today is reported to be 1–4%, a very respectable range. It is commonly used as an alternative to the pill for females who fall into a high risk pill-user group.

The primary advantage of the IUD is that it is convenient. Once it is inserted, it can more or less be forgotten, and fertility is restored when it is removed. Therefore, it has been a contraceptive of choice for people interested in family planning. However, IUDs cause cramps and pain in almost a third of women using them, and frequently they must be removed. Some have also been associated with severe uterine infections, and so they were, for a time, not recommended. Newer designs and safeguards, however, have reduced this risk. They are not recommended for women prone to uterine infections or where there is a risk of other infections (as with multiple sexual contacts) since the IUD can aggravate such problems. Finally, should pregnancy occur, the IUD can create complications leading to sterility.

Chemical Methods One of the most widely used contraceptives today is the **birth control pill.** At present, an estimated 100 million women around the world are using this form of birth control. Its effectiveness rates highest among the various methods—short of surgical sterilization or abortion.

The active substances in the pill are synthetic estrogens and progesterone, used in combination or sequentially. The record of combination pills is slightly better because of its greater hormone content. Both function in the same manner. They inhibit ovulation by overriding the normal rise and fall of estrogens and progesterone produced by the ovary. In a sense, this simulates pregnancy, in which high levels of the two hormones also suppress ovulation (Figure 42.23). To understand this overriding effect, recall the discussion of the interactions of LH, FSH, and estrogens.

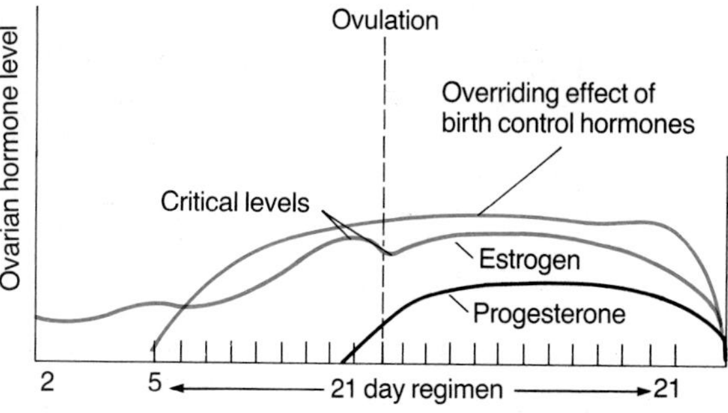

FIGURE 42.23
WORKINGS OF "THE PILL"
The birth control pill, taken for 20 consecutive days (beginning 5 days after menstrual onset), acts to override the normal ebb and flow of reproductive hormones. In particular, it suppresses LH and FSH release, which stops ovulation. The endometrium develops in response to the synthetic hormones of the pill, and when the pill is withheld, the usual menstrual flow soon begins.

Birth control pills receive some bad press occasionally because of limited, but critical, side effects in some women. These include a higher health risk to women who suffer from blood-clotting problems, high blood pressure, diabetes, and family histories of certain other conditions. Further, heavy smokers are under substantially greater risk of side effects. Thus, women should be carefully screened by a physician before using the pill. The debate on side effects is by no means resolved, but one point about the risks stands out very clearly: in all age groups, the risks associated with pregnancy significantly exceed the risks in taking the pill. Deaths from complications in pregnancy are highly age-dependent. For every 100,000 pregnancies in the United States, there are about seven deaths in women between the ages of 15 and 30. This increases steadily to 21 deaths per 100,000 as women approach age 40. Deaths attributed to the pill in women below 35 are one to two per 100,000, increasing to about four as women approach age 40.

The possibility of chemically controlling fertility in males has been repeatedly considered. In spite of the fact that efforts to control male fertility hormonally have not met with great success, the search goes on. Apparently, the regularity of male hormone production and the comparative simplicity of male reproductive physiology don't present an easy target.

In addition to fertility control chemicals, there are several other methods of chemical contraception. They are generally far less effective than hormonal control, however. For the most part, they consist of sperm-killing preparations such as jellies, foams, and suppositories, all used in the vagina. The most effective of these is undoubtedly the aerosol vaginal foam, since it can cover the entire cervix and vaginal lining with the spermicide. (As an aside, spermicides containing *9-nonoxynol* also kill certain of the organisms that cause sexually transmitted diseases.) A more recent chemical device is the **contraceptive sponge**, a device containing spermicides that is inserted into the vagina prior to intercourse. It is reportedly about as effective as the diaphragm.

Surgical Intervention (Sterilization) Male sterilization involves a minor operation known as a **vasectomy** (Figure 42.24a). The procedure can be carried out in a doctor's office with the use of local anesthetic. Essentially, the operation consists of two small incisions in the scrotum and spermatic cords, removal of a small segment of the vas deferens, tying off the cut ends, and sewing up the incisions. As a result of the surgery, sperm travel is blocked just beyond the epididymis. Sperm production continues, but they disintegrate and are cleared away by phagocytes. Other than loss of sperm, semen volume and ejaculation are normal, and many men, their anxieties reduced, report heightened sexual pleasure.

Sterilization in women, **tubal ligation**, is similar in that the surgery involves cutting and tying the oviducts, or just tying them off, thus preventing the sperm and eggs from joining. The surgery can be done through small abdominal incisions or by working through the vaginal wall just below the cervix (Figure 42.24b). While the cutting and tying procedure is essentially irreversible, tying alone can be readily reversed. While almost nothing is known about physiological side effects, sterilization brings many women a new attitude of freedom, greatly enhancing their enjoyment of sex. Because of the difficulties in restoring fertility no one should contemplate complete sterilization

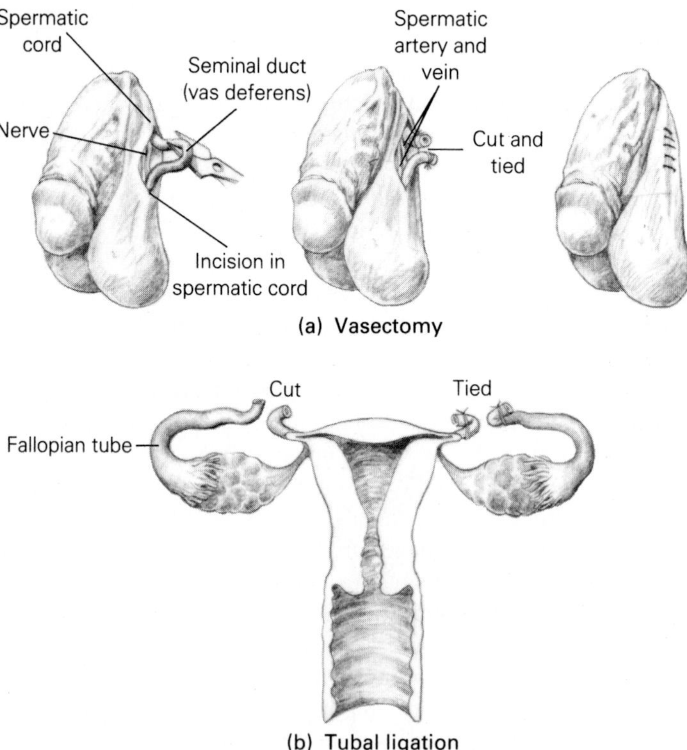

FIGURE 42.24
SURGICAL INTERVENTION
(a) In vasectomy, the vas deferens is opened, a section of the spermatic cord is removed, and the cut ends are tied back. **(b)** In tubal ligation, a portion of the oviduct is removed and the ends tied back.

without thoughtful consideration and preferably counseling. Restoration can be a major operation, requiring full anesthesia and hospitalization, with only about a 50% success rate.

Abortion **Abortion,** the surgical termination of pregnancy, is by far the most controversial means of preventing an unwanted birth. However repugnant or reprehensible it is to many, the facts are that abortion is a widespread practice throughout the developed and developing nations of the world.

In early pregnancies (less than three months), surgical removal of the embryo or fetus is commonly done with a suction or vacuum device, or through **dilation and curettage** ("D&C"). The latter is a bit more severe since a curette (a spoon-shaped surgical knife) is used to scrape the uterine wall, dislodging the embryo.

If the fetus has passed its third month, a saline treatment may be used to induce labor. A 20% saline (NaCl) solution is injected into the uterus. Labor and delivery of the fetus and its supporting placenta follow. Saline treatments are riskier than the other means. There is an occasional problem of salt poisoning and also a greater risk of infection. No one takes saline abortion very lightly. The performance of an abortion after the second trimester can be extremely dangerous to the woman, and is illegal in most places.

The field of reproductive biology, we see, has a number of aspects that have attracted researchers, particularly in recent years. Theorists are interested in the biological advantage of sex. Physiologists are fascinated by the remarkable hormonal control of reproductive processes. Anatomists wonder at animals with two penises. And many of the rest of us are drawn to such discussions because we want to better understand ourselves and to gain more control over our own reproductive activity. The advances and technical information are increasingly hard-won, but each step seems to set up yet another compelling question relating to this complex and critical aspect of life.

APPLICATION OF IDEAS

1. In evolutionary terms, hatching in the precocial state is considered by ornithologists to be primitive, while the altricial state is thought to be advanced. Offer reasons why such a hypothesis might be logical, taking into consideration the evolutionary history of birds and that precocial species are usually ground foragers.
2. Male and female pairing in humans is often described as a matter of "bonding"—that is, the pleasure bond, the pair bond, and so on. Both cultural and biological explanations of bonding have been proposed, and these are often in conflict. In the first, bonding is often considered to be a learned human behavior, approved and reinforced by society. In the second, bonding is considered to be genetic, a product of evolution, and fairly commonplace among higher animal species. Select one or the other of these points of view and offer supporting or opposing arguments. Can either assertion be tested? Can both be correct?

KEY IDEAS

Sexual reproduction promotes genetic diversity in offspring that is not available through asexual reproduction.

MODES OF REPRODUCTION: ADAPTING THEMES

Methods of reproduction vary greatly but follow several themes.

Reproduction Without Sex

1. Asexual reproduction includes **fission** (dividing in two), **budding** (new individuals branching from the adult), **fragmentation** (body breaking into parts and regenerating), **gemmule** formation (clusters of cells that can regenerate the adult), and **parthenogenesis** (development of an unfertilized egg).
2. Parthenogenesis in *Daphnia* occurs when the population is not under pressure, but when unfavorable conditions arise, reproduction through fertilization begins.

Reproduction With Sex

In **monoecious (hermaphroditic)** species, both sexes occur in the same individual. In dioecious species, the sexes occur in different individuals.

In **external fertilization** gametes are released into the surroundings, where fertilization and development occur. Its variations range widely, including random release, timed and simultaneous release, and timed release while in close proximity.

Trends in the evolution of reproduction produced a more efficient use of gametes in some animals. External fertilization and development were replaced by **internal fertilization** and development. This also permitted the transition to land life.

INVERTEBRATE REPRODUCTIVE PATTERNS

1. Sponges release sperm externally, and although they are simple in organization, they make use of internal fertilization. Sperm, released into the sea, are drawn into the sponge body, where they enter the body wall to encounter the eggs. In most animals, internal fertilization occurs through **copulation.**
2. Internal fertilization occurs in hermaphroditic species such as land snails. Most animals are dioecious.
3. Insects have complex reproductive systems, adapted for copulation and internal fertilization. Development occurs externally. In many cases, the copulatory organs are specific to species, which guards against hybridization between similar species.
4. Spiders are often solitary and predatory, which can create mating problems. Since the female is often larger and cannibalistic, mating requires careful approaches and fast escapes.

SEXUAL REPRODUCTION IN VERTEBRATES

Vertebrates utilize both external and internal fertilization.

External Fertilization in Vertebrates

External fertilization occurs in many fishes and may involve the simple, mass shedding of great numbers of gametes. Greater economy occurs through more complex and intimate mating behavior. Salmon migrate to headwaters to pair up and spawn in limited spaces. The male three-spined stickleback displays eleborate courting and mating behavior, restricting actual fertilization to its nest and later protecting the young. Male frogs and toads clasp the female, thereby stimulating egg release. This is followed immediately by fertilization.

Internal Fertilization in Vertebrates

1. The **cloaca** is the common digestive, excretory, and reproductive opening in most vertebrates.
2. In male reptiles, the **penis** is hidden in the cloaca, protruding out during mating. Some have **hemipenes,** paired copulatory organs.
3. Most bird species lack male copulatory organs, and sperm is released as they press their cloacas together. The young are either **precocial** (developed enough to partly fend for themselves) or **altricial** (in a blind, featherless, and helpless state).
4. Nearly all mammals copulate, and subsequently the young develop within the mother's **uterus,** supported there by exchanges provided by the **placenta** or pseudoplacenta. Monotremes (spiny anteaters and duckbilled platypuses) are reproductively primitive egg-layers. Both produce milk but lack nipples.
5. Marsupials are born in a poorly developed state. They move to the pouch, fasten to a nipple and develop there. Females can conceive while the young are developing in the pouch, but development of the new embryo may be slowed.

HUMAN REPRODUCTION

The Male Reproductive System

1. Male genitalia include the penis and **testes,** which occupy the saclike **scrotum.**
2. Most of the testis is made up of highly coiled **seminiferous tubules,** in which a **germinal epithelium** produces sperm, and **interstitial cells** produce testosterone. Maturing sperm are swept by cilia to the **rete testis** and then to the **epididymis** for storage.
3. The path of sperm during ejaculation is from the epididymis to the **vas deferens,** which passes through the **spermatic cord.** Nutritive secretions are received from the **seminal vesicles** as the sperm enter the **urethra.** Next, the **prostate** adds alkaline secretions, and fully formed semen is then forcefully ejected through the urethra.
4. In an **erection** the spongy spaces of the penis, the **corpora cavernosa** and **corpus spongiosum** (which includes the **glans**), become engorged with blood.
5. Sperm cells originate from undifferentiated cells, first becoming **primary spermatocytes,** which then enter meiosis I to form **secondary spermatocytes,** and after meiosis II, **spermatids.** The latter associate with supporting **Sertoli cells,** whereupon they differentiate into **spermatozoa.**

The Female Reproductive System

1. The external female genitalia, or **vulva,** include the **mons veneris,** a fatty mound overlying two folds or lips, the larger **labia majora** and smaller **labia minora.** The minor lips join to produce a hood overlying the **clitoris,** a sensitive, erectile organ tipped by the **glans clitoridis** (homologous to the penis). The labia minora also enclose the **urethral meatus** (opening of the urethra), and the **introitus** (opening of the vagina), the latter of which, in the virgin state, is often partially blocked by a membranous **hymen.**
2. The **vagina** is a distendable, muscular tube containing a squamous epithelial lining, a middle layer of smooth muscle, and an outer layer of fibrous connective tissue. Its inner wall is a moist mucous membrane, and additional moisture and lubrication is provided at the orifice by the **vestibular glands (Bartholin's glands).**
3. The internal female reproductive and anatomy includes the uterus, **oviducts,** and **ovaries.**
4. The uterus opens in a muscular ring called the **cervix.** The uterine wall consists of an inner-most **mucosa,** which produces the vascular **endometrium,** a middle region of smooth muscle, and an outer fibrous connective tissue.

5. The oviducts emerge from the uterus to terminate in fingerlike, movable **fimbriae.** Its ciliated surface sweeps the **oocytes** into the oviduct.
6. The ovaries are supported by the dense ovarian ligament. Each ovary has an outer germinal epithelium that contains the oocytes, all of which formed before birth.

Hormonal Control of Human Reproduction

1. The hormonal control of reproduction involves an interaction between the hypothalamus, pituitary, and gonads in both sexes, with the addition of the placenta in females. Gonadotropin releasing hormones from the hypothalamus prompt the anterior pituitary to release the **gonadotropins,** follicle stimulating hormone (FSH) and luteinizing hormone (LH) (in males, interstitial cell-stimulating hormone or ICSH), which act on the ovaries or testes.
2. In males, FSH stimulates sperm production and works with LH to initiate **testosterone** production. Testosterone levels in the blood create a negative feedback loop back to the hypothalamus. Testosterone influences the development of maleness, the onset of puberty, and sex drive.
3. Puberty in females is associated with FSH secretion, which prompts estrogen production and release by the ovaries. Estrogens influence development in the breasts, reproductive organs, and uterus, and general body growth. Androgens from the adrenal cortex influence the growth of pubic and axillary hair. The onset of fertility begins with the monthly **ovarian** or **menstrual cycle.**
4. In other female mammals, cycles of fertility, **estrous cycles,** occur less frequently, usually annually or semiannually.
5. The menstrual cycle includes the release of an ovum every 28 days, an event that corresponds to growth and thickening of the uterine endometrium, a preparation for the reception of an embryo. The cycle includes the following.
   a. Days 1-14: The **proliferative** or **preovulatory phase** begins with the menses or **menstruation.** The pituitary secretes FSH, which prompts **follicle** development and estrogen secretion. Follicle cells surround the oocyte, supporting its growth into a mature **Graafian follicle.** Under estrogen's influence, the endometrium enlarges and becomes vascularized.
   b. Day 14: Estrogens form a negative feedback loop to the hypothalamus, pituitary FSH secretions slow, and LH increases ("LH surge"). Upon reaching the follicle, LH brings about ovulation—the release of the ovum.
   c. Days 15–28: In the **secretory** or **postovulatory phase,** LH and FSH maintain the **corpus luteum,** which secretes both estrogens and **progesterone,** further influencing endometrial growth. During the days following ovulation, the endometrium is ideally suited to receive the embryo. Terminating events in the cycle are still unclear, but is suspected that ovarian hormones produce a second negative feedback, suppressing LH and FSH secretion. The corpus luteum then fails, and without hormonal support, the endometrium breaks down.
   d. The next cycle may be prompted by the lessening of FSH and LH suppression. But, if fertilization and implantation of an embryo have occurred, cells associated with the embryo secrete **human chorionic gonadotropin,** which supports the corpus luteum for the first two months. Later, progesterone and estrogens from the placenta maintain the uterus.

Contraception

1. Natural methods of **contraception** include **coitus interruptus** (withdrawal prior to ejaculation), **douching** (attempting to flush semen from the vagina), and the **rhythm method,** in which intercourse is avoided during fertile periods. None are very reliable. The variables in the rhythm method include difficulties in determining the fertile period and the longevity of egg and sperm.
2. Mechanical methods include blocking sperm movement with rubber, sheathlike **condoms,** worn over the penis; rubber **diaphragms** inserted with or without spermicide over the cervix; and **intrauterine devices** or **IUDs.** The latter is a device placed in the uterus, where it is thought to prevent implantation, but in ways not understood. Some IUDs contain contraceptive chemicals.
3. Chemical methods of birth control include spermicides and **the pill.** The pill contains synthetic estrogens and progesterone, which are believed to suppress ovulation by overriding the natural rise and fall of fertility hormones. Side effects are noted in certain risk users, and heavy smoking increases the risks. However, in the general population, the risk of pregnancy is greater.
4. Surgical intervention includes the **vasectomy** (cutting and tying the vas deferens) and **tubal ligation** (cutting and tying—or just tying—the oviducts).
5. The most controversial method of birth control is **abortion,** surgically removing an embryo or fetus or inducing its expulsion. Surgical methods in the early months of pregnancy include suction or vacuuming the uterus, and **dilation and curettage** (D&C), scraping the uterus. In later pregnancy, a saline (salt) solution is injected into the uterine cavity to induce labor and expulsion of the fetus.

REVIEW QUESTIONS

1. What appears to be the main adaptive advantage of sexual reproduction? (1014)
2. Distinguish between monoecious and dioecious animals and list examples. (1014)
3. List and briefly explain three means of asexual reproduction in animals. (1013)
4. What seems to trigger alternations between parthenogenesis and fertilization in the *Daphnia*? (1013)
5. Compare external fertilization in the echinoderm with that of the grunion. What is the advantage of such improved efficiency? (1014)
6. What are two important adaptive advantages of internal fertilization? (1014–1015)
7. Describe sperm release and fertilization in sponges. What special characteristic of sponges makes this possible? (1015)
8. Why might it be essential for insects to copulate and fertilize internally? How are some protected from accidental hybridization? (1016)
9. List three classes of vertebrates in which external fertilization is common (also see Chapter 32). (1017–1018)
10. Describe the mating behavior of the three-spined stickleback. (1018)
11. What is a cloaca? In what vertebrate classes does it occur? (1018)
12. Describe copulation in birds. List two exceptions and explain them. (1019)
13. Compare the precocial and altricial states in birds and suggest how these relate to their food requirements. (1019)
14. Briefly summarize how the monotremes and marsupials differ reproductively from the placental mammals. (1019–1020)
15. List the structures and events in the passage of sperm during ejaculation in humans, beginning with the epididymis. (1020–1022)
16. Describe the steps and name the stages in spermatogenesis in the human testis. (1022)
17. In what two ways are the prostate gland secretions essential? (1021)
18. In what way is the erectile tissue of the human penis analogous to the hydrostatic skeleton of some invertebrates? (See Chapter 33.) (1022)
19. Describe the main anatomical structures of a spermatozoan. (1022)
20. List the five structures that make up the human vulva. Which is unrelated to reproduction? (1022, 1026)
21. In what way is the clitoris homologus to the penis? (1026)
22. Describe the makeup of the vagina, starting with the outermost layer of tissue. (1026)
23. Discuss the manner in which the human ovum finds its way into the oviduct, listing the structures involved. (1027)
24. List three structures that play a part in the hormonal control of reproduction, starting with one located in the brain. (1028)
25. Explain how negative feedback works in controlling testosterone production in males. (1029)
26. In what three aspects of life is testosterone important to males? (1028–1029)
27. What two events does the human menstrual cycle closely coordinate? Why is such coordination needed? (1029)
28. Describe events in the ovary, hypothalamus, uterus, and anterior pituitary through the first thirteen days of the human menstrual cycle. Include any negative feedback loops that may form. (1029–1030)
29. Explain how shifting hormones bring on the process of ovulation. (1030)
30. Describe the output of hormones that maintains the uterus through the first part of the postovulatory phase. What events end the cycle? (1030)
31. Once fertilization and implantation have occurred, what is the source of hormonal support during the early and later stages of pregnancy? (1032)
32. Describe the content of one type of birth control pill and suggest how it prevents conception. (1035)
33. List several reasons why the rhythm system of birth control is subject to failure. (1034)
34. List five methods of birth control that are unacceptable from the standpoint of failure. (1033)
35. Discuss the risks of oral contraceptives, but put them in the perspective of normal risks. (1035)
36. List an advantage and a disadvantage of vasectomy and tubal ligation. (1036)
37. Describe the three methods of clinical abortion. Which involves the greatest risks? (1037)

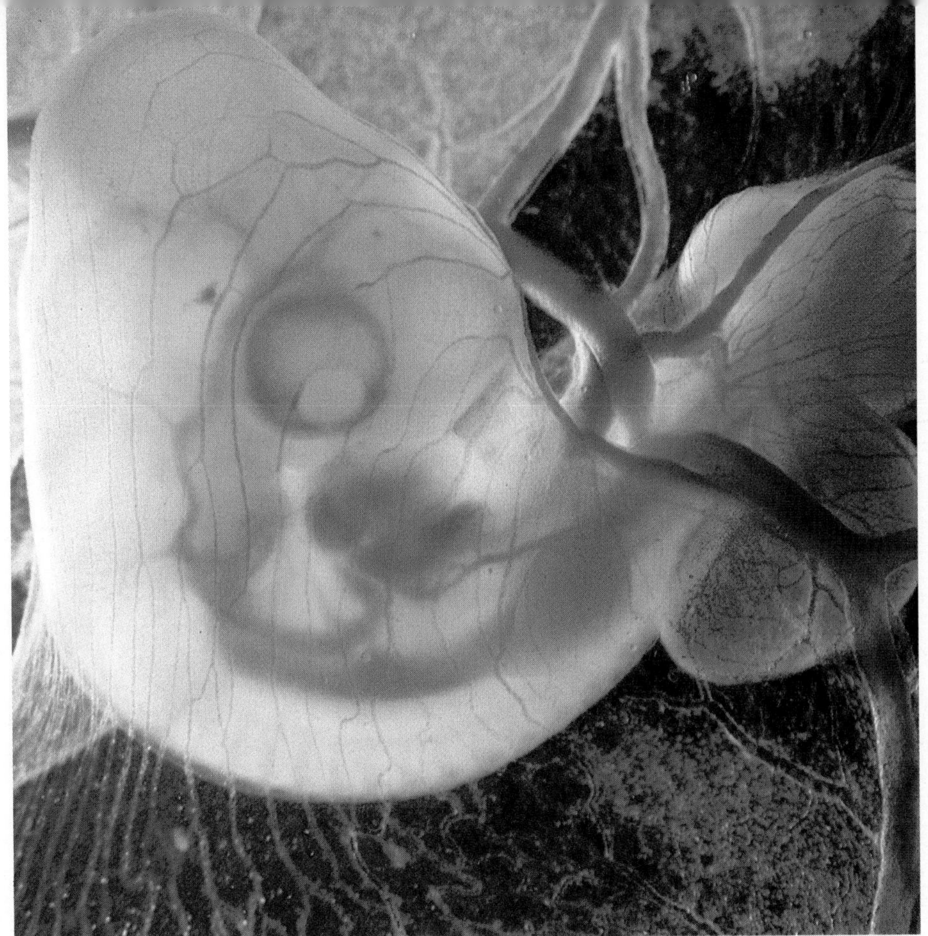

43
ANIMAL DEVELOPMENT

ONE CELL DIVIDES INTO TWO, AND TWO INTO FOUR. THE FOUR AGAIN divide, and the divisions continue until eventually there is a great number of cells. All the cells have stemmed from that one primordial cell, but yet, many divisions later, far down the line, descendant cells may come to be very different from one another. One descendant may become an elongated nerve cell, while another becomes a pigmented cell within the eye's retina, and yet another an unprepossessing liver cell with an impressive battery of biochemical tricks. But since they arise by mitotic divisions from a single fertilized egg, they all carry the same genetic information in their chromosomes.

How does this seeming miracle of development and differentiation occur? This, it turns out, is not a rhetorical question; we don't quite know. It seems that there should be some ready answer, some satisfying explanation.

Part of the problem is that the study of development has two aspects: The descriptive and the experimental—The familiar "what" and "how." The descriptive work—the story of "what" actually can be seen in a developing embryo—is already rather complete. There are serious, unanswered questions in the "how it happened" department, however.

So we will take things a step at a time, beginning with the descriptive aspects of **embryogenesis,** the formation of the embryo. We will include the fascinating areas regarding the forces that direct this development, following various tissues to their specific fates. An overriding consideration will be **determination,** the ongoing genetic commitment of cells and tissues to specific fates. Such commitment may begin very early, continuing until the last touches are made. We will see that the commitment of cells to a certain developmental fate becomes increasingly specific as time passes, until they are completely differentiated (developmentally specialized for a specific role). As various tissues take their place in the developmental pageant of the animal, **morphogenesis,** the development of shape and form, becomes apparent. It is, indeed, a fascinating phenomenon, the development of individuals from those tiny, specialized cells called gametes.

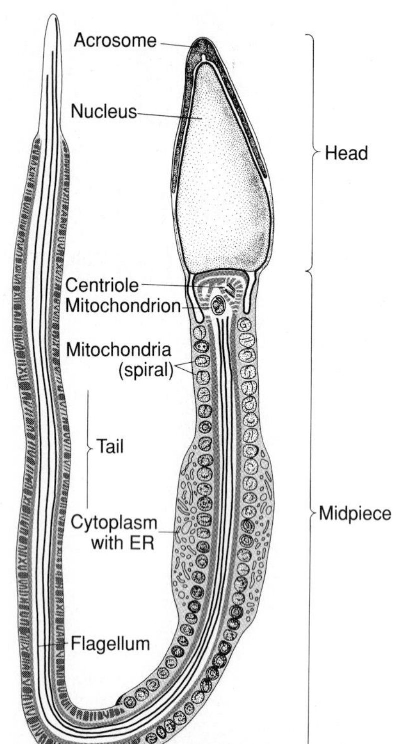

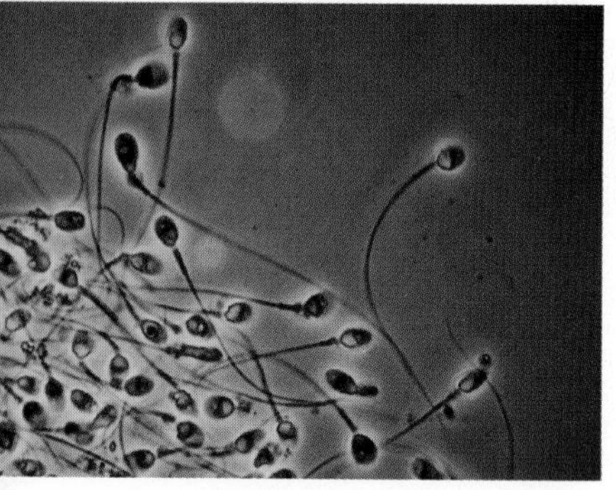

FIGURE 43.1
HUMAN SPERM
The human spermatozoan is highly specialized for its task: fertilizing the ovum. The chromatin-packed sperm head is seen in a side view. Note the broadened shape in the top view seen in some of the cells in the photograph.

GAMETES
The Sperm

The sperm is one of the most highly specialized of cells, and its role is straightforward—it must encounter and penetrate an egg. It thereby delivers its half of the offspring's required chromosome complement. The sperm's entire structure is devoted to this one brief mission.

The three major parts of the sperm are the head, midpiece, and tail (Figure 43.1). The sperm head contains the haploid chromosome complement in the form of highly condensed chromatin. At its tip is found the **acrosome,** a caplike structure that surrounds an enzyme-laden lysosome. If you recall the general nature of lysosomes, you can anticipate its role in penetrating the egg, but we will get to that shortly. The sperm midpiece includes an unusually long, spiralling mitochondrion, which provides the ATP energy needed by the lashing flagellum that propels the sperm along. As with other flagella, it has the familiar 9 + 2 microtubular construction (see Chapter 4). The midpiece also contains other typical organelles in its scant cytoplasm, including a centriole. Upon fertilization, the sperm centriole will join up with the egg centriole, producing the usual paired condition.

The Egg

The **egg,** with its haploid nucleus, contains the other half of the offspring's required chromosome complement. Animal eggs vary considerably in content, organization, and size. For example, whereas a human egg is about the size of the dot over the letter "i," you would be hard put to hide an ostrich egg behind this entire book. Such size differences, we will see, are largely due to differences in the amount of yolk. In the yolky bird and reptile egg, the cytoplasmic region destined to form the embryo, called the **blastodisc** lies perched on the yolk.

The egg cell has the usual cellular organelles but is quite unlike other cells in many respects. In nearly all cases, animal eggs show **polarity.** This means that the cytoplasmic constituents are not equally distributed, but occur in a gradient. The unequal distribution results in essentially two hemispheric divisions. In most kinds of eggs, the metabolic "machinery" (mitochondria, golgi bodies, edoplasmic reticulum, etc.) is concentrated in the more metabolically active **animal hemisphere** and the yolky food reserves in a less active **vegetal hemisphere.** Insect eggs, which differ in many ways from those of other animals, carry organization to the extreme. As we will see later in the chapter, they are so highly organized that the future anterior, posterior, dorsal, and ventral regions of the adult actually can be located in the egg (Figure 43.2a).

The varied yolk content of eggs in the different animal groups, as you might expect, relates closely to the mode of development. The large yolk reserves in the eggs of birds and reptiles will see the embryo through much of its development, which, as you know, occurs independently of the mother. The meager yolk in the mammal egg, on the other hand, is quite sufficient, considering that within a few days after conception the embryo will be fully supported by exchanges across the placenta. (Several animal eggs are compared in Figure 43.2.)

Specialized Surface Structures of the Egg A number of membranes cover the surface of the egg cell. In insects, molluscs, amphibians, and birds, a **vitelline membrane** forms just outside the plasma membrane. Insects produce a tough, thick covering, the chorion, which forms outside the vitelline membrane. Mammals form a **zona pellucida** ("clear zone") around the egg periphery. At ovulation the mammal egg is also surrounded by a cluster of follicle cells called the **corona radiata** ("radiating crown") (Figure 43.3). Echinoderm eggs have a clear **vitelline layer** surrounding the plasma membrane and a dense outer layer of jellylike material called the **jelly coat.** Still other surface structures may be added to egg surfaces as the eggs pass through the oviduct. The leathery egg

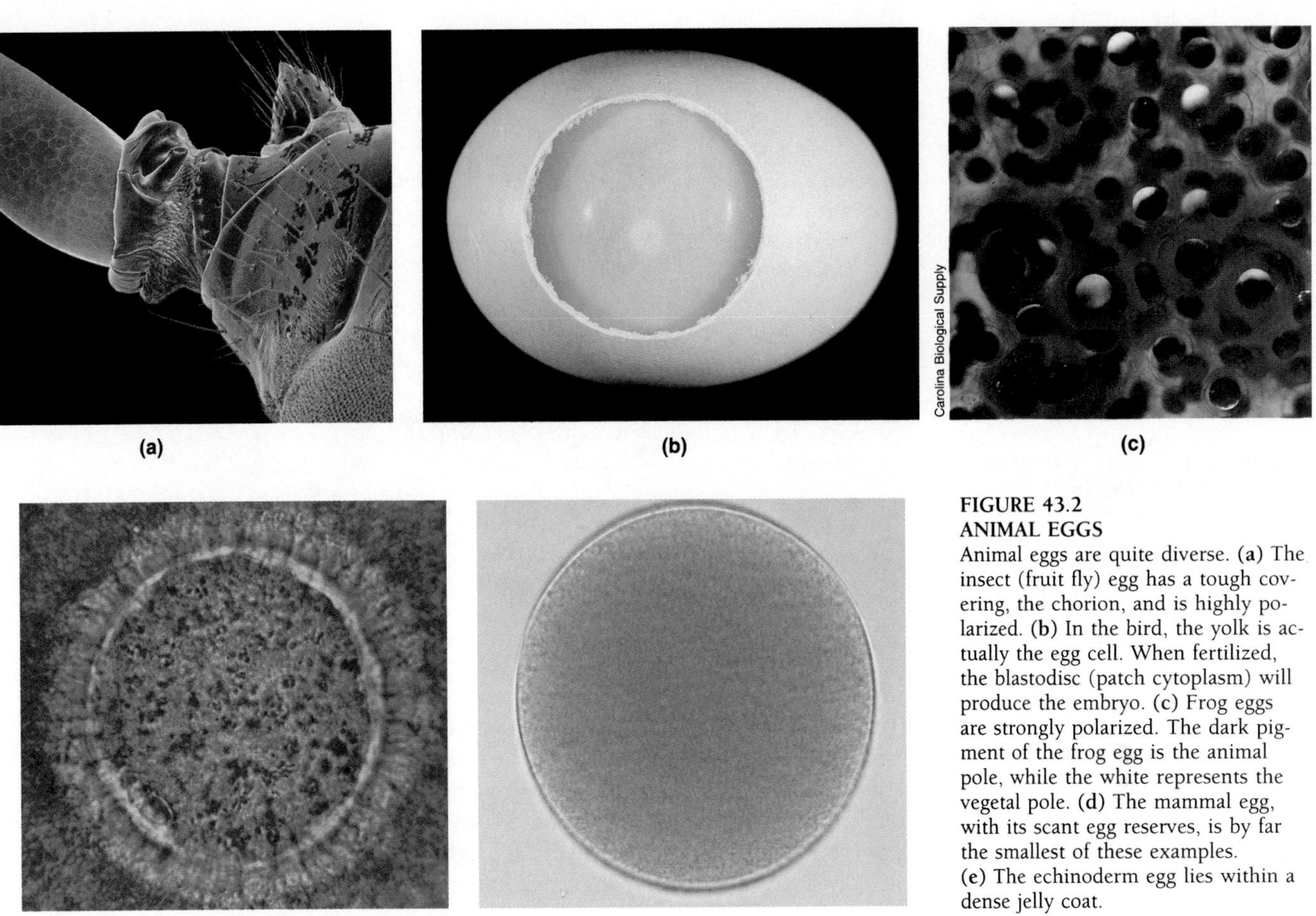

FIGURE 43.2
ANIMAL EGGS
Animal eggs are quite diverse. **(a)** The insect (fruit fly) egg has a tough covering, the chorion, and is highly polarized. **(b)** In the bird, the yolk is actually the egg cell. When fertilized, the blastodisc (patch cytoplasm) will produce the embryo. **(c)** Frog eggs are strongly polarized. The dark pigment of the frog egg is the animal pole, while the white represents the vegetal pole. **(d)** The mammal egg, with its scant egg reserves, is by far the smallest of these examples. **(e)** The echinoderm egg lies within a dense jelly coat.

casing of the shark is one example, and the calcium-hardened shell of the bird egg is another.

FERTILIZATION

Fertilization is the union of a haploid sperm nucleus with a haploid egg nucleus. In the larger sense, however, it includes the events that lead up to this final step. We are beginning to learn a great deal about fertilization in mammals, including humans, but at the moment most of our knowledge comes from studies of other animals, particularly echinoderms (sea urchins, sea stars, and sand dollars), whose hardy gametes are readily available and easily manipulated.

Fertilization in the sea urchin occurs externally, usually on the sea floor, where these slow-moving creatures release their sperm and eggs. Although many sperm approach each egg and a number penetrate the surrounding jelly coat and attach to the vitelline layer (Figure 43.4a), just one usually reaches the actual surface of the egg.

As a sea urchin sperm penetrates the jelly coat, it undergoes changes in its acrosome, the enzyme-laden cap over the sperm head. The **acrosomal reaction,** as these changes are called, has two parts. First, the acrosome ruptures and releases its cargo of enzymes, which then digest a path through the jelly coat and vitelline layer. Then a lengthy **acrosomal process** is extruded. In the successful sperm, the acrosomal process binds to the egg surface (43.4b).

As the acrosomal process binds to the egg's plasma membrane, sweeping changes occur in the egg itself. First, portions of its surface rise up toward the sperm, forming what is called a **fertilization cone.** Some observers report the formation of numerous

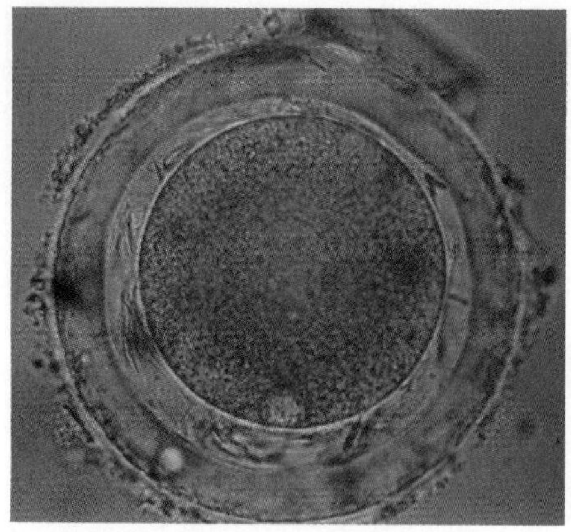

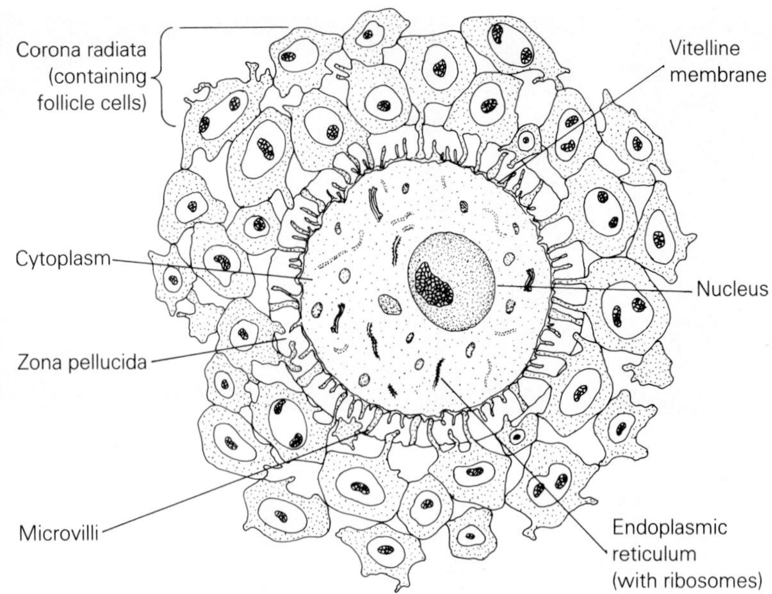

FIGURE 43.3
EGG SPECIALIZATIONS
The human egg, actually still an oocyte after ovulation, is surrounded by cells making up the corona radiata. Within the corona lies the dense zona pelludica, which extends to the egg plasma membrane. All of these structures must be penetrated by a fertilizing sperm. Note the small, oval-shaped polar body off to one side.

fingerlike microvilli, which wrap about the sperm head, actively drawing it into the egg (Figure 43.4b-e). Most agree that the sperm's plasma membrane fuses with the egg's plasma membrane, and the naked sperm head enters the egg cytoplasm. The sperm head detaches from its midpiece and tail and releases its nucleus. Then cytoplasmic microtubules form between the sperm and egg nuclei, or **pronuclei** as they are known, and draw the two together, completing fertilization.

The Cortical Reaction and Sperm Barrier

If more than one sperm penetrates the egg cytoplasm, an event called **polyspermy**, the embryo is usually doomed. However, the echinoderm egg has two kinds of sperm barriers that act to reduce the likelihood of polyspermy. One is almost instantaneous but very temporary, and the other is slower to form but permanent.

Within seconds after a sperm binds to the plasma membrane, the egg undergoes a sudden electrical change, almost like the neural impulse. There is a rapid influx of positive ions, causing the egg cytoplasm to shift from negative to positive. The shift in charges blocks further sperm entry, but the effects are transient, and within 20 seconds the egg's original charges are restored. But in the meantime, the egg will have formed its permanent barrier, the **fertilization membrane.** This process begins with the **cortical reaction** (see Figure 43.4b), whereby numerous enzyme-laden **cortical granules** below the plasma membrane are drawn to the plasma membrane, where they fuse, spilling their enzymes against the vitelline layer. As the enzymes act, the vitelline layer rises up from the plasma membrane and changes into the prominent, impenetrable fertilization membrane. This fertilization membrane appears first as a small mound at the site of sperm entry, but as the cortical reaction spreads, the membrane soon comes to surround the entire egg (Figure 43.4c). The timing must be quite precise, because the new physical barrier must be in place before the electrical barrier disappears.

Fertilization in humans (Figure 43.5) has some similarities to that of sea urchins, but there are critical differences as well. In humans, as well as echinoderms, the egg may encounter a great many sperm. Of the 300 million or so sperm released in a single ejaculation, several thousand usually reach the egg—one of which will fertilize the egg. Although no acrosomal filament forms, human sperm also release enzymes. In this case they digest a path through the corona radiata. (A minimum number of sperm is necessary

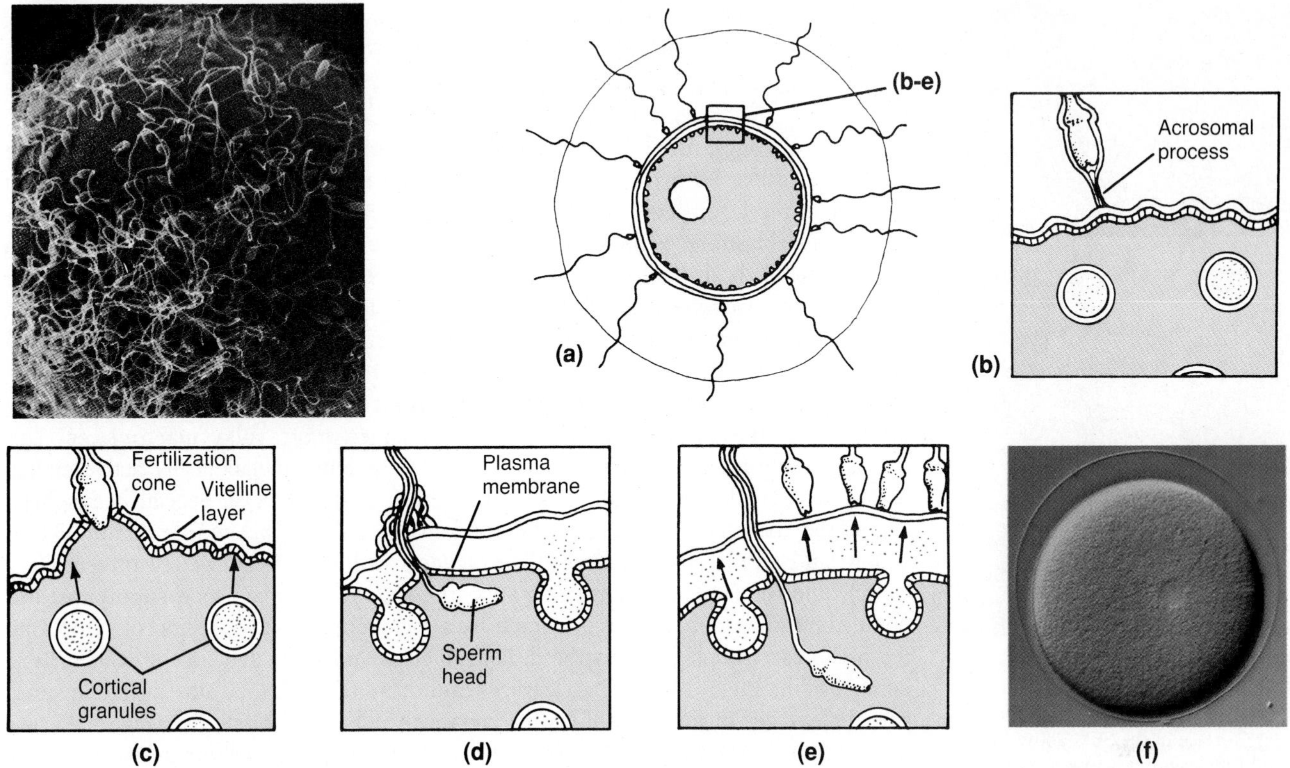

FIGURE 43.4
STUDY OF SEA URCHIN FERTILIZATION
(a) Numerous sperm make their way to the egg surface during fertilization, many penetrating the jelly coat and attaching to the vitelline layer. (b) A successful sperm attaches to the egg plasma membrane via its acrosomal process. (c) As sperm and egg membranes fuse, the egg responds, a cytoplasmic fertilization cone rising up around the sperm head. Cortical granules below have begun to fuse with the plasma membrane. (d) The sperm head has now penetrated the cytoplasm, and some cortical granules have released their contents. (e) In the cortical reaction, the vitelline layer separates from the plasma membrane and rises up to become a sperm barrier, the fertilization membrane. (f) At the completion of the cortical reaction, the fertilization membrane becomes quite prominent.

to break down this barrier, which is why men with low sperm counts, while producing normal sperm, are functionally infertile.) Once through the corona radiata, sperm then bind to the zona pellucida, just as sperm bind to the sea urchin egg's vitelline layer. Reaching the egg surface itself requires the action of still other enzymes. When a sperm finally contacts the egg surface (Figure 43.5), the membranes fuse, the egg is activated, and the sperm is down into its cytoplasm.

Although mammal eggs also form sperm barriers, no visible fertilization membrane forms. One sign of fertilization in humans is the appearance of a second polar body, the result of a second meiotic division. (Recall from Chapter 9 that meiosis, human oogenesis, is not completed unless fertilization has occurred. See Figure 9.14.)

EARLY DEVELOPMENT EVENTS

Fertilization is followed by a period of intense activity, including DNA replication, mitosis, and the first of many cell divisions. Cell division at this time is referred to as **cleavage**, because the entire embryo usually divides in two (is cleaved). During this early development process, the zygote divides again and again, its cells becoming smaller and smaller, finally forming a ball of tiny cells. The resulting cells will provide the raw material from which the embryo will be molded. Let's look into some of the various patterns cleavage takes.

Becoming Multicellular: The First Cleavages

Cleavage takes place in a similar fashion in all animals, but the precise pattern it takes is largely dependent on two factors: yolk content and evolutionary history.

Cleavage patterns in mammals and echinoderms, with their scant yolk supply, are the simplest. The first cleavage divides the zygote into **blastomeres** (early embryonic

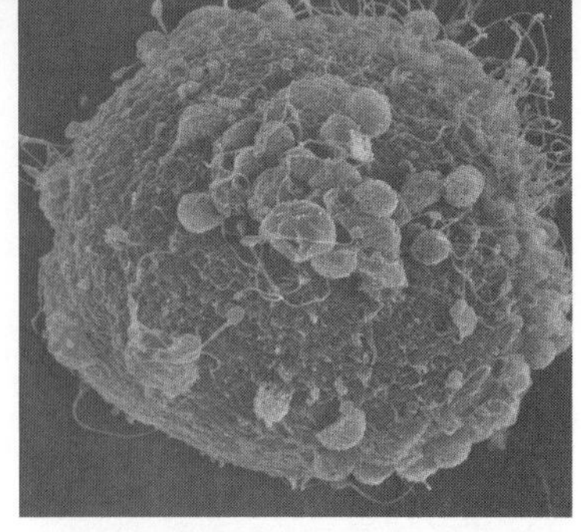

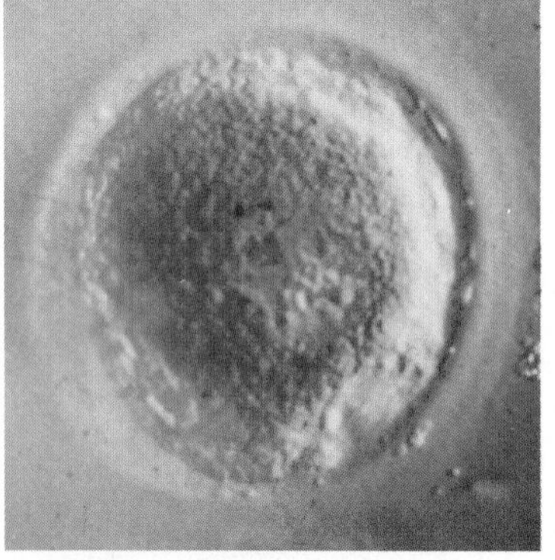

FIGURE 43.5
FERTILIZATION IN HUMANS
Note the sperm tails projecting from the egg surface in the SEM above. In the light micrograph below, a second polar body is taking form at about five o'clock.

daughter cells) of equal size and appearance. The second, third, and subsequent cleavages also produce blastomeres of similar size (Figure 43.6a and b), but with each division the cells become smaller, until the embryo becomes a simple ball of cells just about the size of the original zygote. This stage in mammals is called the **morula** (from "mulberry").

Amphibian eggs, recall, have a substantial amount of yolk, so we can expect cleavage patterns to differ from those of echinoderms and mammals (Figure 43.6c). The first cleavage in the frog zygote is vertical and passes through the animal and vegetal hemispheres. The blastomeres thus contain equal amounts of yolk. The second cleavage again passes through both hemispheres and is at right angles to the first. In this way, four fully equal blastomeres are formed (note how its plane differs from that of the second cleavage in the mammal). The third cleavage is at right angles to the first two, but because it occurs high up in the animal hemisphere (above the "equator"), the earlier symmetry is lost. As a result, four small **micromeres** form, all containing the active cytoplasm of the animal pole. Below are four larger, more yolky **macromeres.** This begins the unique amphibian pattern. Cells of the more metabolically active animal hemisphere divide considerably faster than those of the sluggish vegetal hemisphere, and so they become increasingly smaller relative to the cells below (see Figure 43.6d).

Bird and reptile eggs are large, their contents including clear, watery protein ("egg white") and the large, nutrient-laden yolk (the embryo's primary food supply). At this point, the future embryo is represented by a metabolically active region of cytoplasm, the blastodisc, which contains the diploid nucleus and is located on the yolk surface. Because of the huge yolk region, cleavages cannot cut through the entire zygote. Instead, partial cleavages in the form of simple crevices occur in the blastodisc (Figure 43.6e). Cell division in the partially formed surface cells is completed a short time later, as the

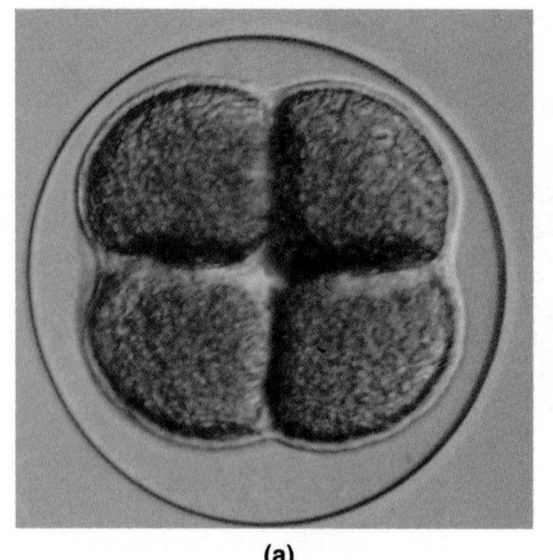

(a)

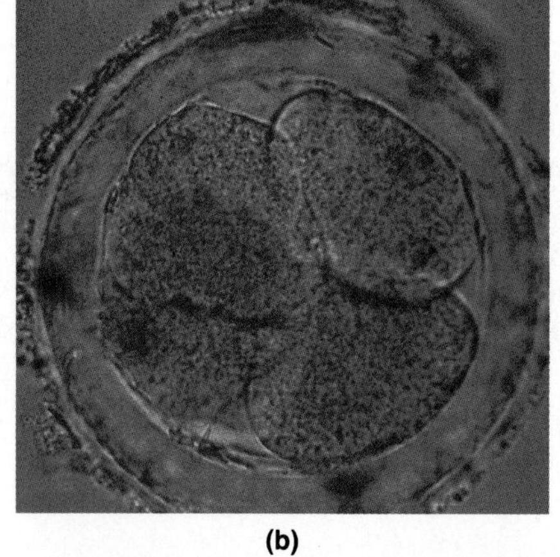

(b)

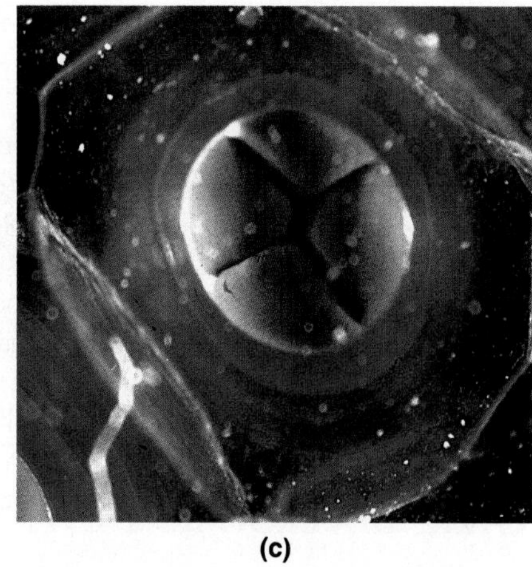

(c)

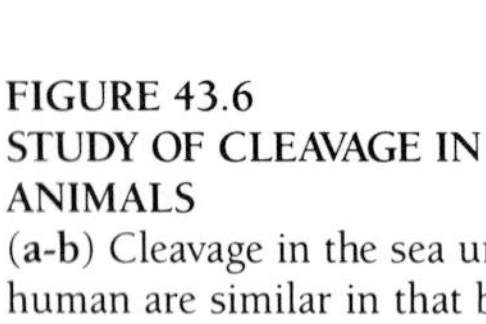

FIGURE 43.6
STUDY OF CLEAVAGE IN ANIMALS
(**a-b**) Cleavage in the sea urchin and human are similar in that blastomeres of equal size appear. (**c**) The first two cleavages in the frog produce equal blastomeres, but from there on micromeres and macromeres emerge (**d**). The immensity of the bird yolk (**e**) results in incomplete cleavages at first, producing blastomeres that bear little resemblance to the other examples.

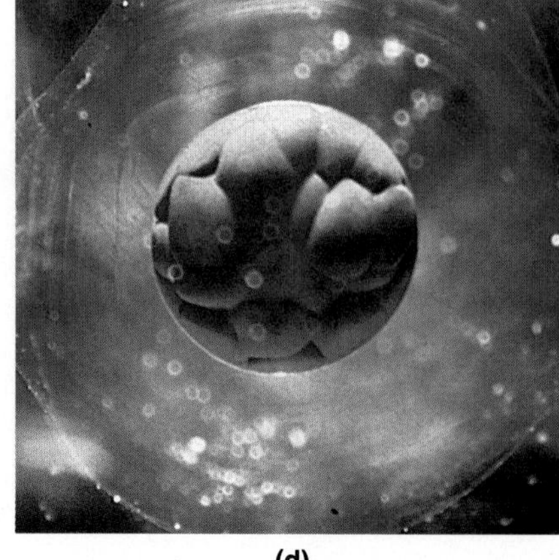

(d)

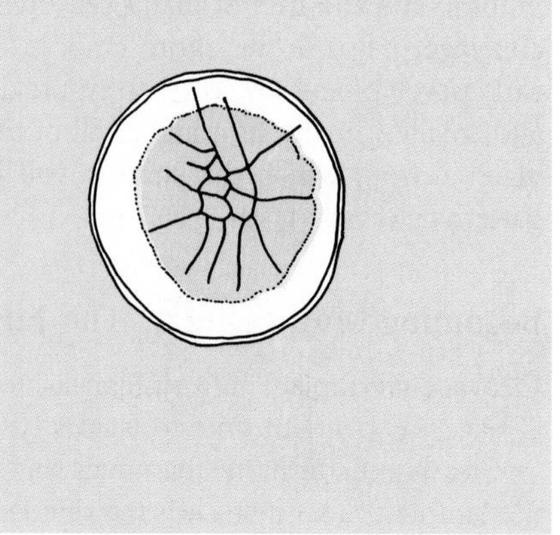

(e)

blastoderm, a flattened two-layered island of cells, arises. Later on in reptile and bird development, as we will see, certain cells of the blastoderm divide rapidly, spreading out and encompassing the yolk as they form the *yolk sac*.

As we have noted, a second factor influencing cleavage patterns in embryos has to do with their past evolutionary history. You may recall from Chapters 30 and 31 that many animals diverged early in their history into two lines, the protostomes (primarily flatworms, molluscs, annelids, and arthropods) and deuterostomes (primarily echinoderms and chordates). One of the embryological distinctions between the two groups has to do with early cleavage patterns. Whereas protostome cleavages follow a spiral pattern, in deuterostomes the pattern is radial (see Essay 31.1).

Insect Cleavage: A Very Different Pattern Insect cleavage differs considerably from that of vertebrates and many other invertebrates as well. The primary difference is due to the fact that the fate of the various parts of an insect embryo are sealed very early in development. In fact, a two-directional polarity (both anterior-posterior and dorsal-ventral) is well established in the egg of the fruit fly (*Drosophila*) well before fertilization. Following fertilization, rapid mitotic activity ensues, but, oddly, no cytoplasmic division occurs (no membranes form between daughter nuclei) until later, so the early embryo is simply multinucleate (Figure 43.7). After hundreds of nuclei have emerged, a few of these begin to migrate to the posterior periphery of the egg, and soon plasma membranes form around these nuclei. Interestingly, the first few cells, called **pole cells** (Figure 43.7d), already have their development fates determined. In the adult, they will give rise to germinal cells, those that will produce sperm and egg.

Nuclear division and migration in the fruit fly embryo continues until about 6000 nuclei are present in a monolayer. Cell membranes then form around each, and the cell layer becomes the insect blastoderm (Figure 43.7f). With this pattern established, the segmented form of the insect larva soon emerges. We'll return to insect development later in the chapter.

The early commitment of embryonic cells in insects is characteristic of what is called **determinate cleavage** (see Essay 31.1), another protostome trait. Protostome embryos have little of the versatility seen in deuterostome embryos, where **indeterminate cleavage** is the rule. As we will see, whereas early deuterostome embryos can be easily manipulated by the experimenter without serious consequences, any such attempts with most protostome embryos, and particularly those of insects, result in serious development problems.

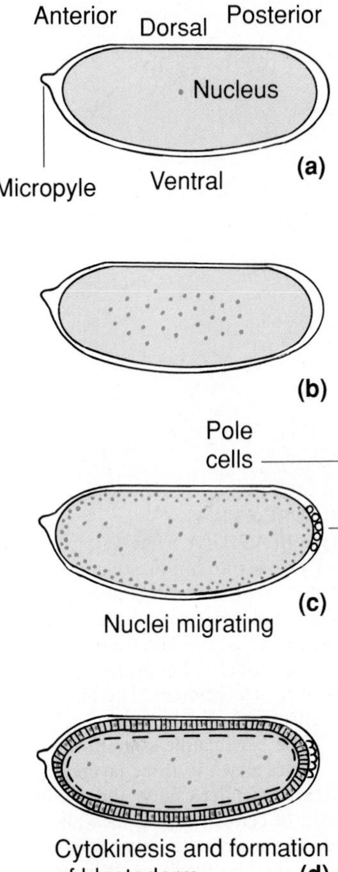

FIGURE 43.7
CLEAVAGES IN THE INSECT EMBRYO
In addition to its pronounced polarity (a), early insect development is unusual in that mitosis (b) is not followed by cytokinesis. Cleavage finally occurs after some 10 rounds of mitosis, but then yielding just a few pole cells (c). By the fourteenth round of mitosis, most of the nuclei have migrated to the egg periphery. (d) Plasma membranes forming about the peripheral nuclei form the insect blastoderm.

The Blastula

As the cells of many embryos continue to divide, they begin to take on a new arrangement, forming a fluid-filled inner cavity or **blastocoel.** At this stage, the embryo is called a **blastula.** In the sea urchin (Figure 43.8a), the outer cells form cilia, and the blastula escapes the fertilization membrane to begin a free-swimming period, spinning along in the sea as its development continues.

The frog blastula (Figure 43.8b) is asymmetrical due to the preponderance of yolk at the vegetal pole. The blastocoel therefore forms near the animal pole, with the pole cells forming a thin rooflike cover over the cavity.

Because of the great amount of yolk, the formation of a simple blastula is not possible in the bird and reptile embryo. Its equivalent here is a flat, platelike, two-layered blastoderm, below which lies a narrow cavity that forms as the blastoderm separates from the yolk (Figure 43.8c).

The blastula stage is represented in mammals by the **blastocyst** which in its earliest period consists of just 32 cells. The blastocyst has three parts, a fluid-filled **blastocyst cavity** and two cellular regions. The **trophoblast** is a thin layer of cells surrounding the cavity, while the **inner cell mass** is an aggregation of cells that protrudes into one side of the cavity (Figure 43.8d). The trophoblast plays a supporting role, whereas the inner cell mass contributes cells to the embryo proper.

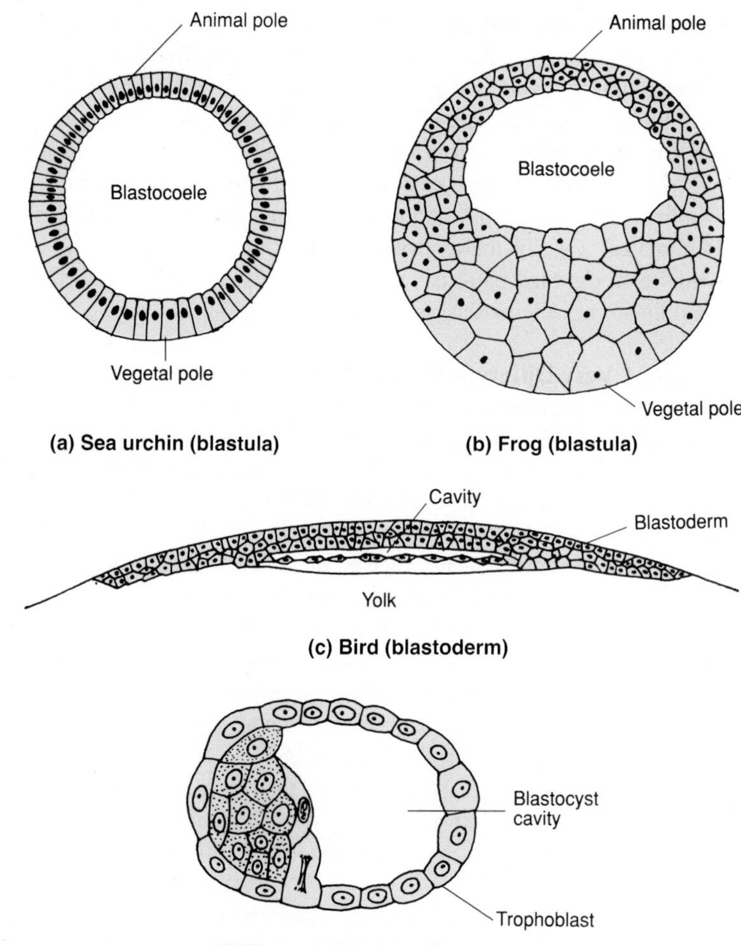

FIGURE 43.8
THE BLASTULA STAGE
(a) The sea urchin blastula is fairly simple, with some vegetal-animal hemisphere orientation and a single layer of cells. (b) The frog blastula is a dense sphere of cells that occur in a definite size gradient. The blastocoel is offset towards the animal pole. (c) The blastulalike stage in the bird includes a two to three-layered blastoderm, a slender cavity, and a thin layer of cells overlying the yolk. (d) The mammalian blastocyst represents the blastula stage. It has three regions: a thin trophoblast, a denser inner cell mass, and the blastocyst cavity.

The blastula and blastocyst represent the end of a preliminary period in development. The embryo has remained simple up to now, but all this will change as it enters **gastrulation**, where extensive changes produce a dramatic rearrangement of cells into specific germ layers marking the **gastrula** stage. This, in turn, paves the way for still later events in which the embryo will take on a more familiar form.

Gastrulation: Organizing the Germ Tissues

"It is not birth, marriage, or death, but gastrulation, which is truly the most important event of your life," said embryologist Lewis Wolpert in 1983. The reason is, nothing can proceed without this critical rearrangement of the embryo. (If your tissues didn't shift, you don't have to worry about the next most important.) Indeed, gastrulation is both fascinating and critical to further development, because it marks the formation of the body's germ layers,—**ectoderm**, **endoderm**, and **mesoderm**. The word *germ*, also *germinal*, refers to beginnings, for it is from these layers that formation of tissues, organs, and organ systems will begin. You may recall that the evolution of the three germ layers (the triploblastic state) in metazoan animals, was one of the key events in animal evolution, paving the way for the formation of true organ systems (see Chapter 30).

The Sea Urchin One of the simplest models for gastrulation is provided by the sea urchin. As we see in Figure 43.9a, the process begins with **invagination**, an inpocketing of the blastula surface (like pushing your fist into a soft beachball). In this way a new cavity, the **archenteron** is created. Its opening is called the **blastopore**. The archenteron, or "primitive gut" as it is also known, later forms the actual gut of the animal, so we see that gastrulation also lays the groundwork for forming the tube-within-a-tube body

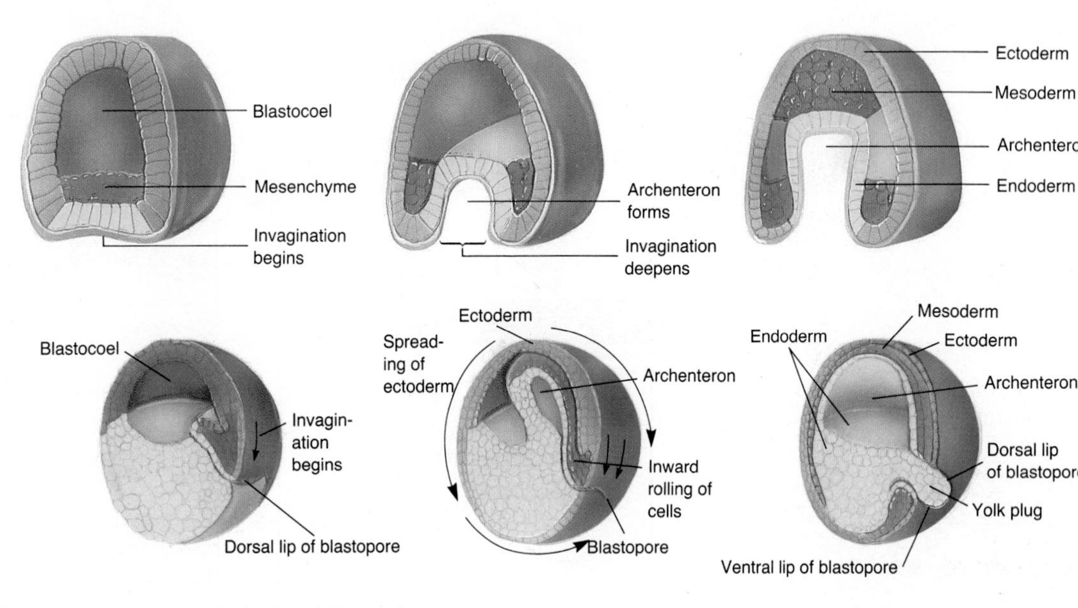

(a) Gastrulation is simplest in the sea urchin, where invagination proceeds from the vegetal hemisphere.

(b) Gastrulation in the frog includes invagination (an inpushing of cells), involution (the inward rolling of cell layers), and epiboly (the flattening and spreading of cells over the spherical gastrula).

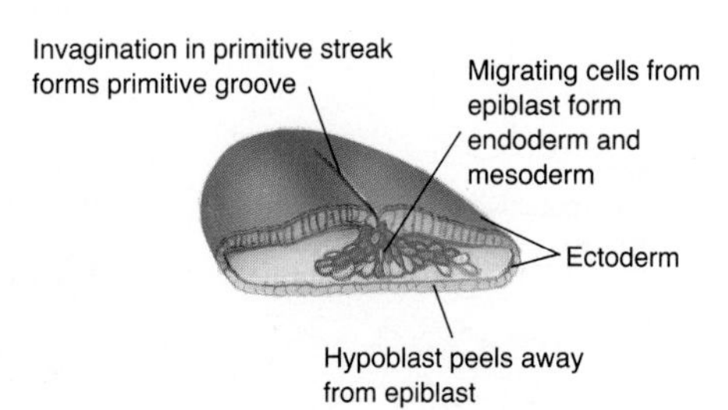

(c) Gastrulation in birds, reptiles, and mammals is similar: following a thickening in the epiblast, insinking cells form the primitive groove. Migrating epiblast contributes to mesoderm and endoderm.

FIGURE 43.9
GASTRULATION
Gastrulation in animals involves forces that bring about a vast relocation of tissues, resulting in the formation of a triploblastic embryo.

plan. Since the sea urchin is a deuterostome, the blastopore region represents the future anus, with a mouth later breaking through at the opposite end (see Essay 31.1). Yet, the ingrowth has a more immediate result: The outer layer is now the ectoderm, the lining of the archenteron is the endoderm, and in between is the mesoderm. The mesoderm is formed from **mesenchyme**, cells that arise from the flattened blastopore region and migrate to the interior of the old blastocoel. From the gastrula, the sea urchin develops into a feeding stage, the **pluteus larva** (see Figure 43.9a), which will later begin the transition into an adult.

The Amphibian We can derive the general principles of gastrulation from the sea urchin, but in some other animals the process can be a bit more complex. Gastrulation in the amphibian (Figure 43.9b) also begins with invagination in the area that will form the blastopore. At first, the insinking involves only a line of mobile cells that forms a kind of curved slit on the surface of the embryo, the **dorsal lip** of the blastopore. With further invagination, though, the edges of the slit are extended they join, forming a circle (the completed blastopore) that surrounds protruding yolky endodermal cell that make up the **yolk plug.** A closeup view at the blastopore (Figure 43.9a, inset) reveals that the cells at this stage change shape, and those on the surface begin to roll under into the indentation caused by invagination. The pocket grows deeper and more rounded, crowding into the old blastocoel until only remnants of the blastocoel remain. It is through this continuing inrolling action that the middle, mesodermal, germ layer becomes established.

During gastrulation, surface cells above the enlarging archenteron also undergo changes. A flattening and spreading process begins in cells at the apex of the animal hemisphere and advances in a wavelike front that spreads entirely over the embryo (Figure 43.9b). So, by late gastrulation the embryo is surrounded by a complete but very thin layer of ectoderm. The embryo is now composed of an outer ectoderm, a middle mesoderm, and an inner endoderm. And with the archenteron in place, the tube-within-a-tube body plan has begun to take form.

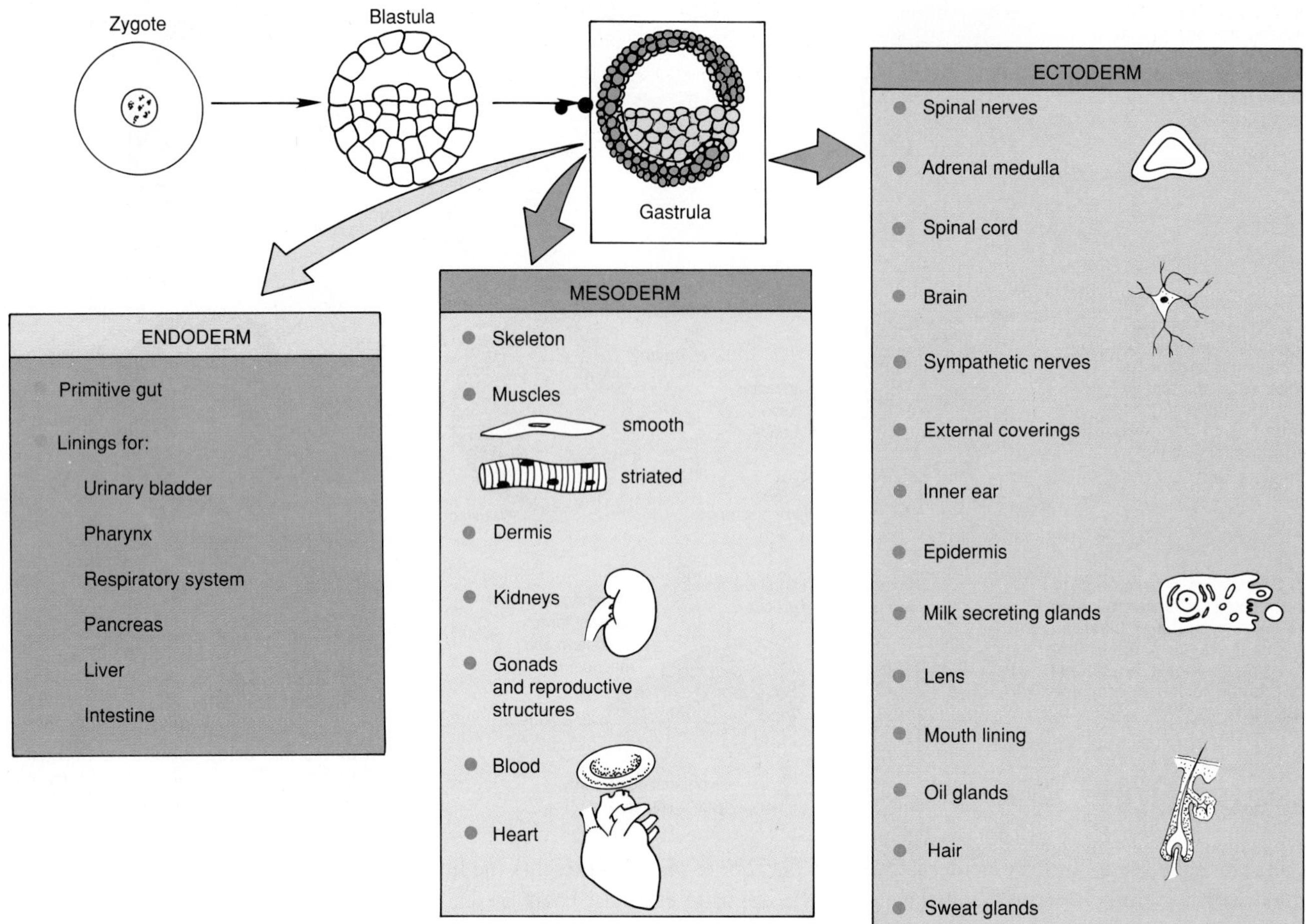

FIGURE 43.10
COMMITMENT OF THE GERM LAYERS
The future role of each of the gastrula's three germ layers—ectoderm, mesoderm, and endoderm—has been generally determined.

Birds, Reptiles, and Mammals Gastrulation in birds and reptiles, as you might expect considering the yolk they must deal with, proceeds a bit differently. As a result of the early cleavages, the embryo has become a flattened blastoderm. Just prior to gastrulation, the blastoderm includes a denser upper region of cells, the **epiblast**, and a thin layer of cells overlying the yolk, the **hypoblast**, the two separated by a narrow, flattened cavity. The epiblast will form the ectoderm and will give rise to the mesoderm and endoderm as its cells are relocated during gastrulation.

The stage is set for gastrulation by the formation of the **primitive streak**, a lengthy thickening of the epiblast. Gastrulation begins with the formation of a crease, the **primitive groove**, along the midline of the streak (Figure 43.9c). The process continues with a deepening of the groove brought about by the inward migration of cells. The effect is somewhat like the invagination that forms the amphibian blastopore described earlier. As the primitive groove deepens, some of the cells turning into the cavity below will become organized into mesoderm, whereas others will displace the older hypoblast, forming endoderm. So gastrulation in the bird, as in the echinoderm and amphibian, produces a three-layered embryo. An archenteron also forms in the bird and reptile gastrula, but in these vertebrates the gut must arise through a number of infoldings in the endoderm, which eventually enclose the cavity. Some of the endoderm and mesoderm will later form a spreading growth that gives rise to the yolk sac. Finally, we should note that the primitive streak also establishes the vertebrate's longitudinal body axis, along which later appears the notochord, dorsal nerve cord, and eventually the vertebral column.

What about mammals? You will probably be pleased to learn that gastrulation in mammals is almost identical to that of birds and reptiles. Mammals also form a two-layered blastoderm, a primitive streak, and primitive groove, and during gastrulation migrating cells from the surface will contribute to mesoderm and endoderm. In view of the vast differences in the egg and other aspects of development, such similarities may be unexpected. But the similarities may be viewed as compelling evidence of the close evolutionary relationship among birds, reptiles, and mammals. It seems that mammals, the last vertebrates to evolve, have retained the early developmental patterns, those deeply ingrained in those ancestors with very yolky eggs.

THE EMBRYO TAKES FORM; ORGANOGENESIS

It is during organogenesis, the formation of organs and organ systems, that we begin to learn the fates of the three germ layers (Figure 43.10). In general, the ectoderm makes its primary contributions to body coverings and neural structure, including the brain and spinal cord. Mesoderm, the middle germ layer, lays out the framework for the skeleton, muscles, blood vessels, and gonads. To do this, mesoderm must migrate throughout the developing embryo. Patches of relocated mesoderm are designated as **mesenchyme** ("middle tissue"). The endoderm specializes in linings, including those of the digestive, respiratory, and excretory systems. Actually, although each germ layer contributes most significantly to specific organs and systems, each organ in the body will eventually contain tissue elements derived from all three germ layers. So, for example, although the brain is principally ectodermal in origin, it also contains tissue elements derived from mesoderm and endoderm.

Establishing the Body Axis

The next events accomplish two things. First, the vertebrate embryonic axis (the anterior-posterior organization) will become clear: that is, distinct head and tail ends will emerge. Second, during the establishment of the axis, the crude outlines of the brain and spinal cord form. Both occur through a fascinating process called **neurulation.** Neurulation begins right after gastrulation and marks the **neurula** stage of the embryo. It proceeds in a similar fashion in most vertebrate groups, so we can concentrate on amphibian development as representative of the process (Figure 43.11).

Neurulation Neurulation begins with the formation of a flattened strip of ectoderm called the **neural plate.** The strip extends part way around the gastrula, extending from the dorsal lip of the blastopore (the future tail region) to a point representing the embryo's head (Figure 43.11a).

As neurulation proceeds, the two edges of the neural plate thicken, forming what are called **neural folds.** Between the folds a depression, the **neural groove**, also takes form. Next, the neural folds rise up along their length, curve inward above the neural groove, and close, thus forming the **neural tube.** In the closing, clusters of cells known as **neural crest cells** become isolated between the neural tube and the overlying ectoderm (Figure 43.11e). We mention these cells because they will later migrate far and wide from this location, contributing to a variety of structures. Finally, we should note that the neural tube is not uniform, but widens out considerably at the end opposite the dorsal lip, forming the crude outlines of an emerging head region.

At the start of neurulation, a second critical structure, the rod-shaped **notochord,** takes form just below the neural plate. The notochord, recall, is present in all chordates at one time or another. Numerous experiments show that the notochord has a profound effect on the overlying neural plate and, in fact, stimulates the formation of the neural tube. In development terminology, the notochord is said to **induce** the ectoderm to form a neural tube. We will look further into such critical influences later in the chapter.

In its later period, the neurula elongates and undergoes many internal changes. With elongation, the archenteron below becomes more tubelike and later will form anal and mouth openings. On either side of the notochord, the mesoderm has organized into paired blocks called **somites.** The somites are forerunners of trunk muscles and the axial skeleton. Their organization reminds us again of the segmental plan so common to all vertebrates.

FIGURE 43.11
NEURULATION
Neurulation in the amphibian is representative of vertebrates in general. The thickened neural plate (**a**) gives rise to neural folds (**b**), which arch upward (**c and d**) over the neural groove until they touch, fusing to form the neural tube (**e**). The neural tube establishes the embryonic axis, its widened portion representing the anterior or head region.

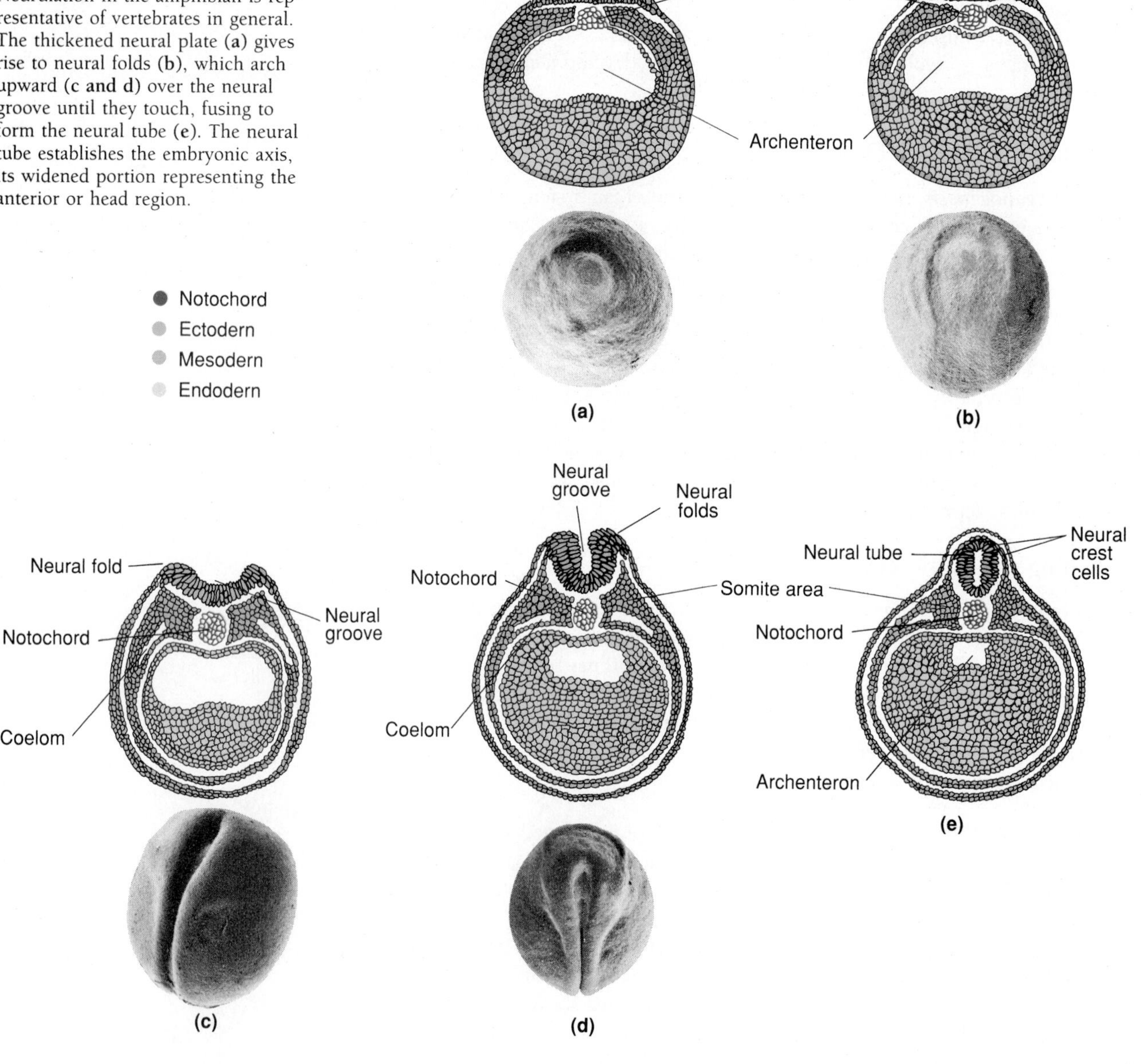

Further Vertebrate Development

Figure 43.12 compares development in our representative vertebrates. Notice in each case the formation of the **gill apparatus**, formed from **gill arches** and **gill slits.** The entire gill apparatus derives from the rudimentary gut in all vertebrates. In fishes, most of the embryonic gill apparatus does just what its name suggests. It forms the gills and their supporting structures. Although terrestrial vertebrates do not produce gills, once the ancestral development pattern of the gill apparatus was laid down it was retained and drastically modified. In terrestrial vertebrates, the gill apparatus contributes to the formation of a number of structures. Part of the first pair of arches contributes to the inner ear, and the next three pairs make contributions to the parathyroid glands, tonsils, and thymus. As you can imagine, the presence of the gill apparatus in the embryo of terrestrial animals has long been cited as strong evidence of evolutionary relatedness among the chordates (see Chapter 32).

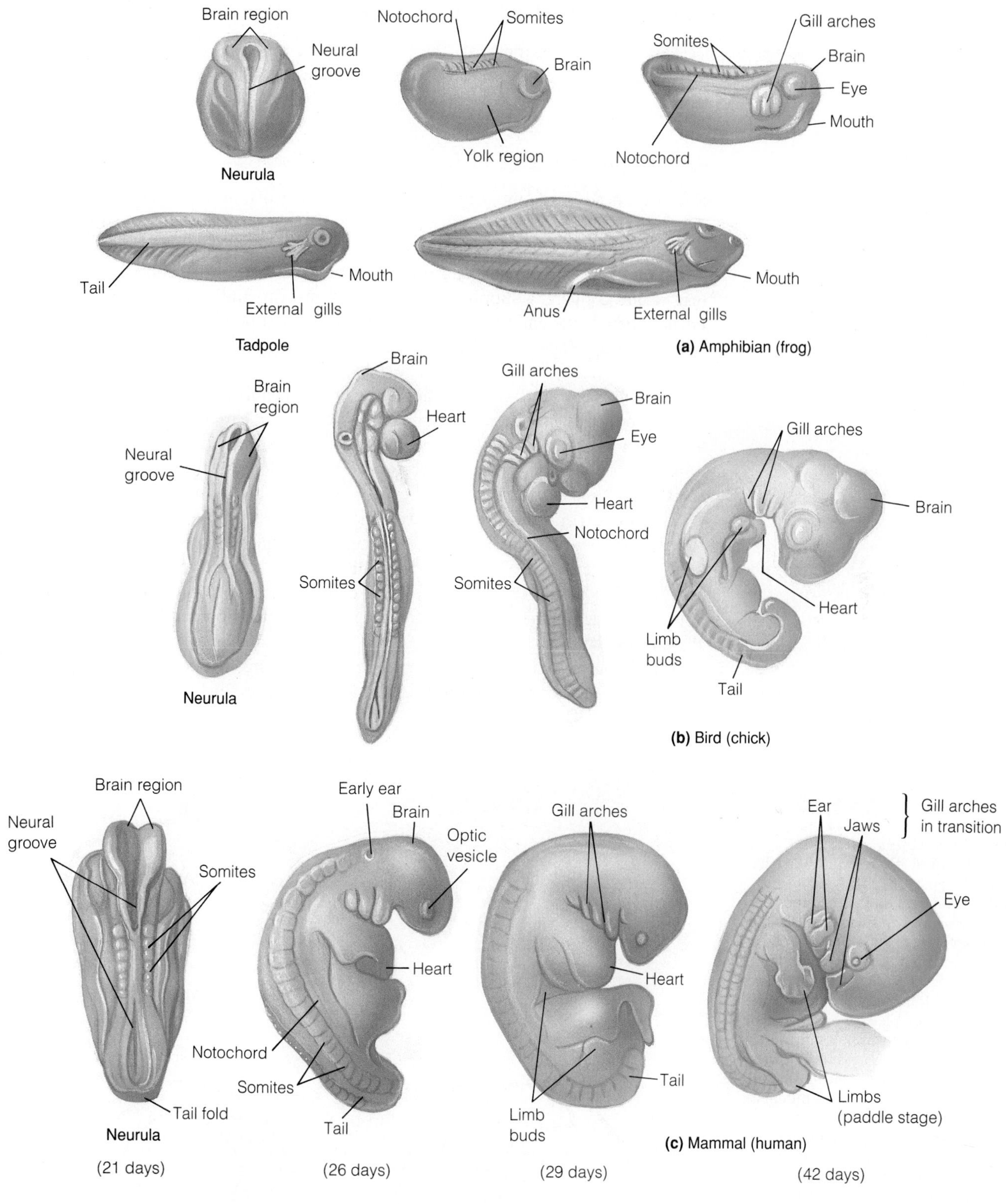

FIGURE 43.12
CONTINUED VERTEBRATE DEVELOPMENT
The general vertebrate developmental plan expresses itself through similarities in representatives of three classes: (a) amphibian, (b) bird, (c) mammal. Such similarities include the anterior-posterior orientation of the axis, the presence of a notochord, formation of the broad outlines of the nervous system, the rise of somites, and the brief appearance of the gill slits. Note the early development of the heart as well as the central nervous system.

VERTEBRATE SUPPORTING STRUCTURES

Development involves not only the developmental processes of the embryo itself, but the appearance of several structures that help sustain and protect the embryo. Such structures, produced by the embryo, will be left behind after birth or hatching.

The Support Systems of Reptiles and Birds

Newly fertilized reptile and bird eggs consist essentially of four parts: shell, yolk, albumen, and embryo. The protective shell itself consists of either a soft leathery covering (reptiles) or a hardened calcareous material (birds). Since the shell is porous, gases readily diffuse in and out. The yolk provides most of the nutrients needed by the embryo. The albumen (egg white) contains a supply of water, stored protein, and some antibacterial agents. Finally, there is the embryo, which will produce four supporting extraembryonic membranes as it develops—the yolk sac, the chorion, the allantois, and the amnion (Figure 43.13).

The **yolk sac** of birds and reptiles is the first extraembryonic membrane to develop. It is a product of endodermal growth and is actually an extension of the primitive gut. As it expands over the surface of the yolk, it becomes suffused with an extensive system of blood vessels. Food substances, absorbed into the blood as it circulates over the yolk's surface, are thus carried into the embryo.

The **amnion** and the **chorion** develop simultaneously, growing up over the embryo shortly after neurulation. The amnion forms a fluid-filled, protective sac that acts as a shock absorber and also keeps the body lubricated so that the growing parts don't fuse together. The chorion continues to grow around the entire egg contents, forming a continuous membrane just under the shell and later fusing with the allantois to form the **chorioallantois.**

The **allantois,** essentially endodermal in origin, begins as a peculiar little pouch, but as it grows it spreads out into a full-fledged membrane, coming to lie against the chorion,

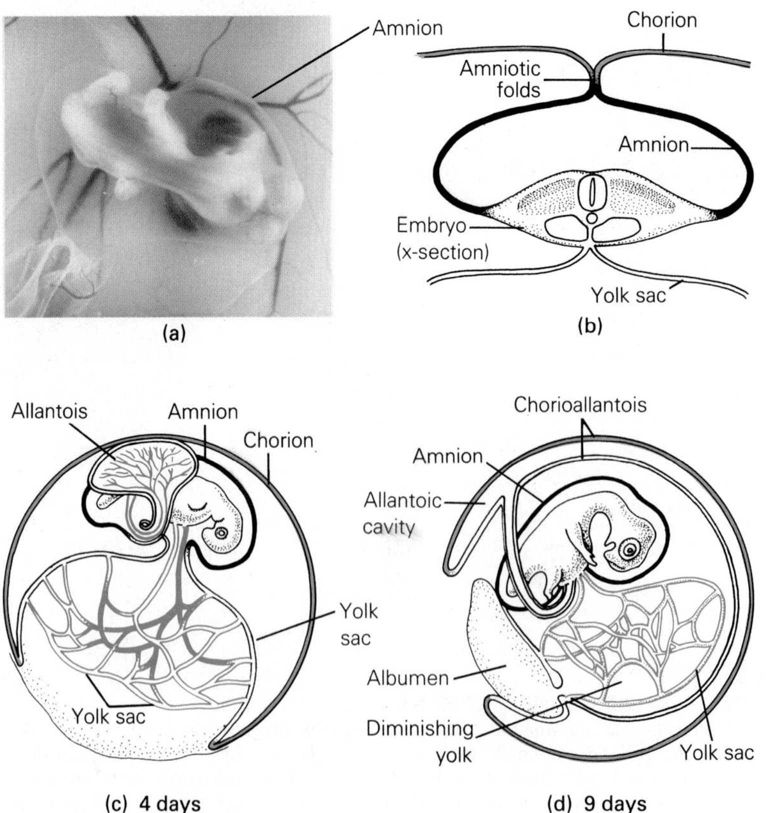

FIGURE 43.13
EXTRAEMBRYONIC MEMBRANES IN BIRDS
(a) The amnion and part of the yolk sac are seen in an intact five-day-old chicken embryo (photo). **(b)** Following neurulation, a fold of mesoderm and ectoderm produces the amniotic folds, which merge together over the embryo to complete the amnion. The same membrane continues outward forming the chorion, while below, a third membrane ventures out to become the yolk sac. **(c)** In a different view, the yolk sac is seen as a vascularized membrane spreading out over the yolk. The allantois begins as a simple sac that follows the path of the chorion as it enlarges. The surrounding chorion is a simple sac that envelops the entire affair. The amnion provides a fluid medium surrounding the delicate embryo. **(d)** As development continues, the yolk sac covers the diminishing yolk; the allantois has grown along the route of the chorion, the two fusing in one area to form the chorioallantois.

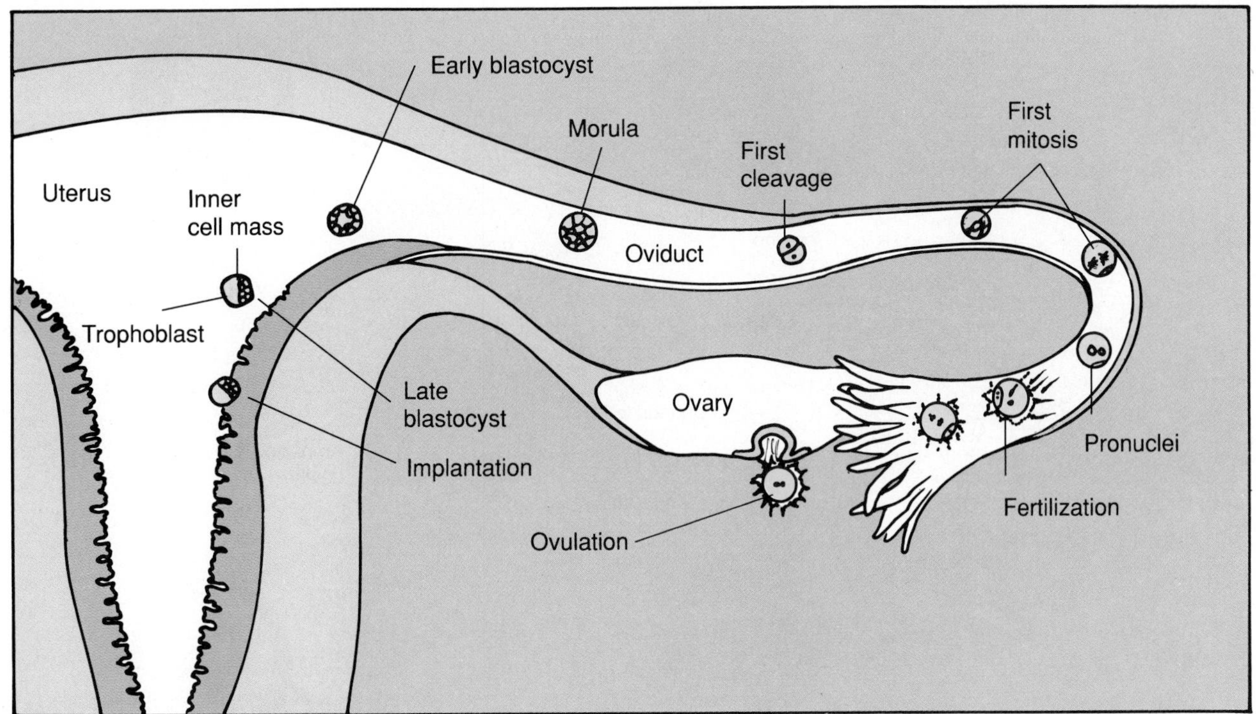

FIGURE 43.14
THE HUMAN EMBRYO: DAYS 1–6
After ovulation, the human oocyte is drawn into the oviduct, where fertilization can occur. From the time of fertilization, about six days are required for the transit to the receptive uterus. The first cleavage occurs about 36 hours after fertilization, but later divisions occur more frequently. When the embryo enters the uterus, it will be a morula. This solid ball of cells will hollow out into a blastocyst, as seen in the photo.

thus setting the stage for their fusion. The network of blood vessels permeating the fused chorioallantois exchanges gases with the air outside the porous shell, while the cavity formed by the allantois functions as a nitrogen waste receptacle.

With the evolution of the reptile and bird egg—the "land egg" as we have called it (see Chapter 32)—the embryo was safely enclosed in a container that retained a watery environment reminiscent of the ancestral condition. But this independent egg, with its supply of food, water, and its supporting extraembryonic membranes, was only one kind of provision for survival in the land environment. The mammals evolved a different system.

The Support System of Mammals

Like the bird and reptile, the mammalian embryo produces four extraembryonic membranes, some with quite distinctive functions. Since the various support systems arise similarly in all the placental mammals, we will use humans as representatives of the group. (In the next section we will trace the development of the human embryo itself.) To follow the formation of support systems in humans, we need to go back to the blastocyst stage, a time when the embryo is about to implant itself in the mother's uterine wall. Figure 43.14 reviews events up to implantation.

Recall that the blastocyst has three features: the fluid-filled cavity, the trophoblast (a thin sphere of cells and the inner cells) and the inner cell mass (a denser region of cells offset to one side). It is the trophoblast that is of interest to us here, since it first makes contact with the uterine wall and is instrumental in establishing the embryo there.

The human embryo reaches the uterus about six or seven days after fertilization, having spent this time slowly moving through the oviduct, where fertilization took place. When it enters the uterus, it contacts the soft endometrium, and implantation begins. The trophoblast secretes enzymes that digest some of the endometrial tissue, and the embryo sinks into the resulting cavity. The cavity becomes filled with nutrient-rich blood (Figure 43.15a), and for a brief time this arrangement will adequately provide for the necessary food and oxygen and the removal of the embryo's meager wastes. But as the embryo grows, its food and oxygen demands increase, and more metabolic wastes are

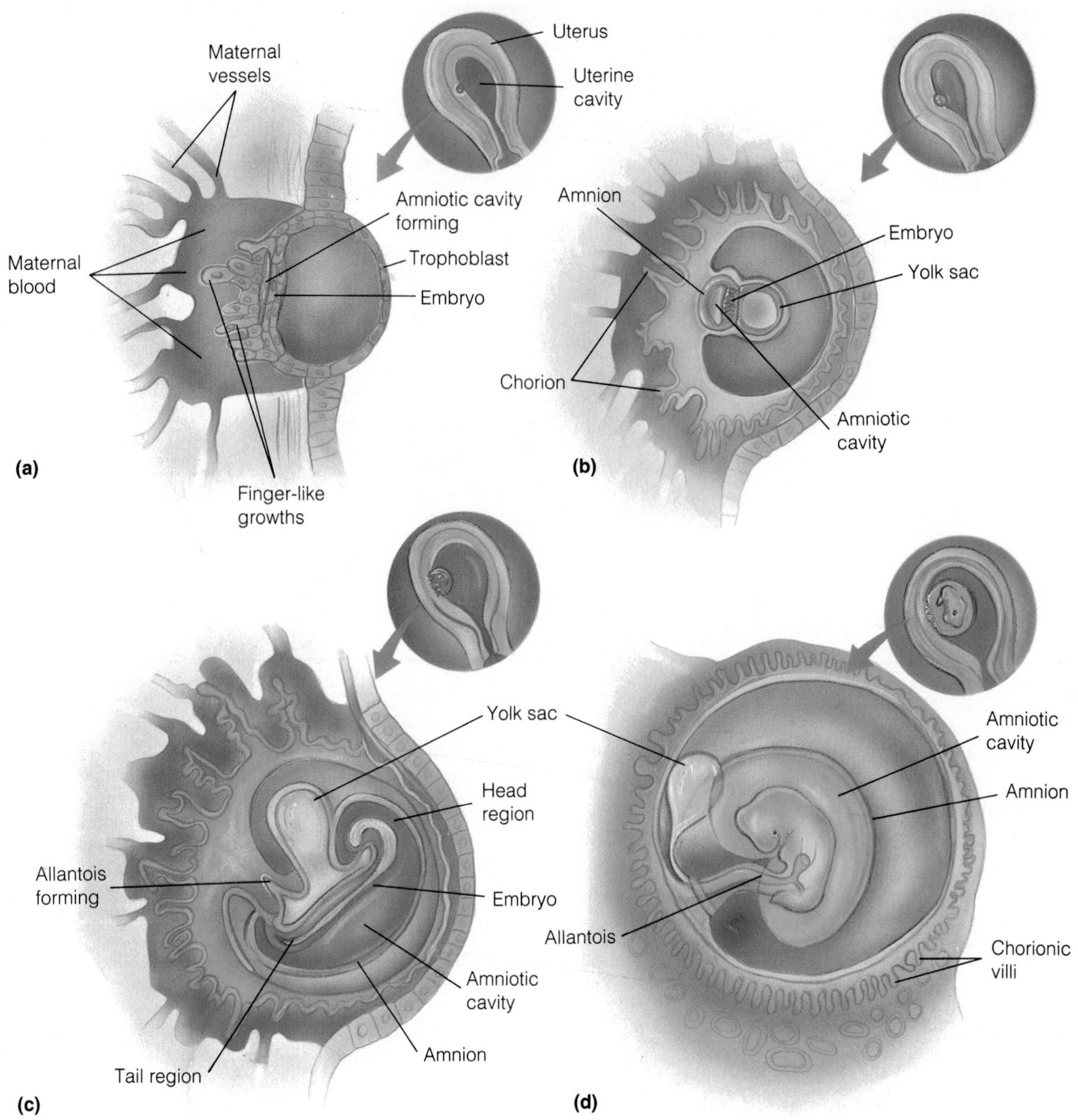

FIGURE 43.15
SUPPORT OF THE HUMAN EMBRYO
(a) Once the blastocyst has penetrated the endometrium, the trophoblast produces spreading cellular fingers that increase the absorptive surface. Spaces forming just above the embryo cell mass announce the emerging amniotic cavity. **(b)** The amnion and yolk arise early, forming cavities above and below the simple, platelike embryo. In addition, the trophoblast gives rise to the chorion, whose early growth produces primary villi that spread into the endometrium, marking the early growth of the placenta. **(c)** As the chorionic villi probe deeper into the endometrium, the allantois arises, first in the form of a slender cavity. **(d)** By five or six weeks, the embryo fakes recognizable human form, and the chorionic villi have increased dramatically. In time, the allantois will form the umbilicus, which houses the all-important umbilical blood vessels.

generated. Thus, the embryo begins to need faster and more efficient means of exchanging materials with the mother's body. This requirement will be met by the formation of the **placenta.**

The continued maintenance of the endometrium is now urgently required by the growing embryo. Should the present menstrual cycle continue (see Chapter 42), hormonal support will soon dwindle, and the endometrium will slough away, carrying the minute embryo with it. However, the trophoblast again comes to the rescue. It begins to secrete **HCG (human chorionic gonadotropin),** which is picked up by the mother's blood and carried to its target, the corpus luteum within the mother's ovary (the same structure that once housed the egg: see Chapter 42). The corpus luteum responds by continuing to secrete progesterone and estrogens, hormones that support the endometrium. Because HCG overrides the ovarian cycle, the embryo is secure. Some HCG is normally present in the mother's urine at this time and can be detected through pregnancy tests (including the much publicized "pregnancy test kits," those designed for home use).

Formation of the Extraembryonic Membranes in Humans Shortly after implantation (during its second week), the embryo forms the amnion, the first of its four extraembryonic membranes. A cavity forms between the trophoblast and the inner cell mass, separating the two and becoming the dome-shaped fluid-filled **amniotic cavity** (Figure 43.15b). Cells of the trophoblast gave rise to the thin amniotic membrane, or amnion, which soon forms a cover over the enlarging amniotic cavity. The floor of the cavity is the embryo itself. At this time the embryo has reached a bilayered stage, quite similar to the blastoderm of the bird (see Figure 43.9). As in the reptile and bird embryos, the fluid of the amniotic cavity will later act as a shock absorber, protecting the embryo from dangerous jolts and acting as a lubricant, keeping the appendages from fusing together as the embryo develops. (At the end of development, the bursting of the amnion [the "bag of waters"] signals the onset of labor.)

Next, the yolk sac emerges, at first outlining a simple cavity below the embryo (see Figure 43.15b). Although hollow and without yolk, the yolk sac serves some functions. It is the site of the first blood cell formation and houses certain lymphoid cells that later migrate into red bone marrow, where they give rise to stem cells (see Chapter 39). In addition, the yolk sac produces cells that migrate to the gonadal area, where they take up residence, later forming the germinal epithelium (which gives rise to sperm and egg cells).

In response to the increasing exchange needs of the embryo, which is now proceeding into gastrulation, cells of the trophoblast lay down the chorion. The chorion will surround the embryo, part of it intimately fusing with the mother's endometrium and forming the placenta. In preparation for this, the chorion produces fingerlike cellular growths called **primary chorionic villi.** These growths break through older cells and deeply penetrate the uterine lining. When joined by blood vessels from the embryo, they become the highly branched chorionic villi (Figure 43.15c-d) and the emerging placenta takes form. Typically, in humans and other primates, the villi come to be surrounded by an extensive blood sinus, which forms as the mother's endometrial blood vessels break down. In this way, the vessels of the chorionic villi are immersed directly in the mother's blood, thereby further facilitating exchanges between mother and embryo (Figure 43.16). Finally, the placenta takes on a disklike form and reaches a diameter of about 20 cm and a thickness of 6 cm (8 in. by 2 1/2 in.).

By the third week of life, neurulation and the elongation of the embryo have begun. At about this time, the allantois (see Figure 43.15 c-d), the fourth extraembryonic membrane, makes its appearance, at first in the form of a simple outgrowth from the rudimentary gut of the embryo. It is reminiscent of its counterpart in the reptile and bird. But here the allantois does not store nitrogen wastes—they are constantly carried away in the mother's blood. In humans and other mammals, the allantois forms the **umbilical cord,** the structure containing the blood vessels that link the circulatory systems of the mother and embryo (see Figure 43.15e).

ORGANOGENESIS IN THE HUMAN

Now that we have traced the embryonic development of various groups of animals and the development and role of their supporting structures, we will continue our focus on

FIGURE 43.16
THE PLACENTA
Eventually, the chorionic villi will have been well-penetrated by blood vessels emerging from the umbilical cord. The villi lie within blood sinuses where the exchanges between mother and fetus go on.

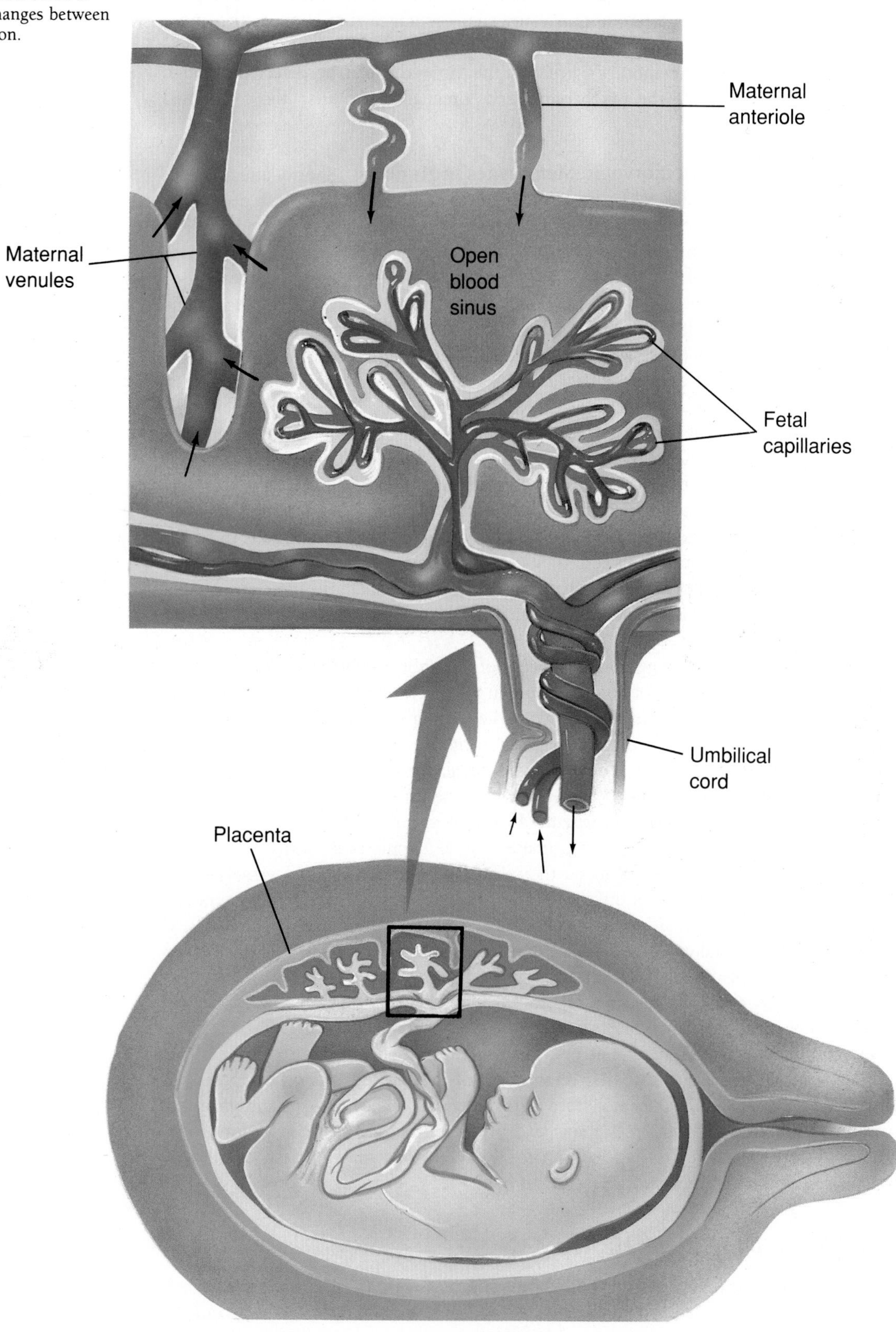

humans. Specifically, we will trace the development of the embryo from organogenesis to birth.

Human development is traditionally divided into three-month sequences called **trimesters,** and so we will review the events of each trimester in turn. Although we will discuss the various systems of the embryo as if they were sequential, we must keep in mind that many of the events are occurring simultaneously.

The First Trimester

The first trimester is marked by a number of sweeping developmental events. Upon the completion of neurulation, organogenesis has begun, and a more familiar form takes shape. It is also the period in which the embryo becomes a fetus (at eight weeks—when it is about 25 mm—only 1 inch— in length).

Nervous System In typical vertebrate fashion, the nervous system develops early. Neurulation begins at about day 18 or 19, closely following gastrulation. As with the amphibian embryo discussed earlier, in humans the neural folds arise on the embryonic disk, reaching upward and then folding together to form the dorsal hollow nerve cord. Its anterior end enlarges to form the vesicles of the brain.

The nervous system continues its rapid progression in the first trimester. While it is the earliest system to begin development, it will not be completed until long after the birth process—perhaps not until the movement of death—since learning is in a sense a development process. By the fourth week, the major regions of the brain and spinal cord are recognizable. When the first trimester ends, these are already well-defined. The still-smooth cerebrum now extends over much of the embryonic brain, and the cerebellum and medulla have become distinct.

Circulatory System As the neural ridges begin to break the contour of the human embryo, the heart and circulatory system makes an early appearance. By day 22, you can make out the first timorous palpitations of the primitive heart. In vertebrates, this great organ is formed as cylinders of mesenchyme converge, producing a single tubelike structure. Within four to five days, the tube will have developed into a fully functional organ; it is crude, but it moves blood. As the blood is pushed along it enters sinuses, forming channels that later become lined with endodermal cells, producing blood vessels. Within another two weeks (or about 40 days from fertilization), the tubular heart will have looped back on itself, paving the way for its four-chambered pattern, which, by now, is nearly completed. An opening between the atrial chambers will remain until some time after birth.

Respiratory and Digestive Systems The respiratory and digestive systems develop fairly rapidly, so that by five to six weeks their basic patterns are clearly established. Once the basic tube has been outlined, a few blind pouches from, then more and more outpocketings, until finally the indistinct outlines of the gut, liver, and pancreas can be seen. The trachea, bronchi, and lungs begin as a small outpocketing in the pharynx. The outpocketing then branches to form the two lung buds, which will give rise to the first individual lung lobes.

Limbs The limbs of humans appear as rounded buds during the fourth week. The arms and legs are distinguishable at six weeks, but fingers and toes require an additional week. The rudimentary hands and feet actually begin as simple webbed paddles that take form through a kind of developmental programmed cell death, as the tissue between the fingers and toes is broken down and absorbed.

Reproductive System The reproductive system begins to develop during the first few weeks, but until the eighth week even a trained observer can't determine the sex of the embryo (without a chromosome test). It is true that before this time the genitals have begun to develop, but the genitals of the two sexes start off in much the same way. The factors controlling sexual differentiation are complex. Certainly, the genetic sex of an individual is determined at fertilization by the chromosomal complement of the sperm cell. Fertilizing sperm cells that carry X chromosomes produce females, while those bearing Y chromosomes produce males. However, this is only a beginning. A number of genes are involved in sexual development, and the control may stem from different levels (such as genetic, hormonal, or environmental). For example, it appears that the Y chromosome of mammals initiates the development of the testes and that the testes take over from there, producing hormones that determine the male's primary sexual characteristics.

In the male embryo, the epididymis and vas deferens (structures of sperm storage and transport, respectively) begin their formation from a pair of embryonic structures, the **Wolffian ducts** (Figure 43.17). But this is only part of a peculiar story. During the indifferent state, all embryos develop a pair of **Mullerian ducts,** which lie alongside the Wolffian ducts. As morphogenesis proceeds, the Mullerian ducts in males degenerate, while the Wolffian ducts form the epididymis and vas deferens as mentioned. In females, however, it is the Wolffian ducts that degenerate, while the Mullerian ducts become the oviducts. The determining factor appears to be testosterone, the male hormone. In its presence, the male structures develop, but in its absence the structures to develop are female. (The role of sex hormones in sexual differentiation is further discussed in Chapter 37 and 42.)

The Second and Third Trimesters

The first trimester is marked by a great deal of activity and sweeping change—so much, in fact, that by its close, most major systems have approached their final form. In contrast, the second trimester is characterized by growth and refinement. Body length increases

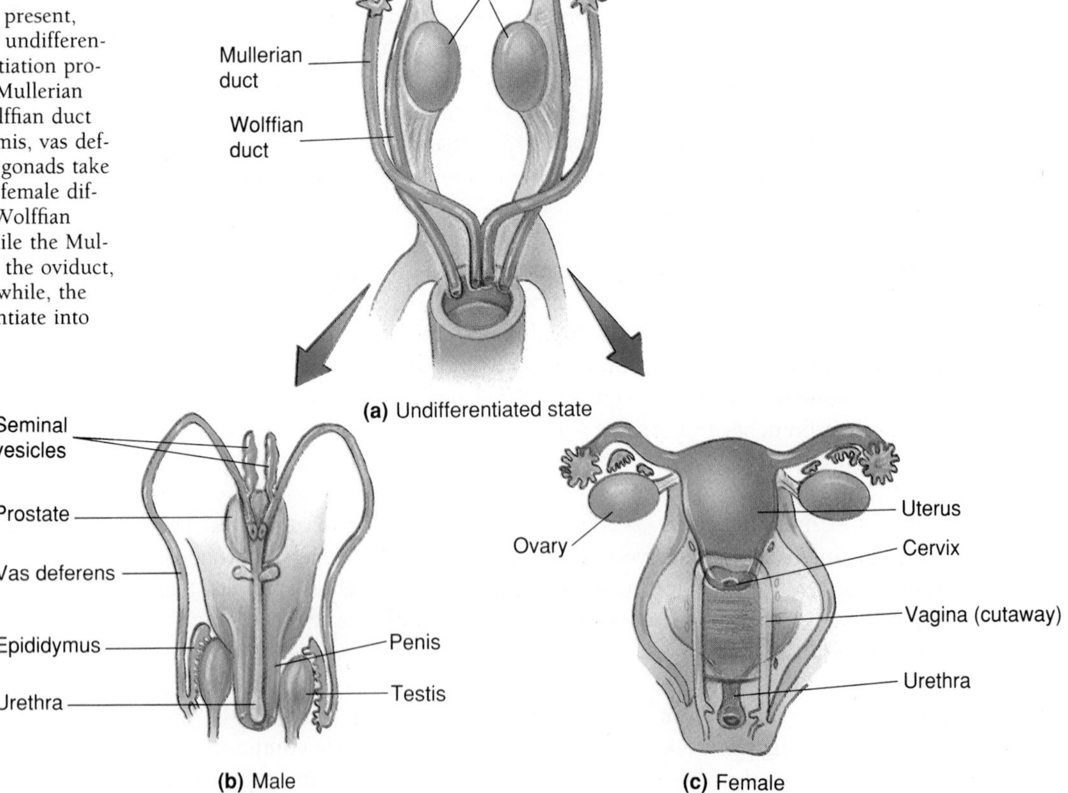

FIGURE 43.17
SEXUAL DIFFERENTIATION: INTERNAL
In the sexually indifferent state (**a**), the embryo will have both Mullerian and Wolffian ducts present, along with the paired but undifferentiated gonads. As differentiation progresses in males (**b**), the Mullerian duct degenerates, the Wolffian duct giving rise to the epididymis, vas deferens, and urethra, while gonads take the form of the testes. In female differentiation (**c**), it is the Wolffian duct that degenerates, while the Mullerian duct contributes to the oviduct, uterus, and vagina. Meanwhile, the indifferent gonads differentiate into the ovaries.

rapidly, catching up to the large head. Systems begin to approach a functional state. The fetus "breathes" but only amniotic fluid enters and leaves the tiny lungs. The digestive system becomes lined by secretory cells that will secrete enzymes. Bile from the liver joins these secretions. The fetus swallows, bringing in fluids and cellular debris that accumulate in the gut as dark, jellylike **meconium.** By 16 weeks, the dermis and epidermis of the skin reach a differentiated state and are penetrated by sweat glands, sebaceous glands, hair follicles, and sensory neurons. If stimulated, the 12-week-old fetus will respond with feeble, uncoordinated movements at first, but these movements soon grow more precise and brisk, and by 16 weeks the mother feels the first fetal movements—usually as a pronounced "kick."

As the third trimester begins, the organ systems begin to be fully functional. A fetus may survive if born now, but its chances are very poor even under the most sophisticated medical supervision. At this time it will measure about 35 cm and weigh around 1000 g (14 in and 2.2 lbs), resembling a very thin and emaciated full-term baby. Growth and refinement continue but, in most cases, at a reduced rate. The fetus, no longer floating free, gradually fills the amniotic sac and uterine cavity. The brain and spinal cord continue to develop rapidly, and the cortex differentiates and begins to form the familiar fissures and convolutions. A layer of fat accumulates beneath the skin. In the skeletal framework, **ossification,** the hardening of bone, which began in the first trimester, increases rapidly. These activities increase the demand for protein and calcium, placing a substantial burden on the mother's reserves. As a final prelude to birth, antibodies cross the placenta during the last month. These will provide the newborn baby with immunity against viral and bacterial infections for a month or two. As the third trimester ends, the fetus, now crowding the mother's abdomen, will measure close to 50 cm and weigh about 3200 g (20 in and 7 lbs).

Birth

In spite of our great interest in the process and the fact that many of us have experienced it, we still don't really understand what brings on the birth process. Currently, considerable research is focused on the changes in the placenta that occur at the end of gestation and the hormonal shifts that bring them about. Since the level of progesterone is important in maintaining the placental tissue, there may be a direct relationship between the aging of this structure and lowered progesterone levels. In addition, at term the uterus contracts more strongly in response to oxytocin.

Recent theories suggest that as pregnancy continues, the number of oxytocin receptors in the uterus increases, possibly brought about by increased estrogen levels and uterine stretching. Thus, while oxytocin levels do not rise appreciably, the uterus becomes increasingly sensitive to the hormone's presence. In addition, prostaglandin secretion by the placenta increases during labor, and prostaglandins are known to enhance the action of oxytocin.

The Stages of Birth In accordance with their penchant for pigeonholing, scientists have divided the birth process into three stages (Figure 43.18). The first is dilation, which involves a softening of the cervix. This is sometimes accompanied by a rupturing of the amnion, often referred to as "breaking the water," although this usually happens in the second stage. The period of dilation is unpredictable and may vary from two to 16 hours. It is accompanied by periodic contractions called labor pains. These contractions increase in frequency through the first stage. When they begin to occur at least every three to four minutes, the fetus will crown, meaning that the head will begin its passage through the cervix and become visible. This is the start of stage two, the fetal expulsion. Expulsion may take just a few minutes, or it may require hours. The mother may have to be given a spinal anesthetic at this time. The level of pain is highly variable, depending on the mother's preparation, her physical condition, her pain threshold, and her emotional state.

The final stage of birth is the placental separation and its expulsion from the uterus

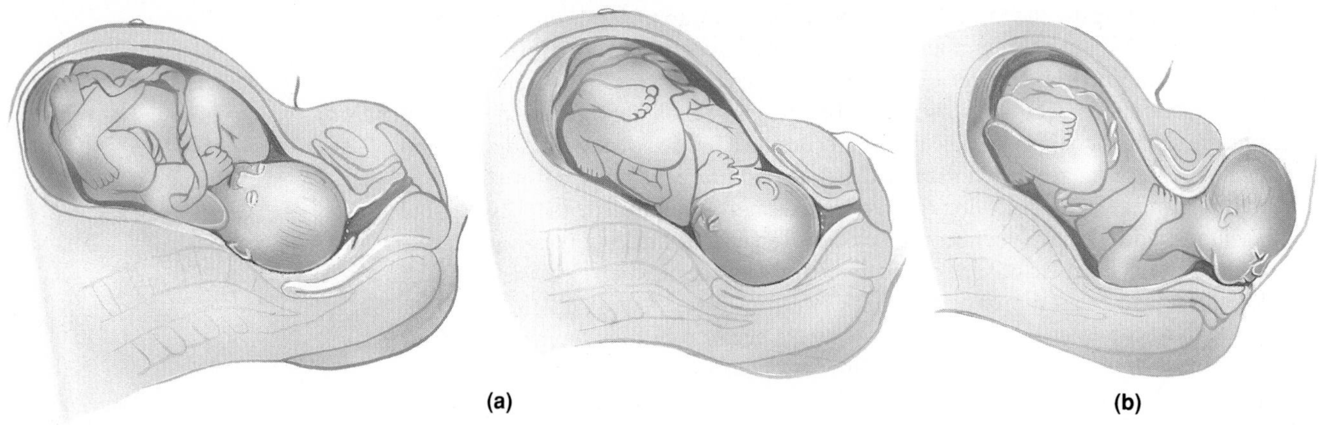

FIGURE 43.18
HUMAN BIRTH
In the first stage of labor, (**a**) dilation of the cervix occurs, and the baby's head eventually crowns. (**b**) The second stage of labor involves the actual expulsion of the fetus. Expulsion of the placenta—the third stage—follows the birth of the baby.

as afterbirth. Following this, the uterus rapidly contracts to its former state, which helps control the bleeding at the site of placental attachment.

Physiological Changes in the Newborn Baby The body systems of the fetus are in a state of readiness prior to birth, for it has been prepared for a new and more threatening kind of existence. For example, the vital exchange of gases has always been provided by the placenta. Now the infant's own respiratory system must function for the first time. There are also changes in the circulatory system (Figure 43.19). Oxygenated blood from the placental circulation had previously entered the fetus through the umbilical vein, proceeding directly into the vena cava and thence to the right side of the heart. A hole in the septum between the fetal atria, the **foramen ovale**, permitted the blood to flow from the right atrium to the left atrium, thus circumventing the route to the still uninflated and functionless embryonic lungs. In addition, a connecting vessel between the pulmonary artery and aorta, the **ductus arteriosus**, permitted blood from the right ventricle to bypass the lungs and go directly into the arterial distribution.

With the baby's first breaths, the ductus arteriosus constricts vigorously, closing that bypass. With the shortcut closed, the blood must move toward the rapidly expanding lungs. As a result of this increased pulmonary flow, the volume of blood returning directly to the left atrium suddenly increases. The foramen ovale is composed of two overlapping flaps of tissue, which are forced closed when the atria contract. Actually, however, the foramen ovale will not seal itself completely until the baby is about a year old.

The first breath of a newborn infant is taken in response to a sudden increase in blood carbon dioxide when the placental exchange is lost. This increase stimulates the respiratory center of the medulla with the usual results, so the baby will breathe with or without the traditional swat on its rear.

Lactation Human mothers, like all mammals, are prepared to nurse their newborn young immediately. The breasts, under the influence of estrogens, and progesterone, have enlarged during pregnancy, and with delivery, **colostrum** secretion begins. Colostrum, a clear yellowish fluid, differs from milk in that it contains more protein, vitamins, and minerals and less sugar and fat. It also contains maternal antibodies that can be important to the infant in its first days.

In a few days, the colostrum is replaced by whiter, thicker milk. Human milk is sufficient in all required vitamins except vitamin D and contains about 700 calories per liter. Interestingly, it is bacteria-free (which is not true of cow's milk, either in grocery stores or at the source). There are additional benefits of breast feeding. Some studies reveal that breast-fed infants are less susceptible to disease, anemia, and vitamin deficiencies.

ESSAY 43.1

THE EMERGENCE OF HUMAN LIFE

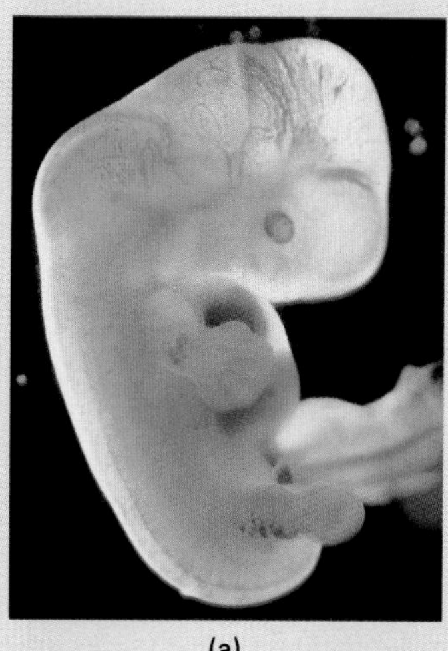
(a)

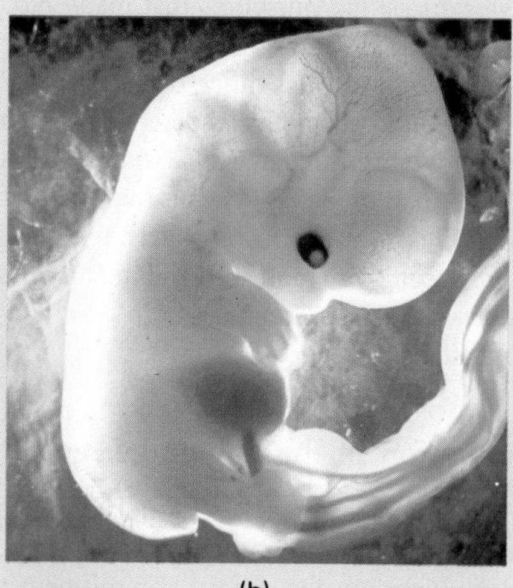
(b)

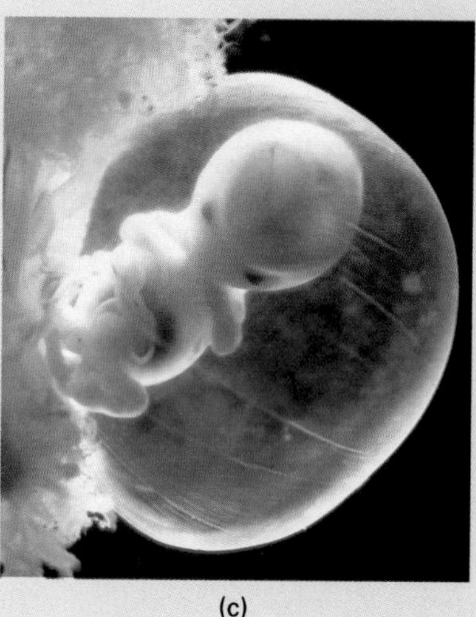
(c)

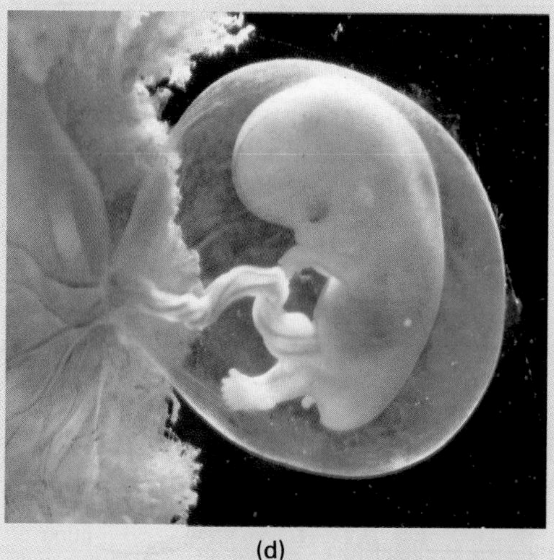
(d)

(a) By the fifth week the human embryo is just ½ inch long. The hands have taken on a paddle shape, and darkened lines mark future centers of bone formation. **(b)** In another few days the limbs have become better defined and blood vessels in the umbilical cord have progressed, as have surface vessels over the head and body. **(c)** At seven weeks, the embryo is 1 inch in length, weighing in at 2 grams. The fleecy outlines of a well-formed placenta are clearly seen. **(d)** The limbs have lengthened in the 9-week fetus and its fingers and toes are clearly defined. The eyes and ears are also taking form. **(e)** At the first trimester's end, all systems in the 2½-inch-long fetus have neared a functional state. **(f)** Although it looks as if it could survive on its own, the 16-week fetus must still undergo a lot of growth and refinement.

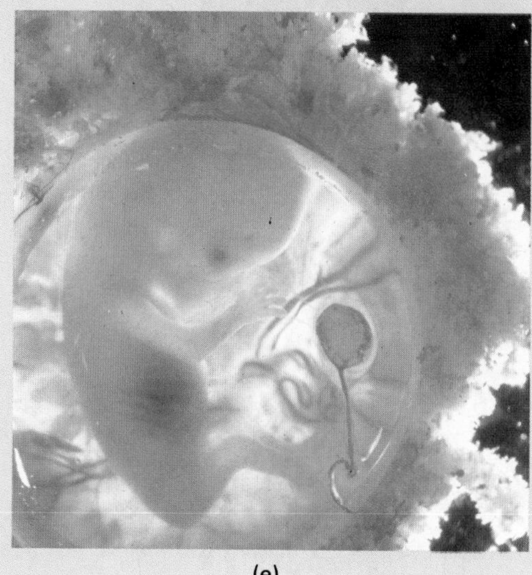
(e)

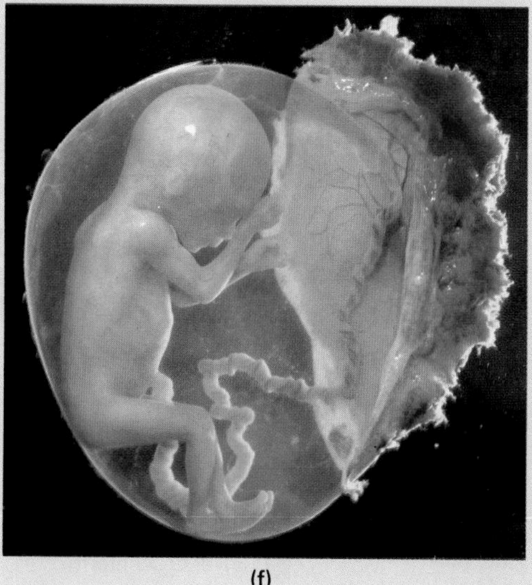
(f)

THE ANALYSIS OF DEVELOPMENT

It is indeed fascinating to watch the changes in a developing embryo. Tissues appear, grow, move, reorganize, and disappear in a genetically choreographed ballet that easily arouses wonder in the most crusty and jaded of biologists. That wonder increases, though when one moves from *when* and *what* happens in development to *why* and *how*. At this point, we start asking questions that many biologists view as the most challenging and exciting in developmental biology.

For example, how does differentiation occur? Why do cells actually change? How are the developmental fates of cells determined? How do genes operate to produce the final phenotype? We are often short on answers to such questions, either because we know too little, or, paradoxically, because we know too much. That is, we may have a great deal of specific information but haven't yet learned what to do with much of it.

We will begin our look into the analytical side of development with a brief return to the polarized egg and zygote, both of which have undergone a considerable amount of organization.

Determination in the Egg and Zygote

Recall that eggs are often highly polarized: that is, their cytoplasmic elements have become organized into a gradient, and that gradient is highly significant to future events. Thus, determination, the genetic commitment to developmental pathways, begins early indeed. In the amphibian egg, the lighter pigmented, metabolically active animal hemisphere is clearly differentiated from the darker vegetal hemisphere. Upon fertilization, the cytoplasm undergoes sweeping changes, becoming even more organized. In one such change, some of the pigment on the side opposite sperm penetration is drawn deeper into animal hemisphere cells, resulting in the formation of a lighter colored, crescent-shaped area called the **gray crescent** (Figure 43.20a). We have noted that the first cleavage cuts across the animal and vegetal hemispheres and in so doing determines

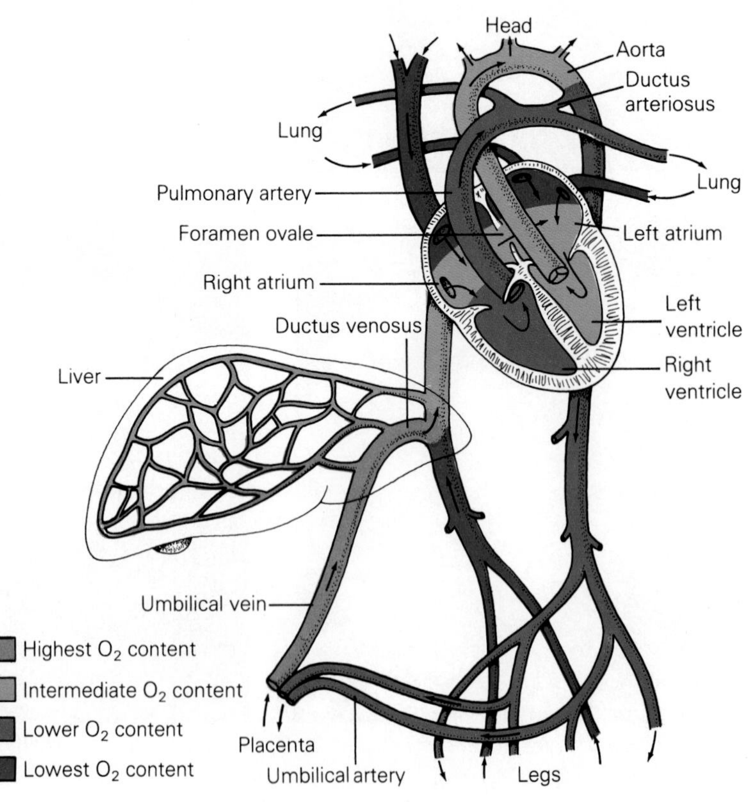

FIGURE 43.19
HUMAN FETAL CIRCULATION
This is the circulatory system of the fetus shortly before birth. Note the foramen ovale, a septum between the right and left atria in the heart, the ductus arteriosus, connecting the pulmonary artery with the aorta and of course, the umbilical circulation.

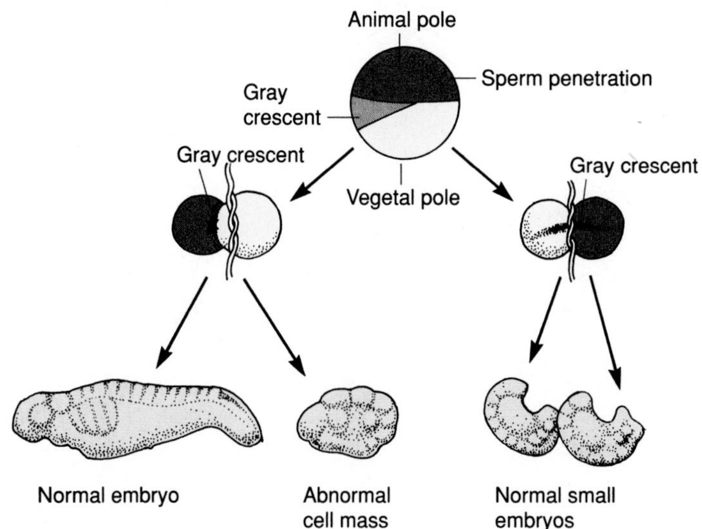

FIGURE 43.20
THE POLARIZED FROG EGG AND EMBRYO
Spemann ligated newt zygotes in two planes. Where the entire gray crescent went to one blastomere, it developed into a normal embryo, but the other blastomere produced only an amorphous mass. Ligating the zygote through the gray crescent resulted in two complete embryos.

the right and left halves of the embryo. That first cleavage also bisects the gray crescent, a momentous developmental event, as we shall see. Keep in mind that the gray crescent marks the location of the future blastopore (which forms at gastrulation) and thus plays a role in determining the future dorsal region of the embryo.

The vital organizational significance of the gray crescent was established about the turn of this century by Nobel laureate Hans Spemann, who tied tiny filaments around newt zygotes, establishing cleavage planes of his choice and partially dividing the zygote. Notice in Figure 43.20b how resulting embryos differ markedly according to where the ties are made. Spemann found that at least part of the gray crescent must be included in the two resulting cells for both to go through development. We should note that development occurs faster in one cell than in the other. At first, only the cell that receives the zygote nucleus undergoes cleavage, but shortly afterwards the second receives a daughter nucleus from an adjacent cell and proceeds through its own development. In spite of many years of study, biologists are still uncertain how the gray crescent exercises its influence.

In the 1930s, embryologist Sven Horstadius found that the urchin egg was likewise in a highly organized state and that, in fact, the future organization of the embryo was well-established even before fertilization. As you see in Figure 43.21, dividing the unfertilized egg cytoplasm in a plane that separated the animal and vegetal hemispheres resulted, upon fertilization, in two abnormal embryos. (We should note that in the artificial cleaving of the egg, one daughter cell lacks a nucleus. It will, upon sperm penetration, develop as a haploid sea urchin.) Thus, we see that determination in the sea urchin begins even before the intervention of the sperm and that much development is set even before it begins.

Totipotency and Sealed Fates Ultimately, the developmental program is genetically based, that is, the developmental program that unfolds during embryogenesis originates in the genome of the animal. Thus, it follows that the fertilized egg contains all of the hereditary material needed to carry out a total developmental program. We say then, that the fertilized egg is **totipotent** ("all powerful"): that is, it can give rise to any kind of cell. But as development continues, more and more cells commit to a specific fate—to becoming certain kinds of specialized cells. Some genes are deactivated, other activated, until only those that are required by the specific kind of cell are active.

But how permanent is this commitment? Can it be reversed? In other words, are the nuclear changes associated with development reversible? The question has been explored in great depth, beginning in the 1950s.

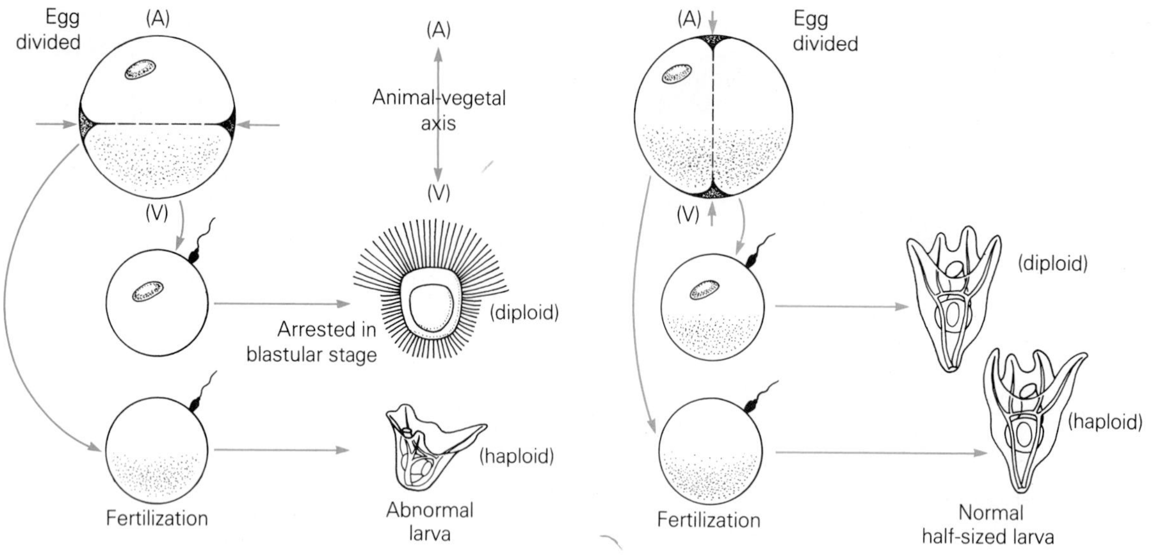

FIGURE 43.21
POLARITY IN THE SEA URCHIN EGG AND EMBRYO
In sea urchins, polarity already exists in the unfertilized egg. When an unfertilized egg is divided in a plane perpendicular to what would be the animal-vegetal axis, and then fertilized, one of the zygotes fails. If, however, the division is made parallel to the animal-vegetal axis, both products successfully develop following fertilization.

Among the first of the great experiments of this era were those by R.W. Briggs and T.J. King, in which cells were enucleated (their nuclei removed) and then replaced by nuclei taken from other cells in various stages of development (Figure 43.22). First, an oocyte (egg cell) from a leopard frog (*Rana pipiens*) was pricked with a glass needle. This sort of stimulation will often induce some kinds of unfertilized eggs to undergo cleavage, in some cases producing a haploid individual. Once the oocyte began showing signs of mitotic activity, the membrane was punctured and the chromosomes removed. The result was an activated and enucleated egg.

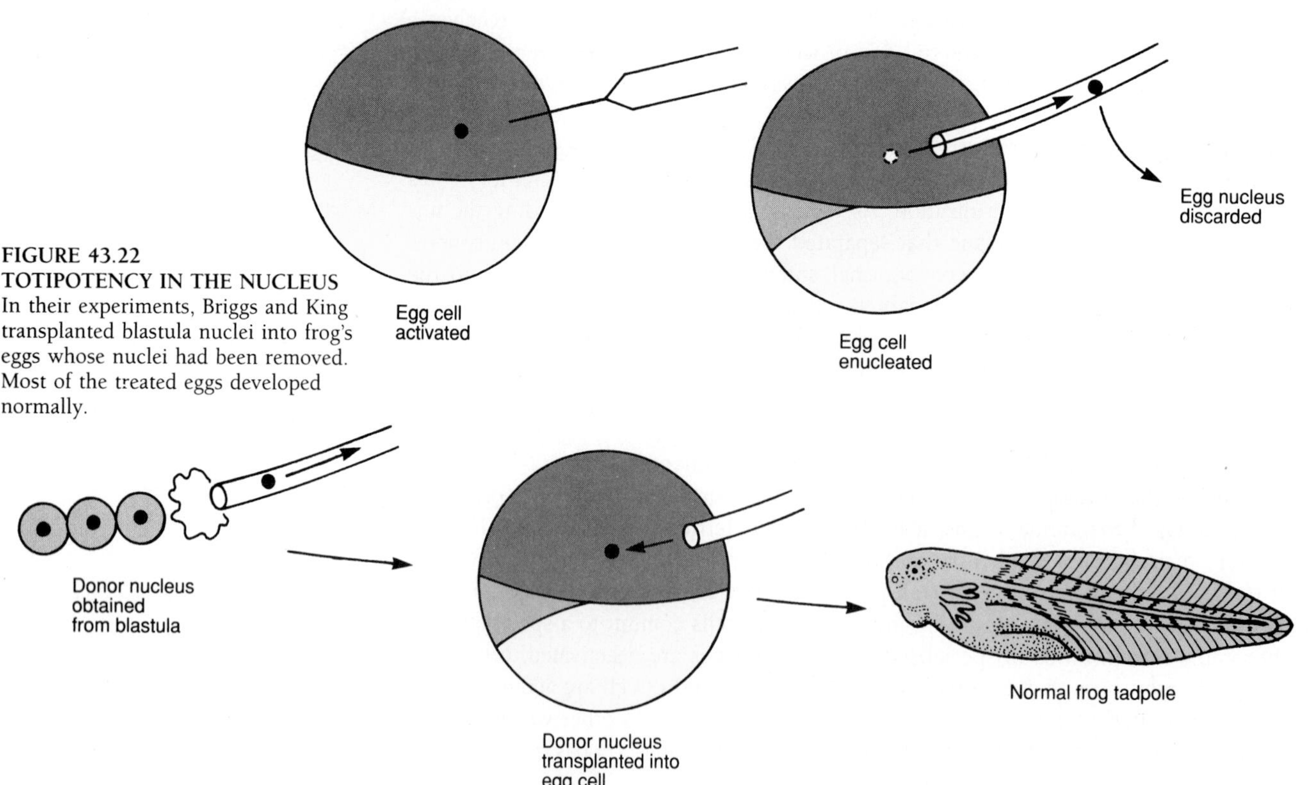

FIGURE 43.22
TOTIPOTENCY IN THE NUCLEUS
In their experiments, Briggs and King transplanted blastula nuclei into frog's eggs whose nuclei had been removed. Most of the treated eggs developed normally.

Then, using a very fine glass pipette, Briggs and King carefully removed nuclei from other frog cells and inserted them into the enucleated eggs. They found that when the nuclei were obtained from cells of the frog blastula stage, most of the recipient eggs went on to develop into normal frogs. Obviously, blastual cell nuclei were still totipotent. Nuclei taken from older cells showed less potency, until nuclei taken from cells at the embryo's tailbud stage would not prompt development in the egg at all. However, nuclei taken from germ cells (those that would have formed gametes) would direct normal development in 40% of the oocytes. Briggs and King then initiated experiments that told us two things. First, cells tend to become increasingly differentiated (losing their ability to form different kinds of cells) as time goes by. Second, certain kinds of cells lose their totipotency before other kinds do.

Regeneration: Setting Back the Development Clock Totipotency can be studied under more natural conditions, such as during **regeneration.** Regeneration is the regrowth of lost structures, such as limbs. Regeneration is common in invertebrates; we've described it in sponges, hydras, sea stars, flatworms, and others. But it is far more limited in vertebrates, the best-known cases involving salamanders.

Following the loss or excision of a limb in the salamander, tissues that were previously committed—having produced skin, bone, connective tissue, and other limb tissues—enter a period of rapid cell division. As these cells divide at the site of the injury, they begin to produce masses of simple cells that form an undifferentiated tissue known as a **blastema.** After a time the rapid cell proliferation ceases, the blastema stops growing, and its cells begin to change. They undergo a rearrangement and reorganization, and they begin to differentiate along different developmental lines. Some cells form bone, others muscle, skin, and nerves, all organized along the same pattern followed in the developing embryo. Finally, the limb is regenerated.

For reasons that are not entirely clear, regeneration in the salamander requires that some nerve tissue remain intact. If large nerves in the area are experimentally isolated and lifted away from the stump, regeneration will not occur, but if they are subsequently restored to their original position, regeneration resumes. Some researchers, experimenting with electrical currents, suggest that the current flow around active neurons is essential to regeneration.

Regeneration is highly limited in most vertebrates and may be altogether absent in mammals. The liver and kidney do regrow tissue following the surgical removal of diseased portions, and this is often cited as an example of regeneration. Actually, however, the liver simply (but dramatically) recovers its lost mass. This occurs through rapid cell division and enlargement, without the events seen in true regeneration.

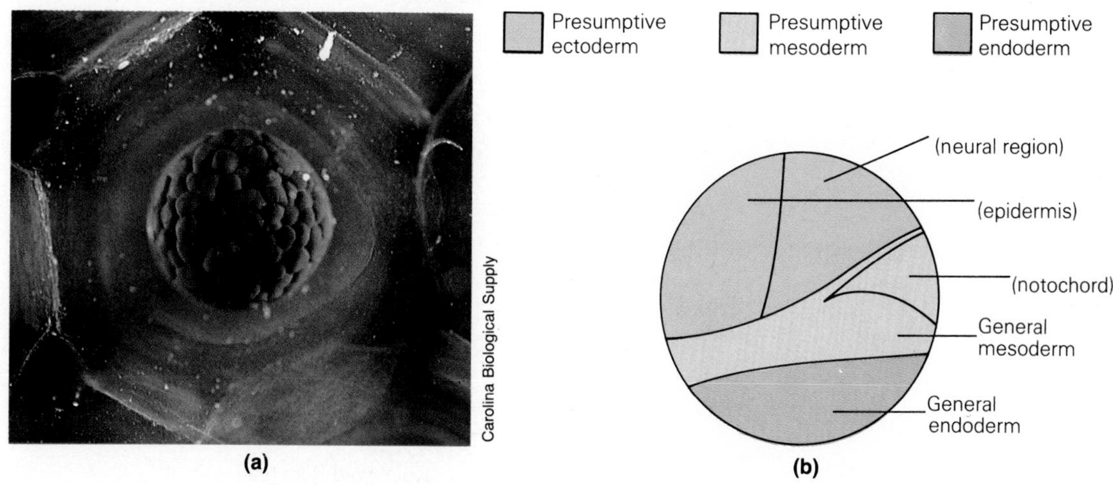

FIGURE 43.23
FATE MAP OF THE AMPHIBIAN BLASTULA
Fate maps indicate the future role of undifferentiated cells. In preparing the amphibian fate map, blastula cells bearing dyes were tracked as gastrulation ensued. Note the special region assigned to the notochord, a structure that appears during neurulation. **(a)** Amphibian blastula. **(b)** Fate map.

The Role of Tissue Interaction in Determination

Fate Mapping Continuing now with determination in the frog embryo, we can note that although the germ layers—ectoderm, endoderm, and mesoderm—first appear in the gastrula, many cells in the blastula are already committed to the formation of the three germ layers and other tissues as well. Such early determination has been well documented by cleverly conceived experiments in which harmless dyes are applied to the blastula's surface and the different-colored regions tracked through subsequent developmental events. From these efforts, embryologists have been able to construct **fate maps** such as the one seen in Figure 43.23. Included in the fate map are regions designated as presumptive ectoderm, presumptive mesoderm, and presumptive endoderm (note also the presumptive notochord). The term "presumptive" indicates the future role of certain cells. With this in mind let's see what happens when the presumptive tissues are experimentally rearranged.

The Primary Organizer In 1921 Spemann and Mangold (mentioned earlier) devised some of the most remarkable experiments in the history of biology. Selecting late frog blastulas as subjects, they transplanted small pieces of presumptive mesoderm (before it rolled under to assume its position in the gastrula) to recipient embryos of similar age. The presumptive mesoderm was placed below the ectoderm of other embryos in a region that would normally form belly skin in the older embryo. The results varied depending on the experiment, but they found, for example, that a recipient embryo would develop a new body axis (a notochord, neural tube, somites, etc.) in that location (Figure 43.24). In some cases, the embryo went on to form a second brain and spinal cord, and in others a second, fully formed embryo emerged (Figure 43.25). Spemann and Mangold concluded that something in the transplanted material was organizing the tissue of the recipient, that is, directing it along different developmental lines. Spemann

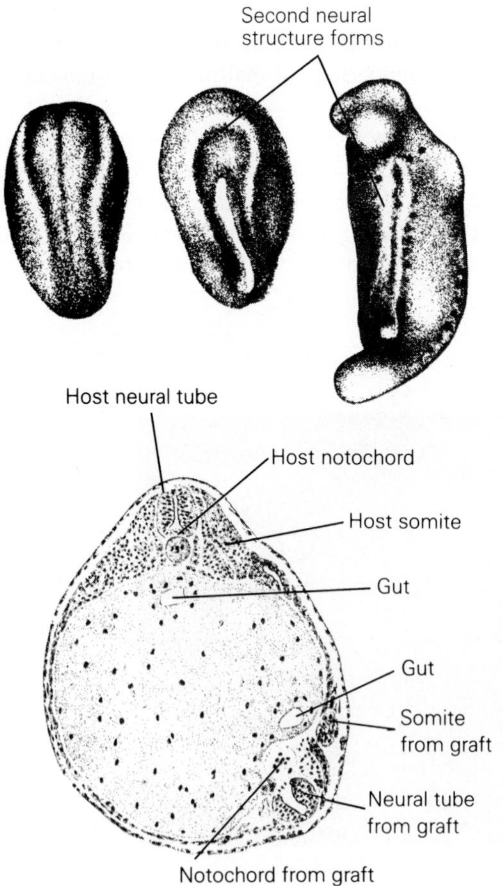

FIGURE 43.24
THE PRIMARY ORGANIZER
Spemann and Mangold transplanted presumptive mesoderm from the dorsal lip of the blastopore to a region of the embryo whose ectoderm would not normally produce a neural tube. The results, seen in whole embryos and in cross sections, reveal a supernumerary neural tube, notochord, and somites.

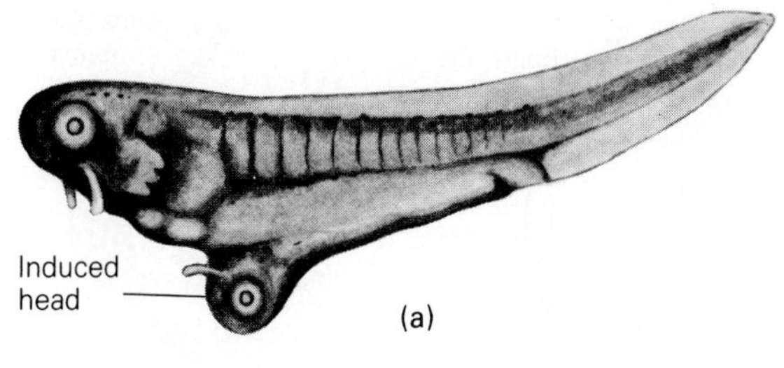

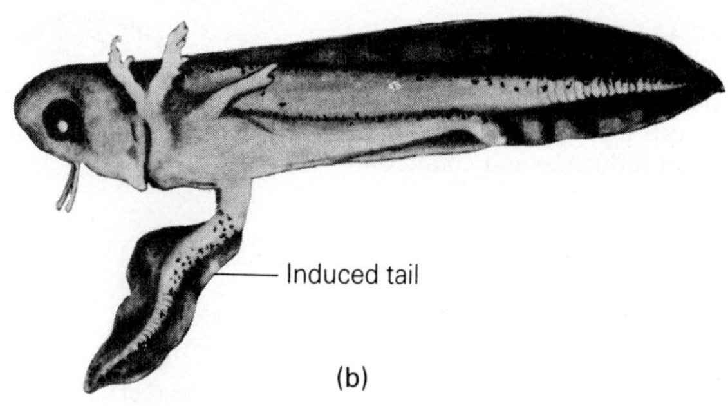

FIGURE 43.25
INDUCTION IN THE OLDER GASTRULA
Spemann and Mangold found that the results of dorsal lip transplants varied with the age of the gastrula. When a source of the graft was an early gastrula, it produced the head region of the embryo **(a)**, while grafts from the dorsal lip of substantially older gastrulas produced only tail regions **(b)**.

labeled the dorsal lip region, from which the tissue was derived, a **primary organizer**, and the interaction between the organizer and the tissue it affected, **embryonic induction.** Induction is now referred to as **tissue interaction**, the ability of one tissue to redirect the developmental fate of another. Subsequent research has shown that such tissue interaction is a normal part of embryonic development.

Scales, Feathers, and Induction Studies carried out in the 1950s and 1960s further illustrated the organizing influence of mesoderm on ectoderm. One such study involved the formation of scales and feathers. In vertebrates, the epidermis contributes to hair, scales, feathers, teeth, and parts of limb buds. Experiments involving chick embryos reveal that the formation of scales and feathers is clearly under the influence of underlying mesenchyme (migrated mesoderm). (Their sensitivity to the same organizer should not be unexpected, since feathers have been shown to be an elaboration on the simpler reptile scale—more evidence of the close evolutionary relationship of reptiles and birds.)

The formation of scales and feathers in the chick embryo begins with an epidermal thickening, an **epidermal placode.** The placode is pushed upward, away from the skin, as mesenchyme below forms a **dermal papilla** that penetrates and elevates the epidermal placode. The placode, under the organizing influence of the dermal papilla, forms a **feather bud** or **scale bud** and begins its differentiation into the feather or scale. It was determined through transplant experimentation that the underlying mesenchyme directed the fate of the overlying epidermis. Researchers switched chick epidermal and mesodermal tissue between the leg and body in various ways. Foot mesenchyme placed beneath body epidermis induced the formation of scales on the body, whereas body mesenchyme implanted below foot epidermis induced feather growth in the foot. However, if the transplanted tissue was epidermis alone, scales and feathers appeared in their usual locations. Clearly, then, the epidermis was passive, with no organizing ability.

Triple Induction in the Vertebrate Eye Among the most complex of tissue interactions is the reciprocal influences that go on as the vertebrate eye takes form. Here, a first tissue organizes a second, which then reciprocates and organizes the first. Furthermore, both of the organizing structures are ectodermal in origin, showing that not *all* induction involves mesoderm organizing ectoderm.

What happens is, the developing brain (an ectodermal structure) forms two bulblike extensions that grow outward and contact the overlying covering of ectoderm (Figure 43.26). The interaction causes the overlying ectoderm to thicken and form a **lens placode**, which will later pinch off to form a lens. The overlying ectoderm becomes the cornea of the eye. (If the brain tissue is experimentally placed under the outer ectoderm anywhere on the body, it will organize the formation of a lens there.) As the lens placode develops, it becomes the organizer, inducing the bulblike extension of the brain to form a two-layered **optic cup** that will, in turn, form the retina of the eye. The optic cup then exerts its influence on the lens placode, inducing it to invaginate, beginning the formation of the lens, as mentioned. Such interactions can be described in coldly analytical ways, but they again show the fascinating nature of life. Those tissue interactions, after all, involve genes, some shutting down, others turning on, in a wonderful interplay

FIGURE 43.26
LENS INDUCTION
Lens induction is a veritable dialog between cell layers, involving three episodes of induction and counterinduction.

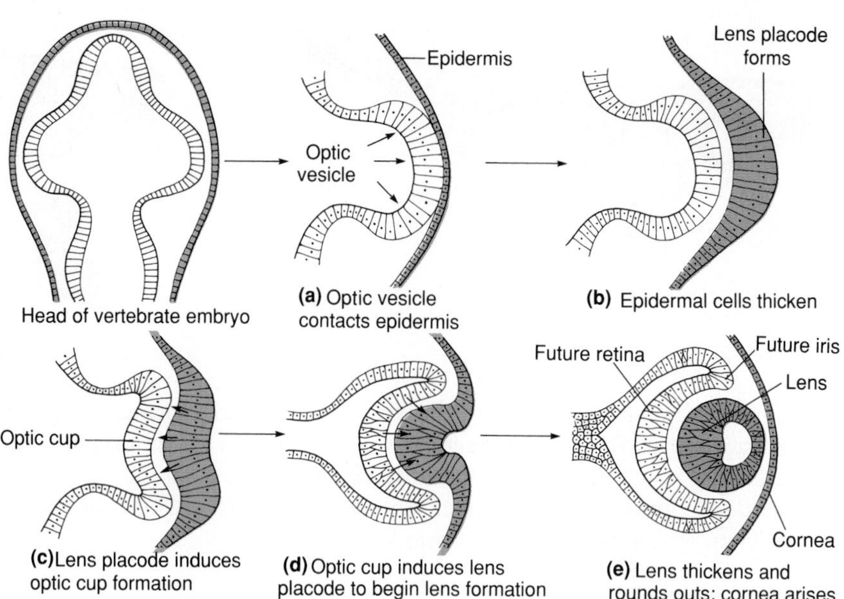

that will cause some cells to focus light and others to flash impulses to the brain as they respond to that light.

Search for the Organizer The obvious question from the tissue interaction research is, what precisely is an organizer? Obviously, it is something that moves from the organizing tissue to the target tissue, but what, exactly, is it? Surprisingly enough, in spite of a search that has lasted nearly 60 years, we still don't know. The organizer is presumed to be molecular in nature, but no such molecule (or molecules) has been found. (Nonetheless, these chemicals have a name waiting for them—**morphogens.**) Many substances have been suggested, from mRNAs to simple inorganic ions, all having had their movements tracked from one cell layer to the next as development progressed. So far, though, none have shown the ability to organize tissue along new developmental pathways.

The Determining Role of Migrating Cells

We've seen how the transplanting of organizing tissues from one region to another alters the usual fate of subservient tissues. Actually, we find a clear analogy to this in the natural migration of cells that commonly occurs in the embryo. Earlier, in our discussion of neurulation, we noted the importance of neural crest cells in the continuing events of development. The neural crest cells, you may recall, form as isolated clusters between the neural tube and the overlying ectoderm (see Figure 43.11). Neural crest cells migrate far and wide in the young embryo, taking on many roles. Some move into the head region, forming the cartilages that are the forerunners of skull bones; others contribute to the adrenal medulla; and still others add to the eye's developing cornea. We've just seen how mesenchyme prompts the epidermis to form scales and feathers. In another of many examples, mesenchyme originating in the blocklike somites of the young chick embryo migrates into the developing limb buds, giving rise to the muscles and skeleton of the limb.

Perhaps the most restless cells of all are the young neurons, whose processes extend into every part of the body. Motor neurons, for example, arise from specific sources in the central nervous system and send their axons out towards target organs. Those that make their connections survive, becoming functional.

Studies of the regeneration of severed salamander limbs (discussed earlier) suggest that neurons get a little help from the target. Cells in the regenerating limb stump secrete a substance called **nerve-growth factor**, which stimulates newly arrived immigrant neurons to sprout branches. In the absence of this stimulus, such as in neurons that reach the wrong target, the cell simply dies. The central nervous system, "covering all bets," sprouts about twice as many motor neurons as are needed, so in this case chaos is averted by cell death.

Some cells are routinely killed because they have migrated to the wrong place. Migrating germinal cells from the chick's yolk sac, those destined for the genital ridges where they help organize the gonads, sometimes go astray. Some of the errant cells end up in inappropriate places—such as the head. The death of such cells helps the animal avoid what could be an interesting problem.

The developmental future of migrating cells often depends upon cues from the new environment in which they find themselves, cues that arise either from resident cells or from the extracellular environment. In one instance, the influential agent is hyaluronic acid, a common constituent of connective tissue. In the formation of the cornea (the tough, transparent capsule over the pupil of the eye), neural crest cells migrating into the extracellular matrix differentiate into connective tissue. However, for the invading process to succeed, cells surrounding the matrix must have first secreted hyaluronic acid. Otherwise, the migrating neural crest cells come to rest short of their goal.

In spite of all we know about animal development, then, we still lack a fully concise, sequential explanation of how the developmental processes are orchestrated. But so far, we have centered our attention on vertebrate development. Perhaps better opportunities lie in another direction, in the mosaic development of insects, where, as we've seen, things are far more regimented. There we can begin to look at what is a well-defined "program of development."

The Fruit Fly Embryo: Determination in Mosaic Development

Developmental commitment in the cells of vertebrate embryos, with their indeterminate cleavage, occurs gradually, the slowest probably occurring in mammals. In fact, until the mammalian embryo reaches the trophoblast stage, its cells remain virtually uncommitted. That is, before this time, any cell can take just about any developmental route. Furthermore, if one cell is damaged or lost, another can fully assume its role. In general, then, if early vertebrate embryos are experimentally altered, a normal individual can still emerge.

At the other end of the commitment spectrum are insects. In this group, the first cells to appear are on their way along a specific development pathway. As we mentioned earlier, cleavage in protostomes is determinate; that is, the earliest cells are already strongly committed. Further, insect embryos undergo mosaic development (each part being critical to the whole), and any tampering with the early embryo, such as is done with vertebrate embryos, will produce an abnormal or dead insect.

As we discussed earlier, insects also pass through development in a manner quite unlike vertebrates. For example, in the formation of the insect blastoderm, there are many mitotic events, but cytokinesis is delayed for a considerable time. The first actual cells to form are the pole cells, and these are already committed to form germinal cells (see Figure 43.7). They will reside in the gonad, a structure that will not appear for several days. Such an early commitment of cells in the insect embryo makes the organism a very useful subject in the ongoing search for clues as to how the developmental program works.

One of the clearest models of developmental determinism is seen in the *Drosophila* larva, where clusters of cells have their adult fate fully determined early in larval development. These clusters, the **imaginal disks** (from *imago,* which means "adult"), enter an inactive state, remaining there until pupation. At that time, each imaginal disk gives rise to a specific adult structure. Other cells in the larva make little or no contribution to adult structure. The larva, then, is a "temporary organism," one dedicated to the nourishment and support of dormant cells set aside for the task of producing the adult. Their activation will occur when a shift in developmental hormones ushers in the events of pupation (see Chapter 37).

FIGURE 43.27
IMAGINAL DISKS
Determination occurs early in *Drosophila* development as the imaginal disks form. Each disk consists of cells that, although inactive in the larva, become active during pupation, each contributing to specific structures in the adult fly.

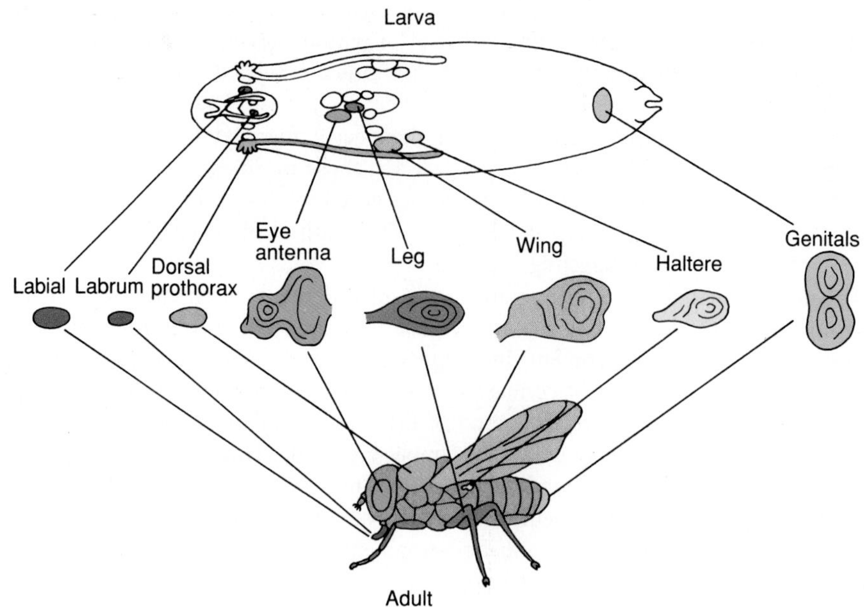

Determination and Imaginal Disks The cells making up the imaginal disks appear to be very simple; there is no visible evidence that they are in a fully determined state. As we see in Figure 43.27, there are several paired imaginal disks and one that is single. The pairs will give rise to antennae, legs, wings, genitals, and other organs. Each disk is in turn subdivided into compartments, and each compartment is made up of cells that contribute to a specific part of the structure to which the disk is committed. In the wing disk, for example, one compartment will contribute cells to the anterior, or leading, region of the wing, and the other to the posterior, or trailing region.

Biologists have taken an intensive interest in the imaginal disks, and their research is finally providing specific clues about how genes really influence development, at least in insects.

Homeotic Mutations A particularly fruitful approach to the question of the relationship of genes and development has been the study of mutations in the imaginal disk cells. These are called **homeotic mutations**, and the genes affected are **homeotic genes**, which, as we saw earlier, are control genes (see Essay 14.1). As you may recall, homeotic genes regulate whole blocks of genes, each of which is responsible for some subpart of the structure assigned to cells of a given imaginal disk.

Because they disturb the regulation of such gene complexes, mutations in homeotic genes profoundly affect the usual developmental sequences. One of the most dramatic examples involves mutations in a large gene complex called *antennapedia* (Figure 43.28a). When active in cells near the eye, this complex normally dictates the formation of an antenna, but when the homeotic gene controlling the complex has mutated, the disk cells may give rise to a completely normal, but seriously misplaced leg. Likewise, homeotic mutations in the so-called *bithorax* gene complex, a block of genes responsible for organizing the adult thorax (including the wings and halteres, or balancing organs), may bring about the production of extra thoracic regions, some bearing a second set of wings (Figure 43.28b).

Normally, cells in a given disk have the potential to produce other structures, but as long as the homeotic genes are functioning normally, they are *inhibited* from doing so. At the moment it is not known just how the homeotic genes inhibit other genes, but molecular biologists suspect that they code for proteins that inhibit gene action by binding certain control regions of DNA. (Not unlike the *lac* operon, as discussed in Chapter 14.) Such a negative control mechanism would itself be affected by cues from

(a) Antennapedia

(b) Bithorax

FIGURE 43.28
HOMEOTIC MUTANTS
Mutations in two gene complexes, (a) antennapedia and (b) bithorax, result in striking miscues in the location of adult structures. Such large complexes of genes, normally held in check by control segments called homeoboxes, express themselves in the wrong places.

the chromosome's surroundings. For example, it is known that when nuclei migrate to the outer region of the fruit fly embryo (see Figure 43.7), cues from the new surroundings bring on responses in certain control genes, which in turn initiate segmentation in the larva. The powerful analytical tools of molecular biology are now being brought to bear on the problem, and what biologists are finding is already casting light on some difficult problems in animal development.

The Homeobox: Molecular Biology of Development

As was mentioned in Chapter 14, molecular biologists have recently discovered that each of the homeotic genes contains a common or similar DNA segment of some 180 nucleotides, a region they called a **homeobox**. The amino acid sequence directed by the homeobox is very similar for the different homeotic gene loci. In *Drosophila,* the homeoboxes are believed to be the key to homeotic gene function, which means that they indirectly determine what the other genes in each cell of the imaginal disk are to do. This greatly simplifies our perception of development, since it follows that perhaps a single cue from the surroundings may have far-reaching effects. We are confident that in the fruit fly, such cues do affect the homeobox, thus activating particular imaginal disks. Such may be the case for control genes in other organisms.

Even greater implications arise from our new knowledge of the homeobox. Molecular biologists, using RNA probes containing the homeobox sequence, have found strikingly similar stretches of nucleotides in other insects. This discovery prompted the search for the sequence in other kinds of animals. As a result, the homeobox was found in the genome of annelid worms (whose ancestors, you may recall, gave rise to the anthropods). Then, to the great surprise of some, the homeobox was found in vertebrates, and a similar sequence has now been discovered in humans. Some theorists are now suggesting that the homeobox occurs in all segmented animals.

How similar are the homeoboxes in different animal groups? The homeobox of the antennapedia gene complex of *Drosophila* is identical in 59 of its 60 amino acids with the homeobox of the frog *Xenopus*. Since the evolutionary lines of fruit flies and frogs must have diverged well over half a billion years ago, it looks as though natural selection has conserved the homeobox almost intact, and to biologists, this suggests that its function is precise and vital.

We've seen that cells in certain embryos pass through increasingly determined states, eventually to be committed to a specific fate in the embryo. We've also seen where determination has a strong genetic basis, involving the selective activation and inactivation of genes, and how specific DNA sequences called homeoboxes may play key roles in activating gene complexes that contribute to the formation of tissues and organs. In addition, we have found that homeoboxes probably react to external cues from their environment and that the cellular environment in the embryo is a constantly shifting, changing entity (Figure 43.29).

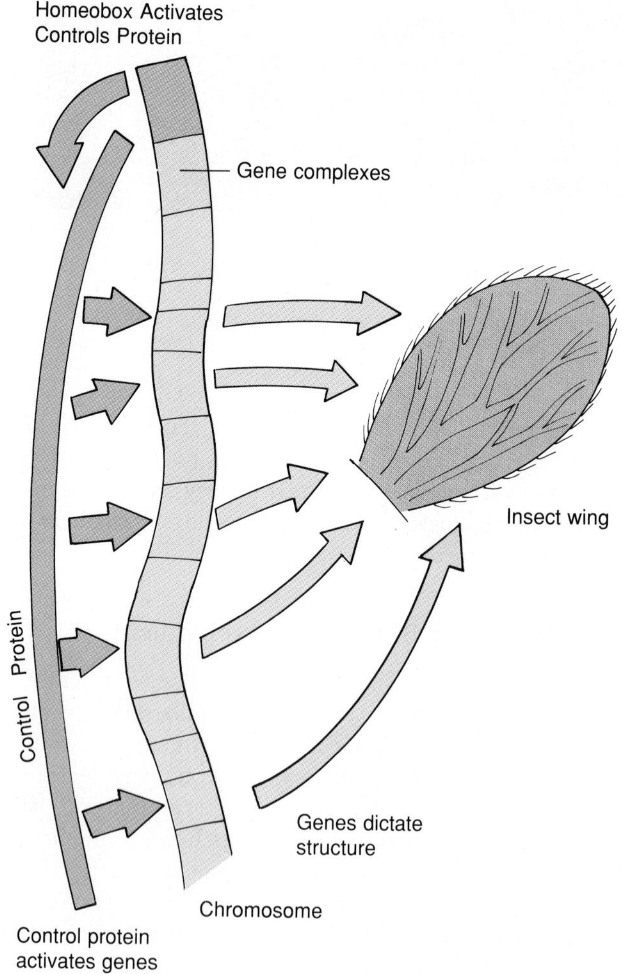

FIGURE 43.29
THE HOMEOBOX
The homeobox is a super-controller, a region of DNA that keeps large gene complexes in check until cues from the surroundings act to lift the inhibition.

The Vertebrate Wing: A Developmental Program

It's intriguing to ponder over the transition of the human hand from its crude paddlelike stage to the complex and versatile structure it becomes. On viewing a newborn baby, we invariably touch a tiny hand, test its tentative grasp, and marvel over the delicate, perfectly formed fingers. How does such a seeming developmental miracle take place? What assures the proper and timely assembly and arrangement of its parts?

For obvious reasons, we have little experimental data on morphogenesis in the human hand, but we are learning a great deal about limb development in other vertebrates, particularly about the formation of the wing. A favorite subject is the chick embryo with its very manageable three-week developmental period, its resilience, and, perhaps most importantly, its easy accessibility. Chick embryos will develop nicely out of the eggshell, in artificial environments where they are available to the researcher. Ongoing studies of chick development are shedding light on some of the basic mechanisms of morphogenesis. As we will see, wing development involves many of the known developmental mechanisms.

The chick wing begins as a simple mound of tissue, the *wing bud,* appearing after some 3½ days of development. The wing bud consists only of a mass of mesenchyme cells (secondary mesoderm) surrounded by simple ectoderm that terminates in a thickening called the **apical ectodermal ridge** (Figure 43.30). But in just three more days, the skeletal elements begin to appear, and the wing takes on a paddlelike shape. After 9½ days, each wing bone is nearly complete and clearly visible. The sequence in which the wing elements appear is highly ordered. Those elements closest to the body (the shoulder joint and upper arm) form first, and those furthest away (the digits) form last.

Many of the wing's connective tissues, including cartilage, bone, muscles, tendons, and dermis, are derived from mesenchyme tissue in the wing bud. The ectoderm, with its prominent apical ridge, is vital to the performance of the mesenchyme. As long as the ectodermal ridge is intact, formation of the wing elements proceeds on schedule. Should the ridge be surgically removed, development will be affected, the degree depending upon the age of the bud. From this fact, it would seem that the apical ectodermal ridge might be in charge, acting in effect as a primary organizer by inducing the mesenchyme to differentiate. This idea has been tested.

As it turns out, the ectoderm isn't in charge at all. Replacing the apical ridge of an early bud with a late apical ridge transplant, or replacing the apical ridge of a late bud with an early ridge has no effect at all. Development proceeds as usual. So, while the presence of an apical ridge is essential, it does not appear to control what happens in differentiating mesenchyme. Determining what controls the mesenchyme required still more experimentation.

A small cluster of mesenchyme cells, the **polarizing region,** located along the posterior margin of the bud just within the ectoderm, appears to be in charge (Figure 43.31). Again, transplant experimentation provides the clues. When a portion of the polarizing region and its overlying apical ridge are removed from a donor and transferred to a location on the *opposite side* of a recipient's limb bud, a wing with two sets of well-formed digits grows (Figure 43.31). But curiously, the digits formed by the transplant grow in a reversed order. Thus, the extra wing tip becomes the mirror image of the original.

Researchers are now trying to understand just how the polarizing region affects the mesenchyme, and, since it must be present, how the apical ridge fits into the picture. We'll concentrate on the polarizing region and come back to the apical ridge later. Observations strongly suggest that the wing bud mesenchyme cells differentiate according to their specific distance from the polarizing region. This suggests that the polarizing region may release an active substance, a *morphogen,* whose subsequent diffusion forms a concentration gradient across the limb bud. The presence of such a substance remains hypothetical, for like other such morphogens, it has not been identified. But to further understand how the distance of mesenchyme from the polarizing region could affect the digit patterns, let's look at the pattern itself.

The relationship between digit formation and distance from the polarizing region

FIGURE 43.30
WING DEVELOPMENTAL PATTERN
Wing development begins in the simple wing bud, which contains undifferentiated mesenchyme and a covering of ectoderm (the ectodermal ridge). Development proceeds from the joint region outward, with the digits, numbered 2, 3, and 4, forming last.

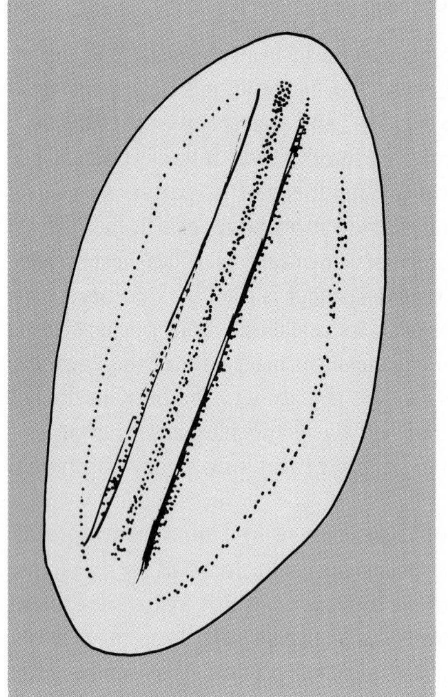

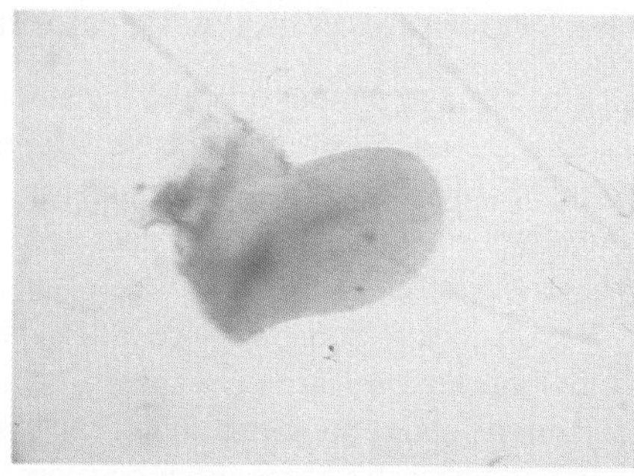

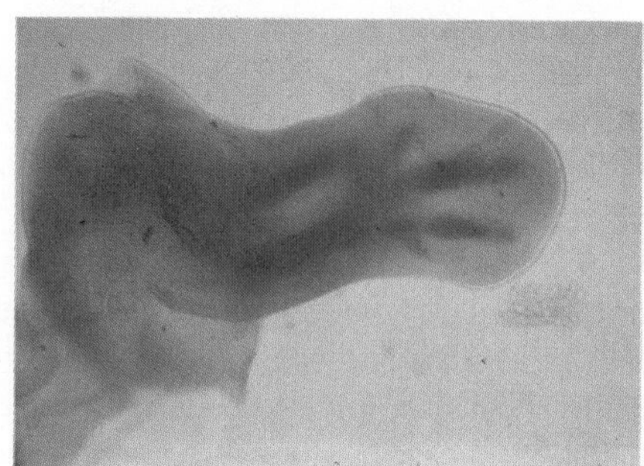

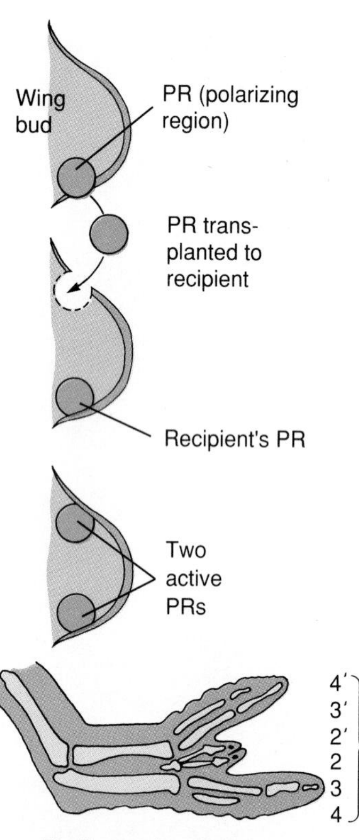

FIGURE 43.31
POLARIZING REGION TRANSPLANT I
When the polarizing region of a donor is implanted opposite the host's polarizing region, a mirror image pattern of digits is established.

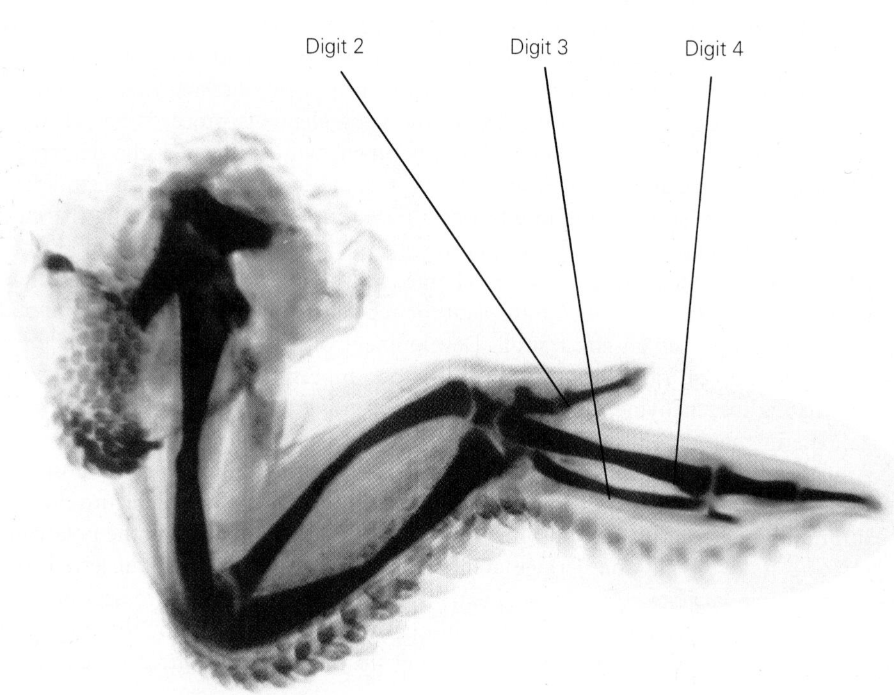

turns out to be very precise. The three digits of birds are homologous to our own second, third, and fourth fingers. (Birds don't have thumbs and pinkies.) Mesenchyme closest to the polarizing region form the fourth digit, those somewhat further away form the third digit, and mesenchyme cells furthest from the polarizing region form the second digit. This is not a guess; the distance factor has been determined through skillful transplant experimentation.

The pattern of digit formation in transplant experiments depends upon where the graft is placed. When polarizing region mesenchyme from a donor is transplanted as far as possible from the host's own polarizing region, both the host and supernumerary

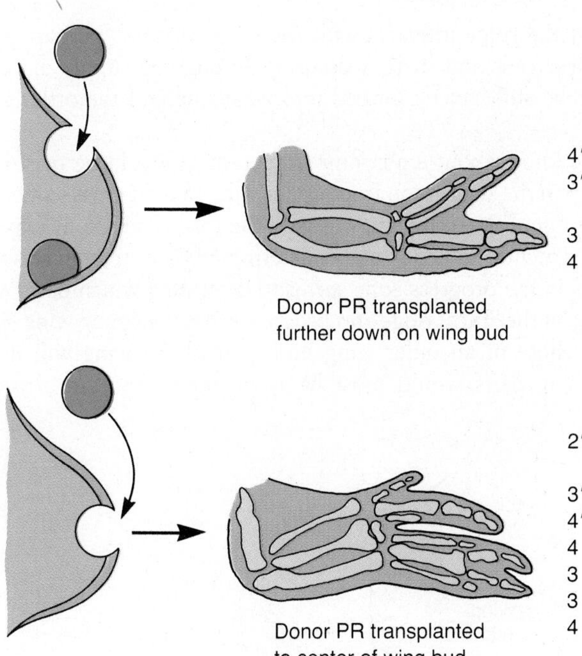

FIGURE 43.32
POLARIZING REGION TRANSPLANT II
The specific effects of polarizing region grafts depend on geography. The location of the transplanted polarizing region determines which digits will appear.

wing tips have the normal number of digits, but as we've said, they occur in reverse order (4–3–2:2′–3′–4′) (see Figure 43.30). But when the transplant is placed nearer the host polarizing region, the resulting decrease in distance (morphogen concentration) inhibits the formation of both second digits, so only the fourth and third digits emerge (4–3:3′–4′) (Figure 43.32). When the transplant is still closer to the host's polarizing region, the extra second digit develops, but the host second digit fails. Further, and stranger still, is the rearrangement in digit order in both wing tips: 4–3–3–4:4′–4′–3′–2′ (see Figure 43.32). This odd result can be attributed to the closeness of the two polarizing regions.

You'll be interested to know that the limb buds of other vertebrates also contain the all-important polarizing region. And, interestingly, should the polarizing zone region of, say, a turtle or mouse or even a human be transplanted into a chick limb bud, the bud will respond as usual. Whatever stimulus the polarizing region provides is apparently similar in some, perhaps all, vertebrates. (Does this bring the homeobox to mind?) Thus, limb development has strong evolutionary overtones. What effect might mutations have on the control genes involved in limb development? Most likely, mutations would produce errors in timing and structure, as seen in the homeotic mutants of *Drosophila,* but then in the long run, the accumulation of some less harmful mutations might just provide the variation upon which natural selection could work. Recall that the limbs of living vertebrates make up a fascinating homologous series (see Figure 18.16) whose evolution awaits explanation. With further study on patterns of limb development and the genetic mechanisms responsible, it seems likely that we may come to understand the evolution of the vertebrate limb.

Earlier, we saw that the formation of the major wing elements—shoulder, arm, and digits—proceeds from the innermost elements to the outermost, the digits forming last. Our final question is, how is this succession determined and controlled? For the answer, we turn to the region of undifferentiated mesenchyme cells just within the apical ridge. These cells make up what is called the **progress zone.** It is the progress zone that produces an ongoing supply of cells for differentiation into limb parts. We mentioned earlier that even though the apical ectodermal ridge has no apparent role in influencing the differentiation of mesenchyme, it is somehow essential to those events. Our best guess now is that the ridge interacts with the progress zone, perhaps suppressing differentiation in these cells and thereby assuring an ongoing supply of versatile, uncommitted cells. But we still haven't looked into what's behind the orderly progression in limb formation.

From what we know about embryonic induction, it would be reasonable to assume that each new step in the succession from upper wing to wing tip is somehow determined by the preceding one, but, strangely, this isn't the case. Transplant experiments clearly indicate that the structure to be formed is determined by a "time in service" factor. That is, the age of cells in the progress zone seems to determine which parts will differentiate next. For example, if the apical ridge and progress zone of a young wing bud are removed and replaced by those of an older wing bud, part of the wing will not form (Figure 43.33). The missing parts would most likely be those that had already begun their

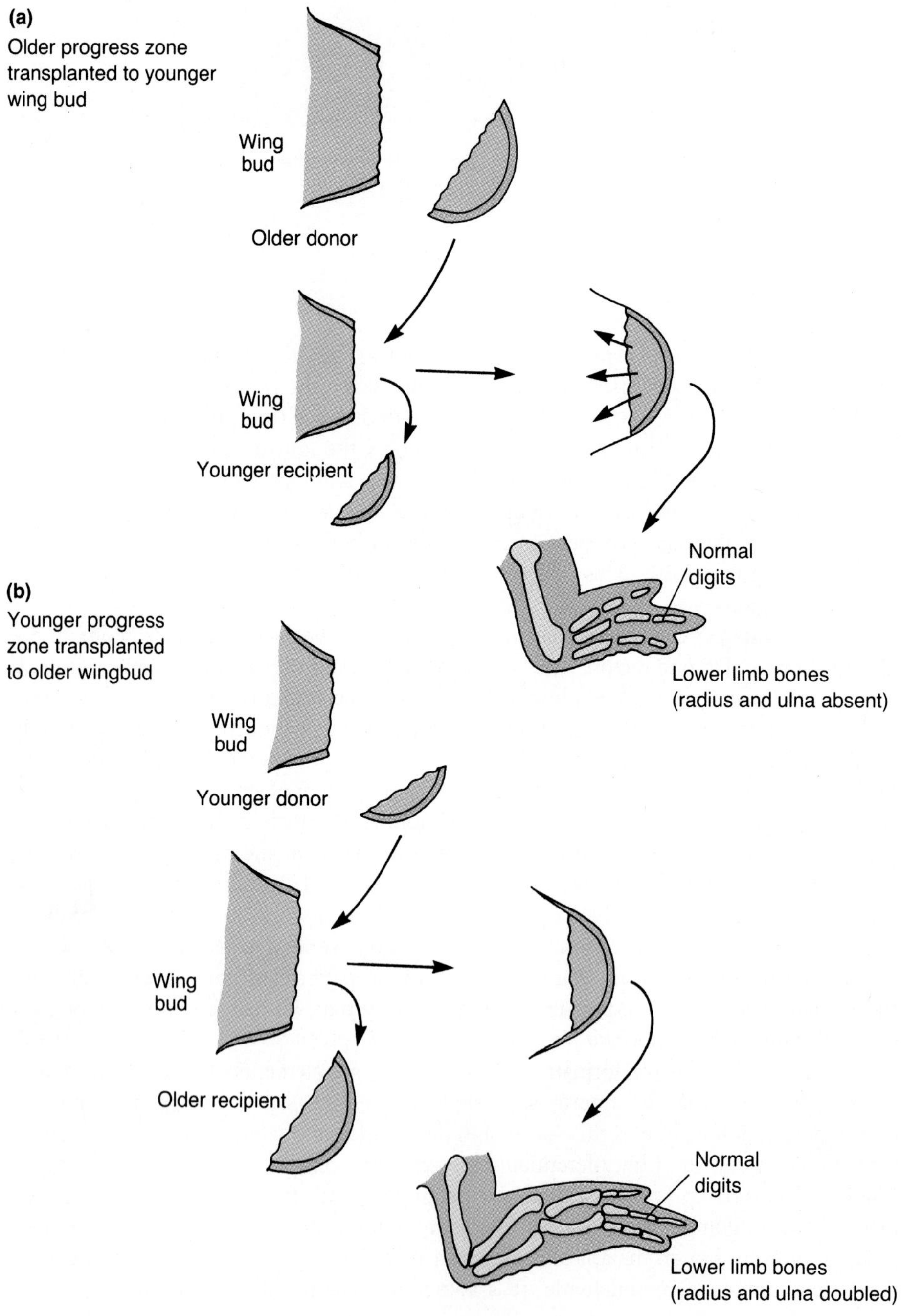

FIGURE 43.33
PROGRESS ZONE TRANSPLANTS
(a) When the late wing bud progress-zone mesenchyme is implanted into an early wing bud, some of the usual wing elements will not form. **(b)** In the reciprocal procedure, transferring early wing bud progress-zone mesenchyme to a late wing bud causes *extra* wing elements to form.

formation in the wing of the donor, say the two lower wing bones (radius and ulna). Once the progress zone has carried out this function, for some reason it cannot repeat it (they lose totipotency, at least in this environment). In the reverse of this, if an older wing bud has its apical ridge and progress zone replaced by those from a younger wing bud, the developing wing will contain some duplicate parts, say two sets of lower wing bones.

So, in the complex interplay of cells that go to form the limbs, we see that some cells determine the specific structures to be produced, whereas other cells determine the organization or pattern such parts will take. Further, the events are carefully orchestrated in a strict but constantly shifting time frame.

blastocyst An early embryonic stage in mammals, just preceding the gastrula, in which the embryo consists of a thin-walled sphere of cells, the trophoblast, and the denser inner cell mass.

blastula In many animals, an early embryonic stage preceding gastrulation that is described as a hollow ball of cells. Its cavity is the blastocoel.

cleavage Cytokinesis (cytoplasmic division) in the early animal embryo.

determinate cleavage Cleavage in which daughter cells have a very early commitment to specific developmental fates—typical of protostomes and particularly evident in insects.

determination The ongoing commitment of cells and tissues in the embryo to specific developmental fates.

differentiation The maturation of cells in which they assume their final, specialized form and function.

embryonic induction (also tissue interaction), An interaction wherein one embryonic tissue influences the developmental fate of a second embryonic tissue.

gastrula An early embryonic stage resulting from the displacement of cells, leading to the emergence of three germ cell layers (ectoderm, mesoderm, and endoderm) and to the formation of a crude gut.

gastrulation The process through which a gastrula emerges from a blastula, involving a rearrangement of surface cells through a spreading process in some cells and an insinking process in others. Gastrulation leads to the emergence of ectoderm, mesoderm, and endoderm, and the formation of a crude gut.

gill apparatus The gill slits and gill arches of the vertebrate embryo derived from the rudimentary gut. Although destined to form gill structures in fishes, the gill apparatus contributes to other structures in terrestrial vertebrates.

homeobox Common nucleotide sequences within different homeotic gene clusters in insects, responsible for activating the homeotic genes in a timely manner. Similar sequences are also seen in the genomes of other animals.

indeterminate cleavage A lack of developmental commitment in newly cleaved cells in the early embryo, resulting for a time in each daughter cell remaining developmentally flexible.

mesenchyme Mesoderm that has migrated to new locations in the embryo. Such mesoderm gives rise to muscle, dermis, gonad, and skeletal tissue.

morphogenesis Changes in shape, structure, and arrangement of cells and tissues during development.

mosaic development The early determination and setting aside of cells for specific developmental roles that may come up much later. Such early commitment makes experimental manipulation of the embryo difficult.

neurulation A developmental process beginning in the gastrula that results in the establishment of the neural tube and body axis.

organogenesis The assembly of cells and tissues into organs and organ systems in the embryo.

placenta In mammals, a highly vascularized, spongelike embryonic structure, arising from the chorion, and providing circulatory exchange between the mother's blood and that of the embryo and fetus.

polarity In the egg cell, an organizational distribution of the egg cytoplasm resulting in a more metabolically active region (the animal hemisphere) and a less metabolically active region opposite (the vegetal hemisphere).

primary organizer An embryonic tissue, often mesoderm, that influences the developmental fate of another embryonic tissue.

totipotent As development proceeds, the retention of the original developmental capabilities (those present in the zygote nucleus) by the nucleus of older cells.

KEY IDEAS

GAMETES

The Sperm

Animal **sperm** cells are made up of a head (condensed chromatin and an **acrosome**), a midpiece (mitochondrion and centriole), and tail (9 + 2 flagellum).

The Egg

1. **Egg** size varies according to yolk quantity, which is related to support needed by the embryo. Reptile and bird eggs are the largest; mammalian eggs the smallest. Much of the larger egg is yolky food supply.
2. Eggs tend to be polarized, with an unequal distribution of cytoplasmic elements. The more metabolically active region is the **animal hemisphere**, the less active region the **vegetal hemisphere.**
3. Many animals produce special surface structures such as a **vitelline membrane, zona pellucida, vitelline layer**, and **jelly coat.**

FERTILIZATION

1. In sea urchins sperm utilize enzymes released from the acrosomal to break through the jelly coat. The **acrosome reaction** includes enzyme release and the formation of an **acrosomal process**, the latter of which binds to the egg surface.
2. On sperm binding in sea urchins, the sperm and egg plasma membranes fuse, the egg forms a **fertilization cone**, and the sperm is drawn in. Egg and sperm nuclei join in fertilization. In sea urchins, electrical changes produce a temporary sperm barrier that prevents **polyspermy**. Following the **cortical reaction,** a more permanent barrier, the fertilization membrane, forms. The membrane forms as a result of the cortical reaction, where **cortical granules** release enzymes into the vitelline layer.
3. The human sperm makes use of acrosomal enzymes to penetrate the **corona radiata** and zona pellucida. No fertilization membrane forms, but the formation of a second polar body indicates fertilization.

EARLY DEVELOPMENTAL EVENTS

Becoming Multicellular: The First Cleavages

1. **Cleavage** patterns in mammals and echinoderms are similar, with equal-sized **blastomeres** appearing through the **morula** stage.
2. In amphibians, the third cleavage produces **micromeres** and **macromeres.** Then, rapid cell division produces small cells in the animal hemisphere, and slower division produces larger, yolk-filled cells at the vegetal hemisphere.
3. Because of the large yolk, cleavage is incomplete at first in reptile and bird embryos. A simple plate of cells, the **blastoderm**, soon replaces the earlier **blastodisc.**
4. Protostome embryos undergo spiral cleavage, whereas deuterostome embryos have radial cleavage.
5. Insect eggs are highly polarized. Mitosis without cytokinesis results in a multinucleate embryo. Some 6000 nuclei will form before much cytokinesis occurs. Then a surrounding cell layer, the blastoderm, appears.

The Blastula

1. A **blastula,** characterized as a hollow ball of cells, is formed by many animals. The sea urchin blastula is ciliated and free-swimming. In amphibians, the blastula is characterized by an offset **blastocoel** underlying the smaller cells of the animal hemisphere.
2. The mammalian version of the blastula is the **blastocyst.** It is made up of a thin-walled **trophoblast** and **inner cell mass,** within which lies the fluid-filled **blastocyst cavity.**

Gastrulation: Organizing the Germ Tissues

1. In **gastrulation,** the inward movement of cells produces a gastrula, a three-layered stage containing an outer **ectoderm,** middle **mesoderm,** and inner **endoderm.** The inpouching forms a new cavity, the **archenteron,** which opens to the outside via the **blastopore.** In sea urchins, the gastrula stage is replaced by the **pluteus larva.**
2. Gastrulation in the amphibian begins with **invagination** at the blastopore, with **dorsal lip** cells rolling inward, giving rise to the archenteron and middle or mesoderm germ layer. Ectodermal cells flatten and spread over the gastrula.
3. In reptiles, birds, and mammals, gastrulation is preceded by the formation of the **primitive streak,** a thickening in the **epiblast** layer. A **primitive groove** forms as cells begin to invade the flattened cavity below, where they give rise to mesoderm and endoderm. The gut arises through infoldings of endoderm.

THE EMBRYO TAKES FORM: ORGANOGENESIS

The germ layers play specific roles in **organogenesis,** the formation of organs. Ectoderm lays down the framework of coverings and neural structures; mesoderm, skeleton, muscle, blood vessels, and gonads; and endoderm forms the linings of the gut and other cavities.

Establishing the Body Axis

1. The formation of a **notochord** and **neural tube** establishes the body axis—the anterior-posterior organization. **Neurulation** begins with the formation of the **neural plate,** whose edges, the neural ridges, rise up and come together, forming the neural tube. **Neural crest cells** gather between the tube and overlying ectoderm.
2. Neural tube formation is prompted by the notochord, a rodlike structure below the ectoderm. Shaping the

neurula occurs through morphogenesis—shape changes—that occur in overlying ectoderm.

3. As the neurula elongates, the archenteron becomes tubelike, and blocklike **somites** of mesoderm form.

Further Vertebrate Development

All vertebrate embryos produce a **gill apparatus,** the **gill arches** and **slits.** They give rise to gills in fishes, but in terrestrial animals they go to form other structures.

VERTEBRATE SUPPORTING STRUCTURES

The embryo produces temporary structures of support.

The Support Systems of Reptiles and Birds

Special egg features include the protective leathery or hardened shell, a water and food supply, and extraembryonic membranes. The latter includes the **amnion** (shock absorber and lubricant), **yolk sac** (for absorbing food), **chorion** (for gas exchange), and the **allantois** (for gas exchange and nitrogen waste storage). The latter two fuse as the **chorioallantois.**

The Support System of Mammals

1. In humans, the blastocyst stage forms one week after fertilization. As it implants in the uterine wall, cells of the trophoblast digest a small, blood-filled cavity in the endometrium. The simple cavity provides the necessary exchanges until fingerlike, cellular extensions of the trophoblast provide greater exchange surfaces.
2. The secretion of **HCG** by trophoblast cells assures endometrial support by prompting the continued secretion of progesterone and estrogens by the corpus luteum.
3. The **amniotic cavity** arises early, followed by growth of the amnion (amniotic membrane). Formation of the yolk sac follows. The amnion functions as it does in birds and reptiles, but the yolk sac has different functions. It produces blood cells, lymphoid cells, and cells that will contribute to the gonads.
4. The chorion and endometrium together form the **placenta.** The chorion, arising from the trophoblast, forms fingerlike **primary chorionic villi** that deeply penetrate the endometrium. When invaded by blood vessels, the fingers become secondary chorionic villi.
5. In humans, the allantois forms the **umbilical cord,** which houses blood vessels that link the embryo to the placenta.

ORGANOGENESIS IN THE HUMAN

The First Trimester

1. Gastrulation and neurulation occur by the third week, followed by organogenesis. By the eighth week, most systems have begun formation. The embryo then becomes a fetus.
2. With neurulation, the crude outlines of the brain arise, and by the fourth week the major regions and the spinal cord are recognizable. The heart begins as a simple tube, which by seven weeks has changed to a four-chambered pump. The digestive and respiratory systems emerge from outpocketings and refinements in the primitive gut. Limb buds appear in the fourth week, with the limbs becoming discernable by six weeks. Fingers and toes go through a paddlelike stage.
3. The reproductive system begins in an indifferent state, with both sexes having the genital-tubercle, —groove, —folds, and swellings. Internally, the presence of **Wolffian** and **Mullerian ducts** precedes sexual differentiation. In the presence of testosterone, the Wolffian duct gives rise to testes and related structures, and male external genitalia emerge. In the hormone's absence, the Mullerian ducts give rise to the ovary and related structures, and female external genitalia emerge.

The Second and Third Trimesters

Organ and organ system refinement continues, and body length and mass rapidly increases. The skin organ, complete with sweat and oil glands, hair follicles, and sensory organs, has formed. By the third **trimester,** all organ systems are fully formed, and most function. The central nervous system takes on its final form, and **ossification** of the cartilaginous skeleton has occurred.

Birth

1. Placental changes related to changing progesterone levels may trigger the birth process. Oxytocin receptors in the uterus may increase, furthering its sensitivity to oxytocin, the hormone that triggers labor contractions.
2. The birth process begins with softening and dilation of the cervix, rupturing of the amnion, and uterine contractions. Crowning and fetal expulsion occur next, followed by placental expulsion.
3. For the respiratory system to function, two shunts ("short circuits") in the fetal circulatory system—the **foramen ovale** (an opening between the atria) and the **ductus arteriosus** (a short vessel connecting the pulmonary artery and aorta)—must close. Their respective closure separates oxygenated and deoxygenated blood and increases pulmonary flow as the fetal lungs become activated.
4. The first breast secretions are **colostrum,** which is clearer and thinner than milk and contains maternal antibodies. Human milk is sufficient in all nutrients except vitamin D and is bacteria-free.

THE ANALYSIS OF DEVELOPMENT

Determination in the Egg and Zygote

1. Some animal eggs are highly polarized, their organization already being critical to future events.

Cleavages must divide the cytoplasmic elements in highly specific ways for development to succeed.

2. The **gray crescent**, which forms at fertilization, is significant to establishing right and left halves, the future blastopore, and the body axis (dorsal region) of the embryo.
3. Spemann tied ligatures in fertilized frog eggs, partially dividing them in various planes. He determined that some gray crescent must be present for a cell to complete development. Using techniques similar to Spemann's, Horstadius determined that cleavages must bisect the animal and vegetal poles of unfertilized sea urchin eggs for normal development to occur.
4. In the **totipotent** state, the cell nucleus can direct the cell or embryo along a complete developmental pathway. Nuclear transplant studies by Briggs and King revealed that the frog nucleus remains totipotent through the blastula stage but becomes increasingly less capable after that.
5. Studies of limb amputation in the salamander reveal that **regeneration** must begin with the formation of a **blastema,** a mass of undifferentiated cells. The cells then follow the developmental pathways similar to those in the embryo. Nerve connections are essential to the process.

The Role of Tissue Interaction in Determination

1. Various tissues in the amphibian blastula are committed to the formation of the three germ cell layers. Prior to gastrulation, they are called presumptive ectoderm, presumptive mesoderm, and presumptive endoderm.
2. Through transplant experiments, Spemann and Mangold determined that presumptive mesoderm in the dorsal lip region acts as a **primary organizer,** capable of orchestrating the formation of the body axis and inducing ectoderm to form a neural tube.
3. Transplant experiments reveal that **mesenchyme** below the chick epidermis is also a primary organizer, capable of inducing the epidermis above to form **feather buds** and **scale buds.**
4. Parts of the vertebrate eye form through a process of triple induction. (a) Brain ectoderm induces overlying ectoderm to thicken, producing a **lens placode.** (b) The placode, in turn, induces the brain ectoderm to form an **optic cup,** which (c) influences the placode to invaginate and form the lens.
5. Whereas the operation of primary organizers is well-established, the molecular basis remains unknown.

The Determining Role of Migrating Cells

Mesenchyme (migrating mesoderm) reaches many parts of the embryo, where it acts as an organizer. Neural crest cells induce the formation of several distant tissues. Somite mesenchyme in the limbs induces muscle and skeletal formation.

1. Neurons emerge from the central nervous system and reach targets throughout the body. Neurons making improper or excessive connections die. Cell death in other tissues also plays an important role in development.
2. Migrating cells take chemical cues from their new surroundings. Hyaluronic acid cues mesenchyme migrating to the eye cells to form connective tissues in the cornea.

The Fruit Fly: Determination in Mosaic Development

1. Strong commitment in the insect occurs in the earliest developmental stages. The first complete cells, pole cells, are already committed to form germinal cells in the adult.
2. **Imaginal disks,** cell clusters in the *Drosophila* larva, are inactive but are destined to form specific adult structures such as antennae, wings, and legs. **Homeotic mutations,** mutations in the **homeotic genes** of the disk cells, can result in misplaced or duplicated organs in the adult (extra legs, double thoracic segments and wings, and others).
3. Homeotic genes are believed to be control genes, those that code for inhibitory proteins that affect entire gene complexes. The homeotic genes may be influenced by cues from their surroundings.

The Homeobox: Molecular Biology of Development

Homeotic genes have common gene sequences called **homeoboxes,** through which homeotic gene control may be exercised. Thus any change in a homeobox drastically affects development. Similar sequences are seen in a great many segmented animals, including humans.

The Vertebrate Wing: A Developmental Program

The vertebrate wing (or any vertebrate forelimb) follows a strictly timed and organized developmental program.

1. The early limb bud consists of mesenchyme and overlying ectoderm, the latter culminating in the **apical ridge** at the leading edge. The presence of the "AR" is essential to future events, but transplants between different age embryos suggest that it does not specifically determine events in the mesenchyme.
2. Within the posterior margin of the AR lies the **polarizing region,** which seems to determine the kinds and numbers of digits that will form. Transplants of the PR indicate that the innermost digits (numbers 2 and 3) form from mesenchyme furthest from the PR, whereas the outermost digit (number 4) forms from closer PR mesenchyme.
3. In the usual developmental events, the succession of limb structures proceeds outward: from upper arm to lower arm and finally to the digits. Transplant experiments indicate that a key factor in this pattern is the age of the **progress zone,** a mesenchyme strip just within the AR. In its earliest form, the PZ directs upper arm formation, and at its oldest period, only the digits. Thus, as the PZ ages, its role in establishing the wing pattern changes.

REVIEW QUESTIONS

1. Prepare concise definitions for the following: embryogenesis, determination, differentiation, organogenesis.
2. Prepare a simple sketch of an animal sperm, labeling the following: sperm head, acrosome, chromatin, midpiece, mitochodrion, centriole, tail, 9 + 2 flagellum.
3. Using the terms *vegetal* and *animal hemispheres,* explain the term *egg polarity*.
4. List the structures that appear outside the cell membrane of sea urchin eggs and human eggs.
5. Describe the events leading up to actual sperm penetration during fertilization in the sea urchin.
6. Describe the formation of two kinds of sea urchin sperm barriers. Why are such barriers essential?
7. List a similarity and a difference in the acrosomal reaction of human and sea urchin sperm.
8. Using simple drawings, compare the patterns of cleavage in the mammal, amphibian, and bird. List two factors that play a determining role in cleavage patterns.
9. Describe the early events in the development of an insect embryo. In what major ways does this differ from the similarly early events in vertebrate development?
10. Briefly distinguish determinate from indeterminate cleavage.
11. Using simple sketches, make comparisons of blastulas, blastocysts, and blastoderms. In what animals does each occur?
12. Using simple sketches if you wish, briefly describe the events of gastrulation in the amphibian.
13. Describe gastrulation in the bird. Explain why the process differs so much from that of amphibians. How do we explain its similarity in mammals?
14. List three important results of gastrulation.
15. Name the primary germ layers and explain their arrangement in the late gastrula.
16. List the several fates of each of the germ layers.
17. Summarize the events of neurulation. What is the apparent role of the notochord in this process?
18. What roles are ascribed to the somites and gill apparatus in the vertebrate embryo?
19. List and state functions for the four extraembryonic membranes of the bird embryo.
20. Outline the events from fertilization through implantation in humans. List three specific supporting acts carried out by the trophoblast.
21. Describe the timing and manner in which the extraembryonic membranes appear in the human embryo.
22. List roles of the yolk sac and allantois that are unique to the mammal.
23. What is the specific role of the chorion? Describe the formation of primary and secondary chorionic villi. What important structure forms from the chorion and endometrium?
24. At what time in the life of a human embryo is its designation changed to "fetus?" Prepare an outline in which the status of each of the following systems is described at that time: nervous, circulatory, digestive, reproductive.
25. What is the role of the Wolffian and Mullerian ducts in producing male and female sexual structures? What appears to be the direct determining factor?
26. Summarize the general changes that occur in the second and third trimesters of human development. At what point in time is the fetus considered viable?
27. How does the presence of the foramen ovale and ductus arteriosus affect circulation in the fetus? What would happen if they remained intact in the newborn baby?
28. Cite the experimental evidence suggesting that the gray crescent is vital to the organization of the amphibian embryo. What actual structures appear to be organized by this region?
29. Describe an experiment whose results suggest that some determination is already present in the sea urchin egg.
30. Summarize the events surrounding limb regeneration in the salamander.
31. How did embryologists determine that much of the blastula's cells had already been committed to specific fates in the gastrula?
32. What is a primary organizer? Briefly explain how Spemann arrived at the notion of a primary organizer.
33. Describe an experiment that supports the idea of embryonic induction in feather and scale formation.
34. Cite two examples of contributions migrating mesoderm, or mesenchyme, makes to vertebrate development.
35. Where do imaginal disks occur? What is their role in the development of adult fruit flies?
36. Where do homeotic genes occur, and what is their role in fruit fly development?
37. Cite two effects of homeotic mutations and explain why the effects are so dramatic.
38. What is the apparent role of the homeobox? Why is this of such great importance to development?
39. Briefly summarize the evidence suggesting that the polarizing region of the chick wing bud, and not the apical ridge, directs wing formation.
40. In what way does distance from the polarizing region in the chick wing bud affect the future of mesenchyme?
41. Review the results of transplant experiments in which older progress region is transferred to young wing bud and younger progress region to older wing bud. What does this indicate about "time in service?"

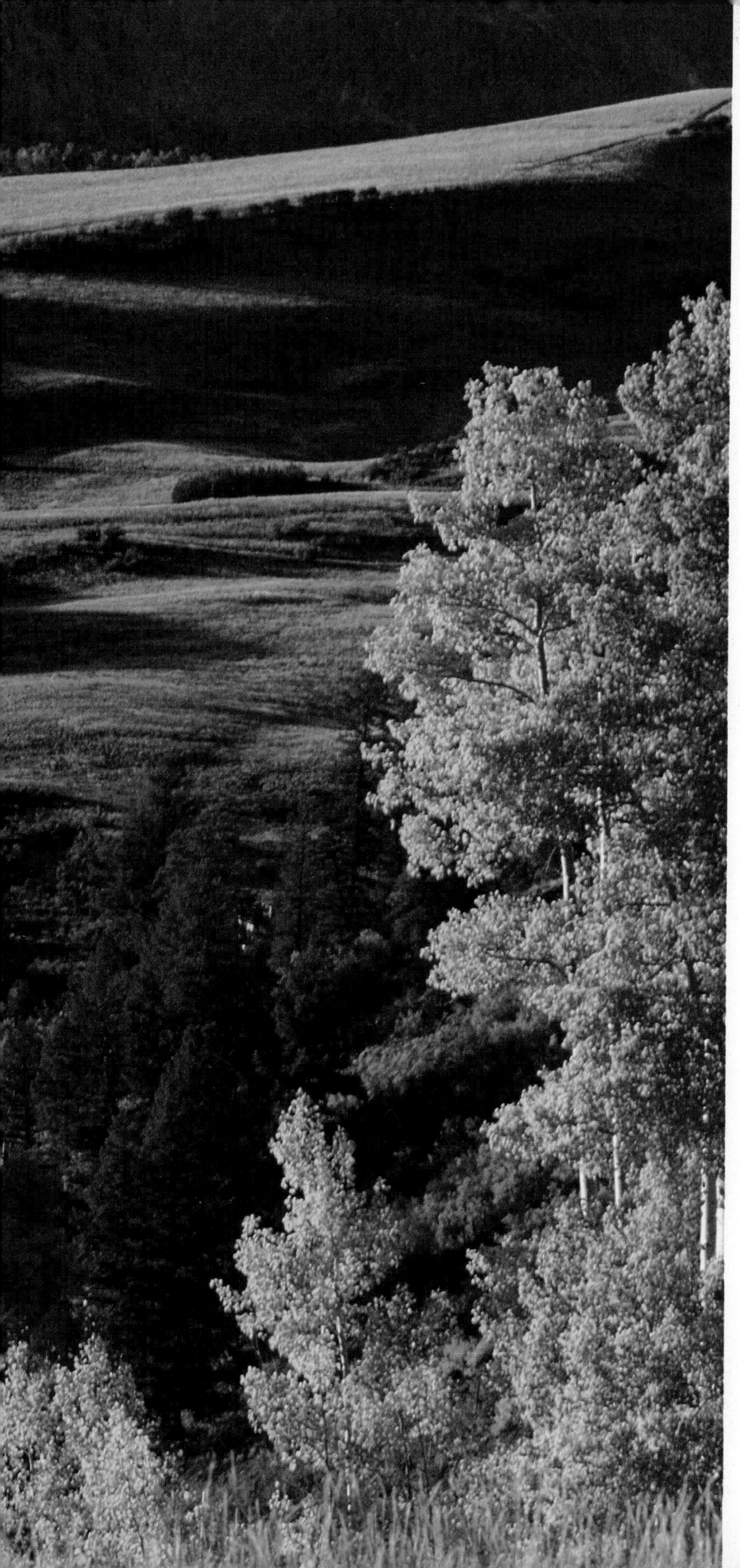

7

BEHAVIOR AND ECOLOGY

44

THE DEVELOPMENT AND STRUCTURE OF ANIMAL BEHAVIOR

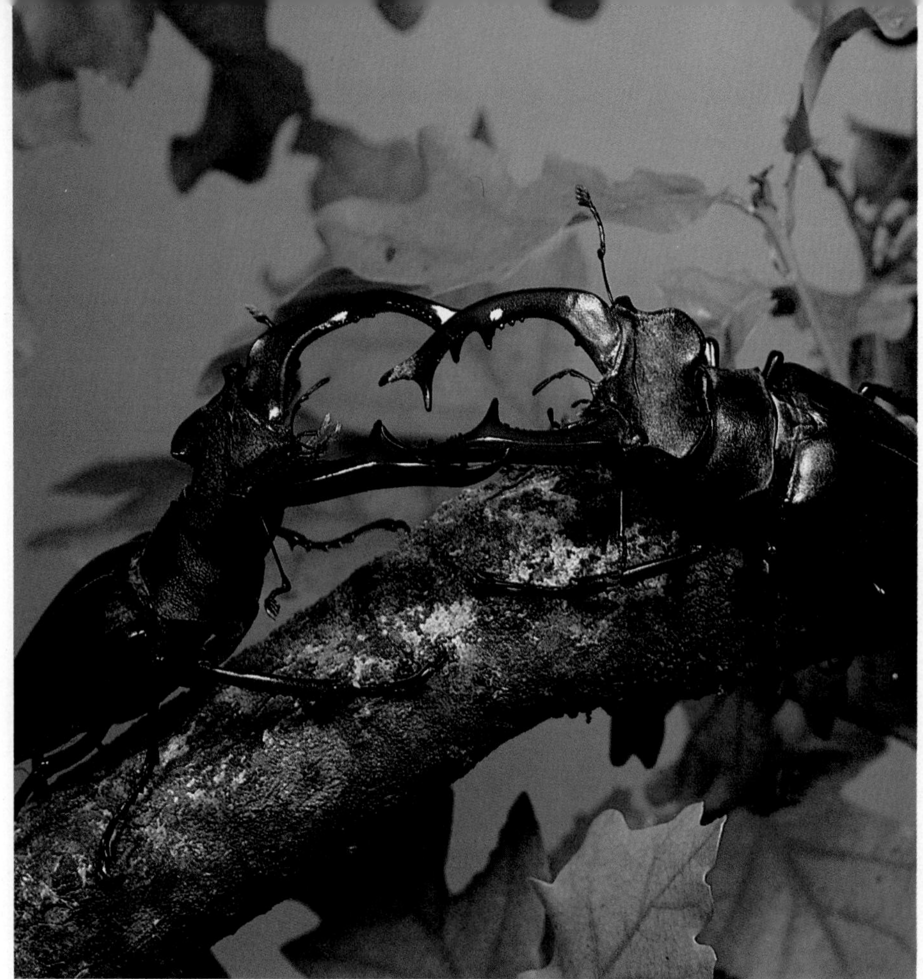

FOR SOME REASON WE HUMANS HAVE A GREAT DEAL OF INTEREST IN OTHER animals. People in small towns all over the country will sit and watch film after TV film of cheetahs running down their prey. Farmers in Arkansas might come in from a day's work and watch something about penguins. (*Anything* they learn about penguins is probably more than they will ever need to know.) The question arises, then, why do we have such an intense interest in the behavior of other species?

There are probably a number of reasons for this interest. The first might be historical. Historically, our lives have been inextricably tied to other species in ways that made our success partly dependent upon what we knew about them. We hunted some of them, we avoided some of them, and in many cases their well-being signaled our well-being (Figure 44.1).

Today, we need to know something about animal behavior for other reasons. Our control of domestic animals, for example, depends on what we know of their behavior. It wouldn't do to put two stallions in the same pen with a group of mares—both the stallions and the mares would be likely to get hurt. Certain kinds of dogs will protect herds of sheep; others will run off with the coyotes. And, of course, every hunter knows that ducks can be drawn to decoys. The purest reason for studying the behavior of other animals, some would say, is simply to learn more about them and, thus, to expand our knowledge of our world. Of course, once we learn something out of simple curiosity, it is not long before someone puts it to some use.

THE DEVELOPMENT OF BEHAVIOR

One of the most intriguing questions in biology is, why does an animal do what it does? How does its behavior develop? A fledgling blue-footed booby (Figure 44.2) on a nest cannot leave the nest yet, but from where it sits the fuzzy youngster picks up nearby twigs and repeatedly tosses them into the air, catching them by one end. This is precisely the motion it will use to handle the fish that it will catch later. It will toss the fish into the air and catch them by their heads so that the spines will be safely flattened for swallowing. One might wonder, then, is the behavior genetically based? Was the bird born with the ability to toss and catch? Is it practicing? If practice results in learning, then can one say the behavior is genetically based? Or can it be an inherited pattern that is improved upon by learning through practice? These are just the sorts of questions that certain kinds of biologists have been working on for years. At one time, these researchers rather clearly (too clearly) took one of two positions (particularly regarding birds and mammals). Some said that behavior is, in general, genetically based, while others believed that almost all behavior is learned (the so-called "nature-nurture controversy"). Fortunately, almost no one takes such rigid positions these days. Instead, researchers ask how behavior develops, fully aware that the development can involve both genetic and learned components.

Perhaps we can best see how certain behaviors develop by noting the influence of genes on behavior and then reviewing the basic concepts of instinct and learning. Finally, we will see how the two can interact to produce an adaptive response—that is, a behavior that better enables the animal to fit into its corner of the world.

FIGURE 44.1
CAVE PAINTING
Humans have been keen observers of other animals, as indicated by these ancient cave paintings found in Altimara, Spain.

Genes and Behavior

It is important to understand that genes do not code for behavior. That is, there is no gene for twig-tossing in blue-footed boobies. Instead, genes code for proteins. These proteins are then put together in such a way as to influence both structure and function within an animal. With this caveat, then, we can proceed to investigate a few lines of evidence that show the effects of genes on behavior.

Inbreeding Experiments One of the best lines of evidence involves inbreeding experiments. Inbred lines, strains created by mating close family members with one another, provide a kind of control in that they hold relatively constant the genetic background of the animals being examined, since inbred lines are homozygous for almost every gene. In this way, one can determine the relative input of genes and environment on behavior. In other words, the behavior of two inbred lines is compared while they interact with the same environment; their behavioral differences are likely to be due to genetic influences.

One of the best-analyzed strain comparisons has been in learning ability, in particular in avoidance learning. (Avoidance learning involves learning to avoid an unpleasantry, such as a mild electric shock.) Avoidance learning in mice can be studied in a shuttle box, a two-compartment apparatus with an electrified floor (Figure 44.3). Mice from two different strains are placed in one compartment, and a light flashes just before the mouse receives a light electrical shock. Some mice soon learn to run through a door into the other compartment to avoid a shock. Some mice never learn to run to safety (as we see in Figure 44.3). Different strains of mice show quite different abilities in avoiding the shocking experience, and the difference is believed to be genetically based.

FIGURE 44.2
BLUE-FOOTED BOOBY
Blue-footed booby chicks will toss sticks in the same manner that they will use to position fish for swallowing.

Artificial Selection Experiments The effects of genetics on behavior have also been shown by artificial selection. For example, pit bulls and fighting bulls have been bred for bravery and aggressiveness. Sled dogs and certain kinds of draft horses have been bred for their willngness to push against a harness when asked. And very specific behaviors have been bred for that seem to have no useful or adaptive qualities at all.

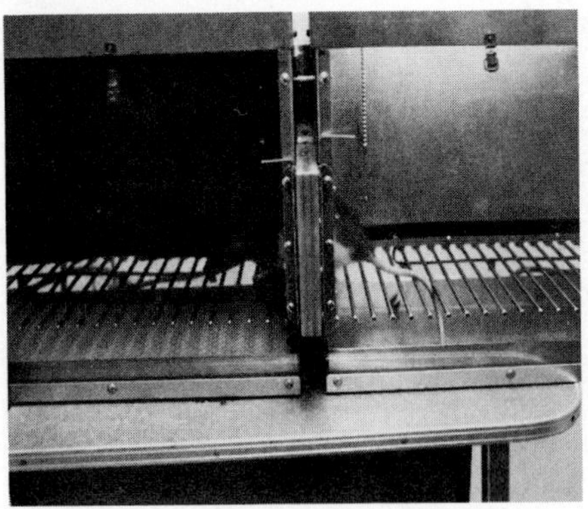

FIGURE 44.3
MICE IN SHUTTLE BOX
The shuttle box is used to test for avoidance learning in mice. Upon a signal they must run through the door to avoid a shock.

FIGURE 44.4
HONEYBEES
Some honeybees are genetically programmed to remove dead larvae; others do not and risk foulbrood infection wiping out the hive.

Hybridization Experiments A third way to demonstrate genetic influences on behavior involves hybridization. The usual procedure is to mate two individuals of strains that exhibit a distinctive type of behavior. Depending on the number of genes involved, the offspring may show one parental type or the other (say, one or two genes) or a blend of the two types (multiple genes).

Behavioral Effects of Few Genes Honeybees, for example, express the hygienic behavior of just one of the parent types. Personal hygiene is usually an important attribute, especially among certain groups of our own species, but for a social insect such as the honeybee *(Apis mellifera)* hive hygiene can be a matter of life and death (Figure 44.4). This is particularly true if the hive has been infected with *Bacillus larvae,* a bacterium that causes American foulbrood. The disease kills larvae and pupae developing within the waxen cells of the comb. If the corpses are not removed, they accumulate and serve as a source of infection that will destroy the colony. Some strains of bees, the Brown and the Squires Resistant, are resistant to American foulbrood mainly because the workers practice good hive hygiene; they open the cells containing dead individuals and remove them. The Van Scoy and Squires Susceptible strains are vulnerable to the disease because their unhygienic workers allow the deceased brood to accumulate and decompose within the hive.

When a hygienic strain is crossed with an unhygienic strain, none of the hybrids remove corpses from the hive. Unhygienic behavior is said to be dominant over the hygienic behavior of removing the bodies of progeny dead from American foulbrood, because it is the form of the behavior that is expressed in the hybrids. Arthur Rothenbuler mated the unhygienic hybrid offspring with homozygous recessive individuals (a backcross) and obtained 4 classes of offspring, each with approximately the same number of colonies:

1. The workers of 9 of the 29 colonies would uncap the cells containing dead bodies but failed to remove the corpses.
2. The workers of 6 of the colonies would remove the dead bodies from cells that were uncapped, but they would not uncap the cells themselves.
3. The workers of 8 colonies were truly unhygienic; they did not uncap cells or remove corpses.
4. The workers of the remaining 6 colonies were hygienic and would uncap cells and remove the dead bodies.

These results support the conclusion that hygienic behavior is controlled by two genes, one for uncapping (the **U** gene) and one for removing dead bodies (the **R** gene). If the hygienic behavior was controlled by a single gene, you would expect a cross of a hybrid (and, therefore, unhygienic) individual with a homozygous recessive (and, therefore, hygienic) individual to result in only two classes of offspring, hygienic and unhygienic, and each class would be expected to have equal numbers of colonies. Four equally sized classes of offspring is what is predicted if two genes are involved in the expression of the trait.

Behavioral Effects of Many Genes Sometimes so many genes are involved that the behavior of hybrids does not resemble that of either parent but is instead intermediate between the parental forms. Consider, for example, the nesting behavior of lovebirds. As their name implies, these members of the parrot family pair while they are still young and remain true to their mate for life. The first part of nesting behavior, obtaining materials, is the same in the Peach-faced lovebird *(Agapornis roseicollis)* and Fisher's lovebird *(Agapornis personata fisheri)*. Members of both species use their bills to clip paper, bark, or leaves into neat little ribbons of nesting material. The Peach-faced lovebird transports these strips by tucking them into the feathers of the lower back (Figure 44.5). The feathers are raised while the strips are tucked in and then lowered to hold them in place. Fisher's lovebird has a more direct method of carrying material to its nest—it

FIGURE 44.5
PEACH-FACED LOVEBIRD
Peach-faced lovebirds carry nest material tucked in their feathers.

holds the strips in its bill. These two species are closely enough related that they will mate in captivity.

The hybrid offspring of crosses between Peach-faced and Fisher's lovebirds display a compromise between the two parental forms of nesting behavior. Since both parental species cut strips of nesting material the same way, hybrids had no trouble with this part, but they were undecided about how to carry the strips. They attempted to tuck the material into the rump feathers, but the strips fell out. About 6% of the time they successfully carried the strips to the nest in their bill. Two months later, they learned not to waste as much time in abortive tucking efforts and carried 41% of their strips to the nest in their bills. This kind of intermediate behavior among hybrids and the difficulty they had learning to use one behavior pattern at the expense of another, then, is evidence that the behavior has a genetic basis and that a number of genes are involved.

Hormones and Behavior

A cat will avoid a German shepherd at all costs—unless, that is, the cat has kittens. A mother cat is likely to attack any large animal that comes near her offspring. The reason is, in part, because certain hormones associated with motherhood compel her to behave protectively. By the same token, mother seals show a retrieval behavior if their young pups wander too near the surging sea before they are ready. The retrieval behavior has been shown to be related to the presence of certain hormones. In fact, if the hormones are altered, the retrieval behavior changes. Hormones, indeed, have important influences on many kinds of behavior. The effects of hormones are basically studied in two ways: by removing the gland and administering hormones, and by correlation analysis.

Gland Removal and Hormone Replacement Rather good evidence of the role of a hormone is produced by removing the gland and simply seeing what happens. For example, castration (the removal of the gonads) is followed by decreased copulation behavior. This result indicates that the decrease is due to the absence of a male hormone (specifically, testosterone) produced by the testes, but the observed results could also be due to the effect of the surgery. To find out if the change in behavior is due to the surgery or to the removal of the source of male hormones, the hormone is injected after the surgery. When this happens, the male's copulatory tendencies reappear, indicating that a complex behavior was, indeed, related to the presence of the hormone.

Correlational Studies Hormonal influences on behavior can also be determined by correlational studies. Here the question is quantitative rather than qualitative—that is, it depends on amounts, not presence. Researchers look for behavioral changes as hormonal level varies.

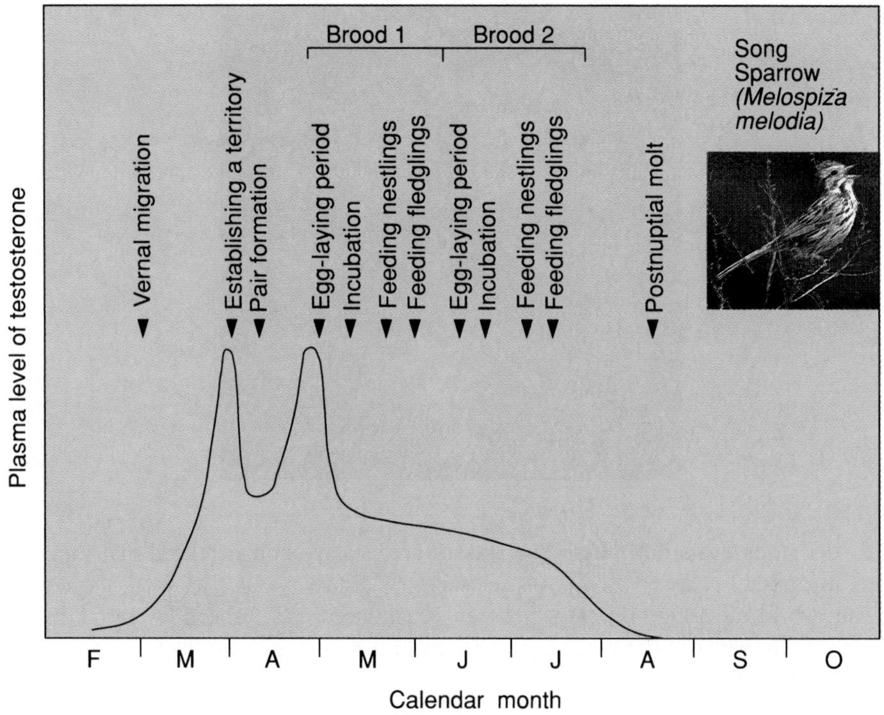

FIGURE 44.6
TESTOSTERONE AND MALE TERRITORIAL BEHAVIOR
Changes in circulating levels of testosterone during the breeding cycle of free-living male song sparrows *(Melospiza melodia)*.

In one such study, John Wingfield analyzed blood samples taken from male song sparrows *(Melospiza melodia)* over the course of a breeding season. He found that testosterone levels were markedly correlated with certain behavioral patterns. Specifically, he found a close correlation between maximum levels of testosterone and male territorial behavior (Figure 44.6).

The Three Ways Hormones Can Influence Behavior

Hormones can operate on behavior at three different levels: by influencing perception, by influencing the development of the central nervous system, and by influencing effectors.

Effects of Hormones on Perception Hormones can influence the ability to detect certain stimuli, sometimes in unexpected ways. For example, the visual ability of women varies with the stages of the menstrual cycle. Visual sensitivity is greatest about the time of ovulation and declines abruptly with menstruation. After menstruation, visual sensitivity begins to increase, again peaking at ovulation.

Hormones can also influence preferences in stimuli. Consider, for example, the migration of the three-spined stickleback *(Gasterosterus aculeatus)*. These fish normally spend the autumn and winter in the sea, migrating in the spring into rivers to breed. The migration depends on the fish preferring fresh water to salt water. This preference is apparently triggered by rising levels of the hormone, thyroxin.

Effects of Hormones on the Development of the Central Nervous System Hormones can influence behavior by altering the morphology, the physiological activity, and the role of neurotransmitters in the central nervous system (the brain and spinal cord). As examples of a morphological change caused by hormones, these messengers can influence the size of the brain, cell number, cell branching, and, interestingly, the percentage of cells sensitive to particular hormones.

Certain brain changes, and the associated physiological activity, caused by hormones are reversible. For example, Fernando Nottebohm has found that male hormones influence the parts of the brain that control singing behavior and that these brain centers change in size in response to the waxing and waning of hormones. Other changes in the brain due to hormones may be more permanent. In the zebra finch (Figure 44.7), sex differences in the brain's song-control centers are established around the time of hatching. At that time the song-control centers become especially sensitive to certain hormones. Females will not respond later to hormonal changes that trigger singing in males, since the appropriate sensitive centers in the females never developed.

Hormonal influences on behavior can be quite complex. For example, the same class of hormones can influence the role of neurotransmitters in a number of ways. To illustrate, corticosteroid hormones affect behavior by influencing both brain excitability and neurotransmitter metabolism.

Effects of Hormones on Effector Mechanisms Hormones can influence behavior by altering effectors, such as muscle and neurons. An example involves the calling behavior of the clawed frog *(Xenopus laevis)* (Figure 44.8). Males of this species attract females by alternating fast and slow metallic "trills." Females, when declining males or when terminating breeding, produce slow, monotonous "clicks." (Sexually receptive females are silent.)

FIGURE 44.7
ZEBRA FINCH

FIGURE 44.8
XENOPUS LAEVIS
THE CLAWED FROG

The differences in the male and female sounds are due to the characteristics of the muscles and neuromuscular junctions of the larynx. Adult males have eight times as many muscles in their larynx. This difference arises under hormonal control metamorphosis, as the frogs leave the tadpole stage. Before that time, the neuromuscular mechanisms of the sexes are identical. Apparently, male hormones (particularly androgen) stimulate the development of additional fibers that, with further specialization, lead to sexual differences in sound production.

It was once thought that any cue could be taught through classical conditioning to act as a conditioned stimulus, whether the unconditioned stimulus was associated with something rewarding (such as food) or something punishing (such as pain or illness). Then John Garcia found that the necessary associations simply could not be made in some cases because of the animal's genetic predisposition. For example, he found that rats could not associate visual or auditory cues with food that made them sick, but that they readily associated scent with such experiences. Quail, he found, could associate colors with tainted food, but not sound or scent. Pigeons could come to associate sounds, but not colors, with danger—and colors, but not sounds, with food.

There were also such genetic strictures in certain kinds of learning. Rats, we know, can easily be taught to press a bar to receive food, but they cannot be taught to press a bar to avoid an electrical shock. Pigeons, on the other hand, can be taught to peck at a certain spot for food, but they cannot easily be taught to hop onto a platform for food. They can be taught to hop onto a platform in order to avoid shock, however.

Animals do, indeed, show genetic propensities to learn certain things. We can assume that they are best able to learn those things that have traditionally been important to the survival of their kind. In some cases, the adaptive nature of the specialization for certain kinds of learning is apparent. For example, rats are nocturnal, so they are unlikely to utilize color in determining the nature of food; scent would be much more important to them. Color would be more important to pigeons, since they have good vision and they feed in the daylight. Pigeon food, however, is not generally noisy. The innate ability to make certain learned associations emphasizes, once again, the close interaction of genes and experience in producing an adaptive response.

THE ETHOLOGISTS' CONCEPT OF INSTINCT

The formal concept of instinct was developed in the 1940s by the Austrian Konrad Lorenz (Fig. 44.9) and refined by the Dutchman Niko Tinbergen in the 1950s. They called themselves **ethologists.** Ethology is concerned with four areas of inquiry: cau-

(a)

(b)

(c)

FIGURE 44.9
ETHOLOGISTS
(a) Konrad Lorenz and gosling. Lorenz, who died in 1989, time and again proved himself to be one of this century's most astute observers and synthesizers in the area of animal behavior. **(b)** Niko Tinbergen, a humanistic and imaginative researcher, shared the Nobel Prize with Lorenz and Karl von Frisch **(c)**, an impeccable experimenter who worked primarily with insects.

sation, development, evolution, and function (which, when preceded by *animal behavior,* can be remembered by a, b, c, d, e, and f—assuming one can remember the alphabet). Lorenz and Tinbergen, with Karl von Frisch (who worked on bee communication), were awarded the Nobel Prize in 1973—the first field biologists to be honored in this way. These pioneers have given rise to new generations of researchers who study animal behavior in the wild as they try to learn just how an animal's behavior helps it adapt to its environment.

Many of the ethological principles are based on instinct being separated into specific facets. Specifically, there are three terms that are central to the concept of instinct: *sign stimulus, innate releasing mechanism,* and *fixed action pattern.*

Sign Stimuli

According to ethologists, instinctive behavior is released by certain very specific signals from the environment. Environmental factors that evoke, or release, instinctive patterns are called **sign stimuli.** The sign stimulus itself may be only a small part of any appropriate situation. For example, fighting behavior may be released in territorial male European robins, not only by the sight of another male, but even by the sight of a tuft of red feathers at a certain height within their territories (Fig. 44.10). Of course, such a response is usually adaptive because tufts of red feathers at that height are normally on the breast of a competitor. The point is that the instinctive act may be triggered by only *certain parts* of the environment.

Innate Releasing Mechanisms

The exact mechanism by which releasers work isn't known, but one theory is that there are certain neural centers called **innate releasing mechanisms (IRMs),** which, when stimulated by impulses set up by the perception of a sign stimulus, trigger a chain of neuromuscular events. It is these events that comprise instinctive behavior. The essence of the IRM is that it is genetically encoded, and once it is triggered by a sign stimulus, it results in the performance of a behavior called the **fixed action pattern.**

Fixed Action Patterns

It can't be denied that animals are born with certain behaviors that are indelibly stamped into their behavioral repertoire. Birds build nests by using peculiar sideways swipes of their heads, with which they jam twigs into the nest mass. And all dogs scratch their

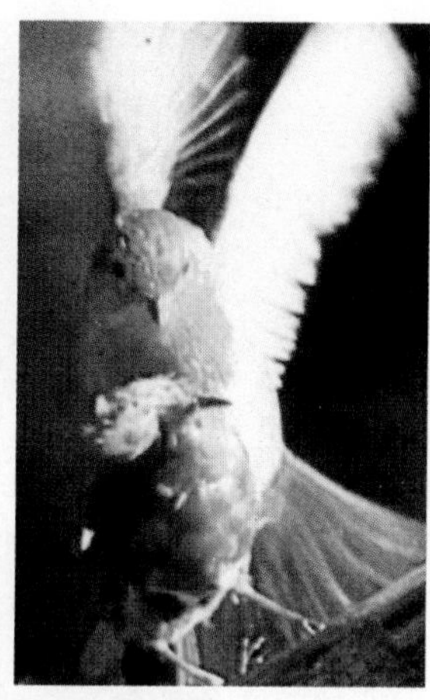

FIGURE 44.10
INSTINCTIVE ATTACK
A male European robin in breeding condition will attack a tuft of red feathers placed in his territory. Since red feathers are usually on the breast of a competitor, it is to his reproductive advantage to behave aggressively at the very sight of them. The phenomenon illustrates that releasers of instinctive behavior need to meet only certain criteria—in other words, need to represent only a specific part of the total situation.

ears the same way, by moving their rear leg outside the foreleg. Such precise and identifiable patterns, which are innate, independent of the environment, and characteristic of a given species, are called fixed action patterns.

Of course, any behavioral pattern, in order to be effective, must be properly coordinated with the environment. The coordination of fixed action patterns with spatial variables in the environment is called **orienting**.

To illustrate the relationship between a fixed action pattern and its orientation, consider the fly-catching movements of a frog (Fig. 44.11). For the fixed action pattern (the tongue flick) to be effective, the frog must carefully orient itself with respect to the fly's position. If the fixed action pattern is performed properly, the frog gains a fly. It is important to realize, however, that in most cases, once the fixed action pattern is initiated it cannot be altered. If the fly should move after the tongue begins its motion, the frog will miss, since the fixed sequence of movements will be completed whether the fly is there or not. Thus the frog first *orients* and then *performs* the fixed action pattern.

We also see that the aggressive threat displays of male three-spined sticklebacks are fixed action patterns (Fig. 44.12). The behaviors are stereotyped, triggered by the same stimuli, and performed in a similar manner by all males of the species.

To sum up the relationship between these three parts of the instinct idea, you might note that a sign stimulus in the environment activates the innate releasing mechanism in the nervous system, which then sends a signal to the muscles that causes a fixed action pattern. (George Barlow has suggested that, partly because these patterns show some flexibility, they be called "modal action patterns." His argument does point out that the behaviors show a certain variability, but this would be expected even if the behaviors are genetically "hard-wired" into the animal.)

Now we will consider the effects of learning on the development of behavior. Although we discuss innate and learned patterns separately, keep in mind that probably in virtually every case, the two kinds of patterns operate together to produce an adaptive behavioral pattern.

FIGURE 44.11
A LEOPARD FROG ORIENTING
The tongue flick, a fixed action pattern, is not released until the central nervous system is stimulated by the sight of an insect in the proper position (close and in the midline).

LEARNING

Learning is the process of developing a behavioral response based on experience. As we consider learning in a wide range of animals, we should keep in mind that the importance of learning varies widely from one species to the next. For example, there are probably very few clever tapeworms. But then, why should they be clever? They live

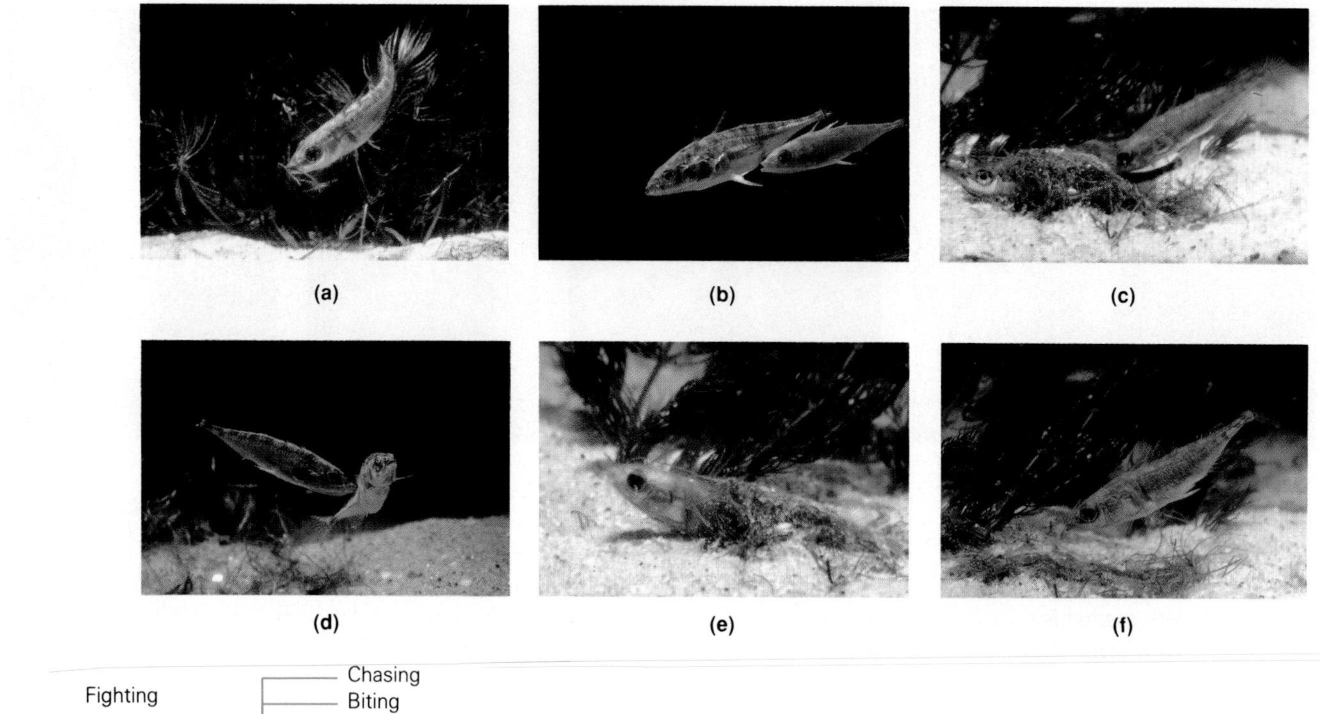

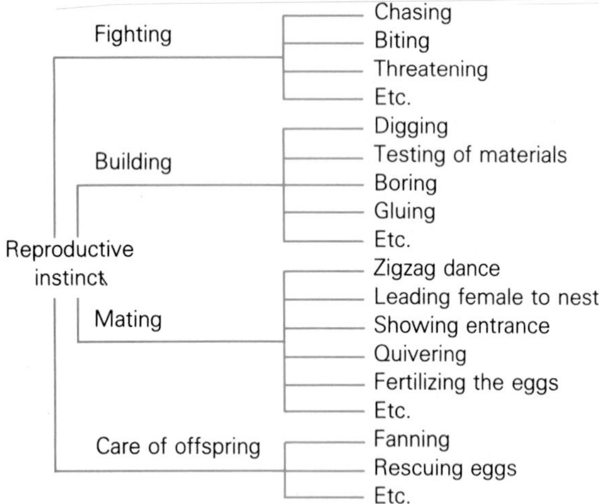

FIGURE 44.12
STICKLEBACK COURTSHIP
The sequence in courtship and mating in the stickleback *Gasterosteus,* a fish found in European ditches and streams. The male is attracted by the sight of the female swollen with eggs. He is torn between chasing her and leading her to the nest he has built **(a)**, and the result is a peculiar zigzag dance. Once she is maneuvered into his "tunnel of love" **(b)**, he pokes her tail with his snout and thus induces her to lay eggs **(c)**. He then promptly chases her off **(d)**, returns to fertilize the eggs **(e)**, and then spends days fanning oxygen-laden water over them until they hatch **(f)**. The hierarchical organization **(g)** of the reproductive instinct of the male stickleback, according to the model devised by Tinbergen in 1942. The overall reproductive instinct is composed of a number of different patterns. The pattern that is expressed is determined by what happens in the environment. These patterns are, in turn, composed of yet more specific ones, and their expression is also environmentally determined.

in an environment that is soft, warm, moist, and filled with food. The matter of leaving offspring is also simplified; they merely lay thousands and thousands of eggs and leave the rest to chance—to sheer blind luck. In contrast, chimpanzees live in changeable and often dangerous environments, and they must learn to cope with a variety of complex conditions. Unlike tapeworms, they are long-lived and highly social, and the young mature slowly enough to give them time to accumulate information. The point is, each kind of animal is able to learn those things that are important to its survival and reproduction.

Kinds of Learning

Interestingly, different species may learn in different ways. However, certain kinds of learning seem to be common to a number of species. Here, we will consider six kinds of learning: habituation, classical conditioning, operant conditioning, latent learning, insight, and imprinting.

Habituation **Habituation** involves learning *not* to respond to a stimulus. In some cases, the first time a stimulus is presented, the response is immediate and vigorous.

But if the stimulus is presented over and over again, the response to it gradually lessens and may disappear altogether. Habituation is not necessarily permanent, however. If the stimulus is withheld for a time after the animal has become habituated to it, the response may reappear when the stimulus is later presented again.

Habituation is important to animals in several ways. For example, a bird must learn not to waste energy by taking flight at the sight of every skittering leaf. A reef fish holding a territory may come to accept and pretty much ignore its neighbors but will immediately drive away a strange fish wandering through the area. The wandering fish is likely to be searching for a territory and therefore be more of a threat.

It may also help explain why animals continue to avoid predators (which they are likely to see only rarely), while ignoring more common, harmless species. Habituation is often ignored in discussions of learning, perhaps because it seems so simple, but it may well be one of the more important learning phenomena in nature.

FIGURE 44.13
PAVLOV
The Russian biologist Ivan Pavlov formulated the notion of *classical conditioning*.

Classical Conditioning Classical conditioning was first described through the well-known experiments of the Russian biologist Ivan Pavlov (Fig. 44.13). In **classical conditioning,** a behavior that is normally released by a certain stimulus comes to be elicited by a substitute stimulus. For example, Pavlov found that a dog would normally salivate at the sight of food. He then experimentally presented a dog with a signal light five seconds before food was dropped into a feeding tray. After every few trials, the light was presented without the food. Pavlov found that the number of drops of saliva elicited by the light alone was in direct proportion to the number of previous trials in which the light had been followed by food.

This conditioning process also worked in reverse. When the conditioned animal was presented repeatedly with a light signal that was not followed by food, the salivary response to the light diminished until it finally was extinguished (Fig. 44.14).

Pavlov's experiments demonstrated two other important properties of classical conditioning: **generalization** and **discrimination.** It was found that if a dog had been conditioned to salivate in response to a green light, it would also respond to a blue or red light. In other words, the dog was able to generalize regarding the qualities of lighted bulbs. However, it was found that with careful conditioning the dog could be taught to respond only to light with certain properties. (Dogs do not see color well and are more likely to be responding to some differences in intensity.) If food was consistently presented only with a green light and never with a red or blue light, the dog would come to respond only to green light, thus demonstrating clearly that it was able to discriminate differences among stimuli (discrimination) as well as the properties they had in common (generalization).

Operant Conditioning **Operant conditioning** involves an animal learning to perform an act in order to receive a reward. Operant conditioning differs from classical conditioning in several important ways. Whereas in classical conditioning the reward (such as food)

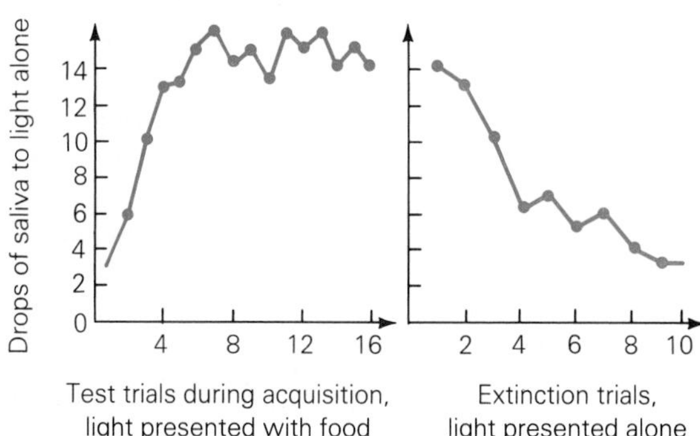

FIGURE 44.14
PAVLOV'S DOG
This apparatus was devised to demonstrate classical conditioning. Upon presentation of a light, meat powder would be blown into the dog's mouth, causing it to salivate at the sight of a light alone. The salivation, then, was *conditional* on the light. Note in the first graph that the dog salivated at maximal levels after only eight trials. When the experiment was reversed, and food no longer followed the light, the dog stopped salivating after only nine more trials.

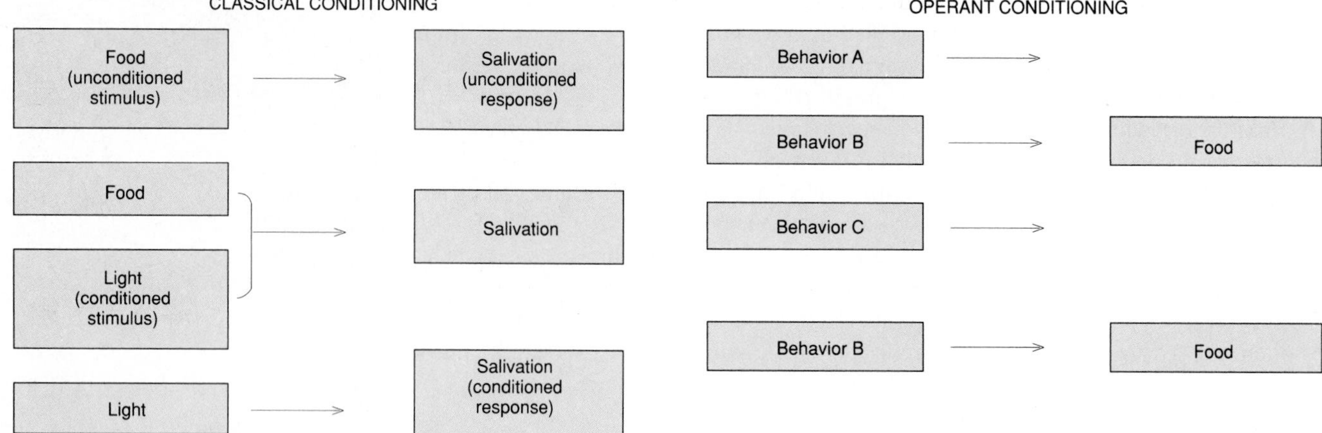

FIGURE 44.15
DIFFERENCE BETWEEN CLASSICAL AND OPERANT CONDITIONING
Note that in classical conditioning, an unconditioned stimulus (such as food) becomes paired with a conditioned stimulus (such as a flash of light). In time, the conditioned stimulus becomes a substitute for the unconditioned stimulus and is then able to produce the response, such as salivation. In operant conditioning, after receiving a conditioned stimulus, there is an opportunity for the animal to respond in various ways. However, only the "correct" response is reinforced. Finally, the stimulus produces a specific response.

follows the stimulus, in operant conditioning the reward follows the behavior (Fig. 44.15). Also, in classical conditioning the experimental animal has no control over the situation. In Pavlov's experiment, all the dog could do was wait for lights to go on and food to appear. There was nothing the dog could do one way or the other to make it happen. In operant conditioning, the animal's own behavior determines whether or not the reward appears.

In the 1930s, B. F. Skinner developed an apparatus that made it possible to demonstrate operant conditioning. This device, now called a **Skinner box,** differed from earlier arrangements involving mazes and boxes from which the animal had to escape in order to reach the food.

Once inside the Skinner box, an animal has to press a small bar in order to receive a pellet of food from an automatic dispenser (Fig. 44.16). When the experimental animal (usually a rat or a hamster) is first placed in the box, it ordinarily responds to hunger with random investigation of its surroundings. When it accidently presses the bar, lo! a food pellet is delivered. The animal doesn't immediately show any signs of associating the two events, bar-pressing and appearance of food, but in time its searching behavior becomes less random. It begins to press the bar more frequently. Eventually, it spends most of its time just sitting and pressing the bar. This sort of learning is based on the principle that if a behavior is rewarded, then the probability of that pattern reappearing is increased.

Latent Learning Usually, learning is thought to be associated with some kind of reward. (In a sense, one learns in order to be rewarded.) However, **latent learning** occurs in the absence of an immediate reward. If rats are allowed to repeatedly run a maze without any sort of reward at the end, they will learn the maze anyway, but they seem to learn it very slowly. However, if after they have been allowed to wander through the maze (apparently without learning much about it) food is placed at the end, these same rats will learn the maze with remarkable swiftness. Apparently, learning had been occurring all along, but it was not evident until it was activated by a patent reward.

Latent learning may be important in the wild, for example, as an animal explores its range, looking into every nook and cranny. The learning that occurs under such circumstances is apparent when a predator appears on the scene and the animal runs for cover, darting into some hole that it had learned was there while wandering through the area.

Insight Learning **Insight learning** involves solving problems, without experience, through processes similar to trial and error. Trial and error, of course, is an important part of certain other kinds of learning (such as operant conditioning), as an animal learns what works and what doesn't through experience. In insight learning, the trials apparently take place mentally. Such an ability is, at present, believed to be a higher order mental function possible only in primates.

In an interesting experiment involving insight learning, researchers hung a stalk of bananas from the ceiling, out of reach of a chimpanzee on the floor. They placed a stick in the room that would reach the bananas, believing that through insight the chimp would use the stick to knock them down. The chimp, however, carefully balanced the stick on end under the bananas, then quickly climbed the stick to the bananas, falling gently to the floor with the entire bunch, and all of them intact.

FIGURE 44.16
RAT IN A SKINNER BOX
Skinner boxes are designed to promote operant conditioning. For example, a hungry rat may move randomly, searching for food, until it accidently presses a bar that delivers a food pellet. Each delivery means a greater probability that the rat will press the bar again, until finally the rat learns simply to press the bar each time it wants a pellet.

Imprinting Some years ago, Konrad Lorenz discovered that newly hatched goslings would follow whatever moving object they saw and that they would continue to identify with this object throughout the rest of their lives. Of course, under most circumstances any such object was likely to be a parent, and in this way they learned to identify their own species. Later, as they approached their first breeding season, they would seek out an individual with the traits of the individual they had followed soon after hatching. If they somehow were exposed to something else at that time, they would focus on whatever resembled that thing when it came time to breed. A group of goslings hatched in an incubator saw Lorenz first, and as they grew up, they would often dutifully fall into single file, following after him as he walked around the farm.

Many animals also learn sexual identification during this critical period. Lorenz once had a tame jackdaw that he had hand-reared, and it would try to "courtship feed" him during mating season. On occasion when Lorenz turned his mouth away, he would receive an earful of worm pulp! The story of Tex, the dancing whooping crane, provides another example of this type of learning (Figure 44.17).

The sort of learning that takes place in such a brief period of a young animal's life has been called **imprinting.** It was once believed to take place through special learning processes, but most researchers now believe that it occurs through the same neural mechanisms that are involved in other types of learning.

Imprinting is especially important in many kinds of bird song. If male white-crowned sparrows are deafened while very young, they will sing disconnected notes, but no real song. The birds apparently must be able to hear themselves in order to learn the song of their species. Normally, these birds must be exposed to the song at about the age of three months. They will not begin to sing, however, until some months later. Then, even a bird that has been experimentally isolated after hearing the song of its group will sing the correct song—not only the basic song of the species, but the variation (dialect) of the particular local population whose song it heard.

FIGURE 44.17
A CRANE IMPRINTED ON HUMANS
Tex, the only female whooping crane at the International Crane Foundation breeding area in 1982, has been hand-reared and therefore had imprinted on humans. She rejected the mate provided for her, but could be enticed to lay eggs (artificially fertilized) by "dancing" with humans. She preferred Caucasian men of average size with dark hair.

HOW INSTINCT AND LEARNING CAN INTERACT

We must keep in mind that a variety of factors are involved in an animal forming an adaptive behavioral pattern. Thus, in most cases, both instinct (genetics) and learning (experience) are involved. As an example of such interaction, consider the development of flight in birds. Flight is a largely innate pattern. A bird must be able to fly pretty well on the first attempt, or it will crash to the ground as surely as would a launched mouse. It was once believed that the little fluttering hops of nestling songbirds were beginning flight movements and that the birds were, in effect, learning to fly before they left the nest. But

FIGURE 44.18
LEARNING IN YOUNG SQUIRRELS
Young squirrels instinctively gnaw at the surfaces of nuts. Later they will learn to bite through the thinnest part of the shell and deftly open the nut.

in a series of experiments, one group of nestlings was allowed to flutter and hop up and down, while another group was reared in boxes that prevented any such movement. At the time when the young birds would normally have begun to fly, both groups were released. Surprise! The restricted birds flew just as well as the ones that had "practiced."

Flight behavior, then, must be an innate, or unlearned, pattern. It is important to realize, though, that generally young birds don't fly as well as adults. Many innate patterns can be improved upon by learning through practice (and strengthening the muscles involved).

In other instances, the innate and learned components of a behavior can be more clearly differentiated. The process by which a red squirrel opens a hazelnut is a rather complex behavior. The squirrels first cut a groove along the growth lines on one or both of the flat sides of the nut, where the shell is thinnest. Then they insert their incisors into the groove and break the shell open. In one study, baby squirrels were reared without any solid food. When they were finally presented with a hazelnut, they correctly performed the gnawing actions and even inserted their incisors into the grooves correctly. Thus, these patterns are apparently innate. The problem was, they tried to gnaw all parts of the shell, even the thickest areas. Eventually they were usually able to break open the shell, but only after great expenditure of energy. In time, however, their performance improved. After they had handled a few nuts, they began to make their grooves at the thinnest part of the shell, and the shell opened more easily. Most of the squirrels soon became proficient at opening hazelnuts, as their innate gnawing pattern was modified through learning to produce the complex adaptive result (Fig. 44.18).

The relative importance of innate and learned components of behavior may vary widely. For example, whereas reproductive behavior in stickleback fish may be largely innate, the nut opening in squirrels is perhaps less so. Furthermore, in some songbird species, the song is largely learned; in others it is almost completely innate.

One of the most interesting examples of the interaction of instinct and learning involves song learning in white-crowned sparrows. In a series of experiments related to those we just discussed regarding imprinting, by Peter Marlen and his colleagues, white-crowned sparrows were reared in sound isolation and then watched to see what kind of song they sang. It turned out that they developed abnormal songs, quite different from the wild stock. Male white-crowned sparrows do not begin to sing until they are several months old, and the researchers found that if they were allowed to hear the normal white-crowned sparrow song, even briefly, between the ages of 10–50 days, when it came time for them to sing they would sing the correct song. They obviously are innately sensitive to the song of their species. Interestingly, if they were allowed to hear the song of other species during their development, they would learn nothing about it. It was as if they were reared in isolation. Also, birds that were deafened after hearing the correct song could never sing it—they had to use that song as a model against which to compare their own sounds (Fig. 44.19). Finally, even if a male heard the correct song after the age of 50 days, he would never be able to sing it. We see then, that white-crowned sparrows can and do learn the correct song of their species, but that the learning is strongly influenced by genetics.

MEMORY AND LEARNING

> Many years had elapsed during which nothing of Cambray had any existence for me, when one day in winter my mother offered me some tea. . . . I raised to my lips a spoonful of the tea in which I had soaked a morsel of cake. No sooner had the warm liquid, and the crumbs with it, touched my palate than a shudder ran through my whole body. . . . And suddenly the memory returns. The taste was that of the little crumb of madeleine which on Sunday mornings at Cambray my Aunt Leonie used to give me, dipping it first in her own cup of real or of

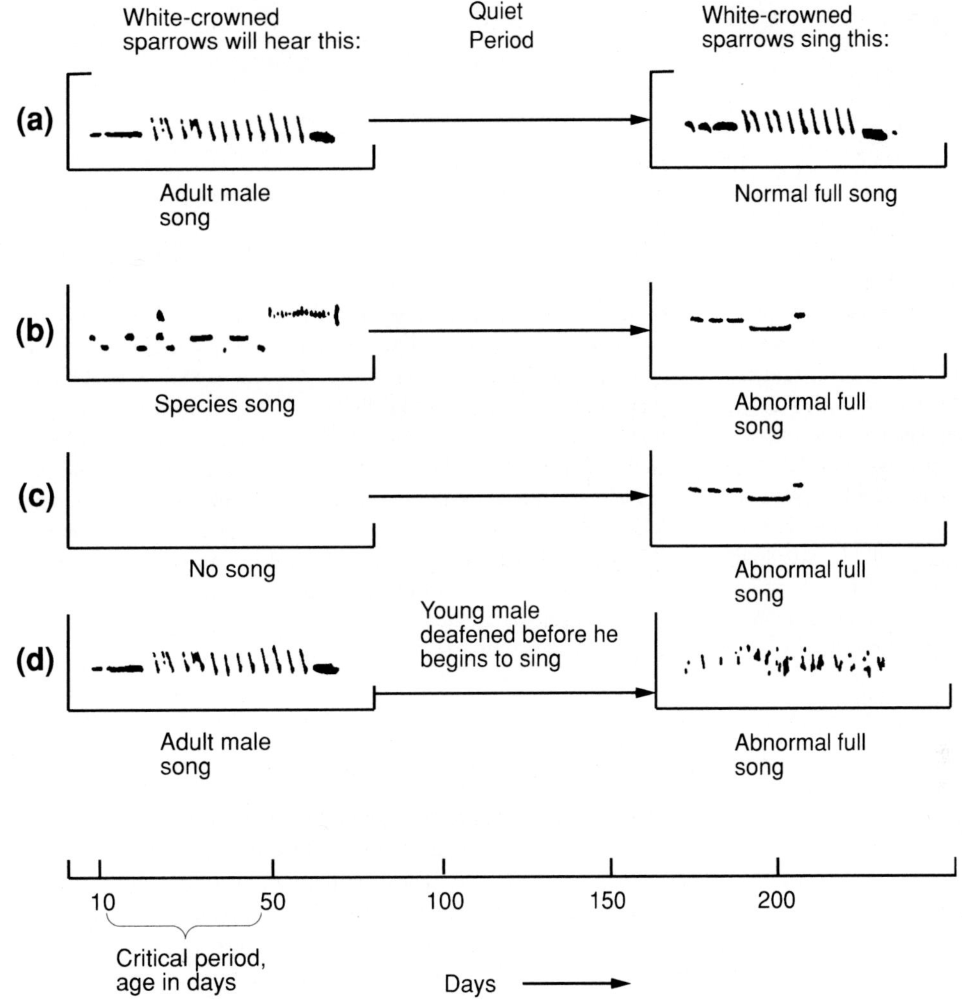

FIGURE 44.19
SONOGRAMS OF SONG-LEARNING IN THE WHITE-CROWNED SPARROW
A sonogram is a visual representation of a sound: **(a)** shows the normal song of a wild male, **(b)** shows that birds reared in isolation produce a different song, **(c)** shows that birds reared in silence sing abnormal songs, and **(d)** indicates that deafened birds sing much more complex songs, but quite unlike the normal song.

lime-flower tea. . . . Once I had recognized the taste, all the flowers in our garden and in M. Swann's park, and all the water lilies on the Vivonne and the good folk of the village and their little dwellings and the parish church and the whole of Cambray sprang into being.

Marcel Proust
Remembrance of Things Past

There are indeed not only bits of information tucked away in our memories, but overall settings, moods, and impressions that can be recalled if properly summoned. In some cases, we can recall things at will, at times we simply cannot, and at other times, remembrances flock to our conscious thought that we would rather have suppressed (Essay 44.1). Obviously, we have not learned to use our memories as well as we might like. But our chances are better if we know more about how they work.

Memory is the storage and retrieval of information, and its characteristics are important in any consideration of learning. Learning is accompanied by changes in the central nervous system. Some change in the neural apparatus produces a more or less permanent record of the learned thing. The physical change that is presumed to occur in the brain when something is learned is called the **engram** (or memory trace).

Memories are often triggered by unexpected stimuli. Just as the tea-soaked cake brought back a rich array of remembrances of the town Cambray for Proust, so can our tucked-away memories be jolted to our consciousness by some small thing that, without the weight of the memory it stirs, would be insignificant. Proust's musing reminds us that memory is more than simple recollection. Each detail we have so neatly stored in our memories is inextricably bound to others in an interwoven fabric that may literally mold our identities, our sense of who we are. We can't pick at a thread without moving the whole cloth.

ESSAY 44.1

LEST YE FORGET

It is suspected that virtually everything we encounter is learned—stored away in the brain. Wilder Penfield and his group, working in Montreal, used electrodes to probe the brains (which has no pain receptors) of conscious patients and asked them to describe their sensations as various parts of the brain were stimulated. The results were startling. Some patients "heard" conversations that had taken place years before, some heard music or seemed to find themselves with old friends, long deceased. They seemed almost to "relive" the experiences, rather than simply remember them. The powers of recall under such circumstances are phenomenal. A hypnotized bricklayer described markings on every brick he had laid in building a wall many years before. His recollections were carefully recorded, and then the wall was found and examined. The markings were there! (Our memory is not infallible, however. We also have a tendency to embellish partially recalled events.) Actually, there may be advantages to forgetting.

Although we may have entered every jot and tittle of our lives into our mind's ledger, we are not consciously able to recall very much of it. Sometimes we can't even recall what we have tried to memorize—as you well know. But apparently the information is there, just the same. The fault lies with our recall techniques. The implications of this finding are enormous. If our research ever enables us to recall or relive earlier events at will, we might "reread" novels on long train trips, and all our exams would be virtually open-book tests. Perhaps terminally ill and pain-racked patients could be stimulated to relive happier, youthful days with loved ones as life dwindles in a changed body.

Theories on Information Storage

It now seems that there are two separate mechanisms by which information is stored, as evidenced in the differences between **long-term memory** and **short-term memory.** The existence of these two types of memory was first noticed in humans with brain concussions who were unable to recall what happened just before an accident but could remember what happened much earlier. On the basis of such findings, researchers postulated two memory systems that could process information independently. Information stored in the long-term memory system is relatively permanent (subject to very slow decay). Once information is processed into the long-term system, it is not easily disrupted and therefore may be recalled with minimum confusion after long periods of time. Short-term memory, on the other hand, is relatively easily disrupted and subject to rather rapid decay. (This is why cramming for exams is not a good idea. If the exam is postponed, all is lost.) Also, there is evidence that the short-term memory can be overloaded, whereas the long-term memory can't be.

Apparently, short-term storage is necessary before long-term storage can occur. In a series of experiments, it was found that if an inedible object is presented to one arm of an octopus, over and over in rapid succession, the arm will come to reject it after learning that it is inedible. If the object is then immediately presented to other arms they will accept it. But if some time is allowed to elapse before the second presentation, all the arms will reject it. This finding suggests that the information acquired by one arm is only locally adaptive. It must filter from a hypothesized short-term center to a long-term system before a general adaptive pattern (total rejection) can appear. The experiment actually only suggests that something filters from neural centers associated with one arm to general centers. The short- and long-term phenomena are postulated because the learning arm acquires no more information than the other arms once the information has become integrated.

In another experiment, a mouse was put into an arena (enclosed area) containing only a box with a hole. If the mouse ran into the hole, its feet received an electric shock, and it never went near that hole again. But if the mouse was given electroconvulsive or chemical shock treatment within about an hour after the unpleasant incident, it would forget the whole experience and helplessly run into the same hole the next day and every day as long as it survived the experiment. On the other hand, if the electroconvulsive

FIGURE 44.20
HOW MEMORIES ARE FORMED
A learning experience causes changes in nerve spines of the hippocampus. At left, a stimulus associated with the experience causes a neuron to release a neurotransmitter that attaches to receptors of the next neuron. This, in turn, causes calcium to enter the cell. The calcium then activates calpain, which begins to break down fodrin. As the fodrin degrades, more receptors begin to appear, allowing more calcium into the cell. The cell membrane weakens, and the cell begins to change shape as new spines develop. These new spines allow new neural connections that are believed to be the basis for memory.

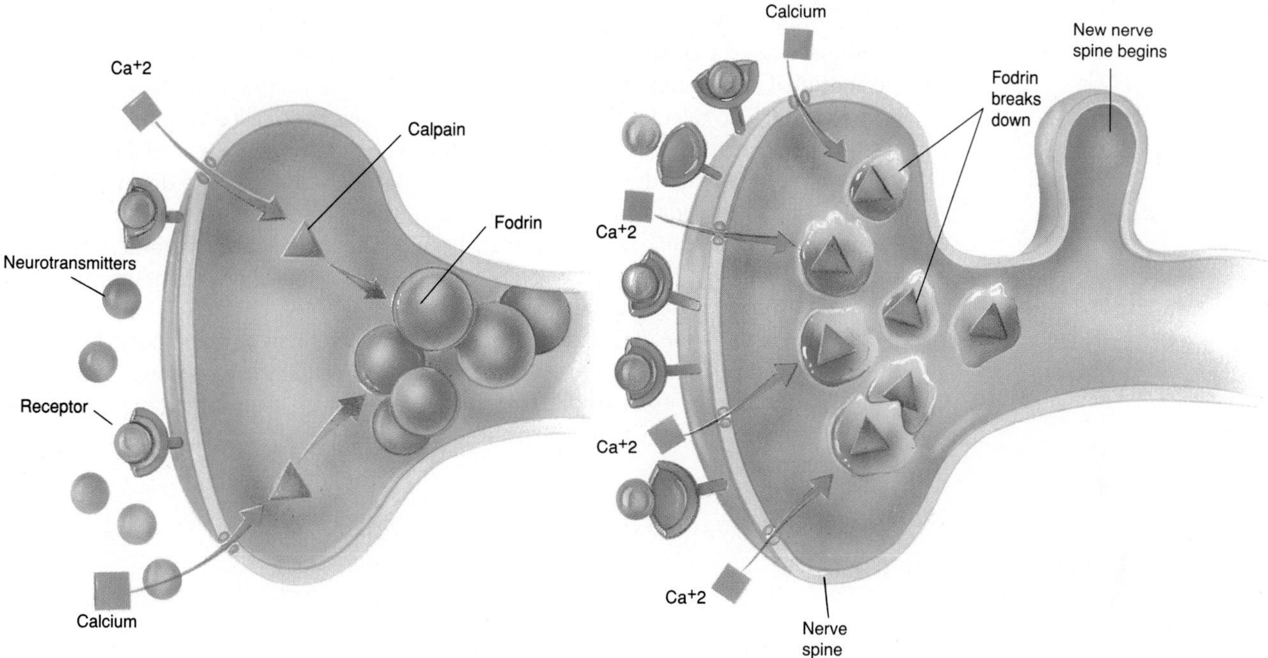

treatment was delayed for two hours, it was not effective, and the mouse did not forget its experience. Mouse memory, then, is labile for up to about an hour and then becomes fixed. We infer that transfer is made from short-term "holding" memory to permanent memory in that time.

The Consolidation Hypothesis Such observations have led to the development of a **consolidation hypothesis,** which states that memory is stored in two ways. First, any registered experience enters a short-term memory system, where its neurological effects last about an hour; then the system returns to normal (one forgets). But while events are still in the short-term system and subject to recall by memory, they are being shifted into the long-term memory system, where they form an engram. The transfer from short-term to long-term is called **consolidation,** and during this process the memory is susceptible to all sorts of disturbances, or even obliteration.

Physiological Differences in Long- and Short-Term Memory Formation

Short-term memory apparently operates on a different physiological basis than does long-term memory. There is no shortage of ideas about how either one works, but the evidence is better for some explanations than for others.

Short-term memory does not form engrams. In other words, there is no physical evidence of short-term memory formation in the brain. It is quickly extinguished, leaving no trace as far as we know. It has been hypothesized that short-term memory is due to temporarily reverberating neural circuits. Such a circuit would produce a feedback loop, so that when an impulse had passed over a circuit, it would be fed back to the receptive area of the circuit (possibly over a "collateral neuron" running the opposite direction), stimulating the memory circuit to fire again. Finally, the stimuli would weaken so that no more impulses could be initiated, and the neural pathway would come to rest, the matter that stimulated it forgotten.

Long-term memory, on the other hand, does leave physical evidence, the engram. This has been clearly evidenced by the formation of new receptors and new growth patterns in neurons as learning (involving memory) takes place. At least some kinds of learning apparently involve an increased likelihood that neural impulses will travel over specific routes. This happens when certain neurons initiate a new direction in growth or when they show an increased sensitivity to signals from the preceding (presynaptic) neuron.

According to one scenario, the sort of learning and memory that takes place with repetition causes a certain neural tract (pathway) to be repeatedly stimulated. Finally, the continual stimulation over that tract causes the postsynaptic neurons to be flooded with calcium. The calcium then activates a dormant enzyme called **calpain,** which now has the ability to break down certain proteins. One such protein, called **fodrin,** lends structural form to the dendrites of neurons. As the calpain breaks down the fodrin (sounds like an episode from *Lord of the Rings*), the reaction exposes new receptors on the postsynaptic neuron, which can then be stimulated by neurotransmitters from the presynaptic cell. With these new receptors exposed, the postsynaptic neuron is much more sensitive to the transmitters (Figure 44.20).

Repeated stimulation of this tract causes the continued breakdown of fodrin by calpain until the protein structure of the postsynaptic neuron actually breaks down, leaving the neuron free to change its shape or to make new connections, thus facilitating the passage of new impulses along the pathway. This new "wiring" of the brain, then, is the physical basis of memory.

APPLICATION OF IDEAS

1. It was once thought that much behavior is either instinctive or learned. We now believe that instinct and learning interact to produce an adaptive behavioral response. What is the evidence?
2. The ethologist's concept of instinct has been said to apply, in its strictest sense, to invertebrates. Why do you suppose this is? Consider the relative influence of cerebral processes in vertebrates and invertebrates.

KEY IDEAS

THE DEVELOPMENT OF BEHAVIOR

At one time researchers believed that behavior was either genetically based or learned. Today researchers ask how behavior develops, admitting both genetic and learned influences.

Genes and Behavior

A number of lines of evidence show the clear effect of genes on behavior.

1. Inbreeding experiments involve mating members of closely related (inbred) lines. As they interact with the same environment, any behavioral differences are considered to be genetically based.
2. Genetic influences can be demonstrated by artificial selection—that is, managed breeding for specific behavioral traits.
3. In hybridization experiments, individuals of closely related strains, but with differing specific behaviors, are bred. If the resulting behaviors are of one parental type or the other, a few genes are believed to be involved. If resulting behavior is a blend of parental types, many genes are believed responsible.

Hormones and Behavior

Hormones can influence behavior, and the effect can be studied in two ways: by gland removal and hormone replacement, and by correlational analysis.

1. In gland removal and hormone replacement, the effects of gland removal are noted. If the effects disappear when the hormone produced by the gland is administered, the gland is considered responsible for the behavior.
2. In correlational studies, the behavior changes proportionally to the level of hormone administered.

The Three Ways Hormones Can Influence Behavior

Hormones can influence behavior by altering perception, development of the nervous system, and effectors.

THE ETHOLOGISTS' CONCEPT OF INSTINCT

The concept of instinct was developed by **ethologists** Lorenz, Tinbergen, and von Frisch. Ethology deals with causation, development, evolution, and function. Three concepts are central to the ethologist's concept of instinct.

Sign Stimuli

Sign stimuli are specific environmental signals that release instinctive behaviors.

Innate Releasing Mechanisms

Innate releasing mechanisms are neural centers that are activated by the perception of sign stimuli.

Fixed Action Patterns

Fixed action patterns are genetically based behaviors that are performed at a signal from the innate releasing mechanism. They are performed independently of environmental influences.

LEARNING

Learning is a process of developing a behavioral response based on experience. **Reinforcement** is the result of an action that increases the probability of the action's being repeated.

Kinds of Learning

1. **Habituation** is learning not to respond to a stimulus (an irrelevant stimulus).
2. **Classical conditioning** involves a behavior that is normally released by one stimulus coming to be released by another stimulus.
3. **Operant conditioning** involves an animal learning to perform an act in order to receive a reward.
4. **Latent learning** occurs in the absence of an immediate reward. Learning occurs, although the reward is delayed.
5. **Insight learning** involves solving problems without resorting to trial and error. The trials take place mentally.
6. **Imprinting**
   Imprinting is a kind of learning that takes place only in a brief period of an animal's early life.

HOW INSTINCT AND LEARNING CAN INTERACT

Most behaviors are based on interactions between instinct (genetically based) and learning (experience-based). The degree to which experience can alter innate patterns depends largely on the species and the specific behavior in question. Most innate patterns can be improved upon by experience. Some of the most intriguing studies of the instinct-learning interaction are on song learning in the white-crowned sparrow.

MEMORY AND LEARNING

Memory is the storage and retrieval of information.

Theories on Information Storage

1. Two separate mechanisms are involved: **long-term memory** and **short-term memory.**
2. Information in long-term memory is slower to decay (forget) and harder to disrupt.
3. Short-term memory must precede long-term memory.
4. The **consolidation hypothesis** states that an effect of experience enters a short-term system. Then one forgets, unless the effect is shifted to long-term memory.

Physiological Differences in Long- and Short-Term Memory Formation

Short-term memory does not form **engrams** and may operate over reverberating neural circuits. Long-term memory does form engrams.

REVIEW QUESTIONS

1. What may be some of the reasons humans are interested in animal behavior? (1086)
2. Do genes code for behavior? State some evidence.
3. What is avoidance learning? (1087)
4. How can one best determine the difference in the effect of a few genes—versus many genes—on a behavior? (1088-1089)
5. What are hygienic bees, and how can they be used to determine how genes affect behavior? (1088)
6. How do crosses between Peach-faced and Fisher's lovebirds carry nest materials? What does this show? (1088-1089)
7. How do hormone replacement studies differ from correlation studies? (1089-1090)
8. When, in a woman's monthly cycle, is visual acuity greatest? What does this show? (1090)
9. Give an example of how hormones can influence preference. (1090-1091)
10. Give an example of how hormones can alter the bird's brain. (1090)
11. What kind of sounds do sexually receptive female clawed frogs produce? (1091)
12. With what four areas of inquiry is ethology concerned? (1092)
13. What are the environmental signals that trigger instinctive actions called? What kinds of neural centers do they stimulate? (1092)
14. Define fixed action patterns and provide two examples. (1092-1093)
15. Define learning. (1093)
16. Define and give an example of reinforcement. (1093-1094)

17. What kind of learning is shown by a bird perched on a familiar scarecrow's shoulder? How is such learning adaptive? (1094)
18. Name and give an example of the kind of learning to which substitute stimuli are integral. (1095)
19. Name and provide an example of the kind of learning where reward follows a stimulus. (1095-1096)
20. What kinds of behavioral sequences are absent in insight learning? (1097)
21. Name and give an example of the kind of learning that involves a brief receptive period, usually very early in life. (1097)
22. Briefly explain the consolidation hypothesis. (1100)
23. Considering the consolidation hypothesis, why might "cramming" be a bad idea? (1100)
24. What happens when calpain breaks down fodrin? (1101)

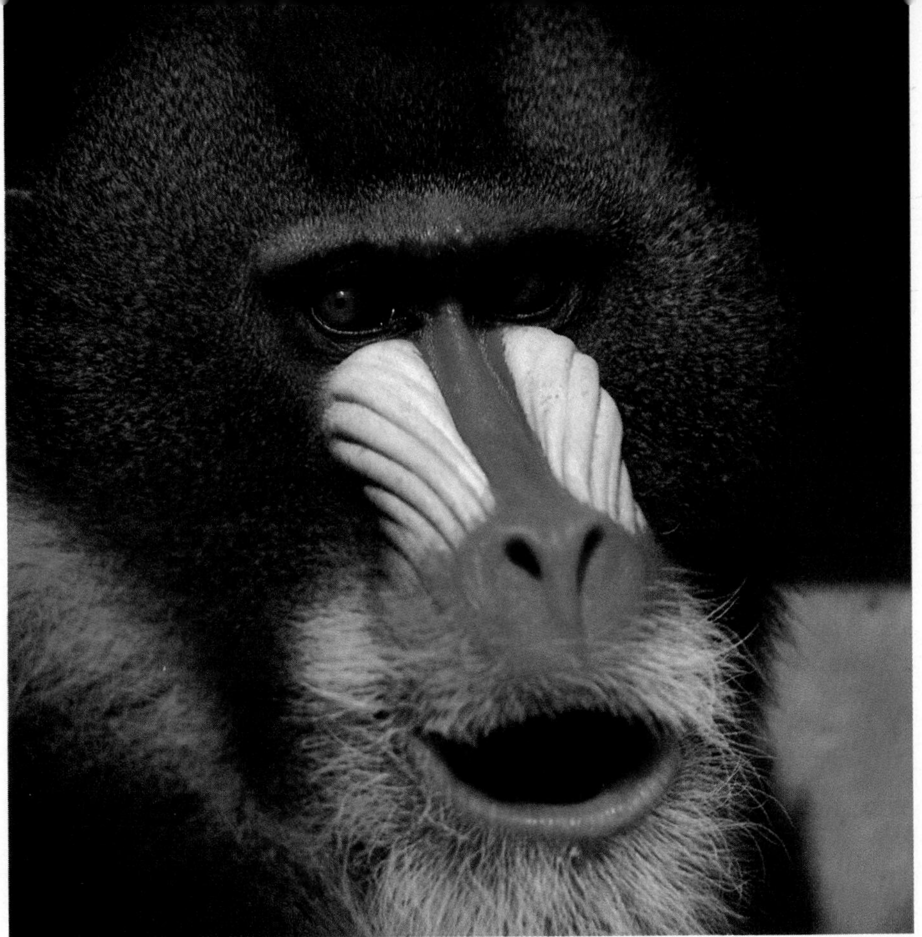

45

ADAPTIVENESS OF BEHAVIOR

ONE OF THE MOST FUNDAMENTAL QUESTIONS IN THE FIELD OF ANIMAL behavior is, why do animals do what they do? Sometimes the answer is obvious. That mouse in your kitchen scurries away when you walk in because it is in its best interest to avoid such large creatures. Tarantulas living in holes in the desert floor avoid drowning by leaving their burrows and climbing into bushes at the first sign of rain. Chickens spend the night on low-lying limbs where foxes can't reach them. We can indeed see the adaptiveness of animals behaving in certain ways.

In other cases the explanations are far less apparent. Why do hammerhead sharks assemble in great numbers and then begin some long migration? How do some woodpecker pairs come to forage close together and others far apart? Why are some species colonial and other territorial? Here, we're getting at causation—the causes of behavior—and we find that questions of causation can be answered at two levels, proximate and ultimate.

PROXIMATE AND ULTIMATE CAUSATION

In the Caribbean, several islands each harbor their own species of woodpecker. The Guadeloupe woodpecker lives on the French resort island of Guadeloupe, the Hispaniolan woodpecker inhabits the island of Hispaniola (Haiti and the Dominican Republic), and the Puerto Rican woodpecker, of course, lives in Puerto Rico. In each of these species the male is larger than the female, and, because of differences in bill size, the members of a pair are able to forage close together, taking largely different foods.

Another species of woodpecker inhabits Jamaica. The Jamaican woodpecker is slightly larger than its Caribbean colleagues, with little sexual difference in body size or bill size (Figure 45.1). They take the same kind of food, and they forage farther apart. The Jamaican woodpecker vocalizes less than the other species, but when it calls, it seems to call more loudly.

FIGURE 45.1
AN ANATOMICAL CORRELATION WITH BEHAVIOR
The sexually monomorphic Jamaican woodpecker has different foraging and social patterns than the sexually dimorphic Puerto Rican woodpecker.

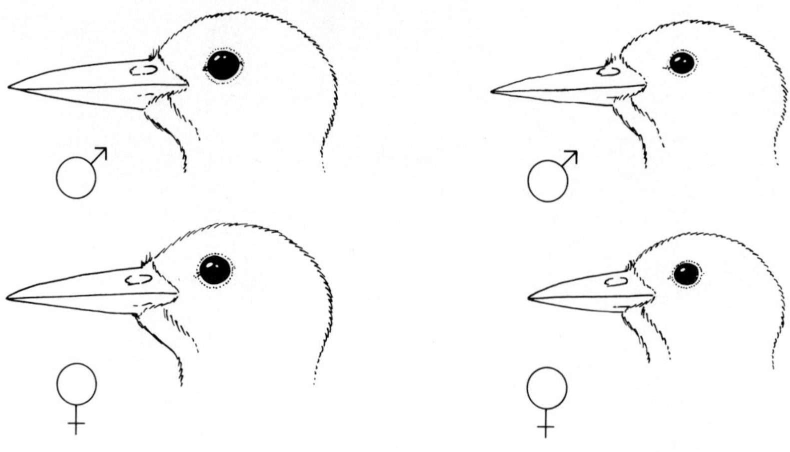

The behavior of the sexually monomorphic ("one form") Jamaican woodpecker and that of the sexually dimorphic ("two forms") species are clearly different. So we may ask about the causes of such differences, and the answers may be formed in two ways.

Proximate causation involves the more immediate bases for a behavior—those usually involving psychological or physiological mechanisms. Proximate questions usually involve *how* the mechanisms of an animal operate to cause it to behave in a certain way. So we might answer the question of how sexually monomorphic species come to forage farther apart by considering whether the sexes have some aversive "feeling" for each other, or if some immediate physiological requirements are fulfilled by the sexes keeping a certain distance. What we would be getting at is just what sorts of internal processes are going on within the body of an animal that result in the performance of a certain behavior.

Ultimate causation involves the evolutionary and adaptive bases for a behavior. Ultimate questions involve *why* an animal behaves in a certain way. In other words, ultimate questions are about why the proximate mechanisms evolved to begin with. Thus, if we ask why sexually monomorphic and sexually dimorphic species forage differently, we can answer in ultimate terms. Perhaps predatory pressures are different on the two types of animals. Maybe sexually monomorphic forms evolved a weaker aggregation tendency because if they foraged closer together, they would be more likely to attract the attention of hawks in the area. Or perhaps by both sexes evolving larger bills, they could each take a larger range of food sizes and thereby effectively increase their kinds of food sources. The behavior, then, could have been the result of predation or historical food pressures—both ultimate factors.

BEHAVIORAL ECOLOGY

Behavioral ecology is the study of how the environment affects behavior. Remember, the environment can be viewed in the historical sense (leading to ultimate causation) or the immediate sense (leading to proximate causation). Behavioral ecology can involve studies of the physical habitat, interactions with other individuals, or less conspicuous influences on behavior, such as weather, oxygen levels, and humidity—any environmental variable that can influence behavior. As examples, we will consider the behavioral ecology of how animals choose a habitat and how they get food.

Habitat Selection

One of the questions regarding habitat selection is, does an animal choose its habitat based on genetic propensities, or through learning which kind of place is best to live in? Stanley Wecker conducted a series of fascinating experiments on prairie deer mice to see to what degree their habitat preferences were genetically influenced. In his ex-

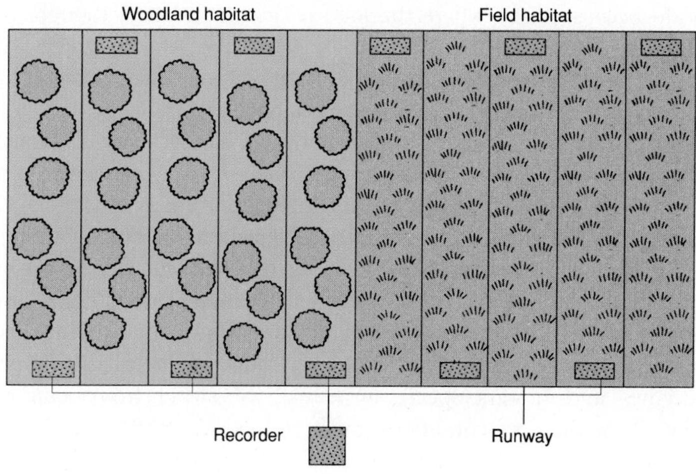

FIGURE 45.2
WECKER'S EXPERIMENTAL SETUP
Mice with different genetic makeups and different experiences were tested to see which kind of habitat they preferred. They were allowed to enter either simulated grassland or forest conditions from a runway. The results are described in the text.

periments, he used two kinds of mice—one of which lives in woodland, the other in grassland. Wecker had found that, in the laboratory, the grassland mice *could* do quite well under simulated forest conditions. He wondered, then, why do they prefer grasslands in the wild? In one of his experiments he tested two groups of grassland mice, one reared in the laboratory. The laboratory group was further subdivided into two groups—one of which had been reared under simulated field conditions, the other under simulated forest conditions.

Not surprisingly, Wecker found that the wild-caught field mice preferred the field-end of the enclosure (Figure 45.2). The "field"-reared laboratory mice made the same choice. However, the "forest"-reared laboratory mice showed no preference whatsoever. Subjecting them to the forest had obliterated any tendency to move toward the field but had not caused them to prefer the forest. Wecker's experiment suggested, then, that habitat preference in prairie deer mice, at least, has a strong genetic component that can be influenced by experience (Table 45.1).

We can also imagine that the field mice would develop certain preferences, while in the field, based primarily on experience. For example, it wouldn't take long to learn not to climb around in thorns; certain kinds of grasses taste bad, while others are quite palatable; and sleeping with ants doesn't work. By experience, then, the mouse would fine-tune its behavior through proximate factors. The principle here might be phrased as: ultimate factors help to set limits, and proximate factors influence the behavior of the animal within those limits.

Foraging Behavior

When asked why he robbed banks, the infamous Willie Sutton is reputed to have replied, "Because that's where the money is." For years, biologists seemed to have tacitly assumed that the same reasoning applies to where animals forage, or seek for food: They forage

TABLE 45.1 THE RESULTS OF WECKER'S EXPERIMENT

NUMBER OF MICE TESTED	HEREDITARY BACKGROUND	EARLY EXPERIENCE	HABITAT PREFERENCE
12	Grassland	Grassland	Grassland
13	Laboratory	Grassland	Grassland
12	Grassland	Laboratory	Grassland
7	Grassland	Forest	Grassland
13	Laboratory	Laboratory	None
9	Laboratory	Forest	None

Source: Data from S.C. Wecker, "The Role of Early Experience in Habitat Selection by the Prairie Deer Mouse, *Perimaniculatus bairdi*," *Ecological Monographs* 33:307–325.

FIGURE 45.3
THE CROW, A FEEDING GENERALIST

where they do because that's where the food is. In recent years, though, that answer has been deemed inadequate. Behavioral ecologists have, in fact, asked why animals forage where they do. (Of course, this is largely an aspect of habitat choice.) Furthermore, they have asked how animals make food choices once there.

Of course, different species forage in different ways. The primary division of foraging styles exists between generalists and specialists. **Generalists** are those species with a broad range of acceptable food items. They are often opportunists and will take advantage of whatever is available, with certain preferences depending on the situation. Crows are an example of feeding generalists (Figure 45.3); they will eat anything from corn to carrion. **Specialists** are those with narrow ranges of acceptable food items. Some species are extremely specialized, such as the Everglade kite (Figure 45.4), which feeds almost exclusively on freshwater snails. There is a wide range of intermediate types between the two extremes, and in some species an animal will switch from being one type to being another depending on conditions, such as food availability or the demands of offspring.

Whatever the strategy, the question is, how does an animal forage most efficiently? More precisely, how does it maximize its gain relative to its expenditure? The gain is measured quite simply as the food value of the item. Expenditure is measured two ways—by the energy the animal spends in searching for food and the energy it expends in handling the food once it has been found.

Obviously, the animal will maximize its foraging success by spending the most time where the most food is. But is this what they do? J.N.M. Smith and H.P.A. Sweatman examined the feeding behavior of the great tit (see Figure 45.5) by setting up a series of grids in an aviary. Each grid contained different densities of mealworms, a favorite food. The food density in each grid could be altered by experimenters. When a certain grid continued to hold the greatest food density for a time, the researchers found that the tits were soon spending almost all their time on that grid. However, the birds continued to hop over and sample from the other grids. The researchers wondered why the birds would waste time in these suboptimal sites, but the answers soon became clear. When the food density was suddenly reduced on the best grid, the birds immediately switched to the second-best grid. They spent little time in making a decision when their primary food source failed. Sampling all the grids, even after the best site had been determined, was adaptive after all.

Another question is, how do animals maximize their foraging efficiency when they are given a choice of food items with different food values? For example, all things being equal, larger food items are more profitable than smaller ones. Thus, the animal will be more successful if it takes larger items. Bluegill sunfish, for example, feed largely on *Daphnia,* the small crustaceans known as water fleas. Researchers found that sunfish will bypass very small water fleas even if they are nearby, in favor of large water fleas further away. In other words, they behave as if the extra travel were worth the extra gain.

The sunfish, of course, don't have to wrestle with the *Daphnia* once they find it, but striped bass may have to consider ease of handling. The bass feed on a variety of foods, including small fish called "shiners" and crayfish. The shiners are taken with a gulp, usually after a short chase, so after the bass catch the shiner, handling it is minimal. The crayfish, on the other hand, is easier to catch but harder to handle. Not only do the pugnacious creatures fight back, requiring careful manipulation by the fish, but much of its body is indigestible chitin, which further reduces its value. So all things being equal, the bass is better off chasing the shiner.

Why, then, are crayfish taken at all? A predator's decision is often based on food availability. It will tend to take the items with the greatest net gain until their numbers are depleted to the degree that less desirable, but more numerous, food items become more attractive. Then it will generally begin to switch to the less desirable items—in which the food value may not be as high, but less time and energy are spent in searching.

Foraging, for most species, then, is a cost-benefit proposition. They tend to maximize the benefit while minimizing the cost to themselves. Since such studies have been undertaken, researchers have often been amazed at what appear to be analytical abilities

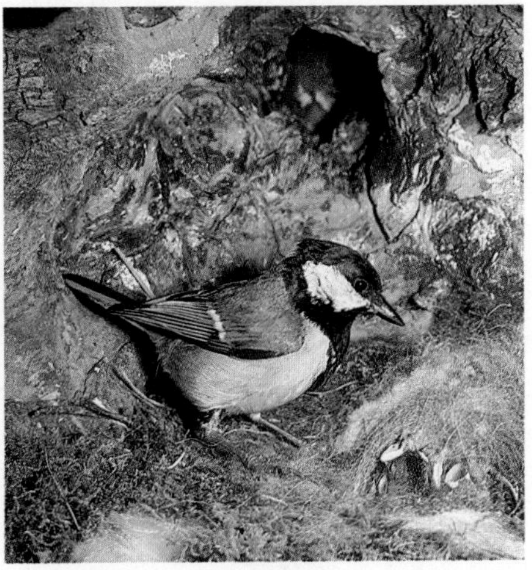

FIGURE 45.4
THE EVERGLADE KITE

FIGURE 45.5
PARUS MAJOR, THE GREAT TIT

of foragers. Of course, any such ability is largely programmed genetically, but learning may also play an important role. For example, young fish are often fairly good optimal foragers, but with age and experience their efficiency improves.

We will now turn our attention to two phenomena that are critical to the adaptation of life on earth. They have to do with how animals adjust with respect to time and place.

BIOLOGICAL CLOCKS

You may have noticed that the small feline carnivore sharing your home becomes restless each evening. The eyes of the cat are dark-adapted, its claws are sheathed in silent pads, its hearing is remarkable, and it has a natural tendency to sneak around. Indeed, if you weren't feeding it, it would have its greatest success in stalking its prey at night. But with the lights on and the TV blaring in your apartment, how does your cat know when it is night outside? After all, you're probably a student (since few people would be reading this for pleasure), and like students everywhere, you probably keep odd hours. You may burn the midnight oil, eat at 3:00 A.M., and turn off the lights and leave at dawn. You might think that such an odd schedule would throw your cat's schedule off, too, but it doesn't. The cat knows when it's night—when to get up and stir around in preparation for the hunt that never happens.

How does the cat "know" when night falls outside its brightly-lit home? Or better yet, how do plants "know" when it is nighttime (see Figure 29.15, p. 651)? (Or, at least, how do they come to behave as if they did?) Although such rhythms were observed some 300 years before Aristotle, we still don't have the answers to some of the most fundamental questions about the timing of living things.

The Adaptiveness of Rhythms

In a cyclic and rhythmic world such as ours, we should expect life to adapt by taking on its own cyclicity to help avail itself of the offerings of each of the earth's phases. One of the longest and most pronounced cycles is the earth's annual rhythm, brought about by the earth's tilt on its axis as it circles the sun in an elliptical orbit. This annual rhythm is expressed as the seasons. In many parts of the world, the seasons are marked by drastic differences. Thus, an animal should begin to lay on fat for the lean, wintry season even while days are long and food is plentiful. The animal's body begins to "tell" the animal to accelerate its feeding schedule, one can say, because that body is the result of generations of ancestors who were subjected to seasonal cyclicity. In time, natural selection would have produced generations whose physiology anticipated the coming winter and who therefore "laid by in store."

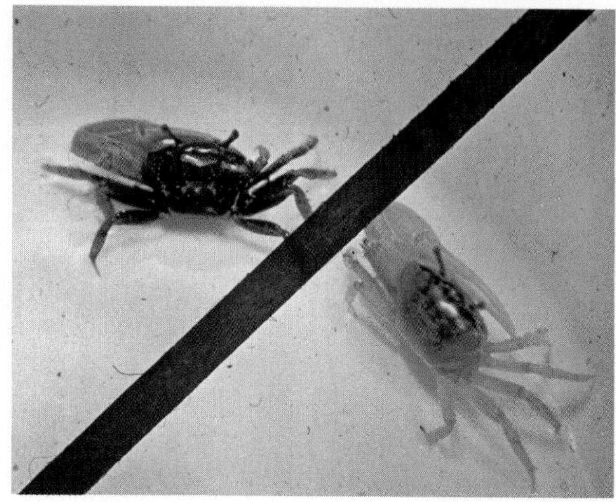

FIGURE 45.6
INTERTIDAL CRABS
They may correlate their physiological and behavioral phases to both diurnal and lunar cycles.

Another cycle to which we earthlings are subjected is that of the moon. Intertidal crabs, for example, may change color on a daily schedule (correlated with sunlight) as a camouflage (Figure 45.6). At the same time, their two daily "running periods" (times of greatest activity) are on a lunar schedule correlated with the rise and fall of the tide. The moon's cycle is about 24 hours, 50 minutes, so the crab's running periods are 50 minutes later each day.

The most obvious of the earth's rhythms is the daily, or **circadian cycle** (*circa,* "about"; *diem,* "day"). The cat starts to prowl at night, about the time that most of us go to sleep. For whatever reason (the evolutionary basis of sleep is unknown), we must lapse into unconsciousness. Since that is the case, we might best do so when we have least use for vision and when we are most difficult for other animals to see. Some of our other daily cycles are less easily explained. For example, many people are drowsy and least effective in early afternoon. This "postprandial" drowsiness is not related to eating lunch.

That circadian rhythms arise internally has been demonstrated by placing organisms under constant conditions of light, temperature, humidity, barometric pressure, and so on. Under such conditions, the organisms will show rhythmic behavior on a roughly 24-hour basis. However, its cycle may drift to 23 or 25 hours. The clock can be reset by exposure to some environmental cue, usually light, but varying from species to species.

Many living things can be "clock-shifted." Clock shifting involves allowing the animal to continue its circadian rhythm, but setting the rhythm ahead or behind of where it would normally be. Most organisms can be clock-shifted somewhat, but beyond a certain point some will not respond and will return to their normal cycling. Clock shifting is done by altering the onset of environmental cues normally associated with some phase of the day. For example, if a light is turned on a little earlier and turned off a little earlier each day in a cage in which an animal is kept, the animal will set its circadian clock according to the onset of the light. If the light is finally ahead of its usual period, the animal will come to be hours ahead of its normal cycle. Clock shifting is important in certain experiments in animal navigation, as we will see shortly.

The Range of Rhythms

Not all biological rhythms are so blatant as the circadian cycles of animals. There are far more subtle clocks measuring time. For example, cell division seems to occur in rather rhythmic spurts (peaking at noon), even in artificial laboratory cultures. If the heart of a hamster is removed and placed in a nutrient medium, it will continue to beat for days, its pace picking up at night, the hamster's normal running period. Even single, excised neurons will exhibit a daily rhythm of firing. Remarkably, sensitivity to drugs varies on a daily cycle. Physicians are beginning to learn about this cyclicity, perhaps spurred on by the finding that certain drugs can be beneficial at one time of day and kill the patient if administered a few hours later. As a final note, most of us were born in the wee hours of the morning, and most of us will breathe our last in these same early hours.

The physiological basis of the clocks remains largely a mystery. Part of the uncertainty arises from contradictory findings. For example, the nucleus of unicellular organisms, such as the green alga *Acetabularia,* apparently controls the cell's timing, but if the nucleus is removed, the cytoplasm can maintain the rhythm on its own. Further confusing the issue, it appears that clocks function through neural mechanisms in some organisms and hormonal mechanisms in others. Further, both neural and hormonal mechanisms may operate within the same individual.

NAVIGATION

On a variable and changing planet such as this, proper positioning can be critical. Animals do, in fact, show a variety of abilities to position themselves and to find their way from one place to another. For example, it is best for certain aquatic insects to move to the protective, darkened bottom as larvae, but as they change to adults, to seek the brightly lit surface that signals a greater abundance of oxygen. Not only do many shorebirds stand so as to face cold winds that would otherwise ruffle their insulating feathers, but some of them also leave those cold winds to travel to warmer climes each fall.

There are undoubtedly people in South America who wonder where their robins go every spring. Those robins, of course, are "our" birds; they merely vacation in the south each winter. If you've been keeping a careful eye on the robins in your neighborhood, you may have noticed that a specific tree is often inhabited by the same bird or pair of birds each year. You may have wondered, since these birds winter thousands of miles away, how they find their way back to that very tree each spring. Such questions have perplexed researchers for years. And the more we learn, it seems, the more strange the story becomes.

Most researchers now believe that there are two primary mechanisms by which birds find their way around. One is a **compass sense,** which enables the animal to know direction. Some birds couple this with an internal clock to form a **time-direction mechanism** with which the bird flies for a certain time in a certain direction. Certain European garden warblers, for example, are born with the ability to reach their wintering grounds in West Africa by flying southwest for a certain time (which takes them to the area of Gibraltar) and then turning and flying southeast for a certain period. Such a mechanism, of course, would be effective for getting the bird into a general area, but not for finding more precise locations.

The second strategy birds apparently use to find their way around is a **map sense.** The map sense is one of the most fascinating and mysterious abilities of animals. An animal with a map sense seems to know its precise longitude and latitude no matter where it is. We will now take a look at two aspects of spatial organization, orientation and navigation, both of which are important to animals on the move.

Orientation is the ability to face in the right direction. **Navigation,** on the other hand, involves starting at point A and finding the way to point B. Much of the work with orientation of birds was triggered in the 1950s when a young German scientist, Gustav Kramer, devised some intriguing experiments on birds and came up with some startling conclusions.

Kramer found that caged migratory birds became very restless at about the time they would normally have begun migration in the wild. Furthermore, he noticed that as they fluttered around in the cage, they usually launched themselves in the direction of their normal migratory route. Kramer devised experiments with caged starlings and found that their orientation was, in fact, in the proper migratory direction—except when the sky was overcast. When they couldn't see the sun, there was no clear direction to their restless movements. Kramer surmised, therefore, that they were orienting by the sun. To test this idea, he blocked their view of the sun and used mirrors to change its apparent position. He found that under these circumstances the birds oriented with respect to the position of the new "suns" (Figure 45.7).

This sort of sun-compass orientation requires that starlings know both the time of day and the normal course of the sun. In other words, incredible as it seems, they apparently know where the sun is supposed to be in the sky at any given time. It seems that this ability is largely innate. In one experiment, a starling reared without ever having seen the sun was able to orient itself fairly well, although not as well as birds that had been exposed to the sun earlier in their lives. (This would appear to be another example of the interaction of innate behavior, maturation, and learning.) Evidence for sun-compass orientation has also been found in a variety of other animals, including insects, fish, and reptiles.

The sun compass can't be the whole answer, of course. Possession of a compass may be necessary for navigation, but it isn't sufficient. Suppose you were blindfolded, driven out into the desert, and released with only a compass. Finding north would be easy; finding Bakersfield wouldn't, even if you had a compass and a map. And, as you sat on a rock, distraught and staring hopelessly at your devices, birds might be passing overhead on their way to a certain tree in Argentina.

As night drew on and you hadn't budged from your rock, other bird species might pass overhead. As you strained to see their dim shapes through teary eyes, you might wonder: How do *they* navigate? Since it is night, they would obviously not be using the sun. Apparently, some birds navigate the way sailors once did—by the stars. In exper-

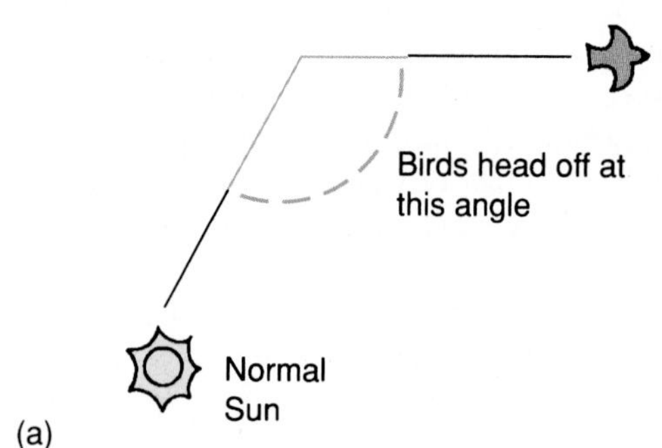

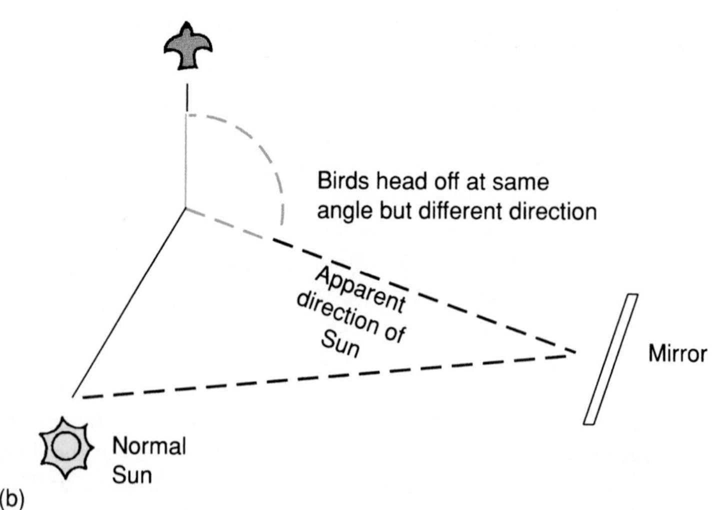

FIGURE 45.7
KRAMER'S SUN-COMPASS ORIENTATION EXPERIMENT
(a) As the sun moves, caged starlings alter the direction of their attempted migratory movement with respect to it, thereby maintaining a constant heading. **(b)** Using mirrors, Kramer altered the apparent position of the sun. The birds shifted their migratory direction by the same angle.

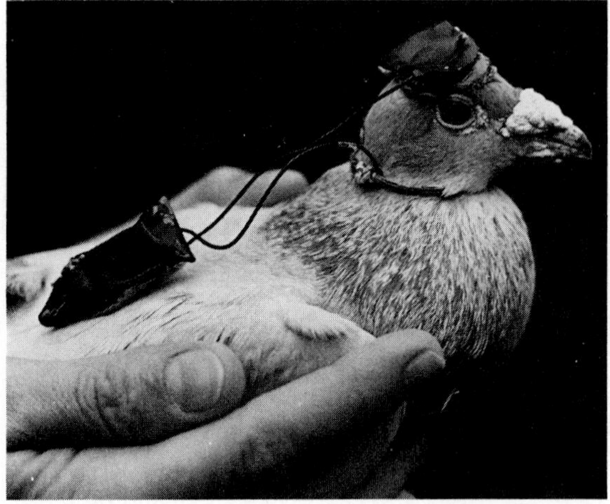

FIGURE 45.8
A FAMOUS HOMING PIGEON
Two hundred fifty American soldiers in World War I owed their lives to this homing pigeon, which carried a distress message to its handler. Reinforcements were sent to rescue the beleaguered battalion.

FIGURE 45.9
HOMING EXPERIMENT
Pigeons carrying coils with reversed polarity on the head tended to home in opposite directions, indicating that pigeons can detect slight magnetic fields and that magnetism can influence their orientation.

iments somewhat analogous to the sun-compass work, night-migrating birds were brought in cages into a planetarium. A planetarium is essentially a theater with a domelike ceiling onto which simulated night skies can be projected. Sure enough, the fly-by-night birds oriented in their cages according to the sky that was projected on the ceiling, even if it was different from the sky outside. In one experiment, birds that normally migrate north and south between western Europe and Africa were shown a simulated sky as it would appear that moment over Siberia. The birds, behaving as though they were thousands of miles off course, oriented toward the "west."

Homing Pigeons

The abilities of homing pigeons have fascinated people for years (Figure 45.8). (All pigeons are technically homing pigeons.) A homing pigeon can be put into a dark box and transported to a distant location by a circuitous route. When it is released, there's a good chance it will soon show up in its home loft. Some make it quickly, others never turn up—there is a great deal of variation in abilities.

Studies with homing pigeons have proven inconclusive but instructive. Researchers have asked a wide range of questions. Do pigeons use a sun compass? If so, how do they home successfully on cloudy days? Do they use landmarks? Pigeons fitted with frosted contact lenses so that they cannot make out landmarks find their way back to the vicinity of their home loft, where they may flutter about directly overhead, or sit on the ground near the loft, unable to see it. Do pigeons use a magnetic compass? Small electromagnets have been attached to the heads of homing pigeons (Figure 45.9). Birds carrying magnets with reversed polarity tended to fly in opposite directions.

Evidently, then, pigeons use a variety of clues, including the sun compass, magnetic force fields, and visual details. Researchers have been foiled in trying to find *the* mechanism. The pigeon apparently has backup systems.

Navigation, as we've noted, requires map sense as well as a compass sense. And pigeons apparently have both. Homing pigeons have further confused researchers by the fact that when they are navigating by the sun, their compass is time-dependent (that is, their directional sense can be changed by artificially altering their clocks). If they are navigating on cloudy days, however, clock-changes do not upset their directional sense.

It should be pointed out that the study of animal navigation and orientation is an extremely vigorous field at present, and new data are appearing almost daily. Any synthesis at this point would be extremely tenuous. However, these examples provide some idea of the value of innate characteristics in orientation, as well as the surprising sensitivity of some species to environmental cues.

FIGURE 45.10a
MIGRATING CARIBOU
On the move to timbered areas farther south.

FIGURE 45.10b
MONARCH BUTTERFLIES OVERWINTERING IN CENTRAL MEXICO

Adaptiveness and Evolution of Migration

Migration involves such costs and risks that one may well wonder how it arose at all. Why would an animal endure such hardships just to be somewhere else? That "somewhere else" must hold great advantages. In some cases the advantages are clear. The great caribou herds feed in summer on the plentiful grasses and lichens on the northern Arctic. In late summer, though, they begin a relentless trek southward to the relative haven of the timbered areas (Figure 45.10). In this way, they extend their food reserves while wintering in a sheltered area.

In other cases, the adaptiveness and evolutionary history of a migratory pattern may not be so apparent. Lincoln Brower has, over years of intensive work, unravelled the story of migration in monarch butterflies. He found that they overwinter in enormous gatherings in a few places, particularly in the highlands of central Mexico where masses hang like great living tapestries from only a few specific trees in about ten separate sheltered valleys (Figure 45.10b). The butterflies are here only to overwinter; they make little demand on the scarce resources of the area. The following spring the great masses break up and the butterflies begin to move northward to take advantage of the abundant, if temporary, milkweed food supply along the Gulf coast, from Texas to Florida. By May they will have laid all of their eggs and played out their energy reserves. The next generation, appearing a short time later, continues on to the Northern United States and Canada. These individuals and their offspring, in turn, lay eggs there and the late August-early September generation begins its long trek back to the sheltered valleys high in the Transvolcanic mountains of Central Mexico.

Brower has determined that this multigenerational migration is a response to the short life span of the monarchs and a spatially variable food supply that proceeds northward from the Gulf coast as the seasons progress. When food is no longer available, the butterflies retreat to their valley where they live off of their reserves until the following spring. The perplexing problem of the adaptiveness and the evolution of monarch migration has now largely been solved. (Just how they navigate, though, remains a mystery.)

There has been a great deal of speculation on the advantages of migration in birds. Some of the adaptive bases are apparent, some are not. The most obvious advantage for example, in a robin migrating southward in the fall, is to escape the severe winter in the north. Not only is it cold up there, but the food supply dwindles in the winter, throwing overwintering species into stronger competition with their neighbors. Some species of food specialists may find that their entire food supply has vanished. So many species escape, not only to warmer climes, but to a better food supply.

If there is so much food in the warmer winter habitats, why don't migratory species just stay there? Why do they return to their summer homes at all? There are certain important advantages to rearing broods in temperate or even polar areas in the spring. For example, days of the northern summer are long, and so the birds' working days can be extended; they can bring more food to their offspring in a given season. Food productivity is high in polar regions in summer because of increased sunlight for photosynthesis, optimal temperatures for photosynthesis, and nutrient-rich upswellings in polar oceans. And since, effectively, more food is available, more young can be raised. It is known that, generally, the farther north from the tropics a species breeds, the larger is its brood. And, of course, with more food, any brood can be raised faster.

Another advantage in returning to the temperate zone to breed is in escaping the high level of competition that exists in the species-packed tropics. The annual flush of life in the temperate zones provides a predictable and otherwise underutilized supply of food that can be exploited readily by mobile species such as birds.

In the polar region, the breeding period is very short, but the brevity of the season may actually work to the advantage of nesting birds that are in danger of falling prey to predators. The short season means that a great number of birds must nest simultaneously; thus, the likelihood of any single individual being taken by a predator is reduced. Also, since the birds come and go from both the summer and winter areas, their predators are denied a stable food supply. By leaving certain geographical areas each year, migratory species deprive many parasites and microorganisms of permanent hosts to which they can closely adapt. In addition, the long harsh winters in the frozen north reduce the numbers of parasites in that area. By the same token, predators that are unable to escape the rigorous northern climes might also be expected to be fewer in number the following spring.

There are three major theories regarding how migration may have evolved. According to one theory, birds evolved in the northern latitudes when the weather was gentler, more benign. Later, as the ice ages advanced, winters would have become increasingly severe, forcing the birds further south. Those evolutionary lines that changed physiologically at a certain time of the year and then were prompted to move south before being caught in the deadly winter weather would have tended to survive. Their inclination to fly northward in the spring is considered an attempt to return to the ancestral breeding ground.

The second theory holds that birds evolved closer to the equator, and as their numbers grew, the increased competition there forced them northward, only to be driven to their ancestral homeland by the severe northern winters.

The third theory is based on continental drift (see Essay 20.1). According to this idea, birds, after originating in the balmy climes of Gondwanaland, followed the separating land masses, exploiting their seasonal offerings, only to return to the southerly areas as each winter drew on.

Let's now turn to another vigorous area in the study of animal behavior, one that also has stimulated a great deal of first-class detective work, but one that is more generally adaptive across a wide range of animals. This area is loosely referred to as communication.

COMMUNICATION

Communication, in its broadest sense, is an action by one animal that influences another animal. Whereas communication was once regarded as a means by which one animal let another animal know what it was going to do, we now see that communication is a "reproductive enabling device" by which one animal manipulates those around it and so enhances its own survival and reproductive output. (As we will see, animals are fully capable of lying to each other.)

Communication may be directly involved with reproductive success as a component of mating behavior or precopulatory displays. It may also indirectly increase reproductive output by helping the offspring avoid danger when parents give warning cries, or by simply helping the reproducing animal live better or longer so that it may successfully mate again. Remember, the charge to all living things is "reproduce, or your genes will be lost." Communication helps animals to carry out that reproductive imperative.

Let's consider a few general methods of communication, and in so doing perhaps we can learn something about the ways in which animals influence the behavior (or the probability of behavior) of other animals. That, after all, is what communication is all about.

Visual Communication

Visual communication is particularly important among certain fish, lizards, birds, and insects—and among some primates as well. Visual messages may be communicated by a variety of means, such as color, posture or shape, or movement and its timing. As was mentioned in Chapter 44, the color red can release territorial behavior in European robins. The female quickly solicits the attention of the male bird by assuming a head-up posture while fluttering her drooping wings. Some female butterflies attract male butterflies by the way they fly. As an example of communication by timing, fireflies are attracted to each other on the basis of their flash intervals, each species having its own frequency. (One predatory species "taps" communication lines by flashing at another species' frequency and then eating whoever comes to call. See Essay 45.1 for other examples of deception.)

ESSAY 45.1

DECEPTION IN THE ANIMAL WORLD

It was once thought that animal communication is essentially a way for one animal to let another know what it is about to do. Now, however, we are aware that this notion is a bit naive. Communication is more realistically thought of as a kind of "enabling device," it enables the communicator to increase its likelihood of success at survival and reproduction. In other words, communication can enable an animal to manipulate its environment more effectively. In some cases, this manipulation involves deception.

For example, a fish may develop an "eyespot" near its tail while concealing its own eye with a stripe; an animal seeking to head it off moves in one direction, while the fish moves off in another. A variety of dangerous species take on common warning coloration such as yellow and black patterns, as we see in the wasps of the genera *Vespula* and *Bembex*. (This kind of convergent evolution is called *Müllerian mimicry*.) The harmless banded king snake mimics the red, yellow, and black stripes of the dangerous coral snake (an

Vespula

Bembex

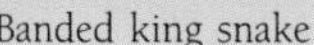

Banded king snake

Coral snake

Because visual signals carry high information content, subtle variations in the message can be conveyed by gradations in the display (Figure 45.11). That is, the display can be given at different levels of intensity. (The same is true for some other means of communication, as we will see.) Of course, **graded displays** are useful only to the species that are sensitive and intelligent enough to be able to recognize such subtleties. Another advantage of the high information load in visual signals is that the same message may be conveyed by more than one means, such as when an aggressive chimpanzee both stares and bares its teeth. Such redundancy may be used as either to modify the message or to emphasize it in order to reduce the chance of error in interpretation.

A visual signal may be a permanent part of the animal, as in the elaborate coloration of the male pheasant and the striking facial marking of the male mandrill (Figure 45.12). These animals advertise their maleness at all times and are continually responded to as males by members of their own species. On appropriate occasions they may emphasize their "machismo" through behavior such as the strutting of the pheasant and the glare of the mandrill. In other cases, a visual sign may be of a more temporary nature, such as the reddened rump areas in female chimpanzees and baboons during estrus. As a more short-term signal, a male baboons may expose his long canine teeth as a threat (Figure 45.13).

Short-term visual signals have the advantage that they can be started or stopped immediately. If a displaying bird suddenly spots a hawk, it can freeze, and its former example of *Batesian mimicry*) and is also avoided by predators.

The orchid *Cryptostylis* emits a scent like that of the female ichneumon wasp. The male wasps thus pollinate the orchids by *pseudocopulation* (shown here). In another example, the *Ophrys* orchids resemble the female black wasp, and they are pollinated as male wasps try to copulate with first one, then another.

The coloration and marking of some animals suggests "I am not here." The leopard, for example, easily blends into its grassy surroundings. Others essentially say "I am something else." A praying mantis captures visiting insects as it appears to be part of a flower.

In some cases, animals deceive others about their conditions. Some birds, such as the common killdeer, feign injury and flutter along the ground in the "broken wing display," drawing predators away from their nesting areas. A great deal of communication, we see, involves living things deceiving each other in a variety of fascinating, unexpected, and bizarre ways.

Ichneumon (male) and *Cryptostylis*

Leopard

Praying mantis

Killdeer

position won't be given away by any lingering images. Also, the recipient of a visual message is usually notified of the exact location of the sender. The recipient can then respond in terms of the sender's precise location, as well as its general presence and behavioral state (aggressive, romantic, or whatever).

Visual signs also have certain disadvantages. For example, the sender must be seen, and all sorts of things can block vision, from mountains and trees to fog and big hats. Visual signals are generally useless at night or in dark places (except for light-producing species). Since visibility weakens with distance, such signals are useless at long distances. Further, as distance increases, the signal must become bolder and simpler; hence it must carry less information.

Sound Communication

Sound plays such an important part in communication in our own species that it may surprise you to learn that it is limited, for the most part, to arthropods and vertebrates. If you are familiar with the songs of the cricket and cicada, you are probably aware that insect sounds are usually produced by some sort of friction, such as rubbing the wings together or rubbing the legs against the wings. The key aspect of these sounds is cadence (timing), whereas with most birds and mammals, pitch or tone is more important.

Vertebrates use a number of different forms of sound communication. Fish may

FIGURE 45.11
GRADED DISPLAY
At right, the bird *Fringilla coelebs* is showing a high-intensity threat. The posture at center is a medium-intensity threat, and that below is a low-intensity threat.

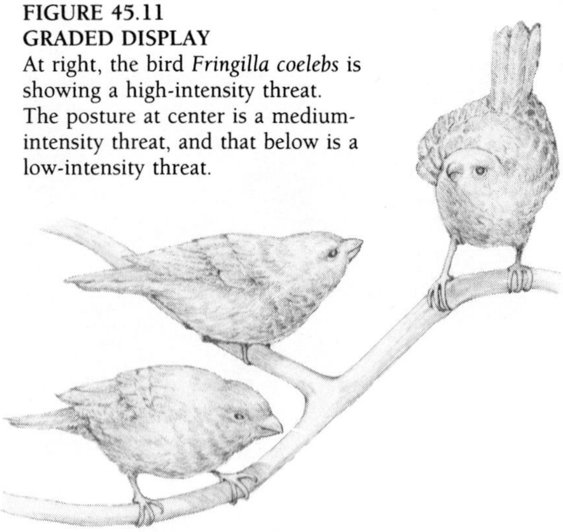

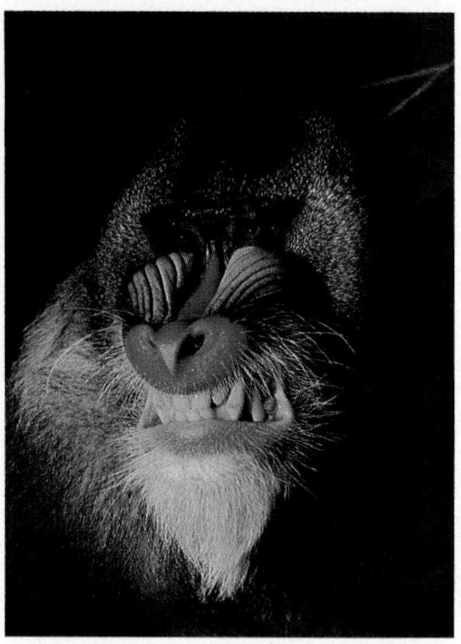

FIGURE 45.12
PERMANENT VERSUS TEMPORARY DISPLAY
A male mandrill has permanent markings that advertise his sex, but his signals of anger are temporary and depend on his mood. It is probably often to his advantage to be perceived as a male, but less frequently advantageous to be perceived as an angry male.

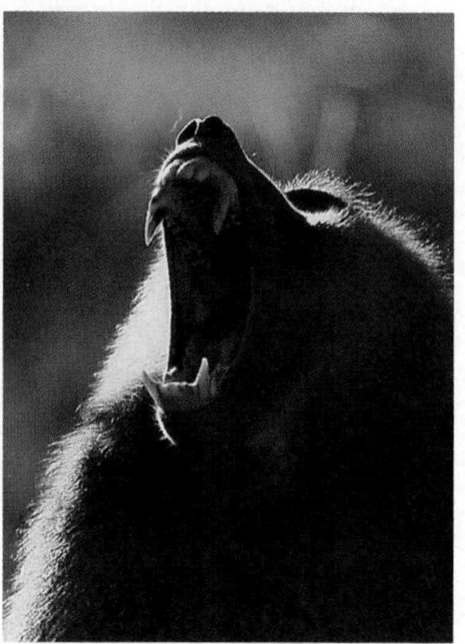

FIGURE 45.13

A male baboon displays his large canine teeth as a threat signal. These animals are powerful and dangerous.

produce sound by means of frictional devices in the head area or by manipulation of the air bladder. Land vertebrates, on the other hand, usually produce sounds by forcing air through vibrating membranes in the respiratory tract. They use sound communication in other ways as well: Rabbits thump the ground, gorillas pound their chests, and woodpeckers hammer hollow trees and drainpipes early on Sunday mornings.

Most of the lower vertebrates rely on signals other than sound, but some species of salamanders can squeak and whistle, and there is even one that barks. Frogs and toads advertise territoriality and choose mates at least partly by sound. Few reptiles communicate by sound, but large territorial bull alligators can be heard roaring in more remote areas of the swamps in the southern United States. Darwin described the roaring and bellowing of mating tortoises when he visited the Galapagos Islands.

Sound may vary in pitch (low and high), in volume, and in tonal quality. The last is apparent as two people hum the same note, yet their voices remain distinguishable. It is possible to show the characteristics of sound graphically by use of a sound spectrogram (facing page). In effect, this is a translation of sound into markings, which makes possible a more precise analysis of the sound.

The function of the message may dictate the characteristics of the sound. For example, Figure 45.14 shows a sound spectrogram of mobbing cries of several bird species. Such calls include brief, low-pitched *chuk* sounds, and their source is easy to pinpoint. When a bird hears the repetitious mobbing call of a member of its own species, it is able to locate the caller quickly and join it in driving away the object of concern—usually a hawk, crow, or other marauder. If an aerial predator is spotted flying overhead, the warning cry of songbirds is usually a high-pitched, extended *tseeeeee,* a sound that is difficult to locate (Figure 45.15). The response of a bird hearing this call is quite different from its response to the mobbing call. The warning cry sends the listener heading for cover, often diving into deep, protective foliage, from which it may also take up the plaintive, hard-to-locate cry.

Sound signals have the advantage or potentially high information load through subtle variations in frequency, volume, timing, and tonal quality. They are distinguishable at low levels, but louder sounds, with their higher energy levels, can carry over greater distances. In addition, sounds are transitory; they don't linger in the environment after they have been emitted. Thus, an animal can cut off a sound signal should its situation suddenly change—for example, with the appearance of a predator. A further advantage is that an animal doesn't ordinarily have to stop what it is doing to produce a sound. And, of course, unlike visual images, can go around or through many kinds of environmental objects.

One disadvantage of sound communication is that it is rather useless in noisy environments. Thus, some sea birds that live on pounding, wave-beaten shorelines rely primarily on visual signaling. Sound also weakens with distance. And the source of a sound is sometimes difficult to locate, especially underwater.

Chemical Communication

You have probably seen ants rushing along single file as they sack your cupboard. You may also have taken perturbingly slow walks with dogs that stop to urinate on every bush. The behavior in both is based on chemical communication. The ants have laid down chemical trails that the others can follow, and the dogs are gloriously advertising their presence. In both cases, the animals are communicating by **pheromones** (from the Greek *pherein,* "to carry"; *horman,* "to excite"), chemicals that are produced by one animal that influence the behavior of another animal of the same species. Pheromones have been found in a number of species but have been most intensively studied in arthropods and mammals.

FIGURE 45.14
MOBBING CALLS
Shown here are sound spectrograms of the mobbing calls of several species of British birds while attacking an owl. The sounds have qualities that make them easy to locate. Such calls are low-pitched *chuk* sounds. Different species of birds have developed similar calls through convergent evolution. That is, their calls serve much the same purpose and were developed under relatively similar conditions; thus, the qualities of the sounds came to be somewhat alike.

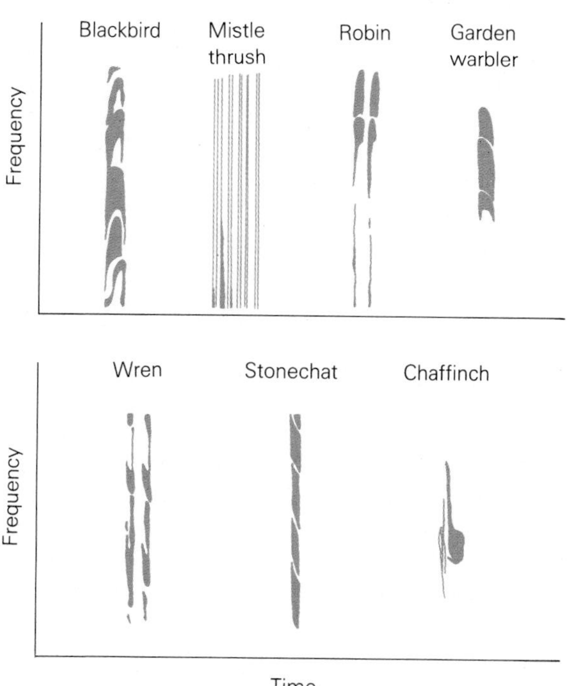

FIGURE 45.15
ALARM CALLS
The sound spectrograms shown here are of five species of British birds when a hawk flies over. Such calls are high-pitched, drawn-out, and difficult to locate. They, too, have achieved their similarity through convergent evolution.

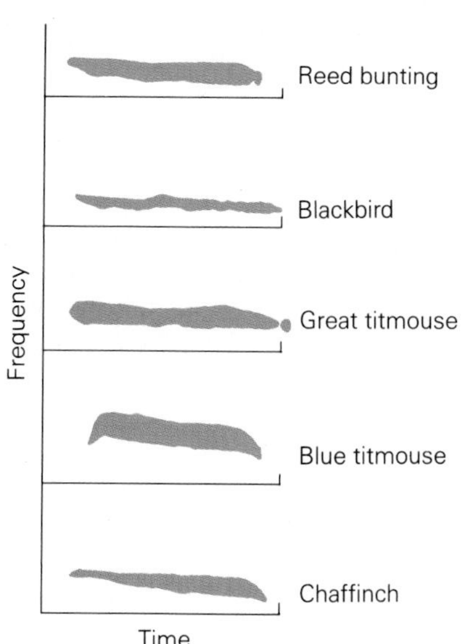

In insects, pheromones may incite very stereotyped behavior. For example, disturbed ants may produce an alarm chemical that causes other ants to drop whatever they are doing and to rush around in an agitated manner, ready to attack an intruder or to help the group in some other way. Insect pheromones, as we saw in Chapter 35, can be very powerful; one molecule of bombykol is enough to excite a male *Bombyx* moth.

Honeybees provide another interesting example of chemical communication in arthropods. The queen produces a "queen substance," which, when fed to developing worker females, inhibits sexual development. When the queen dies and the inhibiting substance is no longer present, new queens begin to develop. (The first one to hatch may massacre the rest.) At this time, the undeveloped ovaries of some adult workers also suddenly begin to mature. Their eggs will produce male drones—royal consorts, one or more of which will mate with the new queen on her nuptial flight.

Pheromones can elicit quite complex behavior in mammals. For example, females of most mammal species elicit sexual responses in males (sometimes coupled with aggression toward other males) by pheromones. In other cases, pheromones can trigger hormonal responses. For example, if pregnant rats of certain species smell the urine of a strange male, some component of that urine will cause them to abort their fetuses and become sexually receptive again.

Careful molecular analysis of several chemical signals has revealed that the backbone of the molecule usually consists of 5 to 20 carbon atoms. The molecular weight of the chains ranges between 80 and 300. It is believed that molecules with at least 5 carbons are needed to provide the variation and specificity for different messages. However, the molecules cannot be too large either, since they are usually carried in the air (lighter molecules could be carried farther) and because more energy is expended in the manufacture of larger molecules.

Chemical signals have the advantage of being extremely potent in very small amounts. Also, because of their persistence in the environment, the sender and receiver do not have to be precisely situated in order to communicate. In addition, chemicals can move around many sorts of environmental obstacles.

The specificity of chemicals, however, limits their information load. Moreover, because chemicals do linger in the environment, they may advertise the signaler to arriving predators as well as to the intended recipient of the message.

COMMUNICATION AND RECOGNITION

An important role of communication among animals involves recognition. That is, how one animal comes to identify another animal, either individually or by groups.

Species Recognition

First, communication may permit one animal to know that another animal is of the same species. This may not seem too important—unless one is interested in reproduction (Essay 45.2). After all, if two animals of different species mate, they cannot normally produce healthy offspring. Many species are very similar, and in the absence of some precise means of identification an animal might waste a lot of time and energy in trying to mate with a member of the wrong species. Therefore, species identification must somehow be quickly established.

For example, the golden-fronted woodpecker and the red-bellied woodpecker often share the same woods along a rather narrow band in Texas (Figure 45.16). The two species look very similar to the casual observer. The most conspicuous difference is that the golden-fronted woodpecker has a small yellow band along the base of its red cap, which is absent on the red-bellied woodpecker. Also, the red-bellied woodpecker has slightly more white in its tail. The habits of the species are much alike, and their calls are very similar, at least to the human ear. Nevertheless, each of these birds is able to recognize its own species. In any case, cross-mating has never been observed, nor have any hybrids. No one knows exactly what cue the birds use, but to them, apparently, the signals are clear.

(a)

(b)

FIGURE 45.16
SIMILAR SPECIES
(a) The golden-fronted woodpecker *(Centurus aurifrons)* is remarkably similar to **(b)** the red-bellied woodpecker *(Centurus carolinus)*. The range of the golden-fronted extends through dry areas from Honduras northward to central Texas, where it rather abruptly stops and that of the red-bellied begins. The red-bellied's range extends to the eastern coast, where it lives primarily in wooded habitats. The birds overlap in a narrow range in Texas, but in spite of strong similarities in appearance, vocalization, and behavior, they don't interbreed. Whereas they don't recognize each other as sexual partners, each does see the other as a competitor, and they mutually exclude each other from their territories as though they were of the same species. Obviously, they use different cues in the recognition of mates and competitors.

Interestingly enough, whereas the two species ignore each other sexually, they have come to treat each other as competitors and will eject a member of the other species from a territory as quickly as they would a member of their own species. Apparently, the stimuli that release sexual and territorial behavior are quite different.

Individual Recognition

It may be important for an individual to be able to recognize specific individuals within its own population. If you have ever watched gulls at the beach, you may have noticed that the adults all look alike. This is because you aren't a gull. (All people may look alike to them.) If you should visit a gullery, the place where they nest, you might see thousands of "identical" gulls. Color banding and recording playback experiments, however, have shown that gulls can recognize their own mates on sight and are able to filter out the calls of their mates from the raucous din of the clamoring gulls wheeling above.

Individual mate recognition is important only in species that establish pairs. In species in which the sexes come together only for copulation, the mate-recognition problem simply becomes one of sexual identification.

One should be able to recognize one's own mate, especially when both parents rear the offspring. Mate recognition ensures that pairs will attend to the nest that harbors their own offspring, thus increasing their chances of reproductive output. This is an essential factor in species that show a high level of coordination while rearing offspring. For example, in the African hunting dog, the male is more likely to regurgitate food to his own mate when she is nursing pups than to any other nursing females. In some species, the recognition on one's own offspring is critical to reproductive success. Female wildebeest learn the scent of their offspring within hours of giving birth and will not allow another calf to nurse. Bank swallows learn the calls of their young and thereby avoid expending energy in feeding unrelated juveniles.

Individual recognition can help reduce conflict by allowing the formation of **dominance hierarchies** in which an individual is recognized and treated in a specific way according to its rank. Once animals in a group know their rank with respect to the others, they will fight less over food or other commodities. In any group where rank is unknown, or not yet established, the incidence of fighting is likely to be high. Rank, obviously, can only be maintained when each animal can recognize the other individually.

Kin Recognition

There is evidence that in some species individuals are able to recognize their relatives, especially among vertebrates. Andrew Blaustein and his colleagues have shown that such

kin recognition can be adaptive on a number of bases. Specifically, it enables an individual to give preferential treatment toward kin (those bearing its kind of genes—see the following discussion of kin selection). Kin recognition also enables individual to avoid competing with kin and, in cannibalistic species, to avoid eating close relatives. Such recognition may also be important in avoiding too much inbreeding.

Kin recognition can occur in three ways. First, if relatives tend to be found together, an individual can simply learn to recognize those around it (familiarity). Second, an individual may learn a specific trait of itself or its relatives (such as an odor or a marking). In this way (and the next), unfamiliar animals can be recognized as kin. Third, "recognition genes" may exist. In this hypothetical case, no learning is involved. Individuals carrying a certain allele would simply be able to recognize others carrying the same allele. It is believed that one or all of these mechanisms could operate in the same individual.

AGONISTIC BEHAVIOR

Agonistic behavior is any behavior that helps to resolve conflict. It usually involves members of the same species. Agonistic behavior includes a wide range of behavioral patterns, such as aggression, threats, and submission. It does *not,* however, include predation. A lion is about as aggressive toward a wild pig as you are toward a hamburger. However, if the fleeing pig should turn around and charge a lion, the lion may, for an instant, show a phase of agonistic behavior—fear. At the other end of the agonistic spectrum is aggression, a behavior that eventually reduces the level of competition between two individuals.

Fighting, a Form of Aggression

Let's consider the most obvious form of aggression—fighting. You can discount the old films you have seen of leopards and pythons battling to the death. Such fights simply aren't likely to happen. What is a python likely to have that a leopard needs badly enough to risk its life for, and vice versa? Although such fighting might occur in the unlikely event that one should try to eat the other, fighting is much more likely to occur between two animals that are competing for the same commodity. The closest competitor is one that uses the same habitat in the same way, and the animal most likely to do this would be a member of the same species. Moreover, if there is a strong competition for mates, fighting is even more likely between members of the same sex within that species. And this is in fact where most fighting occurs—between members of the same sex of a given species.

Of course, fighting between species sometimes occurs. The golden-fronted and red-bellied woodpeckers exclude each other from their respective territories. And lions may attack and kill African cape dogs at the site of a kill. The lions don't eat the dogs; they just exclude them.

FIGURE 45.17
FIGHTING MALE RATTLESNAKES
Each male *(Crotalus horridus)* tries to push the other to the ground, but the deadly poisonous snakes never bite each other.

Animals may fight in a number of ways, but combatants of the same species usually manage to avoid injuring each other (Figure 45.17). There are several apparent benefits to such a system. First, no one is likely to get hurt. The competitor is permitted to continue its existence, it is true, but the possibility of having to compete again entails less risk than does serious fighting. Also, since animals are most likely to breed with the individuals around them, the opponent may be a relative that is carrying some of the same genes; hence, sparing the opponent has a reproductive advantage. In fact, if the population is confined to a small area, the competitor might even be one's own mature offspring; it could also be a prospective mate. So even though the "motive" is selfish, not benevolent, it's best not to hurt each other.

When fighting occurs between potentially dangerous combatants, the fights are usually stylized and relatively harmless. For example, a horned antelope may gore an attacking lion, but when antelopes fight each other, the horns are almost never directed toward the exposed flank of the opponent (Figure 45.18). Such stylized fighting does, however, enable the combatants to establish which is the stronger animal. Once dominance is established, the loser is usually permitted to retreat.

On the other hand, all-out fighting may occur between animals that are unequipped to injure each other seriously, such as hornless female antelope (Figure 45.19), or between animals that are so fast that the loser can usually escape before serious injury, as in house cats.

We might wonder why they don't gore each other. An incurable romantic might assume that they simply don't want to hurt one another. In all likelihood, what the two antelopes "want" has very little to do with it. The fact is, they can't hurt each other. When the system works, an antelope could no more gore an opponent than fly! In terms of the actual mechanism, it may be that the sight of an opponent's exposed flank acts as an inhibitor of butting behavior. Conversely, a facing view, under certain conditions, might serve as a release of very stereotyped fighting behavior.

FIGURE 45.18
MALE PRONGHORN ANTELOPES FIGHTING
Although the horns of these medium-sized antelopes are formidable weapons, neither animal will attack the vulnerable flank of the other. Instead, a harmless pushing contest ensues as the tips of the ridged horns are engaged. These animals effectively employ horns and hoofs against species other than their own, but when confronted with a member of their own species, they are genetically constrained to behave in very circumscribed ways.

But regardless of what we may or may not surmise about animal motivation, such standardized behavior must have evolved because it was beneficial to the ancestors of the present combatants. Thus, we return to the original question: How does the animal benefit by refusing to do serious injury to the opponent?

One benefit is really not that subtle. The antelope might not gore its opponent because if it did so, it might get gored back. The situation is similar to that of two toughs in a barroom brawl. They slug and punch, and each is really trying to win the fight. On the other hand, each has a jackknife in his pocket—and both of them know it. Neither is willing to pull his knife, because then the other would be obliged to retaliate, and someone could get hurt that way. Either brawler would rather risk a loss in a fistfight than a draw in a knife fight.

Animal evolution has proceeded according to the same kind of logical accounting. John Maynard Smith, of the University of Sussex, points out that the antlers of deer are used solely for intraspecific fights and are constructed so as *not* to injure the opponent. The "points," or side branches, cause the antlers to lock in head-to-head combat so that the sharp tips do not actually reach the opponent's flesh. When they are seriously fighting a predator, they use their hooves with dazzling effect. Occasionally, they are male deer with antlers that lack the side branches. When these males engage in what should be a ritual fight, their antlers slip through and gore their opponent. Does this give the mutant deer an advantage? Not at all. The gored opponent may be bigger and stronger and, once gored, may become very angry. Ritual combat then becomes real combat, and the sharp-antlered deer doesn't always win.

Maynard Smith has analyzed fighting strategies with mathematical models and computer simulations. What he looked for was an **evolutionarily stable strategy**, which he defined as an innate behavioral pattern that would outcompete all other behavioral patterns and would be stable against the invasion of a new mutant pattern of behavior. He formed his questions in terms of "hawks" and "doves," asking, is it best never to fight and always retreat? Then a born fighter will always win. Is it better always to fight? No, because there is too much risk of being beaten. Is it best to bluff consistently? No, the bluff will be called. (It's better to bluff inconsistently and unpredictably.) Maynard Smith found two behavioral "strategies" that proved to be the most stable in populations, depending on

FIGURE 45.19
HORNLESS FEMALE ANTELOPES FIGHTING
Hornless females of the Nilgai antelope have no inhibitions against attacking the flank of the competitor, but their butts are quite harmless, at least in the immediate sense. Though the butt itself may not be dangerous, it establishes dominance. A loss, however, usually just means a temporary setback, so it behooves the loser to accept it gracefully and attempt to breed another time. Interestingly, horned males of the same species almost never attack in this way, nor do horned females of other species.

ESSAY 45.2

ANIMAL COURTSHIP

The reproductive success of many species of animals involves persuasion: one animal must persuade another to be its mate. Thus, courtship behavior has evolved, and its array of manifestations never ceases to amaze even the most jaded biologists. The elaborate ceremonies that have resulted are often beautiful and distinctive, but they all serve the same function—to maximize the reproductive success of the participants.

In some cases, courtship involves only species recognition. These patterns are usually relatively brief, and mating occurs between individuals that need never see each other again. In species in which the male is not important in protection or in feeding the young, the sexes join only to mate, and it is in these species that we generally find the larger, more garish, conspicuous males. In such species, *sexual dimorphism* (different appearances of the sexes) is associated with a different strategy. The males maximize their reproductive success by attracting and inseminating as many females as possible. Of course, their appearance and behavior may attract predators and cut their reproductive lives short.

The courtship ritual is more extended in species in which the male will remain with the female. In these cases, ceremonies may last several days. Because the participants tend to remain

Albatrosses may bow and preen and clatter their bills together for days before breeding.

Western grebes go through elaborate rituals of swimming, diving, and "gift-swapping" before they mate.

the circumstances. The first strategy was called the **retaliator strategy**; a retaliator engages only in ritual display and mock battle unless it is seriously attacked, whereupon it will retaliate with just as much seriousness. In a mathematical model, this behavior was always the most successful in the long run, so it should not be surprising to see ritual display and mock battle so common in nature. There's always the threat of real injury to any animal that breaks the rules.

The second strategy, the **bourgeois strategy**, employs quite a different set of rules. The bourgeois approach was even more effective than retaliation, but it required following a peculiar rule. In each encounter between adversaries, one retreats immediately, so that neither wastes time and energy fighting. We find the bourgeois strategy common in territorial encounters. The owner of the territory will attack an intruder, and the intruder almost always quickly retreats. However, the strategy will prevail in a population even if the determination of who concedes is completely arbitrary. Some of our traffic rules work this way. The first car to the intersection has the right of way; if two cars arrive simultaneously, the car on the right has the right of way and the car on the left concedes (but don't count on it). As long as everyone knows the rules, it is advantageous to everyone to follow them.

Can we assume, then, that animals do not fight to the death with their own kind? As a general rule they don't, but there are many exceptions. Accidents may occur in normally harmless fighting, and this can lead to retaliation and escalation. There are also some species that normally engage in dangerous fighting. If a strange rat is placed in a cage with a group of established rats, the group may sniff at the newcomer carefully for a long time, but eventually they will begin to attack it and to do so repeatedly until together for long periods of time, they have more invested in the relationship and therefore must be highly selective. In many of these species (such as the albatross), the males and females may look very much alike (sexually monomorphic). It has been suggested that, in contrast to males in sexually dimorphic species, the drab male coloration may be an adaptation that reduces the possibility of predators. The drab and sexually monomorphic prairie chickens show another strategy. These birds gather at traditional leks (mating areas), where the males whirl and strut and females make their choices. When the dancing stops, however, the males inconspicuously go their way. This, then, is a form of *behavioral dimorphism*.

Hippos mate beneath the water, where the male's great weight can easily be supported. Should the female resist his advances or be unresponsive to his preliminary displays, the male may attack and force her to submit.

A female manatee is physiologically ready to mate before she is behaviorally ready, eluding the cumbersome males for days before accepting one or more of them.

Western grebes not only display to each other in elaborate swimming rituals but may also bring each other gifts of seaweed. In yet other species, the male may bring food to the female. The advantages of this behavior are obvious to a reproductive female. In essence, courtship behavior is a means of finding not only a mate, but a healthy mate that has demonstrated the ability to either recognize the appropriate signals or perform the elaborate and often demanding rituals of the species. In some species, advertising is less important. Male manatees simply follow estrous females until they are accepted, and powerful male hippos may force themselves on the smaller females.

they kill it. If escape is impossible, male guinea pigs and mice often fight to the death. The males of a pride of lions may kill any strange male they find within their hunting area, and a pack of hyenas may kill any of another pack that they can catch. Even gangs of male chimpanzees have been seen to ambush and kill isolated males from other troops.

Is Aggression Instinctive?

Our discussion of aggression so far suggests that aggressive behavior in many animals is largely innate. But there are many who contend that aggressive behavior, especially among mammals, is directly attributable to learning. Others, of course, stress that aggressive behavior has both learned and innate components. The argument has extreme significance for our species. If aggression is socially disruptive in an increasingly crowded world, then we must ask how it can be controlled. As a practical example, does television violence serve to release aggression, and hence dissipate it; or does it serve as a behavioral model for aggression, and hence encourage it? Since there is a tendency in most academic disciplines to focus on aggression as a cultural, or learned, phenomenon, let's consider the evidence that suggests an instinctive component.

In certain species of highly aggressive cichlid fish, the males must fight before they are able to mate. If a male's reproductive state is appropriate and a female is available, the male will frantically seek an opponent upon which to release his fighting behavior. Finding none, he will often attack and kill the female. He is then ready to mate, but of course by then it is too late. The behavior can be described in terms of Tinbergen's hierarchical model. Among mammals, rats will learn mazes in order to be able to kill mice, and it has thus been inferred that the killing is a consummatory act that reduces tension.

Does an animal become aggressive when it is shielded from all opportunities to learn such behavior? Rats and mice have been reared in isolation so that there was no opportunity for them to learn aggression. But when other members of the same species were introduced into their cages, the orphan rodents attacked, showing all the normal threat and fighting patterns. The evidence from all sides is essentially circumstantial, but we would do well to focus more research attention on the roots of aggression in these critical times.

SOCIAL BEHAVIOR

The famous student of animal behavior, Jane Goodall, who has spent much of her life among the chimpanzees of East Africa's Gombe Stream Preserve, once said, "One chimpanzee is no chimpanzee at all." Her point was that researchers should not attempt to study chimpanzee behavior by observing a simple chimpanzee in a cage, because an isolated chimpanzee will behave quite abnormally. Chimpanzees, she noted, are highly social creatures that interact in extremely intensive and complex ways. If you want to know what chimpanzees are like, according to Goodall, then you must watch them when they are with other chimpanzees.

The same statement might be made of any of a number of other creatures, including us. What would a termite be like without other termites? Or a human without other humans? Many of the earth's animals are indeed highly social species, and they interact with each other in subtle and complex ways. Yet, there are some underlying themes. We will look at some of these principles here. In particular, we will consider how animals can cooperate and on what adaptive bases populations may stay together.

Cooperation

Cooperative behavior occurs both within species and between species. As an example of *interspecific* (between species) cooperation, consider the relationship of the rhinoceros and the tickbird. The little birds get free food, while the rhinoceros rids itself of ticks and harbors a wary lookout. Such relationships are well known in nature because of their inherent interest, but the highest levels of cooperation are most likely to exist between members of the same species.

Let's consider a few examples of *intraspecific* (within species) cooperative behavior. Porpoises are air-breathing mammals, much vaunted in the popular press for their intelligence. In fact, certain of their actions support the claim. Groups of porpoises will swim around a female in the throes of birth and will drive away any predatory sharks that might be attracted by the blood. They will also carry a wounded comrade to the surface so that it can breathe. Their behavior in such cases is highly flexible, rather than stereotyped. Such flexibility indicates that their behavior is not solely a blind response to innate genetic influences.

Group cooperation among mammals is probably most common in defensive and hunting behavior. For example, yaks of the Himalayas form a defensive circle around the young at the approach of danger, standing shoulder to shoulder with their massive horns directed outward (Figure 45.20a). This defense is effective against all predators except humans, since it provides no defense against high-powered rifles. Other species, such as birds, wolves, African cape dogs, jackals, and hyenas often hunt in packs and sometimes cooperate in bringing down their prey. In addition, they may bring food to members of the group that were unable to participate in the hunt.

We might expect mammals, with their high intelligence, to cooperate closely. But social behavior and cooperation are most highly developed in certain of the insects (Figure 45.20b). The complex and highly coordinated behavior patterns of insects are usually considered to be genetically programmed, highly stereotyped, and generally not

(a)

(b)

FIGURE 45.20 COOPERATION IN VASTLY DIFFERENT SPECIES
(a) Musk oxen form a defensive circle to protect the females and young. **(b)** These social insects show extremely high levels of cooperation, coordination, and self-sacrifice. The result is a hive in which the watchword is cold, sometimes brutal, efficiency.

influenced by learning. Some of the best examples of insect cooperation are found among the honeybees.

In honeybee colonies, the queen lays the eggs, and all other duties are performed by the workers, which are sterile females. Each worker has a specific job, but that job may change with time. For example, newly emerged workers prepare cells in the hive to receive eggs and food. After a day or so, their brood glands develop, and they begin to feed larvae. Later, they begin to accept nectar from field workers and pack pollen loads into cells. At about this time their wax glands develop, and they begin to build combs. Some of these "house bees" may become guards that patrol the area around the hive. Eventually, each bee becomes a field worker, or forager. She flies afield to collect nectar, pollen, or water, according to the needs of the hive. Apparently, these needs are indicated by the "eagerness" with which the field bees' different loads are accepted by the house bees.

If a large number of bees with a particular duty are removed from the hive, the normal sequence of duties can be altered. Young bees may shorten or omit certain duties and begin to fill in where they are needed. Other bees may revert to a previous job where they are now needed again.

The watchword in a beehive is *efficiency*. In some species, the drones (males) exist only as objects of reproduction. Once the queen has been inseminated, the rest of the drones are quickly killed off by the workers; they are of no further use. The females themselves live only to work. They tend the queen, rear the young, and maintain and defend the hive. When their wings are so torn and battered that they can no longer fly, they either die or are killed by their sisters. But the hive goes on.

Symbiosis

In some cases, members of different species live together in close association, a relationship referred to as **symbiosis.** As a result of symbiotic relationships, one or both individuals can be benefitted (although symbiosis also includes parasitism, when one is harmed.)

Mutualism is an interaction in which all involved species benefit. For example, cattle harbor cellulase-producing bacteria in their complex digestive system. The cattle derive energy-rich breakdown products of cellulose and also vitamins and amino acids from the mutualistic bacteria. The bacteria benefit from the steady food supply ingested by the cattle.

Commensalism is a species interaction in which one species benefits while the other is neither benefitted nor harmed. An example of commensalism is the association between cattle egrets and cattle. The birds, which feed on insects and other small organisms disturbed by moving cattle, benefit because their foraging success is greater near moving cattle than away from them. The cattle appear to be neither helped nor harmed, although observers report an occasional egret picking a parasite from the skin of a bovine associate. Another example is the interaction between oceanic birds, like gulls and terns, and predaceous fish. These birds are attracted to schools of feeding fish that injure more small fish than they can consume during feeding frenzies near the water surface. The birds quickly pick up these easy prey. The predaceous fish do not benefit and are probably not harmed by the opportunistic feeding of the birds.

Commensalism can be more complex and indirect. In the intertidal zone, two species of limpets are subtly involved with a species of chiton. The limpets and chiton have overlapping foods, superficially suggesting the possibility of interspecific competition. The limpets feed on diatoms, and the chiton feeds on both diatoms and kelp. But the limpets are much more abundant in places inhabited by chitons; the limpets appear to benefit from the presence of chitons despite the overlap in foods. This counterintuitive observation results from the chiton's grazing on kelp. Kelp dominates diatoms in the competition for limited space in the intertidal zone. The chiton reduces the impact of kelp on diatoms, allowing both to coexist and in the process benefitting the limpets by indirectly increasing their food supply.

Next we will look at relationships where one animal may be benefitted by another animal performing a behavior, but at a cost to the performer.

ALTRUISM

You probably won't be shattered to learn that most of the "Lassie stories" aren't true. Consider what would happen to the genes of any dog that was given to rushing in front of speeding trains to save baby chickens. The reproductive advantages would be considerable to chickens, but dogs with those tendencies might be selected out of the population by the action of fast trains. In contrast, the genes of a "chicken" dog that spent his energy, not in chivalrous deeds, but in seeking out estrous females, would be expected to increase in the population.

Altruism may be defined as an act by one individual that benefits another, but at the first individual's expense. There are many apparent cases of altruism in the animal world, but our job here will be to ask if these are really altruistic acts, after all, and if so, how the behavior might have evolved.

It is easy to see how certain forms of altruism are maintained in a population. For example, pregnancy, in a sense, is altruistic. The prospective mother is swollen and slowed. Much of her energy goes to the maintenance of the developing fetus. At birth she is not only almost completely incapacitated, but is in marked danger. Pregnancy is clearly detrimental to her. So why do females so willingly take the risk? It may help to understand the enigma if we remember that the population at any time is composed entirely of the offspring of individuals who have made such a sacrifice. Thus, the females in the population are the descendants of generations who did reproduce, so they too are likely to be predisposed to make such a sacrifice. It makes little difference whether the tendency is genetic or learned. The tendency to have offspring or *to do the things that result in having offspring* can be innate (genetically based) or, especially in humans, transmitted culturally.

However, altruism on this basis doesn't explain why a bird may feed the young of another pair, or why an African hunting dog will regurgitate food to almost any puppy in the group. Why, also, would a bird that may have no offspring of its own give a warning cry at the approach of a hawk, alerting other birds at the risk of attracting the hawk's attention to itself? To answer such questions we must look past the answers that first come to mind. It may seem cynical, but we must start with the premise that birds don't give a hoot about each other. A bird that issues a warning call isn't thinking, "I must save the others." At least, there is a simpler explanation of its behavior.

We should note that the biologically "successful" individual is the one that maximizes its reproductive output. One way of accomplishing this is for the organism to leave its own offspring, but another way is through **kin selection.** We need only keep in mind that an individual also shares genes with a cousin, albeit fewer than with a son or a daughter. So it is reproductively beneficial for an individual not only to bear offspring, but to assist relatives, as long as the cost is not too high. In fact, there is theoretically a point at which an individual could increase the success of its genes by saving its nieces and nephews (provided there were enough of them) rather than its own offspring. From the standpoint of effective reproductive output, the organism would be better off leaving 100 nieces than one daughter.

To illustrate, suppose a gene for altruism appears in a population. (Notice that this sets up a mechanism for the continuance of the behavior.) As you can see from Figure 45.21, altruistic behavior would most likely be maintained only in groups in which the individuals are related (that is, have some kinds of genes in common). Altruism might be expected, then, when there is a high probability that proximity indicates kinship, as we see in relatively stationary populations.

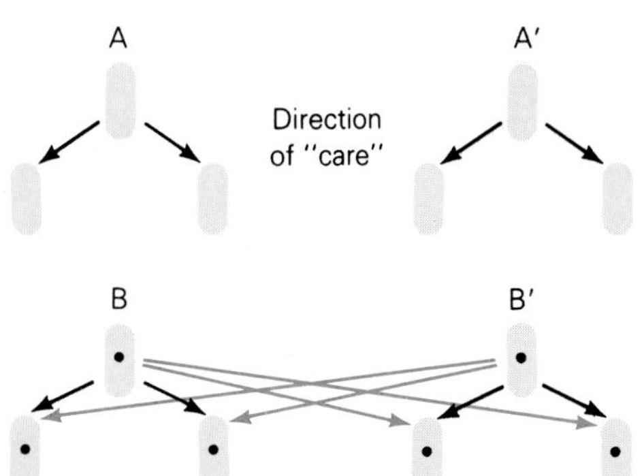

FIGURE 45.21
A GENETIC MECHANISM OF ALTRUISM
In this population, A and A′ are nonaltruists. They behave in such a way as to maximize their own reproductive success but do nothing to benefit the offspring of other individuals. In another segment of the population (B and B′), a gene for altruism has appeared that results in individuals benefiting the offspring of others in some way. It can be seen that, assuming the altruistic behavior is only minimally disadvantageous to the altruist, generations springing from B and B′ are likely to increase in the population over those from A and A′. The altruistic behavior is likely to be greatest where B and B′ are most strongly related, so that B shares the maximum number of genes in common with the offspring of B′ and vice versa. The idea is that B, for example, can increase its own reproductive success by caring for the offspring of a relative with whom it has some genes in common. After all, reproduction is simply a way of continuing one's own kinds of genes.

Keep in mind that no conscious decision on the part of the altruist is necessary. It simply works out that when conditions are right, those individuals that behave altruistically increase their kinds of genes in the population, including the "altruism gene." Nonrelatives would benefit from the behavior of altruists, of course, but there is an increased likelihood that individuals near an altruist are related to it.

It has been determined mathematically that the probability of altruism increasing in a population depends on how closely the altruist and beneficiary are related (as well as on the risk to the individual, of course). In other words, the advantages to the beneficiary must increase as the kinship becomes more remote. For instance, an altruistic act that results in a risk of death of the altruist will be selected for if the net genetic gain to brothers and sisters is more than twice the loss to the altruist; to half-brothers, four times the loss; and so on. To put it another way, an altruistic animal would gain reproductively if it sacrificed its life for more than two brothers, but not for fewer, and so on. Therefore, we can deduce that in highly related groups, such as a small troop of baboons, a male might fight a leopard to the death in defense of the troop.

This model, developed by J. B. S. Haldane in 1932 and expanded by W. D. Hamilton in 1963, helps explain the extreme altruism shown by **eusocial insects** (those with castes—queens, workers, drones, etc.) such as honeybees and paper wasps. Since workers are sterile, their only hope of propagating their own genotype is to maximize the egg-laying output of the queen. In some species, the queen is inseminated only once (by a drone, a haploid male), so all the workers in a hive are sisters and have an average of three-fourths of their genes in common (Figure 45.22). In such a system, then, almost any sacrifice is worth any net gain to the hive and to the queen.

In a brilliant essay, Robert Trivers expanded our understanding of the evolution of altruism by developing the theory of **reciprocal altruism,** which is any altruistic act that depends on the expectation of reciprocation. ("I'll scratch your back if you'll scratch mine.") Reciprocal altruism is an evolutionarily stable strategy with some complex rules. Help (altruistic acts) is given to others—even offered to strangers—when the cost or risk is not too great to the giver. The expectation is that some kind of help will be reciprocated at another time. If the expected reciprocation is not forthcoming, an evolutionarily derived emotion is experienced: moral indignation. The individual that fails to reciprocate is scorned, turned out of the social group, no longer aided. To get ahead, everyone has to play by the rules—or appear to.

Some evidence of intraspecific reciprocal altruism has been reported in troops of social mammals, such as hunting dogs and baboons, but the evidence is not very strong for such behavior in any animals but humans. Trivers implies, in fact, that reciprocal altruism is the key to human evolution. The complexity of such behavior, entailing as it does memory of past actions, the calculation of risk, the foreseeing of the probable consequences of present actions, the possibility of advantageous cheating, and the need to be able to detect such cheating—all require a level of intelligence that is beyond most species. In the opinion of some anthropologists, it is exactly for the management

FIGURE 45.22
RELATEDNESS BETWEEN MOTHER AND OFFSPRING IN EUSOCIAL INSECTS

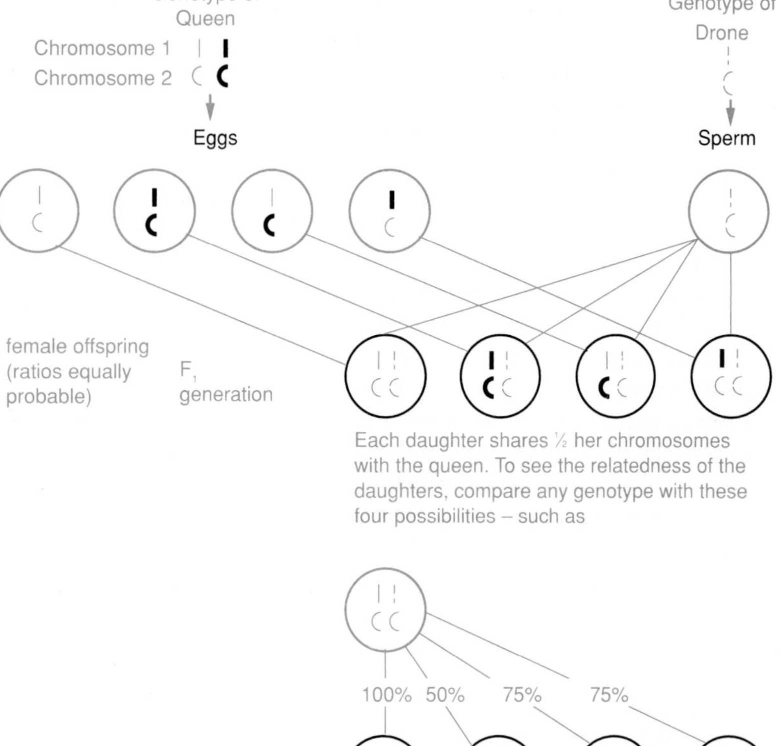
Relatedness Between Mother-Offspring In Eusocial Insects
Genotype of Queen
Chromosome 1
Chromosome 2
Eggs
Genotype of Drone
Sperm
female offspring (ratios equally probable)
F_1 generation
Each daughter shares ½ her chromosomes with the queen. To see the relatedness of the daughters, compare any genotype with these four possibilities – such as
100%
50%
75%
75%
Average: 75%—so daughters are likely to be more strongly related to each other than to their mother. Thus they are likely to sacrifice for each other, based on kin selection.

of these elaborate social interactions that the human brain—and the conscious mind—evolved.

SOCIOBIOLOGY

Sociobiology applies principles of evolution to studies of social behavior in animals. Like other areas of ethology, especially behavioral ecology, it emphasizes the influence of ultimate factors on behavior (including that of humans—see Essay 45.1).

Animals may live together (be social) on a number of adapted bases. We will mention six. (1) **Group foraging.** A group of animals may be more efficient at finding scattered food than a single individual. Of course, once food is found, it must be shared, but overall each individual gets more by searching with others. (2) **Group protection.** A group may offer protection by confusing a predator with sheer numbers. Birds about to be attacked by a falcon will draw closer together. The result is that so many potential targets confound the falcon's attempt to pick out one. (3) The **"selfish herd" effect.** An individual reduces its chance of being caught by a predator by using others as a form of cover. If a predator takes the nearest prey, then it is advantageous to be near other individuals that can act as a shield. (4) **Increased vigilance.** With more eyes and ears around, it is harder for a predator to sneak up (Figure 45.23). (5) **Reproductive coordination.** In a group, reproductive efforts can be coordinated so that offspring are produced together. The advantage for the animals may be to avail themselves of some fleeting commodity, such as a temporary food source, or to control predators on the offspring by overwhelming them. The predators can't take full advantage of a single massive surge in food but would easily wipe the offspring out if they appeared one at a time. (6) **Mutual advantage to parent and offspring.** Offspring may survive and reproduce better eventually if they continue to remain with the parents for at least a time. Wolf packs, for example, are often composed of a single family group in which one male and one female do all the breeding. Younger animals stay around and help with hunting and rearing the pups until their time comes (if it ever does). Nonetheless, their odds are better this way than if they ventured out alone. In some birds, the previous year's offspring may stay around as "helpers." They help defend the nest area and bring food to their younger siblings. The parents gain assistance, and the offspring stand to inherit the territory upon the death of the parents.

We should keep in mind that these advantages are not mutually exclusive. Sociality probably rarely exists on a single adaptive basis but through a variety of adaptive effects.

Sociobiologists, we should add, also study the other end of the social spectrum—those animals that avoid each other. The most prominent examples of such antisocial behavior are found among predators. The adaptive basis is generally believed to be due to the difficulty many carnivores have in finding food. There just isn't enough to share. With predatory animals—such as leopards, jaguars, bears, and weasels—the limits of social behavior seem to be associating during periods of mating and, in some cases, rearing young (Figure 45.24).

Evolution and ecology have obviously played a great role in shaping the social behavior of animals. Just as such forces have molded the shape of an animal's head or the length of its leg, they have also strongly influenced how that animal behaves toward others.

FIGURE 45.23
MEERKATS
These highly social animals are very alert and watchful. As a group they are very difficult for a predator to surprise.

FIGURE 45.24
GRIZZLY FEMALE WITH CUBS
These bears are generally very unsocial, aside from parenting behavior, interacting mainly when they come together at a common feeding area.

APPLICATION OF IDEAS

1. Make an observation of a behavior (either one you've read about or seen personally). Then describe the behavior in terms of probable ultimate and proximate causation.
2. Show how altruism is intimately related to recognition. It would appear more strongly, do you suppose, to which levels of recognition (species, individual, or kin)?

KEY IDEAS

PROXIMATE AND ULTIMATE CAUSATION

1. **Proximate causation** involves the more immediate bases for behavior and describes *how* the animal comes to act in a certain way.
2. **Ultimate causation** involves the evolutionary and adaptive bases for behavior and declares *why* the animal behaves in a certain way.

BEHAVIORAL ECOLOGY

Behavioral ecology is the study of how the environment affects behavior.

Habitat Selection

Wecker's experiments showed both genetic (ultimate) and learned (proximate) elements in habitat selection of mice.

Foraging Behavior

The question of the efficiency of foraging behavior is, how does the animal maximize gain relative to expenditure? In studies of the great tit, it was found that suboptimal sites will be tested as alternate sources of food. Aquatic animals chose food items based on the size and ease of handling of the prey.

BIOLOGICAL CLOCKS

Both plants and animals respond to a day-night cycle, the mechanisms of which are not entirely known.

The Adaptiveness of Rhythms

1. Organisms must adapt to annual, lunar, and daily **(circadian) cycles** (among others).
2. Each time has its own characteristics, of which the animal must take advantage.

NAVIGATION

Navigation involves the directional sense that enables one to get from one place to another.

1. **Compass sense** is the ability to know direction.
2. **Map sense** is the ability to know longitude and latitude.
3. **Time-direction mechanism** is a compass sense coupled with a sense of time.
4. **Orientation** is the ability to face in the right direction.
5. Daytime migrators use the sun as a directional cue.

Homing Pigeons

Homing pigeons use a variety of environmental cues involving both map sense and compass sense.

Adaptiveness and Evolution of Migration

In some cases, the advantages of migration are clear, as in migrating caribou; in other cases, not so clear, as in migrating monarch butterflies.

There are three major theories regarding how bird migration may have evolved. (1) The ice age caused winters in the north to be increasingly severe, (2) competition near the equator forced birds northward to breed, and (3) birds returned each year to the southern land masses where they originated.

COMMUNICATION

Communication is an action by one animal that influences another animal.

Visual Communication, Sound Communication, and Chemical Communication

Visual messages may be sent by color, posture, shape, movement, or timing. Sound messages can be sent by cadence, pitch, or tone. Chemical signals involve molecules called **pheromones.**

COMMUNICATION AND RECOGNITION

Recognition involves identification, either of groups or individuals.

Species Recognition

Species recognition enables one animal to know if another animal is of the same species. An animal recognized as the same species can be treated differently on reproductive and ecological bases.

Individual Recognition

Individual recognition enables one animal to identify specific individuals within its population.

1. Individual recognition is important in mate and offspring identification.
2. Individual recognition is pivotal to the formation of **dominance hierarchies.**

Kin Recognition

Kin recognition, the identification of relatives, occurs in three ways: (1) through familiarity, (2) through a trait borne by such relatives, and (3) through the action of hypothetical "recognition genes."

AGONISTIC BEHAVIOR

Agonistic behavior helps to resolve conflicts among members of the same species.

Fighting, a Form of Aggression

1. Fighting, the most obvious form of aggression, usually occurs between competitors, does not involve injury, and is stylized.
2. Fighting strategies have been analyzed as **evolutionarily stable strategies**, patterns that are selected over mutant variants.

Is Aggression Instinctive?

There is strong evidence that many aspects of aggression are innate. These generally are modulated by learning influences.

SOCIAL BEHAVIOR

Questions of social behavior include how and why (on what adaptive basis) animals stay together.

Cooperation

Cooperation involves animals working together. Intraspecific cooperation occurs between members of the same species.

Symbiosis

Symbiosis involves members of different species living together in close association.

1. In **mutualism**, all species involved benefit.
2. In **commensalism**, one species benefits, while the other is neither benefitted nor harmed.
3. In parasitism, one species benefits, while the other is harmed.

ALTRUISM

Altruism is an act that benefits another, but at the altruist's expense.

1. Altruism may continue in a population through **kin selection**, where relatives are cared for because they bear genes in common with the altruist.
2. The probability of altruism increasing in a population depends on how closely the altruist and beneficiary are related.
3. **Reciprocal altruism** involves performing an altruistic act with some expectation of the favor's being returned.

SOCIOBIOLOGY

Sociobiology applies principles of evolution to studies of social behavior in animals. Animals may be social on several bases: (1) **group foraging**, (2) **group protection**, (3) the **"selfish herd" effect**, (4) **increased vigilance**, (5) **reproductive coordination**, and (6) **mutual advantage to parent and offspring.**

REVIEW QUESTIONS

1. A mother raccoon lies down on a cold day, and her litter of young, following close behind, immediately cuddle up to her belly. Explain the behavior in both ultimate and proximate terms. (1105)
2. Distinguish between proximate and ultimate causation. (1105–1106)
3. In Stanley Wecker's experiment, what did rearing laboratory mice in a forest setting do to their habitat preference? (1106–1107)
4. Describe and give an example of a foraging generalist. (1108)
5. What is a circadian cycle? Name three. (1110)
6. How is clock shifting done? (1110)
7. Distinguish between orientation and navigation. (1111)
8. In a nutshell, what did the planetarium experiments show? (1111–1112)
9. How might competition have helped trigger migration in birds? (1114)
10. Define communication in its broadest sense. (1115)
11. How can visual communication carry such high information content? (1115–1117)
12. What kinds of animals are more likely to use graded displays? (1116)
13. How does sound communication carry such high information content? (1117–1118)
14. Define *pheromone* and provide an example of its workings. (1119–1118)
15. State a disadvantage of chemical communication. (1120)
16. State an advantage of species recognition. (1120)
17. State an advantage of kin recognition in terms of kin selection. (1121–1122)
18. Do you think that agonistic behavior can include retreat? Why? (1122)
19. Define an "evolutionarily stable strategy." (1123)
20. Give one line of evidence suggesting a genetic component to aggression. (1125–1126)
21. What does "One chimpanzee is no chimpanzee at all" mean? (1126)
22. Under what conditions are we most likely to see group cooperation among mammals? (1126–1127)
23. Distinguish between mutualism and commensalism. (1127)
24. Is any harm done in an altruistic act? Explain. (1128–1129)
25. Define sociobiology. (1130)

46

BIOSPHERE AND BIOMES

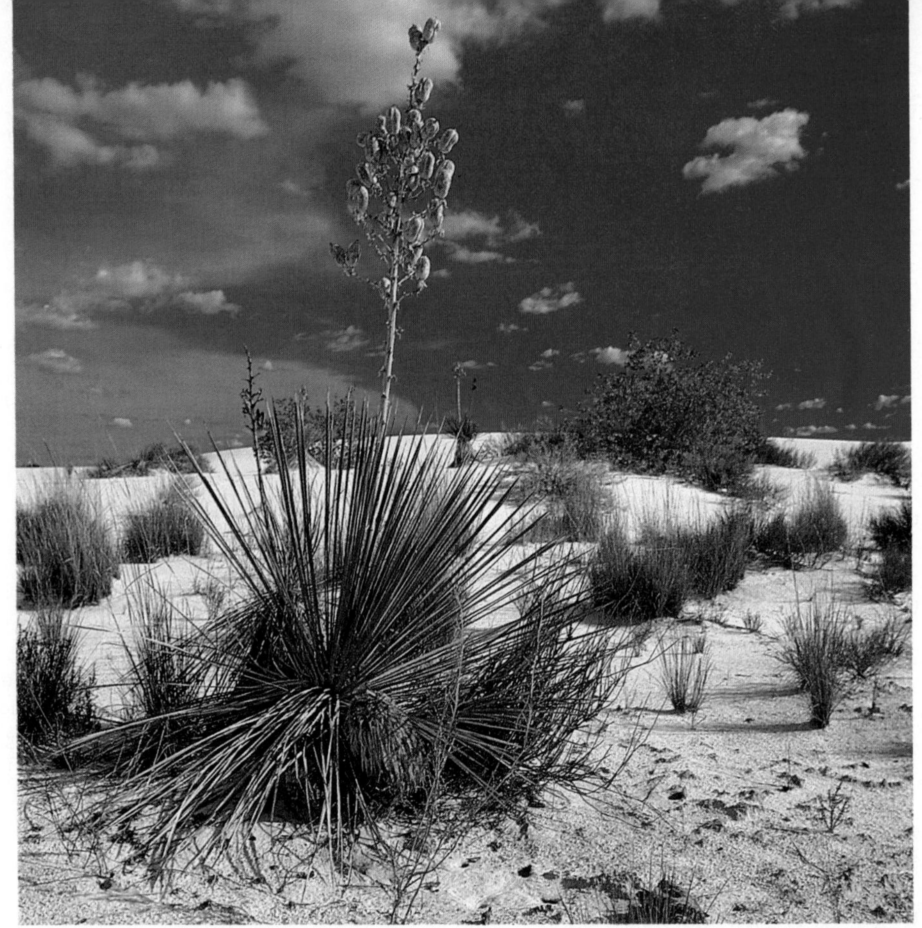

THE WORD "**ECOLOGY**" HAS MADE ITS WAY INTO THE PUBLIC VOCABULARY and has proved to be remarkably resilient there. Often, once a word enters that "great consciousness" it becomes overused, misinterpreted, restructured, and finally battered into uselessness. ("Instinct" is an example.) But *ecology* has weathered the attacks in a remarkable fashion. The reasons, it seems, are twofold. One, it has been freely interchanged with the word **environment** ("What are we doing to the ecology?"), a word with which people seem to be somewhat comfortable, so it has escaped intellectual massage. Two, people have found that it just isn't possible to deal with the idea of ecology in a simplistic fashion. It is a manifold and cumbersome concept, covering too much and touching too much to be handled tidily. We feel comfortable with it only if we don't know much about it.

In this chapter, we certainly want to convey an appreciation of the immensity of the problems associated with ecology. But more than that, we hope to show the great challenge of it all. (Ecologists may one day be recognized as the most important scientists on earth.) What, then, do ecologists study? In the original Greek the root word, *oikos,* means "house." Thus, ecology is "the study of the house"—the place where we live, or the environment. The *environment,* technically, includes all of those factors, both nonliving and living, that affect an organism. Ecology, then, is the study of the interaction between organisms and their environment. The key word is *interaction*.

We saved the study of interaction until late in our discussions of life because a certain amount of basic information about organisms is necessary in order to understand their interactions. Certainly, ecology is where the sciences come together. Let's begin the discussion now by looking at ways in which ecologists have managed to organize and categorize the earth's environment.

THE BIOSPHERE

The **biosphere** is that part of the earth that supports life. It is that thin layer wherein the wondrous properties of light and water interact to permit life, not only in the terrestrial surface, but in the subterrestrial realm and the fresh and marine waters as well. When we consider the biosphere, we find that we're dealing with great expanses, but little depth. Things can't live very far above or below the earth's surface. More precisely, the habitable regions of the earth lie within an amazingly thin layer of approximately 14 miles. This includes the highest mountains and the deepest ocean trenches. If the earth were the size of a basketball, the biosphere would be about the thickness of one coat of paint.

Physical Characteristics of the Biosphere

Conditions within the biosphere are quite special. Our space probes have found no other place in the solar system where these conditions exist. The conditions here result from our distance from the sun, the presence of water, the makeup of our atmosphere, and the earth's solid crust. (Jupiter and Saturn, in fact, are gaseous balls.) Let's look at some of these factors a little more closely and then consider how various life forms are distributed over the earth.

Water You'll recall that much of the biochemistry of life is centered on the peculiar chemical traits of water (see Chapter 2). In addition, you may remember that water has a high specific heat—that is, it requires a relatively great input of energy to raise its temperature. Likewise, water loses energy very slowly (due to the hydrogen bonds between water molecules). Water, then, acts as a great stabilizer of temperature. It absorbs heat slowly and retains it well. Water vapor in the atmosphere is one of the reasons for the earth's comparatively moderate climates. This moisture helps hold heat and slows the radiation of heat from the earth's surface.

The Atmosphere The earth's atmosphere may indeed be wispy and ethereal, but all life depends on this fragile veil. Chemically, it is a protective envelope of gases—78% nitrogen, 21% oxygen, and less than 1% other gases, including small amounts of carbon dioxide and water vapor. Most of the earth's atmosphere clings close to the planet, not extending more than five to seven miles above the surface. In addition to its reservoir of useful gases, the atmosphere is also vital to life in that it screens out much of the dangerous ultraviolet radiation that would otherwise make the earth's surface inhospitable.

As we also have noted, the atmosphere absorbs heat, and in so doing, acts as a great "heat sink," temporarily holding heat close to the earth's surface. In its role as a heat sink, the atmosphere can be compared to a florist's greenhouse. In a greenhouse, light energy is readily admitted through the windows. The absorbed light energy is radiated back as heat energy, but unlike light, the escape of heat is retarded. Thus, greenhouses remain warm without added heating—even in the winter—but only as long as ample sunlight is available. The heat striking the earth is trapped in this manner by atmospheric carbon dioxide and water vapor. (People sometimes learn about the "greenhouse effect" the hard way—when they return from shopping after having left their pets in a closed car.)

It is important to note that while the total energy the earth receives from the sun has historically been equaled by the escape of radiant energy, this equilibrium is being seriously disrupted as humans continue to pollute the atmosphere. But one pollutant we will have difficulty controlling is carbon dioxide. Every form of combustion in which we involve ourselves—from breathing to burning fossil fuels to clearing forests and fields—releases carbon dioxide. The added carbon dioxide in the atmosphere contributes to the earth's greenhouse effect as it traps radiant heat, producing what is believed to be a gradual but inexorable increase in atmospheric temperature. (The greenhouse effect is examined in further detail in the next chapter—see Essay 47.1.)

Solar Energy Only about half of the incoming solar radiation ever reaches the earth's surface. About 30% is reflected back into space by the earth and its atmosphere. Another

FIGURE 46.1
THE EARTH'S ENERGY BUDGET
The relative constancy of conditions in the biosphere depends ultimately on an equilibrium between energy entering and energy leaving. Of the solar energy reaching the upper atmosphere, 30% is immediately reflected back into space. Another 20% is absorbed as heat by water vapor in the atmosphere. The remaining 50% reaches the earth's surface. Most of this energy reenters the atmosphere through evaporation from the earth's waters. The thin arrow represents energy used by organisms.

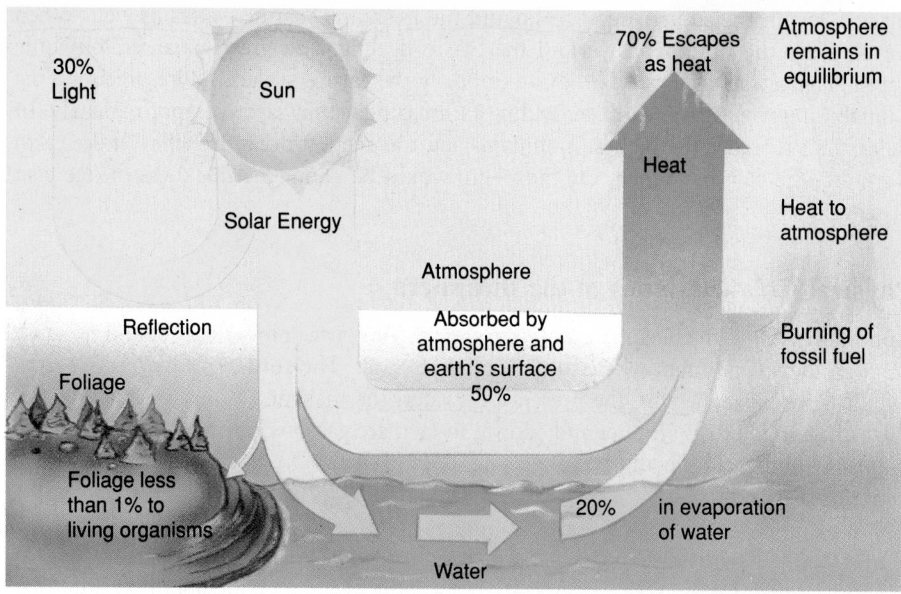

20% is absorbed by gases in the atmosphere. The remaining half reaches the earth's surface as light energy, where it is absorbed by the land and waters, and is then radiated back into space as heat (Figure 46.1). But a great deal of work is accomplished by the 50% reaching the biosphere.

Surprisingly enough, of this light, considerably less than 1% will enter photosynthetic processes. Most of the energy is used in shuffling water around. After all, the heat from solar energy is responsible for most of the evaporation from the oceans, lakes, rivers, and not insignificantly, from the leaves of plants through transpiration. As water absorbs this energy, its molecules gain heat—they move more rapidly—and are finally lifted into the atmosphere as water vapor. There, the heat is lost as the water condenses into its liquid phase, falling as precipitation (such as rain or snow). So the absorption of solar energy is vital in the distributing of water over the earth, which, as we have seen, is also a way of redistributing heat, a factor that moderates our climate. The constant shifting of water between its liquid and gaseous phases over the earth is called the **hydrologic cycle** (Figure 46.2).

Climate in the Biosphere One of the most striking effects of solar energy reaching the earth is seen in the great annual seasons of the planet. The cyclic seasonal changes occur as the earth follows its orbit, exposing different parts of its surface to the direct rays of the sun as it moves. This changing exposure occurs because the earth somehow ended up with a rotational axis that is not perpendicular to the sun's rays, but is 23.5° from vertical (Figure 46.3). Thus, most of the earth's creatures are blessed with fluctuating but moderate surface conditions, and not the alternatives: great heat at the equator and huge expanses of perpetually frozen belts where the temperate zones are now located. This tilted axis also produces less dramatic phenomena, such as the tradewinds and patterns of rainfall. How do such changes arise?

The sun's rays fall more directly on the equator than any other part of the earth, so the equatorial regions are hot. This heat causes great warm air masses to rise, carrying with them large amounts of water vapor. Warmer air holds more moisture than does cooler air. The rising of warm air masses creates a void below, and lower, colder, and drier air masses rush in from the north. These masses, in turn, heat up, become moisture-laden, and rise. Thus, we have air cells both north and south of the equator circulating in opposite directions. Because of the earth's rotational force, the cells are thrown off at an angle. This is called the **Coriolis effect.** These moving air cells produce what are

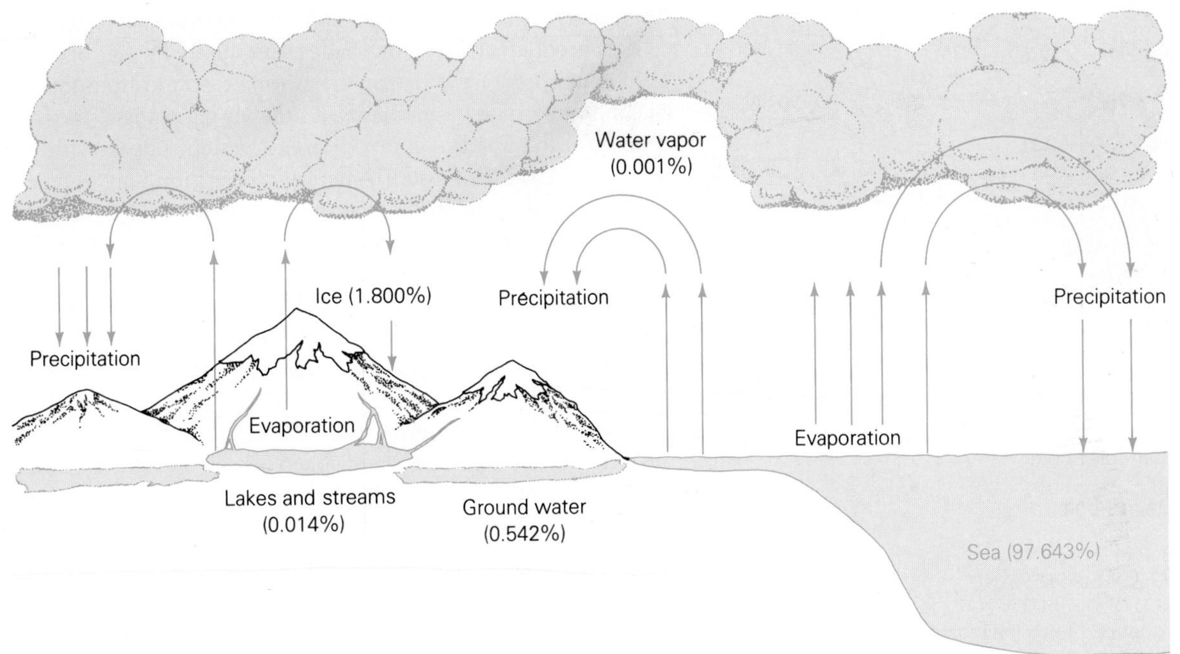

called the tradewinds, the consistent winds so long used by sailing vessels. You can see in Figure 46.4 that there are other air cells in more northerly and southerly latitudes.

The movement of moisture-laden air from the equator creates the equatorial rainfall patterns. The rising air cells cool rapidly and lose most of their water as rainfall near the equator. As the air in the cells moves northward, it becomes increasingly drained of moisture and thus yields less precipitation. Finally, at about 30B latitude, north and south, the lack of rain produces **deserts.**

Other factors may also contribute to the formation of deserts. One is called a **rain shadow**, a region of low rainfall on the lee side of a mountain range. Wherever moisture-laden prevailing winds encounter a high mountain range, most of the precipitation falls on the windward slopes of the mountains. As the air masses move up the mountain slopes, they cool and lose most of their moisture as rain or snow. The drier air mass moves down the leeward slopes (becoming warmer and therefore able to hold more moisture), then scurries over the desert and gathers up whatever moisture there is,

FIGURE 46.2
THE HYDROLOGIC CYCLE
Water is constantly being redistributed about the earth through the ongoing hydrologic cycle. The cycle, driven by solar energy, includes evaporation and precipitation. The distribution of water on the earth is quite uneven, with the greatest percentage—almost 98%—occurring in the oceans. Visible fresh water – that of rivers, lakes, streams, etc.—makes up far less than 1% of the earth's H_2O, an amount that is well exceeded by water locked into the polar ice caps, snow-capped mountains, and glaciers—some 1.8%.

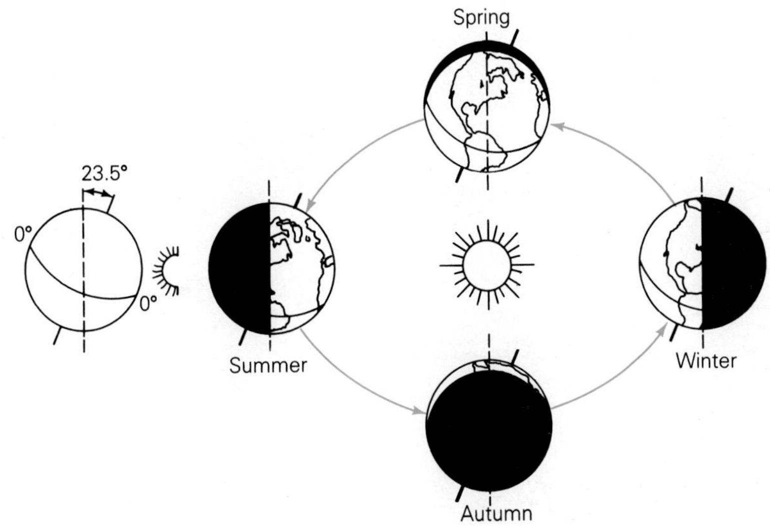

FIGURE 46.3
THE SEASONS IN NORTH AMERICA
Because of its tilted rotational axis, the earth presents a constantly changing face to the sun as it continues its orbit. The tilting produces these seasons in the northern hemisphere.

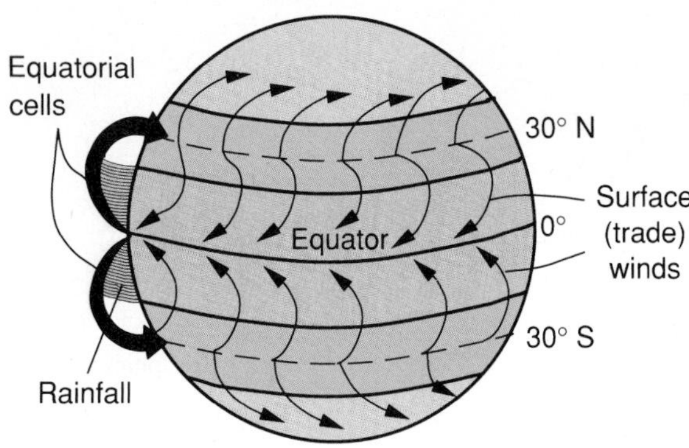

FIGURE 46.4
MOVING AIR MASSES AND RAINFALL DISTRIBUTION
The distribution of precipitation on the earth is determined to a large measure by the formation of several groups of air cells. Air nearest the equator rises, carrying abundant moisture with it. As it rises, cooler, drier air rushes in underneath, and the cell rotates. The rising air mass cools and dumps most of its water in belts north and south of the equator. Rainfall is far more limited just above and below these belts, and here are found some of the earth's great deserts. Note that the great rotating equatorial cells are thrown off center by the spinning of the earth—the Coriolis effect—so the winds do not simply blow north and south, but occur in easterly and westerly directions.

intensifying the situation. As an example, the Great American Desert is a product of the rain shadow produced by the Sierra Nevada mountains. Moisture-laden air, moving eastward from the Pacific, is confronted by the mountains. As the air masses rise, they cool, and most of the water is dumped on the western slopes, with very little to the leeward. In South America, where prevailing winds blow to the west, the rain shadow is produced by the Andes range, and the deserts that form west of the Andes along the coasts of Peru and Chile are some of the driest known.

The moving air masses also help create the ocean currents (Figure 46.5). Ocean currents are, in turn, important because of their effects on the climate of nearby land masses and because they help mix the waters and distribute the nutrients and gases required by aquatic organisms. Major oceanic currents called **gyres** are found in the northern and southern hemisphere. They are chiefly surface phenomena, with the flow extending down just 100 to 200 meters. The deepest flow of a major current is found in the Gulf Stream, extending as far down as 1000 meters. The movement of the major gyres transfers equatorial heat northward and southward, a feat that profoundly affects the climate of coastal regions.

THE DISTRIBUTION OF LIFE: TERRESTRIAL ENVIRONMENT

The earth's surface varies markedly from place to place, as does the life it supports. Vast, distinct, and recognizable associations of life are called **biomes** (Figure 46.6). More precisely, a biome is a particular array of plants and animals within a geographic area brought about by distinctive climatic conditions. Biomes are usually identified more by their plant associations than those of animals, not only because the first is far more obvious, but also because it determines the second. The specific plant associations are, as you would expect, a product of adaptation to several climatic factors, including precipitation, temperature, and light.

FIGURE 46.5
THE OCEAN CURRENTS
The major ocean currents are produced by the earth's winds and modified by its rotational forces. The two great Pacific currents, the Japan Current in the north and the Humboldt Current in the south, carry cold water south and north along the west coasts of North America and South America, respectively. Note that the Atlantic Gulf Stream is quite different, originating in a more tropical region, and carrying warm water north along the east coast of North America before heading out across the Atlantic.

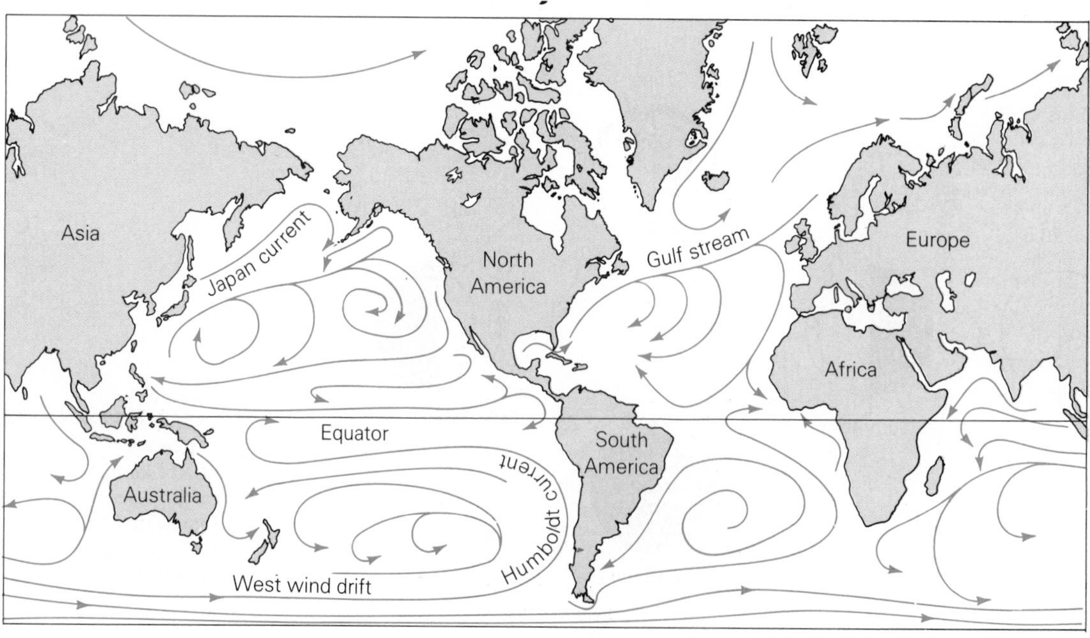

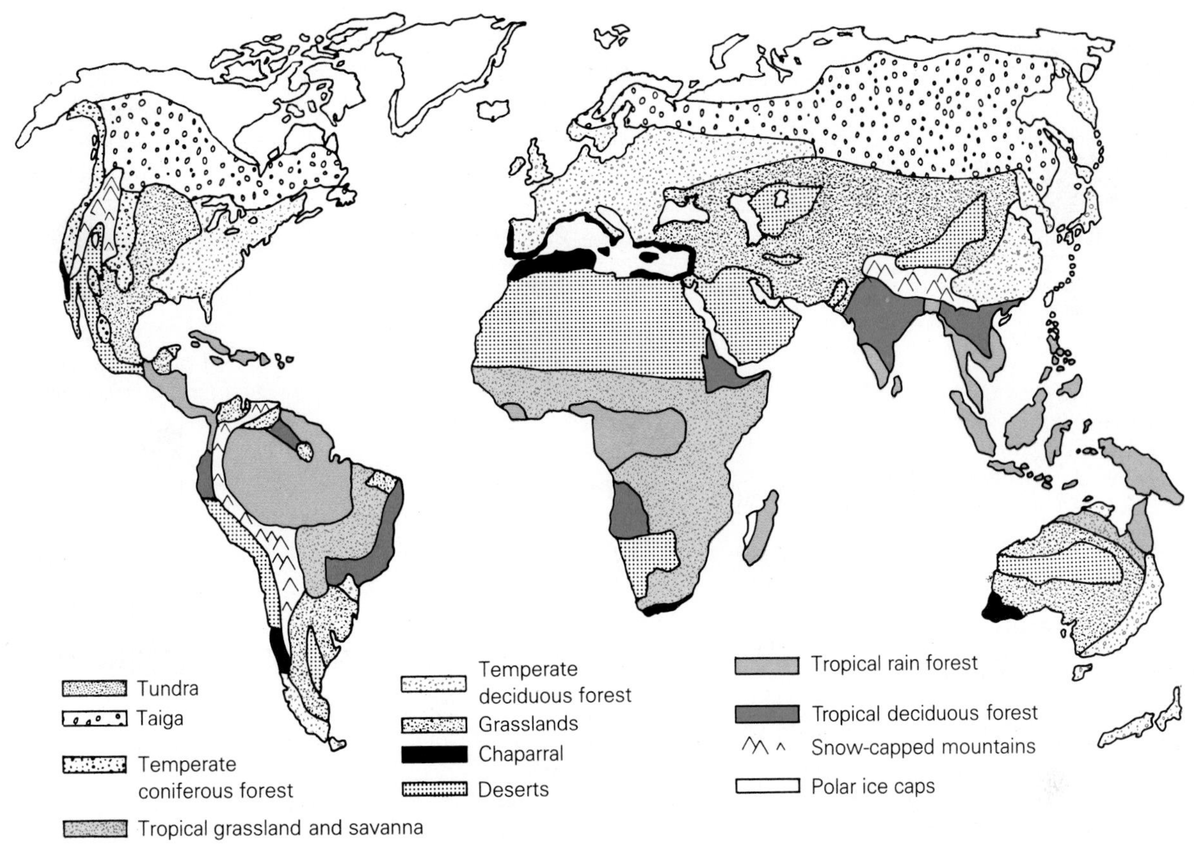

FIGURE 46.6
THE BIOMES
Each of the biomes can be identified by its dominant form of plant life. These forms are adapted to the climatic conditions, including precipitation, the availability of light, and, of course, temperature. Both latitude and altitude affect all these conditions.

Biomes in turn may be subdivided into communities. A **community** is an assemblage of interacting populations of organisms forming an identifiable group within a biome. As with biomes, communities are generally recognized according to their predominating plant species. Several communities may exist within any biome. If you were to begin a walk across a biome, you might notice that although the general climate may remain the same, specific groups of plants and animals change somewhat. Thus, we might find both a mesquite community and a sage community in a desert biome. If you were somehow able to return to that biome hundreds of years later, you would find that biomes also change over time. We will return to community structure and interaction in the next chapter, but for now let's make a few more general observations about biomes and then describe the major ones.

As the biomes change from one place to another, we find some transitions are gradual—forming **transition zones**, regions of overlap between two ecological communities in which species of both communities are found. For example, we find a gradual transition across the United States from the east coast and Appalachian Mountains to the western coastline (Figure 46.7). The moist forests of the Appalachians slowly give way to drier oak-hickory forests and then to forests consisting almost entirely of oak. The forests become less luxuriant as they fade into the great American grasslands. That is, they did before the grasses yielded to the intensive agriculture of "America's breadbasket." The prairies were once seas of tall grasses, which gave way (where there was less precipitation) to shorter and shorter grasses. These yield to the Great American Desert, which is followed by the Sierra Nevada and coastal mountain ranges and finally the Pacific shore. Coniferous forests are common on the mountain slopes, and minor **grasslands** and deserts are found between some ranges. We see, then, that the borders of biomes are usually indistinct, with the mixture of plants seemingly engaged in an endless tug of war over boundaries. The principal determining factors in transitions are precipitation and temperature (see the biome map in Figure 46.7).

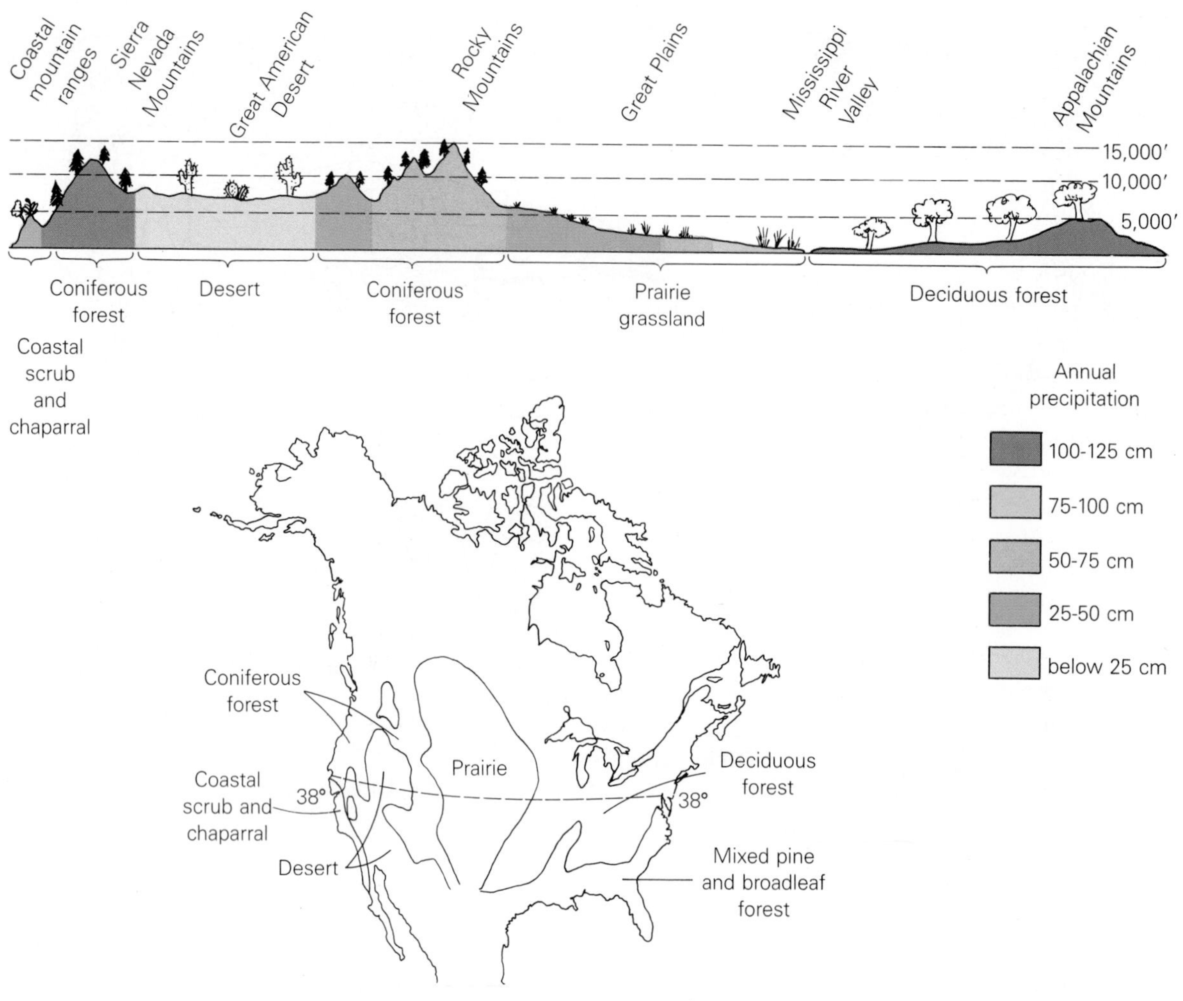

FIGURE 46.7
PROFILE ALONG THE 39TH PARALLEL CROSSING THE UNITED STATES
The line transects several of the major biomes and helps indicate the conditions by which they were produced. For example, the American deserts lie in the rain shadow of the western mountain ranges. Rain and snow fall on the western slopes only. Moisture-laden air from the Atlantic, at the other end of the profile, drops its burden as it moves west, the total rainfall diminishing as it moves inland. Forests occur west to the Mississippi, but from there, prairie extends westward, dwindling into short grasses toward the Rocky Mountains. The Great American Desert extends between the Rocky and Sierra Nevada Mountains.

The distribution of biomes on the earth generally follows latitude, but this pattern is more obvious in the Northern Hemisphere than in the Southern Hemisphere. Going from the equator northward we tend to move from **tropical rain forest** through desert, grassland, **temperate deciduous forest**, the **taiga** (coniferous forest), and finally **tundra** and ice cap. This latitudinal arrangement of biomes is not entirely orderly for several reasons, one of which is the terrain, or topography. Mountain ranges interrupt the orderly distribution of biomes and, as we have seen, are often responsible for the presence of deserts because they can block the movement of moisture.

Mountains also influence biomes in another interesting way: biome distribution is a product of altitude as well as latitude. This is because temperature decreases as altitude increases. Thus, high mountains with permanent snow or ice are found in the tropics, and the usual biome transitions may be represented as well. Further, the *altitudinal* transition may resemble the latitudinal one (grassland into deciduous forest, and so on; see Figure 46.8).

One of the best examples of altitudinal transition is the Ruwenzori mountain range in east central Africa. One can ascend from tropical rain forest through broad-leaved evergreens, to deciduous and coniferous forests, then to alpine meadows, and finally to the barren, snowswept peaks. Once again we see that the factors determining the distribution of life are complex.

Keeping in mind that the dividing lines are often indistinct and arbitrary and that the complexities are greater than any brief discussion can convey, we will now consider the nature of the earth's great biomes, beginning with the driest and perhaps most fabled.

Deserts

Deserts are areas that receive less than 25 cm (10 in) of rain each year. You may be surprised to learn that deserts are not necessarily hot (even in the day) and not necessarily tropical. The largest desert is the Sahara, which, as you can see on our biome map, covers nearly half of the African continent (and, for reasons we will get to shortly, is getting bigger). Other large expanses of desert are found in Australia, Asia, western North America, and South America (in fact, every continent but Europe). Temperatures in these regions undergo dramatic day-night fluctuations and may vary as much as 30° C (54° F) in a 24-hour period. The reason for such extremes is the lack of buffering, heat-retaining moisture in the desert air and soil. The surface heats up rapidly in the daytime, but cools down by evening. In spite of the long periods of drought, much of the actual topography of the desert floor is determined by water. Very seasonal but torrential rains cause flash flooding that continually remakes the face of the desert.

If you have never seen a desert, you might have an image of lifeless regions of drifting sand studded with a few palm-covered oases. Actually, there are such places, but most deserts are alive with plants and animals. Since dry areas are called **xeric** (Greek for "dry"), plants adapted to deserts are called **xerophytes.** Their adaptations can take a number of routes, but they have one common imperative: Save water. Native American perennials, such as cactus, ocotillo, joshua, creosote, and palo verde (Figure 46.9), are adapted to living long periods on what little water they contain, while waiting for rain. Through evolution, the leaves of the cacti have been reduced to spines, thereby conserving water, and photosynthesis occurs mainly in the green stems. The stems are covered with a thick, waxy cuticle, and water is stored in the oversized cells of deeper tissues. The spines of cacti and the thorns of many other desert plants discourage animal browsing and prevent subsequent water loss by the injured stems.

Compared to other plants, desert perennials tend to have fewer and more widely scattered stomata. This helps in conserving water, but the price is a considerably slower growth rate, since CO_2 uptake and photosynthesis are retarded. Other desert perennials have deep root systems able to tap whatever water seeps into the porous soil; very small leathery leaves; and the ability to lie metabolically dormant, growing very little, for long periods. Some desert cacti and other succulents have reversed their gas exchange cycles, taking advantage of the cooler and moister nights. This latter group is composed of the CAM plants discussed in Chapter 28.

The desert annuals have a different strategy for survival. After a spring rain (or *the* spring rain), countless dormant seeds germinate, seedlings quickly sprout and mature, and the desert becomes transformed as tiny flowers of all descriptions erupt into full bloom, a trembling and riotous offering of vibrant color and delicate forms. Their life cycle is short, and they soon die. But their seeds are highly resistant, lying in the sand until subsequent rains propel them into their surge of growth and reproduction.

Just as plants have had to adapt to the rigors of the desert, so have animals. Animals have the advantage of being able to adapt not only anatomically and physiologically, but also behaviorally. The desert is rich in animal life, including a variety of arthropods, some resident birds, and often many seasonal migratory bird species. There are also reptiles and even some mammals. In the hot deserts, the primary problems of all these species include escaping daytime heat and avoiding water loss. Many desert animals beat the heat by simply staying out of the sun and becoming nocturnal. In fact, if you should visit the desert you may see only a few lizards, a bird or two, and a few insects—unless you want to traipse around at night with the snakes.

Desert mammals are largely represented by rodents, animals for which nocturnality has very clear advantages. They tend to lose water quickly because they breathe rapidly and have a large surface area relative to their volume. Thus, most rodents spend the desert days deep in insulating burrows, venturing out only at night to forage. Rodent-hunting predators such as owls and rattlesnakes follow suit, confining most of their activity to the cool of evening or night.

FIGURE 46.8
AN ALTITUDINAL DISTRIBUTION OF BIOMES
The altitudinal variation provided by mountains often mimics latitudinal distribution. In this hypothetical model, a rain forest at sea level gives way to a deciduous forest at a higher elevation. Conifers replace the deciduous trees farther up, which yield to alpine meadows, and finally a rather typical tundra situation near the glacier.

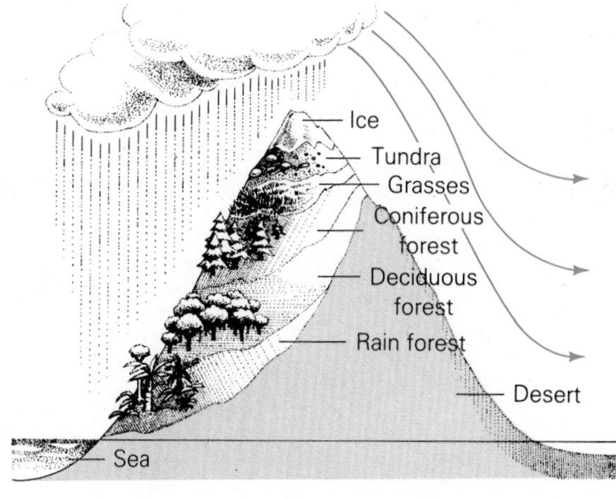

FIGURE 46.9
DESERT PLANTS
The North American desert plants shown here have adapted to limited water; these are called xerophytes. Their thorny epidermis discourages browsing and helps shield the green surfaces from the harsh, direct sunlight. Many xerophytes have pulpy water-storing tissues within. Shown are the Saguaro cactus (near right) and the Joshua tree.

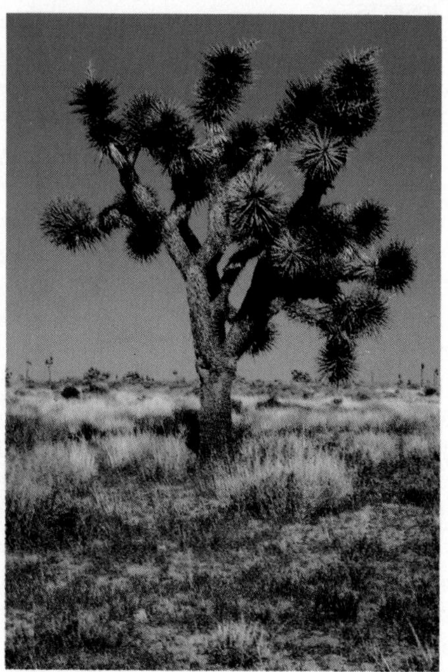

The few daytime animals, such as long-legged and fast-moving lizards, are preyed upon by hawks and roadrunners, which are also adapted to daylight conditions. Even the daytime animals, however, restrict most of their activity to the morning and evening hours (Figure 46.10).

Perhaps we can best gain some insight into the adaptive strategies[1] of desert animals by considering a mammal and, of all things, an amphibian, each of which has made unique adjustments to desert life. These are the versatile kangaroo rat and the spadefoot toad. The kangaroo rat *(Dipodomys deserti)* of the southern California desert is particularly interesting because it doesn't drink. It survives on the water content in its food and supplements this with metabolic water (produced as a waste product in cell respiration; see Chapter 8). The kangaroo rat is also a great water miser. Its remarkably efficient kidney produces only highly concentrated urine (23% urea and 7% salt, compared to 6% urea and 2.2% salt in humans), and very little at that. The feces are also dry and crumbly. Most water loss, in fact, occurs through simple breathing. Even these special physiological features, however, wouldn't permit desert survival were they not coupled with nocturnality. The rat spends its day in a humid, hair-lined burrow, venturing forth only in the cooler evening.

While amphibians are notably scarce in the desert, the fascinating spadefoot toad is a year-round resident of the American deserts. It escapes heat and drying by burrowing, and while entombed it is also capable of **estivation,** a time of dormancy, when buried during long, hot, dry intervals. When sufficient rain occurs, it springs (hops?) into action. For this creature, time is short. Like other toads, the spadefoot must reproduce in water. Finding a temporary pond, the male begins at once to call prospective mates in the briefest of courtships. Eggs, fertilized in the shallow ponds, hatch in a day or two. The young tadpoles complete their metamorphosis and emerge as adults in a few short weeks.

Although deserts seem tough and unyielding, they actually represent one of the more fragile biomes. Simple systems are always the most vulnerable, and deserts are essentially simple places, harboring few species compared to many other places. Because the desert is so vulnerable, the impact of humans can be great.

The recent experience of the region to the south of the Sahara is sobering. People of

[1]"Adaptive strategy" refers to the development of traits in response to an environmental condition and does not imply a purpose on the part of the organism.

the Sahel region long supported themselves and their small herds of cattle by hand-drawn water from a few scattered deep wells. In recent decades, gasoline pumps were introduced, making water suddenly abundant. The result was an explosive increase in the cattle population, followed by a substantial increase in the human population. The cattle placed great pressure on the vegetation of the area until a recent drought—not unlike the many droughts the region has previously survived—killed most of the cattle and many of the people. As a result of such destructive patterns, the Sahara itself has expanded by thousands of square miles, probably permanently, demonstrating that well-intentioned but ill-conceived aid programs can backfire with tragic results.

Grasslands

Grasslands are simply areas dominated by grasses. In the Northern Hemisphere, they exist as huge inland plains and include such areas as the Asian steppes and (in times past) the prairies of North America. Grasslands and deserts are rather similar in important ways and, in fact, many grasslands gradually fade off into deserts. The chief climatic difference between the two is precipitation; grasslands, of course, get more rain—roughly 25 to 75 cm (10 to 30 in)—but the rain is of a seasonal nature.

In the Southern Hemisphere, grasslands (Figure 46.11) are known under various names: the pampas of South America and the veldt and savanna in Africa. (We will consider the savanna as a separate biome because it also harbors scattered groves of trees.) Grassland in Australia is very extensive, occupying over half the continent.

After saying that the dominant plants of grasslands are grasses, we may be able to regain your interest by also noting that grasslands are very different from one place to the next. For example, in the former American prairie, grasses east of the Mississippi reportedly grew 10 feet tall, while those in the west rarely surpassed a foot or two. Again, the difference was principally due to variation in rainfall. But it is important to realize that the grassland biome is also maintained by fire and by the action of grazing

FIGURE 46.10
DAY AND NIGHT DESERT FORAGERS
Animals in the desert forage in two shifts. The day feeders are fewer in number but much more active. Included in their ranks are the African ground squirrel (**a**), roadrunner (**b**), and many insects. But even these hardy foragers often remain in the shadows at midday. At sundown the second shift begins as carnivores like the scorpion (**c**) and the Fennec fox (**d**) go quietly about their deadly business.

FIGURE 46.11
GRASSLANDS
Natural grasslands such as the South American pampas are slowly changing with the encroachment of civilization.

animals. Enormous fires, often started by lightning, periodically sweep through all natural grasslands today. But the rhizomes and extensive root systems survive, and regrowth begins with the next rains. Were it not for fire and grazing animals, the deciduous forest—whose plants are often not fire-adapted—would undoubtedly encroach on many grasslands.

Since rains are usually seasonal in grasslands, the plants have developed important strategies for the dry periods. In lowland regions, root systems may penetrate a permanent water table as far as 3 meters below the surface. More commonly, the grasses rely on vast, spreading diffuse root systems in which no root dominates; the great surface area provided by so many finely branching roots enables the grasses to quickly absorb water from light rains. In addition, grasses readily become dormant, reviving when water is once again available. Some grasses produce underground stems (rhizomes) that remain alive after all the foliage has died. The rhizomes, as well as the aboveground horizontal stems (stolons), form the dense sods that also actively prevent the growth of trees where rainfall would otherwise permit it. This matlike growth prevented agricultural intrusion in the American prairie until there were improvements in plowing implements. (As we mentioned in Chapter 27, the first grassland farmers were dubbed "sodbusters.")

Grasslands are highly efficient at rapidly converting solar energy into the chemical-bond energy of their living matter. They can support larger populations of animals than any other biome on earth. So it is not surprising to find huge herds of grazing animals (Figure 46.12). In fact, grasslands are the habitats of many of the world's large hooved herbivores. The original herbivores of the American prairie—the bison and the pronghorns—have been displaced by cattle and sheep (Figure 46.12), but in the plains of Africa there are still vast herds of wildebeest, zebra, and other natural grazers. These native grazers quickly move on without destroying the grasses, and the plants continue growing from their cropped bases.

Overgrazing, or grazing by herbivores that kill grass by cropping it too short (such as sheep may do), can irreparably harm grassland. The mesquite and cactus-covered wastelands area in Texas (55 million acres) was stable, productive grassland before the cattle barons subjected it to heavy grazing. Much of the Sahara and the deserts of the Middle East, in fact, have been created by domestic grazing in past centuries. The barren Middle East, remember, was once called "the land of milk and honey."

Where wildlife is protected or nurtured in the United States, herds of bison and pronghorn antelope are the conspicuous grazers. But the fields are full of unobtrusive grass eaters, such as jackrabbits, rodents, and prairie dogs, as well as insects and seed-eating birds.

With such a food reserve, the grassland can be expected to support large numbers of predators. Unfortunately, they may prey on domestic herds as well as wild species, and this has presented problems for both us and them. Obviously, we're not going to allow predators to get away with eating herbivores we have raised for profit, so large predators are generally hunted down and shot. These animals include wolves, cougars, coyotes, and even foxes.

Snakes and carnivorous birds range a bit wider, but they don't eat many animals from our domesticated herds. (In spite of the lack of supporting evidence, some ranchers still consider birds of prey to be a threat and thus justify shooting them—including the endangered bald eagle, our national emblem.) One particular and grisly exception is a parrot in New Zealand, the kea *(Nestor notabilis),* which swoops out of the forests, lands on a sheep's back, tears into its flesh, and eats the fat from around the kidneys of the living animal.

Many of the animal species found in grasslands are similar to those of our next group due to convergent evolution, when unrelated groups develop similarities through adapting to similar environments.

Tropical Savannas

FIGURE 46.12
GRASSLAND HERBIVORES
The highly efficient producers of the grassland support an extensive food web, often including incredible numbers of large herbivores. These mammals, in turn, support a sizable predator population.

Tropical savannas are a special kind of grassland that forms at the borders of the tropical rain forest. Unlike other grasslands, the savanna is frequently interrupted by scattered trees or groves (Figure 46.13). And unlike the tropical forests, the savannas have a prolonged dry season contrasting to an annual rainfall of 100 to 150 cm (40 to 60 in). It is during the dry season that the savannas are subject to sweeping fires. The great annual droughts are also responsible for the movements of huge migratory herds of grazers in search of food.

The largest savannas occur in Africa, but savannahs are also found in South America and Australia. While grasses are the dominant form of plant life in the African savanna, the drab landscape is brought to life by palms, colorful acacias, and the strange, misshapen baobab tree, which appears to be growing upside down. The number and variety of hooved animal species exceed that of all other biomes and include the familiar zebras, giraffes, wildebeests, and numerous antelopes of the African plains. The African savanna is also the domain of familiar predators, such as the lion and cheetah. Just as in the grassland, the natural fauna of the savanna is being replaced by domestic grazers, especially cattle.

Tropical Rain Forests

Forests are scattered over much of the world, but only in those places where the water supply is adequate. Rarely is water more available than in the tropical rain forest. These are tropical woodlands with an annual rainfall of at least 250 cm—100 in (and up to 450 cm—180 in)—typically found in lowlands. Rain falls throughout the year but is somewhat heavier during the "rainy season." The largest tropical rain forest is in the Amazon River basin in South America (Figure 46.14). The second largest is in the wilds of the Indonesian archipelago. And then there are those of the Congo basin in Africa, parts of India, Burma, Central America, and the Philippines.

Tropical rain forests are best described as lush, with a very large number of tree species growing to great heights. The floor is dark and wet, and the air is often cool and laden with rich smells. Unlike what we find in the other forest biomes, no single kind of plant dominates the forest. Any tree is likely to be a different species from its neighbors, and nearest trees of the same species are often miles apart. The exceedingly tall trees, ranging in height from 30 to 45 meters (about 100 to 150 ft) form a dense, continuous canopy overhead. The crowns of lesser trees form a subcanopy below their taller neighbors. The trees are invaded by large numbers of vines, and both trees

FIGURE 46.14
THE TROPICAL RAIN FOREST
Tropical rain forests receive enormous amounts of rain throughout the year. There is little seasonal change. No single plant species dominates the terrain, but a number of kinds form a dense canopy over the sodden earth. The humidity on the forest floor can be stifling.

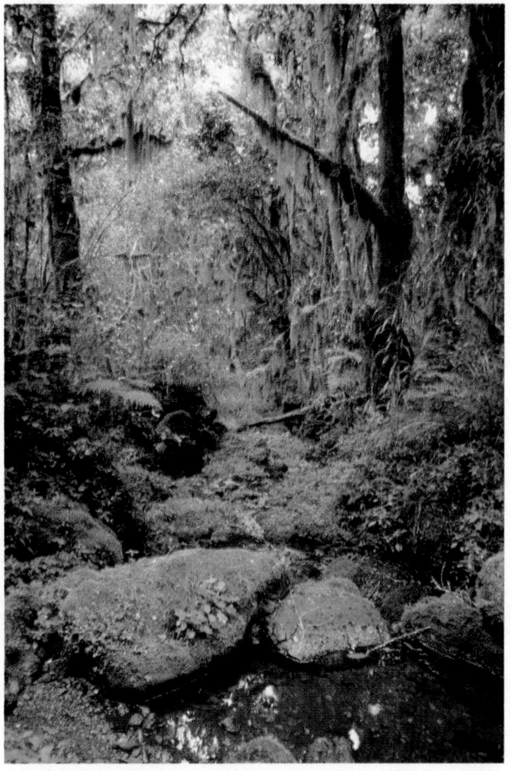

FIGURE 46.13
THE SAVANNA
Large hooved mammals are prominent among the savanna's herbivore population. On the African plains, predators include the large cats.

and vines may be festooned with **epiphytes**, plant species that have evolved ways of joining the keen competition for sunlight. Epiphytes live on the stems and branches of tall trees in the canopy, with no contact with the soil. They absorb water directly from the surrounding humid air. (One species surrounds its roots with a bucketlike base which collects water and insects, the decay of the latter assuring the epiphyte of a continuing supply of nitrogen compounds.) The forest floor may have little to moderate foliage, but it is teeming with fungal and bacterial decomposers and insect scavengers. The darkness, warmth, and blanketing humidity there are ideal for rapid decomposition.

If the rain forest floors often don't have much foliage, then what about those reports of the "impenetrable jungle"? The answer is, jungles are special kinds of tropical forests. Essentially, they contain a low-growing tangle of plants, which is, in fact, almost impenetrable. Jungles arise where light reaches the forest floor. Jungle areas may be scattered through rain forests, but they are particularly common along river banks and steep slopes and in disturbed areas such as clearings and deserted farms. Along river banks the forest is called *wet jungle,* and it abounds with insects and reptiles. The idea that jungles represent tropical rain forests grew from descriptions by river travelers who weren't about to get out of the boat, and so they missed seeing the relatively clear forest floor, often just a few hundred yards from the river bank.

The tropical rain forest harbors a great many animal species, more than any other biome. Insects and birds are particularly abundant, and reptiles, small mammals, and amphibians are common. Many of the animals are arboreal (tree dwellers), and many species are stratified according to the layers established by the plants, becoming specialists at occupying certain levels of the canopy and subcanopy (Figure 46.15). In one study of the Costa Rican rain forest, ecologists found 14 species of ground-foraging birds, 59 species occupying the subcanopy, and 69 in the upper canopy. They further found that about two-thirds of the mammals there were arboreal, as were a number of frogs, lizards, and snakes.

Tropical seasonal forests occupy a considerable portion of the tropical biomes. These differ from true tropical rain forests in that rains are, as you might expect, seasonal. A familiar example is the monsoon forest of Southeast Asia, although such forests exist in other tropical regions. In many instances, the trees there are deciduous, losing their leaves during the dry season. When the monsoons arrive with their torrential rains, the forest takes on some of the characteristics of the tropical rain forest. Some of the more highly valued hardwoods, such as Burmese teak, are found in the tropical seasonal forests.

The tropical rain forests of the world are rapidly disappearing. The Amazon basin rain forest of Brazil, the last really large area of undisturbed forest, is now in the process of being rapidly cleared for timber and farming and being exploited for minerals. The results may prove to be catastrophic for several reasons (see Essay 46.1).

Chaparral

Mediterranean scrub forest, or **chaparral** as it is known in California, is characterized by a dense growth of low evergreen shrubs and trees. As our map shows, it is rather insignificant among the forests of the world. It does, however, have unique characteristics and its own peculiar plant associations. Note, for example, that chaparral is exclusively coastal, found mainly along the Pacific coast of North America and the coastal hills of Chile, the Mediterranean, southernmost Africa, and southern Australia. This forest is unique in that it consists of broad-leaved evergreens, growing in subtropical regions marked by a marine air flow, low rainfall, and a long summer drought. Depending on the altitude, California's coastal chaparral receives between 25 and 75 cm (about 10 to 30 in) of rainfall, almost all of it in the short winter rainy season. The seasonality of the rains mean that the plants experience drought through most of the year.

Plants and animals adapted to chaparral have adopted strategies similar to those of desert dwellers. In the chaparral, plants are chiefly represented by shrubs—mostly with dwarfed, gnarled, and scrubby stems—and scattered succulents (Figure 46.16). Leaves are generally small, with very waxy and tough cuticles. Many plants become dormant

FIGURE 46.15
CANOPY DWELLERS
Each level of vegetation in a rain forest supports specific kinds of animal life. As examples, the dense canopy (the upper layer, broken by occasional taller "emergent" trees) supports monkeys and birds; the subcanopy, composed largely of smaller trees, supports many kinds of insectivorous birds; the smallest trees and bushes are home to many frogs and snakes; and the forest floor supports deer, capybara, and pigs. Ants are found in abundance at all levels.

ESSAY 46.1

THE DESTRUCTION OF TROPICAL RAIN FORESTS

More than half the world's people live in tropical and subtropical areas. About a third of these people are extremely poor. These two facts have set the stage for a potential worldwide disaster. The problem is, these are the sorts of people with direct access to the world's tropical rain forests. Those rain forests can meet some of their immediate needs, and so they are utilized on a pell-mell, devil-take-the-hindmost basis. Locally, the great forests are yielding to the unending search for firewood and new planting and grazing areas. Much of the Central and South American rain forest has been cleared for grazing areas in order to supply North American fast-food places with beef. (Costa Rica is quickly being cleared of its rain forests in order to reduce the price of U.S. fast-food hamburgers by about a nickel. The beef is of relatively poor quality and, in fact, comprises only about 6% of our fast-food needs. Some chains, such as Burger King, no longer import beef from these areas.) Rain forests are also being cut to clear areas for gold mining and for the export of hardwood logs.

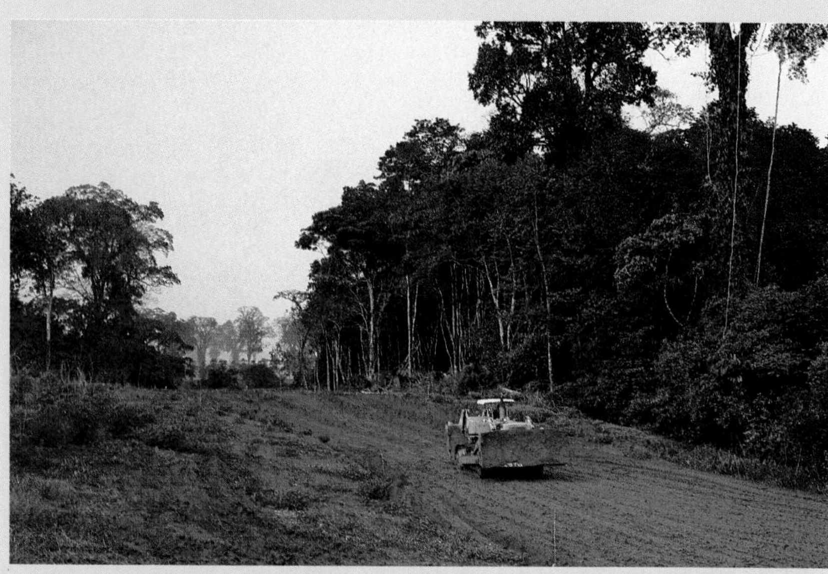

Unfortunately, such logging operations are often highly inefficient. (In some Malaysian forests, over 50% of the trees in an area are cut to obtain 3% that can be sold.) By 1980, almost half the world's tropical forests had been destroyed or severely disturbed. By 2000, almost no tropical rain forests will have escaped such devastation.

The greatest impact occurs through simply clearing the land. About 35 acres of tropical rain forest are cleared each minute. Much of this forest will never be recovered, especially that which is bulldozed, because the thin layer of topsoil with its nutrients is removed. Also, germinating seeds are removed, and watersheds are disrupted (see

after the seasonal rains, remaining that way through the dry summers. Insects abound in the chaparral, including armies of beetles that live in the leaf litter covering the forest floor. Other inhabitants include vertebrates such as mule deer, rabbits, bobcats, rodents, lizards, snakes, and birds. Wrentits and towhees are typical residents, but a large number of birds are migratory.

Chaparral has a peculiar problem related to drought. It is often seared by fast-moving brush fires. Controlling the fires along the California coast is complicated by the rough, hilly terrain. Little of the chaparral escapes the periodic fires, but the plants have adapted to the violence. Some sprout quickly from burned stumps, and others scatter fire-resistant seeds over the burned ground. Ecologists have described the chaparral as a **fire subclimax community.** This means that the chaparral virtually never reaches maturity, and any given stand is always in some stage of recovery. The floor of chaparral is often heavily layered with leaf litter. Decomposition is slow because of the dryness. The accumulation of those leaves (the plants may shed leaves year-round) magnifies the fire problem.

Temperate Deciduous Forests

If you live east of the Mississippi, the **temperate deciduous forests** may dominate your figure). Forests that are simply cut will recover fastest, since seeds, seedlings, and a protective undergrowth are left. Land that has been cleared by burning recovers at an intermediate rate. That is, burned-over land will take about 80 years to recover.

Land is burned in order to clear the trees and release the nutrients they have stored back to the earth through what is called "slash and burn agriculture." Here, the trees are cut over a small area and burned, thereby releasing their stored nutrients back to the soil. The problem is, the land is only fertile for a few years, and then the forest and the crops are both gone.

The extremely heavy rainfall that gave rise to this rain forest in the first place is almost entirely composed of moisture sent aloft from the steaming forest itself. Little additional moisture moves in from the ocean. Cutting down the trees greatly reduces the amount of moisture returned to the atmosphere within the Amazon basin, and this will ultimately reduce the rainfall there by a substantial amount. The rain that does fall into the cleared land will tend to run off into the river system and then to the ocean.

In addition, the tropical rain forests, which are being cleared at an average rate of 1% per year, represent an enormous carbon dioxide "sink." This means that the lush growth absorbs a great deal of carbon dioxide from the atmosphere, locking it up in the carbon compounds of the plants. While death and decay cause a steady turnover, the sudden removal of plants interrupts the normal cycling. Further, since the plants are burned, the carbon stored in their molecules is suddenly released as carbon dioxide. As we will see, atmospheric scientists are particularly alarmed by this steadily increasing carbon dioxide burden.

Finally, you may be surprised to learn that tropical soils are very poor and infertile. The nutrients formed by decomposition are immediately recycled back into plant growth, so no reservoir of humus remains. In addition, the rains continually leach precious nutrients from the porous soil. This fact has become painfully clear to proponents of jungle agriculture. Once the native plants have been removed, the soil may rapidly change into a hard, water-resistant crust known as *laterite*. The term, which means "brick," rather aptly describes the reddened crust.

One of the greatest problems resulting from the destruction of the rain forest is in destroying a form of promise we often didn't even know we had. We know of only about one-sixth of the three million species believed to exist there. Some of these species may have medicinal properties that could remedy specific problems of humankind in ways we can now only imagine. There may be resistant genes in plants related to our vulnerable crops (which could lend their resistance to our crops through genetic recombination). There may be new kinds of food, antibiotics, and therapies awaiting our discovery. But some scientists believe that, at current rates, one million tropical species may be gone by the year 2010. And with them, sadly, go promises untold.

surroundings. Temperate deciduous forests are marked by trees that lose their leaves in winter. Winter is quite dramatic here as the trees rake the gnarly fingers of their bare branches at the winter sky. Each spring they blush anew with fresh faint buds that will change to the deep greens of summer. Temperate deciduous forests, then, are characterized by both leaf fall and changing seasons (Figure 46.17). Such forests extend over much of the eastern United States, northward into southeast Canada. They are also found in the central and northern parts of Europe, including Great Britain, and reach into southern Norway and Sweden. A long finger of the forest pushes into the center of the Soviet Union. In eastern Asia, deciduous forests are found in China, the eastern Soviet Union, Korea, and Japan. Although much less conspicuous in the Southern Hemisphere, they do exist in coastal Brazil, east Africa, the eastern coast of Australia, and across most of New Zealand.

Precipitation in the deciduous forest is rather evenly distributed throughout the year, with rainfall often averaging over 100 cm (39 in)—enough to support a variety of trees. Typically, the larger trees such as oak, maple, beech, birch, and hickory form the canopy. In northerly deciduous forests, communities of beech and maple dominate, while in the south, the oak is common. In times past the oak was joined by the chestnut, but this majestic tree has been all but obliterated by a bark fungus *(Endothia parasitica)*, acci-

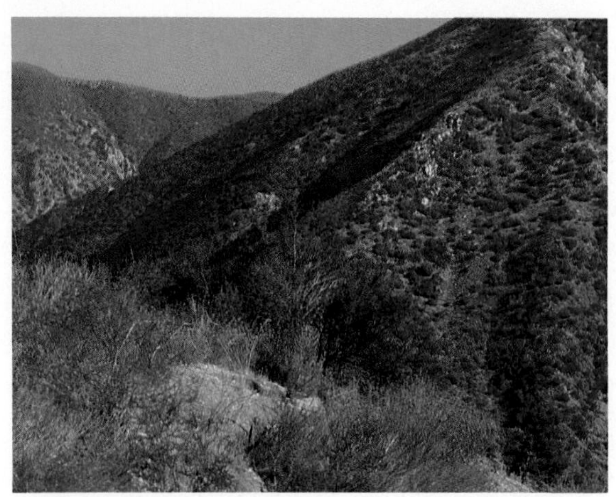

dentally introduced from China at the turn of the century. Interestingly, whereas the adult trees are gone, shoots still emerge from the persistent roots, and some seeds are produced, but further growth is stopped by the fungus.

In temperate deciduous forests, moderate levels of light reach the forest floor and encourage the growth of younger trees, shrubs, ground-hugging plants, and a variety of annuals. The annuals begin to grow in early spring, or even late winter. They rather quickly produce their seeds and die off with the autumn frosts, contributing to the rich leaf litter. The litter and humus harbor an abundance of scavengers and decomposers, especially in the warm summer months. The forest floor is like a giant soft sponge, soaking up rain and contributing to the luxuriant forest growth.

A variety of animal life abounds in deciduous forests (Figure 46.18). Although we have displaced or killed off most of the larger mammals, in some areas they have been protected as game for hunters. The forests were once the home of deer, wolves, bears, foxes, and mountain lions. Mammals in these forests today are largely represented by rabbits, squirrels, raccoons, opossums, and rodents.

FIGURE 46.16
THE CHAPARRAL
The chaparral may lack the lushness of many other forests, but its plants are tenacious and hardy. These scrubby plants resist an annual drought that would discourage most other plants.

The largest predators are likely to be humans, owls, hawks, a few black bears, and occasional bobcats and badgers. In the northern deciduous forests, a few wolves have escaped having their skins trim cheap coats. Finally, as any camper knows, arthropods inhabit the forest in great numbers, and can be found rummaging through the litter and foliage and crawling into tents.

Winter in the deciduous forest is heralded by one of the most beautiful and moving events in nature: autumn. As the abscission layers in the leaf petioles prepare to separate, and the green chlorophyll wanes, other colors come to dominate the woodland scene. The leaves almost seem to celebrate the cool days, as browns, reds, and yellows mix with the greens of conifers. It is too soon over. And the fading light of the shorter days, which changed the leaves, also warns the animals of the approach of winter. Some birds leave; others change color and give up territories. Some mammals, grown heavy with fat, find holes in which to sleep or to hibernate. Others, unable to escape either behaviorally or physiologically, simply face the coming cold; some will not survive. Soon, the stark, lifeless days of winter settle in.

The deciduous forest biome produces trees with strong and flexible cell walls, and much of it has been cleared in order to satisfy not only our need for wood but our extraordinary demand for paper, both of which are made largely from these cell walls. The forests have also fallen simply because we had something else to put in that space. In the American Southeast, native hardwoods have been replaced by faster-growing, more marketable pines, planted in orderly rows. In any case, we have removed so much forest that today almost all American hardwood forests are, at best, second-growth areas. As such, they are in transition, with many regions dominated by scrub oak and other uninspiring invaders.

Taiga and Coniferous Forests

The taiga, or boreal (northern) forest, is almost exclusively confined to the Northern Hemisphere. This is a moist, subarctic forest biome of Europe and North America, dominated by spruces and firs. The taiga forms immense forests across many northern climes, but it is also

FIGURE 46.17
THE CHANGING DECIDUOUS FOREST
The deciduous forest changes its appearance at different times of the year. The lovely green hillside of summer will explode in a riot of colors when autumn arrives. Both are in sharp contrast to the starkness of winter.

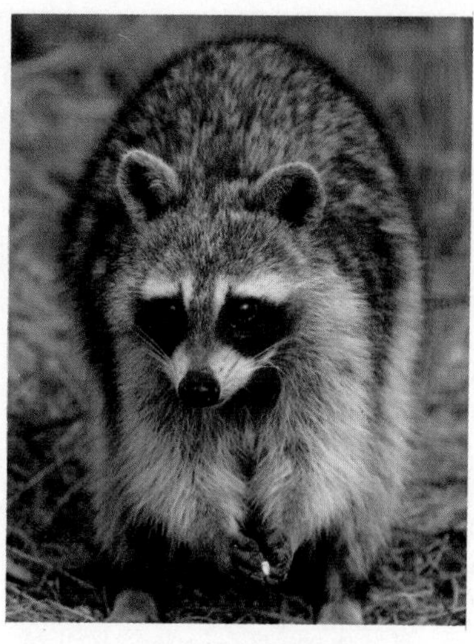

FIGURE 46.18
ANIMALS OF THE DECIDUOUS FOREST
Herbivores of the deciduous forest are numerous, including the cardinal (left) and white-tailed deer (center), in places where they still remain. Typical carnivores today include the raccoon (right).

found at higher elevations in many more southerly mountain ranges (Figure 46.19). The taiga is unmistakable; there is nothing else like it. It is subject to long, cold winters and short summer growing seasons. It is difficult to generalize about rainfall because the taiga is so extensive; some places get a great deal of rain, some don't. But in many parts, there is much less rainfall than in deciduous forests. (In the Canadian interior, for example, rainfall is between 50 and 100 cm—20 to 40 in.) The taiga is interrupted in places by extensive bogs, or **muskegs,** the remnants of large ponds. These wet regions are commonly dominated by spruce, but low-lying shrubs, mosses, and grasses form the spongy ground cover (Figure 46.20). The northernmost taiga is touched by sunlight for only six to eight hours each day through much of the winter, and bathed in sunlight for 19 hours in the summer.

The dominant plants of the taiga are conifers, but occasional communities of hardy, broad-leaved poplar, alder, willow, and birch may be found in disturbed places. Conifer leaves, which are usually needle- or scalelike with dense, waxy surfaces, represent adaptations to the dry conditions common in the taiga. Such adaptations enable conifers to survive the winter with many of their leaves intact; thus, they are sometimes referred to as evergreens. The leaves are adapted for water conservation, since much of the water reaching the taiga may be bound up (and made unavailable) in ice. (One is reminded of the spines and waxy cuticles of desert plants.)

Leisurely strolls can be quite pleasant in a coniferous forest because there is very little underbrush. Those soft needles that feel so good under your boots are hard on other plants. They make the soil notoriously acid (and thus the conifers reduce their

FIGURE 46.20
MUSKEGS
Where bogs and marshes form, the taiga is interrupted. These areas are often referred to as muskegs. They represent a perpetual tug of war between aquatic and terrestrial environment, as plants continually invade the marshes, some failing, others becoming established.

FIGURE 46.19
THE TAIGA
The taiga is an extensive biome, found almost exclusively in the Northern Hemisphere. Its dominant plant is the conifer, although communities of birch, willow, poplar, and alder are not unusual. Since the conditions of the taiga are duplicated in high mountains, similar communities are found there.

FIGURE 46.21
ANIMALS OF THE TAIGA
Herbivores of the taiga include large mammals such as the elk (shown at top) and caribou. Porcupines are very common. The lynx (shown above) and grizzly bear are still found in the taiga, protected somewhat from human intervention by the vastness of the area.

level of competition with other species). The straight trunks and dark, unobstructed forest floor lend a cathedrallike atmosphere to taiga forests, an atmosphere that some find compelling and others forbidding.

The taiga harbors such large herbivores as moose, elk, and deer (Figure 46.21). It is the last refuge of the grizzly bear and black bear, and wolves still roam here, as do lynx and wolverines. While rabbits, porcupines, hares, and rodents abound, insect populations aren't as large as in deciduous forests, although there is an abundance of mosquitoes and flies in boggy regions during the summer.

There are other large coniferous forests that are not considered taiga. A number of temperate coniferous forests are to be found throughout the Northern Hemisphere. For example, the Olympic forest in western Washington is, on its western slopes, a bona fide temperate rain forest with 500 cm (200 in) of rain per year near the glacial peaks. Here, the Sitka spruce reaches its greatest size, approaching that of the famed California redwood. Fingers of the western forests follow the Cascade Mountains into Oregon and California.

Perhaps the most fascinating coniferous forests are the California coastal redwoods. Their gigantic demands for water are not met by rain alone, but also by coastal fogs. On the western slopes of the Sierra Nevada Mountains, we find the magnificent forests of giant sequoias. These awesome and splendid trees are among the world's oldest living things. Finally, not to be ignored, are the expanses of pines that dominate the coastal plain forest in the southeastern United States. These vast forests of longleaf, loblolly, and slash pine may be a temporary invasion of the usual oak-hickory communities that have been disturbed by human intervention and disease.

The coniferous forest is a continuing target of the lumber industry, but its very vastness has protected its more distant regions. The latest and most controversial of foresting methods is called "clear-cutting." Unlike traditional selective foresting, clear-cutting involves clearing *every* tree from the land (Figure 46.22). As this practice continues, the timbering industry is careful to advertise that it is replanting the areas at great expense.

There is a problem with such reforestation, however. Mixed forests of genetically diverse and resilient plants are replaced by artificially developed and genetically homogenous trees that are fast-growing and can quickly be reharvested. Because some kinds of conifers are now being routinely cloned through developments in tissue culture techniques, absolute genetic uniformity in some forests looms just over the horizon. Biologists are quite concerned with the dangers of genetic uniformity in any system, since without variation, destruction by an unforeseen invasion of a mutated fungus, bacterium, or virus could virtually destroy an extensive forest. All of these eventualities aside, where will the jay, bear, and raccoon live while the clearcut forest recovers?

A number of studies have shown that other kinds of damage result from clear-cutting forests. Following clear-cutting, runoff from rainfall increases substantially, as you might expect, but this is accompanied by what ecologists call a "nutrient flush." There is a three- to twentyfold increase in the loss of mineral nutrients from the soil over what is normally recorded in forests. Most dramatic is a substantial loss of critical nitrogen.

Tundra

The tundra is the northernmost land biome, characterized by Arctic plains that support a dense growth of mosses, lichens, and dwarf herbs and shrubs. Except for certain alpine meadows, it has no equivalent in the Southern Hemisphere. Tundra is actually located in a narrow band between northern taiga and the Arctic ice extending from the tip of the Alaskan peninsula around the earth and back to the Bering Sea (see Figure 46.7).

Travellers in the tundra may be struck by the absence of tall trees and shrubs. The annual precipitation is often less than 15 cm (5 in), and much of this occurs as snow. So it's a dry place—at least during the long winter when everything is frozen. However, you wouldn't think it dry if you visited the tundra during its brief, damp summer. When spring and summer finally arrive, the upper few feet of soil thaw, leaving the **permafrost,** perpetually frozen soil, below. The surface thaw produces unusual conditions for a "desert." Ponds begin to form everywhere. Since water cannot percolate down, the plains of the tundra become a veritable bog (Figure 46.23).

The tundra receives less energy from the sun than does any other biome. Because it is near the pole, any sunlight there strikes at an acute angle, losing much of its energy after having traversed diagonally through the atmosphere. The lack of sunlight drastically affects the growing season. During the brief six-to-eight week summer, plants must photosynthesize and store enough food to last through the rest of the year. And they must compress their reproductive period into this brief season, all of which means they can't put much energy into growth. So the tundra is carpeted by low-lying plants. Another factor that dictates that plants hug the ground is the constant high wind, particularly in regions of higher elevation. A tall plant would simply be buffeted to death or even ripped away.

Because of the permafrost, the plants of the tundra can't form deep anchoring root systems; so they form shallow, diffuse roots that often become entangled with the roots of their neighbors to form a continuous mat. In the wet season, any disturbance on hilly slopes can break the entire root mat loose from the wet soil underneath and send a great mass of plant conglomerate sliding down, exposing the barren soil below. Because of the limited growth of its plants, tundra recovery is a very long, very slow process.

Lichens and mosses are common in the tundra, but so are willows and birches. These latter, however, are almost unrecognizable, being only dwarfed symbols of their more southerly relatives. They, with the lichens and mosses, are joined by grasses, rushes, sedges, and other annuals to complete the summer ground cover (Figure 46.24).

Animals life isn't as rare as one might expect in this most northerly biome. Where autotrophs are at work capturing and storing energy there will always be opportunistic heterotrophs ready to harvest it. In fact, year-round residents of the tundra (Figure

FIGURE 46.22
CLEAR-CUTTING
Clear-cutting—a controversial, potentially disastrous foresting practice—involves the widespread clearing of trees in large stretches of land, as can be seen in this view of an Oregon forest.

FIGURE 46.23
THE TUNDRA
In summer, the treeless tundra becomes a veritable marsh as the snow melts. With little runoff, the landscape becomes dotted with innumerable small ponds. Below the water the soils of the tundra are perpetually frozen. The plants dotting the landscape include a number of dwarfed trees, grasses, and abundant reindeer moss.

FIGURE 46.24
PLANTS OF THE TUNDRA
Tundra plants must adapt to drought and strong winds, coupled with shallow soils and seasonal flooding, as we see here.

FIGURE 46.25
ANIMALS OF THE TUNDRA
Because of the paucity of species in the tundra, the food webs may be comparatively simple (and, therefore, easy to disrupt). Common herbivores in the tundra include large animals such as the caribou and small ones such as the lemming. The grizzly bear is one of the many carnivores of the tundra.

46.25) include some rather large herbivores such as, in North America, the caribou and musk oxen, and in Europe and Asia, reindeer. Other browsing animals found here are the ptarmigan, the snowshoe hare, and the ever-present and legendary lemmings. Lemming populations are clear indicators of how good the season is for producers. In good years, lemming populations soar.

Lemmings, in turn, determine the success of a number of predators, including the Arctic fox, lynx, snowy owl, weasel, and Arctic wolf. Also, the jaeger, a migratory bird, travels great distances to feed on the tiny rodents. The duration of the predators' stay, as well as their reproductive success, will depend on the number of lemmings. In years when lemmings are few, many of the predatory birds will migrate early, while the permanent residents will survive by switching to other prey.

A number of waterfowl and shore birds also migrate to the tundra. When they arrive in the spring they must mate and rear their broods quickly, before the brief summer is over. Why would some species travel long distances to breed in such a dismal and risky place? Perhaps because the long days permit extended feeding periods and less competition for food than might be present in the winter feeding grounds. If winters are harsh, resident competitors are few.

While invertebrates are certainly more limited in the frigid tundra, remarkable swarms of mosquitoes and tiny flies abound. They enter a frantic race to complete their reproductive activities before summer ends, since the adults only live for one season. Such insects survive by producing highly resistant immature stages that remain dormant through the long winter.

Winter comes early in the tundra, and with the rapidly shortening days the migratory animals disappear. Some caribou, for example, leave for the forested taiga, where winter food is more plentiful. Those that remain prepare for survival by whatever strategy they have developed. Lemmings retreat to food-laden burrows, while ptarmigans tunnel into snow banks to emerge periodically on foraging expeditions. Since the larger herbivores and predators don't hibernate, they must rove the barren, windswept landscape to feed on mosses, lichens, and each other.

Our own species, with the exception of a few Laplanders in northern Scandinavia and Finland, avoids the tundra. Even the hardy Eskimos prefer the Arctic coastline. For this reason, humans have traditionally had little effect on this most fragile ecosystem. This is fortunate, because the links in the food chains here are few and important. The sudden loss of a single predator or producer through human intervention could be disastrous. We know that recovery from damage is agonizingly slow and costly to the tundra communities. Wheel ruts left by a single wagon that passed over the soil 100 years ago are still clearly visible. Our impact, however, is increasingly disruptive as we search out more oil and other minerals in this delicate biome.

North of the tundra (or in other places at higher altitudes), the vegetation gives way to barren and rocky soil similar to that of the Antarctic. Plant life is sparse and patchy. These dry windswept plains extend to the coastal ice floes and glaciers. Here the marine environment, in sharp contrast, is amazingly rich in life, even in its colder waters. Of course, the perpetual search for more petroleum threatens these rich waters as well.

WATER COMMUNITIES

The Freshwater Province

The waters of the earth are generally not divided into biomes, yet they are similar to those terrestrial realms in that they mold their inhabitants along certain general lines by presenting them with common ecological opportunities and risks. Life had been subjected to the molding effects of water for almost three billion years before it occupied dry land.

Rivers and Streams Estuaries are fed by rivers and, on a smaller scale, streams. Rivers and streams are bodies of water that move in one direction. As we see in Figure 46.26, their headwaters are usually narrow and steep. Since headwaters are often spring-fed and at higher elevations, they may be intensely cold. The falling, turbulent water tends to be mixed and thoroughly aerated. The communities in these areas are composed of relatively few species, such as trout (species that need the cold temperatures and oxygenated waters). A few invertebrates are found among the algae and mosses that cling tenaciously to the rocky banks. Further along, the riverbeds tend to become flatter and the rivers wider and slower. Here, the waters carry more suspended material and may become more murky, relegating the penetration of light to the upper levels and shallow areas where algae and cyanobacteria may thrive. The fish species here need less oxygen, so one may find the slower, less active, bottom-feeding species, such as catfish, carp, and—lurking above them—bass. As the river widens and becomes even slower, suspended materials collected upriver may settle out, forming a nutrient-rich sediment that can support relatively large and complex communities.

Lakes and Ponds As water moves along the channels cut by rivers and streams, it may pause in its inexorable cycle to rest for a time in still bodies called lakes (and on a smaller scale, ponds).

Lakes and ponds differ not only in size, but usually in depth. Ponds are often shallow and therefore can be risky places to live. The shallowness means that light is likely to penetrate the entire pond, encouraging the growth of algae and cyanobacteria, as well as bottom-dwelling plants. Thus, the entire pond may rapidly become clogged with life that can eventually fill in the pond entirely (Figure 46.27). Another problem with shallowness is much more direct: in dry periods, the pond may simply evaporate entirely.

Lakes are grander places, and, compared to other major features of the earth, they may be of relatively recent geological origin. Most occur in northerly or alpine regions as products of the last glacial retreats (10,000 to 12,000 years ago). They are also produced through volcanic activity, as was the famed Crater Lake in Oregon, and through gradual uplifting of land, as were many of the shallow, acid lakes of Florida. The deepest lake in the world is much older than most other lakes: Lake Baikal in the Soviet Union, at 1750 m (5742 ft, or well over a mile) was formed during the Mesozoic Era.

As we will see in more detail in the next chapter, at those depths penetrated by light, photosynthesis stimulates the growth of algae and cyanobacteria and, with the suspended particles washed in by rivers and streams, can give a murkiness to the waters. If the suspended materials settle rapidly, though, the lake may maintain its clarity to some degree. The murkiness that signals productivity can be encouraged by the use of fertilizers

FIGURE 46.27
THE END OF A POND
Many ponds become choked with the growth of water plants. Soon, plants from the shore will encroach and the pond will rapidly disappear.

FIGURE 46.26
THE HEADWATERS
In the headwaters where river systems begin, fast-moving rocky streams are typical.

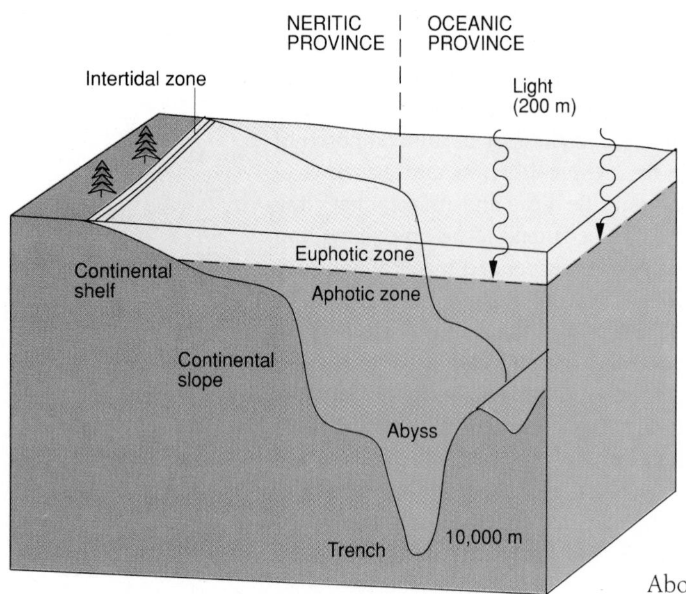

FIGURE 46.28
ORGANIZATION OF THE MARINE ENVIRONMENT
The marine environment is subdivided into the oceanic and neritic provinces, which in turn are divided into lighted (euphotic) and unlighted (aphotic) zones. Most marine life is concentrated in the neritic province, where more abundant mineral nutrients are found.

which, through runoff, find their way into the waters that feed the lake. (Even the private use of lawn fertilizers around Lake Tahoe threatens to turn that wonderfully clear body of water to a flat opaqueness).

Lakes can produce complex communities as small fish and aquatic insects feed on the zooplankton (tiny animal life that feeds on the producers near the surface). Below, larger fish feed on the insects and smaller fish. Around the shore one finds snakes, salamanders, frogs, toads, and birds. Raccoons and opossums may prowl the lake's damp edges, amidst a variety of predatory insects, such as dragonflies, damselflies, and caddis flies.

The Marine Environment

About three-quarters of the earth's surface is covered with water, and most of it is salty (Table 46.1). But not only is the marine environment vast, it's complex as well; it contains many kinds of communities. They fall into two major divisions: those in the relatively shallow areas along continents, or the **neritic province**, and those in the deep-water open sea or **oceanic province** (Figure 46.28). The marine environment can also be subdivided into the **euphotic zone**, through which light penetrates, and the **aphotic zone**, which is in perpetual darkness.

The Oceanic Province Much of the open sea is devoid of life forms. Nonetheless, the oceanic province is the home of the **pelagic** organisms, those that drift in the open sea. Most of these are restricted to the euphotic zone—regions where light can penetrate. Almost all wavelengths of light, depending on the turbidity, are absorbed by the upper 100 m of water, although some shorter wavelengths (blue) may be detected at 300 m. Oceanic waters that are nearly devoid of particles (and thus, of nutrients) are quite blue, while ocean waters with suspended particles (nutrient-rich) may be green, brownish, and sometimes even reddish. In such turbid waters, light may penetrate only 10 meters or so.

Perhaps the greatest mysteries of the sea lie in the **abyssal region**, the deeper waters where depths can vary from 300 to nearly 11,000 m. The famed Marianas Trench is 10,680 m deep (over six miles, and deeper than Mt. Everest is high). Little was known about the ocean abyss until technology permitted its direct exploration. In 1960, the floor of the Marianas Trench was reached by Don Walsh and Jacques Piccard in the bathyscaph, *Trieste*. It was a momentous event, but scientists have visited these depths many times since then, and we are beginning to understand more about what is going on down there.

We know, for instance, that the ocean depths are places of darkness, tremendous pressure, and numbing cold. Nevertheless, these formidable, deep abyssal regions support a surprisingly large number of peculiar **benthic** (bottom-dwelling) scavengers. Tethered cameras focused on bait, miles deep, have photographed primitive hagfish, many species of bony fish, crustaceans, mollusks, echinoderms, and even an occasional shark. For years oceanographers and marine biologists assumed that since producer populations could not exist at such depths, benthic creatures of the aphotic zone had to rely for their energy on the continual "rain" of the remains of pelagic (surface-swimming) organisms as they settled to the ocean floor. But more recently, biologists have discovered that there are unusual communities of organisms living along rifts and vents in the seabed, and that these have producer populations. Obviously, the producers aren't photoautotrophs but are, instead, chemoautotrophic organisms. We will learn more about the chemoautotrophs and the ocean rift community in the next chapter (see Essay 47.2).

The water below the photosynthesizers, but above the ocean floor, harbors some of the most peculiar creatures on earth: the deep-sea fish (Figure 46.29). They are usually small and dark and keen of sight, with big toothy mouths. (They have to be able to handle just about any kind of food they come across.) These predators often produce their own light, with which they signal each other or attract prey. Some species cultivate luminescent patches of bacteria beneath their eyes, which become visible when the fish rolls down a specialized eye covering.

The Neritic Province In the neritic province, the land masses touch the edge of the highly variable **continental shelf,** the shallow waters adjacent to the shores of a continent. The shelf is considered to be a submerged part of the continent. The neritic province ends at the continental slope, where the shelf drops off, the bottom receding to greater depths, often abruptly.

In many shallower areas of the shelf, light penetrates to the ocean bottom. Such regions are continually stirred by waves, winds, and tides, keeping nutrients suspended and supporting many forms of swimming and bottom-dwelling marine life. Giant kelps and other seaweeds form extensive beds, offering hiding places for many fish species (see Figure 23.27). Let's look briefly at the organization of life and the flow of energy in this province.

The seas are home to many tiny species called **plankton,** which drift near the surface of the open ocean. The photosynthetic species are called the **phytoplankton.** Included are algal protists—mainly diatoms and dinoflagellates, all of which are microscopic. Also included are the more recently discovered minute, flagellated **nanoplankton,** which are clearly important ecologically, but poorly understood. Phytoplankton, along with the marine algae—the kelps and seaweeds—are the primary producers of the sea. Thus, energy enters the marine ecosystem via these species. It has been estimated that 80 to 90% of the earth's photosynthetic activity is carried on by marine organisms. Thus, the chain of life in the sea begins with tiny phytoplankton capturing the energy of the sun to build energy-containing compounds within their fragile bodies. But where there is energy, there is likely to be something trying to utilize it, and the tiny algae are fed upon by tiny planktonic animals only a little larger than themselves—the **zooplankton.**

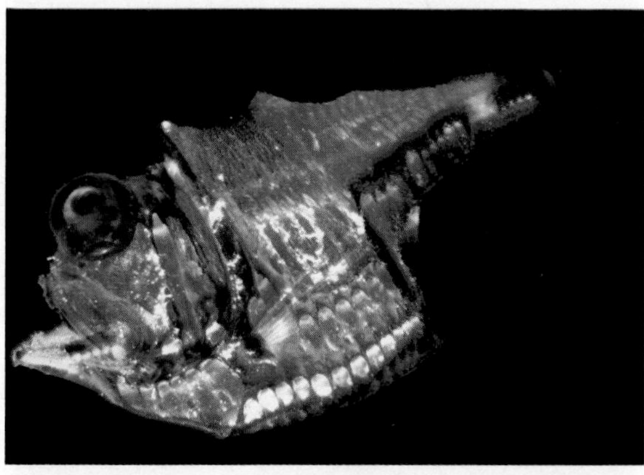

Argyropelecus

Eutaeniophorus

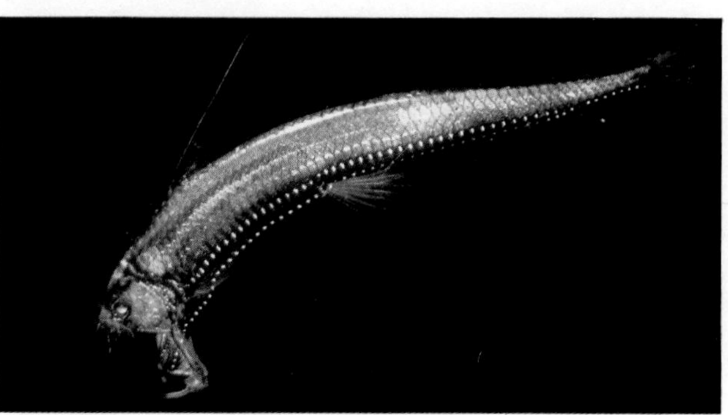

Chauliodus

FIGURE 46.29
FISHES OF THE ABYSS
The deep-sea fishes are among the most bizarre of all creatures. They live where there is no sunlight and very few other animals. Thus, they must be equipped to capitalize on just about any living thing they might come across.

FIGURE 46.30
THE FOOD PYRAMID OF THE OCEAN
The tiny phytoplankton at the base (not drawn to scale) capture the energy of the sun. These are eaten by animals larger than themselves, which are eaten in turn by larger animals. At the top are the largest carnivores of the sea.

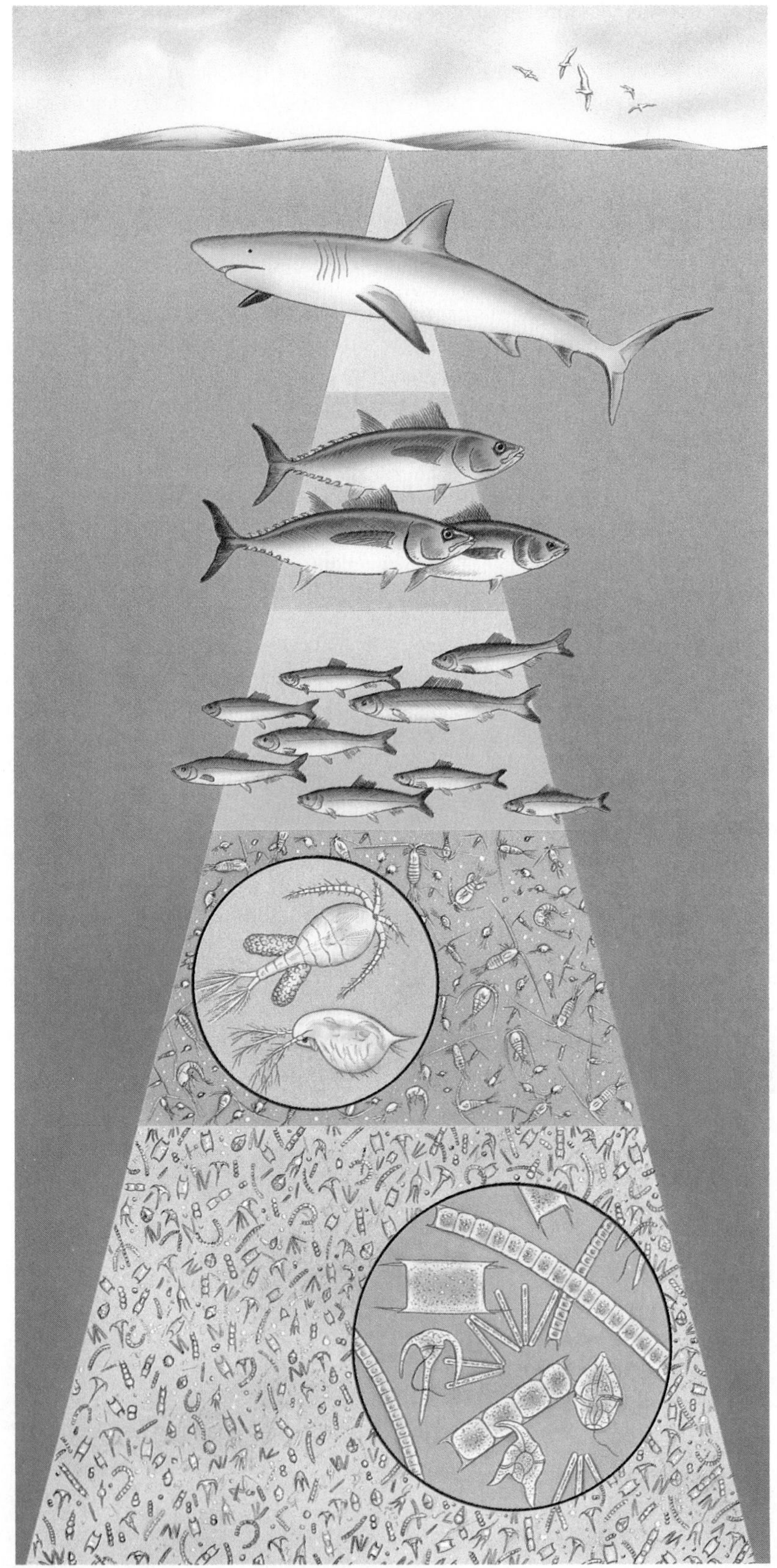

A variety of animals, from tiny fish to the great baleen whales, feed upon plankton of all sorts (Figure 46.30).

Among the more productive regions of the marine environment are the colder offshore waters, where **upwellings** occur. Upwellings are the movement of deep, nutrient-laden colder waters to the surface. While little is known about the vertical movement of water in the open sea, coastal upwellings are better understood. They are generally a seasonal phenomenon, where coastal winds blow either seaward or parallel to the coast, moving the surface layers, which are then replaced by deeper layers. This stirring brings up nutrients that would otherwise be forever locked in the bottom sediments. The nutrients support photosynthetic organisms, which provide the base for marine food chains. The vast anchovy fisheries off the coast of Peru are dependent on such upwellings.

Thus, the ocean communities that are richest in life exist within the neritic province. Unfortunately, the world's neritic fishing grounds are being exploited at a rate that makes the likelihood of their recovery questionable. When increased fishing effort is not accompanied by increased catches, then it is fairly certain that the resource is in danger of depletion. Today's fishing technology is so sophisticated that even in areas where catches have diminished, the last remnants of once-abundant species are being captured.

Coastal communities The coastal communities are extremely diverse. Here we have the **littoral** zone, an area that includes the land along the coast, the waters just offshore, and the **intertidal** zone (the area between high and low tide). Coastal communities include sandy beaches, rocky coasts, bays, estuaries, and tidal mud flats. Such places have one thing in common—they abound with a variety of life. The richness of life in the littoral zone is made possible by the availability of light, shelter (seaweeds, rocks, kelp beds), and nutrient runoff from rivers entering the sea.

TABLE 46.1 OCEAN SALINITY

The ocean averages about 3.5% salts, usually expressed as 35 parts per thousand. Six elements, occurring as ions, comprise 99% of the minerals in sea water. Salinity varies, particularly in surface waters, but also in the depths; the 35 parts per thousand applies to measurements taken at 300 m (about 985 ft).

ELEMENT	% OF TOTAL SALTS	PARTS PER 1000 OF SEA WATER
Chlorine	55.0	18.98
Sodium	30.6	10.56
Magnesium	3.7	1.27
Sulfur (sulfate)	7.7	0.88
Calcium	1.2	0.40
Potassium	1.1	0.38
Less than 1% of salts:		
Bromine		0.065
Carbon		0.028
Strontium		0.013
Boron		0.005
Silicon		0.003
Fluorine		0.001
Nitrogen		0.0007
Aluminum		0.0005
Rubidium		0.0002
Lithium		0.0001
Phosphorus		0.0001
Barium		0.00005
Iodine		0.00005

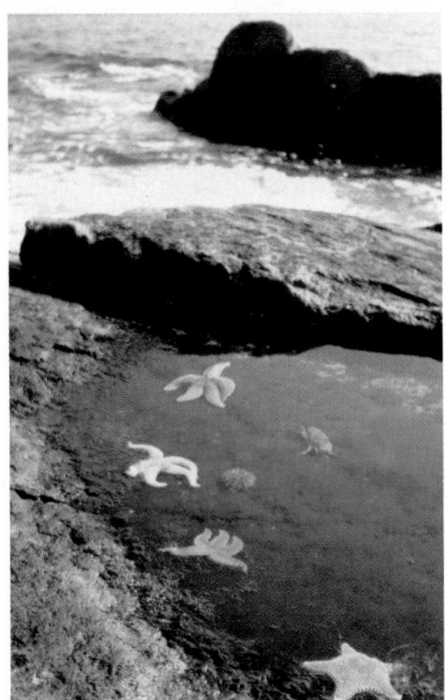

FIGURE 46.31
ROCKY TIDE POOLS
The rocky coast is home for numerous marine animals. Each organism of the rocky tide pool is adapted in some way to withstand both the surging and pounding of waves and intermittent periods of exposure to the air at low tide.

Each region supports its own particular type of plant and animal community. The rocky tidal communities, for example, consist of a large variety of organisms, all adapted in some way for holding on or burrowing in. Their biggest problem is to avoid being swept away or beached by the tremendous force of the waves. If you have dived in such places, you know about the nearly irresistible ebb and flow of the surging water. Like other littoral zone inhabitants, animals of the rocky coast have the added problem of being left "high-and-dry" by low tides, so each must adapt to such intermittent periods.

Visitors to rocky tide pools are usually amazed at the variety of life found there (Figure 46.31). That life may include plants and algae such as seaweeds, eel grass (which is not really grass), and microscopic phytoplankton. Animal life from nearly all phyla is represented, as we mentioned in earlier chapters. Inhabitants include paper-thin flatworms, sponges, bryozoans, tiny but tenacious mollusks, echinoderms, crustaceans, tunicates, and small fishes (often visible only when they dart from one tiny crevice to another). The more sedentary of these species are adapted to regular exposure to the wind and sun as the tides fall.

The greatest productivity in the marine environment is in the estuaries and reefs. Estuaries are the places where rivers run into oceans. It is here that we find salt marshes and mud flats (Figure 46.32). These are often unsightly and smelly areas, and hard to get around in, so they sometimes have not been protected by law as more "desirable" areas have been. This is unfortunate because it is here, where rivers meet oceans, that much of the chain of life in salt water begins. The rivers bring in nutrients washed from the land, and the tides continually cover and then expose the nutrient-rich soil that has been deposited by the rivers, churning up its nutritive riches. Many marine organisms spawn in such places, particularly various species of mollusks, arthropods, and fish. The species that live in such places must be able to tolerate extreme variations in salinity, moisture, and, in some cases, temperature. Because of the great density of life, estuaries are important feeding grounds for amphibians, reptiles, birds, and mammals. Unfortunately, developers have found ways of turning estuaries into valuable waterfront real estate, complete with private boating docks.

Estuaries are fascinating for humans and demanding for marine life because of the special problem of osmoregulation. The problems arise because of the constantly changing salinity, which varies with the changes in the tides and river flow. Estuaries commonly have silty or muddy bottoms, and their backwaters may form mud flats at low tide. The animals inhabiting the mud flats usually survive such drastic changes by burrowing to escape the drying sun. The mud flats are inhabited by polychaete worms, mollusks such as clams and scallops, sea cucumbers, and crustacean arthropods such as crabs. With such rich food sources available, it's no wonder that many species of shore birds visit the mud flats at low tide.

Estuaries are also regions of serious pollution, and many have been drastically altered by humans. As long as rivers are used as dumping places for manufacturing wastes and

FIGURE 46.32
MUD FLATS
Mud flats are saltwater areas in bays and estuaries that appear during intermittent low tides. At such times, their inhabitants are exposed to air, sunlight, and changing salinity, and must adapt to such harsh conditions. Burrowing mollusks and worms are common, and predators from the shore often stalk these areas in search of such prey. The estuarine mud flat may harbor a great variety of organisms if the river has not been heavily polluted.

sewage, the problem will persist. Our policies are often ill-advised because bays and estuaries are rich sources of nutrients and, as we've seen, are the very places where marine animals reproduce. Thus, the young of many marine animals—even those that will eventually move to deeper waters—are frequently endangered by polluted conditions.

Another kind of shallow area that is among the richest of marine habitats is the coral reef, but you haven't seen one unless you have been to tropical regions where the ocean temperatures remain above 20° C. Corals, of course, are cnidarians that live in huge colonies, building heavy exoskeletons of calcium carbonate. As years pass, the mass of exoskeletons grows, with the newer members of the colonies growing on top of the old. Coral atolls, common in the South Pacific, were shown by Darwin to represent coral growth at the top of submerged volcanoes. Barrier reefs, on the other hand, are coral deposits that form along coastlines, usually a short distance from the beach. The largest and most spectacular of these is the Great Barrier Reef, which extends some 1200 miles along the east coast of Queensland, Australia.

Reefs, by their irregular growth, form natural refuges for marine animals and thus set the stage for complex food chains. The complexity is due to the diversity of food types and a multitude of hiding places that forces predators to be able to exploit a number of different kinds of food, specializing in none. Coral reefs usually shelter sponges, encrusting algae, and bryozoans, as well as mollusks such as the octopus. Fishes abound in the reefs, and sharks patrol the deeper waters alongside.

APPLICATION OF IDEAS

1. Contrast the distribution of plant species in the tropical rain forest with that in the taiga and temperate deciduous forest. Suggest reasons for differences.
2. List the ways various animals have adapted to the desert biome. What general rules can be made for desert dwellers? Why are environmentalists more concerned with the human impact on this biome than they might be with most others?
3. Among the greatest environmental concerns today are those associated with the human encroachment on the tropical forests in South America and Asia. What is going on in these places, and how can the effects of this activity possibly be of significant global importance?

KEY IDEAS

Ecology, which has its roots in natural history, is the study of the "house," the interaction of organisms with each other and with their **environment.**

THE BIOSPHERE

The **biosphere** includes those portions of the earth that support life.

Physical Characteristics of the Biosphere

1. Unique characteristics of the earth's biosphere include the presence of water, a gaseous atmosphere, tilted axis, and moderate climate.
2. Water has very significant chemical and physical characteristics that include high specific heat, the resistance to temperature change. Moisture in the atmosphere provides for a moderate climate.
3. The atmosphere contains variable amounts of water and is 78% N_2, 21% O_2, and about 1% rarer gases, including CO_2. The atmosphere screens out harmful ultraviolet radiation. It also acts as a heat sink, producing conditions similar to those in a greenhouse.
4. About 30% of incoming solar energy is reflected from the atmosphere, while 20% is absorbed by the atmosphere, and the rest radiates back into space. Less than 1.0% is captured in photosynthesis, and most becomes involved in the evaporation of water. Evaporation and condensation constitute the **hydrologic cycle.**
5. The earth's changing seasons, and the resulting varying climate, is attributed to its tilted rotational axis (23.5 degrees), which presents a constantly shifting face to the sun.
6. While the equator receives much of the solar radiation, equatorial heat causes moisture-laden air to rise, forming giant rotating air cells that carry moisture and heat northward and southward from the equator. Such air cell movement is thrown westward by rotational forces, producing the tradewinds. Most of the earth's large deserts occur north of the precipitation zone.
7. **Deserts** also form as a result of **rain shadow,** in which high mountains in the path of moisture-laden

air receive the precipitation, while the leeward land masses receive little.

8. The air masses also produce the great ocean currents, including the oceanic **gyres,** the circular surface movements of entire oceans.

THE DISTRIBUTION OF LIFE: TERRESTRIAL ENVIRONMENT

1. **Biomes** are major, recognizable associations of plants and animals maintained by specific climatic conditions.
2. Biomes subdivide into ecological **communities**—interacting populations of organisms—which are also usually identified by dominating plant life.
3. Biome distribution often follows latitude, with **transition zones** from one biome to the next. In a northerly direction from the equator, **tropical rain forests** are followed by desert, **grassland, deciduous forests, taiga,** and **tundra.** Similar distributions are seen in altitudinal transitions, where increasing altitude emulates northerly changes.

Deserts

1. Deserts receive less than 25 cm of highly seasonal precipitation per year and occur in all continents except Europe. Without heat-retaining water, night-day temperature shifts are drastic. Most deserts occur at about 30 degrees north or south latitude, or to the lee of high mountains.
2. Desert-adapted plants are called **xerophytes.** Specific adaptations include water storage, fewer stomata, small leathery leaves, spines, long dormant periods, and specialized cycles as seen in the CAM plants—crassulacean acid metabolism—all of which result in slower growth rates. Many annuals have rapid flowering cycles, and the seeds of some species resist germination unless the quantity of rainfall is enough to assure success.
3. Most animal species are nocturnal, becoming active at night, but some are active in the evening and early morning. Rodents, such as the desert kangaroo rat, survive by remaining in humid burrows. The kangaroo rat has a highly efficient kidney and does not require water. Burrowing desert spadefoot toads undergo **estivation,** remaining dormant through dry periods and rapidly undergoing reproduction when water is available.

Grasslands

1. Major grasslands, which occur in inland plains (steppes, prairies, pampas, and veldts) have more rainfall (25–75 cm/year) than deserts, but it is also seasonal so drought is common. Fires are a significant factor in preventing trees from becoming established.
2. Grasses recover from fires and drought because of their protected stems (underground rhizomes) and extensive **diffuse root systems.** Rhizomes and stolons (surface runners) form sods that prevent the invasion of other plants.
3. Grasslands support huge grazing populations and survive by basal regrowth, although overgrazing has destroyed many marginal grasslands. Most grasslands today have been converted to agriculture.

Tropical Savannas

The **tropical savanna** is characterized by intermittent groves of trees in otherwise typical grassland. Rainfall averages between 100–150 cm/year. Savannas occur in Africa, South America, and Australia. Prolonged dry seasons and fire maintain the grassy state. Huge migratory populations of grazing mammals and large carnivores are common.

Tropical Rain Forests

1. Tropical rain forests receive from 250 to 450 cm of evenly distributed rainfall per year. The largest are in the Amazon, Indonesia, and the Congo. In addition, there are tropical seasonal rain forests, typified by the monsoon forests of Southeast Asia.
2. The foliage of trees in the tropical rain forest forms strata, with an overlying canopy 30 to 45 m high, below which is a subcanopy of smaller trees and vines. Other plants, the **epiphytes,** live entirely within the trees, receiving water mainly from the humid air. Great numbers of species occur, but individuals in each species are widely dispersed.
3. Animal species also occur in great numbers, and each species generally occupies a specific region in the canopy and subcanopy.
4. Clearing of the world's tropical rain forests may produce dramatic changes in climate and soil.

Chaparral

1. The broadleafed evergreens that typify the **chaparral** and Mediterranean scrub are restricted to coastal regions in western South and North America, the Mediterranean, and southernmost Africa and Australia. While rainfall is meager and seasonal (25 to 75 cm), the droughtlike conditions are modified by a moist marine air flow.
2. Plants are typically xerophytic, with well-protected small leaves that are often waxy, leathery, and spined. Decomposition is slow, and ground litter is extensive.
3. Rodents abound, as do migratory birds, but there are few large herbivores or carnivores.
4. The chaparral is a **fire-disclimax community**; that is, it is always in a state of recovery from fires that periodically sweep through.

Temperate Deciduous Forests

1. Deciduous forests, those with trees that shed their leaves seasonally, occur mainly in the north temperate zones of North America, Europe, and Asia. They are all characterized by warm summers and freezing winters, where the principal plant adaptation is leaf-shedding and dormancy. Precipitation averages about 100 cm and is evenly distributed.
2. Typical species are oaks, maples, beeches, birches, and hickory. The forest floor is densely populated by younger trees and many kinds of shrubs and ferns.
3. A rich humus and deep litter support large soil

populations. Larger animals adapt to the seasons by hibernation, migration, or simply facing the cold.

4. Because of human encroachment in much of the deciduous forest biome, large mammalian herbivores and predators have been largely displaced, and only smaller species persist.

Taiga and Coniferous Forests

1. The taiga is the northernmost coniferous forest, extending in a northerly belt around the world. The winters are long and cold. Summers are short. Rainfall is often less than in the deciduous forest. Deciduous communities interrupt the conifers, particularly around boggy **muskegs.** In the northern taiga, light, as well as water, becomes a growth-limiting factor.
2. The needle or scaly conifer leaf with its waxy surface and recessed stomata is well adapted to dryness. The forest floor is relatively clear because of the heavy leaf litter that produces an acidic condition in the soil.
3. Taiga herbivores include moose, elk, deer, rabbits, hares, porcupines, and numerous rodents. Carnivores include bears, wolves, lynx, and wolverines.
4. Temperate coniferous forests in the U.S. are found in the Pacific Northwest, coastal and inland California, and the southeastern Atlantic coast. All are quite accessible and the target of powerful lumbering interests. Clear-cutting, a cost-effective way of lumbering, is apparently a serious ecological threat. Studies reveal that significant soil nutrients are lost by newly increased runoff and that water conditions are drastically changed downstream.

Tundra

1. The northernmost biome is the tundra, an assemblage of dwarfed trees, grasses, and lichens found north of the taiga and at high elevations. Precipitation is less than 15 cm, much of which is snow. Winters are dry, but in the short summer, surface water collects in countless ponds. Below is the **permafrost,** permanently frozen soil.
2. A lack of sunlight results in a short six- to eight-week growing season, so plant growth is retarded. High winds in some regions and the presence of permafrost prevent trees from becoming established. Lichens abound, and plant life includes mosses, grasses, sedges, and dwarf willows and birches.
3. Larger mammals include caribou, musk oxen, and reindeer, along with smaller hares and lemmings. Lemmings make up an important link in the food chain, supporting arctic fox, lynx, snowy owl, weasel, jaeger, and arctic wolf populations. Many waterfowl and shore birds also migrate in to nest in summer, a time when insects also abound.

WATER COMMUNITIES

The Freshwater Province

1. Rivers and streams originate at spring- and snow-fed headwaters, and are clear and oxygen-rich at higher elevations. At lower elevations, river flow slows, and the waters become murky and contain less oxygen. As sediments accumulate further downriver, communities become more complex.
2. Ponds are generally shallow and tend to choke with plant and algal growth.
3. Northern hemisphere lakes tend to be permanent and of glacial or volcanic origin.

The Marine Environment

1. The marine environment is subdivided into the shallower, coastal **neritic province** and the deeper, open-sea **oceanic province.** Each includes a light-penetrable **euphotic zone** and a dark **aphotic zone.**
2. Life is sparse in the nutrient-poor oceanic province and includes floating and drifting **pelagic** organisms, most of which are restricted to the upper 100 to 300 m, the euphotic zone.
3. The perpetually dark **abyssal region** (300 to 11,000 m) supports a few forms of **benthic** (bottom-dwelling) life that are adapted to the cold and immense pressure. Many generate light for signalling and prey capture.
4. The neritic province includes the **continental shelf,** which falls off at the continental slope.
5. Nutrients from upwellings support the phytoplankton—algal protists—and poorly understood **nanoplankton,** the primary producers of the sea. They capture the sun's energy and form the base that makes all other marine life possible. Minute protists and animals (often larval forms) make up the **zooplankton,** upon which many other marine animals feed.
6. The greatest productivity is in colder offshore waters where **upwellings** bring nutrients up to surface waters. Coastal upwellings are brought about by seasonal winds blowing seaward or parallel to the coast.
7. The **littoral zone** of the neritic province contains the coastal communities: sandy beaches, rocky coasts, bays, estuaries, and reefs. Light, shelter, and nutrients abound.
   a. Along rocky coasts, organisms must adapt to wave surge and to exposure at low tide. At low tide numerous tide pools form, each with a mixed population of marine plants, algae, and animals.
   b. Estuaries produce their own problems, including shifts in salinity that require osmoregulatory adaptations. Marine animals in tidal mud flats burrow to escape drying and predatory shore birds. Estuarine life, which often includes immature forms of deeper-water life, is constantly threatened by human encroachment and pollution. Reefs form natural refuges for a variety of marine populations.

REVIEW QUESTIONS

1. Define the term *ecology*. Where does it have its roots? What nonbiological disciplines might be important to ecologists? (1134)
2. Describe the physical extent of the biosphere. (1134-1135)
3. Explain the relationship between water in the biosphere and temperature stability. (1135)
4. Explain what happens to the total amount of solar energy impinging upon the earth's atmosphere. What keeps the earth from continually heating up because of solar energy? (1135-1136)
5. Describe the formation and movement of the equatorial air cells. What do they have to do with the formation and location of the major deserts? (1136-1137)
6. Explain what causes oceanic gyres. How do they contribute to the distribution of heat over the globe? (1138)
7. List the biomes one might cross when moving from the equator to the North Pole. How might this distribution be mimicked by high mountains in the tropics? (1139-1140)
8. Describe the physical conditions of the desert biome, including precipitation and its distribution, day-night temperature changes, and the effects of rainstorms. (1141)
9. List four significant ways in which plants are adapted to desert life. (1141)
10. Briefly discuss three ways in which animals in hot deserts adapt to the heat. (1141-1142)
11. List regions of the world where major grasslands are located and give their regional names. (1143)
12. What characteristics of grasses enable them to recover from both grazing and fire? (1143-1144)
13. Characterize by kind and number the major animal life in grasslands. What does this tell you about the grasses' efficiency of energy capture? (1144)
14. How does the tropical savanna biome differ from the grassland? List some of the major mammals in the African savanna. (1145)
15. Characterize the conditions that support the tropical rain forest of the Amazon basin. In what significant way does the Southeast Asian rain forest differ from the Amazonian forest? (1145-1146)
16. Describe the organization of the tropical rain forest, including numbers of species, vertical stratification, forest floor, and nature of epiphytes. (1146)
17. What are the unique physical conditions of the chaparral? What keeps it from becoming a desert? (1146, 1148)
18. List several climatic adaptations made by chaparral plants. What is the general effect of fire on the chaparral? (1148)
19. List the major countries of the world where deciduous forests are found. What one climatic characteristic do they all share? (1148-1149)
20. What is the major adaptation of broadleafed plants to seasonal changes in the deciduous forest? What three responses do the animals make to seasonal change? (1150)
21. Where is the taiga? Is this the only region of large populations of conifers? Explain. (1150-1151)
22. What are three growth-limiting conditions in the taiga? (1151)
23. Describe the special adaptations of the conifer leaf. (1151)
24. Characterize the physical conditions of the tundra, including precipitation, temperatures, seasons, and special soil conditions. (1152-1153)
25. List five prominent plants of the tundra. What two conditions prevent the growth of tall trees? (1153)
26. Describe the short food chain of the tundra. (1154)
27. Name and describe the two major divisions of the marine environment. In what way is each divided vertically? (1156)
28. What is a pelagic organism? What determines the depths in which these organisms occur? (1156)
29. What are three prominent physical conditions of the abyss? List five phyla of animals represented in the abyssal region. What is their primary food source? (1157)
30. What is an upwelling? How does this phenomenon affect the distribution of marine life? (1159)
31. What two conditions in the rocky coast demand special adaptations by the inhabitants? What are three such adaptations? (1160)

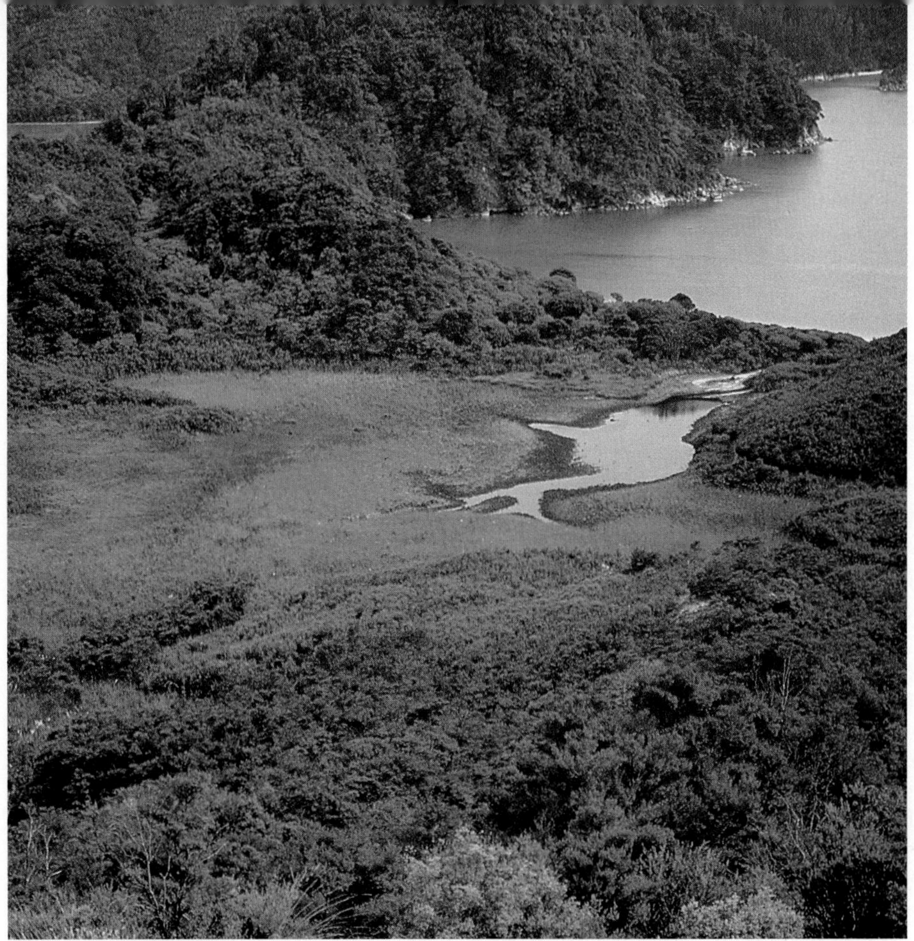

47

ECOSYSTEMS AND COMMUNITIES

THE BEST WAY TO GET SOME IDEA OF THE PROBLEMS FACING ENVIRONmental biologists is simply to look out the window. How is one to make some sense of all that? After all, the view there, complex enough itself, only suggests the array of interwoven factors that play across nature's stage. Not only does the environment vary from place to place, but the view also changes from one time to the next, whether measured in seconds or seasons. The task of making sense of all this is overwhelming, and the scientist has been forced to generalize.

The concept of the biome is one such generalization, but it is a way to organize the living world into general categories that are conceptually somewhat manageable and, one would hope, that indeed reflect something of the nature of the real world. Biomes represent a grand and general view of life in the biosphere, and whereas such conceptual manipulations make our task of understanding the planet somewhat more manageable, they are rather unwieldy divisions, and so we must devise ways of categorizing, separating, and dividing what we see around us. One way of doing this is to try to consider the living world in terms of its interactions. Obviously, here we must focus on smaller ecological units where *interactions* are more easily described.

Biomes are composed of communities. A **community** is defined as an assemblage of interacting plants and animals forming an identifiable group within a biome. The interactions within communities take place in what is called an ecosystem. An **ecosystem** is any community of interacting organisms, including their biotic (living) and abiotic (nonliving) environment, described in terms of the flow of energy through that environment. The ecosystem structure is assumed to occur in all communities, no matter what the makeup of their organisms. We can begin, then, with this system of interaction. Our plan will be first to discuss the common characteristics of ecosystems and then to learn how they apply to communities. A logical place to start is by understanding something about energy flow.

ENERGETICS IN ECOSYSTEMS

Tracing energy flow through a living system sounds like a rather esoteric task, but it also involves such mundane behavior as eating. Since this is the case, we can begin with the synthesis of food and a reminder that the ultimate source of energy for most forms of life on the earth is the visible light from the sun. Of course, energy from visible light enters the living realm of the biosphere through photosynthesis. In the last chapter we saw that about 50% of the solar energy reaching the atmosphere finds its way to the earth's surface. Of this, less than 1% (actually one-tenth of one percent as a worldwide average) is captured by photosynthesizers. But meager as it seems, this is enough to produce 150 to 200 billion tons of dry organic matter each year. (Such estimates are based on the dry weight of organic matter because water is not organic but makes up a substantial part of the weight of living organisms.) Such matter is often referred to as **biomass**—the total weight of organisms per unit of area (often stated as grams or kilograms per square meter [g/m^2, or kg/m^2] and recorded as either dry or wet). On a much smaller scale, this reduces to several kilograms of dry organic matter per year for each square meter. (We should add that chemoautotrophs, such as the organisms using the chemical energy of molecules spewing from vents in the seabed, also contribute to the earth's organic matter. However, this contribution is far less than that of the photosynthesizers.)

Trophic Levels

Energy flowing through the ecosystem is first captured by autotrophs such as photosynthesizers. These, in ecological terms, are the **producers.** From the producers, energy passes to various **consumers,** heterotrophic organisms that cannot utilize external energy and that rely on other organisms for their required organic compounds. In the last chapter we described the passage of food through the marine ecosystem in terms of food chains, where phytoplankton (the primary producers) were fed upon by zooplankton, and these by several levels of larger and larger marine creatures (see Figure 46.30). In the energetics of ecosystems, these steps in energy transfers are called **trophic levels.** Figure 47.1 traces the flow of energy through several trophic levels. The passage of energy, we should emphasize, involves the passage of food molecules. With these points fresh in mind, let's have a closer look at each level.

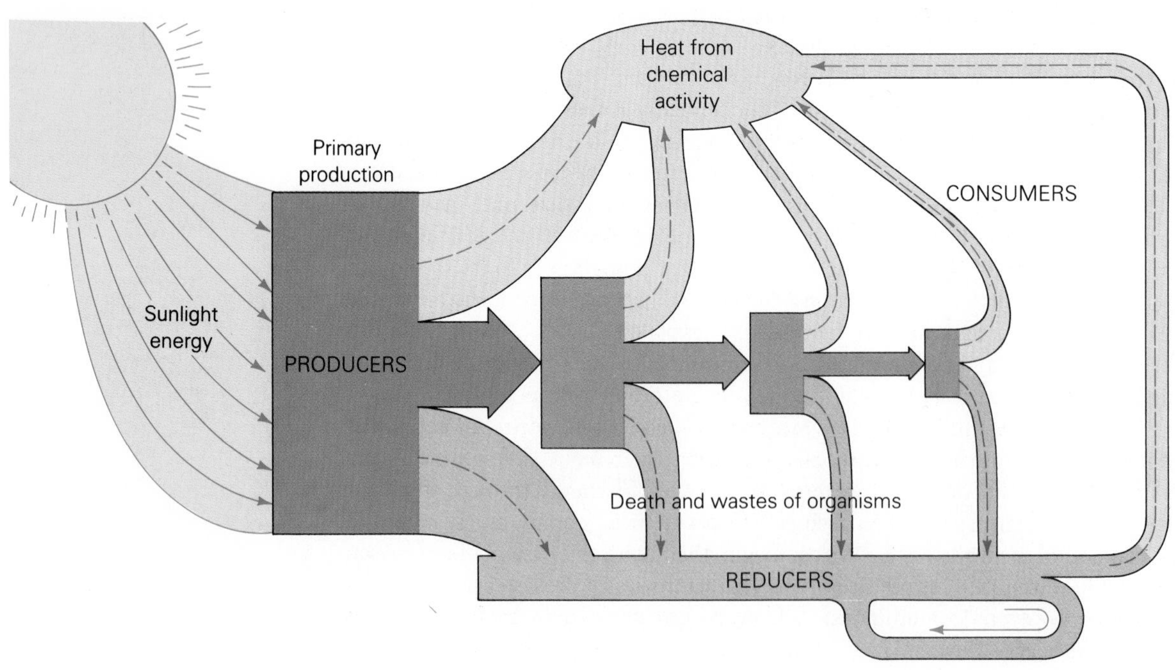

FIGURE 47.1
ENERGY FLOW IN AN ECOSYSTEM
Sunlight energy is first captured by producers during photosynthesis. It then passes, as chemical bond energy in foods, from one consumer to the next and from one trophic level to another. Eventually, the energy passes to decomposers. Each transfer is about 10% efficient. The remaining energy escapes as heat, a product of the respiratory activities that support the organisms.

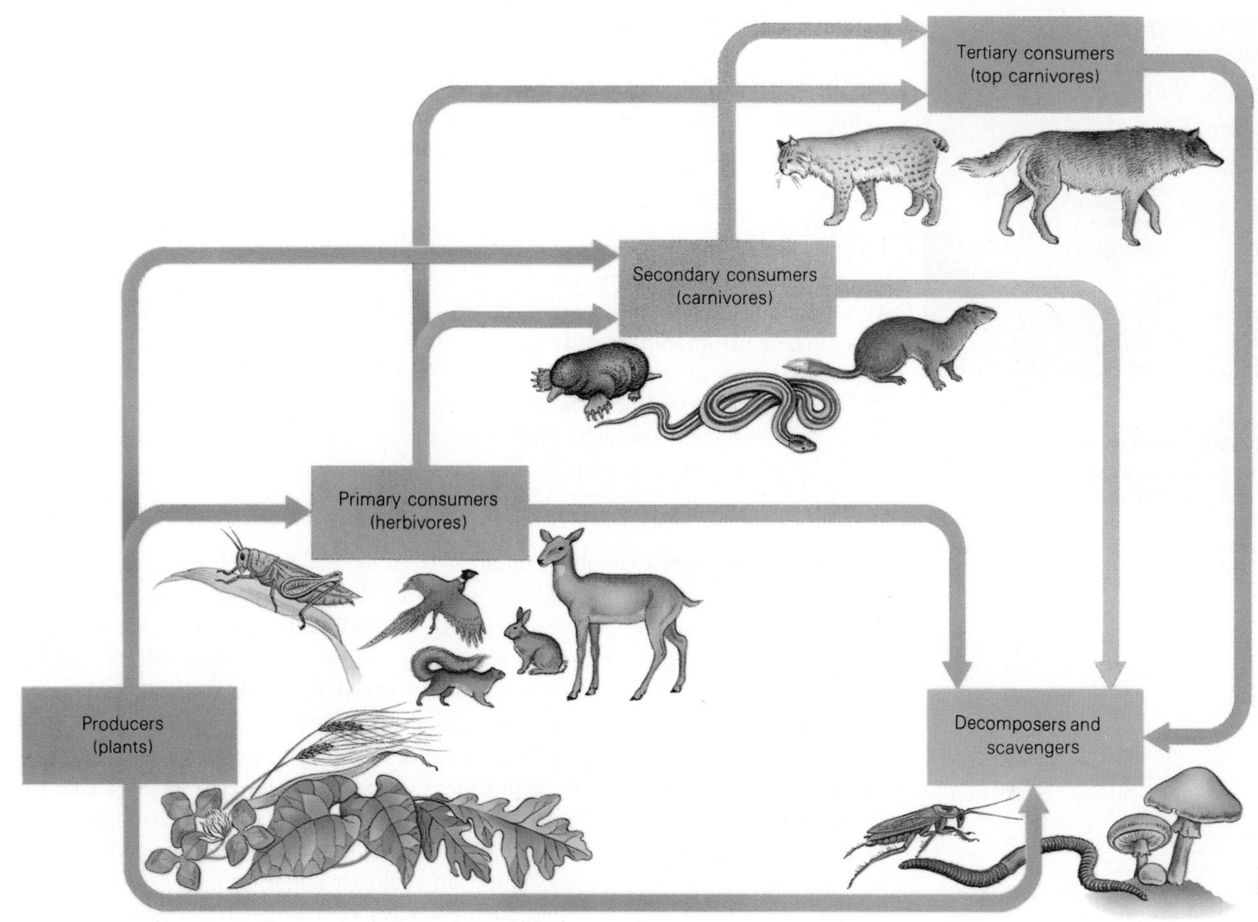

FIGURE 47.2
A SIMPLIFIED FOOD WEB
Food webs reflect the complex feeding patterns in an ecosystem. The arrows indicate the direction of the flow of energy and matter. In nature, some animals feed from more than one trophic level, particularly during shortages when usual food sources become decomposed.

Producers Producers include plants, algal protists, and phototrophic bacteria, all photosynthesizers. As you know, these organisms use light energy, usually captured by chlorophyll pigments in molecular complexes known as photosystems, to produce organic food materials from carbon dioxide, water, and a few minerals (see Chapter 7). The chemoautotrophs include bacteria that obtain energy from inorganic substances in the earth's crust (see Essay 47.2). The biomass of the earth's producers is enormous, about 99% of the total present in the biosphere.

Consumers Consumers include animals, some fungi, many protists, and most bacteria—in a word, heterotrophs. Since some consumers eat producers, while others eat consumers, the flow of energy through the consumers may occur in several steps or trophic levels. Thus, this group has been divided into **primary consumers,** the herbivores that feed directly on producers; **secondary consumers,** carnivores that feed on primary consumers; and so on through **tertiary, quarternary,** and even higher (but rare) consumer levels.

Some consumers—primarily carnivores—are likely to cross trophic levels as they feed. Consider humans as an example. How many trophic levels do *you* occupy? Such complicated feeding patterns in a community are better represented by **food webs,** such as we see in Figure 47.2. However, such diagrams can only suggest the complexity of most food webs; to be accurate, they would have to include every species found in an ecosystem.

Decomposers **Saprobes** feed on the dead. The saprobic fungi and bacteria are called **decomposers.** Decomposers (sometimes called reducers) ensure that essential molecules cycle back to the producer, and so they represent a vital link between consumers and

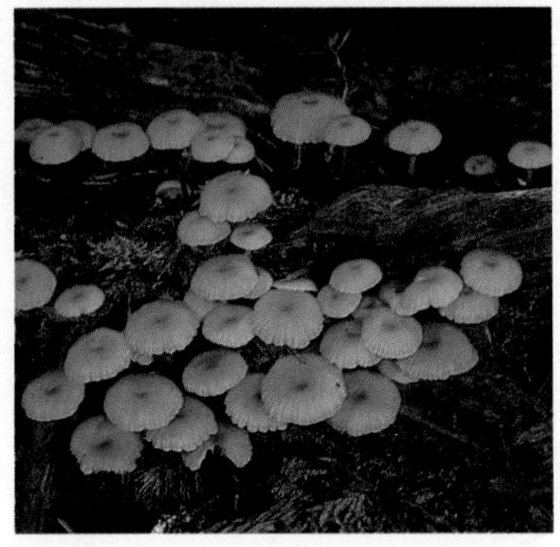

FIGURE 47.3
DECOMPOSER ORGANISMS
Decomposers break down wastes and the remains of plants and animals, making nutrients available to the ecosystem. Most are microscopic in size, but the bracket fungi that live in fallen trees may be quite large.

producers. Essentially, they break down the organic matter, the remains and wastes of organisms, into simple products such as ammonia, sulfates, nitrites, nitrates, phosphates, and the usual carbon dioxide and water. We will look into their activities in more detail shortly, but for now keep in mind that decomposers are the recyclers of the elements of organic nutrients (Figure 47.3), and without them the world would be a far different place.

Trophic Levels as Pyramids The relationships among trophic levels are much easier to visualize if they are represented graphically, something ecologists are prone to do with their data. Pioneering ecologist Charles Elton first conceived of the idea of ecological pyramids, or **Eltonian pyramids** as they are appreciatively known. Ecological pyramids describe changes in some factor at various trophic (feeding) levels.

The producer populations form the base, followed by primary, secondary, tertiary, and higher levels of consumers. Three types of ecological pyramids are commonly used: **pyramids of numbers, pyramids of biomass** (total dry weight of all living matter), and **pyramids of energy.**

Pyramids of Numbers Graphing numbers of individuals at each trophic level in, let's say, a grassland community, produces a typical "stepped" pyramidal shape as seen in Figure 47.4a. But number pyramids can take markedly different forms. For example, if we count trees in a forest and then count the parasites, herbivorous insects, and birds that feed from the trees, a single kind of producer will be supporting many consumers. Thus, the ecological pyramid then takes on an inverted form as seen in Figure 47.4b.

Pyramids of Biomass Graphic depictions of biomass in an ecosystem can be quite revealing. Obviously, one normally cannot weigh a community (or, for that matter, count all of its individuals), so elaborate sampling techniques have been developed to estimate the actual weight or count. For example, an ecologist might estimate biomass by "harvesting" several randomly selected square meters of a much larger area, or perhaps

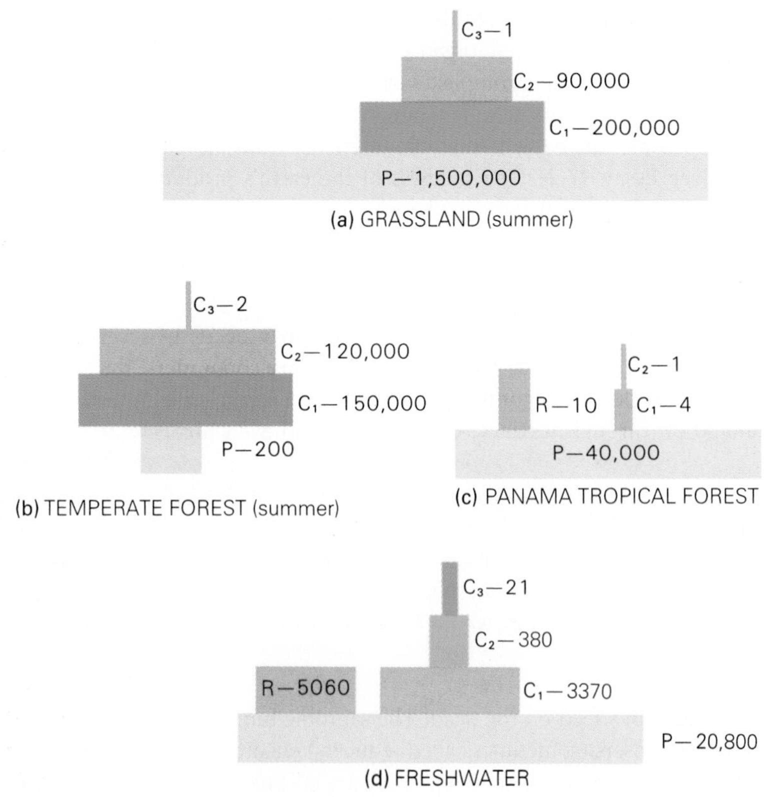

FIGURE 47.4
ECOLOGICAL PYRAMIDS
(a) Pyramid of numbers. Plotting the numbers of organisms in each trophic level of a grassland community produces the stepped pyramid. (b) In a forest community, the pyramid of numbers becomes partly inverted, since most of the producers (P) are large trees and shrubs. Their numbers are far exceeded by the numbers of consumers (C) they support. The numbers represent counts for each 0.1 hectare (quarter acre). (c) Pyramid of biomass. Measurements of biomass (grams per square meter) in a tropical rain forest reveal a great difference between the producer and consumer levels, where 40,000 g/m^2 supports a total consumer mass of only 5 g/m^2. The large decomposer biomass should be expected in a community where nutrient cycling is very rapid. (d) An energy pyramid from a freshwater aquatic community at Silver Springs, Florida. Energy flow here is expressed as kcal/m^2 · year.

harvest in narrow swaths across the designated area. The ecologist is able to weigh all the individuals in these smaller areas and to extrapolate those measurements to the larger area. If the organisms are then sorted according to known trophic levels, dried, and weighed, the data can be plotted. Usually, in such a case, the expected stepped pyramid emerges. Typically, the biomass of the producers is far greater than that of the consumers (Figure 47.4c). As we indicated earlier, 99% of the earth's biomass is to be found in the primary producer level. Not much of this is actually stored following its transfer to the primary consumer level, for several reasons. Of the plant matter eaten by herbivores, a considerable amount is consumed (converted to carbon dioxide, water, nitrogen waste, and so on) during metabolism. In addition, some is not actually absorbed, appearing in the feces. For the same reasons, the primary consumer level can generally support less biomass than does the producer. Thus, less secondary biomass can be sustained, and we see the stepwise narrowing of the pyramid.

Pyramids of Energy Energy pyramids represent the flow of energy from one trophic level to the next and show the loss that occurs in such transfers. That energy is present in the chemical bonds of the molecules taken in as each group of organisms feeds. They have the usual stepped configuration (Figure 47.4d), but they are not as radical in appearance as are biomass pyramids. In this case, the stepped pyramid would be predicted by the second law of thermodynamics, which, as you may recall (Chapter 6), states that energy transfers are never perfect—there is a loss of energy with each transfer. This is true in living as well as nonliving systems. In fact, that loss can be considerable; the energy transfers from one trophic level to the next average about 10% in efficiency. Thus, not much energy handled by any trophic level is available at all to the next level. It turns out that 90% of the energy in the food eaten by consumers is not stored but is used in maintaining the animal (or is lost as waste). Consumers do more than eat, and every act requires energy. (People who measure such things say that about 7% of the chemical bond energy of foods taken in by dairy cattle passes out of their bodies in the chemical bonds of gas!) Eventually, all of the energy entering the earth's ecosystems is released as heat. Heat, of course, is a low grade form of energy, and there is no way for organisms to gather heat energy for useful work. This tremendous reduction in available energy is what ultimately limits the number of trophic levels in nature (to four or so).

Humans and Trophic Levels

Energy pyramids may seem abstract and academic, but they apply dramatically to human populations and suggest some fundamental lessons in economics. (We might point out that economics and ecology share many things other than the prefix, including certain premises and theories.) Humans are basically **omnivores,** that is, capable of feeding at several trophic levels. In large part, the availability of resources and one's economic state determine the trophic level most used. Under poor economic conditions, people generally feed at the primary consumer (herbivore) level. The reason is clear: most of the earth's available energy is stored at the primary level. Feeding at the secondary level or higher involves costly (or wasteful) energy transfers as energy from the primary level is converted, for example, to the bodies of cattle, sheep, hogs, and other animal stocks (Figure 47.5). But in most instances, while affluent people may choose to eat steak and lamb, impoverished people tend to eat cereals and beans and peas. It is also efficient to utilize fishes and other seafood (as do many Asians). Although there are costs in harvesting the high-protein seafood, the energy transfers up the chain are not costly in terms of human effort.

It seems, then, that the earth could support a much larger population if all humans ate only plant food. It's true that more calories would be available without the energy losses that accompany energy transfers from one trophic level to the next. But humans feeding at the herbivore level may face problems with nutritional quality. For example, many common plant foods cannot provide certain essential amino acids (such as tryptophan, methionine, and lysine; see Chapter 38) that are critical to our own synthesis

FIGURE 47.5
HUMANS AS SECOND-LEVEL CONSUMERS
Where humans live as secondary consumers, energy must first flow from a producer to a primary consumer, and the change entails a considerable loss in the energy level. For each 1000 calories stored by a beef-eating human, the producers will have stored nearly 2 million calories.

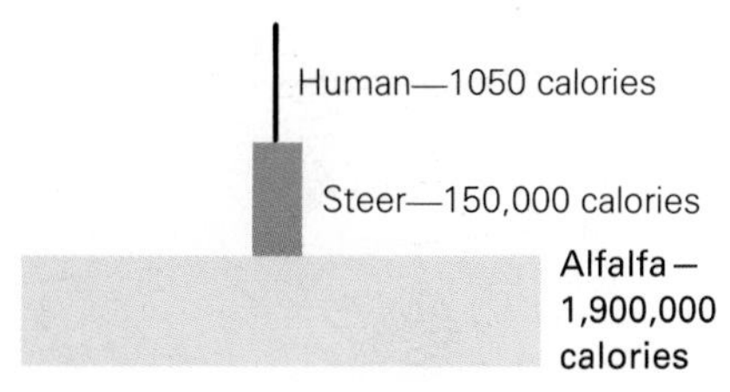

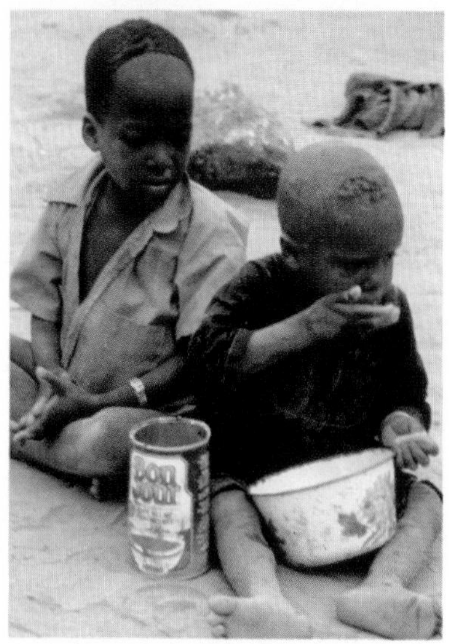

FIGURE 47.6
PROTEIN-DEFICIENT DIET
Kwashiorkor victims suffer from serious dietary amino acid deficiencies, although their calorie intake may be sufficient to provide energy. Symptoms include discolored, often thin, reddish or rust-colored, and brittle hair; wasting of muscles; flaking skin; and puffiness from water retention. Often, simple dietary supplements, such as a daily ration of peas or beans, are all that is required to provide the missing amino acids.

of proteins. The absence of such essential amino acids from the daily diet can lead to such infamous protein deficiency diseases as *kwashiorkor,* a chronic problem in some parts of Africa (Figure 47.6). The critical amino acids can be provided if legumes such as soybeans and peas are included in the diet, but this must be done on a daily basis. Humans do not store amino acids very well, and excesses are rapidly converted to fats or carbohydrates in a daily turnover.

Energy and Productivity

The rate of energy storage in organic substances per unit of space by producers in an ecosystem is referred to as **gross primary productivity (GP).** GP is the total rate at which energy is accumulated by producers over a certain area. GP measurements are useful when comparing one ecosystem or community with another, or making comparisons through some specified time period. However, GP doesn't tell us how much energy the photosynthesizers are actually storing or how much growth is occurring. As we now know, a considerable amount of the captured energy must be expended for synthesis, transport, and numerous other activities. Table 47.1 compares the gross primary productivity—GP—for several major aquatic and terrestrial regions and biomes.

Net primary productivity (NP) takes energy used by the producers into account. It is determined by subtracting the rate of respiration by photosynthesizers (energy utilization) from GP. In other words, NP is the rate of energy stored minus the rate of energy released. NP would be reflected in new growth, seed production, and simple storage of energy-rich compounds such as lipids and carbohydrates. The accumulation of organic matter is reflected as **net community productivity (NCP)**, net primary productivity minus heterotroph respiration.

TABLE 47.1 ESTIMATED GROSS PRIMARY PRODUCTION (ANNUAL BASIS) OF THE BIOSPHERE AND ITS DISTRIBUTION AMONG MAJOR ECOSYSTEMS

ECOSYSTEM	AREA, MILLIONS OF KM	GROSS PRIMARY PRODUCTIVITY KCAL/M^2 · YR	TOTAL GROSS PRODUCTION 10^{16} KCAL · YR
Marine			
Open ocean	326.0	1,000	32.6
Coastal zones	34.0	2,000	6.8
Upwelling zones	0.4	6,000	0.2
Estuaries and reefs	2.0	20,000	4.0
Subtotal	362.4	——	43.6
Terrestrial			
Deserts and tundras	40.0	200	0.8
Grasslands and pastures	42.0	2,500	10.5
Dry forests	9.4	2,500	2.4
Northern coniferous forests	10.0	3,000	3.0
Cultivated lands with little or no energy subsidy	10.0	3,000	3.0
Moist temperate forests	4.9	8,000	3.9
Fuel-subsidized (mechanized) agriculture	4.0	12,000	4.8
Wet tropical and subtropical (broad-leaved evergreen) forests	14.7	20,000	29.0
Subtotal	135.0	——	57.4
Total for biosphere (round figures, not including ice caps)	500.0	2,000	100.0

SOURCE: From *Fundamentals of Ecology*, 3rd Edition, by Eugene P. Odum. Copyright © 1971 by W. B. Saunders Company. Reprinted by permission of Holt, Rinehart and Winston, CBS College Publishing.

Ecologists have devised many techniques for measuring the net productivity in communities, and from this information they can determine the community's status. Is it growing? Has it reached a climax state when the community is stable and self-perpetuating? Is it declining? New productivity accumulates only in communities that are in a growth stage. As they approach the climax state, an equilibrium between the rates of energy assimilation and energy use is established.

NUTRIENT CYCLING IN ECOSYSTEMS

While energy flows *through* an ecosystem, emerging eventually as heat, the nutrients essential to life tend to *cycle* in the usual "eat and be eaten" relationships. Nutrients are taken up by organisms as molecules and ions, and while some remain unchanged, others are incorporated into new molecules and structures, and some are metabolized for their chemical bond energy. But eventually the elements reappear in metabolic wastes, or, when death occurs, in the products of decay. These elements include those of the familiar SPONCH series (Chapter 2)–sulfur, phosphorus, oxygen, nitrogen, carbon, and hydrogen—along with a host of others such as iron, cobalt, sodium, and chlorine. Most of the elements cycle back to the producers as mineral ions, or mineral nutrients as we have called them before (Chapter 28). The pathways of elements as they are taken up and released into the physical environment are called **biogeochemical cycles.** As mentioned earlier, decomposers play a key role in such recycling.

The biogeochemical cycles have three major places where elements are accumulated. First, the elements are integrated into the bodies of living organisms. Second, they may be found in **exchange pools**, the readily available, water-soluble reserves of a mineral nutrient (such as nitrates in soil water that so easily enter plants). In the third, mineral nutrients may be locked away in **reservoirs**, which are less available reserves (such as the elements in animal bones or shells, or atmospheric nitrogen).

Some of the biogeochemical cycles are rather short, involving only a few steps. For example, some of the water taken in from the environment by the plant simply passes through its vascular system to be released back into the environment through transpiration (see Chapter 28). However, the cycling of water may be more complex. It may be used in photosynthesis. Here, the molecules are disrupted, with the oxygen released as a gas into the atmosphere and the hydrogen used in the Calvin cycle to reduce carbon during the synthesis of carbohydrates. Aerobic organisms (including both autotrophs and heterotrophs) then take in the oxygen and use it in cell respiration. There, at the very end of the long mitochondrial process, the oxygen is reunited with the protons and spent electrons, forming water.

In describing biogeochemical cycles, we can again begin with the producers. Once they incorporate simple ions and molecules into their bodies, the substances are then available to be passed from one trophic level to the next as food, eventually reaching the ever-waiting decomposers. (Of course, if the plant dies, its ions and molecules may go *directly* to the decomposer.)

The Nitrogen Cycle

We have mentioned the **nitrogen cycle,** so let's use it as our primary example of biogeochemical cycling (Figure 47.7). Nitrogen, as you know, is highly essential to life, since it is a principal constituent of proteins, nucleic acids, chlorophyll, coenzymes, and many other biomolecules. Like other essential elements, nitrogen is found in both readily available exchange pools and in the less available reservoirs. The largest nitrogen reservoir is the atmosphere, where molecular nitrogen (N_2) accounts for about 78% of the dry atmospheric gases. But, with the exception of certain nitrogen-fixing bacteria, atmospheric nirogen as such is not available to the earth's organisms. Plants and nearly all other producers must have their nitrogen primarily in the form of nitrate ions in order to incorporate it into amino acids. Nitrate ions (NO_3) are produced by soil and water bacteria in two complex pathways: **nitrogen fixation** and **nitrification.** Nitrates in the

FIGURE 47.7
THE NITROGEN CYCLE
The nitrogen cycle can be viewed as two cycles. Blue arrows show the uptake of nitrate by producers and its passage through trophic levels as protein, nucleic acid, and other organic molecules. Several steps at the decomposer level make nitrates available. Red arrows show nitrogen entering an ecosystem through the action of nitrogen fixers, although a minor amount of nitrogen fixation has always occurred during lightning storms and, more recently, through the action of sunlight on air pollutants. Input is roughly balanced by loss through denitrification carried out by anaerobic soil bacteria. In recent years, the extensive human input of synthetic ammonia and nitrates through agriculture has exceeded losses through denitrification. (Dashed lines show loss or potential loss.)

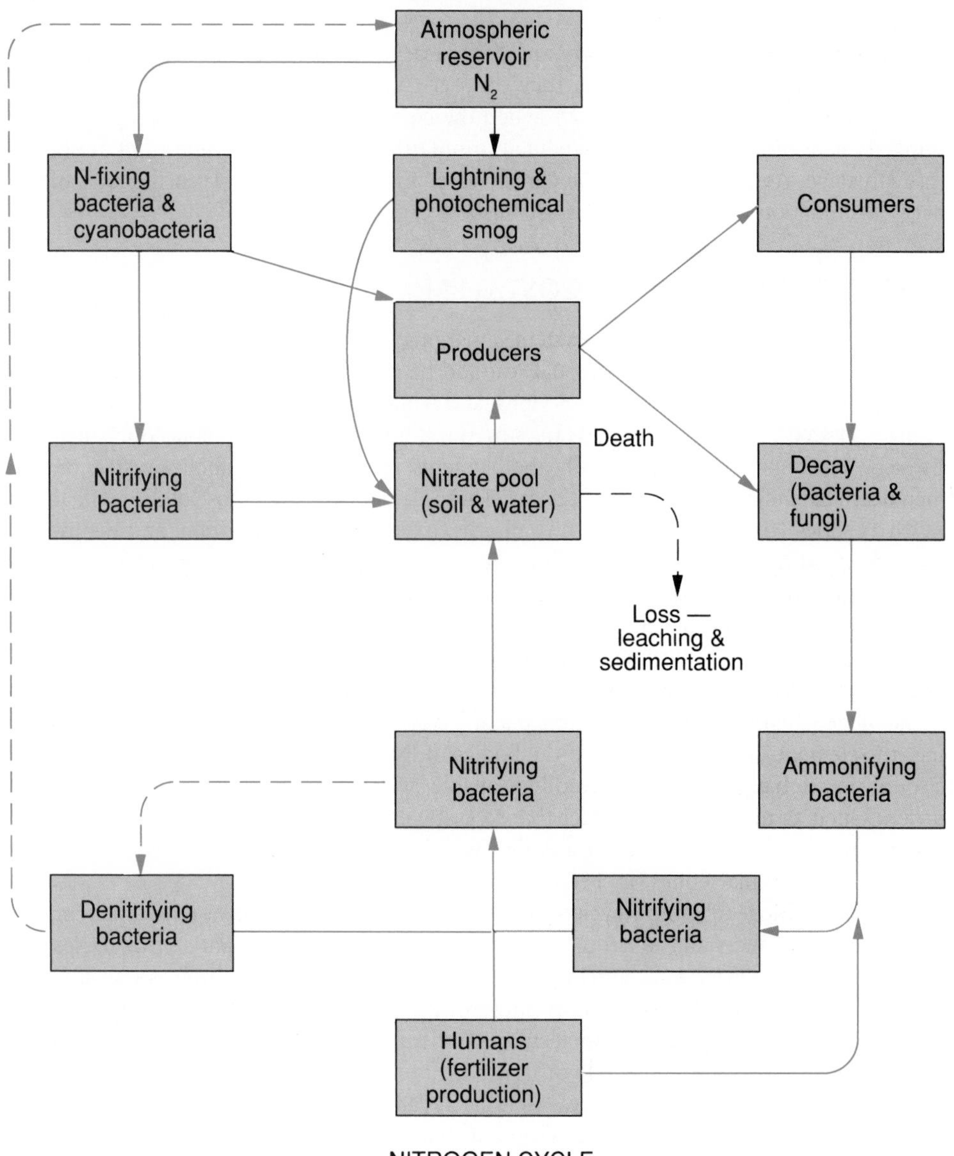

soil—along with other nitrogen-containing ions—make up the readily available exchange pools.

From Producer to Decomposer We can begin, here, to trace the flow of nitrogen from plants to animals to decomposers, and back to plants. First, plants take in nitrate ions, process them, and then incorporate the nitrogen into amino acids, which are finally assembled into plant protein. (For simplicity, we will ignore other nitrogen-containing molecules.) When plants are eaten, the amino acids pass into the consumer levels, where some are used to produce animal protein (and, of course, other nitrogen-containing molecules.) The rest may be metabolized and the nitrogen waste product excreted in various forms (urea, uric acid, or ammonia—see Chapter 36). Eventually, all producers and consumers (and the nitrogen wastes) enter the province of the decomposers.

In the next stages, several populations of microorganisms, each with its role in a multistepped process, recycle the protein and nitrogen wastes. The decomposers include the bacteria and fungi that break down organic wastes into simpler compounds such as ammonia, carbon dioxide, and water. The ammonia readily ionizes in water, forming ammonium ions (NH_4^+). In the next step, nitrification, the ammonium ions are acted upon by bacteria such as the autotroph *Nitrosomonas,* which converts ammonium ions to nitrites (NO_2^-). A second group of nitrifiers, represented by *Nitrobacter,* then converts the nitrites to nitrates (NO_3^-). The nitrates join the exchange pools and cycle back to the plant. Nitrification is vital because ammonium ions are far less available to plants than are nitrates. Although ammonium ions are chemically easier to incorporate into amino acids than nitrate, in the soil these positively charged ions tend to interact with and cling tenaciously to negatively charged clay particles. On the other hand, the negatively charged nitrate ions remain mobile in the soil and are more readily available for uptake by the plant root.

So far, the nitrogen cycle may seem smooth-running and efficient, but there are complications that lead to losses from the exchange pools. For example, the soluble nitrogen-containing ions can be carried out of the producer's reach by leaching—removal by the downward percolation of water. Further, should anaerobic conditions prevail, organisms such as *Pseudomonas denitrificans*—anaerobic soil bacteria that act as *denitrifiers*—can convert the nitrites and nitrates to nitrous oxide (N_2O, "laughing gas") and nitrogen gas (N_2). The two gases escape to enter the atmospheric reservoir. Obviously, denitrifiers can drastically deplete soil fertility. It is for this reason that anaerobic swamps and bogs are notoriously nitrate-poor.

The Nitrogen-Fixers In a balanced ecosystem, the losses through denitrification can be recovered by the gain from nitrogen fixation (see Chapters 22 and 28). Nitrogen-fixing bacteria (including many cyanobacteria) take in atmospheric nitrogen, which they send through a complex biochemical pathway to combine with hydrogen, producing ammonia. Their excesses are released into the soil or water, where nitrifying bacteria convert the ammonia to nitrite and nitrate. As you can see in Figure 47.7, denitrification and nitrogen fixation are not part of the main cycle of nitrification.

Knowledgeable farmers, aware of nitrogen fixation for years, have commonly rotated crops to include periodic alfalfa plantings. ("For years" is an understatement—Greek writings from the third century indicate that legumes were used even then to enrich the soil.) Nitrogen-fixing bacteria of the genus *Rhizobium* invade the roots of alfalfa and other leguminous plants, which respond by forming cystlike nodules around the bacterial colony. In a mutualistic relationship, the bacteria absorb organic nutrients produced by the plant, and the plant gains usable nitrogen fixed by its guest bacteria. In rice paddies and other aquatic ecosystems, much of the nitrogen fixation is carried out by aquatic cyanobacteria. In natural terrestrial ecosystems, nitrogen-fixing bacteria live in mutualistic associations with the roots of wild legumes, alders, buckthorns, and locust. But as far as agriculture is concerned, it is the leguminous plant that is most vital in harboring nitrogen-fixing bacteria.

FIGURE 47.8
NUTRIENT ENRICHMENT THROUGH POLLUTION
A dramatic increase in algal growth occurs when excess nitrate suddenly enriches a body of water. Conditions like this often result in the destruction of other precariously balanced life forms that inhabit the water. Masses of dead and decaying algae will support huge populations of bacteria, which, in turn, will deplete the oxygen supply. The result is the formation of an anaerobic water "desert," with only a few species of anaerobes and highly tolerant aerobes surviving.

Nitrogen may also be fixed the hard way, by lightning, and more gently by the photochemical action of the sun on certain pollutants such as oxides of nitrogen. However, nitrogen fixed through such atmospheric action amounts to less than 10% of that fixed by organisms.

A problem has rather recently arisen regarding our manipulation of nitrogen. It turns out that we are not using up all our nitrate as one might have expected. In fact, we are overloading the environment with nitrogenous products. The problem has only become apparent since we began synthesizing ammonia through industrial means.

Synthetic fertilizers are applied liberally to the soil and thus enter the natural nitrogen cycle. When we add the nitrogen from synthetic fertilizers to the nitrogen compounds produced by natural nitrogen fixation and by automobile exhaust (another major new source), the total amount of available nitrogen is astounding. C.C. Delwiche, at the University of California at Davis, has calculated that there is a net gain of 9 million metric tons of fixed nitrogen to the biosphere each year. Where is the surplus going? In California's central valley, which receives more synthetic fertilizer than any place on earth, the answer is: right into the water table and river systems and from there into San Francisco Bay and on to the Pacific.

How does excess nitrogen affect the water supplied in the soil? What is its effect on river systems and their estuarine life? One visible effect is the sporadic choking of waterways by uncontrolled, runaway algal growth (Figure 47.8). Such rapid growth always accompanies what is called cultural eutrophication, the sudden nutrient enrichment of lakes—to be discussed later. We have yet to measure the effect of the nitrogen load on the marine environment, but there, even more nitrogen is added from sewage dumped into the world's oceans.

Are we altering the nitrogen cycle in other ways? It's difficult to know just what's happening out there, but obviously we need to find out. It is quite conceivable that we can feed the world's burgeoning populations without calling the environment down around us. By coming to understand the nitrogen cycle better, perhaps we will find it vital and even economically feasible to keep our agricultural practices within the limits of natural systems.

The Phosphorus and Calcium Cycles

Now let's briefly consider two more cycles, those of phosphorus (Figure 47.9) and calcium. Their cycles do not involve the atmosphere, but they do involve water. Phosphorus enters the roots of plants as the soluble phosphate ion HPO_4^{2-}. Phosphates are required for the production of many familiar molecules, such as ADP and ATP, phospholipids, nucleic acids, and the coenzymes of photosynthesis and respiration. Since ATP appears in the list, a shortage of phosphates will dramatically affect the energy-requiring activities of plants.

Calcium, another mineral nutrient, is critical to the proper functioning of cell membranes and many enzymatic reactions. It enters living systems through the roots of plants as the cation Ca^{2+}. A shortage of calcium can disrupt transport processes in the plant cells, causing the plant to die. Although it functions in transport systems, the ion itself is rather immobile once in place. When local shortages of other ions occur, they can simply be shifted from other parts of the plant. But not calcium; it must be taken in constantly by the roots.

The cycles of phosphorus and calcium are rather straightforward. Cellular phosphates, for example, are released into the phosphate pool by the action of decomposers. Since they are soluble, they may be recycled at once. Any phosphate incorporated into skeletal,

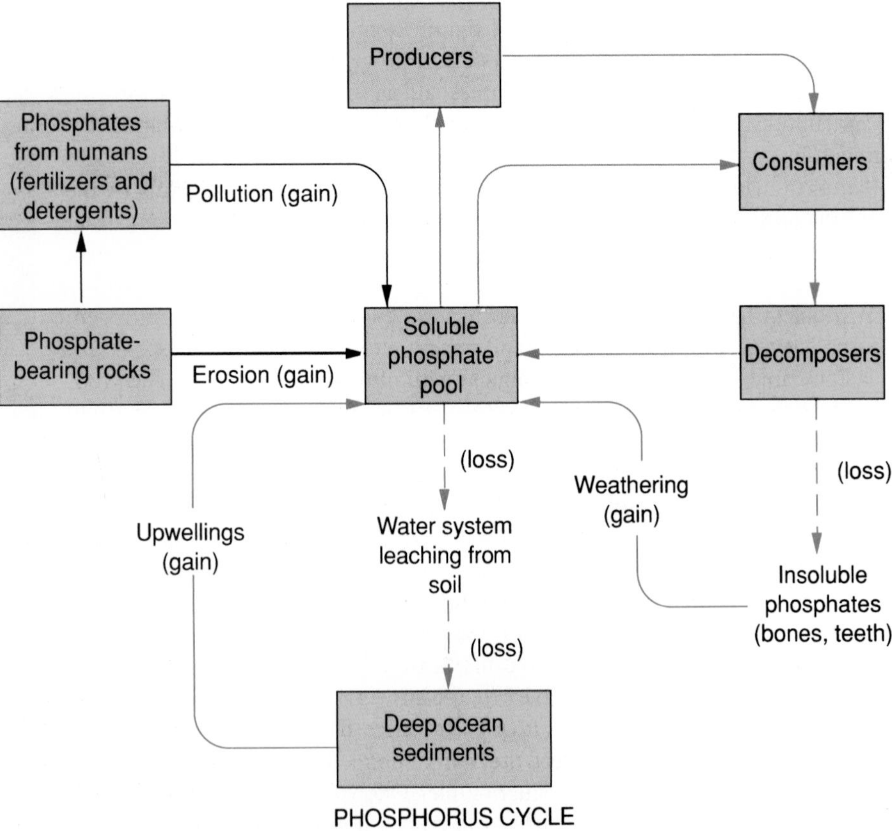

FIGURE 47.9
THE PHOSPHORUS CYCLE
Usable phosphorus in the form of soluble phosphates is found in soil water and in aquatic systems. Some phosphates pass to consumers through the trophic levels, while some are taken in through drinking water. Decomposers make some phosphates available again, but some locked-in animal remains (bones, teeth, shells, etc.) are unavailable for long periods. The loss of phosphates through leaching (from soil water) and through runoff (to the sea) is considerable. Some end up as insoluble phosphorus in deep sediments. A gain in available phosphates occurs through erosion and from pollutants introduced by humans. (Dashed lines show loss or potential loss.)

tooth, or shell material, however, is released very slowly by weathering. Calcium may also cycle very slowly, since it is commonly bound up in skeletons and shells. Calcium from such dense structures is very slowly leached into the soil. In freshwater biomes, the reservoirs of phosphorus and calcium may lie bound in the bottom sediments for long periods of time until currents agitate those murky depths. In marine biomes, occasional upwellings bring the reservoir sediments and dissolved ions to the surface, where they reenter the cycle via the phytoplankton.

The Carbon Cycle

We are aware that life is based on carbon, and as with any key element, its very availability may determine the size of populations. In the terrestrial realm, the main pool of carbon for photosynthesis is atmospheric carbon dioxide (CO_2). In the oceans and other waters, plants and algae use the bicarbonate ion (HCO_3^-) from dissolved carbon dioxide (and carbonate rock) as their principal carbon source.

Paradoxically, as important as atmospheric carbon dioxide is to life, it is present in such small proportions (slightly less than 0.04%) that it can almost be called a "rare gas." The quantity is admittedly small when compared to other gases, but it still represents an enormous amount in absolute terms.

In addition to the carbon found in the atmosphere, the earth has a sizable reservoir in the form of carbonate rock (such as limestone) and fossil fuels (natural gas, oil, coal, peat). However, this reservoir is being steadily altered by human and geological processes. Since the industrial revolution began, CO_2 from fossil fuels has been released into the atmosphere in steadily increasing amounts (Essay 47.1). Today, atmospheric scientists estimate that 5 to 6 billion metric tons of CO_2 from fossil fuels are released into the air annually. Fortunately, as we shall we, about two-thirds of this amount is quickly removed by the oceans and by photosynthesizers. For a look at how CO_2 is shuffled about in its cycle, see Figure 47.10. Since we have already covered critical aspects of the oxygen and water cycles in our discussions of photosynthesis, respiration, and the biosphere, let's move along now to community organization.

THE LAKE AS AN AQUATIC COMMUNITY

The earth is dotted with lakes, some large, some small, and some far more significant in human affairs than others. Here we will focus on those larger lakes that are not immediately giving way to acid, silting in, fertilizers, or other threats. We will consider lakes in their role as viable ecological entities. Specifically, we will concentrate on the communities of life they sustain.

Conditions and Life in the Lake Zones

Limnologists (*limne,* "pool" or "lake"), biologists who study freshwater communities, have divided lakes into zones, each with its own physical features and each harboring a characteristic array of life. These zones are called **littoral, limnetic,** and **profundal** (Figure 47.11). Although some of the terminology is different, the organization is similar to that of the marine environment. The littoral zone includes the shore and adjacent waters in which light penetrates to the lake bottom. Producers in the littoral zone include a variety of free-floating plants, plants that are rooted and submerged, and those rooted and emergent (protruding above the water's surface). The plants form a progression of types as the water deepens. Like marine producers, freshwater producers also include a phytoplankton component—numerous species of photosynthetic bacteria and algal protists. Consumers in the littoral zone include protozoan protists, snails, mussels, aquatic insects, and insect larvae. Salamanders and frogs also prefer the littoral zone, as do both herbivorous and carnivorous fish and turtles, along with a number of birds.

The limnetic zone, like the euphotic zone of the ocean, is defined as the region of open water extending down to the depth of effective light penetration. The actual depth varies considerably with turbidity, but in general the limnetic zone ends where the rate

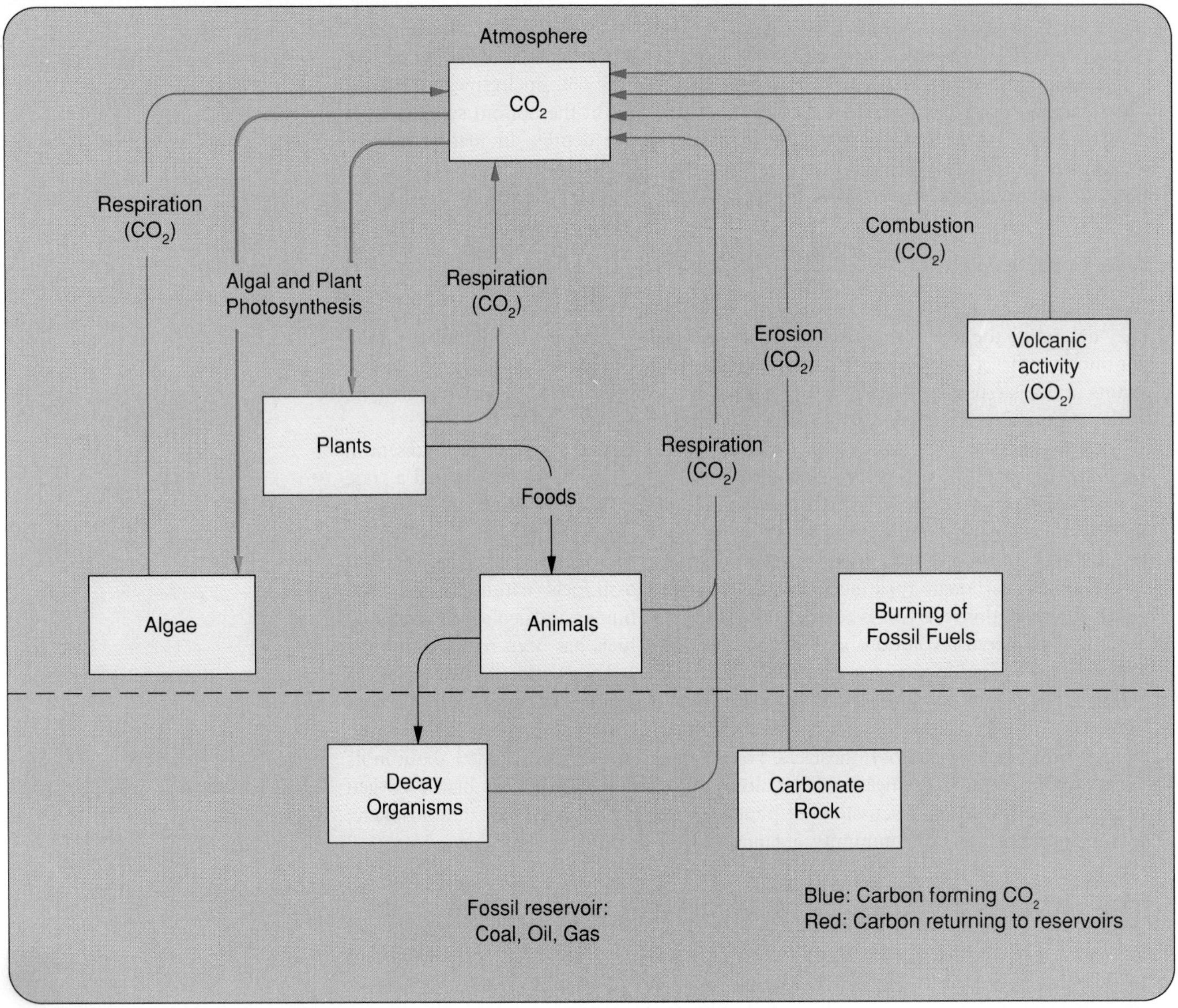

Atmosphere

CO_2

Respiration (CO_2)

Algal and Plant Photosynthesis

Respiration (CO_2)

Combustion (CO_2)

Erosion (CO_2)

Volcanic activity (CO_2)

Plants

Foods

Respiration (CO_2)

Algae

Animals

Burning of Fossil Fuels

Decay Organisms

Carbonate Rock

Fossil reservoir: Coal, Oil, Gas

Blue: Carbon forming CO_2
Red: Carbon returning to reservoirs

FIGURE 47.10
THE CARBON CYCLE
The atmosphere and ocean are the earth's greatest reservoirs of carbon. However, a sizable amount is present in carbonate rock and fossil fuel. The cycling of carbon begins with CO_2 entering plants during photosynthesis and becoming incorporated in carbohydrates. Plants then carry on respiration, releasing some carbon back into the atmosphere and soil as carbon dixoide. Some of the plant carbohydrates enter animals and other heterotrophs, where respiration returns more CO_2 to the atmosphere. The cycle includes one more significant source of CO_2—the burning of fossil fuels by humans.

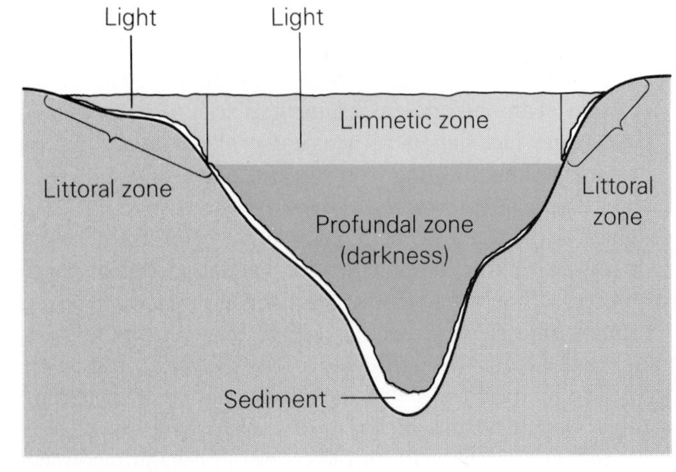

FIGURE 47.11
LAKE ZONATION
Deeper lakes contain three zones, the littoral, limnetic and profundal, each with its own unique physical characteristics and populations of organisms.

ESSAY 47.1

THE CHANGING CARBON CYCLE AND THE GREENHOUSE EFFECT: A DESTABILIZED EQUILIBRIUM

In recent years, both physical and biological scientists have become increasingly concerned with the status of the carbon cycle. In the past 80 years alone, atmospheric carbon dioxide has increased by about 15%, and at the present rate of increase, human activities could easily double the present level over the next 40 years. Recently, a group of aroused scientists from the prestigious National Academy of Sciences informed Congress of the matter and initiated, of course, hearings. The hearings go on and the CO_2 continues to rise, but at least the problem is becoming recognized. The increasingly politically conservative Environmental Protection Agency concurs with the findings of the NAS, and they too are beginning to address the problem. However, there will be no quick fixes to the global problems of increased atmospheric carbon dioxide.

Some of the more hopeful solutions once suggested have not materialized. For example, not long ago some scientists said that the oceans would act as a great buffer by absorbing any overburden of atmospheric CO_2. Unfortunately, the oceans are not very good at absorbing the gas—the mixing of water and gas occurs only in the top 80 meters of the sea.

Distribution of carbon in the biosphere

Plants (550)
Atmosphere (700)
Decomposers and humus (80)
Soil exchange pool (1,100)
Oceans (40,000)

Numbers represent billions of tons

So what does an increasing level of CO_2 mean to us? It means trouble. The trouble arises because carbon dioxide does not absorb or reflect short (ultraviolet) light waves, but it does absorb and reflect the longer (infrared) light waves. Thus, it freely admits solar energy into the biosphere but it slows the escape of heat that radiates from the surface of the earth. Carbon dioxide is thus a factor in the balance between energy entering and leaving the biosphere. So an increase in atmospheric carbon dioxide results in an increase in heat in the biosphere. The principle is similar to that in a greenhouse, which works by letting in the short light waves and retaining the longer waves of heat energy. Thus, the carbon dioxide effect has become known as the *greenhouse effect.*

According to the experts, when this phase of the temperature cycle eventually passes, we may see some severe greenhouse effects. Warming of the earth would result in a warming of the ocean. If the ocean water gets warmer, the solubility of carbon dioxide will decrease. There are much greater reserves of dissolved carbon dioxide in the ocean water than there are in the atmosphere, and an ocean temperature rise of only 1 or 2° C would unload more carbon dioxide into the atmosphere. This in turn would increase the greenhouse effect and further raise the world's temperature, which would cause the release of even more carbon dioxide. To make matters worse, the accelerated melting and subsequent decreases in the size of the polar ice caps would further decrease the amount of solar energy reflected back into space. Again, we see what is known variously as a positive feedback loop, destabilized equilibrium, or vicious circle.

Of particular concern is the Antarctic ice pack, much of which is actually above sea level, either in floating masses or over the Antarctic continent itself. Obviously, the melting of such ice will raise the present sea level. By some estimates, this could be as great as 6 meters (almost 20 ft), but more conservative estimates suggest that the mean sea level will increase by about 3 meters (almost 10 ft) by the year 2100. The most immediate effects will, of course, be felt in coastal regions and low-lying inland plains that have access to the sea.

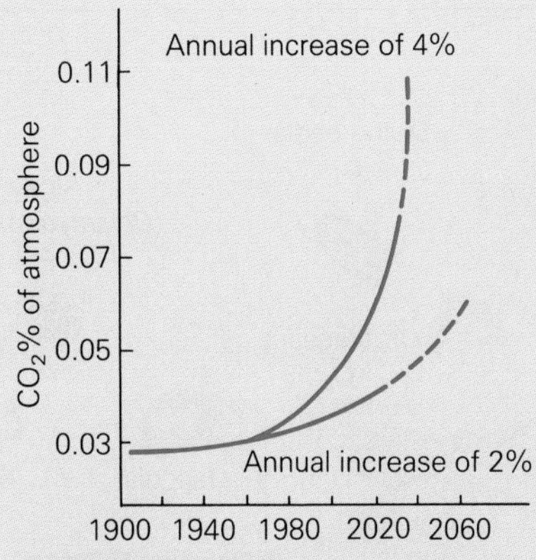

Projections of future carbon dioxide increases

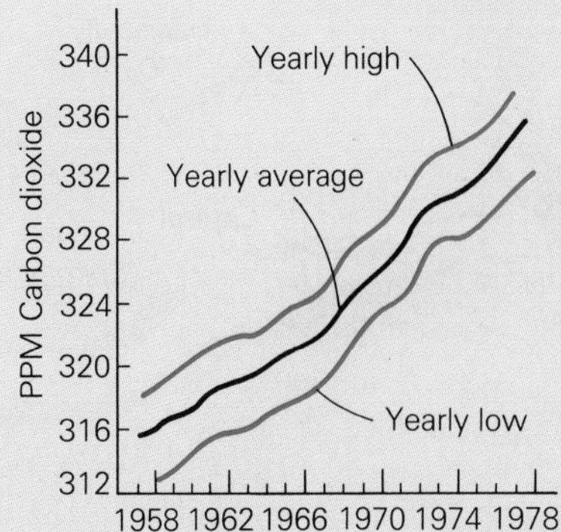

Recorded increases in atmospheric carbon dioxide

But, quite possibly, flooding along the continents is not the most serious consequence to consider. A far greater problem may be the effect of slight increases in the earth's mean temperature on climatic patterns. Studies of model systems help in our predicting patterns of weather change. For example, in the western United States, a change of just a few degrees will alter precipitation enough to reduce the flow of the Colorado River by half. This could be disastrous to the great population centers of the Southwest, which are completely dependent on the Colorado River water for consumption and irrigation of crops. In other parts of the globe, there would be drastically increased river flow—for example, in the Niger and Nile in Africa; the Mekong, Volta, and Tigris-Euphrates in Asia; and the Sao Francisco in Brazil. As far as agriculture *per se* is concerned, Americans have long enjoyed the favorable climatic conditions that make us the world's leading producer of grains. Consider the political and economic ramifications of any significant climatic changes affecting this capability. For example, according to one of the predictions, the rain belt supporting American grain production could well move north, establishing itself in Canada and leaving the American grain belt in a semiarid condition.

of respiration catches up with that of photosynthesis. Producers in the limnetic zone are not obvious, since they are nearly all microscopic. They include many of the phytoplankton found in the littoral zone, along with flagellated algal forms such as *Euglena* and *Volvox*. In northern lakes, phytoplankton populations undergo seasonal blooms, during which their productivity exceeds the plants of the littoral zone. These blooms closely correspond to the availability of nutrients, light, and favorable temperatures and often follow a preceding bloom of nitrogen-fixing cyanobacteria that enrich the waters with excess nitrogen compounds. Zooplankton form large populations of only a few species. Included are minute freshwater crustaceans such as the copepods and cladocerans along with dense blooms of rotifers. These populations rise and fall in response to the numbers of producers. Higher consumer levels are made up principally of the lake fishes: plankton-feeding species such as shad, and the carnivorous species such as bass and pike. The food web of the limnetic zone is often simple and direct. For an overview of life in the lake zones, see Figure 47.12.

The profundal zone begins where the effective penetration of light ceases—or where respiration begins to exceed photosynthesis. Again, turbidity is a determining factor in deep lakes. The profundal zone includes the muddy, sediment-rich lake floor. There is, of course, no producer trophic level here, and life is restricted primarily to bacteria and fungi, decomposers of the lake ecosystem, and a few detritus-feeding clams and wormlike insect larvae. All of the profundal species are adapted to periods of very low oxygen concentrations. As sparse as life is in the profundal zone, the activities of these organisms are critical to the organisms above. As scavengers, they, along with the bacteria, convert a virtual rain of corpses from above into nitrates, phosphates, sulfates, and so on—the usual mineral nutrients. How well these nutrients become distributed in the lake depends, as it did in the marine environment, on the occasional vertical movement of the waters.

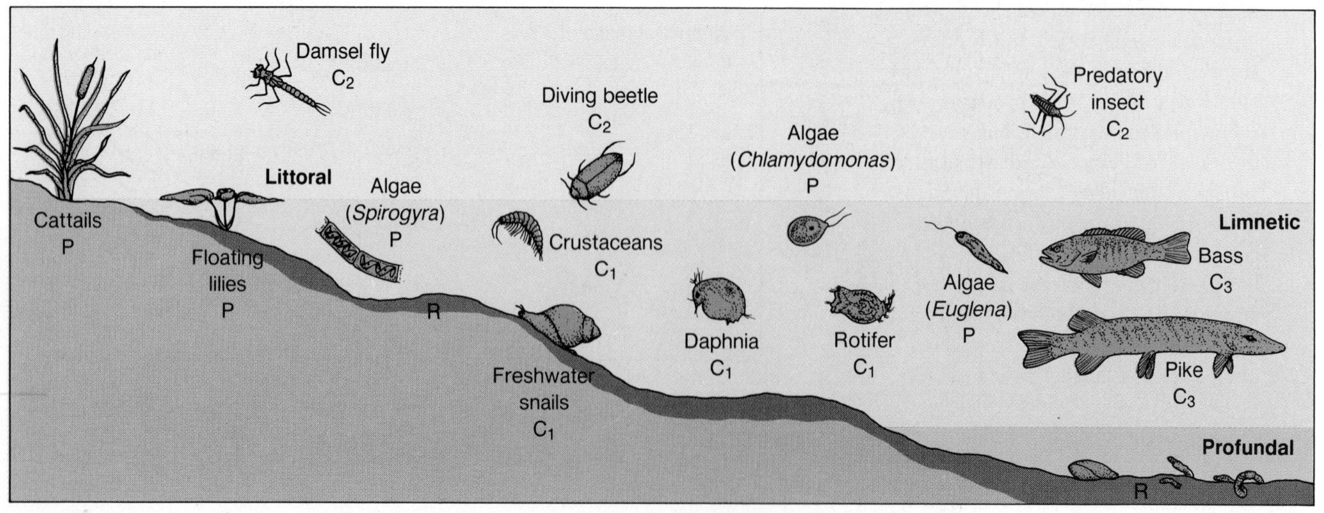

FIGURE 47.12
REPRESENTATIVE LIFE IN LAKE ZONES
Each zone, as you see, supports its own life forms. P = producer, C_1 = 1st order consumer (herbivore), C_2 = 2nd order consumer (carnivore), C_3 = 3rd order consumer (carnivore), R = reducer.

Thermal Overturn and Lake Productivity

The deeper waters of tropical Lake Tanganyika, which is 1450 meters deep (nearly a mile) are devoid of aerobic life, while temperate Lake Baikal, about 300 meters deeper, supports animal life throughout much of its depths. The difference between the two is attributed to the availability of oxygen, which, in turn, is a product of **thermal overturn**, the deep circulation of lake water brought on by seasonal temperature changes. Such changes, as you might expect, are far more significant in the earth's temperature regions where climate extremes are experienced. Thermal overturn carries dissolved oxygen to the lake depths and brings nutrients to the surface. In the absence of seasonal overturns, the lake bottom becomes highly anaerobic, greatly restricting the life there.

The conditions preventing overturn in Lake Tanganyika and some other tropical lakes are similar to those occurring in temperate zone lakes during the summer. In both, a temperature gradient exists between top and bottom waters. The light, warm upper waters may be well mixed by wind action, but they cannot mix with the dense, colder waters below. Many temperate zone lakes have three distinct summer temperature regions—an upper warm region called the **epilimnion** ("upper lake") and a lower, cold-water region called the **hypolimnion** ("lower lake")—each of which experiences little temperature variation throughout its depths. Between the two is a middle region, the **metalimnion,** which reveals a steep **thermocline** (an area where temperature drops most sharply with increasing depth). Typically, oxygen depletion begins just below the thermocline.

Thermal overturn in temperate lakes occurs in the fall and spring, when surface waters undergo temperature changes that profoundly affect their density. If you think back to our discussion of water in Chapter 2, you may recall that cold water has a greater density than warm water, at least until it freezes. Specifically, water reaches its greatest density at 4° C., but paradoxically, it reaches its least dense or lightest state at 0° C. Thus, as autumn temperatures drop, the surface waters of a lake cool until their density exceeds that of the layers below. The dense layers sink, displacing the lighter bottom waters which rise to the top. The sinking surface waters carry oxygen to the bottom and the rising bottom waters carry nutrient rich bottom sediments to the top. This is the **fall overturn.** But, as the cold weather intensifies, the surface waters reach the freezing point, ice forms, and the overturn ceases.

The reverse happens in the spring. The ice thaws and the surface waters begin to warm. When their temperature reaches 4° C, water's greatest density, they sink to the bottom, displacing the waters there and thus bringing about the **spring overturn.** And once again there is a redistribution of oxygen and nutrients. The effects of changing temperatures and density are further explained in Figure 47.13.

So, while all communities are shaped by surrounding climate, we see that it affects the lake communities in unique ways. Rarely do we find that oxygen is a life-limiting factor in the terrestrial communities.

Before leaving the subject of aquatic communities, we should remind ourselves of one of the strangest communities on earth. This is the **oceanic rift community**, composed of the bizarre and recently discovered chemoautotrophs. The inhabitants of this type of community are subjected to the severest conditions, but they manage to thrive in great numbers, as we see in Essay 47.2.

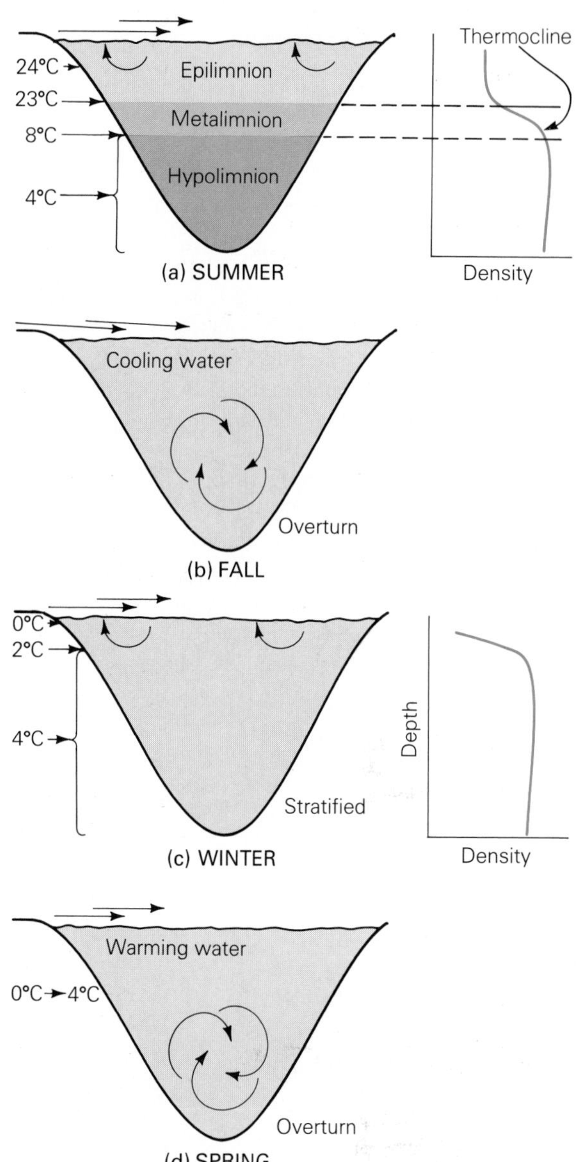

FIGURE 47.13
TEMPERATURE GRADIENTS IN THE LAKE
(a) In summer, water temperature is highest in the epilimnion and lowest in the hypolimnion. A steep temperature gradient, or thermocline, occurs in the metalimnion. (This occurs in deeper lakes only; in shallow lakes, wind mixes the waters and no stratification occurs.) Movement in the epilimnion cannot disturb the denser, cooler layers below, so overturn cannot occur. **(b)** In the fall, cooling surface waters approach 4° C (maximum density), sink to the bottom, and permit a wind-driven overturn. **(c)** In winter, the summer temperature gradient is reversed, with the coldest temperatures at the surface. However, the density gradient is roughly similar to that of summer, preventing overturn. **(d)** In early spring, warming surface waters reach 4° C and sink, starting another wind-driven overturn.

THE FOREST AS A TERRESTRIAL COMMUNITY

In 1969, ecologist George Woodwell and his colleagues completed a decade-long study of productivity in a scrub oak-pine forest community near the Brookhaven National

ESSAY 47.2

AN UNUSUAL COMMUNITY: THE GALAPAGOS RIFT

The Galapagos rift community, a bizarre assemblage of animals and bacteria, centers around the vents of sulfide hot springs in an area of active sea floor spreading some 612 km (380 mi) from Darwin's islands. It was discovered in 1977 by geologists aboard the submarine *Alvin,* which was cruising 2500 m deep (8202 ft) at the time. The geothermal hot springs, spewing boiling-hot solutions of hydrogen sulfide and carbon dioxide, support an entire ecosystem based on the autotrophy called chemosynthesis. It is one of the most dense and productive communities on earth—near the vents, the mass of living tissue approaches 50 to 100 kg per square meter.

Prominent among these denizens are enormous, blood-red tube worms of a previously unknown group, apparently belonging to the pogonophores (an obscure phylum now thought to be related to annelids). Like all pogonophores, they lack all trace of a mouth, anus, or gut. They are chemosynthetic autotrophs, deriving all their energy and carbon needs directly from the oxidation of the hydrogen sulfide and the reduction of the carbon dioxide, probably with the help of intracellular bacterial symbionts. The redness of their flesh is due to heavy concentrations of oxygen-binding hemoglobin; oxygen from the surrounding cold seawater is needed both for the oxidation of H_2S and for the worm's own oxidative metabolism. The tube worms range up to nearly 3 m in length (nearly 10 ft) and are about the girth of a man's wrist.

The entire ecosystem, in fact, is based on hydrogen sulfide chemosynthesis and is entirely cut off from other photosynthesis-based ecosystems. Apart from the tube worms, the primary producers are chemosynthetic bacteria, which swarm in enormous numbers where the often superheated, carbon dioxide- and hydrogen sulfide-laden vent water mixes with the near freezing, oxygenated abyssal seawater. At least 200 bacterial species proliferate near the vents and were visible to observers in *Alvin* as milky clouds. Filter-feeding crabs, clams, mussels, smaller worms (dubbed "spaghetti" by their geologist discoverers), and barnacles live off the bacteria. A little higher on the food chain, larger crabs and a variety of fish

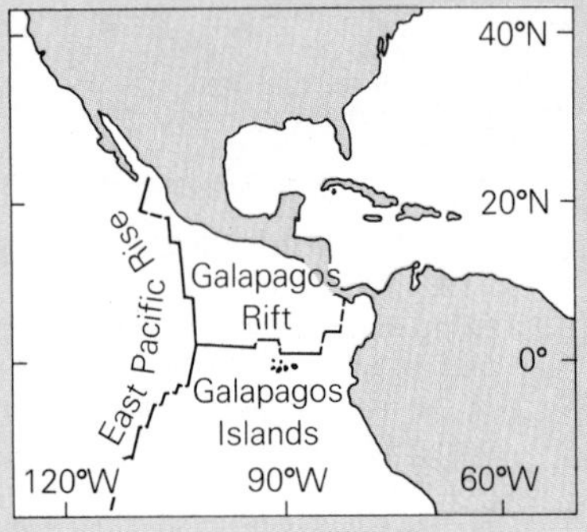

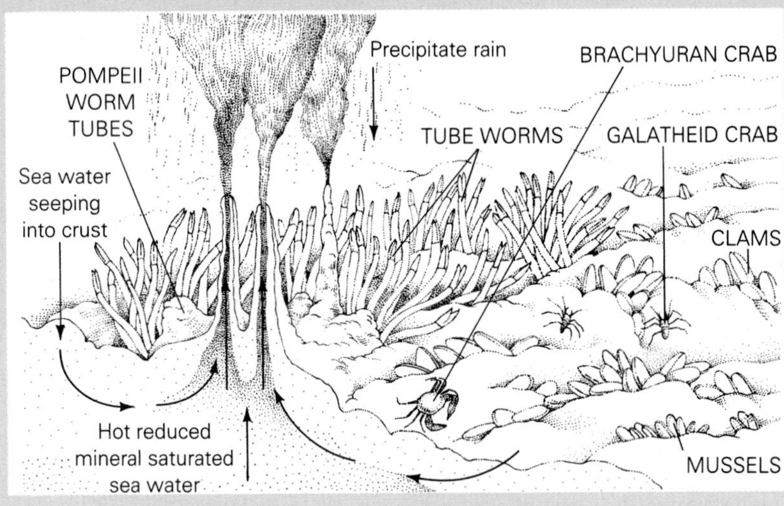

Laboratory in Long Island, New York. Oak-pine communities are common in this area (Figure 47.14), the product of disturbed deciduous forests whose lush elegance once graced the Long Island landscape. Such an undertaking can be enormous in scope and serves as a clear illustration of what studies in community ecology can be like. While productivity is generally expressed in calories or kilocalories (per unit of area per unit of time), in this study the unit used was the gram.

In their preliminary efforts, the ecologists first described the makeup and structure of the community. In addition, they periodically sampled and estimated the total organic scavenge on clumps of bacteria as well as on animal remains. Whelks, leeches, limpets, and miscellaneous worms complete the community. Some species were previously known, but others, such as the tube worms, are new discoveries with uncertain affinities. Similar communities have now been found on the East Pacific Rise Rift, and other deep-sea geothermal communities probably occur wherever there are appropriate sulfide springs. In fact, the rift ecosystems may be so widespread as to constitute a major earth community.

The Galapagos rift community contrasts markedly with the rest of the deep-sea benthic communities, which are characterized by constant cold and a severely limited energy input. The usual benthic communities survive on what little organic material drifts down from the surface waters and are thus based ultimately on energy from distant photosynthesis. The animals of the cold water deep-sea bottom are generally slow-moving and slow-growing, playing a variety of refrigerated waiting games: scavengers patrol listlessly for the chance of a dead fish or a fecal pellet, and predators lie motionless in ambush for wandering scavengers. In the hot springs communities, however, food is virtually unlimited, the temperature is not so uniformly cold, and both growth and metabolism are relatively rapid. The Galapagos rift clams, for instance, grow up to a third of a meter (13 inches) long at a rate of 4 cm per year, some 500 times faster than their smaller cold-water relatives. The flesh of these clams, like that of the giant tube worms, is bright red with hemoglobin. This factor, along with their phenomenal growth rate, indicates a high metabolic rate.

Rich and active as they are, the deep-sea rift communities are ephemeral. The hot springs eventually die down, like volcanoes, leaving behind ghostly communities of empty clam shells. As the earth's crust shifts and new hot springs form, immigrants from established or dying geothermal communities arrive to begin the unusual chemosynthetic ecosystem anew. Rapid ecological succession is one more unusual characteristic of the rift community.

Tube worms

Spaghetti worms

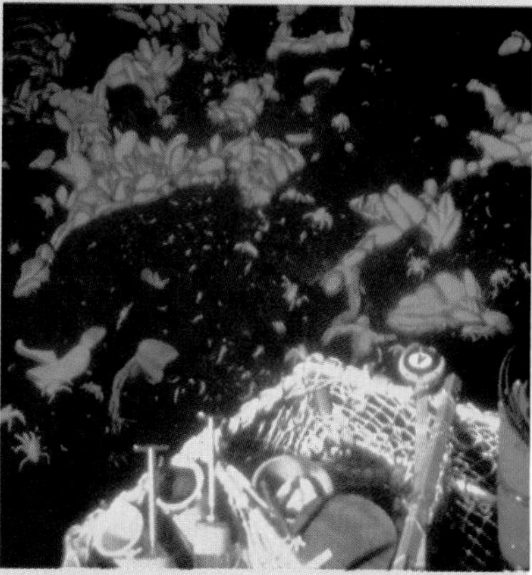

Clam field

matter present. From measurements of the biomass, they determined the gross productivity of the forest. But, as we have seen, the net productivity can only be calculated if the rate of respiration is known. To find this, Woodwell and his associates used a direct indicator: the carbon dioxide output of the forest. As you can imagine, this was easier said than done.

One problem was that such measurements must be made in the dark in order to eliminate the problem of carbon dioxide uptake during photosynthesis. Then, of course, there is the problem of the usual air movement, which would carry away any carbon

FIGURE 47.14
AN OAK-PINE FOREST COMMUNITY
Oak-pine forests are generally in a state of regrowth, since they are periodically ravaged by fires.

FIGURE 47.15
PRODUCTIVITY IN AN OAK-PINE FOREST COMMUNITY
Measurements in this oak-pine forest reveal that the total matter being produced through photosynthesis exceeds the total being metabolized in respiration. Thus, forest growth is on the rise, although the actual matter being stored is only about 20%. The numbers themselves represent a rate—the grams of dry matter per square meter per year. The total amount of dry matter produced by the forest in this time period was 2650 g/m^2. The total metabolized in respiration to meet the energy requirements of life in the forest was 2100 g/m^2. This includes respiration in both autotrophs (R_A = 1450) and in heterotrophs (R_H = 650).

dioxide diffusing from the plants. Fortunately for the study, the region experienced frequent nighttime temperature inversions, such as the ones that produce smoggy days in the Los Angeles area. In an inversion, cool surface air becomes trapped close to the ground by warmer layers of air above. The nighttime inversions in the oak-pine forest prevented the usual vertical movement of air and permitted the accumulation and measurement of respiratory carbon dioxide.

From the lengthy study, it was concluded that the *annual* gross primary productivity of the forest community was 2650 grams (about 5.8 lbs) per square meter, while the annual net primary productivity after the respiration correction was made, was 1200 g/m^2. Further determinations established that the rate at which organic material was accumulating—the net community productivity (NCP) (net production minus heterotroph respiration)—was 550 g/m^2 · year. The forest study is summarized in Figure 47.15.

Compared to some communities, the net productivity of this forest is modest. Annual net productivity in each square meter of some tropical rain forest communities, for example, can reach several thousand grams. However, the greatest annual productivity is not in natural communities, but in agricultural communities. For instance, in tropically grown sugarcane (an efficient C4 plant; see Chapter 7), the annual net productivity can exceed 9000 g/m^2, while grain fields range between 6000 and 10,000 g/m^2. Of course, if we factor in the energy of fossil fuels used to operate farm machinery and the energy used in manufacturing and applying pesticides and fertilizers, the true net yields fall drastically. In fact, on a calorie-for-calorie basis—the caloric yield of food versus the energy needed to produce it—there is often a loss. Table 47.2 compares productivity (expressed in kcal) in six quite different ecosystems.

One more observation—perhaps an obvious one—was made in the Woodwell study: the oak-pine community was growing—it had not reached its ecological climax state. As we mentioned earlier, communities that have reached their climax state have no net productivity. Their respiratory output equals their photosynthetic input. We will look more closely at the climax state and community growth shortly.

COMMUNITY ORGANIZATION AND DYNAMICS

Communities, we have seen, are populations of organisms that interact with each other and *with their physical environment* within a biome. As usual, the key term is *interaction*. Such definitions are admittedly vague, and it is often difficult to determine the boundaries of communities. As with biomes, some communities simply blend gradually into others and for this reason are called **open communities.** Forest communities are like that, as different vegetation types blend together in mixed associations at borders. Open communities are assumed to interact energetically as reserves move relatively freely between

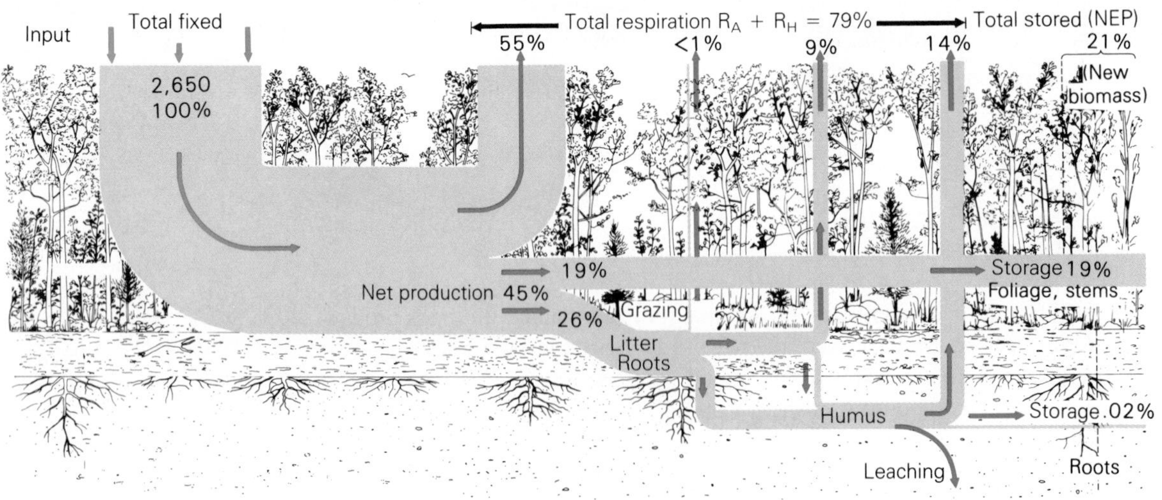

them. **Closed communities** are those with more definite borders. An example might be a spruce forest community bordering Puget Sound, where the presence of salt water ends the forest rather abruptly (Figure 47.16). Caves also tend to contain distinct communities, since they are isolated from their surrounding, and the inhabitants have had to adapt to an entirely different range of conditions.

There is presumed to be little interaction energetically between closed communities. Actually, in terms of energetics, there are no completely closed communities, since there is always some degree of interaction between neighboring communities. The cave community, for instance, must rely on organic materials carried in by streams or rain water, although when an occasional surface animal blunders in, it provides a temporary energy source for a whole host of cave dwellers.

Community organization and structure are the result of a number of interacting ecological factors. We will now consider some of the more important of these. However, we must keep in mind that these are not isolated influences, but a range of intersecting forces that, together, form the communities of life.

The Role of Competition in Community Structure

Competitive interaction can be an important influence on the structure of a community. By definition, **competition** results in some harm coming to the loser of a struggle for any commodity that is in short supply. Ecologists generally recognize two broad types of competition: namely, exploitative competition and interference competition.

Exploitative competition is simply the utilization of a limited resource by two or more organisms, one of which has an advantage that results in the other being harmed by losing the race. Examples are filter-feeding zooplankton that ingest tiny phytoplankton, and fly larvae (maggots) that feed on dead animals. The outcome of exploitative competition is determined by differences in the feeding efficiencies of the competitors. As the resource falls into increasingly short supply, the more efficient competitor will prevail while the other struggles to maintain itself.

Interference competition is any activity that limits another organism's *access* to a necessary resource. Here, individuals may be prevented from consuming resources by the activities of others. Examples are the aggressive behavior of squirrels chasing others away from a food supply; the cannibalism of flour beetle eggs and pupae by flour beetle larvae; among plants, the production of toxic chemicals that inhibit the growth or survival of other individuals; and one wolf shouldering another away from a carcass. As you can see, this form of competition can involve outright killing of the competitor.

Competition exists at two levels, within species (intraspecific competition) and between species (interspecific competition). Intraspecific competition may exist not only for commodities such as food, but for reproductive factors such as mates and nest sites. Generally, the results of interspecific competition are more apparent in the structuring

TABLE 47.2 ANNUAL PRODUCTION AND RESPIRATION AS KCAL/M^2 · YEAR IN GROWING AND CLIMAX* ECOSYSTEMS

	ALFALFA FIELD (USA)	YOUNG PINE PLANTATIONS (ENGLAND)	MEDIUM-AGED OAK-PINE FOREST (NY)	LARGE FLOWING SPRING (SILVER SPRINGS, FL)	MATURE RAIN FOREST (PUERTO RICO)	COASTAL SOUND (LONG ISLAND, NY)
Gross primary production	24,400	12,200	11,500	20,800	45,000	5,700
Autotrophic respiration	9,200	4,700	6,400	12,000	32,000	3,200
Net primary production	15,200	7,500	5,000	8,800	13,000	2,500
Heterotrophic respiration	800	4,600	3,000	6,800	13,000	2,500
Net community production	14,400	2,900	2,000	2,000	Very little or none	Very little or none

*Climax, as we will see later in this chapter, refers to communities of living things that use as much energy as they produce.

SOURCE: Adapted from *Fundamentals of Ecology*, 3rd Edition, by Eugene P. Odum. Copyright © 1971 by W. B. Saunders Company. Reprinted by permission of Holt, Rinehart and Winston, CBS College Publishing.

FIGURE 47.16
BORDER OF A CLOSED COMMUNITY
The spruce forest ends abruptly at the water's edge.

of communities. Such competition can directly influence the species composition of a community. We can best see how this can happen by noting how organisms establish their niches.

The Ecological Niche

The **ecological niche** is an organism's role in nature, involving every aspect of its ecological, physiological, and behavioral interaction with its environment. In describing the niche of a species, ecologists pay special attention to three types of information: the kinds of resources used, the resource characteristics that can affect utilization, and the survival rate when resources are used in the different ways possible by the species. For example, in describing the niche of rainbow trout, we would include foods eaten—for example, insects, crustaceans, fish, and other prey; resource characteristics such as prey sizes; and then the ability to escape or defend themselves. We would also consider the temperature, oxygen concentration, salinity, and pH of the waters where trout feed; and the survival rates of trout when feeding under different conditions, e.g., when feeding on small prey at low temperatures and salinities or on large prey at high temperatures and salinities. Also, the various factors that describe a niche are not necessarily independent of each other. Sometimes, in fact, they are directly correlated (Figure 47.17).

Niches may be thought of in two ways. The **fundamental niche** describes the possible ways that an organism can interact with its surroundings. In other words, the fundamental niche refers to the interactions that an organism is theoretically capable of when not constrained through other factors, such as competition. The **realized niche** reflects the reality of the situation. It describes the part of the fundamental niche that is actually occupied. Because of the array of factors that constrain an organism from doing what it is *able* to, such as competition or predation, an organism occupies only a part of its fundamental niche.

Some niches can be very narrow, while others are quite extensive. For example, some crows are broad-niched. In other words, they interact with their environment in a number of ways. They walk along plowed fields eating newly planted corn, they fly into trees to sample berries, and they don't turn up their beaks at a dead squirrel along the roadside. They may eat the eggs of other birds, and so they are attacked on sight by smaller, quicker nesting birds. They retreat from these tiny tormenters, but they may turn around and mob an owl. Because they interact with their environment in a number of complex ways, they are called broad-niched (see Figure 45.3). On the other hand, the Everglade kite is extremely narrow-niched, as reflected by its diet—almost exclusively freshwater snails. Most microbes are also narrow-niched. Such decomposers as bacteria operate within very rigid and narrow limits. That rotting squirrel was not decomposed by the action of a single kind of bacterium. Instead, a variety of microbes were at work, each utilizing the waste products and by-products of others and each incapable of doing

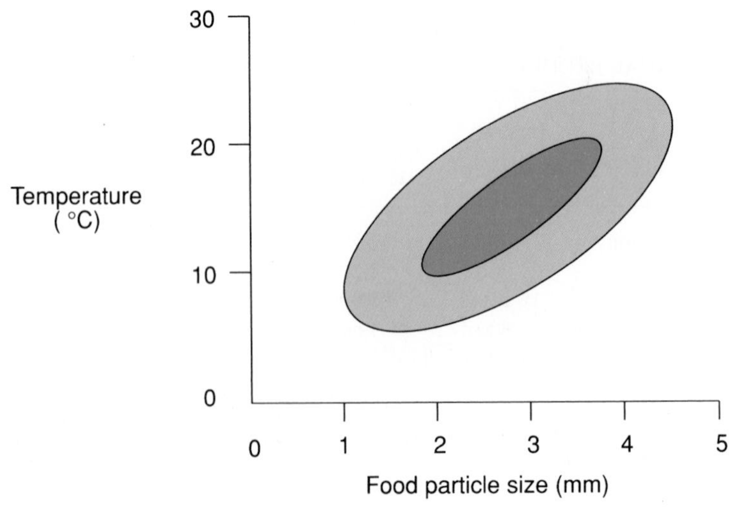

FIGURE 47.17
TWO-DIMENSIONAL NICHE OF TROUT THAT FEED ON DIFFERENT SIZES OF AQUATIC PREY AT DIFFERENT WATER TEMPERATURES
The outer solid line illustrates the actual combinations of temperature and food-particle size used. It shows that the two niche dimensions are not independent and therefore should not be considered separately, because a food-particle size is not necessarily used by the species at all temperatures in the range of tolerance. As we see, larger prey are not taken at the lowest temperatures, and smaller prey are not taken at the highest temperatures. The inner line indicates the portion of the niche associated with greatest survival of the species.

anything else. The decomposition process, then, is complex, each step carried out by a microbial specialist. All specialists are regarded as having narrow niches. Let's turn now to how different species interact when their niches overlap.

Competition leading to out-and-out physical aggression between members of different species is relatively rare. Instead, usually one species is simply and quietly repressed. This repression, in fact, may be so severe that the losing population completely dies out. The classic experiment demonstrating this principle was conducted by the Russian biologist A.F. Gause. According to what is now called Gause's law, or the **principle of competitive exclusion**, two species cannot indefinitely live together and interact with the environment in the same way if resources are limited. The results of the key experiment that led him to this conclusion are shown in Figure 47.18. As you see, the two *Paramecium* species do well alone, but when grown together one—*P. aurelia*—outdoes the other. We are not certain about the specific nature of the competitive edge *P. aurelia* has over *P. caudatum,* but it certainly exists. There is apparently a critical overlap in the niche of the two protists.

As another example, when two species of duckweed, *Lemna gibba* and *Lemna polyrhiza,* were grown alone in a tank, each did well; but when they were grown in the same tank, *L. polyrhiza* died out. The reason is that *L. gibba* is a better competitor for light. It has air-filled sacs that cause it to float higher in the water, blocking light from its competitor. Had the experiment been carried out under a variety of conditions, the experimenters would most likely, sooner or later, have come across a situation in which *L. polyrhiza* was the winner.

Different species are most likely to rely on the same kind of food when that food is plentiful. For example, in the tundra, two predatory birds, snowy owls and jaegers (or skuas as the latter are also known) both feed on lemmings, but only when the lemmings are in abundance. The owls and jaegers switch to alternate prey if lemmings are spare or simply cut short their annual visit. A study of 11 species of Panamanian stream fishes revealed that each species had a very specialized diet during the dry season when food was scarce. But in the wet season, the same abundant food sources were used by all the fishes.

From many such observations grows the hypothesis that competition in the wild rarely results in extermination, but rather it encourages a subdivision of the habitat, each species coming to live where it does best. This point is well illustrated in a classical observation by the noted ecologist R.H. MacArthur. As Figure 47.19 reveals, five species of warblers that utilize the resources and shelter of spruce trees do not actually occupy the same niche. Although some overlap is inevitable, their feeding and nesting habits differ enough so that they are not in serious competition.

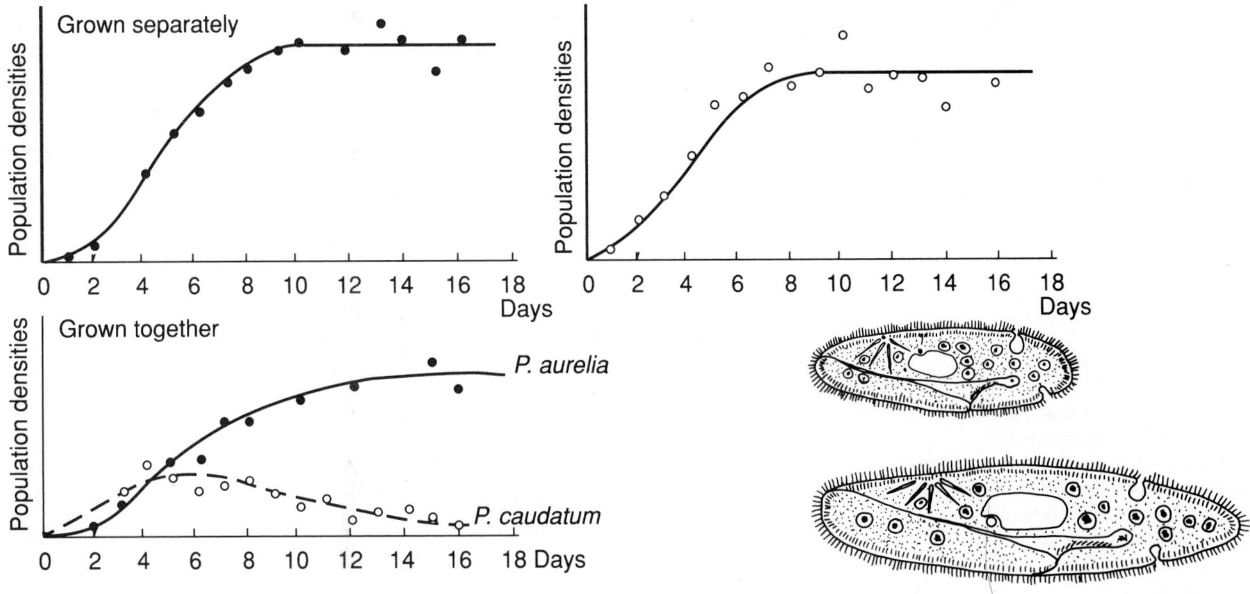

FIGURE 47.18
COMPETITIVE EXCLUSION AT WORK
Part of Gause's evidence for competitive exclusion was the study of population growth in two species of *Paramecium* that utilize the same kinds of foods. When grown separately, the numbers of *P. aurelia* and *P. caudatum* each reached a stable population size. When the two were grown together, *P. caudatum* was displaced by the smaller but apparently more efficient *P. aurelia.*

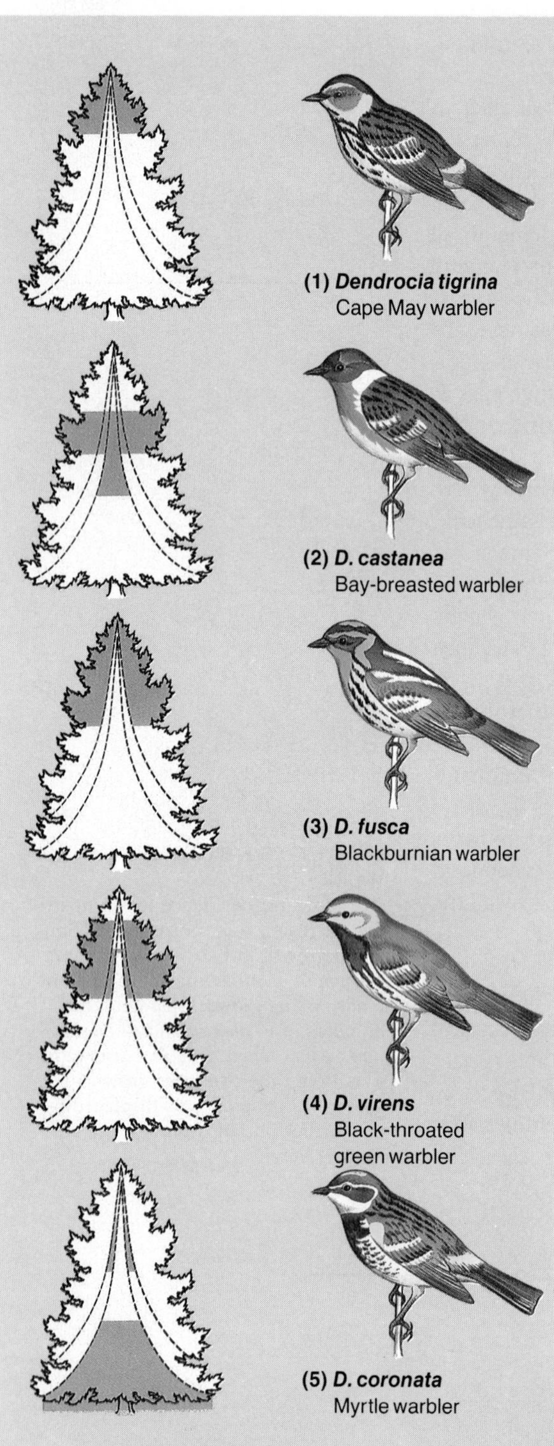

FIGURE 47.19
SUBDIVIDING THE HABITAT
Five species of North American warblers use spruce trees for feeding and nesting, but each has its own zone. The darkened areas indicate where each species spends at least half of its feeding time. By exploiting different parts of the tree, the species avoid direct competition; thus, they can occupy the same habitat.

The Roles of Territoriality and Dominance Hierarchies in Community Structure

In many species, strong and potentially destructive competition between species is held in check by special forms of organization. This organization results in clearer lines between "haves and have nots," but certain individuals in such systems increase their likelihood of surviving and reproducing.

One such organization is the formation of territories. A territory is defined as any defended area. However, territories generally contain some resource (such as food, nest sites, or mates) that must be defended against competitors. If the habitat is variable, some animals will end up with "better" areas than others. If those areas hold more food, then in times of food shortage the individuals holding the best territories will be more likely to survive. In some species of birds, the male must have a good territory in order to attract females. A splendid male holding inferior territory will not find a mate. We see an example of this sort of influence on community structure in certain blackbirds. Those males with peripheral, and hence less protected, nests may not attract mates (Figure 47.20).

Territory size is critical to the reproductive success of the territory holder. Territories that are too large will drain valuable energy and time required in defense; territories that are too small will not provide enough food for rearing offspring. Accordingly, studies show that birds appear to respond to food supply and number of potential trespassers in setting their territory boundaries. Thus territories are larger in lean years than in rich years, and as the number of boundary intrusions grows, territory size dwindles. Such observations suggest that territory size represents a balance between the advantages of a larger territory (more food) and a smaller territory (less defense).

Community structure in some species can be influenced by interactions based on individual recognition. These species are able to form hierarchies (or pecking orders), a ranking of individuals that are then related to according to their rank. The ranking is usually first established early in life through social interactions such as fighting or play. In hierarchies, those higher-ranking animals have freer access to commodities than do others. Hierarchies reduce conflict within groups as an animal acquiesces when confronted by a dominant individual. In times of shortage of critical commodities, then, the higher-ranking animals are more likely to survive, since subordinates give way before them (Figure 47.21). On the other hand, the subordinates are more likely to survive by giving way, rather than being further weakened by a thrashing from a superior individual.

The Roles of Disease and Parasitism in Community Structure

Both disease and parasitism can certainly affect community structure. The more closely that susceptible individuals are packed together, the more opportunity there is for disease transmission. Also, the more individuals there are in a population, the greater will be the number of potential reservoirs in which more virulent mutant strains of the disease microorganisms can develop.

In some cases, disease may interact with predation to depress certain populations and thereby have a profound effect on community structure. For example, a two-week-old caribou fawn can already outsprint a full-grown timber wolf, and healthy caribou seldom fall prey to wolves. However, caribou are subject to a hoof disease that lames them before it affects other parts of the body, and it is these lamed animals that a wolf is likely to cull out of a herd. In areas where the wolves were poisoned in order to protect the migrating caribou, this hoof disease spread unchecked and in a few seasons decimated entire caribou herds. With the caribou herd suffering in this way, opportunities arise for other herbivores that may have competed with the vast herds for grazing opportunities. Do you suppose that a weakened caribou herd would provide more or less food for wolves? Does your answer relate to the short-run or the long-run?

The effect of parasites on community structure is largely dependent on the length of the parasite-host association. Some parasites live in the host for long periods and also reproduce in the host. A long host-life is essential for the maximum fitness of these parasites, and, as might be expected, they have evolved a reduced virulence, minimizing their impact on host survival. Parasites with weaker effects live to reproduce and pass on their genes, but those with stronger effects die with their hosts before reproducing. Good examples are the so-called parasitic castrators, which infest the organs that least influence host survival—the gonads. By feeding on gonadal tissue exclusively, the parasites gain needed resources but have virtually no effect on host survival.

In contrast, other parasites live in or on the host for only brief periods and do not reproduce in the host. A long host-life is not critical to parasite survival and reproduction. In fact, increased virulence may evolve if it is tied to the extraction of large amounts of host resources that speed parasite development. Parasites developing before host death would not be harmed. Appropriate examples are nematodes of the family *Mermithidae*. Juvenile nematodes develop in the body cavity of their insect hosts, but before maturing they bore their way out of the host. This dramatic exit is often fatal to the host, especially if there are a large number of escaping worms. Mermithid nematodes are effective biological weapons in our nonchemical war against economically important insect pests.

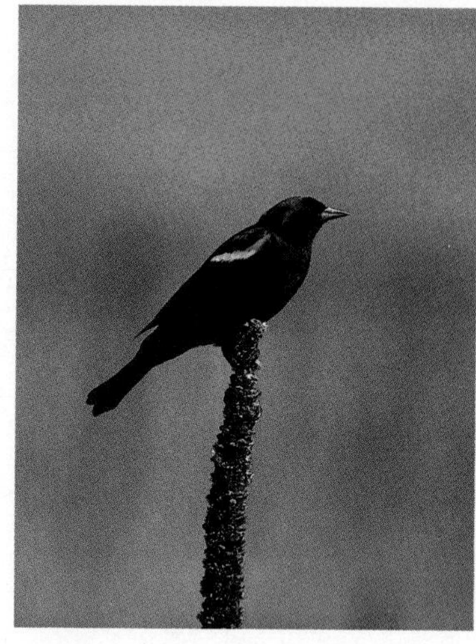

FIGURE 47.20
THE RED-WING, A TERRITORIAL BIRD
Among some birds, such as red-winged blackbirds, males may take territories of varying qualities. Females can assess territories and tend to choose the males with the best real estate. In some cases, a female will choose a male who already has a mate over a bachelor with an inferior status.

The Role of Predation in Community Structure

Predation can be a powerful determinant of community structure. It has a dynamic influence on the numbers and quality of both predator and prey, as it acts as an important agent of natural selection on both groups (see Essay 47.3).

Predators influence the numbers of prey by removing individuals from the prey population, yet they do not normally kill off the prey population. The reason can be explained by examining the classical **Lotka-Volterra theory** (simplified here). Under undisturbed conditions, prey numbers rise steadily, thus providing more and more food for predators. Then the predator numbers begin to rise. Their numbers do not rise immediately, however, since it takes time for the energy from food to be converted into successful reproductive efforts. Because of this time lag, the prey may be well on the road to recovery before the predator population begins to rise. When the predator numbers finally do rise, though, there is increasing pressure on the prey. Then, as the prey begin to be killed off, the predators find themselves with less food, and so their own numbers soon fall off due to starvation or simply a failure to reproduce. Then the prey begin to recover.

FIGURE 47.21
SOCIAL ORGANIZATION IN THE BABOON TROOP
Complex social interaction characterizes the baboon troop, providing a measure of both safety and order. Individuals and subgroups are organized in a hierarchy, with dominant members receiving first priority in feeding and reproduction.

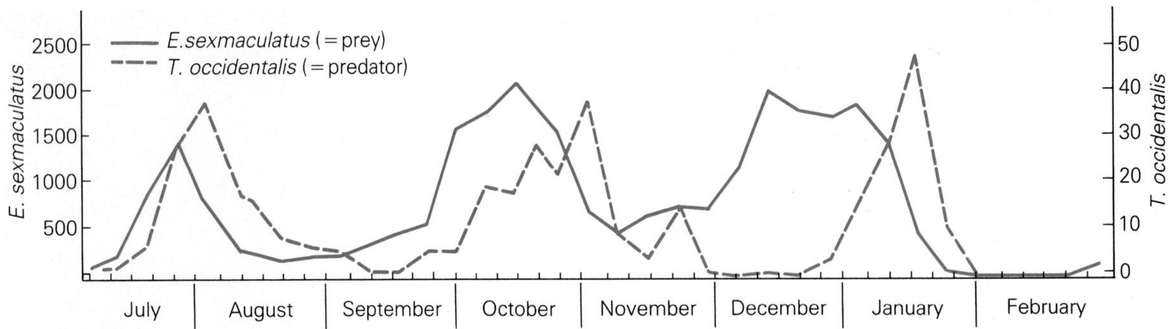

FIGURE 47.22
PREDATOR-PREY INTERACTIONS
The oscillation of population numbers of the predatory mite *(Typhlodromus occidentalis)* and its prey, the six-spotted mite *(Eotetranychus sexmaculatus)*, is almost classic in its characteristics. Prey numbers increase first, followed at once by predator numbers. Then, as the predator increase continues, the prey species diminishes in a population crash. The predator's fate is quite similar as its numbers fall rapidly.

As an example of predator-prey population dynamics, consider the curves produced by the predation of one kind of mite on another in Figure 47.22. We should add, however, that predator-prey relationships are not often as straightforward as this. For example, if the predator is able to hunt more than one prey type, the story can be complicated by prey-switching, in which the predators seek other, more available, food.

By regulating a prey population, predators can affect the level of competition experienced by the prey. Surviving prey often have access to more food and consequently have higher reproductive and growth rates. Also, other species that may have been held in check or even competitively excluded by the prey species may benefit indirectly by the predator's presence. In a classical study of competition between two barnacle species, Connel found that *Balanus* excluded *Chthamalus* from deeper portions of the intertidal zone. However, in the presence of a predatory snail, which preyed selectively on the faster-growing and larger *Balanus,* both barnacle species coexisted in the deeper waters. Similarly, Paine found that in rocky intertidal areas inhabited by a predatory starfish species, many species of barnacles and mussels coexisted. When the starfish was experimentally removed, the level of interspecific competition increased, significantly reducing the number of barnacle and mussel species. A species, such as a predator, whose removal causes the extinction of other species is called a **keystone species.** The activities of a keystone species, then, can indirectly increase the number of species in a community.

The importance of predation and other biological and physical disturbances on the number of species found in a site is recognized in the **intermediate disturbance hypothesis** (Figure 47.23). According to this hypothesis, areas with intermediate levels of disturbance have more species than do areas of lower or higher levels of disturbance. At lower levels, competition is intense, and the resulting exclusion yields only a few

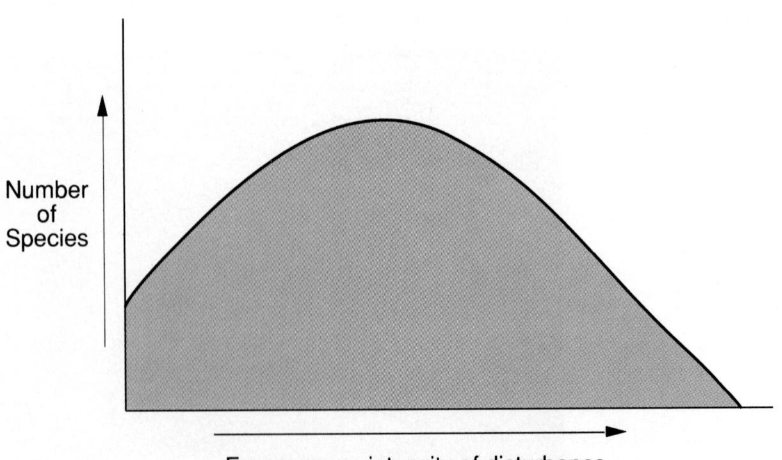

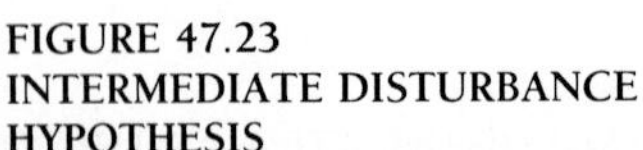

FIGURE 47.23
INTERMEDIATE DISTURBANCE HYPOTHESIS
The number of species is plotted against the frequency or intensity of disturbance. In this diagram illustrating the intermediate disturbance hypothesis, note that the greatest number of species is found at intermediate frequencies or intensities of disturbance.

surviving species. At higher levels, the disturbance itself wipes out all but a few stress-tolerant species. At intermediate levels—not strong enough to kill most species but still strong enough to reduce the competitive impact of dominant species—the number of species is highest because competitively inferior and superior species as well as stress-intolerant and stress-tolerant species survive.

COMMUNITY DEVELOPMENT OVER TIME: ECOLOGICAL SUCCESSION

Our view of biotic communities has for the most part been static—suspended in time—as a convenient way to describe their organizational components. However, they do tend to change as part of their normal development. They change not only in response to climatic and geological forces, but also in response to the activities of their inhabitants. In some cases, the inhabitants will alter the environment, which then influences the community in new ways in a form of feedback loop.

Community change over time is known as **ecological succession.** Where ecological succession is the product of the organisms themselves, it is known as **autogenic succession. Allogenic succession** occurs where outside forces—particularly physical forces such as fire or flood—regularly effect change. In most instances, succession is a result of both autogenic and allogenic factors, although one or the other may have triggered the process.

The entire series of sequential changes involved in succession is a **sere,** and each stage a **seral stage.** Allogenic succession is less predictable than autogenic succession. For example, an orderly progression of species during succession is often interrupted by the sudden bloom of unexpected opportunistic species, such as weeds. In addition, one would not like to leave the impression that one population gracefully gives up its place for the next. On the contrary, species are often quite persistent, seemingly resisting their own displacement.

Ecological succession includes both **primary succession** and **secondary succession.** Primary succession is the establishment of a community where no community previously existed, such as on rocky outcroppings, newly-formed deltas, sand dunes, emerging volcanic islands, and lava flows. (Scientists are now studying the emerging succession on the slopes of Mount St. Helens.) Secondary succession is the establishment of a new community where a previous community has been disrupted. Examples include neglected farms reverting to the wild and forest communities that have been subjected to "clear-cutting," the controversial lumbering practice in which all trees are removed from a stand.

FIGURE 47.24
PIONEER ORGANISMS
Lichens and mosses are able to exist under conditions that discourage most other life. Their success in colonizing bare rock surfaces has earned them the name "pioneer."

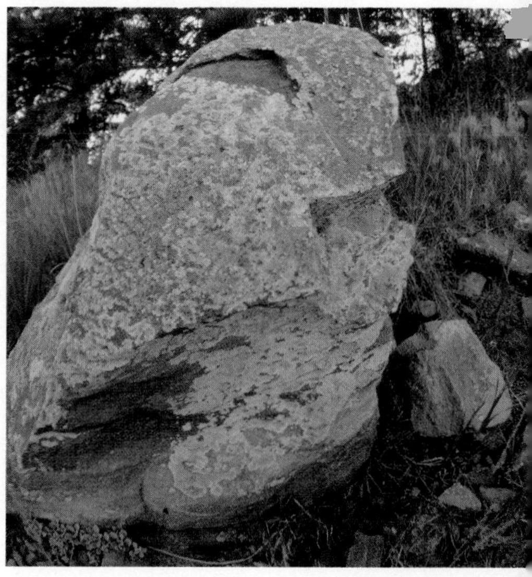

Primary Succession

In primary succession, such as occurs on a rocky outcropping (Fig. 47.24), the sere begins with pioneer organisms, hardy drought-resistant species that can successfully establish themselves and reproduce in places previously not inhabited by their kind. They are specialized for the initial invasion of disturbed or uninhabited areas. Lichens are often the first pioneers to invade rocky outcroppings, where they are held fast by their tenacious, water-seeking, fungal component (see Chapter 24). Lichens are soil builders, producing weak acids that very gradually erode the rock surface. As organic products and sand particles accumulate in tiny fissures, opportunities arise for plants such as grasses and mosses to establish themselves and begin a new seral stage (Figure 47.25).

Plant roots penetrate the rocky crevices, exerting a remarkable turgor pressure, prying at the rocks and gradually widening the fissures. By then, certain insect and decomposer populations will also have established themselves. In time, the lichens that made the penetration of plant roots possible are no longer able to compete for light, water, and minerals, and they give way to the plants. Similarly,

FIGURE 47.25
PRIMARY SUCCESSION
Succession begins here with a bare rock outcropping and ends with a fir-birch-spruce community. Pioneering lichens and mosses begin the soil-building process, followed by the invasion of increasingly larger plants until a more stable long-lived, climax forest community emerges.

these plants will contribute to the soil-building for a time, and then they, too, will be replaced by fast-growing shrubs, and new populations of animals will invade.

The succession on bare rock outcroppings is, at first, an extremely slow process, with a sere often lasting hundreds of years or more. But once soil formation has begun, the process can accelerate. Succession in other sorts of places can also be slow. It is estimated that succession from sand dune to climax forest community on the shores of Lake Michigan took about a thousand years (Figure 47.26).

Secondary Succession

In secondary succession, the principles are similar to those of primary succession, but the seres occur at a more rapid pace. This is possible because the soil is already in place. In a deserted farm, weeds, grasses, shrubs, and saplings are often the first to appear. Weeds are fast-growing, opportunistic plants that quickly invade disturbed communities, but they are often held in check where communities are undisturbed. Weeds are often imported from distant places, where they may be less of a nuisance. For example, in southern California, the large ball-shaped tumbleweed so prevalent there is a native of Russia, and the wild oat and mustard are natives of the Mediterranean region.

As secondary succession progresses, the initial invaders are eventually replaced by plants from the surrounding community. Larger, fast-growing trees such as pines block the sunlight, and a new generation of shade-tolerant shrubs emerges below the canopy. Eventually, there is a general blending with the surrounding community (Figure 47.27). Such a simple transition may require well over 100 years, depending upon the community. Secondary succession in grassland communities, as you might expect, is much faster, taking perhaps 20 to 40 years. At the other extreme, fragile, disturbed tundra may require many hundreds of years to recover, if it ever does.

Succession in the Aquatic Community

Aquatic communities also undergo community development or succession, although such changes may be held in check by shortages of mineral nutrients. Succession in lakes and ponds occurs as a product of **natural eutrophication**, changes brought about by the natural increase in nutrients carried in by streams and runoff from the land.

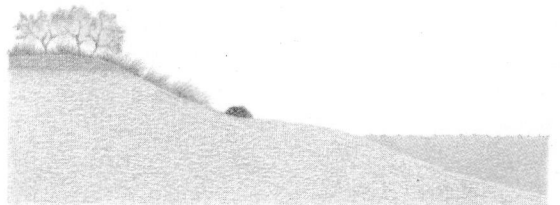

(a) Beach grass and cottonwoods

(b) Oak forest

(c) Oak-hickory-pine forest

(d) Climax beech-maple forest

FIGURE 47.26
SUCCESSION ALONG THE GREAT LAKES
Sand dunes bordering Lake Michigan have long offered ecologists an excellent laboratory for the study of succession. About 1000 years are required for succession to be completed.

Lakes and ponds that are rich in nutrients and high in productivity are called **eutrophic** ("true foods") **lakes,** while those that have limited nutrient supply and little productivity are called **oligotrophic** ("few foods") **lakes.** The general trend in freshwater bodies is toward increased eutrophication and thus increased community growth, but the loss of any of the essential nutrients can reverse the latter trend.

As community growth in a lake progresses, the sediments increase and the depth decreases. Littoral zone plants crowd the shores, extending further and further into the lake, followed by increasing numbers of water-tolerant shore plants (Figure 47.28). Unless the trend is interrupted, the lake will eventually convert to a marsh, and with the invasion of terrestrial plants from the surrounding community, the last traces of the lake will be lost. Interestingly, the ancient and oligotrophic Lake Baikal has shown alarming indications of eutrophication, but not through natural means. The source of nutrient enrichment is the surrounding human community.

Eutrophication is an extremely gradual process in lakes, since nutrients are often whisked away by streams or buried deep in the bottom sediments. Typically, growth is checked by a scarcity of two nutrients, phosphates and nitrates. In recent times, however, such nutrients have become readily available through **cultural eutrophication,** in the form of increased mineral runoff produced by humans as we dump sewage into water systems or permit runoff from heavily fertilized farms, cattle feed-lots, and denuded (clear-cut) forest regions. These practices vastly increase the availability of nitrates.

FIGURE 47.27
SECONDARY SUCCESSION
The stages of secondary succession are revealed in a series of photos taken in the same section of the Bitterroot National Forest in Montana over a 70-year period: **(a)** 1909, **(b)** 1925, **(c)** 1937, and **(d)** 1979.

(a)

(b)

(c)

(d)

(a)

(b)

(c)

FIGURE 47.28
SUCCESSION IN A POND
(a) Early in succession, aquatic plants begin to spread from the edges of the pond. **(b)** Eventually, these plants extend across the open water. **(c)** As the pond's waters disappear, invading marsh grasses, cattails, and sedges replace the floating plants, converting the pond into a marsh.

Furthermore, the addition of phosphates to laundry detergents in recent years has added a heavy phosphate burden to natural water systems, rapidly speeding the succession and aging process of lakes and ponds.

Forces Driving Succession

In some cases of succession, one species will pave the way for the species that follow it. This is especially true in primary succession. The principle has been referred to as the **facilitation model.** In primary succession, for example, the pioneer plants facilitate, or set the stage for, the plants that will follow. They do this by changing the environment in such a way that the habitat is enhanced for the species occupying the next seral stage.

At one time, facilitation was thought to be a primary force in all seral sequences. However, researchers now believe that the facilitation model probably is only valid in primary succession. A number of studies support this idea. For example, it has been shown that nonpioneer plants grow better on plots in which pioneer plants had been experimentally removed. These results suggest that established plants may actually inhibit, not facilitate, the invasion or growth of other plants. Thus, a new model, the **inhibition model,** has been developed. According to the inhibition model, new kinds of plants appear only after established plants have died or been damaged. The established plants, if healthy, inhibit the development of new kinds of plants. Succession occurs despite inhibition because pioneer plants generally have shorter life spans than do nonpioneer plants, so they are replaced more often. As they die, they leave openings for other species to move in while the young pioneer plants are small and vulnerable. The result is that nonpioneer plants will eventually predominate.

Studies of secondary succession in yet other communities suggest that pioneer plants in some cases neither help nor hinder nonpioneer plants. From such findings, the **tolerance model** was developed. According to the tolerance model, succession occurs because of differences in developmental rates and competitive abilities between pioneer and nonpioneer plants. Pioneers develop more quickly and so dominate the early stages of succession. But because pioneers are also likely to be less efficient exploiters of resources, they are eventually replaced by the slower-developing but more efficient nonpioneers.

Climax Communities

Communities in succession tend to produce more organic material than they use, while in **climax communities,** an equilibrium is reached between net production and utilization. In the early stages of succession the exchange rate between organisms and the environment is slow because mineral nutrients are largely stored in environmental reservoirs. But as the climax state is reached, more of the nutrients cycle directly, through exchange pools, between the organisms and the decomposing material. The organisms themselves tend to become more diverse as the community enters the climax state. Concurrently, some ecologists believe, feeding relationships go from a simple chainlike structure to the intricate food web. There is also some evidence that climax communities are much more stable than their transition stages, being less susceptible to external influence such as human intervention.

Trends in Species Richness

If you walk across a desert, you don't see very many kinds of living things. If you crawl among the leaves of a jungle floor, though, you will encounter a lot of species (some with teeth). So some areas are species-poor (with low species density) and some are species-rich (with high species density). The question arises, then, why do some places harbor more species than do other places? There are four primary bases for species diversity.

First, species richness increases with habitat complexity. Deserts have few species, not only because they are hot and dry (although that's probably the main reason), but also because they are ecologically simple places. Grasslands, too, are relatively ecologically simple, and they, too, do not boast great numbers of species relative to complex places, such as hardwood forests and tropical rain forests.

Second, species richness increases with the size of the area. That is, the larger the area, the more species it will hold, all other things being equal. The concept is intuitively apparent, but intuition is not enough in science, so we find that researchers have gone out there and demonstrated the fact, mostly by work on islands of different sizes (Figure 47.29). One probable explanation for this pattern is the greater number of different habitats likely to be included in a larger area. Because different habitats are usually

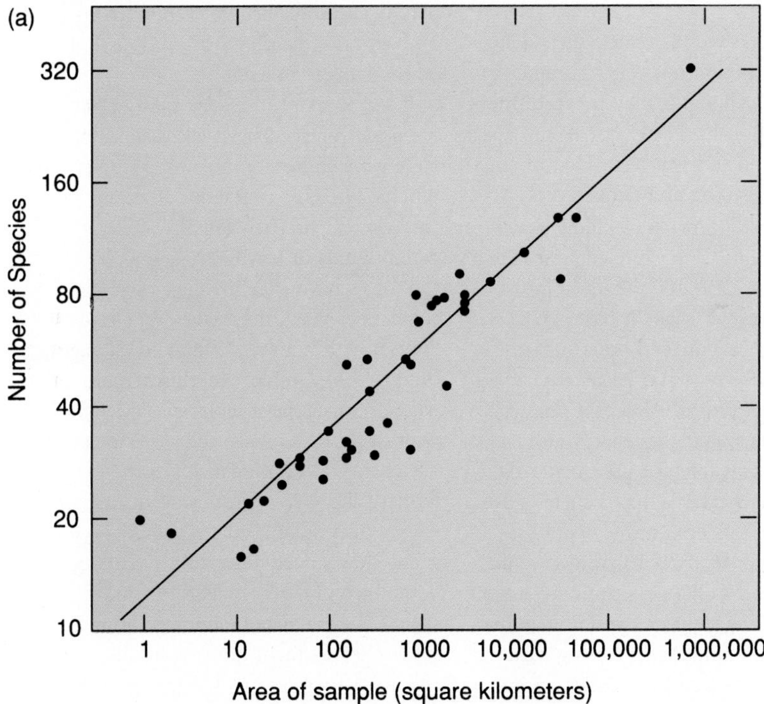

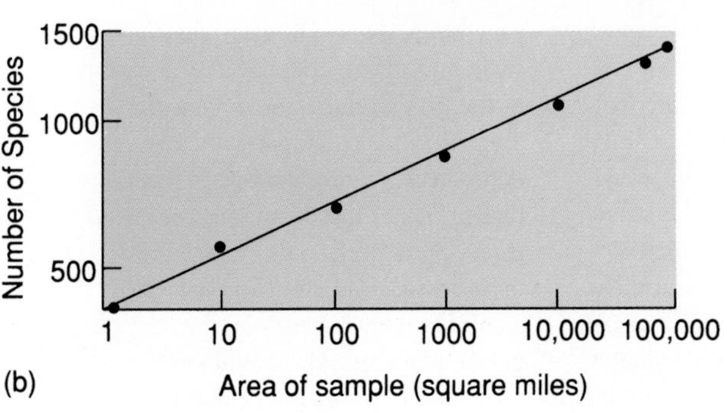

FIGURE 47.29
SPECIES RICHNESS VERSUS AREA
(a) Relationship between number of bird species and island area in the vicinity of New Guinea. (From Ehrlich, P.R., and J.R. Roughgarden, 1987, *The Science of Ecology*. New York: Macmillan.) **(b)** Relationship between number of flowering plant species and sampling area. (From Begon, M., J.L. Harper, and C.R. Townsend, 1986. *Ecology: Individuals, Populations and Communities*. Sunderland, Massachusetts: Sinauer.)

ESSAY 47.3

THE WOLF AS PREDATOR: A CLOSE-UP VIEW

One day I watched a long life of wolves heading along the frozen shore line of Isle Royale in Lake Superior. Suddenly they stopped and faced upwind toward a large moose. After a few seconds the wolves assembled closely, wagged their tails, and touched noses. Then they started upwind single file toward the moose.

L. David Mech
The Wolf (1970)

In recent years, there has been increasing attention to large predators, their life histories, how they kill, and the effects of their predation. For a number of reasons—some scientific, some emotional—North Americans have focused on one of their own, the fabled and mysterious wolf. The hard information, however, was not easy to come by, and so a long-term study by L. David Mech was met with enthusiasm. The story he told was based on many years of watching wolves in the Northern United States, especially on Isle Royale in Lake Superior, and it answered a number of questions. Is the wolf an effective killer? Yes. Does it ever kill more than it can eat? Sometimes. Does it attack humans? No. Is it always successful in its hunts? No. (In fact, fewer than one moose hunting in ten yielded prey.)

One of the things we know from Mech's work and the studies that followed it is that wolves are continually on the hunt. They may attack a prey animal only hours or minutes after a successful kill, and any time they are on the move, they are hunting. When they are successful, they can gorge. Their stomachs can hold up to 20 pounds of food, which is apparently rapidly (or incompletely) digested.

Wolves probably locate prey most often by scent (although they may stumble across prey). Mech found 42 cases in 51 hunts in which he could tell that the wolves were trailing their prey (moose) through the snow by direct scenting. In fact, they once detected a cow and her twins 1.5 miles away. He found that when the lead animals catch wind of the prey, they stop, and all pack members stand alert with eyes, ears, and noses focused toward the prey. Then they carry out a peculiar ceremony, standing nose-to-nose and wagging their tails. If they are in deep snow, they just pile up behind the leader, who then sets out straight for the prey. As they near quarry, they quicken the pace and seem anxious to lunge ahead. But they hold themselves in check, alert and tails wagging. The restrained approach permits the wolves to draw near their prey without sending it into flight.

As soon as wolves realize their quarry has spotted them but is not running, they stop stalking. If they proceed, they do so cautiously. There may be an advantage in this wariness. First of all, a moose is a strong and dangerous animal, easily capable of killing a wolf. If the animal doesn't run, then, it may be out of a spirit of confidence in its own prowess. Perhaps it has successfully dominated other moose—or other wolves—and hasn't developed the habit of retreat. Also, if it doesn't run, it isn't wasting its energy and can therefore probably put up a better fight.

Small or weaker animals such as deer don't have this option; they could certainly not fend off a wolf pack, and so they tend to flee. Interestingly, a wolf may need the sight of a running animal in order to stimulate its closing rush. Mech believes that the sight of a standing animal inhibits the rush response. The wolves may still approach, but they do it more slowly.

If the prey bolts and runs, the wolves give chase. If the quarry is fast enough in its initial run, it usually gets away. Wolves are very perceptive about such matters. If the chase looks hopeless, they stop. If they do give chase, they may run for miles, but this is unusual; they normally chase only a very short distance. Interestingly enough, the prey (whether moose, deer, caribou, or Dall sheep) is also perceptive. It runs no farther than necessary. When the wolve give up, the prey will stop and turn around to watch the wolves. It doesn't waste energy running needlessly.

It is known that wolves attack different animals in different ways. For example, moose are usually bitten on the rump area, with one wolf sometimes clinging to the large, rubbery nose. If the kill is not at first successful, the wounded moose may stiffen and weaken so that it can be brought down some days later. On the other hand, caribou are usually attacked at the shoulder and neck area. When larger prey is not available, wolves will pounce on field mice, landing with all four feet on the hapless rodents.

It is important to realize that wolves are opportunists. They will eat not only large game and mice, but insects, fish, rabbits, birds, or just about anything they can catch. They will also eat certain kinds of berries.

Wolves have changed their diets in some cases when they have come into contact with humans. They raid garbage dumps, for example. More important, where humans have replaced wild game with domestic animals, the wolf's diet has changed in accordance. To the farmer's dismay, whereas wolves almost always dispose of all the remains of wild game, they sometimes eat only parts of domestic animals such as cattle and sheep, but this may be because human activity drives them from such kills.

A lingering question has involved the "sanitizing" effects of wolves on their prey populations. Do they take the weak and infirm and elevate the quality of the prey populations in general? Since they make so many attempts and their success rate is so low, it seems that primarily disadvantaged prey are taken. Such prey could be newborn, inexperienced, malformed, sick, old, wounded, parasitized, starving, crippled, or just plain stupid. Several studies have shown that wolves kill primarily animals less than 1 year old, or those that have lived at least half the usual life span for that species in the wild. In a sample of 93 deer killed by wolves in Minnesota, 59% of the adults were relatively old—at least 4.5 years old. (In contrast, of most deer killed by hunters, only 20% were over 4.5 years old.)

Do wolves control the population of their prey? For a number of reasons it is difficult to know. For example, in many places where such studies have been made, there have been unusually high numbers of ungulates. Possibly, at one time, when ungulate numbers were lower and the number of wolves was higher, wolf activity may have been a major factor in the control of ungulates. But because of the artificial controls on wolves and their prey, it is difficult to know just what sort of population control was imposed by the predators under natural conditions.

As further evidence of the regulatory impact of the wolves, the moose have lived on the island since about the beginning of the century, but the wolves didn't arrive until 1949. In the early 1930s, A. Murie estimated the moose herd at 1000 to 3000 animals. They apparently overbrowsed their area, and their numbers fell drastically through the effects of starvation and disease only a few years later. The herd recovered but began to starve again in the late 1940s. However, for decades since the wolves have arrived, moose numbers were lower and more stable than ever before, and the vegetation recovered. In recent years, the Isle Royale moose population has again been rapidly increasing, this time despite the presence of wolves.

occupied by different species, areas with more habitats will have more species. Another explanation for this pattern applies to isolated areas. Smaller areas are species-poor because they support smaller populations, which are more prone to extinction.

Third, species richness of an area increases with proximity to the geographic source of the species. Again, the principle has been verified by work on islands: the farther an island is from the mainland, the fewer species it will have, all other things being equal (Figure 47.30). This is also an obvious principle; spe-

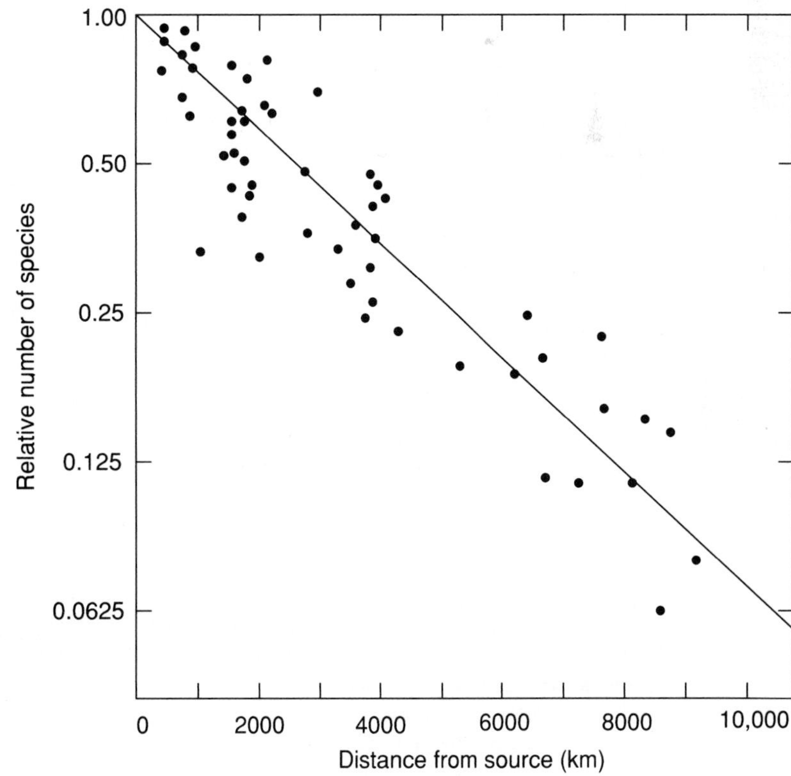

FIGURE 47.30
SPECIES RICHNESS VERSUS DISTANCE FROM SOURCE
Relationship between relative number of bird species on an island and distance of the island from the source of dispersal. Relative number of species is the actual number of species divided by the number of species the island would have if it were very close to the source area. (from Ehrlich, P.R. and J.R. Roughgarden 1987. *The Science of Ecology*. New York: Macmillan.)

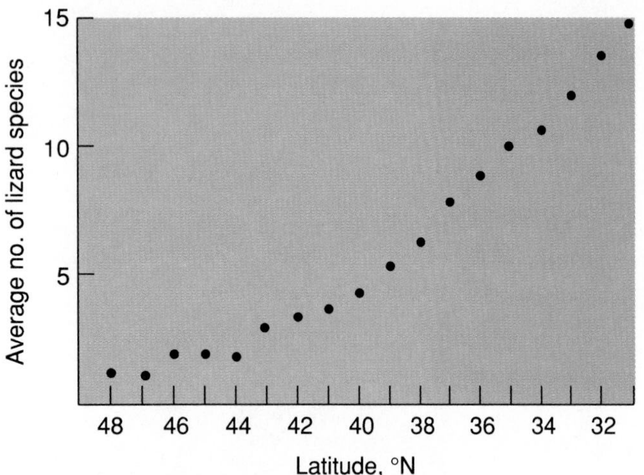

cies have trouble making it out to distant islands.

The fourth relationship is not so obvious. In fact, no one has yet come up with an explanation for it. This is, species richness (in most taxonomic groups) increases at lower latitudes. Thus, there are more species in areas near the equator than in areas of the same size closer to the poles (Figure 47.31). One explanation is that the more benign equatorial weather; without the annual winter disasters, enables a variety of species to survive for long periods and gradually wedge a place for themselves in the complex array of tropical life.

FIGURE 47.31
SPECIES RICHNESS VERSUS LATITUDE
Relationship between average number of lizard species and latitude (degrees north). (from Pianka, E.R. *Evolutionary Ecology*, 4th ed. New York: Harper & Row.)

APPLICATION OF IDEAS

1. While ecological communities may appear vastly different, each has many basic organizational features in common. Name three such different communities and then discuss each in terms of these organizational features. As a start, consider such aspects as energetics, trophic levels, and the cycling of mineral nutrients.
2. Considering how environmentally disruptive some mining and industrial practices can be (shale coal mining, open pit mining, copper smelting, and so on), we can see that there is a vitally important need for industry to reasonably restore the land following operations. In what ways would it be sensible for industry to bring in ecologists for consultation before beginning restorative programs? What specific knowledge might ecologists have to offer, and what kinds of preliminary studies might ecologists make before beginning?

KEY IDEAS

Biomes contain communities of organisms. A complete unit of interaction, regardless of extent, can be called an **ecosystem.** Within ecosystems, the flow of energy and cycling of essential substances occurs through **trophic levels.**

ENERGETICS IN ECOSYSTEMS

About one-tenth of one percent (0.1%) of incoming solar energy is captured in photosynthesis, with 150 to 200 billion tons of dry organic matter or **biomass** produced annually.

Trophic Levels

1. Photosynthesizers and other autotrophs are **producers.** From the producer, energy and molecules go to the **consumer** trophic levels.
2. Photosynthetic producers use chlorophyll and light energy along with carbon dioxide, water, and a few minerals to produce essential organic compounds.
3. Consumer organisms, heterotrophs, occur at several trophic levels and are identified as **primary, secondary, tertiary,** or **quaternary consumers.** It is not uncommon for organisms to live at more than one level. Interaction among trophic levels can be represented by diagrams called **food webs.**
4. **Decomposers** are chiefly fungi and bacteria, organisms that feed on the already dead—also called **saprobes.**
5. Ecological pyramids (or **Eltonian pyramids**) can be used to represent numbers, biomass, or energy flow within communities.
   a. Producers generally represent the great bulk of biomass, with other trophic levels greatly reduced from level to level.
   b. Energy pyramids reveal the total amount of energy stored in each trophic level. Typically, they are radically stepped, with an average transfer of only 10% from one level to the next.

Humans and Trophic Levels

As **omnivores,** humans feed at all trophic levels, but the impoverished tend to live at the primary consumer level. Feeding at higher levels is prohibitive because of the great losses involved in transfers.

Energy and Productivity

1. The rate of energy storage is called **primary productivity,** recorded as **gross primary productivity (GP), net primary productivity (NP),** and net community productivity (NCP).

2. GP is the total rate of energy assimilated by producers, while NP is determined by subtracting the rate of energy utilization by the producers.
3. NCP includes the same measurements, but also takes into account heterotrophs as well as autotrophs, so real community growth is known.
4. Such measurements determine whether a community is declining, growing, or in a nonchanging or **climax state.**

NUTRIENT CYCLING IN ECOSYSTEMS

The movement of mineral ions and molecules in and out of ecosystems occurs through **biogeochemical cycles.** Most ions enter the living realm at the producer level.

The Nitrogen Cycle

1. Outside of life, nitrogen occurs in **exchange pools** and **reservoirs.** The largest reservoir is N_2 in the atmosphere, but it is only available to **nitrogen fixers.** Nitrate ions are made available in soil-exchange pools.
2. Plants incorporate nitrate or ammonium ions into protein. When plants die or are consumed or eaten, the protein goes to the consumer or decomposer level, but eventually all of the incorporated nitrogen goes to the reducers.
   a. **Ammonification:** Bacteria of decomposition reduce the nitrogen to ammonia, which forms ammonium ions.
   b. **Nitrification:** In two steps, ammonium ions are converted to nitrite and then to nitrate, which enters the exchange pools.
3. Where anaerobic conditions prevail, losses from the exchange pools occur through the action of denitrifiers.
4. During **nitrogen fixation,** bacteria (including cyanobacteria) convert atmospheric nitrogen to ammonia, which then undergoes nitrification, forming nitrite and nitrate. Nitrogen fixers include symbionts that live in the roots of leguminous plants.
5. The liberal use of synthetic nitrogen fertilizers in support of crops has produced worldwide soil and water excesses, and the balance between nitrogen fixation and denitrification has been lost. Two products of the nitrogen load are **eutrophication**—nutrient enrichment of waters—and the pollution of soil water supplies.

The Phosphorus and Calcium Cycles

1. Cycles of phosphorus and calcium occur between living organisms and water. The two elements are taken up in soluble phosphate and calcium ions. Phosphates are used in producing ATP, nucleic acids, phospholipids, and tooth and shell materials, while calcium is essential to bone and shell development and in membrane activity.
2. Cellular phosphates cycle directly from living organisms to water and return, while phosphorus in skeletal material and teeth is freed very slowly. Calcium is also freed very slowly from skeletons and shells. Both accumulate in deep ocean and lake bottom sediments until upwellings and overturns redistribute them.

The Carbon Cycle

1. Carbon is an essential part of nearly all the molecules of life. The principal exchange pool on land consists of carbon dioxide gas, while the source in the waters is dissolved carbon dioxide gas and the carbonate ion. A large reservoir occurs in the form of limestone and fossil fuels.
2. Carbon enters the producer level during photosynthesis, where it is used initially to form carbohydrate. Producers, consumers, and decomposers all release carbon during cell respiration.

THE LAKE AS AN AQUATIC COMMUNITY

Conditions and Life in the Lake Zones

Limnologists recognize three lake zones, which are characterized as follows.

a. In the **littoral zone,** light penetrates to the lake bottom. Producers include algae, bacteria, and submerged, emergent, and floating plants. Consumers include protists, snails, mussels, aquatic insects, fishes, frogs and turtles.
b. The **limnetic zone** is open water penetrated by light. Producers are mainly phytoplankton, including green algae and cyanobacteria, whose numbers increase in seasonal "blooms" that correspond to the availability of bottom nutrients. Zooplankton are primary consumers, while lake fishes make up most of the higher-level consumers.
c. The **profundal zone** lacks light, extending to the bottom. Profundal organisms usually tolerate low oxygen concentrations and are made up primarily of decomposers, a few clams, and wormlike insect larvae.

Thermal Overturn and Lake Productivity

1. Oxygen and nutrients are redistributed in lakes that undergo seasonal, wind-driven thermal overturns—the total circulation of lake waters. In summer, temperate zone lakes form a warm, upper **epilimnion,** a cold, lowermost **hypolimnion,** and a **metalimnion** within deeper lakes, a steep **thermocline** between. No mixing occurs below the thermocline because of the greater density of hypolimnion waters.
2. Temperate zone lakes undergo two seasonal thermal overturns when cold surface water sinks, displacing the bottom layers and bringing oxygen down from surface waters and nutrients up from the sediments.

THE FOREST AS A TERRESTRIAL COMMUNITY

1. In determining the productivity of an oak-pine forest community, Woodwell first determined its structure, including a determination of species, organic matter, and biomass. Next, taking advantage of nighttime atmospheric inversions, he determined the output of respiratory carbon dioxide gas.
2. Net community productivity was determined to be 550 $g/m^2/yr$, a modest figure compared to communities that reach 10 to 20 times that figure. The oak-pine community studied was determined to be in a state of growth and not in climax.

COMMUNITY ORGANIZATION AND DYNAMICS

1. **Open communities** have indistinct borders that let adjacent communities interact and blend.
2. **Closed communities** have distinct physical boundaries. Examples include islands, peninsulas, and caves.

The Role of Competition in Community Structure

1. **Competition** is any struggle between two individuals for the same resource.
2. In **exploitative competition**, utilization of a limited resource occurs between two or more organisms.
3. In **interference competition**, one organism blocks another's access to a resource.
4. Competition may occur within or between species, the latter having greater impact on community structure.

The Ecological Niche

1. The **ecological niche** is an organism's role in nature, including its resource utilization.
2. The **fundamental niche** includes all ways an individual can potentially interact with its surroundings. The **realized niche** is more limited, referring to an organism's actual interactions.
3. Niches can vary in breadth from those of very narrow constraints to those with few constraints.
4. Interaction between two species usually leads to repression of one, but often in a subtle way. According to the **principle of competitive exclusion**, two species cannot utilize the same limited resources for long without one species being displaced.

The Roles of Territoriality and Dominance Hierarchies in Community Structure

1. Destructive competition (such as fighting) may be reduced by territorial organization and the ranking of individuals into hierarchies (a dominant-and-subordinate social organization).
2. In times of shortages, social organization can lead to a more equal distribution of resources, but reproduction may be restricted to dominant individuals.

The Roles of Disease and Parasitism in Community Structure

1. Disease and parasitism, by causing selective weakness and death in segments of the population, can be important determinants of community structure.

The Role of Predation in Community Structure

1. **Predation** influences both predator and prey numbers. According to the **Lotka-Volterra theory**, increases and decreases in predator numbers track changes in prey numbers, in predictable oscillations. Thus, the prey number has a negative feedback effect.
2. Predation tends to relieve competition among a prey species, and the loss of a predator species can result in sudden and dramatic increases in prey numbers.

COMMUNITY DEVELOPMENT OVER TIME: ECOLOGICAL SUCCESSION

1. While communities exist for long periods in a state of climax, they change over time under the influence of geological and climatic changes and through biological factors and human intervention. Such change is called **ecological succession**. When brought about by living inhabitants, the process is called **autogenic succession**, while change brought about by outside forces is **allogenic succession.**
2. During community development, the organisms themselves produce changes that bring about their own replacement by other species. Each sequential stage is a **seral stage,** part of a **sere.**
3. Ecological succession includes **primary succession** and **secondary succession.**

Primary Succession

Primary succession occurs where no living organisms have yet become established. Pioneer organisms, such as lichens and mosses, are often first to become established, followed by grasses and sun-tolerant plants. Eventually, the community reaches a climax state.

Secondary Succession

Secondary succession occurs wherever a climax community is disturbed (such as deserted farms, burns, or clearings). It is common for the first invaders to include opportunistic, fast-growing weeds. Initial invaders are soon replaced by fast-growing shrubs and trees, and a gradual blending with the surrounding community occurs.

Succession in the Aquatic Community

1. Succession in aquatic communities is accelerated by **natural eutrophication.** Lakes are rich in nutrients and are described as **eutrophic.** In **oligotrophic lakes,** nutrients are limited, with minute amounts in exchange pools, and succession is extremely slow.
2. Succession in lakes begins with increased nutrient sediments, encroachment by shore plants, and a general increase in both numbers and kinds of organisms. The continued trend will result in the lake filling in and later blending in with the surrounding terrestrial community.
3. Human activities often result in cultural eutrophication, nutrient enrichment through the pollution of lakes.

Forces Driving Succession

1. According to the **facilitation model** of succession, one species or group makes physical changes that favor the establishment of another.
2. In the **inhibition model** of succession, a pioneer species inhibits the establishment of another until it dies off. Some plants do this by producing chemicals that retard others.
3. The **tolerance model** of succession proposes that pioneer plants retard the growth of others by virtue of their faster growth and development. Such plants are likely to be less efficient in other ways, however, and are eventually replaced.

Climax Communities

Climax communities are those whose net production and utilization are in equilibrium, whose cycling of nutrients is direct, and whose organization is complex and stable. Communities in succession or decline have the opposite characteristics.

Trends in Species Richness

1. Species richness increases with habitat complexity.
2. Species richness increases with the size of the area.
3. Species richness increases with proximity to the original source of the species.
4. Species richness increases at lower latitudes.

REVIEW QUESTIONS

1. Carefully distinguish between the terms biosphere, community, ecosystem, and biome. (1165)
2. Approximately what part of incoming solar energy is trapped by producers? In what form is this energy found in the producer? (1166)
3. Describe and provide examples of each of the following: primary producers, primary consumers, secondary consumers, and decomposers. List two foods we eat that place us at the quaternary level of consumption. (1167, 1168)
4. Construct a typical pyramid of biomass or energy. Describe the shape and explain the decrease from one level to the next. (1168, 1169)
5. Approximately what part of the energy moving through trophic levels is stored in each level? Can it ever be 100%? Why? What happens to the rest? (1169)
6. From an energetics viewpoint, which trophic level might support the greatest number of humans? Very specifically, what problems arise when we feed exclusively from this level? (1169–1171)
7. How are the gross primary productivity and net primary productivity determined? Which represents real growth in the producer population? What happens to the difference between them? (1170, 1171)
8. What does net community productivity tell us that GP and NP cannot? If NCP is positive, what does this tell us about the community? (1170, 1171)
9. How does a nutrient reservoir differ from a nutrient exchange pool? (1171)
10. Briefly summarize the events of the "inner nitrogen cycle," including mention of decomposition, ammonification, and nitrification. Why is nitrification desirable when plants can readily incorporate ammonium ions into amino acids? (1171–1174)
11. Briefly describe how nitrogen is constantly lost and gained by the nitrogen cycle. Under what specific conditions do losses occur? How have humans offset the balance between loss and gain? (1172)
12. What are two problems stemming from excessive use of synthetic fertilizers? (1173, 1174)
13. What must happen to phosphates and calcium before they can cycle directly from organisms to exchange pool and back to organisms again? (1174, 1175)
14. Using simple formulas for photosynthesis and respiration, explain how carbon dioxide, oxygen, and water cycle among living things. In addition to carbon dioxide gas, what other sources of carbon are available to life? (1175)
15. Distinguish between an open and a closed community. (1182, 1183)
16. Carefully define the ecological niche and distinguish between the fundamental and realized versions. (1184, 1185)
17. How and why do territoriality and hierarchical organization affect competition in communities? (1186)
18. Why is it that predation does not normally decimate the prey species? What long-range effects on prey species might humans see if they destroy predators such as wolves or mountain lions? (1187–1189)
19. In which of the three lake zones would one expect to find the greatest NCP? Which has the greatest species diversity? The least? (1175–1179)
20. Using a simple diagram, illustrate the organization of the lake into its three zones. What one factor distinguishes the three zones? (1175–1178)
21. Briefly explain why the deep waters of Lake Tanganyika are permanently anaerobic, while the waters of Lake Baikal are aerobic at all depths. (1179)
22. Using the terms *thermocline, epilimnion, metalimnion,* and *hypolimnion,* describe temperature conditions in a temperate zone lake in summer. What effect does such an organization have on the distribution of oxygen and nutrients? (1179)
23. Briefly describe how the density of water changes as it cools. How does such winter cooling in lake waters explain thermal overturn? (1179)
24. Reviewing the Woodwell oak-pine forest study, describe how gross productivity was determined. What was the problem with determining net productivity? How was this overcome? (1179–1182)
25. How did net productivity in the oak-pine community compare with that in the tropical rain forest? A field of grain? (1182)
26. Distinguish between and provide examples of autogenic and allogenic succession. (1189)
27. Suggest several seral stages that might occur during primary succession. How do organisms actually produce the sere? (1189, 1190)
28. List several indicators one can use to distinguish between an oligotrophic and a eutrophic lake. What kinds of human activities tend to increase eutrophication? (1190–1192)
29. List four characteristics of a community in succession.
30. Characterize the productivity of climax communities as opposed to communities in succession. (1192, 1193)

48

POPULATION DYNAMICS

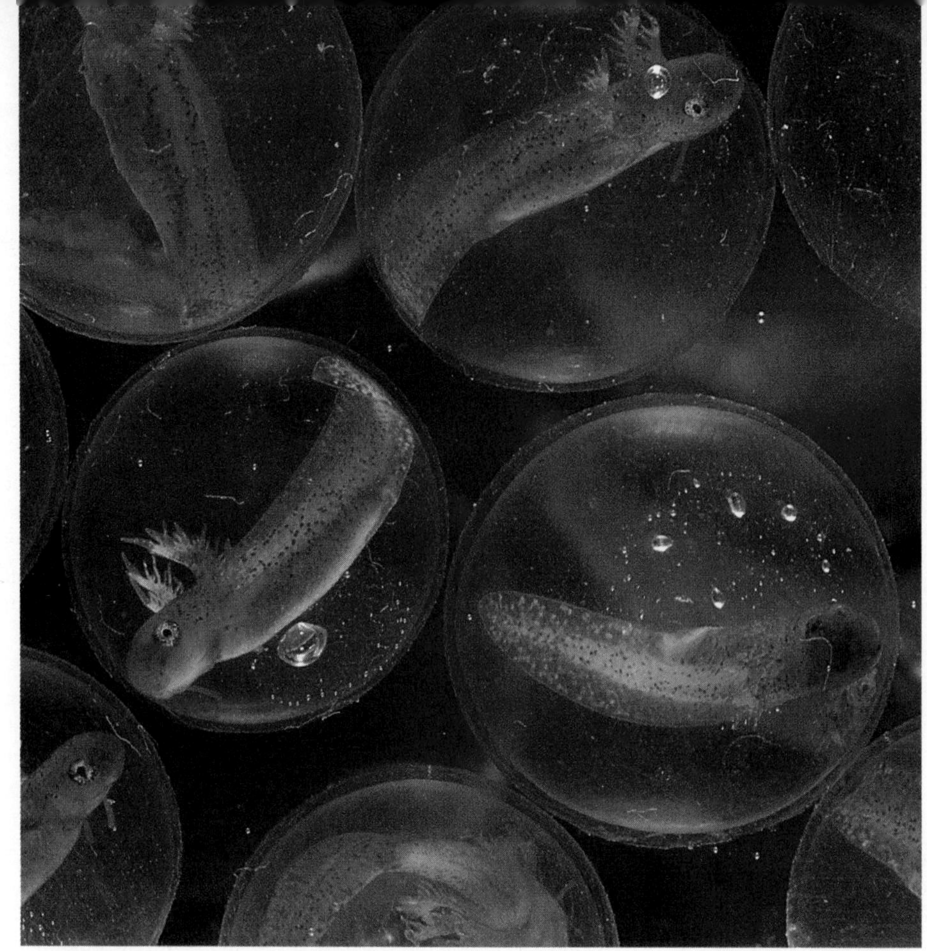

THE HOWLING OF WOLVES ONCE ACCENTED THE NIGHTS OVER MOST OF THE United States. Hunters and hikers regularly came across their tracks, and the remains of their kills were a part of the North American wilderness experience. But now wolves are entirely absent over much of their former range and rare over most of their present range. Today, only a few well-publicized bands precariously persist in scattered areas. What has happened to them? Why have wolf populations declined so precipitously? And what about other species? Why are bluebirds and vultures seen less frequently in Arkansas, just as the beaver is apparently making a comeback there? Sometimes we know the reasons for such population changes; sometimes we don't. Wolves, for example, have traditionally been fair targets for the rifles of farmers and game shooters. But what about places where they are not hunted, for example, protectorates in Michigan and Minnesota? Are the wolf populations in these areas also changing? If so, why? If not, what is keeping them stable? The point is, the numbers of living things change, and often we don't know precisely why.

Because population numbers reflect a myriad of environmental, evolutionary, and physiological effects, most of which are not very well understood, any question about populations tends to be a tough one. A host of researchers are, at this very moment, watching populations of tiny animals in little glass tanks—just trying to fit more small pieces into the great puzzle. Others are traipsing around in fields or spending long hours watching a hole in the ground, waiting for some small face to appear and be counted. A great many people, in fact, are asking very basic questions about how populations change. So our goals here must remain modest. We can only describe, in general terms, a few of the ecological and evolutionary factors that can cause populations to change.

POPULATION CHANGES

Some population changes are short-term or periodic and easily explainable. Grasshoppers, which are plentiful in summer, reproduce and then disappear by winter, leaving their fertilized eggs hidden in the soil. The disappearance of the adults is easy to understand—they die. Some birds disappear from North American forests in the winter as they migrate to more hospitable southern areas. These are familiar and rather predictable phenomena. We will expect the next generation of grasshoppers and the migratory birds to show up in spring.

Other sorts of changes are less predictable and more puzzling. As some species dwindle, we hear of others burgeoning in the form of "plagues" or "invasions." However, the inordinate attention given to such changes may not accurately reflect what is going on. This is because most species seem to be under some sort of control or regulating influence (Figure 48.1). In order to consider any such control, we must first understand something about how populations generally behave. What determines the numbers of a population? What evolutionary and ecological factors are at work? What are the characteristics of population growth? What causes populations to fluctuate, and what sort of stabilizing influences might be at work? Finally, what influences the distribution of populations? We should keep in mind that, technically, a **population** is not just a group of organisms, but a group of the same species that lives in close enough association to be able to interbreed.

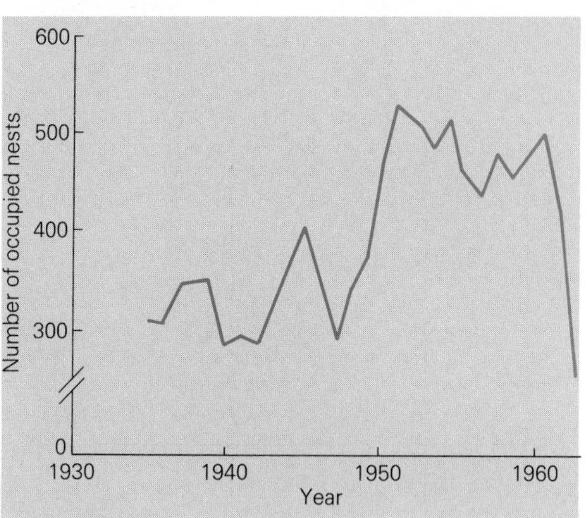

FIGURE 48.1
A STABLE POPULATION
This population of herons has fluctuated around a long-term mean. The data were collected from a study of occupied nests in the area over a period of 30 years.

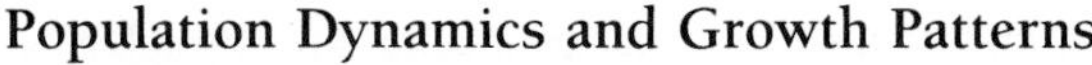

Population Dynamics and Growth Patterns

A good place to begin looking at population change is through an idealized mathematical model. Consider first the simple population equation

$$I = (b - d)N$$

which translates into, "I, the rate of change in the number of individuals in a population, is equal to b, the average birth rate, minus d, the average death rate, times N, the number of individuals in the population."

One can see from the equation that for populations to increase, that is, for I to be positive, the value of b must exceed the value of d—births must exceed deaths. Should they be the same, I will be zero, or should the average death rate, d, exceed the average birth rate, the I will be negative. So, the value of $b - d$ seems to determine what happens to population size (N). The simple difference between the two values we will call r or **realized rate of increase.** The determination of this per-individual rate is simply $r = b - d$. We can now rewrite I to incorporate the new term r. Thus,

$$I = rN$$

or the population rate of change (I) is equal to the per-individual rate of change (r) times the population size (N).

This new equation reminds us of something important in population dynamics. In a growing population, the rate of change in population size is determined not only by r, but by N. For instance, when r is a positive number, N will grow with each new generation of offspring. Thus, I will increase in each generation. The story of population increase and decrease can be summarized in two simple curves, the **J-shaped curve** and the **S-shaped curve.** The letters J and S represent the direction taken by lines when population growth patterns are plotted. The J-shaped curve is the one environmentalists talk about when they try to interest people in the human population growth problem. We will consider this curve first.

Exponential Increase and the J-shaped Curve If a few reproductively active organisms are placed in an idealized environment—one with unlimited resources and space and without danger from disease, predation, or other hazards—they may be expected to reproduce at their maximum physiological rate. This theoretical rate implies maximum gamete production, mating, fertilization, and survival of offspring. (If we allow

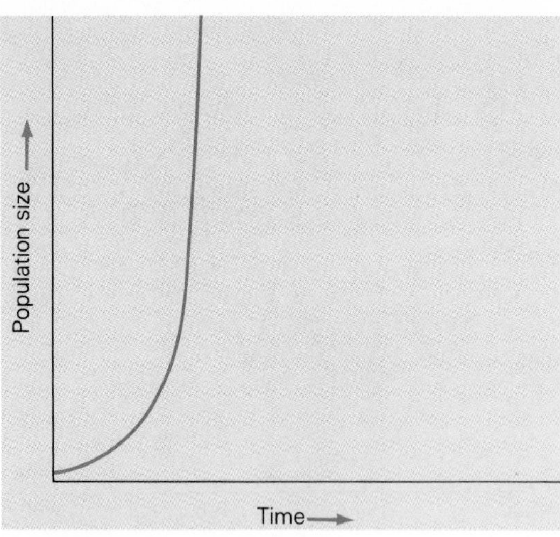

FIGURE 48.2
THE J-SHAPED GROWTH CURVE
The J-shaped or exponential growth curve is characteristic of population growth that occurs when the *r* of a population is constant and positive.

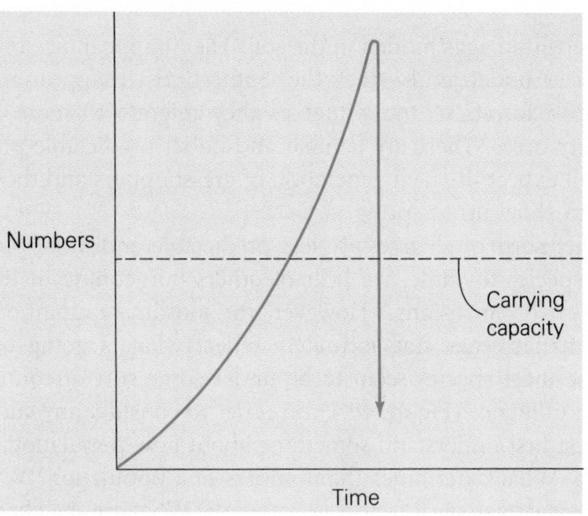

FIGURE 48.3
A POPULATION CRASH
In populations that have undergone a period of unlimited exponential growth, population crashes are not uncommon. The dieback usually places the numbers somewhere well below the ability of the environment to support the population, although in some instances the dieback is complete.

a 30-year reproductive period for humans, then our species' maximum physiological birth rate would be a little over 30 offspring for each female.) This maximum birth rate, together with a physiologically minimum death rate, will produce a maximum value of r, which is symbolized r_m and is called the **intrinsic rate of natural increase.**

Even under r_m, numbers rise slowly at first, simply because we are dealing with a few individuals, say from 2 to 4, to 8, to 16, and so on. In plotting such numbers, at first we see a gently increasing slope. But soon, as the numbers continue to double, the curve arches sharply upward until it seems to approach (but never reaches) the vertical (Figure 48.2). Such increases are called **exponential increases.** They are unlike linear or **arithmetic increases,** those that progress from 1 to 2 to 3, and so on. We have described the exponential growth and J-shaped curve shown by populations growing at their maximum per-individual rate of increase, r_m. (We must add that any constant positive r will also produce a smooth J-shaped exponential growth curve, but with various rises in the curve.)

One of the clearest biological examples of exponential growth is seen when a small number of bacteria are introduced into a rich laboratory culture medium. Under such ideal conditions, the familiar *E. coli,* for example, will divide every 20 minutes. At this rate, after only 24 hours, just one bacterium would give rise to 40 septillion descendants (2^{72}, where the exponent equals the numbers of generations in 24 hours). But as simple as *E. coli's* growth requirements are, such rapid expansion could not be sustained. Very likely, sometime midway through the 24-hour period, the resources that were supporting such phenomenal growth would be reduced to a point where the r_m could no longer be reached. *E. coli* would then have encountered **environmental resistance,** and the rapid rate of increase would begin to slow.

Environmental Resistance and Population Crashes Growth-inhibiting factors, collectively called environmental resistance, can take many forms, but in a laboratory-grown bacterial population, such resistance is commonly in the form of food shortage and the accumulation of toxic waste products. As these things occurred, the population would increase more slowly, and then, for a brief period as the cells died as fast as they were produced, the population would not grow. Finally, the bacterial environment would be so fouled and depleted that the cells would begin to die far more rapidly than they were being produced. Such a sudden decrease in numbers is called a **population crash** (Figure 48.3).

Carrying Capacity and the S-shaped Curve Population crashes may occur when a population exceeds the **carrying capacity** of its environment. The carrying capacity can be defined, in its simplest terms, as the number of individuals of a species that can be supported by the resources of an environment without damage to those resources. The carrying capacity is not fixed; it may change drastically from one time to another. Further, it is constantly affected by many factors, both **biotic** (living) and **abiotic** (nonliving), which we will elaborate upon later. In the case of the bacterial colony, the number of individuals far exceeded the environmental carrying capacity.

The bacterial population crash is an extreme example, since there is no way for such an artificial environment to recover on its own. Typically, however, the resources that maintain a population are renewable: the plants sustaining a population of rabbits will recover; the field mouse population depressed by a population of predatory owls will recover as the owls seek new prey sources, and the rich supply of nitrates supporting a phytoplankton bloom will return at the next season's thermal overturn. It's important to realize that population booms and crashes are the natural way of life for many species.

While population crashes can and do occur under natural conditions, for many species another phenomenon occurs with less drastic results. In such cases, as the population approaches the carrying capacity, growth slows. The nearly vertically rising "*J*" tapers off as a gentler slope, forming the less radical S-shaped curve. Typically, the population numbers then fluctuate around the carrying capacity (Figure 48.4).

Some S-shaped curves are called logistic growth curves, as derived from the logistic model. (In this simple model, environmental resistance is a function only of population size, N, as shown below. Other possible kinds of environmental resistance, such as predation and parasitism, are not included.) Logistic growth curves start out similarly to exponential curves, with the birth rate greatly exceeding the death rate, but then at some point acceleration ceases and deceleration begins. During deceleration, the birth rate generally begins to decrease while the death rate increases. Finally, the two reach equality and $r = 0$, completing the S-shaped curve. The population may then hover about this level.

The carrying capacity is represented in population equations as K, with the following expression or term:

$$\frac{K - N}{K}$$

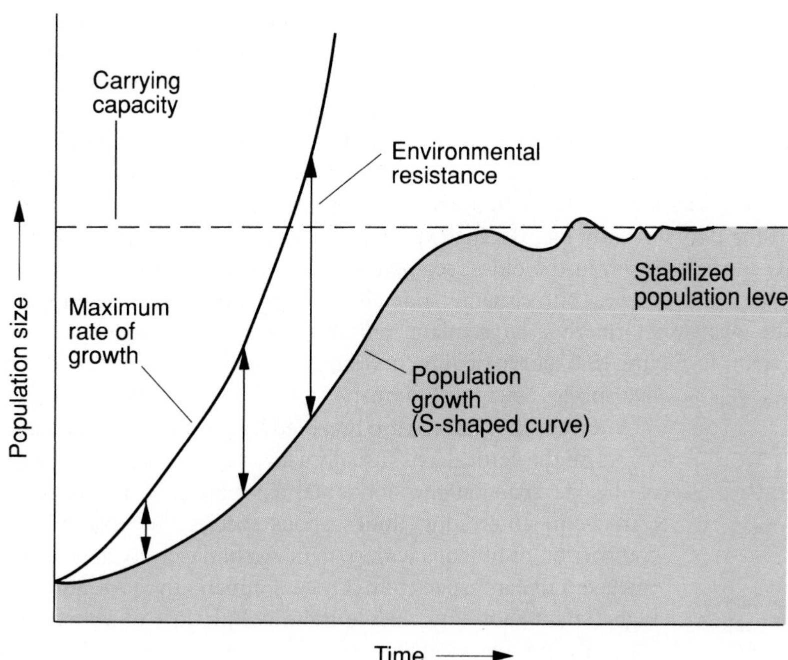

FIGURE 48.4
DYNAMICS OF POPULATION GROWTH
The dynamic factors in population change include intrinsic rate of natural increase, environmental resistance, and carrying capacity. Typically, the interaction of these factors produces the S-shaped curve. Sustained growth at the maximum rate cannot persist because of environmental resistance, which includes limited resources. Such resources help determine the carrying capacity—the numbers any environment can sustain over long periods without serious environmental damage.

To show that population growth is usually not unchecked in nature but is limited as N approaches K, we introduce $(K - N)/K$ into the equation describing growth under uninhibited conditions. (Notice that r_m is used here.)

$$I = r_m\left(\frac{K - N}{K}\right)N$$

Early in the population's growth, the value of N is small compared to K, and subtracting it from K in the numerator has little effect. In fact, at this time the value of $(K - N)/K$ approximates one, and, of course, multiplying the other factors in the equation by the one has no effect at all. As a result, early growth is exponential and rapid. But as the population grows and N becomes larger, the value of $(K - N)/K$ decreases from one and r_mN, now multiplied by a fraction, becomes smaller. In fact, as you may have anticipated, as the value of N reaches K, the new quantity in the numerator becomes zero and, theoretically, no further population increase is possible. To illustrate, simply substitute zero for $(K - N)/K$ and multiply:

$$I = r_m \times 0 \times N = 0$$

In reality, populations often exceed their carrying capacity, and when they do, $(K - N)/K$ becomes a negative number, and we can predict a temporary population decline.

Population growth is certainly more complex than we have portrayed with these theoretical equations, and while populations of some species approach the neat S-shaped, logistic growth curve, most only approximate such idealized conditions (Figure 48.5). Many kinds of populations tend to fluctuate about some density, regulated in part by the apparent negative feedback effects of the carrying capacity. But it is important to keep in mind that the value of K can also change, perhaps due to permanent environmental damage, or cyclically, as seasons change and the availability of resource diminishes or increases. K can also change through the influence of other species, as we will see.

Age, Life Span and the Population A number of factors can influence populations, such as the ages of its members and their longevity. In particular, one might ask: How many individuals within a population are below, within, or beyond the reproductive age? How long is their expected life span? What percentage of individuals complete the life span? As we will find, these are highly significant factors in predicting the future course of human population growth. In 1989, 32% of the human population was below 15 years of age—many having yet to participate in reproduction. How will this affect our future rate of increase? (As an aside, consider this: In Africa 45% of the population is under 15; in the United States, 22%.)

We have seen that the death rate is a vital part of population growth dynamics, but knowing the death rate doesn't give us much information. What is important is knowing the *age groups* that are subjected to the greatest mortality. In many natural populations, we find that death occurs most frequently in the young, and once an individual has survived the rigors of early life, the life expectancy is significantly extended. Obviously, mortality increases again in the oldest segment of the population. This characterization applies to many species, but certainly not all. Modern humans, for example, have overcome what was formerly a large infant mortality and have succeeded in extending the life span. In Figure 48.6, survivorship curves depict the relative average life span of five species, including humans.

FIGURE 48.5
THE S-SHAPED GROWTH CURVE IN NATURE
A long term study of sheep in Tasmania reveals a fairly typical S-shaped curve. Each dot represents an averaging of numbers over a five-year period.

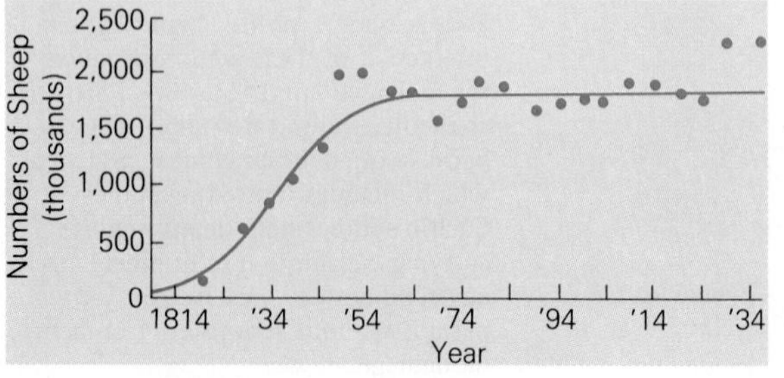

Survivorship curves are prepared by plotting the number of survivors on the vertical axis (usually a logarithmic scale) and the percent of the life span on the horizontal axis. From such curves we can learn some interesting things about species. For example, in the comparison of humans with oysters, we find exaggerated convex and concave curves respectively. Oysters apparently have an extremely high "infant mortality", while there is little loss of young humans. (Actually, it's the oysters' swimming larvae that meet with misfortune, so we have to know something about life history to interpret the curve. With humans, for instance—as with birds, most mammals, and some social insects—parental care accounts for the small loss of offspring.) But at the other end of the spectrum, when humans get old they die off rapidly, while death comes more slowly and steadily to the fewer surviving members of the oyster population. So that's how to tell people from oysters.

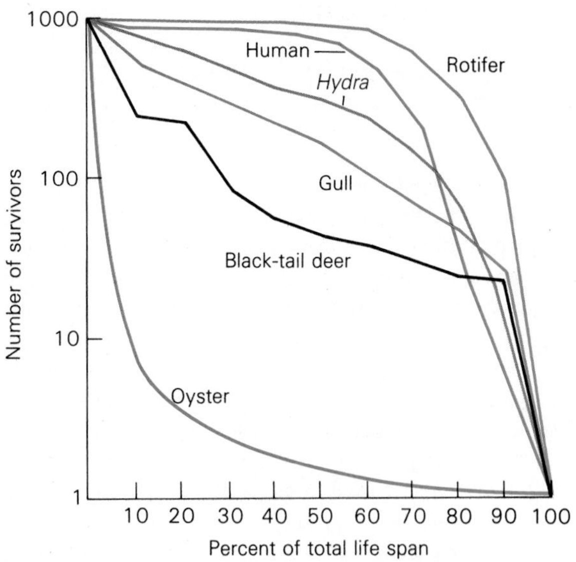

FIGURE 48.6
SURVIVORSHIP CURVES
Survivorship curves for six species, each beginning with a population of 1000 (vertical logarithmic scale). Points along each line represent the percent of the life span reached as the populations age (horizontal axis). In black-tail deer, a high death rate is experienced among the young; the numbers drop off rapidly, with the average individual completing only 6% of its potential life span. Loss during the early period is even more dramatically seen with oysters, where great numbers of offspring are devoured by other marine animals. But most surviving oysters can expect to complete their life span. The average human and rotifer survive the rigors of early life to complete about two-thirds of their potential life span, while the average *Hydra* and gull complete about 37% and 24% respectively.

THE EVOLUTION OF REPRODUCTIVE STRATEGIES

The great array of life on earth at present is here because its forbears reproduced. That's simple enough. And as we look over the diversity among living things, we find that they reproduce in a startling variety of ways. It is important to note, however, that even among such wild diversity, there are common themes that lead to the most efficient (that is, maximal) reproduction. When we say that individuals reproduce maximally, we do not mean they have produced the largest broods (whether litter of young or number of eggs) possible. Instead, they tend to produce that number that leads to the greatest possible number of offspring in the next generation. This is an important distinction.

Natural selection can alter reproductive output in any number of ways, some obvious, some not so obvious. The characteristics with the greatest impact on reproductive output, however, can be grouped into two broad categories. (1) timing of reproduction, and (2) total number of offspring.

Timing of reproduction has two critical aspects, the onset of reproduction and the interval between broods. The onset of reproduction (the time of the first reproductive effort) is important because it is related to generation length, and generally the shorter the generation length, the greater the reproductive output (r). Of two individuals in an ideal environment having the same number of offspring but having different ages at the onset of breeding, the one that reproduces earlier will leave more offspring. The principle is that short generation length means that any offspring produced will make it into the breeding population earlier and be in a position to pass the parents' genes along. In species with multiple broods, the interval between broods is important because the shorter the interval, the greater the opportunity for additional broods.

The total number of offspring is a measure of the brood size and the number of broods produced. All other things being equal, females producing the greatest total number of offspring will have the greatest r (Essay 48.1). This would seem blatantly evident, but in the real world, "all other things" are not equal. For example, if a female produces so many offspring that the effort kills her, she has preempted further opportunities for reproduction. This brings us to our next point, the cost of reproduction.

To assess this cost, we must keep in mind that animals utilize the energy they acquire in three basic ways: maintenance, growth, and reproduction. Maintenance (including such things as homeostasis, muscular contractions, and neural impulses) and growth help an animal to survive. Since there is only so much energy available, any energy expended on maintenance and growth cannot be used in reproduction, and energy expended in reproduction is not available for growth and maintenance. A high reproductive rate, therefore, can decrease survival prospects—and so this is the **cost of reproduction.** Of course, if an animal fails to survive, its reproductive activities are curtailed; so the cost of reproduction can be great, indeed.

What, then, determines the evolution of energy allocation patterns in an organism? How much energy should go toward enhancing reproduction vs. survival? A simple answer is offered: energy should be allocated to maximize fitness, i.e. to maximize r. If, in a particular environment, energy used in reproduction increases fitness more than does the same amount of energy used in maintenance and growth, then genotypes establishing the former pattern of allocation will replace others. If, on the other hand,

ESSAY 48.1

HE EFFECTS OF REPRODUCTIVE RATE AND GENERATION LENGTH

Female population-size projections can show the effects of different reproductive rates (total number of female offspring produced per female) and generation lengths. Individuals of this hypothetical species have a simple reproductive pattern—all reproduce only once and at the same age. All live in an ideal environment, which allows perfect survival until the precise moment after reproduction, when all parents die. If the sex ratio is 1, then total population sizes projected would be twice those listed.

Example (A): This example shows the effects of doubling reproductive rate or halving generation length when the initial reproductive rate is large, i.e. an average of five female offspring per female. The initial generation length is six weeks. Halving generation length produces greater population sizes and thus a greater r than does doubling reproductive rate when reproductive rate is initially large. (Note that in this example and the following, the realized or per-individual rate of increase r is actually equal to r_m, because the hypothetical environment is ideal.)

Example (B): This example shows the effects of doubling reproductive rate or halving generation length when the initial reproductive rate is small, i.e., an average of 1.5 female offspring per female. The initial generation length is six weeks. Doubling reproductive rate produces greater population sizes and thus a greater r than does halving generation length when reproductive rate is initially small. (Expressing

FEMALE POPULATION SIZES, EXAMPLE (A)

WEEK	ORIGINAL POPULATION $r = 0.07$	DOUBLING REPRODUCTIVE RATE $r = 0.18$	HALVING GENERATION LENGTH $r = 0.14$
0	1.00	1.00	1.00
3	1.00	1.00	1.50
6	1.50	3.00	2.25
9	1.50	3.00	3.38
12	2.25	9.00	5.06
15	2.25	9.00	7.59
18	3.38	27.00	11.39
21	3.38	27.00	17.09
24	5.06	81.00	25.63

energy used in maintenance and growth increases fitness more, then genotypes establishing that pattern will replace others. Because species may respond quite differently even to the same environment, they may evolve different reproductive solutions to their environmental problems. Thus, it should not be surprising that some species have evolved high reproductive rates and others low reproductive rates at the same ages. Now should it be surprising that some species reproduce only once but in massive clutches, while others reproduce repeatedly but in smaller clutches.

The Theory of *r* and *K* Selection

One attempt to answer the difficult questions about the evolution of reproductive behavior in populations has led to the theory of **r-selection** and **K-selection.** (The letters *K* and *r* are borrowed from formulas in population statistics, described earlier, where they represent the per-individual rate of increase and the carrying capacity, respectively). The concept is not universally accepted, and many of its adherents have chosen to modify it in various ways to cover special cases, but it can be used to illustrate the range of adaptive options open to species under different conditions. According to the model, K-selected species are those whose population growth curves closely resemble the idealized S-shape. Their numbers hover about the carrying capacity, to which they are highly population sizes in fractions of individuals is one of the advantages of working with a hypothetical species!)

These contrasts highlighted in examples (A) and (B) are caused by the mathematical relationship between r and its determinants. In any species meeting all of the assumptions mentioned for our hypothetical species, r is calculated simply but accurately from the following equation:

$$r = \ln \text{reproductive rate} \Big/ \text{generation length}$$

The logarithmic conversion in the numerator makes changes in reproductive rate less significant than equivalent changes in generation length when the initial reproductive rate is high, but not when it is low.

In species having this simple reproductive habit but living in a less-than-an-ideal environment that causes some prereproductive mortality, r can still be calculated accurately by the similar but more general equation:

$$r = \ln \text{reproductive rate} \Big/ \text{generation length}$$

Replacement rate, which is affected by both reproductive and survival rates, is the number of newborn females produced in a generation per newborn female produced in the previous generation. For example, if the replacement rate were 3.6, then 3.6 newborn females in this generation would replace each newborn female in the last generation.

FEMALE POPULATION SIZES, EXAMPLE (B)

WEEK	ORIGINAL POPULATION $r = 0.07$	DOUBLING REPRODUCTIVE RATE $r = 0.18$	HALVING GENERATION LENGTH $r = 0.14$
0	1.00	1.00	1.00
3	1.00	1.00	1.50
6	1.50	3.00	2.25
9	1.50	3.00	3.38
12	2.25	9.00	5.06
15	2.25	9.00	7.59
18	3.38	27.00	11.39
21	3.38	27.00	17.09
24	5.06	81.00	25.63

responsive. Thus, they have slower rates of population growth. K-selected species tend to be large animals, such as mammals and birds, with long life spans, continuous reproduction—year after year—and lengthy growth periods. At the other extreme are the r-selected species, sometimes referred to as "boom and bust" types. In their adaptation to large population size, they tend to have short life spans, are small in size, mature early, and reproduce only once but with prodigious numbers of offspring—which they generally take no part in raising. Thus, their populations suddenly boom, shooting up beyond the carrying capacity, following the familiar J-shaped growth curve before crashing. While population crashes can be quite dramatic in r-selected species, they are really a routine part of life, and recovery is just as routine and predictable (Table 48.1). You might want to keep in mind, also, that r and K adaptations do not necessarily follow general taxonomic lines—thus, while one might expect insects to be r-selected—and the locusts certainly are—bees are excellent examples of K-selected species. Perhaps we can make the distinction between r- and K-selected species more clear with a couple of examples.

The Tapeworm, an r-Selected Species The tapeworm living so comfortably in the idyllic environs of your intestine is lucky to be there—although its luck is the inverse of yours. It is lucky because the life of a tapeworm is fraught with risk. There are

TABLE 48.1 SOME CHARACTERISTICS OF r- AND K-SELECTION

	r-SELECTION	K-SELECTION
Climate	Variable and/or unpredictable	Fairly constant and/or predictable
Survivorship	High mortality when young; high survivorship afterwards	Either little mortality until a certain age, or constant death rates over a period of time
Intraspecific and interspecific competition	Variable, lax	Usually keen
Selection favors	Rapid development; high rate of population increase; early reproduction; small body size; single reproduction	Slower development; greater competitive ability; delayed reproduction; larger body size; repeated reproductions
Length of life	Usually less than one year	Usually more than one year
Leads to	Productivity	Efficiency

SOURCE: Adapted from Pianka, E.R. 1970. On r- and K-selection. *American Naturalist* 104:592–97 (1970).

incredible odds against its ever finding a host at all. In the case of the beef tapeworm, the egg must pass out with the feces of the previous human host. Then, not only must it survive the elements, but it must also be swallowed by a calf, so already the odds are against it. Once eaten, the larval tapeworm changes it appearance, moves through the intestinal wall of its host, and on through the bloodstream until it comes to lie as a "bladderworm" in the muscle. Here it forms a protective capsule around itself and waits. It waits for a human to eat the beef without cooking it too well. When this happens, the worm attaches to the intestine of its human host, grows into an adult, and begins to lay eggs. Actually, this is a rather simple cycle; some intestinal parasites have three hosts and take different forms in each of these (see Figure 30.14).

There is obviously a strong probability that not all the necessary conditions will be met, and thus only a remarkably small number of tapeworm eggs ever develop into adult tapeworms. So what is the tapeworm's answer to such a demanding life cycle? It has become capable of self-fertilization (although it will also cross-fertilize if another worm is present), and it lays thousands and thousands of eggs. Like other r-selected species, its whole life seems to be devoted to reproduction. It exists only to lay eggs, eggs, EGGS! A few of these will wend their way, by chance, through the complex maze that is the tapeworm's life cycle.

K-Selected Species We can consider the chimpanzee as our representative K-adapted species and then draw on an example from birds. Like other such species, the chimpanzee tends to live in a stable environment, has a long life span, and reproduces only after a prolonged period of development. When the female chimpanzee is sexually receptive, she may copulate with almost any male who signals his desire. Once she has become pregnant, she may not become sexually receptive again for years. Jane Goodall, who spent more than 20 years among the chimps at the Gombe Stream Preserve in Africa, tells us that Flo, the aging but sexy female, did not become sexually receptive again for five years after giving birth. During this time, she attended carefully to her baby.

In the first months, Flo carried her baby everywhere and dilligently guarded it against danger. Later, the baby was permitted brief forays on its own, but it was not allowed to stray far from its mother's vigilant eye. During this time, the youngster would scurry back to Flo at any real or imagined sign of danger. As the young chimpanzee gradually became able to care for itself during those first few years, the association between mother and offspring relaxed, until finally Flo was free to mate again and rear another baby. So, among chimpanzees, the female does not maximize her reproductive success by giving birth to large numbers of offspring. Instead, few offspring are produced, and these receive very careful attention until they become independent (Figure 48.7).

A quick look at some data from birds provides a clearer picture of how the dictates of a particular environment can influence the number of offspring a given species attempts to produce. For example, food supply often influences the number of eggs laid by birds in a particular season. The number will be roughly equivalent to the number of hatchlings they can feed successfully. Here, it is important to note that species can be evolutionarily programmed to take advantage of immediate ecological opportunities. Thus long-run (or "ultimate") factors set the stage for short-run (or "proximate") adaptation.

The number of offspring attempted by some bird species may also reflect the vulnerability of the young to predators. For example, certain species of gulls nest on accessible beaches, where the young are in real danger of being found by prowling foxes and other marauders. These gulls often lay three eggs, although usually only two young can be fed successfully. The odds are that at least one of the young will be eaten, so quite possibly the three-egg nest is an adaptation to high predation. On the other hand, the kittiwake, a species of gull that nests on steep cliffs, normally lays only two eggs. Since any marauding fox would be likely to break its neck trying to climb the cliffs, all the young are safe from foxes (Figure 48.8). Thus, the parents need only produce the number of young that they can feed.

As a final example of an extremely K-selected species, we need only consider the reproductive strategies of our own group.

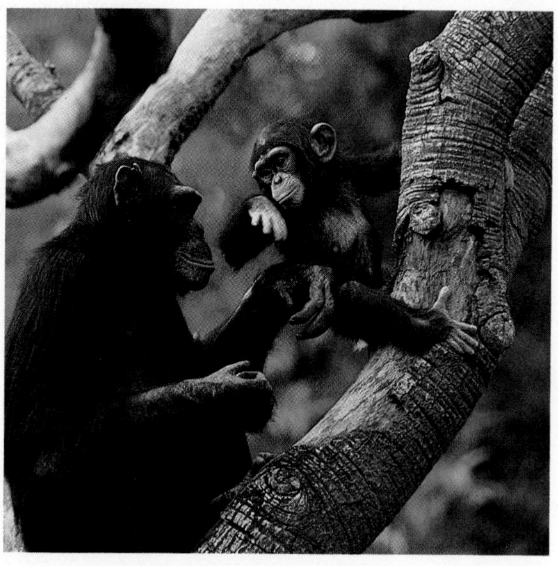

FIGURE 48.7
THE K-SELECTED CHIMPANZEE
Chimpanzee mothers carefully attend to their young. Because of their great investment in each of their offspring, they cannot afford to produce them in great numbers.

The Evolution of Human Reproductive Strategies Since the reproductive strategy of any species is a response to its own evolutionary history and immediate environment, let's take a closer look at the factors that might have determined our own reproductive propensity. The well-nourished human female has the ability to produce one child each year for over 30 years. Does this mean that humans are prepared, either physiologically or psychologically, to rear 30 children? (The record is held by Leontina Albina, an Argentine who at the age of 62 had given birth to 59 children.) Or is our reproductive ability an evolutionary adjustment to a historically rigorous life in which most, or even all, of our offspring were not likely to survive? In other words, has the ability of the human female to produce prolifically evolved in response to a traditionally high death rate among children?

What about today? (See Table 48.2.) A child born in a modern, developed nation such as the United States is very likely to live through its first critical years. In humans today, the mortality rate decreases after the first year and is very low through the teens; then it increases gradually until about age 60, when it rises rather rapidly. (Actually, this sort of curve doesn't apply very well to the poorer countries.)

(a)

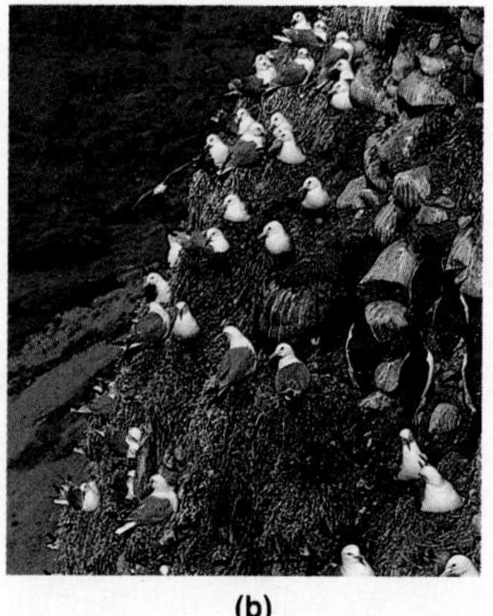

(b)

FIGURE 48.8
PREDATION AND BIRD REPRODUCTION
(a) The ground-nesting gulls, such as the herring gull, have been forced to assume a different reproductive strategy than **(b)** the cliff-nesting kittiwake. The nests of the ground-nesting species are vulnerable to marauding predators, so the birds compensate by laying more eggs than they will be likely to rear. The ground nesters, for example, lay three eggs, whereas they can probably rear only two young. The kittiwake, on the other hand, nests on cliffs safe from predators. Since all their young are likely to survive, they lay only two eggs.

TABLE 48.2 ESTIMATED AVERAGE LIFE SPAN IN HUMAN POPULATIONS

It seems that our life span, as a group, was fixed by evolutionary processes early in our history. Our attempts at increasing our longevity probably have resulted only in our coming closer to reaching the limits set by our physical constitutions. The examination of the remains of our primitive ancestors shows that many of the conditions we associate with old age affected them in much the same way that they affect us today. Also, in classical Rome, a person who lived to the age of 75 was more likely to reach 90 than someone in the United States today. (However, it might be argued that that is because it took a sturdier constitution to reach 75 in those days.)

POPULATION	YEARS
Neanderthal	29
Upper Paleolithic	32
Mesolithic	32
Neolithic Anatolia	38
Austrian Bronze Age	38
Classical Greece	35
Classical Rome	32
United States, 1900–1902	48
United States, 1950	70
United States, 1980	74
United States, 1989	75

SOURCE: Partly adapted from Deevey, E.S. *The Probability of Death*.

The means by which the human species has increased the likelihood of survival are almost universally viewed as good and desirable. For example, we have specific medicines to combat various maladies that, in earlier days, would have proven fatal. We also have therapeutic and corrective devices to aid the sick. If someone in our midst is unable to provide for him- or herself, that person will usually be cared for, however minimally. Our society often provides for those who, in harsher days, would have been selected against. A person doesn't have to be keen of wit and physically agile in order to cross a busy street. He or she simply waits for a light—a light that means it's safe to cross. The result of our social care has been a negation of many of the usual influences of natural selection. However, at the same time that we have reduced selection for swiftness and strength in our species, we have increased the level of variation. Thus we find among us all sorts of interests, talents, tendencies, and appearances. The point is that, with its highly developed social programs, our society is attempting to ensure that every individual will live and reproduce. This raises some important questions. With our reproductive potential so high and so many of the natural curbs removed, are we placing our species in a precarious position? Are we setting the stage for generations of ill-adapted people? Are we psychologically prepared for a great worldwide surge in our numbers?

In essence, then, the reproductive potential of humans has been established through the eons of our development. But although the direction and strength of natural selection have changed as we have altered the environment, we are left with a reproductive capacity better suited to earlier cultures.

POPULATION-REGULATING MECHANISMS

We can now begin to ask what regulates populations. What are the effects of the physical environment? Of other species? In some instances the answers are obvious, while in others they may be quite subtle or completely unknown. Let's begin by listing some such influences. We will see that they fall into two categories, nonliving and living.

Abiotic Control

FIGURE 48.9
DENSITY-INDEPENDENT EFFECTS
Factors such as weather may severely reduce populations without regard to the density of organisms in the population.

Just as seasonal weather changes can alter population numbers, so can irregular or unusual weather. Drought may kill many kinds of plants and animals. Many birds perish in some years because they begin their northward migration in the spring only to be caught by a late cold spell. Such controlling factors of a physical rather than biological nature are called **abiotic population controls** (*a,* "without"; *bios,* "life").

Population-depressing influences such as severe weather are usually **density-independent effects.** (That is, their effects are not influenced by population density—the number of individuals within an area.) In a severe drought, the parching sun doesn't care how many corn plants are struggling in the field below. It kills them all (Figure 48.9). In an area saturated with DDT, most of the insects die—whether there are few or many. Only their individual resistance counts—their numbers means nothing. Thus, mortality brought about by such means is independent of the density of the population.

Nevertheless, there are instances where the severity of the effect of an abiotic factor clearly varies with density—that is, it is **density-dependent.** For instance, if a killing frost strikes, and life or death for a population of insects depends upon the availability of sheltering nooks in their surroundings, then the percentage of the population lost will depend on the density of individuals in the area and the number of available nooks.

Biotic Control

Biotic population controls refer to any influences on a population brought about by a living agent. For example, the organism that causes bubonic plague can reduce populations. So can a tiger. We know how these work, but biotic influences can operate in some subtle ways as well. If a territorial bird drives a competitor into an area where there is less food, when winter comes the underfed competitor may be more likely to succumb to the rigors of the season. Thus, the territory holder has indirectly brought about the reduction of the population.

Unlike abiotic controls, which are usually density-independent, biotic controls on populations are likely to be density-dependent. This means that as population density rises, there will be increasing pressures on it that tend to reduce that density. As the density falls, the pressures will lessen, thus permitting the numbers to increase again. (This is another example of the classical negative feedback loop.)

We have generally been discussing factors that bring about death, but "population regulation" actually refers to the effects of factors that keep population size *within certain limits.* It is a specific term, and intuitively we can see that this sort of regulation is likely to be achieved primarily through density-dependent mechanisms. Density-dependent effects could theoretically keep population sizes within more or less defined ranges, because they generally involve negative feedback control, the familiar stabilizing effect seen so frequently in physiological systems.

The Evolution of Death as a Population Control Mechanism

Have you ever wondered why death occurs at all? Why hasn't natural selection resulted in organisms that simply live forever? If life is better than death, then why is there no marked selection for longevity? Why do humans have so much trouble surviving past their "three score years and ten"? Why death?

All sorts of people have struggled with the "meaning" of death—theologians, poets, novelists, philosophers, drunks, and others of bad habit. Even aboriginal societies deal with the meaning of death in one way or another—some very casually, not fearing it at all but treating it as though it were simply the natural extension of life. So let's join the fray and consider death from a biological point of view.

Programmed Death Annual plants germinate and grow in the early spring, flower in the late spring, undergo seed development in the summer, and disperse their seeds and die in the fall. There are "annuals" among the animals, also. One species of Brazilian fishes lives in temporary ponds that exist only during the rainy season. Shortly before the ponds are due to dry up, the fish spawn and lay cystlike, drought-resistant eggs. The fish die when the ponds dwindle, but the eggs hatch the next year. Strangely, though, if these fish are netted and kept in an aquarium, they continue to follow their natural life cycle. Thus, after the females have laid their eggs and the males have spawned, they die—the fish literally begin to fall apart. The onset of the deterioration is directly associated with the completion of reproduction.

The life spans of squids and octopuses are also intimately associated with their reproductive efforts. Put simply, after they reproduce, they die. The question is, why? As with all such "why" questions in biology, we can answer it in proximate (short-run) terms, or ultimate (long-run, or evolutionary) terms. In the short run, we know that they die at this time because both reproduction and death are under hormonal control. So a surge in certain hormones causes death. If the cephalic gland of a mature octopus is removed, the animal won't engage in sexual activity or reproduce—and also won't die, that is, it won't die on schedule. The experimentally altered animal continues to feed and engage in normal activities. Sooner or later, however, death comes anyway. The altered animals produce tumors and infections, otherwise unknown in the species, and the long-lived but barren animals ultimately succumb.

The question can then be asked in evolutionary terms. Why would their bodies be programmed to self-destruct? Perhaps as a mechanism to avoid parent-offspring competition for resources. Let's explore the notion by asking the question of another species. Do humans have a similar kind of programmed death? There are some reasons for thinking that they do, to some extent at least, although the program is by no means as absolute as that of the squid. Usually between ages 60 and 70 the death rates suddenly begin a precipitous rise and continue to accelerate due to a host of apparently unrelated diseases. It seems that all kinds of systems wear out more or less at the same time. The time of this rapid aging may differ from individual to individual, which smooths out the human death rate curves, but sooner or later it happens, and when you're old, you're old. It has been observed that people who have been saved from cancer soon die of heart disease or some other infirmity of age. In fact, although cancers claim about 25% of all humans, the combined human death rates for all causes increase so rapidly that if we were to eliminate cancer, say at noon today, we would only extend our average life expectancy by less than three years. So why do humans seem to wear out all at once?

Humans are among the longest-lived of all animal species. Our normal life span exceeds that of all other mammals and most vertebrates and invertebrates—possibly only giant tortoises and certain sea anemones regularly outlive us. Correspondingly, we have very effective cancer-suppressing and disease-fighting mechanisms. Our bodies mature very slowly, but they last a long time. We have quite efficient defenses against the wear and tear of time, against infection and our own cancerous cells. We pay a price in physiological energy, in developmental time, and in our rather low rate of reproduction. However, after perhaps 15 to 20 years of maturation, another 25 to 30 years of peak reproduction, and another 20 years to see our youngest children reach their own maturity and independence, our bodies have had it. Once the body has served its allotted time, it wears out—all at once. There doesn't seem to be any strong selection in favor of metabolic systems that protect against the depredations of postreproductive degeneration. From the relentless viewpoint of natural selection, there's just not enough of a payoff involved.

Now, in any system in which the commodities are limited—such as the planet earth—as the numbers of individuals increase, so will the competition for the available commodities, such as food, shelter, land, and natural resources. Consider, then, an organism that finishes its period of reproduction and then hangs around to compete with its offspring. Such an individual could be expected ultimately to leave fewer offspring than a parent that, after having reproduced, died and thus did not interfere with the success of its offspring. Death, then, and its timing may be a means of increasing the individual's own reproductive success.

We've tossed around some important concepts fairly loosely here, as you are aware. However, these are legitimate questions for evolutionary biologists. As scientists, we are not in a position to uncover whole truths, but we can begin to accumulate certain kinds of evidence. In any case, this sort of approach makes as much sense as dealing with the concept in subjective, artistic, abstract, or mystical terms.

Now that we have learned something about the kinds of environments the earth has provided for living things and have reviewed some of the principles of populations and their regulation in nature, we can look at the human situation. We will find that our own species has placed itself in an unheard-of position. We don't yet know the consequences, but we can describe the situation, to some degree.

APPLICATION OF IDEAS

1. If every couple on the planet should, this year, begin to produce only enough children to replace themselves, would the population begin to immediately stabilize? Why?

KEY IDEAS

POPULATION CHANGES

1. In spite of noticeable short-term changes in population size, in the long run the numbers of most species tend to remain about the same. Central questions concern size, growth fluctuations, stabilizing influences, and distribution.

Population Dynamics and Growth Patterns

1. Population changes are represented in the equation $I = (b - d)N$, where I, the rate of change in number of individuals, equals average birth rate minus average death rate times population size. The equation describes unlimited growth where b is maximum and d is minimum.
2. Birth rate minus death rate $(b - d)$ is the realized rate of increase; thus $r = b - d$. Therefore $I = rN$ when r is positive, N grows with each generation, so I increases.
3. Population growth can be characterized with J-shaped curves and S-shaped curves. In the first, the curve rises slowly in the beginning, but increases until it approaches the vertical, due to an **exponential increase** (constantly increasing rate) as opposed to a linear, **arithmetic increase.** The maximum birth rate is symbolized as r_m (**intrinsic rate of natural increase**).
4. Growing populations eventually encounter **environmental resistance,** the total factors that tend to slow such growth, sometimes then undergoing a **population crash.**
5. Crashes generally occur when the environmental **carrying capacity** is drastically exceeded. **Biotic** (living) and **abiotic** (nonliving) factors both determine the carrying capacity—the number that the environment can support. Population booms and crashes are part of the natural cycle of some species.
6. Populations undergoing an S-shaped growth may first follow what seems to be a J-shaped growth that then levels off to fluctuate near the carrying capacity.
7. The carrying capacity is growth-limiting and is represented by K in the expression $(K - N)$. Inserted into the exponential equation, it appears as

$$I = r_m(K - N)N$$

The growth-limiting effects of K increase as the population nears the established carrying capacity, which acts through negative feedback in modulating growth.
8. Age structure and longevity also are important in determining the characteristics of a population. Survivorship curves reveal how aging affects populations. Some species experience a drastic loss of young, with most survivors living out their normal span, while in others the losses occur more equally along the normal span. Still others have light early losses, with few adults reaching the potential life span.

THE EVOLUTION OF REPRODUCTIVE STRATEGIES

1. Reproductive timing and brood size are both important to reproductive success, but reproductive cost is also a vital factor. High reproductive rates are costly and in the long run may lead to reproductive failure (such as when most offspring starve).
2. Reproductive cost or energy can be allocated in various ways. Thus, there are many different successful reproductive strategies.

The Theory of *r* and *K* selection

1. In general, organisms tend to reproduce maximally; that is, they produce the number that assures the greatest survival.
   a. **K-selected species** tend to describe the S-shaped curve, are highly responsive to carrying capacity, and tend to be large, with long life spans and a long reproductive life.
   b. Growth of **r-selected species** ("boom or bust") follows the J-shaped curve. Typically, growth exceeds the carrying capacity, individuals are small with short life spans, and most reproduce but once.
   c. Examples of r-selected species include the tapeworm, a species in which individuals overcome a high mortality rate in their offspring by laying great numbers of eggs.
   d. Birds and mammals are K-selected, producing few offspring but investing much of their energy into assuring survival.
2. The reproductive capacity of humans appears to be a response to a historically high rate of infant mortality. Today, with more assured survival, such a capacity may no longer be adaptive.

POPULATION-REGULATING MECHANISMS

Abiotic Control

1. **Abiotic controls** include nonliving environmental factors such as temperature, rainfall, and drought. Most abiotic control is **density-independent**—its effect has nothing to do with population size and distribution. **Density-dependent control** is modified by the population size and distribution.

Biotic Control

1. **Biotic controls** involve other organisms. They are likely to be density-dependent. They are much more likely to have a stabilizing effect than density-independent factors because they often operate through negative feedback.

The Evolution of Death as a Population Control Mechanism

1. For many species, death closely follows reproduction. Examples include annual plants, certain fishes, and squid. The adaptiveness of such a strategy involves the reduction of competition with one's variable offspring in a system of limited commodities.
2. With humans, the emphasis is on a long period of growth, development, reproduction, and rearing the young, followed by marked deterioration.

REVIEW QUESTIONS

1. What generalization do biologists make about readily visible increases and decreases in natural populations? How do we intuitively suspect that this generalization is valid? (1200)
2. Explain the meaning of the population symbol *r*. How is *r* obtained? Is it a positive number? Explain. (1201)
3. What factors account for the J-shaped growth curve? Can a population actually reach its biotic potential? (1201, 1202)
4. Discuss the meaning of exponential growth. How does this differ from arithmetic growth? (1202)
5. List five factors that might contribute to environmental resistance. (1202)
6. What is the carrying capacity of an environment? What generally happens to a population that exceeds this point, and what happens to the carrying capacity itself? (1203)
7. Is J-shaped growth a rare phenomenon? Where might it be a common occurrence? (1201)
8. Draw three radically different survivorship curves and explain what they mean. Which most closely approximates the human condition? (1204, 1205)
9. Select an example of a K-adapted species and an r-adapted species and discuss the following: growth curves, size, life span, and reproductive lifetime. (1206, 1207)
10. What evidence suggests that the human reproductive capacity was once adaptive but perhaps is no longer so? (1209, 1210)
11. Provide an example of a density-dependent, abiotic population-control factor. Why are most abiotic factors density-independent? (1211)
12. Using an example, describe a way in which density-dependent biotic controls are regulative in nature. (1211)
13. List two examples of programmed death in organisms. Give an example of what happens to such organisms when life is extended experimentally. (1212, 1213)
14. Is human death programmed? In what way(s) might it be adaptive for humans to live far beyond their reproductive years? In the final analysis, how is human death adaptive? (1212, 1213)

49

THE HUMAN IMPACT

WE TURN OUR ATTENTION TO A SET OF NUMBERS THAT STIMULATED A GREAT deal of concern and controversy in recent years: 5.1, 28, 10, 1.7, 40, and 33. You may not at first glance be particularly impressed; they may just look like numbers. However, a demographer will recognize them as the startling figures that represent, respectively, our human population (in billions), our crude birth rate, crude death rate, percentage annual growth, population doubling time (in years), and percentage of people below the age of 15 (see Tables 49.1 and 49.2).

It is just such numbers that have begun to reveal to us how pressing our problems are. For example, we can see that the population of the world will have doubled between the time most of today's college students were born and about the year 2010 (about the time when their children are in college).

The gloom-and-doom statistics are abundant, and anyone who is at all interested has probably heard enough of them by now. But just in case, here is one more: there are 28 more humans living now than there were 10 seconds ago when you began reading this paragraph. By tomorrow at this time, 238,838 more will have been added, and by next year, 87.2 million. The last is approximately the population of Mexico—and most of the newcomers can expect to live in a desperate style similar to that of the average citizen of that struggling nation (Figure 49.1).

We can generate such statements all day, but once we understand the problem, do we wallow in depression? Do we simply look away? Or do we join the ranks of hopeful and determined people who intend to learn as much as they can about the problem and then try to find ways to help? Obviously, we will assume you are in this last group, and so we'll begin by delving into the history of our numbers.

FIG. 49.1
EXPONENTIAL GROWTH—THE HUMAN AFTERMATH
A scene of squalor in Mexico City, the world's fastest growing metropolis. As the population exceeds 20 million, we see drastic unemployment and the continued shortage of dwellings, adequate water, and sewage facilities, all pointing to impending disaster unless heroic efforts to find a solution begin soon.

THE HISTORY OF THE HUMAN POPULATION

The first thing we should know is that throughout most of our 3-to-4-million-year history, populations have remained fairly stable, but in the past million years that stability has been interrupted by three significant growth surges (Figure 49.2). But before that, there were only about 125,000 wandering hominids on the grassy East African plains, and they were not having a great deal of impact on their environment. These early hunters and gatherers were strongly subjected to the same factors that influence most populations of herbivores and carnivores today. Infants and children probably suffered high mortality rates but were quickly replaced in a species where, we believe, fertility was not a seasonal event. The average life span is estimated to have been 30 years, but the high infant mortality hides the possibility that some individuals lived much longer (see Table 48.2). Then, however, the population underwent a sudden, rapid increase.

The First Population Surge

The first growth surge in the human population, having begun about one million years ago, was probably brought about by the development of increasingly efficient tools, leading to more efficient ways of killing game and, in general, easier utilization of whatever was available in the surroundings. In addition, humans—inveterate wanderers—had by then penetrated and established themselves on all of the continents. Thus, by about

FIGURE 49.2
TWO WAYS OF VIEWING HUMAN POPULATION HISTORY
(a) The arithmetic plot of the past 10,000 years clearly reveals the soaring near-vertical rise of the J-shaped curve, a danger signal to most species. Hidden in the gradually rising line preceding this is the earlier population surge brought on by the switch from hunting and gathering to agriculture. The logarithmic plot **(b)**, where time is seemingly compressed, covers human history over the past million years. Here, three growth surges are seen, the products, respectively, of advances in hunting and gathering technology, the rise of agriculture, and the recent era of modern health, industrial, and agricultural practices.

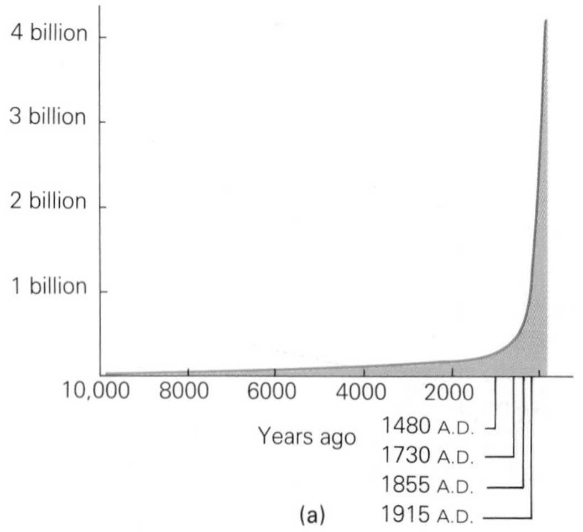

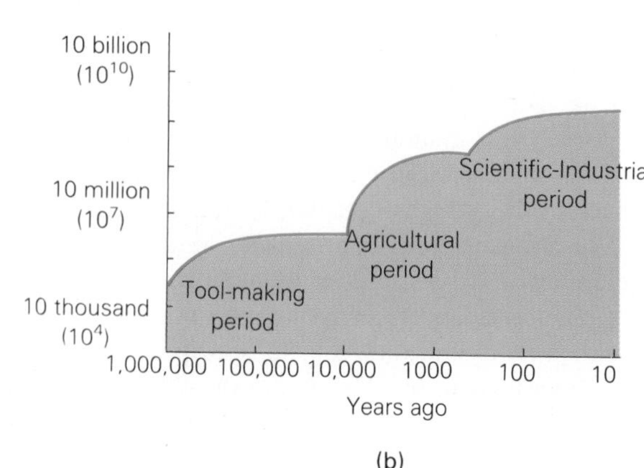

TABLE 49.1 WORLD POPULATION DATA: 1970 AND 1988

REGION	YEAR	TOTAL (MILLIONS)	CRUDE BIRTH RATE	CRUDE DEATH RATE	NATURAL INCREASE (ANNUAL %)	DOUBLING TIME (YEARS)	% BELOW 15 YEARS OF AGE	ESTIMATED POPULATION IN 2020 (MILLIONS)
World	1970	3632	34	14	2.0	35	37	8053
	1988	5128	28	10	1.7	40	33	
Africa	1970	344	47	20	2.6	27	44	1497
	1988	623	44	15	2.9	24	42	
Asia	1970	2045	38	15	2.3	31	40	4629
	1988	2995	28	10	1.8	38	35	
North America	1970	228	18	9	1.1	63	20	327
	1988	272	16	9	0.7	98	21	
Latin America	1970	283	38	9	2.9	24	42	711
	1988	429	29	8	2.2	32	38	
Europe	1970	462	18	10	0.8	88	25	499
	1988	497	13	10	0.4	266	21	
Nations of Special Interest								
United States	1970	205	17.5	9.6	1.0	70	30	297
	1988	246	16.0	9.0	0.7	99	22	
Soviet Union	1970	243	17.9	7.7	1.0	70	28	354
	1988	286	20.0	10.0	1.0	68	26	
People's Republic of China	1970	760	34.0	15.0	1.8	39	?	1404
	1988	1087	21.0	7.0	1.4	49	29	
India	1970	554	42.0	17.0	2.6	27	41	1309
	1988	816	33.0	13.0	2.0	35	38	
Mexico	1970	50.7	44.0	10.0	3.4	21	46	106
	1988	88.5	30.0	6.0	2.4	29	42	

Source of Data: Population Reference Bureau

10,000 years ago, the earth probably supported only about 5 million hunting and gathering humans (about the number in three of New York City's five boroughs), who still had relatively little impact on their surrounding environment.

The Second Population Surge

About 10,000 years ago, the human population began its second growth surge, this time with more authority. With the advent of agriculture and the domestication of animals came increasing densities of local populations. There was less need to roam the countryside in search of food; in fact, there was a great need to stay put and tend the fields and livestock. With surplus food to store, winter and drought no longer exacted such a great toll on human life.

With the increased quantity and dependability of the food supply, humans probably

TABLE 49.2 BASIC POPULATION ARITHMETIC

Crude birth rate = number of births per year per 1000 population

$$\left(\text{determined by: } \frac{\text{total births}}{\text{midyear population}} \times 1000\right)$$

Crude death rate = number of deaths per year per 1000 population

$$\left(\text{determined by: } \frac{\text{total deaths}}{\text{midyear population}} \times 1000\right)$$

Rate of natural increase (or decrease) = crude BR − crude DR

$$\textbf{Percent annual growth} = \frac{\text{rate of natural increase}}{10}$$

$$\underset{\text{(approximately)}}{\textbf{Doubling time}} = \frac{70}{\text{percent annual growth}}$$

(Example: % annual growth in the world in 1986 was 1.7: $\frac{70}{1.7} = 41.18$ years)

$$\textbf{General fertility rate} = \frac{\text{total births}}{\text{total women in reproductive yrs}} \times 1000$$

(Example: In 1983 there were 3,164,000 births per 55,260,000 American women aged 15-44. $\frac{3{,}614{,}000}{55{,}260{,}000} \times 1000 = 65.40$)

experienced a lower death rate, particularly among the young. Furthermore, there was quite possibly an increase in the birth rate because of better nutrition. Large family size may have been encouraged because it meant more hands to till the fields. But life was by no means simple, since the crops were subjected to the inconsistencies of weather and to infestation by insects and other herbivores whose own numbers responded to the novel food supply.

The unprecedented population growth of the early days of agriculture did not continue at its initial soaring rate, but settled into a steadier, more gradual climb. Yet between the advent of agriculture and the time of Christ, the human population rose from 5 million to about 133 million. By 1650 A.D., it had reached an estimated 500 million.

History reveals that on a regional level this growth was interrupted many, many times by the decimating effects of disease, famine, and war. These are largely density-dependent and closely interrelated factors. A drastic example of the population-depressing effects of disease occurred in the fourteenth century, when one-fourth of Europe's population was killed by the "Black Death"—the bubonic plague. Such other diseases as typhus, influenza, and syphilis also took their toll on the crowded and incredibly filthy towns of the medieval period.

Interestingly, the loss of such numbers is insignificant in view of population growth today. At today's population growth rate, for example, the numbers lost to the fourteenth-century bubonic plague could be recouped in one year, while the number killed in all the wars of the last 500 years could be replaced in about six months.

The Third Population Surge

The third population growth surge began in Europe in the mid-seventeenth century, after the unexplained decline of the plague. (Perhaps only those who were naturally immune were left alive.) A number of explanations have been advanced to account for this third surge. For one thing, the crowded populations of Europe expanded into the

New World, with its array of opportunities and unexploited resources. More significantly, the "germ theory" of disease, postulated later by Louis Pasteur and others, quickly led to ways of preventing and combating many dread diseases. By the nineteenth century, public sanitation programs had begun, vaccines were developed, and rapid advances in food storage and transportation technologies led to a marked increase in food supplies. Between 1750 and 1850 the population of Europe doubled; that of the New World increased fivefold.

Populations were surging in other parts of the world as well. In China, the most heavily populated nation at that time, agriculture had made great gains, and a long period of comparative political stability followed the overthrow of the Ming dynasty in 1644. India, however, had known little rest from turmoil and periodic famines. In 1770, the worst famine of all reportedly killed three million people in India. African populations are believed to have remained stable until about 1850, when the impact of imported European medical advances began to depress the death rate. However, recent experiences in Africa, namely the Ethiopian drought and famine, remind us that we are not yet free of ancient hazards (Figure 49.3). Also, we are awaiting the full impact of the more recent threat, AIDS.

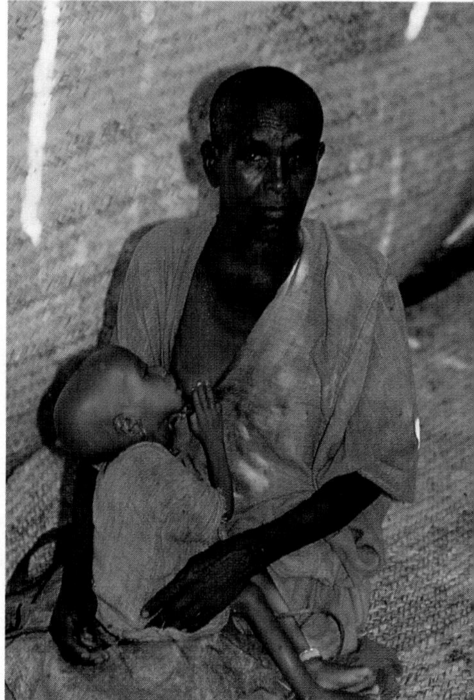

FIGURE 49.3
FAMINE IN ETHIOPIA
In spite of modern achievements, undeveloped and developing nations are still subject to disruptive episodes of famine.

The third surge has continued into modern times, again supported by a host of innovations in industry, agriculture, and public health. Many of these innovations arose in industrialized, developed countries and were exported to heavily populated developing regions.[1] In the developed nations, famine was all but eradicated with the advent of pesticides, chemical fertilizers, and high-yield crops. Potential disease epidemics were routinely controlled by vaccines, antibiotics, and insecticides.

In review, we can note that the human population grew from about 5 million at the dawn of agriculture to 500 million by 1650. By 1850, the world population had doubled, reaching 1 billion. This amazing 200-year **doubling time** was only a hint of things to come. In the 80 years between 1850 and 1930, the numbers doubled again, to 2 billion. The next doubling took only 45 years, so by 1975 world population stood at 4 billion. By 1970, thoroughly alarmed population experts were predicting another doubling, to 8 billion, by the year 2000—a span of only 35 years. (Doubling time is further explored in Table 49.3).

THE HUMAN POPULATION TODAY

To the surprise of nearly everyone, the rate of increase in world population growth slowed toward the end of the 1970s, and that slowing trend continues today (see Table 49.1). (Note the implications of the word *rate*. The occupants of a car approaching a cliff at 30 miles per hour might take little comfort in knowing that its rate is slowing and it will be traveling at only 15 miles per hour by the time it goes over the edge.)

At the end of 1988, demographers estimated the world's population to be over 5.1 billion. The annual growth rate had fallen slightly, and the doubling time had increased to 40 years. These data suggest some headway in our attempts to control our population, which can be at least partially attributed to changing attitudes of women toward their role in society, changes in preferences in family size, advances in methods of birth control, and the liberalization of abortion laws in developed nations.

Growth in the Developing Regions

The new data generated a measure of relief and even a growing optimism. But it turns out that the depressed growth rates occurred primarily in the most developed nations, those that could best support increased populations, such as the United States, Japan, Western Europe, and the Soviet Union. Most poorer, developing nations of the world show few signs of controlling their populations. The annual rate of growth in those regions (excluding China) is ominous (averaging 2.4%), and the doubling time is 30 years (Table 49.3). While the crude birth and death rates (see Table 49.2) in the U.S.

TABLE 49.3 DOUBLING TIMES IN SELECTED NATIONS

DOUBLING TIME	NATION			
	(Africa)	(Asia)	(Latin America)	(North America, Europe)
25 years or less	Algeria, Congo, Egypt, Ghana, Kenya, Libya, Nigeria, Niger	Iran, Iraq, Syria, Philippines, Pakistan	Ecuador, Guatemala, Honduras, Costa Rica, Nicaragua, El Salvador, Paraguay	None
26–35 years	Chad, Ethiopia, Angola, Morocco, South Africa	Afghanistan, Burma, India, Lebanon, Thailand, N. Korea, Vietnam	Bolivia, Brazil, Mexico, Panama, Peru	Albania
36–69 years	Mozambique, Gabon	China, Indonesia, Israel, S. Korea, Taiwan	Argentina, Chile, Cuba, Dominica, Jamaica	None
70+ years	None	Singapore, Japan	Barbados, Uruguay	All except Albania (Belgium; 1034 yrs) (Italy: 3465 yrs)
Negative growth	None	None	None	Denmark, Austria, West Germany, East Germany, Hungary

in 1988 were 16 and 9 per thousand, respectively, those rates in Africa were 44 and 15. Further, the doubling time in Africa is an alarmingly brief 25 years. How do you suppose these shifting population trends will affect political, social, and economic stability in the near future? What effects are they having now? Figure 49.4 illustrates regional population forecasts.

At first glance, the news from Latin America seemed hopeful. In the past few years there has been a decrease in the crude birth rate. But again, the numbers are misleading. In this region, the birth rate is over three times the death rate. Furthermore, many Latin American nations, already troubled by political unrest and dismal economic conditions, face a doubling time shorter than Africa's.

Asia has traditionally troubled demographers because so little is known about it and yet it has enormous reproductive potential. (It is instructive to keep in mind that some three-fifths of the people walking the earth today are Asians, and one in every five humans alive today is Chinese.) However, since 1970 the birth rate in China has been cut nearly 40%, with an equivalent fall in the death rate. Partly because of this shift in China, the doubling time for Asian populations has increased from 31 to 38 years. Massive birth-control programs in China and several other Asian nations have been instrumental to declining birth rates. Similar efforts in India have been somewhat successful, but India's annual increase is still 2%, and the doubling time of this nation of 817 million people is an alarming 35 years.

The successes Asia is currently enjoying, however, may be swamped by another problem, one of momentum. Specifically, 35% of these people are under the age of 15

[1]*Developed* nations are those that have a slow rate of population growth; a stable, industrialized economy; a low percentage of workers employed in the agricultural sector; a high per capita income; and a high degree of literacy. *Developing* nations have the opposite traits. Of course, many countries have intermediate conditions, and some have a combination—a privileged upper class and a poor lower class.

(38% excluding China) and have yet to enter the breeding population. Asia could well be in for another population explosion.

It is difficult if not impossible to predict what the future holds, but there are ways of making educated guesses. Let's gaze into the crystal ball of demography and see how populations are forecast.

Demographic Transition

There is an interesting relationship between a country's developmental progress and its population structure: as nations undergo economic and technological development, their population growth tends to decrease. According to the **theory of demographic transition,** nations go through several developmental phases, the earliest of which is characterized by high birth and death rates and slow growth. As they begin to develop, the birth rate remains high, but the death rate falls. The result is that the population enters a rapid growth phase. Then, as industrialization peaks, the birth rate falls and begins to approximate the death rate. Population growth slows drastically, reaching a phase of very modest growth such as is seen in many European nations today.

One major prediction based on the theory of demographic transition is that population growth in developing parts of Asia, Africa, and Latin America will slow as they become further industrialized. But can the answer to the enigma of world population growth be that simple? Not quite. First, it is the developed nations that must pay the enormous costs of such a massive industrialization program, and this could place a great burden on their own economies. (Also, who needs more competition in the marketplace?) Second, there are severe risks in encouraging developing countries to go through such transitions. For example, if massive development programs stall in the second and most difficult phase, soaring populations will preclude the third stage and send the hopeful developing nation into a deadly population surge and devastating poverty cycle. Thus, any such developmental program must be approached with extreme caution.

THE FUTURE OF THE HUMAN POPULATION

Our track record in predicting population growth is not very impressive; demographers in the U.S. were taken by surprise by the baby boom of the 1950s and they are still trying to explain the latest downward trend. However, increasingly sophisticated and precise calculations show some promise of improving the accuracy of our projections.

One of the more useful statistics in discussing human populations is the **general fertility rate** (GFR). Unlike the birth rate, which reflects the number of babies born per total population in a given year, the GFR reflects the number of births per number of

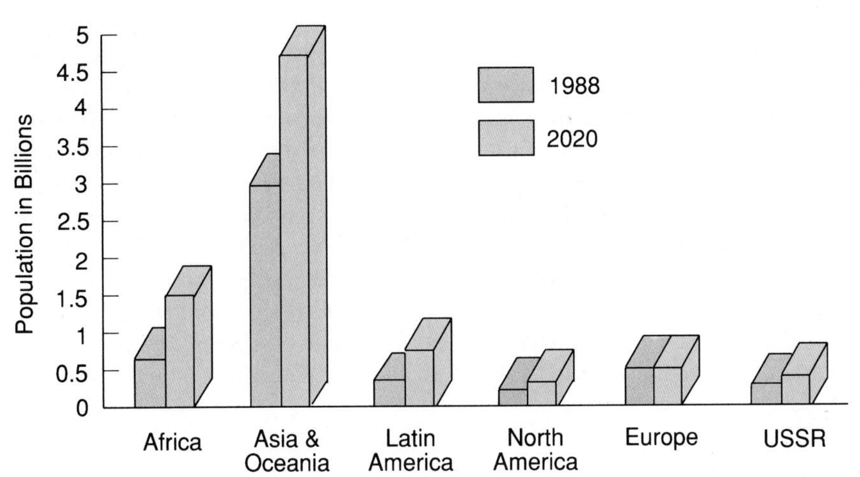

FIGURE 49.4
WORLD POPULATION BY REGION
In 1988 the earth's human population was unevenly distributed over the planet, bringing far greater pressures to bear in some areas than others. Most of the earth's people live in Asia and Africa, and this is not expected to change by 2020. The human population, at 5.1 billion in 1988, is expected to rise to 8.1 billion by 2020. Source: Population Reference Bureau Data Sheet.

women of *reproductive age* per given year. Since the number of women of reproductive age varies from region to region and from nation to nation, comparisons based on the general fertility rate can be considerably more accurate. Reproductive age is somewhat arbitrarily defined as age 15–44, a figure that is most accurate for American women. The calculations for the GFR, together with other population statistics, can be seen in Table 49.2.

One of the highest general fertility rates ever recorded was in Iran: an incredible 200 live births per 1000 women of reproductive age. In that same year, the fertility rate in the Netherlands was 48.

A second useful indicator is the **total fertility rate** (TFR), which is the number of children born to any woman who had conformed to the general fertility rate throughout her reproductive years. More simply, the TFR tells us how many children women are having these days. It is really a prediction of how many children any woman is likely to have if the known general fertility rate remains constant.

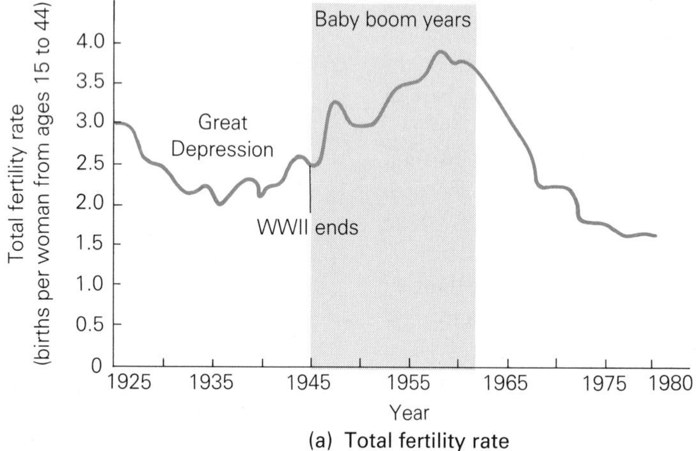

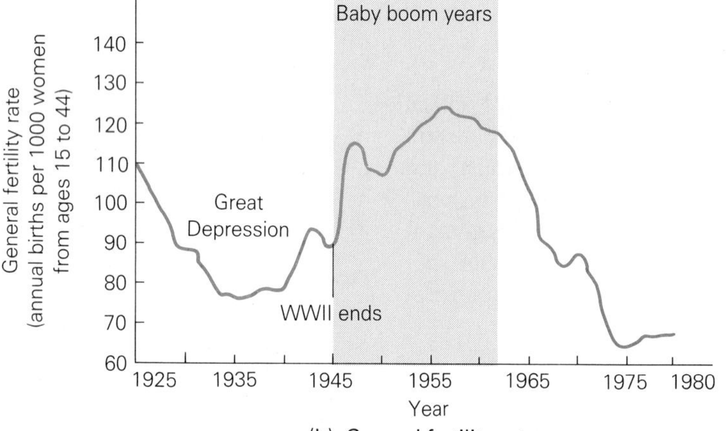

FIGURE 49.5
FERTILITY IN THE U.S.
The two plots include the general fertility rate (top) and total fertility rate (bottom). The Depression years are seen as a valley, followed by a reproductive peak—the "baby boom" years—that followed World War II. The more recent valley is good news to those alarmed by population growth, but keep in mind that attitudes toward family size change, and women born in the baby-boom years are still in the reproductive period of life.

The general fertility rate in the United States and much of Western Europe dipped sharply in the Great Depression years, falling close to the actual replacement level by 1936. The total fertility rate in 1936 was about 2.2. (Replacement level is considered to be 2.1.) This trend was short-lived, however, for by the end of the 1930s the fertility rate had begun to climb, with the GFR peaking in the post-World War II years (1957) at 125 births per 1000 women, with the TFR reaching 3.7 (Figure 49.5). In the United States, this era became known as the "baby-boom" years. This was a time when the "good life" meant a new car, a home in the suburbs, and three or four kids playing in the yard.

Then came the awakening. Environmentally concerned scientists began to spread the alarm about the unprecedented population growth throughout the world. Soon after, reproductive rates in developed countries began to slow. The reasons may never be known, but perhaps people were taking the warnings seriously. By 1975, the total fertility rate in the United States, Japan, and most European nations had fallen to below 2.0. As of the time of this writing, the fertility rates in most of these nations show no signs of beginning a new upward trend.

Demographers have tried to understand what caused such a drastic change in reproductive behavior in the developed world. We can begin to guess at what happened, at least in the United States. It was a matter of attitude. The attitude of American women in their reproductive period toward domestic life and family size changed radically in the 1960s and 1970s (as did that of women in many other developed nations). Women began to assume new roles in our industrialized society. They entered new areas of the work force, challenged male enclaves, demanded rights and privileges previously reserved for men, and placed less emphasis on raising children. Their new attitudes were duly noted by the demographers, who then predicted a reduced rate of population growth. But demographers are only too aware that such trends can change quickly, and if attitude is an important variable in forecasting population sizes, their data must remain current. In any case, measuring attitudes is a very risky business.

Whether the developed nations heeded the demographers' warnings or not, their rate of population growth has slowed. But what about the developing countries, those with poorer, often illiterate populations? After all, these are where most of the earth's teeming billions reside. It turns out, somewhat unexpectedly, that in the developing countries, the rates of increase are also declining, but the effect there is much more difficult to see. The problem is one of population momentum. With a declining death rate (particularly among the young), an increasing life span, and with great numbers of children who have yet to enter their reproductive years, downward trends in fertility will not have an appreciable effect for many years.

Using Mexico as an example once more, let's note that even at the peak of the baby boom in the United States, Mexican women experienced twice as many births per woman. Yet the subsequent dramatic decline seen in the U.S. also occurred in Mexico, and by the mid-1980s the total fertility rate in Mexico had gone from nearly 7 to below 5. Demographers estimate that it will continue to decline, reaching about 2.3 in 2025. Again, the problem is that, because of the large proportion of younger Mexicans in the population, the fertility *decline* will be accompanied by a whopping population *increase* as great numbers of youths enter the breeding population. The Mexican population, by some estimates, will go from 88.5 million in 1988 to 174 million in 2025.

Population Structure

Knowing the age structure of any population is critical to understanding growth patterns and making predictions. One way to portray such data is by **age structure histograms.** In Figure 49.6, you can see how such diagrams are formed. Note the marked differences in the shapes of such histograms between developed and developing nations. In developed areas, recent population increases have been comparatively slow, so the base is not very wide. Also, people tend to live longer and thus to occupy the upper levels in greater proportions. In developing nations, on the other hand, the rate of increase is still expanding, swelling the lower levels. Obviously, these lower levels are important to population forecasting, since they represent future reproducers. Finally, such shapes illustrate that people in developing regions have fewer health advantages and are more likely to die sooner, resulting in a significant decrease in the upper third of these regions' age brackets.

FIGURE 49.6
AGE STRUCTURE HISTOGRAMS
Age structure histograms break the population down into five-year age groups, revealing much about past history and permitting predictions of future trends to be made. *Left,* expanding nation, e.g., Mexico. *Middle,* moderately stable nation, e.g., United States. *Right,* long-term stable nation, e.g., Sweden. Source: Population Reference Bureau.

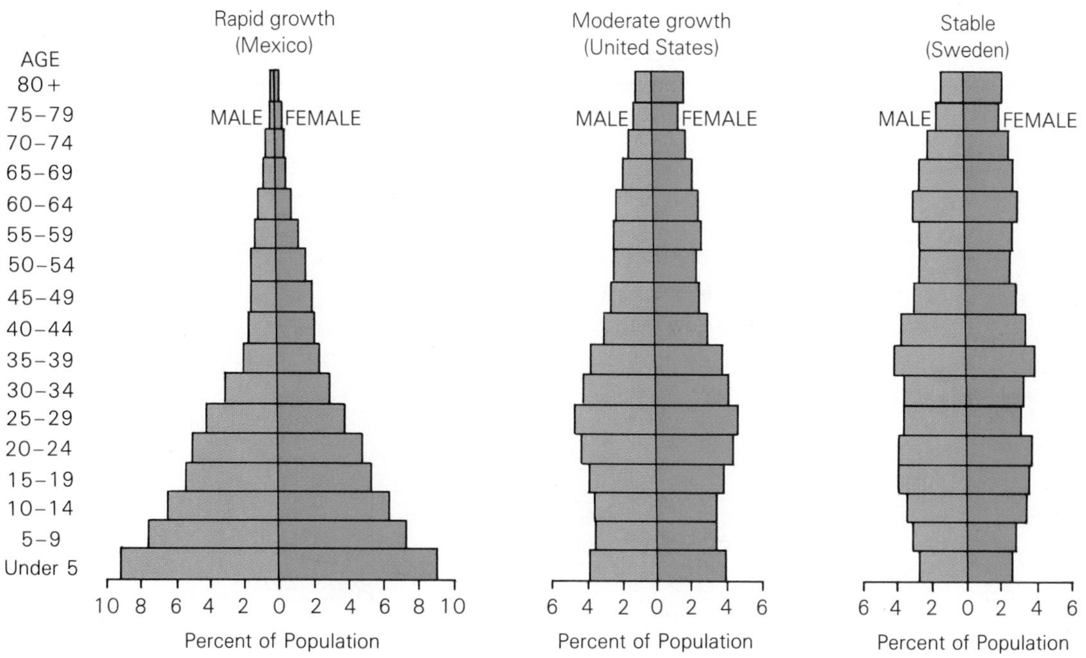

Growth Predictions and the Earth's Carrying Capacity

The fundamental question of how large the human population can become is irrevocably tied to what the earth can support—its human carrying capacity. If we have learned anything from population studies of other species, it is that this capacity cannot be exceeded for long without severe risk—especially, the risk to the environment (see Essays 49.1 and 49.2). Any such damage would lower the environment's carrying capacity and set the stage for a devastating population crash.

The range of estimates of the earth's human carrying capacity is enormous. In other words, the experts cannot agree. Some population biologists believe that we have already exceeded our limits and that our present population represents a drastic overshoot. At the other end of the spectrum are the optimists who believe the human population can increase to 50 billion and still survive easily. (Biologists don't take this latter estimate very seriously.)

Recent estimates by more moderate population experts suggest that the human population could be sustained *temporarily* at 8 to 15 billion. From there, they suggest, our numbers could gradually decrease to new, more stable levels. If we fail to restrain ourselves when we reach the higher numbers, we can expect not a gradual decrease in numbers but a massive increase in our death rate—a dieback or crash. It has been calculated that such a crash might kill 50–80% of the human population. Some population biologists believe that this is probably the way our population will stabilize. They suggest that the dieback will likely be due to a combination of famine, war, disease, and ecological disruption. With the exception of war (which is rare among other species), these are common density-dependent controls.

To leave the subject on a note of cautious optimism, we reiterate that world population growth is slowing down somewhat and that some of the more heavily populated nations have recently joined this trend. Further, it is within the power of the world's family of nations to see that the decreasing trend is continued and accentuated. Our goal, if we are to avoid chaos, is to reach a *replacement level.* Essentially, it means that each couple simply replaces itself, by having only two children. (The popular belief that the United States has achieved this goal is inaccurate.) Because of the burgeoning numbers of young people, the future size of the world's population depends largely upon when that replacement level is reached.

Scenarios portraying a stable world population are admittedly utopian, but they must not be regarded as impossible. Some population experts maintain that many slow-growing developed nations are now reaching the desired level and that others can do so within a few decades. The trend toward family planning and use of birth control is increasing, and one hopes that efforts to stimulate such interest in developing nations will continue. Under the best of circumstances, the world could reach some state of population stability in about 50 years (Figure 49.7).

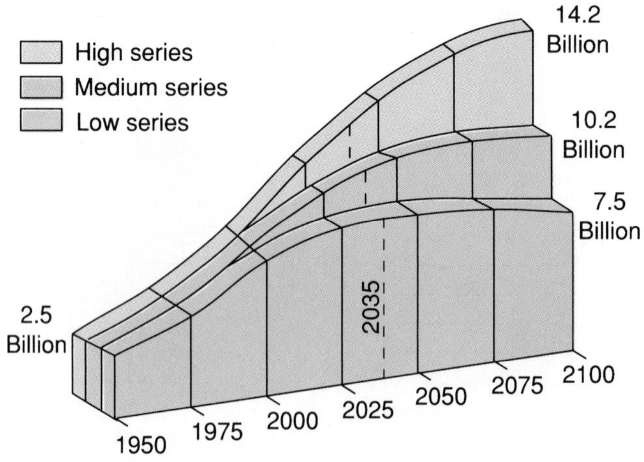

FIG. 49.7
WORLD POPULATION GROWTH, 1950–2100: THREE SCENARIOS
Three projections of world population based on different assumptions. The middle series assumes that, by 2035, each woman will have about two children. Even then, the population will double today's level. If "two-child families" are achieved by 2010, the lower curve results; the higher figure results if the "two-child family" is delayed until 2065. Source: Population Reference Bureau Data Sheet.

ESSAY 49.1

WHAT HAVE THEY DONE TO THE RAIN?

In the 1970s it became clear that the rain was changing. In fact, in some areas the gentle raindrops were downright dangerous. The rain was becoming a dilute mixture of acids. It was first noticed in Scandinavia, then in the northeast United States and southeast Canada, then in Northern Europe and Japan.

Rainwater, of course, had always been slightly acidic because the water dissolved atmospheric carbon dioxide, forming carbonic acids. But now the rain was showing alarming concentrations of the more dangerous sulfuric acid and nitric acid. Where were they coming from? They were the result of accumulations of nitrous oxides and sulfuric oxides in the atmosphere. The nitrous oxides, it turned out, were from power plant and automobile emissions; the sulfuric oxides, mainly from power plants and smelters. Dissolved in the water of cloud formations, they formed nitric acid and sulfuric acid, then fell to earth to bathe our forests and cities and to fill our lakes with the corrosive mix.

The relative proportions of the two acids in rain depends on where one lives. In the northeastern United States the acidity is primarily due to sulfuric acid; in California, to nitric acid. So we do have a choice.

The rain has caused the reduction and even the elimination of fish in many of our lakes. The rain apparently doesn't kill the fish, it just keeps them from reproducing. So no young fish are found as the old ones gradually go the way of all flesh. In fact, about 700 lakes in southern Norway are now *entirely devoid* of fish, and our own northeastern lakes are following one by one. As our Adirondack lakes reach pH levels of 5 (not uncommon), 90 percent have no fish whatever. They are also curiously devoid of frogs and salamanders.

Entire patches of forests world-wide are sickening and dying as ecologists busily try to find out just what effects the rain is having. In fact, such studies have masked inaction by the polluting countries. The Reagan administration (undoubtedly under heavy attack by industrial lobbyists) refused steadfastly for years to take action. Instead, it initiated one "study" after another, finally admitting in 1985 that there was a problem and that it had to do with industrial pollution.

Interestingly, the solution is clear to everyone. We simply need to reduce the levels of our effluent from power plants, smelters, and automobiles. Most of the technology exists, but its implementation would be too expensive for the polluters to willingly bear. Are we willing to pay higher prices for manufactured goods to save our lakes and rivers? The question is a fundamental one and is asked over and over in today's technological world.

Statue eroding in Venice

The effect of acid rain on the Black Forest

ESSAY 49.2

HOLES IN THE SKY

Now let's explore the relationship between underarm deodorants and the death of the oceans. The propellant in underarm sprays is, in many cases, a class of molecules called chlorofluorocarbons (CFC's). These are essentially carbon molecules to which are attached chorine and/or fluorine atoms. Chlorofluorocarbons are used in a variety of manufactured products, such as air conditioning, refrigeration, insulating foams (the type that keep our hamburgers warm in fastfood places), plastics, and industrial solvents.

The problem is, these molecules are very stable. So after you spray under your arms, or after the insulated fastfood box begins to disintegrate, these long-lived little molecules are released into the air. Because they're light, they eventually, perhaps a few years later, end up in the upper atmosphere.

Paradoxically, these molecules would be safer for life if they stayed closer to earth mingling with living things. The truth is, though, they threaten life precisely because they drift upward away from it. The reason is because at an altitude of 15 miles, the CFCs break down the ozone layer. Ozone is O_3, formed by the sun breaking down atmospheric O_2 molecules, allowing them to rejoin as ozone. The chlorine in the CFCs attacks the ozone, breaking it back down into its components. There isn't much ozone up there to begin with. At sea level, all of it together would form a layer over the earth about as deep as a pencil lead is thick.

The ozone, though, is critical to life on earth. Primarily, it functions by blocking destructive ultraviolet light from the sun. Those rays are destructive on three primary bases. First, they increase the risk of skin cancer, particularly among light-skinned people, such as those who invented chlorofluorocarbons. Second, they depress the immune systems of humans, setting the stage for a host of illnesses. Third, they destroy the algae that form the first step in the ocean's food chains.

Ozone depletion was first discovered over the Antarctic. In fact, two thirds of the springtime ozone over the Antarctic is now missing since the British began the measurements some years ago. In the northern latitudes where most people live, the ozone levels have declined by several percent since 1969. Now, an ozone hole is reported developing over the Arctic.

The manufacturers of CFCs have been reluctant to take action to reduce the levels of these chemicals over the earth. In fact, the DuPont Company, which makes about $600 million a year on CFCs, took out ads in newspapers saying that the danger to the ozone layer was improved. The company has now agreed to phase out the manufacture of CFCs, but the phasing out will not be complete until the year 2000. Since the United States makes only 30 percent of the CFC's, the effort will have to be global, demanding more cooperation that one usually finds among industrial nations.

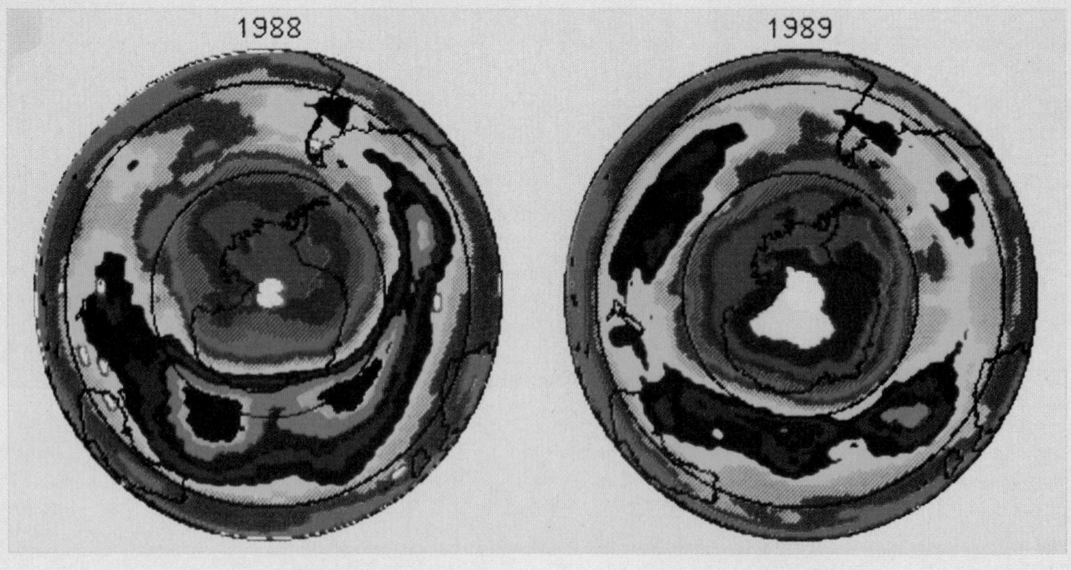

The bottom line is that we must assume our species is not special; it is not exempt from the natural laws that govern population control in other species. Our only special feature is our mental capacity. We have the ability to analyze, predict, imagine, and finally to *choose*. Of course, we can choose by deciding not to choose, not to take a stand, not to be involved. But the time for that luxury is past. We must now learn as much as possible about the nature of overpopulation, apply ourselves to solving the problem, and be ready to stand accountable for our actions when we are judged by future generations.

APPLICATION OF IDEAS

1. Consider what we know about population control in other species, and apply these factors to the future of the human species if voluntary population control measures fail. Considering such control, in what forms might we expect population reduction?
2. Why are the three historical human population growth surges usually understood in terms of decreased death rates?

KEY IDEAS

THE HISTORY OF THE HUMAN POPULATION

1. Human population numbers have remained relatively stable throughout much of our evolutionary history but recently have begun to follow the J-shaped curve.
2. Early hominids lived by hunting and gathering, their numbers regulated by the usual biotic and abiotic factors. Infant mortality was high, but replacement was rapid.

The First Population Surge

In the past 50,000 to 100,000 years, the human population has experienced three significant growth surges. The first growth surge probably occurred with the development of increasingly efficient tools with which people could more effectively modify and exploit the environment.

The Second Population Surge

Around 10,000 years ago, the second growth surge coincided with the development of agriculture and the domestication of animals; humans probably began to experience a lower death rate and possibly a higher birth rate. This growth was interrupted numerous times by a variety of factors such as disease, famine, and war.

The Third Population Surge

In the mid-seventeenth century, the third population surge began, possibly because of expansion into the New World, the introduction of sanitation programs and sophistication of medical practices, advances in the preservation of food, and political stability in some areas. Between 1650 and 1850, the world population doubled (from 500 million to 1 billion). There were 2 billion people by 1930, and 4 billion by 1975; in 1970, some predicted that there would be 8 billion people on earth by the year 2000.

THE HUMAN POPULATION TODAY

The rate of natural increase of the world's population slowed in the late 1970s and is continuing to decelerate.

Growth in the Developing Regions

The doubling time of the human population has increased, primarily in developed nations that could theoretically support increased populations. Developing nations, however, show few signs of controlling their populations. Some developing nations, such as China, have instituted major birth-control propaganda programs to stem their high rates of natural population increase.

Demographic Transition

A country's developmental progress and its population structure are related in an interesting way; as development increases, population growth tends to decrease. Several developmental phases occur, according to the **theory of demographic transition,** with the earliest characterized by high birth and death rates, the next by high birth rates and lower death rates, and the next by low birth and death rates, which reflect a fluctuating equilibrium state. This theory is the basis for the prediction that population growth in developing parts of Asia, Africa, and Latin America will slow as those areas become further industrialized. Programs based on the theory must be approached cautiously, however.

THE FUTURE OF THE HUMAN POPULATION

1. Whereas American couples of the 1950s favored a large family size, attitudes changed abruptly in the 1960s and 1970s, contributing to a sharp decline in population growth.

2. An important population growth indicator is the fertility rate. The **general fertility rate** (GFR) is the number of live births per 1000 women in their reproductive years per year, while the **total fertility rate** (TFR) is a prediction of the average number of children women will bear.
3. Fertility rates in the U.S.A. and Europe dipped during the early depression years but later rose, peaking in the post-WWII "baby boom" years. The increase slowed again by the early 1970s.
4. The recent decline in fertility rates is attributed partly to changes in the attitude of women towards family size and their own roles in society. Other factors include improvements in birth control methods, their increased availability, and increases in abortion.
5. Although declining growth rates have also occurred in developing nations, their effects will not be felt for years because of the large numbers of individuals yet to begin reproducing.

Population Structure

Age structure histograms are useful devices for portraying the age structure of a population. Such information is important in understanding growth patterns and in making predictions.

Growth Predictions and the Earth's Carrying Capacity

The basic issue in human population studies is the carrying capacity of the earth. Estimates vary greatly, and some population biologists believe that humans have already exceeded the carrying capacity and are in a phase of overshoot. A population crash could result in stabilization—or the population could gradually decrease, possibly through zero population growth programs, through which we could reach our replacement level.

REVIEW QUESTIONS

1. Describe the innovations that brought about the first two great population surges in human history. (1216–1218)
2. List the three major causes of death during the Middle Ages. Were these density-dependent or density-independent? Explain. (1218)
3. List three innovations that may explain the most recent human population growth surge, and explain how each may have affected the birth rate and the death rate. (1219)
4. What is the status of human population growth today? Specifically, where is growth most rapid? What are some of the possible political and social ramifications of this? (1219–1221)
5. Briefly predict the future of human population growth according to the theory of demographic transition. Upon what specific observations is this theory based? What are some of the theory's pitfalls? (1221–1224)
6. With simple drawings, depict the population age structure for Mexico, the United States, and Sweden. What future conditions can we predict according to the base of the diagram? According to the top? (1223)
7. What are experts telling us about the earth's carrying capacity? Should human numbers be maintained at our current level? Explain. (1224)

APPENDIX A

SMALL UNITS OF LINEAR MEASUREMENT IN THE METRIC SYSTEM

The United States alone continues to use the peculiar English system of weights and measures in commerce and everyday life, but even in America the times are changing, and there are constant reminders of the switch to metric. Most beverages are sold by the liter these days; track meets now regularly feature the 100-meter dash rather than the 100-yard dash; and the 1500-meter "metric mile" is displacing the traditional mile. (The mile will probably hang on as a special event because of its sentimental value.)

The sciences, including biology, have always used the metric system. Even if you did a good job of memorizing the metric system in grade school, you may well come across some unfamiliar units of measurement in your study of cell structure. You will not be alone. Trained biologists have been having the same problem in recent years because the international metric system itself has been changing. It is becoming even more regular, by decisions of international councils that have authority over such matters. The unit names in the metric system usually consist of a root word and a standard prefix. The root word gives the basic unit for a type of measurement: meter for linear measurement, liter for volume, gram for weight, and so on. The standard prefixes are added to these basic word roots to form larger or smaller secondary units. Each prefix indicates an increase or decrease by a specific positive or negative power of ten. For instance, *micro* means one one-millionth; thus, a microliter is one one-millionth of a liter (10^{-6} liter), a microgram is one one-millionth of a gram (10^{-6} gram), a micrometer is one one-millionth of a meter (10^{-6} meter), and a micromole is one one-millionth of a mole (10^{-6} mole). Another regularity being imposed on the metric system is that each named unit of measurement is one one-thousandth (10^{-3}) of the next larger named unit. The table shows how this works out in practice for units of linear measurement. Note that the familiar centimeter, one one-hundredth of a meter, violates this regularity.

The basic unit of linear measurement is the meter. Other lengths relate directly to it, such as the kilometer (about 0.6 mile) and the centimeter (about 0.39 inch).

UNITS OF LINEAR MEASURE

UNIT NAME	SYMBOL	PORTION OF A METER	EQUIVALENT
kilometer	km	10^3m	1000 m
meter	m	1 m	0.001 km
centimeter	cm	10^{-2} m	0.01 m
millimeter	mm	10^{-3} m	0.001 m
micrometer	μm	10^{-6} m	0.001 mm
nanometer	nm	10^{-9} m	0.001 μm

In addition to the above units, which are now official, three other metric units of linear measurement have been phased out. These units are still common in scientific books and articles, so we'll add them to our table:

UNIT NAME	SYMBOL	PORTION OF A METER	EQUIVALENT
micron	μ	10^{-6} m	1 μm
millimicron	mμ	10^{-9} m	1 nm
angstrom	Å	10^{-10} m	0.1 nm

The change from the name "micron" to "micrometer" means the symbol has changed from μ to μm. The change from "millimicron" (mμ) to "nanometer" (nm) also means a new name and a new symbol for 10^{-9} meter. The angstrom (Å) has been dropped altogether, to be replaced by 0.1 nanometer.

CLASSIFICATION OF ORGANISMS

The classification scheme was drawn from various commonly used sources in microbiology, botany, and zoology, but has been modified to accommodate the most recent ideas. Most taxa mentioned are also described in Chapters 22-25 and 30-32. Notably absent from this scheme are the viruses. Their taxonomic relationships have never been clarified, although some authors include them with the Monera while others have invented "Kingdom Virus."

THE PROKARYOTES

KINGDOM MONERA (also PROKARYOTA): Includes the prokaryotes or bacteria, single-celled and colonial organisms that generally lack membrane-bounded organelles, including an organized nucleus; and whose DNA is organized into a single, circular, main chromosome, sometimes supplemented by minute circular plasmids. Cell division is through fission, and sexual recombination is limited. Where present, the flagellum is tubular and rotating. Spore formation is common, and absorption is the usual mode of feeding by heterotrophs, including the parasites. Autotrophic bacteria include both chemoautotrophs and photoautotrophs. Classification in this kingdom is unsettled and undergoing intensive revision. The newer taxonomies replace Kingdom Monera with two new kingdoms, Archaebacteria and Eubacteria, basing the new distinction in part on ribosomal RNA and cell wall and plasma membrane chemistry.

THE EUKARYOTES

All other kingdoms are eukaryotic. All have cells containing membrane-bounded organelles, their DNA heavily complexed with histone and other proteins, forming chromatin. Mitotic cell division is common, and the eukaryotic flagellum and cilium, where occurring, follows the 9 + 2 microtubular organization. Many undergo meiosis and exchange gametes in sexual reproduction.

KINGDOM PROTISTA: A polyphyletic kingdom with many unresolved taxonomic problems. Includes the protozoans, the fungus-like protists, and the algae. Protists include unicellular, colonial, and multicellular levels of organization. Nutrition is both heterotrophic and photoautotrophic. Sexual reproduction commonly occurs through meiosis and conjugation or the union of gametes, although it has not been observed in many species.

Protozoan Protists

PHYLUM SARCOMASTIGOPHORA: Subphylum Mastiqophora: Flagellated, unicellular heterotrophs that feed by phagocytosis and absorption. Includes parasites such as *Trypanosoma gambiense,* the agent of African sleeping sickness. **Subphylum Sarcodina:** Unicellar heterotrophs with ameboid movement that feed by phagocytosis and absorption. Includes marine radiolarians and foraminiferans and *Entamoeba histolytica,* the parasite of amebic dysentery.

PHYLUM APICOMPLEXA: Unicellar, nonmotile, spore-forming heterotrophs that feed by absorption. Mainly parasitic, including *Plasmodium vivax,* an agent of human malaria.

PHYLUM CILIOPHORA: Ciliates. Unicellular, ciliated heterotrophs that feed by phagocytosis and absorption. Often large, with complex organelles, and sexual reproduction by meiosis and conjugation. Includes *Paramecium.*

Fungus-like Protists

PHYLUM ACRASIOMYCOTA: Cellular slime molds. Individual myxameba that fuse to form a pseudoplasmodium and later compound sporangia. Feed by phagocytosis. *Dictyostelium discoideum.*

PHYLUM MYXOMYCOTA: Acellular slime molds. Unicellular individuals fuse to form a true plasmodium and sporangia. Feed by phagocytosis. Some reproduce sexually. Includes *Physarum polycephalum.*

PHYLUM OOMYCOTA: Water molds. Filamentous with cellulose cell walls. Flagellated stages. Some have large eggs and small nonmotile sperm. Many parasites *(Phytophthora infestans).*

Algal Protists

PHYLUM PYRROPHYTA: Dinoflagellates. Flagellated photoautotrophs with chitinous cell walls and a primitive fission and one-step meiosis. Chlorophylls *a* and *c* and carotenoids, starch storage. Red tide organisms include *Gonyaulax* and *Gymnodinium.*

PHYLUM EUGLENOPHYTA: Euglenoids. Unicellular, flagellated photoautotrophs and heterotrophs with red "eye spot." Chlorophylls *a* and *b* and carotenoids, paramylon starch storage. Includes *Euglena gracilis* and *Phacus.*

PHYLUM CHRYSOPHYTA: Yellow-green and golden-brown algae and diatoms. Unicellular and colonial photoautotrophs with pectin and glassy cell walls. Chlorophylls *a* and *c* and carotenoids and leucosin storage.

PHYLUM RHODOPHYTA: Red algae, seaweeds. Multicellular photoautotrophs with phycocyanin and phycoerythrin, chlorophyll *a,* and carotenoid pigments; floridean or carageenan storage. Sporic cycle. Separate sporophyte and gametophyte with varying dominance. *Polysiphonia, Porphyra.*

PHYLUM PHAEOPHYTA: Brown algae, seaweeds, and kelps. Multicellular phototautotrophs with fucoxanthin, chlorophyll *a,* and carotenoid pigments; laminarin and mannitol storage. Many with sporic cycle. Separate sporophytes and gametophytes in some. Considerable tissue specialization. *Macrocystis, Nereocystis,* and *Sargassum.*

PHYLUM CHLOROPHYTA: Green algae. Unicellular, colonial, and multicellular photoautotrophs, with chlorophylls *a* and *b* and carotenoids, starch storage. *Chlamydomonas, Ulva, Volvox.*

KINGDOM FUNGI: Multicellular heterotrophs with saprobic feeding. Many are parasites. Fungi usually have simple septate or nonseptate vegetative mycelia, but many form complex, specialized, multicellular reproductive structures. The mycelia have chitinous walls. Fungi are primarily zygotic, that is, haploid with brief diploid interludes. Asexual reproduction is by mitotic spore formation, and sexual reproduction is often by conjugation, followed by a lengthy dikaryotic state.

PHYLUM ZYGOMYCOTA: Bread molds. Extensive, simple, nonseptate mycelium, sexual reproduction through conjugation, and zygospore formation. *Rhizopus, Neurospora.*

PHYLUM ASCOMYCOTA: Sac fungi. Extensive, septate mycelium, asexual reproduction by conidiospore, sexual reproduction through conjugation, long dikaryotic state, and fertilization followed by meiosis and the development of ascospores in saclike asci. *Saccharomyces cerevisiae* (baker's yeast), *Claviceps purpurea, Piziza.*

PHYLUM BASIDIOMYCOTA: Club fungi. Extensive septate mycelium, often with a large, raised, fruiting body, the basidiocarp. In sexual reproduction conjugation leads to a long dikaryotic state. Fertilization occurs in the clublike basidia, followed by meiosis and the production of basidiospores. Mushrooms, shelf and bracket fungi, *Puccinia graminis* (wheat rust).

PHYLUM DEUTEROMYCOTA: The Fungi Imperfecti. Those fungi whose sexual reproduction has not been observed and have yet to be classified among the fungi. *Penicillium rocquefortii.*

KINGDOM PLANTAE: A monophyletic kingdom containing multicellular photoautotrophs, most with protected embryos and great tissue specialization. Pigments include chlorophylls *a* and *b* and carotenoids, and the chief storage polysaccharide is starch. Sporic life cycle with the sporophyte clearly dominant in nearly all divisions. (Note: In plant taxonomy, division replaces phylum.)

Nonvascular Plants
(Plants without vascular tissue)

DIVISION BRYOPHYTA: Mosses, liverworts, hornworts. Nonvascular plants with simple tissue organization, little true (vascular) roots, stems, or leaves. Dominant gametophyte, motile sperm, fertilization requires water. Bryophytes include three major classes: Musci (mosses), Hepaticae (liverworts), and Anthocerotae (hornworts).

Vascular Plants (Seedless)
(Vascular tissue, highly dominant sporophyte, embryo develops within separate gametophyte)

DIVISION PSILOPHYTA: Whisk ferns. Primitive group of four species, vascular stem but nonvascular scalelike leaves and rhizoids. Motile sperm; fertilization requires water. *Psilotum nudum.*

DIVISION LYCOPHYTA: Club mosses. Widely distributed; vascular roots, stems, leaves. Motile sperm; fertilization requires water. Homosporous and heterosporous species. *Lycopodium, Selaginella.*

DIVISION SPHENOPHYTA: Horsetails. Limited distribution; leaves highly reduced, slender stems bearing sporangia at tips. Motile sperm; fertilization requires water. *Equisetum.*

DIVISION PTEROPHYTA: Ferns. Widely distributed; foliage often treelike with complex leaves emerging from rhizomes. Motile sperm; fertilization requires water. Tree ferns, bracken ferns, sword ferns, and staghorn ferns.

Vascular Plants with Seeds: Gymnosperms
(Cones rather than flowers, seeds naked—not enclosed in fleshy fruits)

DIVISION GINKGOPHYTA: Maidenhair tree. One species, a large tree with bilobed, fan-shaped leaves. Motile sperm; plant provides fluids for sperm to swim to egg. *Ginkgo biloba.*

DIVISION CYCADOPHYTA: Cycads. Widespread, tropical; palmlike leaves, very prominent cones. Swimming sperm; plant provides fluids for sperm to swim to egg. *Zamia, Cycas.*

DIVISION GNETOPHYTA: Widespread in deserts and warm temperate regions. Advanced, angiosperm characteristics in some: xylem with vessels, phloem with sieve tube members, bladelike leaf, and flowerlike pollen cones. *Gnetum, Welwitschia, Ephedra.*

DIVISION CONIFEROPHYTA: Conifers. Widespread in temperate, subarctic regions and high altitudes; well adapted to dryness. Commonly with needlelike leaves. Nonmotile sperm; no fluids required for fertilization. Pines, spruce hemlock, cedar, larch, fir, juniper, redwood.

Vascular Plants with Seeds: Angiosperms
(Seed plants with flowers, fleshy fruits, and nonmotile sperm)

DIVISION ANTHOPHYTA: Flowering plants. Seeds surrounded by fruit.

> CLASS DICOTYLEDONAE: Dicots. Worldwide distribution; floral parts in fours and fives or their multiples, two cotyledons in embryo, net-veined leaves. Perennials with true secondary growth. Magnolia, oak, willow, maple, apple, rose, cucumber, bean, and pea.
>
> CLASS MONOCOTYLEDONAE: Monocots. Worldwide distribution; floral parts in threes or its multiples, one cotyledon in embryo, parallel-veined leaves. Many soft-bodied, herbaceous. No true secondary growth. Grasses, lily, palm, orchids.

KINGDOM ANIMALIA: A polyphyletic kingdom made up of motile, multicellular heterotrophs, generally with tissue specialization and organ and organ-system organization. Feeding commonly by engulfing and extracellular digestion, but phagocytosis and intracellular digestion is also common. Life cycle is gametic: sexual reproduction by the fusion of haploid gametes that are produced directly through meiosis.

SUBKINGDOM PARAZOA
(Primarily cell-level oranization)

PHYLUM PORIFERA: Sponges. Nonmotile, filter-feeding, solitary or colonial adults with cellular level of organization and few cell types. Asexual reproduction by budding and sexual reproduction by internal fertilization. Swimming planula larva.

> CLASS CALCARIA: Calcium carbonate skeleton. *Leucosolenia.*
>
> CLASS HEXACTINELLIDA: Silicon dioxide skeleton. *Euplectella* (Venus flower basket).
>
> CLASS DEMOSPONGIAE: Skeleton of silicon dioxide or the protein spongin. *Cliona,* bath sponges.

SUBKINGDOM METAZOA
(tissue, organ, and organ-system levels of organization)

Radiate Phyla
(Diploblastic animals with radial symmetry)

PHYLUM CNIDARIA: Radial animals with thin-walled bodies, tentacles armed with stinging cells. Saclike gastrovascular cavity, extracellular and intracellular digestion. Individuals exist as polyps, medusae, or alternating cycles of the two.

CLASS HYDROZOA: Solitary hydroids or those with colonial feeding polyps that bud reproductive medusae in which fertilization occurs. *Obelia, Hydra.*

CLASS SCYPHOZOA: Includes the jellyfish, a medusa in which feeding and sexual reproduction occur. Swimming larva forms a polyp stage that buds off young jellyfish. *Aurelia.*

CLASS ANTHOZOA: Anemones and corals, polyps that feed and reproduce. Development is usually direct from zygote to polyp. *Metridium*, sea fan coral, brain coral.

PHYLUM CTENOPHORA: Radial, marine animals with thin-walled bodies, tentacles armed with glue cells. Saclike bodies like cnidarians, but with eight rows of combs consisting of fused cilia. Comb jelly, Venus girdle.

Bilateral, Protostome, Acoelomate Phyla
(Bilateral symmetry, triploblastic, spiral cleavage, blastopore area gives rise to mouth, solid body lacking a coelom)

PHYLUM PLATYHELMINTHES: Flatworms. Dorsoventrally flattened, solid-bodied; marine, freshwater, and terrestrial worms, most with a highly branched gastrovascular cavity and some organ-system development.

CLASS TURBELLARIA: Free-living planarians, with extensive organ-system development, hermaphroditic. *Dugesia.*

CLASS TREMATODA: Parasitic flukes with oral suckers, often with two or more hosts. *Clonorchis sinensis* (Asian liver fluke), *Schistosoma* (blood flukes).

CLASS CESTODA: Parasitic tapeworms wtih small scolex, numerous proglottids in the mature adults, often with two hosts. *Taenia solium* (pork tapeworm), *Taeniarhynchus* (beef tapeworm).

PHYLUM RHYNCHOCOELA: Proboscis worms. Thin, ribbonlike worms with a complete, one-way digestive tract. Protrusible pharynx with piercing stylet. Marine.

Bilateral, Protostome, Pseudocoelomate Phyla
(Body cavity a false coelom, incompletely lined by mesodermal tissue)

PHYLUM NEMATODA:
Roundworms. Slender, threadlike worms with a fluid-filled body cavity and a complete digestive tract. Terrestrial, free-living predators and internal parasites. Widespread in soil and water—outnumbered only by arthropods. *Rhabditis, Ascaris lumbricoides* (giant roundworm), *Necator americanus* (hookworm), *Enterobias vermicularis* (pinworm).

PHYLUM ROTIFERA: Rotifers or "wheel animals." Complex, free-living, minute, ciliated animals with a complete digestive tract and well-developed organ systems. Sexes separate with parthenogenesis common. *Philodina.*

Bilateral, Protostome, Coelomate Phyla
(Triploblastic animals with a true—mesodermally lined—body cavity)

PHYLA ECTOPROCTA, PHORONIDA, AND BRACHIOPODA: Lophophorate phyla. Ciliated tentacles attached to a ridgelike lophophore. Moss animals, lampshells, and wormlike *Phoronis.*

PHYLUM MOLLUSCA: Mollusks. Highly reduced coelom and segmentation, diversified through adaptive variations in foot, shell, and mantle. Open circulatory system. Chiefly marine and freshwater, some terrestrial.

CLASS POLYPLACOPHORA: Marine, shell in eight parts, lengthy foot surrounded by simple gills and mantle. *Mopalia* (a chiton).

CLASS GASTROPODA: Aquatic and terrestrial, large muscular foot, often retractable foot and spiral shell, feed by radula. *Helix* (land snail), *Haliotis* (an abalone), whelks, limpets, slugs.

CLASS BIVALVIA: Aquatic, hinged shells, digging retractable foot, filter feeders. *Mytilus* (a mussel), *Tagelus* (a clam), oyster, scallop.

CLASS CEPHALOPODA: Marine, enlarged head and brain, image-forming eyes, tentacles, internalized shell in most. *Octopus, Loligo* (a squid), and *Nautilus* (chambered nautilus).

PHYLUM ANNELIDA: Segmented worms. Most with body divided into numerous metameres (segments), well-developed coelom, complete and specialized digestive tract, and closed circulatory system.

CLASS OLIGOCHAETA: Terrestrial and aquatic, free-living and burrowing. Hermaphroditic. *Lumbricus terrestris* (earthworm).

CLASS POLYCHAETA: Marine, free-living burrowing, tube-dwelling and swimming forms. *Neries virens* (clam worm), fanworm, peacock worm.

CLASS HIRUDINEA: Mostly freshwater and terrestrial, parasitic, sucker mouth. *Hirudo* (medicinal leech).

PHYLUM ARTHROPODA: Largest animal phylum. Jointed-legged animals with chitinous exoskeleton, modified segmentation.

Subphylum Chelicerata (no jaws).

CLASSES MEROSTOMATA, ARACHNIDA, AND PYCNOGONIDA: Terrestrial and aquatic, four pairs of legs, fangs rather than mandibles, two body regions and book lungs common. Horseshoe crab, spiders, mites, ticks, scorpions, and sea spiders.

Subphylum Mandibulata (jaws).

CLASS CRUSTACEA: Mainly aquatic, two pairs of antennae, compound eye, hardened exoskeleton. Crabs, lobsters, copepods, barnacles, water fleas.

CLASS INSECTA: Terrestrial and aquatic, three body regions, one pair of antennae, compound eye, three pairs of walking legs, wings are common. Beetles, moths, bees, ants, flies, bugs, grasshoppers, lice, termites, fleas, dragonflies, and others. (For the major insect orders see Table 31.1)

CLASS CHILOPODA AND DIPLOPODA: Centipedes and millipedes.

PHYLUM ONYCHOPHORA: Intermediate between annelids and arthropods, terrestrial with jointed appendages, claws, wormlike body. *Peripatus.*

Bilateral, Deuterostome Phyla
(Radial Cleavage, blastopore area gives rise to anus)

PHYLUM ECHINODERMATA: Spiny-skinned animals. Marine, pentaradial as adults, endoskeleton of calcareous plates, commonly with spines and pedicellaria. All with a water vascular system.

CLASS CRINOIDEA: Cup-shaped body with branched arms, no suckers on tube feet. Sea lily.

CLASS HOLOTHUROIDEA: Cylindrical body, spines reduced, no pedicellaria. *Cucumaria* (a sea cucumber).

CLASS ECHINOIDEA: Spherical or round, flattened body without arms, prominent movable spines, tube feet with suckers. *Lytechinus* (a sea urchin), *Echinarachnius* (a sand dollar).

CLASS ASTEROIDEA: Prominent arms, short, blunt spines, numerous pedicellaria, tube feet with suckers. *Asterias* (a sea star).

CLASS OPHIUROIDEA: Prominent arms, movable spines, tube feet without suckers, no pedicellaria. *Ophiura* (a brittle star), *Gorgonocephalus* (a basket star).

PHYLUM HEMICHORDATA: Hemichordates. Marine, burrowing, acorn-shaped proboscis, gill slits. Larvae similar to echinoderm's. *Saccoglossus* (an acorn worm).

PHYLUM CHORDATA: Chordates. Aquatic and terrestrial, gill slits, notochord, dorsal hollow nerve cord, postanal tail.

Subphylum Urochordata: Marine filter feeders, saclike body with gill slits, bilateral larva with all chordate characteristics. *Ciona* (a tunicate), *Thetys* (a salp).

Subphylum Cephalochordata: Marine filter feeder, fishlike body form, notochord in adults, prominent gill slits and segmented trunk muscles, sensory cirri. *Branchiostoma* (lancelet).

SUBPHYLUM VERTEBRATA: Vertebral column of bone or cartilage, increasing cephalization, ventral heart and dorsal aorta, no more than two pairs of limbs, separate sexes.

CLASS AGNATHA: Fishes with jawless mouth, cartilaginous skeleton. Lamprey and hagfish.

CLASS CHONDRICHTHYES: Marine fishes with cartilaginous skeleton and jaws, placoid scales and rows of replacement teeth, dorsoventrally flattened body. Sharks, rays, chimera.

CLASS OSTEICHTHYES: Marine and freshwater fishes with bony skeleton and jaws. Includes Subclass Crossopterygii (lobe-finned fishes), Subclass Dipneusti (lungfishes), and Subclass Actinopterygii (ray-finned fishes), which include Superorder Chondrostei (sturgeons), Superorder Holostei (gar pike and bowfin), and Superorder Teleostei (perch, bass, tuna, eel, sunfish).

CLASS AMPHIBIA: Freshwater and terrestrial amphibians, chiefly air breathers, moist vascularized skin, three-chambered heart, most requiring water for reproduction. Includes Order Anura (frogs and toads), Order Urodela (salamanders), and Order Apoda (caecilians).

CLASS REPTILIA: Aquatic and terrestrial air breathers with dry, scaly skin and three-chambered hearts in most. Fertilization is internal, and development occurs in the independent, amniotic land egg. Includes Order Chelonia (turtles), Order Squamata (snakes and lizards), Order Crocodilia (crocodiles and alligators), and Order Rhynchocephalia (tuatara).

CLASS AVES: Birds, principally terrestrial and flying; body modifications adapting for flight include lightweight skeleton with fusion in the vertebrae, no teeth, covering of feathers, highly modified forelimbs, four-chambered heart, endothermy, and crosscurrent air-blood flow in lung. Hawks, penguins, peafowl, jays, bluebirds, ducks, turkey.

CLASS MAMMALIA: Mammals, terrestrial and aquatic air-breathers, body covering of hair, glandular skin, milk produced in mammaries; highly specialized, socketed teeth; skull bones highly fused, enlarged cerebrum, and increased learning capacity

Subclass Prototheria: Monotremes, egg-laying mammals. Spiny anteater and duckbilled platypus.

Subclass Metatheria: Marsupial mammals, pseudoplacenta, immature young housed and nourished in marsupium (pouch), nearly all restricted to Australia and New Zealand. Opossums, kangaroos, wallabies, koalas, wombats.

Subclass Eutheria: Placental mammals, true placenta supports embryo development.

Order Insectivora: Smallest mammals, primitive, pointed teeth, tapered snout. Shrew and mole.

Order Dermoptera: Fur-covered patagium, gliding flight. Flying lemur.

Order Chiroptera: Greatly extended fingers with flight membrane, true flight, sonar navigation. Bat.

Order Primates: Enlarged cerebrum, hands and feet with nailed digits, eyes forward, opposable thumb in some. Lemur, monkey, ape, human.

Order Endentata: Toothless or molars only, no enamel. Sloth, anteater.

Order Pholidota: Overlapping horny scales. Scaly anteater.

Order Lagomorpha: Rodentlike, but four upper incisors. Rabbit, hare.

Order Rodentia: Largest order, small, continuous growth in incisors. Squirrel, mouse, gopher, beaver.

Order Cetacea: Forelimbs modified to flippers, hindlimbs absent, lateral tail fluke, nostrils as blowhole, teeth without enamel. Porpoise, whale.

Order Carnivora: Enlarged canines, an adaptation for predation. Bear, cat family, dog family, raccoon, weasel.

Order Pinnipedia: Similar to carnivores, but feet modified for swimming. Seal, sea lion, walrus.

Order Tubulidentata: Piglike snout, no incisors or canines, teeth without enamel Aardvark.

Order Proboscidea: Largest terrestrial mammal, upper lip a prehensile trunk, incisors enlarged to tusks. Elephant.

Order Hyracoidea: small, similar to guinea pig. Hyrax.

Order Sirenia: Forelimbs are paddles, hindlimbs absent, lateral tail flukes, heavy, blunt muzzle. Manatee.

Order Perissodactyla: Hooved feet with odd numbers of toes (tapirs have even-numbered toes on forelimbs, odd on hindlimbs). Horse, ass, zebra, tapir, rhinoceros.

Order Artiodactyla: Hooved feet with even numbers of toes, many with antlers, some ruminants. Pig, hippopotamus, camel, deer, giraffe, cattle.

APPENDIX C

A BIOLOGICAL LEXICON

A list of Greek and Latin prefixes, suffixes, and word roots commonly used in biological terms, alphabetized by the most common combining form in English; with examples illustrating each usage.

a-, an- [Gk. *an-*, not, without, lacking]: anaerobic, abiotic, anorexia, anesthesia, aseptic, asexual

acro- [Gk. *akros,* highest]: acrophobia, acromegaly

ad- [L. *ad-*, toward, to]: adhesion, adrenal, adventitious, adsorption

allo- [Gk. *allos,* other]: allele, allopatric, allosteric, allotetraploid

amphi- [Gk. *amphi-*, two, both, both sides of]: amphibian, Amphineura, amphipod

ana- [Gk. *ana-*, up, up against]: anaphase, anatomy, anabolic

andro- [Gk. *andros,* an old man]: androecium, androgen, androgynous, polyandry

anti- [Gk. *anti-*, against, opposite, opposed to]: antibiotic, antibody, antigen, antidiuretic hormone, antihistamine, antiseptic, antipathy, antiparallel

apo- [Gk., *apo-*, different]: apomoxis, apodeme, aponeurosis

archeo- [Gk. *archaios,* beginning]: archegonium, archenteron

arthro- [Gk. *arthron,* a joint]: arthropod, arthritis

auto- [Gk. *auto-*, self, same]: autoimmune, autotroph, autosome

auxo- [Gk. *aux,* to grow or increase]: auxin, auxospore, auxotroph

bi-, bin- [L. *bis,* twice; *bini,* two-by-two]: binary fission, binocular vision, binomial

bio- [Gk. *bios,* life]: biology, biomass, biome, biosphere, biotic

blasto-, -blast [Gk. *blastos,* sprout; now "pertaining to the embryo"]: blastoderm, blastopore, blastula, trophoblast

brachi- [Gk. *brachion,* arm]: brachiation, brachiopod

brachy- [Gk. *brachys,* short]: brachydactyly, brachycardi

broncho- [Gk. *broncho,* windpipe]: bronchus, bronchiole, bronchitis

carcino- [Gk. *karkin,* a crab, cancer]: carcinogen, carcinoma

cardio- [Gk. *kardia,* heart]: cardiac, cardiology, myocardium, electrocardiogram

cephalo- [Gk. *kephale,* head]: cephalization, cephalochordate, encephalitis, encephalogram

chloro- [Gk. *chloros,* green]: chlorobacterium, chlorophyll, chloroplast, chlorine

chole- [Gk. *chole,* bile; Gk. cholecyst, gall bladder]: cholesterol, cholinesterase, cholecystokinin

chondro- [Gk. *chondros,* cartilage]: achondroplasia, chondroitin, chondroblast

chromo- [Gk. *chroma,* color]: chromosome, chromatophore, chromatin

coelo-, -coel [Gk. *koilos,* hollow, cavity]: coelenteron, coelenterate, coelom, pseudocoelom

com-, con-, col-, cor-, co- [L. *cum,* with, together]: commensal, conjugation, covalent

cranio- [Gk. *kranios,* L. *cranium,* skull]: cranial, craniotomy, cranium

cuti- [L. *cutis,* skin]: cutaneous, cuticle, cutin

cyclo-, -cycle [Gk. *kyklos,* circle, ring, cycle]: cyclosis, cyclostome, pericycle

cyto-, -cyte [Gk. *kytos,* vessel or container; now, "cell"]: cytoplasm, cytology, erythrocyte, cytosine

de- [L. *de-*, away, off; deprivation, removal, separation, negation]: deciduous, decomposer, deoxyribose

derm-, dermato- [Gk. *derma,* skin]: dermatitis, dermis, epidermis

di- [Gk. *dis,* twice]: dikaryon, dicotyledon, diencephalon, diatomic

dia- [Gk. through, passing through, thorough, thoroughly]: diabetes, dialysis, diapause, diaphragm

diplo- [Gk. *diplos,* two-fold]: diplotene, diploid, diploblastic

eco- [Gk. *oikos,* house, home]: ecology, androecium, ecosystem

ecto- [Gk. *ektos,* outside]: ectoderm, ectoplasm, ectoparasite

en- [Gk., L. in, into]: encephalon, encephalitiis, entropy, environment

endo- [Gk. *endon,* within]: endocrine, endoderm, endodermis, endoparasite, endoskeleton

entero- [Gk. *enteron,* intestine]: archenteron, enteric, entameba

epi- [Gk. *epi,* on, upon, over]: epiboly, epicotyl, epidermis, epiphyte, epithelium

eu- [Gk. *eus,* good; *eu,* well; now "true"]: eubacterium, eukaryote

ex-, exo-, ec-, e- [Gk., L. out, out of, from, beyond]: emission, ejaculation, exhale, exocytosis, exoskeleton, tonsillectomy

extra- [L. outside of, beyond]: extracellular, extraembryonic

gam-, gameto- [Gk. *gamos,* marriage; now usually in reference to gametes (sex cells)]: gamete, gametogenesis, isogamete, heterogamete

gastro- [Gk. *gaster,* stomach]: gastric, gastrula, gastrin, gastrovascular cavity

gen- [Gk. *gen,* born, produced by; Gk. *genos,* race, kind; L. *genus, generare,* to beget]: gene, polygenic, genotype, glycogen, florigen, pyrogen

gluco-, glyco- [Gk. *glykys,* sweet; now pertaining to sugar]: glucose, glycogen, glycerol, glycolipid

gyn-, gyno-, gyneco- [Gk. *gyne,* woman]: gynecology, gynoecium, epigyny, hypogyny

hemo-, hemato-, -hemia, -emia [Gk. *haima,* blood]: hematology, hemoglobin, hemorrhoid, hemotoxin, leukemia, toxemia

hepato- [Gk. *hepar, hepat-*, liver]: hepatitis, hepatic

hetero- [Gk. *heteros,* other, different]: heterosexual, heteromorphic, heterozygote

histo- [Gk. *histos,* web of a loom, tissue; now pertaining to biological tissues]: histology, histocompatibility, histamine

homo-, homeo- [Gk. *homos,* same; Gk. *homios,* similar]: homeostasis, homeothermy, homosexual, homology

hydro- [Gk. *hydor,* water; now, confusingly, pertaining either to water or to hydrogen]: dehydration, hydraulic, carbohydrate, hydrophobia

hyper- [Gk. *hyper,* over, above, more than]: hyperacidity, hyperglycemia, hypertension, hypertonic

hypo- [Gk. *hypo,* under, below, beneath, less than]: hypocotyl, hypodermic, hypoglycemia, hypothalamus, hypotonic

inter- [L. *inter,* between, among, together, during]: interbreed, intercellular, interphase, interstitial

intra-, intro- [L. *intra,* within]: intracellular, intravenous, introitus

-itis [L., Gk. *-itis,* inflammation of]: arthritis, bronchitis, dermatitis

leuko-, leuco- [Gk. *leukos,* white]: leukocyte, leukemia, leukoplast

-logue, -logy [Gk. *-logos,* word, language, type of speech]: analogy, homology, dialogue

-logy [Gk. *-logia,* the study of, from *logos,* word]: biology, cytology, embryology

-lysis, lys-, lyso-, -lyze, -lyte [Gk. *lysis,* a loosening, dissolution]: lyse, lysis, lysogeny, hydrolysis, catalysis, catalytic, electrolyte

macro- [Gk. *makro,* now "great," "large"]: macromere, macromolecule, macronucleus

mega-, megalo-, -megaly [Gk. *megas,* large, great, powerful]: megagamete, megalocephalic, megaspore, acromegaly

-mere, -mer, mero- [Gk. *meros,* part]: blastomere, centromere, dimer, polymer, meroblastic, merozoite

meso-, mes- [Gk. *mesos,* middle, in the middle]: mesencephalon, mesentery, mesoderm, mesophyll

meta-, met- [Gk. *meta,* after, beyond; now often denoting change]: metabolism, metacarpal, metamorphosis, metastasis

micro- [Gk. *mikros,* small]: microbe, microbiology, microgametophyte, micropyle

myo- [Gk. *mys,* mouse, muscle]: myasthenia gravis, myocardial infarction, myosin

neuro- [Gk. *neuron,* nerve, sinew, tendon]: neurofibril, neuron, aponeurosis

oligo- [Gk. *oligos,* few, little]: Oligocene, oligochaete, oligotrophic

-oma [Gk. *oma,* tumor, swelling]: carcinoma, glaucoma, hematoma

oo- [Gk. *oion,* egg]: oogenesis, oogonium, oospore

-osis [Gk. *-osis,* a state of being, condition]: arteriosclerosis, cirrhosis, halitosis

osteo-, oss- [Gk. *osteon,* bone; L. *os, ossa,* bone]: ossification, Osteichthyes, teleost

para- [Gk. *para-,* alongside of, beside, beyond]: paramedic, parapatric, paraphyletic, parathyroid

patho-, -pathy, -path [Gk. *pathos,* suffering; now often disease or the treatment of disease]: pathogen, osteopath, homeopathy

peri- [Gk. *peri,* around]: pericarp, pericycle, perinium, periosteum, photoperiod, peristome, peritoneum

phago-, -phage [Gk. *phagein,* to eat]: phagocyte, phage, bacteriophage

plasm-, -plasm, -plast, -plasty [Gk. *plasm,* something molded or formed; Gk. *plassein,* to form or mold]: plasma, plasmid, thromboplastin, cytoplasm, plastid, chloroplast, dermoplasty

-pod [Gk. *pod,* foot]: amphipod, Apoda, arthropod, cephalopod

poly- [Gk. *poly-, polys,* many]: ployandry, polychaete, polydactyly

-rrhea [Gk. *rhoia,* flow]: amenorrea, diarrhea, gonorrhea

septi-, -sepsis, -septic [Gk. *septicos,* rotten, infected]: septic, septicemia, antiseptic

-some, somat- [Gk. *soma,* body; Gk. *somat-,* of the body]: somatic cell, psychosomatic, centrosome, acrosome

-stat, -stasis, stato- [Gk. *stasis,* stand]: statocyst, metastasis, hydrostatic

stoma-, stomato-, -stome [Gk. *stoma,* mouth, opening]: stoma, stomatoplasty, stomodeum, deuterostome

sym-, syn- [Gk. *syn,* with, together]: symbiont, symbiosis, synapsis, synaptinemal complex

taxo-, -taxis [Gk. *taxis,* to arrange, put in order; now often referring to ordered movement]: taxonomy, taxon, chemotaxis, phototaxis

tomo-, -tome, -tomy [Gk. *tome,* a cutting; Gk. *tomos,* slice]: tomography, microtome, anatomy

trich-, tricho-, -trich [Gk. *tricho-* combining form of *thrix,* hair]: trichina, trichocyst, gastrotrich

trop-, tropo-, -tropy, -tropism [Gk. *tropos,* to turn, to turn toward]: tropism, entropy, phototropism

tropho-, -troph, -trophy [Gk. *trope,* nutrition]: trophic level, autotroph, heterotroph

ur-, -uria [Gk. *ouron,* urine]: urea, ureter, uric acid, urine, alkaptonuria, phenylketonuria

uro-, -uran [Gk. *oura,* tail]: urochordate, uropod, anuran

ANSWERS TO SELECTED GENETICS PROBLEMS

CHAPTER 10

1. a. 1/4
   b. 1/2 · 1/2 = 1/4
   c. 1/4 · 1/4 = 1/16
   d. (1/4 · 1/4) + (1/4 · 1/4) = 1/8
   e. In part C, the order of the children is given—so the multiplication law is all that is needed. In part D, no order is specified, so the addition law must be used because here are two possible orders in which the children may occur.
2. a. 1
   b. 0
   c. 0 · 0 = 0
3. a. 0
   b. 1
4. a. Ann—**aa** Mike—**Aa** (how do you know this?)
   Sara—**Aa** Martha—**aa**
   b. 1/2
   c. 1/2 · 1/2 = 1/4
   d. (1/2 · 1/2) + (1/2 · 1/2) = 1/2
5. a. Ralph—**AA**
   b. 1/2 · 1/2 · 1/2 · 1/2 · 1/2 · 1/2 = 1/64
6. a. 1/2 · 1/2 = 1/4
   b. 1/2 · 1/2 = 1/4
   c. 1 · 1/2 = 1/2
   d. 1/2 · 1/2 = 1/4
   e. Genotype would be **A_ B_**; 3/4 · 3/4 = 9/16

CHAPTER 11

1. 9:3:3:1; 1:1:1:1
2. a. Parental strains are **AAbb** and **aaBB**; F_1 hybrid is **AaBb.**
   First backcross progeny: **AaBb**
   AAbb
   AaBb
   Aabb
   Second backcross progeny: **AaBB**
   AaBb
   aaBB
   aaBb
   b. Blue: **aaBB, aaBb, AAbb, Aabb**
   Purple: **AABB, AaBB, AABb, AaBb**
   Scarlet: **aabb**
   c. (1) **AABB** × **aabb**
   (2) Female is **aabb**; male is either **AaBB** or **AABb**
   (3) **AaBb** × **aabb**
   (4) **aaBb** × **Aabb** or vice versa
   (5) **aaBb** × **aaBb** or **Aabb** × **Aabb**

(6) **aaBB** × **aaBB** or **AAbb** × **AAbb**; **aaBB** × **aaBb** and vice versa; **AAbb** × **Aabb** and vice versa

(7) **AAbb** × **aaBB** and vice versa

(8) **aaBb** × **AAbb** and vice versa; **aaBB** × **Aabb** and vice versa

3. a. **B_** = regular **RR** = red
   bb = peloric **Rr** = pink
   rr = white

   BBrr × **bbRR**

   F_1 **Bb Rr**

   F_2 **B_ RR** = regular red
   B_ Rr = regular pink
   B_ rr = regular white
   bb RR = peloric red
   bb Rr = peloric pink
   bb rr = peloric white

				Expected	*Observed*
b. F_2	**B_ RR**	= 3/4 · 1/4 = 3/16	3/16 × 234	43.9	39
	B_ Rr	= 3/4 · 1/2 = 3/8	3/8 × 234	87.8	94
	B_ rr	= 3/4 · 1/4 = 3/16	3/16 × 234	43.9	45
	bb RR	= 1/4 · 1/4 = 1/16	1/16 × 234	14.6	15
	bb Rr	= 1/4 · 1/2 = 1/8	1/8 × 234	29.2	28
	bb rr	= 1/4 · 1/4 = 1/16	1/16 × 234	14.6	13

4. a. X^B = brown spotted teeth
   X^a = hemophilia A
   X^c = colorblindness

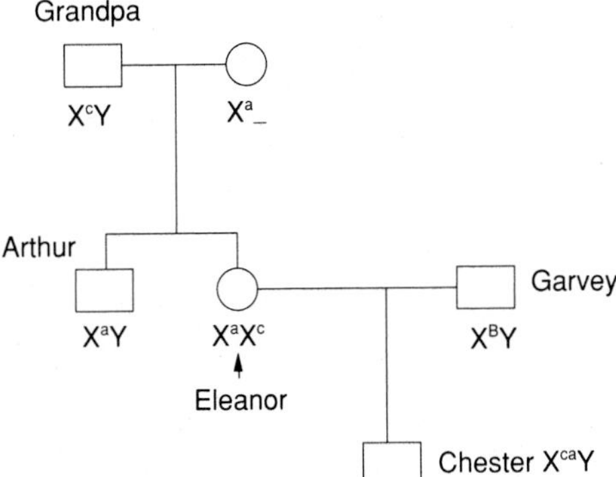

b. She carries X^c and X^a.

c. Chester is also colorblind. Therefore she passed to him an X with both hemophilia and colorblindness. This must have resulted from a crossover between an X^a and an X^c, since she can pass only one X to a son.

d. hemophilic—0
   colorblind—0
   brown teeth—1

e. hemophilic—.5
   colorblind—.5
   brown teeth—0

f. A son cannot be normal, as he would get either an X^c or an X^a from Eleanor, *unless* a crossover occurs to reconstitute a normal X.

   10% crossing over = 5% normal
   5% colorblind and hemophilic
   90% non crossing over = 45% hemophilic
   45% colorblind

5. a. One
   b. Recessive
   c. X chromosome, because F_1 males got trait from mothers (only maternal strain displays spotting trait).
6. a. broad-leaved is dominant
   b. sex linked
   c. F_1: males all narrow-leaved
      females all broad-leaved
      F_2: males 50:50
      females all broad-leaved
7. a. $X^{rb}X^{rb}$ $X^{RB}Y$
      $X^{rb}X^{RB}$ and $X^{rb}Y$
   c. they depart because of recombination between R and B on the X of the F_1 females

SUGGESTED READINGS

PART ONE: MOLECULES TO CELLS

Albert, B., et al. 1989. *Molecular Biology of the Cell.* 2d ed. Garland Publishing, New York. A comprehensive, perhaps vast, "cell book" that covers the field in considerable depth, emphasizing the experiments behind recent discoveries. For the serious student.

Bretscher, Mark S. 1984. "Endocytosis: Relation To Capping and Cell Locomotion." *Science,* vol. 224: pages 681–686.

Darnell, James et al. 1986. *Molecular Cell Biology.* W.H. Freeman, New York. Another comprehensive cell book, and a definite competitor to "Alberts et al." For the serious student.

Darwin, C. 1859. *On the Origin of Species through Natural Selection.* A facsimile of the first edition. Harvard University Press, Cambridge, Mass. We strongly urge all serious biology students to take the time to read it as it was written; Darwin makes much better reading than any of the innumerable books that have since been written about him. Although the writing style is early Victorian, the ideas are fresh and modern and the spirit is infectious.

de Beer, G. 1965. *Charles Darwin: A Scientific Biography.* Doubleday, New York. A sober but intelligent account of Darwin's life and work, with a strong emphasis on the evolution of his scientific thought.

de Duve, Christian. 1984. *A Guided Tour of the Living Cell.* W. H. Freeman, *Scientific American* Library, New York. An authoritative study of the cell, but one written in a delightful manner that is appropriate for layman, student, and teacher.

Dickinson, Robert E., and Ralph J. Cicerone. 1986. "Future Global Warming from Atmospheric Trace Gases." *Nature,* 319:109–115.

Folsome, C. E., ed. 1979. *Life: Origin and Evolution.* W. H. Freeman, *Scientific American* Library, New York.

Gould, S. J. 1977. *Ever Since Darwin: Reflections in Natural History.* W. W. Norton, New York. Essays on evolutionary theory and natural history presented in a lucid and often amusing manner by one of today's most interesting evolutionary theorists.

Griffiths, Gareth, and Kai Simons. 1986. "The *trans* Golgi Network: Sorting at the Exit Site of the Golgi Complex." *Science,* 234:438–443.

Jensen, W. A. 1970. *The Plant Cell.* 2d ed. Wadsworth, Belmont, Calif.

Kennedy, D., ed. 1974. *Cellular and Organismal Biology.* W. H. Freeman, *Scientific American* Library, New York.

Kerr, Richard A. 1988. "Is the Greenhouse Here?" *Science* 239:559–561.

King-Hele, D. 1974. "Erasmus Darwin, Master of Many Crafts." *Nature* 247:87. Charles' wonderful grandfather, poet, inventor, and essayist, made some amazingly clear and sensible statements about evolution and sexual selection. Unlike Charles, he never bothered to collect the facts that would back them up.

Lane, M. Daniel, Peter L. Pedersen, and Albert S. Mildvan. 1986. "The Mitochondrion Updated." *Science,* 234:526–527.

Ledbetter, M. C., and Keith R. Porter. 1970. *Introduction to the Fine Structures of Plant Cells.* Springer-Verlag, New York. A survey of plant cells through the electron microscope, with detailed legends.

Lehninger, A. L. 1975. *Biochemistry.* 2d ed. Worth, New York. The current standard of excellence in biochemistry texts.

________.1982. *Principles of Biochemistry,* Worth, New York. Although writing at a lower, more applied level than *Biochemistry,* Lehninger has succeeded in combining authority with clarity and readability.

Margulis, L., L. To, and D. Chase. 1978. "Microtubules in Prokaryotes." *Science* 200:1118. Margulis found in 1978 just what she had predicted in 1968: some spirochaete bacteria have tubulin-based microtubules.

Moorehead, Alan. 1969. *Darwin and the Beagle.* Harper and Row, New York. We think this book is the most thoroughly enjoyable account of Darwin's seminal years aboard the H.M.S. *Beagle.* Lavishly illustrated.

Oparin, A. I. 1938. *The Origin of Life.* Dover, New York. The ideas of Oparin on the origin of life have spurred an entire field of biology into existence.

Porter, E. 1971. *Galápagos.* Ballantine, New York. All evolutionary biologists dream of making the trip to the Galápagos—and some of them make it. A different world.

Scientific American. 1986. *The Molecules of Life.* W. H. Freeman, New York. This issue of *Scientific American* is devoted entirely to discussions of the important molecules of life: carbohydrate, protein, nucleic acids, and others. An excellent companion to Chapter 3.

Scientific American. 1980. *Molecules to Living Cells.* W. H. Freeman, New York. A collection of *Scientific American* articles that we recommended to accompany Chapters 2–9.

Singer, S. J., and G. Nicolson. 1972. "The Fluid Mosaic Model of the Structure of Cell Membranes." *Science* 175:720. The original description of the now-accepted fluid mosaic model.

Stryer, Lubert. 1988. *Biochemistry.* 3d ed. W. H. Freeman, New York. Although more selective in its coverage than Lehninger's latest biochemistry, Stryer is very readable and includes interesting and colorful graphic aids.

Weissmann, G., and R. Clairborne, eds. 1975. *Cell Membranes: Biochemistry, Cell Biology and Pathology.* HP Publishing, New York. Superbly illustrated articles on all aspects of plasma membranes, incorporating the latest research findings. Notable are E. Racker on the inner mitochondrial membrane and the chemiosmosis hypothesis; W. R. Lowenstein and G. D. Pappas on cell-to-cell communication via membrane junctions; and J. F. Danielli on the Danielli lipid bilayer concept as it has evolved over the years.

***Scientific American* articles. New York: W. H. Freeman Co.**

Allen, Robert Day. 1987. "The Microtubule as an Intracellular Engine." February.

Bretscher, Mark S. "How Animal Cells Move." December.

Dautry-Varsat, Alice, and Harvey F. Lodish. 1984. "How Receptors Bring Proteins and Particles into Cells." May.

Doolittle, Russell F. 1985. "Proteins." October.

Felsenfeld, Gary. 1985. "DNA." October.

Hayflick, L. 1980. "The Cell Biology of Aging." January.

Hinkle, P. C., and R. E. McCarty. 1978. "How Cells Make ATP." March. A thoroughly convincing presentation of the chemiosmosis hypothesis and the evidence for it in both photosynthetic and oxidative phosphorylation. We have drawn heavily on this article in preparing Chapters 6 and 7.

Karplus, Martin, and J. Andrew McCammon. 1986. "The Dynamics of Proteins." April.

Margulis, L. 1971. "Symbiosis and Evolution." August. Still the best presentation of her once startling theory of the origin of eukaryotes.

Porter, K. R., and J. B. Tucker. 1981. "The Ground Substance of the Living Cell." March.

Rothman, James E. 1985. "The Compartmental Organization of the Golgi Apparatus." September.

Satir, P. 1974. "How Cilia Move." October. An intriguing but still speculative model of how sliding microtubules might be responsible for the movement of cilia and flagella.

Schopf, J. W. 1978. "The Evolution of the Earliest Cells." September. Biochemical pathways clearly demonstrate that basic life processes were evolved in an anaerobic world, and that steps involving free molecular oxygen have been "tacked on" to previously evolved systems.

Sharon, Nathan. 1980. "Carbohydrates." November.

Shulman, R. G. 1983. "NMR (nuclear-magnetic-resonance) Spectroscopy of Living Cells." January.

Staehelin, L. Andrew, and Barbara E. Hull. 1978. "Junctions between Living Cells." May.

Unwin, Nigel, and Richard Henderson. 1984. "The Structure of Proteins in Biological Membranes." February.

Weber, Klaus, and Mary Osborn. 1985. "The Molecules of the Cell Matrix." October.

Youvan, Douglas C., and Barry L. Marrs. 1987. "Molecular Mechanisms of Photosynthesis." June.

PART TWO: MOLECULAR BIOLOGY AND HEREDITY

Avery, O. T., C. M. MacLeod, and M. McCarty. 1944. "Studies on the Chemical Nature of the Substance Inducing Transformation of Pneumococcal Types." *Journal of Experimental Medicine* 79:137. The classic experiment that proved to the world—at least in hindsight—that the genetic material is DNA.

Chedd, G. 1981. "Genetic Gibberish in the Code of Life." *Science 81,* November.

Crick, F. H. C. 1979. "Split Genes and RNA Splicing." *Science* 204:264. The codiscover of the structure of DNA was active in the race to decipher the genetic code and has some cogent things to say about introns.

DuPraw, E. J. 1970. *DNA and Chromosomes.* Holt, Rinehart and Winston, New York (paperback).

Goodenough, U. 1984. *Genetics.* 2d ed. Holt, Rinehart and Winston, New York. Of the current general genetics texts, Ursula Goodenough's has the best coverage of recent developments in molecular genetics and is particularly strong in the molecular genetics of higher eukaryotes.

Hamilton, W. D. 1967. "Extraordinary Sex Ratios." *Science* 156:477. Why have some species departed from the usual one-to-one ratio of males to females?

Jacob, F., and J. Monod. 1961. "Genetic Regulatory Mechanisms inthe Synthesis of Proteins." *Journal of Molecular Biology* 33:318. A modern classic and an example of fine scientific writing and reason, this is an account of the revolutionary experiments that demonstrated the existence of operons, the operator, cytoplasmic regulating molecules, and messenger RNA.

Lewin, R. 1983. "A Naturalist of the Genome." *Science* 222:402.

Maynard Smith, J. 1971. "What Use Is Sex?" *Journal of Theoretical Biology* 30:319. One major difference between prokaryotes and eukaryotes is the invention, in the latter, of regular biparental reproduction. Being widespread and of ancient origin, it presumably serves some function, but what is it?

Mendel, G. 1965. "Experiments in Plant Hybridization (1865)." Translated by Eva Sherwood. In *The Origin of Genetics,* eds. C. Stern and E. Sherwood. W. H. Freeman, New York. In addition to the full text of Mendel's classic paper, this volume includes some of Mendel's letters and minor works, the three "rediscovery" papers of 1900, and a fascinating exchange between R. A. Fisher and Sewall Wright on the question of whether or not Mendel drylabbed the whole thing.

Menosky, J. A. 1981. "The Gene Machine." *Science 81,* July/August.

Meselson, M., and F. W. Stahl. 1958. "The Replication of DNA in *E. coli." Proceedings of the National Academy of Sciences* (U.S.) 44:671. The brilliant and influential experiment that proved that DNA unwinds and replicates semiconservatively, as foreseen by Crick. All biology students must study this work in detail at some time or other.

Mourant, A. E., et al. 1978. *The Genetics of the Jews.* Oxford University Press, The Clarendon Press, New York and Oxford. Ashkenazi and Sephardic Jews form a genetically distinct racial group after all, according to blood group and enzyme polymorphisms. Basically Palestinian, they show surprisingly little evidence of past mixing with European groups but rather more (5–10%) negroid admixture, presumably from the time spent in slavery in Egypt.

Okazaki, R. T., et al. 1968. "Mechanism of DNA Chain Growth: Possible Discontinuity and Unusual Secondary Structure of Newly Synthesized Chains." *Proceedings of the National Academy of Sciences* (U.S.). On the Okazaki fragments.

Rensberger, B. "Tinkering with Life." *Science 81,* November.

Sanger, F., et al. 1977. "Nucleotide Sequence of Bacteriophage fx174 DNA." *Nature* 265:687. The first publication of the entire genome of any organism and a *tour de force* of molecular biology.

Scientific American. 1985. *The Molecules of Life.* W. H. Freeman, New York. A beautiful presentation of DNA and gene structure. The set of articles contains wonderful color computer graphic representations of DNA, RNA, and proteins.

Shine, I., and S. Wrobel. 1976. *Thomas Hunt Morgan: Pioneer of Genetics.* University of Kentucky Press, Lexington, Ky. Interesting narrative and lively anecdotes of the early days of genetics in America.

Strickberger, M. W. 1985. *Genetics.* 3d ed. Macmillan, New York. Of the current general genetics texts, Strickberger's has the clearest treatment of classical Mendelian genetics. In addition, we feel that the 120 pages that are included on population genetics and quantitative genetics happen to constitute the best textbook in print on these difficult subjects.

Van Valen, L., and G. W. Mellin. 1967. "Selection in Natural Populations. 7. New York Babies." *Annals of Human Genetics* (London) 31:109. Among newborns, it's better to be average, because small and large babies are both at risk.

Watson, J. D. 1968. *The Double Helix.* Atheneum, New York. Deftly hidden in the narrative of this witty, often hilarious and picaresque account of the personal triumph of a young scientist and an old graduate student is a surprising amount of solid scientific information. Certainly the most enjoyable account of how "the scientific method" actually works in practice.

Watson, J. D., and F. H. C. Crick. 1953. "Molecular Structure of Nucleic Acids. A structure of deoxyribose nucleic acid." *Nature*

171:737. This is the one that started it all: the most influential single page in scientific history.

Watson, J. D., et al. 1987. *Molecular Biology of the Gene*. 4th ed. Benjamin/Cummings, Menlo Park, Calif. This two-volume set is generally recognized as the most comprehensive and accurate treatment of gene structure and function.

***Scientific American* articles. New York: W. H. Freeman Co.**

Aharonowitz, Y., and G. Cohen. 1981. "The Microbiological Production of Pharmaceuticals." September.

Allison, A. C. 1956. "Sickle Cells and Evolution." August. The first demonstration of overdominance: the sickle-cell allele is actually beneficial to carriers in certain environments.

Anderson, W. F., and E. G. Diacumakos. 1981. "Genetic Engineering in Mammalian Cells." July.

Bishop, J. M. 1982. "Oncogenes." March.

Brill, W. J. 1981. "Agricultural Microbiology." September.

Campbell, A. M. 1976. "How Viruses Insert Their DNA into the DNA of the Host Cell." December.

Chambon, P. 1981. "Split Genes." May.

Chilton, M. 1983. "A Vector for Introducing New Genes into Plants." June.

Cohen, S. N., and J. A. Shapiro. 1980. "Transposable Genetic Elements." February.

Crick, F. H. C. 1962. "The Genetic Code." October.

————. 1966. "The Genetic Code III." October.

Dickerson, R. E. 1972. "The Structure and History of an Ancient Protein." April. How chytochrome *c* has evolved its present shape and amino acid sequence over the last 2 billion years.

Friedman, T. 1971. "Prenatal Diagnosis of Genetic Diseases." November. A primer of transabdominal amniocentesis and a valuable discussion of the moral implications involved in the interruption of pregnancy.

Grivell, L. A. 1983. "Mitochondrial DNA." March.

Hopwood, A. 1981. "The Genetic Programming of Industrial Microorganisms." September.

Howard-Flanders, P. 1981. "Inducible Repair of DNA." November.

Hunter, T. 1984. "The Proteins of Oncogenes." August.

Kornberg, R. D., and A. Klug. 1981. "The Nucleosome." February.

Kretchmer, N. 1972. "Lactose and Lactase." 227:70. Milk produces flatulence in adults of Oriental or African ancestry because they lack the enzyme needed to break down milk sugar.

Lake, J. A. 1981. "The Ribosome." August.

Gould, S. J. 1983. *Hens' Teeth and Horses' Toes*. Norton, New York. A collection of essays by a talented philosopher of science.

Gould, S. J., and Eldredge, N. 1977. "Punctuated Equilibria: The Tempo and Mode of Evolution Reconsidered." *Paleobiology*. 3:115. Do the mechanisms of evolution proceed in a slow and orderly manner, or are they rapid and sporadic?

Grant, V. 1963. *The Origin of Adaptations*. Columbia University Press, New York. A balanced view of evolutionary adaptations in plants and animals.

Haldane, J. B. S. 1932. *The Causes of Evolution*. Cornell University Press, Ithaca, N.Y. (paperback). All the difficult math is relegated to the appendix in this rather old—but very wise—discussion of evolutionary genetics. A must for advanced students of population biology.

Joravsky, D. 1970. *The Lysenko Affair*. Harvard University Press, Cambridge, Mass. Fascinating reading about the man whom biologists—especially geneticists—love to hate.

Kettlewell, H. B. D. 1956. "Further Selection Experiments on Industrial Melanisms in the Lepidoptera." *Heredity* 10:287. Here you can read for yourself one of the most often cited experiments in modern evolutionary biology.

King, J. L., and T. H. Jukes. 1969. "Non-Darwinian Evolution." *Science* 164:788. The authors suggest, among other things, that most evolutionary changes on the molecular level may be meaningless noise, the result of mutation and random drift; that most DNA in higher organisms is not genetic material; and that no more than about 1% codes directly for proteins.

Lack, D. 1947. *Darwin's Finches*. Cambridge University Press, New York. Lack investigates the coevolution of competing species on different islands in the Galápagos archipelago.

Levin, D. A. 1979. "The Nature of Plant Species." *Sciences* 204:381. The "biological species concept" quickly breaks down when plant species are considered.

Maynard Smith, J. "Group Selection and Kin Selection." *Nature* 201:1145.

Patterson, C. 1978. *Evolution*. Cornell University Press, Ithaca, N.Y. (paperback). This short textbook makes a useful supplement to a general biology course or stands on its own. It is a good secondary reference for Darwin's finches and industrial melanism.

Scientific American. 1978. *Evolution*. W. H. Freeman, New York. Reprinted from the September 1978 special issue on evolution.

Stebbins, G. L. 1971. *Processes of Organic Evolution*. Prentice-Hall, Englewood Cliffs, N.J. Somewhat dated, but still a superior textbook on evolution.

Whittaker, R. H. 1959. "On the Broad Classification of Organisms." *Quarterly Review of Biology* 34:210. Whittaker's influential revolt against the classic plant-animal dichotomy.

McKusick, V. A. 1965. "The Royal Hemophilia." February. Queen Victoria was heterozygous for the X-linked recessive allele and through political marriages of her daughters managed to pass it on to several of the leading royal families of Europe.

Miller, O. L. 1973. "The Visualization of Genes in Action." March. Some remarkable electron micrographs of transcription and translation, looking almost exactly like diagrams that had originally been made on biochemical evidence alone.

Nomura, M. 1984. "The Control of Ribosome Synthesis." January.

Novick, R. P. 1980. "Plasmids." December.

Pestka, S. 1983. "The Purification and Manufacture of Human Interferons." August.

PART THREE: EVOLUTION

Banks, H. P. 1975. "Early Vascular Plants: Proof and Conjecture." *Bioscience* 25:730. On the origin of one of the five kingdoms.

Darwin, C. 1859, 1966. *On the Origin of Species*. Harvard University Press, Cambridge, Mass. A facsimile of the first edition.

Dawkins, R. 1976. *The Selfish Gene*. Oxford University Press, New York and Oxford. Dawkins is an enthusiastic and often persuasive writer who, in this popular paperback, puts modern evolutionary theory into clear and vivid language. His approach is logical and nonmathematical, with some emphasis on theories of the evolution of animal behavior.

de Beer, G. 1974. "Evolution." In *The New Encyclopedia Brittanica*. 15th ed. 7:7. A concentrated synopsis of evolutionary thought, with an extensive bibliography.

Dobzhansky, T. 1963. "Evolutionary and Population Genetics." *Science* 142:3596.

————. 1970. *Genetics of the Evolutionary Process*. Columbia University Press, New York. A great compendium of observations and interpretations of the genetics of natural populations by an influential evolutionary geneticist.

Dodson, E. O. 1974. "Phylogeny." In *The New Encyclopedia Brittanica.* 15th ed. 14:376. Dodson includes a full historical account of the various one-, two-, three-, four-, and five-kingdom schemes as well as an overview of phylogenetic relationships within the plant and animal kingdoms.

Dodson, E. O., and P. Dodson. 1985. *Evolution: Process and Product.* 3d ed. Prindle, Weber and Schmidt, Boston. An excellent introduction for college students.

Eldredge, N., and S. J. Gould. 1972. "Punctuated Equilibria: An Alternative to Phyletic Gradualism." In *Models in Paleobiology.* Edited by T. J. M. Schopf. W. H. Freeman, New York. The original paper on punctuated equilibrium.

Futuyma, E. J. 1986. *Evolutionary Biology.* 2d ed. Sinauer, Sunderland, Mass. An excellent general text that covers the major points of evolution.

Gilbert, L. E., and P. H. Raven. 1975. *Coevolution of Plants and Animals.* University of Texas Press, Austin, Tex. Such fascinating esoterica as flowers that mimic insects and insects that mimic flowers, as well as such important topics as the coevolution of plants and their herbivores.

***Scientific American* articles. New York: W. H. Freeman Co.**

Bishop, J. A., and Laurence M. Cook. 1975. "Moths, Mechanism and Clean Air." January. How the peppered moth is readapting to the improving atmosphere of postindustrial England.

Eigen, M., et al. 1981. "The Origin of Genetic Information." April.

Gilbert, L. E. 1982. "The Coevolution of a Butterfly and a Vine." August.

Margulis, L. 1971. "Symbiosis and Evolution." August.

Sibley, C. G., and J. F. Ahlquist. 1986. "Reconstructing Bird Phylogeny by Comparing DNAs." February. One of the germinal papers on the evidence of molecular evolution.

Stanley, S. M. 1984. "Mass Extinction in the Ocean." June.

PART FOUR: DIVERSITY AND FUNCTION: MICROORGANISMS AND FUNGI

Ahmadjian, Vernon. 1982. "The Nature of Lichens." *Natural History,* March, pages 30-37.

Alexopoulos, C. J. 1979. *Introduction to Mycology.* Wiley, New York. For enthusiasts of things fungal.

Bold, Harold C., and Michael J. Wynne. 1985. *Introduction to the Algae: Structure and Reproduction.* 2d ed. Prentice-Hall, Englewood Cliffs, N.J. A comprehensive, taxonomically oriented treatment of the algae.

Fox, G. E., et al. 1980. "The Phylogeny of Prokaryotes." *Science* 209:457.

Large, E. C. 1962. *The Advance of the Fungi.* Dover, New York. A close-up, informal look at the historical relationship between humans and fungi.

Lee, J. J., S. H. Hutner, and E. C. Bovee, eds., 1985. *An Illustrated Guide to the Protozoa.* Society of Protozoologists, Lawrence, Kan.

Margulis, Lynn. 1981. *Symbiosis in Cell Evolution: Life and Its Environment on the Early Earth.* W. H. Freeman, New York. Another look at Margulis's novel ideas on the origin of eukaryotic life through serial endosymbiosis.

Miller, S. L. 1935. "Production of Some Organic Compounds under Possible Primitive Earth Conditions." *Journal of the American Chemical Society* 77:2351. The first *experimental* approach to the question of the origin of life on earth.

Nester, Eugene W., et al. 1973. *Microbiology.* 3d ed. Saunders College Publishing, Philadelphia. A popular introductory textbook treating bacteria, viruses, protists, and fungi, with an emphasis on medical and ecological applications.

Raven, P. H. 1970. "A Multiple Origin for Plastids and Mitochondria." *Science* 169:641.

Schopf, J. W., and D. Z. Oehler. 1971. "How Old Are the Eukaryotes?" *Science* 193:47. Perhaps they are not as old as had been thought, but originated only a few hundred million years before the first oldest known fossil animals.

Stanier, R. Y., et al. 1986. *The Microbial World.* 5th ed. Prentice-Hall, Englewood Cliffs, N.J. A highly authoritative, widely respected account of the biology of microorganisms.

Tortora, Gerard J. 1989. *Microbiology. An Introduction.* 3d ed. Benjamin/Cummings, Menlo Park, Calif. A popular introductory text, covering bacteria, viruses, protists, and fungi, with an emphasis on medical, commercial, and ecological applications.

Yates, G. T. 1986. "How Microorganisms Move through Water." *American Scientist.* 74:358–365.

Zinsser, Hans. 1935. *Rats, Lice, and History.* Atlantic Monthly Press/Little, Brown, Boston. A reissue of a classic. A fascinating historical account of those human diseases vectored by animals such as the flea.

***Scientific American* articles. New York: W. H. Freeman Co.**

Barghoorn, E. S. 1971. "The Oldest Fossils." May.

Blakemore, R. P., and R. B. Frankel. 1981. "Magnetic Navigation in Bacteria." December.

Dickerson, R. E. 1978. "Chemical Evolution and the Origin of Life." September. We think that this is the best no-nonsense review of this primarily speculative field.

Friedman, M. J., and W. Trager. 1981. "The Biochemistry of Resistance to Malaria." March.

Gallo, Robert C. 1987. "The AIDS Virus." January.

Groves, D. J., et al. 1981. "An Early Habitat of Life." October. (Stromatolites).

Hogle, J. M., M. Chow, and D. J. Filman. 1987. "The Structure of Poliovirus." March.

Kosikowski, Frank V. 1965. "Cheese." May.

Litten, W. 1975. "The Most Poisonous Mushrooms." March.

Prusiner, S. B. 1984. "Prions." October.

Shapiro, James A. 1988. "Bacteria as Multicellular Organisms." June.

Simons, K., et al. 1982. "How an Animal Virus Gets into and out of Its Host Cell." February.

Strobel, G. A., and G. N. Lanier. 1981. "Dutch Elm Disease." August.

Vidal, G. 1984. "The Oldest Eukaryotic Cells." February.

Woese, C. R. 1981. "Archaebacteria." June.

PART FIVE: DIVERSITY AND FUNCTION: PLANTS

Barth, Friedrich G. 1985. *Insects and Flowers: The Biology of a Partnership.* Princeton University Press, Princeton, N.J. An excellent treatment of the coadaptation of insect and flower, including a look at the diverse flower structures and a very substantial treatment of insect adaptations.

Conard, Henry S., and Paul L. Redfearn, Jr. 1979. *How to know the Mosses and Liverworts.* 2d ed. William C. Brown, Dubuque, Iowa. Just what the beginner needs to identify the common mosses and liverworts and to become familiar with dichotomous keys.

Cutler, David F. 1978. *Applied Plant Anatomy.* Longman, New York. An excellent treatment of plant structure, but with a strong emphasis on applications to everyday life.

Epstein, E. 1972. *Mineral Nutrition of Plants: Principles and Perspectives.* Wiley, New York.

Esau, K. 1977. *Anatomy of Seed Plants.* 2d ed. Wiley, New York. The authoritative sourcebook by one of the world's most distinguished botanists.

Galston, A. W., et al. 1980. *The Life of the Green Plant.* 3d ed. Prentice-Hall, Englewood Cliffs, N.J. For the more advanced student, a comprehensive treatment of botany with a strong physiology emphasis.

Galston, W. W., and P. J. Davies. 1970. *Control Mechanisms in Plant Development.* Prentice-Hall, Englewood Cliffs, N.J.

Heywood, Vernon H., ed. 1985. *Flowering Plants of the World.* Prentice-Hall, Englewood Cliffs, N.J. An excellent guide to plant families. For the advanced student.

Hitch, Charles J. 1982. "Dendrochronology and Serendipity." *American Scientist.* 70:300–305.

Hutchinson, J. 1969. *Evolution and Phylogeny of Flowering Plants.* Academic Press, New York and London.

Lehner, R., and J. Lehner. 1'962. *Folklore and Odysseys of Food and Medicinal Plants.* Tudor, New York. Did you know that crabgrass was deliberately introduced into the United States by Polish immigrants? In nineteenth-century Poland, people made bread from crabgrass seed.

Mulcahy, David L. 1981. "Rise of the Angiosperms." *Natural History* September, pages 30-35.

Norstog, Knut. 1987. "Cycads and the Origin of Insect Pollination." *American Scientist.* 75:270–279.

Raven, P. H., et al. 1986. *Biology of Plants.* 4th ed. Worth, New York. A highly popular general botany with strong evolutionary overtones and a good balance of structure and function.

Sporne, K. R. 1971. *The Mysterious Origin of Flowering Plants.* Carolina Biological Supply Company, Burlington, N.C.

Temple, S. 1977. "The Dodo and the Tambalacoque Tree." *Science* 197:885. On Mauritius there are some geriatric trees whose seeds normally germinate only when passed through the gizzard of a dodo. But the last dodo died in the Seventeenth century.

Wardlaw, I. F. 1974. "Phloem Transport: Physical, Chemical or Impossible?" *Annual Review of Plant Physiology* 25:515. What *does* make sap flow? Although the answer is still not known, at least some hypotheses can be ruled out. Unfortunately, it sometimes seems that all hypotheses can be ruled out.

Went, F. W. 1963. *The Plants.* Time, Inc., Life Nature Library, New York. Although it is now getting along in years, this beautifully illustrated volume remains a superbly readable, informative, and intelligent nontextbook approach to botany.

Zimmermann, Martin H. 1983. *Xylem Structure and the Ascent of Sap.* Springer-Verlag, New York. An excellent, highly readable companion reading for Chapters 27 and 28, where xylem and phloem function is the topic.

Zimmermann, Martin H., and Claude L. Brown. 1975. *Trees: Structure and Function.* Springer-Verlag, New York. Just what you wanted to know about the biology of trees. A nice tie-in between plant structure and function.

***Scientific American* articles. New York: W. H. Freeman Co.**

Albersheim, P. 1975. "The Walls of Growing Plant Cells." April. How cellulose fibers are laid down in careful patterns.

Chilton, Mary-Dell. 1983. "A Vector for Introducing New Genes into Plants." June.

Evans, Michael L., Randy Moore, and Karl-Heinz Hasenstein, 1986. "How Roots Respond to Gravity." December.

Kaplan, D. R. 1983. "The Development of Palm Leaves." July.

Niklas, Karl J. 1987. "Aerodynamics of Wind Pollination." July.

Rosenthal, Gerald A. 1986. "The Chemical Defenses of Higher Plants." January.

Shepart, J. F. 1982. "The Regeneration of Potato Plants from Leaf-Cell Protoplasts." May.

Shigo, Alex L. 1985. "Compartmentalization of Decay in Trees." April.

PART SIX: DIVERSITY AND FUNCTION: ANIMALS

Alexander, T. 1975. "A Revolution Called Plate Tectonics Has Given Us a Whole New Earth." *Smithsonian* 5:30.

Balinsky, B. I. 1974. "Development, animal." In *The New Encyclopedia Brittanica.* 15th ed. 6:625. Although this article is almost entirely devoted to vertebrate embryology, it does that well.

Barnes, Robert D. 1987. *Invertebrate Zoology.* 5th ed. Saunders College/Holt, Rinehart and Winston, Philadelphia. Probably the finest of the more traditional "invert" books, well illustrated and quite comprehensive.

Browder, L. W. 1984. *Developmental Biology.* 2d ed. Saunders College/Holt, Rinehart and Winston, Philadelphia. An excellent, if selective, presentation of development, which explores much of the classical experimental embryology and much of the more recent work as well.

Buchsbaum, Ralph et al. 1987. *Animals without Backbones.* 3d ed. University of Chicago Press, Chicago. Another classic in invertebrate zoology, written for the layman and heavily illustrated with excellent photos and numerous simple drawings.

Cornejo, D. 1982. "Night of the Spadefoot Toad." *Science 82,* September.

Eckert, Roger and David Randall. 1988. *Animal Physiology: Mechanisms and Adaptations.* 3d ed. W. H. Freeman, New York. An advanced-level, but highly readable, animal physiology text. You will find answers to many of your questions in this book.

Golub, Edward S. 1987. *Immunology: A Synthesis.* Sinauer Associates, Sunderland, Mass. For the more advanced student, a thorough grounding in the science of immunology. Quite up to date.

Griffin, D. R., ed. 1974. *Animal Engineering. Readings from Scientific American.* W. H. Freeman, New York. Engineering majors with an interest in biology or biologists with an engineering background will find themselves at home here.

Hood, Leroy E., et al. 1984. *Immunology.* Benjamin/Cummings, Menlo Park, Calif. A survey of modern immunology through an ongoing review of key experiments.

Johanson, D. C., and M. A. Edey. 1981. "Lucy: The Inside Story." *Science 81.* March.

Johanson, D. C., and T. D. White. 1979. "A Systematic Assessment of Early African Hominids." *Science* 203:321. Since our ideas about early hominid relationships are rapidly changing, it pays to keep up with the news. Here the authors take a fresh look at some old bones from East Africa.

Karp, Gerald, and N. J. Berrill. 1981. *Development.* 2d ed. McGraw-Hill, New York. A comprehensive text in animal development, well up to Gerald Karp's usual high standards.

Katchadourian, H. 1974. *Human Sexuality: Sense and Nonsense.* W. H. Freeman, New York. Most of what young adults (and others) need to know about human reproductive anatomy and sexuality. Authoritative, yet quite readable.

Lewin, Roger. 1988. *In the Age of Mankind.* Smithsonian, Washington D.C. A superbly illustrated, well-organized survey of human physical and cultural evolution. A "coffee table" book, but one with authority.

MacArthur, R. H. 1972. *Geographical Ecology: Patterns in the Distribution of Species.* Harper and Row, New York.

McMahon, T. A., and J. T. Bonner. 1985. *One Size and Life,.* W. H. Freeman, New York.

McMenamin, M. A. S. 1982. "A Case for Two Late Proterozoic-Earliest Cambrian Faunal Province Loci." *Geology,* June.

Mayr, E. 1963. *Animal Species and Evolution.* Harvard University Press, Cambridge, Mass. A truly great work uniting biogeography, natural history, evolution, and genetics. Here Mayr introduces and defends "the biological species concept."

Miller, W. H. 1974. "Photoreception." In *The New Encyclopedia Brittanica.* 15th ed. 14:353. In far more detail than found in most *Brittanica* articles, Miller gives a full account of the morphology, physiology, and biochemistry of eyes and other vertebrate and invertebrate photoreceptors.

Money, John, and Anke A. Ehrhardt. 1972. *Man and Woman, Boy and Girl: The Differentiation and Dimorphism of Gender Identify from Conception to Maturity.* Johns Hopkins University Press, Baltimore. From their studies of patients who have been misdiagnosed as to gender because of congenital abnormalities of the genitalia, the authors conclude that personal sexual identity is confirmed by 18 months of age. On the other hand, prenatal sex hormone levels are shown to have a decided influence on personality.

Morton, J. E. 1967. *Guts.* St. Martin's Press, New York.

Nilsson, Lennart, et al. 1986. *A Child Is Born: The Drama of Life before Birth.* Dell, New York.

Oppenheimer, J. H. 1979. "Thyroid Hormone Action at the Cellular Level." *Science* 203:971.

Petit, C., and L. Ehrman. 1969. "Sexual Selection in *Drosophilia.*" *Evolutionary Biology* 3:177. Perhaps you thought those little flies wouldn't *care.* But they do.

Rensberger, B. 1981. "Facing the Past." *Science 81,* October. (Neanderthal Man)

Roitt, Ivan, et al. 1985. *Immunology.* C. V. Mosby, St. Louis. A primer in immunology, organized along a decimal system and undoubtedly the most heavily and colorfully illustrated text on the subject.

Romer, Alfred, and Thomas S. Parsons. 1985. *The Vertebrate Body,* 6th ed. Saunders College/Holt, Rinehart and Winston, Philadelphia. The vertebrate parallel to Barnes's invertebrate zoology. One of the finest vertebrate evolution texts available.

Schmidt-Nielsen, Knut. 1983. *Animal Physiology: Adaptation and Environment.* 3d ed. Cambridge University Press, New York. An intermediate-level animal physiology text, but one written in the usual lively, highly readable style of Schmidt-Nielsen. A companion reader to Chapters 33–43.

Scientific American. 1979. *The Brain.* W. H. Freeman, New York. An entire issue of *Scientific American* devoted to the most recent studies of the brain and related neurobiology. Will relate well to Chapters 34 and 35.

Scientific American. 1985. *Progress in Neuroscience.* W. H. Freeman, New York. Some of the more popular articles on neurobiology from past editions of *Scientific American* magazine have been gathered together in this book.

Scientific American. 1988. *What Science Knows about AIDS.* W. H. Freeman, New York. Certainly required reading for anyone wishing to stay informed on the subject of AIDS. Many highly readable and some more technical articles—but something for everyone.

Shell, E. R. 1982. "The Guinea Pig Town." *Science 82,* December. (Framingham, Mass.)

Shodell, M. 1983. "The Prostaglandin Connection." *Science 83,* March.

Simmons, J. A., M. B. Fenton, and M. J. O'Farrell. 1979. "Echolocation and Pursuit of Prey by Bats." *Science* 203:16

Sperry, R. 1982. "Some Effects of Disconnecting the Cerebral Hemispheres." *Science* 217:1223.

Tanner, J. M. 1974. "Development, human." In *The New Encyclopedia Brittanica.* 15th ed. 6:650. Tanner's emphasis is not on embryology but on the growth and development of human beings between birth and adulthood.

Waddington, C. H. 1974. "Development, biological." In *The New Encyclopedia Brittanica.* 15th ed 6:643: Waddington takes an overview of the *principles* of animal development.

Wallace, R. A. 1980. *How They Do It.* Morrow, New York. Your author's paperback description of how a wide variety of species from bacteria to whales . . . do it.

Weisman, I. L., L. E. Hood, and W. B. Wood. 1978. *Essential Concepts in Immunology.* Benjamin/Cummings, Menlo Park, Calif. There is no way to make the complexities of immunology simple. Advanced students will at least find this paperback a good place to get started.

West, S. 1983. "One Step Behind a Killer." *Science 83,* March. (AIDS).

Wilford, John Noble. 1985. *The Riddle of the Dinosaur.* New York.

Wilson, A. C., S. S. Carlson, and T. J. White. 1977. "Biochemical Evolution." *Annual Review of Biochemistry* 46:573. Each kind of protein appears to evolve (change its amino acid sequence) at a steady rate, making it possible to use protein sequence divergence as an "evolutionary clock."

Woolacott, R. M., and R. L. Zimmer. 1979. *Biology of Bryozoans.* McGraw-Hill, New York.

***Scientific American* articles. New York: W. H. Freeman Co.**

Ada, Gordon L., and Nossal, Gustav. 1987. "The Clonal Selection Theory." August.

Baker, M. A. 1979. "A Brain-Cooling System in Mammals." April.

Beaconsfield, P., et al. 1980. "The Placenta." August.

Berridge, Michael J. 1985. "The Molecular Basis of Communication within the Cell." October.

Bloom, F. E. 1981. "Neuropeptides." October.

Buisseret, P. D. 1982. "Allergy." August.

Cameron, J. N. 1985. "Molting in the Blue Crab," May.

Cantin, Marc, and Jacques Genest. 1980. "The Heart as an Endocrine Gland." February.

Cohen, Irun R. 1988. "The Self, the World and Autoimmunity." April.

Degabriele, R. 1980. "The Physiology of the Koala." July.

Dunant, Yves, and Maurice Israel. 1985. "The Release of Acetylcholine." April.

Eastman, Joseph T., and Arthur L. DeVries. 1986. "Antarctic Fishes." November.

Edelson, Richard L., and Joseph M. Fink. 1985. "The Immunologic Function of Skin." June.

Epel, D. 1978. "The Program of Fertilization." Includes the most dramatic scanning electron micrographs of eggs and sperm that we've seen yet. November.

Feder, Martin E., and Warren W. Burggren. 1985. "Skin Breathing in Vertebrates." November.

Gehring, Walter J. 1985. "The Molecular Basis of Development." October.

Goldstein, Gary W., and A. Lorris Betz. 1986. "The Blood-Brain Barrier." September.

Griffiths, Mervyn. 1988. "The Platypus." May.

Gurdon, J. B. 1968. "Transplanted Nuclei and Cell Differentiation." December.

Hadley, N. F. 1986. "The Arthropod Cuticle." July.

Horner, J. R. 1984. "The Nesting Behavior of Dinosaurs." April.

Hudspeth, A. J. 1983. "The Hair Cells of the Inner Ear." January.

Jarvik. R. K. 1981. "The Total Artificial Heart." January.

Koretz, Jane F., and George H. Handelman. 1988. "How the Human Eye Focuses." July.

Laurence, Jeffrey. 1985. "The Immune System in AIDS." December.

Leder, P. 1982. "The Genetics of Antibody Diversity." May.

Lent, Charles M., and Michael H. Dickinson. 1988. "The Neurobiology of Feeding in Leeches." June.

Lerner, R. A. 1983. "Synthetic Vaccines." February.

Llinas, R. R. 1982. "Calcium in Synaptic Transmission." October.

McMenamin, Mark A. S. 1987. "The Emergence of Animals." April.

Marrack, Philippa, and John Kappler. 1986. "The T Cell and Its Receptor." February.

Milstein, Cesar. 1980. "Monoclonal Antibodies." October.

Mishkin, Mortimer, and Tim Appenzeller. 1987. "The anatomy of Memory." June.

Moog, F. 1981. "The Lining of the Small Intestine." November.

Morrell, P., and W. T. Norton. 1980. "Myelin." May.

Morris, S. C., and H. B. Whittington. 1979. "The Animals of the Burgess Shale." July. An exciting look at an accidentally preserved nearshore community in the early Cambrian era, including primitive members of most major phyla, some members of mysterious and bizarre long-extinct phyla, a marine onychophoran, and one amphioxuslike chordate.

Morrison, A. 1984. "A Window on the Sleeping Brain." April.

Mossman, D., and W. Sarjeant. 1983. "The Footprints of Extinct Animals." January.

Newman, E., and P. Harline. 1982. "The Infrared 'Vision' of Snakes." March.

Philbeam, D. 1984. "The Descent of Hominoids and Hominids." March.

Roper, C., and K. Boss. 1982. "The Giant Squid." April.

Rose, N. R. 1981. "Autoimmune Disease." February.

Rukang, W., and L. Shenglong. 1983. "Peking Man." June.

Russell, D. 1982. "The Mass Extinctions of the Late Mesozoic." January.

Schmidt-Nielsen, K. 1981. "Countercurrent Systems in Animals." May.

Schnapf, Julie L., and Denis A. Baylor. 1987. "How Photoreceptor Cells Respond to Light." April.

Short, R. V. 1984. "Breastfeeding." April.

Snyder, Solomon H. 1985. "The Molecular Basis of Communication between Cells." October.

Stryer, Lubert. 1987. "The Molecules of Visual Excitation." July.

Tonegawa, Susumu. 1985. "The Molecules of the Immune System." October.

Van Dyke, C., and R. Byck. 1982. "Cocaine." March.

Webb, P. W. 1984. "Form and Function in Fish Swimming." July.

Winfree, A. 1983. "Sudden Cardiac Death: A Problem of Topology." May.

Wurtman, R. 1982. "Nutrients That Modify Brain Function." April.

Young, John D., and Zanvil A. Cohn. 1988. "How Killer Cells Kill." January.

Zucker, M. 1980. "The Functioning of Blood Platelets." June.

PART SEVEN: BEHAVIOR AND ECOLOGY

Behavior

Barash, D. P. 1977. *Sociobiology and Behavior*. Elsevier, New York and Amsterdam.

Brown, J. L. 1975. *The Evolution of Behavior*. Norton, New York.

Christian, J. J. 1970. "Social Subordination, Population Density, and Mammalian Evolution." *Science* 168:84. Mammals have marked endocrine responses to crowding and to aggressive encounters.

Darwin, C. 1871. *The Descent of Man, and Selection in Relation to Sex*. Appleton, New York. Darwin has a great deal to say about the evolutionary importance of sexual behavior.

———. 1872. *The Expression of the Emotions in Man and Animals*. Appleton, New York. The first great work of comparative ethology.

Dawkins, R. 1976. *The Selfish Gene*. Oxford University Press, New York and Oxford. The selfish gene concept is applied to animal behavior in the second half of this popular paperback.

Ehrman, L., and P. A. Pasons. 1976. *The Genetics of Behavior*. Sinauer Associates, Sunderland, Mass. A textbook approach.

Gould, J. L. 1982. *Ethology of Behavior*. Norton. New York.

Grier, J. 1984. *Biology of Animal Behavior*. Times Mirror/Mosby, St. Louis. An introduction to general principles.

Jansen, D. H. 1966. "Coevolution of Mutualisms between Ants and Acacias in Central America." *Evolution* 20:249. One of biology's best gee-whiz stories.

Johnsgard, P. A. 1967. "Dawn Rendezvous on the Lek." *Natural History*. 76:16. A *lek* is a special place where males compete with one another for territory and prestige and wait for the arrival of sexually interested females.

Jolly, A. 1985. "The Evolution of Primate Behavior." *American Scientist* 70:230–239. A discussion including the general principles of the evolution of behavior.

Klopfer, P. H., and P. J. Hailman. 1973. *Behavioral Aspects of Ecology*. Prentice-Hall, Englewood Cliffs. N.J.

Maynard Smith, J. 1964. "Group Selection and Kin Selection." *Nature* 201:1145. The problem of group selection and the evolution of altruistic, group-directed behavior.

———. 1974. "The Theory of Games and the Evolution of Animal Conflicts." *Journal of Theoretical Biology* 47:209.

Mech. L. D. 1970. *The Wolf: The Ecology and Behavior of an Endangered Species*. Natural History Press, Garden City, N.Y. Fascinating reading for all you wolf fans.

Oster, G. F., and E. O. Wilson. 1978. *Caste and Ecology in the Social Insects*. Princeton University Press, Princeton, N.J. Hedging bets against rare disasters demands very different allocations of scarce resources from what would be feasible if the world were predictable.

Skinner, B. F. 1938. *The Behavior of Organisms: An Experimental Analysis*. Appleton-Century-Crofts, New York.

Trivers, R. L. 1971. "The Evolution of Reciprocal Altrusim." *Quarterly Review of Biology* 46:35. Frequently behavior that appears to be altruistic is really selfish in the long run: it is insurance for similar help when it might be needed.

Wallace, R. A. 1979. *The Ecology and Evolution of Animal Behavior*. 2d ed. Scott, Foresman, Glenview, Ill. The shorter version of Wallace's behavior book. The previous edition was among the first texts to bring together ecological and evolutionary aspects

of animal behavior with attention to a range of behavior studies.
———. 1979. *Animal Behavior: Its Development, Ecology, and Evolution.* Scott, Foresman, Glenview, Ill. In this treatment, Wallace brings to the field of animal behavior a strong evolutionary viewpoint and a lively style.
———. 1979. *The Genesis Factor.* William Morrow, New York. Wallace considers the evolutionary influences on human behavior in this enjoyable and thought-provoking popular book on the implications of sociobiology.
Wilson, E. O. 1971. *The Insect Societies.* Harvard University Press, Cambridge, Mass.
———. 1975. *Sociobiology, The New Synthesis.* Harvard University Press, Cambridge, Mass.
———. 1976. "Academic Vigilantism and the Political Significance of Sociobiology." *Bioscience* 26:183.
———. 1978. *On Human Nature.* Harvard University Press, Cambridge, Mass. The three books listed above form an "unplanned trilogy" in which Wilson successively considers insect societies, animal social behavior in general, and the application of evolutionary sociobiology to the study of human affairs.

***Scientific American* articles. New York: W. H. Freeman Co.**

Alkon, D. 1983. "Learning in a Marine Snail." *Scientific American,* July.
Heinrich, B. 1981. "The Regulation of Temperature in the Honeybee Swarm." *Scientific American,* June.
Lloyd, J. 1981. "Mimicry in the Sexual Signals of Fireflies." *Scientific American,* July.
Maynard Smith, J. 1978. "The Evolution of Behavior." *Scientific American,* September. More on the key problem of altruism and on the logic of formalized, nonlethal aggressive conflicts.
Wilson, E. O. 1975. "Slavery in Ants." *Scientific American,* June. Some slave-making species have become so specialized that they are no longer capable of feeding themselves.

Ecology

Ayensu, E., ed. 1980. *Jungles.* Crown, New York. Brief descriptions of the flora and fauna of jungles.
Borgstrom, G. 1976. "Never Before Has Humankind Had to Face the Problem of Feeding So Many People with So Little Food." *Smithsonian* 7:70.
Connell, J. H. 1978. "Diversity in Tropical Rain Forests and Coral Reefs." *Science* 199:1302. What makes some habitats so species-rich? Connell suggests that unpredictable natural disasters, such as typhoons, prevent them from ever approaching species equilibrium.
Corliss, J. B., et al. 1979. "Submarine Thermal Springs on the Galápagos Rift." *Science* 203:1073. *Alvin* takes a look at a novel ecosystem built on sulfur-oxidizing chemoautotrophic bacteria.
Egbert, G., et al., eds. 1982. *The Ecology of a Tropical Forest.* Smithsonian, Washington, DC. A fascinating introduction to rainforest ecology.
Gressit, J. L. 1977. "Symbiosis Runs Wild in a Lilliputian 'Forest' Growing on the Back of High-living Weevils in New Guinea." *Smithsonian* 7:135. With a title like that, what more can we say?
Hutchinson, G. E. 1959. "Homage to Santa Rosalia, or Why Are There So Many Kinds of Animals?" *American Naturalist* 93:145. A masterpiece by one of the century's most prestigious ecologists.
Moore, J. A. 1985. "Science as a Way of Knowing." *American Zoologist* 25:1–155. An excellent discussion of the effect of human behavior on the environment.
National Academy of Sciences (U.S.). 1975. *Underexploited Tropical Plants with Promising Economic Value.* Washington D.C. Instead of continuing to grow the standard "world crops" in new tropical nations, why not take a closer look at the native plants?
Odum, E. P. 1983. *Basic Ecology.* Saunders, Philadelphia.
Ricklefs, R. E. 1978. *Ecology,* 2d ed. Chiron Press, Newton, Mass. Our favorite all-around ecology textbook, which also has an excellent coverage of the basics of population genetics.
Schaller, G. B. 1972. *The Serengeti Lion: A Study of Predator—Prey Relations.* University of Chicago Press, Chicago.
Scientific American. 1970. *The Biosphere.* W. H. Freeman, New York.
Southwick, C., ed. 1985. *Global Ecology.* Sinauer Associates, Sunderland, Mass. A collection of readings on human and environmental interaction.
Terborgh, J. 1974. "Preservation of Natural Diversity, the Problem of Extinction-prone Species." *Bioscience* 24:715. How much responsibility should we accept for endangered species—and at what expense?
Wilson, E. O., ed. 1974. *Ecology, Evolution and Population Biology.* Readings from Scientific American. W. H. Freeman, New York.
Wilson, E. O., and W. H. Bossert. 1971. *A Primer of Population Biology.* Sinauer, Sunderland, Mass. (paperback).

***Scientific American* articles. New York: W. H. Freeman Co.**

Bergerud, A. 1984. "Prey Switching in a Simple Ecosystem." December.
Calhoun, J. R. 1962. "Population Density and Social Pathology." February. Rats kept in crowded conditions mistreat their babies and lose interest in sex.
Edmond, J., and K. von Damm. 1984. "Hot Springs on the Ocean Floor." April.
Horn, H. H. 1975. "Forest Succession." May. The succession of trees in a New Jersey forest depends primarily on soil moisture and the geometry of leaves.
Ingersoll, A. 1983. "The Atmosphere." *American,* September.
Mohnen, Volker A. 1988. "The Challenge of Acid Rain." August.
Perry, D. 1984. "The Canopy of the Tropical Rain Forest." November.
Revelle, R. 1982. "Carbon Dioxide and World Climate." July.

GLOSSARY

A band one of the bands of striated muscle, corresponding to the fixed length of the myosin filament.

abiotic 1. characterized by the absence of life. 2. nonbiological; factors independent of living organisms.

abiotic control control of population numbers by nonbiological factors, such as weather.

ABO blood group system a genetically controlled polymorphic cell surface polysaccharide antigen.

abomasum see *ruminant.*

abortion the spontaneous or *induced* expulsion of the human fetus before it is viable.

abscisic acid a plant hormone that suppresses dormancy.

abcission the normal separation, through an abscission zone, of flowers, fruit, and especially leaves from plants.

abscission also, *leaf fall,* the organized (seasonal or periodical) separation of leaves from the stem and the sequence of events involved.

absolute refractory period the brief time period following an action potential and during which repolarization is occurring, when new action potentials cannot be stimulated in that region of the axon.

absorption spectrum a graph indicating the relative light absorption of a molecule as a function of the wave length of light.

abyssal region the lowest depths of the ocean, especially the bottom waters; also, *abysmal region.*

accessory parts in flowers, any parts not directly involved in reproduction, such as the receptacle, calyx, and corolla.

accessory pigment a pigment other than chlorophyll that absorbs energy from light and is capable of transferring that energy to chlorophyll for photosynthesis.

acellular not composed of cells; not divisible into smaller cellular units; also, *noncellular.*

acetyl CoA a key intermediate in metabolism, consisting of an acetyl group covalently bounded to coenzyme A.

acetylcholine a neutrotransmitter released and hydrolyzed in certain synaptic transmission and in the initiation of muscle contraction.

acetylcholinesterase a membrane-bound enzyme that hydrolyzes acetylcholine in the course of synaptic nerve impulse transmission.

achondroplasia a genetic pathology, usually due to a dominant allele, in which the conversion of cartilage into bone is defective, resulting in extremely short appendages, dwarfism, and facial anomalies.

acid a compound capable of neutralizing alkalis and of lowering the pH of aqueous solutions. Acids are proton donors that yield hydronium ions in water solutions.

acid-base buffering system in body fluids such as blood, substances that aid the body in resisting changes in pH.

acidosis an abnormally acidic condition of body fluids.

acoelomate a lower invertebrate animal, one that does not develop a pseudocoelom or coelom.

acrosomal reaction during the fertilization process, the rupture of the sperm acrosome and the physical and chemical changes this produces.

actin a cytoplasmic protein known in both globular and fibrous forms. See *actin filament.*

actin filament also *myofilament,* a cytoplasmic protein fiber found in the cytoskeleton and in muscle contractile units, associated with cellular movement.

action potential during a neural impulse, a brief shift in the voltage potential across the membrane of a neuron from -60 mV to $+40$ mV.

action spectrum a graph relating light-induced biological activity (e.g., carbon fixation or oxygen release in photosynthesis) per photon as a function of wave length.

active immunization the conferring of immunity against a specific disease clinically, that is, through the introduction of weakened or killed disease agents or their antigens.

active site that part of an enzyme directly involved in specific enzymatic activity.

active transport energy-requiring transport of a substance across the cell membrane usually against the concentration gradient.

adaptation 1. an adjusting to conditions. 2. a change that improves function. 3. any alteration in the structure or function of an organism or any of its parts that results from natural selection and by which the organism becomes better able to survive and multiply in its environment.

adaptive radiation the spread of a population into new and differing environments, accompanied by adaptive evolutionary changes.

adenine a purine, one of the nitrogenous bases found in both DNA and RNA as well as in several coenzymes.

adenosine a nucleoside formed of adenine and ribose covalently linked.

adenylate cyclase an enzyme, usually incorporated into the cell membrane, that is capable of transforming ATP into cyclic AMP and pyrophosphate.

adhesion the molecular force of attraction in the area of contact between unlike substances.

ADP adenosine diphosphate, a compound of adenine, ribose, and two phosphate groups.

adrenal also *adrenal gland,* one of a pair of endocrine (ductless) glands located above the kidneys, each consisting of two distinct parts, a central *medulla* and an outer *cortex.*

adrenal cortex the outer portion of the adrenal gland producing steroid hormones.

adrenal medulla the inner portion of the adrenal gland producing epinephrine and norepinephrine (adrenaline and noradrenalin) as its principle hormonal products.

adsorption the adhesion of molecules to surfaces, such as the adsorption of water to cellulose cell walls.

advanced relatively later in evolutionary origin or state, as opposed to *primitive.* See also *derived.*

adventitious in plants, growth of a root or bud in an unusual place; *adventitious root,* a secondary root growing from stem tissue.

aerobe an organism that uses oxygen as a hydrogen acceptor in cellular respiration forming water.

aerobic utilizing oxygen in respiration.

aerobic respiration cell respiration requiring oxygen as an electron acceptor.

afferent vessel any blood vessel carrying blood toward a specific structure such as a gill filament or kidney nephron.

agar also *agar-agar,* a polysaccharide produced by red algae; also, a gel made from this material, used as a moist semisolid base for the experimental growth of microorganisms.

age structure pyramid also *age profile* and *population pyramid,* a pyramidal graph of a population, divided into age groups. Each age group is represented by a horizontal bar, with that of the youngest forming the base.

agglutination the clumping of bacteria, erythrocytes, or other cells by antibody/antigen association.

aggregation a number of independent organisms grouped together either in a casual, temporary way or for more permanent mutual reasons.

aggression hostility, attack, or threat, especially unprovoked, usually against a competitor or potential competitor.

AIDS (acquired immune deficiency syndrome) a recently discovered disease in which the immune system fails.

air sac in birds and insects, any of numerous extensions of the respiratory system as air-filled, membranous sacs into various body parts.

alarm chemicals pheromones released as alarm signals.

albedo effect the reflection of a portion of solar radiation by the atmosphere.

albinism a marked and abnormal, genetically based deficiency in pigmentation.

albumin 1. any of a class of clear, water-soluble plant or animal proteins. 2. *serum albumin,* a clear, water-soluble constituent of blood plasma, thought to serve detoxifying and osmotic functions. 3. *albumen* (egg white protein).

alcohol 1. ethyl alcohol, C_2H_5OH. 2. any of a class of organic compounds analogous to ethyl alcohol in having a —CH_2OH group and lacking aldehyde or carboxyl groups.

aldehyde any organic compound with the reactive group —CHO.

aldosterone a steroid hormone produced by the adrenal cortex, involved in potassium reabsorption by the kidney.

aleurone also *aleurone layer,* a single, outermost layer of cells in the endosperm of cereals.

alga (pl. *algae*), any photosynthetic member of the kingdom *Protista.*

alkali any substance capable of accepting protons and giving off hydroxide ions when in water (also see *base*).

alkaloid any organic, nitrogenous, alkaline, water-soluble, bitter-tasting compound produced by plants, usually with pharmacological effects; e.g., morphine, nicotine, quinine, caffeine.

alkaptonuria a genetic pathology caused by the lack of an essential enzyme and characterized by the excretion of homogentisic acid.

allantois one of the extraembryonic membranes. In birds and reptiles it serves as a repository for nitrogenous wastes of the embryo.

allele a particular form of a gene at a locus.

allele frequency the proportion (relative number) of alleles of a given type at a specific gene locus.

allogenic succession succession brought about by physical forces generally beyond the influence of living organisms.

allopatric speciation the rise of new species from populations that have become geographically isolated.

allosteric control the control of enzyme activity through the binding of small metabolites at one or more secondary (allosteric) binding sites

allosteric site on certain enzymes, a secondary binding site for small metabolites involved in the regulation of enzymatic activity.

allotetraploidy adj., derived from the hybridization of two distinct species and carrying the full diploid chromosome complements of both parental species; n. an allotetraploid organism. Also, *amphidiploid.*

alpha helix the right-handed helical configuration spontaneously formed by certain polymers. See *secondary structure of organization.*

alternation of generations see *sporic cycle.*

altricial of birds, helpless at hatching and requiring parental feeding and care. Compare *precocial.*

altruism behavior that is directly beneficial to others at some cost or risk to the altruistic individual.

Alvarez hypothesis a hypothesis that proposes that much of the massive extinction of life that accompanied the end of the Mesozoic era was produced by the aftereffects of a gigantic asteroid's collision with the earth.

alveolus (pl. *alveoli*), the air cells of the vertebrate lung, formed by the terminal dilation of tiny air passages.

ameba also *amoeba,* 1. any protozoan of the large genus *Amoeba,* characterized by lobose pseudopods and the lack of permanent organelles or supporting structures; 2. any ameboid protist, such as the ameboid stage of a flagellate or sporozoan.

amebocyte in sponges, an amoeboid cell involved in reproduction, digestion, and spicule formation.

ameboid also *amoeboid,* like an ameba in moving or changing shape by protoplasmic flow.

ameboid movement also *amoeboid movement,* or change in the shape of a cell by cytoskeleton deformation, the formation of pseudopods, and protoplasmic flow.

amino acid also *alpha-amino acid,* 1. any organic molecule of the general formula R—CH (NH_2)COOH, having both acidic and basic properties; 2. any of the 20 subunits found as normal constituents of polypeptides.

amino group the univalent group —NH_2, often ionized as —NH_3^+.

aminopeptidase a peptidase (proteolytic enzyme) that attacks peptide bonds adjacent to the amino end of a polypeptide chain.

ammonia 1. a colorless, pungent, poisonous, high soluble gas, NH_3. 2. this compound dissolved in water; ammonium hydroxide.

ammonification the production of ammonia by soil organisms, particularly by the reduction of nitrates or nitrites.

ammonium ion the univalent ion NH_4^+. Ammonia (NH_3) reacts with water to form an ammonium ion and a hydroxyl ion.

amnion the innermost of the extraembryonic membranes of reptiles, birds, and mammals, and thus the sac in which the embryo itself is suspended.

amniote egg as in reptiles and birds, an egg that forms an amnion within the egg case.

amoeba see *ameba.*

AMP (adenosine monophosphate) a molecule consisting of adenine, ribose, and one phosphate group.

amplexus the clasping of the female amphibian by the male during mating.

ampulla (pl. *ampullae*), a dilated portion of a canal or duct, such as the ampullae of the semicircular canals of the inner ear, and the ampullae of the echinoderm water vascular system, which are muscular sacs located above the tube feet.

amylase any enzyme that digests starch.

amylopectin a form of plant starch consisting of branched chains of alpha glucose subunits.

amylose a form of plant starch consisting of unbranched chains of alpha glucose subunits.

anabolic see *anabolism.*

anabolism in cells, synthetic (or building) chemical activity that produces a more highly ordered chemical organization and a higher free energy state. Compare with *catabolism.*

anaerobe an organism that does not require free oxygen to live. *Facultative anerobe,* an organism that can utilize free oxygen when available but does not require it (e.g., yeast). *Obligate anerobe,* an organism that cannot tolerate free oxygen.

anaerobic 1. living or functioning in the absence of oxygen. 2. *anaerobic respiration,* respiration not requiring or utilizing oxygen; glycolysis.

analogous in comparative morphology, similar in form or function but derived from different evolutionary or embryonic precursors (e.g., the wings of insects and birds).

analogy similarity in function, and perhaps appearance, but stemming from different evolutionary origins.

anaphase the stage of mitosis or meiosis in which the centromeres divide and separate and the two daughter chromosomes of each chromosome travel to opposite poles of the spindle; and in which the spindle elongates while the centromeric spindle fibers shorten.

anastomoses a netlike arrangement, as in neural fibers or blood vessels.

androecium a whorl of stamens; the stamens of a flower taken together.

androgen any substance that promotes masculine characteristics; male sex hormone.

anemone *sea anemone,* a sedentary cnidarian having a largish, cylindrical body and one or more whorls of tentacles.

angiosperm a plant in which the seeds are enclosed in an ovary; a flowering plant; a monocot or dicot.

animal a member of the animal kingdom, consisting solely of multicellular forms, almost entirely diploid except for the gamete stage.

animal hemisphere 1. the part of the ovum having the more metabolically active cytoplasm. 2. in the early embryo, the comparable region and the region in which the greatest amount of cell division occurs. Compare *vegetal hemisphere.*

anion a negatively charged ion.

annual yearly; a plant that completes its life cycle within a year; see *perennial.*

annual growth rings the concentric rings seen in the cross section of a woody stem, each ring corresponding to one year's growth.

annulus (pl. *annuli*), ("little ring") 1. a pore in the nuclear membrane. 2. the dense wall of a sporangium, within a sorus (ferns).

antagonist one of a pair of skeletal muscles (or groups of muscles), the actions of which oppose one another.

antenna 1. one of the paired, jointed, movable sensory appendages of the heads of insects or other arthropods. 2. *light antenna,* a cluster of photosynthetic pigments (chlorophyll and accessory pigments) which receives energy from photons and transfers that energy to a single photocenter.

anterior (adj.) toward the front or head end of an organism.

anterior (pituitary) lobe see *pituitary gland.*

anther the pollen-producing organ of the stamen.

antheridium (pl. *antheridia*), in lower plants and fungi, a male reproductive organ containing motile sperm.

antheriodogen a hormone inducing the formation of antheridia.

anthocyanin a class of water-soluble plant pigments, including most of those that give blue and red flowers their colors.

anthropoid 1. resembling a human. 2. *anthropoid ape,* any tailless ape of the family *Pongidae,* comprising the gorillas, chimpanzees, orangutans, gibbons and simians.

antibiotic any of a large number of substances, produced by various microorganisms and fungi, capable of inhibiting or killing bacteria and usually not harmful to higher organisms; e.g., *penicillin, streptomycin.*

antibiotic resistance the ability of an organism to survive in the presence of an antibiotic. Antibiotic resistance is sometimes transmitted from one bacterium to another by means of a plasmid vector.

antibody also *immunoglobulin,* one of a vast number of **Y**-shaped, defensive proteins, secreted by B-cell lymphocytes and capable of binding to matching antigens.

antibody-antigen complex a molecular complex consisting of an antibody bound noncovalently to one or more specific antigens.

anticodon a region of a transfer RNA molecule consisting of three sequential nucleotides capable of antiparallel Watson-Crick pairing with the three sequential nucleotides of a codon.

anticodon loop the region of a transfer RNA molecule containing the anticodon; one of the three or four loops of a transfer RNA molecule.

antidiuretic any substance that helps the body conserve water by increasing water reabsorption in the nephrons.

antidiuretic hormone (ADH) also *vasopressin,* a polypeptide hormone secreted by the posterior pituitary, the action of which is to increase the resorption of fluid from the kidney filtrate.

antigen 1. any large molecule, such as cell-surface protein or carbohydrate, that stimulates the production of specific antibodies, or that binds specifically with such antibodies. 2. any antibody-specific site on such a molecule.

antigenic determinant see *variable region.*

antigen-presenting cell an activated macrophage carrying an antigen that will activate a matching helper T-cell.

antihemophilic factor a blood factor necessary for normal clotting, congenitally lacking in hemophilics. Antihemophilic factor can be isolated from normal blood and administered to hemophilics.

antiparallel (adj.), running parallel but in the opposite direction; specifically, the two strands of DNA, which form parallel interwound helices in which the 5′-to-3′ direction in one strand is the 3′-to-5′ direction in the other.

antiserum (pl. *antisera*), blood serum containing antibodies specific to some particular antigen.

antisprouting factor a local hormone produced by the axons of nerves that inhibits the growth of other nerve axons, in a developmental negative feedback control loop.

anus the opening of the lower end of the alimentary canal (gut), through which solid wastes are voided.

aorta in vertebrates, the principle or largest artery, conveying blood away from the heart.

aortic arch in embryology, one of a series of five paired, curved blood vessels that arise in the embryo from the ventral aorta, pass through the branchial arches, and unite to form the dorsal aorta.

aortic body sensory structures in the aorta that respond to changes in blood CO_2 and pH levels by sending neural impulses to the respiratory control center.

aphotic zone in the aquatic environment, those depths below the penetration of effective light, where respiratory activity exceeds photosynthesis.

apical pertaining to the apex or tip, as of a shoot or root.

apical dominance the uppermost growing tip of a plant stem that hormonally inhibits the upward growth or formation of other branches.

apical meristem the meristem (actively dividing cells) of the growing tip of a root or shoot.

apodeme an ingrowth of the arthropod exoskeleton that serves as the point of attachment for a muscle.

apomixis the production of seeds without the union of gametes.

aponeurosis a flattened sheet of dense connective tissue covering certain muscles.

apoplastic route in roots, the movement of water along cell walls rather than through the cytoplasm, as it makes its way to the stele. (See also *symplastic route.*)

appendicular skeleton the bones of the limbs (two pairs), along with the pelvis and pectoral girdles.

appetitive behavior a variable, nonstereotyped part of instinctive behavior involving searching (for food, water, a mate) for the opportunity to perform.

appetitive stage see *appetitive behavior.*

aqueous pertaining to water; dissolved in water; watery.

arboreal tree-dwelling.

archegonium the female reproductive organs of ferns and bryophytes.

archenteron in embryology, the primitive digestive cavity of the gastrula.

arithmetic increase also *linear increase,* in populations, increases that occur by simple addition, thus 1, 2, 3, 4, 5, 6. 7. etc. Compare *exponential growth curve.*

arteriole a small artery.

arteriosclerosis inelasticity and thickening of the arterial walls.

artery a vessel carrying blood away from the heart and toward a capillary bed.

articulate (v.), to form a joint (with).

artificial selection the deliberate selection for breeding by humans of domesticated animals or plants on the basis of desired characteristics.

ascocarp a cuplike or saclike body in the Ascomytes in which asci are produced.

ascogonium female reproductive organ of ascomycetes which receives the antheridial nuclei in fertilization, and from which dikaryotic hyphae emerge.

ascomycete a sac fungus.

ascospore a haploid spore produced after sexual reproduction within the ascus of a sac fungus.

ascus (pl. *asci*), in ascomycetes, the sac in which meiosis occurs and in which four or eight ascospores are subsequently formed.

aseptic free of microorganisms.

asexual reproduction reproduction not involving the union of genetic material from two sources.

assay system in biology, a situation in which the amount or character of growth or other biological activity is an indicator of the presence or activity of a micronutrient, toxin, hormone, etc.

aster one of a pair of structures, formed in mitosis and meiosis of all animal cells and many other eukaryotic cells, each consisting of *astral rays* (of microtubules and microfilaments) radiating from a centriole.

astrocyte cells of the central nervous system specialized for transporting substances in and out of the capillaries. See also *blood brain barrier.*

atherosclerosis a form of arteriosclerosis characterized by fat deposits (plaques) in the inner lining of the arterial wall.

atlas the first cervical vertebra, which supports the head.

atom the smallest indivisible unit of an element still retaining the element's characteristics.

atomic mass (also, *atomic weight*), the average mass of the atoms of an element, given in daltons; the exact mass of a specific isotope.

atomic nucleus the central region of an atom, consisting of neutrons and positively charged protons, held together by strong nuclear force and constituting most of the atom's mass.

atomic number the number of protons in an atomic nucleus.

atomic weight see atomic mass.

ATP, adenosine triphosphate a ubiquitous small molecule involved in many biological energy exchange reactions, consisting of the nitrogenous base adenine, the sugar ribose, and three phosphate residues.

ATP synthetase an ADP phosphorylating enzyme present in F_1 and CF_1 bodies. Responsible for chemiosmotic phosphorylation (the phosphorylation of ADP, forming ATP).

atrioventricular node a small mass of specialized muscle fibers at the base of the wall between the atria of the heart, conducting impulses to the bundle of His.

atrium also *auricle,* either of the two upper chambers of the heart, each of which receives blood from veins and in turn forces it into the corresponding ventricle.

auditory canal the open, bony canal from the outer ear to the eardrum.

auditory receptor 1. a structure specialized for the reception of sound.

australopithecine pertaining to members of the extinct hominid genus *Australopithecus.*

autogenic succession community development or succession as a result of the activities of the organisms themselves.

autogenous hypothesis hypothetical explanation of the evolutionary origin of certain membrane-bound organelles in the eukaryotic cell, in which their presence is attributed to a series of modifications of the plasma membrane. (Also see *serial endosymbiosis hypothesis.*)

autoimmune disease a disease in which the organism's immune system attacks and destroys one or more of the organism's own tissues.

autoimmune response see *autoimmune disease.*

autoimmunity see *autoimmune disease.*

autonomic learning any learned response mediated by the autonomic nervous system.

autonomic nervous system the system of motor nerves and ganglion which innervates blood vessels, heart, smooth muscles, and glands and controls their involuntary functions. See *sympathetic* and the *parasympathetic divisions.*

autophagy "self-eating," destruction of aging or damaged cellular organelles by powerful hydrolyzing enzymes contained in lysosomes.

autoradiography the production of pictures *(audioradiographs)* revealing the presence of radioactive material in a thin object or section, in which a film or photographic emulsion is laid directly on the object to be tested, is exposed to radiation for a period of time, and is developed photographically.

autosome any chromosome other than the sex (X and Y) chromosomes.

autotetraploid (adj.), having four homologues of each chromosome type, derived from chromosomal duplication without cell division in a diploid organism; n., an autotetraploid organism.

autotroph a microorganism capable of using carbon dioxide as its only source of carbon (and thus receiving its energy from sources other than organic compounds).

auxin a class of natural or artificial substances that act as the principle growth hormone in plants. The naturally occurring auxin is primarily, or solely, *indoleacetic acid.*

axial pertaining to the axis. Axial fruit and flowers grow close to the stem, as opposed to terminal fruit and flowers, which grow on the tips of branches.

axial skeleton in vertebrates, the skull, vertebral column, and bones of the chest, opposed to appendicular skeleton.

axillary bud see *lateral branch bud.*

axon the (usually) long extension of a neuron that conducts nerve impulses away from the body of the cell.

axoneme in cilia and flagella, the active central unit, including the microtubules and their interconnections.

axopod (pl. *axopodia*), in sarcodines, slender, microtubular spines upon which feeding pseudopods move. See also *pseudopod.*

bacillus (pl. *bacilli*), 1. an aerobic, rod-shaped, spore-producing bacterium of the genus *Bacillus.* 2. any rod-shaped bacterium.

backcross a cross between an individual of the first filial generation (i.e., an individual produced by the cross between two true-breeding strains) and an individual of either parental strain; see *testcross.*

bacteriochlorophyll a purple photosynthetic pigment used by various anaerobic photosynthetic bacteria.

bacteriodopsin a light-sensitive pigment used by certain archaebacteria for capturing light energy to be used in chemiosmosis.

bacteriophage a virus that infects and lyses bacteria.

bacterium (pl. *bacteria*), any of numerous prokaryotic organisms.

ball-and-socket joint a joint allowing maximal rotation and flexion, consisting of a ball-like termination on one part, held within a concave, spherical socket on the other; e.g., the hip joint.

ball-and-stick model 1. a solid model of a molecule, in which atomic centers are represented by colored balls and covalent bonds are represented by sticks. 2. a drawing of such a model.

bark the portion of a stem outside the wood (xylem), consisting of cambium, phloem, cortex, epidermis, cork cambium, and cork; everything from the vascular cambium outward.

Barr body a dark-staining feature of the nucleus of the cells of female mammals, representing the condensed X chromosome.

basal body a structure found beneath each eukaryotic flagellum or cilium, consisting of a circle of nine short triplets of microtubules.

basal cell a bulbous cell at the base of the suspensor, believed to anchor and transport food into the embryo.

basal meristem a meristematic reserve in grasses that is below the leaves and stem, and thus is protected from grazing.

basal metabolism rate a measure of oxygen consumption per minute per kilogram of an individual at rest.

base 1. a compound that reacts with an acid to form a salt; a substance that releases hydroxyl ions when dissolved in water. 2. a nitrogenous base (purine or pyrimidine) of nucleic acids.

base deletion a mutation of DNA in which a single nucleotide or nucleotide pair is removed and the phosophate-sugar backbone is rejoined.

base insertion mutation in DNA in which a single nucleotide is inserted into the chain.

base pairing the specific pairing of purines and pyrimidines in DNA, ocurring between adenine and thymine and between guanine and cytosine, and an essential part of replication, transcription, and translation.

base substitution a mutation in DNA in which one nucleotide is replaced by another, or modified into another.

basement membrane in animals, a sheet of collagen that underlies and supports the cells of a tissue; also *basal membrane, basilar membrane.*

basidiocarp the spore-producing organ in the *Basidiomycetes;* a mushroom.

basidiospore an aerial spore produced by meiosis in *Basidiomycetes.*

basidium (pl. *basidia*), the meiotic cell of *Basidiomycetes* that produces basidiospores by budding.

basilar membrane in the vertebrate ear, a membrane that conducts sound waves; also *basement membrane, basal membrane.*

basking to expose one's body to warmth, as the warmth of the sun; a frequent component of *behavioral thermoregulation.*

Batesian mimicry coloration or structural configuration in harmless species, rendering it similar to that of a dangerous species.

B cell a lymphocyte that matures in the bone marrow (mammals) or bursa of Fabricius (birds), and later circulates in the blood; involved in the immune response, especially in the production of free antibodies.

behavior observable activity of an organism; anything an organism does that involves action and/or response to stimulation.

behavioral hierarchy a sequence of fixed action patterns in animals that proceeds from the general to the specific.

behavioral thermoregulation regulating one's internal body temperature by behavioral means, such as *basking,* seeking shade, and taking cold showers.

bell-shaped curve see *normal distribution.*

benthic referring to the benthos, or bottom-dwelling community of organisms.

beta sheet also *beta pleated sheet,* a linear, sheetlike configuration taken by polypeptides as they complete the secondary level of protein structure. Common in connective tissues and their constituents.

bicarbonate ion HCO_3^-.

bilateral symmetry having left and right sides that are approximate mirror images; having a single plane of symmetry.

bile a bitter, highly pigmented, alkaline, fat-emulsifying liquid secreted by the liver, containing bile salts and bile pigments.

bile salts cholesterol and other steroid compounds secreted by the liver and essential for the emulsifying, digestion, and absorption of dietary fats.

binary fission fission (splitting) into two organisms of approximately the same size; cell division (asexual reproduction) in prokaryotes.

binocular having or involving two eyes; *binocular vision,* stereoscopic vision, the ability to perceive depth through the integration of two overlapping fields of vision.

binomial system the tradition, introduced by Linnaeus, that each organism is given a taxonomic name in Latin consisting of a generic term and a specific term.

biochemical pathway a series of enzymatic steps by which an organic molecule is progressively modified.

biochemistry the science dealing with the chemistry of living organisms.

biodegradable capable of being rendered harmless upon exposure to the elements and organisms of the soil or water.

biogeochemical cycle the pathway of elements (e.g., carbon, nitrogen) or compounds (water) as they are taken up and released by organisms into the physical environment.

biogeography the study of the geographical distribution of living things.

biological clock an innate mechanism by

which living organisms are able to perceive the lapse of time.

biomass 1. the total weight of living organisms, usually expressed as dry weight, per unit area of volume in a particular habitat. 2. the weight of organic material produced in a unit time period under specified conditions.

biome a complex of ecological communities characterized by a distinctive type of vegetation, as determined by the climate.

biosphere the entire part of the earth's land, soil, waters, and atmosphere in which the living organisms are found.

biosynthesis see *intermediary metabolism.*

biotic 1. pertaining to life. 2. ecological factors due to the interactions of living organisms, as opposed to abiotic factors as climate.

biotic control population control by living factors, including both intraspecific and interspecific influences. Compare *abiotic control.*

biotic potential the maximum growth rate of a population when it is unrestricted by environmental resistance.

bipedal literally two-footed; walking on two feet, as birds, humans, kangaroos, and some dinosaurs.

bipinnaria larva a larval form in some echinoderms, similar to the tornaria larva of hemichordates.

birth control see *contraception.*

birth control pill an oral steroid contraceptive that inhibits ovulation, fertilization, or implantation, causing temporary infertility in women.

birth rate crude birth rate, the number of births per year per 1000 individuals of all ages in the populaiton. Other birth rates may be expressed in terms of births per year per number of women of a specific age range, the expected lifetime reproduction per female, and so on.

bivalve a mollusk with two shells (valves) hinged together; a pelecypod.

bladder 1. any membranous sac serving as a receptacle for fluid or gas.

bladderworm (technically, *cysticerci*) the encystment stage following the migration of newly hatched tapeworm larvae, often found in skeletal muscle of infected steers and hogs.

blade 1. *leaf blade,* broad part of a leaf, as distinguished from petiole or midrib. 2. any broad, thin part of the thallus (body) of a red, green, or brown alga.

blastocoel the cavity of a blastula.

blastocyst the early preimplantation stage in the mammalian embryo, in which the embryo consists of an *inner cell mass,* a *blastocyst cavity,* and an outer *trophoblast.*

blastoderm the single upper layer of cells of a blastula in embryos that develop in telolecithal eggs (fish, birds, and reptiles).

blastodisc the smallest disc of cytoplasm on the yolk of a bird or reptile egg, containing the egg nucleus, that becomes the early embryo.

blastomere any cell produced during cleavage, through the blastula stage.

blastopore in a gastrula, the opening of the archenteron produced by the invagination of cells during gastrulation.

blastula the early embryonic stage, consisting of a single layer of cells that form a hollow ball enclosing a central cavity, the blastocoel.

blending inheritance a theory of inheritance, which states that parental characters blend to produce an intermediate character in the offspring.

blight invasion of a plant by a parasitic water mold.

blood a circulating fluid in animals that helps distribute gases, nutrients, etc., and often collects cellular wastes.

blood brain barrier in the brain, a state of highly selective permeability to many substances that readily move into or out of other tissues, attributed in part to a lack of the usual looseness of capillary structure.

blood fluke parasitic flatworm of humans and other mammals.

blood type also *blood group,* one of a group of categories into which an individual can be categorized depending on his or her blood cell surface antigens.

Bohr effect the competitive effect of carbon dioxide in reducing the affinity of hemoglobin for oxygen.

bolus a clump of chewed food.

bombykol the sex attractant produced by the female silkworm moth.

bond see *chemical bond, covalent bond, hydrogen bond.*

bone the hard connective tissue forming the skeleton of most vertebrates, consisting primarily of a collagen matrix impregnated with calcium phosphate.

bone marrow see *marrow.*

book lung the respiratory organ of a spider, scorpion, or other terrestrial arachnid, consisting of thin, membranous structures arranged like the leaves of a book; *book gill,* the similar structure in horseshoe crabs.

botanist a scientist who studies plants.

botany the scientific study of plants.

bottle cell during involution in animal gastrulation, cells that take on a peculiar bottlelike appearance when drawn into the embryo.

bottleneck *population bottleneck,* a relatively short period of time during which the size of a population becomes unusually small, resulting in a random change in gene frequencies.

botulism a bacterial disease caused by the presence in foods (usually canned foods) of *Clostridium botulinum,* whose nerve toxins are among the most powerful poisons known.

bound ribosome a ribosome attached to the surface of a membrane, as of the endoplasmic reticulum. See also *endoplasmic reticulum, rough.*

bourgeois strategy in animal behavior, the immediate retreat from an adversary without physical encounter, as though through prearranged rules.

Bowman's capsule one of numerous double-walled membranous capsules in the nephron, each surrounding a glomerulus (ball of capillaries).

brain 1. in vertebrates, the anterior enlargement of the central nervous system, encased in the cranium. 2. in invertebrates, any anterior concentration of neurons more or less corresponding in function to the vertebrate brain.

braincase the primitive bony covering of the vertebrate brain as found in most fish and in the embryos of other vertebrates.

brain hormone (BH), also *prothoracicotrophic hormone (PTTH),* in insects, a hormone originating in the brain that stimulates the prothoracic gland to release ecdysone.

branch primordium see *primordium.*

broad-niched an organism able to live under a wide variety of conditions.

bronchiole a small branch of a bronchus, part of the respiratory tree of the lungs.

bronchus (pl. *bronchi),* either of the two main branches of the trachea.

brown alga any alga of the *Phaeophyta,* usually brown due to fucoxanthin pigment; e.g., kelp.

Brownian motion also *Brownian movement,* the irregular motion of microscopic particles in a liquid or gas, caused by random thermal agitation of molecules in the medium.

bud a small protuberance on the stem of a plant, containing meristematic tissue and covered with overlapping rudimentary foliage.

bud scale an often hairy, waxy, or resinous scale enclosing an immature bud.

bud scale scar external scar marking the extent of each year's growth in a woody stem.

budding 1. asexual cell reproduction with unequal cytoplasmic division, as of yeasts. 2. similar asexual reproduction of a multicellular animal, as in hydras.

buffer a solution of chemical compounds capable of neutralizing both acids and bases, and thus able to maintain an equilibrium pH.

bulk flow the net movement of water brought about by gravity or pressure.

bundle of His, also atrioventricular bundle, a bundle of specialized muscle fibers that conduct impulses along the ventricular septum.

bundle sheath cell cells directly surrounding leaf veins. In C4 plants, such cells carry out the Calvin cycle reactions.

calcareous containing or composed of calcium carbonate.

calcareous sponge a sponge with a skeleton of calcium carbonate spicules.

callus tissue an undifferentiated tissue that forms over wounds in plants.

calorie 1. also *small calorie* (calorie proper), the amount of heat (or equivalent chemical energy) needed to raise the temperature of one gram of water by 1°C. 2. also *large calorie, kilocalorie,* the heat needed to raise the temperature of a kilogram of water by 1°C; 1000 small calories.

Calvin cycle the cycle in C_3 photosynthesis in which NADPH and ATP reduce carbon dioxide to glyceraldehyde phosphate (PGAL).

calyx (pl. calyces) 1. the outermost whorl of floral parts (the sepals), usually green and leaflike. 2. in the kidney, cup-shaped regions that receive urine from the collecting ducts.

cambium undifferentiated meristematic tissue in a plant. See *cork cambium, primary growth, secondary growth, procambium, vascular cambium.*

canaliculus (pl. *canaliculi*), a tiny canal, as in bone, where canals communicate between osteocytes.

canopy the upper leafy area of a tree or especially of a forest.

capillary the smallest blood vessels; the fine channel between the arteriole and venule.

capillary action the tendency of a liquid to rise in a small tube due to adhesion to its inner surfaces and cohesion among water molecules.

capillary bed tissue rich in capillaries; a body of capillaries taken together.

capsid in viruses, the protein coat surrounding the nucleic acid core.

capsomere in viral capsids, an individual protein subunit whose varied arrangements determine shape of different viruses.

capsule 1. a type of fruit that becomes dry and hard before rupturing to release seeds. 2. in animals, any membranous sac or covering.

carapace a hard, shieldlike covering of the dorsal surface of an animal, as a lobster, turtle, armadillo, etc.

carbaminohemoglobin deoxygenated hemoglobin complexed with carbon dioxide.

carbohydrate a class of organic compounds with multiple hydroxyl side groups and an aldehyde or ketone group; sugars, starches, cellulose, and chitin; empirical formula $(CH_2O)n$.

carbon dioxide a colorless gas of the formula CO_2, readily dissolved in water and capable of reacting with water to form a hydroxyl ion (OH^-) and a bicarbonate ion (HCO_3^+).

carbonic anhydrase an enzyme that catalyses the reversible conversion of carbonic acid to carbon dioxide gas and water.

carboxyhemoglobin see *carbaminohemoglobin.*

carboxylation the enzymatic addition of carbon to substrate.

carboxylic acid group an organic acid consisting of carbon, oxygen, and hydrogen and designated —COOH. Upon ionization it becomes a carboxyl ion $-COO^-$

carboxypeptidase an exopeptidase; a protein-digesting enzyme that cleaves peptide bonds of amino acids at the carboxy-terminal end of the chain (see also *aminopeptidase*).

cardiac pertaining to the heart; near or toward the heart.

cardiac circuit the blood vessels of the heart, including the coronary arteries and veins and the capillaries.

cardiac muscle specialized muscle of the heart that is both striated and involuntary.

cardiac output the volume of blood pumped by the heart each minute.

cardiac sphincter the ring of muscle that closes the passageway between the lower esophagus and the stomach.

carnivorous (adj.) flesh-eating.

carotene a red or orange hydrocarbon, $C_{40}H_{56}$, found in most plants as an accessory photosynthetic pigment.

carotenoid any of a group of red, yellow, and orange plant pigments chemically and functionally similar to carotene.

carotid body a mass of cells and nerve endings on either carotid artery that senses blood CO_2 and pH levels and responds by affecting the rate of breathing and the heart beat.

carpel in flowers, a simple pistil, or a single member of a compound pistil; one sector or chamber of a compound fruit.

carposporangium see *carpospore.*

carpospore a diploid spore in members of the genus *Polysiphonia,* formed in the pouchlike carposporangium after fertilization.

carrageenan a polysaccharide produced by Irish moss, a red alga; used as a thickening agent.

carrier a heterozygote for a mutant or rare recessive allele.

carrier molecule a protein that binds a small molecule in facilitated diffusion or active transport.

carrying capacity a property of the environment defined as the size of a population that can be maintained indefinitely.

cartilage a firm, elastic, flexible, translucent type of connective tissue; in development, a precursor of bone formation.

cartilaginous fishes fish having skeletons of cartilage, comprising sharks, rays, and chimeras; one of the eight vertebrate classes.

Casparian strip a waxy strip on cell walls of endoderm that serves as a barrier to the conduction of moisture in roots and stems.

catabolic see *catabolism.*

catabolism in cells, chemical activity that decreases chemical organization and free energy therein. Compare with *anabolism.*

catalyst an agent that causes or accelerates a chemical reaction, while not being permanently altered itself.

cation a positively charged ion.

cecum in vertebrates, a blind pouch or diverticulum of the intestine, at the juncture of the small and large intestines.

cell 1. the structural unit of plant and animal life, consisting of cytoplasm and a nucleus, enclosed in a semipermeable membrane. 2. any similar organization, as that of a protist or prokaryotic organism.

cell body see *neuron.*

cell cycle the cycle of events in the life of a cell, including G_1 (Gap 1, synthesis and growth), **S** (DNA replication), G_2 (Gap 2, synthesis of spindle proteins), and **M** (mitosis and cell division).

cell division the division of a cell into two daughter cells.

cell-mediated response the response of activated cytotoxic T-cells, which includes the identification of, binding to, and lysis of cancerous and virus-infected cells.

cell plate in plant cell division, the forming plasma membrane between nascent daughter cells.

cell respiration the energy-yielding metabolism of foods in which oxygen is used. Also see *respiration* and glycolysis.

cell sap the fluid that fills large vacuoles in plant cells; primarily water with various substances in solution or suspension.

cell streaming see *cyclosis.*

cell-surface antigen any constituent of the cell surface capable of eliciting a specific antibody response.

cell theory the universally accepted proposal that cells are the functional units of organization in living organisms and that all cells today come from preexisting cells.

cellulose an inert, insoluble carbohydrate, a principal constituent of plant cell walls, consisting of unbranched chains of beta glucose.

cell wall the semirigid extracellular encasement of a plant, fungal, algal, cyanophyte, or bacterial cell that gives it a definite shape.

central canal 1. also *Haversian canal.* See *Haversian system.* 2. See *water vascular system.*

central cell or *endosperm mother cell,* in flowering plants, the binucleate cell of the embryo sac, containing two polar nuclei; following fertilization, becomes the triploid primary endosperm nucleus.

central dogma the proposition that all biological information is encoded in DNA, transmitted by DNA replication, transcribed into RNA, and translated into protein; together with several exceptions for certain misbehaving viruses. The term was coined by Fancis Crick.

central lymphoid tissue lymphoid tissue of bone marrow and thymus.

central nervous system the brain and spinal cord.

centriole one of a pair of organelles in animal, protist, fungi, and lower plant cells, each centriole in turn consisting of a pair of short cylinders of nine triplet microtubules.

centromere also *kinetochore,* also *spindle fiber attachment,* the specialized region of a chromosome to which spindle fibers are attached, visible under the light microscope as a light-staining body, and under the electron microscope as a pair of shieldlike plates. A centromere is considering to be a single object until the actual physical separation of daughter centromeres in anaphase.

centromeric spindle fiber any of a group of microtubules attached to each centromere and proceeding to a spindle pole in mitosis or meiosis.

cephalization the evolutionary tendency to concentrate neural and sensory functions in an anterior end.

cephalochordate ("head cord animal"), a lancelet *(Branchiostoma)* of a chordate subphylum in which the permanent notochord extends through what would be the head if it had one, which it doesn't.

cercaria see *human liver fluke.*

cerebellum a portion of the brain serving to coordinate voluntary movement, posture, and balance; it is located behind the cerebrum.

cerebral cortex the outermost region of the cerebrum, the "gray matter," consisting of several dense layers of neural cell bodies and including numerous conscious centers, as well as regions specialized in voluntary movement and sensory reception.

cerebral hemisphere either the right or left half of the cerebrum.

cerebrospinal fluid a cushioning fluid found in the ventricles of the brain and in the central canal of the spinal cord.

cerebrum the anterior, dorsal portion of the vertebrate brain, the largest portion in humans, consisting of two *cerebral hemispheres* and controlling many localized functions, among them voluntary movement, perception, speech, memory and thought.

cervical 1. pertaining to the neck. 2. pertaining to the cervix of the uterus.

cervical cap a birth control device consisting of a small cap that encloses the cervix.

cervical vertebra any vertebra of the neck region.

cervix the necklike base and opening of the uterus.

CF_1 see *CF_0CF_1 complex*. Structure on the outer surface of the thylakoid, the site of chemiosmotic phosphorylation in photosynthesis.

CF_0CF_1complex also called *CF_0CF_1 ATP synthetase,* an ATP-producing structure in the thylakoid membrane of the chloroplast, consisting of a protein-lined channel (CF_0) leading into a spherical region (CF_1) containing the enzyme ATP synthetase.

C4 pathway a CO_2-concentrating adaptation of certain tropical and desert (C4) plants. The C4 pathway originates in leaf mesophyll cells where highly efficient enzymes incorporate CO_2 into four carbon acids that then enter bundle sheath cells, release the CO_2 for use in the Calvin cycle, and recycle to the mesophyll cells.

C4 plant see *C4 pathway*.

chain reaction 1. a reaction which produces a product necessary for the continuation of the reaction. 2. any series of events in which each event causes the next.

chain termination codon any of the three codons of the genetic code causing the termination of polypeptide synthesis; in RNA these are UGA, UAG, and UAA.

chain termination mutation a base substitution mutation in which the new codon created is a chain termination codon.

chaparral a vegetation type common in California, characterized by a dense growth of low evergreen shrubs and trees.

character a classifiable feature, trait, or characteristic of an individual organism; a specific component of a phenotype.

character displacement where similar species share a niche, the tendency for physical differences to become emphasized through natural selection.

character state a specific state of morphological character, defined in terms of presence or absence.

Chargaff's rule in DNA structure, the observation that the quantity of adenine in DNA is equal to the quantity of thymine, while that of guanine is equal to that of cytosine.

charging enzyme any of a group of specific enzymes that covalently attach amino acids to their appropriate tRNA's.

chelicera (pl. *chelicerae*), one of the first pair of appendages in the *Chelicerata;* a spider's fang.

chemical bond any of several forms of attraction between atoms in a molecule.

chemical bond energy the potential energy invested in the formation of a chemical bond or that released upon its dissolution. See also *covalent bond, hydrogen bond.*

chemical communication the transmission of information between individuals by the use of phenomones.

chemical formula 1. representation of a molecule or compound, using chemical symbols and indicating the ratio of one element to another. 2. a mathematical statement including an equation, rule, principle, or answer.

chemically gated channel in a neuron, an ion channel whose gate is activated by chemical means rather than by voltage. See also *ion channel.*

chemical reaction the reciprocal action of chemical agents on one another; chemical change.

chemical synapse snyapses between neurons involving a space, the synaptic cleft, across which neurotransmitters must pass for a neural impulse to begin in the second neuron.

chemiosmosis the process in mitochondria, chloroplasts and aerobic bacteria in which an electron transport system utilizes the energy of photosynthesis or oxidation to pump hydrogen ions across a membrane, resulting in a proton concentration gradient that can be utilized to produce ATP.

chemiosmotic differential also known as *electrochemical proton gradient.* See *chemiosmosis.*

chemiosmotic phosphorylation the production of ATP using the energy of protons passing across a membrane and through F_1 and CF_1 particles.

chemoautotroph an organism capable of utilizing simple inorganic substances as a source of energy and as a source of raw materials for its metabolic activities.

chemoreceptor a neural receptor sensitive to a specific chemical or class of chemicals.

chemosynthesis the synthesis of organic compounds with energy derived from inorganic chemical reactions.

chemotroph see *chemoautotroph.*

chiasma (pl. *chiasmata*), as first seen in diplotene of the first meitoic prophase, the cross- or X-shaped configurations taken by homologous chromatids as repulsion occurs. The regions still fused indicate where crossing over occurred during pachytene.

chief cell a stomach cell that secretes pepsinogen.

chitin a structural carbohydrate that is the principal organic component of arthropod exoskeletons.

chloride ion the anion of the element chlorine (Cl^-).

chloride ion gate see *ion channel.*

chlorophyll a green photosynthetic pigment found in chloroplasts, cyanobacteria, and chloroxybacteria. It occurs in several forms, *chlorophyll a, b,* and *c.*

chloroplast a plastid containing chlorophyll.

chlorosis an abnormally pale or yellow condition of plants caused by lack of chlorophyll, due to disease or mineral deficiency.

choanocyte also *collar cell,* a type of cell in all sponges and certain protists in which a single flagellum is surrounded at its base by a screen of fused cilia (collar) that filters food from the water current created by the flagellum.

cholesterol a common sterol occurring in all animal fats, a vital component of plasma membranes, an important constituent of bile for fat absorption, a precursor of vitamin D, and too much of which is not good for you.

chordae tendineae in the heart, tough, cordlike tendons that prevent backflow by holding the tricuspid and bicuspid (auricular-ventricular) valves in place during systole.

chorioallantois a highly vascular extraembryonic membrane of birds, reptiles, and some mammals, formed by the fusion of the chorion and the allantois.

chorion 1. the tough covering on insect eggs; 2. the outermost of the extraembryonic membranes of birds, reptiles, and mammals, contributing to the formation of the placenta in placental mammals.

chorionic villi small fingerlike processes of the chorion of the early mammalian embryo, especially before the formation of the placenta.

chromatid in a G_2 chromosome, one of the two identical strands of the chromosome following replication and prior to cell division.

chromatin the substance of chromosomes, a molecular complex consisting of DNA, histones, nonhistone chromosomal proteins, and usually some RNA of unknown function.

chromomere one of the beadlike clumps or granules arranged in a linear array of a chromonema, especially when visible in partially condensed prophase chromosomes. Chromomeres may be more or less representative of functional genetic units (genes).

chromosomal mutation a massive spontaneous change in DNA, generally breakage involving a whole chromosome that has not been repaired or has been repaired improperly.

chromosomal replication see *DNA replication.*

chromosome 1. in eukaryotes, an independent nuclear body carrying genetic information in a specific linear order, and consisting of one linear DNA molecule (in G_1) or two DNA molecules (in G_2), one centromere, and associated proteins. 2. in prokaryotes, an analogous circular DNA molecule. 3. the analogous DNA or genetic RNA molecule of a virus.

chromosome puff in polytene chromosomes, an enlargement of one band associated with transcriptional activity (mRNA production).

chrysophyte a yellow-green alga.

chylomicron in digestion, a minute, protein-coated fat droplet formed during lipid transport in the intestinal villi.

chyme churned, semiliquefied food in the stomach.

chymotrypsin a proteolytic enzyme produced by the pancreas.

cilia (sing. *cilium*), fine, hairlike, motile organelles found in groups on the surface of some cells; shorter and more numerous than flagella, but similar in structure, they exhibit coordinated oarlike movement.

ciliary basal body see *basal body*.

circadian rhythm any recurrent sequence of physiological or behavioral activities repeated on a daily basis.

circulatory system the vascular system, consisting of blood-forming organs or tissues, vessels, the heart, and blood.

cisterna slender, membranous channels making up much of the Golgi apparatus.

cistron 1. also *structural gene,* a sequence of DNA specifying the sequence of a polypeptide chain. 2. any continuous genetic unit in which different recessive mutant lesions fail to complement one another in a double heterozygote.

citric acid cycle also *Krebs cycle, tricarboxylic acid cycle,* a cyclic series of chemical tranformations in the mitochondrion by which pyruvate is degraded to carbon dioxide; NAD and FAD are reduced to $NADH_2$ and $FADH_2$; and ATP is generated.

clamp connection upon mitosis in the dikaryotic cells of certain fungi, the formation of a peculiar branching and rejoining of cytoplasm, occurring in such a way as to ensure each daughter cell's receiving both a plus and a minus strain nucleus.

clasper one of a pair of specialized grooved caudal fins used by male sharks and rays as a penis.

class a major taxonomic grouping intermediate between *phylum* (or *division*) and *order*.

classical conditioning see *conditioning, classical*.

cleaning symbiont a usually conspicuous organism that interacts mutualistically with others by removing ectoparasites, necrotic tissue, etc,. receiving nourishment and inhibiting aggressive behavior of the individual being cleaned.

cleavage the total or partial division of a zygote into smaller cells (blastomeres).

climax 1. that stage in ecological succession of plant or animal community that is stable and self-perpetuating. 2. an orgasm.

cline a regular change in allele frequency over geographic space.

clitellum a thickened, ringlike glandular portion of the body wall of earthworms, which secretes mucus to form a cocoon for eggs.

clitoris in female mammals, an erectile, erotically sensitive organ of the vulva, homologous embryologically to most of the penis.

cloaca the common cavity into which the intestinal, urinary, and reproductive canals open in vertebrates other than placental and marsupial mammals.

clonal selection theory in immunology, the proposition that all potential antibody specificity is present in differentiated cells early in development and that the specific immune response consists of inducing appropriately differentiated cells to proliferate clonally.

clone 1. a group of genetically identical organisms derived from a single individual by asexual reproduction. 2 a group of identically differentiated cells derived mitotically from a single differentiated cell.

cloning, gene see *gene cloning*.

closed circulatory system a system in which blood is enclosed within arteries, veins, and capillaries throughout, and is not in direct contact with cells other than those lining these vessels.

closed community biotic communities that are abruptly circumscribed, often because of a confining physical barrier such as a river, bay, or gorge.

cnidoblast in cnidarians, a specialized cell containing a stinging or snaring nematocyst.

cnidocyte in cnidarians, a stinging cell.

coacervate droplets with membranelike surface layers that form spontaneously when certain substances are attracted together in water, a hypothetical step in the earliest formation of cells.

coccus (pl. *cocci*), any spherical bacterium (principally eubacteria), a condition found in many distantly related groups.

cochlea in mammals, a spiral cavity of the inner ear containing fluid, vibrating membranes, and sound-sensitive neural receptors.

cocoon in some invertebrates, a covering used to protect offspring during a stage in development, for example, the silky cocoon of moths and spiders.

codominance the individual expression of both alleles in a heterozygote; see *dominance relationships*.

codon also code group, 1. a series of three nucleotides in mRNA specifying a specific amino acid (or chain termination) in protein synthesis. 2. the colinear, complementary series of three nucleotides or nucleotide pairs in the DNA from which mRNA codon is transcribed.

coelenteron the saclike cavity of cnidarians, which carries out ingestion, digestion, phagocytosis and absorption of food, and also has circulatory functions; also, gastrovascular cavity.

coelom *true coelom,* a principal body cavity, or one of several such cavities between the body wall and gut, entirely lined with mesodermal epithelium; compare *pseudocoelom*.

coelomate having a true coelom.

coenocytic consisting of a multinucleate mass of cytoplasm without subdivision into cells.

coenzyme a small organic molecule required for an enzymatic reaction.

coenzyme Q a mobile electron carrier in the mitochondrion, capable of transporting both electrons and protons.

coevolution evolution of two closely interacting species, such as predator and prey, where changes in the first determine those of the second.

cofactor 1. any organic or inorganic substance, especially an ion, that is required for the function of an enzyme. 2. in blood clotting, any of the many active participating elements.

cohesion the attraction between the molecules of a single substance.

coitus the act of sexual intercourse, especially between people.

coitus interruptus coitus that is intentionally interrupted by withdrawal before ejaculation.

coleorhiza a protective sheath over the radicle in the embryo of grasses and grains.

coleoptile in the embryos and early growth of grasses, a specialized tubular structure, completely enclosing and protecting the pulmule during emergence.

colinearity *principle of colinearity,* the finding that corresponding parts of a structural gene, mRNA, and polypeptide occur in the same linear order.

collagen in animals, a widely distributed fibrous protein of connective tissue that forms much of the structure of tendons and ligaments.

collar cell see *choanocyte*.

collecting duct also *collecting tubule,* the part of a nephron that collects fluids from distal convoluted tubules and discharges it into the renal pelvis.

collenchyma in plants, a strengthening tissue, a modified parenchyma consisting of elongated cells with greatly thickened cellulose walls.

colon in mammals, the large intestine from the cecum to the rectum; including the ascending, transverse, descending, and sigmoid regions, and the rectum.

colonial 1. generally existing in colonies.
2. *colonial organism,* an organism of semidependent parts, derived by asexual reproduction

colony 1. a group of animals or plants of the same kind living in a close semidependent association. 2. an aggregation of bacteria growing together as the descendants of a single individual, usually on a culture plate.

colorblindness the inability to distinguish colors. *Red-green colorblindness,* the inability to distinguish certain shades of red from corresponding shades of green.

colostrum clear, yellowish milk secreted a few days before and after giving birth; colostrum is rich in maternal antibodies.

columnar epithelium epithelium consisting of one *(simple columnar)* or more *(stratified columnar)* layers of elongated, cylindrical cells.

comb a locomotor structure in ctenophorans, one of several slender plates on the body surface, bearing numerous cilia.

comb jelly see *ctenophore*.

combination pill a birth control pill that combines two or more female hormones.

common descent descent of two or more species (or individuals) from a common ancestor; e.g., the similarity in blood chemistry of apes and humans is due to common descent.

community an assemblage of interacting plants and animals forming an identifiable group within a biome, as in salt marsh or sage desert community.

community development also *ecological succes-*

sion and *succession* 1. the process of change in the populations of an area as competing organisms alter their environment. 2. the sequence of identifiable ecological stages or communities occurring over time in the progress of bare rock to a climax community.

compact bone dense, hard bone with spaces of microscopic size.

companion cell in plants, a nucleated cell adjacent to a sieve tube member, believed to assist it in its functions.

comparative anatomy the science of comparing the anatomy of animals, tracing homologies, and drawing evolutionary and phylogenetic inferences.

comparative psychology the study of mental processes (behavior) in animals from a comparative point of view, usually in a laboratory environment.

competition 1. seeking to gain what another is seeking to gain at the same time; a common struggle for the same object. 2. in ecology, the utilization by two or more individuals or species of the same limiting resource.

competitive exclusion principle see *Gause's law.*

competitive inhibition enzyme inhibition involving molecules similar to the substrate that compete for the active site.

complement (n.) 1. a group of blood proteins that interact with antibody-antigen complexes to destroy foreign cells. 2. (v.t.) the production of a normal phenotype by two recessive mutations in a double heterozygote.

complete flower see *flower.*

complete digestive tract a tubular digestive tract with an anal as well as an oral opening.

complete metamorphosis of insects, development includes distinct larval, pupal, and adult stages.

composite flower inflorescence in which the individual flowers are tightly clustered into a disklike head, often with a border of ray flowers that include petals only.

compound in chemistry, a pure substance consisting of two or more elements in a fixed ratio; consisting of a single molecular type.

compound eye an arthropod eye consisting of many simple eyes closely crowded together, each with an individual lens and a restricted field of vision, so that a mosaic image is formed.

compound microscope an optical instrument for forming magnified images of small objects, consisting of an *objective lens* that can be brought close to the object being examined, aligned with an *ocular lens* mounted at the other end of a body tube of fixed length.

concentration gradient 1. a slow, consistent decrease in the concentration of a substance along a line in space. 2. for any spatial difference in concentration, the direction away from the region of greater concentration.

conceptacles in *Fucus,* a strongly heteromorphic brown alga, regions within the blade where the minute gametophyte phase arises.

conclusion a statement, following an experiment or observation, accepting or rejecting a hypothesis. Acceptance may be expressed in provisional terms, indicating the degree of confidence one has in the test.

condensation of chromosomes, the coiling and supercoiling that transfers diffuse chromatin into a compact, discrete body in mitosis.

conditioned response an involuntary response that becomes associated with an arbitrary, previously unrelated stimulus through repeated presentation of the arbitrary stimulus simultaneously with a stimulus normally yielding the response.

conditioning, classical the process by which a conditional response is learned and elicited; compare *operant conditioning.*

condom a thin sheath of rubber or animal membrane worn over the penis during sexual intercourse to prevent conception or venereal infection (after Dr. Condom, an 18th-century English physician); also, *rubber, prophylactic.*

conducting tissue see *vascular tissue.*

conduction 1. the transfer of movement of fluid, heat, ions, impulses, etc. through or along a channel or medium; e.g., the conduction of impulses through the central nervous system. 2. the transfer of heat through a solid, as opposed to *convection* or *radiation.*

cone 1. one of a class of conical photoreceptors in the retina that detect color, consisting of a highly modifed cilium with specialized proteinaceous pigments for detecting red, green, or blue wavelengths. 2. a male or female reproductive structure of conifers, consisting of a cluster of scalelike modified leaves and either pollen or ovules, or the seed.

conidiophore a specialized branch of the mycelium bearing conidia.

conidium (pl. *conidia*) an asexual spore borne on the tip of a fungal hypha; also *conidiospore.*

conifer an evergreen gymnosperm of the order *Coniferales,* bearing ovules and pollen in cones; included are spruce, fir, pine, cedar, and juniper.

conjugated protein a compound of one or more polypeptides with one or more nonprotein substance; e.g., hemoglobin, which consists of four polypeptide chains and four heme groups.

conjugation in ciliates, a temporary cytoplasmic union in pairs, accompanied by meiosis and the exchange of haploid nuclei.

connective tissue a principal type of vertebrate supporting tissue, often with an extracellular matrix of collagen. Included are bone, cartilage, ligaments, and blood.

connective tissue matrix a noncellular, secreted matrix in which the cells of some connective tissues are embedded or immersed, e.g., bone, cartilage, and blood.

conservative replication replication of a molecule in which the original molecule remains intact. Compare *semiconservative replication.*

consolidation the transfer of engrams from short-term memory to long-term memory.

consolidation hypothesis the presumption that memory is first stored in short-term centers, where it is rapidly lost unless it is transferred to long-term centers where its loss is more gradual.

constant region the C-terminal portion of an immunoglobulin (antibody) light or heavy chain, not involved in antigene-specific binding and coded by a constant-region cistron.

consumer one that consumes; in ecology, an animal that feeds on plants (primary consumer) or other animals (secondary consumer).

consummatory behavior the satisfying, stereotyped behavior that completes an instinctive act. Usually preceded by *appetitive behavior.*

consummatory stage see *consummatory behavior.*

contact receptor see *sensory receptor.*

continental drift a theory proposing that today's continents or land masses were once parts of a supercontinent called Pangaea, which started to divide and drift apart some 200 million years ago. Also, *plate tectonics.*

continental shelf the sea bottom of shallow oceanic waters adjacent to the shores of a continent; considered to be a submerged part of the continent.

continuously varying trait see *polygenic trait.*

continuous variation see *polygenic inheritance.*

contraception, also *birth control,* any process or method intended to prevent the sperm from reaching and fertilizing the egg, or preventing ovulation or implantation.

contractile capable of contracting (shortening, drawing in, or becoming smaller).

contractile unit (of muscle), see *sarcomere.*

contractile vacuole an osmoregulatory, water-containing vacuole in protists, capable of filling and emptying through the contraction of microfilaments.

control a standard of comparison in a scientific experiment; a replicate of the experiment in which a possibly crucial factor being studied is omitted.

controlled experiment an experiment involving several replicates, each differing by a single variable factor.

convection the transfer of heat by the movement of a circulating fluid or gas; compare *radiation, conduction.*

convergent evolution the independent evolution of similar structures in distantly related organisms; often found in organisms that have adopted similar ecological niches, as marsupial moles and placental moles.

copulation sexual union or sexual intercourse in animals involving internal fertilization.

coral 1. a colonial anthozoan that secretes a calcareous skeleton. 2. the skeleton itself.

coriolus effect the deflection of a moving body of water or air caused by the rotation of the earth, resulting in clockwise motion in the Northern hemisphere and counterclockwise motion in the Southern hemisphere, as seen in tornadoes, ocean currents, and prevailing winds.

cork 1. in plants, secondary tissue, produced by the *cork cambium,* consisting of cells that become heavily suberized and die at maturity, resistant to the passage of moisture and gasses; the outer layer of bark. 2. such a tissue from the cork oak, used to seal wine bottles.

cork cambium also *phellogen,* in plants, the outermost meristematic layer of the stem of woody plants, from which the outer layer of bark is produced.

corneal lens see *compound eye.*

corolla in flowers, the whorl of petals surrounding the carpels or carpel.

corona radiata ("radiating crown"), an aggregation of follicle cells surrounding the mammalian egg at ovulation.

coronary thrombosis the formation of a blood clot that clings to a coronary vessel.

corpus allatum in insects, one of the paired endocrine bodies that secretes juvenile hormone, which prompts retention of juvenile stages.

corpus callosum a broad, white neural tract that connects the cerebral hemispheres and correlates their activities.

corpus cavernosum (pl. *corpora cavernosa*), in the penis or clitoris, a mass of erectile tissue with large interspaces capable of being distended with blood.

corpus luteum (pl. *corpus lutea*), ("yellow body"), a temporary yellow endocrine body on the surface of the ovary, consisting of secretory cells filling a follicle after ovulation; regressing quickly if the ovum is not fertilized or persisting throughout pregnancy.

cortex 1. *(zoology)* the outer layer or rind of an organ, as *adrenal cortex, kidney cortex.* 2. *(botany)* the portion of stem between the epidermis and the vascular tissue.

cortical (adj.), referring to the cortex.

cortical granules membranous vesicles beneath the surface of unfertilized echinoderm eggs; at fertilization these rupture to form the *fertilization membrane,* during the *cortical reaction.*

cortical reaction during fertilization in echinoderms, the rupture of cortical granules, and subsequent events leading to the formation of a fertilization membrane.

cotransport carrier a membranal carrier that, in each operating cycle, transports two different substances in the same or in opposite directions.

cotyledon also *seed leaf,* a food-storing structure in dicot seeds, sometimes emerging as first leaves; food-digesting organ in most monocot seeds; first leaves in a gymnosperm embryo.

cotylosaur a group of very early Carboniferous reptiles from which most later reptiles evolved.

countercurrent exchange a provision for the efficient uptake or release of heat, molecules, or ions made possible by an opposing flow of fluids between which the exchange is being made. Examples include heat exchangers in many thermoregulators, gas exchangers such as the gill filament of the fish, and the countercurrent sodium exchange loop of the kidney.

countercurrent heat exchanger a heat-retaining mechanism in animals in which warm blood leaving the body core passes cooler blood returning from the extremities in closely paralleled vessels, thus providing the opportunity for heat to pass into the cooler blood along the full extent of the vessels.

covalent bond a relatively strong chemical bond in which an electron pair is shared by two atoms, simultaneously filling the outer electron shells of both.

cranial capacity the volume of the brain vault of the skull and thus the brain, expressed in cubic centimeters, and used in physical anthropology for comparative purposes.

cranial nerve any of the 12 large, paired nerves that emerge directly from the brain to serve both somatic and autonomic nervous systems. Compare *spinal nerves.*

cranium 1. the skull. 2. the part of the skull enclosing the brain.

crash population crash, a sudden die-off or diminution in numbers, particularly after the carrying capacity of the environment has been exceeded and resources have been depleted.

crassulacean acid metabolism (CAM) in desert plants, a means of obtaining carbon dioxide without the risk of opening the stomata during the day. CAM plants utilize a biochemical system in which carbon dioxide is admitted into the plant at night and fixed into organic acids that can be broken back down during daylight, releasing carbon dioxide for the Calvin cycle.

creatine phosphate also *phosphocreatine,* in vertebrate muscles, a source of high-energy phosphate for restoring ATP expended during muscle contraction.

creationism the doctrine that all things, especially plant and animal species, were created fairly recently and substantially as they now exist, by an omnipotent creator and not by gradual evolution.

cretinism a recessive genetic abnormality that results in the inability to produce thyroxine; affected persons are extremely retarded physically and mentally unless thyroxine is administered from early infancy.

Crick strand either of the two strands of DNA, the other being designated the *Watson strand.*

crinoid a type of sessile, stalked echinoderm.

crista (pl. *cristae*), a shelflike fold of the inner mitochondrial membrane into the central cavity of the mitochondrion.

crop also *craw,* 1. an enlargement of the gullet of many birds, which serves as a temporary storage organ. 2. an analogous organ in certain insects and earthworms.

crosscurrent flow in the bird lung, the flow of blood at right angles to the passage of air.

crossing over 1. the exchange of chromatid segments by enzymatic breakage and reunion during meiotic prophase. 2. a specific instance of such an exchange; a crossover.

crossover 1. a crossing over. 2. the result of crossing over. 3. the place on a chromatid where crossing over has occurred.

cryptic coloration coloration that serves to conceal or to render inconspicuous.

cryptogam seedless vascular plants; those that do not produce flowers, cones, or seeds. Simple spores form separate, independent gametophytes in which sexual reproduction occurs (psilophytes, lycophytes, sphenophytes, and pterophytes).

ctenophore also *comb jelly,* any marine swimming invertebrate of the phylum *Ctenophora,* with a gelatinous body and eight rows of plates (combs) of fused cilia.

C-terminal end also *carboxy-terminal end,* the end of a polypeptide chain in which the final amino acid has its primary amino group fixed in a peptide linkage and its primary carboxyl group free; the last part of the chain to be synthesized.

C3 plant a plant in which the initial product of carbon dioxide fixation is a 3-carbon compound of the Calvin cycle. C3 *pathway,* the Calvin cycle.

cuticle a tough, often waterproof, nonliving covering, usually secreted by epidermal cells.

cutin a waxy, water-resistant substance covering the epidermis of leaves and stems.

cutting see *vegetative propagation.*

cuttlebone the skeleton of a cuttlefish, filled with thin, hollow chambers and used as a flotation device.

cyanobacterium formerly *blue-green alga,* any of a large group of blue-green photosynthetic prokaryotes having as photopigments chlorophyll *a,* phycocyanin, and phycoerythrin; and producing oxygen as a photosynthetic waste product.

cycad a plant of an order of palmlike gymnosperms, known from the Triassic to the present.

cyclic AMP (cAMP) adenosine monophosphate in which the phosphate is linked between the 3′ and 5′ carbons of the ribose group; serves as an intracellular gene regulator under a variety of circumstances.

cyclic phosphorylation light reactions employing only photosystem I, during which electrons from P700 reaction center pass through the associated electron transport system and cycle back to P700, thus serving chemiosmosis only.

cyclosis also *cell streaming* or *cytoplasmic streaming,* the movement of cytoplasm within the cell, often in more or less regular, circular pathways.

cyst 1. a bladder, sac, or vesicle. 2. a closed sac, containing fluid, embedded in a tissue. 3. a heavy protective covering of a dormant animal or protist. 4. an encysted organism.

cytochrome 1. any of a group of iron heme enzymes or carrier proteins in the oxidative and photosynthetic electron transport chains. 2. any iron heme protein other than hemoglobin or myoglobin.

cytochrome *c* a mobile, surface electron carrier in the mitochondrion. A widespread molecule, common to nearly all life and whose amino acid content has been the subject of numerous comparative studies.

cytokinesis cytoplasmic cell division; actual division of the cell into two daughter cells.

cytokinin a plant cell hormone, mitogen, and plant tissue culture growth factor, which interacts with other plant hormones in the control of cell differentiation.

cytological map a map locating genes on the physical chromosome by means other than recombination or genetic mapping.
cytoplasm the semisolid, protein-rich matrix of a cell exclusive of the plasma membrane, the nucleus, or other large inclusions.
cytoplasmic binding protein a cytoplasmic protein believed to bind with certain hormones prior to their entry into the nucleus, where the complex activates a gene.
cytoplasmic streaming see *cyclosis.*
cytopyge the fixed position on the surface of a protist from which wastes are discharged by exocytosis; the "anus" of a ciliate protozoan.
cytosine a pyrimadine, one of the four nucleotide bases of DNA, and also one of the four bases of RNA.
cytoskeleton the internal structure of animal cells, composed of microtubules and actin microfilaments; it controls the size, shape, and movement of the cell.
cytotoxic T-cell see *cell-mediated response.*
dalton a unit of atomic and molecular mass, equivalent to one twelfth the mass of an atom of carbon 12.
Danielli model a historically important (and largely validated) conception of the cell membrane as a phospholipid bilayer stabilized by proteins with channels for the passage of water-soluble compounds.
dark clock an unknown mechanism by which plants are able to measure a period of darkness with fair accuracy, and thus produce flowers at the appropriate season.
dark reaction see *light-independent reaction.*
Darwinian fitness see under *fitness.*
daughter cell either of the two cells created when one cell divides.
day neutral plant A plant whose flowering activities are not influenced by the length of day or night.
deamination the removal of an amino group.
death rate 1. in human populations, the number of deaths per 1000 population per year. 2. any analogous rate.
decarboxylation the enzymatic removal of carbon to a substrate.
Compare *carboxylation.*
deciduous plant a perennial plant that seasonally drops its leaves.
decomposer also *reducer,* an organism that breaks down organic wastes and the remains of dead organisms into simpler compounds, such as carbon dioxide, ammonia, and water.
dedifferentiation the return of a differentiated cell to an undifferentiated, plastic state.
deductive reasoning logical progression proceeding from the general to the specific. Making specific deductions based on a larger generalization or premise.
dehydration linkage a covalent bond formed between two compounds by the removal of one oxygen atom and two hydrogen atoms.
dehydration reaction also *dehydration synthesis,* an enzymatic reaction during which water is lost and a covalent bond forms between the reactants. See also *dehydration linkage.*
dehydration synthesis the enzymatic union of molecular subunits involving the removal of water and the forming of a dehydration linkage.
deletion 1. the removal of any segment of a chromosome or gene. 2. the site of such a removal after chromosome healing, considered as a mutation. See also *base deletion.* 3. the deleted chromosome.
deletion map a genetic map constructed from recombination tests between various deletion chromosomes.
demographic transition or *theory of demographic transition,* the proposal that in the development of nations, their population growth has distinct stages: (a) high birth and death rates and slow growth; (b) high birth rate, low death rate, and very rapid growth; and (c) low birth and death rates, and very slow growth.
demography the statistical study of human population (or other) with reference to growth, birth, and death rates, size, density, and migration.
denaturation the alteration of a protein so as to destroy its properties, through heating or chemical treatment (protein denaturation may be reversible or irreversible).
dendrite an extension of a neuron that conducts impulses toward the cell body and axon.
denitrification the conversion of nitrates and nitrites to nitrogen gas; *denitrifying bacteria,* common soil and manure bacteria that are responsible for denitrification.
density-dependent effects factors affecting population parameters (growth, birth, and death rates) in different ways depending on competition and population density, e.g., food availability, nesting site availability, disease, and predation.
density-independent effects those factors affecting population parameters independently of population size, as temperature, salinity, and meteorites.
deoxyhemoglobin hemoglobin not carrying oxygen.
deoxynucleotide any nucleotide of DNA.
deoxyribonucleic acid see DNA.
deoxyribonucleotide a nucleotide in which the sugar portion is deoxyribose rather than ribose.
deoxyribose a 5-carbon sugar identical to ribose except that the 2′ hydroxyl group is replaced by a hydrogen.
depolarizing wave see *neural impulse.*
derived a character that has undergone an evolutionary change in the group being considered; the opposite of *primitive.* See also *advanced.*
dermal (adj.) pertaining to the skin or dermis; *dermal bone,* bone originating in evolution and embryology as flat plates beneath the skin, as certain bones of the cranium.
dermal branchiae "skin gills" of echinoderms, consisting of ciliated outpocketings of the coelom and body wall.
dermal system various plant tissues derived from protoderm, including epidermis, guard cells, leaf hairs, root hairs, and periderm.
dermis in animals, the inner mesodermally-derived layer of the skin, beneath the ectodermally-derived epidermis.
descending colon a part of the large intestine that descends to the rectum.
descent with modification Evolution. Ongoing, continuous change in organisms over time.
desert a region characterized by scanty rainfall (especially less than 25 cm annually).
desiccate (vt.) to dry up, cause to dry up, dehydrate; desiccated; (adj.) dried up, dehydrated.
desmotubule within the plasmodesmata of plant cells, a continuation of the endoplasmic reticulum of one cell with that of the adjacent cell.
determinate cleavage early embryonic cleavages in which daughter cells begin the very early commitment to specific development fates; typical of protostomes and especially of insects.
determination the ongoing commitment of embryonic cells and tissues to specific developmental directions.
deuterostome a bilateral animal with radial cleavage and a mouth that does not arise, developmentally or phylogenetically, from the blastopore; compare *protostome.*
developed nation a nation that is industrialized, has a mechanized labor-free agriculture, high per capita income, and high literacy. Compare *developing nation.*
Developing nation a nation that is characterized by low industrialization, labor-intensive agriculture, low per capita income, and low literacy. Compare *developed nation.*
development 1. the whole process of growth and differentiation by which a zygote, spore, or embryo is transformed into a functioning organism. 2. any part of this process, as the development of the kidney. 3. evolution.
diabetes mellitus a genetic disease of carbohydrate metabolism characterized by abnormally high levels of glucose in the blood and urine, and the inadequate secretion or utilization of insulin.
diakinesis final stage of meiotic prophase, in which the chromosomes are maximally condensed and homologs are attached only at their ends.
diapause in insects, a period of dormancy, interrupting developmental activity of the pupa or other stage, frequently occurring during hibernation or estivation.
diaphragm 1. in mammals, a dome-shaped, muscularized body partition separating the chest and abdominal cavities, involved in breathing movement. 2. a dome-shaped, rubber, contraceptive device fitted over the cervix, often used with a spermicide.
diastole the period of expansion and dilation of the heart during which it fills with blood; the period between forceful contractions (systole) of the heart. See *systole.*
diatom an acellular or colonial yellow-green photosynthetic protist, having a silicon-impregnated cell wall in two parts.
diatomaceous earth geological deposits consisting largely of the cell walls of diatoms.
dicot a flowering plant of the angiosperm class

Dicotyledonae, characterized by producing seeds with two cotyledons; compare *monocot.*

dicotyledon see *dicot* and *cotyledon.*

dictyosome in plants, the Golgi complex. Also see *Golgi complex.*

differentiation in development, the process whereby a cell or cell line becomes morphogically, developmentally, or physiologically specialized. *Terminal differentiation,* differentiation to a form in which the cell will normally not undergo further cell division. See *dedifferentiation.*

diffuse root system also *fibrous root system,* a root system of a plant in which there are many roots all about the same size.

diffusion the random movement of molecules of a gas or solute under thermal agitation, resulting in a net movement from regions of higher initial concentration to regions of lower initial concentration.

digestion the process of making food absorbable by breaking it down into simpler chemical compounds, chiefly through the action of enzymes, acid, and emulsifiers in digestive secretions; see *extra-cellular* and *intracellular digestion.*

digestive vacuole in animals and protists, intracellular vacuoles that arise by the phagocytosis of solid food materials and in which digestive processes occur, nutrient molecules being absorbed and undigested materials eventually being expelled by exocytosis.

digit in terrestrial vertebrates, any of the terminal divisions of the limbs; a finger, thumb, or toe.

dihybrid an organism that is heterozygous at two different loci. *Dihybrid cross,* a cross between two genotypically identical dihybrids.

dikaryotic in many fungi, a condition arising after conjugation, wherein parental plus and minus nuclei remain separated for a time before fusion occurs.

dilation an enlargement or widening; in the human birth process, the initial stage, in which the body of the cervix softens and its opening widens.

dilation and curettage (D and C)a surgical operation in which the cervix is forcefully dilated and the uterine mucosa scraped with a curette; a common method of induced abortion as well as a procedure for removing cysts and polyps.

dimer a molecule consisting of two similar subunits.

dinoflagellate a flagellated, photosynthetic, marine protist of the *Dinoflagellata,* a group considered by various authorities to be an order, class, division, kingdom, or even higher group *(dinokaryotes),* on a level with prokaryotes and eukaryotes.

dinosaur 1. any of a taxon of large, extinct reptiles, widely distributed from the Triassic to the Mesozoic, including thee largest known land animals. 2. any large, extinct reptile.

dioecious 1. in flowering plants, having separate sexes in the sporophyte generation. 2. in algae plants and bryophytes, having separate sexes in the gametophyte generation; that is, producing only eggs or sperm in any one thallus.

dipeptidase an enzyme of the small intestine that hydrolyzes the peptide bond of a dipeptide. See also *dipeptide.*

dipeptide two amino acids united by a peptide linkage; a common intermediate product of digestion. *Dipeptidase,* an enzyme that hydrolyzes dipeptides but not longer polypeptides.

diploblastic a condition in animals in which the adult tissues are derived from just two embryonic germ tissue layers, ectoderm and endoderm (see also *triploblastic*).

diplococcus (pl. *diplococci*), 1. a coccus (spherical) in which the cells are arranged in colonies of two. 2. a bacterium of the genus *Diplococcus,* which includes several serious human pathogens, causing pneumonia and gonorrhea.

diploid having a double set of genes and chromosomes—one set from each parent.

diplotene also *diplonema,* the stage m meiotic prophase during which chiasmata become visible and separation of homologs begins.

direct flight muscles in insects, flight muscles that attach directly to and move the wings (see *indirect flight muscles*).

directional selection selection favoring one extreme of a continuous phenotypic distribution.

disaccharide a carbohydrate consisting of two simple sugar subunits.

discrimination in behavior, the ability to perceive the difference between similar stimuli

disk flower see *composite flower.*

disruptive selection natural selection during which extreme phenotypes receive favorable selection, whereas average phenotypes are selected against.

dissociate to come apart into discrete units; *dissociation constant,* of a chemical reaction, a constant depending on the equilibrium between the combined and dissociated forms of a chemical or chemicals.

distal away from the center of the body, heart, or other reference point.

distal convulted tubule a portion of the nephron between the loop of Henle and the collecting duct.

distance receptor a sensory receptor of stimuli arising at a distance, such as sight and sound.

disulfide bridge see *disulfide linkage.*

disulfide linkage, also *disulfide bond or disulfide bridge,* a covalent link formed by the oxidation of two sulfhydryl groups:

$R_1—SH + HS—R_2 \rightarrow R_1—S—S—R_2 + H_2$

diuretic a chemical agent such as a hormone or drug that slows the recovery of water by the kidneys, thus increasing urine output.

divergence see *divergent evolution.*

divergent evolution following speciation, the continued accumulation of differences between or among species, attributable to adaptive radiation.

diversity 1. variety; variability. 2. the range of types in a major taxon: *plant diversity.* 3. in ecology, a measure of the number of species coexisting in a community.

division a major primary category or taxon of the plant, fungal, and sometimes moneran and protist kingdoms, equivalent to *phylum.*

DNA, deoxyribonucleic acid the genetic material of all organisms (except RNA viruses); in eukaryotes DNA is confined to the nucleus, mitochondria, and plastids.

DNAase an enzyme capable of hydrolyzing DNA sugar-phosphate bonds.

DNA polymerase any of several enzymes or enzyme complexes that catalyze the replication of DNA.

DNA repair system any of several systems of enzymes that detects and excises primary lesions in the Watson or Crick DNA strand, replacing them with a corrected segment.

DNA replication also *replication* or *DNA synthesis,* the semiconservative synthesis of DNA in which the double helix opens, the two strands separate and each is used as a template for producing a new opposing strand.

dominance the phenotypic expression of only one of the two alleles in a heterozygote. See *dominance relationships.*

dominance, behavioral dominance a behavioral relationship between two animals where the *subordinate* individual withdraws or behaves submissively toward the *dominant* individual in any conflict, potential conflict, or interaction.

dominance hierarchy 1. also *pecking order,* behavioral interactions established in a troop, flock, or other species group, in which every individual is dominant to those lower on the order and submissive to those above. 2. in genetics, the relationship among a group of multiple alleles in which each ordered series behaves as a dominant to those below it and as a recessive to those above it.

dominance relationships the interaction between two alleles of a given locus in the development and expression of a specific phenotypic trait. One allele of the pair is *dominant* if its typical phenotype is expressed in both the homozygote and the heterozygote; in this case the other allele, which is expressed only when it is homozygous, is said to be *recessive.* If each of two alleles has a unique phenotypic expression and both are independently expressed in the heterozygote, both are said to be *codominant.* For other dominance relationships consult the text.

dominant 1. a phenothypic trait that is always expressed when a certain allele is present. 2. an allele that expresses a given phenotypic trait whether homozygous or heterozygous. See *dominance relationships.*

dormancy the temporary suspension of growth, development, or other biological activity; suspended animation, *dormant,* in a state of dormancy.

dorsal toward the back or (usually) upper surface of a bilateral animal.

dorsal hollow nerve cord a defining characteristic of chordates, arising in development as a flat dorsal plate of neurectoderm that rounds up and sinks beneath the surface to form, in vertebrates, the *brain* and *spinal cord (central nervous sytem).*

dorsal lip in the animal gastrula, tissue associated with the early blastopore, the site of involution by presumptive mesoderm and endoderm.

dorsoventrally from top to bottom, or back to front, in humans.

double bond a bond between two atoms consisting of two covalent bonds, with a total of four electrons shared.

double fertilization in flowering plants, fertilization of egg cell and central cell by two sperm entering from a pollen tube.

double helix the configuration of the native DNA molecule, which consists of two antiparallel strands wound helically around each other.

doubling time the time it takes a growing population to double in numbers.

Down's syndrome see *trisomy 21*.

downy mildew 1. a protistan parasite of plant leaves. 2. the disease caused by the fungus.

drift *random drift,* also *genetic drift,* 1. the chance fluctuation of allele frequencies from generation to generation in a finite population. 2. the long-term consequences of such fluctuations, such as the loss or fixation of selectively neutral alleles.

drive an urgent, basic, or instinctual need pressing for satisfaction; a physiological or psychological tension, lack, or imbalance that the organism actively seeks to redress, quench, or fulfill.

drive-reduction hypothesis the idea that most behavior aims at (or is rewarded by) the reduction of specific innate drives.

drumstick a small projection (not unlike a chicken leg) of the nucleus of the polymorphonuclear leucocytes of human females, attributed to the condensed second X chromosome.

ductus arteriosis in the mammalian fetus, a short broad vessel conducting blood from the pulmonary artery to the aorta, thus bypassing the fetal lungs.

duodenum the first region of small intestine, just posterior to the stomach, receiving the common bile duct.

dwarf 1. any abnormally short person. 2. also *achondroplastic dwarf,* a genetic abnormality, an abnormally short person with markedly atypical body proportions. 3. any plant or animal unusually small for its species or type. *Pituitary dwarf,* a dwarf whose small stature is due to a lack of growth hormone during development.

dynein arm in cilia and flagella, protein connections between the nine paired groups of microtubules.

ear ossicle one of the bones of the middle ear, in mammals, the malleus, incus, and stapes, that transfers sound from the eardrum to the internal ear.

ecdysis also *molting,* 1. in arthropods, the shedding of the outer, noncellular cuticle. 2. this shedding together with associated changes in size, shape, or function.

ecdysone the molting hormone of insects.

echinoderm any organisms of the marine, coelomate, dueterostome phylum Echinodermata.

ecological community an assemblage of plants, animals, and other organisms forming an interacting unit within a biome, generally identifiable by dominant plant type or types, as seen in a spruce or sage desert community.

ecological niche see *niche.*

ecological succession see *community development.*

ecologist a scientist who studies ecology.

ecology the branch of biology dealing with the relationships between organisms and their environment.

ecosystem the biotic and abiotic factors of an ecological community considered together.

ectoderm in animal development, the outermost of the three primary germ layers of an embryo: the source of all neural tissue, sense organs, the outer cellular layer of the skin and associated organs of the skin; also, any of the tissues derived from embryonic ectoderm; also, the outer cellular layer of a cnidarian.

ectoparasite a parasite that feeds on or attaches to the surface tissue of the host; fleas, lice, ticks, and athlete's foot fungus are human ectoparasites.

ectotherm an organism that makes use of external heat sources to increase its body temperature.

effector also *effector organ,* a bodily organ actively used in behavior, especially in communication—distinguished from *receptor;* e.g., the human eyebrows and associated muscles constitute an effector organ, and a rather effective one at that.

egg cell the larger gamete of an anisogametic organism; in plants, the one haploid cell of the embryo sac that will participate as the female gamete; in anmials, the functional product of meiosis in females.

ejaculation the ejection of semen by involuntary muscular contractions of the vas deferens and urethra in an orgasm.

elastin an extracellular structural protein that forms long, elastic fibers in connective tissue.

elator in liverworts, a springlike structure in the sporangium that ejects the spore outward.

electrical synapse contact between neurons formed by gap junctions, in which an action potential passes directly from one neuron to the next.

electroencephalograph a clinical device for detecting and recording *electroencephalograms* (EEGs) of electrical activity within the brain.

electromagnetic radiation waves of energy propagated by simultaneous electric and magnetic oscillations, including radio waves, microwaves, infrared, visible light, ultraviolet radiation, X rays, and gamma rays. *Electromagnetic spectrum,* the range of electromagnetic radiation from low-energy, low frequency radio waves to high-energy, high frequency gamma rays.

electron one of the three common constituents of an atom, a *lepton* with a mass of 1/1837 of that of a proton and an electrostatic charge of -1.

electron acceptor a molecule that accepts one or more electrons in an oxidation-reduction reaction (and thus becomes reduced).

electron carrier a molecule that behaves cyclically as an electron acceptor and an electron donor.

electron donor a molecule that loses one or more electrons to an electron acceptor in an oxidation-reduction reaction (and thereby becomes oxidized).

electronegativity the relative ability of any element to hold electrons or to attract them from other elements.

electron microscope a device for creating magnified images of small specimens through bombarding it with an electron beam and by subsequent magnetic focusing.

electron orbit also *electron orbital,* 1. the state of an electron as determined by its energy as it moves within an atom. 2. the space within which an electron pair moves within an atom.

electron shell 1. the space occupied by the orbits of a group of electrons of approximately equal energy; 2. an energy level of a group of electrons in an atom.

electron transport see *electron transport system.*

electron transport system also *electron transport chain,* a series of cytochromes and other proteins, bound within a membrane of a thylakoid, mitochondrion, or prokaryotic cell, that passes electrons and/or hydrogen atoms in a series of oxidation-reduction reactions that result in the net movement of hydrogen ions across the membrane.

electrostatic (adj.) pertaining to the attraction or repulsion of electric charges independent of their motion.

element a substance that cannot be separated into simpler substances by purely chemical means.

elongation lengthening; in plant development, the stretching in one direction of stem or root cells under turgor.

Eltonian pyramid also *pyramid of numbers,* a concept in ecology that, because of thermodynamic inefficiency, organisms forming the base of a food chain are numerically abundant and comprise a large total biomass, while organisms of each succeeding level of the chain are successively less abundant and of smaller total biomass, and the top predator is always numerically rare and of relatively small total biomass.

embolism 1. the sudden blockage of a blood vessel by a dislodged blood clot, air bubble or other obstruction. 2. also *embolus,* the clot or obstruction itself.

embryo an animal or seed plant sporophyte in an early state of development. In animals this stage begins with cleavage, includes the laying down of germ layers, organ systems, and basic tissue types, and grades imperceptibly into the *fetal* stage; in seeds the embryo stage lasts from fertilization to germination, the "mature embryo" comprising a rudimentary plant with plumule, radicle, and cotyledons.

embryogenesis in animals, the early period of development when cell division and gastrulation prepare the embryo for morphogenesis.

embryology the scientific study of early development in plants and animals.

embryonic disc 1. also *embryonic shield,* the part of the inner cell mass from which a mammalian embryo develops. 2. *blastodisc.*

embryonic induction see *tissue interaction.*

embryo sac in flowering plants, the mature megagametophyte after division into six haploid cells and one binucleate cell, enclosed in a common cell wall.

endergonic (adj.) of a biochemical reaction the expenditure of energy; moving from a state of lower potential energy to one of higher potential energy; compare *exergonic.*

endocrine gland also *ductless gland,* a discrete gland that secretes hormones into the blood system.

endocrine hormone a hormone that acts at some distance from its source (usually transported in the blood), as opposed to paracrine hormones, which act on neighboring cells, and autocrine hormones, which remain within the producing cell. (For individual hormones see Table 37.1.)

endocrine system the endocrine glands taken together, and their hormonal actions and interactions.

endocytosis the process of taking food or solutes into the cell by engulfment (a form of active transport); see also *phagocytosis, pinocytosis;* compare *exocytosis.*

endoderm also *entoderm,* 1. the innermost of the three primary germ layers of a metazoan embryo and the source of the gut epithelium and its embryonic outpocketings (in vertebrates, the liver, pancreas, lung). 2. any tissue derived from endoderm. 3. in coelenterates, the epithelium of the gastrovascular cavity.

endodermis a single layer of cells around the stele of vascular plant roots that forms a moisture barrier in that the lateral cell walls are closely oppressed and waterproofed in bands, forming the Casparian strip.

endomembranal system within and surrounding the cytoplasm of eukaryotic cells, a number of dynamic, intercovertible membranes, including those surrounding the cell, the nucleus, and those forming the endoplasmic reticulum, Golgi complex, and various vacuoles.

endometrium in mammals, the mucous membrane tissue lining the cavity of the uterus; responds cyclically to ovarian hormones by thickening as preparation for implantation of the blastocyst.

endonuclease a family of enzymes capable of hydrolyzing sugar-phosphate bonds anywhere within DNA strands, an essential enzyme to crossing over. See also *recombination nodule.*

endopeptidase a proteolytic enzyme that cleaves peptide bonds in the middle portions of polypeptides.

endoplasmic reticulum (ER) internal membranes of the cell, usually a site of synthesis; *rough endoplasmic reticulum,* ER without bound ribosomes; the site of synthesis of noncytoplasmic proteins; *smooth endoplasmic reticulum,* ER without bound ribosomes, usually the site of synthesis of nonprotein materials.

endoskeleton a mesodermally derived supporting skeleton inside the organism, surrounded by living tissue, as in vertebrates and echinoderms.

endosperm a nutritive tissue of seeds, formed around the embryo in the embryo sac by the proliferation of the (usually) triploid endosperm nucleus to form a starch-rich mass; the endosperm may persist until germination or be resorbed by the cotyledon(s) during seed maturation.

endosperm mother cell a large, central, usually diploid or binucleate cell formed in the megagametophyte by fusion of haploid nuclei; when fertilized in double fertilization, forms the *endosperm nucleus.*

endospore an asexual resistant spore formed within a bacterial cell; see *spore.*

endosymbiont a mutualistic symbiont that resides within the cells of its symbiotic partner.

endosymbiosis hypothesis see *serial endosymbiosis hypothesis.*

endothelium an epithelial tissue that forms the inner lining of blood and lymph vessels.

endotherm also *homeotherm* and *homoiotherm,* an organism with the ability to metabolically thermoregulate, also called "warm-blooded." See also *ectotherm.*

energetic tendencies the probable behavior of interacting atoms as they proceed to new energetically stable configurations, e.g., the tendency for electrons to pair, for outer shells to fill, and to reach a balance of plus and minus charges.

energy a fundamental concept of physics, either being associated with physical bodies (potential energy, kinetic energy) or with electromagnetic radiation; under some conditions interconvertible with mass by the relation $E = mc^2$ or with entropy but not capable of being created or destroyed.

energy level also *energy shell,* in electrons, a discrete amount of energy that determines the distance any electron will remain from the nucleus.

energy of activation the energy input needed to initiate a chemical reaction.

energy pyramid the Eltonian pyramid with regard to chemical energy rather than numbers or biomass; see *Eltonian pyramid.*

energy shell see *energy level.*

energy state also *energy level,* a quantum state of an electron in an atom, a form of potential energy as the electron changes from a higher energy state to a lower one; a measure of the free energy in a system.

engram (also *memory scar, memory trace*), the physical change that is presumed to occur in the brain when something is learned.

entropy in thermodynamics, the amount of energy in a closed system that is not available for doing work; also defined as a measure of the randomness or disorder of such a system. *Negative entropy,* free energy in the form of organization. See *energy, free energy.*

enucleate without a nucleus.

environment the surrounding conditions, influences, or forces that influence or modify an organism, population, or community.

environmental resistance the sum of environmental factors (e.g., limited resources, drought, disease, predation) that restrict the growth of a population below its biotic potential (maximum possible population size).

enzyme a protein that catalyzes chemical reactions.

enzyme-substrate complex the unit formed by an enzyme bound by non-covalent bonds to its substrate.

ephyra a young, newly budded scyphozoan jellyfish.

epiblast the upper layer of cells in the early bird, reptilian, or mammalian embryo that, following gastrulation, contributes to the three-layered embryo.

epicotyl in the embryo of a seed plant, the part of the stem above the attachment of the cotyledon(s); forms the epicotyl hook in the growth of some seeds.

epidermis 1. in plants, the outer protective cell layer in leaves and in root and stem primary growth, one cell thick and made waterproof with an outer layer of cutin; replaced in secondary growth by *periderm* 2. in animals, the outer epithelial layer, derived from ectoderm; lacking innervation or vascularization in vertebrates.

epididymis in mammals, an elongated soft mass lying alongside each testis, consisting of convoluted tubules; the site of sperm maturation.

epiglottis in the lower pharynx, a flexible cartilaginous flap that folds over the glottis during swallowing.

epiphysis in vertebrates, a part of a bone that ossifies separately and later becomes fused to the rest. *Epiphyseal cartilage,* also *epiphyseal plate,* a plate of cartilage separating the body of a long bone from its terminal epiphysis, being the site of bone growth and elongation.

epiphyte a plant that grows nonparasitically on another plant or sometimes an object.

epistasis the masking of a trait ordinarily determined by one gene locus by the action of a gene or genes at another locus.

epithelial tissue covering a surface or lining a cavity.

epithelium (pl. *epithelia*), a tissue consisting of tightly adjoining cells that cover a surface or line a canal or cavity, and that serves to enclose and protect.

equilibrium frequency in population genetics, the idealized allele frequency at which selection, mutation, and other forces balance, so that there is no expected net change; chance fluctuations may cause oscillations around this idealized equilibrium.

ER see *endoplasmic reticulum.*

era one of the major divisions of geologic time; see the geologic timetable inside the front cover.

erection 1. the becoming stiff and firm (turgid), by dilation under blood pressure, of a

penis, clitoris, or nipple. 2. the stiffened, turgid state of such an organ.
ergot a toxic fungal infective state of rye.
ergotism an illness in humans brought about by eating bread made with ergoted flour, that is, flour milled from grain infected with the fungus *Claviceps purpurea.*
erythroblastosis also *erythroblastosis fetalis,* during Rh incompatibility, the destruction of red blood cells in an Rh^+ fetus or newborn through the action of maternal Rh^- antibodies that have crossed the placenta.
erythrocyte also *red blood cell,* a hemoglobin-filled, oxygen-carrying, circulating blood cell; enucleate in mammals, but having a physiologically inactive, condensed nucleus in other vertebrates.
erythropoietin a hormone that stimulates the production and differentiation of erythrocytes during erythropoiesis.
esophagus the anterior part of the digestive tract; in mammals it is muscularized and leads from the pharynx to the stomach.
essential amino acid one of the amino acids that the body cannot synthesize and thus must be provided by the diet if dietary diseases are to be avoided.
essential fatty acid one of the fatty acids that the body cannot synthesize and thus must be provided by the diet if dietary diseases are to be avoided.
estivation a state of dormancy or torpor induced by heat or dryness; summer dormancy.
estrous cycle the sequence of endocrine and other reproductive system changes that occur from one estrus period to the next.
estrus also *heat,* in the females of most mammalian species, the regularly recurring state of sexual excitement around the time of ovulation; the only time during which the female will accept the male and is capable of conceiving; the term is not applicable to people.
ethology the scientific study of animal behavior as it occurs in the organism's natural environment, or among free-ranging animals in the ethologist's environment.
euchromatin chromatin other than heterochromatin.
eucoelomate also *coelomate,* any animal with a coelom, a body cavity entirely lined by mesodermally derived tissue. See also *coelom.*
eukaryote a nucleated organism.
euphotic zone in the aquatic environment, those depths penetrated by effective light, where photosynthetic activity exceeds respiration.
eutrophication the process in which nutrients increase in a lake, through evaporation and silting, or artifically by pollution.
evolution 1. any gradual process of formation, growth, or change. 2. descent with modification. 3. long-term change and speciation (division into discrete species) of biological entities. 4. the continuous genetic adaptation of organisms or populations through mutation, hybridization, random drift, and natural selection.
evolutionary stable strategy (ESS) an innate behavioral strategy (of conflict, etc.) that confers greater fitness on individuals than can any alternative behavior that might arise by mutation or recombination.
exchange pool in biogeochemical cyles, the readily available reserves of a mineral nutrient, such as a soluble phosphate or nitrate pool in the soil water; compare *reservoir.*
excision repair see *DNA repair system.*
excitatory synapse a synapse in which the secretion of neurotransmitter stimulates neural impulses in the receiving neuron.
excretion the removal of metaboilc wastes from the body.
excretory system an organ system involved in the removal of metabolic wastes; e.g., the kidney, urinary bladder, and associated ducts and valves.
exergonic (adj.) of a biochemical reaction or half-reaction, producing work, heat, or other energy; moving from a higher energy state to a lower energy state; as opposed to *endergonic.*
exocytosis the process of expelling material from the cell through the plasma membrane by the fusion of vacuoles, secretion granules, etc. with the plasma membrane, and their subsequent eversion.
exopeptidase a proteolytic enzyme that hydrolyzes peptide linkages near the two ends of a polypeptide; see *carboxypeptidase* and *aminopeptidase;* compare *dipeptidase, encopeptidase.*
experiment a set of conditions or procedures suggested by a prediction and contrived to test the validity of a hypothesis. The conducting of such a test.
exoskeleton an external skeleton or supportive covering, as in arthropods and armadillos; arthropod exoskeletons are formed by secretions of epidermal cells, largely chitin, proteins, and calcium carbonate.
exponential growth also *geometric growth,* population growth in which the population size increases by a fixed proportion in each time period, successive values forming an exponential series, $P_t = P_0 r^t$ where P_0 is the initial population size, t is the time, and r is the Malthusian parameter.
exponential growth curve also *exponential increase* and *J-shaped curve,* population growth that when plotted takes on a J-shaped curve, growth that approaches the *biotic potential* of a species and is not immediately responsive to *environmental resistance.* Growth increments may be 2, 4, 8, 16, 32, etc. Compare *arithmetic increase.*
expressed sequence or *exon,* the portion of mRNA remaining after the *introns* or *unexpressed sequences* have been removed; see also *intron.*
external ear all parts of the ear external to the eardrum, comprising the external ear canal, the external auditory meatus, and the pinna (pl. *pinnae*—what you wiggle when you wiggle your ears).
external fertilization fertilization outside the body, as in echinoderms and most bony fish.
extinction a coming to an end or dying out of a species or other taxon (group of related organisms).
extracellular outside, between, or among cells.
extracellular digestion digestion outside of cells (usually in the gut).
extraembryonic outside of the embryo proper. *Extraembryonic membrane,* any of several membranes *(amion, chorion, allantois, yolk sac)* produced by the zygote but not part of the embryo proper.
extraembryonic membrane see *extraembryonic.*
extra-Y syndrome see *XYY syndrome.*
extreme halophile an organism, often an archaebacterium, that thrives in an excessively salty environment.
extreme thermophile an organism, often an archaebacterium, that thrives in an excessively hot environment, above 55° C.
eyespot 1. a simple visual organ in many invertebrates, consisting of a pigmented cup with light-sensitive receptors. 2. a simple pigmented visual organelle in various flagellate protists, including euglenoids and *Chlamydomonas.*
facilitated diffusion diffusion of specific molecules across a plasma membrane that is facilitated by reversible association with carrier molecules able to traverse the membrane. Facilitated diffusion differs from active transport in that no energy is expended and net movement follows the concentration gradient.
factor in Mendelian language, the alternative expression of a gene and the equivalent of an allele.
facultative anaerobe an organism such as yeast that can use oxygen when it is available but is also able to live anaerobically.
FAD, $FADH_2$ flavin adenine dinucleotide, a coenzyme and hydrogen carrier in metabolism. FAD is the oxidized form and $FADH_2$ is the reduced form.
fallopian tube see *oviduct.*
family in classification, a grouping smaller than *order* but larger than *genus.*
fascia (pl. *fasciae* or *fascias*), a heavy sheet of connective tissue covering or binding together muscles or other internal structure of the body, often connecting with ligaments or tendons.
fascicle a distinct bundle of muscle fibers, surrounded by connective tissue and carrying nerves and blood vessels.
fat 1. *neural fat,* also *triglyceride,* any of the solid, semisolid, or liquid lipid-soluble compounds consisting of glycerol bound by ester linkages to three fatty acid molecules. 2. also *adipose tissue,* parts of an animal consistently largely of cells distended with triglycerides.
fate map a plan of a zygote, blastula or early embryo indicating the normal developmental fates of the cellular descendants of embryonic ectoderm, mesoderm, and endoderm.
fatty acid an organic acid consisting of a linear hydrocarbon "tail" and one terminal carboxyl group.
feces bodily waste discharged through the anus.
feedback see *negative feedback.*
feedback inhibition see *negative feedback.*

fermentation 1. anaerobic breakdown of glucose to form alcohol and carbon dioxide or lactate. 2. any controlled enzymatic transformation of an organic substrate by microorganisms.

fertility rate the number of births per 1000 women from 15 to 44 years of age; a clearer indicator of reproductive activity in a population than the birth rate.

fertilization the process of the union of gametes or gamete nuclei.

fertlization cone a cone-shaped mound of egg cytoplasm that, upon activation of the egg by a fertilizing sperm, rises to engulf the sperm.

fertilization membrane in echinoderms, a membrane formed upon egg activation by the rupture of cortical granules, preventing further sperm penetration.

fetus in vertebrates, an unborn or unhatched individual past the embryo state; in humans, a developing, unborn individual past the first 8 weeks of pregnancy.

fiber 1. in plants a lengthy, slender, tapering sclerenchyma cell, often thick walled and nonliving at maturity. 2. see *muscle fiber.*

fibril also *myofibril,* cylindrical, striated, contractile units of muscle, containing myofilaments of myosin and actin.

fibrin an insoluble fibrous protein forming blood clots and also contributing to the viscosity of blood; see *fibrinogen.*

fibrinogen a globular blood protein *(globulin)* that is converted into fibrin by the action of thrombin as part of the normal blood clotting process.

fibroblast a cell in the connective tissue group that produces fibers and matrix substances such as collagen.

fibrocartilage slightly compressible cartilage containing thick bundles of collaginous fibers, as in intervertebral disks, the knee joint, and pubic symphysis.

fibrous protein a protein in which the molecules tend to align in linear arrays, in contrast to globular proteins.

fibrous root system also *diffuse root system,* the roots collectively of a plant in which there are many equivalent roots rather than a prominent central root.

fiddlehead a young unfurling frond of a fern, resembling the head of a violin.

filament 1. a long fiber. 2. in flowers, the slender stalk of the stamen, on which the anther is situated.

filamentous algae a group of green algae that tend to form lengthy filaments of cells.

filial generation in Mendelian genetics, 1. *first filial generation,* $\mathbf{F_1}$, the offspring of a cross between two homozygotes for different alleles (or individuals of two true-breeding strains). 2. subsequent filial generations ($\mathbf{F_2}$, $\mathbf{F_3}$ etc.) The progeny of selfing, of inbreeding, or of crosses between individuals of a given filial generation.

filter-feeder an animal that obtains its food by filtering minute organisms from a current of water.

fimbria (pl. *fimbriae*), in vertebrates, a fringe of ciliated tissue surrounding the opening of the oviduct into the peritoneal cavity.

fire-disclimax community an ecological community in continual transition because of recurring fires (e.g., chaparral).

first law of thermodynamics the physical law that states that energy cannot be created or destroyed; later amended to allow for the interconversion of matter and energy.

first messenger a hormone, as distinguished from a *second messenger.*

fission the division of an organism into two (binary fission) or more organisms, as a process of sexual reproduction.

fit in biology, adapted to the environment and thus well able to survive and reproduce.

fitness 1. the state of being adapted or suited (e.g., to the environment). 2. also *relative fitness, Darwinian fitness,* the relative expectation of surviving and reproducing of an individual or of a specific genotype, compared with that of the general population or of a standard genotype. 3. one half the expected number of offspring of a diploid genotype.

five-prime end *(5′ end,)* of a nucleic acid, the end at which the 5′ carbon of ribose or deoxyribose is free to form a phosphate linkage; the first end of the chain in the direction of synthesis.

fixed action pattern a precise and identifiable set of movements, innate and characteristic of a given species; see also *instinctive pattern.*

flagellin the crystalline protein comprising the eubacterial flagellum.

flagellum (pl. flagella), 1. a long, whiplike, motile eukaryotic cell organelle, projecting from the cell surface but enclosed in a membrane continuous with the plasma membrane; longer and fewer in number than cilia, it propels the cell by undulations. 2. an analogous, nonhomologous organelle of bacteria, consisting of a solid, helical protein fiber that passes through the cell wall and propels the cell by rotating like a propeller.

flame bulb see *flame cell system.*

flame cell (see *flame cell system.*

flame cell system in several invertebrate phyla, an osmoregulatory/excretory system consisting of ciliated flame bulbs composed of flame cells, along with associated conducting tubules in which fluids and nitrogen wastes are swept to external pores.

flatworm a platyhelminth, especially a turbellarian such as the common planaria.

floridean starch a storage starch or carbohydrate of the red algae.

florigen a hypothetical flower-inducing hormone.

flower the part of a seed plant that bears the reproductive organs, including pistils, stamens, petals, and sepals.

fluid mosaic model also *Singer model,* a description of the plasma membrane as a phospholipid bilayer stabilized by specifically oriented proteins, with some proteins extending through to both surfaces and other proteins specific for the inner or outer surfaces.

fluke a parasitic trematode flatworm; see also *blood fluke* and *Schistosome.*

food chain a sequence of organisms in an ecological community, each of which is food for the next higher organism, from the primary producer to the top predator.

food pyramid a graphic arrangement of trophic levels with producers forming the base, followed by first, second, third, and subsequent levels of consumers.

food vacuole see *digestive vacuole.*

food web a group of interacting food chains; all of the feeding relations of a community taken together; the flow of chemical energy among organisms.

F-one particle usually $\mathbf{F_1}$ **particle,** see *F_oF_1 complex.*

foramen (pl. *foramina*), an opening or perforation.

foramen magnum the large opening at the base of the skull through which the spinal cord passes into the brain.

foramen ovale in the fetal mammalian heart, an opening in the septum between the atria; it normally closes at the time of birth.

foraminiferan a shelled marine sarcodine protist with one or more nuclei in a single plasma membrane; the carbonate shell having numerous minute openings (foramina) through which slender branching pseudopods are extended.

force filtration in the nephron, the movement of smaller molecules out of the blood and into Bowman's capsule by hydrostatic pressure in the glomerulus.

forebrain 1. the anterior of the three primary divisions of the vertebrate brain. 2. the parts of the brain developed from the embryonic forebrain.

fossil any remains, impression, or trace of an animal or plant of a former geological age, such as mineralized skeleton, a footprint, or a frozen mammoth.

fossil fuel coal, petroleum, or natural gas.

founder effect the population genetic effect of the chance assortment of genes carried by the successful founder or by a few founders of a subsequently large population.

fovea (pl. *foveae*), the most sensitive part of the retina; a small central depression rich in cones and devoid of rods.

$\mathbf{F^+}$ **(fertility-positive)** bacterium containing the plasmid responsible for producing a sex pilus through which a replica of the plasmid can pass to an F^- bacterium.

frame shift mutation a small insertion or deletion in a structural gene such that all mRNA codons downstream are misread in translation.

free energy energy available for work.

free energy of evaporation the energy involved in the evaporation of a liquid.

free radical a highly reactive atom or molecular unit with an unpaired electron.

freeze-etching also *freeze-fracturing,* the slicing of freeze-dried materials for observation through the transmission electron microscope, particularly useful for dividing membranes along their lipid core.

freeze-fracturing see *freeze-etching.*

frequency dependent selection natural selection in which a phenotype is more fit than average when it is numerically rare, and is less fit than average when its relative number surpasses some equilibrium frequency value.

frontal lobe an anterior division of the cerebral hemisphere, believed to be a site of higher cognition.

fruit the seed-bearing ovary of a flowering plant.

fruiting body in fungi, slime molds, and myxobacteria, a structure specialized for the production of spores.

fucoxanthin a brown carotenoid pigment characteristic of brown algae.

functional group in biochemistry, a side group with a characteristic chemical behavior or function.

fundamental niche the potential niche of an organism. Compare *realized niche.*

F_0F_1 complex also *F_0F_1 body* and *F_0F_1 ATP synthetase,* an ATP-producing structure in the mitochondrial inner membrane, consisting of a protein-lined channel (F_0) leading into a spherical region (F_1) containing the enzyme ATP synthetase.

gametic cycle a life cycle with a dominating diploid state, the hapliod state limited to gametes only, and fertilization restoring diploidy (some protists and all animals).

gametocyte also *meiocyte,* the cell that undergoes meiosis to form gametes; a spermatocyte or oocyte.

gametogenesis the process through which haploid gametes (sex cells: eggs and sperm) are produced, commonly including meiosis.

gametophyte in plants with alteration of generations, the haploid form, in which gametes are produced.

gamma radiation electromagnetic radiation of very high energy per photon, beyond X-radiation.

ganglion a mass of nerve tissue containing the cell bodies of neurons.

gap junction dense structures that physically connect plasma membranes of adjacent cells along with channels for cell-to-cell transport.

gas gangrene a bacterial disease caused by *Clostridium perfringins,* the symptoms of which include rotting of the affected flesh.

gas gland in bony fishes, a gas-producing gland associated with the swim bladder and the maintenance of buoyancy. See also *reabsorptive area.*

gastric lipase a fat-digesting enzyme secreted by the stomach lining.

gastrin a hormone, produced by the stomach lining, that induces the secretion of gastric juice (and may be identical with histamine).

gastrointestinal tract 1. the stomach and intestines as a functional unit. 2. the entire passageway from the mouth to and including the anus.

gastrovascular cavity also *coelenteron,* the cavity of cnidarians, ctenophores, and flatworms, which opens to the outside only via the mouth, and serves the functions of a digestive cavity, crude circulatory system, and coelom.

gastrula an early metazoan embryo consisting of a hollow, two-layered cup with an inner cavity (archenteron) opening out through a blastopore.

gastrulation the process usually involving invagination, involution, and epiboly, whereby a blastula becomes a gastrula.

Gause's law also, the *competitive exclusion principle,* the ecological principle that no two genetically isolated kinds of organisms can coexist (exist at the same time and place) while occupying the same ecological niche; the principle that each species of a group of coexisting species must be specialized in some way so as to occupy a unique niche.

gel electrophoresis an analytical procedure in biochemistry, whereby a mixture of biomolecules enters a gel and separates by the differential migration of its constituents in an imposed electrical field.

gemma (pl. *gemmae*), in liverworts, a multicellular bud that becomes detached from the parent thallus in a form of asexual reproduction.

gene (variously defined), 1. the unit of heredity. 2. the unit of heredity transmitted in the chromosome and that, through the interaction with other genes and gene products, controls the development of hereditary character. 3. a continuous length of DNA with a single genetic function. 4. the unit of transcription of DNA. 5. a *cistron* or *structural gene,* that is, the sequence of DNA coding for a single polypeptide sequence.

gene cloning recent techniques whereby pieces of DNA from any source are spliced into plasmid DNA, cultured in growing bacteria, purified, and recovered in quantity. Also *gene splicing, DNA cloning, recombinant DNA.*

gene flow the exchange of genes between populations through migration, pollen dispersal, chance encounters, and the like.

gene frequency see *allele frequency.*

gene pool all the genetic information of a population considered collectively.

gene sequencing determining the specific seuqence of nucleotides in a gene.

gene splicing the use of recombinant DNA techniques to form covalent bonds between DNA of different sources. See *gene cloning.*

general fertility rate see *fertility rate.*

generalization the process whereby a response is made to stimuli similar to, but not identical with, a reference stimulus.

generative cell a cell in the microgametophyte of a seed plant, capable of dividing to form two sperm cells.

generative nucleus one of the haploid nuclei of a pollen grain; when successful pollination occurs, the generative nucleus divides to form two sperm nuclei.

genetic code the relationship by which the 64 possible codons (sequences of three nucleotides in DNA or RNA) each specify an amino acid or chain termination in protein synthesis.

genetic drift see *drift.*

genetic engineering the manipulation of genes through recombinant DNA techniques.

genetic equilibrium the genetic state of a population wherein the frequency of certain alleles remains constant generation after generation.

genetic marker a standard mutant allele used in genetic tests in which it serves only to trace the fate or distribution of a chromosome, chromosome region, or gene locus, especially in recombination studies.

genetic recombination any process, such as crossing over, that leads to new gene combinations on the chromosomes.

genetic variability a broad term indicating the presence of different genetic constitutions in a population or populations.

genetics 1. the science of heredity, dealing with the resemblances and differences of related organisms resulting from the interaction of their genes and the environment. 2. the study of the structure, function, and transmission of genes.

genital swelling in mammalian embryos, a paired swelling beneath the *genital tubercle;* it will form the scrotum in males and the *labia majora* in females.

genital tubercle in the sexually undifferentiated mammalian embryo, three small swellings, the glans and two urogenital folds; in the males these will form the head and shaft of the penis, and in the female they will form the *glans clitoridis* and the *labia majora.*

genitals also *genitalia,* 1. the organs of the reproductive system. 2. also *external genitalia,* the external reproductive organs; the penis and scrotum in human males and the vulva in females.

genome the full haploid complement of genetic information of a diploid organism, usually viewed as a property of the species.

genotype 1. the genetic constitution of an organism with respect to the gene locus or loci under consideration. 2. also *total genotype,* the sum total of genetic information of an organism.

genotype frequency the proportion of individuals in a population having a specified genotype.

genus (pl. *genera*), a group of similar and related species; a taxon smaller than *family* and larger than *species.*

geological era one of five divisions of the earths history (Cenozoic, Mesozoic, Paleozoic, Proterozoic, and Archeozoic);see the geological timetable inside the front cover.

geology the scientific study of the physical history of the earth, the rocks and soil of which it is composed, and the physical changes that the earth has undergone or is undergoing.

geotropism see *gravitropism.*

germ layer any of the three layers of cells formed at gastrulation, the partially differentiated precursors of specific body tissues and organs; see *ectoderm, endoderm,* and *mesoderm.*

germ line *germ line cell,* the gametes and any cell or tissue that will or might give rise to gametocytes, or that otherwise contains the physical precursor of DNA that will be transmitted to offspring. *Germ-line DNA,* the DNA of germ line cells.

germinal adj. 1. pertaining to the germ line. 2.

pertaining to reproduction. 3. giving rise to embryonic structures.

germinal epithelium 1. the epithelium of the gonads, which through mitosis gives rise to gametocytes. 2. also *germinal layer,* the innermost layer of an epithelium, containing mitotic cells.

germinal tissues epithelial tissues specialized for gametogenesis, located in the gonad (testis or ovary) of animals.

germination the process whereby a seed and embryo ends dormancy, the stored materials of the seed being digested, the seed coat ruptured, and the plumule and hypocotyl begin to grow; spore germination, the end of dormancy in a spore.

giant chromosome see *polytene chromosome.*

giantism 1. also *gigantism,* growth to abnormal or unusual size. 2. see *pituitary giant.*

gibberellin any of a family of plant growth hormones that control cell elongation, bud development, differentiation, and other growth effects.

gill 1. an organ for obtaining oxygen from water. 1a. any of the highly vascular, filamentous processes of the pharynxes of fishes, between the gill clefts. 1b. any of the analogous but not homologous organs of invertebrates. 2. the thin, laminar structures on the underside of a mushroom (basidocarp), bearing the spore-forming basidia.

gill arches also *pharyngeal arches,* in fish, a row of bony or cartilaginous curved bars extending vertically between the gill slits on either side of the pharynx, supporting the gills. 2. in land vertebrate embryos, a row of corresponding homologous rudimentary ridges that give rise to jaw, tongue, and ear bones.

gill basket the cartilaginous structure supporting the gills in urochordates, cephalochordates, and larval lampreys, serving as a sieve for filter feeding.

gill chamber in fishes, one of the two chambers housing the gills; covered by a bony flap, the operculum.

gill clefts also *gill slits, pharyngeal clefts,* 1. in fish, the openings between the gills. 2. in vertebrate embryos, the corresponding and homologous grooves in the neck region, between the branchial arches.

gill filament one of the many feathery, leaflike or filamentous structures making up the working surfaces of the gill.

gill pouch also *pharyngeal pouch,* one of the hollowed-out cavities corresponding to the gill clefts in cyclostomes and some sharks.

gill raker in fish, one of the bony processes on the inner side of a gill arch, serving to prevent solids from passing out through the gill clefts.

gill list see *gill cleft.*

gizzard 1. in birds, a thick-walled muscular enlargement of the alimentary canal, filled with small stones and used to grind seeds or other food. 2. any analogous structure in invertebrates, as in earthworms, rotifers, and gastrotrichs.

glandular epithelium simple columnar epithelial tissue that lines glands and the intestine, specializing in synthesis and secretion.

glans (L. *glans,* acorn), 1. also *glans/penis,* the conical, vascular, highly innervated body forming the end of the penis. 2. also *glans clitoridis,* a homologous body forming the tip of the clitoris.

glassy sponge sponges with a skeletal element of silicon dioxide or glass.

glial cell see *neuroglia.*

gliding joint a skeletal joint in which the articulating surfaces glide over one another without twisting.

globulin any of a large class of globular proteins occurring in plant and animal tissues and in blood.

glomerulus (pl. *glomeruli*), the mass or tuft of capillaries within a Bowman's capsule.

glucagon a polypeptide hormone secreted by the pancreatic islets of Langerhans, whose action increases the blood glucose level by stimulating the breakdown of glycogen in the liver.

glottis the opening between the vocal chords in the larynx.

gluconeogenesis in the liver, a biochemical pathway in which, at the expense of ATP, lactate from active muscle is converted back to glucose or glycogen.

glucose also *dextrose, blood sugar, corn sugar, grape sugar,* a 6-carbon sugar, occuring in an open chain form or either of two ring forms; the subunit of which the polysaccharides starch, glycogen, and cellulose are composed, and a constituent of most other polysaccharides and disaccharides.

glucosamine an amino derivative of glucose, a constituent of chitin.

glue cell a glandular, thread-bearing cell, found only in ctenophores; used to capture prey by adhesion.

glycerol formerly *glycerin,* an organic compound composed of a 3-carbon backbone with an alcohol (hydroxyl) group on each carbon; a component of neutral fats and of phospholipids. Pure glycerol is a sweet, sticky, and disgusting liquid.

glycocalyx in plasma membranes, the complex of glycoprotein, glycosaccharides, and/or glycolipids that form a frilly-appearing surface.

glycogen also *animal starch,* a highly branched polysaccharide consisting of alpha glucose subunits; a carbohydrate storage material in the liver, muscle, and other animal tissues.

glycolipid a compounded with lipid and carbohydrate subunits; e.g., cerebrosides and gangliosides.

glycolysis the enzymatic, anaerobic breakdown of glucose in cells, yielding ATP (from ADP), and either lactic acid, or alcohol and CO_2, or pyruvate and NADH (from NAD).

glycolytic pathway the biochemical pathway in which glucose is broken down to pyruvate and in which ATP is synthesized (see *glycolysis*).

glycoprotein a compound containing polypeptide and carbohydrate subunits.

goblet cell a goblet-shaped, mucus-secreting epithelial cell; found in the lining of the nasal cavity, bronchi, and elsewhere.

goiter a visible enlargement of the thyroid gland on the front and sides of the neck resulting from iodine deficiency, associated with normal or subnormal levels of thyroid hormone.

Golgi complex a stack or array of membranous vesicles, formed from endoplasmic reticulum and engaged in the modification and packaging of various protein substances.

Golgi vesicle, also *secretory vesicle* spherical, membrane-bounded bodies formed from the Golgi membranes and containing Golgi-modified substances for transport and for secretion.

gonad in animals, the primary sex gland, with endocrine functions, and the site of meiosis; an ovary or testis.

gonadotropin a hormone that simtulates growth or activity in the gonads; in vertebrates, a specific peptide hormone of the anterior pituitary.

G_1, also *G_1 phase, gap one,* see **cell cycle.**

gonorrhea a sexually transmitted bacterial disease caused by the diplococcus *Neisseria gonorrhoeae,* which thrives in the mucous membranes of the male urethra and genital tract; in women infection occurs in the cervix, often spreading to the fallopian tubes.

graded display a communication pattern in which there exist intermediate states such as a behavioral continuum from low to high levels of motivation.

gradient a continuous change in concentration (or of any other quantity such as temperature). *With the gradient* (in the movement of solutes), in the direction from higher to lower concentration. *Against the gradient,* in the direction from lower to higher concentration, i.e., in the direction opposite to that of expected net diffusion.

gradualism the Darwinian proposal that the pace of evolution is slow but steady with an ongoing accumulation of minor changes leading eventually to the formation of new species. Compare to *punctuated equilibrium.*

Graafian follicle or *follicle,* a fluid-filled body in the ovary during the preovulatory period, containing the maturing oocyte within a cluster of follicular cells.

graft 1. the act of uniting two plants in such a way that they grow together. 2. the resulting growth. 3. v.t., to make a graft.

grafting an artificial method of propagation in which the bud, stem, or shoot of a plant is inserted into a cut in a recipeint stem such that the cambium of both plants comes into contact and will eventually grow together.

grain also *cereal grain,* the hard fruit of a cereal grass, such as wheat, rye, corn, rice, or barley, consisting of a single seed in a tough covering.

Gram stain a solution of iodine and iodide used to stain certain large classes of bacteria. *Gram negative bacterium,* a bacterium that cannot be stained by Gram stain. *Gram positive bacterium,* a bacterium that can be stained with Gram stain.

granum (pl. *grana*), a stack of thylakoid disks

in a chloroplast; seen as a series of minute, multiple green bodies containing most of the chlorophyll of the plant.

granum thylakoid member of a stack of thylakoids, the stack referred to as a *granum*.

gravitropism (formerly *geotropism*) a plant growth response attributed to gravity.

gray crescent in the amphibian embryo, a pigment-free, crescent-shaped region that appears in the egg surface following fertilization.

greenhouse effect the warming principle of greenhouses, in which high-energy solar rays enter easily, while less energetic heat waves are not radiated outward; now especially applied to the analogous effect of increasing atmospheric concentrations of carbon dioxide through the burning of fossil fuels and forest biomass.

gross productivity the amount of biochemical energy captured by photosynthesis in a particular area per unit time.

ground meristem primary plant tissue from which the ground tissues, collenchyma, parenchyma, and sclerenchyma, are derived.

growth curve a graph of population numbers per unit time, especially in a period of rapid initial growth leading into population size stabilization or to a population crash.

growth factor a hormone, usually a mitogen, necessary for the proliferation of cells in tissue culture.

G_2 also *G_2 phase, gap two,* see *cell cycle.*

guanine one of the nitrogenous bases of RNA and DNA.

guanosine a nucleotide of guanine; guanine linked to ribose.

gullet 1. the esophagus. 2. an invagination in the surface of certain protozoan protists, especially when used for the intake of food.

guard cell in leaf and stem epidermis, either of a pair of crescent-shaped cells that with the pore make up of the stoma.

gustation the sense of taste; the act of tasting.

guttation the normal exuding of moisture from the tip of a leaf or stem, presumably due to root pressure.

gynoecium in a flower, the whorl or carpels taken as a group; the pistil or pistils.

habit reversal a test involving the ease with which an animal learns to respond in a manner opposite to that to which it has previously been trained.

habitat the place where a plant or animal species naturally lives and grows.

habituation learning not to respond to environmental stimuli that may have no relevance to the organism.

hair cell see *organ of Croti.*

Haldane-Oparin hypothesis the proposal that life arose spontaneously in the sea following a period of organic synthesis and under conditions that are no longer present on the earth.

half life. 1. of a radioisotope, the time it takes for half of the atoms in a sample to undergo spontaneous decay. 2. the time it takes for half of the amount of an introduced substance (such as a drug) to be eliminated by a natural system, either by excretion or by metabolic breakdown.

haploid having a single set of genes and chromosomes; compare *diploid, polyploid.*

Hardy-Weinberg law a statement of the mathematical expectations of genotype frequencies given allele frequencies in a population of random mating diploid individuals.

haustoria specialized, host-penetrating hyphae of parasitic fungi.

Haversian system a Haversian canal together with its surrounding, concentrically, arranged layers of bone, canaliculi, lacuna, and osteocytes.

heat receptor a sense organ that perceives warmth or radiant heat.

helicase or *unwinding enzyme* see *replication complex.*

heliotropism see *solar tracking.*

heliozoan also *sun animacule,* a protist of a group in the phylum Sarcodina, consisting of free-living, freshwater forms looking rather like tiny suns, with multiple thin, stiff, radiating pseudopods.

heme group an iron-containing, oxygen-binding porphyrin ring present in all hemoglobin chains and in myoglobin; see *porphyrin group.*

hemichordate an animal of the deuterostome phylum *Hemichordata.*

hemipenis either of the paired hemipenes or copulatory organs of a male lizard or snake.

hemizygous a term used to describe the condition of X-linked genes in males, which are neither homozygous nor heterozygous.

hemocoel a body cavity, especially in arthropods and mollusks, formed by expansion of parts of the blood vascular system.

hemocyanin a copper-containing respiratory pigment occurring in solution in the blood plasma of various arthropods and mollusks.

hemoglobin an iron-containing respiratory pigment consisting of one or more polypeptide chains, each associated with a heme group.

hemophilia a genetic tendency to uncontrolled bleeding in humans, due to the lack of a necessary constituent of the blood clotting process; caused by recessive alleles at either of two sex-linked loci, it affects males almost exclusively.

herbivorous plant-eating, *Herbivore,* an animal that feeds only on plants.

heredity the transmission of genetic characters from parents to offspring, and the effects of this transmission. See also *genetics.*

hermaphrodite an animal or plant that is equipped with both male and female reproductive organs.

Herpes simplex see *Type I* and *Type II herpes simplex.*

heterochromatin chromatin that is relatively heavily condensed at times other than mitosis or meiosis, or that condenses early in prophase, and is genetically largely inert.

heterocyst in filamentous cyanophytes, a relatively large, transparent, thick-walled cell specialized for nitrogen fixation.

heterogamete the evolutionarily advanced state in gametes in which a large nonmotile egg and a small motile sperm form.

hetermorphic alternation in alternation of generations, the condition wherein the sporophyte and gametophyte are quite different in appearance, usually with one dominating the life cycle.

heteromorphic chromosomes homologous chromosomes that can be distinguished from one another by some visible difference, as a variable knob or constriction; e.g., X and Y chromosomes.

heterosporous a state in plants in which morphologically distinct spores, microspores, and megaspores are produced.

heterotherm an endothermic animal that thermoregulates only part of the time.

heterotroph a microorganism that requires organic componds as an energy or carbon source.

heterozygous having two different alleles at a specific gene locus on homologous chromosomes of a diploid organism.

Hfr (or *high frequency of recombination*) a strain of *E. coli* in which the F plasmid is incorporated into the main host chromosome, thus increasing the frequency of sexual recombination involving the host chromosome.

hillock or **axonal hillock** a region at the base of an axon in which action potentials arise.

hilum (pl. *hila*), a scar left on a seed, marking the point of separation from the ovary; a bean's belly button.

hindbrain 1. the posterior of the three primary divisions of the embryonic vertebrate brain. 2. the parts of the adult brain derived from the embryonic hindbrain, including the *cerebellum, pons,* and *medulla oblongata.*

hinge joint a joint that moves in one plane, like a door hinge.

hirudin an anticoagulant secreted by leeches.

histone one of a class of small, highly basic proteins that complex with the nuclear DNA of higher eukaryotes, forming nucleosomes that presumably protect the DNA from degradation; one histone (histone I) appears to have a role in chromatin condensation and thus in one level of gene control.

HIV see *human immunodeficiency virus.*

holdfast a rhizoidal bae of a seaweed, serving to anchor the algal thallus to the ocean floor or rock substratum.

homeobox a region within homeotic or control genes, consisting of some 100 nucleotides, whose base sequence is very similar in a variety of organisms. Homeoboxes are thought to play a key role in the activation of control genes.

homeostasis the tendency toward maintaining a stable internal environment in the body of a higher animal through interacting physiological processes involving negative feedback control.

homeostatic mechanism any mechanism involved in maintaining a stable internal equilibrium, especially one involving self-correcting negative feedback.

homeotherm see *endotherm, ectotherm,* and *heterotherm.*

homeotic gene a gene that controls the expression of entire blocks of genes and thus plays an important role in the development of a major structure. In *Drosophilia,* homeotic genes control the activity of the imaginal disks.

hominids a primate group composed of humans and related extinct forms.

hominoid a primate group composed of humans, apes, and related extinct forms.

homologous 1. similar because of a common evolutionary origin. 2. derived by independent evolutionary modification from a corresponding body part of a common ancestor. 3. derived from a common embryological precursor; e.g., the *glans clitoridis* is homologous with the *glans penis.* 4. *homologous chromosomes,* chromosomes which, because of common descent, have the same kind of genes in the same order and which pair in meiosis.

homologue also *homologous chromosome,* either of the two members of each pair of chromosomes in a diploid cell; see *homologous.*

homology similarity due to common descent.

homosporous a state in plants where all spores are the same. See also *heterosporous.*

homozygous having the same allele at a given locus in both homologous chromosomes of a diploid organism.

hormone a chemical messenger transmitted in body fluids or sap from one part of the organism to another, that produces a specific effect on target cells often remote from its point of origin, and that functions to regulate physiology, growth, differentiation, or behavior.

hornwort a nonvascular terrestrial plant allied to mosses and liverworts.

host a living organism that harbors or sustains a parasite, pathogen, or other symbiont.

host-specific a parasite or pathogen that can infect only a single host species.

human chorionic gonadoptropin (HCG) a hormone secreted by the human chorion, prompting the corpus luteum to continue manufacturing progesterone.

Human Immunodeficiency Virus (HIV) the retrovirus responsible for AIDS (acquired immune deficiency syndrome).

human leukocyte-associated antigen (HLA antigen) the major histocompatibility complex of humans. See also *major histocompatibility complex.*

humoral response the manufacture and secretion of specific antibodies by plasma cells (activated B-cells) in response to the presence of antigen or stimulation by helper T-cells.

Huntington's disease also *Huntington's chorea,* a lethal hereditary disease of the nervous system developing in adult life and attributable to a dominant allele.

H-Y cell-surface antigen also *H-YA,* an antigen found on the surface of all cells of all male mammals, but not in female mammals; and also on the cells of all female birds, but not male birds. It is believed to be controlled by a Y-linked gene in mammals, and to be responsible for the differentiation of the mammalian testis.

hyaline cartilage translucent, bluish-white connective tissue consisting of cells embedded in a homogeneous matrix.

hyaluronic acid a viscous mucopolysaccharide occurring chiefly in connective tissues, but also as a cell-binding agent in other tissues, e.g., it holds together the cells of the *corona radiata.*

hybrid the offspring of two animals or plants of different races, breeds, varieties, species, or genera.

hybrid swarm genetically diversified populations produced by introgressive hybridization, that is, the breeding of hybrids with parent stock, thus creating numerous partial hybrids.

hydration shell the layer or layers of water surrounding and loosely bound to anion in solution.

hydrocarbon 1. a chemical compound consisting of hydrogen and carbon only, often in chains or rings: e.g., gasoline, benzene, paraffin; 2. a portion of an organic molecule consisting of hydrogen and carbon only, e.g., the hydrocarbon portion of a fatty acid.

hydrogenation chemically saturating an unsaturated lipid with hydrogen through the elimination of double carbon-to-carbon bonds.

hydrogen bond a weak, noncovalent, usuallyintramolecular electrostatic attraction between the positively polar hydrogen of a side group and a negatively polar oxygen of another side group.

hydrogen carrier a certain membranal element within an electron transport system that accepts both electrons and protons, transporting hydrogen across the membrane, whereupon its proton is released and its electron continues in the system. See *chemiosmosis.*

hydrogen ion 1. a proton. 2. as a convenient fiction, a proton in solution and bound to a water molecule; actually a *hydronium ion* (H_3O^+).

hydroid in cnidarians, a member of class Hydrozoa.

hydrologic cycle also *water cycle,* the cycle of water between its liquid and gaseous phases brought about by heat in the atmosphere.

hydrolysis see *hydrolytic cleavage.*

hydrolytic cleavage also *hydrolysis,* the reaction of a compound with water such that the compound is split into two parts by the greaking of a covalent bond; and water is added in the place of the bond, an —OH group going to one subunit and an —H group going to the other.

hydronium ion a symmetrical charged ion, H_3O^+, that can dissociate into water and a hydrogen ion; most hydrogen ions in solution are actually in the form of hydronium ions.

hydrophilic "water loving," a characteristic of charged molecules in which they readily interact with water molecules.

hydrophobic with regard to a molecule or side group, tending to dissolve readily in organic solvents but not in water; resisting wetting; not containing polar groups or subgroups.

hydrostatic pressure the pressure exerted in all directions within a liquid at rest.

hydrostatic skeleton a supporting and locomotory mechanism involving a fluid confined in a space within layers of muscle and connective tissue; movement occurring when muscle contractions increase or decrease the hydrostatic pressure of the fluids.

hydroxyl group —OH, consisting of an oxygen and hydrogen covalently bonded to the remainder of the molecule; a constituent of alcohols, sugars, glycols, phenols, and other compounds.

hydroxide ion, the anion OH^-, one of the dissociation products of water in the reaction $2H_2O \rightarrow H_3O^+ + OH^-$.

hymen also *maidenhead,* fold of thin mucous membrane closing the orifice of the vagina, especially in virgins.

hyperglycemia an abnormal excess of glucose in the blood.

hyperosmotic pertaining to a solution having a greater concentration of solutes than some reference solution.

hyperpolarization in a neuron, an inhibitory state caused by an influx of negative ions, whereby the threshold of stimulation is greater than during the usual resting state.

hyperthyroidism a condition caused by an excess of circulating thyroid hormone; symptoms in humans include nervousness, sleeplessness, hyperactivity, weight loss, and, if prolonged, a bulging of the eyeballs.

hypertonic (adj.), having a higher osmotic potential (e.g., higher solute concentration) than the cytoplasm of a living cell (or other reference solution).

hyperventilate to breathe deeply deliberately, purging the body of CO_2.

hypha (pl. *hyphae*), one of the individual filaments that make up a fungal mycelium.

hypocotyl the part of a plant embryo below the point of attachment of the cotyledon.

hypocotyl hook in some seedlings, the curved region of the hypocotyl that acts as a bumper as the seedling emerges from the soil.

hypoglycemia 1. an abnormally low level of blood glucose. 2. a medical condition involving recurrent episodes of low blood sugar, due to the overproduction of insulin.

hypoosmotic pertaining to a solution having a lesser concentration of solutes than some reference solution.

hypophysis see *pituitary gland.*

hypothalamus a portion of the floor of the midbrain, containing vital autonomic regulatory centers, and closely associated functionally with the pituitary gland.

hypothesis a proposition set forth as an explanation for a specified group of phenomena, either asserted merely as a provisional conjecture to guide investigation (e.g., working hypothesis) or accepted as highly probably in the light of established facts; see also *theory.*

hypothyroidism a condition caused by abnormally low levels of circulating thyroid hormone; symptoms include physical and mental sluggishness and weight gain.

hypotonic (adj.), having a lower osmotic potential (e.g., a lower concentration of solutes) than the cytoplasm of a living cell (or other reference solution).

hysterectomy the surgical removal of the uterus.

H zone clearer, centermost zone of the relaxed contractile unit consisting of myosin alone.

IAA indoleacetic acid; see *auxin*.

I band one of the striations of striated muscle, of variable width, corresponding to the distance between the ends of the myosin filaments of adjacent units.

imbibition the taking up (absorption) of a fluid (often water) by a hygroscopic gel, colloid, or fibrous matrix; such as the ultramicroscopic spaces in a plant cell wall.

immune response the entire array of physiological and developmental responses involving specific protective actions aginst a foreign substance; including phagocytosis, the production of antibodies, complement fixation, lysis, agglutination, and inflammation.

immune system in vertebrates, widely-dispersed tissues that respond to the presence of the antigens of invading microorganisms or foreign chemical substances.

immunoglobin a protein antibody produced by thymocytes in response to specific foreign substances, consisting of constant regions and two specific antigen binding sites; or a polymer of such molecules.

imperfect flowers lacking either pistils or anthers.

implantation the act of attachment of the mammalian embryo (blastocyst) to the uterine endometrium.

impulse neural impulse, 1. a wave of excitement (transitory membrane depolarization) transmitted through a neuron.

inactivation gate see *sodium ion gate*.

inborm error of metabolism a genetic defect in which an individual lacks one of the enzymes of a biochemical pathway.

incisors in humans and most mammals, the four upper and four lower front teeth, which are chisel-shaped for biting and cutting; in rodents the medial incisors are greatly exaggerated and the lateral incisors are absent.

incomplete flower see under *flower*.

incomplete metamorphosis insect development in which the egg gives rise to nymphs or naiads, which undergo transition directly into the adult state.

indirect flight muscles in insects, flight muscles that attach to and move the thorax, which in turn moves the wings (see also *direct flight muscles*).

indeterminate cleavage a lack of developmental commitment in newly cleaved cells in the early embryo, resulting for a time in daughter cells remaining developmentally flexible.

indoleacetic acid also *IAA,* the naturally-occurring form of auxin, the plant growth hormone; see *auxin*.

induced fit hypothesis the proposition that enzymatic actions proceed because their active site forms an inexact fit with their substrate, producing physical stress.

inducer 1. in molecular genetics, a small molecule that triggers the activity of an inducible enzyme. 2. in embryology, a substance that stimulates the differentiation of cells or the development of a particular structure.

inducible enzyme an enzyme produced by an organism only when induced by an appropriate stimulus. *Inducible operon,* the entire operon involved in the control of a battery of inducible enzymes.

induction the process by which the fate of embryonic cells of tissue is determined, especially as due to tissue interactions.

inductive reasoning logical progression proceeding from the specific to the general. Reaching a conclusion based on a number of observations.

indusium in ferns, a covering over the sorus (spore-forming organ).

inferior situated below or beneath; e.g., *inferior vena cava,* inferior ovary in a flower (epigyny).

inflorescence a clustering of flowers with a specific arrangement.

information load a measure of the level of information that can be communicated in a single display.

ingestion swallowing.

inhibitory block according to the classical definition of instinct, the neurogical inhibitors of behavior that are selectively removed by the perception of the appropriate releaser.

inhibiting hormone any of several hypothalamic neurosecretions, targeted for the adenohypophysis (anterior pituitary), which responds by slowing the release of one of its hormones.

inhibitory synapse a synapse in which the secretion of neutotransmitter increases the threshold voltage requirement of the receiving neuron, thereby inhibiting it.

initiation the start of translation or polypeptide synthesis. See also *initiation complex*.

initiation complex the elements needed to begin initiation or polypeptide synthesis, a process requiring the presence of mRNA, methionine-charged tRNA, and the two ribosomal subunits.

initiator codon or *start condon,* a codon (sequence of three mRNA nucleotides) that initiates translation of a polypeptide; specifically, AUG, which also codes for methionine and N-formyl methionine.

initiator tRNA transfer RNA charged with methionine.

innate inborn, not due to learning.

innate behavioral pattern a genetically programmed behavior pattern.

innate releasing mechanism (IRM) a hypothetical center in the central nervous system that, when stimulated by the proper environmental cue, activates neural pathways that result in an *instinctive behavior pattern*.

inner cell mass in a mammalian blastocyst, the portion that is destined to become the embryo proper.

inner compartment see *matrix*.

inner ear the portion of the ear enclosed within the temporal bone, consisting of a complex fluid-filled labyrinth and the associated auditory nerve; included are the three *semicircular canals,* the *cochlea,* and the *round* and *oval windows;* compare *external ear, middle ear*.

inoculum a small quantity of living cells or organisms introduced into a suitable medium for growth.

inoperative allele see *dominance relationships*.

inorganic molecule one not containing carbon, generally one found occurring outside of living organisms.

inorganic phosphate also P_i, the anion of phosphoric acid; the phosphate group once separated from a nucleotide triphosphate such as ATP or ADP.

insectivore an eater of insects.

insectivorous habitually eating insects; *insectivorous plant,* one with special adaptations for trapping and digesting insects for their nitrogen and mineral content.

insertion 1. in genetics, the addition of extra genetic material into the middle of a chromosome, gene or other DNA sequence; *base insertion,* the insertion of a single base pair into a DNA sequence. Also *muscle insertion,* 2. in anatomy, the distal attachment of a tendon or muscle; compare *origin*.

instinct innate behavior involving appetitive and consummatory phases, the latter usually in response to an environmental releaser but sometimes occurring as vacuum behavior.

instinctive pattern a fixed-action pattern that is coupled with orienting movements.

integument a covering or envelope; in the flower, the covering that encloses the nucellus of an ovule, and later forms part of the seed coat; in animals, the skin, exoskeleton, tunic, cuticle, or other covering.

interbreeding 1. commonly breeding together. 2. breeding together; hybridizing.

intercalated disk in heart muscle, the highly convoluted double plasma membrane between two adjacent cells in a heart muscle fiber.

intercostal muscles the muscles of the rib cage, involved in breathing.

interecdysis also *intermolt, instar,* in arthropods, the period between successive molts.

interferon cellular substance that interferes with replication by viruses, generally produced by virus-infected cells.

intermediary metabolism metabolic activity including the extraction of energy from fuels, its use in the biosynthesis of new substances and in the interconversion of existing substances.

intermediate host in parasitic relationships, a host in which a parasite may undergo asexual reproduction or at least some degree of development.

internal fertilization in animals, fertilization involving copulation, in which sperm release and fertilization occur within the female.

interneuron a neuron typically found in the spinal cord that synapses with sensory and motor neurons and other interneurons.

internode the interval between two nodes, see *nodes*.

interphase all of the cell cycle between successive mitoses; included are G_1, S phase, and G_2.

interstitial pertaining to small spaces between

things or parts, such as the spaces between cells or tissues layers.

interstitial cells cells of the testis that have an endocrine function.

intertidal zone the zone between the mean high tide line and the mean low tide line, intermittently covered by the sea or exposed to air.

intervertebral disk one of the tough, elastic, fibrous disks situated between the centra of adjacent vertebrae.

intestine also *alimentary canal* or *gut,* essentially tubular organ in which the digestion and absorption of foods occur.

intracellular within cells.

intracellular digestion digestion within cells, in digestive vacuoles following phagocytosis.

intraspecific within a species; between members of the same species.

intrauterine device metallic contraceptive device that inhibits implantation when in place in the uterus.

intrinsic rate of increase (*i*) also *instantaneous rate of increase,* the rate of population increase determined by subtracting the average death rate from the average birth rate.

introgression backcrossing of a hybrid with one or the other parental types. Common means of diversification in flowering plants.

introitus the external opening of the vagina.

intron, also *intervening sequence,* a region of DNA separating two parts of a structural gene; transcribed into HnRNA but deleted from mRNA.

invagination the folding inward of a surface or tissue to make a cavity; specifically, the formation of a gastrula by the inward folding of part of the wall of the blastula.

in vitro ("in glass"), in a test tube or other artificial environment; compare *in vivo.*

in vivo in the living body of a plant or animal; compare *in vitro.*

involuntary muscle see *smooth muscle.*

involution an inward roll or curve, in gastrulation, the movement of cells toward the blastopore, over the blastopore lip, and into the archenteron.

ion any electrostatically charged atom or molecule.

ion channel in the neural membrane, sodium and potassium (and chloride) channels that, through the opening and closing of gates, selectively admit or reject ions.

ionic bond a chemical attraction between ions of opposite charge.

ionization 1. the dissociation of a molecule into oppositely charged ions in solution. 2. the creation of ions by the energy of ionizing radiation.

ionizing radiation energetic radiation that produces ions in air and free hydroxyl radicals in water, the latter being responsible for induced mutation and tissue damage; included are X rays, gamma rays, and streams of charged particles from the breakdown of radioactive isotopes.

irregular flower a flower that is bilaterally symmetrical (right and left sides), as opposed to being regular or radially symmetrical.

islets of Langerhans also *islets,* beta, alpha, and delta endocrine cells within the pancreas that secrete the hormones insulin, glucagon, and somatostatin, respectively.

isogametes gametes that are identical in size and appearance, such as the (+) and (−) isogametes of *Chlamydomonas.*

isomorphic alternation in alternation of generations, the condition wherein sporophyte and gametophyte are virtually identical in appearance and no dominance exists.

isoosmotic a solution having the same solute concentration as some reference solution.

isotonic having the same osmotic potential (e.g., the same concentration of solutes) as the cytoplasm of a living cell, or of some other reference fluid.

isotope a particular form of an element in terms of the number of neutrons in the nucleus; *radioactive isotope,* also *radioisotope,* an unstable isotope that spontaneously breaks down with the release of ionizing radiation.

Jacobsons organ a pouchlike olfactory organ in the roof of the mouth of most mammals and other vertebrates; in lizards and snakes it receives the ends of the forked tongue; absent in primates.

jaundice a yellowish pigmentation of the skin and other tissues caused by the deposition of bile pigments, following bile duct obstruction, liver disease, or the excessive breakdown of red blood cells.

jelly coat a layer of transparent substance surrounding the egg of an echinoderm, fish, or amphibian.

joint a point where two bones articulate, often movable.

juvenile hormone (JH), an insect hormone that prevents the differentiation of adult characters; secreted by the corpora allata of larval and subadult insects.

juxtaglomerular complex in the nephron, a physical connection between the afferent arteriole and the distal convoluted tubule, believed to coordinate the reabsorption of sodium.

karyotype 1. a mounted display of enlarged photomicrographs of all the stained chromosomes of an individual, arranged in order of decreasing size. 2. the total chromosome constitution of an individual. 3. the chromosomal makeup of a species.

keel the enlarged breastbone to which the flight muscles of birds are attached.

kelp any of various large brown algal seaweeds.

keratin a cysteine-rich, fibrous, insoluble, intracellular structural protein making up most of the substance of the dead cells of hair, horn, nails, claws, feathers, and the outer epidermis.

kidney 1. in vertebrates, one of a pair of ducted excretory organs situated in the body cavity beneath the dorsal peritoneum, serving to excrete nitrogenous wastes and to regulate the balance of body ions and fluids. 2. any analogous organ in invertebrate metazoans.

kilocalorie see calorie.

kinetic energy the energy of motion.

kinetochore see *centromere.*

kingdom one of the primary divisions of life forms; to the traditional plant and animal kingdoms Whittaker has added fungal, protist, and moneran (prokaryote) kingdoms.

Klinefelter's syndrome in humans, an abnormal condition of males, caused by the chromosomal condition XXY; characterized by increased stature, moderate retardation, small testes, low testosterone, and secondary effects of low testosterone.

Krause corpuscle also *Krause end bulb,* a sensory receptor in the skin that responds to temperature changes.

Krebs cycle see *citric acid cycle.*

krill planktonic crustaceans that constitute the principal food of baleen (toothless) whales.

K-selection population growth characterized as logistic in form, responding to environmental resistance and carrying capacity. Compare *r-selection.*

kwashiorkor a syndrome of severe protein deficiency in human infants and children, including failure to grow, deficiency of melanin pigment, edema, degeneration of the liver, anemia, and retardation.

labia majora (sing. *labium majorum*), in human females, the outer, fatty, often hairy pair of folds bounding the vulva.

labia minora (sing. *labium minorum*), in the human female, the inner, thinner, highly vascular pair of folds surrounding the introitus, enfolding and extending downward from the clitoris.

lac operon (an inducible operon) in the chromosome of *E. coli,* composed of genes that code for three lactose-metabolizing enzymes and the region that controls their transcription. See also *operon.*

lactate fermentation see *fermentation.*

lacteal a lymphatic vessel of the intestinal villi.

lactose intolerance the inability of certain persons to readily digest the milk sugar lactose, which leads to problems in the bowel.

lacuna (pl. *lacunas, lacunae*), a minute cavity in bone or cartilage that holds an osteocyte or chondrocyte.

lagging end in DNA replication, the 5′ end of a DNA strand, where nucleotides are first assembled into Okazaki fragments and then added in using the enzyme ligase.

lamella (pl. *lamellae*), 1. one of the bony concentric layers (ringlike in cross section) that surround a Haversian canal in bone. 2. one of the thin plates composing the gills of a bivalve mollusk. 3. any thin, platelike structure.

lampbrush chromosome a chromosomal region commonly seen in oocytes that becomes temporarily active, spinning out loops of DNA which become active in transcription.

large intestine also colon or bowel, a division of the alimentary canal, primarily functioning in the resorption of water.

larva in animals, an early, active, feeding stage of development during which the offspring may be quite unlike the adult.

larynx in terrestrial vertebrates, the expanded part of the respiratory passage just below the glottis and at the top of the trachea; in mammals in contains the vocal cords and constitutes a resonating *voice box* or *Adam's apple*.

late blight the protisan disease caused by *Phytophthora infestans,* responsible for the Irish potato famine of the 1840s.

latency Darwin's term for the gross phenomenon of recessivity.

lateral branch bud also *axillary bud,* a dormant bud in the leaf axil or above a leaf scar, capable of giving rise to new stems.

lateral line organ in nearly all fish, a sensory organ considered to be responsive to water currents and vibrations; thought to be distantly homologous with the inner ear.

lateral root also *branch root,* roots that arise from the pericycle of the primary root. The branch roots of taproot systems.

laterally from side to side.

law a virtually irrefutable conclusion or explanation of a phenomenon, such as "the law of gravity" or "Mendel's first law."

law of alternate segregation also *Mendel's first law,* the observation, based on the regularity of meiosis, that heterozygous alleles separate in gamete formation, each allele going to approximately half the gametes produced.

law of independent assortment also *Mendel's second law,* the observation, based on the regularity of meiosis, that genetic elements on different chromosomes behave independently in the production of gametes.

laws of thermodynamics in physics, laws governing the interconversions of energy. See also *thermodynamics.*

LDL receptor (low density lipoprotein) a membrane receptor specialized in binding to cholesterol and transporting it into cells where it is stored as fat.

leader sequence see *signal peptide.*

leading end in DNA replication, the 3′ end of a DNA strand, where nucleotides are added one at a time to the growing strand.

leaf in vascular plants, a lateral outgrowth from stem functioning primarily in photosynthesis, arising in regular succession from the apical meristem, consisting typically of flattened blade jointed to the stem by a petiole.

leaf blade the typically flattened, photosynthetic region of a leaf only a few cell layers thick and containing a supporting petiole, midrib, and veins.

leaf hair see *lower epidermis.*

leaf mesophyll the photosynthetic parenchymal cells within a leaf, including an uppermost palisade layer and a lower spongy layer.

leaf hair see *lower epidermis.*

leaf mesophyll the photosynthetic parenchymal cells within a leaf, including an uppermost palisade layer and a lower spongy layer.

leaf mesophyll cell photosynthetic parenchymal cell within the leaf.

leaf primordium see *primordium.*

leaf scale a flattened, photosynthetic, leaflike structure in bryophytes.

leaf scar the mark left on a stem by a fallen leaf.

learning in animal behavior, the process of acquiring a persistent change in a behavioral response as a result of experience.

lenticel a pore in stem of plant for the passage of gasses between the atmosphere and stem tissues.

leptotene (adj. and n.), also *leptonema* (n.), the first stage of meiotic prophase, at which time the still unsynapsed chromosomes appear as elongated, thin threads.

lethal-recessive dominant an allele that has prominent visible effect as a heterozygote, and is lethal as a homozygote; in humans such an allele is also called a *rare dominant* since it may never occur as a homozygote.

leucoplast a colorless plastid; see also *plastid.*

leukocyte also *leucocyte,* a vertebrate white blood cell; including, eosinophils, neutrophils, basophils, monocytes, and lymphocytes.

lichen a combination of a fungus and and an alga growing in a symbiotic relationship.

ligament a tough, flexible, but inelastic band of connective tissue that connects bones, or that supports an organ in place; compare *tendon.*

ligase an enzyme that heals nicks (single-strand breaks) in DNA.

light-harvesting antenna a cluster of photosynthetic pigments (chlorophyll and accessory pigments) which receives energy from photons and transfers that energy to a single *reaction center.*

light-independent reaction that part of photosynthesis not immediately involved in chemiosmosis, specifically the fixation of CO_2 into carbohydrate from the NADPH and ATP produced by the light reaction (Calvin cycle).

light reaction that part of phtosynthesis directly dependent on the capture of photons; specifically the photolysis of water, the thylakoid electron transport system, and the chemiosmotic synthesis of ATP and NADPH.

lignin an amorphous substance that gives wood its rigidity.

limbic system a region of the brain concerned primarily with emotions.

limiting factor any factor in the environment that, by its presence or relative absence, limits the growth of a population; e.g., disease, predation, food supply, mineral nutrient.

limiting resource a resource, such as food, shelter, or mineral nutrient, that by its scarcity limits the growth of a population or determines the carrying capacity of the environment.

limnetic zone in freshwater bodies, the open water deep enough for a photic and an aphotic zone to occur and where food chains begin with phytoplankton.

limnologist an ecologist who specializes in the study of interaction in the freshwaters.

linkage group a group of gene loci shown to be linked by recombination tests; ultimately, a chromosome.

linked (adj.), of two gene loci, not segregating independently.

lipase any fat-digesting enzyme.

lip cell in the fern sporangium, thinner-walled cells that split open through changes in the mature dry annulus, releasing the spores. See also *annulus.*

lipid an organic molecule that tends to be more soluble in nonpolar solvents (such as gasoline) than in polar solvents (such as water).

lipoprotein a conjugated protein with a lipid subgroup.

litter also *leaf litter, duff,* the uppermost slightly decayed layer of organic material on a forest floor.

littoral in reference to the lake or seashore. *Littoral zone,* 1. a coastal region including both the land along the coast, the water along the shore and the intertidal. 2. a similar area in lakes.

liver 1. in vertebrates, a large, glandular, highly vascular organ that serves many metabolic functions including detoxification, the production of blood proteins, food storage, the biochemical alteration of food molecules, and the production of bile. 2. in mollusks, a digestive organ opening into the gut, in which intracellular digestion of food by phagocytosis occurs.

liverwort a kind of ground-hugging bryophyte.

lobe a curved or rounded projection or division, as of the liver or lung.

lobe-finned fish lung fishes and crossopterygians.

locus (pl. *loci*), also *gene locus,* specific place on a chromosome where a gene is located.

logarithmic spiral a form of spiralling curve that keeps its overall shape constant as it increases in length.

logistic growth curve also *logistic increase* and *S-shaped curve,* population growth plot that takes the form of a sigmoid curve, growth that is eventually subject to *environmental resistance* and hovers about the environment's *carrying capacity.*

long-day plant also *short-night plant,* a plant that begins flowering at some specific time before the summer solstice when day length exceeds night, flowering being triggered by a critical, established period of darkness.

long-term memory 1. learning that persists more than a few hours, the memory trace of which is physically located in a different part of the brain that short-term memory. 2. the part of the brain and the general neural function with which such persistent memory traces are associated.

loop of Henle the hairpin loop of the vertebrate nephron, between the proximal and distal convoluted tubules, which leaves the cortex and descends into the medulla, then loops back to the corex; it is involved in water resorption.

lophophorate invertebrate animals bearing the lophophore, a ridgelike feeding structure bearing ciliated tentacles.

lophophore in brachiopods, ectoprocts, and phoronids, a filter-feeding organ consisting of a spiral or horseshoe-shaped ridge surrounding the mouth, and bearing tentacles.

Lotka-Volterra theory the proposal that predator population size rises and falls according to the population size of the prey.

lower epidermis an outer layer of leaf cells, commonly containing numerous stomata and projecting leaf hairs.

lumbar pertaining to the lower trunk; *lumbar region* of the spine, the vertebrae between the thorax and the sacrum.

lumen the cavity or channel of a hollow tubular organ or organelle.

lung in land vertebrates, one of a pair of compound, saccular organs that function in the exchange of gases between the atmosphere and the bloodstream. 2. any of the several analogous organs in invertebrates, e.g., *book lung*.

lymph capillary one of the many tiny, blind endings in the lymphatic system, including the lacteals of the small intestine.

lymph node a rounded, encapsulated mass of lymphoid tissue through which lymph ducts drain, consisting of a fibrous mesh containing numerous lymphocytes and phagocytes.

lymphatic collecting duct larger lymphatic vessels that receive lymph from the lymphatic capillaries.

lymphatic system the system of lymphatic vessels, lymph nodes, lymphocytes, the thoracic duct and the thymus, which together serve to drain body tissues of excess fluids and to combat infections.

lymphatic vessel a thin-walled vessel that conveys lymph, originating as an open intercellular cleft and draining eventually into the *thoracic duct*.

lymphoid tissue tissue in which lymphocytes are activated and aggregate.

lymphocyte any of several varieties of similar-appearing leukocytes involved in the production of antibodies and in other aspects of the immune response; see *B-cell, T-cell*.

Lyon effect in human females, the genetic mosaic created by the random manner in which either the paternal or the maternal X chromosome is selected for permanent inactivation.

lyse (v.t.), to destroy a cell by rupturing the plasma membrane.

lysis (n.), the destruction or lysing of a cell by rupture of the plasma membrane.

lysogenic state a temperate bacteriophage in its quiescent or inactive stage, during which its DNA is incorporated into a host chromosome.

lysogeny the incorporation of a bacteriophage chromosome into the bacterial host chromosome, together with mechanisms preventing further infection and lysis; in later cell generations the incorporated DNA may excise, replicate, and eventually cause cell lysis.

lysosome a small membrane-bounded cytoplasmic organelle, generally containing strong digestive enzymes or other cytootoxic materials.

lysozyme an enzyme that lyses bacteria by dissolving the bacteria cell wall; a natural constituent of eggwhite and tears.

lytic cycle the short cycle following bacteriophage invasion during which viral replication, capsid synthesis, viral assembly, and cell lysis occur, the latter releasing new infective phage particles.

lytic phage any bacteriophage in its reproductive and cell-lysing phase.

macromere a relatively large, yolk-filled blastomere of the lower half (vegetal pole) of a blastula.

macromolecule any large biological polymer, as a protein, carbohydrate, or nucleic acid.

macronucleus the larger of the two nuclei of *Paramecium* and certain other ciliate protozoans, carrying somatic line DNA; compare *micronucleus*.

macronutrient a plant nutrient required in relatively substantial quantities, as nitrogen, phosphorus, potassium sulfur, magnesium, and calcium; compare *micronutrient*.

macrophage a large phagocyte that forms from a monocyte.

madreporite see *water vascular system*.

major histocompatibility complex (MHC) groups of glycopproteins coating the plasma membrane of cells, making individuals biochemically unique.

malaria disease caused by parasitic infection by sporozoan parasites *(Plasmodium)* of the red blood cells; transmitted by *Anopheles* mosquitos.

Malpighian tubule any of numerous blind, hollow, tubular structures that empty into the insect midgut and function as a nitrogenous excretory system.

mammal any vertebrate of the class *Mammalia*.

mammary gland also *milk gland, mamma, mammary,* the organ that, in female mammals, secretes milk for the nourishment of the young.

mandible 1. the vertebrate lower jaw or jaw bone. 2. either of two laterally paired anterior mouth appendages of a mandibulate arthropod, which form strong biting jaws.

mantle a fleshy covering of mollusks that secretes material to form the external shell.

mantle cavity in mollusks, a cavity between the mantle and the body proper, in which the respiratory organs lie; a highly vascularized mantle cavity serves as a lung in pulmonate snails.

marker see *genetic marker*.

marrow *bone marrow,* the soft tissue in the interior cavity of a bone; *blood marrow,* vascularized bone marrow in which blood cells of all types are produced; *fatty marrow,* bone marrow consisting of adipose tissue.

marsupial a mammal of the subclass Methatheria, usually with a pouch *(marsupium)* in the female; included are the kangaroo, wombat, koala, Tasmanian devil, opossum, and wallaby.

masturbation manipulation of one's sexual organs for erotic gratification

mating type in fungi and various protists, a grouping of organisms incapable of sexual reproduction with one another but capable of such reproduction with members of one or more other groups or mating types.

matrix also *inner compartment,* in the mitochondrion, the enzyme-laden region within the highly convoluted inner membrane, the site of the citric acid or Krebs' cycle.

maturation the process of coming to full development.

maxilla (pl. *maxillae*), 1. the upper jaw of a vertebrate, *anterior* to the mandible. 2. in insects, one of the first or second pair of mouth parts *posterior* to the mandibles.

mechanism the theory that everything in the universe is produced by matter in motion through definable physical laws.

mechanoreceptor see *sensory receptor*.

median eye 1. a simple photoreceptive organ, usually lacking a lens, found in lancelets and certain reptiles; apparently used in responses to photoperiodicity. 2. also *pineal eye,* a homologous organ found in birds and various other vertebrates.

medulla 1. the inner portion of a gland or organ; compare *cortex*. 2. also *medulla oblongata,* a part of the brainstem developed from the posterior portion of the hindbrain and tapering into the spinal cord.

medulla oblongata see *medulla*.

medusa the motile, free-swimming jellyfish body type of cnidarian.

megagametophyte the female gametophyte produced from a megaspore in flowering plants; the *embryo sac* consisting of eight haploid nuclei or cells.

megaphyll a large veined leaf, believed to have evolved from primitive branching stems.

megasporangium a sporophyte structure in heterosporous plants in which megaspores are formed through meiosis.

megaspore a single large cell with one haploid nucleus formed after meiosis and a megaspore mother cell in plants, and from which the megagametophyte develops by mitosis.

megaspore mother cell in the ovule, a large diploid cell that will give rise to the megaspore by meiosis and the degeneration of three of the four haploid nuclei.

megasporogenesis the formation of megaspores.

megasporophyll a sporophyte structure specialized for producing megaspores, e.g., the ovules of a flowering plant.

meiosis also *reduction division,* in all sexually reproducing eukaryotes, the process in which a diploid cell or cell nucleus is transformed into four haploid cells or nuclei through one round of chromosome replication and two rounds of nuclear division, the first of which involves the unique pairing of chromosome homologues.

meiospore a haploid spore produced by meiosis.

meiotic division. See *meiosis*.

meiotic interphase the period, which may be prolonged or so brief as to be virtually nonexistent, between telophase I of meiosis and prophase II of meiosis, during which there is no DNA replication.

Meissner's corpuscle in mammals, a small ellipsoidal touch-responsive neural end organ.

melanin the characteristic animal surface pigmentation, a class of polymers of tyrosine or dopa, giving black, brown, orange, or red coloration depending on chemical composition and pigment granule size; also found in plants.

melanism an unusual development of a black or nearly black color in a species not generally so pigmented.

melatonin a hormone produced in the mammalian pineal body that has a role in sexual development and maturation through its effect on the hypothalamus; it also causes color changes in frogs and appears to be related to photoperiodicity responses.

membranal pump any mechanism within the plasma membrane that actively transports molecules or ions into or out of the cell.

membrane 1. any thin, soft, pliable sheet, as of connective tissue. 2. *plasma membrane,* the semipermeable, lipid bilayer, proteinaceous sac enclosing the cytoplasm. 3. any internal cell structure of a similar construction, such as the nucelar membrane, thylakoid membrane, and others; see also *fluid-mosaic model.*

membrane-bound 1. bound to a membrane, e.g., *membrane-bound ribosomes.* 2. also *membrane-bounded,* having an outer boundary consisting of a membrane, e.g., a lysosome is a membrane-bounded organelle.

memory 1. the ability to retain a learned response or to recognize a stimulus previously encountered. 2. the part or function of the brain in which learned responses are maintained. See *engram, long-term memory, short-term memory.*

memory cell a mature, long-lived, B- or T-cell lymphocyte, specialized in retaining specific antigen information.

memory trace also *memory scar,* see *engram.*

menarche in human females, the onset of menstruation; the first menses taken as the beginning of puberty.

meninges tough, protective, connective tissues covering the brain and spinal cord.

menopause in human females 1. the cessation of menstruation, usually occurring between the ages of 45 and 50. 2. the whole group of physical, physiological, and behavioral occurrences and changes associated with the cessation of menstruation.

menses see *menstruation, menstrual cycle.*

menstrual cycle also *ovarian cycle,* the entire cycle of hormonal and physiological events and changes involving the growth of the uterine mucosa, ovulation, and the subsequent breakdown and discharge of the uterine mucosa in menses; normally occurring at three- to five-week periods in nonpregnant women.

menstruation also *menses,* in nonpregnant females of the human species only, the periodic discharge of blood, secretions, and tissue debris resulting from the normal, temporary breakdown of the uterine mucosa in the absence of implantation following ovulation.

meristem also *meristematic tissue,* a plant tissue consisting of small undifferentiated cells that give rise by cell division both to additional meristematic cells and to cells that undergo terminal differentiation; all postembryonic plant growth depends on meristem.

meristematic tissue undifferentiated cells from which the plant produces new tissues.

merozoite a stage of *Plasmodium* that reproduces asexually in a mammalian red blood cell, releasing the poisons that cause malarial symptoms.

mesenchyme patches of mesodermal tissue that has migrated from its original position in the gastrula, such as that which gives rise to muscle, dermis, gonad, and skeletal tissue.

mesentery membranous folds of peritoneum to which the intestines are attached, which, while holding them in their coiled position, permit necessary movement and other changes.

mesoderm the middle of the three primary germ layers of the gastrula, giving rise in development to the skeletal, muscular, vascular, renal, and connective tissues, and to the inner layer of the skin and to the epithelium of the coelom *(peritoneum).*

mesoglea the loose, gelatinous middle layer of the bodies of sponges and cnidarians, between the outer ectoderm and the inner endoderm.

mesohyl in the sponge, a gelatinous matrix in which amebocytes, spicules, and various fibrils are arranged.

mesophyll the chloroplast-rich parenchyma of a leaf, between the upper and lower epidermal cells.

mesosome in many bacteria, an inward extension of the plasma membrane that forms a spherical membranous network; function not resolved.

messenger RNA *mRNA,* 1. in prokaryotes, RNA directly transcribed from an operon or structural gene, containing one or more contiguous regions (cistrons) specifying a polypeptide sequence. 2. in eukaryotes, RNA transcribed from a structural gene, tailored and usually capped and polyadenylated in the nucleus and transported to the cytoplasm; containing a single contiguous region specifying a polypeptide sequence as well as leader and follower sequences.

metabolic defect see *inborn error of metabolism.*

metabolic pathway an orderly series or progression of enzyme-mediated chemical reactions leading to a final product, each step catalyzed by its own specific enzyme.

metabolism the chemical changes and processes of living cells, including but not limited to respiration, the synthesis of biochemicals, and the breakdown of wastes; see also *anabolism, catabolism, basal metabolism, biochemical pathway.*

metabolite 1. a metabolic waste, especially one that is toxic. 2. an intermediate in a biochemical pathway.

metamere segment of a segmented animal. See also *segmentation.*

metamerism see *segmentation.*

metaphase the stage of mitosis or meiosis in which the centromeres of the chromosomes are brought to a well-defined plane in the middle of the mitotic spindle prior to separation in anaphase.

metaphase plate the equatorial plane of the mitotic spindle on which the centromeres are oriented in mitosis.

Metazoa in Whittaker's five-kingdom scheme, all animals other than sponges.

methanogen any archaebacterium from the anaerobic group known to produce methane gas (CH_4) as a metabolic byproduct.

methemoglobinemia, a pathological condition in which methemoglobin polypeptides from which the heme group has been detached are present in the blood (a symptom of nitrite poisoning).

MHC restriction a condition of T-cells in which their ability to recognize other cells is restricted to those with specific Class I or Class II MHC proteins.

mica a mineral that forms thin, clear, waterproof crystals.

microbe also *microorganism,* any very small organism, such as a bacterium, yeast, or protist.

microbial genetics the scientific study of inheritance and DNA function in bacteria, bacteriophage, yeast, fungi, or protist.

microevolution evolutionary change below the species level, including changes in gene frequencies brought about by natural selection and random drift.

microfibril one of the submicroscopic elongated bundles of cellulose in a plant cell wall.

microfilament 1. a submicroscopic filament of the cytoskeleton, involved in cell movement and shape; the principal component is fibers of the protein *actin.* 2. see *myofilament.*

microgametophyte a sperm-producing body developed from a microspore.

micromere in cleavage and in the blastula, a relatively small cell of the animal pole; compare *macromere.*

micrometer 1. symbol μm, one millionth of a meter. 2. an instrument for making very precise or very small measurements.

micronucleus (pl. *micronuclei*), in ciliate protists, the smaller of the two nuclei and the one carrying germ-line DNA; compare *macronucleus.*

micronutrient also *trace element,* an element necessary for plant growth but needed only in vanishingly small quantities.

microorganism also *microbe,* any organism too small to be seen readily without the aid of a microscope; such as bacterium, protist, or yeast.

microphyll a small leaf believed to have evolved from outgrowths of the plant epidermal cells. See also *megaphyll.*

micropyle 1. in insect eggs, a depression or differentiated area through which the sperm enters. 2. in seed plants, a minute opening in the integument of an ovule through which the pollen tube enters.

microsporangium a sporophyte structure in heterosporous plants in which microspores are formed through meiosis.

microspore in seed plants, one of the four haploid cells formed from meiosis of the microspore mother cell, which undergoes mitosis and differentiation to form a pollen grain.

microspore mother cell in the anther of a

flowering plant, the diploid cells that will undergo meiosis to form the haploid microspores.

microsporogenesis in flowering plants the process of producing microspores, which develop into the male gametophyte.

microsporophyll a sporophyte structure specialized for producing microspores, e.g., the anthers of a flowering plant.

microtubule a cytoplasmic hollow tubule composed of spherical molecules of tubulin, found in the cytoskeleton, the spindle, centrioles, basal bodies, cilia, and flagella.

microvilli (sing. *microvillus*), also *brush border,* tiny fingerlike outpocketings of the plasma membrane of various epithelial secretory or absorbing cells, such as those of kidney tubule epithelium and the intestinal epithelium.

midbrain the middle of the three divisions of the vertebrate embryonic brain; the adult structures derived from the embryonic midbrain.

middle ear in mammals, the part of the ear between the eardrum and the oval window, consisting of a chain of three ear ossicles; the *malleus, incus,* and *stapes,* in an air-filled chamber-communicating with the pharynx by way of the eustacian tube.

middle lamella a layer of cementing material between adjacent plant cell walls.

midpiece in a spermatozoan, the segment between the head and the tail, containing one or more mitochondria.

midrib the large central vein of a dicot leaf.

millipede also *myriopod,* nonpoisonous, herbivorous terrestrial arthropod of the class *Diplopoda,* with a long, cylindrical, segmented body, hard integument, and two pairs of legs per segment.

mineral nutrient an inorganic compound, element, or ion needed for normal growth of all organisms.

mineralocorticoid any mineral-regulating steroid hormone of the adrenal cortex.

minimal medium broth or agar gel consisting of glucose, mineral salts, water, and glycine or other nitrogen source; the least complex medium capable of sustaining the growth of a specific microorganism.

minimum mutation tree in systematics, a hypothetical phylogenetic tree selected because it represents the smallest number of evolutionary changes needed to account for known relationships.

missense mutation a base change in a structural gene that alters the coding property of the codon in which it appears, producing an abnormal polypeptide with one amino acid difference.

mitochondrion (pl. *mitochondria*), threadlike, self-replicating, membrane-bounded organelle found in every eukaryotic cell, functioning in oxidative chemiosmotic phosphorylation, apparently derived in evolution from a bacterial endosymbiont.

mitogen a hormone that stimulates cell division.

mitosis cell division or nuclear division in eukaryotes, involving chromosome condensation, spindle formation, precise alignment of centromeres, and the regular segregation of daughter chromosomes to daughter nuclei

mitotic apparatus the centrioles, spindle, spindle fibers, and asters; in plants, the spindle and spindle fibers.

mitotic spindle or *spindle,* a spindle-shaped system of microtubules appearing in the cell during mitosis and meiosis, including both *centromeric spindle fibers* (pole to centrome) and *polar spindle fibers* (pole to midpoint overlapping fibers from opposite pole), and in animals, lower plants, and some protists, the *asters* or *astral rays.*

mitotic spindle apparatus see *mitotic apparatus.*

model a contrived biological mechanism, based on a minimal number of assumptions that, when applied, is expected to yield numerical data consistent with past observations; a biological hypothesis with mathematical predictions, e.g., *laws of thermodynamics, Mendel's first and second laws.*

modifier gene a gene or allele that modifies the expression of genotypes at another gene locus.

molar also *molar tooth,* a cheek tooth adapted for grinding; in human adults, the three most posterior pairs of teeth in each jaw.

mole 1. *gram molecule,* the quantity of a chemical substance that has a mass in grams numerically equal to its molecular mass in daltons; 6.023×10^{23} molecules of a substance.

molecular biology a branch of biology concerned with the ultimate physiochemical organization of living matter; the study of biological systems using biochemical methods.

molecular mass also, *molecular weight,* the mass of a molecule expressed in daltons.

molecule a unit of chemical substance, consisting of atoms bound to one another by covalent bonds.

molt (v.i.) to cast off an outer covering in a periodic process of growth or renewal.

molt (n.) 1. the process of molting; see *ecdysis.* 2. the cast-off covering. 3. the period of a molt; an intermolt in interecdysis.

molting hormone a hormone that initiates or causes molting.

monera the prokaryote kingdom; bacteria.

mongolism an archaic term for *trisomy 21.*

monoclonal antibody a specific immunoglobulin produced by hybrid cells artificially cloned in the laboratory.

monocot a flowering plant of the angiosperm class *Monocotyledonae,* characterized by producing seeds with one cotyledon; included are palms, grasses, orchids, lilies, irises, and others.

monocotyledon see *monocot.*

monoecious 1. of a gametophyte, producing both egg and sperm in the same thallus. 2. of a seed plant sporophyte, producing both ovules and pollen. Compare *dioecious.*

monomer one of the subunits or potential subunits of a polymer.

monophyletic (adj.), of a taxonomic group, deriving entirely from a single ancestral species that possessed the defining characters of the group, while not having given rise to organisms outside of the group.

monosaccharide a sugar not composed of smaller sugar subunits (e.g., glucose, fructose).

monotreme a platypus, echidna, or extinct egg-laying mammal of the order *Monotremata,* subclass *Protheria.*

mons veneris in women, a rounded, usually hairy bulge of fatty tissue over the pubic symphysis and above the vulva.

morph one of the particular forms of an organism that exists in two or more distinct forms in a single population.

morphogen a biologically active substance capable of influencing the developmental fate of embryonic tissue.

morphogenesis the development of form and structure.

morphogenic determinant a developmental event early in the embryo's history that influences later events of development.

mortality death, considered as an aspect of population dynamics.

mortality rate also *death rate,* the number of deaths per unit time occurring among a specified number of individuals (usually per 1000) in a given area or population.

morula in the early embryo, a simple ball preceding the blastula state.

mosaic development the early determination and setting aside of cells for specific developmental roles whose fulfillment may be some time in the future.

mosaic egg an egg in which the developmental fates of different parts of the cytoplasm or surface are fairly rigidly determined prior to fertilization.

motor end plate the terminal branching of the axon of a motor neuron as it forms multiple synapses with a muscle fiber.

motor neuron a neuron that innervates muscle fibers, and impulses from which cause the muscle fibers to contract.

motor unit a motor neuron together with the muscle fibers it innervates, which contract as a unit.

M phase see *cell cycle.*

mRNA *messenger ribonucleic acid, see messenger RNA.*

mucopolysaccharide a polysaccharide that binds water to form a thick, gelatinous material that serves to cement animal cells together; a constituent of mucoproteins, e.g., *chondroitin, hyaluronic acid, heparin.*

mucoprotein a combination of mucopolysaccharide and a polypeptide to form *mucin.*

mucosa also *uterine mucosa, endometrium,* the highly glandular mucous membrane lining of the uterus; inner lining of the gut.

mucous (adj.), covered with or secreting mucus. *mucous membrane.*

mucus a viscid, slippery secretion rich in mucins, that is secreted by mucous membranes and that serves to moisten and protect such membranes.

Mullerian ducts or *paramesonephric ducts,* the embryological origin of oviducts in some female vertebrates.

Mullerian mimicry in a number of dangerous

species, a common coloration or configuration serving as a common warning.
multicellular (adj.), consisting of a number of specialized cells that cooperatively carry out the functions of life.
multinucleate (adj.), having many nuclei within a common cytoplasm.
multiple alleles the alleles of a gene locus when there are more than two alternatives in a population.
multiplicative law in mathematical probability theory, the statement that the probability that all of a group of independent events will occur is equal to the product of their individual probabilities.
muscle a contractile tissue, in vertebrates including skeletal, visceral, and cardiac muscle.
muscle fiber 1. one of the multinucleate cells of striated muscle, which takes the form of a long, unbranched cylinder. 2. cardiac muscle fiber, one of the long, branched, cylindrical subunits of heart muscle, consisting of individual cells joined end to end by intercalated disks.
muscular dystrophy an incurable hereditary disease characterized by the progressive wasting of muscle tissue and eventual death. *Duchenne-type muscular dystrophy,* a common sex-linked variety affecting primarily preteen boys.
muscularis the smooth muscle layer of the wall of a hollow contractile organ such as the uterus, gall bladder, or urinary bladder.
mushroom a basidiocarp, especially but not necessarily an edible one.
muskeg a North American sphagnum bog, particularly in the taiga.
mutagen a chemical or physical agent that causes mutations.
mutant 1. a new or abnormal type of organism produced by a mutation. 2. a mutated gene. 3. an individual that bears and is affected by a mutated gene, 4. (adj.), mutated.
mutate 1. (v.t.), to alter, cause a change, or cause a mutation or DNA change to occur in; to mutagenize. 2. (v.i.), to change in state or genetic condition, to become altered, to undergo a mutation.
mutation 1. any change in DNA. 2. a change in DNA that is not immediately and properly repaired. 3. any abnormal, heritable change in genetic material. 4. the act or process of mutating.
mutation rate the rate at which new mutations occur, generally in terms of mutations per locus per gamete per generation.
mutualism a mutually beneficial association between different kinds of organisms; symbiosis in which both partners gain fitness; compare *symbiosis, parasitism.*
mycelium (pl. *mycelia*), 1. the mass of interwoven hyphae that forms the vegetative body of a fungus. 2. the analogous filaments formed by certain filamentous bacteria.
mycoplasma a tiny, nonmotile, wall-less bacterium of irregular shape, occurring as an intracellular parasite of animals and plants.
mycorrhizae a mutualistic fungus-root association with the fungal mycelium either surrounding or penetrating the roots of a plant.
myelinated having a myelin sheath.
myelin sheath fatty sheath surrounding the axons of many vertebrate neurons; also see *Schwann cell* and *oligodendrocyte.*
myofibril a tubular subunit of muscle fiber structure, consisting of many sarcomeres end-to-end.
myofilament the highly organized microfilaments of striated muscle. *Actin myofilaments,* thin filaments of the protein actin bound to the Z-line of a sarcomere; *myosin myofilament,* thicker filaments of the protein myosin interspersed among the actin myofilaments in a regular hexagonal array.
myosin a protein involved in cell movement and structure; especially muscle cells; see also *myofibril, sarcomere, cytoskeleton.*
myosin bridge the more or less globular head of the generally fibrous protein myosin which forms a movable bridge between a myosin myofilament and an actin myofilament.
myotome 1. the portion of a vertebrate embryonic somite from which skeletal musculature is developed. 2. one of the muscular segments of the body wall of a fish or lancelet.
myxameba the free-living, unicellular, ameboid stage of a slime mold.
myxobacterium a slime or gliding bacterium, many species of which are capable of forming spores in stalked fruiting bodies.
NAD or **nicotine adenine dinucleotide** a coenzyme and hydrogen receptor and carrier in cell metabolism (the oxidized form is NAD^+ and the reduced form, *NADH*).
NADP or nicotine adenine dinucleotide phosphate a coenzyme and hydrogen receptor and carrier in photosynthesis (the oxidized form is $NADP^+$ and the reduced form, *NADPH*).
nanoplankton a newly discovered group of microscopic, flagellated phytoplankton, believed to be important in marine food chains.
nastic response any movement in plants, particularly changes in leaf position, attributed to sudden turgor pressure changes in cells of the petiole.
natural killer (NK) cell a free-roving lymphocyte that identifies, binds to, and lyses cancerous and virus-infected cells as part of the nonspecific immune response.
natural selection the differential survival and reproduction in nature of organisms having different heritable characteristics, resulting in the perpetuation of those charactertistics and/or organisms that are best adapted to a specific environment. Also, *survival of the fittest.*
navigation the ability to locate or maintain reference to a particular place without the use of landmarks.
Neanderthal human an extinct, highly variable type or race of humans (*Homo sapiens neanderthalensis*) of the middle Paleolithic.
nectary a plant gland that secretes a sweet fluid (nectar), most often at the base of a flower, that functions as a reinforcing reward for the visits of pollinating animals.
negative control in a genetic control system, control such that transcription occurs at all times except when prevented by an inhibitor molecule or an inhibitor complex.
negative feedback an automated control mechanism in which an action, brought about by a chemical or physical stimulus, directly or indirectly reduces the stimulus. Such an inhibiting effect constitutes a negative feedback loop.
negative feedback loop see *negative feedback.*
nematocyst one of the minute stinging cells of the cnidarians, consisting of a hollow thread coiled within a capsule, and an external hair trigger.
neoteny the condition in which animals become developmentally arrested in an immature or larval state, while becoming sexually mature and able to reproduce, as in the *Larvaceae.*
nephridiopore the exterior orifice of a nephridium.
nephridium an excretory organ primitive to many coelomate phyla and still found in annelids, brachipods, mollusks and some arthropods, occurring paired in each body segment in segmented animals, and typically consisting of a ciliated funnel (*nephrostome*) draining the coelom through a convuluted, glandular duct to the exterior.
nephron a single excretory unit of a kidney, consisting of a glomerulus, Bowman's capsule, proximal convoluted tubule, loop of Henle, distal convoluted tubule, and collecting duct discharging into the renal pelvis.
neritic province the coastal sea from the low-tide line to a depth of 100 fathoms, generally waters of the continental shelf.
nerve (n.), 1. a filamentous band of nerve cell axons and dendrites, and protective and supporting tissue, that connect parts of the nervous system with other parts of the body. 2. (adj.), pertaining to the nerve or nervous system, e.g., *nerve cell, nerve net, nerve fiber.*
nerve cell see *neuron.*
nerve cord 1. *dorsal hollow nerve cord,* the spinal cord of chordates. 2. *ventral nerve cord,* the solid, paired, segmentally ganglionated, longitudinal central nerve of many invertebrates.
nerve fiber an axon or dendrite.
nerve impulse see under *impulse.*
nerve net in cnidarians, the arrangement of nerve fibers into a netlike pattern with little concentration.
nerve synapse see under *synapse.*
nervous system the brain, spinal cord, nerves, ganglia and the neural parts of receptor organs, considered as an integrated whole.
net community productivity (NCP) the rate of community productivity by autotrophs after their energy utilization through respiration has been subtracted.
net productivity in ecology, the rate of production of energy or biomass by plants, being the gross productivity less the energy used by the plants in their own activities.
neural canal 1. the canal formed by the series

of vertebral neural arches, through which the spinal cord passes. 2. the lumen (cavity) of the spinal cord.

neural folds in early vertebrate embryology, a pair of longitudinal ridges that arise from the neural plate on either side of the neural groove, which fold over and fuse to give rise to the *neural tube*.

neural groove in early vertebrate embryology, the linear depression in the neural plate between the neural folds that will invaginate to form the neural tube.

neural impulse a transient membrane depolarization, followed by immediate repolarization, traveling in a wavelike manner along a neuron.

neural plate in the vertebrate gastrula, a thickened plate of ectoderm along the dorsal midline that gives rise to the neural tube, neural crests, and ultimately to the nervous system.

neural tube 1. in early vertebrate embryology, the hollow dorsal tube formed by the fusion of the neural folds over the neural groove. 2. the spinal cord.

neuroglia also *glia*, supporting cells of the central nervous system.

neuromuscular junction the synapse between a neural motor end plate and a muscle fiber.

neuron also *nerve cell*, a cell specialized for the transmission of nerve impulses, consisting of one or more branched *dendrites*, a *nerve cell body* in which the nucleus resides, and a terminally branched *axon*.

neuropeptide a class of neurotransmitters produced and secreted in the central nervous system and consisting of short chains of amino acids.

neurophysiology the physiology of nerve cells, neural receptors, and other aspects of the nervous system.

neurosecretion 1. a secretion of a nerve cell, especially hormonal secretions into the general circulation, usually produced in the nerve cell body, transmitted along the axon, and released at the terminus of the axon. 2. the process of secretion by neurons.

neurotransmitter a short-lived, hormone-like chemical (e.g., *acetylcholine, norepinephrine*) released from the terminus of an axon into a synaptic cleft, where it stimulates a dendrite in the transmission of a nerve impulse from one cell to another; see also *synapse, synaptic cleft, synaptic knob*.

neurula an early vertebrate embryo slightly older than a gastrula, in which the neural tube has formed.

neurulation the development of a neurula from a gastrula; equivalently, the development of the neural plate, neural folds, and neural tube.

neutral mutation also *selectively neutral mutation*, a mutational change in DNA that has no measurable effect on the fitness of an organism, and which may sometimes by incorporated into the genome of a species by genetic drift.

neutralist a population geneticist or evolutionary theorist who argues that a substantial proportion of protein and DNA changes in evolution have been due to selectively neutral mutations fixed by random drift; compare *selectionist*.

neutrophil the most common mammalian phagocytic leukocyte

neutron one of the two common constituents of an atomic nucleus, a hadron having no charge and no effect on chemical reactions.

nexin in cilia and flagella, protein connections between the nine-paired groups of microtubules.

niche the position or function of an organism in a community of plants and animals. It has been said that if a habitat is an organism's address, a *niche* is its occupation. *Niche* has also been defined as the totality of the adaptations, specializations, tolerance limits, functions, biological interactions, and behavior of a species.

nick 1. (n.), a single-strand break in DNA. 2. (v.t.), to cause a single-strand break in DNA by the hydrolysis of a phosphate linkage.

nicotinic acid a 5-carbon nitrogenous acid, a constituent of NAD and NADP in respiration and photosynthesis; a vitamin (dietary requirement) of animals.

nictitating membrane in many vertebrates, including many mammals and most birds, a thin translucent membrane that can be folded into the inner angle of the eye or drawn across the front of the eyeball; vestigial in humans.

nitrification the chemical oxidation of ammonium salts into nitrites and nitrates by the action of soil bacteria.

nitrogen cycle a biogeochemical cycle in which nitrogen compounds pass from autotrophs (plants and algae) to heterotrophs (animals, fungi, protists, and bacteria) and back to autotrophs. New nitrogen enters the cycle through nitrogen fixation by bacteria, and losses of nitrogen occur through dentrification.

nitrogen fixation the chemical change of nitrogen from the stable, generally unavailable form of atmospheric nitrogen ($N \equiv N$) to soluble and more readily utilized forms such as ammonia, nitrate, or nitrite; usually by the action of cyanophytes or nitrogen-fixing bacteria, but sometimes by lightning, automobile engines, or synthetic fertilizer factories.

nitrogenase an enzyme that catalyzes nitrogen fixation utilizing cell energy.

nitrogenous base 1. a purine or pyrimadine used in the synthesis of nucleic acids; specifically, adenine, cytosine, guanine, thymine, or uracil. 2. any related heterocyclic compound, such as xanthine or uric acid.

noble element also *noble gas*, an element in which the uncharged atoms have all electron shells filled and all electrons paired and which consequently are chemically unreactive.

node 1. in plant stem, any point at which one or more leaves emerge. 2. in a growing stem tip, a region of potential leaf growth, containing meristematic tissue. 3. see *node of Ranvier*. 4. see *lymph node*.

node of Ranvier a constriction in a myelin sheath, corresponding to a gap between successive Schwann cells. See *saltatory propagation*.

nodule 1. any small lump. 2. see *root nodule*.

noncellular see *acellular*.

noncoding strand during transcription, the DNA strand not involved. See also *transcription* and *transcribed strand*.

noncompetitive inhibition enzyme inhibition involving molecules different from the substrate. Such inhibitors alter the active site's shape by binding with some other part of the enzyme.

noncylic phosphorylation light reactions employing both photosystem II and photosystem I, during which electrons from water pass through both photosystems, reducing NADP to NADPH + H^+, thus serving both chemiosmosis and carbohydrate synthesis.

non-Darwinian evolution descent with modification attributable to genetic drift rather than natural selection; the fixation of selectively neutral mutations in evolutionary lineages.

nondisjunction 1. the failure of a pair of homologous chromosomes to segregate to opposite poles in anaphase I of meiosis. 2. the failure of a pair of daughter chromosomes or daughter centromeres to segregate to opposite poles in mitotic anaphase or in anaphase II of meiosis.

nonhistone chromosomal protein a large class of proteins other than histones associated with the chromosomes, constituting about one-third of the substance of chromatin in animals and plants, highly variable from tissue to tissue and known in at least some cases to be involved in gene control.

nonspecific defense any of several fast-acting cellular and chemical responses made by the body against foreign substances, cancerous cells, or invading organisms, but not involving B-cell and T-cell lymphocytes.

nonvascular plant plants lacking tissues specialized for the transport of water and food and providing physical support (bryophytes: mosses, liverworts, and hornworts).

normal distribution also *normal curve, Gaussinan distribution, bell-shaped curve*, the idealized, symmetrical distribution taken by a population of values centering around a mean, when departures from the mean are due to the chance occurrence of a large number of individually small independent effects; often approached in real populations.

notochord in all chordates at some point in development, a turgid, flexible rod running along the back beneath the nerve cord and serving as a skeletal support; replaced during development by the vertebral column in most vertebrates but persistent in adult coelocanths, cyclostomes, lancelets, and larvacean urochordates.

N-terminal end also *amino-terminal end*, the end of a polypeptide chain in which the end amino acid has its primary amino group free (or formylated); the first part of the chain to be synthesized.

nucellus in flowers, a mass of thin-walled parenchymal cells that composes most of the ovule, encloses the embryo sac, and is enclosed by one or more integuments.

nuclear envelope also *nuclear membrane,* the double cellular membrane surrounding the eukaryotic nucleus, the outermost of which is continuous with the endoplasmic reticulum.

nuclear membrane see *nuclear envelope.*

nuclear pore a structure in the nuclear envelope that appears as a hole in electron micrographs although probably filled with a protein mass in life; through which RNA, nuclear proteins, and other large molecules are presumably able to pass. Also *annulus.*

nuclease any enzyme that hydrolyzes a nucleic acid; see also *endonuclease.*

nucleic acid either DNA or RNA, DNA being a double polymer of deoxynucleotides and RNA being a polymer of nucleotides.

nucleolus (pl. *nucleoli*), a dark-staining body of RNA and protein found within the interphase nucleus of a cell, the site of synthesis and storage of ribosomes and ribosomal materials; disperses during mitosis and is reconstituted following nuclear envelope reorganization.

nucleoplasm the viscous fluid matrix of the nucleus, contrasted with *cytoplasm.*

nucleoside triphosphate a nitrogen base, a ribose or deoxyribose, and a chain of three phosphate groups linked to the 5′ carbon of ribose or deoxyribose.

nucleosome in the chromosome, globular bodies of histone about which eukaryotic DNA is wound.

nucleotide 1. a compound consisting of a nitrogenous base and a phosphate group linked to the 1′ and 5′ carbons of ribose respectively; the repeating subunit of RNA. 2. also *deoxynucleotide,* a compound consisting of a nitrogenous base and a phosphate group linked to the 1′ and 5′ carbons of deoxyribose respectively, the repeating subunit of DNA.

nucleus in all eukaryotic cells, a prominent, usually spherical or ellipsoidal membrane-bounded sac containing the chromosomes and providing physical separation between transcription and translation.

nutrient (n.), 1. any substance required for growth and maintenance of an organism. 2. a chemical element or inorganic compound needed for normal growth of a plant, see also *macronutrient, micronutrient, mineral nutrient.* 3. (adj.) furnishing nourishment, e.g., *nutrient broth,* a broth in which microorganisms may readily be grown.

nutrition the process of being nourished, particularly the steps through which an organism obtains food and uses it for bodily processes.

obligate anaerobe an organism, most commonly a bacterium, that cannot utilize oxygen and often is killed by its presence.

occipital lobe the posterior lobe of each cerebral hemisphere.

oceanic province (n.), the open sea as distinguished from the neritic province; *oceanic* (adj.), pertaining to the open sea.

Oceanic rift community a newly discovered kind of community along ocean floor rifts, where a variety of animal life is supported by chemosynthetic bacteria, producers that thrive near vents releasing heated water and hydrogen sulfide gas.

ocellus (pl. *ocelli*), a minute simple eye or eyespot of any organism. 2. one of the elements of an arthropod compound eye; an *ommatidium* 3. one of the three simple eyes that form a triangle dorsally between the compound eyes.

Okazaki fragment in DNA synthesis, the original form of a newly synthesized single strand, being a polynucleotide of some 200–300 bases produced at the lagging ends of each DNA strand and added in by ligase.

olfaction 1. the sense of smell. 2. the process of smelling.

olfactory bulb an extension of the brain that receives neurons from the olfactory epithelium within the nasal cavity.

olfactory receptor the odor-sensitive cells of the mucous membrane of an olfactory organ.

oligodendrite myelin sheath-forming cell of the central nervous system (compare *Schwann cell*).

oligotrophic of a lake, rich in dissolved oxygen and poor in plant nutrients and algal growth, with clear water and no marked stratification.

omasum see *ruminant.*

ommatidium (pl. *ommatidia*), one of the elements of a compound eye, consisting of a corneal lens, crystalline cone, rhabdome, light-sensitive retinula, and sheathing pigment cells.

omnivorous (adj.), feeding on both animal and plant material; literally, eating everything. *omnivore* (n.), an omnivorous animal e.g., pigs, people.

oncogene a cancer-causing gene, often of viral origin.

oocyte an egg cell before maturation; *primary oocyte,* a diploid cell precursor of an egg, before meiosis; *secondary oocyte,* an egg cell after the formation of the first polar body.

oogonium (pl. *oogonia*), a cell that gives rise to oocytes, the large, spherical, unicellular female sex organ of water molds and egg-producing algae in which egg cells are produced.

open circulatory system a circulatory system in which the arterioles end openly into intercellular space, allowing blood to percolate directly through nonvascular tissues.

open communities biotic communities that blend into each other in a gradual transition.

open growth, also *indeterminant growth* in perennial plants the capability of continuous growth.

operant conditioning see *conditioning, operant.*

operator see operon.

operator locus in an operon, the binding site of an inhibitor protein or inhibitor complex.

operculum a body process functioning as a lid or cover, e.g., (a) the horny plate on the foot of certain gastropods, which serves to close off the opening of the shell; (b) the covering flap of a moss spore capsule; (c) the skin-covered bony plates that cover the gills of a fish.

operon in prokaryotes, a region of DNA that includes structural genes and the genes controlling them; transcription may be *inducible,* remaining shut down until activated by an inducer subtance, or *repressible,* remaining active until shut down by a repressor substance. Control regions generally consist of a *promotor region (p),* an *operator region (o),* and a *regulator gene (i),* which produces a *repressor protein.* Also see *lac operon* and *tryptophan operon.*

opsonization the coating of an invading cell by antibodies whose exposed constant regions are keyed to matching receptors on phagocytes, which facilitates phagocytosis.

oral cavity also *mouth cavity,* the cavity between the mouth and the pharynx.

oral-facial-digital syndrome a genetic condition of human females, caused by an allele that is dominant with regard to abnormalities of the mouth, face, and fingers, and is lethal in the hemizygous state.

oral groove in ciliates, a ciliated fold in the body wall, leading into the cytostome or mouth.

orbit 1. see *electron orbit.* 2. in vertebrates, the bony cavity of the eye socket.

order a taxonomic level between class and family.

organ an organized assembly of various tissues performing some major body function; e.g., the heart, brain, and liver.

organelle a functionally and morphologically specialized part of a cell.

organic compound a chemical compound containing carbon.

organic molecule a molecule containing carbon and generally produced by living organisms.

organism 1. a form of life composed of mutually dependent parts that maintain various vital processes. 2. any form of life. 3. an individual plant, animal, or microorganism.

organ of Corti also *spiral organ of Corti,* on the basilar membrane within the cochlea, an organ containing the neural receptors for hearing, including sensory hair cells and associated neural fibers.

organ-system level of organization in organisms, a complex structural organization, including the presence of specific organs performing in a coordinated manner as systems.

orgasm in humans, the climax of sexual excitement typically occurring toward the end of coitus, usually accompanied in men by ejaculation and in women by rhythmic contractions of the cervix.

orientation the directing of bodily position according to the location of a particular stimulus; may be part of an instinctive action.

orienting movements see *orientation.*

origin 1. evolutionary ancestry. 2. the fixed skeletal attachment of a muscle or tendon; compare *insertion.*

osculum an opening in a sponge through which water exits.

osmoconformer an aquatic organism that does not regulate the osmotic potential of its body tissues, but allows it to fluctuate with that of the environment.

osmoregulation in aquatic organisms, the homeostatic regulation of the osmotic potential of body fluids.

osmoregulator an aquatic organism that maintains a constant internal osmotic potential in spite of fluctuations in the salinity of its environment.

osmosis the tendency of water to diffuse through a semipermeable membrane in the net direction from its higher to its lower concentration.

osmotic gradient any difference in the concentration of water molecules across a membrane, the difference between two fluids in terms of osmotic potential.

osmotic potential the tendency or capacity for water to move across a selectively permeable membrane into a second solution, such movement occurring because of the presence of a relatively high solute concentration (or less water) on the other side.

osmotic pressure the actual hydrostatic pressure that builds up in a confined fluid because of osmosis.

ossification the process of bone formation and especially of the replacement of cartilage by bone.

osteoblast a bone-forming cell.

osteoclast a bone-destroying ameboid cell that dissolves calcium phosphate and releases Ca^{2+} into the bloodstream.

osteocyte a bone cell isolated in a lacuna of bone tissue.

osteon see *Haversian system.*

osteoporosis a condition involving the decalcification of bone, producing bone porosity and fragility.

ostracoderm any of a group of extint jawless fish.

outer compartment in the mitochondrion, the region between the outer and inner membranes into which protons are transported.

oval window in the cochlea, a membrane articulating with the stapes that moves in response to its vibration, subsequently creating movement in the fluid perilymph within.

ovarian cycle see *menstrual cycle.*

ovary 1. in animals, the (usually paired) organ in which oogenesis occurs and in which eggs mature. 2. in flowering plants, the enlarged, rounded base of a pistil, consisting of a carpel or several united carpels, in which ovules mature and megasporogeneis occurs.

overdominance if the phenotype of the heterozygote is more fit than those of both homozygotes, both alleles are said to be *overdominant.*

oviduct a tube, usually paired, for the passage of eggs from the ovary toward the exterior or to a uterus, often modified for the secretion of a shell or protective membrane; in humans it is sometimes known as a *fallopian tube.*

oviparity (n.), the condition of being *oviparous.*

oviparous (adj.), producing eggs that develop and hatch outside of the mother's body; compare *ovoviviparous, viviparous.*

ovipositor a female insect organ specialized for the depositing of eggs and often for boring holes in which eggs may be deposited.

ovoviviparity the condition of being ovoviviparous.

ovoviviparous producing eggs that are fertilized internally, develop within the mother's body but without any direct connection with the maternal circulation; the young being released shortly before or after hatching. Compare *viviparous, oviparous.*

ovulate cone also *seed cone,* the female gymnosperm cone, containing megasporangia in which megaspores form, later giving rise to ovules and seeds.

ovulation the release of one or more eggs from an ovary.

ovule 1. in animals, a small egg; an egg in the process of growth and maturation. 2. in seed plants, a rounded outgrowth of the ovary, consisting of the embryo sac surrounded by maternal tissue including a stalk, the nucellus, and one or more integuments.

oxidation 1. the loss of electrons from an element or compound. 2. also dehydrogenation, the loss of hydrogens from a compound.

oxidation-reduction reaction a chemical reaction in which one reactant is oxidized and another is reduced.

oxidative phosphorylation the production of ATP from ADP and phosphate in a process consuming oxygen, as by mitochondria or aerobic bacteria; see also *chemiosmosis, chemiosmotic, phosphorylation, oxidation.*

oxidative respiration the breakdown of biochemicals to produce cellular energy, utilizing oxygen as the final electron acceptor; see *oxidative phosphorylation, respiration.*

oxygen carrier a molecule specialized for the transport or storage of oxygen, e.g., *hemoglobin, myoglobin.*

oxygen debt a state of oxygen depletion after extreme physical exertion; measured by the amount of oxygen required to restore the system to its original state.

oxyhemoglobin hemoglobin carrying four oxygen molecules; the bright red arterial form of hemoglobin.

pacemaker also *sinoatrial* or *SA node,* the portion of the heart in which the impulse for the heartbeat originates and is regulated.

pachytene a phase of meiosis in which crossing over occurs and in which the synapsed chromosomes appear as solid, thickened threads; between zygotene and diplotene.

Pacinian corpuscles oval pressure receptors containing the termini of sensory nerves, especially in the skin of hands and feet.

palate in mammals, the roof of the mouth cavity, separating the mouth cavity from the nasal cavity; consisting of an anterior, bony *hard palate* and a posterior flesh *soft palate.*

paleontology the scientific study of the forms of life existing in former geological periods, as represented by fossil animals, plants, and microorganism.

palisade parenchyma see *leaf mesophyll.*

pampas (sing, *pampa*), the extensive, grassy plains of southern South America.

pancreas a large digestive and endocrine gland of vertebrates, which secretes various digestive enzymes into the duodenum by way of the *pancreatic duct,* and which also contains endocrine tissues in the form of interspersed *islets of Langerhans,* responsible for the production of the hormones *insulin* and *glucagon.*

pancreatic amylase a starch-digesting enzyme secreted by the pancreas. *pancreatic lipase,* a fat-digesting enzyme secreted by the pancreas. *pancreatic proteases,* the protein-digesting enzymes *trypsin, chymotrypsin* and *elastin. pancreatic hormones,* insulin and glucagon.

paradoxical sleep see *REM (rapid eye movement) sleep.*

parallax the apparent displacement of objects seen from different points of view, especially from the slightly different placement of the eyes in binocular vision.

parallel evolution similar evolutionary change occurring simultaneously in separated lines of descent.

parapatric living in separate but adjacent geographic regions not separated by geographical barriers; compare *allopatric, sympatric.*

paraphyletic of a taxonomic group, sharing defining characters because of common descent from a single ancestral species having those characters, but having also given rise to other organisms not now included in the group.

parasite an organism living in or on another living organism from which it obtains its organic sustenance to the detriment of its host; see also *ectoparasite;* compare *symbiont, parasitoid, mutualism.*

parasitism a relationship in which an organism of one kind (the parasite) lives in or on an organism of another kind (the host) at the expense of which it obtains food and shelter, causing some degree of damage but usually not killing the host directly; compare *symbiosis, mutualism.*

parasitoid an organism that invades the body of a host organism and eventually but regularly causes its death; specifically certain insects that lay one or more eggs in a host organism, the resulting larvae developing within and eventually killing the host at about the time that larval development is complete.

parasympathetic division one of the two divisions of the vertebrate autonomic nervous system, the one that utilizes acetylcholine as a neurotransmitter, that increases the activity of smooth muscle and digestive glands, slows the heart, and dilates blood vessels; compare *sympathetic division.*

parasympathetic nerve in the autonomic nervous system, any nerve of the parasympathetic division.

parathyroid glands four small endocrine glands embedded in or adjacent to the thyroid gland and involved in the regulation of calcium ion levels in the blood.

Parazoa a monophyletic group consisting of the sponges, usually included with the Metazoa in the animal kingdom but not directly related to any other animal groups.

parenchyma in higher plants, a tissue consisting of thin-walled living cells that remain capable of cell division even when mature, and function in photosynthesis or food storage.

parietal cells large cells of the stomach lining that secrete hydrochloric acid.

parietal eye a median light-sensitive organ of ancient vertebrates, or its vestige in certain contemporary vertebrates; analogous to the pineal eye but apparently not homologous with it.

parietal lobe the middle division of each cerebral hemisphere.

parotid glands a pair of ducted salivary glands situated below the ears.

parthenogenesis asexual reproduction in which gametes develop without fertilization, either with or without having undergone meiosis; e.g., male hymenopterans are produced parthenogenetically from haploid eggs, and *Daphnia* of either sex may be produced parthenogenetically from unreduced diploid eggs.

partial dominance a phenotype of a heterozygote in which the expressed alleles are halfway between the homozygous phenotypes. Also see *dominance relationships*.

partial pressure 1. the independent pressure exerted by each gas in a mixture of gases. *Total pressure* is equal to the sum of the partial pressures.

passive immunization the conferring of temporary immunity against a specific disease by the injection of antibodies that act against the disease organism.

passive transport movement of fluids, solutes, or other materials without the expenditure of energy, e.g., by diffusion, especially across a membrane.

pasteurization heating of wine or milk to a certain temperature for a given amount of time in order to kill pathogenic or otherwise undesirable bacteria without destroying the quality of the wine or milk.

pathogen an organism that is capable of causing disease in another organism; generally refers to viruses and parasitic bacteria and fungi.

paved teeth also *pavement teeth,* teeth that form a continuous, flat, roughened surface, and that function in crushing or grinding.

pecking order, also *peck order,* in many gregarious animals, a specific linear order of relative dominance and submissiveness; see *dominance hierarchy*.

pecten 1. any comblike organ. 2. a comblike, pigmented, highly vascularized projection from the retina into the vitreous humor of the eyes of most birds; function unknown.

pectin any of a group of complex methylated polysaccharides occurring in plant tissues, especially of fruits and succulent leaves, that bind water and sugar to make viscous solutions or gels.

pectoral girdle the bones or cartilage supporting and articulating with the vertebrate forelimb; in most vertebrates not connected with the axial skeleton, but in humans connecting to the sternum by way of the clavicle and consisting of the clavicle, coracoid process, and scapula.

pelagic (adj.), living in the open sea; *pelagic zone,* the region of the open sea beyond the littoral and neritic zones and extending from the surface to the depth of light penetration; compare *abyssal, benthic, neritic, littoral.*

pellicle 1. the semirigid, proteinaceous integument of many protists. 2. the plasma membrane and associated cytoskeleton of an animal cell.

peloric of a flower, having radial symmetry abnormally in a species in which flowers are normally irregular (bilaterally symmetrical); see also *flower*.

pelvic girdle the bones or cartilage supporting and articulating with the vertebrate hind limbs; in humans consisting of the fused bones of the *pelvis*.

pelvis 1. the bones of the pelvic girdle, consisting of the *sacrum, coccyx* and the paired and fused *ischium, ilium* and *pubis;* 2. *renal pelvis,* the main cavity of the kidney, into which the nephrons discharge urine.

penicillin a group of closely related powerful antibiotics produced by fungi of the genus *Penicillium,* that function by disrupting the synthesis of bacterial cell walls.

penis 1. the copulatory organ of the male in any species in which internal fertilization is achieved by the insertion of a male body part into the female genital tract. 2. the erectile intromittant organ of male mammals, which also serves as a channel for the discharge of urine.

pentaradial symmetry radial symmetry with a five-part organization. See also *radial symmetry*.

PEP cycle also known as *Hatch-Slack Pathway* in C4 plants, a biochemical cycle during which carbon dioxide is first incorporated into 4-carbon compounds in the leaf mesophyll and later released into the Calvin cycle in the bundle sheath cells.

pepsin a proteolytic enzyme secreted as *pepsinogen* by glands of the stomach lining, that is active only at low pH and which acts primarily to reduce complex proteins to simple polypeptides.

pepsinogen the initial, inactive form of pepsin as occurring in and secreted by gastric glands, that readily converts to pepsin in an acid medium.

peptidase a proteolytic enzyme that hydrolyzes peptide bonds; see *aminopeptidase, carboxypeptidase, dipeptidase, endopeptidase, exopeptidase.*

peptide a chain of two or more amino acids linked by peptide bonds, too short to be coagulated by heat or precipitated by saturated ammonium sulfate; most often seen as a partial digestion product or a protein or polypeptide.

peptide bond also *peptide linkage,* the dehydration linkage formed between the carboxyl group of one amino acid and the amino group of another:
R_1—COOH + NH_2—R_2→R_1—CO—NH—R_2 + H_2O.

peptide hormone any hormone consisting of one or more amino acids.

peptidoglycan the primary substance in the cell walls of eubacteria, consisting of N-acetylmuramic acid, N-acetylglucosamine, and certain amino acids.

peptildy transferase during translation, the enzyme involved in the formation of peptide bonds between adjacent amino acids in the growing polypeptide.

perennial (adj.), 1. continuing or lasting for several years. 2. (n.), a plant that lives for an indefinite number of years, as compared with *annual* or *biennial.*

perfect flower see under *flower*.

pericardium 1. the membranous sac surrounding the vertebrate heart. 2. the cavity or hemocoel containing the arthropod heart.

pericarp the ripened and variously modified wall of a plant ovary (fruit) such as a pea pod, seed capsule, berry, or nutshell.

pericycle a layer of parenchyma or sclerenchyma that sheathes the stele of the root, and is associated with the formation of vascular cambium and lateral roots.

periderm in plants, a protective layer of secondary tissue derived from epidermal cells and consisting of cork, cork cambium, and underlying parenchyma.

perinuclear space the space between the two nuclear membranes, continuous with the endoplasmic reticulum.

periosteum tough connective tissue covering of bone.

peripheral lymphoid tissue widely dispersed lymphoid tissues where lymphocytes are activated and aggregate, as in the tonsils, lymph nodes, spleen, and adenoids.

peripheral nervous system nerves and receptors outside the central nervous system, including sensory and motor nerves of the somatic system and autonomic nervous system.

peristalsis successive waves of involuntary contractions passing along the walls of the esophagus, intestine, or other hollow muscularized tube, forcing the contents onward.

peristome. 1. the fringe of teeth around the opening of a moss spore capsule. 2. the lip of a spiral shell. 3. the area surrounding the mouth of an invertebrate.

peritoneum the smooth, transparent membrane lining the abdominal cavity of a mammal. *Peritoneal cavity,* the principal body cavity (abdominal coelom) of a mammal.

peritubular capillaries in the kidney, a capillary bed surrounding the nephron and involved in tubular reabsorption and secretion, related to osmoregulation and excretion.

permafrost the permanently frozen layer of soil and/or subsoil in arctic and subarctic regions.

permeable allowing materials to pass through. *Permeability,* the degree to which materials are able to pass through a substance, membrane, or barrier.

permease an enzymelike carrier that functions in facilitated transport of a specific substrate across a plasma membrane.

peroxisome a cytoplasmic organelle involved in the detoxification of peroxides.

petal one of the usually white or brightly colored leaflike elements of the corolla of a flower.

petiole the small stalk that supports the blade of a leaf; a *leafstalk*.

Petri dish a small shallow dish of thin glass or plastic with a loosely fitting overlapping cover, used for plate cultures in bacteriology.

pH the negative logarithm of the hydronium ion concentration of a solution and a common measure of the acidity or alkalinity of a liquid; pH values of less than 7 indicating acidity, and values greater than 7 indicating alkalinity.

pH differential the difference in acidity between two solutions.

pH scale see *pH*.

phage also *bacteriophage*, a virus that infects and lyses bacteria.

phagocyte 1. any leucocyte that engulfs particulate matter. 2. any cell that characteristically engulfs foreign matter, e.g., cells of the reticuloendothelial system.

phagocytosis taking solid materials into the cell by engulfment and the subsequent pinching off of the plasma membrane to form a digestive vacuole.

phanerogam seed plants; those that produce flowers or cones that bear seeds (gymnosperms and angiosperms).

pharyngeal arch one of several archlike structures in the embryo, some of which form jaws and gill arches in fishes but contribute to a variety of structures in the head and neck in mammals.

pharyngeal gill cleft see *gill cleft*.

pharyngeal pouch in vertebrate embryos, any of the series of outpocketings of the ectoderm of the pharynx opposite the gill clefts, which may or may not break through; in mammals, one of the pairs of pharyngeal pouches becomes the eustachian tubes and the cavities of the middle ears.

pharynx. 1. in fishes, the portion of the alimentary canal in which the gills and gill slits are located. 2. in land vertebrates, the corresponding region, posterior to the oral and nasal cavities and anterior to the esophagus. 3. an analogous region in the alimentary canals of various invertebrates, including some in which it is eversible and toothed.

phenotype the visible or otherwise detectable physical and chemical traits of an organism, as influenced by heredity and by the environment. Compare with *genotype*.

phenotype frequency the proportion of individuals in a population falling into a specific phenotypic category.

phloem a complex vascular tissue of higher plants consisting of sieve tubes, companion cells, and phloem fibers, and functioning in transport of sugars and nutrients.

phloem fiber a fiber cell associated with phloem, with great strength and pliability, e.g., the linen fibers of flax.

phloem parenchyma a parenchymal cell associated with sieve elements and part of the phloem tissue.

phloem ray the part of a vascular ray located in the phloem; see *vascular ray*.

phosphate group phosphate linked by a dehydration linkage to an organic molecule: HPO_4^-.

phosphate ion any of the four ionization steps of phosphate.

phosphate linkage also *phosphate diester*, a phosphate that is covalently linked to two organic residues.

phospholipid also *phosphatide*, any of a class of phosphate-esterified lipids, including *lecithins, cephalins, sphingomyelins;* a major component of cell membranes.

phosphorolysis also *phosphorolytic cleavage*, a reversible reaction analogous to hydrolysis, in which a dehydration linkage is broken by the addition of phosphate:
$R_1—O—R_2 + HPO_4^2 \rightarrow R_1PO_4^2 + HO—R_2$

phosphorylase an enzyme that catalyzes phosphorolysis.

phosphorylation the addition of a phosphate group to a compound, e.g., the addition of phosphate to ADP to produce ATP and water.

photoautotroph a photosynthetic organism; one capable of utilizing light energy and simple inorganic substances in its metabolism.

photoevent the unitary event of the light-dependent reaction of photosynthesis, conceived as the pathway followed by two electrons released by the photolysis of water.

photolysis of water oxidation of water through the removal of hydrogen by highly oxidizing elements of photosystem 680, which in turn are regenerated with the energy of captured photons; the net reaction of a complex series of events being
$2H_2O—4H^+ + 4e^- + O_2\uparrow$.

photon a quantum of electromagnetic radiant energy.

photoperiod the relative lengths of alternating periods of lightness and darkness, which change with the season especially in temperate climates.

photoperiodism the response of an organism to photoperiods, involving sensitivity to the onset of light or darkness and a capacity to measure time.

photoreceptor a receptor of light stimuli.

photorespiration a puzzling phenomenon in which abundant light energy is captured by photosynthesis but little or no net carbon dioxide fixation occurs; common in C_3 plants in bright sunlight on hot days.

photosynthesis the organized capture of light energy and its transformation into usable chemical energy in the synthesis of organic compounds.

photosynthetic pigment any pigment involved in the capture of light; see *chlorophyll, bacteriochlorophyll, accessory pigments*.

photosystem I also *photosystem 700, P700*, the second in the two photosystems in the electron pathway of photosynthesis in cyanobacteria and chloroplasts, and the one involving the reduction of NADP to $NADPH_2$; it may be evolutionarily more ancient than photosystem II.

photosystem II also *photosystem 680, P680*, the first of the two photosystems in the electron pathway of photosynthesis in cyanobacteria and all photosynthetic eukaryotes, and the one involving the photolysis of water; it probably occurred later in evolution than photosystem I.

phototaxis a tendency to move toward light. *Negative phototaxis*, a tendency to move away from light.

phototroph see *photoautotroph;* compare *heterotroph, chemoautotroph*.

phototropism 1. the turning toward light of a growing plant stem. 2. phototaxis. 3. *negative phototropism*, the turning away from light, as of a root tip.

phragmoplast in plants, the cytoplasmic organelle initiating the formation of a cell plate following mitosis.

phycobilin a photosynthetic accessory pigment of cyanobacteria and of red algal chloroplasts (which are apparently derived from saymbiotic cyanobacteria).

phycocyanin a blue-green photosynthetic pigment of cyanobacteria and red algal chloroplasts.

phycoerythrin a red photosynthetic pigment of cyanobacteria and red algal chloroplasts.

phylogenetic tree a graphical representation of the interrelations and evolutionary history of a group of organisms, indicating the relative order of successive divisions of the line of descent, coincident with past speciation events.

phylogeny 1. the evolutionary history of an organism or group of organisms. 2. a phylogenetic tree.

phylum 1. a major taxonomic unit of related, similar classes of animals, e.g. Phylum Annelida. 2. a division of the plant kingdom.

physiological characteristic of an organism's healthy functioning, as contrasted with *pathological* e.g., *physiological pH*, the range of pH found in the tissues of a healthy organism.

physiology 1. a branch of biology dealing with the processes, activities, and phenomena of individual living organisms, organs, tissues, and cells. 2. the normal functioning of an organism.

phytochrome a red-light-sensitive protein complex of certain plant cell membranes, involved in many light-induced phenomena, including phototropism, photoperiodicity, and others.

phytoplankton photosynthesizing planktonic organisms.

phytoplankton photosynthesizing planktonic organisms.

P_i inorganic phosphate.

pigment any chemical substance that absorbs light, whether or not its normal function involves light absorption: e.g., chlorophyll, cytochrome *c*, hemoglobin, melanin.

piloerection the lifting up of hair by tiny involuntary muscles; bristling.

pilus (pl. *pili*), see *sex pilus*.

pineal body also *pineal organ, pineal gland*, a

small body arising from the roof of the third ventricle in all vertebrates, forming a small red cone in most but forming an eyelike photoreceptive organ in larval lampreys, tuatara, and some lizards; directly sensitive to light in reptiles and birds but not in mammals, although it appears to be the center of photoperiod responses in all vertebrates; it secretes the hormone *melatonin.*

pinna in common terms, the "ear."

pinocytosis taking dissolved molecular food materials, such as proteins, into the cell by adhering them to the plasma membrane and invaginating portions of the plasma membrane to form digestive vacuoles; a form of active transport.

pioneer organism 1. an organism that successfully establishes residence and produces offspring in an area not previously inhabited by its kind. 2. a type of organism specialized for the initial invasion of a disturbed area, such as a landslide or burned-out region.

pistil in flowering plants, a unit comprised of one or more ovaries, a style, and a stigma; see also *gynocecium.*

pistillate having pistils but not stamens.

pit also *pit pair,* thin, circular junctions in adjacent plant cell walls, where only the primary cell wall remains.

pith thin-walled parenchymous tissue in the central strand of the primary growth of a stem; the dead remains of such tissue at the center of a woody stem.

pituitary dwarf a small individual resulting from the failure of the anterior pituitary to secrete growth hormone.

pituitary giant an abnormally large individual whose excessive growth is due to excessive secretions of growth hormone, usually because of tumorous overgrowth of the anterior pituitary.

pituitary gland also *hypophysis,* a small double gland lying just below the brain and intimately associated with the hypothalamus in all vertebrates, consisting of an anterior lobe and a functionally distinct posterior lobe; *anterior pituitary,* also *adenohypophysis,* a glandular body communicating with the hypothalamus only by way of a small vascular portal system; secretes growth hormone, prolactin, melanocyte-stimulating hormone, thyrotrophic hormone, ACTH, FSH, and LH; *posterior pituitary,* also *neurohypophysis,* not actually a gland but the enlarged, glandlike termini of axons of cell bodies in the hypothalamus, the hormones being synthesized in the cell bodies and translocated to the posterior pituitary within the axons, to be stored pending release; among these are *oxytocin* and *antidiuretic hormone.*

pivotal joint also *pivot joint,* 1. a joint that permits rotation. 2. a joint that permits only rotation.

placebo a substance having no pharmacological effect but administered as a control in testing experimentally or clinically the efficacy of a biologically active preparation.

placenta 1. in mammals other than monotremes and marsupials, the organ formed by the union of the uterine mucosa with the extraembryonic membranes of the fetus, which provides for the nourishment of the fetus, the elimination of waste products, and the exchange of dissolved gases. 2. in flowering plants, the part of the ovary to which the ovule and seeds attach.

placental mammals also *eutheria,* mammals that form chorioallantoic placentas; apparently a monophyletic group originating about 65 million years ago, and including all living mammals other than marsupials and monotremes.

plankton minute, drifting plant, algal, and animal life in marine and freshwaters.

planula the early, ciliated, free-swimming larva of a cnidarian.

plaque 1. a pathological deposit of lipid, fibrous material and often calcium salts in the inner wall of a blood vessel. 2. a film of bacterial polysaccharide, mucus, and detritus harboring dental decay bacteria.

plasma the fluid matrix of blood tissue, 90% water and 10% various other substances; distinguished from *blood serum* by the ppresence of fibrinogen and the absence of certain platelet-derived hormones.

plasma cell an activated B-cell lymphocyte, specializing in the production and secretion of antibodies.

plasma membrane the external semipermeable limiting layer of the cytoplasm; see also *membrane, fluid-mosaic model.*

plasmid in bacteria, a small ring of DNA that occurs in addition to the main bacterial chromosome and is transferred from host to host by direct contact through pili. *Plasmid plus,* a bacterial strain carrying a plasmid and immune to further transfer. *Plasmid minus,* a bacterial strain not harboring a plasmid and susceptible to plasmid infection.

plasmodesma (pl. *plasmodesmata*), in plants, minute cytoplasmic junctions between cells, occurring at pores in the cell wall through which the plasma membrane of one cell becomes continuous with that of the next. Highly significant to transport.

plasmodium (pl. *plasmodia*), 1. a motile, multinucleate mass of protoplasm produced by the fusion of uninucleate slime mold ameboid cells. 2. also *Plasmodium,* the malarial parasite.

plastid any of several forms of a self-replicating, semiautonomous plant cell organelle, primarily as a *chloroplast* specialized for photosynthesis, a *chromoplast* specialized for pigmentation, or a *leucoplast* specialized for starch storage.

plate (v.t.), to spread a thin suspension of living microorganisms over the surface of an agar gel, so as to be able to isolate and count colonies that result from subsequent growth.

platelet mother cell or **megakaryocyte** see also *platelet.*

platelets minute, cell-like but enucleate, fragile, membrane-bounded cytoplasmic disks present in the vertebrate blood as the result of the programmed fragmentation of thrombocytes; rupture of platelets releases factors that initiate blood clotting, produce or enhance pain, and induce the proliferation of fibroblasts.

plate tectonics the movement of great land and ocean floor masses (plates) on the surface of the earth relative to one another occurring largely in the Cenozoic era; see also *continental drift.*

pleura (pl. *pleurae*), in mammals, the tough, clear, serous connective tissue membrane covering a lung and lining the cavity in which the lung lies.

pleural cavity in mammals, either of two divisions of the coelom constituting the thoracic cavities, harboring the lungs and lined with the pleurae.

plumule 1. the apex of the plant embryo, consisting of immature leaves and an epicotyl that forms the primary stem. 2. a down feather. 3. a feathery extension of certain airborne fruit, such as that of the dandelion.

poikilotherm an animal that does not thermoregulate and whose body temperature hovers about that of its surroundings.

point mutation a small spontaneous change in DNA involving individual nucleotides, such as a single base change, a single base addition or deletion.

polar of a chemical or a part of a molecule, capable of forming hydrogen bonds with water and with other polar molecules.

polar body either of two small cells produced by meiosis of an egg cell: *first polar body,* the recipient of the chromosomes going to one pole in anaphase I; *second polar body,* the recipient of the chromosomes going to one pole in anaphase II of the egg. Neither polar body divides again.

polarity in animal egg cells, the establishment of metabolic gradients that will later reveal themselves as the active animal and sluggish vegetal poles.

polarized state in a neuron, the period when the region outside a neuron is positive and the region inside is negative, producing a membrane voltage potential of -60mv. See also *resting state.*

polar microtubule or **polar spindle fiber** two partially overlapping spindle fibers (not actually continuous), each originating near a pole and extending across the chromosomes without attaching. Also see *mitotic spindle.*

polar nuclei see *central cell.*

pollen cone the male gymnosperm cone, containing microsporangia in which microspores form. Microspores become pollen grains in which the male gametophyte develops.

pollen grain one of the microscopic particles making up pollen, each being the microgametophyte of a seed plant containing a generative nucleus and a tube nucleus.

pollen sac one of two or four chambers in an anther in which pollen develops and is held.

pollen tube a tube that extends from a germinating pollen grain and grows down through style to the embryo sac, into which it releases sperm nuclei.

pollination the transfer of pollen from a stamen to a stigma, preceding fertilization of a flowering plant.

poly-A tail *polyadenylic acid tail,* a string of

about 200 adenosine nucleotides added enzymatically to the 3′ end of HnRNA transcripts in the nucleus during the maturation of mRNA; thought to protect mRNA from degradation by cytoplasmic exonucleases.

polyadenylization the enzymatic addition of a poly-A tail to an RNA transcript.

polydactyly the genetically derived state of having extra fingers or toes.

polygenic inheritance inheritance involving many interacting variable genes, each having a small effect on a specific trait; also, *quantitative inheritance*.

polygenic trait also *continuously varying trait, quantitative trait, metric trait,* a trait in which variation in a population is expressed in continuous increments about a mean, and is attributable to the action and interaction of many variable gene loci and usually also to multiple variable effects of the environment.

polymer a generally large molecule, consisting of chemically bonded subunits, as in a polypeptide or polysaccharide.

polymerase an enzyme that causes polymerization; e.g., *RNA polymerase, DNA polymerase*.

polymorphism 1. the existence within a population of two or more discrete, genetically determined forms other than variations in sex or maturity and apart from rare mutant forms; e.g., black and spotted leopards. 2. the existence within a population of two or more alleles at a locus, at allele frequencies greater than some arbitrary value such as 1% or 5%; e.g., **ABO** and **MN** blood group polymorphisms.

polymorphonuclear leukocyte a large white blood cell with permanently condensed, highly lobed nucleus.

polyp the typical sessile form of a cnidarian; compare *medusa*.

polypeptide a continuous string of amino acids in peptide linkage, longer than a peptide; compare *protein*.

polyphyletic of a taxonomic group, having two or more ancestral lines of origin.

polyploid having more than two complete sets of chromosomes; *autopolyploid,* having three or more complete sets of a specific complement of chromosomes; *allopolyploid,* having two or more full diploid sets of chromosomes derived from different ancestral diploid species.

polyribosome also *polysome,* a number of ribosomes attached to one messenger RNA molecule, each forming the same polypeptide.

polysaccharide a polymer of sugar subunits.

polysome also *polyribosome,* a small group of ribosomes held together by their common attachment to a messenger RNA molecule.

polyspermy an abnormal condition in which two or more sperm successfully penetrate the egg cytoplasm.

polytene chromosome in certain tissues of certain insects, enlarged chromosomes created by chromatin replication without chromosome division, see also *salivary chromosome*.

polyunsaturated of a fatty acid, having more than one carbon-carbon double bond.

pons a broad mass of nerve fibers running across the ventral surface of the mammalian brain.

population 1. (demography) the total number of persons inhabiting a given geographical or political area. 2. (genetics) an aggregate of individuals of one species, interbreeding or closely related through interbreeding and recent common descent, and evolving as a unit. 3. (ecology), the assemblage of plants and/or animals living in a given area; or all of the individuals of one species in a given area. 4. (statistics), a finite or infinite aggregation of individuals under study.

population bottleneck see *bottleneck*.

population genetics the scientific study of genetic variation within populations, of the genetic correlation between related individuals in a population, and of the genetic basis of evolutionary change.

population recognition the ability to recognize members of one's own particular population or subgroup of a species.

porocyte a pore-bearing cell in the body wall of a sponge, specialized for admitting water.

porphyria variegata a rare genetic disease characterized by excess porphyrins in the blood, accompanied by red urine, light sensitivity, and liver damage; its high incidence in Afrikaaners is considered an example of the *founder effect*.

porphyrin group also *porphyrin ring,* a complex organic compound consisting of a flat circle of nitrogenous rings, usually with iron or another metal complexed in the center; a constituent of hemoglobin, myoglobin, chlorophyll, and the cytochromes.

portal circuit an arrangement of blood vessels wherein blood passes through two successive capillary beds (connected by a vein) before its return to the heart.

positive control of a gene or operon, control such that a gene remains inactive until stimulated by an appropriate controlling protein, hormone, or other signal.

posterior 1. in most bilateral organisms, away from the head end of an organism, the opposite of *anterior*. 2. in human anatomy, in view of the upright position of the organism at least in the daytime, *anterior* and *posterior* are often used in place of *ventral* and *dorsal* to mean toward the front and toward the back respectively, while *superior* and *inferior* are used in place of *anterior* and *posterior* meaning toward the head end and toward the tail end respectively.

posterior pituitary in humans, the posterior lobe of the pituitary, a stalked extension of neural tissue extending from the hypothalamus (see *neurohypophysis*).

posterior vena cava in humans *inferior vena cava,* either of two large veins draining the body posterior to the heart; see *vena cava*.

postsynaptic membrane the receptive surface of a dendrite or cell body adjacent to a synapse; see also *synapse*.

posttranscriptional modification the entire series of enzymatic changes made in RNA in the nucleus after transcription, including *capping, polyadenylation,* and *tailoring*.

potassium ion gate see *ion channel*.

potential energy energy stored in chemical bonds, in nonrandom organization, in elastic bodies, in elevated weight, or any other static form in which it can theoretically be transformed into another form or into work.

powdery mildew 1. a fungal plant disease characterized by the presence of powdery conidia. 2. the fungus that causes the disease.

prairie the originally extensive tract of level or rolling treeless grassland of the Mississippi-Missouri valley, characterized by deep fertile soil and tall, coarse perennial grasses.

preadaptation the appearance of some trait that precedes its eventual usefulness.

precapillary sphincter in arterioles, rings of smooth muscle capable of regulating blood flow into capillaries; controlled by the autonomic nervous system.

precocial capable of walking and a high degree of independent activity from hatching or birth; compare *altricial*.

precursor a chemical substance from which another substance is formed, especially in a specific step of a biochemical pathway.

predation 1. the act of catching and eating. 2. being caught and eaten; e.g., *subject to predation*. 3. a mode of life in which food is primarily obtained by killing and eating other animals.

predator an animal that habitually preys on other animals, a carnivorous animal.

prediction in science, a possible outcome logically derived from a known or hypothetical set of circumstances.

prehensile adapted for seizing, grasping, or wrapping around, as the tail of a New World monkey or the upper lip of a rhinoceros.

premolar in humans, any of the eight bicuspid teeth, located in groups of two anterior to the molars.

pressure flow hypothesis a favored explanation of phloem sap movement: the active transport of sugars into the phloem leads to the osmotic uptake of water therein; the resulting increase in hydrostatic pressure provides the force needed to push the sap through the sieve elements.

pressure receptor also *baroreceptor,* see *sensory receptor*.

presumptive in embryology, designating the future fate of a tissue as in *presumptive ectoderm, mesoderm* and *endoderm*.

presynaptic membrane the membrane of an axon or of the synaptic knob of an axon in the region of a synaptic cleft, into which it secretes neurotransmitters in the course of the transmission of a nerve impulse; see *synapse, synaptic cleft*.

prey 1. (n.), an animal hunted or seized for food by a predator. 2. (v.t.), (followed by *upon*), to hunt or seize for food.

prey switching behavior of a predator in response to the relative densities of two prey species, in which the predator ceases hunting one prey and begins to hunt the other.

primary bronchus the left or right division of the trachea. See also *respiratory tree*.

primary consumer a herbivore; see *consumer*.

primary endosperm nucleus in flowering plants, a triploid nucleus resulting from fertilization in the binucleate central or endosperm mother cell.

primary growth of a plant, the initial growth or elongation of a stem or root, resulting primarily in an increase in length and the addition of leaves, buds, and branches. *Primary growth pattern,* the distinctive pattern of xyleme, phloem, and other tissues in primary growth; compare *secondary growth, secondary growth pattern.*

primary host also *definitive host,* in parasitic relationships, a host in which a parasite may undergo sexual reproduction or at least reach a mature state.

primary immune response the slower, initial response against invasion of the body by organisms or foreign molecules, during which immature, inactive lymphocytes are activated into specialized B- and T-cell lymphocytes. See *secondary immune response.*

primary lesion a damaged or mismatched segment of DNA, subject to repair. See *DNA repair system.*

primary meristem any of the three primary plant tissues—protoderm, ground meristem, and procambium—derived from apical meristem.

primary mutation an alteration in DNA subject to repair by repair enzymes, as distinguished from mutations that cannot be repaired.

primary oocyte a diploid cell of egg-producing potential prior to its first meiotic division.

primary phloem phloem developed from apical meristem, that is, the phloem of primary growth.

primary producer in ecology, plants, algae, or other photosynthetic (or chemosynthetic) organisms that produce the food of a food chain.

primary productivity see *gross productivity.*

primary spermatocyte a diploid meiotic cell of sperm-producing potential prior to its first meiotic division.

primary structure or *level of organization,* the linear sequence of amino acids in a polypeptide, or of nucleotides in a nucleic acid.

primary succession the succession of vegetational states that occurs as an area changes from bare earth to a climax community.

primary xylem xylem in primary growth, which is produced by apical meristem rather than vascular cambium.

primitive a character or character state, characteristic of the original condition of the group under consideration; ancestral; not derived; of or like the earliest state within the group considered; old; compare *advanced, derived.*

primitive groove in bird, reptile, and mammalian embryos, a lengthy indentation formed as gastrulation begins.

primitive gut see *archenteron.*

primitive streak in bird, reptile, and mammalian embryos, a thickening in the blastoderm formed by convergence of cells in preparation for gastrulation.

primordium the rudiment or commencement of a part or organ, often appearing as a mass lump of undifferentiated cells; e.g., *leaf primordium, branch primordium, liver primordium.*

principle of colinearity see *colinearity.*

probability 1. the relative likelihood of the occurrence of a specific outcome or event based on the proportion of its occurrence among the number of trials or opportunities for its occurrence, viewed as indefinitely extended. 2. the field of mathematics dealing with probability relationships.

procambium the part of a meristem that gives rise to cambium.

producer see *primary producer.*

product *enzyme product,* one of the resulting compounds of an enzymatic reaction.

productivity see *gross productivity, net productivity.*

profundal zone in freshwater bodies, the dark waters below the photic zone, often having a low oxygen content and limited aerobic life, that consists primarily of reducers and scavengers.

progeny testing determination of an organism's genotype by crossing it with a known recessive homozygous individual and observing the resulting offspring.

proglottid any segment of a mature tapeworm formed by strobilation, containing male and female organs and being shed when full of mature fertilized eggs.

prokaryote also *procaryote,* any organism of the kingdom Monera, having no nucleus and a single circular chromosome of nearly naked DNA; a *eubacterium,* or *archaebacterium.*

proliferation rapid and repeated production of new cells by a succession of cell divisions.

proliferative phase also *preovulatory phase,* in women, that portion of the menstrual cycle prior to ovulation when estrogen levels rise and endometrial growth and repair begins.

promoter a DNA sequence to which RNA polymerase must bind in order for transcription to begin. See *operon.*

pronephros the most primitive form of kidney in vertebrate development, later replaced by the *mesonephros,* which is itself replaced by the *metanephros* from which the final adult kidney forms.

pronucleus the nucleus of either gamete after the entry of sperm into an egg and before nuclear fusion.

prophage a name given to the temperate phage (a virus) when incorporated into a host chromosome.

prophase the first phase of mitosis and meiosis, the events of which include chromosomal condensation, dismantling of the nuclear envelope, organization of the spindle apparatus, and, in meiosis, synapsing and crossing over. See also *mitosis* and *meiosis.*

proplastid an undifferentiated organelle of plant egg cells and embryonic cells. It is precursor of the chloroplast and other plastids.

proprioception the sum of neural mechanisms involved in sensing, integrating, and responding to stimuli from within the body, including the integration of the movement of body parts, balance, and stance.

proprioceptive sensor see *proprioceptor.*

proprioceptor a sensory receptor that is located in internal tissues (as in skeletal, smooth and heart muscle, tendons, carotid body) and responds to conditions of body positions, muscle tension, or internal chemistry.

prop root a type of adventitious root that emerges from the stem, grows down to the soil, and thus "props up" (supports) the plant (typical of corn and certain tropical fig trees).

prostaglandin any of a group of hormonelike substances derived from long-chain fatty acids and produced in most animal tissues by a variety of stimuli including endocrine, neural, and mechanical stimuli as well as inflammation and oxygen deprivation; present in tissues, sometimes transported by the bloodstream but more often found in other body fluids, e.g., prostaglandins in semen, which stimulate uterine contractions, appear to act by adenyl cyclase activation.

prostate gland a pale, firm, partly muscular and partly glandular organ that surrounds and connects with the base of the urethra in male mammals; its viscid, opalescent secretion is a major component of semen.

prosthetic group a part of a functional protein not part of a polypeptide, such as a heme group, coenzyme, carbohydrate, or lipid.

protease a protein-digesting enzyme.

protein a naturally occurring functional macromolecule consisting of one or more polypeptides held together by hydrogen bonds and van der Waals forces and often sulfhydryl linkages, as well frequently including one or more prosthetic groups.

proteinaceous sponge highly complex sponges with skeletal elements formed from the protein spongin.

protein kinase a class of enzymes that activates other enzymes.

proteinoid spherical aggregations of amino acid polymers with a membranelike surface formation, known to grow and divide in a manner reminiscent of cells.

protein-sequencing a determination of the primary structure of proteins or polypeptides—the number, kinds, and order of amino acids—useful in establishing evolutionary relationships.

proteoglycans a class of molecules whose makeup includes protein and polysaccharide components (sugars making up the polysaccharide contain nitrogen).

prothallium also *prothallus,* the fern gametocyte, typically a small, flat, green thallus with rhizoids, bearing numerous antheridia and/or archegonia.

prothoracic gland an endocrine gland located in the prothorax of an insect, the source of the molting hormone *ecdysone.*

prothoracicotrophic hormone or **PTTH** see *brain hormone.*

prothrombin a blood plasma protein that is converted to the protein thrombin during the clotting process.

protocell the earliest form to exhibit characteristics associated with life. A coacervatelike body containing autocatalytic properties that assure faithful replication of specific chemical properties.

protoderm primary plant tissue from which the epidermis is derived.

protoeukaryote a hypothetical predatory microorganism, capable of the engulfment of prey, that by the successive engulfment of a protomitochondrion, a protochloroplast, and perhaps a protocilium evolved into the ancestral eukaryotic cell, perhaps a billion years ago.

Proto-Gondwana see *Proto-Laurasia.*

Proto-Laurasia according to the continental drift theory, one of two land masses (the other, *Proto-Gondwana)* preceding and giving rise to the supercontinent *Pangaea.* The two earlier masses, according to one hypothesis, are the origin of the deuterostomes and protostomes.

proton 1. one of the two hadrons composing the atomic nucleus in ordinary matter, having an electrostatic charge of $+1$ and a mass 1837 times that of an electron, and by its numbers determining the chemical properties of the atom. 2. a hydrogen ion.

proton-motive force free energy or capacity for work of the proton gradient characterizing chemiosmotic systems.

proton pump an active transport system using energy to move hydrogen ions from one side of a membrane to the other against a concentration gradient, as in chemiosmosis.

protonema (pl. *protonemata)* the threadlike and usually transitory stage of a moss or liverwort gametophyte as it emerges from a germinating spore; indistinguishable from a filamentous green alga in structure and appearance.

protoplasm the living substance of cells, including cytoplasm and nuceloplasm.

protoplast 1. a plant or bacterial cell artificially deprived of its cell wall. 2. a plant or bacterial cell, considered apart from its cell wall.

protostome an animal in which the mouth derives (developmentally or phylogenetically) from the blastopore; compare *deuterostome.*

protozoan 1. any of a large group of protists. 2. any nonphotosynthetic protist.

proventriculus in the bird, a glandular region of the stomach between the crop and the gizzard; in the insect foregut, a muscular region fitted with chitinous teeth.

proximal nearer the point of attachment or origin; nearer the center of the body; compare *distal.*

proximal convoluted tubule see *nephron.*

pseudocoelom also *pseudocoel,* "false coelom"; the body cavity in nematodes, rotifers, and certain other invertebrates, between the body wall and the intestine, which is not entirely lined with mesodermal epithelium; in which the gut is entirely endodermal and not muscularized.

pseudoplacenta 1. also *yolk-sac placenta,* the placentalike structure of marsupials. 2. placentalike structure, as in sea snakes, some cartilaginous fishes, and a few insects.

pseudopod also *pseudopodium,* any temporary protrusion of the protoplasm of a cell serving as an organ of locomotion or engulfment; often having a fairly definite filamentous form, and sometimes fusing with others to form a network.

P700 see *photosystem I.*

psilophytes 1. the earliest known vascular plants; simple dichotomously branching plants of the Silurian and Devonian. 2. see *psilopsid.*

psilopsid a simple plant of the tracheophyte subdivision *Psilopsida,* vascular plants lacking roots, cambium, leaves, and leaf traces, and which have a reduced, subterranean, nonphotosynthetic gametophyte.

P680 see *photosystem II.*

pterophyte any member of the plant division sphenophyta, the ferns.

pterosaur also *pterodactyl* any of numerous extinct flying reptiles of the lower Jurassic to the close of the Mesozoic.

pubic symphysis in mammals, the usually semirigid fibrous articulation of the two pubic bones in the midline above the genitalia; subject to considerable hormone-induced loosening toward the end of pregnancy.

pulmonary artery a branching artery that conveys venous (i.e., deoxygenated) blood from the right ventricle to the lungs, consisting of a common pulmonary artery, right and left pulmonary arteries, and further subdivisions.

pulmonary circuit the passage of venous blood from the right side of the heart through the pulmonary arteries to the capillaries of the lung, where it is oxygenated and from which it returns by way of the pulmonary veins to the left atrium of the heart.

pulmonary vein in birds and mammals the only vein that carries arterial (oxygenated) blood, returning it from the lungs to the left atrium.

pulvinus in plants, a cellular complex at the petiole base that is responsible for the turgor changes associated with nastic responses.

punctuated equilibrium a recently proposed theory that evolution does not occur gradually but that life exists over long periods of time with little evolutionary change, interrupted periodically by great changes. Compare to *gradualism.*

pupa a metamorphic insect in a specific quiescent intermolt between the larval and adult stages, during which time extensive body transformations occur, the pupa generally being enclosed in a hardened pupal case or cocoon.

pupil the contractile aperture in the iris of the eye.

purine a heterocyclic nitrogenous base consisting of conjoined 5-membered and 6-membered rings; e.g., adenine, guanine, uric acid.

Purkinje fiber in the heart, specialized conducting fibers that carry contractile impulses from the bundle of His to the outer ventricular walls.

pyloric sphincter also *pyloric valve,* a ring of muscle capable of closing off the opening between the stomach and the small intestine.

pyloric stenosis an inherited but surgically correctable defect of the human gut, characterized by a thickening and restriction in the pyloric orifice (lower valve) of the stomach.

pyramid 1. see *population pyramid.* 2. also *pyramid of numbers:* see *Eltonian pyramid.* 3. any of various anatomical features shaped like the Egyptian pyramids, including certain structures of the brain and kidney.

pyrimidine a nitrogenous base consisting of a six-membered heterocyclic ring; notably cytosine, uracil, and thymine, constituents of nucleic acids.

pyrogen any fever-producing substance or hormone.

pyrophosphate an inorganic ion consisting of two phosphate groups joined by a phosphate-phosphate bond.

quaternary structure or **level of organization** in protein the interaction of two or more polypeptides through disulfide linkages.

radial have parts arranged like rays eminating from a center.

radial canal 1. in a sponge, one of the numerous choanocyte-lined canals radiating from the central cavity. 2. in a jellyfish, one of the gastrovascular canals extending from the central cavity to the marginal circular canal. 3. in an echinoderm water vascular system, a tube extending outward from the ring canal.

radial spoke in cilia and flagella, protein connections between the outer ring of microtubules and the inner sheath surrounding the central pair.

radial symmetry the condition of having similar parts regularly arranged as radii from a central axis, as in a starfish.

radiation 1. the transfer of heat or other energy as particles or waves; compare *convection, conduction.* 2. heat or other energy transmitted as particles or waves; see also *ionizing radiation.*

radiation, electromagnetic see *electromagnetic radiation.*

radiation, ionizing see *ionizing radiation.*

radical 1. also *moiety, residue,* a part or functional group of a molecule. 2. see *free radical.*

radicle the lower portion of the axis of a plant embryo, especially the part that will become the root.

radioactive in elements, a state of instability in which radiation is emitted by the atomic nucleus.

radioactive isotope see *isotope.*

radioactive tracer a radioactive element or a compound containing a radioactive element, that is put into a biological system and transformed or translocated, its eventual fate being determined by physiochemical means.

radioisotope see *isotope.*

radula in all mollusks other than bivalves, a toothed, chitinous band that slides backward and forward over a protrusible prominence from the floor of the mouth, scraping and tearing food and bringing it into the mouth.

rain shadow a region of low rainfall on the prevailing lee side of a mountain or mountain range.

random drift see *drift.*

range the geographical area occupied by a species.
ray see *vascular ray*.
ray flower see *composite flower*.
reabsorptive area in bony fishes, a structure that absorbs gases from the swim bladder, this aiding in the maintenance of buoyancy. Also see *swim bladder* and *gas gland*.
reactant any element, ion, or molecule participating in a chemical reaction.
reaction see *chemical reaction*.
reaction center also *photocenter,* the part of a photosystem in which light-activated chlorophyll *a* transfers an electron to the electron transport system.
reading-frame shift a serious point mutation involving the loss or gain of a single base-pair, whereby all codons of mRNA beyond or "downstream" from the change will be misread.
reafference the presumed reward associated with the performance of a particular movement when an animal is performing an adaptive or instinctive pattern; behavior as its own reward.
realized niche the niche of an organism as it actually is, generally only a portion of the fundamental niche. Compare *fundamental niche* and see *niche*.
receptacle the end of a floral stalk, which forms the base on which the flower parts are borne.
receptor 1. a specialized neuron involved in the detection of environmental stimuli. 2. any sense organ or group of sensitive cells that transmit impulses.
receptor-mediated endocytosis the transport of substances into the cell by vesicles formed from the plasma membrane, following the recognition of and binding of such substances to highly specific membranal receptor sites.
receptor potentials neural signals of the graded type, generated by sensory receptors.
recessive (adj.) of an allele, not expressed in a heterozygote; see *dominance relationships*.
recessivity the failure of one allele of a heterozygote to have any discernible effect on a specific phenotypic trait; see *dominance relationships*.
reciprocal altruism behavior that appears to be altruistic in individual instances but in fact functions to increase the fitness of the individual insofar as it increases the likelihood that the individual will be the recipient of beneficial behavior at another time.
reciprocal cross a cross in which pollen is exchanged between the flowers of two individual plants.
recombinant chromosome a chromosome emerging from meiosis with a combination of alleles not present on the chromosomes entering meiosis.
recombinant DNA a recent general term for the laboratory manipulation of DNA in which DNA molecules or fragments from various sources are severed and combined enzymatically and reinserted into living organisms; see also *gene cloning*.
recombinant DNA technology the techniques used in genetic engineering. See *genetic engineering*.
recombination frequency the proportion of gametes bearing recombinant chromosomes for two specific gene markers that were heterozygous in the parent.
recombination nodule during crossing over in meiosis I, large enzyme complexes within which strands of DNA are exchanged between homologues.
rectum the terminal part of the intestine, used for the temporary storage of feces.
red alga any alga of the division *Rhodophyta*.
red blood cell see *erythrocyte*.
red marrow regions within the ribs, sterum, vertebrae, and hip bones where red blood cells are produced.
redox reaction a coupled reaction with one reactant being oxidized (losing electrons or hydrogens) and the other being reduced (gaining electrons or hydrogens).
red tide seawater discolored by a dinoflagellate bloom in a density fatal to many forms of life.
reducer also *decomposer,* in ecology, a fungus or microorganism that breaks down plant or animal matter into small molecules.
reducing power the relative amount of energy released by a substance for each electron or hydrogen transferred from it in an oxidation-reduction reaction; the relative ability of one substance to transfer electrons or hydrogen atoms to another.
reduction of a substance, the addition of electrons or hydrogen atoms.
reduction division the first division of meiosis.
reductionist in science, one who attempts to divide phenomena and mechanisms to their most elemental parts, isolating as far as possible the effects of individual factors and testing these effects in separate controlled experiments.
reflex arc in the vertebrate nervous system, a simple neural pathway involving as few as two or three neurons that sense and react to a stimulus.
refractory state a brief period when a neuron cannot generate a second impulse.
regeneration following the loss of a limb or other body part, regrowth of that part through dedifferentiation and repetition of the original developmental events.
regular flower see *flower*.
regulator gene see *operon*
releaser in animal behavior, any stimulus that evokes the release of instinctive behavioral patterns.
releasing hormone also *releasing factor,* a chemical messenger released by the hypothalamus that stimulates hormonal release by the pituitary.
REM sleep also *paradoxical sleep,* a normal period of sleep during which the muscles are very relaxed but the eyes move rapidly under closed lids; accompanied by high brain electrical activity.
renal circuit in mammals, the circuit of the branching *renal arteries* arising from the abdominal aorta and continuing through the glomerulus, the capillary beds of the nephron, and the *renal veins* to return to the vena cava.
renal pelvis the cavity of the kidney into which the collecting ducts empty.
repair enzyme any of several different complexes of enzymes that recognize improper base pairing in DNA, excise a region of one of the strands, and rebuild the DNA according to the rules of Watson-Crick pairing; or that otherwise repair mutational damage, including double-strand chromosome breaks.
replacement rate the birth rate (in terms of the lifetime production of offspring per woman) that, at age structure equilibrium, results in a stable population size; e.g., about 2.1 infants per adult woman.
replica plating a screening technique in which an array of colonies on a nutrient-supplemented agar plate are transferred as a group to one or more additional plates of nutrient-deficient agar to test for nutritional mutants.
replication see *DNA replication*.
replication complex during replication, a grouping of the essential enzymes of that process including the unwinding enzyme, helicase, and DNA polymerase.
replication fork the point at which unwinding proteins separate the two DNA strands in the course of DNA replication.
repolarization in the neuron, reestablishment of the resting potential or polarized state following an action potential. See also *neural impulse*.
repressible operon an operon governing a synthetic pathway, and which is generally active but which can be inactivated by the presence of its normal metabolic product in the medium.
repressor molecule in bacterial operons, a small metabolite that combines with an inactive repressor protein to form an active complex that binds with an operator to prevent transcription of an operon.
repressor protein in bacterial operons, a protein that binds the operator and prevents transcription either when bound to an inducing molecule (inducible operon) or when not bound (repressible operon).
reproduction see *sexual reproduction, asexual reproduction*.
reproductive isolation 1. the state of a population in which there is no mating between members of the group and members of other groups, and no immigration or emigration of individuals. 2. the state of a population or species in which successful matings outside of the group are biologically impossible because of physical mating barriers, hybrid inviability, or hybrid sterility.
reservoir in biogeochemical cycles, the less readily available reserves of mineral nutrients, such as atmospheric nitrogen, which is only useful to nitrogen-fixing organisms. Compare *exchange pool*.
residual air or **residual volume** the volume of air remaining in the lungs follow maximum, forceful exhalation.
resolving power the ability of the eye to distinguish objects near each other as distinct

and separate; also the chief factor limiting useful magnification in light microscopes.

respiration 1. breathing. 2. the physical and chemical process by which an organism supplies oxygen to its tissues and removes carbon dioxide. 3. also *aerobic respiration,* any energy-yielding reaction in living matter involving oxygen. 4. the metabolic transformation of food or food storage molecules yielding energy, *aerobic respiration* (e.g, citric acid cycle and electron transport).

respiratory center a region in the medulla oblongata that regulates breathing movements.

respiratory interface a gas exchange surface; generally a moist, thin-walled membrane, such as the skin of an earthworm, the inner lining of the lung, and the lining of a gill filament.

respiratory system in animals, that organ system responsible for the exchange of gases with the environment.

respiratory tree in air-breathing vertebrates, much of the respiratory system, including the trachea, paired bronchi, highly branched tracheoles, and alveoli

resting potential the charge difference across the membrane of a neuron or muscle fiber while not transmitting an impulse.

resting state a state of seeming inactivity in a neuron, but one in which the membrane activity maintains a polarized state in preparation for conduction.

restriction enzyme in bacteria, an enzyme that recognizes and serves a specific, short DNA sequence, thus protecting the cell from all but a few highly adapted, host-specific viruses; such enzymes have proven useful for experimental DNA manipulation.

rete mirabile a small but very dense network of blood vessels in a countercurrent heat exchange.

reticular system also *reticular formation,* a major neural tract in the brainstem containing neural pathways to other parts of the brain and to the *reticular activating system* (RAS), an arousal center.

reticulum see *ruminant.*

retrovirus a single-stranded RNA virus that, after undergoing reverse transcription and producing double-stranded DNA from its RNA coding, inserts into the host chromosome where it is replicated through many host cell generations. It may later escape from the chromosome to enter a period of viral reproduction and cell lysis. (see *HIV* and *AIDS.*)

reverse transcriptase an enzyme of certain RNA viruses that copies RNA sequences into single-stranded and double-stranded DNA sequences.

R-group a chemistry shorthand where R stands for the variable part of an amino acid.

Rh blood group system also *rhesus factor, Rh factor,* a polymorphic blood cell antigen system consisting of three variable antigenic sites controlled by eight alleles at a single locus.

Rh negative (adj.) homozygous for the absence of the most antigenic of the three Rh antigen sites; not responding to anti-Rh antibodies.

rhodopsin a pigment in the rod cells of the retina that, in the presence of light, breaks down into opsin and retinal. The reaction stimulates action potentials in nearby neurons.

Rh positive homozygous or heterozygous for the most antigenic of the three Rh antigen sites; responding with blood cell agglutination to anti-Rh antibodies.

rhizoid 1. a rootlike structure that serves to anchor the gametophyte of a fern or bryophyte to the soil and to absorb water and mineral nutrients. 2. a portion of a fungal mycelium that penetrates its food medium.

rhizome an underground horizontal shoot specialized for food storage.

rhynophyte an extinct plant group thought to represent the first vascular plants.

rhythm method also *natural birth control,* a method of contraception whereby copulation is avoided during periods when conception is likely.

riboflavin a derivative of vitamin B_2 and an active group in the coenzyme FAD, flavin adenine dinucleotide, capable of being reduced during an oxidation-reduction reaction.

ribose a 5-carbon aldose sugar, a constituent of many nucleosides and nucleotides.

ribosomal RNA also *rRNA,* the RNA that forms the matrix of ribosome structure, consisting of a large, intermediate, and two relatively small sequences.

ribosome an ultramicroscopic cytoplasmic organelle or very large molecule consisting of a matrix of RNA and numerous specific proteins, binding a messenger RNA molecule, a charged tRNA, and a growing polypeptide chain during protein synthesis.

RNA also *ribonucleic acid,* a single-stranded nucleic acid macromolecule consisting of adenine, guanine, cytosine, and uracil on a backbone of repeating ribose and phosphate units; divided functionally into rRNA (ribosomal RNA), mRNA (messenger RNA), tRNA (transfer RNA), and viral RNA.

RNAase an enzyme capable of hydrolyzing RNA sugar-phosphate bonds.

RNA polymerase the enzyme or enzyme complex catalyzing transcription.

RNA synthetase see *RNA polymerase.*

rod one of the numerous long, rod-shaped sensory bodies in the vertebrate retina, containing many membrane layers bearing visual pigments, responsive to faint light but not to detail or to variations in color; compare *cone.*

root the portion of a seed plant, originating from the radicle, that functions as an organ of absorption, anchorage, and sometimes food storage, and differs from the stem in lacking nodes, buds, and leaves.

root apical meristem see *apical meristem.*

root cap a protective mass of parenchymal cells that covers the root apical meristem.

root hair one of the many tiny tubular outgrowths of root epidermal cells, especially just behind the root apex, that function in absorption.

root meristem see *apical meristem.*

root nodule one of the multiple swellings on the roots of a leguminous plant, developed in response to a symbiotic nitrogen-fixing bacterium and harboring nourishing colonies of such bacteria.

root pressure the active force by which water rises into the stems from the roots.

root tip in plants, one of the many delicate root endings, containing the apical meristem responsible for primary growth in the root.

rough endoplasmic reticulum see *endoplasmic reticulum.*

round window in the inner ear, a membranous window, similar to the oval window, but at the cochlea's opposite end. It moves in response to perilymph movement started by the oval window. See also *oval window.*

rRNA see *ribosomal RNA*

r-selection a form of natural selection characterized by exponential population growth. Usually characterized by high reproductivity with little parental care.

Ruffini corpuscle sensory receptors of the skin that respond to temperature changes.

rumen see *ruminant.*

ruminant hooved grazing mammals that digest cellulose through the action of microorganisms in a four-part stomach, which includes the *rumen, reticulum, omasum,* and *abomasum.*

runner horizontal stems that grow close to the soil, occasionally taking root and giving rise to new plants (Bermuda and crab grass).

saccule 1. a small sac. 2. the smaller of two chambers of the membranous labyrinth of the ear; compare *utricle.*

sacrum the part of the vertebral column that connects with the pelvis; in people it consists of five fused vertebrae with transverse processes fused into a solid bony mass on either side.

saggital (adj.), 1. the median plane of the body of a bilateral animal. 2. any plane parallel to the median plane.

salinity the concentration of salt in a solution; saltiness.

saliva 1. in land vertebrates, a viscous, colorless, mucoid fluid secreted into the mouth by ducted salivary glands. 2. in invertebrates, any analogous secretion into or from the mouth.

salivary chromosome also *polytene chromosome,* any of the gigantic interphase chromosomes of the salivary glands of larval dipterans (e.g., *Drosophilia).*

salivary gland 1. many of several glands secreting saliva; in snakes the salivary glands are modified into venom glands. 2. in invertebrates, any analogous gland ducting secretions into or from the mouth.

saltatory proceeding by leaps and bounds rather than by gradual transitions. *Saltatory evolution,* also *macroevolution, saltation,* evolution by sudden variation or by periods of active change with intervening periods of morphological stability. *Saltatory propagation,* the skipping movement of an impulse from one node of a myelinated neuron to another.

salting out also *saline abortion,* a method of abortion used in the second trimester, in which the fetus is killed with an injection of a concentrated salt solution into the amniotic cavity.

sap the watery fluid transported by phloem that circulates dissolved gases, sugars, other organic compounds, and mineral nutrients from one part of the plant to another.

saprobe an organism that reduces dead plant and animal matter.

saprobic see *saprobe.*

sarcina an arrangement of coccus (spherical) bacteria in groups of eight.

sarcolemma the membranous sheath enclosing a muscle fibril.

sarcomere the contractile unit of striated muscle bounded by Z-line partitions; consisting of regular hexagonal arrays of actin filaments bound to the Z-line partitions and parallel myosin filaments regularly interspersed between them.

sarcoplasmic reticulum membranous, hollow tubules in the cytoplasm of a muscle fiber, similar to the *endoplasmic reticulum* of other cells.

saturated having accepted as many hydrogens as possible. *Saturated fat,* a triglyceride lacking carbon-carbon double bonds.

savanna a tropical or subtropical grassland with scattered trees or shrubs, usually maintained by such human activities as burning and by foraging for firewood.

scanning electron microscope (SEM) a device for visualizing microscopic objects by scanning them with a moving beam of electrons, recording impulses from scattered electrons, and displaying the image by means of the synchronized scan of an electron beam in a cathode ray (television) tube.

schistosome also *blood fluke,* 1. any of several parasitic trematode flatworms, the primary host of which is a human, the secondary host being a fresh-water snail. 2. any related trematode parasite.

schistosomiasis any of several diseases caused by various schistosomes parasitic on humans.

Schwann cell one of the cells that constitute the myelin sheath, each cell being greatly flattened so as to consist almost entirely of cell membrane, and being repeatedly wrapped around an axon so as to form a sheath many layers thick.

scientific method (variously defined) 1. any method used by scientists to investigate phenomena. 2. any of several more or less formal schemes of research methodology, in which a problem is identified, relevant data are gathered, a hypothesis and its null alternative are formulated from these data, and the null hypothesis is empirically tested for possible rejection.

sclereid short, branched sclerenchyma cells, occurring in seed coats and nut shells.

sclerenchyma a protective or supporting plant tissue composed of cells with greatly thickened, lignified, and often mineralized cell walls.

scolex the hook-bearing head of a larval or adult tapeworm, from which the proglottids are produced by strobilation.

scrotum the external pouch of skin that contains the testes in most adult male mammals.

scutellum in monocots, the cotyledon in its specialized form as a digestive and absorptive organ.

scyphistoma a highly reduced polyp stage in the life cycle of a scyphozoan (jellyfish).

seaweed a multicellular marine alga, of the red, green, or brown algal groups.

secondary cambium or **vascular cambium** meristematic cells derived from procambium as secondary growth ensues; produces secondary xylem and phloem.

secondary consumer a carnivore; see *consumer.*

secondary growth growth in dicot plants that results from the activity of secondary meristem, producing primarily an increase in diameter of stem or root; compare *primary growth.*

secondary growth pattern the arrangement of xylem, cambium, phloem, cork cambium, cork, and other tissues in dicot secondary growth, in which the various tissues form concentric sheaths that appear as rings in cross section; compare *primary growth pattern.*

secondary immune response a rapid response to a second or subsequent invasion of the body by organisms or foreign molecules, during which memory cells quickly produce large numbers of active, specialized B- and T-cell lymphocytes. See also *primary immune response.*

secondary oocyte a developing egg cell after the first polar body has been produced and before the second polar body appears, after meiosis I.

secondary phloem in plants, phloem originating from secondary or vascular cambium.

secondary sex characteristic structural and psychological change that occurs at puberty, or seasonally in seasonal breeders, usually in one sex only, and is not directly part of reproduction; in humans, enlarged breasts and hips in women; beard growth, enlarged larynx, and increased muscle development in men; pubic and axillary hair in both sexes; breeding plumage in adult male birds.

secondary spermatocyte a sperm-forming cell after the first meiotic division and before the second division.

secondary structure or **level of organization** of a macromolecule, the pattern of folding of adjacent residues, usually in the form of helices or sheets.

secondary succession succession occurring in abandoned croplands. See also *community development.*

secondary tissue see *secondary growth.*

secondary xylem in plants, xylem originating from the secondary or vascular cambium.

second law of thermodynamics the statement in physics that all processes occurring spontaneously within a closed system result in an increase of total entropy.

second messenger an intracellular chemical compound transferring a hormonal message from the cell membrane to the nucleus or cytoplasm.

secretory phase or postovulatory phase in women, that portion of the menstrual cycle following ovulation, during which estrogen and progesterone levels rise and the thickening endometrium takes on a glandular appearance.

seed the fertilized and ripened ovule of a seed plant, comprised of an embryo (miniature plant) including one, two, or more cotyledons, and usually a supply of food in a protective seed coat; capable of germinating under proper conditions and developing into a plant.

seed plant a plant of the tracheophyte subdivision *Spermophyta,* consisting of gymnosperms and angiosperms.

segmentation also *metamerism,* 1. the condition of being divided into segments, originally repetitions of nearly identical parts (as still seen in some annelids, chilopods, diplopods, and *Peripatus*), but frequently followed in evolution by the specialization of different segments, as seen in most arthropods, some annelids, and vertebrates. 2. the serial repetition of body parts, such as ganglia or somites.

segmented body plan see *segmentation.*

selectionists biologists that attribute most, if not all, evolutionary change to natural selection; compare to *neutralists.*

selectively permeable in cellular membranes, the characteristic of permitting selected substances to pass through while rejecting others.

self-fertilization fertilization of ovules or eggs effected by pollen or sperm from the same individual.

self-pollination the transfer of pollen from the anther to the stigma of the same flower.

SEM see *scanning electron microscope.*

semen 1. in mammals, a viscous white fluid produced in the male reproductive tract and released by ejaculation, consisting of spermatozoa suspended in the secretions of acccessory glands, principally the prostate and seminal vesicle. 2. any fluid containing sperm and released by male animals.

semicircular canals in vertebrates, a group of three loop-shaped tubular portions of the membranous labyrinth of the inner ear. Serves to sense balance, orientation, and movement.

semiconservative replication replication of a DNA molecule in which the original molecule divides into two complementary parts, both halves being preserved while each half promotes the synthesis of a new complement to itself.

seminal vesicle. 1. in various invertebrates, a pouch in the male reproductive tract serving for the temporary storage of sperm. 2. in male mammals, paired outpocketings of the vas deferens producing much of the fluid substance of semen.

seminiferous tubule any of the coiled, threadlike tubules that make up the bulk of a testis, and are lined with germinal epithelium from which sperm are produced.

semipermeable membrane a membrane that allows some substances to pass freely through it but which restricts or prevents the passage of other substances.

sensillium in the moth antenna, pored, hair-like structures bearing the olfactory neural endings.

sensory 1. pertaining to the senses. 2. receptive of stimuli. 3. conveying nerve impulses from a sense organ to the central nervous system.

sensory cell a peripheral nerve cell that is the primary receptor of a sensory stimulus.

sensory hair an innervated arthropod hair that serves as a sense organ.

sensory neuron also *afferent neuron,* a neuron that conducts impulses carrying sensory information from a receptor to the brain or spinal cord.

sensory palp also *palpus,* a segmented process on either side of the mouth of an arthropod, occurring in insects in pairs on the maxillae and on the labium, having tactile and gustatory functions.

sensory receptor a cell or tissue, specialized in responding to specific kinds of stimuli.

sepals green, leaflike parts of flowers that surround the flower bud before it opens and later form a whorl (the *calyx*) beneath and outside the whorl of petals.

septate divided by septa.

septum (pl. *septa*), any dividing wall or membrane, such as those dividing the transverse partitions dividing the coelom of an annelid.

seral stage one stage in a sere. See also *sere* and *community development.*

sere a recognizable sequence of changes, sometimes predictable, occurring during community development or succession.

serial endosymbiosis hypothesis proposed evolution of the eukaryotic cell through the fusion of several prokaryotic lines, the sequence eventually giving rise to cells with combined phagocytic capabilities, chloroplasts, mitochondria, and flagella.

Sertoli cell also *nurse cell,* one of the elongated cells of the tubules of the testis to which spermatids become attached while they mature.

seta 1. in the moss sporophyte, the stalklike growth that supports the sporangium. 2. the bristlelike, chitinous structure in the body wall of annelids, arthropods, and certain other invertebrates.

sex chromosome an X or Y chromosome, or any chromosome involved in the determination of sex.

sexduction the transfer of bacterial host genes from an F+ bacterium to an F− bacterium following its fortuitous incorporation into the genome of an F episome.

sex-influenced trait a congenital, gene-influenced, or chronic abnormality that can occur in either sex but is more common in one than the other; e.g., breast cancer, ulcers, pyloric stenosis.

sex-limited trait a variable trait that affects members of one sex only; e.g., prostate cancer, pattern baldness, endometriosis.

sex pilus in some prokaryotes, an enlarged pilus (tube) through which a DNA replica can presumably pass from one cell to another.

sexual intercourse also *coitus, copulation,* the entry of the erect penis into the vagina.

sexual recombination genetic recombination as the result of meiosis and biparental reproduction.

sexual reproduction reproduction involving the union of gametes; biparental reproduction.

shelf fungus also *bracket fungus,* a basidiomycete that forms a shelflike basidiocarp on the sides of infected tree trunks.

shell 1. see *electron shell.* 2. a hard, rigid, usually calcareous covering of an animal. 3. the calcareous covering of a bird egg. 4. by extension, the leathery covering of a reptile or monotreme egg. 5. the hard, rigid outer covering of a fruit or seed, e.g., walnut shell, sunflower seed shell.

shoot apical meristem see under *apical meristem.*

short-day plant also *long night plant,* a plant that begins flowering after the summer solstice and in which flowering is triggered by periods of dark longer than some innately determined minimum.

short-term memory see under *memory.*

sickle-cell an abnormally crescent-shaped erythrocyte seen in the disease sickle-cell anemia.

sieve cell the more primitive food-conducting cells in vascular plants other than angiosperms (also see *sieve tube member*).

sieve element includes sieve tube members and sieve cells, the primary food-conducting structures of phloem.

sieve plate. 1. partition or cell wall between sieve tube members, bearing multiple perforations allowing for the continuity of sieve tube cytoplasm. 2. also *madreporite,* the perforated calcareous disk through which the echinoderm water vascular system communicates with the exterior water.

sieve tube a thin-walled tube consisting of an end-to-end series of sieve tube members joined by sieve plates, forming the channels through which sap flows in phloem.

sieve-tube member in angiosperms, a thin-walled phloem cell having no nucleus at maturity and forming a continuous cytoplasm with other such cells to form *sieve tubes,* functioning in the transport of organic solutes, hormones, and mineral nutrients.

signal peptidase in translation by bound ribosomes, an enzyme that cleaves the leader sequence or signal polypeptide from a polypeptide entering the endoplasmic reticulum. See also *signal peptide.*

signal peptide in translation by bound ribosomes, the *leader sequence* of amino acids in the polypeptide being synthesized. Before entering the endoplasmic reticulum, the signal peptide must first interact with a signal recognition protein and then with a *receptor site* on the rough endoplasmic reticulum.

signal recognition protein see *signal peptide.*

silent mutation a change in a DNA codon in which a synonymous codon forms (third letter change), having no effect on the amino acid sequence of the polypeptide.

siliceous containing silica; *siliceous sponge,* a sponge having a skeleton of silica spicules.

silicon one of the most abundant elements in the earth's crust, a major component of rock, its ions found in certain spicules, diatom shells, nettle spines, the outer cell walls of grasses, and elsewhere.

simple eye also *ocellus,* in invertebrates, light receptors resembling single ommatidial units within the compound eye, specialized for detecting changes in light intensity.

Singer model see *fluid-mosaic model.*

sinoatrial node also *SA node, pacemaker,* a small mass of conducting tissue embedded in the musculature of the right atrium of a land vertebrate, representing the vestige of the sinus venosus of ancestral forms, serving as a source of regular contracting impulses to the heart, and transmitting the impulse to the auricles, the atrioventricular node, the bundle of His, and thus to the ventricles.

sink as in "source" and "sink." In plants, a region where a manufactured product such as sucrose or starch is concentrated.

sinus a cavity that forms part of an animal body, e.g., any of the several air-filled, mucous-membrane-lined cavities of the skull; a hemocoel.

sinus venosus in fish and in the embryos of land vertebrates, a distinct chamber of the heart formed by the union of the large systemic veins, opening into the single auricle.

sinusoid a minute, endothelium-lined space or passage for blood in the tissues of an organ, such as the liver or spleen.

siphonous alga coenocytic green algae, an evolutionary line of green algae characterized by a multinucleate cytoplasm, without the usual division by cell walls or membranes, e.g., *Codium.*

skeletal muscle also *voluntary muscle, striated muscle,* muscle attached to the skeleton or in some cases to the dermis, under direct control of the central nervous system, characterized by distinct striations at the subcellular level, with multinucleate unbranched muscle fibers; compare *smooth muscle, cardiac muscle.*

skin-breather any organism in which a significant proportion of the exchange of respiratory gasses occurs through a vascularized moist skin; e.g., most amphibia.

Skinner box a device for investigating operant conditioning, after B. F. Skinner.

skull the skeleton of the head of a vertebrate, being a cartilaginous braincase in cyclostomes, sharks, and the embryos of all forms; in most vertebrates the cartilage is replaced by bone and the structure made more complete by its union with other bones developed in dermal membranes; consisting of the cranium, the bony capsules of the nose, ear, and eye, and the jaws and teeth.

slime mold a funguslike protist possibly related to fungi, consisting of an ameboid, saprophytic *plasmodium* that on maturity coalesces into structured fruiting bodies; two subgroups, the *acellular slime molds* and the *cellular slime molds,* though sharing many features, may also be only distantly related to one another.

small intestine in vertebrates, the region of the alimentary canal and the cecum; the region in which most food absorption occurs.

smooth endoplasmic reticulum see *endoplasmic reticulum.*

smooth muscle also *involuntary muscle,* the muscle tissue of the glands, viscera, iris, piloerection, and other involuntary functions, consisting of masses of uninucleate, unstriated, spindle-shaped cells occurring usually in thin sheets.

smut 1. a fungal disease of plants characterized by black, powdery masses of spores. 2. the fungus that causes the disease.

sodium ion gate either of the two gates controlling sodium ion passage through an ion channel, including an *activation gate* and *inactivation gate.*

sodium/potassium ion exchange pump also *sodium pump,* a poorly understood molecular entity in the plasma membrane, capable of actively transporting sodium out of the cell and potassium in, at a cost of ATP energy.

soft palate in the oval cavity, the membranous, non-bony posterior extension of the hard palate.

solar tracking (formerly *heliotropism*) movement in plants, based on turgor changes, in which the leaves or flowers appear to track the sun throughout the day.

solute something dissolved in solution.

somatic cells cells of the body; any cells other than germinal cells, especially of terminally differentiated tissues.

somatic nervous system the voluntary nervous system, as distinguished from the visceral (autonomic) nervous system.

somatostatin a hormonal secretion of the hypothalamus and delta cells in the *islets of Langerhans.*

somite 1. in the early vertebrate embryo, one of a longitudinal series of paired blocks of tissue that are the forerunners of body muscles and the axial skeleton. 2. one of the segments of any segmented animal.

sorus (pl. *sori*), one of the clusters of sporangia on the underside of a fern frond.

sound spectograph a device for visually displaying the elements of a complex sound; *sound spectrogram,* one such display.

space-filling model a solid model of a molecule, in which the outer orbitals of atoms are represented proportionately by large intersecting spheres.

speciation 1. the division of a species into two or more species. 2. the process or processes whereby new species are formed.

species (pl., *species*), (variously defined), 1. a class of things or individuals having some common characteristics; a distinct type. 2. the major subdivision of a genus or subgenus, regarded as the basic category of biological classification and composed of related individuals that resemble one another through recent common ancestry and that share a single ecological niche. 3. in sexual organisms, a group whose members are the least potentially able to breed with one another but are unable to breed with members of any other group; a reproductively isolated group.

specific defense slow but highly specific cellular and chemical responses made by the B-cell and T-cell lymphocytes against foreign substances, cancerous cells, or invading organisms.

specific heat the amount of heat required to raise the temperature of a substance a specified number of units. With water, the standard reference at standard conditions, the amount of heat required to raise the temperature of one cubic centimeter by 1° C.

specific name the second part of a formal scientific name, indicating species within a genus, always italicized and never capitalized.

sperm 1. a male gamete. 2. a spermatozoan, 3. a spermatozoid. 4. (adj.), pertaining to a male gamete or male gamete function, as in *sperm head, sperm nucleus, sperm cell.*

spermatheca a sac connected with the female reproductive tract of most insects, many other invertebrates, and some vertebrates, that receives and stores spermatozoa after insemination and retains them, often for a long time, until egg maturation and fertilization.

spermatid one of the four haploid cells formed by meiosis of a spermatocyte, prior to maturation into a spermatozoan.

spermatocyte a cell giving rise to spermatozoa; see also *primary spermatocyte, secondary spermatocyte.*

spermatogonium (pl. *spermatogonia*), a diploid testicular cell from which spermatocytes and ultimately sperm are produced; a germ line cell of an adult male.

spermatophore a packet containing sperm, produced by spiders and certain other invertebrates, and by certain salamanders.

spermatozoan (pl. *spermatozoa*), also *spermatozoon, sperm, sperm cell,* a motile male gamete of an animal, produced in great numbers in the male gonad, consisting of a sperm head containing the nucleus and an acrosome, a midpiece and a single flagellum (sperm tail).

S phase *(synthesis phase),* one of the phases of the cell cycle, occurring in interphase between G_1 and G_2 the period of chromosome *replication* and DNA synthesis.

sphenophyte any member of the plant division sphenophyta, the horse tails.

sphincter a ring of muscle surrounding a bodily opening or channel and able to close it off; e.g., *oral sphincter, anal sphincter.*

spicule 1. one of the many small to minute calcareous or siliceous pointed bodies embedded in and serving to stiffen and support the tissues of various invertebrates, including sponges, radiolarians, and sea cucumbers. 2. any small, stiff, spike-like or needle-like body part.

spinal column see *vertebral column.*

spinal cord see *dorsal hollow nerve cord.*

spinal nerve any of the many nerves that enter and leave the spinal cord, including both somatic and autonomic. Compare *cranial nerve.*

spindle a cytoplasmic body present in mitosis and meiosis, in shape resembling the spindle of a primitive loom, and composed of tubulin microtubules and actin microfilaments; serving to segregate daughter chromosomes in anaphase.

spindle apparatus a functional unit consisting of the continuous and centomeric fibers of the spindle, the centromeres, the centrioles (or the analogous *polar ground substance*) and, when present, the asters.

spindle fiber one of the microtubule filaments constituting the mitotic spindle. *Centromeric spindle fiber,* one of the several microtubules attaching firmly to a centromere and extending to a spindle pole. *Fiber,* a microtubule extending from one pole, interlacing with continuous spindle fibers of the other pole, and ending blindly somewhat short of the other spindle pole.

spinneret 1. any organ for producing threads of silk from the secretion of a silk gland. 2. one of the six nipplelike, jointed processes near the end of the spider's abdomen, each bearing numerous minute orifices of silk gland ducts.

spiracle 1. the blowhole of a cetacean. 2. one of the external openings of an insect tracheal system. 3. one of the pair of openings on the upper back part of the head of an elasmobranch or sturgeon, communicating with the pharynx, derived from gill clefts but serving as incurrent respiratory openings in rays.

spiral valve a helical fold of the intestinal wall in the short intestine of a shark, which serves to slow the passage of food and provides additional absorptive surface.

spirillum spiral-shaped, flagellated bacterium with rigid cell walls.

spirochaete spiral-shaped bacterium with a flexible cell wall and locomotion through the movement of axial filaments within the outer membrane.

spleen a discrete organ of the reticuloendothelial system, consisting of lymphoid, reticular, and endothelial tissues with blood supply and red blood cells circulating freely in intercellular spaces; all enclosed in a fibroelastic, muscularized capsule; functions include the scavenging of debris and the maintenance of blood volume.

SPONCH a mnemonic acronym of the six most common elements of living matter, in the order of decreasing atomic mass: Sulfur, Phosphorus, Oxygen, Nitrogen, Carbon, and Hydrogen.

spongin a tough, insoluble protein that makes up the skeleton of a class of sponges including the once commercially important bath sponge.

spongocoel the body cavity of a simple sponge (not a true coelom).

spongy bone bone with a network of thin, hard walls surrounding open, non-bony pockets.

spongy parenchyma see *leaf mesophyll.*

spontaneous mutation a chemical change in DNA not induced by external agents; compare *mutation, mutagen.*

sporangiophore in some protists and most

fungi, an erect growth upon which pore-forming bodies develop.

sporangium a case, cell, or organ in which asexual spores are borne, in algae, fungi, bryophytes, and ferns.

spore a minute unicellular reproductive or resistant body, specialized for dispersal, for surviving unfavorable environmental conditions, and for germinating to produce a new vegetative individual when conditions improve.

sporic cycle a life cycle in which generations alternate between a multicellular, diploid sporophyte and a multicellular haploid gametophyte, the phases sometimes occurring in separate individuals. Meiosis in the sporophyte yields spores from which gametophytes develop. They produce gametes through mitosis, and fertilization restores the sporophyte phase. Either generation may dominate, or they may be equal (some algae, all plants).

sporophyll a modified leaf capable of producing spores. See also *megasporophyll* and *microsporophyll.*

sporophyte in algae and plants having an alternation of generations, a diploid individual capable of producing haploid spores by meiosis; the prominent form of ferns and seed plants; compare *gametophyte.*

sporozoite a small, motile, and elongate infective stage of a sporozoan, a product of a sexual fusion that initiates a new asexual cycle when transmitted to a new host.

squamous epithelium stratified epithelium that consists, at least in its outer layers, of small, flattened, scalelike cells, generated from an underlying germinal layer.

stabilizing selection natural selection that tends to maintain the status quo, selection for the average and against extremes.

stamen the male reproductive structure of the flower, consisting of a pollen-bearing anther and the filament on which it is borne.

staminate having stamens but no pistils; an exclusively male flower.

standard conditions or *standard temperature and pressure,* in chemistry: 0.0°C or 273°K and 1.0 atmosphere of pressure. Conditions to which all gas volumes are measured under the environmental conditions are corrected, permitting one measurement to be accurately compared to another.

staphylococcus a "grape-cluster" arrangement of coccus (spherical) bacteria.

statistics 1. the science that deals with the classification, analysis, and interpretation of numerical data, and that uses mathematical probability theory to seek order from such data. 2. the use of mathematical probability theory in hypothesis testing, specifically in determining whether or not observed correlations or departures from expectation can reasonably be attributed to chance.

statocyst 1. a cell containing one or more statoliths. 2. a sense organ of equilibrium, orientation, and movement consisting of a fluid-filled chamber containing a statolith; widely distributed among invertebrate animals.

statolith 1. the calcareous body within a statocyst which, by its greater specific gravity, tends to move downward or against the direction of acceleration.

stele the cylindrical central portion of the axis of a vascular plant, including pith, xylem, and phloem, surrounded by a pericyle.

stem cell also *hemocytoblast,* generalized cell of red bone marrow from which all blood cells form.

stem reptile also *cotylosaur* a Permian reptile of the group believed to be ancestral to all reptiles.

stereoscopic vision the ability to perceive depth through the integration of two overlapping fields of vision.

sterile jacket cell nonreproductive cell surrounding moss antheridium and archegonium.

steroid any of a class of lipid-soluble compounds on four interlocking saturated hydrocarbon rings; included are all *sterols* (alcoholic steroids), e.g., *cholesterol, estradiol, testosterone, cortisol.*

steroid hormone a class of hormones consisting of the steroid molecule with various side group substitutions, believed to freely pass across the cell membrane and, once bound by a specific carrier protein, interact directly with the chromatin in gene control; included are the vertebrate sex hormones.

stigma the top, slightly enlarged and often sticky end of the style, on which pollen adhere and germinate.

stimulus an aspect of the environment that influences the activity of a living organism or part of an organism, especially through a sense organ.

stimulus generalization the tendency to accept a variety of stimuli that have traits in common with a previously rewarded stimulus.

stipe the stemlike structure in red or brown algae that supports the blades or blade.

stolon 1. in plants, a horizontal branch that produces new plants from buds; also *runner.* 2. a long horizontal fungal hypha that spreads across the surface and periodically sends down a mycelium into the substrate and a group of conidia into the air.

stoma (pl. *stomata*), structure in the epidermis of leaves, stems, and other plant organs, made up of the guard cells and pore, allowing the diffusion of gases into and out of intercellular spaces.

stomach a muscular dilation of the alimentary canal in vertebrates, between the esophagus and the duodenum, that functions in temporary storage, preliminary digestion, sterilization, and physical breakdown of ingested food.

stomatal apparatus see *stoma.*

stone canal a calcified tube of the echinoderm water vascular system, leading from an external sieve plate (madreporite) to the ring canal.

stop codon also *termination codon,* in MRNA, one or more of the codons UAA, UAG, or UGA, signalling the end of polypeptide translation. See also *start codon.*

storage disease a pathological condition characterized by the accumulation within cells of partially metabolized substances that accumulate because of the absence of a critical enzyme.

stratified squamous epithelium multilayered, flattened epithelial cells as seen in the epidermis of the skin, esophagus, and vagina.

strep throat a bacterial disease caused by *Streptococus pyogenes,* characterized by an inflamed throat, which may progress to scarlet fever, kidney inflammation, and rheumatic fever.

streptococcus a "beads-on-a-string" arrangement of coccus (spherical) bacteria.

stretch receptor a proprioceptive sensory receptor that is stimulated by stretching, as in a tendon, muscle, or bladder wall.

striated muscle see *skeletal muscle.*

strobilation asexual reproduction by transverse division of the body into segments that break free as independent organisms; occurring in certain cnidarians and flatworms (tapeworms).

strobilus (pl. *strobili*), a conelike aggregation of sporophylls in a club moss or horsetail.

stroke volume the amount of blood passing through the heart per heartbeat.

stroma 1. the matrix or supporting connective tissue of an organ. 2. the cytoskeleton. 3. the watery proteinaceous matrix surrounding the thylakoids of a chloroplast.

stroma thylakoid a solitary, unstacked thylakoid, usually extending across the stroma from one granum thylakoid to another.

stromatolite a macroscopic geological structure of layered domes of deposited material attributed to the presence of shallow water photosynthetic prokaryotes.

structural formula a two-dimensional representation of the topological relationships of atoms and bonds in a molecule; sometimes including information on spatial relationships as well.

structural protein protein that is incorporated into cellular or extracellular structures.

style the stalk of the pistil in a flower, connecting the stigma with the ovary.

suberin a complex fatty substance of cork cell walls and other waterproofed cell walls.

subspecies a more or less clearly defined, morphologically distinct, named geographic variety of a species; when named, the name forms a third part of the scientific name *(genus, species, subspecies).*

substitution 1. see *base substitution.* 2. *amino acid substitution,* a base substitution that changes the identity of one of the amino acids of a protein.

substrate 1. the base on which an organism lives. 2. a substance acted upon by an enzyme. 3. a nutrient source or medium.

substrate-level phosphorylation the production in one or more steps of ATP from ADP, P_i, and an appropriate organic substrate; the capture of high-energy phosphate bonds directly from metabolic transformations; compare *chemiosmostic phosphorylation.*

succession see *community development.*

sugar diabetes see *diabetes mellitus.*

sulfhydryl group the side group R-SH.

sulfur bacterium 1. *purple sulfur bacterium,* a photosynthetic bacterium utilizing H_2S as a source of hydrogen for photosynthesis; 2. *nonpurple sulfur bacterium,*a nonphotosynthetic chemoautotroph utilizing sulfur compounds as a source of energy.

sun compass an innate mechanism utilizing the angle of the sun and the time of day to compute direction for navigation.

supercoiling bending of a helical linear structure into larger orders of helices, as occurs in the mitotic condensation of chromosomes.

superior vena cava the anterior vena cava in humans; see *vena cava.*

supplemented medium a growth medium providing metabolites not normally required by the organism, and thus able to sustain the growth of certain nutritional mutants that would not be recovered on minimal medium.

suppressor T-cell a specific subpopulation of T-cells whose role is to moderate, slow, and stop the specific immune responses.

surface tension a condition that exists on the free surface of water or other liquid by reason of intermolecular forces unsymmetrically disposed around individual surface molecules, tending to make the surface layer behave in some respects as an elastic membrane.

surface-volume hypothesis the proposal that cells are restricted to a size that assures a surface-volume ratio that provides a sufficient membrane area to support the transport needed to maintain metabolic activity.

survivorship curve a graph with age on the X axis and zero to one on the Y axis, the monotonic curve presenting the probability of surviving to age X for each X.

suspensor in a developing seed, a group or chain of large cells, swollen with water, produced by the zygote and serving to anchor the embryo in place and to push it into contact with the endosperm.

suture in anatomy, immovable joints formed by the articulation of skull bones.

swim bladder also *air bladder,* a gas-filled sac giving controlled buoyancy to most bony fish; homologous with the lungs of land vertebrates and lungfish and probably derived from the primitive lung of an ancestral air-breathing fish.

symbiont a symbiotic organism.

symbiosis 1. the living together in intimate association of two species. 2. *parasitic symbiosis,* also *parasitism,* symbiosis in which one organism gains fitness at some expense to the other. *Commensalism,* symbiosis in which one organism gains at no cost to the other. *Mutualistic symbiosis,* also *mutualism,* symbiosis in which both individuals gain fitness.

sympathetic division see *autonomic nervous system.*

sympathetic nerve in the autonomic nervous system, any nerve of the sympathetic division.

symplastic route in roots, the movement of water through the cytoplasm of adjoining cells (via plasmodesmata) as it makes its way to the stele (see also *apoplastic route*).

synapspe 1. (n.), the place at which nerve impulses pass from the axon or the synaptic knobs of an axon of one neuron to the dendrite or cell body of another. 2. (v.i.), of homologous chromosomes, to pair together, chromomere by chromomere, in zygotene of meiosis.

synapsis the pairing up and fusing of homologous chromosomes during zygotene of the first meiotic prophase, whereby preparations are made for crossing over.

synaptic cleft the minute space between the synaptic knob of one neuron and the dendrite or cell body of another, into which neurotransmitters are released in the transmission of nerve impulses between cells.

synaptic knob one of the multiple bulbous swellings at ends of the branching terminus of an axon, containing neurotransmitters in secretion granules and forming one side of the synaptic cleft.

synaptinemal complex a complex structure composed of protein and RNA, the material of which is first formed between sister strands in leptotene of meiotic prophase and combines with a like structure to form the complex and accomplish the specific zipper-like pairing of homologous chromosomes in zygotene.

synctium 1. a multinucleate mass of protoplasm resulting from the fusion of cells, as in the plasmodium of a slime mold. 2. also *coenocyte,* a multinucleate mass created by repeated nuclear mitosis without cytoplasmic cell division.

synonymous codon in the genetic code, two or more codons that specify the same amino acid.

synovial joint a freely movable joint surrounded by a fibrous capsule lined by a synovial membrane that secretes lubricating synovial fluid.

synthesist in science, one who attempts to deduce broad general principles from widely disparate observations, or who draws upon such observations to substantiate or refute such principles; one who reinterprets established relationships in the support of a new general idea or *synthesis.*

syphilis a sexually transmitted bacterial disease caused by the spirochaete *Treponema pallidum;* symptoms in the primary infection include local chancre on the genitals; in the secondary stages, body rash, mouth lesions, and runny nose, in the tertiary stages, which may occur years later, the widespread infection may damage the blood vessels and eyes and lead to insanity and paralysis.

systole the period of heart contraction, particularly of the ventricles; compare *diastole.*

systolic pressure the highest arterial blood pressure of the cardiac cycle.

systematics 1. the science of classifying organisms or groups of organisms on the basis of their evolutionary relatedness. 2. the science of determining the evolutionary relatedness of groups of organisms, particularly at higher taxonomic levels.

systematist a biologist who tries to determine the evolutionary relationships of organisms.

TACT in plants, an acronym representing the water transporting forces, *t*ranspiration, *a*dhesion, *c*ohesion and *t*ension.

tactile receptor a sensory receptor responsive to light touch; compare *pressure receptor.*

taiga moist, subarctic forest biome of Europe and North America dominated by spruces and firs.

tap root also *taproot* 1. a root having a prominent central portion giving off smaller lateral roots in succession. 2. the central root of such a system, especially when it grows vertically and deep.

tap root system a root system consisting of a large primary root and its secondary and lateral branches.

target cell a cell acted upon by a specific hormone, generally containing or bearing specific hormone receptor proteins not found in other cells.

tarsus 1. the tarsal bones taken together. 2. the vertebrate ankle. 3. the distal portion of an arthropod leg.

taste bud a sensory receptor sensitive to taste, found chiefly in the epithelium of the tongue, being a conical mass of cells made up of supporting cells and *taste cells,* which terminate in modified sensory cilia that project into a porelike space in the overlying epithelium.

taxon 1. any taxonomic category, such as *species, genus, subspecies, phylum,* etc. 2. any named group of related organisms.

taxonomist an individual skilled in identifying, describing, and classifying organisms, usually specializing in a particular group.

taxonomy the science dealing with the identification, naming, and classification of organisms.

Tay-Sachs disease a disease due to homozygosity for a rare recessive autosomal allele, characterized by inclusion bodies in neurons, juvenile idiocy, and early death. The recessive allele and the disease are unusually common in persons of Eastern European Jewish descent.

T-cell a lymphocyte of a variety that matures in the thymus and interacts with invading cells and other cells of the immune system.

T-cell receptor a dual cell surface receptor on the various T-cell lymphocytes, including a highly specific antigen recognition site and an MHC class I or II self-recognition site.

tectorial membrane a membrane of the cochlea, overlying and contacting the hair cells of the organ of Corti.

techoic acid in the cell walls of Gram positive bacteria, an acidic polymer of alternating glycerol and phosphate groups with side-chains such as N-acetyl glucosamine.

telophase the stage of mitosis or meiosis in which new nuclear membranes form around each group of daughter chromosomes, the nucleoli appear, and the chromosomes decondense; at which time the plasma membrane and cytoplasm of the cell usually divide to form two daughter cells.

TEM transmission electron microscope; see *electron microscope*.

temperate deciduous forest a forest biome of the temperate zone, usually in areas of significant winter snowfall, in which the dominant tree species and most other trees are deciduous and are bare in winter months.

temperate rain forest a cool woodland of the temperate zone in an area of heavy rainfall but little or no snow, usually including many different kinds of trees as well as a single dominant tree species.

temperate zone the region of the earth between the Tropic of Cancer and the Arctic Circle together with the region between the Tropic of Capricorn and the Antarctic Circle.

temporal lobe a large lobe on the lateral portion of each cerebral hemisphere.

tendon a touch dense cord of fibrous connective tissue that is attached at one end to a muscle and at the other end to the skeleton, and transmits the force exerted by the muscle in moving the skeleton.

tendril a slender, fast-growing tip of a climbing plant that grows in spirals around an external supporting tree, bush, or trellis.

tensile strength the ability to be pulled without breaking.

tension in the xylem, a pulling force originating through transpiration in the leaf and transferred to water columns in the xylem.

tentacle an elongate, flexible, fleshy, sometimes branched process borne by animals chiefly in groups or pairs on the head or surrounding the mouth, and serving as a tactile and/or prehensile organ.

tenting the expansion of the internal vagina (cul-de-sac) and elevation of the uterus on approaching orgasm.

terminal bud the dormant bud at the stem tip, representing the next season's potential growth.

termination in transcription, a point when the ribosome reaches a stop codon, whereupon the polypeptide is released and the ribosomal subunits separate.

termination signal a DNA sequence that signals the end of a transcription sequence, dislodging RNA polymerase.

terminator codon also *termination codon, stop codon, nonsense codon,* one of the codons UAA, UAG, or UGA in mRNA, or the corresponding codons in DNA, signalling the end of polypeptide transcription.

territorial behavior also *territoriality,* a species-specific pattern of behavior associated with the defense of a territory, in most territorial species by the male, and often including territorial marking, singing, or other displays, threats, attacks on trespassing conspecifics, and patrolling by the defending animal, as well as avoidance and submissive behavior by others when in or around a defended territory.

territory the area defended by a territory animal.

tertiary consumer a consumer at the third level; see *consumer*.

tertiary structure or **level organization** of a protein, the pattern of folding of a polypeptide upon itself, which is generally quite specific for each protein type.

testcross 1. a cross of an individual of unknown genotype with a homozygous recessive to determine zygosity by progeny testing. 2. the cross of a known double heterozygote with a double homozygote to determine linkage relationships.

testis (pl. *testes*), the male gonad, in which spermatozoa are produced by meiosis.

tetanus a bacterial disease characterized by sustained, painful, muscle contraction caused by the toxins of *Clostridium tetani,* which interfere with clearing of the neurotransmitter acetylcholinesterase from neuromuscular junctions.

tetrad 1. a group of four. 2. a group of four equal cells, such as microspores or spermatids, produced by meiosis. 3. also *bivalent,* the synapsed homologues in pachytene through diakinesis, considered as a group of four synapsed chromatids.

tetraploid having four complete genomes in each nucleus; *autotetraploid,* having four genomes (i.e., twice the diploid number) of one normally diploid species; *allotetraploid,* also *amphidiploid,* having two genomes (i.e., the usual diploid number) of each of two different, normally diploid species.

tetrapod also *land vertebrate,* any vertebrate of the classes Amphibia, Reptilia, Mammalia, and Aves, including some that don't have four feet (e.g., birds, people, whales, dugongs).

T-even phage one general type of *E. coli* bacteriophage, with a double-stranded DNA genome and a complex body consisting of a head, tailpiece, end plate, and attachment fibers; arbitrarily designated T_2, T_4, T_6, etc.

thalamus a large subdivision of the diencephalon, consisting of a mass of nuclei in each lateral wall of the centrally-located third ventricle of the brain.

thalassemia also *Cooley's anemia,* a severe congenital anemia due to homozygosity for a variant allele of the beta hemoglobin locus, producing little or no adult hemoglobin. *alpha thalassemia,* an analogous partial deficiency of the hemoglobin alpha chain.

thallose a leafy growth form seen in some liverworts.

thallus a plant body of a multicellular alga, that does not grow from an apical meristem, shows no differentation into distinct tissues, lacks stems, leaves, or roots; may be simple or branched, consisting of filaments, thin plates, or solid bodies of cells, and may include fairly complex blades, bladders, and holdfasts.

theory a proposed explanation whose status is still conjectural, in contrast to well-established propositions that are often regarded as facts. *Theory* and *hypothesis* are terms often often used colloquially to mean an untested idea or opinion. A *theory* properly is a more or less verified explanation accounting for a body of known facts or phenomena, whereas a *hypothesis* is a conjecture put forth as a possible explanation of a specific phenomenon or relationship that serves as a basis for argument or experimentation.

therapsid also *mammallike reptile,* any Permian or Triassic reptile of the extinct order *Therapsida,* with mammallike upright locomotion rather than crawling or bipedal locomotion.

thermal energy 1. energy in the form of heat. 2. the heat equivalent of some other form of energy.

thermogenesis in endothermic animals, the production of body heat through an increase in the metabolic rate, i.e., the release of energy from fuels.

thermoacidophile an organism, often an archaebacterium, that thrives in a hot, acidic environment.

thermodynamics 1. the branch of physics that deals with the interconversions of energy as heat, potential energy, kinetic energy, radiant energy, entropy and work. 2. the processes and phenomena of energy interconversions.

thermoreception the detection of heat, generally by specialized receptors such as mammalian Ruffini and Krause corpuscles.

thermoreceptor a sensory receptor that responds to temperature or to changes in temperature.

thermoregulation. 1. an animal's control over its internal temperature. 2. the physiological mechanisms that maintain a body at a particular temperature in an environment with a fluctuating temperature. *Behavioral thermoregulation,* maintaining a body temperature within acceptable limits by behavioral means, as by basking and by seeking out appropriate microenvironments.

thigmonasty any movement in plants, particularly changes in leaf position, attributed to sudden turgor pressure changes in cells of the petiole and triggered by touch.

thorax 1. in mammals, the part of the body anterior to the diaphragm and posterior to the neck, containing the lungs and heart. 2. the middle of three parts of an insect body, bearing the legs and wings.

threshold of stimulation in neurons or muscles, the lowest level of stimulation that will initiate a response.

threshold voltage The minimum voltage change needed to start an action potential in an axon.

thrombin in the blood clotting reactions, a proteolytic enzyme that catalyzes the conversion of fibrinogen to fibrin by the removal of two short peptide segments, and in turn is produced from *prothrombin* by the action of *thromboplastin*.

thromboplastin a complex phosphoprotein, released from damaged platelets, that catalyzes the conversion of the inactive serum component *prothrombin* into the active enzyme *thrombin* in the initiation of blood clotting.

thylakoid a membranous structure of chloroplasts and cyanophytes consisting of a *thylakoid membrane* containing chlorophyll, accessory pigments, photocenters, and the photosynthetic electron transport chain; an inner lumen that becomes acidic during active photosynthesis; and CF_1 particles attached to the outer sides of the thylakoid membrane, which are the sites of chemiosmotic phosphorylation.

thymine a pyrimidine, one of the four nitrogenous bases of DNA and the only nitrogenous base specific for DNA.

thymus a glandular body above the lungs, believed to stimulate T-cell lymphocyte development through the secretion of thymosin.

thyroid also *thyroid gland,* a large endocrine gland in the lower neck region of all vertebrates, arising as a ventral median outpocketing of the pharynx and believed to be homologous with the endostyle of cephalochordates, lamprey larvae, and urochordates; secretes *throxine* and *calcitonin*.

tissue a group of associated cells, identical in structure and function.

tissue interaction (formerly, *embryonic induction*) the ability of one embryonic tissue to influence the developmental fate of another.

toadstool a mushroom, especially if poisonous, distasteful, or suspicious.

tobacco mosaic virus a common infective agent of tobacco and the first virus to be seen by microscopy.

total fertility rate the predicted general fertility rate, determined through attitudinal studies and other forecasting indicators.

torsion an event of gastropod development in which the upper part of the body is rotated 180 degrees, bringing the anus to a position above the head.

totipotency during development, or throughout life, the retention of complete genetic and developmental potential in the nucleus of a cell.

totipotent of a cell, able to undergo development and differentiation along any of the lines inherently possible for the species; undifferentiated.

touch receptor see *sensory receptor*.

toxin a poisonous substance produced by a biological organism.

trace element any element essential to life, but only in minute or "trace" quantities. See also *micronutrient*.

trachea (pl. *tracheae*), 1. in land vertebrates, the main trunk of the air passage between the lungs and the larynx, usually stiffened with rings of cartilage. 2. one of the air-conveying chitinous tubules comprising the respiratory system of an insect.

tracheal gill in insect nymphs, a feathery apparatus used in exchanging gases with the surrounding water.

tracheal system respiratory system of insects, composed of thin-walled air-conducting tubules opening to *spiracles* and extending to finer, branched *tracheoles,* terminating in *air sacs*.

tracheid a long tubular xylem cell or its lignified, empty cell wall, which functions in support and the conduction of water, distinguished from *xylem vessels* by having tapered, closed ends communicating with other tracheids through pits.

tracheole *see tracheal system*.

tradewind a persistent, directional, global wind created by the rotational displacement of great rising air cells.

transcribed strand the DNA strand of a structural gene that is physically involved in transcription.

transcription the process of RNA synthesis as the RNA nucleotide sequence is directed by specific base pairing with the nucleotide sequence of the transcribed strand of a DNA cistron.

transducer an apparatus that converts nonelectrical forms of energy into electricity. In sensory receptors, the conversion of a physical stimulus such as pressure or heat into a neural impulse.

transduction the transfer of a host DNA fragment from one bacterium to another by a viral particle.

transfer RNA (tRNA) any of a class of relatively short ribonucleotides with a common secondary and tertiary structure consisting of three or four loops and a double-stranded stem, with numerous modified bases in addition to the four normally found in RNA; specific varieties become covalently linked to specific amino acids by specific enzymes, then function in transcription to recognize appropriate mRNA codons and to transfer the amino acid to the growing polypeptide chain.

transformation in a bacterium, the direct incorporation of a DNA fragment from its medium into its own chromosome.

transition zone a region of overlap between two ecological communities in which plants and animals of both communities are found.

translation polypeptide synthesis as it is directed by an mRNA on a ribosome; the transfer of linear information from a nucleotide sequence to an amino acid sequence according to the structures of the genetic code.

translocation 1. the step in protein synthesis in which a transfer RNA molecule that is covalently attached to the growing polypeptide chain is moved (translocated) from one ribosomal tRNA attachment site to the other. 2. a chromosome rearrangement in which a terminal segment of one chromosome replaces a terminal segment of another, nonhomologous chromosome. 3. the directed movement of materials from one part of a plant to another part, especially the directed movement of sugars through phloem.

translocation Down's syndrome also *trisomy 14/21,* a rare inherited form of Down's syndrome where the chromosome number 14 in cells of the carrier are fused to a number 21 chromosome; see also *trisomy 21*.

transmembranal protein plasma membrane protein that extends entirely through the membrane.

transmission electron microscope (TEM) see *electron microscope*.

transpiration. 1. the evaporation of water vapor from leaves, especially through the stomata. 2. the physical effects of such evaporation taken together. *Transpiration pull,* the pulling of water up through the xylem of a plant utilizing the energy of evaporation and the tensile strength of water.

transverse colon see *colon*.

transverse tubule also *T tubule,* a specialized system of tubules in a muscle fiber, transmitting the contractile impulse from the sarcolemma to the sarcomeres.

trichocyst in some ciliates, minute harpoonlike bodies below the pellicle that can be extruded.

triglyceride a compound of glycerol with each of its three hydroxyl side groups bound in an ester linkage to a fatty acid or other residue.

trimester one of three-month periods of the nine months of human gestation, which thus is divided into first, second, and third trimesters.

triploblastic in animals, the existence in the embryo of three primary germ layers, the ectoderm, mesoderm, and endoderm, from which all other tissues are formed.

trisomy 21 also *Down's syndrome,* originally *mongolian idiocy,* a severe human congenital pathology attributable to the presence of three rather than two homologues of chromosome 21 attributable to either nondisjunction or translocation.

tRNA see *transfer RNA*.

tRNA charging enzyme see *charging enzyme*.

trochophore also *trochopore larva,* a rather specific body form of a free-swimming ciliated larva of marine invertebrates of several phyla, including annelids, mollusks, and rotifers.

trophic level a level in a food pyramid.

trophoblast see *blastocyst*.

tropical rain forest a tropical woodland biome that has an annual rainfall of at least 250 cm and often much more, typically restricted to lowland areas, characterized by a mix of many species of tall broad-leaved evergreen trees forming a continuous canopy, with vines and woody epiphytes, and a dark, nearly bare forest floor.

tropism a turning toward or away from a stimulus, usually accomplished by differential growth; see *gravitropism, phototropism*.

tropomyosin a low-molecular-weight filamentous protein that accompanies the globular protein actin in making up actin microfilaments

troponin a protein of low molecular weight that binds calcium in muscle contraction, a specific variant of the ubiquitous calcium-binding molecule *calmodulin*.

true-breeding (adj.), of an organism or strain, when mated with individuals like itself producing offspring like itself; specifically, homozygous for all relevant loci.

trypanosome a parasitic, flagellated protozoan of the genus *Trypanosoma,* infecting mammals and transmitted by insect vectors, responsible for *chagas disease* and *sleeping sickness*.

tryptophan operon (a repressible operon) a region of DNA in *E. coli* that includes structural genes coding for enzymes that synthesize the amino acid tryptophan, and a region that controls their transcription; see also *operon.*

tsetse fly also *tsetse,* a biting fly of Africa that is the carrier of the trypanosome responsible for sleeping sickness.

T tubule see transverse tubule.

tubal ligation sterilization of a female mammal by cutting and tying the oviducts.

tube cell a cell in the microgametophyte of a seed plant, responsible for pollen tube growth.

tube foot one of the numerous specialized, hollow, extensible, flexible organs of locomotion and grasping in echinoderms formed as extensions of the water vascular system, each usually bearing a terminal sucker.

tube nucleus in seed plants, one of the two haploid nuclei of a pollen grain, which controls the subsequent growth and enzyme production of the pollen tube upon germination.

tuber an underground stem typified by the familiar potato.

tubular secretion the active transport of certain wastes from the peritubular capillaries into the nephron.

tubule any slender, elongated channel in an anatomical structure; *kidney tubule,* the tubular part of a nephron, being all of the nephron except Bowman's capsule; see also *proximal convoluted tubule, distal-convoluted tubule.*

tubulin the protein that comprises the spherical subunit of microtubules.

tundra a biome, characterized by level or gently undulating treeless plains of the arctic and subarctic, supporting a dense growth of mosses and lichens and dwarf herbs and shrubs, seasonally covered with snow and underlain with permafrost.

turgor the normal state of turgidity and tension in living plant cells, created by osmosis. *Turgor pressure,* the actual hydrostatic pressure developed by the fluid in a turgid plant cell.

Turner's syndrome also *XO syndrome,* a congenital human abnormality due to the presence of a single X chromosome and no other sex chromosome; resulting in immature female development, short stature, sterility, certain anatomical peculiarities, and normal mental functioning except for a marked deficiency in certain narrowly delimited spatial tasks.

tympanal organ see *tympanum.*

tympanic membrane also *tympanum, eardrum,* a thin, clear, tense double membrane of connective tissue covered with living epithelium, dividing the middle ear from the external auditory canal, vibrating sympathetically with received sound and transmitting the impulses to the inner ear by way of the ossicles of the middle ear; deep within a bony recess in mammals, on the surface in frogs and toads, and intermediate in birds and reptiles.

tympanum 1. see *tympanic membrane.* 2. an analogous thin tense membrane of an organ of hearing in an insect. 3. the membrane of a sound-producing organ that acts as a resonator.

Type I herpes simplex a relatively benign herpes simplex virus that causes cold sores or fever blisters around the mouth.

Type II herpes simplex also *genital herpes,* a highly contagious, sexually transmitted virus, causing intermittent outbreaks of painful, infective blisters on the genitals, followed by a period of remission.

typhlosole 1. in the earthworm, an infolding in the roof of the intestine that greatly increases the digestive and absorptive surface. 2. a similar fold projecting into the gut of certain pelecypod mollusks and starfish.

ultracentrifuge a device capable of creating great rotational force, used to separate fluid-suspended cellular and subcellular materials according to density.

ultraviolet light also *ultraviolet radiation, u.v.,* electromagnetic radiation having a shorter wavelength than visible light and longer than X-rays, including wavelengths normally invisible to humans but visible to bees and hummingbirds; interacts destructively with DNA and thus is mutagenic, carcinogenic, and destructive of skin tissues.

umbilical cord in placental mammals, a vascular cord connecting the fetus with the placenta, containing two *umbilical arteries,* an *umbilical vein* and the blind remnant of the yolk sac, in a connective tissue matrix.

ungulate a mammal that has hoofs, especially of the orders *Perissodactyla* and *Artiodactyla.*

unicellular also *acellular,* 1. consisting of a single cell. 2. having a single cell nucleus.

unsaturated capable of accepting hydrogens; *unsaturated fat,* a triglyceride with one or more carbon-carbon double bonds.

upper epidermis the upper (light-exposed) layer of cutinized epidermal cells in a leaf.

upwelling the coming to the surface of water from the depths of the ocean, which is associated with the introduction of mineral nutrients and a consequent richness in productivity and biomass.

uracil a pyramidine, one of the four nitrogenous bases of RNA and the only one specific to RNA.

urea a highly soluble nitrogenous compound, the principal nitrogenous waste of the urine of mammals; retained in high concentrations in the blood of elasmobranchs and the coelocanth as a means of osmoregulation.

ureter either of the pair of ducts that carry urine from the kidney to the bladder in mammals, or from the kidney to the cloaca in other vertebrates; compare *urethra.*

urethra the canal that carries urine from the mammalian bladder to the exterior, and in males serves also for the transmission of semen.

uric acid a relatively insoluble purine, a principal nitrogenous excretion product of reptiles, birds, and insects; excreted in small quantities as a product of nucleic acid breakdown in mammals; sometimes crystalizing in the tissues of men causing a painful inflammation *(gout).*

urinary bladder see *bladder.*

uterus 1. in female mammals, a muscular, vascularized, mucous-membrane-lined organ for containing and nourishing the developing young prior to birth and for expelling them at the time of birth, consisting of enlargements of the paired oviducts or of a single median enlargement of the fused oviducts. 2. an enlarged section of the oviduct of various vertebrates and invertebrates modified to serve as a place of development of the young or of eggs.

utricle the chamber of the membranous labyrinth of the middle ear into which the semicircular canals open.

uv light *uv radiation,* see *ultraviolet light.*

vacuole a general term for any fluid-filled membrane-bounded body within the cytoplasm of a cell, in higher plants occupying most of the volume of most differentiated living cells, in protozoa performing numerous functions; see *contractile vacuole, digestive vacuole.*

vacuum behavior instinctive actions that are released spontaneously, without the stimulus of a releaser, presumably due to the buildup of action-specific energy that finally demands to be discharged.

vagina 1. an expandable canal that leads from the cervix of the uterus of a female mammal to the external orifice *(introitus),* serving to receive the penis in copulation and as a *birth canal* in parturition. 2. any canal of similar function in other animals.

vagus nerve either of the tenth pair of cranial nerves, with sensory and autonomic motor fibers innervating the heart and viscera.

variability *genetic variability,* the general qualitative term for the presence of genetic differences between individuals in a population.

variable or **experimental variable** the focus of an experiment, to be tested and compared with a control.

variable age of onset of a genetic pathology, not being observed at birth but becoming manifest at some variable time in later development.

variable expressivity of a genetic pathology, having different degrees of abnormality in different individuals with the same or similar genotype.

variable region the portion of an immunoglobulin polypeptide concerned with the binding of an antigen, which varies from antigen to antigen.

vascular (adj.), pertaining to vessels or tissues of conduction, as phloem and xylem in plants or the blood vessels in animals.

vascular bundle a unit of the vascular system of plants, a strand of xylem associated with a strand of phloem and usually a sheath of fibrous sclerenchyma, in the primary growth pattern and in leaves and petioles.

vascular cambium the cylinder of lateral meristem that produces xylem on its inner side and phloem on its outer side in secondary growth, contributing thus to growth in circumference.

vascular plant a plant with xylem and phloem; a tracheophyte.

vascular ray spokelike layers of parenchyma tissue that extend through the wood, vascular cambium, and phloem.

vascular system 1. the circulatory system of an animal. 2. the xylem and phloem of a vascular plant.

vascular tissue plant tissues specialized for conducting water and foods; see also *xylem* and *phloem*.

vas deferens (pl. *vasa deferentia*), the male genital duct or spermatic duct.

vasectomy sterilization of the male by severance and ligature of the vasa deferentia (see vas deferens).

vasoconstriction contraction of sphincters in arterioles resulting in reduced blood flow.

vasodilation relaxation of arteriolar sphincters, allowing increased blood flow.

vector an organism that transmits a parasite, virus, bacterium, or other pathogen from one host to another.

vegetal hemisphere in eggs with polarity established, the less metabolically active, yolky region.

vegetative propagation in plants, the formation of new individuals from fragments or parts of the parental plant. In horticulture such parts would include cuttings.

vegetative reproduction asexual reproduction in plants, in which offspring are produced from portions of the roots, stems, or leaves of the parent.

vein 1. a vessel returning blood toward the heart at relatively lower velocity and pressure; such blood is usually relatively deoxygenated, but in the case of the *pulmonary veins* it is freshly oxygenated. 2. a vascular bundle in a leaf or petiole.

veldt eastern and southern African grassland, usually level, often with scattered trees or shrubs; where the lions live (see *savanna*).

vena cava (pl. *vena cavae*), any one of three large veins by which blood is returned to the right atrium of land vertebrates.

vent in vertebrates the external opening of the cloaca.

ventilation breathing.

ventral (adj.), in bilateral animals, toward the belly; downward, opposite the back; compare *dorsal*. In human anatomy, in view of the peculiar upright stance of the organism, *ventral* is frequently replaced by *anterior* (toward the front).

ventral nerve cord common feature of many invertebrate phyla, the main longitudinal nerve cord of the body, being solid, paired, and ventral, with a series of ganglionic masses.

ventricle a cavity of a body part of organ: 1. one of the large muscular chambers of the four-chambered heart. 2. one of the systems of communicating cavities of the brain, consisting of two *lateral ventricles* and a median *third ventricle*.

venule a small vein.

vertebra (pl. *vertebrae*), one of the bony or cartilaginous elements that together make up the vertebrate spinal column.

vertebral canal the pathway of the spinal cord through the vertebral column.

vertebral column also *spinal column*, the articulated series of vertebrae connected by ligaments and separated by intervertebral disks that in vertebrates forms the supporting axis of the body, and of the tail in most forms.

vesicle a small cavity, fluid-filled sac, blister, cyst, vacuole, etc.

vessel 1. a tube or canal in which a body fluid is contained or circulated within the interior, e.g., *blood vessel, lymph vessel*. 2. a water conducting tube in an angiosperm formed in the xylem by the end-to-end fusion of a series of *vessel elements*. Compare *tracheid*.

vestibular apparatus that portion of the inner ear involved in sensing position and movement of the head, containing the semicircular canals, saccule, and utricle.

vestibular gland also *gland of Bartholin*, a pair of oval, ducted glands lying on either side of the lower part of the vagina, secreting a lubricating mucus, especially on appropriate occasions.

vibrio curved or comma-shaped bacilli (rodlike bacteria).

villus (pl. *villi*), a small, slender, vascular, fingerlike process: 1. one of the minute processes that cover and give a velvety appearance to the inner surface of the small intestine, that contain blood vessels and a lacteal, are covered with microvilli, and serve in the absorption of nutrients. 2. one of the branching processes of the surface of the implanted blastodermic vesicle of most mammals.

virulence 1. disease-producing capability. 2. the capacity of an infective organism to overcome host defenses.

virus a noncellular organism consisting of DNA or RNA enclosed in a protein coat, often together with a few enzymes; replicating only within a host cell, utilizing host ribosomes, enzymes, and energy.

visceral muscle also *smooth muscle* **or** *involuntary muscle*, muscle of the gut, uterus, and blood vessels. **visible light** electromagnetic wavelengths longer than about 400 nm and shorter than about 750 nm, which can serve as visual stimuli to most photoreceptive organisms.

visual receptor a sense receptor stimulated by light or by patterns of light.

vital capacity the maximum volume of air that can be forcefully exhaled.

vitalism a doctrine in biology that ascribes the functions of a living organism to a *vital principle* or *vital force* distinct from chemical and physical forces; the doctrine is no longer taken seriously by practicing biologists.

vitamin any organic substance that is essential to the nutrition of an organism, usually by supplying part of a coenzyme.

viviparity the condition of producing live young.

viviparous (adj.), producing live young from within the body of the female, following development of the young within a uterus in intimate association with the tissues of the mother; compare *ovoviviparous, oviparous*.

vocal cord in many vertebrates, one of the paired mucous membranes stretched across the glottis, capable of producing varied sounds when vibrated by air.

voice box see *larynx*.

voluntary muscle see *skeletal muscle*.

vulva the external genitalia of a woman.

water mold an aquatic protist of the oomycetes.

water potential the potential energy of water to move, as a result of concentration, gravity, pressure, or solute content.

water vascular system a system of vessels in echinoderms, containing sea water, used in the movement of tentacles and tube feet and possibly in excretion and respiration.

Watson strand either of the two strands of DNA, the other being designated the *Crick strand*.

wavelength the physical distance from one point of maximum intensity to the next in a propagated wave, especially of electromagnetic radiation in a vacuum; inversely proportional to *frequency*, the number of waves passing a point per second, and also inversely proportional to the energy per photon.

wax 1. a dense, hard, lipid-soluble ester of a long-chain alcohol and a fatty acid. 2. beeswax, one such compound. 3. a natural mixture of such monohydroxy esters and other lipid substances.

Wernicke's area a region of the cerebrum concerned with deciphering speech.

wheat rust 1. any of three distinct fungal diseases of wheat. 2. a rust fungus responsible for such a disease.

whorl 1. a group of parts repeated in a circle. 2. any of the four basic radially repeated groups of flower parts; see *gynoecium, androecium, corolla, calyx*. 3. a fingerprint in which the ridges form concentric circles.

wild type 1. of a phenotype, normal or typical in appearance; not mutant, rare or abnormal. 2. of an allele, normal and common in the population; not mutant; giving rise to a typical phenotype when homozygous.

wilt (v.i.), to lose turgor, especially because of an inadequate supply of water to a plant or plant part.

wood the hard, fibrous xylem of secondary growth, especially that of the central stem of a tree, consisting of lignified cellulose cell walls.

woody containing wood or wood fibers; consisting largely of secondary growth with hard lignified tissues.

work in physics, the transfer of energy resulting in the movement of an object formerly at rest.

X chromosome 1. one of the two heteromorphic sex chromosomes of mammals, flies, and certain other insects, such that the possession of two X chromosomes without a Y chromosome results in the development of a normal female, while the possession of one X and one Y chromosome results in a normal male, and other combinations are rare and abnormal. 2. the sex chromsome of grasshoppers and certain other insects such that the possession of two X chromosomes results in female development while the possession of only one results in male development. 3. see *Z chromosome*.

xeric (adj.), of a dry environment (e.g., desert).

xeroderma pigmentosum a rare autosomal recessive condition of humans in which the DNA repair system specific for u.v.-induced damage is defunct, resulting in a pathological sensitivity to sunlight that usually leads to multiple skin cancers.

xerophyte a plant adapted to dry areas.

X-linked also *sex-linked,* the condition, common to many genes of functions unrelated to sex, of being present on the X chromosome; thus males have only one copy rather than the normal diploid two, and recessive X-linked alleles are always expressed in males, thus radically affecting patterns of inheritance.

X-organ an endocrine gland, found in the eyestalk of crayfish, that produces *molt-inhibiting hormone;* compare *Y-organ.*

X-ray a highly energetic form of electromagnetic radiation with wavelengths of .01 nm to 10 nm, between gamma radiation and u.v. radiation.

X-ray crystallography a procedure of directing X rays at a crystal of a protein or other large molecule, recording the pattern produced as the rays are bent by the regular, repeating molecular structures within the crystal, and using this pattern to reconstruct the 3-dimensional structure of the molecule.

XXX syndrome see *Klinefelter's Syndrome.*

xylem one of the two complex tissues in the vascular system of vascular plants, consisting of the dead cell walls of vessels, tracheids, or both, often together with sclerenchyma and parenchyma cells, functioning chiefly in the conduction of water and in support, but sometimes also in food storage; the substance of wood in secondary growth as well as part of the vascular bundles of primary growth; see *tracheid, vessel, lignin;* compare *phloem.*

xylem vessel water-conducting tissue in plants, consisting of vessel elements arranged end to end.

XYY syndrome also called *extra-Y syndrome,* an abnormal genetic condition of males due to nondisjunction, characterized by markedly excessive height, a tendency to severe acne, and mild mental retardation that in itself tends to get affected individuals into trouble of various kinds.

Y chromosome 1. one of the two heteromorphic sex chromosomes of mammals, flies, and certain other insects such that XY individuals are male and XX individuals are female; in mammals carrying the primary male sex determinant for the H-Y antigen, a determinant for increased height, and a few other genes; in general nearly devoid of genes not concerned with maleness; compare *X chromosome.*

yeast a unicellular sac fungus, especially *Saccharomyces cerevisiae,* which is used in the production of bread, beer, wine, and raised doughnuts.

yellow marrow yellow, fatty material within the central cavity of long bones.

yolk the material stored in an ovum that supplies food material to the developing embryo, consisting chiefly of *vitellin* (a lecithin-containing phosphoprotein), nucleoproteins, other proteins, lecithin (a phospholipid), and cholesterol.

yolk sac 1. one of the extraembryonic membranes of a bird, reptile, or mammal, that in birds and reptiles grows over and encloses the yolk mass and in placental mammals encloses a fluid-filled space; the first site of blood cell and circulatory system formation. 2. an analogous structure in elasmobranchs and cephalochordates.

Y-organ one of the endocrine glands of a crayfish eyestalk, secreting *crustecdysone,* the crustacean molt-inducing hormone.

Z chromosome one of the two heteromorphic sex chromosomes of birds and butterflies, such that ZZ individuals are male and ZW individuals are female; sometimes considered to be an X chromsome; compare *X chromosome, Y chromosome.*

zeatin a plant cytokinin hormone and growth factor extracted from corn.

zero population growth (ZPG) a point in population dynamics where the birth rate equals the death rate.

Z line in striated muscle, the partition between adjacent contractile units to which actin filaments are anchored.

zona pellucida a thick, transparent, elastic membrane or envelop secreted around an ovum by follicle cells.

zooplankton animal life drifting at or near the surface of the open sea.

zoospore 1. any independently motile spore. 2. the flagellated gamete of a foraminiferan.

Z scheme a graphic presentation of the oxidation-reduction reactions occurring in the light reaction and electron transport chain in photosynthesis.

zygospore a diploid fungal or algal spore formed by the union of two similar sexual cells, having a thickened and usually ornamented wall, that serves as a resistant resting spore.

zygote a cell formed by the union of two gametes; a fertilized egg.

zygotene the stage in meiotic prophase in which homologous chromosomes synapse; see *synapse, synaptinemal complex.*

zygotic cycle a primitive life cycle with a dominating haploid state and one in which gametes form through mitosis. Upon fertilization, a brief diploid state is followed by a meiosis and a return to the haploid state (some protists, all fungi).

ACKNOWLEDGMENTS

Unless otherwise acknowledged, all photos are the property of HarperCollins Publishers Inc.

FRONT COVER
© Gerard Lacz/Peter Arnold, Inc.

BACK COVER
© Luiz Claudio Marigo/Peter Arnold, Inc.

CONTENTS
p. xiii, Dr. E.R. Degginger; **p. xv**, Jeff Rotman/ Peter Arnold, Inc.; **p. xvii**, Robert Lyons/Visuals Unlimited; **p. xvii**, R. Langridge/Dan McCoy/ Rainbow; **p. xviii**, K.G. Murti/Visuals Unlimited; **p. xix**, Dwight R. Kuhn; **p. xx**, Farrell Grehan/Science Source/Photo Researchers; **p. xxi**, John D. Cunningham/ Visuals Unlimited; **p. xxii**, Fred Bavendam/Peter Arnold, Inc.; **p. xxiii**, Robert Noonan/; **p. xxiv**, Joseph Van Wormer/Bruce Coleman Inc.; **p. xxiv**, Bruce Davidson/ANIMALS ANIMALS; **p. xxvi** T. Kitchin/Tom Stack & Associates; **p. xxvii**, Guhl/Photographic Resources, Inc.; **p. xxviii**, From THE INCREDIBLE MACHINE, © Lennart Nilsson, National Geographic Society; **p. xxviii**, Peter J. Bryant/ Biological Photo Service; **p. xxx**, Stanley Flegler/ Visuals Unlimited; **p. xxx**, Adrian Davis/Bruce Coleman, Inc. **p. xxxi**, Werner H. Muller/Peter Arnold, Inc.;

PART ONE, CHAPTER ONE
p. xxxii-1, Grant Heilman Photography; **p. 2**, Susan Pierres/Peter Arnold, Inc.; **p. 3**, "Charles Darwin, Age 30" by George Richmond, By permission of the Darwin Museum Down House; **p. 4**, The Bettmann Archive; **p. 5**, Jen and Des Bartlett/Bruce Coleman Inc.; **p. 7**, Tim Rock/ Earth Scenes; **p. 8**, George Harrison; **p. 8**, Elliott Varner Smith; **p. 8**, Elliott Varner Smith; **p. 8**, David M. Stone; **p. 8**, David M. Stone; **p.9**, Elliott Varner Smith; **p.9**, George Harrison; **p. 9**, Dr. E.R. Degginger; **p. 15**, Adrienne T. Gibson/ ANIMALS ANIMALS; **p. 15**, Barry L. Runk from Grant Heilman Photography; **p. 15**, Barry L. Runk from Grant Heilman Photography; **p. 16**, Doug Wechsler; **p. 16**, By permission of the Darwin Museum, Down House; **p. 17**, The Bettmann Archive; **p. 17**, Brian Seed/TSW/Click Chicago; **p. 17**, Lynn M. Stone/Bruce Coleman Inc.; **p. 20**, Dwight R. Kuhn; **p. 20**, Dr. E.R. Degginger; **p. 20**, Dwight R. Kuhn; **p. 20**, Rajesh Bedi/Life Magazine © Time Inc.; **p. 21**, Baron Hugo van Lawick © National Geographic Society; **p. 21**, Stephen J. Krasemann/DRK Photo; **p. 21**, Manfred Kage/Peter Arnold, Inc.; **p. 21**, Dr. Frank M. Carpenter; **p. 21**, Dwight R. Kuhn; **p. 22**, Runk/Schoenberger from Grant Heilman Photography; **p. 22**, NASA

CHAPTER TWO
p. 25, Randall M. Feenstra and J.A. Stroscio, Courtesy of IBM; **p. 26**, IBM Research/Visuals Unlimited; **p. 33**, Dr. E.R. Degginger; **p. 37**, William E. Ferguson Photography; **p. 39** Rod Planck/Tom Stack & Associates; **p. 39** Ken W. Davis/Tom Stack & Associates; **p. 39**, Dwight R. Kuhn; **p. 39**, Dwight R. Kuhn; **p. 39**, Jeanette Thomas/Visuals Unlimited; **p. 42**, Patrice Ceisel/ Visuals Unlimited; **p. 42**, William Amos/Bruce Coleman Inc.

CHAPTER THREE
p. 50, Runk/Schoenberger from Grant Heilman Photography; **p. 55**, Eric V. Grave; **p. 56**, Alfred Owczarzak/Taurus Photos, Inc.; **p. 57**, Brown, R.M. and J.H.M. Willson, 1977. In INTERNATIONAL CELL BIOLOGY 1976-1977, ed., B.R. Brinkley and K.R. Porter, pp. 267-283. © 1977 by The Rockefeller University Press; **p. 58**, Jane Burton/Bruce Coleman Inc.; **p. 62**, Ed Reschke

CHAPTER FOUR
p. 76, Stanley Flegler/Visuals Unlimited; **p. 79**, From TISSUES AND ORGANS: A TEXT-ATLAS OF SCANNING ELECTRON MICROSCOPY, by Richard G. Kessel and Randy H. Kardon, W.H. Freeman and Company © 1979; **p. 81**, University of California; **p. 82**, Dr. Daniel Branton; **p. 83**, Dr. Kwang Shing Kim, NYU Medical Center/Peter Arnold, Inc.; **p. 86**, Dr. Eva Frei and Professor R.D. Preston; **p. 88**, K.G. Murti/Visuals Unlimited; **p. 89**, From BEHOLD MAN, © Lennart Nilsson, National Geographic Society, Inc.; **p. 90**, P.R. Burton, Univ. of Kansas/Biological Photo Service; **p. 90**, P.R. Burton, Univ. of Kansas/Biological Photo Service; **p. 90**, L.E. Roth/Biological Photo Service; **p. 91**, Dr. Don Fawcett/Photo Researchers; **p. 91**, Dr. David Branton; **p. 92**, Dr. John Taylor; **p. 93**, Richard Chao; **p. 96**, Frederick, S.E., and E.H. Newcomb, 1969. J. CELL BIOLOGY 43: 343; **p. 97**, G.F. Leedale/ Science Source/Photo Researchers; **p. 98**, Runk/ Schoenberger from Grant Heilman Photography; **p. 98**, Runk/Schoenberger from Grant Heilman Photography; **p. 102**, Photo Researchers; **p. 102**, R. Bhatnagar/Visuals Unlimited; **p. 103**, McGill et al. 1976. J. Ultrastruct Res. 57:43-53. Micrograph courtesy Dr. B.R. Brinkley; **p. 103**, Dr. William E. Barstow; **p. 104**, Dr. Karl Aufderheide; **p. 105**, Bouck, G.B. 1971. J. CELL BIOLOGY, 50: 362-384. Reprinted by copyright permission of the Rockefeller University Press; **p. 105**, Dr. William E. Barstow

CHAPTER FIVE
p. 110, Peter Parks/Oxford Scientific Films/ ANIMALS ANIMALS; **p. 112**, Dr. Henry C. Aldrich; **p. 113**, Fawcett/Photo Researchers; **p. 125**, M.M. Perry and A.B. Gilbert, Agricultural Research Council's Poultry Research Centre, Edinburgh

CHAPTER SIX
p. 129, Tom Ives; **p. 130**, Janet Thiede/ ANIMALS ANIMALS; **p. 130**, Tony Stone Worldwide/Click Chicago; **p. 131**, Adrian Davies/Bruce Coleman Inc.; **p. 136**, Regents of the University of California

CHAPTER SEVEN
p. 152, Adrian Davies/Bruce Coleman Inc.; **p. 153**, Robert Frerck/Odyssey Productions, Chicago; **p. 157**, Hugh Spencer; **p. 157**, Courtesy, William E. Barstow; **p. 172**, Wayne A. Bladholm; **p. 172**, Alan Pitcairn from Grant Heilman Photography

CHAPTER EIGHT
p. 177, Bruce Coleman, Inc.; **p. 197**, Dan Cabe/ Photo Researchers

PART TWO, CHAPTER NINE
p. 202-203, Dan McCoy/Rainbow; **p. 204**, Guhl/Photographic Resources, Inc.; **p. 205**, Lester V. Bergman & Associates; **p. 206**, Du Praw, E.J. and G.F. Bahr, 1969 ACTA CYTOLOGICA 13:188. Science Printers & Pub. Inc.; **p. 209**, Ed Reschke; **p. 210**, J. Pickett-Heaps/Science Source/Photo Researchers; **p. 210**, Biological Photo Service; **p. 212**, Kessel, R.G., and C.Y. Shih. 1974. SCANNING ELECTRON MICROSCOPY IN BIOLOGY: A STUDENT'S ATLAS ON BIOLOGICAL ORGANIZATION. New York, Heidelberg, Berlin: Springer-Verlag © 1974; **p. 213**, Gimenez-Martin, G.C. de la Torre, and J.F. Lopez-Saez. In MECHANISMS AND CONTROL OF CELL

DIVISION, ed. T.L. Rost and E.M. Gifford, Jr., pp. 267-283; **p. 214**, Runk/Schoenberger from Grant Heilman Photography; **p. 214**, Courtesy, F.R. Turner and Anthony P. Mahowald, from CELLS, by David M. Prescott.; **p. 215**, Biological Photo Service; **p. 215**, Hasenkampf 1986/ Biological Photo Service; **p. 217**, J. Kezer, In Novitski, E., 1982 HUMAN GENETICS, 2nd ed. New York, Macmillan Publishing Co.; **p. 218**, Professor M. Westergaard. In DNA CHROMATIN AND CHROMOSOMES by E.M. Bradbury, N. Maclean and H.C. Matthews © 1981 by Blackwell Scientific Publications.; **p. 222**, L.B. Shettles/Science Source/Photo Researchers; **p. 227**, Robert Lyons/Visuals Unlimited; **p. 229**, The Bettmann Archive

CHAPTER TEN
p. 229, The Bettmann Archive

CHAPTER ELEVEN
p. 246, J. Bryant/Biological Photo Service; **p. 247**, Jan Halaska/Photo Researchers; **p. 248**, Carolina Biological Supply; **p. 249**, Dr. E.R. Degginger; **p. 250**, Runk/Schoenberger from Grant Heilman Photography; **p. 250**, Milwaukee Journal Photo; **p. 251**, Dr. E.R. Degginger; **p. 251**, Runk/Schoenberger from Grant Heilman Photography; **p. 253**, Ed Reschke; **p. 254**, G.R. Bishop/Visual Innovations; **p. 254**, Peter J. Bryant/Biological Photo Service; **p. 255**, Lester V. Bergman & Associates; **p. 256**, Runk/ Schoenberger from Grant Heilman Photography; **p. 263**, Dr. Murray L. Barr; **p. 264**, Dr. Henry L. Nadler, Genetics Dept., Children's Memorial Hospital; **p. 265**, Courtesy MacMillan Science Co., Inc.; **p. 266**, The Bettmann Archive

CHAPTER TWELVE
p. 272, R. Langridge/Dan McCoy/Rainbow; **p. 276**, Lee D. Simon/Science Source/Photo Researchers; **p. 278**, x-ray difference of DNA by Rosalind Franklin. In Watson, J.D. 1968. THE DOUBLE HELIX. New York: Atheneum, p. 215 © 1968 by J.D.Watson; **p. 279**, Watson and Crick in front of the DNA model. Photographer: A.C. Barrington Brown. In Watson, J.D. 1968 THE DOUBLE HELIX. New York: Atheneum, p. 215 © 1968 by J.D. Watson; **p. 280**, Regents of the University of California; **p. 288**, National Institute of Health; **p. 291**, Omikron/Science Source/Photo Researchers

CHAPTER THIRTEEN
p. 295, K.G. Murti/Visuals Unlimited; **p. 298**, Miller, O.L. Jr. and B.R. Beatty, 1969. SCIENCE 164: 955-957; **p. 309**, Miller, O.L., Jr., B.A. Hamkalo, and C.A. Thomas, Jr. 1970. SCIENCE 169:392-395)

CHAPTER FOURTEEN
p. 313, Courtesy, Peter Lichter and David C. Ward, Yale University; **p. 322**, Edstrom, J., and W. Beermann. 1962 J. CELL BIOLOGY 14:374. Reprinted by copyright permission of The Rockefeller University Press

CHAPTER FIFTEEN
p. 326, K.G. Murti/Visuals Unlimited; **p. 327**, Tom Broker/Rainbow; **p. 327**, Lee D. Simon/ Photo Researchers; **p. 328**, Bruce Iverson; **p. 332**, Charles C. Brinton, Jr., and Judith Carnahan

CHAPTER SIXTEEN
p. 341, Copyright © 1988 New England Biolabs Inc. All rights reserved. The re-use or reproduction of any of the information, design or layout contained in this catalog without permission of NEBI is prohibited. **p. 346**, Courtesy, Robert J. Ferl, University of Florida; **p. 347**, Dan McCoy/Rainbow; **p. 348**, Robert J. Ferl, University of Florida; **p. 350**, Jon Gordon/ Phototake; **p. 353**, Dan McCoy/Rainbow

CHAPTER SEVENTEEN
p. 366, G.R. Roberts

PART THREE, CHAPTER EIGHTEEN
p. 372, Joseph Van Wormer/Bruce Coleman Inc.; **p. 373**, The Bettmann Archive; **p. 373**, Norman Myers/Bruce Coleman Inc.; **p. 376**, Hans Reinhard/Bruce Coleman Inc.; **p. 376**, Hans Reinhard/Bruce Coleman Inc.; **p. 378**, Hans Reinhard/Bruce Coleman Inc.; **p. 379**, Erwin & Peggy Bauer/Bruce Coleman Inc.; **p. 381**, David T. Horsley, University of Nottingham; **p. 383**, Dr. H.B.D. Kettleweil; **p. 384**, David C. Houston/Bruce Coleman Inc.

CHAPTER NINETEEN
p. 395, Sandved and Coleman Photography; **p. 402**, Milwaukee Journal photo; **p. 406**, Bill Field

CHAPTER TWENTY
p. 410, Derek Fell; **p. 411**, Kjell B. Sandved/ Bruce Coleman Inc.; **p. 411**, Robert P. Carr; **p.411**, Stephen J. Krasemann/DRK Photo; **p. 412**, J. Van Wommer/Bruce Coleman Inc.; **p. 412**, Bob and Clara Calhoun/Bruce Coleman Inc.; **p. 412**, Russ Kinne/National Audubon Society/Photo Researchers; **p. 412**, Mark S. Carlson/Tom Stack & Associates; **p. 412**, Leonard Lee Rue III/Bruce Coleman Inc.; **p. 412**, Jen and Des Bartlett/National Audubon Society/Photo Researchers; **p. 412**, Clem Haagner/Bruce Coleman Inc.; **p. 412**, Warren Garst/Tom Stack & Associates; **p. 413**, Tom McHugh/National Audubon Society/Photo Researchers; **p. 413**, Stephen Maslowski/ National Audubon Society/Photo Researchers; **p. 415**, Richard Ellis; **p. 415**, Dr. Merlin D. Tuttle/ Bat Conservation International; **p. 415**, Bullaty and Lomeo; **p. 422**, Raymond A. Mendez/ ANIMALS ANIMALS; **p. 422**, Allan Root/Bruce Coleman Inc.; **p. 425**, U.S. Dept. of Agriculture; **p. 426**, Jungle Larry's Safari Land, Inc.; **p. 427**, C.C. Lockwood/Bruce Coleman Inc.; **p. 427**, Robert P. Carr/Bruce Coleman Inc.; **p. 428**, Michael Habicht/ANIMALS ANIMALS; **p. 430**, Sandved and Coleman Photography

CHAPTER TWENTY-ONE
p. 437, Farrell Grehan/Science Source/Photo Researchers; **p. 439**, Roger Ressmeyer/Starlight; **p. 444**, Courtesy Sidney W. Fox in Fox, S.W. and K. Kose, 1977. MOLECULAR EVOLUTION AND THE ORIGINS OF LIFE, 2d ed. New York Marcel Dekker Inc.

PART FOUR, CHAPTER TWENTY-TWO
p. 452-453, Robert Noonan; **p. 454**, David M. Phillips/Visuals Unlimited; **p. 455**, Rick Smolan; **p. 456**, Biological Photo Service; **p. 458**, Manfred Kage/Peter Arnold, Inc.; **p. 458**, Centers for Disease Control, Atlanta; **p. 458**, Dr. G. de Haller, Courtesy of Dr. C.F. Robinow; **p. 459**, Centers for Disease Control, Atlanta; **p. 459**, Clay Adams; **p. 459**, Martin M. Rotker/ Taurus Photos, Inc.; **p. 459**, Manfred Kage/Peter Arnold, Inc.; **p. 459**, Courtesy Dr. C.F. Robinow; **p. 459**, Ed Reschke; **p. 459**, Biological Photo Service; **p. 460**, Courtesy Dr. R. Wyckoff/ National Institutes of Health; **p. 461**, Ames Research Center/NASA; **p. 463**, Lester V. Bergman & Associates; **p. 464**, Lester V. Bergman & Associates; **p. 464**, Lester V. Bergman & Associates; **p. 465**, D.A. Glawe; **p. 466**, Sinclair Stammers/Science Source/Photo Researchers; **p. 466**, J.R. Waaland, University of Washington/Biological Photo Service; **p. 466**, Sherman Thomson/Visuals Unlimited; **p. 467**, Courtesy of D.L. Findley, P.L. Walne and R.W. Holton, University of Tennessee, Knoxville. From J. PSYCHOLOGY 6: 182-188, 1970; **p. 471**, Biological Photo Service; **p. 471**, Gene M. Milbrath, PhD.; **p. 473**, Dennis Kunkel/ Phototake; **p. 473**, Phototake; **p. 473**, Valentine, R.C., H.G. Pereira, 1965. © by Academic Press J. MOLECULAR BIOLOGY 13:13-20; **p. 473**, Martin M. Rotker/Phototake; **p. 473**, Dr. Michael Moody; **p. 477**, Courtesy Sally Hensen, Pediatrics Dept. and John Hay, Microbiology Dept., U.S.U.H.S., Bethesda, MD; **p. 477**, Simons, K., H. Garoff, and A. Helenius, 1982. Scientific American 246: 58-66; **p. 479**, Dr. Erskine Palmer/Alyne Harrison/Taurus Photos, Inc. Science Photo Library International

CHAPTER TWENTY-THREE
p. 485, John D. Cunningham/Visuals Unlimited; **p. 486**, The Granger Collection, New York; **p. 486**, The Granger Collection, New York; **p. 490**, Lester V. Bergman & Associates; **p. 491**, Steven T. Brentano, University of Iowa; **p. 492**, John D. Cunningham/Visuals Unlimited; **p. 493**, M. Schliwa/Visuals Unlimited; **p. 493**, Manfred Kage/Peter Arnold, Inc.; **p. 493**, Dr. Rudolf Rottger; **p. 494**, Don and Pat Valenti; **p. 494**, Ed Reschke; **p. 495**, Dr. E.R. Degginger; **p. 495**, Eric V. Grave; **p. 495**, Eric V. Grave; **p. 497**, Ed Reschke; **p. 499**, Dr. E.R. Degginger; **p. 499**, John Shaw; **p. 500**, Dr. James W. Richardson; **p. 501**, Kevin Schafer/Tom Stack & Associates; **p. 502**, Eric V. Grave; **p. 502**, Dr. Paul E. Hargraves; **p. 502**, J. Pickett-Heaps/Science Source/Photo Researchers; **p. 502**, Gordon Leedale Biophoto Associates/Photo Researchers (Science Source); **p. 503**, Runk/Schoenberger from Grant Heilman Photography; **p. 503**, Jan Hinsch/Science Photo Library/Photo Researchers; **p. 503**, Burmeister/Visuals Unlimited; **p. 503**, Brian Parker/Tom Stack & Associates; **p. 505**, John D. Cunningham/Visuals Unlimited; **p. 506**, Runk/Schoenberger/Grant Heilman Photography; **p. 506**, Jeff Foot/Bruce Coleman Ltd.; **p. 506**, F. Stuart Westmoreland/ Tom Stack & Associates; **p. 508**, Doug Wechsler; **p. 510**, Kim Taylor/Bruce Coleman Ltd.; **p. 510**, Runk/Schoenberger from Grant Heilman Photography; **p. 510**, Dr. E.R. Degginger; **p. 511**, Grant Heilman from Grant Heilman Photography

CHAPTER TWENTY-FOUR
p. 517, Robert Noonan; **p. 518**, Grant Heilman Photography; **p. 519**, Alec Duncan/Taurus Photos, Inc.; **p. 519**, Charles Marden Fitch; **p. 522**, Runk/Schoenberger from Grant Heilman Photography; **p. 524**, 1984. DISCOVER THE INVISIBLE, Eric V. Grave, Englewood Cliffs, N.J. Prentice-Hall, Inc. Used by permission; **p. 525**,

Stanley Flegler/Visuals Unlimited; **p. 525**, Don and Pat Valenti; **p. 525**, Dwight R. Kuhn; **p. 526**, Dr. James W. Richardson; **p. 526**, James L. Castner; **p. 527**, Eric V. Grave/Phototake; **p. 528**, Dr. E.R. Degginger; **p. 528**, Peter Ward/Bruce Coleman Ltd.; **p. 528**, Doug Wechsler; **p. 528**, Don and Pat Valenti; **p. 531**, N. Allin and G.L. Barron. Courtesy NSERC Canada. Can. J. Bot. 57:187-193

PART FIVE, CHAPTER TWENTY-FIVE

p. 536537, J. Vermer/Visuals Unlimited; **p. 538**, Jacques Jangoux/Peter Arnold, Inc.; **p. 542**, Runk/Schoenberger from Grant Heilman Photography; **p. 542**, John Cunningham/Visuals Unlimited; **p. 544**, Runk/Schoenberger from Grant Heilman Photography; **p. 544**, Dwight R. Kuhn; **p. 544**, Dr. James W. Richardson; **p. 544**, Runk/Schoenberger from Grant Heilman Photography; **p. 547**, Dr. Linda E. Graham University of Wisconsin; **p. 550**, The Peabody Museum of Archaeology and Ethnology, Yale University; **p. 551**, Doug Wechsler; **p. 552**, Runk/Schoenberger from Grant Heilman Photography; **p. 554**, William E. Ferguson Photography; **p. 555**, Rod Planck/Tom Stack & Associates; **p. 555**, Ed Cooper; **p. 555**, Doug Wechsler; **p. 555**, Dr. E.R. Degginger; **p. 556**, Runk/Schoenberger from Grant Heilman Photography; **p. 556**, Runk/Schoenberger from Grant Heilman Photography; **p. 557**, Dr. E.R. Degginger; **p. 558**, Runk/Schoenberger from Grant Heilman Photography; **p. 558**, Lennart Nilsson/Boehringer Ingelheim Zentrale GmbH; **p. 559**, William H. Allen, Jr.; **p. 559**, William E. Ferguson Photography; **p. 560**, Brian Parker/Tom Stack & Associates; **p. 561**, Brian Parker/Tom Stack & Associates

CHAPTER TWENTY-SIX

p. 572, Frank Oberle/Photographic Resources, Inc.; **p. 574**, Barry L. Runk from Grant Heilman Photography; **p. 574**, John Gerlach/Visuals Unlimited; **p. 574**, Bruce Coleman Inc.; **p. 574**, Runk/Schoenberger from Grant Heilman Photography; **p. 574**, Alfred B. Thomas/Earth Scenes; **p. 574**, W.H. Hodge/Peter Arnold, Inc.; **p. 574**, Derek Fell; **p. 576**, William E. Ferguson Photography; **p. 576**, Doug Wechsler; **p. 577**, Dr. Thomas Eisner; **p. 577**, Eric Crichton/Bruce Coleman Ltd.; **p. 577**, Frieder Sauer/Bruce Coleman Ltd.; **p. 577**, Lester V. Bergman & Associates; **p. 578**, William E. Ferguson Photography; **p. 578**, Rod Planck; **p. 579**, G.R. Roberts; **p. 580**, James L. Castner; **p. 580; p. Jack M. Bostrack**; **p. 584**, G.R. Roberts; **p. 585**, G.R. Roberts; **p. 586**, G.R. Roberts; **p. 586**, John Ebeline; **p. 586**, G.R. Roberts; **p. 586**, John Shaw; **p. 586**, Robert P. Carr; **p. 590**, Lester V. Bergman & Associates; **p. 590**, Doug Wechsler

CHAPTER TWENTY-SEVEN

p. 595, David M. Dennis/Tom Stack & Associates; **p. 598**, U.S. Forestry Service; **p. 603**, From LIVING IMAGES by Gene Shih and Richard G. Kessel. Reprinted courtesy of Jones and Bartlett Publishers, Inc., Boston.; **p. 603**, Dr. Alice Belling; **p. 603**, Suny College of Environmental Science and Forestry; **p. 603**, Dr. Jeremy Burgess/Science Photo Library/Photo Researchers; **p. 605**, Lester V. Bergman & Associates; **p. 607**, Ed Reschke; **p. 607**, Omikron/Science Source/Photo Researchers; **p. 608**, John D. Cunningham/Visuals Unlimited; **p. 609**, James L. Castner; **p. 610**, Ed Reschke; **p. 611**, Runk/Schoenberger from Grant Heilman Photography; **p. 612**, David Newman/Visual Unlimited; **p. 612**, Ed Reschke; **p. 612**, Lott/Biological Photo Service; **p. 612**, Lott/Biological Photo Service; **p. 614**, Ed Reschke

CHAPTER TWENTY-EIGHT

p. 618, John D. Cunningham/Visuals Unlimited; **p. 619**, William H. Mullins; **p. 626**, Runk/Schoenberger from Grant Heilman Photography; **p. 626**, E. Zeiger and N. Burstein, Stanford University/Biological Photo Service; **p. 631**, Lynn M. Stone; **p. 633**, S.A. Wilde; **p. 634**, Dwight R. Kuhn; **p. 634**, Runk/Schoenberger from Grant Heilman Photography

CHAPTER TWENTY-NINE

p. 638, John Cancalosi/Tom Stack & Associates; **p. 641**, Lester V. Bergman & Associates; **p. 642**, Robert E. Lyons/Visuals Unlimited; **p. 647**, Lynn M. Stone; **p. 647**, Eric Crichton/Bruce Coleman Ltd.; **p. 647**, Rod Planck; **p. 650**, R. Calentine; **p. 650**, William E. Ferguson Photography; **p. 651**, Lillian N. Bolstad/Peter Arnold, Inc.; **p. 652**, Runk/Schoenberger from Grant Heilman Photography; **p. 653**, Eric Crichton/Bruce Coleman Ltd.; **p. 653**, Wayne A. Bladholm; **p. 653**, Eric Crichton/Bruce Coleman Ltd.

PART SIX, CHAPTER THIRTY

p. 659, Jerry Cooke/ANIMALS ANIMALS; **p. 660**, Jeff Rotman/Peter Arnold, Inc.; **p. 662**, Christopher Newbert/Bruce Coleman Inc.; **p. 671**, Ed Reschke; **p. 672**, Kim Taylor/Bruce Coleman Ltd.; **p. 673**, Nancy Sefton; **p. 674**, Nancy Sefton; **p. 674**, Doug Wechsler; **p. 675**, Jeffrey L. Rotman; **p. 676**, Ed Reschke; **p. 679**, Michael Klein/Peter Arnold, Inc.; **p. 682**, T.E. Adams/TSW/Click Chicago; **p. 682**, Tom Adams; **p. 682**, Larry Jensen/Visuals Unlimited; **p. 683**, Russ Kinne/Science Source/Photo Researchers; **p. 683**, Roland Birke/Peter Arnold, Inc.

CHAPTER THIRTY-ONE

p. 688, Fred Bavendam/Peter Arnold, Inc.; **p. 690**, Frieder Sauer/Bruce Coleman Ltd.; **p. 690**, Fred Bavendam/Peter Arnold, Inc.; **p. 690**, H. Chaumeton/Nature; **p. 691**, Kjell B. Sandved/Photo Researchers; **p. 691**, James H. Carmichael/Photo Researchers; **p. 691**, Carolina Biological Supply; **p. 691**, Tom McHugh/Steinhart Aquarium/Photo Researchers; **p. 692**, Dr. E.R. Degginger/ANIMALS ANIMALS; **p. 696**, Doug Wechsler; **p. 696**, Alex Kerstitch/Sea of Cortez Enterprises; **p. 696**, Bob Evans/Peter Arnold, Inc.; **p. 697**, C.W. Bernard/ANIMALS ANIMALS; **p. 698**, Jan Taylor/Bruce Coleman Ltd.; **p. 699**, Jeff Goodman/NHPA; **p. 699**, Oxford Scientific Films/ANIMALS ANIMALS; **p. 701**, John Cancalosi/Peter Arnold, Inc.; **p. 703**, Alex Kerstitch/Sea of Cortez Enterprises; **p. 703**, James L. Castner; **p. 703**, Alex Kerstitch/Sea of Cortez Enterprises; **p. 703**, James L. Castner; **p. 703**, James L. Castner; **p. 703**, Carolina Biological Supply; **p. 704**, Robert P. Carr; **p. 706**, Wardene Weisser/Bruce Coleman Inc.; **p. 706**, Wardene Weisser/Bruce Coleman Inc.; **p. 706**, Wardene Weisser/Bruce Coleman Inc.; **p. 706**, Wardene Weisser/Bruce Coleman Inc.; **p. 706**, R.A. Mendez/ANIMALS ANIMALS; **p. 706**, C.W. Perkins/ANIMALS ANIMALS; **p. 706**, Jack Wilburn/ANIMALS ANIMALS; **p. 708**, Dwight R. Kuhn; **p. 708**, Kim Taylor/Bruce Coleman Ltd.; **p. 709**, Fred Bavendam/Peter Arnold, Inc.; **p. 709**, William E. Ferguson Photography; **p. 709**, C.B. and D.W. Frith/Bruce Coleman Ltd.; **p. 710**, Carolina Biological Supply; **p. 710**, Carolina Biological Supply; **p. 710**, Robert Dunne/Photo Researchers; **p. 710**, Gilbert Grant/Photo Researchers; **p. 710**, Chesher/Photo Researchers; **p. 710**, Jeffrey L. Rotman; **p. 713**, Jeffrey L. Rotman; **p. 713**, H. Chaumeton/Nature

CHAPTER THIRTY-TWO

p. 718, Bruce Davidson/ANIMALS ANIMALS; **p. 720**, Nancy Sefton; **p. 720**, Mike Newman/Photo Researchers; **p. 721**, Heather Angel/Biofotos; **p. 721**, Carolina Biological Supply; **p. 722**, Dr. Giuseppe Mazza; **p. 722**, Heather Angel/Biofotos; **p. 722**, Steinhart Aquarium/National Audubon Society/Photo Researchers; **p. 723**, Tom McHugh/Photo Researchers; **p. 723**, Wolf H. Fahrenbach; **p. 725**, Alex Kerstitch/Sea of Cortez Enterprises; **p. 725**, Jeffrey L. Rotman; **p. 725**, Peter David/Planet Earth Pictures—Seaphot; **p. 726**, From TISSUES AND ORGANS: A TEXT-ATLAS OF SCANNING ELECTRON MICROSCOPY, by Richard G. Kessel and Randy H. Kardon. W.H. Freeman and Company, copyright © 1979.; **p. 727**, Tom McHugh/Stinehart Aquarium/Photo Researchers; **p. 727**, S.W. Ross/Visuals Unlimited; **p. 728**, Dr. E.R. Degginger/Bruce Coleman Inc.; **p. 728**, Wardene Weisser/Bruce Coleman Inc.; **p. 728**, M. Fogden/Bruce Coleman Inc.; **p. 729**, Peter Scoones/Planet Earth Pictures—Seaphot; **p. 729**, C. Allan Morgan; **p. 729**, Dr. Edward S. Ross; **p. 730**, Joe McDonald/Bruce Coleman Inc.; **p. 731**, Tom McHugh/Photo Researchers; **p. 736**, Palaontaologisches Museum, Museum; **p. 736**, Courtesy Dr. Carl Welty; **p. 737**, G.R. Roberts; **p. 737**, William E. Ferguson Photography; **p. 737**, Dwight R. Kuhn; **p. 738**, The Photographic Library of Australia Pty. Ltd., Sydney; **p. 738**, S. J. Krasemann/Peter Arnold, Inc.; **p. 738**, Andrew Rakoczy/Bruce Coleman Inc.; **p. 742**, Bruce Coleman Ltd.; **p. 742**, David Agee/Anthro-Photo; **p. 743**, Jen & Des Bartlett/Bruce Coleman Inc.; **p. 744**, Carolina Biological Supply; **p. 747**, Copyright © 1986 Jay H. Mattermes, Courtesy SCIENCE 81

CHAPTER THIRTY-THREE

p. 752, T. Kitchin/Tom Stack & Associates; **p. 753**, James Carmichael/NHPA; **p. 753**, Carolina Biological Supply; **p. 755**, Dr. Giuseppe Mazza; **p. 756**, Dr. E.R. Degginger; **p. 756**, Jack Wilburn/ANIMALS ANIMALS; **p. 758**, Stephen Dalton/ANIMALS ANIMALS; **p. 758**, Stephen Dalton/NHPA; **p. 759**, Dr. Giuseppe Mazza; **p. 759**, Russ Kinne/Natl. Audubon Society/Photo Researchers; **p. 759**, Kathie Atkinson/Oxford Scientific Films/ANIMALS ANIMALS; **p. 759**, Carolina Biological Supply; **p. 759**, Hans Pfletschinger/Peter Arnold, Inc.; **p. 760**, Dr. Ralph Buchsbaum; **p. 761**, Carolina Biological Supply; **p. 761**, Ed Reschke; **p. 762**, Ed Reschke; **p. 762**, Rod Planck; **p. 762**, Hans Pfletschinger/Peter Arnold, Inc.; **p. 763**, From BEHOLD MAN, © Lennart Nilsson, National Geographic Society; **p. 763**, John Watney Photo Library; **p. 766**, Glenn Oliver/Visuals Unlimited; **p. 766**, Ken Lucas/Biological Photo Service; **p. 771**, Carolina Biological Supply; **p. 775**,

From TISSUES AND ORGANS: A TEXT-ATLAS OF SCANNING ELECTRON MICROSCOPY, by Richard G. Kessel and Randy H. Kardon, W.H. Freeman and Company. Copyright © 1979.; **p. 775**, Biophoto Associates/Photo Researchers; **p. 776**, Biophoto Associates; **p. 777**, Omikron/Science Source/Photo Researchers

CHAPTER THIRTY-FOUR
p. 784, from THE INCREDIBLE MACHINE, © Lennart Nilsson, National Geographic Society; **p. 785**, Irwin and Peggy Bauer/Bruce Coleman Inc.; **p. 787**, Dr. Cedric S. Raine, Dept. of Neuropathology, Albert Einstein College of Medicine; **p. 788**, From TISSUES AND ORGANS: A TEXT-ATLAS OF SCANNING MICROSCOPY, by Richard G. Kessel and Randy H. Kardon, W.H. Freeman and Company, copyright © 1979; **p. 798**, Dr. John Heuser

CHAPTER THIRTY-FIVE
p. 804, Breck P. Kent/ANIMALS ANIMALS; **p. 806**, Carolina Biological Supply; **p. 810**, Martin M. Rotker/Taurus Photos, Inc.; **p. 810**, Biophoto Associates/Photo Researchers; **p. 823**, Dan McCoy/Rainbow; **p. 823**, Scientific Catalog, © J. Allan/Courtesy DREAMSTAGE Hobson and Hoffmann-LaRoche Inc.; **p. 827**, Charlton Photos; **p. 827**, Carolina Biological Supply; **p. 829**, Larry West/Bruce Coleman Inc.; **p. 830**, Michael P. Gadomski/Bruce Coleman Inc.; **p. 830**, Michael P. Gadomski/Bruce Coleman Inc.; **p. 834**, Stephen Dalton/Oxford Scientific Films/ANIMALS ANIMALS; **p. 837**, Rod Planck; **p. 837**, Biophoto Associates/Science Source/Photo Researchers; **p. 839**, Omikron/Science Source/Photo Researchers

CHAPTER THIRTY-SIX
p. 845, Frans Lanting; **p. 846**, Bruce Coleman Inc.; **p. 848**, Eric & David Hosking/Photo Researchers; **p. 849**, John Gerlach/Tom Stack & Associates; **p. 853**, Bruce Coleman Inc.; **p. 853**, Larry West/Bruce Coleman Inc.; **p. 854**, Dan Guravion/Photo Researchers; **p. 857**, George Holton/Photo Researchers; **p. 858**, William E. Ferguson Photography; **p. 858**, Lynn M. Stone; **p. 858**, Rod Planck

CHAPTER THIRTY-SEVEN
p. 876, John D. Cunningham/Visuals Unlimited; **p. 877**, Oxford Scientific Films/ANIMALS ANIMALS; **p. 886**, Bob Cranston; **p. 894**, The Bettmann Archive; **p. 894**, AP/Wide World; **p. 895**, Mac E. Hadley, University of Arizona; **p. 897**, Paul Almasy/Ominkron/Science Source/Photo Researchers; **p. 897**, AFIP/Science Source/Photo Researchers; **p. 899**, Ed Reschke; **p. 903**, G.R. Roberts

CHAPTER THIRTY-EIGHT
p. 909, Dwight R. Kuhn; **p. 910**, Leonard Lee Rue III/Photo Researchers; **p. 913**, Jeff Rotman/Peter Arnold, Inc.; **p. 914**, Fritz Polking GDT/Peter Arnold, Inc.; **p. 915**, Leonard Lee Rue III/Photo Researchers; **p. 915**, Bruce Coleman Inc.; **p. 915**, Ken Miyata/Photo Researchers; **p. 915**, Tom McHugh/Photo Researchers; **p. 915**, Frans Lanting/Photo Researchers; **p. 919**, Omikron/Science Source/Photo Researchers; **p. 933**, From TISSUES AND ORGANS: A TEXT-ATLAS OF SCANNING ELECTRON MICROSCOPY, by Richard G. Kessel and Randy H. Kardon, W.H. Freeman and Company, copyright © 1979.

CHAPTER THIRTY-NINE
p. 937, Dr. Tony Brain/Science Photo Library/Photo Researchers; **p. 938**, Carolina Biological Supply; **p. 943**, Carolina Biological Supply; **p. 943**, Biophoto Associates/Photo Researchers; **p. 945**, From BEHOLD MAN, © Lennart Nilsson, National Geographic Society; **p. 950**, Biophoto Associates/Science Source/Photo Researchers; **p. 950**, D. W. Fawcett/Science Source/Photo Researchers; **p. 952**, CNRI/Science Photo Library/Photo Researchers; **p. 954**, Ed Reschke

CHAPTER FORTY
p. 959, David M. Dennis/Tom Stack & Associates; **p. 960**, Taurus Photos, Inc.; **p. 960**, D. Wilder; **p. 960**, G.R. Roberts; **p. 960**, William Boehm; **p. 960**, Sally Faulkner Collection/Douglas Faulkner; **p. 960**, Thomas Kitchin/Tom Stack & Associates; **p. 968**, B.F. King, University of California/Biological Photo Service; **p. 969**, Netland, P.A. and B.R. Zetter, 1984 SCIENCE; Cover © 1984 by the AAAS

CHAPTER FORTY-ONE
p. 980, From THE INCREDIBLE MACHINE, © Lennart Nilsson, National Geographic Society; **p. 985**, John D. Cunningham/Visuals Unlimited; **p. 985**, Alfred Owczarzak/Taurus Photos, Inc.; **p. 985**, Alfred Owczarzak/Science Photo Library Int'l/Photo Researchers; **p. 986**, Biology Media/Science Source/Photo Researchers; **p. 987**, Zucker-Franklin, D., M.F. Greaves, C.E. Grossi, and A.M. Marmont. 1981 ATLAS OF BLOOD CELLS: FUNCTION AND PATHOLOGY, vol. 2. Philadelphia: Lea & Febiger, © 1981 by Edi. Ermes s.r.l.-Milan, Italy; **p. 1006**, Jane Schreibman/Photo Researchers

CHAPTER FORTY-TWO
p. 1012, Leonard Lee Rue III/Photo Researchers; **p. 1013**, Biophoto Associates/Science Source/Photo Researchers; **p. 1013**, C. Allan Morgan/Peter Arnold, Inc.; **p. 1013**, Carolina Biological Supply; **p. 1014**, Susan Ernst, Department of Biology, Tufts University; **p. 1015**, Tom McHugh/Photo Researchers; **p. 1015**, Hans Pfletsching/Peter Arnold, Inc.; **p. 1015**, Nancy Sefton/Natl. Audubon Society/Photo Researchers; **p. 1016**, P. Lorne-Jacana/Photo Researchers; **p. 1017**, Carolina Biological Supply; **p. 1018**, Carolina Biological Supply; **p. 1018**, Fletcher W.K./Photo Researchers; **p. 1018**, R.M. Meadows/Peter Arnold, Inc.; **p. 1019**, Marty Snyderman; **p. 1019**, Phillipa Scott/Photo Researchers; **p. 1020**, Tom McHugh/Photo Researchers; **p. 1020**, Jack Green/The Photo Library of Australia; **p. 1020**, Ken M. Higheill/Photo Researchers; **p. 1023**, Dr. G. Schatten/Photo Reserachers; **p. 1023**, From TISSUES AND ORGANS: A TEXT-ATLAS OF SCANNING ELECTRON MICROSCOPY, by Richard G. Kessel and Randy H. Kardon, W.H. Freeman and Company, copyright © 1979.; **p. 1034**, Ray Ellis/Photo Researchers

CHAPTER FORTY-THREE
p. 1041, Oxford Scientific Films/ANIMALS ANIMALS; **p. 1042**, John Walsh/Science Source/Photo Researchers; **p. 1043**, David Scharf/Peter Arnold, Inc.; **p. 1043**, Oxford Scientific Films/ANIMALS ANIMALS; **p. 1043**, Carolina Biological Supply; **p. 1043**, Petit Format/Nestle/Science Source/Photo Researchers; **p. 1043**, Carolina Biological Supply; **p. 1044**, Petit Format/Nestle/Science Source/Photo Researchers; **p. 1045**, Tegler, M.J., and D. Epel, 1973. SCIENCE 179:685-688 1973 by the AAAs; **p. 1045**, Dr. G. Schatten/Science Photo Library/Photo Researchers; **p. 1046**, From BEHOLD MAN, © Lennart Nilsson, National Geographic Society; **p. 1046**, Sea Studios, Inc./Peter Arnold, Inc.; **p. 1046**, Petit Format/Nestle/Science Source/Photo Researchers; **p. 1046**, Photo Researchers; **p. 1046**, Carolina Biological/Photo Researchers; **p. 1046**, From BEHOLD MAN, © Lennart Nilsson National Geographic Society; **p. 1054**, Oxford Scientific Films/ANIMALS ANIMALS; **p. 1063**, After Mangold and Tiedemann, from Balinsky, 1975 Sum. Fig. p. 980. Nilsson L. 1977. A CHILD IS BORN, New York: Dell Publishing Co., Inc.; **p. 1067**, Carolina Biological Supply; **p. 1073**, David Scharf/Peter Arnold, Inc.; **p. 1073**, David Scharf/Peter Arnold, Inc.; **p. 1076**, Courtesy, Professor Dennis Summerbell, National Insititue for Medical Research, London, U.K.; **p. 1084**, Galen Rowell/Peter Arnold, Inc.

PART SEVEN; CHAPTER FORTY-FOUR
p. 1086, Hans Reinhard/Bruce Coleman Inc.; **p. 1087**, Robert Frerck/Odyssey Productions, Chicago; **p. 1087**, George Holton/Ocelot/Photo Researchers; **p. 1088**, Natl. Audubon Society/Photo Researchers; **p. 1088**, Dr. Philip G. Zimbardo; **p. 1089**, J.M. Labat/Photo Researchers; **p. 1090**, Photo Researchers; **p. 1091**, Dr. E.R. Degginger; **p. 1092**, AP/Wide World; **p. 1098**, Dr. I. Eibl-Eibesfeldt

CHAPTER FORTY-FIVE
p. 1105, Werner H. Muller/Peter Arnold, Inc.; **p. 1108**, John Shaw/Tom Stack & Associates; **p. 1109**, Neil Bromhall/Genesis Films Ltd./Oxford Scientific Films/ANIMALS ANIMALS; **p. 1109**, Patricia Caulfield/ANIMALS ANIMALS; **p. 1110**, Courtesy, Guidance Associates, Mt. Kisco, N.Y., from the film strip RHYTHMS OF LIFE; **p. 1113**, The Bettmann Archive; **p. 1113**, Charles Walcott; **p. 1114**, Warren Garst/Tom Stack & Associates; **p. 1114**, Frans Lanting/Minden Pictures; **p. 1116**, Dr. Edward S. Ross; **p. 1116**, Dr. E.R. Degginger; **p. 1117**, Oxford Scientific Films/ANIMALS ANIMALS; **p. 1117**, Charles G. Summers, Jr./Charles G. and Rita Summers; **p. 1117**, William E. Ferguson Photography; **p. 1117**, Lynn M. Stone; **p. 1118**, George Harrison; **p. 1118**, M.P. Kahl; **p. 1121**, Robert P. Carr; **p. 1121**, Lynn M. Stone; **p. 1122**, Gordon Wiltsie/Bruce Coleman Inc.; **p. 1123**, Charles G. Summers, Jr./Charles G. and Rita Summers; **p. 1124**, Candace Bayer/Wolfgang Bayer Productions; **p. 1124**, Robert C. Fields/ANIMALS ANIMALS; **p. 1125**, Mrs. Lorrimer Armstrong; **p. 1125**, Fred Bavendam/Peter Arnold, Inc.; **p. 1127**, Stephen J. Krasemann/DRK Photo; **p. 1127**, Stephen Dalton/NHPA; **p. 1131**, David MacDonald/Oxford Scientific Films/ANIMALS ANIMALS; **p. 1131**, Jeff Foott/Bruce Coleman Inc.

CHAPTER FORTY-SIX
p. 1134, Dr. E.R. Degginger; **p. 1142**, Rod Planck; **p. 1143**, K.G. Preston-Matham/ANIMALS ANIMALS; **p. 1143**, Sullivan and Rogers/Bruce Coleman Inc.; **p. 1143**, Alex Kerstitch/Sea of Cortez Enterprises; **p. 1143**,

E.P.I./Nancy Adams; **p. 1144**, Clem Haagner/Bruce Coleman Inc.; **p. 1145**, Peter Johnson/NHPA; **p. 1145**, David C. Fritts; **p. 1145**, David C. Fritts/ANIMALS ANIMALS; **p. 1148**, Jack Swenson/Tom Stack & Associates; **p. 1150**, Tom McHugh/Photo Researchers; **p. 1150**, William E. Ferguson Photography; **p. 1150**, Rod Planck; **p. 1151**, John Shaw; **p. 1151**, Rita Summers/Charles G. and Rita Summers; **p. 1151**, Don and Pat Valenti; **p. 1151**, David C. Fritts; **p. 1151**, G.R. Roberts; **p. 1152**, Charles G. and Rita Summers; **p. 1152**, Dr. E.R. Degginger; **p. 1153**, David R. Frazier Photolibrary; **p. 1153**, David C. Fritts; **p. 1153**, John Shaw/Tom Stack & Associates; **p. 1154**, David C. Fritts; **p. 1154**, Steve McCutcheon; **p. 1154**, Stephen J. Kraseman/DRK Photo; **p. 1155**, Doug Sokell/Tom Stack & Associates; **p. 1155**, Marty Stouffer/Earth Scenes; **p. 1157**, H. Chaumeton/Nature; **p. 1157**, Peter David/Plant Earth Pictures—Seaphot; **p. 1157**, J.M. Bassot/H. Chaumeton/Nature; **p. 1160**, Jeffrey L. Rotman; **p. 1160**, Steve McCutcheon; **p. 1160**, Dr. E.R. Degginger

CHAPTER FORTY-SEVEN

p. 1165, G.R. Roberts; **p. 1168**, G.R. Roberts; **p. 1170**, Joseph P. Shapiro/U.S. News & World Report; **p. 1173**, William E. Ferguson Photography; **p. 1181**, Dudley Foster/Woods Hole Oceanographic Institution; **p. 1181**, James Childress/Woods Hole Oceanographic Institution; **p. 1181**, Alvin External Camera/Woods Hole Oceanographic Institution; **p. 1182**, John D. Cunningham/Visuals Unlimited; **p. 1184**, Breck P. Kent/Earth Scenes; **p. 1187**, John Shaw/Tom Stack & Associates; **p. 1187**, Tim Davis/Photo Researchers; **p. 1189**, Don and Pat Valenti; **p. 1191**, U.S. Forestry Service; **p. 1192**, John Ebeling; **p. 1195**, Tom Stack & Associates

CHAPTER FORTY-EIGHT

p. 1200, Dr. E.R. Degginger; **p. 1206**, Don Cabe/Photo Researchers; **p. 1209**, Tom McHugh/Photo Researchers; **p. 1209**, Dr. E.R. Degginger; **p. 1211**, Dr. E.R. Degginger; **p. 1215**, Norman Myers/Bruce Coleman Inc.

CHAPTER FORTY-NINE

p. 1216, Sergio Dorantes/Gamma-Liaison; **p. 1219**, Thomas S. England/Photo Researchers; **p. 1225**, Gary Milburn/Tom Stack & Associates; **p. 1225**, Stern/Black Star; **p. 1226**, NASA

ILLUSTRATION CREDITS

Jean Helmer, Floyd Hosmer, JAK Graphis LTD, Sandra McMahon, Pagecrafters, Inc., Precision Graphics, R.S.R. Associates, Nadine B. Sokol

INDEX